Astronomical Data

Earth

Mass	5.98×10^{24} kg
Mean radius	6.378×10^{6} m
Mean density	5.5×10^{3} kg/m^3
Surface gravitational field strength	9.80 N/kg
Mean earth–sun distance	1.5×10^{11} m

Moon

Mass	7.35×10^{22} kg
Mean radius	1.74×10^{6} m
Mean density	3.3×10^{3} kg/m^3
Surface gravitational field strength	1.62 N/kg
Mean earth–moon distance	3.85×10^{8} m

Sun

Mass	1.99×10^{30} kg
Mean radius	6.96×10^{8} m
Mean density	1.4×10^{3} kg/m^3
Surface gravitational field strength	274 N/kg

Planets of the Solar System

	EQUATORIAL RADIUS (km)	MASS (10^{24} kg)	SURFACE GRAVITY (Earth = 1)	PERIOD OF REVOLUTION	ORBITAL PERIOD (Years)
Mercury	2440	0.33	0.38	58.65 days	0.241
Venus	6050	4.87	0.91	243.01 days	0.615
Earth	6378	5.98	1.00	23 h 56 min 4.1 s	1.000
Mars	3394	0.64	0.39	24 h 37 min 22.6 s	1.88
Jupiter	71 400	1900	2.74	9 h 50.5 min	11.86
Saturn	60 000	569	1.17	10 h 14 min	29.46
Uranus	25 050	87	0.94	17 h	84.01
Neptune	27 700	103	1.15	17 h 50 min	164.8
Pluto	1100	0.01	0.03	6.39 days	248.6

Greek Letters

Alpha	A	α	Nu	N	ν
Beta	B	β	Xi	Ξ	ξ
Gamma	Γ	γ	Omicron	O	o
Delta	Δ	δ	Pi	Π	π
Epsilon	E	ϵ	Rho	P	ρ
Zeta	Z	ζ	Sigma	Σ	σ
Eta	H	η	Tau	T	τ
Theta	Θ	θ	Upsilon	Υ	υ
Iota	I	ι	Phi	Φ	ϕ
Kappa	K	κ	Chi	X	χ
Lambda	Λ	λ	Psi	Ψ	ψ
Mu	M	μ	Omega	Ω	ω

UNIVERSITY PHYSICS

Models and
Applications

UNIVERSITY PHYSICS

Models and Applications

William P. Crummett
Centre College

Arthur B. Western
Rose-Hulman Institute of Technology

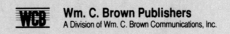
Wm. C. Brown Publishers
A Division of Wm. C. Brown Communications, Inc.

Book Team

Executive Editor   *Jeffrey L. Hahn*
Editor   *Jane Ducham*
Publishing Services Coordinator   *Julie Avery Kennedy*

Wm. C. Brown Publishers
A Division of Wm. C. Brown Communications, Inc.

Vice President and General Manager   *Beverly Kolz*
Vice President, Publisher   *Earl McPeek*
Vice President, Director of Sales and Marketing   *Virginia S. Moffat*
National Sales Manager   *Douglas J. DiNardo*
Marketing Manager   *Amy Schmitz*
Advertising Manager   *Janelle Keeffer*
Director of Production   *Colleen A. Yonda*
Publishing Services Manager   *Karen J. Slaght*
Permissions/Records Manager   *Connie Allendorf*

Wm. C. Brown Communications, Inc.

President and Chief Executive Officer   *G. Franklin Lewis*
Corporate Senior Vice President, President of WCB Manufacturing   *Roger Meyer*
Corporate Senior Vice President and Chief Financial Officer   *Robert Chesterman*

Cover photo by Karl Weatherly, Tony Stone Worldwide

Copyediting, permissions, and production by York Production Services

Cover and interior design by York Production Services

Composition by York Graphic Services

The credits section for this book begins on page 1222 and is
considered an extension of the copyright page.

Library of Congress Catalog Card Number: 93–72117

ISBN 0–697–11199–7

Printed in the United States of America by Wm. C. Brown Communications, Inc.,
2460 Kerper Boulevard, Dubuque, IA 52001

10 9 8 7 6 5 4 3 2 1

CONTENTS

Chapters marked with a ★ contain numerical applications and spreadsheet problems. Special numerical sections of the text are in **boldface.**

Preface   xix

PART I: MECHANICS   xxiv

1   PHYSICS AND MEASUREMENT   2
1.1 Physics and Models   3
1.2 Systems of Units   5
Length   5
Time   7
Mass   8
The Metric System   10
1.3 Unit Conversion and Dimensional Analysis   10
1.4 Significant Figures   13
1.5 Summary   15
Problems   16

★2   VECTOR ALGEBRA   18
2.1 Coordinate Systems   19
2.2 Scalars and Vectors   21
2.3 Vector Addition   23
Multiplication by a Scalar   25
Unit Vectors   25
Vector Components   26
Vector Addition: Component Method   26
2.4 Vector Multiplication   30
The Scalar Product   31
The Vector Product   31
Scalar and Vector Products in the Unit-Vector Representation   33
2.5 Summary   35
Problems   36

★3   MOTION ALONG A STRAIGHT LINE   40
3.1 Position, Velocity, and Acceleration   41
Average Velocity and Average Speed   42
Average Acceleration   43
Instantaneous Velocity   44
Instantaneous Acceleration   45
3.2 Graphical Interpretation of Position, Velocity, and Acceleration   46
Average Velocity   46
Instantaneous Velocity   47

Average and Instantaneous Acceleration 48
Position from Velocity Graphs 49
Velocity from Acceleration Graphs 51

3.3 The Constant-Acceleration Model 52
The Sign of Acceleration 55
Freely Falling Objects 57

3.4 Applying Integral Calculus to Find Position and Velocity (Optional) 60

3.5 Handling Real Data (Optional) 61
Computing Areas 61
Taking Slopes 64
Computing Acceleration from Velocity 66
Computing Acceleration from Position 66

3.6 Summary 68
Problems 69

★**4 MOTION IN TWO DIMENSIONS 76**
4.1 Velocity and Acceleration Vectors 77
Average Speed 78

4.2 Projectile Motion 79
4.3 Circular Motion 85
Tangential and Normal Components of Acceleration in Two Dimensions 89

4.4 Numerical Methods in Two Dimensions (Optional) 90
4.5 Summary 93
Problems 94

5 FORCES: NEWTON'S THREE LAWS OF MOTION 102
5.1 Newton's First Law: Inertia 103
5.2 Newton's Second Law 104
5.3 Some Common Forces 105
The Weight Force 106
Spring Forces 108
Normal Forces 109
Frictional Forces 109
Tension 109

5.4 Simple Applications of Newton's Second Law 110
5.5 Newton's Third Law 117
5.6 Gravitational Field Strength, Falling Objects, and Mass (Optional) 119
5.7 Summary 120
Problems 121

★**6 ADDITIONAL FORCE MODELS AND CIRCULAR MOTION 126**
6.1 The Coefficient-of-Friction Model 127
6.2 Circular Motion 130
6.3 Newton's Universal Law of Gravitation 136
The Cavendish Balance 138
"Weighing" the Sun 139
Weight and Universal Gravitation 140

6.4 Models for Motion through a Resistive Medium (Optional) 142

6.5 Summary of Force Models   146

6.6 Numerical Methods for Newton's Second Law (Optional)   146

6.7 Summary   152

Guest Essay: Physics and Racing at the Indianapolis 500   160

★7   WORK AND KINETIC ENERGY   163

7.1 Work (Constant Force, Constant Direction)   164

7.2 Work (Variable Force, Constant Direction)   165

7.3 Work (Variable Force, Variable Direction)   169

7.4 Work and Kinetic Energy   170

7.5 Power   174

7.6 Numerical Calculations of Work (Optional)   176

7.7 Summary   177

Problems   178

★8   CONSERVATION OF ENERGY   186

8.1 Conservative Forces and Potential Energy   187

8.2 Potential Energy: The Negative of the Integral of a Conservative Force over Distance   188

Gravitational Potential Energy near the Earth's Surface   188

Potential Energy of a Hooke's-Law Spring   190

8.3 Conservation of Mechanical Energy   190

Energy Conservation and Isolated Systems   195

8.4 Systems with Nonconservative Forces   196

8.5 Conservative Forces: The Negative of the Derivative of Potential Energy   199

Equilibrium   201

8.6 Potential Energy for Newtonian Gravity (Optional)   202

8.7 Conservative Forces in Two and Three Dimensions (Optional)   206

Motion in Two Dimensions   207

Motion in Three Dimensions   208

8.8 Summary   210

Problems   212

★9   IMPULSE AND LINEAR MOMENTUM   219

9.1 Impulse   220

9.2 Impulse and Momentum   221

9.3 The Momentum Statement of Newton's Second Law   223

9.4 Conservation of Momentum   226

Recoil   227

Relative Motion   230

9.5 Collisions in One Dimension   232

Perfectly Inelastic Collisions   233

Perfectly Elastic Collisions   234

Partially Elastic Collisions   236

9.6 Collisions in Two Dimensions   238
   Perfectly Inelastic Collisions   239
   Perfectly Elastic Collisions   240
   Partially Elastic Collisions   241
9.7 Summary   241
Problems   242

★10   MOMENTUM, ENERGY, AND THE CENTER OF MASS   249
10.1 The Big Picture   250
10.2 The Ballistic Pendulum and Its Friends   250
10.3 The Center of Mass   252
   Center of Mass for a Collection of Point-Objects   253
   Center of Mass for Symmetrical Solid Objects   254
   Center of Mass for Solid Objects by Integration (Optional)   255
10.4 Center of Mass: Energy, Momentum, and Newton's Second Law   257
   Gravitational Potential Energy and the Center of Mass   258
   Newton's Second Law and the Center of Mass   258
   Momentum and the Center of Mass   259
   Kinetic Energy and the Center of Mass   262
10.5 Systems of Variable Mass (Optional)   263
10.6 Summary   268
Problems   269

★11   ROTATION ABOUT A FIXED AXIS   276
11.1 Kinematics of Rotational Motion   277
   The Constant-Acceleration Model   282
11.2 Kinetic Energy and Rotational Inertia   283
11.3 Torque   292
   Work Done by a Torque   295
11.4 Newton's Second Law for Rotation about a Fixed Axis   296
11.5 Angular Momentum for Rotation about a Fixed Axis   299
   Conservation of Angular Momentum   300
11.6 Summary   303
Problems   304

★12   STATIC EQUILIBRIUM AND ROLLING OBJECTS   311
12.1 Static Equilibrium   312
   Torques Due to the Weight Force   312
   Force and Torque in Static Equilibrium   313
12.2 Rotation and Translation with No Slipping (Optional)   316
   Energy Analysis   317
   Force Analysis   319
12.3 Rotation and Translation with Slipping (Optional)   324
   Angular Impulse–Angular Momentum Theorem   324
   Center of Percussion   326
12.4 Summary   327
Problems   328
Guest Essay: Biomechanical Loading of the Human Body   335

★**13   VECTOR DESCRIPTIONS OF ROTATIONAL MOTION   339**

13.1 The Torque Vector   340

13.2 The Angular Velocity and Angular Acceleration Vectors   341

13.3 The Angular Momentum Vector   344

Newton's Second Law for Angular Momentum   345

Conservation of Vector Angular Momentum   346

Precession of a Gyroscope (Optional)   346

Precession of the Earth's Axis of Rotation (Optional)   348

13.4 Vector Relations between **L** and **ω** (Optional)   349

Single Point-Masses   350

Multiple Point-Masses   351

13.5 Summary   352

Problems   353

★**14   OSCILLATIONS   356**

14.1 Kinematics of Simple Harmonic Motion   357

Velocity and Acceleration for Simple Harmonic Motion   361

14.2 The Dynamics of Simple Harmonic Motion   363

Mass Attached to a Spring   363

The Simple Pendulum   366

The Physical Pendulum   368

The Torsion Pendulum   369

A Two-Mass System   370

14.3 Energy of a Simple Harmonic Oscillator   371

14.4 Uniform Circular Motion and Simple Harmonic Motion   374

14.5 Damped and Driven Oscillations   375

Damped Oscillations   375

Driven Oscillations   377

14.6 **Numerical Calculations for Periodic Motion (Optional)   378**

Equation of Motion for Simple Harmonic Motion   378

Velocity and Acceleration for Simple Harmonic Motion   379

Large-Amplitude Pendulum Oscillations   380

14.7 Summary   380

Problems   381

★**15   ONE-DIMENSIONAL WAVES   387**

15.1 Introduction   388

Particle Motion with Respect to Wave Direction   388

Wave Dimension   389

Particle Behavior in Time   389

15.2 Waves Traveling on a String   390

Wave Pulses   390

Sinusoidal Wave Train   392

Longitudinal Waves   394

15.3 Wave Velocity on a String   395

15.4 Energy Transported by Sinusoidal Waves   396

15.5 Superposition and Interference of Waves   397

Wave Interference   398

Adding Waves That Differ in Phase Only: Interference   398

Adding Waves That Differ in Frequency Only: Beats   399

Adding Waves That Differ in Direction Only: Standing Waves   400
Wave Reflection   402
Fourier's Theorem (Optional)   403

15.6 The Wave Equation (Optional)   404
Wave Velocity: The Wave Equation Analysis   406

15.7 Spreadsheet Calculations for Superposition of One-Dimensional Waves (Optional)   407

15.8 Summary   408

Problems   410

★16 **SOLIDS, LIQUIDS, AND GASES   414**

16.1 States of Matter   415

16.2 Stress, Strain, and the Elastic Moduli   416
Young's Modulus   418
Shear Modulus   419
Bulk Modulus   419

16.3 Density and Pressure   421

16.4 Fluid Statics   423
Variation of Pressure with Depth   423
The Incompressible-Fluid Model   424
Pascal's Principle   426
Archimedes' Principle   427
A Compressible-Fluid Model   428

16.5 Pressure-Measuring Devices   430
The Manometer   430
The Mercury Barometer   431

16.6 Fluid Dynamics   432
The Ideal-Fluid Model   432
The Equation of Continuity   432
Bernoulli's Equation   434
Qualitative Applications of Bernoulli's Equation   438

16.7 Dynamic Viscosity (Optional)   438

16.8 Summary   440

Problems   441

Guest Essay: The Continuity Equation in Everyday Life   446

★17 **SOUND   450**

17.1 Models for Sound Waves in a Gas   451

17.2 The Velocity of Sound   452
Other Distortion Waves (Optional)   455

17.3 Harmonic Waves in Air   457

17.4 Sound Intensity and Sound Intensity Level   458
The Decibel Scale   460

17.5 Sources of Sound   463
Vibrating Strings   463
Air Columns   466
Resonance and Beats   467

17.6 The Doppler Effect   468
Shock Waves   472

17.7 Summary   473

Problems   473

PART II: THERMODYNAMICS AND KINETIC THEORY 482

★18 TEMPERATURE, HEAT, AND THE EQUATION OF STATE 484

18.1 Temperature 485
  The Celsius and Fahrenheit Scales 485
  Constant-Volume Gas Thermometer 486

18.2 Thermal Expansion 488
  Linear Expansion 490
  Area Expansion 491
  Volume Expansion 491

18.3 Heat and Energy Transfer Mechanisms 493
  Conduction 494
  Building Insulation (Optional) 497
  Convection and Radiation 498

18.4 Heat Capacity and Latent Heat 500
  Heat of Transformation 502

18.5 The Equation of State 504
  The van der Waals Gas (Optional) 507

18.6 Summary 508
Problems 509

★19 THERMODYNAMICS I: PROCESSES AND THE FIRST LAW 513

19.1 Equilibrium, the Zeroth Law of Thermodynamics, and Processes 514
  The Zeroth Law of Thermodynamics 514

19.2 Work 518
19.3 The First Law of Thermodynamics 520
19.4 Specific Thermodynamic Processes 522
  Isochoric Processes 523
  Isobaric Processes 523
  Isothermal Processes 524
  Adiabatic Processes 525

19.5 Cyclic Processes 528
19.6 Summary 532
Problems 533

20 THERMODYNAMICS II: THE SECOND LAW 538

20.1 The Second Law of Thermodynamics and Heat Engines 540
20.2 The Carnot Cycle 542
20.3 Refrigerators and Heat Pumps 546
  The Curious Fraction Q/T 548

20.4 The Absolute Temperature Scale and the Third Law of
       Thermodynamics 548
20.5 General Cyclic Processes 549
  Efficiencies of Real Engines: The Clausius Inequality 551
  The Curious Q/T Result for Arbitrary Cycles 553

20.6 A Formal Definition of State Variables 554
20.7 Entropy: A State Variable 555
20.8 Entropy Changes for Irreversible Processes 556
  The Principle of Increasing Entropy 556

20.9 Entropy and Disorder 559

20.10 Summary 559

Problems 560

★21 MICROSCOPIC CONNECTIONS TO THERMODYNAMICS 564

21.1 The Kinetic Theory of an Ideal Gas 565
The Ideal-Gas Model 565
Pressure and Molecular Motion 565
Temperature and Molecular Motion 568

21.2 The Equipartition-of-Energy Theorem 570
Ideal Polyatomic-Gas Models 571

21.3 Another Look at Specific Heats 572
Monatomic Ideal Gases 572
Diatomic Ideal Gases 573
Quantum Mechanical Effects 574
Specific Heat Capacity of Solids 575

21.4 Distribution of Molecular Speeds 576
The Maxwell Speed-Distribution Function 576
Mean Free Path 577

21.5 Summary 578

Problems 579

PART III: ELECTRICITY AND MAGNETISM 582

★22 STATIONARY CHARGES AND THE ELECTRIC FIELD 584

22.1 Electric Charge 585
Continuous Charge Distributions 588

22.2 Coulomb's Law 590

22.3 The Electric Field 595
Point-Charge Distributions 596
Continuous Charge Distributions 601
Electric Field Lines 608

22.4 The Electric Dipole in a Uniform Electric Field 611

22.5 The Electric Dipole in Nonuniform Fields 614

22.6 Summary 615

Problems 617

★23 GAUSS'S LAW 622

23.1 Electric Flux 623

23.2 Gauss's Law 626
Gaussian Surfaces 626
Coulomb's Law from Gauss's Law 627

23.3 The Ideal-Conductor Model in Electrostatic Equilibrium 633

23.4 Summary 638

Problems 639

★24 ELECTRIC POTENTIAL 642

24.1 Electric Potential Energy 643

24.2 The Electric Potential 646

Electric Potential Change Due to a Uniform Electric Field 649

Electric Potential Change in a Nonuniform Electric Field 652

24.3 Computing the Electric Potential from a Charge Distribution 655

Point-Charge Distributions 655

Continuous Charge Distributions 656

24.4 Computing the Electric Field from the Electric Potential 658

The Relation Between V and E in Three Dimensions (Optional) 660

Why Some of This Should Seem Familiar 662

24.5 Equipotential Surfaces 662

24.6 The Big Picture 666

24.7 Numerical Methods: Relaxation (Optional) 666

24.8 Summary 671

Problems 672

★25 CAPACITORS AND DIELECTRICS 678

25.1 Capacitance 680

25.2 Combinations of Capacitors 685

Capacitors in Parallel 685

Capacitors in Series 688

25.3 Energy Storage in a Capacitor 691

25.4 Dielectrics 695

25.5 Gauss's Law and the Electric Field Vectors 703

Polarization and the Displacement Field (Optional) 704

Gauss's Law for the Displacement Field (Optional) 706

25.6 A Numerical Application (Optional) 707

25.7 Summary 709

Problems 710

★26 ELECTRIC CURRENT AND RESISTANCE 716

26.1 Electric Current and Current Density 717

26.2 Resistivity and Resistance 722

Resistance 723

Temperature Dependence of Resistivity (Optional) 726

26.3 Energy Dissipation 727

26.4 Microscopic Models of Resistance (Optional) 729

Valence-Bonding Model of Conduction 729

The Band Model of Conductivity 732

Superconductors 735

26.5 Summary 737

Problems 738

★27 DIRECT-CURRENT CIRCUITS 742

27.1 Electromotive Force 743

27.2 Combinations of Resistors 745

Resistors in Series 745

Resistors in Parallel 746

27.3 Multiple-Loop Circuits: Kirchhoff's Rules   749

27.4 Potential Difference, Current, and Resistance Measurements   753
    Voltmeters   754
    Ammeters   755
    The Wheatstone Bridge   756
    The Potentiometer   756

27.5 RC Circuits   757
    Electronic Calculus: Differentiating and Integrating Circuits (Optional)   761

27.6 Digital Voltmeters (Optional)   762

27.7 A Matrix Method for Complex Circuits (Optional)   763

27.8 Summary   766

Problems   767

★28 **THE MAGNETIC FIELD   774**

28.1 The Magnetic Field   775
    Moving Charges in Uniform Magnetic Fields   779
    Moving Charges in Nonuniform Magnetic Fields   781
    The Hall Effect (Optional)   783
    Magnetohydrodynamics (Optional)   785

28.2 Force on a Current-Carrying Conductor   787

28.3 Current-Carrying Loops in a Uniform Magnetic Field   789
    The Magnetic Dipole Moment   790

28.4 Summary   793

Problems   794

★29 **SOURCES OF MAGNETIC FIELDS   810**

29.1 The Biot-Savart Law   811

29.2 Parallel Wires, Amperes, and Coulombs   818

29.3 Ampère's Law   820
    Infinite Current Sheet   821
    The Solenoid   822
    The Toroid   825

29.4 Magnetic Flux and Gauss's Law for Magnetism   826

29.5 Field on the Axis of a Solenoid: A Numerical Application (Optional)   827

29.6 Summary   828

Problems   829

★30 **FARADAY'S LAW AND INDUCTION   826**

30.1 The Laws of Faraday and Lenz   827
    Faraday's Law   830
    Lenz's Law   831

30.2 Induced Electric Fields   836
    Eddy Currents   838

30.3 The Displacement Current and Maxwell's Equations   839
    Maxwell's Equations   841

30.4 Summary   842

Problems   843

★31 INDUCTANCE 848

31.1 Induction 849

31.2 LR Circuits 853

31.3 Energy and the Magnetic Field 854

31.4 LC Circuits (Optional) 857

  The RLC Circuit 859

31.5 Summary 861

Problems 863

★32 MAGNETIC PROPERTIES OF MATERIALS (OPTIONAL) 867

32.1 Overview of the Magnetic Properties of Matter 868

32.2 The Source of Magnetism in Materials 870

32.3 The Magnetic Field Vectors 872

32.4 Diamagnetic, Paramagnetic, and Ferromagnetic Phases 876

  Diamagnetism 876

  Paramagnetism 878

  Ferromagnetism 878

32.5 Summary 883

Problems 884

Guest Essay: Geophysical Applications of Electromagnetic Induction 886

★33 ALTERNATING-CURRENT CIRCUITS 890

33.1 Circuit Elements in AC Circuits 891

  A Resistor in an AC Circuit 892

  A Capacitor in an AC Circuit 894

  An Inductor in an AC Circuit 896

33.2 RLC Circuits 900

33.3 The Root-Mean-Square Potential and Current 907

33.4 Power and Resonance in AC Circuits 910

33.5 Transformers (Optional) 916

33.6 Filter Circuits (Optional) 921

33.7 Summary 923

Problems 924

Guest Essay: Magnetic Resonance 929

34 ELECTROMAGNETIC WAVES 932

34.1 The Prediction of Waves from Maxwell's Equations 933

34.2 Sinusoidal Electromagnetic Waves 937

34.3 Energy Transport by Electromagnetic Waves 938

34.4 Radiation Pressure 941

34.5 Sources of Electromagnetic Waves 942

34.6 The Electromagnetic Spectrum 945

34.7 Summary 946

Problems 948

PART IV: OPTICS   950

★35   REFLECTION, REFRACTION, AND POLARIZATION OF LIGHT   952

35.1   Particles and Waves: A Tale of Two Models   953

35.2   Properties of the Wave Model of Light   954

Wavefronts and Rays   954

Huygens' Principle   955

The Ray Model   956

35.3   Reflection   956

35.4   Refraction   957

Total Internal Reflection   960

Optical Fibers (Optional)   962

Dispersion   964

Prism Geometry (Optional)   965

35.5   Polarization   966

Polarization by Scattering   967

Polarization by Reflection   968

Wire-Grid Polarizers   970

Birefringence (Optional)   972

Half-Wave and Quarter-Wave Plates (Optional)   974

Optics of the Compact-Disc Player (Optional)   975

35.6   Summary   976

Problem   977

★36   GEOMETRICAL OPTICS   982

36.1   Images   983

36.2   Images Formed by Plane Mirrors   984

36.3   Images Formed by Curved Mirrors   986

Concave Mirrors   988

Convex Mirrors   994

36.4   Images Formed by Refracting Surfaces   995

Refraction at Curved Surfaces   996

Apparent Depth   997

36.5   Images Formed by Lenses   999

The Lens-Maker's Formula   999

The Thin-Lens Formula   1001

36.6   The Eye and Simple Optical Instruments (Optional)   1008

The Eye   1008

The Simple Magnifying Glass   1009

The Compound Microscope   1011

Telescopes   1012

36.7   Optical Aberrations (Optional)   1013

Chromatic Aberration   1014

Third-Order Aberrations   1014

36.8   Numerical Methods for Paraxial Ray Tracing (Optional)   1015

A Spreadsheet Template for the ABCD Parameters   1017

The ABCD Matrix (Optional)   1018

36.9   Summary   1019

Problems   1021

★**37   INTERFERENCE OF LIGHT   1028**

37.1   Interference   1029
Two-Source Interference   1029
Thin-Film Interference   1031

37.2   Irradiance for Simple Interference Patterns   1034
Coherence   1036
Optical Beats (Optional)   1037
Thin-Film Interference   1038
High-Reflectance Result for Thin-Film Interference (Optional)   1039
Two-Source Interference   1040

37.3   Multiple-Source Interference   1044
Phasors   1044

37.4   Summary   1046
Problems   1048

★**38   DIFFRACTION OF LIGHT   1052**

38.1   Single-Slit Diffraction   1053
Diffraction-Limited Optics   1055
Resolution   1056

38.2   Effect of Finite Slit Width on Double-Slit Interference Patterns
(Optional)   1058

38.3   Diffraction Gratings (Optional)   1060
Resolving Power of a Grating   1060
Free Spectral Range (Optional)   1062

38.4   X-Ray Diffraction (Optional)   1064
Bragg Diffraction   1065

38.5   Summary   1070
Problems   1071

PART V: MODERN PHYSICS   1076

★**39   SPECIAL THEORY OF RELATIVITY   1078**

39.1   The Speed of Light   1079
Luminiferous Ether   1079
Michelson Interferometer   1080
The Experiment of Michelson and Morley   1081

39.2   Postulates of Special Relativity   1084
The Downfall of Simultaneity   1085

39.3   Relativistic Kinematics   1085
Lorentz Transformations   1086
Time Dilation   1088
Length Contraction   1089
Experimental Tests of Relativity   1090
Relativistic Doppler Shift   1092
The Twin Paradox   1093
Velocity Transformations   1096

39.4   Relativistic Dynamics   1098
Relativistic Momentum   1099
Kinetic Energy   1100

39.5 Binding Energy and Mass   1101
39.6 Summary   1103
Problems   1105

★**40   THE BIRTH OF QUANTUM PHYSICS   1109**

40.1 The Particle Model for Light Revisited   1110
  Blackbody Radiation   1110
  Heat Capacity of Solids   1114
  Photoelectric Effect   1115
  Compton Scattering   1119

40.2 The Wave Model for Particles   1121
  The Davisson-Germer Experiment   1122
  The Experiment of G. P. Thomson   1123

40.3 The Spectra of Atoms   1125
40.4 The Rutherford-Bohr Model of the Atom   1127
  Rutherford Scattering   1127
  Bohr Model of the Atom   1128

40.5 Summary   1133
Problems   1134

APPENDIXES   1139

1 **MATHEMATICS SUMMARY   1139**

2 **INTRODUCTION TO NUMERICAL METHODS USEFUL IN PHYSICS   1144**

3 **LIST OF SYMBOLS   1147**

4 **GREEK LETTERS   1149**

5 **CONVERSION FACTORS   1149**

6 **PERIODIC TABLE AND ATOMIC MASSES   1151**

7 **PHYSICAL CONSTANTS   1155**

8 **ASTRONOMICAL TABLES   1156**

9 **PROPERTIES OF SPORTING BALLS AND RELATED INFORMATION   1157**

10 **INTRODUCTION TO SPREADSHEETS   1158**

11 **LIST OF TABLES   1162**

12 **NOBEL PRIZE WINNERS IN PHYSICS   1163**

13 **ANSWERS TO ODD-NUMBERED PROBLEMS   1170**

Never judge a book by its preface. The only part of this text that was composed for instructors is this preface. The remainder is intended for students enrolled in a college-level calculus-based introductory physics course of one to one-and-one-half year's duration. We have tried carefully to honor this ideal throughout the text while bringing a modern flavor to traditional introductory material. By "modern" we mean more than contemporary examples. Indeed, we incorporated much of the fruits of recent research in physics education. Here are the important features of this text:

A TEXT FOR STUDENTS

This is an armchair text not a podium text. We write in an informal and conversational tone to and for students, not their instructors. We take special care to discuss underlying, often unstated, assumptions and concepts that can act as barriers to student understanding.

PHENOMENA, MODELS, EQUATIONS

We begin the discussion of new concepts by using descriptions of related phenomena with which students have had direct experience. Next, we proceed to describe explicitly models that extract essential features of each phenomenon. Only after these steps are completed are equations derived from an analysis of the relevant model.

EXAMPLES FROM MODERN TECHNOLOGY

We use many examples drawn from up-to-date modern technology relevant to today's student. For example we include a more detailed analysis than most texts of the principles behind such topics as computerized motion analysis, digital voltmeters, optical fibers, and compact-disk players.

TEACHING THROUGH EXAMPLES

Solutions to example problems are often longer than those in most texts. We take the time to discuss the logic behind the procedures as well as possible pitfalls. Sometimes bad examples using incorrect solution methods are shown and discussed.

CALCULUS AS A PREREQUISITE OR A COREQUISITE

In the ideal case students will have had one-term of calculus prior to starting this physics course. The text may, however, be used in courses in which students are taking their first term of calculus concurrently with their first term of physics. By omitting optional sections in the early chapters, the introduction of calculus may be delayed until students have encountered elementary concepts in their calculus course. Derivatives are introduced in Chapter 3, but are used sparingly throughout the next several chapters. For those who wish to use the full power of the calculus early in the course, the integral relations between acceleration, velocity, and position are presented in an optional section of Chapter 3. However, the first necessary use of the integral does not occur until the discussion of work and kinetic energy in Chapter 7.

COMPUTERS, SPREADSHEETS, AND REAL DATA

In addition to conventional algebraic techniques, methods of solution that allow students to take advantage of the power of modern hand-held scientific calculators are presented for selected cases, for example, iterative techniques for solving complex equations in mechanics and the matrix solution of simultaneous equations in DC circuit problems. Many examples and problems make use of real data obtained from actual student experiments or published articles. Suggestions for the use of spreadsheets to analyze data and aid in problem solutions are given.

NUMERICAL METHODS: A TEXT WITHIN A TEXT

Numerical techniques available to today's students with access to desktop computers are presented in optional sections throughout the text. Computer programs (source code) and spreadsheet templates are provided so that students can concentrate on the physics rather than generating software. Students may be introduced to numerical integration and differentiation, numerical solutions to differential equations, relaxation methods for potential problems, and spreadsheet methods for paraxial ray tracing in optics. The optional numerical sections are fully integrated with the physics of the chapters in which they appear; however, *no reference to any numerical method or solution is made from sections within the conventional parts of the text.* The sections on numerical methods can be used as an integral part of the course, reserved as independent study projects for interested students, or combined and presented in a separate minicourse on numerical techniques in physics.

THE DISKETTE ACCOMPANYING THIS TEXT . . .

A computer diskette accompanies this text. It contains 56 programs, data files, and spreadsheet templates used in examples and problems found throughout this text. (Both IBM-compatible and Macintosh-compatible diskettes are available.)

AND ALL THE PEDAGOGICAL FEATURES YOU EXPECT

In addition, the text contains all the pedagogical features you've come to expect in a modern textbook: four-color concept-consistent line art, boldface type highlighting new vocabulary words and concepts, boxed problem-solving strategies, concept questions interspersed with the text material, lots of worked examples, chapter summaries, a variety of problems of varying difficulty, and guest essays for enrichment. Photographs have been selected for their educational value, either to illustrate a specific phenomenon or to suggest an application of a principle under study.

Historical anecdotes that we have found of interest to students have been included where they do not detract from the flow of the text. Portraits and short biographies of a few of the great contributors to science are also presented to remind us of the human side of the endeavor we call physics.

ACKNOWLEDGMENTS

We gratefully thank the following persons for their thoughtful suggestions, assistance, and patience throughout the preparation of this textbook. First, our thanks to the students who reviewed early versions of the text and helped to provide coherent, workable problems, especially: John Buetow, John Thompson, and Arthur Usher, IV.

A special debt is owed to many reviewers whose constructive critiques helped us to avoid many pitfalls and improve both the pedagogy and the physics: Don C. Hopkins, South Dakota School of Mines/Technology; Edward H. Carlson, Michigan State University; Ronald M. Cosby, Ball State University; John W. Jewett, Jr., Cal. State Polytechnic University; Kirby W. Kemper, Florida State University; Richard D. Haracz, Drexel University; Ralph V. McGrew, Broome Community College; Thurman R. Kremser, Albright College; Philip N. Parks, Michigan Technological University; and John A. Gilreath, Clemson University.

We are indebted to out editors at William C. Brown Company: Jeffrey Hahn, who started us down the path, Lynne Meyers, who still vows to get even, and Jane Ducham and Julie Kennedy, who saw the project to completion.

We are also indebted to Mary Jo Gregory and her colleagues at York Graphic Services for their aid during production. Special mention is due John Robson, Director of the Rose-Hulman Library, for assistance with biographical research, Bruce Danner, Director of The Waters Computer Center at Rose-Hulman, for assistance converting spreadsheet files, and Phil Lockett and Marshall Wilt of Centre College for helpful critiques. We are particularly indebted to John Gilreath of Clemson University for his assistance with problem statements and solutions.

Color Key

Mechanics

Cartesian Unit Vectors

Position / Displacement / Distance

Velocity

Acceleration

Force

Linear Momentum

Gravitational Field

Torque

Angular Momentum

Angular Velocity

Work and Energy

Thermodynamics

Fluid

High Temperature

Low Temperature

Optics

Glass Tint

Wavefront

Light Ray

Mirror

Electricity and Magnetism

Area Vector

Positive Charge

Negative Charge

Metal

Electric Field

Electric Dipole Moment

Electric Potential

Polarization

Displacement Field

Dielectric

Current

Magnetic Induction Field

Magnetic Dipole Moment

Magnetization

H-Field

Poynting Vector

Impedance

General

Construction Lines / Labels / Arrows

MECHANICS

The first seventeen chapters of this text are about **classical mechanics,** which is sometimes called Newtonian mechanics and often shortened to just plain "mechanics." Classical mechanics is the study of the motion of everyday objects and includes the special case of *no* motion. Mechanics plays a fundamental role in many areas of physics and engineering. When you compute the trajectory of a rocket, determine the rotational speed of a spinning ice skater, or even design a bridge, you are "doing mechanics." Our objective is to develop a way of thinking that is sufficiently general to allow us to study the motions of macroscopic objects. The term *macroscopic object* is often used somewhat vaguely, but here we mean objects that are considerably bigger than molecules: bullets, basketballs, and planets, for example. Moreover, we will consider only objects whose speeds are much slower than the speed of light.

In a rough sense, our study of mechanics can be divided into five areas. After some fundamental definitions and mathematical preliminaries (Chapters 1–2), we will begin by studying *translational motion* (Chapters 3–10). Later (Chapters 11–13), we will find that many of the concepts of translational motion have analogs in our second area of study, *rotational motion.* We will then take up two rather special types of motion: *oscillatory* and *wave motion* (Chapters 14 and 15, respectively). Next (Chapter 16), we will examine the mechanics of *solids, liquids,* and *gases.* Finally, we will apply the mechanical concepts we have learned to *sound* (Chapter 17). These divisions are somewhat fuzzy, and the topics are not mutually exclusive. For example, in Chapter 4 we must provide a few definitions that pertain to rotations, and in Chapter 12 we will study the general motions of objects that both rotate and translate.

The study of mechanics is a great place to begin to learn about the science of physics because so many of the concepts, such as momentum and energy, play important roles in other branches of physics. A firm understanding and appreciation of the fundamental ideas of mechanics will be a valuable tool for your study of other areas, such as thermodynamics, electromagnetism, relativity, and quantum physics. And, of course, all of these disciplines are important to other areas of science and engineering.

CHAPTER

1

Physics and Measurement

In this chapter you should learn

- about the structure of physical theories and their range of applicability.
- about the *Système Internationale* and other unit systems.
- how to convert from one system of units to another.
- about significant figures.

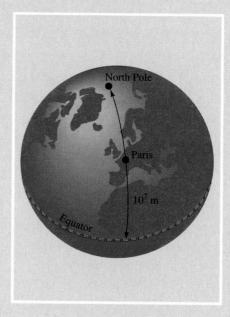

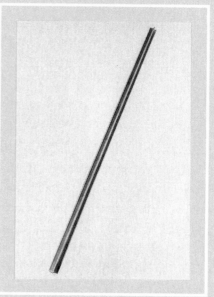

$$c \equiv 299\ 792\ 458 \text{ m/s}$$

> Physics is a game. Any number may play. The object of the game is to discover the Rules of Nature. The playing field is the entire universe. Any device, physical, conceptual, or computational may be used. Players may score points but can never win. The game is never over. Players score points by discovering a Rule of Nature. The greater the number of phenomena explained by a proposed Rule of Nature, the greater the number of points awarded. Bonus points are awarded if the proposed rule predicts previously unobserved, and especially unexpected, phenomena. An untested candidate for the status of Rule of Nature shall be called a *hypothesis*. When a hypothesis has described many phenomena, it may achieve the status of *theory*. Especially well-tested theories that explain much may, by agreement of the players, be awarded the exulted status of *law*. Any hypothesis, theory, or law may be challenged by any player at any time. All disputes will be settled by experiments, mutually agreed on by the players. The decisions of Nature, as revealed through experiments, are final.

Most of us are born accidental physicists. Through many, often unintentional and, at best, poorly planned experiments, we quickly learn what seem to be Rules of Nature. For example, one of the earliest rules we discover is that "When I drop a heavy object, it falls down."[1] Sometime later, without realizing that we are "Doing Physics," we may begin to find *relations* between *variables* when we realize "the greater the distance the object falls, the greater its speed at impact." As we grow, we develop a storehouse of these private rules, some of which may, unfortunately, be incorrect. Physics is a public collection of such rules. The rules are stated in a quantitative fashion so they can be *tested*. Only when the rules have been tested by a number of experiments are they admitted to the body of knowledge called physics. But first and foremost, Doing Physics is a human endeavor, and its primary activity is to bring order and understanding to what we observe as the fundamental rules of nature. And, yes, mistakes have been made. Rules that passed test after test for many years were later proved inaccurate when applied to extreme circumstances never dreamed of by their creators. (Such rules are seldom discarded altogether. However, their reputation is somewhat tarnished by the realization that they are only "special cases" after all.)

Physicists attempt to describe nature with as much economy as possible. One standard they like to apply to measure the success of a theory is to count the number of basic principles or assumptions; the fewer, the better. Indeed, one feature that makes physics so attractive to those who become physicists is the vast number of phenomena that can be understood in terms of a very few basic concepts, principles, and equations.

In this chapter we lay important groundwork for much of the remainder of this text. In addition to defining some commonly used terms, we will provide you with a small sketch of what we call the "big picture" so that you can better see where the physics you learn from this text fits into the larger scheme.

1.1 Physics and Models

Physics is founded on experimental observations. In its infancy a physical **theory** begins when generalizations are made from these observations; fundamental laws are proposed and predictions are made based on these propositions. More experiments are performed to test these predictions, and if they are verified, the theory is supported but never proved. If the predictions are *not* verified, the theory must be abandoned or at least modified. Over the past several hundred years theories have evolved that have become very successful in describing diverse aspects of the physical universe. These theories are based on a remarkably small number of basic principles.

---

[1] Many children explore this law in some detail, usually between the ages of 1 and 2 while seated in a high chair.

Most of the theories with which we are concerned in this text constitute **classical physics** and include the *macroscopic* mechanical, thermodynamic, electromagnetic, and optical behavior of objects in our universe. We will also briefly describe two areas that are known as **modern physics,** although we prefer the name **twentieth-century physics.** These areas include **quantum physics** and **relativistic physics.** The former describes the behavior of subatomic particles, atoms, and molecules, as well as many of the microscopic properties of solids, liquids, and gases. Relativistic physics is necessary for the description of objects with speeds comparable to that of light. The objects may be either microscopic or macroscopic, but the predictions of relativistic physics reduce to those of classical physics when the speeds of the objects under consideration are much smaller than that of light.

It is important to know a theory's range of validity. The laws of motion that we will study in the early chapters of this text are part of classical physics and are suitable for objects traveling at speeds far less than the speed of light. When objects travel with speeds close to the speed of light, the more general relations of the special theory of relativity are required. The theory of relativity is applicable to all objects. However, as we mentioned above, in the limit of much smaller speeds, relativity's predictions reduce to those of classical physics. This feature of reducibility is required of any theory that proposes to be a more general theory than the one it is designed to replace; a new theory must encompass the successes of old theories as well as explain phenomena the old one does not. Quantum theory was developed to explain phenomena the classical theories could not. When quantum theory is applied to atoms and molecules, discontinuities in various physical parameters appear that are not predicted by classical physics. In the limit that the system becomes large enough that the size of these discontinuities becomes negligible, the predictions of quantum theory approach those of classical physics.

A useful model of a softball's motion might neglect gravitational force variations, the Earth's rotation, the tidal effects of the sun and moon, and perhaps even air friction.

A **model** is an approximation of reality. For a softball lobbed from first base to the pitcher's mound, reality includes air friction, gravitational force variations, the rotation of the earth, and the tidal effects of the sun and moon. However, we can learn some important characteristics about the ball's motion from a primitive model containing *only* the most important effects, the constant downward pull of gravity for example. When we notice that the softball's true behavior is generally as we predict but not exactly, we *may* want to improve our model by including other effects, perhaps air resistance, for instance. On the other hand, for rockets that travel from the earth to the moon, our model *must* also include the variation of the gravitational force to get anything close to the correct description. Throughout this text we will stress how scientists use models. It is important for you to keep in mind what the assumptions are for a particular model so you will know the model's range of validity.

In some sense, the art of Doing Physics is contained in the construction of models. The most elegant model is one that includes *just enough* of the complications to give answers to the required accuracy for the purpose at hand. Often, the most fundamental laws are best revealed from the simplest models. Thus, in many of the early chapters, we will reduce a great deal of the beauty and diversity of nature to single particles that can be described as mathematical points. Basketballs, baseballs, and ballerinas all must follow the same basic law of gravitation. While we are trying to understand that basic underlying law, it is best not to be too concerned with the seams in the cover of the baseball or whether the ballerina is wearing toe shoes.

1.2 Systems of Units

Many concepts of physics are most compactly expressed in the language of mathematics. It is important, therefore, that the symbols used in any mathematical expression be precisely defined. Indeed, terms, such as *position, force, work,* and *energy,* that are used casually in everyday language are given formal and exact definitions in physics. Associated with each of the symbols representing physical quantities is a number. For a particular number to be useful, however, we must all agree on what it means. If we told you that we walked 3.0, you would naturally ask: "Three what?" Three meters? Three miles? Three furlongs? Or even, three seconds? It is always important to write *units* with a number.

In physics, there are a very limited number of **basic units** from which the units of all quantities in which we have an interest can be derived. In mechanics, we only need the basic units of length, mass, and time. At the Fourteenth General Conference on Weights and Measures in 1971, **standard units** for these and four other quantities were chosen. This collection of units is called the *Système Internationale* and is abbreviated **SI.** The standard units for length, mass, and time were chosen to be the meter, the kilogram, and the second, respectively. Below, one at a time, we answer the questions: What distance is a meter? How much is a kilogram? How long is a second? The SI units used in mechanics are summarized in Table 1.1.

TABLE 1.1   **SI Units Used in Mechanics**

PROPERTY	NAME	ABBREVIATION
Length	meter	m
Mass	kilogram	kg
Time	second	s

The other standard units chosen in 1971 include the mole (which we also define below), the unit of temperature called the kelvin, the unit of electric current called the ampere, and the unit of luminous intensity called the candela. The units of all other physical quantities can be expressed as a combination of these fundamental units. For mechanics, the units of physical quantities other than length, mass, and time can be expressed in terms of these three basic units. For example, velocity has the unit of length per time and in SI this unit is the meter per second (m/s). In Chapter 5 we discover that the SI unit of force is kilogram meter per second squared ($kg \cdot m/s^2$). The **density** of a substance is defined as its *mass per unit volume,* so in SI the unit of density is kilograms per cubic meter (kg/m^3).

Length

The original standard of length was established in France in 1792. At that time the **meter** was defined to be one ten-millionth of the distance you travel if you walk due South from the North Pole to the equator along a path through Paris! This definition may sound impractical to you, and indeed because it was impractical this standard was changed to be the distance between two scratches ruled on a particular platinum–iridium bar stored at the International Bureau of Weights and Measures in Sèvres, France. Other secondary standard bars were constructed by comparison with this bar and distributed throughout the world so that copies of the copies could be made and everyone who needed a standard could have one.

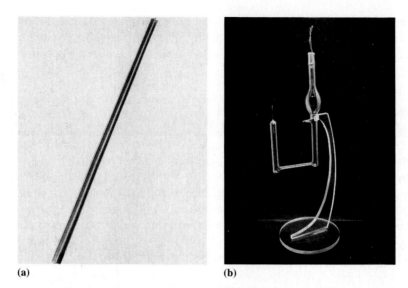

(a)   (b)

(a) The standard meter bar, made from a platinum-iridium alloy and kept at the National Institute of Standards and Technology, was the U.S. standard from 1893 until 1960. (b) A krypton 86 lamp provided light for a standard meter definition from 1960 to 1983. (Courtesy of the National Institute of Standards and Technology.)

By 1960, length measurements could be made the accuracy of which far exceeded the uncertainty inherent in the distance between the two scratches on the original platinum–iridium bar. Moreover, some uncertainty in length is always introduced in any copy of the original bar, and errors compound when copies of copies are constructed. In 1960, a length of one meter was redefined to be the distance of 1 650 763.73 wavelengths of a particular orange-red colored light (the $2p_{10} - 5d_5$ spectral line) emitted by the atoms of krypton-86 in a gas discharge tube.[2] This peculiar number was chosen so that the length of a meter came out to be close to the old standard. This new definition of length had the advantage of portability without loss of any accuracy. Although considerable technological skills are required to reproduce the standard, anyone possessing such skills can recreate the standard even if the original is lost or destroyed. Moreover, one need not travel to France!

In 1983, at the Seventeenth General Conference on Weights and Measures, the meter was redefined with even greater precision.

The Système Internationale recommends the use of a space to separate each sequence of three digits in long numbers such as 299 792 458. The custom in the United States is to use commas for this purpose, as in 299,792,458. The U.S. practice can cause confusion because much of the rest of the world uses a comma to designate the location of the decimal. We shall follow the SI convention.

In SI, the length of one meter is the distance traveled by light in a vacuum during a time interval of 1/299 792 458 s.

An equivalent statement is that the **speed of light** c in a vacuum is, *by definition,*

$$c \equiv 299\ 792\ 458\ \text{m/s}$$

The United States is almost the only country whose population does not use the meter for its length unit. The common U.S. length unit is the **foot,** which is defined as one-third of a **yard.** But whether we Americans like it or not, we are inherently tied to the metric system! The yard is related, by definition, to the standard meter:

$$1\ \text{yd} = 0.9144\ \text{m (exactly)}$$

so that

$$1\ \text{ft} = 1/3\ \text{yd} = 0.3048\ \text{m (exactly)}$$

Of course, a foot is made up of 12 in, and the definition above makes an inch equal to 0.0254 m (exactly). Table 1.2 provides some approximate lengths.

---

[2] The number 86 refers to one of the five forms, or *isotopes,* of the element krypton found in nature. The atoms of these isotopes have different atomic masses.

TABLE 1.2   Some Approximate Lengths

DISTANCE	LENGTH (m)
To the Andromeda galaxy	2×10^{22}
From the sun to nearest star	4×10^{16}
From the earth to the sun	1.5×10^{11}
Radius of the earth	6.4×10^{6}
North Pole to the equator (not just through Paris!)	1×10^{7}
Length of a football field	1×10^{2}
You are here → Height of a typical person	1.7
Thickness of a hand calculator	$\sim 1 \times 10^{-2}$
Paper thickness	$\sim 1 \times 10^{-4}$
Diameter of a hydrogen atom	1×10^{-10}
Diameter of the nucleus of an atom	$\sim 1 \times 10^{-14}$

Throughout this text we employ exponential notation to designate large and small numbers, as illustrated below. Most calculators and computers employ a similar E notation, which is also shown for comparison. (Some computers use a D notation to signify the exponent when double-precision arithmetic is employed.)

$$1000 = 10^{3} = \text{E}3$$
$$100\ 000 = 10^{5} = \text{E}5$$
$$0.001 = 10^{-3} = \text{E} - 3$$
$$0.000\ 01 = 10^{-5} = \text{E} - 5$$

Time

If we are going to use the distance light travels in a certain amount of time (*1/299 792 458 second*) to define the length standard of a meter, we had better have a good standard for time. The original definition of time was also related to the rotational motion of the earth. Prior to 1960, a **second** was defined to be $(\frac{1}{60})(\frac{1}{60})(\frac{1}{24})$ of the average time between successive high noons for the year 1900. We now know that the rotation time for the earth varies significantly; so in 1967 the second was redefined in terms of the time for cesium atoms to make a certain atomic transition.

One second is the time required for 9 192 631 770 periods of the light wave emitted by cesium-133 atoms making a particular atomic transition.

Cesium clocks are very accurate. The uncertainty gained over time by such a device is less than 1 s in 30 thousand years! Clocks based upon hydrogen atoms have been built that are even more accurate (1 s in 30 million years), but to date the cesium clock is still the

A cesium clock at the National Institute of Standards and Technology in Boulder, Colorado. (Courtesy of the National Institute of Standards and Technology.)

standard. Although expensive and somewhat intricate to build, cesium clocks (sometimes called *atomic clocks*) in principle provide a portable standard. As was the case with the length standard, we don't have to run off to Sèvres, France or Boulder, Colorado to copy a time standard. We need only build a cesium clock! Some time intervals are given in Table 1.3.

TABLE 1.3   Some Approximate Time Intervals

INTERVAL	TIME (s)
Age of the universe	5×10^{17}
Time since dinosaurs became extinct	2×10^{15}
Age of the Grand Canyon	3×10^{14}
You are here → Age of a typical college student	6.4×10^{8}
One year	3.2×10^{7}
One hour	3.6×10^{3}
Time for light to travel from the earth to the moon	1.3
Time for guitar A-string to make one vibration	2×10^{-3}
Typical time for one cycle of a wave received by a radio	1×10^{-6}
Time for light to travel across a hydrogen atom	3×10^{-19}

Mass

We have to go back to France to find a standard for mass. The SI unit of mass is the **kilogram,** a standard that was established in 1887 and one that has not been modified since that time.

> The kilogram is defined as the mass of a certain platinum–iridium cylinder located at the International Bureau of Weights and Measures in Sèvres, France.

The United States owns a copy of the standard mass, which it keeps at the National Institute for Standards and Technology in Gaithersburg, Maryland. It is about 0.039 m in both height and diameter. The cylinder is removed only once a year for the purpose of comparing other standards to it. In the past century it has been returned to France only twice for comparison with *the* standard. Some common masses are given in Table 1.4. Note the wide range of values given in this table.

The U.S. standard mass, a copy of the international standard kilogram, kept at the National Institute of Standards and Technology in Gaithersburg, Maryland. (Courtesy of the National Institute of Standards and Technology.)

TABLE **1.4**   Some Approximate Masses

	OBJECT	MASS (kg)
	Milky Way galaxy	2×10^{41}
	Earth	6×10^{24}
	Boeing 747 aircraft	2×10^{8}
	Pick-up truck	2×10^{3}
You are here →	Average human	7×10^{1}
	Basketball	0.6
	Dust particle	1×10^{-9}
	Copper atom	1×10^{-23}
	Proton	2×10^{-27}
	Electron	9×10^{-31}

There is an important additional standard unit of mass. When we are dealing with atoms, either in experiment or in theory, the kilogram is a quite inconvenient unit. Therefore, a **unified atomic mass unit,** abbreviated u, was established. By international agreement, this unit is defined based on a carbon-12 atom having a mass of exactly 12 u. In terms of the standard kilogram

$$1 \text{ u} = 1.660\ 540\ 2 \times 10^{-27} \text{ kg}$$

The masses of other atoms can be stated in terms of the unified atomic mass unit. Table 1.5 provides some examples.

TABLE **1.5**   Several Atomic Masses

ELEMENT	MASS (u)
Hydrogen-1	1.007 825
Helium-4	4.002 603
Carbon-12	12.000 000 0
Copper-64	63.929 766
Gold-197	196.966 56
Cesium-133	132.905 43
Oxygen-16	15.994 915
Nitrogen-14	14.003 074
Uranium-238	238.050 786

The **mole** is also a standard unit. One mole of a substance contains **Avogadro's number** of molecules of that substance. Avogadro's number N_A is defined so that exactly 0.012 kg of carbon-12 contains exactly one mole of carbon atoms. This number is

$$N_A = 6.022 \times 10^{23} \text{ particles}$$

One mole of any element contains precisely this same number of atoms. The mass of one mole of substance is called the **molecular mass** M. The molecular mass of carbon is, of course, exactly 0.012 kg. Each molecule of nitrogen gas N_2 has two nitrogen atoms, and, therefore, from Table 1.5 we see that the molecular mass of nitrogen gas is $2(0.014 \text{ kg}) = 2.8 \times 10^{-2}$ kg. That is, 2.8×10^{-2} kg of nitrogen gas contain N_A molecules of nitrogen.

The Metric System

Sometimes, the standards of length (the meter), mass (the kilogram), and time (the second) are too small or too large to be convenient when used to describe certain quantities. For this reason prefixes defined in powers of 10 may be associated with the standard units. For example, the prefix *milli* means 10^{-3}, so that a *milli*meter is just 10^{-3} meter (1 mm = 10^{-3} m). Similarly, because *micro* means 10^{-6}, a *micro*second means 10^{-6} s. Finally, a *kilo*meter is 10^3 m. Table 1.6 lists the prefixes defined for powers of 10. Any system of units based upon multiples of 10 is called a **metric system.** Sometimes, SI is referred to as the **mks** system (for meter-kilogram-second). However, another commonly referred to metric system is **cgs,** for centimeter-gram-second. A gram is one-thousandth of a kilogram. Finally, we occasionally refer to the **British engineering** system in which the units of length, mass, and time are the foot, slug, and second. Because SI units are used almost universally throughout the world in science and engineering, we concentrate on this system throughout this text.

TABLE 1.6   Powers of 10 Prefixes

PREFIX	POWER OF 10	EXAMPLE (m)
femto	10^{-15}	fm
pico	10^{-12}	pm
nano	10^{-9}	nm
micro	10^{-6}	μm
milli	10^{-3}	mm
centi	10^{-2}	cm
deci	10^{-1}	dm
deka	10^{1}	dam
hecto	10^{2}	hm
kilo	10^{3}	km
mega	10^{6}	Mm
giga	10^{9}	Gm
tera	10^{12}	Tm
petra	10^{15}	Pm
exa	10^{18}	Em

1.3 Unit Conversion and Dimensional Analysis

At the beginning of the last section we emphasized how important it is to always include units with a number that is used to describe a physical quantity. Sometimes it is convenient to change from one system of units to another. In this case, the physical quantity does not change, but the number associated with its units might. For example, suppose you measure the distance between your physics lecture hall and your physics laboratory and

find it to be 45.0 yd, but your instructor asks for the distance in meters. The relationship between meters and yards is given by

$$0.9144 \text{ m} = 1.00 \text{ yd}$$

Dividing both sides of this expression by 1 yd we have

$$\frac{0.9144 \text{ m}}{1.00 \text{ yd}} = 1$$

Recall the mathematical principle of identity that allows us to multiply a value by 1 and leave it unchanged. Hence, to change 45.0 yd to the equivalent distance in meters we write

$$(45.0 \text{ yd})\left(\frac{0.9144 \text{ m}}{1.00 \text{ yd}}\right) = 41.1 \text{ m}$$

Notice that we canceled the units of yards, just as we do for algebraic symbols.

Equations such as 0.9144 m = 1 yd, 1 kg = 10^3 g, 60 s = 1 min, and 660 ft = 1 furlong, can be written as the ratio of two quantities equal to 1. The ratios themselves are called **conversion factors.** These ratios make conversions between systems of units quick and efficient, so that we need only insert appropriate conversion factors in such a way that all the units cancel, except those that we desire. This technique is illustrated in Example 1.1. A list of useful conversion factors is given in Appendix 5.

EXAMPLE 1.1 *That's Moving!*

Some horse enthusiasts believe that Secretariat was the greatest thoroughbred of all time. During his 1973 Kentucky Derby win, Secretariat averaged a speed of just over 5.00 furlongs/min. What is this speed in meters per second?

SOLUTION: The required conversion factors are

$$1.00 \text{ min} = 60.0 \text{ s}$$

$$1.00 \text{ furlong} = 660. \text{ ft}$$

$$3.00 \text{ ft} = 1.00 \text{ yd}$$

$$1.00 \text{ yd} = 0.9144 \text{ m}$$

Applying these to the speed of 5.00 furlongs/min we have

$$\left(\frac{5.00 \text{ furlongs}}{1.00 \text{ min}}\right)\left(\frac{1.00 \text{ min}}{60.0 \text{ s}}\right)\left(\frac{660. \text{ ft}}{1.00 \text{ furlong}}\right)\left(\frac{1.00 \text{ yd}}{3.00 \text{ ft}}\right)\left(\frac{0.9144 \text{ m}}{1.00 \text{ yd}}\right) = 16.8 \text{ m/s}$$

◀

Secretariat in the homestretch of the 1973 Kentucky Derby.

When we use symbols to represent physical quantities in mathematical equations we often do not need to be concerned with the units of the symbols until we actually substitute numbers into an equation. However, it is sometimes useful to refer to the **dimension** of a quantity. The nature of a physical quantity is described by its dimension. For example, the distance between the goals on a soccer field is described by the dimension of length; meters, centimeters, feet, and inches, all are particular *units* with which the *dimension* length may be measured. The duration of your physics class lecture has the dimension of *time,* but may be measured in *units* of seconds, minutes, hours, or (heaven forbid) days. Physical quantities that are not fundamental are described by combinations of dimensions. For example, the dimensions of speed are *length/time* and the volume occupied by an object has dimensions (*length*)^3.

Dimensional analysis can help us check our derivations of mathematical equations. As we have already noted, when the magnitudes of two or more physical quantities are multiplied, their units should be treated in the same manner as ordinary algebraic symbols; we may cancel identical units in the numerator and denominator. The same is true for the dimensions of a physical quantity. The magnitudes of physical quantities may be added together only if they have the same dimensions. Similarly, the physical quantities represented by symbols on both sides of a mathematical equation must have the same dimensions. Hence, if we derive an equation for the length of an object, regardless of the symbols appearing in the original mathematical relation, when all the individual dimensions are simplified (all cancellations completed), the remaining dimension must be that of length. If we derive an equation for speed, the dimensions on both sides of our equation must, when simplified, be length per time. If we ever find that the dimensions of two terms that are added (or subtracted) differ, we have made a mistake somewhere in our derivation. Similarly, if we find that the dimensions of the left side of an equation differ from those of the right side, we can be sure the equation is wrong! Regrettably, the converse is not true. Dimensional consistency does not guarantee correct equations. In addition, the arguments of the special functions, such as the sine, cosine, tangent, logarithm, and exponent, must be dimensionless.

EXAMPLE 1.2　*They Never Went into This Dimension on "Star Trek."*

Under certain circumstances an equation for the distance x a particular object travels after time t is given by the equation

$$x = v_o t + \tfrac{1}{2}at^2$$

where the dimensions of the symbols are as follows:

SYMBOL	DIMENSION
x	length
v_o	length/time
t	time
a	length/time^2
$\tfrac{1}{2}$	a pure number, no dimensions

Verify that these dimensions are consistent.

SOLUTION:　We substitute these dimensions into the equation:

$$\text{length} \overset{?}{=} \left(\frac{\text{length}}{\text{time}}\right)(\text{time}) + \left(\frac{\text{length}}{\text{time}^2}\right)(\text{time})^2$$

$$\text{length} = \text{length} + \text{length}$$

Each term on the right-hand side of this "equation" has the same dimension; namely, length. Moreover, the left-hand side also has this dimension. All is well! ◀

As a final note regarding SI units, notice that the shorter designations for units such as kg, m, and s are **symbols,** not abbreviations; thus they are *not* followed by a period. The unit names are never capitalized. The unit symbols are capitalized only if the unit is named after a person. Thus, the symbol for the unit ampere is A, and the symbol for the unit joule is J. The single exception is L, which is the symbol for the unit liter. This exception is made to avoid confusion of the lowercase letter l with the Arabic numeral 1.

1.4 Significant Figures

Most calculators can provide their users with ten or more digits. We have noticed that some students have a tendency to write answers that include all these digits, but a number written in this manner is misleading. Usually, the number is not really known to the accuracy implied by all ten digits. For example, if you measure the length of a racket-ball court as $L = 12.2$ m, there are three significant digits in this measurement. The digit to the right of the decimal place is the least well known, and to say that the court is 12.2 m long means that the true length is between 12.15 m and 12.25 m. More succinctly, we could write $L = 12.20 \pm 0.05$ m. Suppose your friend finds that the width of the court is $W = 6.1$ m. There are two significant digits in this result and the uncertainty in its value is ±0.05 m. We can compute the area of the court by multiplying its length by its width. If you multiply 12.2 m times 6.1 m, your calculator gives you $LW = 74.42$ m^2. This answer, however, is misleading since it implies a precision we cannot in fact claim. Because the numerical value of the width (6.1 m) has only *two* significant digits, we must write that the area of the court is 74 m^2.

Worst-Case Analysis of Uncertainty

Consider the multiplication of $L = 12.20 \pm 0.05$ m by $W = 6.10 \pm 0.05$ m. To find the *extreme* limits of the uncertainty, we first multiply the two lowest values of L and W to obtain

$$L_{min}W_{min} = (12.15 \text{ m})(6.05 \text{ m}) = 73.5075 \text{ m}^2$$

If L and W both have their maximum values, we have

$$L_{max}W_{max} = (12.25 \text{ m})(6.15 \text{ m}) = 75.3375 \text{ m}^2$$

(continued on next page)

We can now see the folly of quoting the product of the experimentally measured L and W values as 74.42 m². The truth of the matter is that we're not even completely sure of the 4 in 74. To suggest that the final "2" is experimentally significant is stretching the truth much too far.

A More Optimistic Analysis

An alternative rule that is often applied to determine the uncertainty of the product of two (or more) experimental values is founded upon probability. This rule is based on the assumption that not even physicists are so unlucky as to have uncertainties combine to produce the worst case every time. If, instead, we assume that uncertainties combine randomly, the following rule results: When two or more experimentally obtained numbers are multiplied, the percentage uncertainty of the answer is equal to the square root of the sum of the squares of the percentage uncertainties of the original numbers.

In the preceding example for instance, we have

$$L = 12.20 \pm 0.05 \text{ m} = 12.20 \text{ m} \pm 0.4\%$$

and

$$W = 6.10 \pm 0.05 \text{ m} = 6.10 \pm 0.8\%$$

Following the square-root-of-the-sum-of-the-squares rule we find

$$LW = 74.42 \pm \sqrt{(0.4\%)^2 + (0.8\%)^2} = 74.42 \pm 0.9\% = 74.42 \pm 0.7$$

This result leads us to quote the answer as

$$LW = 74.4 \pm 0.7 \text{ m}^2$$

This approach justifies the approximate rule of thumb for significant figures as given in the following paragraphs.

There is a popular saying regarding computer computations: "Garbage in. Garbage out." We might paraphrase this quotation for the results of calculations involving experimentally measured quantities: "Uncertainty in. Uncertainty out." Physicists and mathematicians have developed an extensive set of rules to predict how uncertainties in the measurement of several input variables propagate through a calculation to the uncertainty in the answer. You may study these rules in the laboratory portion of this course. However, for purposes of this text we employ a popular approximate rule that serves pretty well to ensure that we do not claim greater precision in our answers than they really deserve. Here is the approximate rule of thumb:

When multiplying or dividing two or more values, the number of significant digits in the answer can be no greater than the least number of significant digits in any of the values multiplied or divided.

We must be careful if a number has leading zeros (such as 0.0000023 kg, which has only *two* significant figures) or trailing zeros (such as 1500 m). For example, a race track may be $L = 1500$ m in length. How many significant figures are there in this number? We can only be sure of two, but we suspect the race officials would not be satisfied if L were as long as 1540 m or as short as 1450 m; when rounded to two significant digits, these numbers are both 1500 m. If we write the distance as $L = 1500.$ m, the decimal point implies that we have measured L to the nearest half-meter and the statement of its length with its uncertainty can be written $L = 1500.0 \pm 0.5$ m. Because of the possibility of ambiguities, it is common in science to use **scientific notation** to write values that have leading or trailing zeros. In this case, when we write $L = 1.500 \times 10^3$ m, it is clear that there are four significant figures. On the other hand, a mass $M = 2.3 \times 10^{-6}$ kg has only two significant figures.

Note that exact integers, such as 2, are assumed to have no uncertainty. Thus, if we know some physical quantity x to four significant figures, we know $2x$ to four (or perhaps five) significant figures, not just one significant figure. For example, if $x = 567.3$, then $2x = 1134.6$ and all five digits are significant. This result follows because $2x$ is equivalent to $x + x$, and when physical quantities are added (or subtracted), the uncertainties combine in a different fashion as we show next.

When we add or subtract values we must pay careful attention to the *number of decimal places* rather than the number of *significant figures*. The "number of decimal places" means the number of *significant* digits to the right of the decimal when all numbers to be added (or subtracted) are written in exponential notation *with the same power of 10 multiplier.*

> When adding or subtracting two or more numbers, the number of decimal places in the answer is equal to the smallest number of decimal places in any of the values added or subtracted.

For example, when we perform the sum $S = 54.45 \times 10^3$ kg $+ 16.3 \times 10^3$ kg $- 8.63 \times 10^3$ kg, our calculator yields 62.12×10^3 kg. However, the answer must be written 62.1×10^3 kg because the second term in the sum S has only one place to the right of the decimal. When we subtract 1.0082 u from 1.0087 u, we obtain 0.0005 u, a result with four decimal places but only one significant figure. There are times when retaining the proper number of decimal places can lead to an increase in the number of significant figures. Suppose we add the numbers 1.0073 u and 0.0005 u to obtain 1.0078 u. The answer has five significant digits whereas one of the original terms (0.0005 u) has only one significant digit.

In the examples given throughout this text we use three significant figures. Our purpose in example problems is to illustrate physical concepts and problem-solving strategies and to use a fewer number of significant figures may, on occasion, mask a possible computational error. This approach occasionally leads us to pose problems in which data with three significant figures seem a bit improbable. We hope you forgive this small bit of impracticality in the interest of clarity. Finally, for sequential calculations, we carry all the digits in our calculator even when we quote an intermediate result to fewer significant figures. In other words, if we have $c = ab$ and $d = cf$, we calculate d from $(ab)f$, not from a rounded value of c times f.

(margin note) $1500 \rightarrow 1500 \pm 50$ has only two significant figures and is written 1.5×10^3

(margin note) $1500. \rightarrow 1500.0 \pm 0.5$ has four significant figures and is written 1.500×10^3

1.5 Summary

The basic units employed in the study of mechanics define **standards** for length, mass, time, and the quantity of matter. In SI, these units are the **meter, kilogram, second,** and **mole.** The number of atoms or molecules in one mole of a substance is called **Avogadro's number** N_A and its value is 6.022×10^{23}.

When physics is applied to a particular phenomenon it is often possible to neglect certain aspects of that phenomenon. In such cases the resulting application is called a **model** for that phenomenon. It is important to know the range of validity of a model.

The **dimensions** of length, mass, time, and combinations of these dimensions describe the nature of physical quantities. We may use dimensional analysis to check our derivation of an equation.

PROBLEMS

1.2 Systems of Units

1. Apply the proper prefix from Table 1.6 to rename the following quantities: (a) 10^{-6} phones, (b) 10^6 phones, (c) 2×10^3 mocking-birds, (d) 10^{-12} boos, (e) 10^{12} bulls.
2. Through what distance does light travel in 1.00 ns?
3. A pulse of light from a Q-switched ruby laser lasts 30.0 ns. How far in space does the pulse extend from beginning to end?
4. From the data in Table 1.5 compute the mass in kilograms for an atom of (a) hydrogen, (b) carbon, and (c) gold.
5. The mass of a certain penny is 3.00×10^{-3} kg. How many copper atoms would be in this penny if it were made of pure copper?
6. Estimate the number of molecules in 1.00 m^3 of air by assuming that it has a density of 1.21 kg/m^3 and that it is entirely composed of nitrogen molecules.

1.3 Unit Conversion and Dimensional Analysis

7. The mean radius of the earth is 6370 km. (a) What is the area of the earth's surface in square meters? (b) What is the area of the earth's surface in square centimeters? (c) What is the area of the earth's surface in square yards?
8. A *cord* of wood occupies a space 8.0 ft long, 4.0 ft wide, and 4.0 ft high. What is the volume of a cord in cubic meters?
9. The area of the paper on a roll commonly used to print newspapers is 1.00×10^4 yd^2. (a) Express this area in square meters. (b) Express this area in square centimeters. (c) Express this area in square miles.
10. One *acre* is 43 560 ft^2. What is this area in square meters?
11. A *section* is a 1.00 mile by 1.00 mile area. What is the area of one section in square kilometers?
12. A **light-year** is the *distance* traveled by light in one year. (a) How many kilometers are in a light-year? (b) How many miles are in a light-year?
13. The density of water at 4°C is 1.00 g/cm^3. Convert this density to kilograms per cubic meter.
14. Show that the dimensions of the equation $v^2 = v_o^2 + 2ax$ are consistent if the dimensions of x, v, and a are length, length per time, and length per time squared, respectively.
15. Determine the units of the constants α and β in each of the following equations if x is in meters, t is in seconds, and v is in meters per second:
    (a) $x = \alpha t^2 + \beta t^3$ (b) $v = \alpha + \beta t$
    (c) $x = \alpha \cos(\beta x)$ (d) $x = \alpha e^{-\beta t^2}$
    (e) $x = \alpha \sin(\beta t)$ (f) $v = \alpha \ln(\beta t^2)$
16. Verify that if the constants α and β in the equation $x = \alpha t + \beta t^4$ have the correct units, then α and β also have the correct units in the expression for dx/dt, the derivative of the function x with respect to the variable t if x is in meters, t is in seconds, and v is in meters per second.
17. Verify that if the constants α and β in the equation $v = \alpha \cos(\beta t)$ have the correct units, then α and β also have the correct units in the expression for dv/dt, the derivative of v with respect to the variable t if x is in meters, t is in seconds, and v is in meters per second.

1.4 Significant Figures

18. State the number of significant figures in each of the following measured quantities: (a) 5.634 m, (b) 5600 ft, (c) 6.02×10^{23} molecules, (d) 1.20×10^{-1} mol.
19. State the number of significant figures in each of the following measured quantities: (a) 9.700×10^4 cm, (b) 5.000 02 in, (c) 1.01×10^5 $kg/(m \cdot s^2)$, (d) 2.240×10^{-2} m^3.
20. State the number of significant figures in each of the following measurements of a mass m: (a) 120 g, (b) 120. g, (c) 0.12 kg, (d) 120 000 mg, (e) 120.00 g, (f) 1.2×10^2 g, (g) 1.20×10^2 g.
21. State the number of significant figures in each of the following measurements of a distance d:
    (a) 0.0044 km,   (b) 4.4 m,
    (c) 440.0 cm,   (d) 4400 mm,
    (e) 4.40×10^6 μm,   (f) 4 400 000 μm.
22. (a) Compute the circumference of a circle that is 2.56 cm in radius. (b) Compute the radius of a circle that is 2.560 cm in circumference.
23. The speed of light in a vacuum is 2.998×10^8 m/s. Determine the speed of light in miles per hour, if 5280. ft = 1.000 mile.
24. The length and width of a dormitory room were measured to be 3.10 ± 0.03 m and 3.50 ± 0.03 m, respectively. Find the area of the room and estimate its uncertainty.
25. A fraternity purchased a cylindrical hot tub with a diameter of 2.44 ± 0.01 m and a depth of 1.22 ± 0.05 m. (a) Compute the volume of water required to fill the hot tub. (b) Estimate the uncertainty in this volume.
26. A prankster fills a balloon with water. If the balloon may be modeled as a sphere with a diameter $d = 12.40 \pm 0.05$ cm, (a) compute the volume of the balloon, and (b) estimate the uncertainty in the balloon's volume.

General Problems

27. Compute the number of water molecules in 1.00 L of water (1.00 L $= 10^{-3}$ m^3). How many moles of water molecules (H_2O) are there in a liter of water? How many hydrogen atoms are in this volume?
28. Assume that all the air in a 3.00-m by 15.0-m by 20.0-m classroom is made up of gaseous nitrogen of density 1.21 kg/m^3. If this nitrogen is condensed to a liquid of density 804 kg/m^3, what volume does the nitrogen occupy?
29. The label on a gallon container of a particular brand of paint makes the claim that its contents can cover 420. ft^2. (a) Compute its coverage in square meters per liter (1 L $= 10^{-3}$ m^3). (b) How thick is the paint layer?
30. One *astronomical unit* AU is the distance between the earth and the sun. A *parsec* pc is equal to 206 265 AU. (a) Express the distance between the earth and the sun in parsecs. (b) If the distance between the earth and the sun is 9.29×10^7 miles, find the conversion factor between parsecs and miles. (c) What is the conversion factor between astronomical units and kilometers?
31. Determine the values of m and n in each of the following equations if x, t, and v have the units of centimeters, seconds, and centimeters per second, respectively:
    (a) $x = (1.00$ cm$)t^m$ (b) $v^2 = (2.00$ cm$)x^m t^n$
    (c) $v = (3.50$ $m^{-2})x^m t^n$ (d) $x = (9.80$ $s^{-1})vt^m$

32. Compute the following, assuming each number represents a measured physical quantity the uncertainty of which is correctly represented by the number of significant figures given
(a) $321.3 + 12.22 + 9.09$ (b) $18 - 9.2 + 36.55$
(c) $(132.5)(27)$ (d) $\pi(7.56 \times 10^4)$
(e) $(1.434 \times 10^3) +$ (f) $(6.38 \times 10^2)/(0.218)$
(9.56×10^{-4})

33. Compute the following, assuming each number represents a measured physical quantity the uncertainty of which is correctly represented by the number of significant figures given:
(a) $(4.53 \times 10^{-5}) + (2.93 \times 10^{-6})$
(b) $\pi(3.87 \times 10^2)^2$

(c) $(9.38 \times 10^4) + (1.046 \times 10^3)$
(d) $12.4 + 0.565 - 6.22$
(e) $123 + 12.3 + 1.23$
(f) $5(142)$

34. A simple pendulum consists of a spherical ball suspended from the ceiling by a length of string. A student needs to know the distance L from the upper end of the string to the center of the ball. The student measures the string's length with a meter stick and finds it to be 76.2 cm. With a micrometer the student measures the diameter of the ball to be 3.452 cm. What distance should the student quote for the length L?

CHAPTER

2

Vector Algebra

In this chapter you should learn

- how to designate points in space with Cartesian and polar coordinates.

- the definitions of a scalar and a vector.

- how to apply the rules for vector addition both graphically and analytically.

- how to represent vectors with the unit-vector notation.

- how to compute the scalar and vector products of two vectors.

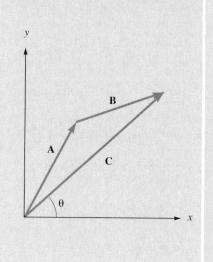

$$C_x = A_x + B_x$$
$$C_y = A_y + B_y$$
$$C = \sqrt{C_x^2 + C_y^2}$$
$$\theta = \arctan\left(\frac{C_y}{C_x}\right)$$

Vectors play a central role in physics. In this chapter we review the definition of a vector, describe the rules for vector algebra, and provide some examples to illustrate the basic vector operations needed in this text. Some of the material in this chapter is used almost immediately (Chapter 4), whereas that presented in Sections 2 through 4 will not be needed until Chapter 7. Your instructor may choose to cover all the material at the beginning of the course or leave some of it until later. Whatever the approach, ultimately you need to master the material in this chapter. You may find you already know much of it. If so, fine. Nevertheless, we urge you to read the material carefully—first as a useful review but also to become accustomed to the notation used throughout this text.

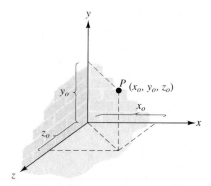

FIGURE 2.1

The three-dimensional Cartesian coordinate system.

2.1 Coordinate Systems

Much of physics is about motion. In order to describe movement, we must be able to specify where an object has been, where it is, and where it's going; we need a **coordinate system.** One of the most convenient coordinate systems is the **Cartesian coordinate system,** which consists of three mutually perpendicular lines, called **axes.** These axes, designated as x-axis, y-axis, and z-axis, intersect at a point, called the **origin** where the real number line of each axis has its zero. The location of any point in the three-dimensional space in which we live can be designated by three numbers known as **coordinates.** As shown in Figure 2.1, these coordinates are distances measured along each of the axes from the origin. These coordinates are usually written as an ordered triple (x,y,z). The x-coordinate of point P can be found by passing a plane parallel to the yz-plane through point P. The point at which this plane intersects the x-axis is the x-coordinate. The y- and z-coordinates can be found in an analogous fashion.

For many of our applications, quantities of interest are confined to a plane. For these cases we require a Cartesian coordinate system with only two axes, such as that shown in Figure 2.2. Positions with positive x-coordinates are located to the right of the y-axis, and negative x-values are located to the left of the y-axis. Positions designated with positive y-values lie above the x-axis, and negative values of y are below the x-axis. If you like, think of the positive z-axis as in Figure 2.2 as directed out of the page. In this case, points with positive values of z are located above the page, and negative z-values lie below it. By the way, sometimes Cartesian coordinates are called **rectangular coordinates.**

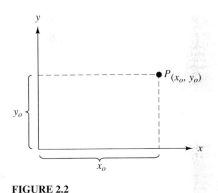

FIGURE 2.2

A two-dimensional Cartesian coordinate system.

We often refer to a **frame of reference.** A frame of reference is just another name for the particular coordinate system with respect to which we are making measurements. For example, often we refer to the ground as our frame of reference. What we mean by this is that the origin of our coordinate system is fixed at some specified place on the ground. It may be convenient to specify other frames of reference as attached to a table top or on a moving object, such as a car, train, or plane. If we want to discuss the motion of the earth, it may be advantageous to establish a frame of reference by imagining a coordinate system attached to the sun.

We occasionally find **polar coordinates** useful. In a polar coordinate system spatial positions in a plane are designated by a length r from the origin, and an angle θ usually measured from the positive x-axis. The polar coordinates (r, θ) for the point designated by the Cartesian coordinates (x, y) are shown in Figure 2.3. From simple trigonometry we see that the relationships between the polar coordinates and the Cartesian coordinates are

$$x = r\cos(\theta) \quad \text{and} \quad y = r\sin(\theta) \tag{2.1}$$

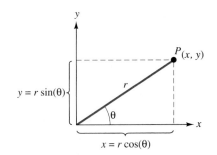

FIGURE 2.3

Polar coordinates r and θ.

Hence, if we are given the polar coordinates (r, θ) and we wish to find the Cartesian coordinates (x, y), we need only employ Equation (2.1). On the other hand, if we are given x and y, we can find r and θ. From Figure 2.3 you should verify that

$$r = \sqrt{x^2 + y^2} \qquad \text{and} \qquad \theta = \arctan\left(\frac{y}{x}\right)$$

(2.2)

Be careful to apply Equation (2.1) only when the angle θ is defined from the positive x-axis. If θ is defined as an angle from any other axis, you must examine the figure in order to determine the relationship between the polar coordinates (r, θ) and the Cartesian coordinates (x, y). Additional care is required when calculating θ using the arctangent of Equation (2.2). By drawing two different r-lines from the origin to points with the (x, y) coordinates $(-1, 2)$ and $(1, -2)$, you should verify that the arctan $(-1/2) \neq$ arctan $(1/-2)$. For this reason you need to be wary of an angle that is provided by the arctangent function of your calculator. We illustrate the reason for this caution in Example 2.1. Don't skip it!

EXAMPLE 2.1   *Floating Out of the First and Fourth Quadrant*

An aviator's most prized balloon takes off prematurely. While clinging to a rope the other end of which is held by a friend on the ground, the balloonist rises to an altitude of 12.0 m above his friend. It's high noon; so at this particular latitude the sun is directly overhead. As illustrated in Figure 2.4(a), the distance between the shadows of the balloon and the person on the ground is 9.00 m. For the coordinate system shown, find the polar coordinates of the balloon end of the rope.

SOLUTION   The x- and y-coordinates of the balloon end of the rope are

$$x = -9.00 \text{ m}$$

$$y = +12.0 \text{ m}$$

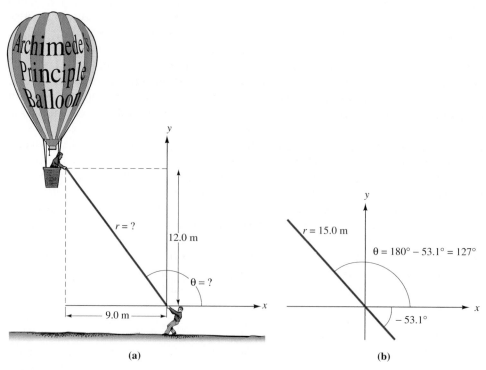

(a)   (b)

FIGURE 2.4

(a) From the x and y values, we compute the polar coordinates r and θ. (b) We must add 180° to the angle provided by the calculator to obtain the correct angle θ.

Hence,

$$r = \sqrt{x^2 + y^2} = \sqrt{(-9.00 \text{ m})^2 + (12.0 \text{ m})^2} = 15.0 \text{ m}$$

We compute the angle from

$$\theta = \arctan\left(\frac{y}{x}\right) = \arctan\left(\frac{12.0 \text{ m}}{-9.00 \text{ m}}\right)$$

If you take the inverse tangent with your calculator, you will find θ to be $-53.1°$. From Figure 2.4, however, we expect θ to be greater than 90°. The problem is not with our mathematics, but rather with your calculator (and ours too, for that matter). The angles provided from the inverse tangent function by most calculators are given for the first and fourth quadrants only. Hence, if our position actually lies in the second or third quadrant we must always be careful to convert the calculator angle to the proper value. As shown in Figure 2.4(b), the actual polar coordinate angle is

$$\theta = 180.0° - 53.1° = 126.9°$$

Usually, such angle conversions are not difficult if we pay attention to our coordinate system. You should investigate your calculator and determine the quadrants for which the inverse sine and inverse cosine functions are given.   ◀

2.2 Scalars and Vectors

Often in physics and engineering, a single number and its units are not sufficient to provide a complete description of a physical quantity. For example, if you walk three kilometers East, your position is much different than if you had walked three kilometers West. The change in position of an object is called **displacement,** and we shall have much more to say about this quantity in the next chapter. A displacement is an example of a **vector.** A vector is a quantity that requires both a **magnitude** and a **direction** for a complete description. Velocity is another example of a vector. Traveling 5.00 m/s North is different from traveling 5.00 m/s South. Many quantities in physics, including force, acceleration, and momentum, require both a magnitude and a direction for a complete description.

In diagrams, we designate a vector by a *directed line segment*. By choosing some suitable scale, the length of the line is made to represent the magnitude of the vector. This length may represent a displacement, a velocity, or any other vector quantity. Therefore, the units associated with the directed line segment are not necessarily length. An arrow head is placed at the end of the line to indicate the vector's direction. Figure 2.5 shows *one* vector located at three different positions. Since each arrow has the same magnitude and direction, the three arrows are three symbols for the same vector. It's rather like the symbol "2" for the integer two. If we print the symbol 2   2   and   2 at different places on this page and then ask, "How many numbers are there in this sentence?" the answer is, "Only one number; namely, two." Of course, the numeral "2," a symbol for the number two, appeared three times. The point of all this is that the *location* of a vector is not part of its definition. All that counts is the vector's magnitude and direction.

Some quantities have no direction and require only a single number for a complete description. Such quantities are called **scalars.** Mass, volume, density, pressure, and temperature are all examples of scalars. The mathematics of scalar quantities is the ordinary algebra with which you are very familiar. However, when dealing mathematically with

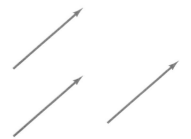

FIGURE 2.5

Three representations of the same vector. The vector is represented by directed line segments. The lengths of the lines are the same, indicating the same magnitude. The directions of the vectors are also the same.

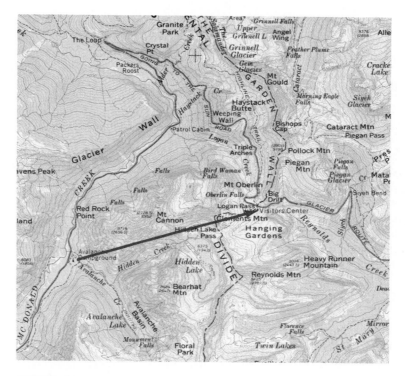

FIGURE 2.6

The displacement vector from the Avalanche Campground to the Logan Pass Visitors' Center is designated on this topographical map. The displacement vector tells us nothing about the actual path, which is the road indicated by the dark curve.

vectors we must conform to the rules of **vector algebra.**[1] These rules are significantly different from those of ordinary algebra. We will spend the remainder of this chapter illustrating vector algebra.

We want to give you a little more information about the displacement vector since it is convenient to use it in some examples below. The magnitude of this vector is determined by the straight-line distance from the original position of an object to its final position. The direction of the displacement vector, as indicated by the arrow head, points *from* the original location (we'll call it the *tail* of the vector) *to* the final location (we'll call it the *tip*). For example, Figure 2.6 shows the displacement vector from the Avalanche Campground to the Logan Pass Visitors' Center in Glacier National Park. If you drive your car from the campground to the visitors' center, you must follow the road indicated by the dark line. However, after the trip, your displacement vector is represented by the directed line segment shown in the figure. The displacement vector indicates nothing about the path, and depends only upon the locations of the original and final positions.

When we write about vector quantities it is convenient to use a symbol that is distinctly different from symbols we might use for a scalar. A long-standing custom in

---

[1] The actual *definition* of a vector describes the properties of its components under transformations of the coordinate system. Here we list *properties* that follow from the formal definition. For a more complete discussion see Banish Hoffman, *About Vectors* (Englewood Cliffs, N.J., Prentice-Hall, 1966).

physics and engineering lecture halls is to draw a small arrow over the symbol that represents a vector. Thus, $\vec{s}$ might be written on the chalkboard or overhead projector to represent a displacement vector, or the symbol $\vec{v}$ might be used to designate a velocity vector. It is important to realize that when someone writes $\vec{s}$ or $\vec{v}$, we really don't know anything about $\vec{s}$ or $\vec{v}$, other than the fact that they are vectors. If we want to know their values, we need two pieces of information about each: a magnitude and a direction. Another custom is to use the vector's symbol without the arrow to designate the magnitude only of the vector. Hence, s might represent the magnitude of the displacement $\vec{s}$ and it might have units, such as meters, kilometers, or feet. Similarly, v may stand for the magnitude of the velocity vector $\vec{v}$, appropriate units being m/s, ft/s, mi/h, or some other combination of length divided by time. Occasionally, for emphasis we use the absolute value to indicate the magnitude of the vector. For example, $s = |\vec{s}|$ and $v = |\vec{v}|$. In each of these cases, knowledge of the magnitude supplies us with only part of the information about each vector. We can't emphasize enough that to fully describe a vector you must also provide a direction.

There is another long-standing custom involving vectors. In textbooks, rather than using an arrow over the symbol $\vec{s}$, a bold character is employed. Thus, the displacement vector is written **s,** and the velocity vector is given by **v.** We follow this convention throughout this text. The magnitude of a vector is still designated by the nonbold symbol, or the absolute value notation.

2.3 Vector Addition

It is often necessary to add two or more vectors. For example, we may want to know the total displacement of an object that has undergone two successive displacements. The operation performed when the word "add" is applied to vectors is more involved than its counterpart in ordinary algebra. When there is both a magnitude and a direction associated with an addition process, special rules apply. Many of the vector additions we need to perform can be carried out in two dimensions. In this case, one way to add the vectors is to simply draw them. Suppose we want to add two vectors represented by **A** and **B.** We begin by selecting a suitable scale so that we can fit both vectors on our drawing. This scale must be the same for each vector. For example, 1.00 cm on our diagram may actually represent 2.00 km of displacement. While paying careful attention to its orientation, we draw a directed line segment to represent the vector **A.** Next, we draw a directed line segment representing vector **B,** but in such a way so that its tail is placed at the tip of

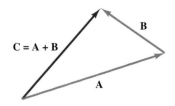

FIGURE 2.7

The vector **C** is the resultant from adding vector **B** to vector **A**. Vector **C** runs from the tail of vector **A** to the tip of vector **B**.

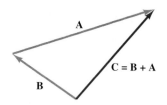

FIGURE 2.8

When vector **A** is added to vector **B**, the resultant **C** is exactly the same as for the process shown in Figure 2.7, in which vector **B** is added to vector **A**.

vector **A**. As illustrated in Figure 2.7, the **resultant vector C** is represented by the directed line segment from the tail of vector **A** to the tip of vector **B**. That is, since **C** is to be equivalent to **A** + **B**, it must have the same start and the same finish as **A** + **B** when they are linked tip to tail. In equation form we write

$$\mathbf{C} = \mathbf{A} + \mathbf{B} \tag{2.3}$$

Equation (2.3) says that vector **C** is the sum of the vectors **A** and **B**. Above, we have used the displacement as an example of a vector, but Equation (2.3), as well as all the rules given below, are applicable to any collection of vectors, be they displacements, velocities, forces, or other vector quantities.

As is the case of ordinary algebra, it makes sense to add vectors only when they have the same units. It is very difficult to figure out what you have if you add a displacement to a velocity; whatever it is, it's of no value to us and really doesn't make any sense! So, be just as careful with the units of a vector as you are with those of a scalar.

Vector addition is **commutative.** That is, the order of the addition process does not matter.

$$\mathbf{A} + \mathbf{B} = \mathbf{B} + \mathbf{A} \tag{2.4}$$

In Figure 2.7 we found the resultant vector **C** by first drawing vector **A** and then adding vector **B** to it. In Figure 2.8 we show the process of adding vector **A** to vector **B**. The resultant vector **C** has the same magnitude and direction as that shown in Figure 2.7. The order of addition has been reversed, but the answer is the same.

Vector addition is **associative.** Hence, if we add three vectors **A, B,** and **C,** we are free to add **A** to **B,** and then add the vector **C** to the resultant (**A** + **B**) as in Figures 2.9(a) and (b). Or, we can begin by adding the vectors **B** and **C,** and then add the vector **A** to the resultant (**B** + **C**) as in Figures 2.9(c) and (d). That is

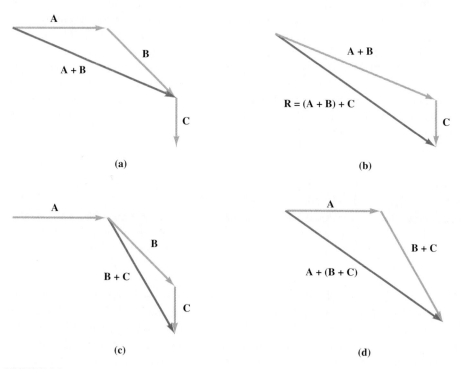

FIGURE 2.9

The vectors **A** and **B** and **C** can be added in any order. In (a) and (b) we add **C** to (**A** + **B**), and in (c) and (d) we add **A** to (**B** + **C**). The resultant **R** is the same for each case.

$$(A + B) + C = A + (B + C) \qquad (2.5)$$

We may also want to subtract one vector from another. To subtract a vector means to add the vector's additive inverse. The additive inverse of a vector (denoted an opposite sign $-$) is defined to be equal in magnitude to the original vector but antiparallel in direction:

$$A - B \equiv A + (-B) \qquad (2.6)$$

The graphical process of subtracting two vectors is illustrated in Figure 2.10.

Multiplication by a Scalar

When dealing with scalars we take the meaning of $3a$ to be $(a + a + a)$. In exactly the same way we take $3A = A + A + A$. Thus, the result of multiplying a vector by a scalar follows immediately from the rules for addition of vectors.

When we multiply a vector A by a scalar s, then $|sA| = |s| \cdot |A|$, whereas the direction of the resultant vector depends upon the sign of s. If s is positive, the direction of sA is the same as that of the vector A. If s is negative, then sA points in the opposite direction from A. Dividing a vector by a scalar is equivalent to multiplying the vector by the reciprocal of the scalar. The same rules just stated apply for the direction of a vector formed from division by a scalar.

You should always be careful with the units when you multiply a vector by a scalar. Very often the scalar has units, and these units are usually different from those of the vector. Never forget to perform any necessary multiplication of units.

Using the rules for addition, subtraction, and multiplication of vectors by a scalar, we are now able to construct **linear combinations of vectors** such as $2A - 3B + C/2$. Constructing linear combinations of vectors graphically provides a pictorial approach that is useful for visualizing the process. However, in order to find the magnitude and direction of the resultant, we must lay a ruler and a protractor down on the figure and actually measure the length and angle of the resultant. Then, using our chosen scale, we must convert the magnitude back to its proper value (and units). Needless to say, this process is not only cumbersome but of limited accuracy. A more convenient analytical technique used to add vectors is based on our ability to resolve vectors into components and use what are called unit vectors.

Unit Vectors

Unit vectors are a convenient shorthand used to specify a particular direction in space. They have a magnitude of unity and no physical units. Unit vectors are indicated by placing a "hat," or circumflex, over a symbol. Thus, a unit vector in the direction of some particular velocity is written $\hat{v}$. In Cartesian coordinate systems it is convenient to define three special unit vectors whose directions are parallel to the coordinate axes. These unit vectors are labeled $\hat{i}, \hat{j}$, and $\hat{k}$, and are directed along the x-, y-, and z-axes, respectively.[2] These three unit vectors are shown in Figure 2.11.

The most significant property of the set of unit vectors we have described is that any vector in our three-dimensional space can be written as a linear combination of these unit vectors. In general we can describe any vector A through a linear combination of unit vectors

$$A = A_x\hat{i} + A_y\hat{j} + A_z\hat{k} \qquad (2.7)$$

[2] Verbally, we refer to these unit vectors as "i-hat, j-hat, and k-hat."

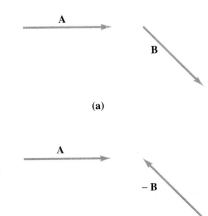

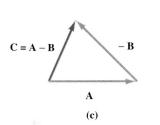

FIGURE 2.10

Subtracting B from A. (a) Vectors A and B. (b) Vectors A and $-B$. (c) The vector addition of A and $-B$.

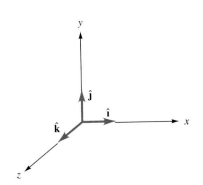

FIGURE 2.11

The unit vectors $\hat{i}, \hat{j}$, and $\hat{k}$ for Cartesian coordinates. These vectors have unit magnitude, no dimensions, and are directed along the positive x, y, and z axes, respectively.

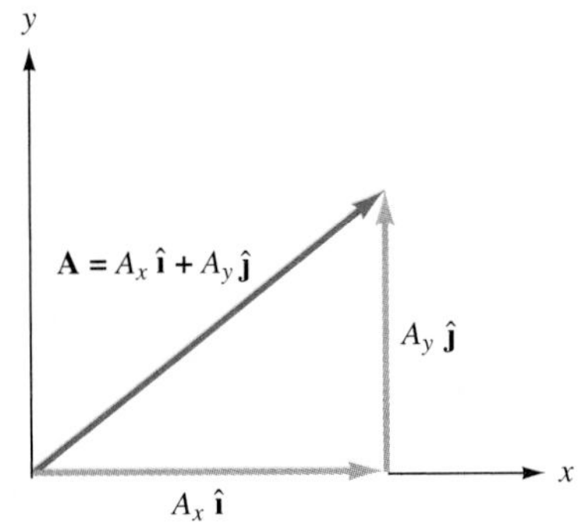

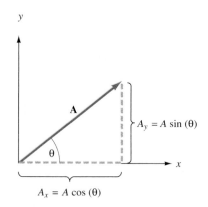

FIGURE 2.12

The vector $\mathbf{A} = A_x\hat{\mathbf{i}} + A_y\hat{\mathbf{j}}$ is the vector sum of the component vectors $A_x\hat{\mathbf{i}}$ and $A_y\hat{\mathbf{j}}$.

The numbers A_x, A_y, and A_z are the **Cartesian components** of vector **A.** You need to distinguish three closely related terms: The vector $A_x\hat{\mathbf{i}}$ is known as a **component vector,** whereas A_x is a **component,** and the *magnitude* of the component vector is given by the absolute value of the component A_x (written $|A_x|$). When A_x is positive, $A_x\hat{\mathbf{i}}$ is directed along the positive x-axis. However, if A_x is negative, then $A_x\hat{\mathbf{i}}$ points along the negative x-axis. Similar statements hold for $A_y\hat{\mathbf{j}}$ and $A_z\hat{\mathbf{k}}.$ As stated by Equation (2.7) and illustrated in Figure 2.12, the vector **A** is simply the vector sum of its component vectors.

Vector Components

In order to write a vector in terms of the unit vectors $\hat{\mathbf{i}}$, $\hat{\mathbf{j}}$, and $\hat{\mathbf{k}},$ we must be able to calculate the vector's components. For simplicity we illustrate this process using a vector lying in the xy-plane. Figure 2.13 shows a vector **A** drawn with its tail located at the origin of a Cartesian coordinate system. We have indicated the magnitude of the vector as A, and the angle between the vector and the positive x-axis as θ. We have also drawn a line from the tip of **A** down to the x-axis in such a way that this line strikes the x-axis at 90°. (Some math professors we know call this process *dropping a perpendicular;* we suppose that is all right if the floor is carpeted.) The right-triangle in this figure is obvious, so it should be easy for you to verify that the lengths of the sides of this triangle are

$$A_x = A\cos(\theta) \quad \text{and} \quad A_y = A\sin(\theta) \tag{2.8}$$

The quantities A_x and A_y are exactly the Cartesian components of the vector **A** we desire. The process of determining these components is known as **resolving** the vector into components. Also note that, in general, the component of a vector along *any* direction in space can be calculated in a similar fashion.

Once we choose a coordinate system, the components are uniquely determined. However, it is clear that the components of **A** depend on how we orient our coordinate system. For example, if we tilt the x-axis so that it runs along the direction of **A,** then the angle θ is zero and $A_x = A$, whereas $A_y = 0$. Therefore, it is always important to make some statement about the orientation of the coordinate system you are using to resolve a vector into its components. Often, a simple diagram is sufficient.

If we are given the components A_x and A_y of a vector, we may find the magnitude A and direction as indicated by θ. We apply the Pythagorean theorem to the right triangle of Figure 2.13 and also the definition of the arctangent:

$$A = \sqrt{A_x^2 + A_y^2} \quad \text{and} \quad \theta = \arctan\left(\frac{A_y}{A_x}\right) \tag{2.9}$$

You have probably noticed that Equations (2.8) and (2.9) are the same relationships we found in the last section, which relate Cartesian coordinates to polar coordinates. Consequently, the caution in Example 2.1 regarding the angle computed by the arctangent function of your calculator applies equally for Equation (2.9).

In three dimensions it is necessary to consider three components of a vector. In this case, the components are determined by an extension of the process described above. Because we most often need to consider vectors in a plane, we only briefly mention the third component in this chapter, leaving the details until we encounter cases where the third component is required.

Vector Addition: Component Method

Suppose a vector **C** is the resultant of adding vectors **A** and **B.** Then

$$\mathbf{C} = \mathbf{A} + \mathbf{B}$$

FIGURE 2.13

Resolving vector **A** into its x and y components A_x and A_y, respectively.

Now, two vectors are equal when the vectors have both the same magnitude and direction. This can be the case only if both **C** and (**A + B**) have the same components. Thus,

$$C_x\widehat{\mathbf{i}} + C_y\widehat{\mathbf{j}} = (A_x\widehat{\mathbf{i}} + A_y\widehat{\mathbf{j}}) + (B_x\widehat{\mathbf{i}} + B_y\widehat{\mathbf{j}})$$
$$C_x\widehat{\mathbf{i}} + C_y\widehat{\mathbf{j}} = (A_x + B_x)\widehat{\mathbf{i}} + (A_y + B_y)\widehat{\mathbf{j}} \tag{2.10}$$

We see that

$$C_x = A_x + B_x \quad \text{and} \quad C_y = A_y + B_y \tag{2.11}$$

If the vectors **A** and **B** have three nonzero components, their sum is computed in a similar fashion. In particular, if A_z and B_z are the z-components of vectors **A** and **B**, then

$$\mathbf{A} + \mathbf{B} = (A_x\widehat{\mathbf{i}} + A_y\widehat{\mathbf{j}} + A_z\widehat{\mathbf{k}}) + (B_x\widehat{\mathbf{i}} + B_y\widehat{\mathbf{j}} + B_z\widehat{\mathbf{k}})$$
$$= (A_x + B_x)\widehat{\mathbf{i}} + (A_y + B_y)\widehat{\mathbf{j}} + (A_z + B_z)\widehat{\mathbf{k}} \tag{2.12}$$

We now have a simple analytic procedure for adding two or more vectors:

1. Draw a convenient coordinate system and sketch the vectors.
2. Resolve each vector into component vectors parallel to the coordinate axes, taking care to associate the proper sign with each component.
3. Form the components of the resultant vector by summing the individual components associated with each coordinate axis.
4. Apply the Pythagorean theorem to compute the magnitude of the resultant.
5. Use the arctangent function or other suitable inverse trigonometric function to find the direction of the resultant. (*Don't omit this step or you will almost assuredly lose points on a quiz!*)

We illustrate this procedure with Examples 2.2 through 2.4.

EXAMPLE 2.2   *The Team Now Has a ¼ Back, a ½ Back, and a √2 Back*

Much to the dismay of the fans, a physics teacher takes on the position of football coach. During a game that coach instructs a receiver to run 8.00 m at an angle of 30.0° north of east, turn, and then run 4.00 m 70.0° west of north. What is the receiver's displacement after running this route?

SOLUTION

STEP 1: The receiver's displacement vectors, labeled **A** and **B,** are shown in Figure 2.14(a). We are comfortable drawing the vectors in this manner because they map the actual path taken by the player. We have converted 70.0° W of N to the equivalent angle of 160.0° as measured from the direction parallel to the positive x-axis.

STEP 2: In Figure 2.14(b) and Figure 2.14(c) we show the vectors **A** and **B** drawn on their own coordinate system. We can easily compute the x- and y-components of these vectors. Notice that B_x is negative because it is directed along the negative x-axis:

$$A_x = A\cos(30.0°) = (8.00 \text{ m})\cos(30.0°) = 6.93 \text{ m}$$

$$A_y = A\sin(30.0°) = (8.00 \text{ m})\sin(30.0°) = 4.00 \text{ m}$$

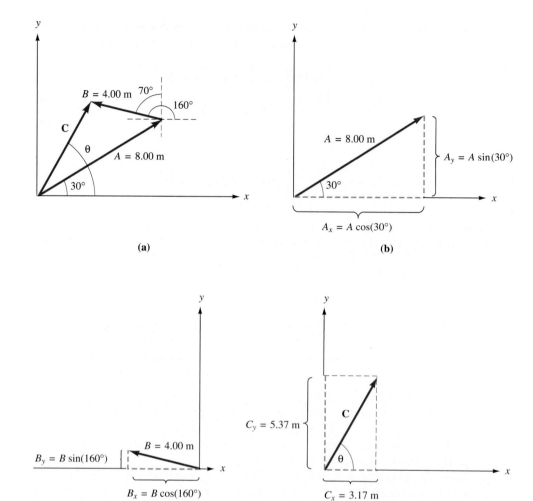

FIGURE 2.14

(a) Displacement vectors of the receiver in Example 2.2. (b) Vector **A** and (c) vector **B** drawn on their own coordinate system. (d) The Cartesian components of the resultant.

$$B_x = B \cos(160.0°) = (4.00 \text{ m}) \cos(160.0°) = -3.76 \text{ m}$$

$$B_y = B \sin(160.0°) = (4.00 \text{ m}) \sin(160.0°) = 1.37 \text{ m}$$

STEP 3: The components of the resultant vector **C** are

$$C_x = A_x + B_x = 6.93 \text{ m} - 3.76 \text{ m} = 3.17 \text{ m}$$

$$C_y = A_y + B_y = 4.00 \text{ m} + 1.37 \text{ m} = 5.37 \text{ m}$$

In unit vector form we write the resultant

$$\mathbf{C} = (3.17 \text{ m})\widehat{\mathbf{i}} + (5.37 \text{ m})\widehat{\mathbf{j}}$$

STEP 4: The magnitude of the resultant is

$$C = \sqrt{C_x^2 + C_y^2} = \sqrt{(3.17 \text{ m})^2 + (5.37 \text{ m})^2} = 6.23 \text{ m}$$

STEP 5: The angle the resultant makes with the positive x-axis is

$$\theta = \arctan\left(\frac{C_y}{C_x}\right) = \arctan\left(\frac{5.37 \text{ m}}{3.17 \text{ m}}\right) = 59.4°$$

The polar form of the answer can be written: **C** is 6.23 m at an angle of 59.4° North of East (see Fig. 2.14(d)). ◄

EXAMPLE 2.3 *Stepping Through Vector Addition*

Find the sum of three vectors whose magnitudes and directions are:

Vector **A:** 5.00 units at an angle of 25.0° from the $+x$-axis
Vector **B:** 3.00 units at an angle of $-50.0°$ from the $+x$-axis
Vector **C:** 4.00 units at an angle of 60.0° above the $-x$-axis

SOLUTION

STEP 1: As long as we maintain the correct magnitudes and directions, we are free to slide a vector around to any convenient location. (Formally, this process is called **vector translation.**) As shown in Figure 2.15(a), for this example it is convenient to place the tails of all three vectors at the origin of the coordinate system. Moreover, if we pay careful attention to signs, we can use the angles provided for us to compute the components of each vector. This approach allows us to avoid referencing all angles to the positive x-axis.

STEP 2: We resolve each vector into its component vectors. Any component vector directed along a positive axis has a positive component, but any component vector whose direction is along a negative axis has a minus sign associated with the corresponding component.

$$A_x = A \cos(25.0°) = (5.00 \text{ units}) \cos(25.0°) = 4.53 \text{ units}$$

$$A_y = A \sin(25.0°) = (5.00 \text{ units}) \sin(25.0°) = 2.11 \text{ units}$$

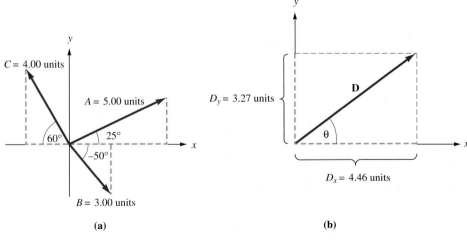

(a) (b)

FIGURE 2.15

(a) Vector translation showing all vector tails at the origin of the coordinate system. (b) The resultant vector **D.**

$$B_x = B \cos(-50.0°) = (3.00 \text{ units}) \cos(50.0°) = 1.93 \text{ units}$$

$$B_y = B \sin(-50.0°) = -(3.00 \text{ units}) \sin(50.0°) = -2.30 \text{ units}$$

$$C_x = -C \cos(60.0°) = -(4.00 \text{ units}) \cos(60.0°) = -2.00 \text{ units}$$

$$C_y = C \sin(60.0°) = (4.00 \text{ units}) \sin(60.0°) = 3.46 \text{ units}$$

STEP 3: We compute the components of the resultant **D**:

$$D_x = A_x + B_x + C_x = (4.53 + 1.93 - 2.00) \text{ units} = 4.46 \text{ units}$$

$$D_y = A_y + B_y + C_y = (2.11 - 2.30 + 3.46) \text{ units} = 3.27 \text{ units}$$

STEP 4: The magnitude of the resultant is

$$D = |\mathbf{D}| = \sqrt{D_x^2 + D_y^2} = \sqrt{(4.46 \text{ units})^2 + (3.27 \text{ units})^2} = 5.53 \text{ units}$$

STEP 5: From Figure 2.15(b) we find the direction of **D**:

$$\theta = \arctan\left(\frac{D_y}{D_x}\right) = \arctan\left(\frac{3.27}{4.46}\right) = 36.2°$$

The resultant vector **D** has a magnitude of 5.53 units directed at an angle of 36.2° from the positive x-axis.

EXAMPLE 2.4

Three vectors are given by $\mathbf{A} = (2.00 \text{ km})\hat{\mathbf{i}} + (3.00 \text{ km})\hat{\mathbf{j}}$, $\mathbf{B} = (1.00 \text{ km})\hat{\mathbf{i}} - (2.00 \text{ km})\hat{\mathbf{j}}$, and $\mathbf{C} = -(6.00 \text{ km})\hat{\mathbf{i}}$. Find the resultant vector **D**.

SOLUTION

$$\begin{aligned}\mathbf{D} &= \mathbf{A} + \mathbf{B} + \mathbf{C} \\ &= [(2.00 \text{ km})\hat{\mathbf{i}} + (3.00 \text{ km})\hat{\mathbf{j}}] + [(1.00 \text{ km})\hat{\mathbf{i}} - (2.00 \text{ km})\hat{\mathbf{j}}] - (6.00 \text{ km})\hat{\mathbf{i}} \\ &= [(2.00 + 1.00 - 6.00)\text{km}]\hat{\mathbf{i}} + [(3.00 - 2.00 + 0)\text{km}]\hat{\mathbf{j}} \\ &= -(3.00 \text{ km})\hat{\mathbf{i}} + (1.00 \text{ km})\hat{\mathbf{j}}\end{aligned}$$

Although this expression for **D** is complete, let's also find its magnitude and direction, just for practice!

$$|\mathbf{D}| = D = \sqrt{D_x^2 + D_y^2} = \sqrt{(-3.00 \text{ km})^2 + (1.00 \text{ km})^2} = 3.16 \text{ km}$$

$$\theta = \arctan\left(\frac{D_y}{D_x}\right) = \arctan\left(\frac{1.0 \text{ km}}{-3.0 \text{ km}}\right) = -18.4°$$

Measured from the positive x-axis the direction of **D** is $180° - 18° = 162°$. The vectors **A, B, C,** and their resultant **D** are shown in Figure 2.16.

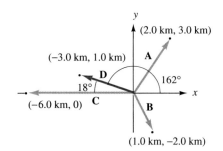

FIGURE 2.16

The vectors **A, B, C,** and their vector sum **D**.

2.4 Vector Multiplication

How are vectors multiplied? We have already seen how to multiply a vector by a scalar. Now we look at two ways to multiply vectors together.

The Scalar Product

Figure 2.17(a) shows two vectors **A** and **B** that have been drawn with their "tails" (that is, their starting points) at a common location. The smaller of the two angles between **A** and **B** is labeled θ. The **scalar product** (or **inner product**) of these two vectors is labeled **A · B** and is defined as the product of their magnitudes and the cosine of the smaller angle between them. That is,

$$\mathbf{A} \cdot \mathbf{B} = AB \cos(\theta) \tag{2.13}$$

The notation **A · B** is spoken "A dot B." For this reason the scalar product is also referred to as the **dot product.**

Both A and B are scalars and so is $\cos(\theta)$. Hence, as you probably suspect from the name, the result of a scalar product is a scalar. It should be clear that the scalar product is commutative:

$$\mathbf{A} \cdot \mathbf{B} = \mathbf{B} \cdot \mathbf{A} \tag{2.14}$$

Graphically, we may interpret the scalar product of two vectors as the product of (1) the magnitude of the first vector and (2) the component of the second vector resolved along the direction of the first vector. Figure 2.17(a) shows the projection of vector **A** onto the direction of vector **B**. This projection has a value $A \cos(\theta)$. The product of $A \cos(\theta)$ and B is $AB \cos(\theta) = \mathbf{A} \cdot \mathbf{B}$. Similarly, in Figure 2.17(b) we show that $B \cos(\theta)$ is the projection of vector **B** along the direction of vector **A.** In this case the product of this projection and A is also $AB \cos(\theta) = \mathbf{A} \cdot \mathbf{B}$. Note that this product may be negative if the projection of one vector is actually antiparallel to the other. If **A** and **B** are perpendicular to each other then $\mathbf{A} \cdot \mathbf{B} = 0$ (see also Problem 46).

Finally, we want to mention that after performing the dot product operation we are left with a scalar, that is, a number and units. We have no direction associated with this number and, therefore, nothing left that is appropriate to draw on the diagram of vectors **A** and **B**; you can't draw the result of a scalar product.

The Vector Product

The **vector product,** or **outer product,** of two vectors **A** and **B** is written **A × B.** The result of this operation is another vector **C:**

$$\mathbf{C} = \mathbf{A} \times \mathbf{B} \tag{2.15}$$

This combination is spoken "A cross B," and thus sometimes called the **cross product.** If θ is the smaller angle between **A** and **B,** the magnitude of **C** is given by

$$C = |\mathbf{A} \times \mathbf{B}| = AB \sin(\theta) \tag{2.16}$$

The direction of $\mathbf{C} = \mathbf{A} \times \mathbf{B}$ is perpendicular to the plane containing the vectors **A** and **B,** in the sense given by what is termed the **right-hand rule.** The application of the right-hand rule is illustrated in Figure 2.18, where we show the vectors **A** and **B** drawn with their tails at a common point. To apply the right-hand rule, the fingers of the right hand are placed along the direction of vector **A** in such a manner that they can be turned at

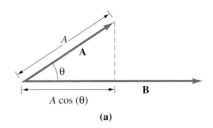

(a)

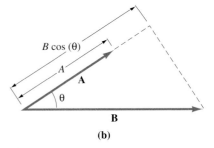

(b)

FIGURE 2.17

Graphically, the scalar product of two vectors **A** and **B** can be viewed as the product of the magnitude of one of the vectors and the component of the other vector that is parallel to the direction of the first. (a) $\mathbf{A} \cdot \mathbf{B} = (A \cos(\theta))B = AB \cos(\theta)$. (b) $\mathbf{A} \cdot \mathbf{B} = (B \cos(\theta))A = AB \cos(\theta)$.

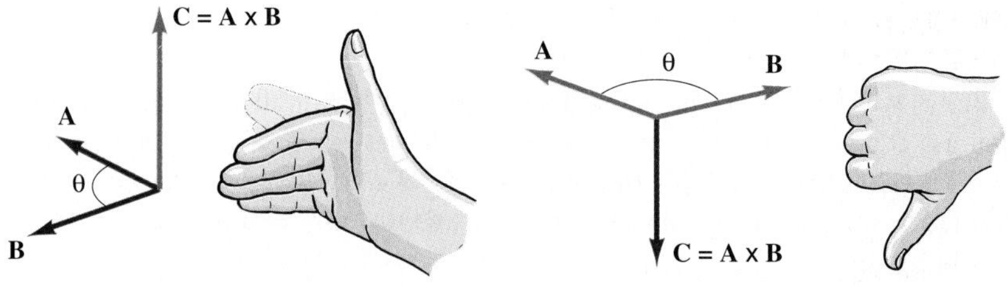

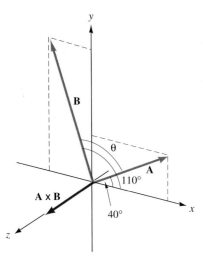

FIGURE 2.18

(a) The direction of the vector product $\mathbf{A} \times \mathbf{B}$ is determined by the right-hand rule. The fingers of the right hand are placed along the direction of vector $\mathbf{A}$ and turned toward the direction of vector $\mathbf{B}$. (b) The extended thumb points along the direction of $\mathbf{C} = \mathbf{A} \times \mathbf{B}$.

the knuckles toward the direction of vector $\mathbf{B}$. The extended thumb of the right hand provides the direction of $\mathbf{A} \times \mathbf{B}$. (It is easy to determine if you are turning your fingers in the correct sense. If you hear loud popping sounds and experience more than usual pain while doing physics problems, you've got it backward.)

Although their magnitudes are equal, the vector $\mathbf{A} \times \mathbf{B}$ is not equivalent to the vector $\mathbf{B} \times \mathbf{A}$. In fact, you should verify by the right-hand rule that their directions are opposite, so that

$$\mathbf{B} \times \mathbf{A} = -\mathbf{A} \times \mathbf{B} \tag{2.17}$$

Formally, Equation (2.17) states that the vector product does not satisfy the commutative property. There are other important properties of the cross product. If s is a scalar

$$\mathbf{A} \times (s\mathbf{B}) = (s\mathbf{A}) \times \mathbf{B} = s(\mathbf{A} \times \mathbf{B}) \tag{2.18}$$

In the special case where $\mathbf{A}$ and $\mathbf{B}$ are perpendicular, the magnitude of the cross product is

$$|\mathbf{A} \times \mathbf{B}| = AB \qquad (\mathbf{A} \perp \mathbf{B})$$

and if $\mathbf{A}$ and $\mathbf{B}$ are parallel

$$\mathbf{A} \times \mathbf{B} = 0 \qquad (\mathbf{A} \| \mathbf{B})$$

The vector cross product obeys the **distributive law.** That is, for the three vectors $\mathbf{A}$, $\mathbf{B}$, and $\mathbf{C}$,

$$\mathbf{A} \times (\mathbf{B} + \mathbf{C}) = \mathbf{A} \times \mathbf{B} + \mathbf{A} \times \mathbf{C} \tag{2.19}$$

FIGURE 2.19

The right-hand rule gives the direction of the vector cross product $\mathbf{A} \times \mathbf{B}$.

EXAMPLE 2.5   *Thumbthing to Think About*

Two vectors each lie in the *xy*-plane. The vector $\mathbf{A}$ has a magnitude of 4.00 units and makes an angle of $40.0°$ with the positive *x*-axis, whereas vector $\mathbf{B}$ is 6.00 units in magnitude and makes an angle of $110.0°$ with the positive *x*-axis. (a) Compute $\mathbf{A} \cdot \mathbf{B}$. (b) Compute the vectors $\mathbf{A} \times \mathbf{B}$ and $\mathbf{B} \times \mathbf{A}$.

SOLUTION   (a) The vectors $\mathbf{A}$ and $\mathbf{B}$ are shown in Figure 2.19. The angle θ between the vectors is

$$\theta = 110.0° - 40.0° = 70.0°$$

From Equation (2.13) we obtain the dot product

$$\mathbf{A} \cdot \mathbf{B} = AB\cos(\theta) = (4.00)(6.00)\cos(70.0°) = 8.21$$

(b) The magnitude of the vector cross product is given by Equation (2.16):

$$|\mathbf{A} \times \mathbf{B}| = AB\sin(\theta) = (4.00)(6.00)\sin(70.0°) = 22.6$$

The direction of the $\mathbf{A} \times \mathbf{B}$ is given by the right-hand rule. As shown in Figure 2.19, this direction is along the positive z-axis; if you place the fingers of your right hand along the direction of $\mathbf{A}$ and curl them toward $\mathbf{B}$, your thumb points along the positive z-axis. Thus, we are led to write

$$\mathbf{A} \times \mathbf{B} = (22.6)\hat{\mathbf{k}}$$

The magnitude of the vector $\mathbf{B} \times \mathbf{A}$ is the same as that of $\mathbf{A} \times \mathbf{B}$:

$$|\mathbf{B} \times \mathbf{A}| = BA\sin(\theta) = 22.6$$

However, the direction of $\mathbf{B} \times \mathbf{A}$, also given by the right-hand rule, is along the negative z-axis; if you place the fingers of your right hand along the direction of $\mathbf{B}$ and curl them toward $\mathbf{A}$, your thumb points along the negative z-axis. ◀

Scalar and Vector Products in the Unit-Vector Representation

When the vectors $\mathbf{A}$ and $\mathbf{B}$ are represented with the unit vectors $\hat{\mathbf{i}}, \hat{\mathbf{j}},$ and $\hat{\mathbf{k}}$, we find some very convenient expressions for the vector multiplication definitions given above. For multiplication by a scalar, the result is simple:

$$s\mathbf{A} = s(A_x\hat{\mathbf{i}} + A_y\hat{\mathbf{j}} + A_z\hat{\mathbf{k}}) = (sA_x)\hat{\mathbf{i}} + (sA_y)\hat{\mathbf{j}} + (sA_z)\hat{\mathbf{k}} \qquad (2.20)$$

For the vector dot product we must first compute the dot product for every possible combination of unit vectors. For example $\hat{\mathbf{i}} \cdot \hat{\mathbf{i}} = (1)(1)\cos(0°) = 1$ and $\hat{\mathbf{i}} \cdot \hat{\mathbf{j}} = (1)(1)\cos(90°) = 0$. You should verify that

$$\hat{\mathbf{i}} \cdot \hat{\mathbf{i}} = \hat{\mathbf{j}} \cdot \hat{\mathbf{j}} = \hat{\mathbf{k}} \cdot \hat{\mathbf{k}} = 1 \qquad (2.21a)$$
$$\hat{\mathbf{i}} \cdot \hat{\mathbf{j}} = \hat{\mathbf{j}} \cdot \hat{\mathbf{k}} = \hat{\mathbf{k}} \cdot \hat{\mathbf{i}} = 0 \qquad (2.21b)$$

Applying these relations term by term to the vector dot product of $\mathbf{A}$ and $\mathbf{B}$, we find

$$\mathbf{A} \cdot \mathbf{B} = (A_x\hat{\mathbf{i}} + A_y\hat{\mathbf{j}} + A_z\hat{\mathbf{k}}) \cdot (B_x\hat{\mathbf{i}} + B_y\hat{\mathbf{j}} + B_z\hat{\mathbf{k}})$$

$$\mathbf{A} \cdot \mathbf{B} = A_xB_x + A_yB_y + A_zB_z \qquad (2.22)$$

For the particular case where $\mathbf{B} = \mathbf{A}$,

$$\mathbf{A} \cdot \mathbf{A} = A^2 = A_x^2 + A_y^2 + A_z^2 \qquad (2.23)$$

That is,

$$A = |\mathbf{A}| = \sqrt{\mathbf{A} \cdot \mathbf{A}} = \sqrt{A_x^2 + A_y^2 + A_z^2} \qquad (2.24)$$

To find the vector cross product we must evaluate the cross products of the unit vectors. You should verify that

$$\hat{i} \times \hat{i} = \hat{j} \times \hat{j} = \hat{k} \times \hat{k} = 0 \tag{2.25a}$$

$$\hat{i} \times \hat{j} = -\hat{j} \times \hat{i} = \hat{k} \tag{2.25b}$$

$$\hat{j} \times \hat{k} = -\hat{k} \times \hat{j} = \hat{i} \tag{2.25c}$$

$$\hat{k} \times \hat{i} = -\hat{i} \times \hat{k} = \hat{j} \tag{2.25d}$$

You don't need to remember all the combinations in Equations (2.25b), (c), and (d). There is a pattern in each equation, such as $\hat{k} \times \hat{i} = \hat{j}$. To see the pattern, first note that all three letters always appear. Next, regardless of where it appears, begin with $\hat{i}$ and read the symbols left to right, jumping over the $\times$ and $=$ symbols. When you get to the end of the equation, loop back to the first symbol of the equation. If you have read the letters i, j, and k in alphabetical order, this sequence is known as a **cyclic permutation,** and the result of the cross product is a positive unit vector. Otherwise, the permutation of $\hat{i}$, $\hat{j}$, and $\hat{k}$ is **noncyclic,** and the cross product is negative. When we apply this procedure to the equation $\hat{k} \times \hat{i} = \hat{j}$ we read "$i\,j\,k$" and there is a + sign in front of the $\hat{j}$. On the other hand, for the equation $\hat{k} \times \hat{j} \overset{?}{=} \hat{i}$ we read "$i\,k\,j$," which is not alphabetical in order and we therefore place a − sign in front of $\hat{i}$ to obtain the correct relation $\hat{k} \times \hat{j} = -\hat{i}.$

In order to determine the cross product of the vectors $\mathbf{A} = A_x\hat{i} + A_y\hat{j} + A_z\hat{k}$ and $\mathbf{B} = B_x\hat{i} + B_y\hat{j} + B_z\hat{k}$, we must compute the cross product of each term. We leave it as a homework problem for you to apply Equation (2.25) to show that[3]

$$\mathbf{A} \times \mathbf{B} = (A_yB_z - A_zB_y)\hat{i} + (A_zB_x - A_xB_z)\hat{j} + (A_xB_y - A_yB_x)\hat{k} \tag{2.26}$$

Equation (2.26) can be expressed in the form of a determinant:

$$\mathbf{A} \times \mathbf{B} = \hat{i}\begin{vmatrix} A_y & A_z \\ B_y & B_z \end{vmatrix} - \hat{j}\begin{vmatrix} A_x & A_z \\ B_x & B_z \end{vmatrix} + \hat{k}\begin{vmatrix} A_x & A_y \\ B_x & B_y \end{vmatrix}$$

$$\mathbf{A} \times \mathbf{B} = \begin{vmatrix} \hat{i} & \hat{j} & \hat{k} \\ A_x & A_y & A_z \\ B_x & B_y & B_z \end{vmatrix} \tag{2.27}$$

You should verify that the expansion of the determinant in Equation (2.27) does indeed yield Equation (2.26).

EXAMPLE 2.6   *Doing Your ABC's*

For $\mathbf{A} = (2.0)\hat{i} - (5.0)\hat{j}$ and $\mathbf{B} = (4.0)\hat{i} + (3.0)\hat{j}$ compute (a) the scalar product and (b) the vector product for these two vectors.

SOLUTION   (a) We apply Equation (2.22) to compute the dot product:

---

[3] Many modern scientific calculators have built-in functions for calculating the dot and cross products. Check the user's manual for your calculator. If your calculator has such a function, it is well worth half an hour to learn how to use it for cross products.

$$\mathbf{A} \cdot \mathbf{B} = [(2.0)\hat{\mathbf{i}} - (5.0)\hat{\mathbf{j}}] \cdot [(4.0)\hat{\mathbf{i}} + (3.0)\hat{\mathbf{j}}]$$

$$= (2.0)(4.0) + (-5.0)(3.0) = -7.0$$

(b) The cross product is given by Equation (2.28), where the components A_z and B_z are zero:

$$\mathbf{A} \times \mathbf{B} = \begin{vmatrix} \hat{\mathbf{i}} & \hat{\mathbf{j}} & \hat{\mathbf{k}} \\ 2.0 & -5.0 & 0 \\ 4.0 & 3.0 & 0 \end{vmatrix}$$

$$= [(-5.0)(0) - (0)(3.0)]\hat{\mathbf{i}} + [(0)(4.0) - (2.0)(0)]\hat{\mathbf{j}}$$
$$+ [(2.0)(3.0) - (-5.0)(4.0)]\hat{\mathbf{k}} = 26\hat{\mathbf{k}}$$

Can you verify by direct computation that $\mathbf{B} \times \mathbf{A} = -\mathbf{A} \times \mathbf{B}$? Establish the determinant for $\mathbf{B} \times \mathbf{A}$ and show that it yields $-26\hat{\mathbf{k}}$. ◀

2.5 Summary

Positions in space are designated relative to coordinate systems. The **Cartesian coordinate system** is a particularly convenient coordinate system in which positions are designated by distances (x, y, z) along three perpendicular axes that intersect at a point called the origin. In a plane, the polar coordinates (r, θ) are also useful. When we describe a situation relative to some coordinate system, that coordinate system is our **frame of reference.**

For a complete description, a **scalar** quantity requires only a magnitude to be given. **Vectors,** however, require both a magnitude and a direction for a complete description. Vectors may be represented graphically as a directed line segment. The length of the line represents the vector's magnitude and the line's angle with respect to some coordinate system specifies the vector's direction.

Two vectors $\mathbf{A}$ and $\mathbf{B}$ can be added graphically by choosing a suitable scale and placing the tail of vector $\mathbf{B}$ at the tip of vector $\mathbf{A}$. The resultant vector runs from the tail $\mathbf{A}$ to the tip of $\mathbf{B}$. A vector $-\mathbf{A}$ has the same magnitude as vector $\mathbf{A},$ but opposite direction. Vector addition is commutative and associative.

The components of a vector $\mathbf{A}$ are the projections A_x and A_y on the x- and y-axes. If θ is the angle between the x-axis and the vector $\mathbf{A},$ then

$$A_x = A \cos(\theta) \qquad \text{and} \qquad A_y = A \sin(\theta)$$

where A is the magnitude of $\mathbf{A}$. From the components A_x and A_y the magnitude and angle θ of vector $\mathbf{A}$ are given by

$$A = \sqrt{A_x^2 + A_y^2} \qquad \text{and} \qquad \theta = \arctan\left(\frac{A_y}{A_x}\right) \tag{2.8}$$

Interpreting the meaning of angle θ requires knowledge of the quadrant in which the vector lies.

If A_x and A_y are the components of vector $\mathbf{A},$ whereas B_x and B_y are the components of vector $\mathbf{B},$ the components of the vector sum $\mathbf{C} = \mathbf{A} + \mathbf{B}$ are

$$C_x = A_x + B_x \qquad \text{and} \qquad C_y = A_y + B_y \tag{2.9}$$

Unit vectors $\hat{\mathbf{i}}, \hat{\mathbf{j}},$ and $\hat{\mathbf{k}}$ are defined to have unit magnitude and are directed along the x-, y-, and z-axes, respectively. In terms of these unit vectors in the xy-plane, a vector $\mathbf{A}$ with components A_x and A_y may be written.

$$\mathbf{A} = A_x\hat{\mathbf{i}} + A_y\hat{\mathbf{j}}$$

The **scalar product** of two vectors is given by

$$\mathbf{A} \cdot \mathbf{B} = AB\cos(\theta) \tag{2.13}$$

where θ is the smaller angle between the vectors $\mathbf{A}$ and $\mathbf{B}$. The magnitude of the **vector product** of vectors $\mathbf{A}$ and $\mathbf{B}$ is

$$|\mathbf{A} \times \mathbf{B}| = AB\sin(\theta) \tag{2.16}$$

where, again, θ is the smaller angle between $\mathbf{A}$ and $\mathbf{B}$. The direction of $\mathbf{A} \times \mathbf{B}$ is given by the **right-hand rule.** Some properties of the vector product are

$$\mathbf{B} \times \mathbf{A} = -\mathbf{A} \times \mathbf{B} \tag{2.17}$$

$$\mathbf{A} \times (s\mathbf{B}) = (s\mathbf{A}) \times \mathbf{B} = s(\mathbf{A} \times \mathbf{B}) \tag{2.18}$$

where s is a scalar.

$$\mathbf{A} \times (\mathbf{B} + \mathbf{C}) = \mathbf{A} \times \mathbf{B} + \mathbf{A} \times \mathbf{C} \tag{2.19}$$

For vectors represented by the unit-vector notation

$$\mathbf{A} \cdot \mathbf{B} = A_xB_x + A_yB_y + A_zB_z \tag{2.22}$$

$$\mathbf{A} \times \mathbf{B} = (A_yB_z - A_zB_y)\hat{\mathbf{i}} + (A_zB_x - A_xB_z)\hat{\mathbf{j}} + (A_xB_y - A_yB_x)\hat{\mathbf{k}} \tag{2.26}$$

or

$$\mathbf{A} \times \mathbf{B} = \begin{vmatrix} \hat{\mathbf{i}} & \hat{\mathbf{j}} & \hat{\mathbf{k}} \\ A_x & A_y & A_z \\ B_x & B_y & B_z \end{vmatrix} \tag{2.27}$$

PROBLEMS

2.1 Coordinate Systems

1. A point in the xy-plane is designated by the Cartesian coordinates (3.50 m, 5.60 m). What are the polar coordinates of this point?
2. A point in the xy-plane is designated by the Cartesian coordinates (−4.00 m, 5.50 m). What are the polar coordinates of this point?
3. A point in the xy-plane is designated by the polar coordinates (12.5 m, 230.°). What are the Cartesian coordinates of this point?
4. Two points in the xy-plane are given by the Cartesian coordinates (2.30 cm, 8.60 cm) and (−4.40 cm, 1.20 cm). (a) What is the distance between these two points? (b) What are the polar coordinates of the point (−4.40 cm, 1.20 cm) from a new origin placed at the point (2.30 cm, 8.60 cm)?
5. Two points in the xy-plane have polar coordinates (5.00 m, 36.9°) and (7.00 m, 136°). (a) Compute the Cartesian coordinates (x, y) for each of these points. (b) What is the distance between these

two points? (c) What are the polar coordinates of the point (7.00 m, 136°) from a new origin placed at the position of (5.00, 36.9°)?

2.3 Vector Addition

6. A car is driven 36.0 km due North and then 18.0 km at an angle of 30.0° West of North. Apply the graphical method to determine the car's total displacement.
7. A child walks 24.0 m due east, turns through a 40.0° angle toward the north and walks 12.0 m. Use the graphical method to determine the child's displacement.
8. Vector $\mathbf{A}$ has a magnitude of 8.00 units and makes an angle of 20.0° with the positive x-axis. Vector $\mathbf{B}$ has a magnitude of 6.00 units and makes an angle of 115° with the positive x-axis. Finally, vector $\mathbf{C}$ has a magnitude of 12.0 units and makes an angle of 210.° with the positive x-axis. Apply the graphical method to find the resultant of these three vectors.

9. What is the resultant (magnitude and direction) of the vector sum **A + B + C** shown in Figure 2.P1?

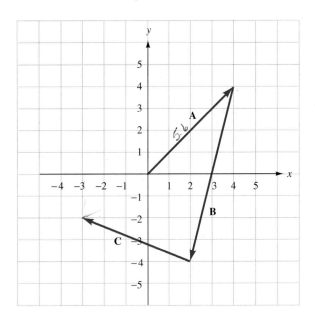

FIGURE 2.P1   Problems 9, 17

10. A wildlife biologist tracks an elk over flat terrain by following the emissions from a radio collar worn by the animal. If the elk travels 3.00 km at an angle of 30.0° E of N, followed by a 2.00 km displacement at an angle of 40.0° W of N and, finally, a 4.00 km displacement at 20.0° S of W, apply the graphical method to find the elk's total displacement.

11. Compute the components of a vector whose magnitude is 18.6 m/s at an angle of 150.° from the positive x-axis.

12. Compute the components of a vector whose magnitude is 35.0 kg-m/s^2 at an angle of 53.1° below the positive x-axis.

13. What are the components of a vector whose magnitude is 25.0 m and whose direction is 245° from the positive x-axis?

14. The x-component of a vector is 4.60 kg-m/s, and the y-component is −8.80 kg-m/s. What are the magnitude and direction of this vector?

15. The x-component of a displacement vector is −7.20 m, and the y-component is 3.80 m. Compute the magnitude and the direction of this vector.

16. The vector **A** has components $A_x = 18.0$ units and $A_y = -6.00$ units. Vector **B** has components $B_x = -3.00$ units and $B_y = -7.00$ units. Find the magnitude and direction of the vector **C** such that $\mathbf{A} + 2\mathbf{B} - 3\mathbf{C} = 0$.

17. What are the components of each vector **A, B,** and **C** and the resultant **A + B + C** shown in Figure 2.P1?

18. A pilot flies an airplane from city A to city B along a straight-line path that is 136 km in length and at a bearing of 65.0° E of N. He then flies at an angle of 40.0° W of N for 78.0 km, also in a straight line. (a) Resolve both displacement vectors into their x- and y-components. (b) Compute the magnitude and direction of the pilot's total displacement.

19. The vector **A** has components $A_x = -72.0$ units and $A_y = 48.0$ units. Vector **B** has components $B_x = 48.0$ units and $B_y = 72.0$ units. Compute the components of (a) **A + B**, (b) **A − B**, and (c) **2A + B**.

20. The magnitudes of the vectors shown in Figure 2.P2 are $A = 36.0$ units and $B = 24.0$ units, whereas $\theta_1 = 37.0°$ and $\theta_2 = 124°$. (a) Compute the components of vectors **A** and **B**. (b) What are the components of **A + B**? (c) What are the magnitude and direction of **A + B**?

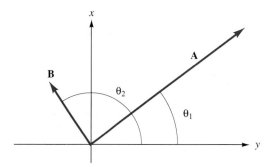

FIGURE 2.P2   Problems 20, 21

21. The magnitudes of the vectors shown in Figure 2.P2 are $A = 84.0$ units and $B = 60.0$ units, whereas $\theta_1 = 40.0°$ and $\theta_2 = 118°$. (a) Compute the components of vectors **A** and **B**. (b) What are the components of **A − 3B**? (c) What are the magnitude and direction of **A − 3B**?

22. The magnitudes of the vectors shown in Figure 2.P3 are $A = 16.0$ m, $B = 13.0$ m, and $C = 9.80$ m. The angles are $\theta_1 = 30.0°$, $\theta_2 = 130.°$, and $\theta_3 = 250.°$ (a) Find the components of each vector. (b) Find the components of the resultant **A + B + C**. (c) What are the magnitude and direction of the resultant **A + B + C**?

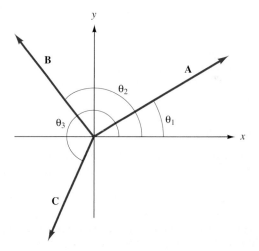

FIGURE 2.P3   Problems 22, 23

23. The magnitudes of the vectors shown in Figure 2.P3 are $A = 12.5$ m, $B = 10.0$ m, and $C = 7.50$ m. The angles are $\theta_1 = 36.0°$, $\theta_2 = 125°$, and $\theta_3 = 255°$. (a) Find the components of each vector. (b) Find the components of the vector **D = 2A − B + C**. (c) What are the magnitude and direction of **D**?

24. (a) Compute the vector sum in unit-vector notation of the two vectors $\mathbf{A} = 2.00\widehat{\mathbf{i}} + 5.00\widehat{\mathbf{j}}$ and $\mathbf{B} = 7.00\widehat{\mathbf{i}} - 3.00\widehat{\mathbf{j}}$. (b) What are the magnitude and direction of the resultant you found in part (a)?

25. If $\mathbf{A} = (3.60 \text{ m})\widehat{\mathbf{i}} - (2.40 \text{ m})\widehat{\mathbf{j}}$ and $\mathbf{B} = (-8.40 \text{ m})\widehat{\mathbf{i}} - (1.20 \text{ m})\widehat{\mathbf{j}}$, compute the components, magnitudes, and directions for (a) **A + B**, (b) **A − 2B**, and (c) **−2A + B**.

26. In Figure 2.P4 the magnitudes of vectors **A** and **B** are 9.00 units and 7.00 units, respectively. The angle $\theta_1 = 55.0°$, and $\theta_2 = 115°$. (a) Write vectors **A** and **B** in unit-vector notation. (b) Compute the resultant **C** = **A** + **B** in unit-vector notation. (c) What are the magnitude and direction of **C**?

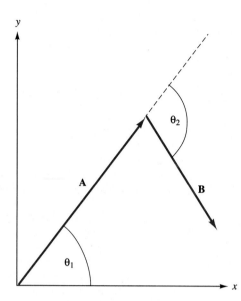

FIGURE 2.P4   Problem 26

27. If **A** = $12.0\hat{\mathbf{i}} - 8.00\hat{\mathbf{j}} + 9.00\hat{\mathbf{k}}$ and **B** = $7.00\hat{\mathbf{i}} + 3.00\hat{\mathbf{j}} - 14.0\hat{\mathbf{k}}$, compute in unit-vector notation (a) **A** + **B**, (b) **A** − 2**B**, and (c) −2**A** + **B**.

28. For the two vectors **A** = $8.00\hat{\mathbf{i}} + 3.00\hat{\mathbf{j}} - 6.00\hat{\mathbf{k}}$ and **B** = $-4.00\hat{\mathbf{i}} + 2.00\hat{\mathbf{j}} + 5.00\hat{\mathbf{k}}$, (a) find the unit-vector form of the vector **C** such that 3**A** + 4**B** − 2**C** = 0. (b) What are the magnitudes of **A**, **B**, and **C**?

2.4 Vector Multiplication

29. (a) Compute the scalar product of the two vectors **A** = $3.00\hat{\mathbf{i}} - 5.00\hat{\mathbf{j}}$ and **B** = $6.00\hat{\mathbf{i}} + 2.00\hat{\mathbf{j}}$. (b) What is the smaller angle between **A** and **B**?

30. The magnitude of **A** is 12.5 units and the magnitude of **B** is 8.00 units. If the angle between **A** and **B** is 32.0°, evaluate **A** · **B**.

31. The magnitude of **A** is 16.0 units and the magnitude of **B** is 12.0 units. What must be the angle between **A** and **B** for their scalar product to be −81.0?

32. For **A** = $3.00\hat{\mathbf{i}} + 2.00\hat{\mathbf{j}} - 5.00\hat{\mathbf{k}}$, **B** = $9.00\hat{\mathbf{i}} - 7.00\hat{\mathbf{j}} - 2.00\hat{\mathbf{k}}$, and **C** = $2.00\hat{\mathbf{i}} + 3.00\hat{\mathbf{j}} + 1.00\hat{\mathbf{k}}$, evaluate (a) **A** · (**B** + **C**), (b) **A** · (**B** − 2**C**), (c) **B** · (2**A** + **C**).

33. Find the angle between each of the following vectors: (a) $2.00\hat{\mathbf{i}} - 4.00\hat{\mathbf{j}}$ and $-3.00\hat{\mathbf{i}} + 3.00\hat{\mathbf{j}}$, (b) $4.00\hat{\mathbf{i}} - 5.00\hat{\mathbf{j}}$ and $-2.00\hat{\mathbf{i}} - 3.00\hat{\mathbf{j}}$, (c) $3.00\hat{\mathbf{i}} - 2.00\hat{\mathbf{j}}$ and $-3.00\hat{\mathbf{i}} + 2.00\hat{\mathbf{j}}$.

34. Use the definition of the scalar product to find the angle between the vectors **A** = $5.00\hat{\mathbf{i}} + 2.00\hat{\mathbf{j}} - 5.00\hat{\mathbf{k}}$ and **B** = $-6.00\hat{\mathbf{i}} - 4.00\hat{\mathbf{j}} + 3.00\hat{\mathbf{k}}$.

35. The sum of two vectors **A** and **B**, which have an angle θ between them when placed tail to tail, can be written in symbolic form as **C** = **A** + **B**. Take the dot product of this equation with itself and thereby prove the law of cosines: $C^2 = A^2 + B^2 - 2AB\cos(\phi)$,

where $\phi = 180° - \theta$ is the angle between the vectors when added tip to tail.

36. Find the angle between vector **A** = $3.00\hat{\mathbf{i}} - 2.00\hat{\mathbf{j}} + 4.00\hat{\mathbf{k}}$ and y-axis.

37. For the vectors **A** = $A_x\hat{\mathbf{i}} + A_y\hat{\mathbf{j}} + A_z\hat{\mathbf{k}}$ and **B** = $B_x\hat{\mathbf{i}} + B_y\hat{\mathbf{j}} + B_z\hat{\mathbf{k}}$, verify by direct computation that **A** · **B** is given by Equation (2.22).

38. By direct computation, apply Equation (2.26) to show that **A** × **B** is given by Equation (2.27).

39. For the vectors **A** = $4.00\hat{\mathbf{i}} + 3.00\hat{\mathbf{j}}$ and **B** = $2.00\hat{\mathbf{i}} - 5.00\hat{\mathbf{j}}$ compute (a) **A** × **B**, (b) **B** × **A**, (c) **A** × (2**B**), and (d) 2**A** × **B**.

40. For the vectors **A** = $3.00\hat{\mathbf{i}} - 2.00\hat{\mathbf{j}} + 4.00\hat{\mathbf{k}}$ and **B** = $1.00\hat{\mathbf{i}} + 2.00\hat{\mathbf{j}} + 3.00\hat{\mathbf{k}}$, compute (a) **A** × **B**, (b) **B** × **A**, (c) **A** × (2**B**), and (d) 2**A** × **B**.

41. Show that |**A** × **B**| is the area of the parallelogram that has **A** and **B** for two of its sides.

42. Compute the angle(s) between **A** and **B** if |**A** × **B**| = **A** · **B**.

43. Show that the scalar **A** · **B** × **C** is equal to the volume of the parallelepiped whose edges are defined by the vectors **A**, **B**, and **C**, as shown in Figure 2.P5.

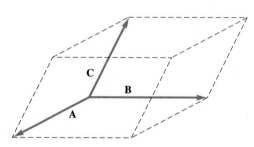

FIGURE 2.P5   Problem 43

44. Show that $|\mathbf{A} \times \mathbf{B}|^2 = |\mathbf{A}|^2|\mathbf{B}|^2 - (\mathbf{A} \cdot \mathbf{B})^2$.

45. Vector **A** lies in the xy-plane and is 5.00 units long and makes an angle of +36.87° with the positive x-axis. Vector **B** = $+2.00\hat{\mathbf{i}} - 1.50\hat{\mathbf{j}} + 2.50\hat{\mathbf{k}}$. (a) Find **A** + **B** and (b) the angle between **A** and **B** when they are placed tail to tail.

46. (a) Prove that if **A** · **B** = 0 then **A** is perpendicular to **B**. (b) If **A** × **B** = 0, what can be said of **A** and **B**? Prove your assertion. (Did you get both cases?)

Problems for Scientific Calculators

47. Use one of the vector functions of your calculator to calculate the magnitude of a vector with components [17.0, 35.0, −22.0].

48. Practice using your calculator's vector functions by verifying the famous "BAC − CAB" rule **A** × (**B** × **C**) = **B**(**A** · **C**) − **C**(**A** · **B**) for the special case of the vectors given in Problem 49.

49. Prove to yourself you can use your calculator's cross product function by calculating **A** × (**B** × (**C** × **D**)) for **A** = $3.00\hat{\mathbf{i}} + 2.00\hat{\mathbf{j}} - 5.00\hat{\mathbf{k}}$, **B** = $9.00\hat{\mathbf{i}} - 7.00\hat{\mathbf{j}} - 2.00\hat{\mathbf{k}}$, **C** = $2.00\hat{\mathbf{i}} + 3.00\hat{\mathbf{j}} + 1.00\hat{\mathbf{k}}$, and **D** = $1.00\hat{\mathbf{i}} + 0.500\hat{\mathbf{k}}$.

50. Use your calculator to verify the relation stated in Problem 44 for the special case of the vectors listed in Problem 49.

51. Use the determinant function of your calculator to evaluate **A** · (**B** × **C**) for the vectors given in Problem 49. [Hint: see Problem 61.] Compare your answer with that obtained using the cross and dot product functions.

General Problems

52. A surveying student uses the base of a flagpole in front of a library as an origin. She finds that the front door of the library is 8.50 m due East of the flagpole. She finds that the entrance the physics building is 36.0 m at an angle of 55.0° W of N from the pole, and the entrance to the mathematics building is 60.0° S of W and 24.0 m from the pole. Use the graphical method to find a student's displacement if she travels (a) from the math building to the physics building, (b) from the physics building to the math building, (c) from the library to the physics building, and (d) from the math building to the library.

53. A child runs 4.00 m due North, 3.00 m directly Northeast, and then 2.00 m directly Southeast. (a) What are the components of the child's total displacement? (b) What are the magnitude and direction of the child's displacement vector?

54. Three vectors have the following components: $(A_x, A_y) = (4.00, -5.00)$, $(B_x, B_y) = (-6.00, 8.00)$, and $(C_x, C_y) = (-2.00, -6.00)$. (a) Find the components of the resultant for these three vectors. (b) What are the magnitude and direction of the resultant?

55. With chalk, a physics student draws an xy-coordinate system on an empty parking lot. From the origin, she walks 14.0 m at an angle of 30.0° measured anticlockwise from the positive x-axis, turns and walks 18.0 m at an angle of 60.0° below a line parallel to the positive x-axis (Fig. 2.P6). What must the magnitude and direction of her next displacement be for her net displacement from the origin to be 16.0 m at an angle of 50.0° below the positive x-axis?

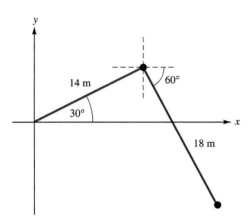

FIGURE 2.P6   Problem 55

56. For the three vectors, $\mathbf{A} = 2.00\hat{\mathbf{i}} - 3.00\hat{\mathbf{j}}$, $\mathbf{B} = -3.00\hat{\mathbf{i}} + 4.00\hat{\mathbf{j}}$, and $\mathbf{C} = -3.00\hat{\mathbf{i}} - 5.00\hat{\mathbf{j}}$, compute in unit vector notation and also find the magnitude and direction of (a) $\mathbf{A} + \mathbf{B}$, (b) $3\mathbf{A} + \mathbf{B} - \mathbf{C}$, and (c) $\mathbf{A} - 2\mathbf{B} + \mathbf{C}$.

57. For vectors $\mathbf{A} = -2.00\hat{\mathbf{i}} + 5.00\hat{\mathbf{j}} + 2.00\hat{\mathbf{k}}$, $\mathbf{B} = 4.00\hat{\mathbf{i}} + 1.00\hat{\mathbf{j}} - 2.00\hat{\mathbf{k}}$, $\mathbf{C} = 2.00\hat{\mathbf{i}} + 3.00\hat{\mathbf{j}} - 1.00\hat{\mathbf{k}}$, and $\mathbf{D} = 1.00\hat{\mathbf{i}} - 1.00\hat{\mathbf{j}} + 1.00\hat{\mathbf{k}}$ find scalars l, m, and n such that $l\mathbf{A} + m\mathbf{B} + n\mathbf{C} = \mathbf{D}$.

58. A vector $\mathbf{A} = A_x\hat{\mathbf{i}} + A_y\hat{\mathbf{j}} + A_z\hat{\mathbf{k}}$ makes angles α, β, and γ with the x-, y-, and z-axes, respectively. (a) Apply the definition of the scalar product to the vector $\mathbf{A}$ and the unit vectors $\hat{\mathbf{i}}$, $\hat{\mathbf{j}}$, and $\hat{\mathbf{k}}$, individually to show that

$$\cos(\alpha) = \frac{A_x}{\sqrt{A_x^2 + A_y^2 + A_z^2}}, \quad \cos(\beta) = \frac{A_y}{\sqrt{A_x^2 + A_y^2 + A_z^2}},$$

$$\cos(\gamma) = \frac{A_z}{\sqrt{A_x^2 + A_y^2 + A_z^2}}$$

(b) Show that $\cos^2(\alpha) + \cos^2(\beta) + \cos^2(\gamma) = 1$.

59. For the vectors $\mathbf{A} = 2.00\hat{\mathbf{i}} - 4.00\hat{\mathbf{j}}$, $\mathbf{B} = 3.00\hat{\mathbf{i}} - 2.00\hat{\mathbf{j}}$, and $\mathbf{C} = -1.00\hat{\mathbf{i}} + 2.00\hat{\mathbf{j}}$ find (a) $\mathbf{A} \times \mathbf{B}$, (b) $\mathbf{A} \cdot (\mathbf{B} \times \mathbf{C})$, (c) $\mathbf{A} \times (\mathbf{B} \times \mathbf{C})$.

60. For the vectors $\mathbf{A} = 1.00\hat{\mathbf{i}} - 2.00\hat{\mathbf{j}} + 1.00\hat{\mathbf{k}}$, $\mathbf{B} = 2.00\hat{\mathbf{i}} + 2.00\hat{\mathbf{j}}$, and $\mathbf{C} = -1.00\hat{\mathbf{i}}$ find (a) $\mathbf{A} \times \mathbf{B}$, (b) $\mathbf{A} \times \mathbf{C}$, (c) $\mathbf{C} \cdot (\mathbf{B} \times \mathbf{A})$.

61. Show that if $\mathbf{A} = A_x\hat{\mathbf{i}} + A_y\hat{\mathbf{j}} + A_z\hat{\mathbf{k}}$, $\mathbf{B} = B_x\hat{\mathbf{i}} + B_y\hat{\mathbf{j}} + B_z\hat{\mathbf{k}}$, and $\mathbf{C} = C_x\hat{\mathbf{i}} + C_y\hat{\mathbf{j}} + C_z\hat{\mathbf{k}}$ then

$$\mathbf{A} \cdot (\mathbf{B} \times \mathbf{C}) = \begin{vmatrix} A_x & A_y & A_z \\ B_x & B_y & B_z \\ C_x & C_y & C_z \end{vmatrix}$$

62. Consider vectors $\mathbf{A} = 3.00\hat{\mathbf{i}} - 5.00\hat{\mathbf{j}}$ and $\mathbf{B} = -6.00\hat{\mathbf{i}} + 2.00\hat{\mathbf{j}}$. (a) Calculate the magnitude of the cross product $\mathbf{A} \times \mathbf{B}$. (b) Show that solving the magnitude equation $C = AB \sin(\theta)$ for θ and using the arcsine function of your calculator does *not* lead to the correct value for θ, the angle between the two vectors, when they are tail to tail.

CHAPTER

3

Motion along a Straight Line

In this chapter you should learn

- the definitions of displacement, speed, average velocity, and average acceleration.

- the definitions of instantaneous velocity and instantaneous acceleration.

- how to apply the displacement, velocity, and acceleration descriptions of motion.

- the relations between the graphical and analytic representations of motion descriptions.

- about the constant-acceleration model and how it can be applied to freely falling objects.

- how the calculus can be used to efficiently describe motion.

- (optional) about numerical methods for handling position, velocity, and acceleration data.

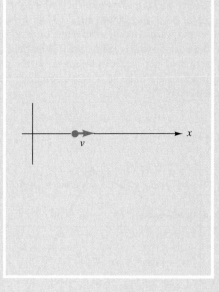

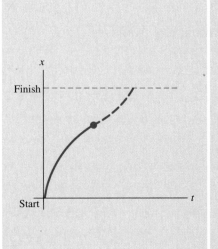

Much of physics is about motion. So to "do physics," you need to be able to describe the motion of objects very precisely. So precisely in fact, that if you took a trip, someone else using your description could duplicate the trip, taking the same route, traveling with the same speeds, doing everything exactly as you did in your original trip.

You can probably think of several ways to describe your trip so that it can be duplicated. One obvious way is to create a position record. That is, to record exactly where you were at each moment. It would be impossible, of course, to list every moment of time. But, if you described where your feet were at the end of each minute during your summer vacation, you could be pretty sure someone could reproduce your trip very accurately!

An alternative way to record your trip is to keep a record of your velocity. That is, you can describe how fast and in what direction you were moving at each moment. Some sports-car rallies are run in this manner. Each driver receives a list of instructions that describes how fast to drive in certain directions and how long to keep each speed before changing. The winner of the rally drives the car that most accurately reproduces the route, arriving at designated check points at exactly the right times.

A third record, a description of your acceleration, although used much less frequently in day-to-day conversation, is employed quite often in physics. With this method, you describe how your motion *changed* each moment: how you increased your speed, slowed down, or changed your direction. This technique is similar to a description of how hard you pressed the accelerator or the brake at each moment and when and how far you turned the steering wheel.

There are other possibilities for records of motion, but the three we use are the position, velocity, and acceleration. The three illustrations provided above are examples of each of these three types of records. However, we need more precise definitions than those given in these descriptions, and that is the task of the first section of this chapter.

By now, you might suspect that because each of the three records—position, velocity, and acceleration—contains a complete description of motion, they must be somehow equivalent to one another. This is indeed the case. In fact, you will soon learn how to start with any one of the three descriptions and produce the other two.

Exactly how you go back and forth between the descriptions depends on the mathematical form of the motion record. The form may be a graph, a function, or just a list of numbers. In the case of graphs you use simple geometry; for functions you use algebra or calculus; and for numbers you can use a calculator or computer. We hope you will notice that in each case you are really doing the same thing, merely with different tools.

All of this probably sounds like a lot to learn. Please be assured that it is not all that bad. To make it a little easier, in this chapter we restrict ourselves to motion along a straight line. After you get the hang of that, we consider motions that have changes in direction (Chapter 4). But, before we start, we need to get our definitions pinned down.

3.1 Position, Velocity, and Acceleration

The first term we need to define is **position.** Position is specified by using a Cartesian coordinate system: three perpendicular straight lines that intersect at a point. The three lines are known as the x-, y-, and z-axes, and the point of intersection is the origin. By specifying distances along each line we can designate the position of a point in our three-dimensional space. We promised to use just one dimension in this chapter, so for now we confine ourselves to objects that move only along the x-axis. Thus, to describe an object's position we have only to specify one number; namely, how far the object is from the origin along the x-axis. We refer to this number as the coordinate of the object. It is a positive number for positions to the right of the origin and a negative number for positions to the left.

Suppose an object is small enough that we can use the particle model described at the end of Section 1.1. In this case the object is treated as a mathematical point moving along the x-axis. We label the particle's initial and final position as shown in Figure 3.1.

The object's **displacement** is defined as the difference between its final position and its original position. We represent the displacement with a Δx, which is read "delta x."

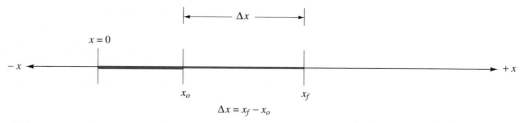

FIGURE 3.1

The displacement Δx is computed by subtracting the original position x_o from the final position x_f.

Note, this is not delta times x; the delta x is just one quantity. The notation we use is

$$\Delta x = x_f - x_o$$

The subscripts o and f refer to the original and final positions. These are *not* necessarily the positions from which the particle starts its motion nor where its motion ceases. The o and f designate the particular original and final positions we are considering out of the entire motion of the object. Note the order: final position minus original position. Whenever we calculate "delta" anything, we always take the final value minus the original value. Hence, the time $\Delta t = t_f - t_o$ may designate the change in time during which the particle traveled through displacement Δx.

Average Velocity and Average Speed

It is almost impossible to talk about motion without using words, such as *velocity* and *speed*. We start by defining exactly what we mean by average velocity and average speed.

> The **average velocity** of an object traveling along the x-axis is defined as the ratio of its displacement to the change in time for that displacement.

Stated algebraically,

$$v_{av} \equiv \frac{\Delta x}{\Delta t} = \frac{x_f - x_o}{t_f - t_o} \tag{3.1}$$

The units of average velocity are meters per second (m/s) in the SI system and feet per second (ft/s) in the British engineering system. Other units you may encounter include miles per hour (mi/h or mph) and kilometers per hour (km/h). A related idea is **average speed:**

$$Average\ speed = \frac{Total\ distance\ traveled}{\Delta t}$$

Notice that for our purposes *velocity* and *speed* have different meanings. The difference between the two definitions can be illustrated by means of a simple example.

EXAMPLE 3.1   "Speed" and "Velocity" Have Different Meanings

Consider the following short trip down a straight hallway. Start from your physics classroom door, which we use as the origin of our coordinate system. Walk 20.0 m down the corridor, then stop, turn around, and walk 5.0 m back toward the classroom and stop 15.0 m from the door. All of this takes 25.0 s. Compute your average velocity and average speed.

SOLUTION   A sketch of your path is shown in Figure 3.2. According to our definition, your average velocity is given by

$$v_{av} = \frac{\Delta x}{\Delta t} = \frac{x_f - x_o}{t_f - t_o} = \frac{15.0 \text{ m} - 0.0 \text{ m}}{25.0 \text{ s} - 0.0 \text{ s}} = 0.600 \text{ m/s}$$

On the other hand, your average speed is

$$Speed_{av} = \frac{Total \ distance \ traveled}{\Delta t} = \frac{20.0 \text{ m} + 5.0 \text{ m}}{25.0 \text{ s}} = 1.00 \text{ m/s} \quad \blacktriangleleft$$

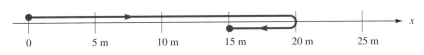

0	5 m	10 m	15 m	20 m	25 m

FIGURE 3.2

You walk 20.0 m down a hallway, turn, and walk back 5.0 m.

From Example 3.1 you can see the distinction we are making between average velocity and average speed. The average velocity depends only on the starting point x_o and the end point x_f of the motion. The average speed depends on the length of the total path traveled between these two points.

Notice that the average velocity can be a negative quantity. Negative average velocity merely means that on our coordinate axis the final position is to the left of the original position.

Concept Question 3.1
For any time interval during its motion is it possible for an object to have zero average velocity but a nonzero average speed? Explain.

Bad Idea!

Another way that people sometimes try to calculate average velocity is by just adding two velocities and dividing by two. *This procedure is wrong*, and maybe this famous trick question will help you to remember that it's wrong: A college student drives a car 1 mi at 30 mph. How fast must the student drive a second mile in order to average 60 mph for the 2-mi trip. Answer: It can't be done; there's no way the student can average 60 mph for the trip! Here's why: Sixty mph is 1 mi/min. In order to average 60 mph for 2 mi, the trip must be driven in 2 min. But going the first mile at 30 mph takes 2 min, so the driver has no time left at all to go the second mile. The moral of the story is this: Use the *definition* of average velocity given by Equation (3.1).

Average Acceleration

When we speak of acceleration we mean the time rate of change of velocity. It is appropriate to use the same word *acceleration* whether we are speeding up or slowing down. However, if the object is slowing down it is sometimes described as decelerating. (We want to caution you not to associate the word "deceleration" with a negative value of *a*. See *The Sign of Acceleration* in Section 3.3.) **Average acceleration** is defined by the relation

$$a_{av} = \frac{\Delta v}{\Delta t} = \frac{v_f - v_o}{t_f - t_o}$$

(3.2)

Because acceleration is a velocity change divided by a time change, it has units, such as miles per hour per second (mi/h/s or mph/s) and meters per second per second ((m/s)/s). You sometimes see meters per second per second (m/s)/s written as meters per second squared (m/s^2). It is clearer that (m/s)/s is a velocity unit divided by a time unit, but m/s^2 is mathematically equivalent and somewhat easier to say.

EXAMPLE 3.2   *Yes, But What's a Square Second?*

You are driving East through a town at 35.0 mph. The speed limit increases to 55.0 mph, so you increase your speed to the new limit in 10.0 s. Your average acceleration is calculated

$$a_{av} = \frac{v_f - v_o}{t_f - t_o} = \frac{55.0 \text{ mi/h} - 35.0 \text{ mi/h}}{10.0 \text{ s}} = 2.00 \frac{\text{mi/h}}{\text{s}}$$

That is, you were increasing your velocity at a rate of 2.00 mph each second. Since 1.00 mph = 0.447 m/s, we can write this result as

$$a_{av} = \left(2.00 \frac{\text{mi/h}}{\text{s}}\right)\left(\frac{0.447 \text{ m/s}}{1.00 \text{ mi/h}}\right) = 0.894 \frac{\text{m/s}}{\text{s}} = 0.894 \text{ m/s}^2 \quad \blacktriangleleft$$

Instantaneous Velocity

Next, we attach a meaning to the concept of velocity at a single instant of time. By *instant of time* we mean one mathematical point in time with no duration at all. When we have explored what this concept means, we shall title our definition **instantaneous velocity.** How meaning can be attached to the concept of velocity at an instant in time is a question that has puzzled humankind for centuries. Zeno of Elea, a philosopher from ancient Greece, presented the problem this way. A mathematical point in time has no extent. Thus, nothing can move from one place to another *during* an instant, since the instant has no duration. But, all of time is merely a collection of points of time. So, if nothing can move at any one of them, and all time is but a collection of such points, nothing can move! Depressing thought, isn't it? You are trapped here reading your physics text for all eternity! Perhaps you should stop and get up to get a drink of water to reassure yourself that this can't be right. When you get back we will try to resolve Zeno's paradox.

The fact of the matter is that we cannot directly *measure* the velocity of an object at a single point in time. Every method you can imagine actually measures an average velocity. This restriction doesn't mean that the situation is hopeless, only that we must be more subtle. Suppose as you are speeding up on a highway, a patrolman wants to know your instantaneous velocity. How can he attempt to measure it? One possibility might be for him to fly in a helicopter and use a stop watch to measure the time it takes your car to travel 100. m between two white lines drawn on the highway. This approach provides only an average velocity over a time interval Δt of, perhaps, 5.00 s. Another patrolman might lay two electronic cables across the road 10.0 m apart. A clock could be made to start when your front tire hits the first cord and stop when your front tire hits the second. This method also gives an average velocity, but the Δt is shorter, say, 0.480 s. In an extreme case we could imagine a fanatical officer with a pulsed laser who could detect your motion over the distance of 2.00 m during a time of only 9.52 hundredths of a second. But, alas, this technique too merely provides an average velocity, albeit over a very short time interval.

If all three of the officers in our story above were to make their measurements together, that is, with the cables stretched across the road somewhere between painted

An officer explains the difference between average velocity and instantaneous velocity.

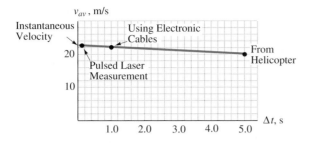

FIGURE 3.3

The instantaneous velocity is found by extending a smooth curve through the points on a v_{av}-versus-Δt graph to the vertical axis where $\Delta t = 0$. The value at this point is the instantaneous velocity.

white lines, and the laser pulses striking the car while its front wheels were between the cables, the officers could compare notes. If they were to plot their results on a graph, it might look like Figure 3.3. In this figure the average velocities are plotted on the vertical axis, and the duration of the Δt intervals along the horizontal axis. Because you are speeding up, each officer measures a different average velocity. The point we want on our graph is a point *exactly* on the vertical axis where Δt equals zero. But we can't get there; we can only come close. Nonetheless, you can see in Figure 3.3 that a line drawn through all three measurements *approaches* the vertical axis in a smooth way. It appears that if we extrapolate our measurements of average velocity to smaller and smaller Δt-values, the values actually approach a unique point on the vertical axis. That value of average velocity that we approach as Δt goes to zero is what we shall mean by *instantaneous velocity*.

This process of choosing smaller and smaller time intervals is actually an application of the mathematical technique of taking a limit. We formally define the instantaneous velocity $v(t)$ using the mathematical notation for this limiting process:

$$v(t) \equiv \lim_{\Delta t \to 0} v_{av} = \lim_{\Delta t \to 0} \frac{\Delta x}{\Delta t} \qquad (3.3)$$

Equation (3.3) is the definition of the derivative of the position function with respect to time. Therefore, we can write this definition in its complete mathematical form:

$$v(t) = \frac{dx(t)}{dt} \qquad (3.4)$$

The instantaneous velocity function is the derivative with respect to the time of the position function.

Instantaneous Acceleration

Instantaneous acceleration is defined in a fashion analogous to the method for defining instantaneous velocity. That is, **instantaneous acceleration** is the value approached by the average acceleration as the time interval for the measurement becomes closer and closer to zero.

$$a(t) \equiv \lim_{\Delta t \to 0} a_{av} = \lim_{\Delta t \to 0} \frac{\Delta v}{\Delta t} \qquad (3.5)$$

Equation (3.5) is the definition of the derivative of the velocity function with respect to time; therefore, we can write the definition of instantaneous acceleration in its complete mathematical form:

$$a(t) = \frac{dv(t)}{dt} \qquad (3.6)$$

The instantaneous acceleration function is the derivative with respect to time of the velocity function.

To summarize, the ability to perform the derivative operation permits us to find the velocity from the position function and the acceleration from the velocity function. The technique is demonstrated in Example 3.3.

EXAMPLE 3.3 Using the Derivative

After attaining a velocity of 2.00 m/s a cyclist starts peddling harder for 5.00 s, achieving the following velocity function:

$$v(t) = (0.600 \text{ m/s}^3)t^2 + 2.00 \text{ m/s} \qquad \text{for } 0.0 \le t \le 5.00 \text{ s}$$

Find the cyclist's acceleration function during this interval.

SOLUTION The acceleration function is calculated from Equation (3.6):

$$a(t) = \frac{dv(t)}{dt} = \frac{d}{dt}[(0.600 \text{ m/s}^3)t^2 + 2.00 \text{ m/s}] = 2(0.600 \text{ m/s}^3)t + 0$$

$$a(t) = (1.20 \text{ m/s}^3)t$$

This is a linear function of t with a slope of 1.20 m/s^3. ◀

We have now completed our first major task, that of introducing the formal definitions of kinematic quantities used to describe motion. Using these definitions we can progress from a position record to a velocity record to an acceleration record. It is possible to reverse this process and progress from an acceleration record to a velocity record to a position record. This reverse process can be accomplished using graphical means or integral calculus. In the latter portion of Section 3.2 we describe graphical methods. Techniques using calculus are included in optional Section 3.4 for students who are familiar with integral calculus.

3.2 Graphical Interpretation of Position, Velocity, and Acceleration

As mentioned in the introduction to this chapter, we are concerned with three different descriptions of motion: position, velocity, and acceleration. Our goal is to learn how, given any one of these records, to obtain the other two. We now describe how to do this when we are given motion records in graphical form.

Average Velocity

We begin with a graphical record of position and see how to obtain the velocity record. Figure 3.4(a) shows a graph of position versus time. The graph describes the motion of a person walking along a straight line. The person started at $x = 2.0$ m when $t = 0.0$ s and walked out to $x = 10.0$ m, arriving there at $t = 15.0$ s. Then the person turned around and walked back to $x = 5.0$ m arriving when $t = 20.0$ s.

Suppose we wish to calculate the average velocity between $t = 5.0$ s and $t = 15.0$ s. From the graph we see that at $t_o = 5.0$ s the position is $x_o = 7.0$ m, and at $t_f = 15.0$ s the position is $x_f = 10.0$ m. We substitute these values into our definition:

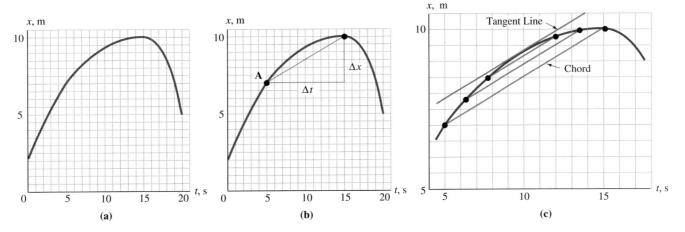

FIGURE 3.4

(a) A position-versus-time record. (b) The average velocity from $t = 5$ s to $t = 15$ s is the slope of the straight line connecting points A and B. (c) The instantaneous velocity of $t = 10$ s is the slope of the line tangent to the position curve at this time.

$$v_{av} = \frac{x_f - x_o}{t_f - t_o} = \frac{10.0 \text{ m} - 7.0 \text{ m}}{15.0 \text{ s} - 5.0 \text{ s}} = 0.30 \text{ m/s}$$

Let's look at this calculation in terms of the graph. In Figure 3.4(b) we identify the points A (t_o, x_o) and B (t_f, x_f) with large dots. We join the dots with a straight line and use this line as the hypotenuse of a right triangle. You can see that the vertical leg of the right triangle is equal in length to $(x_f - x_o)$, whereas the horizontal leg is $(t_f - t_o)$. (We speak of the "length" of this triangle leg, although it is not a length in the ordinary sense because its unit is time, not length. Nonetheless, it is easier to talk this way. No harm is done so long as you remember to append time as the unit whenever you use this "length.") The ratio of the length of the vertical leg to the horizontal leg is, by definition, the geometric **slope** of the line AB. But this ratio also matches our definition of average velocity. We are led to conclude:

> The average velocity between two points in time can be obtained from a position-versus-time graph by computing the slope of the straight line joining the coordinates of the two points.

Instantaneous Velocity

How does the instantaneous velocity look on a position-versus-time graph? Our definition (Eq. (3.3)) says to take the average velocity over smaller and smaller Δt intervals to determine the limit of such a process. We do this for our previous example by calculating the instantaneous velocity at $t = 10.0$ s.

In Figure 3.4(c) the region around 10 s has been enlarged. We know that the average velocity over a given time interval is the slope of the line joining the end points of the interval. Therefore, in Figure 3.4(c) we draw a series of chords the end points of which become closer and closer to $t = 10.0$ s. You can see what is happening as Δt becomes smaller and smaller. The slope of the chord does approach a limit; this limit is the slope of what is called the **tangent line.** This tangent line touches the curve at only one point, but its slope is defined by the limit of the slopes of the chords as Δt becomes smaller and smaller.

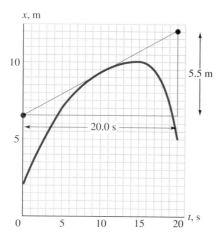

FIGURE 3.5

The instantaneous velocity at $t = 10.0$ s is the slope of the tangent to the curve at $t = 10.0$ s.

For position-versus-time graphs:

> When you take the slope of a line connecting two points on a tangent line, you are computing the *instantaneous velocity at the point of tangency.*
> When you take the slope of a line connecting two points on the curve itself, you are computing the *average velocity between the two points.*

Concept Question 3.2
Is it possible during the same time interval for both the position and velocity graphs of an object to consist of horizontal, straight lines? Explain.

> The instantaneous velocity at time t is the slope of the tangent line drawn to the position-versus-time graph at that time.

Fortunately, you don't have to try to take the slope from those very tiny chords. Once you draw the tangent, you can take its slope over as large a triangle as you wish since the slope of a straight line is the same everywhere. This explanation of the instantaneous velocity is consistent with the graphical interpretation of the velocity function given by Equation (3.4): The derivative of a function at a point is the slope of the function at that point.

EXAMPLE 3.4   *Going Off on a Tangent*

For the instant $t = 10.0$ s, find the instantaneous velocity of the particle the position-versus-time graph of which is given in Figure 3.5.

SOLUTION   First draw the tangent line that just touches the curve at $t = 10.0$ s. (If you have trouble judging the tangent, put two small dots on the curve just on either side of the $t = 10.0$ s point. Then draw a straight line that passes through both of them. Your line should be quite close to the tangent.) Now pick any two convenient times and put points *on the tangent line* at these times. In Figure 3.5 we use $t = 0.0$ s and $t = 20.0$ s. Next draw a right triangle with the tangent as the hypotenuse and the end points as vertices. Finally, calculate the slope using

$$v_{\text{inst}} = \left(\frac{x_f - x_o}{t_f - t_o}\right)_{\text{on tangent line}} = \frac{11.5 \text{ m} - 7.5 \text{ m}}{20.0 \text{ s} - 0.0 \text{ s}} = 0.20 \text{ m/s}$$

The instantaneous velocity at $t = 10.0$ s is 0.20 m/s and is written $v(10.0 \text{ s}) = 0.20$ m/s.   ◀

We now know how to get the *entire* velocity graph from the position graph. At each point of interest, we must draw a tangent to the curve of the position graph. The velocity of the object at the point in time where the tangent is drawn is given by the slope of the tangent. A velocity-versus-time graph may be constructed by plotting the results of this tangent-taking process for the range of time being graphed. Of course, if we are fortunate enough to know the position *function*, we need only compute the time derivative of this function to obtain the velocity function (which we can then graph).

Average and Instantaneous Acceleration

Now you're in for some good news. In order to compute the acceleration we need only apply this same procedure of taking slopes. This time, however, we start with a velocity graph instead of a position graph.

> The average acceleration between two points in time is equal to the slope of the chord connecting the points on a velocity-versus-time graph.

The instantaneous acceleration may be computed similarly to the instantaneous velocity, but this time from a velocity-versus-time graph:

> The instantaneous acceleration at a time t is the slope of the tangent drawn to the velocity-versus-time graph at that time.

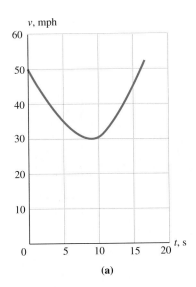

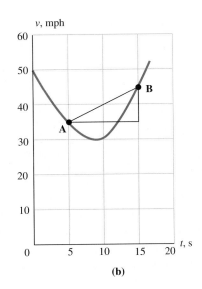

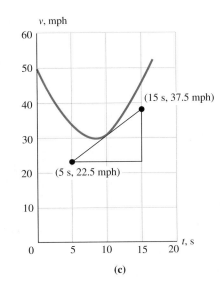

FIGURE 3.6

For Example 3.5. (a) Velocity versus time. (b) The slope of the line $\overline{AB}$ is the average acceleration a_{av}. (c) The slope of the tangent at $t = 10.0$ s is the instantaneous acceleration a_{inst}.

Again, this statement is consistent with the graphical interpretation of the derivative of a function: The derivative of the velocity function is just the slope of the velocity curve. We now tie all this together in Example 3.5.

EXAMPLE 3.5   *Averages and Instants: Chords and Tangents*

From the velocity-versus-time graph shown in Figure 3.6(a) find (a) the average acceleration between times $t = 5.0$ s and $t = 15.0$ s, and (b) the instantaneous acceleration at $t = 10.0$ s.

SOLUTION   (a) As shown in Figure 3.6(b), we connect the points A (5.0 s, 35.0 mph) and B (15.0 s, 45.0 mph) with a straight line. The slope of this line is the average acceleration:

$$a_{av} = \frac{\Delta v}{\Delta t} = \frac{v_f - v_o}{t_f - t_o} = \frac{45.0 \text{ mph} - 35.0 \text{ mph}}{15.0 \text{ s} - 5.0 \text{ s}} = 1.00 \text{ mph/s}$$

$$= (1.00 \text{ mph/s})\left(\frac{0.447 \text{ m/s}}{1.00 \text{ mph}}\right) = 0.447 \text{ m/s}^2$$

(b) In Figure 3.6(c), we draw a tangent to the curve at $t = 10.0$ s then calculate the slope of the tangent between $t = 5.0$ s and $t = 15.0$ s.

$$a_{inst} = \left(\frac{\Delta v}{\Delta t}\right)_{tangent} = \left(\frac{v_f - v_o}{t_f - t_o}\right)_{on \text{ tangent line}} = \frac{37.5 \text{ mph} - 22.5 \text{ mph}}{15.0 \text{ s} - 5.0 \text{ s}}$$

$$= 1.50 \text{ mph/s} = 0.670 \text{ m/s}^2$$

Notice that at $t = 9.0$ s the tangent is horizontal, the slope is zero and, therefore, the acceleration is zero. At that *instant* the velocity is not changing.   ◀

Position from Velocity Graphs

Given graphical records, we now know how to progress from position to velocity and from velocity to acceleration. What about the reverse process? Given the velocity record, how can we calculate displacement? For a hint on how to proceed, look at the special case

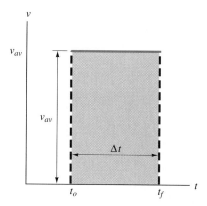

FIGURE 3.7

Because v_{av} is a constant, the "area" under this v-versus-t curve is the displacement $\Delta x = v_{av} \Delta t$.

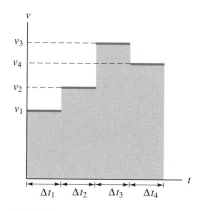

FIGURE 3.8

For each segment of the motion, the velocity is a different constant. The displacement Δx_i during the ith interval is the area $v_i \, \Delta t_i$, so that the total displacement is $\Delta x = v_1 \, \Delta t_1 + v_2 \, \Delta t_2 + v_3 \, \Delta t_3 + v_4 \, \Delta t_4$.

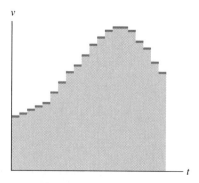

FIGURE 3.9

When the velocity changes in a complex fashion, we may approximate the velocity function by breaking it up into many segments, each with a constant velocity. The total displacement is the net area under the curve (shaded region).

of constant velocity. For constant velocity the average velocity and the instantaneous velocity are the same. We can rearrange Equation (3.1) to read

$$\Delta x = v_{av} \, \Delta t$$

Now look at the velocity-versus-time graph in Figure 3.7. What is the product of v_{av} and Δt? It is simply the "area" of the rectangle with height v_{av} and width Δt. Once again we have strange units for something we call an "area." One "length" has the unit of meters per second, and the other has the unit of seconds. The product of these two "lengths" has meters as the unit—fine for displacement Δx, but strange for area. The area of the rectangle in Figure 3.7 is the displacement $\Delta x = v_{av} \, \Delta t$.

Next consider the more complex motion graphed in Figure 3.8, made up of several segments in which the velocity is constant but different in each of the segments. For each interval the displacement is the rectangular area under the constant-velocity segment. Hence, for the combined trip the displacement is the area under the entire curve. For an even more complex trip we could continue in this fashion by breaking the trip into many short segments, each with a constant velocity. Figure 3.9 shows an example. Here again the total distance traveled is the area under the curve.

By now you have probably figured out where we are going with these ideas. If the "area-under-the-curve" argument holds for curves like Figure 3.9, why not for smooth curves like Figure 3.10? It turns out that indeed it does hold. We will not prove this here, but rather content ourselves with the intuitive argument presented above. A rigorous proof can be found in most introductory texts on calculus. Our rule can be stated as follows:

> Given a velocity-versus-time graph, the displacement during an interval between times t_o and t_f is the "area" bounded by the velocity curve, the $v = 0$ axis, and the two vertical lines $t = t_o$ and $t = t_f$.

For semantic reasons this rule is sometimes shortened:

> The displacement is the area under the velocity curve.

This statement provides an easy way to remember the rule, but don't take it too literally. Sometimes the area *under* the curve is actually *above* the curve as we see in Example 3.6.

EXAMPLE 3.6   *If You Draw a Negative Area Does Your Pencil Grow Longer?*

Consider the object the velocity-versus-time graph of which is given in Figure 3.11. Calculate the displacement during each 5.0-s interval and reconstruct the position history of the object.

SOLUTION   (a) From $t = 0.0$ s to $t = 5.0$ s the area under the curve is approximately a rectangle the area formula of which is *Area = Height × Base*. (We ignore the slight curvature of the graphs where the slope of the lines changes.) Thus, the displacement Δx_1 during the first 5.0-s interval is

$$\Delta x_1 = (6.0 \text{ m/s}) \times (5.0 \text{ s}) = 30. \text{ m}$$

(b) From 5.0 s to 10.0 s the area is approximately a trapezoid. The area of a trapezoid is *area = ½(Height₁ + Height₂) × (Base)*. Thus,

$$\Delta x_2 = \frac{1}{2}(6.0 \text{ m/s} + 10.0 \text{ m/s})(5.0 \text{ s}) = 40. \text{ m}$$

(c) From 10.0 s to 15.0 s the area is a triangle, *area* = $\frac{1}{2}$(*Height* × *Base*). Thus,

$$\Delta x_3 = \frac{1}{2}(10.0 \text{ m/s})(5.0 \text{ s}) = 25 \text{ m}$$

(d) Here we must be careful! Although the area between the curve and the $v = 0$ axis is a rectangle, the height of the rectangle is *negative* 5.0 m/s. Thus,

$$\Delta x_4 = (-5.0 \text{ m/s})(5.0 \text{ s}) = -25 \text{ m}$$

We obtain the total displacement of the particle from $t = 0$ s to any time t by summing all of the displacements up to that time. Assuming that this particle started from $x = 0$, we have for the *total* displacement at time t:

Position History

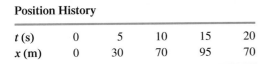

t (s)	0	5	10	15	20
x (m)	0	30	70	95	70

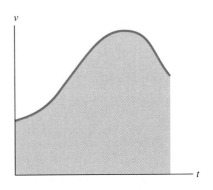

FIGURE 3.10

Smooth version of "curve" in Figure 3.9.

Velocity from Acceleration Graphs

Now for some more good news. As before, we find that the rules connecting acceleration to velocity are the same as those connecting velocity to displacement:

> Given an acceleration-versus-time graph, the change in velocity between $t = t_o$ and $t = t_f$ is the "area" bounded by the acceleration curve, the $a = 0$ axis, and the vertical lines $t = t_o$ and $t = t_f$.

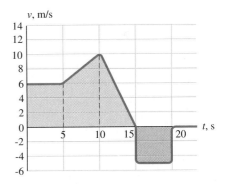

FIGURE 3.11

Velocity versus time.

EXAMPLE 3.7   *Another Strange Area of Physics*

An object is traveling with a velocity of 8.0 m/s when it experiences an acceleration the graph of which is shown in Figure 3.12(a). Calculate the velocity of the object at 1-s intervals and plot a velocity-versus-time graph for the object.

SOLUTION   The "areas" for each 1-s interval represent the *change* in velocity during that interval:

INTERVAL	Δv
$0.0 \text{ s} \le t \le 1.0 \text{ s}$	$(1.0 \text{ s})(-2.0 \text{ m/s}^2) = -2.0 \text{ m/s}$
$1.0 \text{ s} \le t \le 2.0 \text{ s}$	$(1.0 \text{ s})(-2.0 \text{ m/s}^2) = -2.0 \text{ m/s}$
$2.0 \text{ s} \le t \le 3.0 \text{ s}$	$(1.0 \text{ s})(-2.0 \text{ m/s}^2) = -2.0 \text{ m/s}$
$3.0 \text{ s} \le t \le 4.0 \text{ s}$	$(1.0 \text{ s})(+3.0 \text{ m/s}^2) = +3.0 \text{ m/s}$
$4.0 \text{ s} \le t \le 5.0 \text{ s}$	$(1.0 \text{ s})(+3.0 \text{ m/s}^2) = +3.0 \text{ m/s}$
$5.0 \text{ s} \le t \le 6.0 \text{ s}$	$(1.0 \text{ s}) \frac{1}{2}(3.0 \text{ m/s}^2 + 4.0 \text{ m/s}^2) = +3.5 \text{ m/s}$
$6.0 \text{ s} \le t \le 7.0 \text{ s}$	$(1.0 \text{ s}) \frac{1}{2}(4.0 \text{ m/s}^2 + 5.0 \text{ m/s}^2) = +4.5 \text{ m/s}$
$7.0 \text{ s} \le t \le 8.0 \text{ s}$	$(1.0 \text{ s}) \frac{1}{2}(5.0 \text{ m/s}^2 + 6.0 \text{ m/s}^2) = +5.5 \text{ m/s}$

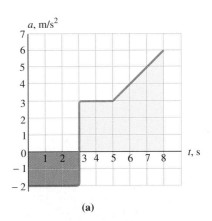

(a)

FIGURE 3.12

(a) Acceleration versus time.

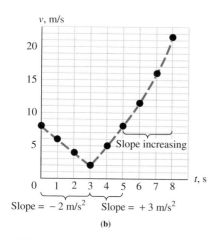

FIGURE 3.12

(b) Velocity versus time.

We next construct a table of *total* velocity versus time:

Velocity History

t (s)	0	1	2	3	4	5	6	7	8
v (m/s)	8.0	6.0	4.0	2.0	5.0	8.0	11.5	16.0	21.5

A graph of these points is shown in Figure 3.12(b). Notice that for the intervals $0.0 \leq t \leq 3.0$ s and 3.0 s $\leq t \leq 5.0$ s the data points fall on straight lines. This result should come as no surprise because the slope of the velocity-versus-time curve is the acceleration, and in each of these intervals the acceleration is constant (see Fig. 3.12(a)). In the interval 5.0 s $\leq t \leq 8.0$ s the acceleration is not constant, and a smooth curve drawn through the velocity-versus-time curve in this interval appears to have a slope that is increasing, as it must from the data in Figure 3.12(a). ◀

We have now completed our second major task. We know how to go between position, velocity, and acceleration using graphical records. To deduce velocity from position graphs or acceleration from velocity graphs, we take slopes. To reverse the process and deduce velocity changes from acceleration graphs or position changes from velocity graphs, we calculate areas.

3.3 The Constant-Acceleration Model

Most motion in the world is quite complex. However, we can often obtain a quite reasonable approximation of complex motions by using simplified models that retain only the most essential features of the motion. One such useful model is that of constant acceleration. This model is attractive to use (and tempting to overuse!) because the results are relatively simple, and answers can be calculated using only algebra.

We use the results from the previous section on graphical records to find these algebraic relations. A graph of constant acceleration is shown in Figure 3.13(a). The velocity change from $t = 0$ to some final time t is the area under the acceleration curve:

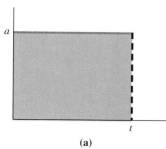

$$\Delta v = v_f - v_o = at \tag{3.7}$$

For a constant acceleration, the slope of the velocity-versus-time curve is a constant. A graph of velocity versus time with a constant slope (acceleration) is shown in Figure 3.13(b). The displacement is the area under this curve

$$\Delta x = x_f - x_o = \frac{1}{2}(v_f + v_o)t \tag{3.8}$$

To simplify the notation we write v_f as v, dropping the subscript. Equations (3.7) and (3.8) can then be written

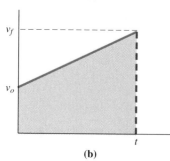

$$v = v_o + at \tag{3.9}$$

FIGURE 3.13

(a) Constant acceleration. (b) Velocity versus time with constant acceleration.

$$x = x_o + \frac{1}{2}(v + v_o)t \tag{3.10}$$

We challenge you to eliminate the time t between Equations (3.9) and (3.10) to show that

$$v^2 = v_o^2 + 2a(x - x_o) \qquad (3.11)$$

If you substitute for the velocity v from Equation (3.9) into Equation (3.10), you can show that

$$x = x_o + v_o t + \frac{1}{2}at^2 \qquad (3.12)$$

These last three equations relate the six variables: x, x_o, v, v_o, a, and t for constant accelerations. If for one of these equations you have all the values except one, you can solve for the unknown variable. These observations suggest the following procedure for problem solving:

Problem-Solving Strategy

1. Draw a picture of the situation described. We can't overemphasize the importance of this step.
2. Draw a coordinate axis on your picture and label the origin $x = 0$.
3. Translate the words describing the problem into equalities using the letters x, x_o, v, v_o, a, or t. Express all numerical values in a consistent unit system (preferably SI).
4. If you don't have four variables the values of which are specified in step 3, look closely at the problem to identify a *special condition* regarding the object at the point in question. There may be some information implied in the problem. In any case you can't proceed without having four known quantities! An exception to this rule is the case in which three known quantities and the unknown all appear in Equation (3.9), in which case Equation (3.9) may be solved directly for the unknown.
5. Identify the letter that corresponds to the unknown quantity you are trying to find.
6. Examine the five letters for the four quantities you know plus the unknown. Pick the one of the four kinematic equations that has these five letters in it.
7. Solve the equation algebraically for the unknown. At this point keep only letters in your equation.
8. After the equation is solved, put in numbers and calculate an answer.
9. Check for algebra errors by putting only units in your equation. Reduce the units to the simplest form. Do they make sense? If the answer is supposed to be time and the units are seconds, that's good. If the units are meters, check your work!
10. Finally, think about your answer. Does it make sense? Birds don't fly at thousands of meters per second; it takes seconds, not years, for automobiles to accelerate to reasonable speeds.

> I don't know where to start! What do I do first?

> How do I decide which equation to use?

Let's apply these steps to Example 3.8.

EXAMPLE 3.8   *Coming Up to Speed Using the Constant-Acceleration Model*

An automobile accelerates from 54.0 km/h to 90.0 km/h. How far does it travel in this time if the acceleration is a constant 2.52 (km/h)/s?

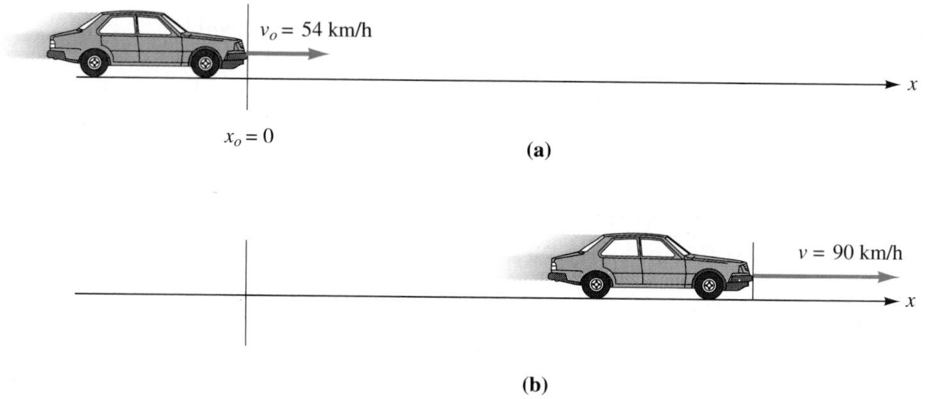

FIGURE 3.14

Strategy step 1 of Example 3.8.

SOLUTION   Strategy steps (1) and (2) are shown in Figure 3.14. Results for steps (3), (4), and (5) are summarized in the table below:

$$v_o = 54.0 \text{ km/h} = 15.0 \text{ m/s}$$

$$v = 90.0 \text{ km/h} = 25.0 \text{ m/s}$$

$$a = 2.52 \text{ km/h/s} = 0.700 \text{ m/s}^2$$

$$x_o = 0.0 \text{ m}$$

$$x = ?$$

In the first three lines of our table we identified the initial velocity, the final velocity, and the acceleration, all of which can be identified from the problem statement. We made the choice for x_o back in strategy step (2) when we placed our coordinate system on the drawing. Any choice would have been correct, but choosing $x_o = 0$ makes things a bit simpler.

Based on the list in our table, we can now choose an equation. We are looking for an equation in which the time does not appear. Apparently, Equation (3.11) is the one:

$$v^2 = v_o^2 + 2a(x - x_o)$$

Applying strategy step (7) we solve for x:

$$x = x_o + \frac{v^2 - v_o^2}{2a}$$

We follow strategy step (8) and put in numbers,

$$x = \frac{(25.0 \text{ m/s})^2 - (15.0 \text{ m/s})^2}{2 \ (0.700 \text{ m/s}^2)} = 286 \text{ m}$$

Checking the units as in strategy step (9), we find

$$[x] = \frac{(\text{m/s})^2}{\text{m/s}^2} = \text{m}$$

Concept Question 3.3
Can an automobile be in motion toward the east and have an acceleration the direction of which is toward the west? Explain.

Finally (strategy step (10)), we think about the answer. Yes, 286 m does sound reasonable based on our driving experience.   ◀

The Sign of Acceleration

Before going further, we must raise an extremely important point regarding the algebraic sign (+ or −) of various kinematic quantities, acceleration in particular. If you know that an object's coordinate x is positive, you know it is located to the right of the origin. If x is negative, the particle is to the left of the origin.

If you know the sign of the velocity, what do you know? If the velocity is positive, you know the object is moving toward the right; if the velocity is negative, you know the object is moving toward the left. Notice, you cannot tell from the sign of the velocity alone *where* the object is located, only the direction of its motion.

Now what about the sign of the acceleration? *You may be tempted to say* that if the acceleration is positive the object is speeding up, and if the acceleration is negative the object is slowing down. Unfortunately, this simple interpretation is *wrong*. An object with positive acceleration can be slowing down, and an object with negative acceleration can be speeding up!

Acceleration tells us about *changes* in velocity. If your acceleration is positive, your velocity is changing to be more toward the right. But there are *two* ways this can happen. If you are already moving to the right and your change in velocity is toward the right, then you are speeding up. On the other hand, if you are moving to the left (v is negative) and your change in velocity is to the right, you are slowing down.

Conversely, if your acceleration is negative and your velocity is positive, you are slowing down. But if your velocity is negative and your acceleration is negative, then the acceleration is in the same direction as your velocity and you are speeding up (to the left). Thus, we are led to the following correct interpretation of the sign of the acceleration:

If velocity and acceleration have the same sign, the object is speeding up. If velocity and acceleration have opposite signs, the object is slowing down. In either case, the sign of the acceleration indicates the direction in space in which velocity changes are pointing.

Getting these signs correct is extremely important. Example 3.9 shows how to handle the signs of accelerations.

Concept Question 3.4
Can an object maintain its motion in the same direction while reversing the direction of its acceleration? Explain.

EXAMPLE 3.9   *The Sign of Acceleration*

Two cars are approaching each other along a two-lane highway. Car A is traveling east at 72.0 km/h, and car B is traveling west at 54.0 km/h. When the cars are 50.0 m apart, the drivers recognize each other and both begin to slow down at 10.8 (km/h)/s so they can exchange greetings as they pass. When and where will the cars pass?

SOLUTION   Our strategy is to describe both cars with the general equations for position when acceleration is constant. Following our problem-solving procedure, we draw a picture and put a coordinate system on it as shown in Figure 3.15. This time, our table has two columns, one for each car. Note carefully that we have converted km/h to m/s using the methods illustrated in Example 1.1.

CAR A	CAR B
$x_{Ao} = 0.0$	$x_{Bo} = +50.0$ m
$v_{Ao} = +20.0$ m/s	$v_{Bo} = -15.0$ m/s
$a_A = -3.00$ m/s^2	$a_B = +3.00$ m/s^2

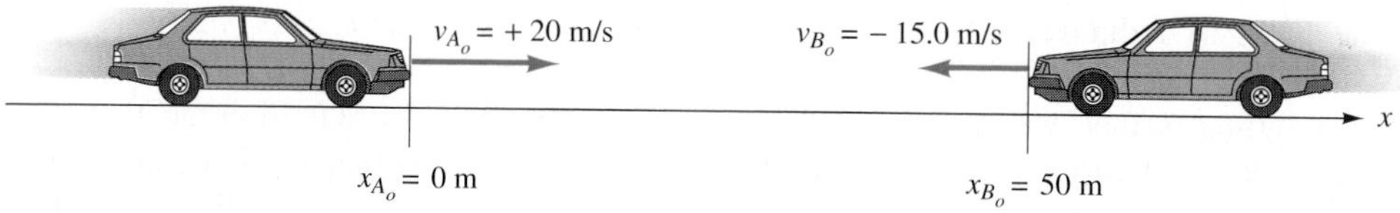

FIGURE 3.15

Two cars approaching each other. See Example 3.9.

The initial positions are simply the coordinates of the two cars on the coordinate system shown in Figure 3.15. The initial velocity for car A is positive because it moves toward the right. Similarly, the velocity for car B is negative because its motion is to the left.

The sign of acceleration for car A seems quite natural; because car A is moving to the right and slowing down, its acceleration is negative. The acceleration for car B is perhaps less obvious. To slow down car B's motion to the left requires an acceleration to the right; that is, a *positive* acceleration.

Continuing, we note that we have only three pieces of information for each car, and we know we need four. Step (4) of our problem-solving strategy suggests that we look for a special condition. Rereading the problem we realize that the coordinate of interest is the point " . . . when the cars pass each other." This statement provides a special condition:

$$x_A = x_B$$

So if we write

$$x_A = x_{Ao} + v_{Ao}t + \frac{1}{2}a_A t^2 = (20.0 \text{ m/s})t - (1.50 \text{ m/s}^2)t^2$$

$$x_B = x_{Bo} + v_{Bo}t + \frac{1}{2}a_B t^2 = (50.0 \text{ m}) - (15.0 \text{ m/s})t + (1.50 \text{ m/s}^2)t^2$$

Then equating x_A to x_B, we can solve for the time when the special condition is satisfied:

$$(20.0 \text{ m/s})t - (1.50 \text{ m/s}^2)t^2 = (50.0 \text{ m}) - (15.0 \text{ m/s})t + (1.50 \text{ m/s}^2)t^2$$

$$(3.00 \text{ m/s}^2)t^2 - (35.0 \text{ m/s})t + 50.0 \text{ m} = 0$$

This is a quadratic equation of the form $At^2 + Bt + C = 0$, with solutions

$$t = \frac{-B \pm \sqrt{B^2 - 4AC}}{2A}$$

$$= \frac{-(-35.0 \text{ m/s}) \pm \sqrt{(-35.0 \text{ m/s})^2 - 4(3.00 \text{ m/s}^2)(50.0 \text{ m})}}{2(3.00 \text{ m/s}^2)}$$

$$= 1.67 \text{ s and } 10.0 \text{ s}$$

The first root is the solution we are seeking: 1.67 s after the cars start braking they pass. What does the second root ($t = 10.0$ s) mean? If each car continues with its acceleration (-3.00 m/s^2 for car A and $+3.00$ m/s^2 for car B), they eventually come momentarily to rest. If by shifting into reverse they continue to maintain these same accelerations, their velocities change directions and they once again meet at $t = 10.0$ s after their initial braking.

To find the position at which the cars pass, we can solve for either x_A or x_B. Choosing the simpler, we have

$$x_A = x_{Ao} + v_{Ao}t + \frac{1}{2}a_A t^2 = (20.0 \text{ m/s})(1.67 \text{ s}) - (1.50 \text{ m/s}^2)(1.67 \text{ s})^2 = 29.2 \text{ m}$$

◄

Freely Falling Objects

In 1638, Italian physicist Galileo Galilei Linceo put forth the definition of constant acceleration in his book *Dialogues and Mathematical Demonstrations Concerning Two New Sciences Pertaining to Mechanics and Local Motion.* In the *Dialogues* Galileo not only defined constant acceleration but boldly asserted that objects really do fall with constant acceleration and furthermore this acceleration is the same for all objects.

Galileo's claim was hotly debated for years, many people arguing that heavier objects obviously fall faster than lighter ones. Legend has it that Galileo tried to prove his point by dropping two stones of different weights from the leaning tower of Pisa. If he did so, observers would have seen the two stones arriving at very nearly the same instant. Any discrepancy would have been caused by air friction, but Galileo would have had a difficult time convincing his detractors. Galileo wanted his readers to accept the following model for the motion of *freely falling* objects.

> All objects near the surface of the earth, moving under the influence of gravity only, accelerate toward the earth with constant acceleration.

Galileo's model is an accurate description of objects traveling vertically at low speeds near the surface of the earth. Here is an important feature of this model you should not miss: The acceleration of an object near the earth's surface is downward and constant, whether the object is rising, falling, or stopped at the top of its trajectory.

If we were to idealize Galileo's experiment by dropping an object in a vacuum and finding its instantaneous velocity at one-tenth of a second intervals, we would obtain data as shown in the left two columns of Table 3.1.

TABLE 3.1 **Data from "Galileo's experiment"**

TIME (s)	VELOCITY (m/s)	CHANGE IN VELOCITY (m/s)	ACCELERATION (m/s²)
0.000	0.000		
0.100	−0.980	−0.980	−9.80
0.200	−1.960	−0.980	−9.80
0.300	−2.940	−0.980	−9.80
0.400	−3.920	−0.980	−9.80
0.500	−4.900	−0.980	−9.80

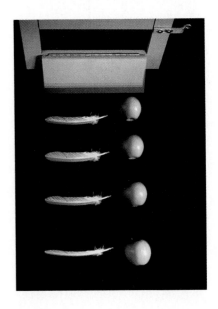

In the third column of Table 3.1 is the change in velocity occurring during each time interval. The last column is the average acceleration during the interval calculated from $a_{av} = \Delta v / \Delta t$. Because the velocity changes by the same amount (namely, −0.980 m/s) during each one-tenth second interval, the rate of change of velocity (the acceleration) is a constant: $(-0.980 \text{ m/s})/(0.100 \text{ s}) = -9.80 \text{ m/s}^2$. The *magnitude* of this **acceleration due to gravity** is designated a_g.

Our four equations for constant acceleration (Eqs. (3.9) through (3.12)) can be customized to describe the motion of objects that rise and fall under the influence of gravity alone. This case is often called **free fall** even though objects are sometimes rising!

$$v_y = v_{o_y} - a_g t \tag{3.13}$$

$$y = y_o + \frac{1}{2}(v_y + v_{o_y})t \tag{3.14}$$

(Freely falling objects)

$$v_y^2 = v_{o_y}^2 - 2a_g(y - y_o) \tag{3.15}$$

$$y = y_o + v_{o_y}t - \frac{1}{2}a_g t^2 \tag{3.16}$$

Notice that in the Equations (3.13) through (3.16) we changed the position variable to y (instead of x) and that the upward direction is taken as positive. This choice makes the acceleration $a = -a_g$. Furthermore, we emphasize that velocity is parallel to the y-axis by adding a y-subscript to v and v_o.

Example 3.10 illustrates how these equations may be used to compute the position and velocity of an object undergoing vertical motion near the surface of the earth. But remember, these equations work only for models that permit us to neglect air friction.

EXAMPLE 3.10   Next Time, Use a Rope!

A climber on a high ledge above the base of a cliff attempts to throw a sandwich to his climbing partner located above him. When the sandwich leaves the lower climber's hand it is 40.0 m above the base of the cliff and traveling upward with a velocity of 11.0 m/s. The upper climber holds his hand 5.00 m above the point at which the sandwich is released, but just misses catching it as it goes by. (a) What is the sandwich's velocity as it travels upward past the upper climber's outstretched hand? (b) How long does it take the sandwich to reach its maximum height? (c) To what height above the upper climber's hand does the sandwich reach? (d) How long after the lower climber throws the sandwich does it hit the bottom of the cliff?

SOLUTION   The origin of the coordinate system is placed at the point of release of the sandwich as shown in Figure 3.16. From the problem statement we deduce that

$$y_o = 0$$
$$v_{o_y} = +11.0 \text{ m/s}$$
$$a = -a_g = -9.80 \text{ m/s}^2$$

(a) In question (a) the special condition is "... past the upper climber's outstretched hand." That is, $y = +5.00$ m. The unknown is v_y. Thus, we look among Equations (3.13) through (3.16) for the equation with t missing; it is Equation (3.15):

$$v_y^2 = v_{o_y}^2 - 2a_g(y - y_o)$$
$$= (-11.0 \text{ m/s})^2 - 2(9.80 \text{ m/s}^2)(5.00 \text{ m} - 0)$$
$$v_y = 4.80 \text{ m/s}$$

(b) The point we seek is determined by the phrase "... maximum height." We need to realize that at its highest point the sandwich stops for an instant as it reverses direction, and thus $v_y = 0$ is the special condition occurring at maximum height. We are asked "how long," so that the unknown is t. We look for an equation without the variable y; it is Equation (3.13):

$$v_y = v_{o_y} - a_g t$$

$$t = \frac{v_{o_y} - v_y}{a_g} = \frac{4.80 \text{ m/s} - 0.0 \text{ m/s}}{9.80 \text{ m/s}^2} = 1.12 \text{ s}$$

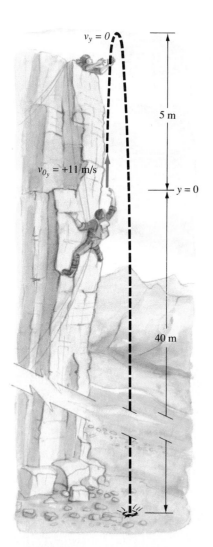

FIGURE 3.16

The climbers and the falling sandwich of Example 3.10.

(c) To find the height of this extreme point we substitute the time from part (b) into Equation (3.14).

$$y = y_o + v_{o_y}t - \frac{1}{2}a_g t^2$$

$$= 0.0 + (11.0 \text{ m/s})(1.12 \text{ s}) - \frac{1}{2}(9.80 \text{ m/s}^2)(1.12 \text{ s})^2$$

$$= 1.17 \text{ m}$$

Notice that this position indicates the distance above the *lower* climber's hand where we placed the origin of the coordinate system. Because we were asked to compute the height above the *upper* climber's hand, we must subtract 5.00 m:

Height above the upper climber's hand = 11.5 m − 5.00 m = 6.5 m

(d) The bottom of the cliff is located at $y = -40.0$ m on our chosen coordinate system. Thus, we again use Equation (3.16), but this time we solve for t.

$$\frac{1}{2}a_g t^2 - v_{o_y}t + (y - y_o) = 0$$

This is a quadratic equation of the form $At^2 + Bt + C = 0$. The general solutions are

$$t = \frac{-B \pm \sqrt{B^2 - 4AC}}{2A}$$

$$= \frac{-(-v_{o_y}) \pm \sqrt{(-v_{o_y})^2 - 4\left(\frac{1}{2}a_g\right)(y - y_o)}}{2\left(\frac{1}{2}a_g\right)}$$

$$= \frac{+v_{o_y} \pm \sqrt{v_{o_y}^2 - 2a_g(y - y_o)}}{a_g}$$

For the present case

$$t = \frac{(11.0 \text{ m/s}) \pm \sqrt{(11.0 \text{ m/s})^2 - 2(9.80 \text{ m/s}^2)[(-40.0 \text{ m}) - (0.0 \text{ m})]}}{(9.80 \text{ m/s}^2)}$$

$$= -1.95 \text{ s and } 4.19 \text{ s}$$

In the present example the negative time has no physical meaning, so the answer is 4.19 s. ◄

Concept Question 3.5
At a particular instant during its motion an object's instantaneous velocity is zero. Can it have a nonzero acceleration at this same instant? Explain.

Did you notice in part (d) of Example 3.10 that we used the position of the base of the cliff as the special condition? If you were tempted to write the special condition as $v_y = 0$, *don't do it!* We freely admit that after the sandwich hits the ground, its velocity is zero. (At least what's left of it has zero velocity!) But during the stopping process, the acceleration of the sandwich was definitely *not* constant. Thus, the constant acceleration equations do not apply once the object actually touches the ground. Many problems use phrases such as " . . . *when* the object hits the ground." However, what is really meant is " . . . the instant *just before* the object hits the ground."

3.4 Applying Integral Calculus to Find Position and Velocity (Optional)

This section is included for students who have preceded this physics course with a course in the calculus. If you are currently enrolled in your first quarter or first semester of calculus, your physics instructor may advise you to return and read this section after you have been introduced to integral calculus techniques.

In order to find the change in the position, we learned that it was necessary to compute the area bounded by the velocity curve and the $v = 0$ axis. An area computation was also required to find the change in the velocity from the acceleration curve. The area under a curve may be computed by the calculus procedure of integration.

The displacement can be calculated using the definition of the definite integral and is usually written

$$\Delta x = x - x_o = \int_{t_o}^{t_f} v(t)\, dt \qquad (3.17)$$

The change in position is the integral of the velocity function.

If we know the velocity function, we can compute the change in the position by applying Equation (3.17) to the velocity. We can repeat this process for the area under the velocity curve. Recall that the change in the velocity is the area under the acceleration curve and, therefore,

$$\Delta v = v - v_o = \int_{t_o}^{t_f} a(t)\, dt \qquad (3.18)$$

The change in the velocity is the integral of the acceleration function.

We demonstrate the power of integration in Example 3.11.

EXAMPLE 3.11   Using the Integral

Consider again the cyclist we first encountered in Example 3.3, whose velocity function remains

$$v(t) = (0.600 \text{ m/s}^3)t^2 + 2.00 \text{ m/s} \qquad \text{for } 0.0 \le t \le 5.0 \text{ s}$$

Find the cyclist's position function during this interval if the bicycle crossed the $x = 10.0$ m point at $t = 2.00$ s.

SOLUTION   The change in position $\Delta x = x - x_o$ is the integral of the velocity function.

$$x(t) - x_o = \int_{t_o=0}^{t_f=t} [(0.600 \text{ m/s}^3)t^2 + 2.00 \text{ m/s}]\, dt$$

After integrating we have

$$x(t) = (0.200 \text{ m/s}^3)t^3 + (2.00 \text{ m/s})t + x_o$$

To determine the initial position x_o we must satisfy the condition $x = 10.0$ m when $t = 2.00$ s:

$$10.0 \text{ m} = (0.200 \text{ m/s}^3)(2.00 \text{ s})^3 + (2.00 \text{ m/s})(2.00 \text{ s}) + x_o$$

$$x_o = 4.4 \text{ m}$$

Finally, we can write the complete position function:

$$x(t) = (0.200 \text{ m/s}^3)t^3 + (2.00 \text{ m/s})t + 4.4 \text{ m}$$

With this function we can predict the position of the cyclist at any time t during the time interval $0 \le t \le 5$ s. ◀

We end this section by applying Equations (3.17) and (3.18) to the constant-acceleration model. For an object that has velocity v_o at time $t_o = 0$ and a constant acceleration a, Equation (3.18) becomes

$$v - v_o = \int_0^t a \, dt = at$$

or

$$v(t) = v_o + at$$

which is Equation (3.9).

Substituting $v(t) = v_o + at$ into Equation (3.17) and taking the initial position to be x_o at time $t_o = 0$, we have

$$x - x_o = \int_0^t (v_o + at) \, dt = v_o t + \frac{1}{2}at^2$$

or

$$x(t) = x_o + v_o t + \frac{1}{2}at^2$$

This result is Equation (3.12). We may derive the other two kinematic equations for the constant acceleration model (Eqs. (3.10) and (3.11)) from these expressions for $v(t)$ and $x(t)$. The integral calculus provides a powerful technique to produce a velocity record from an acceleration record and a position record from a velocity record.

3.5 Handling Real Data (Optional)

The graphical representations of motion have permitted us to come a long way toward understanding the relationships between position, velocity, and acceleration. However, in many situations position, velocity, or acceleration may be handed to us in tabular form: a list of numbers! It is then up to us to figure out how to move from one description of motion to another. In this section we describe some common methods for doing just that.

Computing Areas

We have learned already that if we have the velocity history of an object's motion, we can obtain its position history by finding the "area" under the velocity graph. Let's now consider how we can compute this area if we are given the history as a set of data that lists

TABLE 3.2   Trained Hammerfist
             Strike Data[1]

T (s)	V (m/s)
0.0	0
0.01	0.5
0.02	1.1
0.03	2.2
0.04	2.4
0.05	2.2
0.06	2.0
0.07	2.6
0.08	3.2
0.09	3.8
0.10	4.5
0.11	5.3
0.12	6.9
0.13	8.1
0.14	9.3
0.15	10.5
0.16	11.1

[1] Data taken from the article S. R. Wilk, R. E. McNair, and M. S. Feld, "The physics of karate," *American Journal of Physics,* 51 (1983):783.

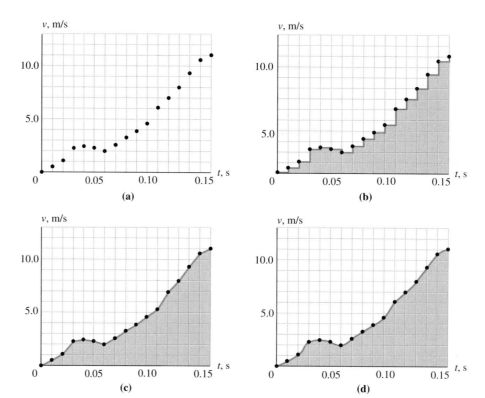

FIGURE 3.17

(a) Velocity-versus-time data for a hand executing a hammerfist strike. (b) A crude method used to estimate the "area" under the velocity-versus-time curve. (c) The "area" under the velocity-versus-time curve by the trapezoidal method. (d) The "area" under the velocity-versus-time curve by Simpson's rule.

the velocity of the object at a number of instants of time. Table 3.2 provides an example. It lists the velocity of a point on a fist during a karate hammerfist strike. A graph of these data is shown in Figure 3.17(a).

If we want to determine the area under this curve, one easy (but crude) method is simply to draw rectangles of width equal to the time between points and heights equal to the velocity at the start of the time interval. The area that we calculate in this manner is shown in Figure 3.17(b). The formula for the distance that this "area" represents is

$$\Delta x = v_1\,\Delta t_1 + v_2\,\Delta t_2 + v_3\,\Delta t_3 + \cdots + v_{16}\,\Delta t_{16}$$

We can shorten this equation by using the summation notation

$$\Delta x = \sum_{i=1}^{16} v_i\,\Delta t_i$$

where Σ means to sum up the $v_i\,\Delta t_i$ terms with the i first equal to 1 and then equal to 2, then 3, and so on, up to a maximum of 16 in this case. Notice we can't use the 17*th* point.

We think you would agree that a better approximation is obtained by using trapezoids instead of rectangles as shown in Figure 3.17(c). In case you've forgotten, the area of a trapezoid is the average of the length of the two parallel sides times the distance between them. So for our example the "area" is

$$\Delta x = \frac{1}{2}(v_1 + v_2)\,\Delta t_1 + \frac{1}{2}(v_2 + v_3)\,\Delta t_2 + \cdots + \frac{1}{2}(v_{16} + v_{17})\,\Delta t_{16}$$

or more compactly

$$\Delta x = \frac{1}{2} \sum_{i=1}^{16} (v_i + v_{i+1}) \, \Delta t_i \qquad (3.19)$$

Note that the 17*th* point is now used even though the sum goes up to only 16.

An even better approximation is made if we use a smoother curve to join adjacent points. For example, we can use a parabola to fit each three adjacent points, as in Figure 3.17(d). If you are handy with calculus, you can prove that the area under a parabola going through three equally spaced points (t_1, v_1), (t_2, v_2), and (t_3, v_3) is

$$\Delta x = Area = \frac{1}{3}(v_1 + 4v_2 + v_3) \, \Delta t$$

where $\Delta t = (t_3 - t_2) = (t_2 - t_1)$ are the equal time intervals. So for our entire graph in Figure 3.17(d) we obtain

$$\Delta x = \frac{\Delta t}{3}[(v_1 + 4v_2 + v_3) + (v_3 + 4v_4 + v_5) + \cdots + (v_{15} + 4v_{16} + v_{17})] \quad (3.20)$$

or, in the summation notation,

$$\Delta x = \frac{\Delta t}{3} \sum_{i=1}^{8} (v_{2i-1} + 4v_{2i} + v_{2i+1}) \qquad (3.21)$$

You should substitute several values of i into Equation (3.21) and convince yourself that it reproduces Equation (3.20). Notice that it is necessary to have an odd number of data points to use this method and that the times must be equally spaced. The expression given by Equation (3.21) is known as **Simpson's rule,** and for N data points the rule can be written

$$\Delta x = \frac{\Delta t}{3} \sum_{i=1}^{(N-1)/2} (v_{2i-1} + 4v_{2i} + v_{2i+1}) \qquad (3.22)$$

EXAMPLE 3.12 *A BASIC Example*

Write a simple program in BASIC to compute the displacement-versus-time record of the fist the velocity-versus-time record for which is given in Table 3.2, using the trapezoidal rule of Equation (3.19).

SOLUTION (See Figure 3.18) After you load the program and type RUN you should see three columns printed; the final X value should be X = 0.701 m. Line 50 sets the maximum number of data pairs for V and T to 20. Line 70 reads the number of data pairs from the DATA statement in line 1000. Lines 80 through 110 constitute a loop to read the data pairs T and V from DATA statements 1010 through 1040 and convert the time value to seconds. Line 130 initializes the X variable so that $x_o = 0$. Lines 140 through 160 perform the summation indicated in Equation (3.19) for the trapezoidal rule. Lines 180 through 210 print T, V, and X for each increment of time. Line 220 marks the end of executable statements.

A copy of the source code for this program is located in the BASIC directory of the diskette accompanying this text. You will need your own copy of a BASIC interpreter or compiler to run the program. In Problem 57 you are asked to modify this program to use Simpson's rule to calculate $x(t)$. Which do you think is more accurate, the trapezoidal rule or Simpson's rule? ◀

```
40    REM ********************************************************************
50    DIM V(20), T(20), X(20)
60    REM *********************** READ  INPUT DATA  *********************
70    READ N                           'READ NUMBER OF DATA POINTS
80    FOR I=1 TO N                     'LOOP TO READ DATA
90     READ T(I), V(I)
100    T(I)=T(I) *0.01                  'CONVERT TIMES TO SECONDS
110   NEXT I
120   REM ******************** TRAPEZOIDAL RULE ****************************
130   X(1)=0                           'INITIAL POSITION
140   FOR I=2 TO N                     'LOOP TO CALCULATE POSITION
150    X(I)=X(I-1)+(T(I)-T(I-1))*(V(I)+V(I-1))/2
160   NEXT I
170   REM ******************** SCREEN PRINT ********************************
180   PRINT ''   T'';''   V'';''  X''   'PRINT OUTPUT HEADER
190   FOR I=2 TO N
200    PRINT USING ''###.###''; T(I), V(I), X(I)
210   NEXT I
220   STOP
230   REM ******************** INPUT DATA **********************************
1000  DATA 17
1010  DATA 0,0.0,1,0.5,2,1.1,3,2.2,4,2.4
1020  DATA 5,2.2,6,2.0,7,2.6,8,3.2,9,3.8
1030  DATA 10,4.5,11,5.3,12,6.9,13,8.1,14,9.3
1040  DATA 15,10.5,16,11.1
```

FIGURE 3.18

Program in BASIC computing the displacement-versus-time record of the hammerfist.

EXAMPLE 3.13 *Spread the News*

Use a spreadsheet to implement the trapezoidal rule to compute the displacement record from the velocity-versus-time data in Table 3.2.

SOLUTION (See Figure 3.19) After entering the data in columns A and B (starting with row 11), type the formula $+(A12 - A11)*(B12 + B11)/2$ into cell D12. Then use your spreadsheet's copy command to replicate the formula into cells D13 through D27. Typing the formula @SUM(D11..D27) into cell D29 then sums all the Δx's and provides you with the total displacement. To obtain the complete record of x versus t you can type 0 (zero) into cell E11. Then type $+E11 + D12$ into cell E12. Use the copy command to replicate cell E12 into cells E13 through E27.

A copy of the spreadsheet TRAPZ.WK1 is in the SS directory of the disk supplied with this text. In Problem 58 you are asked to modify this spreadsheet to use Simpson's rule. ◄

Until this point we have computed the position changes from a velocity record. We hope you have already guessed that precisely the same procedure allows you to deduce the velocity record if you are given a list of the accelerations.

Taking Slopes

Now that you know how to go from acceleration to velocity or velocity to position, using the trapezoidal rule or Simpson's rule, it is natural to ask, "How do I go the other way?" That is, how do we obtain a velocity record from a set of position data or an acceleration

	A	B	C	D	E	F
5	========		=	========	========	
6	Original data		!	Calculated values		
7	========		=	========	========	
8	Time	Velocity	!	Delta x	Position	
9	(s)	(m/s)	!	(m)	(m)	
10	========		=	========	========	
11	0	0	!		0	+(A12−A11)*(B12+B11)/2
12	0.01	0.5	!	0.0025	0.0025	
13	0.02	1.1	!	0.008	0.0105	
14	0.03	2.2	!	0.0165	0.027	+E11+D12
15	0.04	2.4	!	0.023	0.05	
16	0.05	2.2	!	0.023	0.073	
17	0.06	2	!	0.021	0.094	
18	0.07	2.6	!	0.023	0.117	
19	0.08	3.2	!	0.029	0.146	
20	0.09	3.8	!	0.035	0.181	
21	0.1	4.5	!	0.0415	0.2225	
22	0.11	5.3	!	0.049	0.2715	
23	0.12	6.9	!	0.061	0.3325	
24	0.13	8.1	!	0.075	0.4075	
25	0.14	9.3	!	0.087	0.4945	
26	0.15	10.5	!	0.099	0.5935	
27	0.16	11.1	!	0.108	0.7015	@sum(D11..D27)
28	========			========	========	
29	Total distance traveled 0.7015					

FIGURE 3.19

Spreadsheet for implementing the trapezoidal rule for the data in Table 3.2.

record from a set of velocity data? We found the answers to these questions back in Section 3.2; namely, we should draw a smooth curve through the data points and then draw tangents to that curve at the points of interest.

The question then is "How can we obtain the *best* value for the slope at a given point?" To answer this question, let's focus our attention on just three points in a position record and, as is usually the case, take the points to be equally spaced in time. Three such points are shown in Figure 3.20. The point of interest is the center point, we'll call it point i or the ith point. The earlier point to the left, we'll call $i - 1$, the later point to the right we'll call $i + 1$.

In Figure 3.20 we connected the three points with a parabola. (Remember, two points determine a unique straight line; three points determine a unique parabola.) Our question then becomes "What is the slope of a parabola at its midpoint?" The answer is remarkably simple: The slope of a parabola at its midpoint is the same as the slope of the chord connecting the end points. (You can prove this to yourself by doing Problem 76.) This result can be stated algebraically as

$$v_i = \frac{x_{i+1} - x_{i-1}}{t_{i+1} - t_{i-1}} \tag{3.23}$$

Connecting the three points in question with a parabola is the same as assuming that the acceleration is constant for the $2\Delta t$-long time interval. Hence, our approximation should be reasonable as long as the acceleration does not change too dramatically during this interval.

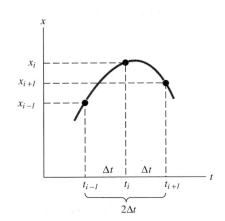

FIGURE 3.20

Three points describing a unique parabola. The slope of the parabola at its midpoint equals the slope of the line connecting the end points.

Computing Acceleration from Velocity

Once again the good news is that the same procedure we use to calculate velocity from position can be used to compute acceleration from velocity. Thus, if we are given velocity data, the acceleration can be calculated using the following rule:

$$a_i = \frac{v_{i+1} - v_{i-1}}{t_{i+1} - t_{i-1}} \tag{3.24}$$

Computing Acceleration from Position

In order to obtain acceleration directly from position data, you might be tempted to apply Equation (3.23), followed by Equation (3.24). This approach works, but the answers are not the best we could obtain from the same data. Using Equations (3.23) and (3.24) one after another is equivalent to using two parabolas to connect five points. (The first three points are connected with one parabola; the last three points with a second parabola. The middle point is shared by both parabolas.) What we would really like is for all five points to be connected with a single smooth curve: a fourth-order polynomial.

The rule for finding the curvature at the midpoint of a fourth-order polynomial drawn through five equally spaced points is a little more complicated than that for a parabola. We content ourselves with just giving the answer. If you are interested in additional details, see References cited in Appendix 2.

$$a_i = \frac{1}{12(\Delta t)^2} \{-y_{i+2} + 16y_{i+1} - 30y_i + 16y_{i-1} - y_{i-2}\} \tag{3.25}$$

TABLE 3.3   Author's Hammerfist Strike Data

TIME (s)	POSITION (m)
0.0167	0.4417
0.0333	0.4603
0.0500	0.4667
0.0667	0.4760
0.0833	0.4883
0.1000	0.5037
0.1167	0.5190
0.1333	0.5497
0.1500	0.5960
0.1667	0.6640
0.1833	0.7290
0.2000	0.7937
0.2167	0.8553
0.2333	0.9230
0.2500	0.9910
0.2667	1.0467
0.2833	1.0747
0.3000	1.0807
0.3167	1.0650
0.3333	1.0650

EXAMPLE 3.14   *It's Healing Nicely, Thank You*

One of the authors, who prefers not to identify himself, tried to duplicate the hammerfist strike data from Examples 3.12 and 3.13. Table 3.3 gives the position and time data for this author's fist. Calculate the velocity-versus-time record for the author's hand using a computer program or spreadsheet and compare the maximum speed of the author's fist to that of the karate expert in Table 3.2.

SOLUTION 1   A BASIC program
(See Figure 3.21) Use the same program structure as that in Example 3.12. The calculation loop is based on Equation (3.23).
Of course, you must change the DATA statements to reflect the values in Table 3.3. The complete implementation of this example is in the file AFIST.BAS found on the diskette accompanying this text.

SOLUTION 2   A spreadsheet template
(See Figure 3.22) A complete template is provided as AFIST.WK1 on the diskette accompanying this text.   ◀

```
7     REM *********************************************************
10    DIM V(20), T(20), X(20)
15    REM *************** READ INPUT DATA *************************
20    READ N                        'NUMBER OF DATA POINTS
30    FOR I=1 TO N                   'LOOP TO READ DATA
40     READ T(I), X(I)
50     T(I)=T(I)/60                  'CONVERT TIMES TO SECONDS
60    NEXT I
70    REM ****************** CALCULATE DIFFERENCES ***************
90    FOR I=2 TO N-1
100    V(I)=(X(I+1)-X(I-1))/(T(I+1)-T(I-1))
110   NEXT I
120   REM ****************** OUTPUT TO SCREEN *******************
130   PRINT''    T'';''   V'';''   X''
140   FOR I=2 TO N-1
150    PRINT USING ''###.###'';T(I), V(I), X(I)
160   NEXT I
170   STOP
200   REM ******************** INPUT DATA **********************
1000  DATA 20
1010  DATA  1,0.4417, 2,0.4603, 3,0.4667, 4,0.4760, 5,0.4883
1020  DATA  6,0.5037, 7,0.5190, 8,0.5497, 9,0.5960,10,0.6640
1030  DATA 11,0.7237,12,0.7937,13,0.8553,14,0.9230,15,0.9910
1040  DATA 16,1.0467,17,1.0747,18,1.0807,19,1.0650,20,1.0650
```

FIGURE 3.21

A portion of the BASIC program AFIST.BAS.

	A	B	C	D	E	F
5						
6	Smoothed data			!	Calculated values	
7						
8	I	Time	Position	!	Velocity	
9	number	(s)	(m)	!	(m/s)	
10						
11	+A11+1 1	0.0167	0.4417	!		
12	2	0.0333	0.4603	!	0.750	
13	+A11/60 3	0.0500	0.4667	!	0.470	(C13-C11)*30
14	4	0.0667	0.4760	!	0.650	
15	5	0.0833	0.4883	!	0.830	
16	6	0.1000	0.5037	!	0.920	
17	7	0.1167	0.5190	!	1.380	
18	8	0.1333	0.5497	!	2.310	
19	9	0.1500	0.5960	!	3.430	
20	10	0.1667	0.6640	!	3.990	
21	11	0.1833	0.7290	!	3.890	
22	12	0.2000	0.7937	!	3.790	
23	13	0.2167	0.8553	!	3.880	
24	14	0.2333	0.9230	!	4.070	
25	15	0.2500	0.9910	!	3.710	
26	16	0.2667	1.0467	!	2.510	
27	17	0.2833	1.0747	!	1.020	
28	18	0.3000	1.0807	!	-0.290	
29	19	0.3167	1.0650	!	-0.470	
30	20	0.3333	1.0650	!		

FIGURE 3.22

A portion of the spreadsheet template AFIST.WK1.

3.6 Summary

Fundamental Definitions   In this chapter we defined, for the case of one dimension, the terms used to describe motion:

position: a coordinate specified along a Cartesian axis

displacement: a change in position, $\Delta x = x_f - x_o$

average velocity: $v_{av} = \dfrac{\Delta x}{\Delta t}$

average speed: $\dfrac{Total\ distance\ traveled}{\Delta t}$

average acceleration: $a_{av} = \dfrac{\Delta v}{\Delta t}$

instantaneous velocity: the limit approached by the average velocity as the time interval of measurement goes to zero. In the language of the calculus, the instantaneous velocity is the time derivative of the position function:

$$v(t) = \frac{dx(t)}{dt} \tag{3.4}$$

instantaneous acceleration: the limit approached by the average acceleration as the time interval of measurement goes to zero. In the language of the calculus, the instantaneous acceleration is the time derivative of the velocity function:

$$a(t) = \frac{dv(t)}{dt} \tag{3.6}$$

Graphical Methods   Velocity is the slope of the position-versus-time graph. The average velocity between two points is the slope of the chord joining the coordinates of the two points. Instantaneous velocity at a point is the slope of the tangent at that point.

Acceleration is the slope of the velocity-versus-time graph. The average acceleration between two points is the slope of the chord joining the coordinates of the two points. Instantaneous acceleration at a point is the slope of the tangent to that point.

The velocity change between two times is the area bounded by the acceleration curve, the $a = 0$ axis, and vertical lines through those two times. This rule is often remembered in the simple form: velocity change is the area "under" the acceleration curve.

The position change between two times is the area bounded by the velocity curve, the $v = 0$ axis, and vertical lines through those two times. This rule is often remembered in the simple form: displacement is the area "under" the velocity curve.

The Constant-Acceleration Model   Four kinematic equations relate position, velocity, and acceleration when the acceleration is constant.

$$v = v_o + at \tag{3.9}$$

$$x = x_o + \frac{1}{2}(v + v_o)t \tag{3.10}$$

$$v^2 = v_o^2 + 2a(x - x_o) \tag{3.11}$$

$$x = x_o + v_o t + \frac{1}{2}at^2 \tag{3.12}$$

The acceleration of objects falling freely at low speeds near the surface of the earth is constant with a magnitude a_g approximately equal to 9.80 m/s².

Integral Calculus (Optional)   Displacement is the integral of the velocity function:

$$\Delta x = x - x_o = \int_{t_o}^{t_f} v(t)\, dt \tag{3.17}$$

Velocity change is the integral of the acceleration function:

$$\Delta v = v - v_o = \int_{t_o}^{t_f} a(t)\, dt \tag{3.18}$$

Numerical Methods (Optional)   Computers can be used to automate the process of taking slopes or calculating areas using simple programs or spreadsheet software. When N data points are equally spaced in time, each three adjacent points can be considered on a parabola. The area formula for parabolas is Simpson's rule:

$$\Delta x = \frac{1}{3}\Delta t \sum_{i=1}^{(N-1)/2} (v_{2i-1} + 4v_{2i} + v_{2i+1}) \tag{3.22}$$

A slope rule for first derivatives based on three points is

$$v_i = \frac{x_{i+1} - x_{i-1}}{t_{i+1} - t_{i-1}} \tag{3.23}$$

PROBLEMS

3.1 Position, Velocity, and Acceleration

1. A brisk walking speed is 3.00 mph (mi/h). Convert this speed to kilometers per hour and meters per second.
2. During part of a ''fun run'' one participant jogs along at 2.00 m/s for 100. m, then sprints at 6.00 m/s for another 100. m. What is the participant's average velocity for the 200. m?
3. A driver in a hurry decides to take a freeway that cuts straight across town. The freeway entrance and exit are 4.00 mi apart. By weaving dangerously in and out of traffic the driver manages to average 75.0 mi/h from entrance to exit. How much time did the driver save by averaging 75.0 mi/h instead of 65.0 mi/h? Express your answer in seconds.
4. A family drives 60.0 mi/h for 3.00 h. They then stop at a rest stop for 15.0 min. Finally, the family drives the last 150. mi in 3.50 h. What is the magnitude of their average velocity for the trip?
5. A hiker walks 3.00 mi along a straight trail in a high mountain valley in 1.25 h. Discovering that she has dropped her map, she leaves her pack and jogs back along her original route. She finds her map 0.50 mi from where she discovered it missing, picks it up, and returns to the spot where she left her pack arriving 15.0 min after she first left it. (a) What is the hiker's average velocity from the start of her trip until she returns to her pack? (b) What is the hiker's average speed at this same point?
6. A pass defender intercepts a football on his own 20.0-yd line and runs straight down the sideline toward the opposition goal for 25.0 yd. An opponent then strips the ball from his hands and runs back in the opposite direction 45.0 yd, scoring a touchdown. The play lasted 10.0 s from the interception to the touchdown.

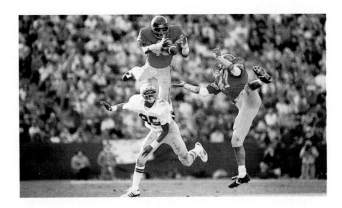

(a) What was the average velocity of the football during this 10.0 s? (b) What was the average speed of the football during this time?

7. An automobile braking hard can slow down at a rate of 15.0 mi/h/s. Convert this rate to kilometers per hour per second and meters per second per second.
8. A driver brakes a minivan from 30.0 mi/h to 20.0 mi/h in 1.50 s. What is the car's average acceleration in miles per hour per second? Express your answer in meters per second per second.
9. A Taurus SHO accelerates from 0.0 to 90.0 mph in 15.0 s. What is the average acceleration in miles per hour per second? Express your answer in meters per second per second.
10. A sprinter accelerates from 0.00 to 9.50 m/s in 2.50 s. Calculate the sprinter's average acceleration. Express your answer in meters per second per second and in miles per hour per second.

11. You desire to know the instantaneous velocity of a point particle at $t = 10.00$ s. From a record of the particle's position versus time, displacements are measured for shorter and shorter Δt, around the point $t = 10.00$ s. From the table below calculate the average velocity for each Δt. Make a plot of average velocity versus Δt, and from the plot determine the instantaneous velocity at $t = 10.00$ s.

DISPLACEMENT (cm)	ΔT (s)
10.0	2.5
7.6	2.0
5.4	1.5
3.4	1.0
1.6	0.5

12. Repeat the analysis of Problem 11 for the particle the motion of which is described by the table below.

DISPLACEMENT (m)	ΔT (s)
144	6
125	5
104	4
81	3
56	2
29	1

13. The "●" symbols in Figure 3.P2 represent the position of a particle at 0.10-s intervals. The object of this exercise is to calculate the instantaneous velocity of this particle at the point marked by the symbol ×. Use a centimeter ruler to measure the displacement for several values of Δt surrounding the × point. Construct a table and graph like that in Problem 11 for this particle and then calculate the instantaneous velocity of the particle at the point marked ×.

FIGURE 3.P2   Problem 13

14. Using the technique described in Problem 13, calculate the instantaneous velocity of the object the position record of which is shown in Figure 3.P3 at the point marked by the symbol ×. Assume that the time interval between ● marks is 0.50 s.

FIGURE 3.P3   Problem 14

15. A particle moving along the x-axis has a position function given by $x(t) = (4.00 \text{ m/s}^3)t^3 - (5.00 \text{ m/s})t + 6.00 \text{ m}$. (a) Find the velocity function of the particle. (b) Find the acceleration function of the particle. (c) How fast and in what direction is the particle moving at $t = 0.00$ s? (d) What is the average velocity of the particle between $t = 1.00$ s and $t = 2.00$ s?

16. Answer the questions posed in Problem 15 for a particle the position of which is given by $x(t) = (5.00 \text{ m/s}^3)t^3 + (3.00 \text{ m/s}^2)t^2$.

17. The velocity function of an object moving along a straight line is given by $v(t) = At^2 + Bt$. (a) What are the units of the constants A and B? (b) If at $t = 2.00$ s the function's velocity is 3.00 m/s and its acceleration is 0.500 m/s², find the values of the constants A and B. (c) What is the object's acceleration at $t = 6.00$ s?

18. An object can be modeled as a particle moving along a straight line with a velocity function $v(t) = (3.00 \text{ m/s}^3)(t - 2.00 \text{ s})^2 + 4.00 \text{ m/s}$. (a) What is the average acceleration for the object between $t = 3.00$ s and $t = 5.00$ s? (b) Compute the acceleration function for this particle. (c) What is the object's instantaneous acceleration at $t = 4.00$ s?

3.2 Graphical Interpretation of Position, Velocity, and Acceleration

19. Assume that Figure 3.P4 shows the velocity-versus-time history of two cars moving along the same road. If both cars started at the same point together, at what later time do they have the same (a) position? (b) speed? (c) acceleration?

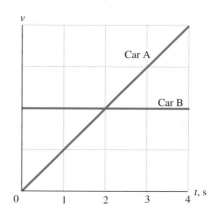

FIGURE 3.P4   Problem 19

20. Assume that Figure 3.P5 shows the position-versus-time history of two cars on the same road. When do the two cars have the same (a) position? (b) velocity? (c) acceleration?

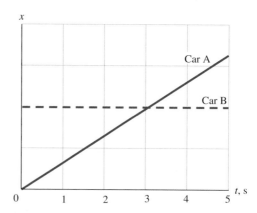

FIGURE 3.P5   Problem 20

21. Land to the west of the San Andreas fault in California is moving northwest relative to the land on the east side of the fault. Figure 3.P6 is a graph of the displacement of the west side relative to the east side in Stone Canyon, California. (a) What was the average speed of the slip along this segment of the San Andreas fault between July 1, 1968, and October 31, 1968? Express your

answer in meters per second and in centimeters per year. (b) If this slip continues at the 1968 rate, what total displacement occurs between October 31, 1968, and October 1, 1998?

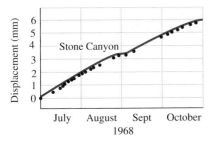

FIGURE 3.P6   Problem 21

22. An *instant* is one single point in time, for example, $t = 2.00$ s. An *interval* of time is an extended period of time, for example, 1.50 s $\leq t \leq 4.00$ s. Consider the motion of an object represented by the graph in Figure 3.P7. (a) List times when the object was stopped only for an instant. (b) List the extended periods of time when the object was at rest.

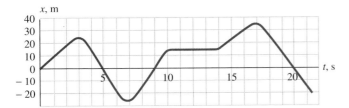

FIGURE 3.P7   Problem 22

23. Answer the questions in Problem 22 for the graph in Figure 3.P8.

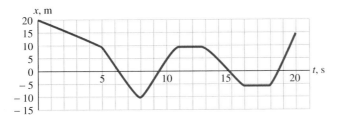

FIGURE 3.P8   Problem 23

24. Answer the questions in Problem 22 for the object the velocity-versus-time graph of which is given in Figure 3.P9.

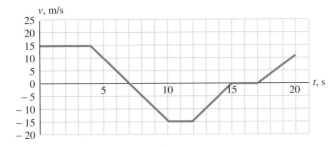

FIGURE 3.P9   Problem 24

25. Answer the questions in Problem 22 for the object the velocity-versus-time graph of which is given in Figure 3.P10.

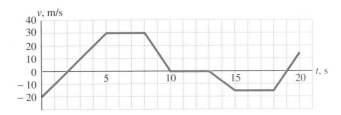

FIGURE 3.P10   Problem 25

26. (a) What is the average velocity between $t = 0.0$ and $t = 10.0$ s for the object the position history of which is shown in Figure 3.P7? (b) What is the average speed of the object in Figure 3.P7 between $t = 0.0$ and $t = 5.0$ s?

27. Answer the questions in Problem 26 for the object the position history of which is given in Figure 3.P8.

28. Answer the questions in Problem 26 for the object the velocity history of which is given in Figure 3.P9.

29. Answer the questions in Problem 26 for the object the velocity history of which is given in Figure 3.P10.

30. One way to check to see if you need new shock absorbers on your car is to stand on the rear bumper, then quickly jump off. If the bumper motion looks like that depicted by the graph in Figure 3.P11, you need new shocks. (a) Sketch the velocity-versus-time curve for the automobile bumper the vertical position-versus-time graph of which is given in Figure 3.P11. (b) Sketch the acceleration-versus-time curve for the bumper.

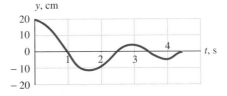

FIGURE 3.P11   Problem 30

31. Figure 3.P12 is a graph of velocity-versus-time for a Lamborghini Miuras. (a) Estimate from the graph how long it takes the Lamborghini to complete a quarter mile starting from rest. (b) In what gear is the acceleration the greatest? (c) Estimate the magnitude of the greatest acceleration. Express your answer in g's (the number of times greater than 9.80 m/s^2).

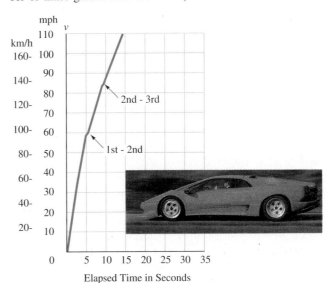

FIGURE 3.P12   Problem 31

32. Figure 3.P13 is a graph of velocity versus time for a Mazda MPV with an automatic transmission. (a) Estimate from the graph how long it takes the Mazda MPV to complete a quarter mile starting from rest. (b) What is the instantaneous acceleration of the Mazda at the 5.0-s mark? (c) What is the average velocity of the Mazda during the first 35.0 s? (d) What is the average acceleration of the Mazda during the first 35.0 s? (e) At what time is the instantaneous acceleration equal to the average acceleration found in part (d)?

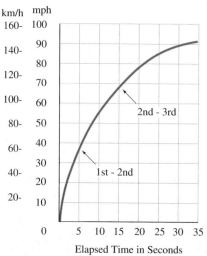

FIGURE 3.P13   Problem 32

★**33.** A car with an initial velocity of 10.0 m/s accelerates at a rate of $\mathscr{A}$ m/s² for 3.00 s, then decelerates at a rate of $-\mathscr{A}$ m/s² for 3.00 s. Sketch the velocity-versus-time graph for the motion described and from it calculate the value of $\mathscr{A}$, such that the total distance traveled during the 6.00-s time interval is equal to 75.0 m.

3.3   The Constant-Acceleration Model

34. Carry out the algebra required to derive Equations (3.9) and (3.10) from Equations (3.7) and (3.8).

35. The position function of a particle traveling along the x-axis is given by $x(t) = 3.00$ m $+ (2.50$ m/s$)t + (4.00$ m/s²$)t^2$. (a) What is the acceleration of the particle? (b) What is the particle's velocity at $t = 0.0$ s? (c) At what time is the particle located at $x = 4.00$ m?

36. The position function of an object traveling along the x-axis is given by $x(t) = 12.0$ m $- (2.6$ m/s²$)t^2$. (a) What is the acceleration of the particle? (b) What is the object's velocity at $t = 0.0$ s? (c) What is the object's velocity at $t = 4.00$ s? (d) What is the object's position at $t = 4.00$ s?

37. A hypervelocity cannon is to launch a projectile at 10.0 km/s. The maximum acceleration the projectile can survive is 100 000g (100 000 times 9.80 m/s²). Use the constant-acceleration model to calculate the minimum length of the barrel.

38. An arrow has a velocity of 45.0 m/s when it reaches a straw target. (a) If its tip comes to rest at a depth of 20.0 cm into the straw, use the constant-acceleration model to calculate its average acceleration while coming to rest. (b) How long after first touching the straw does it take for the arrow to come to rest?

39. A child on a bicycle increases her velocity from 3.00 m/s to 5.00 m/s in a distance of 4.00 m. (a) Find her acceleration, assuming it was constant. (b) Find the time it took her to go this distance.

40. An electron with an initial velocity of 5.00×10^5 m/s decelerates uniformly and comes to rest in 4.00 ms (1 ms = 10^{-3} s). (a) Determine the acceleration of the electron. (b) How far does the electron travel in this time?

41. A karatekas can attain a velocity of 32.3 mph during a karate strike while moving the hand through a displacement of 0.750 m. If the acceleration is constant, what is the hand's average acceleration?

42. A passenger airplane must reach a speed of 90.0 m/s (about 200 mph) for lift-off. Use the constant-acceleration model with $a = 0.400g$ (0.400 times 9.80 m/s²) to estimate the minimum length of runway required.

43. A laser scans a line of information initially at a rate of 1.25 m/s. The scan rate decreases uniformly to 1.05 m/s in 1.50 h. What is the total length of the line of information scanned by the laser? (These data describe the scan rate of the laser in a typical compact-disk player. Your answer is the length of the spiral of information on the surface of the compact disk.)

44. Equation (3.10) describes the position at any time t of an object moving with constant acceleration. (a) Write this equation for three particular times: 0, T, and $2T$. (b) Use the equations you wrote in part (a) to calculate the average velocity of the object between 0 and $2T$. (c) Show that the average velocity from 0 to $2T$ is the instantaneous velocity at the midpoint of the interval, that is, at $t = T$.

45. Your vehicle and a car you trail by 20.0 m are both traveling 45.0 mph (20.1 m/s). You decide to pass. You accelerate at 5.00 mph/s (2.35 m/s²) up to 65.0 mph (29.0 m/s), after which you hold your speed constant. When you are 5.00 m ahead of the other car you can safely pull back into the right-hand lane. (a) On a single graph, plot the velocity versus time for both cars. (You may want to make a rough sketch first, then answer the questions below before making your final graph.) (b) How long does it take you to reach 65.0 mph? How far does each car travel from the instant you started to accelerate until you reached a speed of 65 mph? What is the relative position of your car to the other car? (c) How much more time elapses before you can pull into the right-hand lane? (d) What total distance do you travel while passing?

46. A group of freshmen students decide to "lake" their resident advisor (RA). The RA discovers their plot when the freshmen are 4.00 m away, and he starts running at a constant speed (born of fear) of 8.00 m/s. The fastest freshman sprints after him initially at 10.0 m/s, but (no doubt due to fatigue from studying physics) slows down at a rate of 0.500 m/s². Does the freshman catch the RA? If he does, tell where and when; if not, what is the closest he gets?

47. A popular test of reaction time is to bet a friend that he can't catch a dollar bill. The friend places the side of his hand on the edge of a table and extends his index finger and thumb about an inch-and-a-half apart as if to pinch. You hold a dollar bill lengthwise just above the space between his fingers. The bet is that your friend can have the money if he can catch it. Use a relatively new bill and stiffen it by folding it lengthwise, then straightening it, leaving a slight crease. The bill falls without appreciable air resistance for its entire length. How long does your friend have to catch the bill? (The length of a dollar bill is 15.6 cm.)

48. A baton is thrown vertically into the air with a speed of 24.0 m/s. It is released from a height of 2.00 m above the ground. (a) How high does it rise? (b) How fast is it traveling when it returns to waist height (1.00 m above the ground)? (c) How long does it remain in the catchable range from 2.00 m to 0.50 m above the ground?

49. The average velocity of an object thrown vertically upward and caught at the point of release is zero. Calculate the object's average speed in terms of its initial velocity upward.

3.4 Applying Integral Calculus to Find Position and Velocity

50. An object moving along the x-axis is described by the acceleration function $a(t) = (32.0 \text{ m/s}^3)t - (6.00 \text{ m/s}^4)t^2$ and has velocity 2.00 m/s at $t = 1.00$ s and position $x = 4.00$ m at $t = 2.00$ s. (a) Find the velocity function for this particle. (b) Find the position function for this particle.

51. Answer the questions posed in Problem 50 for a particle described by the acceleration function $a(t) = (12.0 \text{ m/s}^3)t + 5.00$ m/s^2, which is located at $x = -3.00$ m when $t = 0.00$ s and moving at 5.00 m/s when $t = 1.00$ s.

52. The velocity of a particle moving along the x-axis is given by $v(t) = (3.00 \text{ m/s}^3)t^2 + 5.00 \text{ m/s}$ and the particle is located at $x = 7.00$ m when $t = 1.00$ s. (a) Find the position function of the particle. (b) For this particle find the average velocity between $t = 0.00$ s and $t = 2.00$ s. (c) Find the instantaneous acceleration of the particle at $t = 3.00$ s.

53. Answer the questions posed in Problem 52 for a particle moving along the x-axis with velocity given by $v(t) = (12.0 \text{ m/s}^3)t^2 - (8.00 \text{ m/s}^2)t + 4.00$ m/s if the particle is located at $x = -3.00$ m when $t = 2.00$ s.

54. Consider the velocity function $v(t) = (108 \text{ m/s}^2)t - (3.00 \text{ m/s}^3)t^2$ for an object moving along the x-axis. (a) For $t > 0$, when is the first time at which the object is instantaneously at rest? (b) In what direction is the object traveling when $t = 9.00$ s? (c) What is the total distance traveled by the object during the first 9.00 s of its motion? (By "total distance traveled" we mean the total distance moved in the odometer sense, and *not* the total displacement.) (d) What is the average speed of the object during the first 9.00 s? (e) What is the average velocity of the object during the first 9.00 s?

55. Answer the questions in Problem 54 for an object described by the velocity function $v(t) = 10.0 \text{ m/s} - (0.200 \text{ m/s}^3)(t - 5.00 \text{ s})^2$.

3.5 Numerical Methods for Handling Real Data

56. Explore the capabilities of your favorite spreadsheet program by importing the template CNSTA.WK1 found on the diskette accompanying your text. (a) Load the template and view the graph. (b) Change some of the values for XO, VO, or A and examine the changes in the graph. (c) Add a second graph of the same equation $(X = XO + VO*T + (0.500)*A*T^2)$ using different parameters (perhaps labeling them XXO, VVO, and AA). If you have trouble, take a peek at CNSTAHLP.WK1. (d) Adjust the values for your two equations so that the two objects collide. Hand in a printed graph.

57. Modify the program used in Example 3.12 to calculate the area under the curve using Simpson's rule instead of the trapezoidal rule. (The program from Example 3.12 is on the diskette supplied with your text. It is called TRAPZ.BAS.)

58. Modify the spreadsheet template of Example 3.13 to calculate the area under the curve using Simpson's rule instead of the trapezoidal rule. (The template for Example 3.13 is on the diskette supplied with your text. It is called TRAPZ.WK1.)

59. Data describing acceleration versus time for a point particle is located in the spreadsheet template V__OF__T.WK1. Calculate the velocity-versus-time record for this particle using both the trapezoidal rule and Simpson's rule. Compare both results with the expression $v(t) = (4.00 \text{ m/s})t - (3.00 \text{ m/s}^2)t^2$, which provides the exact answer.

60. Perform the calculations described in Problem 59 using the BASIC data statements in the file V__OF__T.BAS.

61. The data in the table below describe position versus time for an arrow fired from a compound bow. Use a computer, pocket calculator, or spreadsheet to calculate the velocity-versus-time record for the arrow until it leaves the bow, assuming it starts from rest. (The data below is on the diskette accompanying this text. The spreadsheet version is ARROW.WK1; for BASIC programs use the DATA statements found in ARROW.BAS.)

TIME (s)	POSITION (cm)
0	0
0.001	0.12
0.002	0.43
0.003	0.99
0.004	1.87
0.005	3.11
0.006	4.77
0.007	6.88
0.008	9.47
0.009	12.49
0.010	15.89
0.011	19.61
0.012	23.59
0.013	27.83
0.014	32.32
0.015	37.04
0.016	41.96
0.017	47.04
0.018	52.11

62. Use the data in the table in Problem 61 to compute the acceleration-versus-time history for the arrow while it is in the bow, (a) by applying Equation (3.23) followed by Equation (3.24), and (b) by applying Equation (3.25) to the position data directly. How do the results compare?

General Problems

63. Figure 3.P14 shows two sets of three graphs each. Each column contains three graphs for the same object: position (x) versus time, velocity (v) versus time, and acceleration (a) versus time. For each column, one of the three graphs is given, and the other two are unfinished. Complete the missing graphs for each of the two objects. Note that initial positions and velocities are marked on the uncompleted graphs where needed.

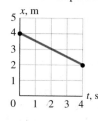
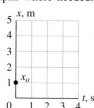

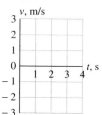
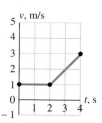

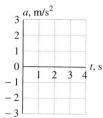
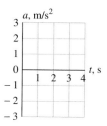

FIGURE 3.P14   Problem 63

64. Complete the missing graphs in Figure 3.P15 as described in the instructions for Problem 63.

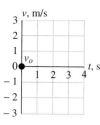
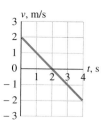

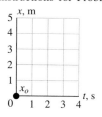
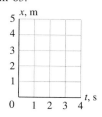

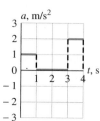
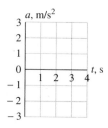

FIGURE 3.P15   Problem 64

65. (a) For the Lamborghini described in Problem 31 calculate the average acceleration between $t = 0.0$ s and $t = 15.0$ s. (b) If you were to use the acceleration calculated in part (a) as if it were the actual (constant) acceleration between $t = 0.0$ s and $t = 15.0$ s, what would be the largest amount by which your calculated velocity would be in error? (c) Estimate the total distance error that results from using this constant-acceleration approximation.

★66. A point particle with an initial velocity v_o accelerates with constant acceleration $\mathcal{A}$ for 4.00 s during which time it travels through a displacement of 12.0 m. After this interval the particle continues with constant velocity for an additional 4.00 s, covering an additional 20.0 m. (a) Sketch a velocity-versus-time graph for this motion. (b) Use the graph to help you calculate the value of v_o and $\mathcal{A}$.

67. A dragster completes the quarter mile in exactly 12.00 s. (a) Use the constant-acceleration model to calculate the average required acceleration. Express your answer in mph/s and in g's. (One $g = 9.80$ m/s^2.) (b) How fast is the dragster going as it crosses the quarter-mile mark? (c) If the dragster brakes from top speed with an acceleration of $1.20g$, how far does it go before stopping?

68. A race car starts from rest with a constant acceleration. After a short time its velocity is 40.0 ft/s. After traveling a distance of 215 ft farther, its velocity is 90.0 ft/s. (a) Calculate the car's acceleration. (b) How much time does the car take to travel the 215 ft? (c) How much time does the car take to reach 40.0 ft/s from rest? (d) How far does the car travel to reach 40.0 ft/s starting from rest?

69. A car traveling 12.0 m/s passes a motorcyclist starting his ignition. At 5.00 s after the car passes the cycle, the motorcyclist begins to accelerate in the same direction as the car, with a constant acceleration of 2.00 m/s^2. (a) How much time elapses after the cyclist starts moving until he catches up to the car? (b) How far does each travel between the time the car passes the motorcycle and the motorcycle catches up to the car? (c) If the speed limit is 55 mph, should the motorcyclist be concerned about a speed trap?

70. A three-year-old child is standing up in the front seat of a pickup truck. The truck travels 30.0 mph (13.4 m/s) down a straight road. The child's head is 0.70 m from the front windshield. To avoid a dog that runs into the road, the driver brakes moderately hard, decelerating the truck 10.0 mph/s (4.47 m/s^2). If not caught, the child continues forward with constant velocity (30.0 mph) until striking the windshield. How long does a parent have to recognize the danger, react, and move to catch the child before the child's head strikes the windshield? What is the moral of this story?

71. Some engineering students are trying to construct a soft container within which an egg can survive a 10.0-m fall from the top of a dormitory. (a) As a preliminary trial a net is placed 2.00 m above the ground. How long after release does the container hit the net? (b) A physics student carrying a basket of laundry the top of which is 1.00 m above the ground walks with a speed of 1.50 m/s out the door of the dormitory. How far from the impact point should the basket be in order to catch the container when it is released at a signal from a student on the ground? (c) Correct your answer to (b) by assuming there is a 0.20-s delay in release due to the reaction time of the dropper.

72. A pitching machine can throw a baseball at speeds in excess of 80 mph. (a) If the machine throws the ball straight upward at 80.0 mph (35.8 m/s), neglecting air resistance, how high does it rise? Assume the point of release is 1.00 m above the ground. (b) Suppose the player tries to catch this ball but loses it in the sun. If he spots the ball when it is 20.0 m above the ground, how long does he have before it reaches his glove, which he holds at a height of 2.00 m?

★73. A jewel thief steals a valuable golden egg from an exhibit on the third floor of a multistory office building. Police quickly enter the building and the thief runs to hide on one of the upper floors of the building. To be free of the evidence, the thief drops the egg out the window. An alert detective on the third floor sees the egg falling and times it as it passes the lower half of her window. The floors of the building are located 4.00 m apart. The egg took 0.101 s to fall 1.50 m past the lower part of the detective's window. The bottom of the detective's window is 8.00 m above the ground. On what floor is the thief hiding?

74. A rock is held just under the surface of the water and released from rest. Its position is given as a function of time by $y(t) = -(0.500 \text{ m/s})t + (0.500 \text{ m})e^{-t/\tau}$, where $\tau = 2.00$ s. (a) Find the velocity function for the rock. (b) Find the velocity of the rock at $t = 1.00$, 2.00, 3.00, and 4.00 s. (c) What is the "terminal velocity" of the rock (i.e., the velocity of the rock after a very long time)?

75. A particle moving along the x-axis has a velocity function given by $v(t) = (2.00 \text{ m/s})e^{-t/\tau}$, where $\tau = 5.00$ s. The particle is lo-cated at $x = 0.00$ when $t = 0.00$ s. (a) Find the position function of the particle. (b) Show that although the particle never stops, it does not go beyond the position $x = 10.0$ m.

76. (a) Show that the parabola given by the following formula passes through the points (x_1, y_1), (x_2, y_2), (x_3, y_3)

$$y = \frac{(x - x_1)(x - x_2)}{(x_3 - x_1)(x_3 - x_2)}y_3 + \frac{(x - x_1)(x - x_3)}{(x_2 - x_1)(x_2 - x_3)}y_2$$
$$+ \frac{(x - x_2)(x - x_3)}{(x_1 - x_2)(x_1 - x_3)}y_1$$

(b) Find the derivative of the function in part (a). Leave the answer as the sum of six fractions. Do not simplify.

(c) Assume that the x-coordinates are equally spaced a distance h apart so that $(x_2 - x_1) = (x_3 - x_2) = h$. Show that the derivative in part (b) reduces to $(y_3 - y_1)/2h$. This result shows that the slope at the midpoint of a section of a parabola is equal to the slope of the chord joining the end points.

Motion in Two Dimensions

In this chapter you should learn

- to apply the vector definitions of displacement, velocity, and acceleration.

- the relation between speed and the velocity vector.

- to apply the kinematic equations for constant acceleration to projectile motion problems.

- to calculate the velocity and acceleration for uniform circular motion.

- to estimate the tangential and radial components of acceleration by looking at the record of a particle's trajectory in two dimensions.

- (optional) to apply numerical techniques to analyze the photographic record of an object moving in two dimensions.

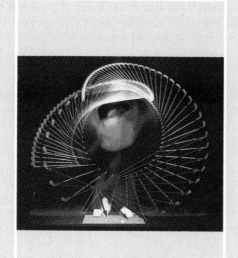

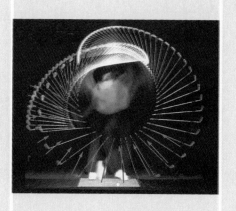

In the previous chapter we examined the relationships among position, velocity, and acceleration for an object moving in a straight line. This study is known as **kinematics.** But, of course, real objects don't always move in straight lines. In order to study the trajectories of basketballs and the orbits of satellites we must extend our study of kinematics to at least two dimensions. For the most part we simply repeat the processes learned in the previous chapter, applying the same procedures to vectors, component by component. There is perhaps one surprise in this chapter: objects accelerate when they change direction even if they don't change speed. Watch for this development in Section 4.3. We must start once again, however, with some definitions.

4.1 Velocity and Acceleration Vectors

The **position** of a particle can be expressed with vector notation as

$$\mathbf{r} = x\hat{\mathbf{i}} + y\hat{\mathbf{j}} + z\hat{\mathbf{k}} \tag{4.1}$$

In the previous chapter we restricted our discussion to objects moving in one dimension. Each case involved changes in only one of the components in Equation (4.1), designated by $\hat{\mathbf{i}}$ or $\hat{\mathbf{j}}$ or $\hat{\mathbf{k}}$. Nonetheless, the definitions of kinematic quantities used to describe motion in one dimension apply equally well to the three-dimensional case. To extend any definition to the three-dimensional case, we merely perform all indicated operations on each of the three vector components *independently*. For example, **displacement** is now

$$\Delta\mathbf{r} = \mathbf{r}_f - \mathbf{r}_o$$
$$= (x_f\hat{\mathbf{i}} + y_f\hat{\mathbf{j}} + z_f\hat{\mathbf{k}}) - (x_o\hat{\mathbf{i}} + y_o\hat{\mathbf{j}} + z_o\hat{\mathbf{k}})$$
$$= (x_f - x_o)\hat{\mathbf{i}} + (y_f - y_o)\hat{\mathbf{j}} + (z_f - z_o)\hat{\mathbf{k}}$$

or, in short,

$$\Delta\mathbf{r} = \Delta x\hat{\mathbf{i}} + \Delta y\hat{\mathbf{j}} + \Delta z\hat{\mathbf{k}} \tag{4.2}$$

Equation (4.2) tells us that the displacement is the vector sum of its component vectors. As a consequence, the **average velocity vector** is just

$$\mathbf{v}_{av} = \frac{\Delta\mathbf{r}}{\Delta t} = \frac{\Delta x}{\Delta t}\hat{\mathbf{i}} + \frac{\Delta y}{\Delta t}\hat{\mathbf{j}} + \frac{\Delta z}{\Delta t}\hat{\mathbf{k}}$$

The definition for **instantaneous velocity** is written similarly:

$$\mathbf{v} = \frac{d\mathbf{r}}{dt} = \frac{dx}{dt}\hat{\mathbf{i}} + \frac{dy}{dt}\hat{\mathbf{j}} + \frac{dz}{dt}\hat{\mathbf{k}} \tag{4.3}$$

where $dx/dt = v_x$ is called the **x-component** of the velocity, $dy/dt = v_y$ is the **y-component** of the velocity, and similarly $dz/dt = v_z$ is the **z-component.**

The **average acceleration** vector is

$$\mathbf{a}_{av} = \frac{\Delta \mathbf{v}}{\Delta t} = \frac{\Delta v_x}{\Delta t}\hat{\mathbf{i}} + \frac{\Delta v_y}{\Delta t}\hat{\mathbf{j}} + \frac{\Delta v_z}{\Delta t}\hat{\mathbf{k}}$$

In a similar fashion, the **instantaneous acceleration** vector is defined

$$\mathbf{a} = \frac{d\mathbf{v}}{dt} = \frac{dv_x}{dt}\hat{\mathbf{i}} + \frac{dv_y}{dt}\hat{\mathbf{j}} + \frac{dv_z}{dt}\hat{\mathbf{k}}$$ (4.4)

Average Speed

Just as in Chapter 3 we use the word *speed* to mean something different from *velocity*. For average speed we keep exactly the same definition as in the previous chapter; namely,

$$Average\ speed = \frac{Total\ distance\ traveled}{Time\ taken}$$

Also as in the previous chapter, this definition means that although the average velocity depends only on the positions of the end points of a trip, the average speed depends upon the path taken.

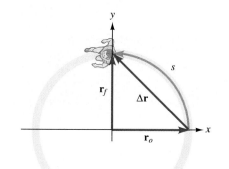

FIGURE 4.1

Average speed depends on total distance traveled, but average velocity depends only on the location of the initial and final points. See Example 4.1.

Concept Question 1
Concept Question 1
If a runner completes a lap around a track 400 m in circumference in 60 s, (a) what is the runner's average velocity? (b) What is the runner's average speed?

EXAMPLE 4.1   Running in Circles

A runner jogs a quarter of the way around a circular path of radius 25.0 m in 10.0 s. As shown in Figure 4.1 his initial position is given by $\mathbf{r}_o = (25.0 \text{ m})\hat{\mathbf{i}}$, and his final position is $\mathbf{r}_f = (25.0 \text{ m})\hat{\mathbf{j}}$. (a) Compute the runner's displacement $\Delta \mathbf{r}$ and his average velocity $\mathbf{v}_{av}$. (b) Compute the magnitude of the runner's average velocity and his average speed.

SOLUTION   (a) The displacement is given by

$$\Delta \mathbf{r} = \mathbf{r}_f - \mathbf{r}_o$$
$$= (25.0 \text{ m})\hat{\mathbf{j}} - (25.0 \text{ m})\hat{\mathbf{i}}$$
$$= -(25.0 \text{ m})\hat{\mathbf{i}} + (25.0 \text{ m})\hat{\mathbf{j}}$$

and the average velocity is

$$v_{av} = \frac{(-25.0\hat{\mathbf{i}} + 25.0\hat{\mathbf{j}}) \text{ m}}{10.0 \text{ s}} = (-2.50\hat{\mathbf{i}} + 2.50\hat{\mathbf{j}}) \text{ m/s}$$

(b) The magnitude of the runner's average velocity vector is

$$|v_{av}| = v_{av} = \sqrt{(2.50 \text{ m/s})^2 + (2.50 \text{ m/s})^2} = 3.54 \text{ m/s}$$

On the other hand, the average speed is determined by the length of the path traveled by the runner. In this case, the path length is one quarter the circumference $(2\pi R)$ of the circle:

$$Average\ speed = \frac{\frac{1}{4}(2\pi \times 25.0 \text{ m})}{10.0 \text{ s}} = 3.93 \text{ m/s} \qquad \blacktriangleleft$$

As you can see from Example 4.1 the average speed and the magnitude of the average velocity vector are not necessarily the same. However, if the starting and ending points in this example are brought closer and closer together, the length of the path becomes shorter. Moreover, the average speed and the magnitude of the average velocity become more nearly equal as illustrated in Figure 4.2. In the limit as Δt goes to zero they become equal. That is, the *instantaneous* speed is equal to the magnitude of the *instantaneous* velocity vector.

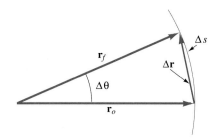

FIGURE 4.2

For small angles the arc length Δs approaches the chord length Δr.

4.2 Projectile Motion

In this section we examine the motion of objects, such as basketballs and horseshoes, that are thrown with an initial velocity and continue in their motion essentially under the influence of gravity only. We refer to such objects as **projectiles.** First we develop a simple model for such motion and then use the model to predict the shape of the trajectory for projectiles. Finally, we develop a set of procedures that allow us to predict the location of a projectile at any time based on its initial velocity.

The statement that velocity and acceleration both behave like vectors has some interesting implications. In Chapter 3 we discussed "freely falling" objects. We modeled objects that fall near the surface of the earth as having a constant downward acceleration. If velocity really does behave like a vector, then this acceleration affects only the vertical ($\hat{\mathbf{j}}$) component of the motion. That is, if the object is moving in the horizontal ($\hat{\mathbf{i}}$) direction, this motion is not altered by the downward fall. Furthermore, the horizontal motion does not affect the downward acceleration.

In Figure 4.3 you can see one of the many tests of this remarkable fact. The two balls in the picture were released at the same instant. Their positions were recorded simultaneously at *equally spaced instants of time* by means of a flashing light. One ball had no horizontal velocity and merely fell straight down. The other ball had an initial velocity that was entirely horizontal the instant it was released. The significant thing to notice about the picture is that the vertical positions of both balls are identical at each instant of time. The ball with the horizontal initial velocity fell with the same downward acceleration as the ball that was not moving horizontally. Moreover, if you measure the horizontal velocity of the second ball, you can tell that its velocity was constant because the ball traveled an equal horizontal distance during each interval of time.

Although the x- and y-components of a projectile's motion are independent in the sense just described, both are dependent on time. This mutual time dependence can be used to relate the projectile's vertical and horizontal positions. That is, we can determine the shape of its trajectory. The motion of the projectile with the horizontal motion in Figure 4.3 can be easily described using the constant-acceleration model. In Chapter 3 we wrote the general expression for an object traveling in one dimension as

$$x = x_o + v_o t + \frac{1}{2} a t^2 \qquad (4.5)$$

Taking $x_o = 0$ and using $a = 0$ in the x-direction, we obtain

$$x = v_o t \qquad (4.6)$$

For the vertical motion x should be replaced by y:

$$y = y_o + v_0 t + \frac{1}{2} a t^2$$

Picking $y_o = 0$ and recognizing that the original velocity in the y-direction was zero, the expression for y becomes

$$y = -\frac{1}{2} a_g t^2 \qquad (4.7)$$

FIGURE 4.3

A multiple-flash picture of two balls released at the same instant. Notice that the forward motion of the second ball does not alter its vertical acceleration from that of the ball falling straight down. If you carefully measure the horizontal motion of the projected ball, you find that it moves forward an equal amount during each of the equal time intervals between flashes.

Concept Question 2
A steward pours a cup of coffee on an airplane traveling at 150 m/s (336 mph). During the 0.10 s that the coffee falls from the pot to the level of the cup, the cup moves forward 15 m! Why is it possible to pour coffee on airplanes?

Were it not for the effect of air resistance, water droplets from this fountain would follow a parabolic trajectory.

If we solve Equation (4.6) for t and substitute the resulting expression (x/v_o) into Equation (4.7) we obtain

$$y = -\frac{1}{2}\left(\frac{a_g}{v_o^2}\right)x^2 \tag{4.8}$$

If you focus your attention on how the vertical height y depends on horizontal position x, you may recognize Equation (4.8) as the equation of a parabola that opens downward. In the *freely-falling-body* model, the path of a projectile is always a parabola.

Now let's consider a projectile launched so that the initial velocity vector makes a nonzero angle with the horizontal. Before the two-dimensional problem can be reduced to two, one-dimensional problems, the initial velocity vector must be resolved into its horizontal and vertical components. As shown in Figure 4.4, this resolution is accomplished by forming a right triangle with the original velocity vector as the hypotenuse. The sides of the right triangle are parallel to the x- and y-axes. Therefore, we label the horizontal component v_{o_x} and the vertical component v_{o_y}. Note that the component vectors must have directions such that they add vectorially to give the original vector.

We recall the definition of the sine and use the triangle formed by the vectors in Figure 4.4, $\sin(\theta) = $ (*Side opposite/Hypotenuse*) $= v_{o_y}/v_o$. Thus, we obtain the magnitude of v_{o_y} from

$$v_{o_y} = v_o \sin(\theta) \tag{4.9}$$

Making use of the cosine definition we find that

$$v_{o_x} = v_o \cos(\theta) \tag{4.10}$$

With Equations (4.9) and (4.10) we have completed the task of dividing the problem into two component problems: one for the x-motion and one for the y-motion.

In Table 4.1 we show the final results of this decomposition. We start with the equations for our constant-acceleration model and then specialize them for horizontal and vertical components. We write each of these specialized equations in a separate column, one for x-components, another for y-components. We make this separation to emphasize the importance of not intermingling x- and y-components.

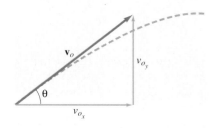

FIGURE 4.4

The initial velocity $\mathbf{v}_o$ must be resolved into x- and y-components such that $v_{o_x}\hat{\mathbf{i}} + v_{o_y}\hat{\mathbf{j}} = \mathbf{v}_o$.

TABLE **4.1**   Kinematic Equations Specialized
for Projectile Motion

EQUATIONS FOR ANY DIMENSION

$$s = s_o + v_o t + \frac{1}{2}at^2$$

$$v = v_o + at$$

$$v^2 = v_o^2 + 2a(s - s_o)$$

EQUATIONS FOR PROJECTILE MOTION

x-component	*y-component*

$(s_o = x_o)$	$(s_o = y_o)$
$(v_o = v_{o_x})$	$(v_o = v_{o_y})$ Substitutions
$(a = 0)$	$(a = -a_g)$
$x = x_o + v_{o_x}t$	$y = y_o + v_{o_y}t - \frac{1}{2}a_g t^2$
$v_x = v_{o_x}$	$v_y = v_{o_y} - a_g t$
	$v_y^2 = v_{o_y}^2 - 2a_g(y - y_o)$

We recommend you do *not* memorize the specialized equations summarized in Table 4.1. They are merely special cases of the one-dimensional constant-acceleration equations.

Before tying all this theory together with an example, we give you our suggestions for a general method for attacking projectile-motion problems.

Problem-Solving Strategy

How to attack projectile-motion problems with the constant-acceleration model:

1. Draw a picture showing the object and anticipated trajectory.
2. Place an *xy*-coordinate system on your picture. Make the *y*-axis parallel to the acceleration of gravity. Where you place the origin is arbitrary, but make it convenient.
3. If the initial velocity is not parallel to either the *x*- or the *y*-axis, then resolve it into components by using Equations (4.9) and (4.10).
4. Make two columns (one for the *x*-components and one for the *y*-components) that summarize the initial conditions of the problem. Include all items (x_o, v_{o_x}, a_x, y_o, v_{o_y}, and a_y), even if some are unknown.
5. Identify the unknown and place its symbol in the appropriate column.
6. Identify the ''special condition'' that describes some property of the object's position or velocity at the point of interest. Put the algebraic statement of the special condition in the *x* or *y* column as appropriate. (If this step is unclear to you now, don't worry. We discuss it further as we work through some examples below.)
7. **(a)** If your unknown is in a column with four known quantities, you can solve for it immediately from one of the constant-acceleration equations.
   (b) If your unknown is in a column with only three known quantities, but the other column has four known quantities, use one of the constant-acceleration equations on the side with the four known quantities to solve for the time. Then use this time as the fourth known in the other column.
   (c) If you don't have four known quantities on at least one side, go back to steps 4 and 5 and see if you've left something out.
   (d) If the unknown is either v_o or θ, write Equation (4.5) for both sides and solve simultaneously.
8. After solving the equation algebraically, that is with letters, substitute numerical values. (You may prefer to substitute numbers earlier, but we encourage you to wait until the end. Sometimes seemingly impossible problems can be solved this way when an unknown cancels out of the problem!)
9. Substitute units into your equation and check for dimensional consistency.
10. Check your answer to see if it is reasonable.

Now that we have set out our plan, let's apply it to a couple of examples to see exactly how it works.

EXAMPLE 4.2   *Splat!*

A spring-loaded toy gun fires a small, sticky ball horizontally toward a vertical wall located 2.00 m away. If the "muzzle velocity" of the ball is 5.00 m/s, how far below the horizontal position of the gun does the ball hit the wall?

SOLUTION   Figure 4.5 shows steps 1 and 2 of the problem-solving strategy. *Note:* We have put the origin of the coordinate system at the tip of the gun. Because the initial velocity is horizontal, we do not have to perform step 3. (Or, from another point of view, the angle θ is zero so that $v_{o_x} = v_o \cos(0°) = v_o$, and $v_{o_y} = v_o \sin(0°) = 0$.) For step 5 we note that the unknown in the problem is y, because we are asked for the vertical position on the wall where the ball hits.

For this problem the "special condition" of step 6 stems from the phrase ". . . hit the wall." In order to hit the wall, the ball must reach $x = 2.00$ m on our coordinate system. Hence, "$x = 2.00$ m" is the special condition. (Be careful here. You might be tempted to use $v_x = 0$ as the special condition because you know the ball stops there. However, after the ball actually contacts the wall it is no longer described by our equations for the constant-acceleration model. Once it really touches, all bets are off! When we say ". . . hit the wall," we really mean ". . . the instant just before it hits the wall.")

With steps 1 through 6 completed we can fill in our table as follows:

x-COMPONENTS		*y*-COMPONENTS	
$x_o = 0$		$y_o = 0$	
$v_{o_x} = 5.00$ m/s		$v_{o_y} = 0$	
$v_x = 0$		$a_y = -a_g = -9.80$ m/s^2	
$x = 2.00$ m	(special condition)	$y = ?$	(the unknown)

As you look at the table above you can see that we are in the situation described in step 7b of our problem-solving strategy. That is, we want y, but we know only three quantities on the y-side of the table. (Remember, none of the kinematic equations involving position can be solved unless we know four of the variables.) The only quantity common to both the x- and y-components of the motion is time. Hence, we follow the suggestion from step 7b and solve for t from the x-side.

Before actually proceeding let's specialize the general x- and y-equations of motion by substituting from their respective columns:

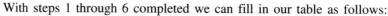

$$x = x_o + v_{o_x}t + \frac{1}{2}a_x t^2 \qquad \Big| \qquad y = y_o + v_{o_y}t + \frac{1}{2}a_y t^2$$

$$x = \qquad v_o t \qquad \Big| \qquad y = \qquad -\frac{1}{2}a_g t^2$$

We can now solve the x-equation for $t = x/v_o$ and substitute this fraction into the y-equation, obtaining

$$y = -\frac{1}{2}a_g\left(\frac{x^2}{v_o^2}\right)$$

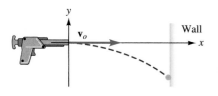

FIGURE 4.5

A ball fired horizontally follows a parabolic path to a vertical wall. See Example 4.2.

We now substitute numbers and units

$$y = -\frac{1}{2}(9.80 \text{ m/s}^2)\left(\frac{(2.00 \text{ m})^2}{(5.00 \text{ m/s})^2}\right)$$

$$= -0.784 \text{ m/s}^2 \frac{\text{m}^2}{\text{m}^2/\text{s}^2}$$

$$= -0.784 \text{ m}$$

The physical interpretation of our answer is that after the ball has traveled a horizontal distance of 2.00 m (and ''hits the wall''), it is 0.784 m below the point labeled $y = 0$ on our coordinate system. Referring to Figure 4.5, we see that this location is 0.784 m below the level of the toy gun. Needless to say, if the gun is only, say, 0.500 m above the floor, the ball hits the floor before hitting the wall. As step 10 suggests, you should always check your answer to see if it is physically reasonable. ◀

Let's now look at a somewhat more involved problem.

EXAMPLE 4.3   A Bad Bunt!

A baseball player tries to bunt a pitch down the third-base line. When the ball leaves the bat it is 1.20 m above the ground, traveling 10.0 m/s at a 36.87° angle above the horizontal. (a) How high does the ball rise? (b) Where does the ball first hit the ground?

SOLUTION   Figure 4.6 illustrates the ball's parabolic trajectory. Superposed on the trajectory is our choice of a coordinate system (steps 1 and 2). This time, the original velocity vector must be resolved into x- and y-components by using Equations (4.9) and (4.10). The table of initial conditions becomes

x-COMPONENTS	y-COMPONENTS
$x_o = 0$	$y_o = +1.20$ m
$v_{o_x} = +v_o \cos(36.87°)$	$v_{o_y} = +v_o \sin(36.87°)$
$\quad = (10.0 \text{ m/s})(0.800) = 8.00$ m/s	$\quad = (10.0 \text{ m/s})(0.600) = 6.00$ m/s
$a_x = 0$	$a_y = -a_g = -9.80 \text{ m/s}^2$

(a) How high does the ball rise? The unknown here is y. We must identify the special condition that describes the ball's highest point. The unique feature at the ball's highest point is that the ball stops going up and starts going down; at that one instant of time it is doing neither. That is, the y-component of the ball's velocity is zero. The special condition is, therefore, $v_y = 0$. Consequently, we add to our table:

$$y = ? \qquad \text{(the unknown)}$$

$$v_y = 0 \qquad \text{(the special condition)}$$

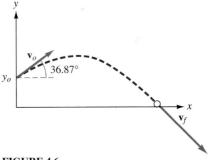

FIGURE 4.6

A baseball is bunted from a height y_o and follows a parabolic trajectory to ground level where $y = 0$. See Example 4.3.

Notice both of these entries are in the y-column. We now have the situation described in step 7a of our procedure; namely, four known quantities in the same column with the unknown. Thus, we can solve for y immediately using the appropriate constant-acceleration equation. Because t does not appear in the y-column, we look for the equation with t missing:

$$v^2 = v_o^2 + 2a(s - s_o)$$

We specialize this equation for the y-motion

$$v_y^2 = v_{o_y}^2 - 2a_g(y - y_o)$$

and solve for the unknown y:

$$y = y_o + \frac{v_{o_y}^2 - v_y^2}{2a_g}$$

Substituting the numbers we obtain

$$y = 1.20 \text{ m} + \frac{(6.00 \text{ m/s})^2 - 0^2}{2(9.80 \text{ m/s}^2)}$$

$$= 1.20 \text{ m} + 1.84 \text{ m} = 3.04 \text{ m}$$

Because we designated $y = 0$ as ground level, the result for y indicates that the bunt reaches 3.04 m above the ground.

(b) Where does the ball first hit the ground? The unknown "where" is x. In Example 4.2 we learned that the special condition for an object hitting a surface should be expressed in terms of the object's position (not velocity). Hence, the special condition here is $y = 0$. Placing the unknown and the special condition in the table we have:

$$x = ? \qquad\qquad | \qquad\qquad y = 0$$

The situation is now the one described in step 7b of our problem-solving strategy. Thus, we know that we should solve the y-equations for t and then use this value of t in one of the x-equations. Equation (4.5) contains the variables necessary to solve for t:

$$s = s_o + v_o t + \frac{1}{2}at^2$$

which becomes for y-motion

$$y = y_o + v_{o_y} t - \frac{1}{2}a_g t^2$$

Or, because $y = 0$,

$$0 = y_o + v_{o_y} t - \frac{1}{2}a_g t^2$$

This relation is a quadratic equation of the form

$$At^2 + Bt + C = 0$$

with solutions

$$t = \frac{-B \pm \sqrt{B^2 - 4AC}}{2A}$$

In our case $A = -\frac{1}{2}a_g$, $B = v_{o_y}$, and $C = y_o$. Substituting the values for these quantities, we find $t = -0.175$ s and $t = 1.40$ s. We take the positive root as the physically meaningful one for this case. This value for t can now be substituted into the x-equation:

$$x = x_o + v_{o_x}t + \frac{1}{2}a_x t^2$$

$$= v_{o_x}t$$

$$= (8.00 \text{ m/s})(1.40 \text{ s}) = 11.2 \text{ m}$$

What about the physical interpretation? You be the judge: It is 27.4 m from home plate to third base. If the third baseman has 1.40 s to reach the ball and can maintain an *average speed* of 6.00 m/s, how far can he be from the point of the ball's impact from the ground and expect to just catch it? Answer: 8.40 m. ◀

In Examples 4.2 and 4.3 you saw the importance of identifying the "special condition" in order to solve the problem. In those examples three common, special conditions were introduced. Table 4.2 summarizes those special conditions plus another that commonly occurs. You should add your own to the list.

TABLE **4.2**   Special Conditions in Projectile-Motion Problems

WORD DESCRIPTION IN PROBLEM	SPECIAL CONDITION
". . . hits the (vertical) wall . . ."	$x = x_{\text{wall}}$
". . . hits the ground . . ."	$y = y_{\text{ground}}$
". . . at the top (of its trajectory) . . ."	$v_y = 0$
". . . hits some point . . ."	$x = x_{\text{point}}$ and $y = y_{\text{point}}$

4.3 Circular Motion

As is evident from our analysis of projectile motion, when we look at motion in more than one dimension the true vector nature of velocity and acceleration becomes apparent. However, an even more dramatic demonstration of just what it means for a quantity to be a vector is provided by the case of uniform circular motion.

By **uniform circular motion** we mean motion in a circle with *constant radius and constant speed.* Figure 4.7 shows such motion for a person in an amusement park ride. At the instant shown, the person has traveled through an arc length s and is located at an angle θ measured anticlockwise from the x-axis. In mathematics and physics it is usually convenient to express the angle θ in **radians.** The symbol for radians is **rad.** The definition is

$$Angle \ (in \ radians) = \frac{Arc \ length}{Radius \ length}$$

or

$$\theta = \frac{s}{r} \tag{4.11}$$

Concept Question 3
If a ball is thrown upward at an angle $\theta > 0$ above the horizontal, what is the ball's acceleration at the top of its trajectory? What is the ball's acceleration when it is halfway (vertically) to the top of its trajectory?

Concept Question 4
Where in its parabolic path does the magnitude of a kicked football's velocity have its minimum value? Its maximum value?

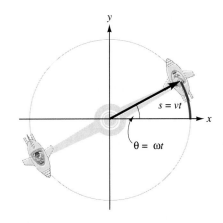

FIGURE 4.7

In uniform circular motion both the angle and the arc length increase uniformly in time.

Notice that the dimensions of θ are (length/length). Hence, an angle expressed in radian units really has no units at all! We write *rad* to remind ourselves that the number that precedes this label is an angle, whose definition is given by Equation (4.11).

Returning to the amusement park and our discussion of the person moving with a constant speed in a circle, we note that the angle θ does not stay constant but rather increases linearly in time. A linear equation for θ at time t is

$$\theta = \omega t$$

where we have assumed that $\theta = 0$ rad at $t = 0$ s. The slope of this linear equation ω (omega) is a constant that tells us the amount by which the angle increases during each second. This constant ω is known as the **angular velocity.** It is best written in units of radians per second. You may also see it expressed in rpm (revolution per minute) or even degrees per second. If you do see these other units, we suggest you change them to radians per second as fast as you can turn on your calculator! From here on out we assume θ is in radians and ω is in radians per second.

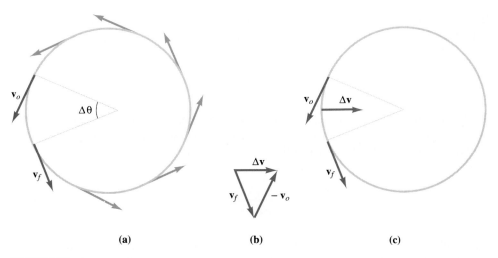

FIGURE 4.8

(a) Velocity vectors tangent to the circle represent the direction an object takes if there is no centripetal acceleration. (b) Two velocity vectors are subtracted to find the change in velocity $\Delta\mathbf{v} = \mathbf{v}_f - \mathbf{v}_o$. (c) The change $\Delta\mathbf{v}$ between initial and final velocity vectors, when placed at the average position, points toward the center of the circular motion. This $\Delta\mathbf{v}$ vector when divided by Δt becomes the average acceleration, which also points toward the center.

Now, let's look at the acceleration and velocity vectors of our person undergoing uniform circular motion. Acceleration is defined as the time rate of change of the velocity *vector*. Therefore, we can produce an acceleration either by changing our speed (the length of the velocity vector) or by changing the *direction* of the velocity vector. That's right, if we change the direction of the vector and leave its magnitude constant, we still have a nonzero acceleration! Let's see how this situation is possible.

Figure 4.8(a) illustrates uniform circular motion and shows that the length of the velocity vector never changes, even though its direction does vary constantly. This constant change in direction produces an acceleration that is just as real as the acceleration that changes speed.

In uniform circular motion the direction of the *instantaneous* acceleration is toward the center of the circle. We can see how this direction comes about by looking at the *average* acceleration *over a short time interval*. In Figure 4.8(b) two of the velocity vectors in Figure 4.8(a) are subtracted to compute $\Delta\mathbf{v} = \mathbf{v}_f - \mathbf{v}_o$. Because $\mathbf{a}_{av} = \Delta\mathbf{v}/\Delta t$, the direction of the average acceleration is in the direction of $\Delta\mathbf{v}$. In Figure 4.8(c) $\Delta\mathbf{v}$ is redrawn on the original circular path, halfway between $\mathbf{v}_f$ and $\mathbf{v}_o$. Note that it points inward toward the center of the circle. This acceleration is denoted by $\mathbf{a}_c$, and is known as the **centripetal acceleration.** (The word *centripetal* was coined by Isaac Newton from the Latin words for "center seeking" as opposed to *centrifugal*, which he coined to indicate "center fleeing.")

The magnitude of the centripetal acceleration can be found by combining the powerful techniques of calculus with vector notation. In Figure 4.9 the radius vector $\mathbf{r}$ points to a particle in uniform circular motion. The expression for $\mathbf{r}$ in rectangular coordinates is

$$\mathbf{r} = x\hat{\mathbf{i}} + y\hat{\mathbf{j}}$$

$$\mathbf{r} = R_o \cos(\theta)\hat{\mathbf{i}} + R_o \sin(\theta)\hat{\mathbf{j}} \tag{4.12}$$

Using $\theta = \omega t$, the position vector in Equation (4.12) can now be written as an explicit function of time:

$$\mathbf{r}(t) = R_o \cos(\omega t)\hat{\mathbf{i}} + R_o \sin(\omega t)\hat{\mathbf{j}} \tag{4.13}$$

Concept Question 5
Explain how an object can have a nonzero acceleration while its velocity magnitude remains constant.

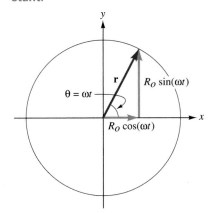

FIGURE 4.9

The position vector $\mathbf{r}$ has length $|\mathbf{r}| = R_o$. The angle θ increases linearly with time.

It is easy to find the velocity vector as a function of time by differentiating Equation (4.13) with respect to time. Remember that R_o, ω, $\hat{\mathbf{i}}$, and $\hat{\mathbf{j}}$ are all constants. The result is

$$\mathbf{v}(t) = \frac{d\mathbf{r}}{dt} = -R_o\omega\sin(\omega t)\hat{\mathbf{i}} + R_o\omega\cos(\omega t)\hat{\mathbf{j}} \qquad (4.14)$$

We leave it to you to prove that vectors $\mathbf{v}$ and $\mathbf{r}$ are perpendicular (Problem 27). This condition means that the velocity is tangent to the circular path of the particle. The magnitude of the velocity, found in the usual manner (see Problem 27), is

$$v = \omega R_o \qquad (4.15)$$

Note that this equation holds only when ω is expressed in radians per unit time.

One more derivative, this time of Equation (4.14), takes us to the acceleration vector

$$\mathbf{a}(t) = \frac{d\mathbf{v}(t)}{dt} = -R_o\omega^2\cos(\omega t)\hat{\mathbf{i}} - R_o\omega^2\sin(\omega t)\hat{\mathbf{j}} \qquad (4.16)$$

If you haven't fallen asleep by now you probably noticed that Equation (4.16) for the acceleration looks remarkably like Equation (4.13). In fact,

$$\mathbf{a}(t) = -\omega^2\mathbf{r}(t) \qquad (4.17)$$

There are two important points revealed by Equation (4.17). First, this equation confirms our expectation that the centripetal acceleration points toward the center of the circle. Because Equation (4.17) is a vector equation, the directions of the two vectors on either side of this equation must be identical. The minus sign indicates that although vector $\mathbf{a}$ and vector $\mathbf{r}$ are parallel, they point in opposite directions. (This geometry is sometimes called antiparallel.) Second, Equation (4.17) tells us how the magnitudes of the two vectors are related; namely, that

$$a \equiv a_c = \omega^2 R_o \qquad (4.18)$$

Note that this equation holds only when ω is expressed in radians per unit time. From Equation (4.15) we see that $\omega R_o = v$, and, therefore, Equation (4.18) can also be written

$$a_c = \frac{v^2}{R_o} \qquad (4.19)$$

To summarize, an object moving in uniform circular motion at radius R_o and constant angular speed ω has

1. a constant linear speed $v = R_o\omega$ tangent to the circle, and
2. a centripetal acceleration $a_c = \omega^2 R_o = v^2/R_o$ directed inward toward the center of the circle.

EXAMPLE 4.4   *Now That's Taking a Spin*

You are driving along the interstate highway at 25.0 m/s (56 mph) beside a truck traveling at the same speed. Looking out the window you observe a tire on the truck. The tire has a radius of 0.700 m. (a) What is the angular velocity of the tire? (b) What is the centripetal acceleration of a point on the tire's rim?

SOLUTION   (a) From your point of view the center of the tire is standing still and the road is moving backward past the tire's bottom edge at 25.0 m/s. Because the tire is not slipping on the road, at any instant the tire's bottom edge must also have a speed of 25.0 m/s. Thus, we have as our given quantities

$$v = 25.0 \text{ m/s}$$

$$R_o = 0.700 \text{ m}$$

From Equation (4.15) we have

$$\omega = \frac{v}{R_o} = \frac{25.0 \text{ m/s}}{0.700 \text{ m}} = 35.7 \text{ rad/s}$$

Notice we have changed the units from 1/s to rad/s in order to make it clear that the angular units are radians.
    (b) From Equation (4.19),

$$a_c = \frac{v^2}{R_o} = \frac{(25.0 \text{ m/s})^2}{0.700 \text{ m}} = 893 \text{ m/s}^2$$

It is sometimes revealing to compare such accelerations to that due to gravity. The resulting ratio a/a_g is traditionally referred to as the number of g's for that acceleration. We note that $a/a_g = (893 \text{ m/s}^2)/(9.80 \text{ m/s}^2) = 91.1$ That is, the tread on the edge of the tire must hold against 91 g's. No wonder a lot of rubber from truck-tire retreads appears along the highway!   ◀

Tangential and Normal Components of Acceleration in Two Dimensions

Consider a figure skater gliding in graceful arcs across the surface of an ice rink. Even if the skater is modeled as a point particle traveling along a winding path with a changing speed, a detailed description of the skater's velocity and acceleration is complicated. Nonetheless, some general features of such complex motion can still be easily described.

1. The velocity vector is always tangent to the path.
2. The acceleration vector may have two components: one tangent to the path and one perpendicular to the path.

   (a) The component of the acceleration parallel to the path is due to changes in speed. When the speed is increasing, the tangential component a_t points in the same direction as the velocity; when the speed is decreasing, the tangential component points opposite the velocity.

   (b) When the path of an object curves, there is a component of the acceleration perpendicular to the velocity. This component of the acceleration a_c points toward the inside of the curve.

   (c) The total acceleration is the vector sum of the tangential and centripetal components.

The techniques illustrated in the following example show how we can sometimes simplify our view of an object's motion by separating acceleration into components parallel and perpendicular to the object's path.

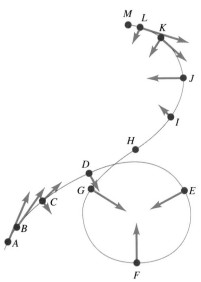

FIGURE 4.10

Tangential and normal components of the acceleration for a figure skater.

Concept Question 6
If you are traveling in a circle with nonzero speed, (a) can you have zero tangential acceleration? (b) Can you have zero centripetal acceleration? (c) Can your tangential acceleration ever be greater in magnitude than your centripetal acceleration? (d) How does your speed have to change so that your total acceleration is nearly antiparallel to your velocity?

EXAMPLE 4.5   Cutting a Fine Line

An ice skater travels along the path shown in Figure 4.10. The heavy dots along the path indicate the position of the skater every 2.00 s. On the curve are approximate directions for the tangential and centripetal acceleration vectors for several points. Explain these acceleration components.

SOLUTION   Point *B*: From points *A* to *C* the skater's speed increases. We know this because the dots are spaced farther and farther apart, and the time interval between each is the same (2.00 s). Therefore, at point *B* we show an acceleration component tangent to the path in the forward direction. Also, from *C* to *D* there is a slight change in direction. Thus, we show a small acceleration component perpendicular to the path.

Points *E*, *F*, and *G*: From point *D* to point *G* the skater's speed is nearly constant. We know this because the dots are equally spaced along the path. Thus, we show no tangential acceleration component during this interval. At points *E*, *F*, and *G*, however, the skater is changing direction rapidly, traveling nearly in a circle at a high speed. Consequently, we show a fairly large centripetal acceleration (toward the center of the circle) at points *E*, *F*, and *G*.

Point *H*: At point *H* the skater is traveling at constant speed along a straight line. There is no acceleration of any type.

Point *J*: Along *IJK* the skater moves in a larger circle at constant speed. For this portion of the path the centripetal acceleration points toward the center of the curve.

Point *L*: Starting at point *K*, the skater slows down. The tangential acceleration is directed opposite the skater's velocity. The centripetal component shrinks due to the smaller velocity at point *L*.   ◄

4.4 Numerical Techniques in Two Dimensions (Optional)

In this chapter we have analyzed two simple models that are sometimes useful approximations of the motion of actual objects; namely, idealized projectile motion and uniform circular motion. The constant-acceleration model for projectiles can break down for any number of reasons: air resistance, the curvature of the earth, and changes in the acceleration of gravity with altitude, to name just a few. We postpone until Chapter 6 the discussion of models that better describe projectile motion when these effects are important. Instead, let's look at the motion of an object moving in a nearly circular path but with a nonconstant speed.

Because we are going to need a couple of important results from Chapter 2 about vectors, we repeat them here to refresh your memory. First, recall how to create a unit vector in a particular direction. It's easy; just find *any* vector in the required direction, then divide that vector by its own magnitude. For example, we are going to need a unit vector parallel to the velocity vector, so we'll divide $\mathbf{v}$ by $|\mathbf{v}|$.

We also need to find a vector perpendicular to the velocity vector. You may recall that when two vectors are perpendicular their dot product is zero. In two dimensions this property means that vector $\mathbf{B}$ is perpendicular to $\mathbf{A}$ if $\mathbf{A} \cdot \mathbf{B} = A_x B_x + A_y B_y = 0$ or, in other words, if $B_y/B_x = -A_x/A_y$. This result tells us that to make a vector perpendicular to $\mathbf{A}$, we have only to interchange the *x*- and *y*-components of $\mathbf{A}$ and take the opposite sign for one of them. For example, if $\hat{\mathbf{T}} = T_x\hat{\mathbf{i}} + T_y\hat{\mathbf{j}}$ is a *unit* vector, then $\hat{\mathbf{N}}$ is a unit vector perpendicular to $\hat{\mathbf{T}}$ if $\hat{\mathbf{N}} = -T_y\hat{\mathbf{i}} + T_x\hat{\mathbf{j}}$. Not convinced? Take the dot product of $\hat{\mathbf{T}}$ and $\hat{\mathbf{N}}$, and then compare the magnitude of $\hat{\mathbf{T}}$ to the magnitude of $\hat{\mathbf{N}}$.

Now that we've had our quick review of vectors, we're ready to proceed with a practical example.

EXAMPLE 4.6 *Analysis of a Golf Swing*

Figure 4.11 shows a multiflash photograph of a golfer's swing taken by Harold Edgerton, one of the pioneers of high-speed photography. The position of the head of the club in the plane of its swing is estimated from the photograph. (Club head coordinates are corrected for some, but not all, distortions due to camera magnification.) Use the position data to calculate the velocity and acceleration of the club head.

SOLUTION In columns A through C of the spreadsheet GOLFSWNG.WK1, we record the time and the x- and y-positions of the golf club head for forty of the images on the photograph. In column E we calculate the v_x-components of velocity from the x-positions just as we did in Section 3.5 for the one-dimensional case. Similarly, we calculate values for v_y in column F. The magnitude of the velocity is easily obtained from the v_x- and v_y-components: $v = \sqrt{v_x^2 + v_y^2}$. These values of v are shown in column G. Next, in columns H and I we calculate the a_x-component of the acceleration from the change in v_x, and the a_y-component of acceleration from the change v_y. We used the simple difference method:

$$a_x(i) = \frac{v_x(i+1) - v_x(i-1)}{2\,\Delta t}$$

At this point our spreadsheet looks like Figure 4.12.

FIGURE 4.11

Multiflash photograph of a golfer's swing taken by Harold Edgerton, a pioneer in high-speed flash photography.

	A	B	C	D	E	F	G	H	I	J
5										
6	t	x	y	‖	vx	vy	v	ax	ay	a
7	(0.01	(cm)	(cm)	‖	(cm/s)	(cm/s)	(cm/s)	(cm/s^2)	(cm/s^2)	(cm/s^2)
8										
9	1	−48.89	69.03	‖						
10	2	−64.24	58.47	‖	−1535	−1421	2092			
11	3	−79.59	40.61	‖	−1450	−2030	2495	1.85E+04	−6.29E+04	6.56E+04
12	4	−93.23	17.87	‖	−1165	−2680	2922	4.55E+04	−6.90E+04	8.27E+04
13	5	−102.90	−12.99	‖	−540	−3411	3453	8.24E+04	−6.09E+04	1.02E+05
14	6	−104.04	−50.35	‖	483	−3898	3928	1.17E+05	−2.03E+04	1.18E+05
15	7	−93.23	−90.96	‖	1791	−3817	4216	1.34E+05	4.26E+04	1.40E+05
16	8	−68.22	−126.69	‖	3155	−3046	4385	1.07E+05	1.16E+05	1.57E+05
17	9	−30.13	−151.87	‖	3923	−1502	4201	2.13E+04	1.64E+05	1.66E+05
18	10	10.23	−156.74	‖	3582	244	3590	−4.83E+04	1.50E+05	1.58E+05
19	11	41.50	−147.00	‖	2956	1502	3316	−5.54E+04	1.12E+05	1.25E+05
20	12	69.36	−126.69	‖	2473	2477	3500	−5.83E+04	7.72E+04	9.67E+04

FIGURE 4.12

Part of a spreadsheet template for calculating velocity and acceleration in two dimensions.

The total acceleration $\sqrt{a_x^2 + a_y^2}$, is calculated in column J.

Everything so far has been pretty "straightforward" as physicists like to say. Now we are ready to change the x- and y-components into components that are parallel and perpendicular to the path, respectively. This calculation makes the power of vectors really apparent. Let's call the unit vector tangent to the path $\hat{\mathbf{T}}$. Remember $\hat{\mathbf{T}} = \mathbf{v}/v$. To obtain a_t, the component of acceleration in the direction of $\hat{\mathbf{T}}$, we compute the dot product of $\hat{\mathbf{T}}$ with $\mathbf{a}$. In vector notation the process is symbolized:

$$a_t = \mathbf{a} \cdot \hat{\mathbf{T}} \tag{4.20}$$

A side view of a golfer's swing indicates the true plane of motion.

In component form this dot product appears as

$$a_t = a_x T_x + a_y T_y \tag{4.21}$$

In cells K11 and L11 of our spreadsheet we calculate T_x and T_y from v_x/v and v_y/v, respectively. Then the formula for a_t in cell M11 is

$$M11: \quad (H11*K11) + (I11*L11) \tag{4.22}$$

The component perpendicular to $\hat{\mathbf{T}}$ is just as easy to obtain. Let's call $\hat{\mathbf{N}}$ the unit vector perpendicular to $\hat{\mathbf{T}}$. Then we can write

$$a_n = \mathbf{a} \cdot \hat{\mathbf{N}} \tag{4.23}$$

Remember how the components of $\hat{\mathbf{N}}$ and $\hat{\mathbf{T}}$ are related? To write $\hat{\mathbf{N}}$ in terms of $\hat{\mathbf{T}}$ we use our simple rule: switch the x- and y-components and change the sign of one of them. (We arbitrarily change the sign of the x-component so that a_n is positive.)

$$a_n = a_x N_x + a_y N_y = -a_x T_y + a_y T_x \tag{4.24}$$

Therefore, cell N11 in the spreadsheet contains the formula

$$N11: \quad (-H11*L11) + (I11*K11) \tag{4.25}$$

Thus, the final columns of the spreadsheet appear as in Figure 4.13.

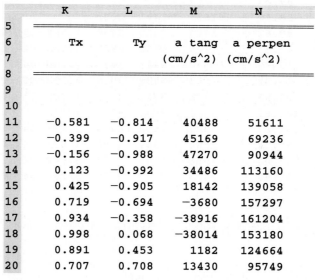

	K	L	M	N
	Tx	Ty	a tang (cm/s^2)	a perpen (cm/s^2)
11	−0.581	−0.814	40488	51611
12	−0.399	−0.917	45169	69236
13	−0.156	−0.988	47270	90944
14	0.123	−0.992	34486	113160
15	0.425	−0.905	18142	139058
16	0.719	−0.694	−3680	157297
17	0.934	−0.358	−38916	161204
18	0.998	0.068	−38014	153180
19	0.891	0.453	1182	124664
20	0.707	0.708	13430	95749

FIGURE 4.13

Continuation of the spreadsheet of Figure 4.12.
The portion shown here shows the calculation
of the normal and tangential components of acceleration.

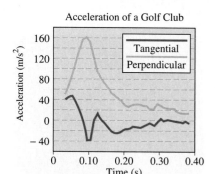

FIGURE 4.14

Estimates of the tangential and
perpendicular components of
acceleration for the head of the golf
club in Figure 4.11.

In Figure 4.14 we show the fruits of our calculation. The component of the acceleration perpendicular to the path is largest at the bottom of the golfer's swing where the velocity is greatest. Here the club head is traveling in a nearly perfect circle. During this time the perpendicular ($\hat{\mathbf{N}}$) component of the acceleration points toward the center of that circle and may properly be called the centripetal acceleration.

During the down swing, the tangential acceleration is positive. That is, it is in the same direction as the velocity. This acceleration reaches a maximum value of about 470 m/s^2! Note that the peak speed of 44 m/s is reached at the point where the tangential acceleration is zero. (Think about why this result must be so!) After contact with the ball the acceleration quickly becomes negative, that is, antiparallel to the velocity vector. (Because we calculate velocities over two intervals and assign them to the middle image, the club appears to slow down in image 9.) In Figure 4.15 we show the position, velocity, and acceleration vectors on the original photograph. ◀

(a)

In this section, the vector methods of Chapter 2 were combined with the numerical techniques of Chapter 3 to analyze motion in two dimensions. Be assured that this procedure is no "academic exercise." These techniques are used every day. Entire companies are devoted to the rapid analysis of high-speed photography. These techniques have been applied to many subjects ranging from the training of athletes to the mechanical design of high-speed assembly lines.

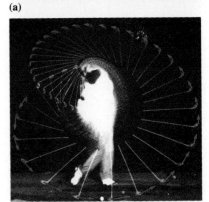

4.5 Summary

Definitions of kinematic variables were extended to three dimensions:

(b)

position:
$$\mathbf{r} = x\hat{\mathbf{i}} + y\hat{\mathbf{j}} + z\hat{\mathbf{k}}$$

displacement:
$$\Delta\mathbf{r} = \Delta x\hat{\mathbf{i}} + \Delta y\hat{\mathbf{j}} + \Delta z\hat{\mathbf{k}}$$

average velocity:
$$\mathbf{v}_{av} = \frac{\Delta\mathbf{r}}{\Delta t} = \frac{\Delta x}{\Delta t}\hat{\mathbf{i}} + \frac{\Delta y}{\Delta t}\hat{\mathbf{j}} + \frac{\Delta z}{\Delta t}\hat{\mathbf{k}}$$

instantaneous velocity:
$$\mathbf{v} = \frac{d\mathbf{r}}{dt} = \frac{dx}{dt}\hat{\mathbf{i}} + \frac{dy}{dt}\hat{\mathbf{j}} + \frac{dz}{dt}\hat{\mathbf{k}}$$

average acceleration:
$$\mathbf{a}_{av} = \frac{\Delta\mathbf{v}}{\Delta t} = \frac{\Delta v_x}{\Delta t}\hat{\mathbf{i}} + \frac{\Delta v_y}{\Delta t}\hat{\mathbf{j}} + \frac{\Delta v_z}{\Delta t}\hat{\mathbf{k}}$$

instantaneous acceleration:
$$\mathbf{a} = \frac{d\mathbf{v}}{dt} = \frac{dv_x}{dt}\hat{\mathbf{i}} + \frac{dv_y}{dt}\hat{\mathbf{j}} + \frac{dv_z}{dt}\hat{\mathbf{k}}$$

average speed:
$$v_{av} = \frac{Total\ path\ length}{\Delta t}$$

instantaneous speed:
$$v = |\mathbf{v}| = Magnitude\ of\ velocity\ vector$$

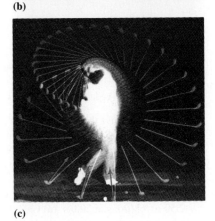

(c)

FIGURE 4.15

(a) Position vectors of the golf club head. (b) Velocity vectors are obtained from differences in position vectors. (c) Acceleration vectors indicate the direction of changes in the velocity vectors.

Projectile Motion   The freely-falling-body model may be extended to include projectile motion, and a systematic problem-solving strategy for projectile motion can be applied to determine the projectile's motion. The main emphasis of this strategy is on the decomposition of a two-dimensional problem into two, one-dimensional problems.

The general equations for constant acceleration were specialized for x- and y-components of projectile motion:

x-COMPONENTS	y-COMPONENTS
$x = x_o + v_{o_x}t$	$y = y_o + v_{o_y}t - \frac{1}{2}a_g t^2$
$v_x = v_{o_x}$	$v_y = v_{o_y} - a_g t$
	$v_y^2 = v_{o_y}^2 - 2a_g(y - y_o)$

Uniform Circular Motion   A convenient means for designating an angle is by the radian measure: if s is the arc length subtended by the angle at radius r, then in radians the angle θ is defined by

$$\theta = \frac{s}{r}$$

When an object exhibits uniform circular motion, it travels in a circle of constant radius with a constant speed. If ω is the angular velocity in units of rad/s, the angle through which an object rotates in time t (given in radians) is $\theta = \omega t$. The speed of the object exhibiting uniform circular motion at radius R with angular velocity ω is

$$v = R\omega$$

Moreover, the object has an acceleration toward the center of its circular path called the centripetal acceleration a_c. The magnitude of this acceleration is given by

$$a_c = \frac{v^2}{R} = R\omega^2$$

It is sometimes useful to think of the acceleration along a curved path as having two components, one tangential to the path due to changes in speed, and the other perpendicular to the path due to changes in direction.

Optional: Numerical methods can be employed to compute values for an object's velocity and acceleration functions in two dimensions.

PROBLEMS

4.1 Velocity and Acceleration Vectors

1. A San Francisco trolley car rounds a corner along a track that forms a circular arc of radius 9.50 m. The trolley takes 20.0 s to move around the quarter circle. (a) What is the magnitude of the trolley's average velocity vector? (b) What is the average speed of the trolley?

2. A child on a merry-go-round with a diameter of 5.00 m makes 2.50 revolutions in 22.0 s. (a) Calculate the magnitude and direction of the child's average velocity. (b) Calculate the child's average speed.

3. An ROTC student marches 40.0 m due east, makes a right turn and marches another 30.0 m. These two segments of the march together take 60.0 s to complete. (a) What is the student's average velocity vector? What is the magnitude of this vector? (b) What is the student's average speed?

4. While practicing for a race, a yacht sails 1.20 km due east, then 1.60 km NW (45° W of N) and finally 0.80 km due south. If all this takes place in 15.0 min, calculate (a) the average velocity of the yacht and (b) the average speed of the yacht. Express both answers in meters per second.

5. The position of a particle is described by the vector equation $\mathbf{r} = (2.00 \text{ m/s})t\hat{\mathbf{i}} - (3.00 \text{ m/s}^2)t^2\hat{\mathbf{j}}$. (a) Determine the particle's velocity function. (b) Determine the particle's acceleration function. (c) What is the particle's speed at $t = 2.00$ s? (d) What is the magnitude of the particle's acceleration at $t = 2.00$ s?

6. From an origin located at the top of an incline, the position vector of a ball that rolls down the incline is given by $\mathbf{r} = (4.24 \text{ m/s}^2)t^2\hat{\mathbf{i}} - (2.45 \text{ m/s}^2)t^2\hat{\mathbf{j}}$. (a) What is the magnitude of the ball's displacement at $t = 2.50$ s? (b) What is the magnitude of the ball's velocity at $t = 2.50$ s? (c) Determine the ball's acceleration function, and the magnitude of its acceleration at $t = 2.50$ s.

7. An object that can be modeled as a point particle moves with a position function given by $\mathbf{r} = (3.00 \text{ m/s}^3)t^3\hat{\mathbf{i}} - [8.00 \text{ m} - (2.00 \text{ m/s}^2)t^2]\hat{\mathbf{j}}$. (a) Compute the velocity function for the object. (b) What is the magnitude of the object's velocity at $t = 2.50$ s? (c) What is the object's acceleration at $t = 2.50$ s?

8. From an origin located at the base of the ladder, the position of a child who slides from the top of a playground-slide that reaches the ground is given by the position vector $\mathbf{r} = (1.50 \text{ m/s})t\hat{\mathbf{i}} + [4.00 \text{ m} - (1.00 \text{ m/s}^2)t^2]\hat{\mathbf{j}}$. (a) How much time does it take the child to reach the ground? (b) What is the magnitude of the child's velocity at the instant she reaches the ground? (c) What horizontal distance does the child travel during her slide?

4.2 Projectile Motion

9. A child throws a snowball from a porch. When the snowball leaves her hand it is 3.00 m above the ground and has an initial velocity of 12.0 m/s at an angle of 25.0° above the horizontal. (a) Draw a picture of the snowball's path and include all the information given above. Also include a coordinate system with its origin at the point where the snowball leaves the child's hand, y-axis vertical. (b) Resolve the initial velocity of the ball into x- and y-components. (c) Summarize the initial conditions of the problem in x and y columns and include "$t = ?$" in the y-column. (d) Solve for the ball's time of flight from one of the constant-

acceleration equations in the y-column. (e) Compute the horizontal distance the ball travels.

10. A child throws a ball onto a roof that is sloped at 22.5°. The ball rolls down the roof and leaves it with a velocity of 12.0 m/s. The ball hits the ground 5.00 m from the bottom of the wall of the house. Find the height of the roof edge: (a) Draw a picture of the ball's motion. Include a coordinate system with the origin at the point where the ball leaves the roof, y-axis vertical. Be sure to show all the information you know. (b) Resolve the initial velocity of the ball into x- and y-components. (Be careful with the sign of v_{o_y}!) (c) Summarize the initial conditions of the problem in x- and y-component columns and include ''y = ?'' in the y-column. (d) In equation form, solve for t from a constant-acceleration equation in the x-column (Why?). (e) Substitute this result into a constant-acceleration equation from the y-column to find the maximum height y. (f) Compute the velocity (magnitude and direction) of the ball at the instant it hits the ground.

11. A movie stuntman is participating in an exciting rooftop chase. During the pursuit he is to run horizontally off the edge of one rooftop and land on a second roof that is 1.50 m away (horizontally) and 1.00 m lower. Find the minimum speed the stuntman should have by modeling him as a point particle.

12. In a children's party game pennies are flicked horizontally off the edge of a table toward a wastebasket on the floor. Suppose that the table top is 1.00 m high, and the top edge of the wastebasket is 0.45 m high and has a diameter at the top of 0.30 m. If the near edge of the wastebasket is 1.50 m (horizontally) from the edge of the table, what range of velocities can a penny have if it is to land in the basket?

13. A baseball is bunted down the third-base line leaving the bat from a height of 0.900 m with an initial velocity of 3.00 m/s at an angle of 20.0° above the horizontal. (a) How far does the ball travel before hitting the ground? (b) What is the speed of the baseball just before it hits the ground?

14. A soccer ball is kicked at an angle of 53.13° above the horizontal with a speed of 11.3 m/s toward the center of a soccer goal 11.0 m away. The top crossbar of the goal is at a height of 2.44 m. The goalie is drawn to the side of the goal, but can deflect any shot less than 1.00 m above ground by diving. Does the shot score?

15. A novice bowler ''lofts'' the ball down the alley. The ball leaves the bowler's hand at a height of 1.10 m above the alley directed at an angle of 15.0° above the horizontal with a speed of 3.20 m/s. (a) How far in front of the bowler's hand does the ball hit the alley? (b) What is the speed of the ball just before it hits the alley?

16. A volleyball is served underhand from a height of 1.20 m above the ground at an angle of 40.0° above the horizontal with an initial speed of 11.5 m/s. The net is 2.44 m high at a distance of 9.50 m away. The out-of-bounds line on the opposite court is 18.6 m away. Is the serve legal? That is, does it both clear the net and stay in bounds?

17. A popular carnival game is to throw quarters at a set of glasses on a table a few meters behind a counter. A quarter is thrown from a height of 1.10 m at an angle of 53.13° above the horizontal. (a) At what speed must the quarter be thrown to land in the center of a glass 3.20 m away, if the mouth of the glass is 0.50 m above the ground? (b) If the mouth of the glass is 5.00 cm wide, what range of velocities results in the center of the quarter falling within the diameter of the glass? Assume the launch angle is kept constant.

18. A tightly crumpled wad of paper is tossed toward a wastebasket sitting against a wall. The paper is thrown toward the basket from a height of 1.25 m at an angle of 10.0° below the horizontal with a speed of 2.00 m/s. The wastebasket has a height of 30.0 cm and a diameter of 25.0 cm, and its center is located a distance of 75.0 cm from the point of release of the paper. Where does the paper land?

19. The end of a water slide is 2.00 m above the surface of the pool. For the last several meters the slide is inclined downward at 20.0°. If a swimmer tucked in a ''cannonball'' position leaves the end of the slide at 3.50 m/s, (a) how far (horizontally) from the end of the slide does the swimmer hit the water? (b) How fast is the swimmer traveling just before hitting the water?

20. A student wishes to hit his roommate with a water balloon. His roommate, who is 2.00 m tall, is standing outside the dorm 4.00 m from the dorm wall. (a) If the student throws the balloon horizontally out of the window 15.0 m above the ground with a speed of 3.30 m/s, does he hit his roommate? (Don't forget: anything from head-to-toe counts as a hit.) (b) Calculate the speed of the water balloon when it is a height of 2.00 m from the ground. Express your answer in meters per second and miles per hour. (A student at one college lost sight in one eye after being hit by a water balloon in an incident such as that described above. *Never* drop anything on a person.)

21. To simulate weightlessness for astronaut training (Fig. 4.P1), a pilot flies a cargo plane in a parabolic trajectory with a constant horizontal speed and a downward acceleration of 9.80 m/s². (a) If the plane starts this maneuver at 30 000 ft and terminates the simulation at 5000 ft, how many seconds of weightlessness can be simulated? (Neglect the change in gravity over this altitude change.) (b) If the plane's horizontal velocity is 300 mph (134 m/s) how far forward does the plane travel during this time?

FIGURE 4.P1 Astronauts aboard the NASA training aircraft (known as the vomit comet) achieve ''weightlessness'' when the plane flies in a parabolic arc. (See Problem 21.)

22. Calculate the horizontal velocity of the golf ball during the first bounce in Figure 4.P2. [*Hints*: (1) The diameter of a golf ball is 4.26 cm. (2) The picture was taken on earth where $a_g = 9.80$ m/s². (3) You do not have to be able to count all the images near the top of the bounce to know how many are there. Think about the ball's horizontal motion!]

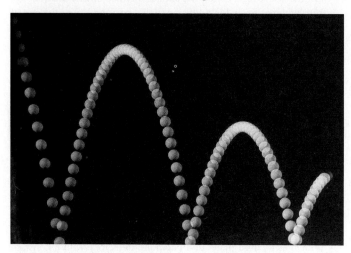

FIGURE 4.P2   Can you calculate the time interval between flashes from this photo? See Problem 22.

23. (a) Show that, when fired over level ground ($y_o = y_f$), the range R of a projectile launched with speed v_o at an angle θ above the horizontal is given by $R = (v_o^2/a_g) \sin(2\theta)$. (b) What is the maximum range and at what angle does it occur? (c) Derive an expression for the maximum height of the projectile.

24. Galileo in the *Two New Sciences* argues that for projectiles with launch angles that "... exceed or fall short of 45° by equal amounts, the ranges are equal." Start with the range formula derived in Problem 23, set $\theta = (\pi/4) \pm \Delta\theta$, and use trigonometric identities to prove Galileo's assertion for the constant-acceleration model.

25. A golf ball rolls horizontally into the top of an empty elevator shaft. The shaft is 2.00 m across and 25.0 high. As the ball falls it bounces back and forth between the two opposite walls. Each time the ball strikes the side, the horizontal component of its velocity reverses, but its vertical velocity is unchanged from its value just prior to the collision. If the original horizontal velocity of the ball was 10.0 m/s, (a) how many times does it hit a wall before reaching the bottom of the shaft and (b) where (relative to the side it rolled in from) does it strike the bottom of the shaft?

★26. A woman is standing 15.0 ft from a point directly below the basket when she shoots a free throw. As the ball leaves her hand it is 6.00 ft above the ground traveling at an angle of 54.0° above the horizontal. If the basket is 10.00 ft above the floor, find the veloc-

ity she must give the ball to "make the basket" without hitting the rim or backboard.

4.3 Circular Motion

27. (a) Prove that the vectors $\mathbf{r} = R_o[\cos(\omega t)\widehat{\mathbf{i}} + \sin(\omega t)\widehat{\mathbf{j}}]$ and $\mathbf{v} = \omega R_o[-\sin(\omega t)\widehat{\mathbf{i}} + \cos(\omega t)\widehat{\mathbf{j}}]$ are perpendicular by showing that their scalar product is zero. (b) Show that the magnitude of vector $\mathbf{v}$ is ωR_o.

28. An ultracentrifuge spins test tubes in circles with a rotation rate of 1.00×10^4 rpm. What is the centripetal acceleration of a point in the test tube 25.0 cm from the center of rotation? How many times the acceleration of gravity is this?

29. A bobsled traveling 85.0 km/h rounds a circular curve of radius 12.0 m. (a) Calculate the centripetal acceleration of the sled. How many times the acceleration of gravity is this? (b) What is the angular speed of the sled around the curve?

30. Take the moon's orbit around the earth as circular. (a) Using data from Appendix 8, calculate the centripetal acceleration of the moon in its orbit around the earth. (b) What is the angular speed of the moon around the earth?

31. (a) What is the centripetal acceleration of a point on the earth's equator due to the earth's rotation? (b) What is the centripetal acceleration of the earth in its orbit around the sun? (You may find data in Appendix 8 helpful.)

32. Our sun rotates about the center of the Milky Way galaxy with a speed of 220 km/s at a distance of 8.50 kpc (kiloparsecs). (a) Calculate the time it takes the sun to make one full orbit around the galactic center. (b) What is the centripetal acceleration caused by this motion? (1 pc $= 3.086 \times 10^{16}$ m.)

33. In the Bohr model of an atom an electron revolves around the nucleus in a circular orbit of radius 0.0529 nm with a velocity of 2.19×10^6 m/s. (a) How long does it take the electron to orbit the nucleus in Bohr's model? (b) What is the centripetal acceleration of the electron? (c) What is the angular frequency of the electron?

34. A driver rounds a circular corner of radius 12.0 m. If she wishes passengers to experience a maximum centripetal acceleration of 0.200 "*g*," what maximum speed can the car have as it rounds the corner? (An acceleration of 0.200 "*g*" means 0.200 times 9.80 m/s².)

35. A fighter pilot flies a jet into a horizontal 3 "*g*" turn, that is, the jet plane travels in a circular path such that the centripetal acceleration is 3.00 times 9.80 m/s². If the plane is flying at 4.00×10^2 m/s, what is the radius of the plane's circular path?

36. After defeating the faculty in a basketball game (by a particularly large margin), the intramural team members "rub it in." One player spins the basketball about a vertical axis while balancing it on her finger. A second estimates that the ball turns 4.00 rev/s about a vertical axis and has a radius of 12.0 cm. (a) The players inform the (still gasping for breath) faculty of the value for the maximum centripetal acceleration of a point on the ball. What is this value? (b) What is the centripetal acceleration of a point on the sphere located at an angle of 45.0° from the vertical axis? Make a sketch showing the direction of that acceleration.

37. Horizontal acceleration can be measured by observing the surface of a liquid. The tangent of the angle between the liquid surface and the horizontal is equal to a/a_g. A student makes a simple accelerometer using this principle by bending a transparent tube into a **U** shape and taping it to a stiff piece of cardboard with graph paper glued to it. The student fixes the accelerometer to the dashboard of the car so that the plane of the cardboard is perpendicular to the car's forward motion. The student drives at a constant speed of 25.0 mph around a circular curve while a passenger records the angle made by the fluid level between the two ends of

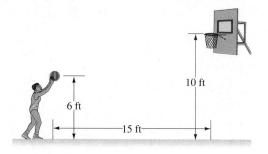

FIGURE 4.P3   Problem 26

the tube. If the angle of the liquid is 12.0°, what is the radius of the curve around which the students are driving?

38. A particle is initially traveling at 7.00 m/s around a circular path of radius 3.00 m. At $t = 0.00$ s it begins to decelerate at a constant rate of 2.00 m/s². (a) Calculate the particle's centripetal acceleration at $t = 1.00$ s and $t = 3.00$ s. (b) Calculate the particle's total acceleration at $t = 1.00$ s and at $t = 3.00$ s. (c) What is the particle's speed when the magnitudes of the centripetal acceleration and the tangential acceleration are equal?

39. A race driver accelerates a car from 0.00 m/s at a constant rate of 3.00 m/s² around a circular track of radius 50.0 m. (a) How much time elapses before the car's centripetal acceleration equals its tangential acceleration? (b) How far does the car travel during this time? (c) Through what angle around the track does the car travel? (d) What is the magnitude of the car's total acceleration at that instant?

40. Some children are trying to hit a soft-drink can with a stone tied to the end of a long string. One child swings the rock in a horizontal circle of radius 1.20 m at a height of 1.40 m above the ground as shown in Figure 4.P4. When the rock is circling at a rate of 1.50 rev/s, the child releases the string and the rock flies forward. How far in front of the child does the rock hit the ground?

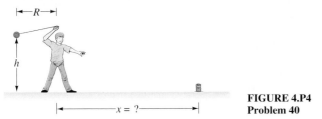

FIGURE 4.P4
Problem 40

41. Figure 4.P5 shows the position of a race car driving at constant speed on a circular track at three equally spaced instants of time. (a) Make a copy of the figure and draw arrows on your diagram to show the direction of the centripetal acceleration at each of the three points. Scale your drawing so your acceleration vectors are about half the length of the radius of the circle. (b) On a second copy of the figure show three positions of a car that is continually slowing down as it travels around the circle. At each position draw an arrow to represent the total acceleration of the particle. Make the tangential acceleration vector half the length of the initial centripetal acceleration vector. (c) Repeat part (b) except with the car increasing in speed.

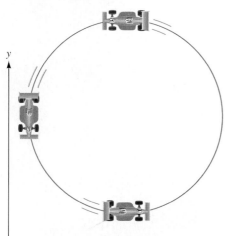

FIGURE 4.P5
Problem 41

42. Figure 4.P6 shows the path driven by an automobile. The positions of the car at four different equally spaced times are marked by A, B, C, and D. Copy this path onto a sheet of paper three

times. (a) On the first copy of the figure draw position vectors from the origin to the position of the car at each of the four times marked A, B, C, and D. (b) On your second figure draw vectors indicating the car's velocity at each of the four points. (c) On your third figure draw vectors indicating the acceleration of the car at each of the four points.

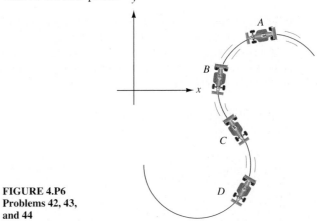

FIGURE 4.P6
Problems 42, 43, and 44

43. Reread Problem 42 for an explanation of Figure 4.P6. Make three careful copies of Figure 4.P6 except for the position of the car at points B, C, and D. Assume a car is traveling the path in the figure and has the same velocity at point A as the car in Problem 42, except this time the car is constantly increasing its speed. (a) On your first copy of the figure show, using position vectors, where this car is located for the same times as the car in Problem 42. (b) On your second drawing show velocity vectors for these same instants of time. (c) On your third drawing indicate acceleration vectors at the same four points in time.

44. Repeat the exercise described in Problem 43 except this time assume the car continually decreases its speed after it reaches point **A.**

45. Make a simple tracing of Figure 4.P7. (a) On your sketch draw arrows to indicate the estimated sizes and approximate directions

FIGURE 4.P7   Problem 45

of the velocity and acceleration vectors for the diver's feet. (b) Indicate the direction of the velocity and acceleration of the diver's head for the last five images.

46. Consider a particle moving in a circular path with radius 0.750 m. At a particular instant the particle has a speed of 3.00 m/s and is accelerating at 6.00 m/s². (a) Calculate the particle's centripetal acceleration. (b) Draw a circle and indicate the particle with a point. Draw both the tangential acceleration vector and the centripetal acceleration vector of the particle on your diagram. (c) Calculate the total acceleration of the particle. (d) Answer parts (a) through (c) for a particle with the same initial speed but decelerating at 6.00 m/s².

4.4 Numerical Techniques in Two Dimensions (Optional)

47. Construct a spreadsheet to plot ideal projectile motion. Make a column for t, x, and y. Let x_o, y_o, v_{o_x}, v_{o_y}, a_x, and a_y be "input" variables. Use your spreadsheet to verify the answer to Problem 11 found in the back of the book.

48. The position of a point on an epicycle (see Fig. 4.P13, Problem 65) can be written $\mathbf{r} = [R_1 \cos(\omega_1 t) + R_2 \cos(\omega_2 t)]\hat{\mathbf{i}} + [R_1 \sin(\omega_1 t) + R_2 \sin(\omega_2 t)]\hat{\mathbf{j}}$. (a) Determine the expression for the acceleration of this point. (b) Assume numerical values of $R_1 = 4.00$ m and $R_2 = 2.00$ m for the radii and $T_1 = 5.50$ s and $T_2 = 8.00$ s as the times for complete cycles of the two circular motions. Use a spreadsheet or computer program to plot the magnitude of the total acceleration versus time for this motion. Make certain you plot the motion for a sufficiently large range of the variable t to observe the full range of behavior.

49. Table 4.P1 gives position versus time for a golf ball swinging on the end of a string. Perform an analysis similar to that in Example 4.6. Calculate the maximum tangential acceleration and the maximum centripetal acceleration of the golf ball. Where do these two maxima occur? (These data are contained in the data file GOLFBALL.DAT on the disk accompanying this text. Numerical values are separated by commas and may be imported into a spreadsheet or read by a computer program.)

TABLE 4.P1   x- and y-Positions of a Golf Ball on the End of a String at 1/30-s Intervals[1]

x	y	x	y
2.25	7.70	27.15	0.10
3.20	6.80	30.10	0.75
4.65	6.00	32.75	1.20
6.10	5.00	35.60	2.00
8.10	4.00	37.90	2.90
10.20	2.90	40.00	3.90
12.75	2.10	41.90	4.85
15.50	1.20	43.60	5.70
18.25	0.85	44.75	6.70
21.15	0.25	45.50	7.25
24.25	0.01	46.25	7.80

[1] Multiply by 10/6 to convert readings to centimeters. See Problem 49. (Data courtesy Dana Gossert and Michael McInerney, Rose-Hulman Institute of Technology.)

50. Consider the volleyball service described in Problem 16. Create a spreadsheet template or computer program to calculate the ball's trajectory. Use your software to determine (a) the range of angles the server can use to make a legal serve if the initial velocity

remains constant at 11.5 m/s. (b) Suppose the angle stays fixed at 40°, what range of initial velocities results in a legal serve?

★51. (a) Extend Problem 50 to produce a map indicating the locus of all pairs of v_o and θ for which the server can make a legal serve from the starting point given in Problem 16. Present your answer in the form of a plot with v_o on the vertical axis and θ on the horizontal axis with the allowable region(s) shaded. [Consider only initial speeds less than 20 m/s.] (b) Now change the initial height of the serve to 2.50 m (an overhead serve) and once again map the region of velocity-angle space that results in a legal serve. (c) Which serve, underhand (the original case) or overhand (the second case), leaves less margin for error on the server's part?

52. Table 4.P2 gives the x-, y-, and z-coordinates of the heel of a ballet dancer's foot during a portion of a jump from the toe. (See target 4 of Fig. 4.P8.) (a) As a first approximation assume that the heel is moving only in the xz-plane, and use a spreadsheet or computer program to plot its trajectory. (b) Again ignoring the y-motion, calculate and plot the velocity as a function of time for this interval. [Note: you can expect some noise here. This is real data.] Draw a few velocity vectors on your plot of the trajectory of the heel. (c) Create an acceleration-versus-time graph. Estimate the magnitude of the largest acceleration of the heel. Express this value in "g's" [These data may also be found in the comma-separated file HEEL.DAT on the diskette accompanying this text.]

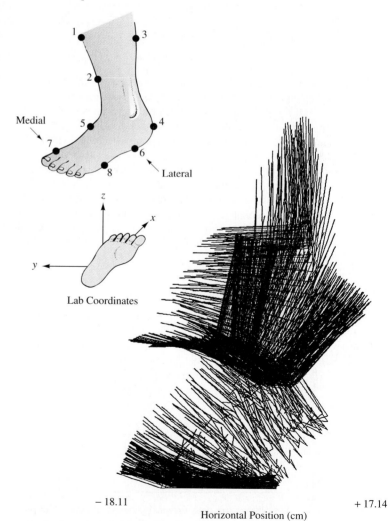

FIGURE 4.P8   Problem 52

TABLE **4.P2**   Position of left calcaneous for ballet dancer. Lateral view, jump from toe. File kb10s.ted, frame interval 1/60 s.[1]

FRAME NUMBER	TARGET 4			TARGET 3		
	x (cm)	y (cm)	z (cm)	x (cm)	y (cm)	z (cm)
47	−17.0682	1.4736	2.1652	−6.8642	2.6294	19.4448
48	−16.9964	1.3502	2.4774	−6.5020	2.5110	19.5821
49	−16.9167	1.2028	2.7905	−6.1747	2.4050	19.6891
50	−16.8244	1.0524	3.1058	−5.8651	2.3430	19.7851
51	−16.7177	0.9126	3.4081	−5.5649	2.3528	19.8881
52	−16.6011	0.8052	3.6815	−5.2991	2.4291	20.0072
53	−16.4882	0.7603	3.9292	−5.1011	2.5638	20.1435
54	−16.3927	0.7838	4.1572	−4.9787	2.7525	20.2877
55	−16.3157	0.8593	4.3636	−4.9316	2.9830	20.4250
56	−16.2525	0.9678	4.5356	−4.9553	3.2386	20.5484
57	−16.2065	1.0856	4.6496	−5.0691	3.4850	20.6555
58	−16.1807	1.1938	4.6935	−5.3156	3.6899	20.7468
59	−16.1722	1.3035	4.6875	−5.7143	3.8502	20.8434
60	−16.1565	1.4445	4.6845	−6.2492	3.9852	20.9805
61	−16.0897	1.6484	4.7606	−6.8853	4.1054	21.2038
62	−15.9188	1.9176	5.0110	−7.5743	4.2138	21.5710
63	−15.5758	2.1706	5.5415	−8.2665	4.2879	22.1402
64	−14.9596	2.2323	6.4464	−8.9127	4.2568	22.9412
65	−13.9465	1.9559	7.7882	−9.4173	4.0252	23.9649
66	−12.4902	1.3605	9.5488	−9.6297	3.5341	25.1894
67	−10.7221	0.6379	11.5991	−9.4513	2.8451	26.5732
68	−8.9290	0.0307	13.7303	−8.9195	2.1480	28.0525
69	−7.3858	−0.3169	15.7216	−8.1744	1.6407	29.5383
70	−6.1841	−0.4338	17.4218	−7.3775	1.3903	30.9173
71	−5 2186	−0.4741	18.7804	−6.6282	1.3018	32.0913
72	−4.3251	−0.5767	19.8090	−5.9334	1.2033	33.0161
73	−3.4512	−0.7881	20.5258	−5.2778	0.9939	33.7007
74	−2.6813	−1.0920	20.9445	−4.6854	0.7010	34.1560
75	−2.1222	−1.4470	21.0628	−4.1993	0.4042	34.3732
76	−1.8070	−1.8106	20.8601	−3.8516	0.1681	34.3390

[1] Data courtesy of Robert Soutas-Little, Biomechanics Evaluation Laboratory, St. Lawrence Health Science Pavilion, East Lansing, Michigan.

53. See Problem 52 for a description of the data in Table 4.P2. (a) Use the data to calculate the distance between the heel (target 4) and the posterior shank (target 3) of the dancer's leg as a function of time. (b) Calculate and plot the rate at which the distance between the targets is changing as a function of time. What is the maximum rate of contraction of the tendons and muscles between these two targets? Express your answer in meters per second and miles per hour.

General Problems

54. The position of an object, which may be modeled as a particle, is described by the vector equation $\mathbf{r} = r_o \exp(-t/\tau)\hat{\mathbf{i}} + Vt\hat{\mathbf{j}}$ where $r_o = 8.00$ m, $\tau = 10.0$ s, and $V = 4.00$ m/s. (a) Determine the object's velocity function. (b) Determine the particle's acceleration function. (c) What is the object's speed at $t = 5.00$ s? (d) What is the magnitude of the object's acceleration at $t = 5.00$ s?

55. A sports car traveling along an S-curve follows a path described by $\mathbf{r} = Vt\hat{\mathbf{i}} + r_o \cos(\omega t)\hat{\mathbf{j}}$ where $V = 10.0$ m/s, $r_o = 20.0$ m, and $\omega = 0.500$ rad/s. (a) Determine the car's velocity function. (b) Determine the car's acceleration function. (c) Find the car's speed at $t = 1.00$ s and at $t = 3.1416$ s. (d) What direction (with respect to the x-axis) is the car moving at $t = 1.00$ s? At $t = 3.1416$ s?

56. An electron in a cloud chamber follows a path described by $\mathbf{r} = (r_o - Vt)[\sin(\omega t)\hat{\mathbf{i}} + \cos(\omega t)\hat{\mathbf{j}}]$ where $r_o = 10.0$ cm and $V = 2.00 \times 10^2$ cm/s. (a) Determine the electron's velocity function. (b) If $\omega = 4.00 \times 10^3$ rad/s, find a unit vector parallel to the velocity vector at $t = 1.00$ ms.

57. A particle travels a path described by

$$\mathbf{r} = [v_o t + R_o \cos(\omega t)]\hat{\mathbf{i}} + [R_o + R_o \sin(\omega t)]\hat{\mathbf{j}}$$

where R_o, v_o, and ω are constants. (a) Sketch its path. (b) Derive the particle's velocity function. (c) What is the particle's maximum speed? On your sketch indicate where the particle attains this maximum speed. (d) What is the particle's minimum speed? On your sketch indicate where the particle attains this speed. (e) Describe how such motion might be made to occur. [*Hint:* See Fig. 4.P9.]

FIGURE 4.P9   Problem 57

58. A particle's trajectory is described by the equation

$$\mathbf{r} = v_o t \widehat{\mathbf{i}} + R_o \cos(\omega t) \widehat{\mathbf{j}}$$

(a) Determine the particle's velocity function. (b) Determine the particle's acceleration function. (c) Sketch the particle's trajectory. (d) Determine the particle's minimum speed. Draw an arrow on your sketch of the particle trajectory to show the direction of the velocity vector when the particle has this minimum speed. (e) What is the particle's minimum acceleration? Where is this minimum attained?

★59. One child is trying to hide from another who is throwing snowballs. The first child hides by crouching behind a 2.50-m high fence that is 4.00 m from the other child's snow fort. The second child tries to hit the hiding child by lobbing snowballs over the fence as shown in Figure 4.P10. If the maximum speed with which the second child can throw a snowball is 10.0 m/s, what is the closest distance to the backside of the fence that he can get a snowball to land? (Assume the snowball is released at a height equal to that of the base of the fence.)

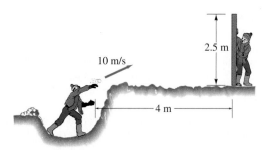

FIGURE 4.P10   Problem 59

★60. An automatic launcher accelerates a tennis ball along a tube 1.20 m long, inclined at 15.0°. While in the tube the ball accelerates from rest at a rate of 200. m/s². (a) Find the maximum height above the end of the tube that the ball reaches. (b) If the base of

the tube is 0.80 m above the ground, what horizontal distance does the ball travel before hitting the ground?

61. A tennis player at the net tries a drop shot. The tennis ball leaves her racquet at a height of 0.850 m above the ground with a speed of 4.55 m/s at an angle of 55.0° above the horizontal. The net is 0.910 m high at a point 1.75 m away directly in front of the ball as shown in Figure 4.P11. (a) By what amount does the ball clear the net? (b) How long does her opponent have from the time the ball leaves her racquet until it lands in the opposite court?

FIGURE 4.P11   Problem 61

62. Show that for $y_o = y_f$ there is a unique angle of projection such that the maximum height attained by a projectile is equal to its horizontal range regardless of initial velocity.

63. Galileo described constant acceleration by saying that the distances fallen in equal time increments stand to each other as the ratio of odd integers. Prove Galileo's assertion. [*Hint:* Make a series of numbers attained by adding consecutive odd integers.]

64. Just before the instant of release, a (very) novice hammer thrower rotates at a rate of 1.80 rev/s (Figure 4.P12). He wields the ham-

FIGURE 4.P12   The trained athlete in this photo throws the hammer much farther than the novice described in Problem 64.

mer in a horizontal plane 1.30 m above the ground. If the distance from his rotation axis to the sphere is 1.50 m, how far does the hammer travel from its point of release before it hits the ground?

★65. Certain amusement park rides consist of two circular motions superimposed as shown in Figure 4.P13. Assume the passenger travels in a circle of radius R_1 with constant angular speed ω_1 about a center point that itself is traveling around a circle of radius R_2 with constant angular speed ω_2. (a) Write an expression for the maximum speed attained by the passenger. Draw a sketch showing where this maximum occurs. (b) What is the minimum speed of the passenger? Where does it occur? (c) What is the maximum acceleration experienced by the passenger? (d) Show that the effective instantaneous radius of curvature of the arc in which the passenger travels at the moment of experiencing maximum acceleration is $R_{eff} = (\omega_1 R_1 + \omega_2 R_2)^2/(\omega_1^2 R_1 + \omega_2^2 R_2)$.

★66. Because water is a nearly ideal liquid, each point on the surface of a liquid has a slope equal to a/a_g, where a is the horizontal acceleration. Use this fact to prove that if a pan of water is placed in the center of a turntable and rotated, the surface of the water forms a parabola along any vertical plane through the center of rotation.

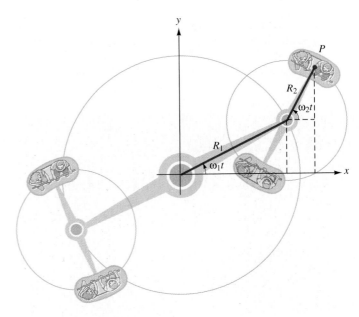

FIGURE 4.P13   Problems 48 and 65

Forces: Newton's Three Laws of Motion

In this chapter you should learn:

- Newton's three laws of motion.
- to recognize five common types of forces that affect the motion of objects.
- how to apply Newton's second law to compute the accelerations of objects subject to unbalanced forces.

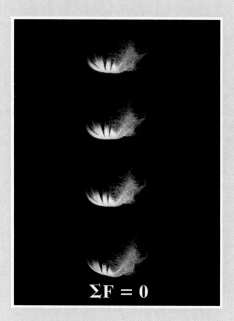

$$\Sigma F = 0$$

$$\Sigma F = m\mathbf{a}$$

$$\mathbf{F}_{12} \qquad -\mathbf{F}_{21}$$

$$\mathbf{F}_{12} = -\mathbf{F}_{21}$$

From a practical standpoint the next two chapters are the most important in this book. A thorough understanding of the classical treatment of forces underlies virtually every branch of engineering and applied science. The force concepts central to these chapters have nearly universal application. It is true that these concepts require modification when applied to detailed studies of the internal structure of atoms and nuclei or the enormous expanses of the universe. Nonetheless, for typical objects around us, our bridges, buildings, bowling balls, and bodies, the laws of classical mechanics formulated by Isaac Newton in the seventeenth century are the laws we live by.

In Chapters 3 and 4 we discussed the mathematical description of motion. This branch of mechanics is called kinematics. While doing kinematics we never asked (nor answered!) questions, such as ''*Why* does the object move that way?'' The business of asking about the causes of motion is called **dynamics** and is the subject of these next two chapters. In this chapter we lay the groundwork. You will be happy to learn there are only three new formulas in this chapter. As measured by mathematical complexity, this chapter is one of the simplest in this book. However, there are a number of new ideas that you must understand completely in order to be successful in the remainder of your physics course.

You no doubt already have some idea of what it takes to get things moving. Unless you're an extremely rare individual, you have exerted pushes or pulls on many objects ranging from desks to dumbbells in order to move them. These pushes and pulls are examples of what we mean by forces.

5.1 Newton's First Law: Inertia

People who study how students learn physics say that the ''law of inertia'' as formulated by Newton is one of the most difficult ideas of introductory physics. The reason for this difficulty is that you may already have an idea about inertia, and it may be wrong! To get this matter into the open, we state the law straight out so that we can explore its consequences.

> Newton's first law:
>
> When an object moves with a constant velocity (magnitude and direction), the total force on it is zero.

The reason this law may be difficult to remember is that it seems to contradict our everyday experience. We all know that if you want to move a desk from one side of the room to the other you have to push it. If you stop pushing, the desk stops moving. Thus, it seems that if you want something to move with a constant velocity, you should push with a constant force. But this notion is in apparent contradiction to Newton's first law!

To reconcile our everyday experiences with Newton's first law we need only read the law carefully. It says, that when an object moves with constant velocity, the *total* force on the object is zero. It turns out that as you push the desk across the floor there are many (at least four!) different forces on the desk. When the desk moves with constant velocity, the vector *sum* of all these forces is zero. Confusion occurs if we think that our force is the only one acting. This view is an easy trap to fall into since our push may be to us the most obvious force acting on the desk.

You must train yourself to remember a simple and powerful fact: When an object moves with constant velocity, the *total* force on the object is *zero*. When a car travels down a straight and level road at 55 mph the total of all forces from gravity, the road, and wind, is zero. When a parachutist falls with constant velocity downward, the total force from parachute, gravity, and wind is zero. A heavy desk that you push across the floor at constant velocity has a total force of zero acting on it.

We should mention in passing that a velocity that is always zero is certainly constant and, therefore, a stationary object has zero total force acting on it. However, the important point is not that the velocity is zero, but that it is *constant*.

Concept Question 1
If a mass does not accelerate, is it necessarily the case that there are no forces acting on the mass? Explain.

Concept Question 2
For which of the following is the total force acting on the object zero (ignore small effects due to the motion of the earth itself): (a) a physics book at rest on your desk (b) an automobile traveling with constant speed along a long, straight highway (c) a parachutist falling straight down at a constant velocity (d) a child sliding down a straight incline at constant speed (e) an apple falling from a tree to the ground (f) a model train engine traveling around a circular track at constant speed?

Newton's first law is often called the **law of inertia.** *Inertia* is just a term used to describe the fact that objects obey Newton's first law: all objects subject to zero net force move with a constant velocity. To say that a "body has inertia" means that the body has a natural tendency to move with a constant velocity; we must apply a force to change its velocity.

As a final point we need to emphasize how to add up the forces on an object: *forces always add as vectors.* This is another experimental fact that probably doesn't surprise you. This point is illustrated in Example 5.1. For now, don't be concerned about the units of force. We will discover them quite naturally in the next section.

EXAMPLE 5.1   How to Get Nothing from Something

An object traveling at a constant velocity is acted on by three forces. One force has a magnitude of 4.00 units directly east. A second force has magnitude 6.00 units directed 60.0° north of east. What is the magnitude and direction of the third force?

SOLUTION   Because the object is traveling at constant velocity, we know the total of all three forces is zero. Hence, the force vectors form a closed triangle when all are added together. (A closed triangle means the resultant vector has zero length.) Thus, the two known forces and the one unknown force form a triangle as shown in Figure 5.1.

Because $\mathbf{F}_1 + \mathbf{F}_2 + \mathbf{F}_3 = 0$, we know $\mathbf{F}_3 = -\mathbf{F}_1 - \mathbf{F}_2$. Resolving $\mathbf{F}_1$ and $\mathbf{F}_2$ into components

$$\mathbf{F}_1 = 4.00\hat{\mathbf{i}}$$

and

$$\mathbf{F}_2 = 6.00\cos(60.0°)\hat{\mathbf{i}} + 6.00\sin(60.0°)\hat{\mathbf{j}} = 3.00\hat{\mathbf{i}} + 5.20\hat{\mathbf{j}}$$

Combining these results gives us

$$\mathbf{F}_3 = -\mathbf{F}_1 - \mathbf{F}_2$$
$$= -(4.00\hat{\mathbf{i}}) - (3.00\hat{\mathbf{i}} + 5.20\hat{\mathbf{j}})$$
$$= -7.00\hat{\mathbf{i}} - 5.20\hat{\mathbf{j}}$$

The magnitude of $\mathbf{F}_3$ is $F_3 = \sqrt{(F_{3_x})^2 + (F_{3_y})^2} = \sqrt{(-7.00)^2 + (-5.20)^2} = 8.72$ units of force. The reference angle, α, for $\mathbf{F}_3$ can be found from $\tan(\alpha) = 5.20/7.00 = 0.743$, giving $\alpha = 36.6°$. Thus, $\mathbf{F}_3$ points 36.6° south of west. ◀

We close this section with a note about vocabulary. Several phrases are sometimes used to express the fact that the total force on an object is zero. The (vector) total force is often called the **net force** and written $\mathbf{F}_{net}$. Thus, Newton's first law is often written: $\mathbf{v}$ is constant when $\mathbf{F}_{net} = 0$. Because a total force of zero implies that all the forces are "balanced," $\mathbf{F}_{net}$ is also often called the **unbalanced force.** Writing $\mathbf{F}_{net} = 0$ is a way of saying that there is no unbalanced force acting on the object under consideration. Finally to emphasize that the net force is not itself a kind of force, but rather the sum of other forces, $\mathbf{F}_{net}$ is sometimes symbolized as $\Sigma\mathbf{F}$. We use all of these designations at various times throughout the remainder of this book.

5.2 Newton's Second Law

We now know what happens when there is no unbalanced force on an object: its velocity remains constant. It is natural (at least for physicists) to ask "What happens when there *is* an unbalanced force on an object?" For the answer, we must again turn to experiment.

Concept Question 3
When you are seated in a car that quickly accelerates forward, why do you feel pressed backward against the seat?

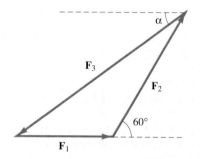

FIGURE 5.1

Vectors $\mathbf{F}_1 + \mathbf{F}_2 + \mathbf{F}_3$ form a closed triangle indicating that their vector sum is zero.

The result from numerous experiments turns out to be remarkably simple. When there is a constant unbalanced force on an object, the object moves with constant acceleration. Furthermore, if the force changes, the acceleration changes in direct proportion, with larger forces producing larger accelerations: Twice the force produces twice the acceleration; three times the force produces three times the acceleration, and so on.

The magnitude of the acceleration produced depends on the quantity of matter being pushed. Identical forces on a golf ball and a bowling ball result in considerably different accelerations. This quantity of matter is referred to as the **inertial mass** and is designated by m. Inertial mass measures the quantity of matter by its ''sluggishness.'' If you have two chunks of iron, one with twice as many atoms in it, the larger one has twice the inertial mass. It is twice as difficult to get the larger mass moving in the sense that, if the two chunks have equal forces acting on them, the larger one has only half the acceleration.

The relation between the total force and the inertial mass is neatly symbolized by the simple algebraic statement of Newton's second law.

Newton's second law:

$$\Sigma \mathbf{F} = m\mathbf{a} \qquad (5.1)$$

Note that this is a vector equation. That is, the direction of the acceleration is in the direction of the net force.

To make Newton's second law useful we must assign units to inertial mass and force. The mass unit is the **kilogram.** A force sufficient to impart a constant acceleration of one meter per second per second to a kilogram mass is called a **newton.** The unit newton is abbreviated N.

The SI unit of force is the newton:
$$1 \text{ N} = 1 \text{ kg} \cdot \text{m/s}^2$$

EXAMPLE 5.2   *Units in Newton's Second Law*

A mass of 3.35 kg is acted on by an unbalanced force of 5.00 N. What is the magnitude of acceleration of the mass?

SOLUTION   From Equation (5.1) we have

$$a = \frac{F_{\text{net}}}{m} = \frac{5.00 \text{ N}}{3.35 \text{ kg}} = 1.49 \text{ N/kg} = 1.49 \text{ m/s}^2 \qquad \blacktriangleleft$$

Newton's second law, Equation (5.1), is the focus of the remainder of this chapter and all of the next. Our problem-solving strategy is the same throughout. We search for forces acting on objects and put their total on the left-hand side of Equation (5.1). Then the object's mass is multiplied by the acceleration that describes the object's motion and this product is placed on the right-hand side. The resulting equations are then solved for any unknowns. We think you will be amazed that so much information can come from such a simple process.

In order to proceed, we need some forces to put into Newton's second law. In the next section we describe several forces commonly encountered in our day-to-day experience.

Concept Question 4
If a single force acts on a body that is already in motion, can its velocity ever be zero? Explain.

5.3 Some Common Forces

In this section we describe five forces we commonly encounter in our everyday activities: weight, spring forces, ''normal'' forces, friction, and tension. We discuss friction only briefly, leaving a detailed treatment until the next chapter. Weight is an example of what we shall call a body force, whereas the others are contact forces. The distinction is explained in the following subsections.

The barbell is attracted to the earth by the force of gravity. We call this force the barbell's weight.

The Weight Force

One of the most common forces we encounter is weight *W*. Briefly, an object's **weight** is the force of the earth's gravitational attraction, which pulls the object toward the earth's center. Weight is a body force. That is to say, it is a force that acts on every particle throughout an object's body. Consider a bag of sand. Every grain of sand has a weight. You can think of the bag of sand as having thousands of little weight forces pointing downward, one from each grain of sand. If we are interested in only the total weight of the sand, we can ignore the fact that the weight is distributed over all the grains and simply assign a single total weight vector to the whole bag. This weight vector must be placed at a position called the *center of gravity* of the bag of sand. In Chapter 10 we will discuss how to find this location. For the time being, it won't matter because we continue to use our point–object model. That is, until we get to Chapter 10 we need not be concerned about the spatial extent of objects. Everything from trunks to trains are treated as if they were simply a single point. When we take this extreme point of view it doesn't matter where on the object the force acts because the object is just a single point.

The magnitude of the weight force depends on the strength of the earth's gravitational attraction at the point where the mass is placed. In this chapter we are interested only in the gravitational force near the earth's surface. We will delay until Chapter 6 a discussion of how weight changes with distance from the center of the earth. The weight of an object also depends on the amount of matter in that object. Thinking back to the bag of sand, it is perhaps obvious that, if the grains of sand are identical, then a bag with twice as many grains has twice the weight. If, instead of grains, we measure the quantity of sand in terms of its mass in kilograms, then the proportion between the weight force *W* and mass can be written

$$W = mg$$

where *g* is known as the local **gravitational field strength.**

Over the surface of the earth, the gravitational field is fairly constant. At the National Institute of Standards and Technology in Gaithersburg, Maryland, the value is $g = 9.80$ N/kg, and we shall use this value throughout this text.[1]

From the relation $W = mg$ it is clear that mass *m* and weight *W* are two different quantities; weight is a pull (a force), whereas mass is a measure of a quantity of matter ("muchness," if you will). Because we are accustomed to living in a nearly constant gravitational field and because weight is proportional to mass, it is perhaps not surprising that we often confuse the two concepts, weight and mass. But they are measures of two *very* different physical properties. The difference between the two is also readily apparent when we look at units: mass is measured in kilograms and weight measured in newtons.

Despite the "obvious" difference between weight and mass, our culture is littered with confusing misuses of these words. Canned foods have such labels as "Net weight 326 grams." We often speak of "weighing" an object, yet report the answer in kilograms. It is vital that you keep mass and weight separate and not confuse the two concepts, despite what you may hear in the supermarket.

The confusion between mass and weight may go back to earlier common unit systems, which are listed in Appendix 5. Two other systems you may yet encounter (which we fervently hope will soon die out as we "go metric") are listed in Table 5.1 along with their standard SI counterparts. We use the British engineering system only enough to give you confidence in your ability to handle these units should you encounter them in the future.

---

[1] The fact that the magnitude of *g* is numerically equal to the acceleration of a freely falling body is not a simple coincidence. This equality is related to the, perhaps surprising, fact that the earth pulls on one kilogram of sand with the same gravitational force with which it pulls on one kilogram of lead, or water, or, indeed, one kilogram of anything. The importance of this result is discussed in Section 5.6 for those who are interested.

A Message to Earthlings

Throughout this text we have used the notation a_g to represent the kinematic acceleration of a freely falling object and g to represent the local gravitational field strength near the surface of the earth. The equality of these two quantities tells us something important about the nature of matter. (See Section 5.6.) For those readers who live on the surface of the earth, we should point out a subtlety in convention for quoting the value of a_g (and therefore g). Because the earth is rotating about its axis, the value of g actually measured by an apparatus at rest on the surface of the earth is smaller than the value which would be obtained were the earth not rotating. Experimental values of g such as those in Appendix 7 are smaller than the g values which would be measured on a nonrotating earth.

As a practical matter, when we wish to know the *effective* weight of a person on the surface of the earth or solve a projectile-motion problem such as the trajectory of a softball, it is the experimentally measured g value including the effects of the earth's rotation that should be used. If we were more pure of heart, we might adapt the notation $W = mg_{eff}$ to reinforce the fact that it is the effective g that should be used for experiments on the surface of the rotating earth. However, the numerical difference is quite small, and the notation is awkward. Instead, we ask you to remember that in this text for problems describing objects fixed on the surface of the earth, the symbol g means the earth's gravitational field strength minus a small correction for the earth's rotation. When objects are not fixed to the surface of the earth, we denote the force due to gravity alone as F_g (rather than W) and write $F_g = mg$. In this latter context g means the gravitational field strength *not* including any rotational effect.

TABLE 5.1   Units of Force, Mass, and Acceleration in Several Systems[1]

SYSTEM	FORCE	MASS	ACCELERATION
SI	newton (N)	kilogram (kg)	m/s^2
cgs	dyne	gram (g)	cm/s^2
British Engineering	pound (lbf)	slug	ft/s^2

[1] In each case the force unit equals the mass unit times the corresponding acceleration unit.

EXAMPLE 5.3   Weight and Mass

(a) According to its label, the "net weight" of a jar of peanut butter is 1134 g. What is the actual weight of the jar? (b) If a 2.00-lbf force is the only force that acts on a 5.00-lbf object, what is the object's acceleration?

SOLUTION   (a) The units indicate that we have been given the mass of the peanut butter. Since we want the peanut butter's weight, we convert using $W = mg$. The result is

$$W = mg = (1.134 \text{ kg})(9.80 \text{ N/kg}) = 11.1 \text{ N}$$

The earth pulls on the peanut butter with a force of 11.1 N.

(b) We plan to use $F_{net} = ma$ to solve the problem, but notice that we do not know m. We are given only the object's weight ($W = 5.00$ lbf). Thus, we first find the object's mass using $W = mg$. The gravitational field strength at the surface of the earth in the British Engineering system is 32.2 lbf/slug. Solving for m we obtain

$$m = \frac{W}{g} = \frac{5.00 \text{ lbf}}{32.2 \text{ lbf/slug}} = 0.155 \text{ slug}$$

The spring exerts a force opposing its deformation.

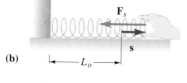

(a) $\longmapsto L_o \longmapsto$

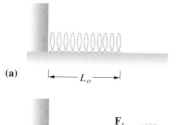

(b) $\longmapsto L_o \longmapsto$ F_s s

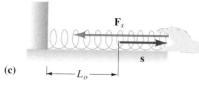

(c) $\longmapsto L_o \longmapsto$ F_s s

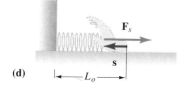

(d) $\longmapsto L_o \longmapsto$ F_s s

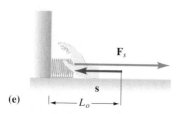

(e) $\longmapsto L_o \longmapsto$ F_s s

FIGURE 5.2

(a) A spring with natural length L_o. (b), (c) When the spring is stretched, it exerts a restoring force in direct proportion to the amount of stretch. (d), (e) When the spring is compressed, the restoring force increases in direct proportion to the amount of compression.

Now that we have the net force and the object's mass we can use $F_{net} = ma$ to find the acceleration

$$a = \frac{F_{net}}{m} = \frac{2.00\ \text{lbf}}{0.155\ \text{slug}} = 12.9\ \frac{\text{ft/s}}{\text{s}}$$

The 2.00 lbf force accelerates the 0.155-slug mass at a rate of 12.9 ft/s². ◀

We leave this section with two final words of advice:

1. It is vital that you understand the difference between mass and weight.
2. Never mix unit systems.

Spring Forces

A second type of force you may encounter is that due to springs. As you may have discovered for yourself, springs resist attempts to change their length. In fact, the more you alter a spring's length, the harder it resists. One simple model of this behavior is to write the force exerted by a spring as

$$\mathbf{F}_s = -k\mathbf{s} \qquad (5.2)$$

where s is the length change caused by the stretch (or squish) of the spring from its natural unstretched length, and k is called the **stiffness constant** of the spring or, simply, the **spring constant.** Values for the spring constant are usually quoted in units of N/m. Equation (5.2) is known as **Hooke's law.** This model for the behavior of a spring is accurate only if s doesn't become too large. (As anyone who has ever owned a Slinky can testify, the force law for springs that have been stretched too far is strongly dependent on the history of those overstretches!)

The minus sign in Hooke's law reminds us that the direction of the force exerted by the spring is opposite the displacement (stretch or squish) that produces it. Suppose the left end of a spring is fixed as shown in Figure 5.2. If you pull the free end of the spring to the right, the spring pulls back to the left. On the other hand, if you compress this spring by moving its free end toward the left, it pushes back to the right.

We will have much more to say about springs in the coming chapters. We introduce them here primarily to remind you that there are objects that "push back" in proportion to the amount we deform them.

EXAMPLE 5.4   *Working Out with Springs*

A student exercises by pulling on a handle grip attached to the free end of a long spring the opposite end of which is fixed to a wall. To stretch the spring by 0.800 m the student must exert a force of 85.0 N. What is the spring's stiffness constant? (Assume the spring obeys Hooke's law.)

SOLUTION   We use Hooke's law $\mathbf{F}_s = -k\mathbf{s}$, but recognize that the force exerted by the student is equal in magnitude (and opposite in direction) to the force exerted by the spring. Thus, $|\mathbf{F}_{student}| = +ks$, and we solve for k as follows:

$$k = \frac{F_{student}}{s} = \frac{85.0\ \text{N}}{0.800\ \text{m}} = 106\ \text{N/m}$$

That is, the spring increases its pull by 106 N for every meter it is stretched. ◀

Normal Forces

Whenever two surfaces are in contact they exert forces on each other. Such forces are, appropriately enough, known as **contact forces.** We usually find it convenient to resolve these contact forces into components, one parallel to the contact surface, the other perpendicular to that surface. Figure 5.3 illustrates the contact force on your finger by a tabletop as you slide your finger toward the left along the rough surface. The **normal force** is the component of the contact force that is perpendicular to the surface.

The normal force is a measure of how strongly the surfaces in contact are pressed together. Push your hand straight down on a table or desk. The force you feel resisting your push is the normal force of the table pushing up on your hand. This push by the table on your hand is, in fact, a spring force of sorts. You push on the table and it "gives" a little under your force. Like a spring, when the table deforms, it "pushes back."

Structures, such as concrete walls, have very large "spring constants." That is, they exert very large forces when deformed by imperceptibly small amounts. This effect sometimes makes it difficult to visualize the source of the contact force. (Or worse yet, we have a tendency to even forget that the force is there.) If desktops had the springiness of mattresses we would be less likely to forget the normal force. If you set your book down on such a desk, the book would sink in until the springs' push balanced the weight of the book. The same thing really does happen when you lay your book on an actual desk; it's just that the springs are much stiffer, and the deformation is quite small. (*Note:* We are not advocating sleeping on your desk!)

As a final point, if you still have doubt that inanimate objects, such as walls and floors, exert forces, go to the nearest brick wall. Standing next to it, repeatedly bang your head against it with moderate vigor. If you continue to do this long enough, we guarantee that eventually you will concede that the wall does exert a force on your head.

Frictional Forces

In Figure 5.3 the contact force between two surfaces was resolved into two components. The component perpendicular to the contact surface we called the normal force. The component of the contact force parallel to the contact surface is generally called the **frictional force.** The direction of the frictional force is opposite from the relative motion (or attempted motion) of the two surfaces in contact. Frictional forces are in general quite complicated. In the next chapter we will present a simple model that attempts to describe the major features of the behavior of frictional forces. For the present it is enough for you to know that the frictional force is (1) a contact force, (2) parallel to the contact surface, and (3) directed so as to oppose relative motion (or attempted motion) of the surfaces in contact.

Tension

The final kind of force we wish to introduce in this chapter is the force exerted by strings, ropes, and wires when they are stretched tight. We use the word *string* in the following discussion, but you can substitute *rope* or *wire* anyplace we use *string*. The force exerted by the end of a taut string is called **tension.** The direction of the force is such that the string pulls the object to which it is attached. (It's difficult to push with a string!) If the string is not accelerating, then the tension is the same on both ends of the string.

If the string is accelerating, the tension is not the same on both ends of the string. The difference in tension is the unbalanced force that accelerates the string itself. However, if the string's mass is much, much smaller than the masses to which its ends are attached, then the difference in tensions is quite small, and it is a good approximation to say that the tension is the same on both ends of the string.

In many of our exercises and examples we state that the string is "light." In this case you should take the string tension as constant even when the string accelerates. We show examples of how to calculate tension for both "light" and heavy strings in the following section.

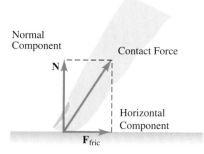

FIGURE 5.3

The contact force between two surfaces can be regarded as the vector sum of two components. The component normal to the surfaces in contact is called the *normal* force N. The component parallel to the contact surface is the *frictional force* F_{fric}.

The tension force pulls on objects attached to the end of the rope.

5.4 Simple Applications of Newton's Second Law

In Section 5.3 we introduced five common forces that we encounter on a daily basis: weight, spring forces, normal forces, friction, and tension. These forces[2] are the ingredients that, when added together, become the "net force" on the left-hand side of Newton's second law, $\Sigma \mathbf{F} = m\mathbf{a}$. We are going to follow the same procedure in all of our force analyses from the simplest to the most complex.

Before beginning our examples, we need to call your attention to a very important point concerning the way we use + and − signs in this text. Forces are always labeled with letters like W, F_s, and N. *These italic letters which are not in boldface type are magnitudes.* Our convention is to treat these as positive quantities, and indicate their direction with explicit + or − signs. Thus, $W (= +mg)$ is a positive quantity (and g itself is positive). If we are using a coordinate system with its y-axis positive upward, then we indicate the weight of an object as $-W$. Similarly, we write $F_s = +ks$ for the *magnitude* of the spring force.

Because we do not always know which way a system accelerates, we treat the acceleration a somewhat differently. We always *assume* that the acceleration is positive. That is, the symbol a is treated as if the acceleration were in the direction of the positive coordinate axis. If the acceleration is actually in the opposite direction, a comes out equal to a negative number.

Since we are going to use one standard procedure, we list it here at the start so that you can follow it in the examples ahead.

Problem Solving Strategy: Using Newton's Second Law

1. Make it clear in your mind what object you are considering. This sounds incredibly simple, but a lot of mistakes are made by overlooking this obvious point. When analyzing a chair, don't include forces on the floor or the table! **Make a simple sketch showing the object under consideration.**

2. Ask yourself "What are the forces?" acting on the object you are analyzing. Until you've had a lot of practice, try using the five forces we've described previously as a check list. **Draw arrows on your sketch to show the direction of each force acting on the object.**

3. Next, it is very helpful to **redraw the forces making a *free-body diagram*.** In this diagram you reduce the object of interest to a single point. On this point place each of the forces you identified in step 2, with its "tail" on the point.

4. **Choose a coordinate system for the problem and sketch it on your free-body diagram.** Here's a hint about how to keep the algebra simple in the latter part of the problem: always choose one axis of your coordinate system parallel to the object's *acceleration*. For most of the problems you can probably guess the direction of the acceleration even before you start. Pick this direction for one of your coordinate axes.

5. **If any force is not parallel to the x- and y-axes of the coordinate system you chose, resolve that force into components that are parallel to those axes.**

6. **Write Newton's second law for each of the coordinate directions.** That is, write $\Sigma \mathbf{F}_x = m\mathbf{a}_x$ and $\Sigma \mathbf{F}_y = m\mathbf{a}_y$ (and $\Sigma \mathbf{F}_z = m\mathbf{a}_z$ if necessary). Put only forces on the left-hand side, and place the proper component of the acceleration (multiplied by the mass) on the right.

7. **Solve the resulting set of equations for any unknowns.** Often you need additional equations that give the magnitudes of special forces (for example, $W = mg$). In addition, you should be alert for any geometric relations between the variables in the problem, including the acceleration.

8. **After solving for the unknown in letters, substitute numbers as required.** Substitute units as a final check of your algebra.

---

[2] These forces are representative. Other types will be added to our list later.

Now let's look at Examples 5.5 through 5.8 to see how all this fits together.

EXAMPLE 5.5 *Apparent Weight*

(a) A 60.0-kg student stands on a vertical spring that has a stiffness constant of 2.00×10^4 N/m. The spring is in an elevator moving upward with a constant velocity. How far does the spring compress? (b) Close to the top of the elevator's motion, its velocity decreases by 1.50 m/s^2 as it comes to rest. What is the compression of the spring as the elevator slows down?

SOLUTION "What am I analyzing?" We are going to analyze the student shown in Figure 5.4(a), so we look at forces on her only.

"What are the forces?" We look for weight, spring, normal, friction, and tension forces. Weight is present because we assume the elevator is near the surface of the earth. The weight force points down. The spring force is clearly also a part of the problem. Since the spring is compressed downward, it exerts an upward force on the student. The student is not touching the floor of the elevator or the walls, hence normal and frictional contact forces are absent. Finally, there are no strings attached, so no tension is present. The only two forces acting are the spring force and the weight force. Figure 5.4(b) shows the free-body diagram.

Since both forces are vertical, it seems only sensible to pick a coordinate system with the y-axis vertical. Both forces are already parallel to our coordinate axis, so there is no need to resolve forces; we may skip step 5.

We are now ready to write Newton's second law for the y-direction. On the left-hand side we sum the forces, using $+$ and $-$ signs to indicate their directions.

$$\Sigma \mathbf{F}_y = m\mathbf{a}_y$$
$$+F_s\hat{\mathbf{j}} - W\hat{\mathbf{j}} = ma_y\hat{\mathbf{j}}$$
$$+F_s - W = ma_y \qquad (5.3)$$

Because the elevator is moving with constant velocity, $a_y = 0$. We complete step 6 by substituting $a_y = 0$ into Equation (5.3) and obtain

$$F_s - W = 0$$

As indicated in step 7 we substitute $F_s = ks$ and $W = mg$, and solve for s

$$ks - mg = 0$$

$$s = +\frac{mg}{k} = \frac{(60.0 \text{ kg})(9.80 \text{ N/kg})}{2.00 \times 10^4 \text{ N/m}} = 0.0294 \text{ m}$$

Thus, we find the spring compresses by 2.94 cm.

(b) The same type forces, weight and spring, act on the block as the elevator slows down. In fact the only difference between part (a) and part (b) of this example is that in (b) the acceleration of the student is no longer zero. Consequently, Equation (5.3) becomes

$$ks - mg = ma_y$$

which we once again solve for s to obtain

$$s = \frac{m(a_y + g)}{k} = \frac{(60.0 \text{ kg})(-1.50 \text{ m/s}^2 + 9.80 \text{ N/kg})}{2.00 \times 10^4 \text{ N/m}}$$

$$= \frac{(60.0 \text{ kg})(8.30 \text{ N/kg})}{2.00 \times 10^4 \text{ N/m}} = 0.0249 \text{ m}$$

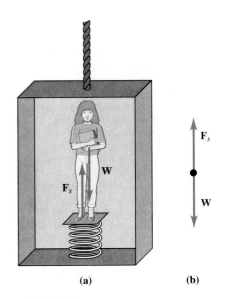

(a)   **(b)**

FIGURE 5.4

(a) The two forces acting on the student are the weight force W and the spring force F_s. (b) A free-body diagram using the point–mass model for the student.

Concept Question 6
(a) In Example 5.5, what happens to the student's apparent weight when the elevator accelerates from rest and starts moving upward again? (b) What do the magnitude and direction of the elevator's acceleration have to be to make the student's apparent weight equal to zero?

Concept Question 7
If, while riding on an airplane, you hang a large mass on the end of a spring of small stiffness constant and it does not stretch, what do you conclude? (Parachutes are located behind the seats!)

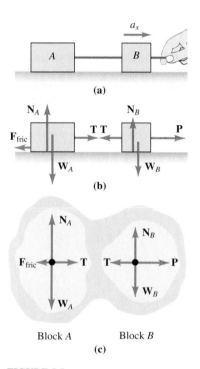

(a)

(b)

(c)

Block *A*   Block *B*

FIGURE 5.5

(a) Blocks *A* and *B* are pulled along a horizontal surface by the applied force *P*. (b) The forces acting on blocks *A* and *B*. (c) A free-body diagram for blocks *A* and *B* using the point–mass model.

Did you notice the negative value used for a_y ($\mathbf{a}_y = (-1.5 \text{ m/s}^2)\hat{\mathbf{j}}$)? The motion of the elevator is up (positive) and, because it is slowing down, the acceleration must have the opposite sign. Reread the section entitled "The sign of acceleration" in Section 3.3 of Chapter 3 if it is not clear to you why a_y is negative. Also note that g is positive in accordance with our remarks at the beginning of this section.

Finally you should notice the answer. During the few moments while the elevator was stopping, the spring was compressed 0.45 cm less than while the elevator traveled with constant velocity. As you know, springs are often used to indicate an object's weight. Had the spring in our problem been the spring inside of, say, a bathroom scale, we would have seen the scale indicate a decrease in weight for a moment as the elevator came to rest. Of course, the student's weight did not actually change; it stayed fixed at mg. Only her "apparent weight," as measured by the compression of a spring, changed.   ◀

EXAMPLE 5.6   Connected Masses

A block of mass $M_A = 5.00$ kg is connected to another block $M_B = 3.00$ kg by a light string. A second string is attached to the 3.00-kg mass and used to pull the two masses along a level table by applying a horizontal pull of 17.0 N as shown in Figure 5.5(a). The bottom of the 5.00-kg block has been roughened so that a constant frictional force of 1.00 N acts on it. The smoother 3.00-kg block slides without friction. Calculate the acceleration of the two blocks and the tension in the string that connects them.

SOLUTION   What system are we analyzing? We are analyzing two blocks in this case. In Figure 5.5(b) we draw the forces acting on each block separately. Each has a weight due to the pull of the earth. There are no spring forces on either block. Each block is in contact with the table top, hence there are normal forces on each block indicating that the table top is pushing up on each. (Notice we *don't* draw the downward force of the blocks on the table. We are analyzing the blocks, not the table!) There is also a frictional force acting on the larger block. The frictional force is shown parallel to the contact surface and in a direction opposing the motion. Finally, tension forces are shown. The string connecting the blocks pulls on both blocks. The string pulls the 5.00-kg block forward, but tends to hold the 3.00-kg block back. The other tension force of 17.0 N applied to pull the 3.00-kg mass to the right is labeled P.

In Figure 5.5(c) we show separate free-body diagrams of the two blocks. Only forces acting on each block are drawn; we are treating each block as a separate system. We choose the x-axis to be parallel to the expected acceleration, that is, horizontal. Because all forces are parallel to one or the other coordinate axis, there is no need to resolve forces.

We are now ready to apply step 6 of our problem-solving strategy by writing $\Sigma \mathbf{F} = m\mathbf{a}$ four times! We should write $\Sigma \mathbf{F}_y = m\mathbf{a}_y$ for each block and $\Sigma \mathbf{F}_x = m\mathbf{a}_x$ for each block. Because the y equations are essentially the same for both blocks and because we find them of little use anyway, we write Newton's second law for only one of them, say block A:

$$N_A \hat{\mathbf{j}} - W_A \hat{\mathbf{j}} = m_A a_y \hat{\mathbf{j}}$$

leading to

$$N_A - W_A = m_A a_y$$

Note that the acceleration in the y-direction is zero since the block has a constant y-velocity (namely, zero). Thus, we can find the normal force easily as follows:

$$N_A - W_A = 0$$

$$N_A = W_A = m_A g = (5.00 \text{ kg})(9.80 \text{ N/kg}) = 49.0 \text{ N}$$

The normal force on block B can, of course, be found in an analogous fashion.

Let's proceed, however, directly to the more interesting task of analyzing the horizontal motion of the blocks. Because the method is the same for each, we'll write equations for blocks A and B side by side.

BLOCK A	BLOCK B	
$\Sigma \mathbf{F}_{A_x} = m_A \mathbf{a}_{A_x}$	$\Sigma \mathbf{F}_{B_x} = m_B \mathbf{a}_{B_x}$	
$T\hat{\mathbf{i}} - F_{\text{fric}}\hat{\mathbf{i}} = m_A a_{A_x}\hat{\mathbf{i}}$	$P\hat{\mathbf{i}} - T\hat{\mathbf{i}} = m_B a_{B_x}\hat{\mathbf{i}}$	
$T - F_{\text{fric}} = m_A a_x$	$P - T = m_B a_x$	(5.4)

Notice that the tension force T has the sign that indicates its direction for the block on which it is acting. Also (if the string doesn't break), the acceleration is the same for both masses, so we have set $a_{A_x} = a_{B_x} = a_x$.

Equations 5.4 are two equations with two unknowns, T and a_x. The equations can be solved "simultaneously" for example, by solving the block-A equation for T, and substituting the resulting expression into the block-B equation:

$$T = F_{\text{fric}} + m_A a_x \qquad (5.5)$$

$$P - (F_{\text{fric}} + m_A a_x) = m_B a_x$$

With the unknown T eliminated, we can solve for the acceleration

$$P - F_{\text{fric}} = (m_A + m_B)a_x$$

$$a_x = \frac{P - F_{\text{fric}}}{m_A + m_B} = \frac{17.0\ \text{N} - 1.00\ \text{N}}{8.00\ \text{kg}} = +2.00\ \text{m/s}^2$$

Having found a_x, we can now substitute back into Equation (5.5) to find the tension:

$$T = 1.00\ \text{N} + (5.00\ \text{kg})(2.00\ \text{m/s}^2) = 11.0\ \text{N}$$

As a final check we substitute the known values for T and a_x into the block-B side of Equation (5.4) to make sure it is satisfied.

$$P - T \overset{?}{=} m_B a_x$$

$$17.0\ \text{N} - 11.0\ \text{N} = (3.00\ \text{kg})(2.00\ \text{m/s}^2)$$

$$6.00\ \text{N} = 6.00\ \text{N}$$

Our problem checks. ◀

Incidentally, don't try to memorize any of the equations that result from our examples. They are unlikely to be valid in any other problem. Instead, you should learn how to apply the *procedures* listed in the problem-solving strategy. The only fundamental law we have encountered thus far is Newton's second law. This law, along with a few helper expressions like $W = mg$ and $F_s = ks$, are all you really need to remember.

Concept Question 8
A 1.0-kg mass is attached to the ceiling by a light string but also has an identical light string attached to its bottom side. When a steadily increasing force is applied downward to the bottom string, the upper string between the mass and the ceiling eventually breaks. However, when the lower string is jerked swiftly downward, it is the lower string that breaks. Explain these experimental results. [*Hint:* It is essential that the upper string be able to stretch slightly. Try imagining the upper string to be a rubber band.]

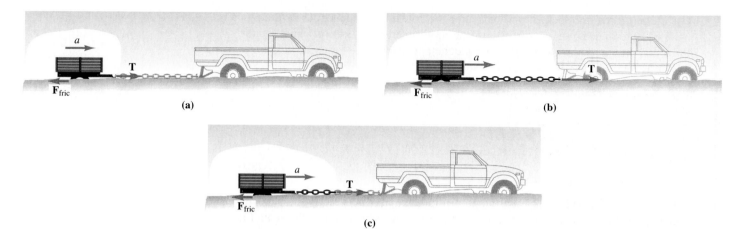

(a)

(b)

(c)

FIGURE 5.6

A utility trailer is pulled from the mud by a heavy chain. (a) To find the tension at the point where the chain is attached to the trailer, we treat the trailer as the system and look at the tension as an applied force. (b) To find the tension in the chain at the point where it is attached to the truck, we treat the trailer and the chain as the system. (c) To find the tension at the midpoint of the chain, we treat the trailer and one-half of the chain as the system.

EXAMPLE 5.7   *When Tension Is Not Constant*

A heavy towing-chain attached to a four-wheel-drive truck is used to pull a light utility-trailer from the mud as shown in Figure 5.6(a). The trailer has a mass $m_T = 85.0$ kg, and the chain has a mass $m_C = 15.0$ kg. Assume the mud exerts a constant frictional force of 400. N on the trailer. Calculate the tension in the chain at both ends and the center during the moments when the trailer is accelerating from rest at 0.500 m/s^2.

SOLUTION   We first analyze the trailer alone. The two horizontal forces are tension and friction. Writing Newton's second law for horizontal forces we find

$$\Sigma \mathbf{F}_x = m_T \mathbf{a}$$

$$T - F_{\text{fric}} = m_T a$$

$$T = F_{\text{fric}} + m_T a$$

$$= (400.\ \text{N}) + (85.0\ \text{kg})(0.500\ \text{m/s}^2) = 443\ \text{N}$$

Thus, at the point where the chain is attached to the trailer the tension is 443 N.

We now analyze a system made up of the trailer plus the chain as shown in Figure 5.6(b). The forces are of the same type, but the mass is now that of the trailer plus the chain. Thus, proceeding as before, we now find

$$T - F_{\text{fric}} = (m_T + m_C)a$$

$$T = F_{\text{fric}} + (m_T + m_C)a$$

$$= 400.\ \text{N} + (85.0\ \text{kg} + 15.0\ \text{kg})(0.500\ \text{m/s}^2) = 450\ \text{N}$$

The tension in the towing chain at the point where it is attached to the truck is 450 N.

To find the tension in the center of the chain we consider the system to be the trailer plus one half of the chain with forces acting as shown in Figure 5.6(c). This time we have

$$T - F_{\text{fric}} = \left(m_T + \frac{m_C}{2}\right)a$$

$$T = F_{\text{fric}} + \left(m_T + \frac{m_C}{2}\right)a$$

$$= 400.\,\text{N} + (85.0\,\text{kg} + 7.50\,\text{kg})(0.500\,\text{m/s}^2) = 446\,\text{N}$$

as the tension at the center of the chain.

◀

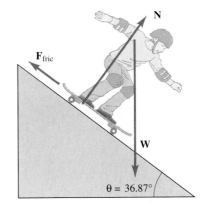

EXAMPLE 5.8   *Resolving Forces*

A 65.0-kg skate-boarder rides down an inclined ramp the surface of which makes a 36.87° angle with the horizontal. If a constant frictional force of 75.0 N acts on the skate-boarder as he rides, what is his acceleration?

SOLUTION   We are analyzing the skate-boarder shown in Figure 5.7(a). The forces acting on the skate-boarder are weight, normal, and friction. In Figure 5.7(b) we show the free-body diagram for the skate-boarder. The direction of the acceleration for the skate-boarder is certainly going to be along the surface of the incline. As suggested in problem-solving strategy step 4, we pick the direction of one of the coordinate axes, the x-axis in this case, to be parallel to the acceleration. As a result of our choice of coordinate axes, we see in Figure 5.7(b) that although N and F_{fric} are parallel to the y- and x-axes, respectively, the weight force W is parallel to neither. Hence, we must carry out step 5 of our problem-solving strategy by resolving W into its x- and y-components. We can accomplish this resolution by using the usual trigonometric identities and the construction shown in Figure 5.7(b). (To see how we determined which angle of the force triangle was equal to the incline angle θ, read the problem-solving hint: "Picking the angle" in the boxed section following this example.)

$$|W_x| = W\sin(\theta); \qquad |W_y| = W\cos(\theta)$$

At this point, at least until you've had more practice, you may wish to redraw the free-body diagram and replace the weight force by its two components as in Figure 5.7(c).

Newton's second law may now be written for the x- and y-directions separately:

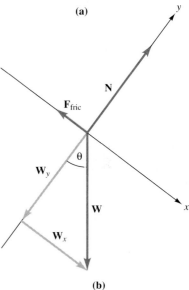

(b)

x-COMPONENTS	y-COMPONENTS
$\Sigma F_x = ma_x$	$\Sigma F_y = ma_y$
$W\sin(\theta) - F_{\text{fric}} = ma_x$	$N - W\cos(\theta) = ma_y$

We now note that the y-acceleration-component is zero since the skate-boarder's velocity perpendicular to the plane is zero (a constant). Thus, the x-equation can be solved for a_x and the y-equation for N:

$$a_x = \frac{W\sin(\theta) - F_{\text{fric}}}{m}$$

$$= \frac{mg\sin(\theta) - F_{\text{fric}}}{m}$$

$$= g\sin(\theta) - \frac{F_{\text{fric}}}{m}$$

$$= (9.80\,\text{N/kg})(0.600) - \frac{75.0\,\text{N}}{65.0\,\text{kg}}$$

$$= 4.73\,\text{m/s}^2$$

$$N = W\cos(\theta)$$

$$= mg\cos(\theta)$$

$$= (65.0\,\text{kg})(9.80\,\text{N/kg})(0.800)$$

$$= 510.\,\text{N}$$

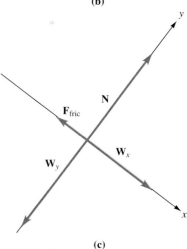

(c)

FIGURE 5.7

(a) The forces acting on a skate-boarder on an inclined plane. (b) A free-body diagram showing the resolution of the weight force W into two components. We know that the angle labeled θ is the same as the incline angle because its two rays are perpendicular to the two rays of the incline angle left to left and right to right. (c) Free-body diagram with the weight force replaced by its components W_x and W_y.

We note that in this case there is no need to solve for the normal force N; it is shown here for the sake of completeness.   ◀

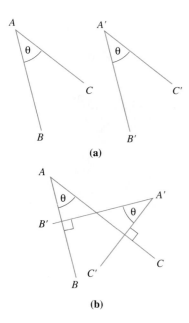

FIGURE 5.8

(a) Two angles are equal if their rays are parallel left to left and right to right. (b) Two angles are equal if their rays are perpendicular left to left and right to right.

Problem-Solving Hint: Picking the Angle

When resolving forces into components it is usually necessary to determine which acute angle of the force triangle (see Fig. 5.7(b), for example) is equal to an angle given in the problem (the incline angle θ in Fig. 5.7(a), for example). The easiest way to make this determination is to make use of the following two theorems from plane geometry:

1. Two angles are equal if their rays are parallel, left to left and right to right.
2. Two angles are equal if their rays are perpendicular, left to left and right to right.

We will not prove these theorems here, but their truth is apparent from Figure 5.8.

Which of these two theorems is applicable for angle θ in Example 5.8? What is true of two angles if one pair of rays is parallel and the other pair is perpendicular?

EXAMPLE 5.9   The Atwood Machine

Two masses are connected by a light string draped over a frictionless dowel. Mass A on the left in Figure 5.9(a) is 3.00 kg, and mass B on the right is 5.00 kg. The system is released from rest. Calculate the acceleration of the masses and the tension in the string.

SOLUTION   What are we analyzing? We are analyzing each of the two blocks A and B. Of the five forces (weight, spring, normal, friction, and tension), only weight and tension act on the blocks in question. The directions of the forces are shown in Figure 5.9(a). Free-body diagrams of each block appear in Figure 5.9(b).

We begin by choosing the standard "obvious" coordinate system for the problem. As we progress we will find that a different choice makes thinking about the problem somewhat simpler. We expect you may wish to adopt the second coordinate system in your own work, but for now let's begin as usual: We pick the coordinate system to have the y-axis in the vertical direction with the positive direction upward as indicated in Figure 5.9(b).

As usual, we write $\Sigma \mathbf{F}_y = m\mathbf{a}_y$ for each block separately:

BLOCK A	BLOCK B	
$\Sigma \mathbf{F}_{A_y} = m_A \mathbf{a}_{A_y}$	$\Sigma \mathbf{F}_{B_y} = m_B \mathbf{a}_{B_y}$	
$T\hat{\mathbf{j}} - W_A \hat{\mathbf{j}} = m_A a_{A_y} \hat{\mathbf{j}}$	$T\hat{\mathbf{j}} - W_B \hat{\mathbf{j}} = m_B a_{B_y} \hat{\mathbf{j}}$	
$T - W_A = m_A a_{A_y}$	$T - W_B = m_B a_{B_y}$	(5.6)

We note that the magnitudes of a_{A_y} and a_{B_y} are the same, but that their directions are opposite, so that $a_{A_y} = -a_{B_y}$. If we make this substitution in Equations (5.6), they become

$$T - W_A = m_A a_{A_y} \qquad \qquad T - W_B = -m_B a_{A_y}$$

or

$$T - W_A = m_A a_{A_y} \qquad \qquad W_B - T = m_B a_{A_y} \qquad (5.7)$$

Equations (5.7) can be solved simultaneously for the two unknowns T and a_{A_y}. Before we solve them, however, we would like to introduce a new way of interpreting these equations. We like to think of Equations (5.7) as if they were written using a strange coordinate system in which the y-axis started on the side of block A and then was bent around the rod at the top so that the y-axis was positive downward on the B side. Figure 5.9(c) shows how you might draw such a coordinate axis.

In this interpretation, mass A and mass B are in a kind of tug-of-war. Any force that tends to help B win tries to make the system move clockwise and is treated as a positive force, because it points in the positive sense on our folded y-axis. On the other hand, any force that tends to help A win tries to make the system move anti-clockwise and is regarded as negative. The advantage of this way of thinking is that if the system moves clockwise, both masses have a positive acceleration so you don't need to distinguish between a_{A_y} and a_{B_y}. In fact you can call the acceleration just plain a without any ambiguity. When the acceleration a is positive, mass A moves upward in Figure 5.9(c), and mass B moves downward.

In Figure 5.9(d) we show still another way of thinking about the competition between mass A and mass B. In this figure the "bent" y-axis is straightened out. We are careful to keep the forces parallel to the axis as it is straightened. In Figure 5.9(d) the $+$ and $-$ signs for the various forces in Equations (5.7) are completely consistent with an ordinary one-dimensional sign convention.

We have now shown you three different points of view that lead you to Equations (5.7). You may have a fourth that makes sense to you. However, whatever point of view you take, you need a *consistent* set of signs to get to the pair of Equations (5.7).

With these matters out of the way, we can now finish solving Equations (5.7). The easiest way is to add the two equations and solve for a_{A_y}, which we now call simply a. (This step is equivalent to solving one of Equations (5.7) for T and substituting the result into the other equation.)

$$W_B - W_A = m_A a + m_B a$$

$$a = \frac{W_B - W_A}{m_A + m_B} = \frac{m_B - m_A}{m_A - m_B} g$$

$$= \frac{2.00 \text{ kg}}{8.00 \text{ kg}} (9.80 \text{ N/kg}) = 2.45 \text{ m/s}^2$$

This value for acceleration can then be substituted into either of Equations (5.7) to obtain the tension:

$$T = W_A + m_A a = m_A(g + a) = (3.00 \text{ kg})(9.80 \text{ m/s}^2 + 2.45 \text{ m/s}^2) = 36.8 \text{ N}$$

A parting note of caution: Should the acceleration a turn out to be a negative quantity, be sure to substitute its value into the tension equation as a negative number. You must preserve the algebraic integrity of the variables. ◄

Examples 5.5 through 5.9 have been provided to show, in a relatively simple context, several useful techniques for analyzing the forces on objects and their resulting motion. In more complicated problems you may need to apply several of the techniques at the same time. For example, a problem might require you to resolve forces into components and then solve simultaneous equations. Although each problem is different, each one makes use of the same general procedural steps given in the problem-solving strategy. What you should learn here is how to attack problems using the strategy.

5.5 Newton's Third Law

The last of what are referred to as Newton's three laws of motion is often called the law of "action and reaction." It is easiest to give a more precise statement of the law if we first discuss a popular convention for describing forces.

We believe that it is desirable to train yourself to always describe a force using the following pattern. Tell what the force is exerted *on*, and then, what the force is exerted *by*. For example, if you wish to describe the upward normal force on the bottom of your feet,

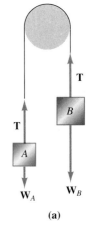

(a)

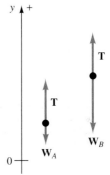

(b)

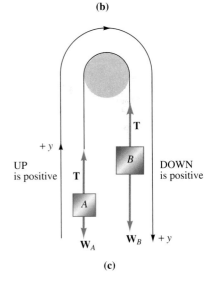

(c)

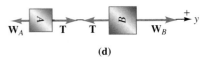

(d)

FIGURE 5.9

(a) Two masses connected by a light string that rides without friction over a dowel rod. (b) Free-body diagram with conventional y-axis. (c) An alternative folded coordinate system. (d) The folded y-axis is straightened showing a possible interpretation of the signs in Equation (5.7).

say "*N* is the force exerted *on* my feet *by* the floor." This way, there can be no ambiguity regarding which force you are talking about. Using this jargon you describe your weight as "The gravitational force exerted *on* my body *by* the earth."

Using the "*on . . . by . . .*" description of forces, it is easy to state Newton's third law.

> **Newton's third law:**
>
> If a force is exerted on body 1 by body 2, then a force is also exerted on body 2 by body 1. These two forces are equal in magnitude, but opposite in direction.

Concept Question 9
While seated waiting for class to start a student incorrectly tells his classmate that the Newton's-third-law reaction force to the earth's downward gravitational force on him (his weight) is the upward normal force of the floor on the feet of his chair. Help correct this student's misconception by describing the actual Newton's-third-law reaction force for each of the two forces he has mentioned.

Newton's third law says that forces always come in equal and oppositely directed pairs. If you know one force and wish to discover what the reaction force is, it is a simple matter of reversing the nouns in the "*on . . . by . . .*" description. Thus, the upward normal force *on* your feet *by* the floor is paired with the downward normal force *on* the floor *by* your feet. Your downward weight is the gravitational pull *on* your body *by* the earth. Therefore, there must be an equal upward gravitational pull *on* the earth *by* your body.

Sometimes subscripts are used to denote the "*on . . . by . . .*" relationship of forces, with the first subscript designating the "on" object and the second subscript the "by" object. Thus, the force exerted on the wall by your hand is written $\mathbf{F}_{wh}$, and the force on your hand by the wall is $\mathbf{F}_{hw}$. Using this notation, Newton's third law is written succinctly as

$$\mathbf{F}_{wh} = -\mathbf{F}_{hw} \tag{5.8}$$

To illustrate one final point regarding Newton's third law, we would like to challenge you and your class to a tug-of-war. That's right, the two of us against your entire class. We don't care how large the class is, we'll take you all on! We'll use the standard tug-of-war arrangement, you and your classmates on one end of a rope, us on the other.

Because we can't actually be with you today we have to decide the winner by reasoning. We hereby declare the battle a draw! Here is our (incorrect) line of reasoning. In order to win, you have to figure out what's wrong with our argument.

We claim that because of Newton's third law, whatever force your class exerts on us, we exert an equal and opposite force on your class. This is, after all, what Newton's third law declares. Since these forces are equal and opposite, they cancel. Thus, although we can't win, neither can you (unless you figure out what's wrong with this line of argument). No, don't try to win by weakening the third law. It's really true. Newton didn't restrict it at all; the law doesn't mean the magnitudes of the forces are *nearly* equal, it means the magnitudes are *exactly* equal. The law doesn't mean the forces are equal and opposite unless there is an acceleration. The law means exactly what it says, the forces are exactly equal and opposite, *all the time!*

Give up? If you need a hint, go back and read step 1 of our problem-solving strategy again. The message in step 1 is this: if you want to know whether *we* are going to move, you should look at forces on *us.* What is wrong with our argument is that we tried to convince you that a force *on you* and your classmates could affect *our* motion. In Figure 5.10 we show our fictitious tug-of-war with all the forces acting on us. From the figure it should be clear that to cause us to accelerate, all that is necessary is that the rope tension be larger than the frictional force of the ground on our feet. It doesn't matter to *our* motion that the rope pulls back *on the class* with the exact same tension *T*. In one sense then, the real tug-of-war is between the frictional force on our feet, and the frictional force on your class's feet. You win!

The moral of the tug-of-war story is that when analyzing a particular object's motion, be sure you look at the forces on that object only. The paired forces of Newton's third law always act on different objects. Sometimes, however, the two objects referred to in Newton's third law are merely different parts of a larger object. In that case the forces, called **internal forces,** do cancel when we consider the motion of the larger object. This of

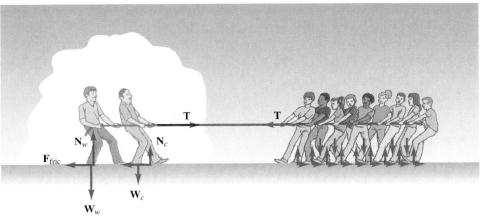

FIGURE 5.10

The authors challenge your class to a tug-of-war. You can defeat the authors only if you understand Newton's third law.

course is nothing more than a formal statement that you can't lift yourself off the ground by pulling on your own feet. The upward force on your feet by your hands is equal and opposite to the downward force on your hands by your feet.

5.6 Gravitational Field Strength, Falling Objects, and Mass (Optional)

In Section 5.4 we claimed that the gravitational attraction for an object near the earth's surface is 9.80 N/kg. You probably noticed that this field strength is numerically equal to the downward acceleration of freely falling bodies near the earth's surface. Not only are the numbers equal but the units are also, because the units newtons per kilogram is equivalent to meters per second squared. (Check this by substituting $N = kg \cdot m/s^2$.) The equivalence between the acceleration due to gravity and the gravitational field strength is especially intriguing to physicists. We may get to the heart of their interest by pointing out that there are actually two types of mass! One type is inertial mass, the other is gravitational mass. Inertial mass m_i is a measure of the "sluggishness" of an object and is the mass we have used in our applications of Newton's second law $\mathbf{a} = \mathbf{F}_{net}/m_i$. This mass is a measure of a body's inertia (resistance to change in motion). Larger inertial masses m_i result in smaller accelerations $\mathbf{a}$ for a given force.

The other mass used in the equation $W = mg$ is a measure of "attractiveness." It is this type of mass that determines the strength of the pull of the earth on a given object. The strength of this pull has nothing whatsoever to do with motion. The expression we should have written in Section 5.4 for the weight of an object is $W = m_g g$, where m_g is the **gravitational mass.**

There is no obvious reason why the two types of mass m_i and m_g should be equivalent. Why should the attraction of two bits of matter for one another have anything to do with how difficult it is to change their motion? A great many experiments have been performed to measure a difference between these two seemingly different properties of matter. So far, no experiment has been able to conclusively measure any difference. Many physicists think that there are fundamental reasons to believe the two types of mass must be the same. Indeed, the exact equality of these two masses lies at the heart of Einstein's general theory of relativity (which is itself still being challenged!). As a practical matter, if there is a difference, it is exceedingly small. We therefore follow the convention of using the same symbol m and same name, mass, for both.

Concept Question 10
A car tows a small trailer behind it. If the forward pull of the car on the trailer is equal to the backward pull of the trailer on the car, how can either accelerate?

When we accept the equivalence of gravitational mass and inertial mass, it is easy to see how g is equal to a_g. Let's calculate the acceleration of an object falling freely near the surface of the earth. We use Newton's second law, $\Sigma \mathbf{F} = m_i \mathbf{a}$, putting forces on the left and $m_i a$ on the right. The only force on the falling object is gravity, that is, its weight. Thus, $\Sigma \mathbf{F} = m_i \mathbf{a}$ becomes

$$W = m_i a$$

Because $W = m_g g$, we have

$$m_g g = m_i a$$

When we equate m_g and m_i, we may cancel them to find

$$g = a \qquad\qquad (5.9)$$

Equation (5.9) tells us that the kinematic downward acceleration a of a freely falling body is numerically equal to the local gravitational field strength. That is, $a = 9.80$ N/kg $= 9.80$ m/s^2. This, of course, is the 9.80 m/s^2 that we encountered in Chapter 3 where we referred to it as a_g. We now see that the acceleration due to gravity alone is a measure of the gravitational field strength. The acceleration due to gravity is constant only if the gravitational field is constant. In Chapter 6 the gravitational force is treated in more detail.

5.7 Summary

Newton's three laws of motion may be stated as follows:

1. Law of inertia: The total force on an object moving at constant velocity is zero. Total force means the *vector sum* of all forces.
2. $\Sigma \mathbf{F} = m\mathbf{a}$, where the net force $\Sigma \mathbf{F}$ equals the *vector sum* of all forces acting on the object of mass m.
3. $\mathbf{F}_{AB} = -\mathbf{F}_{BA}$, that is, if body A exerts a force on body B, then body B exerts an equal and opposite force on body A.

The SI unit of force is the newton (N) and the SI unit of mass is the kilogram (kg). In the British Engineering system the corresponding units are the pound force (lbf) and the slug. Five different forces commonly encountered in classical physics problems have been introduced. They are

1. Weight, $W = mg$
2. the Hooke's-law spring force, $\mathbf{F}_s = -k\mathbf{s}$
3. the normal force N, a contact force perpendicular to the surfaces in contact
4. the frictional force F_{fric}, a contact force parallel to the surfaces in contact and directed so as to oppose the relative motion or attempted relative motion of the surfaces
5. the tension T, the force exerted by a taut string or rope

(Optional) There are actually two types of mass: inertial mass and gravitational mass. Experiments imply that these two are equivalent.

PROBLEMS

5.1 Newton's First Law: Inertia

1. Three forces act on an object located at the origin of a Cartesian coordinate system. Force A is described by $\mathbf{F}_A = (3.00 \text{ N})\hat{\mathbf{i}} + (4.00 \text{ N})\hat{\mathbf{j}}$. Force B lies in the xy-plane, has magnitude 8.00 N, and is directed so that it makes an angle of $110°$ (measured anticlockwise) with the positive x-axis. (a) Find a third force $\mathbf{F}_C$ such that the sum of the three forces is zero. Give your answer in both polar and unit vector form. (b) If these are the only three forces acting on the object, is it necessarily at rest?

2. Three puppies tug at an old sock. Two pull to the right with equal forces of 10.0 N separated by an angle of $45.0°$. (a) With what force does the third pup have to pull to hold the sock stationary? (b) How would your answer change if the sock was moving at a constant velocity of 0.0500 m/s in the direction of the third pup?

3. Prove that if three forces total zero, the forces must be coplanar. To keep your proof as general as possible, use only the fact that $\mathbf{F}_1 + \mathbf{F}_2 + \mathbf{F}_3 = 0$. [*Hint:* See Problem 2.43.]

4. Three forces act on a 2.00-kg mass as shown in Figure 5.P1. If $F_1 = 30.0$ N, calculate the magnitude of F_2 and F_3, (a) if the mass is stationary, (b) if the mass is moving with a constant velocity of 2.00 m/s to the right.

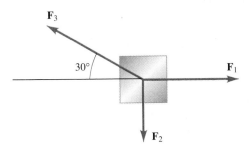

FIGURE 5.P1   Problem 4

5.2 Newton's Second Law

5. A 2.00-kg mass has forces F_1 and F_2 exerted on it as shown in Figure 5.P2. (a) If the mass is moving at a constant velocity to the left, what is the magnitude of the third force F_3? (b) If the third force is reduced to half the value calculated in part (a), what is the acceleration of the mass?

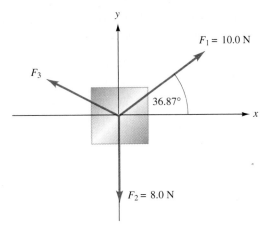

FIGURE 5.P2   Problem 5

6. (a) A force of $4.00 \text{ N}\hat{\mathbf{i}} + 3.00 \text{ N}\hat{\mathbf{j}}$ is exerted on a 2.00-kg mass. If this is the only force acting, find the magnitude of the mass's acceleration. (b) If an additional force of $-2.00 \text{ N}\hat{\mathbf{i}} + 2.00 \text{ N}\hat{\mathbf{j}}$ acts on the mass, what is the magnitude of the mass's acceleration?

7. Two forces are available, one of 10.0 N, the other of 15.0 N. If both forces are applied to a 6.00-kg mass, (a) what is the largest acceleration the mass can have? (b) What is the smallest acceleration the mass can have? (c) What should be the angle between the two forces to accelerate the mass at exactly 3.00 m/s²?

8. Force F_1 gives mass A an acceleration of a_o. Force F_2 gives mass A an acceleration of 2.00 m/s². Force F_1 gives mass B an acceleration of $3a_o$. A 10.0-N force gives mass B an acceleration of $4a_o$, and F_2 acting on mass B produces $a_o/2$. Find the magnitude of F_1, F_2, a_o, m_A, and m_B.

9. A 4.00-kg mass is acted on by a 12.0-N force parallel to the x-axis. However, the mass exhibits an acceleration at a constant angle of $36.87°$ above the positive x-axis. (a) What is the magnitude of the other force acting on the mass if this force is directed parallel to the positive y-axis? (b) What is the magnitude of the resulting acceleration?

10. If a constant unbalanced force of 125 N is applied to a 7.25-kg shotput for 0.450 s, what is the velocity of the shot at the end of this time interval?

5.3 Some Common Forces

11. (a) What is the mass of a person who weighs 100. lbf on earth? Express your answer in both kilograms and slugs. (b) What is the weight of a 5.00-kg mass on earth? Express your answer in both pounds and newtons.

12. On the surface of the moon the gravitational field strength is about one-sixth that on the surface of the earth. (a) If there were a supermarket on the surface of the moon where you could buy 2.00 lbf of hotdogs, how many hotdogs would you get? (A 2.00-lb hotdog-package on earth contains 10 hotdogs.) (b) If you purchased 2.00 kg of hotdogs on the moon, how would the number of hotdogs compare to a 2.00-kg purchase on earth?

13. (a) A 16.0-lbf shotput is pushed by an unbalanced force of 70.0 lbf. What is the acceleration of the shotput during the push? (b) A ball accelerates at 20.0 m/s² when pushed with an unbalanced 60.0-N force. What is the weight of the ball on earth?

14. (a) If you ask for 1.00 slug of hamburger at the grocery store, how much hamburger do you receive (expressed in pounds)? (b) If you ask for 1.00 N of hamburger how much do you get (expressed in kilograms)? How much is this expressed in pounds?

15. One of the authors of this text has an earth weight of 588 N (except during holiday season). (a) What is the author's mass? (b) If this author were to take his bathroom scale to Mars, what weight, in newtons, would it indicate? [*Hint:* You may find the data in Appendix 8 helpful.]

16. Consider the following thought experiment. Hang a bowling ball from the ceiling by a strong wire. Imagine that you butt your head into the bowling ball. It hurts! Your head is very sore. Now imagine that you are in your spaceship in the farthest reaches of the cosmos where the gravitational field strength is essentially zero. "Hang up" the bowling ball in your spaceship. (No wire is needed this time!) Once again you butt your head into the bowling ball with the same speed as before. How does this new collision compare with the one on earth? What property of the bowling ball (weight or mass) is important here?

17. A student discovers an old spring in the physics stockroom and decides to measure its stiffness constant. The student finds that it takes a pull of 15.0 N to stretch the spring 0.250 m. (a) What is the stiffness constant of the spring? (b) What force is required to stretch the spring to 0.450 m?

18. To calibrate a spring to use as a simple scale a physics professor attaches a cut-out of an arrow on the lower end of a spring and places a sheet of cardboard behind the spring. When the professor hangs a 1.50-kg mass from the end of the spring, it stretches by 0.370 m. (a) What is the spring's stiffness constant? (b) How far apart should the professor make the 1.00-N marks on the cardboard scale?

5.4 Simple Applications of Newton's Second Law

19. Draw a diagram of the following situations showing the forces acting on the object in question. Make your forces of approximately correct relative length. Label them as W, N, F_{fric}, T, F_s, or P as appropriate. (Use P as a push or pull applied by an external agent, such as a person.) (a) a book resting on a horizontal desk (b) a book pushed by a horizontal force along a level table at a constant speed (c) a book being held stationary against a vertical wall by someone's hand exerting a horizontal force on the book (d) a child sliding down a rough incline at constant speed (e) an apple falling freely toward the ground.

20. Draw a diagram of the following situations showing the forces acting on the object in question. Make your forces of approximately correct relative length. Label them as W, N, F_{fric}, T, F_s, or P as appropriate. (Use P as a push or pull applied by an external agent, such as a person.) (a) a barbell being held aloft by a weight lifter (b) a small (stubborn) dog being pulled by its leash (Have the leash make an angle of about 45° with the horizontal.) (c) a ball resting in a deep V-shaped trough (d) a lamp, which is attached to the ceiling by a cord but is pulled to one side by a horizontal force (e) a coin that has been tossed upward, at a moment when it is free of the hand that threw it but still moving upward.

21. A mass of 1.50 kg is dragged at a constant velocity along the top of a table by means of a horizontal spring with stiffness constant $k = 200.$ N/m. (a) If the spring is extended by 5.00 cm as the block is pulled, what is the force of friction on the block? (b) Assuming the frictional force stays constant, by what amount is the spring extended if the block is accelerated at 2.50 m/s²?

22. A spring with a 200.-N/m stiffness constant is placed between a person's hand and a 2.50-kg mass. When the person pushes the mass across a rough surface, the spring is compressed 10.0 cm. (a) If the acceleration of the mass is 4.00 m/s², what frictional force acts on the mass? (b) How much must the person compress the spring for the mass to move at a constant speed across the surface?

23. A spring is used to pull a 6.00-kg mass up a frictionless 35.0° incline. Assume the spring is parallel to the incline. (a) If the spring stretches 15.0 cm from its equilibrium length while the mass accelerates at 3.00 m/s² up the incline, what is the stiffness constant of the spring? (b) What length must the spring be stretched to cause the mass to accelerate up the incline at 4.50 m/s²?

24. A 2.00-kg mass hangs on a spring connected to the top of an elevator car. While the elevator is at rest this mass causes the spring to stretch by 20.0 cm. (a) Calculate the stiffness constant of the spring. (b) If the elevator accelerates from rest to 5.00 m/s in 2.00 s and the acceleration is constant, what is the magnitude of the elevator's acceleration? (c) If that acceleration is upward, by how much is the spring stretched (from its unstretched length)

during that acceleration? (d) By what amount is the spring stretched if the elevator's acceleration is downward?

25. An elevator initially moving upward with a speed of 3.00 m/s comes to rest in 2.00 s. Attached to the ceiling of the elevator is a spring with a 3.00-kg mass connected to its lower end. The stiffness constant of the spring is 300. N/m. (a) What is the elevator's acceleration? (b) How much is the spring extended from its natural length during this 2.00 s? (c) With what acceleration and in what direction must the elevator accelerate so that the spring is in an unstretched state?

26. When a certain archery bow is at full draw the archer must exert a force of 45.0 lbf straight back on the bow string to keep it in place. The two halves of the string each make an angle of 36.87° with the horizontal. (a) Draw a free-body diagram showing the forces on the archer's fingertips. (b) What is the tension in each string?

27. In a high-wire circus act a 400.-N performer carrying a 10.0-kg balance pole walks out onto the center of a tightly stretched wire. The wire sags 5.00° from the horizontal. (a) Draw a free-body diagram of the forces on the wire beneath the performer's feet. (b) What is the tension in the wire?

28. A hot-air balloon weighing 500. N is held down by two tethers on opposite sides of the balloon. One rope makes an angle of 53.13° with the ground and has a tension of 800. N. If there is a buoyant force of 1750 N directly upward on the balloon, what is the tension in the other tether rope, and what angle does it make with the ground?

29. A 50.0-kg mass is suspended by a wire. The wire is held to the side by a second wire attached as shown in Figure 5.P3. Calculate the tension in each of the wires.

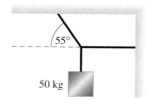

FIGURE 5.P3   Problem 29

30. A 5.00-kg mass is held at rest on a frictionless incline by means of a horizontal force. If the incline makes an angle of 53.13° with the horizontal, (a) what is the magnitude of the required horizontal force? (b) What is the magnitude of the normal force?

31. A 2.50-kg block is at rest on a 36.87° frictionless incline. The block is held in place by a spring as shown in Figure 5.P4. (a) If the spring has a stiffness constant of 50.0 N/m, how far is it compressed? (b) If the incline is not frictionless and the maximum static frictional force is 15.0 N, by how much additional distance can the spring be compressed so that the block remains at rest? (c) By how much can the spring be stretched and still permit the block to be in equilibrium?

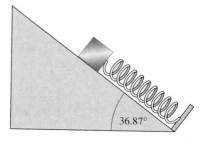

FIGURE 5.P4   Problem 31

32. A 3.00-kg block slides down an incline that makes a 30.0° angle with the horizontal. If a constant frictional force of 10.0 N resists its motion, (a) calculate the block's acceleration. (b) What is the magnitude of the normal force of the plane on the block? (c) If the block was sliding up the incline, what would be its acceleration?

33. A long-haul truck is pulling two empty trailers behind it. The mass of the first trailer is 2.40×10^3 kg; the mass of the second is 2.00×10^3 kg. The truck pulls on the first trailer with a force of 3.00×10^3 N. Assuming a constant frictional force of 300. N acts on each of the two trailers, calculate the acceleration of the system and the tension in the hitch connecting the first and second trailers.

34. Four 1.00-kg masses hang vertically below one another, connected by light strings. An upward force of 66.0 N is exerted on the uppermost mass. (a) Calculate the acceleration of the system and (b) the tension in each string.

35. A miniature train used to transport passengers around a zoo has an engine and three cars. Each car has a mass of 300. kg and is acted upon by a constant frictional force of 60.0 N. (a) With what force must the engine pull on the first car to accelerate the train at 0.500 m/s²? (b) What is the tension in the hitch between the first and the second cars? (c) between the second and the third?

36. A 3.00-kg mass (A) and a 1.50-kg mass (B) are connected by a light string. The 3.00-kg mass slides without friction along a level tabletop, and the 1.50-kg mass hangs off the edge of the table as shown in Figure 5.P5. Assume there is no friction between the string and the smooth dowel. Calculate the acceleration of the masses and the tension in the string.

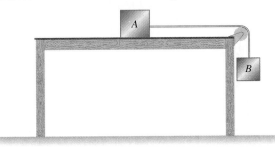

FIGURE 5.P5   Problems 36, 37

37. Two blocks are connected as in Figure 5.P5. Although the dowel over which the string slides is frictionless, the table top is rough and a constant frictional force of 5.00 N is exerted on block A ($m_A = 3.00$ kg). (a) Calculate the acceleration of the masses and the tension in the string if block B ($m_B = 1.50$ kg) is initially moving downward. (b) If someone gives block A a shove so that block B is initially moving upward, what is the acceleration of the system and the tension in the string right after the person lets go?

38. Three masses of $m_A = 2.00$ kg, $m_B = 3.00$ kg, and $m_C = 4.00$ kg are connected by a light string. The string is draped over a smooth dowel so that it slides without friction. As shown in Figure 5.P6, the 2.00-kg and 3.00-kg masses hang on one side of the dowel, and the 4.00-kg mass hangs on the other. Calculate the acceleration of this system of masses and the tension in each of the two string segments.

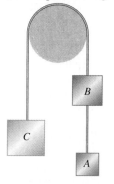

FIGURE 5.P6   Problem 38

39. Three masses are connected as shown in Figure 5.P7. The masses of m_1, m_2, and m_3 are 5.00 kg, 2.00 kg, and 7.00 kg, respectively. The mass on the tabletop experiences a constant frictional force of 40.0 N. Assuming the string slides without friction over the dowel at the edge of the table, calculate the acceleration of the system and the tension in each of the two string segments.

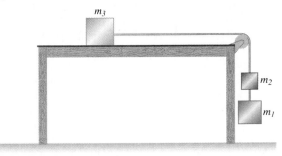

FIGURE 5.P7   Problem 39

40. Three masses are connected as shown in Figure 5.P8. The masses of m_A, m_B, and m_C are 2.00 kg, 4.00 kg, and 6.00 kg, respectively. Mass A experiences a frictional force of 10.0 N, and mass B experiences a frictional force of 15.0 N. Assuming the string rides without friction over the rod at the table's edge, calculate the acceleration of the system of masses and the tension in each of the two string segments.

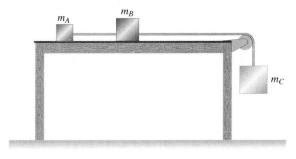

FIGURE 5.P8   Problem 40

5.5   Newton's Third Law

41. When a 1.00-kg mass hangs from a spring scale as shown in Figure 5.P9(a), the scale reads 9.80 N. (a) If a string from which the mass hangs is draped over a perfectly frictionless pulley as in Figure 5.P9(b), what is the reading on the spring scale? (b) If the upper end of the spring scale is attached to a second string that is, in turn, draped over a second perfect pulley and attached to a second 1.00-kg mass as shown in Figure 5.P9(c), what is the reading on the spring scale?

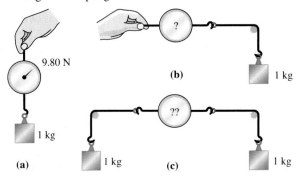

FIGURE 5.P9   Problem 41

42. Two football players compare their strength by each pulling a rope attached to a spring scale that is firmly attached to an immobile tree. One player is able to generate a tension of 190 lbf, and the other independently is able to create a tension of 230 lbf. The spring scale is now attached to two ropes, one on each end of the scale, and the two football players pull on the ropes in opposite directions, but the scale does not move. (a) What is the maximum possible reading on the spring scale now? (b) Draw a sketch of each of the three situations described. On the sketches draw vectors of approximately correct scale showing the forces on the football players in each case.

43. For each of the forces that follow, state the reaction force required by Newton's third law using the "on ... by ..." designation of Section 5.5. Also, classify the force as gravitational, spring, normal, friction, or tension. (a) The upward force on a barbell by a weight lifter's hands, (b) the horizontal force on the road by a car tire, (c) the forward pull on a following trailer by a car bumper, (d) the force on the floor surface by the heel of your shoe, (e) your weight.

44. For each of the forces that follow, state the reaction force required by Newton's third law using the "on ... by .." designation of Section 5.5. Also, classify the force as gravitational, spring, normal, friction, or tension. (a) The downward force of a book on the desk top on which it rests, (b) the rearward push of a canoe paddle on the water, (c) the gravitational pull of the earth on the moon, (d) the push of a rocket engine on the exhaust gases, (e) the leftward horizontal frictional force on the bottom of a trunk sliding to the right across a floor, (f) the upward force of a rope on a rock climber.

45. Two masses are pushed along a smooth tabletop by a horizontal force $F_P = 6.00$ N as shown in Figure 5.P10. Mass A is 4.00 kg and mass B is 2.00 kg, so the two, of course, accelerate forward at 1.00 m/s². (a) Draw a free-body diagram of the forces on block A only. What is the magnitude of the horizontal force causing A to accelerate? What kind of force is it: weight, spring, normal, friction, or tension? (b) Draw a free-body diagram of block B. Because B pushes A forward, A must push B backward, so don't forget to include this force on your diagram. Verify that Newton's second law is satisfied for block B.

FIGURE 5.P10   Problem 45

46. Two blocks are set on a horizontal table as shown in Figure 5.P11. The upper block, $m_A = 3.00$ kg, is fastened to a fixed wall by a string, and the lower block, $m_B = 5.00$ kg, is pulled from beneath by a force $P = 10.0$ N. The tabletop is frictionless, but there is a frictional force between the two blocks. (a) Draw separate free-body diagrams for block A and block B. (b) If block B accelerates at 1.50 m/s², what is the tension in the string that holds block A?

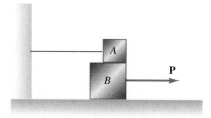

FIGURE 5.P11   Problem 46

General Problems

47. A 75.0-kg student steps horizontally off a 1.00-m-high crate onto the ground. (a) How long before the student's feet hit the ground? (b) What is the student's weight? (c) Newton's third law implies that the student attracts the earth gravitationally. What is the magnitude of the upward pull of the student on the earth? (d) What is the magnitude of the earth's upward acceleration as it rises to meet the student? (e) How far does the earth move upward while the student is falling downward?

48. A spring of stiffness constant 150. N/m is used to pull a 4.00-kg mass up a frictionless 30° incline. Assume the spring is parallel to the incline. (a) If the spring stretches 16.0 cm from its equilibrium length, what is the mass's acceleration up the incline? (b) What length must the spring be stretched for the mass to move up the incline at a constant speed? (c) If the spring in part (b) was replaced by a light string, what must be the tension in the string for the mass to move up the incline at a constant speed?

★49. One end of an unstretched spring of stiffness constant k_1 is attached to the ceiling of an elevator, and a 1.50-kg mass is connected to the other end. The ends of another unstretched spring of stiffness constant k_2 are attached to the bottom of the mass and to the floor of the elevator as shown in Figure 5.P12. When the mass is released, the top spring stretches 40.0 cm and the bottom spring is compressed by this same amount. When the elevator accelerates upward, the top spring stretches an additional 8.00 cm. (a) What is the elevator's acceleration? (b) What must the elevator's acceleration be to stretch the springs to a total length of 52.0 cm from their unstretched lengths?

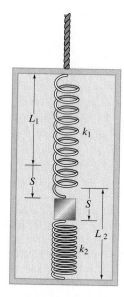

FIGURE 5.P12   Problem 49

50. A spring is used to pull a 6.00-kg mass down a rough 36.87° incline. The spring has a stiffness constant of 250. N/m, and a frictional force of 13.0 N opposes the motion down the incline. What is the acceleration of the mass when the spring is stretched by 10.0 cm?

51. A 2.50-kg block is held stationary on a frictionless incline by means of a string at the top of the incline and a spring at the bottom of the incline as shown in Figure 5.P13. The spring has stiffness constant $k = 75.0$ N/m and is stretched by 0.200 m. If the incline makes an angle of 60.0° from the horizontal, what is the tension in the string?

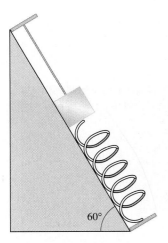

FIGURE 5.P13   Problem 51

52. Two 5.00-kg masses are connected by a heavy (1.00-kg!) cable. The two masses are pulled along a smooth horizontal surface by an external force $P = 22.0$ N as shown in Figure 5.P14. Calculate the tension in the cable (a) where it is attached to the first mass, (b) where it is attached to the second mass, and (c) at its midpoint.

FIGURE 5.P14   Problem 52

53. The three masses connected as shown in Figure 5.P15 have masses of $m_1 = 1.00$ kg, $m_2 = 2.00$ kg, $m_3 = 2.00$ kg. (a) If the tabletop is frictionless, calculate the acceleration of the system of masses and the tension in each of the two strings. (b) If the block on the table experiences a constant frictional force of 5.00 N, what is the acceleration of the system and the tension in each of the two strings if the system is initially moving to the left. (c) If the system is initially moving to the right, what is its acceleration and the tension in each string?

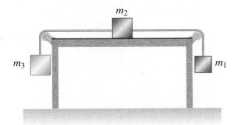

FIGURE 5.P15   Problem 53

54. Two blocks are connected by a light string as shown in Figure 5.P16. Block A has mass 4.00 kg, and block B has mass 5.00 kg, and the incline angle θ is 36.87°. If the incline is frictionless, find the acceleration of this system of masses and the tension in the string.

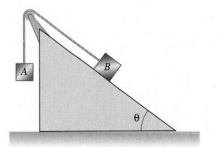

FIGURE 5.P16   Problem 54

55. Two blocks are connected by a light string as shown in Figure 5.P17. Block A has mass 3.50 kg, and block B has mass 5.00 kg. If both sides of the incline are frictionless, calculate the acceleration of this system of masses and the tension in the string.

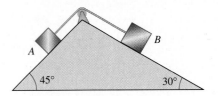

FIGURE 5.P17   Problem 55

CHAPTER

6

Additional Force Models and Circular Motion

In this chapter you should learn:

- to use the coefficient-of-friction model to approximate the frictional force on sliding objects.

- to apply Newton's second law to objects moving in uniform circular motion.

- to compute the gravitational force between spherical objects using Newton's universal law of gravitation.

- (optional) to calculate Reynolds number and from it determine an appropriate model for viscous drag.

- (optional) to understand terminal velocity and calculate its value for objects falling through air.

- (optional) to apply a numerical technique to compute the motion of objects subject to very general forces.

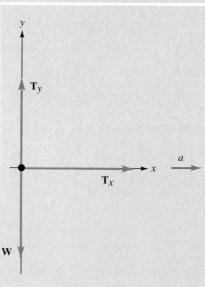

126

In the previous chapter we examined Newton's laws of motion in a preliminary way. We introduced several force models including weight, spring, normal, friction, and tension. Using these models you gained experience solving for the acceleration of objects to which these different types of forces are applied. In this chapter we present a more detailed model for friction by introducing the concept of a coefficient of friction, which characterizes the frictional force between two surfaces. Following the introduction of this model we apply Newton's second law to the important special case of uniform circular motion. In the latter half of the chapter we introduce additional force models required to expand the range of phenomena we can consider: We broaden our concept of weight to include the general gravitational attraction between objects throughout the universe. We introduce two models that can be used to describe the friction on moving objects due to air or other resistive media. Finally, we introduce a numerical method for deducing the motion of an object experiencing quite general forces.

6.1 The Coefficient-of-Friction Model

The frictional forces between two surfaces can be quite complicated. In general, these forces depend on the surfaces' temperature, preparation, relative speed, and even recent history. At this point, to describe the frictional force in great detail is not terribly useful. Instead, we introduce a relatively simple model that includes several of the major phenomena related to friction. First, we discuss the frictional force between two surfaces at rest relative to one another, and then describe the case in which the surfaces slide past one another.

Generally, frictional forces change their size in response to applied forces. For example, if you lay a book on a level table, the frictional force is zero. If you push horizontally on the book, but gently enough so that the book still does not move, the frictional force on the book increases to balance your push. The harder you push, the greater the frictional force, but it grows large enough only to exactly balance your push. The net force on the book remains zero. If you continue increasing the size of your push, eventually the book starts to accelerate. At the instant just before motion begins, the frictional force reaches a maximum value beyond which it can no longer grow. At this point any additional push by you becomes an unbalanced force. It is this maximum force of friction, just before motion occurs, that we wish to characterize.

In general, the harder two surfaces are pressed together, the larger the maximum force of friction between them. This observation is no surprise. You know it's more difficult to slide a box of heavy books across the floor than it is to slide only the empty box. One way of looking at this effect is to say that when the box is full of books its bottom surface is pressed more tightly against the floor. A simple approximation that describes this effect is to assume that the maximum static frictional force increases in direct proportion to the normal force. The proportionality constant between the maximum static frictional force $F_{\text{fric}_{\text{max}}}$ and the normal force N is known as the **coefficient of static friction** and is designated by μ_s. Thus, one can write

$$F_{\text{fric}_{\text{max}}} = \mu_s N$$

This approximation works fairly well for many surfaces and is simple enough to make its application straightforward.

The second property of frictional force is one you may have noticed if you have moved much furniture. You start by pushing quite hard on the object, then it seems to "break loose" and start sliding. After motion begins, a much smaller applied force is necessary to balance the frictional force and keep the object moving at a constant speed. In order to build this property of friction into our model, we need to introduce a second coefficient of friction μ_k for the sliding (kinetic) case. Because our experience suggests that the maximum static frictional force is greater than the kinetic frictional force, we expect $\mu_s > \mu_k$.

To summarize our model of frictional forces, we state the following:

$$F_{\text{fric}} = \mu_k N \qquad \text{(For kinetic friction)} \qquad (6.1)$$

$$F_{\text{fric}} \leq \mu_s N \qquad \text{(For static friction)}$$

Concept Question 1
If a tire rolls without slipping over a road surface, which is more appropriate to describe the frictional force between the tire and the road—static friction or kinetic friction?

Some *typical* values for the coefficients of friction between several different surfaces are given in Table 6.1.[1] These coefficients of friction allow us more accurately to model situations where frictional forces take on nonconstant values. One such application of this model is illustrated in Example 6.1.

TABLE 6.1 Typical Coefficients of Static and Kinetic Friction

SURFACES	STATIC	KINETIC
Steel on steel	0.7	0.6
Aluminum on steel	0.6	0.5
Copper on steel	0.5	0.4
Teflon on steel	0.04	0.04
Rubber on concrete	1.	0.8
Wood on wood	0.4	0.2
Teflon on Teflon	0.04	0.04
Brake lining on cast iron	0.4	0.3
Hemp rope on metal	0.3	0.2
Metal on ice	0.02	0.02
Rubber tires on smooth pavement (dry)	0.9	0.8

EXAMPLE 6.1   *An Adventure in Physics*

A 60.0-kg dresser is pushed up an inclined ramp into the back of a moving van by means of a horizontal push force **P** as shown in Figure 6.1(a). The coefficients of friction between the ramp and dresser are $\mu_s = 0.600$ and $\mu_k = 0.400$, and the ramp makes an angle of 20.0° with the horizontal. (a) What value of **P** is required to move the dresser at a constant speed of 0.200 m/s along the ramp? (b) If the person pushing should stop for a rest and let **P** = 0, does the dresser slide back down the ramp?

SOLUTION   (a) We continue to follow the problem-solving strategies of Chapter 5. We are analyzing the dresser; it is the "system" under consideration. The forces present are weight, normal, friction, and, of course, **P**. These forces are shown in Figure 6.1(b). Figure 6.1(c) shows the forces redrawn on a free-body diagram for the dresser. We expect any acceleration to be parallel to the incline (unless the ramp breaks!). Therefore, we take the *x*-axis parallel to the incline.

The two forces **W** and **P**, neither of which are along the *x*- or the *y*-axis, must both be resolved into components that are parallel to these axes. Following procedures similar to those of Example 5.8, we find the magnitudes of the components:

$$|P_x| = P\cos(\theta) \qquad |W_x| = W\sin(\theta)$$
$$|P_y| = P\sin(\theta) \qquad |W_y| = W\cos(\theta)$$

These components are shown in the free-body diagram in Figure 6.1(d).

We are now ready to proceed to step 6 of our strategy and write Newton's second law for both the *x*- and *y*-directions.

---

[1] Because the frictional forces between surfaces depend on the many factors listed previously (surface preparation, history, etc.) the values listed in Table 6.1 are only approximate and may vary as much as 20% from those listed.

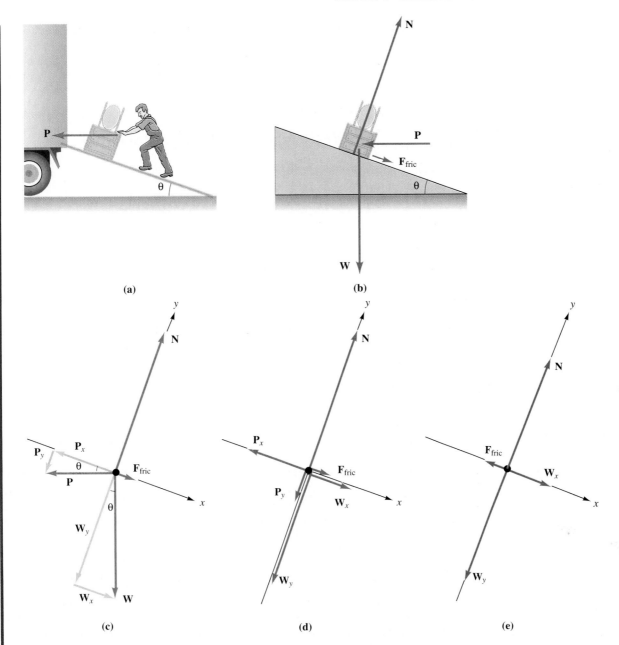

FIGURE 6.1

(a) A dresser is pushed up an inclined plane by a horizontal push force P. (b) Forces acting on the dresser as it moves up the rough incline. (c) A free-body diagram. The x-axis of the coordinate system is chosen parallel to the acceleration along the incline. Forces P and W are resolved into components parallel to the coordinate axes. (d) A free-body diagram showing all resolved forces. (e) The force P is reduced to zero. The frictional force now opposes the attempted motion down the plane.

$$x\text{-components:} \qquad F_{\text{fric}} + W\sin(\theta) - P\cos(\theta) = ma_x \qquad \textbf{(6.2)}$$

$$y\text{-components:} \qquad N - W\cos(\theta) - P\sin(\theta) = ma_y \qquad \textbf{(6.3)}$$

Because the velocity along the ramp is constant, $a_x = 0$. (The exact value $v_x = -0.200$ m/s is irrelevant; what matters is that v_x is a *constant*.) Similarly, because v_y is also a constant (namely, 0), $a_y = 0$.

Equations (6.2) and (6.3) contain three unknowns: N, F_{fric}, and P. Therefore, we need a third equation before we can proceed. Equation (6.1) relates F_{fric} and N, and provides our third equation. In this case the dresser is sliding, so we write $F_{\text{fric}} = \mu_k N$. We can now solve the three equations for the three unknowns. Our plan is first to solve Equation (6.3) for N, and then replace F_{fric} in Equation (6.2) with $\mu_k N$. From Equation (6.3) we obtain the normal force,

$$N = W\cos(\theta) + P\sin(\theta) \qquad\qquad (6.4)$$

and from Equation (6.2)

$$\mu_k[W\cos(\theta) + P\sin(\theta)] + W\sin(\theta) - P\cos(\theta) = 0$$

Collecting terms in P we have

$$\mu_k P\sin(\theta) - P\cos(\theta) = -[\mu_k\, W\cos(\theta) + W\sin(\theta)].$$

From here we factor out P and W

$$P[\mu_k\sin(\theta) - \cos(\theta)] = -W[\mu_k\cos(\theta) + \sin(\theta)]$$

and then divide by the coefficient of P to obtain

$$P = \frac{\sin(\theta) + \mu_k\cos(\theta)}{\cos(\theta) - \mu_k\sin(\theta)}\, W$$

$$P = \frac{\sin(20.0°) + (0.400)\cos(20.0°)}{\cos(20.0°) - (0.400)\sin(20.0°)}(60.0\ \text{kg})(9.80\ \text{N/kg}) = 526\ \text{N}$$

(b) If the force P is reduced to zero, the dresser would, in the absence of friction, slide down the incline. Hence, the *attempted* motion is down the incline. Because friction opposes the relative motion (or attempted motion) of the surfaces, the frictional force acts up the incline. Figure 6.1(e) shows the revised free-body diagram for $P = 0$. The component of the gravitational force trying to move the dresser down the incline is just $W_x = W\sin(\theta) = (60.0\ \text{kg})(9.80\ \text{N/kg})\sin(20.0°) = 201\ \text{N}$. Now we must answer the question, "Is this force larger or smaller than the maximum possible frictional force?"

The normal force can be obtained from Equation (6.4) by setting $P = 0$. Thus, the maximum available static frictional force is $F_{\text{fric}_{\text{max}}} = \mu_s N = \mu_s W\cos(\theta) = (0.600)(60.0\ \text{kg})(9.80\ \text{m/s}^2)\cos(20.0°) = 332\ \text{N}$. The *maximum* possible frictional force (332 N) exceeds the force trying to move the dresser down the incline (201 N). The *actual* frictional force is not 332 N. In a way it would be nice if it were, for then the dresser would crawl up the incline under its own (frictional) power. But, alas, friction doesn't work that way. The actual frictional force simply grows until it exactly balances the 201-N force acting down the plane, and then it grows no more. But at least the dresser won't slide back down the incline. ◀

6.2 Circular Motion

In Chapter 4 we saw that objects traveling in a circular path with a constant speed also have a constant acceleration toward the center of the circle. The centripetal acceleration changes the direction of the object's velocity vector. From Newton's second law it necessarily follows that this acceleration is caused by an unbalanced force acting toward the center of the circle. This centripetal force is *not a new kind of force*. The cause of the centripetal acceleration must be some identifiable force, such as those we have already introduced: weight, spring, normal, friction, or tension.

Every time you see anything moving in a circle, a force is keeping it from flying off in a straight line. When a massive hammer on the end of a chain is whirled in a circle by an Olympic hammer thrower, the chain is pulling the hammer inward toward the center of the circle. When you drive a car around a curve, the frictional force on your tires pulls the

Concept Question 2
According to the coefficient-of-friction model, is the frictional force on a brick being pulled across a table larger when the brick lays on its largest face or on its smallest face? Explain.

The hammer throw. The athlete must exert a large inward force on the mass to hold it in a circular path.

car toward the center of the curve. If you are on the end of a chain of people in a roller rink playing "crack the whip," the person holding your hand is pulling you toward the center of rotation. If that person lets go, you head off in a straight line.

There are no new force laws in this section. We are merely going to apply Newton's second law to a number of situations where we find an unbalanced force giving rise to a centripetal acceleration.

EXAMPLE 6.2 A Swinging Time

As shown in Figure 6.2(a), a child whirls a 0.100-kg stone on the end of a light string in a circle around her head. The length L of the string is 0.500 m, and the string makes an angle $\theta = 36.87°$ with the vertical. Find (a) the time for one rotation of the stone and (b) the tension in the string.

SOLUTION   (a) The system we are analyzing is the stone. Once again following our problem-solving strategy, we draw a free-body diagram like that in Figure 6.2(b). There are only two forces acting on the rock: weight and tension. There are no other forces (if we ignore the very small air resistance). In particular, there is no "extra" force toward the center of the circle. (What could this extra force be? There is no normal, friction, or spring force.) Furthermore, there is no force into or out of the page that "keeps the rock moving." No force is necessary to keep the rock moving. Forces are needed only to *change* motion. (Remember, we are ignoring the effect of air resistance, which slows the rock down slightly with each rotation.)

The next step is to choose coordinate axes. As always, our advice is to choose one axis of the coordinate system parallel to the acceleration. In this case, the acceleration is toward the center of the horizontal circle in which the rock is moving. As shown in Figure 6.2(c), we deliberately draw the acceleration in a different color to emphasize that it is *not* a force.

The tension force is not parallel to either coordinate axis and, therefore, must be resolved in the usual manner:

$$|T_x| = T \sin(\theta) \qquad |T_y| = T \cos(\theta)$$

We are now ready to write the x- and y-components of Newton's second law:

$$\Sigma \mathbf{F_x} = m\mathbf{a}_x \qquad \Sigma \mathbf{F_y} = m\mathbf{a}_y$$

$$T \sin(\theta) = ma_x \qquad T \cos(\theta) - W = ma_y$$

Because the stone's motion is entirely in the horizontal plane, v_y is always a constant (0), thus $a_y = 0$. On the other hand, a_x is the centripetal acceleration v^2/r. With these substitutions our x- and y-equations become

$$T \sin(\theta) = m \frac{v^2}{r} \qquad T \cos(\theta) = mg \qquad \textbf{(6.5)}$$

where we have also substituted $W = mg$. Equations (6.5) are two equations in two unknowns, T and v. One easy way to solve for v is to divide the left equation by the right equation to obtain

$$\tan(\theta) = \frac{v^2}{rg}$$

This result leads to

$$v = \sqrt{rg \tan(\theta)} \qquad \textbf{(6.6)}$$

When we substitute values into Equation (6.6) we must remember that r is the radius of the circle in which the rock is traveling, and we are given only the length L

Concept Question 3
Describe what force is responsible for the centripetal acceleration in each of these cases: (a) a rock on a string whirled in a circle around a child's head, (b) a roller-coaster car at the bottom of a circular loop-the-loop, (c) a car turning a corner on a level road, (d) the moon orbiting the earth.

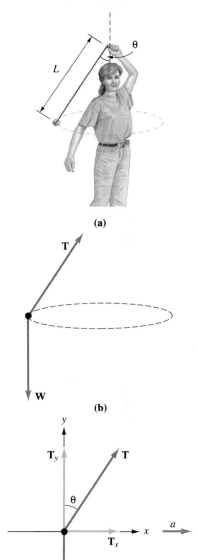

(a)

(b)

(c)

FIGURE 6.2

(a) A student whirls a stone in a horizontal circle. (b) There are only two forces acting on the stone as it travels in its circular arc. (c) The centripetal acceleration points toward the center of the horizontal circle in which the stone travels. The x-axis is chosen parallel to the acceleration. The component T_x is an unbalanced force causing the centripetal acceleration.

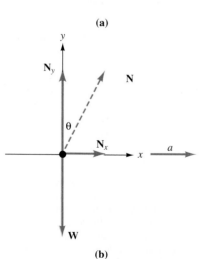

(a)

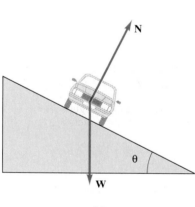

(b)

of the string. However, from the geometry of Figure 6.2(a) it is easy to see that $r = L \sin(\theta) = 0.300$ m. Thus, we obtain

$$v = \sqrt{(0.300 \text{ m})(9.80 \text{ N/kg}) \tan(36.87°)} = 1.48 \text{ m/s}$$

The total distance the stone must travel to make one circle is $2\pi r = 2\pi(0.300 \text{ m}) = 1.88$ m. The time t to travel this distance is

$$t = \frac{d}{v} = \frac{1.88 \text{ m}}{1.48 \text{ m/s}} = 1.27 \text{ s}$$

(b) Solving for the tension in the string is much simpler. We need to use only the y-components of Equations (6.5):

$$T = \frac{mg}{\cos(\theta)} = \frac{(0.100 \text{ kg})(9.80 \text{ N/kg})}{\cos(36.87°)} = 1.22 \text{ N} \quad \blacktriangleleft$$

The coefficient-of-friction model can be combined with the analysis of circular motion to describe the behavior of automobiles rounding banked curves. Example 6.3 illustrates such an analysis.

EXAMPLE 6.3   *Friction Keeps Us on the Right Track*

A section of a cloverleaf highway exit is to be laid out as a circle with radius 25.0 m. (a) What should be the banking angle θ of the roadbed so that cars traveling at 11.2 m/s (about 25 mph) need no frictional force from the tires to negotiate the turn? (b) The coefficients of friction between the tires and the road are $\mu_s = 0.900$ and $\mu_k = 0.800$. At what maximum speed can a car enter the curve without sliding toward the top edge of the banked turn?

SOLUTION   In Figure 6.3(a) we see a head-on view of a car traveling around the banked curve. The normal and weight forces are indicated. We are assuming there is no friction, so these are the only forces in the plane of the figure. (The only other forces acting on the car are perpendicular to the plane of the drawing. These forces all add to exactly zero, so that the car travels with a *constant speed.* We therefore ignore these forces while analyzing the car's motion and concentrate on only the normal and weight forces.) Figure 6.3(b) shows a free-body diagram for the car. We have indicated the forces, the acceleration (which is not a force!), and our choice of coordinate system with the x-axis parallel to the acceleration.

This time, it is the normal force which must be resolved into components with magnitudes

$$|N_x| = N \sin(\theta) \qquad |N_y| = N \cos(\theta)$$

Now we write Newton's second law in component form

$$\Sigma \mathbf{F}_x = m\mathbf{a}_x \qquad\qquad \Sigma \mathbf{F}_y = m\mathbf{a}_y$$
$$N \sin(\theta)\hat{\mathbf{i}} = ma_x\hat{\mathbf{i}} \qquad\qquad N \cos(\theta)\hat{\mathbf{j}} - W\hat{\mathbf{j}} = ma_y\hat{\mathbf{j}}$$
$$N \sin(\theta) = ma_x \qquad\qquad N \cos(\theta) - W = ma_y$$

Because the car is to travel in a horizontal circle, $a_y = 0$ and $a_x = v^2/r$. These substitutions lead to

$$N \sin(\theta) = m\frac{v^2}{r} \qquad\qquad N \cos(\theta) = mg$$

Dividing the *x*-equation by the *y*-equation, we can solve for the desired banking angle θ,

$$\tan(\theta) = \frac{v^2}{rg} = \frac{(11.2 \text{ m/s})^2}{(25.0 \text{ m})(9.80 \text{ m/s}^2)} = 0.512$$

$$\theta = \arctan(0.512) = 27.1°$$

(b) Up to now we have assumed that no friction aids the car as it rounds the turn. If the driver goes faster than the design speed of 11.2 m/s, a frictional force must act parallel to the road and inward toward the center of the turn. This situation is shown in Figure 6.3(c). The frictional force must be resolved into components. As you can see from Figure 6.3(d) we have,

$$\left|F_{\text{fric}_x}\right| = F_{\text{fric}} \cos(\theta) \qquad \left|F_{\text{fric}_y}\right| = F_{\text{fric}} \sin(\theta)$$

Next, these force components must be included when we write Newton's second law in component form:

$$\Sigma \mathbf{F}_x = m\mathbf{a}_x \qquad\qquad \Sigma \mathbf{F}_y = m\mathbf{a}_y$$

$$N \sin(\theta) + F_{\text{fric}} \cos(\theta) = m\frac{v^2}{r} \qquad N \cos(\theta) - F_{\text{fric}} \sin(\theta) - W = 0$$

For a tire rolling without slipping, there is no relative motion between the road and that part of the tire in contact with the road. Hence, it is the *static* frictional force we need to evaluate. Since we are looking for the maximum speed with which we can round the corner, we set $F_{\text{fric}} = F_{\text{fric}_{\text{max}}} = \mu_s N$ to obtain

$$N[\sin(\theta) + \mu_s \cos(\theta)] = m\frac{v^2}{r} \qquad N[\cos(\theta) - \mu_s \sin(\theta)] = mg$$

Once again, we can eliminate *N* by dividing the *x*-equation by the *y*-equation to obtain

$$\frac{\sin(\theta) + \mu_s \cos(\theta)}{\cos(\theta) - \mu_s \sin(\theta)} = \frac{v^2}{rg}$$

which gives, for *v*

$$v = \sqrt{\frac{\sin(\theta) + \mu_s \cos(\theta)}{\cos(\theta) - \mu_s \sin(\theta)} rg}$$

$$= \sqrt{\frac{\sin(27.1°) + (0.900) \cos(27.1°)}{\cos(27.1°) - (0.900) \sin(27.1°)}(25.0 \text{ m})(9.80 \text{ m/s}^2)}$$

$$= 25.3 \text{ m/s} \approx 57 \text{ mph}$$

We leave it as an exercise for you to calculate the slowest speed with which a car can round this curve without sliding down, toward the center of the curve. (See Problem 17.) ◀

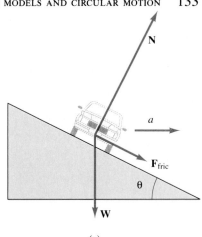

(c)

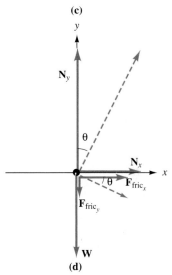

(d)

FIGURE 6.3

A car travels around a banked curve. (a) If the car travels at just the right speed, only weight and normal forces act in the plane of the figure. The normal force shown is actually the sum of four normal forces acting on the tires. Here we model the car as a single point–object. (b) The *x*-axis is chosen to be parallel to the centripetal acceleration, which points toward the center of the horizontal circle in which the car travels. The normal force *N* is resolved into components parallel to the coordinate axes. (c) When the car travels faster than the design speed for the given banking angle, a frictional force acts to hold the car in its circular path. (d) Both frictional force and the normal force must be resolved into components.

EXAMPLE 6.4   *Putt-Putt Physics*

A certain par-three hole at the local miniature golf course starts when the 0.0460-kg ball must be sent around a double vertical loop with radius 0.600 m as shown in Figure 6.4(a). Ignore the small frictional force between the track and the ball, and apply the point-particle model to determine (a) the normal force between the track and the ball at the top of the first loop where the ball is traveling 3.00 m/s. (b) What is the speed the ball must have at the top of the second loop so it is just barely in contact with the surface? (c) Find the force between the track and the ball at the bottom of the first loop where the ball's speed is 5.50 m/s.

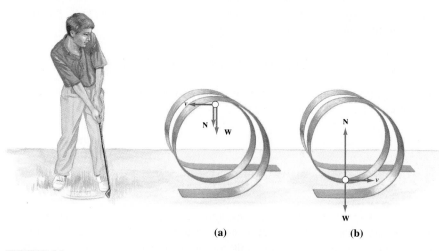

(a)   (b)

FIGURE 6.4

A golf ball is to travel around a double, circular loop. (a) At the top of the loop the normal force of the track and the weight force are in the same direction. (b) At the bottom the normal force and the weight force on the ball are oppositely directed.

SOLUTION   Although the motion is not *uniform* circular motion, we can always apply Newton's second law. In general there are two components to the ball's acceleration—one normal to its trajectory and one tangent to its trajectory. The acceleration component normal to the trajectory is always v^2/r. As the ball moves upward in the arc, it slows, and thus the tangential acceleration is directed antiparallel to the ball's velocity. As the ball speeds up in the downward arc, the tangential acceleration is parallel to the velocity vector. At the top and bottom the tangential acceleration is zero.

(a) Figure 6.4(a) shows the ball and the forces acting on it when it is at the top of the first loop. These forces are the weight force W and the normal force N. Note that the direction of the normal force is downward. We are analyzing the *ball,* and the force on the ball by the track is downward. Ignoring friction, these are the only forces acting on the ball. As you can see from Figure 6.4(a), at the top of the circular motion the weight force and the normal force act in the same direction. Hence, we write Newton's second law as

$$\Sigma \mathbf{F} = m\mathbf{a}$$

$$-W - N = ma$$

But a is the centripetal acceleration, which at this moment points downward and thus can be written $a_c = -v^2/r$. With this substitution, we may solve for the normal force:

$$-W - N = m\left(-\frac{v^2}{r}\right)$$

$$N = m\frac{v^2}{r} - W$$

$$= 0.0460 \text{ kg} \frac{(3.00 \text{ m/s})^2}{0.600 \text{ m}} - (0.0460 \text{ kg})(9.80 \text{ m/s}^2)$$

$$= 0.239 \text{ N}$$

(b) We interpret the phrase ''just barely in contact'' to mean that the normal contact force reduces to zero. Thus, the expression

$$N = m\frac{v^2}{r} - W$$

developed in part (a) implies

$$0 = m\frac{v^2}{r} - mg$$

so that

$$\frac{v^2}{r} = g$$

and

$$v = \sqrt{rg}$$
$$= \sqrt{(0.600 \text{ m})(9.80 \text{ m/s}^2)} = 2.42 \text{ m/s}$$

(c) As shown in Figure 6.4(b), at the bottom of the ball's motion the normal force is upward, whereas the weight force is still downward. When the ball is in its lowest position, the centripetal acceleration $a_c = v^2/r$ toward the center of the circular motion is upward and therefore positive on our coordinate system. For this new geometry Newton's second law gives

$$+N - W = ma = m\left(+\frac{v^2}{r}\right)$$

so that

$$N = m\frac{v^2}{r} + W$$

$$= m\left(\frac{v^2}{r} + g\right)$$

$$= (0.0460 \text{ kg})\left(\frac{(5.50 \text{ m/s})^2}{0.600 \text{ m}} + 9.80 \text{ m/s}^2\right)$$

$$= 2.77 \text{ N} \qquad \blacktriangleleft$$

We end this section by reiterating an important point. When you solve problems for objects traveling in a circle *don't invent a new force* to be the centripetal force. Just choose from the list of old standbys and realize that whichever you use, these forces must add up so as to leave an unbalanced force toward the center of the circular motion.

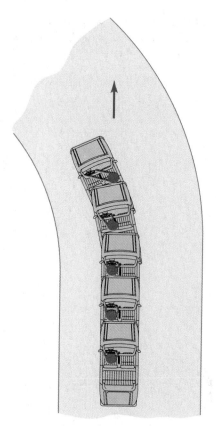

FIGURE 6.5

A bird's-eye view of a car rounding a corner at constant speed. The driver travels in a straight line but thinks he is being thrown to the outside of the curve.

Feeling the Physics: Pseudo Forces in Accelerated Frames of Reference

You may be having trouble reconciling the net *inward* force required for circular motion as discussed in the previous section with an outward (centrifugal) "force" you have felt, for example, while traveling in a car rounding a sharp corner. We all know that when the car goes around a corner, it certainly *feels* like there is an outward force on us. If we take a "bird's-eye view" of the situation as illustrated in Figure 6.5, we see things differently. From this top view we see that you (represented by the dot) are trying to move in accordance with Newton's first law. That is, you are trying to move in a straight line with constant speed. When the wheels of the car turn, however, the frictional force of the road pushes on them and the car turns so that its path is a circle. Now, if you are moving in a straight line and you continue to move in a straight line while the car turns to the left, you soon find yourself on the right side of the car! Newton's first law says that you travel in a straight line until some force pushes or pulls you into the circular arc. For example, if you are driving, you may hold on to the steering wheel and let it pull you toward the left; that is, toward the center of the circle. If you are a passenger on the right side of the car, then the car door may push you to the left.

This explanation is completely consistent with our analysis of circular motion and the required net inward force. But it's easy to forget this when you are riding inside a car. There, the car is your frame of reference. When you find yourself sliding toward the right door, you tend to think of yourself as having been pushed against it. It's hard to imagine that the door has moved over and is pushing on you. Doors don't usually do things like that!

The whole problem, as you by now realize, is that you are judging your motion in a frame of reference (the car) that is itself accelerating toward the center of its curved path. You, in the car, believe that it is you who is accelerating—in the opposite direction. Hence, you think there is a force on you. Often, this fictitious force is called the centrifugal force.

The feeling of a fictitious force occurs in any accelerating frame of reference. When the car accelerates forward, you feel pressed backward. If you wanted, you could also invent a force to explain this pressing effect. These invented, or "pseudo," forces that arise in accelerated frames of reference feel very much like an extra gravitational force. In fact, one of the tenets of Einstein's general theory of relativity is that you cannot possibly tell the difference between being in an accelerated frame of reference and being in a gravitational field. Engineers have proposed that by spinning space stations, "artificial gravity" can be introduced into environments in which astronauts will spend long periods of time. (See Problem 22.)

It is not uncommon, particularly in engineering classes, to put oneself in an accelerated reference system and use the fictitious forces as if they were real. Despite Einstein's admonition, it is traditional in most physics classes not to do this; and we advise that you not do it either. The danger is that you might start introducing fictitious forces into force diagrams when you are *not* in an accelerating frame of reference. Then you're in trouble. (Nonetheless, we couldn't restrain ourselves from including a couple of problems for which it is terribly convenient to go into the accelerating frame and consider an artificial gravity in the opposite direction of the centripetal acceleration. (See Problems 21(b) and 22.)

6.3 Newton's Universal Law of Gravitation

Up to this point, we have looked at the weight force from a rather parochial point of view; that is, from the view of creatures who live on the surface of the earth. In this section we expand our outlook and consider the gravitational force in a more general context. The central point is the introduction of a "universal" law of gravitation, first deduced by

Newton. With it, and his three laws of motion introduced previously, Newton was able to do something rather remarkable for his time. He was able to describe the motion of the moon and the planets with the very same laws of physics that are used to describe the motion of objects on the earth's surface.

You have probably noticed that in this chapter, as well as the last, we always return to a single problem-solving method and a single, central equation: Newton's second law. Scientists in general, and perhaps physicists in particular, delight in this kind of economy of thought. The ability to explain as much of the universe as possible from as few starting principles as possible is a cultural value held by every physicist we know. It is no surprise then that Newton's success in uniting heaven and earth with a single set of physical laws is heralded as one of the greatest intellectual achievements in history. In this brief section we can only hint at the history behind the **Great Newtonian Synthesis.**

In 1665, when the black plague swept through England, Newton left Trinity College of Cambridge University where he was a student. He returned to his boyhood home in Woolsthorpe, Lincolnshire. It was probably there that he ". . . began to think of gravity extending to the orb of the moon, and . . . compared the force requisite to keep the moon in her orb with the force of gravity at the surface of the Earth, and found them to answer pretty nearly." [2] Let's use the physics you learned in the last few chapters to discover what Newton was talking about and, moreover, lead us to his universal law of gravitation.

For simplicity, we'll take the moon's orbit to be a circle around the earth. (It's actually an ellipse, but its deviation from a circle is very small.) We also need some astronomical data, such as the radius of the earth and the distance from the earth to the moon. You can find these and similar data in Appendix 8. These distances were known "pretty nearly," even in 1665. Perhaps Newton's line of reasoning went something like this. If the moon travels in a circle, then it has a centripetal acceleration pointing toward the center of its orbit. This centripetal acceleration is easy to calculate from the expression

$$a_c = \frac{v^2}{r}$$

where $v = 2\pi r/T$ and $r = 3.84 \times 10^5$ km, and $T = 27.3$ days $= 2.36 \times 10^6$ s. Substituting these values we obtain $a_c = 0.00272$ m/s^2. Now, the gravitational field strength near the surface of the earth is 9.80 N/kg = 9.80 m/s^2. Clearly this is much too large compared with 0.00272 m/s^2. If gravity does extend "to the orb of the moon," it must grow weaker on the way. The radius of the moon's orbit is about 60 times the earth's radius. Suppose we guess that gravity grows weaker in inverse proportion to distance. Then the gravitational field strength is (9.80 m/s^2)/60 = 0.16 m/s^2, still much too large compared with 0.00272 m/s^2. We have to make gravity even weaker, so let's try an inverse square: (9.80 m/s^2)/(60)^2 = 0.00272 m/s^2. Awesome! Surely, this result is no accident.

We now can guess how gravitation depends on distance; namely, as an inverse square. We also know that our weight is proportional to our mass. On the other hand, we learned from Newton's third law that we are also attracting the earth with a force of equal magnitude. Therefore, it seems that the gravitational force of attraction is proportional to the earth's mass too. These dependencies can be written as an equation if we introduce a proportionality constant:

$$F_{\text{grav}} = G \frac{mM}{R^2} \tag{6.7}$$

The proportionality constant G is a universal constant 6.67×10^{-11} N $\cdot$ m^2/kg^2. Don't confuse G with g. Big G is a universal constant; its value is the same everywhere in the

Isaac Newton (1642–1727) wearing the wig and robes of a student at Trinity College, Cambridge, which he entered in 1661 at the age of eighteen. In 1672, he published *Theory of Light and Color.* A bitter controversy surrounding his theories left him unwilling to publish again. Despite this decision his friend, astronomer Edmond Halley, persuaded Newton to publish his works on mechanics (Halley even assumed responsibility for publication costs). The subsequent publication of the *Principia* in 1687 established Newton as the greatest scientist of his time, indeed perhaps all time. Newton was also well versed in the chemistry of metals and served as Master of the British mint. He was knighted in 1705 by Queen Anne.

---

[2] Sir Isaac Newton, as quoted in F. J. Rutherford, G. Holton, and F. G. Watson, *The Project Physics Course* (New York: Holt, Rinehart and Winston, 1970) Unit II: 86.

Concept Question 5
If the distance between the centers of two spherical masses is tripled, by what factor is the gravitational attraction between these masses changed?

Concept Question 6
(a) If the average density of the earth stays constant, by what factor must the radius of the earth increase so that the mass of the earth increases by a factor of 8? (b) If this change occurred, by what factor would the surface gravity of the earth change?

universe. Little g is just the local gravitational field strength here on the earth's surface. We discuss the relation between g and G a little later in this section.

Equation (6.7) is a statement of **Newton's universal law of gravitation** (NULG). Newton asserted that every object in the universe attracts every other object in the universe with a force that can be calculated using NULG. If the objects are spherical, the distance between them should be measured from their *centers*. If the objects are not spherical, the situation is more complex. In order to compute the net force of one mass distribution due to another, we must perform an integration over each pair of points that make up the two objects. (See Problem 25.)

Unless specifically stated otherwise, we continue to use our point-particle model of the objects we consider. However, such a model is really valid only if the dimensions of the actual object are small compared to the distance between the objects. For example, if we consider the gravitational force between our bodies and the earth, the proper distance in Equation (6.7) is from the center of the earth to some point on our body. Because the radius of the earth is quite large compared to our size, the value of R does not change noticeably regardless of whether we take the distance to our head or to our feet.

EXAMPLE 6.5   *So That's Why I Missed That Strike!*

Calculate the gravitational attraction between two 16.0-lb bowling balls that are touching. Take the circumference of the bowling balls to equal the legal maximum of 27.0 in.

SOLUTION   Using $2\pi r$ for the circumference, we find the radius of a bowling ball to be 0.109 m. The mass is found from

$$m = \frac{W}{g} = \frac{16.0 \text{ lbf}}{32.2 \text{ lbf/slug}} = 0.497 \text{ slug} = 7.25 \text{ kg}$$

Remembering that the R in NULG is the distance between centers, we substitute into Equation (6.14) to obtain

$$F_{\text{grav}} = G\frac{mM}{R^2} = (6.67 \times 10^{-11})\frac{(7.25 \text{ kg})(7.25 \text{ kg})}{(0.218 \text{ m})^2} = 7.38 \times 10^{-8} \text{ N}$$

A newton is about the weight of a Big Mac, therefore, it seems unimaginable that one could ever measure the force of gravitational attraction between two bowling balls. This measurement has, however, been made as described in the following section. ◀

The Cavendish Balance

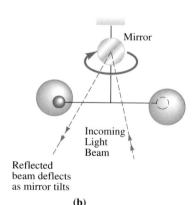

(a)

(b)

FIGURE 6.6

Cavendish balance. The forces of attraction drawing the smaller masses (m) toward the larger masses (M) cause the supporting fiber to twist. The mirror tilts, causing the reflected light beam to deflect.

In order to determine the magnitude of the universal constant G, it is necessary to measure all the other quantities in Newton's universal law of gravitation (NULG). Hence, we must measure the gravitational attraction between laboratory-sized objects. We illustrated in Example 6.5 that the force between these objects is quite small, and, therefore, measuring G requires special experimental techniques. The gravitational force was first measured by Lord Henry Cavendish in 1798, using what is now called a **Cavendish balance** (illustrated in Figure 6.6). It consists of two small lead balls attached to the ends of a light rod to form a dumbbell. This dumbbell is suspended from a thin quartz fiber to which a small mirror is attached. A pencil-wide beam of light is reflected off the mirror to a distant wall, where the bright spot is observed. A second pair of massive spheres is brought near the suspended pair in such a way that the gravitational attraction of the second pair on the first pair twists the quartz fiber. This twist causes the mirror to rotate, and the position of the reflected light spot moves.

The twist of the quartz fiber is similar to the spring force we described in the previous chapter. When a spring is stretched, it exerts a force that tends to return the spring to its original length. When the fiber is twisted, it exerts forces that tend to return the fiber to its untwisted condition. The extent to which the fiber can exert a restoring twist is characterized by a twist constant (torsional constant, in the official jargon), which can be measured beforehand. Thus, from the angle turned by the mirror, the gravitational force can be measured. Consequently all the variables in NULG are known except G, which we can now calculate.

"Weighing" the Sun

The tables in Appendix 8 include such data as the masses of the sun, the earth, and the other planets. Did you ever wonder where these numbers come from? How do we know the mass of the sun? It certainly wasn't measured with a pan balance! Actually, in a real sense, it was measured with a Cavendish balance, because now that we know G, we can calculate the mass of the sun. Indeed, we may find the mass of any object with an orbiting satellite.

To calculate the mass of the sun we (once again) use Newton's second law, this time to analyze the earth in its orbit around the sun.

$$\Sigma \mathbf{F} = m_{\text{earth}}\mathbf{a} \tag{6.8}$$

The net force in this case is the gravitational force of the sun on the earth, and is given by NULG. The acceleration is the centripetal acceleration, $v^2/R = 4\pi^2 R/T^2$, where R is the radius of the earth's orbit, and T is 1 year. This time T for a planet to make one complete revolution around the sun is called the **period** of the planet's orbit. Thus, Equation (6.8) can be written

$$G\frac{m_{\text{earth}}M_{\text{sun}}}{R^2} = m_{\text{earth}}\frac{4\pi^2 R}{T^2} \tag{6.9}$$

Canceling the mass of the earth, which appears on both sides of Equation (6.9), we find

$$\frac{T^2}{R^3} = \frac{4\pi^2}{GM_{\text{sun}}} \tag{6.10}$$

or, solving for the mass of the sun,

$$M_{\text{sun}} = \frac{4\pi^2 R^3}{GT^2} \tag{6.11}$$

From Equation (6.11) we can obtain the mass of the sun if we know the universal constant G, the period of the earth's orbit, and the radius of the earth's orbit around the sun.

You should realize that although we derived Equation (6.11) for the sun's mass based on the earth's orbit, a completely analogous equation holds for any body with an orbiting satellite. That is, we can find the earth's mass by using the orbit of the moon (or any of our numerous artificial satellites). We can find the mass of Jupiter by analyzing its moons, and so forth.

Equation (6.10) is interesting from a historical viewpoint. Although we derived it for the earth's orbit around the sun, the mass of the earth canceled out. Indeed, we would have obtained precisely Equation (6.10) had we been analyzing the orbit of Mars or Saturn or any of the sun's satellites. In other words, for every planet in our solar system the quantity T^2/R^3 has the same value, a result first published by Johannes Kepler in 1619. The statement, that T^2/R^3 is a constant, is known as Kepler's law. (Oftentimes, it is called **Kepler's third law.**) Kepler discovered this law essentially by trial and error. He used data of planetary positions carefully observed and recorded by Danish astronomer Tycho Brahe. For Kepler, the right-hand side of Equation (6.10) was just a number. He did not under-

Concept Question 7
An artificial satellite is in a circular orbit around the earth such that it takes 1 hour to complete its orbit. (a) If it is propelled into an orbit with a radius twice as great, does it take the satellite more or less time to complete an orbit? (b) By what factor does the time for one orbit change?

stand why this rule should hold. Such a law, which describes a regularity in nature but has no known theoretical explanation, is called an **empirical law.** Newton was aware of Kepler's third law but was able to deduce this law independently from his universal law of gravitation. This derivation no doubt convinced Newton of the correctness of his inverse-square law. The interplay between Brahe's observations, Kepler's empirical law, and Newton's theory is typical of the way science has operated for the past 400 years.

Weight and Universal Gravitation

We now have two alternative equations from which we can compute the weight of an object on the surface of the earth; we may use mg or Newton's universal law of gravitation (NULG). Because on a nonrotating earth these two equations must produce equivalent results, we may equate the expressions for each. From the result we learn what determines the local gravitational field strength. Setting mg equal to the force from NULG, we have

$$mg = G \frac{M_{earth} m}{(R_{earth})^2}$$

which leads to

$$g = G \frac{M_{earth}}{(R_{earth})^2} \tag{6.12}$$

From Equation (6.12) we can observe the factors that determine the gravitational field strength here on the earth's surface; namely, G, M_{earth}, and R_{earth}. (Remember that the value of g being discussed here is that due to gravity alone. The effective g at the earth's surface includes a small correction due to the earth's rotation.) We also see that gravity grows weaker as the distance from the earth's surface increases. Therefore, g is weaker on the tops of mountains than at their bases. To see how quickly gravity falls off with distance from the surface of the earth, we need to find the rate of change of g with increasing R. If you think that sounds like a calculus problem, you're exactly right. The derivative we need is

$$\left. \frac{dg}{dR} \right|_{R=R_{earth}} = \frac{d}{dR} \left(\frac{GM_{earth}}{R^2} \right)_{R=R_{earth}} = -\frac{2GM_{earth}}{(R_{earth})^3} \tag{6.13}$$

which can be rewritten

$$\left. \frac{dg}{dR} \right|_{R=R_{earth}} = -\frac{2g}{R_{earth}} \tag{6.14}$$

For small changes in R (that is, small with respect to the radius of the earth), the corresponding change in g can be calculated from

$$\Delta g = \frac{dg}{dR} \Delta R$$

Therefore, near the surface of the earth

$$\Delta g = -\frac{2g}{R_{earth}} \Delta R \tag{6.15}$$

We show how to apply this result in Example 6.6.

EXAMPLE 6.6   *A Sure-Fire Weight-Loss Program*

Calculate the change in gravitational field strength with a 1.00-km increase in altitude. How much weight does a 100.-lb person lose by going to this altitude?

SOLUTION   We find the change in g from Equation (6.15):

$$\Delta g = -\frac{2g}{R_{\text{earth}}} \Delta R$$

$$= -2\left(\frac{9.80 \text{ N/kg}}{6.38 \times 10^6 \text{ m}}\right)(1.00 \times 10^3 \text{ m}) = -0.00307 \text{ N/kg}$$

Because weight is mg, the fractional weight loss is equal to the fractional change in g (which is also equal to twice the magnitude of the fractional change in R, as shown by Equation (6.15)).

$$\text{Fractional weight change} = \frac{\Delta g}{g} = \frac{-0.00307}{9.80} = -3.13 \times 10^{-4}$$

A person with a low-altitude weight of 100 lb thus weighs 0.031% less with an elevation increase of 1.00 km, that is, 0.031 lb less. But, of course, that person's mass does not decrease at all.   ◀

You no doubt have seen movies that show astronauts in orbit around the earth as they float about their capsules in a "weightless" state. You may have read about many experiments performed on the Space Shuttle in "zero gravity." In Example 6.7 we check to see if orbiting astronauts are really free of the earth's gravity. You will see that, like the projectile in Figure 6.7, the astronauts are not free of gravity, but rather in a state of continuous free fall toward the earth!

EXAMPLE 6.7   *How Can You Be Weightless in Gravity?*

A typical satellite orbit may have an altitude of 1.00×10^6 m (620 miles) above the earth's surface. Calculate the value of g at this altitude and reconcile the answer to claims of the "weightlessness" of astronauts in orbit.

SOLUTION   The satellite is less than one earth radius above the earth, but this altitude is not small enough to use the differential approximation of Equation (6.15). We must instead use Equation (6.12) with the radius of the earth (6.38×10^6 m) replaced by the distance of the satellite from the center of the earth: $R = R_{\text{earth}} + \textit{Altitude above earth} = 7.27 \times 10^6$ m.

$$g = G \frac{M_{\text{earth}}}{R^2}$$

$$= \frac{(6.67 \times 10^{-11} \text{ N m}^2/\text{kg}^2)(5.97 \times 10^{24} \text{ kg})}{(7.38 \times 10^6 \text{ m})^2} = 7.31 \text{ N/kg}$$

We see that although the gravitational field strength is 25% lower than at the earth's surface, it is far from zero! Why then do astronauts at these altitudes float about the cabin and appear to be weightless?

The astronauts are not "weightless" because the weight force mg acting on them is not zero. However, they do have zero *apparent weight* in the sense of Example 5.5 in the previous chapter. If you drop your physics book, it accelerates to the ground at 9.80 m/s². But suppose both you and your physics book are in a very long elevator that is itself accelerating downward at 9.80 m/s². If you release the book, it does not move relative to your hand. Your book, your hand, and the floor are accelerating downward at the same rate. Hence, there is no relative motion. If your book is

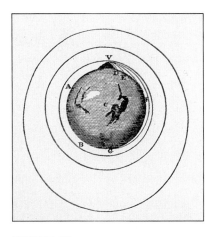

FIGURE 6.7

A drawing from Newton's *The System of the World* showing the trajectory of projectiles fired horizontally with different speeds. When the launch speed is large enough the projectile "falls" continuously into a closed orbit.

Orbiting astronauts with zero "apparent weight" may have a large weight force mg acting on them but be in a continuous state of falling toward the earth.

attached to your hand by a spring, the spring does not need to stretch to "hold" the book stationary relative to your hand. Your book (and you) appear weightless. In fact, astronauts can practice weightlessness in airplanes that are "flying" in a free-fall trajectory. (See Problem 4.21.)

The astronauts have zero apparent weight because they are in a state of continual free fall toward the earth. This seems to raise another paradox. If they're in free fall, why don't they get closer to the earth? The answer is that their forward orbital motion carries them tangentially away from the earth, but the gravitational force causes them to "fall" toward the center of the earth. It is this combination of competing effects that results in circular motion; as the astronauts orbit the earth, the amount they fall is just matched by the curvature of the earth.   ◄

6.4 Models for Motion Through a Resistive Medium (Optional)

Until now, our problems and examples ignored the effect of air resistance on the motion of objects. We deliberately picked situations in which these effects are small, talking about basketballs and bunts, not badminton and bullets. In succeeding paragraphs we describe the drag forces on simple objects, such as spheres. These drag forces turn out to be relatively complicated, but we find two classes of motion in which the drag can be described with fairly simple models. We use these two models in two examples to estimate the size of the drag effects.

The magnitude of the **drag force** D on an object traveling through a resistive medium can be described by the expression

$$D = \frac{1}{2}\, C_D A \rho v^2 \qquad (6.16)$$

Here A is the cross-sectional area normal to the object's path, ρ is the density of the medium through which the object is traveling, v is the object's speed, and C_D is the drag coefficient. If C_D were constant in the above expression, drag force calculations would be fairly simple. The bad news is that C_D is, in general, not constant. It depends in a complicated way on the object's shape and velocity as well as the density and dynamic **viscosity** of the medium through which the object is moving. If you are not familiar with the term *viscosity,* you need not be overly concerned. We will introduce a careful definition of viscosity in Chapter 16. For the time being it is sufficient to think of viscosity η (eta) as a number that characterizes the internal friction of the medium. Cold honey is quite viscous; warm honey is much less viscous. A short table of viscosities is given here for convenience. More complete data are located in Chapter 16.

TABLE 6.2 Dynamic Viscosity of Common Substances at 24°C and Atmospheric Pressure

SUBSTANCE	DYNAMIC VISCOSITY $(N \cdot s/m^2)$
Hydrogen	9.6×10^{-6}
Air	1.8×10^{-5}
Gasoline	2.9×10^{-4}
Water	9.6×10^{-4}
Mercury	1.5×10^{-3}
Glycerine	9.5×10^{-1}

The good news is that the drag coefficient depends on the variables mentioned only in a certain dimensionless combination. This combination is known as the **Reynolds number** after the British engineer Osborne Reynolds (1842–1912). The Reynolds number Re is defined as

$$Re = \frac{L\rho v}{\eta} \qquad (6.17)$$

where L is some characteristic dimension of the object, for example, the diameter in the case of a sphere. A small value for the Reynolds number means slow motion through a stiff medium; a large Reynolds number indicates rapid motion through a slippery medium.

Figure 6.8 shows how the drag coefficient for a sphere depends on the Reynolds number. Note that the behavior is quite complex overall. There are, however, two regions in which the behavior is relatively simple. One case is for $Re < 1$, in which the drag coefficient is nearly linear in $1/Re$. The other is for the region $1000 < Re < 300\,000$, in which the coefficient is nearly constant at about 0.5. We develop different models for these two regions in the following paragraphs.

For $Re < 1$, it can be shown from Figure 6.8 that $C_D = 24/Re$. Substituting this result into the expression for the drag force we get

$$D = 6\pi\eta r v \qquad (6.18)$$

where r is the radius of the sphere. Equation (6.18) is known as **Stokes' law** after Sir George G. Stokes (1819–1903) who first derived it. It is applicable in situations of slow motion through viscous fluids.

For spheres moving with Reynolds number between 1000 and 300 000, we can use Equation (6.16) as it stands and take $C_D = 0.500$. We refer to this as the **quadratic model** of fluid resistance. Objects that are not spheres have different values of the drag coefficient C_D. In addition, the range of Reynolds numbers over which the quadratic model is appropriate varies with shape. Example 6.8 illustrates how to determine which of these two models to use.

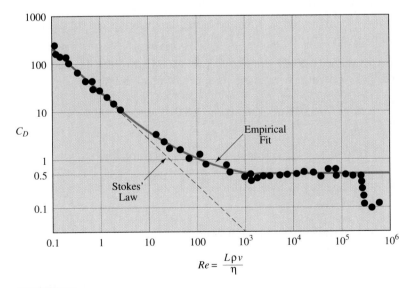

FIGURE 6.8

The drag coefficient of a smooth sphere as a function of Reynolds number. See Problem 6.45 for the equation of the empirical fit denoted by the solid line. Note the use of logarithmic scales for the axes.

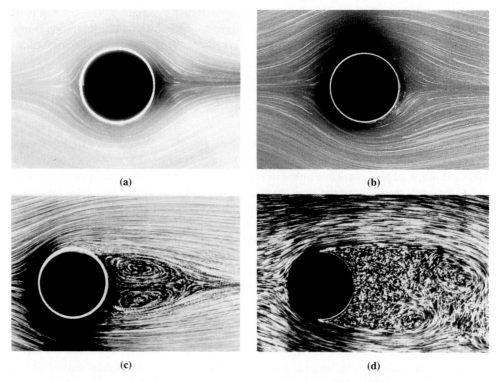

(a) Uniform flow past a circular cylinder at $Re = 0.16$. The flow pattern is similar to that of a smooth sphere. At low Reynolds number it is difficult to tell that the flow is from left to right. This is the region where Stokes' law is valid. (b) Flow past a circular cylinder at $Re = 1.54$. The flow is no longer symmetric fore and aft. Streamlines are made visible by aluminum powder in water. (c) Flow past a circular cylinder at $Re = 26$. Standing eddies form around $Re = 9$ and grow longer with increasing Reynolds number until the flow becomes unstable around $Re = 40$. (d) Circular cylinder at $Re = 2000$. At this Reynolds number the boundary layer is laminar over the front, separates, and breaks up into a turbulent wake. The drag coefficient is approximately constant until $Re \approx 300\,000$ when the boundary layer becomes turbulent at the separation point.

EXAMPLE 6.8   *Calculating Reynolds Number*

Calculate the Reynolds number for (a) a marble falling in glycerine and (b) a softball on the way to home plate.

SOLUTION   (a) The density of glycerine at 24°C is 1260 kg/m^3. For the marble we take the diameter L as 1.5 cm and estimate its speed at about 1.0 cm/s:

$$Re = \frac{L\rho v}{\eta} = \frac{(0.015 \text{ m})(1260 \text{ kg/m}^3)(0.010 \text{ m/s})}{0.95 \text{ N} \cdot \text{s/m}^2}$$

$$= 0.20 < 1 \qquad \text{We would use Stokes' law for drag.}$$

(b) For a 9.5-cm diameter softball with a velocity of, say, 30 m/s in air ($\rho = 1.2$ kg/m^3), we calculate

$$Re = \frac{(0.095 \text{ m})(1.2 \text{ kg/m}^3)(30 \text{ m/s})}{1.8 \times 10^{-5} \text{ N} \cdot \text{s/m}^2} = 1.9 \times 10^5$$

Because $1000 < Re < 3 \times 10^5$, the quadratic model is applicable to the softball's motion.   ◀

For macroscopic objects traveling in air, Example 6.8 shows that the quadratic model is usually the model of choice, unless the object's dimensions and speed are extremely small. We present the solution for free fall in the one-dimensional case here. In Section 6.7

we present numerical techniques that can be used to correct a projectile's trajectory for air resistance using the quadratic model. (See Problem 47.)

For the case of free fall we continue to ignore the small buoyant effect of air (see Chapter 16) and model air friction by using the quadratic model. Taking a coordinate system with the upward direction as positive and assuming the object is to fall from rest, the net force on the object is

$$\Sigma F = -mg + \frac{1}{2} C_D A \rho v^2 \qquad (6.19)$$

The first thing to notice about Equation (6.19) is that, if v increases sufficiently, a speed is reached at which the weight force is balanced by the force of the air resistance, making the total force zero. After this speed is attained, there is no further increase in speed; the object has reached its **terminal velocity** v_T. We can find this terminal velocity by setting the net force in Equation (6.19) to zero and solving for v_T. The result is

$$v_T = \sqrt{\frac{2mg}{C_D A \rho}} \qquad (6.20)$$

After a sufficiently long time, the falling object has a velocity very close to $v = -v_T$. The acceleration can be written compactly in terms of the terminal velocity by substituting its definition from Equation (6.20) into Equation (6.19) and dividing both sides of the resulting equation by the object's mass m:

$$a = g\left(-1 + \left(\frac{v}{v_T}\right)^2\right) \qquad (6.21)$$

Since $a = dv/dt$, Equation (6.21) involves both v and its derivative. Such an equation is known as a **differential equation.** When we assume that $v = 0$ at $t = 0$, the solution to this differential equation is

$$v = -v_T \tanh\left(\frac{gt}{v_T}\right) \qquad (6.22)$$

where

$$\tanh(x) = \frac{e^x - e^{-x}}{e^x + e^{-x}}$$

The function $\tanh(x)$ is the **hyperbolic tangent,** a function included on most modern scientific calculators.

Concept Question 8
(a) If two raindrops, one with a large diameter and the other with a small diameter, fall from a cloud, which has the greater terminal velocity? (b) If both raindrops have the same speed, which has the larger acceleration?

EXAMPLE 6.9 *Correcting for Air Resistance*

A spherical water balloon of radius 10.0 cm is dropped from the top of a tall building. (a) Calculate the balloon's terminal velocity. (b) Calculate the velocity of the balloon 2.00 s after release. (c) Compare the answer in part (b) to that obtained by using the constant-acceleration model.

SOLUTION   (a) For a sphere we take $C_D = 0.500$. The density of dry air is 1.20 kg/m³ at 20°C and atmospheric pressure. The mass of the water can be calculated from the balloon's volume and the density ρ_w of water: $m = \rho_w(4/3)\pi r^3 = (1.00 \times 10^3 \text{ kg/m}^3)(4/3)\pi(0.100 \text{ m})^3 = 4.19 \text{ kg}$.

$$v_T = \sqrt{\frac{2mg}{C_D A \rho}} = \sqrt{\frac{2(4.19 \text{ kg})(9.80 \text{ N/kg})}{(0.5)(0.0314 \text{ m}^2)(1.20 \text{ kg/m}^3)}} = 66.0 \text{ m/s}$$

(b) The velocity at any time t can be calculated directly from Equation (6.22):

$$v = -v_T \tanh\left(\frac{gt}{v_T}\right)$$

$$= -(66.0 \text{ m/s}) \tanh\left(\frac{(9.80 \text{ N/kg})(2.00 \text{ s})}{66.0 \text{ m/s}}\right) = -19.0 \text{ m/s}$$

(c) The constant-acceleration model predicts that after 2.00 s the velocity is

$$v = v_0 - a_g t = 0 - (9.80 \text{ m/s}^2)(2.00 \text{ s}) = -19.6 \text{ m/s}$$

This speed is about 3% larger than the answer in part (b), which takes air resistance into account.

You may have noticed that we used the quadratic model for an object starting from rest. Certainly for a short time after release, when the velocity was quite small, the Reynolds number was less than 1000. Hence, the quadratic model was not valid for this portion of the fall. We leave it as a problem for you to estimate the size of the resulting error. (See Problem 40.)   ◀

6.5 Summary of Force Models

In this chapter and the last we introduced a number of simple models for the forces we commonly encounter. Table 6.3 summarizes these models.

TABLE 6.3   Force Models

NAME	FORMULA	COMMENTS
Weight	$W = mg$	body force; $g = 9.80$ N/kg near earth's surface
Normal	N, solve for each time	contact force; perpendicular to contact surface
Friction	$F_{\text{fric}} = \mu_k N$	kinetic friction; parallel to contact surface, opposes relative motion
	$F_{\text{fric}} \leq \mu_s N$	static friction, $\mu_s > \mu_k$; parallel to contact surface, opposes attempted relative motion between surfaces
Fluid drag	$D = 6\pi\eta r v$	Stokes' law; for $Re < 1$
	$D = \dfrac{1}{2} C_D A\rho v^2$	quadratic model; for $1000 < Re < 300\ 000$; for spheres $C_D \simeq 0.5$
Springs	$F_s = ks$	ideal Hooke's law spring only; force is opposed to extension ($\mathbf{F}_s = -k\mathbf{s}$)
Tension	T, solve for each time	you can't push with a rope
Gravity	$F = G\dfrac{mM}{R^2}$	Newtonian gravity, universal law

6.6 Numerical Methods for Newton's Second Law (Optional)

We hope that by now you realize the central role that Newton's second law plays in mechanics. From the forces on an object we may find the acceleration. From this acceleration and the initial conditions, the entire motion record, including velocity and position, can be deduced. We have introduced simple models for many of the common forces and summarized them in Table 6.3. With these models and simple algebra, we are able to describe a great many phenomena, from frictional effects on sliding objects to the circular orbits of satellites. Still, there are many problems we haven't touched on. How can you calculate a satellite's orbit if it isn't circular? What is the effect of air resistance on the range of real projectiles? In principle, we know how to solve these problems: select appropriate models for the forces and apply Newton's second law. However, sometimes this approach generates equations we may not be able to solve. When this happens the calculating power of the digital computer can come to our aid.

In the paragraphs that follow, we introduce one method for determining a particle's motion when the forces acting on it are known. Officially, it is known as the **fourth-order Runge-Kutta** method. We do not attempt to justify the mathematics entirely, although we hope to make the result plausible. We present a computer program for implementing this method.

The idea behind this method is quite simple. We assume that at the outset we know the particle's location and velocity. We assume too that the forces on the particle can be calculated from this knowledge of position and velocity. If the force is known, then the acceleration can be obtained from Newton's second law. With the acceleration known we can begin to "step through" the particle's motion by advancing time in very small steps. At each step we can approximate the small changes in **v** and **r** using derivatives.

Let's think about the one-dimensional case first. The simplest guess we might make for the "new" velocity after some small time Δt is

$$v_{\text{new}} = v_{\text{old}} + a_{\text{old}}\Delta t$$

Similarly, we can obtain the new position using

$$x_{\text{new}} = x_{\text{old}} + v_{\text{old}}\Delta t$$

Once we obtain this "new" velocity and position, we can again evaluate the force and from this new force compute a new acceleration. Then we can take the "new" values as "old" values and advance the solution another step to some new "new" values, and so on.

The plan is a good one, but the bad news is that our approximations are not particularly accurate. In fact, you may recognize that the x-equation is not exactly correct even for something as simple as constant acceleration. (What is missing of course is the $\frac{1}{2}a_{\text{old}}\Delta t^2$ term that we derived back when studying the constant-acceleration model.) In the jargon of numerical analysis one says that the x-equation is accurate only to first order in Δt. (The second-order term Δt^2 is missing.) Thus, we have to make the Δt steps very small to obtain reasonable accuracy as the solution marches along. Such a solution takes a long time to calculate very far into the future and we have to keep a large number of digits in the calculation to obtain reasonable accuracy.

The Runge-Kutta method employs the same type of strategy as the one we just described. That is, it starts with known values of position and velocity and then computes new ones at some time Δt later. The Runge-Kutta method is more accurate and "robust" as a numerical technique because it does not depend on the values of the force only at the initial point. Instead it "looks ahead" and samples the force at three additional points along the trajectory before actually taking the step!

As just described, the Runge-Kutta method looks at a total of four points before deciding the result of an individual step. One point is the initial point. As the second point, it takes only half a step and samples the force halfway along. Based on this "test force" it then takes another trial half-step assuming that the force from the first test was constant. Finally, it takes a trial full step based on these last two and samples the force near the end of the proposed step. The actual full step is then calculated based on a judicious average of the four trial steps. We won't prove it here, but the method is accurate to the fourth order. Figure 6.9 shows the positions where the fourth-order Runge-Kutta samples the force along the actual trajectory.

So you will recognize the six equations in the Runge-Kutta calculation when you see them in a program, we write them out here. Trajectories that can be computed using this recipe are those for which the force(s) depend on time, position, or velocity of the particle (or any combination of these).

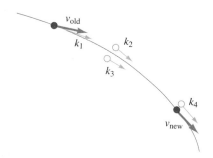

FIGURE 6.9

The Runge-Kutta method "looks ahead" and samples the force at four points before calculating a new position and new velocity from the old position and old velocity.

In Equations (6.23) the $F(t, x, v)$ stands for a force function that can be evaluated from knowledge of the time t, position x, and velocity v. You supply the force function that applies to your own problem. The k's are velocities at the four trial points, and m is the mass of the particle.

$$k_1 = F(t_{old}, x_{old}, v_{old}) \, \Delta t/m \qquad\qquad (6.23)$$

$$k_2 = F\left(t_{old} + \frac{\Delta t}{2}, x_{old} + \frac{v_{old}\,\Delta t}{2} + \frac{k_1\,\Delta t}{8}, v_{old} + \frac{k_1}{2}\right) \Delta t/m$$

$$k_3 = F\left(t_{old} + \frac{\Delta t}{2}, x_{old} + \frac{v_{old}\,\Delta t}{2} + \frac{k_1\,\Delta t}{8}, v_{old} + \frac{k_2}{2}\right) \Delta t/m$$

$$k_4 = F\left(t_{old} + \Delta t, x_{old} + v_{old}\,\Delta t + \frac{k_3\,\Delta t}{2}, v_{old} + k_3\right) \Delta t/m$$

$$x_{new} = x_{old} + \left(v_{old} + \frac{k_1 + k_2 + k_3}{6}\right) \Delta t$$

$$v_{new} = v_{old} + \frac{k_1 + 2k_2 + 2k_3 + k_4}{6}$$

These equations may look imposing, but we've got good news. You have programs and spreadsheets with the Runge-Kutta method already written out; they are on the diskette accompanying this text. Use them as a tool in your own computer "laboratory" to explore the results of some "what if" problems. The only thing you have to do is answer our perennial question, "What are the forces?" Put some forces into our programs and see for yourself how air friction affects real trajectories. Try putting in a $1/r^2$ force and see if you can launch a satellite into stable orbit. Invent your own strange forces and explore their effect on masses.

EXAMPLE 6.10   *Spring into Runge-Kutta*

As an example of the application of the Runge-Kutta method to one-dimensional motion, create a spreadsheet template to determine the motion of a mass connected to a Hooke's law spring with an (optional) drag force proportional to velocity.

SOLUTION   The solution based on Equations (6.23) may be found in RK1D.WK1 on the diskette accompanying this text. You can explore the behavior of this system simply by typing your own values of k, m, or the drag parameter b directly into the appropriate cells in column B as shown in Figure 6.10. The resulting position-versus-time graph is already prepared for viewing.

While you are experimenting keep a careful eye on your graphs. If you make the mass too light, or the spring too stiff, you may see some very strange results. If things start to look weird, you are probably learning something about numerical analysis but not very much about physics: the step size (Δt) in the Runge-Kutta calculations must result in small changes in the object's motion. If you are in doubt about the validity of your result, reduce the step size. If this changes the answer, reduce Δt again. Don't trust any result until making Δt smaller doesn't cause changes in the object's behavior! (If you keep the same number of steps you, of course, see less of the motion when you make Δt smaller.)

It is much more important for you to learn how to put your own force function into the spreadsheet than to play with our example. The critical cell is H10, which holds the first copy of the force function. In our case it reads

```
H10:   -$B$17*E10-$B$20*F10
```

If you check the cell references from Figure 6.10, you will see that the force described in H10 is

	A	B	C	D	E	F	G	H
1	Fourth-order Runge-Kutta solution of F = ma							
2	Formula 25.5.20 Handbook of mathematical functions, Abramowitz and Stegun							
3	Hooke's law spring with linear damping							
4	Constants		Time	Position	Velocity	Accel.	Force	k1
5			0	0.0500	1	−12.5	−25	−0.125
6	Delta t =	0.01	0.01	0.0593	0.8630	−14.833	−29.667	−0.1483
7	XO =	0.05	0.02	0.0671	0.7045	−16.797	−33.594	−0.1679
8	VO =	1	0.03	0.0733	0.5284	−18.341	−36.683	−0.1834
9	mass =	2	0.04	0.0777	0.3392	−19.428	−38.857	−0.1942
10			0.05	0.0801	0.1415	−20.031	−40.061	−0.2003
11			0.06	0.0805	−0.0597	−20.133	−40.267	−0.2013
12	Spring k=	500	0.07	0.0789	−0.2594	−19.733	−39.467	−0.1973
13			0.08	0.0753	−0.4527	−18.841	−37.683	−0.1884
14			0.09	0.0699	−0.6347	−17.479	−34.958	−0.1747
15	drag b=	0	0.1	0.0627	−0.8008	−15.681	−31.362	−0.1568
16			0.11	0.0539	−0.9470	−13.491	−26.983	−0.1349
17			0.12	0.0438	−1.0695	−10.965	−21.931	−0.1096

FIGURE 6.10

A portion of the spreadsheet used to apply the fourth-order Runge-Kutta method to one-dimensional motion.

$$F = -kx - bv$$

This equation is Hooke's law with a resistive term proportional to velocity. Once your own force formula is typed into H10, it must be copied to other cells in the spreadsheet. Follow the steps below to copy your formula. (You can make your spreadsheet as many rows deep as your computer memory allows. We assume 60 here.)

1. Copy H10 to H11 through H60.
2. Copy H10 through H60 to N10 through N60.
3. Copy H10 through H60 to T10 through T60.
4. Copy H10 through H60 to Z10 through Z60.

That's all there is to it. You can even write a "macro" to perform all the copying if you're handy with your spreadsheet.

We have suggested some calculations you might wish to try in the problem section, but we hope you think of others on your own. ◀

The spreadsheet template used in Example 6.10 was designed to make the introduction of new force functions as simple as possible rather than to conserve computer memory. For calculations in two dimensions we recommend a computer program similar to the one described in the next example.

EXAMPLE 6.11   *The Shot Seen Around the World*

Write a computer program to implement the fourth-order Runge-Kutta equations to solve for motion in two dimensions.

SOLUTION   The BASIC computer program RK2D.BAS implements a two-dimensional version of the Runge-Kutta method. The user (that's you) needs to supply the force function and the initial conditions as part of the source code.

```
30      REM  ****************************************************************************
50      REM  ****************************************************************************
60      REM  *   VARIABLES:INITIALIZATION AND DEFINITIONS                               *
70      REM  ****************************************************************************
100     RAD = 3          'Radius of planet, if any
110     RMIN = RAD       'Minimum radius, calculation stops for R < RMIN
120     T = 0            'Initialize time variable
130     X = 0            'Initial X-coordinate
135     Y=RAD +.5        'Initial Y-coordinate
140     XMAX=11          'Use to set aspect ratio for video monitor,
141                      ' 11 for EGA and Herculus mono
145     YMAX=8           'Use to set aspect ratio for video monitor,
146                      ' 8 for EGA and Hercules mono
150     VX=8.5           'X-component of initial velocity (arbitrary units)
155     VY=0             'Y-component of initial velocity (arbitrary units)
160     N=1000           'Maximum number of steps in orbit plot
170     H=0.05           'Time step for each new point
180     M=0.5            'Projectile mass (arbitrary units)
200     REM  ****************************************************************************
210     REM  *          FORCE FUNCTION DEFINITION                                       *
220     REM  ****************************************************************************
230     MP = 1.0         'Planet mass (arbitrary units)
240     G = 200          'Universal Constant (arbitrary units)
300     DEF FNFORCEX(T,X,Y,VX,VY) = -G*M*MP*(X/(X^2+y^2)^1.5)
310     DEF FNFORCEY(T,X,Y,VX,VY) = -G*M*MP*(Y/(X^2+y^2)^1.5)
400     REM  ****************************************************************************
410     REM  *           MAIN PROGRAM                                                   *
420     REM  ****************************************************************************
440     GOSUB 1100    'INITIALIZE GRAPHICS
450     GOSUB 1300    'DRAW CIRCLE
500     FOR I = 1 TO N
540       GOSUB 1000 'ADVANCE ONE STEP
550       GOSUB 1200 'PLOT (X,Y)
560     IF (INKEY$<>"")   THEN STOP        'Stop if user hits any key
570     IF (SQR(X^2+Y^2)<RMIN) THEN STOP   'Stop if too close to origin
580     T = T + H 'Advance time
590     NEXT I
999     STOP
1000    REM  ****************************************************************************
1010    REM  *         SUBROUTINE TO ADVANCE STEP                                       *
1020    REM  ****************************************************************************
1030    KX1=H*FNFORCEX(T,X,Y,VX,VY)/M
1032    KY1=H*FNFORCEY(T,X,Y,VX,VY)/M
1034    KX2=H*FNFORCEX(T+H/2,X+(H/2)*VX+(H/8)*KX1,Y+(H/2)*VY+(H/8)*KY1,VX+KX1/2,VY+KY1/2)/M
1036    KY2=H*FNFORCEY(T+H/2,X+(H/2)*VX+(H/8)*KX1,Y+(H/2)*VY+(H/8)*KY1,VX+KX1/2,VY+KY1/2)/M
1038    KX3=H*FNFORCEX(T+H/2,X+(H/2)*VX+(H/8)*KX1,Y+(H/2)*VY+(H/8)*KY1,VX+KX2/2,VY+KY2/2)/M
1040    KY3=H*FNFORCEY(T+H/2,X+(H/2)*VX+(H/8)*KX1,Y+(H/2)*VY+(H/8)*KY1,VX+KX2/2,VY+KY2/2)/M
1042    KX4=H*FNFORCEX(T+H,X+H*VX+(H/2)*KX3,Y+H*VY+(H/2)*KY3,VX+KX3,VY+KY3)/M
1044    KY4=H*FNFORCEY(T+H,X+H*VX+(H/2)*KX3,Y+H*VY+(H/2)*KY3,VX+KX3,VY+KY3)/M
1046    X=X+H*(VX+(KX1+KX2+KX3)/6)
1048    VX=VX+(KX1+2*KX2+2*KX3+KX4)/6
1050    Y=Y+H*(VY+(KY1+KY2+KY3)/6)
1052    VY=VY+(KY1+2*KY2+2*KY3+KY4)/6
1090    RETURN
```

(Continued on the next page)

```
1100   REM ********************************************************************************
1102   REM *              INITIALIZE GRAPICS                                              *
1104   REM ********************************************************************************
1110   SCREEN 2
1120   WINDOW (XMAX,YMAX)-(-XMAX,-YMAX)
1190   RETURN
1200   REM ********************************************************************************
1210   REM *    PLOT POINT                                                                *
1220   REM ********************************************************************************
1230   PSET(X,Y)
1290   RETURN
1300   REM ********************************************************************************
1302   REM *     DRAW CIRCLE IN WORLD COORDINATES                                         *
1303   REM ********************************************************************************
1310   FOR J = 0 TO 40
1312     THETA = (3.1415/80)*J
1315     XR = RAD*SIN(THETA)
1320     YR = SQR(RAD^2-XR^2)
1330     PSET (XR,YR):PSET(-XR,YR):PSET(XR,-YR):PSET(-XR,-YR)
1340   NEXT J
1350   RETURN
```

FIGURE 6.11

A portion of a BASIC program RK2D.BAS used to predict the motion of an object subject to nonconstant forces in two dimensions.

The particle is launched at time T from a point X and Y with velocity components VX and VY. These initial conditions are written in lines 130, 135, 150, and 155, respectively. Line 160 contains the number of steps to take (N) and line 170 is the step size (H).

In addition to the initial conditions you must, of course, answer the question, "What are the forces?" by defining a force function (in x- and y-component form) in lines 300 and 310. Here you must supply the right-hand side of the statements

```
300 DEF FNFORCEX(T,X,Y,VX,VY)=
```

and

```
310 DEF FNFORCEY(T,X,Y,VX,VY)=
```

The expression following the equal sign may contain any of the variables T, X, Y, VX, or VY.

As a specific example, in RK2D.BAS we show how to launch a projectile from the surface of a fictitious planet. Because this planet has no atmosphere, the force is due to Newton's universal law of gravitation only. In a Cartesian coordinate system the x-component of force is given by

$$F_x = \frac{GmM}{R^2}\cos(\theta) = \frac{GmM}{R^2}\frac{x}{R}$$

$$= GmM\left(\frac{x}{(x^2 + y^2)^{3/2}}\right)$$

Thus, in line 300 we write the force function for x as

```
300 DEF FNFORCEX(T,X,Y,VX,VY)=-G*M*MP*(X/(X^2+Y^2)^1.5)
```

Similarly, for the y-component we enter

```
310 DEF FNFORCEY(T,X,Y,VX,VY)=-G*M*MP*(Y/(X^2+Y^2)^1.5)
```

Notice we used arbitrary units throughout the program. In RK2D.BAS we added a few extra touches, such as drawing a circle of radius RAD to represent our planet. If the circle isn't round on your computer screen, change the values of XMAX and YMAX (lines 140 and 145) until your circle is round so that your orbits plot with a true aspect ratio.

We also added a step (570) that stops the calculation when the projectile "hits" the planet surface. If you eliminate this step and shrink RAD to say 0.2, you can simulate larger satellite orbits on your screen.

When you use the program to simulate satellite orbits be skeptical of its results if the satellite gets too close to $R = 0$. If the force is too large, you learn a lot about the convergence of Runge-Kutta calculations but nothing about physics. (We've even had one of our satellites fall through $R = 0$ then rocket off the screen as if propelled at incredible speeds.)

More sophisticated Runge-Kutta programs use an "adaptive step size," which means that when the force becomes large the step size automatically shrinks. This approach makes calculations faster and mitigates the problem of large forces giving too large a velocity change in one step. We have used a fixed step-size. One advantage of fixing the step-size is that the apparent speed of the point on your computer screen is proportional to the actual particle velocity. Thus, when your satellite speeds up as it approaches nearer to the source of gravity, you can see how satellites actually behave in real time! (This velocity behavior was discovered by Kepler and is described in more detail in Chapter 13.)   ◀

6.7 Summary

In this chapter we introduced the coefficient-of-friction model:

$$F_{\text{fric}} = \mu_k N \qquad \text{(For kinetic friction)}$$

$$F_{\text{fric}} \leq \mu_s N \qquad \text{(For static friction)}$$

The constants μ_s and μ_k are the coefficients of static and kinetic friction, respectively. In general, $\mu_s > \mu_k$.

Objects traveling in uniform circular motion must have an unbalanced force acting on them in the direction of the center of the circle. The magnitude of this net force may be set equal to mv^2/R.

Newton's universal law of gravitation (NULG) describes the gravitational attractive force between every two masses in the universe. The magnitude of the force is given by

$$F = G\frac{mM}{R^2}$$

Fictitious "pseudo" forces arise in accelerated reference frames. Sometimes it is convenient to treat the fictitious force as a pseudo gravitational force. The magnitude of the corresponding fictitious gravitational field is equal to the magnitude of the acceleration of the coordinate system. The direction of the fictitious gravitational field is opposite that of the coordinate system's acceleration.

(Optional) Stokes' law and the quadratic model are two frequently employed models used to describe the effect of viscous drag:

$$D = 6\pi\eta\tau v \qquad \text{Stokes' law}$$

$$D = \frac{1}{2}C_D A \rho v^2 \qquad \text{Quadratic model}$$

(Optional) The Runge-Kutta technique for the solution to $\Sigma \mathbf{F} = m\mathbf{a}$ may be applied to determine the motion of objects subjected to complicated forces. A spreadsheet template for problems in one dimension and BASIC programs for one- and two-dimensional problems are provided with this text.

PROBLEMS

6.1 The Coefficient-of-Friction Model

1. A tractor exerts a constant force of 7.85×10^3 N on a 1.00×10^3-kg load moving it forward at a constant speed of 0.550 m/s. What is the coefficient of kinetic friction between the bottom of the load and the ground?

2. A 3.00-kg block rests on a horizontal table. The coefficient of static friction between the block and the table is 0.750. A spring with stiffness constant $k = 2.50 \times 10^2$ N/m lies on the tabletop with one end fixed and the other attached to the block. If the block is moved along a straight line compressing or stretching the spring, over what range of distances can it be moved and still remain at rest when released?

3. (a) Show that a block resting on an incline remains at rest as long as the incline angle does not exceed $\arctan(\mu_s)$. (b) Such a block just starts to slip when the incline angle is 36.87°. It then accelerates, traveling 0.800 m along the plane in 2.00 s. What was the acceleration of the block down the plane? (c) What is the coefficient of kinetic friction between the block and the plane?

4. Two blocks ($m_A = 2.00$ kg and $m_B = 6.00$ kg) are pulled by horizontal forces P and T as shown in Figure 6.P1. The coefficient of kinetic friction between the lower block and the table is 0.333 and between the two blocks is 0.600. (a) Draw a free-body diagram showing all forces for each block separately. (b) If the forces are $P = 50.0$ N and $T = 20.0$ N, determine the acceleration of each block.

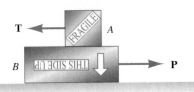

FIGURE 6.P1   Problem 4

5. A 100.-kg trunk is riding in the center of the back of a pickup truck. The coefficient of friction between the trunk and the pickup bed is 0.650. What is the greatest deceleration the truck can have without the trunk sliding forward toward the back of the cab?

6. Two blocks are pushed along a frictionless table as shown in Figure 6.P2(a). (a) Write an expression for the magnitude of the normal force between the two blocks as a function of the masses m_A and m_B, and the push force P. (b) Suppose that block B is now raised off the table as shown in Figure 6.P2(b). If the surfaces between the blocks are rough, then block B can be prevented from slipping back down by the force of friction. Determine an expression for the minimum force P that prevents block B from falling if the coefficient of static friction between the block faces is μ_s.

(a)   (b)

FIGURE 6.P2   Problem 6

7. A block of mass 2.00 kg is given an initial velocity of 3.50 m/s up a rough plane inclined at 30.0° with respect to the horizontal. If the coefficients of friction between the block and the plane are $\mu_k = 0.150$ and $\mu_s = 0.600$, (a) how far up the incline does the block travel before coming to rest? (b) Does the block remain at rest or does it slide back down the incline?

8. Two workers move a 50.0-kg crate along a rough floor. One pushes with a force of 200. N at an angle $\theta = 36.87°$ below the horizontal. The other worker pulls with a 100.-N force upward at $\phi = 30°$ as shown in Figure 6.P3. If the coefficient of friction between the block and the floor is 0.250, what is the acceleration of the block?

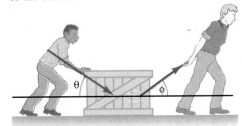

FIGURE 6.P3   Problem 8

9. A trunk weighing 5.00×10^2 N is to be pushed up a rough incline by an applied horizontal force **P**. The incline makes an angle $\theta = 36.87°$ from the horizontal as shown in Figure 6.P4. If a push force of 1.00×10^3 N is sufficient to move the trunk at a constant velocity of 0.200 m/s, what is the coefficient of kinetic friction between the trunk and the incline?

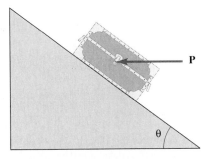

FIGURE 6.P4   Problem 9

10. A trunk of mass 50.0 kg is to be pulled along a rough horizontal floor by an applied force **P**. This force is applied at an angle $\theta = 53.13°$ from the vertical as shown in Figure 6.P5. If the coefficient of kinetic friction between the trunk and the floor is $\mu_k = 0.500$, what must the magnitude of the force **P** be to move the trunk at a constant velocity of 0.200 m/s?

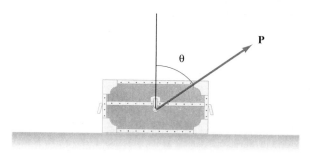

FIGURE 6.P5   Problem 10

11. Two blocks slide down an inclined plane as shown in Figure 6.P6. The coefficient of kinetic friction μ_A between block A and the plane is greater than μ_B, the coefficient of kinetic friction between block B and the plane. (a) Draw separate free-body diagrams for blocks A and B. (b) Determine an expression for the contact force between the two blocks as a function of m_A, m_B, μ_A, μ_B, θ, and g.

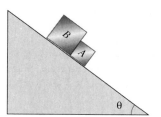

FIGURE 6.P6   Problem 11

12. Three masses ($m_A = 3.00$ kg, $m_B = 5.00$ kg, $m_C = 2.00$ kg) are connected by a light string. The mass B slides on a horizontal tabletop, and masses A and B hang off opposite sides of the table as shown in Figure 6.P7. The strings slide without friction over rods near the edge, but the coefficient of kinetic friction between the tabletop and block B is 0.500. If the system of masses is given an initial velocity to the right, calculate the acceleration of the system and the tension in each of the two strings.

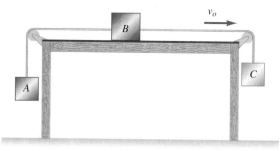

FIGURE 6.P7   Problem 12

6.2 Circular Motion

13. A phonograph turntable rotates at $33\frac{1}{3}$ rpm. If the coefficient of friction between the turntable and a nickel is 0.0300, how close to the center can the nickel be placed and still ride without slipping?

14. A block of mass m is attached to a spring of stiffness constant k and unstretched length l. One end of the spring is attached to a pivot in the center of a horizontal frictionless table. The mass is set into circular motion around the pivot. What rotation rate is necessary to stretch the spring to twice its unstretched length?

15. A mass of 0.200 kg hangs on a 0.500-m light string from the center of the roof in a van. The van is traveling at 20.0 m/s around a curve with radius 50.0 m. What angle does the string make with the vertical? [You may ignore the length of the string compared to the radius of the circle in which the mass travels.]

16. In an amusement park ride known as the rotor, brave participants stand around the inside wall of a circular platform that begins to rotate. When the platform is rotating sufficiently fast, the bottom falls away and passengers are held in place by only the force of friction between themselves and the wall, as shown in Figure 6.P8. (a) Draw a free-body diagram of a passenger. Carefully show weight, the normal force of the wall on the passenger's back, and the frictional force that holds the passenger up. (b) If the platform has a radius of 2.50 m and is rotating at 3.50 rad/s, what is the minimum coefficient of friction necessary to hold the passengers in place?

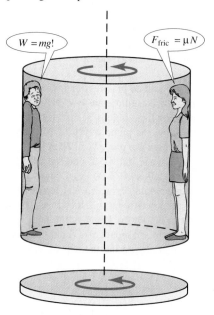

FIGURE 6.P8   Problem 16

17. Compute the slowest speed the car in Example 6.3 could have and still round the curve without sliding toward the center.

18. A soup bowl is formed with a flat bottom and sides that are inclined at $\theta = 36.87°$ with the bottom. An ice cube is set sliding at constant speed in a horizontal circle around the inside of the bowl as shown in Figure 6.P9. (a) Draw a free-body diagram for the ice cube. (b) If the ice cube slides without friction around the bowl in a horizontal circle of radius 0.100 m, what is its speed?

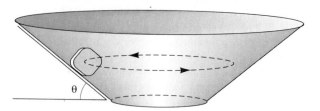

FIGURE 6.P9   Problem 18

19. A 1.50-kg mass on the end of a string is made to undergo uniform circular motion with a radius of 1.20 m in a vertical plane. (a) Draw free-body diagrams for the mass when it is at the top and when it is at the bottom of its circular path. (b) What must be the mass's speed for the tension in the string to just be zero at the top of the circular path? (c) What is the tension in the string at the bottom of the circle?

20. A quarter lies on the seat of a roller coaster traveling rapidly over a series of vertical waves as shown in Figure 6.10. (a) If the radius of curvature R of the wave is 12.5 m, how fast can the car travel over the crest of the wave before the quarter just loses contact with the seat, that is, before the normal force on the quarter becomes zero? (b) If the car travels the same speed and encounters a dip with the same radius of curvature, what is the normal force N on the quarter at the bottom of the dip? State your answer for N as a multiple of the quarter's weight.

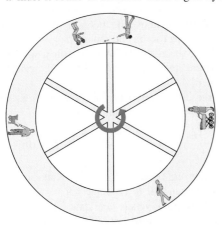

FIGURE 6.P10   Problem 20

21. Consider a mass at rest on a spring scale at a point on the earth's equator. Because of the earth's rotation, the spring scale does not read a true measure of the weight force mg. (a) Calculate the apparent weight as measured by the spring-scale force on a 1.000-kg mass, if the local gravitational field strength at this point is 9.800 N/kg. What is the percentage difference in the apparent weight and mg? (b) What is the measured acceleration a_g of a freely falling object at this location?

22. Engineers have proposed building space stations in a shape resembling a hollow donut. In order to produce "artificial gravity," the space station is to rotate about an axis through the center as in Figure 6.P11. This artificial gravitational field is equal and opposite to the centripetal acceleration of the passenger. If the space station has an outer radius of 200. m, with what angular velocity ω must it rotate to simulate earth's gravity?

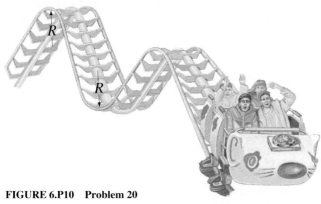

FIGURE 6.P11   Problem 22

6.3 Newton's Universal Law of Gravitation

23. In a hydrogen atom the electron and proton are separated by 5.30×10^{-11} m. (a) Calculate the gravitational attraction between these particles. (b) If the electron orbits the proton with a speed of 2.20×10^6 m/s in a circular orbit with radius equal to the distance in part (a), is the gravitational attraction sufficient to hold it in its orbit? Explain.

24. Calculate the gravitational attraction between two 5.00-kg spheres of pure lead that just touch one another.

25. Determine an equation for the force of attraction between a steel bar of mass M and a lead sphere of mass m arranged as in Figure 6.P12. (*Hint:* Consider the bar as made up of small chunks of length dx each with mass $dm = (M/L)\, dx$. The force on each chunk can be written $dF = G\left(\dfrac{m\, dm}{R^2}\right)$, where $R = a + x$. The total force is obtained by integrating dF as x ranges from 0 to L.)

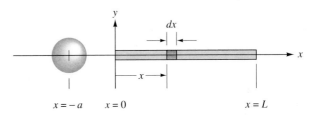

FIGURE 6.P12   Problem 25

26. There is a point between the earth and the moon where a rocket would experience an equal gravitational pull from each. How far from the center of the earth is this point of neutral gravity?

27. Imagine a satellite orbiting 0.500 m above a perfectly spherical earth. Ignoring mountains, trees, air resistance, and a few billion knees, calculate the period of such a satellite.

28. Jupiter's four brightest moons orbit this planet with the following parameters:

	ORBITAL PERIOD (EARTH DAYS)	AVERAGE RADIUS OF ORBIT (10^3 km)
Io	1.769	422
Europa	3.55	671
Ganymede	7.15	1070
Callisto	16.69	1883

(a) Show that T^2/R^3 is constant for these satellites. (b) Calculate the mass of Jupiter based on these orbits.

29. Using data from Appendix 8, calculate the "surface" gravity of the sun. How much would a 1.00-kg mass weigh there?

30. (a) Calculate the gravitational field strength on the surface of the moon. [See Appendix 8 for required data.] (b) If you jump upward on earth, what initial velocity must you have in order to raise your center of mass to a maximum height of 0.250 m? (c) If you give yourself the same initial velocity on the moon, how high do you jump?

31. Show that the pull of the sun on the moon is always greater than the pull of the earth on the moon. This result explains Figure 6.P13, which shows that the moon's orbit is never concave away from the sun.

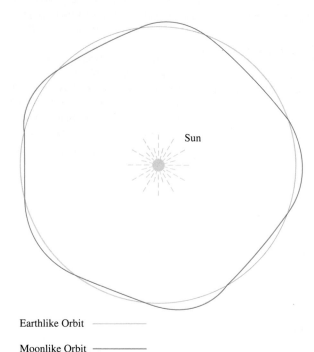

Earthlike Orbit ————

Moonlike Orbit ————

FIGURE 6.P13   The moon's orbit around the sun is similar to the dark line in this figure in that it is always concave toward the sun. (See Problem 31.) The distance from the earth to the moon has been exaggerated for clarity. This exaggeration causes it to appear as if the moon travels around the earth only five times each year.

32. Geophysicists routinely measure changes in the earth's surface gravity as small as 0.100 milliGal (1 Gal = 10^{-2} N/kg). Suppose you were standing over a buried geologic structure that could be modeled as a sphere of radius 0.500 km with its center 0.700 km below the surface of the earth. If the spherical structure has a density of $\rho_s = 2.80$ gm/cm^3 and the surrounding earth has a density of $\rho_e = 2.5$ gm/cm^3, the buried sphere appears to be an excess mass with density $(\rho_s - \rho_e)$. Using this model, calculate the increase in the gravitational field g directly over the buried sphere. Express your answer in milliGals.

6.4 Models for Motion Through a Resistive Medium (Optional)

33. Calculate the Reynolds number of (a) a 25.0-cm diameter soccer ball traveling at 3.00 m/s in air, (b) a 60.0-cm diameter spherical depth charge falling 0.500 m/s in water, (c) a 0.300-mm grain of sand falling 1.00 cm/s in water.

34. Over what range of speeds can a 1.00-cm diameter spherical projectile travel through the air and have the drag force described reasonably well by the quadratic model?

35. Show that $v = -v_T \tanh (gt/v_T)$ is a solution to Equation (6.21). (*Hint:* Substitute v and dv/dt into Equation (6.21) and show that the equation reduces to an identity.)

36. By use of reasoning similar to that leading from Equation (6.19) to Equation (6.20), determine the expression for the terminal velocity of an object in free fall when the drag coefficient is given by Stokes' law.

37. (a) Estimate the Reynolds number for a rain drop falling in air. (b) Estimate the terminal velocity of a rain drop.

38. (a) Estimate the magnitude of the air-drag force on a bicyclist traveling at 10.0 mph by treating the cyclist as a sphere of radius 0.300 m. (b) What happens to the drag if the cyclist's speed increases to 15.0 mph?

39. Robert Millikan measured the charge of an electron by measuring electric forces on very small oil droplets. To accomplish this measurement he needed to know the weight of individual droplets. In order to determine a droplet's weight he timed its fall while watching through a microscope. If the density of the oil was 0.950×10^3 kg/m^3 and the terminal velocity of the droplet was 1.54 mm/s, (a) what was the radius of the droplet? (b) What was the weight of the droplet? (*Hint:* Estimate the Reynolds number of the droplet before choosing an air-drag model.)

40. Review Example 6.9, where it was shown that the velocity of a freely falling water balloon after 2.00 s is 19.6 m/s when air resistance is negligible, but 19.0 m/s when the quadratic model for air resistance is used. Early in the balloon's fall (while $Re < 1000$) there is less air drag on the balloon than predicted by the quadratic model. In order to estimate the magnitude of the error made by using the quadratic model for the entire fall, (a) calculate the length of time after release before the balloon reaches $Re = 1000$. (A simple free-fall estimate is all that is required here.) (b) What difference in speed is predicted by the free-fall model and the quadratic model for an object falling for the length of time calculated in part (a)? Is this difference in speed significant? (See Problem 45 for a more accurate model for the entire range of Re.)

6.6 Numerical Methods for Newton's Second Law (optional)

41. Adjust the initial velocity in RK2D.BAS to simulate the projectile motions from Newton's *Systems of The World* as shown in Figure 6.7. (a) What is the minimum speed (in the units of this program) that allows your projectile to travel around the planet? (b) Can you launch the projectile so that it hits the planet after traveling over halfway around? Explain.

42. Adjust the damping constant b in the spreadsheet template of Example 6.10 so that the spring returns to equilibrium in the shortest time interval. More specifically, determine the value of b that results in the shortest time for the absolute value of y to fall below 0.950 of the initial y-value (and never exceed this value again). A value of b within 10% of the optimum is adequate.

43. Modify the spreadsheet template of Example 6.10 to produce a y-versus-t graph of an object "dropped" toward the moon from a distance of three moon radii above the moon's surface. How long does it take such an object to hit the moon?

44. Modify the BASIC program RK1D.BAS (or write a similar program for a programmable calculator) to perform the calculation described in Problem 43.

45. The behavior of the drag coefficient for a sphere as shown in Figure 6.8 can be approximated by the equation

$$C_D = \frac{24}{Re} + \frac{6}{1 + \sqrt{Re}} + 0.4, \qquad 0 \le Re \le 2 \times 10^5$$

Use this equation and the fourth-order Runge-Kutta method to obtain a more accurate answer to Example 6.9.

46. Table 6.P1 gives the position-versus-time data for a falling badminton shuttlecock released from rest. Which fluid-drag model better fits the data: Stokes' law or the quadratic model?

TABLE 6.P1   Position versus Time for a Falling Shuttlecock[1]

$-y$ (m)	t (s)
0.61	0.347
1.00	0.470
1.22	0.519
1.52	0.582
1.83	0.650
2.00	0.674
2.13	0.717
2.44	0.766
2.74	0.823
3.00	0.870
4.00	1.031
5.00	1.193
6.00	1.354
7.00	1.501
8.50	1.726
9.50	1.873

[1] Mark Peastrel, Rosemary Lynch, and Angelo Armenti, Jr., "Terminal Velocity of a Shuttlecock in Vertical Fall," *American Journal of Physics* 48 (1980): 511–513.

★47. Write a computer program (or modify RK2D.BAS) to use the fourth-order Runge-Kutta method to calculate the trajectory of a 7.30-cm diameter smooth spherical projectile ($m = 0.145$ kg) launched with an initial velocity of 45.0 m/s at an angle 40.0° from horizontal in a uniform gravitational field $g = 9.80$ N/kg over level ground. Compare the range and maximum height to that obtained using the constant-acceleration model without air resistance. [*Hint:* Use the quadratic model to obtain the magnitude of the drag force D. This drag force is directed antiparallel to the velocity vector so that $\mathbf{D} = -D\ (\mathbf{v}/v)$. Thus, use $D_x = -D(v_x/v)$ to obtain the x-component for the drag force. An analogous expression gives D_y.]

General Problems

48. Sacks of flour fall from a stationary hopper straight down onto a horizontal conveyor belt traveling at 0.350 m/s. (a) If the coefficient of kinetic friction between the belt and the flour sacks is 0.235, how long does it take a sack to stop sliding relative to the belt? (b) How far (as measured on the belt) is it from the point where a sack hits the belt to the point where the sack stops moving relative to the belt? (c) How far is the flour sack from the hopper when it stops moving relative to the belt?

49. The block B in Figure 6.P14 is pulled by an applied force $P = 5.00$ N. The block B slides on the table without friction, but the coefficient of sliding friction between block A and block B is 0.250. For $m_A = 1.00$ kg and $m_B = 2.00$ kg, compute the acceleration of the blocks and the tension in the string connecting them. Assume the pulley is massless and frictionless.

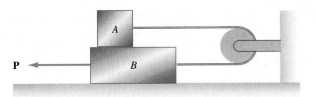

FIGURE 6.P14   Problem 49

50. A 1.50-kg block is pushed upward along a rough, vertical wall. The coefficient of kinetic friction between the wall and the block is 0.200. An external force P is applied to the block at an angle $\theta = 36.87°$ from the vertical (Figure 6.P15). What should the magnitude of P be so that the block moves at a constant velocity of 0.100 m/s up the wall?

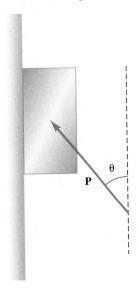

FIGURE 6.P15   Problem 50

★51. A block of granite with weight W is pulled along the ground by a force P acting at an angle θ above the horizontal as shown in Figure 6.P16. If the coefficient of friction between the block and the floor is μ_k, determine the angle θ that minimizes the force P required to pull the trunk at a constant velocity.

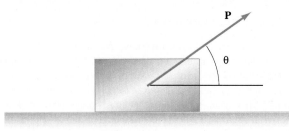

FIGURE 6.P16   Problem 51

52. A playful puppy ($m_A = 3.00$ kg) jumps up, grabs the edge of a tablecloth, and hangs off the edge of the table as shown in Figure 6.P17. A birthday cake ($m_B = 1.50$ kg) and a punch bowl ($m_C = 5.00$ kg) are at rest on the tablecloth. The friction between the tablecloth and the bowl and between the tablecloth and the cake plate is such that the bowl and plate do not move relative to the tablecloth. The tablecloth, however, slides across the polished table with a coefficient of sliding friction equal to 0.200. Model this potential disaster as a system of three connected masses (in

the spirit of Figures 6.P17 and 6.P18), ignoring the friction between the tablecloth and the table where the cloth passes over the rounded table edge. Calculate the acceleration of the system and the tension in the tablecloth between the puppy and the cake and between the cake and the punch bowl.

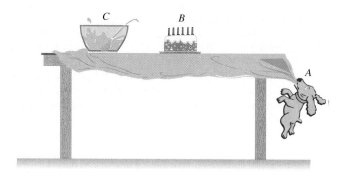

FIGURE 6.P17   Problem 52

53. Three masses ($m_A = 2.00$ kg, $m_B = 2.00$ kg, $m_C = 1.00$ kg) are connected by a light string. One mass slides on the top of a table, and the other two hang over the edge of the table as shown in Figure 6.P18. The string slides over a frictionless rod near the edge of the table, which is rough with coefficient of kinetic friction $\mu_k = 0.400$ between it and mass A. (a) If the masses are placed into motion such that mass A is moving to the left, find the acceleration of the system and the tension in each of the strings. (b) If the system is moving such that mass A moves to the right, calculate the acceleration and the tension in each of the two strings.

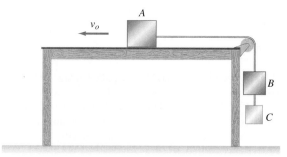

FIGURE 6.P18   Problem 53(a)

54. A physics experiment you may have performed involves threading a light string through a soda straw and then attaching a mass to each end of the string. The straw is then held vertical in one hand over your head, and one mass is set into circular motion around your head while the other mass hangs straight down as shown in Figure 6.P19. If the straw edge is flared slightly at the top, the string slides almost without friction so that the tension is the same in both segments of the string. If $L = 30.0$ cm, the hanging mass $m_A = 0.500$ kg, and the orbiting mass $m_B = 0.250$ kg, find the angle θ the string makes with the vertical and the rotation rate (in radians per second) of the circling mass. (b) Explain why this experiment cannot be performed with two equal masses.

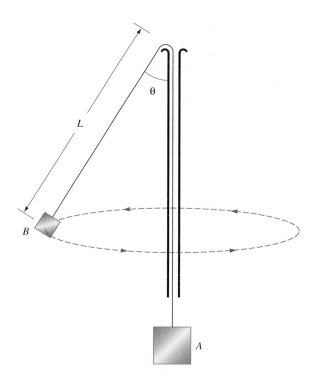

FIGURE 6.P19   Problem 54

55. One part of a roller coaster ride takes passengers around the inside of a vertical circle (Figure 6.P20). (a) If the loop has a radius of 20.0 m and the cars are traveling at 17.0 m/s at the top, what is the magnitude of the normal force of the seat on a 50.0-kg passenger? (b) The track maintains its curvature at the bottom as the cars head up the next hill. If the cars travel 25.0 m/s at the bottom, what is the normal force of the seat on the passenger?

FIGURE 6.P20   Problem 55

56. A pilot is flying a plane in a horizontal circle of radius 1.00 km at a speed of 200. km/h. To passengers in the plane the turn feels like a small, extra gravitational field (pseudo-g) directed outward with a magnitude equal to the centripetal acceleration of the plane. (See ''Feeling the physics'' at the end of Section 6.2.) The passengers ride most comfortably if the total effective gravitational field (the vector sum of actual g and pseudo-g) is perpendicular to their chair seats. At what angle should the pilot bank the plane in this turn to give the passengers a comfortable ride?

57. Geosynchronous satellites link major communication networks in the United States. These satellites orbit the earth above the equator and have a period equal to exactly 24.0 h. This feature permits them to stay ''stationary'' relative to the surface of the earth. Calculate the altitude above the surface of the earth that all geosynchronous satellites must have.

58. Consider two satellites each in circular orbit around the earth. Satellite A is one earth radius R_e above the surface of the earth. Satellite B is $2R_e$ above the surface. (a) Calculate the speed of each satellite. (b) Calculate the period of each satellite (the time to make one revolution).

59. Saturn's moon Tethys orbits Saturn in a nearly perfect circle of radius 294.66×10^3 km. Calculate the orbital period of Tethys.

60. Consider a simplified ''satellite-docking'' problem where you want to dock your Space Shuttle with a satellite. Assume all orbits are circular and a satellite can change from one orbit to another instantly (but only along a radius). You and the satellite are in orbits 1.00×10^6 m above the earth's surface, but you are trailing the satellite by $90.0°$ (you are a quarter circle behind the satellite). In order to catch up with the satellite you *decrease* your speed by 25%. (a) What is your new orbital radius? (b) What is your orbital period? (c) How long does it take you to catch up with the satellite?

Physics and Racing at the Indianapolis 500

by RUTH H. HOWES

Advances in the design and fine-tuning of racing engines for Indy-500 cars have produced cars the speed capability of which is limited by their ability to negotiate the oval of the Indianapolis Motor Speedway. Modern racing engines use computers to control fuel flow over the wide range of revolutions per minute required in the race and to balance the power generated in each engine cylinder. Cars carry a "black box," an onboard data collection system that can be down-loaded into a computer during pit stops and analyzed by the pit crew, who then make minute adjustments during the next stop.

In the interest of safety, the United States Auto Club, USAC, limits the power these engines produce by such measures as requiring pop-off valves on turbo chargers that hold engine pressure at a fixed level and mandating that cars must burn methanol. Nevertheless, modern Indycar engines propel the cars at speeds that compel racing teams to apply physics in order to hold the cars on the track, accelerate them around the corners of the oval, and protect the driver in the event of an accident.

The fastest lap speeds exceed 225 mph. Lap speed is an average of speeds along the straights and the slower speeds through the corners. Thus, Indy cars travel down the straight legs of the speedway at speeds over 250 mph. At such speeds, the fast-moving air passing over the car exerts considerably less pressure than stationary air, which can be trapped under the car; consequently the Bernoulli effect tends to lift the car off the track. The overall shape of the car with a curved top and flat bottom tends to increase this effect. Not only can the cars literally take off from the track, but even a little lift can reduce the critical lateral adhesion between the tires and the track. Because the lateral adhesion provides a significant amount of the centripetal force needed to accelerate the car around the corners of the oval, any loss in adhesion can and does cause serious accidents.

Designers use two mechanisms to hold Indycars on the track at very high racing speeds. In a modern ground-effects design, the body panels of the cars are designed to cause a fast-moving airstream under the car while a slower airstream passes over the car. The front wings of the car are shaped like inverted airfoils and force the car down onto the track. The space between the wheels of the car is occupied by two inverted airfoils, one on each side, which channel air into a Venturi, a path under the car in which

RUTH HOWES

Ball State University

Ruth Howes holds a B.A. in physics from Mount Holyoke College. She earned an M.A. in physics and a Ph.D. in nuclear physics from Columbia University. She is currently the George and Frances Ball Distinguished Professor of Physics and Astronomy at Ball State University. Like all Hoosiers, she takes a keen interest in the annual running of the Indianapolis-500. In addition to teaching physics and astronomy, she studies physics applications in such policy areas as energy and the verification of arms control agreements.

the air's channel for flow is compressed so that the air moves faster. The Bernoulli effect forces the car down onto the track towards the faster moving air. Ducting for cooling air also contributes to the fast-moving airstream under the car. Many cars continue the tunnels formed by the airfoils beyond the rear wheels of the car, and the air expands rapidly behind the car. The expanding air creates turbulence, which eliminates the advantage of drafting in Indianapolis racing.

The car in the figure illustrates the shape of a modern racing chassis. In addition to using the Bernoulli effect to hold the car onto the track, the car's body is carefully constructed to promote laminar flow of air around the car. Turbulence can reduce the forces holding the car to the track, increase the force of air resistance, and make the car hard to handle. Ground effects are the subjects of complex computer models and tests in air tunnels for racing teams that can afford them. All parts of a chassis are adjustable. Skilled racing crews and drivers make minute adjustments to tune the car for particular track conditions. Adjustments as tiny as 0.1° in the angle of a front wing can gain a fraction of a second each lap, which may allow the driver to win the race.

The second mechanism holding the car to the track is the wing mounted behind and above the car. The airstream strikes the flat surface of the wing and exerts a force on it. Particles of air are reflected from the wing, and the angle of incidence equals the angle of reflection. The resulting force is toward the track and opposing the motion of the car. The force toward the track helps to hold the car down onto the track, but the backward force slows the car. By changing the angle of the wing, crews can increase or decrease the size of this force. If a car has a more powerful engine, its crews can increase the downward force on the car to permit higher speeds. Of course the ability of the car to accelerate down the straightaway is reduced as the backward force on the car increases. The crew must choose between acceleration on this tangent and speed in the corners. Increasing the downward force on the car also increases its lateral adhesion to the track and allows the driver to maintain higher speeds through the corners. A modern Indycar typically generates nearly four times its weight in down force because of its ground-effects design. Lateral forces in corners frequently top three times the car's weight.

The track at the Indianapolis Motor Speedway is an oval 2.5 mi around. In each circuit, race cars must make four sharp turns. Drivers run almost straight into the turns up the hill of the banked track so that most of the car's weight slows the car. Just before touching the wall, they turn the car sharply and accelerate down the banked track, using a component of their weight to gain speed on the straight. The centripetal force pulling the car around the curve depends on the banking of the track and the lateral adhesion between the racing tires and the track. If the driver turns at a speed greater than that allowed by the centripetal force on the car, he can brush the wall, a sure ticket to a dangerous and expensive accident. On the other hand, taking the turns too slowly costs seconds of precious time and loses races.

The banking of the turns is fixed by the design of the track and cannot be changed by the drivers. The banking causes part of the car's weight to act outward from the center of the banked curve normal to the track. The reaction force of the track on the car provides part of the centripetal force needed to pull the car around the turn.

The lateral adhesion between the tires and the surface of the speedway depends on the surfaces of the tires and the track as well as on the downward force acting on the car. As racing speeds have increased, the surface of the track has changed. The first racing surface was actually paved with bricks (the track is colloquially known as the brickyard). Of course one early race car lost 104 gal of oil during the 500-mi race, and an average car lost more than 20 gal. By the end of the race, the cars were literally skating on oil slicks. Today, Indy racing stops as soon as rain wets the track, and the yellow caution flag flies to slow traffic whenever oil or fuel spills on the track.

The tires in Indycars are more than twice as wide as normal tires. They are treadless and made of soft rubber, which adheres to the track. Drivers zig zag before the race during warm-up laps to heat their tires and make the rubber sticky. A racing tire is hot to the touch after the car has been driven at high speeds. A layer of rubber comes off the tires onto the track and forms a black track around the speedway known as the "groove." The sticky rubber on the track in turn sticks to the rubber in the tires of trailing cars. During the race, drivers generally stay in the groove around corners. The adhesion of the tires to the track also increases the force the tires can exert on the track as the wheels roll forward, thereby increasing the force of the track on the tires and accelerating the cars forward.

Adjustments as tiny as 0.1° . . . can gain a fraction of a second each lap . . .

Because the rubber of the racing tires is designed to come off on the track, tires wear out quickly, and the expense of new tires is a major factor in the cost of racing. So USAC rules limit the car's fuel cell to 40 gal, cars must stop at least six times during the race. Tires can be changed nearly as fast as fuel can be pumped into the tank. A car uses four sets of tires during a racing day and two during a test day. To run the Indianapolis 500 from practice to the race itself requires around a dozen sets of tires. Goodyear Rubber Company, which supplies all tires, provides each team with four sets of tires for free and charges at least $1000 a set for the rest. All teams use the same rubber compound so that no one has an advantage, and the tires are numbered and collected by the company after they have been used. The tires are carefully engineered to wear evenly.

Because Indycars turn in only one direction, the tires on the outside of the car, the driver's right, are actually a little larger than those on the inside to make turning easier. Tolerances are

on the order of 0.1 in under the enormous down and lateral forces on Indycars. Indy tires are inflated with nitrogen so that as they heat and tend to expand they expand less than they would if filled with air. In spite of all precautions, tire performance changes with the surface temperature of the track and the wear on the tires produced by as little as one lap of running at racing speeds.

With increasing speeds, designers and USAC have constantly attempted to prevent serious driver injuries. The driver sits towards the front of the car with a 40-gal fuel cell behind him. Methanol burns with a colorless flame, and several early accidents involved bad burns because crews didn't realize that flames were present. Today's drivers wear heavy suits of flame-retardant Nomex and helmets that protect their faces. Unfortunately the temperatures in the cockpits of the cars climb to 150°F. Drivers must function in the Nomex suits while fighting the jolting steering wheel and controlling the car for at least 3 h during the race.

The fuel cell consists of a thick steel or aluminum box filled with foam. The fuel is held by a thick rubber bladder inside the foam lining. In the event that an accident breaks the metal shell, the bladder holds the fuel. Should the bladder break open, the foam absorbs the fuel and prevents it from soaking the driver and crew or forming an explosive mist.

Cars running at the Indianapolis 500 must weight at least 1550 lb. Designers have nevertheless taken advantage of modern light, strong, composite materials. The one-piece chassis and frame is made of aircraft-quality honeycomb aluminum reinforced with Kevlar and carbon fiber materials. The underwings are also composite materials, although the rear wing is generally aluminum. Wheels remain magnesium. Engines are currently aluminum although teams have begun to experiment with new high-temperature ceramic components. Thus, real limits on the speeds that cars can achieve come from limiting the down force on the car rather than the power of the engine. In this effort, officials check such properties as the car's ground clearance and its width. These factors will control speeds at the Indianapolis 500, at least until engineers devise a new technique for increasing speeds again.

Drivers are strapped tightly into their seats. Straps are, of course, designed for quick release in the event of an accident. Helmets have restraints to help the driver support the huge lateral forces encountered in the corners. These large lateral forces push the driver against the side of the car and the restraints that hold him in the cockpit. Safety crews, fire trucks, and an ambulance are present whenever a car runs on the Indianapolis Motor Speedway. In the event of an accident, the crew's first concern is to prevent fire. They spray the car and anyone in the vicinity with foam. Their second mission is to render first aid to the driver or anyone else who needs it. The yellow caution flag slows other drivers. If debris must be removed from the track, the race may be stopped.

As a final precaution, the bodies of race cars are designed to break apart in a collision. The kinetic energy of the car is carried in part by the pieces of the car, which travel out of the collision and are transformed into deformational energy as the chassis bends. Thus, a relatively minor accident can destroy the body of a car, but the driver walks away unharmed. The law of conservation of energy has its uses in preserving the safety of race drivers.

Despite USAC's concern with driver safety, Indianapolis racing remains dangerous. The increasing use of high-technology materials, computer monitoring of racing, electronic engine controls, and air tunnels to check ground effects has increased the cost of Indianapolis racing. Teams are rarely able to make a profit even when they win generous purses. IndyCar's governing board recently limited such features as the use of ceramics and titanium in engines as well as advanced electronic systems. Teams are allowed only ten testing days during the racing season. Electronics and new materials are two of the most expensive new technologies in racing. Track time at the Indianapolis Motor Speedway costs $3000 a day just for safety personnel, not including such expenses as wear and tear on the engines, which have to be rebuilt every 500 miles or so, and racing tires.

Thus, the future of Indianapolis racing may well see limits on the application of new technologies developed for aircraft and the space program. It is unlikely that budgets or safety rules will ever curb the clever applications of basic physics to increase the speed and excitement of the Indianapolis 500-Mile Race.

Questions for Thought

1. Friction does not depend on cross-sectional area but racing tires are much wider than normal tires. Why do you think this is so? Why do you think that racing tires are treadless?
2. Formula-1 race cars run on tracks that curve sharply in both directions. For example, the Grand Prix of Monte Carlo is run through city streets. What design changes would you make in an Indycar for such a race?
3. The previous essay described some of the measures that USAC officials have taken to slow lap speeds at the Indianapolis 500. What are some other properties of the car that might be regulated?
4. What changes in the body of the car should be made if the car is to be turned into a drag racer for running short distances on a straight track?

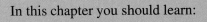

$$W = \int \mathbf{F} \cdot d\mathbf{s}$$

CHAPTER

7

Work and Kinetic Energy

In this chapter you should learn:

- to calculate the work done by a force.
- to define kinetic energy.
- to apply the work–energy theorem.
- to calculate power as the rate of performing work.
- (optional) to apply Simpson's rule to compute the work done by a variable force.

Newton's three laws of motion allow us to deduce how objects move from knowledge of the forces acting on them: Newton's second law relates the total force acting on an object to the object's acceleration. Once the acceleration is known, the velocity and position records can be deduced using algebra, calculus, graphs, or digital computers, depending on circumstances. Sometimes, however, we are not interested in such an exhaustive record of the object's motion. If you shoot an arrow from a bow, you may not care about the details of the arrow's acceleration. From a practical point of view, perhaps all you really want to know is how fast the arrow is traveling when it leaves the bow. What seems desirable then is a simple way to relate the final velocity of an object to the forces that act on it *without having to wrestle with the details of how the object came to that final velocity*. In the next several chapters of this text (Chapters 7 through 10) we will study two procedures for doing exactly that.

7.1 Work (Constant Force, Constant Direction)

Let's examine a task that we might all agree is work—mowing a lawn, for example. As you push on the handle, the mower moves forward. As the mower moves, the point where you are applying the force moves with it. Thus, the force you apply acts on the object through some distance. When a force acts over a distance, we say **work** is done. The formal definition of work is, as usual, mathematical. Because the work concept is somewhat novel, we introduce several expressions for work valid for special cases before tackling the general definition.

Consider first a simple case (as in our mowing example above) where a *constant force* **F** acts on an object moving in a *constant direction* through displacement $\Delta \mathbf{s}$. For this special case the work $\mathcal{W}$ done on the object by the force is given by

$$\mathcal{W} = \mathbf{F} \cdot \Delta \mathbf{s} \qquad \text{(Constant force, constant direction)} \qquad (7.1)$$

The vertical force by students on their books does no work as the books move horizontally. When students walk up the stairs they do positive work on their books; when they descend the stairs students perform negative work on their books.

The indicated vector multiplication is the scalar, or "dot," product. (If you don't recall how to take the dot product of two vectors, reread Section 2.4.) This definition of work is hardly intuitive. Only later in the chapter does it become clear why anyone would define such a concept.

Work, as calculated from Equation (7.1), bears some resemblance to our common notion of what work is, but the differences between Equation (7.1) and our everyday language are probably more striking than the similarities. For one thing, notice that if the object doesn't move, no work is performed. That is, if you hold your physics book out at arm's length and don't move it, you are not doing any work. In short order your muscles begin to ache, your arm may even tremble from the effort, but you are doing no work on the book in the sense defined by Equation (7.1)!

Suppose you carry a heavy physics book across a level floor. If you move the book only horizontally, you still do no work. To see why, recall that the dot product in Equation (7.1) can be written in terms of the angle ϕ between the force **F** and the displacement $\Delta \mathbf{s}$:

$$\mathcal{W} = F \, \Delta s \, \cos(\phi) \qquad (7.2)$$

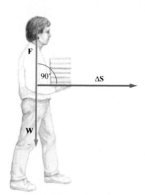

FIGURE 7.1

The angle between the force and the displacement is 90°. Since $\cos(90°) = 0$, no work is done.

The force *F* you exert on the book is upward, but the distance you travel is horizontal. Thus, as shown in Figure 7.1, the angle ϕ is 90°. Because the cosine of 90° is zero, $\mathcal{W} = 0$.

Stranger still is the result if **F** and $\Delta \mathbf{s}$ are in *opposite* directions. Such is the case, for example, when you lower a book down from a shelf. You must exert a force upward to keep the book from falling, yet as you lower your hand, the book travels downward. In this case the angle between the vector that represents the force on the book by your hand and the vector that represents the book's displacement is 180°. Since the cosine of 180° is −1, the work done is negative! Clearly, the physics definition of work differs from our everyday use of the word. (This is probably a good thing. How would you feel if you worked all week unloading heavy cartons from high truck-beds to a lower conveyer belt, then at the end of the week you received a *bill* from your boss because you had been doing negative work?)

The previous remarks may be summarized by the following statement:

It is only the component of the force acting in the direction of the displacement that contributes to the work done.

The unit of work in the SI system is the joule (1 J = 1 N · m) after James Prescott Joule, who showed the equivalence of many different forms of work. In the British Engineering system the unit of work is a pound force times a foot distance and known, logically enough, as a foot-pound (ft · lbf).

EXAMPLE 7.1 *Getting the Work Done*

A physics student on an inner tube slides down an inclined plane. Suppose the incline makes an angle $\theta = 36.87°$ with the horizontal as shown in Figure 7.2(a). If the student weighs 5.00×10^2 N and a frictional force of 1.50×10^2 N acts on the inner tube as it slides downward, find the work done on the student by each of the forces as she slides through a distance of 2.00 m.

SOLUTION Figure 7.2(b) shows the three forces acting on the student as she slides down the incline. The work done by each of the forces is calculated in the same manner; namely, using

$$\mathcal{W} = \mathbf{F} \cdot \Delta \mathbf{s} = |\mathbf{F}||\Delta \mathbf{s}| \cos(\phi)$$

where ϕ is the angle between the force vector and the displacement vector when they are placed tail to tail as shown in Figure 7.2(b).

The work done by the weight force $\mathbf{W}$ is

$$\mathcal{W} = W \Delta s \cos(\phi_1) = (5.00 \times 10^2 \text{ N})(2.00 \text{ m}) \cos(90° - \theta)$$

$$= (1.00 \times 10^3 \text{ N} \cdot \text{m}) \cos(53.13°) = 600. \text{ J}$$

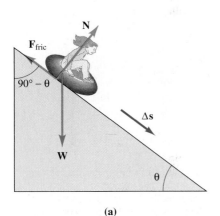

The work done by the frictional force $\mathbf{F}_{\text{fric}}$ is

$$\mathcal{W} = F_{\text{fric}} \Delta s \cos(\phi_2) = (150. \text{ N})(2.00 \text{ m}) \cos(180°)$$

$$= (300. \text{ N} \cdot \text{m})(-1) = -300. \text{ J}$$

Finally, for the normal force $\mathbf{N}$ we run into a bit of luck. The work done by this force may be computed without finding the magnitude of N because the angle between this force and the displacement is 90°:

$$\mathcal{W} = N \Delta s \cos(\phi_3) = N (2.00 \text{ M}) \cos(90°)$$

$$= N (2.00 \text{ m})(0) = 0 \text{ J}$$

◀

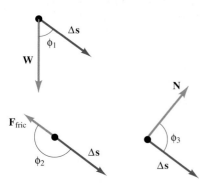

7.2 Work (Variable Force, Constant Direction)

Next, we consider the work performed when both the direction of the force and the direction of the object's displacement are constant, but the *magnitude* of the force changes. Before allowing the magnitude of the force to vary, however, let's examine the

FIGURE 7.2

(a) The forces acting while a person slides down an inclined plane. (b) In each case the work is calculated using the angle ϕ between the force and the displacement when the vectors are placed tail to tail.

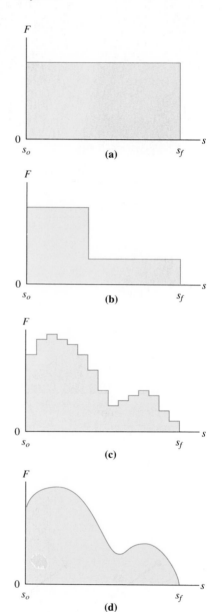

FIGURE 7.3

When the force is parallel to the displacement, the work done is the area between the force curve and the $F = 0$ axis.

graphical significance of work. When the constant force **F** moves an object through a displacement $\Delta \mathbf{s}$ and is parallel to this displacement, Equation (7.1) gives

$$\mathcal{W} = \mathbf{F} \cdot \Delta \mathbf{s} = F \, \Delta s \cos(0°) = F \, \Delta s$$

Figure 7.3(a) shows the graph of a constant force F versus the distance s. Notice that the work done is the area bounded by the force curve and the $F = 0$ axis between s_o and s_f. If the force were to change abruptly once during the push, the graph might appear as in Figure 7.3(b). Here, the total work done is the sum of the work performed during each of the two segments and is still equal to the area under the force curve. Figure 7.3(c) shows a still more complicated process with the force changing abruptly many times. In this case the total work done is again the sum of the contributions during each of the fifteen Δs intervals when the force was constant:

$$\mathcal{W}_{\text{total}} = \sum_{i=1}^{15} F_i \, \Delta s_i$$

When the force changes smoothly with distance as in Figure 7.3(d), we imagine dividing the interval up into N subintervals, each of very small width Δs, then letting N become infinite as the interval width approaches zero. This process is symbolized

$$\lim_{\substack{N \to \infty \\ \Delta s_i \to 0}} \sum_{i=1}^{N} F_i \, \Delta s_i \qquad (7.3)$$

We recognize Equation (7.3) as the definition of the integral. Thus, a more general definition of work, when force and displacement are parallel, is

$$\mathcal{W} = \int_{s_o}^{s_f} F \, ds \qquad \text{(Force parallel to displacement)} \qquad (7.4)$$

Note that the work done is still the area between the F curve and the $F = 0$ axis. If the force is not parallel to the displacement (but the angle ϕ between them remains constant), then only the component of F parallel to the displacement contributes to the work. Because this component is $F \cos(\phi)$, the integral in Equation (7.4) must be multiplied by the factor $\cos(\phi)$:

$$\mathcal{W} = \cos(\phi) \int_{s_o}^{s_f} F \, ds \qquad (\cos(\phi) \text{ constant}) \qquad (7.5)$$

An important example of a force that changes with distance is that of the Hooke's-law spring introduced in Section 5.3. The *magnitude* of the force exerted by an ideal spring when it is stretched or compressed from its natural length by a distance s is $F_{sp} = ks$. If the distance s represents an extension, the spring is in **tension**, that is, it *pulls* on objects attached to its ends. If s represents a shortening of the spring, the spring is in **compression**, meaning it *pushes* on objects in contact with its ends. Consider such a spring lying parallel to the x-axis with one end fixed and its free end resting at the origin. To stretch this spring from its natural length by an additional amount x, we must exert $\mathbf{F} = +kx\hat{\mathbf{i}}$ on the spring. Moreover, this applied force is parallel to the displacement $dx\hat{\mathbf{i}}$ as shown in Figure 7.4. The work *we* do to stretch the spring to some final extension s is thus

$$\mathcal{W} = \int_0^s kx \, dx$$

$$= \frac{1}{2}kx^2 \Big|_0^s$$

$$\mathcal{W} = \frac{1}{2}ks^2 \qquad \text{(Work by external agent stretching Hooke's-law spring from 0 to } s) \qquad (7.6)$$

This result can also be obtained graphically. In Figure 7.5 we show a plot of the applied force $F_{appl} = +kx$ versus the extension x. The area under this force curve for total extension s is

$$\mathcal{W} = \frac{1}{2}(Base)(Height) = \frac{1}{2}(s)ks = \frac{1}{2}ks^2$$

During this stretching process the work performed *by the spring* force is *negative* because the spring pulls back along the $-\hat{\mathbf{i}}$ direction. In this case, ϕ in Equation (7.5) is 180°, whereas the magnitude of F_{sp} is still kx. Therefore,

$$\mathcal{W}_{sp} = \cos(180°)\int_0^s kx\,dx$$

$$= -\frac{1}{2}ks^2 \qquad \text{(Work } by \text{ a Hooke's-law spring when stretched from 0 to } s)$$

Can you determine the signs of the work when the spring is returned from its stretched configuration to its natural length? If we slowly return the spring to its natural length, we do $-\frac{1}{2}ks^2$ work, while the spring does $+\frac{1}{2}ks^2$ work. Do not try to remember all these cases! Simply look to see whether the force is parallel to the displacement ($\cos(\phi) = \cos(0°) = +1$) or antiparallel ($\cos(\phi) = \cos(180°) = -1$).

Concept Question 3
Sketch a continuation of the force function shown in Figure 7.3(d) such that the force becomes negative when extended past s_f. (a) What is the physical interpretation of the negative force? (b) Shade the area corresponding to the work done by this force. (c) What is the algebraic sign of the work done by the negative force?

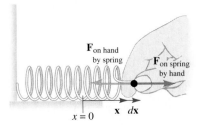

$\mathbf{F}_{\text{on hand}}$ by spring $\mathbf{F}_{\text{on spring}}$ by hand

$x = 0$ **x** $d\mathbf{x}$

FIGURE 7.4

The spring shown has been stretched by a distance x. For this case, the work done *by* the spring is negative, but the work done *on* the spring is positive.

EXAMPLE 7.2   ¡Un Ejemplo Lleno de Dulces!

A spring of stiffness constant $k = 2.00 \times 10^2$ N/m has a natural length of 1.00 m when suspended vertically from the ceiling. A 1.50-kg piñata is attached to the spring and raised, compressing the spring until its end is 0.500 m from the ceiling. The piñata is then released and falls downward. Between the time the piñata is released and the time that it falls through a point 1.200 m below the ceiling (a) how much work is done on the piñata by gravity? (b) How much work is done on the piñata by the spring during this same time interval? (c) What is the total work done on the piñata?

SOLUTION   The forces on the piñata as it moves from its original position to its final position are shown in Figure 7.6.

(a) As the piñata falls the force due to gravity is constant and in the same direction as the displacement. Thus, the work done by the weight force is

$$\mathcal{W} = F\,\Delta s\,\cos(\phi) = mg\,\Delta y\,\cos(0°)$$

$$= (1.50\ \text{kg})(9.80\ \text{N/m})(0.700\ \text{m})(1) = +10.3\ \text{J}$$

(b) To calculate the work done by the spring we must be more careful. During the first 0.500 m of the piñata's descent the spring is in a compressed state and, therefore, it pushes downward on the piñata as it moves downward. That is, the spring force and the displacement are in the same direction. Thus, the work done by the spring on the piñata during this portion of the fall is positive and given by Equation (7.6):

$$\mathcal{W} = \frac{1}{2}ks^2 = \frac{1}{2}(200.\ \text{N/m})(0.500\ \text{m})^2 = +25.0\ \text{J}$$

However, during the next 0.200 m of the 0.700 m fall, the spring is stretched and, therefore, exerts an *upward* force on the *downward* moving piñata. Thus, the work

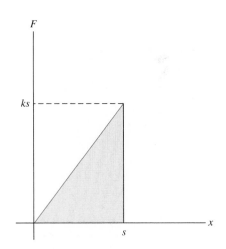

F

ks

s

x

FIGURE 7.5

The force required to stretch a Hooke's-law spring. The work done is the area under the curve and can be calculated from simple geometry as $\frac{1}{2}ks^2$.

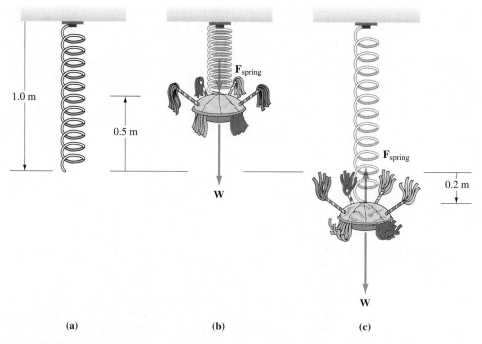

FIGURE 7.6

(a) A spring with natural (unstretched) length 1.00 m hangs from the ceiling. (b) A piñata is attached to the spring and raised a distance 0.50 m by an external agent (not shown). (c) The external agent releases the piñata and allows it to fall downward. (See Example 7.2.)

done by the spring during this portion of the motion is negative:

$$\mathcal{W} = -\frac{1}{2}ks^2 = -\frac{1}{2}(200.\ \text{N/m})(0.200\ \text{m})^2 = -4.00\ \text{J}$$

The total work done by the spring during the entire 0.700-m fall is

$$\mathcal{W} = +25.0\ \text{J} - 4.00\ \text{J} = 21.0\ \text{J}$$

(c) The "net work" done on the piñata means the algebraic sum of the work done on the piñata by all forces. (By "algebraic sum" we mean that if a work value is negative you "add" it as a negative number.) The net work done in this case is

$$\mathcal{W}_{\text{net}} = +10.3\ \text{J} + 21.0\ \text{J} = 31.3\ \text{J} \qquad \blacktriangleleft$$

EXAMPLE 7.3   *What Area Do You Work In?*

The force-versus-extension curve for a non-Hooke's-law spring fixed at one end but free to move at the other is shown in Figure 7.7(a). Compute the work done by an *external* force that displaces the end of the spring from $x = 0.000$ to $x = 0.400$ m.

SOLUTION   As shown in Figure 7.7(a), when this spring is extended to the right $(s > 0)$, it still pulls back to the left $(F < 0)$ similar to a Hooke's-law force. However, for $s > 0.3$ m the restoring force increases faster than predicted by Hooke's law. At each location during the stretch of this spring an external agent must apply a force of the same magnitude as that shown in Figure 7.7(a), but in the positive direction. Such a force is graphed by the red line in Figure 7.7(b). Notice that this force is just the negative of that plotted in Figure 7.7(a). The work done by the *external force* to

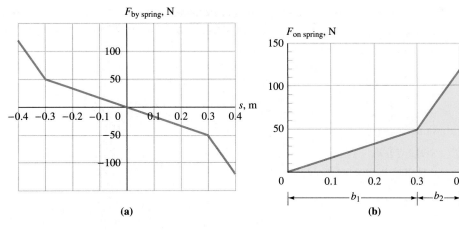

FIGURE 7.7

The force curve of a spring not obeying Hooke's law. (a) The force *by* the spring. (b) The (equal and opposite) force *on* the spring.

stretch the spring from $x = 0$ to $x = 0.400$ m is positive and equal to the shaded area between the force curve and the $F = 0$ axis in Figure 7.7(b). This area can be calculated easily by dividing it into two triangles and a rectangle, and then using elementary formulas from geometry:

$$Work = Shaded\ area = \frac{1}{2}b_1h_1 + b_2h_1 + \frac{1}{2}b_2h_2$$

$$= \frac{1}{2}(0.300\ \text{m})(50.0\ \text{N}) + (0.100\ \text{m})(50.0\ \text{N}) + \frac{1}{2}(0.100\ \text{m})(70.0\ \text{N})$$

$$= +7.5\ \text{J} + 5.0\ \text{J} + 3.5\ \text{J} = +16.0\ \text{J} \qquad \blacktriangleleft$$

7.3 Work (Variable Force, Variable Direction)

In the last two sections we set the stage for a formal and complete definition of work by illustrating the work concept for some special cases. By now we hope you realize that (1) the force must act through some displacement for work to be performed, and (2) the angle between the force and the displacement is important. We are now ready to state the general definition of work. This definition is applicable when the force varies and the angle between the force and the displacement also changes through some path from the original location designated by position vector $\mathbf{s}_o$ to the final location designated by $\mathbf{s}_f$.

$$\mathcal{W} = \int_{s_o}^{s_f} \mathbf{F} \cdot d\mathbf{s} = \int_{s_o}^{s_f} F\cos(\phi)\,ds \qquad (7.7)$$

Evaluation of this integral when both F and ϕ change can be complicated and may even require a numerical technique. The following example shows a simple case where only the angle changes.

(a)

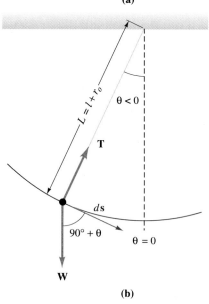

(b)

FIGURE 7.8

(a) A student prepares to swing across the river. (b) The forces acting on the student during his swing. Note that the angle θ is negative when directed to the left of vertical.

EXAMPLE 7.4   *Physics Students Are Real Swingers*

Physics student Sam, of mass m, holds the end of a long rope of length l tied at the other end to the branch of a tree that extends over a lake. Sam, whose height is $2r_o$, stands on the bank above the water in such a way that the rope makes an angle θ_o with the vertical. Sam then steps off the bank and swings out over the water. (a) Calculate the work done by the weight force as Sam moves from the initial position to the point where the rope is vertical. (b) Calculate the work done by the rope tension for this same interval.

SOLUTION   We model this problem by treating Sam as a point-mass (This is *not* how instructors usually think of their students!) attached to a massless rope of length $L = l + r_o$. (If the distance r_o is large compared to the length of the rope, the point–mass model is inappropriate and the problem must be treated by methods described in Chapter 13.) Figure 7.8(b) is a picture of our model showing the (arbitrary) origin of our coordinate system to be at the lowest point along Sam's path.

When the rope makes an angle θ with the vertical, the weight force **W** makes an angle of $(90° + \theta)$ with the instantaneous displacement vector $d\mathbf{s}$, the latter being tangent to the circular arc in which the student travels. (You may be bothered by our use of $(90° + \theta)$ as the angle, but remember that the θ shown in Figure 7.8(b) is a negative number. Try drawing the picture when Sam is on his way back up on the other side of vertical to convince yourself that the $+$ sign is correct.) Thus, the incremental work done is

$$dW = |F||ds|\cos(\phi) = (mg)(ds)\cos(90° + \theta)$$

Now the arc length s is related to the angle θ in radians by $s = L\theta$, so $ds = L\,d\theta$. Moreover, $\cos(90° + \theta) = -\sin(\theta)$. With these substitutions we have

$$dW = -(mg)(L\,d\theta)\sin(\theta)$$

The total work done when the rope moves from θ_o to 0 is

$$\mathcal{W} = -\int_{\theta_o}^{0} mgL\sin(\theta)\,d\theta$$

$$= mgL\int_{0}^{\theta_o}\sin(\theta)\,d\theta$$

$$= mgL\left(-\cos(\theta)\right)\Big|_{0}^{\theta_o}$$

$$= mgL[1 - \cos(\theta_o)]$$

(b) The tension force **T** is along the radius and therefore at any location perpendicular to $d\mathbf{s}$. (The direction of $d\mathbf{s}$ is tangent to the circular arc.) Hence, the work differential is

$$dW = T\,ds\cos(90°) = 0$$

Thus, the tension force does no work on the student.   ◀

7.4 Work and Kinetic Energy

In the previous sections we described several techniques for computing the work done on an object by various forces. In this section we wish to relate the work done on an object to that object's motion. It should come as no surprise to you that we find this relation by applying Newton's second law to the total force acting on the object.

We calculate the work done by the total force $\mathbf{F}_{net}$ along some general path from s_o to s_f:

$$\mathcal{W}_{net} = \int_{s_o}^{s_f} \mathbf{F}_{net} \cdot d\mathbf{s} \qquad (7.8)$$

In Equation (7.8) we replace $\mathbf{F}_{net}$ with $m\mathbf{a}$ (Newton's second law), in which m is the mass of the object experiencing the net unbalanced force $\mathbf{F}_{net}$. For simplicity we take $\mathbf{a}$ parallel to $d\mathbf{s}$ so that the dot product yields $\mathbf{F}_{net} \cdot d\mathbf{s} = (ma) \cos(0°)(ds) = ma\, ds$. Equation (7.8) can then be written

$$\mathcal{W}_{net} = \int_{s_o}^{s_f} ma\, ds \qquad (7.9)$$

The acceleration in Equation (7.9) may be replaced by dv/dt. Using the chain rule from calculus, dv/dt may be written

$$\frac{dv}{dt} = \frac{dv}{ds}\frac{ds}{dt}$$

However, ds/dt is just the definition of speed v, so that

$$\frac{dv}{dt} = \frac{dv}{ds}v$$

Hence, the integral in Equation (7.9) becomes

$$\mathcal{W}_{net} = \int_{s_o}^{s_f} m\left(v\frac{dv}{ds}\right)ds = \int_{s_o}^{s_f} mv\left(\frac{dv}{ds}ds\right) \qquad (7.10)$$

The quantity $(dv/ds)ds$ is the definition of the differential dv. Thus, the integral can be changed to an integration over v, providing we remember to change the position limits to velocity limits; namely, the original velocity v_o at position s_o and final velocity v_f at s_f:

$$\mathcal{W}_{net} = \int_{v_o}^{v_f} mv\, dv$$

This last integral is easy to perform for constant mass m and results in

$$\mathcal{W}_{net} = \frac{1}{2}mv_f^2 - \frac{1}{2}mv_o^2 \qquad (7.11)$$

Equation (7.11) relates the work done on an object to the change in its speed. The rather unlikely combination $\frac{1}{2}mv^2$ occurs so often in physics, and is of such significance, that it is given its own name: **kinetic energy.**

$$K = \frac{1}{2}mv^2 \qquad \text{Definition of kinetic energy} \qquad (7.12)$$

As a consequence, Equation (7.11) is known as the **work–energy theorem** and is sometimes written

$$\mathcal{W}_{net} = K_f - K_o = \Delta K \qquad \text{Work–energy theorem} \qquad (7.13)$$

Concept Question 4
Can the kinetic energy of an object ever be negative? Explain.

Concept Question 5
A ball tossed upward begins with a speed v_o when it leaves your hand. At the top of its flight its speed is zero. Apply the work–energy theorem to explain the ball's change in kinetic energy during this portion of its trajectory.

This theorem is the most important result in this chapter, and its utility is illustrated in Examples 7.5 through 7.7.

EXAMPLE 7.5   *This Must Be What the English Call "High Tea"*

The force of a certain defective toaster on a 28.3-g slice of toast can be modeled using Hooke's law with a spring constant of 75.0 N/m. The spring is compressed by 7.20 cm when starting the toaster. When the toast pops up, what is the speed of the toast as it passes through the point where the spring reaches its uncompressed state?

SOLUTION   As the bread is *raised* 7.20 cm, the two forces acting on it are the *downward* gravitational pull of the earth and the upward push of the spring. We calculate the work done by each,

$$Work\ by\ gravity = (mg)(\Delta y)\cos(180°) = -mg\,\Delta y$$

$$= -(28.3 \times 10^{-3}\ \text{kg})(9.80\ \text{N/kg})(7.20 \times 10^{-2}\ \text{m})$$

$$= -20.0 \times 10^{-3}\ \text{J} = -20.0\ \text{mJ}$$

$$Work\ done\ by\ spring = +\frac{1}{2}ks^2 = +\frac{1}{2}k(\Delta y)^2$$

$$= \frac{1}{2}(75.0\ \text{N/m})(7.20 \times 10^{-2}\ \text{m})^2 = 194\ \text{mJ}$$

Thus, the total work done on the slice of toast is

$$\mathcal{W}_{net} = Work\ by\ gravity + Work\ done\ by\ spring = 194\ \text{mJ} - 20.0\ \text{mJ} = 174\ \text{mJ}$$

According to the work–energy theorem

$$\mathcal{W}_{net} = K_f - K_o = \frac{1}{2}mv_f^2 - \frac{1}{2}mv_o^2$$

Because in this case $v_o = 0$, when we solve for v_f we obtain

$$v_f = \sqrt{\frac{2\mathcal{W}_{net}}{m}}$$

$$= \sqrt{\frac{2(174 \times 10^{-3}\ \text{J})}{28.3 \times 10^{-3}\ \text{kg}}} = 3.51\ \text{m/s}$$ ◀

EXAMPLE 7.6   *Look Out Below!*

If the student riding the inner tube described in Example 7.1 started down the slope with a speed of 0.500 m/s, (a) calculate her kinetic energy after traveling the 2.00 m down the incline. (b) Calculate her speed at this point.

SOLUTION   In Example 7.1 we found the work done by each of the forces that act on the tuber. (See Fig. 7.2.) Summing these, we find the net work done on the tuber:

$$Work\ by\ weight\ force = +600.\ \text{J}$$

$$Work\ by\ friction \qquad = -300.\ \text{J}$$

$$Work\ by\ normal\ force = \qquad 0\ \text{J}$$

$$\overline{Net\ work\ done \quad \mathcal{W}_{net} = +300.\ \text{J}}$$

(a) We rearrange the work–energy theorem to find the final kinetic energy K_f

$$\mathcal{W}_{net} = K_f - K_o$$

$$K_f = K_o + \mathcal{W}_{net}$$

$$= \frac{1}{2}mv_o^2 + \mathcal{W}_{net}$$

$$= \frac{1}{2}\left(\frac{500. \text{ N}}{9.80 \text{ N/kg}}\right)(0.500 \text{ m/s})^2 + 300. \text{ J}$$

$$= 6.38 \text{ J} + 300. \text{ J} = 306 \text{ J}$$

Note that we calculated the mass of the student from $m = W/g$.

(b) Now that the final kinetic energy is known, the final speed can be computed easily:

$$K_f = \frac{1}{2}mv_f^2$$

$$v_f = \sqrt{\frac{2K_f}{m}}$$

$$= \sqrt{\frac{2(306 \text{ J})}{\left(\frac{500 \text{ N}}{9.80 \text{ N/kg}}\right)}} = 3.47 \text{ m/s}$$

Note that the quantity v is the speed, hence only the positive value of the square root is considered. ◄

EXAMPLE 7.7 *But It's Not as Much Fun Traveling in a Straight Line*

If the student holding onto the end of the rope in Example 7.4 starts on the bank from rest, how fast does he travel at the bottom of the arc?

SOLUTION In Example 7.4 we found that the only work performed on the student was that done by gravity:

$$\mathcal{W}_{net} = mgL[1 - \cos(\theta_o)]$$

From the information that $v_o = 0$ and the work–energy theorem we can calculate the final speed of the student:

$$\mathcal{W}_{net} = \Delta K = K_f - K_o = K_f - 0 = \frac{1}{2}mv_f^2$$

Solving for v_f we obtain

$$v_f = \sqrt{2gL[1 - \cos(\theta_o)]}$$

You should notice that the term $L[1 - \cos(\theta_o)] = L - L\cos(\theta_o)$ is the change in vertical height of the student. If we call this height h, then our answer becomes $v_f = \sqrt{2gh}$. This answer is exactly the value we obtain from the constant-acceleration model for objects in free fall when they descend a vertical distance h after starting from rest. Here is the point you should not miss: When the only work done is that performed by gravity, a mass attains the same final speed *regardless of the path it takes* between the two heights. This point foreshadows an important general principle that we will explore in more detail in Chapter 8. ◄

7.5 Power

Our discussion of work would not be complete without mention of the related concept of power. **Power** is the time rate of energy transfer. For example, when a net force F accelerates a mass m, kinetic energy is transferred to the mass. The amount of kinetic energy transferred to the mass is equal to the work done by the net force. If the net force does work W during time Δt, the average power is defined as

$$\mathcal{P}_{av} = \frac{\Delta W}{\Delta t} \tag{7.14}$$

If the amount of work done can be written as an explicit function of time, then it is possible to calculate the instantaneous power from the time derivative of the work:

$$\mathcal{P} = \frac{dW}{dt} \tag{7.15}$$

We can find an alternative expression for the instantaneous power by writing $\mathcal{P}\,dt = dW$ and $dW = \mathbf{F} \cdot d\mathbf{s}$ so that

$$\mathcal{P} = \frac{(\mathbf{F} \cdot d\mathbf{s})}{dt} = \mathbf{F} \cdot \frac{d\mathbf{s}}{dt}$$

$$= \mathbf{F} \cdot \mathbf{v} \tag{7.16}$$

Equation (7.16) provides the instantaneous power delivered by a force $\mathbf{F}$ to an object moving at velocity $\mathbf{v}$.

As we proceed through the text we will encounter many forms of energy in addition to the kinetic energy. The definition of power may be extended to include the rate at which any form of energy is transferred.

In the SI system, power is expressed in joules per second, which is defined as a **watt**, after James Watt, the Scottish inventor. From the definition of the watt it is clear that a watt · second is a joule. As a larger practical unit, electric companies usually bill customers for energy in units of kilowatt · hour (kW · h). Substitution of 3600 s in place of hours should quickly convince you that

$$1 \text{ kW} \cdot \text{h} = 3\,600\,000 \text{ J}$$

Quite a bargain, considering 3.6 million joules still costs only a few cents!

The work unit in the British Engineering system is the foot pound (ft · lbf), and the power unit is the foot pound per second (ft · lbf/s). However, another unit, the **horse power (hp)**, is commonly used and is equal to 550 ft · lbf/s. As you might guess, a bit of interesting history goes along with this definition.

James Watt, the Scotsman for whom the power unit is named, introduced a major improvement in the design of steam engines in 1765. Steam engines work by heating water until it becomes steam, then injecting this high-pressure steam into a cylinder where it pushes a piston forward. The steam is then cooled to condense it back into water, creating a partial vacuum into which the piston is pushed by external air pressure. Before Watt, the piston cylinder itself was cooled to condense the steam during each stroke. Watt realized that a lot of fuel was wasted by repeatedly reheating the cylinder each time new steam was introduced. Instead, Watt used not only a separate boiler to heat the steam, but also a separate condenser to cool it, with valves to control the flow of the steam to the cylinder. Thus (as they used to say in one 1980s hamburger commercial), the hot side stayed hot and the cold side stayed cold. This improvement resulted in doubling the amount of work a steam engine could do for the same amount of fuel. The industrial revolution was off and running!

Concept Question 6
Can the average and instantaneous power ever be equal? Explain.

Concept Question 7
One often speaks of paying an electric bill sent by the "power company." When billed for a certain number of kilowatt · hours, are you really being billed for the *power* used? Explain.

The original definition of the unit horsepower was based on the rate at which horses can raise water from the depths of coal mines.

In order to market his improved steam engine, Watt had to show not only that it was better than other steam engines, but also that it was better than the other main existing competition: horses. One of the major uses for steam engines in those days was pumping water out of coal mines. Watt found that a strong horse working at a typical rate could raise 150 lbf at over 3.6 ft/s. Thus was born the definition of one horse power as 550 ft · lbf/s. Still used in engineering today, the horse power is defined in terms of the watt as

$$1 \text{ hp} = 746 \text{ W (exactly)}$$

EXAMPLE 7.8   *This Is a True Story!*

A former professional football player was helping train players on a college team. He wanted to quantify the power of some of the players and used himself for an example. He picked up two of the lighter players and put one on each shoulder. He then climbed two flights of stairs carrying the players. The players he carried had a total weight of 360. lbf, the height of the stairs he climbed was 24.0 ft. It took him 17.0 s to climb the stairs. (He apologetically mentioned being tired by the time he reached the top.) Calculate this trainer's average power output to raise the players through this height in this time.

SOLUTION   The work done by the trainer is computed from the product of the force he applied and the distance through which he applied this force:

$$\Delta W = F \, \Delta y = (360. \text{ lbf})(24.0 \text{ ft}) = 8640 \text{ ft} \cdot \text{lbf}$$

The power is the average rate of doing work:

$$\mathcal{P}_{av} = \frac{\Delta W}{\Delta t} = \frac{8640 \text{ ft} \cdot \text{lbf}}{17.0 \text{ s}} = 508 \text{ ft} \cdot \text{lbf/s}$$

In horse power this is

$$\mathcal{P}_{av} = \frac{508 \text{ (ft} \cdot \text{lbf/s)}}{550 \text{ (ft} \cdot \text{lbf/s)/hp}} = 0.924 \text{ hp}$$

a remarkable athletic feat in any units!

James Watt (1736–1819). Son of a Scottish carpenter, Watt became skilled in the use of tools in his father's workshop. He studied the construction and repair of machines in London. At the age of 21 he returned to Scotland where he obtained a position as an instrument maker at the University of Glasgow. He is shown here contemplating a model steam engine used for lecture demonstrations. His analysis of the energy wasted in reheating the cylinder walls with every stroke of the piston led to his invention of the separate steam condenser. He was 27 years old at the time.

The archer does work on the bow string while pulling it back. The string does work on the arrow when the string is released.

7.6 Numerical Calculations of Work (Optional)

In Section 7.4 we saw that work can be calculated as the integral of a force acting over a distance. In Section 3.5 of this text we discussed numerical methods for integration of functions of a single variable. In particular, we presented formulas for Simpson's rule. Exactly the same techniques can be used to calculate the work, providing we know the magnitude of the force acting at each position.

As a simple illustration, in Example 7.9 we calculate the work necessary to pull back a compound bow.

EXAMPLE 7.9   . . . with Thanks to McMillan's Sports for Assistance Taking Data

The data in Figure 7.9 show the force versus extension necessary to pull back the string of a compound bow. Calculate the work done to bring the string back 40.0 cm.

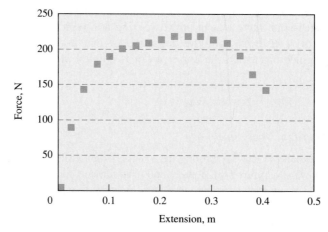

FIGURE 7.9

The force-versus-extension curve for an actual compound bow.

SOLUTION   We use Simpson's rule for integration as introduced in Chapter 3 and summarized in Appendix 2. Changing variables to match the work integral, Simpson's rule is written

$$W = \frac{\Delta x}{3} \sum_{i=1}^{8} (F_{2i-1} + 4F_{2i} + F_{2i+1})$$

for the seventeen force values labeled F_j, where $j = 1$ to 17. In our experiment the Δx interval is 0.0254 m because the original displacements were measured in inches. In the spreadsheet template CMPDBOW.WK1 we list distances in column A, forces in column B, and the first term of the sum

$$\left(\frac{\Delta x}{3}\right)\left(F_{2i-1} + 4 \cdot F_{2i} + F_{2i+1}\right)$$

is calculated in cell C11 using the formula

$$\text{C11:} \quad (0.0254/3)*(B10 + 4*B11 + B12)$$

The cell C12 is left blank. The pair of cells (C11 and C12) is then copied into cells C13 to C14 so that the odd cells contain the formula and the even cells are blank. This copying procedure is repeated through cells C25 and C26. The total area under the curve shown in Figure 7.9 is found by summing column C. The result is

$$\mathcal{W} = 74.8 \text{ J} \qquad \blacktriangleleft$$

	A	B	C	D	E	F
5	Calculated values (MKS):				Original data:	
6						
7	Distance	Force	Simpson-rule terms		Distance	Force
8	(m)	(N)	(see text)		(inches)	(lbf)
9						
10	0.000	0			0	0
11	0.025	89	4.2		1	20
12	0.051	142			2	32
13	0.076	178	8.8		3	40
14	0.102	189			4	42.5
15	0.127	200	10.1		5	45
16	0.152	205			6	46
17	0.178	209	10.6		7	47
18	0.203	214			8	48
19	0.229	218	11.0		9	49
20	0.254	218			10	49
21	0.279	218	11.0		11	49
22	0.305	214			12	48
23	0.330	209	10.5		13	47
24	0.356	191			14	43
25	0.381	165	8.4		15	37
26	0.406	142			16	32
27						
28	Total column C:		74.8	Nm = total work		

FIGURE 7.10

A spreadsheet template to calculate the area under the curve using Simpson's rule.

7.7 Summary

Work is defined as

$$\mathcal{W} = \int_{s_o}^{s_f} \mathbf{F} \cdot d\mathbf{s} = \int_{s_o}^{s_f} F \cos(\phi) \, ds \qquad (7.7)$$

where ϕ is the angle between the force $\mathbf{F}$ and the displacement $d\mathbf{s}$. If the force and angle are constant this definition reduces to

$$\mathcal{W} = F \, \Delta s \cos(\phi) \qquad \text{(Constant } F \text{ and } \phi)$$

Kinetic energy is defined as

$$K = \frac{1}{2}mv^2 \tag{7.12}$$

The work–energy theorem relates the net work done to the change in kinetic energy:

$$\mathcal{W}_{net} = K_f - K_o = \Delta K \tag{7.13}$$

Power is the time rate of energy transfer. When this power transfer is due to a net force doing work on an object, the power is related to the work through the following expressions.

Average power: $\mathcal{P}_{av} = \dfrac{\Delta W}{\Delta t}$ \hfill (7.14)

Instantaneous power: $\mathcal{P} = \dfrac{dW}{dt}$ \hfill (7.15)

Instantaneous power delivered by a moving object: $\mathcal{P} \doteq \mathbf{F} \cdot \mathbf{v}$ \hfill (7.16)

(Optional) The work integral can be calculated using numerical integration techniques, such as Simpson's rule as described in Chapter 3.

PROBLEMS

7.1 Work (Constant Force, Constant Direction)

1. A 50.0-N force acts at an angle of 36.87° below the horizontal (Fig. 7.P1) while the object that the force is acting on travels 4.00 m horizontally. Calculate the work done by this force.

FIGURE 7.P1   Problem 1

2. An unknown force F acts at 40.0° above the horizontal as shown in Figure 7.P2. If the force F does 50.0 J of work while the object moves forward 4.00 m horizontally, what is the magnitude of F?

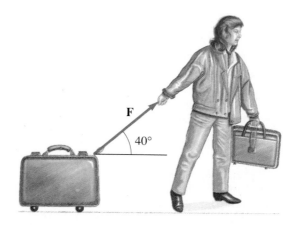

FIGURE 7.P2   Problem 2

3. You raise a 5.00-kg mass vertically 1.50 m at constant speed. You then translate the mass horizontally 2.00 m, again at constant speed. Finally, you lower the mass 0.500 m at constant speed. (a) Calculate the work *you* do for each of the three segments of the motion and the total work you do for the complete trip. (b) Calculate the work done by gravity on each of the three legs of the trip and the total work done by gravity for the entire trip.

4. (a) If you bench-press 150. lbf, raising the weights 16.0 in. with each lift, how much work do you do raising the weights? Express your answer in foot pounds and joules. (b) How much work do you do each time you lower the weights? (c) If you do ten complete repetitions, what is the total work done?

5. A constant horizontal force of 6.00 N $\widehat{\mathbf{i}}$ acts on a 2.00-kg mass as it moves along the path shown in Figure 7.P3. Calculate the work done by this force along each segment of the path and the total work done by the 6.00-N force along the entire path.

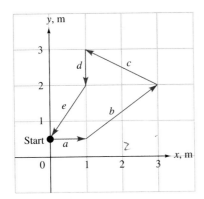

FIGURE 7.P3 Problem 5

6. Three constant forces act on a mass as it traverses the path shown in Figure 7.P4. (a) Calculate the work done by each of the three forces along this path. (b) Calculate the net work done on the mass by all three forces.

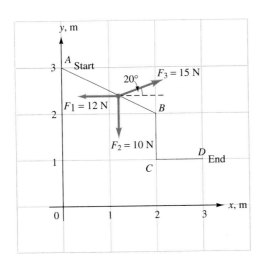

FIGURE 7.P4 Problem 6

7. Three constant forces act on a 5.00-kg mass as it traverses the path shown in Figure 7.P5. If the net work done by the three forces is 750. J, calculate the magnitude of the force P.

8. A force given by $\mathbf{F} = 8.00$ N $\widehat{\mathbf{i}} + 4.00$ N $\widehat{\mathbf{j}}$ acts on an object as it moves through a displacement $\Delta \mathbf{s} = 3.00$ m $\widehat{\mathbf{i}} + 2.00 \,\widehat{\mathbf{j}}$. What is the work done by this force?

9. Evaluate the work done by each of the following forces acting over the displacement given: (a) $\mathbf{F} = 3.00$ N $\widehat{\mathbf{i}} - 2.00$ N $\widehat{\mathbf{j}} + 4.00$

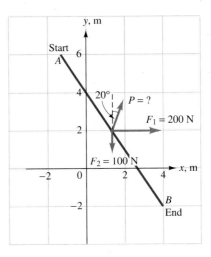

FIGURE 7.P5 Problem 7

N $\widehat{\mathbf{k}}$, $\Delta \mathbf{s} = 8.00$ m $\widehat{\mathbf{i}} + 2.00$ m $\widehat{\mathbf{j}} - 3.00$ m $\widehat{\mathbf{k}}$. (b) In the xy-plane $\mathbf{F} = 4.00$ N acting at 12.0° anticlockwise from the positive x-axis, $\Delta s = 8.00$ m along a line making an angle of 65.0°, also anticlockwise from the x-axis.

10. (a) If a force given by $\mathbf{F} = 8.00$ N $\widehat{\mathbf{i}} - 4.00$ N $\widehat{\mathbf{j}}$ acts on an object moving along the displacement vector $\Delta \mathbf{s} = 3.00$ m $\widehat{\mathbf{i}} + 2.00$ m $\widehat{\mathbf{j}}$, how much work does the force do on the object? (b) How much work is done by a force $\mathbf{F} = 2.00$ N $\widehat{\mathbf{i}} + 3.00$ N $\widehat{\mathbf{j}}$ when it moves a 3.00-kg mass through a displacement of $\Delta \mathbf{s} = 6.00$ m $\widehat{\mathbf{i}} + 2.00$ m $\widehat{\mathbf{j}} + 1.50$ m $\widehat{\mathbf{k}}$?

7.2 Work (Variable Forces, Constant Direction)

11. A Hooke's-law spring with stiffness constant $k = 30.0$ N/m has an unstretched length of 40.0 cm. How much work must you do on the spring to change its length from 50.0 cm to 60.0 cm?

12. A Hooke's-law spring has a spring constant of $k = 250.$ N/m. Calculate the work done by the spring as its length is changed from being compressed by 0.200 m to being stretched by 0.500 m.

13. A Hooke's-law spring can be stretched 0.150 m beyond its equilibrium length by a force of 25.5 N. (a) How much work do you do if you change the spring from being stretched by 0.200 m to being compressed by 0.300 m? (b) How much work does the spring do on you in the case described in part (a)?

14. An electrical engineer designs a servosystem so that a robotic arm exerts a restoring force given by $F = -(5.00 \times 10^2 \text{ N/m}^3)x^3$ if the arm is displaced a distance x from its set point. How much work does the arm do when it moves a mass to its set point, if it starts from a displacement of 5.00 cm?

15. A non-Hooke's-law spring has a force function described by $F = -as - bs^3$, where a and b are constants and s is the distance the spring is stretched or compressed from its natural length. (a) How much work must be done to stretch this spring by an amount x? (b) What are the units of the constants a and b in SI? If $a = 100$ and $b = 10$ in SI units, calculate the total work done to stretch the spring 0.300 m.

16. The attractive force in newtons between two small magnets can be approximated as $F = A/(a + x)^2$, where A and a are constants and x is the distance in meters between the magnet faces. (a) What are the SI units of A and a? (b) How much work must you do to

separate two touching magnets to positions infinitely far apart? (c) Show that your answer has the units of work.

17. A loosely wound non-Hooke's-law spring exerts a restoring force given by $F = -A\sqrt{s}$ where $A = 30.0 \text{ N}/\sqrt{\text{m}}$. Calculate the work done by this spring if it is compressed by 0.250 m.

18. The force on an atom bound to a molecule can be crudely modeled as shown in Figure 7.P6. How much work must be done to move the atom from $x = a$ to $x \gg d$?

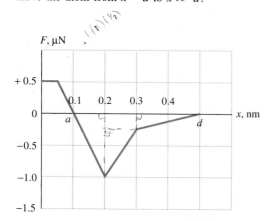

FIGURE 7.P6   Problem 18

19. Consider the force function in Figure 7.P7. (a) In what x-regions is the force repulsive (i.e., directed away from the origin)? (b) In what regions is the force attractive? (Is this an attractive or repulsive problem?) (c) What is the total work done by this force as it acts on an object moving from $x = 0$ to $x = 5.00$ m? (d) What is the total work done by this force if it acts on an object moving from $x = 5.00$ m to $x = 0$?

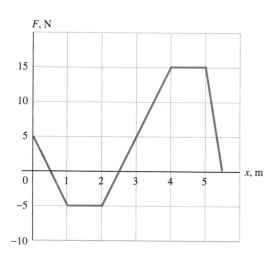

FIGURE 7.P7   Problem 19

20. Answer the questions posed in Problem 19 for the force function shown in Figure 7.P8.

21. A force function is given by the expression

$$\mathbf{F}(x) = (Ax^3 - Bx)\,\widehat{\mathbf{i}}$$

where $A = 2.00 \text{ N/m}^3$ and $B = 2.00 \text{ N/m}$. (a) Sketch this function. (b) Calculate the work done by this force if it acts on an object as it is moved from $x = 1.00$ m to $x = 3.00$ m.

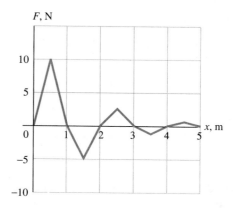

FIGURE 7.P8   Problem 20

7.3 Work (Variable Forces, Variable Direction)

★22. A particle moves in the xy-plane subject to a force $F = x(\widehat{\mathbf{i}} + \widehat{\mathbf{j}})$. If the particle is constrained to move along the path $y = 4x^2$, how much work is done by the given force on the particle as it moves from $(0, 0)$ to $(0, 4)$? [*Hint*: Take $d\mathbf{s} = dx\,\widehat{\mathbf{i}} + dy\,\widehat{\mathbf{j}}$ and perform the dot product with **F**. Note that before you can perform the integrations, each integral must be written as a function of a single variable. The path equation must be used to relate x and y.] See Problem 61 for a generalization of this result.

23. Rework Example 7.4 using the hint given in Problem 7.22. Take as the object a simple pendulum of length L; as shown in Figure 7.P9 the path is described by $y = -\sqrt{L^2 - x^2}$.

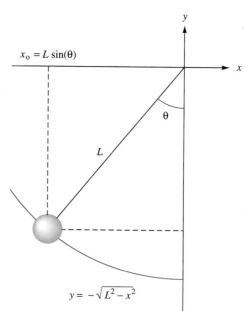

FIGURE 7.P9   Problem 23

24. A particle travels a path in the xy-plane by following straight line segments from $(0.00, 0.00)$ to $(0.00, 0.30$ m$)$ to $(0.40$ m$, 0.30$ m$)$ to $(0.40$ m$, 0.40$ m$)$. As it travels this path, a force given by $\mathbf{F} = 0.500 \text{ N}\,\widehat{\mathbf{i}} + 0.200 \text{ N}\,\widehat{\mathbf{j}}$ acts on the particle. (a) How much work does this force do on the particle as it travels the three legs of the path? (b) If instead of following the path described in part (a), the particle moves directly from $(0, 0)$ to $(0.40$ m$, 0.40$ m$)$ along a straight path, how much work does the force do?

7.4 Work and Kinetic Energy

25. What is the kinetic energy of (a) a 0.142-kg baseball traveling 35.0 m/s? (b) a 16.0-lbf bowling ball traveling 25.0 ft/s?

26. Calculate the kinetic energy in joules of (a) a 9.70×10^{-3}-kg bullet traveling 500. m/s and (b) a 190.-lbf fullback running 10.0 m/s.

27. How fast must a 0.0100-g mosquito fly to have the same kinetic energy as a proton moving 3.00×10^6 m/s?

28. How fast must a 110.-lb golfer move to have the same kinetic energy as a 4.60×10^{-2}-kg golf ball traveling 50.0 m/s?

29. When a 0.190-kg softball is thrown toward you it reaches your hand with a speed of 12.0 m/s. (a) What is the ball's kinetic energy? (b) You catch it bare-handed. If you keep your arms stiff you can stop the ball in a distance of 1.00 cm. What average force must your hands exert to bring the ball to rest? (c) If you bring your hands 25.0 cm backward smoothly as you catch the ball, what average force does the ball exert on your hands?

30. The work–energy theorem is quite general and requires no restriction on the acceleration. It should thus contain the rules for constant acceleration as a special case. Show that the constant-acceleration kinematic equation $v^2 = v_o^2 - 2a\,\Delta s$ follows from the work–energy theorem.

31. A 4.00-kg block is traveling 15.0 m/s. (a) How fast is the block traveling if 2.00×10^2 J of work is done on the block? (b) If -2.00×10^2 J of work were done on the block instead, what would be its final speed?

32. A 750.-kg automobile is traveling 20.0 m/s. (a) What is the automobile's new speed if 150. kJ net work is done on the car? (b) How much work must be done on the car to bring it to rest (after the work in part (a) is added)?

33. A graph of the force versus extension for a slingshot is shown in Figure 7.P10. (a) Calculate the work done by someone pulling the slingshot from $s = 0.00$ to $s = 0.80$ m. (b) If a 50.0-g ball bearing is fired from the slingshot, how fast is it traveling when it leaves the slingshot? (Assume the force on the ball bearing is identical to the force curve in the figure. The force is actually somewhat smaller since the bands themselves must be accelerated.)

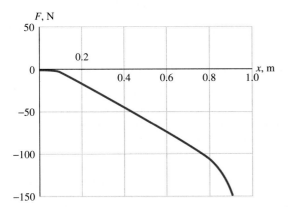

FIGURE 7.P10   Problem 33

34. A 110.-g clay pigeon is launched horizontally by a Hooke's-law spring with an effective spring constant of 90.0 N/m. If the spring is compressed by 0.500 m, how fast is the pigeon moving when it leaves the thrower?

35. A 2.50-kg mass is pressed against an ideal massless horizontal spring compressing it 0.500 m. The stiffness constant of the spring is 1.00×10^2 N/m. The mass is released, and the spring pushes the block forward along a rough horizontal surface. If the coefficient of friction between the block and the surface is $\mu_k = 0.400$, how fast is the block going when it is 1.50 m beyond the point of release?

36. A block is pushed up an inclined plane by a horizontal force **P** as shown in Figure 7.P11. Attached to the upper end of the incline is a spring with stiffness constant $k = 320.$ N/m. The horizontal force continues to act as the block moves an additional 0.500 m along the incline. Assume the following magnitudes for the forces acting on the block as it moves the 0.500 m after contacting the spring: $P = 300.$ N, $N = 260.$ N, $W = 100.$ N, and $F_{\text{fric}} = 70.0$ N. Calculate the work done on the block by each of the following forces for the 0.500-m distance: (a) the push force **P**, (b) the normal force **N**, (c) the weight force **W**, (d) the frictional force $\mathbf{F}_{\text{fric}}$, (e) the spring. (f) What is the net work done on the block over the 0.500-m distance? (g) If the block had a speed of 1.00 m/s when it first contacted the spring, what is its speed after traveling the 0.500 m as described above?

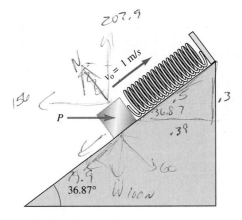

FIGURE 7.P11   Problem 36

★37. An ideal Hooke's-law spring ($k = 1.00 \times 10^2$ N/m) is attached to the upper end of an inclined plane as shown in Figure 7.P12(a). The region of the incline on which the spring rests is frictionless. However, the portion of the incline beyond the end of the spring is rough, with a coefficient of kinetic friction of 0.333 between it and a 5.00-kg block. The block is permanently attached to the spring and the spring is compressed 0.200 m as shown in Figure 7.P12(b). The block is then released and moves down the

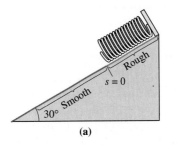
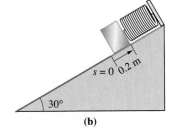

FIGURE 7.P12   Problem 37

incline to a point 0.400 m beyond the point where the spring is at its natural length (Fig. 7.12(c)). Calculate the work done by (a) the spring, (b) the normal force, (c) the weight force, and (d) friction, as the block moves from its release point to the 0.400-m point. (e) Finally, find the speed of the block at the 0.400-m point.

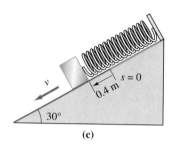

(c)

FIGURE 7.P12   Problem 37 (Continued)

38. An 8.00-kg block is permanently attached to a spring on an incline as shown in Figure 7.P13(a). The block initially compresses the spring as shown in Figure 7.P13(b), and the block is then released from rest. Questions (a) through (e) concern the work done on the block as it moves from the position shown in Figure 7.P13(b) and a position 0.250 m below the normal end of the spring as shown in Figure 7.P13(c). Calculate the work done by (a) the normal force, (b) the gravitational force, (c) friction, and (d) the spring. (e) What is the net work done on the block? (f) Find the speed of the block as it passes the point indicated in Figure 7.P13(c).

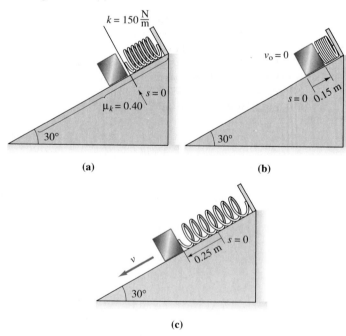

FIGURE 7.P13   Problem 38

39. An experimental rail-gun exerts a force on a 10.0-g wax plug as the plug proceeds down a straight rail. The magnitude of the force is given by

$$F = A - B(x - 2.00 \text{ m})^2, \qquad 0 \le x \le 4.00 \text{ m}$$

and is equal to zero elsewhere. (a) Take $A = 4.00$ kN and $B = 1.00$ kN/m^2 and sketch the force-versus-distance curve. (b) Find

the total work done on the projectile as it moves from $x = 0$ to $x = 4.00$ m. (c) Find the speed of the projectile at $x = 4.00$ m if it started from rest at $x = 0$.

40. In drivers' training courses instructors often warn students that braking distance increases with the square of velocity. For example, an automobile traveling 50 mph takes four times as far to stop as one traveling 25 mph. Use the coefficient-of-friction model and the work–energy theorem to verify this claim.

7.5 Power

41. A student earns a few extra dollars by cutting lawns. A yard is 20.0 m long and 10.0 m wide. The student can cut a strip of grass 0.45 m wide with each pass. If the student exerts a force of 35.0 N at an angle of 35.0° from the horizontal when pushing the mower, (a) how much total work does the student do? (b) If it takes the student 20.0 min to complete the job, what is the average power expended by the student? Answer in horsepower and in watts.

42. For bicyclists the main source of resistance is air drag. (a) Show that the quadratic model for air resistance $D = \frac{1}{2}C_D A \rho v^2$ (see Section 6.3) implies that the power required to maintain a constant speed v is proportional to v^3. (b) Estimate the power required for a cyclist with a cross-sectional area $A = 0.750$ m^2 and a drag coefficient $C_D = 1.40$ to maintain a speed of 40.0 km/h. (c) If the drag coefficient could be cut in half, how much faster could the cyclist go using the same power as in part (b)?

43. One month, the electric utility bill for a home was $61.19 for 870 kW · h. (a) What is the average cost of a joule of work? (b) If all the electrical work that month went to raise a 10 metric ton (10 000 kg) weight, how high was the weight raised?

44. While rafting, the oarsman pulls with a constant force of 45.0 N on each of the two oars during a stroke of length 0.800 m. What is the average power output, if the oarsman makes ten strokes in 30.0 s?

45. A small, pulsed ruby-laser delivers 30.0 mJ of light energy in 20.0 ns. (a) What is the average optical output power of the laser during firing? (b) The laser can be fired at a maximum repetition rate of one shot every 1.50 s. What is the average optical power output of the laser when firing at the maximum repetition rate?

46. A cuckoo clock is powered by a mass of 0.800 kg hanging from a light chain. The mass falls a distance of 1.30 m during 24.0 h. What is the average power supplied by the falling weight?

7.6 Numerical Calculations of Work (Optional)

47. Force-versus-deformation data for a tennis ball in compression is shown in Figures 7.P14 and 7.P15 and included in the file TENNISB.WK1 on the diskette accompanying this text. For a given deformation the force is greater when the ball is being compressed than when it is being released. (a) Use Simpson's rule implemented on a spreadsheet or computer program to calculate the work done on the ball as it is being compressed and the work done by the ball as it is expanding back to its normal state. (b) Determine what fraction of the work done compressing the ball is not recovered when the ball flexes back. (c) If a 65.0-g ball was subjected to these forces when it was dropped onto a hard surface, use the work–energy theorem to calculate the velocity of the ball just before and just after the collision. (d) From your results in part (c) calculate the coefficient of restitution of the tennis ball. The coefficient of restitution is defined as $\epsilon = $ (Speed of separation)/(Speed of approach) and is discussed in more detail in Chapter 9.

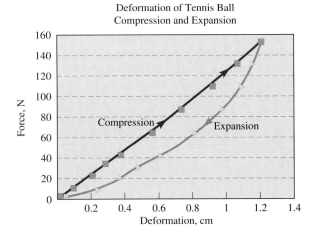

Deformation of Tennis Ball
Compression and Expansion

FIGURE 7.P14   Problem 47. The force required to statically deform a tennis ball, one side of which is in a hemispherical cup. (Adapted from data by H. Brody, "Physics of the Tennis Racket," *American Journal of Physics*, 47 (1979): 482–487.)

Problem 7.47: Deformation of a tennis ball

Plot Coordin. (cm)	Compression: Position (cm)	Force (N)	Expansion: Position (cm)	Force (N)
0	0	0		
0.05			0.05	0
0.10	0.1	10		
0.21	0.21	24		
0.24			0.24	10
0.29	0.29	35		
0.37			0.37	22
0.38	0.38	44		
0.48			0.48	32
0.57	0.57	65		
0.61			0.61	43
0.74	0.74	88		
0.82			0.82	65
0.93	0.93	110		
0.98			0.98	88
1.08	1.08	132		
1.10			1.10	109
1.16			1.16	131
1.22	1.22	153	1.22	153

(b)

48. The strings of a tennis racket were deformed by the force shown in Figure 7.P16, which acted over a circular area of 12.0 cm². Data is also contained in TRACKET.WK1 on the diskette accompanying this text. (a) Use a numerical integration technique to determine the total work done in deforming the strings by 1.4 cm. (b) Use the least-square regression capabilities of your spreadsheet software, or other technique, to determine the slope of the force-versus-deformation graph between $s = 0.0$ and $s = 1.0$ cm. This slope is the stiffness constant of the equivalent Hooke's-law spring.

Problem 7.48:

String Deformation (cm)	Force (N)
0.00	0
0.10	30
0.20	55
0.30	75
0.33	95
0.42	115
0.47	140
0.55	150
0.62	175
0.65	195
0.75	215
0.83	235
0.83	240
0.87	250
0.88	255
0.97	285
1.00	300
1.07	320
1.13	340
1.18	370
1.27	390
1.31	405
1.35	425
1.40	440

FIGURE 7.P15   The work done compressing a tennis ball and deforming the racket strings is described in Problems 47 and 48.

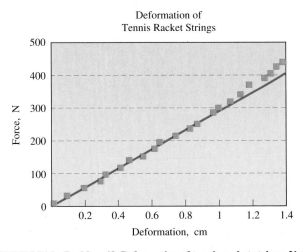

Deformation of Tennis Racket Strings

FIGURE 7.P16   Problem 48. Deformation of tennis racket strings. Note that the Hooke's law approximation is quite good. (Adapted from data by H. Brody, "Physics of the Tennis Racket," *American Journal of Physics*, 47 (1979): 482–487.)

49. Write a computer program to calculate the integral given in part (a) of Problem 61. Test your program by using the path and force given in part (b) of that problem.

General Problems

50. A 3.60-kg mass with a hole through the center slides without friction along a wire bent in the shape shown in Figure 7.P17. During this motion a constant force -8.00 N $\hat{\mathbf{j}}$ acts on the mass. (a) Calculate the work done by this force along each segment of the path and the total work done by the 8.00-N force along the entire path. (b) If the mass leaves the origin with a speed of 2.00 m/s, what is its speed at the end point?

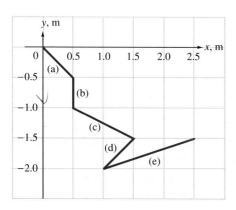

FIGURE 7.P17   Problem 50

51. If a hole were bored through the center of a planet with uniform density, the force on a mass m dropped into such a hole would be given by $F = mgr/R$, where r is the (variable) distance of the mass from the center of the planet, R is the (constant) radius of the planet, and g is the gravitational field strength at the surface of the planet. If the mass were dropped from rest at the surface, what would be the speed of the mass m as it passed through the center of the planet? (Assume the planet has no atmosphere.)

52. Figure 7.P18 shows the force acting on a 125-g ball bearing as it is fired from a slingshot. What is the speed of the ball bearing as it leaves the slingshot at $x = 75.0$ cm?

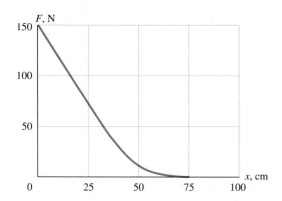

FIGURE 7.P18   Problem 52

53. When a certain 250.-g plastic ball collides with a wall, the force on it behaves as shown in Figure 7.P19. The upper line shows the force as the ball is compressed; the lower line shows the restoring force as the ball bounces away. (a) What was the speed of the ball just before touching the wall? (b) What is the speed of the ball as it leaves the wall?

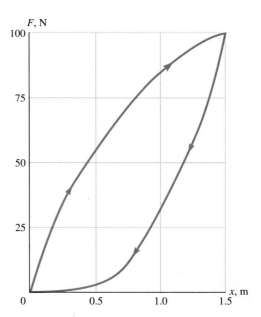

FIGURE 7.P19   Problem 53

54. As a major-league pitcher throws a 0.145-kg baseball the force acting along the ball's path varies as shown in Figure 7.P20. What is the speed of the ball as it leaves the pitcher's hand?

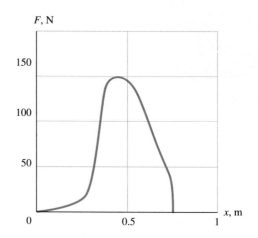

FIGURE 7.P20   Problem 54

55. When a 3.50-kg mass is acted on by the force shown in Figure 7.P21, its speed changes from 1.50 m/s at $x = 0.000$ m to 2.50 m/s at $x = 0.500$ m. If the force shown is the only force acting on the mass, what is the magnitude of F_{max}?

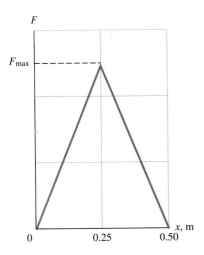

FIGURE 7.P21   Problem 55

56. Consider a mass of weight W. (a) How much work does an external agent perform while lifting the mass a vertical distance h at constant speed? (b) Suppose an ideal (frictionless, massless) pulley was used to lift the weight as shown in Figure 7.P22. What is the tension T in the string? Through what distance does the agent pulling on the free end of the string have to move to raise the weight a height h? How much work does the external agent perform while raising the weight by a height h?

FIGURE 7.P22   Problem 56

57. Two forces act on a mass of 0.674 kg. The forces are $\mathbf{F}_1 = (0.175\,\hat{\mathbf{i}} + 0.440\,\hat{\mathbf{j}})$ N and $\mathbf{F}_2 = (0.250\,\hat{\mathbf{i}} - 0.250\,\hat{\mathbf{j}})$ N. The mass undergoes a displacement of $\Delta\mathbf{r} = (0.675\,\hat{\mathbf{i}} + 0.750\,\hat{\mathbf{j}})$ m along a straight line. (a) Calculate the net work on the mass. (b) If the mass initially had a speed of 1.50 m/s, what is its speed at the end of the displacement?

58. A constant force of $F = 2.50$ N is the only force acting on a 15.0-kg mass that is initially at rest. (a) What is the instantaneous power delivered to the mass at $t = 5.00$ s? at $t = 10.0$ s? (b) What is the total work done on the block during the first 10.0 s? (c) What is the average power delivered during the first 10.0 s?

59. Suppose that the force F that a motor can exert on a certain load decreases as the load moves faster according to the linear relation $F = A(v_{max} - v)$ for speeds $0 < v < v_{max}$. (a) Write an expression for the power delivered to the load. Sketch a graph of the power delivered as a function of v. (b) At what speed should the load move for the motor to deliver maximum power?

★60. It has been suggested[1] that the force exerted by a rubber band can be modeled as

$$F = F_o\left(\frac{l_o + s}{l_o} - \frac{l_o^2}{(l_o + s)^2}\right)$$

where l_o is the unstretched length of the rubber band, s is the distance the rubber band is stretched beyond its natural length, and F_o is a (temperature-dependent) force characterizing the strength of the rubber band. Calculate the work required to double the length of the rubber band.

★61. If a path in the xy-plane is described by the function $y = g(x)$, then the differential distance ds along the path is

$$d\mathbf{s} = dx\,\hat{\mathbf{i}} + \left(\frac{dg}{dx}\right)dx\,\hat{\mathbf{j}}$$

(a) Show that the work done by a force $\mathbf{F}(x, y) = F_x(x, y)\hat{\mathbf{i}} + F_y(x, y)\hat{\mathbf{j}}$ along the path $y = g(x)$ can be written

$$\mathcal{W} = \int_{x_o}^{x_f} F_x(x, g(x))\,dx + \int_{x_o}^{x_f} F_y(x, g(x))\frac{dg}{dx}dx$$

(b) If a child slides down a slide shaped like the parabola $y = \frac{3}{2}(x - 2)^2$, from $x = 0.00$ m to $x = 2.00$ m, how much work is done by the gravitational force $-mg\,\hat{\mathbf{j}}$?

62. Find the expression that gives the instantaneous power delivered by the gravitational field to the student Sam in Example 7.4.

---

[1] Richard Wolfson and Jay Pasaschoff, *Physics, Extended with Modern Physics* (Glenview, IL: Scott, Foresman and Co., 1990), p. 144.

CHAPTER

8

Conservation of Energy

In this chapter you should learn

- to distinguish between conservative and nonconservative forces.

- to calculate potential energy functions from conservative force functions.

- to apply conservation of mechanical energy to conservative systems to solve problems.

- to apply a generalization of the conservation of mechanical energy principle that is applicable to systems in which nonconservative forces act.

- (optional) to calculate the gradient of a potential energy function.

- (optional) to calculate the gravitational potential energy of a pair of spherical masses when they are separated by large distances.

- (optional) to test three-dimensional force functions to determine if they are conservative.

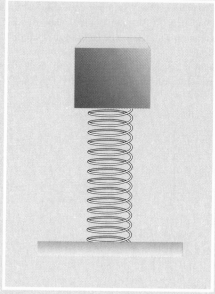

$$\frac{\begin{array}{c} K_o \\ +U_o \\ +W_{noncon} \end{array}}{K_f + U_f}$$

8.1 Conservative Forces and Potential Energy

In Chapter 7 we found that the work–energy theorem allows us to compute the change in an object's speed from the work done on the object. When we experiment with objects by applying forces of many different types, we notice two distinctly different types of behavior due to two distinctly different classes of force. To illustrate this difference let's think about two familiar forces that are representative of the two different classes: gravity and friction.

If you toss your textbook vertically into the air, you give it some initial velocity and, therefore, some initial kinetic energy. As the book travels upward, gravity does negative work on the book, thereby reducing its kinetic energy until it is zero at the instant the book stops. But, of course, the story doesn't end there; the book falls. And as it falls, gravity does positive work on the book, increasing its kinetic energy. When your text reaches the original point of release, it has regained all of the kinetic energy you initially gave it. (We are ignoring air resistance here, of course.) The point of this illustration is that the kinetic energy disappeared for a time, but then came back again. The earth's gravitational attraction provided a means by which the kinetic energy was "stored" for a short time.

In contrast, consider friction. Take the same book and give it a shove so that it begins to slide across a horizontal table. If you like, give it the same initial kinetic energy as in the previous illustration. As the book slides, the frictional force does negative work on the book, decreasing its kinetic energy to zero as before. This time, however, the book won't come back to your hand no matter how long you wait! The frictional force "used up" the kinetic energy in a nonreversible way.

In general, forces, such as gravity in the first of our two examples, are called **conservative** forces. Forces, such as friction, that do not act in a reversible fashion are called **nonconservative** forces. Conservative forces act so as to guarantee that when an object returns to a particular position it has the same kinetic energy each time it reaches that position. In some sense then, any kinetic energy temporarily missing during the trip is not really gone, it is always potentially available. For this reason it is both convenient and useful to change the way we describe the work done by conservative forces. Instead of repeatedly talking about the "negative work done that diminishes the kinetic energy, which we can always get back," it is easier to simply say that the kinetic energy is converted to **potential energy.**

It is much simpler to describe what happens to the textbook after it is tossed into the air if we use the potential energy concept: First, we give the book an initial kinetic energy. Then on the way up, the book loses kinetic energy but gains (gravitational) potential energy. Finally, as the book falls, the potential energy is converted back to kinetic energy. As a specific example, suppose that the work done by gravity on the book during the upward portion of its motion was $-5.0\,\text{J}$. Then we say that the book has gained $+5.0\,\text{J}$ of potential energy. In other words, we took the opposite of the work done as the change in potential energy. Symbolically this procedure is written

$$\Delta U = -\mathcal{W}_{\text{con}} \qquad (8.1)$$

where ΔU is the change in the potential energy U. The subscript "con" reminds us that this is a useful definition only when the work $\mathcal{W}$ is done by a *con*servative force.

Later in this chapter we will find that the potential energy concept plays a central role in determining the behavior of many physical systems. For this reason it is important that you be able to distinguish conservative forces from nonconservative ones. We have already given some characteristics of each that should help you to distinguish intuitively between them. Now we provide a formal definition of conservative forces.

Concept Question 1
While playing with a toy spring a young child stretches it so much that its shape becomes permanently changed. During the portion of the stretch when the deformation took place was the force of the spring conservative?

Kinetic energy and potential energy are continuously interchanged during juggling.

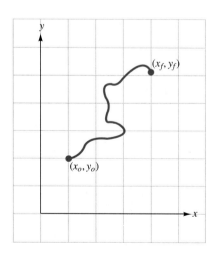

FIGURE 8.1

When a mass is moved along an arbitrary path from a height y_o to y_f, the work done by gravity depends only on the initial and final heights.

Conservative force: A force is conservative if the work done by that force when moving an object from point P_o to point P_f is independent of the path taken between P_o and P_f.

(This definition is equivalent to the statement that the work done by the force around *every* closed path totals zero.)

The previous definition, although concise and exact, is often difficult to apply as a test to determine if a force is, or is not, conservative. Note that we must show that the work done along *every* path is the same. Thus, we could not merely pick two specific paths and show that the work was the same along both. We are happy to report, however, that for problems involving only one dimension, it is quite simple to identify conservative forces.

Conservative forces—one dimension: In a one-dimensional system a force is conservative if it is a function of position only.

For cases involving two or more dimensions there is a mathematical test that is based on the derivatives of the force components, which we can use to identify conservative forces. A description of this test is given in optional Section 8.7.

8.2 Potential Energy: The Negative of the Integral of a Conservative Force over Distance

For conservative forces we defined the change in potential energy as

$$\Delta U = -\mathcal{W}_{con}$$

which, employing the definition of work, becomes

$$\Delta U = -\int_o^f \mathbf{F} \cdot d\mathbf{s} \qquad (8.2)$$

Note that the force $\mathbf{F}$ in Equation (8.2) is the conservative force. We illustrate the application of Equation (8.2) by deriving the potential energy changes for two familiar conservative forces: gravity near the earth's surface and a Hooke's-law spring.

Gravitational Potential Energy Near the Earth's Surface

Let's consider what happens when we move a mass m along an arbitrary path from a height y_o to y_f as shown in Figure 8.1. We wish to calculate the change in gravitational potential energy. The force in Equation (8.2) is therefore the weight force $\mathbf{W} = -mg\,\hat{\mathbf{j}}$. The displacement $d\mathbf{s}$ along any portion of the path is $d\mathbf{s} = dx\,\hat{\mathbf{i}} + dy\,\hat{\mathbf{j}} + dz\,\hat{\mathbf{k}}$. With these substitutions the incremental change in potential energy is

$$dU = \mathbf{F} \cdot d\mathbf{s} = (-mg\,\hat{\mathbf{j}}) \cdot (dx\,\hat{\mathbf{i}} + dy\,\hat{\mathbf{j}} + dz\,\hat{\mathbf{k}}) = -mg\,dy$$

With this substitution we have

$$\Delta U = -\int_{y_o}^{y_f} (-mg)\,dy$$

If we restrict ourselves to relatively small y-values (near the surface of the earth), mg is a constant that can be taken outside the integration, making the integral trivial

Only changes in potential energy are important. Here the ground below has been lowered by centuries of erosion. The gravitational potential energy of the balanced rock has increased relative to the ground below.

Thomas Jefferson's 7-day clock in his home, *Monticello*. Jefferson's clock is powered by falling masses connected by cables running over pulleys. The clock did not run for a full 7 days when Jefferson first built it. His solution was to redefine the zero of potential energy by cutting holes in the floor. Days of the week are indicated by the weights passing the dark strips on the wall. Friday and Saturday are in the basement!

$$\Delta U = mg \int_{y_o}^{y_f} dy$$
$$= mgy_f - mgy_o$$

Because ΔU means $U_f - U_o$, we have

$$U_f - U_o = mgy_f - mgy_o$$

and it is natural to set $U_f = mgy_f$ and $U_o = mgy_o$. However, you probably already realize that this is not the only possibility. If $U_f = mgy_f + C$ and $U_o = mgy_o + C$, where C is any constant, we obtain the same result for ΔU:

$$U_f - U_o = (mgy_f + C) - (mgy_o + C)$$
$$= mgy_f - mgy_o \qquad (8.3)$$

This result is a quite general feature of potential energy functions:

> Only *changes* in potential energy functions are significant, and the actual value of the function at any point is arbitrary to within an additive constant.

That is, the position you choose for the $U = 0$ point is arbitrary. For the gravitational potential energy function the simplest choice is to make $U = 0$, where $y = 0$. (And, of course, where you pick $y = 0$ is also up to you!)

To apply this choice to the gravitational potential energy function (near the earth's surface) we specify $U_o = 0$ when $y_o = 0$, and drop the unneeded f subscripts on the remaining terms of Equation (8.3) to obtain

$$\boxed{U = mgy \qquad \text{(Near the earth's surface)}} \qquad (8.4)$$

Note that if you choose $y = 0$ to be on the top of a lab table, you wind up with negative potential energy for masses positioned below the tabletop as shown in Figure (8.2). There is nothing wrong or mysterious about this. Negative potential energy simply means less potential energy than at the zero point. If you are extremely bothered by such negative numbers you can simply move the $y = 0$ to the floor so that all potential energies are positive (as long as there's no basement under you).

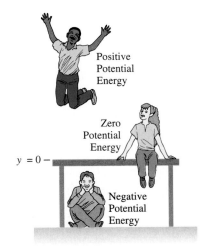

FIGURE 8.2

The actual value of potential energy is arbitrary. If $y = 0$ is chosen to be at ground level, all three students have positive potential energy.

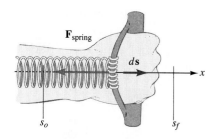

FIGURE 8.3

The elastic potential energy of the spring increases as its end is moved from s_o to s_f.

A final point we wish to emphasize is that, even though the path taken was quite arbitrary, only the *vertical* component of displacement contributes to a change in the gravitational potential energy of the mass m. Hence, if a child slides down a curved playground slide, we need to know only her change in height to find her change in potential energy.

Potential Energy of a Hooke's-Law Spring

The force from a Hooke's-law spring acts along a single direction. Let's fix one end of the spring and place an x-axis along the length of the spring so that the spring's free end is at $x = 0$. Then when the free end is displaced by x, the restoring force is given by

$$\mathbf{F} = -kx\,\hat{\mathbf{i}}$$

and hence is a function of position only. According to the criteria of the previous section, this characteristic means that the Hooke's-law spring force is conservative. Let's calculate the potential energy function for such a spring when we stretch it from s_o to s_f as shown in Figure 8.3. The force exerted by the spring is $\mathbf{F} = -kx\hat{\mathbf{i}}$, provided we choose $x = 0$ at the point where the end of the spring rests when unstretched. The displacement vector is $d\mathbf{s} = +\,dx\hat{\mathbf{i}}$. The integral defining the change in potential energy is easily calculated:

$$\Delta U = -\mathcal{W}_{\text{spring}}$$

$$= -\int_{s_o}^{s_f} (-kx\,\hat{\mathbf{i}}) \cdot (dx\,\hat{\mathbf{i}})$$

$$= +\int_{s_o}^{s_f} kx\,dx$$

$$= +k\int_{s_o}^{s_f} x\,dx$$

$$U_f - U_o = +\frac{1}{2}ks_f^2 - \frac{1}{2}ks_o^2$$

Once again, we can deduce the potential energy function. This time $U = \frac{1}{2}ks^2 + C$, for which any value of the constant C gives the correct value for ΔU. It is certainly convenient in this case to pick $C = 0$ so that we assign zero potential energy to the spring in its unstretched state. If we perform the above calculation for a compression of the spring, we obtain the same expression for the potential energy. Thus, for a Hooke's-law spring the potential energy can be written

$$U = \frac{1}{2}ks^2 \tag{8.5}$$

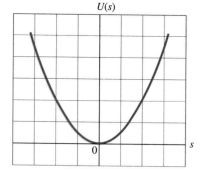

FIGURE 8.4

The graph of the potential energy function of a Hooke's-law spring is a parabola.

where s is the distance by which the spring is stretched or compressed from its natural length. The potential energy function for such a spring is thus a parabola like that shown in Figure 8.4.

8.3 Conservation of Mechanical Energy

In Section 8.1 we found it useful to redefine the work performed by a conservative force in terms of the change in potential energy:

$$\Delta U = -\mathcal{W}_{\text{con}}$$

where U is the potential energy function and $\mathcal{W}_{con}$ is the work done by a conservative force. In the previous chapter we found that the net work done on a mass can be related to that mass's change in speed through the work–energy theorem:

$$\mathcal{W}_{net} = \Delta K$$

where $\mathcal{W}_{net}$ represents the sum of the work done by all of the forces acting on the mass. For N different forces we may simply write

$$\mathcal{W}_1 + \mathcal{W}_2 + \mathcal{W}_3 + \cdots + \mathcal{W}_N = \Delta K$$

If each of these N forces is conservative, each can be replaced by the negative of its respective potential energy change:

$$-\Delta U_1 - \Delta U_2 - \Delta U_3 - \cdots - \Delta U_N = \Delta K$$

Or we may write

$$\Delta K + \Delta U_1 + \Delta U_2 + \Delta U_3 + \cdots + \Delta U_N = 0$$

For the sake of brevity we write the sum of the changes in all the potential energy terms from 1 to N simply as ΔU. Then,

$$\Delta K + \Delta U = 0 \qquad (8.6)$$

Equation (8.6) tells us that the total change in kinetic energy plus the total change in potential energy is zero. That is, there is no change in the sum of K plus U. Stated mathematically

$$\Delta(K + U) = 0 \qquad (8.7)$$

The quantity $K + U$ is called the **total mechanical energy.** Equation (8.7) thus says that when only conservative forces act, the change in total mechanical energy of a system is zero. This statement is a powerful theorem.

If only conservative forces perform work on and within a system of masses, the total mechanical energy of the system is conserved.	**CONSERVATION OF MECHANICAL ENERGY**

To further emphasize the meaning of the conservation of mechanical energy theorem, we write it in another form that is often more convenient for problem solving. We merely need to recall the meaning of the Δ symbol. Recall $\Delta\mathcal{X}$ (where $\mathcal{X}$ is anything) means $\mathcal{X}_{final} - \mathcal{X}_{original}$. Thus, Equation (8.7) can be written

$$(K_f + U_f) - (K_o + U_o) = 0$$

or, in other words

$$K_o + U_o = K_f + U_f \qquad (8.8)$$

Equation (8.8) suggests how we can exploit the conservation of mechanical energy to solve for the mechanical state of a system of masses: we treat this number called mechanical energy as a useful "bookkeeping" quantity. Our first step is to look at a picture showing the initial configuration of the system and then compute the total mechanical energy present in this initial state. The system is then allowed to evolve. If only conserva-

Concept Question 3
Three students throw water balloons with the *same speed* from the *same height* off the edge of a building roof. If one student throws his balloon straight up, another horizontally, and the third throws his straight down, what are the relative speeds of the balloons when they each strike the ground? (Ignore any effect of air friction.)

tive forces act, we are guaranteed that when the system reaches any later configuration, the total mechanical energy is unchanged, although the distribution of energy between kinetic and potential terms may be different. Therefore, a mathematical expression for the total mechanical energy $K_f + U_f$ must be equal to the same number we computed for the initial state $K_o + U_o$. In Example 8.1 we show how to apply this strategy.

Problem-Solving Strategy for the Conservation of Mechanical Energy

1. Draw some pictures of the system. At a minimum draw (I) the initial configuration of masses (and springs if any) and (II) the final configuration. (When a spring is present, we like to first draw a zeroth picture (0) showing what the system looks like when the spring is at its normal (unstretched) length.)

2. Put a coordinate system on your pictures. Be sure to use the *same coordinate system on all pictures*. (If a spring is present, we strongly recommend that you put the zero of gravitational potential energy ($y = 0$) at the point where the movable end of the spring is located when the spring is at its natural length.)

3. Look at picture I and ask yourself, "What forms of energy are present?" Your check list is fairly small at this point in your study: kinetic energy, gravitational potential energy, and spring potential energy.

If the object is moving include	$\frac{1}{2}mv_o^2$
If the mass is not located at $y = 0$, include	mgy_o
If the spring is stretched or squashed, include	$\frac{1}{2}ks_o^2$

As you learn of other types of potential energy (e.g., electrical), add them to this list.

4. Look at picture II and ask yourself, "What forms of energy are present?"

If the object is moving include	$\frac{1}{2}mv_f^2$
If the mass is not located at $y = 0$, include	mgy_f
If the spring is stretched or squashed, include	$\frac{1}{2}ks_f^2$

As you learn of other types of potential energy (e.g., electrical), add them to this list.

5. Now simply equate the total energy in picture I to the total in picture II. Check to see how many unknowns you have. You may often find there is only one unknown, and you can solve for it immediately. If you have two or more unknowns, you need to search for additional equations that relate these unknown variables. Quite often these equations come from geometric relationships between y (the vertical coordinate) and s (the spring distortion).

6. Check your result to be sure your answer is reasonable and has the proper units. If you had to solve a quadratic equation, choose the root consistent with the equations written in step 4.

Although the total mechanical energy of a conservative system remains constant, the distribution of energy to kinetic and potential forms ($K, U_1, U_2, \ldots, U_N$) may change. An analogy might be leaving the house with a ten dollar bill, but spending none of it. During the day you may change the ten for a five, four ones, and a dollar in change, but if you don't spend anything, when you get back home you still have ten dollars. (Unless you have great willpower your finances are likely to resemble the nonconservative systems discussed in the next section more than the conservative systems in this section.)

In Section 8.2 we derived two important potential energy functions, which we repeat here for convenience

$$U = mgy \qquad \text{(Gravitational potential energy near surface of earth)}$$

$$U = \frac{1}{2}ks^2 \qquad \text{(Potential energy for a Hooke's-law spring)}$$

We begin our illustrations of the conservation of mechanical energy principle with Example 8.1, which uses only the gravitational potential energy function. In Example 8.2 we use both the gravitational and spring potential energy functions.

EXAMPLE 8.1 *No Parachutes Allowed*

The mechanical engineering department sponsors an egg-dropping contest in which students design protective packaging for raw eggs. A student drops a test package out of his dorm window from a height of 12.0 m above the ground. The package is dropped from rest. Use the conservation of energy theorem to find the speed with which the package hits the ground.

SOLUTION Following step 1 in our problem-solving strategy, Figure 8.5 shows the package just before the drop and just as it impacts the ground. We place our coordinate system so that the origin is at ground level (step 2). Now for step 3: At the initial height $y_o = 12.0$ m, the package has only gravitational potential energy, since it is initially at rest. Thus,

$$K_o = 0 \qquad \text{and} \qquad U_o = mgy_o$$

At the instant the container contacts the ground it has velocity v and its height is zero, so

$$K_f = \frac{1}{2}mv^2 \qquad \text{and} \qquad U_f = 0$$

Substituting these results into the conversation of energy theorem,

$$K_o + U_o = K_f + U_f$$

we have

$$\frac{1}{2}mv^2 + 0 = 0 + mgy_o$$

Canceling the mass m and solving for v, we obtain

$$v = \sqrt{2gy_o}$$

We have now obtained this same result by applying three different techniques: the simple kinematic equations for falling objects, the work–energy theorem of the previous chapter (see Example 7.7), and now the conservation-of-mechanical-energy approach. Substituting $y_o = 12.0$ m we find

$$v = \sqrt{2(9.80 \text{ N/kg})(12.0 \text{ m})} = 15.3 \text{ m/s} \qquad \blacktriangleleft$$

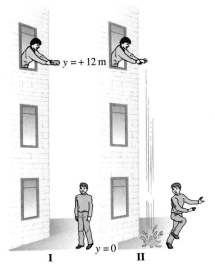

FIGURE 8.5

The speed of the egg carton just before it hits the ground can be found from its initial potential energy. (See Example 8.1.)

In order to compare two solution methods we reanalyze the spring–mass system of Example 7.2. Here, however, instead of using the work–energy theorem as in Chapter 7, we apply the conservation of mechanical energy so we can emphasize this different point of view.

EXAMPLE 8.2   *But Do We Get the Same Answer?*

A spring of stiffness constant $k = 200.$ N/m has a natural length of 1.000 m when suspended from the ceiling. A 1.50-kg piñata is attached to the spring and raised, compressing the spring until its end is 0.500 m from the ceiling. The piñata is then released and falls downward. What is the speed of the piñata when the end of the spring passes a point 1.200 m from the ceiling?

SOLUTION   As indicated in step 1 of our problem-solving strategy, Figure 8.6 shows three pictures important for the development of our solution. Picture 0 shows the natural length of the spring without the piñata attached. As suggested in step 2 of our problem-solving strategy, we choose the bottom end point of the spring as the $y = 0$ point for this problem.

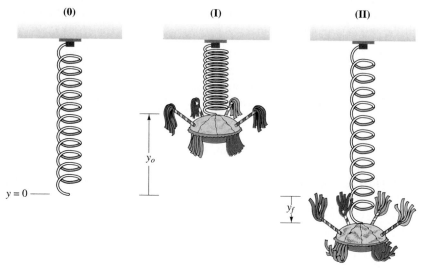

FIGURE 8.6

A piñata is attached to a spring fastened to the ceiling. As the piñata falls the total mechanical energy is conserved. (See Example 8.2.)

Picture I shows the initial state for the problem, with the piñata attached and held (by an unshown hand) in the original position described in the problem. Following step 3 of our strategy, in picture I we look for three possible types of energy: kinetic energy, gravitational potential energy, and spring potential energy. The mass is stationary, so $v = 0$ and, therefore, $K_o = 0$. The spring is squashed, so there is spring potential $U_o = \frac{1}{2}ky_o^2$. The mass is not located at $y = 0$. Hence, there is gravitational potential $U = mgy_o$. The total energy in picture I is, therefore,

$$K_o + U_o = 0 + \frac{1}{2}ky_o^2 + mgy_o$$

Notice that we picked the zero of gravitational potential energy for the mass to be the point $y = 0$ where the top of the piñata is located.

We proceed to step 4 of the problem-solving strategy. From picture II we see that the piñata is in motion, the spring is stretched, and the mass is not at $y = 0$. Thus, all three forms of mechanical energy are present:

$$K_f + U_f = \frac{1}{2}mv_f^2 + \frac{1}{2}ky_f^2 + mgy_f$$

Note carefully the sign of $+mgy_f$. When the piñata is 1.20 m below the ceiling, its y location is 0.200 m *below* the $y = 0$ position; that is, $y = -0.200$ m. Substituting

$y_f = -0.200$ m, we find $+mgy_f = +(1.50 \text{ kg})(+9.80 \text{ N/kg})(-0.200 \text{ m}) = -2.94$ J, a negative potential energy as we expect.

Remember that the invisible agent holding the piñata in picture I is assumed to be instantly removed when the problem starts so that it does no work. Thus, only conservative forces act as the piñata moves between the configurations in picture I and picture II. Consequently, as anticipated in strategy step 5, we can equate the total mechanical energy in pictures I and II.

$$K_I + U_I = K_{II} + U_{II}$$

$$\frac{1}{2}ky_o^2 + mgy_o = \frac{1}{2}mv_f^2 + \frac{1}{2}ky_f^2 + mgy_f$$

For this case there is only one unknown (v_f), so we may proceed with step 6 and immediately solve for v_f:

$$v_f = \sqrt{\frac{2[mg(y_o - y_f) + \frac{1}{2}k(y_o^2 - y_f^2)]}{m}}$$

$$= \sqrt{\frac{2\{(1.50 \text{ kg})(9.80 \text{ m/s}^2)[(+0.500 \text{ m}) - (-0.200 \text{ m})] + \frac{1}{2}(200. \text{ N/m})[(0.500 \text{ m})^2 - (-0.200 \text{ m})^2]\}}{1.50 \text{ kg}}}$$

$$= 6.46 \text{ m/s}$$

Remember, this is the speed. The velocity can be either positive or negative depending on whether the mass is moving upward or downward. ◀

Energy Conservation and Isolated Systems

We close this section on the conservation of mechanical energy with a discussion of a subtle but terribly important point. In the statement of the law of conservation of mechanical energy we asserted that

> if only conservative forces perform work within an isolated system of masses, the total mechanical energy of the system is conserved.

However, we did not define exactly what we meant by *isolated system*. We intend to rectify that omission right now. One implication of the phrase *isolated system* should be fairly obvious; namely, whatever system we consider in our "before" picture, we had better consider exactly the same system in our "after" picture. If we first consider four masses and two springs, when we finish we better still be considering the same four masses and same two springs, and furthermore *these masses and springs must not interact with any other masses or springs (or anything else) during the time interval under study!*

It is in this last italicized expression that the subtlety lies. For it is all too easy to leave out relevant masses. Moreover, sometimes even relevant masses are intentionally left out as an approximation without explicit reference to the fact that an approximation has been made. In fact, we have already done this ourselves in both of the last two examples! (We will explain the error—we prefer to call it an "approximation"—in later paragraphs and show why we obtained the "right" answer anyway. Before you continue, you might like to reread Examples 8.1 and 8.2 to see if you can spot what on earth we left out.)

The object we left out of Examples 8.1 and 8.2 was the earth! The egg container *and the earth* constitute the isolated system in Example 8.1. The piñata we analyzed in Example 8.2 is coupled to the earth by both gravity and the spring (through the roof supports, attached to the walls, which are supported by the earth). In fact, it is a general property of potential energy functions that they are *not* a property of an individual object, but always express a relationship between two or more objects. It is the system of the piñata, spring, and earth that constitutes the isolated system in Example 8.2. Strictly speaking, the conservation of mechanical energy holds only if we consider that entire system.

Let's examine the egg-container example more carefully. In Example 8.1, when we set $mgy_o = \frac{1}{2}mv^2$, we were actually making an approximation. As the egg container falls toward the earth, the earth also falls toward the egg container. Thus, strictly speaking, we must include both of these objects in our energy expression:

$$mgy_o = \frac{1}{2}mv^2 + \frac{1}{2}m_{earth}v_{earth}^2$$

Note that mgy_o is the potential energy of the combined earth–container system. (The mass of the earth is hidden within the factor g as shown in Equation (6.12).) As you no doubt already realize, the kinetic energy gained by the earth is much smaller than the energy gained by the container. In fact it turns out that the kinetic energy gained by the earth is smaller than that gained by the container by a factor of m/m_{earth}. (See Problem 61.) Because the kinetic energy gained by the earth is immeasurably small, we simply left it out of the problem altogether without warning you that we were making an approximation. Now that we have cleared our conscience on this matter, we continue to leave out the earth's kinetic energy when solving similar problems with masses that are small compared to the earth.

Here then is the bottom line on isolated systems. A system of masses is isolated only if the masses do not interact with other masses outside the system. You can sometimes apply the conservation of mechanical energy to nonisolated systems if (1) the interacting masses outside the system are so large that their energies do not change appreciably or (2) the forces connecting the system to the outside masses are so small that no appreciable energy changes are brought about by them.

8.4 Systems with Nonconservative Forces

When only conservative forces act we found it useful to recast the work–energy theorem

$$\mathcal{W}_{net} = \Delta K$$

into the conservation-of-mechanical-energy theorem

$$K_o + U_o = K_f + U_f$$

In Section 8.3 we discovered that the potential energy function for a conservative force can be found from the negative of the work done by that force between two arbitrary points (Eq. (8.2)). There is no ambiguity within this definition because the work done along any path between the two arbitrary points is the same for conservative forces. This feature is not true for nonconservative forces, such as friction. Therefore,

it is ***not*** possible to define a potential energy function for nonconservative forces.

Please don't ever try to define a potential energy for a nonconservative force. It will almost assuredly lead to disaster—well, at least to an incorrect result. To see the effect of nonconservative forces on the conservation-of-mechanical-energy theorem, let's repeat our derivation of this theorem. We again start from the work–energy theorem. This time, however, we split the net work into two pieces: one due to conservative forces ($\mathcal{W}_{con}$) and one due to nonconservative forces ($\mathcal{W}_{noncon}$). In this notation the work–energy theorem

$$\mathscr{W}_{net} = \Delta K$$

becomes

$$\mathscr{W}_{con} + \mathscr{W}_{noncon} = \Delta K$$

For the work done by conservative forces we can always compute the potential energy change $\Delta U = -\mathscr{W}_{con}$, or

$$\mathscr{W}_{con} = -\Delta U$$

so that we have

$$-\Delta U + \mathscr{W}_{noncon} = \Delta K$$

or

$$\mathscr{W}_{noncon} = \Delta K + \Delta U$$

Because the sum of the changes is the change in the sum, we have

$$\Delta(K + U) = \mathscr{W}_{noncon} \tag{8.9}$$

In words, Equation (8.9) says that the change in the total mechanical energy of the system is equal to the work done by the nonconservative force(s).

As an example, consider friction, the archetypical nonconservative force. We know that the work done by kinetic friction is usually negative; in such cases $\Delta(K + U)$ is negative. That is, the total mechanical energy decreases when friction does negative work. (We should mention here that nonconservative forces, including friction, can do positive work. One example is an external applied force, such as a person pushing in the direction of the motion. In such a case the mechanical energy of the system increases.)

If we expand the Δ notation in Equation 8.9, it can be recast into the "before-and-after" form similar to that which we found useful for conservative systems. The result is the following theorem

$$K_o + U_o + \mathscr{W}_{noncon} = K_f + U_f \tag{8.10}$$

In this form, $\mathscr{W}_{noncon}$ should be interpreted as the work done by all nonconservative forces as the system evolves from the configuration labeled o to the configuration labeled f. Don't forget that *when the work done by kinetic friction is a negative number*, and you "add" its value, you actually reduce the total energy on the left-hand side of Equation (8.10). Also remember that in our notation $U (= U_1 + U_2 + \cdots + U_N)$ stands for the sum of all the forms of potential energy present.

EXAMPLE 8.3   *One Way to Combine Physics and Chemistry*

With the pond in front of the dorm safely frozen over, students decide to see how far they can propel an old chemistry book across the pond using an inner tube stretched between two trees as a slingshot. Model the slingshot as a Hooke's-law spring with force constant $k = 300.$ N/m and assume the coefficient of friction between the book and ice is 0.220. The mass of the chemistry book is 1.60 kg. If the "spring" is compressed by 0.750 m (a) how fast is the book traveling when it is 1.50 m in front of the end of the unstretched spring? (b) What distance does the book travel from where the inner tube is released to where the book finally stops?

(0)

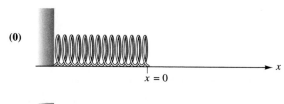

$x = 0$

(I)

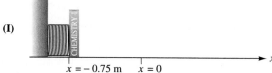

$x = -0.75\ \text{m}$ $x = 0$

(II)

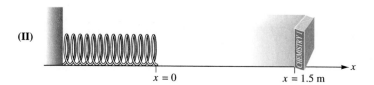

$x = 0$ $x = 1.5\ \text{m}$

(III)

$x = 0$

FIGURE 8.7

A book is compressed against a Hooke's-law ''spring'' and is then released. The work done by the nonconservative force of friction eventually reduces the total mechanical energy to zero.

SOLUTION In Figure 8.7, picture 0 shows the position of the end of the unperturbed spring, which we take as $x = 0$. Picture I shows the situation just as the book is released from rest. Picture II shows the book as it sails past the point 1.50 m in front of the spring. Finally, picture III shows the final stopping point of the book.

(a) Following steps 3 and 4 of the problem-solving strategy, we see that there is no kinetic energy and no gravitational potential energy in picture I. There is, however, spring potential energy in the amount $\frac{1}{2}kx_o^2$. In picture II there is no energy associated with the spring, but there is now the kinetic energy $\frac{1}{2}mv_f^2$. The energy in these two pictures is related by the modified conservation of mechanical energy Equation (8.10):

$$K_o + U_o + \mathcal{W}_{\text{noncon}} = K_f + U_f$$

$$\frac{1}{2}kx_o^2 + \mathcal{W}_{\text{noncon}} \text{ (from I to II)} = \frac{1}{2}mv_f^2$$

The work done by kinetic friction between picture I and picture II is

$$\mathcal{W}_{\text{fric}} = (F_{\text{fric}})(\Delta x)\cos(180°) = (\mu_k mg)(x_f - x_o)(-1)$$
$$= -\mu_k mg(x_f - x_o)$$

Substituting this expression for $\mathcal{W}_{\text{fric}}$, we have

$$\frac{1}{2}kx_o^2 - \mu_k mg(x_f - x_o) = \frac{1}{2}mv_f^2$$

This equation is easily solved for v_f

$$v_f = \sqrt{\frac{kx_o^2}{m} - 2\mu_k g(x_f - x_o)}$$

$$= \sqrt{\frac{(300.\ \text{N/m})(-0.750\ \text{m})^2}{(1.60\ \text{kg})} - 2(0.220)(9.80\ \text{N/kg})[1.50\ \text{m} - (-0.750\ \text{m})]}$$

$$= 9.79\ \text{m/s} \qquad \text{(Which is definitely fast-track chemistry!)}$$

Notice that Δx is, as always, $x_{\text{final}} - x_{\text{original}}$ and leads to the correct expression (1.50 m + 0.75 m) for the distance traveled by the mass between picture I and picture II.

(b) To find how far the book ultimately travels, we note that in picture III there is no kinetic energy left because $v_f = 0$. The equation that takes us from picture I directly to picture III is:

$$\frac{1}{2}kx_o^2 + \mathcal{W}_{\text{noncon}} \ (\text{from I to III}) = \frac{1}{2}mv_f^2 = 0$$

or

$$\frac{1}{2}kx_o^2 - \mu_k mg(x_f - x_o) = 0$$

which we can solve for x_f:

$$x_f = x_o + \frac{kx_o^2}{2\mu_k mg}$$

$$= -0.750\ \text{m} + \frac{(300.\ \text{N/m})(-0.750\ \text{m})^2}{2(0.220)(1.60\ \text{kg})(9.80\ \text{N/kg})} = 23.7\ \text{m} \qquad \blacktriangleleft$$

8.5 Conservative Forces: The Negative Derivative of Potential Energy

Thus far we have emphasized how to obtain the potential energy function by integrating the conservative force function (Eq. (8.2)). In some cases it turns out to be easier to do the reverse. That is, it may be easier to model the potential energy and then deduce the corresponding force. As you no doubt have already guessed, since we obtain the potential energy from the *negative integral* of the force function, we obtain the conservative force from the *negative derivative* of the potential energy function.

In one dimension it is particularly easy to obtain the force from the potential energy. For example, in the case of the gravitational force near the surface of the earth we have seen that the potential energy is given by

$$U = mgy$$

The component of the force in the y-direction can be found by

$$F_y = -\frac{d}{dy}(mgy) = -mg$$

This component is negative because the force is downward, whereas in the potential energy expression $U = mgy$ the positive y-direction is upward. We may write the component vector as

$$F_y \hat{\mathbf{j}} = -mg\hat{\mathbf{j}}$$

Similarly, if a Hooke's-law spring with one end fixed rests on a horizontal table, its potential energy function can be written

$$U = \frac{1}{2}kx^2$$

where x is the displacement of the free end of the spring from its unstretched position. The force exerted by the spring when its end is displaced is found from

$$F_x = -\frac{d}{dx}\left(\frac{1}{2}kx^2\right)$$

$$F_x = -kx$$

Or, in vector form,

$$F_x \hat{\mathbf{i}} = -kx\,\hat{\mathbf{i}}$$

In general, if the *one-dimensional* potential energy function U is a function of position s, then the conservative force $F(s)$ is found from

$$F(s) = -\frac{dU}{ds} \qquad (8.11)$$

For potential energy functions that depend on more than one spatial coordinate, such as $U(x, y)$ and $U(x, y, z)$, a generalization of Equation (8.11) is required. We investigate these cases in the optional Section 8.7.

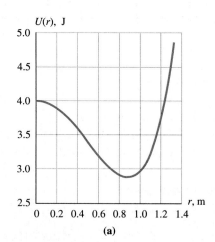

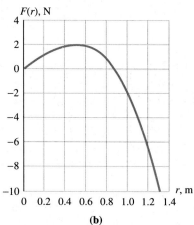

FIGURE 8.8

(a) A graph of the potential energy function described in Example 8.4.
(b) The force can be deduced from the negative slope of the potential energy function.

EXAMPLE 8.4   Can You Find the Equilibrium Points by Inspection of the Potential Energy Function?

In the region $r \geq 0$ the particle model can be used to describe the motion of an object that has a potential energy function $U(r) = Ar^4 - Br^2 + C$ where $A = 2.00$ J/m^4, $B = 3.00$ J/m^2, and $C = 4.00$ J. Find the conservative force $F(r)$ and the locations where $F(r) = 0$.

SOLUTION   From Equation (8.11) we have

$$F(r) = -\frac{dU(r)}{dr} = -\frac{d}{dr}(Ar^4 - Br^2 + C)$$

$$= -4Ar^3 + 2Br - 0$$

We find the locations where $F = 0$ by setting our result equal to zero and solving for r:

$$0 = -4Ar^3 + 2Br$$

$$r = 0 \qquad \text{and} \qquad r = \sqrt{\frac{B}{2A}}$$

Upon substitution of numerical values, we have

$$r = 0 \qquad \text{and} \qquad r = \frac{\sqrt{3}}{2}\,\text{m}$$

Plots of $U(r)$ and $F(r)$ are shown in Figure 8.8. You should verify that the graph of $F(r)$ is the slope of the $U(r)$ graph. ◀

Because of the relation between the force and the potential energy function expressed by Equation (8.11), it is possible to get a good qualitative "feel" for how an object is going to behave just by looking at a graph of the potential energy function. Think of the potential energy curve as an elevation map and the object on which the conservative force acts as an ice cube sliding without friction over the landscape.

As an example, consider the one-dimensional potential energy curve shown in Figure 8.9, where we use the particle model of an object. In regions where the slope is positive, there is a negative force (that is, a force directed toward the left) on the particle. In regions where the slope is negative, the positive conservative force accelerates the particle to the right. Hence, as the particle travels along a range of x-values for which the $U(x)$ curve appears "uphill" to the particle, it slows down. When the particle travels along x in such a direction and in a region where $U(x)$ appears "downhill," the magnitude of the particle's velocity increases.

All these observations are nothing more than manifestations of the conservation-of-energy theorem. To understand the connection, look again at Figure 8.9 and keep in mind that the total mechanical energy must be constant. In regions where the potential energy $U(x)$ is large, the kinetic energy must be small. Conversely, when $U(x)$ is small, the particle must have a large kinetic energy and, therefore, a large velocity magnitude. By drawing a horizontal line at an energy corresponding to the total mechanical energy of the particle E_{total}, we can determine the spatial limits of the particle's motion.

As another illustration, consider the potential energy function shown in Figure 8.10. If our ice cube is released from point A in Figure 8.10, you could no doubt predict with confidence that it gains speed sliding down, slips over the top of point D, travels down into the next valley, stops for an instant at point G, and then begins a return journey back toward point A.

Equilibrium

Figure 8.10 provides another illustration of how we can predict a great deal about an object's motion from its potential energy function without performing any computations. A particle is in equilibrium when the total force on it is zero. Points B, D, and F are points of equilibrium because the slopes of the $U(x)$ curve are zero at these points. It should be clear from this figure, however, that there is a considerable difference between what happens to an ice cube placed at these points. Point B is a point of **stable equilibrium.** If the ice cube moves slightly from B, a restoring force arises pushing it back toward B. If we give the ice cube a little shove when it is at point B, it oscillates back and forth about B like the mass on the end of a spring. On the other hand, D is a point of **unstable equilibrium.** A small displacement of a stationary ice cube from D results in a force pushing the ice cube even farther from D. When you try to balance a pencil vertically on its rounded eraser you are trying to establish it in a point of unstable equilibrium. Finally, F is a point of **neutral equilibrium** where a small displacement from F results in no unbalanced force.

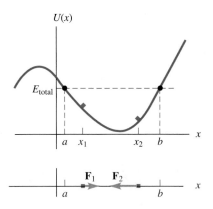

FIGURE 8.9

When looking at potential energy curves you must remember that the actual particle motion is along a straight line. The points a and b are called turning points. A particle with total mechanical energy E_{total} is confined to the region $a \le x \le b$.

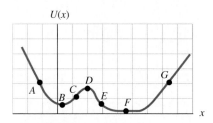

FIGURE 8.10

A particle released from rest at point A can travel to point G but no farther.

Concept Question 6
Describe three orientations of a right circular cone that correspond to (a) unstable equilibrium, (b) stable equilibrium, and (c) neutral equilibrium.

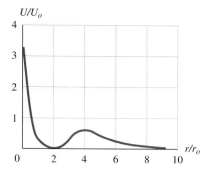

U/U_o

FIGURE 8.11

A charged particle inside the nucleus of an atom might have a potential energy function similar in form to this one. (See Example 8.5.)

Stars are formed when the gravitational potential energy of hydrogen gas is converted into kinetic energy. When the kinetic energy of the hydrogen nuclei is large enough to permit the nuclei to collide, the process of nuclear fusion starts.

EXAMPLE 8.5   *If Only We Could Tunnel Out of Here*

Consider the potential function $U(r)$ shown in Figure 8.11. (a) Identify any points of stable, unstable, and neutral equilibrium. (b) Identify the points of strongest attraction toward $r = 0$ and strongest repulsion from $r = 0$. (c) If a particle is released from rest at $r/r_o = 1.0$ over what range of values (r/r_o) does your intuition tell you the particle should be able to travel?

SOLUTION   We use our mental picture of an ice cube sliding without friction along the potential energy curve. (a) We see that $r/r_o = 2.0$ is a point of stable equilibrium, and that $r/r_o = 4.0$ is an unstable equilibrium.

(b) Points near $r/r_o = 3.0$ have large positive slopes indicating the strongest attraction toward the origin. Clearly, there is a very strong repulsion as we get near $r/r_o = 0$. At points for which $r/r_o > 4.0$ the force is repulsive.

(c) If the particle is released from rest at $r/r_o = 1.0$, then its total energy is equal to its potential energy. Drawing a horizontal line at this energy, we see that the particle again has zero kinetic energy at $r/r_o = 3.0$.   ◄

8.6 Potential Energy for Newtonian Gravity (Optional)

As we discussed in Section 6.3, Newton's universal law of gravitation (NULG) describes an attractive force between two point–masses according to the relation

$$F = G\frac{mM}{R^2} \tag{8.12}$$

where R is the distance between the point–masses. Using the calculus, which he invented for this purpose, Newton was able to show that Equation (8.12) can also be used for spherical masses if the R is taken as the distance between the centers of the masses. If nonspherical masses are far enough apart so their dimensions are small compared to the distance between them, the masses may be modeled as points. In such cases Equation (8.12) provides an approximately correct result. To obtain accurate answers for the forces between two irregularly shaped masses, the objects must first be subdivided (mentally at least) into differential mass elements. The force between every pair of elements must be calculated and then summed vectorially. (See Problem 6.25, for example.)

Although this Newtonian model for gravity has been superseded by more recent theories due to Einstein and others, it is sufficiently accurate for most practical purposes, which include putting satellites into orbit and people on the moon. Consequently, it is of interest to examine the potential energy function associated with the Newtonian gravitational model.

Consider two spherical masses separated by some distance. We want to calculate the change in potential energy if the masses are pulled farther apart. Using the techniques in optional Section 8.7 it can be shown that the gravitational force described by Equation (8.12) is conservative. (See Problem 46.) Thus, the gravitational potential energy is equal to the negative of the work done by the gravitational force as the masses are moved apart. Because gravity is a conservative force, the path along which we compute the work does not matter. Let's make it easy on ourselves and take a path straight out along a radius placed parallel to the x-axis as shown in Figure 8.12.

We need to calculate

$$\Delta U = -\mathcal{W}_{\text{grav}} = -\int_{R_o}^{R_f} \mathbf{F} \cdot d\mathbf{s} \tag{8.13}$$

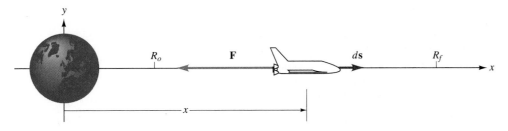

FIGURE 8.12

As the rocket moves from R_o to R_f, its gravitational potential energy increases. Newton's universal law of gravitation must be used to calculate the change in potential energy.

The path of integration is from R_o to R_f. Thus, $d\mathbf{s} = dx\,\hat{\mathbf{i}}$. Because the force of gravity is attractive, the force on our test mass m is

$$\mathbf{F} = -G\frac{Mm}{x^2}\,\hat{\mathbf{i}}$$

Substituting this force equation into Equation (8.13) we have, after taking the indicated dot product,

$$\Delta U = +GMm \int_{R_o}^{R_f} x^{-2}\,dx$$

This integral is easily performed using $\int x^n\,dx = x^{(n+1)}/(n+1)$ with $n = -2$:

$$\Delta U = -GMm(x^{-1})_{R_o}^{R_f}$$

When we substitute the limits, we have

$$\Delta U = \left(-G\frac{Mm}{R_f}\right) - \left(-G\frac{Mm}{R_o}\right)$$

Since $\Delta U = U_f - U_o$, the form of the above equation helps us to identify candidate functions for U. Clearly, a function of the form

$$U = -G\frac{Mm}{R} + C$$

where C is any constant, gives the correct result. As usual, we can pick any value for C we wish. The simple choice $C = 0$ is often taken.

$$U = -G\frac{Mm}{R} \qquad \text{Potential energy for Newtonian gravity} \qquad (8.14)$$

The choice $C = 0$ does, however, have some consequences which take a bit of getting used to. For example, if you look for the R-value where $U = 0$, you will discover it's very far away. In fact, $U = 0$ only as R becomes infinite! Moreover, our choice of $C = 0$ means that the potential energy of any two masses separated by a *finite* distance is negative.

Concept Question 7
Spaceships *A* and *B* are identical. Because of the earth, spaceship *A* has gravitational potential energy − 1 MJ, and spaceship *B* has potential energy − 2 MJ. (a) Which ship has the greater potential energy? (b) Which ship is farther from the earth?

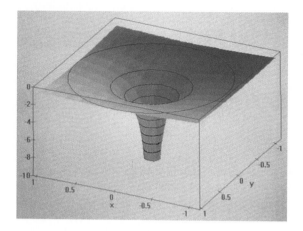

FIGURE 8.13

The potential energy of a mass m in the gravitational field of a planet of mass M.

Remember, there is nothing strange about a negative potential energy. Having negative potential energy merely means that you have less potential energy than you do at the zero point. Perhaps the following analogy will help. Think of the gravitational force as a kind of "rubber band" trying to hold two masses together. Now, the farther apart we pull the masses, the larger the potential energy stored in the gravitational "rubber band." If we label the largest potential energy stored in the fully stretched rubber band as zero, then, when it has less potential energy, we must use negative numbers to describe the energy.

The potential energy function described by Equation (8.14) is graphed in Figure 8.13. If one of the masses is fixed (or so large compared to the other that the larger mass's motion is negligible), then we can picture the second mass as moving in a potential well as shown in Figure 8.13. Remember, the actual motion is in a plane, and that the third dimension is a representation of the *potential energy,* not displacement. We can visualize the smaller mass's path by looking at the motion of the object when the picture is viewed from above.

We now have two expressions that model the gravitational potential energy. We must be careful to use them appropriately. The expression mgy was derived for a constant gravitational field g. Thus, it is useful over limited heights above the surface of the earth (or other celestial body if g is suitably modified). We refer to "$U = mgy$" as the **local model.** The expression "$U = -GMm/R$" is a **universal model,** and it is useful over large distances for objects that can be modeled as uniform spheres. Notice also that the two expressions have $U = 0$ at different locations. Because of these differences, you can never, ever use both expressions in the same problem. To see how we can use the universal model of potential energy to solve problems, look at Example 8.6.

EXAMPLE 8.6   *Only a Stone's Throw Away*

If a lunar landing module is launched upward from the surface of the moon with a speed of 1.00×10^3 m/s, how far above the moon's surface does it rise?

INCORRECT SOLUTION   Since it is usually easier to use the local model, let's try it first. Let the surface of the moon be the $y = 0$ point, so that $U = 0$ at this location. Then we have only kinetic energy on the "before" side of the conservation of mechanical energy, Equation (8.8). At the highest point the module reaches above the moon's surface, the kinetic energy is zero, having turned entirely into gravitational potential energy mgy_{max}:

$$K_o + U_o = K_f + U_f$$

$$\frac{1}{2}mv_o^2 + 0 = 0 + mg_{\text{moon}} y_{\text{max}}$$

canceling the mass m, and solving for y_{max} we obtain, with $g_{\text{moon}} = 1.62$ N/kg,

$$y_{\text{max}} = \frac{v_o^2}{2g_{\text{moon}}} = \frac{(1000.\ \text{m/s})^2}{2(1.62\ \text{N/kg})}$$

$$= 309\ \text{km}$$

Because the radius of the moon is only 1740 km, y_{max} is a sizable fraction of R_{moon}. Our "local" model for gravity is suspect!

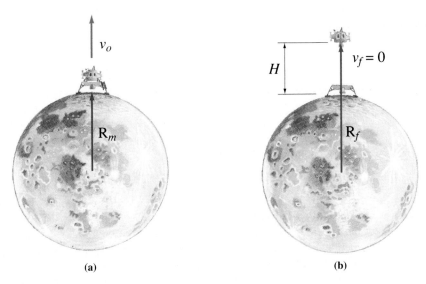

(a)   (b)

FIGURE 8.14

When applying the conservation of mechanical energy to an object that moves far from the surface of the planet, the universal form of the gravitational potential energy function must be used and distances measured from the center of the planet. (See Example 8.6.)

CORRECT SOLUTION   Using the universal model for gravitational potential energy, the landing module's potential energy at the moon's surface is *not* zero. We must make use of the fact that the module starts at a distance R_{moon} from the center of the moon as shown in Figure 8.14(a). The final position of the module is R_f as shown in Figure 8.14(b). Thus, the conservation of mechanical energy becomes

$$K_o + U_o = K_f + U_f$$

$$\frac{1}{2}mv_o^2 - \frac{GmM}{R_o} = 0 - \frac{GmM}{R_f}$$

Canceling the mass of the lunar module m and solving for R_f, we obtain

$$\frac{1}{R_f} = \frac{1}{R_o} - \frac{v_o^2}{2GM}$$

$$= \frac{1}{1.738 \times 10^6\ \text{m}} - \frac{(1.00 \times 10^3\ \text{m/s})^2}{2(6.67 \times 10^{-11}\ \text{N m}^2/\text{kg}^2)(7.36 \times 10^{22}\ \text{kg})}$$

$$= 4.70 \times 10^{-7}\ \text{m}^{-1}$$

which gives $R_f = 2130$ km as the distance from the center of the moon. To obtain the module's height above the moon we must subtract the moon's radius $R_m = 1740$ km:

$$H = R_f - R_{\text{moon}} = 390 \text{ km}$$

This result, as we might expect, is larger than our previous (incorrect) solution because the strength of the moon's gravitational attraction decreases with height above the moon's surface. ◄

An interesting question arises when we consider problems like Example 8.6. We might well ask how fast a projectile needs to be launched from the surface of a planet to reach an arbitrarily distant point. The only way to ensure that it reaches *any* point is to assume that this destination is located at infinity ($R = \infty$ in Eq. (8.14)). At this distance we assume the projectile has no remaining kinetic energy. Because its gravitational potential energy also approaches zero at infinity, the total energy on the right-hand side of our "before and after" conservation-of-energy equation is zero:

$$K_o + U_o = K_f + U_f$$

$$\frac{1}{2}mv^2 - \frac{GmM}{R_o} = 0 + 0$$

Solving for v we obtain

$$v = \sqrt{\frac{2GM}{R_o}} \qquad \text{(Escape velocity)} \qquad (8.15)$$

The velocity v in Equation (8.15) is known as the **escape velocity.** It is the minimum velocity an object must have to escape completely the "gravitational bonds" that hold it to another mass. If an object's velocity is less than the escape velocity, its kinetic energy is less than the absolute value of the gravitational potential energy and, therefore, its total energy is negative. Thus, objects that are gravitationally bound to each other have negative total energy. The (negative) gravitational potential energy may be thought of as a kind of "energy debt" that must be paid before the object can be free. The earth's velocity is less than that necessary to escape the sun's gravitation, thus it is bound in its orbit around the sun.

8.7 Conservative Forces in Two and Three Dimensions (Optional)

In Section 8.5 we found that for one-dimensional systems, the conservative force F can be calculated from the negative derivative of potential energy function:

$$F = -\frac{d}{ds}U(s)$$

The position label s might represent a displacement along the x-axis, a height y, a radial position r, or any other coordinate necessary to specify the single dimension. This relation between U and F is just a special case of a more general rule that allows the conservative force to be computed from the potential energy function. The rule is

$$\mathbf{F} = -\left(\left(\frac{\partial U}{\partial x}\right)\hat{\mathbf{i}} + \left(\frac{\partial U}{\partial y}\right)\hat{\mathbf{j}} + \left(\frac{\partial U}{\partial z}\right)\hat{\mathbf{k}}\right) \qquad (8.16)$$

The combination of partial derivatives and unit vectors occurring in Equation (8.16) is known as the **gradient** and is often symbolized by the "del" operator

$$\mathbf{\nabla} = \hat{\mathbf{i}}\frac{\partial}{\partial x} + \hat{\mathbf{j}}\frac{\partial}{\partial y} + \hat{\mathbf{k}}\frac{\partial}{\partial z}$$

Partial derivatives are reviewed in Appendix 1. However, all you really need to know is that when you take the partial derivative with respect to one variable, treat the others as if they were *constants*. (Hence, $\partial U/\partial x$ means take the derivative with respect to x and pretend y and z are constants.) Making use of the del operator, Equation (8.16) can be written compactly as

$$\mathbf{F} = -\mathbf{\nabla}U = -\mathbf{grad}\ U$$

The name "gradient" is well chosen for this operator because it does indeed tell us about the "steepness" of the function on which it operates.

Motion in Two Dimensions

In Section 8.5 we found it helpful to view the potential energy function as a sort of "roller coaster" map over which an object modeled as a particle could travel without friction. This kind of pictorial reasoning can be extended to motion in two dimensions. Figure 8.15(a) is a diagram of the potential energy function that an alpha particle experiences when it is launched at a gold nucleus. Now you may not have any idea how an alpha particle behaves in such circumstances (you may not even remember what an alpha particle is), but you can surely guess how an ice cube moves if it is "launched" up the side of this "potential hill." Although this type of mental picture is helpful, it is important to remember that the particle is not really going up and down, but rather moving only in a two-dimensional plane. The actual track of the alpha particle is that observed from directly overhead as in Figure 8.15(b).

Equation (8.16) is nothing more than a detailed mathematical prescription for how to obtain the steepness of a potential hill at any point (x, y, z). Its application is illustrated in Example 8.7.

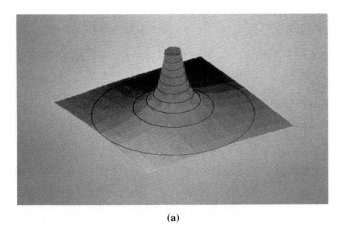

(a)

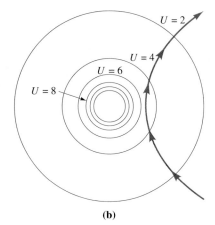

(b)

FIGURE 8.15

(a) The potential energy function of a positively charged alpha particle a distance R from the nucleus of a gold atom. (b) If launched toward the gold nucleus, the alpha particle follows a hyperbolic trajectory similar to the one shown. Circles represent contour lines along which the potential energy is constant.

EXAMPLE 8.7   *Climbing Mt. Coulomb*

For motion restricted to the *xy*-plane, the potential energy function (the sides of the potential hill) shown in Figure 8.15 can be described by

$$U(x, y) = \frac{A}{\sqrt{(x^2 + y^2)}}$$

where *A* is a constant. Show that forces calculated from Equation (8.16) agree with your intuition. Specifically, show that the force always points away from the origin and becomes larger as you get closer to the origin.

SOLUTION   Recall that the partial derivative operator $(\partial/\partial x)$ means that we differentiate with respect to *x*, treating the other variables (e.g., *y* and *z*) as constants. Thus,

$$\frac{\partial U}{\partial x} = \frac{\partial}{\partial x}[A(x^2 + y^2)^{-1/2}] = -\frac{1}{2}A(x^2 + y^2)^{-3/2}(2x) = -A\frac{x}{r^3}$$

where $r = \sqrt{x^2 + y^2}$ is the radial distance from the origin.
In a similar manner it is easy to show that

$$\frac{\partial U}{\partial y} = -A\frac{y}{r^3} \qquad \text{and} \qquad \frac{\partial U}{\partial z} = 0$$

Substituting these values into Equation (8.16), we find

$$\mathbf{F} = -\left(\left(\frac{\partial U}{\partial x}\right)\hat{\mathbf{i}} + \left(\frac{\partial U}{\partial y}\right)\hat{\mathbf{j}} + \left(\frac{\partial U}{\partial z}\right)\hat{\mathbf{k}}\right)$$

$$= +A\frac{x\hat{\mathbf{i}} + y\hat{\mathbf{j}}}{r^3} = A\frac{\mathbf{r}}{r^3}$$

In this final form we see that **F** does indeed fit our intuitive sense of how the force should behave. That is, the magnitude *F* is proportional to $1/r^2$ so that as *r* becomes small the force increases rapidly. Furthermore, since the direction of **F** is parallel to the direction of **r**, the force points directly ''downhill'' as we expect looking at Figure 8.15.   ◀

Motion in Three Dimensions

If a force is a function of position only, it can be written

$$\mathbf{F} = F_x(x, y, z)\hat{\mathbf{i}} + F_y(x, y, z)\hat{\mathbf{j}} + F_z(x, y, z)\hat{\mathbf{k}}$$

A force with this form is conservative if the following three equalities are all true:

$$\frac{\partial F_x}{\partial y} = \frac{\partial F_y}{\partial x}, \qquad \frac{\partial F_x}{\partial z} = \frac{\partial F_z}{\partial x}, \qquad \frac{\partial F_y}{\partial z} = \frac{\partial F_z}{\partial y}$$

We refer to the process by which we check to see if these three equalities hold as the **partial derivative test.**

EXAMPLE 8.8 *Does It Work on Southern Democrats and Northern Republicans?*

Determine whether or not the following force function describes a conservative force.

$$\mathbf{F} = (1 \text{ N/m}^5)xyz^3\hat{\mathbf{i}} + \left(\frac{1}{2}\text{N/m}^5\right)x^2z^3\hat{\mathbf{j}} + \left(\frac{3}{2}\text{N/m}^5\right)x^2yz^2\hat{\mathbf{k}}$$

SOLUTION   The units in the previous equation are required for dimensional consistency. However, they are cumbersome to repeat while performing the derivatives. We omit them and perform the partial derivative test on the three components of **F** written as

$$F_x = xyz^3, \qquad F_y = \frac{x^2z^3}{2}, \qquad F_z = \frac{3x^2yz^2}{2}$$

The first pair of derivatives yields

$$\frac{\partial F_x}{\partial y} = \frac{\partial(xyz^3)}{\partial y} = xz^3, \qquad \frac{\partial F_y}{\partial x} = \frac{\partial}{\partial x}\left(\frac{x^2z^3}{2}\right) = xz^3$$

Because these results are equal, F passes the first one-third of the test. The second pair of derivatives also check:

$$\frac{\partial F_x}{\partial z} = \frac{\partial}{\partial z}(xyz^3) = 3xyz^2, \qquad \frac{\partial F_z}{\partial x} = \frac{\partial}{\partial x}\left(\frac{3x^2yz^2}{2}\right) = 3xyz^2$$

The final pair of derivatives

$$\frac{\partial F_y}{\partial z} = \frac{\partial}{\partial z}\left(\frac{x^2z^3}{2}\right) = \frac{3x^2z^2}{2}, \qquad \frac{\partial F_z}{\partial y} = \frac{\partial}{\partial y}\left(\frac{3x^2yz^2}{2}\right) = \frac{3x^2z^2}{2}$$

are also equal, and the force function passes all three parts of the test. The force is conservative. ◀

In the preceding paragraphs we described a partial derivative test for conservative forces that may have appeared to you to come out of thin air. That is, we made no attempt to justify the processes of this test. In the following paragraphs we justify the test by showing that it reduces to a mathematical identity for forces that are the gradient of some potential function.

For a force to pass the test, three sets of partial derivatives have to be equal:

$$\frac{\partial F_x}{\partial y} = \frac{\partial F_y}{\partial x}, \qquad \frac{\partial F_x}{\partial z} = \frac{\partial F_z}{\partial x}, \qquad \frac{\partial F_y}{\partial z} = \frac{\partial F_z}{\partial y}$$

We know from our previous discussion that the conservative force F is the negative gradient of a potential energy function (Eq. (8.11)). Therefore, F_x itself is a derivative; namely,

$$F_x = -\frac{\partial U}{\partial x}$$

Thus, the first derivative of F_x is actually the negative second derivative of U:

$$\frac{\partial F_x}{\partial y} = -\frac{\partial^2 U}{\partial y\, \partial x} \tag{8.17}$$

In similar fashion, since

$$F_y = -\frac{\partial U}{\partial y}$$

we find

$$\frac{\partial F_y}{\partial x} = -\frac{\partial^2 U}{\partial x\, \partial y} \tag{8.18}$$

If we substitute Equations (8.17) and (8.18) into the first of the three partial derivative tests, it becomes, after canceling the minus signs on both sides,

$$\frac{\partial^2 U}{\partial y\, \partial x} = \frac{\partial^2 U}{\partial x\, \partial y}$$

On the left-hand side of the equation above we have U twice differentiated, first with respect to x, then with respect to y. On the right-hand side, U is differentiated first with respect to y then with respect to x. In short, the two derivatives are identical except for the order of the x- and y-differentiations. It is customary in calculus courses to prove that for "reasonably well behaved" functions these two derivatives are always equal. That is, the order of the differentiation doesn't matter. It is left as an exercise for you to show that a similar result holds true for the other two parts of the derivative test.

The following two statements summarize our results:

If a force is the gradient of a potential energy function, it is a conservative force.

If a force satisfies the "three derivative test," it is conservative and therefore can be written as the gradient of some potential energy function.

8.8 Summary

A force is conservative if the work done by that force when moving an object from point P_o to point P_f is independent of the path taken between P_o and P_f. (This is equivalent to the statement that the work done by the force around every closed path totals zero.)

For forces that act in only one dimension, the force is conservative if it is a function of position only.

For conservative forces it is useful to define the change in potential energy function to be equal to the negative of the work done by the conservative force. The location where a potential energy function is zero is arbitrary. Two useful potential energy functions are:

$$U = mgy \qquad \text{(Gravitation in a constant field of strength } g\text{)}$$

$$U = \frac{1}{2}ks^2 \qquad \text{(Hooke's-law spring stretched or compressed by } s\text{)}$$

In an isolated system with only conservative forces performing work, mechanical energy is conserved:

$$K_o + U_o = K_f + U_f$$

If nonconservative forces act in an isolated system, the work done by the nonconservative forces is equal to the change in the total mechanical energy of the system:

$$K_o + U_o + W_{\text{noncon}} = K_f + U_f \qquad (8.10)$$

The force function associated with a one-dimensional potential energy function $U(s)$ can be obtained from the negative of the derivative of the potential energy:

$$F(s) = -\frac{dU}{ds} \qquad (8.11)$$

(Optional) Potential energy for Newtonian gravity is given by

$$U = -G\frac{Mm}{R} \qquad (8.14)$$

where R is the distance between the centers of spherical masses.

(Optional) In three dimensions the conservative force is the negative gradient of the potential energy function $U(x, y, z)$:

$$\mathbf{F} = -\left(\left(\frac{\partial U}{\partial x}\right)\hat{\mathbf{i}} + \left(\frac{\partial U}{\partial y}\right)\hat{\mathbf{j}} + \left(\frac{\partial U}{\partial z}\right)\hat{\mathbf{k}}\right) \qquad (8.16)$$

$$= -\boldsymbol{\nabla} U$$

where $\boldsymbol{\nabla}$ is the gradient operator

$$\boldsymbol{\nabla} = \hat{\mathbf{i}}\frac{\partial}{\partial x} + \hat{\mathbf{j}}\frac{\partial}{\partial y} + \hat{\mathbf{k}}\frac{\partial}{\partial z}$$

(Optional) A three-dimensional force is conservative if it passes the partial derivative test

$$\frac{\partial F_x}{\partial y} = \frac{\partial F_y}{\partial x}, \qquad \frac{\partial F_x}{\partial z} = \frac{\partial F_z}{\partial x}, \qquad \frac{\partial F_y}{\partial z} = \frac{\partial F_z}{\partial y}$$

PROBLEMS

8.1 Conservative Forces and Potential Energy

1. In Figure 8.P1 the work (in joules) done by a force **F** along the five paths A, B, C, D, and E is $+3.00$, $+5.00$, $+10.00$, $+2.00$, and -5.00, respectively. Can you tell from these data if the force **F** is conservative?

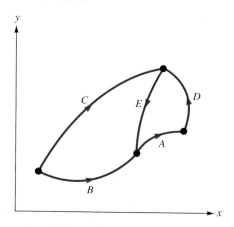

FIGURE 8.P1   Problem 1

2. Determine if the following forces are conservative, giving reasons for your answers: (a) the force of air resistance, (b) an ideal Hooke's-law spring, (c) the force from a tennis ball being compressed (see the graph in Problem 7.47), (d) gravity.

3. Calculate the work done by the force $\mathbf{F} = (2.00\ \text{N/m})x\,\widehat{\mathbf{i}} - 3.00\ \text{N}\,\widehat{\mathbf{j}}$, in Figure 8.P2 (a) along path A and (b) along path B. (c) Does the fact that the work done is the same in both cases prove that the force is conservative?

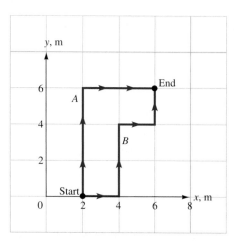

FIGURE 8.P2   Problems 3 and 4

4. Calculate the work done by the force $\mathbf{F} = (0.500\ \text{N/m}^2)y^2\,\widehat{\mathbf{i}} + (1.00\ \text{N/m})x\,\widehat{\mathbf{j}}$ in Figure 8.P2 (a) along path A and (b) along path B. (c) Is this force conservative?

5. Calculate the work done by the force $\mathbf{F} = (3.00\ \text{N})\,\widehat{\mathbf{i}} + (4.00\ \text{N/m})y\,\widehat{\mathbf{j}}$ in Figure 8.P3 (a) along path A, (b) along path B, and (c) along path C. [*Hint:* Along path C, $d\mathbf{s} = dx\,\widehat{\mathbf{i}} + dy\,\widehat{\mathbf{j}}$.] (d) Does the equality of these three work values prove that **F** is a conservative force?

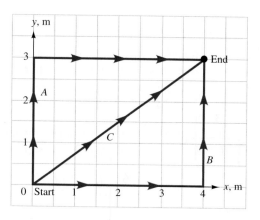

FIGURE 8.P3   Problem 5

6. Calculate the work done by the force $\mathbf{F} = (1.00\ \text{N/m})x\,\widehat{\mathbf{i}} + (1.00\ \text{N/m}^2)xy\,\widehat{\mathbf{j}}$ along paths A, B, and C in Figure 8.P3. (See Problem 5 for a hint.) (d) Does the inequality of these three work values prove that **F** is nonconservative?

8.2 Potential Energy: The Negative Integral of Conservative Force over Distance

7. A 2.50-kg book rests on top of a desk 0.850 m above the floor. If the origin of the y-coordinate system is on the floor, (a) what is the gravitational potential energy of the book as it rests on the desk? If the book falls from the desk to the floor, what is the book's change in potential energy? (b) Answer the two questions in part (a) except this time pick the tabletop as the $y = 0$ point. (c) Repeat the calculations of part (a) taking the ceiling 2.50 m above the floor as the $y = 0$ point.

8. (a) Calculate the increase in potential energy of a Hooke's-law spring of stiffness constant $k = 200.\ \text{N/m}$ when the spring is stretched from s $= 0.000$ to s $= 0.500$ m. (b) What is the increase in potential energy when the spring is stretched from 0.500 m to 1.00 m? (c) from 1.00 m to 1.50 m?

9. (a) Calculate the work done when a conservative force described by $F = -B/x^2$ acts on a particle that moves from $x = x_o$ to $x = x_f$ ($B =$ constant). (b) What form does your answer to part (a) suggest for a potential energy function? (c) Why can't the $x = 0$ point be taken as the zero of potential energy? (d) What points can be taken as the zero for potential energy?

10. A force function often used to model intermolecular forces is given by

$$F = \left(\frac{A}{12}\right) r^{-11} - \left(\frac{B}{6}\right) r^{-5}$$

where A and B are constants. (a) Calculate the work done if a particle is moved from $r = x_o$ to $r = x_f$ by this force. (b) What potential energy function is suggested by your answer to part (a)? (c) Where is the "natural" zero for this potential energy function? (d) Sketch the potential energy function.

11. A one-dimensional conservative force depends on distance as shown in Figure 8.P4. Sketch this force's potential energy function.

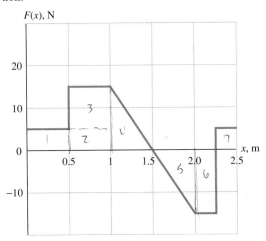

FIGURE 8.P4   Problem 11

12. A one-dimensional conservative force depends on distance as shown in Figure 8.P5. Sketch this force function's potential energy graph (a) taking $x = 0.00$ as the point of zero potential energy and (b) taking $x = 5.00$ m as the point of zero potential energy.

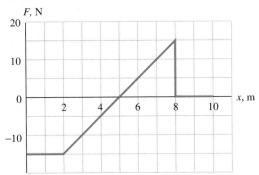

FIGURE 8.P5   Problem 12

13. Near an abrupt pn semiconductor junction, the conservative electric force on positive charge carriers (called "holes") has the form shown in Figure 8.P6. Sketch the shape of the potential energy function for holes near the pn junction.

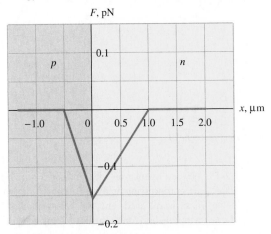

FIGURE 8.P6   Problem 13

14. Near the base region of a forward-biased pnp transistor, the (conservative) electric force on holes (see Problem 13) can be approximated by the function graphed in Figure 8.P7. Sketch the potential energy function of holes in the transistor.

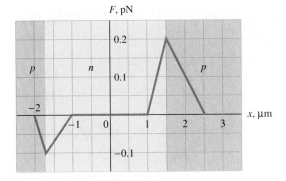

FIGURE 8.P7   Problem 14

15. Figure 8.P8 shows a certain one-dimensional conservative force as a function of position along the positive x-axis. (a) Make a quantitative graph of the potential energy as a function of position. Take $U = 0$ at $x = 0$. (b) Describe exactly what changes to your graph result if the $U = 0$ point is taken at $x = 7.00$ m instead of $x = 0.00$.

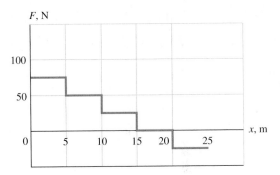

FIGURE 8.P8   Problem 15

16. A one-dimensional force function is described by $F = (3.00 \text{ N/m})|x|$. (a) Sketch the force function over the domain $-2.00 \text{ m} < x < 2.00 \text{ m}$. (b) Sketch the corresponding potential energy function over the same domain.

8.3 Conservation of Mechanical Energy

17. A 2.00-kg mass is released from rest at point A in Figure 8.P9. The track on which it slides is frictionless. (a) What is the block's velocity at point B? (b) What is the maximum compression of the spring, which has stiffness constant 30.0 N/m? (Assume the spring to be massless.)

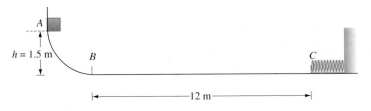

FIGURE 8.P9   Problem 17

18. When a 2.00-kg block is 0.500 m directly above the end of a spring, it is traveling downward with a velocity of 3.00 m/s. If the spring has a stiffness constant of $k = 160.$ N/m, how far does the block compress the spring before coming momentarily to rest?

FIGURE 8.P10   A skater turns kinetic energy into gravitational potential energy—and back again.

19. A 0.250-kg mass is traveling to the right with a speed of 5.60 m/s at point A on the frictionless track shown in Figure 8.P11. (a) What is the maximum speed of the mass? (b) What is the maximum compression of the spring of stiffness constant $k = 100.$ N/m? (Assume the spring has zero mass.)

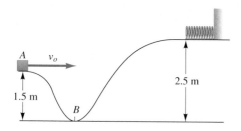

FIGURE 8.P11   Problem 19

20. Continue with Example 8.2 and find (a) the lowest point to which the piñata falls and (b) the maximum speed the piñata attains.

21. A spring-loaded launcher fires a 155-g ball bearing off a horizontal tabletop that is 1.01 m high. The ball strikes the floor 2.34 m horizontally from the point of release, as shown in Figure 8.P12. If the spring is compressed by 4.50 cm, what is its spring constant?

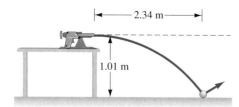

FIGURE 8.P12   Problem 21

22. An Atwood machine consists of two masses ($m_A = 2.75$ kg and $m_B = 1.50$ kg) tied together by a light string draped over a frictionless rod. (See Fig. 8.P13.) If the masses are released from rest, use conservation of energy to determine how fast they are moving when m_A has fallen 1.60 m.

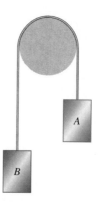

FIGURE 8.P13   Problem 22

8.4 Systems with Nonconservative Forces

23. Three blocks ($m_A = 3.50$ kg, $m_B = 1.50$ kg, and $m_C = 2.00$ kg) are connected as shown in Figure 8.P14. The strings slide over the pegs without friction. However, the coefficient of sliding friction between block B and the tabletop is $\mu_k = 0.250$. Use the generalization of the conservation of mechanical energy theorem to determine the speed of the blocks after mass A has fallen 0.500 m.

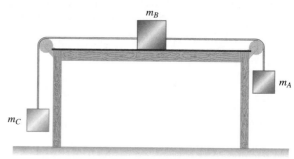

FIGURE 8.P14   Problem 23

24. A 0.250-kg bead slides on a curved wire as shown in Figure 8.P15. The segment from A to B is rough, but the segment from B to C is frictionless. (a) If the bead slides from rest at point A and is observed to have velocity 5.40 m/s at point B, how much work was done by friction from A to B? (b) To what height does the bead rise between B and C?

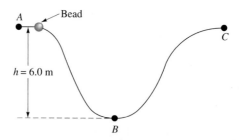

FIGURE 8.P15   Problem 24

25. A 0.250-kg mass is pressed against an ideal massless horizontal spring ($k = 100.$ N/m) compressing it by 0.500 m as shown in Figure 8.P16. The mass is released, and the spring pushes the block forward over a rough horizontal surface. How fast is the block moving when it is a distance of 1.50 m beyond the spring if the coefficient of friction between the block and the surface is $\mu_k = 0.400$?

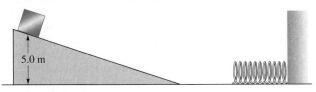

FIGURE 8.P16   Problem 25

26. A 2.00-kg mass slides down an incline as shown in Figure 8.P17. At the bottom of the slope, it glides without friction along a horizontal surface and collides with a Hooke's-law spring with stiffness constant $k = 392$ N/m. (a) What is the potential energy of the mass at the top of the incline (relative to the bottom of the incline)? (b) If the speed of the block at the bottom of the incline is 7.00 m/s, how much work did friction do on the block as it traveled down the incline? (c) How far does the block compress the spring? (Assume the spring is massless.)

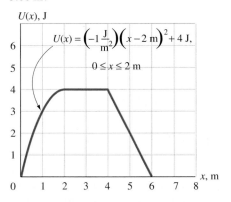

FIGURE 8.P17   Problem 26

★27. A 0.250-kg mass is pressed against a horizontal spring ($k = 100.$ N/m), compressing it by 0.500 m. The mass is released, and the spring pushes it forward along a horizontal frictionless surface. (a) Use energy methods to find the speed of the spring as it leaves the end of the spring. (b) As the block slides along it encounters a rough inclined plane tilted 30.0° above the horizontal. If the coefficient of friction between the block and the incline is $\mu_k = 0.1178$, how far up the incline, as measured along the incline, does the block travel before its speed is reduced to 2.00 m/s?

8.5 Conservative Force: The Negative Derivative of Potential Energy

28. Consider the potential energy function $U(x)$ shown in Figure 8.P18. (a) Carefully sketch the force function corresponding to this potential function. (b) If a 2.00-kg mass is moving to the right at 2.00 m/s at $x = 3.00$ m, what is the block's speed at $x = 5.00$ m?

$U(x)$, J

$$U(x) = \left(-1\frac{\text{J}}{\text{m}^2}\right)(x - 2 \text{ m})^2 + 4 \text{ J},$$

$$0 \le x \le 2 \text{ m}$$

FIGURE 8.P18   Problem 28

29. Consider the potential energy function of a force that acts in only one dimension: $U = Ax^2 - Bx^{-1}$, where $A = 2.00$ J·m^2 and $B = 25.0$ J·m. (a) Sketch this potential function for $0 < x \le 1.5$ m. (b) Determine the force function for this potential. (c) Where is the point of equilibrium? (d) Is the equilibrium stable or unstable?

30. The Yukawa potential, given by

$$U = -U_o \frac{r_o}{r} \exp\left(-\frac{r}{r_o}\right)$$

where r_o and U_o are constants, is used to model the strong attractive force between the heavy particles (protons and neutrons) inside the nucleus of atoms. Because the potential is spherically symmetric, the problem can be treated as if it were one-dimensional in the variable r. (a) Sketch the Yukawa potential as a function of r for $U_o = 1.00 \times 10^{17}$ J and $r_o = 1.00 \times 10^{-15}$ m. (b) Calculate the force function associated with the Yukawa potential. (c) Find the equilibrium point in terms of r_o. Is the equilibrium stable, unstable, or neutral?

31. Calculate the force function for which $U = A[(x^2 + y^2)z]$ is the potential energy function.

32. Calculate the force function for which $U = A(xy + yz + xz)$ is the potential energy function.

33. Calculate the force function for which $U = Axy \cos(z/a)$ is the potential function.

34. Calculate the force function for which $U = A/(x^2 + y^2 + z^2)$ is the potential function.

35. The potential energy function in a certain region of space can be described by

$$U(x) = \begin{cases} -1.50 \text{ J} + (3.00 \text{ J/m})x, & 0 \le x \le 2.00 \\ +6.00 \text{ J} - (1.50 \text{ J/m}^2)(x - 3.00 \text{ m})^2, \\ & 2.00 \le x \le 4.00 \text{ m} \end{cases}$$

as shown in Figure 8.P19. Sketch a graph of the corresponding force function.

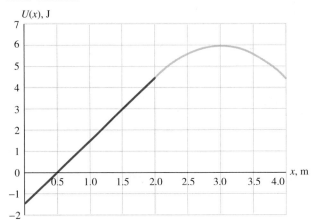

FIGURE 8.P19   Problem 35

36. The potential energy function for electrons near a semiconductor *pn*-graded junction can be modeled in one dimension by $U(x)$, where

$$U(x) = -A\frac{W^3}{12} \qquad\qquad x \le -\frac{W}{2}$$

$$U(x) = A\left(\left(\frac{W}{2}\right)^2 x - \frac{x^3}{3}\right), \qquad -\frac{W}{2} < x < \frac{W}{2}$$

$$U(x) = A\frac{W^3}{12} \qquad\qquad x \ge \frac{W}{2}$$

as shown in Figure 8.P20. (a) Carefully sketch the graph of the corresponding force function. (b) In what direction is the force on an electron located at $x = 0$?

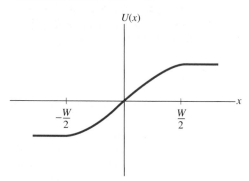

FIGURE 8.P20   Problem 36

8.6 Potential Energy for Newtonian Gravity (Optional)

37. Calculate the escape velocity required for an object to escape (a) the sun, (b) the earth, (c) the moon.
38. Is the moon more tightly bound (in terms of gravitational potential energy debt) to the earth or to the sun?
39. One model suggests that a black hole occurs when a star's mass becomes compacted into a sphere so small that the escape velocity becomes greater than the speed of light c. The radius at which this occurs is called the Schwarzschild radius. (a) Derive an expression for the Schwarzschild radius of a star of mass M based upon Newtonian gravity. (b) Show that a mass m has an energy debt equal to $mc^2/2$ when it is at the Schwarzschild radius away from the center of a black hole.
40. Show that if a rocket is given a speed equal to one-half of the escape velocity from a planet of radius R, it rises a distance of $\frac{1}{3}R$ above the planet's surface. (Ignore any resistance due to the planet's atmosphere.)
41. If a rocket is given a velocity equal to the escape velocity while it is on a planet's surface, how far from the surface is it when its velocity is equal to half the escape velocity? (Ignore frictional effects of the planet's atmosphere.)
★42. Show that the kinetic energy of a satellite in orbit around a planet is equal to one-half the absolute value of the satellite's gravitational potential energy. (*Hint:* You will need equations for circular motion and gravitational force in addition to energy equations.)

8.7 Conservative Forces in Two and Three Dimensions (Optional)

43. Use the partial derivative test to determine if the following are conservative forces:
    (a) $\mathbf{F} = (3.00 \text{ N/m})x\,\hat{\mathbf{i}} + (3.00 \text{ N/m})y\,\hat{\mathbf{j}} + (3.00 \text{ N/m})z\,\hat{\mathbf{k}}$,
    (b) $\mathbf{F} = (1.00 \text{ N/m}^2)y^2\,\hat{\mathbf{i}} + (1.00 \text{ N/m}^2)x^2\,\hat{\mathbf{j}} + (1.00 \text{ N/m}^2)z^2\,\hat{\mathbf{k}}$
44. Use the partial derivative test to determine if the following forces are conservative:
    (a) $\mathbf{F} = (1.00 \text{ N/m}^5)y^2z^3\,\hat{\mathbf{i}} + (2.00 \text{ N/m}^5)xyz^3\,\hat{\mathbf{j}} + (3.00 \text{ N/m}^5)xy^2z^2\,\hat{\mathbf{k}}$
    (b) $\mathbf{F} = (1.00 \text{ N/m})z\cos(y/a)\,\hat{\mathbf{i}} - (1.00 \text{ N/m})\dfrac{xz}{a}\sin(y/a)\,\hat{\mathbf{j}} + (1.00 \text{ N/m})x\cos(y/a)\,\hat{\mathbf{k}}$, where a is a constant. What are the units of the parameter a?
★45. Determine a force component F_z such that the following force vector is conservative: $\mathbf{F} = (2 \text{ N/m}^2)xz\,\hat{\mathbf{i}} + (2 \text{ N/m}^2)yz\,\hat{\mathbf{j}} + F_z\,\hat{\mathbf{k}}$

46. Show that the Newtonian model for gravity

$$\mathbf{F} = GMm\left(\frac{x\,\hat{\mathbf{i}} + y\,\hat{\mathbf{j}} + z\,\hat{\mathbf{k}}}{(x^2 + y^2 + z^2)^{3/2}}\right)$$

is a conservative force. *Hint:* Can you make a convincing argument about the equality of the derivatives in the derivative test based upon symmetry?

Numerical Methods and Computer Applications

47. Use the symbolic differentiation capabilities of your calculator or computer to calculate the required derivatives for Problem 46.
48. Construct a spreadsheet template to analyze the piñata in Example 8.2. Make columns for y, mgy, $\frac{1}{2}ks^2$, and $\frac{1}{2}mv^2$. Calculate each of these quantities at 2.0-cm intervals from $y = 0.500$ m to $y = -0.500$ m. (a) Make a graph showing how each of these quantities changes with y. (b) Use your spreadsheet to answer the questions posed in Problem 20.
49. Use a spreadsheet to produce a plot of the Yukawa potential described in Problem 30. Plot values of r from 0.10×10^{-15} to 5.00×10^{-15} m. Display three plots on the same graph each with a different value of r_o: 1×10^{-15}, 2×10^{-15}, and 3×10^{-15}.
50. The force required to hold a mousetrap spring at different angles is shown in Table 8.P1. As shown in Figure 8.P21, in each case the force was applied tangentially to the circular arc in which the bar travels. The radius of this arc was 3.4 cm. (a) Use the trapezoidal rule to numerically integrate the force function to produce the potential energy function. (b) What is the maximum potential energy stored in the fully compressed mousetrap? *Hint:* The distance traveled by the bar is $s = r\theta$, where θ must be in radians. (c) If the 26-g mousetrap is upside down and the trap goes off, what is the maximum height to which the trap can rise?

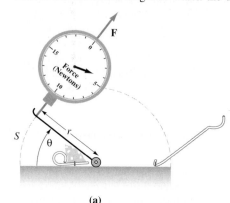

(a)

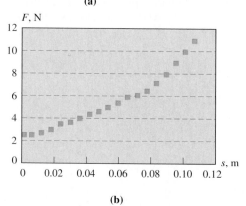

(b)

FIGURE 8.P21   Problem 50. (a) As the mousetrap is set, the force is always kept parallel to the motion along the semicircle. (b) The tangential force required to set a mousetrap.

TABLE **8.P1**   Force to Set a
Mousetrap[1]

$\theta(°)$	$F(N)$
10	2.5
20	2.7
30	3.0
40	3.5
50	3.7
60	4.0
70	4.4
80	4.6
90	5.0
100	5.4
110	5.9
120	6.1
130	6.5
140	7.2
150	8.0
160	9.0
170	10
180	11

[1] These data are in the spreadsheet template
MOUSETRP.WK1 on the diskette accompanying this text.

General Problems

51. Tarzan is swinging from a tall tree on the end of a vine of length
    L. The other end of the vine is fastened to a tree limb at the same
    elevation as Tarzan. As shown in Fig. 8.P22 there is a broken
    limb a distance D below the point where the vine is secured. The
    broken limb snags the vine as Tarzan swings past his lowest
    point. (a) Assume Tarzan leaves his perch with zero velocity. Use
    the conservation of mechanical energy to write an expression for
    the velocity of Tarzan when he reaches the highest point in his
    swing (i.e., when the vine between the snag and Tarzan is verti-
    cal). (b) Show that if $D < 3/5\ L$ then Tarzan cannot circle com-
    pletely around the snag.

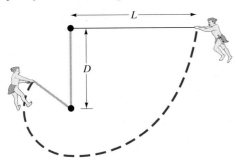

FIGURE 8.P22   Problem 51

52. Try designing a spring-powered car. You might use heavy springs
    like those in the rear suspension of a car. (a) You can estimate
    their stiffness constant by noticing that if a 75.0-kg person stands
    on the rear bumper of a car, the two rear springs are each com-
    pressed by about 20.0 cm. (b) A gallon of gasoline has the equiva-
    lent of 1.3×10^8 J of energy, but a car is only about 15% efficient
    in turning this chemical energy into mechanical energy. Assum-
    ing your spring-powered car to be 100% efficient, how far does
    the spring, with stiffness constant as found in (a), need to be
    compressed to provide the same mechanical energy as a gallon of
    gas? (c) What force is needed to hold the spring compressed?

★53. Put a small wet ice cube on the top of a basketball and watch it
    slide off. You notice that it leaves the surface of the basketball
    before reaching the equator. Model the ice cube as a point–mass
    sliding on a frictionless hemisphere of radius R as shown in Fig-
    ure 8.P23. Show that the ice cube leaves the sphere's surface at a
    height $R/3$ below the top of the basketball.

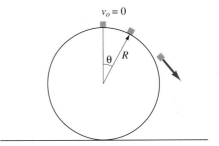

FIGURE 8.P23   Problem 53

★54. In all of the problems and examples in this text we modeled the
    springs as "massless." In doing this we ignored the kinetic en-
    ergy of the moving coils of the spring itself. A somewhat better
    (but still approximate) model is to assume, for a spring with one
    end fixed and the other end moving with velocity v_e, that the
    velocity of intermediate points changes linearly with distance
    from the fixed end, that is, $v = (v_e/L)x$, where L is the *total* length
    of the stretched spring as shown in Figure 8.P24. (a) Calculate the
    total kinetic energy of all mass elements $dm = (m/L)\ dx$ in the
    spring of mass m. (b) Show that the total kinetic energy of the
    spring–mass system is the same as that of a massless spring with
    mass $M + (m/3)$ attached to its free end. (c) Why is this model
    only approximate?

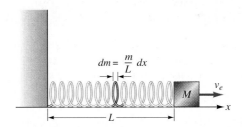

FIGURE 8.P24   Problem 54

55. Consider a long uniform rod of mass M and length L. The rod has
    a linear mass density of $\lambda = M/L$. When the rod is held vertically
    with the lower end at $y = 0$, each small section of thickness dy
    has a mass $dm = \lambda\ dy$. This mass element dm has a gravitational
    potential energy $dU = gy\ dm$. (a) Find the total gravitational po-
    tential energy of the rod by summing dU from all of the mass
    elements between $y = 0$ to $y = L$. (b) If one wished to calculate
    the potential energy found in (a) by treating the entire mass $M =
    \lambda L$ as if it were located at a single point, what would be the y
    coordinate of that point?

56. A slingshot has a force-versus-extension curve as shown in Fig-
    ure 8.P25. If a 210.-g ball bearing is fired at a 45° angle above the
    horizontal from the slingshot when it is pulled back to 0.650 m,
    where should a wastepaper basket be placed to catch the projec-
    tile if the mouth of the wastebasket is level with the point of
    release of the ball bearing? (We have actually performed this ex-
    ercise as a classroom demonstration. Determine the force-versus-

distance curve with a spring scale. It is helpful to cut a small hole in the leather pouch that holds the ball bearing and hold the bearing with an electromagnet on the back side. Breaking the electric circuit of the electromagnet ensures a clean release.)

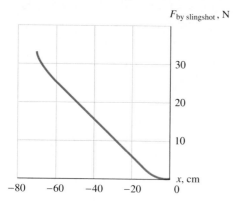

FIGURE 8.P25 Problem 56

★**57.** A 2.00-kg block is released from point A as shown in Figure 8.P26. The curved portion of the track is frictionless. The horizontal track, including the portion under the spring, has kinetic coefficient of friction $\mu_k = 0.100$. (a) What is the block's velocity at point B? (b) What is the block's velocity at point C where the spring's length is equal to its natural length? (c) How much does the block compress the spring ($k = 30.0$ N/m) before coming momentarily to rest? (d) Where does the block ultimately come to rest?

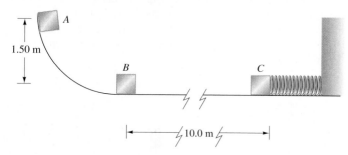

FIGURE 8.P26 Problem 57

58. Two masses ($m_A = 10.0$ kg and $m_B = 5.00$ kg) are connected to a Hooke's-law spring ($k = 200$. N/m) as shown in Figure 8.P27. Mass B rides without friction along the tabletop. The system is released from rest from a position where the spring is at its natural unstretched length. (a) How fast is block A moving after it has fallen 0.500 m? (b) What is the greatest distance that mass A falls before coming momentarily to rest? (c) What is the maximum speed attained by the masses?

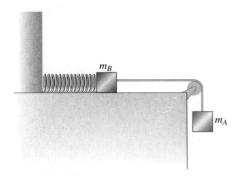

FIGURE 8.P27 Problems 58 and 59

59. Consider the two masses connected to a Hooke's-law spring as shown in Figure 8.P27. Mass A is 10.0 kg, $m_B = 5.00$ kg, the spring constant is $k = 200$. N/m, and the coefficient of friction between block B and the tabletop is $\mu_k = \mu_s = 0.250$. The masses are released from rest from a point where the spring is at its natural, unstretched length. (a) How far downward from the point of release does mass A fall before starting upward again? (b) How close to the point of release does mass A rise on the return bounce? (c) How far below the point of release does the mass fall the second time it reaches a lowest point? (d) Through what approximate total distance does mass B travel before the system comes to rest? Why is this answer only approximate?

60. An ice cube starting from rest slides without friction down an incline and into a vertical circular loop-the-loop of radius R as shown in Figure 8.P28. If the ice cube is to remain in contact with the track at the top of the loop, what is the minimum height h above the bottom of the loop from which the ice cube must start?

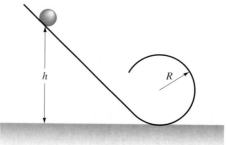

FIGURE 8.P28 Problem 60

61. When an apple of mass m falls to the ground, the earth also "falls" toward the apple. (a) What is the ratio of the magnitudes of the force of the apple on the earth to the force of the earth on the apple? (b) What is the ratio of the acceleration of the apple to that of the earth? (c) What is the ratio of the distance traveled by the apple to that traveled by the earth? (d) Show that the ratio of the kinetic energy gained by the apple to that gained by the earth is equal to m_{earth}/m.

62. A popular toy consists of a small (8.75 g) rubber ball attached to a thin elastic band of unstretched length 70.0 cm. The other end of the elastic is connected to the center of a flat wooden paddle. When stretched, the elastic band behaves as a Hooke's-law spring of force constant $k = 0.250$ N/m. Because the elastic band is quite thin, when the ball is closer than 50.0 cm to the paddle, the elastic acts like a limp string and exerts no force on the ball. If the ball is batted vertically upward so that it leaves the stationary paddle face with a velocity of 7.50 m/s, how high does the ball rise before coming momentarily to rest?

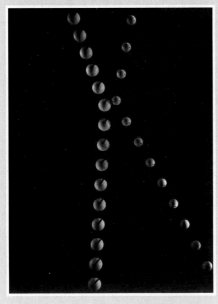

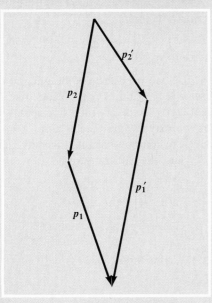

Impulse and Linear Momentum

In this chapter you should learn

- to compute the impulse caused by a force from either the analytical or graphical description of that force.
- to compute an object's momentum from its mass and velocity.
- to compute an object's change in momentum from the applied impulse and vice versa.
- to write Newton's second law in terms of momentum changes.
- to apply the law of conservation of momentum to solve recoil problems.
- to apply the law of conservation of momentum to one- and two-dimensional collisions.

In the previous two chapters we found that the concept of kinetic energy provides a useful means to help us solve motion problems. However, kinetic energy is a scalar quantity, and if you have ever made a bank shot while playing pool, knocked a tennis ball into the net, or become lost while driving in a new town, you know that *direction* is a critical part of motion. Thus, although kinetic energy is important, it cannot be the last word in a description of motion. In this chapter and the next we discuss momentum, which also describes the "quantity" of motion. A major distinction between this new concept and that of kinetic energy is that *momentum is a vector.* As the change in kinetic energy was related to the integral of force with respect to *distance,* we show that the change in momentum is related to the integral of force with respect to *time.*

When you hit a softball, spike a volleyball, or shoot a basketball, you push on an object for a short time, and that push changes the object's motion. Intuitively, you know that the harder you push on a free object, the greater is the change in that object's motion. You also can guess that the longer you push on the object, the greater its change in motion. The concept of *impulse,* which we discuss in the next section, formalizes these ideas.

9.1 Impulse

We begin this chapter with another definition and (once again) ask your indulgence until later when we will show that the definition is a useful one. **Impulse** is defined as the integral of force with respect to time:

$$\mathbf{J} = \int_{t_o}^{t_f} \mathbf{F}\, dt \tag{9.1}$$

Because force is a vector and time is a scalar, the result of the integral in Equation (9.1) is a vector. If the force is constant (remember constant vectors must have constant magnitude and constant direction), it may be removed from the integral so that the impulse is simply

$$\mathbf{J} = \mathbf{F}\, \Delta t$$

If the force is given as an explicit function of time, the impulse in Equation (9.1) can be calculated by straightforward integration. If instead, the time dependence of each force component is given in graphical form, the impulse is the area between the force curve and the $F = 0$ axis. The SI unit of impulse is a newton-second (N · s).

Concept Question 1
When a large force acts on an object, does it always mean that the resulting impulse is large? Explain.

EXAMPLE 9.1   *How Do You Integrate a Vector?*

Calculate the impulse due to the force $\mathbf{F} = A\hat{\mathbf{i}} + Bt\hat{\mathbf{j}}$ where $A = 2.00$ N and $B = 4.00$ N/s, if this force acts from $t_o = 0.000$ s to $t_f = 0.300$ s.

SOLUTION   By the definition Equation (9.1) we need to take the integral of the force vector with respect to time. This vector is the sum of two vectors: one in the x-direction ($A\hat{\mathbf{i}}$), the other in the y-direction ($Bt\hat{\mathbf{j}}$). Elementary rules of calculus tell us that, when taking the integral of the sum, we may integrate each term of the sum separately. Recall that the unit vectors $\hat{\mathbf{i}}$ and $\hat{\mathbf{j}}$ are constant vectors. Like any constant they may be taken outside of the integral. Thus, the integral reduces to

$$\mathbf{J} = \int_{t_o}^{t_f} \mathbf{F}\, dt = A\hat{\mathbf{i}} \int_{t_o}^{t_f} dt + B\hat{\mathbf{j}} \int_{t_o}^{t_f} t\, dt$$

$$= A\hat{\mathbf{i}}(t)\big|_{t_o}^{t_f} + B\hat{\mathbf{j}}(t^2/2)\big|_{t_o}^{t_f}$$

$$= (0.600\ \text{N} \cdot \text{s})\hat{\mathbf{i}} + (0.180\ \text{N} \cdot \text{s})\hat{\mathbf{j}}$$

Note that the answer is a vector.   ◀

EXAMPLE 9.2   *Rack 'Em Up!*

A model for the force on a billiard ball when a cue strikes it "head on" is shown in Figure 9.1. Calculate the impulse of this force.

SOLUTION   Because all of the motion is parallel to the *x*-axis only, we need not employ the unit-vector notation. The impulse $J = \int F \, dt$ is the area under the curve that consists of two triangles and a rectangle as shown by the shading in Figure 9.1.

$$J = \tfrac{1}{2}(500.\ \text{N})(1.00 \times 10^{-3}\ \text{s}) + (500.\ \text{N})(1.00 \times 10^{-3}\ \text{s})$$

$$+ \tfrac{1}{2}(500.\ \text{N})(1.00 \times 10^{-3}\ \text{s})$$

$$= 1.00\ \text{N} \cdot \text{s}\quad \text{in the direction of the force } \mathbf{F}.$$

◀

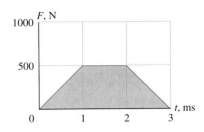

FIGURE 9.1

The impulse is the area between the force curve and the $F = 0$ axis.

9.2 Impulse and Momentum

The concept of impulse is useful because it relates an object's initial state of motion to its final state of motion. We can discover the exact form of this relationship by substituting the net force into the definition of the impulse. Because our vector equation is true for each component, we simplify the notation and look at only one component:

$$J_{\text{net}} = \int_{t_o}^{t_f} F_{\text{net}} \, dt$$

The net force, according to Newton's second law, is equal to the product of mass and acceleration.

$$J_{\text{net}} = \int_{t_o}^{t_f} ma \, dt$$

A short impulse applied to the cue ball imparts momentum to the cue ball, which it soon shares with the other balls.

We can substitute the definition of acceleration $a = dv/dt$ into the integral and make use of the definition of the differential $dv = (dv/dt) \, dt$, leading us to write

$$J_{\text{net}} = \int_{t_o}^{t_f} \left(m \frac{dv}{dt} \right) dt = \int_{v_o}^{v_f} (m) \, dv$$

Notice when we change the variable of integration to v, we must also change the time limits of the integral to the corresponding limits of v. Up to this point our derivation has been quite general. We now consider the special case for the motion of an object with a constant mass. The constant m can now be factored from the integral. (In Section 10.4 we will discuss what happens if we do not make this restriction.) For constant m

$$J_{\text{net}} = m \int_{v_o}^{v_f} dv = m(v_f - v_o) = mv_f - mv_o$$

Recall that the symbol Δ is reserved to indicate the change in any quantity. This change is the difference between the final value of a quantity and its original value. Hence, $mv_f - mv_o = \Delta(mv)$. This relation is true for each vector component, thus the net impulse on an object of constant mass is related to that object's motion by the equation

$$\mathbf{J}_{\text{net}} = \Delta(m\mathbf{v})$$

The quantity $m\mathbf{v}$ is of such significance in physics and engineering that it is given its own name and symbol.

Concept Question 2
The total energy of an object is zero. Must its linear momentum also be zero?

Concept Question 3
If the translational kinetic energy of an object is reduced by a factor of 4, by what factor does its linear momentum change?

Concept Question 4
A constant force **F** acts at different times on two different particles. One particle has mass m and the other has mass $2m$. If the force acts on mass m for twice the length of time that it acts on mass $2m$, what is the ratio of the velocity of the particle with mass m to that of particle with mass $2m$?

> Definition of momentum: momentum = mass × velocity
>
> $$\mathbf{p} = m\mathbf{v}$$

The SI unit of momentum is kilogram-meters per second (kg · m/s). You should be certain to note that momentum is a *vector*. The impulse of a force can now be related to an object's motion in a concise fashion:

$$\mathbf{J}_{\text{net}} = \Delta\mathbf{p} \tag{9.2}$$

Equation (9.2) is often referred to as the **impulse-momentum theorem.** In words, Equation (9.2) says that "the net impulse on an object is equal to the object's change in momentum." The implications of this theorem are far reaching. To unfold them takes the rest of this chapter and most of the next.

EXAMPLE 9.3   *That's Some Break!*

If a 0.165-kg cue ball is hit dead center by the impulse described in Example 9.2, calculate its approximate speed immediately after impact.

SOLUTION   In our solution we ignore the frictional force of the tabletop on the ball (and assume no "English" is applied, that is, the ball is struck horizontally at the midline). The error introduced by ignoring friction is discussed in the boxed section at the end of this example. When the frictional force of the tabletop is ignored, the net impulse on the cue ball is due entirely to the cue. The magnitude of this impulse was found in Example 9.2 to be 1.00 N · s. To find the speed change of the ball, we solve Equation (9.2) for Δv

$$J_{\text{net}} = \Delta p$$
$$= m\,\Delta v$$

Solving for Δv, we have

$$\Delta v = \frac{J_{\text{net}}}{m} = \frac{1.00\ \text{N} \cdot \text{s}}{0.165\ \text{kg}} = 6.06\ \text{m/s}$$

The direction of $\Delta \mathbf{v}$ is the same as the direction of $\mathbf{F}_{\text{cue}}$.   ◀

EXAMPLE 9.4   *How Do I Handle the Vectors?*

A falling volleyball with velocity $\mathbf{v}_o = -(0.650\ \text{m/s})\hat{\mathbf{i}} - (0.350\ \text{m/s})\hat{\mathbf{j}}$ is subjected to the net impulse described in Example 9.1 and shown in Figure 9.2. If the volleyball has a mass of 0.275 kg, calculate its velocity immediately following the impulse.

SOLUTION   We solve Equation (9.2) for the velocity change

$$\mathbf{J}_{\text{net}} = \Delta\mathbf{p} = m\,\Delta\mathbf{v}$$
$$\Delta\mathbf{v} = \frac{\mathbf{J}_{\text{net}}}{m}$$

Recalling that $\Delta\mathbf{v} = \mathbf{v}_f - \mathbf{v}_o$, we solve for the final velocity

The volleyball player changes the ball's momentum by applying an impulse.

$$\mathbf{v}_f = \mathbf{v}_o + \frac{\mathbf{J}_{net}}{m}$$

where $\mathbf{J}_{net}$ was found in Example 9.1. Substituting the known values we obtain

$$\mathbf{v}_f = -(0.650 \text{ m/s})\widehat{\mathbf{i}} - (0.350 \text{ m/s})\widehat{\mathbf{j}} + \frac{(0.600 \text{ N} \cdot \text{s } \widehat{\mathbf{i}} + 0.180 \text{ N} \cdot \text{s } \widehat{\mathbf{j}})}{(0.275 \text{ kg})}$$

We combine the vectors component by component, as usual:

$$\mathbf{v}_f = (-0.650 \text{ m/s} + 2.18 \text{ m/s})\widehat{\mathbf{i}} + (-0.350 \text{ m/s} + 0.655 \text{ m/s})\widehat{\mathbf{j}}$$
$$= (1.53 \text{ m/s})\widehat{\mathbf{i}} + (0.305 \text{ m/s})\widehat{\mathbf{j}} \qquad \blacktriangleleft$$

The Impulse Approximation

In Example 9.3 we calculated the speed of the billiard ball using the impulse-momentum theorem of Equation (9.2). To use this theorem we really need to consider the *net* force on the ball. In this case the net force is the vector sum of four forces: weight, normal, friction, and the cue. If the pool table is level, the weight force and normal force are equal and opposite; hence, their vector sum is zero. However, the horizontal frictional force of the tabletop on the ball is quite complicated. This force begins as static friction with zero magnitude. As the cue force on the ball increases in magnitude, the static frictional force increases to a maximum of $\mu_s N$. When the cue force exceeds the maximum static frictional force the ball begins to accelerate. The frictional force then rapidly decreases to $\mu_k N$ as the ball begins to slide. In our example we ignored all of this and omitted friction entirely! To be sure that it is a reasonable approximation, we must estimate the size of the error we introduce. To perform this estimate, let's examine the maximum force of friction, which is $\mu_s N$. If we can ignore the maximum frictional force, we can surely ignore the entire frictional force. The normal force is equal to the weight of the ball, $mg = 1.62$ N. Our experience with the felt on the top of pool tables allows us to estimate the coefficient of static friction as less than 1. Therefore, the largest frictional force we might encounter is less than 1.62 N. Because this force is much less than the force due to the cue during nearly the entire impulse (see Fig. 9.1), we can safely ignore it. (We leave as a problem for you to estimate the magnitude of the error. See Problem 69.) It is a fairly common practice to ignore small external forces during impulses due to large impact-forces. We make this approximation often and will not burden you with such detail again. However, you should always think about what forces you are ignoring and be sure the approximation is reasonable for the particular case you are considering.

9.3 The Momentum Statement of Newton's Second Law

A popular picnic game is egg-toss. Pairs of participants toss a raw egg back and forth over larger and larger distances. The pair who can successfully toss the egg when farthest apart wins. It doesn't take long to figure out the way to keep the egg from breaking: make initial contact with the egg as far in front of your body as possible, and then move your hands backward to bring the egg slowly to rest. The physics of the game is simple. When the egg reaches you it has some momentum. To stop the egg you must reduce the momentum to

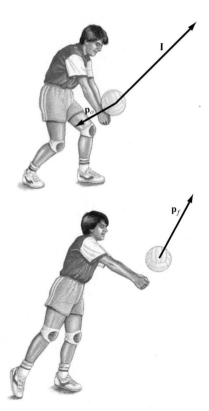

FIGURE 9.2

The final momentum of the volleyball is the vector sum of its original momentum plus the applied impulse. See Example 9.4.

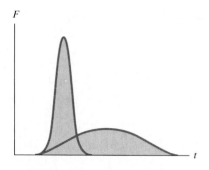

F

t

FIGURE 9.3

A large force acting for a short time and a small force acting for a long time may result in the same impulse, as indicated by the equal areas under the force-versus-time curves.

zero by applying an impulse. That is, you must have a certain area under the force-versus-time curve. As shown in Figure 9.3, you have a choice. You can exert a large force for a short time or a small force for a long time. One way or another you must generate an area under the force-versus-time curve equal to Δp. In the paragraphs below we describe how to apply this idea quantitatively.

The impulse momentum theorem of Section 9.2 was derived directly from Newton's second law by integration. It comes as no surprise then that we can recast Newton's second law in terms of momentum. One simple way to "work backward" is to replace the impulse in Equation (9.2) with its integral definition

$$\mathbf{J}_{net} = \Delta \mathbf{p}$$

$$\int_{t_o}^{t_f} \mathbf{F}_{net} \, dt = \Delta \mathbf{p}$$

and then let the force be constant so it can be removed from the integral:

$$\mathbf{F}_{net} \int_{t_o}^{t_f} dt = \Delta \mathbf{p}$$

$$\mathbf{F}_{net} \, \Delta t = \Delta \mathbf{p}$$

For a constant force we can solve for $\mathbf{F}_{net}$ exactly. Usually, however, the force is not constant, and we can find only the average force from

$$\mathbf{F}_{net \, av} = \frac{\Delta \mathbf{p}}{\Delta t} \tag{9.3}$$

In this form $\mathbf{F}_{net \, av}$ is the average net force that acts during time Δt. For short intervals $\Delta \mathbf{p}/\Delta t$ approximates the instantaneous force. In the limit as Δt approaches zero the fraction $\Delta \mathbf{p}/\Delta t$ becomes the derivative

$$\mathbf{F}_{net} = \frac{d\mathbf{p}}{dt} \tag{9.4}$$

Equation (9.4) is the form in which Newton cast his second law. When the mass is a constant, it can be moved in front of the differentiation symbol, and the familiar $\mathbf{F}_{net} = m\mathbf{a}$ is recovered:

$$\mathbf{F}_{net} = \frac{d(m\mathbf{v})}{dt}$$

$$= m \frac{d\mathbf{v}}{dt} = m\mathbf{a}$$

Using Equation (9.3), we can calculate the average net force that acts during an impact if the time duration of the impact can be estimated.

EXAMPLE 9.5   Serve and Volley

A tennis ball ($m = 5.70 \times 10^{-2}$ kg) is traveling with velocity $-(15.0$ m/s$)\hat{\mathbf{i}}$. The ball is struck by a racket so that its velocity immediately after impact is $+(10.0$ m/s$)\hat{\mathbf{i}}$. If the ball is in contact with the racket for 6.00 ms (1 ms $= 10^{-3}$ s), calculate the average force of the racket on the ball.

SOLUTION   We intend to use $\mathbf{F}_{net\ av} = \Delta\mathbf{p}/\Delta t$ and must, therefore, determine the ball's change in momentum.

$$\Delta\mathbf{p} = \mathbf{p}_f - \mathbf{p}_o = m\mathbf{v}_f - m\mathbf{v}_o = m(\mathbf{v}_f - \mathbf{v}_o)$$
$$= (5.70 \times 10^{-2}\text{ kg})[(+10.0\text{ m/s})\hat{\mathbf{i}} - (-15.0\text{ m/s})\hat{\mathbf{i}}]$$
$$= (5.70 \times 10^{-2}\text{ kg})(+25.0\text{ m/s }\hat{\mathbf{i}})$$
$$= +(1.42\text{ kg}\cdot\text{m/s})\hat{\mathbf{i}}$$

Note very carefully that the change in velocity is 25 m/s, *not* 5 m/s. Make sure you understand why. The average force on the ball may now be calculated:

$$\mathbf{F}_{net\ av} = \frac{\Delta\mathbf{p}}{\Delta t}$$
$$= \frac{(+1.42\text{ kg}\cdot\text{m/s})\hat{\mathbf{i}}}{6.00 \times 10^{-3}\text{ s}} = 238\text{ N }\hat{\mathbf{i}}$$

From Newton's third law we also know that the average force *on the racket by the ball* is the opposite of the force *on the ball by the racket,* and thus is given by -238 N $\hat{\mathbf{i}}$. ◀

The magnitude of the forces during a brief impulse may be quite large as indicated by the deformation of the tennis ball.

Equation (9.3) provides the average force that acts to change the momentum of a single object (which we have been modeling as a particle). From Newton's third law, we know that the magnitude of this force is also the average force exerted by the particle on whatever it has collided with. However, this equation can also be interpreted as the average force due to a large number of particles colliding with some object. For example, suppose your class started throwing tennis balls at the door to the classroom in a continuous but random fashion. Let's also suppose your class can make an average of 15 balls per second strike the door head on. To make life easy, let's further suppose that each tennis ball has a momentum change equal to that found in Example 9.5, namely, 1.43 kg m/s. Then, in one second, the total momentum change is 15 times this amount, or 21.5 kg m/s. The average force is this momentum change divided by the time taken, which in this case is 1.0 s. Thus,

$$F_{av} = \frac{21.5\text{ kg}\cdot\text{m/s}}{1\text{ s}} = 21.5\text{ N}$$

We can summarize the reasoning used here in a simple formula:

$$\mathbf{F}_{net\ av} = (\Delta\mathbf{p}\ for\ one\ collision) \times \left(\frac{number\ collisions}{\Delta t}\right)$$

The extension of this idea to a continuous stream of matter is illustrated in Example 9.6.

EXAMPLE 9.6   *Water, Water, Everywhere*

A fire hose is directed at a vertical wall as shown in Figure 9.4. The water is traveling at 8.00 m/s. If 15.0 gal/s strikes the wall, what force does the water exert on the wall? (Assume that the water's velocity is zero after it makes contact with the wall.)

FIGURE 9.4

The stream of water exerts a steady force against the wall. Example 9.6 illustrates how this force may be calculated.

SOLUTION   Let's take one gallon as our ''particle'' unit. Using the density of water, we find that each gallon has a mass of 3.785 kg. The change in momentum of one gallon is thus,

$$\Delta \mathbf{p} = m(\mathbf{v}_f - \mathbf{v}_o) = (3.785 \text{ kg})[(0.00 \text{ m/s})\hat{\mathbf{i}} - (8.00 \text{ m/s})\hat{\mathbf{i}}] = -30.3 \text{ kg} \cdot \text{m/s } \hat{\mathbf{i}}$$

Now, 15 of these gallons of water are stopped by the wall in 1.00 s, hence

$$\mathbf{F}_{\text{net av}} = (\Delta \mathbf{p} \text{ for one collision}) \times \left(\frac{\text{number collisions}}{\Delta t} \right)$$

$$= (-30.3 \text{ kg} \cdot \text{m/s})\hat{\mathbf{i}} \times \left(\frac{15.0}{1.00 \text{ s}} \right) = -454 \text{ N } \hat{\mathbf{i}}$$

This result is the force exerted *by* the wall *on* the water, so the average force *by* the water *on* the wall is $+454 \text{ N } \hat{\mathbf{i}}$ (or 102 lbf!).

9.4 Conservation of Momentum

In this section we wish to emphasize the summation process used to calculate the net force on an object. Consequently, we use the symbol $\Sigma \mathbf{F}$ for the net force $\mathbf{F}_{\text{net}}$ throughout this section. When the sum of the forces on an object is zero, the equation $\Sigma \mathbf{F} = d\mathbf{p}/dt$ tells us that the time derivative of momentum is zero. As you know, if the derivative of a quantity is zero, that quantity is a constant. Consequently, we can state a conservation law for momentum: when the net force on an object is zero, its momentum is constant. This law is hardly any surprise. In fact, it is equivalent to Newton's first law, the law of inertia we discussed in Chapter 5.

The real utility of the momentum conservation concept comes about when it is applied to a collection of objects. This collection might be a group of discrete objects, such as billiard balls on a pool table. Alternatively, the collection might be a group of connected particles that make up an extended object the internal structure of which is too complicated for it to usefully be modeled as a point–particle; a rocket, for example. In either case, we refer to the particles of interest as a *system* of particles.

When we consider a system of particles, the total momentum of the system is simply the *vector sum* of the momentum of each of the particles in the system. We use an uppercase **P** to represent the total momentum of a system of N particles.

$$\mathbf{P} = \sum_{i=1}^{N} \mathbf{p}_i$$

Now consider the net force on a system of particles. There are two kinds of forces: (1) *internal forces,* resulting from the forces between the particles within the system and (2) *external forces,* arising between particles in the system and objects outside the system. For example, imagine a system constructed by tying two one-kilogram masses together with a spring. Next imagine hurling the combination through the air. The force of the spring on each of the two masses is internal to the system. However, gravity and air resistance are external forces on the system.

When we calculate the net force on a system of particles by performing the vector addition implied by $\Sigma\mathbf{F}$, all of the *internal* forces cancel! This follows immediately from Newton's third law of motion, which tells us that forces arise in pairs of equal magnitude but opposite directions. This result is, of course, just the old "You-can't-pick-yourself-up-by-your-own-bootstraps" rule. Because the internal forces cancel in pairs, the vector sum in $\Sigma\mathbf{F}$ over all forces can be replaced by a sum over the *external* forces only.

$$\sum\mathbf{F}_{\text{external}} = \frac{d\mathbf{P}}{dt}$$

Now, if the summation of the external forces is zero, then $d\mathbf{P}/dt$ is zero, which tells us that the total momentum of the system is constant for all time. This condition permits us to make a concise statement of the **law of conservation of momentum:**

In the absence of a *net* external force, the momentum of a closed system is constant.

We have italicized the word *net* in this momentum conservation statement because this condition is easy to forget. For momentum to be conserved it is absolutely necessary that the external forces total to zero. Be certain this condition is satisfied before you apply the law of conservation of momentum!

The conservation of momentum law can be used to relate the initial motion of objects within a system to the motion of those same objects sometime later. This use prompts us to emphasize the equality of momentum "before" and "after" something happens within the system. Thus, it is often useful to write this conservation law as

$$\sum_{i=1}^{N} \mathbf{p}_i \text{ (before)} = \sum_{i=1}^{N} \mathbf{p}_i' \text{ (after)} \qquad \text{(provided } \sum\mathbf{F}_{\text{external}} = 0) \qquad (9.5)$$

We denote the value of variables in the "after" case with primes (').

Recoil

The first type of problem to which we apply the conservation of momentum is **recoil.** If you have ever fired a shotgun or big-game rifle, you are already familiar with the phenomenon. Typically, the problem starts with a system of two or more parts with no relative velocity. Then some type of stored energy is released causing the parts of the system to fly

Concept Question 5
If at some time all the particles constituting an isolated system are at rest, under what condition can the particles of the same isolated system have motion?

apart. In the case of a shotgun, for example, the chemical energy stored in the gun powder is released by burning. Pressure from the resulting gases causes the steel shot to be propelled from the muzzle of the gun. At the same time, the gun attains a velocity in the opposite direction; the gun is said to recoil. Examples 9.7 through 9.9 show three different aspects of the application of the conservation of momentum.

EXAMPLE 9.7   *Always Carry a Pencil When You Travel in Space*

A 60.0-kg astronaut is in a spacecraft that has a constant-velocity trajectory toward Mars. At one point in the trip she finds herself suspended in the middle of the spacecraft with zero velocity (relative to the spacecraft) and unable to reach any wall. In order to rescue herself, she throws a 0.00550-kg pencil at 5.00 m/s (relative to the spacecraft) toward the rear of the craft. How does this help?

SOLUTION   The system we consider is the astronaut plus the pencil as shown in Figure 9.5. Because the system is traveling with constant velocity with respect to the background of the stars, the net external force is zero. We can thus apply the conservation of momentum to this system. Let's use a coordinate system at rest with respect to the spacecraft and choose the x-axis along the direction of the pencil's motion. We use m_a and v_a as the astronaut mass and velocity before the pencil is thrown and m_p and v_p to designate those quantities for the pencil. After the pencil is thrown we write v'_a and v'_p for the new velocities of the astronaut and pencil, respectively. From Equation (9.5),

$$\sum_{i=1}^{2} \mathbf{p}_i = \sum_{i=1}^{2} \mathbf{p}'_i$$

$$m_a \mathbf{v}_a + m_p \mathbf{v}_p = m_a \mathbf{v}'_a + m_p \mathbf{v}'_p$$

where, because the astronaut and pencil are initially at rest with respect to the spacecraft, $v_a = v_p = 0$. Because the pencil's final velocity is parallel to the x-axis, $\mathbf{v}_p$ may be written $v'_p\,\widehat{\mathbf{i}}$. Thus, we write

$$0 + 0 = m_a \mathbf{v}'_a + m_p v'_p\,\widehat{\mathbf{i}}$$

Solving for the astronaut's velocity after she has thrown the pencil, we find

$$\mathbf{v}'_a = -\frac{m_p v'_p\,\widehat{\mathbf{i}}}{m_a} = -\frac{(0.00550\text{ kg})(5.00\text{ m/s})\widehat{\mathbf{i}}}{(60.0\text{ kg})}$$

$$= -4.58 \times 10^{-4}\text{ m/s}\,\widehat{\mathbf{i}} = -0.458\text{ mm/s}\,\widehat{\mathbf{i}}$$

Throwing the pencil helps the astronaut because she recoils in the direction opposite to the pencil and makes contact (eventually!) with the spacecraft wall.   ◄

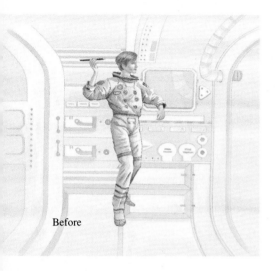

Before

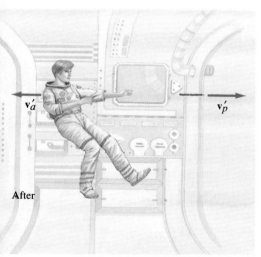

v'_a v'_p

After

FIGURE 9.5

The total momentum of the astronaut plus that of the pencil is zero before and after the pencil is thrown.

Having carefully warned you not to apply conservation of momentum unless the net external force is zero, we must admit to the following real-world practicality. When the internal forces are large compared to the external forces, the conservation of momentum law can be applied to obtain good approximate answers even if the net external force is not precisely zero. We illustrate this point in Example 9.8.

EXAMPLE 9.8 *How Do I Use Two-Dimensional Vectors?*

Three students (Teresa, 50.0 kg; Sara, 60.0 kg; and Doug, 80.0 kg) decide to go to the roller rink instead of studying physics. In order to salve their consciences, they perform a quick experiment at the rink. While standing close together in a small circle, each student presses the palms of his or her hands against those of the other two students. Then, they simultaneously push off from each other. Doug rolls off at 0.650 m/s directly toward the seats. Sara moves away at 0.500 m/s at an angle of 110° anticlockwise from Doug's line of motion as shown in Figure 9.6. How fast and in what direction is Teresa pushed?

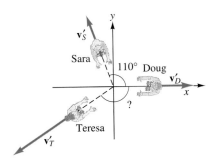

FIGURE 9.6

The total momentum of the three skaters in Example 9.8 remains zero even after they push off from one another.

SOLUTION Each skater's weight is balanced by the normal force of the floor on his or her skates. Thus, the only external force on our system is friction. If we are willing to ignore small frictional forces during the brief push, we can obtain an approximate solution using the conservation of momentum.

As shown in Figure 9.6, we take Doug's direction as the x-axis and write $\mathbf{v}'_D = 0.650$ m/s$\hat{\mathbf{i}}$. This choice makes Sara's velocity after the push $\mathbf{v}'_S = (-0.171\hat{\mathbf{i}} + 0.470\hat{\mathbf{j}})$ m/s. We solve the conservation of momentum condition (Eq. (9.5)) for v'_T:

$$\sum_{i=1}^{2} \mathbf{p}_i = \sum_{i=1}^{2} \mathbf{p}'_i$$

$$m_D\mathbf{v}_D + m_S\mathbf{v}_S + m_T\mathbf{v}_T = m_D\mathbf{v}'_D + m_S\mathbf{v}'_S + m_T\mathbf{v}'_T$$

Because all three skaters start from rest, each has zero initial momentum:

$$0 + 0 + 0 = m_D\mathbf{v}'_D + m_S\mathbf{v}'_S + m_T\mathbf{v}'_T$$

We can immediately solve for Teresa's velocity:

$$\mathbf{v}'_T = -\frac{m_D\mathbf{v}'_D + m_S\mathbf{v}'_S}{m_T}$$

$$= -\frac{(80.0 \text{ kg})(0.650\hat{\mathbf{i}} \text{ m/s}) + (60.0 \text{ kg})[(-0.171\hat{\mathbf{i}} + 0.470\hat{\mathbf{j}}) \text{ m/s}]}{(50.0 \text{ kg})}$$

$$= -\frac{[52.0\hat{\mathbf{i}} + (-10.3\hat{\mathbf{i}} + 28.2\hat{\mathbf{j}})] \text{ kg} \cdot \text{m/s}}{50.0 \text{ kg}}$$

$$= -(0.835\hat{\mathbf{i}} + 0.564\hat{\mathbf{j}}) \text{ m/s}$$

The magnitude of Teresa's velocity is $\sqrt{v_x'^2 + v_y'^2} = 1.01$ m/s. The direction is 146° from Doug's left hand as shown in Figure 9.6.

ALTERNATIVE SOLUTION The problem above can also be solved with the vectors in polar form. Because the total momentum before the push is zero, the total afterward is also zero. The three vectors $\mathbf{p}'_D$, $\mathbf{p}'_S$, and $\mathbf{p}'_T$ must, therefore, form a closed triangle so that the result of the vector addition is zero. The magnitudes of $\mathbf{p}'_D$ and $\mathbf{p}'_S$ are calculated in the usual manner:

$$p'_D = m_D v'_D = (80.0 \text{ kg})(0.650 \text{ m/s}) = 52.0 \text{ kg} \cdot \text{m/s}$$

$$p'_S = m_S v'_S = (60.0 \text{ kg})(0.500 \text{ m/s}) = 30.0 \text{ kg} \cdot \text{m/s}$$

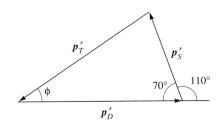

FIGURE 9.7

That these three momentum vectors have a vector sum of zero is indicated by the fact that they form a closed triangle when added together.

These two vectors are added as shown in Figure 9.7. Because the total momentum is zero, the unknown vector $\mathbf{p}'_T$ can now be constructed by drawing a line that forms a closed triangle with the previous two vectors, as shown in Figure 9.7. The direction of $\mathbf{p}'_T$ is that which causes all three vectors to form a "one-way street." The magnitude of $\mathbf{p}'_T$ can now be found using the law of cosines

$$C^2 = A^2 + B^2 - 2AB\cos(\theta)$$

$$p'^2_T = p'^2_D + p'^2_S - 2\,p'_D p'_S \cos(\theta)$$

$$p'_T = \sqrt{(52.0\ \text{kg}\cdot\text{m/s})^2 + (30.0\ \text{kg}\cdot\text{m/s})^2 - 2\,(52.0\ \text{kg}\cdot\text{m/s})(30.0\ \text{kg}\cdot\text{m/s})\cos(70°)}$$

$$= 50.4\ \text{kg}\cdot\text{m/s}$$

Teresa's speed is then $v'_T = p'_T/m_T = (50.4\ \text{kg}\cdot\text{m/s})/50\ \text{kg} = 1.01\ \text{m/s}$ as before. The angle ϕ in Figure 9.7 can be found from the law of sines

$$\frac{\sin(A)}{a} = \frac{\sin(B)}{b}$$

$$\frac{\sin(\phi)}{p'_D} = \frac{\sin(\theta)}{p'_T}$$

$$\sin(\phi) = \left(\frac{p'_D}{p'_T}\right)\sin(\theta)$$

$$= \left(\frac{30.0\ \text{kg}\cdot\text{m/s}}{50.0\ \text{kg}\cdot\text{m/s}}\right)\sin(70°) = 0.564$$

$$\phi = \sin^{-1}(0.564) = 34.3°$$

This result is the angle between Teresa's momentum vector and the negative x-axis and confirms the answer we obtained above when we used vectors in rectangular form.   ◀

Relative Motion

Concept Question 6
When a steel ball is released in a vertical vacuum tube, the ball's momentum increases as it approaches the earth. How is momentum conserved in this case?

In Examples 9.7 and 9.8 the velocities of the objects were given relative to some external coordinate system (the spacecraft in Example 9.7 and the roller rink in Example 9.8). This type of assignment may not always be possible. Often, only the *relative* velocity of two objects is known. For example, an Olympic target rifle might fire a bullet with a speed of 750 m/s when the rifle is mounted rigidly. If the rifle is allowed to recoil freely when the bullet is fired, the bullet still leaves the barrel with a speed of 750 m/s *relative to the backward recoiling rifle*. However, the bullet's velocity with respect to the ground is less than 750 m/s. We may still apply the conservation of momentum condition but must be careful to specify all velocities relative to a single, inertial coordinate system.

Let's see how to combine relative velocities so that they may be easily written with respect to one frame of reference. Suppose you are riding in the back of a large truck that is traveling 50 mph. If you throw a tennis ball forward at 20 mph in the same direction in which the truck is moving, the ball will be traveling at 70 mph with respect to the road. If you throw the tennis ball at 20 mph in the direction opposite to the truck's velocity, the ball will be traveling 30 mph forward with respect to the road. The ball's velocity with respect to the road in either case can be calculated from the vector expression

$$\mathbf{v}_{\text{ball(road)}} = \mathbf{v}_{\text{ball(truck)}} + \mathbf{v}_{\text{truck(road)}}$$

where $\mathbf{v}_{\text{ball(road)}}$ and $\mathbf{v}_{\text{truck(road)}}$ are the velocities of the ball and the truck, each measured with respect to the road, and $\mathbf{v}_{\text{ball(truck)}}$ is the velocity of the ball relative to the truck. To make the equation above a little more general we can write

$$\mathbf{v}_{AC} = \mathbf{v}_{AB} + \mathbf{v}_{BC} \tag{9.6}$$

In Equation (9.6) we use two subscripts. The first subscript denotes the object we are describing, and the second subscript denotes the frame of reference. Thus, $\mathbf{v}_{AB}$ means the velocity of object A relative to B. For example, the velocity of the tennis ball relative to the road is written $\mathbf{v}_{\text{ball road}}$. Equation (9.6) is easy to remember if you allow yourself a few liberties with the subscripts, namely, treating them like fractions. Think of $\mathbf{v}_{AB}$ as $\mathbf{v}_{A/B}$ and $\mathbf{v}_{BC}$ as $\mathbf{v}_{B/C}$. Then when you add the velocities, think of the B's as "canceling": $\mathbf{v}_{A/B} + \mathbf{v}_{B/C} = \mathbf{v}_{A/C}$. Note, this is simply a mnemonic device to help you remember the order of the subscripts in Equation (9.6) and has no mathematical significance.

The point of all this is that when you apply the conservation of momentum condition, you must *be sure that all velocities are relative to the same coordinate system,* as we illustrate in Example 9.9.

EXAMPLE 9.9 *Using Relative Velocities and the Conservation of Momentum*

When standing on shore, a 75.0-kg boater can throw a 10.0-kg anchor horizontally with a speed of 2.00 m/s. If this boater stands in a 50.0-kg boat (which is stationary in the water) and heaves the anchor with the same effort, how fast will boater and boat recoil?

(a) Before

(b) After

$\mathbf{v}'_{BW}$ $\mathbf{v}'_{AW}$

FIGURE 9.8

(a) A bass fisherman is about to throw a heavy anchor. (b) When the anchor is thrown forward, the boat and fisherman recoil backward. When only the relative velocity between the anchor and the fisherman is known, special care must be taken when applying the conservation of momentum. The required steps are illustrated in Example 9.9.

SOLUTION We treat the boat and boater as a single object with mass $m_B = 125.$ kg. We denote the velocity of the boat and boater relative to the water as $\mathbf{v}_{BW}$. The velocity of the anchor relative to the water is $\mathbf{v}_{AW}$. The given velocity 2.00 m/s is the velocity of the anchor with respect to the boater. Hence, we denote it as $\mathbf{v}_{AB}$. We ignore the frictional effect of the water's surface during the short time the anchor is being thrown and can thus apply the law of conservation of momentum.

$$\sum \mathbf{p}_i = \sum \mathbf{p}'_i$$

$$m_B \mathbf{v}_{BW} + m_A \mathbf{v}_{AW} = m_B \mathbf{v}'_{BW} + m_A \mathbf{v}'_{AW}$$

Note that in the conservation of momentum equation above, all velocities are given relative to one frame of reference, the water in this case. Because both boat and anchor are stationary to begin with, we have

$$0 + 0 = m_B \mathbf{v}'_{BW} + m_A \mathbf{v}'_{AW} \tag{9.7}$$

All motion is along a straight line, so we drop the explicit vector notation and indicate direction with + and − signs. Equation 9.7 has two unknowns v'_{BW} and v'_{AW}. We need a second equation and a way of introducing the given quantity v'_{AB}. We can satisfy both these necessities by adapting the relative velocity expression from Equation (9.6) to the problem at hand:

$$v'_{AW} = v'_{AB} + v'_{BW}$$

This expression for v'_{AW} may now be substituted into Equation (9.7), which can then be solved for $\mathbf{v}'_{BW}$.

$$0 = m_B v'_{BW} + m_A (v'_{AB} + v'_{BW})$$

$$v'_{BW} = -\left(\frac{m_A}{m_B + m_A}\right) v'_{AB}$$

$$= -\left(\frac{10.0 \text{ kg}}{125. \text{ kg} + 10.0 \text{ kg}}\right)(2.00 \text{ m/s}) = -0.148 \text{ m/s}$$

For the sake of completeness, we note that the anchor travels with velocity

$$v'_{AW} = v'_{AB} + v'_{BW} = 2.00 \text{ m/s} + (-0.148 \text{ m/s}) = 1.85 \text{ m/s}$$

relative to the water.   ◀

9.5 Collisions in One Dimension

The conservation of momentum is extremely helpful in the analysis of **collisions.** You probably think of collisions as brief events between objects that contact each other, such as the collision of a bowling ball with a bowling pin. We define collisions more broadly: The interaction between two or more objects is called a **collision** if there exists three identifiable stages to this interaction: before, during, and after. In the before and after regimes the interaction forces are zero or approach zero asymptotically. Between these two stages the interaction forces are large and often the dominant forces governing the objects' motions. For example, a moving airtrack glider with a bumper formed from a magnet might approach an identical stationary glider to within a centimeter or so. The repulsive magnetic forces between the gliders cause the former to slow down and the latter to speed up. After the magnets separate by a few tenths of a meter the force between them drops to essentially zero, leaving both gliders with different velocities after separation. In such a case we say the gliders have "collided" even though they have not actually touched. In the same spirit, two widely separated galaxies coasting through space may pass close enough to each other that their mutual gravitational attraction causes them both to alter course. After the galaxies are sufficiently far apart, the gravitational attraction is once again negligible, but their velocities through space have been permanently altered. In the case of colliding galaxies the collision may take millions of years! The salient point defining collisions is that the masses go through three distinct phases: a noninteracting phase before collision, an interacting phase (the collision itself), and a return to a noninteracting phase. The conservation of momentum statement is useful for relating the initial velocities before the interaction to the final velocities after the interaction *without* requiring a detailed knowledge of the interaction forces.

When two objects collide and the initial and final velocities of both are parallel (or antiparallel), the collision is said to be one-dimensional. We start our discussion of collisions with this simple case. In addition, we restrict ourselves to collisions where the net external force on the colliding objects is zero. Many real-world collisions can be modeled as if they satisfy this condition. For example, during an automobile collision, the collision forces are likely to be enormous compared to the frictional forces between the road and the tires. Consequently, during the collision we can treat the cars as if the net external force were zero; that is, as if they were sliding on a frictionless surface. We shall call the model in which we ignore the outside forces while calculating the result of a collision the **collision approximation.** In the collision approximation momentum is conserved.

In general, there is a change in total kinetic energy whenever two objects collide. The exceptional instance where there is no change in kinetic energy is a very special case called a **perfectly elastic** collision. In most collisions the kinetic energy of the colliding objects is reduced during the collision. Such collisions are called **inelastic** collisions. In this case the kinetic energy is converted into other forms of energy or work is performed deforming the objects. There is an upper bound to the amount by which the kinetic energy can be reduced. When the collision is such that the kinetic energy reduction is a maximum, the objects stick together and the collision is called **perfectly inelastic.** Although rare, **hyperelastic collisions** can occur in which the kinetic energy is larger after the collision than before. In such cases the kinetic energy must come from some potential energy source, for example, from the chemical energy released by an explosion occurring at the time of the collision.

During a collision the collision forces are often much larger than any other force in the problem. For example, the frictional force on the car tires can be ignored during the collision.

Perfectly Inelastic Collisions

When objects stick together after a collision, the collision is said to be perfectly inelastic. In football, when a linebacker tackles a running back, he hopes the collision will be inelastic. Let's first think simply of two masses m_A and m_B. The velocities of m_A and m_B are denoted as $\mathbf{v}_A$ and $\mathbf{v}_B$ before the collision and $\mathbf{v}_A'$ and $\mathbf{v}_B'$ after the collision. Because the objects have the same velocity after a perfectly inelastic collision, we call the final velocity $\mathbf{v}'$ without a subscript. The conservation of momentum can be applied to the system composed of m_A and m_B:

$$m_A\mathbf{v}_A + m_B\mathbf{v}_B = (m_A + m_B)\mathbf{v}'$$

There are five variables in this equation. If four of them are known, the equation can be solved for the fifth.

EXAMPLE 9.10   Over the Top

A 96.0-kg running back attempts to score a touchdown from the one-yard line by leaping over the forward line. He is met in midair by a 105.-kg linebacker who grabs hold of the ball carrier. When they meet, the horizontal velocity of the ball carrier is 5.22 m/s, and the linebacker has a horizontal speed of 4.93 m/s in the opposite direction. Is the momentum of the ball carrier enough to put him over the goal line?

SOLUTION   We can apply the conservation of momentum to the horizontal motion of two football players because the net external horizontal force is temporarily zero while they are not in contact with the ground (or their fellow players). For the one-dimensional case we drop the explicit vector motion, representing direction by + and − signs.

$$m_A v_A + m_B v_B = (m_A + m_B)v'$$

$$v' = \frac{m_A v_A + m_B v_B}{m_A + m_B}$$

$$= \frac{(96.0 \text{ kg})(5.22 \text{ m/s}) + (105. \text{ kg})(-4.93 \text{ m/s})}{96.0 \text{ kg} + 105. \text{ kg}}$$

$$= -8.22 \times 10^{-2} \text{ m/s} = -8.22 \text{ cm/s}$$

The defender hopes to make an inelastic collision with the ball carrier.

The linebacker has successfully stopped the ball carrier's forward progress and even causes a small backward velocity. The outcome of the play critically depends on what happens in the next few seconds when the players make contact with the ground and other players.   ◀

Perfectly Elastic Collisions

A perfectly elastic collision is defined as a collision in which kinetic energy is conserved. If the net external force on the system of masses is zero so that momentum is also conserved, for one-dimensional collisions we have two equations that can be solved for two unknowns.

$$m_A v_A + m_B v_B = m_A v'_A + m_B v'_B$$

$$\frac{1}{2} m_A v_A^2 + \frac{1}{2} m_B v_B^2 = \frac{1}{2} m_A v'^2_A + \frac{1}{2} m_B v'^2_B$$

Concept Question 7
If two objects have equal kinetic energies, can you conclude that their momenta are equal? Explain.

These two equations can be solved simultaneously as they stand. However, we can obtain a set of simpler "working equations" by performing a little algebra. The notation is becoming a bit cumbersome, so let's simplify it before doing the algebra. Let's divide the conservation of momentum and the conservation of kinetic energy equations by m_A and call the ratio of masses $m_B/m_A = \mathcal{M}$. While we're at it we may as well multiply the conservation of energy equation by 2 to get rid of the pesky $\frac{1}{2}$. With these algebraic steps completed the conservation of momentum and the conservation of kinetic energy become:

$$v_A + \mathcal{M} v_B = v'_A + \mathcal{M} v'_B$$

$$v_A^2 + \mathcal{M} v_B^2 = v'^2_A + \mathcal{M} v'^2_B$$

From the first of these two equations we see that

$$v_A - v'_A = \mathcal{M}(v'_B - v_B) \tag{9.8}$$

The second equation we can write as

$$v_A^2 - v'^2_A = \mathcal{M}(v'^2_B - v_B^2)$$

The difference of two squares can be factored into sums and differences so that the last equation becomes

$$(v_A + v'_A)(v_A - v'_A) = \mathcal{M}(v'_B - v_B)(v'_B + v_B)$$

On the left-hand side of this equation we can substitute the expression for $v_A - v'_A$ found in Equation (9.8). The result is

$$(v_A + v'_A)\mathcal{M}(v'_B - v_B) = \mathcal{M}(v'_B - v_B)(v'_B + v_B)$$

Providing $v_B \neq v'_B$, we can now cancel $\mathcal{M}(v'_B - v_B)$ from both sides of the above equation, leaving

$$v_A + v'_A = v'_B + v_B$$

or, as we prefer to write,

$$(v_A - v_B) = (v'_B - v'_A) \tag{9.9}$$

The quantity $v_A - v_B$ is the speed of approach of the two objects. Similarly, $v_B' - v_A'$ (note the reverse order of the subscripts) is the speed of separation. Equation (9.9) tells us that, for perfectly elastic collisions, the speed of approach is equal to the speed of separation.

The conservation of momentum statement and Equation (9.9) now give us two *linear,* working equations that can be solved for two unknowns. Working with these linear equations is equivalent to solving the equations for conservation of momentum and kinetic energy directly. Algebraically, however, the working equations are much simpler to handle.

EXAMPLE 9.11 *Gliding Through Physics*

Two gliders collide on an airtrack during a physics laboratory experiment. The target glider of mass $m_B = 0.287$ kg is at rest when struck by an incoming glider of mass $m_A = 0.189$ kg. The velocity of the target glider after the collision was measured to be $v_B' = 0.250$ m/s in the same direction that the first glider had before the collision. By using magnetic "bumpers" aligned so that the gliders repel one another but never actually touch, the collision is made to be essentially perfectly elastic. Calculate the velocity of glider A before and after the collision.

SOLUTION Using the labels shown in Figure 9.9, we write the two working equations for perfectly elastic collisions

$$v_A + \mathcal{M} v_B = v_A' + \mathcal{M} v_B'$$

$$v_A - v_B = v_B' - v_A'$$

with $v_B = 0$ and $\mathcal{M} = m_B/m_A = 1.52$. By adding our two working equations, we obtain

$$2v_A + (\mathcal{M} - 1)v_B = (1 + \mathcal{M})v_B'$$

or

$$v_A = \frac{(1 - \mathcal{M})v_B + (1 + \mathcal{M})v_B'}{2} = \frac{(1 - 1.52)(0 \text{ m/s}) + (1 + 1.52)(0.250 \text{ m/s})}{2}$$

$$= 0.315 \text{ m/s}$$

Subtracting the two working equations, we find

$$(1 + \mathcal{M})v_B = 2v_A' + (\mathcal{M} - 1)v_B'$$

or

$$v_A' = \frac{(1 + \mathcal{M})v_B + (1 - \mathcal{M})v_B'}{2} = \frac{(1 + 1.52)(0 \text{ m/s}) + (1 - 1.52)(0.250 \text{ m/s})}{2}$$

$$= -0.0650 \text{ m/s}$$

We can make a quick check of our algebra and arithmetic by comparing the speed of approach with the speed of separation.

$$\text{speed of approach} = v_A - v_B = (0.315 \text{ m/s}) - (0.000 \text{ m/s}) = 0.315 \text{ m/s}$$

$$\text{speed of separation} = v_B' - v_A' = (0.250 \text{ m/s}) - (-0.065 \text{ m/s}) = 0.315 \text{ m/s}$$

As expected, for this perfectly elastic collision, the speed of approach equals the speed of separation. ◀

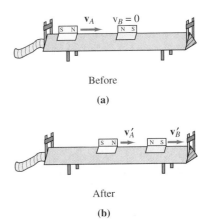

Before

(a)

After

(b)

FIGURE 9.9

A near perfectly elastic collision using magnets mounted on airtrack gliders.

Partially Elastic Collisions

Most real collisions fall somewhere between perfectly elastic and perfectly inelastic. In *partially elastic* collisions kinetic energy is reduced, and the speed of separation, although nonzero, is less than the speed of approach. A model that describes such a collision may be constructed by assigning a **coefficient of restitution** to the collision process. The coefficient of restitution is symbolized by the Greek letter epsilon ϵ and defined

$$\epsilon = \frac{\textit{Speed of separation}}{\textit{Speed of approach}}$$

$$\epsilon = \frac{(v'_B - v'_A)}{(v_A - v_B)} \tag{9.10}$$

This definition allows us to use working equations for partially elastic collisions that are similar to those equations for perfectly elastic collisions:

$$v_A + \mathcal{M}v_B = v'_A + \mathcal{M}v'_B \tag{9.11}$$

$$\epsilon(v_A - v_B) = (v'_B - v'_A) \tag{9.12}$$

TABLE 9.1   Coefficient of Restitution Model

Perfectly elastic	$\epsilon = 1$
Partially elastic	$0 < \epsilon < 1$
Perfectly inelastic	$\epsilon = 0$
Hyperelastic	$\epsilon > 1$

In fact, both perfectly inelastic and perfectly elastic collisions can be considered special cases of the equations above. When $\epsilon = 1$ we have the perfectly elastic case. Perfectly inelastic collisions have $\epsilon = 0$. For partially elastic collisions $0 < \epsilon < 1$. Hyperelastic collisions have $\epsilon > 1$.

EXAMPLE 9.12   *How Can I Measure the Coefficient of Restitution?*

A "legal" racquetball must rebound between 68.0 and 72.0 in. when dropped from a height of 100. in. onto the court floor. What are the limits of the coefficient of restitution for a legal racquetball?

SOLUTION   We can calculate the velocity of the ball just before it hits the floor by using the simple kinematics of Chapter 3.

$$v_f^2 = v_o^2 + 2a(y - y_o)$$
$$v_f = -\sqrt{v_o^2 + 2a(y - y_o)} = -\sqrt{2a_g y_o}$$

Similarly, we find the velocity immediately after rebound by using the maximum height $y_{max} = (y - y_o)$:

$$v'_f = +\sqrt{2a_g y_{max}}$$

Taking the floor's velocity as zero both before and after the collision, the coefficient of restitution is

$$\epsilon = \frac{(0 - \sqrt{2a_g y_{max}})}{(-\sqrt{2a_g y_o} - 0)} = \sqrt{\frac{y_{max}}{y_o}}$$

Thus, the acceptable limits of the coefficient of restitution are

$$\sqrt{\frac{68.0}{100.}} \leq \epsilon \leq \sqrt{\frac{72.0}{100.}}$$

$$0.825 \leq \epsilon \leq 0.849$$

◀

EXAMPLE 9.13 Boing!

If the airtrack gliders described in Example 9.11 had spring bumpers characterized by a coefficient of restitution $\epsilon = 0.850$, what would be the outcome of a collision in which the two gliders approached one another, each with a speed of 0.500 m/s?

SOLUTION The masses are $m_A = 0.189$ kg and $m_B = 0.287$ kg, so the mass ratio $\mathcal{M} = 1.52$. Adding our two working Equations (9.11) and (9.12), we find

$$(\epsilon + 1)v_A + (\mathcal{M} - \epsilon)v_B = (\mathcal{M} + 1)v_B'$$

or

$$v_B' = \frac{\epsilon + 1}{\mathcal{M} + 1}v_A + \frac{\mathcal{M} - \epsilon}{\mathcal{M} + 1}v_B$$

In this special case, $v_B = -v_A$, so we have

$$v_B' = \frac{2\epsilon - \mathcal{M} + 1}{\mathcal{M} + 1}v_A$$

$$= \frac{2(0.850) - 1.52 + 1}{1 + 1.52}(0.500 \text{ m/s})$$

$$= \left(\frac{1.18}{2.52}\right)(0.500 \text{ m/s}) = 0.234 \text{ m/s}$$

From the second of our working equations,

$$\epsilon(v_A - v_B) = (v_B' - v_A')$$

which can be solved for v_A'

$$v_A' = v_B' - \epsilon(v_A - v_B) = 0.234 \text{ m/s} - (0.85)[0.500 \text{ m/s} - (-0.500)] \text{ m/s}$$

$$= -0.616 \text{ m/s} \blacktriangleleft$$

It is interesting to look at the range of values for the final velocities of various targets when the values of the mass ratio $\mathcal{M}$ and the coefficient of restitution ϵ are changed. Let's consider the special case of a stationary target mass, so that $v_B = 0$ in our working equations. Then it is easy to show (see Problem 36) that

$$v_B' = \frac{1 + \epsilon}{1 + \mathcal{M}}v_A$$

Consider the perfectly elastic case where $\epsilon = 1$ for a very light target. A very light target means $\mathcal{M} \ll 1$, so we can ignore $\mathcal{M}$ compared to 1, and the above equation becomes

$$v_B' \approx 2v_A$$

This result indicates that the target gains a speed approximately twice that of the incoming projectile.

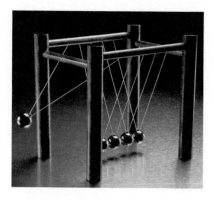

FIGURE 9.10

In 1668, Christiaan Huygens explained the action of this famous toy using the conservation of momentum and the conservation of kinetic energy.

EXAMPLE 9.14   *I Always Wondered Why That Worked*

A well-known desktop toy consists of five identical hard spheres that just touch when each hangs freely from a bifilar support. When one sphere is drawn to the side and released, it collides with the row of four stationary spheres (Fig. 9.10). As a result of this collision the sphere on the opposite end (and that one only) is knocked forward. If two spheres are pulled aside and released initially, two (and only two) spheres are knocked out from the far end. Relate these observations to the conservation of momentum and kinetic energy.

SOLUTION   We model the system as perfectly elastic and write the conservation of momentum and conservation of kinetic energy conditions for the spheres. We use m for the common mass and v_A, v_B, v_C, v_D, and v_E for the velocities.

$$mv_A + mv_B + mv_C + mv_D + mv_E = mv'_A + mv'_B + mv'_C + mv'_D + mv'_E$$

$$\frac{1}{2}mv_A^2 + \cdots + \frac{1}{2}mv_E^2 = \frac{1}{2}mv'^2_A + \cdots + \frac{1}{2}mv'^2_E$$

We can, of course, divide both equations by m, and multiply the second equation by 2. Let's take the case of a single sphere moving initially with speed $v_A = 1.00$. (This is not really restrictive, v_A can be any velocity; we just change the velocity units to make A's velocity equal to 1 velocity unit.) The equations then become

$$1 = v'_A + v'_B + v'_C + v'_D + v'_E$$

$$1 = v'^2_A + v'^2_B + v'^2_C + v'^2_D + v'^2_E$$

We also note that physically the spheres can't change order, so that $v'_A < v'_B < v'_C < v'_D < v'_E$. However, this observation doesn't help much because we still have five unknowns and only two equations. From watching the experiment we know one answer, namely, $v'_E = 1$ and $v'_A = v'_B = v'_C = v'_D = 0$. You should try a few others like $v'_D = v'_E = \frac{1}{2}$ and $v'_A = v'_B = v'_C = 0$, which satisfy the first equation. You will quickly discover that such solutions fail to satisfy the second equation. After you give up trying to find alternative solutions, you might want to look up the "triangular equality" in a math book and ponder its relationship to this example.

You may be interested to know that in 1668 this "toy" was the subject of three papers presented to the *Royal Society of London for Improving of Natural Knowledge*.[1] Of the three presenters, John Wallis, Christopher Wren, and Christiaan Huygens, only Huygens analyzed the problem in complete detail. Huygens showed that the explanation depended on both the conservation of momentum and the conservation of *vis viva*, a quantity equal to twice what we now call kinetic energy. The necessity of the conservation of both momentum and *vis viva* in elastic collisions was significant because there had been a long, sometimes spirited, debate about which of the two was really conserved. René Descartes (1596–1650) proposed in his *Principles of Philosophy* (1644) that it was momentum that was the properly conserved quantity of motion. In 1680, Gottfried Wilhelm Leibniz (1646–1716) still argued that Descartes' physics was defective. He was convinced that mv^2 was the properly conserved quantity. So you can see you need not be discouraged if you occasionally have difficulty with these matters. It has taken some of the greatest minds since the Renaissance to puzzle out the laws governing colliding objects.   ◄

9.6 Collisions in Two Dimensions

By *collisions in two dimensions* we mean that all of the objects in the collision move in the same two-dimensional plane before and after the collision. We continue to describe colli-

---

[1] Founded in 1662, one of the earliest and perhaps the most famous of scientific societies. Newton was president of the society from 1703 until his death in 1727.

sions between only objects that can be modeled as point–particles. We also continue to use the collision model that allows us to apply the conservation of momentum to our collisions.

Because momentum is a vector, each of its components must be separately conserved. Thus, for collisions in two dimensions the conservation of momentum results in *two* independent equations. If we pick the *xy*-plane, for example, we have

$$\Sigma \mathbf{p}_x \text{ (before)} = \Sigma \mathbf{p}'_x \text{ (after)}$$

$$\Sigma \mathbf{p}_y \text{ (before)} = \Sigma \mathbf{p}'_y \text{ (after)}$$

As in the case for one-dimensional collisions, we discuss the perfectly inelastic, perfectly elastic, and partially elastic cases separately.

Perfectly Inelastic Collisions

Two-dimensional collisions are really quite simple. As with other vector problems, we have merely to decompose the two-dimensional problem into two one-dimensional problems and solve each of them, exactly as in the previous section. When finished, we can recombine components to find vector magnitudes and directions.

In the collision approximation the outcome of the collision between hockey players can be predicted using the conservation of momentum.

EXAMPLE 9.15   *"Checking" Your Momentum Vectors*

Two hockey players race for a free puck. Before reaching the puck they collide, becoming entangled. The red-team player has a mass of 75.0 kg and was traveling with a velocity of 11.0 m/s straight up the ice (say along the *x*-axis). The green-team player (mass 68.0 kg) was traveling 8.50 m/s parallel to the blue line (*y*-axis) as shown in Figure 9.11(a). How fast and in what direction are the pair traveling immediately after the collision?

SOLUTION   The momenta before the collision are

$$\mathbf{p}_{red} = m_{red}\mathbf{v}_{red} = (75.0 \text{ kg})(11.0 \text{ m/s})\widehat{\mathbf{i}} = 825 \text{ kg} \cdot \text{m/s } \widehat{\mathbf{i}}$$

$$\mathbf{p}_{green} = m_{green}\mathbf{v}_{green} = (68.0 \text{ kg})(8.50 \text{ m/s})\widehat{\mathbf{j}} = 578 \text{ kg} \cdot \text{m/s } \widehat{\mathbf{j}}$$

Using the conservation of momentum, the final momentum is equal to the initial momentum

$$\Sigma \mathbf{p}' \text{ (after)} = \Sigma \mathbf{p} \text{ (before)}$$

$$\Sigma \mathbf{p}' = \mathbf{p}_{red} + \mathbf{p}_{green} = (825\widehat{\mathbf{i}} + 578\widehat{\mathbf{j}}) \text{ kg} \cdot \text{m/s}$$

The magnitude of the final momentum is

$$p' = \sqrt{(p_x'^2 + p_x'^2)} = \sqrt{(825 \text{ kg} \cdot \text{m/s})^2 + (578 \text{ kg} \cdot \text{m/s})^2} = 1010 \text{ kg} \cdot \text{m/s}$$

The speed of the combined masses after the collision is

$$v' = \frac{p'}{m_{tot}} = \frac{\sqrt{(825 \text{ kg} \cdot \text{m/s})^2 + (578 \text{ kg} \cdot \text{m/s})^2}}{143 \text{ kg}} = 7.04 \text{ m/s}$$

The direction of the pair is found from

$$\theta = \arctan\left(\frac{p'_y}{p'_x}\right) = \arctan\left(\frac{578}{825}\right) = 35.0°$$

from the original direction of the red player's travel as shown in Figure 9.11(b).

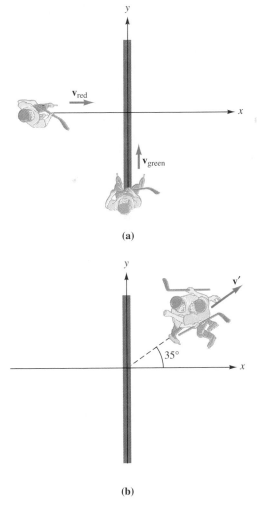

FIGURE 9.11

(a) The two hockey players of Example 9.15 approach along right angles. (b) After an inelastic collision, their final trajectory can be predicted using the conservation of momentum.

Concept Question 8
Figure 9.12(a) shows the momentum vectors of two particles A and B before they collide. Figure 9.12(b) shows the momentum vector for particle A after the collision. Sketch the momentum vector for particle B after the collision assuming momentum is conserved.

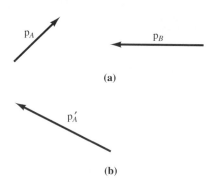

(a)

(b)

FIGURE 9.12

(a) The momentum vectors of particles A and B before the particles collide.
(b) The momentum vector of particle A after the collision. Can you draw the momentum vector for particle B following the collision?

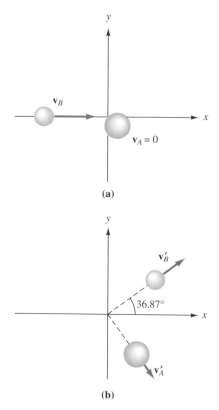

(a)

(b)

FIGURE 9.13

Two unequal masses collide while sliding over a frictionless surface. Vector momentum is conserved, but kinetic energy may not be.

Perfectly Elastic Collisions

In the case of perfectly elastic collisions kinetic energy is conserved. This condition provides a third equation to be satisfied by the colliding particles. Given the initial speeds and directions of the two approaching particles, three equations are insufficient to solve for the final speeds and directions of both particles after collision because there are four unknowns after the collision: two speeds and two directions.

It makes physical sense that we cannot determine the result of a two-dimensional collision from the initial momenta only, because the results of such a collision must depend not only on the original velocities, but also on the location on the objects where they come into contact. If you have played pool or shuffleboard, you know that the result of a two-dimensional collision depends strongly on the point of contact of the objects. Two-dimensional collisions can range anywhere from head on to a barely glancing impact. In addition, the details of how the collision force depends on the separation of the particles during the collision influences the result.

The realm of atomic physics is one in which perfectly elastic collisions do occur. The molecules of a gas may collide in perfectly elastic collisions. (Such collisions may also be partially elastic if the molecules are excited to higher internal energy levels. See Chapter 21 for more details.) If sufficient information can be obtained concerning the direction of particles after a collision, their momenta can be deduced.

EXAMPLE 9.16   Now for Some Air Hockey

On a frictionless air-hockey table two pucks collide in what is, for all practical purposes, a perfectly elastic collision. Puck A, which has a mass twice that of puck B, is initially at rest. Before the collision, puck B moves with an initial velocity of $v_A = 5.00$ m/s along the x-axis. After the pucks collide, puck B is observed to travel at an angle $\theta = 36.87°$ above the positive x-axis. Find the speeds of both pucks and the direction of travel for puck A after the collision.

SOLUTION   The masses are unknown, so we let one puck have mass m and the other mass $2m$. Because no net external force acts in the horizontal plane, momentum is conserved in this plane. Therefore, we expect to obtain two equations, one from equating the total x-momentum before the collision to the total x-momentum after the collision, and a second from equating the total y-momentum before the collision to the total y-momentum after the collision.

Momentum before = Momentum after

x-momentum: $mv_A = mv_B' \cos(\theta) + 2mv_{A_x}'$

y-momentum: $0 = mv_B' \sin(\theta) + 2mv_{A_y}'$

Because we are modeling the collision as if it were perfectly elastic, we may also equate the kinetic energy before and after the collision:

kinetic energy: $\dfrac{1}{2}mv_A^2 = \dfrac{1}{2}mv_B'^2 + \dfrac{1}{2}(2m)v_A'^2$

We may simplify the above equations by dividing each by m. The kinetic energy equation can be multiplied by 2. We also note that $v_A'^2 = v_{A_x}'^2 + v_{A_y}'^2$. The resulting three equations have three unknowns:

$$v_A = \cos(\theta)\, v_B' + 2v_{A_x}' \tag{9.13}$$

$$0 = \sin(\theta)\, v_B' + 2v_{A_y}' \tag{9.14}$$

$$v_A^2 = v_B'^2 + 2v_{A_x}'^2 + 2v_{A_y}'^2 \tag{9.15}$$

We solve the first two equations in this last set for v_{A_x}' and v_{A_y}', respectively, then substitute the results into the third equation to obtain

$$v_A^2 = v_B'^2 + (v_A - \cos(\theta) \, v_B')^2 + (\sin(\theta) \, v_B')^2$$

When numerical values are substituted, the above equation reduces to

$$(3.00)v_B'^2 - (8.00 \text{ m/s})v_B' - (25.00 \text{ m}^2/\text{s}^2) = 0,$$

which we solve using the quadratic formula to obtain

$$v_B' = \begin{cases} +4.51 \\ -1.85 \end{cases} \text{m/s}$$

Because v_B' is the magnitude of the vector $\mathbf{v}_B'$, it must be a positive quantity; thus, we choose the positive root. Substituting $v_B' = 4.51$ m/s into Equations (9.13) and (9.14), we obtain $v_{A_x}' = 0.695$ m/s and $v_{A_y}' = -1.35$ m/s. Prudence dictates that we should substitute these three values into Equation (9.15) to check our solution: $25.00 \stackrel{?}{=} (4.51)^2 + 2(0.695)^2 + 2(-1.35)^2 = 24.95$, which checks well enough.

Finally we can find the angle made by puck A and the x-axis from

$$\phi = \arctan\left(\frac{v_{A_y}'}{v_{A_x}'}\right) = \arctan\left(\frac{-1.35}{0.695}\right) = -62.8°$$

◀

Concept Question 9
If you can hear the collision between two objects, can the collision be perfectly elastic? Explain.

Partially Elastic Collisions

In the case of partially elastic collisions, kinetic energy is not conserved. Therefore, we have only the conservation of momentum condition to apply to our problem. In two dimensions this means two equations. Thus, we can solve problems with no more than two unknowns. Typically you might know everything about all the participants in the collision except the final velocity of one. You could then solve for the magnitude and direction of the velocity for that one object. We do not show a detailed example for this case. Look at the first steps in Examples 9.15 and 9.16 and you will see how to write the conservation of momentum condition for a collision in two dimensions.

9.7 Summary

Impulse is defined as integral of force with respect to time.

$$\mathbf{J} = \int_{t_o}^{t_f} \mathbf{F} \, dt \tag{9.1}$$

Momentum is mass times velocity, $\mathbf{p} = m\mathbf{v}$. The **impulse–momentum theorem** states that the net impulse on an object with a constant mass is equal to the object's change in momentum.

$$\mathbf{J}_{\text{net}} = \Delta \mathbf{p} \tag{9.2}$$

Newton's second law can be written in terms of the time rate of change of momentum

$$\mathbf{F}_{\text{net}} = \frac{d\mathbf{p}}{dt} \tag{9.4}$$

The relation between the **relative velocities** of three objects can be written

$$\mathbf{v}_{AC} = \mathbf{v}_{AB} + \mathbf{v}_{BC} \tag{9.6}$$

The **conservation of momentum** law states that when the net external force on a system is zero, the total momentum of the system is constant. When this condition is satisfied, we may apply the conservation of momentum to **recoil** and **collisions.** During some collisions, the collision forces are substantially greater than the external forces, and the conservation of momentum theorem can be applied to obtain good, but approximate, results even when the system is not completely isolated. In the **collision approximation,** outside forces that act during the collision are ignored.

Collisions may be classified according to their degree of elasticity by the **coefficient of restitution ϵ:**

$$\epsilon = \frac{Speed\ of\ separation}{Speed\ of\ approach}$$

$$\epsilon = \frac{(v'_B - v'_A)}{(v_A - v_B)} \tag{9.10}$$

In terms of the coefficient of restitution we classify collisions as outlined in the following table.

ELASTICITY	KINETIC ENERGY	ONE-DIMENSIONAL COEFFICIENT OF RESTITUTION
Perfect	Conserved	$\epsilon = 1$
Partially elastic Partially inelastic Inelastic	Not conserved	$0 < \epsilon < 1$
Perfectly inelastic	Not conserved (maximum possible lost)	$\epsilon = 0$
Hyperelastic	Not conserved	$\epsilon > 1$

PROBLEMS

9.1 Impulse

1. Determine the impulse delivered to an object by the following forces: (a) A 5.00-N force acting in a direction parallel to the x-axis for 3.20 s. (b) The force of gravity acting on a freely falling mass of 0.500 kg as it falls 1.00 m starting from rest. (c) A 2.00-N force acting parallel to the y-axis for 2.00 s and then changing to a 3.00-N force acting parallel to the x-axis for 1.00 s.

2. Determine the impulse delivered by the following forces: (a) The force of gravity acting on a 1.50-kg mass falling freely (from rest at $y = 0.00$ m) as the mass falls between $y = -1.00$ m and $y = -2.00$ m. (b) The force graphed in Figure 9.P1. (c) The force $\mathbf{F} = 0.600\ \mathrm{N}\,\widehat{\mathbf{i}} + 0.250\ \mathrm{N}\,\widehat{\mathbf{j}}$ acting for 5.50 s.

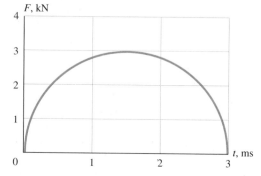

FIGURE 9.P1   Problem 2

3. Determine the impulse delivered by the force $\mathbf{F} = (3.00\ \mathrm{N/s^2})\,t^2\widehat{\mathbf{i}} - (2.50\ \mathrm{N/s})\,t\widehat{\mathbf{j}}$ acting from $t = 0.00$ s to $t = 2.00$ s.

4. A force given by $\mathbf{F} = F_o \exp(-at)\,\widehat{\mathbf{i}}$ acts on an interstellar rocket for an essentially infinite length of time starting from $t = 0.00$. (a) For $F_o = 5.00 \times 10^4$ N and $a = 1.00 \times 10^{-4}\ \mathrm{s^{-1}}$, calculate the net impulse delivered to the rocket. (b) How much time elapses until one-quarter of the total impulse is delivered? (c) How long does it take before the second quarter of the impulse is delivered?

9.2 Impulse and Momentum

5. (a) Write the kinetic energy of a particle in terms of momentum and mass only. (b) How fast would a 5.50-kg bowling ball have to travel to have the same momentum as a 45.9-g golf ball traveling 50.0 m/s?

6. The force-versus-time curve for a club striking a golf ball is modeled by the curve shown in Figure 9.P2. (a) What is the impulse on the golf ball? (b) If the 45.9-g ball starts from rest, what is its speed after being acted upon by the impulse?

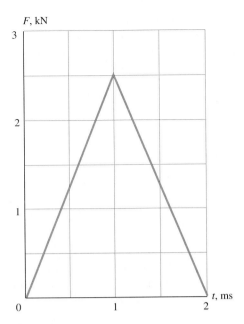

FIGURE 9.P2   Problem 6

7. A robotic braking system supplies a braking force as shown in Figure 9.P3, which is described by $F = (5.00 \text{ N/s})t$. (a) How long does this braking system have to act to stop a 25.0-kg mass traveling at 3.00 m/s? (b) What maximum force is reached? (c) What is the average force acting?

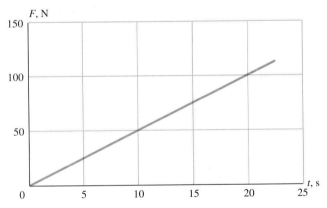

FIGURE 9.P3   Problem 7

8. A constant net force of 48.0 N acts on a 0.0459-kg golf ball, increasing its speed from 1.20 m/s to 6.72 m/s. (a) What impulse acts on the golf ball? (b) What is the time interval Δt during which the force acts?

9. A 1.10×10^3 kg car traveling at a constant speed of 12.0 m/s rounds a 90° corner in 4.50 s. (a) What is the magnitude of the net impulse on the car? (b) What is the average force on the car? (c) Draw a sketch showing the direction of the net impulse relative to the initial and final directions of the car.

10. A Dudley SB-12LND cork-center regulation slow-pitch softball of 0.189-kg mass is acted on by a gravitational force of $-1.95 \text{ N} \hat{\mathbf{j}}$ and a force due to air resistance of $-0.350 \text{ N} \hat{\mathbf{i}} + 0.120 \text{ N} \hat{\mathbf{j}}$ for a time interval of 0.100 s. (a) Calculate the net impulse on the soft-

ball. (b) If the initial velocity of the softball was $\mathbf{v}_o = +13.6 \text{ m/s} \hat{\mathbf{i}} - 8.3 \text{ m/s} \hat{\mathbf{j}}$, find the velocity of the ball at the end of the 0.100 s.

9.3 Momentum and Newton's Second Law

11. A 65.0-kg diver does a "cannon ball" dive into the water. Just before striking the water the diver's speed is 5.50 m/s straight down. At a time 0.750 s after hitting the water, the diver's downward velocity is 1.25 m/s. What is the average force acting on the diver during entry into the water?

12. A volleyball ($m = 0.275$ kg) is traveling with a velocity of 12.5 m/s when a player returns it directly along its original path (but in the opposite direction) with a speed of 8.50 m/s. (a) What impulse did the ball receive from the player? (b) If the ball was in contact with the player's hand for 80.0 ms, what average force did she apply to the volleyball?

13. A 56.0-g tennis ball traveling at 9.00 m/s strikes a vertical wall at an angle of 36.87° from a normal to the wall. The ball rebounds at a speed of 5.00 m/s at the same angle on the other side of the normal as shown in Figure 9.P4. (a) Calculate the magnitude and direction of the ball's change in momentum. (b) If the collision lasted 2.50×10^{-3} s, find the average force exerted on the wall.

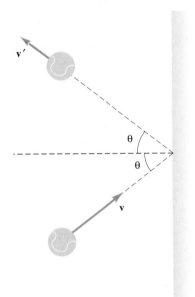

FIGURE 9.P4   Problem 13

14. A basketball of mass 0.595 kg is bounce-passed from player to player. Just before striking the floor it has a velocity of 7.75 m/s at an angle of 50.0° with a normal to the floor. It rebounds from the floor with a speed of 5.40 m/s, making an angle of 35.0° on the other side of the normal. (The change of angle is due to back spin.) If the ball is in contact with the floor for 35.0 m/s, find the average force of the floor on the ball.   , 035 s

15. During a downpour, rain is falling with an average intensity of 1.00 drop/cm²/s. Take the terminal velocity of all the drops to be 10.0 m/s. If the mass of each drop is 0.0450 g, what is the average force exerted on each square meter of a horizontal roof due to the impact of the rain? (Assume each drop has zero velocity after the impact. Also ignore the weight of the water.)

16. Suppose that 4.20×10^{22} molecules of a gas, each with mass 1.93×10^{-23} kg and traveling 600. m/s collide with a 10.0 cm ×

10.0 cm wall area each second. If the molecules all strike the wall along a normal and rebound with the same speed they had before impact, what is the average force of the molecules on the wall?

9.4   Conservation of Momentum

17. A 3.00-kg target rifle fires a 150-grain (9.72 g) bullet with a speed of 2500 ft/s (762 m/s). If the rifle is allowed to recoil freely backward, what is its recoil velocity?

18. A 12.5-kg dog jumps off a stationary 2.40-kg sled, which is at rest on a frozen pond. If the dog hits the ice traveling at 0.350 m/s, how fast will the sled be moving and in what direction?

19. Two skateboarders are on a smooth surface between two fences that are 5.00 m apart as shown in Figure 9.P5. One skater has a mass of 60.0 kg, the other a mass of 75.0 kg. Where should they place their skateboards initially so that they each reach the fence to their backs at the same moment?

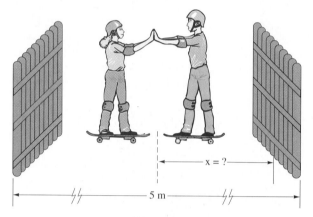

FIGURE 9.P5   Problem 19

20. A 255-kg payload is to be separated from its 575-kg third-stage booster rocket in such a way that the booster rocket has zero velocity, relative to the earth, after separation. If the combined payload and booster have a speed of 2.65 km/s just before separation, what is the speed of the payload after separation?

21. A 5.50-kg mass and a 1.75-kg mass are connected together with a small explosive charge between them. When the explosive charge is set off, the 5.50-kg mass flies to the right with a speed of 15.0 m/s. (a) What is the velocity of the 1.75-kg mass? (b) How much kinetic energy was transferred to the masses by the explosion?

22. A light Hooke's-law spring of stiffness constant $k = 225$ N/m is compressed 0.100 m and held by a nylon string. The spring is placed between two airtrack gliders of masses 200. g and 300. g. When the string is burned into two segments, the spring expands. Assume all of the energy of the compressed spring goes into the gliders, and calculate the velocity of each after the spring expands.

23. A 55.0-kg physics student is at rest on a 5.00-kg sled that also holds two chunks of ice each with a mass of 1.50 kg. The student throws each of them, one at a time, in the same horizontal direction with a speed of 12.0 m/s *relative to the student*. (a) If the sled slides over a frozen pond frictionlessly, how fast are the sled and student traveling after throwing the first chunk of ice? (b) How fast are they traveling after throwing both chunks?

24. A 50.0-kg astronaut finds herself at rest relative to her space station 1.00×10^3 m away. She holds four 2.00-kg masses and decides to propel herself toward the station by throwing the masses

in the direction away from the station. Because of the encumbrance of the space suit, she can push the masses away at a speed of 3.00 m/s *relative to herself* regardless of how many masses she shoves away at any given time. (a) If she throws all four masses simultaneously, what speed does she gain? (b) If she throws the masses two at a time, what final speed does she gain from all four? (c) If she throws the masses one at a time, what is her final speed? (See Problem 49 for an extension of this problem.)

9.5   Collisions in One Dimension

25. A 6.00-kg mass traveling 5.00 m/s collides with a stationary 10.0-kg mass. (a) If the masses stick together after the collision, what is the final velocity of the combined masses? (b) By what fraction is the kinetic energy reduced by the collision?

26. A loaded boxcar of 2.50×10^4 kg mass is rolling freely down a level railroad track at a speed of 0.600 m/s. The boxcar collides with a stationary boxcar and couples to it. (a) If the two cars move off together at 0.400 m/s, what is the mass of the second boxcar? (b) By what fraction is the kinetic energy reduced by the collision?

27. A neutron with a velocity of 3.00×10^6 m/s makes a perfectly elastic head-on collision with a stationary alpha particle. (a) What is the mass ratio $\mathcal{M} = m_\alpha/m_n$? (b) Find the velocity of both particles after the collision. (The neutron and alpha particle are particles encountered often in subatomic physics. The mass of an alpha particle is $m_\alpha = 6.64 \times 10^{-27}$ kg and the neutron mass can be found in Appendix 7.) (c) What fraction of the neutron's kinetic energy is transferred to the alpha particle?

28. Solve Problem 27 with the roles of the alpha particle and the neutron reversed.

29. A 2.00-kg mass traveling 15.0 m/s collides head on with a stationary 1.00-kg mass. (a) If the collision is characterized by a coefficient of restitution $\epsilon = 0.750$, calculate the final velocities of the masses after the collision. (b) By what fraction is the kinetic energy reduced by the collision?

30. A 6.00-kg mass traveling 2.50 m/s collides head on with a stationary 4.00-kg mass. After the collision the 6.00-kg mass travels in its original direction with a speed of 1.00 m/s. (a) What is the velocity of the 4.00-kg mass after the collision? (b) What is the coefficient of restitution for this collision? (c) What is the fractional change of the kinetic energy for this collision?

31. A 4.00-kg mass is traveling to the right with a velocity of $+12.0$ m/s. This mass makes a head-on collision with a stationary 2.00-kg mass. (a) If the coefficient of restitution between the two blocks is $\epsilon = 0.250$, calculate the final velocities of the two masses. (b) By what fraction is the kinetic energy reduced by the collision? (c) If the contact between the two blocks lasted 0.333 s, what was the average force on the 4.00-kg mass?

32. A 16.0-lb bowling ball traveling at 5.00 m/s collides with a single 3.50-lb bowling pin. If the coefficient of restitution between the ball and the pin is 0.800, use the collision approximation to estimate the speed of the pin immediately after the collision.

33. Two masses, *A* and *B,* collide head on. Before the collision their velocities are $v_A = 2.00$ m/s and $v_B = -3.00$ m/s. After the collision the velocities are $v'_A = -1.50$ m/s and $v'_B = +0.50$ m/s. Find the ratio $\mathcal{M} = m_B/m_A$ of the masses and the coefficient of restitution for the collision.

34. A space module is at rest in space relative to the background of the stars. It is struck by a small meteorite. (a) Without doing any calculations, guess which of the following causes the space module to gain greater speed: the meteorite makes a perfectly inelastic

collision with the module or the meteorite makes a perfectly elastic collision with the module. (b) Now test your intuitive answer to part (a) by calculating the final speed of the module if the module has a mass of 100. kg and the meteorite has mass of 10.0 kg and an original speed of 1.00 km/s.

35. Model the batting of a baseball using the collision approximation. Take the mass of a baseball as 0.142 kg and the mass of the bat as 1.00 kg. Let the initial speed of the ball be 40.0 m/s and suppose the bat moves with a velocity of −15.0 m/s. Assume the coefficient of restitution is 0.400. (a) Calculate the velocity of the ball and the bat after the collision. (b) Are your answers reasonable? Comment on the validity of the collision approximation.

36. (a) Show that the working Equations (9.11) and (9.12) can be solved in general for the final velocities, yielding

$$v'_A = \frac{1 - \mathcal{M}\epsilon}{1 + \mathcal{M}}v_A + \frac{\mathcal{M}(1 + \epsilon)}{1 + \mathcal{M}}v_B$$

$$v'_B = \frac{1 + \epsilon}{1 + \mathcal{M}}v_A + \frac{\mathcal{M} - \epsilon}{1 + \mathcal{M}}v_B$$

(b) What do these equations predict for the result of a perfectly elastic collision between equal masses?

9.6 Collisions in Two Dimensions

37. A croquet ball traveling with a velocity of 4.80 m/s collides with an identical ball initially at rest. Immediately after the collision the original ball is observed to have a speed of 3.60 m/s at an angle of 28.0° from its original direction. What is the velocity of the other croquet ball immediately after the collision?

38. A particle of mass m traveling with a 6.80 cm/s velocity collides elastically with a mass $2m$ initially at rest. If the mass $2m$ is observed to have a speed of 1.80 cm/s at an angle of 18.0° from the direction of mass m's original velocity, what is the velocity (magnitude and direction) of mass m after the collision?

39. In one strategy for the Strategic Defense Initiative proposed in the late 1980s, small high-velocity "pebbles" were to be fired at incoming targets, which might be actual nuclear warheads or simply decoys. Officials hoped that the heavy warheads could be distinguished from lighter decoys from observations of the different collision results for the pebbles colliding with each. Model the warheads as 250.-kg masses and the decoys as 20.0-kg masses. Assume the pebbles have a mass of 0.500 kg and that they strike the targets at right angles. Take the initial speed of the targets as 2.00 km/s and the speed of the pebbles as 7.00 km/s. If the collisions are perfectly inelastic, calculate the difference in deflection angles between the warheads and the decoys after being hit with a pebble.

40. Two cars collide at an intersection and become entangled. Car A has a mass of 785 kg and was traveling North at 45.0 km/h. Car B has a mass of 700. kg and was traveling due East. Skid marks indicate that after the collision both cars moved together at an angle of 60.0° North of East immediately after the collision. Use the collision approximation to estimate the velocity of car B before the collision.

41. Figure 9.P6 shows a multiple-flash photograph of two colliding spheres. Assuming that momentum is conserved, calculate the mass ratio of the two spheres. [Hint: You do not need to know the absolute distance or time scales. Measure the angles on the photograph with a protractor.]

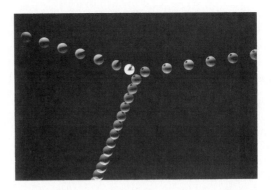

FIGURE 9.P6   Problem 41. The sphere with the dot struck the stationary, unmarked sphere.

42. Figure 9.P7 shows a multiple-flash photograph of the collision between two spheres. The time interval between exposures is 0.0100 s. (a) Make a scale drawing showing the velocity change of each disk due to the collision. (b) If momentum is conserved, what is the mass ratio of the two disks? (c) Does the time interval really matter? (d) Can you tell the direction of the original motion from the photographs?

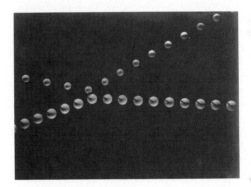

FIGURE 9.P7   Problem 42

43. Two spheres collide as shown in Figure 9.P8. Initially, sphere A travels along the x-axis with a velocity of 20.0 m/s, and the target sphere B is at rest. After the collision sphere A travels 10.0 m/s at an angle of 30° above the x-axis. Sphere A has a mass of 0.200 kg, and sphere B has a mass of 0.500 kg. (a) Calculate the velocity (magnitude and direction as an angle from the positive x-axis) of the target sphere immediately after the collision. (b) Draw vectors showing (1) the vector addition whose resultant is the total momentum before the collision and (2) the vector addition leading to the total momentum after the collision. (c) Is the collision inelastic, perfectly elastic, or hyperelastic?

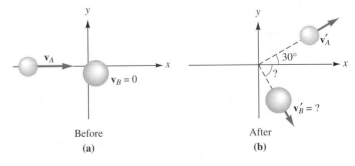

FIGURE 9.P8   Problem 43

44. A 2.00-kg object travels along the x-axis at a speed of 5.00 m/s and collides with a stationary 3.00-kg target. After the collision the 2.00-kg object is seen to travel at 1.00 m/s at an angle of 36.87° below the x-axis as shown in Figure 9.P9. (a) Find the magnitude and direction (as an angle from the positive x-axis) of the 1.00-kg mass immediately after the collision. (b) Is the collision perfectly elastic? Explain.

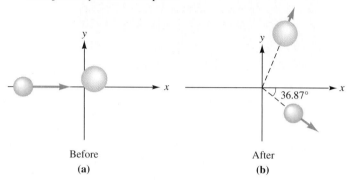

Before
(a)

After
(b)

FIGURE 9.P9   Problem 44

45. Figure 9.P10 shows a vector statement of the conservation of momentum in two dimensions for a stationary target object B and an object A with momentum p_A before the collision. Vectors p_B' and p_A' are the vectors after the collision. Because momentum is conserved, the triangle formed by the vectors must be closed. Use Figure 9.P10 to show that for the special case of equal masses and a perfectly elastic collision, angle θ is 90.0°. (*Hint:* Write the kinetic energy of each object as $p^2/2m$.)

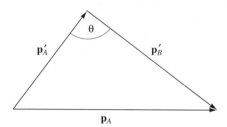

FIGURE 9.P10   Problem 45

46. A 55.0-kg chunk of debris drifts in space at 2.00 km/s. (a) If the debris is hit at right angles by a meteorite of 0.600-kg mass traveling 30.0 km/s, at what angle from its original direction is the debris deflected if the collision is perfectly inelastic? (b) What angle is the debris deflected through from its original direction if the collision is perfectly elastic? (c) Compare your answers to (a) and (b) and qualitatively explain the results in terms of the conservation of momentum.

Numerical Methods and Computer Applications

47. (a) Use Simpson's rule (see Chapter 3, Section 5) to calculate the impulse of the force described by Figure 9.P11 and the data in Table 9.P1. (b) If this impulse is applied to a stationary 46.0-g golf ball, what is the ball's speed as it leaves the tee? These data may also be found in the spreadsheet template GOLFBALL.WK1 on the diskette accompanying this text.

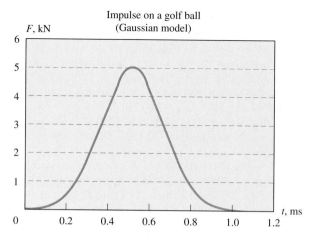

FIGURE 9.P11   Problem 47. See data in Table 9.P1 and spreadsheet template GOLFBALL.WK1.

TABLE 9.P1

TIME (ms)	FORCE (kN)	TIME (ms)	FORCE (kN)
0.00	0.01	0.60	3.89
0.05	0.03	0.65	2.85
0.10	0.09	0.70	1.84
0.15	0.23	0.75	1.05
0.20	0.53	0.80	0.53
0.25	1.05	0.85	0.23
0.30	1.84	0.90	0.09
0.35	2.85	0.95	0.03
0.40	3.89	1.00	0.01
0.45	4.70	1.05	0.00
0.50	5.00	1.10	0.00
0.55	4.70		

TABLE 9.P2   Incomplete Lab Data

DATA POINT NUMBER	TIME (??)	FORCE (N)
0		0.00
1		0.05
2		0.10
3		0.15
4		0.22
5		0.30
6		0.40
7		0.48
8		0.46
9		0.25
10		0.00

48. Table 9.P2 gives the force-versus-time record for a laboratory experiment. The instantaneous force was recorded at equally separated instants of time, but, unfortunately, no one in the lab group recorded the size of the time increments. It is known however that a 1.50-kg cart, which started at rest, had a speed of 0.750 m/s after it was acted on by the impulse. Determine the time interval between points when the force was recorded. These data may also be found in the spreadsheet template LOSTTIME.WK1 on the diskette accompanying this text.

49. Consider again Problem 24. Suppose that the original 8.00-kg mass can be divided into an arbitrary number of pieces. Construct a spreadsheet (or computer program) that calculates the final speed of the astronaut after throwing N chunks of mass, each with mass (8.00 kg)/N for values of N ranging from 1 to 25. Present your results as a graph of final velocity versus N.

50. Data files GLIDERS1.WK1 contains the position-versus-time history of two gliders colliding on an airtrack. (a) Use a spreadsheet to make a graph of the position-versus-time history of both gliders on the same plot. (b) From your graph identify, for each glider, regions of constant velocity just before and just after the collision. Then use the "regression" capability of your spreadsheet to calculate the slopes for each of these four line segments. These slopes are, of course, the velocities of the gliders. (c) Use these four velocities to find the mass ratio of the gliders and the coefficient of restitution of the collision.

51. Repeat the analysis described in Problem 50 for the data file GLIDERS2.WK1.

52. Repeat the analysis described in Problem 50 for the data file GLIDERS3.WK1.

53. Repeat the analysis described in Problem 50 for the data file GLIDERS4.WK1.

54. Repeat the analysis described in Problem 50 for the data file GLIDERS5.WK1.

55. Repeat the analysis described in Problem 50 for the data file GLIDERS6.WK1.

56. Repeat the analysis described in Problem 50 for the data file GLIDERS7.WK1.

57. Using a spreadsheet or computer program, plot a diagram showing the results of one-dimensional collisions for various values of the mass ratio $\mathcal{M} = m_B/m_A$ and coefficient of restitution ϵ. Use the general result of Problem 36 with the projectile traveling with initial velocity $v_A = 1$ and the target at rest, that is, $v_B = 0$. On the vertical axis plot v'_B; on the horizontal axis plot v'_A. For $\epsilon = 0$ choose 10 mass ratios $\mathcal{M} = 0.02, 0.04, 0.08, 0.16, 0.32, 0.64, 1.28, 2.56, 5.12,$ and 10.24. For each mass ratio plot a point with coordinates (v'_A, v'_B) on your graph. Connect these points with a line. Now repeat the entire process for each of the following values of ϵ: 0.2, 0.4, 0.6, 0.8, 1.0. Finally connect each of the sets of points with the same mass ratio but different ϵ. Label your graph clearly so that it can be used to predict the outcome of any collision.

58. Consider a series of N individual airtrack gliders on a very long track. The first glider has mass M, the second mass fM, the third mass f^2M, and so on to the Nth, which has mass $f^{N-1}M$. The first glider is set into motion with a velocity of, say, 1.00 m/s, toward the others. The first glider collides with the second, which is then set in motion and collides with the third, and so on. (a) Using the working equations for one-dimensional collisions, prepare a spreadsheet that calculates the final speed of each glider after it has been hit and before it hits the next glider. (The first and last gliders, of course, suffer only one collision.) Use $M = 2.00$ kg, $N = 25$, $\epsilon = 1.00$, and $f = 0.95$ and create a plot showing the speed of each glider on the vertical axis, with the glider number used as the horizontal coordinate. (b) On the same graph, plot four additional curves showing the results for $f = 0.90, 0.85,$ 0.80, and 0.75. (c) What fraction of the original kinetic energy of the first glider is transmitted to the Nth in each case? Compare these results to the case where the first mass hits the Nth mass directly.

59. Refer to Problem 58. (a) Perform a similar analysis, except make $f = 0.90$, and on the same graph produce plots for $\epsilon = 1.00, 0.95, 0.90, 0.85,$ and 0.80. (b) What fraction of the total kinetic energy is delivered to the Nth glider in each case? Compare these cases to the result obtained if the first mass hits the Nth mass directly.

General Problems

60. A force function described by $F = At(1.00 \text{ s} - t)$ acts on a 0.750-kg mass for 1.00 s. (a) What is the net impulse delivered to the mass? (b) If the mass started from rest and had a final velocity of 3.50 m/s after the impulse acted, what is the numerical value (and units!) of A?

61. Model the impulse of the collision between a 0.145-kg baseball and a bat using the force function $F = Bt^2(t_f - t)$ for times $0 < t < t_f$. Assume the ball had an initial speed (relative to the ground) of 35.0 m/s and that, after the collision, its speed in the opposite direction was 30.0 m/s. (a) If the contact time $t_f = 2.00$ ms, what is the magnitude of the constant B? (b) What is the maximum force on the ball? (c) What is the average force on the ball?

62. (a) Please graph the force-versus-time curve for the data described in Table 9.P3. Estimate (or use numerical integration to calculate) the impulse delivered by the shot-putter who exerted this force (Fig. 9.P12). (b) If this impulse is delivered to a 16.0-lb shot released from a height of 2.00 m at an angle of 40.0° above the horizontal, how far does the shot travel?

TABLE 9.P3   Shot Put (Model Data)

TIME (s)	FORCE (N)	TIME (s)	FORCE (N)
0.000	0	0.275	265
0.025	0	0.300	260
0.050	0	0.325	260
0.075	0	0.350	250
0.100	0	0.375	200
0.125	5	0.400	190
0.150	5	0.425	190
0.175	30	0.450	180
0.200	107	0.475	100
0.225	220	0.500	0
0.250	260		

FIGURE 9.P12   An impulse is delivered to a shot.

63. In his novel *Chesapeake* James Michener describes one of the long guns used by early watermen to hunt ducks on the Chesapeake Bay in the 1800s. The gun is described as "eleven feet six inches long, about a hundred and ten pounds in weight." It was reportedly loaded with 1.5 lb (plus a fistful) of number-six shot and charged with three-quarters of a pound of black powder. (a) If the shot achieved a modest muzzle velocity of 800 ft/s when the gun was fired, what would be the recoil velocity of a 125-lb skiff carrying the gun, a 120-lb hunter, and a 75-lb retriever? (Ignore the momentum of any unburnt powder and combustion gases.) (b) In a subsequent anecdote this gun was loaded with double the shot and powder and fired. As Michener tells it, the gun ". . . burst out of the back of Jake's skiff, knocked him unconscious and threw him a good twenty yards aft into the dark and icy waters." Assume Jake was launched at 45°, and calculate the initial speed required to launch him that far. What percentage of the recoil momentum would this require? Is this reasonable?

64. A 10.0-g projectile is fired at 100. m/s toward a stationary 10.00-kg target. When struck, the target breaks into two pieces with masses 9.80 kg and 0.20 kg. The original projectile and the smaller part of the target fly off along the original direction of the projectile at speeds of 60.0 m/s and 20.0 m/s, respectively. Calculate the velocity (including direction) of the larger piece of the target after the collision. (The results may be surprising, but they are quite real. See "A physicist examines the Kennedy assassination film," Luis W. Alvares, *American Journal of Physics, 44*(9), 813–827 (1976).)

65. The circumference of a basketball must be between 75 and 78 cm. When dropped from a height such that the *bottom* of the basketball is 1.80 m from the floor, the *top* of the ball must rebound to a height of 1.20 to 1.40 m. What are the legal limits of the coefficient of restitution of a basketball?

66. A small rubber ball can be made to rebound to surprising heights by holding it on top of a heavier ball, then dropping the two together as shown in Figure 9.P13. Model the result of dropping a 50.0-g rubber ball and a 600.-g basketball together. Take the original height of the bottom of the basketball to be $h = 1.50$ m from the floor and the coefficient of restitution to be 0.800 for all collisions. First disregard the smaller ball and consider only the collision of the basketball with the floor. Then let the downward moving rubber ball collide with the upward moving basketball, which has a diameter of 24.0 cm. How high does the small rubber ball rise?

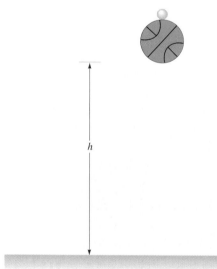

FIGURE 9.P13   Problem 66

★67. A 5.00-kg mass slides along a horizontal frictionless track with a speed of 4.00 m/s. This mass collides with a stationary 2.00-kg mass, which has a light spring ($k = 1.60 \times 10^3$ N/m) attached as shown in Figure 9.P14. As the masses collide the 5.00-kg mass slows down, the 2.00-kg mass speeds up, and the spring compresses. At the moment of maximum spring compression, the collision can be modeled as perfectly inelastic because both masses are moving at the same speed. The "missing" kinetic energy at this moment has been converted into spring potential energy. (a) Calculate the maximum compression of the spring. (b) After the spring forces the two blocks to separate, what is the final velocity of each mass?

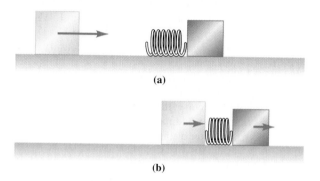

(a)

(b)

FIGURE 9.P14   Problem 67

★68. A 0.200-kg mass is sliding over a frictionless tabletop with a velocity of 4.00 m/s when it encounters a frictionless ramp as shown in Figure 9.P15. The ramp itself (initially at rest) has a mass of 0.500 kg and can slide frictionlessly over the tabletop. (a) What maximum vertical distance up the ramp does the 2.00-kg mass climb during the "collision"? (b) After the 0.200-kg mass slides back down the ramp, what is the final velocity of the block and the ramp?

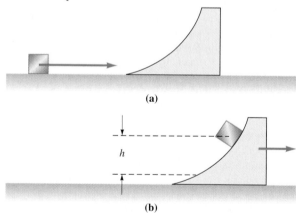

(a)

h

(b)

FIGURE 9.P15   Problem 68

69. Calculate the fractional error in the final velocity of the cue ball of Example 9.3 if the coefficient of kinetic friction between ball and tabletop is 0.400. This error is a result of using the impulse approximation in which the frictional force of the tabletop on the cue ball is ignored.

70. A particle of mass m collides elastically with a particle of mass $4m$. After the collision, mass m is observed to have a speed of 6.78 m/s at an angle of $+22.0°$ from its original direction. Mass $4m$ has a speed of 2.55 m/s at an angle of $-14.4°$ from the original direction of mass m. Find the initial velocities of the masses and the angle at which mass $4m$ approached mass m before the collision.

10

Momentum, Energy, and the Center of Mass

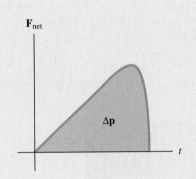

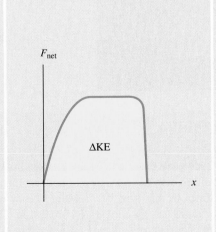

In this chapter you should learn

- to determine when conservation of momentum and conservation of energy can be applied to different parts of the same problem.

- to calculate the coordinates and the velocity of the center of mass of a collection of point–particles.

- (optional) to calculate the center of mass of extended objects.

- to write a simple expression for the gravitational potential energy in terms of the center of mass of a collection of objects or an extended object.

- to write the kinetic energy of a system of particles in terms of the velocity of the center of mass and velocities relative to the center of mass.

- (optional) to analyze systems with variable mass.

10.1 The Big Picture

In Chapters 7, 8, and 9 we found that integrals of the net force on an object can be related to changes in that object's motion. These relations culminated in two of the cornerstones of classical mechanics: the conservation of mechanical energy and the conservation of linear momentum. The developments were quite similar but had important differences. Before proceeding further, let's step back and look at a side-by-side summary of these two related ideas:

TABLE **10.1**   **The Big Picture**

CHAPTERS 7, 8	CHAPTER 9
Work: $\mathcal{W} = \int \mathbf{F} \cdot \mathbf{ds}$	Impulse: $\mathbf{J} = \int \mathbf{F}\, dt$
Kinetic energy: $K = \frac{1}{2}mv^2$	Momentum: $\mathbf{p} = m\mathbf{v}$
Work-energy theorem: $\mathcal{W}_{\text{net}} = \Delta K$	Impulse-momentum theorem: $\mathbf{J}_{\text{net}} = \Delta \mathbf{p}$
Conservation of mechanical energy:	Conservation of momentum:
If only conservative forces act, $\Sigma(U + K) = \Sigma(U' + K')$	If net external force is zero, $\Sigma \mathbf{p} = \Sigma \mathbf{p}'$

The similarity in logical development is rather obvious from the comparisons shown in Table 10.1; both work and impulse are defined as integrals of force, these integrals represent changes in kinetic energy and momentum, respectively, and under precise conditions, each concept leads to a conservation condition. The differences are equally clear: (1) Energy is related to force and *distance;* momentum is related to force and *time.* (2) Energy is a scalar; momentum is a vector. (3) Conservation of mechanical energy holds when only conservative forces act; conservation of momentum holds for all types of forces provided the total external force is zero.

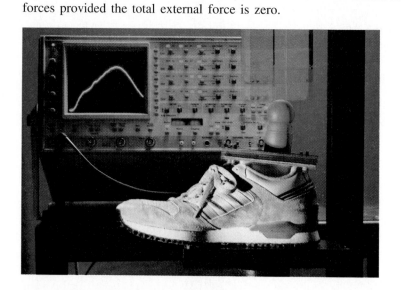

Scientists at Battelle Laboratories test the cushioning effect of running shoes by analyzing the force-versus-time graph when a large mass is dropped into the shoe. (Photo courtesy of Battelle Laboratories)

10.2 The Ballistic Pendulum and Its Friends

The ballistic pendulum and problems similar to it involve conservation of both momentum and mechanical energy but not at the same time. In this type of problem one or more collisions are followed (or preceded) by one or more energy conversion processes. These problems provide a good way for you to test your knowledge of when to apply the conservation of momentum and when to apply the conservation of mechanical energy.

The classic problem of this sort is the ballistic pendulum itself.

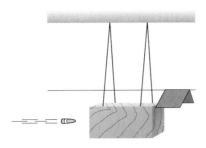

EXAMPLE 10.1 *Name Three Classics. Gone with the Wind, the 1964½ Mustang, and the Ballistic Pendulum*

In order to measure the muzzle velocity of a certain rifle, a bullet is fired into a 1.25-kg block of wood supported by a bifilar suspension as shown in Figure 10.1. The bullet lodges in the wood, and the combined wood and bullet rise a vertical distance $h = 2.40$ cm. If the bullet has a mass of 2.00×10^{-3} kg, what was its original velocity?

FIGURE 10.1

A simple ballistic pendulum comprised of a block of wood into which a bullet is fired. A light piece of folded paper hangs over a wire to indicate the maximum forward distance the block moves. From the forward distance the change in height of the block can be calculated. (See Example 10.1.)

SOLUTION First we'll tell you what *not* to do! At first it may appear like a simple conservation of energy problem. You may be *tempted* to simply equate the original kinetic energy of the bullet to the final potential energy of the block and bullet. WRONG, WRONG, WRONG! PLEASE DON'T DO IT! Mechanical energy is not conserved from beginning to end. The collision is inelastic and considerable kinetic energy is "lost," that is, converted to sound and thermal energy.

What you *should do* is to think of the problem as two problems, as illustrated in Figure 10.2. Picture I in Figure 10.2 shows the bullet and block before the bullet strikes. Picture II illustrates the bullet and block just after the bullet has come to rest relative to the (now moving) block. Picture III shows the block at its highest point, where it momentarily stops before swinging back.

When we go from picture I to picture II we have a simple, perfectly inelastic collision for which momentum is conserved, but kinetic energy definitely is not conserved. From picture II to picture III mechanical energy is conserved because only the conservative force of gravity does work. (Remember, the radially directed tension force does no work because it is perpendicular to the displacement along the circular arc of the block's path.)

You may realize that the division of the problem into two completely uncoupled problems is an approximation. We act as if it were possible for the block to suddenly acquire a velocity without moving in the process. We will learn more about this "sudden approximation" in Example 10.2.

Finishing the present example is really quite simple when the problem is divided into two parts. Using the notation of Figure 10.2 we have a projectile of mass m and initial velocity v_o striking a block of mass M. Immediately after the collision the block and embedded projectile travel with speed V'. Thus, we may apply our two conservation laws in succession:

Conservation of momentum from picture I to picture II:

$$mv_o = (M + m)V'$$

Conservation of mechanical energy from picture II to picture III:

$$\frac{1}{2}(M + m)V'^2 = (M + m)gh$$

We solve the second equation for V' and substitute the resulting expression into the first equation to obtain

$$v_o = \left(\frac{M + m}{m}\right)\sqrt{2gh}$$

$$= \left(\frac{1.25 \text{ kg} + 2.00 \times 10^{-3} \text{ kg}}{2.00 \times 10^{-3} \text{ kg}}\right)\sqrt{2(9.80 \text{ N/kg})(2.40 \times 10^{-2} \text{ m})} = 429 \text{ m/s}$$

◄

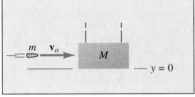

I

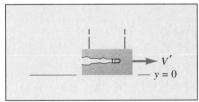

II

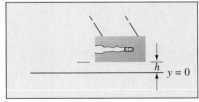

III

FIGURE 10.2

From picture I to picture II only momentum is conserved. From picture II to picture III only mechanical energy is conserved. (See Example 10.1.)

Concept Question 3
Decide if the conservation of momentum and/or the conservation of mechanical energy can be applied to the following occurrences: (a) Two clay balls traveling in outer space collide and stick together. (b) A child slides down a frictionless slide. (c) Two helium atoms collide leaving both in their original electronic state. (d) Two snowballs collide in midair, breaking into 37 different pieces.

Concept Question 4
Several sequential processes are described. In each case tell to which part of the sequence the conservation of momentum may be applied and to which part the conservation of mechanical energy (or a generalization thereof) may be applied: (a) A clay ball attached to a long string is pulled to one side and allowed to swing downward in a circular arc. At the bottom of the arc the ball strikes a second clay ball hanging on a string of equal length. The two balls stick together and move upward in a circular arc until they come momentarily to rest. (b) A block rests on a vertically mounted Hooke's-law spring. A second stationary block is released from a distance h above the first block. The block falls, collides with and sticks to the first block, and then the pair move downward together compressing the spring until they both stop momentarily.

EXAMPLE 10.2   A Critical Look at the "Sudden Approximation"

If you read Example 10.1 critically, and we hope you did, you should be worried about the claim that the problem could be split neatly into two independent problems. You might have thought ". . . but the block starts moving forward before the bullet stops, so some of the kinetic energy is converted into potential energy even before picture II." You are correct, of course. The separation of the problem into two totally distinct subproblems is an approximation. We refer to it as the **sudden approximation.** This approximation is that, although the block acquires a velocity V from the impact, the collision is so quick that the block does not have time to move any distance due to its acquired velocity. Let's see how well this argument holds up when we check the actual numerical values.

Let's "guesstimate" that the bullet burrowed into the block a distance of about 10 cm. Because a rough estimate is all we need, let's model the stopping process with a constant acceleration. For a constant acceleration we can use

$$d = \frac{(v_o + v_f)}{2} t$$

where $v_f = 0$ (relative to the block). This result gives us an (admittedly crude) estimate of the time for the bullet to lodge in the wood:

$$t = \frac{2d}{v_o} = \frac{0.2 \text{ m}}{430 \text{ m/s}} = 0.5 \times 10^{-3} \text{ s} = 0.5 \text{ ms}$$

We can work backward from the block's final height to find its speed immediately after the collision by using $\sqrt{2gh} = 0.7$ m/s. For the block let's assume a constant acceleration from rest so that during the time the bullet burrows its way in, the block moves.

$$x = \frac{(v_o + v_f)}{2} t = \left(\frac{0 + 0.7 \text{ m/s}}{2} \right)(0.5 \text{ ms}) = 0.2 \text{ mm}$$

Thus, we estimate that the block moved forward about 0.2 mm during the brief collision. Suppose the block were suspended by, say, 1.00 m long strings. Then, after being hit by the bullet, the block moves forward by nearly 22 cm before coming to rest. As you can see, the distance moved by the block during the collision is less than 0.1% of the total distance moved by the block. We are led to conclude that the detailed behavior of the bullet is not really important. It hardly matters whether the block moved 0.1 mm or 0.4 mm. The point is that it was a *very* small distance and completely ignorable given the number of significant figures in the problem. Thus, we have reason to have confidence in our answer when we apply the *sudden approximation.* ◀

10.3 Center of Mass

Thus far in our study of mechanics we have made extensive use of the point–mass model. We have treated everything from elephants to elementary particles as simple point–objects. In Chapters 11 and 12 we will have to abandon this simple model to describe the motion of extended objects that rotate as they move. We will treat these extended objects as collections of point–particles. In this section we begin a transition to this point of view by looking for laws that pertain to collections of point–objects. We find that there is one special point that has some interesting and simple properties no matter how complicated the system. This point is called the **center of mass.** We begin with a definition and then look at some of its interesting properties.

Center of Mass for a Collection of Point–Objects

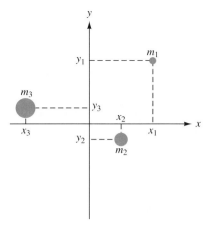

If you have ever tried to carry a tray full of glasses and a pitcher of soft drinks across a room, you have some notion of the importance of the center of mass. In a uniform gravitational field, the center of mass is a balance point. In Chapter 14 we provide a more complete explanation of the nature of a balance point and its relation to the center of mass. For the time being we rely on your intuition and assume that you know from experience that there is a point associated with any mass distribution through which you can provide an upward force to balance the object so that it doesn't rotate or fall to the ground. For a collection of point–objects in a single plane, such as those shown in Figure 10.3, the coordinates of the center of mass are calculated from

$$x_{CM} = \frac{\sum_{i=1}^{N} m_i x_i}{\sum_{i=1}^{N} m_i} \tag{10.1}$$

$$y_{CM} = \frac{\sum_{i=1}^{N} m_i y_i}{\sum_{i=1}^{N} m_i} \tag{10.2}$$

FIGURE 10.3

Geometry for calculating the center of mass.

Concept Question 5
Two children are on a teeter-totter. One child is slightly heavier than the other. The heavier child causes his end of the teeter-totter to touch the ground, raising the end with the lighter child into the air. The heavier child threatens never to let the lighter child back to the ground. What can the lighter child do to cause her end to move to the ground?

where mass m_i is located at coordinates (x_i, y_i). If the mass–points are spread out in three dimensions, a similar equation must be written for z_{CM}. From these equations you can see that the center of mass is just a weighted average of the positions of all the masses. The average is said to be "weighted" because all objects are not treated equally. A 5-kg mass is five times more important than a 1-kg mass. To illustrate this idea we begin with Example 10.3 in which the masses are arranged along a single line. Example 10.4 then discusses a case involving a two-dimensional mass distribution.

EXAMPLE 10.3 *Balancing a Molecule*

In equilibrium the potassium and iodine atoms of KI are separated by 2.79×10^{-10} m. Potassium has a mass of 39.1 u and iodine has a mass of 127 u. (The u stands for unified mass units. See Appendix 5 for conversion factors to SI.) How far from the K atom is the center of mass of this molecule located?

SOLUTION As shown in Figure 10.4 we place the origin of a coordinate system at the position of the potassium atom and then place the iodine atom at $x = 2.79 \times 10^{-10}$ m. We need only apply Equation 10.1 to find x_{CM}:

$$x_{CM} = \frac{\sum_{i=1}^{2} m_i x_i}{\sum_{i=1}^{2} m_i} = \frac{(39.1 \text{ u})(0.00 \text{ m}) + (127 \text{ u})(2.79 \times 10^{-10} \text{ m})}{39.1 \text{ u} + 127 \text{ u}}$$

$$= 2.13 \times 10^{-10} \text{ m}$$

As you may have guessed, the average mass position (the balance point) is closest to the more massive iodine atom. ◄

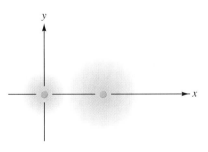

FIGURE 10.4

Conceptual picture of KI molecule. Nearly all of each atom's mass is located at its nucleus, which we model as a point. (See Example 10.3.)

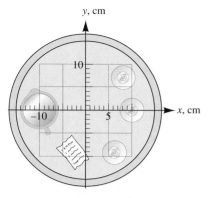

FIGURE 10.5

Where is the balance point of this tray? (See Example 10.4.)

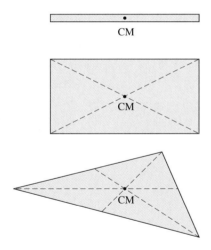

FIGURE 10.6

The center of mass for uniform planar objects is easily found using geometry. For a triangle the center of mass is located at the intersection of the three medians.

Concept Question 6

Without actually carrying out the process, describe in words how to calculate the center of mass of a flat object comprising a rectangle with a triangle sitting on its upper edge.

EXAMPLE 10.4   I Think I'll Still Use Two Hands!

A 2.200-kg pitcher of soft drink and three glasses, each with mass $m = 0.200$ kg, are loaded on a light tray at the positions shown in Figure 10.5. If you wanted to carry this tray with one hand, under what point should you place your hand?

SOLUTION   You must place your hand under the center of mass for this arrangement of masses. The pitcher is centered at (x, y) coordinates $(-10.0$ cm, 0.00 cm$)$, and the glasses are located at $(8.00$ cm, 7.00 cm$)$, $(10.00$ cm, 0.00 cm$)$, and $(6.00$ cm, -9.00 cm$)$. We calculate the x-coordinate of the center of mass from Equation (10.1)

$$x_{CM} = \frac{\sum\limits_{i=1}^{4} m_i x_i}{\sum\limits_{i=1}^{4} m_i}$$

$$= \frac{[(2.200)(-10.0) + (0.200)(8.00) + (0.200)(10.0) + (0.200)(6.00)]\text{kg} \cdot \text{cm}}{(2.200 + 0.200 + 0.200 + 0.200)\ \text{kg}}$$

$$= \frac{-17.2\ \text{kg} \cdot \text{cm}}{2.800\ \text{kg}} = -6.14\ \text{cm}$$

In an analogous fashion from Equation (10.2)

$$y_{CM} = \frac{\sum\limits_{i=1}^{N} m_i y_i}{\sum\limits_{i=1}^{N} m_i}$$

$$= \frac{[(2.200)(0.00) + (0.200)(7.00) + (0.200)(0.00) + (0.200)(-9.00)]\text{kg} \cdot \text{cm}}{(2.200 + 0.200 + 0.200 + 0.200)\ \text{kg}}$$

$$= \frac{-0.40\ \text{kg} \cdot \text{cm}}{2.800\ \text{kg}} = -0.14\ \text{cm}$$

The center of mass for this distribution of pitcher and glasses is $(-6.14$ cm, -0.14 cm$)$. ◀

Center of Mass for Symmetrical Solid Objects

For symmetrical solid objects we can use our intuition about balance points to locate the center of mass easily. A uniform rod, for example, balances when supported in the center. A thin rectangle of uniform density balances at the center point where the diagonals cross, as in Figure 10.6. If each vertex of a plane triangle is connected to the midpoint of the opposite side, these median lines cross at a single point two thirds of the distance along the median from each vertex. This crossing point is the center of mass.

If the object is not symmetrical, it is possible to find the center of mass experimentally by hanging the object from a string several times. Each time the object is hung the string is attached to a different point on the object's edge. The line of the string is then extended down through the object as shown in Figure 10.7(a). The extended lines from every hanging cross at the single point, the center of mass in Figure 10.7(b).

Center of Mass for Solid Objects by Integration (Optional)

For nonsymmetrical solid objects (as well as symmetrical ones) we can calculate the coordinates of the center of mass by using a method like the one we used for a collection of point–masses. To do this, we must (mentally!) divide the solid object up into a collection of small mass elements, then we perform the summations in Equations (10.1) and (10.2). As the chunks into which we divide the object become smaller and smaller, their number becomes larger and larger. As the limit that each m_i approaches zero and the number of mass chunks approaches infinity, the summations in Equations (10.1) and (10.2) approach the integrals

$$x_{CM} = \frac{\int x \, dm}{M} \tag{10.3}$$

$$y_{CM} = \frac{\int y \, dm}{M} \tag{10.4}$$

where $M = \int dm$ is the total mass of the object. If the object is linear in shape, the integral is simply a single integral in one dimension. If the object is planar, the integral becomes two integrals over the object's area. If the object is fully three-dimensional, three integrals over the volume must be performed.

To actually perform the integration, the chunk of mass dm must be written in terms of the spatial coordinates x, y, and z. This is normally done through a density function. You may encounter three types of density functions depending on whether the object may be considered linear, planar (flat), or fully three-dimensional. The three density functions are:

$$Linear \ mass \ density = \lambda = \frac{Mass}{Length}$$

$$Area \ mass \ density = \sigma = \frac{Mass}{Area}$$

$$Volume \ density = \rho = \frac{Mass}{Volume}$$

If the object is sticklike and lies along the x-axis, we can write that

$$dm = \lambda \, dx$$

If the object is essentially flat and lies in the xy-plane, we can write in rectangular coordinates

$$dm = \sigma \, dx \, dy$$

If the object is fully three-dimensional, we can write in rectangular coordinates

$$dm = \rho \, dx \, dy \, dz$$

These three possibilities are illustrated in Figure 10.8. The expressions for dm for the two- and the three-dimensional cases imply that two and three integrals are required, respectively. You will be pleased to hear that this is not always the case. Often, even in the two- and three-dimensional cases the expression for dm may be written in terms of a single integration variable by appropriate choice for the geometry of the mass element. See Example 10.6 for an illustration of such a two-dimensional case.

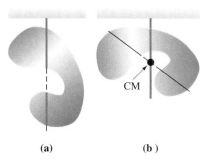

FIGURE 10.7

The center of mass (CM) of an object can be located experimentally by hanging the mass several times from a string attached to the mass at different points. The intersection of lines formed by extending the string indicates the CM. Note that the CM may lie outside the object itself.

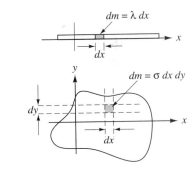

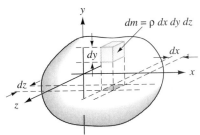

FIGURE 10.8

Line, area, or volume densities may be encountered depending on the dimensionality of the object being considered.

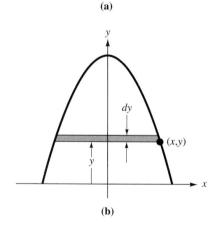

(a)

y

dy

(x,y)

y

x

(b)

FIGURE 10.9

(a) The outline of the vertical back of this homecoming float is a parabola. (b) The shaded mass element has an area $dA = (2x)y\,dy$. All of the mass in this strip is located the same distance y from the x-axis. (See Example 10.6.)

EXAMPLE 10.5   *When They're Not Biting, Try Balancing the Fly Rod*

Suppose a long slender rod of length L has a linear mass density given by

$$\lambda(x) = Cx$$

where C is a constant. The rod lies along the x-axis, and its end with the smaller mass density is placed at $x = 0$. Find the center of mass of the rod.

SOLUTION   To find the center of mass we must calculate two integrals: one to find the mass of the rod, the other to calculate the numerator of Equation (10.3). We first calculate the mass using

$$M = \int_0^L dm$$

Because the problem is one-dimensional, $dm = \lambda\,dx$ and, of course, here $\lambda = \lambda(x)$.

$$M = \int_0^L \lambda\,dx = \int_0^L (Cx)dx = C\int_0^L x\,dx$$

$$M = C\left(\frac{x^2}{2}\right)_0^L = \frac{CL^2}{2}$$

We now have to perform the integral in the numerator of Equation (10.3), which is known as I_1 the *first moment* of the mass distribution

$$I_1 = \int_0^L x\,dm = \int_0^L x\,\lambda\,dx = C\int_0^L x^2\,dx$$

$$= C\left(\frac{x^3}{3}\right)_0^L = \frac{CL^3}{3}$$

Thus, we can now find the center of mass

$$x_{CM} = \frac{I_1}{M} = \frac{CL^3/3}{CL^2/2} = \frac{2}{3}L$$

The answer seems reasonable because we expect the center of mass to be closer to the heavier end.   ◀

EXAMPLE 10.6   *It Must Be the ΣΠΣ Float!*

A college group decides to build a huge homecoming float with a large plywood backdrop in the shape of a parabola. It is necessary to locate the center of mass of the plywood so that a brace can be attached where wind pressure will not cause tilting. Find the center of mass of a uniform piece of plywood cut as in Figure 10.9 so that the top edge follows the curve $y = 4.00 - x^2$ and the bottom edge is the x-axis.

SOLUTION   We plot the curves carefully and note that the parabola cuts the x-axis at ± 2.00, and the y-axis at $y = 4.00$. From symmetry we can see that $x_{CM} = 0$, so the real problem is to find y_{CM}. We mentally divide the parabolic section into horizontal strips of height dy as shown in Figure 10.9(b). Each of the strips is $2x$ wide and has mass $dm = \sigma\,dA = \sigma(2x\,dy)$. Because all of the mass in any given strip is the same distance y from the x-axis, we can write the first moment about the x-axis as

$$I_1 = \int_0^L y \, dm = \int_0^4 y(2\sigma x \, dy) = 2\sigma \int_0^4 xy \, dy$$

The x-value at the right end of each strip is obtained from the equation of the parabola. Solving $y = 4.00 - x^2$, for x we have

$$x = \sqrt{4.00 - y}$$

Thus, the integral becomes

$$I_1 = 2\sigma \int_0^4 \sqrt{4.00 - y} \; y \, dy$$

which can be integrated using the substitution $u = 4.00 - y$ with $du = -dy$

$$I_1 = -2\sigma \int_4^0 u^{1/2}(4.00 - u) \, du$$

Note that the limits of the definite integral have been changed to u limits as required by the variable substitution. The integral can now be easily evaluated

$$I_1 = +2\sigma \int_0^4 [(4.00)u^{1/2} - u^{3/2}] \cdot du = \sigma\left(\frac{256}{15}\right)$$

We must also find the mass

$$M = \int_0^4 \sigma \, dA = \int_0^4 2\sigma x \, dy = 2\sigma \int_0^4 x \, dy = 2\sigma \int_0^4 \sqrt{4.00 - y} \, dy$$

The integral can be evaluated in a fashion similar to the previous one with the result

$$M = \sigma\left(\frac{32}{3}\right)$$

Thus, the center of mass is located at

$$y_{CM} = \frac{I_1}{M} = \frac{\sigma\left(\dfrac{256}{15}\right)}{\sigma\left(\dfrac{32}{3}\right)} = \frac{8}{5} = 1.60 \text{ m}$$

For the sake of clarity we have not observed our usual practice of carrying units explicitly on all numerical quantities. It is an instructive exercise to do so. (See Problem 61.) ◀

10.4 Center of Mass: Energy, Momentum, and Newton's Second Law

At the beginning of the previous section we defined the center of mass and promised that its definition would simplify the way we view the motion of systems that could not be modeled as a point–particle. In this section we show you why knowledge of the center of mass of an extended object simplifies the analysis of that object's motion. We begin with an important illustration and then derive more general results. These general results not only justify many of the point–particle models we have used thus far but also provide us with insight into ways to attack the more complex problems we shall eventually encounter.

Gravitational Potential Energy and the Center of Mass

Potential energy is a scalar quantity, so to compute the gravitational potential energy of a collection of N objects we need only sum the potential energies of the separate objects. Let's suppose that the mass of each object is sufficiently small such that we may ignore the gravitational attraction between each pair of the objects and consider only the potential energy of each due to the earth's uniform gravitational field. In this case, a collection of N objects near the earth's surface has potential energy

$$U = \sum_{i=1}^{N} m_i g y_i$$

where y_i is the height of mass m_i above some reference level. Factoring out the constant g, we can write this as

$$U = g \sum_{i=1}^{N} m_i y_i$$

The summation in the equation above is the same as the summation in numerator on the right-hand side of Equation (10.2). As we have shown, this summation can be replaced by $M y_{CM}$, leading to

$$U = M g y_{CM} \qquad (10.5)$$

Hence, we see that the potential energy of a collection of objects or an extended object can be calculated by treating the entire mass as if it were all located at the center of mass. This result is just one example of how knowledge of the center of mass of an object (or collection of objects) can help simplify an analysis. Let's be more general.

Newton's Second Law and the Center of Mass

The three equations for the x-, y-, and z-coordinates of the center of mass can be compactly written in vector form as

$$\mathbf{r}_{CM} = \frac{\sum_{i=1}^{N} m_i \mathbf{r}_i}{M} \qquad (10.6)$$

The first time derivative of this equation yields an equation for the velocity of the center of mass

$$\mathbf{v}_{CM} = \frac{\sum_{i=1}^{N} m_i \mathbf{v}_i}{M}$$

Concept Question 7
(a) If you start with your arms at your sides and then raise them above your head, what happens to the location of your center of mass relative to your torso?
(b) How does your gravitational potential energy change?

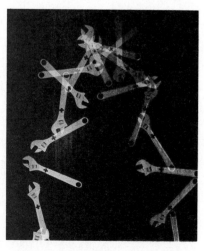

Despite the complicated motion the center of mass of the wrench travels as the point–particle model predicts.

Rearranging this result, we have

$$Mv_{CM} = \sum_{i=1}^{N} m_i v_i \qquad (10.7)$$

The right side of Equation (10.7) is the total momentum of the system of N masses. The left side of this equation is the momentum of a point–particle with mass equal to the total system mass M and velocity equal to that of the center of mass. Thus, we may treat the momentum of a collection of masses as if these masses were a single point mass located at the center of mass and traveling with the velocity v_{CM}.

We continue by taking a time derivative of Equation (10.7) to obtain

$$Ma_{CM} = \sum_{i=1}^{N} m_i a_i$$

You probably recognize the terms $m_i a_i$ as the net force on each mass m_i. The summation is thus the vector sum of all of the forces on the particles in our system. As we noted before, when summed, the forces acting between the particles within the system itself all cancel in pairs in accordance with Newton's third law. Thus, the sum of all forces can be replaced by the sum of all the external forces. If $F_{i\ \text{external}}$ is the net external force on mass m_i, our last equation becomes

$$Ma_{CM} = \sum_{i=1}^{N} F_{i\ \text{external}} \qquad (10.8)$$

This result, of course, is Newton's second law. But this time it is telling us something new. This remarkable equation says that no matter how complicated the system of particles, there is one point that behaves just as our simple particle model predicts. That point is none other than the center of mass. If you take an extended object (a cat, for example) and gently toss it into the air, regardless of how it twists and turns in flight, so long as air resistance is negligible, its center of mass travels in a simple parabola.

You can change the location of your center of mass relative to your torso by bending your body. Athletes often take advantage of this ability. A high jumper's center of mass travels in a parabola once he leaves the ground. As shown in Figure 10.10, by bending over backward as he goes over the crossbar, the jumper can actually squirm over the bar while his center of gravity travels in a parabolic arc under the bar!

Ballet dancers are noted for graceful jumps in which they seem to hang in the air, momentarily defying gravity. In actual fact, the dancer's center of mass must, of course, follow a parabola. However, on the way up, the dancer raises both legs, thus raising her center of gravity. As she begins to descend, she quickly lowers her legs, and thus her center of gravity moves downward relative to her torso. Because she has moved her center of mass up and down by moving her legs only, her head moves in a nearly level path instead of a parabola. Because we normally watch the dancer's head, it appears in Figure 10.11 as if she has floated horizontally rather than following a parabolic trajectory.

Momentum and the Center of Mass

We have seen in the paragraphs above that the center of mass of a collection of objects moves as a point–particle with mass equal to the total system mass. Often, the motion of the system appears simpler if we put ourselves into a frame of reference moving with the center of mass. The analysis is particularly simple if the total external force on the system is zero so that the center of mass moves with a constant velocity. Below, we view several applications of momentum conservation from a frame of reference that coincides with the center of mass.

The simplest case one can imagine is that of a stationary center of mass. If no external forces act and the system is initially at rest, the center of mass remains fixed in space. The parts of the system may rearrange themselves, but the center of mass remains at the same location.

FIGURE 10.10

The jumper's body is never entirely above the bar. Correct form may allow the jumper's center of mass to pass under the bar even though the jumper passes his body over the bar one part at a time.

FIGURE 10.11

The dancer raises and lowers her center of mass relative to her torso by changing leg and arm position. This allows her head to travel in a nearly horizontal path while her center of mass travels in a parabola.

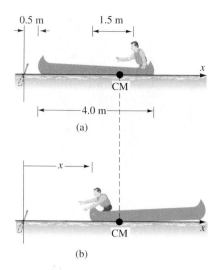

0.5 m   1.5 m

CM

4.0 m

(a)

x

CM

x

(b)

FIGURE 10.12

The position of the center of mass remains fixed unless there is a net external force to change the zero velocity of the center of mass. (See Example 10.7.)

EXAMPLE 10.7   Maybe I'll Hope for a Wind

A lone canoeist sitting in the stern of a canoe unknowingly drops the only paddle in the water. By the time the canoeist notices the problem, the paddle is 0.50 m in front of the bow. The canoeist walks to the bow to retrieve the paddle. After reaching the bow how far is the canoeist from the paddle? Assume the canoeist has a mass of 50.0 kg, the canoe a mass of 30.0 kg, that the canoe is 4.00 m long, and that the canoeist starts from the stern 1.50 m from the center of the canoe.

SOLUTION   We take a coordinate system fixed relative to the water with $x = 0.00$ cm at the position of the paddle as shown in Figure 10.12(a). We assume that the water exerts no horizontal force during the movement of the canoeist. Because no unbalanced external force acts, the center of mass is at the same location before and after the move. The position of the center of mass before the canoeist moves is calculated as usual

$$x_{CM} = \frac{m_A x_A + m_B x_B}{m_A + m_B} = \frac{(30.0 \text{ kg})(2.50 \text{ m}) + (50.0 \text{ kg})(4.00 \text{ m})}{30.0 \text{ kg} + 50.0 \text{ kg}} = 3.44 \text{ m}$$

After the canoeist moves, the situation is as shown in Figure 10.12(b); notice that we do not know the position x of the bow of the canoe. Nonetheless, the position of the center of mass can be calculated as

$$x_{CM} = \frac{m_A x'_A + m_B x'_B}{m_A + m_B} = \frac{(30.0 \text{ kg})(x + 2.00 \text{ m}) + (50.0 \text{ kg})(x)}{80.0 \text{ kg}} = x + 0.75 \text{ m}$$

Because the position of the center of mass is unchanged, we may equate its coordinate before and after the canoeist's movement

$$3.44 = x + 0.75$$

$$x = 2.69 \text{ m}$$

Thus, the maneuver is, in some sense, self-defeating: as the canoeist moves to the bow, the bow must move away from the paddle. Clearly the canoeist cannot avoid the need for an external force to move the center of mass.   ◄

When the center of mass of a system is not stationary, its velocity can be found from Equation (10.7), which was obtained from the time derivative of the equation for the position of the center of mass. This equation is restated here, rearranged slightly for convenience.

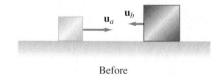

u_a u_b

Before

u'_a u'_b

After

FIGURE 10.13

In the center-of-mass frame of reference the total momentum is zero before and after the collision.

$$\mathbf{v}_{CM} = \frac{\sum_{i=1}^{N} m_i \mathbf{v}_i}{M} \tag{10.9}$$

Once the velocity of the center of mass is calculated, it is easy to "jump into" the moving center-of-mass frame of reference simply by *subtracting* $\mathbf{v}_{CM}$ *from the velocity of every particle in the system.* Once you are in the center-of-mass frame, you discover that the total momentum of the system is zero. (See Problem 59.)

As an example of the simplicity that results in the center-of-mass frame, let's look again at the one-dimensional collision of two objects. (Our convention is to denote velocity relative to the center of mass with $\mathbf{u}$ rather than $\mathbf{v}$.) The two objects now must approach each other as in Figure 10.13, so that one object's negative momentum cancels the other object's positive momentum and gives zero total momentum. Because the motion is one dimension we drop the vector notation and write

$$m_a u_a + m_b u_b = 0$$

or

$$\frac{u_a}{u_b} = -\frac{m_b}{m_a}$$

However, if no external forces act on the masses, the total momentum is zero after the collision as well. Thus, the same equation must hold true after the collision, so that

$$\frac{u_a'}{u_b'} = -\frac{m_b}{m_a}$$

In other words, the *ratio* of the two objects' velocities must be the same before and after the collision. If this result is true, the only thing that can happen because of the collision is that both objects' velocities are multiplied by the same factor. So you can see that solving any one-dimensional collision is trivial in the center-of-mass frame: you obtain the new velocities by multiplying the old velocities by ± this factor, which we call ϵ. This factor ϵ is the same *coefficient of restitution* introduced in Chapter 9. Thus, we write for example,

$$u_a' = -\epsilon u_a \tag{10.10}$$

$$u_b' = -\epsilon u_b \tag{10.11}$$

The negative signs in Equations (10.10) and (10.11) are chosen so that the objects rebound from one another. This choice is not required; positive signs result if the objects somehow pass through one another as a bullet might pass through a soft clay target.

Jumping into the center-of-mass frame, solving the collision problem using Equations (10.10) and (10.11), and then jumping back out of the center-of-mass frame (by *adding* v_{CM} to the velocity of each mass) provides an alternative method for solving one-dimensional collisions, which some students prefer to the algebraic methods of Chapter 9. We show how to apply this technique in Example 10.8.

EXAMPLE 10.8   *Knocking Around in the CM Frame*

A 4.00-kg mass traveling at 6.00 m/s along the x-axis collides with a stationary 2.00-kg mass in a head-on collision characterized by a coefficient of restitution $\epsilon = 0.800$. Find the final velocities of both masses.

SOLUTION   This situation takes place in what is usually referred to as the "lab" frame of reference and is shown in Figure 10.14(a). We must first find the velocity of the center of mass, using Equation (10.7):

$$v_{CM} = \frac{m_a v_a + m_b v_b}{m_a + m_b}$$

$$= \frac{(4.00 \text{ kg})(6.00 \text{ m/s}) + (2.00 \text{ kg})(0.00 \text{ m/s})}{6.00 \text{ kg}} = 4.00 \text{ m/s}$$

We now "jump into" the center-of-mass (CM) frame by subtracting v_{CM} from both lab velocities. We designate the velocities in the CM frame by u.

$$u_a = v_a - v_{CM} = 6.00 \text{ m/s} - 4.00 \text{ m/s} = +2.00 \text{ m/s}$$

$$u_b = v_b - v_{CM} = 0.00 \text{ m/s} - 4.00 \text{ m/s} = -4.00 \text{ m/s}$$

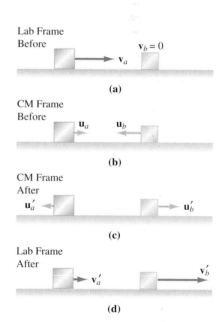

FIGURE 10.14

Four steps to solve collision problems using the center-of-mass frame of reference. (See Example 10.8.)

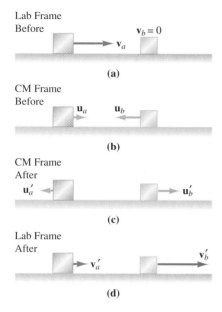

Lab Frame
Before

(a)

CM Frame
Before

(b)

CM Frame
After

(c)

Lab Frame
After

(d)

FIGURE 10.14

The impending collision as viewed in the CM frame is shown in Figure 10.14(b). Note that we can check our work so far by seeing if we are indeed in a zero momentum frame:

$$4.00 \text{ kg}(2.00 \text{ m/s}) + 2.00 \text{ kg}(-4.00 \text{ m/s}) = 0$$

We now "solve" the collision problem by using Equations (10.10) and (10.11).

$$u'_a = -\epsilon u_a = -(0.800)(+2.00 \text{ m/s}) = -1.60 \text{ m/s}$$
$$u'_b = -\epsilon u_b = -(0.800)(-4.00 \text{ m/s}) = +3.20 \text{ m/s}$$

The velocities of the two masses in the CM frame after the collision are shown in Figure 10.14(c). To complete the problem we have merely to "jump back" to the original laboratory frame of reference. To find the velocities relative to the CM frame we subtracted v_{CM} from each lab-frame velocity. To return to the lab frame we need only add v_{CM} back to each CM velocity:

$$v'_a = u'_a + v_{CM} = -1.60 \text{ m/s} + 4.00 \text{ m/s} = +2.40 \text{ m/s}$$
$$v'_b = u'_b + v_{CM} = +3.20 \text{ m/s} + 4.00 \text{ m/s} = +7.20 \text{ m/s}$$

as shown in Figure 10.14(d).

As a final check on our work we can verify that

Speed of separation $= \epsilon$ (*Speed of approach*)

$$v'_b - v'_a \stackrel{?}{=} \epsilon (v_a - v_b)$$

$$7.20 \text{ m/s} - 2.40 \text{ m/s} \stackrel{?}{=} 0.800(6.00 \text{ m/s} - 0.00 \text{ m/s})$$

$$4.80 \text{ m/s} = 4.80 \text{ m/s} \quad \text{(It works!)} \quad \blacktriangleleft$$

Kinetic Energy and the Center of Mass

By now you have noticed that collision analysis is simplified when velocities are referred to the center of mass. We can also gain new insights into the kinetic energy of a system of particles by viewing the particles from a coordinate system moving with their center of mass. If we have N particles, their total kinetic energy as measured relative to the lab frame of reference is

$$K = \sum_{i=1}^{N} \frac{1}{2} m_i \mathbf{v}_i \cdot \mathbf{v}_i$$

In the expression above for kinetic energy we have calculated v_i^2 by taking the dot product of $\mathbf{v}_i$ with itself.

Recall that the relation between the lab-frame velocities v and the CM-frame velocities u is

$$\mathbf{v} = \mathbf{v}_{CM} + \mathbf{u} \tag{10.12}$$

Substituting this relation into the expression for kinetic energy, we find that

$$K = \sum_{i=1}^{N} \frac{1}{2} m_i (\mathbf{v}_{CM} + \mathbf{u}_i) \cdot (\mathbf{v}_{CM} + \mathbf{u}_i)$$

Performing the indicated dot product in the above summation, we obtain three terms, each of which can be summed separately

$$K = \sum_{i=1}^{N} \frac{1}{2} m_i v_{CM}^2 + \sum_{i=1}^{N} m_i(\mathbf{v}_{CM} \cdot \mathbf{u}_i) + \sum_{i=1}^{N} \frac{1}{2} m_i u_i^2 \qquad (10.13)$$

In the first summation $\frac{1}{2}v_{CM}^2$ can be factored out of the sum, leaving $\sum_{i=1}^{N} m_i$, which is just the total system mass M. Thus, the first sum is $\frac{1}{2}Mv_{CM}^2$. The third sum in Equation (10.13) cannot be further simplified. It represents the total kinetic energy of the particles *as seen in the CM frame*.

The middle summation of Equation (10.13) can be rearranged by factoring out the constant vector $\mathbf{v}_{CM}$ and performing the dot product after the summation

$$\mathbf{v}_{CM} \cdot \sum_{i=1}^{N} m_i \mathbf{u}_i$$

Take a careful look at this last summation. Note that $\sum_{i=1}^{N} m_i \mathbf{u}_i$ is just the total momentum in the CM frame. But we have already seen that this is zero! Consequently, the middle term is zero, and Equation (10.13) can be written

$$K = \frac{1}{2} M v_{CM}^2 + \sum_{i=1}^{N} \frac{1}{2} m_i u_i^2 \qquad (10.14)$$

This equation tells us something extremely important and useful: The kinetic energy of a system is the sum of two parts. The first part may be obtained by treating the whole system as a single particle of mass M moving with a velocity v_{CM}. The second part of the kinetic energy is the energy as seen in the center-of-mass frame. We can say more succinctly that

> The kinetic energy of a system of particles is the sum of the kinetic energy *of* the center of mass plus the kinetic energy *relative* to the center of mass.

When particles collide, conservation of momentum dictates that only the kinetic energy described by the second term of Equation (10.14) can be altered. When particle physicists accelerate subatomic particles to enormous speeds and slam them together, only the center-of-mass kinetic energy is available to create new particles. The kinetic energy of the center of mass is useless because that energy cannot change. This restriction is the reason that physicists design "supercolliders" within which colliding particles are accelerated in opposite directions. In these designs the lab frame is made to be a center-of-mass frame, hence all the energy spent accelerating the particles is available to make the reaction "go."

10.5 Systems of Variable Mass (Optional)

In all of our discussions of momentum conservation to this point we have considered only *closed systems*. By a closed system we mean that the mass of our system stays constant. We draw a mental boundary around the objects of interest and let nothing pass through it; no mass enters or leaves the system during the process we are analyzing. If there is no net external force acting on this system, the total momentum of the system enclosed by the boundary is conserved. That is, the center of mass of the system has constant velocity.

Concept Question 8
A system is composed of two 1.0-kg masses. These masses move toward each other with a speed of 2.0 m/s. (a) What is the total momentum of the system? (b) What is the total kinetic energy associated with the motion of the center of mass of the system? (c) What is the kinetic energy of the masses relative to the center of mass of the system? (d) Answer questions (a)–(c) for the case where the two masses move in the same direction.

The velocity of the center of mass of a system of particles is calculated from Equation (10.9), which we repeat here for convenience

$$\mathbf{v}_{CM} = \frac{\sum\limits_{i=1}^{N} m_i \mathbf{v}_i}{M}$$

When an unbalanced external force acts on the system, the velocity of the center of mass changes in accordance with Newton's second law expressed as in Equation (10.8):

$$\sum_{i=1}^{N} \mathbf{F}_{i \; \text{external}} = M\mathbf{a}_{CM}$$

If mass enters or leaves the system under consideration, the velocity of the center of mass as calculated in Equation (10.9) must change.[1] Thus, similar to the presence of a net external force, a change in the mass of the system also produces a change in the velocity of the center of mass. Therefore, it is not surprising that when we write the equation of motion for a system that is gaining or losing mass, there appears to be an additional force acting on the object. This additional force is a direct effect of the change of system mass.

Probably the most famous example of a system with a changing mass is rocket propulsion. A rocket propels itself forward by expelling exhaust gases backward at a high rate. We find that the forward recoil of the rocket due to expulsion of mass to the rear can be described as an additional forward force on the rocket. In the case of a rocket this forward force is called *thrust*.

Let's start out thinking about the rocket and its fuel-oxidizer mixture as a closed system. For the sake of argument, we begin with the rocket in interstellar space at rest with respect to the background of the stars, which we take as our reference frame. Now suppose the rocket turns on its engines and starts expelling exhaust gases at 100 m/s *relative to the rocket*. The first bit of exhaust travels backward through the vacuum of space at 100 m/s. The rocket recoils and gains speed in the forward direction. A little later, when the rocket is moving forward at 40 m/s, the exhaust it spews out is traveling backward 100 m/s relative to the rocket but at only 60 m/s *relative to the stars*. At the moment the rocket itself is moving forward at 100 m/s the exhaust it expels is actually stationary in space! Even more thought-provoking is the fact that when the rocket is traveling forward 175 m/s, its exhaust is going *forward* at 75 m/s.

If we now look at the rocket and all the exhaust it has ever put out, we see a strange sight. The exhaust gases are strewn throughout space, some going backward, some nearly stationary, and still other portions chasing the rocket. As long as we keep track of all the exhaust gas and the rocket with its remaining fuel, our system is closed. In fact the total momentum of the whole mess is zero for the case we described. But who wants to try to keep track of all that exhaust gas? All we're interested in is the rocket. Thus, in the following paragraphs we derive an expression for the acceleration of the rocket, paying special attention to the mass of the exhaust gas, which we allow to leave the system.

In rocket problems it is the speed of the exhaust *relative to the rocket* that is generally known. The use of relative velocities in the conservation of momentum takes some special care. Before proceeding you should quickly reread *Relative Motion* in Section 9.4, especially Example 9.9.

The rocket is propelled forward by the recoil from the exhaust gas, which it expels to the rear. We analyze this recoil using notation similar to the notation in Example 9.9. That is, each velocity has two subscripts. The first subscript indicates the object being described, and the second subscript indicates the frame of reference in which the object's

The mass of the rocket is not constant. The effect of the exhaust mass leaving the rocket can be treated as an additional external force on the rocket. This force is called *thrust*.

---

[1] There is one exception to this statement. When mass leaves the system with a velocity precisely equal to v_{CM} and is not interacting with the remaining portion of the system through any forces, the velocity of the center of mass does not change.

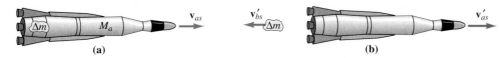

FIGURE 10.15

The rocket expels mass Δm and recoils in the opposite direction.

velocity is measured. For example, in Figure 10.15(a) we show mass $M_a + \Delta m$ traveling at velocity $\mathbf{v}_{as}$ relative to the background of the stars. After the mass Δm is expelled backward with velocity $\mathbf{v}'_{bs}$ relative to the stars, mass M_a moves forward at $\mathbf{v}'_{as}$ as shown in Figure 10.15(b).

The momentum of the system comprising $M_a + \Delta m$ in Figure 10.15(a) is

$$\mathbf{p}_{\text{tot}} = (M_a + \Delta m)\mathbf{v}_{as}$$

After the exhaust is expelled the momentum of the closed system as pictured in Figure 10.15(b) is

$$\mathbf{p}'_{\text{tot}} = M_a\mathbf{v}'_{as} + \Delta m\,\mathbf{v}'_{bs}$$

The quantity that is usually known in these circumstances is not $\mathbf{v}'_{bs}$, the velocity of the exhaust relative to the stars, but rather $\mathbf{v}'_{ba}$, the velocity of the exhaust relative to the rocket. Using our relative velocity subscript rule from Chapter 9, we can write $\mathbf{v}'_{bs} = \mathbf{v}'_{ba} + \mathbf{v}'_{as}$. Thus, the momentum after the expulsion of the exhaust can be written

$$\mathbf{p}'_{\text{tot}} = M_a\mathbf{v}'_{as} + \Delta m(\mathbf{v}'_{ba} + \mathbf{v}'_{as})$$

or

$$\mathbf{p}'_{\text{tot}} = (M_a + \Delta m)\mathbf{v}'_{as} + \Delta m\,\mathbf{v}'_{ba}$$

Now we can calculate the change in momentum of the closed system

$$\Delta\mathbf{p}_{\text{tot}} = \mathbf{p}'_{\text{tot}} - \mathbf{p}_{\text{tot}}$$
$$= (M_a + \Delta m)(\mathbf{v}'_{as} - \mathbf{v}_{as}) + \Delta m\,\mathbf{v}'_{ba}$$
$$= (M_a + \Delta m)(\Delta\mathbf{v}_{as}) + \Delta m\,\mathbf{v}'_{ba}$$

Next, we divide the equation above by Δt, the time interval during which Δm was expelled.

$$\frac{\Delta\mathbf{p}_{\text{tot}}}{\Delta t} = \frac{(M_a + \Delta m)(\Delta\mathbf{v}_{as})}{\Delta t} + \frac{\Delta m}{\Delta t}\,\mathbf{v}'_{ba}$$

In the limit as Δt approaches zero, Δm also approaches zero. However, three of the fractions approach finite limits:

$$\lim_{\Delta t \to 0}\frac{\Delta\mathbf{p}_{\text{tot}}}{\Delta t} = \frac{d\mathbf{p}_{\text{tot}}}{dt}$$

$$\lim_{\Delta t \to 0}\frac{\Delta\mathbf{v}_{as}}{\Delta t} = \mathbf{a}_{as}$$

$$\lim_{\Delta t \to 0}\frac{\Delta m}{\Delta t} = +\frac{dm}{dt} = -\frac{dM_a}{dt}$$

To establish the third result we have observed that, as the exhaust part of the system (small m) increases in mass, the rocket (large M) decreases in mass by an equal amount. Thus, dM_a/dt is itself a negative number, so we must set the negative of it equal to dm/dt.

With these substitutions we have

$$\frac{d\mathbf{p}_{\text{tot}}}{dt} = M_a \mathbf{a}_{as} - \left(\frac{dM_a}{dt}\right)\mathbf{v}'_{ba}$$

According to Newton's second law, the total momentum of a closed system can be changed only by *external* forces:

$$\sum_{i=1}^{N} \mathbf{F}_{i \text{ external}} = \frac{d\mathbf{p}_{\text{tot}}}{dt}$$

Making use of this statement of Newton's second law, we can write for our closed system

$$\sum_{i=1}^{N} \mathbf{F}'_{i \text{ external}} = M_a \mathbf{a}_{as} - \left(\frac{dM_a}{dt}\right)\mathbf{v}'_{ba} \tag{10.15}$$

As you can see this isn't just $F_{\text{net}} = ma$! Actually, we shouldn't have expected to obtain the simple form of Newton's second law. When you write $F_{\text{net}} = ma$ the "m" represents the total mass. In Equation (10.15) the mass M_a is just part of the total system mass. We may recover something that *looks* like our old friend $F_{\text{net}} = ma$ by the simple expedient of adding $(dM_a/dt)\mathbf{v}'_{ba}$ to both sides of Equation (10.15):

$$\sum_{i=1}^{N} \mathbf{F}_{i \text{ external}} + \left(\frac{dM_a}{dt}\right)\mathbf{v}'_{ba} = M_a \mathbf{a}_{as} \tag{10.16}$$

In this form we see that if we wish to focus our attention on the rocket only, we can treat the effect of the departing exhaust mass as if it were an external force. When treated in this manner the term $(dM_a/dt)\mathbf{v}'_{ba}$ is called the **thrust** of the rocket engine. The velocity $\mathbf{v}'_{ba}$ is just the velocity of the exhaust relative to the rocket and is often written simply $\mathbf{v}_{\text{rel}}$.

When no external forces act on the rocket and the exhaust velocity is constant, Equation (10.16) can be integrated to show the relation of final rocket speed to the ratio of the fuel mass to payload mass. We let the constant velocity $\mathbf{v}'_{ba} = \mathbf{v}_{\text{rel}}$ be directed in the negative-x direction and drop subscripts, which are no longer required for clarity. We can then write Equation (10.16) for motion along the x-axis as

$$-v_{\text{rel}}\frac{dM}{dt} = M\frac{dv}{dt}$$

Separating variables, we have

$$dv = -v_{\text{rel}}\frac{dM}{M}$$

This expression can be integrated directly from some initial velocity v_o when the mass of the rocket is $M = M_o$ to a point where the rocket is traveling at speed v_f with mass M_f.

$$\int_{v_o}^{v_f} dv = -v_{\text{rel}} \int_{M_o}^{M_f} \frac{dM}{M}$$

$$v_f - v_o = +v_{\text{rel}} \ln\left(\frac{M_o}{M_f}\right) \tag{10.17}$$

This equation tells us that the increase in rocket speed is directly proportional to the velocity of the exhaust gases. It comes as no surprise that the more fuel the rocket can carry, the higher final speed it can achieve. A look at Figure 10.16, however, shows that the benefits of increased fuel decrease rapidly after M_o/M_f exceeds 3 or 4. This is an indication of the need for multistage rockets. (See Problems 38 and 39.)

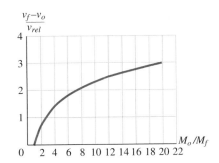

FIGURE 10.16

Doubling M_o/M_f from 2 to 4 increases the final speed by a factor of 2. Doubling M_o/M_f from 10 to 20 increases the final speed by only 30%. Multistage rockets give higher final speeds for the same fuel expenditure.

EXAMPLE 10.9   A Problem for Trekkies

The third stage of an intergalactic probe is traveling in interstellar space at 1.30×10^3 m/s. Its "dry weight" (i.e., mass without fuel) is 1.25×10^3 kg, and it carries 2.00×10^3 kg of fuel. The rocket engine can exhaust fuel at 120. kg/s with an exhaust velocity of 3.00×10^3 m/s. (a) What is the initial acceleration of the rocket? (b) Is that acceleration constant? (c) What is the final speed of the rocket after the fuel is spent?

SOLUTION   (a) The magnitude of the thrust T of the rocket is

$$T = \frac{dM}{dt} v_{\text{rel}} = (120. \text{ kg/s})(3.00 \times 10^3 \text{ m/s}) = 3.60 \times 10^5 \text{ N}$$

The initial acceleration of the rocket plus fuel is the force divided by the total mass:

$$a = \frac{T}{M + m} = \frac{3.60 \times 10^5 \text{ N}}{3.25 \times 10^3 \text{ kg}} = 111 \text{ m/s}^2$$

(b) As the fuel burns, the thrust remains constant, but it acts on a smaller and smaller mass. Therefore, the acceleration is *not* constant.

(c) Because the acceleration is not constant, we cannot use the kinematic equations for constant acceleration from Chapter 3. Instead we use Equation (10.17)

$$v_f - v_o = +v_{\text{rel}} \ln\left(\frac{M_o}{M_f}\right) = (3.00 \times 10^3 \text{ m/s}) \ln\left(\frac{3.25}{1.25}\right)$$

$$= (3.00 \times 10^3 \text{ m/s})(0.956) = 2.87 \times 10^3 \text{ m/s}$$

Thus, the final speed of the third stage is

$$v_f = v_o + 2.87 \times 10^3 \text{ m/s} = 4.17 \text{ km/s}$$ ◀

EXAMPLE 10.10   A Problem for Mechanical Engineers

Conveyer belts provide one of the most useful means for moving loose materials, such as mineral ore, from one place to another. In addition, they provide another example of a system with constantly varying mass. (a) Calculate an expression for the force required to keep a conveyer belt moving at constant speed when it is continually being loaded with additional ore mass as shown in Figure 10.17. (b) Compare the power that must be delivered to the belt with the rate at which ore is gaining kinetic energy.

SOLUTION   (a) We start with our expression Equation (10.16)

$$\sum_{i=1}^{N} \mathbf{F}_{i \text{ external}} + \left(\frac{dM_a}{dt}\right)\mathbf{v}'_{ba} = M_a\mathbf{a}_{as}$$

which acts as the equation of motion for our belt-plus-moving-ore system with changing mass M_a. For simplicity, we call mass M_a simply M. The acceleration $\mathbf{a}_{as}$ is the acceleration of the system we are analyzing, namely the belt with its ore load. Be-

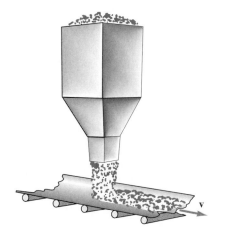

FIGURE 10.17

Mineral ore is deposited on a continuously moving conveyer belt.

cause we desire the belt to move at constant velocity, $a_{as} = 0$. In our original derivation of Equation (10.16), $\mathbf{v}'_{ba}$ was the velocity of the exhaust relative to the rocket. Here it is the velocity of the ore *relative to the belt,* which we write as $-v\hat{\mathbf{i}}$. The term (dM_a/dt) is the rate at which the belt is changing mass, a positive quantity in this case, which we denote by $\dot{M}$. This symbol is read as "M dot," the dot representing differentiation with respect to time. Finally, if we ignore the friction between the belt and its rollers, the only horizontal external force is the applied force for which we are solving. With these substitutions our equation becomes

$$\mathbf{F} + \dot{M}(-v\hat{\mathbf{i}}) = 0$$

or simply

$$\mathbf{F} = \dot{M}v\hat{\mathbf{i}}$$

(b) The rate at which $\mathbf{F}$ does work to keep the belt moving is the power $\mathcal{P}$

$$\mathcal{P} = \mathbf{F} \cdot \mathbf{v} = \dot{M}v^2$$

It is interesting to compare this to the rate at which ore is gaining kinetic energy

$$\frac{dK}{dt} = \frac{d}{dt}\left(\frac{1}{2}mv^2\right) = \frac{1}{2}\dot{M}v^2$$

We notice that only half the power supplied by the motor to keep the belt moving is going to the kinetic energy of the ore. So where does the rest go? To answer this question, you have only to think about the mechanism by which the ore acquires its horizontal velocity after hitting the belt. The answer is that friction ultimately drags the ore up to speed. While the ore is slipping across the surface of the belt, mechanical energy is dissipated. What is perhaps surprising is that the result is completely independent of the magnitude of the frictional force. This matter is explored further in Problem 56.   ◀

10.6 Summary

The **sudden approximation** allows many problems (e.g., the ballistic pendulum) to be subdivided into smaller problems within which mechanical energy or momentum is conserved, even though neither is conserved in the overall problem.

The position, velocity, and acceleration of the **center of mass** of a system of particles can be calculated from

$$\mathbf{r}_{CM} = \frac{\displaystyle\sum_{i=1}^{N} m_i \mathbf{r}_i}{M} \tag{10.6}$$

$$\mathbf{v}_{CM} = \frac{\displaystyle\sum_{i=1}^{N} m_i \mathbf{v}_i}{M}$$

$$\mathbf{a}_{CM} = \frac{\displaystyle\sum_{i=1}^{N} m_i \mathbf{a}_i}{M}$$

For a continuous distribution of masses these equations become integrals, such as

$$\mathbf{r}_{CM} = \frac{\int \mathbf{r} \, dm}{M}$$

where M is the total mass of the system.

The **gravitational potential energy of a system of particles** near the earth's surface can be calculated as if the entire system mass were located at the center of mass.

The **center of mass of a system of particles moves as if it were a point–particle** of mass M acted upon by $\Sigma_{i=1}^{N} \mathbf{F}_{i \text{ external}}$. That is,

$$\sum_{i=1}^{N} \mathbf{F}_{i \text{ external}} = M\mathbf{a}_{CM} \qquad (10.8)$$

The **kinetic energy of a system of particles** is the sum of the kinetic energy of total system mass M moving at v_{CM} plus that of each particle's kinetic energy computed relative to the center of mass.

$$K = \frac{1}{2} M v_{CM}^2 + \sum_{i=1}^{N} \frac{1}{2} m_i u_i^2 \qquad (10.14)$$

(Optional) Newton's second law may be written for an open system with changing mass M_a in the form

$$\sum_{i=1}^{N} \mathbf{F}_{i \text{ external}} + \frac{dM_a}{dt} \mathbf{v}_{\text{rel}} = M_a \mathbf{a}_{as} \qquad (10.16)$$

The final velocity of a rocket with initial mass M_o and final mass M_f and constant exhaust velocity v_{rel} is

$$v_f - v_o = +v_{\text{rel}} \ln\left(\frac{M_o}{M_f}\right) \qquad (10.17)$$

PROBLEMS

10.1 The Big Picture

1. (a) If two objects with different masses have the same momentum, which has the larger kinetic energy, the greater or lesser mass? (b) If two objects have the same kinetic energy, which has the greater momentum, the greater mass or the lesser mass? (c) True or false: If the total momentum of an isolated system of particles is conserved, the total kinetic energy must also be conserved. (d) True or false: If the total kinetic energy of an isolated system of particles is conserved, the momentum must also be conserved.

2. Which, if either, of the following is true near the surface of the earth when air resistance is ignored? (a) All masses falling from the same height have the same kinetic energy. (b) All masses falling from the same height have the same momentum. Explain your answers in terms of the work–energy theorem and the impulse–momentum theorem.

3. Consider a 1.00-kg mass and a 2.00-kg mass. Compare their momentum and their kinetic energy if they start from rest and are each pushed by an unbalanced 1.00-N force (a) for the same time, say, $t = 5.00$ s and (b) for the same distance, say, 2.00 m.

4. Two stationary blocks of masses 0.500 kg and 1.500 kg are each pushed along a frictionless horizontal surface by a net horizontal force of 2.00 N for 2.50 s. They then encounter a frictionless 30° incline. (a) How far does each block travel upward along the incline? (b) Reconcile this result with the work–energy theorem by calculating the distance each travels while being pushed on by the 2.00-N force.

5. A 0.500-kg block and a 1.50-kg block originally at rest are each pushed over a horizontal frictionless surface by a net horizontal force of 2.00 N for a distance of 1.50 m. Subsequently each block collides with a different stationary 50.0-kg mass in a perfectly elastic collision. (a) Which block imparts the greater velocity to the 50.0-kg mass? (b) Reconcile this result with the impulse–

momentum theorem by calculating the length of time the 2.00-N force acts on each mass.

6. A 0.500-kg block and a 1.500-kg block initially at rest are both pushed in opposite directions along a frictionless horizontal surface by a 2.00-N force for 3.50 s. (a) Subsequently, the blocks collide with each other in a perfectly *in*elastic collision. What is the velocity of the pair after collision? (b) Suppose the two blocks starting from rest are pushed in opposite directions by the 2.00-N force for the same 0.500-m distance before they collide. What is their final velocity after they collide inelastically?

10.2 The Ballistic Pendulum and Its Friends

7. A 5.00-kg mass sliding along a horizontal frictionless surface at 2.00 m/s collides with a 1.00-kg stationary mass. As shown in Figure 10.P1, after the collision the two masses stick together and contact a light Hooke's-law spring of stiffness constant $k = 100.$ N/m. What is the maximum compression of the spring?

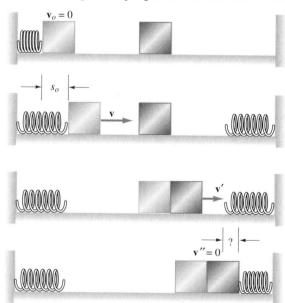

FIGURE 10.P1   Problem 7

8. Two gliders each of mass 0.500 kg are free to slide without friction along a laboratory airtrack. At one end of the track the first glider is held against a spring of stiffness constant $k = 5.00$ N/m,

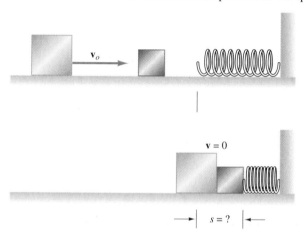

FIGURE 10.P2   Problem 8

compressing it by 0.200 m. The first glider is released and acquires a velocity from the compressed spring. This glider then strikes the second glider and sticks to it. The combined gliders then strike a second identical spring on the other end of the track. (See Fig. 10.P2.) How far does the second spring compress before the gliders come (momentarily) to rest?

9. A 0.500-kg mass slides without friction along a horizontal surface at a rate of $v_o = 2.50$ m/s. This mass collides with a stationary mass $M = 2.00$ kg attached to the end of a light, unstretched spring with stiffness constant $k = 200$ N/m. Following the collision, the lighter mass rebounds straight back with a speed of $v_f = 0.500$ m/s. (See Fig. 10.P3.) What is the maximum compression of the spring?

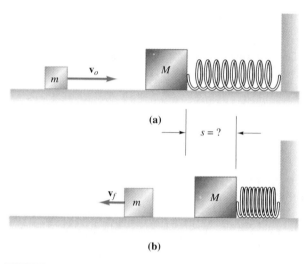

FIGURE 10.P3   Problem 9

10. A 2.00-kg mass is sliding along a tabletop. Just as its speed is 10.0 m/s, it collides with a stationary 8.00-kg block. Immediately after the collision the 2.00-kg block rebounds with a speed of 1.50 m/s in the opposite direction. If the kinetic coefficient of friction between the tabletop and the 8.00-kg mass is $\mu_k = 0.450$, how far does the 8.00-kg mass slide before coming to rest?

11. A sphere of mass $m_a = 100.$ g is attached to a rigid support with a light string. As shown in Figure 10.P4(a), the sphere is held out to the side so that the string is horizontal and taut and then released from rest. When at its lowest point, the first sphere collides with a similarly suspended second sphere of mass $m_b = 250.$ g. If the collision is characterized by a coefficient of restitution $\epsilon = 0.900$, what maximum angle θ with the vertical does the string holding the 100.-g mass make when it comes momentarily to rest?

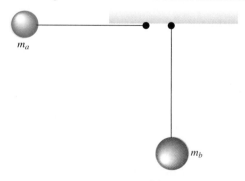

(a)

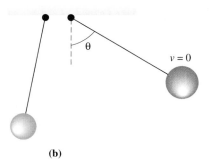

FIGURE 10.P4   Problem 11

12. Two gliders each of mass 0.500 kg are free to slide without friction along a horizontal airtrack. At one end of the track the first glider is held against a Hooke's-law spring, compressing it 0.200 m. The first glider is released and acquires a velocity from the compressed spring. This first glider then collides with the stationary second glider in a collision characterized by a coefficient of restitution $\epsilon = 0.650$. The second glider subsequently strikes the second spring at the far end of the track. How far is the second spring compressed before the second glider comes (momentarily) to rest? The springs both have stiffness constant $k = 5.00$ N/m.

10.3 Center of Mass

13. If three point–particles of 1.50 kg, 2.50 kg, and 2.00 kg are located in a line at $x = -5.00$ cm, -1.00 cm, and 3.00 cm, respectively, where is the center of mass of this system of particles?

14. (a) Treat the earth and the moon as point–particles, and find the location of the center of mass of the pair. (b) Compare this answer with the radius of the earth. (Data for masses and distances can be found in Appendix 8.)

15. Calculate the center of mass of four point–particles with the following masses and positions:

MASS (kg)	(x, y) (cm)
0.500	$(-10.0, 50.0)$
0.400	$(0.00, -10.0)$
0.300	$(5.00, 0.00)$
0.200	$(10.0, -15.0)$

16. Suppose the uniform circular tray in Example 10.4 has a mass of 0.50 kg. Where is the new center of mass of the tray, pitcher, and glasses?

17. Calculate the location of the center of mass of each of the three objects shown in Figure 10.P5.

18. Calculate the center of mass of a rectangular sheet of plywood with a circle cut out of it as shown in Figure 10.P6. [*Hint:* Treat the problem as made up of a whole rectangular sheet of plywood and a circular piece of plywood with a negative mass!]

19. Calculate (a) the total mass and (b) the center of mass of a thin rod of length L if, when the rod lies along the x-axis with one end at $x = 0$, the linear mass density of the rod is given by $\lambda = \lambda_o[2 - (x/L)^2]$, where λ_o is a constant.

20. Calculate (a) the total mass and (b) the center of mass of an infinitely long rod with linear mass density $\lambda = \lambda_o e^{-x/L}$, where λ_o and L are constants. (c) We know there are no infinitely long rods in your physics laboratory. How long must a real rod with

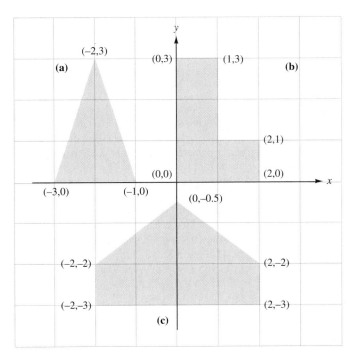

FIGURE 10.P5   Problem 17

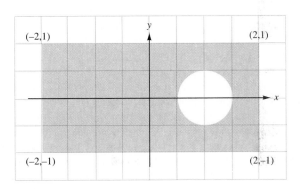

FIGURE 10.P6   Problem 18. The radius of the circle of 0.50 units.

this density function be for the error to be less than 0.1% between the mass of real truncated rod and that of the infinitely long model? Give your answer as a multiple of L, a distance that scales the problem answer.

21. Find the center of mass of a semicircle of radius R as shown in Figure 10.P7.

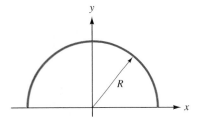

FIGURE 10.P7   Problem 21

22. Find the center of mass of an area bounded by $y^2 = 4x$, and $2x - y = 4$.

10.4 Center of Mass: Energy, Momentum, and Newton's Second Law

23. Solve Problem 9.25 by jumping into the center-of-mass frame of reference and using the method of Example 10.8.

24. A loaded boxcar of 2.50×10^4 kg mass is rolling freely down a level railroad track at a speed of 0.600 m/s. The boxcar collides with a stationary boxcar and couples to it. The two cars move off together at 0.400 m/s. Use the definition of the velocity of the center of mass to find the mass of the second boxcar.

25. A proton with a velocity of 3.00×10^6 m/s makes a perfectly elastic head-on collision with a stationary alpha particle. Using the method of Example 10.8, find the velocities of each particle after the collision. (The proton and alpha particle are particles encountered often in subatomic physics. The mass of an alpha particle is $m_\alpha = 6.64 \times 10^{-27}$ kg, and the proton mass can be found in Appendix 7.)

26. Solve Problem 25 with the roles of the proton and the alpha particle reversed.

27. A 2.50-kg mass traveling 12.0 m/s collides head on with a stationary 3.50-kg mass. (a) If the collision is characterized by a coefficient of restitution $\epsilon = 0.750$, calculate the final velocities of the masses after the collision using the method shown in Example 10.8. (b) Calculate the total kinetic energy before the collision as measured in the laboratory frame. (c) Calculate the total kinetic energy before the collision as measured in the center-of-mass frame. (d) Calculate the kinetic energy of a fictitious particle traveling with the velocity of the center of mass. (e) Relate the three kinetic energies calculated in parts (b) through (d).

28. Solve Problem 9.30 by jumping into the center-of-mass frame of reference and using the method of Example 10.8.

29. A 4.00-kg mass is traveling to the right with a velocity of 12.0 m/s. This mass makes an in-line collision with a 2.00-kg mass traveling at 3.00 m/s also to the right. (a) If the coefficient of restitution between the two blocks is $\epsilon = 0.250$, calculate the final velocities of the two masses using the method illustrated in Example 10.8. (b) What is the speed of approach of the two blocks? What is their speed of separation? Relate these two speeds to the coefficient of restitution.

30. A 6.00-kg mass traveling 3.00 m/s toward the right collides head on with a 4.00-kg mass traveling 2.00 m/s toward the left. The coefficient of restitution for the collision is 0.800. (a) Use the method of Example 10.8 to determine the final velocity of each mass. (b) What is the speed of approach of the two masses? What is their speed of separation? Relate these two speeds to the coefficient of restitution.

31. Solve Problem 9.32 by jumping into the center-of-mass frame of reference and using the method of Example 10.8.

32. A 5.00-kg mass and a 10.00-kg mass travel toward each other each with a speed of 15.0 m/s. (a) For what value of the coefficient of restitution between 0.00 and 1.00 does the 10-kg mass have the greatest final speed? (b) The least final speed?

33. (a) Using the conservation of momentum, show that if all the particles in a closed system collide and stick together that the single resulting superparticle has velocity v_{CM}. What is the kinetic energy of the resulting superparticle? (b) In the text it is shown that the total kinetic energy of a system of particles can be written

$$K = \frac{1}{2} M v_{CM}^2 + \frac{1}{2} \sum_{i=1}^{N} m_i u_i^2$$

where the u_i are the velocities of the particles in the system as measured relative to the center of mass. What do you conclude about the maximum reduction in kinetic energy that can occur in an inelastic collision?

34. Three particles with masses 1.00 kg, 2.00 kg, and 3.00 kg are moving with velocities of 2.50 m/s $\hat{\mathbf{i}}$, 1.50 m/s $\hat{\mathbf{j}}$, and 0.50 m/s $\hat{\mathbf{i}}$ + 0.75 m/s $\hat{\mathbf{j}}$, respectively. (a) Calculate the total kinetic energy of the system. (b) Calculate the velocity (vector) of the center of mass. (c) Calculate the velocity of each particle as seen in the center-of-mass frame of reference by subtracting the center-of-mass velocity from each of the three particles' velocities. (c) Calculate the kinetic energy as seen in the center-of-mass frame. (d) Calculate the kinetic energy of a fictitious point–particle the mass of which is equal to the total mass of the three particles if this fictitious mass travels with velocity v_{CM}. (e) Compare the answer to (a) with the sum of the energies you calculated in parts (c) and (d).

10.5 Systems of Variable Mass

35. A toy rocket uses compressed air to squirt water out of a nozzle. The plastic rocket has a "dry weight" of 75.0 g and carries 250. g of water. Assume the water is expelled at a uniform velocity of 7.00 m/s during a 0.500-s time interval. (a) What is the thrust on the rocket? How does this thrust compare with the rocket's weight? (b) What is the initial acceleration of the rocket if it starts at the surface of the earth? (c) If the rocket were used in outer space, starting from rest, what final velocity would it reach?

36. When watching the space shuttle lift off the launch pad (Fig. 10.P8), we notice that during the first few seconds the rocket appears to slowly rise at a nearly constant velocity. Knowing that the initial mass of the rocket at lift-off is approximately 3.0×10^6 kg and that the exhaust velocity is about 2.5 km/s, estimate the rate at which fuel and oxidizer are being burned.

FIGURE 10.P8   Problem 36

37. In a now infamous editorial, the *New York Times* on January 13, 1920 sarcastically criticized Robert Goddard when he proposed that a rocket could be used to travel to the moon. In essence, the editorial said that anyone with "common sense" knows that the rocket exhaust has nothing to push against in the vacuum of space, hence it cannot propel a rocket as it does near the surface of the earth where there is air to push against. The belief was based on the old "you can't pick yourself up by your bootstraps" logic. (a) Detail exactly what is wrong with this commonsense argument by explaining what the rocket is pushing against. (b) Like all good physics arguments, this one was settled by an experiment. The January 1975 issue of *Scientific American* magazine describes how Goddard tried to persuade his critics by firing a revolver loaded with a blank cartridge in a vacuum chamber. What do you expect happened? How is this relevant to the rocket argument?

38. Examine the advantages of a two-stage rocket by analyzing the following models. Assume you have a 250-kg payload, 5500 kg of fuel–oxidizer, and 500-kg overhead of rocket engine and structural materials. Furthermore, take the speed of the exhaust gases relative to the rocket to be 1000 m/s and assume the rocket starts from rest from a point in space where gravity can be ignored. (a) Use the rocket equation to predict the final speed of the payload (plus overhead) if a single-stage rocket is used, that is, take $M_o = 6250$ kg and $M_f = 750$ kg. (b) Now split the fuel–oxidizer and its engine overhead, into two equally massive stages. Use the first stage ($M_o = 6250$ kg, $M_f = 3500$ kg) to boost the second stage to a velocity v_1. Then jettison the first-stage rocket (250 kg) and burn the fuel in the second stage ($M_o = 3250$ kg, $M_f = 500$ kg) to add additional velocity. Compare the resulting final velocity to that of part (a). (See Problem 42 for an extension of this problem.)

39. Repeat the analysis of Problem 38, except divide combined fuel–oxidizer and overhead masses into three stages with one-half of their total in the first stage, and one-fourth in each of the following two stages. The flight plan is summarized in Table 10.P1.

TABLE 10.P1

		PAYLOAD (kg)	FUEL (kg)	OVERHEAD (kg)
Stage 1	start	250	5500	500
	end	250	2750	500
Stage 2	start	250	2750	250
	end	250	1375	250
Stage 3	start	250	1375	125
	end	250	0	125

40. A conveyer belt is used to move 750. kg of ore per minute horizontally to a concentrator. (a) Ignoring friction between the belt and its rollers, what power is required to drive the belt at a constant speed of 0.500 m/s? (b) How much kinetic energy is added to the ore mass each second? (c) Why are the answers to (a) and (b) different?

Numerical Methods and Computer Applications

41. (a) If you have not already worked Problem 9.49, work it now. (b) Compare the results of Problem 9.49 with the final velocity predicted by the rocket equation when the mass is expelled continuously.

42. Pursue the model of Problem 38 further by changing the ratio of mass assigned to the two stages of the rocket to values other than 50/50. Assume that the engine mass overhead is proportional to the fuel. That is, if you assign 90% of the fuel to the first stage also assign 90% of the 500-kg overhead to the first stage. Let f be the fraction of the fuel assigned to the first stage, and $(1 - f)$ be the fraction of the fuel in the second stage. Vary f from 0.05 to 0.95 in steps of 0.05, and produce a plot of final payload velocity versus f.

43. Spreadsheet template GLIDERS1.WK1 described in Problem 9.50 contains the position-versus-time data for two gliders colliding on an airtrack. (a) Calculate the velocity of the center of mass of the two gliders before and after the collision. Do they agree within experimental uncertainty? (b) Modify the position-versus-time record so that it describes the collision as it would appear in the center-of-mass frame of reference. Prepare a graph showing x versus t for both gliders as seen from the center-of-mass frame of reference.

44. Carry out the calculations described in Problem 43 for the spreadsheet template GLIDERS2.WK1, which is described in Problem 9.51.

45. Table 10.P2 contains a list of particle masses along with their velocity components. Using a spreadsheet or computer program, find (a) the total kinetic energy of the system, (b) the velocity of the center of mass, (c) the kinetic energy of the system of particles *as measured in the center-of-mass frame of reference*, and (d) the kinetic energy *of the center of mass*. (e) Compare the answers to parts (c) and (d) with the answer to part (a).

TABLE 10.P2

MASS (g)	v_x (cm/s)	v_y (cm/s)	v_z (cm/s)
12	12	-5	2
21	-21	-3	-9
5	3	4	-11
25	-4	15	20
16	9	0	5

★46. Modify the Runge-Kutta program from Chapter 5 to solve the rocket Equation (10.15) with an external force equal to the (continuously changing!) weight of the rocket. Apply your program to the toy rocket described in Problem 35 to (a) calculate the speed of the rocket at "burn out" when all the water is gone and (b) to find the maximum altitude of the rocket.

★47. Add in air resistance using the quadratic model of Chapter 5 ($D = \frac{1}{2}C_D\rho Av^2$) to your analysis of the rocket in Problem 46. Take $C_D = 0.5$ and $A = 4.0$ cm^2.

★48. Perform the calculation described in Problem 46 using a modification of the Runge-Kutta spreadsheet from Chapter 5.

★49. Perform the calculation described in Problem 47 using a modification of the Runge-Kutta spreadsheet from Chapter 5.

★50. Two physics students are arguing about automobile accidents. One says that if you know you are going to be hit from behind, you should take your foot off the brake because then the collision occurs over a greater distance and the peak forces are smaller. The other says you should push hard on the brake so that you gain less momentum and the net impulse and peak force are smaller. Model such a collision by taking the mass of each automobile as 1000. kg. Let car A be moving at 15.0 km/h and collide with car B, which is at rest. Model the interaction force between the two as a Hooke's-law spring, so that when $\Delta x < s_o$ the force on each car has magnitude $F = |k(s_o - \Delta x)|$, where $\Delta x = |x_B - x_A|$, $s_o = 1.00$ m, and $k = 1.00 \times 10^5$ N/m. Start the model with $x_B = 1.00$ m and $x_A = 0.00$. Calculate x and a for each car as they collide using the Runge-Kutta method. (Note this requires substantial modification of the single particle Runge-Kutta programs or spreadsheets presented previously.) Continue the calculation up to the point of maximum acceleration of car B. Perform the calculation twice, once with no frictional force on car B and again with a frictional force due to a coefficient of friction of 0.95.

General Problems

51. What is the gravitational potential energy of three 8.00 ft long 2 × 4's nailed together in the form of the letter "H" standing upright if each 2 × 4 has a mass of 3.50 kg? (Assume the crossbar is connected at the midpoint of the two verticals, and ignore the mass of the nails.)

52. For distances small compared with the radius of the earth, the earth's gravitational field decreases 3.1×10^{-6} N/kg for each meter distance above the surface of the earth. Calculate the difference in distance between the center of mass and the center of gravity of a system composed of two 1.00-kg masses, one at the surface of the earth, the other at a distance of 10.0 km directly above the first. (The **center of gravity** is defined as $y_{cg} = (\Sigma_{i=1}^{N} m_i g_i y_i)/(\Sigma_{i=1}^{N} m_i g_i)$.)

53. A 6.50-g bullet is fired at a 1.50-kg block of wood suspended by strings to form a ballistic pendulum. As shown in Figure 10.P9, the bullet penetrates completely through the wood and strikes a second block of wood of mass 2.00 kg. The 1.50-kg mass rises 3.35 cm above its original height, and the 2.00-kg block is knocked straight back by the impact 5.50 cm across a horizontal surface. If the coefficient of friction between the 4.00-kg block and the surface is $\mu_k = 0.850$, what was the original speed of the bullet?

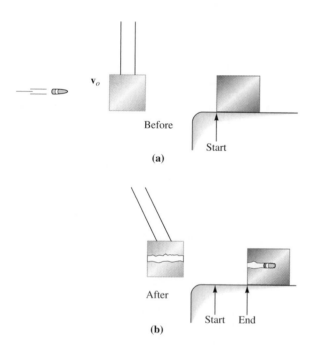

Before

(a)

After

Start   End

(b)

FIGURE 10.P9   Problem 53

54. During the filming of an adventure movie, the 55.0-kg heroine makes a flying leap, grabbing her 70.0-kg partner as he stands on the edge of a building. The building is 20.0 m high, and after the collision the pair fall together off the building with an initial velocity that is horizontal. In the stunt, the pair are to fall into a swimming pool below, as shown in Figure 10.P10. The near edge of the pool is 3.00 m from the wall of the building and the far edge is 15.0 m from the wall. (a) What speed must the heroine have in order for the pair to land in the center of the pool? (b) What are the limits of her speed resulting in a reasonably safe landing, say, 1.00 m from either edge? (c) Are the speeds reasonable? Compare them, for example, with a very fast sprinter who can run 100. m in 10.0 s.

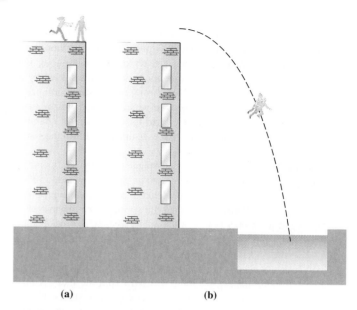

(a)   (b)

FIGURE 10.P10   Problem 54

55. A 100.-grain (6.48 g) bullet traveling 2600. ft/s collides with a 2.00-kg block of wood, which slides along a horizontal surface after being hit. (a) If the coefficient of friction between the surface and the block is $\mu_k = 0.885$, how far does the block slide before coming to rest? Assume the bullet remains within the wooden block. (b) If a 200.-grain bullet with the same momentum hits the block, how far does the block slide? (c) If a 200.-grain bullet with the same kinetic energy as the original 100.-grain bullet hits the block, how far does the block slide?

56. A 2.00-kg chunk of ore is dropped vertically onto a conveyer belt moving horizontally with a constant speed of 5.00 m/s. (a) If the coefficient of kinetic friction between the belt and the ore is $\mu_k = 0.60$, how long after hitting the belt does the ore stop slipping? (b) How far does the belt travel during this time? (c) How far does the ore travel relative to the stationary floor during this time? (d) Along what length of the belt does the ore slide before coming to rest relative to the belt? (e) How much work is done by the frictional force? How much kinetic energy does the ore gain? How much work does the belt perform during this time? (f) Reconcile the work done by the belt with the energy gained by the ore, and relate the result to Example 10.10. (g) If the coefficient of friction is half of the given value, does it affect the energy balance? Explain.

57. An open truck is loaded with 2.00×10^3 kg of corn from a hopper as shown in Figure 10.P11. If the truck travels at 3.00 m/s

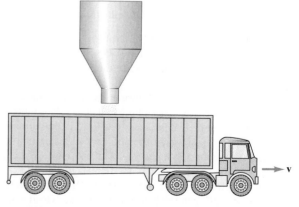

FIGURE 10.P11   Problem 57

under the hopper during the 4.00 s it takes to load the corn, (a) what average force is required to keep the truck moving forward at the constant speed? (b) How many joules of work does this force perform? (c) What is the change in kinetic energy of the corn? (d) Why are the answers to (b) and (c) different?

58. Whether a collision is elastic or inelastic often depends on the kinetic energy available to be transformed to other energy types. For example, a slow-moving electron may collide with an atom perfectly elastically, whereas a faster moving electron makes a partially elastic collision, leaving the atom in an excited state. Because of the need to conserve momentum, the only kinetic energy *available to excite the atom* is the system kinetic energy as measured in the center-of-mass frame of reference, that is, $\sum_{i=1}^{N} \frac{1}{2} m_i u_i^2$. An electron may make an inelastic collision with a hydrogen atom in its ground state if the available kinetic energy is greater than 1.64×10^{-18} J. (a) What is the minimum velocity of an electron that can make an inelastic collision with a stationary hydrogen atom? (b) What kinetic energy does such an electron have?

59. Show that, given any system of N particles with masses m_i and velocities $\mathbf{v}_i$, the total momentum in the center-of-mass frame $\sum_{i=1}^{N} m_i \mathbf{u}_i$ is zero.

60. Combine the individual expressions, such as Equations (10.1) and (10.2), for x_{CM}, y_{CM}, and z_{CM} to show that Equation (10.6) is indeed a correct expression for $\mathbf{r}_{CM}$.

61. Recopy each equation and variable substitution in Example 10.6, showing explicit units on all numerical quantities. For example, start with the equation $y = 4.00$ m $- (1.00$ m$^{-1})x^2$.

62. Three particles with masses 1.50 kg, 2.00 kg, and 2.50 kg are moving with velocities of 2.50 m/s $\hat{\mathbf{i}}$, 2.50 m/s $\hat{\mathbf{j}}$, and 0.75 m/s $\hat{\mathbf{i}}$ − 0.60 m/s $\hat{\mathbf{j}}$, respectively. (a) Calculate the total kinetic energy of the system. (b) Calculate the velocity (vector) of the center of mass. (c) Calculate the kinetic energy as seen in the center-of-mass frame. (d) Calculate the kinetic energy of a fictitious point–particle whose mass is equal to the total mass of the three particles if this fictitious mass travels with velocity v_{CM}. (e) Compare the answer to (a) with the sum of the energies you calculated in parts (c) and (d).

★63. Suppose a long slender rod of length L has a linear mass density given by

$$\lambda(x) = \lambda_o \exp\left(\frac{x}{L}\right)$$

where λ_o is a constant. The rod lies along the x-axis, and its end with the smaller mass density is placed at $x = 0$. Find the center of mass of the rod.

Rotation about a Fixed Axis

In this chapter you should learn (for objects rotating about a fixed axis):

- the analogy between the kinematics of rotational motion and linear motion.
- how to calculate the energy of a rotating object.
- how to calculate the rotational inertia of a rotating object.
- how to apply the rotational analog of Newton's second law.
- how to calculate the angular momentum of a rotating object.

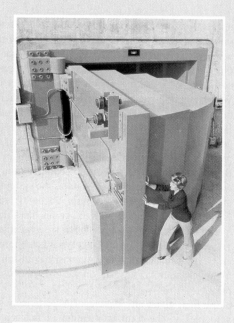

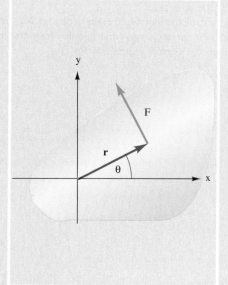

$$\tau = I\alpha$$

In the previous chapters of this text we have treated everything from people to piñatas as dimensionless point-particles. In Chapter 10 we discovered that this point-particle model is not an unreasonable view because for all objects there is in fact one point, the center of mass, that behaves just as this model predicts. On the other hand, it is clear that many features of an object's motion have been omitted. Think of two divers jumping upward off a three-meter board into a swimming pool. One diver might merely tuck up into a ball and plummet downward resulting in nothing more interesting than a big splash. The other diver might execute a graceful one-and-a-half-gainer with a half-twist and win a gold medal. If both divers leave the board at the same angle and speed, the center of mass of each follows the same parabolic trajectory. This is certainly an important bit of information, but clearly, many interesting details of the divers' motions are not described by the point-particle model. What is missing, of course, is information about rotation.

In this chapter we begin to explore the mechanics of rotating objects. Rotational effects are important in many diverse applications from diving to ice skating, from the height of the cushion above a pool table to the inertial guidance systems for ships and airplanes. We begin by studying the rotation of simple shapes about fixed axes[1] to reveal the underlying physical laws. We will meet many familiar concepts: For example, the kinematic relations among position, velocity, and acceleration have analogs in rotational motion. Not surprisingly, the rotational equivalent to Newton's second law is a central concept. We tinker with the conservation of mechanical energy to include rotational motion. Finally, the law of conservation of momentum has a rotational counterpart. So you see it's all stuff you've seen before with just a little different twist. . . .

11.1 Kinematics of Rotational Motion

Recall that kinematics is the description of motion. While "doing kinematics" we don't ask, "Why does the object move that way?" We merely seek to describe the motion. As shown in Figure 11.1, we once again use a Cartesian coordinate system. For the time being, we assume that whatever object is rotating does so about a fixed rotational axis that coincides with the z-axis of the coordinate system. Because we want to describe the object's rotation about this axis, we need to watch a particular point, call it P, on the object. For simplicity we take P to lie in the xy-plane and designate the radius vector from the origin to point P as **R**. Following the usual convention, we denote the angle measured anticlockwise from the positive x-axis to the radius vector **R** as a positive angle θ. Now, the single variable θ called the **angular position** is sufficient to describe the orientation of our object. A change in the angular orientation is described by the **angular displacement** and denoted by Δθ.

The angle θ could be measured in degrees, but we generally find it useful (and often imperative!) to measure θ in *radians*. Consequently, we once again remind you of the definition of radian measure:

$$\theta = \frac{s}{R} \qquad \text{definition of radian measure}$$

Here, s is the arc length, and R is the radius of the circle along which the arc length is measured. Note that we can determine the actual distance the point P moves through space when our object rotates by a known angle from the relation $s = r\theta$, but *only* if θ is measured in radians.

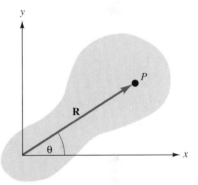

FIGURE 11.1

The object rotates about a fixed axle along the z-axis (out of the page). The angle θ is sufficient to specify the orientation of the object.

---

[1] Throughout this chapter we continually qualify our results as applicable to rotation about a fixed axis only. This axis must be fixed in an inertial frame of reference. In addition, as we will show in Chapter 12, these results may also be applied to the case of a moving axis provided that (a) the axis is an axis of symmetry passing through the center of mass and (b) that the axis moves only parallel to itself. For the sake of brevity we continue to say simply "fixed axis" to indicate the above conditions.

EXAMPLE 11.1   Bit by Bit I'm Getting the Hang of It

The read head of a floppy disk drive is stationed 4.00 cm from the center of rotation of the diskette. If the diskette rotates through 45.0°, what is the length of the diskette track that passes under the head?

SOLUTION   We must convert the angular displacement $\Delta\theta$ from degrees to radians. Since 2π rad $= 360°$, we can write

$$\Delta\theta(\text{in radians}) = \Delta\theta(\text{in degrees})\left(\frac{2\pi}{360°}\right) = 45°\left(\frac{2\pi}{360°}\right) = \frac{\pi}{4} \text{ rad}$$

Now we may use $\Delta s = R\,\Delta\theta$ to find the length of the arc that passes below the read head of the disk drive in Figure 11.2:

$$\Delta s = R\,\Delta\theta = (4.00 \text{ cm})\left(\frac{\pi}{4} \text{ rad}\right) = 3.14 \text{ cm}$$

Recall that the unit symbol *rad* is a place holder indicating the convention used to express the angular measure and is itself unitless.   ◄

We can describe a rotating object's motion by quoting the rate at which θ is changing. The rate of change of position we called velocity. The rate of change of angular position we call **angular velocity** and designate it with the lowercase Greek letter omega ω. Both average and instantaneous angular velocity may be defined analogously to their translational counterparts:

$$\omega_{av} = \frac{\Delta\theta}{\Delta t} \qquad \text{Average angular velocity}$$

$$\omega = \frac{d\theta}{dt} \qquad \text{Instantaneous angular velocity (fixed axis)}$$

There are many units for angular velocity. One of the most common is rpm, which stands for revolutions per minute (rev/min). In principle, any consistent set of units works for kinematics, but as soon as we begin to do dynamics we absolutely must use radians per second (rad/s). Our advice is start *now* to get yourself in the habit of immediately changing all angular velocities to radians per second as the very first thing you do when you start a problem. For kinematic problems you may waste a few calculator strokes converting between revolution, degrees, and radians as required by the problem, but we earnestly believe you will save yourself much irritation later by starting this habit now.

You have probably anticipated the next step; namely, we look at the time rate of change of angular velocity. As you might expect, this quantity is called **angular acceleration.** Designated by the lowercase Greek letter alpha α, the preferred units are (rad/s)/s or more succinctly rad/s^2.

$$\alpha_{av} = \frac{\Delta\omega}{\Delta t} \qquad \text{Average angular acceleration}$$

$$\alpha = \frac{d\omega}{dt} = \frac{d^2\theta}{dt^2} \qquad \text{Instantaneous angular acceleration (fixed axis)}$$

If you have driven a car with a tachometer, you know that when you step down on the accelerator the indicator dial shows your "rpms" increasing as your engine turns faster and faster. You might, for example, be able to increase the tachometer reading by 50 rpm

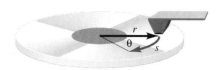

FIGURE 11.2

The arc length s traced out by the read head of the disk drive is given by $s = r\theta$. See Example 11.1.

during each second. This process is an angular acceleration of 50 rpm/s. Such a unit is convenient and helps us to understand that angular acceleration means increasing (or decreasing) the rotation rate of an object. However, as convenient as it might be, we once again recommend you change these units to the less intelligible rad/s^2 as quickly as your calculator allows.

Given θ as a function of t, we can proceed from $\theta(t)$ to $\omega(t)$ to $\alpha(t)$ by taking derivatives. If instead we are given $\alpha(t)$, we can reverse the process, moving from $\alpha(t)$ to $\omega(t)$ to $\theta(t)$ by the inverse process of integration. In particular

$$\omega(t) = \int \alpha(t)\, dt + C \tag{11.1}$$

$$\theta(t) = \int \omega(t)\, dt + C' \tag{11.2}$$

where C and C' are integration constants. As shown in Table 11.1, these relationships for the rotational motion of an extended object about a fixed axis are identical to those of Chapter 3 for particle motion along a straight line. In Example 11.2 we demonstrate that our problem-solving approach is also analogous.

TABLE 11.1   **Analogy Between Rotational Kinematics and Straight-Line Kinematics**

	LINEAR	(UNITS)	ANGULAR	(UNITS)
position	x	(m)	θ	(rad)
displacement	Δx	(m)	$\Delta\theta$	(rad)
velocity	$v \equiv dx/dt$	(m/s)	$\omega \equiv d\theta/dt$	(rad/s)
acceleration	$a \equiv dv/dt$	(m/s^2)	$\alpha \equiv d\omega/dt$	(rad/s^2)
	$v = \int a\, dt + C$		$\omega = \int \alpha\, dt + C$	
	$x = \int v\, dt + C'$		$\theta = \int \omega\, dt + C'$	

EXAMPLE 11.2   *The One I Had When I Was a Kid Never Spun This Long*

A navigation-quality gryroscope is spinning at $\omega_o = 3.60 \times 10^3$ rpm. The power is shut off, and the gyroscope begins to slow down with its angular velocity described by

$$\omega = \omega_o e^{-t/\tau}$$

with $\tau = 1.00 \times 10^4$ s. (a) How long does it take for the gryoscope to reach half its original angular velocity? (b) What is the gryoscope's original angular acceleration? (c) What is the gyroscope's average acceleration between the time the power is cut off and the instant it reaches half speed? (d) Through how many revolutions does the gyroscope turn during this time?

SOLUTION   First we clean up the units:

$$\omega_o = \left(3600\ \frac{\text{rev}}{\text{min}}\right)\left(\frac{\text{min}}{60\ \text{s}}\right)\left(\frac{2\pi\ \text{rad}}{\text{rev}}\right) = 120\pi\ \text{rad/s}$$

(a) This question asks "*When* does $\omega = \omega_o/2$?" We must therefore solve $\omega = \omega_o\, e^{-t/\tau}$ for t. We divide both sides of this equation by ω_o and then take the natural logarithm of both sides, leading to

$$\ln\left(\frac{\omega}{\omega_o}\right) = \ln\left(e^{-t/\tau}\right) = -\frac{t}{\tau}$$

Solving for t, we find

$$t = -\tau \ln\left(\frac{\omega}{\omega_o}\right)$$

or

$$t = +\tau \ln\left(\frac{\omega_o}{\omega}\right)$$

Because $\omega = \omega_o/2$, the value for t is

$$t = \tau \ln(2) = (1.00 \times 10^4 \text{ s})(0.693) = 6930 \text{ s} = 1.93 \text{ h}$$

(b) The angular acceleration function is the time derivative of the angular velocity function

$$\alpha(t) = \frac{d\omega}{dt} = \frac{d}{dt}\left(\omega_o e^{-t/\tau}\right) = -\frac{\omega_o}{\tau} e^{-t/\tau}$$

The original acceleration α_o at $t = 0.00$ s is

$$\alpha_o = -\frac{\omega_o}{\tau} e^{-0/\tau} = -\frac{\omega_o}{\tau}$$

$$= -\frac{120\pi \text{ rad/s}}{1.00 \times 10^4 \text{ s}} = -1.20\pi \times 10^{-2} \text{ rad/s}^2 = -3.77 \times 10^{-2} \text{ rad/s}^2$$

The negative value of angular acceleration indicates that the angular acceleration is in the clockwise sense. In the present case, since the flywheel is rotating anticlockwise (ω is positive), the flywheel is slowing down. Beware! A negative angular acceleration *does not always mean the object is slowing down*. When ω and α have the same sign, the rotation rate is increasing; when ω and α have opposite signs, the rotation rate is decreasing. (Recall our similar rule in Chapter 3 for the velocity and acceleration of translational motion.)

(c) Our definition of average angular acceleration is

$$\alpha_{av} = \frac{\Delta\omega}{\Delta t} = \frac{\omega_f - \omega_o}{t_f - t_o} = \frac{(\omega_o/2) - \omega_o}{t_f - 0} = -\frac{\omega_o}{2t_f} = -\frac{120\pi \text{ rad/s}}{2(6930 \text{ s})}$$

$$= -2.72 \times 10^{-2} \text{ rad/s}^2$$

(d) The angular displacement function is the integral of the angular velocity

$$\theta(t) = \int \omega(t)\, dt + C$$

$$= \int \omega_o e^{-t/\tau}\, dt + C$$

$$= -\omega_o \tau e^{-t/\tau} + C$$

Following the usual custom, we let $\theta = 0$ at $t = 0$, so that

$$0 = -\omega_o \tau + C$$

leading to $C = +\omega_o \tau$. Consequently, the angular position function is

$$\theta(t) = \omega_o \tau (1 - e^{-t/\tau})$$

We next evaluate $\theta(t)$ at the time the gyroscope's angular velocity reaches half its initial value, which was found in part (a) to be $t = \ln(2)\tau$.

$$\theta\,(\ln(2)\tau) = \omega_o \tau (1 - e^{-(\ln(2)\tau)/\tau}) = \omega_o \tau \left(1 - \frac{1}{2}\right)$$

$$= \frac{(120\pi \text{ rad/s})(1.00 \times 10^4 \text{ s})}{2}$$

$$= 60.0 \times 10^4 \pi \text{ rad} = 30.0 \times 10^4 \text{ revolutions} \qquad \blacktriangleleft$$

By now you cannot have failed to note that the definitions of angular position, angular displacement, angular velocity, and angular acceleration parallel their Chapter 3 counterparts: position, displacement, velocity, and acceleration. In fact the definitions are so close, it is possible to confuse them. For this reason it is traditional to state explicitly *linear* displacement, *linear* velocity, and *linear* acceleration when referring to the quantities **r**, **v**, and **a** from Chapter 3. The phrase *linear* velocity is meant to distinguish ds/dt measured in meters per second from *angular* velocity measured in radians per second. It takes a while to become accustomed to saying that an object going in a circle has a *linear* velocity because we often associate ''linear'' with a straight line. But of course, an arc is just as much a line as is a straight segment. If you were driving a car around a circular track, you might have an *angular* velocity of 100 rev/h. While performing this circular motion, the reading on your speedometer is your *linear* velocity, that is, the rate at which you are hurtling along the track. Having made all these comments, we are ready to relate the angular quantities to the linear quantities for motion in a circle, where the radius can be treated as a constant. We have only to differentiate the expression for arc length, which results from the definition of radian measure.

$$s = R\theta \qquad \text{(radian units only)}$$

$$\frac{ds}{dt} = \frac{d}{dt}(R\theta) = R\frac{d\theta}{dt}$$

According to our definitions, this result means

$$v_t = R\omega \qquad \text{(radian units only)}$$

where we have added a subscript t to the linear velocity to remind ourselves that the direction of the linear velocity vector is tangent to the circle. Differentiating again, we find the connection between linear and angular acceleration.

$$\frac{dv_t}{dt} = R\frac{d\omega}{dt}$$

$$a_t = R\alpha \qquad \text{(radian units only)}$$

Here too, the t subscript reminds us that the component of linear acceleration caused by changes in rotation rate is tangent to the circle. If the object in question is increasing its rotation rate, a_t is parallel to v_t. If the rotation rate is slowing, a_t is antiparallel to v_t.

Concept Question 1
(a) Can two race cars driving on the same wide circular track have the same angular speed but different linear speeds? Explain.
(b) If a clock were designed so that the tip of the second-hand and the tip of the minute-hand both had the same linear speed, what would the clock look like?

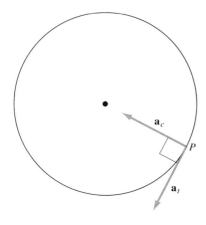

As you may recall from Chapter 4, there is another component of linear acceleration associated with circular motion *which exists independent of* α. This is the *centripetal* acceleration

$$a_c = \frac{v_t^2}{R}$$

directed radially inward. Using the relation $v_t = R\omega$, it is easy to show that a_c depends on ω, not α.

$$a_c = R\omega^2 \qquad \text{(Radian units only)}$$

Because the tangential a_t and the radial a_c are at right angles, the magnitude of the total linear acceleration can be found from the square root of the sum of the squares of each:

$$a_{\text{tot}} = \sqrt{a_t^2 + a_c^2} = \sqrt{(R\alpha)^2 + (R\omega^2)^2} = R\sqrt{\alpha^2 + \omega^4}$$

The vectors associated with the linear accelerations are shown in Figure 11.3.

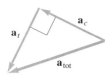

FIGURE 11.3

Because the radial and tangential components of the acceleration are perpendicular, the total acceleration of the point P can be found by the Pythagorean theorem.

The Constant-Acceleration Model

In Chapter 3 we introduced the constant-acceleration model for *translational* motion and we subsequently found this model useful on numerous occasions. The constant-acceleration model for *rotational* motion also has important applications. When we take α to be a constant, Equation (11.1) may be integrated

$$\omega = \int \alpha \, dt + C$$

$$= \alpha t + C$$

TABLE 11.2   Relations Between Angular and Linear Quantities for Constant Radius R

$$s = R\theta$$
$$v_t = R\omega$$
$$a_t = R\alpha$$
$$a_c = R\omega^2$$
$$a_{\text{tot}} = R\sqrt{\alpha^2 + \omega^4}$$

These expressions are valid only when θ, ω, and α are expressed in radian measure.

Concept Question 2
(a) Is it possible for an object traveling in a circular path to have zero tangential acceleration and nonzero angular acceleration? (b) Is it possible for an object traveling in a circular path to have zero centripetal acceleration and nonzero angular acceleration?

If we assume the angular velocity is ω_o at $t = 0$, we find that the constant $C = \omega_o$ and obtain the result

$$\omega = \omega_o + \alpha t$$

We now substitute this expression for ω into Equation (11.2) and perform the resulting integration. With the initial condition that $\theta = \theta_o$ at $t = 0$, we find

$$\theta(t) = \theta_o + \omega_o t + \frac{1}{2}\alpha t^2$$

If you eliminate t from these last two equations, you will find

$$\omega^2 = \omega_o^2 + 2\alpha(\theta - \theta_o)$$

Similarly, if you eliminate α instead of t, you will obtain

$$\theta = \theta_o + \frac{(\omega + \omega_o)}{2}t$$

These equations have the same form as the equations for one-dimensional linear motion that we developed in Chapter 3. This result is hardly surprising because we began with parallel definitions and made the same special case assumption.

TABLE **11.3**   Comparison of Linear and
Rotational Equations for the
Constant-Acceleration Model

$\omega = \omega_o + \alpha t$	$v = v_o + a_o t$
$\theta = \theta_o + \omega_o t + \dfrac{1}{2}\alpha t^2$	$s = s_o + v_o t + \dfrac{1}{2}at^2$
$\omega^2 = \omega_o^2 + 2\alpha(\theta - \theta_o)$	$v^2 = v_o^2 + 2a(s - s_o)$
$\theta = \theta_o + \dfrac{\omega + \omega_o}{2}t$	$s = s_o + \dfrac{v + v_o}{2}t$

The similarity between linear and rotational kinematics for the constant-acceleration model also extends to the problem-solving strategies. You may wish to review the problem-solving strategy for the constant-acceleration model as given in Chapter 3.

EXAMPLE 11.3   *Enough to Make Your Head Spin*

An electric motor is initially rotating at 500. rpm. It slows its rotation at a constant rate to 200. rpm in 25.0 s. (a) Find the angular acceleration and (b) the number of revolutions the motor makes while slowing down between these two rates.

SOLUTION   We write out the "given" quantities, changing units as we go

$$\theta_o = 0.00 \quad \text{(this is a free choice)}$$

$$\omega_o = 500. \text{ rpm} = 500. \text{ rev/min} = 52.36 \text{ rad/s}$$

$$\omega_f = 200. \text{ rpm} = 20.94 \text{ rad/s}$$

$$t = 25.0 \text{ s}$$

In part (a) we wish to find α. The equation that contains our given quantities and the one we desire is $\omega = \omega_o + \alpha t$, which we immediately solve for α.

$$\alpha = \frac{\omega - \omega_o}{t} = \frac{20.94 \text{ rad/s} - 52.36 \text{ rad/s}}{25.0 \text{ s}} = -1.26 \text{ rad/s}^2$$

To find the number of revolutions we must find θ. The simplest choice is

$$\theta = \theta_o + \frac{\omega + \omega_o}{2}t = \frac{20.94 \text{ rad/s} + 52.36 \text{ rad/s}}{2}(25.0 \text{ s}) = 916 \text{ rad}$$

$$= 916 \text{ rad} \times \frac{1 \text{ rev}}{2\pi \text{ rad}} = 146 \text{ rev} \qquad \blacktriangleleft$$

11.2 Kinetic Energy and Rotational Inertia

We now look at the familiar concept of kinetic energy as it applies to a rotating object. To begin simply, let's look at a collection of N point-masses arranged in a straight line. If you like, think of them as connected by a massless rigid rod. Let this rod be fixed at one end and free to rotate about that end so that the masses all move in the xy-plane. Such a collection of masses, which maintain their relative position while rotating, is known as a **rigid body.**

As shown in Figure 11.4 for $N = 4$, we denote each mass as m_i and its distance from the pivot point by r_i. When our rod is rotating with angular velocity ω about the pivot,

Concept Question 3
Make up (but do not solve) a rotational motion problem that is analogous to the following linear motion problem: "A car traveling 30 m/s brakes and comes to rest in a time of 8.00 s. How far did the car travel during this interval?"

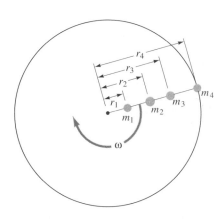

FIGURE 11.4

Four masses in a straight line rotate with a common angular velocity ω about a fixed axis perpendicular to the page. The linear velocity of each mass is $v = r\omega$.

Calvin and Hobbes   by Bill Watterson

each mass is moving with a different *linear* velocity $v_i = \omega r_i$. The total kinetic energy of all N masses can then be written

$$K = \sum_{i=1}^{N} \frac{1}{2} m_i v_i^2 \tag{11.3}$$

$$K = \sum_{i=1}^{N} \frac{1}{2} m_i \omega_i^2 r_i^2$$

Notice that each mass travels through exactly the same angle in the same amount of time. Hence, the *angular* velocity is the same for each mass, $\omega_i = \omega$ for all i. The total kinetic energy is thus

$$K = \frac{1}{2} \left(\sum_{i=1}^{N} m_i r_i^2 \right) \omega^2$$

We can cast this expression in a simple form that resembles our expression $\frac{1}{2}mv^2$ by writing

$$K = \frac{1}{2} I \omega^2 \qquad \text{Kinetic energy of rotation} \tag{11.4}$$

As you no doubt realize, we obtained this simple form for the total kinetic energy of the rotating mass-points by hiding a messy summation behind the innocent looking symbol I. The quantity I is known as the **rotational inertia** (or **moment of inertia**). The bad news is that the rotational inertia is somewhat tedious to calculate, although not particularly difficult. The good news is that, once it is calculated, it's just a quantity that stays fixed throughout the problem *so long as both the mass distribution and the axis of rotation stay the same.* Formally, we define the rotational inertia for a collection of N mass-points as

$$I = \sum_{i=1}^{N} m_i r_i^2 \qquad \text{Rotational inertia of } N \text{ point–masses} \tag{11.5}$$

where r_i is the shortest straight-line distance between the point-mass m_i and the axis of rotation. The SI units of the rotational inertia are $\text{kg} \cdot \text{m}^2$.

EXAMPLE 11.4   A Collection of Points in Three Dimensions about a Fixed Axis

Fourteen point-masses, each of mass $m = 10.0$ g, are located on the vertices and face centers of a cube 2.00 cm on a side as shown in Figure 11.5. Calculate the rotational inertia of this system of masses about the y-axis, assuming the point-particles maintain their relative position as they rotate.

SOLUTION   The important distances r_i in the expression $I = \Sigma_{i=1}^{N} m_i r_i^2$ are the shortest distances between the mass–points and the axis of rotation. For the eight masses located at the vertices of the cube, these distances are the same, and $r = \sqrt{2.00}$ cm. For the four masses at the center of the side cube faces the distance $r = 1.0$ cm. For the two masses at the center of the top and bottom faces, $r = 0$. Consequently, we can calculate

$$I = \sum_{i=1}^{14} m_i r_i^2 = m \sum_{i=1}^{14} r_i^2$$

$$= (10.0 \text{ g})[8(\sqrt{2.00} \text{ cm})^2 + 4(1.00 \text{ cm})^2 + 2(0.00 \text{ cm})^2] = (10.0 \text{ g})(20.0 \text{ cm}^2)$$

$$= 200. \text{ g} \cdot \text{cm}^2 = 2.00 \times 10^{-5} \text{ kg} \cdot \text{m}^2$$

FIGURE 11.5

A face-centered cubic array of point–particles. See Example 11.4.

Of course, seldom does a collection of isolated mass-points make up a rigid body. Usually, rigid bodies by their very nature are composed of a collection of contiguous mass-points. To determine the rotational inertia of a solid object we mentally divide the object into small mass chunks Δm and perform the summation

$$I \approx \sum_{i=1}^{N} \Delta m_i r_i^2$$

In the limit as the number of mass-points increases without bound while each mass Δm_i approaches zero, the summation becomes the integral

$$I = \int r^2 \, dm \qquad \text{rotational inertia for solid masses} \qquad (11.6)$$

where r is the shortest straight-line distance between each dm mass element and the axis of rotation. The integration is understood to include the entire mass distribution, thus the single integral above is symbolic of one, two, or three integrals depending on the dimensionality of the problem. You may wish to review the definitions of linear, area, and volume mass densities (λ, σ, and ρ, respectively) in Section 10.3 before proceeding with the example calculations of I.

Concept Question 4
Three students are arguing about how to calculate the kinetic energy of two small blocks, each of mass m located on opposite ends of a massless rod that is rotating with angular velocity ω about an axis through and perpendicular to its center. Student *A* says that you must find the v of each mass and then double the quantity $(1/2)mv^2$. Student *B* says that you must calculate the moment of inertia of each of the two masses and then sum $(1/2)I\omega^2$. Student *C* claims that you must perform calculations suggested by both students *A* and *B* and then add the answers to obtain the total kinetic energy. Who's right here?

Moments of Mass Distributions

The integrals of dm that we have encountered in Chapters 10 and 11 are collectively known as *moments* of the mass distribution. We have encountered many already.

$$I_0 = \int dm \qquad \text{the zeroth moment (the total system mass)}$$

$$I_1 = \int r \, dm \qquad \text{the first moment (used to calculate the center of mass)}$$

$$I_2 = \int r^2 \, dm \qquad \text{the second moment (rotational inertia)}$$

$$I_N = \int r^N \, dm \qquad \text{the } N\text{th moment of a mass distribution (useful in advanced mechanics)}$$

The rotational inertia is the second moment of the mass distribution and is sometimes called simply the **moment of inertia.**

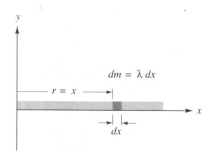

FIGURE 11.6

The rotational inertia of a thin rod is the sum of the rotational inertia of each of the mass elements making up the rod. See Example 11.5.

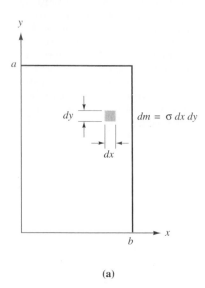

(a)

FIGURE 11.7

(a) The usual differential area unit in rectangular coordinates has dimensions $dx\,dy$.

EXAMPLE 11.5   *Getting into the Swing of It*

Calculate the rotational inertia of a slender rod of length L and uniform density λ_o if it is pivoted about one end.

SOLUTION   We orient the rod along the x-axis with the pivoted end at the origin as shown in Figure 11.6. A typical mass element $dm = \lambda \, dx$ is located a distance $r = x$ from the pivot. Thus, the definition of I becomes

$$I = \int r^2 \, dm = \int_0^L x^2 \lambda_o \, dx = \lambda_o \int_0^L x^2 \, dx = \lambda_o \left[\frac{x^3}{3} \right]_0^L = \lambda_o \frac{L^3}{3}$$

It is traditional to write I in terms of the total mass M. In this case $M = \lambda_o L$, so that

$$I = \frac{1}{3} ML^2 \qquad \blacktriangleleft$$

EXAMPLE 11.6   *Don't Slam That Door*

Calculate the rotational inertia of a door of height a, width b, and uniform area density σ_o when it is pivoted about an axis parallel to one edge.

SOLUTION   As shown in Figure 11.7(a) we locate the lower left-hand corner of the door at the origin and pivot the door around the y-axis. We choose a rectangular mass element $dm = \sigma_o \, dx \, dy$. You can see from Figure 11.7(a) that the shortest distance between dm and the y-axis is the distance $r = x$. Integrating over the entire door requires two integrals

$$I = \int_{\text{door}} r^2 dm = \int_0^b \int_0^a x^2 \sigma_o \, dy \, dx$$

We first perform the y-integration

$$I = \int_0^b x^2 \sigma_o \left(\int_0^a dy \right) dx = \int_0^b x^2 \sigma_o (a) \, dx$$

As you may have already observed, we could have reached this last expression directly by choosing mass $dm = \sigma_o\, dA = \sigma_o(a\, dx)$ as shown as Figure 11.7(b). In this case, the area element is a rectangle of height a and width dx. Every point within this thin strip is the same distance x (to within dx) from the axis of rotation. Although not necessary, when you can identify strips of mass dm all located *the same distance from the axis of rotation* and use them as dm elements, you can save time and reduce the amount of computation.

However you arrive at the previous expression, the x-integration can now be performed

$$I = \int_0^b x^2 \sigma_o a\, dx = \sigma_o a \int_0^b x^2\, dx = \sigma_o a \left[\frac{x^3}{3}\right]_0^b = \frac{1}{3}\, \sigma_o a b^3$$

Because the area density is constant, the mass of the door is simply $M = \sigma_o A = \sigma_o a b$. Thus, we can write the moment of inertia as

$$I = \frac{1}{3}\, M b^2$$

You should not miss the fact that this result is the same as that for the thin rod pivoted about one end, as derived in Example 11.5. The rotational inertia is not affected by the distribution of the mass parallel to the axis of rotation. Only the distance from the mass to the axis of rotation is important. ◀

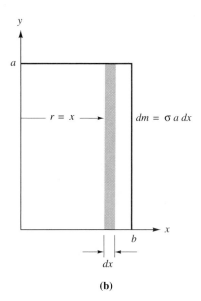

(b)

FIGURE 11.7

(b) Because all of the mass in the shaded rectangle is the same distance from the rotation axis, a larger area element $a\, dx$ may be taken. See Example 11.6.

EXAMPLE 11.7   *I Warned You Not to Slam the Door*

Calculate the rotational inertia of the door in the problem above but this time take the axis to be perpendicular to the door and through its center.

SOLUTION   The geometry for our calculation is shown in Figure 11.8 where we take the rotation axis to be the z-axis and align the edges of the door parallel to the x- and y-axes. Once again we take the mass element to be $dm = \sigma_o\, dx\, dy$. This time the shortest distance from the mass element dm to the axis of rotation is given by $r = \sqrt{x^2 + y^2}$. With these substitutions the moment of inertia integral becomes

$$I = \int_{\text{door}} r^2\, dm = \int_{-b/2}^{b/2} \int_{-a/2}^{a/2} (x^2 + y^2)\sigma_o\, dy\, dx$$

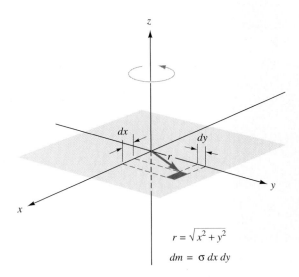

FIGURE 11.8

A flat rectangular mass rotating about an axis perpendicular to its plane. If the axis of rotation is the z-axis, each mass element is located at a distance $r = \sqrt{x^2 + y^2}$. See Example 11.7.

We perform the y-integration first, treating x as a constant,

$$I = \int_{-b/2}^{b/2} \left(x^2 y + \frac{y^3}{3} \right) \Bigg|_{-a/2}^{a/2} \sigma_o \, dx = \sigma_o \int_{-b/2}^{b/2} \left(x^2 a + \frac{a^3}{12} \right) dx$$

Next, we perform the x-integration:

$$I = \sigma_o \left(\frac{x^3}{3} a + \frac{a^3}{12} x \right) \Bigg|_{-b/2}^{b/2} = \sigma_o \left(\frac{b^3}{12} a + \frac{a^3}{12} b \right)$$

The total mass of the door is $M = \sigma_o ab$, so we may write the moment of inertia as

$$I = \frac{1}{12} M(a^2 + b^2)$$

Note that this rotational inertia is for the same door as in Example 11.6, but the result is quite different. The point you should not miss here is that the moment of inertia depends both on the object *and the axis of rotation.* ◄

Table 11.4 provides the rotational inertia for some common shapes. The equations given in this figure were derived by applications of Equation (11.6) for the particular geometry of each object and rotation axis. More complicated shapes can be built up from combinations of these shapes. The moments of inertia are additive, provided, of course, that the axes of rotation are one and the same.

Notice that objects that have mass distributions with most of their mass located at large r-values have large rotational inertia. Objects with mass concentrated at small distances r from the rotation axis have small I-values. We can observe these features in Table 11.4. The rotational inertia for a solid disk is $I_{\text{disk}} = \frac{1}{2}MR^2$, whereas a hoop with the same mass and radius has $I_{\text{hoop}} = MR^2$. The hoop has all its mass located at $r = R$, whereas the disk has a substantial amount of its mass located much closer to the axis where $r \ll R$. Hence, we expect the rotational inertia of the solid disk to be smaller than that of the hoop with the same mass and radius. Moreover, we see that for a hoop and a solid disk each with identical mass, radius, and angular velocity, the hoop has a larger kinetic energy (Eq. (11.4)) because its rotational inertia is larger.

Torques and flywheels find application in the world's largest steam hoist at the Quincy Mine located in Michigan.

TABLE 11.4   Rotational Inertia for Common Shapes

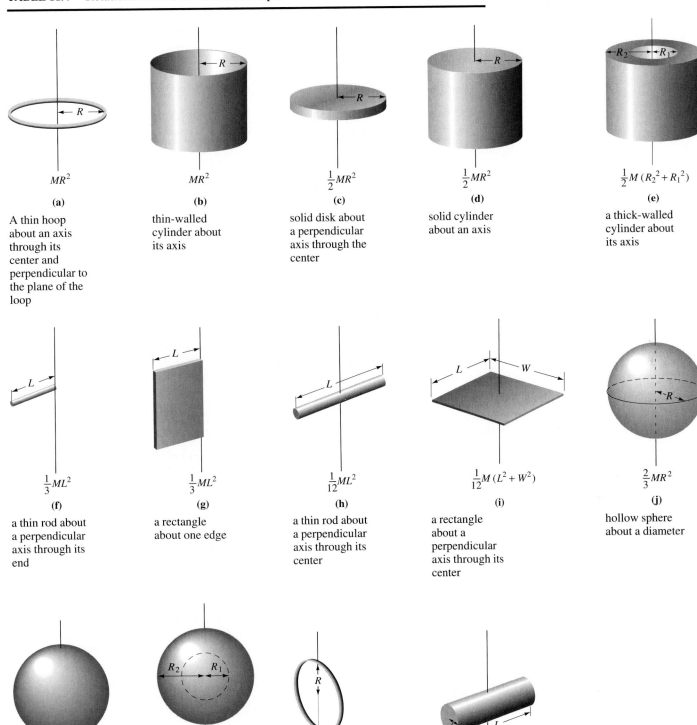

MR^2

(a)

A thin hoop about an axis through its center and perpendicular to the plane of the loop

MR^2

(b)

thin-walled cylinder about its axis

$\frac{1}{2}MR^2$

(c)

solid disk about a perpendicular axis through the center

$\frac{1}{2}MR^2$

(d)

solid cylinder about an axis

$\frac{1}{2}M(R_2^2 + R_1^2)$

(e)

a thick-walled cylinder about its axis

$\frac{1}{3}ML^2$

(f)

a thin rod about a perpendicular axis through its end

$\frac{1}{3}ML^2$

(g)

a rectangle about one edge

$\frac{1}{12}ML^2$

(h)

a thin rod about a perpendicular axis through its center

$\frac{1}{12}M(L^2 + W^2)$

(i)

a rectangle about a perpendicular axis through its center

$\frac{2}{3}MR^2$

(j)

hollow sphere about a diameter

$\frac{2}{5}MR^2$

(k)

a solid sphere about its diameter

$\frac{2}{5}M\left(\dfrac{R_2^5 - R_1^5}{R_2^3 - R_1^3}\right)$

(l)

a thick-walled hollow sphere about a diameter

$\frac{1}{2}MR^2$

(m)

a thin hoop about a diameter

$\frac{1}{4}MR^2 + \frac{1}{12}ML^2$

(n)

a solid cylinder about a perpendicular axis though its center.

The Parallel-Axis Theorem and the Perpendicular-Axis Theorem

The parallel-axis theorem and the perpendicular-axis theorem can be used to extend the utility of Table 11.4. We state the theorems here and leave the proofs as exercises.

The Parallel-Axis Theorem

Let the moment of inertia of an object through its center of mass be I_{CM}. The moment of inertia of the mass through any other axis *parallel to the first* is given by

$$I_{\text{parallel}} = I_{CM} + Ml^2$$

where M is the total mass of the object and l is the perpendicular distance between the two parallel axes. You should compare the answer to Example 11.5 for the moment of inertia of a thin rod to the entry in Table 11.4 for a thin rod rotated about its center. Convince yourself that they are indeed related by the parallel-axis theorem.

Perpendicular-Axis Theorem

By a **laminar body** we mean an object that can be treated as two-dimensional, such as a large figure cut from a thin sheet of plywood. Let the z-axis be perpendicular to a laminar body and run through the body's center of mass. Call the moment of inertia of the body about that axis I_z. Let the x- and y-axes lie in the plane of the body with the origin at the center of mass as shown in Figure 11.9. Define the moment of

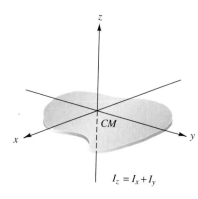

$$I_z = I_x + I_y$$

FIGURE 11.9

Let I_x be the rotational inertia of a laminar body about the x-axis lying in the plane of the body and passing though the body's center of mass. Let I_y be the rotational inertia about a second axis passing through the center of mass, perpendicular to the first and also lying in the plane of the object. Then the rotational inertia I_z of the laminar object about a third axis normal to its plane and passing through the center of mass is given by $I_z = I_x + I_y$.

inertia about the x- and y-axes to be I_x and I_y. The perpendicular-axis theorem states that

$$I_z = I_x + I_y$$

For example, if you double the expression for the moment of inertia for the circular hoop rotating about its diameter (part (m) from Table 11.4), you find the result equal to the rotational inertia of the hoop rotating as in part (a) in Table 11.4.

Once the moment of inertia has been calculated, we can incorporate the kinetic energy of rotation into problems involving the conservation of mechanical energy. We once again use the "before" and "after" approach of Chapter 8. You may wish to review the problem-solving steps from that chapter before reading Examples 11.8 and 11.9. Note that the rotational kinetic energy $\frac{1}{2}I\omega^2$ is not a new form of energy but merely a convenient way to calculate the total energy of motion of the many particles that make up the rotating object.

EXAMPLE 11.8 *Who Was Supposed to Hold the Wall?*

One wall of a house was framed using 8.00 ft long 2x4's spaced on 24.0-in centers as shown in Figure 11.10(a). (a) Calculate the moment of inertia of the wall for rotation about its lower edge. Take the mass of each 2x4 to be 3.50 kg. (b) If the wall starts from vertical with zero speed and falls over, how fast is the top edge traveling just before it hits the ground?

SOLUTION   (a) We model the wall as consisting of seven thin rods each of mass m and length l. Five of the rods rotate about their ends. From Table 11.4 we find each rod has a moment of inertia $\frac{1}{3}ml^2$. The header (the top 2x4) has all of its mass located at distance l from the axis of rotation and hence has a moment of inertia of ml^2. The footer (the bottom 2x4) has all its mass located at such a small distance from the rotation axis that its moment of inertia about this axis is negligible. Thus, the total moment of inertia of our model is

$$I = 5 \times \left(\frac{1}{3}ml^2\right) + ml^2 = \frac{8}{3}ml^2$$

(b) When the wall is vertical all of its energy is gravitational potential energy. In Chapter 10 we saw that the gravitational potential energy of an extended object can be calculated from Mgh_{CM}, where h_{CM} is the height of the center of mass. By inspection, we see that the center of mass of the wall is located at a height of $l/2$ above the rotation axis.

If the wall simply rotates about its base, the initial potential energy turns entirely into kinetic energy just before the wall slams into the ground. Thus, the conservation of mechanical energy

$$U_o + K_o = U_f + K_f$$

becomes

$$Mgh_{CM} + 0 = 0 + \frac{1}{2}I\omega^2$$

$$(7m)g\,\frac{1}{2} = \frac{1}{2}\left(\frac{8}{3}ml^2\right)\omega^2$$

$$\omega = \sqrt{\frac{21}{8}\frac{g}{l}}$$

The header is traveling in a circular arc of radius $r = l$. We find its tangential velocity using

$$v_t = R\omega = l\sqrt{\frac{21}{8}\frac{g}{l}} = \sqrt{\frac{21}{8}gl}$$

$$= \sqrt{\frac{21}{8}(9.80 \text{ N/kg})(2.44 \text{ m})} = 7.92 \text{ m/s}$$

Note we had to substitute $M = 7m$. You might be tempted to leave out the mass of the footer since it does not change height and had zero moment of inertia. However, we must include the footer mass, otherwise, the center-of-mass height would not be $l/2$.

(a) Before

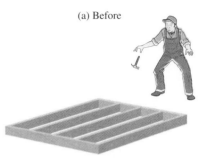

(b) After

FIGURE 11.10

A wall framed of 2x4's falls over starting from rest. See Example 11.8.

EXAMPLE 11.9   No More "Massless-Pulley" Models

A certain mass M is free to slide without friction across a horizontal table. This mass is connected by a light thread to a mass m that hangs over the edge of the table. (See Figure 11.11.) The connecting thread passes over a frictionless pulley in the shape of a disk with radius R and mass m_p. Calculate the velocity of the string after the hanging mass has fallen a distance y_o starting from rest.

SOLUTION   Because the pulley is a disk, we use part (d) in Table 11.4 to write its moment of inertia as $I = \frac{1}{2}m_p R^2$. We again employ the conservation of mechanical energy. The potential energy of mass m is going to be changed into the kinetic energy due to the translation of both masses plus the kinetic energy due to the rotation of the

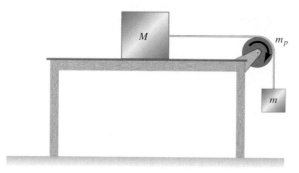

FIGURE 11.11

The potential energy of the falling mass is converted into kinetic energy of all three masses. The kinetic energy of the pulley is due to its rotation.

pulley. From the conservation of mechanical energy theorem we may write

$$U_o + K_o = U_f + K_f$$

$$mgy_o = \frac{1}{2}mv^2 + \frac{1}{2}Mv^2 + \frac{1}{2}I\omega^2$$

where v is the translational velocity of the blocks and ω is the rotational velocity of the pulley. Note that, because mass M does not change height, only m appears in the potential energy term. At any instant the string moves with the same velocity v, as do the masses. Now, *provided that the string does not slip* on the pulley edge, the tangential velocity of the edge of the pulley is also v. Thus, we can write $v = v_t = R\omega$ and substitute $\omega = v/R$. If at the same time we substitute the expression for I of a uniform disk as found in Table 11.4, we obtain

$$mgy_0 = \frac{1}{2}mv^2 + \frac{1}{2}Mv^2 + \frac{1}{2}\left(\frac{1}{2}m_p R^2\right)\left(\frac{v}{R}\right)^2$$

In this expression v is the only unknown, so we may solve immediately:

$$v = \sqrt{\frac{2mgy_o}{m + M + m_p/2}}$$ ◀

11.3 Torque

We are well on our way to finding the rotational analog of Newton's second law $F_{net} = ma$. We have already discovered that the analog of a is α and the analog of m is I. It remains for us to find the analog of force. Forces are necessary to change an object's linear motion. Therefore, it seems natural to ask, "What is necessary to change an object's rotational motion?" or, more simply, "What is it that makes things start to rotate?" Obviously, we need a force, but there's more to it than that, as we can see by examining a simple wrench. Figure 11.12 shows three wrenches all with the same force applied. Which

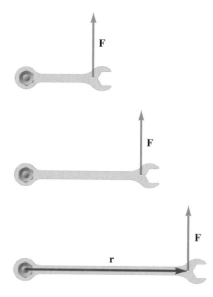

FIGURE 11.12

The farther the point of application of force **F** from the nut, the greater turning effect it has.

one has the greatest turning effect on the nut? You're right. The one with the longest handle, of course. Many a "frozen" automobile head bolt has been loosened by slipping a pipe over the end of the wrench to provide more leverage. We conclude that the length of **r**, the vector from the pivot to the point of application of the force, is an important factor.

But distance and force alone cannot be the whole story. Look at Figures 11.13(a) and 13(b). You can push or pull on the end of a wrench for a long time, and it will never turn the nut! Clearly, any force component parallel or antiparallel to **r** will not turn the wrench. As shown in Figure 11.14, if a force **F** is applied at an arbitrary angle with respect to the direction of **r**, the force can be resolved into two components: one parallel to **r**, and one perpendicular to **r**. Only the component perpendicular to **r** contributes to the turning effect. If the vectors **r** and **F** form an angle ϕ when placed tail to tail, as in Figure 11.14, then the magnitude of the component of **F** perpendicular to **r** is $F \sin(\phi)$.

We now have the factors to create rotational motion; namely, $rF \sin(\phi)$. What remains is to give them a name. The product of these turning factors is called **torque** and is symbolized by the lowercase Greek letter τ (tau). Thus, we write[2]

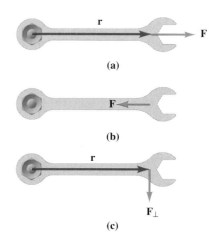

FIGURE 11.13

Forces such as those in (a) and (b) are parallel to the vector **r**, which joins the pivot point to the point of application of the force **F**. Such forces have no turning effect.

$$\tau = rF \sin(\phi) \qquad \text{Magnitude of torque} \qquad (11.7)$$
$$\text{(} \mathbf{r} \text{ and } \mathbf{F} \text{ in a plane perpendicular}$$
$$\text{to the axis of rotation)}$$

In SI, the units of torque are newton-meters (N · m). Torque wrenches using the British Engineering system are labeled in foot-pounds (ft · lbs) or ounce-inches (oz · in).

We have interpreted $F \sin(\phi)$ in the torque definition to be the component of **F** perpendicular to **r**. In engineering circles it is conventional to associate the $\sin(\phi)$ term with r and think of $r \sin(\phi)$ as the component of **r** perpendicular to **F**. With this interpretation $r \sin(\phi)$ is known as the **moment arm,** or **lever arm,** of the force **F**. The moment arm is sometimes designated as $r_\perp$; when this is done the magnitude of the torque may be written $\tau = r_\perp F$. The moment arm of a force is easy to find. As shown in Figure 11.15, to find the lever arm you draw a line parallel to the force **F** through the point where this force is applied. This line is sometimes called the **line of action** of the force **F**. The perpendicular distance from the pivot to the line of action is the moment arm.

In its fullest definition torque is a vector. For the special cases we consider in this chapter and the next, the direction of the torque vector is parallel to the axis of rotation. To determine in which direction parallel to the axis the torque vector points, you have only to wrap the fingers of your right hand around the rotation axis in the same sense in which the torque tries to rotate the object. Your thumb then points along the axis in the direction of the torque. Thus, for example, in Figure 11.15 the force **F** turns the wrench anticlockwise. If you wrap the fingers of your right hand around the nut in an anticlockwise sense, your right thumb points upward out of the page. This vector designation may seem a rather bizarre convention at this point in your study, but it does allow us to conveniently distin-

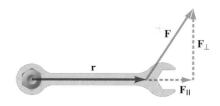
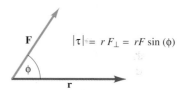

$$|\tau| = rF_\perp = rF \sin(\phi)$$

FIGURE 11.14

When a force acts in a plane perpendicular to the axis of rotation, the magnitude of the torque is found from $\tau = rF \sin(\phi)$.

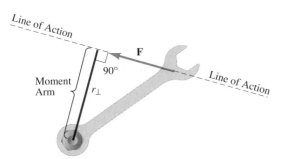

FIGURE 11.15

The *line of action* of a force is a line parallel to the force that passes through the point of application of the force. The *moment arm* $r_\perp$ is the length of the shortest line segment connecting the pivot and the line of action of the force. The magnitude of the torque about the pivot is given by $r_\perp F$. See Example 11.10.

---

[2] We have assumed that **F** and **r** are in a plane perpendicular to the rotational axis. In Chapter 13 we will expand this definition to include a vector convention for the direction of torque.

guish between torques that cause clockwise motion and torques that cause anticlockwise motion. Later, in Chapter 13, a more complete definition of the torque vector will be presented. In that chapter we study the motion of gyroscopes and other rotating systems with axes whose directions in space change. Only then does the beauty and power of the vector convention for torque become apparent. However, you will find it helpful later if you begin now practicing the vector designation of torque direction.

EXAMPLE 11.10   One Good Turn Deserves Another

Find the net torque about point P produced by the two forces shown in Figure 11.16, if $F_1 = 60.0$ N, $r_1 = 0.100$ m, $\theta_1 = 40.0°$, $F_2 = 45.0$ N, $r_2 = 0.300$ m, and $\theta_2 = 30.0°$.

SOLUTION   We can calculate the magnitude of the torque caused by F_1 directly from the definition Equation (11.7):

$$\tau_1 = F_1 r_1 \sin(\phi_1)$$

Note very carefully that the angle ϕ in our definition is the angle between r and F when they are placed *tail to tail*. For $\mathbf{F}_1$ and $\mathbf{r}_1$ the angle $\phi_1 = 180.0° - 40.0° = 140.0°$.

$$\tau_1 = (60.0 \text{ N})(0.100 \text{ m}) \sin(140.0°) = 3.86 \text{ N} \cdot \text{m}$$

The torque τ_1 acting alone causes the wheel to rotate anticlockwise, and thus the direction of this torque is out of the page.

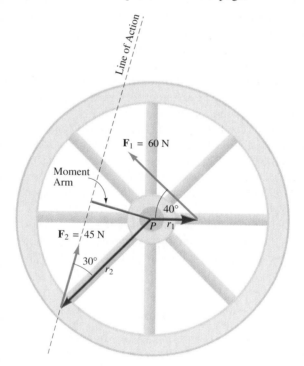

FIGURE 11.16

We could calculate the magnitude of τ_2 in the same manner. However, to demonstrate the use of the "moment-arm method" we extend the line of force vector $\mathbf{F}_2$ in Figure 11.16 in both directions to indicate the line of action of $\mathbf{F}_2$. The moment arm is a line segment extending from the pivot point to the line of action and meeting the line of action at a right angle. The length of the moment arm can be calculated from trigonometry

$$\textit{Moment arm} = r_\perp = r_2 \sin(\theta) = (0.300 \text{ m}) \sin(30.0°) = 0.150 \text{ m}$$

The torque is then computed from

$$\tau_2 = r_\perp F_2 = (0.150 \text{ m})(45.0 \text{ N}) = 6.75 \text{ N} \cdot \text{m}$$

Note that τ_2 acting alone causes the wheel to rotate in the clockwise sense and thus is represented by a vector directed into the page.

The two torques attempt to rotate the wheel in opposite senses (τ_1 tries to rotate the wheel anticlockwise, whereas τ_2 tries to rotate the wheel clockwise). Thus, the two torques tend to counteract one another. We represent this competition with $+$ and $-$ signs, in this instance taking $+$ as anticlockwise. Consequently, the net torque is

$$\tau_\text{net} = \tau_1(\text{out of the page}) + \tau_2(\text{into the page})$$

$$= +3.86 \text{ N} \cdot \text{m} - 6.75 \text{ N} \cdot \text{m} = -2.89 \text{ N} \cdot \text{m}$$

The minus in this case indicates that the net effect of the combined torques is to rotate the wheel clockwise and is represented by a vector into the page. ◀

Concept Question 5
You have available only a short box-end wrench with which you are unable to loosen a nut when you pull with all your might. Explain why sliding a long pipe over the end of the wrench will help.

Work Done by a Torque

In Chapter 7 we defined the increment of work done by force **F** on an object that moved an incremental distance ds as $dW = \mathbf{F} \cdot d\mathbf{s} = F \, ds \cos(\theta)$. We interpreted this equation to mean that only the component of **F** parallel to the displacement ds does work, that is, $dW = F_\parallel ds$. Let's look at this interpretation in the context of the tangential force **F**, which applies a torque to a rotating body as shown in Figure 11.17. Point P is moving in a circle

FIGURE 11.17

The motorist applies a tangential force as her hand turns the jack handle in circular motion.

of radius R. Thus, the distance ds is tangent to the circle and equal to $R \, d\theta$. Now, because $F_\parallel$ is tangent to the circle, it is perpendicular to R. Consequently, $F_\parallel$ can be written as $F \sin(\phi)$. Putting these substitutions together, we can write

$$dW = F \sin(\phi) \, R \, d\theta$$

You no doubt recognize the combination $F \sin(\phi) \, R$ in the above expression as our definition of torque. Thus, we can write

$$dW = \tau \, d\theta$$

If the torque remains constant, this expression can be integrated to produce a simple expression for the work done

$$\Delta W = \tau \, \Delta\theta \qquad \text{Work done by a constant torque on an object rotating about a fixed axis}$$

Recalling the definition of power $\mathscr{P} = dW/dt$ from Chapter 7, it immediately follows that

$$\mathscr{P} = \frac{dW}{dt} = \lim_{\Delta t \to 0} \frac{\Delta W}{\Delta t} = \lim_{\Delta t \to 0} \frac{\tau \Delta \theta}{\Delta t} = \tau \lim_{\Delta t \to 0} \frac{\Delta \theta}{\Delta t} = \tau \frac{d\theta}{dt}$$

or

$$\mathscr{P} = \tau \omega \qquad \text{Instantaneous power expended by a torque on an object rotating about a fixed axis}$$

11.4 Newton's Second Law for Rotation about a Fixed Axis

We can now determine the form of Newton's second law for rotations about a fixed axis by combining the results of the previous two sections with the work–energy theorem of Chapter 7. The three results we combine are

$$K = \frac{1}{2} I\omega^2 \qquad \text{Kinetic energy of rotation}$$

$$\mathscr{P} = \tau \omega \qquad \text{Power applied to a rotating object}$$

and

$$\mathscr{W}_{\text{net}} = \Delta K = K - K_o \qquad \text{Work–energy theorem}$$

where K_o, the initial kinetic energy of the object, is a constant, and K is the final kinetic energy after the work $\mathscr{W}_{\text{net}}$ is done. Taking the time derivative of the work–energy theorem and recalling that the time rate of performing work is power, we obtain

$$\frac{d}{dt}(\mathscr{W}_{\text{net}}) = \frac{d}{dt}\left(\frac{1}{2} I\omega^2 - \frac{1}{2} I\omega_o^2\right)$$

$$\mathscr{P}_{\text{net}} = I\omega \frac{d\omega}{dt} + 0 = I\omega\alpha$$

Now substituting $\mathscr{P}_{\text{net}} = \tau_{\text{net}}\omega$, we obtain:

$$\tau_{\text{net}}\omega = I\omega\alpha$$

and dividing both sides by ω leads to

$$\tau_{\text{net}} = I\alpha \qquad \text{Newton's second law for an object rotating about a fixed axis} \qquad (11.8)$$

With Newton's second law for rotational motion we are able to solve a wide variety of rotational motion problems. We may also obtain a little more insight into the meaning of the moment of inertia I. When we solve Equation (11.8) for the angular acceleration α, we obtain

$$\alpha = \frac{\tau_{\text{net}}}{I}$$

This result tells us that for a given torque τ_{net} the angular acceleration is small when the moment of inertia is large and, conversely, the angular acceleration is large when the moment of inertia is small. Hence, we see that the moment of inertia really does play an inertial role, just as the mass does for translational motion. It is important to remember, as we pointed out in Section 11.2, that mass alone does not determine the rotational inertia; the mass's location relative to the axis of rotation is also critical. For example, we have found that the moment of inertia of a hoop is larger than that of a solid disk with the same mass and radius. Hence, more net torque is required to change the rotational state of the hoop than that of the disk. Objects with large moments of inertia require more torque to change their rotation state than do those with small moments of inertia.

As a specific example Equation (11.8) permits us to include more realistic pulley models into our force-analysis problems. You may find it helpful to take a few minutes to review the problem-solving strategy for force-analysis problems, which we introduced in Chapter 5. A quick reading of Example 5.7 may also help bring to mind strategies for connected-mass type problems. Below we list only the new features you will need to keep in mind when you model pulleys with nonzero moments of inertia.

Special Precautions for Problems Containing Objects with Nonzero Rotational Inertia

1. In connected-mass problems the tensions in strings on either side of a pulley are no longer necessarily equal. Denote the tension in all string segments with different symbols.
2. For each object that translates only, write $\mathbf{F}_{net} = m\mathbf{a}$ as described in the problem-solving strategy of Chapter 5. For each object that rotates only, write $\tau_{net} = I\alpha$. (Objects that both translate and rotate are treated in Chapter 12.)
3. Check the sign convention used for each a and α in the equations you have written to ensure they are consistent.
4. Solve the equations generated by steps 2 and 3 for all unknowns.

Omission of Step 3 in the procedure outlined above is responsible for more student errors than any other cause. In Example 11.11 we show details of how to check for this sign consistency. We urge you to never omit this step in your work.

EXAMPLE 11.11   Here We Go 'Round Again . . .

Using the connected masses described in Example 11.9, determine expressions for the acceleration of the system and tension in the two segments of the string. Find numerical values for these unknowns for the case $M = 0.500$ kg, $m = 0.400$ kg, and the pulley mass $m_p = 0.200$ kg.

SOLUTION   Figure 11.18 shows the connected masses with the tension T_1 between M and the pulley, and the tension T_2 between the pulley and mass m. As the figure

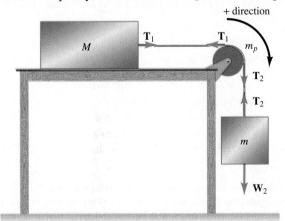

FIGURE 11.18

Connected masses with a pulley with significant rotational inertia. Only forces that move the system are shown.

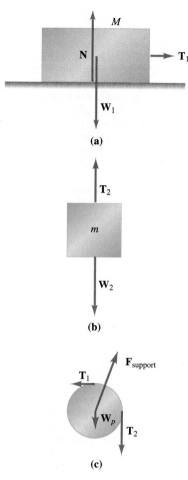

FIGURE 11.19

(a) The weight force $\mathbf{W}_1$ and the normal force $\mathbf{N}$ cancel. The only unbalanced force on M is the horizontal tension T_1. (b) The downward acceleration of m is equal to the rightward acceleration of M. Both accelerations are positive according to the "folded-axis" sign convention of Figure 11.19. (c) The force $\mathbf{F}_{\text{support}}$ of mount supporting the pulley axle must balance the weight $\mathbf{W}_p$ of the pulley and both tensions. However, neither $\mathbf{F}_{\text{support}}$ nor $\mathbf{W}_p$ exert a torque about the pulley axis.

also indicates, we use the "folded-axis" sign convention first introduced in Example 5.9. That is, we take rightward motion of mass M as positive. Consistent with this same sense of motion, we take *downward* motion of mass m as positive. Thus, any force, torque, or acceleration consistent with rightward motion of M or downward motion of m, is taken as positive. Our game plan is to write $\mathbf{F}_{\text{net}} = m\mathbf{a}$ for each mass and $\tau_{\text{net}} = I\alpha$ for the pulley, and then solve the three equations for the unknowns.

$\mathbf{F}_{\text{net}} = M\mathbf{a}$ **for mass M**   Figure 11.19(a) shows the free-body diagram for mass M. We ignore the vertical forces $\mathbf{N}$ and $\mathbf{W}_1$, which cancel. The net horizontal force on M (indeed, the only horizontal force on M) is T_1. We thus write simply

$$+T_1 = Ma$$

$\mathbf{F}_{\text{net}} = m\mathbf{a}$ **for mass m**   Figure 11.19(b) shows the free-body diagram for mass m. In accord with our sign convention we write

$$W_2 - T_2 = ma$$

$\tau_{\text{net}} = I\alpha$ **for the pulley**   Figure 11.19(c) shows an isolated view of the pulley with the tensions T_1 and T_2 acting (remember strings always pull!). The string always leaves the circular pulley along a tangent line. Consequently, $\mathbf{r}$ and $\mathbf{T}$ are always perpendicular, making $\phi = 90°$. Thus, the *magnitudes* of the two torques are

$$|\tau_1| = rT_1 \qquad \text{and} \qquad |\tau_2| = rT_2$$

These two torques tend to turn the pulley in opposite directions so we must give them opposite signs. These signs *must be chosen in accord with our adopted sign convention.* Our convention suggests that anything that tends to make our system move in a clockwise sense should be counted as positive. Thus, we *must* write

$$+rT_2 - rT_1 = I\alpha$$

We now have three equations of motion for our systems:

$$T_1 = Ma$$
$$W_2 - T_2 = ma$$
$$rT_2 - rT_1 = I\alpha$$

Let's stop now and double check that we have followed our original sign convention as we so cautioned in step 3. Compare each equation with its respective free-body diagram and verify that each force and torque is consistent with the sign convention in Figure 11.18.

Our three equations still contain four unknowns: T_1, T_2, a, and α. If the string does not slip on the edge of the pulley, then the tangential acceleration of the pulley edge is equal to a and we may write

$$\alpha = \frac{a_t}{r} = \frac{a}{r} \qquad \text{(No-slip condition)}$$

With this expression for α, the third equation becomes

$$rT_2 - rT_1 = I\frac{a}{r}$$

We now have three equations in three unknowns. It is best to adopt some systematic procedure for solving these equations. We choose the following strategy: solve the force equations for the tensions, then substitute these expressions for the tensions into the torque equation. This procedure results in one equation for the acceleration. Here we go:

$$T_1 = Ma \qquad \text{(First equation already solved for } T_1\text{)}$$

$$T_2 = W_2 - ma \qquad \text{(Second equation solved for } T_2\text{)}$$

$$T_2 - T_1 = I\frac{a}{r^2} \qquad \text{(Torque equation solved for } T\text{'s)}$$

$$(W_2 - ma) - (Ma) = I\frac{a}{r^2} \qquad \text{(Expressions for } T_1 \text{ and } T_2 \text{ substituted)}$$

$$W_2 = \left(M + m + \frac{I}{r^2}\right)a$$

$$a = \frac{W_2}{\left(M + m + \dfrac{I}{r^2}\right)}$$

Recalling that $I = \frac{1}{2}m_p r^2$ and that $W_2 = mg$, we can finally write

$$a = \frac{m}{M + m + \dfrac{1}{2}m_p}g = \frac{0.400\ \text{kg}}{0.500\ \text{kg} + 0.400\ \text{kg} + 0.100\ \text{kg}}g$$

$$= 0.400g = 0.400(+9.80\ \text{N/kg}) = 3.92\ \text{m/s}^2$$

The positive value for this acceleration indicates that the motion is to the right, as we expect. This value of a can now be substituted into our expressions for T_1 and T_2:

$$T_1 = Ma = (0.500\ \text{kg})(0.400\ g) = 1.96\ \text{N}$$

$$T_2 = W_2 - ma = m(g - a) = (0.400\ \text{kg})(g - 0.400g)$$

$$= (0.400\ \text{kg})(0.600g) = (0.400\ \text{kg})(0.600)(+9.80\ \text{N/kg}) = 2.35\ \text{N}$$

Note that all tensions must turn out to be positive values because T's represent magnitudes only. (We put in the tension directions using explicit $+$ and $-$ signs.) In other problems, the acceleration a may turn out to be a negative number, indicating an acceleration in the direction opposite to that which was taken as positive. If a is negative, be sure to substitute it as a negative value into the tension equations. ◀

11.5 Angular Momentum for Rotation about a Fixed Axis

As is the case for the rest of the equations developed in this chapter, the relations we develop next are restricted to objects that rotate about a fixed axis. In what follows we develop the rotational analog to the conservation of momentum theorem. After establishing this new law, we present a series of examples that show applications of this powerful tool.

Let's first consider a point-particle rotating about a fixed axis. Our consideration is strictly limited to the special case of particles traveling in a plane to which the axis of rotation is normal. For such a case we define the **angular momentum** as

$$l = rmv \sin(\phi)$$ Magnitude of the angular momentum of a   (11.9)
point–mass moving in a plane perpendicular
to axis of rotation

Here r is the distance of the mass m from the axis of rotation, v is the speed of the mass, and ϕ is the angle between the **r** and **v** when the vectors are placed tail to tail. This definition remains valid even if the particle is not traveling in a circle about the axis. In fact, the particle can be traveling in a straight line and we still say that it has an angular momentum about the chosen axis. The only restriction associated with this definition is that the plane of the particle's motion must be perpendicular to the axis. We leave it for you to show that the angular momentum of a particle traveling in a straight line is a constant (Problem 63).

If the particle *is* traveling in a circle of constant radius r, then ϕ, the angle between **r** and **v** is 90°, and we may substitute $v = r\omega$ to rewrite the expression for angular momentum as

$$l = mr^2\omega$$ (point-particle moving in a circle, radian measure)

As you may realize, mr^2 is the moment of inertia of a point-particle located a fixed distance r from the axis of rotation. For a collection of N particles *all in the same plane perpendicular to the rotation axis,* the total angular momentum L is

$$L = \left(\sum_{i=1}^{N} m_i r_i^2\right)\omega$$

The quantity in brackets is, as you probably anticipated, the rotational inertia of the N particles. Thus, we may use the following expression for the magnitude of angular momentum under a limited set of circumstances

$$L = I\omega$$ Magnitude of the angular momentum for   (11.10)
1. a laminar body rotating about a fixed axis
perpendicular to the plane of the body, or
2. a symmetrical body rotating about an axis
of symmetry through its center of mass.

Angular momentum, like torque, is properly defined as a vector. For the special cases listed in Equation (11.10), the angular momentum vector is parallel to the axle. The direction of the angular momentum vector can be determined by wrapping the fingers of your right hand around the axle in the direction of the rotation. Your extended thumb then indicates the direction parallel to the axle along which the angular momentum vector points. The full vector definition of angular momentum is presented in Chapter 13.

Conservation of Angular Momentum

In the case of linear motion, the absence of a net external force led directly to the law of conservation of linear momentum. The case of rotation follows the same pattern. When $\tau_{\text{net}} = 0$, we know that the angular acceleration α is zero, which implies that ω is a constant. For the special case of a rotation about a fixed axle through an axis of symmetry, a constant ω implies a constant L so that there is no change in the angular momentum. This law of the **conservation of angular momentum** is much more profound in its vector form discussed in Chapter 13. Nonetheless, it is a powerful theorem even in this restricted version. As in the case of linear motion, conservation of angular momentum allows us to predict the outcome of collisions, such as the one examined in Example 11.12.

EXAMPLE 11.12   *I Wonder If Those Little Kids Realize They're Doing Physics Experiments*

A 25.0-kg child runs at 5.00 m/s along a line tangent to the circumference of a stationary merry-go-round of radius 2.00 m. If the merry-go-round has a rotational inertia of $4.00 \times 10^2 \ kg \cdot m^2$, what is the rotational rate of the merry-go-round if the child grabs hold and jumps on while going by?

SOLUTION   This example sounds like a collision problem, so we are led from experience to look for conserved quantities. Candidates are: mechanical energy, linear momentum, and angular momentum. The event is an inelastic collision (the child "sticks" to the merry-go-round) and our experience reminds us that mechanical

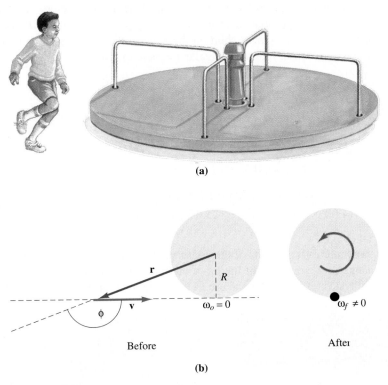

FIGURE 11.20

(a) A child runs along a path that is tangent to the edge of a playground merry-go-round. (b) We model the child's action as an inelastic collision that conserves angular momentum. See Example 11.12.

energy is not conserved in such cases. We reject linear momentum because we realize that the system is not isolated; large horizontal forces from the earth are transmitted to the merry-go-round via the post in the center. We can, however, apply the conservation of angular momentum because the large horizontal forces at the center of the merry-go-round do not exert torques around the post because they act at the rotation axis. We must, however, be willing to ignore the frictional torque about the post. We assume that the merry-go-round is constructed from bearings with good lubrication so that this frictional torque is small. Moreover, we expect that the frictional torque is small because the moment arm from the center of the post to the point where the friction acts is small. When we ignore the small frictional torque during the collision, we can employ the rotational equivalent of the "collision approximation" introduced in Chapter 9.

Having settled on our strategy, we now write the conservation of angular momentum

$$l_{\text{child}} + L_{\text{mgr}} = l'_{\text{child}} + L'_{\text{mgr}}$$

where, as usual, quantities before the collision are unprimed and quantities after the collision are primed. We model the child as a point–particle (poor kid) and calculate l_{child} using Equation (11.9). From Figure 11.20(b) it is clear that $r \sin(\phi) = R$ for all values of r and ϕ along the tangent line that indicates the child's path. This observation leads to a simple expression for the child's angular momentum

$$l_{\text{child}} = mrv \sin(\phi) = mRv$$

The angular momentum of the merry-go-round we write as $L_{\text{mgr}} = I_{\text{mgr}}\omega$. Because $\omega = 0$ before the collision, the initial angular momentum of the merry-go-round is zero.

After the collision, it is convenient to think of the child-plus-merry-go-round as a single compound object with moment of inertia $I' = I_{\text{mgr}} + mR^2$, where mR^2 is the moment of inertia of our point–mass child on the edge of the merry-go-round. With these substitutions our conservation of angular momentum equation becomes

$$mRv = I'\omega'$$

so that

$$\omega' = \frac{mRv}{I'} = \frac{mRv}{I_{\text{mgr}} + mR^2}$$

$$= \frac{(25.0 \text{ kg})(2.00 \text{ m})(5.00 \text{ m/s})}{(4.00 \times 10^2 \text{ kg} \cdot \text{m}^2) + (25.0 \text{ kg})(2.00 \text{ m})^2} = 0.500 \text{ rad/s} \quad \blacktriangleleft$$

Some of the most interesting results of the conservation of angular momentum occur when there is a change in the moment of inertia of an object as it rotates. Figure skaters, ballet dancers, and physics teachers all make use of this phenomenon, albeit with differing

An ice skater executing a spin employs the conservation of angular momentum. (a) The skater starts with a large rotational inertia and small rotation rate. (b) The skater ends with a small rotational inertia and large angular speed.

degrees of grace. If you have watched figure skaters, you have perhaps noted their ability to spin on the point of one skate with remarkably high angular velocity. The next time you see such a performance, notice that the skater starts his or her turn in a very open stance, often with arms and a leg held far outward. After starting the turn, the skater then draws arms and legs as close to the body as possible. As this happens, the rotation rate increases. The frictional torque on the point of one skate is small, and the angular momentum $I\omega$ is nearly constant. At the beginning of the rotation the skater's I is large and ω is small. As the skater reduces I, ω must increase to keep the product $I\omega$ constant. We examine this phenomenon quantitatively in Example 11.13.

EXAMPLE 11.13   *Taking the Teacher for a Spin*

A physics teacher sits on a stool that is free to rotate nearly without friction about a vertical axis. (See Figure 11.21.) Her outstretched hands each hold a large mass so that her rotational inertia is $12.00 \ \text{kg} \cdot \text{m}^2$. By pulling her arms in close to her body she is able to reduce her rotational inertia to $6.00 \ \text{kg} \cdot \text{m}^2$. If her students start her spinning at $0.500 \ \text{rad/s}$, what is her rotational speed after she draws her arms in?

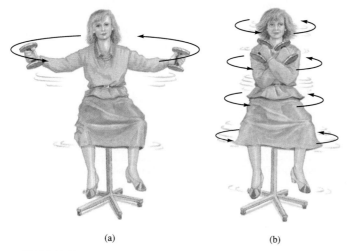

(a)   (b)

FIGURE 11.21

By reducing her moment of inertia the physics professor can increase her angular velocity because $I\omega$ is conserved.

SOLUTION   In the absence of external torques her angular momentum stays constant so that

$$I\omega = I'\omega'$$

$$\omega' = \frac{I\omega}{I'} = \frac{(12.00 \ \text{kg} \cdot \text{m}^2)(0.500 \ \text{rad/s})}{6.00 \ \text{kg} \cdot \text{m}^2} = 1.00 \ \text{rad/s}$$

When her rotational inertia halves, her angular velocity doubles.   ◀

11.6 Summary

A system of equations has been developed to describe rotation about a fixed axis. These equations are analogous to the equations for mechanics in one dimension. We have listed the major results in tabular form below. The quantities in the two columns are *not*, in general, equal. The column title should prefix the labels in the leftmost column. For example, the correct relation for the row labeled "velocity" is "*linear* velocity v is analogous to *rotational* velocity ω."

	LINEAR	(UNITS)	ROTATIONAL	(UNITS)
Kinematic Quantities				
position	x	(m)	θ	(rad)
displacement	Δx	(m)	$\Delta\theta$	(rad)
velocity	$v \equiv dx/dt$	(m/s)	$\omega \equiv d\theta/dt$	(rad/s)
acceleration	$a \equiv dv/dt$	(m/s²)	$\alpha \equiv d\omega/dt$	(rad/s²)
	$v = \int a\,dt + C$		$\omega = \int \alpha\,dt + C$	
	$x = \int v\,dt + C'$		$\theta = \int \omega\,dt + C'$	
Dynamic Quantities				
force/torque	F	(N)	$\tau = rF\sin(\phi)$	(N·m)
inertia	m	(kg)	$I = \sum_{i=1}^{n} m_i r_i^2$	(kg·m²)
kinetic energy*	$K = \frac{1}{2}mv^2$	(J)	$K = \frac{1}{2}I\omega^2$	(J)
work*	$\mathcal{W} = F\,\Delta s$	(J)	$\mathcal{W} = \tau\,\Delta\theta$	(J)
power*	$\mathcal{P} = Fv$	(W)	$\mathcal{P} = \tau\omega$	(W)
momentum	$p = mv$	(kg·m/s)	$l = mrv\sin(\phi)$ or $L = I\omega$	(kg·m²/s)
Physical Laws				
Newton's second law	$F_{net} = ma$		$\tau_{net} = I\alpha$	
Work–Energy theorem*	$\mathcal{W}_{net} = \Delta K$		$\mathcal{W}_{net} = \Delta K$	
Conservation of momentum:				
	$F_{net} = 0$ implies $P =$ constant		$\tau_{net} = 0$ implies $L =$ constant	

*These quantities are not merely analogous. They are alternative expressions for the exact same physical quantity.

PROBLEMS

11.1 Kinematics of Rotational Motion

1. State whether each of the following is an angle ($\Delta\theta$), angular speed (ω), or angular acceleration (α) and convert each quantity to radian-based units: (a) 36.87°, (b) $33\frac{1}{3}$ rpm, (c) 25.0 revolutions, (d) one revolution in 24.0 h, (e) 5.00 rpm/h.

2. State whether each of the following is an angle ($\Delta\theta$), angular speed (ω), or angular acceleration (α) and convert each to radian-based units: (a) 120°/s per day (b) 100 rpm (c) 1.00° (d) π revolutions (e) reduce one revolution in 365.25 days.

3. The angular acceleration of a flywheel is given by $\alpha = \alpha_o e^{-t/\tau}$. (a) Determine $\omega(t)$ and $\theta(t)$ if the flywheel starts from rest. (b) If $\alpha_o = 2.00$ rpm/s and $\tau = 5.00$ s, what is the average angular acceleration during the first minute? (c) How many revolutions did the wheel turn in the first minute? (d) What was the average angular velocity during the first minute?

4. The angular acceleration of a disk is described by $\alpha = 8.00$ rad/s² $- (0.0500$ rad/s⁴$)\ t^2$ for $0.0 \leq t \leq 16.0$ s. Assuming the disk started from rest, for the period of time between 0.00 s and 16.0 s find (a) the average angular acceleration, (b) the total number of revolutions, and (c) the final angular velocity.

5. A graph of the angular velocity versus time for a certain rotating object is shown in Figure 11.P1. (a) Calculate the total number of revolutions made by the disk between $t = 5.00$ s and $t = 15.00$ s.

(b) What is the angular acceleration of the object at $t = 3.00$ s?

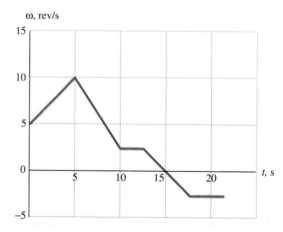

FIGURE 11.P1   Problem 5

6. The angular velocity versus time for a particular compact disk (they're not all the same!) is shown in Figure 11.P2. (a) Estimate the total number of revolutions made by the disk. (b) Determine the initial and final angular accelerations. (The rather complicated kinematics of compact disks is discussed in "Kinematics of the

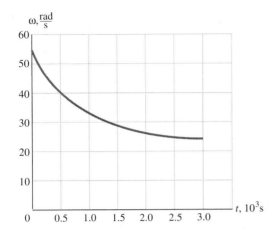

FIGURE 11.P2   Problem 6

Compact Disk/Digital Audio System,'' J. P. McKelvey, *American Journal of Physics*, **54** (12), December 1985, pp. 1160–1165.)

7. A race car is traveling 127 mph around a circular track of circumference 1.00 mi. Calculate the car's angular velocity around the track.

8. From the data in Appendix 8 determine the velocity of the earth through space due to its orbital motion around the sun.

9. The tip of the minute hand of a certain clock moves through a distance of 50.0 cm in 45.0 min. (a) What is the length of the minute hand? (b) What is the angular velocity of the minute hand? (c) What is the linear velocity of the tip of the minute hand?

10. (a) What is the angular velocity due to the earth's rotation of a person standing on the equator? (b) Using data from Appendix 8, determine the linear speed of that person due to the earth's rotation. (c) Calculate the centripetal acceleration of this person.

11. As viewed from above, a merry-go-round of radius 3.00 m is rotating clockwise with a rotation rate of 6.00 rpm relative to the ground. The operator walks along the edge in the direction opposite to the rotation of the merry-go-round so that he has a *clockwise* rotation rate relative to the ground of only 4.00 rpm. (a) What is the velocity (relative to the earth) of the edge of the merry-go-round? (b) What is the velocity (relative to the earth) of the operator? (c) What is the velocity of the operator relative to the merry-go-round surface?

12. A wheel of radius 20.0 cm starts to turn from rest with an angular acceleration of 1.00 rad/s^2 for 10.0 s, after which its angular velocity is constant. Make graphs of the velocity versus time, the tangential acceleration versus time, and the centripetal acceleration versus time for $0.0 \leq t \leq 15.0$ s. Position the graphs directly below one another and use the same time scale on each.

13. A wheel of radius R starts from rest and maintains a constant angular acceleration α. At what time t are the centripetal and tangential accelerations of a point on the rim equal?

14. A bicyclist starts from rest and accelerates at a constant rate for 15.0 s. After 10.0 s the bicycle's front wheel is turning at a rate of 2.50 rev/s. (a) Through how many revolutions does the wheel turn in the first 10.0 s? (b) If the front wheel has a diameter of 0.800 m, what is the linear speed of a piece of mud stuck on the outside edge of the tire when $t = 10.0$ s? (c) What is the total linear acceleration of the piece of mud when $t = 2.00$ s?

15. A high-performance flywheel is rotating at 8.00×10^4 rpm when the power is disconnected and it is allowed to turn freely. After 48 h it is rotating at 2.00×10^4 rpm. (a) If the angular accelera-

tion is constant, how much more time passes before the flywheel stops? (b) Through how many revolutions does it turn from the instant it is disconnected until its speed is zero? (c) What is its angular acceleration in radians per second squared?

16. Use the constant-acceleration model to calculate the angular acceleration of a compact disk, if the disk accelerates from rest to a rotation rate of 50.0 rad/s in a quarter turn.

17. When a certain ceiling fan is switched from low speed to high speed, it increases its rotation rate from 1.00 rad/s to 3.00 rad/s in 15.0 s. If the angular acceleration is constant, find (a) its magnitude and (b) the number of revolutions of the fan while speeding up.

18. Starting at $t = t_o$ an automobile engine steadily increases its rotation speed from 100. to 1000. rpm in 1.00 s. It maintains this rotation rate for 3.00 s, and then finally returns to 100. rpm at t_f by decreasing its rotational speed at a constant rate of 300. rpm/s. Find the total number of rotations made by the engine between t_o and t_f.

★19. A disk hangs horizontally by a light thread attached to the ceiling. A second disk is suspended below the first by a ribbon that joins their centers. At $t = 0.00$ s the ribbon has 100. complete anticlockwise twists in it. (That is, the ribbon is twisted by keeping the upper disk fixed and rotating the lower disk anticlockwise when viewed from above as shown in Figure 11.P3.) At $t = 0.00$ s the upper disk starts to rotate with constant $\omega_u = 2.00$ rpm anticlockwise, and the lower disk, starting from rest, begins to rotate with a clockwise angular acceleration with a magnitude of 1.00 rpm/s. (a) At what time t_f does the ribbon become (momentarily) untwisted? (b) How many revolutions does each disk make between $t = 0$ and $t = t_f$? (*Hint:* What is the equivalent linear motion problem?)

FIGURE 11.P3   Problem 19

★20. Two disks are suspended with a ribbon between them as described in Problem 19. At $t = 0.00$ s the ribbon connecting them is untwisted, but the upper disk is rotating clockwise with a constant angular velocity of 15.0 rpm. At $t = 0.00$ s the lower disk is stationary but begins to rotate clockwise with constant angular acceleration of 2.00 rpm/s. (a) At what time is the ribbon once again (momentarily) untwisted? (b) What is the angular velocity of each disk at that moment?

11.2 Kinetic Energy and Rotational Inertia

21. (a) Calculate the rotational inertia of two 0.250-kg masses located a fixed distance of 0.500 m on opposite sides of an axis of rotation perpendicular to the line joining the masses. (b) Calculate the rotational inertia of a uniform slender rod of length 1.00 m and mass 0.500 kg when it rotates about an axis perpendicular to the rod and through its center. (c) Calculate the moment of inertia of a thin hoop of radius 0.500 m and total mass 0.500 kg if it rotates about an axis perpendicular to the loop and passing through its center. (d) Why are the answers to (a) and (c) the same, but different from (b)?

22. Compute the moment of inertia of a wagon wheel composed of six spokes each of length l and mass m, with a circular hoop of mass M around them. (Assume the hoop is thin.)

23. A rectangle of mass 6.00 kg has a perimeter of 3.00 m. The moment of inertia of the rectangle when rotated about an axis perpendicular to it and through its center is $0.625 \text{ kg} \cdot \text{m}^2$. What are the dimensions of the rectangle?

24. A 0.600-kg plastic tube that is 2.00 m long, has a small diameter compared to its length. It is known that two 0.50-kg masses are located inside the tube at equal distances on either side of the center. When rotated about a perpendicular axis through the center of the tube, the rotational inertia of the tube and hidden masses is $0.260 \text{ kg} \cdot \text{m}^2$. Where are the hidden masses located?

25. Three 8.00 ft long 2x4's each of mass 3.00 kg are nailed together to form an H. (The crossbar is nailed to the midpoints of the two vertical bars.) What is the moment of inertia of the H about an axis perpendicular to the plane of the H and passing through its center? (*Hint:* Use the parallel-axis theorem to find the moment of inertia of the sides of the H.)

26. Two thin rods of mass m and length L are crossed at right angles to make an X shape. The rods are free to rotate about a pivot perpendicular to the plane of the X and through its center. On the end of each rod is a thin hoop of mass M and radius R. The edge of the hoop is attached to the end of the rod so that the plane of each hoop is the same as the plane of the X and the center of the hoop is aligned with the rod. (See Fig. 11.P4.) (a) What is the moment of inertia of the composite object? (b) Suppose that the hoops are rotated so that their planes are perpendicular to the plane of the X. What is the moment of inertia of the new composite object?

27. A thin rod of length L is located along the x-axis with one end at the origin. If the density of the rod is given by $\lambda(x) = \lambda_o(x/L)^2$, where λ_o is a constant, find (a) the total mass M of the rod and (b) the moment of inertia of the rod when rotated about the y-axis. Write your answer to (b) in terms of M, not λ_o.

28. A thin rod of length $2L$ lies along the y-axis with the origin at its midpoint. If the density of the rod is given by $\lambda = \lambda_o(L - y)/L$, where λ_o is a constant, calculate (a) the total mass M of the rod and (b) the moment of inertia of the rod about the x-axis. Express the moment of inertia in terms of M, not λ_o.

29. By direct integration, calculate the moment of inertia of a uniform circular disk about an axis through its center and perpendicular to the plane of the disk.

30. A circular disk of radius R rotates about an axis through its center and perpendicular to the plane of the disk. If the area density of the disk varies as $\sigma = \sigma_o r/R$, where σ_o is a constant, find (a) the mass of the disk and (b) the moment of inertia of the disk. Express the rotational inertia in terms of R and M only.

31. Model the earth as a uniform solid sphere using data from Appendix 8. (a) What is the rotational kinetic energy of the earth due to its rotation about its axis? (b) Compare this energy with the kinetic energy of the earth due to its orbital motion around the sun.

32. At what rate does a stationary bowling ball of mass m have to rotate to have the same kinetic energy as it would if it were moving in a straight line (without rotation) with velocity v?

33. Model a Frisbee as a uniform disk of mass m and diameter 28.0 cm rotating at 5.00 rev/s. (a) How fast must it travel through the air so that its kinetic energy due to translational motion is equal to its kinetic energy of rotation? (b) For real Frisbees the lip around the outer edge causes a larger percentage of the Frisbee's mass to be concentrated about the edge than the uniform-disk model implies. Is your answer to (a) larger or smaller for the case of a real Frisbee?

34. A uniform disk of mass 2.50 kg and radius 15.0 cm is free to rotate without friction around a fixed horizontal axle passing through its center. A light thread is attached to the edge of the disk and wrapped around its circumference several times. The loose end of the thread is attached to a 0.50-kg mass that hangs freely from the thread. The disk and the hanging mass start from rest when the mass is released. Use the conservation of energy to determine the velocity of the mass after it has fallen 1.25 m.

35. The Atwood machine shown in Figure 11.P5 consists of a mass $M = 500.$ g and a mass $m = 300.$ g connected by a light string that passes over a pulley in the shape of a disk of mass 750. g and radius $r = 10.0$ cm. The string does not slip on the edge of the pulley, which turns without friction about its center. Use the conservation of energy to calculate the velocity of the 500-g mass after it has fallen 1.00 m if the system starts from rest.

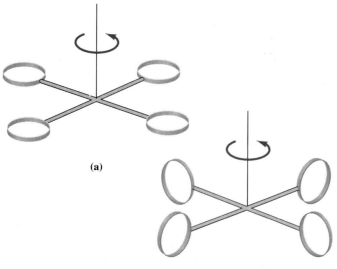

(a)

(b)

FIGURE 11.P4   Problem 26

FIGURE 11.P5   Problem 35

36. A modification of the Atwood machine shown in Figure 11.P6 uses two disks of different diameters. The disks are locked together so that they must turn with the same angular velocity. The larger disk has a radius $R = 20.0$ cm and a mass 0.500 kg, and a mass $M = 0.600$ kg is attached to the string wound around its edge. The smaller disk has a radius $r = 15.0$ cm and a mass 0.250 kg, and a mass $m = 0.400$ kg is attached by another string. If the system starts from rest, find the velocity of the 0.600-kg mass after it has fallen 1.00 m.

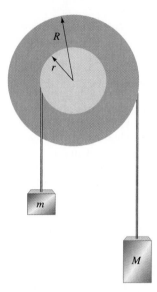

FIGURE 11.P6   Problem 36

37. A thin uniform hoop of diameter 1.10 m and mass 150. g is balanced upright on its edge. If the hoop falls over, how fast is its top edge traveling the instant before it hits the ground?

38. Two rods of length L and mass M are connected to form the letter T. The T is originally in the upright position and has zero angular velocity about an axis through its base and parallel to the upper rod of the T. The T falls forward and swings around the axis. Find the expression for the angular velocity of the T at the moment when it is inverted.

11.3 Torque

39. Calculate the net torque on the rod in Figure 11.P7 about the pivot P shown.

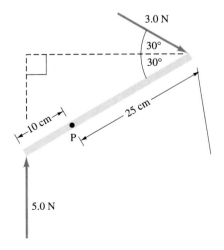

FIGURE 11.P7   Problem 39

40. Calculate the net torque about the pivot point P on the object shown in Figure 11.P8.

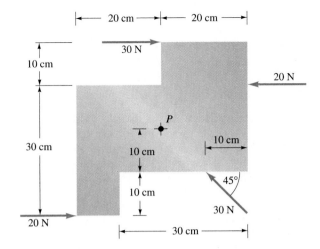

FIGURE 11.P8   Problem 40

41. The two forces shown in Figure 11.P9 are known as a *couple*. Show that the net torque about the center of the bar is given by $LF \sin(\phi)$.

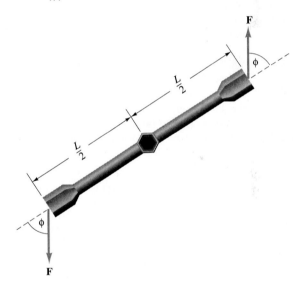

FIGURE 11.P9   Problem 41

42. Calculate the net torque on the wheel shown in Figure 11.P10.

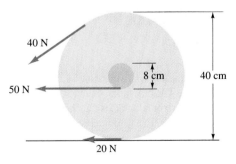

FIGURE 11.P10   Problem 42

43. A toy gyroscope has a moment of inertia of 3.50×10^3 g·cm². A 70.0-mm long string is wrapped without overlapping around the gyroscope axis, which has a radius of 2.00 mm. If a child pulls the string off the gyroscope while keeping the tension in the string at 25.0 N, what is the final angular speed of the gyroscope? (*Hint:* Use the work–energy theorem in rotational form.)

44. A wheel of moment of inertia of 60.0 kg · m² and radius 40.0 cm is spinning at a rate of 500. rpm. A braking force of 300. N is applied tangential to the rim of the wheel while it makes 2.00 rev. What is the angular speed of the wheel after the force is applied?

11.4 Newton's Second Law for Rotation about a Fixed Axis

45. A mass m is constrained to move in a circle of radius R. A tangential force F acts as the particle moves. Start by writing both the radial and tangential components of Newton's second law for the mass. Multiply both sides of the equation for tangential motion by R. Now using the definitions of torque, and moment of inertia, transform this equation to $\tau = I\alpha$.

46. Use the force method to calculate the acceleration of the mass and the tension in the string for the disk–mass system described in Problem 34.

47. Use the force analysis method to calculate the acceleration of the 500.-g mass and the tension in each of the two segments of the string for the Atwood machine described in Problem 35.

48. Find the tension in each of the two strings and the acceleration of the 0.600-kg mass for the modified Atwood machine described in Problem 36.

★49. Two masses are connected by a light string that passes over a pulley as shown in Figure 11.P11. The mass on the incline is $M = 1.50$ kg. The surface of the frictionless incline makes an angle $\theta = 36.87°$ with the horizontal. The falling mass is $m = 1.00$ kg, and the frictionless pulley has a rotational inertia $I = 5.00 \times 10^{-3}$ kg · m² and a radius of 0.100 m. (a) If the string rides without slipping on the pulley, find the tensions in the string segments and the acceleration a of the system. (b) What are the values of a and the tensions if the pulley experiences a constant frictional torque $\tau_f = 0.200$ N · m?

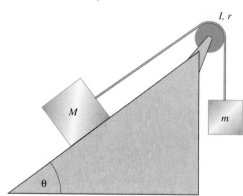

FIGURE 11.P11   Problem 49

★50. A rod of radius r and mass m_r is firmly fitted in the center of a disk of radius R and mass M_R. A block of mass M, which slides without friction over a horizontal table, is attached to the rod via a light string. A mass m hangs freely from a second string, which is wrapped around the circumference of the disk as shown in Figure 11.P12. Find the expression for the acceleration of m.

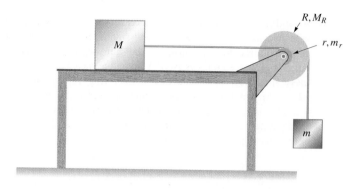

FIGURE 11.P12   Problem 50

11.5 Angular Momentum about a Fixed Axis

51. In a college physics lab, students measure the moment of inertia of a disk by dropping a sticky clay ball of known mass on a horizontal rotating disk and observing the decrease in rotation rate (Fig. 11.P13). If the disk is originally rotating at 30.0 rpm and then slows to 20.0 rpm when a small 100. g mass is dropped on the disk 12.0 cm from the axis of rotation, what is the moment of inertia of the disk?

FIGURE 11.P13   Problem 51

52. Two disks are rotating independently about the same axis. Disk A has a moment of inertia of 0.400 kg · m² and an initial angular velocity of 2.50 rad/s clockwise. The second disk B is rotating anticlockwise at 1.50 rad/s. The two disks are then linked together (without application of an external torque) so that they rotate together with a common angular velocity of 0.500 rad/s anticlockwise. Find the moment of inertia of disk B.

53. A carnival game is played by throwing cloth-covered balls at a Velcro-covered target. The target is shaped as a horizontal rod free to spin about a vertical axis through its center. Each end of the rod has a circular target attached to the end of the rod as shown in Figure 11.P14. The rod has a total length of $L =$

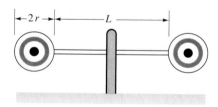

FIGURE 11.P14   Problem 53

50.0 cm and a mass of 60.0 g. Each circle has mass of 100. g and a radius $r = 10.0$ cm. A player throws a ball of mass 50.0 g at a speed of 10.0 m/s, striking the right target circle dead center. If the ball sticks to the target, what is the angular speed of the target and ball after the collision?

54. A target at a shooting gallery is made of a 250.-g metal disk (diameter = 20.0 cm) that is free to spin about an axis running along one diameter. A shooter shoots a 1.00-g pellet at the stationary target, striking it 8.00 cm from the axis of rotation. Before colliding with the disk the pellet has a velocity of 200. m/s. After the collision the pellet continues along the same straight-line trajectory as before the collision but with velocity of 50.0 m/s. Find the speed of rotation of the target after being struck by the pellet.

Numerical Methods and Computer Applications

55. Write a computer program or devise a spreadsheet template to calculate the moment of inertia about the y-axis of N mass–points, each of mass m/N equally spaced from 0 to L along the x-axis. Let N vary from 1 to 50. For $N = 1$ let all of the mass be at distance $x = L$; for $N = 2$, let two masses of $m/2$ each be located at $x = L/2$ and $x = L$; for $N = 3$, let the masses be $m/3$ each and the distances be $x = L/3, 2L/3$, and L; and so on. (a) Hand in a graph of I versus N for $m = 2.00$ kg and $L = 1.00$ m. (b) What is the significance of the horizontal asymptote of your graph as $N \to \infty$?

56. Model a uniform rod as a collection of 50 equal point–masses spread out along the horizontal axis. Devise a spreadsheet template or computer program to calculate the moment of inertia of the rod using any one of the mass–points as the position of an axis of rotation perpendicular to the rod. Calculate the moment of inertia for the rod for each of 50 possible axis positions and produce a plot of I versus axis position. Comment on the axis position for maximum and minimum values of I.

57. The circumference of a 0.825 kg size-3 metal baseball bat varies with distance from the end as shown in Table 11.P1.

TABLE 11.P1   The Circumference of a Size-3 Metal Baseball Bat

DISTANCE FROM END (cm)	CIRCUMFERENCE (cm)
2	15
4	8
6	8
⋮	⋮
28	8
30	8
32	9
34	10
36	11
38	12
40	13
42	14
44	15
46	16
48	17
50	18
52	18
⋮	⋮
78	18
80	18

Assume the mass of the hollow bat varies in direct proportion to its circumference, and calculate (a) the center of mass of the bat and (b) the moment of inertia of the bat about an axis 15.0 cm from the smaller end.

58. Calculate the moment of inertia of the bat in Problem 57 about each of 40 axes separated by 2.00-cm intervals from the end of the bat to the tip. Produce a graph of I versus axis position and locate the point of minimum moment of inertia.

59. Explore the effect of the pulley inertia on the connected-mass problem of Example 11.11. In particular, produce a single graph that shows the value of both string tensions as the mass of the pulley is varied, such that $0 \le m_p \le 10M$ in steps of $0.500\ M$. Discuss the significance of the $m_p \to \infty$ asymptotic values of the tensions.

60. A flat, 4.00-kg disk rotates about an axis perpendicular to the plane of the disk and through its center of mass. The radius of the disk is 25.0 cm. A tangential torque τ acts on the edge of the disk in such a manner that the angular momentum of the disk is given by $L(t) = (3.00\ \text{N} \cdot \text{m})\, te^{-t/\tau}$, where $\tau = 2.00$ s. (a) Plot the angular momentum function in steps of 0.200 s from $t = 0.00$ to $t = 4.00$ s. (b) Compute the torque at each time by taking the appropriate numerical derivative, and graph your results. (c) Compute the torque function $\tau = dL/dt$ directly from $L(t)$, plot the numerical values of τ for the interval $0.00\ \text{s} \le t \le 4.00$ s, and compare your graph with that of part (b).

General Problems

61. As a string rides without slipping on the edge of a 2.00-cm diameter pulley, the string accelerates from rest with acceleration 0.500 m/s². After 3.00 s what is (a) the tangential acceleration of the pulley edge? (b) the centripetal acceleration of the pulley edge? (c) the angular acceleration of the pulley? (d) the angular velocity of the pulley?

62. A bicycle chain is wrapped around the front crank sprocket of diameter D and a rear sprocket of diameter d as shown in Figure 11.P15. (a) When the front sprocket rotates with an angular velocity Ω, what is the angular velocity of the rear sprocket? (b) If torque τ is applied to the front sprocket while it turns at Ω, and the input power is all transferred to the rear sprocket, what torque τ is transferred to the rear sprocket? (c) If $D > d$, what do you gain? What do you lose?

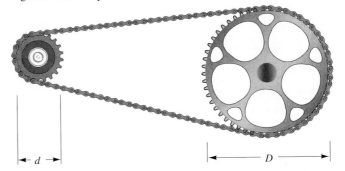

FIGURE 11.P15   Problem 62

63. Prove that a particle traveling with constant velocity has a constant angular momentum about an arbitrary point P.

★**64.** Prove the parallel-axis theorem.

★**65.** Prove the perpendicular-axis theorem.

66. An 80.0-g meterstick is pivoted about one end so that it may rotate in a plane perpendicular to an axis through the pivot. The meterstick begins to rotate and its angular momentum function is

given by the function $L(t) = (3.00 \text{ N} \cdot \text{m/s})t^2 - (2.00 \text{ N} \cdot \text{m})t$.
(a) Evaluate the torque τ that acts on the meterstick at $t = 0.500$ s.
(b) What is the angular acceleration of the meterstick at $t = 0.500$ s?

67. A 0.250-kg uniform rod is 1.20 m long and is pivoted about a point through its center of mass. The rod begins to rotate in a plane perpendicular to the axis through the pivot and its angular momentum function is given by $L(t) = (0.500 \text{ N} \cdot \text{m/s}^2)t^3 + (2.00 \text{ N} \cdot \text{m/s})t^2$. (a) Evaluate the torque τ that acts on the rod at $t = 1.50$ s. (b) What is the angular acceleration of the rod at $t = 1.50$ s?

68. A cord passes through a small hole in a frictionless, horizontal surface and is attached to a small mass m as shown in Figure 11.P16. The mass is initially traveling in a circle of radius R_o with velocity v_o when the cord is pulled slowly from below so that the radius of the circle decreases to R. (a) What is the velocity of the mass when the radius of the circle is R? (b) How much work is required to change the radius from R_o to R? (c) Find an expression for the magnitude of the force F as a function of R.

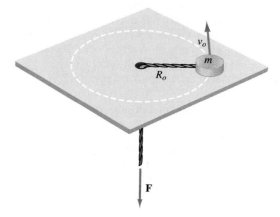

FIGURE 11.P16   Problem 68

Static Equilibrium and Rolling Objects

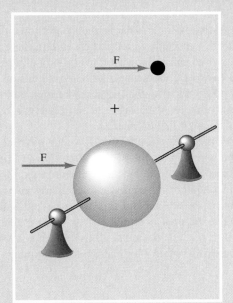

In this chapter you should learn:

- how to calculate the torque due to the weight of an extended object.

- how to analyze simple rigid objects in static equilibrium.

- how to calculate the kinetic energy of an object that both rotates and translates.

- (optional) the rotational impulse–angular momentum theorem.

- (optional) how to perform a force analysis on an object that both rotates and translates.

$$\Sigma F_{net} = ma_{cm}$$

$$\Sigma \tau_{net} = I_{cm}\alpha$$

Since the invention of the wheel, objects that roll have played a central role in civilization. Everywhere you look, from automobile tires on the highway to golf balls on the putting green, objects are rolling as they move. Whether one is designing roller bearings for machinery or putting back spin on the cue ball, an understanding of the physics of combined rotation and translation makes the job easier. Paradoxically, objects that neither translate nor rotate are also critical to modern life. We like buildings and billboards that stay put and don't fall over. In this chapter you will discover that the unity of physics allows us to analyze both stationary structures and rolling objects using the same framework, namely the point–particle model and rotation about a fixed axis. Because you have already studied the basic physics of both these cases, you should find this chapter a good place to use your newfound knowledge on some interesting applications.

12.1 Static Equilibrium

When we used the point–particle model we defined a particle to be in equilibrium when the total force on it was zero. Although it took some getting used to, we came to realize that a particle traveling with *any* constant velocity was in equilibrium. When that constant velocity happens to be zero in our frame of reference, we speak of the object as being in **static equilibrium.**

In Chapter 5 we mentioned that weight is a body force. That is, the weight force does not act at only one point on an object. Instead, each individual particle within the body experiences a small weight force due to its own mass. However, when modeling real objects as point–particles we did not need to worry about the distributed nature of the weight force. If you model an elephant as a point–particle, the fact that its trunk, tail, and tusks each have their individual weight doesn't matter. We assign the entire weight to the idealized single point that the elephant has become in our model.

When we began to consider extended objects we found that it suddenly became important to analyze the effect of the distributed nature of the weight force. For example, when we calculated the gravitational potential energy of an extended object in a uniform gravitational field for the first time, we summed up the entire gravitational energy of every single mass–point in a generic extended mass. When we did this we were pleased to discover that it is possible to calculate the gravitational potential energy of an extended object by pretending that the entire mass of the object is located at a single point known as the *center of mass.* We now need to take an equally careful look at the total torque on an object caused by its own weight.

Torques Due to the Weight Force

In order to properly calculate the torque on an object due to its own weight, we must find the torque due to each individual mass–point and sum these torques. It turns out that for any axis this summation process produces a remarkably simple result:

> In a uniform gravitational field, the torque about any axis produced by the weight of an extended object is equal to the torque produced by the total mass acting as if it were concentrated at the center of mass.

Thus, we discover yet another convenient property of this wonderful point called the center of mass. In the following paragraphs we give a short proof of this property for a laminar object with an axis perpendicular to its plane. A more general proof must await the vector treatment of torque in Chapter 13.

Let's calculate the torque due to the weight of the mass M about the Z-axis of the XYZ-frame of reference shown in Figure 12.1. Let a second coordinate system (xyz) have its origin at the center of mass of the extended object. The xyz-coordinate system origin is located at the point $(X_{CM}, Y_{CM}, 0)$ in the XYZ-system. We imagine dividing the extended object into a large number of smaller masses, each with weight $g \, dm$. To find the total

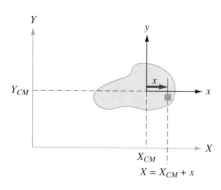

FIGURE 12.1

The relationship between the coordinates of a mass point in two different coordinate systems.

torque about point P due to the weight of the extended mass, we find the torque due to the weight of each dm using the moment-arm method and sum the resulting torques:

$$\tau = \int R_\perp \, dF = \int_{\text{object}} (\textit{Moment-arm})(g \, dm)$$

The line of action of the weight force of each mass–point is straight down. Consequently, the moment arm of the mass dm is just X, the X-coordinate of the mass–point. As you can see from Figure 12.1, $X = X_{CM} + x$. The integral thus becomes

$$\tau = \int_{\text{object}} (X_{CM} + x) g \, dm$$

$$\tau = X_{CM} g \int_{\text{object}} dm + g \int_{\text{object}} x \, dm$$

The first of the two integrals on the right-hand side of the previous equation is just the total mass M of the object. The second integral is also familiar; it is the numerator of the fraction we used to calculate the x-coordinate of the center of mass: $x_{CM} = \int x \, dm/M$. Because $x_{CM} = 0$ in the xyz-coordinate system, the numerator of the fraction (that is to say the second integral above) is also zero. Hence, the total torque due to the weight of our object is

$$\tau = MgX_{CM}$$

which is just the torque of a single point-mass M located at the center of mass of the extended object.

Force and Torque in Static Equilibrium

The analysis of static structures, such as bridges and buildings, is vital in many professional fields such as civil and mechanical engineering. Although it may not be immediately obvious, the approach we found useful in the *force analysis* of objects that translate only and the *torque analysis* used to describe objects that rotate only can be combined to study extended objects that neither translate nor rotate. That is, we first invoke the point–particle model and apply Newton's second law $F_{\text{net}} = ma_{CM}$, ignoring the possibility of rotation. Following that, we apply a rotational analysis $\tau_{\text{net}} = I\alpha$, ignoring the possibility of translation. For the *static equilibrium* we desire both a_{CM} and α to be zero. Thus, we require that

$$\mathbf{F}_{\text{net}} = 0$$

and

$$\tau_{\text{net}} = 0$$

The two previous equations actually represent six equations. You already know that Newton's second law is a vector equation so that $\mathbf{F}_{\text{net}} = 0$ means that $\Sigma \, \mathbf{F}_x = 0$, $\Sigma \, \mathbf{F}_y = 0$, and $\Sigma \, \mathbf{F}_z = 0$. The vector nature of the torque equation is not yet evident because we have deliberately put off introducing the full vector-nature of the torque concept until the next chapter. Nonetheless, it should be clear that a structure could, in principle at least, rotate about the x-, y-, or z-axes. If the structure is to be static, it must not rotate about *any* of these directions. Thus, the expression for zero net torque could stand for as many as three equations. A general problem in static equilibrium, then, might involve the simultaneous solution of six equations in six unknowns! The good news is that we have time for only an introductory look at the field of statics and, hence, we restrict ourselves to structures with

Concept Question 1
(a) If the velocity vector for the center of mass of an extended object is constant, is that object necessarily in equilibrium? (b) If the velocity vector for the center of mass of an extended object is zero, is that object necessarily in static equilibrium?

only two dimensions. This simplification results in at most two force equations and one torque equation. Rest assured that there are a sufficient number of problems that may be modeled in two dimensions to give you the flavor of the subject of statics.

Again because we are not introducing any new concepts, we proceed directly to a set of examples. Before beginning you may wish to review the general methods for computing torques, particularly the moment-arm method from Section 11.3.

EXAMPLE 12.1   *Getting the Short End of the Stick*

After a day of white-water rafting Tom and Art were dragging their raft back to the pickup (Fig. 12.2). One end of the drag rope was attached to the raft and the other attached to a stick of length $l = 70.0$ cm. Tom and Art each pulled on opposite ends of the stick. Under the cover of darkness, Art moved the point where the rope was attached from the center of the stick to a point 30.0 cm from Tom's end of the stick. If the tension in the rope was 350. N while the raft was dragged with a constant velocity, how much force did each of the rafters exert on the stick?

FIGURE 12.2

The two rafters do not share equally in pulling the raft unless the pull rope is tied to the center of the crossbar.

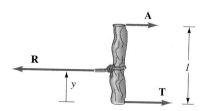

FIGURE 12.3

Forces acting on the stick used to pull the raft.

SOLUTION   As shown in Figure 12.3, all three forces are parallel to the *x*-axis. We represent the forces of Art, Tom, and the rope tension on the stick as *A*, *T*, and *R*, respectively. Because the stick is moving with a constant velocity, according to Newton's second law the total force on the stick is zero.

$$T + A - R = 0 \qquad \text{(Force equation)}$$

We now wish to sum the torques and set them equal to zero. However, the question immediately arises, "Where should we place the axis of rotation?" The answer is easy, "Anyplace you like!" You can't go wrong. The fact of the matter is that we do not want the stick to start rotating about *any* point, so the torques must sum to zero about *every* axis you choose. We do have some advice, however, about how to make your life a little simpler by choosing your axis position cleverly. The magnitude of the torque is $\tau = rF \sin(\phi)$. The factor of *r* guarantees that any force that acts at the point where the rotation axis is located exerts no torque about that point. Therefore, we can eliminate unknown forces from our equations by locating the rotation axis at their point of application. For example, if we take Tom's hand as the pivot point, the force *T* exerts no torque about this pivot. Then, instead of having two unknowns in the resulting torque equation, there is only one.

We proceed with the example by locating the axis of rotation at Tom's hand. The total torque about this point must sum to zero, so that

$$-yR - lA = 0 \qquad \text{(Torque equation)}$$

We have arbitrarily taken anticlockwise torques as negative. We do not need to be consistent in our sign convention for translational and rotational motion because linear and angular acceleration are both zero. Because of our clever choice of pivot point, the torque equation has only one unknown, namely, A. Hence, we solve for it immediately.

$$A = \frac{yR}{l} = \frac{(30.0 \text{ cm})(350. \text{ N})}{70.0 \text{ cm}} = 150. \text{ N}$$

This value for A may now be substituted into the force equation, and we may solve for T:

$$T = R - A = 350. \text{ N} - 150. \text{ N} = 200. \text{ N}$$

◀ **Concept Question 2**
(a) In Example 12.1, exactly what is it that is being treated as an object in equilibrium? (b) Is Example 12.1 a case of static or dynamic equilibrium?

EXAMPLE 12.2   *Just Hangin' Around*

A sign is supported by a rod that is pivoted freely by a hinge attached to a building. As shown in Figure 12.4 the support rod is held horizontal by a guy wire attached to its end. The wire makes a 60.0° angle with the building. If the sign and support rod weigh 50.0 lbf and 20.0 lbf, respectively, find the tension in the guy wire and the force at the hinge.

SOLUTION   In Figure 12.5 we show a free-body diagram of the horizontal rod with all of the forces acting on it. Note that we have already solved a small subproblem leading to the forces shown. That is, we have reasoned that the sign is supported symmetrically by two chains and, hence, both chains should have equal tensions. Because the sign weighs 50.0 lbf, each chain must support 25.0 lbf. This 25.0 lbf is transmitted by the chain tension to the horizontal rod as shown in Figure 12.5. Notice, too, that the weight of the rod itself has been placed at its center of mass.

Finally, note that we have put the unknown force of the hinge on the end of the rod into the problem as two forces **H** and **V**. These symbols represent the horizontal and vertical components of the single force of the hinges on the rod. We have anticipated the need to resolve this force into *x*- and *y*-components. Don't worry if you are not sure of the direction of these forces; **V,** for example, could be up or down. Just pick a direction. If the answer comes out negative, it means you guessed wrong, but the magnitude of your answer is still correct. Incidentally, you should always put in both horizontal and vertical components. Often, one of the aims of good engineering design is to ensure that the force at the hinge is directed along the rod so that it is in compression only. However, you cannot make this assumption at the start of the problem.

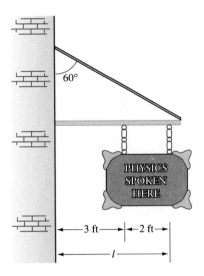

FIGURE 12.4

The horizontal rod holding the sign is in static equilibrium because it neither translates nor rotates. See Example 12.2.

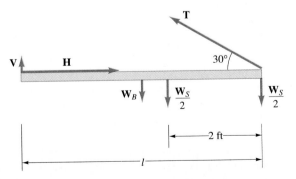

FIGURE 12.5

Forces acting on the horizontal rod of Figure 12.4. Note that the force of the wall acting on the end of the rod has been resolved into components **H** and **V** for convenience.

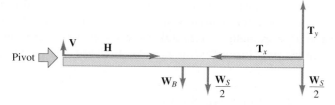

FIGURE 12.6

All forces acting on the horizontal rod have been resolved into x- and y-components for easy summation of forces and torques.

Now that we understand the free-body diagram we can proceed with the force analysis. The guy-wire tension force must be resolved into x- and y-components. The magnitudes of these components are: $|\mathbf{T}_y| = T\sin(30.0°)$ and $|\mathbf{T}_x| = T\cos(30.0°)$. After the tension force is resolved we have a free-body diagram as shown in Figure 12.6. All of the forces are resolved so that both force and torque equations are simple to write:

$$+V + T_y - W_B - \frac{W_S}{2} - \frac{W_S}{2} = 0 \qquad (y\text{-component force equation})$$

$$-H + T_x = 0 \qquad (x\text{-component force equation})$$

Before taking torques we must decide on a pivot point. We note that by choosing the hinge as a pivot, we can eliminate *two* unknowns from the resulting equation. How can we resist? The torques about the hinge are

$$-W_B\frac{l}{2} - \frac{W_S}{2}(l - 2\text{ ft}) - \frac{W_S}{2}l + T_y l = 0 \qquad (\text{torques about hinge})$$

where we have taken torques tending to rotate the rod anticlockwise as positive torques. Because of our choice of pivot, the torque equation has only one unknown and may be solved immediately for T after substituting $|T_y| = T\sin(30.0°)$

$$T = \frac{W_B l + 2W_S(l - 1\text{ ft})}{2l\sin(30.0°)} = \frac{(20.0\text{ lbf})(5.00\text{ ft}) + 2(50.0\text{ lbf})(4.00\text{ ft})}{2(5.00\text{ ft})(0.500)} = 100.\text{ lbf}$$

Values for H and V can now be obtained by substituting T into the force equations

$$H = T_x = T\cos(30.0°) = (100.\text{ lbf})(0.866) = 866\text{ lbf}$$

$$V = -T_y + W_B + \frac{W_S}{2} + \frac{W_S}{2} = -T\sin(30.0°) + W_B + W_S = 20.0\text{ lbf}$$

In this case both H and V are positive values, indicating that our choice for their directions was correct in Figure 12.5. ◀

Concept Question 3
True or False. (a) If an extended object is in rotational equilibrium about some particular axis, then it must be in rotational equilibrium about all parallel axes. (b) If an object is in rotational equilibrium about some particular axis, then it must be in rotational equilibrium about all axes.

12.2 Rotation and Translation with No Slipping (Optional)

The equations of motion for objects that are both translating and rotating can be found by an application of either an energy analysis or a force analysis. In this section we examine both of these powerful techniques for the case of objects that roll without slipping.

Energy Analysis

We begin with the technique of energy analysis that employs two properties of the center of mass first deduced in Chapter 10:

1.
$$K = \frac{1}{2}MV_{CM}^2 + \sum_{i=1}^{N}\frac{1}{2}m_i u_i^2$$

The kinetic energy of a system of particles can be divided into the kinetic energy *of* the center of mass and the kinetic energy *relative* to the center of mass.

2.
$$U = Mgy_{CM}$$

The gravitational potential energy of a system of particles in a uniform gravitational field can be calculated as if the entire mass were concentrated at the center of mass.

Focusing our attention on item (1) above, let's think about what we mean by the kinetic energy *relative* to the center of mass for a *rigid* body. Because the object can't change shape, the only way individual parts of the object can move relative to the center of mass is if the entire object rotates as a whole around some axis passing through the center of mass. Thus, for rigid-body motion the relationship in (1) becomes

$$K = \frac{1}{2}MV_{CM}^2 + \frac{1}{2}I_{CM}\omega^2 \qquad (12.1)$$

where I_{CM} is the rotational inertia of our object about a particular axis through the center of mass. Equation (12.1) suggests a remarkably simple way to view the motion of a rigid object that is both rotating and translating. The kinetic energy due to rotational motion (about an axis through the center mass) can simply be added to the kinetic energy due to the center-of-mass motion.

One of the most interesting applications of Equation (12.1) is to an object of circular cross section rolling on a flat surface. Especially useful is the case when the center-of-mass velocity V_{CM} is equal to the tangential velocity $R\omega$ of the edge of the object. As Figure 12.7 shows, the motion of each particle is the result of two causes. Part of the motion of any given particle is due to the translation of the entire object through space. Combined with this whole-body motion is the motion of each point due to rotation about the center of mass. The actual velocity of each point is then the vector sum of these two individual velocities. Notice that in the special case where $V_{CM} = R\omega$, the velocity of the point at the bottom of the object is zero! This result is precisely what occurs when a wheel rolls along the ground without slipping. For this reason we refer to $V_{CM} = R\omega$ as the **no-slip condition.** There is no relative motion of the bottom of each tire and the surface of the road when your tires don't slip. When your tires are spinning either too slow or too fast for your forward motion, there is relative motion between the road surface and your tires, and you hear a squeal (and burn a lot of expensive rubber). In the paragraphs that follow we show how this simple division of the kinetic energy can be used to solve problems.

Let's now put this information to use by looking at the classic problem of an object of circular cross section rolling without slipping down an incline. Such an object is shown in Figure 12.8. If you look back to Figure 11.9 you will discover that the moments of inertia for objects we might want to roll down the incline all have the form $I_{object} = kMR^2$; for a sphere $k = \frac{2}{5}$, for a disk $k = \frac{1}{2}$, and for a hoop $k = 1$. We can find the object's final velocity in terms of k, M, R, and the vertical drop y of the incline.

We use the conservation of mechanical energy, noting that the object's initial gravitational potential energy is converted to kinetic energy as it rolls down the incline.[1]

---

[1] In the next section we will point out that it is necessary for friction to act at the point of contact between the object and the floor. Because there is no relative motion between the object and the floor at this point, the frictional force does no work. Consequently, the conservation of mechanical energy may be applied even in the presence of a nonconservative force.

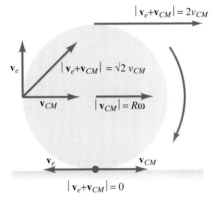

FIGURE 12.7

The total velocity of any point on an object that rotates and translates is the sum of the velocity of the center of mass of the object $\mathbf{v}_{CM}$ and the velocity relative to the center of mass $\mathbf{v}_e$.

$$V_{CM} = R\omega \qquad \text{no-slip condition}$$

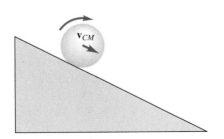

FIGURE 12.8

An object with circular cross section rolls without slipping down an inclined plane.

$$U_o + K_o = U + K$$

$$Mgy_o = \frac{1}{2}MV_{CM}^2 + \frac{1}{2}I_{CM}\omega^2$$

From the *no-slip condition* we have

$$V_{CM} = R\omega \qquad \text{(No-slip condition)}$$

which allows us to replace ω with V_{CM}/R.

$$Mgy_o = \frac{1}{2}MV_{CM}^2 + \frac{1}{2}I_{CM}\frac{V_{CM}^2}{R^2}$$

We now replace I_{CM} with kMR^2 to produce

$$Mgy_o = \frac{1}{2}MV_{CM}^2 + \frac{1}{2}kMV_{CM}^2 \tag{12.2}$$

$$V_{CM} = \sqrt{\frac{2gy_o}{1+k}} \tag{12.3}$$

It is well worth our while to inspect this answer carefully. First we note that for $k = 0$ we obtain the familiar result $V_{CM} = \sqrt{2gy_o}$, which is just the velocity we have previously obtained (many times) for the point–mass model when rotational effects are not present. The parameter k tells us how important rotational inertia is to the object's rolling motion. The right-hand side of Equation (12.2) is the total kinetic energy of the object at the bottom of the incline:

$$K = \frac{1}{2}MV_{CM}^2 + \frac{1}{2}kMV_{CM}^2$$

$$K = (1+k)\left(\frac{1}{2}MV_{CM}^2\right) \tag{12.4}$$

In Equation (12.4) the parameter k tells us what fraction of the translational kinetic energy is stored in the rotational kinetic energy. Because the total kinetic energy is K and rotational kinetic energy is $\frac{1}{2}kMV_{CM}^2$, the fraction f of the original potential energy that must become rotational energy is

$$f = \frac{\frac{1}{2}kMV_{CM}^2}{K} = \frac{\frac{1}{2}kMV_{CM}^2}{(1+k)\left(\frac{1}{2}MV_{CM}^2\right)} = \frac{k}{1+k}$$

For a hoop, $k = 1$ and the rotational inertia is very important. Every bit of the mass must roll in a circle of radius R; one-half of the available energy ($f = 1/(1 + 1)$) goes into making the hoop rotate. For a cylinder, $k = \frac{1}{2}$ and one-third of the potential energy is used for rotation. For a solid sphere even more mass is closer to the axis of rotation so that k is smaller ($\frac{2}{5}$), and only two-sevenths of the energy is locked up in the rotational motion. These results are summarized in Table 12.1.

TABLE 12.1 Energy Division between Translational Motion of the Center of Mass and Rotational Motion about the Center of Mass for Objects that Roll without Slipping

OBJECT	$I_{CM} = kMR^2$ k	FRACTION OF TOTAL KINETIC ENERGY THAT IS ROTATIONAL $k/(k + 1)$
Hoop	1	$\dfrac{1}{2}$
Disk	$\dfrac{1}{2}$	$\dfrac{2}{3}$
Hollow sphere	$\dfrac{2}{3}$	$\dfrac{2}{5}$
Solid sphere	$\dfrac{2}{5}$	$\dfrac{2}{7}$

Using the results of Equation (12.3) and Table 12.1, we can predict the outcome of a downhill race between objects of different shape. If we let the objects described in Table 12.1 roll downhill starting from rest at the top of the incline, the outcome is quite predictable: the solid sphere reaches the bottom first! In fact, regardless of the mass and radius of any participant, the solid sphere always wins. That the mass of the participants does not matter is perhaps not surprising. We have seen before that the gravitational force increases in direct proportion to the mass so that the effect of inertia and weight cancel. It is more surprising that the outcome does not depend on the radius. If we look back into the steps leading to Equation (12.3), we can see where the radius cancels. The rotational inertia increases as R^2. On the other hand, during one rotation of the object its center of mass moves forward $s = 2\pi R$. Thus, for a given forward velocity, a larger R results in a smaller ω. Because the kinetic energy of rotation is proportional to ω^2, the two effects related to the size of R cancel.

Force Analysis

To analyze the forces that act on an object when it rolls without slipping we need three important results that have already been introduced. They are repeated here for convenience:

1. In a uniform gravitational field, the torque about any axis produced by the weight of an extended object is equal to the torque produced by the total mass acting as if it were concentrated at the center of mass. (See Section 12.1.)

2. $$\mathbf{F}_{net} = M\mathbf{a}_{CM}$$

   The center of mass of a system of particles follows a trajectory as if all external forces were acting on a single (fictitious) particle with a mass equal to the total mass of the system. (See Section 10.4.)

3. $$\tau_{net} = I_{CM}\alpha_{CM}$$

   The total external torque about the center of mass of a rigid body produces an angular acceleration about an axis through its center of mass, provided the axis moves parallel to itself. (See below.)

This last result is yet another property of this wonderful point called the center of mass. To appreciate its origin recall from Chapter 11, Equation (11.8), that the form of Newton's second law for rotational motion about a fixed axis is $\tau_{net} = I\alpha$. We derived this expression for rigid bodies rotating about a fixed axis in an inertial coordinate system. There are a number of special cases, which we will explore more fully in Section 13.4, to which this equation may be applied even if the axis of rotation is not fixed in an inertial

Concept Question 4
Why is it possible to apply the conservation of mechanical energy to objects that roll down inclined planes when it is clear that a frictional force must act to make the object roll rather than slide?

Concept Question 5
You are given a paper plate and six quarters. The quarters are to be taped to the plate so that when rolled downhill the plate has as much energy as possible tied up in its rotational motion. Where should the quarters be taped to the plate?

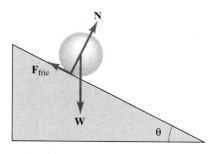

FIGURE 12.9

Forces acting on an object rolling down an inclined plane. The initial point of each vector represents the point where the force is applied to the rolling object.

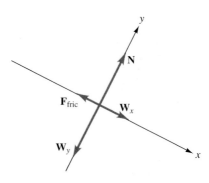

FIGURE 12.10

Free-body diagram of the forces acting on a circular object rolling down an incline. The vertical weight force has been replaced by its components parallel to and perpendicular to the incline. We have modeled the object as a point-mass; hence, the point of application of the forces is irrelevant for the force analysis.

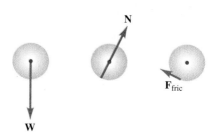

FIGURE 12.11

Diagrams showing each of the three forces acting on a circular object rolling down an incline. We are analyzing the torques about the center of mass, hence, the point of application of each force (indicated by the tail of each force vector) is important.

frame. Of particular importance is the *special case* with the following two restrictions: (i) the object rotates about an axis of symmetry through the center of mass, which is fixed relative to the body, and (ii) the axis translates through space keeping parallel to its original orientation. For such a case, the unique properties of the center of mass allow us to apply this powerful theorem.

Taken together properties (1), (2), and (3) tell us that our force analysis of an object that both rotates and translates can be divided into two problems, both of which we already know how to solve! First, *ignore the fact that the object is rotating,* and write $\mathbf{F}_{net} = M\mathbf{a}_{CM}$, using the point–particle model for the object. Next, *ignore the fact that the object is translating* and write $\tau_{net} = I_{CM}\alpha$ as if the object were merely rotating about the center of mass.[2] When the object rolls without slipping, the two analyses can be connected with a *no-slip condition* relating the accelerations.

As an illustration, let's again examine an object of cylindrical symmetry rolling down an incline. We begin with our familiar question from Chapters 5 and 6, "What are the forces?" In Figure 12.9 we show the familiar forces $\mathbf{W}$, $\mathbf{N}$, and $\mathbf{F}_{fric}$. The frictional force in rotational motion requires extra attention. We discuss a way to determine its presence and direction in a special section "Frictional Forces in Rotational Motion." You may wish to skip the special section now, particularly if the presence and direction of $\mathbf{F}_{fric}$ in Figure 12.9 looks correct and natural to you. If later you discover some frictional forces that don't seem obvious, you may want to come back and look in the box.

We now proceed to follow the problem-solving strategy for Newton's second law as introduced in Chapter 5. Figure 12.10 shows our choice of coordinate system. Resolving the weight force into components parallel to the x- and y-coordinates, we find $|\mathbf{W}_x| = W\sin(\theta)$ and $|\mathbf{W}_y| = W\cos(\theta)$. Next we write $\mathbf{F}_{net} = M\mathbf{a}_{CM}$ in component form

X-COMPONENTS	Y-COMPONENTS
$W\sin(\theta) - F_{fric} = Ma_{CM}$	$N - W\cos(\theta) = Ma_y = 0$

As usual, we have set the y-component of acceleration equal to zero in the above expression. Now comes the first major change from our previous analysis using the point–mass model: We may *not* substitute $F_{fric} = \mu_k N$ for the frictional force. Part of the reason is that, as we have shown earlier, if there is no slipping, the bottom edge of the wheel is *stationary* relative to the incline's surface. Furthermore, you may *not* use $F_{fric} = \mu_s N$ either! The reason is that $\mu_s N$ is the *maximum* static frictional force and there is no reason to believe that the maximum frictional force is acting. The point is that F_{fric} is now an independent unknown, and that our coefficient-of-friction model can, at best, put bounds on its value: $0 \le |\mathbf{F}_{fric}| \le \mu_s N$.

We now switch our point of view and consider the wheel as if it were merely rotating about a fixed axis through the center of mass. We alter our favorite question slightly by asking, "What are the torques?" We must examine each of the three forces in Figure 12.11 to determine the torque it exerts *about the center of mass* of the wheel. Recall that our torque expression is $\tau = rF\sin(\phi)$. For the weight force $\mathbf{W}$, the effective point of application is the center of mass. Hence, the r in our torque expression is zero; the gravitational force exerts no torque. The normal force $\mathbf{N}$ is applied at the edge of the wheel, where $r = R$. However, for this force the angle $\phi = 180°$, which makes the torque zero. The frictional force does exert a nonzero torque about the center of mass. The angle between the tangential frictional force and the $\mathbf{r}$ vector from the axle to the point of application of the force is $\phi = 90°$. Consequently, the magnitude of the frictional torque is

---

[2] You may very well be surprised that a law such as $\tau_{net} = I_{CM}\alpha_{CM}$ can be applied about an axis that is *accelerating,* that is, in a noninertial coordinate system! As you know, in accelerating coordinate systems, even Newton's second law does not hold unless we are willing to invent fictitious (inertial) forces. As discussed in Section 5.7, these inertial forces can be treated like an extra gravitational force due to an added gravitational field of magnitude $\mathbf{g}_i = -\mathbf{a}_{sys}$, where a_{sys} is the acceleration of the system. As shown in Section 12.1, the torque on an object due to the weight force caused by a uniform gravitational field may be calculated as if the entire weight of the body were concentrated at the object's center of mass. For exactly the same reason, the torque due to inertial forces can be calculated as if the inertial force were concentrated at the center of mass. Consequently, the inertial force, like the true weight force, exerts no torque *about the center of mass*. Thus, we can apply $\tau_{net} = I_{CM}\alpha_{CM}$ even if the center of mass is accelerating.

Frictional Forces in Rotational Motion

It is sometimes difficult, when objects both translate and rotate, to determine the role played by friction and, in particular, to determine the direction of the frictional force. We've found one of the best ways to determine what friction is doing is to "think it away." By trying to imagine what happens without friction, you can sometimes see the effect friction has when you, mentally, turn it back on. For example, think about a disk rolling down an incline. Suppose the disk and incline are made of ice and the surfaces of both are wet, making them really slippery. What happens when the disk is placed on the incline? You set the disk down and it starts to *slide* down the incline with no tendency to roll! Now let's see if we can think about how friction must act. At the risk of rubbing you the wrong way, imagine that the bottom edge of the ice disk, where it is in sliding contact with the incline, is the end of your nose sliding down the incline (Fig. 12.12). Now turn friction back on by imagining that the incline is suddenly sandpaper instead of ice. Ouch!! We know it hurts, but ask yourself which way the sandpaper is dragging your poor nose. Why, up the incline, of course. And "up the incline" is the direction of the frictional force when the sphere rolls down the hill.

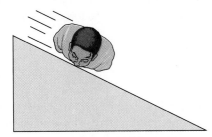

FIGURE 12.12

Imagine sliding your nose down a smooth incline that suddenly turns to sandpaper. Your nose knows the direction of the frictional force.

Let's look at another example as depicted in Figure 12.13. Which way does friction act on your car tires when you accelerate forward from rest, say to the right? Again, think away friction and imagine what happens without it. You're correct: The tires turn clockwise while the car stays stationary. Once again imagine that the bottom edge of the tire is replaced by your nose. Imagine your nose moving to the left across the ice as your head rotates clockwise. Suddenly the ice turns to sandpaper, and what do you feel? The skin on the end of your nose is scraped to the right in Figure 12.13. This forward frictional force is in fact the external force necessary to move your car forward.

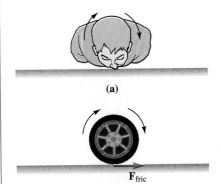

(a)

(b)

$\mathbf{F}_{fric}$

FIGURE 12.13

Rotate your head as if it were a tire spinning on perfectly smooth ice. If the ice turns to sandpaper, your nose knows the direction of the frictional force.

The method, we admit, may seem a bit silly, but over the years we have found that one's nose is an excellent friction sensor and seldom leads you wrong; just follow your nose.

Concept Question 6
A bicycle wheel is caused to spin clockwise as seen from the side. The wheel is then lowered to the ground, which is covered with loose gravel. (a) In what direction is the frictional force on the edge of the wheel? (b) Relate the force described in part (a) to the direction in which gravel is thrown.

simply $|\tau| = RF_{\text{fric}}$. Thus, we can now write $\tau = I_{CM}\alpha$, specialized for this case, as

$$+RF_{\text{fric}} = I_{CM}\alpha$$

The algebraic sign of the torque (+ in this case) is critical. You *must* choose the sign convention for torques consistently with the sign convention for the translational motion. Here's how:

1. Imagine that the torque in question is the only torque acting on the object. Determine which way the object would rotate due to this torque.
2. Determine which way the object would try to translate, *if* it did rotate as determined in (1).
3. If the translation determined in (2) is in the positive translation direction, sign the torque positive. If the torque in question, acting alone, tends to roll the object in the negative direction, sign it negative.

We must choose the sign conventions in the force equations(s) and the torque equation(s) consistently because we intend to solve these equations simultaneously. We cannot overemphasize the care you must take with these signs because a wrong choice leads to an erroneous result and often confuses the interpretation of the answer. Let's look further at our two equations, one from the x-component of force, the other from the torque equation:

$$W\sin(\theta) - F_{\text{fric}} = Ma_{CM}$$

$$+RF_{\text{fric}} = I_{CM}\alpha$$

In *this case* the frictional force contributes a negative force and a positive torque. Please do not memorize this result; the signs do not always turn out this way. Think about the sign of each torque as carefully as you think about the sign of each force.

Having finished haranguing you about signs, let's return to the problem at hand. We now have two equations with *three* unknowns, namely a_{CM}, α, and F_{fric}. We need a third equation before we can proceed. This third equation is provided by the *no-slip condition* $a_{CM} = R\alpha$. Substituting this expression into the torque equation, we have

$$+RF_{\text{fric}} = I_{CM}\frac{a_{CM}}{R}$$

or

$$F_{\text{fric}} = \frac{I_{CM}}{R^2}a_{CM}$$

If we, once again, let $I_{CM} = kMR^2$, the expression for F_{fric} becomes

$$F_{\text{fric}} = kMa_{CM}$$

This expression may now be substituted into the x-component force equation to yield

$$W\sin(\theta) - kMa_{CM} = Ma_{CM}$$

Letting $W = Mg$ and solving for the acceleration, we obtain

$$a_{CM} = \frac{1}{1 + k}g\sin(\theta)$$

With this value of a_{CM}, the expression for friction becomes

$$F_{\text{fric}} = \frac{k}{1 + k}Mg\sin(\theta)$$

Once again we see the importance of the mass distribution as reflected by the parameter k. The farther from the axis the mass is located, the larger is k, and the smaller the acceleration of the center of mass.

Although the results we have just obtained are interesting, they are by no means general and should not be memorized. The purpose of the preceding calculation was to illustrate the application of the force method to objects that both roll and translate. What is important is the *method,* which we reiterate here:

Summary of the Problem-Solving Strategy for Force Analysis of Objects that both Rotate and Translate

---

1. Write $\mathbf{F}_{net} = Ma_{CM}$ for the object as if it were a point–mass, that is, ignoring rotation.
2. Write $\tau = I_{CM}\alpha$ as if the object were only rotating about the center of mass, that is, ignoring translation.
3. Use the no-slip condition $a_{CM} = R\alpha$, if appropriate.
4. Solve the resulting equations simultaneously for any unknowns. Cautions:

  a. In general, it is not the case that $F_{fric} = \mu N$.
  b. Be certain that the sign convention for forces and torques are consistent.

EXAMPLE 12.3   Nonslipped Disk

A solid disk of mass M has a thin string wrapped several times around its circumference. If the free end of the string is held stationary while the disk is released from rest, find the acceleration of the disk and the tension in the string.

SOLUTION We first treat the disk as a point–mass and write Newton's second law. As shown in Figure 12.14, the only forces are along the y-axis, hence we write

$$T - W = Ma_{CM} \qquad \text{(Force equation)}$$

We now calculate the torque about the center of mass from each of these forces. As usual the weight force $\mathbf{W}$ exerts no torque about the center of mass. The torque caused by the tension force has magnitude $\tau = RT$. We note that if the tension force $\mathbf{T}$ were the only force acting on the disk, it would tend to rotate anticlockwise. If the disk did rotate anticlockwise, it would naturally tend to roll downward. In our force equation we called downward the negative direction, hence, we must write

$$-RT = I_{CM}\alpha \qquad \text{(Torque equation)}$$

We now employ the no-slip condition and replace α with a_{CM}/R.

$$T = -\frac{I_{CM}}{R^2}a_{CM} = -\frac{1}{2}Ma_{CM}$$

where we also used $I_{CM} = \frac{1}{2}MR^2$ for the moment of inertia of the solid disk about its center. This expression for T can now be substituted into the force equation to give

$$-\frac{1}{2}Ma_{CM} - W = Ma_{CM}$$

which can be solved for a_{CM} to yield

$$a_{CM} = -\frac{2}{3}g$$

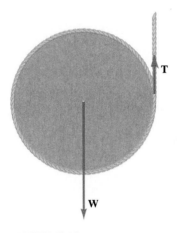

FIGURE 12.14

A cylinder with a string wrapped around its circumference is released. Only two forces act on the cylinder as it falls. See Example 12.3.

Recall that g is the *magnitude* of the gravitational field strength and is, therefore, positive. Hence, the acceleration is negative (downward) as expected. This value for a_{CM} can be substituted into the expression for T to obtain the tension

$$T = +\frac{1}{3}Mg$$

Thus, the string tension is only $\frac{1}{3}$ the weight of the disk.   ◀

12.3 Rotation and Translation with Slipping (Optional)

Some of the more interesting examples of combined rotation and translation occur when slipping does take place. Perhaps you have taken a Ping-Pong ball and, with your finger tips, pinched the edge of the ball against a smooth table. When released just right, the ball leaps forward with a great amount of back spin, and considerable slippage occurs between the edge of the ball and the tabletop. If the table is long enough, eventually friction causes the slipping to stop, and the ball rolls back to your hand. It is possible to relate the final state of motion (the velocity after the ball stops slipping) to the initial state of motion of the ball. That is, given $V_{o_{CM}}$ and ω_o, it is possible to find $V_{f_{CM}}$. To find this relation we adopt the same basic strategy as in our force analysis. That is, we analyze the motion *of* the center of mass, ignoring rotation, and then analyze the rotation *about* the center of mass, ignoring translation. This time, however, instead of using Newton's second law directly, we employ the impulse–momentum theorem and its rotational counterpart. We call the rotational analog of the impulse–momentum theorem the rotational impulse–angular momentum theorem.

Angular Impulse–Angular Momentum Theorem

By integrating $\tau_{net} = I\alpha$ with respect to time and using the expression $L = I\omega$ for angular momentum, we can arrive at an angular analog of the impulse–momentum theorem of Chapter 9. In analogy with the linear impulse $J = \int F\,dt$, we define the **rotational impulse** of a torque as

$$\mathcal{J} = \int_{t_o}^{t_f} \tau\,dt$$

Following the procedure similar to that used for the linear case, we calculate the rotational impulse caused by the net torque. In the resulting integral we substitute $I\alpha$ for the net torque, and perform a change of variable

$$\mathcal{J}_{net} = \int_{t_o}^{t_f} \tau_{net}\,dt = \int_{t_o}^{t_f} I\alpha\,dt = \int_{t_o}^{t_f} I\frac{d\omega}{dt}\,dt = \int_{\omega_o}^{\omega_f} I\,d\omega = I(\omega_f - \omega_o)$$

Equating the first and last expressions, we have

$$\mathcal{J}_{net} = \Delta L \qquad \text{Scalar version of rotational impulse–angular momentum theorem (see limitations for Eq. (11.10))}$$

Concept Question 7
Why is mechanical energy not conserved when a solid disk slips while rolling?

We show how to employ both the linear and angular versions of the impulse–momentum theorem in Example 12.4.

EXAMPLE 12.4   *Scratchin' Off*

A bicycle wheel has its tire replaced by a smooth, heavy, metal rim. The wheel is held by a handle on its axle and given an initial angular velocity ω_o. The wheel is then placed in contact with the ground. At first the wheel remains stationary, spinning in place. After a short time it begins to move forward and eventually reaches the point where it rolls without slipping. Find the final velocity of the wheel in terms of the initial angular velocity ω_o.

SOLUTION   We assume the wheel is initially turning clockwise and take the positive x-direction to the right as shown in Figure 12.15(a). Proceeding with the force analysis first, we note that the vertical forces, **W** and **N**, cancel. The important direction here is along the x-axis. There is only one horizontal force, namely, $\mathbf{F}_{\text{fric}}$ directed to the right as shown in Figure 12.15(b). (If the direction of $\mathbf{F}_{\text{fric}}$ is not clear, read the box "Frictional Forces in Rotational Motion" in Section 12.2 concerning the frictional force.) We write the impulse–momentum theorem for the x-component using the subscript "*o*" to denote initial values and unsubscripted variables for the final values when the wheel begins to roll without slipping. We also drop the explicit "*CM*" subscript on the velocity of the center of mass.

$$J_{\text{net}} = \Delta p \qquad \text{(Linear impulse–momentum theorem)}$$

becomes for the present case

$$+F_{\text{fric}}\Delta t = MV - MV_o \qquad \text{(Translational motion)}$$

where Δt is the interval between the time when the wheel is set on the floor and when it begins to roll without slipping.

We now look at the torque about the center of mass imposed by all three forces. The torques about the center of mass from both **N** and **W** are zero. The only torque is that due to F_{fric}. We note that, if $\mathbf{F}_{\text{fric}}$ were the only force acting, it would tend to make the wheel rotate anticlockwise. If the wheel did rotate anticlockwise, it would roll to the left, that is, in the negative x-direction. Consequently, we take the torque due to $\mathbf{F}_{\text{fric}}$ as negative. With the sign convention decided we now apply the rotational analog of the impulse–momentum theorem:

$$\mathcal{J}_{\text{net}} = \Delta L \qquad \text{(Rotational impulse–angular momentum theorem)}$$

$$-RF_{\text{fric}}\Delta t = I\omega - I\omega_o$$

The left-hand side of this expression is the rotational impulse, that is, the torque $(-RF_{\text{fric}})$ multiplied by the time Δt during which it acts. We can solve this last equation for the impulse $F_{\text{fric}}\Delta t$

$$F_{\text{fric}}\Delta t = \frac{I\omega - I\omega_o}{-R}$$

and substitute the resulting value into the impulse equation for translational motion

$$+\frac{I\omega_o - I\omega}{R} = MV - MV_o$$

In the present case we were given that $V_o = 0$. Hence,

$$MV = \frac{I\omega}{R} - \frac{I\omega_o}{R}$$

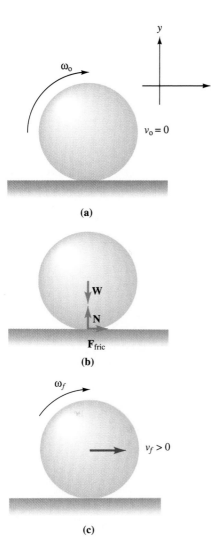

FIGURE 12.15

(a) A spinning object is placed on a horizontal surface without giving the object any translational velocity. (b) The forces acting on the object while it is slipping. (c) After a short time, friction causes the object to begin to roll without slipping. See Example 12.4.

For the final velocity (only!) the no-slip condition $\omega = V/R$ may be used:

$$MV = \frac{I\omega_o}{R} - \frac{IV}{R^2}$$

Once again we let $I = kMR^2$ and solve for V:

$$V = \left(\frac{k}{1+k}\right)R\omega_o$$

An interesting way to think about this answer is to recall that $R\omega_o$ is the initial tangential velocity given to the edge of the wheel. Our result then tells us that the final translational velocity of the wheel is the fraction $k/(1+k)$ of this initial tangential velocity. This fraction is, in fact, the same one that occurs in Table 12.1.

In this particular example, we could use the coefficient-of-friction model to calculate $F_{fric} = \mu_k N$. (We use μ_k because the edge of the wheel is sliding across the floor when there is slippage.) Thus, we can actually calculate the duration Δt of the impulse. Note, however, that it is not necessary to do so. It is important to also recognize that the details of the impulse itself might never be known. This observation is particularly important for nonconstant impulse forces, such as that of a cue striking a pool ball.   ◄

Center of Percussion

Objects do not necessarily have to roll in order for the technique of combined linear and rotational analysis to be useful. If you have played baseball you may have had the unpleasant sensation of striking the ball and feeling a sharp sting when the ball contacts the bat either near the end or too far down the handle. We can understand this phenomenon by modeling the bat as if it were a free body during the collision with the ball.[3]

Once again we first analyze the motion of the center of mass and then the rotation about the center of mass. To simplify the analysis we assume the bat to be stationary and let the bat be struck by an impulse J at a distance y from the center of mass as shown in Figure 12.16. (If you like, think of yourself in the frame of reference of a bat moving

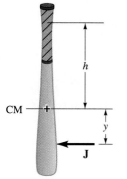

FIGURE 12.16

A freely suspended bat is struck by an impulse.

parallel to it at a high velocity.) The impulse–momentum theorem for linear motion, $J_{net} = \Delta p$, tells us that, as a result of the impulse, the center of mass of the bat moves with velocity v_{CM} given by

$$J_{net} = mv_{CM}$$

We also know that the torque exerted by this impulse causes the bat to rotate about the center of mass in accordance with the rotational impulse–angular momentum theorem, $\mathcal{J}_{net} = \Delta L$. In the present case, the *angular* impulse $\mathcal{J}_{net}$ is related to the *linear* impulse

[3] For a discussion of the validity of this model see Howard Brody, ''Models of baseball bats,'' *American Journal of Physics*, **58**(8), 756–758 (1990).

J_{net} by $\mathcal{J}_{net} = yJ_{net}$ because the impulsive torque τ is related to the impulsive force F by $\tau = yF$. Because $\Delta L = I_{CM}\omega$, we have

$$yJ_{net} = I_{CM}\omega$$

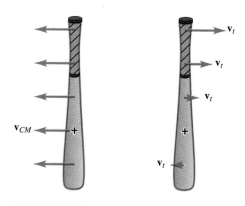

Now imagine your hand to be located at a point that is distance h from the center of mass as shown in Figure 12.16. This point has a net velocity that is the sum of two velocities: the center of mass velocity v_{CM} and the tangential velocity v_t caused by the bat's rotation about the center of mass. As you can see from Figure 12.17, these two velocities are in opposite directions. If they are of equal magnitude, the point h acquires no net velocity due to the impulse; that is, no sting! We can equate the magnitude of the two velocities, substitute their values from the two impulse expressions, and solve for the resulting requirement:

FIGURE 12.17

The resulting motion of a freely suspended bat struck by an impulse can be thought of as the superposition of translational motion of the entire bat and rotation of the bat around its center of mass.

$$v_{CM} = v_t$$

$$v_{CM} = h\omega$$

$$\frac{J_{net}}{m} = h\frac{yJ_{net}}{I}$$

$$y = \frac{I_{CM}}{mh} \tag{12.5}$$

The point whose position is given by y is known officially as the **center of percussion,** although many of us still call it the "sweet spot" on the bat.

12.4 Summary

When calculating the torque due to the weight of an extended object, the weight of the entire mass may be treated as if it acted on the center of mass of the object.

An extended object is in **static equilibrium** if $\mathbf{F}_{net} = 0$ and $\tau_{net} = 0$ for torques about all axes.

When the object rolls without slipping, two **no-slip conditions** apply:

$$V_{CM} = R\omega$$

$$a_{CM} = R\alpha$$

The kinetic energy of an object that both rotates and translates can be calculated by adding the kinetic energy due to the linear velocity *of* the center of mass to the kinetic energy of the object's rotation *about* the center of mass:

$$K = \frac{1}{2}MV_{CM}^2 + \frac{1}{2}I_{CM}\omega^2$$

(Optional) When analyzing the forces on a cylindrically symmetric object that translates and rotates about an axis that itself moves parallel to its original position, we may divide the analysis of the object's motion into a point–particle model analysis of the center-of-mass motion, and rotation about the center of mass.

$$\mathbf{F}_{net} = M\mathbf{a}_{CM}$$

$$\tau_{net} = I_{CM}\alpha$$

(Optional) The list of analogies between linear and rotational motion from Chapter 11 may be extended to include **rotational impulse** and the **rotational impulse–momentum theorem.**

	LINEAR	(UNITS)	ROTATIONAL	(UNITS)
Kinematic Quantities				
position	x	(m)	θ	(rad)
displacement	Δx	(m)	$\Delta \theta$	(rad)
velocity	$v \equiv dx/dt$	(m/s)	$\omega \equiv d\theta/dt$	(rad/s)
acceleration	$a \equiv dv/dt$	(m/s²)	$\alpha \equiv d\omega/dt$	(rad/s²)
	$v = \int a\,dt + C$		$\omega = \int \alpha\,dt + C$	
	$x = \int v\,dt + C'$		$\theta = \int \omega\,dt + C'$	
Dynamical Quantities				
force and torque	F	(N)	$\tau = rF\sin(\phi)$	(N·m)
inertia and moment of inertia	m	(kg)	$I = \sum_{i=1}^{N} m_i r_i^2$	(kg·m²)
kinetic energy*	$K = \frac{1}{2}mv^2$	(J)	$K = \frac{1}{2}I\omega^2$	(J)
work*	$\mathcal{W} = F\,\Delta s$	(J)	$\mathcal{W} = \tau\,\Delta\theta$	(J)
power*	$\mathcal{P} = Fv$	(W)	$\mathcal{P} = \tau\omega$	(W)
momentum	$p = mv$	(kg·m/s)	$l = mrv\sin(\phi)$ or $L = I\omega$	(kg·m²/s)
impulse	$J = F\,\Delta t$	(N·s)	$\mathcal{J} = \tau\,\Delta t$	(N·m·s)
Physical Laws				
Newton's second law	$F_{net} = ma$		$\tau_{net} = I\alpha$	
Work–energy theorem*	$\mathcal{W}_{net} = \Delta K$		$\mathcal{W}_{net} = \Delta K$	
Impulse–momentum theorem	$J_{net} = \Delta p$		$\mathcal{J}_{net} = \Delta L$	

*These quantities are not merely analogous; they are alternative expressions for the exact same physical quantity.

PROBLEMS

12.1 Static Equilibrium

1. Four masses of 100. g, 200. g, 300. g, and 150. g are located along the x-axis at 10.0 cm, 15.0 cm, 25.0 cm, and 30.0 cm, respectively. Take the force of gravity as acting straight down parallel to the y-axis. (a) Calculate the torque about the origin caused by the weight of each of the four masses. (b) Find the center of mass of the four masses and calculate the torque about the origin due to the weight of a single mass of 750. g placed at the center of mass. (c) Compare the total of the torques calculated in part (a) to the answer in part (b).

2. (a) Using the calculus, compute the total torque about the origin due to the weight of a uniform rod of length L and constant linear mass density λ if the rod has one end at $x = a$ and the other at $x = L + a$. (b) Show that this result is identical to the torque due to the weight of a point–mass $m = \lambda L$ located at $x = a + L/2$. (*Hint*: In (a) divide the rod into segments dm, calculate the torque caused by each segment, and integrate the resulting expression for $d\tau$.)

3. A 70.0-kg painter stands 1.00 m from the left end of a 3.00-m long uniform plank that is supported by a cable at each end. If the plank itself weighs 30.0 kg, find the tension in each cable.

4. A 65.0-kg person stands at the end of a uniform diving board of length $L = 3.50$ m as shown in Figure 12.P1. The board itself has a mass of 25.0 kg and is bolted to two supports: one located at the far end of the board and the other at $S = 1.20$ m from the far end. Calculate the force on the board at each of the supports.

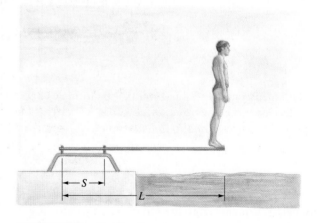

FIGURE 12.P1   Problem 4

5. A physics professor wishes to find the mass of an 8.00-ft long 2x4 for a book he is writing but he has no scale and only a 1.00-kg mass as a reference. He places the 1.00-kg mass on one end of the 2x4 and is able to balance the board on a knife edge placed 3.00 ft from the 1.00-kg mass. What is the mass of the 2x4?

6. Model the forces acting on the forearm when a 25.0-lbf weight is being curled as shown in Figure 12.P3. Treat the biceps as you

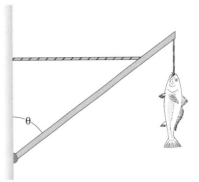

FIGURE 12.P2   The human musculoskeletal system is a collection of levers.

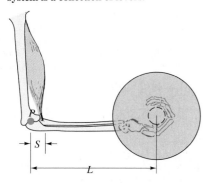

FIGURE 12.P3   Problem 6

would a vertical guy wire. Take the weight of the forearm itself to be 4.00 lbf and assume the center of mass of the forearm to be located 14.0 cm from the "hinge" P at the elbow. Take $L = 32.0$ cm and $S = 4.50$ cm. (a) Calculate the tension in the biceps and the force exerted on the forearm at the elbow joint P. Express the answer in both newtons and pounds. (b) Calculate these same forces after the forearm has been raised until it is 30.0° above the horizontal, assuming that the tension force from the biceps continues to act at right angles to the forearm. [Don't forget both the **H** and **V** forces at the pivot.]

7. A 59.0-kg person is standing on one end of a uniform horizontal plank of length $L = 4.00$ m with mass 20.0 kg. The end of the plank extends $S = 2.50$ m beyond the support surface as shown in Figure 12.P4. How far from the edge of the support can an 80.0-kg person walk before the plank begins to tip?

8. Suppose the guy wire in Example 12.2 is moved to the midpoint of the horizontal rod. Assuming the point where the guy wire is attached to the wall stays fixed, calculate the new values for H, V, and T.

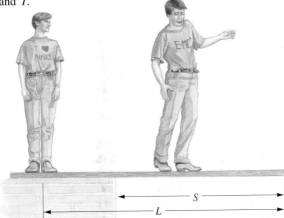

FIGURE 12.P4   Problem 7

9. A 110.-lbf fish hangs from the end of a 8.00-ft long, 20.0-lbf uniform pole, which is pivoted at one end and held stationary by a horizontal guy wire attached three-quarters of the way from the pivot to the free end. The pole makes an angle $\theta = 53.13°$ with the vertical as shown in Figure 12.P5. Calculate the tension in the guy wire and the force of the pivot on the end of the pole.

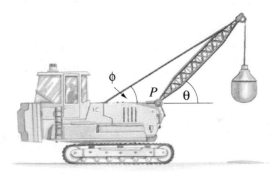

FIGURE 12.P5   Problem 9

10. A wrecking ball of mass 500. kg is held by a uniform crane boom of mass 300. kg as shown in Figure 12.P6. The boom is attached to a support cable at a point seven-eighths of the length of the boom from the pivot. If the boom and the cable make angles $\theta = 45.0°$ and $\phi = 30.0°$ with the horizontal, calculate the tension in the support cable and the force at the boom hinge P.

FIGURE 12.P6   Problem 10

11. A rock climber is rappelling down the side of a vertical cliff. Assume the rope is looped to a carabiner located at the climber's

FIGURE 12.P7   One of the authors (ABW) checks his answer for Problem 11. (Photo by Craig Zaspel, Western Montana College.)

center of mass as shown in Figure 12.P8. If the rope is attached to the climber at $l = 1.00$ m horizontally from the wall and the climber's feet are $L = 3.00$ m from the top of the cliff, find the tension in the rope and the force exerted by the wall on the climber's feet.

FIGURE 12.P8   Problem 11

12. A light bookshelf, $L = 25.0$ cm deep, is supported by loosely hinged braces as shown in an end view in Figure 12.P9. At the particular brace shown, there is a force $F = 20.0$ N downward due to books on the shelf at an effective distance of $S = 10.0$ cm from the wall. The brace is attached to the wall at point Q, 30.0 cm below point P. Find the horizontal and vertical components of the force at the point P where the shelf is attached to the wall. (*Hint*: You may assume that the brace tension is parallel to the brace. This is a special case due to the loose hinges and the fact that we ignore the weight of the brace.)

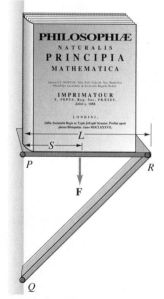

FIGURE 12.P9   Problem 12

13. A ladder of length L leans against a wall at an angle θ from the vertical as shown in Figure 12.P10. The wall is smooth so that only a horizontal force N is exerted on the top of the ladder. A fireman of mass M climbs a distance d upward as measured along the ladder. (a) Assuming the ladder stays in equilibrium, calculate H and V, the horizontal and vertical force components on the foot of the ladder as a function of M, d, and θ. (Ignore the mass of the ladder itself.) (b) If the maximum horizontal frictional force H is related to V through $H = \mu V$, what is the minimum coefficient of friction μ required to prevent the ladder from slipping.

FIGURE 12.P10   Problems 13 and 14

14. Repeat the calculation of Problem 13 but this time include the mass m of the ladder of uniform linear mass density. Comment on the effect of the ladder mass: Is a heavy ladder more or less safe for a given angle and coefficient of friction?

12.2 Rotation and Translation with No Slipping

15. A solid disk of radius 10.0 cm and mass 0.500 kg, a thin hoop of radius 4.00 cm and mass 0.100 kg, and a hollow sphere of radius 15.0 cm and mass 0.333 kg are released at the same instant from a point 1.00 m (measured along the incline) from the bottom of a ramp that is inclined 15.0° from the horizontal. (a) In what order do the objects reach the bottom? (b) How much sooner does the winner reach the bottom than the last place finisher? [*Hint*: The acceleration of the center of mass is constant.]

16. A thick-walled hollow cylinder and a wheel composed of a thin hoop and six spokes are placed side-by-side on an incline and released from rest. The cylinder of outer radius $R_2 = 8.00$ cm and inner radius $R_1 = 5.00$ cm has a mass of 1.25 kg. The wheel's rim has a mass of 0.350 kg and a radius of 7.50 cm. Each of the six spokes has length 7.50 cm and mass 0.150 kg. Calculate the speed of both the hollow cylinder and the wheel if they roll without slipping down a ramp of length 2.00 m that is inclined at 15.0° from the horizontal.

17. A wheel of radius R is made from a thin hoop of mass M and two thin rods of length $2R$ and mass m. The rods cross each other at right angles. The wheel is rolling without slipping along a horizontal table when it encounters an upward-sloping incline. If the initial speed of the wheel center is v_o, determine the maximum vertical height that the center of mass attains.

18. A wheel of radius 2.54 cm and mass 150. g starts from rest and rolls down a 1.50-m long incline without slipping. If the incline makes an angle of 30.0° from the horizontal and the velocity of the center of mass of the wheel is 3.00 m/s at the bottom of the incline, what is the moment of inertia of the wheel?

19. A block of mass $M = 2.00$ kg travels over a frictionless horizontal surface at 3.00 m/s. The block encounters a frictionless incline as shown in Figure 12.P11. (a) What vertical distance up the incline does the block travel before coming momentarily to rest

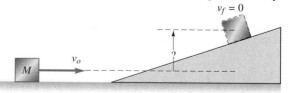

(b) Suppose a solid sphere of identical mass is rolling without slipping on a rough surface and encounters a rough incline. If the velocity of the center of mass of the sphere is the same as that of the block, how far up the incline does it roll?

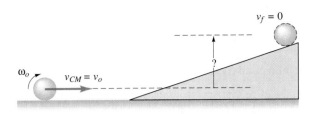

FIGURE 12.P11   Problem 19

20. A toy car is powered by a wind-up spring of negligible mass. The car is constructed from a 100.-g body and four disk-shaped wheels, each of mass 20.0 g and radius 2.00 cm. The spring is wound tightly, storing 10.0 J of energy. If 90.0% of this energy is converted to the kinetic energy of the car, how fast is the car moving when the spring comes unwound? (Assume the wheels roll without slipping.)

21. The axle of a solid cylinder compresses two Hooke's-law springs by 0.200 m as shown in Figure 12.P12. The cylinder has mass $M = 0.600$ kg and radius $r = 8.00$ cm. If the spring constant $k = 10.0$ N/m and the cylinder always rolls without slipping, what is the final speed of the cylinder after it leaves the springs?

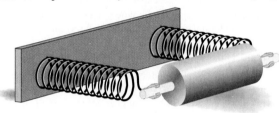

Perspective View

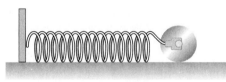

End View

FIGURE 12.P12   Problem 21

22. A cart has mass $m = 2.00$ kg including four wheels of radius $r = 3.00$ cm. The cart is traveling over a level surface at 2.00 m/s, its wheels rolling without slipping. When the cart collides with a light Hooke's-law spring of stiffness constant $k = 150.$ N/m, the spring is compressed 30.0 cm before the cart comes momentarily to rest. Calculate the moment of inertia of one of the wheels.

23. Show that for a sphere of radius r rolling inside of a sphere of radius $R > r$ as shown in Figure 12.P13, the no-slip condition implies $v_{CM} = \omega r(R - r)/R$, where ω is the angular velocity of the small sphere about its center of mass.

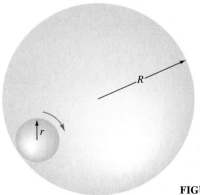

FIGURE 12.P13   Problem 23

24. A small solid sphere of radius r rolls without slipping on the inside of a sphere of radius R. If the small sphere is originally at rest at a position where the radius of the large sphere makes an angle θ with the vertical, use the results of Problem 23 to find the velocity of the center of mass of the small sphere as it passes through the lowest point.

25. What is the minimum coefficient of friction required to allow a wheel of mass M, radius R, and moment of inertia $I = kMR^2$ to roll without friction down an incline that makes an angle θ with the horizontal?

26. A yo-yo is composed of two disks each of radius 4.0 cm and mass $M = 100.$ g connected by a rod of radius 0.250 cm and mass 5.00 g. A light, thin string is wound around the rod without overlap. The free end of the string is held stationary and the yo-yo is released from rest. Find the tension in the string and the downward acceleration of the center of mass of the yo-yo.

27. A soil compacter in the shape of a large solid cylinder is pulled up a 10.0° incline as shown in Figure 12.P14. If the cylinder has a diameter of 0.800 m and a mass of 1.20×10^3 kg, (a) what force must be applied to the yoke to make the cylinder accelerate 0.100 m/s² up the incline, and (b) what is the magnitude of the frictional force that acts on the curved surface of the cylinder in contact with the incline?

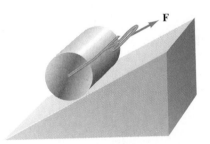

FIGURE 12.P14   Problem 27

28. A toy much favored by toddlers is constructed from a hollow sphere that is rolled along the ground by means of a long handle attached to a yoke which, in turn, holds the axle of the sphere (Fig. 12.P15). The sphere has a radius of 8.00 cm and a mass of 0.500 kg. (a) What is the acceleration of the sphere if the child pushes with a force of 5.00 N at an angle of 30.0° from horizontal? (Ignore the mass of the handle and yoke.) (b) What is the magnitude of the frictional force that acts on the sphere at its point of contact with the ground?

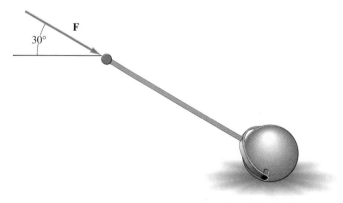

FIGURE 12.P15   Problem 28

29. Show that if any cylindrically symmetric object with rotational inertia $I = kMR^2$ is pulled by a yoke connected to its axle as shown in Figure 12.P16, the acceleration of the object is

$$a_{CM} = \left(\frac{1}{1 + k}\right)\frac{F}{M}.$$

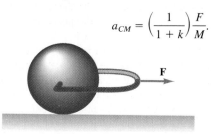

FIGURE 12.P16   Problem 29

30. Consider, as in Figure 12.P17, two methods of applying a force F to a heavy disk that is to be rolled up an incline. In the first instance the force is applied to the disk axle (by applying $F/2$ to each side of the axle to ensure the disk goes straight up the incline). In the second method a rope is wound around the circumference of the disk, so that the force F is applied at the top edge of the disk. (a) Show that the frictional forces between the incline and the disk's edges act in opposite directions for the two cases. Indicate the directions on sketches and explain your reasoning. (b) Show for the first case that the no-slip condition implies the velocity of the tip of the rope v_{rope} is given by $v_{rope} = R\omega$, but for the second case $v_{rope} = 2R\omega$.

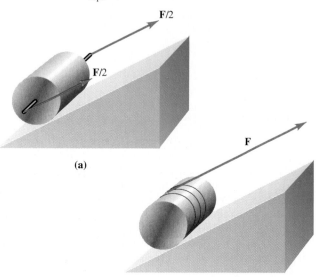

(a)

(b)

FIGURE 12.P17   Problem 30

★31. Calculate the acceleration of mass m in Figure 12.P18 in terms of m, M, R, θ, g, and k, where $I = kMR^2$. Ignore the pulley and assume no slipping between the wheel and the incline. (*Hint:* See Problem 30.)

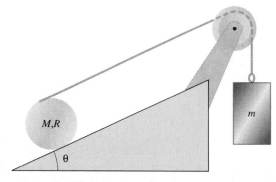

FIGURE 12.P18   Problem 31

★32. Calculate the minimum coefficient of friction required to prevent the wheel in Problem 31 from slipping.

12.3 Rotation and Translation with Slipping (Optional)

33. A force of 5.00 N acts tangentially on the edge of a disk with mass 0.800 kg and radius 20.0 cm for a time of 3.00 s. The disk is free to turn about an axis perpendicular to the plane of the disk and through its center. If the disk starts at rest, what is its angular speed after the force acts?

34. A 25-cent coin has a diameter of 2.40 cm and a mass of 5.70 g. It is held on edge and briefly tapped with a flick of the finger to set it spinning. After being struck the quarter is observed to spin with angular velocity of 5.00 rev/s. (a) What is the magnitude of the angular impulse $\mathcal{J}$? (b) If the collision of finger and coin is estimated to last about 10.0 ms and the force is directed at a right angle to the plane of the quarter, what is the magnitude of the average force acting on the coin?

35. A bicycle wheel of mass 1.10 kg and radius 35.0 cm is free to turn about a fixed axle through its center. A force of 10.0 N is applied tangentially to the edge of a wheel, which was originally stationary. After the force is applied the wheel is observed to be rotating at 0.800 rev/s. Model the wheel as a thin hoop and estimate the duration of the force.

36. A flywheel with moment of inertia $150.\ kg \cdot m^2$ is rotating at 100. rpm. A friction brake applies a force of 500. N tangentially at a point 0.500 m from the center of the wheel. If the brake is applied for 2.00 s, what is the angular velocity of the flywheel after the brake has been applied?

37. Work the inverse problem of Example 12.4. That is, let the wheel start with $V_o \neq 0$ and $\omega_o = 0$. Determine the final velocity of the wheel after it starts to roll with no slipping.

38. If the coefficient of friction between the wheel and the floor in Problem 37 is μ_k, how far does the wheel move before it begins to roll without slipping?

39. A bowling ball is thrown with a speed of 8.00 m/s and no initial spin. After the ball has moved down the alley a distance of 15.0 m, it begins to roll without slipping. Calculate the kinetic coefficient of friction between the ball and the alley surface.

40. A Ping-Pong ball is given an initial forward velocity V_o along the surface of a table. At the instant of release, it also has a *backspin* with angular velocity ω_o. (a) Find the velocity of the Ping-Pong ball when it starts to roll without slipping. (b) What condition must be fulfilled so that the Ping-Pong ball rolls back toward the point of release?

41. A pool cue strikes a ball of radius R at height h above the center of the ball as shown in Figure 12.P19. Show that if the ball is to begin motion with no slipping, then $h = \frac{2}{5}R$.

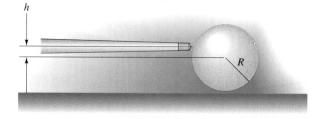

FIGURE 12.P19   Problem 41

42. A child is playing with a thin hoop of diameter 1.00 m and mass 0.250 kg. With the plane of the hoop vertical and the bottom edge resting on the ground, the child starts the hoop moving parallel to its own plane by hitting the hoop and thus imparting a horizontal impulse directed toward the center of the hoop. If the magnitude

of the impulse is 0.200 N · s, use the collision approximation and calculate (a) the initial velocity of the hoop, and (b) the final velocity of the hoop after it begins to roll without slipping.

43. (a) Show that if any cylindrically symmetric object with moment of inertia $I = kMR^2$ is given an initial angular velocity ω_o about its axis and then placed on a rough horizontal surface, its final angular velocity when it begins to roll without slipping is given by

$$\omega_f = \left(\frac{k}{1+k}\right)\omega_o.$$

(b) Show that the fractional change in kinetic energy is $1/(1+k)$.

44. A camper finds a long, thin, straight branch on the ground and decides to use it for firewood. The branch is too long, so he decides to break it by hitting it over a rock as shown in Figure 12.P20. If the camper holds the branch by one end, at what point on the stick should he hit the rock so that he feels no "sting" from the impact?

FIGURE 12.P20   Problem 44

Numerical Methods and Computer Applications

45. Consider Example 12.2 again. This time, let the point of attachment of the guy wire vary from the point on the end of the rod to a point 0.10 m from the wall, keeping the length of the guy wire constant. Plot a graph of the behavior of T, H, and V as a function of the point of attachment.

46. Explore the results of Problem 13. Use a spreadsheet to plot the minimum required μ as a function of d/L for six θ values between 10° and 60° in 10° increments. Produce all the graphs on a single plot and comment on your results.

General Problems

47. As shown in Figure 12.P21 a garden roller of mass M and radius R is pulled horizontally by the yoke over a curb of height h (where $h < R$). Find the minimum force F necessary to pull the roller over the curb in terms of M, R, and h.

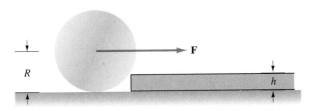

FIGURE 12.P21   Problem 47

48. A uniform disk ($M = 2.50$ kg) of radius $R = 20.0$ cm rests against a vertical wall and is held in place by two light cords that make 30° angles with the wall. As shown in Figure 12.P22 the cords are attached to the disk a horizontal distance of $\frac{2}{3}R$ from the disk center. (Only one cord is visible in the figure. The second cord is directly behind the first on the back side of the disk.) (a) Find the tensions in the cords, and the horizontal and vertical components of the force exerted on the disk by the wall at the point of its contact with the disk. (b) What is the minimum coefficient of static friction between the wall and the disk for this system to remain in equilibrium?

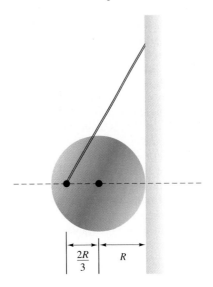

$\frac{2R}{3}$ R

FIGURE 12.P22   Problem 48

49. Suppose the wall in Problem 48 were polished to the point that it becomes frictionless. (a) What angle would the cords make with the vertical wall when the disk is in its new position of equilibrium? (b) Compute the new tension in the cords and the normal force of the wall on the disk.

50. As shown in Figure 12.P23 a cylindrical container of radius R contains two spheres of radii r where $R < 2r < 2R$. If each sphere has a weight W, find the force of the cylinder base on the bottom sphere and the force of the cylinder walls on the spheres. (Assume the wall of the cylinder is smooth.)

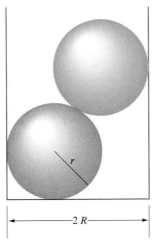

r

$2R$

FIGURE 12.P23   Problem 50

51. A disk, a solid sphere, and a hoop each with the same mass and radius are all set into rotation with the same initial angular velocity ω. All are then gently lowered to the floor and given no initial translational velocity. After all three have reached the point where

they begin to roll without slipping, which has the greatest horizontal velocity? Which has the least?

52. Two students carry a trunk of weight $W = 500.$ N up the dormitory stairs using handles on both ends. Assume the trunk is uniformly packed so that its center of mass is midway on a line joining the two handles, which are separated by distance $L = 1.20$ m as shown in Figure 12.P24(a). Also assume that each student applies only a vertical force to the trunk. (a) At a moment when the trunk is stationary and inclined by an angle of 36.87°, how much force is exerted by each of the two students? (b) If the center of mass of the trunk is a distance $d = 0.300$ m below the midpoint of the line joining the trunk handles as in Figure 12.P24(b), how much force is exerted by each student?

★53. Consider the face of a rectangular dam of width L and total depth D. The force dF on a horizontal strip of the dam increases with the depth y of the water according to $dF = \rho Lgy\, dy$, where ρ is the density of the water. (a) What is the magnitude of the total force on the dam face? (b) What is the total torque on the face of the dam about an axis along the top of the dam? (c) If the total force of the water on the dam is modeled as acting at a single point, how far below the surface should this point be located in order for this force to cause the correct total torque?

★54. (a) Show that if an object is in equilibrium under the influence of two forces only, the two forces must be equal and opposite and have the same line of action. (b) Show that if an object subject to only three forces is in equilibrium that (i) either the lines of action are all parallel or (ii) they all meet at a single point. (Forces acting such as those in case (ii) are said to be **concurrent.**)

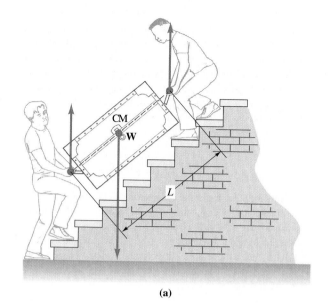

(a)

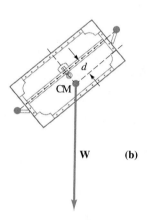

W   (b)

FIGURE 12.P24   Problem 52

GUEST ESSAY

Biomechanical Loading of the Human Body

by ROBERT SOUTAS-LITTLE

The branch of physics known as classical mechanics quantitatively analyzes and predicts the motion, position, forces, and deformations of materials or bodies. For this purpose, many new areas of mathematics were developed: calculus, vector analysis, and tensors, among others. Clinical medicine, on the other hand, is descriptive by nature, having few quantitative measurements. It has been observed that most of the major advances in medicine have come from the introduction of new tools. Physics and mathematics are now playing the part of "a new tool." **Biomechanics** as a part of bioengineering is helping bring the worlds of engineering and medicine

together. Quantitative measurements are made by those using classical mechanics, and these data are then put to use by those in the clinical field.

Biomechanics is a large field and one which has made major advances during the past 30 years. The field is closely aligned with medical specialties, such as orthopaedics, rehabilitation, sports medicine, and many others. One application of mechanics is to study the loading and function of the musculoskeletal system. This information is useful in planning surgery for crippled children, designing artificial limbs or joints, and understanding injury, among many applications.

It is astonishing to most people to learn the amount of force that parts of the human body must withstand dur-

ing daily activities or during exercise and sport. The bones in our legs and arms serve as levers and can be accurately modeled as such. Before beginning to examine how the musculoskeletal system functions, let us first examine a simple lever as shown in Figure 1.

The required geometry of the lever is shown by specifying the distance from the fulcrum to the ends where loads are placed. The loads are labeled F_A and F_B and are resisted by the compression at the fulcrum, or pivot point, R. In engineering, the sketch with loads shown is called a **free-body diagram** and is the starting point of all analyses. Now, if the lever is in balance, or equilibrium, there is no movement and no imbalance in

ROBERT SOUTAS-LITTLE
Michigan State University

Robert Soutas-Little received his B.S. in Mechanical Engineering from Duke University and M.S. and Ph.D. degrees in Mechanics with minors in Physics and Mathematics from the University of Wisconsin. He taught theoretical mechanics at Marquette University, the University of Wisconsin, Oklahoma State University, and Michigan State University. In addition, he has been a guest Professor at Lawrence Livermore Laboratory, Cambridge University, England, and The Technion in Israel. Soutas-Little is the winner of the 1970–71 Western Electric Award for Teaching Excellence in Engineering and the author of over 100 articles and presentations. Currently, he is Director of the Biomechanics Evaluation Laboratory of Michigan State University.

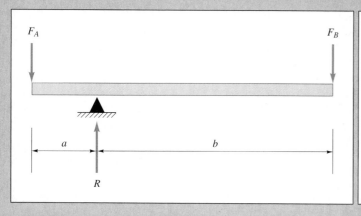

FIGURE 1

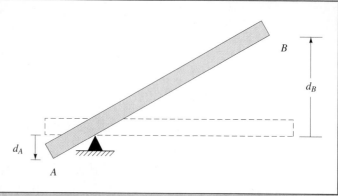

FIGURE 2

forces or torques. We can express this in a mathematical form as follows:

$$R = F_A + F_B \quad (1)$$

$$F_A a = F_B b \quad (2)$$

If we assume that the distance b is greater than a,

$$F_A = \frac{bF_B}{a} \quad (3)$$

and

$$R = \frac{(b + a)}{(a)} F_B \quad (4)$$

F_B has the greatest mechanical advantage and, therefore, is lower in magnitude than F_A. If, for example, b is 10 in and a is 2 in, F_A has to be five times greater than F_B. If 10 lbf are placed at B, then 50 lbf are required at A to balance the lever, and the compression at the fulcrum would be 60 lbf.

Now, let us examine movement of our lever (see Figure 2). By examination, $d_B = (b/a) d_A$, or with the dimensions given, the movement at B is five times greater than at A. Now, conceptually, we can see the way levers work; the point where the movement is greatest has the lower force, and where the movement is low, the force is higher.

Let us model the forearm and hand as a lever as shown in Figure 3. This lever is slightly different from the previous one because the fulcrum is back at the elbow. The force M generated by the biceps muscle has a very small lever arm, 2 cm, and the weight W in the hand has a lever arm of

40 cm. If we balance torques about the fulcrum at the elbow, we obtain:

$$2M - 40W = 0 \quad (5)$$

$$M = 20W \quad (6)$$

If we hold a 20-lbf weight in our hand, the biceps must balance it with a force of 400 lbf. At most points in our body, the muscle has a short lever arm (it gets the short end of the stick). This is the major reason that the muscle must generate large forces. There are physiological bases for this type of lever system because although muscles make only small changes in length when they contract, the limbs have a large range of motion. Therefore, to produce large movements with small contractions, the muscle insertion point must be close to the fulcrum of the lever and the muscle must be able to generate large forces.

We can demand more of the muscle if we increase the effective length of the arm, an example of which is a tennis player who doubles his lever arm by using a racket to hit a tennis ball as hard as he can. The medical condition that results from stress at the fulcrum is tennis elbow. The loading at the elbow can be approximated using these models in such activities as shoveling, hammering, or using other tools.

Let us look at another free-body model, which, although similar, is harder to conceptualize. We attempt to determine loads on the lower back, or lumbar spine, by modeling the upper part of the body as a free body (Fig. 4).

Viewed as loading on a simple lever, this free-body model appears as shown in Figure 5.

We identify these loads as follows. P is the load held in the hands; W is the weight of the upper body,

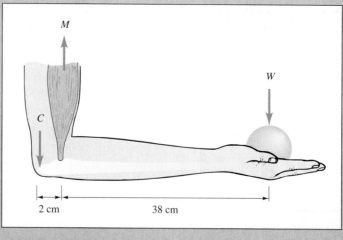

FIGURE 3

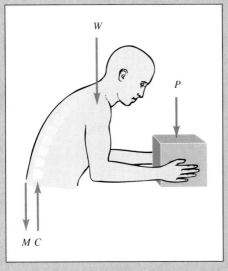

FIGURE 4

approximately two-thirds of body weight; M is the muscle force posterior to the spine; and C is the compression on the spine (the disk and vertebrae). The lever arms are designated by a, b, and l. If we balance forces and torques at the spine (which is the fulcrum), we obtain

$$M = \frac{bW + lP}{a} \qquad (7)$$

$$C = M + W + P \qquad (8)$$

Now, let us examine what loads we can place on our spine while performing a task of lifting up a box. In this example, suppose a 150-lbf man is lifting a box that weighs 30 lbf. Realistic values of the lever arms for this case are: $l = 25''$, $b = 15''$, and $a = 1''$. The overturning torque caused by the 30-

lbf box and the 100 lbf of upper body weight is 2250 in · lbf and requires the muscle to generate a force of 2250 lbf to resist with only a 1-in lever arm. The resulting compression on the spine is 2380 lbf, more than a ton. It is difficult to believe that the body can support loads of this magnitude, but bioengineering tests have shown the accuracy of these calculations.

The previous examples assume static equilibrium, that is, that no parts of the body are accelerating. This is not generally the case when we are performing even the simplest daily activity. For example, during walking, the center of mass of our body rises and falls (accelerates up and down) during each step. Biomechanists measure the forces between the foot and the ground with an instrument called a force plate or force platform. Far more sophisticated than an expensive bathroom scale, a typical force plate measures the vertical force, the force forward and backward, and the force side to side. In addition, the moments about these axes are also measured (see Fig. 6).

The forces are called the **ground reaction forces**, and the three torque measurements allow researchers to calculate where the ground reaction force acts on the interface of the foot with the plate. This point moves as the walker touches first with his heel, then has his foot flat and, finally, pushes off. These six quantities can be measured a thousand times a second (1000 Hz). Although this may seem to be taking more measures than necessary (every millisecond), the average walker has his foot in contact with the ground for only 600–800 ms, and high

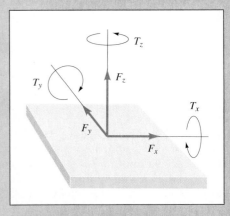

FIGURE 6

heel-strike forces can occur during the first 30–50 ms. During running, the contact time can be only 200 ms, with significant changes in position and force happing every 10–20 ms. A three-dimensional computer plot of the ground reaction force at its point of interface for a typical walking pace is shown in Figure 7.

As the right foot first touches the ground, the transfer of weight from the left foot to the right foot begins. This phase is termed **double-stance**, meaning both feet are on the ground. After the left foot toes-off, or pushes off with the toes, single-stance occurs until the left foot is swung forward and the transfer of weight is initiated again. Examination of Figure 7 shows a loading shape that looks like the letter "M." The two peaks are due to the dynamics of transfer of the weight from one foot to the other and occur during double-stance. Note that the value of these peak forces is between 1.25 and 1.5 times body weight during normal walking. Because both feet are on the ground at the time of weight transfer, each leg bears this load, and

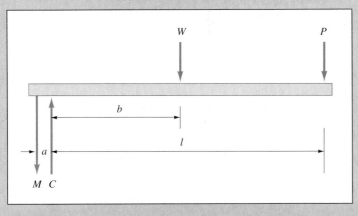

FIGURE 5

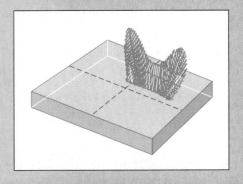

FIGURE 7

an upward force of 2.5–3 times body weight is exerted on the body. This force is necessary to decelerate the weight on the front leg and accelerate it off the rear limb. During single-stance, when the body is supported by only one leg, the ground reaction force drops to 0.7 to 0.8 times body weight. In a dynamic sense, the individual is not supporting his weight and is accelerating down.

Sports activities produce higher ground reaction forces than simple walking. During running, the forces on each leg reach values of 2.5–3 times body weight. High-impact aerobic exercise develops forces of 5–6 times body weight. The landing from the "slam dunk" generates 12–14 times body weight. Recently, architects and engineers have begun to take human dynamic loads into consideration in structural design of sports stadiums and music halls. Obviously, such buildings cannot be designed based only on attendance; the activity of the spectators must be considered.

The landing from the "slam dunk" generates 12–14 times body weight.

Now let us examine some of the functional aspects of forces and torques on the body. In particular, biomechanists are concerned with the loads and torques at the body joints. As mentioned in the previous examples, these joint forces and torques are caused by external loads but are resisted by muscles, ligaments, and compression in the joint. To illustrate, let's consider the knee joint when a man squats down to do exercises, lifts

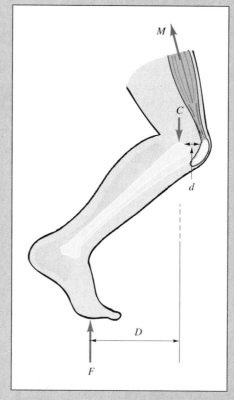

FIGURE 8

weights, or performs normal activities. Again, begin by modeling the lower part of the leg as a free body as shown in Figure 8.

The ground reaction force F is equal to more than one-half the body weight because of dynamic effects, C is the joint compression, and M is the muscle force generated by the muscles on the front of your leg, the quadriceps. Again, the required muscle force is greater than F because of the difference in lever arms.

$$M = \frac{DF}{d} \qquad (9)$$

The applied torque FD is trying to flex the knee joint while the muscle force M is trying to extend the knee. If the woman extends her knee (standing up), then the muscle torque is greater and the muscle shortens and is said to be in **concentric contraction**. Another way to express this is that the muscle

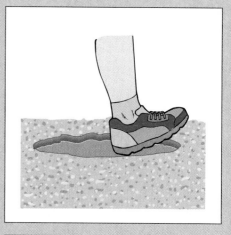

FIGURE 9

is producing motion against the resistance of the body weight. If, however, the woman is squatting down or flexing her knee, the torque due to the ground reaction force is greater than the muscle torque and the muscle is braking against the external force. The muscle is stretching during this process and is said to be in **eccentric contraction**.

Most injuries occur when the muscles are trying to resist external loads placed on the body and are in eccentric contraction. During these phases, the person is not in control of the loads on her body. Modeling and understanding body loading in the workplace has led to the development of the field of ergonomics and has helped in the reduction of injuries. This essay provides just a small insight into biomechanics—the application of physics to the human body.

Questions for Thought

1. A runner unexpectedly lands on the far side of a pothole as shown in Figure 9. Explain why this could result in a tear of his Achilles tendon.
2. Explain why obesity or pregnancy can cause back pain.
3. Discuss the differences between pushing and pulling a large object.
4. Model the loading at the knee when kicking a soccer ball.

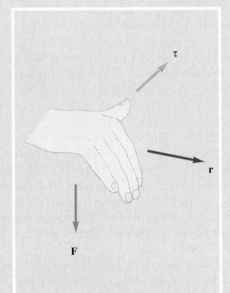

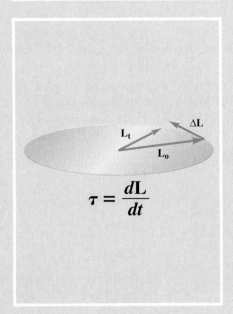

$$\tau = \frac{d\mathbf{L}}{dt}$$

13

Vector Descriptions of Rotational Motion

In this chapter you should learn

- the vector definitions of torque, angular velocity, and angular momentum.

- the vector form of Newton's second law for rotational motion.

- (optional) the cause of the precession of gyroscopes.

- (optional) when you may use the simplified theory of rotation presented in Chapters 11 and 12.

Up to this point in our study of rotational motion we have looked at a restricted subset of objects that includes those that rotate about a fixed axis of symmetry (or perhaps an axis moving parallel to its original position). You have developed, we trust, some intuitive "feel" for rotational quantities, such as torque, rotational inertia, and angular momentum. Armed with this background, we are now ready to examine more general rotations and cast many of the laws we have discussed in their general vector form. We make considerable use of vector algebra (particularly the vector product) introduced in Chapter 2. You may find it helpful to review this material before proceeding.

13.1 The Torque Vector

In Chapter 11 we defined the torque about a rotational axis as $\tau = rF\sin(\phi)$ with the restriction that **r** and **F** both lie in a plane perpendicular to the axis of rotation. In general, torque is defined not about an axis, but rather about a *point:*

$$\tau = \mathbf{r} \times \mathbf{F} \qquad \text{General definition of the torque about a point} \qquad (13.1)$$

This definition assumes that the point about which torques are to be taken is the origin, and **r** is the vector from the origin to the point of application of the force.

Let's compare this definition with our previous version. First, we note that when **r** and **F** are in a plane containing the origin and perpendicular to the axis of rotation, the magnitude of the cross product $|\mathbf{r} \times \mathbf{F}| = rF\sin(\phi)$ is exactly the same as our scalar definition. The new feature is that torque is now a vector. In general, because it is the cross product of **r** with **F**, the torque vector is perpendicular to the plane defined by **r** and **F**. Thus, the torque vector itself indicates the orientation of the axis about which **F** is *trying* to cause turning motion. This result is very convenient indeed. In Chapters 11 and 12 we described the action of the torque as either clockwise or anticlockwise. This kind of information is now encoded in the direction of the torque vector as given by the right-hand rule of Chapter 2. As shown in Figure 13.1, when the torque vector is directed toward you, you see the torque as trying to turn an object anticlockwise. Put another way, if you wrap the fingers of your right hand around a screwdriver so that the screw advances, your thumb points in the direction of the torque.

What if an object is constrained to rotate about an axis that is fixed in space, but the torque vector given by Equation (13.1) points in a different direction? In such a case, only the component of the torque vector that is parallel to the fixed rotation axis changes the object's rotation. Example 13.1 shows in detail what we mean.

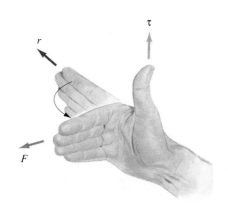

FIGURE 13.1

The direction of the torque can be found from the right-hand rule. Start with your right hand flat and fingers pointing in the **r** direction. Rotate your fingers (by bending them at the knuckle) until they point in the direction of **F**. Your outstretched thumb indicates the direction of the torque.

Concept Question 1
If you hold your book straight out in front of you, in what direction is the torque vector about your shoulder due to the book's weight?

EXAMPLE 13.1   *The Torque Vector Depends on the Choice of Coordinate System Origin*

A child stands beside a small merry-go-round. In order to start it rotating she applies a constant force to its railing as shown in Figure 13.2. In relation to the coordinate system shown in the figure the force $\mathbf{F} = 20.0\,\text{N}\,\hat{\mathbf{i}} + 15.0\,\text{N}\,\hat{\mathbf{k}}$ at the railing position

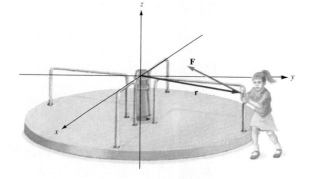

FIGURE 13.2

A small child exerts a force on the bar of a playground merry-go-round. See Example 13.1.

$\mathbf{r} = 0.30\ \mathrm{m}\,\hat{\mathbf{i}} + 1.20\ \mathrm{m}\,\hat{\mathbf{j}}$. (a) Find the torque on the merry-go-round about the origin. (b) Find the torque about the z-axis. (c) How do the answers to parts (a) and (b) change if the origin of the coordinate system is moved upward by a meters?

SOLUTION   (a) The torque is calculated from the cross product, which we expand as a determinant:

$$\boldsymbol{\tau} = \mathbf{r} \times \mathbf{F} = \begin{vmatrix} \hat{\mathbf{i}} & \hat{\mathbf{j}} & \hat{\mathbf{k}} \\ 0.30 & 1.20 & 0.00 \\ 20.0 & 0.0 & 15.0 \end{vmatrix} \mathrm{N}\cdot\mathrm{m} = (18.0\,\hat{\mathbf{i}} - 4.5\,\hat{\mathbf{j}} - 24.0\,\hat{\mathbf{k}})\ \mathrm{N}\cdot\mathrm{m}$$

This torque has magnitude $|\boldsymbol{\tau}| = \tau = \sqrt{(18.0)^2 + (-4.5)^2 + (-24.0)^2}\ \mathrm{N}\cdot\mathrm{m} = 30.3\ \mathrm{N}\cdot\mathrm{m}$ (about the origin).

(b) In general, to find the *component* of a torque that acts about an axis parallel to the unit vector $\hat{\mathbf{n}}$, we need to compute $\tau_n = \boldsymbol{\tau}\cdot\hat{\mathbf{n}}$. Therefore, when the axis is the z-axis, we need to compute

$$\tau_z = \boldsymbol{\tau}\cdot\hat{\mathbf{k}} = -24.0\ \mathrm{N}\cdot\mathrm{m}$$

Because the merry-go-round is constrained to rotate about the z-axis, it is this component of the torque that begins the rotational motion of the merry-go-round.

(c) If the origin of the coordinate system used is moved upward by a meters, then the $\mathbf{r}$ indicating the point of application of the force in this new system becomes $\mathbf{r} = 0.30\ \mathrm{m}\,\hat{\mathbf{i}} + 1.20\ \mathrm{m}\,\hat{\mathbf{j}} - a\ \mathrm{m}\,\hat{\mathbf{k}}$. Thus, the torque as measured in this new coordinate system is

$$\boldsymbol{\tau} = \mathbf{r} \times \mathbf{F} = \begin{vmatrix} \hat{\mathbf{i}} & \hat{\mathbf{j}} & \hat{\mathbf{k}} \\ 0.30 & 1.20 & -a \\ 20.0 & 0.0 & 15.0 \end{vmatrix} \mathrm{N}\cdot\mathrm{m} = [18.0\,\hat{\mathbf{i}} + (-4.5 - 20a)\hat{\mathbf{j}} - 24.0\,\hat{\mathbf{k}}]\ \mathrm{N}\cdot\mathrm{m}$$

This result is indeed a different torque. However, we would be utterly surprised if the physical effect of the child's push were different just because we changed coordinate systems! Notice that if we calculate the component of this new torque that is available to turn the merry-go-round about the z-axis, we arrive at the same answer as in part (b):

$$\tau_z = \boldsymbol{\tau}\cdot\hat{\mathbf{k}} = -24.0\ \mathrm{N}\cdot\mathrm{m} \qquad \blacktriangleleft$$

13.2 The Angular Velocity and Angular Acceleration Vectors

We might be tempted to take all of our scalar quantities from Chapter 11 and convert them into vectors. If we attempted this conversion for rotations through finite angles, we would fail. The effort is doomed because we cannot make θ a vector merely by asking the publisher to set the symbol θ in boldface. There's more to being a vector than wearing a bold face. In order to be a vector, the quantity must obey the laws of vector arithmetic; for example, the commutative law.[1] Let's symbolize two rotations of a disk by $\boldsymbol{\theta}_x$ and $\boldsymbol{\theta}_y$. Here, we let $\boldsymbol{\theta}_x$ symbolize a rotation of the disk by $90°$ about an axis along the North–South direction, and $\boldsymbol{\theta}_y$ symbolize a rotation through $90°$ about an axis along the East–West direction. Now, if the vectors commute, we should get the same final orientation of a disk regardless of the order in which we perform the rotations. That is, we expect $\boldsymbol{\theta}_x + \boldsymbol{\theta}_y = \boldsymbol{\theta}_y + \boldsymbol{\theta}_x$. As shown in Figure 13.3, this is not the case. Finite rotations do not commute and, therefore, are not vectors.

---

[1] The ultimate tests of a quantity's vector nature are that quantity's properties under a rotation of coordinate system.

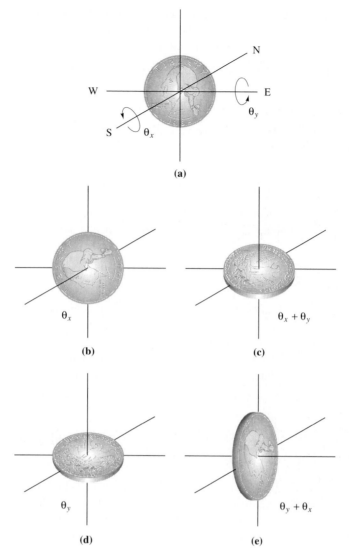

FIGURE 13.3

The two rotations θ_x and θ_y, when performed in reverse order, do not result in the same final orientation of the quarter.

Now you may be surprised to learn that, although *finite* rotations are not vectors, *infinitesimal* rotations *are* vectors. To make this seeming contradiction plausible, repeat the experiment described above wherein a disk is rotated about two perpendicular axes. This time, however, make the amount of each rotation very small. You find that the orientation of the disk is nearly the same regardless of the order in which the two rotations are performed. In the limit as the size of the angular displacement $\Delta\theta$ approaches zero, it becomes a vector.

Now it may seem that a quantity, such as an infinitesimally small rotation, is of little practical value, whether it is a vector or not. However, recall that a very useful quantity has just such a vanishingly small rotation as part of its definition. In particular the angular velocity ω is

$$\omega = \frac{d\theta}{dt} = \lim_{\Delta t \to 0} \frac{\Delta\theta}{\Delta t}$$

In the limit as Δt approaches zero $\Delta\theta/\Delta t$ approaches a finite limit, and that limit has the properties necessary to be a vector. What direction can we usefully assign to the angular velocity? A clue can be found by recalling the relation between the tangential linear velocity v_t and ω,

$$v_t = \omega r$$

where r is the radius of the circular motion. We wish to define the direction of ω so the equation above becomes a useful vector equation. We can make such a useful vector definition for the case that the particle is traveling in a circular arc[2] of radius r. We see that if we choose the direction of ω perpendicular to both $\mathbf{r}$ and $\mathbf{v}_t$ in the sense shown in Figure 13.4, we can write

$$\mathbf{v}_t = \boldsymbol{\omega} \times \mathbf{r} \qquad (13.2)$$

Note carefully that the $\mathbf{r}$ in this definition differs significantly from the $\mathbf{r}$ in the definition of torque (and, later, angular momentum) in that in Equation (13.2) the origin of vector $\mathbf{r}$ is not arbitrary. Here $\mathbf{r}$ must originate along some axis that passes through the center of the particle's circular motion. Furthermore, that axis must be normal to the plane of the circle.

Let's examine this definition of $\boldsymbol{\omega}$ for a moment. In the simplest case, shown in Figure 13.5(a), we see that $\boldsymbol{\omega}$, as implicitly defined above, is a vector perpendicular to the plane of the circular motion. That is, Equation (13.2) provides the correct direction for $\mathbf{v}_t$ when $\boldsymbol{\omega}$ is along the indicated direction. Moreover, the magnitude of ω is v_t/r. When the circular orbit does not contain the origin of the coordinate system as in Figure 13.5(b), the magnitude of ω is given by v_t/R, where R is the radius of the circle as measured in the plane of this circle. From the figure it is clear that $R = r \sin(\phi)$, so that when we look at the magnitude of v_t implied by our cross product definition $\mathbf{v}_t = \boldsymbol{\omega} \times \mathbf{r}$, we obtain $v_t = r\omega \sin(\phi) = R\omega$, the correct relationship between omega and tangential velocity!

To tie the vector definition of angular velocity together with our earlier study of circular motion, consider a particle moving in a circle of constant radius but not necessarily changing speed. By differentiating Equation (13.2) with respect to time we can find an expression for the acceleration. When we take the derivative of a vector cross product, we use the product rule from calculus but must be careful to preserve the order of the members:

$$\frac{d}{dt}(\mathbf{v}_t) = \frac{d}{dt}(\boldsymbol{\omega} \times \mathbf{r})$$

$$\mathbf{a} = \frac{d\boldsymbol{\omega}}{dt} \times \mathbf{r} + \boldsymbol{\omega} \times \frac{d\mathbf{r}}{dt}$$

In the case we are considering, the magnitude of r is constant so that $d\mathbf{r}/dt = \mathbf{v}_t$, the tangential velocity. Therefore, we may write

$$\mathbf{a} = \boldsymbol{\alpha} \times \mathbf{r} + \boldsymbol{\omega} \times \mathbf{v}_t \qquad (13.3)$$

where we have defined the angular acceleration vector $\boldsymbol{\alpha}$ as $d\boldsymbol{\omega}/dt$. For the case of a particle traveling in a circle of constant radius, $\boldsymbol{\omega}$ may change in magnitude but not direction; consequently, $\boldsymbol{\alpha}$ is either parallel to $\boldsymbol{\omega}$ when the particle is increasing speed or antiparallel to $\boldsymbol{\omega}$ when the particle is slowing down. A quick check of the directions in the cross product should convince you that the first term in Equation (13.3) is tangent to the circular motion, and, thus, we identify it with the tangential acceleration a_t. Similarly, the second term in Equation (13.3) is directed radially inward and is our old friend, the centripetal acceleration a_c.

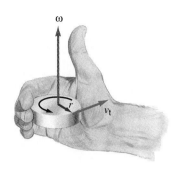

FIGURE 13.4

When you wrap your right hand around a rotating object in the direction of rotation, your extended thumb points in the direction of the angular velocity.

Concept Question 2
As you stare at a ceiling fan from below you see that its blades rotate anticlockwise. In what direction is the angular velocity vector of the fan?

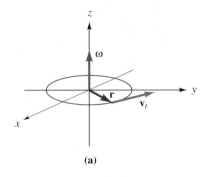

(a)

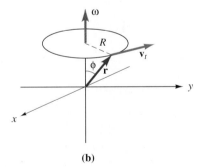

(b)

FIGURE 13.5

Geometry relating the angular velocity vector, the tangential linear velocity, and the position vector from the origin of the chosen coordinate system.

---

[2] The particle need not actually travel in a complete circle. Any point in a general trajectory can be considered on the arc of *some* circle of appropriate radius. In such cases one speaks of the "instantaneous axis of rotation," which changes during the particle's motion. We confine our discussion to particles that travel in circular paths of constant radius.

Equations (13.2) and (13.3) are, of course, vector statements that relate the linear quantities ($\mathbf{v}_t$, $\mathbf{a}_t$, and $\mathbf{a}_c$) to the rotational quantities ($\boldsymbol{\omega}$ and $\boldsymbol{\alpha}$), the scalar forms of which we first encountered in Chapter 11. We leave as an exercise (Problem 9) for you to verify that these expressions give the correct magnitudes.

13.3 The Angular Momentum Vector

We next provide a vector definition for angular momentum. As is the case for torque, angular momentum is defined about a point rather than an axis. You may by now have guessed that angular momentum, like torque, is defined as a cross product. For a single point–particle we write

$$\mathbf{l} = \mathbf{r} \times \mathbf{p} \qquad (13.4)$$

for the angular momentum about the origin of a particle with linear momentum $\mathbf{p}$ located at position $\mathbf{r}$ from the origin. Like the radius vector in the torque equation, $\mathbf{r}$ may be defined from any origin. Thus, we may speak of a particle's angular momentum about any point. The magnitude $|\mathbf{l}| = rp \sin(\phi)$ is consistent with our definition in Chapter 11. The added feature that we now associate with angular momentum is a direction. *The direction of the angular momentum vector depends on the point with respect to which it is defined.* Consequently, when specifying the angular momentum, we must speak of the angular momentum *about some point.* As shown in Figure 13.6(a), for a point-particle rotating in a circle with the origin at its center, the angular momentum is parallel to the axis of rotation. If the plane of the circle does not contain the origin, as in Figure 13.6(b), the angular momentum of the particle with respect to this origin is not parallel to the axis of rotation.

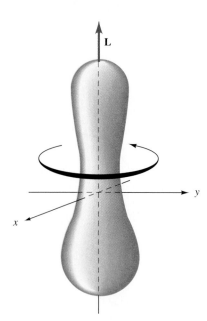

FIGURE 13.7

For an object rotating about an axis of symmetry, you can find the direction of the angular momentum vector by wrapping the fingers of your right hand around the object in the sense of its rotation. Your extended thumb then indicates the direction of the angular momentum.

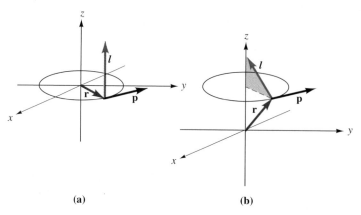

(a)   (b)

FIGURE 13.6

The angular momentum vector $\mathbf{l}$ depends on the coordinate system used.

We think of extended objects as a collection of point-particles. Because angular momentum is a vector, the total angular momentum of an extended object is the vector sum of the angular momenta of the individual mass-points that make up the object. As before, we use an uppercase $\mathbf{L}$ to designate the angular momentum of an extended object so that we have $\mathbf{L} = \Sigma_{i=1}^{N} \mathbf{l}_i$. For the simple case of a symmetrical object rotating about an axis of symmetry, this total angular momentum is parallel to the axis of rotation and directed in the sense given by the right-hand rule as shown in Figure 13.7. It is important to recognize that the angular momentum is *not always* parallel to the rotation axis for all objects and axes! We will discuss the circumstances in which the two are parallel in Section 13.4.

Newton's Second Law for Angular Momentum

In Chapter 9 we found that the vector form of Newton's second law relates the net force on a mass to the time rate of change of its linear momentum, $\mathbf{F}_{net} = d\mathbf{p}/dt$. We wish now to develop the rotational analog of this vector relation. We confine our development to the case in which torques and angular momentum are taken with respect to a point fixed in an inertial coordinate system. Let's start by looking at the time derivative of the angular momentum of a point-mass m. We are again careful to preserve the order of the cross product members:

$$\frac{d\mathbf{l}}{dt} = \frac{d}{dt}(\mathbf{r} \times \mathbf{p}) = \frac{d\mathbf{r}}{dt} \times \mathbf{p} + \mathbf{r} \times \frac{d\mathbf{p}}{dt}$$

The term $d\mathbf{r}/dt \times \mathbf{p}$ in the expression above is zero because any vector (in this case $d\mathbf{r}/dt = \mathbf{v}$) crossed with itself is zero:

$$\frac{d\mathbf{r}}{dt} \times \mathbf{p} = \mathbf{v} \times \mathbf{p} = \mathbf{v} \times (m\mathbf{v}) = m(\mathbf{v} \times \mathbf{v}) = 0$$

The term $\mathbf{r} \times d\mathbf{p}/dt$ is the net torque because, according to Newton's second law, $d\mathbf{p}/dt$ is the net force:

$$\mathbf{r} \times \frac{d\mathbf{p}}{dt} = \mathbf{r} \times \mathbf{F}_{net} = \boldsymbol{\tau}_{net}$$

Therefore, we have shown *for a single particle* that

$$\boldsymbol{\tau}_{net} = \frac{d\mathbf{l}}{dt}$$

For a composite object, we must sum the torques and angular momenta of all of the constituent particles. For N particles we have

$$\sum_{i=1}^{N} \boldsymbol{\tau}_{net_i} = \frac{d}{dt} \sum_{i=1}^{N} \mathbf{l}_i$$

The net torque on the ith particle may be the resultant of many forces, some internal, some external. When we are simply summing forces, we have already seen that Newton's third law leads to a cancellation of the internal forces. Let's see if the internal torques also cancel. In Figure 13.8 we show a general object with two generic internal forces $\mathbf{F}_{12} = -\mathbf{F}_{21}$. These forces have different $\mathbf{r}$ vectors. However, since the forces point *directly*

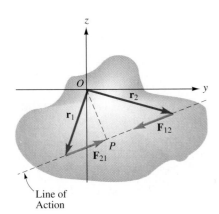

FIGURE 13.8

The lever arm *OP* is the same for both forces $\mathbf{F}_{12}$ and $\mathbf{F}_{21}$. Thus, since the forces are equal and opposite, the torques are equal and opposite and, therefore, cancel.

toward one another,[3] they have a common line of action, and, hence, the same lever arm. Thus, the torques are equal in magnitude and opposite in direction and, therefore, cancel. Hence, the expression for the sum of the torques on all particles that make up the object needs only include external torques (i.e., torques due to external forces), and we can write finally

$$\sum \boldsymbol{\tau}_{\text{ext}} = \frac{d\mathbf{L}}{dt} \qquad \text{Newton's second law for rotational motion (vector form)} \qquad (13.5)$$

This expression is not the most general one possible. We restricted our derivation by measuring torques and angular momenta about a pivot fixed in an inertial coordinate system. If the pivot accelerates, the expression is, as you would expect, more complicated:[4]

$$\sum \boldsymbol{\tau}_{\text{ext}} = \frac{d\mathbf{L}}{dt} + m\,(\mathbf{r}_{CM} - \mathbf{r}_o) \times \mathbf{a}_o \qquad (13.6)$$

where $\mathbf{r}_o$ and $\mathbf{a}_o$ are the position and acceleration of the pivot point itself, and $\mathbf{r}_{CM}$ is the position of the center of mass, all with respect to an inertial frame of reference. The most significant point to note about this expression is that, when torques and angular momenta are calculated about the center of mass, then $\mathbf{r}_o = \mathbf{r}_{CM}$, so that Equation (13.6) reduces to the simpler Equation (13.5). (See footnote 2 of Chapter 12 on page 320.) By examining the cross product in Equation (13.6), you may be able to discover other special cases where the simpler Equation (13.5) may be applied.[5] We suggest you forget them as soon as possible after you discover them. The only two cases of sufficient generality to warrant memorizing are (i) a pivot fixed in an inertial coordinate system and (ii) a pivot taken at the center of mass.

Conservation of Vector Angular Momentum

In the absence of external torques Equation (13.6) tells us that $d\mathbf{L}/dt = 0$. Thus, we find that $\mathbf{L}$ is constant in time, both in magnitude *and direction*. This condition gives a new dimension (well, actually three) to the law of conservation of angular momentum. The conservation of angular momentum is one of the most basic and fundamental physical laws. No violation has ever been found.[6]

Precession of a Gyroscope (Optional)

To illustrate the descriptive power of the vector relation between torque and angular momentum, let's consider a gyroscope. If you have ever observed a gyroscope, chances are you were fascinated by its seeming ability to defy gravity. When held horizontally by one end of its axle, as in Figure 13.9, a spinning gyroscope does not fall downward as we

Concept Question 5
Suppose you were at the earth's equator and set a very good gyroscope to turning with its axis horizontal and parallel to an East–West line. Let the gyroscope be mounted in an excellent gimbal mount so that no external torques act on it. To help us keep track of the orientation of the axis let the East end of the gyroscope axis be painted red. You and the gyroscope are stationary relative to the earth. (a) Describe how the gyroscope looks to you 6 h later, (b) 12 h later, (c) 18 h later, (d) 24 h later. [*Hint:* It may be helpful to draw a picture of you and your gyroscope as you would appear to an imaginary observer looking down from a point directly above the North Pole. The earth rotates anticlockwise as seen from the North Pole.]

FIGURE 13.9

A spinning gyroscope seems to defy gravity as it precesses.

---

[3] This is the so-called strong form of Newton's third law.

[4] See for example K. R. Symon, *Mechanics,* 3rd Ed. (Addison-Wesley, Reading, MA, 1971), pp. 163–164.

[5] For example see F. R. Zypman, ''Moments to remember. . . ,'' *American Journal of Physics,* **58**(1), 41–42 (1990).

[6] Interestingly, the conservation of angular momentum holds even when our derivation of Equation (13.6) does not. In some instances, electromagnetic forces do not obey the strong form of Newton's third law, that is, the equal and opposite forces are not directed along the same line of action. This situation is a nice reminder that physics is fundamentally an *inductive* science. Although we may deduce the consequences of our physical laws, we always arrive at the laws, themselves, by induction—sometimes accompanied by a great leap of faith.

might expect; rather, its free end lazily traces out a horizontal circle, executing a motion known as precession. We might at first have said that we expect the gyroscope to "fall." However, when we examine the forces on the gyroscope as shown in Figure 13.10, we see that $N = W$ is the only condition required to keep the gyroscope from falling. If $N = W$, then the vector sum of the forces is zero, so the center of mass does not accelerate, in other words, the gyroscope does not fall. This explanation, we admit, is not at all satisfying. However, it does help us to realize that we are not expecting the gyroscope to *fall* but rather to *twist*. To describe the twisting motion we use the rotational form of Newton's second law as derived above.

The important result from Newton's second law for rotation is that torques cause the angular momentum to change. Let's start then by looking at the angular momentum vector. As shown in Figure 13.10, **L** is essentially parallel to the gyroscope axis.[7] Next, let's calculate the torque on the gyroscope using the support point as our pivot. The only force exerting a torque about the pivot is the gyroscope's weight. The direction of this torque is easy to find using the right-hand rule and Figure 13.10. The resultant torque is horizontal and at a right angle to the gyroscope axis. Because Equation (13.6) is a vector equation, the direction of the torque is also the direction of the change in **L**. In Figure 13.10 we see that the result of adding $\Delta \mathbf{L}$ to $\mathbf{L}_o$ is to create a new vector $\mathbf{L}_f$ displaced anticlockwise (when viewed from above) from $\mathbf{L}_o$. This movement of the gyroscope axis occurs continuously, and the gyroscope is said to process.

We may also obtain a quantitative expression for the precession rate based on the magnitudes of the quantities involved in the precession. The gravitational torque is $\tau = rF \sin(\phi) = hmg \sin(90°) = hmg$, where h is the distance from the pivot to the center of mass of the gyroscope. In one complete precession turn, the tip of the **L** vector traces out a distance (in momentum space) of $2\pi L$. Thus, the magnitude of the rate of change of L is

$$\frac{\Delta L}{\Delta t} = \frac{2\pi L}{T_p} = \Omega_p L$$

where T_p is the precession period and Ω_p is the precessional angular frequency. Substituting these values for torque and $d\mathbf{L}/dt$ into Equation (13.6), we find

$$hmg = \Omega_p L$$

from which we see

$$\Omega_p = \frac{hmg}{L} \qquad (13.7)$$

Notice that this expression holds only when $\Omega_p \ll \omega$. In such cases, if the moment of inertia and rotation rate of the gyroscope about its axis are known, the precession rate can be found. Interestingly, if the axis of the gyroscope is not horizontal, the expression for the precession frequency remains unchanged. (See Problem 17.)

EXAMPLE 13.2 *How Long Have You Been Going Around Together?*

A gyroscope with an axis 14.0 cm long has a mass of 0.150 kg. The vast majority of the mass is in a circular ring of mean radius 5.00 cm. The gyroscope is set rotating at a rate of 20.0 rev/s. If it is held horizontally by one end of its axis, what is its rate of precession?

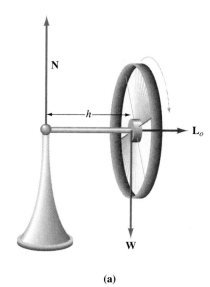

(a)

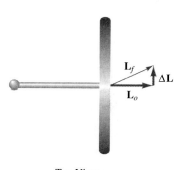

Top View

(b)

FIGURE 13.10

(a) The torque due to the gyroscope's weight causes the angular momentum to vector $\mathbf{L}_o$ to rotate into the page without changing length. (b) A *top view* showing the direction of the change in angular momentum, which is also the direction of the torque due to the gyroscope's weight W acting at distance h from the pivot.

---

[7] Notice that **L** is not exactly parallel to the gyroscope axis because the slow precessional motion contributes a small vertical component to **L**.

SOLUTION   We model the gyroscope as a thin ring and calculate its moment of inertia as

$$I = MR^2 = (0.150 \text{ kg})(5.00 \times 10^{-2} \text{ m})^2 = 3.75 \times 10^{-4} \text{ kg} \cdot \text{m}^2$$

The rotation rate is

$$\omega = 20.0 \times 2\pi \text{ rad/s} = 40.0\pi \text{ rad/s}$$

Consequently,

$$L = I\omega = 4.71 \times 10^{-2} \text{ kg} \cdot \text{m}^2/\text{s}$$

Employing Equation (13.7), we have

$$\Omega_p = \frac{dmg}{L} = \frac{(0.070 \text{ m})(0.150 \text{ kg})(9.80 \text{ N/kg})}{4.71 \times 10^{-2} \text{ kg} \cdot \text{m}^2/\text{s}} = 2.18 \text{ rad/s}$$

$$T_p = \frac{2\pi}{\Omega_p} = 2.88 \text{ s}$$

When no external torques act on an object, the direction of its angular momentum remains constant. Satellites are often set spinning when placed in their final orbit to stabilize their orientation in space. Many satellites contain three small gyroscopes oriented in orthogonal directions so that the satellite orientation can be changed. When an electrically powered gyroscope starts spinning, the satellite body must spin oppositely so that angular momentum is conserved. When the body reaches the desired orientation, the gyroscope is stopped. (See Problem 20.) You can observe a similar effect by sitting on a rotatable stool and holding a bicycle wheel vertically by its axle. If you start the wheel rotating, the stool "recoils" with the opposite rotation. (What happens if you turn the wheel's axle horizontal while the wheel is rotating?)

Precession of the Earth's Axis of Rotation (Optional)

As you know, the earth spins on its axis and, therefore, acts like a giant gyroscope. As it travels around the sun, its axis of rotation remains directed at a fixed point in space near the North Star. This axis of revolution is inclined with respect to the plane of the earth's orbital motion by 66.5° as shown in Figure 13.11. In actual fact, however, the earth's

As the earth rotates on its axis the stars appear to make circles about the celestial north pole.

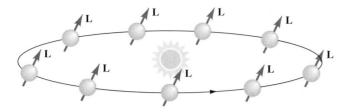

FIGURE 13.11

As the earth orbits the sun during the course of a year, its axis of rotation remains pointed in the direction relative to the background of the stars.

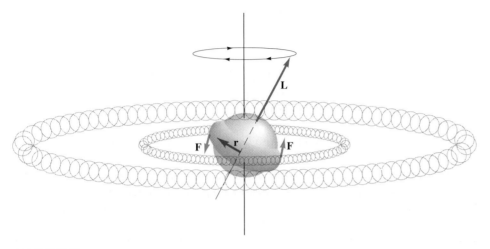

FIGURE 13.12

On the average, during the course of many centuries, the mass of the sun and moon, as seen from the earth, appears to be smeared out in doughnut shapes. The small equatorial bulge at the earth's equator is pulled by gravity toward the plane of the "doughnuts".

rotation axis is not *perfectly* stationary with respect to the stars. The axis of the earth's rotation precesses with a period of about 26 000 years. Explaining this precession was another triumph for—you guessed it—Isaac Newton. We will not present a detailed account here, but the fundamental cause is easy to see. If the axis precesses, there must be an external torque. This torque comes from the gravitational attraction of the sun and the moon. Compared to the precession period of 26 000 years, the positions of the sun and moon relative to the earth vary rapidly. In Figure 13.12 we imagine the mass of the sun and moon smeared out in doughnut shapes around the earth. These doughnut models suggest average mass distributions of the sun and moon over many years as observed from the earth. We must also realize that the earth is not a perfect sphere. Because it spins on its axis, there is a bulge at its equator. We can model this bulge as a belt of extra mass as shown in Figure 13.12. The average mass distribution of the sun and moon attract the earth's equatorial bulge. A quick check using the right-hand rule should convince you that the torque generated on this earth model is out of the page in Figure 13.12, and, therefore, this torque is perpendicular to the earth's rotation axis. The earth's precession is caused by this torque, just as the downward force of gravity causes our simple gyroscope to precess.

13.4 Vector Relations Between **L** and ω (Optional)

It is sometimes possible to reduce Newton's second law written in the form $\tau_{net} = d\mathbf{L}/dt$ to the simple form $\tau_{net} = I\alpha$. Let's investigate what is required to make this simplification so as to clearly understand the limitations of the simpler form. The derivation is quite simple, having only three steps, as follows:

$$\frac{d\mathbf{L}}{dt} = \frac{d(I\boldsymbol{\omega})}{dt} = I\frac{d\boldsymbol{\omega}}{dt} = I\boldsymbol{\alpha}$$

The derivation is valid only *if I* is a constant, and *if* we can write $\mathbf{L} = I\boldsymbol{\omega}$. Recall that *I* depends not only on the body but also on the axis of rotation. Thus, for *I* to be constant, *the rotation axis must remain fixed relative to the object.* To learn what the requirement $\mathbf{L} = I\boldsymbol{\omega}$ means let's return to our defining equations, concentrating first on a single point-mass, for which

$$\mathbf{l} = \mathbf{r} \times m\mathbf{v}$$

and

$$\mathbf{v} = \boldsymbol{\omega} \times \mathbf{r}$$

Substituting **v** from the second expression into the first, we obtain

$$\mathbf{l} = \mathbf{r} \times m\,(\boldsymbol{\omega} \times \mathbf{r}) = m\mathbf{r} \times (\boldsymbol{\omega} \times \mathbf{r}) = m[\mathbf{r} \times (\mathbf{v} \times \mathbf{r})]$$

When expanding the vector cross product like the one above, the order of the parentheses is important. We use the "BAC-CAB" rule to simplify the double cross product. (See Problem 12.26.)

$$\mathbf{A} \times (\mathbf{B} \times \mathbf{C}) = \mathbf{B}(\mathbf{A} \cdot \mathbf{C}) - \mathbf{C}(\mathbf{A} \cdot \mathbf{B})$$

Applying this expansion to our expression for angular momentum, we find

$$\mathbf{l} = m[(r^2)\boldsymbol{\omega} - (\mathbf{r} \cdot \boldsymbol{\omega})\mathbf{r}] \tag{13.8}$$

We use Equation (13.8) to explore the circumstances under which $\mathbf{L} = I\boldsymbol{\omega}$ holds. We look first at a single point-mass and then at a combination of point-masses. The latter leads us to extended masses.

Single Point-Masses

From Equation (13.8) we see that **l** is parallel to $\boldsymbol{\omega}$ when the coefficient of **r** is zero. This situation occurs when $\mathbf{r} \cdot \boldsymbol{\omega}$ is zero, that is, when **r** and $\boldsymbol{\omega}$ are perpendicular. This case is shown in Figure 13.13(a) where we see that the condition is fulfilled when the origin of

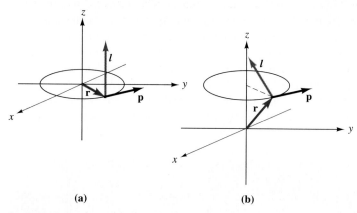

(a)   (b)

FIGURE 13.13

The angular momentum vector depends on the position of the particle's orbit relative to the coordinate system. The angular momentum is not necessarily parallel to the angular velocity.

the coordinate system used is in the plane of the particle's circular motion. In this case $l = mr^2\boldsymbol{\omega}$. We recognize mr^2 as the moment of inertia of a point-particle, so $l = I\boldsymbol{\omega}$ is valid. In Figure 13.13(b) we see that l is not parallel to $\boldsymbol{\omega}$ when the origin is not in the plane of the circle. This situation may at first seem strange because the particle motion is exactly the same. Remember, however, that both the torque and the angular momentum depend on the choice of coordinate system.

Multiple Point-Masses

We have already observed that l is parallel to $\boldsymbol{\omega}$ when the origin of the coordinate system is located in the plane of a single particle's circular motion. The total angular momentum of a collection of particles is just the vector sum of the individual angular momenta. It immediately follows that the total angular momentum $\mathbf{L}$ of any collection of particles all rotating in the same plane is parallel to $\boldsymbol{\omega}$ if the origin of the coordinate system is in that plane. Thus, we have shown that $\mathbf{L} = I\boldsymbol{\omega}$ holds for *any laminar body rotating about an axis perpendicular to the plane of the lamina if* $\mathbf{L}$ *and* $\boldsymbol{\omega}$ *are defined with respect to an origin in the plane of the lamina.*

Let's now look at two points located such that the origin is not in the plane of either's rotation. Because l is not parallel to $\boldsymbol{\omega}$ for either of the points separately, the sum of the two l's in general is not parallel to $\boldsymbol{\omega}$. However, for the special case of two oppositely located, identical masses as shown in Figure 13.14, an interesting cancellation occurs.

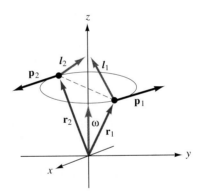

FIGURE 13.14

For two particles located on opposite sides of the axis of rotation, the *total* angular momentum is parallel to the axis of rotation.

Let's examine Equation (13.8) carefully for these two particles.

$$\mathbf{L} = \mathbf{l}_1 + \mathbf{l}_2 = m_1[(r_1^2)\boldsymbol{\omega} - (\mathbf{r}_1 \cdot \boldsymbol{\omega})\mathbf{r}_1] + m_2[(r_2^2)\boldsymbol{\omega} - (\mathbf{r}_2 \cdot \boldsymbol{\omega})\mathbf{r}_2]$$

Call the common masses $m_1 = m_2 = m$ and the common *magnitude* of the radii $|\mathbf{r}_1| = |\mathbf{r}_2| = r$. Then each of the indicated dot products has the value $r\omega \sin(\phi)$. The total angular momentum thus becomes

$$\mathbf{L} = 2mr^2\boldsymbol{\omega} - mr\omega \cos(\phi)(\mathbf{r}_1 + \mathbf{r}_2)$$

The significant feature of this expression for the vectors in Figure 13.14 is that the direction of the sum $\mathbf{r}_1 + \mathbf{r}_2$ is parallel to $\boldsymbol{\omega}$! In fact we can write $\mathbf{r}_1 + \mathbf{r}_2 = 2r \cos(\phi)\,\hat{\mathbf{n}}$, where $\hat{\mathbf{n}}$ is a unit vector parallel to $\boldsymbol{\omega}$. Because $\omega\hat{\mathbf{n}} = \boldsymbol{\omega}$, we can write the angular momentum

$$\mathbf{L} = 2mr^2\boldsymbol{\omega}[1 - \cos^2(\phi)] = 2mr^2\boldsymbol{\omega} \sin^2(\phi) = 2mR^2\boldsymbol{\omega} = I\boldsymbol{\omega}$$

Now that we know $\mathbf{L} = I\boldsymbol{\omega}$ holds for oppositely located point-masses, we can construct a great number of objects by assembling pairs of point-masses. Several examples are shown in Figure 13.15. You can extrapolate this process to explore the kind of symmetry required. In a problem at the end of this chapter we suggest another pair of points that lead to other possibilities. (See Problem 24.)

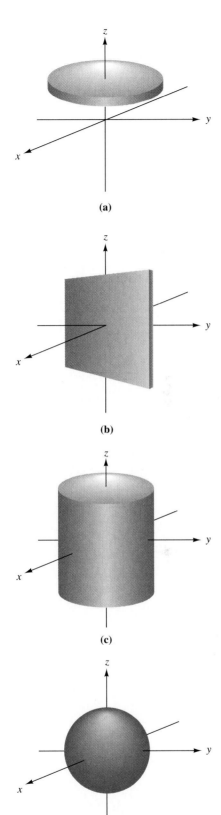

(a)

(b)

(c)

(d)

FIGURE 13.15

Symmetrical objects that can be formed by adding combinations of oppositely paired points as in Figure 13.14.

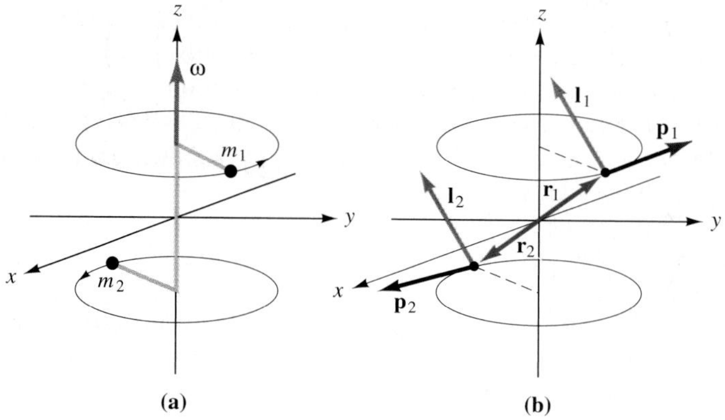

FIGURE 13.16

For the asymmetrical rotor the total angular momentum is not parallel to the angular velocity.

EXAMPLE 13.3   *Just to Convince You* $\mathbf{L} = I\boldsymbol{\omega}$ *Isn't Always True!*

Show, for the asymmetrical rotor in Figure 13.16, that $\mathbf{L}$ is not parallel to $\boldsymbol{\omega}$.

SOLUTION   We again sum the individual momenta of the two particles

$$\mathbf{L} = \mathbf{l}_1 + \mathbf{l}_2 = m_1[(r_1^2)\boldsymbol{\omega} - (\mathbf{r}_1 \cdot \boldsymbol{\omega})\mathbf{r}_1] + m_2[(r_2^2)\boldsymbol{\omega} - (\mathbf{r}_2 \cdot \boldsymbol{\omega})\mathbf{r}_2]$$

Once again the masses and the magnitudes of the r vectors are equal. This time, however, $\mathbf{r}_1 \cdot \boldsymbol{\omega} = -\mathbf{r}_2 \cdot \boldsymbol{\omega}$, so that we obtain

$$\mathbf{L} = 2mr^2\boldsymbol{\omega} - mr\omega \cos(\phi)(\mathbf{r}_1 - \mathbf{r}_2)$$

The difference $\mathbf{r}_1 - \mathbf{r}_2$ is *perpendicular* to $\boldsymbol{\omega}$, so $\mathbf{L}$ cannot be parallel to $\boldsymbol{\omega}$ unless $\phi = \pi/2$.   ◀

We conclude this section by mentioning the general result regarding the parallelism of $\mathbf{L}$ and $\boldsymbol{\omega}$. It is not at all obvious from our previous remarks, but it turns out that for all objects, regardless of shape, there are always three, mutually perpendicular axes related to the body for which $\mathbf{L} = I\boldsymbol{\omega}$. These three axes are known as the **principal axes** of that body. Such matters are the subject of more advanced courses in mechanics.

13.5 Summary

Torque, angular velocity, and angular momentum may be defined as vector quantities:

$\boldsymbol{\tau} = \mathbf{r} \times \mathbf{F}$ definition of torque
$\mathbf{v} = \boldsymbol{\omega} \times \mathbf{r}$ relationship between angular velocity and tangential velocity for circular motion
$\mathbf{l} = \mathbf{r} \times \mathbf{p}$ definition of angular momentum of a particle

Torque and angular momentum may be calculated about any point. On the other hand, the origin of the $\mathbf{r}$ vector in the implicit definition of angular velocity must have the center of the particle's circular trajectory as its origin.

The vector form of the rotational equivalent of Newton's second law of motion for an inertial coordinate system is

$$\boldsymbol{\tau}_{\text{net}} = \frac{d\mathbf{L}}{dt}$$

This expression holds when $\boldsymbol{\tau}$ and $\mathbf{L}$ are measured relative to either (1) a point fixed in an inertial coordinate system or (2) the center of mass of the rotating object.

The law of the **conservation of angular momentum** states that when no net external torques act on an object, both the magnitude and direction of its angular momentum are constant.

When a rapidly rotating body is acted on by an external torque the axis of rotation precesses through space with an angular frequency given by $\Omega_p = hmg/L$, where m is the gyroscope's mass and h is the distance from the pivot point to the gyroscope's center of mass.

The expressions $\mathbf{L} = I\boldsymbol{\omega}$ and $\boldsymbol{\tau} = I\boldsymbol{\alpha}$ hold for restricted cases. Two of these cases are (1) a laminar body rotating about an axis perpendicular to itself and containing the origin, and (2) an object that is rotationally symmetric and rotating about its axis of symmetry. Other examples are discussed in the text. For all objects these statements are true for three mutually perpendicular *principal axes*.

PROBLEMS

13.1 The Torque Vector

1. (a) Calculate the torque about the origin exerted by a force $\mathbf{F} = (2.00\,\hat{\mathbf{i}} + 3.00\,\hat{\mathbf{j}} - 1.00\,\hat{\mathbf{k}})$ N that acts at the point $\mathbf{R} = (3.00\,\hat{\mathbf{i}} + 1.00\,\hat{\mathbf{j}} + 2.00\,\hat{\mathbf{k}})$ m. (b) What is the component of this torque that acts about an axis through the origin parallel to the unit vector $\hat{\mathbf{n}} = (\hat{\mathbf{i}} + \hat{\mathbf{j}})/\sqrt{2}$?

2. (a) Calculate the torque exerted about the origin by the force $\mathbf{F} = (10.0\,\hat{\mathbf{i}} - 5.0\,\hat{\mathbf{j}} + 10.0\,\hat{\mathbf{k}})$ N that acts at the point $\mathbf{R} = (0.50\,\hat{\mathbf{i}} + 2.00\,\hat{\mathbf{k}})$ m. (b) What is the component of this torque that acts along an axis through the origin parallel to the unit vector $(\hat{\mathbf{i}} + \hat{\mathbf{j}} + \hat{\mathbf{k}})/\sqrt{3}$?

3. (a) A force $\mathbf{F} = (3.00\,\hat{\mathbf{i}} + 2.00\,\hat{\mathbf{j}} + 3.00\,\hat{\mathbf{k}})$ N acts at the point A with coordinates $(0.00, -1.00, 3.00)$ m. Calculate the torque about the point P with coordinates $(1.00, 2.00, 2.00)$ m that results from this force. (b) What is the component of this torque about an axis through P and parallel to the z-axis?

4. The force on a rigid object is given by $\mathbf{F} = (15.0\,\hat{\mathbf{i}} + 10.0\,\hat{\mathbf{k}})$ N and acts at a point A with coordinates of $(0.200, 0.500, 0.600)$ m. (a) Calculate the torque about the point P that has coordinates $(0.500, 0.500, 0.500)$ m. (b) What is the component of the torque about an axis through P and parallel to the unit vector $\hat{\mathbf{n}} = (\hat{\mathbf{j}} + \hat{\mathbf{k}})/\sqrt{2}$?

5. (a) Compute the torque about the origin caused by the force $\mathbf{F} = 10.0\,\mathrm{N}\,\hat{\mathbf{j}}$ acting at point A in Figure 13.P1 if $a = 0.500$ m, $b = 0.600$ m, and $c = 0.800$ m. (b) Calculate the torque caused by $\mathbf{F}$ about point P that has coordinates $(-0.400, 0.300, 0.000)$ m.

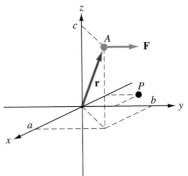

FIGURE 13.P1   Problem 5

6. Calculate the net torque about the origin caused by forces $\mathbf{F}_1$ and $\mathbf{F}_2$ in Figure 13.P2 if $a = b = 20.0$ cm and $\mathbf{F}_1 = 4.00\,\mathrm{N}\,\hat{\mathbf{i}}$ and $\mathbf{F}_2 = -5.00\,\mathrm{N}\,\hat{\mathbf{j}}$.

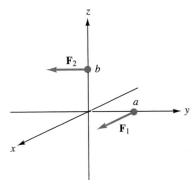

FIGURE 13.P2   Problem 6

7. Show that if $\mathbf{r}$ and $\mathbf{F}$ lie in the xy-plane, the torque is normal to that plane.

8. A force $(20.0, 35.0, -10.0)$ N is applied to a rigid object at the point $(40.0, 60.0, 40.0)$ cm. What force can be applied at the point $(-10.0, -30.0, 80.0)$ cm in a direction perpendicular to the corresponding $\mathbf{r}$ vector so that the total torque about the origin is zero?

13.2 The Angular Velocity and Angular Acceleration Vectors

9. Show that Equations (13.2) and (13.3) give the correct magnitudes for v_t, a_t, and a_c.

10. Consider a circular disk rolling without slipping at speed v_o as shown in Figure 13.P3. (a) Consider the motion to be composed of center of mass translation and rotation about the center of mass. Take the center of the disk as the pivot and calculate the tangential velocity of points A, B, and C due to the rotation at angular speed $\omega_o = v_o/R$. Point B is located midway between points A and O. Find the total velocity vector for each of these three points by vectorially adding the velocity due to rotation only to the velocity due to the disk's translational motion $\mathbf{v} = v_o\hat{\mathbf{i}}$. (b) Now consider the motion as rotation only about an (instantaneous) axis of rotation located at the point of contact between the

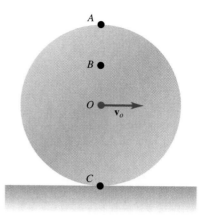

FIGURE 13.P3   Problem 10

ground and the disk. Taking the rotation rate as ω_o calculate the velocity of the same three points using $\mathbf{v} = \boldsymbol{\omega} \times \mathbf{r}$. (c) Compare your results from (a) and (b), and comment.

11. Show that the equation $\mathbf{v} = \boldsymbol{\omega} \times \mathbf{r}$ can be solved for a unique value of $\boldsymbol{\omega}$ if $\boldsymbol{\omega}$ and $\mathbf{r}$ are perpendicular. [*Hint:* Operate on both sides of the equation with $\mathbf{r} \times$, then expand the right-hand side using the BAC − CAB rule. (See Problem 26.)]

12. Show that (a) the equation $\mathbf{v} = \boldsymbol{\omega} \times \mathbf{r}$ leads to the same expression for $\mathbf{v}$ when applied to both Figure 13.5(a) and 13.5(b), but (b) that an attempt to define $\boldsymbol{\omega}$ directly using $\boldsymbol{\omega} = (\mathbf{v} \times \mathbf{r})/r^2$ gives different results for $\boldsymbol{\omega}$ for the two figures.

13.3 The Angular Momentum Vector

13. A 0.0100-kg particle is located at a point with (x, y, z) coordinates $(15.0, 20.0, -5.0)$ cm and travels with a velocity $\mathbf{v} = (2.00\,\hat{\mathbf{i}} + 3.00\,\hat{\mathbf{j}} - 1.00\,\hat{\mathbf{k}})$ m/s. (a) Calculate this particle's angular momentum about the origin. (b) Calculate this particle's angular momentum about the point $(10.0, -10.0, 10.0)$ cm.

14. Compute the total angular momentum about the origin of the three independent particles with masses, positions, and velocities as follows: 0.200 kg, $(20.0, -30.0, 50.0)$ cm, $(-1.50, 0.00, 2.00)$ m/s; 0.150 kg, $(-10.0, 20.0, 25.0)$ cm, $(2.00, -0.50, 1.00)$ m/s; 0.250 kg, $(10.0, 50.0, 40.0)$ cm, $(-1.00, -1.50, -2.00)$ m/s.

15. Calculate the angular momentum about the origin of the particles shown in Figure 13.P4.

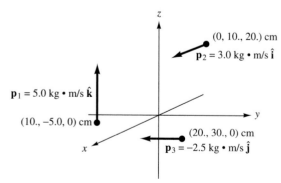

FIGURE 13.P4   Problem 15

16. Calculate the total angular momentum about the origin of the three particles shown in Figure 13.P5.

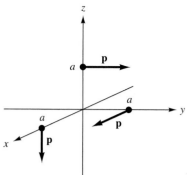

FIGURE 13.P5   Problem 16

17. (a) Using arguments similar to those in Section 13.3, show that if the gyroscope axis makes an angle ϕ with the vertical the precession frequency is still given by $\omega_p = mgh/L$. (b) Show that the vector equation $\boldsymbol{\tau} = \boldsymbol{\Omega}_p \times \mathbf{L}_o$ gives the correct relation for magnitudes and the correct direction for $\boldsymbol{\Omega}_p$ when $\mathbf{L}_o$ is the angular momentum due to the rotation about the main gyroscope axis.

18. A bicycle wheel has been modified so that the tire is replaced by a strip of heavy metal giving the wheel a mass of 3.50 kg, effectively located in a thin hoop of radius 30.0 cm. A string is attached to one end of the axle at a point 15.0 cm from the center of the wheel. If the wheel is set into rapid rotation with the axle horizontal, a person can hold up the wheel by the free end of the string, which remains vertical as the axle remains horizontal and slowly precesses about the string. Calculate the precession rate of the wheel if the wheel is spinning at a rate of 4.50 rev/s.

19. A student sits on a stool that is free to rotate about a vertical axis. The student holds a stationary bicycle wheel with its axis vertical. The student and chair have rotational inertia 2.00 kg · m² about the vertical axis and the wheel has $I = 0.300$ kg · m². (a) The student starts the wheel spinning at 3.00 rev/s with $\boldsymbol{\omega}$ pointing directly upward. What is the magnitude and direction of the student's resulting rotation? (b) Now suppose that the student is stationary and the wheel is spinning at 3.00 rev/s with $\boldsymbol{\omega}$ again upward. The student takes the handles on either side of the wheel, one in each hand, and turns the wheel so that its axle is horizontal with $\boldsymbol{\omega}$ pointing toward the student's right. What is the magnitude and direction of the student's rotation? (c) In part (b) the system composed of the student, the chair, and the bicycle wheel gained a component of angular momentum in the horizontal direction. This could not happen unless an *external* torque acts on the system. In what direction must the external torque act? Draw a simple sketch and show the additional forces (over and above the usual normal forces) on the feet of the stool that could cause a torque in the required direction.

20. A satellite contains three gyroscopes with orthogonal axes as shown in Figure 13.P6. Each gyroscope has a rotational inertia I about its axis, and the rotational inertia of the entire satellite (including all three gyroscopes) about the x-, y-, or z-axis is $100 I$. (a) If gyroscope B starts turning at a rate of 200. rpm due to torques supplied by a magnetic motor attached to the satellite body, what happens to the body of the satellite? Quantify your answer. (b) Starting from the position shown in Figure 13.P6, suppose you wish to realign the satellite so that the z-axis points toward the star S_2. Which gyroscope should you activate? (Angle θ is in the yz-plane.) In which direction should it turn? If angle θ is 15.0° and the gyroscope revolves at 200. rpm, how long should you let it run?

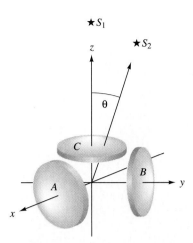

FIGURE 13.P6 Problem 20

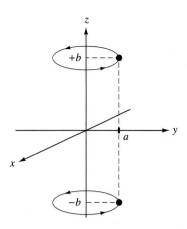

FIGURE 13.P7 Problem 24

21. Show that for a point-mass m, if $\mathbf{r}$ and $\mathbf{v}$ are in the xy-plane, $\mathbf{l}$ is normal to that plane.

22. Prove that angular momentum is constant for any particle that moves under the influence of a central force, that is, such that $\mathbf{F}_{net} = k\hat{\mathbf{r}}$, where k is a constant.

13.4 Vector Relations between L and ω (Optional)

23. In Example 13.3 it was shown that for the asymmetrical rotor of Figure 13.16, $\mathbf{L}$ is not parallel to $\boldsymbol{\omega}$. (a) Draw a sketch similar to Figure 13.16 and show the total $\mathbf{L}$ vector. Indicate on your sketch what happens to the $\mathbf{L}$ vector during one complete rotation of the rotor. For convenience, keep the "tail" of $\mathbf{L}$ located at the origin of the coordinate system. (b) On your sketch draw two vectors $\mathbf{L}_o$ and $\mathbf{L}_f$ for positions of the rotor before and after a rotation of just a few degrees. Next, draw the $\Delta \mathbf{L}$ vector between $\mathbf{L}_o$ and $\mathbf{L}_f$ such that $\mathbf{L}_f = \mathbf{L}_o + \Delta \mathbf{L}$. (c) Bearings are located at points A and B of Figure 13.16. Draw equal and opposite forces at these bearing points that create the torque required to produce the $\Delta \mathbf{L}$ vector you found in part (b).

24. (a) Using arguments similar to those in Sections 13.4, show that for two point–particles located as shown in Figure 13.P7, $L = I\omega$. (b) Draw three examples of laminar bodies that can be constructed using collections of such paired points. What general symmetry rule can you write for this type of laminar body? (c) Draw at least two, three-dimensional (i.e., nonlaminar) solid objects that can be generated using collections of paired points like those of Figure 13.P7.

Numerical Methods and Computer Applications

25. Design a spreadsheet template or computer program to calculate the cross product of two vectors.

General Problem

26. Verify the BAC–CAB rule introduced in Section 13.4, by explicitly expanding both sides of the equation

$$\mathbf{A} \times (\mathbf{B} \times \mathbf{C}) = \mathbf{B}(\mathbf{A} \cdot \mathbf{C}) - \mathbf{C}(\mathbf{A} \cdot \mathbf{B})$$

14

Oscillations

In this chapter you should learn to

- recognize the condition for simple harmonic motion.
- compute the frequency and period for simple harmonic motion.
- determine $x(t)$, $v(t)$, and $a(t)$ for many cases of simple harmonic motion.
- compute the kinetic energy and potential energy of a simple harmonic oscillator.

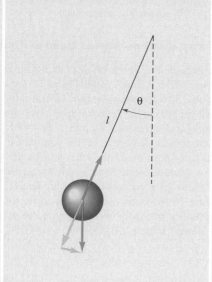

$$\theta = \theta_o \cos(\omega t + \phi)$$

$$\omega = \sqrt{\frac{\kappa}{I}}$$

When an object undergoes periodic motion its position, velocity, and acceleration all repeat themselves in equal time intervals. Periodic motion is important because so many objects exhibit this type of behavior: the atoms or molecules in solids, a child on a swing, the string of a guitar, the pistons in your car, and even your car itself after you hit a bump. The orbit of the moon around the earth and that of the earth around the sun are also examples of this type of motion. Periodic motion is not confined to mechanical objects. In Chapter 33 we will discover that electric currents can behave in a periodic manner and that electromagnetic waves are periodically varying electric and magnetic fields.

Many mechanical vibrations are not truly periodic because they die out when left to themselves. Friction causes mechanical energy to dissipate, and the motion is said to be damped. On the other hand, when a periodic driving force is applied to the system, the oscillations continue and often reach large amplitude; the motion is said to be driven. We will study a damped and driven system toward the end of this chapter. In our applications until then we consider only models in which friction is negligible.

14.1 Kinematics of Simple Harmonic Motion

One very important type of periodic motion is called **simple harmonic motion** (SHM). A great many oscillations can be modeled as SHM. Indeed, virtually all oscillations around a stable equilibrium position are SHM if the displacements from equilibrium are small enough. Applications range from the rattle of an automobile fender to the vibration of molecules. Simple harmonic motion is motion for which the displacement can be expressed as a simple sine or cosine function. Thus, the equation of motion for an object executing SHM may be written

$$x(t) = A \cos(\omega t + \phi) \tag{14.1}$$

where A, ω, and ϕ are all constants. In the paragraphs that follow we describe the physical meaning of these three parameters. The coefficient A is called the **amplitude** of the SHM. The cosine function is bounded by ± 1, so A sets a limit on the object's displacement from the $x = 0$ position (called **equilibrium**): the object may move between $+A$ and $-A$. The position function given by Equation (14.1) is illustrated in Figure 14.1.

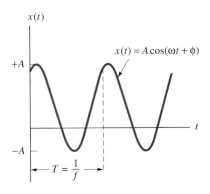

FIGURE 14.1

The position function $x(t)$ for an object exhibiting simple harmonic motion. The time for one complete cycle is the period T.

The *Foucault pendulum* at the Smithsonian Museum, Washington, D.C. The plane in space of the pendulum's swing remains fixed. However, to observers on the rotating earth the plane of rotation appears to rotate.

The constant ω is called the **angular frequency** of the simple harmonic motion. The units of ωt must be radians, so ω has the unit of radians per second. As in Chapter 11, we urge you to always use radians for angular measurements. Moreover, we again caution you to be certain your calculator knows you are conversing in radians. Probably the most common mistake students make with the material in this chapter is to forget to tell their calculator that an angle is in radians!

Let's investigate the angular frequency ω a little further. Although ω is the parameter that most often appears in the position equation $x(t)$, there are two other quantities, the period and frequency, that are related to how quickly the object repeats its motion. The **period** T is the length of time for one complete cycle of any periodic motion. We can relate the period to the angular frequency ω by noting that the cosine repeats its values each time its argument $(\omega t + \phi)$ changes by 2π. The time required for the argument to increase by 2π is *one period*. Therefore, when t increases by T, the change in $(\omega t + \phi)$ must be 2π:

$$\Delta(\omega t + \phi) = 2\pi$$

$$[\omega(t_o + T) + \phi] - (\omega t_o + \phi) = 2\pi$$

or,

$$\omega T = 2\pi$$

Solving for the period we find

$$T = \frac{2\pi}{\omega}$$

The period T is the number of seconds required for each complete cycle of motion. This time is shown in Figure 14.1. As an example, the period of the second hand on a mechanical watch is $T = 60$ s; this amount of time is required for one complete revolution of that hand. The reciprocal of T is called the **frequency** f and is the number of cycles in a fixed interval of time:

$$f = \frac{1}{T} \tag{14.2}$$

The units of f can be thought of as cycles per second. The SI unit of frequency is **hertz** (Hz), named after the physicist Heinrich Hertz.

$$1 \text{ Hz} = 1\frac{\text{cycle}}{\text{second}}$$

The frequency of a watch's second hand is $f = 1/T = 1/60$ cycle/s $= 0.0167$ Hz. (The *frequency* f is different from the *angular frequency* ω.) Let's summarize the relations between the period T, the angular frequency ω, and the frequency f:

$$\omega = 2\pi f = \frac{2\pi}{T} \tag{14.3}$$

Let's now turn our attention to ϕ, the last parameter in the SHM equation of motion, Equation (14.1). This quantity is called the **phase constant,** and it is used to designate where within the period of the cosine function the object starts its motion at time $t = 0$ s.

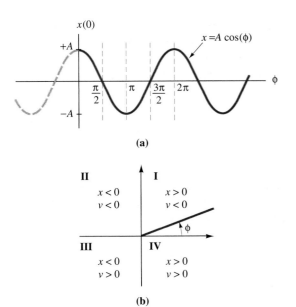

$x(0)$

$x = A \cos(\phi)$

$+A$

$-A$

$\frac{\pi}{2}$ π $\frac{3\pi}{2}$ 2π

ϕ

(a)

II	I
$x < 0$ $v < 0$	$x > 0$ $v < 0$
III	IV
$x < 0$ $v > 0$	$x > 0$ $v > 0$

ϕ

(b)

FIGURE 14.2

(a) A generic cosine curve as a function of angle ϕ. (b) Sign of x and v for each quadrant.

Figure 14.2(a) shows a generic cosine curve as a function of argument angle ϕ. We have drawn vertical lines dividing the domain of the cosine into four quadrants. In each quadrant focus your attention on the x-value and on the slope of the curve. Notice that each quadrant is different. In quadrant I the displacement x is positive but the slope is negative. In quadrant II x is negative and the slope is negative. In quadrant III x remains negative but the slope becomes positive, whereas in quadrant IV both x and the slope are positive. These results are summarized in Figure 14.2(b), and Figure 14.3 shows the range of ϕ for all possible initial conditions for positive ϕ. The point here is that ϕ depends on initial *position* and initial *velocity*.

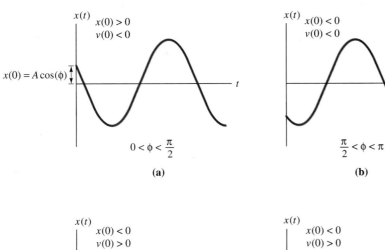

$x(t)$

$x(0) > 0$
$v(0) < 0$

$x(0) = A\cos(\phi)$

t

$0 < \phi < \frac{\pi}{2}$

(a)

$x(t)$

$x(0) < 0$
$v(0) < 0$

t

$\frac{\pi}{2} < \phi < \pi$

(b)

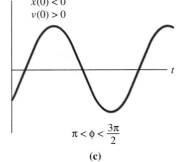

$x(t)$

$x(0) < 0$
$v(0) > 0$

t

$\pi < \phi < \frac{3\pi}{2}$

(c)

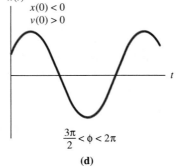

$x(t)$

$x(0) < 0$
$v(0) > 0$

t

$\frac{3\pi}{2} < \phi < 2\pi$

(d)

FIGURE 14.3

The value of the phase constant ϕ depends on the sign of the initial velocity as well as the initial position.

If we are given the initial displacement x_o, we might at first think that we can find ϕ from Equation (14.1) with $t = 0$:

$$x_o = A \cos[\omega(0) + \phi]$$

Solving this equation for ϕ, we have

$$\phi = \cos^{-1}\left(\frac{x_o}{A}\right)$$

However, it is not quite that simple. The problem is that your calculator's arccosine function returns values of ϕ in the first and second quadrant only. For these two quadrants the initial velocity is negative. What do we do if the initial velocity is in fact positive? Inspection of Figure 14.2(b) provides the answer. There we see that for a given initial displacement, positive initial velocities can be obtained by using negative ϕ values. That is, if the initial velocity is positive take the negative of the value obtained from the arccosine. A handy way to remember this rule is to write

$$\phi = -[\text{sign}(v_o)] \cos^{-1}\left(\frac{x_o}{A}\right) \tag{14.4}$$

By the function $\text{sign}(v_o)$ we mean that you should insert the algebraic sign ($\pm$) of the initial velocity. We illustrate the use of this rule in Examples 14.1 and 14.2. Problem 73 describes how to find the phase angle when the initial velocity rather than the amplitude is known.

You should also learn to sketch the SHM curve if the value of ϕ is given. This process is quite simple. Start with a quick sketch of a cosine curve with its maximum value at the origin of the t-axis ($t = 0$ s). Next, compute the time $t' = \phi/\omega$. Find this time on the time axis and relabel it $t = 0$. Adjust all other times accordingly. (If you prefer, you can leave the time axis stationary and slide the cosine curve instead. If ϕ is negative, slide the cosine curve to the right by $|t'|$. If ϕ is positive, slide the curve to the left.)

EXAMPLE 14.1   A Leap into Simple Harmonic Motion

After an athlete leaves a diving board the end of the board oscillates in SHM with a frequency of 2.50 Hz and an amplitude of 15.0 cm. At $t = 0.00$ s it is located at $y = +5.00$ cm and is traveling toward the positive y direction. (a) Find the equation of motion of a point on the end of the board. (b) Where is the point at $t = 0.300$ s?

SOLUTION   (a) Because the amplitude is 15.0 cm, we begin with the equation

$$y(t) = (15.0 \text{ cm}) \cos(\omega t + \phi)$$

Now $\omega = 2\pi f = 2\pi(2.50 \text{ Hz}) = 5.00\pi$ rad/s and

$$y(t) = (15.0 \text{ cm}) \cos[(5.00\pi \text{ rad/s})t + \phi]$$

To find the phase constant ϕ we apply Equation (14.4). We note that the initial position $y_o = 5.00$ cm, and the amplitude is $A = 15.0$ cm. Moreover, the initial velocity is positive (upward, along $+y$) so that $[\text{sign}(v_o)] = [+]$. Hence,

$$\phi = -[\text{sign}(v_o)] \cos^{-1}\left(\frac{x_o}{A}\right) = -[+]\cos^{-1}\left(\frac{5.00 \text{ cm}}{15.0 \text{ cm}}\right) = -1.23 \text{ rad}$$

Finally, the position equation becomes

$$y(t) = (15.0 \text{ cm}) \cos[(5.00\pi \text{ rad/s})t - 1.23 \text{ rad}]$$

A graph of this function is shown in Figure 14.4. The period is $T = 1/f = 1/(2.50$ Hz$) = 0.400$ s.

(b) At $t = 0.300$ s,

$$y = (15.0 \text{ cm}) \cos[(5.00\pi \text{ rad/s})(0.300 \text{ s}) - 1.23 \text{ rad}]$$

$$= (15.0 \text{ cm})(-0.943) = -14.1 \text{ cm}$$

Notice that we have been careful to take the cosine of the angle in radians! ◀

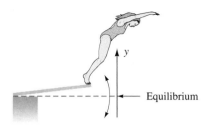

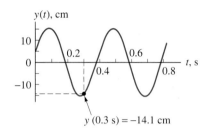

FIGURE 14.4

See Example 14.1.

Velocity and Acceleration for Simple Harmonic Motion

The velocity and acceleration functions of a particle executing SHM are found by computing the first and second time derivatives of $x(t)$. From Equation (14.1)

$$v(t) = \frac{dx}{dt} = -A\omega \sin(\omega t + \phi) \qquad (14.5)$$

and

$$a(t) = \frac{d^2x}{dt^2} = -A\omega^2 \cos(\omega t + \phi) \qquad (14.6)$$

Notice that the maximum values of the velocity and acceleration are given by

$$v_{\text{max}} = \omega A$$

$$a_{\text{max}} = \omega^2 A$$

Now, $x(t) = A \cos(\omega t + \phi)$, so Equation (14.6) may be rewritten

$$a(t) = -\omega^2 x(t)$$

That is,

the acceleration is proportional to, but in the opposite direction from, the displacement.

This statement is a characteristic of all SHM and should be good news to you for the following reason: If ever you find that this condition is satisfied for an object, you immediately know that the position function for that object is given by Equation (14.1). Furthermore, Equations (14.5) and (14.6) provide the object's velocity and acceleration functions. You need only determine A, ω, and ϕ for your particular simple harmonic oscillator.

In addition to the familiar derivative relations between $x(t)$, $v(t)$, and $a(t)$ given previously, there are graphical connections between these functions for SHM. These relationships are easiest to demonstrate if, for the moment, we suppose that $\phi = 0$. Then

$$x(t) = A \cos(\omega t) \qquad (14.7a)$$

$$v(t) = -\omega A \sin(\omega t) \qquad (14.7b)$$

$$a(t) = -\omega^2 A \cos(\omega t) \qquad (14.7c)$$

Now, you may not remember the following trigonometric identities, but using Appendix 1 you can verify that for any angle α

$$-\sin(\alpha) = \cos\left(\alpha + \frac{\pi}{2}\right)$$

and

$$-\cos(\alpha) = \cos(\alpha + \pi)$$

Making these substitutions in Equations (14.7b,c), we have

$$x(t) = A\cos(\omega t) \tag{14.8a}$$

$$v(t) = \omega A\cos\left(\omega t + \frac{\pi}{2}\right) \tag{14.8b}$$

$$a(t) = \omega^2 A\cos(\omega t + \pi) \tag{14.8c}$$

These equations are illustrated in Figure 14.5. Notice that the acceleration is always opposite in direction but proportional in magnitude to the displacement.

Before we end this section with an example, we want to caution you against memorizing too many equations. You should know Equations (14.1), (14.3), and (14.4). You can always find the expressions for $v(t)$ and $a(t)$ by differentiating Equation (14.1). Don't clutter up your mind with a bunch of equations you know how to quickly compute!

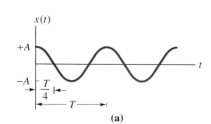

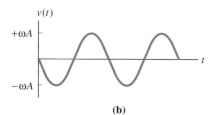

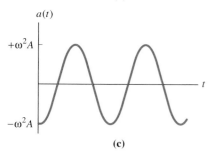

FIGURE 14.5

The phase relations between $x(t)$, $v(t)$, and $a(t)$. Notice that not until one-fourth period ($T/4$) has elapsed does the $x(t)$ curve cross the $x = 0$ position, whereas the velocity function crosses zero at $t = 0$ s. Therefore, the velocity is $\pi/2$ rad out of phase with the position function. Similarly, the acceleration function is π rad out of phase with the position function.

EXAMPLE 14.2 *Phasing in the Physics*

An object that is modeled as a particle oscillating in SHM with an amplitude of 20.0 cm repeats its motion every 4.00 s. At time $t = 0$ s it is located at $x = -10.0$ cm and its velocity is -27.2 cm/s. (a) Find the particle's equation of motion. (b) What is the particle's velocity and acceleration at $t = 1.30$ s?

SOLUTION (a) The angular velocity is $\omega = 2\pi/T = 2\pi/(4.00 \text{ s}) = \pi/2$ rad/s. Therefore,

$$x(t) = (20.0 \text{ cm})\cos\left[\left(\frac{\pi}{2}\text{ rad/s}\right)t + \phi\right]$$

To find the phase constant we apply Equation (14.4):

$$\phi = -[\text{sign}(v_o)]\cos^{-1}\left(\frac{x_o}{A}\right)$$

$$= -[-]\cos^{-1}\left(\frac{-10.0 \text{ cm}}{20.0 \text{ cm}}\right)$$

$$= +2.09 \text{ rad} = +\frac{2\pi}{3}\text{ rad}$$

The equation of motion is

$$x(t) = (20.0 \text{ cm})\cos\left[\left(\frac{\pi}{2}\text{ rad/s}\right)t + \frac{2}{3}\pi\text{ rad}\right]$$

(b) The velocity function is given by dx/dt:

$$v(t) = \frac{d}{dt}\left\{(20.0 \text{ cm})\cos\left[\left(\frac{\pi}{2}\text{ rad/s}\right)t + \frac{2}{3}\pi\text{ rad}\right]\right\}$$

$$= -\left(\frac{\pi}{2} \text{ rad/s}\right)(20.0 \text{ cm}) \sin\left[\left(\frac{\pi}{2} \text{ rad/s}\right)t + \frac{2}{3}\pi \text{ rad}\right]$$

$$= -(31.4 \text{ cm/s}) \sin\left[\left(\frac{\pi}{2} \text{ rad/s}\right)t + \frac{2}{3}\pi \text{ rad}\right]$$

At $t = 1.30$ s,

$$v(1.30 \text{ s}) = -(31.4 \text{ cm/s}) \sin\left[\left(\frac{\pi}{2} \text{ rad/s}\right)(1.30 \text{ s}) + \frac{2}{3}\pi \text{ rad}\right]$$

$$= +26.3 \text{ cm/s}$$

The acceleration function is given by dv/dt.

$$a(t) = \frac{d}{dt}\left\{-(31.4 \text{ cm/s}) \sin\left[\left(\frac{\pi}{2} \text{ rad/s}\right)(1.30 \text{ s}) + \frac{2}{3}\pi \text{ rad}\right]\right\}$$

$$= -(49.3 \text{ m/s}^2) \cos\left[\left(\frac{\pi}{2} \text{ rad/s}\right)t + \frac{2}{3}\pi \text{ rad}\right]$$

At $t = 1.30$ s,

$$a = -(49.3 \text{ cm/s}^2) \cos\left[\left(\frac{\pi}{2} \text{ rad/s}\right)(1.30 \text{ s}) + \frac{2}{3}\pi \text{ rad}\right]$$

$$= +26.8 \text{ cm/s}^2$$

14.2 The Dynamics of Simple Harmonic Motion

In this section we examine several common examples of SHM. Each of these illustrations are sometimes used to model some other physical system. For example, from our model of a mass on the end of a spring we may gain insight into how an atom or molecule might behave when bound in some region between other, heavier molecules within a solid. In each case we apply the Newton's second law technique described in Chapters 5 and 6: add the forces acting on the mass and set the sum equal to "ma." We find in each case that the acceleration is indeed proportional to, but in the opposite direction from, the displacement.

Mass Attached to a Spring

We begin by looking at the frictionless, spring–mass system shown in Figure 14.6. As we saw in Chapter 5, when mass m is given a displacement x from equilibrium it experiences a restoring force $F_{sp} = -kx$. Newton's second law gives

$$\Sigma F = ma \tag{14.9}$$

$$-kx = ma$$

or

$$a = -\left(\frac{k}{m}\right)x \tag{14.10}$$

Since both k and m are constants, Equation (14.10) tells us that the acceleration of the

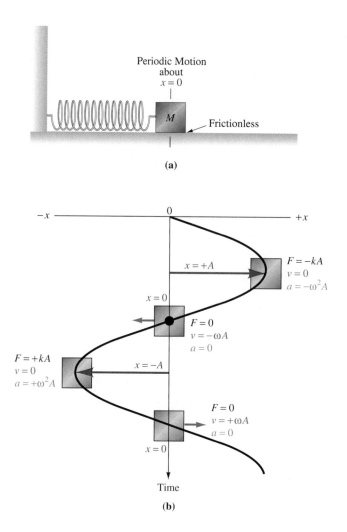

FIGURE 14.6

A simple harmonic oscillator consisting of a mass m attached to a spring of constant k. The other end of the spring is fixed.

mass m is proportional to, but in the opposite direction (note the minus sign) from, the displacement x. Therefore, the condition for SHM is satisfied.

Both the acceleration a and the position x vary in time. When we substitute the definition of acceleration $a = d^2x(t)/dt^2$ into Equation (14.10), we obtain

$$\frac{d^2x(t)}{dt^2} + \left(\frac{k}{m}\right)x(t) = 0 \qquad \textbf{(14.11)}$$

This equation is called a **second-order differential equation.** (Actually, your math professor probably calls it a "homogeneous second-order linear differential equation with constant coefficients.") What you and I want to know about this equation is simply, "What is $x(t)$?" Equation (14.11) says if you take the second derivative of $x(t)$ with respect to t, and add (k/m) times $x(t)$ to it, the result all cancels out to give zero. Thus, $x(t)$ must have a special form. That is, $x(t)$ must be a function that, after taking the second derivative, has been turned back into itself multiplied by the constant, $-(k/m)$.

You probably already suspect that the solution to Equation (14.11) is given by Equation (14.1). Indeed, both the cosine and sine functions have the properties we just described. Let's see if Equation (14.1) is indeed a solution to the differential Equation (14.11). The first derivative of $x(t)$ is

$$\frac{dx}{dt} = \frac{d}{dt}[A\cos(\omega t + \phi)]$$

$$= \omega A \sin(\omega t + \phi)$$

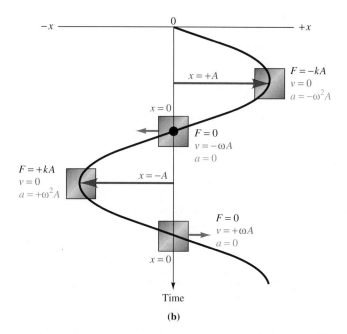

and the second derivative is

$$\frac{d^2x}{dt^2} = -\omega^2 A \cos(\omega t + \phi)$$

Substituting this result and Equation (14.1) into Equation (14.11), we have

$$-\omega^2 A \cos(\omega t + \phi) + \left(\frac{k}{m}\right) A \cos(\omega t + \phi) = 0$$

The term $A \cos(\omega t + \phi)$ is common to each side of this equation, so canceling it and solving for ω, we have

$$\omega = \sqrt{\frac{k}{m}} \qquad \qquad \textbf{(14.12)}$$

That is, Equation (14.1) is indeed a solution to Equation (14.11), as long as ω is given by Equation (14.12). Let's see what Equation (14.12) means. Using Equation (14.3), we have

$$f = \frac{\omega}{2\pi} = \frac{1}{2\pi} \sqrt{\frac{k}{m}} \qquad \qquad \textbf{(14.13)}$$

Two springs with the same size masses but different spring constants k have different frequencies of vibration. The stiffer spring has a larger k value, and Equation (14.13) tells us that it also has a larger frequency of vibration. On the other hand, if we have two identical springs but two different masses, Equation (14.13) says that the system with the larger mass has the smaller frequency.

One last point: We may substitute $\omega = \sqrt{k/m}$ into the differential Equation (14.11) to obtain

$$\frac{d^2x}{dt^2} + \omega^2 x = 0 \qquad \qquad \textbf{(14.14)}$$

Anytime you stumble on this form of a differential equation you know its solution: $x(t) = A \cos(\omega t + \phi)$! Often, $x(t)$ may be disguised as a $y(t)$ or even $\theta(t)$, and ω^2 is almost always a factor consisting of several constants that we'll call "junk:"

$$\frac{d^2x}{dt^2} + (\text{junk})x = 0$$

The point is, however, that once you obtain this expression, you know the motion is simple harmonic. Furthermore, you know the equation of motion (it's Eq. (14.1)), and you know ω is just the square root of the "junk." Simple harmonic motion is simple, isn't it?

EXAMPLE 14.3   A Spring–Mass System (Are There Summer–Mass Systems?)

A spring of 200. N/m constant is fixed at one end, and a 2.00-kg mass is attached to the other end. The mass is pulled 10.0 cm from equilibrium and released. As the mass first passes through the $x = 0$ position a stopwatch is started. (a) What are the angular frequency and period of the mass's motion? (b) What is the equation of motion $x(t)$ for the mass? (c) What is the mass's velocity and acceleration at $t = 1.50$ s?

SOLUTION   (a) The angular frequency is given by

$$\omega = \sqrt{\frac{k}{m}} = \sqrt{\frac{200.\ \text{N/m}}{2.00\ \text{kg}}} = 10.0\ \text{rad/s}$$

Concept Question 1
It has been said that the process of aging is one in which a person's natural frequency becomes lower, because both relevant constants go in the wrong direction. What is meant by this tongue-in-cheek statement?

Concept Question 2
How does the fact that a real spring has mass affect its oscillation period compared to the ideal model in which the spring is massless?

and the period is

$$T = \frac{2\pi}{\omega} = \frac{2\pi}{10.0 \text{ rad/s}} = 0.628 \text{ s}$$

(b) The amplitude of the motion is 10.0 cm, so the general equation of motion is

$$x(t) = (10.0 \text{ cm}) \cos(\omega t + \phi)$$

The initial position when the watch is started is $x_o = 0$, and the velocity is negative because it is directed toward the negative x-axis. Hence,

$$\phi = -[\text{sign}(v_o)] \cos^{-1}\left(\frac{x_o}{A}\right) = -[-] \cos^{-1}(0) = \frac{\pi}{2} \text{ rad}$$

Therefore,

$$x(t) = (10.0 \text{ cm}) \cos\left[(10.0 \text{ rad/s})t + \frac{\pi}{2}\right]$$

However, $\cos(\alpha + \pi/2) = -\sin(\alpha)$ for any angle α, and $x(t)$ becomes

$$x(t) = -(10.0 \text{ cm}) \sin[(10.0 \text{ rad/s})t]$$

(c) The velocity function is given by $v(t) = dx(t)/dt$:

$$v(t) = \frac{d}{dt}\{-(10.0 \text{ cm}) \sin[(10.0 \text{ rad/s})t]\}$$

$$= -(100. \text{ cm/s}) \cos[(10.0 \text{ rad/s})t]$$

At $t = 1.50$ s we find

$$v = -(100. \text{ cm/s}) \cos[(10.0 \text{ rad/s})(1.50 \text{ s})] = 76.0 \text{ cm/s}$$

The acceleration function is

$$a(t) = \frac{d}{dt}\{-(100. \text{ cm/s}) \cos[(10.0 \text{ rad/s})t]\}$$

$$= +(1.00 \times 10^3 \text{ cm/s}^2) \sin[(10.0 \text{ rad/s})t]$$

At $t = 1.50$ s we have

$$a = 650. \text{ cm/s}^2 \qquad \blacktriangleleft$$

The Simple Pendulum

We next examine a model known as the "simple pendulum." This model consists of a point-mass attached to the end of a light cord. The other end of the cord is fixed so that in equilibrium the cord is vertical. The mass is displaced from equilibrium through an arc length s and released as shown in Figure 14.7. When applying the simple pendulum model the length of the cord is the distance from the point of suspension to the center of mass of the object attached to the cord. This model is accurate only when used to describe small objects hanging on the end of long strings. The reason for these restrictions is that any real object attached to the end of the string rotates (through the same angle as the pendulum string) as the pendulum swings back and forth. In the simple pendulum model we ignore

Concept Question 3
What effect does doubling the amplitude of a spring–mass oscillator have on the maximum momentum of the mass?

Concept Question 4
In a spring–mass oscillator, what effect does doubling the spring constant have on the maximum force that acts on the mass?

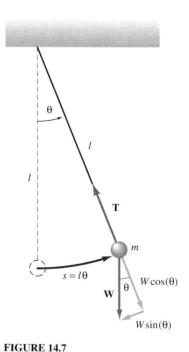

FIGURE 14.7

A simple pendulum.

the rotation of the object and consider only the object's translational motion. When the object is large compared to the length of the string, you must use the *physical pendulum* model described later.

After the point-mass in Figure 14.7 is released, both s and θ vary in time. In fact, the definition of θ in radians is $\theta(t) = s(t)/l$. The unbalanced force on mass m is $\omega \sin(\theta) = mg \sin(\theta)$. However, this force acts in the direction opposite that of the displacement $s(t)$, so we include a minus sign in the expression for the net force on m:

$$F_{\text{net}} = -mg \sin(\theta)$$

So Newton's second law ($\Sigma \mathbf{F} = m\mathbf{a}$) implies

$$ma = -mg \sin(\theta)$$

Solving this equation for the acceleration, we get

$$a = -g \sin(\theta) \tag{14.15}$$

The acceleration a is *not* proportional to angular displacement θ as required for SHM. However, for small angles, $\sin(\theta) \approx \theta$. (Try this on your calculator, but remember: θ is in radians!) With this approximation Equation (14.15) becomes

$$a = -g\theta \tag{14.16}$$

and the condition for SHM is satisfied as long as we agree to *keep θ small*; the acceleration a is proportional to, but in the opposite direction from, the angular displacement θ. Now, $a = d^2s/dt^2$ and $s = l\theta$, so that $a = l\, d^2\theta/dt^2$. Substituting this result into Equation (14.16), we have

$$\frac{d^2\theta}{dt^2} + \left(\frac{g}{l}\right)\theta = 0 \tag{14.17}$$

Compare Equation (14.17) with Equation (14.14). They're the same! Sure, the letters are different, but the form of the differential equation is the same! The variable θ has taken the place of x, and $\sqrt{g/l}$ has appeared in place of ω (note that g/l is the "junk"). We now know the equation of motion for a pendulum with small amplitude (small θ). We simply make the above replacements in Equation (14.1):

$$\theta(t) = \theta_{\max} \cos(\omega t + \phi) \tag{14.18}$$

where $\omega = \sqrt{\text{junk}} = \sqrt{g/l}$, and we have also taken the liberty of replacing A by another amplitude symbol $\theta_{\max}$. As was the case for the spring and mass system, both ϕ and $\theta_{\max}$ must be determined for a particular problem.

We may also determine the period or frequency of the simple pendulum motion. Because $\omega = \sqrt{g/l}$,

$$f = \frac{1}{T} = \frac{\omega}{2\pi} = \frac{1}{2\pi}\sqrt{\frac{g}{l}} \tag{14.19}$$

The interesting feature of this result is that the frequency of vibrations for the simple pendulum is independent of the mass m. Equation (14.19) depends only on g and the length of the pendulum l. Two pendulums of equal length, one on the moon and the other on earth's surface, have different frequencies.

Finally, we comment on the approximation we made in order to obtain the differential Equation (14.17) from Equation (14.15): We had to assume θ is small. When the amplitude of the motion is not small, the motion is not SHM. Numerical techniques for the large-amplitude case are described in Section 14.6.

Concept Question 5
How does the period of a simple pendulum change in an elevator that is accelerating (a) upward and (b) downward?

EXAMPLE 14.4 *A One-Second Example*

Find the length of a pendulum that takes 1.00 s to swing in each direction.

SOLUTION   The period of such a pendulum is 2.00 s. From Equation (14.19)

$$T = 2\pi\sqrt{\frac{l}{g}}$$

so

$$l = g\left(\frac{T}{2\pi}\right)^2$$

$$= (9.80 \text{ N/kg})\left(\frac{2.00 \text{ s}}{2\pi}\right)^2 = 0.993 \text{ m} \quad \blacktriangleleft$$

The Physical Pendulum

Any rigid body suspended from a fixed point and displaced from equilibrium can undergo oscillatory motion. We consider only objects that are sufficiently thin that we can model them as planar or objects that have a plane of symmetry passing through the center of mass and perpendicular to the axis of rotation. Figure 14.8 shows such a rigid mass pivoted about point P and displaced from equilibrium through angle θ. If the distance from P to the object's center of mass is l, then gravity provides a restoring torque of magnitude $wl\sin(\theta) = mgl\sin(\theta)$ about point P. That is,

$$\tau_{\text{net}} = -mgl\sin(\theta)$$

Employing the rotational form of Newton's second law ($\Sigma\tau = I\alpha$), we have

$$-mgl\sin(\theta) = I\alpha$$

where I is the moment of inertia about pivot point P for the rigid body. If we again restrict the motion to small θ, $\sin(\theta) \approx \theta$ (in radians!), we have

$$\alpha = -\left(\frac{mgl}{I}\right)\theta$$

Once again, the angular acceleration is proportional to, but in the opposite direction from, the displacement θ. The substitution $\alpha = d^2\theta/dt^2$ yields

$$\frac{d^2\theta}{dt^2} + \left(\frac{mgl}{I}\right)\theta = 0$$

Once again we have an equation of the form

$$\frac{d^2x}{dt^2} + (\text{junk})x = 0$$

Hence, the equation of motion is again $\theta(t) = \theta_{\text{max}}\cos(\omega t + \phi)$, and the angular velocity is $\omega = \sqrt{\text{junk}} = \sqrt{mgl/I}$. The frequency of oscillation is

$$f = \frac{1}{T} = \frac{\omega}{2\pi} = \frac{1}{2\pi}\sqrt{\frac{mgl}{I}} \quad\quad (14.20)$$

Notice that when the object is a simple pendulum, $I = ml^2$ and Equation (14.20) reduces to Equation (14.19). Finally, we can use Equation (14.20) to find the moment of

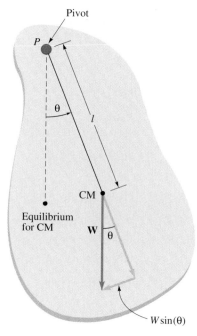

FIGURE 14.8

A physical pendulum. In Chapter 12 we learned that to find the gravitational torque acting on a mass we can treat the mass as a point-particle located at its center of mass (CM).

inertia about the rotation axis for an irregularly shaped but planar object. We simply measure f and solve Equation (14.20) for I.

EXAMPLE 14.5   The Hoop and the Pendulum

A child suspends a circular hoop of mass M and of radius R from a nail so that it hangs in a vertical plane as shown in Figure 14.9. Find the period of this physical pendulum for small oscillations about equilibrium.

SOLUTION   The period can be computed from Equation (14.20). The moment of inertia for a hoop about its axis is $I_{CM} = MR^2$. The pivot point, however, is not along the axis of the hoop, so we must apply the parallel axis theorem (see Section 11.2):

$$I = I_{CM} + Md^2$$

In our case, the distance between the axis through the center of mass and the pivot point is just the radius of the hoop R. Therefore,

$$I = MR^2 + MR^2 = 2MR^2$$

The distance between the pivot and the center of mass is $l = R$, so Equation (14.20) gives us

$$T = 2\pi\sqrt{\frac{I}{MgR}} = 2\pi\sqrt{\frac{2MR^2}{MgR}}$$
$$= 2\pi\sqrt{\frac{2R}{g}}$$

This period is equal to that of a simple pendulum with length $2R$.   ◀

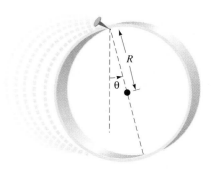

FIGURE 14.9

A circular hoop hangs on a nail and swings back and forth in SHM. See Example 14.5.

The Torsion Pendulum

Figure 14.10 shows a flexible rod fixed to the ceiling at one end with the other end attached to the axis of a large disk. When the disk is twisted through a small angle θ, the rod provides a restoring torque with magnitude proportional to θ. That is, the motion of the disk obeys a rotational Hooke's law with the restoring torque given by

$$\tau_{\text{net}} = -\kappa\theta \tag{14.21}$$

The torsion constant κ has units of newton-meters per radian. Following the techniques of Chapter 12, we apply Newton's second law for rotational motion ($\Sigma\tau = I\alpha$) with the net torque given by Equation (14.21):

$$-\kappa\theta = I\alpha$$

Solving for the angular acceleration, we obtain

$$\alpha = -\left(\frac{\kappa}{I}\right)\theta$$

where I is the moment of inertia of the disk. This equation indicates that the angular acceleration is proportional to angular displacement but in the opposite direction. Therefore, we expect to observe SHM. Continuing with $\alpha = d^2\theta/dt^2$, we have

$$\frac{d^2\theta}{dt^2} + \left(\frac{\kappa}{I}\right)\theta = 0$$

FIGURE 14.10

A torsion pendulum consisting of a solid disk attached to a flexible rod.

We recognize this differential equation as just another form of $d^2x/dt^2 + (\text{junk})x = 0$ and, therefore, we know the solution to be

$$\theta(t) = \theta_{max} \cos(\omega t + \phi) \qquad (14.22)$$

where now $\omega = \sqrt{\text{junk}} = \sqrt{\kappa/I}$. The frequency of the disk's motion is

$$f = \frac{1}{T} = \frac{\omega}{2\pi} = \frac{1}{2\pi}\sqrt{\frac{\kappa}{I}}$$

Concept Question 6
What effect does taking a torsion pendulum to the planet Mars have on its frequency of oscillations? Explain.

A Two-Mass System

As a final example of SHM let's look at the two masses connected by a single spring shown in Figure 14.11. Although it may seem to be a rather crude approximation, this system is actually a useful model for a diatomic molecule; the spring represents the bonds between the two atoms. The model accurately describes molecular vibrations at ordinary gas temperatures because, when the atoms are near their equilibrium position, the potential energy function for the diatomic system has a minimum. Near this minimum the potential energy function is U-shaped. When the vibrational amplitudes of the atoms are sufficiently small, the bottom of the U is shaped like a parabola. Consequently, when x is the displacement of the atom-to-atom distance from equilibrium, the potential energy function is $\frac{1}{2}kx^2$, where the k value describes the bond strength. In fact, this model is useful in Chapter 21 when we study a gas composed of diatomic molecules.

We begin by establishing the coordinate system also shown in Figure 14.11. Let the separation between the masses be l when the spring is unstretched. If x_1 and x_2 are the positions of the masses m_1 and m_2 on the x-axis, we can define a new coordinate

$$x = (x_2 - x_1) - l \qquad (14.23)$$

You should convince yourself that x represents the stretch length of the spring when positive and the compression of the spring when negative. When x is positive, the spring is in tension. Therefore, the spring force on m_1 is to the right, that is, the positive direction in Figure 14.11. Consequently, we must write Hooke's law as $F_1 = +kx$. However, if F_2 is the force on mass m_2, Newton's third law says that $F_2 = -F_1$, so that $F_2 = -kx$.

With these thoughts in mind we apply Newton's second law to each of the masses:

$$\Sigma F = ma$$

$$+kx = m_1 \frac{d^2x_1}{dt^2}$$

and

$$-kx = m_2 \frac{d^2x_2}{dt^2}$$

We multiply the first of these equations by m_2 and the second by m_1, and subtract the first equation from the second to obtain

$$-k(m_1 + m_2)x = m_1 m_2 \frac{d^2(x_2 - x_1)}{dt^2} \qquad (14.24)$$

Because l is a constant in our definition of x, the derivative on the right-hand side of the above equation can be written

$$\frac{d^2(x_2 - x_1)}{dt^2} = \frac{d^2[(x_2 - x_1) - l]}{dt^2} = \frac{d^2x}{dt^2}$$

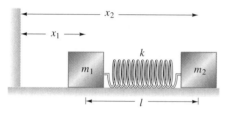

FIGURE 14.11

Two masses connected by a light spring. When the spring is unstretched the separation between the masses is l.

After some rearrangement, Equation (14.24) can be written

$$\left(\frac{m_1 m_2}{m_1 + m_2}\right)\frac{d^2x}{dt^2} = -kx \tag{14.25}$$

We define the **reduced mass** $\mu = m_1 m_2/(m_1 + m_2)$ and rewrite Equation (14.25) to obtain

$$\frac{d^2x}{dt^2} + \frac{k}{\mu}x = 0 \tag{14.26}$$

Equation (14.26) is just the, by now familiar, differential equation $d^2x/dt^2 + (\text{junk})x = 0$. Therefore, the spring-distortion distance x exhibits SHM with an angular frequency $\omega = \sqrt{\text{junk}} = \sqrt{k/\mu}$. The reduced mass is so called because its value is always less than either m_1 or m_2.

All this means that when we have two masses coupled together by a spring of stiffness constant k, we can treat the system as if it were composed of a single mass μ on the end of the same spring that is fixed at the other end. The distortion length x is the displacement of the hypothetical mass μ from equilibrium but is not a real position of either of the actual masses.

To continue the analysis, we can find an expression for the relative velocity and acceleration between the masses by differentiating Equation (14.23) with respect to time:

$$x = (x_2 - x_1) - l$$

$$v = \frac{dx}{dt} = v_2 - v_1$$

$$a = \frac{dv}{dt} = a_2 - a_1$$

This last analysis concludes our illustrations of SHM. There are many other examples in which the acceleration of an object is proportional to the negative of the displacement. We will leave some of them for you to work out in the problems at the end of this chapter.

14.3 Energy of a Simple Harmonic Oscillator

The total mechanical energy of a simple harmonic oscillator consists of the sum of its kinetic and potential energies. By using Equation (14.5) for $v(t)$, we find that the kinetic energy can be written

$$K = \frac{1}{2}mv^2 = \frac{1}{2}m\omega^2 A^2 \sin^2(\omega t + \phi)$$

In the first section of Chapter 8 we found that the potential energy of a spring stretched or compressed from equilibrium could be written $U = \frac{1}{2}kx^2$. Thus,

$$U = \frac{1}{2}kx^2 = \frac{1}{2}kA^2 \cos^2(\omega t + \phi)$$

where we have used Equation (14.1) for $x(t)$. Now $\omega^2 = k/m$, so the total mechanical energy $E = K + U$ becomes

$$E = \frac{1}{2}kA^2[\sin^2(\omega t + \phi) + \cos^2(\omega t + \phi)]$$

However, $\sin^2(\alpha) + \cos^2(\alpha) = 1$ for any angle α, so the total mechanical energy of the oscillating system is

$$E = \frac{1}{2}kA^2$$ (14.27a)

or, with $k = m\omega^2$,

$$E = \frac{1}{2}m\omega^2A^2$$ (14.27b)

Equations (14.27a,b) indicate that the total mechanical energy of an undamped harmonic oscillator is proportional to the square of the amplitude and is constant in time. Figure 14.12 illustrates how energy is exchanged between the kinetic and potential terms. As m passes through the equilibrium position its potential energy is zero and the entire energy ($\frac{1}{2}kA^2$) is kinetic. At $x = \pm A$ the velocity is zero and the energy is entirely potential. As the mass oscillates *between* $\pm A$ there is an exchange between the kinetic energy of the mass and the potential energy of the spring. At every instant of time, however, the sum of $K + U$ is equal $\frac{1}{2}kA^2$.

Finally, note that for $E = K + U$ we can write

$$\frac{1}{2}kA^2 = \frac{1}{2}mv^2 + \frac{1}{2}kx^2$$

Solving this expression for the speed of the mass at position x, we have

$$v = \sqrt{\frac{k}{m}(A^2 - x^2)}$$ (14.28)

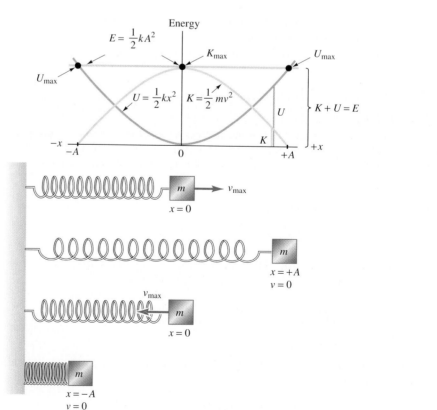

FIGURE 14.12

The kinetic, potential, and total energy of a simple harmonic oscillator as a function of position. Total energy $E = K + U$ is constant at all positions and times.

Probably the most important feature you should recognize about this equation is that the velocity is *not* proportional to the displacement. (Remember that the magnitude of the acceleration is proportional to the displacement.) *You cannot use proportions to find the velocity of the mass at position x!* This approach is a common error students make when they attempt to find v. You must use Equation (14.28) (a result of the conservation of energy), or if you know the time, apply Equation (14.5). The equations of SHM are not linear; don't use proportions!

Concept Question 7
By how much does the total energy of a spring–mass oscillator change when (a) the amplitude is doubled, (b) the mass is doubled, and (c) the spring constant is doubled?

EXAMPLE 14.6   *Energy and Simple Harmonic Motion*

A 2.00-kg mass is set into SHM on the end of a spring, with a 30.0-cm amplitude. The period is found to be 3.00 s. (a) Find the total energy of the mass–spring system. (b) What is the speed of the mass when it is 20.0 cm from equilibrium? (c) Compute the potential and kinetic energies of the system when the mass is 10.0 cm from equilibrium.

SOLUTION   (a) The total energy is $E = \frac{1}{2}kA^2$, so we must find the spring constant k. Now, because $\omega = \sqrt{k/m}$,

$$k = m\omega^2 = m\left(\frac{2\pi}{T}\right)^2$$

$$= (2.00 \text{ kg})\left(\frac{2\pi}{3.00 \text{ s}}\right)^2 = 8.77 \text{ N/m}$$

Therefore,

$$E = \frac{1}{2}kA^2 = \frac{1}{2}(8.77 \text{ N/m})(0.300 \text{ m})^2 = 0.395 \text{ J}$$

(b) To find the speed of the mass when $x = 20.0$ cm, we apply Equation (14.28):

$$v = \sqrt{\frac{k}{m}(A^2 - x^2)}$$

$$= \sqrt{\left(\frac{8.77 \text{ N/m}}{2.00 \text{ kg}}\right)[(0.300 \text{ m})^2 - (0.200 \text{ m})^2]}$$

$$= 0.468 \text{ m/s}$$

(c) The potential energy at $x = 10.0$ cm is

$$U = \frac{1}{2}kx^2 = \frac{1}{2}(8.77 \text{ N/m})(0.10 \text{ m})^2 = 0.0439 \text{ J}$$

We can find the kinetic energy from

$$K = E - U = 0.395 \text{ J} - 0.0439 \text{ J} = 0.351 \text{ J}$$

Or, we can compute v again from Equation (14.25)

$$v = \sqrt{\left(\frac{8.77 \text{ N/m}}{2.00 \text{ kg}}\right)[(0.300 \text{ m})^2 - (0.100 \text{ m})^2]}$$

$$= 0.592 \text{ m/s}$$

and use

$$K = \frac{1}{2}mv^2 = \frac{1}{2}(2.00 \text{ kg})(0.592 \text{ m/s})^2 = 0.351 \text{ J}$$

14.4 Uniform Circular Motion and Simple Harmonic Motion

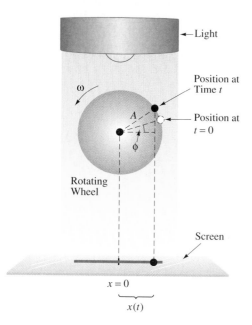

FIGURE 14.13

The projection of uniform circular motion onto a plane is SHM.

There is a relationship between uniform circular motion and SHM that provides a convenient way to visualize SHM. Consider Figure 14.13 in which we show a small mass attached to the rim of a wheel of radius A. The wheel is undergoing uniform circular motion with *constant angular velocity* ω. The wheel is placed between a light source and a screen so that a shadow of the wheel and the mass is projected onto the screen. The plane of the wheel is perpendicular to the plane of the screen. When the line connecting the center of the wheel to the mass is vertical, we designate the mass's shadow as $x = 0$ on the screen. If we start a clock (time $t = 0$) when the wheel has turned through angle ϕ from the horizontal position, then at time t later the total angle through which the mass has rotated is $\omega t + \phi$. However, the horizontal displacement of the mass's shadow from its equilibrium position is $x = A \cos(\omega t + \phi)$. Comparing this result with Equation (14.1), we see

the projection of uniform circular motion along a diameter is simple harmonic motion.

Perhaps now it makes more sense to you why ω is called the *angular* frequency; ω is the constant angular velocity with which a point moves so that its projection onto a plane exhibits SHM. The velocity $v(t) = -\omega A \sin(\omega t + \phi)$ and acceleration $a(t) = -\omega^2 A \cos(\omega t + \phi)$ of a mass undergoing SHM can also be visualized as projections of the velocity and acceleration vectors of a point undergoing uniform circular motion. The SHM position, velocity, and acceleration relationships to those of uniform circular motion are illustrated in Figure 14.14 in which we show the projections on the horizontal x-axis.

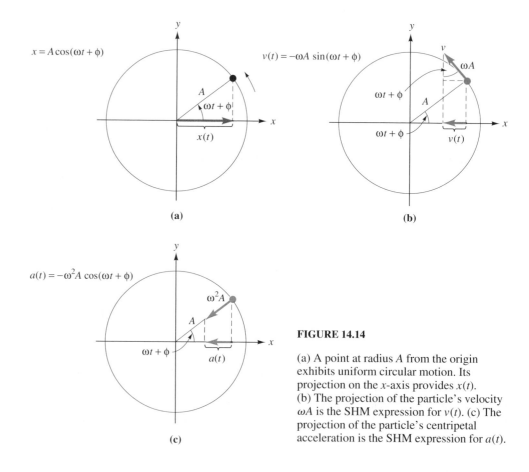

FIGURE 14.14

(a) A point at radius A from the origin exhibits uniform circular motion. Its projection on the x-axis provides $x(t)$. (b) The projection of the particle's velocity ωA is the SHM expression for $v(t)$. (c) The projection of the particle's centripetal acceleration is the SHM expression for $a(t)$.

14.5 Damped and Driven Oscillations

Damped Oscillations

In all of the oscillatory motions we have considered thus far we have neglected friction. Often the frictional force may be modeled as proportional to the magnitude of the velocity (translational velocity or rotational velocity, depending on the type of motion). Of course, the direction of this frictional force always opposes the motion. Let's use the spring–mass system to investigate the effects of this type of friction. Figure 14.15 shows a possible arrangement for our oscillating system with a frictional force provided by a fluid. The frictional force is $F_f = -bv$, where b is called the damping constant. The net force on our mass m with displacement x and velocity $v = dx/dt$ is

$$F_{net} = -kx - bv$$

Thus, Newton's second law gives

$$-kx - bv = ma$$

That is,

$$-kx - b\frac{dx}{dt} = m\frac{d^2x}{dt^2}$$

or, in standard form,

$$m\frac{d^2x}{dt^2} + b\frac{dx}{dt} + kx = 0 \qquad (14.29)$$

This differential equation is a little more complicated than Equation (14.11) although it reduces to Equation (14.11) when $b = 0$. Because the solution to Equation (14.29) does not lend itself to a quick analysis and we don't want you to fall asleep this close to the end of this chapter, we simply inform you of the solution and let you look forward to a course in differential equations!

The solution depends on the relative sizes of the spring constant k and the damping constant b. There are three possibilities:

I. Underdamped Motion: Small enough b so that $b/2m < \sqrt{k/m}$ It is convenient to define $\omega_o = \sqrt{k/m}$, the angular velocity for the case with no damping. Then the solution can be written

$$x(t) = Ae^{-(b/2m)t}\cos(\omega t + \phi) \qquad (14.30)$$

where now

$$\omega = \sqrt{\omega_o^2 - \left(\frac{b}{2m}\right)^2}$$

and ϕ is still our familiar phase constant.

Notice when $b = 0$, $\omega = \omega_o$. Graphs of $x(t)$ with no damping and $x(t)$ for the underdamped case are shown in Figure 14.16(a), (b), and (c). The motion is oscillatory, but the amplitude decreases exponentially. In fact, one way to view Equation (14.30) is that the SHM has a new angular frequency ω and an amplitude $Ae^{-(b/2m)t}$ that decays exponentially in time. In Figure 14.16(c) the damping constant b is larger than that in graph (a), yet still small enough so that $b/2m < \omega_o$.

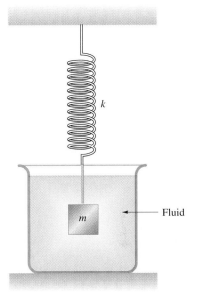

FIGURE 14.15

A spring–mass system damped by oscillating in a fluid. The magnitude of the frictional force provided by the fluid is proportional to the velocity of m.

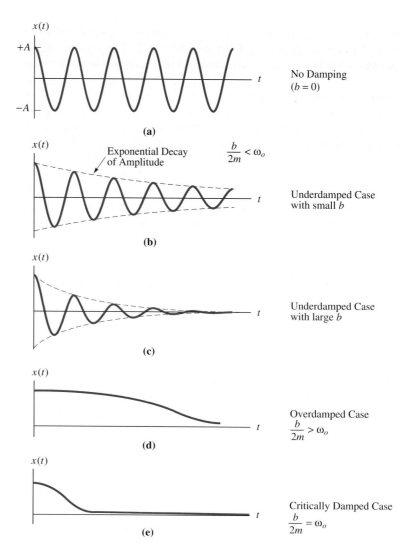

FIGURE 14.16

Graphs of $x(t)$ for a spring–mass system with different damping constants b.

II. Overdamped Case: $b/2m > \omega_o$ In the case of a relatively large damping constant the motion is *not* oscillatory. The $x(t)$ function is given by

$$x(t) = (Ae^{\omega't} + Be^{-\omega't})e^{-(b/2m)t}$$

where

$$\omega' = \sqrt{\left(\frac{b}{2m}\right)^2 - \omega_o^2}$$

and, A and B are constants the values of which depend on the initial position and velocity of the mass. A graph that displays one form of this solution is given in Figure 14.16(d).

III. Critically Damped Motion: $b/2m = \omega_o$ The solution in this case is given by

$$x(t) = (C + Dt)e^{-(b/2m)t}$$

where C and D are constants the values of which depend on the initial position and velocity of the mass. Figure 14.16(d) shows a plot of $x(t)$ for a critically damped case.

Note that the decay of the motion toward $x = 0$ is more rapid in the critically damped case (Fig. 14.16(e)) than for the overdamped case (Fig. 14.16(d)). Sometimes it is important to design an oscillating system to be critically damped. For example, you do not want the frictional retarding force of a swinging grocery store door to permit underdamped motion, particularly in cold weather. Much heat is lost if the door is allowed to swing back and forth in underdamped harmonic motion. Similarly, a slowly closing (overdamped) door is also undesirable. A good mechanical engineer knows how to compute the proper damping constant for a variety of mechanically oscillating systems.

Driven Oscillations

Suppose in addition to the spring and damping forces, the mass m is acted on by a periodic force

$$F(t) = F_o \cos(\omega_d t)$$

where F_o is a constant and ω_d is the angular velocity of this driving force. The differential equation for this motion is

$$m \frac{d^2 x}{dt^2} + b \frac{dx}{dt} + kx = F_o \cos(\omega_d t)$$

After a sufficiently long time (which once again depends on the relative sizes of the constants k and b), this driving force is performing work on the mass at the same rate at which energy is dissipated by friction. In this case, we reach what is called the **steady-state condition,** and $x(t)$ settles down to a simple harmonic oscillation described by

$$x(t) = A \cos(\omega_d t - \phi_o)$$

where

$$A = \frac{F_o/m}{\sqrt{(\omega_d^2 - \omega_o^2)^2 + (b\omega_d/m)^2}} \qquad (14.31)$$

and

$$\phi_o = \arctan\left(\frac{b\omega_d/m}{\omega_o^2 - \omega_d^2}\right)$$

This angle specifies how far the displacement lags behind the driving force.

The important parameter in this solution is the amplitude A given by Equation (14.31). A graph of A versus driving frequency ω_d is shown in Figure 14.17. The steady-state oscillation amplitude depends on the driving frequency ω_d. As the driving frequency is changed the amplitude peaks dramatically at a particular value of ω_d known as the *resonance frequency*. For a large frictional force (large b) the peak is broad and low. When the damping constant is small, the peak is high and centered closer to ω_o. If there were no friction at all, the amplitude would grow without bound as the driving angular frequency approaches ω_o. The angular frequency ω_o is called the **natural frequency,** and the dramatic increase in amplitude near this frequency is called **resonance.**

There are many examples of resonance phenomena. In machines, such as automobiles (or even airplanes), the vibrations in one part of the system may be near the resonant frequency of another part of the machine. Whereas in the least significant case this can produce an annoying rattle, in other situations the resonating piece may even fly off (a phenomenon that has been known to occur in an old truck driven by one of the authors)! A more serious situation occurred in 1850 in France when a column of marching soldiers attempted to cross a swinging bridge. The frequency of their march set the bridge into resonance, and several hundred men were killed as the bridge broke apart.

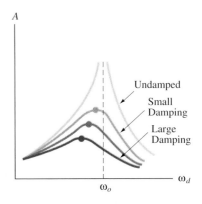

FIGURE 14.17

The steady-state amplitude of a driven harmonic oscillator as a function of driving frequency ω_d. For large damping the resonance curve is broad. As the damping constant is made smaller the resonant frequency shifts closer to ω_o and the peak sharpens.

FIGURE 14.18

Tacoma Narrows Bridge: resonant mode and subsequent collapse.

Another dramatic example of resonance is that of the famous "Tacoma Narrows Bridge Disaster." On July 1, 1940, winds reaching speeds of 40–45 mph produced a resonant torsional vibration of the bridge shown in Figure 14.18. As the wind continued to blow for a couple of hours, the bridge oscillated in various torsional modes until the frequency reached about 14 vibrations/min. The bridge then collapsed and fell 190 ft into the water.

Resonance is also an important concept for the construction of musical instruments. As we will see in Chapter 17, the shape and size of an instrument must be such that resonance can occur for certain, desired frequencies. Resonance also has important applications in physics, chemistry, and engineering. Much can be learned about atomic and intermolecular structure and bonding by discovering the values of the resonant frequencies in atoms, molecules, and solids. Recall that the natural frequency of a spring–mass system is given by $\omega_o = 2\pi f = \sqrt{k/m}$. In real systems the spring constant k "plays the part" of a bond; that is, its value characterizes the strength of a particular bond. If we begin by measuring the natural vibrational frequencies in an atom, molecule, or those atoms and molecules that constitute a solid, and if we know the masses involved, we can compute the spring constant k. For real substances, the number and nature of the spring constants can be large and complicated. However, the basic idea is just as we have stated it: measure a certain frequency and compute a related bond strength k. One modern approach is to use electromagnetic radiation to provide the driving force (that is, light waves or microwaves).

14.6 Numerical Calculations for Periodic Motion (Optional)

Equation of Motion for Simple Harmonic Motion

If you want to practice your understanding of the effects of different phase constants on the initial positions of the object, you can use the spreadsheet SHMX(T).WK1 provided on the disk accompanying this text. This spreadsheet template permits you to plot Equation (14.1) for any ω and ϕ. See if you can verify the rules for determining ϕ as described in Section 14.1.

Velocity and Acceleration for Simple Harmonic Motion

Spreadsheets provide a quick and convenient method by which you can verify your solutions for object position, velocity, and acceleration in SHM. As illustrated in the following example, Equations (14.8) can easily be implemented in a spreadsheet template.

EXAMPLE 14.7   *Spreading Out Simple Harmonic Motion*

Construct a spreadsheet to show the phase relations between $x(t)$, $v(t)$, and $a(t)$. Compute values for x, v, and a for the first two periods of an oscillating object. Test the spreadsheet for an amplitude of 2.00 m and a period 4.00 s.

SOLUTION   A possible spreadsheet template is shown in Figure 14.19. We placed data to be supplied (amplitude, period, and phase constant) in column B, and the calculated values (ω and frequency) are in cells E5 and E6. The initial time ($t = 0$) is placed in cell B11.

	A	B	C	D	E	F
3	SIMPLE HARMONIC MOTION - X(t), V(t), A(t) PHASE RELATIONS					
4						
5	Motion Amplitude A (m) =	2.00	! Omega (rad/s) =		1.5708	!
6	Period T (s) =	4.00	!Frequency (Hz) =		0.2500	!
7	Phase Constant (RAD) =	0.6280	!			!
8						
9		Times (s)	!	X(t)	V(t)	
10						
11		0	!	1.6184	−1.84571772	
12		0.2	!	1.1761	−2.54092147	
13		0.4	!	0.6187	−2.98741666	
14		0.6	!	0.0007	−3.14149981	
15		0.8	!	−0.6174	−2.98808906	
16		1	!	−1.1750	−2.54220044	
17		1.2	!	−1.6176	−1.84747808	
18		1.4	!	−1.9019	−0.97192228	
19		1.6	!	−2.0000	−0.00123352	
20		1.8	!	−1.9024	0.969575979	

FIGURE 14.19

A spreadsheet for $x(t)$, $v(t)$, and $a(t)$ for Example 14.7.

We can compute 20 values for the first two periods by placing the formula +B11+$B$6/20 in cell B12, and copying it to cells B13 through B51. The functions $x(t)$, $v(t)$, and $a(t)$ may be computed by entering Equations (14.1), (14.5), and (14.6) as the following formulas:

CELL	FORMULA
D11	$B$5*@COS($E$5*B11+$B$7)
E11	−$E$5*$B$5*@SIN($E$5*B11+$B$7)
G11	−(($E$5)^2)*$B$5*@COS($E$5*B11+$B$7)

These formulas should each be copied to rows 12 through 51 of their respective columns. A graph of the values for $x(t)$, $v(t)$, and $a(t)$ for this spreadsheet is shown in Figure 14.20. The file SHMXVA.WK1 on the disk accompanying this text contains the spreadsheet template of this example. ◄

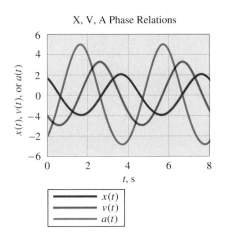

FIGURE 14.20

Graphs of $x(t)$, $v(t)$, and $a(t)$ for Example 14.7.

Large-Amplitude Pendulum Oscillations

In Chapter 6 (Section 6) you learned that the Runge-Kutta method can be applied to find the motion for complicated forces. The spreadsheet LARGPEND.WK1 contains a modification of the Runge-Kutta spreadsheet you may have used with Chapter 6. We can apply this numerical technique to discover the differences between the solution provided by the position function for simple harmonic motion and the more "complicated" position function with a large amplitude.

In LARGPEND.WK1 we followed the steps outlined in Section 6.9 to change the force to $F_{net} = -mg \sin(\theta)$. The graphical result from this spreadsheet for a pendulum of 1-m length released from an initial angle of $\pi/3$ rad (1.05 rad) is shown in Figure 14.21. Notice that the period of the large-amplitude pendulum is larger than that predicted by the SHM model. When the initial angle is reduced to a small angle, both curves coincide. That is, as the amplitude is made smaller, the green curve shifts toward the blue SHM curve. Verify this feature for yourself with LARGPEND.WK1.

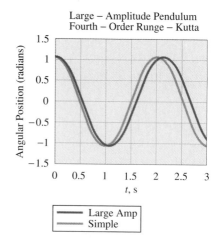

Large – Amplitude Pendulum
Fourth – Order Runge – Kutta

FIGURE 14.21

Comparison of $x(t)$ for large- and small-amplitude pendulums.

14.7 Summary

Objects with accelerations proportional to, but in the opposite direction from, their displacements, exhibit simple harmonic motion (SHM). The differential equation of such motion is of the form

$$\frac{d^2 x(t)}{dt^2} + \omega^2 x(t) = 0 \qquad (14.14)$$

and the solution is

$$x(t) = A \cos(\omega t + \phi) \qquad (14.1)$$

The amplitude A determines the maximum displacement from the equilibrium position $x = 0$. The phase constant ϕ is determined from the initial position and velocity. The angular frequency ω is related to the frequency f and period T:

$$\omega = 2\pi f = \frac{2\pi}{T} \qquad (14.3)$$

The velocity and acceleration functions for SHM are given by

$$v(t) = \frac{dx}{dt} = -A\omega \sin(\omega t + \phi) \qquad (14.5)$$

$$a(t) = \frac{d^2 x}{dt^2} = -A\omega^2 \cos(\omega t + \phi) \qquad (14.6)$$

The maximum velocity is ωA, and the maximum acceleration is $\omega^2 A$.

For some common systems the expressions for ω are as follows:

SYSTEM	ω	TERM DEFINITIONS
Spring–mass system	$\sqrt{k/m}$	k = spring constant
		m = mass of particle
Simple pendulum	$\sqrt{g/l}$	l = pendulum length
(small amplitude only)		g = acceleration of gravity
Torsion pendulum	$\sqrt{\kappa/I}$	κ = torsion constant
		I = moment of inertia about axis through pivot
Physical pendulum	$\sqrt{mgl/I}$	l = distance from pivot to center of mass

The kinetic and potential energies of a simple harmonic oscillator are given by

$$K = \frac{1}{2} mv^2 = \frac{1}{2} m\omega^2 A^2 \sin^2(\omega t + \phi)$$

$$U = \frac{1}{2} kx^2 = \frac{1}{2} kA^2 \cos^2(\omega t + \phi)$$

In the absence of damping, the total mechanical energy $E = K + U$ is constant:

$$E = \frac{1}{2} kA^2 \qquad\qquad \textbf{(14.27a)}$$

When frictional forces are present, harmonic motion is damped. For weak damping the amplitude decreases exponentially in time. Driven oscillations exhibit resonance when the driving frequency matches the natural frequency of the system. At resonance the amplitude of the oscillations is a maximum.

PROBLEMS

14.1 Kinematics of Simple Harmonic Motion

1. The piston in an engine takes 144 ms to undergo six complete vibrations. Find the period, frequency, and angular frequency of the vibrations.

2. An oscillating mass requires 1.60 s before it begins to repeat its motion. Find the frequency and angular frequency of the mass's motion.

3. A point-mass undergoes simple harmonic motion with a frequency of 25.0 Hz and an amplitude of 2.50 cm. At $t = 0$ s the mass is located at $x = -1.75$ cm and travels toward the positive x-direction. (a) What is the period and angular frequency of the simple harmonic motion? (b) Find the equation of motion $x(t)$ for the mass and make a plot of it for the first two cycles. (c) What is the mass's position at $t = 0.118$ s?

4. The end of a spring oscillates up and down in simple harmonic motion with an amplitude of 40.0 cm. It takes the spring 1.50 s to make one complete cycle up and down. If a stopwatch is started when the end of the spring is 15.0 cm below the equilibrium position and traveling toward the negative y-direction, (a) write the equation of motion $y(t)$ for the end of the spring. (b) Plot $y(t)$ from $t = 0.00$ s to $t = 4.00$ s. (c) What is the position of the end of the spring at $t = 2.75$ s?

5. A small child bounces up and down while staying in contact with a trampoline. She undergoes SHM and repeats her motion every 1.75 s. The height of each bounce above the equilibrium position of the trampoline is 35.0 cm. (a) What is the amplitude and angular frequency of the child's motion? (b) Write an equation of motion $y(t)$ for the child's motion if timing is to begin at the highest point of the child's motion. (c) What is the child's position $t = 4.50$ s?

6. A 0.250-kg particle oscillating in simple harmonic motion has a period of 5.00 s and an amplitude of 30.0 cm. At $t = 2.87$ s its position is 23.5 cm and it is traveling along the x-axis in the direction of decreasing x. Find the position of the particle at 3.00 s.

7. A small speck of dust afloat in an organ pipe oscillates back and forth in simple harmonic motion. Its motion is described by the equation $x(t) = (0.150$ mm$) \cos[(1000\pi$ rad/s$)t - 1.42$ rad$]$. (a) What are the frequency and period of the speck's motion? (b) What is its position at $t = 1.23$ s? (c) What is the speck's velocity at $t = 1.23$ s? (d) Plot the speck's position and velocity functions from $t = 0.00$ s to $t = 5.00 \times 10^{-3}$ s.

8. A particle oscillating in SHM with an amplitude of 15.0 cm repeats its motion every 6.00 s. At time $t = 0.00$ s it is located at $x = -7.50$ cm and its velocity is $+13.6$ cm/s. (a) Find the particle's equation of motion $x(t)$. (b) What is the particle's velocity and acceleration when $t = 1.80$ s?

9. The end of a tuning fork vibrates in SHM with a frequency of 440. Hz and an amplitude of 0.600 mm. At $t = 0.00$ s its position is 0.300 mm and its velocity is 750. mm/s. (a) Find the position, velocity, and acceleration functions for the end of the tuning fork. (b) What are the position, velocity, and acceleration of the end of the tuning fork at $t = 3.00 \times 10^{-3}$ s?

10. A mass oscillates in simple harmonic motion with an amplitude of 35.0 cm and a period of 2.50 s. The initial position $x_o = -25.0$ cm and the initial velocity $v_o = -50.0$ cm/s. (a) Find $x(t)$, $v(t)$, and $a(t)$ for the mass. (b) Compute the values of $x(t)$, $v(t)$, and $a(t)$ at $t = 1.00$ s. (c) Plot $x(t)$, $v(t)$, and $a(t)$ from $t = 0.00$ s to $t = 5.00$ s.

11. A particle executes SHM with an equation of motion given by

$$x(t) = (20.0 \text{ cm}) \cos\left[(12\pi \text{ rad/s})t - \frac{\pi}{3} \text{ rad}\right]$$

(a) What are the amplitude, period, and frequency of the particle's motion? (b) What are its maximum velocity and acceleration? (c) What is the particle's velocity at $t = 0.150$ s?

12. The position function of an object undergoing SHM is given by the expression

$$x(t) = (1.50 \text{ m}) \sin\left[(2 \text{ rad/s})t + \frac{5\pi}{6} \text{ rad}\right]$$

(a) What are the amplitude and period of the object's motion? (b) What are the velocity and acceleration of the object at $t = 3.00$ s?

13. Figure 14.P1 shows the position function for a 0.500-kg mass executing SHM at the end of a spring. (a) What is the amplitude, period, and angular frequency for this SHM? (b) What is the spring constant k? (c) Find the position function $x(t)$ for this mass. (d) What is the mass's velocity and acceleration at $t = 5.00$ s?

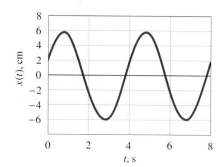

FIGURE 14.P1
Problem 13

14. Figure 14.P2 shows the position function for a 200.-g mass exhibiting SHM at the end of a spring. (a) What is the amplitude, period, and angular frequency for this SHM? (b) What is the spring constant k? (c) Find the position function $x(t)$ for this mass. (d) What is the mass's velocity and acceleration at $t = 0.400$ s?

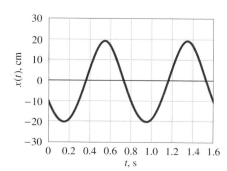

FIGURE 14.P2
Problem 14

14.2 The Dynamics of Simple Harmonic Motion

15. A uniform spring of constant k is cut into three segments of equal length. What is the spring constant of each segment?

16. A 16.0-kg mass on the end of a spring oscillates on a frictionless, horizontal surface with the equation of motion

$$x(t) = (5.00 \text{ cm}) \cos\left[(24.0 \text{ rad/s})t - \frac{\pi}{8}\right]$$

(a) Find the maximum velocity and acceleration of the mass. (b) What is the spring constant k? (c) What are the velocity and acceleration when $x = 2.50$ cm?

17. One end of a spring of negligible mass is attached to the ceiling. When a 250.-g mass is placed on the free end (without stretching the spring) and then released, the mass descends 20.0 cm before it changes direction and begins to ascend. (a) What is the spring constant k? (b) Find the frequency of the mass's oscillations.

18. A 60.0-g mass oscillates with a frequency of 2.00 Hz up and down on the end of a vertical spring. The maximum speed of the mass is 4.00 m/s. Find (a) the spring constant k, (b) the amplitude of the motion, and (c) the maximum force of the spring on the mass.

19. A mass on the end of a spring oscillates in SHM with an amplitude of 15.0 cm on a horizontal, frictionless surface. The maximum force of the spring on the mass is 12.0 N, and the period of the motion is 0.500 s. (a) What is the mass? (b) What is the spring constant?

20. An unknown mass m is placed on the end of a spring and set into oscillation. The time for 10.0 vibrations is found to be 24.0 s. Next, a 0.500-kg mass is added to the unknown mass so that both hang on the spring. The new time for 10.0 vibrations is found to be 36.0 s. (a) What is the unknown mass? (b) What is the spring constant?

21. The shock absorbers of the truck mentioned at the end of Section 14.5 are completely worn out. When a physics student pushes down on the truck she finds that it oscillates up and down with a period of 1.25 s. If the mass of the truck is 1130 kg, find the spring constant for each of the four identical springs that form what is left of the suspension system.

22. Two springs (with constants k_1 and k_2) and a mass m are attached as shown in Figure 14.P3. If the surface on which they oscillate is frictionless, find the frequency f of the oscillations.

FIGURE 14.P3   Problem 22

23. A mass m is placed between two springs of constants k_1 and k_2 as shown in Figure 14.P4. If the surface on which they oscillate is frictionless, find the period of the mass's oscillations.

FIGURE 14.P4   Problem 23

24. A simple pendulum has a period of 6.50 s. (a) What is its length? (b) How much must it be shortened to change its period to 6.00 s?

25. A simple pendulum has a frequency of 0.400 Hz. What is its period on the moon if the gravitational field strength there is one-sixth of that on the earth's surface?

26. The effective gravitational field strength g is 9.801 N/kg in Washington, D.C., and 9.809 N/kg in Sevres, France. If a pendulum has a period of exactly 5.000 s in Washington, what is its period in Sevres?

27. A simple pendulum 1.50 m in length is attached to the ceiling of an elevator. If the period of the pendulum is found to be 2.75 s, what is the magnitude and direction of the elevator's acceleration?

28. A simple pendulum is constructed from a light string 80.0 cm long with a small 150.-g ball attached to its free end. The pendulum is pulled to one side through an angle of 10.0° and released. A watch is started just as the ball passes through the equilibrium point for the first time. Write an equation of motion $\theta(t)$ for the angle (in radians) of the string from the vertical.

29. A simple pendulum consists of a 60.0-g mass at the end of a 1.20-m long string. The mass is pulled to one side through an angle of 5.00° and released just as a stopwatch is started. Write an equation of motion $\theta(t)$ for angle of the pendulum in radians.

30. The disk in Figure 14.10 has a mass of 1.50 kg and a 20.0-cm diameter. A torque of 5.00 N · m is required to twist the disk through an angle of 2.00°. (a) What is the torsion constant κ? (b) With what frequency does the disk oscillate when it is released?

31. The disk in Figure 14.10 has a radius of 12.0 cm and an unknown mass. When it is rotated through a small angle and released, its period of oscillation is found to be 0.200 s. The unknown mass is

removed from the rod and another disk with the same radius but with a known mass of 1.50 kg is placed on top of the unknown mass. The combination is then reattached to the vertical rod and set into rotational oscillation. The new period is found to be 0.500 s. What is the unknown mass?

32. Two small masses, each of 50.0 g, are fastened to the ends of a bar of negligible mass. The bar is 20.0 cm in length and is suspended at its midpoint by a wire, as shown in Figure 14.P5. When the masses are rotated so that the wire twists through a small angle, the bar is found to oscillate with a period of 2.00 s. Find the torsion constant κ.

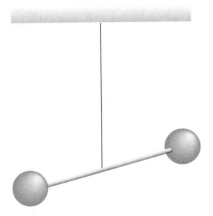

FIGURE 14.P5   Problems 32 and 33

33. The rod of negligible mass that connects the two small masses in Problem 32 is replaced with one that has a mass of 250. g. Find the new period of small oscillations.

34. A 3.20-kg planar body is pivoted 25.0 cm from its center of mass to form a physical pendulum. It is found to oscillate with a frequency of 2.40 Hz. What is the moment of inertia of the body about the axis of its pivot?

35. Two small masses m_1 and m_2 are attached to the ends of a rod of negligible mass as shown in Figure 14.P6. The rod is pivoted so that the masses and rod form a physical pendulum. Find an expression for the period of oscillation. Express your answer in terms of the parameters given in Figure 14.P6.

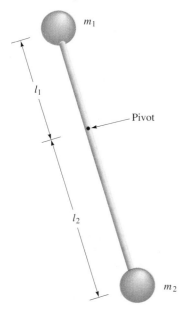

FIGURE 14.P6   Problem 35

36. A solid, 0.800-kg disk has a radius $R = 15.0$ cm. It is pivoted at point P a distance $r = 10.0$ cm from its center so that it can oscillate in the plane of the disk as a physical pendulum (Fig. 14.P7). Find the frequency of small oscillations.

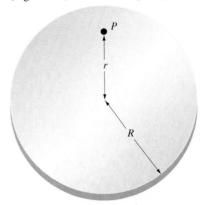

FIGURE 14.P7   Problems 36 and 40

37. A child turns a bicycle upside-down to play with the wheels, each of which has a 0.125-kg · m^2 moment of inertia and a radius of 32.0 cm. When the child attaches a small mass to the edge of one of the wheels and rotates the wheel through a small angle, it is found to oscillate with a small amplitude about equilibrium with a period of 2.00 s. How much mass did the child place on the wheel edge?

38. A small 40.0-g mass is attached to the end of an 80.0-g uniform rod of 1.20-m length. The rod hangs from a pivot at the other end. Compute the frequency of small oscillations for this system.

39. The pendulum of a grandfather clock consists of a 2.50-kg solid sphere (radius = 10.0 cm) attached to the end of a 1.50-kg rod, 0.800 m in length (Fig. 14.P8). What is the period of this pendulum?

FIGURE 14.P8   Problem 39

40. The solid disk of mass M and radius R shown in Figure 14.P7 is pivoted at point P. The point P is a distance r from the center of the disk. (a) Find an expression for the period of small oscillations in terms of R, r, and M. (b) Find the position of r for the period to be a minimum.

41. A uniform rod of length l and mass m is pivoted a distance d from its center of mass. (a) Find an expression for the period of small oscillations. (b) Find the distance from one end of the rod at which it should be pivoted for the period to be minimum.

42. When a 90.0-lbf child walks to the end of a lightweight diving board, the end is lowered by 6.00 in. If the child and board are modeled as a spring–mass system, what is the frequency of the oscillation of the board?

14.3 Energy of a Simple Harmonic Oscillator

43. A 0.500-kg mass oscillates in SHM at the end of a spring of constant 150. N/m. The total energy of the system is 2.50 J. (a) What is the amplitude of the motion? (b) What is the maximum speed of the mass? (c) What is the period of the oscillations?

44. A 1.50-kg mass attached to the end of a spring oscillates with a period of 2.00 s and an amplitude of 18.0 cm. (a) What is the total energy of the system? (b) What is the kinetic energy of the mass when its displacement is 6.00 cm? (c) What is the mass's displacement when its velocity is half the maximum value?

45. A 25.0-g shotgun slug travels with a velocity of 272 m/s horizontally into a 10.0-kg block where it remains. The block rests on a surface of negligible friction and is also attached to a spring. The other end of the spring is fixed to a wall. (a) If the amplitude of the oscillations is 15.0 cm, find the spring constant k. (b) What is the maximum kinetic energy of the mass–bullet combination?

46. When the 2.00-kg mass of a spring–mass system has a displacement of 5.00 cm its velocity is 12.0 cm/s, and when its displacement is 10.0 cm its velocity is 6.00 cm/s. (a) What is the spring constant k? (b) What is the total energy of the system?

47. In terms of the amplitude A, the spring constant k, and the mass m, find the displacement of the mass in a spring–mass system when the potential energy is equal to the kinetic energy.

48. Find the kinetic and potential energies of a spring–mass system when the displacement is half the amplitude.

14.5 Damped and Driven Oscillations

49. A 0.500-kg mass is attached to a spring of constant 150. N/m. A driving force $F(t) = (12.0 \text{ N}) \cos(\omega_d t)$ is applied to the mass, and the damping coefficient b is 6.00 N · s/m. (a) What is the resonant frequency of this system? (b) What is the amplitude of the steady-state motion when the driving frequency is half the resonant frequency? (c) What is the amplitude at resonance?

50. Find an expression for the driving frequency ω_d at which the amplitude A given by Equation (14.31) is a maximum. Your answer should be expressed in terms of m, b, and ω_o.

14.6 Numerical Calculations for Periodic Motion

51. Modify the spreadsheet template SHMX(T).WK1 to compute and plot $x(t)$, $v(t)$, and $a(t)$ when the amplitude A, frequency f, $x(0)$, and $v(0)$ are given.

52. Write a computer program that computes 20 values of $x(t)$ over the first two periods of SHM when the amplitude, frequency, x_o, and v_o are given. Be sure to include statements that provide the phase constant with the proper sign.

53. Develop a spreadsheet template that computes and plots 20 values of $x(t)$ over the first two periods of SHM when the amplitude, frequency, x_o, and v_o are given. Pay careful attention that the proper sign of the phase constant is used.

54. Construct a spreadsheet template that provides graphs of the potential energy function $U(x)$, the kinetic energy function $K(x) = $ $\frac{1}{2}mv^2$, and the total energy $E = K + U$ for the spring–mass system. Input should consist of the amplitude, frequency, mass, and spring constant.

55. Develop a spreadsheet template or a computer program that plots $x(t)$ for the underdamped harmonic oscillator when the amplitude, period, phase constant, and damping constant are provided.

56. Write a computer program or construct a spreadsheet template to plot the solution $x(t)$ for the critically damped oscillator and the overdamped oscillator. In addition to the mass m and the damping constant b, input should also include the constants A and B for the overdamped case and C and D for the critically damped case.

57. Write a program or construct a spreadsheet template to plot the amplitude of the driven harmonic oscillator A as a function of the driving frequency ω_d (Eq. (14.31)). Input should consist of the frequency f, damping constant b, mass m, and driving force amplitude F_o.

58. Modify the spreadsheet template LARGPEND.WK1 to include a damping force proportional to the angular velocity of the pendulum. Plot a curve for a 0.500-kg simple pendulum of length 2.00 m with $b = 0.30$ N · s/m starting from rest at $\theta = 60°$ from the vertical.

59. Figure 14.P9 shows a cross section of a simple model for a gasoline engine. As the wheel turns anticlockwise the piston moves with periodic motion that is not SHM. (a) Derive an expression for $x(t)$ from an equilibrium position where x_o is shown in Fig. 14.P9(b). It is helpful to apply the laws of the cosines and sines to Figure 14.P9(c) to write an expression for x in terms of angle $\theta = \omega t$, the length l, and the radius r. (b) Write a program

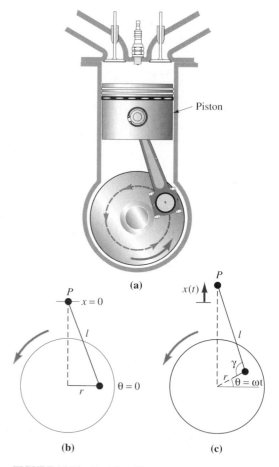

FIGURE 14.P9   Problem 59

or construct a spreadsheet template to plot two periods of the motion $x(t)$ for various values of l, r, and ω. Graphically compare your results with a simple sine function to show that the motion is not SHM.

60. (a) Use the Runge-Kutta method to produce a graph showing the time evolution of the first five oscillations of a driven damped oscillator with parameters $k = 550.$ N/m, $m = 335$ g, $b = 15.0$ N · s/m, $F_o = 25.0$ N, and driving frequency 10.0 Hz. Take $x_o = 0.00$ cm and $v_o = 0.00$ m/s. (b) What is the steady-state amplitude of these oscillations?

General Problems

61. A 250.-g mass is attached to a spring with constant $k = 16.0$ N/m. The mass is pulled 20.0 cm from equilibrium so that the spring is stretched. It is released, and 0.300 s later a clock is started. (a) Find the equation of motion $x(t)$ for the mass, where t is the time on the clock. (b) What is the mass's momentum when the clock reads $t = 0.250$ s? (c) Compute the mass's kinetic energy when its displacement from equilibrium is one-fourth its amplitude.

62. A particle oscillates with simple harmonic motion. The maximum displacements from $y = 0.00$ m are ± 0.500 m, and the frequency of the motion is 10.0 Hz. At $t = 0.00$ s the particle is located at 0.150 m and has a velocity of 24.0 m/s toward the negative x-axis. (a) Find the position, velocity, and acceleration functions for the particle. (b) Make a plot of $y(t)$, $v(t)$, and $a(t)$ for the first two periods of the particle's motion.

63. A mass oscillates in simple harmonic motion with an amplitude of 10.0 cm and a frequency of 50.0 Hz. At $t = 2.50$ ms its position is -6.00 cm and its velocity is positive. How much time elapses (from $t = 2.50$ ms) until the mass reaches $x = 0.00$?

64. A 27.0-kg child jumps up and down on a pogo stick with an amplitude of 15.0 cm. If the child's foot just becomes separated from the pogo stick pedal at the top of each jump, how many jumps does the child make in 10.0 s?

65. A light spring of constant k is fixed at one end to the top of a frictionless inclined plane as shown in Figure 14.P10. The other end is attached to a mass m. (a) What length does the spring stretch when the mass is in equilibrium? (b) Find an expression for the period when the mass oscillates back and forth along the plane.

FIGURE 14.P10   Problem 65

66. The disk shown in Figure 14.10 has a mass of 2.00 kg and a radius of 12.0 cm. When oscillating, it is found to have a period of 0.800 s. How far from the center of the disk must a 600.-g point–mass be placed to increase the period to 0.900 s?

67. The flywheel in a mechanical watch has a mass of 0.0500 g, a moment of inertia of 3.61×10^{-2} g · cm^2, and takes 10.0 s to complete exactly 28.0 complete rotational oscillations. (a) Find the torsion constant κ. (b) If the watch gains 2.00 min/day, by

how much should the moment of inertia of the flywheel be changed to keep the correct time?

68. A hole is drilled through a meterstick at the 20.0-cm mark. The meterstick is then pivoted at the hole by placing it on a nail in a wall. The bottom of the meterstick is displaced through an angle of $\theta = +5.00°$ and released at the same instant a stopwatch is started. (a) What is the period and angular frequency of the meterstick's oscillations? (b) What is the equation of motion $\theta(t)$ for the meterstick? (c) What is the speed of a point at the lower end of the meterstick at $t = 1.26$ s?

69. Two light, elastic strings of equal length l are attached to a mass m as shown in Figure 14.P11. When the mass oscillates horizontally with a small amplitude x, the tension T in the strings is approximately constant. (a) Show that the restoring force on mass m is $-(2Tl/l)x$. (b) Show that the period of oscillation is $2\pi\sqrt{ml/2T}$. (c) Compute the kinetic energy of the mass when it passes through a point that is one-fourth the distance from equilibrium to its maximum displacement.

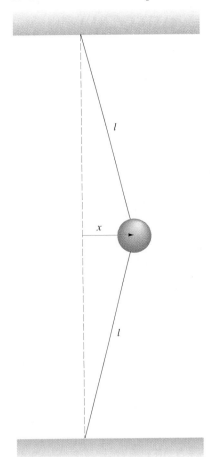

FIGURE 14.P11   Problem 69

70. A uniform bar of length l and mass m is pivoted at its center. It is attached to two unstretched springs of equal constants k as shown in Figure 14.P12. When the bar is rotated through a small angle and released, it exhibits simple harmonic motion. (a) Show that the angular frequency of its oscillations is $\sqrt{6k/m}$. (b) If the bar is released from rest at an angle θ_o, find an expression for the kinetic energy of the bar when it makes an angle of $\theta_o/3$ from the equilibrium (horizontal) position.

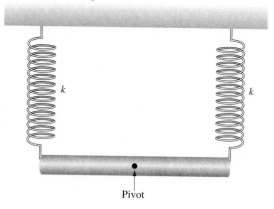

Pivot

FIGURE 14.P12   Problem 70

71. A cart consists of a body of mass m and two wheels, each of mass M and radius R. The cart is attached to a spring of constant k. The other end of the spring is fixed to a wall as shown in Figure 14.P13. Find an expression for the horizontal oscillatory motion of the cart.

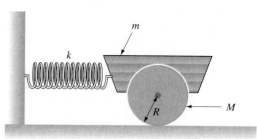

FIGURE 14.P13   Problem 71

72. A mass $m_1 = 2.00$ kg sits on a frictionless, horizontal surface and is attached to a spring of constant $k = 400.$ N/m. A second mass $m_2 = 3.00$ kg is placed on top of the 2.00-kg mass as shown in Figure 14.P14. (a) If the top mass does not slip, find the oscillation frequency for the masses. (b) What must the coefficient of static friction μ_s be between the blocks for the oscillations to have an amplitude of 7.50 cm?

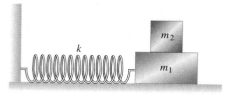

FIGURE 14.P14   Problem 72

73. (a) By computing the ratio of Equation (14.5) to Equation (14.1), show that the phase constant ϕ in Equation (14.1) is given in terms of the initial position x_o and velocity v_o by $\phi = \tan^{-1}(-v_o/\omega x_o)$. (b) What rule must you apply to the sign of the angle ϕ given by this relation to allow for the fact that most calculators return a value of the arctangent between $-\pi/2$ and $+\pi/2$?

74. A small mass is placed 8.00 cm from the center of the disk of the torsion pendulum shown in Figure 14.10. The disk oscillates with a period of 0.750 s. If the coefficient of friction between the mass and the disk is $\mu_s = 0.600$, what maximum amplitude can the oscillations have and the mass not slip?

★75. A 9.72-g bullet traveling horizontally collides with a 10.0-kg block in which it becomes embedded. The block rests on a surface of negligible friction and is attached to a Hooke's-law spring. The other end of the spring is fixed to a wall. As a result of the impact the block begins simple harmonic motion with a frequency of 0.775 Hz and an amplitude of 7.50 cm. What was the magnitude of the bullet's initial velocity before hitting the block?

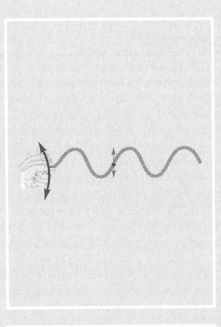

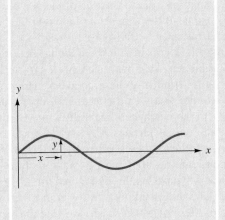

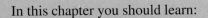

$$y(x,t) =$$

$$A \sin(kx \pm \omega t + \phi)$$

$$\mathscr{P} = \tfrac{1}{2}\,\mu\omega^2 A^2 v$$

One-Dimensional Waves

In this chapter you should learn:

- the characteristic features of wave motion.
- to determine and apply the position equation for waves.
- how to determine the rate at which sinusoidal waves transport energy.
- the consequences of wave superposition.

15.1 Introduction

Without a doubt you have witnessed many types of wave motion. The waves you have seen may include ripples in the water caused by a rock tossed into a pond, or the bouncing motion of a telephone wire in a gusty breeze. Perhaps you have observed the vibrations of the strings on a guitar. Examples of wave motion are all around us. You've probably also read that sounds, radio and TV signals, and even visible light pulses travel as waves.

If we stay within the realm of classical physics (and defer until Chapter 40 some exciting new ways of looking at the world from an atomic view), there are two basic types of waves: mechanical and electromagnetic. Radio and TV stations broadcast signals in the form of electromagnetic waves. Light, whether emitted from distant stars or laboratory lasers, can be modeled as electromagnetic waves. These waves can travel through empty space. Mechanical waves, on the other hand, require an elastic medium in which to propagate. Water waves, waves on a string, and sound are all examples of mechanical waves.

The characteristic that all mechanical waves have in common is *the transport of energy and momentum without any bulk movement of matter from one place to another.* Think about the beach. (But don't daydream!) Water waves from many miles out in the sea travel to the beach where they deposit their energy by crashing onto the shore. (The waves also occasionally do work on surfers, imparting a small fraction of the wave's total energy to adventurous humans.) But of course, none of the water from far out in the ocean actually travels to the shore; only the wave disturbance is transported toward the beach. (It's a good thing all that water doesn't end up on the beach or there would be a pretty good size flood in Malibu!) Very near the beach deep ocean waves change character, and there is an actual transport of water to the shore. Ironically these breakers, which most people are accustomed to calling "waves," are not waves in the technical sense that we are describing. This usage is another case where there is a significant difference between the everyday language and the technical definitions of physics.

We begin with some good news. Many of the concepts we used in the previous chapter to describe simple harmonic motion are directly applicable to wave motion. One such concept is amplitude. When a mechanical wave passes through a medium, the particles that make up the medium are displaced. The **amplitude** of a wave is just the maximum displacement from equilibrium of one of these particles. Later, we find that terms, such as frequency, angular frequency, and period, also have applications to wave motion.

There are three wave properties traditionally used to classify mechanical waves: (1) the direction in which individual particles move when a wave goes by, (2) the dimensionality of the wave propagation, and (3) the behavior of the wave in time. Each of these properties is described in detail in the following paragraphs.

Particle Motion with Respect to Wave Direction

Transverse Waves: If the *particle* motion of the medium in which the wave is traveling is *perpendicular* to the direction of wave propagation, the wave is called **transverse.**

Figure 15.1(a) shows a transverse wave on a string. When you move the end of a string up and down, the disturbance you produce travels along the string until friction damps it out. That is, the direction of the wave's velocity is *along* the string, but the medium (the string) moves up and down. Suppose you use a dark pen and mark a dot on the string and focus your attention on the dot as the wave passes. The dot moves up and down, perpendicular to the direction of the wave velocity.

Longitudinal Waves: If the particle motion of the medium is *parallel* to the direction of wave propagation, the wave is said to be **longitudinal.**

Concept Question 1
When wind blows across a wheat field, one can see "amber waves of grain." Are these disturbances true waves? Explain?

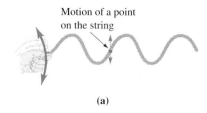

Motion of a point on the string

(a)

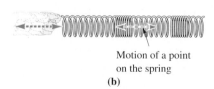

Motion of a point on the spring

(b)

FIGURE 15.1

(a) A transverse wave on a string. (b) A longitudinal wave on a spring.

Figure 15.1(b) shows a long and very elastic spring. When we repeatedly push and pull on the end of the spring, the compressions and stretches of the spring travel along the spring direction. A dot marked on the spring travels back and forth, parallel and antiparallel to the direction of the wave velocity.

For some types of waves the motion of particles in the medium is neither purely transverse nor purely longitudinal. Instead, a point in the medium travels in a small ellipse as the wave passes through. Such motion often takes place when the wave propagates along the surface of a medium. For example, ocean waves and Rayleigh waves have both longitudinal and transverse components. (A Rayleigh wave is a particular type of deformation wave that travels along the surface of the earth after an earthquake or an explosion.)

Wave Dimension

One-Dimensional Waves: Waves confined to travel either to the right or left along a straight line are one-dimensional waves.

Waves on a string or other flexible cord are good examples of one-dimensional waves. A potentially confusing feature of one-dimensional waves is that they actually require a second parameter for their description. That second parameter may be a spatial coordinate. So, although we need only one dimension to designate the direction in which a wave moves along a string, we need a second coordinate to specify the direction in which the string moves as the wave passes. Usually, we specify the wave's position by a distance along the x-axis and the displacement from equilibrium at that x-value by the coordinate y; however, we classify such a wave as one-dimensional. In Figure 15.2 we show the x- and y-coordinates necessary to describe a transverse wave. Remember, it is only the disturbance that propagates; there is no bulk movement of matter.

Two-Dimensional Waves: Waves that propagate over a surface are two-dimensional waves.

The vibration of the surface of a drum head (Fig. 15.3) is a good example of a two-dimensional wave. Notice that although the wave is classified as a two-dimensional wave, three spatial dimensions are required for its description: two to specify a particular position on the drum head and a third to designate the surface displacement at that position.

Three-Dimensional Waves: Three-dimensional waves propagate in all directions.

A sound wave is an example of a three-dimensional wave. As we will see in Chapter 17, sound waves in air consist of rapid increases and decreases in air pressure. When you "snap" your fingers a sound wave travels radially outward in all directions from your hand. Light waves from a small lamp are an example of three-dimensional electromagnetic waves. Here again, in three dimensions an additional (fourth) parameter is needed to describe the magnitude of the disturbance displacement from equilibrium.

Particle Behavior in Time

The previous two classifications have pertained to the spatial description of a wave. The last classification describes a wave's behavior in time.

Wave Pulse: If the motion of a particle within a restoring medium follows a time sequence that consists of equilibrium (no motion) followed in time by some type of motion, and finally a return to equilibrium, the wave is said to consist of a **pulse.**

White particles suspended in water reveal that each point in the water travels in an ellipse as a wave passes by. Particles near the surface travel in circles, while near the bottom the motion is almost purely longitudinal.

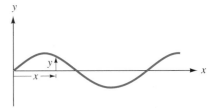

FIGURE 15.2

The description of a one-dimensional wave requires two spatial dimensions: (1) position along the wave and (2) the displacement from equilibrium.

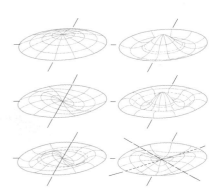

FIGURE 15.3

Two-dimensional waves on the surface of a drum.

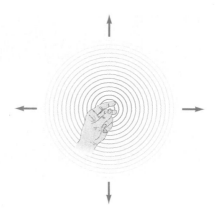

FIGURE 15.4

The sound wave produced by "snapping" your fingers is a three-dimensional, spherical wave.

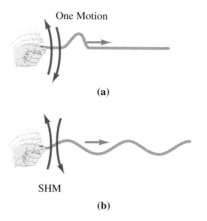

One Motion

(a)

SHM

(b)

FIGURE 15.5

(a) A wave pulse. (b) A wave train.

The sound emitted by the snap of your fingers is an example of a three-dimensional wave pulse. You can generate a transverse wave pulse on a string by rapidly displacing one end of the string up and down, but just once. As the displacement pulse travels along the string each particle in the string begins at rest, experiences a displacement as the pulse passes through it, and then returns to equilibrium (Fig. 15.5(a)).

Wave Train: In a wave train the wave travels through the medium, and the medium's particles undergo periodic motions. If the periodic motions are simple harmonic oscillations, the disturbance is called a **sinusoidal wave train.**

You can generate a wave train on a stretched string by continuously moving the end of the string up and down in simple harmonic motion (Fig. 15.5(b)). A wave train need not be sinusoidal; any continuous succession of pulses constitutes a wave train.

In this chapter we consider waves that propagate along only one dimension. Many of the concepts and some of our results are directly applicable to three-dimensional waves, particularly sound waves, which we will study in Chapter 17. Although our analysis is confined to transverse waves, the equations we generate are also applicable to one-dimensional longitudinal waves, such as a compression wave traveling along the length of a spring.

We begin our study of wave motion by considering a transverse wave traveling along a string of mass per unit length μ. In SI the unit of μ is kilograms per meter and in the cgs system it is grams per centimeter. There are several effects that quickly complicate the motion of a wave on a string. Friction, as you have probably guessed, is a very important complication. Friction may be either internal to the string or caused by air resistance. In both cases the result is a loss of wave amplitude as mechanical energy is dissipated by friction. Also, if the restoring force on the string does not exactly obey Hooke's law, the width of any pulse increases and its amplitude decreases as the disturbance travels along the string. This property is known as **dispersion** and is often present for large-amplitude waves.

We adopt a model for wave motion in which the amplitude of the wave pulse (or wave train) is small, and the tension F in the string is constant.

15.2 Waves Traveling on a String

Before we begin to find the position equation for a one-dimensional wave, it is helpful to remind you of the translation theorem from algebra: If $f(x)$ is some function, then $f(x - d)$ is just that same function translated to the right by an amount d. Similarly, $f(x + d)$ looks like the function $f(x)$ but translated to the left by an amount d. These features of function translations are illustrated in Figure 15.6.

Wave Pulses

Let's begin by considering a wave pulse traveling along a string. Figure 15.7 shows a pulse moving along the positive x-axis. The vertical position of the string at any point x and time t is the distance $y(x, t)$ where $y = 0$ is the equilibrium position. At time $t = 0$ the shape of the string may be represented by some function $f(x)$. Let's suppose we know the shape of the pulse, that is, the function $f(x)$. Furthermore, suppose we also know the velocity v with which the pulse is traveling along the x-axis. This velocity is sometimes called the **phase velocity** because it represents the speed of a particular point on the wave. Because the pulse travels a distance $d = vt$ during time t, we can predict the shape of the wave on the string at time t by simply replacing x by $x - vt$:

$$y(x, t) = f(x - vt) \qquad \text{(Wave pulse traveling to the right)} \qquad \textbf{(15.1a)}$$

Or, if the wave pulse is traveling to the left we replace x by $x + vt$:

$$y(x, t) = f(x + vt) \qquad \text{(Wave pulse traveling to the left)} \qquad \textbf{(15.1b)}$$

Notice that y is a function of both position x and time t. It is important to realize the meaning of $y(x, t)$: it is the *vertical displacement from equilibrium at time t of the point on the string located at horizontal position x*. The position function for the wave is represented by $y(x, t)$ and is sometimes called the **wave function.** If you substitute a particular time t_o into $y(x, t)$, then $y(x, t_o)$ is a function of x only. A graph of $y(x, t_o)$ is the shape of the entire string at time t_o.

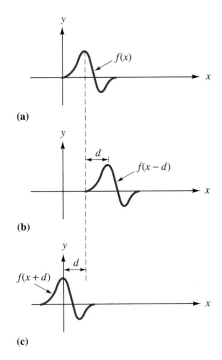

(a)

(b)

(c)

FIGURE 15.6

When x in the function $f(x)$ is replaced by $(x - d)$, the function is translated to the right by an amount d. When x is replaced by $(x + d)$, the function is translated to the left by an amount d.

EXAMPLE 15.1   *Checking Your Pulse*

The shape of a wave pulse at time $t = 0$ is given by the function

$$f(x) = \frac{2x}{1 + ax^2}$$

where $a = 1.00 \text{ cm}^{-2}$. If the wave travels along the positive x-axis with a phase velocity of 4.00 cm/s, plot the wave function at times $t = 0.00$ s, 2.00 s, and 3.00 s.

SOLUTION   The wave function is found by replacing x by $x - vt$:

$$y(x, t) = f(x - vt) = \frac{2(x - vt)}{1 + a(x - vt)^2}$$

At $t = 0.00$ s,

$$y(x, 0 \text{ s}) = \frac{2x}{[1 + (1.00 \text{ cm}^{-2})x^2]}$$

A graph of this function is shown in Figure 15.8(a).
    At $t = 2.0$ s,

$$y(x, 2.00 \text{ s}) = \frac{2[x - (4.00 \text{ cm/s})(2.00 \text{ s})]}{\{1 + (1.00 \text{ cm}^{-2})[x - (4.00 \text{ cm/s})(2.00 \text{ s})]^2\}}$$

$$= \frac{2(x - 8.00 \text{ cm})}{[1 + (1.00 \text{ cm}^{-2})(x - 8.00 \text{ cm})^2]}$$

Figure 15.8(b) shows that this function is just that of Figure 15.8(a) translated to the right by $vt = (4.00 \text{ cm/s})(2.00 \text{s}) = 8.00$ cm.

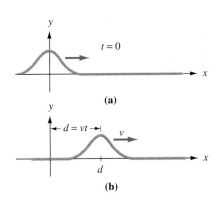

(a)

(b)

FIGURE 15.7

A wave pulse traveling with velocity v along a string.

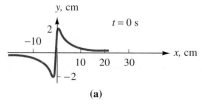

(a)

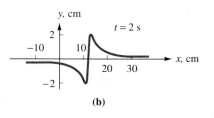

(b)

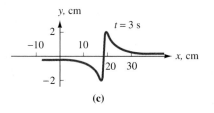

(c)

FIGURE 15.8

Finally, at $t = 4.00$ s

$$y(x, 4.00 \text{ s}) = \frac{2(x - 12.0 \text{ cm})}{[1 + (1.00 \text{ cm}^{-2})(x - 12.0 \text{ cm})^2]}$$

This wave function is shown in Figure 15.8(c).[1]

Sinusoidal Wave Train

Next, let's generate a sinusoidal wave train. Suppose you hold on to one end of a string and begin to move it up and down continuously in simple harmonic motion (SHM). After your hand has moved up, down, and back up again to its original position, the string appears as in Figure 15.9(a). The length of the wave shown in the figure is called one wavelength λ. In the region $0 \le x \le \lambda$ the general mathematical description of this function at this instant is

$$y(x) = A \sin\left(\frac{2\pi}{\lambda}x\right) \tag{15.2}$$

Once again, keep in mind the meaning of $y(x)$: it is the string's displacement from equilibrium at position x. You can verify that Equation (15.2) works by substituting $x = 0$, $x = \lambda/2$, and $x = \lambda$; in each case $y = 0$. Similarly, if you substitute $x = \frac{1}{4}\lambda$ or $\frac{3}{4}\lambda$ you obtain $x = +A$ and $-A$, respectively.

Equation (15.2) is not the whole story, however, because the shape (or profile) of this function travels to the right with phase velocity v. In fact, at a time t later this same profile is located a distance $d = vt$ further down the x-axis (Fig. 15.9(b)). Once again, from the translation theorem that we stated at the beginning of this section we know that we can obtain the functional form of this wave by simply replacing x by $x - vt$. Therefore, the position equation for our sinusoidal wave train is

$$y(x, t) = A \sin\left[\frac{2\pi}{\lambda}(x - vt)\right] \qquad \text{(Wave moving toward the right)} \tag{15.3}$$

Similarly, if we were to generate a wave at the other end of the string and send it toward the left, in time t the wave profile would be translated a distance $d = vt$ to the left. Equation (15.2) would again describe the initial profile, and at time t later the sinusoidal wave would have the form

$$y(x, t) = A \sin\left[\frac{2\pi}{\lambda}(x + vt)\right] \qquad \text{(Wave moving toward the left)} \tag{15.4}$$

We can obtain a more compact form of Equations (15.3) and (15.4) by using our SHM definitions for the period, frequency, and angular frequency. These definitions, together with a general definition of wavelength, are summarized below.

> **Period T:** the time required for a point on the string to complete one cycle of its periodic motion. This is, of course, exactly the same time that it takes for one wavelength to pass the point
>
> **Frequency f:** the number of complete vibrations of a point on the string that occur in one second or, equivalently, the number of wavelengths that pass a given point in one second

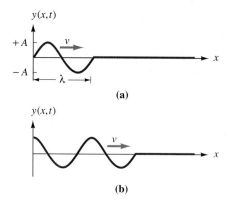

(a)

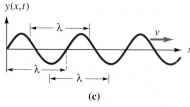

(b)

(c)

FIGURE 15.9

A sinusoidal wave train.

---

[1] See Section 15.7 for a spreadsheet template for this example.

Angular frequency ω: $2\pi f$

Wavelength λ: the shortest straight-line distance measured along the direction of wave propagation between two points on the wave that have the same displacement and slope

The relations between T, f, and ω are the same as those in Chapter 14:

$$\omega = 2\pi f = \frac{2\pi}{T}$$

Finally, notice that the term $2\pi/\lambda$ appears in both Equations (15.3) and (15.4). It is convenient to define this term as the **wave number k:**

$$k = \frac{2\pi}{\lambda} \tag{15.5}$$

The dimension of k is $(\text{length})^{-1}$.

If the period T is the time for a complete wavelength λ to pass a point on the string, then the wave velocity with which the wave travels along the x-axis must be

$$v = \frac{\textit{Distance traveled}}{\textit{Time to travel this distance}} = \frac{\lambda}{T} \tag{15.6a}$$

or

$$v = \lambda f \tag{15.6b}$$

The wave velocity may also be written

$$v = \frac{\omega}{k} \tag{15.7}$$

The equality of these two expressions for v is easily shown because the ratio ω/k is just

$$v = \frac{\omega}{k} = \frac{2\pi f}{2\pi/\lambda} = \lambda f$$

We are now ready to write Equations (15.3) and (15.4) in a compact form. We have

$$y(x, t) = A \sin\left[\frac{2\pi}{\lambda}(x \pm vt)\right]$$

$$= A \sin(kx \pm kvt)$$

But as Equation (15.7) shows, $kv = \omega$, so that, finally,

$$y(x, t) = A \sin(kx - \omega t) \qquad \text{(Sinusoidal wave traveling} \qquad \tag{15.8a}$$
$$\text{in the direction of}$$
$$\text{increasing } x)$$

$$y(x, t) = A \sin(kx + \omega t) \qquad \text{(Sinusoidal wave traveling} \qquad \textbf{(15.8b)}$$
$$\text{in the direction of}$$
$$\text{decreasing } x)$$

These functions are the equations of motion (or the wave functions) for a transverse, one-dimensional wave, traveling with amplitude A, wave number k, and angular frequency ω. Note that, in general, the argument of the sine function may also contain a phase constant ϕ if the wave does not start with zero amplitude and positive slope exactly at $x = 0$ when $t = 0$. Thus, in general,

$$y(x, t) = A \sin(kx \pm \omega t + \phi) \qquad \textbf{(15.8c)}$$

Longitudinal Waves

Equations (15.8) are also the equations of motion for one-dimensional *longitudinal* waves. In the longitudinal case, however, $y(x, t)$ is the displacement of the vibrating medium from equilibrium in the direction *parallel* to the wave propagation.

EXAMPLE 15.2   A Sinusoidal Traveling Wave

A sinusoidal transverse wave traveling along the x-axis has the wave function

$$y(x, t) = (1.50 \text{ cm}) \sin[(15.0 \text{ cm}^{-1})x - (3\pi \text{ rad/s})t]$$

(a) What are the frequency, period, and wavelength of this wave? (b) With what velocity does the wave travel along the x-axis? (c) What is the transverse velocity of a point on the string located at $x = 1.25$ m at $t = 1.60$ s?

SOLUTION   (a) By comparing $y(x, t)$ to Equation (15.8a), we identify the wave number and the angular frequency:

$$k = 15.0 \text{ cm}^{-1}$$

$$\omega = 3\pi \text{ rad/s}$$

The frequency is given by

$$f = \frac{\omega}{2\pi} = \frac{3\pi \text{ rad/s}}{2\pi} = 1.50 \text{ Hz}$$

and the period is

$$T = \frac{1}{f} = \frac{1}{1.50 \text{ Hz}} = 0.667 \text{ s}$$

The wavelength is given by

$$\lambda = \frac{2\pi}{k} = \frac{2\pi}{15.0 \text{ cm}^{-1}} = 0.419 \text{ cm}$$

(b) The phase velocity is

$$v = \frac{\omega}{k} = \frac{3\pi \text{ rad/s}}{0.419 \text{ cm}^{-1}} = 22.5 \text{ cm/s}$$

(c) The transverse velocity of a point on the string at a fixed location x is the derivative dy/dt with x held constant. Such a derivative is called a **partial derivative** and is symbolized $\partial y/\partial t$. You take this derivative with respect to t in the usual manner, but treat x as a constant.

$$v_t(x, t) = \frac{\partial y}{\partial t} = -(3\pi \text{ rad/s})(1.50 \text{ cm}) \cos[(15.0 \text{ cm}^{-1})x - (3\pi \text{ rad/s})t]$$

$$= -(14.1 \text{ cm/s}) \cos[(15.0 \text{ cm}^{-1})x - (3\pi \text{ rad/s})t]$$

At $x = 1.25$ m and $t = 1.60$ s

$$v_t(1.25 \text{ m}, 1.60 \text{ s}) = -(14.1 \text{ cm/s}) \cos[(15.0 \text{ cm}^{-1})(1.25 \text{ cm}) - (3\pi \text{ rad/s})(1.60 \text{ s})]$$

$$= 12.2 \text{ cm/s} \blacktriangleleft$$

15.3 Wave Velocity on a String

As we show in the following paragraphs, the phase velocity of the wave, given by ω/k, depends on the tension F in the string and its mass per length μ. In order to find v in terms of F and μ we can apply Newton's second law to a small segment of a circular wave pulse traveling with speed v. For this analysis it is convenient to choose a frame of reference that is traveling along with the wave at velocity v. In this frame of reference the string continually flows past us, and as each segment reaches our position it rises up and over, following a circular path for a very small distance at the top. It's all right to do this because Newton's laws are valid in any frame of reference that moves at a constant velocity. We continue to use our small-amplitude, constant-tension model.

Figure 15.10(a) shows a small string segment of mass Δm that subtends an arc length $\Delta l = R(2\theta)$. In Figure 15.10(b) we expand the drawing of the arc length and show the tension forces F that act on this segment. The horizontal components $F \cos(\theta)$ cancel. However, in the limit that Δl becomes very small, the vertical components $F \sin(\theta)$ both act toward the center of curvature and provide the centripetal acceleration that drives the mass Δm back toward the equilibrium position. Therefore, the total centripetal force is $F_c = 2F \sin(\theta)$. For a small amplitude, θ is small and $\sin(\theta) \approx \theta$, so $F_c \approx 2F\theta$. That is,

$$F_{net} = ma$$

$$2F\theta = \frac{\Delta m v^2}{R}$$

But $\Delta m = \mu \Delta l = \mu R(2\theta)$, so we have

$$2F\theta = \frac{2\mu R \theta v^2}{R}$$

Solving for the wave speed v, we find

$$v = \sqrt{\frac{F}{\mu}} \qquad (15.9)$$

The velocity of a sinusoidal wave is proportional to the square root of the tension F and inversely proportional to the square root of the mass per length μ. Although we assumed a circular shape for the pulse in the above derivation, Equation (15.9) is valid for wave pulses and trains of any shape as long as the other assumptions of our model are satisfied.

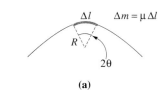

(a)

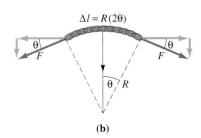

(b)

FIGURE 15.10

The forces acting on a small segment of a string in wave motion.

Concept Question 2
Explain why the wave speed changes when a wave is sent up a long rope attached to a high ceiling.

EXAMPLE 15.3

A sinusoidal wave with a wavelength of 20.0 cm, an amplitude of 3.00 cm, and a velocity of 2.00 m/s travels on a string ($\mu = 0.250$ g/cm) to the left along the *x*-axis. (a) What are the frequency and period of the wave motion? (b) What is the tension in the string? (c) What is the position function for the wave?

SOLUTION   (a) Because $v = f\lambda$, the frequency is given by

$$f = \frac{v}{\lambda} = \frac{200.\ \text{cm/s}}{20.0\ \text{cm}} = 10.0\ \text{Hz}$$

and the period is

$$T = \frac{1}{f} = \frac{1}{10.0\ \text{Hz}} = 0.100\ \text{s}$$

(b) We can find the tension F from Equation (15.9):

$$F = \mu v^2 = (0.250\ \text{g/cm})(200.\ \text{cm/s})^2 = 1.00 \times 10^4\ \text{dyn} = 1.00 \times 10^{-3}\ \text{N}$$

(c) In order to write the equation describing the wave's propagation we must find k and ω:

$$k = \frac{2\pi}{\lambda} = \frac{2\pi}{20.0\ \text{cm}} = \frac{\pi}{10.0}\ \text{cm}^{-1} = 0.314\ \text{cm}^{-1}$$

$$\omega = 2\pi f = 2\pi(10.0\ \text{Hz}) = 20.0\pi\ \text{rad/s} = 62.8\ \text{rad/s}$$

For motion to the left the propagation equation is given by Equation (15.8b):

$$y(x, t) = (3.00\ \text{cm}) \sin[(0.314\ \text{cm}^{-1})x + (62.8\ \text{rad/s})t] \quad \blacktriangleleft$$

15.4 Energy Transported by Sinusoidal Waves

When we move the end of a string up and down to generate a sinusoidal wave, we perform work on the string. If there are no frictional losses, the work we do becomes the energy of the wave. In this section we compute the rate at which this energy is transported by the wave.

We begin by looking at a small mass element *dm* of the string shown in Figure 15.11. If μ is the mass per unit length of the string, then the mass of this element can be written $dm = \mu\ dx$, where *dx* is the length of this element. As the wave travels along the string, *dm* oscillates up and down in SHM. It obtains its energy from work done on it by the mass element immediately to its left. Similarly, *dm* performs work on the mass element to its right, thereby transferring energy along the string. Let's use our knowledge of the dynamics of a mass *m* that oscillates in SHM with an angular frequency ω and amplitude *A*. In Chapter 14 we found that the total energy of such a mass is equal to the maximum kinetic energy $E = \frac{1}{2}mv_{max}^2 = \frac{1}{2}m\omega^2 A^2$. Then, the energy of a small mass element *dm* as it oscillates in SHM and passes this energy to the adjacent *dm* across distance *dx* is

$$dE = \frac{1}{2}(dm)\ \omega^2 A^2 = \frac{1}{2}(\mu\ dx)\omega^2 A^2$$

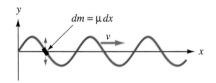

FIGURE 15.11

A small mass element *dm* oscillates in SHM when a sinusoidal wave passes through it.

The rate at which energy is transferred along the string is dE/dt, the power:

$$\mathcal{P} = \frac{dE}{dt} = \frac{1}{2}\mu \frac{dx}{dt}\,\omega^2 A^2$$

However, dx/dt is the velocity v of the wave:

$$\mathcal{P} = \frac{1}{2}\mu\omega^2 A^2 v \qquad\qquad (15.10)$$

The rate at which energy is transferred by a sinusoidal wave on a string is proportional to the square of both the frequency and amplitude and is also proportional to the velocity of the wave. Although we obtained these results for sinusoidal waves on a string, *the power transmitted by any mechanical sinusoidal wave is always proportional to the square of the frequency and the square of the amplitude.*

EXAMPLE 15.4 *Pass the Energy, Please*

A 120.-Hz simple harmonic oscillator is attached to the end of a cord that has a mass density of 20.0 g/m and is under a tension of 12.5 N. What power must the oscillator provide the cord to generate sinusoidal waves with an amplitude of 5.00 cm?

SOLUTION The expression for power given by Equation (15.10) requires us to find both the angular frequency and the velocity of the wave. The angular frequency is

$$\omega = 2\pi f = 2\pi(120.\text{ Hz}) = 240\pi \text{ rad/s} = 754 \text{ rad/s}$$

The velocity is given by Equation (15.9):

$$v = \sqrt{\frac{F}{\mu}} = \sqrt{\frac{12.5 \text{ N}}{0.0200 \text{ kg/m}}} = 25.0 \text{ m/s}$$

Finally, the required power is

$$\mathcal{P} = \frac{1}{2}\mu\omega^2 A^2 v$$

$$= \frac{1}{2}(0.0200 \text{ kg/m})(754 \text{ rad/s})^2(0.0500 \text{ m})^2(25.0 \text{ m/s})$$

$$= 355 \text{ W} \qquad\blacktriangleleft$$

15.5 Superposition and Interference of Waves

When the amplitude of two waves traveling through the same elastic medium is small, a remarkable thing happens: the instantaneous displacement of each particle of the medium is the vector sum of the displacements due to each wave. This property is called the **principle of superposition.** It means that the waves do not alter one another and each propagates through the medium as if the other were not there.[2]

---

[2] The principle of superposition follows mathematically from the wave equation presented in optional Section 15.6.

Concept Question 3
A harmonic oscillator of fixed frequency is sending transverse waves along a horizontal cord. By what factor must the power output of the oscillator change to double the amplitude while the tension in the cord remains constant?

Concept Question 4
If the tension in the cord carrying a transverse wave is doubled while the amplitude of the wave remains constant, by what factor does the power output of the oscillator change?

Concept Question 5
If the amplitude and wavelength of a wave traveling on a string remain constant but the frequency of the vibrator attached to a string is doubled, by what factor does the transmitted power change?

The principle of superposition holds only for waves propagating through media in which the restoring forces obey Hooke's law. For mechanical waves this restriction usually means the principle applies to small amplitude waves only. The principle accurately describes low-amplitude waves on a string and sound waves that are not too loud. When large-amplitude waves travel on a string, the tension in the string is not constant, and the superposition principle does not apply. Similarly, shock waves in air, such as the "sonic boom" created by an aircraft traveling faster than the speed of sound in air, result in such large air-pressure fluctuations that the superposition principle again does not hold. We consider only small-amplitude waves for which we can discover the consequences of the superposition principle. We also continue to use the wave on a string as our model. You should keep in mind, however, that the results we obtain are applicable to any mechanical wave of small amplitude traveling through an elastic medium obeying Hooke's law.

Wave Interference

Let's begin by considering two waves both traveling on the same string. When we look at the position function for either of the waves, $y(x, t) = A \sin(kx \pm \omega t + \phi)$, it is clear that there are many ways that the two waves can differ. The waves can have opposite directions, different amplitudes A, different wavelengths $\lambda = 2\pi/k$, different frequencies $f = \omega/2\pi$, or one wave may differ in phase from the other by an amount ϕ. We examine several of these possibilities in the paragraphs that follow. Let's examine this last situation first.

Adding Waves That Differ in Phase Only: Interference

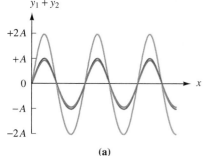

Suppose two waves traveling on the same string with the same amplitude, wavelength, and frequency are described by the equations

$$y_1(x, t) = A \sin(kx - \omega t)$$

$$y_2(x, t) = A \sin(kx - \omega t - \phi)$$

Notice that the phase constant ϕ specifies the extent to which the second wave is shifted along the x-axis relative to the first wave. When the two waves are both present at the same time, the principle of superposition allows us to find the equation for the resulting wave by simple addition of the two separate wave descriptions.

$$y(x, t) = y_1 + y_2 = A \left[\sin(kx - \omega t) + \sin(kx - \omega t - \phi) \right] \quad \text{(15.11)}$$

We will find the following trigonometric identity useful several times in the remainder of this chapter:

$$\sin(\alpha) + \sin(\beta) = 2 \sin\left(\frac{\alpha + \beta}{2}\right) \cos\left(\frac{\alpha - \beta}{2}\right) \quad \text{(15.12)}$$

When we apply this identity to Equation (15.11), we obtain

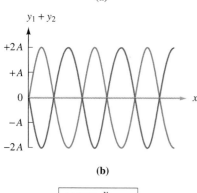

$$y(x, t) = \left[2A \cos\left(\frac{\phi}{2}\right) \right] \sin\left(kx - \omega t - \frac{\phi}{2} \right) \quad \text{(15.13)}$$

FIGURE 15.12

(a) Complete constructive interference between two waves with the same amplitude, frequency, and wavelength: $\phi = 0$ rad. (b) Complete destructive interference between the same two waves: $\phi = \pi$ rad.

The argument of the sine term in Equation (15.13) indicates that we still have a wave with the same frequency $f = \omega/2\pi$ and wavelength $\lambda = 2\pi/k$. However, the wave is phase-shifted by $\phi/2$ relative to wave 1. The term $2A \cos(\phi/2)$ multiplying the sine function may be interpreted as the new amplitude of our wave. This new amplitude depends critically on the phase constant ϕ. Figure 15.12(a) illustrates the situation for $\phi = 0$ rad. This case is called **constructive interference,** and both waves are in phase: the positions of the crests of each wave coincide, as do the positions of the troughs. When $\phi = \pi$ rad, as shown in Figure 15.12(b), the waves are completely out of phase and cancel each other entirely. This case is called **destructive interference.** If the two waves have different amplitudes, complete destructive interference does not take place even if the phases differ by π rad.

Adding Waves That Differ in Frequency Only: Beats

Next, let's investigate what happens when we combine two waves with the same amplitude and phase but slightly different frequencies f_1 and f_2 and, therefore, different wavelengths and wave numbers: $f_1 = v/\lambda_1 = (k_1/2\pi)v$ and $f_2 = v/\lambda_2 = (k_2/2\pi)v$. The superposition of these waves is

$$y(x, t) = A \sin(k_1 x - \omega_1 t) + A \sin(k_2 x - \omega_2 t)$$

Applying Equation (15.12), the previous equation becomes

$$y(x, t) = \left\{ 2A \cos\left[\frac{1}{2}(k_2 - k_1)x - \frac{1}{2}(\omega_2 - \omega_1)t \right] \right\} \sin\left[\frac{1}{2}(k_2 + k_1)x - \frac{1}{2}(\omega_2 + \omega_1)t \right]$$

(15.14)

In order to interpret Equation (15.14) let's apply our "snapshot" technique and see how this wave looks at $t = 0$ s. Also, let's define the average wave number

$$k_{av} = \frac{1}{2}(k_1 + k_2)$$

and the wave number difference

$$\Delta k = k_2 - k_1$$

With these substitutions at $t = 0$ s, Equation (15.14) becomes

$$y(x, 0) = \left[2A \cos\left(\frac{\Delta k\, x}{2} \right) \right] \sin(k_{av} x) \qquad \textbf{(15.15)}$$

Because we assume the frequency difference to be small, Δk is small compared to both k_1 and k_2. We can interpret Equation (15.15) as the equation of a sinusoidal wave with a wave number equal to the average of the two component waves. Moreover, the amplitude of this wave is determined by the factor $2A \cos(\Delta kx/2)$. This amplitude *varies with position x.* At $x = 0$ it is a maximum $2A$, at $x_{\min} = \pi/\Delta k$ it decreases to zero, and at $x_{\max} = 2\pi/\Delta k$ it again reaches a maximum. Figure 15.13 is a graphical representation of

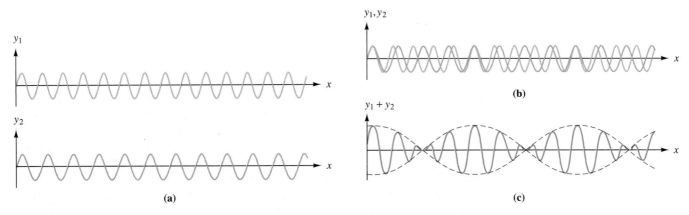

FIGURE 15.13

(a) Two waves of slightly different frequencies. (b) The graphs from part (a) plotted on the same graph. (c) The superposition of the two waves.

how these waves combine. A wave with an amplitude that varies is said to be **amplitude-modulated.**

As t increases from zero in Equation (15.14) the whole pattern shown in Figure 15.13(c) moves along the x-axis with the wave velocity. At any point in space along the wave path the wave amplitude oscillates in time from a maximum of $2A$ to a minimum of zero. These pulsations of the amplitude are called **beats.** The number of amplitude pulsations per second is called the **beat frequency.** The time between amplitude maximums for a wave traveling with velocity v is

$$T_{\text{beat}} = \frac{x_{\max}}{v} = \frac{2\pi/\Delta k}{v}$$

So the beat frequency is

$$f_{\text{beat}} = \frac{v\Delta k}{2\pi} = \frac{vk_1}{2\pi} - \frac{vk_2}{2\pi}$$

$$= \frac{v}{\lambda_2} - \frac{v}{\lambda_1}$$

$$\boxed{f_{\text{beat}} = f_1 - f_2} \tag{15.16}$$

In Chapter 17 we apply Equation (15.16) to sound waves.

Adding Waves That Differ in Direction Only: Standing Waves

In our analysis to this point we have investigated how differing phases and frequencies of two superposed waves affect the sum. Let's now see the result of combining two waves y_1 and y_2 that have the same amplitude, wavelength, and frequency but travel in opposite directions:

$$y_1 = A \sin(kx - \omega t)$$
$$y_2 = A \sin(kx + \omega t)$$
$$y(x, t) = y_1 + y_2 = A \left[\sin(kx - \omega t) + \sin(kx + \omega t)\right]$$

Applying the trigonometric identity Equation (15.12), we rewrite this expression as

$$y(x, t) = [2A \sin(kx)] \cos(\omega t) \tag{15.17}$$

By now you know to recognize the term $2A \sin(kx)$ as the amplitude of the y-oscillation of the point on the string located at position x. When we choose this interpretation, the second factor, $\cos(\omega t)$, indicates to us that the string oscillates up and down in SHM; every point x along the string oscillates with the same angular frequency ω. Another important feature of Equation (15.17) pertains to the periodic spatial variation of the amplitude. There are x-positions for which the amplitude of the y-motion is always zero, and other x-positions for which the amplitude of the y-motion is always $2A$. A graph of this wave at several different times t is shown in Figure 15.14. This type of wave is called a **standing wave** because it *appears* to travel neither to the left nor to the right. The zero amplitude positions as well as those of the amplitude maximums do not change.

Any position for which (kx) is a multiple of π yields a zero for the amplitude factor $2A \sin(kx)$. These positions are called **nodes:**

$$kx = 0,\ \pi,\ 2\pi,\ 3\pi,\ \cdots$$

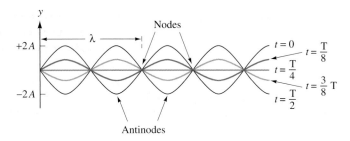

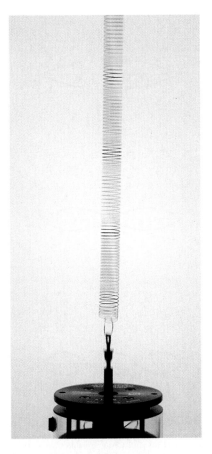

FIGURE 15.14

A standing wave at different instants of time (T = one period = $2\pi/\omega$).

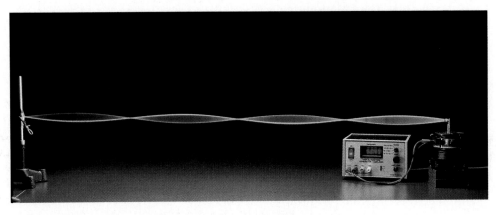

A vibrating rod creates transverse standing waves on a string.

Longitudinal standing waves on a spring. Notice the alternating regions of compression and rarefaction.

When we make the substitution $k = 2\pi/\lambda$, we find the node locations to be

$$x_n = 0, \frac{\lambda}{2}, \frac{2\lambda}{2}, \frac{3\lambda}{2}, \cdots$$

These particular node locations are correct only for the special case in which both constituent waves are described by sine functions with zero phase angle. For other cases the nodes may have other positions; however, as shown in Figure 15.15, the distance between nodes is always $\lambda/2$.

Spaced halfway between the nodes are the **antinodes.** At these positions the amplitude $2A\sin(kx)$ is a maximum. The sine term in the amplitude factor is a maximum when

$$kx = \frac{\pi}{2}, \frac{3\pi}{2}, \frac{5\pi}{2}, \cdots$$

or

$$x_a = \frac{\lambda}{4}, \frac{3\lambda}{4}, \frac{5\lambda}{4}, \cdots$$

Here again these particular antinode positions also depend on the assumed form of the original waves.

The sounds from stringed musical instruments are generated by combinations of standing waves. An important characteristic of the waves on these instruments is that both ends of the string are fixed either by the instrument itself (such as on a piano or harp) or by the instrument at one end and a finger at the other end (instruments, such as a guitar or

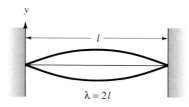

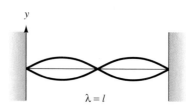

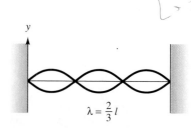

FIGURE 15.15

The first three normal modes of a string fixed at both ends.

Concept Question 6
What net energy is transported past a single point on an ideal string that exhibits standing waves?

Concept Question 7
A student watching standing waves claims that at regular intervals the string is completely straight for an instant. Can this student be correct?

Concept Question 8
Draw a sketch of two waves of equal amplitude and wavelength overlapping on the same axis. (a) How does the superposition of the two waves appear? (b) Suppose the waves move in opposite directions, each by one-quarter wavelength. Again draw the picture of both waves and the result of their superposition. (c) Advance each wave one-quarter wavelength again. How does the resultant wave appear?

violin). Conditions such as these, which place restrictions on the behavior of a wave at particular positions, are called **boundary conditions.** For stringed instruments the boundary conditions are that the end points be nodes. In Figure 15.15 we show the three lowest frequency standing waves for a string of length L with both ends fixed. The standing wave patterns possible on the string are called the **normal modes.** From Figure 15.15 we see that the wavelengths of the normal modes must be

$$\lambda = 2L, \; L, \; \frac{2}{3}L, \; \frac{1}{2}L, \; \cdots$$

In general, then

$$\lambda = \frac{2L}{n} \qquad \text{where } n = 1, 2, 3, \cdots \qquad (15.18)$$

The relation between the frequency of a wave and its wavelength is given by Equation (15.6b), $\lambda f = v$ the wave velocity. Therefore, the frequencies with which a string fixed at both ends may vibrate can be found by substituting $\lambda = v/f$ in Equation (15.26) to give

$$f = \frac{v}{2L}, \; \frac{2v}{2L}, \; \frac{3v}{2L}, \; \frac{4v}{2L}, \; \cdots$$

That is,

$$f = n\left(\frac{v}{2L}\right) \qquad \text{where } n = 1, 2, 3, \cdots \qquad (15.19)$$

These frequencies are called the **normal frequencies** of the string. The first frequency ($v/2L$) is called the **fundamental** (or the *first harmonic*). The second frequency ($n = 2$) is called the **first overtone** (or *second harmonic*), the third frequency is called the **second overtone** (or *third harmonic*), and so on.

The general motion of a string fixed at both ends consists of the fundamental plus a number of the overtones. The amplitude of most of the overtones depends on the method of excitation of the string. We will have more to say about the sounds created by stringed instruments in Chapter 17.

Wave Reflection

Suppose we send a wave pulse down a string that is fixed at the other end, as illustrated in Figure 15.16(a). When the wave reaches the fixed point it exerts an upward pull on the end. By Newton's third law the fixed point exerts an equal and opposite force downward on the string. The result is a phase change of π rad for the reflected wave; as it travels back

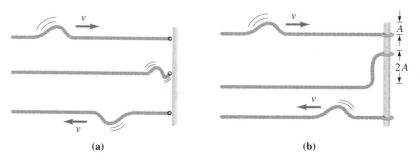

(a)   (b)

FIGURE 15.16

(a) A pulse reflected from a fixed end undergoes a phase change of π radians. (b) A wave reflected from a free end undergoes no phase change.

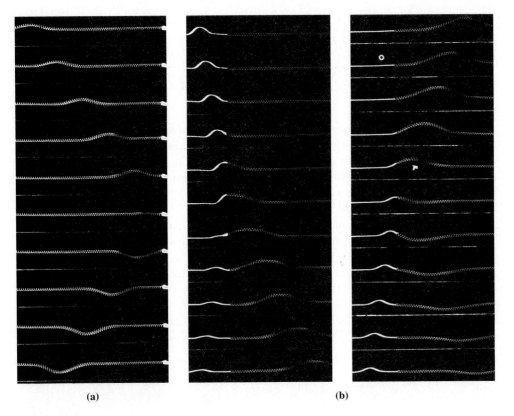

(a)   (b)

FIGURE 15.17

(a) Waves reflected from a fixed end. (b) Wave reflection and transmission at a junction between springs of different mass densities.

toward us the wave is inverted. This case is shown in Figure 15.17(a) for a wave traveling down a string.

On the other hand, if the end of the string is free to move vertically (Figure 15.16(b)), the pulse overshoots to an amplitude twice the normal amount, and the reflected pulse is not inverted.

When two strings of different mass densities are joined and a pulse sent down one of the strings, both reflected and transmitted pulses are generated at the junction. When a wave pulse travels from a medium of high mass density to one of lower mass density, the reflected wave pulse is not inverted. However, when the pulse travels from a medium of low mass density to higher mass density, the reflected pulse is inverted. These effects are shown in Figure 15.17(b). Notice the junction is between a light and a heavy spring. The fraction of energy received by each pulse at the junction depends on the relative magnitudes of the two mass densities.

Fourier's Theorem (Optional)

In this section we have been investigating some of the consequences of the superposition principle. One of the most important theorems associated with wave superposition was discovered by the French mathematician Jean Baptiste Joseph Fourier (1768–1830). Fourier's theorem states that any arbitrarily shaped, periodic wave can be constructed by the superposition of a sufficiently large number of sinusoidal waves.[3]

---

[3] There are other restrictions requiring that the wave (over the length L defined in the text) have a finite number of discontinuities and that its amplitude be bounded. However, for physical systems (such as waves on a string) we usually don't have to worry about these restrictions, so if you don't understand them, ignore this footnote.

For an arbitrary wave, at time $t = 0$, one general expression for the Fourier series representation of the wave is

$$y(x) = \sum_{n=0}^{\infty} \left[A_n \sin\left(n\frac{2\pi x}{L}\right) + B_n \cos\left(n\frac{2\pi x}{L}\right)\right]$$

where L is the length in space over which the periodic wave extends before it begins to repeat. Also, A_n and B_n are constants that represent the amplitudes of the various component waves that superpose to form the arbitrary wave. There are equations for A_n and B_n; however, a presentation of Fourier analysis is, as we authors sometimes like to say, "beyond the scope of this text." We simply point out that for any periodic wave shape, no matter how complex, A_n and B_n may be computed.

Usually, an infinite number of A_n and B_n are required for a completely accurate representation of an arbitrary wave. However, often just a few terms are sufficient to begin to closely approximate the wave. Figure 15.18(a) shows a square wave (again at time $t = 0$ s), a shape very different from the smooth curves of the sine or cosine functions. The Fourier series for this wave turns out to be

$$y(x) = \left(\frac{4A}{\pi}\right)\sin\left(\frac{2\pi x}{L}\right) + \left(\frac{4A}{3\pi}\right)\sin\left(\frac{6\pi x}{L}\right) + \left(\frac{4A}{5\pi}\right)\sin\left(\frac{10\pi x}{L}\right) + \cdots$$

$$= \left(\frac{4A}{\pi}\right)\sum_{n=1,3,5\cdots}^{\infty} \frac{1}{n}\sin\left(\frac{2n\pi x}{L}\right) \qquad \text{(All } n \text{ values are odd)} \qquad \textbf{(15.20)}$$

When we use only the first six terms of this series we obtain the rough approximation to the square wave shown in Figure 15.18(b). The fact that any periodic wave can be decomposed into a superposition of sinusoidal waves gives added importance to the study of sinusoidal waves.

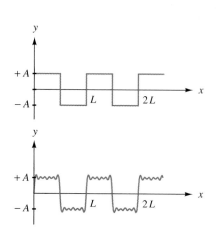

FIGURE 15.18

(a) A square wave. (b) A six-term Fourier series approximation of a square wave.

15.6 The Wave Equation[4] (Optional)

In this section we apply Newton's second law to a segment of a string through which a wave is passing. We discover that this time Newton's second law leads us to a second-order partial differential equation known as the **wave equation.** Figure 15.19(a) shows a wave pulse traveling along a string. We have exaggerated the amplitude so that the x- and y-coordinates are clear to you. The important thing to realize about Figure 15.19(a) is that it represents a "snapshot" of the wave motion. That is, the displacement y at position x is as drawn in this figure at only one particular instant of time. As the wave continues to move along the string the displacement y at position x changes in time. Therefore, y depends not only on the position x but also on the time t. Hence, we write y as a function of both x and t: $y = y(x, t)$. It is this function $y(x, t)$ we wish to determine. Once we have $y(x, t)$ we have the position function for the wave: we know the displacement of the string y for any location x, at any time t.

In Figure 15.19(b) we show a small segment of the string located at position x. The length of this segment is Δx so its mass is $\Delta m = \mu\, \Delta x$. The magnitude of the tension force at both ends of the segment is F. However, the angle that **F** makes with the horizontal changes from θ at position x to $\theta + \Delta\theta$ at position $x + \Delta x$. A free-body diagram for the components of **F** that act on Δm is shown in Figure 15.19(c). For small values of θ, $\cos(\theta)$ and $\cos(\theta + \Delta\theta)$ are both almost equal to one. Hence, the x-component of force on the segment Δm is very close to zero, a result we expect because there is no net horizontal

---

[4] Your instructor may choose to skip this section, and it may be omitted without loss of continuity. However, it contains the physics and mathematics of what may be the founding problem of mathematical physics, so we hope you will take the time to study it. (See G. F. Wheeler and W. P. Crummett, "The vibrating string controversy," *American Journal of Physics*, **55**, 33–37 (1987)).

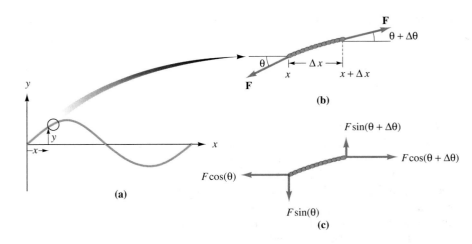

FIGURE 15.19

Application of Newton's second law to a segment of a string undergoing wave motion.

displacement of the string. However, there is a net force along the y-direction, and it is given by

$$\Delta F_y = F \sin(\theta + \Delta\theta) - F \sin(\theta) \tag{15.21}$$

If the angle θ is sufficiently small, we can use the approximation

$$\sin(\theta) \approx \tan(\theta) = \frac{dy}{dx}\bigg|_{t=\text{const}} = \frac{\partial y}{\partial x} \tag{15.22}$$

where $\partial y/\partial x$ is the partial derivative of y with respect to x. (Partial derivatives are reviewed in Appendix 1.)

Now $\partial y/\partial x$ and $\partial y/\partial t$ have very different meanings. The slope of the string at position x is $\partial y/\partial x$, whereas $\partial y/\partial t$ is the *transverse* velocity of the string at position x and time t; that is, the up or down velocity of a particular point on the string as the wave passes through it. Don't confuse the transverse velocity with the phase velocity that a wave pulse or train has as it travels along the string.

Let's get back to the application of Newton's second law to our string segment. When we substitute Equation (15.22) into Equation (15.21), we have

$$\Delta F_y = F\left[\left(\frac{\partial y}{\partial x}\right)_{x+\Delta x} - \left(\frac{\partial y}{\partial x}\right)_x\right] \tag{15.23}$$

But $\Delta F_y = \Delta m a_y$ (Newton's second law) and $a_y = \partial^2 y/\partial t^2$, so

$$\Delta F_y = \mu\,\Delta x\frac{\partial^2 y}{\partial t^2}$$

Substituting this result for the left-hand side of Equation (15.23), we have

$$\frac{\partial^2 y}{\partial t^2} = \left(\frac{F}{\mu}\right)\left[\frac{\left(\frac{\partial y}{\partial x}\right)_{x+\Delta x} - \left(\frac{\partial y}{\partial x}\right)_x}{\Delta x}\right] \tag{15.24}$$

Next, let's take the limit as $\Delta x \to 0$. The term on the right-hand side of Equation (15.24) that is affected by this limit is

$$\lim_{\Delta x \to 0}\left[\frac{\left(\frac{\partial y}{\partial x}\right)_{x+\Delta x} - \left(\frac{\partial y}{\partial x}\right)_x}{\Delta x}\right] \tag{15.25}$$

In Equation (15.25) the term $\partial y/\partial x$ is a partial derivative and therefore must still be some function $f(x, t)$. Moreover, $(\partial y/\partial x)_x$ is that function evaluated at position x. Similarly, $(\partial y/\partial x)_{x+\Delta x}$ is this same function evaluated at $x + \Delta x$: $f(x + \Delta x, t)$. Therefore, Equation (15.25) is just the definition of the partial derivative of $f(x, t)$ with respect to x. That is,

$$\lim_{\Delta x \to 0} \frac{f(x + \Delta x) - f(x)}{\Delta x} = \left(\frac{df(x, t)}{dx}\right)_{t=\text{const}} = \frac{\partial f(x, t)}{\partial x} = \frac{\partial}{\partial x}\left(\frac{\partial y}{\partial x}\right) = \frac{\partial^2 y}{\partial x^2}$$

Substituting this result into Equation (15.24), we have

$$\frac{\partial^2 y(x, t)}{\partial t^2} = \left(\frac{F}{\mu}\right)\frac{\partial^2 y(x, t)}{\partial x^2} \tag{15.26}$$

Equation (15.26) is a second-order partial differential equation and is called the **wave equation.** The solution $y(x, t)$ is the position function for any one-dimensional wave that travels through a frictionless medium with elastic properties that obey Hooke's law (no dispersion).

Wave Velocity: The Wave Equation Analysis

The position function $y(x, t)$ for one-dimensional waves must be a solution to the wave equation. Let's see if Equation (15.8a) is indeed a solution to Equation (15.26). We substitute $y(x, t)$ for a wave traveling along the positive x-axis and allow you to check it for the wave traveling toward decreasing values of x. We need the second derivatives with respect to both time and position:

$$y(x, t) = A \sin(kx - \omega t)$$

TIME DERIVATIVES	SPATIAL DERIVATIVES
$\dfrac{\partial y}{\partial t} = -\omega A \cos(kx - \omega t)$	$\dfrac{\partial y}{\partial x} = -kA \cos(kx - \omega t)$
$\dfrac{\partial^2 y}{\partial t^2} = -\omega^2 A \sin(kx - \omega t)$	$\dfrac{\partial^2 y}{\partial x^2} = -k^2 A \sin(kx - \omega t)$

Substituting these results into Equation (15.26) yields

$$-\omega^2 A \sin(kx - \omega t) = -\left(\frac{F}{\mu}\right)k^2 A \sin(kx - \omega t)$$

Canceling terms common to both sides, we have

$$\left(\frac{\omega}{k}\right)^2 = \frac{F}{\mu} \tag{15.27}$$

However, $\omega/k = v$, and Equation (15.27) becomes

$$v = \sqrt{\frac{F}{\mu}} \tag{15.28}$$

This result is just Equation (15.9), the expression for the wave velocity in terms of the physical characteristics of the string. This last derivation is analogous to another we performed in Chapter 14. There, we found an *ordinary* differential equation for SHM (Eq. (14.11)), and when we substituted a solution we had proposed on physical grounds (Eq. (14.1)) we discovered a bonus: an expression for the constant ω. In our present derivation we substituted the proposed solution Equation (15.8) into the *partial* differential wave Equation (15.26) and found that the velocity of the wave is given by Equation (15.28). It appears that when we are consistent with our approach to a problem nice things happen!

Concept Question 9
When we discussed simple harmonic motion of a mass on the end of a spring, we found two competing factors (the spring's stiffness and the mass on the end) that determined the frequency of oscillation. (a) Why do we call these "competing factors"? (b) Describe how the wave velocity on a string is determined by similar competing factors.

15.7 Spreadsheet Calculations for Superposition of One-Dimensional Waves (Optional)

In Example 15.1 we examined a particular wave pulse $y(x, t)$. The file WAVEPULS.WK1 on the disk accompanying this text provides a spreadsheet with the function $y(x, t)$ from this example. Try increasing values of t_o in this spreadsheet to see how the wave pulse moves to the right. Modify the spreadsheet so that the pulse travels to the left. Finally, try different velocities v to confirm that the wave speed changes as you would expect.

The spreadsheet SUPERPOS.WK1 on the disk accompanying this text permits you to see the superposition of two waves on the same string. We examine the physics of this spreadsheet and its application in the following example.

EXAMPLE 15.5   *Spreadsheet Superposition*

Construct a spreadsheet that provides a graph of two different waves and their sum. Allow different amplitudes, wavelengths, frequencies, and phase constants. Plot y versus x at a given t_o.

SOLUTION   The file SUPERPOS.WK1 on the disk accompanying this text contains such a spreadsheet and its template is shown in Figure 15.20. We are to supply required information in columns B and D. Cells B7 through B10 contain the amplitude, wavelength, frequency, and phase constant for wave 1. Cells D7, D8, and D10 contain the corresponding parameters for wave 2.

	A	B	C	D	E	F	G	H	I	J	K	L
4												!
5		Wave #1	!	Wave #2	!		Calculated Value					!
6												!
7	Amplitude =	1.000	!	3.000	!	Frequency-2 =		2.000	!			!
8	Wavelength =	3.000	!	3.000	!	Omega-1 =		12.566	!	Wv Number-1 = 2.094		!
9	Frequency-1 =	2.000	!		!	Omega-2 =		12.566	!	Wv Number-2 = 2.094		!
10	Phase Constant =	0.000	!	0.000	!				!			!
11												!
12	TIME =	0.000	!	X	!	Y1(x,t)	!	Y2(x,t)	!	Y1 + Y2		!
13												!
14			!	0.000	!	0.000	!	0.000	!	0.000		
15			!	0.140	!	0.288	!	0.865	!	1.153		
16			!	0.279	!	0.552	!	1.656	!	2.208		
17			!	0.419	!	0.769	!	2.307	!	3.076		
18			!	0.559	!	0.921	!	2.762	!	3.683		
19			!	0.698	!	0.994	!	2.982	!	3.976		
20			!	0.838	!	0.983	!	2.949	!	3.933		

FIGURE 15.20

A spreadsheet template for SUPERPOS.WK1.

Superposition of Two Waves
Snapshot at Time t_o

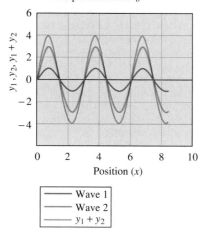

FIGURE 15.21

Graph of complete constructive interference between two waves of equal wavelength and frequency and zero phase constants but amplitudes of 1 unit and 3 units.

Superposition of Two Waves
Snapshot at Time t_o

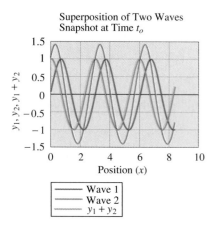

FIGURE 15.22

Graph of two waves of equal amplitude, wavelength, and frequency, but wave 2 is phase-shifted by $\pi/2$ rad.

You may be wondering why we are not allowed to designate the frequency of wave 2, or more likely, you already know: Both waves are traveling through the same medium (perhaps a string). Therefore, the phase velocities of each must be equal: $v_1 = v_2$. That is, $f_1\lambda_1 = f_2\lambda_2$, or

$$f_2 = \frac{f_1\lambda_1}{\lambda_2}$$

This equation has been entered in cell H7. The other calculated parameters include the angular velocities and wave numbers. The equations for these quantities are in columns H and K.

Column D also provides values for the x-axis. You may modify this any way you like. For now, 60 points are calculated, and the step size is based on an average value for the wave number. This approach gives you reasonable graphs as long as you don't let the wavelengths become too different.

Columns F and H contain the wave functions

$$y(x, t) = A_i \sin(k_i x - \omega_i t + \phi)$$

where $i = 1$ for wave 1 and $i = 2$ for wave 2. In cell F14, for instance, we have

```
$B$7*@SIN($K$8*$D14−$H$8*$B$12+$B$10)
```

Column J contains the superposition of the two waves. That is, the sum of columns F and H.

For complete constructive interference between two waves of equal wavelength and frequency and zero phase constants but amplitudes of 1 unit and 3 units, we obtain the graph shown in Figure 15.21. For two waves of equal amplitude (1 unit), wavelength, and frequency, but with wave 2 phase-shifted by $\pi/2$ rad, we obtain the graph shown in Figure 15.22. The resultant, as we predicted earlier, is just another wave phase-shifted by $\frac{1}{2}(\pi/2)$ rad = $\pi/4$ rad.

Use this spreadsheet to investigate additional applications of the superposition principle. In particular, you may want to apply it as you review this chapter or while you read Chapter 17 and investigate sound waves. ◄

15.8 Summary

Mechanical waves can propagate through elastic media, transporting energy without the bulk movement of matter. **Longitudinal waves** have particle motions parallel to the direction of wave propagation. **Transverse waves** have particle motions perpendicular to the wave direction.

If the function $y = f(x)$ represents the shape of a wave pulse at time $t = 0$ s, then $y = f(x \pm vt)$ provides the displacement from equilibrium of the particle located at position x at time t as the pulse travels with velocity v to the right (minus sign) or left (plus sign). A **sinusoidal wave** is described by the equation

$$y(x, t) = A \sin(kx \pm \omega t + \phi)$$

where A is the **amplitude**, k is the **wave number**, and ω is the **angular frequency**. The wave number is defined as

$$k = \frac{2\pi}{\lambda} \tag{15.5}$$

where λ is the wavelength of the wave. The angular frequency ω is related to the frequency f and the period T of the wave by

$$\omega = 2\pi f = \frac{2\pi}{T}$$

The speed of a wave on a string is determined by the tension in the string F and the mass density of the string μ and is given by

$$v = \sqrt{\frac{F}{\mu}} \qquad (15.9)$$

When the amplitudes are not too large the displacement of a particle at a point where two waves pass at the same time is the vector sum of the individual wave displacements. This statement is called the **superposition principle.**

The power transmitted by a sinusoidal wave along a string is

$$\mathcal{P} = \frac{1}{2}\mu\omega^2 A^2 v \qquad (15.12)$$

Two sinusoidal waves traveling in the same medium with the same amplitude and frequency (and wavelength) but phase-shifted by ϕ from one another superpose to yield another sinusoidal wave with a phase constant $\phi/2$. The amplitude of the composite wave also depends on the phase constant and is given by $2A\cos(\phi/2)$.

Two waves, each with the same amplitude A and phase ϕ but slightly different frequencies f_1 and f_2, produce a wave with an amplitude that, at any instant, changes periodically with position from 0 to $2A$ and back to 0 again. Such waves are said to **beat;** the frequency of the beats is given by

$$f_{\text{beat}} = f_1 - f_2 \qquad (15.14)$$

Standing waves are created by two waves of equal amplitudes and frequencies, one traveling to the right and the other to the left. The displacements of a standing wave at the nodes is zero, and $2A$ at the antinodes. For a string clamped at $x = 0$ carrying a wave of wavelength λ, **nodes** are located at

$$x_n = 0, \frac{\lambda}{2}, \frac{2\lambda}{2}, \frac{3\lambda}{2}, \cdots$$

and **antinodes** are found at

$$x_a = \frac{\lambda}{4}, \frac{3\lambda}{4}, \frac{5\lambda}{4}, \cdots$$

The possible frequencies of a standing wave are called the **normal frequencies.** The normal frequencies are determined by the boundary conditions.

(Optional) Any periodic wave may be represented by an infinite series of sinusoidal waves called a **Fourier series.**

(Optional) All one-dimensional waves of small amplitude satisfy the **one-dimensional wave equation**

$$\frac{\partial^2 y(x, t)}{\partial t^2} = \left(\frac{F}{\mu}\right)\frac{\partial^2 y(x, t)}{\partial x^2} \qquad (15.26)$$

If $f(x)$ is any function with a finite second derivative, the $f(x \pm vt)$ satisfies the wave equation.

PROBLEMS

15.2 Waves Traveling on a String

15.3 Wave Velocity on a String

1. At time $t = 0$ s a wave pulse has the shape given by the wave function

$$y(x) = \frac{a}{4 + bx^2}$$

where $a = 3.00$ cm and $b = 2.00$ cm^{-2}. (a) If the wave travels along the positive x-axis with a phase velocity of 1.50 m/s, write the wave function that describes this traveling wave pulse. (b) Make a plot of this wave function at times $t = 0.000$ s and $t = 0.500$ s.

2. At time $t = 0.000$ s a wave pulse has the shape given by the wave function

$$y(x) = \frac{ax}{4 + bx^2}$$

where $a = 4.00$ cm and $b = 1.00$ cm^{-2}. (a) If the wave travels along the negative x-axis with a phase velocity of 2.50 m/s, write the wave function that describes this traveling wave pulse. (b) Make a plot of this wave function at times $t = 0.000$ s and $t = 0.300$ s.

3. Figure 15.P1 shows a graph of a sinusoidal wave traveling to the right at three different times. From the graph determine (a) the amplitude, (b) the wavelength, (c) the minimum wave speed, and (d) the frequency of the wave.

y, cm

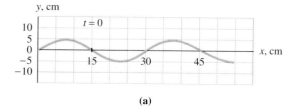

(a)

y, cm

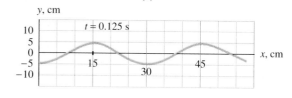

(b)

y, cm

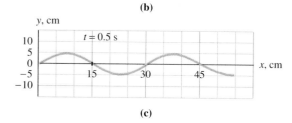

(c)

FIGURE 15.P1   Problem 3

4. Ocean waves pass a stationary cruise ship at a rate of 10.0 waves per minute. If the waves are traveling with a speed of 15.0 ft/s what is the distance between wave crests?

5. A sinusoidal wave travels along a rope with a velocity of 12.0 m/s. If 24.0 waves pass through a point on the rope in 8.00 s, find the wavelength.

6. The distance between successive maxima of a sinusoidal wave traveling along a string is 60.0 cm. Eighty such maxima pass through a point on the string in 10.0 s. Find the phase velocity of the wave.

7. A position function for a wave traveling on a string is given by

$$y(x, t) = (4.00 \text{ cm}) \sin[(\pi \text{ cm}^{-1})x + (12.0 \text{ rad/s})t]$$

(a) Find the amplitude, frequency, period, and wavelength of the wave. (b) Compute the displacement of the string at $x = 1.00$ m when $t = 1.80$ s. (c) What is the phase velocity of the wave? (d) What is the transverse velocity of the string at $x = 1.00$ m and $t = 1.80$ s?

8. Plot y versus t for the wave given in Problem 7 for (a) $x = 0.00$, and (b) $x = 1.50$ cm.

9. Plot y versus x for the wave given in Problem 7 for (a) $t = 0.000$ s and (b) $t = 0.250$ s.

10. The wave function for a wave traveling along a string is given by

$$y(x, t) = (1.20 \text{ cm}) \sin[(5\pi \text{ cm}^{-1})x - (72.0 \text{ rad/s})t]$$

(a) Find the amplitude, frequency, period, and wavelength for this wave. (b) With what velocity does the wave travel along the string? (c) Find the displacement from equilibrium and the transverse velocity of a point on the string located at $x = 3.50$ m at $t = 4.00$ s.

11. (a) Write the position function for a wave traveling on a string to the left with an amplitude of 6.00 cm, a wavelength of 1.80 m, and a velocity of 24.0 m/s. (b) At $t = 2.50$ s compute the displacement, velocity, and acceleration of a point on the string located at $x = 2.00$ m.

12. The amplitude of a wave traveling in the direction of decreasing x on a string is 3.20 cm, its wavelength is 30.0 cm, and its frequency is 30.0 Hz. (a) Write the position function for this wave, assuming the phase constant $\phi = 0$. (b) Compute the displacement and transverse velocity of a point on the string at $x = 30.0$ cm at $t = 0.667$ s.

13. A wave on a string travels in the direction of increasing x with a phase velocity of 3.600 m/s and wavelength of 1.200 m. The amplitude of the wave is 2.400 cm. (a) Compute the frequency and wave number for the wave. (b) Write the wave function for this wave, assuming the phase constant $\phi = 0$. (c) At $t = 1.40$ s find the displacement from equilibrium and the transverse velocity of the wave at $x = 2.60$ m.

14. Compute the maximum transverse velocity and acceleration of the wave described in Problem 13.

15. The amplitude of a wave traveling on a string is 1.50 cm. The wave has a frequency of 90.0 Hz and is traveling in the direction of increasing x with a velocity of 16.0 m/s. (a) Find the wavelength and position function for this wave. (b) For $t = 2.00$ s, assuming a phase constant of zero, compute the displacement and transverse velocity of a point on the string located at $x = 1.50$ m.

16. Find the maximum transverse velocity and acceleration of the wave described in Problem 15.

17. A longitudinal wave is propagating along a spring. The distance between successive regions of maximum compression is 30.0 cm, and the frequency of the wave is 15.0 Hz. The maximum displacement from equilibrium of a point on the spring is 3.20 cm.

(a) Compute the wave's phase velocity. (b) If the wave propagates in the direction of increasing x, write the position function for the wave. Assume the displacement is zero at $x = 0$ and $t = 0$.

18. The velocity of a wave pulse on a 20.0-m long cord is found to be 3.60 m/s. If the tension in the cord is 60.0 N, find the cord's total mass.

19. What tension must be applied to a 12.0-m long string of 280.-g total mass for a wave to travel on the string with a velocity of 10.0 m/s?

20. The velocity of a wave on a certain string is 15.0 m/s when a tension of 10.0 N is applied to the string. What tension is required if the wave velocity is to be increased to 60.0 m/s?

21. One end of a 20.0-g string, 4.00 m in length, is attached to an oscillator that vibrates at 120. Hz. What tension must be applied to the string for waves with a wave number of 10π m^{-1} to travel along the string?

22. A 1.50-g thread, 2.00 m in length, is stretched horizontally. One end is attached to a simple harmonic oscillator, and the other is passed over a light pulley and attached to a 400.-g mass. The frequency of the oscillator is 60.0 Hz. (a) When the oscillator is turned on, with what velocity do waves travel along the thread? (b) What is the wavelength of the waves?

23. Figure 15.P2 is the graph of a wave function at $t = 0$ s for a wave on a string traveling to the right. The tension in the string is 80.0 N, and its linear mass density is 20.0 g/m. (a) Determine the amplitude, wavelength, and frequency of the wave. (b) Find the wave function for this wave.

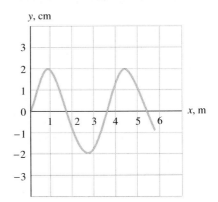

y, cm

FIGURE 15.P2   Problem 23

24. A sinusoidal wave is propagating along a cord toward the left (in the direction of decreasing x). Figure 15.P3 is a graph of the wave function at $t = 0.00$ s. The string is under a tension of 3.60 N and has a linear mass density of 5.00 g/cm. (a) Find the amplitude, wavelength, and velocity of the wave. (b) Using appropriate numerical values, write an equation in the form $y(x, t) = \sin(kx \pm \omega t \pm \phi)$ to describe this wave. (*Hint:* A phase constant is required because the displacement is not zero at $t = 0$ s.)

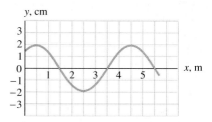

y, cm

FIGURE 15.P3   Problem 24

15.4 Energy Transport by Sinusoidal Waves

25. A 20.0-m rope has a mass of 2.00 kg and is under a tension of 100. N. What power must a 50.0-Hz vibrator supply to the rope to produce waves with an amplitude of 10.0 cm?

26. A sinusoidal wave travels on a string of 2.40×10^{-2} kg/m mass density under a tension of 80.0 N. The frequency of the wave is 40.0 Hz, and the amplitude is 5.00 cm. At what rate is energy transported by the wave?

27. A 30.0-Hz harmonic oscillator supplies 100. W of power to a string along which waves are traveling with an amplitude of 3.00 cm. If the distance between successive wave maxima (crests) is 65.0 cm, what is the mass density μ of the string?

28. A cord with a linear mass density of 36.0 g/cm is attached to a harmonic oscillator that has a maximum power output of 160. W. (a) What is the minimum wavelength that can be generated on this cord if a constant tension of 30.0 N is applied to it and the wave amplitude is 0.800 cm? (b) What is the power output of the oscillator if the wavelength remains unchanged and the tension is reduced to 20.0 N?

29. A wave traveling along a string ($\mu = 16.0$ g/m) is described by the wave function

$$y(x, t) = (2.48 \text{ cm}) \sin[(7.85 \text{ m}^{-1})x - (118 \text{ rad/s})t]$$

Find (a) the wave velocity, (b) the wave frequency, and (c) the power transmitted by the wave.

30. A wave traveling along a string under a tension of 7.78 N is described by the wave function

$$y(x, t) = (1.75 \text{ cm}) \sin[(5.24 \text{ m}^{-1})x - (94.3 \text{ rad/s})t]$$

Find (a) the wave velocity, (b) the wave frequency, and (c) the power transmitted by the wave.

31. An oscillator attached to a string provides 480. mW of power, which is transmitted by the string. The string is under a tension of 12.4 N, and its waves have a 1.80-cm amplitude and a wavelength of 1.15 m. With what velocity do the waves travel along the string?

15.5 Superposition and Interference of Waves

32. Two sinusoidal waves, each with an amplitude of 3.50 cm and the same wavelength, travel along a string toward the right. If one wave is phase-shifted by $\pi/3$ rad from the other, what is the amplitude of the composite wave?

33. Two sinusoidal waves of equal amplitudes and wavelengths travel along a string in the same direction. One wave is phase-shifted by $\pi/6$ rad from the other. If the composite wave is observed to have an amplitude of 2.60 cm, what are the amplitudes of the component waves?

34. The following two waves travel on the same string:

$$y_1(x, t) = (2.00 \text{ cm}) \sin[(3\pi \text{ m}^{-1})x - (600. \text{ rad/s})t]$$

$$y_2(x, t) = (2.00 \text{ cm}) \sin[(3\pi \text{ m}^{-1})x - (600. \text{ rad/s})t - \pi/3]$$

(a) Find the amplitude and frequency of the resultant wave. (b) At $t = 1.30$ s, what is the displacement and transverse velocity of the point on the string located at $x = 2.50$ m?

35. Two sinusoidal waves travel on the same string. Their wave functions are

$$y_1(x, t) = (5.00 \text{ cm}) \sin[(12.0 \text{ m}^{-1})x - (180. \text{ rad/s})t - \pi/4]$$

$$y_2(x, t) = (5.00 \text{ cm}) \sin[(12.0 \text{ m}^{-1})x - (180. \text{ rad/s})t + \pi/3]$$

(a) What is the phase difference between these two waves? (b) Apply Equation (15.12) to find the wave function for the superposition of these two waves. (c) What is the displacement of the string at $x = 2.00$ m at $t = 1.70$ s?

36. A string vibrates according to the position function

$$y(x, t) = (5.00 \text{ cm}) \sin[(\pi/3 \text{ cm}^{-1})x] \cos[(40\pi \text{ rad/s})t]$$

(a) What are the amplitudes and velocities of the component waves that give rise to this standing wave? (b) What is the distance between nodes?

37. The position function for a standing wave is given by

$$y(x, t) = (3.60 \text{ cm}) \sin[(35.0 \text{ m}^{-1})x] \cos[(1.60 \times 10^3 \text{ rad/s})t]$$

Determine the amplitude, velocity, frequency, and wavelength of the two interfering waves.

38. The position function for a standing wave is given by

$$y(x, t) = (4.00 \text{ cm}) \sin[(52.0 \text{ m}^{-1})x] \cos[(804. \text{ rad/s})t]$$

Find the position function for the two interfering waves.

39. The wave function for a sinusoidal wave on a string is

$$y(x, t) = (6.00 \text{ cm}) \sin[(5\pi \text{ m}^{-1})x - (314. \text{ rad/s})t]$$

Find the position function for a wave of the same amplitude, frequency, and wavelength that when added to this wave produces a wave with an amplitude of 4.00 cm.

40. The graph of Figure 15.P4 shows the maximum displacements for a standing wave on a string. (a) What is the wave number for each of the two waves that combine to form this standing wave? (b) If the velocity of one of the component waves is 12.0 m/s, what is the position function for the standing wave? (c) What is the transverse velocity of a point on the string at $x = 3.00$ m at $t = 8.00$ s?

FIGURE 15.P4   Problem 40

41. A string 60.0 cm in length has a linear mass density of 20.0 g/m. What tension must be applied to the string to establish a fundamental frequency of 60.0 Hz?

42. A 120.-Hz oscillator vibrates a 1.80-m string, fixed at both ends, such that there are five standing wave segments. Determine (a) the wavelength of the standing waves and (b) the fundamental frequency of a string this length.

43. A string of 0.400 g/cm linear mass density is under a tension of 500. N. When both ends are fixed, the fundamental mode is observed to have a frequency of 224 Hz. What is the string's length?

44. A string fixed at both ends has a length of 6.00 m and a mass of 36.0 g and is under a tension of 36.0 N. Determine the positions of the nodes and antinodes for the second overtone.

45. A string fixed at both ends is to be able to oscillate with wavelengths of 48.0 cm or 56.0 cm. What is the minimum length of the string?

46. A string fixed at both ends is replaced by another with a linear mass density that is a factor of 4 greater than the first. By what factor does the fundamental frequency change?

47. The tension of a guitar string is increased by 40.0%. By what factor does the fundamental frequency of vibration change?

48. A violin string is vibrating with a frequency of 440. Hz and has a maximum amplitude of 0.800 mm. Find the maximum acceleration of a point on the string.

15.6   The Wave Equation (Optional)

49. Show that the function $y(x, t) = Ae^{i(kx-\omega t)}$ is a solution to the one-dimensional wave equation, where $i = \sqrt{-1}$.

50. Show that the function $A(x + vt)^2$ (where A and v are constants) is a solution to the one-dimensional wave equation.

51. Show that, if f is any function possessing a second derivative, then $f(x \pm vt)$ is a solution of the one-dimensional wave equation.

52. Verify that the wave function for standing waves, Equation (15.17), is a solution to the wave equation.

53. Show that if $y_1(x, t)$ and $y_2(x, t)$ are each solutions to the wave equation (Eq. (15.26)) that $y(x, t) = y_1(x, t) + y_2(x, t)$ is also a solution.

15.7   Spreadsheet Calculations for Superposition of One-Dimensional Waves (Optional)

54. Modify the spreadsheet WAVEPULS.WK1 so that a graph for the pulse in Problem 1 is plotted.

55. Modify the spreadsheet WAVEPULS.WK1 so that a graph for the pulse in Problem 2 is plotted.

56. Construct a spreadsheet to plot an approximation to the square wave. Utilize the first 10 terms in Equation (15.20).

57. The Fourier series for a "sawtooth" wave at $t = 0$ s is given by

$$y(x, 0) =$$
$$-\frac{1}{\pi} \sin(kx) - \frac{1}{2\pi} \sin(2kx) - \frac{1}{3\pi} \sin(3kx) - \frac{1}{4\pi} \sin(4kx) -$$
$$\cdots$$

Construct a spreadsheet or a computer program to plot the contribution of each of the first five terms to $y(x, 0)$ and their sum.

General Problems

58. A wave travels along a string with the wave function

$$y(x, t) = (2.50 \text{ cm}) \sin[(1.05 \text{ cm}^{-1})x - (2\pi \text{ rad/s})t]$$

(a) Find the amplitude, frequency, period, and wavelength of the wave. (b) What is the phase velocity of the wave? (c) For the time $t = 1.20$ s compute the displacement from equilibrium and transverse velocity of a point on the string located at $x = 1.50$ m.

59. Plot y versus t for the wave given in Problem 10 for (a) $x = 0.00$, and (b) $x = 3.00$ cm.

60. Plot y versus x for the wave given in Problem 10 for (a) $t = 0.000$ s and (b) $t = 0.250$ s.

61. A wave on a string travels along the x-axis in the direction of decreasing x on a cable of mass density $\mu = 50.0$ g/cm. The frequency of the wave is 20.0 Hz, and it is under a tension of 60.0 N. (a) Find the wavelength and period of this wave. (b) If the amplitude of the wave is 1.50 cm, find the position function for this wave, assuming a phase constant of $\phi = 0$. (c) Compute the displacement and transverse velocity of a point on the cord at $x = 1.50$ m and $t = 2.00$ s. (d) At what rate is power transmitted by this wave?

62. A 4.00-m rope has a mass of 480. g and is under a tension of 50.0 N. An oscillator sends waves along the rope with a frequency of 40.0 Hz. (a) Compute the wavelength of the waves. (b) Write the equation of motion for the waves if they travel to the right with an amplitude of 1.20 cm (take $\phi = 0$). (c) Find the displacement and velocity of a point on the rope located at $x = 3.00$ m, at $t = 2.00$ s. (d) What must the amplitude of the rope be changed to for it to transmit energy at a rate of 15.0 W?

63. A string fixed at both ends vibrates at its fundamental frequency of 220. Hz. If the string were half as long, with nine times the tension and twice the mass density, what would be its new fundamental frequency?

64. A 120.-cm long string has a mass density of 20.0 g/cm. What tension must be applied to the string, fixed at both ends, to establish the second harmonic standing wave with a frequency of 512 Hz?

65. A string fixed at both ends is under a tension of 384 N and has a mass density of 0.600 g/cm. A normal mode is observed to have the frequency of 50.0 Hz. The next higher normal mode is found to have a frequency of 75.0 Hz. What is the length of the string?

66. When properly tuned, the D string of a violin vibrates with a frequency of 294 Hz. When the tension in the string is 64.0 N, the frequency is found to be 10.0 Hz too low. By how much must the tension be increased for the violin to be tuned at the correct frequency?

67. Two strings of different mass densities are joined together and stretched tightly, and the free ends are fixed in place. One string is 3.00 m in length and has a mass density of 2.00 g/m, and the other is 1.00 m in length and has a mass density of 0.500 g/m. If the tension in both strings is 4.80 N, find the lowest frequency that permits standing waves in both strings with a node at the junction.

CHAPTER

16

Solids, Liquids, and Gases

In this chapter you should learn

- the definitions of stress and strain for several types of deformations.

- to apply the definitions of Young's modulus, shear modulus, and bulk modulus to compute the deformation of objects subject to a net stress.

- the characteristics of the ideal fluid model.

- how pressure varies with depth in a static fluid.

- the principles of Pascal and Archimedes for static fluids.

- the equation of continuity and Bernoulli's equation for moving fluids.

- the definition of dynamic viscosity.

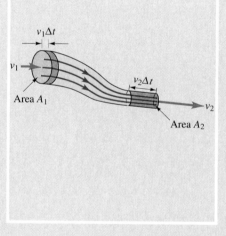

$$P + \tfrac{1}{2}\rho v^2 + \rho g h = \text{Constant}$$

Calvin and Hobbes
by Bill Watterson

From Bill Waterson, *The Authoritative Calvin and Hobbes*. (Andrews and McMeel, a Universal Press Syndicate Company, Kansas City, MO, 1990), p. 190. Reprinted with the permission of Universal Press Syndicate.

Concept Question 1
Calvin has his physics wrong. The air pressure inside the balloon is always greater than the pressure outside. So why does the balloon break? Explain.

This chapter is devoted to the mechanical properties of matter. Sensible people make springs from steel not lead. They do so because steel is hard and "springy," whereas lead is soft and plastic. Air can be squashed with relative ease, so that a room full of air can be compressed into a small cylinder. Water, on the other hand, stubbornly refuses to have its volume changed by more than a minuscule amount, even at enormous pressure. Such properties as the elasticity of metals and the compressibility of gases are known as **mechanical properties.** The mechanical behavior of materials depends on the chemical makeup and atomic structure of the substance, but we are not concerned with these details. Rather we concentrate on the macroscopic description of the deformation of bulk matter when forces are applied.

With the exception of the Hooke's-law spring model, all the solids we have dealt with have been modeled as *rigid* bodies, that is, objects that do not change their shape. Real objects, however, deform to some extent when an external force is applied to them. In Chapter 4 we found that the "normal force" originates from such deformations, but our description was merely qualitative. In this chapter we formulate some systematic ways to describe quantitatively the deformation of solids that are subjected to applied forces.

16.1 The States of Matter

At the beginning of the twentieth century, matter was classified as one of three states: solid, liquid, or gas. We now know that this classification scheme is much too limited. Newly discovered states of matter, such as liquid crystals, plasmas, and microclusters, simply don't fit nicely into such a scheme. Nonetheless, the simple three-state scheme does adequately describe the vast majority of matter we encounter daily. Thus, we continue to use it, knowing full well that some "stuff" just won't fit.

The present chapter is concerned with how matter deforms under applied forces. The deformations differ dramatically depending on whether the matter is solid, liquid, or gas. In the next section we define three convenient **elastic moduli** to describe deformations. Right now, however, we begin with a qualitative definition of each type of modulus and see how they differ for the three states of matter.

Young's Modulus: A measure of a material's resistance to a change in its length

Shear Modulus: A measure of the resistance to motion between parallel layers within a material

Bulk Modulus: A measure of the resistance of a material to a change in its volume

From these three qualitative definitions it is possible to devise a simplified scheme from which we can choose whether to model a quantity of matter as a solid, a liquid, or a gas. Table 16.1 provides this scheme. Gases are distinguished by their small bulk modulus. It is relatively easy to change the volume of a gas by compressing it. On the other hand it requires extraordinary effort to change the volume of a solid or a liquid, and both these states have large bulk moduli. Liquids and gases provide very little resistance to relative motions between the molecules that compose them. Hence, wind and ocean currents are possible. Liquids are distinguished from solids by their zero shear modulus. The zero shear modulus results in the liquid's ability to take on the shape of its container.

TABLE 16.1   **A Classification Scheme for the States of Matter**

STATE	SHEAR MODULUS	BULK MODULUS
Solid	Large	Large
Liquid	Zero	Large
Gas	Zero	Small

Both liquids and gases have the ability to flow and are called **fluids.** This ability to flow, however, leads directly to an illustration of why the classification scheme of Table 16.1 is only a qualitative instrument and, in fact, does not always work very well. In particular, some forms of matter are not easily classified into any of these categories. Take glass for example. We commonly think of a windowpane as a solid. However, over very long periods (often years), glass has been observed to flow. Some old stained-glass windows are thicker at the bottom than at the top! We avoid such complications (interesting as they are) and consider only materials that can unambiguously be modeled as a solid, liquid, or gas in the sense shown in Table 16.1.

16.2 Stress, Strain, and the Elastic Moduli

To begin our more detailed look at elastic properties let's first direct our attention to solids. If you squeeze a round blob of clay in your hand, it takes on the new shape and stays that way. This type of deformation is called **plastic.** That is, the clay makes no

Even objects that appear to be as solid as rock deform from large forces applied over long periods. (Photo by Debra Hanneman.)

attempt to restore itself to its original shape. On the other hand, if you bend a diving board, after you release the board it rebounds to its original shape. This kind of response to an applied force is called **elastic.**

The ways that an elastic body may change shape are as numerous as the ways we push, pull, twist, or bend that body. Therefore, it is clear that we need some systematic way to describe both the manner in which an applied force acts and the resulting change in shape. Let's first consider the applied force.

There are two independent ways a force may be applied to the face of an object: (1) perpendicular to the surface and (2) parallel to the surface. In Figure 16.1(a), a hammer blow produces a force perpendicular to its area of contact. A force applied across the top of your physics text is parallel to this surface, as shown in Figure 16.1(b). A force applied at an angle between these two directions may be resolved into perpendicular and parallel components. Either of these two force components applied over an area is associated with a closely related quantity known as **stress.** The perpendicular component of force produces a **normal stress;** the parallel component produces a **shear stress.** The magnitude of either type of stress is defined as

$$Stress \equiv \frac{Force}{Area} = \frac{F}{A} \tag{16.1}$$

Notice that to introduce the stress concept we had to change our mental model of how forces are applied. In earlier chapters we thought of forces as being applied at a point. Indeed we used such phrases as ". . . the *point* of application of the force." When we modeled objects as rigid bodies, this force-at-a-point model was adequate. Now that we wish to consider the deformations of real objects, it is no longer so. We must admit that a person's weight applied to the area under a spiked heel has a much different effect on a wooden floor than does that same weight applied to a flat-bottomed tennis shoe.

With these observations about how forces can be applied to an object, we are ready to consider deformations. The **strain** is a dimensionless measure of the fractional change in

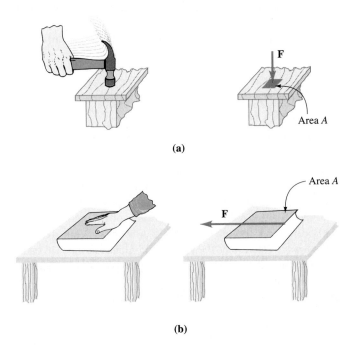

(a)

(b)

FIGURE 16.1

Independent stresses may be applied either (a) perpendicular to a surface or (b) parallel across a surface.

an object's shape when stress is applied. There are several types of strain. In the paragraphs that follow we give three different strain definitions that are useful to characterize these different deformations.

Prior to this chapter the only deformable object we considered was the spring. Let's recall the details of this model so that we can use it as a guide for useful definitions of other types of deformations. In the spring system, the deformation s (the stretch or squish) is proportional to the applied force. This relation is Hooke's law. When F is the *applied* force the relation between s and F is

$$F = ks$$

or

$$s = \frac{1}{k}F \tag{16.2}$$

There is no minus sign in Equation (16.2) because F is the force *applied to* the spring rather than the restoring force provided *by* the spring. We write Hooke's law in the form of Equation (16.2) to emphasize that this model predicts that the amount of distortion s is proportional to the applied force F. It's also important for you to recognize that the spring constant k characterizes how easy it is to deform the spring. For a fixed force F, a large k means a small deformation s, and small k implies a large deformation.

As anyone who has ever owned a Slinky toy spring knows, Hooke's law is valid only as long as the spring is not overstretched. When stretched beyond the *elastic limit* the spring will not return to its original length when the applied force is removed. As long as the stress is not too great, we can apply a relation similar to Hooke's law to solid materials. For such elastic materials this generalization of Hooke's law simply states that the *strain is proportional to the applied stress,* an idea that can be expressed mathematically as

$$\text{Strain} \propto \text{Stress} \quad \text{(Elastic materials)} \tag{16.3}$$

The proportionality constant depends on the shape of the material and the type of stress applied (remember, F can be perpendicular or parallel to the surface over which it acts). Just as in Equation (16.2) for a Hooke's-law spring, the proportionality constant in Equation (16.3) can be written as the inverse of another constant called a **modulus.** In other words,

$$\text{Strain} = \frac{1}{\text{Modulus}} \text{Stress} \tag{16.4}$$

where the stress is given by Equation (16.1). Equation (16.4) simply means that a large modulus implies a stiff material because the strain (deformation) is relatively small for a given stress.

The Hooke's-law spring constant depends not only on the material from which the spring is made but also on details of the spring's construction, such as the wire size, the diameter of the turns, and the number of turns. Several springs all constructed from the same type of steel can have vastly different Hooke's-law spring constants. What we seek now is an *intrinsic* elastic constant that characterizes the steel itself. Consequently, in each of the strain definitions you will see that dimensional factors are divided out to leave a modulus that is characteristic of the material only.

Now let's see how the strain is defined for three important cases.

Young's Modulus

The easiest stress to visualize is that applied so as to stretch or shorten the length of an object that is fixed in place at one end. In this case the **tensile stress** is the force acting perpendicular to the cross-sectional area of the object, as shown in Figure 16.2. The strain is defined as the change in length ΔL divided by the original length L_o, and the corresponding modulus is called **Young's modulus Y.** Equation (16.4) becomes

$$\frac{\Delta L}{L_o} = \frac{1}{Y}\frac{F}{A} \tag{16.5}$$

Because the strain $\Delta L/L_o$ is dimensionless, Young's modulus has units of force per unit area.

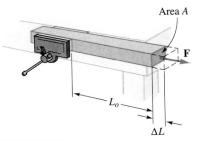

FIGURE 16.2

Material is lengthened by the tensile stress F/A.

Shear Modulus

Suppose we apply a tangential force across the surface of an object while a parallel surface is held fixed. For example, the next time the cafeteria serves you one of those rectangular blocks of Jello, try pushing on its top surface with the back of a fork similar to the illustration in Figure 16.1(b). In doing so you exert a **shear stress** on the Jello. The magnitude of the stress is again F/A, where A is the area to which the force is applied. We can define a quantitative measure of the deformation from the parameters shown in Figure 16.3. The amount by which a point on the edge of the sheared object is moved laterally increases with perpendicular distance from the fixed surface on the opposite side. It therefore makes sense to define the strain as the ratio of the lateral displacement x to the perpendicular distance y. The shear strain is thus x/y and the stress–strain relation, Equation (16.4), becomes

$$\frac{x}{y} = \frac{1}{S}\frac{F}{A} \tag{16.6}$$

where S is the **shear modulus.** The shear modulus is large for stiff materials and has units of force per unit area.

Bulk Modulus

Suppose we apply a force uniformly over all the surfaces of an object and in a manner such that the force is perpendicular to every surface (Fig. 16.4). This type of stress is called **pressure,** and for an elastic body we observe a decrease in its volume. The strain definition for this type of process is written as the ratio of the change in volume $\Delta \mathcal{V}$ to the original volume $\mathcal{V}_o$. The modulus that characterizes how much the material's volume changes is called the **bulk modulus** B. The stress–strain relation given by Equation (16.4) becomes

$$\frac{\Delta \mathcal{V}}{\mathcal{V}_o} = -\frac{1}{B}\frac{F}{A} \tag{16.7}$$

Application of a pressure F/A results in a volume *decrease*, so $\Delta \mathcal{V}$ is a negative number. Thus, it is necessary to introduce a minus sign to keep B positive. Substances with large bulk moduli are difficult to compress but they need not be stiff. For example, think of water. Sometimes it is convenient to work with the inverse of the bulk modulus called the **compressibility.** Materials with large compressibilities have small bulk moduli and are, therefore, easily compressed.

Concept Question 2
(a) If one end of a wire is fastened rigidly to a wall and the other end is pulled by a 10-N force, what is the magnitude of the force that should be substituted into the equation for stress? (b) If a 10-N force is applied to both ends of a wire so as to stretch the wire, what is the magnitude of the force that should be substituted into the stress equation?

Concept Question 3
Two rods of circular cross section, each fixed in place at one end, are made of the same material and have the same original length. One rod has half the diameter of the other, and the same force is applied to the free end of each. If the larger diameter rod undergoes a 1% change in length, what is the percentage change in length of the other rod?

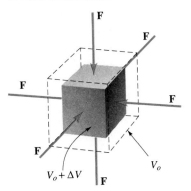

FIGURE 16.3

A shear deformation.

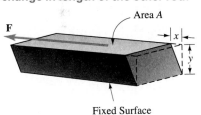

FIGURE 16.4

A cube under a uniform pressure reduces volume by ΔV.

Table 16.2 provides values for the moduli of various substances. The SI unit of stress newtons per square meter (N/m²) is the Pascal (Pa):

$$1 \text{ N/m}^2 = 1 \text{ Pa}$$

TABLE 16.2   Elastic moduli for various substances

SUBSTANCE	YOUNG'S MODULUS (GPa)	SHEAR MODULUS (GPa) (1 Gigapascal = 10^9 Pa)	BULK MODULUS (GPa)
Aluminum	69.8	24	72.2
Brass (Cu–Zn alloy)	82.7–117	36	61
Copper	112	42	137
Lead	13.8	5.6	43
Steel	172–226	84	160
Tungsten	340	150	323
Diamond	1120	450	540
Glass	55	25	450
Water	0	0	2.2
Mercury	0	0	27

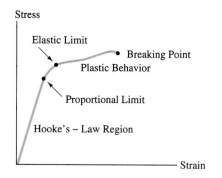

Stress

Elastic Limit

Breaking Point

Plastic Behavior

Proportional Limit

Hooke's – Law Region

Strain

FIGURE 16.5

A stress–strain diagram for a metal.

What happens if a material is stressed beyond the point to which Hooke's law applies? A typical stress–strain graph for a stretching metal wire is shown in Figure 16.5. As long as the stress does not exceed the **proportional limit,** the linear relation between stress and strain defined by Equation (16.4) is valid. If the stress does exceed the proportional limit but is less than that of the **elastic limit,** the material still returns to its original shape when the stress is removed. However, the linear relation of Equation (16.4) is not valid in the region between the proportional limit and the elastic limit. Beyond the elastic limit the material does not return to its original shape when the stress is removed. The material is said to exhibit **plastic** behavior in the region between the elastic limit and the breaking point. Beyond the elastic limit the restoring forces within the material are not conservative because they are a function of the material's history. In the plastic region additional work by some external agent is required to return the material to its original shape. If we continue to increase the stress beyond the elastic limit, eventually the material reaches the **breaking point** at which regions of the object begin to separate from each other.

EXAMPLE 16.1   *That's Probably a Safe Margin!*

One of the steel cables on the aileron-rudder system of a small aircraft is 15.0 ft long and has a 0.1875-in diameter. For proper operation, the maintenance manual states that the cable should be tensioned with 2.50 lbf of force. What force must be applied to the cable to cause its length to change by 1.00%? Young's modulus for the cable is 29.0×10^6 lb/in².

SOLUTION   Many cables, such as those used in light aircraft, small-boat steering linkages, and bicycle hand brakes, are actually composed of many fine strands of wire braided together. This process provides a more flexible cable and one more immune to surface damage (which, in turn, reduces the breaking point of the wire). There may also be some initial stretch in the cable due merely to the cable fibers tightening together. However, we ignore braiding in all our applications of Young's modulus. The cross-sectional area of the cable is πr^2 where $r = 0.1875/2$ in $= 0.09375$ in. The change in cable length for a 1.00% increase in length is

$$\Delta L = (0.0100)L_o$$

or

$$\frac{\Delta L}{L_o} = 0.0100$$

The force necessary for this percentage change in length may be found from Equation (16.5):

$$F = Y\frac{\Delta L}{L_o}A = Y\frac{\Delta L}{L_o}(\pi r^2) = (29.0 \times 10^6 \text{ lb/in}^2)(0.0100)\pi(0.09375 \text{ in})^2$$

$$= 8.01 \times 10^3 \text{ lbf}$$

The required 2.50 lbf causes the cable to change length by an insignificant amount. Can you compute the change in length for this applied force? (Answer: 4.68×10^{-5} in) ◄

EXAMPLE 16.2 *Jello Is a Shear Delight*

A certain college cafeteria's raspberry Jello servings are 10.0 cm × 10.0 cm × 5.00 cm thick. When a force of 0.150 N is applied across the top as shown in Figure 16.3, the top edge is displaced horizontally 1.80 cm from the fixed, parallel bottom. Compute the shear modulus of the Jello.

SOLUTION We are given $F = 0.150$ N and $A = (0.100 \text{ m})(0.100 \text{ m}) = 0.0100 \text{ m}^2$. Therefore, the stress is

$$\frac{F}{A} = \frac{0.150 \text{ N}}{0.0100 \text{ m}^2} = 15.0 \text{ N/m}^2$$

The strain parameters, defined in Figure 16.3, are given as $x = 0.0180$ m and $y = 0.0500$ m, resulting in a strain of

$$\frac{x}{y} = \frac{0.0180 \text{ m}}{0.0500 \text{ m}} = 0.360$$

From Equation (16.6) the shear modulus is

$$S = \frac{Stress}{Strain} = \frac{F/A}{x/y} = \frac{15.0 \text{ N/m}^2}{0.360} = 41.7 \text{ N/m}^2 \quad ◄$$

16.3 Density and Pressure

If you hold a styrofoam cup in one hand and a glass mug of identical size in the other, you quickly notice one of the more obvious properties of matter, namely, density. Styrofoam is probably the lowest density ''solid'' most of us encounter. Lead is probably one of the highest density solids we are likely to hold (at least in any quantity). Ask your instructor if the physics department has a lead brick that you can try lifting; you'll be amazed at its weight. If the brick were platinum it would weigh nearly twice as much! (The failure of movie directors to realize that a 5 cm × 6 cm × 25 cm gold bar weighs 32 lbf has ruined more than one good adventure movie for us.)

TABLE 16.3 Densities of some substances

SUBSTANCE*	DENSITY (kg/m^3)
Air	1.29
Aluminum	2.70×10^3
Copper	8.96×10^3
Ethyl alcohol	0.79×10^3
Gasoline	0.7×10^3
Glycerin (0°C)	1.26×10^3
Gold	19.3×10^3
Helium	1.77×10^{-1}
Hydrogen	8.99×10^{-2}
Ice (0°C)	0.917×10^3
Iron	7.87×10^3
Lead	11.4×10^3
Mercury	13.6×10^3
Oxygen	1.43
Platinum	21.5×10^3
Silver	10.5×10^3
Water (4°C)	1.000×10^3
(20°C)	0.998×10^3
Seawater (15°C)	1.025×10^3

*At 20°C and 1 atm of pressure, unless otherwise noted

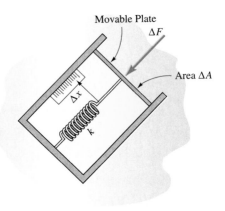

FIGURE 16.6

A simple pressure-measuring device consisting of a spring-loaded piston.

The **density** ρ of a substance is defined as mass per unit volume of a sample of the substance. That is, if a small mass element Δm occupies a volume $\Delta \mathcal{V}$, the density is given by

$$\rho = \frac{\Delta m}{\Delta \mathcal{V}} \tag{16.8}$$

In general, the density of an object depends on position, so that ρ is a function of x, y, and z: $\rho(x, y, z)$. However, if the object is **homogeneous,** its physical parameters (such as density) do not change with position throughout its volume. Therefore, if an object of mass M and volume $\mathcal{V}$ is homogeneous, Equation (16.8) reduces to

$$\rho = \frac{M}{\mathcal{V}}$$

The units of density are kilograms per cubic meter in SI and grams per cubic centimeter in the cgs system. Because all materials change volume when their temperature changes, ρ is also a function of temperature. However, this variation is small enough that, for our purposes within this chapter, we adopt a model that ignores the temperature dependence of ρ. We do, however, have opportunities to consider a case in which ρ varies with position. Table 16.3 provides some common densities of substances at 20°C.

Sometimes the density of a substance is given in terms of its **specific gravity,** or relative density. The use of this parameter is not part of a sinister plot intended to confuse students (although it sometimes turns out that way)! Specific gravity is simply the ratio of the density of the substance to that of water at 4°C. At this temperature water has a density of 1.00×10^3 kg/m³. Therefore, to convert specific gravity to density, we need only multiply by 1.00×10^3 kg/m³. For example, the specific gravity of mercury is 13.6. Hence, its density is 13.6×10^3 kg/m³.

The **pressure** exerted by a fluid is defined as the force per unit area at a point within the fluid. We can illustrate the concept of pressure by examining the pressure-measuring device shown in Figure 16.6. When the fluid surrounding the pressure-measuring device exerts an external force ΔF on the piston area ΔA, we can measure the spring compression Δx and from it compute the fluid force $\Delta F = k \Delta x$ applied to this area. The *average pressure* in the fluid at the position of our measuring device is then given by

$$P_{av} = \frac{\Delta F}{\Delta A}$$

At least conceptually, we may find the pressure at a single point by using pistons of smaller and smaller cross-sectional area ΔA and extrapolating the ratio $\Delta F/\Delta A$ to its value at which ΔA is zero. Although this procedure is obviously cumbersome from a practical standpoint, the formal definition of pressure is given as the force per unit area *at a point:*

$$P = \lim_{\Delta A \to 0} \frac{\Delta F}{\Delta A} = \frac{dF}{dA} \tag{16.9}$$

When the force F is constant over the surface A, this definition reduces to the simple form

$$P = \frac{F}{A} \tag{16.10}$$

In the previous section we saw that the SI unit N/m² is called a pascal. This is also the SI pressure unit: 1 N/m² = 1 Pa. Two other common pressure units are the **atmosphere** and the **bar:**

$$1 \text{ atm} = 1.013\,25 \times 10^5 \text{ Pa}$$
$$1 \text{ bar} = 1.000\,00 \times 10^5 \text{ Pa}$$

There is yet another way to designate pressure that is convenient if a mercury barometer is employed. The corresponding unit (millimeters of mercury) is defined in Section 16.5.

Imagine a static fluid, and consider a small cubic element of it deep within the fluid. We know that the fluid enclosed by this imagined boundary is in equilibrium because it is at rest. Because the pressure is the force per area $P = F/A$, and the areas of each face are equal, by symmetry the pressures must also be the same on each of the lateral faces (otherwise the cube of fluid would accelerate laterally). Moreover, in the limit that the cube volume (and therefore its weight) shrinks to zero, the forces (and therefore the pressures) on the top and bottom surfaces become equal. Thus, the pressure exerted by a fluid at a point is the same in all directions. Because a fluid cannot support a shear stress, the force exerted by a fluid pressure must also be perpendicular to the surface of the container that holds it. All these observations simply mean that we need not designate the pressure as a vector; the force due to static fluid pressure always acts perpendicular to *any* surface within the fluid, real or imagined.

A diamond anvil cell used to generate pressures comparable to those in the earth's mantle.

16.4 Fluid Statics

In this section we investigate some properties of fluids at rest. We discover how pressure changes with depth and also learn the principles of Pascal and Archimedes that help guide our understanding of how fluids interact with applied forces. You may be surprised to learn what enormous pressure can be exerted by small volumes of a fluid.

Variation of Pressure with Depth

In Figure 16.7 we designate an infinitesimal mass Δm of fluid by surrounding it with an imaginary boundary. Although the fluid shown on the left in Figure 16.7(a) happens to be in a container, the result we obtain can also be applied to uncontained fluids, such as our atmosphere. We place the origin of our coordinate system at the *bottom* of the container so that the bottom of the fluid element Δm has coordinate y and its top is located at $y + \Delta y$. The fluid contained within this imaginary boundary is in equilibrium (it doesn't accelerate), so the net force on the fluid slab must be zero. In Figure 16.7(b) we see the fluid forces on opposite sides of the vertical boundary are equal in magnitude but opposite in

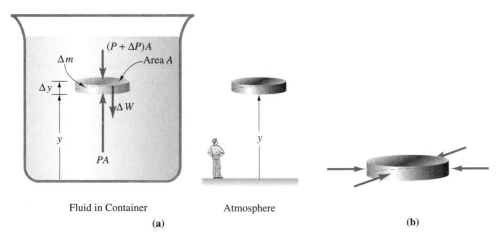

FIGURE 16.7

(a) The forces on a small mass element of fluid located at a height y from the origin. (b) The forces on the sides of the element cancel.

direction and, therefore, cancel. The other forces acting on the fluid contained within the imaginary boundary are the weight force $\Delta W = (\Delta m)g$ and forces due to fluid pressure $P(y)$ at height y. These latter forces may be computed by solving Equation (16.10) for F. In particular, the upward force due to pressure $P(y)$ on the slab's bottom is $F = P(y)A$, and the magnitude of the downward force due to the pressure at the top of the slab is $P(y + \Delta y)A$. For equilibrium, the vector sum of the forces on mass Δm must total zero:

$$P(y)A - P(y + \Delta y)A - (\Delta m)\,g = 0$$

or

$$[P(y) - P(y + \Delta y)]\,A - (\Delta m)\,g = 0$$

Now $[P(y) - P(y + \Delta y)] = -\Delta P$, and from Equation (16.8) $\Delta m = \rho\,\Delta \mathcal{V}$. However, the volume of our small fluid element is $\Delta \mathcal{V} = A\,\Delta y$, so our last equation becomes

$$-A\,\Delta P - \rho(A\,\Delta y)\,g = 0$$

or

$$\frac{\Delta P}{\Delta y} = -\rho g$$

In the limit as Δy approaches zero, $\Delta P/\Delta y$ becomes

$$\frac{dP}{dy} = -\rho g \tag{16.11}$$

Equation (16.11) indicates that on a graph, the slope of P versus y is negative. That is, the pressure P decreases with height y from the bottom of the fluid. This result is also true of our atmosphere; as we travel to higher altitudes, the air pressure decreases. (This pressure decrease is the source of the discomfort we sometimes feel in our ears when making rapid altitude changes.)

The Incompressible-Fluid Model

When the density of the fluid ρ is a constant, it is easy to integrate Equation (16.11) to find the pressure P at any point in the fluid. As shown in Figure 16.8 we designate the pressure at height y from the bottom of the fluid as P and the particular pressure at the surface of the fluid ($y = y_f$) as P_o. Then Equation (16.11) can be written

$$dP = \left(\frac{dP}{dy}\right) dy$$

$$dP = -\rho g\,dy$$

$$\int_P^{P_o} dP = -\rho g \int_{y_o}^{y_f} dy$$

Notice we have removed the constant factor ρg from the integral. Integrating and substituting the limits, we have

$$P_o - P = -\rho g(y_f - y)$$

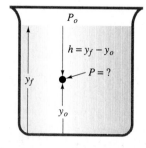

FIGURE 16.8

Computing the pressure at depth h.

or

$$P - P_o = +\rho g(y_f - y)$$

However, $(y_f - y) = h$, the depth to point y_o, so

$$P = P_o + \rho gh \qquad (\rho = \text{constant}) \qquad \textbf{(16.12)}$$

where h is a positive distance measured downward from the surface.

Equation (16.12) allows us to find the pressure at any depth h within an *incompressible* fluid. There is a rather amazing consequence of this equation. Notice that the pressure depends only on depth and not on the cross-sectional area of the fluid body. Therefore, two different dams securing two lakes of equal depth at the dam wall, must be constructed of equivalent strengths, even if one lake is only a few meters in length behind the dam, but the other extends many kilometers to its headwaters!

EXAMPLE 16.3   *Just Another Dam-Physics Problem*

The water behind a dam of width L has a depth H. Compute the total force of the water on the dam.

SOLUTION   The pressure on an area $dA = L\,dy$ is shown in Figure 16.9. At depth $h = H - y$ this pressure is, from Equation (16.12),

$$P = \rho_w gh = \rho_w g(H - y)$$

where ρ_w is the density of water. We omitted atmospheric pressure P_o because it acts on both sides of the dam. From Equation (16.12) we can find the force on the strip dA:

$$dF = P\,dA = \rho_w g(H - y)\,dA$$
$$= \rho_w g(H - y)L\,dy$$

Therefore, the total force on the dam is

$$F = \int_0^H \rho_w g(H - y)L\,dy$$
$$= \frac{1}{2}\rho_w gLH^2$$

Norris Dam is in the TVA system near Knoxville, Tennessee. Its length is 567 m, and its front side rises 81 m, although the water depth at the dam is about 64 m (in water-skiing season). With these data we find that the total force of the water on the dam is 1.1×10^{10} N! Check our answer. How many pounds of force is this? ◄

The French scientist Blaise Pascal (1623–1662) devised an experiment to show that pressure is independent of cross-sectional area (Eq. (16.12)). As shown in Figure 16.10, he filled a barrel with water and then snugly secured a tall pipe to a hole in its top. When he filled the pipe with water to a height h, the pressure within the barrel increased by ρgh at all points, and for only moderate heights the barrel was not sufficiently strong to withstand the additional pressure. The amazing feature of this experiment is that the cross-sectional area of the pipe makes no difference; a small-diameter pipe causes the same increase in pressure as a pipe the size of the barrel top! This experiment takes advantage of the "ρgh" term in Equation (16.12) to increase P. Can you see another way to increase P?

Because pressure increases with depth, the bands are spaced closer together near the base of this silo. (Photo by Tim Thornberry.)

Concept Question 4
Explain why silos are constructed to have many, closely spaced reinforcement bands around their bases but fewer toward the top?

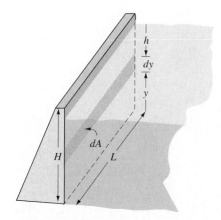

FIGURE 16.9

An area segment $dA = L\,dy$.

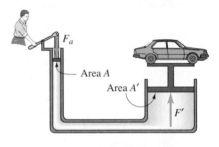

FIGURE 16.10

Pascal's experiment.

FIGURE 16.11

Pascal's principle applied to a hydraulic car lift.

Pascal's Principle

Equation (16.12) indicates that we can raise the pressure at every depth h in a fluid by increasing the pressure P_o at the surface. Pascal recognized a consequence of this fact that we now call **Pascal's principle:**

A pressure applied to a confined fluid at rest is transmitted throughout that fluid.

Anyone who has ever had serious car trouble has probably observed Pascal's principle in action as the mechanic raises the car on a hydraulic lift. Figure 16.11 shows a U-shaped pipe with a cross-sectional area that varies from a relatively small value to a much larger one. If, at the small-area end, we apply a force F_a, this force produces an additional applied pressure $P_a = F_a/A$ at all points within the fluid. Hence, at the large end of the pipe this additional applied pressure $P' = F'/A'$ must be equal to P_a:

$$P' = P_a$$

$$\frac{F'}{A'} = \frac{F_a}{A}$$

Or,

$$F' = \frac{A'}{A} F_a$$

If we make A' larger than A, we are able to multiply the applied force F_a by the ratio A'/A. For example, if $A = 5.0 \text{ cm}^2$ and $A' = 300 \text{ cm}^2$, an applied force of 400 N (90 lbf) results in a lifting force of $(300 \text{ cm}^2/5.0 \text{ cm}^2)(400 \text{ N}) = 24\,000$ N, plenty to raise most cars! We pay a penalty for this force multiplication. The applied force F_a must act through a large distance to cause F' to be applied through a small distance. This requirement has to do with conservation of energy and is explored in homework Problem 26 at the end of this chapter.

Archimedes' Principle

You have probably noticed that submerged objects have smaller apparent weight. This effect can also be understood from Equation (16.12). Because pressure increases with depth, the force on the bottom of a submerged object is greater than that on its top. Therefore, the fluid exerts a net upward **buoyant force** on the object. There is a short statement, known as **Archimedes' principle,** that can be derived from Equation (16.12). From this principle we can easily compute the magnitude of the buoyant force on an object.

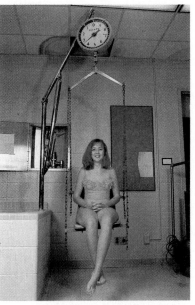

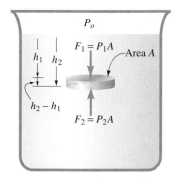

FIGURE 16.12

The buoyant force on a submerged object is a result of the increase in pressure with depth.

We derive Archimedes' principle for the cylinder shown in Figure 16.12, although the result is applicable to objects of any shape. The top of the cylinder is at depth h_1, and its bottom is at h_2. At any depth between h_1 and h_2 the horizontal forces on the cylinder cancel because for each force on one side there is a force of equal magnitude but opposite direction on the other. (See Fig. 16.7(b).) The cross-sectional area of the cylinder is A, so the net force on the cylinder top is $F_1 = P_1A$, where $P_1 = P_o + \rho gh_1$. Similarly, the fluid force on the cylinder bottom is $F_2 = P_2A$, where $P_2 = P_o + \rho gh_2$. Therefore, the net buoyant (upward) force F_b on the cylinder due only to the fluid is

$$F_b = F_2 - F_1 = P_2A - P_1A$$
$$= (P_o + \rho gh_2)A - (P_o + \rho gh_1)A$$
$$= \rho gA(h_2 - h_1) \qquad (16.13)$$

But $A(h_2 - h_1)$ is just the volume of the cylinder, which we can also call the *volume of the fluid displaced* by the cylinder $\mathcal{V}$. Equation (16.13) can, therefore, be written

$$F_b = (\rho \mathcal{V})g$$

However, $\rho \mathcal{V}$ is just the *mass of the fluid displaced,* and we have

$$F_b = mg$$

Finally, mg is the weight of the *fluid* displaced, and we are led to the following statement first presented by Archimedes and named after him. **Archimedes' principle** states

> The buoyant force acting on a submerged volume is equal to the weight of an equivalent volume of the fluid displaced.

This statement means for submerged objects we have another force to add to our list (weight, normal, spring, friction, etc.) of possible forces acting on a body. For example, we must include the buoyant force on objects due to the "ocean of air" we call our atmosphere. In highly accurate mass measurements, this effect must be taken into account.

"Hydrostatic weighing." The scale reading is smaller when the swimmer is submerged. (Photos by Tim Thornberry.)

Concept Question 5
A student sitting in a rowboat drops the anchor into the pond, whereupon it sinks to the bottom. Does the level of the pond rise, fall, or remain the same? Explain.

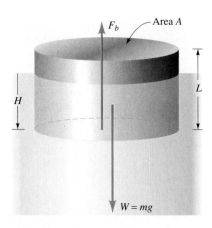

FIGURE 16.13

A cylinder of length L, cross-sectional area A, and density ρ_c floating in water.

EXAMPLE 16.4   *Just Bob, Bob, Bobbin' Along*

A cylinder of length L, cross-sectional area A, and density ρ_c floats in water as shown in Figure 16.13. (a) Find the buoyant force on the cylinder. (b) To what depth H does the cylinder sink?

SOLUTION   (a) When the cylinder is in equilibrium, the upward buoyant force of the fluid F_b is equal in magnitude to the downward force of gravity W:

$$F_b = W = M_c g = (\rho_c \mathcal{V}_c)g$$

$$= \rho_c (LA)g$$

(b) By Archimedes' principle the buoyant force F_b is equal to the weight of the fluid displaced:

$$F_b = W_w = m_w g$$

where m_w is the mass of the water *displaced* by the cylinder. This mass is $\rho_w \mathcal{V}_d$, for which the volume of the water displaced is $\mathcal{V}_d = AH$, so $m_w = \rho_w AH$. Therefore,

$$F_b = \rho_w AHg$$

Setting this expression equal to that of F_b found in part (a), we have

$$\rho_w AHg = \rho_c LAg$$

or,

$$H = \frac{\rho_c}{\rho_w} L$$

$$= s_c L$$

where we have recognized ρ_c/ρ_w as the specific gravity of the cylinder s_c. This direct proportion between the specific gravity and the depth at which an object floats forms the basis for the calibrated floats used to measure the specific gravity of many liquids from maple syrup to fermenting beer. ◀

A Compressible-Fluid Model

For gases, the constant density assumed in the incompressible-fluid model is often not adequate. However, an alternative simplifying assumption can be made, namely, that density is proportional to pressure. As a specific application, let's apply this model to our atmosphere. When we say "the density is proportional to the pressure," we mean that there is a constant k such that $\rho = kP$. In addition, we assume that we are given a reference density ρ_o at a particular gas pressure P_o. For our atmosphere we take our reference at the earth's surface, where $\rho = \rho_o$ and $P = P_o$, so

$$\rho_o = kP_o$$

or

$$k = \frac{\rho_o}{P_o}$$

The general equation for the density at other pressures is, therefore,

$$\rho = \frac{\rho_o}{P_o} P$$

We rearrange Equation (16.11) to obtain

$$dP = -\rho g \, dy$$

and substituting for ρ, we have

$$dP = -\frac{\rho_o}{P_o} Pg \, dy$$

or

$$\frac{dP}{P} = -g\frac{\rho_o}{P_o} dy$$

To integrate this expression we recognize that the pressure is P_o when the altitude y is zero and the pressure is P when the altitude is h. Hence, the integrals complete with their limits are

$$\int_{P_o}^{P} \frac{dP}{P} = -g\frac{\rho_o}{P_o} \int_0^h dy$$

Performing the integration and substituting the limits, we have

$$\ln(P) - \ln(P_o) = -g\frac{\rho_o}{P_o} h$$

$$\ln(P/P_o) = -g\frac{\rho_o}{P_o} h$$

$$P = P_o \exp\left\{-\frac{h}{[P_o/(g\rho_o)]}\right\} \qquad (16.14)$$

or

$$P = P_o \exp\left(-\frac{h}{H_o}\right)$$

Instead of a linear decrease in pressure with increasing altitude, as in the case of an incompressible fluid, our current model leads to the exponentially decreasing pressure, as shown in Figure 16.14. The quantity $H_o = P_o/(g\rho_o)$ has the units of length. This length is a characteristic distance in the sense that, when h increases by this distance, the pressure drops by $1/e$, that is, to about 37% of its original value. As it turns out, this model describes reasonably well the way our atmosphere actually behaves.

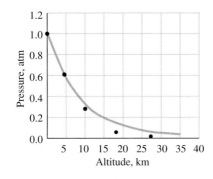

FIGURE 16.14

Comparison between the compressible fluid model and actual atmospheric pressure. The model assumes air density is proportional to pressure and that the acceleration due to gravity g does not change with altitude h. Data points indicate experimental measurements, and the solid line represents the model prediction.

EXAMPLE 16.5   *No Wonder I Ran Out of Breath on This Climb!*

By assuming atmospheric density to be proportional to pressure, find the air pressure at the summit of Granite Peak, Montana, elevation 3900 m above sea level. Take the sea-level density of air ρ_o as 1.29 kg/m^3 and its pressure P_o to be 1.01×10^5 Pa.

SOLUTION   This example is a direct application of the simple compressibility model in the previous paragraph. We begin by calculating the length scale factor in Equation (16.14):

$$H_o = \frac{P_o}{g\rho_o} = \frac{1.01 \times 10^5 \text{ Pa}}{(9.80 \text{ m/s}^2)(1.29 \text{ kg/m}^3)} = 7.99 \text{ km}$$

If the atmosphere exactly obeyed our model, air pressure would decrease by a factor of 0.37 every time the altitude increased by 7.99 km. Using this value for the characteristic length, we write

$$P = P_o \exp\left(-\frac{h}{H_o}\right)$$

When $h = 3.90$ km, this yields

$$P = (1.01 \times 10^5 \text{ Pa}) \exp\left(-\frac{3.90 \text{ km}}{7.99 \text{ km}}\right) = (1.01 \times 10^5 \text{ Pa}) \exp(-0.488)$$

$$= 0.620 \times 10^5 \text{ Pa} = 0.620 \text{ atm} \quad \blacktriangleleft$$

16.5 Pressure-Measuring Devices

Despite many instrumental innovations, two old but reliable devices are routinely employed to measure the pressure of a confined fluid and the atmosphere. In this section we describe these two instruments.

The Manometer

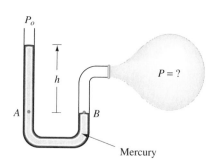

FIGURE 16.15

Measuring the pressure in a balloon with a manometer.

We may construct a **manometer** from a tube open at both ends, bent into the shape of a "U," and partially filled with mercury (Fig. 16.15). When one end of the tube is subjected to an unknown pressure P, the mercury level drops on that side of the tube and rises on the other so that the difference in mercury levels is h. Because points A and B are both the same height above the bottom of the U-tube, the pressures at these points are equal. In Figure 16.15 it is easy to see that the pressure at A is

$$P = P_o + \rho_{Hg}gh \tag{16.15}$$

where ρ_{Hg} is the density of mercury. Equation (16.15) is, therefore, also the downward pressure on the mercury surface at point B. The total pressure P at point A is the sum of the atmospheric pressure P_o and the so-called gauge pressure $\rho_{Hg}gh$.

Gauge pressure is the difference between the total **absolute** pressure and the local atmospheric pressure. When you measure your car's tire pressure, the gauge you are using measures the gauge pressure (hence the name). Even when your tire is flat, it still has one atmosphere (absolute) of pressure in it. Unfortunately, this pressure won't help keep your tire inflated because there is also one atmosphere of pressure exerting a force on the outside of the tire.

EXAMPLE 16.6   So Where Did the Air Go?

A motorist about to leave Denver (the "mile-high city") fills the car's tires to 220. kPa (gauge pressure). Upon reaching sea level the motorist rechecks the (cold) tire pressure. There has been a noticeable change in gauge pressure! Which way did the gauge pressure change and by how much?

SOLUTION   We can estimate the atmospheric pressure in Denver using the method of Example 16.5 and taking 1.60 km (roughly 1 mile) as the approximate elevation of Denver above sea level.

$$P = (101 \text{ kPa}) \exp(-h/7.99 \text{ km}) = (101 \text{ kPa}) \exp\left(-\frac{1.60 \text{ km}}{7.99 \text{ km}}\right) = 82.7 \text{ kPa}$$

Thus, the total pressure in the tire in Denver was

$$P(\text{absolute}) = P(\text{gauge}) + P(\text{atm}) = 220. \text{ kPa} + 82.7 \text{ kPa} = 303 \text{ kPa}$$

We assume no air leaked from the tire, the tire's volume does not change, and its gauge pressure was remeasured at the same temperature as at the higher elevation. In this case the tire's *absolute* pressure at the lower elevation is the same. The gauge pressure, however, depends on the new (higher) atmospheric pressure. So at the lower elevation

$$P(\text{gauge}) = P(\text{absolute}) - P(\text{atm}) = 303 \text{ kPa} - 101 \text{ kPa} = 202 \text{ kPa}$$

Thus, the motorist notices a drop of about 18 kPa (about 2.5 psi). No air has escaped from the tire, but the motorist should add more air to compensate the increased pressure of the atmosphere. ◄

Suppose we are using a mercury manometer to measure some unknown pressure, and the atmospheric pressure has already been determined. (Below, we explain how the latter can be measured.) Because in Equation (16.15) $\rho_{Hg}g$ is a constant, we need only know the height h to be able to determine the pressure. For this reason it has become customary to specify pressures in **millimeters of mercury (mm of Hg).** For 1.00 mm Hg

$$\rho_{Hg}gh = (13.6 \times 10^3 \text{ kg/m}^3)(9.80 \text{ N/kg})(1.00 \times 10^{-3} \text{ m}) = 1.33 \times 10^2 \text{ Pa}$$

That is, the conversion factor between millimeters of mercury and SI pressure units is

$$\frac{1.33 \times 10^2 \text{ Pa}}{1.00 \text{ mm Hg}} = 1$$

With this convention in hand, let's next describe a common device used to measure atmospheric pressure.

The Mercury Barometer

This time, we take a straight glass tube closed at one end and completely filled with mercury. We invert it into a dish also filled with mercury, taking care that no air enters the tube (Fig. 16.16). Atmospheric pressure supports the column of mercury in the tube to a height h. The pressure between the closed end of the tube and the column of mercury is zero, so $P_o = 0$ in this case. Therefore, the pressures at points A and B are equal, and

$$P = 0 + \rho_{Hg}gh$$

At sea level, atmospheric pressure P_o can support a column of mercury about 76.0 cm in height. Hence,

$$P_o = (760 \text{ mm Hg})\left(\frac{1.33 \times 10^2 \text{ Pa}}{1.00 \text{ mm Hg}}\right) = 1.01 \times 10^5 \text{ Pa}$$

EXAMPLE 16.6 *Now That's a Straw!*

What must be the length of a barometer tube used to measure atmospheric pressure if we are to use water instead of mercury?

SOLUTION From Equation (16.12),

$$P = P_o + \rho_w gh = 0 + \rho_w gh$$

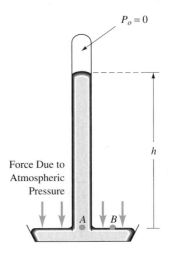

FIGURE 16.16

The barometer.

Solving for h, we obtain

$$h = \frac{P}{\rho_w g} = \frac{1.01 \times 10^5 \text{ Pa}}{(1.00 \times 10^3 \text{ kg/m}^3)(9.80 \text{ m/s}^2)} = 10.3 \text{ m} \qquad \text{(about 33 ft!)}$$

This result means that even if you have the strongest mouth this side of the Mississippi River, you couldn't drink a glass of water with a straw any longer than 10.3 m. When you do drink through a straw, you lower the air pressure at one end while the atmosphere continues to provide a constant pressure on the surface of the water in your glass. This pressure imbalance therefore results in a net force that *pushes* the water up the straw. Even when there is a perfect vacuum at one end of the straw, the atmosphere can push a column of water only 10.3 m high. ◀

16.6 Fluid Dynamics

When a fluid is in motion, the net force on a small element of the fluid may or may not be zero, depending on whether it is moving with a constant velocity or accelerating. Therefore, even for the incompressible-fluid model, Equation (16.12) is usually not sufficiently general to describe the pressure variation if the fluid is moving. In this section we correct this situation by deriving two important relations that may be used to describe pressure and velocity variations within moving fluids.

We want to keep the derivations as simple as possible but we also want our results to be sufficiently useful that we may have some confidence in their application. Therefore, we consider a fluid model with the characteristics described below.

The Ideal-Fluid Model

1. **The fluid is incompressible.** Thus, the density ρ is independent of pressure.
2. **The flow is *steady*.** That is, the fluid velocity at any point within the fluid is a constant. As we move from one point to the next within the fluid the velocity may change, but at a given location the velocity of the fluid that flows past that point is always the same.
3. **The flow is not turbulent.** This requirement means that no small region of the fluid (sometimes called a *fluid element*) has any angular velocity. Whirlpools are definitely not allowed.
4. **The fluid is nonviscous.** We neglect any friction between adjacent regions of a fluid.

Although the flow of real fluids to some degree violates one or even all of these assumptions, the ideal-fluid model is often applied to obtain at least a first approximation to the true properties of fluid flow.

The Equation of Continuity

In our second assumption we defined steady flow; at any location all the fluid passing through that point has the same velocity. Rather than focusing our attention on a particular location, we trace out the sequence of points through which one of the fluid particles flows. In fluid dynamics this path is called a **streamline.** For steady flow, streamlines do not change position in time. Figure 16.17 shows how streamlines can be used to illustrate steady fluid flow. Photographs of streamlines generated from smoke trails in a wind tunnel are shown in Figure 16.18. Streamlines cannot cross because to do so would imply two possible paths and, therefore, two different velocities, a violation of the steady-flow assumption.

You may have heard the proverb, "Still waters run deep." A person more interested in kayaking may have preferred the statement, "Shallow waters run rapid." Both of these

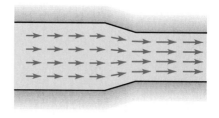

(a)

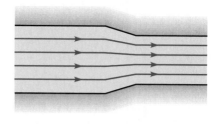

(b)

FIGURE 16.17

(a) Velocity vectors for steady fluid flow through a constricted pipe.
(b) Streamlines for the same fluid flow.

FIGURE 16.18

Smoke-generated streamline over an aerodynamically designed vehicle.

statements are illustrations of the equation we are about to derive. Figure 16.19 shows a fluid moving through a tube with a cross-sectional area that changes from A_1 to A_2 along its length. There is at least one thing we know for sure about the fluid flow through this pipe: whatever amount of fluid-mass enters the pipe from the left, this same amount must exit at the right. That is, *fluid–mass is conserved.* Another way to make this statement is to say that the rate of mass-flow past any point is a constant. Suppose the fluid's velocity and density are v_1 and ρ_1 at the left end, and v_2 and ρ_2 at the right end. In time Δt the fluid at the left end travels through a distance $v_1\,\Delta t$, and the fluid-mass Δm_1 contained within volume $(v_1\,\Delta t)\,A_1$ is simply the density multiplied by this volume:

$$\Delta m_1 = \rho_1 v_1 \,\Delta t\, A_1$$

or, the rate at which mass flows through the left side is

$$\frac{\Delta m_1}{\Delta t} = \rho_1 v_1 A_1$$

At the right end, the fluid density is ρ_2, and its velocity is v_2. In the same time interval Δt the mass Δm_2 travels through a cross-sectional area A_2 so that it occupies a volume $v_2\,\Delta t\, A_2$. The mass contained within this volume is

$$\Delta m_2 = \rho_2 v_2 \,\Delta t\, A_2$$

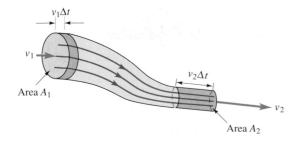

FIGURE 16.19

Deriving the equation of continuity. In time Δt a fluid volume $v_1\,\Delta t\, A_1$ enters at the left and a volume $v_2\,\Delta t\, A_2$ leaves at the right.

And the rate of mass flow at the right side is

$$\frac{\Delta m_2}{\Delta t} = \rho_2 v_2 A_2$$

For mass to be conserved (that is, no mass is piling up or disappearing between the left and right ends of the pipe), the rates $\Delta m_1/\Delta t$ and $\Delta m_2/\Delta t$ must be equal. Therefore,

$$\rho_1 v_1 A_1 = \rho_2 v_2 A_2 \qquad (16.16)$$

or

$$\rho v A = \text{constant}$$

Equation (16.16) is called the **equation of continuity,** and it is a consequence of the conservation of mass. For our incompressible-fluid model $\rho_1 = \rho_2$, and Equation (16.16) becomes

$$v_1 A_1 = v_2 A_2 \qquad (16.17)$$

or

$$v A = \text{constant}$$

Equation (16.17) is the reason "Still waters run deep!" For the product vA to remain constant while the cross-sectional area increases, the velocity must decrease. Similarly, the nozzle of your garden hose is constructed to have a small cross-sectional area so that the water velocity increases as it emerges. The units of vA are volume per time, and this quantity is called the **flow rate,** or **volume flux.**

Bernoulli's Equation

In the previous paragraphs we found that mass conservation leads to an important relation between fluid velocity and the cross-sectional area through which it flows. We next apply the *conservation of energy* to a moving fluid with elevation and cross-sectional area that may change. Our result is the generalization of Equation (16.12) we seek. The path of such a flow for an ideal fluid is shown in Figure 16.20. Remember, we are still using the ideal-fluid model. Look back at our assumptions if you have forgotten what this means!

To apply the conservation of energy theorem $W = \Delta K + \Delta U$ we first compute the work performed on the fluid to move it from the configuration at the lower left side of the pipe to that at the upper right in Figure 16.20. The fluid to the left of that shown exerts a pressure P_1 on cross-sectional area A_1. The resulting force is $F_1 = P_1 A_1$. This force moves the fluid through a distance Δx_1 and therefore does work $W_1 = F_1 \Delta x_1 = P_1 A_1 \Delta x_1$ on the fluid.

At the same time, an equivalent volume of fluid at the upper right end is displaced through Δx_2. However, the force $F_2 = P_2 A_2$ on this fluid volume is toward the left so the

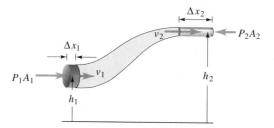

FIGURE 16.20

Ideal fluid flow through a constricted pipe. The fluid at height y_1 flows with velocity v_1 through area A_1 and at height y_2 it flows with velocity v_2 through area A_2.

work done on this fluid volume is $W_2 = -F_2 \, \Delta x_2 = -P_2 A_2 \, \Delta x_2$. The work is negative because the force is in the opposite direction from the displacement. Equivalently, work is performed *by* the fluid that moves through displacement Δx_2. The total work performed on the entire fluid element shown is

$$W = W_1 + W_2 = P_1 A_1 \, \Delta x_1 - P_2 A_2 \, \Delta x_2 \qquad (16.18)$$

As noted above, the left and right volume elements are equivalent:

$$A_1 \, \Delta x_1 = A_2 \, \Delta x_2 = \Delta \mathcal{V}$$

Equation (16.18) becomes

$$W = (P_1 - P_2) \, \Delta \mathcal{V} \qquad (16.19)$$

Next, let's investigate the kinetic and potential energy of the moving fluid element. If Δm is the mass of the fluid contained within volume $\Delta \mathcal{V}$, the change in its kinetic energy between the two ends is

$$\Delta K = \frac{1}{2} \Delta m \, v_2^2 - \frac{1}{2} \Delta m \, v_1^2 \qquad (16.20)$$

and the change in gravitational potential energy is

$$\Delta U = \Delta m \, g h_2 - \Delta m \, g h_1 \qquad (16.21)$$

When we substitute Equations (16.19), (16.20), and (16.21) into the energy conservation condition $W = \Delta K + \Delta U$ we have

$$(P_1 - P_2) \, \Delta \mathcal{V} = \left(\frac{1}{2} \Delta m \, v_2^2 - \frac{1}{2} \Delta m \, v_1^2 \right) + (\Delta m \, g h_2 - \Delta m \, g h_1)$$

We divide this equation through by $\Delta \mathcal{V}$ and recognize the density as $\rho = \Delta m / \Delta \mathcal{V}$. Rearranging, we find

$$P_1 + \frac{1}{2}\rho v_1^2 + \rho g h_1 = P_2 + \frac{1}{2}\rho v_2^2 + \rho g h_2 \qquad (16.22)$$

or

$$P + \frac{1}{2}\rho v^2 + \rho g h = \text{constant}$$

Equation (16.22) is called **Bernoulli's equation.** The next two examples illustrate how the equation of continuity and Bernoulli's equation may be applied to fluids in motion. In the paragraphs following these examples we provide some qualitative illustrations that can be understood from Bernoulli's equation.

EXAMPLE 16.7 *Garden Physics*

A garden hose has an inside cross-sectional area of 3.60 cm², and the opening in the nozzle is 0.250 cm². If the water velocity is 50.0 cm/s in a segment of the hose that lies on the ground (a) with what velocity does the water leave the nozzle when it is held 1.50 m above the ground, and (b) what is the water pressure in the hose on the ground?

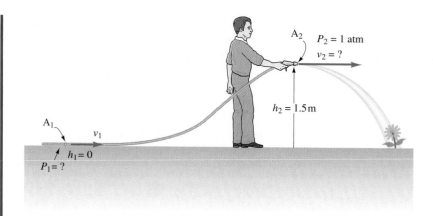

FIGURE 16.21

Application of the equation of continuity and Bernoulli's equation.

SOLUTION   (a) The problem is illustrated in Figure 16.21. We first apply the equation of continuity, Equation (16.17), to find the velocity of the fluid at the nozzle:

$$v_2 = \frac{A_1}{A_2}v_1 = \left(\frac{3.60 \text{ cm}^2}{0.250 \text{ cm}^2}\right)(50.0 \text{ cm/s}) = 720. \text{ cm/s} = 7.20 \text{ m/s}$$

(b) We next apply Bernoulli's equation to find the pressure P_1. From Figure 16.21 we see the height $h_1 = 0.00$ m and $h_2 = 1.50$ m. The pressure at the nozzle is atmospheric pressure $P_2 = 1.01 \times 10^5$ Pa. Solving for P_1 and using the density of water $\rho = 1.00 \times 10^3$ kg/m^3, we have

$$P_1 = P_2 + \frac{1}{2}\rho(v_2^2 - v_1^2) + \rho g(h_2 - h_1)$$

$$= (1.01 \times 10^5 \text{ Pa}) + \frac{1}{2}(1.00 \times 10^3 \text{ kg/m}^3)[(7.20 \text{ m/s})^2 - (0.50 \text{ m/s})^2]$$

$$+ (1.00 \times 10^3 \text{ kg/m}^3)(9.80 \text{ m/s}^2)(1.50 \text{ m} - 0.00 \text{ m})$$

$$= 1.41 \times 10^5 \text{ Pa} \qquad \blacktriangleleft$$

EXAMPLE 16.8   *The Pitot Tube*

The Pitot tube shown in Figure 16.22 is a device used to measure the velocity of moving fluids. Determine the velocity v of the fluid in terms of its density ρ_f, the density of the fluid in the manometer ρ, and the height h.

SOLUTION   Let's begin by stating Bernoulli's law:

$$P_1 + \frac{1}{2}\rho v_1^2 + \rho g h_1 = P_2 + \frac{1}{2}\rho v_2^2 + \rho g h_2$$

We apply the left side of this equation to the fluid flowing past the Pitot tube and use the right side for the gas within the tube. The average height of the fluid flowing past the openings at points A is the same as the height of the point B. Therefore, $h_1 = h_2$, and Bernoulli's equation becomes

$$P_1 + \frac{1}{2}\rho v_1^2 = P_2 + \frac{1}{2}\rho v_2^2$$

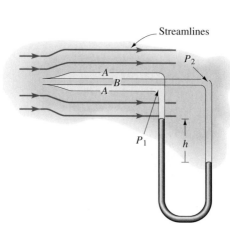

FIGURE 16.22

The Pitot tube.

The openings at points A are located far enough along the tube that the pressure in the moving fluid at these points is the same as that on the other side of these openings, within the Pitot tube. Therefore, the pressure at the left side of the manometer is the same as the fluid pressure P_f; that is, $P_1 = P_f$. The velocity v_1 is the fluid velocity v_f. In equilibrium, the velocity v_2 of the fluid at point B is zero, and the pressure on the right side of the manometer is P_2, the so-called stagnation pressure. Thus,

$$P_f + \frac{1}{2}\rho_f v_f^2 = P_2 \qquad (16.23)$$

Because ρ is the density of the manometer fluid, we can apply Equation (16.12) to find a relation between P_1 and P_2:

$$P_2 = P_1 + \rho gh$$

or

$$P_2 = P_f + \rho gh$$

Substituting this result into Equation (16.23), we have

$$\frac{1}{2}\rho_f v_f^2 = \rho gh$$

or

$$v_f = \sqrt{\frac{2\rho gh}{\rho_f}}$$

Often, Pitot tubes are calibrated for specific fluids with densities ρ_f (air, for example) so that the fluid velocity may be read directly from the manometer. ◀

EXAMPLE 16.9   Torricelli's Law

A tank filled to a height y_2 with a fluid of density ρ has a small hole in its side at a height y_1 (Fig. 16.23). If the pressure at the top of the fluid is P_t, determine the velocity with which the fluid leaves the hole. Assume the cross-sectional area of the tank is very large compared to that of the hole.

SOLUTION   We can apply the equation of continuity, Equation (16.17), to discover what the assumption $A_1 \ll A_2$ means. Solving Equation (16.17) for v_2, we have

$$v_2 = \frac{A_1}{A_2}v_1$$

Clearly, because $A_1 \ll A_2$, v_2 must be very small compared to the velocity we seek, v_1. Our approximation is, therefore, $v_2 \approx 0$.

The fluid emerging from the hole is at atmospheric pressure P_o. Applying the left and right sides of Bernoulli's equation to the hole and the top of the fluid, respectively, we have from Equation (16.22),

$$P_o + \frac{1}{2}\rho v_1^2 + \rho gy_1 = P_t + \rho gy_2$$

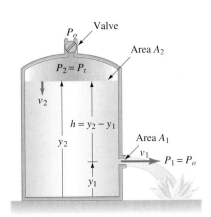

FIGURE 16.23

Torricelli's law.

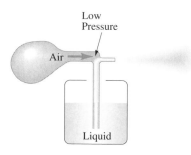

FIGURE 16.24

An atomizer.

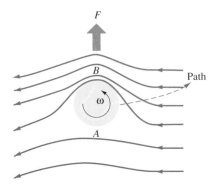

FIGURE 16.25

Part of the reason a rotating baseball veers toward side B is because the pressure on side A is greater than that on side B. Momentum conservation also plays an important role. Notice that the streamlines indicate a change in the air direction after it flows past the ball.

Concept Question 6
Why do golf balls have dimpled surfaces?

Concept Question 7
Airplane wings are designed so that air flows slower under the wing than over it. Use Bernoulli's law to explain how this design causes a net upward force on the wing.

Concept Question 8
Explain why, as you drive along a highway, you might feel your car pulled sideways toward a passing truck.

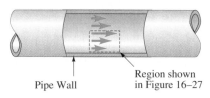

FIGURE 16.26

Flow of a real fluid through a pipe. The fluid velocity increases with perpendicular distance from the pipe wall.

Solving for v_1, we obtain

$$v_1 = \sqrt{\frac{2(P_t - P_o)}{\rho} + 2gh}$$

When the valve at the top of the tank is opened so that the pressure $P_t = P_o$, the velocity is

$$v_1 = \sqrt{2gh}$$

This result is sometimes called **Torricelli's law.**

On the other hand, if a pump is attached to the open valve so that the pressure P_t is made large enough that $2(P_t - P_o)/\rho \gg \rho gh$, then $v_1 \approx \sqrt{2(P_t - P_o)/\rho}$. That is, the velocity of the emerging water depends primarily on the magnitude of the applied pressure P_t. ◄

Qualitative Applications of Bernoulli's Equation

Atomizers are used in perfume dispensers and bug sprays and even, on rare occasions, to paint college dormitories. A simple atomizer consists of a vertical tube extending into the liquid to be dispensed, and a mechanism by which a stream of air can be made to pass over the top of the tube (Fig. 16.24). You may be surprised to learn that for modest pressures and flow velocities even air can be successfully modeled as an incompressible fluid. From Bernoulli's equation we know that the pressure at the tube top must decrease because the velocity has increased. (That is, for a constant h, $P + \frac{1}{2}\rho v^2 + \rho gh = $ constant.) The excess pressure on the fluid surface within the container forces it up the tube where it mixes with the airstream.

It probably won't help your curve ball to know how it works, but Bernoulli's equation helps to explain what makes the ball veer from a straight trajectory. Figure 16.25 shows streamlines around a rotating baseball. The ball's rough surface and stitching cause a small layer of air close to its surface to rotate with the ball. Therefore, the air velocity on side A is lower than that on side B, and Bernoulli's law tells us that the pressure on side A is greater than that on side B. This pressure difference contributes a net force on the ball so that it curves along the path shown. Also notice that the second effect of the ball's rotation is to deflect the air in a direction opposite to the ball's deflection. Thus, momentum conservation also plays a significant role in causing the ball to veer.

There are numerous other phenomena for which we can at least obtain a qualitative understanding from Bernoulli's law. Try your hand at an explanation of how a chimney works, or perhaps why gophers build a mound around one of the two entrances to their burrows.

16.7 Dynamic Viscosity (Optional)

Although fluids cannot support a shear stress, they do offer some degree of resistance to a shear type of deformation. This resistance, called dynamic **viscosity,** causes some of the mechanical energy of the fluid flow to be dissipated. Figure 16.26 shows the velocity vectors for a viscous liquid flowing through a pipe. The fluid close to the pipe wall moves slower than that near the center of the pipe. Thus, there is a difference in the rate of fluid displacement as layers of the fluid slide past each other. You can think of viscosity as arising from internal friction between these infinitesimally separated layers within the fluid.

Let's carefully examine a small flow region of the viscous fluid shown in Figure 16.26. Figure 16.27 shows such a region. The blue arrows in this figure represent fluid *displacement* during a small time Δt. The fluid in the layer immediately above the region shown in Figure 16.27 produces a shear stress F/A on an area A of the fluid. The shear

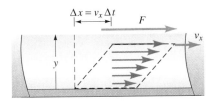

FIGURE 16.27

The definition of strain in a viscous fluid. The flow of the fluid layer above that shown causes a shear stress on the fluid. The blue vectors represent fluid displacement.

strain is defined just as it was in Section 16.2, namely, the lateral displacement Δx per perpendicular distance y.

$$Shear\ strain = \frac{\Delta x}{y}$$

However, because the fluid at the top has a velocity v_x,

$$\Delta x = v_x\, \Delta t$$

and

$$Shear\ strain = \frac{v_x\, \Delta t}{y}$$

The *shear strain per unit time* is

$$\frac{Shear\ strain}{\Delta t} = \frac{v_x}{y}$$

That is, the time rate of change of the shear strain is v_x/y. Moreover, we expect this rate of change to be proportional to the shear stress:

$$\frac{v_x}{y} \propto \frac{F}{A}$$

Just as in Section 16.2, the proportionality constant is written as the inverse of another constant, this time called the **coefficient of dynamic viscosity** η:

$$\frac{v_x}{y} = \frac{1}{\eta} \frac{F}{A} \tag{16.24}$$

or,

$$\eta = \frac{F/A}{v_x/y} \tag{16.25}$$

Now, it is probably no surprise to you that there is at least one additional way the fluid flow can be more complicated. Often, the fluid velocity v_x does not change *linearly* with the vertical distance y. That is, v_x/y may not be a constant. In this case v_x/y is replaced with the **velocity gradient** dv_x/dy, and the definition Equation (16.25) becomes

$$\eta = \frac{F/A}{dv_x/dy} \tag{16.26}$$

The dynamic viscosities of some common fluids are given in Table 16.4. A discussion of the motion of objects through a viscous medium is given in Chapter 6, Section 4.

TABLE 16.4   Dynamic viscosities of some fluids

FLUID	TEMPERATURE (°C)	VISCOSITY η (10^{-3} N · s/m^2)
Air	24	0.018
Water	24	0.96
Water	100	0.3
Glycerine	24	950
Mercury	24	1.5
Gasoline	24	0.29
Motor oil (10W-30)	24	100
Crude oil	24	7

16.8 Summary

The stress is the force per area acting on a material, and the strain is a measure of the material's deformation. The elastic moduli are defined from the stress–strain relation

$$Strain = \frac{1}{Modulus} \, Stress \tag{16.4}$$

Young's modulus is a measure of the material's resistance to a change in length under a tensional or compressional force. The **shear modulus** is a measure of the material's resistance to a deformation caused by a force parallel to a plane defined within the material. A material's resistance to a change in its volume caused by confining forces is characterized by the **bulk modulus.**

The **density** of a homogeneous substance with mass M occupying volume $\mathcal{V}$ is

$$\rho = \frac{M}{\mathcal{V}}$$

The **pressure** P is the force per area: $P = F/A$. In SI, the unit of pressure is N/m^2 and 1 N/m^2 = 1 Pa (pascal). In the **incompressible-fluid model** the pressure of a static fluid increases with depth h in a fluid of density ρ according to the equation

$$P = P_o + \rho gh \tag{16.12}$$

where P_o is atmospheric pressure at the fluid surface.

Pascal's principle states that when a pressure is applied to a confined fluid at rest, the pressure is transmitted throughout that fluid. An object submerged in a fluid appears to weigh less because of an upward **buoyant force** on the object. **Archimedes' principle** states that the buoyant force acting on a submerged object is equal to the weight of the fluid displaced.

In the **ideal-fluid model** the fluid is incompressible, nonviscous, and has a steady flow with no turbulence. Two important results obtained from this model are the **equation of continuity**

$$v_1 A_1 = v_2 A_2 \tag{16.17}$$

and **Bernoulli's equation**

$$P + \frac{1}{2}\rho v^2 + \rho gh = \text{constant}$$

The **dynamic viscosity** of a substance is a measure of its resistance to dynamic shear.

PROBLEMS

16.2 Stress, Strain, and the Elastic Moduli

1. A 250.-N force is applied to one end of a 6.00-m long, 3.00-mm diameter aluminum wire that is fixed at the other end. Compute (a) the stress in the wire and (b) the change in its length.
2. A copper wire 3.00 m in length has a cross-sectional area of 2.50 mm². When one end is fixed, what mass must be hung from the other end of the wire to change its length by 1.00 mm?
3. Power companies routinely apply a 2670-N tension to steel support cables they string between power poles. If the original length of such a cable is 34.0 m and the cable is 1.60 cm in diameter, by how much does the cable stretch?
4. A steel pipe is used as a post to support part of a building. The inner and outer diameters of the pipe are 10.2 cm and 12.1 cm, respectively. If the unloaded length of the pipe is 2.75 m, what is the length of the pipe when a compressional load of 1.25×10^4 N is applied to it?
5. Show that the work necessary to stretch a wire of Young's modulus Y by an amount ΔL is $YA(\Delta L)^2/(2L_o)$, where L_o is the original length of the wire and A is its cross-sectional area.
6. A tractor is attached by a 2.00-m steel cable to a broken-down piece of machinery. When the tractor begins to pull, it applies an initial 1650-N force to the cable before the machine begins to move. If the diameter of the cable is 1.27 cm, find the maximum amount the cable stretches.
7. One end of a 3.00-m long copper wire of 2.00-mm² cross-sectional area is bonded to the end of a 2.00-m long aluminum wire with a cross-sectional area of 1.50 mm². What tension must be applied to this 5.00-m combination to change the total length by 1.00 cm?
8. The bottom face of a copper cube 10.0 cm along an edge is fixed in place while a shearing force of 2.20×10^3 N is applied to the top face. (a) What is the shear strain of the cube? (b) Through what horizontal distance is the sheared face displaced?
9. A toy elastic block is 8.00 cm on an edge. As shown in Figure 16.P1 when a force of 12.5 lbf is applied across the top of the block, its side tilts at an angle of 6.00°. Find the shear modulus of the block.

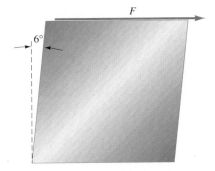

FIGURE 16.P1   Problem 9

10. A solid brass sphere is submerged in the ocean to a depth where the pressure is 1.80×10^7 N/m². What is the percentage change in the sphere's radius?

16.3 Density and Pressure

11. A solid sphere, 6.00 cm in diameter, has a 1.01-kg mass. Identify the substance from Table 16.3.

12. One end of a string is tied to the ceiling. (a) What volume of lead must be hung on the other end if the tension in the string is to be 44.0 N? (b) What volume of aluminum must be hung in addition to the lead to increase the tension to 50.0 N?
13. Neutron stars are believed to have densities so high that only neutrons can exist within them. (a) If a neutron star has a mass of 3.00×10^{28} kg and a radius of 10.0 km, what is its density? (b) What would be the weight of 1.00 cm³ of this material at the earth's surface?
14. A 45.0-kg gymnast balances on one hand. If the area of her hand is 38.0 cm², what average pressure does she exert on the floor?
15. The total area of the United States is 3.54×10^6 mi². Estimate the total force of the atmosphere on the United States.

16.4 Fluid Statics

16. Compute the total pressure at the bottom of a swimming pool 3.00 m in depth.
17. A certain ocean trench is 5.00 km deep. (a) What pressure does the constant-density model predict for this depth? (b) Is the actual pressure larger or smaller than the value you found in part (a)? Explain.
18. Mercury is poured into a beaker until its depth is 20.0 cm. (a) What is the gauge pressure of the mercury at the bottom of the beaker? (b) How tall of a beaker of water is necessary to achieve the same pressure at the bottom?
19. If the earth's atmosphere were of constant density (1.30 kg/m³), to what height must it extend for the pressure at the earth's surface to be 1.00 atm?
20. The water in a swimming pool is 2.50 m deep, and the flat-bottomed pool is 8.00 m long and 6.00 m wide. (a) Find the total force on the pool bottom due only to the water in the pool. (b) Find the total force on the side of the pool that is 8.00 m long. (c) What is the total force on the 6.00-m long side?
21. When the manometer shown in Figure 16.15 is suddenly detached from the pressure source, the mercury level is observed to rise and fall in simple harmonic motion. If the total length of the mercury in the tube is L, show that the frequency of the oscillations is $1/\pi \sqrt{g/2L}$.
22. Two liquids that do not mix are poured into a U-shaped tube as shown in Figure 16.P2. Show that the difference in fluid heights is $H = (\rho_1 - \rho_2)h/\rho_2$.

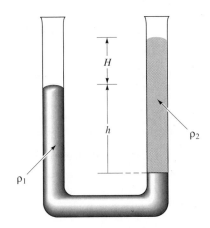

FIGURE 16.P2   Problem 22

23. Two liquids that do not mix have densities ρ_1 and ρ_2. They are poured into a cylindrical container such that their respective heights are h_1 and h_2. If P_o is the atmospheric pressure at the top of the cylinder, prove that the pressure at the bottom is $P_o + \rho_1 g h_1 + \rho_2 g h_2$.

24. In the ideal-gas model (see Chapter 18) the density of a gas is given by the expression $\rho = P/RT$, where P is the gas pressure, R is a constant, and T is the temperature (in kelvins). If at altitude y_o the pressure is P_o, show from Equation (16.11) that the pressure at altitude y is

$$P = P_o \exp\left[-\left(\frac{g}{RT}\right)(y - y_o)\right]$$

This expression is sometimes used to compute the pressure variation in the stratosphere.

25. Figure 16.P3 shows a hydraulic apparatus with pistons initially at the same level. An unstretched spring ($k = 285$ N/cm) is attached to the smaller piston, and a 70.0-kg mass is placed on the other. How much is the spring compressed?

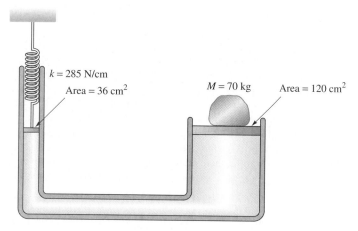

$k = 285$ N/cm
Area = 36 cm^2
$M = 70$ kg
Area = 120 cm^2

FIGURE 16.P3 Problem 25

26. One side of a U-shaped container similar to that shown in Figure 16.11 has a small piston of cross-sectional area equal to 30.0 cm^2. The area of the other piston is 120. cm^2. (a) If no energy is lost to friction, how much work is performed by a 200.-N force when it moves the small piston downward through a distance of 75.0 cm? (b) Through what distance is the large piston moved?

27. (a) Compute the buoyant force when a chunk of aluminum with volume 0.250 m^3 is submerged in water. (b) What volume of lead experiences this same buoyant force when it is submerged in water?

28. A weight lifter is preparing to hoist a 110.0-kg set of iron weights. How much help can he expect to receive from the atmosphere? That is, what is the buoyant force of the atmosphere on this volume of iron?

29. A cylindrical buoy 2.00 m in length and 0.750 m in diameter is constructed of a material with a specific gravity of 0.600. One end of a light cord is attached to the bottom of the buoy, and the other is fixed to an anchor that rests at the bottom of the waterway. If only 0.400 m of the buoy remains above water, what is the tension in the cord?

30. A hot-air balloon has a volume of 2.40×10^3 m^3. Its hot air has a density of 0.950 kg/m^3, whereas that of the surrounding air is 1.29 kg/m^3. What is the maximum load the balloon can lift (including its own weight)?

31. A solid object has a volume of 434 cm^3. It is attached by a string to a spring scale. When the object is submerged in water, the scale indicates an apparent weight of 7.23 N. Use Table 16.3 to identify the material from which the object is made. (There is a legend that Archimedes used this technique to determine if the king's crown was really gold!)

32. If seawater has a specific gravity of 1.03 and that of ice is 0.920, show that about 89% of an iceberg is below the surface.

33. When a cube of wood floats in water, 60.0% of its volume is submerged. When the same cube floats in an unknown fluid, 85.0% of its volume is submerged. Find the densities of the wood and the unknown fluid.

34. An air-filled beach ball is 46.0 cm in diameter and has a negligible mass. (a) What mass of lead must be attached to the bottom of the ball for half of it to be submerged in water? (b) If the lead could be "balanced" on top of the ball, what mass of lead would be required to again half submerge the ball?

35. A detective wants to determine if a piece of lead contains an empty cavity. In air it weighs 55.4 N, and when it is submerged in water its apparent weight is 48.0 N. Does the lead object contain a cavity? If so, what is the cavity's volume?

16.6 Fluid Dynamics

36. Water flows through a horizontal pipe of radius r_o with a velocity of 3.60 m/s and a pressure of 2.50×10^5 Pa. Find the velocity and pressure within the pipe at a point where the diameter is $2r_o$.

37. The water tunnel through a dam has a circular opening with a diameter of 2.00 m at a depth of 20.0 m below the water's surface. The exit end of the tunnel is 50.0 m below the water's surface and has a 1.00-m diameter. (a) With what velocity does the water exit the tunnel? (b) What is the water pressure as it enters the tunnel?

38. Water enters a building at street level through a 10.0-cm diameter pipe. It rises through pipes to a height of 30.0 m, where it emerges with a velocity of 1.50 m/s from a faucet 1.25 cm in diameter. Assume this faucet is the only one turned on within the building. (a) What is the water's velocity at street level? (b) What is the pressure of the water at street level?

39. A water storage tank has the shape of a sphere 5.00 m in radius. As shown in Figure 16.P4 its center is 10.0 m above the ground.

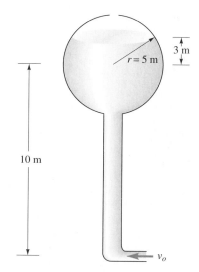

$r = 5$ m
3 m
10 m
v_o

FIGURE 16.P4 Problem 39

The vertical pipe has a radius of 0.25 m. A pump at ground level delivers water at a velocity $v_o = 16.0$ m/s up the vertical supply pipe. (a) What is the speed of the water at the surface when this surface is 3.00 m above the tank's center? (b) What is the pressure of the water at the bottom of the vertical pipe?

40. The funnel-shaped reservoir shown in Figure 16.P5 is being filled with water from the bottom. The top diameter $D = 10.0$ m and at the bottom diameter $d = 2.00$ m. The height H of the funnel is 8.00 m. Just as the tank is filled the water is entering the bottom at a rate of 6.00 m³/s. At this instant: (a) What is the velocity of the water entering the reservoir? (b) What is the velocity of the water at the top of the reservoir? (c) What is the pressure at the bottom of the reservoir?

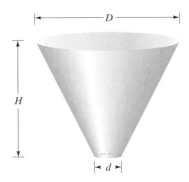

FIGURE 16.P5   Problem 40

41. A sliding piston forces water out of the tube-shaped container shown in Figure 16.P6. The diameters d_1 and d_2 are 8.00 cm and 2.00 cm, respectively. If $v = 8.00$ m/s, (a) find the velocity of the piston. (b) What force F is applied to the piston? Assume the water leaving the container enters air at 1.00 atm of pressure.

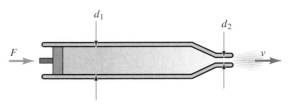

FIGURE 16.P6   Problem 41

42. The tank shown in Figure 16.23 contains water, and the opening is 12.0 m below the water's surface. The valve is opened so that the pressure at the top of the tank is one atmosphere. Water from the hole is observed to fill a 0.500-m³ tank in 30.0 min. (a) With what speed does the water leave the hole? (b) What is the cross-sectional area of the hole?

43. Water emerging from the hole in the tank shown in Figure 16.23 hits the ground 2.00 m below the opening and 1.30 m away (horizontally) from the tank. (a) What is the water height above the hole? (b) If the diameter of the hole is 2.00 cm, how long does it take to fill a 0.500-m³ tank placed in the path of the water stream?

44. A Pitot tube is used to measure the speed of air in a wind tunnel. The Pitot tube manometer is filled with alcohol of 0.800×10^3 kg/m³ density. If the level difference of the alcohol in the two arms of the manometer is 4.20 cm, what is the wind velocity?

45. A light aircraft has a total wing area of 16.5 m² and a mass of 1270 kg. As the plane flies horizontally, the wind speed *over* the wings is 56.0 m/s and the speed of the air under the wing is 52.0 m/s, compute the net upward force on the plane. Is this lift sufficient to raise the plane, or is some additional mechanism required?

Numerical Methods and Computer Applications

46. Write a program or develop a spreadsheet to compute and plot the variation of pressure in the stratosphere given in Problem 24. Use $P_o = 2.3 \times 10^4$ Pa at $y_o = 10^4$ m and $T = 220$ K ($= -53°C = -63°F$). Compute the pressure in the range from 10 000 m to 20 000 m.

47. Write a program or develop a spreadsheet to compute and plot the variation of pressure in the troposphere given in Problem 52. Use $P_o = 1.01 \times 10^5$ Pa at $y_o = 0$ m, $T_o = 293$ K ($= 20°C = 68°F$), $R = 287$ J/kg · K, and $a = 6.5 \times 10^{-3}$ K/m.

General Problems

48. A steel transition cable for the aileron-rudder system of a small aircraft is 30.0 in long with a diameter of 0.1875 in. Young's modulus for this wire is 30.0×10^6 lb/in². The maintenance manual states that the cable should be tensioned to 30.0 ± 10.0 lbf. (a) What are the minimum and maximum stresses permitted by these instructions? (b) What range is possible for the length changes for this cable if it is tensioned to the minimum and maximum limits?

49. Many materials exhibit unusual properties (such as changes in electrical or magnetic behavior or rearrangements of atomic structure) when subjected to large pressures. In some solid-state physics experiments it is not unusual to apply pressures of 1.00×10^9 N/m². If an aluminum cube 0.500 cm on a side is subjected to this pressure (a) what is its volume at this pressure? (b) What is the new length of a side of the cube?

50. The pressure-measuring instrument shown in Figure 16.6 is evacuated and immersed in a water reservoir so that the piston surface maintains an upward orientation. The cross-sectional area of the piston is 20.0 cm², and the spring constant is 2.40×10^4 N/m. With this orientation the spring is compressed 1.20 cm from its equilibrium length. When the device is inverted at this same depth so that the piston face is straight down, the spring stretches 0.500 mm from its previous position. What is the mass of the piston and the instrument's depth?

★51. Compute the net torque due to the water on the dam shown in Figure 16.9 about an axis that runs along the front, bottom edge of the dam.

52. In one model for the variation of air pressure in the troposphere the density of air is given by $\rho = P/RT$, where P is the pressure, R is a constant, and T is the temperature in kelvins (see Chapter 18). The temperature is given by the expression $T = T_o - a(y - y_o)$, where T_o is the temperature at an altitude at which the pressure is known, and a is a constant called the *lapse rate*. If at altitude y_o the pressure is P_o, show from Equation (16.11) that the pressure in the troposphere is given by

$$P = P_o\left(\frac{T_o - a(y - y_o)}{T_o}\right)^{(g/aR)}$$

53. Suppose the cylinder of Example 16.4 is pushed down a distance x and released. (a) Show that it exhibits simple harmonic motion. (b) Find the period of its oscillations.

54. A 300.-cm³ wooden block with specific gravity 0.700 is attached by a light string to a lead anchor. Both are placed in a barrel of water so that the anchor rests on the bottom and the block is completely submerged. The barrel is then placed on an elevator. (a) What is the tension in the string when the elevator ascends at a constant velocity? (b) What is the tension while the elevator is accelerating downward at 2.00 m/s²?

55. In one model (called isostasy) of the earth the continents are assumed to float on layers of molten rock below. Take the average specific gravity of the continents to be $s_c = 2.67$. Assume that they are supported hydrostatically by the underlying material with an average specific gravity $s_o = 3.27$. (a) If the continents are modeled as blocks of thickness t, show that the depth of the continent is $(s_c/s_o)t$. (b) If the average elevation of a continent were 3.00 km, what thickness would this model predict for that continent?

56. Water flows from an open faucet as shown in Figure 16.P7. The diameter of the faucet is 2.50 cm, and at a distance of 35.0 cm below the faucet the stream has a diameter of 1.50 cm. The flow rate is such as to fill a 1.00-L beaker in 15.0 s. (a) Find the velocity of the water as it leaves the faucet. (b) What is the velocity of the water 35.0 cm below the faucet?

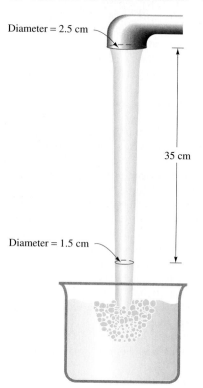

Diameter = 2.5 cm

35 cm

Diameter = 1.5 cm

FIGURE 16.P7   Problem 56

57. A small stream of water emerges from a very small hole in a very large, open tank as shown in Figure 16.P8. If the water rises to a height $y = 1.60$ m above the opening, how far below the water's surface is the opening?

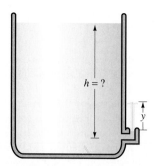

$h = ?$

y

FIGURE 16.P8   Problem 57

58. A Venturi tube, shown in Figure 16.P9, is a device used to measure the velocity of a fluid through a pipe. The cross-sectional area of the pipe at position a is A_1, and the area at b is A_2. The manometer may be used to find the pressure difference $P_a - P_b$ between points a and b. Show that the velocity at point a is

$$v_a = \sqrt{\frac{2(P_a - P_b)}{\rho[(A_1/A_2)^2 - 1]}}$$

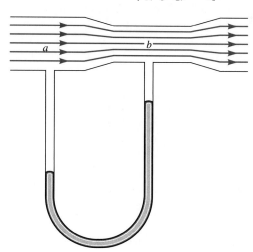

a b

FIGURE 16.P9   Problems 58 and 59

59. The manometer is removed from the constricted pipe shown in Figure 16.P9, and the pipe carries water. The diameter at point a is 4.00 cm, and the pressure at this point is 2.50×10^5 Pa. The diameter at point b is 1.00 cm. Compute the velocities at points a and b such that the pressure at b is zero. This phenomenon is called *cavitation,* and the water vaporizes to form bubbles at point b.

60. A *siphon* (Fig. 16.P10) is sometimes used to remove fluid from a container. The tube must initially be filled with fluid but then operates until the fluid level reaches the submerged end of the tube. Show that the speed of the fluid leaving the tube is $\sqrt{2g(h + y)}$.

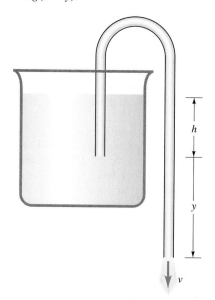

h

y

v

FIGURE 16.P10   Problem 60

61. The U-shaped tube shown in Figure 16.P11 rotates in still air with angular velocity ω about the vertical axis AB. Compute the difference in fluid level h in terms of ω, the radii r_1 and r_2, and the fluid density ρ.

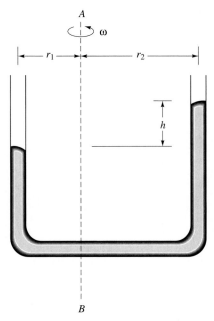

FIGURE 16.P11   Problem 61

★**62.** The fluid-filled cylindrical container shown in Figure 16.P12 is rotating with constant angular velocity ω. By computing the centripetal force due to the pressure difference on a small fluid element of width Δx, show that the fluid surface is parabolic and given by $y = (\omega^2/2\rho g)x^2 + y_o$, where ρ is the density of the fluid.

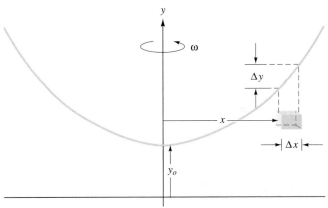

FIGURE 16.P12   Problem 62

GUEST ESSAY

The Continuity Equation in Everyday Life

by ALBERT A. BARTLETT

The continuity equation is a simple expression that represents conservation of matter. In steady-state flow, the number of kilograms of matter entering a pipe or similar container per second must equal the total number of kilograms per second that flow (or leak) out of the pipe or container each second.

Suppose a pipe has ends designated as 1 and 2, with cross-sectional areas A_1 and A_2. Imagine that matter flows in end 1 with a speed v_1 and density ρ_1 and that the matter flows out of end 2 with a speed v_2 and density ρ_2. The requirement that matter be conserved is expressed by equating the number of kilograms per second in at end 1 ($A_1 v_1 \rho_1$) to the number of kilograms per second that come out of end 2 ($A_2 v_2 \rho_2$). We describe this situation by saying that this flow conserves the quantity ($A v \rho$). Operationally, this means that

$$A_1 v_1 \rho_1 = A_2 v_2 \rho_2$$

If the pipe is rectangular in cross section, $A = hw$, then the flow conserves the quantity ($hwv\rho$). If the matter is incompressible (as is approximately the case with many liquids), then the density ρ does not change in the flow so the conserved quantity is Av or hwv.

The Flow of Persons or Vehicles

If 82 persons/min flow steadily into one end of a corridor, the law of conservation of persons leads us to expect that a total of 82 persons/min flow out of the various outlets from the corridor. Let us start by looking at the flow of persons through a door of height h and width w. We first note two things about the relation of h and w to the flow of persons through the door. Let us restrict our discussion to cases in which $w > 1$ m and $h > 2$ m. Thus, we exclude from our consideration widths so small that people can not get through the door, and we limit

ALBERT A. BARTLETT
University of Colorado, Boulder

Albert A. Bartlett is a retired Professor of Physics at the University of Colorado in Boulder. He has a B.A. in physics from Colgate University (1944) and a Ph.D. in nuclear physics from Harvard University (1951). He has been a member of the University of Colorado faculty since 1950. His interests center on teaching physics and on educating people about the arithmetic of growth. He has lectured on this subject over a thousand times.

He was president of the American Association of Physics Teachers in 1978. In 1991 he received the association's Melba Newill Phillips Award for his contributions to physics education.

our consideration to a range of widths within which increasing w increases the number of persons per second that can flow through the door. Thus, w continues to play a part in the continuity equation. In contrast, increasing h beyond 2 m does not increase the rate of flow of persons through the door so that for $h > 2$ m the height of the door plays no role in the door's ability to transmit people. As a consequence, h is not needed in the continuity equation. So, with these restrictions on h and w, the conserved quantity in the flow is the product ($wv\rho$), and the continuity equation becomes

$$w_1 v_1 \rho_1 = w_2 v_2 \rho_2$$

Qualitatively, the equation indicates that where the width of a roadway, door, or sidewalk is large, the speed and density of the flow of persons or vehicles is low, and vice versa. Let us look now at some examples of the application of this equation.

Checkout Lines at a Large Store or Supermarket

Let us imagine a store that is operating with a capacity crowd of customers. (See Figure 1.) As they prepare to leave the store, the customers have to wait in line at one of a dozen or more checkout lines. These lines we call situation 1. The customers then flow out through the front door of the store, which is situation 2. In situation 1 the customers are waiting in line and therefore their speed v_1 is very low; consequently their density ρ_1 and flow width w_1 must be very large. This necessarily large width w_1 is achieved by having as many as a dozen cash registers in parallel giving a large width w_1 to the flow. As soon as the customers leave the cash registers, their speed increases as they head for the front door, where $v_2 \gg v_1$. The large value of v_2 means that the product $w_2 \rho_2$ can be small. One or two ordinary doors are usually enough to accommodate the output of a dozen checkout lines.

Toll Gates on Highways

Let us look at two lanes of a highway ($w_1 = 2$) that is crowded to capacity with cars all traveling in the same direction at the speed limit ($v_1 =$

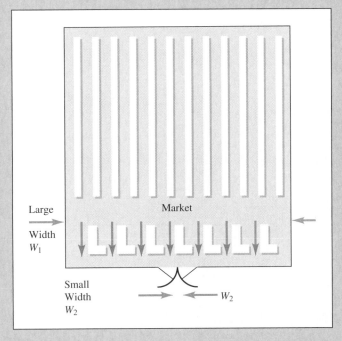

FIGURE 1

Checkout lines at a supermarket. Checkout speed is low: $V_1 \approx 0$; checkout density is high. As a consequence, w_1 is large.

25 m/s). At this speed, the cars are spaced about 10–15 car lengths apart in each lane, so the density ρ_1 is low. When the cars approach a toll gate, they must slow down to a very low speed (v_2 can be approximately 1 m/s) to pay the toll. The cars then are bumper to bumper, so that the density ρ_2 is approximately ten times ρ_1. We can use these values to estimate the width w_2 that is needed at the toll gate.

$$w_2 \approx w_1 \left(\frac{v_1}{v_2} \right) \left(\frac{\rho_1}{\rho_2} \right)$$

$$= (2 \text{ lanes}) \left(\frac{25}{1} \right) \left(\frac{1}{10} \right) \approx 5 \text{ lanes}$$

Thus, the two-lane highway must increase in width to about five lanes with five parallel toll gates in which the cars slow to pay the toll. (See Figure 2.)

Corridors and Stairs

Imagine a corridor of width w_1 crowded with people so that ρ_1 is large. The people are moving with a speed v_1 toward a staircase of width w_2 that is at the end of the corridor. In

Indeed, much of what is described here might be thought of as "common sense."

this crowded situation, the people are approximately incompressible, so ρ_2 on the staircase is approximately the same as ρ_1 in the corridor. Because people move more slowly on stairs than they do in a level corridor, v_2 tends to be less than v_1. These observations, along with the continuity equation, tell us that w_2 should be greater than w_1. If the two widths are the same, then one has a jam, and the incompressibility of the people means that v_1 in the corridor is forcibly reduced so it has the same value as v_2. If, as is often observed, w_2 is less than w_1, then v_2 is reduced to a value less than v_1.

■■■■ FIGURE 2

Toll plaza at the San Francisco Bay Bridge. This large toll plaza is an excellent example of the continuity equation. Where car velocities are small (stopping to pay tolls), the width of the roadway is very large. Where the velocity is high (before or after the toll plaza), the width of the roadway is reduced.

escalator to move quickly away from the end and to spread out so that jams are avoided.

Railroad Yards

When a freight train comes into the area of a railroad junction with a speed v_1, it must stop so that its cars can be separated and sorted into new trains that go to various destinations. The average speed v_2 of the cars in the sorting process is very low. The densities of the cars (cars per unit length of track) in the sorting process is only a little less than the density of the cars in the incoming train. The continuity equation tells us the consequence of this reduction in velocity; w_2 must be much greater than w_1. One can see this in a railroad classification yard where the trains arriving on a single track ($w_1 = 1$) are broken up into groups of cars that are distributed over a couple of dozen tracks ($w_2 = 24$) and the cars are assembled into new trains going to several new destinations. (See Figure 3.)

■■■■ FIGURE 3

Bailey Yard, Union Pacific Railroad, North Platte, Nebraska. Single lead tracks branch out into multiple parallel assembly tracks which then converge to a few outgoing tracks.

However, in some instances we see staircases that are much wider than the corridors to which they connect. The monumental outside staircases at the entrances of large public buildings are often many times as wide as the entry doors at the top of the stairs. These wide stairs can accommodate comfortably all of the people who move rapidly through the doors and who then move slowly down the stairs.

A related problem is seen at the exit ends of escalators. The escalators deliver a constant stream of people, but for the system to work smoothly the people leaving the escalator must move away from the end with at least the same speed as that of the escalator that delivered them. However, one often sees people step off the escalator and stop to look around or to orient themselves. This stopping would require a very large w, but generally this increased width is not available. When a crowded escalator is delivering a steady stream of people, the continuity equation indicates that it is necessary for the people stepping off the

Merging Highway Traffic

We can see interesting examples of applications of the continuity equation as the morning rush-hour traffic from many on ramps merges into the traffic on multilane arterial roads. Very commonly the traffic from an on ramp has only a very short distance along the arterial in which to speed up so that it can merge smoothly with the traffic on the arterial. The merging traffic generally does not cause the arterial traffic to lower its speed. The width of the arterial generally does not increase at every on ramp, so the result of the added traffic from the on ramp is to increase the traffic density on the arterial. This is the origin of traffic jams. When the traffic density becomes too high, some drivers tend to slow down, thereby increasing the severity of the jams.

However, sometimes a sign on the on ramp says "Continuous Lane Ahead." This means that the w of the arterial is being increased by one lane so that the arterial traffic and the merging traffic can all maintain their speeds with no major change in the traffic density. The reader can see that it is very convenient for the driver to have a continuous lane ahead. However, it is not possible to add very many lanes to the arterials that enter a city. For example, if a two-lane freeway has 30 on ramps as it enters a large metropolitan area and if each of the 30 has a continuous lane ahead, then the downtown freeway would be an expensive, awkward, and an improbable 32 lanes wide.

Conclusion

The reader can see that the continuity equation has many applications in our everyday lives. Indeed, much of what is described here might be thought of as "common sense." Yet it is important that we see these common sense things as being applications of a very important equation, because when we understand the connection between the equation and the things that we see, we can see how to calculate widths, speeds, and densities in new cases. Yet, in spite of the common sense aspect of the continuity equation, we often see cases where designs reflect no understanding at all of this important fundamental relation.

CHAPTER

17

Sound

In this chapter you should learn

- about the nature of sound waves.
- what factors determine the speed of sound waves.
- how to determine the intensity and sound intensity level of sound waves.
- the basic physics of string and wind musical instruments.
- how the frequency of sound waves appears to change if the source, observer, or both are in motion.

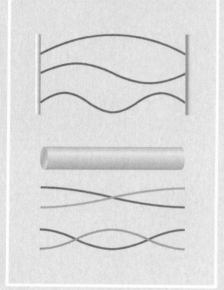

$$f = \frac{n}{2\,l} \sqrt{\frac{F}{\mu}}$$

$$f = n\,\frac{v}{2\,l}$$

After an earthquake, seismic waves can travel through the earth, reaching a quarter of the way around the world in less than an hour. A submarine can locate unseen underwater objects by ''pinging'' with its sonar and listening for the returning echoes. Grasp a long metal rod in the middle between your index finger and thumb, then rap the end of the rod smartly on the floor. The rod vibrates, sometimes for minutes, with a surprisingly pure tone. You know the rod is vibrating because you hear sound waves in air created by the rod's vibrations.

All of the phenomena described in the previous paragraph are examples of distortion waves propagating in elastic media. In this chapter we discuss primarily sound waves in a gas. However, many of the phenomena we describe also hold true for distortion waves in other media, and we point out similarities and differences with other types of waves from time to time.

17.1 Models for Sound Waves in a Gas

Before we can describe sound waves in a gas, we must first make a few comments regarding the models used to describe the gas itself. One model is called the **kinetic theory of gases** and is described in detail in Chapter 21. For now, it is sufficient to remind you that in a cubic meter of gas, such as nitrogen, at standard temperature and pressure, there are an enormous number of molecules (about 26×10^{24}). However, the molecules themselves are so small that, if they are condensed into a liquid, they would fill only $1/640th$ of a cubic meter and easily fit into two soft drink cans. Thus, most of the space ''occupied'' by a gas is in fact empty. The gas pressure (and ''springiness'' we feel from blown-up balloons) is due to the constant bombardment of surfaces by billions and billions of molecules each second. This springiness suggests a second model in which we describe air as composed of small, but finite, regions that we call *mass elements*. We imagine these mass elements to be compressible cells that expand and contract as the sound wave passes. The average position of these cells, however, does not change. Please be alert to the fact that we alternate between the particle model and the cell model as necessary to emphasize different features of sound waves.

With the particle model in mind, let's now apply the ''snapshot'' view of a wave that was useful in Chapter 15, this time to visualize a sound wave at one particular instant. In this frozen moment, a sound wave appears to consist of alternating regions where the air molecules are compressed slightly closer together than their average equilibrium distance and regions where they are slightly more separated than in equilibrium. These regions are called **compressions** and **rarefactions,** respectively, and are illustrated in Figure 17.1. In regions where the density is greater than average, the pressure is also above the normal value. Similarly, the pressure is below normal in the rarefaction regions.

Let's now consider the cell model in which we ignore the molecular motion and consider the gas to be made up of a large number of small, discrete mass elements. In the absence of a sound wave the pressure is the same at all points, and each mass element is at rest and occupies a volume $\mathcal{V}$. As a sound wave propagates in the gas, each element is set into motion and oscillates in some fashion about its rest (equilibrium) position. The pressure varies both with time and position, and as a result, the volume of each mass element varies with time. In gases (and liquids) the motion of the mass element is parallel to the direction of propagation, and for this reason sound waves are classified as longitudinal waves. If the motion of each mass element is simple harmonic, the wave is called a harmonic wave. When the mass elements in any plane at right angles to the direction of propagation of the wave are all in phase with each other, the wave is said to be a **plane wave.** This type of wave is of considerable importance, because at a large distance from almost any sound source the waves radiated by the source may be considered to be plane waves.

Another important type of sound wave is a **spherical wave.** If the surface of a sphere is made to expand and contract periodically, sound energy is radiated uniformly in all directions (Fig. 17.2). As the disturbance created by the motion of the sphere propagates outward, the mass elements of air are set in motion, each vibrating along a radial direction. Because the disturbance arrives simultaneously at all mass elements located the same

FIGURE 17.1

Sound waves consist of alternating regions of compressions and rarefactions.

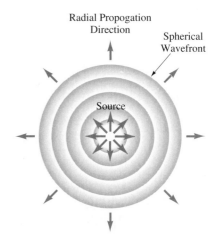

FIGURE 17.2

Spherical sound waves emitted from a point source.

Plane Wavefronts

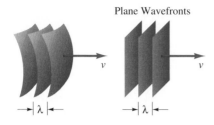

FIGURE 17.3

At large distances from the source, spherical wavefronts approximate plane waves.

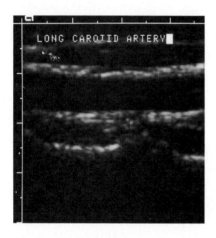

An ultrasound image of a carotid artery. (Image courtesy of Charles Cyr.)

distance from the center of the sphere, it should be evident that the motions of all the elements located on the same spherical surface are in phase with each other. Such surfaces are called **wavefronts.** As shown in Figure 17.3, at large distances from a spherically symmetrical source, the spherical wavefronts approximate plane wavefronts.

Our ears are sensitive to both the amplitude and the frequency of sound waves. We recognize sound waves with different frequencies as having different pitches; frequencies between about 20 Hz and 20 kHz are audible to the human ear. Sound waves with frequencies below 20 Hz are called **infrasonic** waves, and those above 20 kHz are called **ultrasonic** waves. Infrasonic waves are created by earthquakes, volcanic eruptions, ocean waves, and even some animals, such as whales and elephants. Ultrasonic waves are created by and audible to many animals, including bats, porpoises, birds, and insects. The short wavelengths of ultrasonic waves have made them useful for medical imaging. The short wavelength permits detailed features of different tissue types to be detected. Sensitive detectors allow small-amplitude ultrasound waves to be observed; the small amplitude prevents tissue damage.

Frequency of audible ranges for various sources.

17.2 The Velocity of Sound

Mathematically, the analysis of plane waves and spherical waves is somewhat more complex than that of one-dimensional waves. Before tackling three-dimensional waves, we invent a simple model by which we can take a one-dimensional look at a sound wave. Later, when we consider how the intensity of a sound wave varies with distance from the source, we necessarily return to the three-dimensional nature of the wave.

One way to generate sound waves is to use a speaker from your stereo. In order to find an expression for the speed of sound in a gas, such as air, let's consider a very crude speaker system. Suppose we have a piston that can oscillate back and forth in an air-filled tube, thereby generating the compressions and rarefactions of a sound wave. Such a system is shown in Figure 17.4. Obviously, this device is not the speaker system you would consider purchasing for your stereo, but it helps us understand what factors are important for determining the velocity of sound.

Piston motion

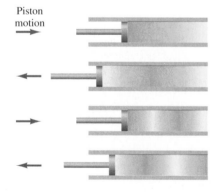

FIGURE 17.4

An oscillating piston sends a sound wave down a gas-filled tube.

In order to see the effect of the advancing piston on the gas more easily, let's simplify the molecular-gas model introduced at the beginning of Section 17.1 by pretending that the gas molecules travel only parallel to the x-axis and, furthermore, that they all travel with the same speed v_m. This view is, of course, a gross oversimplification, but it makes it easier to grasp the essence of what's going on. We take the speed of our molecules as the average molecule speed in actual gases, which is on the order of 400 m/s! We assume that the piston's speed of advance is a much smaller value v_p. Before the piston moves, on average our gas molecules don't go anywhere. They simply collide with the piston wall and then with each other. When they collide with each other in elastic collisions, each molecule simply turns around; at any given moment, half are moving to the left, half to the right, and thus the average gas velocity is zero. However, when the piston starts to move, the distribution of molecular speeds changes.

To see how the molecules' speeds change when the piston starts to move, we need to solve two simple collision problems, one for molecules striking the piston, the second for molecules striking each other. If you have forgotten the details of elastic collisions, you may want to review Section 9.5 before reading further. For molecules striking the piston, the speed of approach is $v_p + v_m$. If the molecular collision is perfectly elastic, the speed of separation is also $v_p + v_m$. Because the piston continues to advance at v_p, the rebounding molecules must move off at $v_m + 2v_p$ after striking the piston. As shown in Figure 17.5, each molecule that strikes the moving piston has the quantity $2v_p$ added to its speed. Thus, kinetic energy is added to the gas.

The second collision problem we need to think about is a molecule moving at v_m toward the piston that strikes head-on another molecule moving away from the piston with speed $v_m + 2v_p$. Assuming again that the collision is perfectly elastic, you can quickly determine, using the methods of Chapters 9 or 10, that the two equal-mass molecules merely exchange velocities; the one originally moving toward the piston with velocity v_m now moves away with velocity $v_m + 2v_p$, and the other molecule heads back toward the piston with speed v_m. The net result of all this banging around is that the excess velocity $2v_p$ is passed on from one molecule to the next. The rightmost molecule with the extra velocity is at the leading edge of a propagating disturbance as shown in Figure 17.5. If we look within the disturbed region, at any given moment we can expect that half the molecules have velocity $-v_m$ toward the piston and half have velocity $+v_m + 2v_p$ away from the piston. *Thus, the gas in the disturbed region acquires an average velocity $v_{av} = (-v_m + v_m + 2v_p)/2 = v_p$.* When the piston stops moving, the molecules returning to it with speed v_m simply have their velocities reversed, and we are back where we started, except for the propagating disturbance, which is now traveling down the tube with a speed of $v_m + 2v_p \approx v_m$. Be sure to get this straight. In the disturbed region the gas molecules have excess average speed v_p, but that region of excess speed propagates with much higher speed v_m.

We can take three important points from our model: (1) The gas in the region between the piston and the leading edge of the disturbance has acquired an average velocity v_p to the right, but (2) each gas molecule hit by the piston gains velocity $2v_p$ in the x-direction, thus increasing the kinetic energy of the gas. Furthermore, this "extra" kinetic energy is passed on from molecule to molecule *at roughly the speed v_m,* which is much faster than the piston speed v_p. (3) The molecules that were pushed out of the way by the piston also create a region of increased density that itself also propagates down the tube along with the excess kinetic energy.

When we introduce into our model the full, three-dimensional nature of the molecules' trajectories and permit the molecules to have a distribution of speeds, the details of our analysis become more complicated. However, the basic mechanism in our previous description is still at work. Rather than pursuing this detailed microscopic picture further, we now treat the gas as if it were a continuous "fluid" medium. Nonetheless, we urge you to keep our kinetic model in mind; it should then be clear why we make the assumptions we do in the compressible-cell model described next.

Let's take a detailed look at what happens when one stroke of the piston sends a compression along the tube. Our game plan is to apply Newton's second law to the disturbed gas. As in Chapter 9, it is convenient to apply the second law in the form of the impulse–momentum theorem, $F_{net} \Delta t = \Delta p$, where p is momentum, not pressure! We assume P_o and ρ to be the equilibrium pressure and density of the gas, respectively. When the piston shown in Figure 17.6 moves to the right with velocity v_p, our microscopic model predicts two results. First, the entire column of gas between the piston and the leading edge of the disturbance moves to the right with an average velocity v_p. Second, the front edge of the compression moves to the right with the sound-wave speed, which is much greater than the piston speed.

In time Δt the wavefront of the advancing disturbance moves a distance $\Delta x = v \Delta t$ to the right, where v is the velocity of sound. The net force on the compressed region between the piston and this advancing wavefront disturbance is

$$F_{net} = (P + \Delta P) A - PA = \Delta P A$$

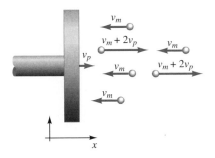

FIGURE 17.5

Molecular collisions with an advancing piston. Only the *magnitudes* of the velocities are designated. The topmost molecule moves toward the piston with *relative* velocity $v_m + v_p$. The second row of molecules are about to have an elastic collision. The third row of molecules shows the velocity transferred after the collision.

Concept Question 1
If the cone on a speaker suddenly moves forward creating a compression pulse in the air in front of it, does that pulse move through the gas with a speed that is close to (a) the speed of the speaker cone, or (b) the average speed of the gas molecules?

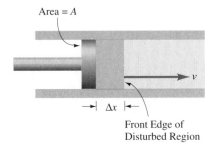

FIGURE 17.6

The piston oscillating in SHM imparts energy to a mass Δm and causes it to oscillate.

where P is the pressure just ahead of the disturbance, $P + \Delta P$ is the pressure of the piston on the compressed region, and A is the cross-sectional area of the piston. Knowing the net force on the compressed region, we can calculate the impulse delivered to this region in time Δt

$$F_{net}\, \Delta t = \Delta P A\, \Delta t \tag{17.1}$$

To find the change in the momentum of the compressed region we note that the mass Δm of the air in the compressed region is its density ρ times its volume $\Delta \mathcal{V} = A\, \Delta x$:

$$\Delta m = \rho A\, \Delta x = \rho A v\, \Delta t$$

Because the initial average velocity of the air is zero (before the piston moves) and after the piston's motion the compressed air column moves with speed v_p, the *change* in velocity of the cylinder of air is v_p. Thus, the change in the momentum of the air in the compressed region is

$$\Delta p = \Delta m\, v_p = \rho A v\, \Delta t\, v_p \tag{17.2}$$

By the impulse–momentum theorem ($F_{net}\, \Delta t = \Delta p$), we can equate Equations (17.1) and (17.2)

$$\Delta P A\, \Delta t = \rho A v\, \Delta t\, v_p$$

and solve for the **acoustic pressure**

$$\Delta P = \rho v_p v \tag{17.3}$$

Our last step is to relate this change in pressure to the properties of the gas. In Chapter 16 we saw that changes in the pressure of a gas are related to changes in its volume by a quantity called the bulk modulus, which was defined as

$$B = -\frac{\Delta P}{(\Delta \mathcal{V}/\mathcal{V})} \tag{17.4}$$

The original volume of the region we are analyzing was $Av\, \Delta t$ before the piston moved. The piston's motion made this volume smaller by

$$\Delta \mathcal{V} = -Av_p\, \Delta t \tag{17.5}$$

Substituting this expression for $\Delta \mathcal{V}$ and the expression for ΔP from Equation (17.3) into the definition of the bulk modulus and solving for v, we have

$$v = \sqrt{\frac{B_{ad}}{\rho}} \tag{17.6}$$

The numerical value of the bulk modulus of a gas depends on whether or not the gas is kept at constant temperature when it is squeezed. As we noted when we discussed our microscopic model, kinetic energy is added to the gas by the moving piston, and this is tantamount to raising its temperature in the compressed region. Under these conditions the appropriate bulk modulus is known as the **adiabatic bulk modulus** B_{ad}, as opposed to the **isothermal** (constant-temperature) bulk modulus B discussed in Chapter 16. The distinction between adiabatic and isothermal processes is explained in detail in Chapter 19. For liquids and solids the difference is often too small to be of practical importance. Although we have derived Equation (17.6) by supposing the material in the tube to be air, this expression provides the velocity of sound for any liquid or gas of adiabatic bulk modulus B_{ad} and density ρ.

The velocity of sound in air and some other materials is given in Table 17.1.

TABLE **17.1**   **The Velocity of Sound in Various Materials**

MATERIAL	VELOCITY (m/s)
Gases	
Air (0°C)	331
(20°C)	343
Helium (0°C)	970
Nitrogen (0°C)	334
Oxygen (0°C)	316
Liquids **(25°C)**	
Water (distilled)	1498
Seawater	1531
Ethanol	1270
Mercury	1450
Solids **(longitudinal waves in thin rods)**	
Aluminum	5100
Copper	4760
Gold	3240
Lead	2160
Stainless steel	5790
Other	
Granite	6000
Glass	
Flint	3980
Crown	5100
Pyrex	5640
Wood	
Ash	4670
Beech	3340
Oak	3850

Other Distortion Waves (Optional)

Table 17.2 provides formulas for the propagation speeds of several types of distortion waves. Although the formulas are all different in detail, notice that each has the form of the square root of a fraction. In every case the numerator of the fraction is a factor that relates the magnitude of the restoring force on a mass element in the medium to the displacement of that element from equilibrium. In the denominator of each fraction is a mass density that describes the inertia of the displaced mass element. This fraction represents a competition between a ''springiness'' factor and a ''sluggishness'' factor.

Although we do not intend to derive (or even use) all of the formulas in Table 17.2, we cannot resist commenting on them, because they give us insight into the rich diversity of the mechanisms that drive elastic waves. In the case of a thin rod the distortion may be a compression pulse that causes a bulge to propagate along the rod (or perhaps an extension pulse that results in the propagation of a thin region). Thus, it comes as no surprise that the restoring force is related to the Young's modulus of the material, because, as shown in Chapter 16, it is the Young's modulus that relates longitudinal stress to longitudinal strain.

For a compression wave traveling within an extended material (as opposed to a thin rod), the surface distortion is not a factor. As the wave passes through a particular region

TABLE 17.2   **Distortion-Wave Velocities**

WAVE TYPE	VELOCITY EXPRESSION
Longitudinal Waves	
Thin solid rod	$\sqrt{\dfrac{Y}{\rho}}$
Extended solid	$\sqrt{\dfrac{B + \frac{4}{3}S}{\rho}}$
Liquid	$\sqrt{\dfrac{B}{\rho}}$
Gas	$\sqrt{\dfrac{B_{ad}}{\rho}} = \sqrt{\dfrac{\gamma P}{\rho}} = \sqrt{\dfrac{\gamma RT}{m_o}}$
Transverse Waves	
Extended solid	$\sqrt{\dfrac{S}{\rho}}$
String	$\sqrt{\dfrac{F}{\mu}}$

Springiness Terms
B = isothermal bulk modulus
B_{ad} = adiabatic bulk modulus
F = string tension
S = shear modulus
Y = Young's modulus

Inertial Terms
ρ = mass density
μ = linear mass density

within the solid, the material in that region is compressed, so perhaps you are not surprised to see the bulk modulus B appearing in the numerator. However, finding the shear modulus in the formula for the speed of a ''pure'' compression wave is surprising indeed. The shear modulus appears, because in an extended solid a ''pure'' compression wave is not possible. The material must be in shear over at least part of the wavefront.

In the case of a liquid, the shear modulus is zero. Hence, we have simply $\sqrt{B/\rho}$. Even when the liquid is sheared, no shear stress results.

The longitudinal waves in a gas are especially interesting. As derived in the previous paragraphs, their speed is given by $\sqrt{B_{ad}/\rho}$, the same as for liquids. We have cheated a little and used some results that do not appear in this text until Chapter 21; we have rewritten the equation for the sound velocity in gases as $\sqrt{\gamma RT/m_o}$. In this expression R is a universal constant, and m_o is the molecular weight of the gas. Gamma is a constant with a value that is characteristic of the particular gas and at most a weak function of temperature. This expression means that the speed of sound waves in a given gas is a function of temperature only! This temperature dependence occurs because the compressions and rarefactions in gases are transmitted by collisions between widely separated gas molecules. In Chapter 21 we will learn that the average speed of the molecules is directly related to the temperature of the gas; the hotter the gas, the faster the molecules move. The faster the molecules move, the more often they collide. The more often they collide, the faster a wave disturbance is passed along.

We should also point out that the models of sound waves we are using lead us to conclude that the phase velocity of sound waves in pipes does not depend on wavelength. This result is not completely accurate. If you clap your hands together in front of a very long hollow pipe, the returning echo is not a sharp clap. Instead, the echo is a rather drawn out buzz-zing similar to the artificial sound often generated by video games when you shoot the bad guys. This effect is known as **wave-guide dispersion** and is due to waves of different wavelength traveling at different speeds. The effect, though fascinating, is ''beyond the scope of this text.'' (You can hear a similar effect in a metallic ''waveguide'' by stretching a Slinky across the room. Hold one end to your ear and have a friend tap the other end with a pencil. If you listen carefully to the sound, you will be able to tell whether the high-frequency sound waves travel faster or slower than the low-frequency sound waves.)

Notice that the formula for the speed of transverse waves on a string is significantly different from the other expressions in Table 17.2. The term related to the restoring force appearing in the numerator is not the elastic modulus of the string. In fact, the elastic property of the string used in the derivation of $v = \sqrt{F/\mu}$ is that the string is soft. That is, it has zero Young's modulus. The tension is a *force,* not an elastic modulus. This formula reminds us that the restoring force in the case of the string comes from components of the impressed tension, *not* the elastic properties of the medium.

Before resuming our discussion of sound waves, we want to mention that although sound waves are longitudinal waves, deformation waves in solids can be either longitudinal or transverse. Many of the properties we describe for sound waves are also true of transverse (shear) deformation waves in elastic media. Indeed, low-frequency **seismic** waves propagating through the earth after an earthquake consist of both longitudinal components (P waves) and transverse components (S waves). These two wave types travel with different velocities. Moreover, the S waves usually do more damage to surface structures than do the P waves.

17.3 Harmonic Waves in Air

Let's return to our "speaker system" made up of a piston moving inside a cylinder first discussed at the beginning of Section 17.2 and illustrated in Figure 17.4. This time, instead of generating a single pulse, we move the piston in simple harmonic motion to create a harmonic wave train. This wave train can be described using the general wave functions developed in Chapter 15.

Imagine the gas in front of the piston divided up into a series of thin mass elements shaped in this case as layers parallel to the face of the piston. Actually, we can characterize the wave in two different ways. We can describe the gas pressure in each of the thin layers, or we can describe the displacement of a given layer from its equilibrium position. We use the notation $X(x, t)$ to represent the *displacement* X from equilibrium of a layer with equilibrium position x. Similarly, $P(x, t)$ represents the pressure at time t in the layer located at point x.

In the graphs in Figure 17.7, we show the variation of both the air molecule displacement and the pressure change along the length of the tube, at the instant when the piston is located in the final figure. It is important to recognize that the pressure wave is the *variation of pressure* ΔP relative to atmospheric pressure. As illustrated in Figure 17.7 the pressure and displacement waves are $\pi/2$ rad out of phase from each other. Both functions, the pressure variation $\Delta P(x, t)$ and the displacement $X(x, t)$, are sinusoidal waves and therefore may be written

$$\Delta P = \Delta P_o \sin(kx - \omega t) \tag{17.7}$$

and

$$X(x, t) = X_o \sin\left(kx - \omega t + \frac{\pi}{2}\right)$$

$$X(x, t) = X_o \cos(kx - \omega t) \tag{17.8}$$

Equations (17.7) and (17.8) describe the pressure variation and gas-layer displacement for the sound waves we produced in air-filled tubes. They also correctly describe sound waves *with plane wavefronts* in air but are not correct for spherical wavefronts. The latter have amplitudes that decrease in magnitude with distance from the source. That is, for spherical waves $X_o = X_o(x, y, z)$. But before we worry about spherical waves, let's first discover how the amplitude of the pressure waves is related to the amplitude of the displacement waves.

The amplitude ΔP_o of the pressure wave in Equation (17.7) is related to the amplitude X_o of the displacement wave in Equation (17.8). We can find the relation between them by

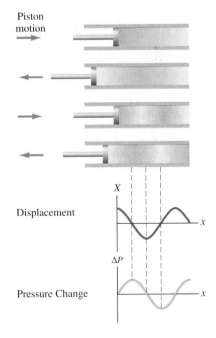

FIGURE 17.7

An oscillating piston sends a sound wave down a gas-filled tube. The displacement and pressure variation graphs display the harmonic nature of the sound wave.

again using the definition of the bulk modulus, Equation (17.4). Referring back to Figure 17.6, we recall that the volume enclosed by the cross-sectional area A and the length Δx is $A\,\Delta x$. At any instant t, the amount X by which air molecules are displaced from equilibrium changes with x, the distance along the tube; that is, X is a function of x as well as t: $X(x, t)$. When air molecules undergo a displacement $\Delta X = X(x + \Delta x, t) - X(x, t)$, the volume change associated with this displacement is $\Delta \mathcal{V} = A\,\Delta X$. We can solve Equation (17.4) for the pressure change:

$$\Delta P = -B_{ad}\frac{\Delta \mathcal{V}}{\mathcal{V}} = -B_{ad}\frac{A\,\Delta X}{A\,\Delta x}$$

$$= -B_{ad}\frac{\Delta X}{\Delta x}$$

In the limit as Δx approaches zero $\Delta X/\Delta x$ becomes $\partial X/\partial x$, a partial derivative because t is held constant. (Partial derivatives are reviewed in Appendix 1.)

$$\Delta P = -B_{ad}\frac{\partial X}{\partial x}$$

Substituting for $X(x, t)$ from Equation (17.8), we have

$$\Delta P = -B_{ad}\frac{\partial}{\partial x}[X_o \cos(kx - \omega t)]$$

$$= B_{ad}X_o k \sin(kx - \omega t)$$

The bulk modulus is related to the velocity through Equation (17.6), $B_{ad} = \rho v^2$, so

$$\Delta P = \rho v^2 X_o k \sin(kx - \omega t)$$

In Chapter 15 we found that the velocity of a sinusoidal wave is given by $v = \omega/k$, so we can rewrite this last equation as

$$\Delta P = \rho \omega v X_o \sin(kx - \omega t)$$

Comparing this result with Equation (17.7), we find the pressure amplitude to be related to the displacement amplitude by

$$\Delta P_o = \rho \omega v X_o \tag{17.9}$$

In the next section, we use Equation (17.9) to discuss the intensity of sound.

Concept Question 2
Because the amplitude of the pressure wave ΔP_o is related to the amplitude of the displacement wave X_o by the relation $\Delta P_o = \rho \omega v X_o$, must the pressure and displacement reach their respective maximums at the same time? Explain.

17.4 Sound Intensity and Sound Intensity Level

The **intensity** I of sound is defined as the average rate at which sound energy passes through a unit area perpendicular to the direction of propagation. To see how this intensity changes with distance let's consider a small (point) source that is radiating spherical waves. If $\mathcal{P}_{av}$ is the rate at which the source is radiating sound energy (called the **sound power** of the source), then assuming none of the radiated energy is absorbed, all of it must eventually pass through any closed spherical surface surrounding the source. If the source has been operating for some time, then the rate at which sound energy passes through any such spherical surface must equal the rate $\mathcal{P}_{av}$ at which energy is radiated by the source. It follows that the intensity at a distance r_o from the source is given by the average power divided by the surface area of a sphere of radius r_o:

$$I_o = \frac{\mathscr{P}_{av}}{4\pi r_o^2}$$

In the absence of absorbers $\mathscr{P}_{av}$ cannot change once it is emitted from the source. Thus, at a distance $r > r_o$ the intensity I is

$$I = \frac{\mathscr{P}_{av}}{4\pi r^2} \tag{17.10}$$

Eliminating $\mathscr{P}_{av}$ between these last two equations, we find

$$I = I_o\left(\frac{r_o^2}{r^2}\right) \tag{17.11}$$

That is, *the intensity of a sound wave dies off inversely as the square of the distance from the source.*

The intensity of a sound wave is related to the pressure wave amplitude. To find how the two are related we once again return to our one-dimensional approach and consider a small mass Δm of air that is driven by our oscillating piston in a tube (Fig. 17.6, for the last time, we promise!). In Chapter 14, Equation (14.27b), we found that the total energy of a mass m oscillating in SHM can be calculated through knowledge of the amplitude A_o of the oscillation as $E = \frac{1}{2}m\omega^2 A_o^2$. Similarly, for a sound wave traveling through air, the energy ΔE of a small air mass $\Delta m = \rho(A\,\Delta x)$ is related to the maximum displacement X_o of the particles through

$$\Delta E = \frac{1}{2}\Delta m\omega^2 X_o^2 = \frac{1}{2}\rho A\,\Delta x\,(\omega X_o)^2$$

The power is the time rate of energy transfer:

$$Power = \mathscr{P} = \frac{\Delta E}{\Delta t} = \frac{1}{2}\rho v\frac{\Delta x}{\Delta t}(\omega X_o)^2$$

However, $\Delta x/\Delta t = v$, the velocity of the sound wave, so we can write

$$\mathscr{P} = \frac{1}{2}\rho Av(\omega X_o)^2$$

The intensity is the power transferred per unit area,

$$I = \frac{\mathscr{P}}{A} = \frac{1}{2}\rho v(\omega X_o)^2 \tag{17.12}$$

Finally, we can solve Equation (17.9) for the displacement amplitude X_o and substitute this result into Equation (17.12) to obtain an expression for the intensity in terms of the pressure amplitude ΔP_o:

$$I = \frac{\Delta P_o^2}{2\rho v} \tag{17.13}$$

Concept Question 3
If you increase your distance from a certain sound source so that its amplitude has been reduced to one-fourth its original value, by what factor does its intensity change? By what factor does the amplitude of the pressure wave change?

EXAMPLE 17.1   Is the Threshold of Pain for "Rap" Different from That for "Country and Western"?

The most intense sounds a human ear can tolerate at a frequency of 1.00 kHz have an intensity of about 1.00 W/m². Compute the pressure and displacement amplitudes for a sound wave of this intensity.

SOLUTION   We begin by solving Equation (17.13) for the pressure amplitude:

$$\Delta P_o = \sqrt{2\rho v I}$$

Using $v = 343$ m/s for sound velocity and taking the density of air to be $\rho = 1.29$ kg/m³, we have

$$\Delta P_o = \sqrt{2(1.29 \text{ kg/m}^3)(343 \text{ m/s})(1.00 \text{ W/m}^2)} = 29.7 \text{ N/m}^2$$

This result is the magnitude of the maximum pressure fluctuation above and below normal atmospheric pressure. The variation is about 300 parts per million.

The displacement amplitude is related to the pressure amplitude by Equation (17.9). Solving this equation for X_o, we have

$$X_o = \frac{\Delta P_o}{\rho \omega v} = \frac{29.7 \text{ N/m}^2}{(1.29 \text{ kg/m}^3)(2\pi \times 10^3 \text{ Hz})(343 \text{ m/s})}$$

$$= 1.07 \times 10^{-5} \text{ m} = 0.0107 \text{ mm}$$

This is a fairly small displacement. However, you may be surprised to learn how small a displacement your ear can detect! This topic is the subject of Problem 28 at the end of this chapter.

◀

The Decibel Scale

Our ears can detect a very wide range of sound wave intensities. In fact, this range extends from a barely audible level of 10^{-12} W/m² to a painful 10^0 W/m² = 1 W/m². These levels also depend on the frequency of the sound, as illustrated in Figure 17.8. The range of intensities a human can sense is so vast that it is convenient to express the intensities using a logarithmic scale. This scale defines the **sound intensity level** and is given by the

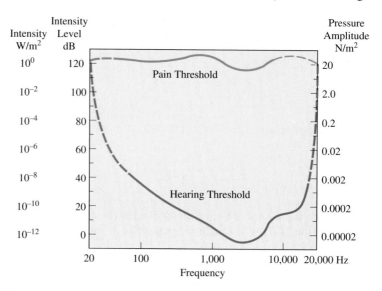

FIGURE 17.8

The average hearing range for the human ear is frequency dependent. Notice that the frequency scale is logarithmic.

TABLE 17.3 Sound Intensity Levels for Some Common Sources

SOUND SOURCE	INTENSITY LEVEL
Threshold of hearing	0 dB
Rustling leaves (no students rolling in them)	10 dB
Whisper	20 dB
Rustling leaves with giggles	30 dB
Classroom lecturer	60 dB
Classroom lecturer (the day the test is returned)	65 dB
Busy city street	70 dB
Lawnmower with a bad muffler (from where you push)	100 dB
Threshold of pain	120 dB
Jet engine (from passenger left on runway)	130 dB

(Note that the decibel scale does not correspond exactly to our perception of loudness. Few people would say that the sound of a classroom lecture is half the threshold of pain. Well, maybe physics lectures, but only in a metaphoric sense.)

relation

$$L_l = (10 \text{ dB}) \log_{10} \left(\frac{I}{I_o} \right)$$ (17.14)

We hasten to make several comments about this definition. First, the logarithm is to base-10 and is called a **common logarithm.** It is *not* the natural (base-*e*) logarithm. Second, the reference intensity I_o is taken to be the threshold of hearing for a ''standard'' human:

$$I_o = 10^{-12} \text{ W/m}^2$$ (17.15)

Finally, the unit of L_l is the decibel (dB), and its size is one-tenth of a bel. To help you get your bearings with it we have included Table 17.3, which gives approximate sound intensity levels for some common sounds. (This unit of intensity, bel, is named after Alexander Graham Bell, and we have no idea where the other ''l'' in Bell went!)

Sound intensity is measured in watts per square meter (W/m^2).

Sound intensity *level* is measured in decibels (dB).

EXAMPLE 17.2 *Easy as Falling off a Log*

Verify the sound intensity levels for the threshold of hearing ($I = I_o$), normal conversation ($I = 10^{-6}$ W/m^2), and the threshold of pain ($I = 1$ W/m^2) given in Table 17.3.

SOLUTION We apply Equation (17.14), first for the threshold of hearing:

$$L_l = (10 \text{ dB}) \log_{10} \left(\frac{I}{I_o} \right) = (10 \text{ dB}) \log_{10} \left(\frac{I_o}{I_o} \right) = (10 \text{ dB}) \log_{10} (1) = 0 \text{ dB}$$

Next, for normal conversation,

$$L_l = (10 \text{ dB}) \log_{10} \left(\frac{10^{-6} \text{ W/m}^2}{10^{-12} \text{ W/m}^2} \right) = (10 \text{ dB}) \log_{10} (10^{+6})$$

$$= (10 \text{ dB})(6) = 60 \text{ dB}$$

Finally, for $I = 1$ W/m^2,

$$L_l = (10 \text{ dB}) \log_{10} \left(\frac{1 \text{ W/m}^2}{10^{-12} \text{ W/m}^2} \right) = (10 \text{ dB}) \log_{10} (10^{+12})$$

$$= (10 \text{ dB})(12) = 120 \text{ dB}$$

◀

EXAMPLE 17.3 *Did You Realize the Volume Control Turns in Both Directions?*

An outdoor concert speaker emits music at a level such that the total radiated audio power is 20.0 W. If we approximate this speaker as a point source, (a) what is the sound intensity level at a point 4.00 m away? (b) How far away from the speaker does the sound intensity level drop to half the level that it has at 4.00 m?

SOLUTION   (a) In order to apply Equation (17.14) we must first compute the intensity at a radial distance of 4.00 m from the source. We can find this intensity from Equation (17.10):

$$I = \frac{\mathscr{P}_{av}}{4\pi r^2} = \frac{20.0 \text{ W}}{4\pi(4.00 \text{ m})^2} = 9.95 \times 10^{-2} \text{ W/m}^2$$

From Equation (17.14) we find

$$L_I = (10 \text{ dB}) \log_{10}\left(\frac{9.95 \times 10^{-2} \text{ W/m}^2}{10^{-12} \text{ W/m}^2}\right)$$

$$= (10 \text{ dB})(11.0) = 110 \text{ dB}$$

(b) Half the sound intensity level from part (a) is $L_I' = (110 \text{ dB})/2 = 55$ dB. We solve

$$L_I' = (10 \text{ dB}) \log_{10}\left(\frac{I'}{I_o}\right)$$

for the new intensity I':

$$\log_{10}\left(\frac{I'}{I_o}\right) = \frac{L_I'}{10 \text{ dB}}$$

$$I' = I_o \times 10^{L_I'/10 \text{ dB}}$$

But $I' = \mathscr{P}_{av}/4\pi(r')^2$, so

$$\frac{\mathscr{P}_{av}}{4\pi(r')^2} = I_o \times 10^{L_I'/10 \text{ dB}}$$

or

$$r' = \sqrt{\frac{\mathscr{P}_{av}}{4\pi I_o(10^{L_I'/10 \text{ dB}})}}$$

Now, if we assume there is no dissipation of energy (due to sound absorption by trees, buildings, ears, etc.), the total power $\mathscr{P}_{av}$ being radiated remains at 20.0 W/m² and

$$r' = \sqrt{\frac{20.0 \text{ W/m}^2}{4\pi(1.00 \times 10^{-12} \text{ W/m}^2)(10^{(55 \text{ dB}/10 \text{ dB})})}}$$

$$= 2.24 \times 10^3 \text{ m} = 2.24 \text{ km}$$

Comparing the sound intensity levels in this problem with those of Table 17.3, we find that if no sound were absorbed the sound intensity level from the speaker over 2 km away is comparable to listening to a person in the same room. It should be clear from this example that sound intensity levels do not die off rapidly with distance from the source, an oversight that has caused more than a few fraternities to anger college deans! ◀

Concept Question 4
When the power emitted from a speaker doubles from 5 W to 10 W, the sound intensity level at a certain point in front of the speaker increases by 3 dB. What must the speaker power be to cause an additional 3 dB increase?

17.5 Sources of Sound

In this section we study the physics of musical instruments. You should be aware that there is much more to the construction of a good musical instrument than what we have to say (although we will tell you *why* there is more to it). We begin with stringed instruments because we have already developed much of the required physics.

Vibrating Strings

In Chapter 15 we found that the boundary conditions for a string fixed at both ends restricts the string's motion; an integral number of half-wavelengths must equal the length of the string (review Fig. 15.15). This requirement leads to the following expression for the frequencies with which the string can vibrate, Equation (15.19):

$$f = n\frac{v}{2l} \qquad n = 1, 2, 3, \ldots \tag{17.16}$$

where $n = 1$ for the first harmonic, $n = 2$ for the second harmonic, and so on. In Equation (17.16), v is the velocity of the wave *on the string,* and is given by $\sqrt{F/\mu}$. The tension in the string is F, and μ is the string's linear mass density. Please note, therefore, that Equation (17.16) can also be written

$$f = \frac{n}{2l}\sqrt{\frac{F}{\mu}} \qquad n = 1, 2, 3, \ldots \tag{17.17}$$

As a string vibrates with one of these frequencies, the air surrounding it is alternately compressed and rarefied by the back-and-forth motion of the string. Therefore, the frequency of a sound wave emitted by a vibrating string is the same as that of the string, Equation (17.17). However, *the wavelength of the sound wave is different from that of the string* because the velocity of sound in air is different from the velocity of the wave on the string. If v is the velocity of sound in air, then the wavelength of the sound wave is

$$\lambda = \frac{v}{f} = \frac{2l}{n}\sqrt{\frac{\mu}{F}}\,v \qquad n = 1, 2, 3, \ldots \tag{17.18}$$

When a stringed instrument is plucked or bowed, frequencies other than the first harmonic (the fundamental) are generally excited. However, only frequencies described by Equation (17.18) are created. These higher harmonics may remain as long as the fundamental itself (although often they are damped more rapidly than the fundamental by velocity-dependent frictional effects). Figure 17.9 shows how a string might look at various times during an oscillation when only the first two harmonics are present.

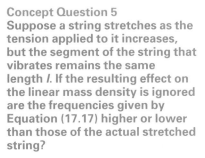

Concept Question 5
Suppose a string stretches as the tension applied to it increases, but the segment of the string that vibrates remains the same length *l*. If the resulting effect on the linear mass density is ignored are the frequencies given by Equation (17.17) higher or lower than those of the actual stretched string?

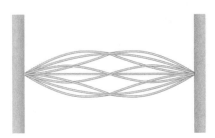

FIGURE 17.9

A string vibrating with its first two harmonics. The string position is shown at nine different times.

One reason a guitar sounds different from a banjo has to do with the number of harmonics present and their relative amplitudes. The great variety of musical sounds that various instruments can produce results from the superposition of selected harmonics. Musical instruments are designed with various shapes so that certain harmonics are reinforced and others quickly damp out. The best way to describe the relative amplitudes of

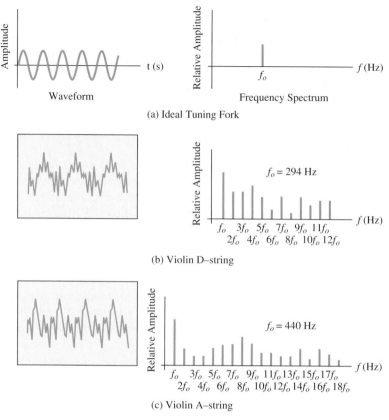

FIGURE 17.10

Waveforms and frequency spectra for (a) a tuning fork, (b) a violin D-string, and (c) a violin A-string.

the harmonics present for a particular instrument is by a **frequency spectrum.** A frequency spectrum is a graph with a frequency scale along the horizontal axis and an amplitude scale along the vertical axis. Figure 17.10(a) shows the waveform and the frequency spectrum for an ideal tuning fork. The sound produced by an ideal tuning fork vibrating at constant amplitude has only one frequency f_o, and, therefore, the frequency spectrum has only a single spike. (The vertical scale is somewhat arbitrary, but all that is really important to us is the *relative* amplitudes of the various component frequencies that make up the sound from an instrument.) In Figures 17.10(b) and (c) we show the waveforms and frequency spectra for the D- and A-strings of a violin. Notice that in addition to the fundamental there are numerous overtones present. Figure 17.11 shows the measured waveforms and measured frequency spectra for a flute, a trumpet, and a saxophone, all playing the same 440-Hz note. Clearly, different harmonics are present with different amplitudes in these instruments. Each instrument is playing the same note, but the variation in amplitude for different harmonics is one important reason that different instruments playing the same note sound different to us. Following Example 17.4 we describe how sound is produced in air columns, the basis for the wind instruments.

Concept Question 6
Three young children antagonize their babysitter by blowing whistles at the same time. If the whistles each emit sound waves with the same amplitude but with frequencies of 480 Hz, 512 Hz, and 536 Hz, draw a frequency spectrum of the "music" heard by the babysitter.

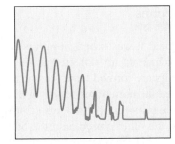

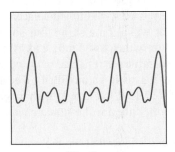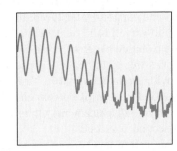

Waveform   Frequency Spectrum

(a)

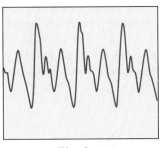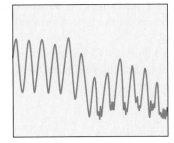

Waveform   Frequency Spectrum

(b)

Waveform   Frequency Spectrum

(c)

FIGURE 17.11

Waveforms and frequency spectra for (a) a flute, (b) a trumpet, and (c) a saxophone. Each play the same note with f = 440 Hz.

EXAMPLE 17.4 *Fiddling with Physics*

The fundamental mode of the D-string of a violin has a frequency of 294 Hz. What are the frequencies of the first and second overtones?

SOLUTION   The fundamental frequency f_1 is given by Equation (17.17) with n = 1. Therefore,

$$f_1 = \frac{1}{2l} \sqrt{\frac{F}{\mu}} = 294 \text{ Hz}$$

The first and second overtones are also given by Equation (17.17), with n = 2 and 3, respectively:

$$f_2 = \frac{2}{2l} \sqrt{\frac{F}{\mu}} = 2f_1 = 2(294 \text{ Hz}) = 588 \text{ Hz}$$

$$f_3 = \frac{3}{2l} \sqrt{\frac{F}{\mu}} = 3f_1 = 3(294 \text{ Hz}) = 882 \text{ Hz}$$

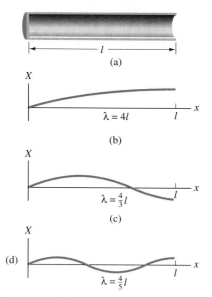

FIGURE 17.12

Standing waves in a tube closed at one end. The air molecule displacement amplitudes are shown for (a) the fundamental, (b) the first overtone, and (c) the second overtone.

Concept Question 7
For harmonic sound waves the pressure wave and the displacement wave are 90° out of phase. Explain in your own words the physical conditions that suggest that it is reasonable for the open end of a resonating pipe to be a pressure node and a displacement antinode.

Air Columns

Longitudinal waves propagating inside an air-filled tube, such as an organ pipe or other wind instrument, are reflected at the ends of the tube. The sounds we hear from such instruments are formed from standing waves in these tubes. Just as in the case for waves on a string, the normal modes of vibration are determined by the *boundary conditions*. We will consider two simple models of wind instruments, one model with one end of the tube open and the other closed, and the other model with both ends open. Let's see if we can determine what the boundary conditions must be for each of these models and from them find the frequencies of the normal modes.

Figure 17.12(a) shows a tube of length l with one closed end. Because the air next to the closed end cannot be displaced, this position must correspond to a displacement node of the sound wave in the tube. Just as in the case of waves on a string, our sound wave undergoes a phase change of π rad as it is reflected from the closed tube-end. At the open end the air is free to move, and consequently the sound wave reflected at an open end is in phase with the incident wave. Thus, for standing waves in an air-filled tube, the open end corresponds to a displacement antinode. Figures 17.12(b), (c), and (d) show graphs of the air-molecule displacements along the length of the tube for the fundamental and the first and second overtones.

A more careful analysis of air vibrations in tubes shows that the antinode boundary condition at the open end is only approximately correct and works best if the radius of the tube is small compared to the wavelength of the sound wave. In fact, if R is the inside radius of the tube, then the actual antinode occurs at a distance of approximately $0.6R$ beyond the end of the tube. Our applications of the boundary conditions ignore this small correction.

From Figure 17.12 we see that we can satisfy the boundary conditions by placing $\frac{1}{4}\lambda$, $\frac{3}{4}\lambda$, $\frac{5}{4}\lambda$, . . . in the length. Hence, the wavelengths of the normal modes in our tube with one end closed are given by

$$\lambda = 4l, \frac{4l}{3}, \frac{4l}{5}, \ldots$$

$$= \frac{4l}{n_o} \qquad n_o = 1, 3, 5, 7, \ldots \qquad (17.19)$$

The frequencies are given by $f = v/\lambda$, where v is the *velocity of sound in air*:

$$f = n_o \frac{v}{4l} \qquad n_o = 1, 3, 5, \ldots \qquad (17.20)$$

These frequencies differ from the normal mode frequencies of a string, Equation (17.17), because the boundary conditions are different.

A tube of length l open at both ends is shown in Figure 17.13(a). The boundary conditions for this model place an antinode at both ends. The displacement waves for the first three normal modes are shown in Figures 17.13(b), (c), and (d). From this figure we

FIGURE 17.13

Standing waves in a tube open at both ends. Air molecule displacement amplitudes are shown for (a) the fundamental, (b) the first overtone, and (c) the second overtone.

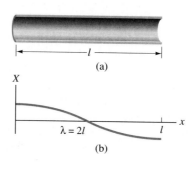

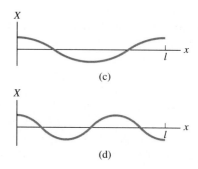

see that the wavelengths of the normal modes are

$$\lambda = 2l, \frac{2l}{2}, \frac{2l}{3}, \dots$$

$$\lambda = \frac{2l}{n} \qquad n = 1, 2, 3, \dots \qquad \text{(17.21)}$$

The frequencies of the normal modes are given by $f = v/\lambda$:

$$f = n\frac{v}{2l} \qquad n = 1, 2, 3, \dots \qquad \text{(17.22)}$$

where again, v is the velocity of sound in air.

Just as in the case of the stringed instruments, the variety of sounds emitted by different wind instruments is a result of the reinforcement of different harmonics by the different instrument shapes. (Look back at Fig. 17.11.) The standing wave in a wind instrument is excited by a stream of air crossing an open end. The process by which this stream gives rise to a standing wave is complex and involves a succession of air vortices (small air "whirlpools") that are released into the tube. If this sounds like a difficult problem, it is, and you will be glad to hear that we are going to drop the issue now and say no more about it, not even in the problems at the chapter end!

Resonance and Beats

Recall the meaning of **resonance** (Section 14.5): When an oscillating system experiences a periodic force with a frequency that is the natural frequency of the system, the amplitude of its motion is much larger than that for a driving frequency at some other value. The frequencies of the normal modes for any vibrating system, such as a string or an air column in a tube, are the **resonant frequencies** for that system. A mechanical oscillator tied to the end of a stretched string that is fixed at the other end creates standing waves only when it vibrates at one of the resonant frequencies, given by Equation (17.17). It is important, therefore, that the shape of a musical instrument be designed so as to emphasize the desired harmonics for the particular sound of that instrument.

Two identical tuning forks, each fixed to the top of hollow boxes, may be used to illustrate resonance (Fig. 17.14). Each box resonates at the tuning fork's frequency providing larger vibrating surfaces that allow the vibrational energy of the tuning fork to be transferred to the air more efficiently. When we strike one tuning fork, causing it to vibrate at its natural frequency, the sound waves travel to the other tuning fork. The frequency of these sound waves are precisely at the natural frequency of the second fork, and therefore it is driven into motion. If by touching it we stop the motion of the first tuning fork, we find that the second fork is vibrating and emitting sound waves at the same frequency. When we repeat this experiment but slightly change the natural frequency of the second fork (perhaps by winding a small rubber band on one of the tips), we find that this second fork does not resonate.

The two tuning forks in this example may be used to illustrate the phenomenon of beats discussed in Chapter 15. Suppose we set both tuning forks into vibration by simultaneously striking each. (Remember, we slightly changed the frequency of one of the forks!) In Section 15.5 we found that the superposition of two waves with slightly different frequencies causes the amplitude to be modulated. That is, the amplitude of the composite sound wave varies periodically. The frequency f with which the composite amplitude varies is given by Equation (15.16): $f_{\text{beat}} = f_1 - f_2$, the difference between the frequencies of the two tuning forks. Also, the frequency of the combined wave is the average frequency of the two forks. (See Problem 65.) For example, if the two tuning forks vibrate at 220 Hz and 216 Hz, respectively, the resultant sound wave has a frequency of 218 Hz, and its amplitude varies in intensity with a beat frequency of 4 Hz.

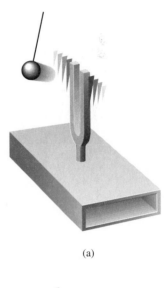

(a)

(b)

FIGURE 17.14

An example of resonance: When tuning fork (a) is set into vibration, the identical tuning fork (b) is driven to vibrate at the same frequency.

Christian Johann Doppler (1803–1853)
Austrian mathematician and physicist.

17.6 The Doppler Effect

According to Table 17.1, the speed of sound waves emitted from a source in still air at 20°C is 343 m/s. However, if an observer moves toward or away from the sound source, the apparent speed of the sound waves must increase or decrease with respect to the observer. That is, the rate at which sound wavefronts impinge on the observer's ear increases if the observer is in motion toward the source or decreases if she moves away from the source. The rate at which wavefronts reach the observer's ear is just the frequency she notes for that sound wave. Therefore, a moving observer notices a change in the frequency of the sound source.

Similarly, a stationary observer measures a change in the frequency if the source is in motion. For example, if your friend drives her car toward you while sounding the horn, you will notice the frequency of the horn to be higher than the frequency you would hear if the car were not moving. After the car passes and is moving away from you, the frequency of the horn will seem lower to you. The frequency change due to motion of the source, observer, or both is called the **Doppler effect** after Christian Doppler (1803–1853) who first derived the formulas that describe this effect.

Let's begin to compute the Doppler frequency shift by first thinking about what happens when neither the source nor the observer is in motion. Figure 17.15 shows a source and an observer separated by a distance d and the sound wavefronts traveling from

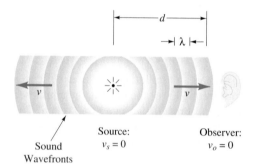

Source:
$v_s = 0$

Observer:
$v_o = 0$

Sound
Wavefronts

FIGURE 17.15

Sound waves emitted by a stationary source and observed by a stationary observer.

the source to the observer with speed v. The speed of sound is given by $v = f\lambda$. In time t the number N of wavefronts reaching the observer is

$$N = \frac{Distance\ traveled\ in\ time\ t}{Wavelength} = \frac{d}{\lambda} = \frac{vt}{\lambda}$$

Therefore, the frequency f' of the sound waves at the observer position is

$$f' = Frequency\ at\ observer = \frac{N}{t}$$

$$= \frac{vt/\lambda}{t} = \frac{v}{\lambda} = f$$

$$f' = f$$

This result is no great surprise; the frequency of the observed waves is exactly equal to the frequency of wavefronts emitted.

Now let's see what happens when the observer is in motion toward or away from the source. If the observer moves *toward* the source with speed v_o (Fig. 17.16), the wavefronts have a relative speed of $v + v_o$ toward the observer. Similarly, if the observer were moving away from the source, he would see wavefronts traveling toward him with the reduced speed of $v - v_o$. If we express the relative speed of the sound waves and the

Source:
$v_s = 0$

Observer:
v_o Toward Left

FIGURE 17.16

Sound waves emitted by a stationary source and heard by an observer moving toward the source.

observer as $v \pm v_o$, the number of waves reaching the observer's ear in time t is

$$N = \frac{(v \pm v_o)t}{\lambda}$$

and the frequency of the sound waves at the observer is

$$f' = \frac{N}{t} = \frac{\left[\dfrac{(v \pm v_o)t}{\lambda}\right]}{t}$$

$$= \frac{v \pm v_o}{\lambda} = \frac{v \pm v_o}{v/f}$$

or finally

$$f' = f\frac{v \pm v_o}{v} \qquad \text{(Use: + for motion toward source, and} \qquad \textbf{(17.23)}$$
$$\text{− for motion away from source)}$$

Suppose next that the observer is at rest and the source is in motion with speed v_s. This situation is shown in Figure 17.17. Notice that we place stationary observers both to the left and right of the source. The source is in motion to the right. From this figure we see that the wavelength of the sound waves decreases for motion toward an observer and increases for motion away from an observer. Let's first consider motion toward the observer (the right side of Fig. 17.17).

During the time between wavefront emissions the source has moved to the right, decreasing the distance between wavefronts by just the distance it travels during this time. Therefore,

$$\lambda' = \lambda - (\textit{Distance the source moves between wavefront emissions})$$

$$= \lambda - (v_s \times \textit{Time between emissions})$$

Now, the time between wavefront emissions is just the period $T = 1/f$ of the wave, so

$$\lambda' = \lambda - v_s T = \lambda - \frac{v_s}{f}$$

$$\frac{v}{f'} = \frac{v}{f} - \frac{v_s}{f} = \frac{v - v_s}{f}$$

$$f' = f\frac{v}{v - v_s} \qquad \text{(Source moving toward the observer)} \qquad \textbf{(17.24)}$$

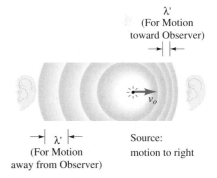

λ'
(For Motion
toward Observer)

λ'
(For Motion
away from Observer)

Source:
motion to right

FIGURE 17.17

Sound waves emitted by a source in motion toward the right. For an observer on the right the wavelength appears to be shortened, whereas an observer on the left receives wavelengths that are lengthened.

When the source moves away from the observer, as illustrated on the left in Figure 17.17, the distance between wavefronts is increased by $v_s T$. Therefore,

$$\lambda' = \lambda + v_s T$$

$$= \lambda + \frac{v_s}{f}$$

Replacing each occurrence of λ by the appropriate v/f this expression becomes

$$\frac{v}{f'} = \frac{v}{f} + \frac{v_s}{f} = \frac{v + v_s}{f}$$

Upon solving for f' we find

$$f' = f\frac{v}{v + v_s} \qquad \text{(Source moving away from the observer)} \qquad \textbf{(17.25)}$$

We can combine the results of Equations (17.24) and (17.25) to obtain

$$f' = f\frac{v}{v \mp v_s} \qquad \begin{array}{l}\text{(Use: + for motion away from observer}\\ \quad\quad\; - \text{ for motion toward observer)}\end{array} \qquad \textbf{(17.26)}$$

In fact, if both source and observer are in motion, the general expression for the observed frequency is

$$f' = f\frac{v \pm v_o}{v \mp v_s} \qquad\qquad \textbf{(17.27)}$$

Table 17.4 summarizes the various combinations of sign choices for all applications of Equation (17.27). Finally, we wish to emphasize that the sound velocity appearing in Equation (17.27) is invariably measured relative to the air through which the sound is moving. Thus, all other speeds must be measured relative to the air.

TABLE 17.4   Doppler frequencies (f')

	SOURCE STATIONARY	SOURCE TOWARD OBSERVER	SOURCE AWAY FROM OBSERVER
Observer stationary	f	$f\left(\dfrac{v}{v - v_s}\right)$	$f\left(\dfrac{v}{v + v_s}\right)$
Observer toward source	$f\left(\dfrac{v + v_o}{v}\right)$	$f\left(\dfrac{v + v_o}{v - v_s}\right)$	$f\left(\dfrac{v + v_o}{v + v_s}\right)$
Observer away from source	$f\left(\dfrac{v - v_o}{v}\right)$	$f\left(\dfrac{v - v_o}{v - v_s}\right)$	$f\left(\dfrac{v - v_o}{v + v_s}\right)$

v = sound speed; v_s = source speed; v_o = observer speed

EXAMPLE 17.5 *The Doppler Shift*

A student finds a battery-powered horn in the physics stockroom. When connected to the battery, the horn has a frequency of 800. Hz. Determine the frequency another student (the observer) hears if (a) the source is stationary while the observer rides a bicycle toward it at a speed of 15.0 m/s, (b) the source moves at a speed of 15.0 m/s toward the stationary observer, and (c) both the source and observer move toward each other, each with a speed of 15.0 m/s.

SOLUTION (a) When the source is stationary $v_s = 0$ and when the observer moves toward the source at $v_o = 15.0$ m/s, we must use the plus sign in the numerator of Equation (17.27):

$$f' = f\frac{v + v_o}{v} = (800.\ \text{Hz})\left(\frac{343\ \text{m/s} + 15.0\ \text{m/s}}{343\ \text{m/s}}\right)$$

$$= 835\ \text{Hz}$$

(b) In this case $v_o = 0$ and $v_s = 15.0$ m/s, and we use the negative sign in the denominator of Equation (17.27):

$$f' = f\frac{v}{v - v_s} = (800.\ \text{Hz})\left(\frac{343\ \text{m/s}}{343\ \text{m/s} - 15.0\ \text{m/s}}\right)$$

$$= 836\ \text{Hz}$$

(c) When both source and observer are moving toward each other, Equation (17.27) gives

$$f' = f\frac{v + v_o}{v - v_s} = (800.\ \text{Hz})\left(\frac{343\ \text{m/s} + 15.0\ \text{m/s}}{343\ \text{m/s} - 15.0\ \text{m/s}}\right)$$

$$= 873\ \text{Hz}$$

◀

The Doppler effect can occur for any type of periodic wave. Although we have confined our attention to sound waves, the effect can be observed for light waves. As we shall see in Chapter 34, light travels with a speed of 3.00×10^8 m/s. In astronomy, the Doppler effect is employed to determine the rate at which a star is moving away from us. Light waves with the lowest frequencies visible to our eyes appear red to us. If a star emits light of a higher frequency, such as blue, then, because of its motion away from us, the frequency of its light appears "red-shifted" toward lower frequencies. However, the same Doppler Equation (17.27) *cannot* be used to find the velocity of the star. Instead, we must use a modification of Equation (17.27) that arises from the effects of the theory of special relativity. For electromagnetic waves there is no "special" frame of reference, such as air. Light travels at 3.00×10^8 m/s in vacuum relative to all frames of reference, and it is impossible to tell whether it is the source that is moving or the observer. When the effects of special relativity are taken into account, the relativistic Doppler equation becomes

$$f' = f\sqrt{\frac{1 \pm (v/c)}{1 \mp (v/c)}}$$

where v is the *relative* speed of the source and observer and c is the velocity of light in vacuum. The upper signs are used for approaching source and observer, and the lower signs are for receding source and observer.

Shock Waves

In our derivation of the Doppler shift (Eq. (17.27)) we did not consider cases in which the source velocity v_s exceeds the velocity of sound v. When $v_s > v$, a shock wave results as illustrated in Figure 17.18. The source is traveling to the right with velocity v_s and, just as in Figures 17.15, 16, and 17, the circles represent spherical wavefronts emitted at a regular

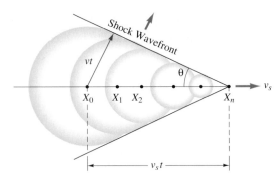

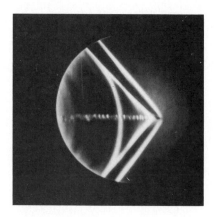

Shock waves are produced when a projectile travels faster than the speed of sound in air.

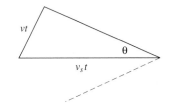

FIGURE 17.18

A sound source moves with velocity v_s *greater* than the speed of sound v in the medium. The envelope of spherical wavefront forms the conical shock wave.

sequence of times. The air in front of the source is undisturbed because the source travels at a speed greater than that of sound. The source is located at position X_o at time $t = 0$ and at position X_n at time t later. During this time t, the wavefront emitted at X_o has traveled a distance vt from this position while the source has traveled a distance of $v_s t$. The line from X_n to the wavefront emitted from X_o is tangent to all the wavefronts emitted at the regular time intervals from $t = 0$ to time t. It is this superposition of wavefronts that forms the conically shaped shock wave. From Figure 17.18 we see that the path length $v_s t$ and the X_o wavefront radius vt form a right triangle. Therefore, we can compute the angle of the shock-wave cone:

$$\sin(\theta) = \frac{vt}{v_s t} = \frac{v}{v_s}$$

The ratio $v_s/v = 1/\sin(\theta)$ is called the **Mach number.** The shock waves produced by aircraft traveling at supersonic speeds carry a great amount of energy. The air pressure variation associated with this type of shock wave produces the unpleasant "sonic boom" you hear when in the vicinity of an aircraft traveling with such a speed. There are other examples of shock waves: The V-shaped wave emanating from the bow of a boat is a kind of "water shock wave" produced when the boat travels faster than the speed of the water waves. When charged particles move through water at a speed greater than the speed of light in water, the light waves they emit are a type of shock wave called Cerenkov radiation.

Changing the fuel element of the High Flux Isotope Reactor (HFIR) at Oak Ridge National Laboratory. The blue glow in the water surrounding the fuel element is Cerenkov radiation. (Photo courtesy of Oak Ridge National Laboratory.)

17.7 Summary

Sound waves are three-dimensional longitudinal waves that consist of compressions and rarefactions of the medium. The speed of sound in a fluid of adiabatic bulk modulus B_{ad} and density ρ is

$$v = \sqrt{\frac{B_{ad}}{\rho}} \qquad (17.6)$$

The intensity of a sound wave is the power per area that is distributed over a wavefront and is given by the expression

$$I = \frac{\mathcal{P}}{A} = \frac{1}{2}\rho v(\omega X_o)^2 \qquad (17.12)$$

where X_o is the wave amplitude, ρ is the medium density, and v is the wave velocity. For spherical waves the intensity decreases inversely as the square of the distance from the sound source. The sound intensity level L_I is measured in decibels and is computed from the sound intensity I:

$$L_I = (10 \text{ dB}) \log_{10}\left(\frac{I}{I_o}\right) \qquad (17.14)$$

where I_o is 10^{-12} W/m^2.

Standing sound waves create musical notes in string and wind instruments. The frequency of the sound waves emitted by a particular instrument is determined by the boundary conditions for the waves. String instruments have nodes at the fixed ends of the string. In the tube model of a wind instrument both ends are displacement antinodes if the tube is open at both ends. If one end of the tube is closed, that end becomes a displacement node.

If a source of sound or an observer moves with velocity v_s or v_o, respectively, the observed frequency of the sound f' is changed. This Doppler shift in frequency is given by

$$f' = f\frac{v \pm v_o}{v \mp v_s} \qquad (17.27)$$

where f is the source frequency and v is the speed of sound. The signs are chosen so that the observed frequency f' is higher for relative motion toward the observer and lower for relative motion away from the observer.

If a sound source moves through a medium with a speed greater than the speed of sound through that medium, a shock wave results.

PROBLEMS

Unless otherwise noted, use the velocity of sound in air as 343 m/s and the density of air as 1.29 kg/m^3.

17.2 The Velocity of Sound

1. From the data in Tables 16.2 and 16.3 compute the speed of sound in copper and lead.
2. When a gauge pressure of 2.00 atm is applied to a certain material, its volume decreases by 5.00 parts in 10^6. If the density of the material is 6.00 g/cm^3, what is the speed of sound in this material?
3. Show that the right-hand side of Equation (17.6) has the units of velocity.
4. If the density of mercury is 13.6 g/cm^3, use the data in Table 17.1 to compute the bulk modulus of mercury.
5. What are the minimum and maximum wavelengths of sound audible to the average human ear?
6. Bats can emit ultrasonic waves with frequencies as high as $100. \times 10^3$ Hz (100. kHz). What is the wavelength of such a wave?
7. A rock is dropped from a bridge into the river below. The splash is heard 2.00 s after the rock is released. What is the height of the bridge above the water?

8. A sound wave travels from the top of a long, cylindrical container filled with distilled water. It is reflected by the container's bottom and returns to the top of the container. The total time for the sound wave to travel to the bottom and back is 0.0200 s. What length of container is required for a sound wave to travel the same path in the same time if the container is filled with seawater?

9. In many land regions where oil exploration is carried out there is a surface layer consisting of an inhomogeneous region several hundred meters thick. Seismic waves generated at the earth's surface (by a dynamite blast or other source) travel through this layer and are reflected by the boundary between the inhomogeneous layer and underlying rock. Suppose such a layer is 300. m thick and overlays a layer of granite. When a dynamite source is detonated, a detector located on the surface 800. m from the source receives a signal from a reflected sound wave 0.220 s after the detonation. What is the average sound velocity in the inhomogeneous layer?

17.3 Harmonic Waves in Air

10. Find the displacement amplitude of a 5.00-kHz sound wave that has a pressure wave amplitude of 1.30×10^{-2} N/m^2.

11. Find the pressure amplitude of a 1.00-kHz sound wave with a displacement amplitude of 1.00×10^{-10} m in air.

12. A sound wave in air has a displacement amplitude of 1.20×10^{-8} m and a pressure wave amplitude of 0.900 Pa. Is this sound wave audible to the human ear?

13. The wavelength of a sound wave in air is 2.00 m. What is the pressure amplitude of this wave if the displacement amplitude is 2.20×10^{-7} m?

14. The pressure amplitude of a 1.00-kHz sound wave is 2.00 Pa. At $t = 0.00$ s the pressure is a maximum at position $x = 0.00$. (a) Write the pressure wave function for this sound wave. (b) What is the pressure variation at $x = 1.60$ m and $t = 0.500$ s?

15. The displacement amplitude of a 5000.-Hz sound wave is 1.20×10^{-10} m. At $t = 0.00$ s and $x = 0.00$ m the displacement is zero. (a) Write the displacement wave function for this sound wave. (b) What is the displacement at $x = 0.500$ m and $t = 1.20$ s?

16. The displacement function of a sound wave is given by the wave equation

$$X(x, t) = (2.40 \times 10^{-10} \text{ m}) \sin\{2\pi[(2.00 \text{ m}^{-1})x - (686 \text{ s}^{-1})t]\}$$

(a) What are the frequency and the wavelength of the wave? (b) What is the displacement of the wave at $x = 2.50$ m and $t = 1.50$ s?

17. A plane sound wave propagates through still air with a frequency of 512 Hz and a pressure variation amplitude of 4.00 Pa. Find the (a) wavelength and (b) the displacement amplitude.

18. Write the wave function for the pressure variation of a sound wave that has a wavelength of 5.00 cm and a displacement maximum 1.10×10^{-7} m.

19. Write the displacement wave function for a sound wave with a frequency of 2.00 kHz and a maximum pressure variation of 2.00 Pa.

17.4 Sound Intensity and Sound Intensity Level

20. A 60.0-W point source emits sound waves into still air. Compute the intensity of the sound at a distance of 2.00 m from the source.

21. Find the sound intensity level (in decibels) at a position 20.0 m from a source that emits sound at a power of 50.0 W. Assume the wavefronts are spherical.

22. At a distance of 2.00 m from a source the sound level is 120.0 dB. At what distance is the sound level (a) 60.0 dB and (b) 30.0 dB?

23. Compute the sound intensity levels (in decibels) of a sound wave that has an intensity of (a) 0.200×10^{-3} W/m^2, (b) 2.00×10^{-3} W/m^2, (c) 20.0×10^{-3} W/m^2, and (d) 0.200 W/m^2.

24. (a) Compute the sound intensity level (in decibels) at a position 10.0 m from a source that emits sound at a power of 20.0 W. Assume the sound waves are spherical. (b) At what distance from the source is the sound intensity level reduced to half its value at the 10.0-m position?

25. At what power must a sound wave be produced if it is to be barely audible from a distance of 6.00 m from the source? Assume the sound waves are spherical.

26. A 150.-W source emits spherical sound waves. How far from the source must one stand for the sound intensity level to be 120.0 dB?

27. Compute the pressure wave amplitude for a sound wave with a sound intensity level of (a) 60.0 dB and (b) 120.0 dB.

28. The faintest sound intensity the human ear can detect at a frequency of 1.00 kHz is about 1.00×10^{-12} W/m^2. (a) Determine the pressure variation amplitude for a wave of this intensity. (b) Find the corresponding displacement amplitude and compare it with the average diameter of an air molecule, 1.00×10^{-10} m.

29. A 10.0-W point source emits spherical sound waves with a frequency of 2.00 kHz. (a) Compute the intensity and displacement amplitude of the sound waves at a distance of 1.00 m from the source. (b) Compute the intensity and displacement amplitude at a distance of 10.0 m from the source.

30. A 15.0-W point source emits spherical sound waves with a wavelength of 70.0 cm. Find the intensity, sound intensity level, and displacement amplitude of the sound wave at a distance of 100. m from the source.

17.5 Sources of Sound

31. A violin string 33.0 cm in length has a mass density of 2.00×10^{-3} kg/m. If the tension in the string is 30.0 N, what is the highest frequency overtone that can be heard by a human ear?

32. A musical note whose frequency is a factor of 2.00 above another note is said to be an *octave* above that note. By how much must the tension in a cello string be increased for its fundamental frequency to be increased by an octave?

33. A guitar string is plucked so as to produce its fundamental frequency. The wavelength of the fundamental sound wave is 87.5 cm. What are the frequencies of the first and second overtones for this string?

34. A 65.0-cm guitar string has a mass of 2.60 g and is plucked so as to produce its fundamental frequency. The sound wave emitted has a wavelength of 1.17 m. (a) What is the tension in the string? (b) What is the wavelength of the wave on the string?

35. A piano wire 1.20 m in length is set into vibration so that between the wire ends there are three antinodes and two nodes. What is the wavelength of the sound wave emitted if the string has a mass of 1.80 g and experiences a tension of 600. N?

36. A violin string 32.6 cm in length has a mass of 2.00 g. The sound wavelength of the fundamental mode is 1.75 m. Where must a finger be placed on this same string to play a note with a frequency of 262 Hz?

37. Find the frequencies of the first three harmonics for a pipe 60.0 cm in length that is (a) open at both ends, and (b) closed at one end.

38. Find the minimum length of a pipe that has a fundamental frequency of 256 Hz and is (a) open at both ends, and (b) closed at one end.

39. The air in a tube 1.20 m in length vibrates in the second overtone with a frequency of 352 Hz. If the tube is closed at one end, what is the speed of sound in the air of the tube?

40. The frequency of a certain harmonic in a 1.80-m long tube closed at one end is 338 Hz. The next higher harmonic has a frequency of 435 Hz. Find the velocity of sound in the tube.

41. The air in a tube 1.80 m in length and closed at one end vibrates with a frequency of 138 Hz. If the frequency of the next higher harmonic is 230. Hz, what is the velocity of sound in the tube?

42. A thin-walled tube has a diameter of 4.00 cm and is closed at one end. (a) What length should the tube be for a fundamental frequency of 256 Hz if we ignore any correction for the tube's diameter? (b) If we now take the tube's diameter into account, what is the fundamental frequency for the tube?

43. A thin-walled tube 1.20 m in length has a diameter of 6.00 cm and is open at both ends. (a) By taking into account the correction for the tube's diameter, compute the frequencies of the fundamental and the second harmonic for standing waves in the tube. (b) By how much does your answer to part (a) differ from the frequencies you computed by ignoring the effect of the tube's diameter?

44. To produce the note A (220. Hz) on the E-string of a guitar a finger is placed behind the fifth fret while the E-string is plucked. When a properly tuned E-string is set into vibration in this manner and the A-string is also plucked, 4 beats per second are heard. What are the possible frequencies of the A-string?

45. Two identical tuning forks vibrate at a frequency of 512 Hz. A mischievous physics professor shaves the forked end of one of the tuning forks so that its mass is smaller than that of the other fork. When the two forks are set into vibration, 6.00 beats/s are heard. What is the frequency of the damaged fork? (Warning: Do *not* attempt this exercise on your own. It should be performed only by qualified physics instructors with a sufficient budget!)

17.6 The Doppler Effect

Unless otherwise stated, in Problems 46 through 55 assume the air through which the sound waves travel is at rest.

46. A car travels down the freeway in still air with a speed of 60.0 mph, and the driver sounds the horn as he approaches a friend standing by the road. If the horn has a frequency of 360. Hz, (a) what frequency does his friend hear? (b) What frequency does his friend hear if the driver continues to sound the horn after he has passed and is traveling away at 60.0 mph?

47. A trumpet player on an open train-car plays a steady F-note (349 Hz). If the train is traveling with a speed of 30.0 mph, what frequencies do observers standing by the track hear as the train approaches and recedes? (This type of experiment—trumpets, trains, and observers—is precisely the one by which Buys Ballot first attempted, in 1845, to verify Doppler's equation.)

48. A stationary motorist sounds his horn while a child on a bicycle rides away from her at a speed of 60.0 m/s. (a) If the horn has a frequency of 300. Hz, what frequency does the child hear? (b) What velocity must the child have to hear a frequency of 295 Hz?

49. Two children with identical horns on their bicycles ride toward each other. Their speeds are 5.00 m/s and 7.00 m/s, respectively. If the horns emit sound waves at a frequency of 340. Hz, what frequencies do the children hear?

50. A sound source emits an 800.-Hz tone while traveling toward a stationary object with a speed of 12.0 m/s. The sound waves from the tone are reflected from the object back to the source. What frequency is observed at the source position?

51. A car travels along a street at a speed of 8.00 m/s while the driver sounds a horn with a frequency of 310. Hz. The driver of a parked car sounds its identical horn. How many beats per second do the drivers hear if the cars move toward each other?

52. Two motorists are traveling at 55.0 mph in the same direction on a freeway, and each is equipped with a horn that has a frequency of 350. Hz. (a) If the front motorist sounds her horn, what frequency does the rear motorist hear? (b) If the rear motorist sounds his horn, what frequency does the front motorist hear? (c) Repeat parts (a) and (b) but this time take the velocity of the leading car to be 60.0 mph.

53. When the Concorde SST travels at Mach 1.80, (a) with what speed does it travel and (b) what is the half-angle of the shock-wave cone? (Assume the speed of sound is 320. m/s.)

54. What must be the speed of a supersonic aircraft for the half-angle of the shock-wave cone to be 45.0°?

55. A bullet has a speed of 600. m/s in air. What is the angle of the shock-wave cone?

Numerical Methods and Computer Applications

56. In Section 17.4 the sound intensity level was defined as $L_l = 10 \log_{10}(I/I_o)$, where $I_o = 1.00 \times 10^{-12}$ W/m². Write a program or construct a spreadsheet template that provides a table of decibel values for intensities that range from 1.00×10^{-13} W/m² to 100. W/m². Increase I by factors of 10 (i.e., $I_1 = 10^{-13}$ W/m², $I_2 = 10^{-12}$ W/m², . . . , $I_{16} = 100$ W/m²). Why is a linear graph of L_l versus I not useful?

57. (a) From the definition of sound intensity level given in Section 17.4 show that $\Delta I = (\Delta L_l/4.34)I$. (b) Write a program or construct a spreadsheet template to compute ΔI for I-values that range from 1.00×10^{-5} W/m² to 100. W/m². Increase I by factors of 10 (i.e., $I_1 = 10^{-5}$ W/m², $I_2 = 10^{-4}$ W/m², . . . , $I_8 = 100.$ W/m²). Verify that $\Delta I \propto I$ by plotting ΔI versus I. (You may find it necessary to make several plots, each with no more than four points. Why?)

58. Equation (17.17) provides the frequencies of a sound wave emitted by a vibrating string that is fixed at both ends. Write a program or construct a spreadsheet template to plot the frequency f versus the string tension F for $0 \le F \le 100$ N for *two different* strings. Input should consist of the string length l, the harmonic number n, and the mass density μ. Try various values for the parameters and observe how the frequency changes with the tension F. Is it possible for the curves to cross at some point?

General Problems

59. A boy stands 400. m in front of a high wall, and a girl stands between the boy and the wall. The girl hears the boy shout her name and 0.500 s later hears her name echoed from the wall. How far from the wall is the girl standing?

60. The pressure amplitude of a sound wave is given by the wave function

$$\Delta P(x, t) = (0.500 \text{ Pa}) \cos\{2\pi[(0.500 \text{ m}^{-1})x - (166 \text{ s}^{-1})t]\}$$

(a) What are the frequency, wavelength, and velocity of this wave? (b) What is the pressure variation at $x = 0.850$ m and $t = 0.0250$ s?

61. A speaker system emits sound at a level such that at a particular distance from the speaker the sound intensity level is 100. dB. How many additional, identical speakers located at the same position are required for the sound level to reach 120. dB at the same distance from the speaker?

62. What must be the wavelength in still air of a 10.0-W sound wave for the displacement amplitude to be 2.00×10^{-11} m at a distance of 50.0 m from the source?

63. An organ pipe open at both ends is to be constructed so as to resonate at a fundamental frequency of 262 Hz (middle C) when the temperature is such that the speed of sound is 343 m/s. (a) Compute the required length of the pipe. (b) If the temperature changes so that the speed of sound falls to 333 m/s, what is the new fundamental frequency of this pipe?

64. A simple device used to demonstrate resonance in a tube closed at one end is shown in Figure 17.P1. The device consists of a long

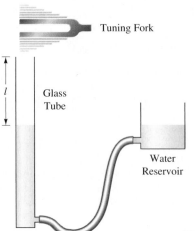

Tuning Fork

l Glass
     Tube

Water
Reservoir

FIGURE 17.P1   Problem 64

glass tube connected to a water reservoir through a flexible hose. By lowering or raising the reservoir, the water level in the tube can be changed, thereby changing the length of the tube. When a vibrating tuning fork is placed above the open end of the tube, the level of the water can be adjusted until the sound waves from the fork are reinforced and a resonance occurs. If the tuning fork has a frequency of 440. Hz, at what values of l are the first, third, and fifth harmonics heard?

65. Show that the superposition of two waves $y_1(t) = A \sin(\omega_1 t)$ and $y_2(t) = A \sin(\omega_2 t)$ results in a wave with a frequency that is the average of those of y_1 and y_2.

66. An amateur radio operator is attempting to adjust the oscillator in his transmitter by comparing its frequency with that of a standard oscillator having a frequency of 5.0000 MHz (1 MHz = 10^6 Hz). If he wants his frequency to be within one-hundredth of a percent (0.0100%) of the standard, what is the maximum beat frequency he can permit between the two oscillators?

67. As a train passes, an observer standing near the track hears the whistle blow with a frequency of 558 Hz as it approaches, and 452 Hz as it recedes. What is the frequency of the whistle and the speed of the train?

68. A car travels with a speed of 20.0 m/s toward a stationary observer. The car's horn is sounded with a frequency of 320. Hz. (a) What frequency does the observer hear if there is no wind? (b) If there is a 10.0 m/s wind blowing from the source toward the observer, what frequency does the observer hear? (c) If this wind is blowing from the observer toward the source what frequency does the observer hear?

69. Two students decide to use their saxophones to measure the speed of a car. The saxophones are both tuned to a frequency of 262 Hz, and each student plays this note while one rides in the car and the other remains stationary. When the car approaches the stationary student, 6.00 beats/s are heard by this student. What is the speed of the car?

70. Atoms of hydrogen in a star emit light with a characteristic frequency of 4.570×10^{14} Hz. If the measured frequency from the hydrogen in a distant star is 4.560×10^{14} Hz, with what speed does this star move away from earth?

71. The speed of light in a certain medium is 2.30×10^8 m/s. If the half-angle between the direction of a particle's motion and the Cerenkov radiation shock-wave cone is 60.0°, what is the particle's speed?

GUEST ESSAY

Musical Acoustics

by THOMAS D. ROSSING

Acoustics is the science of sound. It is a broad and interesting field that encompasses the academic disciplines of physics, psychology, engineering, speech, audiology, architecture, and music. Applications of acoustics are everywhere in our society. We use sound waves to communicate, to diagnose and heal illness, to explore the oceans, to transmit music, to remove dirt and corrosion from delicate surfaces, and for many other purposes.

Among the many branches of acoustics are physical, architectural, engineering, musical, psychological, and electroacoustics; noise control; shock and vibration; speech; underwater acoustics; and bioacoustics. Each of these is a fascinating area that embodies many interesting principles and applications of physics. This essay discusses only one of these areas: musical acoustics.

Musical acoustics deals with:

1. The production of musical sound.

2. The transmission of musical sound to the listener (via the concert hall or through recording and reproduction of sound).
3. The perception of musical sound by the listener.

Musical performance depends on all of these, and all three are based on principles of physics. We will try to show how some of these physical principles apply to musical acoustics.

THOMAS D. ROSSING
Northern Illinois University

Thomas Rossing is a Distinguished Research Professor of Physics at Northern Illinois University. He is the author of over 200 publications mainly in acoustics, magnetism, environmental noise control, and physics education. His current research interests are in musical acoustics and magnetic levitation using superconductors.

Tom is a graduate of Luther College and received his PhD in physics from Iowa State University. He began his professional career with the UNIVAC Division of Sperry Rand, but has been a professor of physics since 1957. He taught at St. Olaf College from 1957 to 1971 and also chaired the Physics Department. He has done research at Stanford University, MIT, Oxford University (England), the Royal Institute of Technology (Sweden), the Institute of Perception Research (The Netherlands), and the Physikalisch-Technische Bundesanstalt (Germany). For the past 4 years he has served as scientist-in-residence at Argonne National Laboratory.

He has been an active member of several scientific societies. He served as president of the American Association of Physics Teachers, and has served on the Governing Board of the American Institute of Physics. In 1992, he was awarded the Silver Medal in Musical Acoustics by the Acoustical Society of America. He served as a National Lecturer for Sigma Xi, and has organized several symposia for the American Association for the Advancement of Science.

The Production of Musical Sound

The oldest musical instrument of all is the human voice. Early in human history, however, other types of instruments were developed to supplement the voice. First came percussion instruments, then primitive wind and string instruments. Throughout history, hundreds of different types of musical instruments have enriched the many cultures of the world.

Musical instruments can be classified according to the nature of their primary vibrating system. **String instruments** employ metal, nylon, or gut strings that are set into vibration by plucking, bowing, or striking. By tuning various strings to different frequencies, a wide range of musical tones can be produced. Because vibrating strings themselves radiate sound very weakly, string instruments nearly always include a soundboard that is driven by the vibrating string.

The primary vibrating element in most **wind instruments** is a column of air, generally enclosed in a pipe. In many instruments, the air column is set into vibration by a vibrating reed, the player's lips, or an air jet. The reed of a woodwind instrument and the lips of a brass instrument player act as pressure-controlled valves, letting puffs of air flow into the air column at just the right time to sustain oscillations. This feedback mechanism is a rather complex nonlinear process and has been the object of considerable research in recent years.

Percussion instruments most commonly depend on some type of vibrating bar, membrane, or plate to produce sound when struck by a drumstick or mallet. Because the vibrational modes of these vibrators are not harmonics of a fundamental frequency (as are the vibrational modes of strings and air columns), the sounds of percussion instruments do not, in general, have harmonic partials. Percussion instruments, therefore, have a rather distinctive quality, or timbre (pronounced "tam-ber"). Some percussion instruments (e.g., timpani, marimba, xylophone, bells, gongs) produce musical sounds having a distinct pitch, and in these cases the vibrating elements have been specially shaped or loaded so that several prominent partials do, in fact, have

frequencies that are close to being harmonics of a fundamental frequency. Other percussion instruments (e.g., bass drum, snare drum, cymbals, triangle) do not convey a sense of pitch, and therefore can be played, without change, with music written in any key.

In our Acoustics Laboratory at Northern Illinois University, we have studied the vibrational modes of several percussion instruments as a way of understanding their acoustical behavior. One of the ways we do this is to use holographic interferometry. This interesting technique displays interference patterns in which the nodes (or pivot lines) appear as bright lines and the antinodes (points of maximum motion) appear as "bull's eyes," as shown in Figure 1. In between, the interference fringes create a sort of contour map that tells us the vibrational amplitude of each point on the vibrator.

Other methods for studying the

◼◼◼ FIGURE 1

Time-average holographic interferograms showing modes of vibration of a Chinese two-tone bell. This bell, copied from 2000-year-old bells found in an ancient Chinese tomb, sounds two distinctly different notes depending on where it is struck. The modes of vibration occur in pairs, which we have labeled as *a* and *b*. (From T. D. Rossing and J. Tsai, "Vibrational modes of a modern Chinese two-tone bell," *Acoustics Australia,* **19,** 73–74 (1992). Reprinted with permission.

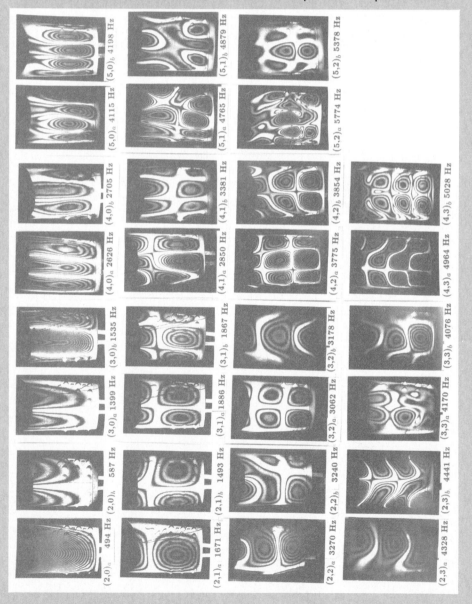

vibrational modes of musical instruments are Chladni patterns, computer-aided modal analysis, and using a small microphone to probe the sound field close to the vibrating surface. By combining the results of all these different methods, we can obtain a fairly accurate picture of how these musical instruments vibrate and produce sound.

Transmission of Musical Sound to the Listener

At one time, studying the transmission of musical sound from the source to the listener was synonymous with the study of concert hall acoustics. That is still a very important part of musical acoustics. Nowadays, however, more music is heard via loudspeakers and headphones than is heard in live performances, and so understanding the recording and reproduction of sound is also a very important aspect of musical acoustics (as well as engineering acoustics). In this essay, we briefly discuss both concert hall acoustics and sound recording.

Concert Halls

In a concert hall, it is important to consider direct, early, and reverberant sound. Sound waves travel at about 344 m/s (1130 ft/s); the **direct** sound may reach the listener after 0.02–0.2 s, depending on the distance from the source to the listener. A short time later, the same sound reaches the listener from various reflecting surfaces, mainly the ceiling and the walls. The first group of reflections that reaches the listener within about 50 ms of the direct sound is generally considered the **early** sound. After the first group of reflections, the reflected sounds arrive fast and furious from all directions, gradually becoming smaller and closer together and blending into what is called the **reverberant** sound. The sound pressure at a particular seat in a concert hall due to an impulsive sound (such as breaking a balloon) on stage might resemble that shown in Figure 2.

Modern research in sound perception (*psychoacoustics*) has pointed out the important contributions to the listeners' experience from each of

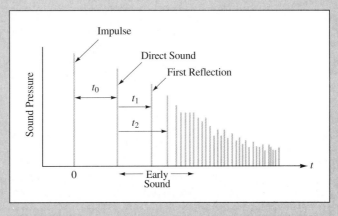

FIGURE 2

Sound pressure at a particular seat in a concert hall due to an impulsive sound on stage. The direct sound first reaches the listener, followed (after time $t_1, t_2, \ldots$) by early reflections, and later by the many reflections from all directions that make up the reverberant sound.

these three types of sound. For example, we judge the direction of the sound source almost entirely on the basis of the first sound to reach our ears, which is ordinarily the direct sound. When a musician attacks a note or a speaker utters a word, our ears are provided with several repetitions of this sound, due to room reflections, that closely follow the direct sound. Provided they arrive within about 35 ms of the direct sound, we do not hear them as separate sounds, but rather as a reinforcement of the first sound. Even if these early reflections carry more energy to our ears than the direct sound, we ignore them in judging the direction of the sound source. This remarkable characteristic of our auditory system is called the **precedence effect**, and it has considerable significance in our perception of stereophonic sound reproduction as well as our perception of sound in a live performance in the concert hall. It accounts for the confusion that can result when a loudspeaker is located closer to the listener than the direct sound source (as in some poorly designed sound systems in concert halls or churches).

Several features of the early reflected sound are important in determining whether a concert hall has "good" or "bad" acoustics. For example, two qualities that are often mentioned by critics are "intimacy" and "spaciousness," and both of these depend on the early reflected sound. A hall is generally considered to have

intimacy when the time delay between the direct and the first reflected sound is less than 20 ms. This is generally the case for listeners seated within 4–5 m of a side wall who receive their first reflection from such a wall, but other listeners who are far away from both side walls and the ceiling are not within this range. For this reason, suspended reflectors are often hung below the ceiling to provide early reflections within this time interval.

A feeling of spaciousness, on the other hand, results from a predominance of lateral reflections in the early sound, so that the early sound is slightly different at the two ears. Many outstanding concert halls are quite narrow so that strong lateral reflections from the side walls are available to most listeners. Other halls include specially designed ceilings that direct sound toward the side walls rather than directly downward to the listeners.

The characteristic of concert halls most familiar to laypersons is the reverberation time. Reverberation time is generally defined as the time for the sound level to decrease 60 dB (in other words, for the sound intensity to fall to one-millionth of its initial value). In a bare room in which all surfaces absorb the same fraction of the sound that reaches them, the reverberation time is proportional to the ratio of room volume to surface area:

$$RT = K \frac{volume}{area}$$

The constant of proportionality K is determined by the nature of the floor, ceiling, and walls of the room.

In practice, of course, concert halls are not bare rooms, and calculating the reverberation time can be fairly complicated. It is important to consider how sound is absorbed by people and seats as well as by the irregular surfaces in the hall. Often the initial sound decay rate is different from the later rate and dominates the judgments of the listeners. The optimum reverberation time depends on the size of the concert hall, the type of music performed, and the listeners' tastes. The reverberation time for low-frequency sounds should be somewhat longer than that of high-frequency sounds. A hall with a short reverberation time is said to be "dry," and a hall with too much reverberation is said to lack "clarity."

Many other considerations enter into the listeners' judgment of whether a hall has good or bad acoustics. It should be emphasized that no two listeners in a concert hall hear exactly the same sounds. A concert hall should never be judged for quality until a listener has sat in a number of different locations.

Sound Recording and Reproduction

The main media for recording sound and music are phonograph discs, magnetic tapes, and compact discs. Phonograph discs record an analog representation of sound; compact discs record a digital representation; magnetic tapes use either one.

Most phonograph records are vinyl plastic discs with circular grooves, with an average density of about 100 grooves per centimeter (250 grooves per inch) and designed to revolve at $33\frac{1}{3}$ rpm. When a record is played, a sharp stylus rides in the groove, following its undulations. The horizontal and vertical motions of the stylus generate electric signals that can be amplified electronically. These signals are usually generated by electromagnetic induction in the phonograph pickup.

Compact disc digital audio is the most exciting development in high-fidelity sound reproduction in recent years. Only 12 cm in diameter, a compact disc can store more than 6 billion bits of binary data to be read out by a laser—equivalent to storing 782 megabytes or more than the capacity of 500 floppy discs. Used for digital audio, a compact disc stores 74 min of digitally encoded music that can be reproduced over the full audible range of 20 to 20 000 Hz. The signal-to-noise ratio and the dynamic range can both exceed 90 dB, and the sound is virtually unaffected by dust, scratches, or fingerprints on the disc, or by small variations in turntable speed (defects that are so annoying in a phonograph).

Digital data, whether in a computer or on a compact disc, is stored in a binary representation, that is, as a series of 1's and 0's. Binary encoding is accomplished by sampling the electrical waveform representing the sound at regular intervals and expressing the samples as binary numbers. The conversion from voltage samples to binary numbers is done by an analog-to-digital converter (ADC).

Recorded information on a compact disc is stored in pits, 0.5 μm wide and 0.11 μm deep, in an aluminized surface. The track, which spirals from the inside out, is about 3 mi in length. This track of pits is recorded and read at a constant 1.25 m/s, so the rotation rate of the disc must change from about 8 to 3.5 rev/s as the spiral track changes its diameter. The optical pickup in a compact disc player embodies some very sophisticated electronic and optical devices, including an aluminum-gallium-arsenide semiconductor laser. The physics of compact disc digital audio was described in an article by the author in *The Physics Teacher* (December 1987).[1]

Modern recording tape consists of a thin coating of magnetic particles (typically 0.5 μm in length) on a plastic base. The coating may be thought of as having a large number of tiny bar magnets that are in random alignment in the unmagnetized state. When a magnetic field is applied by means of the record head, the magnets align themselves in the direction of the field. (The magnetic particles do not move or rotate; the "motion" involved in alignment of the magnets is inside the atoms themselves.)

In a typical tape recorder, the tape passes three heads in succession. First the *erase* head applies a rapidly oscillating magnetic field to erase the old information (demagnetize the tape); then the *record* head magnetizes the tape in the desired pattern; and finally the *playback* head reads the recorded pattern and generates an output voltage as the tape passes by. Some tape recorders use a single head for the record and playback functions, but this is generally less satisfactory than using separate heads.

Digital audio tape recorders (DATs) may employ either stationary heads (S-DAT) or rotating heads (R-DAT). R-DATs use a rotating head design and helical scanning quite similar to that used in video tape recorders. The head rotates at 2000 rpm, and the tape passes it at an angle of 6°23' and a speed of 8.15 mm/s. With a sampling rate of 44.1 kHz, recording 16-bit samples requires a bit rate of 705 600 bits/s. Allowing for error detection, synchronization, and so on, pulse frequencies of 1 MHz and higher are required, which is comparable to the frequency bandwidth required for video recording. S-DATs use many parallel tracks (as many as 22) to record each audio channel, and thus make it possible to

> *. . . no two listeners in a concert hall hear exactly the same sounds.*

record a high bit rate at a modest tape speed.

Recording and playback of music by magnetic and optical devices have led to many interesting applications of physics. As a physicist, I am always curious about how things work. As an amateur musician, I find that understanding the physics of sound recording and reproduction can be useful in improving my performance. I hope this brief discussion has helped motivate some readers to investigate the subject further.

Perception of Sound and Music

Psychoacoustics is the science that deals with the perception of sound. This interdisciplinary field overlaps the academic disciplines of physics, biology, psychology, music, audiology, and engineering and utilizes principles from each of them. Our understanding of sound perception has increased substantially in recent years.

Loudness, pitch, and *timbre* are three attributes used to describe sound, especially musical sound. These subjective attributes depend, in a rather complex way, on measurable quantities, such as sound pressure, frequency, spectrum of partials, envelope, and duration. Relating the subjective attributes of sound to physical quantities is the central problem of psychoacoustics. We illustrate this with a couple of examples.

The sound pressure level (SPL) at a given location can be measured with an instrument called a sound-level meter, which consists of a microphone, an amplifier, and a suitably calibrated voltmeter. We express the sound pressure level in decibels above a reference sound pressure $p_o = 2 \times 10^{-5}$ N/m²:

$$\text{SPL (dB)} = 20 \log \frac{p}{p_o},$$

where p is the sound pressure at the microphone. Note that SPL is proportional to the logarithm of sound pressure. Thus, a tenfold increase in sound pressure results in a 20-dB increase in SPL, whether the pressure changes from 0.0001 N/m² to 0.001 N/m² or from 0.01 N/m² to 0.1 N/m². Doubling the sound pressure results in a 6-dB increase in SPL (because 20 log 2 = 6).

The subjective **loudness** of a sound depends on the SPL at our ears, but it also depends on such quantities as frequency, spectrum of partials, duration, and envelope (i.e., how the amplitude varies with time). Figure 3 shows how the subjective loudness (expressed in *sones*) depends on frequency and sound pressure level for pure tones (i.e., sine waves) and for "musical tones" with five harmonics. The peaks in the pure tone response at about 3000 Hz and 9000 Hz are due to

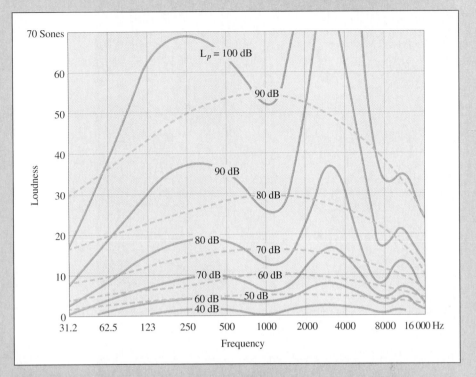

FIGURE 3

Subjective loudness of pure tones (solid curves) and "musical" tones having five harmonics (dashed curves) as a function of frequency and sound pressure level. Loudness is expressed in subjective units called *sones*. These curves represent "average" judgments of subjective loudness, and the variation from person to person is substantial. (From T. D. Rossing, *The Science of Sound,* 2nd ed. (Addison-Wesley, Reading, MA, 1990), p. 94.

resonances of the ear canal in a "typical" individual; the resonance frequencies differ from individual to individual, of course.

In a similar way, the subjective **pitch** of a sound is found to depend mainly on the fundamental frequency, but it also depends on sound level, spectrum, envelope, and the presence of other sounds. The auditory system has a remarkable ability to recognize the fundamental frequency of a complex tone, even when the fundamental is totally missing from the sound spectrum.

In this brief essay, I have attempted to describe the fascinating science of musical acoustics and how understanding the physics of sound can enrich our appreciation of musical sound. It is yet another example of how physics applies to the arts.

References and Further Reading

1. T. D. Rossing, "The compact disc digital audio system," *The Physics Teacher* **25**, 556–562 (1987).

2. D. E. Hall, *Musical Acoustics,* 2nd ed. (Brooks/Cole, Pacific Grove, CA, 1991).

3. K. C. Pohlmann, *Principles of Digital Audio* (H. W. Sams, Indianapolis, 1985).

4. T. D. Rossing, *Musical Acoustics: Selected Reprints* (American Association of Physics Teachers, College Park, MD, 1988).

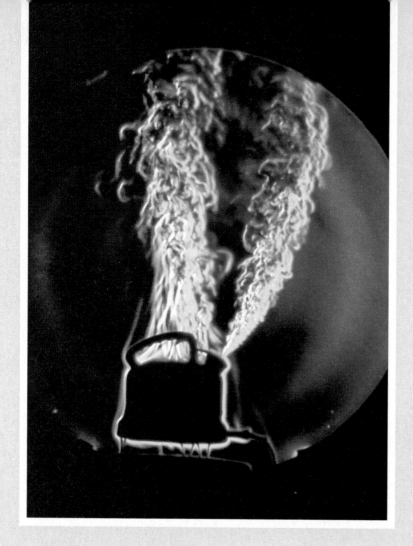

PART

II

THERMODYNAMICS AND KINETIC THEORY

The science of thermodynamics origically evolved as a separate and distinct branch of physics. However, as scientists probed for a deeper understanding of heat, new links with the branch of physics called mechanics were discovered. The links once again revealed an underlying unity behind seemingly different phenomena: mechanics and thermodynamics. Thermodynamics is an experimental science based on empirical laws. It is concerned with the macroscopic behavior of matter when it is subject to energy transfer by two mechanisms: work and heat flow. In these chapters we answer such questions as: What happens to the kinetic energy of a dropped object when it comes to rest on the ground? How much work can a particular engine process perform? When cold and hot fluids are poured into a common container, why don't we ever observe them to im-mediately acquire their former temperatures when poured back into their original containers?

In Chapter 18 we will state a working definition of temperature and establish some other important definitions that allow us to describe the thermodynamic properties of matter. In Chapters 19 and 20 we will examine the consequences of the empirical laws of thermodynamics and study several important thermodynamic processes.

In Chapter 21 we will apply the methods of classical mechanics to a *microscopic* model of a gas and examine some additional consequences of more advanced classical treatments. We obtain a microscopic interpretation of two important macroscopic variables, temperature and pressure, and discover important limitations of our classical model.

CHAPTER

18

Temperature, Heat, and the Equation of State

In this chapter you should learn

- how to compute the change in length, area, and volume of objects when their temperature is changed.

- the definition of heat and how to compute its rate of flow through a material.

- how to compute the temperature change of an object when heat is added to it.

- about state variables that describe a system and the relations between them.

- about Ernest Hemingway:

"About what time do you think I'm going to die?" he asked.

"What?"

"About how long will it be before I die?"

"You aren't going to die. What's the matter with you."

"Oh, yes, I am. I heard him say a hundred and two."

"People don't die with a fever of a hundred and two. That's a silly way to talk."

"I know they do. At school in France the boys told me you can't live with forty-four degrees. I've got a hundred and two."

"You poor Schatz," I said. "Poor old Schatz. It's like miles and kilometers. You aren't going to die. That's a different thermometer. On that thermometer thirty-seven is normal. On this kind it's ninety-eight."[1]

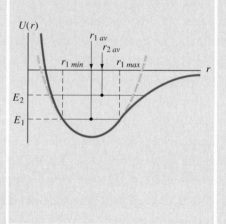

$$\Delta L = \alpha L \, \Delta T$$

---

[1] From Ernest Hemingway, *Winner Take Nothing.* Reprinted with permission of Charles Scribner's Sons, an imprint of Macmillan Publishing Company, Copyright 1933 by Charles Scribner's Sons; renewal copyright © 1961 by Mary Hemingway.

If you grab the metal rack out of a hot oven with your bare hands, you will receive a painful and severe burn. Yet these days it is quite fashionable for brave people (even physics teachers) to walk across a bed of glowing coals without burning their feet. How can this feat be possible? The concepts that allow us to understand this and other fascinating and puzzling questions about hot and cold are the subjects of this chapter.

There are actually two different physical concepts required to answer the question just posed. You probably have some intuitive feel for these concepts, but we need to define each carefully if we are going to base a physical theory on them. These two concepts are temperature and heat. When we have established working definitions that quantify temperature and heat, we will be able to analyze our examples in more detail.

18.1 Temperature

When we studied mechanics, we were able to describe all physical phenomena in terms of only three basic units: length, time, and mass. Concepts, such as velocity, force, kinetic energy, and pressure, all have units that can be reduced to combinations of these three quantities. For the study of thermodynamics we must introduce a new fundamental property: **temperature.** Although our senses can give us an intuitive feel for temperature differences, they are neither reliable nor sufficiently accurate to be useful for a study of the thermal properties of matter. Temperature can be given a precise definition based on the second law of thermodynamics. However, it is sufficient for us at this point to focus our attention on the definition of a **temperature scale** based on some property of a material that changes with temperature.

The Celsius and Fahrenheit Scales

Many materials expand when their temperature is raised. We can use this expansion as a property to distinguish one temperature T from another. As an example, let's define a temperature scale in which T is a linear function of L, the length of mercury in an evacuated glass tube.[2]

$$T = aL + b \qquad (18.1)$$

By assigning particular numbers to T for two different L-values on our scale, we can find the slope a and the intercept b of the linear relation in Equation (18.1). One common and useful assignment is the Celsius scale shown in Figure 18.1. The two **fixed points** of this scale are the ice point and the steam point. The **ice point** is chosen to be 0°C for a mixture of water and ice in equilibrium at one atmosphere of pressure, and the **steam point,** chosen to be 100°C for a mixture of water and water vapor (steam) at the boiling point under one atmosphere of pressure. The following example provides the details of how to compute a and b.

> ### EXAMPLE 18.1 Say "Ahhh"
>
> A thermometer is made by evacuating all the air from a glass tube, partially filling it with mercury, then sealing the ends. The length of a mercury column in the glass tube is 2.40 cm when the tube is immersed in a bath of water and ice at one atmosphere of pressure. The length is 20.8 cm when the tube is placed in a steam bath, also at one atmosphere of pressure. According to this thermometer, what is the temperature of a liquid for which the mercury is 12.5 cm in length?

Physicist John Taylor (University of Colorado) walks across 1200°F glowing embers. **Do not attempt this feat yourself!** An improperly prepared bed of coals can result in severe burns that lead to permanent handicaps. Heat transferred to the feet depends on the heat capacity and thermal conductivity of the embers as well as the time spent in contact with the feet. (See John R. Taylor, "Fire walking: A lesson in physics," *The Physics Teacher* **23** (1989) p. 166.)

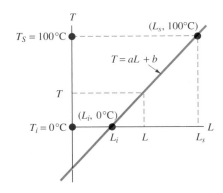

FIGURE 18.1

The Celsius temperature scale for a mercury-in-glass thermometer. The two points $(L_i, 0°C)$ and $(L_s, 100°C)$ are used to find the slope and intercept of the line $T = aL + b$. When the mercury is of length L the thermometer temperature is T.

---

[2] We might just as easily define our scale as a linear relation between T and the length L of alcohol in an evacuated tube, or even the length of an aluminum rod.

SOLUTION We know by the definition of our temperature scale that

$$T = aL + b$$

We are given $L = 2.40$ cm when $T = 0.0°C$ and $L = 20.8$ cm when $T = 100.0°C$. Substituting each into our linear equation, we have

$$T_i = aL_i + b$$

$$T_s = aL_s + b$$

Solving these two simultaneous equations for a and b, we find

$$a = \frac{T_s - T_i}{L_s - L_i} = \frac{100.0°C}{18.4 \text{ cm}} = 5.43°C/\text{cm},$$

$$b = \frac{1}{2}\left(\frac{T_i - aL_i}{T_s - aL_s}\right) = -13.0°C$$

So,

$$T(°C) = (5.43°C/\text{cm})L - 13.0°C$$

This linear equation permits us to find an unknown temperature T from the length of the mercury column in the glass tube. For $L = 12.5$ cm,

$$T = (5.43°C/\text{cm})(12.5 \text{ cm}) - 13.0°C$$

$$= 54.9°C$$

The Fahrenheit scale is another common temperature scale. The ice point is assigned a value of 32°F and the steam point is taken to be 212°F. A linear equation of the same form as Equation (18.1) can be constructed for the Fahrenheit scale in the same manner as in the previous example. Because both the Celsius and Fahrenheit scales are each linear, there is a linear relation between a temperature given in degrees Fahrenheit (T_F) and that given in degrees Celsius (T_C): $T_F = AT_C + B$. We leave as an exercise (Problem 7) for you to show that $T_F = \frac{9}{5}T_C + 32°F$.

In Example 18.1 we used a column of mercury to establish a temperature-measuring device for the Celsius scale. However, a problem arises anytime we try to define a temperature scale by using a particular material. Suppose we construct two glass thermometers, one containing mercury and the other alcohol. We find that even when the two are calibrated so as to agree at both the ice and steam points, they do not exactly agree when brought in contact with an object at some other temperature. The reason for this discrepancy is that the expansion of real materials (mercury and alcohol, in our case) is not exactly linear over all temperature ranges. If the ratio of changes in length for the two substances does not stay constant, then a linear temperature scale based on one of the substances is not linear for the other.

This problem is not unique to mercury and alcohol thermometers. The difficulty is that we have tried to define a fundamental quantity (temperature) in terms of a physical property of a specific substance that behaves in a unique way. Dilute gases are far less sensitive to this problem because they all behave in a nearly identical fashion. For this reason thermometers that employ dilute gases are useful.

Constant-Volume Gas Thermometer

The gas thermometer consists of a gas-filled bulb connected to an open-ended manometer (Figure 18.2). The U-shaped portion of the manometer is a flexible hose. The left arm of the hose is connected to the bulb and has a calibration mark so that during a temperature

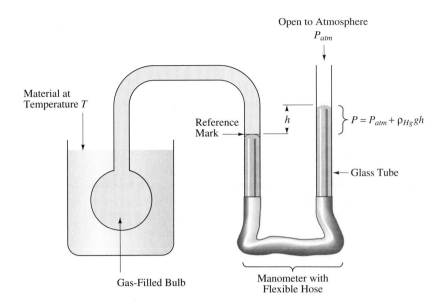

FIGURE 18.2

A constant-volume gas thermometer.

measurement the volume of the gas enclosed in the bulb portion of the device may be kept constant. A constant volume is maintained by raising or lowering the right arm of the manometer tube. That is, every time the temperature of the bulb changes, the height of the right arm of the manometer is adjusted so that the volume of the enclosed gas remains unchanged. The bulb may be made of ordinary glass, porcelain, quartz, or platinum, depending on the range of temperatures to be measured.

Instead of the length L of mercury in a glass tube, the pressure P of the constant volume of gas is the physical property used to establish the temperature scale. We again choose a linear relation, but this time between T and P, and use the ice and steam points to establish a Celsius scale: $T = aP + b$ (Fig. 18.3a). Just as in Example 18.1, the slope and intercept of this relation are computed from the values of the measured physical property at the ice and steam points, this time pressure P_i and P_s, instead of L_i and L_s.

You may be surprised at what happens when we try different gases in the bulb. The phenomenon is shown in Figure 18.3(b). When we extend the lines of our temperature scales to $P = 0$, they all intersect at the same point: $T = -273.15°C$. This result is one bit of evidence that *all gases tend toward the same behavior at low pressures.* We will return to this idea later in this chapter; however, for now we are content to point out that a new temperature scale may be devised that has the advantage of requiring only *one* fixed point. We define the **gas temperature scale,** or the **Kelvin scale:**

$$T_K = T_C + 273.15 \text{ K}$$

The units of this new scale are called **kelvins** (symbol K) and Figure 18.3(c) shows the result of this definition. The graph of the linear relation between T and P passes through the origin so that only one point is required to fix the slope of the line. By international agreement, the fixed point is chosen to be the **triple point** of water. At this temperature and pressure, all three phases of water (solid, liquid, and gas) exist in equilibrium, no one phase growing in mass at the expense of either of the others. This phenomenon happens only when the water, ice, and water vapor are all at a pressure of 4.58 mm Hg. The temperature of this mixture at the triple point is assigned a value of 273.16 K. Suppose we were able to immerse the bulb of our constant-volume gas thermometer into a mixture of water, ice, and water vapor all at the triple point. Moreover, suppose we were to find that the pressure in the bulb was some value P_o. Then the coordinates $(P_o, 273.16 \text{ K})$ would fix the slope of the line of the T-versus-P graph of Figure 18.3(c). It is important to recognize that the pressure P_o in this figure is that of the gas in the bulb of the constant-volume gas thermometer, and *not* 4.58 mm of Hg, the pressure of the water, ice, and water vapor at the triple point.

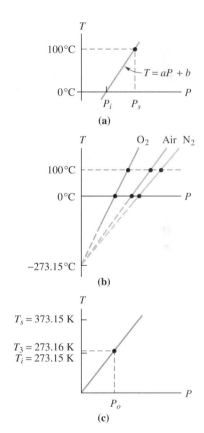

FIGURE 18.3

(a) A linear relation to establish a temperature scale based on the pressure of a constant volume of gas. (b) Linear relations for different gases all have the same intercept. (c) The kelvin temperature scale.

The constant-volume gas thermometer is a useful device because the temperature scales it establishes are far less sensitive to the physical properties of the particular gas used in the bulb. However, this approach still uses a physical property of a substance (the pressure of a gas) to define temperature. A means by which we can define a true thermodynamic temperature is given in Chapter 20. By the way, we don't use the "degree" symbol to designate a kelvin temperature; a temperature of 300 K is read "three hundred kelvins."

18.2 Thermal Expansion

Concept Question 1
If nitrogen were used in a constant-volume gas thermometer what would limit the temperature range over which the thermometer could be used?

Most materials expand when their temperature is raised. There are exceptions to this statement, however, and the most notable is water, which contracts when its temperature is increased from 0°C to 3.98°C. Thus, water has its minimum volume (and hence maximum density) at 3.98°C.

We can understand the thermal expansion of solids by examining the potential energy function of the atoms (or molecules) that compose the material. We can model an individual molecule's motion as that of a point–particle oscillating in a potential well caused by the interatomic forces. In the simplest case we imagine the potential well to be parabolic such as in the case of a Hooke's-law spring ($U(x) = \frac{1}{2}kx^2$). For that case we found that the mass oscillates in simple harmonic motion between maximum and minimum positions, and its *average* position is located at the minimum of the symmetrical potential energy function.

To understand the expansion of solids we must realize that the actual interatomic potential is not symmetrical. A typical potential energy function for a particle in a material

William Thompson (Lord Kelvin) (1824–1907). Sometimes called the father of British physics, Thompson worked in geology, electricity, hydrodynamics, and thermodynamics. His contributions ranged from theoretical to practical and he held many patents. He is reported to have been the first in Scotland to install electric lights. Thompson was knighted Lord Kelvin in 1892 and is buried next to Isaac Newton in Westminster Abbey.

Thermal expansion caused these railroad rails to buckle.

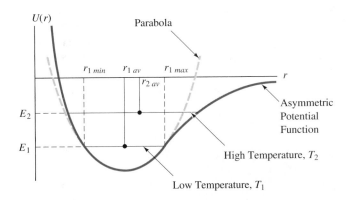

FIGURE 18.4

The potential energy function for an atom or molecule within a material (solid line) and the symmetrical parabolic curve (dashed line). At temperature T_1 the average distance to its nearest neighbor is $r_{1_{av}}$ and at a higher temperature T_2 this distance has increased to $r_{2_{av}}$.

is shown in Figure 18.4. The variable r is the separation distance between this particle and its nearest neighbor. The shape of this $U(r)$ curve is the result of all interactions with the other particles in the material. In Chapter 19 we will find that the temperature is related to the energy of the atoms or molecules composing the material. At temperature T_1 the particle has a total energy E_1, and its separation from its nearest neighbor oscillates between $r_{1_{min}}$ and $r_{1_{max}}$. The particle's average distance from its nearest neighbor is $r_{1_{av}}$. The important difference between this curve and that of the harmonic oscillator is the shape: $U(r)$ is *not symmetrical*. The potential energy curve is flatter to the right at larger r-values. For this reason, at higher temperatures the particle spends more time at r-values toward the less steep portion of the curve. In Figure 18.4, the average position of the particle for a higher temperature T_2, corresponding to the higher energy E_2, is $r_{2_{av}}$. Because $r_{2_{av}} > r_{1_{av}}$, the average separation of the atoms or molecules in the solid increases with increasing temperature.

The previous arguments describe why substances expand when their temperature is raised. But what about the contraction of water when T increases from 0°C to 3.98°C? Substances that contract when their temperature is raised undergo changes in the overall form of their intermolecular potential energy function. That is, the *shape* of the $U(r)$ curve in Figure 18.4 changes with temperature. A much more complex analysis is required to describe this process, but the final result is, of course, that the average separation distance between the atoms or molecules decreases with increasing temperature.

To avoid buckling, bridge surfaces employ expansion joints to permit the highway surface to expand and contract with temperature changes.

Linear Expansion

Figure 18.5 shows a concrete road median that has buckled on a hot day. This phenomenon is just one of many examples in which objects expand when their temperature is raised. The amount of expansion depends, as you might expect, on the particular material.

FIGURE 18.5

Thermal expansion has caused this concrete road median to buckle.

Figure 18.6 illustrates that a rod changes length when we raise its temperature. Experimentally, we find that if we don't change the temperature of a solid by too much, the change in length of the solid is proportional to (1) the original length l, and (2) the change in temperature ΔT. That is,

$$\Delta l = \alpha l \, \Delta T \qquad (18.2)$$

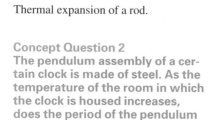

FIGURE 18.6

Thermal expansion of a rod.

Concept Question 2
The pendulum assembly of a certain clock is made of steel. As the temperature of the room in which the clock is housed increases, does the period of the pendulum increase, decrease, or remain the same?

The proportionality constant α is called the **coefficient of thermal expansion.** Alpha itself depends somewhat on temperature, but if the change in temperature ΔT is not too great (often, several degrees or sometimes even several tens of degrees is all right) α can be taken as a constant. However, if you are going to investigate expansion properties over a large temperature range, or if you need a very accurate result, you should check a handbook for how α changes in the temperature region of your interest. Table 18.1 lists coefficients of expansion for some common materials.

TABLE 18.1   Expansion Coefficients for Some Materials

MATERIAL	LINEAR EXPANSION ($°C^{-1}$)	VOLUME EXPANSION ($°C^{-1}$)
Aluminum	24×10^{-6}	
Brass	19×10^{-6}	
Concrete	12×10^{-6}	
Copper	17×10^{-6}	
Glass (ordinary)	9×10^{-6}	
Glass (Pyrex)	3×10^{-6}	
Invar (36% Ni, 64% Fe)	0.9×10^{-6}	
Lead	29×10^{-6}	
Quartz	0.4×10^{-6}	
Steel	11×10^{-6}	
Air		3.7×10^{-3}
Alcohol (Ethyl)		1.0×10^{-3}
Gasoline		9.6×10^{-4}
Helium		3.7×10^{-4}
Mercury		1.8×10^{-4}

EXAMPLE 18.2   *The Long and Short of It*

The road median shown in Figure 18.5 is to be replaced by another concrete median, this time in sections each 40.0 m in length. How much spacing must be left between adjacent sections when the temperature is 15.0°C to prevent buckling up to a temperature of 50.0°C?

SOLUTION   We apply Equation (18.2):

$$\Delta l = \alpha l\, \Delta T$$

$$= (12 \times 10^{-6}\text{°C}^{-1})(40.0\ \text{m})(50.0\text{°C} - 15.0\text{°C})$$

$$= 0.017\ \text{m}$$

The construction crew should leave at least 1.7 cm between each section. ◄

Area Expansion

Objects that can be modeled as two dimensional, such as a thin metallic plate, change area when their temperature is raised or lowered. For an isotropic material, the area expansion may be easily expressed in terms of the *linear* expansion coefficient α. Suppose the rectangular plate shown in Figure 18.7 has an original area A. If its sides change length by an amount Δl and width by Δw when the plate's temperature is changed by ΔT, the new area is

$$A + \Delta A = (l + \Delta l)(w + \Delta w) = (l + \alpha l\, \Delta T)(w + \alpha w\, \Delta T)$$

$$= lw(1 + \alpha\, \Delta T)^2 = A(1 + 2\alpha\, \Delta T + \alpha^2\, \Delta T^2)$$

$$= A + 2A(\alpha\, \Delta T) + A(\alpha\, \Delta T)^2$$

$$\Delta A = A[2(\alpha\, \Delta T) + (\alpha\, \Delta T)^2]$$

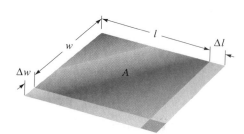

FIGURE 18.7

The area expansion of a flat plate.

The data in Table 18.1 indicates that the size of α is around 10^{-6}°C^{-1}, so the term $(\alpha\, \Delta T)^2$ is about 10^{-12}, one millionth the size of $(\alpha\, \Delta T)$, and may therefore be ignored. Hence,

$$\boxed{\Delta A = A(2a)\, \Delta T} \qquad (18.3)$$

Low T

This equation has the same form as Equation (18.2), with α replaced with 2α.

Figure 18.8 illustrates what happens to a cavity within a flat plate when its temperature is raised: *the cavity expands as would the missing material.*

Volume Expansion

Most objects with expansion properties that are important to us are three-dimensional. Thus we are interested in how their volume changes with temperature. Analogous to the definition of linear expansion, the **coefficient of volume expansion** β is defined as

$$\beta = \frac{1}{\mathcal{V}}\frac{\Delta \mathcal{V}}{\Delta T}$$

or

$$\boxed{\Delta \mathcal{V} = \beta \mathcal{V} \Delta T} \qquad (18.4)$$

High T

FIGURE 18.8

The thermal expansion of a heated flat plate with a cavity. All dimensions increase when the temperature increases, including that of the cavity.

Like the coefficient of linear expansion, β is also a function of temperature. The constant values given for β in Table 18.1 can be used for many applications over reasonable temperature ranges, but the precise definition of β is given by the limiting case of Equation (18.4) as ΔT approaches zero:

$$\beta = \frac{1}{\mathcal{V}} \frac{d\mathcal{V}}{dT} \tag{18.5}$$

Concept Question 3
When a mercury thermometer is placed in a fluid that is initially hotter than the thermometer, the level of the mercury first falls a small amount before rising. Explain this phenomenon.

where it is understood that *pressure is held constant.* We leave it as a problem for you to show that for an isotropic material $\beta = 3\alpha$ (see Problem 22).

Finally, just as in the case of area expansion, a cavity in a three-dimensional object also *expands* as would the missing material.

EXAMPLE 18.3   *Not at These Prices!*

A 4.00×10^3 cm^3 steel can is filled to the brim with gasoline at 20.0°C. How much gasoline overflows the can through the loose cap if while sitting in the sun the temperature of the can and gasoline increases to 35.0°C?

SOLUTION   From Equation (18.4) we write

$$\Delta \mathcal{V} = \beta \mathcal{V} \Delta T$$

The volume of the can (i.e., the cavity) expands as

$$\begin{aligned}
\Delta \mathcal{V}_{can} &= \beta_{can} \mathcal{V}_{can} \Delta T = (3\alpha_{steel}) \mathcal{V}_{can} \Delta T \\
&= 3(11 \times 10^{-6}\,°C^{-1})(4.00 \times 10^3\,cm^3)(15.0°C) \\
&= 2.0\,cm^3
\end{aligned}$$

The volume change of the gasoline is

$$\begin{aligned}
\Delta \mathcal{V}_{gas} &= \beta_{gas} \mathcal{V}_{gas} \Delta T \\
&= (9.6 \times 10^{-4}\,°C^{-1})(4.00 \times 10^3\,cm^3)(15.0°C) \\
&= 58\,cm^3
\end{aligned}$$

The overflow is $\Delta \mathcal{V}_{gas} - \Delta \mathcal{V}_{can} = (58 - 2.0)\,cm^3 = 56\,cm^3!$ ◄

EXAMPLE 18.4   *Is It Enough to Make an Extra Penny?*

For temperatures between 500. K and 1000. K, the coefficient of volume expansion for copper is given by the expression $\beta(T) = AT + B$, where $A = 2.3 \times 10^{-8}$ K^{-2} and $B = 4.4 \times 10^{-5}$ K^{-1}. How much does a 1.00-cm^3 cube of copper change in volume between these temperatures?

SOLUTION   From Equation (18.5)

$$\beta = \frac{1}{\mathcal{V}} \frac{d\mathcal{V}}{dT}$$

so that

$$\beta\,dT = \left(\frac{1}{\mathcal{V}} \frac{d\mathcal{V}}{dT} \right) dT = \frac{1}{\mathcal{V}} \left(\frac{d\mathcal{V}}{dT} dT \right) = \frac{1}{\mathcal{V}} d\mathcal{V}$$

Substituting the expression for $\beta(T)$, this expression becomes

$$\frac{d\mathcal{V}}{\mathcal{V}} = \beta(T)\,dT = (AT + B)\,dT$$

Now, we integrate from the initial configuration to the final

$$\int_{\mathcal{V}_o}^{\mathcal{V}_f} \frac{d\mathcal{V}}{\mathcal{V}} = \int_{T_o}^{T_f} (AT + B)\,dT$$

$$\ln(\mathcal{V})\Big|_{\mathcal{V}_o}^{\mathcal{V}_f} = \left(\frac{AT^2}{2} + BT\right)_{T_o}^{T_f}$$

$$\ln\left(\frac{\mathcal{V}_f}{\mathcal{V}_o}\right) = \frac{A}{2}(T_f^2 - T_o^2) + B(T_f - T_o)$$

$$\mathcal{V}_f = \mathcal{V}_o \exp\left[\frac{A}{2}(T_f^2 - T_o^2) + B(T_f - T_o)\right]$$

$$= (1.00\text{ cm}^3)\exp\left\{\frac{(2.3 \times 10^{-8}\text{ K}^{-2})}{2}[(1000.\text{ K})^2 - (500.\text{ K})^2]\right.$$
$$\left. + (4.4 \times 10^{-5}\text{ K}^{-1})(1000.\text{ K} - 500.\text{ K})\right\}$$
$$= 1.03\text{ cm}^3,$$

Therefore, $\Delta\mathcal{V} = 0.03\text{ cm}^3$.

18.3 Heat and Energy Transfer Mechanisms

In the first two sections of this chapter we presented the concepts of temperature and thermal expansion. We described three temperature scales to qualify the concept of relative hotness or coldness. Then, thermal expansion was shown to be a result of the asymmetry in the intermolecular potential energy: molecules with more energy are located at a greater average distance from their neighbors. This type of energy is associated with the microscopic states of an enormous number of individual molecules. From our macroscopic view, we denote it as the **internal energy** (or **thermal energy**):

> Internal energy (or thermal energy): Kinetic and potential energy due to random motions of the atoms or molecules in matter

Internal energy is transferred between two adjacent bodies when their temperatures are different. The energy transferred is called **heat:**

> Heat: Energy transferred between two chunks of matter due only to a temperature difference between the two

Because heat is transferred energy, its units are the same as those for energy: joules (J) in SI and ergs in the cgs system.

Another common heat unit, the **calorie,** has its origin in the days before heat was recognized to be transferred energy. It was originally thought that heat was an actual fluid called *caloric* that was transferred between objects. This fluid model predicted that the total quantity of heat was constant. However, in 1798, an American named Benjamin Thompson (1753–1814) who later became Count Rumford of Bavaria, found that heat

(a)

(b)

(a) A light dusting of snow has not melted over buried railroad tracks where the insulating wood prevented transfer of ground heat. (b) Wooden railroad ties are in fact covered by 5–15 mm of sand. (From A. B. Western, ''Snow on a thermal pattern,'' *The Physics Teacher*, **22,** 29 (1984).)

could be continuously created in ever-increasing amounts by doing work on a system. The particular system he observed was a cannon as it was being drilled; a horse provided the mechanical work. Shortly after Thompson's discovery it was recognized that the word *heat* was being used to describe the amount of internal energy transferred to or from an object.

James Prescott Joule (a British brewer for whom the SI energy unit is named) was among the first to attempt a measurement of the relationship between heat and energy. Joule performed an enormous number of experiments in which he converted numerous types of work into heat. For example, he measured the temperature change of water when stirred with paddles run by descending weights. Joule's value for the mechanical energy equivalent of heat has been superseded by the value described in the paragraph below.

The unit of heat called the *calorie* was originally defined to be the amount of heat necessary to raise the temperature of one gram of water from 14.5°C to 15.5°C at one atmosphere of pressure. Now, however, the definition is given in terms of the SI system:[3]

$$1 \text{ cal} = 4.184 \text{ J} \quad \text{(exactly!)}$$

Now that we've made this statement we better warn you that there are other definitions of the calorie around! This one is that of the United States National Institute of Standards and Technology, NIST (formerly called the National Bureau of Standards). The 1929 International Steam Table Conference in London defined the calorie as 1/860 of something called the *international watt-hour*. This definition results in a value of 4.186 J for a calorie. There is more to the story, but it's not worth going into. The main point is this: if you wish to get the answers given in the back of this text, you must use the calorie that is equivalent to 4.184 J!

In the British engineering system the unit of heat is the **British thermal unit (BTU).** One BTU is the amount of heat required to raise the temperature of one pound of water from 63°F to 64°F. The BTU is equal to 1055 J.

As we have stated above, heat is energy transferred between two pieces of matter that have different temperatures. In the following subsections we describe how this transfer of internal energy can take place. These transfers can usually be classified into one of three categories: conduction, convection, and radiation.

Conduction

Suppose a mass is placed between (and in good physical contact with) two bodies with different temperatures. As Figure 18.9 illustrates, heat flows from the hotter body, through this central mass, to the cooler body. On a microscopic scale this heat flow, called *conduction,* may be viewed as a transfer of kinetic energy from molecules with more kinetic energy to those with less. In a gas, the transfer takes place through collisions. In a solid, the molecules (or atoms) are bound together much more closely than those of a gas. When the temperature of one end of a solid rod is raised, the molecules at that end vibrate with larger and larger amplitudes. These vibrations are coupled to the neighboring molecules (through the bonds that hold the solid together), and this kinetic energy of vibration is

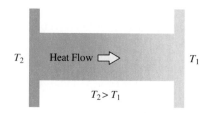

T_2 Heat Flow T_1

$T_2 > T_1$

FIGURE 18.9

Thermal conduction through a mass whose ends are fixed at temperatures T_1 and T_2.

---

[3] If you are a ''weight watcher,'' you need to be aware that the unit nutritionists define as a Calorie (Cal, with a capital C) is 1000 times that of physicists' definition of a calorie (cal, with a small c). So a 500-Cal milkshake is a 500 000-cal (500-kcal) milkshake to a physicist!

transferred from neighbor to neighbor. By this process internal energy is transferred through the solid.

In general, solids are better thermal conductors than gases. One reason is, of course, the closer spacing of the molecules in the solid. Metals are usually very good thermal conductors because in addition to the atomic vibration mechanism for energy transfer, electrons are free to travel throughout the metal and transfer energy. Nonmetallic solids have fewer free electrons and are generally poorer thermal conductors.

The ability of a particular substance to conduct heat is characterized by a constant called the **thermal conductivity k.** Consider the flow of internal energy illustrated in Figure 18.10 in which a bar is conducting heat from a high-temperature region (on the left) to a low-temperature region (on the right). The amount of heat transferred $\bar{d}Q$ through a slab of material in time dt is proportional to the cross-sectional area A of the slab, and the temperature difference across it $\Delta T = T_1 - T_2$ and the duration dt of the contact. After we finish this derivation we will explain the reason we draw a bar through the differential in the symbol $\bar{d}Q$. The heat transfer $\bar{d}Q$ is also inversely proportional to the thickness Δx of the slab: heat is conducted more readily through a thin slab than a thick one. Putting these observations together, we have

$$\bar{d}Q \propto \frac{A \, \Delta T \, dt}{\Delta x}$$

The proportionality constant is k, the thermal conductivity. Thus, the rate at which heat is transferred through the slab is given by

$$\frac{\bar{d}Q}{dt} = -kA \frac{\Delta T}{\Delta x}$$

In the limit of small distances Δx, this relation becomes

$$\frac{\bar{d}Q}{dt} = -kA \frac{dT}{dx} \qquad (18.6)$$

This equation expresses the rate of heat flow through an area A when there is a **temperature gradient** dT/dx across the material. Because heat flows from higher to lower

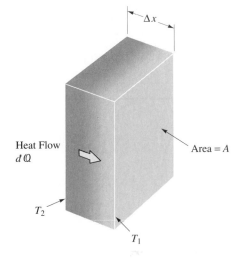

FIGURE 18.10

Heat transfer through a slab of area A and thickness Δx. The opposite faces are maintained at temperatures T_1 and T_2.

Different temperatures are shown by different colors in this liquid-crystal thermometer. The gradation of color from the hot end to the cold end gives visual indication of the temperature gradient.

temperatures, dT/dx is negative, and the minus sign in Equation (18.6) makes dQ/dt positive when heat flows from hot to cold. Table 18.2 provides the values of k for some common solids and gases. We must caution you that the symbol dQ means a small quantity of heat and does *not* represent a differential in the calculus sense. (Hence, we draw a bar through the differential.) Furthermore, the ratio dQ/dt is not a true derivative but merely represents the rate of heat flow.

TABLE 18.2   Thermal Conductivities

SUBSTANCE	THERMAL CONDUCTIVITY, k	
	$\left(\dfrac{\text{cal}}{\text{s}\cdot{}^\circ\text{C}\cdot\text{m}}\right)$	$\left(\dfrac{\text{J}}{\text{s}\cdot{}^\circ\text{C}\cdot\text{m}}\right)$
Metals		
Aluminum	55	230
Brass	29	120
Copper	96	400
Iron	20	84
Lead	8.4	35
Silver	100	420
Gases		
Air	5.7×10^{-3}	2.4×10^{-2}
Hydrogen	3.3×10^{-2}	0.14
Oxygen	5.5×10^{-3}	2.3×10^{-2}
Other		
Concrete	0.33	1.4
Corkboard	9.6×10^{-3}	4×10^{-2}
Glass	0.19	0.8
Ice	0.48	2.0
Polystyrene	2×10^{-3}	1×10^{-2}
Wood	1.9×10^{-2}	8×10^{-2}

EXAMPLE 18.5   *Putting on an Extra Blanket*

A compound slab consists of two materials of conductivities k_1 and k_2 mounted in good thermal contact (Figure 18.11). Each has an area A, and the thicknesses of the slabs are l_1 and l_2, respectively. If the temperatures of the outer surfaces are T_c and T_h, find the rate of heat transfer through the slabs in the steady state.

SOLUTION   In order to apply Equation (18.6) to either of the materials, we must know the temperature of the interface between them. Let's assume it is some (yet to be determined) value T. Then, for material 1,

$$\frac{dQ_1}{dt} = -k_1 A \frac{(T - T_h)}{l_1}$$

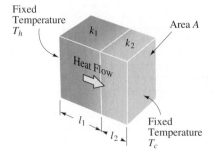

Fixed Temperature T_h

k_1

k_2

Area A

Heat Flow

l_1

l_2

Fixed Temperature T_c

FIGURE 18.11

Heat conduction through a compound slab. See Example 18.5.

And for material 2,

$$\frac{dQ_2}{dt} = -k_2 A \frac{(T_c - T)}{l_2}$$

In the steady state, the rate of heat flow through each material must be identical:

$$\frac{dQ_1}{dt} = \frac{dQ_2}{dt} = \frac{dQ}{dt}$$

From which we conclude

$$k_1 A \frac{(T - T_h)}{l_1} = k_2 A \frac{(T_c - T)}{l_2}$$

Solving for T, we obtain

$$T = \frac{k_1 T_h l_2 + k_2 T_c l_1}{k_1 l_2 + k_2 l_1}$$

Substituting this result into either of the expressions for dQ_1/dt or dQ_2/dt yields

$$\frac{dQ}{dt} = \frac{A(T_h - T_c)}{(l_1/k_1) + (l_2/k_2)} \qquad (18.7)$$

◀

Building Insulation (Optional)

A generalization of Equation (18.6) can be used to compute the rate of heat loss from a home or other building. A generalization is necessary because most walls and ceilings consist of several layers of different materials. Example 18.5 provided us with an expression for the rate of heat flow through a compound slab, so the modification of Equation (18.6) for a two-layer wall or ceiling is just Equation (18.7). For an n-layer boundary of area A,

$$\frac{dQ}{dt} = \frac{A(T_h - T_c)}{\sum_{i=1}^{n} l_i/k_i} \qquad (18.8)$$

where i runs from 1 to the number of layers n. Equation (18.8) is the generalization required for calculating heat loss through multiple-layered walls and roofs.

The term within the sum is the heat path-length divided by the thermal conductivity for one of the building materials. This quantity is called the *R-value* for that material, and Equation (18.8) may be rewritten as

$$\frac{dQ}{dt} = \frac{A(T_h - T_c)}{\sum_{i=1}^{n} R_i} \qquad (18.9)$$

The R-values for various materials at commercial building supply stores are given in units of $ft^2 \cdot °F \cdot h/BTU$. Table 18.3 provides a list of common building materials in both these units and SI. In addition, this table also contains a typical R-value for a thin layer of stagnant air film that is always present on any wall. The actual thickness (and R-value) of this layer is reduced by wind, so the heat loss is greater on breezy days than on calm days.

Concept Question 4
Explain why air "feels" cooler when there is a wind present than when the air is still.

TABLE 18.3   R-Values for Some Typical Building Materials

MATERIAL	CONVENTIONAL UNITS $(\text{ft}^2 \cdot {}^\circ\text{F} \cdot \text{h}/\text{BTU})$	CONVERTED TO SI $(\text{m}^2 \cdot {}^\circ\text{C} \cdot \text{s}/\text{J})$
Air film	0.25	0.04
Brick (4 in thick)	0.8	0.14
Concrete blocks (filled cores)	1.93	0.34
Fiberglass blanket (3.5 in thick)	12	2.1
Fiberglass board (1 in thick)	4.00	0.70
Glass		
Single	0.96	0.17
Insulated (0.25 in air space)	1.64	0.29
Gypsum (0.5 in)	0.45	0.08
Hardwood (1 in thick)	0.91	0.16
Particle board (1 in thick)	1.31	0.23
Polystyrene (1 in thick)	5	0.9
Shingles		
Asphalt	0.44	0.08
Wood	0.87	0.15
Stucco (1 in thick)	0.20	0.035
Vermiculite (1 in fill)	2.13	0.38
Vertical air space (3.5 in thick)	1.01	0.18

EXAMPLE 18.6   Thank Goodness for Fiberglass

(a) Compute the R-value (in both units) of a wall consisting of brick, 1.00-in thick particle board, 3.5-in fiberglass blanket, and half-inch gypsum. (b) Compute the heat loss rate in watts for a 3.00 m × 5.00 m wall when the inside temperature is 20.0°C and outside it is 5.00°C.

SOLUTION   (a) From Table 18.3 we find

	$(\text{ft}^2 \cdot {}^\circ\text{F} \cdot \text{h}/\text{BTU})$	$(\text{m}^2 \cdot {}^\circ\text{C} \cdot \text{s}/\text{J})$
R1 (outside air film)	0.25	0.04
R2 (brick)	0.8	0.14
R3 (particle board)	1.31	0.23
R4 (fiberglass)	12	2.1
R5 (gypsum)	0.45	0.08
R6 (inside air film)	0.25	0.04
R (total)	15	2.6

It should be clear that most of the wall's insulating effectiveness comes from the fiberglass layer. You may be interested to learn that trapped air pockets within the fiberglass actually provide the insulation and not the glass fibers themselves.
  (b) From Equation (18.9)

$$\frac{dQ}{dt} = \frac{A(T_2 - T_1)}{R_{\text{tot}}} = \frac{(15.0 \text{ m}^2)(20.0^\circ\text{C} - 5.0^\circ\text{C})}{2.6 \text{ m}^2 \cdot {}^\circ\text{C} \cdot \text{s}/\text{J}} = 86 \text{ W} \quad \blacktriangleleft$$

Convection and Radiation

When heat is transferred from one part of a fluid to another by movement of the fluid itself, the process is called **convection.** Convection is driven by gravity. When one region of a fluid is heated it expands and has a lower density than the surrounding fluid. Archime-

New Lithosphere Forms at Spreading Center

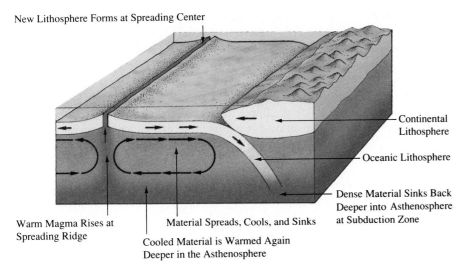

Warm Magma Rises at
Spreading Ridge

Material Spreads, Cools, and Sinks

Cooled Material is Warmed Again
Deeper in the Asthenosphere

Continental
Lithosphere

Oceanic Lithosphere

Dense Material Sinks Back
Deeper into Asthenosphere
at Subduction Zone

Convection of the molten asthenosphere may provide the mechanism for opening ocean ridges and continental motion. (From Carla W. Montgomery, *Physical Geology,* 2nd ed. (William C. Brown, Dubuque, IA, 1990) p. 184. Reprinted with permission.)

des' principle (Chapter 16) says that this less-dense fluid rises. This process is called *natural convection* and is exhibited by air masses in the atmosphere, water masses in the oceans, and even the molten mantle below the solid crust of the earth's surface.

In the heating systems of many homes or buildings, air or water is first warmed through a conduction process and then moved by a fan or pump. This latter type of energy transfer is called *forced convection.* Strictly speaking, this thermal energy transferred by forced convection should not be called "heat" because the transfer is not caused by a temperature difference.

The last form of energy transfer we consider is called **radiation.** Because a complete description of this phenomenon requires some background in electromagnetism, we summarize only the basics. It turns out that whenever an electric charge is accelerated, it emits electromagnetic waves. When an object is hot enough, our eyes can sense these electromagnetic waves and we call them light. However, even when an object is merely warm, it still gives off electromagnetic waves although they are not visible to the unaided eye. Nonetheless, this "radiant energy" travels at the speed of light. When electromagnetic waves impinge on matter, they transfer energy to it by doing work on the electric charges of that matter. The warmth you feel from a campfire is mostly the result of electromagnetic radiation. Two objects that are at the same temperature continually exchange energy by radiation; however, the net transfer of energy between them is zero. Only the net energy exchanged because of a temperature difference is called *heat.*

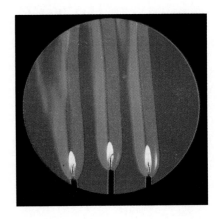

The speeds with which energy can be transferred are much smaller for the processes of convection and conduction than for radiation. The actual movement of hot material in the convection process is typically only several centimeters per second. The conduction of heat through a solid is limited by the rate at which electrons and atomic vibrations can propagate from atom to atom through the solid. The speed of sound in the solid sets an upper limit to the rate of vibrational energy transfer, about 10^4 m/s. In practice, the speed with which heat is conducted in solids is almost always much smaller than this value, due to scattering of the lattice vibrations. On the other hand, radiant energy travels at the speed of light: 3×10^8 m/s in vacuum!

18.4 Heat Capacity and Latent Heat

Generally, when a body absorbs heat, its temperature increases.[4] However, different bodies require different amounts of heat to change their temperatures by the same amount. For example, 1 cal of heat is required to raise the temperature of 1 g of water by 1°C, but we need to add only about 0.1 cal to 1 g of copper to raise its temperature by the same 1°C. The quantity that describes how much heat is necessary to change the temperature of an object is called the object's **heat capacity** C'.

> Heat Capacity: The amount of heat required to raise the temperature of an object by one degree Celsius

If Q is the amount of heat required to raise the temperature of an object by ΔT, then the heat capacity C' is given by

$$C' = \frac{Q}{\Delta T} \tag{18.10}$$

The **specific heat** is the heat capacity per unit mass, $C = C'/m$:

$$C = \frac{1}{m}\frac{Q}{\Delta T} \tag{18.11}$$

Specific heat is a property of materials rather than particular objects. Table 18.4 provides the specific heats of some common substances.

Sometimes it is more convenient to specify the amount of matter present by giving the number of moles present rather than the number of grams. You probably recall from Chapter 1 of this text that a mole of a substance contains Avogadro's number of molecules of that substance, $N_A = 6.022 \times 10^{23}$ particles/mol. The **molar heat capacity** c is the heat capacity per mole of substance, C'/n, where n is the number of moles:

$$c = \frac{1}{n}\frac{Q}{\Delta T} \tag{8.12}$$

All these definitions assume that the heat capacity does not change very much over the temperature range ΔT. If this approximation does not hold, then the proper definition of the heat capacity is $C' = dQ/dT$. Furthermore, $Q/\Delta T$ in Equations (18.11) and (18.12) must also be replaced by the fraction[5] dQ/dT. For example, the specific heat is given by

$$C(T) = \frac{1}{m}\frac{dQ}{dT} \tag{18.13}$$

where we have explicitly indicated the temperature dependence of $C(T)$.

When the specific heat is not constant with temperature, we find the total heat required for a given temperature change from T_o to T_f by integration of the relation ex-

---

[4] An exception to this statement occurs when the body undergoes a phase transition. We consider transitions between solid and liquid and between liquid and vapor phases in more detail later in this chapter.

[5] We remind you once again that dQ/dT is not a derivative but merely represents a ratio.

TABLE 18.4   Specific Heats and Molar Specific Heat Capacities of Some Common Substances (Room temperature, unless otherwise noted)

SUBSTANCE	(cal/g/°C)	(cal/mol/°C)
Aluminum	0.215	5.82
Copper	0.093	5.84
Iron	0.113	5.99
Lead	0.031	6.36
Gold	0.031	6.08
Silver	0.056	6.06
Mercury	0.033	6.69
Glass	0.2	—
Water (15°C)	1.000	34.015
Ice (-10°C to 0°C)	0.55	18.71
Water Vapor (100°C to 120°C)	0.48	16.33
Wood	0.33	—

pressed in Equation (18.13):

$$dQ = mC(T)\, dT$$

and integrating

$$Q = m \int_{T_o}^{T_f} C(T)\, dT \tag{18.14}$$

Only when $C(T)$ is constant over the temperature range from T_o to T_f, can C be taken out of the integral. In that case Equation (18.14) reduces to

$$Q = mC\Delta T \tag{18.15}$$

which is consistent with Equation (18.11). Let's do some examples.

EXAMPLE 18.7   *I Think I'll Stick with My Microwave*

A 40.0-g aluminum spoon at 60.0°C is placed in a thermos container with 30.0 g of soup at 10.0°C. If the heat absorbed by the container is small enough to be ignored, find the final temperature of the soup and spoon. Approximate the specific heat of the soup with that of water.

SOLUTION   If no energy is transferred from outside sources then heat neither enters nor leaves the system. Thus, the total heat transfer Q_{tot} for the system is zero and this problem may be summarized by the following statement:

$$Q_{\text{tot}} = Q_{Al} + Q_{\text{soup}} = 0$$

The heat transfer of both the spoon and the soup is given by Equation (18.15):

$$Q = mC\Delta T$$

Applying this equation to our $Q_{\text{tot}} = 0$ statement, we obtain

$$m_{Al} C_{Al}(T_f - T_{Al}) + m_s C_s(T_f - T_{s_o}) = 0$$

Notice the order of the temperatures in the ΔT terms. The change in temperature is given by "final" − "original" just as all "delta" quantities are in this text. Solving for T_f we obtain

$$T_f = \frac{m_{Al} C_{Al} T_{Al} + m_s C_s T_{s_o}}{m_{Al} C_{Al} + m_s C_s}$$

Substituting numerical values for the masses and the specific heats from Table 18.4, we obtain

$$T_f = \frac{(40.0 \text{ g})(0.215 \text{ cal/g/°C})(60.0°C) + (30.0 \text{ g})(1.00 \text{ cal/g/°C})(10.0°C)}{(40.0 \text{ g})(0.215 \text{ cal/g/°C}) + (30.0 \text{ g})(1.00 \text{ cal/g/°C})}$$

$$= 21.1°C$$

◀

EXAMPLE 18.8   *Just When I Thought I Was Done with Calculus*

The specific heat of carbon dioxide in the temperature range $300 \text{ K} \le T \le 500 \text{ K}$ is given by the expression

$$C(T) = A - B\frac{1}{T} + C\frac{1}{T^2}$$

where $A = (0.105 \text{ kcal/kg/K})$, $B = (76.2 \text{ kcal/kg})$, and $C = (2.96 \times 10^4 \text{ kcal} \cdot \text{K/kg})$. Compute the amount of heat required to raise the temperature of 0.250 kg of CO_2 from 300. K to 500. K.

SOLUTION   Equation (18.14) may be applied to $C(T)$ to compute ΔQ:

$$Q = m \int_{T_o}^{T_f} C(T) \, dT$$

Substituting for $C(T)$, we obtain

$$Q = m \int_{T_o}^{T_f} \left(A - B\frac{1}{T} + C\frac{1}{T^2} \right) dT$$

$$= m \left[AT - B \ln(T) - C\frac{1}{T} \right]_{T_o = 300 \text{ K}}^{T_f = 500 \text{ K}}$$

Concept Question 5
Explain why the numerical value of the specific heat capacity given with the unit of cal/g/°C is exactly equal to the value with a unit cal/g/K.

Substituting numerical values of the constants and the integration limits, we find

$$Q = 5.38 \text{ kcal}$$

◀

Heat of Transformation

When a substance undergoes a change of phase, for example from a solid to a liquid or a liquid to a gas, its temperature does not change as heat is added to the substance. Figure 18.12 shows the temperature variation of one gram of ice initially at −10°C when heat is slowly added to it until it finally reaches a temperature of 110°C as steam. While in route to this final state, the ice undergoes two phase changes. During a phase change the internal energy transferred to the substance does not increase the average kinetic energy of the molecules. Instead, it is expended as work to overcome intermolecular attractive forces so that the molecules may separate from their neighbors. The amount of heat required to change the phase of a substance is called the **heat of transformation.** The heat required to change the phase of a unit mass of a substance is called the **latent heat** L. Therefore, the

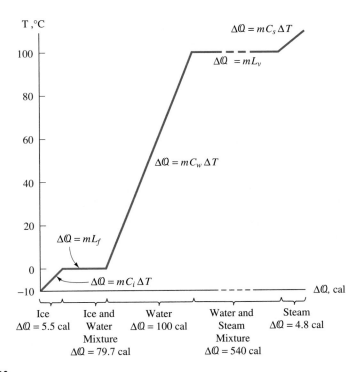

FIGURE 18.12

When heat is added to 1 g of ice at $-10°C$, its temperature rises to 0°C. As more heat is added the ice's temperature remains at 0°C while it changes phase to water at 0°C. When additional heat is added, the water temperature increases to 100°C where it again remains fixed until all the water changes phase to steam at 100°C.

amount of heat necessary to change the phase of a mass m is

$$Q = mL \tag{18.16}$$

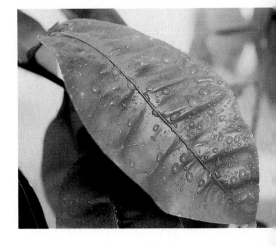

The **latent heat of fusion** L_f is used to compute the heat transferred if the phase change is from a solid to a liquid. If the phase change is from a liquid to a gas, the **latent heat of vaporization** L_v is used. The latent heat of fusion for the ice-to-water transition at 0°C is 79.7 cal/g, and the latent heat of vaporization for the water-to-steam transition at 100°C and atmospheric pressure is 540. cal/g. When the transition takes place in the other direction, heat is liberated by the substance rather than absorbed by it: When water vapor condenses to liquid, each gram of steam gives up 540. cal of heat. Similarly, 79.7 cal of heat are given up by each gram of water that solidifies to ice.

When an early morning frost is expected, gardeners often spray water on their plants. The temperature of the water in contact with the plant does not go below 0°C until large quantities of heat from this water are lost to the cool air as the water freezes. If the air remains below freezing for only a short while, there may be insufficient time to remove enough heat to freeze the water, and the plants will be saved.

EXAMPLE 18.9   *This Sounds like Spring in the Midwest*

How much heat must be added to 40.0 g of ice at 0.0°C to convert it all to steam at 100.°C?

SOLUTION   There are three steps when ice at 0°C is changed to steam at 100°C:

1. ice changes phase to water at 0°C;
2. the temperature of the water is raised from 0°C to 100°C, and
3. the water changes to steam at 100°C.

Heat is required for each of these steps; the total heat required to convert the ice to steam is the sum of the heat increments required for each step:

$$Q = mL_f + mC_w \, \Delta T + mL_v$$

$$= m(L_f + C_w \, \Delta T + L_v)$$

$$= (40.0 \text{ g})[79.7 \text{ cal/g} + (1.00 \text{ cal/g°C})(100.\text{°C} - 0.0\text{°C}) + 540. \text{ cal/g}]$$

$$= 28.8 \text{ kcal}$$

◀

18.5 The Equation of State

In the science of thermodynamics, a **system** refers to a collection of matter that is in some way bounded. This boundary may be real, such as that of an actual container, or it may be the imaginary boundary we place around a particular portion of matter on which we wish to focus our attention. One system we will often refer to is that of a gas enclosed in a cylinder equipped with a movable piston (Fig. 18.13). The cylinder walls and the piston form the boundary; the enclosed gas is the system.

The **state** of a thermodynamic system is designated by giving the values of certain parameters. These parameters, called **state variables,** describe the physical properties of the system and can be experimentally verified. For a gas, appropriate state variables are the pressure P, the volume $\mathcal{V}$, and the temperature T. (We find that P, $\mathcal{V}$, and T are not the only state variables that can be used to describe the state of a gas, but they are sufficient to uniquely specify a gas's state. In later chapters we will meet other state variables, such as internal energy.) Systems other than gases (e.g., electrical and magnetic systems) may require a completely different set of state variables for their description.

Only a certain number of state variables can be independently controlled. When magnitudes are imposed on these variables, the values of the remaining state variables are determined by the nature of the matter in the system. Let's consider a gas as an example. We confine a certain amount of gas to the cylindrical piston shown in Figure 18.13. Experimentally, we find that when we fix the temperature and volume of the gas, the pressure is determined: we cannot choose an arbitrary value for the pressure once the temperature and volume are assigned values for a given amount of gas. Similarly, if we choose the pressure and volume, the nature of the gas fixes the temperature.

The mathematical relation between the state variables of a system is called the **equation of state.** In general, this equation can be very complicated. However, for a gas at low pressures (or, equivalently, low densities), the relation is fairly simple. If you perform experiments on the fixed amount of gas in our piston system (be sure to keep the pressure low!), you find that at a fixed temperature the product of the pressure and volume is a constant. Robert Boyle (1627–1691) was the first to discover this fact about 1660. Continuing to experiment, you find that when the pressure is kept constant, the volume is proportional to the kelvin temperature. This is the law of Charles and Gay-Lussac, discovered in 1802. These results may be combined into the **equation of state for an ideal gas:**

$$P\mathcal{V} = nRT \qquad (18.17)$$

In Equation (18.17), n is the number of moles of the gas, and R is a constant called the **universal gas constant.** Its value is

$$R = 8.314 \text{ J/mol/K} = 0.0821 \text{ L} \cdot \text{atm/mol/K} = 1.986 \text{ cal/mol/K}$$

The ideal-gas equation of state, Equation (18.17), is often called the **ideal-gas law.** An ideal gas is a fictitious gas that is exactly described by Equation (18.17). The ideal gas law describes the behavior exhibited by all real gases when their pressure is sufficiently

FIGURE 18.13

A cylindrical container with a movable piston.

low. This equation works best for monatomic gases (such as helium or argon). For these noble gases the ideal-gas law is applicable even for pressures up to a few atmospheres.

The ideal-gas law contains the three thermodynamic variables P, $\mathcal{V}$, and T. We can see how any two of these quantities are related by plotting their variations while holding the third fixed. Figure 18.14(a) is a graph of P versus $\mathcal{V}$ for several different T-values. The form of the equation $P\mathcal{V} = constant$ is a hyperbola. For larger values of the constant the hyperbolas are farther from the origin. Because the constant is nRT, higher temperatures have $P\mathcal{V}$-hyperbolas farther from the origin. The constant temperature lines are called **isotherms.**

Figure 18.14(b) is a graph of P versus T for several different constant $\mathcal{V}$-values. We can solve Equation (18.17) for P and obtain $P = (nR/\mathcal{V})T$. In the PT-plane this relation is linear with slope $nR/\mathcal{V}$. For smaller $\mathcal{V}$-values, the slope of the P-versus-T line is large. The lines of constant-$\mathcal{V}$ are called **isochors.** The results are similar for the graph of $\mathcal{V}$ versus T when P is held constant; this behavior is shown in Figure 18.14(c). The ideal-gas equation gives $\mathcal{V} = (nR/P)T$. A graph of $\mathcal{V}$ versus T is a straight line with a slope nR/P. Small P-values yield steeper slopes. Constant-P lines are called **isobars.**

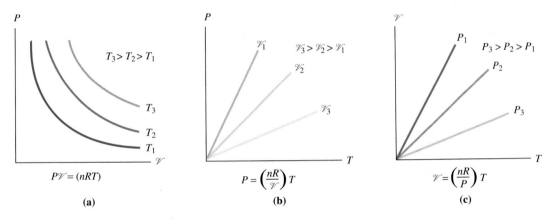

FIGURE 18.14

Graphs of the ideal-gas law. (a) P versus $\mathcal{V}$ for three isotherms, (b) P versus T for three isochors, and (c) $\mathcal{V}$ versus T for three isobars.

We can combine all these two-dimensional plots into a single three-dimensional graph. The result is the $P\mathcal{V}T$-surface shown in Figure 18.15. Points on this surface designate states of an ideal-gas system. An ideal gas may exist with P-, $\mathcal{V}$-, or T-values that lie on this surface.

Real substances have $P\mathcal{V}T$-surfaces that are much more complex than that depicted in Figure 18.15, because all three phases (solid, liquid, and gas) are represented on these

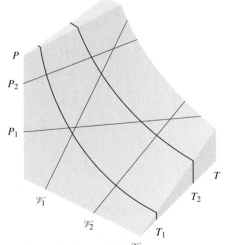

FIGURE 18.15

$P\mathcal{V}T$-surface for an ideal gas. Isotherms, isochors, and isobars from Figure 18.14 are included. Note the isochors and isobars are straight lines, but the isotherms are hyperbolas.

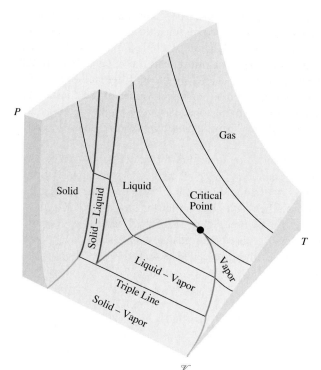

FIGURE 18.16

$P\mathcal{V}T$-surface for a real substance
that contracts on solidification
from the liquid phase.

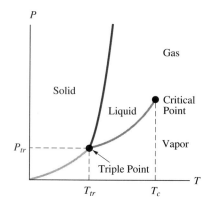

FIGURE 18.17

PT-plane projection of Figure 18.16.
The lines that separate the three phases
intersect at the triple point.

diagrams. One such surface, for a substance that contracts when it freezes, is shown in
Figure 18.16. The projection of this surface onto the PT-plane is shown in Figure 18.17.
The **triple point** is defined by the temperature and pressure at which all three phases of the
substance coexist. This point is the projection of the triple line shown in Figure 18.16. For
a given substance there is only one triple-point coordinate (P_{tr}, T_{tr}). The value of $\mathcal{V}$ at this
point depends on the particular mass of the substance that happens to be present.

The liquid–gas portion of a $P\mathcal{V}$-plot for a real substance is shown in Figure 18.18. At
high temperatures the isotherms approach the shape of those for an ideal gas. Above the
critical temperature T_c the gas does not have a discontinuous change from a vapor state

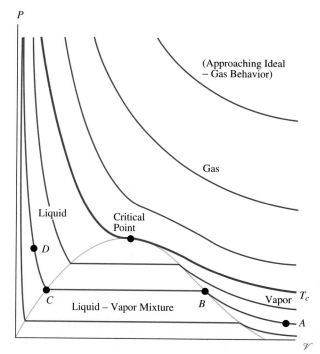

FIGURE 18.18

The liquid–gas portion of a $P\mathcal{V}$
plot from Figure 18.16.

to a liquid state. Indeed, for $T > T_c$ it is a matter of convenience which term, *liquid* or *gas,* we want to use. Usually, if the ideal-gas law is approximately obeyed, the term *gas* is applied. At a temperature below T_c it is traditional to use the term *vapor* for the gas phase. If the volume of the gas is decreased along the isotherm *A,* it begins to liquefy at point *B.* The pressure remains constant from *B* to *C,* and the state of the substance consists of a mixture of liquid and vapor. At point *C* the substance is entirely in the liquid phase. The isotherms are very steep in the liquid region *D* because a liquid requires a very large increase in pressure to decrease its volume by a small amount; liquids are only slightly compressible.

The van der Waals Gas (Optional)

Concept Question 6
Two tanks contain the same number of moles of the same kind of gas. If one tank has twice the volume of the other, but the gas pressures are the same, what can you say about the temperature of the gas in the smaller tank relative to that of the larger tank?

The isotherms of Figure 18.18 in the region above, but not too far from, the critical point have shapes distorted from those of an ideal gas. From this figure we can see that the distortion is due to the gas pressure and temperature coming near the region where liquefication may begin. The distortion of these isotherms is therefore due to molecular attractions that become strong in this $P\mathcal{V}$-region and are not accounted for within the ideal-gas model. An improvement on the ideal-gas law may be made by taking into account these intermolecular forces. Such a gas model that also accounts for the volume actually occupied by the gas molecules (and therefore not available for the molecules to move around within) is called the **van der Waals model.** The form of the van der Waals equation of state is

$$\left(P + \frac{n^2 a}{\mathcal{V}^2}\right)(\mathcal{V} - nb) = nRT \tag{18.18}$$

The constant a is related to the forces between the molecules, and b is related to the volume they occupy. Both constants a and b are chosen to give the best fit to experimental data. Table 18.5 provides these constants for some common gases. Notice that helium has a relatively small correction for intermolecular forces: a is small. However, both fluorobenzene and propane have very large values for the intermolecular force constant and thus require only modest pressures for liquefication to begin.

TABLE 18.5 Van der Waals Constants for Some Gases

GAS	a $(m^6 \cdot Pa/mol^2)$	b $(10^{-5}\ m^3/mol)$
Argon	13.62	3.219
Carbon dioxide	36.39	4.267
Chlorine	65.77	5.622
Fluorobenzene	201.9	12.86
Helium	0.3456	2.370
Hydrogen	2.476	2.661
Methane	22.82	4.278
Nitrogen	14.08	3.913
Oxygen	13.78	3.183
Propane	87.77	8.445
Water	55.35	3.049

When the pressure and temperature are reduced to values within a region sufficiently close to the critical point, the van der Waals equation of state also deviates from the observed behavior of real gases. The graph for the spreadsheet VWAALS.WK1 on the disk accompanying this text provides a comparison of isotherms for a van der Waals gas and an ideal gas. You may change the van der Waals parameters as well as the temperature to see the difference between these two gas models. There are other equations of state that improve Equation (18.18). However, because intermolecular forces are very complex, no simple equation describes with complete accuracy the properties of a real gas in all PVT-regions.

18.6 Summary

Thermodynamics is concerned with macroscopic relations between quantities, such as temperature, heat, and work.

Although temperature can be given a substance–independent definition, empirical temperature scales may be devised that are based on some property of a substance that changes with temperature. Two common scales are established by assigning temperature values to the freezing point and steam point of water at one atmosphere of pressure.

Celsius Scale: $T_f = 0.0°C$ $T_s = 100.0°C$

Fahrenheit Scale: $T_f = 32.0°F$ $T_s = 212.0°F$

The Kelvin scale is related to the Celsius scale by

$$T_K = T_C + 273.15\ K$$

Thermal expansion is a result of the change in average separation between the molecules or atoms of a substance. For many solids this phenomenon may be understood in terms of the shape of the interatomic potential energy function.

Linear Expansion: $\Delta l = \alpha l\, \Delta T$ (18.2)

where α is the **coefficient of linear expansion.**

Area Expansion (isotropic substance): $\Delta A = A(2\alpha)\, \Delta T$ (18.3)

Volume Expansion: $\Delta V = \beta V \Delta T$ (18.4)

For isotropic substances, $\beta = 3\alpha$.

Energy transferred between two masses due to a temperature difference between them is called **heat. Internal energy (thermal energy)** is the kinetic and potential energy associated with the random motions of atoms and molecules in matter.

The rate of one-dimensional heat flow through a uniform material is given in terms of the **thermal conductivity** k of the material

$$\frac{dQ}{dt} = -kA\frac{dT}{dx}$$ (18.6)

where A is the cross-sectional area, and dT/dx is the temperature gradient across the thermal conductor.

The amount of heat required to raise the temperature of an object by one degree Celsius is called the **heat capacity** of that object.

Heat capacity: $C' = \dfrac{Q}{\Delta T}$ (18.10)

The **specific heat capacity** is the heat capacity per unit mass, $C = C'/m$:

$$C = \frac{1}{m}\frac{Q}{\Delta T}$$ (18.11)

The **molar heat capacity** is defined as

$$c = \frac{1}{n}\frac{Q}{\Delta T}$$

(18.12)

where n = number of moles.

When a substance undergoes a change of phase, heat is absorbed or liberated. For a mass m that undergoes a change of phase between a solid and liquid the **latent heat** of fusion is given by

$$Q = mL_f$$

When mass m undergoes a change of phase between the liquid and gaseous phases, the latent heat of vaporization is

$$Q = mL_v$$

The quantities that are uniquely specified by the thermodynamic state of a system are called **state variables.** The **equation of state** is the mathematical relation between the state variables of a system.

Ideal-gas equation of state: $$PV = nRT$$ (18.17)

The van der Waals equation of state takes into account intermolecular forces and the volume occupied by the gas:

$$\left(P + \frac{n^2a}{V^2}\right)(V - nb) = nRT$$ (18.18)

where a and b are constants for a particular gas.

PROBLEMS

18.1 Temperature

1. In 1989, the temperature in Fairbanks, Alaska, dropped to $-86°F$. Express this temperature in (a) degrees Celsius, (b) in kelvins.
2. Normal human body temperature is $98.6°F$. Express this temperature in degrees Celsius.
3. The boiling point of helium is $-268.98°C$, whereas that of oxygen is $-182.97°C$. The freezing point of copper is $1083°C$ and that of cesium is only $28.5°C$. Express all these temperatures (a) in degrees Fahrenheit, (b) in kelvins.
4. Fahrenheit and Celsius thermometers are both immersed in a fluid. The Fahrenheit temperature is 1.60 times the Celsius temperature. Find the temperature on each scale.
5. A new temperature scale is devised in which $30.0°$ is assigned to the ice point and $150.0°$ is assigned to the steam point. Derive an equation that relates this temperature scale to (a) the Celsius scale, (b) the Fahrenheit scale.
6. For the temperature scale devised in Problem 5, what temperature corresponds to zero on the Kelvin scale?
7. Show that the linear relationship between the Fahrenheit and Celsius temperature scales is $T(°F) = \frac{9}{5}T(°C) + 32°F$. At what temperature is the numerical value of each scale the same?
8. The Rankin temperature scale, sometimes used in engineering, is related to the Kelvin scale by $T_R = \frac{9}{5}T_K$. (a) What are the ice and steam points in degrees Rankin ($°R$)? (b) What is the relation between degrees Fahrenheit and degrees Rankin?
9. Platinum resistance thermometers are sometimes employed in experimental physics and engineering. The resistance of such a thermometer is given by the equation $R = A + BT + CT^2$, where A, B, and C are constants. What minimum number of standard temperatures must be measured to calibrate this type of thermometer?
10. A constant-volume gas thermometer reads a pressure of 15.6 cm Hg at $0.00°C$. What is the temperature of the gas when its pressure is 21.4 cm Hg?
11. A constant-volume gas thermometer registers a pressure of 3.00 cm Hg at the triple point of water. What pressure does it register at (a) the boiling point of water, and (b) the melting point of lead ($327.3°C$)?
12. A gas thermometer registers a pressure of 0.800 atm at a temperature of $100.0°C$. What is the temperature when the pressure is 0.600 atm?

18.2 Thermal Expansion

13. An aluminum flagpole is 12.00 m long at $15°C$. What is its length at $40°C$?

14. A steel section of the Alaskan pipeline is 50.0 m long when installed at 70.0°F. What is its length at −60.0°F?

15. What is the fractional change in length of a 500.0-m length of steel railroad track if its temperature is changed from 0.00°C to 35.0°C?

16. The Ni–Fe alloy Invar has a particularly small expansion coefficient ($0.90 \times 10^{-6}°C^{-1}$) and is used to make measuring tapes that have very small temperature-change errors. (a) By how much does the distance between the 10.00-cm and 90.00-cm marks change when the temperature of the tape increases by 50.0°C? (b) How much does the distance between these same two marks change on a tape made from aluminum?

17. At 20.0°C a copper wire 50.00 m in length is supported between two telephone poles. What is its length at 40.0°C?

18. The inner radius of a thick-walled hollow aluminum ball is 5.00 cm, and the outer radius is 8.00 cm. By how much does the wall thickness increase when the temperature is changed by 100.0°C?

19. An ordinary glass windowpane is 1.80 m wide and 1.20 m tall. What change in area must the installer permit for temperature changes from −10.0°F to 100.0°F?

20. A telescope at the former Soviet Special Astrophysical Observatory has a mirror that is 6.00 m in diameter. If the mirror is made of Pyrex, compute the change in its area for a temperature change of 50.0°C.

21. A brass sphere has a radius of 4.00 cm. Find the change in surface area of the sphere when its temperature is raised from 25.0°C to 100.0°C.

22. Show for an isotropic material of length l, width w, and height h, the coefficient of volume expansion β is 3α.

23. A cylindrical copper kettle has a diameter of 10.00 cm and is 15.00 cm high. If the kettle is initially at room temperature (20.0°C), find the change in volume of the kettle when it is filled with boiling water.

24. A fishing sinker made from lead has a volume of 0.200 cm³ at 40.0°C. By how much does its volume change when used for ice fishing? Assume the water temperature is 0.00°C immediately under the ice.

25. Show that the coefficient of volume expansion β may be expressed in terms of the density ρ as

$$\beta = -\frac{1}{\rho}\frac{d\rho}{dT}$$

26. Compute the coefficient of volume expansion for an ideal gas by applying Equation 18.5 to the ideal-gas law. (Remember, P is held constant in this definition.)

18.3 Heat and Energy Transfer Mechanisms

27. A 50.0-kg woman runs up many flights of stairs in a building. If the vertical distance she travels through is 60.0 m, how many (physicist) calories does she use up?

28. A 2.00-kg iron block is at the top of an 8.00-m high inclined plane. The block is given an initial velocity of 2.00 m/s down the incline. Because of friction the block comes to rest just as it reaches the bottom of the incline. If 60.0% of the block's lost mechanical energy is absorbed by the block, what is its temperature increase?

29. A 10.0-g lead bullet is fired into a 1.00-kg wood block, initially at rest. The velocity of the bullet just before impact is 256 m/s. If the initial temperature of both the bullet and block is 20.0°C, what is the final temperature of the block–bullet combination? (Assume no heat is lost from the block-bullet system.)

30. What must be the radius of a cylindrical copper rod for it to conduct heat at a rate of 500. cal/s if the rod is 5.00 cm long and the temperature gradient along the cylinder axis is 1.50°C/cm? (Ignore any heat lost along the sides of the rod.)

31. Figure 18.P1 shows a lock on the inside of a door leading outside a classroom somewhere in the northwestern United States. If the lock shaft is 9.00 cm long and the temperature gradient across the lock is 3.7°C/cm, what is the outside temperature? *Hint*: Note the drop of water forming at the bottom of the keyhole. (From W. P. Crummett, "Frost inside a classroom," *The Physics Teacher* **23**, 363 (1985)).

FIGURE 18.P1   Problem 31

32. An air-filled brass pipe is surrounded by insulation and has an inner diameter of 2.00 cm and an outer diameter of 4.00 cm. The pipe is 30.0 cm long, and one end is placed in good thermal contact with a heat source at 80.0°C. If the other end is maintained at a temperature of 5.00°C, find the rate of heat flow from the 80.0°C end to the 5.00°C end.

33. A glass windowpane is 2.00 m wide and 1.50 m high. How much heat flows through the window each hour when the inside temperature is 72.0°F and that outside is 20.0°F? Assume the windowpane thickness is 3.00 mm.

34. Three cylindrical rods made of aluminum, brass, and iron are each 12.0 cm long and have diameters of 2.00 cm. They are placed end to end in this order and surrounded by insulation. The left end of the aluminum rod is placed in contact with a heat source at 100.0°C, while the right end of the iron rod is maintained at 20.0°C. Find the temperatures at the ends of the brass rod and the rate of heat flow through the rods.

35. Thermopane windows consist of two glass plates separated by a thin air layer. If 1.20 m × 1.50 m × 4.00 mm glass plates are separated by 6.00 mm of air in one such window, at what rate is heat lost through the window when the temperature difference across it is 30.0°C?

36. Two materials of different thermal conductivities k_1 and k_2 form a compound slab shaped as a cube. As shown in Figure 18.P2 the interface between the two materials is parallel to one of the cubic faces. When two opposite faces of the cube are placed between a temperature gradient, which direction maximizes the heat flow: parallel to the interface or perpendicular to it?

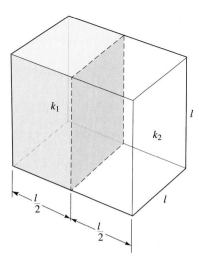

FIGURE 18.P2   Problem 36

18.4 Heat Capacity and Latent Heat

37. How much does the temperature of 60.0 g of gold rise when 225 cal of heat are added to it?

38. How much heat is required to raise the temperature of 0.500 kg of a lead radiation shield from 20.0°C to 100.°C?

39. What is the final temperature when 245 g of water at 90.0°C are poured into an insulating vessel containing 350. g of water at 20.0°C? (Ignore the heat capacity of the vessel.)

40. How much water at 60.0°C must be added to a glass containing 150. g of ice at 0.00°C for the final temperature to be 15.0°C? Ignore any heat absorbed by the glass.

41. A 750.-g block of copper is heated to 250.°C. It is placed in a 350.-g aluminum vessel that contains 200. g of water at 5.00°C. What is the final temperature of the copper, water, and aluminum vessel?

42. A 300.-g cylinder of unknown material is heated to 100.°C and placed in a 200.-g glass vessel wrapped in insulating material. The vessel initially contained 250. g of water at 10.0°C. The final temperature of the vessel, water, and cylinder is 19.4°C. From an inspection of Table 18.4, what material do you guess the cylinder is made from?

43. What is the final temperature when 30.0 g of cream at 10°C are added to 200.0 g of coffee at 65.0°C in a 300.-g glass cup, also initially at 65.0°C. (Approximate specific heat capacities of cream and coffee are the same as that of water.)

44. How much heat is necessary to convert 20.0 g of ice at −10.0°C entirely to water at +10.0°C?

45. How much heat is required to convert 10.0 g of water at 20.0°C entirely to steam at 120.°C?

46. What is the final temperature when 200.0 g of ice at −8.00°C is mixed with 25.0 g of steam at 105.0°C?

18.5 The Equation of State

47. One-half mole of an ideal gas is maintained at a constant volume of 1.2 L. Its initial temperature is 15.0°C. What is its final temperature if its pressure is decreased by 30.0%?

48. An ideal gas is confined to a cylinder with a movable piston. When the temperature is 25.0°C, the pressure and volume are 1.50 atm and 2.80 L, respectively. What is the final pressure if the gas is compressed to a volume of 1.20 L while the temperature increases to 40.0°C?

49. One-half mole of an ideal gas is confined to a cylinder with a movable piston. Initially, the temperature of the gas is 10.0°C, and the volume it occupies is 3.20 L. What is the temperature of the gas when the pressure is doubled and the volume is halved?

50. The pressure of a gas is held constant while its temperature is raised from 15.0°C to 215°C. By what factor must the volume of the gas change?

51. Show that 1.00 mol of any gas at a pressure of 1.01×10^5 Pa and at a temperature of 0.00°C occupies a volume of 22.4 L.

52. The remaining gas in a vacuum system is under a pressure of 10^{-9} mm of Hg. (Recall that 1 atm = 760 mm Hg.) How many molecules are there in 1.00 L of gas if its temperature is 0.00°C?

Numerical Methods and Computer Applications

53. Although controversial, the equation used to compute the so-called wind chill factor is

$$T_{\text{chill}} = 33°F - \{10.45 + [10(\text{s/m})^{1/2}\sqrt{v} - (1 \text{ s/m})v]\}$$
$$\times \left(\frac{33°F - T}{22}\right)$$

where T is the air temperature, and v is its speed. The idea is that T_{chill} is the temperature at which a person feels just as cold in still air as he does at temperature T in wind with speed v. Write a program or construct a spreadsheet template to make a table and a graph of T_{chill} for one-degree increments from $T = 0°F$ to 100°F, for wind speeds of 5, 10, 15, and 20 mph. (See H. Richard Crane, "Brrrr! The origin of the wind chill factor" in *The Physics Teacher* **27**, 59–61 (Jan. 1989).)

54. Write a computer program or spreadsheet template to compute the amount of heat required to raise the temperature of a substance with a specific heat capacity that depends on the temperature. Input should include the initial and final temperatures T_o and T_f, the mass of the substance, and the increment size ΔT for the integration. Use a subroutine for the $C(T)$ function so you can try it out for (a) a constant: $C = 0.22$ cal/g°C, and (b) the function $C(T) = (7.31 \text{ cal/g} \cdot K)(T/\theta)^3$, where $\theta = 343 \text{ K}^{-1}$. (This equation is that of $C(T)$ for copper when T is far below room temperature.)

55. Write a computer program that yields the final temperature when a mass m_1 at temperature T_1 is brought into contact with a mass m_2 at temperature T_2. Input should also include the specific heats C_1 and C_2 for the masses.

56. Using the data of Table 18.3, develop a computer program or a spreadsheet template to compute the R-value of an n-layer wall. Expand the program or spreadsheet to compute the rate of heat loss through the wall at a given temperature gradient across the wall.

57. Construct a spreadsheet template that computes the P-values for a range of V-values for a van der Waals gas. From these data graphically compare the shapes of the isotherms for several of the gases given in Table 18.5.

General Problems

58. The length of colored alcohol in an evacuated glass tube is 3.00 cm when the tube is immersed in a bath of water and ice at 1.00 atm of pressure. The length is 28.0 cm when the tube is placed in a steam bath, also at 1.00 atm of pressure. (a) Find the linear relation that allows you to use this device as a thermometer. (b) When the thermometer is immersed in another liquid, the length of the alcohol column is 22.4 cm. What is the temperature of the liquid?

59. A steel wire 20.0 m in length supports a 225-lbf Foucault pendulum bob temporarily at rest and at a temperature of 40.0°C. The wire diameter is 0.500 mm. If the temperature drops to 15.0°C, how much additional weight must be applied to the bob to return the length to 20.0 m? Assume Young's modulus for the steel wire is $200. \times 10^9$ N/m².

60. The length of an object is l_o at a temperature T_o. If the coefficient of linear expansion is a function of temperature $\alpha(T)$, show that the length of the object at temperature T is

$$l_f = l_o\left(1 + \int_{T_o}^{T_f} \alpha(T)\, dT\right)$$

61. The inner diameter of the glass tube for a mercury-in-glass barometer is 0.500 cm. The barometer is calibrated to read 1.00 atm when the length of mercury is 760.0 mm at 0.00°C. If the expansion of the glass can be ignored, by how much does the height of mercury change when the temperature of the barometer is raised to 20.0°C?

62. The mercury in a thermometer just reaches the top of the bulb at temperature T. The volume of the bulb at this temperature is $\mathcal{V}_T$ and the cross-sectional area of the capillary is A_T. If the change in the area A_T can be ignored, show that the length of mercury in the capillary when the temperature is changed by ΔT is

$$l = \frac{\mathcal{V}_T}{A_T}(\beta_{Hg} - 3\alpha_{glass})\,\Delta T$$

63. If a volume of water at 0.00°C is $\mathcal{V}_o$, then in the temperature range from 0°C to 33°C the approximate volume is given by the expression

$$\mathcal{V}(T) = \mathcal{V}_o[1 - AT + BT^2]$$

where $A = 6.427 \times 10^{-5}\,°C^{-1}$ and $B = 8.505 \times 10^{-6}\,°C^{-2}$. Compute the coefficient of volume expansion β for water at 10.0°C.

64. A cylindrical steam pipe is of length L, and the inner and outer radii are R_1 and R_2, respectively (Figure 18.P3). The pipe con-

FIGURE 18.P3 Problem 64

ducts heat radially outward at a constant rate dQ/dt. Show that the temperature difference between the inner and outer radii is given by

$$\frac{1}{2\pi kL}\left(\frac{dQ}{dt}\right)\ln\left(\frac{R_2}{R_1}\right)$$

where k is the thermal conductivity of the pipe material.

65. A hollow sphere conducts heat radially outward at a rate dQ/dt. If the inner and outer radii of the sphere are r_1 and r_2, respectively, show that the temperature difference between the inner and outer surfaces is given by

$$\frac{1}{4\pi k}\left(\frac{dQ}{dt}\right)\left(\frac{1}{r_1} - \frac{1}{r_2}\right)$$

where k is the thermal conductivity of the material composing the sphere.

66. The specific heat of nitrogen in the temperature range $300\ \text{K} \leq T \leq 5000\ \text{K}$ is given by the expression

$$C(T) = (0.096\ \text{kcal/kg K}) - (63.7\ \text{kcal/kg})\frac{1}{T}$$
$$+ (3.83 \times 10^4\ \text{kcal K/kg})\frac{1}{T^2}$$

Compute the amount of heat required to raise the temperature of 0.100 kg of N_2 from 500. K to 2500 K.

67. The specific heat of air in the temperature range $500\ \text{K} \leq T \leq 700\ \text{K}$ is given by the expression

$$C(T) = (0.062\ \text{kcal/kg K}) - (5.42 \times 10^{-6}\ \text{kcal/kg K}^2)T$$
$$+ (2.58 \times 10^{-10}\ \text{kcal/kg K}^3)T^2$$

Compute the amount of heat required to raise the temperature of 0.200 kg of air from 500. K to 700. K.

68. A Styrofoam ice chest has dimensions of 45.0 cm × 45.0 cm × 60.0 cm, and its wall thickness is 2.00 cm. It is filled with 6.00 kg of ice at 0.00°C on a day when the outside temperature is 32.0°C. The ice melts to water at 0.00°C in 13.0 h. What is the average thermal conductivity of the Styrofoam?

Thermodynamics I: Processes and the First Law

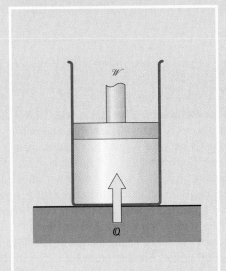

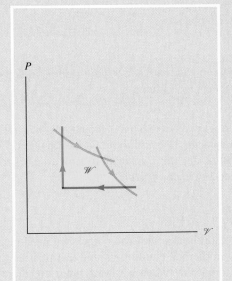

In this chapter you should learn

- the zeroth law of thermodynamics.
- the first law of thermodynamics.
- the relationship between internal energy, heat, and work.
- to analyze isochoric, isobaric, isothermal, and adiabatic processes.
- to compute the work performed and the heat transferred during cyclic processes.

Cooling towers provide an efficient mechanism for heat to be transferred to the low-temperature reservoir of the atmosphere.

Modern civilization as we know it would not be possible without the heat engine. Automobiles, airplanes, and turbine engines all convert thermal energy into mechanical energy. All of these engines share certain properties: each has a source of internal energy at high temperature, an outlet of internal energy at low temperature, and a ''working fluid'' with a temperature that drops as mechanical work is extracted from it. In an effort to expose the fundamental limitations that nature places on such engines we avoid the interesting, but nonessential, details of internal combustion engines or gas turbines. Instead we construct admittedly abstract models of the types of processes occurring in actual engines. Keep in mind however, that this is very ''real-world'' stuff. No one can understand the economic, social, and environmental implications of an industrialized world who does not understand the laws of thermodynamics.

The previous chapter dealt with several of the most obvious effects of temperature changes and differences. You learned to compute some useful quantities, such as thermal expansion, the rate of heat loss from an object, and the temperature changes when two bodies of different temperatures are brought into contact. In this chapter you can look forward to learning how to find the amount of work you might expect to produce from a system if you change its state. (You may wish to review what the term *state* means, by looking back to Section 18.5.) You should also learn several different ways to change the state of a system. And last, but certainly not least, you will learn the first of two important restrictions that nature imposes on us during these changes of state: the first law of thermodynamics.

19.1 Equilibrium, the Zeroth Law of Thermodynamics, and Processes

Consider the following experiment. A metal spoon is supported so that one end is in a candle flame and the other rests on an ice cube. If we take several tiny thermometers and place them along the length of the spoon, we find that after a short time each of the thermometers reaches some *constant* temperature, but of course they do not all read the *same* temperature. When the temperature reading on each thermometer remains constant in time, the spoon is in a **steady state.** If we remove the candle and the ice cube, soon all the tiny thermometers along the spoon show the same temperature. We say that the spoon is now in **equilibrium.** We consider fairly simple systems, such as a gas in a cylinder or a small amount of material in the liquid or solid phase. For such cases it is most convenient to use the following definitions for a steady state and thermodynamic equilibrium:

> Steady State: A thermodynamic system is in a steady state if all the state variables are constant in time at each point within the system.

> Thermodynamic Equilibrium: A system is in thermodynamic equilibrium if every state variable has the same value throughout the system.

With the definition of equilibrium of a single object in hand it should now be straightforward to determine when two different objects are in equilibrium with each other. Let's stick to a single state variable, say temperature, for this explanation. However, what we say about temperature must also be true for any state variable.

The Zeroth Law of Thermodynamics

To see whether two objects are in thermal equilibrium with each other, we need merely determine that both objects are individually in thermal equilibrium and that their equilibrium temperatures are the same. Then we know that when we bring them together, the combined system is in equilibrium and nothing happens. But this simple argument is more subtle than it may seem at first. In this experiment it is really necessary to have a *third* object present. What is the third object? It is a thermometer, of course. You need the thermometer to tell when the temperatures are the same. If you missed the subtlety, don't be embarrassed, a lot of people do. That's why the generalization of this experiment is known as the **zeroth law of thermodynamics.** The logical necessity of this law was recognized only after the first and second laws had been established, but a logical and systematic development of thermodynamics dictates that the more fundamental law be placed prior to the other two. That leaves only ''the zeroth law'' for its name. We hope we've got them all now. It would be awkward to have to refer to ''negative laws'' of thermodynamics!

> Zeroth Law of Thermodynamics: If two systems are in thermodynamic equilibrium with a third, they are in equilibrium with each other.

This result may seem obvious but only because you have experienced it in some way or another on numerous occasions and is, moreover, precisely the reason it is referred to as a law; it is the result of many observations. (Recall that thermodynamics is an empirical science. However, the game physicists play is to see how few basic postulates they can use and still logically explain all the known observations. Thermodynamics has four such basic postulates. This law is one of them.)

An important consequence of the zeroth law is that the *state variables can be used to determine if two systems are in equilibrium with each other.* If two systems' state variables are all the same, then we know that the two systems are in equilibrium when placed in thermal contact, and no heat flows from one to the other. This law is not as obvious as it may first appear. There are many examples in which this type of reasoning doesn't work. For example, if you love your dog Dudley and you love your cat Clyde, it does not necessarily follow that Dudley even likes Clyde. Or, to be more physical, gasoline and water both mix with alcohol, but gasoline and water remain separated when placed in a vessel together.

We are going to take an approach to thermodynamics that is not unlike others we have followed in this text. When we began to study motion on an inclined plane (bring back fond memories?) we first ignored friction. We found this simplification convenient because it allowed us to temporarily ignore one complication and permitted us to concentrate on the physics of the problem in a simple context. We also took this approach when we studied the harmonic oscillator: during our first pass we ignored friction. Later, when we had the physics of the frictionless system well in hand, we introduced damping.

In thermodynamics too, we can conceive of some ideal circumstances in which we ignore such effects as friction, heat loss, or finite temperature changes. In reality, we can sometimes come close to these models, but we can never really achieve them. One important idealization is that of a heat reservoir:

> Heat Reservoir: An infinite source or sink of heat the temperature of which does not change no matter how much heat is added or withdrawn.

A large lake often approximates a heat reservoir. When we throw an ice cube into a lake, heat flows from the lake into the ice, causing the cube to melt. Yet, for all practical purposes, the temperature of the lake does not change. In thermodynamic examples, a heat reservoir is often represented by a very large mass with a constant temperature (Fig. 19.1).

Much of what we do in this chapter is concerned with what happens in a system when we take it from one state to another. We call this transition between states a process:

> Process: A sequence of events that takes a system from one thermodynamic state to another.

Concept Question 1
Give some similar examples for which a law analogous to that of the zeroth law of thermodynamics does not hold.

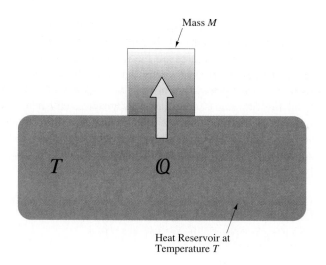

FIGURE 19.1

A heat reservoir is often represented in diagrams by a very large mass that remains at a constant temperature T. As depicted here, the temperature of the reservoir does not change when heat Q flows from it into the mass m.

There are two important types of processes in thermodynamics: reversible and irreversible. It is very important that you have a clear picture of each.

> Reversible Process: A process during which the equation of state applies at all times.

To say that a process is reversible means that the system is taken through a series of equilibrium states and, therefore, the process can be shown as a path on the $P\mathcal{V}T$-surface of the substance. Such a process is illustrated in Figure 19.2 for an ideal gas. Reversible processes are often called **quasi-static processes** because an equilibrium state is a static state: the thermodynamic variables of the system are constant (static) throughout the system. Clearly, however, if this is the case, the system does not change and a process does not take place. That's where the "quasi" comes in. The idea is that during a reversible process the system can deviate from equilibrium by only an *infinitesimal* amount. Now, there is a distinction between what we can do mathematically to get from one state to another, and what we have to do in reality to travel between those states. Mathemati-

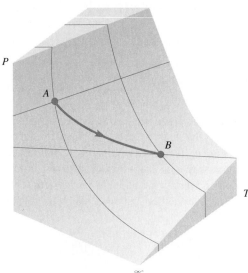

FIGURE 19.2

A reversible process for an ideal gas. During the change of state from A to B the equation of state applies at all times. The process, therefore, is a succession of equilibrium (or quasi-static) states.

cally, we require the equation of state to apply at all times during the process. In reality, we approach this situation by carrying out the process very slowly.

A pan of water at temperature T_o is heated quasi-statically (reversibly) by placing it in contact with a heat reservoir at temperature $T_o + dT$. Then the pan of water is placed in contact with another heat reservoir at temperature $T_o + 2(dT)$, and so on until it finally reaches a temperature T_f (Fig. 19.3). Notice by this approach that if we make an infinitesimal change in the temperature T (in the reverse direction: change T by $-dT$), we reverse the direction of the process: hence, the name *reversible process*. Obviously, we can't really construct an infinite series of heat reservoirs. Therefore, we can only approximate a reversible process, and the processes we carry out in real life are called irreversible processes:

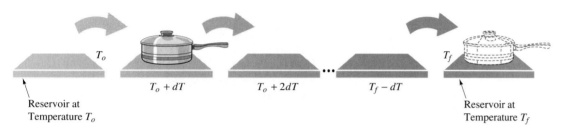

FIGURE 19.3

Heating a pan of water reversibly with an infinite number of heat reservoirs.

> Irreversible Process: A process which occurs naturally in only one direction, toward an equilibrium state.

An infinitesimal change in a thermodynamic variable such as temperature does not render a process irreversible, but a sudden *finite* change in the variable does. Furthermore, *the equation of state does not apply during an irreversible process*. The thermodynamic variables are not constant throughout the system, and therefore, by definition, the system is not in equilibrium. When you begin to heat a pan of water on the stove, the water next to the heat source is at a higher temperature than that near the water's surface.

As we stated above, all real processes are irreversible. However, this restriction does not mean that a system cannot be returned to its original state. Our pan of hot water may be made cool again by placing it in contact with a cool reservoir: a refrigerator or a cool lake.

You may be wondering what good it is to even consider the reversible-process model if all real processes are irreversible. Actually, the concept of a reversible process is very useful when we want to determine if an as yet untried real (irreversible) process occurs at all. But this is the subject of the second law of thermodynamics, which we take up in the next chapter.

We must define two final terms. First, a perfect **insulator** is a boundary that permits no heat to flow in or out of a system. Next, processes that take place with no heat loss or gain to the system under study are called **adiabatic** processes. We can sometimes come very close to a true adiabatic process by carrying out the process very quickly so that there is little time for heat to be lost. However, by our previous arguments, such a fast process must necessarily be irreversible. A reversible, adiabatic process must be carried out slowly, and we must rely on an ideal insulator to prevent heat from flowing out of or into the system.

Concept Question 2
Give some examples of irreversible processes that occur in nature.

19.2 Work

In the previous chapter (Section 18.5) we found it convenient to consider as our system a cylinder with a movable piston. Such a system is again shown in Figure 19.4. This time the enclosed gas, by virtue of its pressure P, has moved the piston through a small displacement dx. Because the cross-sectional area of the piston is A, and dx is small enough that P is constant during this displacement, the force on the piston is $F = PA$.

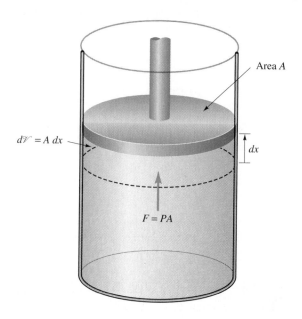

FIGURE 19.4

The work done by pressure P to change the volume of a fixed quantity of gas by $d\mathcal{V}$.

Therefore, the infinitesimal work $d\mathcal{W}$ is

$$d\mathcal{W} = F\,dx = PA\,dx = P(A\,dx) = P\,d\mathcal{V}$$

where we recognize the change in the gas volume as $A\,dx = d\mathcal{V}$. The work done by pressure P when the volume of a gas changes from $\mathcal{V}_o$ to $\mathcal{V}_f$ is

$$\mathcal{W} = \int_{\mathcal{V}_0}^{\mathcal{V}_f} P\,d\mathcal{V}$$

(19.1)

Hence, the work done *by* a gas is the area under the P-versus-$\mathcal{V}$ curve. There are two important points regarding this definition of work that you should not miss:

1. *The work depends on the* path *between the original and final states.* This point should be clear because the area under the P-versus-$\mathcal{V}$ curve depends on the shape of the curve. One special case occurs when there is no change in volume. Such processes are called **isochoric** processes. Because $d\mathcal{V} = 0$, the work is zero.
2. *The distinction between work done* by *the system and that done* on *the system is critical.* We chose to define work so that when the system does work on the environment, the value calculated is positive. When the environment does work on the system, the quantity computed by Equation (19.1) is negative.

Let's illustrate these points with the following examples.

EXAMPLE 19.1   A Gas That Works

A gas expands from a volume of 100. L to 250. L while the pressure changes according to the relation $P = a\mathcal{V}^2$, where $a = 2.00$ N/m^8. How much work is done by the gas during the expansion?

SOLUTION   The graph of the expansion is shown in Figure 19.5. From Equation (19.1) the work is

$$\mathcal{W} = \int_{\mathcal{V}_o}^{\mathcal{V}_f} a\mathcal{V}^2 \, d\mathcal{V} = a\left(\frac{\mathcal{V}^3}{3}\right)_{\mathcal{V}_o}^{\mathcal{V}_f} = \frac{a}{3}(\mathcal{V}_f^3 - \mathcal{V}_o^3)$$

$$= \left(\frac{2.00 \text{ N/m}^8}{3}\right)[(0.250 \text{ m}^3)^3 - (0.100 \text{ m}^3)^3]$$

$$= 9.75 \times 10^{-3} \text{ J}$$

This result is represented by the "area" bounded by the curve and the $P = 0$ axis between $\mathcal{V}_o$ and $\mathcal{V}_f$. ◀

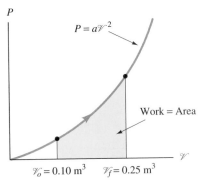

FIGURE 19.5

A graph of the expansion of a gas from 100. L to 250. L when the pressure changes according to $P = a\mathcal{V}^2$, where $a = 2.00$ N/m^8. See Example 19.1.

EXAMPLE 19.2   Working in Another Area of Physics

A thermodynamic system is taken from an original state A to an intermediate state B by the linear process shown in Figure 19.6. Its volume is then reduced to the original value from B to C by an isobaric process. Compute the total work done by the gas from A to B to C.

SOLUTION   The work performed by the gas is the area bounded by the curve representing the process and the $\mathcal{V}$-axis. From A to B the total area is the sum of the area of the triangle ABC and the rectangle below CB:

$$\mathcal{W}_{AB} = \frac{1}{2}[(5.00 \times 10^{-3} \text{ m}^3) - (2.00 \times 10^{-3} \text{ m}^3)](600. \text{ Pa} - 300. \text{ Pa})$$

$$+ [(5.00 \times 10^{-3} \text{ m}^3) - (2.00 \times 10^{-3} \text{ m}^3)](300. \text{ Pa})$$

$$= 1.35 \text{ J}$$

The work from B to C is the area below the rectangle BC. Note carefully, however, that the volume is decreasing. In the following equation, as always, $\Delta\mathcal{V}$ is calculated as $\mathcal{V}_f - \mathcal{V}_o$.

$$\mathcal{W}_{BC} = P \, \Delta\mathcal{V} = P_B(\mathcal{V}_C - \mathcal{V}_B) = (300. \text{ Pa})[(2.00 \times 10^{-3} \text{ m}^3) - (5.00 \times 10^{-3} \text{ m}^3)]$$

$$= -0.90 \text{ J}$$

As we expected, because of the decreasing volume, the work done as the system evolved from B to C is negative. That is, work is done *on* the system. As a consequence the net work done *by* the system is

$$\mathcal{W}_{\text{tot}} = \mathcal{W}_{AB} + \mathcal{W}_{BC} = 1.35 \text{ J} + (-0.90 \text{ J}) = 0.45 \text{ J}$$ ◀

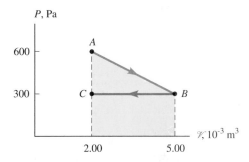

FIGURE 19.6

A linear process of the change in a thermodynamic system from original state A to intermediate state B to final state C.

Perhaps you have begun to realize that we are using two types of quantities to describe the thermodynamics of a system. On one hand, we have parameters, such as pressure, temperature, and volume. These are quantities that depend only on the state of the system, and to emphasize this fact they are called *state variables* (Section 18.5). On the other hand, both work and the amount of heat transferred depend on the particular *path* of the process, so we might call those *path variables*. Although it makes perfectly good sense to refer to the temperature, pressure, or volume of a system (as long as it is in an equilibrium state), to speak of the ''heat in a system'' or the ''work in a system'' is nonsense. Don't use these phrases; they always lead to incorrect thinking. We emphasize the fact that heat and work are not state variables by using script letters to designate them:

> Both the work $\mathcal{W}$ done by (or on) a system and the heat $\mathcal{Q}$ transferred into (or out of) a system depend on the path between the original and final states. Therefore, *work and heat are not state variables*.

We are now ready to introduce a new state variable.

19.3 The First Law of Thermodynamics

When a thermodynamic system is taken through a process in which heat $\mathcal{Q}$ is added to the system and work $\mathcal{W}$ is performed by the system, we find a surprising result. For all processes between the same original and final states, although the numerical value of $\mathcal{Q}$ and the numerical value of $\mathcal{W}$ may be different for each process, the value of $\mathcal{Q} - \mathcal{W}$ is always the same! This point is illustrated in Figure 19.7 in which, even though we follow different paths between states A and C, $\mathcal{Q} - \mathcal{W}$ always turns out to be the same number. This result means that the difference $\mathcal{Q} - \mathcal{W}$ depends only on the original and final states of the system. We conclude, therefore, that $\mathcal{Q} - \mathcal{W}$ is itself a state variable. This new state variable is called the **internal energy** U. The quantity $\mathcal{Q} - \mathcal{W}$ is the change in the internal energy between the original and final states of the process:

$$U_f - U_o = \mathcal{Q} - \mathcal{W} \tag{19.2}$$

Equation (19.2) is called the **first law of thermodynamics,** and it implicitly includes the assertion that this new state variable U (the internal energy) exists. Several comments are appropriate about this law. The first law of thermodynamics is based on many experiments, and no process has been discovered that violates this law. That's comforting because Equation (19.2) looks very much like an expression of the conservation of energy, which is exactly what it is! Now, however, we don't have to make any assumptions about the true nature of ''internal energy.'' Thermodynamics makes the quite general statement that there is such a quantity, and the value of its change is always the same for all processes with the same beginning point and the same ending point. In Chapter 21 we show that this thermodynamic internal energy is in fact the same as that defined in Chapter 18; namely, the sum of the kinetic and potential energies of the individual atoms or molecules that make up the system.

Our second comment regarding the first law of thermodynamics is related to the signs of $\mathcal{Q}$ and $\mathcal{W}$. When you apply Equation (19.2), pay careful attention to the following conventions. We defined $\mathcal{Q}$ as the heat added to the system. Therefore, $\mathcal{Q}$ is positive when heat is transferred *into* the system and negative when heat is removed *from* the system. The work $\mathcal{W}$ is positive when work is done *by* the system and negative when work is done *on* the system. (Watch out for this latter sign convention in other courses. It is convenient for chemists to define work with the opposite sign.) For most applications of the first law in physics, the sign conventions of Equation (19.2) are the most useful.

It is often convenient to consider infinitesimal changes in the internal energy, heat transferred, and work performed. In this case we use a differential representation of Equation (19.2). However, we emphasize that heat transfer and work are of a completely different nature from a change in internal energy. It doesn't hurt to say it again: the internal energy change is independent of path, whereas heat transfer and work performed depend on path. The differential form of $\mathcal{Q}$ and $\mathcal{W}$ are, as we noted in Chapter 18, often written with a bar through the differential d as in $đ\mathcal{Q}$ and $đ\mathcal{W}$ to emphasize that they are

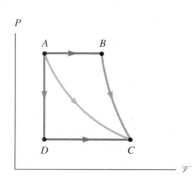

FIGURE 19.7

A surprising experimental result: the value of $\mathcal{Q} - \mathcal{W}$ is independent of path and depends only on the initial and final states $(P_A, \mathcal{V}_A)$ and $(P_C, \mathcal{V}_C)$.

$\mathcal{Q} > 0$ when heat is added *to the* system

$\mathcal{Q} < 0$ when heat is removed *from* the system

$\mathcal{W} > 0$ when work is done *by* the system

$\mathcal{W} < 0$ when work is done *on* the system

not exact differentials of some function but merely represent small amounts of heat and work. In this case, Equation (19.2) becomes

$$dU = đQ - đW \qquad \text{(19.3)}$$

Finally, you should note that Equation (19.2) defines only *changes* in the internal energy. Therefore, we can assign an arbitrary reference value for the internal energy of a particular state. This arbitrariness is the same as that for all potential energy functions. For example, recall the gravitational potential energy mgy near the earth's surface. Depending on the particular problem, we found it useful at different times to assign a zero potential energy to floor level, a tabletop, or some other convenient position. The same is true for the internal energy; we can pick a convenient state and assign it a convenient reference value U_o.

Concept Question 3
Apply the first law of thermodynamics to demonstrate that the total internal energy of an isolated system remains constant.

EXAMPLE 19.3   *Balancing Your Energy Checkbook*

A thermodynamic system is taken from an original state A to another state B and then from B to state C along the path shown in Figure 19.8. During the process AB, 3.60 J of heat are added to the system. The internal energy decreases from B to C by 1.40 J. Find (a) the change in internal energy from A to C, and (b) the amount of heat added to the system from B to C.

SOLUTION   In this type of problem, for which we are called on to apply the first law of thermodynamics, it is advisable to make a table of each term, ΔU, Q, and W, for each part of the process. We suggest you list these terms in the order in which they appear in the law and immediately fill in values given in the problem statement. For our problem we know $Q_{AB} = 3.60$ J and $\Delta U_{BC} = -1.40$ J. Hence, our table first appears as

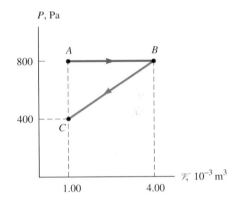

FIGURE 19.8

A thermodynamic process for which $Q_{AB} = 3.60$ J and $\Delta U_{BC} = -1.40$ J. See Example 19.3.

PROCESS	ΔU	Q	W
$A \rightarrow B$		3.60 J	
$B \rightarrow C$	-1.40 J		

In order to compute ΔU_{AB} from the first law $\Delta U_{AB} = Q_{AB} - W_{AB}$, we must know W_{AB}. The work performed from A to B is the area bounded by the curve AB and the $P = 0$ axis:

$$W_{AB} = (800.\ \text{Pa})[(4.00 \times 10^{-3}) - (1.0 \times 10^{-3})]\text{m}^3 = 2.40\ \text{J}$$

Therefore, $\Delta U_{AB} = Q_{AB} - W_{AB} = 3.60$ J $- 2.40$ J $= 1.20$ J. The first row of our box summary may be filled in as follows:

PROCESS	ΔU	Q	W
$A \rightarrow B$	1.20 J	3.60 J	2.40 J
$B \rightarrow C$	-1.40 J		

This second box summary makes it easy to check our result. We see that the value in the ΔU cell of the first row is indeed the difference of the values in the Q and W cells for the AB-process.

For the process BC we are given $\Delta U = -1.40$ J, so to compute Q_{BC} we must find W_{BC}. The area under the BC-curve is -1.80 J because the volume decreases during this process. (See Example 19.2 to review negative areas.) Then $Q_{BC} = \Delta U_{BC} +$

$\mathcal{W}_{BC} = -1.40\ \text{J} - 1.80\ \text{J} = -3.20\ \text{J}$. The final form of our table is

PROCESS	ΔU	Q	$\mathcal{W}$
$A \to B$	1.20 J	3.60 J	2.40 J
$B \to C$	−1.40 J	−3.20 J	−1.80 J

Concept Question 4
Under what condition can a certain amount of internal energy be completely converted into mechanical work?

Although our summary table may appear to be little more than extra work for the purpose of verifying our results, we assure you that this approach becomes increasingly useful in future problems. When a complete thermodynamic process consists of several individual processes, this method can become almost indispensable. We *strongly* urge you to become practiced with this method so that it will be a familiar tool when you encounter more complicated processes.   ◀

19.4 Specific Thermodynamic Processes

In this section we examine the changes in the internal energy, the amount of heat transferred, and the work performed by a system during some specific processes. We use an ideal gas as our model, so most (but not all) of the results we obtain apply only to ideal gases. Where appropriate, we mention the results for a van der Waals gas or other model. How we arrive at a result is as important as the result itself. In essence, we illustrate the techniques of thermodynamics by using the vehicle of the ideal-gas model. But before we begin, a complication has arisen that we must resolve.

The molar specific heat c was defined by Equation (18.12), which may be rewritten

$$Q = nc\ \Delta T$$

where n is the number of moles of the substance under consideration. You may have already recognized a shortcoming of this definition. The heat Q transferred into or out of the system depends on the particular process. Therefore, the equation $Q = nc\ \Delta T$ depends on the path. That is, the magnitude of Q may be different for different paths between the same two temperatures. But n and ΔT remain the same, so it must be c that is path-dependent. The only way around this complication is to always state the process for which c is defined. Actually, it turns out to be useful to define only two particular processes, one a constant-volume process and the other a constant-pressure process. During a constant-volume process, the amount of heat transferred is

$$Q = nc_v\ \Delta T \qquad (19.4)$$

where c_v is the **molar specific heat at constant volume.** For an isobaric process

$$Q = nc_p\ \Delta T \qquad (19.5)$$

and c_p is the **molar specific heat at constant pressure.**

We find that many other important processes can be described in terms of combinations of constant pressure and constant volume processes. Remember, the formal results of thermodynamics can be deduced from basic postulates, but this science is fundamentally empirical. If we are to obtain anything more than relationships between quantities, we must find some way to introduce numerical values. The quantities c_p and c_v provide us with such a mechanism. By experimentally measuring c_p and c_v we can compute numerical results from the relations we derive in this section. Measurements of c_p and c_v for helium (which behaves very much like an ideal gas) yield

$$\left. \begin{array}{l} c_p = 20.8\ \text{J/mol/K} \\ c_v = 12.5\ \text{J/mol/K} \end{array} \right\} \text{Helium}$$

Isochoric Processes

During an **isochoric** process the volume is held constant so that $d\mathcal{V} = 0$ and the work done by the system is zero:

$$\mathcal{W} = \int P\, d\mathcal{V} = 0$$

The first law of thermodynamics therefore reduces in this special case to

$$\Delta U = Q \tag{19.6}$$

However, for a constant-volume process, $Q = nc_v\, \Delta T$ and Equation (19.6) becomes

$$\Delta U = nc_v\, \Delta T = nc_v(T_f - T_o) \tag{19.7}$$

Equation (19.7) is an important result because the left side of this equation is the change in a state variable and, therefore, is independent of the path from T_o to T_f. Hence, ΔU *must always be equal to* $nc_v\, \Delta T$, for any process. This result is of great practical significance. It enables you to calculate the change in internal energy for *any* process taking a system between temperatures T_o and T_f by finding the change in energy for an isochoric process between the same two temperatures. Equation (19.7) also suggests that the temperature of a system is closely related to the internal energy of the system; ΔU depends on the temperature change.

Because the internal energy U is a state variable, Equation (19.7) can be written in differential form:

$$dU = nc_v\, dT$$

or, because $dU = (dU/dt)dt$,

$$c_v = \frac{1}{n}\frac{dU}{dT} \tag{19.8}$$

This expression is helpful when we examine a gas from a microscopic view. It permits us to tie down the relationship between internal energy and temperature. For now, we simply note that *for an ideal gas when there is no change in temperature, there is no change in internal energy.* That is, *isotherms of an ideal gas are constant-energy curves.* These statements are true for real gases only under conditions in which they behave as ideal gases; that is, at low pressures and far from the vicinity of a phase transition. Near phase transitions the internal energy can depend critically on the pressure and the volume per mole of the gas, and the relation $Q = nc_v\, \Delta T$ is not applicable (see Eq. (18.16)).

Isobaric Processes

When the pressure is held constant during a process, the process is said to be **isobaric.** The work performed during a constant-pressure process is

$$\mathcal{W}_P = \int_{\mathcal{V}_o}^{\mathcal{V}_f} P\, d\mathcal{V} = P\int_{\mathcal{V}_o}^{\mathcal{V}_f} d\mathcal{V} = P(\mathcal{V}_f - \mathcal{V}_o) = P\,\Delta\mathcal{V}$$

For this special case the first law of thermodynamics can be written

$$\Delta U = Q - P\,\Delta\mathcal{V} \tag{19.9}$$

We now transform Equation (19.9) by making three substitutions: (1) the process is isobaric so we can write $Q = nc_p \Delta T$, (2) Equation (19.8) allows us to write $\Delta U = nc_v \Delta T$, and (3) the ideal-gas equation of state gives $P \Delta \mathcal{V} = nR \Delta T$. With these three substitutions, Equation (19.9) becomes

$$nc_v \Delta T = nc_p \Delta T - nR \Delta T$$

Canceling $n \Delta T$, we have $c_v = c_p - R$, or

$$\boxed{c_p - c_v = R} \qquad \text{(Ideal gases only)} \qquad (19.10)$$

Equation (19.10) is an important and useful result. However, you should remember that we used the ideal-gas equation of state to derive it, so Equation (19.10) is only applicable to ideal gases. The equivalent expression for a van der Waals gas is somewhat more complex, and the value for $c_p - c_v$ turns out to depend on the volume $\mathcal{V}$ for that gas model.

Equation (19.10) says that the *specific heat at constant pressure is greater than the specific heat at constant volume.* Can you think of a reasonable explanation for this result? Recall the definition of the molar specific heat capacity: it is the amount of heat necessary to raise the temperature of one mole of a substance by one temperature degree. During a constant-pressure process some of the heat transferred to the substance must go to perform work on the container as its volume changes. During a constant-volume process, however, no heat goes into work and all of it can contribute to the temperature increases. Thus, we see that the extra heat required in the constant-pressure case is used to perform work. This result foreshadows the extremely important application of thermodynamics: the design of engines to convert heat into work. Because in the constant-volume process the first law is $\Delta U = Q - 0$, our explanation again suggests that the temperature is related to the internal energy.

Isothermal Processes

A process during which the temperature is held constant is called **isothermal.** In order to compute the work performed by a substance during a constant-temperature process, it is necessary to know the equation of state. That is, we need to know how P varies with T in Equation (19.1). For a substance that obeys the ideal-gas equation of state ($P\mathcal{V} = nRT$), we can write

$$\mathcal{W}_T = \int_{\mathcal{V}_o}^{\mathcal{V}_f} P \, d\mathcal{V} = nRT\int_{\mathcal{V}_o}^{\mathcal{V}_f} \frac{d\mathcal{V}}{\mathcal{V}} = nRT \ln(\mathcal{V}) \Big|_{\mathcal{V}_o}^{\mathcal{V}_f}$$

$$= nRT[\ln(\mathcal{V}_f) - \ln(\mathcal{V}_o)]$$

$$\mathcal{W}_T = nRT \ln\left(\frac{\mathcal{V}_f}{\mathcal{V}_o}\right) \qquad \text{(Ideal gas only)} \qquad (19.11)$$

The work given by Equation (19.11) is represented by the area bounded by the $P = 0$ axis and the isotherm shown in Figure 19.9.

We can find the amount of heat required for this process by the first law, $\Delta U = Q - \mathcal{W}$ in combination with our earlier result $\Delta U = nc_v \Delta T$. In the isothermal case $\Delta U = nc_v \Delta T = nc_v(0) = 0$. So from the first law we find $Q = \mathcal{W}$:

$$Q_T = nRT \ln\left(\frac{\mathcal{V}_f}{\mathcal{V}_o}\right) \qquad \text{(Ideal gas only)} \qquad (19.12)$$

where we have used a T-subscript to remind ourselves that the process is isothermal.

We have again relied on the ideal-gas equation of state to obtain these results. So be sure to remember that Equations (19.11) and (19.12) are appropriate only for ideal gases.

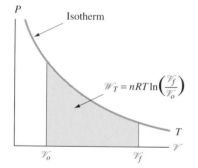

$$\mathcal{W}_T = nRT \ln\left(\frac{\mathcal{V}_f}{\mathcal{V}_o}\right)$$

FIGURE 19.9

Work done during an isothermal expansion of an ideal gas from volume $\mathcal{V}_o$ to volume $\mathcal{V}_f$.

Heat transfer during an isothermal process

Adiabatic Processes

The last specific process we consider in detail is an **adiabatic** process defined by the condition $Q = 0$. This type of process might be carried out in a well-insulated container, so that no heat flows across the container boundary. It is not necessary, however, to have a system surrounded by a good insulator for a process to come very close to being adiabatic. If a process is carried out very rapidly, there is little time for heat transfer.

With $Q = 0$, the first law of thermodynamics becomes $\Delta U = -\mathcal{W}$, or $\mathcal{W} = -\Delta U$. Hence, the work performed by a system during an adiabatic process is done completely at the expense of the internal energy. Because $\Delta U = nc_v \, \Delta T$, the work performed during an adiabatic process is

$$\mathcal{W}_Q = -nc_v \, \Delta T \qquad (19.13)$$

Work done during an adiabatic process

We can find the equation in the $P\mathcal{V}$-plane for an adiabat of an ideal gas from the differential form of the first law $dU = \overline{d}Q - \overline{d}\mathcal{W}$, or

$$dU = -\overline{d}\mathcal{W} = -P \, d\mathcal{V}$$

The last two equations taken together imply

$$nc_v \, dT = -P \, d\mathcal{V} \qquad (19.14)$$

In order to eliminate T from this expression we must be somewhat clever. Let's take the differential of the equation of state $P\mathcal{V} = nRT$:

$$d(P\mathcal{V}) = d(nRT)$$

$$P \, d\mathcal{V} + \mathcal{V} \, dP = nR \, dT$$

where on the left-hand side we used the so called "product rule." Solving this last expression for $n \, dT = (P \, d\mathcal{V} + \mathcal{V} \, dP)/R$ and substituting into Equation (19.14), we have

$$\frac{c_v(P \, d\mathcal{V} + \mathcal{V} \, dP)}{R} = -P \, d\mathcal{V}$$

Rearranging this result, we have

$$\frac{(c_v + R) \, d\mathcal{V}}{\mathcal{V}} + c_v \frac{dP}{P} = 0 \qquad (19.15)$$

But $c_v + R = c_p$, so that Equation (19.15) can be written

$$\left(\frac{c_p}{c_v}\right)\frac{d\mathcal{V}}{\mathcal{V}} + \frac{dP}{P} = 0$$

Next, we integrate this expression and write the integration constant on the right side of the equality as the natural logarithm of another constant:

$$\left(\frac{c_p}{c_v}\right)\ln(\mathcal{V}) + \ln(P) = \ln(\text{constant})$$

It is customary to define the ratio of molar heat capacities as the lowercase Greek letter gamma so that $c_p/c_v = \gamma$. With this final substitution, we have

$$\ln(\mathcal{V}^\gamma) + \ln(P) = \ln(\text{constant})$$

$$\ln(P\mathcal{V}^\gamma) = \ln(\text{constant})$$

$P\mathcal{V}$-relation along an adiabat for an ideal gas

or more simply,

$$P\mathcal{V}^{\gamma} = \text{constant} \qquad \text{(Ideal gas only)} \qquad (19.16)$$

Equation (19.16) is the mathematical expression for the curve in the $P\mathcal{V}$-plane that represents an adiabat for an ideal gas. The value of the constant is determined when P and $\mathcal{V}$ are known at any point along the adiabatic process. Because $c_p > c_v$, γ is greater than 1 and, therefore, the P-versus-$\mathcal{V}$ graph of "$P\mathcal{V}^{\gamma}$ = constant" is a steeper curve than $P\mathcal{V}$ = constant. That is, on a $P\mathcal{V}$-diagram an adiabat passing through a given point has a negative slope of greater absolute value than an isotherm passing through that same point. This result is illustrated in Figure 19.10. We may apply Equation (19.16) to the adiabat shown

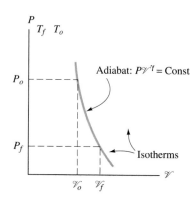

FIGURE 19.10

Adiabats are steeper than isotherms. The work done during an adiabatic expansion from volume $\mathcal{V}_o$ to volume $\mathcal{V}_f$ is the area between $\mathcal{V}_o$ and $\mathcal{V}_f$ bounded by the adiabat and the $P = 0$ axis. Its numerical value is $-n\,C_v\,(T_f - T_o)$.

in this figure to obtain

$$P_o\mathcal{V}_o^{\gamma} = P_f\mathcal{V}_f^{\gamma} \qquad (19.17)$$

We have now completed an analysis of several of the more important thermodynamic processes. Before looking at an example, let's summarize these processes, emphasizing what is true in general about each and the specific results we obtained for an ideal gas. Such a summary is shown in Table 19.1.

TABLE 19.1   Summary of Thermodynamic Processes

PROCESS AND GENERAL RESULT	RESULT FOR IDEAL GAS
Isochoric	
$\mathcal{W} = 0$	
$\Delta U = nc_v\,\Delta T$ (True also for nonisochoric processes)	
$Q = \Delta U$	
Isobaric	
$\mathcal{W} = P\,\Delta\mathcal{V}$	
$\Delta U = nc_v\,\Delta T = Q - \mathcal{W}$	
$Q = nc_p\,\Delta T$	$c_p - c_v = R$
Isothermal	
$\mathcal{W} = \displaystyle\int P\,d\mathcal{V}$	$\mathcal{W} = Q = nRT\ln\!\left(\dfrac{\mathcal{V}_f}{\mathcal{V}_o}\right)$
$Q = \Delta U + \mathcal{W}$	$\Delta U = 0$
Adiabatic	
$\mathcal{W} = -\Delta U = -nc_v\,\Delta T$	$P\mathcal{V}^{\gamma} = \text{constant} \qquad (\gamma = c_p/c_v)$
$Q = 0$	

Concept Question 5
When you let air out of a tire that is very warm from a long drive, is the temperature of the escaping air higher, lower, or the same as that in the tire? Use a $P\mathcal{V}$-diagram to explain your answer.

EXAMPLE 19.4   A Thermodynamic Two-Step (Isobaric and Isothermal Processes)

A vessel containing 0.0600 mol of an ideal gas is taken through the process $A \to B \to C$ shown in Figure 19.11. Compute the work performed by the gas, the heat transferred, and the change in internal energy for the isobaric process $A \to B$ and the isothermal process $B \to C$.

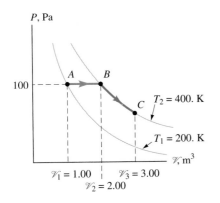

FIGURE 19.11

An isobaric process followed by an isothermal process. See Example 19.4.

SOLUTION   We again construct a table for our results. We can immediately place a zero under ΔU for the $B \to C$ process because there is no change in internal energy along an isotherm of an ideal gas.

PROCESS	ΔU	Q	W
$A \to B$			
$B \to C$	0		
Total			

The process occurring between A and B is isobaric, so the work done is

$$W_{AB} = \int_{V_1}^{V_2} P\, dV = P \int_{V_1}^{V_2} dV = P(V_2 - V_1) = (100.\ \text{Pa})(2.00 - 1.00)\text{m}^3$$

$$= 100.\ \text{J}$$

The change in internal energy from A to B is

$$\Delta U = nc_v\, \Delta T = (0.0600\ \text{mol})(12.5\ \text{J/mol/K})(400.\ \text{K} - 200.\ \text{K}) = 150.\ \text{J}$$

The heat transferred to the gas during $A \to B$ can be computed two ways. First,

$$Q = nc_p\, \Delta T = (0.0600\ \text{mol})(20.8\ \text{J/mol/K})(400.\ \text{K} - 200.\ \text{K}) = 250.\ \text{J}$$

Or, from the first law, $Q = \Delta U + W = 150\ \text{J} + 100\ \text{J} = 250\ \text{J}$. We are now able to fill in the first row of our table:

PROCESS	ΔU	Q	W
$A \to B$	150 J	250 J	100 J
$B \to C$	0		
Total	0		

The work done by the gas along the isotherm $B \to C$ is

$$W_{BC} = \int_{V_2}^{V_3} P\, dV = nRT \int_{V_2}^{V_3} \frac{dV}{V} = nRT \ln\left(\frac{V_3}{V_2}\right)$$

$$= (0.0600 \text{ mol})(8.314 \text{ J/mol/K})(400.\text{ K})\ln\left(\frac{3.00 \text{ m}^3}{2.00 \text{ m}^3}\right) = 80.9 \text{ J}$$

Because $\Delta U = 0$ along the isotherm, $Q_{BC} = W_{BC} = 80.9$ J. We can now complete our summary table of the process:

PROCESS	ΔU	Q	W
$A \to B$	150 J	250 J	100 J
$B \to C$	0	80.9 J	80.9 J
Total	150 J	331 J	181 J

19.5 Cyclic Processes

In this section we study the basic thermodynamic principles behind the operation of heat engines, such as those powered by gasoline, diesel fuel, coal, and nuclear energy. Engines are constructed in such a manner as to produce periodic, mechanical motion, and they perform a certain amount of mechanical work during each cycle of the motion. We consider engines from a purely thermodynamic standpoint and do not concern ourselves with the details of the sources (or sinks) of energy; we continue to rely on the heat-reservoir model. You should realize that a reservoir may in fact be a burning fuel or a cooling tower. Let's begin with a definition.

Processes during which the thermodynamic variables periodically return to their original values are called **cyclic processes.** However, we must pay a price for the work we get, and that price is determined, of course, by the first law of thermodynamics. At the end of a complete cycle, all state variables return to their original values, including the internal energy: for a complete cycle $\Delta U_{\text{cycle}} = 0$. Therefore, $Q - W = 0$, and *the work we obtain from a reversible cyclic process is equal to the total heat transferred into the system.*

Radioactive decay processes provide the high-temperature reservoir for the engine of this nuclear-powered submarine. The cool ocean water is the low-temperature reservoir.

A general cyclic process is illustrated in Figure 19.12. During the portion of the cycle from A to B the work performed by the system is positive; this work is represented by the area under the curve from A to B. During the return from B to A the work is negative because the volume is decreasing. When this latter negative work is combined with the positive work of the A to B part of the cycle, *the net work performed is the area enclosed by the PV-diagram of the cycle.*

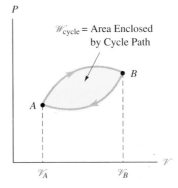

FIGURE 19.12

A cyclic thermodynamic process. The work done during the cycle is the area enclosed by the path representing the cycle.

Problem-Solving Strategy for Thermodynamic Cycles

Finding all the parameters associated with a thermodynamic cycle is a little like solving a crossword puzzle; a detailed recipe cannot be set out for every case. However, the following general strategy usually brings success:

I. First sketch the process in the PV-plane and make a table of ΔU, Q, and W for each segment of the cycle as well as the total for these quantities.

   A. Fill in any given information in the appropriate cells of the table.

   B. From your knowledge that $\Delta U = Q - W$ for each segment and that $\Delta U_{cycle} = 0$ fill in as many of the empty cells in the table as you can.

   C. If you cannot complete the table apply the strategy in II below.

II. Find the temperature, pressure, and volume at each vertex of the cycle.

   A. If two of the variables are known, the equation of state can be used to find the third.

   B. If one variable is known at a given vertex, find a second by relating the values at a neighboring vertex to that at the vertex in question using relations valid along the connecting path.

     i. If the connecting path is an isotherm, an isochor, or an isobar, use relations implied by the equation of state. For an ideal gas these are $PV = constant$, $P/T = constant$, or $V/T = constant$, respectively.

     ii. If the connecting path is an adiabat $PV^\gamma = constant$ may be used for an ideal gas.

III. Use the relations summarized in Table 19.1 to calculate ΔU, Q, and W for each path.

When analyzing a cyclic process it is very helpful to make a table of ΔU, Q, and W just as we did in Examples 19.3 and 19.4. It is also helpful to include the total for each of these quantities. That way, if ΔU_{cycle} does not come out to be zero, you know something is wrong! In Example 19.5 we illustrate how to use this method with a specific cycle.

EXAMPLE 19.5   The Physics-Teachers' Cycle

One mole of helium is enclosed in a cylinder with a movable piston. By placing the cylinder in contact with various reservoirs and also insulating it at the proper times, the helium performs the cycle shown in Figure 19.13. Process AB is isothermal, BC is

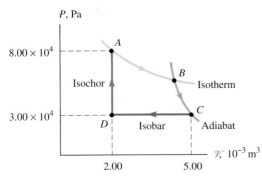

FIGURE 19.13

A cyclic process consisting of an isothermal expansion (AB), an adiabatic expansion (BC), an isobaric compression (CD), and an isochoric return (DA) to the original state. See Example 19.5.

adiabatic, *CD* is isobaric, and *DA* is isochoric. Compute the internal energy change, heat transferred, and work performed for each portion of the cycle and the total amount of each of these quantities for the entire cycle. Assume helium to be an ideal gas. (To help you visualize the cycle, Figure 19.14 illustrates each step.)

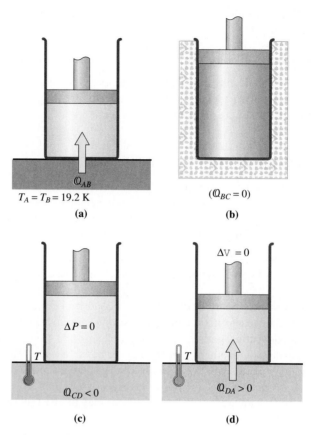

FIGURE 19.14

(a) Heat is added to the gas while it expands and does work from *A* to *B*. (b) The cylinder is isolated from *B* to *C* so that no heat is transferred as work is performed. (c) Work is done on the gas from *C* to *D*. Heat is removed as the gas is compressed at constant pressure back to its original volume. (d) The volume remains unchanged as heat is added and the temperature is raised to the original value. See Example 19.5.

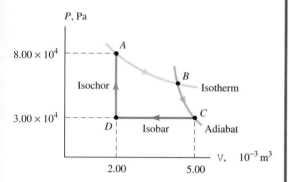

FIGURE 19.13   Repeat

SOLUTION   We begin by computing some useful information, such as the temperatures $T_A = T_B$, T_C, T_D; the volume $\mathcal{V}_B$; and the pressure P_B. The temperatures can be obtained from the equation of state for an ideal gas.

$$T_A = \frac{P_A \mathcal{V}_A}{nR} = \frac{(8.00 \times 10^4 \text{ Pa})(2.00 \times 10^{-3} \text{ m}^3)}{(1 \text{ mole})(8.314 \text{ J/mole/K})} = 19.2 \text{ K}$$

$$T_C = \frac{P_C \mathcal{V}_C}{nR} = 18.0 \text{ K}$$

$$T_D = \frac{P_D \mathcal{V}_D}{nR} = 7.22 \text{ K}$$

In order to compute the pressure and volume at point *B* we use Equation (19.17) for the adiabat *BC*:

$$P_C \mathcal{V}_C^\gamma = P_B \mathcal{V}_B^\gamma$$

For helium we have

$$\gamma = \frac{c_p}{c_v} = \frac{(20.8 \text{ J/mol/K})}{(12.5 \text{ J/mol/K})} = 1.66$$

However, we are faced with two unknowns: both P_B and $\mathcal{V}_B$. Therefore, we need another equation. We also know

$$P_B \mathcal{V}_B = nRT_B$$

Eliminating P_B from these last two equations, we find

$$P_C \mathcal{V}_C^{\gamma} = P_B \mathcal{V}_B^{\gamma} = P_B \mathcal{V}_B (\mathcal{V}_B)^{\gamma-1} = nRT_B (\mathcal{V}_B)^{\gamma-1}$$

Upon solving for $\mathcal{V}_B$, we find

$$\mathcal{V}_B = \left(\frac{P_C \mathcal{V}_C^{\gamma}}{nRT_B}\right)^{1/(\gamma-1)} = 4.55 \times 10^{-3} \text{ m}^3$$

And returning to $P_B \mathcal{V}_B = nRT_B$, we have

$$P_B = \frac{nRT_B}{\mathcal{V}_B} = 3.51 \times 10^4 \text{ Pa}$$

We now have the pressure, volume, and temperature at each vertex of the cycle. Thus, we finally begin the calculation of the heat, work, and internal energy along each path of the cycle in order to fill in a summary table.

ISOTHERM $A \rightarrow B$

$$\mathcal{W}_{AB} = nRT_A \ln\left(\frac{\mathcal{V}_B}{\mathcal{V}_A}\right)$$

$$= (1 \text{ mol})(8.314 \text{ J/mol/K})(19.2 \text{ K}) \ln\left(\frac{4.55 \times 10^{-3} \text{ m}^3}{2.00 \times 10^{-3} \text{ m}^3}\right) = 132 \text{ J}$$

Along the isotherm AB we know $\Delta U_{AB} = 0$, so $\mathcal{Q}_{AB} = \mathcal{W}_{AB} = 132$ J.

ADIABAT $B \rightarrow C$

$$\mathcal{Q}_{BC} = 0$$

Because $\mathcal{Q}_{BC}$ is zero, the work can be found from the change in internal energy. Recalling that the change in energy can always be found from $\Delta U = nc_v \Delta T$ even when the process is not actually isochoric, we have

$$\mathcal{W}_{BC} = -\Delta U_{BC} = -nc_v \Delta T$$

$$= -(1 \text{ mol})(12.5 \text{ J/mol/K})(18.0 \text{ K} - 19.2 \text{ K})$$

$$= 15.0 \text{ J}$$

and,

$$\Delta U_{BC} = -\mathcal{W}_{BC} = -15.0 \text{ J}$$

The work done by the gas during this portion of the cycle is at the expense of the internal energy because ΔU_{BC} decreases and $\mathcal{Q}_{BC} = 0$.

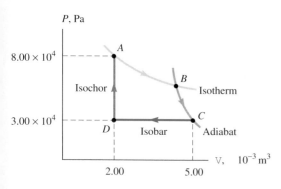

FIGURE 19.13   Repeat

ISOBAR C → D   During this isobaric process

$$\mathcal{W}_{CD} = P\,\Delta\mathcal{V} = (3.00 \times 10^4 \text{ Pa})[(2.00 \times 10^{-3} \text{ m}^3) - (5.00 \times 10^{-3} \text{ m}^3)] = -90.0 \text{ J}$$

$$\mathcal{Q}_{CD} = nc_p\,\Delta T = (1 \text{ mol})(20.8 \text{ J/mol} \cdot \text{K})(7.2 \text{ K} - 18.0 \text{ K}) = -225 \text{ J}$$

$$\Delta U_{CD} = \mathcal{Q}_{CD} - \mathcal{W}_{CD} = (-225 \text{ J}) - (-90 \text{ J}) = -135 \text{ J}$$

(We could also have used $\Delta U_{CD} = nc_v\,\Delta T_{CD}$ to compute this last value. See if you can obtain $\Delta U_{CD} = -135$ J by this approach.)

ISOCHOR D → A   This is a constant-volume process, so

$$\mathcal{W}_{DA} = 0$$

$$\Delta U_{DA} = \mathcal{Q}_{DA} = nc_v\,\Delta T_{DA} = (1 \text{ mol})(12.5 \text{ J/mol/K})(19.2 \text{ K} - 7.2 \text{ K}) = 150. \text{ J}$$

We can now fill in our summary table, complete with the cycle totals:

PROCESS	ΔU	$\mathcal{Q}$	$\mathcal{W}$
$A \rightarrow B$	0	132 J	132 J
$B \rightarrow C$	−15 J	0	15 J
$C \rightarrow D$	−135 J	−225 J	−90 J
$D \rightarrow A$	150 J	150 J	0
Total	0	57 J	57 J

During each complete cycle 132 J + 150 J = 282 J of heat are removed from high-temperature reservoirs. Of that, 57 J of heat are converted into 57 J of work, and 225 J of heat are moved to the low-temperature reservoir.   ◀

19.6 Summary

A simple thermodynamic system is in equilibrium when its thermodynamic variables are constant throughout the system. The **zeroth law of thermodynamics** states that two systems are in equilibrium with each other when they are each in equilibrium with a third.

A **process** is a sequence of events that takes a system from one state to another. In a **reversible process,** the equation of state applies at all times. In an **irreversible process** the state variables of the system may not have uniform values throughout the system, and, therefore, the equation of state does not apply to the system. All natural processes are irreversible. Volume is fixed during an **isochoric process.** Pressure is maintained constant during an **isobaric process.** Temperature is constant during an **isothermal process.** An **adiabatic process** is one in which there is no transfer of heat into or out of the thermodynamic system.

Quantities that depend only on the state of a system are called **state variables.** The change in a state variable is independent of path and depends only on the original and final states of the system. Pressure, volume, temperature, and internal energy are examples of state variables.

The work performed by a thermodynamic system that undergoes a volume change from $\mathcal{V}_o$ to $\mathcal{V}_f$ is

$$\mathcal{W} = \int_{\mathcal{V}_o}^{\mathcal{V}_f} P\, d\mathcal{V} \qquad\qquad (19.1)$$

Work is not a state variable. Heat is not a state variable either because the amount of heat absorbed by a system depends on the particular process. The heat capacity of a material depends on the conditions under which the heat is added. Two useful, special cases are the molar heat capacity at constant volume and constant pressure, c_v and c_p, respectively.

The **first law of thermodynamics** is a statement of the law of conservation of energy: $\Delta U = Q - \mathcal{W}$, where ΔU is the change in **internal energy** of the system, Q is the heat added to the system, and $\mathcal{W}$ is the work performed by the system. In differential form this law is $dU = \bar{d}Q - \bar{d}\mathcal{W}$. The differentials with a bar through them remind us that heat and work are not state variables. The internal energy is a state variable. For cyclic processes $\Delta U = 0$.

The change in internal energy during an isothermal process of an ideal gas is zero.

Table 19.1 provides a summary of some important thermodynamic processes. A **cyclic process** is one in which the thermodynamic variables periodically return to their original values.

PROBLEMS

19.2 Work

1. Figure 19.P1 shows two thermodynamic processes in the $P\mathcal{V}$-plane for which $P_A = 2.00 \times 10^5$ Pa, $P_B = 4.00 \times 10^5$ Pa, $\mathcal{V}_A = 1.00$ L, and $\mathcal{V}_B = 3.00$ L. (a) Compute the work done by the system along the path AB. (b) Compute the work done by the system along the path ACB.

2. Figure 19.P1 shows two thermodynamic processes in the $P\mathcal{V}$-plane for which $P_A = 3.30 \times 10^5$ Pa, $P_B = 7.50 \times 10^5$ Pa, $\mathcal{V}_A = 2.50$ L, and $\mathcal{V}_B = 5.00$ L. (a) Compute the work done by the system along the path AB. (b) Compute the work done by the system along the path ACB.

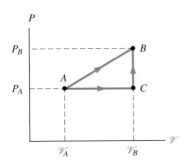

FIGURE 19.P1   Problems 1, 2, 18 and 19

3. For the thermodynamic processes shown in Figure 19.P2, 440 J of work are performed by the system along the diagonal path AC, and 320 J of work are done by the system along path ADC. How much work does the system do along the path ABC?

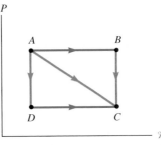

FIGURE 19.P2   Problems 3 and 20

4. For the thermodynamic processes shown in Figure 19.P3, $P_A = 1.00 \times 10^5$ Pa, $P_B = 0.30 \times 10^5$ Pa, $P_D = 0.60 \times 10^5$ Pa, $\mathcal{V}_A = 0.20$ L, and $\mathcal{V}_D = 1.30$ L. (a) Compute the work performed by the system along path AD. (b) If the total work done by the system along the path ADC is 85.0 J, find the volume at point C. (c) How much work is performed by the system along the path CDA?

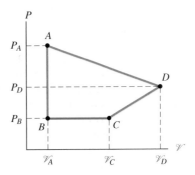

FIGURE 19.P3   Problem 4

5. In Figure 19.P4, $V_o = 0.100$ m³, $V_f = 0.250$ m³, $a = 250.$ Pa/m⁶, and $b = 500.$ Pa/m⁹. (a) Find the work done by the system along the path ABC. (b) Compute the work done by the system along the path $ABCDA$.

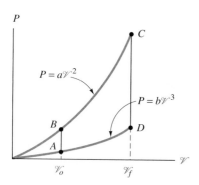

FIGURE 19.P4   Problems 5 and 21

6. The equation of an adiabatic process for an ideal gas is $P = AV^{-\gamma}$, where A and γ are constants. (a) Show that the work performed by the gas from state (P_o, V_o) to state (P_f, V_f) is given by

$$W_{ad} = \frac{1}{1 - \gamma}(P_f V_f - P_o V_o)$$

(b) Show the result of part (a) is equivalent to

$$W_{ad} = \frac{nR}{1 - \gamma}(T_f - T_o)$$

7. One mole of an ideal gas initially at a pressure of 1.20 atm and a temperature of 24.0 K doubles its volume while the temperature remains constant. Compute the work done by the gas.

8. The pressure of an ideal gas is 1.50 atm, and its volume is 2.00 L. The gas expands isothermally to a final volume of 6.00 L. How much work does the gas perform?

9. An ideal gas, initially at a pressure of 2.00 atm with a volume of 4.00 L is compressed isothermally until its pressure is 3.00 atm. How much work is done on the gas?

10. A pressure of 0.800 atm is maintained on a piston of a pump, and this pressure causes the piston to travel a distance of 20.0 cm along a cylindrical container. If the cross-sectional area of the piston is 3.00×10^3 cm², compute the work done on the gas within the piston.

11. During a constant-pressure process a piston moves through a distance of 18.0 cm along a cylindrical container while the volume of the gas increases. If the cross-sectional area of the piston is 1.20×10^3 cm² and the gas performs 1.25 kJ of work, what is the gas pressure during the process?

19.3 The First Law of Thermodynamics

12. An ideal gas's volume is decreased from 2.00 L to 1.00 L at a constant pressure of 1.50 atm. At the same time, the internal energy of the system is reduced by 54.0 cal. How much heat was removed from the system?

13. The volume of an ideal gas increases from 2.00 L to 5.00 L while the pressure remains constant at 0.800 atm. During this process 600. J of heat are added to the gas. What is the change in internal energy during the process?

14. The volume of an ideal gas increases from 1.00 L to 3.00 L while the pressure remains constant at 1.20 atm. During this process the internal energy increases by 150. J. How much heat is added to the system?

15. During an isothermal process, 0.500 mol of an ideal gas is compressed from an initial volume of 6.00 L to a final volume of 2.00 L. The initial pressure of the gas was 1.40 atm. How much heat is removed from the system?

16. A certain gas expands its volume isothermally from 1.50 L to 6.00 L. The initial temperature and pressure of the gas were 27.0°C and 1.00 atm, respectively. How much heat is removed from the system during the process if it can be modeled as an ideal gas?

★17. Figure 19.P5 shows various processes in the PV-plane for an ideal gas. The process ADC is an adiabat, AB is an isotherm, and BC is a constant-volume process. The heat added to the gas along AB is 400. cal. The change in internal energy from C to A is $+1000.$ cal, and the shaded area is 150. cal. (a) How much work is done by the gas from A to B? (b) How much heat is added to the system from B to C? (c) What is the change in internal energy from C to D? (d) How much work is done by the gas from D to A?

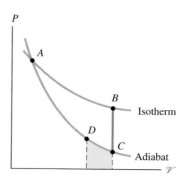

FIGURE 19.P5   Problem 17

18. For the processes shown in Figure 19.P1 and described in Problem 1, $Q_{AB} = 2.10 \times 10^3$ J of heat are added to the system along the path AB. (a) What is the change in internal energy along the path AB? (b) How much heat is added to the system along the indirect path ACB?

19. For the processes shown in Figure 19.P1 and described in Problem 2, 3.60×10^3 J of heat are added to the system along the path ACB. (a) What is the change in internal energy along the path ACB? (b) How much heat is added to the system along the direct path AB?

20. The change in internal energy along the process ABC shown in Figure 19.P2 and described in Problem 3 is -1.00×10^3 J. (a) How much heat is removed from the system along the path ABC? (b) How much heat is removed along the path ADC? (c) How much heat is added to the system during the AC process?

21. Along the path ABC of Figure 19.P4 (see Problem 5), 1.67 J of heat are added to the system. (a) What is the change in internal energy of the system from point A to point C? (b) How much heat is added to the system along the path ADC?

19.4 Specific Thermodynamic Processes

22. One mole of an ideal gas expands isothermally at a temperature of 0.00°C until its volume is doubled. (a) How much work is done by the gas? (b) How much heat is transferred to the gas during the expansion?

23. The volume of 2.00 mol of helium (c_p = 20.8 J/mol/K and c_v = 12.5 J/mol/K) increases from 1.00 L to 2.00 L at a constant pressure of 1.00×10^5 Pa. The pressure of the helium then decreases isothermally until its volume is 3.00 L. (a) Sketch a $P\mathcal{V}$-diagram for this process. (b) How much heat is added to the gas during the isobaric expansion? (c) What is the change in internal energy during the isobaric expansion? (d) What are the changes in internal energy and the heat transfer during the isothermal expansion?

24. One mole of an ideal gas for which c_p = 7.00 cal/mol/K and c_v = 5.00 cal/mol/K undergoes an isochoric increase in pressure from 0.50 atm to 1.50 atm at a volume of 2.00 L. This process is followed by an isobaric increase in volume to 4.00 L. (a) Sketch a $P\mathcal{V}$-diagram for this process. (b) What are the change in internal energy and the heat added to the gas for the isochoric and isobaric processes?

25. Two moles of a gas, for which c_p = 7.00 cal/mol/K and c_v = 5.00 cal/mol/K, undergo an isothermal expansion from a volume of 3.00 L at 1.00 atm to a volume of 6.00 L at 0.50 atm. The volume is next returned to its original value at constant pressure. (a) Sketch a $P\mathcal{V}$-diagram for this process. (b) How much heat is added to the system during the isothermal expansion? (c) What are the change in internal energy and the heat added to the gas during the isobaric decrease in volume?

26. One mole of helium gas expands adiabatically from a pressure of 1.20 atm and a volume of 1.00 L until its volume is doubled. (a) What are the final pressure and temperature of the helium? (b) What is the change in internal energy of the helium? (c) How much work was performed by the helium during the expansion?

27. Two moles of helium at 250. K are expanded adiabatically from a volume of 1.50 L to 4.50 L. (a) What are the final pressure and temperature of the gas? (b) What is the change in internal energy of the gas?

19.5 Cyclic Processes

28. An ideal gas is used as the working substance for the engine cycle shown in Figure 19.P6. Process AB is isobaric, BC is isochoric, and CA is isothermal. During each cycle 80.0 J of work are performed by the engine. During AB while 35.0 J of heat is moved from C to A and 140. J of heat is added to the gas. (a) How much work is done on the system from C to A? (b) What is the change in internal energy from A to B? (c) How much heat is removed from B to C? Assume the gas is ideal with c_p = 20.8 J/mol/K and c_v = 12.5 J/mol/K.

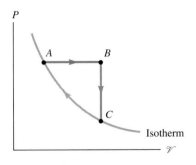

FIGURE 19.P6 Problems 28 and 38

29. One mole of gas is taken through the cycle shown in Figure 19.P7. Segment CA is an adiabat and $\mathcal{V}_A$ = 2.00 L, $\mathcal{V}_B$ = 3.00 L, and P_C = 1.00×10^5 Pa. For each segment of the cycle compute the change in internal energy, the heat transferred to the

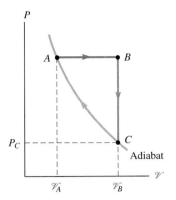

FIGURE 19.P7 Problem 29

system, the work performed by the system, and the total per cycle for each of these quantities. Place the results of your computations in a table as illustrated in Example 19.5. Assume the gas is ideal with c_p = 20.8 J/mol/K and c_v = 12.5 J/mol/K.

30. Figure 19.P8 shows a cyclic process in the $T\mathcal{V}$-plane. The segment CA is an adiabat, and $\mathcal{V}_A$ = 2.00 L, $\mathcal{V}_B$ = 3.00 L, and T_A = 300. K. The working substance can be modeled as 1.00 mol of an ideal gas, and 60.0 J of work are performed during each cycle. From B to C, 30.0 J of heat are removed. (a) Plot the cycle in the $P\mathcal{V}$-plane. (b) For each segment of the cycle compute the change in internal energy, the heat transferred to the system, the work performed by the system, and the total per cycle for each of these quantities. Place the results of your computations in a table like that in Example 19.5. Assume the gas is ideal with c_p = 20.8 J/mol/K and c_v = 12.5 J/mol/K.

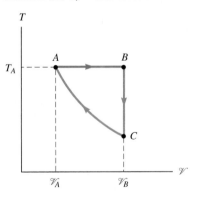

FIGURE 19.P8 Problem 30

31. Figure 19.P9 shows an isothermal expansion from A to B for 0.150 mol of helium at 10.0°C. Assume P_A = 2.00×10^5 Pa and

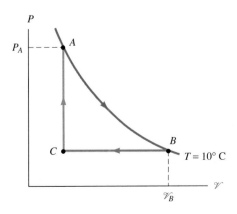

FIGURE 19.P9 Problem 31

$V_B = 5.00$ L. The expansion is followed by an isobaric reduction in volume from B to C. Finally, the gas is returned to its original state at constant volume. (a) What is the pressure and temperature at point C? (b) Compute the change in internal energy, the heat transferred to the system, the work performed by the system for each segment of the cycle. (c) Find the total per cycle for each of the quantities in part (b). Assume the gas is ideal with $c_p = 20.8$ J/mol/K and $c_v = 12.5$ J/mol/K.

32. In a cyclic thermodynamic process an ideal gas is first compressed at constant volume while its internal energy is increased by 350. J. It next expands adiabatically until it returns to its original pressure. Finally, the gas is returned isobarically to its original state when 60.0 J of work are performed on the gas while its internal energy decreases by 200. J. (a) How much heat is added to the gas during the initial compression at constant volume? (b) What is the change in internal energy during the adiabatic expansion? (c) How much heat is added to the gas during the final volume reduction at constant pressure? Assume the gas is ideal with $c_p = 20.8$ J/mol/K and $c_v = 12.5$ J/mol/K.

Numerical Methods and Computer Applications

33. Write a program or construct a spreadsheet template to plot ideal-gas adiabats and isotherms that originate at the same pressure and volume (P_o, V_o) and terminate at pressure V_f. Try out your program or spreadsheet for 1 mol of the ideal gas with $P_o = 12.0 \times 10^5$ Pa, $V_o = 0.020$ L, and $V_f = 0.025$ L. Verify that adiabats are steeper than isotherms.

34. Write a program or construct a spreadsheet template that fills in the table in Example 19.5 given any set of data (P_A, V_A), (P_B, V_B), (P_C, V_C), and (P_D, V_D) for n moles of an ideal gas. Try out your program or spreadsheet for the data given in Example 19.5.

35. The pressure and volume data given in Table 19.P1 describe a cyclic thermodynamic process. Apply a numerical integration technique to compute the work done during each cycle. [The data in Table 19.2 may also be found in the file PV1.DAT on the disk accompanying this text.]

TABLE 19.P1   PV1.DAT

VOLUME (m³)	PRESSURE (10⁶ Pa)
0.0785	1.400
0.0752	1.486
0.0720	1.581
0.0687	1.688
0.0654	1.807
0.0621	1.942
0.0589	2.094
0.0556	2.269
0.0523	2.470
0.0491	2.703
0.0458	2.977
0.0425	3.303
0.0392	3.695
0.0425	3.695
0.0458	3.695
0.0491	3.695
0.0523	3.695

TABLE 19.P1   PV1.DAT

VOLUME (m³)	PRESSURE (10⁶ Pa)
0.0556	3.695
0.0589	3.410
0.0621	3.162
0.0654	2.943
0.0687	2.748
0.0720	2.575
0.0752	2.420
0.0785	2.280
0.0785	2.104
0.0785	1.928
0.0785	1.752
0.0785	1.576
0.0785	1.400

General Problems

36. A certain engine cycle can be modeled to consist of three segments: an isobaric expansion followed by an isochoric decrease in pressure and, finally, a return to the original state along an adiabat. During the isobaric expansion 220. J of heat are added to the gas, and 90.0 J of heat are removed along the isochor. Also 60.0 J of work are performed on the gas during the adiabatic portion of the cycle. (a) Sketch the general shape of this cycle in the PV-plane. (b) What is the change in internal energy of the gas during the adiabatic process? (c) Compute the change in internal energy during the isobaric expansion. (d) How much work does the engine do each cycle?

37. An ideal-gas engine cycle consists of an isobaric expansion followed by an isochoric decrease in pressure and, finally, a return to the original state along an adiabat. During the isobaric expansion 296 J of heat are added to the gas while 228 J of work are performed. The internal energy increases by 84.0 J during the adiabatic portion of the cycle. (a) Sketch the general shape of this cycle in the PV-plane. (b) How much work does the engine do each cycle? (c) What is the change in internal energy of the gas along the isobaric expansion? (d) How much heat is transferred to the gas during the isochoric expansion?

38. An ideal gas is used as the working substance for the engine cycle shown in Figure 19.P6. Process AB is isobaric, BC is isochoric, and CA is isothermal. During each cycle 190. J of heat are added to the engine. During process BC, 65.0 J of heat are removed, while 360. J of work are done during process AB. (a) How much work does the engine do during each cycle? (b) How much heat is added to the engine during process AB? (c) How much heat is removed from the engine during process CA?

39. Consider 0.100 mol of an ideal gas taken through the cyclic thermodynamic process shown in Figure 19.P10. For this process $V_A = 1.00$ L, $V_B = 4.00$ L, $P_A = 4.00 \times 10^4$ Pa, and $P_C = 1.00 \times 10^4$ Pa. For each segment of the cycle compute the change in internal energy, the heat transferred to the system, the work performed by the system, and the total per cycle for each of these quantities. Place the results of your computations in a table like that in Example 19.5. (Assume the gas is ideal with $c_p = 20.8$ J/mol/K and $c_v = 12.5$ J/mol/K.)

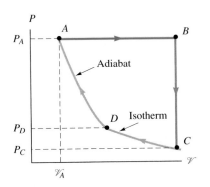

FIGURE 19.P10   Problem 39

40. In a cyclic thermodynamic process 0.0500 mol of an ideal gas has an initial volume of 0.500 L and a temperature of 120. K. The gas expands isobarically to a volume of 2.50 L and is then compressed at constant volume until its temperature is again 120. K. Finally, the gas is compressed isothermally to its original volume. (a) Plot the cycle in the PV-plane. (b) Compute the change in internal energy, the heat added to the gas, and the work performed by the gas for each segment of the cycle. (c) Compute the total internal energy change, the total heat added, and the net work performed by the gas during the cycle. (Assume the gas is ideal with $c_p = 20.8$ J/mol/K and $c_v = 12.5$ J/mol/K.)

41. Suppose 0.100 mol of an ideal gas is taken through a cyclic thermodynamic process as follows. From an initial state $(P_A, V_A) = (12.6$ bar, 2.50 L) the gas expands adiabatically to a state (P_B, V_B), and then expands isothermally to $(P_C, V_C) = (4.50$ bar, 6.80 L). The gas is next compressed isobarically to its original volume (2.50 L) and, finally, undergoes a pressure increase at constant volume to its original state. (a) Plot the cycle in the PV-plane. (b) Compute the change in internal energy, the heat added to the gas, and the work performed by the gas for each segment of the cycle. (c) Compute the total internal energy change, the total heat added, and the net work performed by the gas during the cycle. (Assume the gas is ideal with $c_p = 20.8$ J/mol/K and $c_v = 12.5$ J/mol/K.)

42. An ideal gas consisting of $n = 0.150$ mol is taken through the cycle shown in Figure 19.P11. For this process $P_A = 8.60 \times 10^5$

Pa, $P_D = 2.20 \times 10^5$ Pa, $P_C = 0.80 \times 10^5$ Pa, and $V_A = 0.150$ L. (a) Compute the change in internal energy, the heat added to the gas, and the work performed by the gas for each segment of the cycle. (b) Compute the total internal energy change, the total heat added, and the net work performed by the gas during the cycle. (Assume the gas is ideal with $c_p = 20.8$ J/mol/K and $c_v = 12.5$ J/mol/K.)

43. Two samples of 1.00 mol each of an ideal gas both have initial state variables V_o, P_o, and T_o. The gases are allowed to expand, one isothermally, the other isobarically, to a final volume V_f. (a) Determine the expression for the work done by each gas in terms of V_o, P_o, and V_f only. (b) For the limiting case where $V_f \gg V_o$, which process results in the greater amount of work done by the gas?

44. A 2.000-L piston chamber is filled with helium gas ($\gamma = 5/3$) at a temperature of 20.0°C and pressure of 1.00 atm. The gas is compressed adiabatically to a volume of 0.400 L. Find the new pressure and temperature of the gas.

45. Three samples of 1.00 mol of an ideal gas are each initially at a temperature of 127°C and a pressure of 1.00 atm. Each of the three samples is caused to expand to three times its initial volume by one of the following processes: (a) adiabatically, (b) isobarically, and (c) isothermally. Calculate the heat absorbed by the gas and the work done by the gas in each case. (Assume $c_p = 20.8$ J/mol/K and $c_v = 12.5$ J/mol/K.)

★46. (a) Show that for an ideal gas undergoing an adiabatic expansion from P_o, T_o, V_o to P_f, T_f, V_f the initial and final state variables are related by

$$\frac{T_f}{T_o} = \left(\frac{V_o}{V_f}\right)^{\gamma-1} = \left(\frac{P_f}{P_o}\right)^{(\gamma-1)/\gamma}$$

and (b) that the work done is given by

$$W = \frac{nRT_o}{\gamma - 1}\left[1 - \left(\frac{P_f}{P_o}\right)^{(\gamma-1)/\gamma}\right]$$

47. Completely analyze the following thermodynamic cycle performed on 0.100 mol of an ideal diatomic gas ($c_p = 29.1$ J/mol/K and $c_v = 20.8$ J/mol/K). At the starting point A the gas has a volume of 2.24 L at a pressure of 2.00 atm. The gas then expands isothermally to B at which point the pressure is 1.00 atm. Next the gas is cooled to a pressure of 0.500 atm while the volume is kept constant (point C). Next the gas is compressed adiabatically to point D at which the pressure is again 2.00 atm. Finally heat is added while the gas is held at constant pressure until the volume is again 2.24 L, returning the gas to point A. (a) Determine the pressure, temperature, and volume at each of the points A, B, C, and D. (b) Sketch the cycle on a PV-diagram. (c) Determine the heat added to the system, the work done by the system, and the change in the internal energy of the gas for each of the four processes. Express all your answers in tabular form.

FIGURE 19.P11   Problem 42

CHAPTER

20

Thermodynamics II: The Second Law

In this chapter you should learn

- to quote two descriptive and one mathematical statement of the second law of thermodynamics.
- to describe the Carnot cycle.
- to compute the maximum and the actual efficiencies of heat engines.
- to define temperature in thermodynamic terms.
- to define entropy.
- to compute the change in entropy for reversible and irreversible processes.

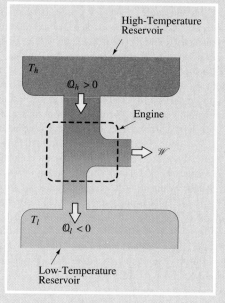

T_h

$Q_h > 0$

High-Temperature Reservoir

Engine

W

T_l

$Q_l < 0$

Low-Temperature Reservoir

$$e = \frac{W_{net}}{Q_h}$$

The final example in the previous chapter (Example 19.5) described how helium gas could be used to move a piston back and forth by alternately heating and cooling the gas using a high-temperature and a low-temperature reservoir. Although admittedly artificial, it is important to realize that this process can serve as a generic example of how modern industrialized nations get work done. Whether it is coal-fired or nuclear-powered electrical generator or a gasoline-powered automobile engine, the thermodynamic process is similar. Fuel is ''burned'' to produce a high-temperature heat source. The heat from the high-temperature reservoir is then transferred to some low-temperature heat sink, typically the air, or a nearby river or ocean. As the heat flows from hot to cold, part of the heat is turned into mechanical work, for example turning the armature of an electrical generator.

Because the extraction of work from heat is so fundamental to modern society, we do well to examine the process carefully. (Remember as you read this chapter that the high-temperature heat comes from burning fuel, which means it costs you money.) The desirable result of Example 19.5 is that 57 J of heat was transformed into 57 J of mechanical work. That sounds pretty good; you get what you pay for. The bad news is that, while this heat is being converted there was another effect, namely, an additional 225 J of heat was removed from the high-temperature reservoir and delivered to the low-temperature reservoir. You paid for this 225 J, but it did you no good at all. Indeed very often the heat delivered to the low-temperature reservoir winds up being released to the environment, sometimes with undesirable side effects.

This secondary effect, dumping heat to a low-temperature reservoir, is not an artifact of our example. Rather it is intrinsic to the process of converting heat into mechanical energy. Whether your intended career is science, engineering, medicine, law, politics, or something else, it's important that you understand the limitation that nature imposes on us when we seek to obtain useful work from heat. This chapter is the description of those limitations.

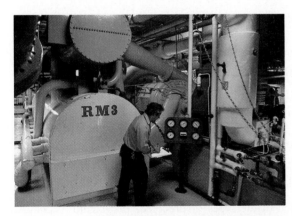

Modern industrialized nations use millions of devices designed to convert heat into work or use work to move heat from low-temperature reservoirs to high-temperature reservoirs.

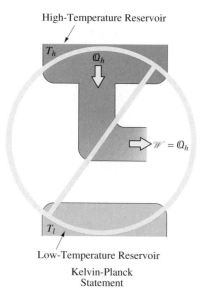

Kelvin-Planck
Statement

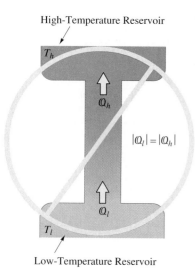

Clausius
Statement

FIGURE 20.1

Processes forbidden by the second law of thermodynamics.

Concept Question 1
Can a given amount of mechanical energy be completely converted into thermal energy? Explain.

20.1 The Second Law of Thermodynamics and Heat Engines

We have, thus far, two postulates from which we have been able to derive many of the thermodynamic properties of matter: the zeroth and first laws of thermodynamics. There are two common examples that help us recognize why another postulate is necessary. First, consider two glasses containing equal amounts of water, one hot and the other cold. When the two are mixed in a pitcher the result is fairly obvious: we obtain warm water. Now, when you pour the warm water from the pitcher back into the glasses, you never (ever!) are able to pour the water back in such a way that the hot water returns to one glass and the cold to the other. However remarkable it may sound, this unlikely result could actually happen without violating the first law of thermodynamics because energy would be conserved.

Consider another process. You turn your bicycle over and give the wheel a spin. After a few seconds the wheel comes to rest and you notice that the bearings have warmed up. Next consider the reverse process: Begin with cool bearings and heat them with a torch. As the bearings cool, it would not violate the first law if the internal energy of the hot bearings was spontaneously converted into mechanical energy. But do you expect to see the wheel start to rotate as the bearings cool? Never. There must be another law that says something about the direction of natural processes. This new law is the second law of thermodynamics.

There are at least four good statements of the second law. All, of course, are equivalent. Two we state immediately; the other two will have to wait for some additional definitions. The first two statements of the second law are called the Kelvin-Planck statement and the Clausius statement:

> **Kelvin-Planck Statement:** It is impossible for any cyclic process during one cycle to convert internal energy into work with no other effect.
>
> **Clausius Statement:** It is impossible for any cyclic process during one cycle to remove an amount of heat from a low-temperature reservoir and transfer an equivalent amount of heat to a high-temperature reservoir with no other effect.

The Kelvin-Planck and Clausius statements of the second law of thermodynamics are equivalent. We will not take time to prove their equivalence, although the proof is not difficult. The usual approach is to assume that one expression of the law can be violated and then show that this leads to a violation of the other. As a matter of fact, this last sentence sounds like an excellent homework problem! So if you're motivated, check out Problem 38; it includes helpful hints.

Figure 20.1 provides an illustration of the types of processes that cannot occur. A process that can occur is shown in Figure 20.2. This second schematic illustrates the idea of a **heat engine.** This device converts internal energy into mechanical energy by absorbing heat Q_h at a high temperature, performing an amount of work W, and expelling heat Q_l to a low-temperature reservoir. There are two important features to recognize about this schematic. First, it is intended to represent only the flow of heat and the work performed during some type of real engine process. The actual engine is not shown because its type may vary. It may be helpful for you to visualize the engine as something rather crude like an ideal gas in our old and familiar cylinder, rather than the intricate system in that Porsche you might really want to own. If you prefer, you can think of the engine process as that of Example 19.5. We will provide another important engine process later in this chapter. The problems for this chapter also contain some useful models for real engines.

The second important feature of this schematic is that it represents all effects resulting from one *complete cycle.* At the end of a cycle the internal energy has returned to its original value, so $\Delta U_{\text{cycle}} = 0$. There is no need, therefore, to show ΔU in Figure 20.2. The first law says $\Delta U = Q - W$ for any process. When the first law is written for a complete cycle, we have $\Delta U_{\text{cycle}} = Q_{\text{net}} - W_{\text{net}}$, where Q_{net} is the net heat added to the engine system and W_{net} is the net work done by the system during the cycle. If we continue our sign convention that heat added to a system (the engine) is positive and heat removed from a system is negative, then Q_h is positive and Q_l is negative so that $Q_{\text{net}} = Q_h + Q_l$. Because $\Delta U_{\text{cycle}} = 0$, the first law implies that $Q_{\text{net}} = W_{\text{net}}$. Therefore, the work performed by the engine during a single cycle is

$$W = Q_h + Q_l = Q_h - |Q_l| \tag{20.1}$$

A natural question to ask about any engine is "How much work can I get out of this thing for a given amount of fuel?" The **efficiency** e of a heat engine is defined as the ratio of the work we obtain to the heat we put into the engine:

$$e = \frac{Work\ done\ per\ cycle}{Heat\ input\ per\ cycle} = \frac{W}{Q_h} \qquad \text{Efficiency of a heat engine} \tag{20.2}$$

We can substitute the expression for W from Equation (20.1) into Equation (20.2) to obtain

$$e = \frac{Q_h + Q_l}{Q_h} = 1 - \frac{|Q_l|}{Q_h} \tag{20.3}$$

where we have recognized that $Q_l < 0$. For instance, in Example 19.5 we found that during each cycle the heat transferred *into* the system (132 J during the isothermal process and 150 J during the isobaric process) was $132\ \text{J} + 150\ \text{J} = 282\ \text{J}$. The net work done during the cycle was 57.0 J. Hence, the efficiency of an engine operating with this particular cycle is $e = (57.0\ \text{J})/(282\ \text{J}) = 0.202$, or 20.2%.

In the next section we will examine a particular cyclic process with which you need to become familiar. Even though it is a theoretical process it is very important from a practical standpoint.

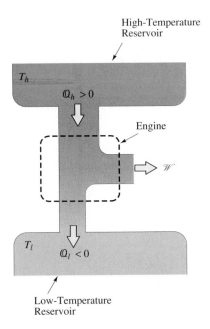

FIGURE 20.2

A heat engine. This schematic represents one cycle of a heat engine. Heat Q_h is removed from the high-temperature reservoir, heat Q_l is transferred to the low-temperature reservoir, and work W is performed during each cycle. The change in internal energy ΔU is zero.

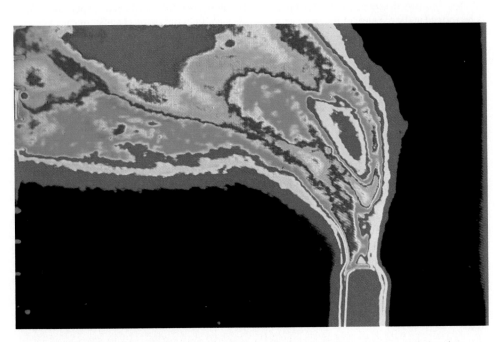

Infrared thermography records the emissions from a smokestack. The inevitable result of turning thermal energy into mechanical energy is the deposition of low-temperature heat into the environment. Here thermal energy is ejected in the form of hot air.

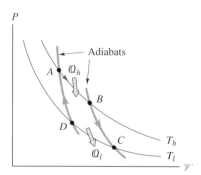

FIGURE 20.3

The Carnot cycle. Processes AB and CD are isothermal, and BC and DA are adiabatic. During process AB heat Q_h is added to the system and during CD heat Q_R is removed.

20.2 The Carnot Cycle

In 1824, a French engineer named Nicolas Léonard Sadi Carnot (1796–1832) invented a cyclic process consisting of two isotherms and two adiabats. The sequence of processes that make up the cycle bearing Carnot's name is illustrated in Figure 20.3. The engine we imagine that carries a working substance through this cycle is called a **Carnot engine.** It is convenient to think of the engine as consisting of our piston with insulating walls and plunger. The base, however, may be placed in contact with either additional insulation or with a heat reservoir. When the base is in contact with the reservoir, heat may enter or leave the system depending on the temperatures of the system and the reservoir. A pictorial representation of one cycle of this engine is shown in Figure 20.4. During each cycle heat Q_h is provided to the engine by a high-temperature reservoir, and heat Q_l is removed from the system and deposited into the low-temperature reservoir. Therefore, according to our sign convention, $Q_h > 0$ and $Q_l < 0$ for the Carnot engine.

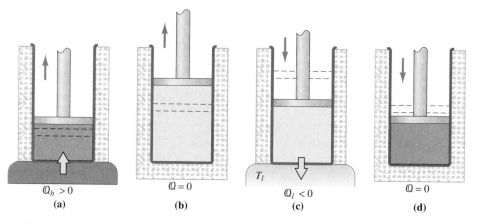

FIGURE 20.4

The Carnot cycle applied to a piston in a cylindrical chamber. The walls of the chamber are insulating, and the base can be placed in contact with either a heat reservoir or additional insulation. (a) Heat Q_h flows into the chamber during an isothermal process. (b) An adiabatic expansion: Work is performed at the expense of internal energy. (c) Heat Q_l flows out of the chamber during an isothermal compression. (d) The system is returned to its original state via an adiabatic compression.

N. L. Sadi Carnot (1796–1832) French engineer and scientist. Son of Lazare Carnot, a noted scientist and politician of the Napoleonic era, Sadi Carnot carried on his father's interest in engines. His most famous essay, *Réflexions sur la puissance mortice de feu* (Reflections on the Motive Power of Fire), was published in Paris in 1824. Sadi Carnot died at the age of 36 of cholera.

Our goal for this section is to develop an expression for the efficiency of a Carnot engine. We therefore need to calculate the work done and the heat added (or extracted) along each of the four paths making up the Carnot cycle. We follow the general procedure described in the problem-solving strategy for analyzing thermodynamic cycles as described in Chapter 19. That is, we first determine the values of P, V, and T at each vertex on the PV-diagram and then calculate the heat and work along each path of the cycle.

Although almost anything may be used as the working substance in a Carnot cycle, for simplicity we will use an ideal gas. However, we do not use particular values of P, V, and T at each vertex. So instead of finding a numerical value for each as we usually do, we find relationships between the volumes at the four thermodynamic coordinates for the process: (P_A, V_A), (P_B, V_B), (P_C, V_C), (P_D, V_D). Referring to Figure 20.3, we recall that when we have numbers, we relate pressures and volumes for an isothermal process using

$$(A \rightarrow B) \qquad\qquad P_A V_A = P_B V_B \qquad\qquad (20.4)$$

and

$$(C \rightarrow D) \qquad\qquad P_C V_C = P_D V_D \qquad\qquad (20.5)$$

Whereas for the adiabatic processes, we employ

$(B \rightarrow C)$
$$P_C \mathcal{V}_C^{\gamma} = P_B \mathcal{V}_B^{\gamma}$$
(20.6)

$(D \rightarrow A)$
$$P_A \mathcal{V}_A^{\gamma} = P_D \mathcal{V}_D^{\gamma}$$
(20.7)

When we divide Equation (20.4) by Equation (20.7) and Equation (20.5) by Equation (20.6), we obtain two different expressions for the ratio P_B/P_D:

$$\frac{P_D}{P_B} = \frac{1}{\mathcal{V}_A^{(1-\gamma)}} \frac{\mathcal{V}_B}{\mathcal{V}_D^{\gamma}}$$

$$\frac{P_D}{P_B} = \mathcal{V}_C^{(1-\gamma)} \cdot \frac{\mathcal{V}_B^{\gamma}}{\mathcal{V}_D}$$

Equating these expressions, we (eventually) find the surprisingly simple result

$$\frac{\mathcal{V}_A}{\mathcal{V}_B} = \frac{\mathcal{V}_D}{\mathcal{V}_C}$$
(20.8)

Equation (20.8) relates the volumes at the corners of the cycle, which we will need later.

We are now ready to find the work done, heat transferred, and internal energy change along each path of the Carnot cycle. Before beginning let's review the results for both isothermal and adiabatic processes. For a process that takes a thermodynamic system from $(P_o, \mathcal{V}_o, T_o)$, to $(P_f, \mathcal{V}_f, T_f)$, we learned in Chapter 19 that

$$\mathcal{W}_{\text{isotherm}} = nRT \ln\left(\frac{\mathcal{V}_f}{\mathcal{V}_o}\right)$$
(20.9)

$$\mathcal{W}_{\text{adiabat}} = -nc_v(T_f - T_o) \qquad (= -\Delta U_{\text{adiabat}})$$
(20.10)

$$Q_{\text{isotherm}} = nRT \ln\left(\frac{\mathcal{V}_f}{\mathcal{V}_o}\right) \qquad (= \mathcal{W}_{\text{isotherm}})$$
(20.11)

$$Q_{\text{adiabat}} = 0$$
(20.12)

$$\Delta U_{\text{isotherm}} = 0$$
(20.13)

$$\Delta U_{\text{adiabat}} = nc_v(T_f - T_o)$$
(20.14)

The stage is now set for our analysis of one cycle of the Carnot engine.

We follow the tabular procedure introduced in Chapter 19. The results of these calculations are shown in Table 20.1. Notice that in the $C \rightarrow D$ portion of the cycle there are two alternative forms for Q and $\mathcal{W}$. The second form comes from taking the natural logarithm of both sides of Equation (20.8), which results in $\ln(\mathcal{V}_D/\mathcal{V}_C) = \ln(\mathcal{V}_A/\mathcal{V}_B)$ and then employing the identity $\ln(\mathcal{V}_A/\mathcal{V}_B) = -\ln(\mathcal{V}_B/\mathcal{V}_A)$. The row of totals for ΔU, Q, and $\mathcal{W}$ are obtained by using the $-\ln(\mathcal{V}_B/\mathcal{V}_A)$ form from the $C \rightarrow D$ row.

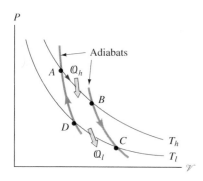

FIGURE 20.3   Repeat

TABLE 20.1   Carnot Cycle

PROCESS	ΔU	Q	W
$A \rightarrow B$ (isotherm)	0	$nRT_h \ln\left(\dfrac{V_B}{V_A}\right)$	$nRT_h \ln\left(\dfrac{V_B}{V_A}\right)$
$B \rightarrow C$ (adiabat)	$nc_v(T_l - T_h)$	0	$-nc_v(T_l - T_h)$
$C \rightarrow D$ (isotherm)	0	$nRT_l \ln\left(\dfrac{V_D}{V_C}\right)$	$nRT_l \ln\left(\dfrac{V_D}{V_C}\right)$
		$\downarrow$	$\downarrow$
[alternative forms]		$-nRT_l \ln\left(\dfrac{V_B}{V_A}\right)$	$-nRT_l \ln\left(\dfrac{V_B}{V_A}\right)$
$D \rightarrow A$ (adiabat)	$nc_v(T_h - T_l)$	0	$nc_v(T_h - T_l)$
TOTAL	0	$nR(T_h - T_l) \ln\left(\dfrac{V_B}{V_A}\right)$	$nR(T_h - T_l) \ln\left(\dfrac{V_B}{V_A}\right)$

We are now ready to complete the final phase of our calculation to find the efficiency of the Carnot cycle. From Table 20.1 we can find the net work obtained from one complete Carnot cycle operating with an ideal gas. We notice that, as predicted by the first law with $\Delta U_{\text{cycle}} = 0$, this work has the same value as the net heat transferred to the engine during this cycle:

$$W_{\text{net}} = Q_{\text{net}} = nR(T_h - T_l) \ln\left(\frac{V_B}{V_A}\right) \tag{20.15}$$

Recalling our expression for the net heat transferred for the Carnot engine, we write $W_{\text{net}} = Q_h + Q_l$ (remember that Q_l is a negative number). To obtain the efficiency we divide the net work done by Q_h, the total heat extracted from the high-temperature reservoir. Recall that the efficiency of any engine is

$$e = \frac{W}{Q_h} = \frac{Q_h + Q_l}{Q_h} = 1 + \frac{Q_l}{Q_h} \tag{20.16}$$

For the case of the Carnot engine we can write this efficiency in terms of the temperatures of the two reservoirs by using our results for Q_l and Q_h from Table 20.1:

$$Q_l = Q_{CD} = -nRT_l \ln\left(\frac{V_B}{V_A}\right) \quad \text{and} \quad Q_h = Q_{AB} = nRT_h \ln\left(\frac{V_B}{V_A}\right) \tag{20.17}$$

so that

$$e = 1 - \frac{nRT_l \ln\left(\dfrac{V_B}{V_A}\right)}{nRT_h \ln\left(\dfrac{V_B}{V_A}\right)}$$

Which immediately simplifies to

$$e_{\text{Carnot}} = 1 - \frac{T_l}{T_h} \qquad \text{Ideal Carnot efficiency} \tag{20.18}$$

Equation (20.18) is the expression for Carnot efficiency that we set out to obtain.

Often, efficiencies are quoted as percentages. To convert the decimal value of the efficiency *e* to a percentage we need only multiply *e* by 100%. Although we derived Equation (20.18) for a Carnot engine that uses an ideal gas as a working substance, this relation turns out to be true, in general. That is, for any working substance in the Carnot engine, we discover the remarkable fact that the efficiency depends *only* on the temperatures of the two reservoirs and is given by Equation (20.18).

The efficiency of the Carnot engine is important because, as we will show in the next section,

> No real engine operating between temperatures T_l and T_h can have an efficiency greater than the efficiency of a Carnot engine operating between those same temperatures.

This statement, often called **Carnot's theorem,** places an absolute upper limit on the efficiency of *any* engine operating between two thermal reservoirs with temperatures T_l and T_h. No matter how clever we are, no matter how much money we spend, no matter what legislation governments pass, we can never do better for engine efficiency than that of a Carnot engine. In practice, because of such factors as friction and heat loss through imperfect insulating boundaries, most real engines operate at significantly lower efficiencies than the limit set by the ideal Carnot engine.

EXAMPLE 20.1   *Extracting Work from Heat When It Flows "Downhill"*

What operating characteristics might we expect from a Carnot engine if we operate it between a reservoir of water at the boiling point (100°C) and another reservoir composed of an ice water bath (0°C)? (a) Compute the theoretical efficiency of such a Carnot engine. (b) What work might we expect to obtain from such an engine if we extract 12.0 kJ of heat from the high-temperature reservoir? (c) If we in fact obtain 1.50 kJ of work for each 12.0 kJ of heat extracted from the high-temperature reservoir, what is the actual efficiency of the engine?

SOLUTION   (a) We obtain the theoretical efficiency from Equation (20.18)

$$e_{\text{Carnot}} = 1 - \frac{T_l}{T_h} = 1 - \frac{273\text{ K}}{373\text{ K}} = 0.268$$

That is, we should expect a maximum efficiency of 26.8%. Notice that we converted temperatures from degrees Celsius to kelvins. Don't forget to do this or you will lose points on the exam! The *T*'s in all our equations are in kelvins.

(b) The general definition of efficiency is given by Equation (20.2), $e = \mathcal{W}_{\text{net}}/Q_h$; so using the theoretical efficiency, we expect to extract

$$\mathcal{W}_{\text{net}} = eQ_h = (0.268)(12.0\text{ kJ}) = 3.22\text{ kJ}$$

(c) In fact only 1.50 kJ was obtained, so the *actual* efficiency is

$$e = \frac{\mathcal{W}_{\text{net}}}{Q_h} = \frac{1.50\text{ kJ}}{12.0\text{ kJ}} = 0.125$$

The actual efficiency of this engine is 12.5%.

Concept Question 2
If you want to increase the efficiency of a Carnot engine by the easiest method but can only change the temperature of one reservoir by a small amount ΔT, do you decrease T_l or increase T_h?

Conversion of thermal energy into mechanical energy. The pressurized liquid carbon dioxide at room temperature inside the cylinder has large thermal energy and low mechanical energy. When the valve is opened, the escaping crystals of carbon dioxide snow have high mechanical energy and low thermal energy. Why does this process not violate Clausius' statement of the second law of thermodynamics?

20.3 Refrigerators and Heat Pumps

In the first two sections of this chapter we introduced engine systems that function because some heat from a high-temperature reservoir is converted to work done *by* the system, while the remainder of the heat is transferred to a low-temperature reservoir. In this section we consider refrigeration processes in which work is done *on* a system so that heat may be removed from a low-temperature reservoir and transferred to a high-temperature reservoir.

We can run a Carnot engine in reverse as a **Carnot refrigerator.** Figure 20.5 illustrates such a process. Heat Q_l is extracted from the low-temperature reservoir and transferred to the high-temperature reservoir. However, this process requires the input of mechanical energy, and we must perform work *on* the system. In order to characterize the effectiveness of a particular refrigeration process, we define a **coefficient of performance** μ_r:

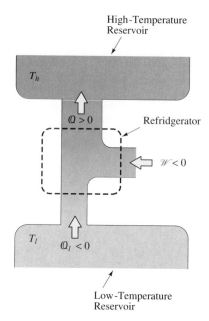

High-Temperature
Reservoir

T_h

$Q > 0$

Refridgerator

$W < 0$

T_l $Q_l < 0$

Low-Temperature
Reservoir

FIGURE 20.5

The Carnot refrigerator.

$$\mu_r = \frac{\textit{Heat removed from the low-temperature reservoir}}{\textit{Magnitude of the net work performed on the system}} = \frac{Q_l}{|W_{net}|} \quad \textbf{(20.19)}$$

The signs for the net work and heat flows for a refrigerator are reversed from those for an engine. That is, $W_{net} < 0$, $Q_h < 0$, and $Q_l > 0$. However, these quantities are related as before: $W_{net} = Q_h + Q_l$, so that

$$\mu_r = \frac{Q_l}{|Q_h + Q_l|} = \frac{Q_l}{|Q_h| - Q_l} = \frac{1}{|Q_h|/|Q_l| - 1} \quad \textbf{(20.20)}$$

Good refrigerators have μ_r values much greater than one. For a Carnot refrigerator we have from Equations (2.17)

$$\left|\frac{Q_h}{Q_l}\right| = \frac{nRT_h \ln\left(\frac{V_B}{V_A}\right)}{nRT_l \ln\left(\frac{V_B}{V_A}\right)} = \frac{T_h}{T_l}$$

Thus, Equation (20.20) gives

$$\mu_{r_{Carnot}} = \frac{1}{(T_h/T_l) - 1} = \frac{T_l}{T_h - T_l} \quad \textbf{(20.21)}$$

An important application of the refrigeration cycle is the **heat pump.** This device is often used to remove heat from the air on a cool day and transfer it to the warmer interior of a building. A heat pump and a refrigerator are the same device, but for a refrigerator the important quantity is the heat *removed* from the low-temperature reservoir. For the heat pump the important quantity is the heat transferred to the high-temperature reservoir. Consequently, the coefficient of performance μ_{hp} of a heat pump is defined differently from that of a refrigerator:

$$\mu_{hp} = \frac{\left|\textit{Heat transferred to the high-temperature reservoir}\right|}{\left|\textit{Work performed on the system}\right|} = \frac{|Q_h|}{|W_{net}|} \quad \textbf{(20.22)}$$

For the Carnot heat pump both the heat Q_l we extract from the low-temperature reservoir and the mechanical work W_{net} we add to cause the extraction are ultimately deposited as heat Q_h in the high-temperature reservoir, so that $|W_{net}| + Q_l = |Q_h|$. Thus, the work required is $|W_{net}| = |Q_h| - Q_l$, and Equation (20.22) can be rewritten

$$\mu_{hp_{Carnot}} = \frac{T_h}{T_h - T_l} \qquad (20.23)$$

In practice, the work W_{net} necessary to run a heat pump depends on the efficiency of the mechanical components of the engine. Therefore, although Equation (20.23) can be used to find the maximum theoretical (Carnot) coefficient of performance, Equation (20.22) must be employed to compute the actual value. Sometimes, well water or even the earth itself is used as a heat source. For T_h values at normal room temperatures, actual coefficients of performance can drop to less than unity for outside temperatures T_l below the midteens (°F).

On warm days a heat pump can be operated in reverse as an air-conditioner; heat is removed from the building and transferred to the warmer, outside air. We explore this feature of a heat pump in Example 20.2. Commercially available heat pumps are often rated by the amount of heat they can transfer in British thermal units per minute (BTU/min). The **British Thermal Unit** (BTU) is a unit of work and corresponds to 778.169 ft-lbs. One BTU is equal to 1.056 kJ.

EXAMPLE 20.2   Pumping Heat Back "Uphill"

A Carnot refrigeration cycle is to be used to cool a house. It is to remove heat from the air inside the home by means of a cooling coil at 10°C (50°F) and transfer it outside at an exhaust temperature of 36.0°C (97°F). (a) Compute the ideal coefficient of performance. (b) If the refrigerator cycle actually performed this well, how much heat would be transferred to the outside air when 1000 BTU of heat are removed from the inside air? (c) What is the minimum amount of power we should expect from the refrigerator motor if it is to remove 1000 BTU/min?

SOLUTION   (a) Converting the temperatures to kelvins, we have $T_l = 283$ K and $T_h = 309$ K. The Carnot coefficient of performance for a refrigerator is given by Equation (20.21):

$$\mu_r = \frac{T_l}{T_h - T_l} = \frac{283 \text{ K}}{309 \text{ K} - 283 \text{ K}} = 10.9$$

(b) From Equation (20.19), $\mu_r = Q_l/|W_{net}|$. However, $|W_{net}| = |Q_{net}| = |Q_h| - |Q_l|$. Thus,

$$\mu_r = \frac{Q_l}{|Q_h| - |Q_l|}$$

Solving for $|Q_h|$ and recalling that the problem statement tells us $Q_l = 1000$ BTU, we obtain

$$|Q_h| = \frac{(1 + \mu_r)Q_l}{\mu_r} = \frac{(1 + 10.9)(1000 \text{ BTU})}{10.9} = 1090 \text{ BTU}$$

However,

$$Q_h = -1090 \text{ BTU}$$

where the minus sign is required by our sign convention that heat leaving the system (the refrigerator in this case) is negative.

(c) The work required by the refrigerator for each 1000 BTU removed from the home is

$$|W_{net}| = |Q_h| - |Q_l| = 1090 \text{ BTU} - 1000 \text{ BTU} = 90 \text{ BTU}$$

Concept Question 3
When a heat pump is used as an air-conditioner, what coefficient of performance should be used?

Concept Question 4
Can a kitchen be cooled by leaving the refrigerator door open? Explain.

Concept Question 5
Can a heat-pump have a coefficient of performance less than 1? Explain.

We must, however, realize that $W_{net} = -90$ BTU. The negative sign is a result of work being performed on the system by the motor. That is, ideally the motor must do $(90 \text{ BTU})(1.056 \text{ kJ/BTU}) \approx 93$ kJ of work for each 1000 BTU of heat removed. This amount of work is required each minute if 1000 BTU/min are to be transferred. The required minimum power is, therefore,

$$\text{Power} = \frac{93 \times 10^3 \text{ J}}{60.0 \text{ s}} = 1600 \text{ W} \qquad (=2.1 \text{ hp})$$

We better plan to use a significantly larger motor, however, because we did not account for friction nor heat loss due to imperfect insulation!   ◀

The Curious Fraction Q/T

Before we end our discussion of the Carnot cycle there is one more result we wish to derive for future use. Equation (20.17) can be solved for the ratios Q_l/T_l and Q_h/T_h:

$$\frac{Q_l}{T_l} = -nR \ln\left(\frac{V_B}{V_A}\right)$$

$$\frac{Q_h}{T_h} = nR \ln\left(\frac{V_B}{V_A}\right)$$

Adding these two equations, we find for the reversible processes of the Carnot cycle

$$\frac{Q_l}{T_l} + \frac{Q_h}{T_h} = 0 \qquad\qquad (20.24)$$

Although we obtained this result using an ideal gas as the working substance in our Carnot engine, Equation (20.24) is a general result, true for any working substance. This result has major significance, which we will explore later.

20.4 The Absolute Temperature Scale and the Third Law of Thermodynamics

The curious result for the ratio of Q/T that we obtained at the end of the previous section can be used to once and for all define a temperature scale that is independent of the working substance. The scale is called the **absolute temperature scale,** and it is equivalent to the Kelvin scale. Let's solve Equation (20.24) for the ratio of the heat removed from a Carnot engine to that transferred to the engine for a complete cycle:

$$\left|\frac{Q_l}{Q_h}\right| = \frac{T_l}{T_h} \qquad\qquad (20.25)$$

We can determine the unknown temperature T of any reservoir by simply operating a Carnot engine between it and another reservoir at a reference temperature T_{fix}. If we measure the ratio of the heats transferred to and from the engine and take $T_h = T_{fix}$, we can solve for the unknown temperature $T_l = T$ from Equation (20.25):

$$T = \left|\frac{Q}{Q_{fix}}\right| T_{fix}$$

where Q is the heat transferred to (or from) the reservoir of unknown temperature, and Q_{fix} is the heat transferred from (or to) the reservoir with a temperature that is defined as T_{fix}.

The absolute temperature scale is defined by choosing the fixed reservoir to be that of the triple point of water, and T_{fix} is assigned a value of 273.16 K (Fig. 20.6). Then, any other temperature is given by

$$T = \left| \frac{Q}{Q_{tp}} \right| (273.16 \text{ K}) \qquad \qquad (20.26)$$

where Q_{tp} is the heat transferred to or from the reservoir at the triple point.

Temperatures as low as 10^{-5} K have been reached with great difficulty, and scientists specializing in the production of low temperatures have found that the degree (no pun intended) of difficulty in reducing the temperature of a system increases enormously as lower temperatures are sought. Based on this experimental evidence, another law has been postulated, and it is known as the **third law of thermodynamics.**

It is impossible to reduce the temperature of a system to absolute zero.

The third law of thermodynamics

20.5 General Cyclic Processes

All real processes take place with finite (as opposed to infinitesimal) changes in the thermodynamic variables (P, V, T, etc.) during each portion of a cycle. Therefore, all real processes are irreversible. It is possible, however, to obtain useful information about many real engine processes by analyzing them as if they were reversible. In particular, we can find their theoretical maximum efficiency. That way engineers can judge the actual performance of a real engine against some theoretical maximum instead of the impossible goal of 100%.

In principle, any reversible cyclic process can be analyzed (for work output, heat transfer, efficiency, and so on) as a combination of Carnot cycles. (We warned you that the Carnot cycle is important!) Figure 20.7 shows a reversible, but otherwise arbitrary, cycle and how it can be broken up into a family of Carnot cycles. If we are careful to travel along each Carnot segment in the same direction, the interior segments cancel and we have a kind of jagged approximation to the true cycle. By allowing the number of Carnot cycles to become very large, we may obtain as accurate an approximation to the true cycle as we wish. Equation (20.24) holds for each of the cycles, so for the entire family of

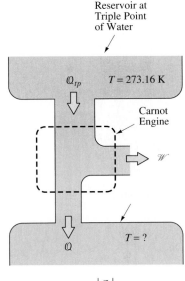

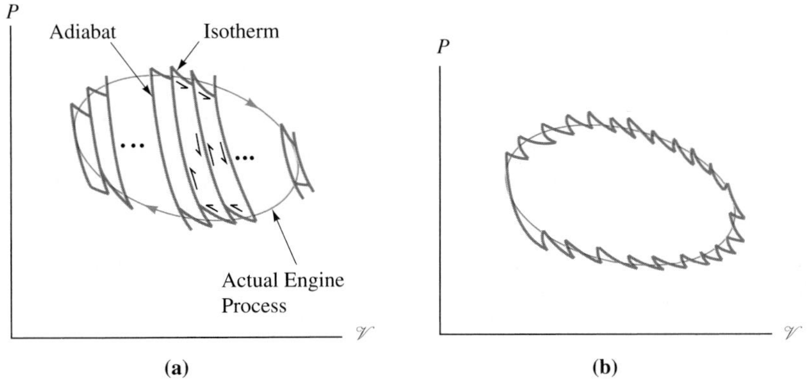

$$T = \frac{|Q|}{|Q_{tp}|} (273.16 \text{ K})$$

FIGURE 20.6

Defining the absolute temperature scale.

FIGURE 20.7

(a) An arbitrary engine cycle may be analyzed as a large number of Carnot cycles. (b) Because interior paths cancel, we have an approximation to the true cycle. This approximation can be improved by choosing a larger number of individual Carnot cycles.

Carnot cycles

$$\sum_{j=1}^{N} \left(\frac{Q_{h_j}}{T_{h_j}} + \frac{Q_{l_j}}{T_{l_j}} \right) = 0, \qquad N = \text{number of Carnot cycles} \qquad \textbf{(20.27a)}$$

If we count processes instead of cycles, we can simplify the notation to

$$\sum_{i=1}^{N} \frac{Q_i}{T_i} = 0, \qquad N = \text{number of processes in the family of Carnot cycles} \qquad \textbf{(20.27b)}$$

As an example of a real process we analyze the gasoline engine, a schematic of which is shown in Figure 20.8. A completely accurate $P\mathcal{V}$-diagram for a gasoline engine can be quite complicated. However, a useful model that approximates the gasoline engine cycle is called the **Otto cycle** (Fig. 20.9). Another engine process, the Diesel cycle, is a subject of Problem 20.

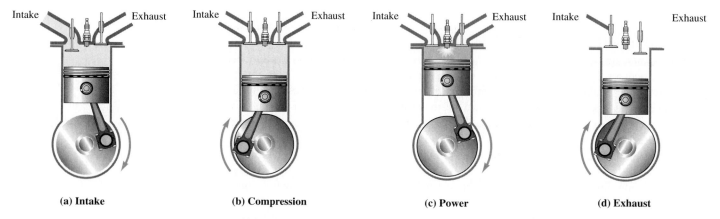

(a) Intake   **(b) Compression**   **(c) Power**   **(d) Exhaust**

FIGURE 20.8

The four processes of a gasoline engine: (a) Air and fuel are mixed in the intake stroke. (b) After the intake valve is closed the fuel and air mixture is compressed by the piston in the compression stroke. (c) The spark plug ignites the mixture in the power stroke. (d) The spent combustion gases are exhausted.

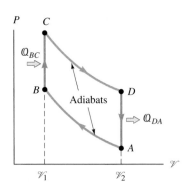

FIGURE 20.9

The Otto cycle: The compression and power strokes are adiabatic processes. The combustion and exhaust strokes are isochoric processes.

EXAMPLE 20.3   *Modeling a Gasoline Engine*

Compute the efficiency of an Otto cycle employing an ideal gas as the working fluid.

SOLUTION   The efficiency of a heat engine is given by Equation (20.2). Heat is transferred into the Otto cycle during the combustion stroke ($Q_h = Q_{BC}$). For a complete cycle $\Delta U_{\text{cycle}} = 0$, and the work performed is $W_{\text{net}} = Q_{\text{net}} = Q_{BC} + Q_{DA}$. Therefore,

$$e = \frac{W_{\text{net}}}{Q_h} = \frac{Q_{BC} + Q_{DA}}{Q_{BC}} = 1 + \frac{Q_{DA}}{Q_{BC}}$$

Now along the isochors, $Q_{BC} = nc_v(T_C - T_B)$ and $Q_{DA} = nc_v(T_A - T_D)$. Substituting these expressions on the right-hand side of the previous expression, the efficiency can be written

$$e = 1 + \frac{T_A - T_D}{T_C - T_B}$$

The next step is to replace the temperatures with volumes. To do this we start with the realization that processes AB and CD are both adiabatic. Thus, we can write for the AB process

$$P_A \mathcal{V}_2^\gamma = P_B \mathcal{V}_1^\gamma \qquad \text{or} \qquad (P_A \mathcal{V}_2) \mathcal{V}_2^{(\gamma-1)} = (P_B \mathcal{V}_1) \mathcal{V}_1^{(\gamma-1)}$$

But for an ideal gas

$$P_A \mathcal{V}_2 = nRT_A \qquad \text{and} \qquad P_B \mathcal{V}_1 = nRT_B$$

so we have

$$T_A \mathcal{V}_2^{(\gamma-1)} = T_B \mathcal{V}_1^{(\gamma-1)}$$

or

$$\frac{T_B}{T_A} = \left(\frac{\mathcal{V}_2}{\mathcal{V}_1}\right)^{(\gamma-1)}$$

This last result is a useful relationship between the temperatures and volumes of an ideal gas for any adiabatic process. You may find this relationship helpful in the problem set at the end of this chapter. It is important, however, that you develop the skill to derive such relations, so try your hand at computing the similar result for the CD portion of the Otto cycle. The answer is

$$\frac{T_C}{T_D} = \left(\frac{\mathcal{V}_2}{\mathcal{V}_1}\right)^{(\gamma-1)}$$

Substituting these last two expressions into the last equation for the efficiency e, we (eventually) find

$$e = 1 - \frac{1}{(\mathcal{V}_2/\mathcal{V}_1)^{(\gamma-1)}}$$

The term $\mathcal{V}_2/\mathcal{V}_1$ is called the **compression ratio.** For a compression ratio of 7 and $\gamma = 1.4$ (nominal values for gasoline engines and the gases they burn), we find

$$e = 1 - \frac{1}{(2.18)} = 0.541 \qquad \text{or} \qquad 54.1\%$$

True gasoline efficiencies are much smaller than this value, usually ranging between 15% and 20%. Such factors as incomplete combustion, heat loss, and friction serve to reduce the actual efficiency from the ideal value.   ◀

Efficiencies of Real Engines: The Clausius Inequality

In Section 20.2 we asserted that no engine (not even an ideal one) can have a greater efficiency than the corresponding Carnot engine. This result is often called **Carnot's theorem.** The word *corresponding* means, in the simplest case, that both engines (alternative and Carnot) operate between the same two heat reservoirs. In more complicated situations, such as the Otto cycle described in Example 20.3, the two engines operate between the same series of reservoirs.

In order to prove Carnot's theorem, let's first assume we have an engine that *does* have an efficiency greater than that of a Carnot engine operating between the same two heat reservoirs and then show that this leads to a contradiction. We use primed quantities to represent the proposed super-engine. Therefore, if e_C is the efficiency of the Carnot

Concept Question 6
Why are automobile engines less efficient in the winter than in the summer?

engine, our claim is

$$e' > e_C$$

or

$$\frac{\mathcal{W}'_{\text{net}}}{\mathcal{Q}'_h} > \frac{\mathcal{W}_{\text{net}}}{\mathcal{Q}_h} \qquad\qquad \textbf{(20.28)}$$

Now, all's fair in love and proofs (as long as the mathematics is correct!) so we run our Carnot engine in reverse as a refrigerator. Furthermore, we throttle this refrigerator so that the amount of heat $|\mathcal{Q}_h|$ it transfers *to* the high-temperature reservoir is exactly equal to $\mathcal{Q}'_h$, the heat removed *from* the same reservoir by our "better" super-engine. A schematic of this process is shown in Figure 20.10. Some extra care is now needed with signs because

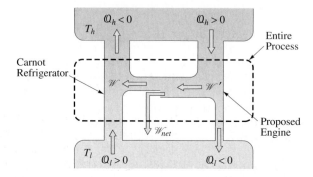

FIGURE 20.10

Proof of Carnot's theorem.

both $\mathcal{W}_{\text{net}}$ and $\mathcal{Q}_h$ are negative for the Carnot refrigerator. Thus, Equation (20.28) must be written

$$\frac{\mathcal{W}'_{\text{net}}}{\mathcal{Q}'_h} > \frac{|\mathcal{W}_{\text{net}}|}{|\mathcal{Q}_h|} \qquad\qquad \textbf{(20.28a)}$$

We have conspired to make $\mathcal{Q}'_h = |\mathcal{Q}_h|$, and as a result Equation (20.28a) becomes

$$\mathcal{W}'_{\text{net}} > |\mathcal{W}_{\text{net}}|$$

Because of the relative size of these magnitudes, after one complete cycle of both the engine and the refrigerator there is actually an overall output of useful work available from the combined system

$$\mathcal{W}_{\text{tot}} = \mathcal{W}'_{\text{net}} + \mathcal{W}_{\text{net}} = \mathcal{W}'_{\text{net}} - |\mathcal{W}_{\text{net}}| > 0$$

Where does the energy for this work come from? The heat withdrawn from the high-temperature reservoir by the super-engine has been returned there by the refrigerator. Consequently, the energy source for the work must be the low-temperature reservoir. However, this result is a violation of the second law of thermodynamics! The overall effect of one cycle of the combined super-engine–refrigerator is to remove an amount of heat $(\mathcal{Q}_l - |\mathcal{Q}'_l|)$ from one reservoir and convert it entirely into work $\mathcal{W}_{\text{tot}}$ with no other effect: a blatant contradiction of the Kelvin-Planck statement of the second law. Therefore, we must conclude that our initial supposition regarding the efficiency of our conjectured super-engine is wrong and its efficiency e' is not greater than that of a Carnot engine operating between the same two temperatures. That is,

$$e' \le e$$

which is to say that no engine can have an efficiency greater than the corresponding Carnot engine.

Previously we showed that any reversible cycle can be modeled to arbitrary precision as a series of Carnot cycles. Thus, the efficiency of any reversible engine is equal to the Carnot efficiency. In other words, *all* reversible engines have the same efficiency when they operate between the same two temperatures.

This concludes our discussion of Carnot's theorem. However, one more result follows that will be useful later.

The Curious Q/T Result for Arbitrary Cycles

For the Carnot cycle we previously found (Eq. (20.24))

$$\frac{Q_l}{T_l} + \frac{Q_h}{T_h} = 0$$

We now wish to obtain a similar result valid for arbitrary cycles that, as argued previously, can have efficiencies e' less than or equal to the ideal Carnot efficiency e_C:

$$e' \leq e_C$$

On the left-hand side we substitute the definition of efficiency, and on the right-hand side we substitute the expression we derived for the Carnot efficiency in terms of the reservoir temperatures.

$$\frac{W'_{net}}{Q'_h} \leq 1 - \frac{T_l}{T_h}$$

For one complete cycle of the arbitrary engine $\Delta U'_{cycle} = 0$, so that $W'_{net} = Q'_h + Q'_l$. With this substitution and some rearrangement of the right-hand side, the previous inequality can be written

$$\frac{Q'_h + Q'_l}{Q'_h} \leq \frac{T_h - T_l}{T_h}$$

Multiplying both sides of the inequality by the positive quantity $Q'_h T_h$, we find

$$Q'_h T_h + Q'_l T_h \leq Q'_h T_h - Q'_h T_l$$

After subtracting the common term $Q'_h T_h$ from both sides and rearranging, we are left with

$$\frac{Q'_l}{T_l} + \frac{Q'_h}{T_h} \leq 0$$

In practice, such cycles need not consist of only two stages during which heat is transferred. When many processes are involved this result can be generalized to

$$\sum_{i=1}^{N} \frac{Q_i}{T_i} \leq 0, \qquad N = \text{number of processes in a complete cycle} \qquad \textbf{(20.29a)}$$

where the equality holds only when all processes are reversible. We abbreviate this cumbersome notation as

$$\sum_{cycle} \frac{Q}{T} \leq 0 \qquad\qquad \textbf{(20.29b)}$$

Equation (20.29) is called the **Clausius inequality.** It has a fairly simple meaning: when we add up the quotient of the heat transferred Q and temperature T for each reservoir involved in a cyclic process, the sum of Q/T must be less than, or at most equal to, zero. For more complex processes, such as those analyzed with a large number of cycles, each of which is infinitesimally separated in temperature from its neighboring cycle, the sum becomes the integral

$$\oint \frac{dQ}{T} \leq 0 \qquad (20.30)$$

The circle on the integral sign reminds us that this integration is to be performed over a complete cycle. In either Equation (20.29) or (20.30) the equality applies to reversible processes, whereas the inequality must be used for irreversible (real) processes. Equation (20.30) will be helpful in Section 20.7.

20.6 A Formal Definition of State Variables

We previously described state variables as quantities that characterize the thermodynamic "state" of a system. We noted that heat Q and work W are not state variables because their values depend on the history of the system. On the other hand the internal energy U is a state variable. In this section we wish to generalize these ideas and present a mathematical criterion for elevating a variable to the status of "state variable." We start with the internal energy as an example because you are already familiar with this concept.

Recall that the internal energy U of a system depends only on the state of that system. Furthermore, for cyclic processes we found that the net change in internal energy is zero. That is,

$$\sum_{\text{cycle}} \Delta U = 0$$

If the cycle consists of an infinite number of infinitesimal changes in U, the sum becomes an integral over the cycle:

$$\oint dU = 0 \qquad (20.31)$$

Any state variable must have this same property because a state variable depends only on the state of the system. Conversely, if a variable X exhibits the property

$$\oint dX = 0 \qquad \text{(Condition for } X \text{ to be a state variable)} \qquad (20.32)$$

for all closed paths, then X is a state variable.

In order to prove that a variable is a state variable we have to show that its integral around *all* closed paths is zero. On the other hand, to show that a variable is not a state variable we need find only one closed path where the integral is not zero. For instance, we have seen in several examples that the net work performed by a system and the net heat added to the system do depend on the path so that:

$$\oint dW \neq 0$$

$$\oint dQ \neq 0$$

The results of such examples prove that these two quantities are not state variables.

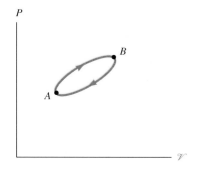

FIGURE 20.11

In cyclic processes the total change in internal energy for a complete cycle is zero.

20.7 Entropy: A State Variable

In the Section 20.5 we found for reversible processes

$$\oint \frac{dQ}{T} = 0$$

where we use the equality in the Clausius inequality (Eq. (20.30)) because the process is reversible. By our definition Equation (20.32) in Section 20.6, the quantity dQ/T must be a state variable. We designate this new variable S and call it **entropy.** That is,

$$dS = \frac{dQ}{T}$$

The change in entropy between any two states (*not* a cycle) is

$$\Delta S = \int_o^f \frac{dQ}{T} \qquad (20.33)$$

Processes for which the entropy does not change are called **isentropic.** That the entropy is a state variable and depends only on the original and final states and not on the path taken between the states is illustrated in the following example.

EXAMPLE 20.4   *Entropy Changes During a Reversible Process for an Ideal Monatomic Gas*

Compute the change in entropy of an ideal gas when it is taken from state (V_o, T_o) to (V_f, T_f) by an unspecified, but reversible, path. Apply the result to find the change in entropy of 1.00 mol of an ideal gas when its volume is slowly doubled, isothermally.

SOLUTION   From the first law of thermodynamics

$$dU = dQ - dW \qquad \text{or} \qquad dQ = dU + dW$$

But $dU = nc_v\, dT$ and $dW = P\, dV$, so $dQ = nc_v\, dT + P\, dV$. Then, with $P = nRT/V$, we have, after dividing by T

$$\frac{dQ}{T} = nc_v \frac{dT}{T} + nR \frac{dV}{V}$$

The change in entropy is

$$\Delta S = \int_o^f \frac{dQ}{T} = nc_v \int_{T_o}^{T_f} \frac{dT}{T} + nR \int_{V_o}^{V_f} \frac{dV}{V}$$

$$= nc_v \ln\!\left(\frac{T_f}{T_o}\right) + nR \ln\!\left(\frac{V_f}{V_o}\right)$$

Note that we were able to compute ΔS without any consideration of the path between the original and final states. Our expression for ΔS depends only on the values of the volumes and temperatures at the original and final states (V_o, T_o) and (V_f, T_f).

For a mole of a monatomic ideal gas $n = 1.00$. Furthermore, when its volume is doubled while the temperature remains constant $V_f = 2V_o$ and $T_f = T_o$, so $\ln(V_f/V_o) = \ln(2)$ and $\ln(T_f/T_o) = 0$. Our expression for the change in entropy becomes

$$\Delta S = nR \ln(2) = (1.00\ \text{mol})(8.314\ \text{J/mol} \cdot \text{K}) \ln(2) = 5.76\ \text{J/K} \qquad \blacktriangleleft$$

EXAMPLE 20.5   *Entropy Change During a Reversible Change of Phase*

Compute the change in entropy of 1.00 kg of water at 100.°C when it is reversibly converted to steam at 100.°C.

SOLUTION   We apply Equation (20.33):

$$\Delta S = \int_o^f \frac{dQ}{T}$$

In our case, the temperature is a constant, so

$$\Delta S = \frac{1}{T} \int_o^f dQ = \frac{Q}{T}$$

During a transition from the liquid phase to the vapor phase $Q = mL_f$. Therefore,

$$\Delta S = \frac{mL_f}{T} = \frac{(1.00 \text{ kg})(540. \text{ kcal/kg})}{(373. \text{ K})} = 1.45 \text{ kcal/K} \quad \blacktriangleleft$$

20.8 Entropy Changes for Irreversible Processes

The processes illustrated in Examples 20.4 and 20.5 are both reversible. You might be wondering how to compute the entropy change for irreversible processes. More likely, however, you are wondering why anyone would want to compute the change in entropy in the first place. After all, what could dQ/T possibly mean? In this section we show that the concept of entropy provides a powerful, mathematical expression for the second law of thermodynamics, from which we can decide if a process occurs naturally, or not. This result assuredly has important applications to chemistry, engineer and, of course, physics.

Before we derive this mathematical expression for the second law, let's first clear up the mystery of how you compute entropy changes for irreversible processes. It's easy! Recall that entropy is a state variable and, therefore, the entropy change is independent of the path between the original and final states. Because we can always construct a reversible path between the original and final states of an irreversible process, the entropy change must be the same for both processes, the irreversible process as well as the imaginary process along the reversible path. We illustrate this procedure with an example at the end of this section.

The Principle of Increasing Entropy

The Clausius inequality (Eq. (20.30)) is a consequence of the second law of thermodynamics. We arrived at this result in Section 20.5 by applying the second law to a real (irreversible) engine process. Let's now apply Equation (20.30) to an irreversible process of an *isolated* system. In order to do so we must make a complete cycle out of the process because Equation (20.30) is valid only for cycles. The cycle consists of our irreversible process of the isolated system, followed by a reversible return to the original state. This cycle is illustrated in Figure 20.12. We use a hashed line for the irreversible process because nonequilibrium states cannot actually be represented by a curve on a PVT-surface. (Perhaps a specific example will help keep the arguments below clear. You can heat a mass of water irreversibly by placing it in contact with a heat reservoir at a much higher temperature than the water. The water and the reservoir form the isolated system.

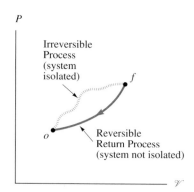

FIGURE 20.12

An irreversible process of an isolated system made into a cyclic process. The reversible return to the original state is not in isolation.

The water may be returned to its original state by sequentially placing it in contact with heat reservoirs infinitesimally separated in temperature. In this latter case the system is not isolated.)

Because a portion of the cycle $(o \rightarrow f)$ is irreversible, we must use the inequality in Equation (20.30):

$$\oint \frac{đQ}{T} < 0$$

This integration may be rewritten in terms of each segment of the cycle,

$$\oint \frac{đQ}{T} = \int_o^f \frac{đQ}{T} + \int_f^o \frac{đQ}{T} < 0$$

However, the first integral on the right side of the equal sign is zero because for the isolated process no heat flows into or out of the system (both the water and the high-temperature reservoir, in our example) and $đQ = 0$. Therefore, we have

$$\int_f^o \frac{đQ}{T} < 0$$

During the return portion of the cycle the system is not isolated, and a *reversible* path is followed from state f to state o. Therefore, $đQ/T = dS$ along this path. That is to say, we can always integrate $đQ/T$ along this path, and one way to express this fact is to write $đQ/T$ as dS. Then

$$\int_f^o dS < 0$$

or

$$S_o - S_f < 0$$

This result may be rewritten

$$S_f > S_o$$

This result is remarkable! It says that the entropy of the final state of the irreversible process for our isolated system is always greater than that of the initial state. This result is an important enough consequence of the second law of thermodynamics that we emphasize it:

The entropy of an isolated system increases for all natural processes.

What we have here is a significantly different kind of law than we've discussed thus far in this text. You are probably more comfortable with such laws as the *conservation* of energy, and the *conservation* of linear and angular momentum. Now we are confronted with a "nonconservation" condition. Entropy is a strange concept in the sense that all real processes are irreversible and, therefore, the entropy of the universe continues to grow and grow. However, as we mentioned above, it is a very useful concept. In Example 20.6 we illustrate just how the mathematics works out so that the entropy increases for a process that is clearly irreversible.

EXAMPLE 20.6 *Death, Taxes, and Entropy Increase*

What is the change in entropy of the universe when 1.00 kg of water at $T_1 = 10.0°C$ is mixed with 2.50 kg of water at $T_2 = 80.0°C$?

SOLUTION   We must first compute the final temperature of the mixture:

$$Q_{net} = 0$$

$$m_1 C_w(T_f - T_2) + m_2 C_w(T_f - T_1) = 0$$

$$T_f = \frac{m_1 C_w T_2 + m_2 C_w T_1}{m_1 C_2 + m_2 C_w} = \frac{m_1 T_2 + m_2 T_1}{m_1 + m_2}$$

$$= \frac{(2.50 \text{ kg})(80.0°C) + (1.00 \text{ kg})(10.0°C)}{2.50 \text{ kg} + 1.00 \text{ kg}}$$

$$= 60.0°C$$

Next, we must convert all temperatures to kelvins by adding 273 K:

$$T_1 = 10°C = 283 \text{ K}$$

$$T_2 = 80.0°C = 353 \text{ K}$$

$$T_f = 60.0°C = 333 \text{ K}$$

If we assume that heat is exchanged only between the two masses of water (none lost to the surrounding environment), then the total change in entropy is ΔS_1 (the change in entropy of the cold water) plus ΔS_2 (the change in entropy of the warm water):

$$\Delta S = \Delta S_1 + \Delta S_2 = \int_{T_1}^{T_f} \frac{dQ}{T} + \int_{T_2}^{T_f} \frac{dQ}{T}$$

But $dQ = mC_w \, dT$, so

$$\Delta S = m_1 C_w \int_{T_1}^{T_f} \frac{dT}{T} + m_2 C_w \int_{T_2}^{T_f} \frac{dT}{T}$$

$$= C_w \left[m_1 \ln\left(\frac{T_f}{T_1}\right) + m_2 \ln\left(\frac{T_f}{T_2}\right) \right]$$

$$= (1.00 \text{ kcal/kg/K}) \left[(1.00 \text{ kg}) \ln\left(\frac{333 \text{ K}}{283 \text{ K}}\right) + (2.50 \text{ kg}) \ln\left(\frac{333 \text{ K}}{353 \text{ K}}\right) \right]$$

$$= (1.00 \text{ kcal/kg/K})[(1.00 \text{ kg})(0.163) + (2.50 \text{ kg})(-0.0583)]$$

$$= 16.9 \text{ cal/K}$$

This example illustrates the striking contrast between the conservation of energy (used to find T_f) and the nonconservation of entropy; the entropy of the universe has increased as a result of mixing cold and hot water.

Work an example on your own to see that even in "symmetrical" cases entropy increases. Consider a mixture of *equal amounts* of hot and cold water, say 1.00 kg of each. If $T_1 = 10.0°C$ and $T_2 = 80.0°C$, it should be easy for you to find that T_f is 45.0°C. By following the same method as that shown above you should be able to show that $\Delta S = 12.2$ cal/K. Try it for the practice!   ◀

20.9 Entropy and Disorder

In Sections 20.7 and 20.8 we had quite a lot to say about how to calculate entropy changes. But perhaps you had the uneasy notion that although you can calculate numbers for entropy, you still don't know what it is! This is the question we address in this section.

In the first section of this chapter we considered two illustrations from which we realized that there must be a second law of thermodynamics. In the first thought experiment we mixed a glass of hot water and a glass of cold water to obtain a pitcher of warm water. In the second illustration we decided that the bearing of our bicycle wheel became hot as friction caused the wheel to come to rest. There is another way to view these two natural processes: disorder has been increased. When you have a glass of hot water and a glass of cold water, in a sense your system has a degree of order to it; cold water is segregated from hot water. After mixing, you have no such order. In the case of the bicycle wheel, orderly circular motion is changed to the disorderly random thermal motions of the atoms in the bearing.

Entropy is related to the disorder of a system. The disorder usually manifests itself as an increase in the randomness of the atoms or molecules of a system, such as in our examples above. It is the business of the science called **statistical mechanics** to predict the most probable behavior of large numbers of atoms or molecules by combining the mathematics of statistics with the physics of mechanics. An important result of statistical mechanics is that the quantity we have defined as entropy is indeed a measure of the disorder of a system. Moreover, the statistical mechanical statement of the second law of thermodynamics is:

All isolated systems tend toward a state of greater disorder.

If a particular system under consideration is not isolated, the entire system consists of the particular system and its surroundings (perhaps the rest of the universe). Because the entropy is a measure of the disorder, another statement of the second law is:

In all natural processes the entropy of the universe increases.

Because entropy increases for all naturally occurring processes, the entropy of the universe must continue to increase until it reaches some maximum value. Presumably, at that time all matter in the universe will be at the same temperature. There will then be no heat reservoirs at different temperatures, and no process, whether it be thermodynamic, chemical, or biological, can take place. That day, however, will not come before the end of this term, so you should continue to do your homework! In fact, our sun will have long reached its final stages before the so-called "heat death" of the universe takes place.

20.10 Summary

There are two descriptive and equivalent statements of the **second law of thermodynamics.**

The Kelvin-Planck Statement: It is impossible for any cyclic process during one cycle to convert internal energy into work with no other effect.

The Clausius Statement: It is impossible for any cyclic process during one cycle to remove an amount of heat from a low-temperature reservoir and transfer an equivalent amount of heat to a high-temperature reservoir with no other effect.

Concept Question 7
If during a thermodynamic process a certain system's entropy change is negative, what can you conclude about the change in entropy of the system's surroundings?

Concept Question 8
Plants change disordered molecules of nitrogen, oxygen, water, and carbon dioxide into highly ordered complex molecules. From where does the negative entropy come?

Concept Question 9
It has been said that teenagers' bedrooms are proof positive of the second law of thermodynamics. How would you explain this joke to someone who has not taken a physics course?

The **Carnot cycle** consists of two adiabats and two isotherms. The **efficiency** of a heat engine is defined as:

$$e = \frac{\textit{Work done per cycle}}{\textit{Heat input per cycle}} = \frac{W}{Q_h} \tag{20.2}$$

The efficiency of a Carnot engine is

$$e_{\text{Carnot}} = 1 - \frac{T_l}{T_h} \tag{20.18}$$

where T_1 and T_2 are the temperatures of the low- and high-temperature reservoirs, respectively. **Carnot's theorem** states that no engine can have a greater efficiency than the corresponding Carnot engine.

The **coefficient of performance** μ_r of a refrigeration cycle is given by:

$$\mu_r = \frac{\textit{Heat removed from the low-temperature reservoir}}{\textit{Magnitude of the work performed on the system}} = \frac{Q_l}{|W_{\text{net}}|} \tag{20.19}$$

The absolute temperature scale is defined in terms of the amount of heat transferred during a Carnot cycle. It coincides with the Kelvin scale.

The **third law of thermodynamics** states that it is impossible for a system to reach a temperature of absolute zero.

The **entropy** S of a system is a state variable. The change in entropy for reversible processes is given by

$$\Delta S = \int_o^f \frac{dQ}{T} \tag{20.33}$$

This same expression can be used to compute the entropy change for an irreversible process by connecting the beginning and end states of the irreversible process by a reversible path and then calculating ΔS along the reversible path.

The entropy statement of the second law of thermodynamics is: The entropy of an isolated system increases for all natural processes. Entropy is a measure of disorder. The entropy of the universe increases for all irreversible processes.

PROBLEMS

20.1 The Second Law of Thermodynamics and Heat Engines
20.2 The Carnot Cycle
20.3 Refrigerators and Heat Pumps

1. An engine is to operate between heat reservoirs at 200.°C and 400.°C. (a) Compute the Carnot efficiency of the engine. (b) If we extract 15.0 kJ of heat from the high-temperature reservoir, what work should we expect to obtain from the engine? (c) If we only obtain 1.20 kJ of work for each 15.0 kJ of heat extracted from the high-temperature reservoir, what is the actual efficiency of the engine?

2. A Carnot engine is to provide 50.0 kJ of work while 300. kJ of heat enters the engine each cycle. Determine the efficiency of the cycle and the amount of heat rejected to a low-temperature reservoir.

3. A Carnot engine performs 25.0 kJ of work and rejects 10.0 kJ of heat to a low-temperature reservoir during each cycle. (a) How much heat enters the engine from a high-temperature reservoir?

(b) What is the efficiency of the engine? (c) If the low-temperature reservoir has a temperature of 120.°C, what is the temperature of the other heat reservoir?

4. An engine performs at 80.0% efficiency of a Carnot engine operating between the same two temperatures as the engine. The low-temperature reservoir is at 150.°C. The engine performs 800. J of work while 1400. J of heat are absorbed from the high-temperature reservoir. (a) What is the temperature of the other reservoir? (b) What is the efficiency of the engine?

5. Is it possible for an engine cycle to produce 1000. J of work if it receives 2800. J of heat at 500.°C and rejects heat at 250.°C? Explain.

6. One type of heat engine has established a record for operating at about 60.0% of the Carnot efficiency when the low-temperature reservoir is within the temperature range of 50.0°C to 150.°C. (a) What is the range of the actual efficiency if a high-temperature reservoir of 600.°C is provided? (b) How much heat must be provided by the 600.°C reservoir if 1.00 kJ of work is to be per-

formed by the engine while heat is rejected at 50.0°C? (c) Repeat part (b) for heat rejection at 150.°C.

7. A Carnot refrigerator is to remove 1000. J of heat from air at 20.°C and discard heat at 100.°C. (a) What is the coefficient of performance for this refrigerator? (b) How much work is necessary to remove the heat from the air? (c) How much heat is discarded at 100.°C?

8. A Carnot refrigerator is to remove heat from 2.50 kg of water at 0.00°C until it all changes phase to ice at 0.00°C. The heat is to be rejected to air at 25.0°C. (a) How much heat is transferred to the air? (b) How much work is required?

9. A refrigerator is to remove 2500. J from water at 15.0°C and discard heat to air at 30.0°C. The refrigerator operates with a coefficient of performance that is 80.0% of that for the Carnot refrigerator operating between these same two temperatures. (a) How much work is actually required? (b) How much heat is rejected to the air?

10. A building is to be heated by a reverse Carnot cycle. When the outside air is 36.0°F, the rate of heat flow required to keep the building interior at 68.0°F is 500. kW. Compute the power required for the motor and μ_{hp}.

11. A refrigerator is to operate between 5.00°C and 40.0°C. It is to remove heat at a rate of 1.80 kW while operating at an actual coefficient of performance that is three-fourths that of the Carnot refrigerator operating between these same two temperatures. (a) What work must be performed by the refrigerator motor each second? (b) What is the rate of heat transfer at the 40.0°C temperature?

20.5 General Cyclic Processes

12. Compute the efficiency for the engine process of Example 19.5.

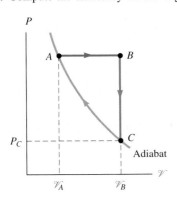

FIGURE 20.P1   Problems 13 and 35

13. Find the efficiency of an engine that carries 1.00 mol of a monatomic ideal gas (c_p = 20.8 J/mol/K and c_v = 12.5 J/mol/K) through the cyclic processes shown in Figure 20.P1. Use the following numerical values: n = 1.00 mol, P_C = 1.00 × 10⁵ Pa, V_A = 2.00 L, and V_B = 3.00 L.

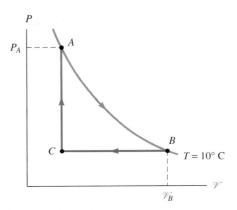

FIGURE 20.P2   Problems 14 and 36

14. Compute the efficiency of an engine that carries 0.150 mol of a monatomic ideal gas (c_p = 20.8 J/mol/K and c_v = 12.5 J/mol/K) through the cyclic processes shown in Figure 20.P2. Use the following data: n = 0.150 mol, P_A = 2.00 × 10⁵ Pa, and V_B = 5.00 L.

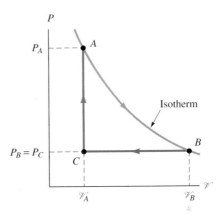

FIGURE 20.P3   Problems 15 and 37

15. Find the efficiency of an engine that carries 0.100 mol of a monatomic ideal gas (c_p = 20.8 J/mol/K and c_v = 12.5 J/mol/K) through the cyclic processes shown in Figure 20.P3. Use the following data: n = 0.100 mol, P_A = 4.00 × 10⁴ Pa, P_B = 1.00 × 10⁴ Pa, V_A = 1.00 L, and V_B = 4.00 L.

16. An Otto cycle operates with a compression ratio of 6.00 and has minimum and maximum temperatures of 100.°C and 550.°C, respectively. Assume c_p = 29.1 J/mol/K and c_v = 20.8 J/mol/K for the gas used in this engine. (a) Find the efficiency of the cycle and compare it to that of a Carnot engine operating between these temperature extremes. (b) Determine the heat added and rejected per mole of gas for the constant-volume segments of the cycle. (c) Determine the work per mole of gas for each cycle.

17. The compression ratio of an Otto cycle is 7.00, and 60.0 kJ of heat are transferred to the engine during each cycle. The lowest temperature is 25.0°C, and the pressure at this point is 2.00 × 10⁴ Pa. For 1.00 mol of a gas as an operating fluid (a) determine the maximum operating temperature. Assume c_p = 29.1 J/mol/K and c_v = 20.8 J/mol/K for the gas used in this engine. (b) Compute the work done each cycle.

18. The heat input to an Otto cycle operating with 1.00 mol of a gas is 150. kJ, and the compression ratio is 8.00. The lowest temperature and pressure are 80.0°C and 1.00 × 10⁵ Pa, respectively, and c_p = 29.1 J/mol/K and c_v = 20.8 J/mol/K for the gas. (a) Compute the efficiency of the engine and the work performed for the

cycle. (b) Find the temperature and pressure at each vertex of the cycle.

19. The heat input to an Otto cycle operating with 1.00 mol of gas is 100. kJ, and the compression ratio is 7.00. The lowest temperature and pressure are 100.°C and 1.50×10^5 Pa, respectively, and $c_p = 29.1$ J/mol/K and $c_v = 20.8$ J/mol/K for the gas. (a) Compute the efficiency of the engine and the work performed for the cycle. (b) Find the temperature and pressure at each vertex in the cycle.

20. The diesel cycle is shown in Figure 20.P4. It consists of an adiabatic compression, an isobaric and an adiabatic expansion, and finally an isochoric rejection of heat. Show that the efficiency of the diesel cycle is

$$e_D = 1 - \frac{1}{\gamma}\left(\frac{T_d - T_a}{T_c - T_b}\right)$$

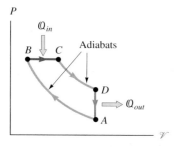

FIGURE 20.P4   The diesel cycle. See Problem 20.

21. The Stirling cycle is shown in Figure 20.P5. It consists of two isothermal and two isochoric processes. Show that the efficiency of the cycle is the same as that of a Carnot engine operating between the same two temperatures.

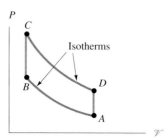

FIGURE 20.P5   The Stirling cycle. See Problem 21.

20.7 Entropy: A State Variable

22. Compute the change in entropy of one mole of an ideal, monatomic gas ($c_p = 20.8$ J/mol/K and $c_v = 12.5$ J/mol/K) when its temperature is increased from 0.00°C to 100.°C, and its volume remains constant.

23. One mole of an ideal, diatomic gas ($c_p = 29.1$ J/mol/K and $c_v = 20.8$ J/mol/K) expands isothermally to 3.00 times its original volume. Compute its change in entropy.

24. Two moles of an ideal, monatomic gas ($c_p = 20.8$ J/mol/K and $c_v = 12.5$ J/mol/K) expand to 4.00 times their original volume while the temperature decreases to 0.500 its initial value. What is the change in entropy of the gas?

25. Two moles of an ideal, monatomic gas ($c_p = 20.8$ J/mol/K and $c_v = 12.5$ J/mol/K) expand to 2.00 times the original volume while the pressure drops to 0.250 times its initial value. (a) Compute the change in entropy of the gas. (b) Could the system have been isolated while this process occurred?

26. An experimenter proposes a process in which 1.00 mol of an ideal, diatomic gas ($c_p = 29.1$ J/mol/K and $c_v = 20.8$ J/mol/K) undergoes a tripling of its volume and a halving of its temperature. Can this process occur naturally, or must it take place in conjunction with other processes? Explain.

27. A 200.-g block of aluminum at 70.0°C is placed in a thermos bottle that contains 400. g of water at 10.0°C and is allowed to come to equilibrium. Assume a negligible amount of heat is absorbed by the thermos. (a) Compute the change in entropy of the aluminum. (b) Compute the change in entropy of the water. (c) What is the change in entropy of the universe for this process?

28. Compute the change in entropy of the universe when 1.50 kg of ice at 0.00°C is converted to water at 0.00°C during a reversible process.

29. Compute the change in entropy of the universe when 0.500 kg of water at 5.00°C is mixed with 3.00 kg of water at 20.0°C in an insulated container.

30. A vessel with 2.00 mol of an ideal, diatomic gas ($c_p = 29.1$ J/mol/K and $c_v = 20.8$ J/mol/K) at 200. K is placed in good thermal contact with a vessel containing 1.00 mol of an ideal, diatomic gas at 400. K. If the only heat transferred is that between the two gases, what is the change in entropy of the universe when the gases reach equilibrium?

31. Compute the change in entropy of the universe when 2.00 kg of ice at -10.0°C is mixed with 10.0 kg of water at 80.0°C in an insulated container.

32. Compute the entropy change when 20.0 g of steam at 110.°C is mixed with 500. g of water at 10.0°C. (Use $c_{steam} = 0.480$ cal/g/°C.)

33. By application of the first law of thermodynamics show that the work done by a system during a reversible process is $\int T \, dS$.

34. Find the change in entropy for each segment of the cyclic process in Example 19.5.

35. Compute the change in entropy for each segment of the cyclic process shown in Figure 20.P1. Use the numerical values given in Problem 13.

36. Find the change in entropy for each segment of the cyclic process shown in Figure 20.P2. Use the numerical values given in Problem 14.

37. Compute the change in entropy for each segment of the cyclic process shown in Figure 20.P3. Use the numerical values given in Problem 15.

General Problems

38. Prove that the Kelvin-Planck and Clausius statements of the second law of thermodynamics are equivalent. Assume that you can violate the Clausius statement of the law. Use the heat you transfer to the high-temperature reservoir to be that which is extracted from this same reservoir to run a heat engine. Show that the net result of one cycle of such a two-component system violates the Kelvin-Planck statement.

39. When a certain Carnot engine operates between the temperatures of 40.0°C and 100°C, it generates 350. W of power output. Compute (a) the rate of heat input and (b) the rate of heat exhausted from the engine.

40. A heat pump is to remove heat from the air at 25.0°C from inside a home and transfer it outside where the temperature is 40.0°C. (a) Compute the ideal coefficient of performance μ_{hp}. (b) If the heat pump were actually to perform this well, how much heat would be removed from the home air when 1200. BTU of heat are transferred to the outside air? (c) What is the minimum amount of

power we should expect from the heat pump motor if it is to remove 1200. BTU/min?

41. Two Carnot engines operate from the same high-temperature reservoir, but different low-temperature reservoirs. Show that the engine with the largest temperature difference is the most efficient.

42. The coefficient of performance of a certain heat pump is 3.60. If the outside air is $-10.0°C$, and the occupants of a building want to maintain the inside air at $24.0°C$, at what rate must the heat pump do work to provide 1.50×10^6 cal of heat to the building each hour?

43. Derive a relationship between the efficiency of a Carnot engine and the coefficient of performance of a Carnot refrigerator that operates between the same two reservoirs as the Carnot engine.

44. Show that $\mu_{hp} = 1 + \mu_r$.

45. Compute the efficiency of an engine that carries 0.150 mol of a monatomic ideal gas ($c_p = 20.8$ J/mol/K and $c_v = 12.5$ J/mol/K) through the cyclic processes shown in Figure 20.P6.

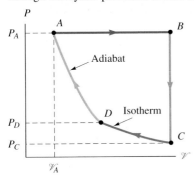

FIGURE 20.P6 Problems 45 and 53

46. A gas undergoes the reversible cyclic process shown in Figure 20.P7. (a) If $\mathcal{V}_B = 4\mathcal{V}_A$, then in terms of P_A, $\mathcal{V}_A$, T_A, and the ratio of specific heats $\gamma = c_p/c_v$ find P_B, P_C, T_B, and T_C. (b) Find the work $\mathcal{W}_{net}$, the net heat Q_{net} added per cycle, ΔU_{cycle}, and ΔS_{cycle} for 1.00 mol of this gas.

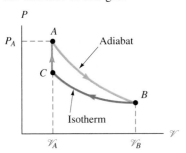

FIGURE 20.P7 Problem 46

47. The Brayton cycle (shown in Fig. 20.P8) is sometimes used to analyze the gas turbine engine. It consists of an adiabatic com-

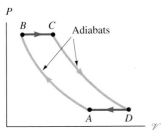

FIGURE 20.P8 The Brayton cycle. See Problem 47.

pression, an isobaric and an adiabatic expansion, and finally an isobaric return to the original state. Show for this cycle that $T_C/T_B = T_D/T_A$. [*Hint:* For this reversible cycle $\Delta S_{cycle} = 0$.]

48. The Brayton cycle is described in Problem 47. Show that the efficiency of this cycle is

$$e_B = 1 - \frac{T_A}{T_B}$$

[*Hint:* Use the result of Problem 47.]

49. Show that the efficiency of an Otto cycle may be written $e_{Otto} = 1 - T_A/T_B = 1 - T_D/T_C$. Use this result to show that this efficiency is less than that of a Carnot engine operating between the maximum temperature extremes of an Otto cycle.

50. An ideal gas increases its entropy by 72.0 J/K during an isothermal expansion at 212°C. How much heat was absorbed by the gas?

51. Compute the increase in entropy of 1.00 mol of an ideal, monatomic gas when it undergoes an isothermal expansion that doubles its volume.

52. Before it became illegal, a 60.0-kg physics instructor jumped from a height of 4.00 m into a Yellowstone Park hot-spring. (He checked out the temperature and depth before the jump!) Assume that the water, air, and instructor were all at a temperature of 98.6°F. Compute the change in entropy of the universe caused by this instructor's jump.

53. Compute the change in entropy for each segment of the cyclic process shown in Figure 20.P6.

54. For the cycle shown in Figure 20.P9, $n = 1.00$ mol, $P_A = 1.00$ atm, $P_B = 3.00$ atm, $\mathcal{V}_A = 1.00$ L, and $\mathcal{V}_C = 3.00$ L. For each cycle find (a) the work done, (b) the net heat added, (c) the engine efficiency, and (d) the efficiency of a Carnot engine operating between the highest and lowest temperatures of the cycle. Assume $c_p = 20.8$ J/mol/K and $c_v = 12.5$ J/mol/K.

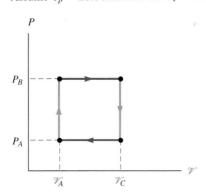

FIGURE 20.P9 Problem 54

55. A refrigerator is to remove heat from 3.00 kg of water at 10.0°C until it all changes phase to ice at 0.0°C. The refrigerator operates at a low temperature of 0.00°C, and heat is rejected to air at 20.0°C. (a) How much heat is removed from the water? (b) If the actual μ_r value is 60% of the Carnot refrigerator value, how much heat is transferred to the 20.0°C air? (c) If the water is to be completely converted to ice in 2.00 h, what is the minimum power required from the refrigerator motor?

Microscopic Connections to Thermodynamics

In this chapter you should learn

- the microscopic interpretations of pressure and temperature.

- how the internal energy of a substance is distributed among its molecules.

- a microscopic theory for the specific heat capacities of gases and solids.

- how the various speeds are distributed among a gas's molecules.

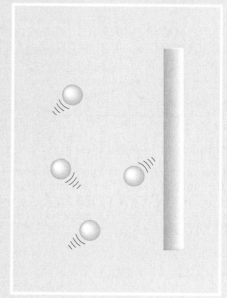

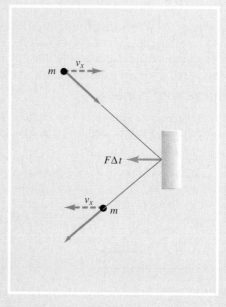

Most of Chapters 18, 19, and 20 dealt with the macroscopic view of solids, liquids, and gases. In this chapter we will take a brief look at a microscopic model of a system by applying the methods of mechanics to the random motions of individual molecules. This approach will provide important microscopic interpretations of two macroscopic quantities: pressure and temperature. We will also provide you with some results of an analysis that ignores the details of the molecular motions but takes advantage of the large number of molecules within a system. In order to keep the mathematics reasonable, we will restrict our attention mostly (but not entirely) to the ideal-gas model.

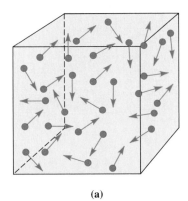

21.1 The Kinetic Theory of an Ideal Gas

The objective for this section is to relate the microscopic motion of the molecules of a gas to the macroscopically measured thermodynamic variables of pressure and temperature. To do this we apply the laws of Newtonian mechanics to the molecules in a container of gas. In order to simplify the calculations we employ a simplified model of the gas.

The Ideal-Gas Model

At ordinary temperatures and pressures the molecules in a gas are widely separated; the gas molecules themselves take up very little of the volume occupied by the gas. Gas pressure arises from the collisions of the gas molecules with the container walls. In order to model this situation we make the following assumptions, which serve to define the **microscopic model of an ideal gas:**

1. Molecules are point-particles that take up no space. The number of them in an ordinary sized container is enormous.
2. The molecules move in a random fashion with high velocities. The molecules undergo elastic collisions with the container walls. Momentum and kinetic energy are therefore conserved during the collisions.
3. The forces between particles and the forces between the particles and the container walls act only during collisions.

Pressure and Molecular Motion

Our first objective for this section is to compute the pressure that an ideal gas exerts on the walls of its container based on the laws of mechanics applied to gas molecules. Let's begin our computation of the pressure of the gas in the container shown in Figure 21.1(a). Focus your attention on a volume of the gas immediately in front of an area A of the container wall. The volume in question is indicated in Figure 21.1(b). There are billions of gas molecules in this region, of which we have shown only a select few. In fact we have drawn only that small fraction of molecules which happen to have one particular magnitude (v_{x1}) for the x-component of their velocity. Of these, half are moving toward the wall and half away from the wall. In order for a particle with velocity component $+v_{x1}$ to collide with the container wall within a time Δt, it must be within a perpendicular distance of $v_{x1} \Delta t$ from the wall. Suppose there are n_1 such particles *per unit volume*,[1] that is $n_1 =$ the number of particles per volume with x-component equal to v_{x1}.

We can predict the number of particles of this type that collide with the wall during the interval Δt by multiplying the number per unit volume by the size of the volume within the critical distance from the wall.

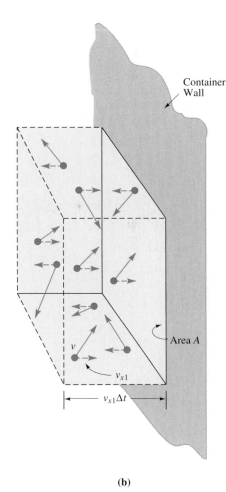

FIGURE 21.1

(a) A model of gas molecules in a container. (b) A volume $v_{x1} \Delta t A$ bounding a region of a gas. Only the gas molecules that have the particular x-component of their velocity magnitudes v_{x1} are shown.

---

[1] Particles outside this volume that also have velocity components v_{x1} and collide with the area A in time Δt are compensated for by those that move out of the volume before colliding with A.

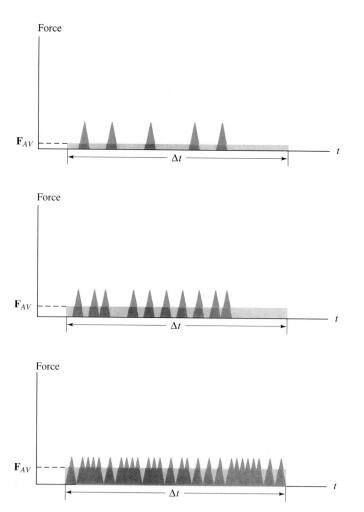

The average force for all the molecules with velocity component v_{x1} is defined so that the area $F_{AV} \Delta t$ is equal to the total area under all the individual impulses.

$$\left(\begin{array}{c} \textit{Number of collisions} \\ \textit{with area A during time } \Delta t \end{array}\right) = \frac{1}{2}\left(\begin{array}{c} \textit{Number of particles} \\ \textit{per volume} \end{array}\right) \times \left(\begin{array}{c} \textit{Total volume within} \\ \textit{distance } (v_{x1} \, \Delta t) \textit{ of area A} \end{array}\right)$$

$$= \frac{1}{2}(n_1)(A v_{x1} \, \Delta t) = \frac{n_1 v_{x1} A \, \Delta t}{2} \qquad (21.1)$$

When a particle of mass m undergoes an elastic collision with the wall, the particle exerts a tiny force on the wall for a brief interval of time much, much shorter than Δt. However, the combined effect of billions of such collisions each second is a nearly steady macroscopic average force of the particles on the wall. From Newton's third law we know that the average force $\mathbf{F}_1$ on the wall is equal in magnitude and opposite in direction to the average force $\mathbf{F}_1'$ exerted on the stream of molecules hitting the wall. We can calculate the latter force from the momentum statement of Newton's second law:

$$\mathbf{F}_1' = \frac{\Delta \mathbf{p}}{\Delta t}$$

as written in Section 9.3 (for instance see Example 9.6)

$$\mathbf{F}_1' = (\Delta \mathbf{p} \textit{ for one collision}) \times \left(\frac{\textit{Number of collisions}}{\Delta t}\right) \qquad (21.2)$$

For the perfectly elastic collisions our particles make with the container wall, $\mathbf{v}_o = v_{x1}\hat{\mathbf{i}}$ and $\mathbf{v}_f = -v_{x1}\hat{\mathbf{i}}$. This collision process is illustrated in Figure 21.2. Notice that the vertical component of the velocity is not affected by the collision. Only the x-component contributes to the pressure of the gas on the wall. Thus, for one collision the change in momentum of one molecule is $\Delta\mathbf{p} = m\mathbf{v}_f - m\mathbf{v}_o = -2m\mathbf{v}_{x1}$. Using the calculated value for $\Delta\mathbf{p}$ and Equation (21.1) for the number of collisions, Equation (21.2) becomes

$$\mathbf{F}_1' = (-2m\mathbf{v}_{x1}) \times \left(\frac{n_1 v_{x1} A \,\Delta t/2}{\Delta t}\right) = -m\mathbf{v}_{x1} n_1 v_{x1} A$$

and so the magnitude of the average force of the gas on the wall is

$$F_1 = n_1 m v_{x1}^2 A$$

The average pressure on the wall is given by $P_1 = F_1/A$. Using the preceding expression for the force, we obtain

$$P_1 = n_1 m v_{x1}^2$$

This pressure P_1 is that due to only those molecules with x-component of velocity v_{x1}. For the total pressure from all the particles we must sum the pressures from particles with x-components v_{x2}, v_{x3}, v_{x4}, . . . until we have contributions from all the particles within the gas:

$$P = P_1 + P_2 + P_3 + \cdots = m(n_1 v_{x1}^2 + n_2 v_{x2}^2 + n_3 v_{x3}^2 + \cdots) \qquad (21.3)$$

It may seem like this is about as far as we can go. However, the term within the parentheses of Equation (21.3) is related to the average value of v_x^2 for all the particles. To see this, let's recall how we take an average. If N_1 is the number of particles with velocity v_{x1}^2, N_2 the number with v_{x2}^2, N_3 the number with v_{x3}^2, . . . , the average value of v_x^2 (written $<v_x^2>_{av}$) is

$$<v_x^2>_{av} = \frac{N_1 v_{x1}^2 + N_2 v_{x2}^2 + N_3 v_{x3}^2 + \cdots}{N_1 + N_2 + N_3 + \cdots} = \frac{N_1 v_{x1}^2 + N_2 v_{x2}^2 + N_3 v_{x3}^2 + \cdots}{N} \qquad (21.4)$$

where N is the total number of particles in the container of gas under consideration.

When the numerator and denominator of the final fraction on the right-hand side of Equation (21.4) are each divided by $\mathcal{V}$, the volume occupied by the gas, and we make the substitutions $\mathcal{N} = N/\mathcal{V}$, $n_1 = N_1/\mathcal{V}$, $n_2 = N_2/\mathcal{V}$, . . . , we have

$$<v_x^2>_{av} = \frac{n_1 v_{x1}^2 + n_2 v_{x2}^2 + n_3 v_{x3}^2 + \cdots}{\mathcal{N}}$$

Upon rearranging, this equation becomes

$$n_1 v_{x1}^2 + n_2 v_{x2}^2 + n_3 v_{x3}^2 + \cdots = \mathcal{N}<v_x^2>_{av}$$

where $\mathcal{N}$ is called the *number density*. Substituting this result into Equation (21.3), we have

$$P = \mathcal{N}m<v_x^2>_{av} \qquad (21.5)$$

Let's now examine the average $<v_x^2>_{av}$ in more detail. For every velocity vector $\mathbf{v}$ it is true that $v^2 = v_x^2 + v_y^2 + v_z^2$, so this equality must certainly hold on the average:

$$<v^2>_{av} = <v_x^2>_{av} + <v_y^2>_{av} + <v_z^2>_{av}$$

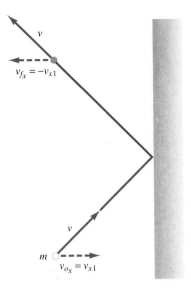

FIGURE 21.2

An elastic molecular collision with the container wall.

Concept Question 1
The time interval Δt in Equation (21.1) establishes the distance $v_{x1}\,\Delta t$ within which n_1 particles per volume might collide with the wall, whereas the time Δt in Equation (21.2) is related to the time interval over which the average force F_1 acts to give impulse $F_1\,\Delta t$. Explain how these two time intervals are the same.

Furthermore, because the particles move with random motions, the average values of the three components v_x^2, v_y^2, and v_z^2 are equal and therefore their total is three times any one of them

$$<v^2>_{av} = 3<v_x^2>_{av}$$

or

$$<v_x^2>_{av} = \frac{1}{3}<v^2>_{av}$$

With this substitution Equation (21.5) becomes

$$P = \frac{1}{3}\mathcal{N}m<v^2>_{av}$$

Now who among us can look at the quantity $m<v^2>_{av}$ without an irresistible urge to put a 1/2 in front? Doing so and recalling that $\mathcal{N} = N/\mathcal{V}$, we have

$$P = \frac{2}{3}\left[\frac{N}{\mathcal{V}}\left(\frac{m<v^2>_{av}}{2}\right)\right]$$

$$P = \frac{2}{3}\frac{N}{\mathcal{V}}<K_{trans}>_{av} \tag{21.6}$$

We have now accomplished our first objective for this section. We have related the macroscopic variable pressure to a property characteristic of the motion of individual gas molecules, the average translational kinetic energy $<K_{trans}>_{av}$. The following sentence summarizes what Equation (21.6) is telling us:

> The pressure of the gas is proportional to the number of particles per volume and the average translational kinetic energy of the molecules.

Temperature and Molecular Motion

We are now ready to take the second important step and relate the macroscopic concept of temperature to the microscopic motion of the molecules. We proceed by solving Equation (21.6) for the product $P\mathcal{V}$,

$$P\mathcal{V} = \frac{2}{3}N<K_{trans}>_{av}$$

Combining this equation with the ideal-gas equation of state, $P\mathcal{V} = nRT$, we have

$$nRT = \frac{2}{3}N<K_{trans}>_{av}$$

The number of particles in the system N may be expressed in terms of the number of moles n through $N = nN_A$ (N_A = Avogadro's number = 6.02×10^{23} particles/mol):

$$nRT = \frac{2}{3}nN_A<K_{trans}>_{av}$$

$$\left(\frac{R}{N_A}\right)T = \frac{2}{3}<K_{trans}>_{av}$$

The constant R/N_A is known as **Boltzmann's constant** $k_B = R/N_A = 1.38 \times 10^{-23}$ J/K. Using the definition of k_B this last result can then be rewritten

$$<K_{\text{trans}}>_{av} = \frac{3}{2} k_B T \qquad (21.7)$$

Equation (21.7) is the second major result of this chapter:

> The Kelvin temperature is a direct measure of the average translational kinetic energy of the ideal-gas molecules.

In our ideal-gas model the particles are noninteracting, thus there is no interparticle potential energy. Consequently, the total energy for N particles is the sum of their individual kinetic energies. This sum is N times the average translational kinetic energy

$$U = N<K_{\text{trans}}>_{av} = \frac{3}{2} N k_B T \qquad (21.8)$$

As foreshadowed in the previous chapters, the internal energy for an *ideal* gas is a function of temperature only.

We may obtain an idea of how fast the molecules of our ideal gas are moving by substituting $<K_{\text{trans}}>_{av} = m<v^2>_{av}/2$ into Equation (21.7) and solving for $<v^2>_{av}$. Upon taking the square root, we obtain a representative speed v_{rms} known as the **root-mean-square speed** of the collection of particles:

$$v_{rms} = \sqrt{<v^2>_{av}} \qquad (21.9a)$$

For our monatomic ideal gas we obtain

$$v_{rms} = \sqrt{<v^2>_{av}} = \sqrt{\frac{3k_B T}{m}} \qquad (21.9b)$$

Concept Question 2
In the derivations of Equations (21.6) and (21.7) we ignored the gravitational potential energy of the gas molecules within the container. What is the justification for this omission?

Concept Question 3
In the derivation of Equation (21.7) we assumed that there are a very large number of molecules. However, we also used the ideal-gas equation of state, which holds for gases with low densities. Reconcile the inconsistency of this approach.

EXAMPLE 21.1   *Faster Than a Speeding Bullet*

A pressurized tank at 25.0°C contains 4.00 mol of argon. (a) What is the root-mean-square speed of the argon atoms? (b) What is the total energy of the gas?

SOLUTION   (a) We can find v_{rms} from Equation (21.9b) if we can first find the mass of an argon molecule. From Appendix 6 we have the molecular weight of argon (39.96 u), so that

$$m = (39.96 \text{ g/mol})\left(\frac{1.00 \text{ mol}}{6.02 \times 10^{23} \text{ molecules}}\right)$$

$$= 6.64 \times 10^{-23} \text{ g} = 6.64 \times 10^{-26} \text{ kg}$$

Then, Equation (21.9b) gives

$$v_{rms} = \sqrt{\frac{3k_B T}{m}} = \sqrt{\frac{3(1.38 \times 10^{-23} \text{ J/K})(298 \text{ K})}{6.64 \times 10^{-26} \text{ kg}}}$$

$$= 431 \text{ m/s} \ (= 1550 \text{ km/h!})$$

(b) From Equation (21.8) we have $U = \frac{3}{2}Nk_BT$. But $Nk_B = nR$ and

$$U = \frac{3}{2}nRT = \frac{3}{2}(4.00 \text{ mol})(8.314 \text{ J/mol/K})(298 \text{ K}) = 1.49 \times 10^4 \text{ J}$$

Concept Question 4
Using the concepts of the kinetic theory, explain why a tied balloon stays inflated. In particular, mention what factor causes the pressure inside to be greater than the pressure outside when inner and outer gases are at the same temperature.

The total energy must also be equal to $N<K_{\text{trans}}>_{av} = nN_A(\frac{1}{2}m<v_{rms}^2>_{av})$:

$$U = (4.00 \text{ mol})(6.02 \times 10^{23} \text{ molecules/mol})\left[\frac{1}{2}(6.64 \times 10^{-26} \text{ kg})(431 \text{ m/s})^2\right]$$

$$= 1.49 \times 10^4 \text{ J} \quad \text{(Our answer checks!)} \blacktriangleleft$$

21.2 The Equipartition-of-Energy Theorem

The result $U = \frac{3}{2}k_BT$ for the internal energy of an ideal monatomic gas as derived in the previous section can be viewed as a special case of a still more fundamental theorem. This more basic result is known as the **equipartition-of-energy theorem,** and it has been derived from an approach called statistical mechanics. In this branch of mechanics one does not try to keep track of individual molecular motions but rather calculates statistical averages for the properties of the very large number of molecules within a system. (Our argument in the previous section that $<v_x^2>_{av} = <v_y^2>_{av} = <v_z^2>_{av}$ for gas particles in equilibrium is an example of statistical mechanics.) Let's build some background before we actually state the equipartition theorem.

The "3" on the right-hand side of Equation (21.7) comes from the substitution we made in Equation (21.5), $<v^2>_{av} = 3<v_x^2>_{av}$, to obtain Equation (21.6). If we work backward, substituting this result into Equation (21.7), we have

$$\frac{m<v_x^2>_{av}}{2} = \frac{1}{2}k_BT \tag{21.10a}$$

This equation says that an energy of $\frac{1}{2}k_BT$ is associated with the x-component of the translational kinetic energy of the particle. Because the average x-, y-, and z-components are equal, we have the same result for these components:

$$\frac{m<v_y^2>_{av}}{2} = \frac{1}{2}k_BT \tag{21.10b}$$

$$\frac{m<v_z^2>_{av}}{2} = \frac{1}{2}k_BT \tag{21.10c}$$

Evidently, there is $\frac{1}{2}k_BT$ of energy associated with each component of the velocity. When we add together all three of Equations (21.10a,b,c), we obtain Equation (21.7):

$$\frac{m<v_x^2>_{av}}{2} + \frac{m<v_y^2>_{av}}{2} + \frac{m<v_z^2>_{av}}{2} = \frac{3}{2}k_BT$$

The equipartition-of-energy theorem is a generalization of this result. Each independent variable necessary to describe the energy of a molecule corresponds to a **degree of freedom.** For point-particles we have three degrees of freedom: only v_x, v_y, and v_z are required to write an expression for the total energy. Molecules with more complex structure may also rotate and vibrate and, therefore, have more degrees of freedom. We consider a couple of these cases below. With this definition of degrees of freedom we can state the theorem:

> **Equipartition-of-Energy Theorem:** There is associated an average energy of $\frac{1}{2}k_B T$ per molecule with each degree of freedom of the molecules in a system.

The point-particle picture of our ideal gas may be called a monatomic ideal-gas model. The energy U_m of each molecule consists only of its kinetic energy, $U = \frac{1}{2}m(v_x^2 + v_y^2 + v_z^2)$. As noted above, we had to use *three* variable v_x, v_y, and v_z to write this expression. Therefore, according to the equipartition-of-energy theorem, the average energy per molecule is $3(\frac{1}{2}k_B T)$. For a gas of N molecules, the total energy is

$$U = \frac{3}{2}Nk_B T \qquad \text{(Monatomic ideal gas)}$$

in agreement with our previous results.

Ideal Polyatomic-Gas Models

Many real gases, such as O_2 and H_2, consist of two tightly bound atoms. We can model such gases as rigid diatomic molecules and use the equipartition-of-energy theorem to compute the total energy of the gas at temperature T. In addition to its translational kinetic energy, a molecule of such a gas may rotate as illustrated in Figure 21.3. Normally, we need three perpendicular axes to describe all the rotations that the molecule could have. However, because we are modeling the individual atoms as point-masses, the moment of inertia about the symmetry axis through the line joining the masses is zero, thus we ignore rotations about this axis. However, we still need two angular velocities to describe rotations about any two other axes. So, if M is the total mass of the diatomic molecule, the expression for the total kinetic energy of a molecule is

$$U_m = \frac{1}{2}M(v_x^2 + v_y^2 + v_z^2) + \frac{1}{2}(I_1\omega_1^2 + I_2\omega_2^2)$$

There are five variables in this equation: v_x, v_y, v_z, ω_1, and ω_2. The average energy per molecule is, therefore, $5(\frac{1}{2}k_B T)$. For a rigid diatomic gas of N molecules the total energy is

$$U = \frac{5}{2}Nk_B T \qquad \text{(Rigid diatomic ideal gas)} \qquad \textbf{(21.11)}$$

Some diatomic molecules, such as NO and Cl_2, are not very rigid even at room temperature. The equilibrium separation oscillates about an average distance, an effect that is often modeled by a spring interaction as shown in Figure 21.4. In addition to the five variables necessary to describe its translational and rotational kinetic energies, we need two more: one to designate the velocity of one of the molecules relative to the other, and a relative separation coordinate necessary to describe the potential energy between the two atoms. Therefore, the total number of degrees of freedom is 7 and the average energy of a molecule is $7(\frac{1}{2}k_B T)$. For N molecules in the gas,

$$U = \frac{7}{2}Nk_B T \qquad \text{(Vibrating diatomic ideal gas)} \qquad \textbf{(21.12)}$$

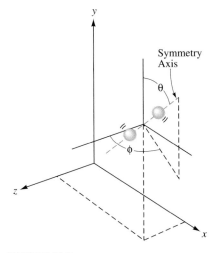

FIGURE 21.3

When the moment of inertia about the symmetry axis is negligible, a rigid diatomic molecule has 5 degrees of freedom. The angular velocities about the axes shown are $\omega_1 = d\phi/dt$ and $\omega_2 = d\theta/dt$.

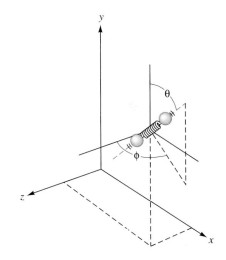

FIGURE 21.4

The motion of a vibrating diatomic molecule. There are two new vibrational degrees of freedom in addition to the five necessary to describe the rigid diatomic molecule.

21.3 Another Look at Specific Heats

We can use the expressions for the internal energy derived in the previous section to arrive at a result that should seem rather extraordinary to you. We compute the specific heat capacity for each of the three ideal-gas models. Why might our result be unusual? Just watch the outcome! The molar specific heat at constant volume c_v is given by Equation (19.8),

$$c_v = \frac{1}{n}\frac{dU}{dT} \tag{21.13}$$

and c_p can be computed from Equation (19.10):

$$c_p = c_v + R \tag{21.14}$$

Let's apply these equations to a monatomic ideal gas.

Monatomic Ideal Gases

Recall that $k_B = R/N_A$ and $N = nN_A$. Therefore,

$$Nk_B = (nN_A)\left(\frac{R}{N_a}\right) = nR$$

and Equation (21.8) for the total energy of the monatomic gas becomes

$$U = \frac{3}{2}Nk_BT = \frac{3}{2}nRT \tag{21.15}$$

Substituting this result into Equation (21.13), we have

$$c_v = \frac{1}{n}\frac{d}{dT}\left(\frac{3}{2}nRT\right) = \frac{3}{2}R = \left(\frac{3}{2}\right)(8.314 \text{ J/mol/K})$$

$$c_v = 12.5 \text{ J/mol/K} \quad \text{(Monatomic ideal gas)} \tag{21.16}$$

Now, isn't this result remarkable? We have a number! Equation (21.16) states that it takes 12.5 J of energy to raise the temperature of 1.00 mol of a monatomic ideal gas by 1.00 K. This number is precisely the experimental value for the specific heat quoted in Section 19.4 for helium and other inert gases in the homework problems in Chapter 19! If you don't find this remarkable, reflect on this: We have been able to sit down with a paper and pencil and predict a numerical value for the specific heat of a real gas just by thinking! (We do, of course, need to know the *experimental* value of one of nature's fundamental constants: the gas constant R.) Let's review how we obtained this result. By considering how gas particles bounce off container walls, in the first section of this chapter, we computed the pressure on the walls. Combining this relation with the ideal-gas equation of state we derived Equation (21.8). And finally from Equation (21.13) we found a value for c_v. From Equation (21.14) we can now find c_p:

$$c_p = c_v + R = \frac{3}{2}R + R = \frac{5}{2}R = \frac{5}{2}(8.314 \text{ J/mol/K})$$

$$c_p = 20.8 \text{ J/mol/K} \quad \text{(Monatomic ideal gas)} \tag{21.17}$$

Diatomic Ideal Gases

While we are on a roll, let's go ahead and use the values of U we found from the equipartition-of-energy theorem to compute c_v and c_p for the diatomic ideal gases. In the case of the rigid diatomic molecule we have

$$U = \frac{5}{2} N k_B T = \frac{5}{2} nRT$$

so that

$$c_v = \frac{1}{n}\frac{d}{dT}\left(\frac{5}{2}nRT\right) = \frac{5}{2}R = \left(\frac{5}{2}\right)(8.314 \text{ J/mol/K})$$

$$c_v = 20.8 \text{ J/mol/K} \qquad \text{(Rigid diatomic ideal gas)} \qquad \textbf{(21.18)}$$

And

$$c_p = c_v + R = \frac{5}{2}R + R = \frac{7}{2}R = \frac{7}{2}\,(8.314 \text{ J/mol/K})$$

$$c_p = 29.1 \text{ J/mol/K} \qquad \text{(Rigid diatomic ideal gas)} \qquad \textbf{(21.19)}$$

For the vibrating diatomic gas, the internal energy is

$$U = \frac{7}{2} N k_B T = \frac{7}{2} nRT$$

so that

$$c_v = \frac{1}{n}\frac{d}{dT}\left(\frac{7}{2}nRT\right) = \frac{7}{2}R = \left(\frac{7}{2}\right)(8.314 \text{ J/mol/K})$$

$$c_v = 29.1 \text{ J/mol/K} \qquad \text{(Vibrating diatomic ideal gas)} \qquad \textbf{(21.20)}$$

In addition

$$c_p = c_v + R = \frac{7}{2}R + R = \frac{9}{2}R = \frac{9}{2}(8.314 \text{ J/mol/K})$$

$$c_p = 37.4 \text{ J/mol/K} \qquad \text{(Vibrating diatomic ideal gas)} \qquad \textbf{(21.21)}$$

Now that we have all these predictions for the specific heat capacities of these three types of gases, it would be nice to see how well our theory agrees with experiment. Table 21.1 provides the experimental data for several gases of each type. In addition to c_v and c_p, the difference $c_p - c_v$ is provided. For our models this value should be $R = 8.314$ J/mol/K. The experimental specific heat values agree very well for the monatomic gases in Table 21.1, and the agreement for the rigid diatomic molecules is reasonably close. However, there is a real problem with vibrating diatomic molecules. Our classical model is just not sufficient to predict the specific heats for these types of gases. Quantum theory is required for an accurate description of molecules with structures more complex than that of the rigid diatomic molecules.

TABLE 21.1   Specific Heat Capacities for Selected Gases at 15°C

GAS	c_p (J/mol/K)	c_v (J/mol/K)	$c_p - c_v$ (J/mol/K)	γ
Monatomic				
Model	20.8	12.5	8.31	5/3
Argon	20.95	12.56	8.39	1.67
Helium	20.80	12.53	8.27	1.66
Rigid Diatomic				
Model	29.1	20.8	8.31	7/5
H_2	28.56	20.25	8.31	1.41
N_2	29.02	20.67	8.35	1.40
O_2	29.15	20.81	8.34	1.40
Vibrating Diatomic				
Model	37.4	29.1	8.31	9/7
Cl_2	34.09	25.15	8.94	1.36
NO	29.85	21.0	8.85	1.43

Quantum Mechanical Effects

The classical theory presented previously predicts that the specific heat of gases is independent of temperature (see Eqs. 21.17 to 21.21). However, experimentally we find that specific heats do depend on temperature (see Eq. (18.13) and Example 18.8). Our classical model is inadequate to explain the observed temperature dependence of the specific heat of gases.

In order to account for the temperature dependence of the specific heat we must employ a quantum mechanical model of gas molecules. For our purposes, the important difference between the quantum mechanical model and the classical model of molecules is this: the rotational and vibrational energy of the molecules is **quantized.** That is, a molecule is permitted to have only discrete values of energy.

If the full impact of that last sentence did not hit you, read it again. It is saying that molecules cannot rotate with any angular velocity they (or we) desire, but only certain allowed values! For example, in order to change its angular velocity the molecule must absorb enough energy to make the jump from one energy level to the next. The molecule is not allowed to save up energy for the jump; the energy must be delivered all at once. For the rigid diatomic molecule at sufficiently low temperatures the energy difference between the rotational energy levels is sufficiently large that usually the rotational state of the molecule cannot be changed by a collision between molecules.

A critical parameter that determines whether the gas behaves classically or quantum mechanically is the magnitude of the average energy of the gas molecules compared to the size of the steps between adjacent energy levels of the molecule. For *translational motion* the size of the steps is very small compared to the average energy of the molecules; we can treat the translational motion of the molecules classically. For *rotational motion* the energy steps are quite a bit larger. These steps are large enough in fact that at low temperatures virtually no molecules have sufficient energy to start to rotate. The classical equipartition theorem breaks down, and energy is not shared equally among all the degrees of freedom. The rotational motion is ''frozen out.'' As the temperature is increased, the average kinetic energy of the molecules eventually becomes large with respect to the rotational energy steps. At that point the rotational degrees of freedom share equally in the energy as predicted by the equipartition theorem.

For diatomic molecules the energy steps between *vibrational* energy levels is even larger than the steps for rotational motion. However, when a sufficiently high temperature

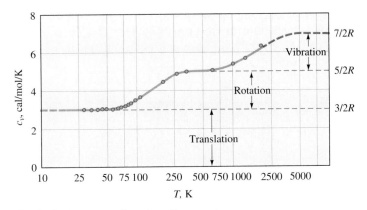

The behavior of c_v for hydrogen as a function of temperature.

is reached, a molecule may gain enough energy to start to vibrate at higher energy levels. In fact, even the diatomic molecules that are rigid at 15°C (see Table 21.1) can start to vibrate at higher temperatures. The quantization of vibrational energy levels is the reason for the disagreement between the theoretical values and the experimental values of c_p and c_v in Table 21.1 for the vibrating diatomic molecules. When quantum theory is applied to these systems, the agreement of the new theory with experiment is excellent.

Specific Heat Capacity of Solids

The molar heat capacity c_v for a typical solid is shown in Figure 21.5. The change in c_v with temperature is a result of the quantization of energy levels. As we saw in the previous section this quantization is important at low temperatures. Quantum theory predicts that as the absolute temperature T becomes small, c_v varies as $(T/T_D)^3$. The parameter T_D is called the **Debye temperature** and it has a characteristic value for a particular solid. This result is known as the **Debye T-cubed law.** The law works best for temperatures at which $T/T_D < 0.1$. From Figure 21.5 we can see that at high temperatures, c_v approaches a value of $3R$. This phenomenon is known as the **law of Dulong and Petit,** and we can understand this high-temperature behavior from the classical equipartition-of-energy theorem. Each atom in the solid executes simple harmonic motion in three directions about its equilibrium position. Its total energy is, therefore,

$$U_m = \frac{1}{2}m(v_x^2 + v_y^2 + v_z^2) + \frac{1}{2}(k_x x^2 + k_y y^2 + k_z z^2)$$

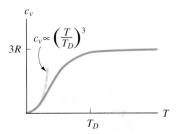

FIGURE 21.5

The molar heat capacity for a typical solid. At low temperatures c_v approaches zero as $(T/T_D)^3$. At high temperatures (above the Debye temperature T_D) c_v approaches a value of $3R$.

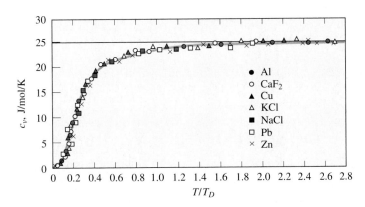

The molar heat capacity of various solids.

where k_x, k_y, and k_z are spring constants that represent bond strengths for the three directions. There are six variables in this expression, hence 6 degrees of freedom, so the equipartition theorem predicts that, when in equilibrium at temperature T, the total internal energy of a solid of N atoms is

$$U = 6N\left(\frac{1}{2}k_B T\right) = 3Nk_B T = 3nRT$$

Concept Question 5
What can account for the observation that some elements in the solid state have molar heat capacities greater than 3R?

and

$$c_v = \frac{1}{n}\frac{d}{dT}\left(3nRT\right) = 3R$$

21.4 Distribution of Molecular Speeds

Because the motions of molecules in a gas are random, at any instant not all molecules have the same speed or energy. We have already seen that the characteristic speed v_{rms} depends upon temperature. A much more difficult question is to ask about the *distribution* of speeds among the various particles and how this distribution changes with temperature. The concept that embodies this description is known as the **speed-distribution function.** In essence the speed-distribution function describes the number of particles that have speeds in the neighborhood of some particular speed.

The Maxwell Speed-Distribution Function

James Clark Maxwell (1831–1879) was the first to derive an expression for the speed-distribution function $N_v(v)$ for molecular speeds (1860). This function gives the number of molecules N_v with speeds within a specified speed interval Δv. That is to say $N_v(v)\,\Delta v$ is the number of molecules with speeds between v and $v + \Delta v$. The total number of molecules is N, so

$$N = \lim_{\Delta v \to 0} \sum_{v=0}^{\infty} N_v(v)\,\Delta v = \int_0^{\infty} N_v(v)\,dv$$

The expression Maxwell derived for $N_v(v)$ is

$$N_v(v) = 4\pi N\left(\frac{m}{2\pi k_B T}\right)^{3/2} v^2 e^{-(mv^2/2k_B T)} \qquad \textbf{(21.22)}$$

A plot of this function is shown in Figure 21.6. The important feature to note in Equation (21.22) is that for a particular gas the distribution function depends only on the absolute temperature T. Once the temperature is given, the shape of the $N_v(v)$-versus-v curve is

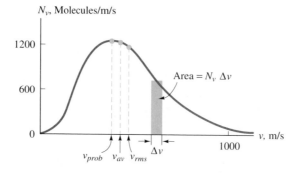

FIGURE 21.6

The distribution function $N_v(v)$ for about 600 000 molecules (10^{-18} mol) of oxygen at 300 K. The number of atoms with speeds between v and $v + \Delta v$ is $N_v(v)\Delta v$, and the total area under the curve is N, the total number of molecules. For this temperature $v_{prob} = 393$ m/s, $v_{av} = 447$ m/s, and $v_{rms} = 494$ m/s.

determined. From Figure 21.6 we see that there are only a few molecules with very large or very small speeds. Also note that the curve is asymmetric. Therefore, the average speed v_{av} is different from the most probable speed v_{prob} that corresponds to the speed at which $N_v(v)$ is a maximum. Another consequence of the asymmetry is that v_{av} is somewhat lower than v_{rms}. In fact, these three characteristic speeds are related by

$$v_{av} = \int_0^\infty v N_v(v)\, dv = 1.60 \sqrt{\frac{k_B T}{m}}$$

$$v_{rms} = \sqrt{\int_0^\infty v^2 N_v(v)\, dv} = 1.77 \sqrt{\frac{k_B T}{m}}$$

$$v_{prob} = 1.41 \sqrt{\frac{k_B T}{m}}$$

We leave it as an exercise for you to compute these values (Problems 22 and 23).

The effect of temperature on the velocity distribution function is shown in Figure 21.7.[2] Because average molecular speed increases as the temperature is raised, the curve shifts toward higher v-values for larger T. Furthermore, as the temperature is increased, the curve broadens so that the range of speeds increases.

The mass m of a gas molecule appears in the numerator of the negative exponential term in Equation (21.22). Thus, for a given temperature the larger a molecule's mass, the more unlikely it is to have a high speed. Therefore, light molecules, such as He and H_2, are more likely to escape from the earth's atmosphere than are heavier molecules. There are more light molecules that have speeds close to the escape velocity.

The distribution function for the molecular speeds of liquids is very similar to that of Equation (21.22). Evaporation takes place when a molecule near the surface of a liquid acquires (through collision) a velocity near the higher end of the distribution function. At higher temperatures there are more molecules with sufficient speed to escape the bonds that attract the liquid molecules together.

Mean Free Path

Gas molecules travel in a random fashion, undergoing collisions with each other as illustrated in Figure 21.8. The average distance a gas molecule travels between collisions is called the **mean free path** ℓ. Intuitively, we expect this path length to be related to both the density of the gas and the diameter of the molecules.

It is possible to estimate the mean free path if we adopt a model in which the gas molecules are spheres of diameter d. Figure 21.9(a) shows that in order for two molecules to collide, the path directions of their centers must pass within a distance d from each other. We can construct an equivalent description by replacing one of the molecules with a sphere of diameter $2d$ and the other molecule with a point-particle (see Fig. 21.9(b)). If during time Δt the average velocity of the $2d$-diameter molecule is v_{av}, the distance it travels is $v_{av} \Delta t$. During this time it sweeps out a cylinder of volume $\pi d^2 (v_{av} \Delta t)$. If the number of molecules per unit volume of the gas is n_v, then the number of molecules within this cylinder is $n_v(\pi d^2 v_{av} \Delta t)$. The $2d$-diameter molecule must collide with each of these molecules. Hence, the total number of collisions in time Δt is $n_v v_{av} \Delta t$. However, the mean free path ℓ is just the distance $v_{av} \Delta t$ the molecule travels divided by the number of collisions it has along this distance:

$$\ell = \frac{v_{av} \Delta t}{n_v \pi d^2 v_{av} \Delta t} = \frac{1}{\pi d^2 n_v} \tag{21.23}$$

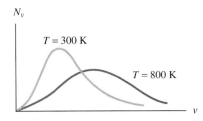

FIGURE 21.7

The Maxwell's speed distribution function for oxygen at 300 K and at 800 K. The curve shifts toward higher speeds at higher temperatures.

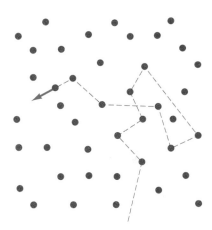

FIGURE 21.8

Random collisions of a gas molecule.

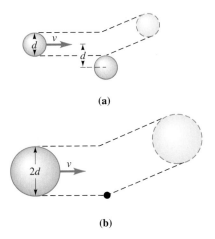

FIGURE 21.9

(a) The path directions of the centers of two spherical molecules must pass within a distance $2d$ from each other for a collision to take place. (b) An equivalent collision between a sphere of radius $2d$ and a point–particle.

Concept Question 6
If hot air rises, why is it often cooler at the top of mountains than at their bases?

Concept Question 7
From a microscopic view, what parameters are important for the determination of the speed of sound in a gas?

This result was obtained by assuming the point-particles were at rest when the $2d$-diameter molecule collided with them. When the motions of the other molecules are taken into account, the effect is to reduce the mean free path. The correct expression is

$$\ell = \frac{1}{\sqrt{2}\pi d^2 n_v} \tag{21.24}$$

EXAMPLE 21.2   Gases Are Mostly Empty Space

A tank at 25.0°C contains oxygen at 1.00 atm of pressure. If the molecular diameter of oxygen is about 2.50×10^{-10} m, find the number of molecular radii an oxygen molecule typically travels between collisions at this temperature.

SOLUTION   In order to use Equation (21.24) we must compute n_v. From the ideal-gas equation of state, $P\mathcal{V} = Nk_BT$

$$n_v = \frac{N}{\mathcal{V}} = \frac{P}{k_BT} = \frac{1.01 \times 10^5 \text{ Pa}}{(1.38 \times 10^{-23} \text{ J/K})(298 \text{ K})} = 2.46 \times 10^{25} \text{ molecules/m}^3$$

Then,

$$\ell = \frac{1}{\sqrt{2}\pi d^2 n_v} = \frac{1}{\sqrt{2}\pi(2.50 \times 10^{-10} \text{ m})^2(2.46 \times 10^{25} \text{ molecules/m}^3)}$$

$$= 1.47 \times 10^{-7} \text{ m}$$

The number of molecular diameters in this distance is

$$\frac{\ell}{d} = \frac{1.47 \times 10^{-7} \text{ m}}{2.50 \times 10^{-10} \text{ m}} = 587$$

◀

21.5 Summary

In this chapter we applied the laws of mechanics to a model of an ideal gas. The pressure of a gas was found to be proportional to the number of particles per volume and the average translational kinetic energy of the particles:

$$P = \frac{2}{3}\left[\frac{N}{\mathcal{V}}\left(\frac{m<v^2>_{av}}{2}\right)\right] = \frac{2}{3}\frac{N}{\mathcal{V}}<K_{trans}>_{av} \tag{21.6}$$

The absolute temperature of the gas was found to be related to the average translational kinetic energy of the gas molecules:

$$\left(\frac{m<v^2>_{av}}{2}\right) = \frac{3}{2}k_BT \tag{21.7}$$

The **equipartition-of-energy theorem** states that there is an average energy of $\frac{1}{2}k_BT$ per molecule associated with each degree of freedom of a system. From this theorem we can compute the total internal energy and molar heat capacities for monatomic and diatomic models of an ideal gas:

MODEL	INTERNAL ENERGY	c_p	c_v
Monatomic	$\dfrac{3}{2}nRT$	$\dfrac{5}{2}R$	$\dfrac{3}{2}R$
Diatomic (rigid)	$\dfrac{5}{2}nRT$	$\dfrac{7}{2}R$	$\dfrac{5}{2}R$
Diatomic (vibrating)	$\dfrac{7}{2}nRT$	$\dfrac{9}{2}R$	$\dfrac{7}{2}R$

The model works well for monatomic and rigid diatomic molecules but does not correctly predict the molar heat capacities for vibrating diatomic molecules or more complex structures. For these latter two types of molecules quantum theory is required.

The **Maxwell speed-distribution function** gives the number of molecules in a Δv speed interval near speed v when the gas is in equilibrium at temperature T:

$$N_v(v) = 4\pi N\left(\frac{m}{2\pi k_B T}\right)^{3/2} v^2 e^{-(mv^2/2k_B T)} \qquad (21.22)$$

From a plot of this function at different temperatures we find that the average speed of gas molecules increases when the temperature is raised.

The **mean free path** is the average distance a molecule travels between collisions:

$$\ell = \frac{1}{\sqrt{2}\pi d^2 n_v} \qquad (21.24)$$

PROBLEMS

21.1 The Kinetic Theory for an Ideal Gas
21.2 The Equipartition-of-Energy Theorem

1. An automatic launcher fires tennis balls ($m = 58.5$ g) toward a wall so that they hit it at an angle of 45°. The balls have a velocity of 15.0 m/s when they strike the wall, and they rebound elastically. If 15 such launchers are aimed so that tennis balls hit a 4.00-m^2 area uniformly at a rate of 20.0 balls every 5.00 s, find the average pressure on the wall.

2. During a time of 0.100 s, 2.00×10^{23} oxygen molecules strike a 10.0-cm^2 area of a wall. If the average velocity component perpendicular to the wall is 250. m/s, find the pressure on this area due to these molecules.

3. The speeds of 12 molecules in meters per second are: 1.00, 2.00, 2.00, 4.00, 5.00, 6.00, 6.00, 6.00, 7.00, 8.00, 9.00, and 9.00. Compute the (a) average speed, and (b) the root-mean-square speed.

4. The speeds of ten molecules in meters per second are: 2.00, 3.00, 4.00, 4.00, 5.00, 6.00, 7.00, 8.00, 8.00, and 9.00. Compute the (a) average speed, and (b) the root-mean-square speed.

5. (a) Find the root-mean-square speed of the molecules in a cylinder of hydrogen gas in equilibrium at 300. K. (b) If the cylinder contains pure oxygen at this temperature what is the root-mean-square speed of these molecules?

6. The surface of the sun consists mainly of hydrogen atoms. What is the root-mean-square speed of the hydrogen atoms if the temperature at the sun's surface is 6.00×10^3 K?

7. Compute the total kinetic energy of 1.00 mol of nitrogen molecules at 300. K.

8. At what temperature do the atoms in a container filled with helium gas have a root-mean-square speed of 200. m/s?

9. What is the average translational kinetic energy of water molecules in water vapor in thermal equilibrium at 100.°C? What is the root-mean-square speed of these molecules?

10. A 10.0-L vessel contains 2.00 mol of an ideal gas at 1.50 atm of pressure. What is the average translational kinetic energy of a single molecule?

11. A 3.00-L vessel contains an ideal gas at 0.500 atm of pressure at a temperature of 84.0°C. What is the average translational kinetic energy of a single molecule?

21.3 Another Look at Specific Heats

12. Compute c_v for an ideal gas in units of calories per mole per degree Celsius and compare it with that of water.

13. An insulated vessel contains 1.00 mol of helium at 10.0°C on one side of a removable partition and 1.00 mol of oxygen at 90.0°C on the other side, both at 1.00 atm of pressure. What is the final temperature when the partition is removed and the mixture of gases is allowed to come to equilibrium?

14. (a) How many moles of helium are necessary so that the helium gas has the same temperature change as 1.00 mol of nitrogen when the same amount of heat is added to each gas during a constant-volume process? (b) How many moles of helium are required if the process is carried out at constant pressure? Assume no heat is lost to the surroundings.

15. Two identical cylinders with movable pistons are both at 20.0°C. One cylinder contains 1.00 mol of argon and the other is filled with 1.00 mol of hydrogen. Each is initially at 1.00 atm of pressure. If the volumes of both gases are increased adiabatically to twice their initial value, what is the change in temperature of each gas?

16. A cylinder with a movable piston contains 1.00 mol of a monatomic ideal gas at 1.50 atm of pressure. Suppose 750. J of heat are slowly added to the gas with an initial volume of 1.00 L. As its volume increases the gas pressure remains constant. (a) What is the final temperature of the gas? (b) What is the final volume of the gas?

17. A 2.00-L vessel contains 6.00×10^{-2} mol of an ideal gas consisting of rigid diatomic molecules. The temperature of the gas is 200. K. What is the final pressure when 45.0 cal of heat are absorbed by the gas?

18. An ideal gas consists of 4.00×10^{-2} mol of rigid diatomic molecules at a temperature of 77.0°C. The gas is enclosed in a container of fixed volume at a pressure of 1.50 atm. How much heat must be removed from the gas to reduce its pressure to 0.500 atm?

19. (a) Convert the specific heats of the metals listed in Table 18.4 into molecular specific heats in units of joules per mole per kelvin and then write each in the form aR, where R is the ideal-gas constant and a is a number. (b) Compare these values to the predictions of the classical equipartition theorem.

21.4 Distribution of Molecular Speeds

20. An experimental arrangement used to measure the distribution function of molecular speeds is shown in Figure 21.P1. The disks rotate with an angular velocity ω and are separated by a length d. (a) If the angular off-set between the centers of the two slots is θ, show that molecules with velocity $v = \omega d/\theta$ reach the detector. (b) Suppose $d = 0.500$ m, the angular velocity $\omega = 200.$ rev/s, and $\theta = 60.0°$. What average speed is observed at the detector?

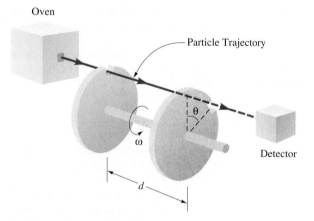

FIGURE 21.P1 Problem 20

21. Show that the Maxwell distribution function satisfies the condition

$$N = \int_0^\infty N_v(v)\, dv$$

[*Hint:* This is not an elementary integral. You will need to transform the integral into a form, such as $\int_0^\infty x^{2m} e^{-ax^2} dx$, and look it up in a good integral table or use symbolic mathematics software, such as Mathematica or Maple.]

22. Verify by integration that

$$v_{av} = \int_0^\infty v N_v(v)\, dv = 1.60 \sqrt{\frac{k_B T}{m}}$$

and

$$v_{rms} = \sqrt{\int_0^\infty v^2 N_v(v)\, dv} = 1.77 \sqrt{\frac{k_B T}{m}}$$

[*Hint:* These are not elementary integrals. You will need to transform them into forms, such as $\int_0^\infty x^m e^{-ax^2} dx$, and look them up in a good integral table or use symbolic mathematics software, such as Mathematica or Maple.]

23. Show that the most probable velocity of the Maxwell distribution is

$$v_{prob} = 1.41 \sqrt{\frac{k_B T}{m}}$$

24. Find the mean free path of air molecules at 300. K and 1.00 atm of pressure by assuming that air is composed of nitrogen molecules with a diameter of 1.80×10^{-10} m.

25. It has been estimated that in "empty" space there is about one hydrogen molecule per cubic meter. Assume the diameter of a hydrogen molecule is 1.00×10^{-10} m and find its mean free path in empty space.

Numerical Methods and Computer Applications

26. (a) Write a computer program to compute $N_v(v)$ for several molecular speeds within a specified range of speeds from the Maxwell distribution function. Input should consist of the minimum and maximum speeds, the speed increment, the molecular mass, and kelvin temperature. (b) Modify the program to provide a graph of the distribution function (either on screen or in printed format).

27. Modify the spreadsheet MAXWLDIS.WK1 on the disk accompanying this text so that graphs of $N_v(v)$ are generated for gases with two different masses, both at the same temperature.

28. (a) Use the spreadsheet template MAXWLDIS.WK1 or a suitable computer program to determine the total fraction of molecules in a gas that have speeds greater than twice v_{rms} when the temperature is 300. K. (b) Does this fraction change if the temperature is doubled?

29. Use numerical integration to perform the integrals

$$I_0 = \int_0^\infty N_v(v)\, dv$$

$$I_1 = \int_0^\infty v N_v(v)\, dv$$

and

$$I_2 = \int_0^\infty v^2 N_v(v)\, dv$$

for 1.00 mol of helium at 300 K. Check your work using the relations described in Problems 21 and 22.

General Problems

30. (a) Compute the temperature at which helium atoms have a root-mean-square speed equal to 0.100 times the escape velocity from a height of 600. km above the earth's surface. (b) Repeat part (a) for a helium atom on the surface of the moon ($g_{moon} = 0.16 g_{earth}$).

31. What are the ratios of the root-mean-square velocities of ^{235}U and ^{238}U atoms? (Assume the mass of ^{235}U is 235 g/mol and ^{238}U is 238 g/mol. This difference is the basis for the gaseous diffusion method of separating these two isotopes.)

32. Neutrons with average kinetic energies equal to that of a gas molecule at about 300. K are called **thermal neutrons.** What is the root-mean-square velocity of a thermal neutron? (The neutron mass is 1.67×10^{-27} kg.)

33. The erratic motion of very small (but still macroscopic) objects, such as smoke particles in air or pollen in water, is called **Brownian motion.** (It was discovered by an English botanist Robert Brown in 1827.) This behavior is due to statistical fluctuations in the bombardment of the particles by gas or water molecules. If a smoke particle is observed to have a root-mean-square speed of 5.00 mm/s when in thermal equilibrium with a gas at 300. K, what is the particle's mass?

34. From Equation 20.6 show that the pressure of an ideal gas is given by $P = \frac{1}{3}\rho v_{rms}^2$, where ρ is the mass density of the gas. Use this result to compute the pressure exerted by 10.0 g of helium at 300. K in a cubical container 1.00 m^2 on a side.

35. A gas is enclosed in a cylindrical container with a movable piston. The initial 4.00-L volume contains 0.500 mol of the gas. (a) If 480. cal of heat must be added to raise the temperature of the gas from 80.0 K to 273. K at a constant pressure, do you think the gas is monatomic or diatomic? Justify your answer. (b) What is the final volume of the gas? (c) By how much did the internal energy change during the heat-absorption process?

36. Show that Equation 20.24 for the mean free path may be written in the form

$$\ell = \frac{k_B T}{4\pi\sqrt{2}d^2 P}$$

where T and P are the gas temperature and pressure, respectively.

37. In order to collide with the phosphors and produce a bright spot on the screen, a beam of electrons must travel the length of a television tube. The mean free path of the electrons must, therefore, be at least as long as the length of the tube. Compute the pressure of the gas within the tube that results in a mean free path of 0.500 m. Assume the temperature of the gas is 300. K and that a collision occurs if an electron comes within 3.00 nm from the center of a molecule.

ELECTRICITY AND MAGNETISM

By now you are familiar with the way physicists like to view nature: We prefer to begin with a few fundamental principles that summarize the essence of the outcome of many experiments. From these basic ideas we build models to describe more complex phenomena. This adventure has evolved into one in which we try to find as few principles as possible from which we can explain all our observations.

Physicists are particularly happy when the fundamental principles of a theory can be expressed cleanly and compactly in the form of mathematical relations. The theory of electricity and magnetism is one in which physicists take special delight. You see, they have this whole branch of physics condensed into just four equations! They are called *Maxwell's equations,* and they are the fruit of experiments performed over many years by many scientists. James Clerk Maxwell was one of these scientists and the first to write down these equations in their complete form. Maxwell's equations are so robust that all classical electromagnetic phenomena are described by them. Additional restrictions are required for phenomena that involve processes that take place on an atomic scale. In order to apply Maxwell's equations, however, you must clearly understand the meanings of each term and symbol within them! In fact, it is really a good idea to build up to Maxwell's equations by working through some of the evidence for their existence. This approach will help us to establish a mathematical framework for the statement of each of these equations and also give us some experience with the mathematical machinery necessary for their application.

It will be one of our objectives as we journey through the following chapters to provide the evidence for each of Maxwell's equations. At the proper times, when we have established enough background, we will state each equation and show how it works.

Our first step toward Maxwell's marvelous equations must necessarily be the introduction of a property of matter that we have not previously mentioned. Through many experiments, scientists have found that matter acts in such a way that is most efficiently described by associating with it a property called charge. In Chapter 22 we will tell you what we have come to believe about this fundamental quantity, how it is related to some of the more basic constituents of matter, and how its effects can be conveniently described by the field concepts we have already found useful for gravitational phenomena. In Chapters 22 through 25 we will describe what is called *electrostatics,* that is, the features of electrical phenomena for which these fields are generated by charges that are fixed in position. We consider steady currents and the circuits they constitute in Chapters 26 and 27. Magnetism is described in Chapters 28 through 32, and we end this section of the text with introductions to alternating current circuits (Chapter 33) and electromagnetic waves (Chapter 34).

CHAPTER

22

Stationary Charges and the Electric Field

In this chapter you should learn

- to apply Coulomb's law to point charges and continuous charge distributions.
- to compute the electric field from various, high-symmetry charge distributions.
- about the electric dipole.
- to compute the motion of charges in a uniform electric field.

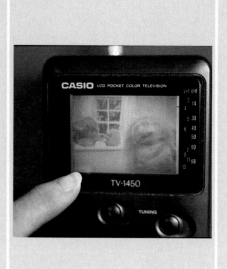

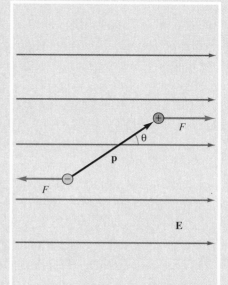

$$\tau = p \times E$$

$$U = -p \cdot E$$

For some people, the first introduction to electricity comes as a shock. You may have had the experience of walking across a rug and then casually reaching for the doorknob, only to be greeted by a painful spark that jumps between your hand and the doorknob. Other encounters are more benign. We have all had the experience of running a comb through our hair and finding that the hair is attracted to the comb instead of lying flat where we had hoped to put it. And who among us can resist the temptation to rub an air-filled balloon across a sleeve to see if it sticks to the wall? These are familiar phenomena associated with what is known as **static electricity.**

Static electricity is sometimes a nuisance, although the multimillion-dollar industry that produces ''fabric softeners'' for the laundry dryer to reduce ''static cling'' might not think so. At other times this same electrostatic attraction is exploited for commercial applications. For example, electrostatic attraction is at the heart of the familiar copy machine found in virtually every office. This same force, which attracts dust particles to the screen of our TV set, is used in the smokestack ''scrubbers'' of many industrial plants to remove fine particulates from the smoke plume.

On a more fundamental level, the strength of the electric force determines the size of atoms. Electric forces determine whether a substance is a solid, liquid, or a gas at a given temperature. These forces also govern membrane transport and nerve impulses in biological systems. Electric forces are used to accelerate elementary particles to velocities close to the speed of light, providing high-energy projectiles to hurl at other particles in the attempt to unravel the fundamental building blocks of the universe.

Because of the wide range of phenomena governed by electric forces, an understanding of them is of great importance to every scientist and engineer. We begin this exciting study with the notion of charge.

22.1 Electric Charge

The study of the physical property we now call *charge* began at least several hundred years BC when people noticed that resinous materials, when rubbed with cloth, attracted feathers or pieces of straw. Since that time many scientists have formulated theories about the origin of this phenomenon. Some early models suggested that electricity was a fluid. It was thought that when a resinous material and cloth are rubbed together, the electric fluid was transferred, leaving one material with an excess of the fluid and the other with a deficit. A rival model suggested that there were two fluids that were evenly mixed when an object was electrically neutral. Today's charge model is closer to the two-fluid theory, but different in two important aspects. First, electricity is not a substance in and of itself but rather an attribute called **charge** that can be assigned to some kinds of fundamental particles. Second, this attribute is not fluidlike, because there appears to be graininess to it, that is, a smallest size in which the attribute is distributed.

The American scientist Benjamin Franklin made important studies of electricity, and in addition to flying his famous kite, he introduced the plus (+) and minus (−) notation we use today for the two types of charge. We might mention here that the + and − notation is a bit of a holdover from the one–fluid model. There is nothing intrinsically different in the two types of charge, making one ''better'' than the other. Franklin could equally well have called the two types of charge red and blue or sweet and sour. Franklin is also credited with the statement of charge conservation we make below. We will not take time to mention any more of the fascinating history of electricity. Rather, we content ourselves with a summary of the properties of charge and its place in the atomic model.

First of all, there are two types of charge, which are called **positive** and **negative.** Only two types of charge are necessary to explain the observed forces between charged bits of matter. (In Problem 1 we suggest the kinds of observations that would make it necessary to ''invent'' a third kind of charge.) One aspect of the force between charged particles is easy to state:

Unlike charges attract each other, and like charges repel each other.

That is, a positive charge attracts a negative charge, and at the same time a negative charge attracts a positive charge. However, two positive charges repel each other as do two negative charges.

Charge also comes only in discrete bundles and, therefore, is said to be **quantized.** It is a long-standing tradition to label the magnitude of the smallest observed charge with the letter e. If q is any quantity of charge, the charge quantization condition may be written

$$q = \pm ne, \qquad n = 0, 1, 2, 3, 4, \ldots$$

You are probably familiar with all these properties of charge, but bear with us. We want to present a complete and coherent picture. While we are reviewing, let's summarize the main features of the atom. In this model, the center of the atom is called the nucleus and it contains **protons,** each of charge $q = +e$, and neutral particles ($q = 0$) called **neutrons.** Much less massive **electrons** with charge $q = -e$ move about the nucleus in patterns that depend on the electron's energy. In Chapter 40 we will describe this model in more detail. For now, we note that our present microscopic model of ordinary matter asserts that all solids, liquids, and gases are composed of atoms. An atom of any element consists of equal numbers of electrons and protons and hence has zero net charge. The atoms of different elements contain different numbers of electrons and protons, the simplest being hydrogen with one proton as a nucleus and one electron traveling about it. All other atoms also contain neutrons in the nucleus, but the neutrons do not upset the charge neutrality of the atom.

Scanning tunneling microscope image, 7×7 nm, of a Cs zig-zag chain (red) on the GaAs(110) surface (blue) recorded at a sample bias of -3 V. From L. J. Whitman, J. A. Stroscio, R. A. Dragoset, and R. J. Celotta, Phys. Rev. Lett. **66,** 1338 (1991).

You may be wondering how the magnitude of e, the smallest charge bundle, compares with the size of electric charges you may have already encountered—for example, when rubbing a balloon and sticking it to the wall. The answer is that e is very small by these standards. Billions of elementary charges are involved when electrostatic forces are large enough for us to notice on a macroscopic scale. The standard SI unit of charge turns out to represent an enormous quantity of charge. In SI the unit of charge is the **coulomb** (C), named after the discoverer of the force law between two charges. For now we take a pragmatic approach and simply state that, to seven significant figures,

$$e = 1.602\ 189 \times 10^{-19}\ \text{C}$$

In Chapter 28 we will describe an operational definition of the coulomb. We promise that in Chapter 29 we will provide you with a precise prescription from which e can be clearly defined.

Table 22.1 provides the charge and mass of the three basic particles that are the building blocks of the atoms of ordinary matter. Physicists have discovered many fundamental particles, and the list continues to grow. One feature of each of the particles discovered to date is that they have only three possible values of charge: $-e$, 0, or $+e$. For our purposes we need not consider other properties of these particles. However, it is interesting to note that one popular model of elementary particles, called the *standard model*, proposes particles called **quarks,** which have charges of $\frac{1}{3}e$ and $\frac{2}{3}e$. To date, no fractional charge of e has been unambiguously detected, and, in fact, some quark models suggest we may never be able to see these fractional charges even if they "exist."

TABLE 22.1   **Charge and Mass of the Proton, Electron, and Neutron**

PARTICLE	CHARGE	MASS
Proton	$+e$	$1.672\ 623 \times 10^{-27}\ \text{kg}$
Neutron	0	$1.674\ 929 \times 10^{-27}\ \text{kg}$
Electron	$-e$	$9.109\ 390 \times 10^{-31}\ \text{kg}$

In the following chapters we will have much to say about the electrical properties of matter. For now, however, it is both convenient and sufficient to qualitatively define three classes of materials: conductors, insulators, and semiconductors. In a **conductor** most of the electrons are bound to the atoms that compose the material. However, there is about one electron per atom that is essentially free to travel within the conductor, occasionally colliding with impurities and irregularities in the conductor. In fact, within very pure metals these electrons can often travel hundreds of atomic spacings before colliding with imperfections. For this reason they are often called **free electrons.** Metals, electrolytes (liquids with dissociated ions), and moist earth are examples of conductors.

Virtually all of the electrons within a material classified as an **insulator** (also called a **dielectric**) are bound to the atoms of this type of material and are not free to move around. Glass, plastics, and rubber are good insulators. Most gases are insulators unless the atoms of the gas become ionized, for example, by high temperatures or large electric forces. These ionized gases, known as plasmas, may be highly conducting. Intermediate between insulators and conductors are **semiconductors.** At room temperature these materials, such as silicon and germanium, normally have only about one mobile charge per 10^{12} atoms. By adding the proper impurity atoms to these semiconductors, a process called doping, the number of mobile charge carriers can be raised to levels as high as one per 10 000 atoms.

At the surface of a substance the electrical properties may be different from those well within the material. The actual mechanism by which electrons can be removed from the surface of one insulator when it is in contact with another is not well understood. However, we do know that the surfaces of two different types of insulators bind their electrons with different strengths. When the two materials are brought into contact, electrons from

Concept Question 1
A ribbon of light aluminum foil is draped over a pencil so that equal lengths hang off each side. When a charged object is brought near the foil where it passes over the pencil, the two ends of the foil separate apart. Explain this phenomenon. What happens when the charged object touches the foil and is then removed?

Concept Question 2
An uncharged Ping-Pong ball is coated with a thin layer of metal and suspended on a string (like a pendulum) in the narrow region between two metallic and oppositely charged plates. If the ball accidentally touches the positively charged plate for an instant, describe the subsequent motion of the Ping-Pong ball.

the material with weaker surface bonds are transferred to the material with the stronger surface bonds. Although not essential, rubbing the materials together may increase the charge transfer by increasing the surface area in contact. The material that wins the tug-of-war for electrons acquires a net negative charge, and the material from which the electrons are removed is left with a deficiency of electrons, and, therefore, is positive. Hence, when we say an object is negatively charged, we mean it has an excess of electrons. If the object is positively charged, it has a deficiency of electrons.

Continuous Charge Distributions

We have already mentioned that charge is quantized. Let's now compute the number of fundamental particles (electrons or protons) that carry a total charge magnitude of 1.00 C. Because the *magnitude* of the charge of an electron or proton is 1.60×10^{-19} C, the number of electrons or protons per coulomb is

$$\frac{1}{1.60 \times 10^{-19} \text{ C/particle}} = 6.25 \times 10^{18} \text{ particles/C}$$

Although this number of particles may seem enormous, it is, in fact, roughly the number of electrons that flows through a 100-W light bulb in only one second! Even one microcoulomb (1 μC = 10^{-6} C) contains 6.25×10^{12} charge carriers. If we were to take 1 μC of charge and spread it out uniformly over the length of 1 km, the distance between adjacent charges would be only 1.60×10^{-7} mm! Consequently, we often ignore the discrete nature of charge, and model charge distributions as if they were completely continuous. (Shades of the fluid model!) This approach works well as long as the distances in which we are interested are large compared to the distances between individual charged particles.

Although there are an infinite number of ways we can spread a continuous charge distribution over a region of space, there are three means by which we can describe all types of distributions. Charges may be spread out along a line (curved or straight) like a row of dominoes, over a surface like new paint on a car, or throughout a volume like chalk dust particles in air when two chalk erasers are clapped together. With these possibilities in mind we define three different **charge densities:**

SYMBOL	DEFINITION	SI UNITS
λ =	charge per length	C/m
σ =	charge per area	C/m^2
ρ =	charge per volume	C/m^3

If a total charge q is distributed *uniformly* along a curve of length l, over a surface area A, or throughout a volume $\mathscr{V}$, we can calculate these densities from

$$\lambda = \frac{q}{l}$$

$$\sigma = \frac{q}{A}$$

$$\rho = \frac{q}{\mathscr{V}}$$

In general, λ, σ, and ρ may not be uniform but instead change as a function of position within the charge distribution. In either case, the amount of charge dq that is within a small element dl, dA, or $d\mathscr{V}$ is

$$dq = \lambda \, dl \qquad \text{(Charge distributed along a curve)} \qquad (22.1a)$$

$$dq = \sigma \, dA \qquad \text{(Charge distributed over a surface)} \qquad (22.1b)$$

$$dq = \rho \, d\mathcal{V} \qquad \text{(Charge distributed throughout a volume)} \qquad (22.1c)$$

In Examples 22.1 and 22.2 we illustrate how two of these charge distributions may be used.

EXAMPLE 22.1   *A Nonuniform Linear Charge Distribution*

Charge is distributed along the x-axis from $x = 0.0$ to $x = L = 50.0$ cm in such a way that its linear charge density is given by

$$\lambda = ax^2$$

where $a = (18.0 \; \mu C/m^3)$. Compute the total charge in the region $0 \le x \le L$.

SOLUTION   We apply Equation (22.1a) where $dl = dx$ because, as shown in Figure 22.1, the charge lies along the x-axis:

$$dq = \lambda \, dx \qquad \text{or} \qquad q = \int_0^L \lambda \, dx$$

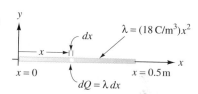

FIGURE 22.1

A nonuniform, linear charge distribution. See Example 22.1.

Substituting for λ and integrating we find

$$q = \int_0^L ax^2 \, dx = \left. \frac{ax^3}{3} \right|_0^L = \frac{aL^3}{3} = \frac{(18.0 \; \mu C/m^3)(0.500 \; m)^3}{3} = 0.750 \; \mu C \qquad \blacktriangleleft$$

EXAMPLE 22.2   *A Nonuniform Volume Charge Distribution*

Charge is distributed throughout a spherical region of space in such a manner that its volume charge density is given by

$$\rho = br, \qquad 0 \le r \le R$$

where $b = (4.00 \; \mu C/m^4)$ and $R = 0.500$ m. Find the total charge within the sphere.

SOLUTION   For a three-dimensional charge distribution we must apply Equation (22.1c):

$$dq = \rho \, d\mathcal{V} \qquad \text{or} \qquad q = \int \rho \, d\mathcal{V}$$

With ρ given, we have only to write $d\mathcal{V}$, a small volume element within the sphere. We can save some effort if we are clever when we write $d\mathcal{V}$. First, the charge density ρ depends only on r, the distance from the center of the sphere. Therefore, we choose a spherical coordinate system and place our origin at the center of the sphere. Moreover, because the charge distribution is spherically symmetric, ρ is approximately constant within a thin spherical shell of radius r and thickness dr. Thus, we can think of the volume of the sphere as made up of concentric shells like the layers of an onion (Fig. 22.2). Because the area of an entire sphere is $4\pi r^2$, we can take our volume element to be

$$d\mathcal{V} = 4\pi r^2 \, dr$$

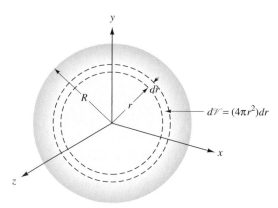

FIGURE 22.2

The spherical shell volume element. See Example 22.2.

and write

$$q = \int \rho \, d\mathcal{V} = \int_0^R \rho (4\pi r^2 \, dr)$$

Substituting for ρ, we have

$$q = \int_0^R br(4\pi r^2 \, dr) = 4\pi b \int_0^R r^3 \, dr = 4\pi b \frac{r^4}{4}\Big|_0^R = \pi(4.00 \ \mu C/m^4)(0.500 \ m)^4$$

$$= 0.785 \ \mu C$$

◀

22.2 Coulomb's Law

The first quantitative studies of the electrostatic force were performed by Charles Coulomb (1736–1806). In 1785, he invented the torsion balance shown in Figure 22.3. The primary component of this device is a beam suspended horizontally from a vertical fiber. On one end of the beam is a sphere that can be charged by bringing it into contact with a charged object (perhaps a rubbed piece of glass or rubber). The beam is balanced by a counterweight on the other end. When a second charged sphere is brought near the first as shown in Figure 22.3, the electrostatic force causes the beam to rotate. This device can be calibrated independently of the Coulomb force so that the force necessary to cause a given rotation is known. With this device Coulomb was able to firmly establish the following properties for the electrostatic force between stationary charges:

1. The magnitude of the force is proportional to the product of the magnitudes of the charges q_1 and q_2.
2. The magnitude of the force is inversely proportional to the square of the distance separating the charges and is directed along the line connecting the two charges.
3. The force is repulsive if the charges have the same sign and attractive if their signs are opposite.

Because there are two charges present in each of these observations, it is helpful to adopt the "on . . . by" subscript notation first introduced in Chapter 5: F_{12} is the magnitude of the force *on* charge q_1 *by* charge q_2. Mathematically, Coulomb's law can then be written

$$F_{12} = k \frac{q_1 q_2}{r^2}$$

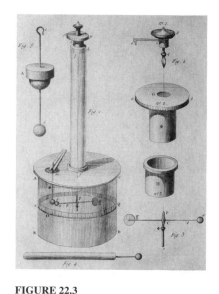

FIGURE 22.3

Coulomb's torsion balance.

where the proportionality constant k is called **Coulomb's constant.** In SI,

$$k = 8.9875 \times 10^9 \ N \cdot m^2/C^2$$

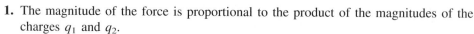
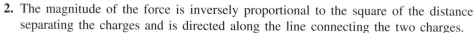

The constant k is often written in terms of another constant ϵ_o:

$$k = \frac{1}{4\pi\epsilon_o}$$

This new constant ϵ_o is called the **permittivity of free space,** and its value is

$$\epsilon_o = 8.8542 \times 10^{-12} \frac{\text{C}}{\text{N} \cdot \text{m}^2}$$

For now, the reason anyone might want to introduce this second constant remains as mysterious as its name. Although it appears to make Coulomb's law look more complicated, it eventually has the effect of making other equations (Maxwell's, in particular!) appear less complex. We'll get around to explaining the meaning of "permittivity of free space" in Chapter 25. For now, however, we often shorten the name of ϵ_o to the **permittivity constant.**

With this relation between k and ϵ_o, Coulomb's law can be written

$$F_{12} = \frac{1}{4\pi\epsilon_o} \frac{q_1 q_2}{r^2} \qquad (22.2)$$

You must be careful with the sign of F_{12} and the direction of this force. Notice when the charges are of the same sign, F_{12} is positive, and the charge q_1 is repelled by charge q_2. When the charges are of opposite signs, F_{12} is negative. We interpret this negative value to mean that the force is attractive; that is, it is directed so as to draw the charge q_1 toward q_2. Now, for some important advice that you should read carefully:

When you draw a figure that includes the Coulomb force vector, never, never, label F_{12} (or F_{21}) with a minus sign even if F_{12} is a negative number.

A minus sign that results from Equation (22.2) tells you that the force F_{12} is attractive. Use this knowledge to orient the vector $\mathbf{F}_{12}$ of your figure in the correct direction. Then use the *magnitude* of F_{12} (with units) to label this vector. If you put a minus sign with this label, we can almost guarantee that you are going to become confused about its direction.

The absolute value of Equation (22.2) provides the magnitude of the Coulomb force. As we have just stated, the sign merely indicates whether the charges are attracted toward each other or repelled away from each other. It is possible to remove any ambiguity about the direction of F_{12}. All that we need to do is remember that force is a vector and define a *unit vector* $\hat{\mathbf{r}}_{12}$ the magnitude of which is 1 and the direction of which is *toward q_1 from q_2*. In this case

$$\mathbf{F}_{12} = \frac{1}{4\pi\epsilon_o} \frac{q_1 q_2}{r^2} \hat{\mathbf{r}}_{12} \qquad (22.3)$$

This form of Coulomb's law is illustrated in Figure 22.4 for three different point-charge combinations and orientations. If you apply Equation (22.3) carefully and recognize that $\hat{\mathbf{r}}_{12} = -\hat{\mathbf{r}}_{21}$, you will always obtain the correct direction for $\mathbf{F}_{12}$ and $\mathbf{F}_{21}$. Notice that Coulomb's law obeys Newton's third law. That is,

$$\mathbf{F}_{12} = \frac{1}{4\pi\epsilon_o} \frac{q_1 q_2}{r^2} \hat{\mathbf{r}}_{12} = \frac{1}{4\pi\epsilon_o} \frac{q_2 q_1}{r^2} (-\hat{\mathbf{r}}_{21}) = -\frac{1}{4\pi\epsilon_o} \frac{q_2 q_1}{r^2} \hat{\mathbf{r}}_{21} = -\mathbf{F_{21}}$$

What happens when more than two charges are present? By experimenting, it has been discovered that the net force on one of the charges is the vector sum of the individual

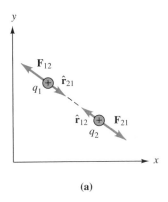

(a)

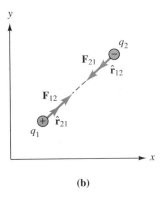

(b)

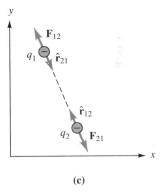

(c)

FIGURE 22.4

The vector form of Coulomb's law applied to three different pairs of point-charges. In each case $|\mathbf{F}_{12}| = |\mathbf{F}_{21}| = (1/4\pi\epsilon_o)(|q_1 q_2|/r^2)$.

Subscript convention:

$$F_{\text{on,by}}$$

$$r_{\text{to,from}}$$

Coulomb forces. For N charges, the total force on the charge labeled 1 due to the other $N - 1$ charges is

$$\mathbf{F}_1 = \mathbf{F}_{12} + \mathbf{F}_{13} + \mathbf{F}_{14} + \cdots + \mathbf{F}_{1N} = \sum_{i=2}^{N} \mathbf{F}_{1i}$$

This result is called the **principle of superposition.** How to apply Coulomb's force law for two point-charges is illustrated in Example 22.3, and the principle of superposition is used in Example 22.4.

EXAMPLE 22.3 *The Hydrogen Atom*

Compute the attractive force of the proton on the electron in the hydrogen atom. Compare the magnitude of this force with that of the gravitational force between these two particles. The distance between the two particles is 5.29×10^{-11} m.

SOLUTION The charges, forces, and instantaneous positions are shown in Figure 22.5. The proton has charge $q_1 = +e$, and the electron has charge $q_2 = -e$. Therefore, the force *on* the electron *by* the proton is $\mathbf{F}_{21}$:

$$\mathbf{F}_{21} = \frac{1}{4\pi\epsilon_o} \frac{(+e)(-e)}{r^2} \hat{\mathbf{r}}_{21} = -\frac{1}{4\pi\epsilon_o} \frac{e^2}{r^2} \hat{\mathbf{r}}_{21}$$

$$= -\frac{1}{4\pi\epsilon_o} \frac{e^2}{r^2}(-\hat{\mathbf{r}}_{12})$$

$$= +\frac{1}{4\pi\epsilon_o} \frac{e^2}{r^2} \hat{\mathbf{r}}_{12}$$

The unit vector $\hat{\mathbf{r}}_{12}$ points to the left (toward q_1 from q_2), so the force $\mathbf{F}_{21}$ is directed to the left, and its magnitude is

$$F_{21} = \frac{1}{4\pi\epsilon_o} \frac{e^2}{r^2} = \frac{1}{4\pi[8.854 \times 10^{-12} \text{ C}^2/(\text{N} \cdot \text{m}^2)]} \cdot \frac{(1.60 \times 10^{-19} \text{ C})^2}{(5.29 \times 10^{-11} \text{ m})^2}$$

$$= 8.22 \times 10^{-8} \text{ N}$$

The magnitude of the gravitational force is

$$F_{G_{21}} = G\frac{M_e M_p}{r^2} = (6.67 \times 10^{-11} \text{ N} \cdot \text{m}^2/\text{kg}^2)\frac{(1.67 \times 10^{-27} \text{ kg})(9.11 \times 10^{-31} \text{ kg})}{(5.29 \times 10^{-11} \text{ m})^2}$$

$$= 3.63 \times 10^{-47} \text{ N}$$

Clearly, the electrostatic force is much stronger than the gravitational force. In fact, the ratio of the electrostatic force to the gravitational force $F_{21}/F_{G_{21}}$ is about 2×10^{39}. ◀

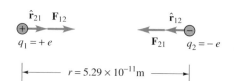

FIGURE 22.5

The hydrogen atom.

Concept Question 3
In what ways are Newton's universal law of gravitation and Coulomb's law similar? In what ways are they different?

EXAMPLE 22.4 *An Electrifying Review of Vector Addition*

Three charges $q_1 = -e$, $q_2 = +2e$, and $q_3 = +e$ are fixed in the xy-plane at positions $(0, a)$, $(0, -a)$, and $(b, 0)$, respectively. Find the net electrostatic force on q_3.

SOLUTION The charges and the resulting electrostatic forces $\mathbf{F}_{31}$ and $\mathbf{F}_{32}$ are shown in Figure 22.6. Application of Equation (22.2) gives

$$F_{31} = \frac{1}{4\pi\epsilon_o} \frac{q_3 q_1}{R^2} = -\frac{1}{4\pi\epsilon_o} \frac{e^2}{R^2} = -\frac{1}{4\pi\epsilon_o} \frac{e^2}{(a^2 + b^2)}$$

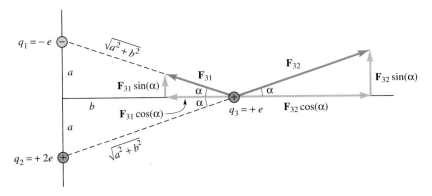

FIGURE 22.6

Computing the net force on q_3 by q_1 and q_2. See Example 22.4.

and

$$F_{32} = \frac{1}{4\pi\epsilon_o} \frac{q_3 q_2}{R^2} = \frac{1}{4\pi\epsilon_o} \frac{2e^2}{R^2} = \frac{1}{4\pi\epsilon_o} \frac{2e^2}{(a^2 + b^2)}$$

We must be careful to realize that the minus sign out in front of the expression for F_{31} means that q_3 is attracted toward q_1, as shown in Figure 22.6. The magnitude of the vector F_{31} is the absolute value of this expression. We cannot emphasize too strongly that the total force on q_3 is the *vector* sum of $\mathbf{F}_{31}$ and $\mathbf{F}_{32}$. We add these two forces component by component.

The x-component of the resultant force F_3 is

$$F_{3_x} = |F_{32}| \cos(\alpha) - |F_{31}| \cos(\alpha) = (|F_{32}| - |F_{31}|) \cos(\alpha)$$

From the figure we see

$$\cos(\alpha) = \frac{b}{\sqrt{a^2 + b^2}}$$

So that

$$F_{3_x} = \left(\frac{1}{4\pi\epsilon_o} \frac{2e^2}{(a^2 + b^2)} - \frac{1}{4\pi\epsilon_o} \frac{e^2}{(a^2 + b^2)} \right) \frac{b}{\sqrt{a^2 + b^2}}$$

$$F_{3_x} = \frac{1}{4\pi\epsilon_o} \frac{e^2 b}{(a^2 + b^2)^{3/2}}$$

The y-component of the resultant force is

$$F_{3_y} = |F_{31}| \sin(\alpha) + |F_{32}| \sin(\alpha) = (|F_{31}| + |F_{32}|) \sin(\alpha)$$

where $\sin(\alpha) = a/\sqrt{a^2 + b^2}$, so that

$$F_{3_y} = \left(\frac{1}{4\pi\epsilon_o} \frac{e^2}{(a^2 + b^2)} + \frac{1}{4\pi\epsilon_o} \frac{2e^2}{(a^2 + b^2)} \right) \frac{a}{\sqrt{a^2 + b^2}}$$

$$= \frac{1}{4\pi\epsilon_o} \frac{3e^2 a}{(a^2 + b^2)^{3/2}}$$

The magnitude of the resultant force is

$$F_3 = \sqrt{(F_{3_x})^2 + (F_{3_y})^2}$$

Substituting the expressions previously found for each component, we find after simplifying

$$F_3 = \frac{e^2}{4\pi\epsilon_o} \frac{\sqrt{9a^2 + b^2}}{(a^2 + b^2)^{3/2}}$$

The resultant force is directed along an angle with the x-axis given by

$$\theta = \arctan\left(\frac{F_{3_y}}{F_{3_x}}\right) = \arctan\left(\frac{3a}{b}\right)$$

◀

EXAMPLE 22.5   *Enough to Make Your Hair Stand on End!*

A linear charge distribution $\lambda = +0.400\ \mu C/m$ lies along a straight line 10.0 cm long. A point-charge $q_o = +0.200\ \mu C$ is placed a distance $d = 15.0$ cm from the near end of the charge distribution and along the same line. Find the force of the charge distribution on charge q_o.

SOLUTION   We place the x-axis along the line of charge as shown in Figure 22.7. Because both the charge distribution and q_o are positive, the direction of the force on q_o is toward the right. The magnitude of the force ΔF from a small charge Δq within the distribution is given by the magnitude of Equation (22.2):

$$\Delta F = \frac{1}{4\pi\epsilon_o} \frac{q_o\,\Delta q}{r^2}$$

If we make the charge infinitesimally small $\Delta q \to dq$ and we have

$$dF = \frac{1}{4\pi\epsilon_o} \frac{q_o\,dq}{r^2}$$

To obtain the total force F we sum the dF contributions from all the charge distributed along the line of charge. That is, we integrate over dq:

$$F = \int dF = \frac{q_o}{4\pi\epsilon_o} \int \frac{dq}{r^2}$$

Notice that we removed all constants from the integral. The distance r is not a constant because it is the distance from dq to q_o and, therefore, depends on the location of dq. Now

$$dq = \lambda\,dx$$

and to include the entire line we must integrate x from $x = 0$ to $x = L = 10$ cm. From Figure 22.7 we see $r = d + L - x$.

This Van de Graaff generator can produce an electric potential difference of as much as 100 000 V. Is this young woman scared, or is there another explanation for her hairdo?

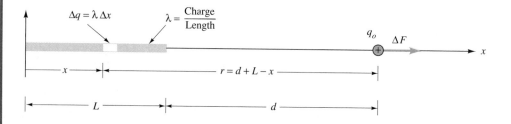

FIGURE 22.7

The x-axis lies along the line of the charge.

$$F = \frac{q_o}{4\pi\epsilon_o} \int_0^L \frac{\lambda \, dx}{(d + L - x)^2}$$

If we make the substitution $u = d + L - x$, $du = -dx$, and are careful to change the integration limits in accord with the substitution, we have

$$F = -\frac{q_o\lambda}{4\pi\epsilon_o} \int_{d+L}^d \frac{du}{u^2} = +\frac{q_o\lambda}{4\pi\epsilon_o}\left(\frac{1}{u}\right)_{d+L}^d = +\frac{q_o\lambda}{4\pi\epsilon_o}\left(\frac{1}{d} - \frac{1}{d+L}\right)$$

$$F = \frac{(0.200 \times 10^{-6}\ \text{C})(0.400 \times 10^{-6}\ \text{C/m})}{4\pi[8.854 \times 10^{-12}\ \text{C}^2/(\text{N}\cdot\text{m}^2)]}\left(\frac{1}{0.150\ \text{m}} - \frac{1}{0.250\ \text{m}}\right)$$

$$= 1.92 \times 10^{-3}\ \text{N} = 1.92\ \text{mN}$$

The magnitude of the force of the point–charge on the line of charge is also F. This example could serve as a model for the electrostatic interaction of a brush (the point) attracting a charged strand of hair. The model, although admittedly crude, is accurate enough to indicate the order of magnitude of static charges we encounter in such cases.   ◀

22.3 The Electric Field

The concept of a field has been found to be quite useful in physics and engineering. In Chapter 5 (Section 3) we first encountered the idea of a gravitational field; g was defined as the force *per unit mass* exerted by the earth on an object. There are other circumstances in nature that can be conveniently described with this concept, including the effects of electric and magnetic forces. But it doesn't stop there. Scientists and engineers have employed the field concept to describe many other phenomena, including the strains in materials and even the motion of fluids.

Before we begin to explain the idea of an electric field we think you should take a quiz. You can never get enough practice before a physics exam, so take out a sheet of scrap paper and turn on your calculator. We're not fooling, get with it! Here's the problem: A charge of 8.00×10^{-10} C is attached to the origin of a coordinate system. (a) What is the force on a 1.00-C charge placed at $x = 2.00$ m? (We are starting out easy to help you gain confidence. But there are going to be several parts to this quiz!)

To answer this question you probably performed the following computation:

FIGURE 22.8

Pop quiz!

$$F = \frac{1}{4\pi\epsilon_o}\frac{q_1 q_2}{r^2} = \frac{1}{4\pi[8.854 \times 10^{-12}\ \text{C}^2/(\text{N}\cdot\text{m}^2)]} \cdot \frac{(8.00 \times 10^{-10}\ \text{C})(1.00\ \text{C})}{(2.00\ \text{m})^2}$$

$$= 1.80\ \text{N}$$

If you did, congratulations. You are beginning to master the material in this chapter. The force of the 8.00×10^{-10} C charge on the 1.00-C charge is 1.80 N. Now for parts (b) and

GARFIELD® by Jim Davis

GARFIELD reprinted by permission of UFS, Inc.

(c): (b) Quickly, if the 1.00-C charge is replaced by a 2.00-C charge, what is the force on this new 2.00-C charge, and (c) what is the force on a 4.00-C charge if it replaces the 2.00-C charge?

Were you able to rapidly answer parts (b) and (c) as 3.60 N and 7.20 N, respectively, *without* repeating the use of Coulomb's law? Hopefully, you realized that your first computation provided you with the force on a 1.00-C charge, so it was easy to find the force on the 2.00-C or 4.00-C charges; you simply multiply the first answer by 2 or by 4. Why does this work? It works because your first answer (1.80 N) is the *force per unit charge*. That is, the 8.00×10^{-10} C charge provides a force of 1.80 *newtons per coulomb* (N/C) on any charge placed 2.00 m away. This "would-be force" is the whole idea of the electric field! If we know the force on one unit charge at a certain point in space, we can compute the force on *any* charge by simply multiplying the force that would exist on one unit charge by the size of the actual charge. In the usual jargon, the force that would exist on one unit of charge is simply called the force *per* unit charge. Let's try to formulate a precise definition of the electric field (without any more quizzes).

Suppose way over there (down the hall and around the corner), there is a glob of charge, tacked to the bulletin board. If we place a small positive charge q_{test} (called a **test charge**) at some other position in space (perhaps here in front of you) and this test charge experiences a force **F,** the electric field **E** at the position of q_{test} is defined as

$$\mathbf{E} = \frac{\mathbf{F}}{q_{test}} \tag{22.4}$$

Notice that because q_{test} is a positive scalar, **E** must be parallel to **F**. In other words,

the electric field at a position in space is in the direction of the force on a positive test charge placed at that position.

We must be careful to apply a small enough test charge so that the charge distribution causing the electric field (the glob on the bulletin board) is not affected by the presence of q_{test}. A more proper definition is one in which we measure **F** for smaller and smaller q_{test} and define **E** as

$$\mathbf{E} = \lim_{q_{test} \to 0} \frac{\mathbf{F}}{q_{test}} \tag{22.5}$$

But charge is quantized, so we can never make q_{test} smaller than e. In practice, however, it is usually possible to choose a small enough q_{test} so that **E** can be measured to sufficient accuracy. By the way, we are never really going to ask you to compute a limit. But, we do want you to understand that the electric field as defined by Equations (22.4) and (22.5) is that of charge distributions and *not* that of the test charge q_{test}.

Point–Charge Distributions

The simplest charge distribution is that of a single point-charge $+q$. If we place a test charge q_{test} a distance r away from q, the force on q_{test} by q is, if we shorten the notation q_{test} to q_t,

$$\mathbf{F} = \frac{1}{4\pi\epsilon_o} \frac{qq_t}{r^2} \hat{\mathbf{r}}$$

Hence, the electric field produced by q at position r is $\mathbf{E} = \mathbf{F}/q_t$:

$$\mathbf{E} = \frac{1}{4\pi\epsilon_o} \frac{q}{r^2} \hat{\mathbf{r}} \tag{22.6}$$

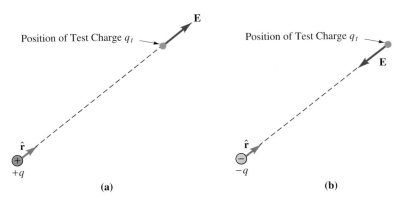

FIGURE 22.9

The electric field at point P a distance r from a point-charge. The direction of the field is determined by the direction of the force of a test charge q_t. (a) The electric field of the charge $+q_o$ points radially outward. (b) The electric field of the charge $-q_o$ points radially inward.

where the unit vector, by our convention, points toward the position of q_t from q. That is, the electric field of a positive charge q points radially outward from q, as shown in Figure 22.9(a). On the other hand, as is illustrated in Figure 22.9(b), if q is negative the field is directed inward toward q.

It's worthwhile to examine Equation (22.6) for just another moment. Notice q_t does *not* appear in this equation. The charge q_t is absent because the electric field is a property of the charge q only and has nothing whatsoever to do with the test charge q_t. By writing Equation (22.6) we are taking the view that charge q establishes about itself a field $\mathbf{E}$ that is described by this equation. If you want to find the force caused by this field on some other charge q', located at position r, use the definition of the electric field: $\mathbf{E}$ is the *force per unit charge,* so the force by q on q' is $\mathbf{F}' = q'\mathbf{E}$. (The force per unit charge times the charge equals the force.) We hope you do not miss the fact that this approach is exactly parallel to the way we think about a gravitational field. If you want to calculate the force on a mass m' by the earth of mass m, you can do so by finding the gravitational field $\mathbf{g}$ created by the earth near its surface, which is the *force per unit mass,* 9.80 N/kg. Then the force on any other mass m' placed in this field is $\mathbf{F}' = m'\mathbf{g}$.

$\mathbf{F}' = q'\mathbf{E}$ the force on a charge q' by a field $\mathbf{E}$ created by other charges

$\mathbf{F}' = m'\mathbf{g}$ the force on a mass m' by a field $\mathbf{g}$ created by other masses

Concept Question 4
An electron and a proton both experience the same electric field *E*. Which is subject to the larger force? Which experiences the larger acceleration?

To compute the electric field due to several point-charges we must apply the principle of superposition. This process is shown in Figure 22.10 where we want to find the total electric field at point P due to the charges shown. We begin by applying Equation (22.6) to find the field at P from each charge. Then, we must compute the *vector* sum of the individual field vectors. In general, for N charges

$$\mathbf{E} = \mathbf{E}_1 + \mathbf{E}_2 + \mathbf{E}_3 + \cdots + \mathbf{E}_N = \sum_{i=1}^{N} \mathbf{E}_i \qquad (22.7)$$

where

$$\mathbf{E}_i = \frac{1}{4\pi\epsilon_o} \frac{q_i}{r_i^2}\hat{\mathbf{r}}_i \qquad (22.8)$$

The distance r_i is that from the charge q_i to the point P, and $\hat{\mathbf{r}}_i$ is a unit vector directed from q_i to P. Note that we dropped the t subscript from $\hat{\mathbf{r}}_{it}$ because we have found that the field is independent of the test charge. This technique is illustrated in Examples 22.6 and 22.7. Both are very important, so work through each carefully! The first, Example 22.6, introduces a model that is the basis for the charge distributions of many molecules. Example 22.7 is important because the charges are distributed so that the resultant electric field has a symmetry feature we want you to be sure to understand.

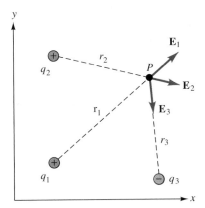

FIGURE 22.10

The total electric field at a point in space is the vector sum of the fields from each point-charge.

Before beginning, let's summarize our game plan for calculating the electric field from multiple sources:

Problem-Solving Strategy: Calculating the Total Electric Field from Multiple Charges

0. Remember that the electric field is a vector.

1. Draw a picture! Locate each charge on a Cartesian coordinate system. Locate the point P where you wish to know the electric field; label it clearly.

2. For each charge draw an electric field vector *at point P* representing the field contribution from the charge. The field vector is parallel to the line joining the charge and point P. Determine the direction of each field vector by imagining a *positive* test charge at point P and using the "opposites attract, likes repel" rule.

3. Determine the magnitude of each of the electric field contributions using

$$E = \frac{1}{4\pi\epsilon_o} \frac{|q|}{r^2}$$

4. Find the x- and y-components of each of the electric field values computed above.

5. Add all the x-components together. Add all the y-components together. You now have the components of the total E-field at point P. The magnitude is, as for any vector, the square root of the sum of the squares of the components.

6. Remember that the electric field is a vector.

Concept Question 5
Two point-charges equal in magnitude but opposite in sign are separated by a distance *L*. How many locations along the line are there where the magnitudes of the fields from each charge are equal? At how many locations along the line on which they lie is the resultant *E*-field zero? Explain.

EXAMPLE 22.6   *The Electric Dipole*

The pair of point-charges shown in Figure 22.11 is called an **electric dipole.** It consists of positive and negative charges of equal magnitude q separated by a distance d. It is convenient to introduce a quantity called the **electric dipole moment** as a vector of magnitude p equal to the product of the magnitude of either charge and the distance between the charges: $p = qd$. Find the electric field E at a distance x along the perpendicular bisector of the straight line connecting $-q$ to $+q$.

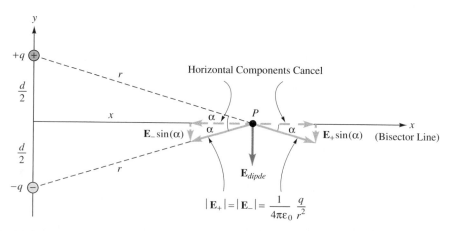

FIGURE 22.11

The electric dipole. See Example 22.6.

SOLUTION Figure 22.11 contains the elements of steps (1) and (2) in our problem-solving strategy. From this figure we see that the E-field contributions at the point P from both charges make an angle α with the perpendicular bisector of the dipole. However, the field from the positive charge (we call it E_+) points away from $+q$, whereas that of the negative charge (called E_-) points toward $-q$. Now on to step 3: The magnitude of both fields is the same and is

$$E_+ = E_- = \frac{1}{4\pi\epsilon_o}\frac{q}{r^2}$$

However, both E_+ and E_- must be resolved into components parallel and perpendicular to the bisector line as described in step (4) of the problem-solving strategy. From the figure it is easy to see that the components parallel to the x-axis add to zero, and those perpendicular to the x-axis combine to give a total field that points parallel to the y-axis:

$$E_{\text{dipole}} = E_+ \sin(\alpha) + E_- \sin(\alpha) = 2E_+ \sin(\alpha) = \frac{1}{2\pi\epsilon_o}\frac{q}{r^2}\sin(\alpha)$$

From the geometry of Figure 22.11 we see that

$$\sin(\alpha) = \frac{(d/2)}{r} \quad \text{and} \quad r = \sqrt{x^2 + (d/2)^2} \quad \text{so}$$

$$E_{\text{dipole}} = \frac{1}{2\pi\epsilon_o}\frac{q}{[x^2 + (d/2)^2]}\frac{(d/2)}{\sqrt{x^2 + (d/2)^2}} = \frac{1}{4\pi\epsilon_o}\frac{qd}{[x^2 + (d/2)^2]^{3/2}}$$

If we use the definition $p = qd$, we have

$$E_{\text{dipole}} = \frac{1}{4\pi\epsilon_o}\frac{p}{[x^2 + (d/2)^2]^{3/2}}$$

The electric dipole is an important charge distribution because it provides a useful model for many types of molecules. We will have much more to say about how a dipole behaves in an externally applied electric field in Section 22.4. For now, we want to end this example by mentioning that there are many situations in which the distance x is large compared to the charge separation distance d. In this case, $x \gg d/2$ and we can ignore this $d/2$ term in the denominator of E_{dipole} to obtain

$$E_{\text{dipole}} \approx \frac{p}{4\pi\epsilon_o x^3}$$

Note that this field strength decreases as $1/x^3$, whereas the magnitude of the E-field of a point-charge decreases more slowly as $1/r^2$. We expect the field of a dipole to drop off more quickly because its net charge is zero. ◀

EXAMPLE 22.7 *If You Can Do Two, You Can Do Five*

Five point-charges, $+e$ each, are distributed at equal intervals along the x-axis at locations $(-2a, 0)$, $(-a, 0)$, $(0, 0)$, $(a, 0)$, and $(+2a, 0)$. Compute the resultant electric field at a point P on the y-axis with coordinates $(0, b)$, if $a = 2.00$ cm and $b = 5.00$ cm.

SOLUTION Following problem-solving steps (1) and (2) produces the charge distribution and the E-field vectors shown in Figure 22.12. As we proceed with step (3) we note from the symmetry of the charge locations, that the field strengths from

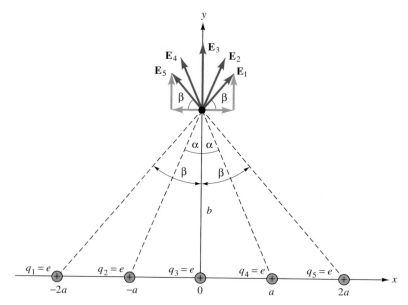

FIGURE 22.12

The electric field due to five point charges. See Example 22.7.

charges q_1 and q_5 are equal, as are the field strengths of charges q_2 and q_4. In particular

$$E_1 = E_5 = \frac{1}{4\pi\epsilon_o} \frac{e}{(2a)^2 + b^2}$$

and

$$E_2 = E_4 = \frac{1}{4\pi\epsilon_o} \frac{e}{a^2 + b^2}$$

The field strength from charge q_3 is

$$E_3 = \frac{1}{4\pi\epsilon_o} \frac{e}{b^2}$$

In order to find the total field at point P we must add these five contributions together *vectorially* as emphasized in problem-solving step (4). That is to say, we must resolve each electric field contribution into its horizontal and vertical components and then add the components. The simplifying feature about these fields is that the *x-components* of E_1 and E_5 cancel, as do the x-components of E_2 and E_4. (We have only shown the components of E_1 and E_5 in Figure 22.12. We think you can visualize the analogous components of E_2 and E_4.) Hence, only the y-components of all the fields contribute to the total field:

$$E_{\text{tot}_x} = 0$$

$$E_{\text{tot}_y} = E_1 \sin(\beta) + E_2 \sin(\alpha) + E_3 + E_4 \sin(\alpha) + E_5 \sin(\beta)$$

$$E_{\text{tot}} = E_{\text{tot}_y} = 2E_1 \sin(\beta) + 2E_2 \sin(\alpha) + E_3$$

From the geometry of Figure 22.12 we see that

$$\sin(\alpha) = \frac{a}{\sqrt{a^2 + b^2}} \qquad \text{and} \qquad \sin(\beta) = \frac{2a}{\sqrt{(2a)^2 + b^2}}$$

Substituting for E_1, E_2, E_3, $\sin(\alpha)$, and $\sin(\beta)$, we have

$$E = 2\left(\frac{1}{4\pi\epsilon_o}\frac{e}{(2a)^2 + b^2}\right)\frac{2a}{\sqrt{(2a)^2 + b^2}} + 2\left(\frac{1}{4\pi\epsilon_o}\frac{e}{a^2 + b^2}\right)\frac{a}{\sqrt{a^2 + b^2}}$$
$$+ \frac{1}{4\pi\epsilon_o}\frac{e}{b^2}$$

$$E = \frac{e}{4\pi\epsilon_o}\left(\frac{4a}{[(2a)^2 + b^2]^{3/2}} + \frac{4a}{[a^2 + b^2]^{3/2}} + \frac{1}{b^2}\right)$$

Substituting $a = 0.0200$ m and $b = 0.0500$ m yields $E = 1.75 \times 10^{-6}$ N/C. We re-emphasize two important features of this problem: (1) Electric fields must be added vectorially and (2) the charges were arranged symmetrically so that the horizontal components of the E-field from each charge canceled. Keep your eyes open for similar features in some of the *continuous* charge distributions to come later! ◀

Continuous Charge Distributions

How do we compute the electric field from a *continuous* charge distribution? We begin by dividing the distribution up into small elements each of which contains a charge Δq as shown in Figure 22.13. From Equation (22.8), the field ΔE from just one of these charges Δq is

$$\Delta \mathbf{E} = \frac{1}{4\pi\epsilon_o}\frac{\Delta q}{r^2}\hat{\mathbf{r}} \qquad (22.9)$$

where $\hat{\mathbf{r}}$ is the unit vector directed *from the location of Δq toward the point in space where we want to compute* $\mathbf{E}$. To obtain the net electric field from the entire distribution, we apply the principle of superposition. The vector sum of the fields from all the charges that make up the distribution is

$$\mathbf{E} = \frac{1}{4\pi\epsilon_o}\sum_i \frac{\Delta q_i}{r_i^2}\hat{\mathbf{r}}_i$$

In the limit as $\Delta q_i \to 0$ we have

$$\mathbf{E} = \frac{1}{4\pi\epsilon_o}\lim_{\Delta q \to 0}\sum_i \frac{\Delta q_i}{r_i^2}\hat{\mathbf{r}}_i$$
$$= \frac{1}{4\pi\epsilon_o}\int \frac{dq}{r^2}\hat{\mathbf{r}} \qquad (22.10)$$

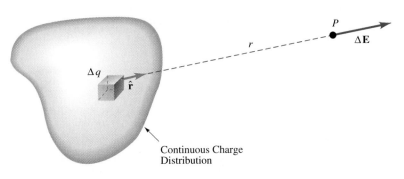

FIGURE 22.13

The total electric field at point P from a continuous charge distribution is the vector sum of the E-field contributions from each Δq of the distribution.

There are two features of Equation (22.10) that need your special attention. First, Equation (22.10) is a *vector* integration. Therefore, you cannot really use Equation (22.10) as it stands. Rather, it is a signal to you to resolve the vector *E*-field into components. *Only* after you have done this step can you perform the summations indicated by the integral of Equation (22.10). The good news is that often times the charge distribution will be symmetrical, so that you will be able to perform some of the integrals in your head! Look back at Example 22.7. The symmetrical charge locations were helpful because we did not have to compute the horizontal components of the fields; we could see that they all canceled. The same thing can happen when we consider charge distributions that are long, straight lines, or disks, or spheres, and so forth.

After you have resolved the *E*-fields of Equation (22.10) into components, you then have to perform an integration over the source charge distribution. We know how much you enjoy that process, particularly if it is over a volume! However, symmetrical charge distributions have the added bonus that the integrals turn out to be pretty easy ones to compute. Moreover, we often want to know the field only at positions where the required integrals are routine in nature. Not convinced? Well, to further bolster your confidence we give you steps we think will help you establish a systematic attack on these types of problems:

Problem-Solving Strategy: Calculating the Total Electric Field from Continuous Charge Distributions

0. Remember that the electric field is a vector.

1. Identify the type of charge distribution and, if necessary, compute the charge density λ, σ, or ρ.

2. Draw a picture showing the location of the source charge distribution and the point P where you wish to calculate the field. Mark a generic chunk of charge dq on the source charge distribution. *Don't pick the exact middle or an end!* Draw at point P the $d\mathbf{E}$ vector produced by the chunk of charge dq.

3. Write down an expression for the magnitude of $d\mathbf{E}$ using

$$dE = \frac{1}{4\pi\epsilon_o}\frac{dq}{r^2}$$

4. Resolve the $d\mathbf{E}$ vector into its components.

4 (a). Identify any special symmetry features of the problem and use them to identify components that sum to zero.

4 (b). Write an expression for the component(s) of the field that are not canceled by other components.

5. Write the distance r and any trigonometric factors in terms of given coordinates and parameters.

6. Perform the indicated integrations. *Remember that the limits of the integration should be such that the region integrated is the one that contains the source charges.*

7. Remember that the electric field is a vector.

Step (6) in the continuous charge problem-solving strategy is particularly important because sometimes r, as well as sines and cosines in the expression for dE, depend on the location of dq and *cannot* be removed from the integral for E.

We urge you to look carefully at Examples 22.8 and 22.9. You will eventually receive a substantial "pay-back" if you learn to apply our steps. After you have diligently studied these examples, cover the solutions and see if you can rework each. But don't just memorize; remember the steps and know why you perform each one.

EXAMPLE 22.8 *If You Can Do Five, You Can Do an Infinite Number*

Compute the electric field E a distance y from an infinite line of charge along the x-axis if the charge per length is constant and given by λ.

SOLUTION

0. Glance back at Example 22.7. Make sure you understand it. We sincerely believe you will not understand the following solution unless you are 100% comfortable with Example 22.7.
1. In this example a linear charge density λ is given, so we expect to integrate over a single dimension.
2. We choose an xy-coordinate system as shown in Figure 22.14 and designate a small charge element $dq = \lambda\, dx$. We wish to compute the E-field at point P, a *fixed* distance y from the line of charge.

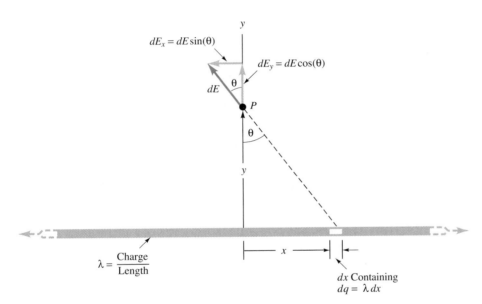

FIGURE 22.14

Computing the E-field at point P, a distance y from a uniform line of charge.

3. The magnitude of the field dE is given by

$$dE = \frac{1}{4\pi\epsilon_o}\,\frac{dq}{r^2} = \frac{1}{4\pi\epsilon_o}\,\frac{\lambda\, dx}{r^2}$$

and dE makes an angle θ with the vertical (y-axis).

4. The field dE can be resolved into components parallel and perpendicular to the y-axis:

$$dE_y = dE\cos(\theta)$$

$$dE_x = dE\sin(\theta)$$

4 (a). From symmetry we see that the combined components of dE_x cancel; for the x-component of the E-field from every dq on the left side of the line of charge there is an identical x-component of the E-field from a dq on the right side. These components dE_x, however, point in opposite directions and, therefore, cancel.

4 (b). The components dE_y are all parallel and simply add. Hence, the total electric field at point P is found by integrating the expression

$$dE_{\text{line}} = dE \cos(\theta) = \frac{1}{4\pi\epsilon_o} \frac{\lambda \, dx}{r^2} \cos(\theta)$$

5. The direct approach to solve for E_{line} is to write both r and θ in terms of x and y, and then integrate x from $-\infty$ to $+\infty$. If we follow this plan we find ourselves with an integral that could be solved using a trigonometric substitution or an integral table.

We are going to use an alternative approach and recognize that another way to integrate dx over the entire line is to let the angle θ go from $-\pi/2$ to $+\pi/2$ radians. In order to take this approach we have to write the variables x and r (and the differential dx) in terms of θ. We begin by noting

$$\tan(\theta) = \frac{x}{y} \quad \text{or} \quad x = y \tan(\theta)$$

so that

$$dx = y \sec^2(\theta) \, d\theta = \frac{y}{\cos^2(\theta)} \, d\theta$$

Furthermore,

$$\cos(\theta) = \frac{y}{r} \quad \text{or} \quad \frac{1}{r^2} = \frac{\cos^2(\theta)}{y^2}$$

Making these substitutions for dx and $1/r^2$ into our expression for E_{line}, we have

$$dE_{\text{line}} = \frac{\lambda}{4\pi\epsilon_o} \left(\frac{\cos^2(\theta)}{y^2} \right) \left(\frac{y}{\cos^2(\theta)} \, d\theta \right) \cos(\theta)$$

6.
$$E_{\text{line}} = \frac{1}{4\pi\epsilon_o} \frac{\lambda}{y} \int_{-\pi/2}^{+\pi/2} \cos(\theta) \, d\theta$$

$$= \frac{1}{4\pi\epsilon_o} \frac{\lambda}{y} [\sin(\pi/2) - \sin(-\pi/2)]$$

$$E_{\text{line}} = \frac{\lambda}{2\pi\epsilon_o y}$$

Notice that the direction of E_{line} is perpendicular to and away from the line of charge. Now, it might occur to you to wonder why anyone is interested in the electric field from an infinitely long, straight, charge distribution. After all, any real, linear charge distribution must be finite in length! However, for positions that are close to the charged line where y is small compared to the total length, our result works very well as long as we are not close to the ends of the charged line. Geiger-Müller tubes used to detect moving charged particles, such as ions and electrons, have a long charged wire running down the center of a cylindrical housing. Also, charged wires are employed as part of the scrubbing mechanism to clean air on some smokestacks. ◀

EXAMPLE 22.9   *Charge a Ring*

A total charge $+q$ is uniformly distributed around a ring of radius y, as shown in Figure 22.15. Compute the electric field at a point on the axis of the ring a distance x from its center.

SOLUTION

1. Because the charge is distributed along a circle, we use the linear charge density λ. For a circle of radius y, the charge per length around the ring is

$$\lambda = +\frac{q}{2\pi y}$$

2. We wish to determine the field dE at the point P. This differential field element is produced by the infinitesimal charge $dq = \lambda\, dl$ shown in Figure 22.15.

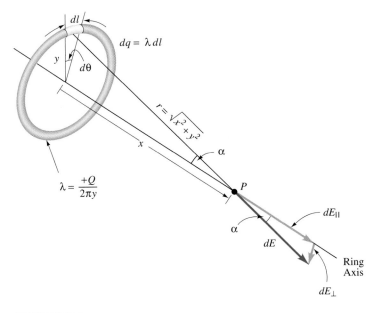

FIGURE 22.15

Computing the E-field on the axis of a charged ring. The field dE is produced by the charge element $dq = \lambda\, dl$.

3. The magnitude of the differential form of Equation (22.9) is

$$dE = \frac{1}{4\pi\epsilon_o}\frac{dq}{r^2} = \frac{1}{4\pi\epsilon_o}\frac{\lambda\, dl}{r^2}$$

4. In our problem, the field dE has components both parallel ($dE_{\parallel}$) and perpendicular ($dE_{\perp}$) to the x-axis:

$$dE_{\parallel} = dE \cos(\alpha)$$

$$dE_{\perp} = dE \sin(\alpha)$$

4 (a). When we sum the contributions from all the charge elements dq around the ring, the perpendicular components $dE_{\perp}$ cancel. That is, the perpendicular component from a dq on one side of the ring is exactly canceled by the perpendicular component from a dq on the opposite side of the ring.

4 (b). The components parallel to the x-axis simply add to a nonzero value. Hence, the total electric field on the ring axis is found by integrating the expression

$$dE_{\text{ring}} = dE \cos(\alpha) = \frac{1}{4\pi\epsilon_o} \frac{\lambda \, dl}{r^2} \cos(\alpha)$$

5. We must now write r and $\cos(\alpha)$ in terms of the given variables x and y. From the figure we see

$$r = \sqrt{x^2 + y^2}$$

and

$$\cos(\alpha) = \frac{x}{r} = \frac{x}{\sqrt{x^2 + y^2}}$$

Moreover, $dl = y \, d\theta$, and our expression for dE becomes

$$dE_{\text{ring}} = \frac{1}{4\pi\epsilon_o} \lambda \frac{y \, d\theta}{(x^2 + y^2)^2} \frac{x}{\sqrt{x^2 + y^2}}$$

$$E_{\text{ring}} = \frac{1}{4\pi\epsilon_o} \int \frac{\lambda xy \, d\theta}{(x^2 + y^2)^{3/2}}$$

We have a bit of a luck in this problem because neither of the parameters x nor y depend on θ. Therefore, all factors can be removed from the integral, and

$$E_{\text{ring}} = \frac{\lambda}{4\pi\epsilon_o} \frac{xy}{(x^2 + y^2)^{3/2}} \int_0^{2\pi} d\theta$$

6.
$$E_{\text{ring}} = \frac{\lambda}{2\epsilon_o} \frac{xy}{(x^2 + y^2)^{3/2}}$$

Finally, we may substitute for the $\lambda = q/2\pi y$ to obtain

$$E_{\text{ring}} = \frac{q}{4\pi\epsilon_o} \frac{x}{(x^2 + y^2)^{3/2}}$$

Once again we want to look at this solution for some special cases. Note if $x = 0$ that $E = 0$ exactly as we would expect from the symmetry. This result is a partial check on our work; certainly if E wasn't zero at $x = 0$ we would have known our answer was wrong. Notice, too, what happens when the distance x becomes very large compared to the ring diameter y. For the position $x \gg y$ we can ignore y in the denominator of our result so that

$$E_{\text{ring}} \rightarrow \frac{q}{4\pi\epsilon_o} \frac{x}{(x^2)^{3/2}} = \frac{1}{4\pi\epsilon_o} \frac{q}{x^2}$$

This result is just the electric field of a point-charge q. When we are at such a large distance from the ring that its shape is not distinguishable from a point, we simply observe the field from a point-charge q. This is another check on our solution. ◀

EXAMPLE 22.10 *Making a Ring into a Disk into a Plane*

A charge $+q$ is uniformly distributed over the surface of a disk with radius R. (a) Find the electric field E on the axis of the disk at a distance x from its center. (b) Use your result to find the electric field from an infinite plane of charge with a charge per unit area that is given by σ.

SOLUTION (a) Sometimes we get lucky and can use the results of a previous E-field computation to bypass some of our six steps. In this example we recognize that one way to construct a disk of charge is to make the disk out of a series of rings, as shown in Figure 22.16. In this figure we show a ring of radius y and thickness dy.

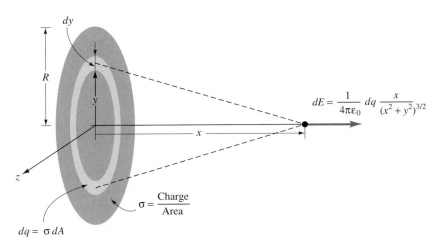

$$dE = \frac{1}{4\pi\varepsilon_0}\, dq\, \frac{x}{(x^2 + y^2)^{3/2}}$$

$$\sigma = \frac{\text{Charge}}{\text{Area}}$$

$$dq = \sigma\, dA$$

FIGURE 22.16

A disk of charge can be constructed from many concentric charged rings. The charge dq is distributed over a ring of width dy and length $2\pi y$.

We are thus led to look back at the answer to Example 22.9. When we compare that answer to the current case we think of the charge q on the ring as only part of the charge and now call it dq. Similarly, the final answer for E is no longer the total field but only a contribution dE from a single ring that is now one of many:

$$E_{\text{ring}} = \frac{q}{4\pi\epsilon_o}\frac{x}{(x^2 + y^2)^{3/2}} \qquad \text{(Answer for one ring)}$$

$$dE_{\text{ring}} = \frac{dq}{4\pi\epsilon_o}\frac{x}{(x^2 + y^2)^{3/2}} \qquad \text{(One ring's contribution to a disk)}$$

We obtain the field of the disk by summing all the rings to form a disk. That is,

$$E_{\text{disk}} = \int dE_{\text{ring}}$$

Notice from Figure 22.16 that we don't have to worry about any components when we use this expression because each dE_{ring} is parallel to the ring axis. Because the charge q is spread over a surface, we use the area charge density $dq = \sigma\, dA$ where

$$\sigma = +\frac{q}{\pi R^2} \qquad \text{and} \qquad dA = (2\pi y)\, dy$$

so that $dq = \sigma\, dA = \sigma(2\pi y)\, dy$. Using this expression we can jump directly to step

(5), writing

$$dE_{\text{disk}} = \frac{\sigma}{4\pi\epsilon_o}(2\pi y\, dy)\frac{x}{(x^2 + y^2)^{3/2}}$$

Now we must sum these dE contributions from each of the source charge rings. All of these rings lie in the yz-plane between $y = 0$ and $y = R$. Note that x remains constant for all the source rings. Thus, we write the integral over the source charge as

$$E_{\text{disk}} = \frac{\sigma x}{2\epsilon_o}\int_0^R \frac{y\, dy}{(x^2 + y^2)^{3/2}}$$

Step 6. The integral may be solved using the "u substitution": $u = (x^2 + y^2)$ with $du = 2y\, dy$ because x is a constant. We must also, of course, change the limits to u limits, so our integral is

$$E_{\text{disk}} = \frac{\sigma x}{2\epsilon_o}\frac{1}{2}\int_{x^2}^{x^2+R^2}\frac{du}{u^{3/2}} = \frac{\sigma x}{2\epsilon_o}\frac{1}{2}\left(\frac{u^{-1/2}}{-\frac{1}{2}}\right)_{x^2}^{x^2+R^2}$$

$$E_{\text{disk}} = \frac{\sigma}{2\epsilon_o}\left(1 - \frac{x}{\sqrt{x^2 + R^2}}\right)$$

(b) We can make our disk into an infinite plane by allowing $R \to \infty$. In this case, it is easy to see that the second term in the expression for E_{disk} goes to zero, and we are left with

$$E_{\text{plane}} = E_{\text{disk}}\bigg|_{R\to\infty} = \frac{\sigma}{2\epsilon_o}$$

This result is independent of x, so the field strength E_{plane} is a constant at all points in space (but off the plane). Although $E_{\text{plane}} = \sigma/2\epsilon_o$ strictly applies to planar charge distributions of infinite expanse, we can apply this relation to planes of finite extent provided we remain near enough to the plane that our distance from it is small compared to the distance across the plane. Of course we must also stay away from the edges and be sure the charge distribution is uniform. ◀

Electric Field Lines

A convenient way to graphically visualize the electric field is with **electric field lines,** which are also called **lines of force.** This model has the following characteristics:

1. The electric field line is tangent to the electric field vectors at each point in space.
2. The number of lines per area that pass through a surface perpendicular to the electric field lines is proportional to the strength of the field in that region.

The second characteristic means that in regions where the electric field lines are relatively close together the field strength is large. On the other hand, if the electric field lines are far apart, the field strength is small. These features are illustrated in Figures 22.17 through 22.21. In these figures we show the field lines in the plane of the page only. However, the electric field really extends in all directions from a charge distribution.

There are three rules that help you draw electric field lines correctly:

1. No field lines can cross.
2. Field lines must begin on (point outward from) positive charges and end on (point inward to) negative charges. (Field lines may also start or end at infinity, which for our purposes is often the edge of the page.)

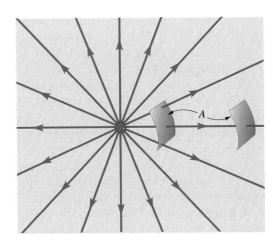

FIGURE 22.17

Electric field lines from a point-charge. Close to the charge the field strength is large and there is a relatively large number of field lines passing through the surface area A. At a distance farther from the charge the field is weaker and fewer lines pass through the same area A.

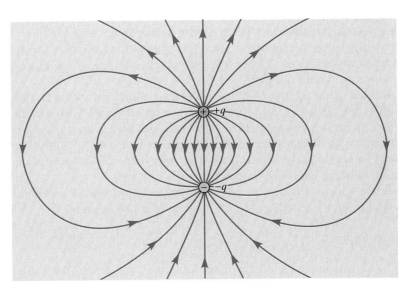

FIGURE 22.18

Electric field lines for two point-charges of equal magnitude but opposite sign: an electric dipole.

FIGURE 22.19

Electric field lines for two positive point-charges of equal magnitude.

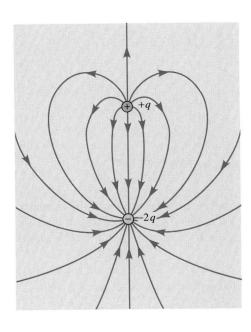

FIGURE 22.20

Electric field lines of two unequal charges of opposite sign.

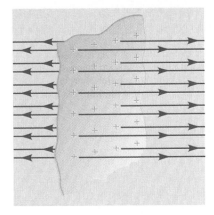

FIGURE 22.21

Electric field lines of an infinite plane of charge. Only a small portion of the plane is shown.

3. The number of field lines starting from, or ending at, a charge is proportional to the magnitude of that charge.

You should verify these rules for Figures 22.17 through 22.21.

We can confirm for a point-charge that this mechanism for visualizing the electric field is consistent with Coulomb's law. What follows is a very good exercise because it provides an opportunity to picture the three-dimensional features of the electric field. This is an important drill for the chapters to come; you should get used to visualizing the electric field in all three dimensions.

Suppose we surround an isolated positive charge with an imaginary spherical surface with a center that coincides with the position of the point-charge. Visualize the point-charge as a sea urchin with sharp spines protruding in every direction. Imagine the sphere as a thin sheet wrapped around the sea urchin in a perfect sphere. When the radius of this sphere is r, its area is $4\pi r^2$. If N is the total number of field lines drawn outward from the charge q, the number of *lines per unit area* passing through the spherical surface is $N/4\pi r^2$. At this point you should be visualizing a spherical surface with many small holes poked through its surface. These holes are the intersections of the field lines (sea urchin spines) with the spherical surface. Because the number of holes per unit area is $N/4\pi r^2$, the electric field strength is proportional to $1/r^2$, precisely the result we obtained for the field of a point-charge (see Eq. 22.6). Notice that if Coulomb's law were not inverse square our field line picture would not work.

There are two features of the electric field line model about which we should caution you. First, it is possible to draw only a very limited number of field lines, however, *the field still exists in the regions between the field lines.* Second, you must keep in mind that the electric field always exists in *three* dimensions. In many of the figures we present we are able to show field lines only in the plane of the page. Sometimes the behavior of the field outside the plane of the page is also important.

We end this section with an example illustrating how the knowledge of a uniform electric field strength can be used to determine the motion of a point-charge.

Concept Question 6
Explain why electric field lines do not cross.

The elastic fibers of a "cush ball" point radially outward analogous to the E-field lines of a point-charge. However, the E-field lines extend to infinity.

EXAMPLE 22.11   *This Looks Like Chapter 4!*

A point-charge q of mass m travels with an initial velocity $\mathbf{v}_o = v_o\hat{\mathbf{i}}$ along the x-axis. As it crosses the origin it encounters a uniform electric field $\mathbf{E} = -E_o\hat{\mathbf{j}}$ as shown in Figure 22.22. If the field E_o is sufficiently large that gravity can be ignored, find the equations of motion $x(t)$ and $y(t)$ for the charge.

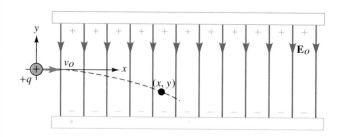

FIGURE 22.22

A charge $+q$ with velocity v_o enters a region of uniform electric field.

SOLUTION   Figure 22.22 shows how we can visualize the constant E-field created by two parallel planes of charge, as long as we ignore any nonuniformity in the field near the planes' edges. The force on the charge is $\mathbf{F} = q\mathbf{E} = -qE_o\hat{\mathbf{j}}$. From Newton's second law, $m\mathbf{a} = -qE_o\hat{\mathbf{j}}$, or

$$\mathbf{a} = -\frac{qE_o}{m}\hat{\mathbf{j}}$$

Because q, E_o, and m are all constants, we can use the constant-acceleration model in which the positions are given by $s(t) = s_o + v_o t + \frac{1}{2}at^2$. In our case

X-COMPONENT	Y-COMPONENT
$a_x = 0,$	$a_y = -qE_o/m,$
$v_{xo} = v_o$	$v_{yo} = 0,$
$x_o = 0$	$y_o = 0$
$x(t) = v_o t$	$y(t) = -\dfrac{1}{2}(qE_o/m)t^2$

You no doubt recognize the familiar equations for a parabolic trajectory. ◀

22.4 The Electric Dipole in a Uniform Electric Field

The electric dipole consists of positive and negative charges of equal magnitude, separated by a distance d. In Example 22.6 we found it convenient to define the magnitude of the dipole moment as the product

$$p = qd \tag{22.11}$$

For reasons we shall discover below it is also helpful to define the electric dipole moment as a *vector* directed *from the negative charge to the positive charge* (Fig. 22.23).

Figure 22.24 shows an electric dipole within a uniform electric field **E**. Even when the dipole is oriented at an arbitrary angle θ with respect to the field direction, the net force on the dipole is zero:

$$\mathbf{F}_{\text{net}} = (+q)\mathbf{E} + (-q)\mathbf{E} = 0$$

Hence, there is no *translational* acceleration of the dipole. However, as Figure 22.24 also shows, both forces $\mathbf{F} = +q\mathbf{E}$ and $\mathbf{F} = -q\mathbf{E}$ tend to rotate the dipole in the same direction about point O. Therefore, the dipole does experience a net torque τ that tends to align the dipole moment **p** along the direction of the E-field. We can resolve the forces on each of the charges $\pm q$ into components parallel and perpendicular to the dipole axis. The magnitude of the net torque on the dipole is

$$\tau = 2\left(\frac{d}{2}\, F \sin(\theta)\right) = d(qE)\sin(\theta) = (qd)E\sin(\theta) = pE\sin(\theta)$$

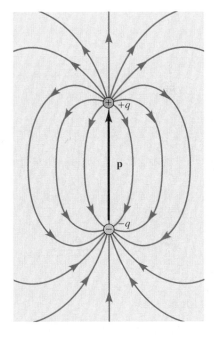

FIGURE 22.23

The electric dipole vector **p**.

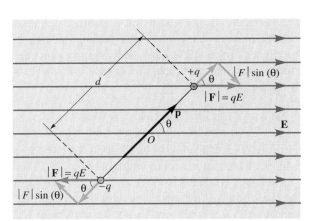

FIGURE 22.24

An electric dipole in a uniform electric field. When the dipole moment **p** is at a nonzero angle θ with respect to the E-field direction there is a torque τ on the dipole.

Now, a quick application of the right-hand rule to Figure 22.24 shows us that the direction of the torque is into the page. Moreover, θ is the angle between the vectors $\mathbf{p}$ and $\mathbf{E}$. Therefore, we can represent the torque $\boldsymbol{\tau}$ by the vector cross product:

$$\boldsymbol{\tau} = \mathbf{p} \times \mathbf{E} \qquad (22.12)$$

We must perform work on the dipole to change its orientation. Moreover, we can represent this work as a change in the potential energy of the dipole and E-field system. In Chapter 11 we found that the work $d\mathcal{W}$ done by a torque τ is $\tau\, d\theta$. For the angle θ as defined in Figure 22.24 the torque $\tau = pE \sin(\theta)$ tends to *decrease* the angle $d\theta$. Thus, as the electric field does positive work to align the dipole, $d\theta$ is a negative number. As a consequence we must include a minus sign in the expression for work to ensure that the work done by the field is a positive number:

$$d\mathcal{W} = -pE \sin(\theta)\, d\theta$$

Moreover, the change in potential energy is the negative of the work done (see Eq. 8.1) so that

$$dU = -d\mathcal{W} = +pE \sin(\theta)\, d\theta$$

The change in potential energy between any two angles θ_o and θ is thus

$$U - U_o = \int_{\theta_o}^{\theta} pE \sin(\theta)\, d\theta = pE[-\cos(\theta)]_{\theta_o}^{\theta}$$

$$U - U_o = [-pE \cos(\theta)] - [-pE \cos(\theta_o)]$$

This form suggests a potential function $U = -pE \cos(\theta) + C$. The usual convention is to take the constant $C = 0$. This choice is not without consequences, and we discuss them in the next paragraph. But first we note that, because $\mathbf{p}$ and $\mathbf{E}$ are both vectors and θ is the angle between them, we can represent the potential energy U as the dot product of $\mathbf{p}$ and $\mathbf{E}$:

$$U = -\mathbf{p} \cdot \mathbf{E} \qquad (22.13)$$

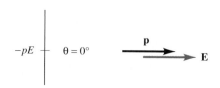

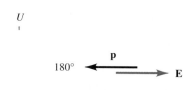

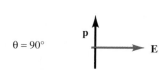

FIGURE 22.25

The dipole potential energy for three orientations. Whereas the maximum potential energy is pE, the maximum *change* in potential energy is $2pE$.

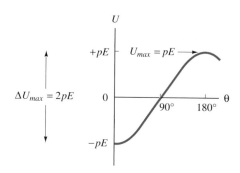

FIGURE 22.26

U as a function of θ for an electric dipole in a uniform E-field.

This potential energy function has caused enough student turmoil that we think you better take a close look at Figure 22.25. In this figure we represent the potential energy of the dipole in the field $\mathbf{E}$ for three important orientations: $0°$, $90°$, and $180°$. Orientations between those shown are easy to visualize. Notice that the maximum and minimum potential energies are $+pE$ and $-pE$, respectively, and by Equation (22.13), zero potential energy occurs at $\theta = 90°$. But here's the catch: Even though the maximum potential energy is $+pE$, the difference between the maximum and minimum potential energy is $2pE$. That is, if the dipole is originally aligned antiparallel to the field where its potential energy is $U_o = +pE$, and then it is rotated to an orientation parallel to the field where its potential energy is $U_f = -pE$, its *change* in potential energy is

$$\Delta U = U_f - U_o = -pE - (+pE) = -2pE$$

Figure 22.26 shows the potential energy U as a function of θ. The slope of this curve is the magnitude of the torque on the dipole. Note that in accord with Equation 22.12, the torque is maximum at $\theta = 90°$ and zero at $\theta = 0°$ and $\theta = 180°$. As you can see from Figure 22.26, $\theta = 180°$ is an unstable equilibrium. Any small deviation from $\theta = 180°$ results in a torque that tries to align the dipole at $\theta = 0°$. By looking at the maximum and minimum energy values, you can easily see why the maximum *change* in potential energy is $-2pE$.

Many molecules exhibit a permanent dipole moment and are said to be **polar.** The moments arise from asymmetric charge arrangements of the ions and electron distributions that form the bonds between the ions. For example, there is an angle of about 105° between the hydrogen and oxygen bonds of a water molecule as shown in Figure 22.27. Because the oxygen ion has a net negative charge and the hydrogen ions are each positive, the molecule can be modeled as negative and positive charges of equal magnitude located along a line through the oxygen atom center that also bisects the hydrogen atoms. Even molecules with symmetric charge distributions may acquire a dipole moment when placed in an electric field. The field pulls the positive charges along the direction of the field lines while the negative charges are moved in the opposite direction. The result is an **induced dipole moment.**

Dipoles have many practical applications. Chances are good that you will use many of them as you work homework problems for this chapter: these days most calculator displays use liquid crystals. These liquid crystals displays (LCDs) are composed of dipolar molecules with strong optical properties. The electric charge acts as a handle on these molecules, allowing them to be rotated by electric forces, thereby changing the polarization of the light that reflects from them. To say much more about the optical properties takes us too far ahead of ourselves; we will return to this topic in Chapter 35. However, we can appreciate that liquid crystals are so popular in calculator displays and miniature TVs because they rely on *electrostatic* fields. Displays that actually light up (you can see them in a dark room) draw a continuous flow of charges from a battery and cause it to ''go dead,'' often fairly quickly. In the case of LCDs, virtually no continuous flow of charge is required for the display. All the battery need supply is a very small amount of charge to align the dipoles of the liquid crystals.

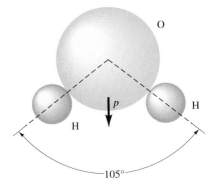

FIGURE 22.27

A model of a water molecule. The asymmetric charge distribution may be represented by an electric dipole moment **p.**

TABLE 22.2 Electric Dipole Moments
A convenient unit of electric dipole moment is the debye (D).
$1 D = 3.33564 \times 10^{-30} C \cdot m$.

COMPOUND*	FORMULA	MOMENT (D)
Silver chloride	AgCl	5.73
Potassium bromide	BrK	10.41
Water	H_2O	1.85
Ozone	O_3	0.53
Carbon dioxide	CO_2	0 (s)†
Benzene	C_6H_6	0 (s)

*Gas phase.
†(s) Symmetrical molecules.

Liquid crystal displays take advantage of the small amount of energy necessary to align microscopic dipoles.

EXAMPLE 22.12 *Bottoms Up!*

The dipole moment of a water molecule is 6.23×10^{-30} C · m. One mole of water molecules (6.02×10^{23} molecules) is placed in a uniform E-field of 3.00×10^5 N/C. (a) What is the magnitude of the torque on a single molecule when it is oriented at an angle of 60.0° from the field? (b) How much work must be done on this quantity of water to rotate all the dipoles from alignment with the field ($\theta_o = 0°$) to $\theta_f = 135°$?

SOLUTION   (a) From Equation (22.11)

$$|\tau| = |\mathbf{p} \times \mathbf{E}| = pE \sin(60.0°) = (6.23 \times 10^{-30} \text{ C} \cdot \text{m})(3.00 \times 10^5 \text{ N/C}) \sin(60°)$$

$$= 1.62 \times 10^{-24} \text{ N} \cdot \text{m}$$

(b) The work required from an external agent to rotate a *single* water molecule from 0° to 135° can be found from its change in potential energy:

$$\Delta U_{\text{molecule}} = U(\theta_f) - U(\theta_o) = [-pE\cos(\theta_f)] - [-pE\cos(\theta_o)]$$

$$= pE[\cos(\theta_o) - \cos(\theta_f)]$$

$$= (6.23 \times 10^{-30}\,\text{C}\cdot\text{m})(3.00 \times 10^5\,\text{N/C})[\cos(0°) - \cos(135°)]$$

$$= 3.19 \times 10^{-24}\,\text{J}$$

The total work required to change the potential energy of 1 mol of molecules is therefore

$$\mathcal{W}_{\text{tot}} = (6.02 \times 10^{23}\ \text{molecules/mol})(3.19 \times 10^{-24}\ \text{J/molecule}) = 1.92\ \text{J} \quad \blacktriangleleft$$

22.5 The Electric Dipole in Nonuniform Fields

The behavior of dipoles in a nonuniform field is responsible for phenomena, such as the dust accumulation on your TV screen and the dissolving power of water. As in the uniform field case discussed above, the dipoles tend to align themselves with the external field. Unlike the uniform field case, however, the net force on the dipole is not zero. As you can see from Figure 22.28, when the dipole is aligned with the field, it is always attracted to regions of higher field strength, regardless of the field's direction.

Sometime (when no one is watching) tear a bit of paper into tiny fragments and lay them on the desk. Run a comb through your hair briskly a few times and then bring it near the bits of paper. If the humidity is not too high, you should be able to get the paper bits to jump up to the comb. (If anyone sees you, just claim to be doing your physics homework.) There are two things happening here. First you are creating *induced dipoles* out of the bits of paper. That is, the paper is neutral overall but the electrons on the paper are driven away from the negatively charged comb as shown in Figure 22.29. Next, this induced dipole is attracted to the region of higher electric field, which is, of course, near the comb. This is the same mechanism that attracts dust to a TV or computer terminal screen. The screen is bombarded by electrons and builds up net negative charge. Then dust particles become induced dipoles and are attracted to the high-field region of the screen face.

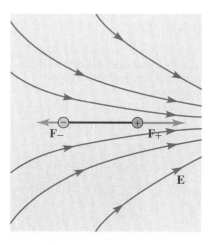

FIGURE 22.28

Because the force is stronger in the region of higher *E*-field intensity, a dipole experiences a net force in a nonuniform field.

FIGURE 22.29

Bits of paper become induced dipoles and are attracted to the region of higher electric field provided by the comb.

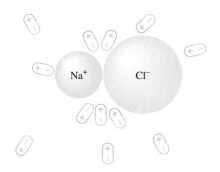

FIGURE 22.30

The nonuniform *E*-field about sodium chloride molecules attracts polar water molecules that eventually separate the ions.

The permanent dipoles of the water molecule are responsible for water's ability to dissolve many ionic solids. Salt for example is composed of Na^+ and Cl^- ions bound together by strong electrostatic attractive forces. These same electric forces attract the dipolar water molecules because the *E*-field from the Na^+ and Cl^- "point–charges" is very nonuniform. The ions become surrounded by the dipoles as shown in Figure 22.30. The ions are thus forced farther apart, and the electric force that previously bound them is greatly reduced because of the $1/r^2$ nature of the Coulomb attraction.

Concept Question 7
When you rub a rubber balloon (an insulator) on your sweater it acquires a net negative charge. Explain the movement of charges that results in such a balloon being attracted to (and perhaps sticking to) an electrically neutral wall.

22.6 Summary

The property of matter called **charge** has the following characteristics:

1. There are two types of charge labeled positive and negative. Like charges repel each other, and unlike charges attract each other.
2. Charge is conserved.
3. Charge is quantized in multiples of the fundamental amount $\pm e = \pm 1.60 \times 10^{-19}$ C.

The force on a point-charge q_1 by a point-charge q_2, separated by distance r is given by **Coulomb's law**

$$F_{12} = \frac{1}{4\pi\epsilon_o} \frac{q_1 q_2}{r^2} \tag{22.2}$$

where ϵ_o is 8.8542×10^{-9} C^2/(N·m^2), and is called the **permittivity of free space.**
The **electric field E** at a point in space is defined as the force **F** per unit charge q_{test} at that location:

$$\mathbf{E} = \frac{\mathbf{F}}{q_{test}}$$

The electric field at a distance r from a point-charge q is

$$\mathbf{E} = \frac{1}{4\pi\epsilon_o} \frac{q}{r^2} \hat{\mathbf{r}} \tag{22.6}$$

where $\hat{\mathbf{r}}$ is a unit vector that points from the location of q to the point designated by $\mathbf{r}$. The electric field from a collection of point-charges is the vector sum of the fields from each charge. For a continuous charge distribution, the electric field is

$$\mathbf{E} = \frac{1}{4\pi\epsilon_o} \int \frac{dq}{r^2}\hat{\mathbf{r}} \tag{22.10}$$

where dq is an infinitesimal charge element within the distribution. For a linear charge distribution $\lambda =$ **charge per length,** the charge contained in a segment dl is $dq = \lambda\, dl$; for an area charge distribution with $\sigma =$ **charge per area,** the charge contained in an area element dA is $dq = \sigma\, dA$; for volume charge density with $\rho =$ **charge per volume,** the charge contained in volume element $d\mathcal{V}$ is $dq = \rho\, d\mathcal{V}$.

Electric field lines can be used pictorially to model the electric field. The direction of the lines is along the direction of a force on a positive test charge, and the number of lines per area through a surface perpendicular to the field lines is proportional to the field strength.

Some E-field magnitudes calculated in this chapter are summarized in Table 22.3.

TABLE 22.3   *E-field Magnitudes for Various Symmetric Geometries*

GEOMETRY	FIELD	PARAMETERS
Point–charge	$E_{pt} = \dfrac{1}{4\pi\epsilon_o}\dfrac{q}{r^2}$	$q =$ charge $r =$ distance from point-charge
Infinite line of charge	$E_{\text{line}} = \dfrac{\lambda}{2\pi\epsilon_o y}$	$\lambda =$ charge/length $y = \perp$ distance from line
Axis of a charged ring	$E_{\text{ring}} = \dfrac{q}{4\pi\epsilon_o}\dfrac{x}{(x^2 + y^2)^{3/2}}$	$q =$ total charge on ring $y =$ ring radius $x =$ distance along axis from ring center
Axis of a charged disk	$E_{\text{disk}} = \dfrac{\sigma}{2\epsilon_o}\left(1 - \dfrac{x}{\sqrt{x^2 + R^2}}\right)$	$\sigma =$ charge per area $R =$ disk radius $x =$ distance along axis from disk
Infinite plane of charge	$E_{\text{plane}} = \dfrac{\sigma}{2\epsilon_o}$	$\sigma =$ charge per area

Two charges of equal magnitude but opposite sign, separated by a distance d form an **electric dipole.** The electric dipole moment $\mathbf{p}$ is a vector that points from the negative charge toward the positive charge, and its magnitude is

$$p = qd \tag{22.11}$$

The torque on an electric dipole in a uniform electric field E is

$$\boldsymbol{\tau} = \mathbf{p} \times \mathbf{E} \tag{22.12}$$

and its potential energy U is

$$U = -\mathbf{p} \cdot \mathbf{E} \tag{22.13}$$

Dipoles (both permanent and induced) are attracted into regions of higher field strength from regions of lower field strength.

PROBLEMS

22.1 Electric Charge

1. How do we know there are only two types of charge? Answer this question by proposing a third kind of charge; call it $*$. Your new charge must interact with existing charges in a unique way. Describe how your charge works by filling in the interaction table below in which R means repulsion and A means attraction.

	$+$	$-$	$*$
$+$	R	A	
$-$	A	R	
$*$			

   Be sure that $*$ has force rules that make it different from $+$ and $-$, and that the columns and rows are self-consistent. Now go find a particle that behaves the way your $*$ charge does. No one has found a charge obeying your new rule yet, but who knows? (P.S. If you find one, be sure to invite us to the party when you celebrate winning the Nobel Prize.)

2. How many electrons are in an ordinary copper penny ($m = 2.65$ g)?

3. **Faraday's constant** is the electric charge of 1.00 mol of protons. What is the magnitude of this constant?

4. Based on the results of alpha-particle experiments performed by Geiger, Marsden, and himself, Lord Rutherford first proposed an atomic model. In these experiments an alpha particle (charge = $+2e$) approaches a gold nucleus (charge = $+79e$). What is the force of the gold nucleus on the alpha particle when they are separated by (a) 2.00×10^{-13} m and (b) 1.00×10^{-13} m?

5. A nonuniform charge distribution lies along the x-axis between $x = 0.0$ and $x = 20.0$ cm and is given by $\lambda = -(3.00\ \mu C/m)\sin[(5\pi\ rad/m)x]$ in this region. (a) Compute the total charge in the region $0 \le x \le 20.0$ cm. (b) How much charge is contained within the region between $x = 5.0$ cm and $x = 15.0$ cm?

6. Charge is distributed along a semicircle of 4.00-cm radius according to the charge density $\lambda(\theta) = (24.0\ \mu C/rad)[1 - \cos(\theta)]$. What is the total charge located between $\theta = 0$ rad and $\theta = \pi$ rad?

7. The two hemispheres that compose a spherical surface (radius = 16.0 cm) are covered with different, but uniform, charge distributions. The charge density on one-half of the sphere is $2.40\ \mu C/m^2$, and that on the other is $-3.60\ \mu C/m^2$. What is the total charge on each hemisphere?

8. The ends and sides of a closed cylinder are covered with different uniform surface-charge distributions. The ends of the cylinder each carry densities $\sigma_{end} = -2.40\ \mu C/m^2$, whereas the curved surface has a charge density $\sigma_{side} = +1.20\ \mu C/m^2$. If the cylinder radius is 8.00 cm and its length is 64.0 cm, find the total charge on the cylinder.

9. Charge is distributed over the surface of a disk of 10.0-cm radius. The resulting charge density is given by $\sigma(r) = (40.0\ \mu C/m^3)r$ for $0 \le r \le 10.0$ cm. What is the total charge on the disk?

10. A cylindrically shaped charge distribution is oriented with its axis along the y-axis. The cylinder is 25.0 cm long and has a radius of 12.0 cm. The charge density increases with distance r from the cylinder axis and is given by

$$\rho(r) = (12.0\ \mu C/m^5)r^2, \qquad 0 \le r \le 12.0\ cm$$

   (a) Compute the total charge within the cylinder. (b) How much charge is contained within a smaller cylindrical region bounded by $r = 6.00$ cm?

11. Charge is distributed throughout a spherical region of space in such a manner that its volume charge density is given by $\rho(r) = (8.00\ \mu C/m^6)r^2$, where $0 \le r \le 0.360$ m. (a) Compute the total charge within this sphere of 36.0-cm radius. (b) How much charge is contained within the spherical region bounded by $r = 0.180$ m?

22.2 Coulomb's Law

12. The natural state of hydrogen at a temperature of $-269°$ is a solid. The closest distance between the protons in this state is 3.75×10^{-10} m. What is the magnitude of the electrostatic force on a proton due to a single nearest-neighbor proton?

13. What must be the separation between two electrons for the Coulomb force between them to be 1.00 dyn?

14. A charge $q_1 = 2.40\ \mu C$ is located in the xy-plane with coordinates (0.500 m, 1.50 m), and a second charge $q_2 = -3.60\ \mu C$ is located at (2.00 m, 1.00 m). (a) Write an expression for a unit vector $\hat{n}$ directed from q_1 to q_2 in terms of $\hat{i}$ and $\hat{j}$. (b) Compute (in unit-vector notation) the force on q_2 by q_1. (c) What is the force on q_1 by q_2?

15. Three charges are placed along the x-axis. A $+3.60$-μC charge is located at $x = 0.50$ m, and charges of $-1.80\ \mu C$ and $+2.40\ \mu C$ are located at $x = 2.00$ m and $x = 3.00$ m, respectively. (a) Find the net electrostatic force on the -1.80-μC charge. (b) Find the net electrostatic force on the $+2.40$-μC charge.

16. Three point–charges lie along a straight line. Charge $q_1 = +4.00\ \mu C$ and is located at 4.00 m from charge $q_2 = +2.00\ \mu C$. Where along the line must the third charge $q_3 = -2.00\ \mu C$ be placed so that the net force on it is zero?

17. The base of a pyramid forms a square of side length a. The top of the pyramid is a distance b above the center of the base. At each of the corners of the pyramid base there is a charge $+Q$. (a) Find an expression for the resultant force on a charge q placed at the top vertex. (b) If $a = 10.0$ cm, $b = 15.0$ cm, and $Q = 2.40\ \mu C$, find the force on a 3.60-μC charge at the top of the pyramid.

18. Three charges are located at the vertices of an equilateral triangle of sides l as shown in Figure 22.P1. If $q_1 = +2.00\ \mu C$, $q_2 =$

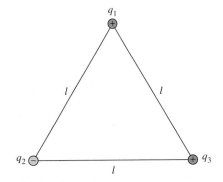

FIGURE 22.P1   Problems 18 and 25

$-3.00 \ \mu C$, $q_3 = +6.00 \ \mu C$, and $l = 8.00$ cm, find the resultant force on q_3.

19. Four point–charges are located at the corners of a square of side l as shown in Figure 22.P2. If $q_1 = -q_2 = e$, $q_3 = -q_4 = -2e$, and $l = 25.0$ cm, what is the resultant electrostatic force on q_2?

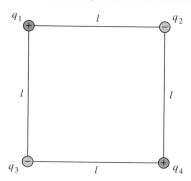

FIGURE 22.P2   Problems 19 and 26

20. A linear charge distribution $\lambda(x) = +(2.00 \ \mu C/m^3)x^2$ lies along the x-axis from $x = 0.00$ to $x = 5.00$ cm. A point-charge $q = +4.00 \ \mu C$ is placed at $x = 15.0$ cm. Find the force of the charge distribution on q.

22.3 The Electric Field

21. The Coulomb force on a proton located at a particular point in space is 6.40×10^{-14} N. What is the magnitude of the electric field at this location?

22. A point-charge of $6.40 \ \mu C$ is located at the origin. Find the magnitude and direction (the angle from the direction of the positive x-axis) of the electric field at (a) $(x, y) = (0.00 \text{ m}, 2.00 \text{ m})$, and (b) $(x, y) = (4.00 \text{ m}, -6.00 \text{ m})$.

23. In his experiment to determine the fundamental charge e, R. M. Millikan balanced the gravitational force on a charged oil droplet by an electric force from an applied, uniform electric field. If an oil drop contained 18 excess electrons, what must its mass have been to be balanced in an electric field of magnitude $E = 3.18 \times 10^5$ N/C?

24. Two point-charges, $q_1 = 4.00 \ \mu C$ and $q_2 = -2.00 \ \mu C$, are located along the x-axis at $x = 1.00$ m and $x = 5.00$ m, respectively. (a) Find the electric field at the origin. (b) Find the electric field at $x = 7.00$ m. (c) Find the electric field at $x = 5.00$ m. (d) Sketch the field lines.

25. Three charges are located at the vertices of an equilateral triangle of sides l as shown in Figure 22.P1. (a) If $q_1 = +4.00 \ \mu C$, $q_2 = -6.00 \ \mu C$, and $l = 12.0$ cm, find the electric field at the position of q_3. (b) If $q_3 = -2.00 \ \mu C$, what is the force on it from the resultant electric field? (c) Sketch the field lines due to q_1 and q_2 only.

26. In the charge arrangement shown in Figure 22.P2, $l = 40.0$ cm, $q_2 = q_3 = -8.00 \ \mu C$, and $q_1 = 12.0 \ \mu C$. (a) Find the electric field at the location of charge q_4. (b) If $q_4 = 3.00 \ \mu C$, what is the magnitude of the Coulomb force on q_4?

27. Two point-charges, $q_1 = +3e$ and $q_2 = -2e$, are located along the y-axis at $y = 0.00$ m and $y = 4.00$ m, respectively. (a) Sketch the resultant field. (b) Find the location on the y-axis where the resultant electric field is zero.

28. At a temperature of $-195°C$ atoms of lithium occupy the corners and the center of a cube 3.49×10^{-10} m along an edge. What is the electric field strength at the position of one of the corners due to the electric field of the eight lithium nuclei located at the other

corners and the cube center? There are three protons in the lithium nucleus.

★29. A charge distribution consisting of three point-charges q, $-2q$, and q located along the y-axis as shown in Figure 22.P3 is called an **electric quadrupole**. (a) Find an expression for the electric field E at a point on the x-axis. (b) Show for large x ($x \gg d$) that

$$\mathbf{E} = -\frac{1}{4\pi\epsilon_o} \frac{3q \, d^2}{x^4} \hat{\mathbf{i}}$$

The quantity $3q \, d^2$ is called the **quadrupole moment**.

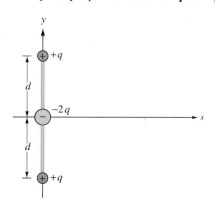

FIGURE 22.P3   Problem 29

30. A linear charge distribution $\lambda = -12.0 \ \mu C/m$ lies along the x-axis from $x = 0.0$ cm to $x = 24.0$ cm. Find the magnitude and direction of the electric field at $x = 36.0$ cm.

31. A linear charge distribution $\lambda(x) = +(2.00 \ \mu C/m^2)x$ lies along the x-axis from $x = 0.0$ cm to $x = 10.0$ cm. (a) Find the electric field at a point on the x-axis located at $x = 15.0$ cm. (b) Sketch the field lines.

32. A total charge Q is distributed uniformly along a rod of length L. Find the electric field E at a distance y along the perpendicular bisector of the rod (Fig. 22.P4).

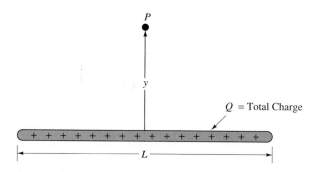

FIGURE 22.P4   Problem 32

33. Charge is distributed uniformly along the entire positive x-axis (from $x = 0$ to $x \to \infty$) with a constant charge density λ. Find the electric field at a point located a distance y from the origin and along the y-axis.

34. A total charge $Q = 20.0 \ \mu C$ is distributed uniformly along a rod bent in the shape of a semicircle of radius $r = 5.00$ cm (Fig. 22.P5). Find the electric field at point P, the center of the semicircle.

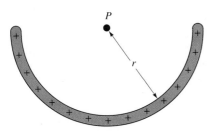

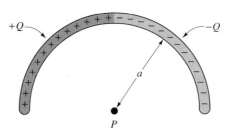

FIGURE 22.P5   Problem 34

FIGURE 22.P8   Problem 38

35. A rod is charged with a uniform charge density λ and is bent into an arc of radius a. If the arc subtends an angle 2θ as shown in Figure 22.P6, find the electric field at point P.

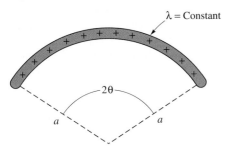

FIGURE 22.P6   Problem 35

36. A linear charge distribution along the y-axis from $y = 0.0$ cm to $y = 10.0$ cm is given by the linear charge density

$$\lambda(y) = -(160.\ \mu C/m^2)y + 8.00\ \mu C/m$$

The charge is confined to this region, and there are no other charges outside the region $0 \le y \le 10.0$ cm. (a) Make a graph of $\lambda(y)$. (b) Compute the electric field strength at $y = 15.0$ cm.

37. Two uniformly charged rods (each with total charge $Q = -8.00\ \mu C$ and length $L = 10.0$ cm) are arranged as shown in Figure 22.P7. Find the total electric field at point P if $a = 5.00$ cm.

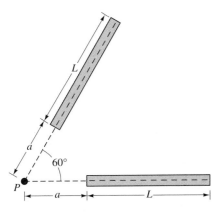

FIGURE 22.P7   Problem 37

38. Charge is distributed uniformly along a semicircular rod of radius a such that half of the semicircle is charged with $+Q$ and the other half is charged with $-Q$ (Figure 22.P8). Find the electric field at point P.

★39. Charge is distributed along a semicircle of radius a with constant charge density λ. The semicircle is located in the yz-plane as shown in Figure 22.P9. Find the electric field at point P a distance x along the x-axis.

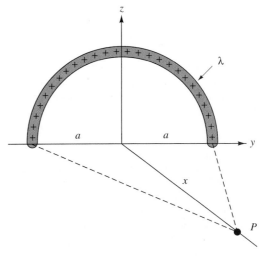

FIGURE 22.P9   Problem 39

22.4 The Electric Dipole in a Uniform Field

40. An electric dipole with a moment $\mathbf{p} = (3.00\ \mu C \cdot m)\hat{\mathbf{i}} - (1.00\ \mu C \cdot m)\hat{\mathbf{j}}$ is placed in the uniform field $\mathbf{E} = (2.00 \times 10^5\ N/C)\hat{\mathbf{i}} + (3.00 \times 10^5\ N/C)\hat{\mathbf{j}}$. (a) What is the magnitude of the torque that $\mathbf{E}$ exerts on $\mathbf{p}$? (b) What is the potential energy of the dipole? (c) What is the angle between $\mathbf{p}$ and $\mathbf{E}$?

41. An electric dipole moment $\mathbf{p} = (2.00\ \mu C \cdot m)\hat{\mathbf{i}} + (3.00\ \mu C \cdot m)\hat{\mathbf{j}} - (1.00\ \mu C \cdot m)\hat{\mathbf{k}}$ is placed in the uniform field $\mathbf{E} = (3.00 \times 10^5\ N/C)\hat{\mathbf{i}} + (2.00 \times 10^5\ N/C)\hat{\mathbf{k}}$. (a) Compute the magnitude of the torque that $\mathbf{E}$ exerts on $\mathbf{p}$. (b) What is the potential energy of the dipole moment $\mathbf{p}$? (c) If the dipole were to be rotated in the field $\mathbf{E}$ what is the maximum potential energy of the dipole? (d) How much work does the E-field perform on the dipole if it rotates it from an antiparallel alignment to a parallel alignment with $\mathbf{E}$?

42. An electric dipole consists of charges with a magnitude of 2.00×10^{-12} C separated by 1.00 mm. The dipole is placed in a uniform E-field of magnitude 3.00×10^6 N/C, where the dipole and field point in the same direction. (a) What is the potential energy of the dipole in this orientation? (b) What is the potential energy of the dipole after it has been rotated to an orientation antiparallel to the field? (c) How much work was required to rotate the dipole from its initial alignment with the field to its orientation in part (b)?

43. Two identical dipoles of moment p lie along a straight line and are oriented in the same direction along the line. If their centers are a distance R apart, where R is much greater than the size of the individual dipoles, show that the force by one dipole on the other is $3p^2/(2\pi\epsilon_o R^4)$.

44. An electric dipole consists of two point-charges $\pm q$, each of mass m, separated by distance d. The dipole is placed in a uniform electric field of magnitude E_o. Show that the dipole exhibits simple harmonic motion with period $2\pi\sqrt{md/2qE_o}$ when it is rotated through a *small* angle θ and released.

Numerical Methods and Computer Applications

45. An electric dipole is oriented along the y-axis with its charges $+q$ and $-q$ located at $(0, d/2)$ and $(0, -d/2)$, respectively. Write a computer program or develop a spreadsheet template to compute and plot the resulting electric field magnitude from $x = -10d$ to $x = +10d$ along lines parallel to the x-axis given by an input value for y. Use the program or spreadsheet template for $q = 2.00\ \mu C$, $d = 1.00$ mm, and (a) $y = 0.25$ mm, (b) $y = 0.75$ mm, (c) $y = 2.0$ mm and (d) $y = 10.$ mm. Compare and explain the results of each plot.

46. In Example 22.6 we derived an expression for the E-field on the axis of a dipole. An approximate result was also obtained that we expect to work for distances $x \gg d$. Write a program or develop a spreadsheet template to compute and plot the magnitude of the electric field along the x-axis for both results. Try out the program for $q = 5.00\ \mu C$ and $d = 0.500$ mm. Use your program or spreadsheet template to find the distance from the dipole where the exact expression for E and the approximate one agree to within 1%.

47. Write a program or develop a spreadsheet template to compute the force on a point-charge q_o located at position x_o due to other charges located along the x-axis. (a) In the first case let there be two other charges each of magnitude $Q/2$, one of which is located at the $x = 0$ origin and the other at $x = L < x_o$. (b) Modify the program or spreadsheet template to compute the force on q_o from four charges each of magnitude $Q/4$, equally spaced from the origin to location $x = L$. (c) Modify this program or spreadsheet template to compute the force on q_o from 24 charges each of magnitude $Q/24$. Assume the charges are equally spaced between the origin and $x = L$. Take $q_o = 0.200\ \mu C$, $L = 10.0$ cm, and $x_o = 25.0$ cm. Select Q so that the total charge is the same as that on the rod in Example 22.5 and compare the force from this example to the results you obtain from your program or spreadsheet.

48. Modify the Runge-Kutta spreadsheet template RK1D.WK1 of Chapter 6 to calculate the position versus time of a point-charge q_o moving with initial velocity v_o directly toward a fixed point-charge q. That is, use the Coulomb force law to compute and plot r versus t for the charge q_o.

49. Modify the Runge-Kutta program RK2D.BAS (or write your own program) to plot the trajectory of a point-charge q_o that is fired at a stationary charge q located at the origin. Let the initial velocity of q_o be $v = v_o\hat{\mathbf{i}}$ and its initial position be (x_o, y_o). Choose x_o so that the particle starts at the far left-hand side of the screen. The initial aiming error y_0 is known as the **impact parameter**. The paths you are simulating here are the kind of trajectories that alpha particles take when scattering from gold nuclei in the famous Rutherford scattering experiments.

50. A uniform distribution of constant linear charge density λ lies along the x-axis from $x = 0$ to $x = L$. (a) Derive an expression for the *magnitude* only of the electric field strength at a general point (X_o, Y_o) in the xy-plane. (b) Develop a spreadsheet or write a program to compute this magnitude at 51 points at intervals of ΔX along a line parallel to the line of charge at a fixed distance Y_f from $X = -25\ \Delta X$ to $X = +25\ \Delta X$. (c) Test the program or spreadsheet for the following parameters: $\lambda = 24.0$ pC/m, $L = 60.0$ cm, $Y_o = 2.00$ cm, and $\Delta X = 3.00$ cm.

General Problems

51. Three parallel, infinite planes carry different charge densities and are parallel to the yz-plane. One plane passes through the origin and has a charge density $\sigma_1 = -2.00\ \mu C/m^2$. Another plane passes through $x = 1.00$ m and has $\sigma_2 = -1.00\ \mu C/m^2$. Finally, a plane passing through $x = 3.00$ m has $\sigma_3 = +3.00\ \mu C/m^2$. What is the electric field at (a) $x = -1.00$ m, (b) $x = 0.500$ m, (c) $x = 2.00$ m, and (d) $x = 4.00$ m?

52. (a) Sketch the E-field in the neighborhood of a uniformly charged disk. (b) Show that the E-field on the axis of a uniformly charged disk approaches the E-field of a point-charge for large distances from the disk. [*Hint:* Using E_{disk} from Example 22.10 show first that $x(x^2 + R^2)^{-1/2} = (1 + R^2/x^2)^{-1/2}$, and then apply the binomial expansion where $(1 + u)^n \approx 1 + nu$ when $u \ll 1$.]

★53. A point-charge $-q_o$ is placed at the center of a uniformly charged ring of total charge $+Q$ and radius R. The point-charge is confined to move only along the axis of the ring. The point-charge is displaced a small distance along the axis $(x \ll R)$ and released. Show that the point-charge executes simple harmonic motion with period $2\pi\sqrt{(4\pi\epsilon_o mR^3)/(qQ)}$ where m is the mass of the point-charge.

54. A small mass $m = 2.00$ g carrying a charge $Q = +6.00\ \mu C$ is attached to the end of a light string. The other end of the string is fixed to a vertical, infinite plane of charge density σ. As shown in Figure 22.P10 the string makes an angle $\theta = 36.87°$ with the vertical. Find the charge density σ and the tension in the string.

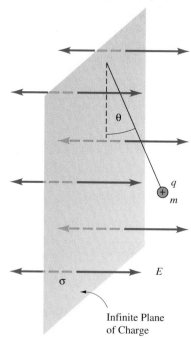

FIGURE 22.P10   Problem 54

55. Two small spheres carry charges of equal magnitude but opposite signs and are attached to strings of length $l = 20.0$ cm as shown in Figure 22.P11. The mass of each sphere is 5.00 g. A horizontal uniform electric field $E = 1.75 \times 10^5$ N/C exists in this region of space. The angle θ is 25.0°. Find the magnitude of the charge q.

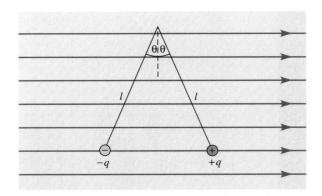

FIGURE 22.P11   Problem 55

56. What must be the electric field strength E to cause a charge q of mass m to accelerate *upward* with an acceleration magnitude $a_g = 9.80 \text{ m/s}^2$?

57. A uniform electric field $E = 3200.$ N/C is applied to a proton initially at rest. (a) What is the magnitude of the proton's acceleration? (b) How far has the proton traveled when its velocity is 2.50×10^5 m/s? (c) How much time has it taken to travel this distance?

58. An alpha particle is a helium nucleus (see Problem 4) and therefore consists of two protons and two neutrons. Such a particle has an initial velocity of $+1.60 \times 10^3$ m/s $\hat{\mathbf{i}}$ and enters a region where the electric field is $\mathbf{E} = -500.$ N/C $\hat{\mathbf{i}}$. (a) How much time passes before the alpha particle comes to rest? (b) How far has the alpha particle traveled during this time?

59. An electron enters a region of space where there is an electric field of 200. N/C directed downward parallel to the y-axis. The velocity of the electron as it passes the origin is 5.00×10^4 m/s at an angle of $36.87°$ below the positive x-axis. (a) How much time elapses before the y-coordinate of the electron is $+10.0$ cm? (b) What is the x-coordinate when $y = +10.0$ cm? (c) What is the electron's velocity when $y = +10.0$ cm?

60. A proton traveling with velocity 3.60×10^4 m/s $\hat{\mathbf{i}}$ enters a region of space where there is an electric field $\mathbf{E} = -5.0 \times 10^6$ N/C $\hat{\mathbf{i}}$. (a) What is the proton's acceleration? (b) After it enters the electric field, how much time passes before the proton comes to rest? (c) How far does it travel in the field before coming to rest?

61. An electron is traveling with a speed of 2.40×10^5 m/s along the x-axis in the direction of increasing x when it enters a region of space where the electric field has a magnitude of 8.00 N/C. The field is directed downward parallel to the negative y-axis. (a) Determine the proton's y-coordinate after it has traveled into the field a distance of 20.0 cm as measured parallel to the x-axis. (b) What is the proton's velocity at this instant?

★62. In a cathode-ray oscilloscope an electron with an initial velocity v_o enters a uniform electric field E as shown in Figure 22.P12. As it passes through the region of length l where an electric field exists, it is deflected a total vertical distance h from its original horizontal trajectory. (a) Show that the deflection distance h is given by

$$h = \left(\frac{eE}{2mv_o{}^2}\right)l^2$$

(b) Find an expression for the total deflection distance H at a vertical screen that is a distance L from the region of the uniform field. (Assume the transition from full field to no field is abrupt.)

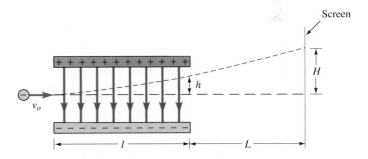

FIGURE 22.P12   Problem 62

Gauss's Law

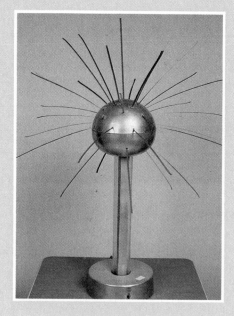

In this chapter you should learn

- the definition of electric flux.
- Gauss's law and how to apply it to compute the electric field for three symmetrical charge distributions.
- some properties of the ideal-conductor model.

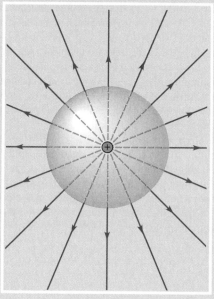

$$\oint \mathbf{E} \cdot d\mathbf{A} = \frac{q_{\text{encl}}}{\epsilon_o}$$

In Chapter 22 we introduced the concept of an electric field, and you learned how to compute the magnitude and direction of the *E*-field for a variety of point and continuous charge distributions. With the field concept well in hand and some experience with its application, we are almost ready to present the first of Maxwell's equations that were foreshadowed in the introduction to this part of the text (Part III). However, the form of this equation that will be most useful to us involves yet one more new idea, that of flux. We devote the next section to the definition and properties of electric flux. The idea of flux will be useful not only to describe electric fields in this chapter but also magnetic fields in later chapters.

23.1 Electric Flux

In Section 3 of Chapter 22 we mentioned that it is important to visualize electric field lines in three dimensions. Moreover, we hinted that you must also be able to picture three-dimensional surfaces surrounding a charge distribution. In this section we bring both of these representations together. Although we argued that the number of lines per unit area that pass perpendicular through a surface is *proportional* to the electric field strength, thus far we have used electric field lines primarily to provide a *qualitative* representation of the electric field.

What we want to try to establish now is a *quantitative* way to describe the number of such lines passing through an area. This quantitative measure is called the **electric flux** Φ_E.

> The electric flux Φ_E is a measure of the number of electric field lines passing perpendicularly through a surface.

Let's begin by visualizing field lines passing at right angles through a rectangular surface, as in Figure 23.1(a). Coincident with one of the field lines, we have also drawn a unit vector $\hat{\mathbf{n}}$ that is used to keep track of the surface orientation. We will maintain $\hat{\mathbf{n}}$ perpendicular to the surface area A, no matter which way we tilt the surface.

Because the number of field lines *per area* is proportional to the electric field strength E, the total number of field lines passing through area A of Figure 23.1(a) is proportional to the product EA. That is, stronger *E*-fields have more lines per area. Similarly, for a constant field strength E, larger areas A have more field lines passing through them. Hence, a quantitative measure of the number of field lines that pass through a surface A is

$$\Phi_{E\perp} = EA \qquad (23.1)$$

In Figure 23.1(b) we have tilted the surface area A so that the normal $\hat{\mathbf{n}}$ makes an angle θ with the field lines. Clearly, there are fewer lines passing through the surface (some of them now miss the surface). In fact, when $\theta = 90°$ *no* field lines pass *through* the surface.

Figure 23.1(c) shows a projection of the tilted area A onto a plane perpendicular to the *E*-field. The size of the projected area is $A \cos(\theta)$. The number of field lines passing through the area $A \cos(\theta)$ is the same as the number passing through the tilted area A. Hence, in this case the total number of field lines is proportional to the product of E and $A \cos(\theta)$, and we may write the electric flux as

$$\Phi_E = EA \cos(\theta) \qquad (23.2)$$

Notice when $\theta = 0°$, Equation (23.2) becomes Equation (23.1). Equation (23.2) is the quantitative measure we seek for the number of field lines passing through a surface, as long as *the electric field is constant and the surface area is flat*.

There is another way to view Equation (23.2). Notice that we can rewrite it as $\Phi_E = [E \cos(\theta)]A$. Now, $E \cos(\theta)$ is the component of the electric field perpendicular to the

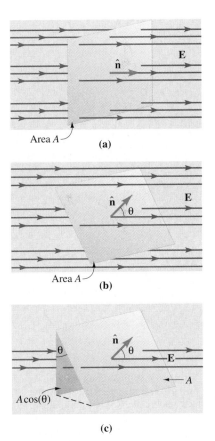

FIGURE 23.1

(a) *E*-field lines passing perpendicular through a surface area A. The unit vector $\hat{\mathbf{n}}$ is perpendicular to the surface A. (b) When the surface is tilted at angle θ, fewer lines pass through it. (c) The number of field lines passing perpendicular through the projected area $A\cos(\theta)$ is equal to the number of lines passing through the tilted area A.

surface area A. We call that component $E_\perp$. Consequently, another interpretation of the flux definition in Equation (23.2) is that, for a given surface area A, Φ_E includes only the perpendicular component of E through the (untilted) area A. We will remind you of this useful point of view at the end of this section.

Before we generalize Equation (23.2) for variable E-fields and other surface shapes, we resurrect the vector dot product notation in order to put this equation into a more convenient form. That is, we want to make $EA\cos(\theta)$ into a dot product. To do this we make the area A into a vector with direction parallel to $\hat{\mathbf{n}}$. The magnitude of $\mathbf{A}$ is just the size of the surface area A. All this may seem a little strange to you now, but we promise that it is useful and you will quickly realize the convenience of such a convention. To summarize this new definition,

$\mathbf{A}$ is the area vector; its magnitude is the area A (units of m^2 in SI) and its direction is perpendicular to the surface it represents.

Now, we realize that there are two sides to a surface, and we suspect that you are probably wondering what determines the side on which you should place the vector $\mathbf{A}$. For the present, it doesn't matter because we will always designate our choice. Later in this section when we consider *closed* surfaces we provide a convention for designating the direction of $\mathbf{A}$. In later chapters additional significance of the choice of area direction will be given.

With this definition of the area vector we can write Equation (23.2) as

$$\Phi_E = \mathbf{E} \cdot \mathbf{A} \tag{23.3}$$

We now extend the definition of electric flux to include surfaces of more general shape over which the electric field can vary in both magnitude and direction. Keep in mind that Φ_E is just a measure of the number of field lines that pass through a surface at right angles to that surface. In Figure 23.2 we show an arbitrarily shaped surface through which many field lines can be imagined to pass, although we only show a few such lines. We can obtain an approximate value for the total flux through the entire surface area by dividing the surface into small area elements ΔA. These area elements are chosen to be small enough so that over each of them $\mathbf{E}$ is approximately constant. If we number these area elements from $i = 1$ to N, the flux Φ_{E_i} through the ith area element ΔA_i is

$$\Phi_{E_i} = E_i\,\Delta A_i \cos(\theta) = \mathbf{E}_i \cdot \Delta\mathbf{A}_i$$

and the total electric flux through all N area elements is approximately

$$\Phi_E \approx \sum_{i=1}^{N} \mathbf{E}_i \cdot \Delta\mathbf{A}_i$$

We obtain the exact expression for Φ_E by taking the limit as the number of area elements grows to infinity and the size of each approaches zero:

$$\Phi_E = \lim_{\substack{N \to \infty \\ \Delta A_i \to 0}} \sum_{i=1}^{N} \mathbf{E}_i \cdot \Delta\mathbf{A}_i$$

$$\Phi_E = \int \mathbf{E} \cdot d\mathbf{A} \tag{23.4}$$

The integral on the right-hand side of Equation (23.4) is known as a **surface integral**, and you must always pay careful attention to the shape of the surface over which it is evaluated. Also, as we mentioned previously, $\mathbf{E}$ can vary over the surface; only in certain high-symmetry cases does $\mathbf{E}$ turn out to be a constant on a surface. In this latter case,

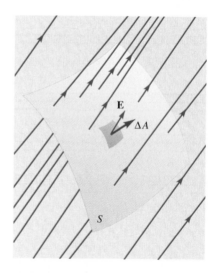

FIGURE 23.2

The number of field lines per area is not constant over the surface S. However, the electric field $\mathbf{E}$ is approximately constant over a small surface element $\Delta\mathbf{A}$. The net electric flow through the entire surface is the sum of the flux through all area elements, which together constitute the surface S.

however, **E** can then be removed from the integral. We are happy to report that you will encounter many of these fortunate situations!

We are on a collision course with one of Maxwell's equations, and that equation contains a term that is the flux through a *closed* surface. Such a surface is shown in Figure 23.3. For a closed surface we must reach some agreement about the direction of the area vector. We adopt the convention (consistent with the rest of the world) that area vectors always point *outward* from the closed surface. The directions of the area vector at several different locations are shown on the closed surface in Figure 23.3. Notice that $d\mathbf{A}$ does indeed change direction with position on the surface. Every $d\mathbf{A}$ must be perpendicular to the particular surface element it represents. We use the symbol $\oint$ to represent an integral over a *closed* surface. In this case

$$\Phi_{E_{closed}} = \oint \mathbf{E} \cdot d\mathbf{A} \qquad (23.5a)$$

As we mentioned previously, $E\cos(\theta)$ is just the component of the electric field $E_\perp$ that is perpendicular to the surface dA. When we write out the dot product in Equation (23.5a) the integral becomes $\oint E\cos(\theta)dA = \oint E_\perp\,dA$. Therefore, we can also write the electric flux definition for a closed surface as

$$\Phi_{E_{closed}} = \oint E_\perp\,dA \qquad (23.5b)$$

We illustrate the concept to electric flux in Example 23.1.

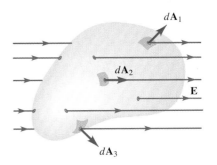

FIGURE 23.3

Electric field lines passing through a closed surface. The area vectors are perpendicular and outward from the surface at every point on the surface.

EXAMPLE 23.1 *What Goes In, Must Come Out (at Least for This Example)*

In a region of space the electric field is uniform with a magnitude E_o. Compute the electric flux through a closed cylindrical surface of radius R if the axis of the cylinder is parallel to the field direction.

SOLUTION The cylindrical surface and several E-field lines with associated surface elements are shown in Figure 23.4. We must compute the flux through both ends of the cylinder and the curved portion of its surface. Hence, the closed integral given in Equation (23.5) consists of the sum of three terms:

$$\Phi_{E_{closed}} = \oint \mathbf{E} \cdot d\mathbf{A} = \int_{\substack{left\\end}} \mathbf{E} \cdot d\mathbf{A} + \int_{\substack{right\\end}} \mathbf{E} \cdot d\mathbf{A} + \int_{\substack{curved\\surf}} \mathbf{E} \cdot d\mathbf{A}$$

As Figure 23.4 shows, for all the surface elements on the left end the angle between **E** and $d\mathbf{A}$ is 180°, so

$$\int_{\substack{left\\end}} \mathbf{E} \cdot d\mathbf{A} = \int_{\substack{left\\end}} E_o\cos(180°)\,dA = -E_o\int_{\substack{left\\end}} dA = -E_o\pi R^2$$

where we have removed E_o from the integral because it is a constant and $\int_{left\,end} dA$ is just the area of the circle that makes up the left end.

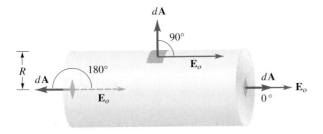

FIGURE 23.4

The electric flux through the surface of a closed cylinder. See Example 23.1.

All the surface elements on the right end of the cylinder are parallel to E_o, so

$$\int_{\substack{\text{right} \\ \text{end}}} \mathbf{E} \cdot d\mathbf{A} = \int_{\substack{\text{right} \\ \text{end}}} E_o \cos(0°) \, dA = +E_o \int_{\substack{\text{right} \\ \text{end}}} dA = E_o \pi R^2$$

Finally, at every point on the curved surface the area vectors are perpendicular to the direction of the electric field. Thus,

$$\int_{\substack{\text{curved} \\ \text{surf}}} \mathbf{E} \cdot d\mathbf{A} = \int_{\substack{\text{curved} \\ \text{surf}}} E_o \cos(90°) \, dA = 0$$

Combining these three results, we have

$$\Phi_{E_{\text{closed}}} = -E_o \pi R^2 + E_o \pi R^2 + 0 = 0$$

We have shown that the total flux through a closed cylindrical surface from a uniform electric field parallel to the cylinder's axis is zero. This result foreshadows an important law discussed in the following section. ◄

23.2 Gauss's Law

We are now in a position to state the first of Maxwell's equations, called **Gauss's law**. This equation encompasses many experimental results, including Coulomb's law. In fact, it is possible to derive Coulomb's law from Gauss's law. To view the two laws as equivalent, however, is a mistake. Coulomb's law expresses the mutual force between two *static* charges. When charges are in relative motion Coulomb's law does not provide the correct force of one charge on another. However, Gauss's law remains valid for moving charges and is, therefore, more general than Coulomb's law.

Gauss's law provides a relation between the total electric flux through any closed surface and the charge enclosed within that surface. The law simply states

$$\Phi_{E_{\text{closed}}} = \frac{q_{\text{encl}}}{\epsilon_o} \tag{23.6}$$

where q_{encl} is the net charge enclosed by the surface over which the integral is performed. This surface need not be real and, in fact, for most applications of Gauss's law it is an imagined, closed boundary called a **Gaussian surface**.

In principle, Gauss's law may be used to compute the electric field for any charge geometry, including both point-charge collections and continuous charge distributions. However, the form of the law given by Equation (23.6) is most useful for certain symmetrical charge distributions. In these cases a Gaussian surface may be found over which the electric field is constant and $\mathbf{E}$ can be removed from the integral in Equation (23.6). Before we apply Gauss's law to compute electric fields we want to illustrate how the law applies to two general cases.

Gaussian Surfaces

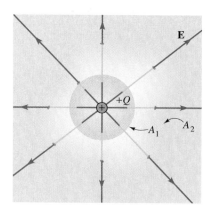

FIGURE 23.5

A spherical symmetric charge distribution surrounded by two concentric Gaussian surfaces. The inner Gaussian surface A_1 is spherical, whereas the outer surface A_2 is arbitrarily shaped.

Figure 23.5 shows a uniform ball of charge surrounded by two concentric, Gaussian surfaces. The inner surface is spherical with its center coinciding with that of the charge. The outer surface is nonspherical. Gauss's law states that the electric flux through both these surfaces is exactly the same, namely, Q/ϵ_o. How can this be? Although the surface area A_2 is larger than A_1, it is located farther away from the charge so the field strength is less at all points on the surface A_2. According to Gauss's law, this decrease in the E-field strength just compensates for the increase in surface area so that Φ_E remains equal to Q/ϵ_o.

When no charge is enclosed by the Gaussian surface, as illustrated in Figure 23.6, the net flux through the surface is zero. The total flux entering the surface is equal in magnitude but opposite in sign to that leaving the surface. You should now understand why the total flux through the cylindrical surface in Example 23.1 turned out to be zero; the charge enclosed by the cylinder was zero. Even when there are charges within a closed surface, if they are both positive and negative so that the total charge is zero, Gauss's law tells us that the net flux through the surface is zero.

Coulomb's Law from Gauss's Law

Every result of the previous chapter may be obtained from Gauss's law because Coulomb's law can be derived from Gauss's law. In fact, let's do just that. The derivation of Coulomb's law from Gauss's law is an excellent starting point for you to see how to apply Gauss's law to certain high-symmetry cases.

We begin by surrounding a point-charge $+q$ by a spherical Gaussian surface of radius r, as shown in Figure 23.7. We chose the Gaussian surface to be a sphere centered on the point-charge for two very important reasons. First, from symmetry, the electric field strength E *is constant* on the surface of the sphere. In the previous sentence, the words *from symmetry* mean that it doesn't matter where you are on the surface, the charge q looks the same. Moreover, any point on the sphere is the same distance from this charge, so **E** must have the same magnitude at all points on this Gaussian surface.[1] These observations are important because they mean we can remove **E** from the surface integral in Equation (23.6). The second reason we chose a spherical Gaussian surface is that at every point on this surface the electric field **E** is parallel to the area vector $d\mathbf{A}$. This point is illustrated for one of the E-field lines shown in Figure 23.7. Hence, Gauss's law gives

$$\oint \mathbf{E} \cdot d\mathbf{A} = \frac{q_{\text{encl}}}{\epsilon_o}$$

$$\oint E \cos(0°)\, dA = \oint E\, dA = \frac{q}{\epsilon_o}$$

But, as we said, E is constant over the spherical surface and, therefore, can be taken out of the integral:

$$E \oint dA = \frac{q}{\epsilon_o}$$

Now, the area of a sphere is $4\pi r^2$ so

$$E(4\pi r^2) = \frac{q}{\epsilon_o}$$

or

$$E = \frac{1}{4\pi\epsilon_o}\frac{q}{r^2}$$

With this result we can establish all the physics we developed in the previous chapter, including Coulomb's law: From our definition of the electric field, the force on a point-

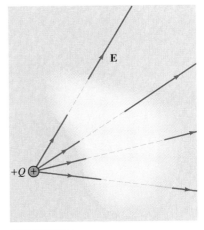

FIGURE 23.6

When no charge is enclosed by the Gaussian surface, the net electric flux through the surface is zero. The total number of E-field lines entering the surface equals the number leaving it.

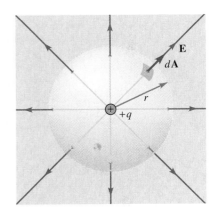

FIGURE 23.7

Derivation of Coulomb's law from Gauss's law. A point-charge q is surrounded by a spherical Gaussian surface with a center that coincides with the position of q.

---

[1] This equality is the case only if the electric field squirts outward from the charge evenly in every direction. This assumption is equivalent to saying that all directions in space are the same. If anything makes one direction special, the symmetry is said to be "broken." For example, motion of the charge in a particular direction could, in principle, break this symmetry, so perhaps we should not be surprised if the result (Coulomb's law) does not hold for moving charges.

charge q_o located a distance r from the charge q is $F = q_o E$. Therefore,

$$F = \frac{1}{4\pi\epsilon_o} \frac{qq_o}{r^2}$$

the result we were seeking.

As we mentioned toward the beginning of this section, Gauss's law is most useful when the charge distribution has a high degree of symmetry. There is a systematic means by which we can apply Gauss's law to these types of problems. Below we list the steps we apply to the examples that follow.

Problem-Solving Strategy for Gauss's Law

Calculate the flux:

1. Observe the symmetry of the electric field produced by the charge distribution. If it's spherical, select a spherical Gaussian surface. If it's linear, cylindrical, or planar, select a cylindrical Gaussian surface.
2. Orient the Gaussian surface so that it surrounds all or a portion of the charge in such a manner that the electric field is either constant over all parts of the surface or is perpendicular to the surface area vector.
3. Evaluate the flux through each surface element of the Gaussian surface. Be sure to note the angle between **E** and the area vector pointing outward from the Gaussian surface.

Calculate the charge:

4. Compute the charge enclosed within the surface and divide it by ϵ_o.

Apply Gauss's law:

5. Write Gauss's law. Replace the flux $\Phi_{E_{closed}}$ with the expression you derived in steps (1)–(3) and replace q_{encl}/ϵ_o with your result from step (4). Solve the resulting equation for the electric field.

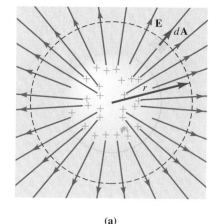

(a)

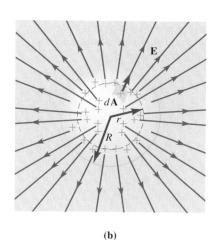

(b)

FIGURE 23.8

(a) A spherical Gaussian surface surrounds a ball of charge. (b) A spherical Gaussian surface within the ball of charge. See Example 23.2.

This is a good time for you to look at a few examples. Example 23.2 points out an extremely important result for the special case of spherical charge distribution. Examples 23.3 and 23.4 reexamine some of the examples of the previous chapter.

EXAMPLE 23.2   Everything Looks Like a Point When You're Far Away, But Spheres Look Like Points Even When You're Close. Get the Point?

A total charge Q is distributed uniformly throughout a sphere of radius R so that the charge density ρ is a constant. Compute the electric field strength at a point (a) outside the sphere, and (b) inside the sphere.

SOLUTION   (a) We first calculate the flux. Because the charge distribution is spherically symmetrical, step (1) of the problem-solving strategy tells us to select a spherical Gaussian surface. Following step (2) we note that if we center the Gaussian surface at the center of the ball of charge (Fig. 23.8(a)), E must have the same magnitude at all points on the surface of the sphere. Moreover, E is also parallel to every area vector $d\mathbf{A}$ of the surface. Therefore, at all points on the Gaussian surface $\mathbf{E} \cdot d\mathbf{A} = E\, dA$, and the expression for flux reduces to

$$\Phi_E = \oint \mathbf{E} \cdot d\mathbf{A} = E \oint dA = E(4\pi r^2)$$

where in the final step we used the fact that the area of the sphere is $4\pi r^2$. We have now finished steps (1)–(3). In step (4) we calculate the charge inside of the Gaussian surface. For part (a) of this example this is trivial: We are told the total charge is Q and all of this charge is inside the Gaussian surface. Finally (step (5)) we write Gauss's law

$$\Phi_{E_{\text{closed}}} = \frac{q_{\text{encl}}}{\epsilon_o}$$

Into this law we substitute the flux $E(4\pi r^2)$ on the left-hand side and the charge Q on the right-hand side:

$$E(4\pi r^2) = \frac{Q}{\epsilon_o}$$

Solving for E, we have

$$E_{\text{out}} = \frac{1}{4\pi\epsilon_o}\frac{Q}{r^2} \qquad (r \geq R)$$

We subscripted this E field as E_{out} to emphasize that this expression for the value of the electric field is restricted to positions outside the sphere of charge. This is so because for values of $r < R$ not all of the charge Q is inside the Gaussian surface. Notice too that this expression is the same as the expression for the field of a point-charge Q located at the center of the sphere. Hence, we conclude that for a uniform, spherical charge distribution, the electric field *outside* the charge distribution is equivalent to that of a point-charge located at the sphere's center. This result turns out to be true for *any* spherically symmetrical charge distribution (see Problem 38 at the end of this chapter).

(b) Steps 1 and 2: For a point within the sphere we also select a spherical Gaussian surface, as shown in Figure 23.8(b), but note that this time the radius of that imaginary sphere r is less than R. The electric field is again constant in magnitude on the imaginary surface and is parallel to all of the outward directed area vectors $d\mathbf{A}$. Therefore, we can again calculate electric flux through the spherical surface just as before. In fact we obtain the exact same expression for the flux:

$$\Phi_E = \oint \mathbf{E} \cdot d\mathbf{A} = E(4\pi r^2)$$

When we apply step (4) we must be careful to use only the charge *enclosed within the Gaussian surface*. The charge enclosed is the constant-charge density multiplied by the volume *enclosed by the Gaussian surface*. Denoting the density of the charge as ρ, the charge enclosed by a sphere of radius $r < R$ is

$$q_{\text{encl}} = \rho\left(\frac{4}{3}\pi r^3\right)$$

We proceed with step (5), using these new expressions for the flux and q_{encl}. Gauss's law becomes

$$E(4\pi r^2) = \frac{\rho}{\epsilon_o}\left(\frac{4}{3}\pi r^3\right)$$

and solving for the electric field, we find

$$E_{\text{in}} = \left(\frac{\rho}{3\epsilon_o}\right)r$$

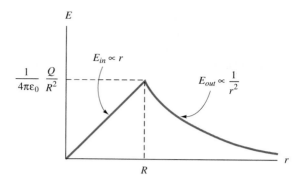

FIGURE 23.9

The electric field inside and outside a uniform ball of charge with radius R. Notice that $E_{\text{in}} = E_{\text{out}}$ at $r = R$.

Because the charge density is uniform and $\rho = Q(\frac{4}{3}\pi R^3)^{-1}$, we can rewrite this result as

$$E_{\text{in}} = \frac{1}{4\pi\epsilon_o}\frac{Q}{R^3}r \qquad (r \leq R)$$

The important feature of this result is that E increases *linearly* in r in the region $0 \leq r \leq R$. Also notice that $E_{\text{in}} = 0$ at $r = 0$.

Notice that at the boundary $r = R$ between the charge sphere and empty region outside the two solutions E_{out} and E_{in} agree. This is not too surprising because we know that electric field lines start and end only on charges. Imagine the Gaussian surface starting at $r > R$ and then shrinking. As the radius of the Gaussian surface shrinks past the $r = R$ boundary, at first only a few charges are excluded and the E-field decreases gradually. This is an example of the general rule that the electric field must be continuous in the absence of a **surface charge distribution**. In a surface charge distribution a great many charges are spread out in a nearly zero-thickness layer. In such a case the E-field jumps abruptly. (In the next section we will show that E may be discontinuous at the surface of a conductor.) You should verify that E_{out} and E_{in} are indeed equivalent at $r = R$. The solutions for the two regions are graphed in Figure 23.9. Notice that as r increases there is a linear growth of E_{in} and a $1/r^2$ decay of E_{out}.

Concept Question 1
What is the total flux through a Gaussian surface that surrounds an electric dipole?

EXAMPLE 23.3   *Impress Your Friends! Confound Your Enemies! Do Vector Surface Integrals in Your Head!*

Compute the electric field at distance y from an infinite line of charge if the charge per length is equal to the constant λ.

SOLUTION

1. The charge distribution is linear so we use a cylindrical Gaussian surface.
2. A uniform line looks the same from the top, bottom, front, and back. If you were in a universe where only the line existed, there would be only two special directions: parallel to the line and perpendicular to the line. In the direction parallel to the line, right is no different from left; nature has no way to choose one over the other. Thus, symmetry dictates that a test charge placed a distance y from the infinite line of charge experiences an equal force to the right and to the left; which is to say no net force at all parallel to the wire. From this symmetry we see that the electric field lines must be perpendicular to the line of charge. Moreover, as we move around at a constant distance y from the line of charge the strength of

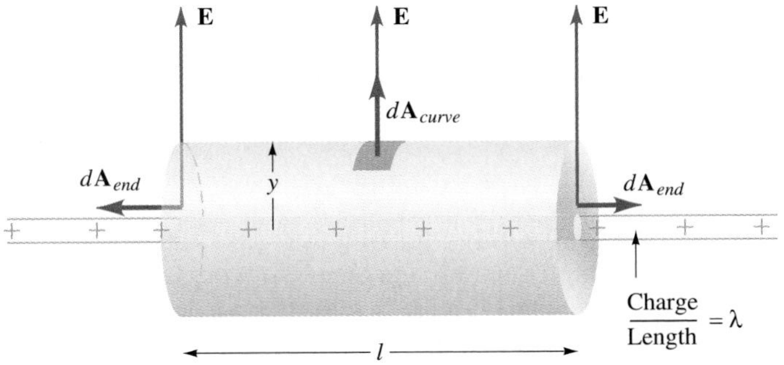

FIGURE 23.10

A cylindrical Gaussian surface surrounding a segment of an infinite line of charge. See Example 23.3.

the E-field is the same in all directions. We therefore surround a segment of the charged line with our cylindrical Gaussian surface of radius y, and length l as shown in Figure 23.10.

3. We must be sure to include all the cylindrical surfaces when we compute the flux:

$$\Phi_E = \oint \mathbf{E} \cdot d\mathbf{A} = \int_{\substack{\text{left}\\\text{end}}} \mathbf{E} \cdot d\mathbf{A} + \int_{\substack{\text{right}\\\text{end}}} \mathbf{E} \cdot d\mathbf{A} + \int_{\substack{\text{curved}\\\text{surf}}} \mathbf{E} \cdot d\mathbf{A}$$

On both ends of the cylinder, the radial $\mathbf{E}$ is perpendicular to the axial area vectors $d\mathbf{A}$. Hence, in the first two integrals $\mathbf{E} \cdot d\mathbf{A} = 0$. On the curved surface $\mathbf{E}$ is always parallel to the area vector so that in the second integral $\mathbf{E} \cdot d\mathbf{A} = E\,dA$. Moreover, E is constant on the curved surface, so the flux becomes

$$\Phi_E = 0 + 0 + E\int_{\substack{\text{curved}\\\text{surf}}} dA = E(2\pi yl)$$

where in the previous step we used the fact that the lateral surface area of a cylinder is $2\pi yl$. Don't miss the fact that we just completed an intimidating bit of vector calculus (the surface integral) by merely multiplying the constant E times an area. (Would that all calculus were this easy!)

4. The total charge enclosed by the Gaussian surface is the product of the charge per length times the length: $q_{\text{encl}} = \lambda l$.

5. We write Gauss's law, and substitute our expressions for flux and charge:

$$\Phi_{E_{\text{closed}}} = \frac{q_{\text{encl}}}{\epsilon_o}$$

$$E(2\pi yl) = \frac{\lambda l}{\epsilon_o}$$

Finally, we solve the resulting equation for E:

$$E = \frac{\lambda}{2\pi\epsilon_o y}$$

This result is precisely that which we obtained in Example 22.8, but this time the integration was as simple as multiplying E and A! ◀

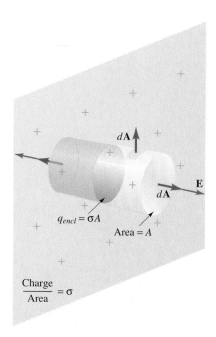

FIGURE 23.11

A cylindrical Gaussian surface passing perpendicular through an infinite plane of charge. See Example 23.4.

Concept Question 2
If there are fewer *E*-field lines leaving a Gaussian surface than entering, what can you conclude about the net charge enclosed by the surface?

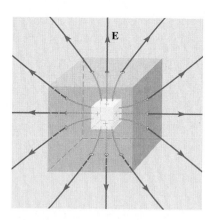

FIGURE 23.12

The electric field is not constant over a cubic Gaussian surface that surrounds a uniform cube of charge.

EXAMPLE 23.4   *Look Ma, No Integral Table!*

Use Gauss's law to find the electric field strength of an infinite plane carrying a uniform charge density σ.

SOLUTION

1. For a planar charge surface we can use a cylindrical Gaussian surface.
2. We place the cylinder so that it passes through the plane with its axis perpendicular to the plane as shown in Figure 23.11. If only this plane existed in empty space, there would be only two special directions: parallel to the plane and perpendicular to it. From symmetry, we know that the electric field lines can only be perpendicular to the plane of charge if they exist at all. That is, if we place a small, positive test charge *anywhere* off the plane (on either side), it experiences a force directly away from the positively charged plane.
3. Calculating the flux for the cylindrical Gaussian surface in Figure 23.11, we have

$$\Phi_E = \oint \mathbf{E} \cdot d\mathbf{A} = \int_{\substack{\text{left}\\\text{end}}} \mathbf{E} \cdot d\mathbf{A} + \int_{\substack{\text{right}\\\text{end}}} \mathbf{E} \cdot d\mathbf{A} + \int_{\text{curve}} \mathbf{E} \cdot d\mathbf{A}$$

At every point on the curved surface, $\mathbf{E}$ is perpendicular to $d\mathbf{A}$. As a consequence the dot product in the final integral above is zero. However, on both ends of the cylinder $\mathbf{E}$ and $d\mathbf{A}$ are parallel. If the area of each cylinder end is A, the flux becomes

$$\Phi_E = EA + EA + 0 = 2EA$$

4. The area of the charged surface inside the Gaussian cylinder is also A. Since the area surface charge σ is constant, the charge enclosed by the surface is $q_{\text{encl}} = \sigma A$.
5. We write Gauss's law and substitute for both flux and charge from our previous calculations

$$\Phi_{E_{\text{closed}}} = \frac{q_{\text{encl}}}{\epsilon_o}$$

$$2EA = \frac{\sigma A}{\epsilon_o}$$

Upon solving for E we find

$$E = \frac{\sigma}{2\epsilon_o}$$

This is the same result we obtained in Example 22.10.   ◀

You must always be careful to select a Gaussian surface over which the electric field is constant if you wish to calculate the surface integral simply. The following illustration will help you visualize a pitfall if you are not cautious with this step. Imagine a cube filled with a uniform charge distribution. We might be tempted to surround the charge with a cubic Gaussian surface. However, the electric field produced by a cube of charge is not sufficiently symmetric for us to be able to apply Gauss's law. As Figure 23.12 shows, the electric field is *not* constant over this cubic Gaussian surface. Indeed, the field produced by a cube of charge is difficult to evaluate over this surface and we will not attempt to perform this integration.

It is important for you to realize that, even though the flux is difficult to calculate, Gauss's law is still valid for the Gaussian surface shown in Figure 23.12. If we add up the normal component of the electric field over the entire cubic Gaussian surface, the result is in fact equal to the total charge enclosed divided by ϵ_o.

Concept Question 3
Explain why a simple application of Gauss's law cannot be used to compute the electric field of a charged ring.

23.3 The Ideal-Conductor Model in Electrostatic Equilibrium

In this section we describe a simple model of a metallic conductor. In the **ideal metallic-conductor model** we assume there are completely free electrons that can move around within the conductor and on its surface. From this model we establish some properties of a conductor when it is placed in an external electric field or when charge is placed on the conductor. In each case we express our result for a state of **electrostatic equilibrium**. By electrostatic equilibrium we mean that the charge distribution is static and we model the charge distribution as if the mobile charges were completely stationary.[2]

Let's first consider what happens when we place an ideal conductor in an applied electric field **E**. As illustrated in Figure 23.13, the free electrons experience a force in the opposite direction of the field and migrate to one side of the conductor (the left side in Fig. 23.13). The accumulation of electrons on this side of the conductor leaves the opposite side positively charged. Moreover, this charge buildup creates an electric field with a direction opposite to the applied field. Eventually, the field created within the conductor by the accumulated charges exactly cancels the applied field, and the net field within the conductor is zero. In most good conductors, such as copper, the process we described is amazingly fast, usually taking place in less than 10^{-10} s. For our purposes, this is instantaneous! Thus, for our model it is quite reasonable to say:

In electrostatic equilibrium, the electric field inside an ideal conductor is zero.

We next use this result together with Gauss's law to see what happens when we add excess charge to a conductor. We can charge the conductor negative by adding more electrons to it, or we can make it positive by removing electrons from it. Figure 23.14 shows an isolated conductor in which we have placed a Gaussian surface just below the ideal-conductor surface, that is, an infinitesimal distance below the surface. Our previous result says that the electric field is zero at every point on this Gaussian surface because it is inside the conductor. Gauss's law then implies that the net charge contained within the Gaussian surface is zero. Because the excess charge cannot be within this surface it must reside *on* the conductor's surface. (Remember, the Gaussian surface is arbitrarily close to the conductor's surface.) Therefore, in our model we will assume:

In electrostatic equilibrium, any excess charge on an ideal, isolated conductor must reside on the conductor's surface.

Now that we are on a roll, let's investigate the magnitude and direction of an electric field just outside a conductor. We do not require the conductor to be isolated, but we insist that it be in electrostatic equilibrium. Furthermore, let's assume that either the conductor is charged or it is placed in an applied electric field. In either case, there are regions of the

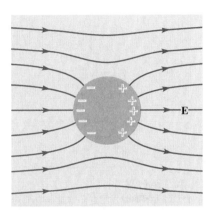

FIGURE 23.13

An ideal conductor placed in an external electric field. In electrostatic equilibrium, the E-field inside the conductor is zero.

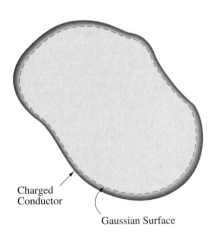

FIGURE 23.14

An isolated, charged conductor with a Gaussian surface just inside its surface.

---

[2] Conduction electrons always undergo random thermal motions making collisions with impurities and imperfections. Thus, even in a state of electrostatic equilibrium the average speed of any given electron is nonzero. However, for any fixed area element we wish to choose, the net flux of electrons through that area is zero for reasonably long times (e.g., a microsecond or longer).

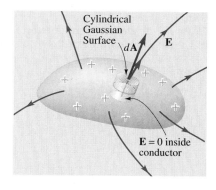

FIGURE 23.15

A cylindrical Gaussian surface projected through the surface of a charged conductor.

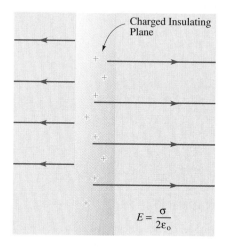

(a)

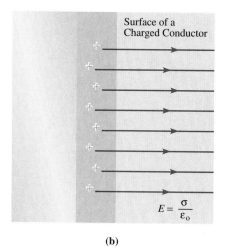

(b)

FIGURE 23.16

Each charge on a charged surface gives rise to the same number of E-field lines. (a) For an insulating surface, field lines are directed in both directions, perpendicular to the surface. (b) For a charged conducting surface all field lines leave the surface in the same direction.

conducting surface that hold excess charge. One such region is shown in Figure 23.15. We first note that:

The electric field lines intercept the conducting surface perpendicular to the surface.

If there were a component of the electric field parallel to the surface, the free electrons would move along the surface under the influence of this field to positions where they would establish an opposing field. In equilibrium, this opposing field would just cancel the parallel component of the applied field. Consequently, in electrostatic equilibrium, there can be no component of the applied field parallel to the conductor surface. (This argument should sound familiar because it is the same we applied to establish that the electric field is zero inside the conductor.)

It is possible to relate the magnitude of the electric field at a location on the conductor surface to the surface charge density at that location only. We apply Gauss's law to the small cylindrical Gaussian surface shown in Figure 23.15. The length of this cylinder is made small enough so that outside the conductor $\mathbf{E}$ is parallel to the cylinder axis and, therefore, $\mathbf{E} \cdot d\mathbf{A} = 0$ on the curved portion of this Gaussian surface. Moreover, $E = 0$ inside the conductor, so the only contribution to the flux term in Gauss's law comes from the end of the cylinder that is outside the conductor. Hence,

$$\oint \mathbf{E} \cdot d\mathbf{A} = \frac{q_{\text{encl}}}{\epsilon_o}$$

becomes

$$EA = \frac{\sigma A}{\epsilon_o}$$

from which we find

$$E = \frac{\sigma}{\epsilon_o} \qquad (23.7)$$

Equation (23.7) is the electric field strength just outside a conductor with surface charge density σ. Inside the conductor $E = 0$. Hence, there is a discontinuity in the electric field at the surface. Moreover, this discontinuity is related to the surface charge density: $E_{\text{out}} - E_{\text{in}} = \sigma/\epsilon_o$.

You should carefully compare the value for E given by Equation (23.7) with the field just outside an infinite sheet of charge on a thin *insulator*. For that case we found in Example 23.4 that $E = \sigma/2\epsilon_o$, exactly half of the value for a conductor. There is nothing mysterious about this result. You can think of each charge producing the same number of field lines in both cases. For the conductor, the field lines must all leave from the same side (the outer surface) of the plane of charge. On the other hand, for the insulator the field lines leave the surface equally *on both sides*, which makes the field just half as strong on each side compared with the field outside the conductor. These two cases are illustrated in Figure 23.16.

EXAMPLE 23.5 *A Problem for Party Animals!*

We've come a long way in our understanding of electrostatic phenomena. Let's now apply our knowledge to an important practical problem at which we hinted toward the beginning of Chapter 22. How much charge *does* it take to stick a balloon on the ceiling by rubbing the balloon on your sleeve?

SOLUTION   To solve this problem we have to build a reasonable (but not necessarily great) physical model. Let's assume the balloon is perfectly spherical and covered with a uniform layer of charge. For reasons you'll see in a minute, let's make the

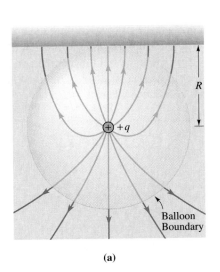

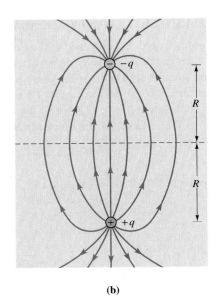

(a)   (b)

FIGURE 23.17

(a) A charged balloon in contact with the ceiling is modeled as a point-charge in front of a conductor. (b) The E-field lines are equivalent to those of the charge Q and an image charge $-Q$.

ceiling a conductor. We use two important results from this and the previous chapters: (1) The electric field outside a uniform spherical charge distribution can be calculated by pretending that all the charge is concentrated at the center of the sphere. (2) The electric field lines from the point-charge that represents the balloon intercept the conducting ceiling at right angles as shown in Figure 23.17(a). When we compare the resulting electric field pattern with that of two oppositely charged point-charges as shown in Figure 23.17(b), we notice something *very* interesting. From the point of view of the "balloon charge" there is no difference! (Cover the upper portion of both figures with a sheet of paper and you can't tell which is which.) Therefore, we can calculate the force on the balloon by calculating the force between our fictitious balloon charge and its even more fictitious "mirror-image charge" on the other side of the ceiling.

Now that the model is complete, the actual solution is easy. For a balloon of radius R, the balloon charge and its mirror image are a distance $r = 2R$ apart, leading to a Coulomb force of

$$F = \frac{1}{4\pi\epsilon_o} \frac{q^2}{(2R)^2}$$

which must at least balance the weight mg of the balloon. (We ignore the small buoyant force of air.) Thus, we require

$$\frac{1}{4\pi\epsilon_o} \frac{q^2}{4R^2} \geq mg$$

which leads to

$$q \geq \sqrt{16\pi\epsilon_o mg}\, R$$

For a 1-g balloon of radius $R = 0.1$ m, a charge of

$$q \geq 0.2\ \mu C$$

is required. Try working this result into the conversation the next time you're at a party!

Concept Question 4
How does the E-field vary inside a spherical rubber balloon that carries a uniform charge distribution on its surface?

EXAMPLE 23.6   *You Can Run, But You Can't Hide*

A point charge $+Q$ is placed at the center of an uncharged spherical conducting shell of inner radius A and outer radius B as shown in Figure 23.18(a). (a) Find the electric field strength within the shell where $r < A$. (b) What is the magnitude and sign of the induced charge q' on the inner shell surface? (c) What is the field inside the conducting shell where $A < r < B$? (d) What is the electric field at points $r > B$? (e) What is the surface charge on the outer surface of the conductor where $r = B$?

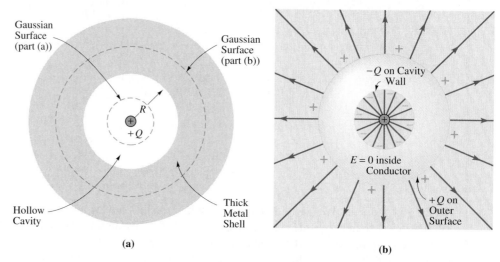

(a)   **(b)**

FIGURE 23.18

A point-charge $+Q$ located at the center of a spherical conducting shell. The cavity is surrounded by a spherical, Gaussian surface, the center of which coincides with Q.

SOLUTION   (a) We place a spherical Gaussian surface of radius $r < R$ inside the cavity, centered on the charge $+Q$. Applying our problem-solving strategy, Gauss's law

$$\oint \mathbf{E} \cdot d\mathbf{A} = \frac{q_{encl}}{\epsilon_o}$$

becomes

$$E(4\pi r^2) = \frac{+Q}{\epsilon_o}$$

from which we find the electric field to be

$$E = \frac{1}{4\pi\epsilon_o} \frac{Q}{r^2}$$

This result is the same as that of a point-charge in empty space!

(b) We enclose the entire cavity (and some of the conducting material) with another spherical Gaussian surface. We don't know ahead of time if there is a charge on the inside surface of the conductor or not, so we'll play it safe and say there is, calling it q'. If q' is zero, Gauss's law will tell us so. Because the imaginary Gaussian surface is inside the conductor, the electric field is everywhere zero. As a consequence the flux integral on the left-hand side of Gauss's law is zero, leading to

$$\oint \mathbf{E} \cdot d\mathbf{A} = \frac{q_{encl}}{\epsilon_o}$$

$$0 = \frac{Q + q'}{\epsilon_o}$$

which implies

$$q' = -Q$$

We discover there *is* a charge on the inside surface of the conductor. In fact, the total charge induced on the inside surface of the cavity is the negative of the charge placed at its center.

(c) Hah! We can spot a trick question as well as the next person. The field inside a conductor in electrostatic equilibrium is always zero. You think we didn't read the chapter or something?

(d) Because we want the field at a point located a distance r from the center, we take a spherical Gaussian surface of radius r and apply Gauss's law one more time. The surface integral on the left-hand side has the same form as before, so the flux is simply the product of the new field E and the new area $4\pi r^2$. On the right-hand side of Gauss's law we need the charge enclosed by our new Gaussian surface. In the problem statement we were told that the conducting sphere has no net charge. Consequently, the total charge inside our Gaussian surface is just Q. Thus, Gauss's law

$$\oint \mathbf{E} \cdot d\mathbf{A} = \frac{q_{encl}}{\epsilon_o}$$

becomes

$$E(4\pi r^2) = \frac{+Q}{\epsilon_o}$$

Solving for E, we find

$$E = \frac{1}{4\pi\epsilon_o}\frac{Q}{r^2}$$

Here we are again! Once more we have simply the Coulomb field of the point-charge Q. The conducting sphere had no shielding effect at all. (However, it is interesting to note that such a conducting shield *does* prevent electrostatic fields from charges outside the shell from entering it!)

(e) The conducting shell has no net charge, yet there is a surface charge $-Q$ pinned to its inner surface. Because the net charge on the shell is zero and no charge can be internal to the conductor, there must be $+Q$ on the outer surface of the conductor. Examine Figure 23.18(b) carefully, and relate the field picture and charge distributions to these results. ◄

Concept Question 5
When the point-charge in Example 23.6 is moved laterally from the center, does the net charge induced on the inside surface change? Explain.

Concept Question 6
Draw the new *E*-field lines when the point-charge in Example 23.6 is moved laterally from the center toward the inside surface of the hollow conductor.

23.4 Summary

The **electric flux** is a measure of the number of electric field lines that pass perpendicularly through a surface. The general definition of electric flux is

$$\Phi_E = \int \mathbf{E} \cdot d\mathbf{A} \tag{23.4}$$

Gauss's law relates the electric flux through a *closed* surface to the charge enclosed by the surface:

$$\Phi_{E_{\text{closed}}} = \oint \mathbf{E} \cdot d\mathbf{A} = \frac{q_{\text{encl}}}{\epsilon_o}$$

This law is one of Maxwell's equations. Coulomb's law may be viewed as a consequence of Gauss's law. For certain, high-symmetry charge distributions Gauss's law can provide a simple way to compute the electric field strength.

In an **ideal conductor** large numbers of electrons are completely free to move. In **electrostatic equilibrium** all charges can be modeled as if stationary. For an ideal conductor in electrostatic equilibrium: (1) the electric field is zero at all points within the conducting material, (2) any excess charge (due to an electron excess or deficit) resides on the surface of the conductor, and (3) the electric field just outside a conductor intercepts the conductor at right angles to its surface and has magnitude $\sigma/(2\epsilon_o)$.

PROBLEMS

23.1 Electric Flux

1. Sketch, on a three-dimensional Cartesian coordinate system, rectangles that have the following unit vectors perpendicular to their surface: (a) $\hat{\mathbf{n}} = \hat{\mathbf{i}}$, (b) $\hat{\mathbf{n}} = \hat{\mathbf{k}}$, (c) $\hat{\mathbf{n}} = (\hat{\mathbf{i}} + \hat{\mathbf{j}})/\sqrt{2}$

2. For each of the following, draw a representation of a three-dimensional Cartesian coordinate system and on it (1) sketch the electric field (which is uniform and fills all space), (2) sketch a rectangular area that could represent the orientation of the area **A**, and (3) calculate the electric flux through that area: (a) $\mathbf{E} = 5.00 \text{ N/C}\,\hat{\mathbf{i}} + 3.00 \text{ N/C}\,\hat{\mathbf{j}}$, $\mathbf{A} = 4.00 \text{ m}^2\,(0.600\,\hat{\mathbf{i}} + 0.800\,\hat{\mathbf{j}})$, (b) $\mathbf{E} = 8.00 \text{ N/C}\,\hat{\mathbf{j}}$, $\mathbf{A} = 0.250 \text{ m}^2\,\hat{\mathbf{n}}$, $\hat{\mathbf{n}} = (\hat{\mathbf{i}} + \hat{\mathbf{j}} + \hat{\mathbf{k}})/\sqrt{3}$.

3. A uniform electric field is given by $\mathbf{E} = (2.00 \text{ N/C})\,\hat{\mathbf{i}} + (4.00 \text{ N/C})\,\hat{\mathbf{j}}$. (a) Compute the magnitude of the electric flux through a rectangular surface perpendicular to the x-axis with an area of 1.50 m^2. (b) Compute the magnitude of the electric flux through a rectangular surface perpendicular to the y-axis with an area of 1.50 m^2.

4. A uniform electric field of 2.00×10^5 N/C parallel to the positive z-axis fills all of space. Compute the magnitude of the electric flux through a rectangle of 4.80-m^2 area if the rectangle is placed (a) in the xy-plane, (b) in the yz-plane, and (c) so that the unit vector $\hat{\mathbf{n}}$, which is perpendicular to the plane of the rectangle, lies in the yz-plane and makes a 30° angle with the z-axis.

5. The direction of a uniform electric field is perpendicular to the z-axis and makes a 50.0° angle with the positive x-axis. A rectangular surface is bounded by the x- and z-axes and the lines $x = 0.800$ m and $z = 1.20$ m. If the magnitude of the flux through the rectangle is $350. \text{ N} \cdot \text{m}^2/\text{C}$, (a) what is the magnitude of the electric field, and (b) through what angle from the x-axis must the rectangular surface be tilted to reduce the total flux through the rectangle to $100. \text{ N} \cdot \text{m}^2/\text{C}$?

6. Calculate the electric flux through an imaginary spherical surface 1.00 m in radius if there is a charge of 1.00 μC at the center of the sphere.

7. Use the result of Example 22.8 to find the electric flux through a closed cylindrical surface 25.0 cm in radius and 1.50 m in length if a line with charge density 2.40 μC/m coincides with the axis of the cylinder and extends far beyond the cylinder ends.

8. A uniform electric field of magnitude E_o is directed along the positive y-axis. By direct computation (that is, do not use Gauss's law) find the net electric flux through each of the six walls of a closed rectangular box with faces passing through the origin and $(a,0,0)$ $(0,b,0)$, and $(0,0,c)$. What is the net flux through the box? (Fig. 23.P1.)

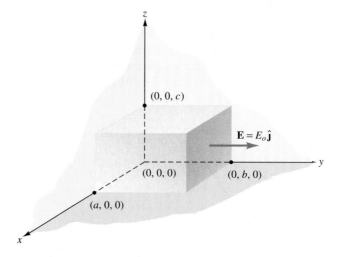

FIGURE 23.P1   Problem 8

23.2 Gauss's Law

9. Three charges $q_1 = +2.00\ \mu C$, $q_2 = -4.00\ \mu C$, and $q_3 = +6.00\ \mu C$ are surrounded by a spherical Gaussian surface of 3.00-m radius. What is the net electric flux through this Gaussian surface?

10. A charge of $+12.0\ \mu C$ is placed in the center of a cube 24.0 cm on an edge. (a) What is the total electric flux through the cubic surface? (b) What is the electric flux through one of the square surfaces of the cube?

11. A uniform electric field $E = 500$. N/C passes through a hemispherical surface of radius $R = 12.0$ cm as shown in Figure 23.P2. Find the net electric flux through the hemispherical surface only.

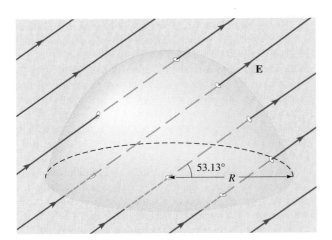

FIGURE 23.P2   Problem 11

★12. A nonuniform electric field extends throughout all space and is given by $\mathbf{E} = \{[3.00\ \text{N/(C} \cdot \text{m)}]x + [1.00\ \text{N/(C} \cdot \text{m}^2)]x^2\}\hat{\mathbf{j}}$. (a) Compute the net electric flux through the surface of a cube, three sides of which are coincident with the xy-, yz-, and xz-planes so that one corner of the cube passes through the origin, and the diagonally opposite corner is located at the point (0.500 m, 0.500 m, 0.500 m). (b) What is the total charge enclosed within the cube?

13. Two horizontal, infinite, and parallel planes carry uniform surface charge densities of $+\sigma$ and $-\sigma$. Compute the electric field strength (a) between the two planes and (b) below the bottom plane.

14. Two horizontal, infinite, and parallel planes carry identical, uniform surface charge densities of $+\sigma$. Compute the electric field strength (a) between the two planes and (b) below the bottom plane.

15. A total charge of 100. μC is distributed uniformly throughout a spherical region of space with a radius of 20.0 cm. Find the electric field strength at (a) $r = 10.0$ cm from the sphere's center and (b) 30.0 cm from the sphere's center.

16. A spherical balloon with a 6.00-cm radius has charge uniformly distributed over its surface. At a distance of 50.0 cm from the center of the balloon the electric field strength is 7.20×10^4 N/C. (a) What is the magnitude of the charge on the balloon? (b) What is the electric field strength 18.0 cm from the center of the balloon?

17. A total charge of $Q = 0.360\ \mu C$ is distributed uniformly over the surface of a spherical balloon. The balloon has a 24.0-cm radius.

(a) Compute the electric field at a point inside the balloon. (b) Compute the field strength at a point 12.0 cm above the surface of the balloon. (c) Compute the field strength at a distance of 0.500 cm above the balloon surface. (d) Use the result of Equation (23.7) to estimate the field strength 0.500 cm above the balloon surface and compare the result to that in part (c).

18. A uniform, spherical charge distribution of total charge $+Q$ is surrounded by a nonconducting spherical shell of total charge $-Q$ as shown in Figure 23.P3. If $Q = 50.0\ \mu C$, $R_1 = 5.00$ cm, $R_2 = 12.0$ cm, and $R_3 = 16.0$ cm, (a) compute the electric field strength at $r = 10.0$ cm from the center of the distribution. (b) Compute the electric field strength at $r = 14.0$ cm and (c) at $r = 20.0$ cm.

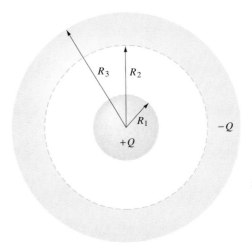

FIGURE 23.P3   Problem 18

19. A hollow, spherical charge distribution of inner radius R_1 and outer radius R_2 has a uniform density ρ (Fig. 23.P4). (a) Compute the electric field strength in the region $r \le R_1$. (b) Compute the electric field strength in the region $R_1 \le r \le R_2$. (c) Compute the electric field strength in the region $r \ge R_2$. (d) Show that the answers to parts (a) and (b) agree at $r = R_1$, and parts (b) and (c) agree at $r = R_2$.

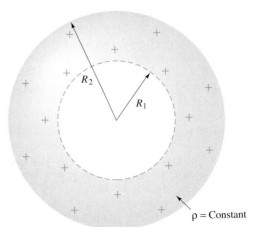

FIGURE 23.P4   Problem 19

20. The electric field a distance of 1.50 m from a very long line of charge is 2.40×10^4 N/C. (a) Find the linear charge density λ for this charge distribution. (b) What is the electric field a distance of 0.750 m from the line?

21. A very long, cylindrical shell carries a uniform charge density ρ as shown in Figure 23.P5. The inner radius of the shell is R_1, and its outer radius is R_2. Find the electric field strength in the regions (a) $0 \le r \le R_1$, (b) $R_1 \le r \le R_2$, and (c) $r \ge R_2$.

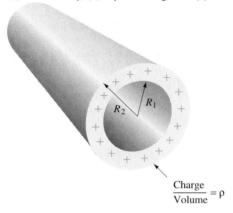

$$\frac{\text{Charge}}{\text{Volume}} = \rho$$

FIGURE 23.P5   Problem 21

22. A very long cylindrical charge distribution of radius R has a constant charge density ρ. Compute the electric field strength in the regions (a) inside the cylinder ($r \le R$), and (b) outside the cylinder ($r \ge R$). (c) Show that your answers to parts (a) and (b) agree at $r = R$.

23.3 The Ideal-Conductor Model in Electrostatic Equilibrium

23. The electric field near the earth's surface on a clear day is about 100. N/C, and is directed downward toward the earth's center. Compute the net charge on the surface of the earth by modeling it as an ideal-conducting sphere of 6370-km radius.

24. A point-charge $Q = -30.0 \ \mu C$ is placed at the center of a hollow, conducting spherical shell of inner radius $R_1 = 12.0$ cm and outer radius $R_2 = 16.0$ cm. A charge of $+10.0 \ \mu C$ is transferred from an external source to the conducting shell. Compute the electric field at the following distances from the sphere's center: (a) $r = 8.00$ cm, (b) $r = 14.0$ cm, and (c) $r = 20.0$ cm. (d) What is the magnitude and sign of the induced charge on the outer surface of the sphere?

25. A charge Q is placed at the center of a charged, spherical, conducting shell of inner radius $R_1 = 8.00$ cm and outer radius $R_2 = 12.0$ cm. The induced charge on the inner surface of the shell is $-4.00 \ \mu C$, and at a distance of 16.0 cm from the center of the shell the electric field is 3.50×10^6 N/C directed radially outward. What are the magnitude and sign of the additional charge on the conducting shell?

26. A thin, conducting, spherical shell of 6.00-cm radius carries an unknown charge spread uniformly over its surface. Concentric with the shell is a 4.00-cm thick, conducting shell with an inner radius of 10.0 cm. The outer shell carries an excess charge of unknown sign. At a distance $r = 8.00$ cm from the center of the shells the electric field magnitude is 9.00×10^7 N/C and is directed toward the spheres' centers. When $r = 18.0$ cm, the electric field strength is 2.20×10^7 N/C and is directed radially outward. What are the magnitudes and signs of the charges on the inner and outer surfaces of the spherical shells?

27. A very long wire carries a uniform charge along its length with linear density $\lambda = +3.20 \ \mu C/m$. The wire is surrounded by a conducting cylinder with a 4.00-cm inner radius and a 6.00-cm outer radius. The cylinder's axis coincides with the wire, and the shell is charged so that along its length its charge density is $-9.60 \ \mu C/m$. (a) Compute the electric field strength at $r = 2.00$ cm, $r = 5.00$ cm, and $r = 8.00$ cm. (b) What is the linear charge density induced on the inner surface of the cylinder?

28. Two square, copper plates are 2.00 m on a side, parallel to each other, and separated by a small distance. Each plate carries a charge of equal magnitude but opposite sign. If the electric field between the plates is 5.00×10^4 N/C and we ignore edge effects, what is the magnitude of the total charge on each plate?

Numerical Methods and Computer Applications

29. The first 20 rows and 10 columns of cells in the spreadsheet template FLUX.WK1 contain numbers that are the vertical components of an electric field that we imagine to extend through the cells (toward you). By placing a ruler on your computer screen, find the area of a cell. (a) From this area compute the net electric flux through the 20×10 cell region on the screen. (b) Reduce the width of each cell by three character lengths and recompute the electric flux.

30. Write a program to compute the approximate electric flux from a point-charge q through a square, planar area: (a) Assume the charge is located a distance z_o along the z-axis and the square is in the xy-plane bounded by $-1 \le x \le +1$ and $-1 \le y \le +1$. Divide the square into 16, equal-sized area elements. (b) Modify the program from part (a) for charge locations off the z-axis at positions (x_o, y_o, z_o).

31. A very long cylindrical charge distribution of radius $R = 4.00$ cm has a constant charge density $\lambda = 20.0 \ \mu C/m$. Develop a computer program or implement a spreadsheet to compute and graph the electric field strength in the region from the center of the cylinder to a distance of 24.0 cm from the cylinder axis.

32. A very long cylindrical charge distribution of radius $R = 2.00$ cm has a charge density that increases linearly with distance from the axis of the cylinder. The charge density is given by $\rho(r) = 2.00 \ \mu C/m^3 + (50.0 \ \mu C/m^4)r$. Develop a program or implement a spreadsheet to compute and graph the electric field strength in the region from the center of the cylinder to a distance of 16.0 cm from the cylinder axis.

33. Charge is distributed uniformly throughout a spherical region of space with a radius of 4.00 cm. The charge density is $6.00 \ \mu C/m^3$. Develop a program or implement a spreadsheet to compute and graph the electric field strength in the region from the center of the sphere to a distance of 16.0 cm from its center.

34. Charge is distributed throughout a spherical region of space with radius $R = 6.00$ cm so that the charge density increases linearly from the center of the sphere and is given by $\rho(r) = (4.00 \ \mu C/m^4)r$. Develop a program or implement a spreadsheet to compute and graph the electric field strength in the region from the center of the sphere to a distance of 20.0 cm from its center.

General Problems

35. Charge is uniformly distributed throughout a large flat, platelike volume of thickness d. If the density of the charge is ρ and the plate is assumed to be infinite in lateral extent, show that the electric field a distance x from the center of the plate along a line perpendicular to the face of the plate is $(\rho/\epsilon_o)x$, where $|x| < d/2$.

36. Charge is distributed throughout a spherical region of space with radius R such that within the sphere the charge density increases linearly from the center of the sphere and is given by $\rho(r) = Cr$, where C is a constant. (a) Compute the electric field at a position $r \ge R$. (b) Compute the electric field strength at a position $r \le R$. (c) Show that the answers to parts (a) and (b) are equal at $r = R$.

37. Charge is distributed throughout a spherical region of space with radius R so that the charge density increases from the center of the sphere and is given by $\rho(r) = Cr^2$, where C is a constant. (a) Compute the electric field strength at a position $r \geq R$. (b) Compute the electric field strength at a position $r \leq R$. (c) Show that the answers to parts (a) and (b) are equal at $r = R$.

38. Prove that the electric field outside any spherical charge distribution for which the charge density depends only on the radial distance r (that is, $\rho = \rho(r)$), is the same as that of a point-charge equal to the total charge enclosed within the sphere.

39. Gauss's law for gravity relates the gravitational flux to the net mass enclosed in a Gaussian surface. If M_{encl} is the enclosed mass and G is the gravitational constant, Gauss's law for gravity is

$$\oint \mathbf{g} \cdot d\mathbf{A} = -4\pi G M_{\text{encl}}$$

Use this law and the fact that $\mathbf{F} = m\mathbf{g}$ to derive Newton's universal law of gravitation.

40. A very long cylindrical charge distribution of radius R has a charge density that increases linearly with distance from the axis of the cylinder. The charge density is given by $\rho(r) = A + Br$, where A and B are constants. (a) Compute the electric field strength inside the cylinder ($r \leq R$). (b) Compute the electric field strength outside the cylinder ($r \geq R$).

★41. A very long cylinder of radius R_1 and uniform charge density ρ has a cylinder hole cut along its entire length so that the axes of the cylinder and the hole are parallel and separated by a distance d (Fig. 23.P6). The radius of the hole is $R_2 < R_1$. Compute the electric field strength along the line AB in the following regions: (a) $r \leq d - R_2$, (b) $d - R_2 \leq r \leq d + R_2$, (c) $d + R_2 \leq r \leq R_1$, and (d) $r > R_1$.

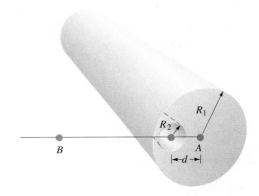

FIGURE 23.P6   Problem 41

42. A conducting sphere of radius r carries a charge q. Show that Equation (23.7) gives the correct electric field just above the conducting surface.

43. A very long, solid, cylindrical charge distribution with a constant-charge density ρ has a radius R_1. Concentric with the long cylinder is an uncharged conducting cylindrical shell of inner radius R_2 and outer radius R_3. (a) For a position far from the ends of the cylinder, compute the electric field in the following regions: $r < R_1$, $R_1 \leq r \leq R_2$, $R_2 \leq r \leq R_3$, and $r \geq R_3$. (b) What is the induced charge per length on the inner and outer surfaces of the conducting cylinder?

44. A square, conducting horizontal plate 36.0 cm on a side is charged so that at a distance of 0.500 cm above the surface of the center of the plate the electric field strength is 1.40×10^7 N/C. (a) Estimate the total charge on the plate. (b) Estimate the electric field strength 12.0 m above the center of the plate.

24

Electric Potential

In this chapter you should learn

- to compute the electric potential energy of two point-charges.
- the definition of electric potential.
- to compute the electric potential from the electric field.
- to compute the electric potential from a charge distribution.
- to compute the electric field from the electric potential.
- the definition of equipotential surface.
- (optional) the relaxation method of solving two-dimensional potential problems numerically.

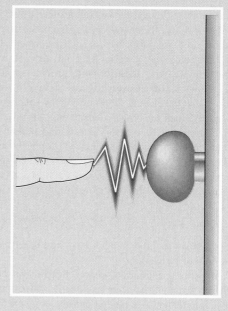

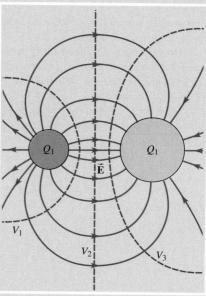

$$V = -\int \mathbf{E} \cdot d\mathbf{s} + C$$

In Chapters 7 and 8 we discovered a useful technique that allowed us to avoid working directly with such conservative forces as gravity, springs, and an assortment of other position-dependent forces. We introduced the concept of potential energy, which led to the conservation-of-energy theorem, and we employed this theorem to solve many problems. The solutions of some of these problems would have been quite complex had we used Newton's second law. The energy concept is also useful in electricity and magnetism, and is so because the Coulomb force is conservative. We begin this chapter by computing the potential energy between two point-charges. Rather quickly, we will discover that the field concept we have come to know and "love" (you might prefer "respect") can be united with the potential energy concept.

There is a delightful correspondence between the physics developed in this chapter and that of Chapters 7 and 8. In fact, it is worth reminding you about the view we established in those chapters. Recall that, if you are given a conservative force, you can find the potential energy function (to within an arbitrary constant) from the negative of the work done by the force:

$$U(x,y,z) = -\int \mathbf{F} \cdot d\mathbf{s} + U_o \qquad (24.1)$$

On the other hand, for the one-dimensional case the conservative force $F(x)$ can be derived from the potential energy function $U(x)$ by using

$$F(x) = -\frac{dU}{dx} \qquad (24.2)$$

Later in an optional section of this chapter we will describe an extension of Equation (24.2) by which we can determine the force $\mathbf{F}$ from the potential energy function for the three-dimensional case.

The picture we develop in this chapter builds on procedures described by Equations (24.1) and (24.2). If you're a little rusty in the application of these equations, it will be worth an hour of your time to review the material in Chapters 7 and 8, particularly the examples in Sections 7.3 and 8.5.

24.1 Electric Potential Energy

We begin by computing the expression for the potential energy of two point-charges q and q_t. Following the view that evolved in the previous two chapters, we envision the charge q as establishing an electric field $\mathbf{E}$ in which we place the test charge q_t. The force of this field on q_t is $q_t\mathbf{E}$, and we can compute the change in potential energy of charge q_t as it is moved in the E-field of the charge q from Equation (24.1):

$$U_f - U_o = -\int_o^f \mathbf{F} \cdot d\mathbf{s} = -\int_o^f q_t\mathbf{E} \cdot d\mathbf{s}$$

$$\Delta U = U_f - U_o = -q_o\int_o^f \mathbf{E} \cdot d\mathbf{s} \qquad (24.3)$$

The electric field from a point-charge q is

$$\mathbf{E} = \frac{1}{4\pi\epsilon_o}\frac{q}{r^2}\hat{\mathbf{r}}$$

and Equation (24.3) becomes

$$U_f - U_o = -\frac{qq_t}{4\pi\epsilon_o}\int_o^f \left(\frac{1}{r^2}\right)\hat{\mathbf{r}}\cdot d\mathbf{s}$$

$$= -\frac{qq_t}{4\pi\epsilon_o}\int_o^f \left(\frac{1}{r^2}\right)\cos(\theta)\,ds \qquad \textbf{(24.4)}$$

where θ is the angle between the radial direction from the origin to each integration point and the vector $d\mathbf{s}$ that is tangent to the path of integration at each integration point. Figure 24.1 shows the charges q and q_t, and a general path between two points o and f. Because the force of q on q_t is conservative, the path we take to evaluate Equation (24.4) need not be the general one shown; ΔU must be the same for *any* path between o and f. It is easy to evaluate Equation (24.4) if we choose the two-segment path $o \to o' \to f$ also shown in Figure 24.1. Along the path $o \to o'$ the angle between $d\mathbf{s}$ and the radial direction is 90°, whereas along the path $o' \to f$, $\theta = 0°$ because $d\mathbf{s}$ is along the radial direction. Moreover, $ds = dr$ along $o' \to f$. Hence, Equation (24.4) becomes

$$U_f - U_o = -\frac{qq_t}{4\pi\epsilon_o}\left[\int_o^{o'} \frac{\cos(90°)}{r^2}ds + \int_{o'}^f \frac{\cos(0°)}{r^2}ds\right]$$

$$= -\frac{qq_t}{4\pi\epsilon_o}\left(0 + \int_{r_o}^{r_f} \frac{1}{r^2}dr\right)$$

where we recognize that $r = r_o$ at point o' and $r = r_f$ at point f. Evaluating this last integral, we have

$$U_f - U_o = -\frac{qq_t}{4\pi\epsilon_o}\left(-\frac{1}{r}\right)_{r_o}^{r_f}$$

$$U_f - U_o = \frac{1}{4\pi\epsilon_o}\frac{qq_t}{r_f} - \frac{1}{4\pi\epsilon_o}\frac{qq_t}{r_o} \qquad \textbf{(24.5)}$$

The previous equation suggests a potential function of the form

$$U = \frac{1}{4\pi\epsilon_o}\frac{qq_t}{r} + C$$

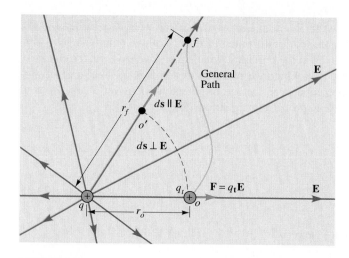

FIGURE 24.1

The work done on charge q_t as it moves from point o to point f is independent of path. Along $o \to o'$, $d\mathbf{s}$ is perpendicular to $\mathbf{E}$. Along $o' \to f$, $d\mathbf{s}$ is parallel to $\mathbf{E}$.

If we choose $C = 0$ so that $U = 0$ at $r = \infty$, our potential energy function becomes

$$U(r) = \frac{1}{4\pi\epsilon_o} \frac{qq_t}{r} \qquad (24.6)$$

Equation (24.6) is the potential energy of the system composed of the two charges q_t and q. The selection that the potential energy is zero at $r = \infty$ seems a natural choice when we remember that only at an infinite separation are the forces on the charges zero so that they do not interact at this distance. If the charges are both of the same sign, the potential energy, as described by Equation (24.6), is zero at infinity and grows more positive as the charges are brought closer together. This result seems reasonable because like charges repel, and we therefore expect the potential energy to increase as the charges approach one another.

When the charges are of opposite sign, their maximum potential energy is zero, corresponding again to infinite separation. As the charges are brought closer together $U(r)$ becomes more negative, and the potential energy decreases. Don't allow this negative potential energy to bother you; remember, potential energy means the ability to do work for us. Because the force between two unlike charges is attractive, the unlike charges have little potential energy when they are very close and more potential energy when their separation is increased. This is precisely what happens when $U(r) = 0$ at $r = \infty$; $U(r)$ becomes more negative for smaller r.

In Chapter 8 we found that the work done on a system by external agents is equal to the change in the sum of the kinetic and potential energies ($\mathcal{W}_{ext} = \Delta K + \Delta U$). Thus, the electric potential energy given by Equation (24.5) is just the amount of work required to bring the two charges q_1 and q_2 from a state of rest at an infinite separation to a state of rest in which their separation distance is r.

When more than two charges are present in some region of space, the total potential energy is the algebraic sum of the potential energy stored by each pair of charges. For three charges

$$U = \frac{1}{4\pi\epsilon_o}\left(\frac{q_1 q_2}{r_{12}} + \frac{q_1 q_3}{r_{13}} + \frac{q_2 q_3}{r_{23}} \right)$$

where r_{12} is the separation between charges q_1 and q_2, r_{13} is the separation between q_1 and q_3, and r_{23} is the separation between q_2 and q_3 (Fig. 24.2).

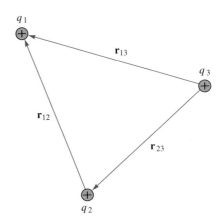

FIGURE 24.2

Three point-charges fixed in place with separations r_{12}, r_{13}, and r_{23}.

EXAMPLE 24.1   *It Takes Work to Bring Them Together*—Henry Kissinger

(a) What must be the separation of two 1.00-μC charges for the potential energy of the pair to be 1.00 J? (b) How much work must be done to bring the separation distance between the two charges to one-fourth the separation found in part (a)? Assume the charges are at rest at both the original and final positions.

SOLUTION   (a) We have, from Equation (24.6)

$$U(r) = \frac{1}{4\pi\epsilon_o} \frac{qq_t}{r}$$

or

$$r = \frac{qq_t}{4\pi\epsilon_o U} = \frac{(1.00 \times 10^{-6}\ \text{C})^2}{4\pi(8.854 \times 10^{-12}\ \text{C}^2/(\text{N}\cdot\text{m}^2))(1.00\ \text{J})}$$

$$= 8.99 \times 10^{-3}\ \text{m} = 0.899\ \text{cm}$$

(b) The final separation is $r_f = (8.99 \times 10^{-3}\ \text{m})/4 = 2.25 \times 10^{-3}\ \text{m}$. The work–energy theorem of Chapter 8 tells us that the work we must do on the charges is equal to the change in their total energy, kinetic plus potential. Because both the original and final kinetic energies are zero, the work done on the charges equals the change in their potential energy:

$$\mathcal{W} = \Delta U = U_f - U_o = \frac{1}{4\pi\epsilon_o}\left(\frac{qq_t}{r_f} - \frac{qq_t}{r_o}\right) = \frac{qq_t}{4\pi\epsilon_o}\left(\frac{1}{r_f} - \frac{1}{r_o}\right)$$

$$= \frac{(1.00 \times 10^{-6}\ \text{C})^2}{4\pi[8.854 \times 10^{-12}\ \text{C}^2/(\text{N}\cdot\text{m}^2)]}\left(\frac{1}{2.25 \times 10^{-3}\ \text{m}} - \frac{1}{8.99 \times 10^{-3}\ \text{m}}\right) = 3.00\ \text{J}$$

◀

24.2 The Electric Potential

It will be helpful if we review our incentive for defining the electric field. In Chapter 22, we found that the force of a charge q on another charge q_t is given by Coulomb's law:

$$\mathbf{F} = \frac{1}{4\pi\epsilon_o}\frac{qq_t}{r^2}\hat{\mathbf{r}}$$

We noticed that this force $\mathbf{F}$ *is proportional to* q_t, and this observation motivated us to define the electric field as the force per charge, $\mathbf{E} = \mathbf{F}/q_t$. For a point-charge we found

$$\mathbf{E} = \frac{1}{4\pi\epsilon_o}\frac{q}{r^2}\hat{\mathbf{r}}$$

From this point (pun optional) we went on to find expressions for the electric field produced by many charge geometries. We are in a very similar circumstance right now: Equation (24.6) tells us that the potential energy of the charge q_t in the electric field of charge q *is proportional to* q_t. We are thus inspired to define the **electric potential** V (as opposed to the electric potential *energy*):

> The electric potential V is the potential energy per unit positive charge.

For a point-charge we can compute the electric potential from Equation (24.6):

$$V(r) = \frac{U(r)}{q_o} = \frac{1}{4\pi\epsilon_o}\frac{q}{r} \tag{24.7}$$

We hope you can see from this expression that the electric potential is itself a field; in this case a scalar field. The potential function $V(r)$ has a value at every point in space (except at $r = 0$, which is not allowed because q is already there). For a given charge q, you supply an r-value, and Equation (24.7) provides you with a number that is the potential energy per unit charge at that position. What good is this number? If you know the potential energy *per unit charge*, you need only multiply this result by the size of any charge q' you place at position r, and the product is the potential energy of system made up of q' and q.

The SI unit for potential energy is the joule and the unit of charge is the coulomb. In SI, the unit of electric potential is the **volt** (V). Hence,

$$1\ \text{V} = 1\ \frac{\text{J}}{\text{C}}$$

Concept Question 1
When an electron moves in an electric field parallel to an *E*-field line, does its potential energy increase or decrease? Explain.

Now, the electric potential function given by Equation (24.7) is only applicable to a single point-charge. We are usually interested in more extensive charge distributions, so we must make our definition of electric potential more general.

We begin by noting that the potential energy is defined only to within an arbitrary constant.[1] That is, only the *change* in potential energy is defined by Equation (24.3). Therefore, in a formal definition only the *change* in the electric potential is designated. In equation form our definition of electric potential is

$$\Delta V = \frac{\Delta U}{q_t} \tag{24.8}$$

but $\Delta U = U - U_o$ is given by Equation (24.1), so

$$\Delta V = -\frac{1}{q_t} \int_o^f \mathbf{F} \cdot d\mathbf{s}$$

Now, for any charge q_t in a field $\mathbf{E}$, $\mathbf{F} = q_t\mathbf{E}$, so

$$\Delta V = -\frac{1}{q_t} \int_o^f q_t\mathbf{E} \cdot d\mathbf{s}$$

$$= V_f - V_o = -\int_o^f \mathbf{E} \cdot d\mathbf{s} \tag{24.9}$$

Equation (24.9) permits us to find the electric potential difference between points o and f if we know the electric field $\mathbf{E}$ between these two points. As we mentioned above, the value V_o is arbitrary, but, just as in the case of potential energy, there are usually convenient values to assign to V_o at certain locations.

The definite integral of Equation (24.9) allows calculation of the potential difference between two points in space due to an electric field $\mathbf{E}$. The same potential change results from any potential function $V(x, y, z)$ that satisfies

$$\boxed{V(x, y, z) = -\int \mathbf{E} \cdot d\mathbf{s} + C} \tag{24.10}$$

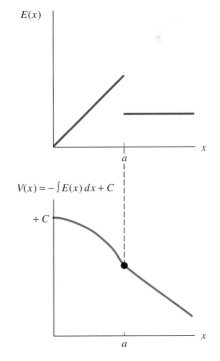

where C is a constant. We refer to this relation as our *master equation* for computing the electric potential function $V(x, y, z)$ from a known electric field $\mathbf{E}(x, y, z)$. Equation (24.7) can be viewed as a special case of the master equation for a point-charge in which both fields are functions of r only and $C = 0$.

Note that $V(x, y, z)$ must be continuous because $V(x, y, z)$ is the integral of the electric field $E(x, y, z)$, and the electric field must always be finite (otherwise, we have the possibility of an infinite force $\mathbf{F} = q\mathbf{E}$). At worst, the E-field may have a jump discontinuity due to a sheet of charge, for example at the surface of a conductor. In this case V has a kink where the integral of E changes slope but remains continuous (Fig. 24.3). Hence, there can be no abrupt changes (discontinuities) in the electric potential function:

The electric potential V must be a continuous function.

As we stated before, the electric potential V, like the electric field $\mathbf{E}$ is a field quantity. That is to say, V has a value at all points (x, y, z) in space. There is, however, a significant difference between the $\mathbf{E}$ and V fields. Although the electric field is a *vector field*, we must remember:

The electric potential is a **scalar** field.

FIGURE 24.3

A discontinuity in $E(x)$ at $x = a$ yields a kink in $V(x)$ at $x = a$.

[1] For two point-charges Equation (24.7) shows that this constant was chosen so that $V = 0$ at $r = \infty$. This step was actually performed when we chose the form of the potential energy function, Equation (24.6).

This feature makes perfectly good sense when we remember that the electric potential is the potential energy per charge, and the potential energy is a scalar. This distinction means that, for the same charge distribution, the calculation of the electric potential is usually much less complicated than the calculation of the electric field.

There is one more point we need to make before we apply Equation (24.10) in some examples. The change in electric potential is the change in potential energy per unit charge (Eq. (24.8)), and by Equation (24.1) the change in potential energy is the negative of the work performed *by the conservative force* of the electric field:

$$\Delta U = -\int_o^f \mathbf{F} \cdot d\mathbf{s} = -\mathcal{W}_{of}$$

From Equation (24.8) we therefore have

$$\Delta V = \frac{\Delta U}{q_t} = -\frac{\mathcal{W}_{of}}{q_t} \tag{24.11}$$

Now, $\mathcal{W}_{of}$ is the work done *by the force from the electric field* in some region of space when the charge q_t moves from location o to location f. If *we* want to move the charge q_t from o to f, we must do work $\mathcal{W}_{ext} = -\mathcal{W}_{of}$. Hence, Equation (24.11) can be written

$$\Delta V = +\frac{\mathcal{W}_{ext}}{q_f} \tag{24.12}$$

Here's the point: Equation (24.12) means that we can interpret the change in electric potential as the *work per unit charge we must do* to move a unit charge from location o to location f. This result is helpful when we want to know how much work we must perform to move a charge through a potential difference of a certain number of volts. We can rearrange Equation (24.12) to obtain

$$\mathcal{W}_{ext} = q_t \, \Delta V$$

When $q_t = 1$ C and $\Delta V = 1$ V, this equation says:

> We must perform one joule of work to carry a one coulomb positive charge through an increase in potential difference of one volt.

Be wary of the signs. If *we* do work on a positive charge to move it in an electric field, then we are moving it to a region of higher potential. On the other hand, if the charge is negative *it* does work for us as it moves into regions of higher potential. That is, the sign of the potential change is defined by what happens to a *positive* test charge. For negative charges, when the electric *potential* (V) goes up, the negative charge's potential *energy* (U) goes down. This occurs naturally if you put the correct sign of the charge in $\Delta U = q \, \Delta V$. Another helpful rule of thumb is that *electric field lines point toward regions of lower potential*.

There is a non-SI unit of energy that is particularly useful in atomic and nuclear physics. This unit is the **electron volt** (abbreviated eV):

> One electron volt is the kinetic energy acquired by an electron or a proton when it is accelerated through a potential difference of one volt.

Now, the magnitude of the charge on an electron is 1.60×10^{-19} C and one volt is 1 J/C, so applying Equation (24.8), we have for the energy of one electron volt:

$$\Delta U = q \, \Delta V = (1.60 \times 10^{-19}\text{C})(1.00 \text{ J/C}) = 1.60 \times 10^{-19} \text{ J}$$

That is,

$$1 \text{ eV} = 1.60 \times 10^{-19} \text{ J}$$

For example, an electron in a TV picture tube might be accelerated by a potential difference of 6.8 kV. The energy this electron acquires from this acceleration is 6.8×10^3 eV because for each volt of potential difference the electron is accelerated through, it gains 1 eV of energy. In joules this amount of energy is $(6.8 \times 10^3 \text{ eV})(1.6 \times 10^{-19} \text{ J/eV}) = 1.1 \times 10^{-15}$ J. This might not seem like much energy, but remember that an electron is small; its mass is only 9.1×10^{-31} kg so with this energy its velocity is about 5×10^7 m/s, nearly fast enough so that corrections for special relativity are needed!

EXAMPLE 24.2   *Electric Potential of an Isolated Point-Charge*

(a) Find the potential difference between points a and b located at distances of $r_a = 2.00$ m and $r_b = 1.00$ m from an isolated 1.00-μC charge. (b) How much work is required to move a 36.0-μC charge from a to b?

SOLUTION   (a) We have already derived Equation (24.7) for the potential field $V(r)$ of a point-charge q. The potential difference between two locations r_a and r_b is thus

$$\Delta V = V(r_b) - V(r_a) = \frac{q}{4\pi\epsilon_o}\left(\frac{1}{r_b} - \frac{1}{r_a}\right)$$

Substituting $r_a = 2.00$ m, $r_b = 1.00$ m, and $q = 1.00$ μC, we have

$$\Delta V = \frac{(1.00 \text{ } \mu\text{C})}{4\pi[8.854 \times 10^{-12} \text{ C}^2/(\text{N} \cdot \text{m}^2)]}\left(\frac{1}{1.00 \text{ m}} - \frac{1}{2.00 \text{ m}}\right) = 4490 \text{ V}$$

(b) From Equation (24.12)

$$\mathcal{W}_{ext} = q_t \, \Delta V = (36.0 \times 10^{-6} \text{ C})(4490 \text{ V}) = 0.162 \text{ J} \quad \blacktriangleleft$$

Electric Potential Change Due to a Uniform Electric Field

In this and the next subsection we illustrate how to apply Equation (24.9) to compute the potential difference between two points in a region of space where **E** is known. Figure 24.4 shows a region in which there is a uniform electric field **E**. Our first objective is to compute the change in potential ΔV between two points o and f that are situated at different locations along a line parallel to the uniform electric field. We apply Equation (24.9):

$$\Delta V = V_f - V_o = -\int_o^f \mathbf{E} \cdot d\mathbf{s} = -\int_o^f E \cos(180°) \, ds$$

The angle between **E** and $d\mathbf{s}$ is 180° because according to our integration limits, $d\mathbf{s}$ goes from location o to location f, and **E** is directed opposite to this direction. Also, E is a constant and can be removed from the integral.[2] Hence, because the distance from point o to point f is d, we have

$$\Delta V = +E\int_o^f ds$$

$$\Delta V = E \, d \qquad (24.13)$$

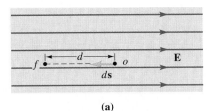

(a)

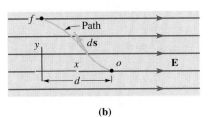

(b)

FIGURE 24.4

Computing the change in electric potential when the original and final points are along a line (a) parallel to the E-field lines and (b) when they are not along a line parallel to the E-field lines.

---

[2] Remember that in general **E** is not constant along field lines. The magnitude of **E** depends on the distance between field lines.

From this result we may find the change in potential *energy* of a charge q_t when it is moved from point o to point f. Applying Equation (24.8), we obtain

$$\Delta U = q_t \, \Delta V$$

or

$$\Delta U = q_t \, E \, d \qquad \qquad (24.14)$$

When the charge q_t is positive, there is an increase in potential energy as q_t is moved from o to f. This result agrees with our previous observations about the sign of ΔU because the charge q_t experiences a force from the electric field toward the right in Figure 24.4(a). When q_t is positive we must perform work on this charge, thereby increasing its potential energy when we move it in a direction opposite to the E-field. If the charge q_t is negative, Equation (24.14) says ΔU is negative; that is, its potential energy *decreases* when q_t is moved from o to f.

We next show that if the points o and f do not lie along a line parallel to a field line, we still obtain the same result, Equation (24.13). Figure 24.4(b) shows the same region of space where there is a uniform electric field **E,** but this time the points o and f define a line not parallel to the field. We again apply Equation (24.9):

$$\Delta V = V_f - V_o = -\int_o^f \mathbf{E} \cdot d\mathbf{s}$$

This time we draw a coordinate system with the x-axis parallel to the E-field direction. Thus,

$$\mathbf{E} = E\hat{\mathbf{i}} \qquad \text{and} \qquad d\mathbf{s} = dx\hat{\mathbf{i}} + dy\hat{\mathbf{j}}$$

Our integral becomes

$$\Delta V = -\int_o^f (E\hat{\mathbf{i}}) \cdot (dx\hat{\mathbf{i}} + dy\hat{\mathbf{j}}) = -\int_d^0 E \, dx$$

where we have recognized that the x-value of the original coordinate is $x = d$, and the final coordinate is $x = 0$. Performing the final integration, we find

$$\Delta V = -E\,x \Big|_d^0 = +E\,d$$

which is Equation (24.13).

We can obtain an important result from Equation (24.13). Solving for E, we have

$$E = \frac{\Delta V}{d}$$

In Chapter 22 we found that the unit of electric field is the newton per coulomb. However, in Equation (24.13), ΔV has the unit of volt and d is measured in meters. Therefore, the electric field unit can also be written volt per meter. That is,

$$1\,\frac{\text{N}}{\text{C}} = 1\,\frac{\text{V}}{\text{m}},$$

a result that can also be deduced from the more general equation, Equation (24.9).

Before we present an example we want to mention a device that can help us to establish a region of space across which there is a known potential difference. In

Chapter 26 we will give a more complete description of what a **battery** does, but for now it is sufficient to know that this device is capable (by chemical means) of establishing a potential difference ΔV between its terminals. During the life of the battery the positive terminal is maintained at a higher electric potential than the negative terminal. When you purchase a 1.5-V battery from the local discount store, you are armed with the ability to raise or lower electric potentials by 1.5 V. As we have said, this ability is handy on various occasions throughout this and the next several chapters.

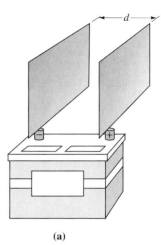

(a)

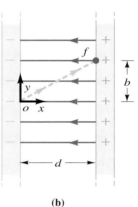

(b)

EXAMPLE 24.3   *This Field of Physics Has a Lot of Potential*

The terminals of a 12.0-V battery are connected to two parallel conducting plates that are separated by a distance $d = 10.0$ cm as shown in Figure 24.5(a). In the region near the center of the plates this arrangement produces a uniform electric field (see Fig. 24.5(b)). (a) What is the electric field strength in the region between the center of the plates? (b) Use the result of part (a) and Equation (24.9) to show that ΔV along the diagonal path in Figure 24.5(b) is 12.0 V.

SOLUTION   (a) From Equation (24.13) the electric field strength between the plates is

$$E = \frac{\Delta V}{d} = \frac{12.0 \text{ V}}{0.100 \text{ m}} = 120. \text{ V/m}$$

(b) In Figure 24.5(b) we place the origin of a coordinate system at point o, so that the coordinate of the point f is (d, b). The electric field is along the negative x-axis, so $\mathbf{E} = -E\hat{\mathbf{i}}$. In the xy-plane $d\mathbf{s} = dx\hat{\mathbf{i}} + dy\hat{\mathbf{j}}$, so Equation (24.9) gives

$$\Delta V = V_f - V_o = -\int_o^f \mathbf{E} \cdot d\mathbf{s} = -\int_o^f (-E\hat{\mathbf{i}}) \cdot (dx\hat{\mathbf{i}} + dy\hat{\mathbf{j}}) = +\int_0^d E \, dx$$

$$= E \, d = (120. \text{ V/m})(0.100 \text{ m}) = 12.0 \text{ V} \qquad \blacktriangleleft$$

FIGURE 24.5

A 12.0-V battery. (a) The terminals are connected to two parallel conducting plates. (b) The uniform electric field between the plates.

EXAMPLE 24.4   *A Potentially Confusing Example*

A 6.00-V battery is used to establish a uniform electric field between two parallel conducting plates separated by 12.0 cm. (a) If an electron is released from rest at the negative plate, what is its speed when it reaches the positive plate?

SOLUTION   The change in potential energy of the electron as it travels from the negative plate to the positive plate can be found by applying Equation (24.8):

$$\Delta U = q_t \, \Delta V = -e \, \Delta V$$

or

$$U_f - U_o = -e(V_f - V_o)$$

Recall the strange result when dealing with negative charge: the potential energy is low when the electric potential is high! We set up the problem so that the electron's potential energy is lowest when it reaches the positive plate. Hence, we choose $U_f = 0$ at the positive plate where $V_f = +6.00$ V (Fig. 24.6). Substituting this choice into the previous equation we can find the value of the initial potential energy U_o when the electron is at its initial position on the plate with zero potential ($V_o = 0$).

$$U_o = +eV_f$$

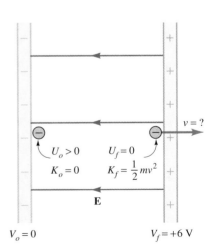

FIGURE 24.6

The potential energy of the negative charge is low when the electric potential is high: when $V_f = +6.00$ V, $U_f = 0$.

Because the electric field is conservative, we apply the conservation of mechanical energy

$$U_o + K_o = U_f + K_f$$

$$eV_f + 0 = 0 + \frac{1}{2} m_e v^2$$

We see that the potential energy is converted entirely into kinetic energy at the positive plate. Solving for the final velocity, we have

$$v = \sqrt{\frac{2eV_f}{m_e}} = \sqrt{\frac{2(1.60 \times 10^{-19} \text{ C})(6.00 \text{ V})}{9.11 \times 10^{-31} \text{ kg}}} = 1.45 \times 10^6 \text{ m/s} \qquad \blacktriangleleft$$

Electric Potential Change in a Nonuniform Electric Field

We already computed the electric potential function for a point-charge q, and it is given by Equation (24.7). In Example 24.2 we used this expression to compute the potential difference between two points at different radial distances from the charge. We next illustrate how to apply Equation (24.10), the master equation, to compute ΔV for nonuniform fields. The charge distribution in the example below is also spherically symmetric. However, this simple distribution is sufficient to illustrate the caution you must take when you apply Equation (24.10). In particular, the dot product must be performed with care. You should attack problems with cylindrical symmetry with the same approach as that illustrated in Example 24.5.

EXAMPLE 24.5   The Constants Are Players to Be Named Later

A total charge $+Q$ is spread uniformly throughout a spherical region of space with radius R. (a) Compute the electric potential in region I, where $0 \leq r \leq R$. (b) Compute the electric potential in region II, where $r \geq R$, making certain that $V \to 0$ at $r \to \infty$. Require that the potential in region II match the potential in region I at the surface of the sphere where $r = R$.

SOLUTION   We really have two problems here. The first is to find the *functional form* of the potential function in regions I and II. The second problem is to match the two solutions where they meet and to satisfy the boundary condition $V(r \to \infty) = 0$. Let's tackle each problem separately.

STEP 1: Find the form of the potential function in regions I and II

Our master equation is

$$V(r) = -\int \mathbf{E} \cdot d\mathbf{s} + C$$

We will proceed from the inside out, being careful always to integrate in the direction of increasing x in order to avoid potential confusion over the direction of $d\mathbf{s}$ and the order of the limits of the integration. We identify the form of the potential energy function by integrating between two arbitrary points along the x-axis within the appropriate region.
Region I:
    For the given charge distribution we found in Example 23.2 that the electric field along the radial direction is given by

$$E_{\text{in}} = \frac{1}{4\pi\epsilon_o} \frac{Q}{R^3} r \qquad (r \leq R)$$

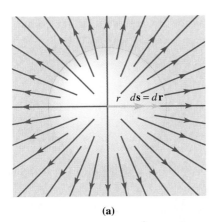

(a)

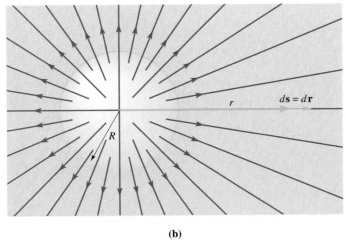

(b)

FIGURE 24.7

Computing the electric potential (a) inside a spherical distribution of charge with uniform charge density and (b) outside the sphere. See Example 24.5.

Because **E** is parallel to $d\mathbf{s}$, we have $\mathbf{E} \cdot d\mathbf{s} = E_{\text{in}}\, ds = E_{\text{in}}\, dr$, so that the master equation becomes

$$V_{\text{I}}(r) = -\int \frac{1}{4\pi\epsilon_o} \frac{Q}{R^3} r\, dr \qquad (0 < r < R)$$

Integrating, we have in region I

$$V_{\text{I}}(r) = -\frac{1}{4\pi\epsilon_o} \frac{Q}{R^3} \frac{r^2}{2} + C_{\text{I}}$$

where C_{I} is a constant to be determined.

Region II:

We found in Example 23.2 that along the x-axis the field in region II outside the sphere is given by

$$E_{\text{out}} = \frac{1}{4\pi\epsilon_o} \frac{Q}{r^2} \qquad (r \geq R)$$

Again **E** is parallel to $d\mathbf{s}$, so that $\mathbf{E} \cdot d\mathbf{s} = E_{\text{in}}\, ds = E_{\text{in}}\, dr$, giving

$$V_{\text{II}}(r) = -\int \frac{1}{4\pi\epsilon_o} \frac{Q}{r^2}\, dr + C_{\text{II}} \qquad (R < r < \infty)$$

Integrating, we find that the potential function for region II has the form

$$V_{II}(r) = \frac{1}{4\pi\epsilon_o}\frac{Q}{r} + C_{II}$$

where C_{II} is to be determined.

STEP 2: Satisfy the boundary conditions and match the solutions at $r = R$.

We now have the *form* of the potential function in each of our two regions:

$$V_{I}(r) = -\frac{1}{4\pi\epsilon_o}\frac{Q}{R^3}\frac{r^2}{2} + C_{I}$$

$$V_{II}(r) = \frac{1}{4\pi\epsilon_o}\frac{Q}{r} + C_{II}$$

It remains for us to find the values for the two constants C_I and C_{II}. We must choose these two constants to meet the remaining conditions of the problem: make the potential approach 0 as $r \to \infty$, and make the two functions match at $r = R$.

The first condition requires $V_{II}(\infty) = 0$, which is satisfied if

$$C_{II} = 0$$

The second condition requires $V_I(R) = V_{II}(R)$:

$$-\frac{1}{4\pi\epsilon_o}\frac{Q}{R^3}\frac{R^2}{2} + C_I = \frac{1}{4\pi\epsilon_o}\frac{Q}{R}$$

which leads to

$$C_I = \frac{3}{2}\frac{1}{4\pi\epsilon_o}\frac{Q}{R}$$

With these values for the constants our final solution is

$$V_{I}(r) = \frac{1}{4\pi\epsilon_o}\frac{Q}{R^3}\left(\frac{3R^2 - r^2}{2}\right), \qquad (r \leq R)$$

Concept Question 2
If you know the electric field **E** at a particular *point*, can you find the electric potential *V* at that point? Explain.

$$V_{II}(r) = \frac{1}{4\pi\epsilon_o}\frac{Q}{r}, \qquad (r \geq R)$$

A graph of $V(r)$ is shown in Figure 24.8.   ◀

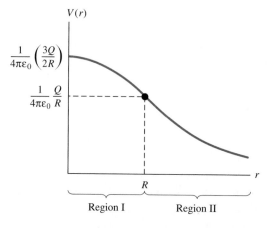

FIGURE 24.8

A graph of $V(r)$. See Example 24.5.

24.3 Computing the Electric Potential from a Charge Distribution

Equation (24.9) and our master equation allow us to compute the electric potential in some region of space if we already know the electric field **E** in that region. It is possible, and indeed often desirable, to calculate the potential directly from the charge distribution. In this section we describe how to find $V(r)$ for both point-charge and continuous charge distributions.

Point-Charge Distributions

From Equation (24.7) we know that the electric potential at distance r_i from a point-charge q_i is given by

$$V_i = \frac{1}{4\pi\epsilon_o} \frac{q_i}{r_i} \tag{24.15}$$

Again we want to remind you that we are using V_i and not ΔV_i in Equation (24.15) because we have already selected V_o to be zero at an infinite distance from the charge q_i. To obtain the potential from a collection of point-charges we use the principle of superposition. For N point-charges, the electric potential is

$$V = \frac{1}{4\pi\epsilon_o} \sum_{i=1}^{N} \frac{q_i}{r_i} \tag{24.16}$$

The most important feature to realize about Equation (24.16) is that *the distances r_i are from the charge q_i to the point in space where V is to be evaluated.* We illustrate this point in the following examples.

EXAMPLE 24.6 *The Potential on the Axis of an Electric Dipole*

Two point-charges $+q$ and $-q$ are located at $(d/2, 0)$ and $(-d/2, 0)$, respectively. Find the electric potential at distance x along the x-axis.

SOLUTION Figure 24.9 shows charges $q_1 = -q$ and $q_2 = +q$, and distances r_1 and r_2. Note the happy circumstance in Figure 24.9 that there are *no vectors* drawn at point P; the electric potential is a *scalar* field. Hence, we need only compute its net value at point P. From Equation (24.16) we have

$$V(x) = \frac{1}{4\pi\epsilon_o}\left(\frac{q}{r_1} - \frac{q}{r_2}\right) = \frac{q}{4\pi\epsilon_o}\left(\frac{1}{x + d/2} - \frac{1}{x - d/2}\right)$$

$$= \frac{qd}{4\pi\epsilon_o}\left[\frac{1}{x^2 - (d/2)^2}\right] = \frac{p}{4\pi\epsilon_o}\left[\frac{1}{x^2 - (d/2)^2}\right]$$

where in the last step we recognize $p = qd$ as the electric dipole moment of this charge arrangement. For distances that are large compared to the charge separation, $x \gg d$, this expression becomes

$$V(x) \approx \frac{1}{4\pi\epsilon_o} \frac{p}{x^2}$$

The potential dies off proportional to $1/x^2$. It is not surprising that the potential decreases more quickly than that of a point-charge (the dependence is $1/r$ in Eq. (24.7)); the net charge of a dipole is zero. (Recall the similar arguments in Example 22.6 for the electric field dependence of the dipole.) ◀

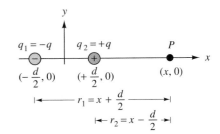

FIGURE 24.9

Computing the electric potential on the axis of an electric dipole. (See Example 24.6.)

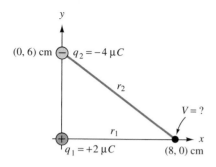

FIGURE 24.10

The locations of charges q_1 and q_2. See Example 24.7.

Concept Question 3
Concept Question 3
In Example 24.7 which charge has its potential energy raised by 0.300 J? Carefully explain your answer.

EXAMPLE 24.7   *Electrostatic Potential Energy*

Two point-charges $q_1 = +2.00 \ \mu C$ and $q_2 = -4.00 \ \mu C$ are located at (x, y) coordinates $(0.00, 0.00)$ and $(0.00, 6.00)$ cm, respectively. (a) Compute the electric potential at $(8.00, 0.00)$ cm. (b) Find the increase in electrostatic potential energy if the system of a -5.00-μC charge is placed at $(8.00, 0.00)$ cm.

SOLUTION   Figure 24.10 shows the locations of charges q_1 and q_2. From Equation (24.16) we have

$$V = \frac{1}{4\pi\epsilon_o}\left(\frac{q_1}{r_1} + \frac{q_2}{r_2}\right)$$

Notice again that we have no vector components to worry about! Now,

$$r_2 = \sqrt{(8.00 \text{ cm})^2 + (6.00 \text{ cm})^2} = 10.0 \text{ cm}$$

Hence, the electric potential at $(8.00, 0.00)$ cm is

$$V = \frac{1}{4\pi[8.854 \times 10^{-12} \text{ C}^2/(\text{N} \cdot \text{m}^2)]}\left(\frac{2.00 \times 10^{-6} \text{ C}}{0.0600 \text{ m}} - \frac{4.00 \times 10^{-6} \text{ C}}{0.100 \text{ m}}\right)$$
$$= -5.99 \times 10^4 \text{ V}$$

(b) The potential energy is related to the electric potential by Equation (24.8), $\Delta V = \Delta U/q_t$. In our case, we know V to be -5.99×10^4 V, so

$$U = q_t V = (-5.00 \times 10^{-6} \text{ C})(-5.99 \times 10^4 \text{ J/C}) = 0.300 \text{ J} \quad \blacktriangleleft$$

Continuous Charge Distributions

To compute the electric potential of a continuous charge distribution we mentally break up the distribution into a large number of small charge elements dq as illustrated in Figure 24.11. For one of these small charge elements, Equation (24.7) tells us that the contribution to the potential at point P is

$$dV = \frac{1}{4\pi\epsilon_o}\frac{dq}{r}$$

where dq is $\lambda \, ds$, $\sigma \, dA$, or $\rho \, dV$, depending on the dimensionality of the charge distribution. The total electric potential at point P is obtained by integrating over the entire charge distribution:

$$V = \frac{1}{4\pi\epsilon_o}\int \frac{dq}{r} \qquad (V = 0 \quad \text{at} \quad r = \infty) \qquad \textbf{(24.17)}$$

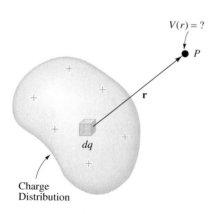

FIGURE 24.11

The electric potential at point P is computed by summing the contributions from all charges dq that compose the distribution.

You must be sure to realize two features of Equation (24.17). First, this equation assumes $V = 0$ at $r = \infty$. Second, note that the r in Equation (24.17) depends on the location of the particular charge element dq and, therefore, usually cannot be removed from the integral. With these cautions in mind we think you will find the following examples to be more attractive than those in Chapter 22 because, again, they are not vector integrations.

EXAMPLE 24.8   *A Kinder, Gentler Integration*

A linear charge distribution $\lambda = +0.400 \ \mu C/m$ lies along a straight line $L = 10.0$ cm long. Compute the electric potential at a point along the line located at a distance $d = 15.0$ cm from one end of the distribution.

SOLUTION   This charge distribution is the same as that in Example 22.5 in which we computed the force on a point-charge placed 15.0 cm from the end of the distribution. In the present case, however, we do not have to be concerned with any vectors. As shown in Figure 24.12 we place the line of charge along the x-axis with the left

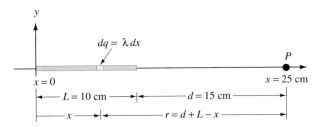

FIGURE 24.12

The line of charge lies along the x-axis with the left end at the origin. See Example 24.8.

end at the origin. We want to compute the electric potential at point P, 15.0 cm from the right end of the distribution. Applying Equation (24.17)

$$V = \frac{1}{4\pi\epsilon_o}\int \frac{dq}{r}$$

where in our case $dq = \lambda\, dx$. To include the entire line of charge we integrate dx from $x = 0$ to $x = 10.0$ cm. From the figure we see $r = d + L - x$. Therefore,

$$V = \frac{\lambda}{4\pi\epsilon_o}\int_0^L \frac{dx}{(d+L)-x} = \frac{\lambda}{4\pi\epsilon_o}\left\{-\ln[(d+L)-x]\right\}_0^L = \frac{\lambda}{4\pi\epsilon_o}\ln\left(\frac{d+L}{d}\right)$$

$$= \frac{(0.400 \times 10^{-6}\,\text{C/m})}{4\pi[8.854 \times 10^{-12}\,\text{C}^2/(\text{N}\cdot\text{m}^2)]}\ln\left(\frac{0.25\,\text{m}}{0.15\,\text{m}}\right) = 1.84 \times 10^3\,\text{V} \quad \blacktriangleleft$$

EXAMPLE 24.9   *The Potential of a Charged Disk*

Find the electric potential on the axis of a disk of radius R that carries a uniform charge density σ.

SOLUTION   As Figure 24.13 shows, all the charge enclosed in a ring of radius y and width dy is located a constant distance $r = \sqrt{x^2 + y^2}$ from point P, which is on the disk axis. The area of this ring is $(2\pi y)\,dy$, so the charge of the ring is

$$dq = \sigma\, dA = 2\pi\sigma y\, dy$$

Equation (24.17) for the electric potential is, therefore,

$$V = \frac{1}{4\pi\epsilon_o}\int \frac{dq}{r} = \frac{1}{4\pi\epsilon_o}\int_0^R \frac{2\pi\sigma y\, dy}{\sqrt{x^2 + y^2}} = \frac{\sigma}{2\epsilon_o}\int_0^R \frac{y\, dy}{\sqrt{x^2 + y^2}}$$

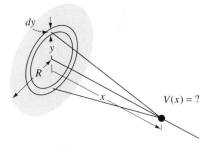

FIGURE 24.13

All the charge enclosed in a ring of radius y and width dy is located at a constant distance $r = \sqrt{x^2 + y^2}$ from point P. See Example 24.9.

The integral may be simplified by the "u substitution" $u = x^2 + y^2$ and, because x is a constant, $du = 2y\, dy$. Making this substitution and keeping track of the integration limits, our expression for the potential becomes

$$V = \frac{\sigma}{4\epsilon_o}\int_{x^2}^{x^2+R^2} \frac{du}{u^{1/2}}$$

The integral is of the form

$$\int u^n \, du = \frac{u^{n+1}}{(n+1)}$$

so

$$V(x) = \frac{\sigma}{4\epsilon_o} \left(\frac{u^{1/2}}{1/2} \right)_{x^2}^{x^2+R^2}$$

$$= \frac{\sigma}{2\epsilon_o} (\sqrt{x^2 + R^2} - x)$$

You should verify that for large x (that is, $x \gg R$) this expression satisfies the condition that $V(x \rightarrow \infty) = 0$. Moreover, you should also verify the units of this expression for $V(x)$ at $x = 0$.   ◀

24.4 Computing the Electric Field from the Electric Potential

We hope that the examples of the previous section convinced you that it is often easier to compute the electric potential than the electric field. For a given charge arrangement, both computations require either a summation over discrete charges or an integration over a charge distribution. However, in the case of the electric potential, neither the sum nor the integration involves vectors. It would be nice, therefore, if there were an easy way to compute the electric field from the electric potential.

As we're sure you have already guessed, there is indeed a relatively easy procedure to compute $\mathbf{E}$ from V. The short version of the story (good only for the one-dimensional field or fields of spherical or cylindrical symmetry) is this: to find the electric field from the electric potential we must take a derivative (more accurately, a *negative* derivative). We don't know about you, but we would rather take a derivative any day and leave a vector integration to someone else.

We first find the relation between V and $\mathbf{E}$ for one-dimensional fields. That is, we consider electric fields parallel to the x-axis so that $\mathbf{E} = E(x)\hat{\mathbf{i}}$ and $V = V(x)$. (You are always free to orient your coordinate system so that the x-axis coincides with the direction of the uniform field.) The differential form of Equation (24.9) is

$$dV = -\mathbf{E} \cdot d\mathbf{s}$$

So, with $d\mathbf{s} = dx\hat{\mathbf{i}} + dy\hat{\mathbf{j}} + dz\hat{\mathbf{k}}$, we have

$$dV = -E(x)\hat{\mathbf{i}} \cdot (dx\hat{\mathbf{i}} + dy\hat{\mathbf{j}} + dz\hat{\mathbf{k}}) = -E(x) \, dx$$

Because the differentials dV and dx are related by the derivative through $dV = (dV/dx) \, dx$, we have

$$E(x) = -\frac{dV}{dx} \tag{24.18}$$

In one dimension, the electric field is the negative derivative of the electric potential. On a graph of the electric potential function, the electric field strength is the negative slope of the electric potential function.

We illustrate how E may be computed from knowledge of V in the following examples.

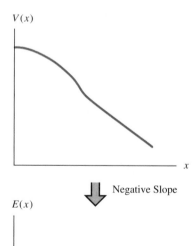

$V(x)$

Negative Slope

$E(x)$

FIGURE 24.14

The electric field is the negative slope of V versus x. (Look back at Fig. 24.3 for a surprise!)

EXAMPLE 24.10   *Obtaining the Electric Field from the Electric Potential (One Dimension)*

Use the electric potential function derived in Example 24.9 to find the electric field on the axis of a charged disk.

SOLUTION   The electric potential we derived in Example 24.9 is

$$V(x) = \frac{\sigma}{2\epsilon_o}(\sqrt{x^2 + R^2} - x)$$

Applying Equation (24.18), we have

$$E(x) = -\frac{d}{dx}\left[\frac{\sigma}{2\epsilon_o}(\sqrt{x^2 + R^2} - x)\right]$$

$$= \frac{\sigma}{2\epsilon_o}\left(1 - \frac{x}{\sqrt{x^2 + R^2}}\right)$$

This is precisely the answer we obtained in Chapter 22 for E_{disk}; however, this time we did not have to perform the vector integrations of Examples 22.9 and 22.10. We emphasize that this result is valid for axial positions only. To obtain the off-axis field you must use the methods of the next section. ◀

EXAMPLE 24.11   *Undoing a Previous Example*

(a) Find the electric potential on the axis of a uniformly charged ring at distance x from its center. The ring carries a total charge $+Q$. (b) Find the electric field strength on the axis of the ring at distance x from its center.

SOLUTION   (a) Figure 24.15 shows the charge distribution and designates point P at distance x along the ring axis. The charge $dq = \lambda\, ds$, where $\lambda = Q/2\pi R$ and $ds = R\, d\theta$. The electric potential is given by

$$V(x) = \frac{1}{4\pi\epsilon_o}\int_{\text{ring}}\frac{dq}{r} = \frac{1}{4\pi\epsilon_o}\int_0^{2\pi}\frac{\lambda R\, d\theta}{\sqrt{x^2 + R^2}} = \frac{1}{4\pi\epsilon_o}\frac{\lambda R}{\sqrt{x^2 + R^2}}\int_0^{2\pi} d\theta$$

$$= \frac{\lambda}{2\epsilon_o}\frac{R}{\sqrt{x^2 + R^2}}$$

Substituting $\lambda = Q/2\pi R$, we obtain

$$V(x) = \frac{1}{4\pi\epsilon_o}\frac{Q}{\sqrt{x^2 + R^2}}$$

(b) From the symmetry of the problem, it is clear that **E** has only an x-component as long as we stay on the x-axis. Consequently, we can use our one-dimensional approach. (For an off-axis point we need to use the methods described in the next section.) We apply Equation (24.18) to find the electric field at point P:

$$E(x) = -\frac{dV}{dx} = -\frac{Q}{4\pi\epsilon_o}\frac{d}{dx}\left(\frac{1}{\sqrt{x^2 + R^2}}\right) = -\frac{Q}{4\pi\epsilon_o}\left(-\frac{1}{2}\right)(x^2 + R^2)^{-3/2}(2x)$$

$$= \frac{Q}{4\pi\epsilon_o}\frac{x}{(x^2 + R^2)^{3/2}}$$

This is the same result we obtained by a direct, vector integration in Example 22.9. (In the present example the ring radius is $y = R$.) ◀

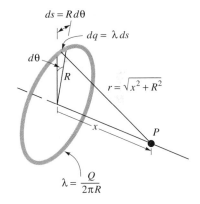

$$ds = R\, d\theta$$
$$dq = \lambda\, ds$$
$$r = \sqrt{x^2 + R^2}$$
$$\lambda = \frac{Q}{2\pi R}$$

FIGURE 24.15

Computing the electric potential and field on the axis of a charged ring. See Example 24.11.

EXAMPLE 24.12   *Now We're Back to the Beginning Again*

From the electric potential function of a point-charge q (Eq. (24.7)) find the electric field of the point-charge.

SOLUTION   The potential function for a point-charge is

$$V(r) = \frac{1}{4\pi\epsilon_o}\frac{q}{r}$$

Although the electric potential field is not one-dimensional, there is only one variable r in the potential function. Hence, we can write Equation (24.17) as a derivative with respect to r, rather than x, and

$$E(r) = -\frac{dV(r)}{dr} = -\frac{d}{dr}\left(\frac{1}{4\pi\epsilon_o}\frac{q}{r}\right) = \frac{1}{4\pi\epsilon_o}\frac{q}{r^2}$$

a familiar result!   ◄

Concept Question 4
If you know the electric potential *V* at a particular *point,* can you find the electric field **E** at that point? Explain.

Concept Question 5
If the electric field in some region of space is zero, does it mean that the electric potential is also zero in this same region? Explain.

The Relation Between *V* and *E* in Three Dimensions (Optional)

We begin again with the differential form of Equation (24.9):

$$dV = -\mathbf{E}\cdot d\mathbf{s} \qquad\qquad (24.19)$$

In general, the potential function V depends on all three coordinates: $V = V(x, y, z)$. From the calculus of multivariable functions

$$dV = \frac{\partial V}{\partial x}\,dx + \frac{\partial V}{\partial y}\,dy + \frac{\partial V}{\partial z}\,dz$$

Next, writing $\mathbf{E} = E_x\hat{\mathbf{i}} + E_y\hat{\mathbf{j}} + E_z\hat{\mathbf{k}}$ and $d\mathbf{s} = dx\hat{\mathbf{i}} + dy\hat{\mathbf{j}} + dz\hat{\mathbf{k}}$, we have

$$\mathbf{E}\cdot d\mathbf{s} = (E_x\hat{\mathbf{i}} + E_y\hat{\mathbf{j}} + E_z\hat{\mathbf{k}})\cdot(dx\hat{\mathbf{i}} + dy\hat{\mathbf{j}} + dz\hat{\mathbf{k}})$$
$$= E_x\,dx + E_y\,dy + E_z\,dz$$

Using these expressions for dV and $\mathbf{E}\cdot d\mathbf{s}$ in Equation (24.19) gives us

$$\frac{\partial V}{\partial x}\,dx + \frac{\partial V}{\partial y}\,dy + \frac{\partial V}{\partial z}\,dz = -(E_x\,dx + E_y\,dy + E_z\,dz)$$

Equating the coefficients of dx, dy, and dz on both sides of this equation results in the following expressions for the components of the electric field **E**:

$$E_x = -\frac{\partial V}{\partial x}, \qquad E_y = -\frac{\partial V}{\partial y}, \qquad E_z = -\frac{\partial V}{\partial z}$$

Because $\mathbf{E} = E_x\hat{\mathbf{i}} + E_y\hat{\mathbf{j}} + E_z\hat{\mathbf{k}}$, we have

$$\mathbf{E} = -\left(\frac{\partial V}{\partial x}\hat{\mathbf{i}} + \frac{\partial V}{\partial y}\hat{\mathbf{j}} + \frac{\partial V}{\partial z}\hat{\mathbf{k}}\right)$$

$$= -\left(\frac{\partial}{\partial x}\hat{\mathbf{i}} + \frac{\partial}{\partial y}\hat{\mathbf{j}} + \frac{\partial}{\partial z}\hat{\mathbf{k}}\right)V(x, y, z)$$

$$\mathbf{E}(x, y, z) = -\boldsymbol{\nabla}V(x, y, z) \qquad\qquad (24.20)$$

where $\boldsymbol{\nabla}$ is the gradient operator we first encountered in Chapter 8.

Combined with the general expression for the electric potential, Equation (24.16), our previous result, Equation (24.20), provides a powerful mechanism for computing the electric field of a given charge distribution: we first apply Equation (24.16) to compute the electric potential by a scalar integration. We then find the electric field by applying Equation (24.20). Once we know both the potential field $V(x, y, z)$ and electric field $\mathbf{E}(x, y, z)$, we are in a position to compute the potential energy $U = q_t V$ and the force $\mathbf{F} = q_t \mathbf{E}$ on any charge q_t placed at point (x, y, z) in space.

Knowledge of V and $\mathbf{E}$ permits us to do much more than just predict the behavior of point-charges bouncing around electric fields (although some of us do get very excited about this type of behavior, particularly when we can learn something about nature from the outcome). From a practical standpoint, knowledge of V and $\mathbf{E}$ helps scientists to build electron microscopes for biological as well as physical applications; mechanical engineers can design electrostatic precipitators to clean the air; and electrical engineers can design a myriad of electronic devices for an ever-growing list of applications.

EXAMPLE 24.13   *Using the Gradient Operator*

The electric potential in a region of space is given by the function

$$V(x, y, z) = ax + by^2 - cz^3$$

where $a = (3.00 \text{ V/m})$, $b = (2.00 \text{ V/m}^2)$, and $c = (1.00 \text{ V/m}^3)$. (a) Compute the potential difference between the points P_o and P_f with (x, y, z) coordinates $(0.00, 1.00, 2.00)$ m and $(1.00, 2.00, 1.00)$ m, respectively. (b) What is the magnitude of the electric field at point A: $(3.00, 1.00, 2.00)$ m? (c) What is the magnitude of the acceleration of an electron placed at point A?

SOLUTION   (a) The potential difference is

$$\Delta V = V_f(1.00, 2.00, 1.00) - V_o(0.00, 1.00, 2.00)$$

$$= [(3.00 \text{ V/m})(1.00 \text{ m}) + (2.00 \text{ V/m}^2)(2.00 \text{ m})^2 - (1.00 \text{ V/m}^3)(1.00 \text{ m})^3]$$

$$- [(3.00 \text{ V/m})(0) + (2.00 \text{ V/m}^2)(1.00 \text{ m})^2 - (1.00 \text{ V/m}^3)(2.00 \text{ m})^3]$$

$$= 10.0 \text{ V} - (-6.00 \text{ V}) = 16.0 \text{ V}$$

(b) To compute the electric field we use Equation (24.20):

$$\mathbf{E}(x, y, z) = -\boldsymbol{\nabla} V(x, y, z) = -\left(\frac{\partial}{\partial x}\widehat{\mathbf{i}} + \frac{\partial}{\partial y}\widehat{\mathbf{j}} + \frac{\partial}{\partial z}\widehat{\mathbf{k}} \right)(ax + by^2 - cz^3)$$

$$= -a\widehat{\mathbf{i}} - 2by\widehat{\mathbf{j}} + 3cz^2\widehat{\mathbf{k}}$$

At point A,

$$\mathbf{E}(3.00 \text{ m}, 1.00 \text{ m}, 2.00 \text{ m}) = -(3.00 \text{ V/m})\widehat{\mathbf{i}} - 2(2.00 \text{ V/m}^2)(1.00 \text{ m})\widehat{\mathbf{j}}$$

$$+ (1.00 \text{ V/m}^3)(2.00 \text{ m})^2\widehat{\mathbf{k}}$$

$$= -(3.00 \text{ V/m})\widehat{\mathbf{i}} - (4.00 \text{ V/m})\widehat{\mathbf{j}} + (12.0 \text{ V/m})\widehat{\mathbf{k}}$$

The magnitude of $\mathbf{E}$ at this point is

$$E = \sqrt{E_x^2 + E_y^2 + E_z^2} = \sqrt{(-3.00 \text{ V/m})^2 + (-4.00 \text{ V/m})^2 + (12.0 \text{ V/m})^2}$$

$$= 13.0 \text{ V/m}$$

(c) The magnitude of the force on an electron is $F = eE$, so by Newton's second law $eE = m_e a$, where m_e is the electron's mass. Therefore,

$$a = \frac{e}{m_e}E = \frac{(1.60 \times 10^{-19} \text{ C})}{(9.11 \times 10^{-31} \text{ kg})}(13.0 \text{ N/C}) = 2.28 \times 10^{12} \text{ m/s}^2 \quad \blacktriangleleft$$

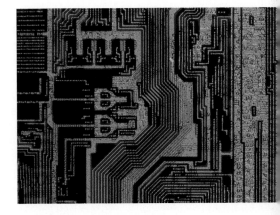

Microchip circuits have many applications in modern technology.

Why Some of This Should Seem Familiar

There is a clear correspondence between the relation of the electric field to the electric potential that has evolved thus far in this chapter and the relation of the conservative force to the potential energy that we developed in Chapters 7 and 8. This correspondence should come as no surprise because the electric field is defined as the force per unit charge and the electric potential is defined as the potential energy per unit charge. That is, in order to obtain the electric field we measure or compute the electric force on a unit charge ($\mathbf{E} = \mathbf{F}/q_t$). To find the electric potential (to within an arbitrary constant) we measure or compute the electric potential energy of a unit charge ($\Delta V = \Delta U/q_t$).

Table 24.1 illustrates the relations between the force and potential energy concepts of Chapters 7 and 8 and the electric field and potential ideas of this chapter. In Chapters 7 and 8 we found the relations between the conservative force and the potential energy: the potential energy is the negative line integral of the conservative force, and the conservative force is the negative derivative of the potential energy. (Those of you who studied optional Section 8.7 should recognize that in three dimensions the conservative force is the negative gradient of the potential energy function.) In this chapter we found the relations between the electric field and the electric potential: the electric potential is the negative line integral of the electric field, and the electric field is the negative derivative (or, in three dimensions, the negative gradient) of the electric potential.

TABLE 24.1   Correspondence of Force and Potential Energy Relations to Electric Field and Electric Potential Relations

FORCE AND POTENTIAL ENERGY	ELECTRIC FIELD AND POTENTIAL RELATIONS
$\Delta U = -\int \mathbf{F} \cdot d\mathbf{s}$	$\Delta V = -\int \mathbf{E} \cdot d\mathbf{s}$
$F = -\dfrac{dU}{dx}$ (one dimension)	$E = -\dfrac{dV}{dx}$ (one dimension)
$\left(\mathbf{F} = -\nabla U\right.$ (from optional Section 8.7)	$\mathbf{E} = -\nabla V$ (from optional material in this section)$\left.\right)$

CONNECTING EQUATIONS

$$\mathbf{F} = q_t\mathbf{E}$$
$$\Delta U = q_t\,\Delta V$$

24.5 Equipotential Surfaces

The collection of points that all have the same electric potential constitute what is called an **equipotential surface.** We have seen that the potential function for a point-charge, Equation (24.7), depends only on the distance from the charge. Hence, the equipotential surfaces of a point-charge are concentric spheres, perpendicular to the electric field lines (Fig. 24.16(a)). For a region of space in which there is a uniform electric field, the equipotential surfaces are planes that are also perpendicular to the direction of the E-field (Fig. 24.16(b)). In fact, it is easy to see that equipotential surfaces must always be perpendicular to E-field lines. From Equation (24.12) we know that the work we must do in order to move a charge q_t between two points on an equipotential surface is

$$\mathcal{W}_{\text{ext}} = q_t\,\Delta V = 0$$

because ΔV is zero on the surface. This observation means that there can be no component of the electric field $E_\parallel$ along the surface; otherwise, we would have to perform work to overcome the force $F = q_t E_\parallel$. Hence, $\mathbf{E}$ must be perpendicular to the equipotential surface. Equipotential surfaces for the field of a dipole are shown in Figure 24.16(c).

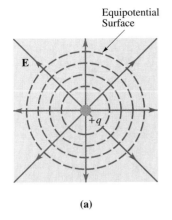

Equipotential Surface

E

+q

(a)

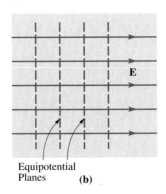

E

Equipotential Planes

(b)

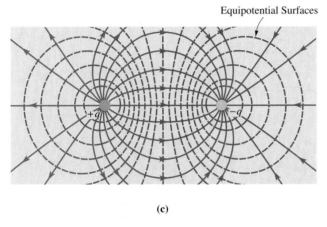

Equipotential Surfaces

+q   -q

(c)

FIGURE 24.16

Equipotential surfaces. In each case only a planar cut through the surface is actually shown (dashed lines). (a) Spherical equipotential surfaces for a point-charge. (b) Equipotential planes for a uniform E-field. (c) Equipotential surfaces for a dipole.

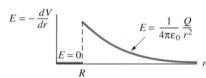

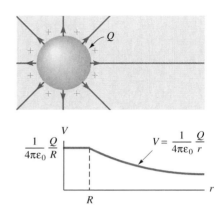

$$\frac{1}{4\pi\varepsilon_0}\frac{Q}{R}$$ $$V = \frac{1}{4\pi\varepsilon_0}\frac{Q}{r}$$

$$E = -\frac{dV}{dr}$$ $$E = \frac{1}{4\pi\varepsilon_0}\frac{Q}{r^2}$$

$E = 0$

FIGURE 24.17

The electric potential and field inside and outside a solid conducting sphere. The potential is a continuous function and, therefore, is a constant inside the sphere. The electric field is the negative slope of the electric potential function.

In Chapter 23 we saw that in equilibrium E-field lines must always intercept the surface of a conductor, charged or uncharged, at right angles. It is not difficult to see that:

The surface of a conductor is an equipotential surface.

Consider two points on the surface of a conductor. If these two points are not at the same potential, free electrons tend to move from the lower potential point to the higher potential point (because they are negative charges). But this is a violation of our electrostatic equilibrium assumption. Hence, in equilibrium all points on the surface of a conductor must be at the same potential. A similar argument holds for points inside the conductor; all points inside and on the surface of a conductor must be at the same potential.

Figure 24.17 shows graphs of the electric potential and field inside and outside a charged, spherical conductor. Outside the sphere both the electric potential and electric field are that of a point-charge Q located at the sphere's center. But inside the sphere the potential is constant and equal to its value at the surface. The electric field inside the sphere is the negative derivative of the potential, which is zero because the potential is a constant.

When a conductor is placed in a uniform electric field, the E-field lines are perturbed so that they intercept the conducting surface at right angles. Figure 24.18 shows a planar section of the E-field lines that were computed numerically for an irregularly shaped conductor placed in a uniform E-field.

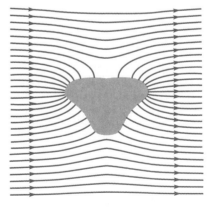

FIGURE 24.18

The result of a numerical calculation of the E-field lines around an irregularly shaped conductor placed in a uniform E-field. (From R. L. Spencer, "Electrical field lines near an oddly shaped conductor in a uniform electric field," *American Journal of Physics*, **56** (6), 512 (1988). Reprinted with permission.)

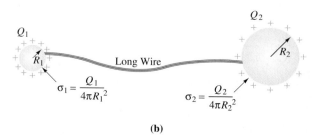

(a)

(b)

FIGURE 24.19

(a) The surface charge density and electric field strength of a charged conductor are greatest in regions of small curvature. (b) A model of the conductor shown in part (a).

The charge distribution on an irregularly shaped conductor is not uniform. As shown in Figure 24.19(a), the surface charge density σ is relatively large in regions where the curvature of the conductor is large (small radius R), and σ is small where the curvature is small (large radius R). Hence, the electric field around a charged conductor can be relatively high in regions where the conductor forms a sharp point. This result is the reason your fingertips—the sharp point—can easily receive a "shock" from a door knob after you have stored charge on yourself by walking across a carpet.

We can understand why charge density is greater at sharper corners by modeling the conductor in Figure 24.19(a) with two spheres of different radii joined together by a long wire. As shown in Figure 24.19(b), the spheres are separated by a large enough distance that in the regions near each we can approximate the electric potential as that of an isolated, charged sphere. The connecting wire makes the arrangement all one conductor and ensures that both spheres are at the same potential. We designate the charge and radius of sphere 1 as Q_1 and R_1, and those of sphere 2 as Q_2 and R_2. The potential at the surface of each is

$$V_1 = \frac{1}{4\pi\epsilon_o}\frac{Q_1}{R_1} \quad \text{and} \quad V_2 = \frac{1}{4\pi\epsilon_o}\frac{Q_2}{R_2}$$

But because the spheres are connected by a conductor, in equilibrium $V_1 = V_2$, so

$$\frac{1}{4\pi\epsilon_o}\frac{Q_1}{R_1} = \frac{1}{4\pi\epsilon_o}\frac{Q_2}{R_2}$$

or

$$\frac{Q_1}{Q_2} = \frac{R_1}{R_2}$$

Concept Question 6
Describe equipotential surfaces for an infinite line of charge.

Concept Question 7
Explain why two equipotential surfaces cannot cross.

This equation tells us that the larger sphere holds more charge, but let's look at the surface charge densities σ_1 and σ_2 of the two spheres. The charge on each sphere is the product of its area and the charge density. Thus, $Q_1 = 4\pi R_1^2 \sigma_1$ and $Q_2 = 4\pi R_2^2 \sigma_2$, so our last result can be written

$$\frac{4\pi R_1^2 \sigma_1}{4\pi R_2^2 \sigma_2} = \frac{R_1}{R_2}$$

or

$$\sigma_1 = \frac{R_2}{R_1}\sigma_2 \qquad (24.21)$$

When the radius $R_1 < R_2$, Equation (24.21) tells us that $\sigma_1 > \sigma_2$ and we have proved our assertion that the charge density is greater in regions of smaller curvature.

EXAMPLE 24.14 *A Shocking Example!*

It has been found experimentally that an electric field strength of 3.0×10^6 V/m can cause a spark in ordinary air. Suppose you shuffle across a rug so that your body becomes negatively charged. You then reach toward an uncharged doorknob, but a spark jumps between your index finger and the doorknob when they are separated by 0.50 cm. Model the electric field between your fingertip and the doorknob as the field between two infinite planes and estimate (a) the potential difference between your finger and the doorknob and (b) the charge density on the end of your finger.

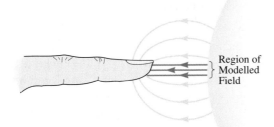

Region of Modelled Field

FIGURE 24.20

$\Delta V = 15\,000$ V. Ouch!!
See Example 24.14.

SOLUTION (a) Our parallel-plate model implies that the electric field is constant between the two planes. The potential difference is therefore $\Delta V = Ed$, where d is the separation of the planes. Because $E \geq 3.0 \times 10^6$, $\Delta V \geq (3.0 \times 10^6 \text{ V/m})(5.0 \times 10^{-3} \text{ m}) = 15\,000$ V!! Ouch, no wonder that smarts!!

(b) In Chapter 23 we found that when the field lines all leave a plane surface from the same side of the surface, the electric field just outside that surface is given by Equation (23.7)

$$E = \frac{\sigma}{\epsilon_o}$$

So, for the present case,

$$\sigma = \epsilon_o E = (8.854 \times 10^{-12} \text{ C}^2/(\text{N}\cdot\text{m}^2))(3.0 \times 10^6 \text{ N/C}) = 27 \ \mu\text{C/m}^2$$

From this result we find that on the roughly 1-cm^2 area of the end of your finger there is something like 0.003 μC of excess charge. The fact that the quantity of charge is small is what prevents your finger from becoming charred! There's plenty of voltage, but the total energy involved, when this quantity of charge moves between your finger and the doorknob, is only $U = q \, \Delta V = (0.003 \ \mu\text{C})(15 \text{ kV}) = 0.05 \text{ mJ}$. This is about the same amount of kinetic energy that a 1-g mass acquires after falling 0.5 cm from rest near the surface of the earth. ◀

24.6 The Big Picture

We are now aware of three ways to describe an electrostatic configuration. We can tell where the charges are located. That is, we can give the charge density $\rho(\mathbf{r})$ at every point in space; we can describe the electric field $\mathbf{E}(r)$ at every point in space; or we can describe the electric potential $V(r)$ at every point. As you no doubt realize by now, these three descriptions are not independent; in fact, any one of them is complete in the sense that given one, the other two can be deduced. Figure 24.21 summarizes the relationships among these three fields and shows the pathways we have developed to get between them. The letter Q symbolizes knowledge of the charge distribution, which may actually be given in terms of q, ρ, σ, or λ. The indicated integrals may be summations over discrete charges rather than integrations over continuous charge distributions.

You may use the "electrifying" triangle in Figure 24.21 to plan your problem-solving strategy *before* striking off into what admittedly can be a bewildering array of electrostatic formulas. When you start a problem, the "given" information determines from which corner you should begin. The information the problem seeks determines the corner at which you need to finish. You should then be able to map out several possible strategies for getting from start to finish along the sides of the triangle. Determining which path is easiest takes experience. We suggest that you go back over several examples in Chapters 22 through 24 as well as problems you have already worked and map out the starting point, end point, and path taken.

Incidentally, you may feel that the triangle is incomplete, there being no paths shown from $\mathbf{E}$ or V back to Q. Rest assured that the required relationships do exist although, as they say, "the mathematics is beyond the level of this text." If you're interested, the equation connecting $\mathbf{E}$ with Q is called "the differential form of Gauss's law" and written $\nabla \cdot \mathbf{E} = \rho/\epsilon_o$. Another connection between V and Q (which can be used to move in either direction) is called Poisson's equation: $\nabla^2 V = \rho/\epsilon_o$. The mathematics can be found in standard texts on vector calculus.

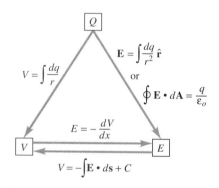

FIGURE 24.21

The relationships between charge density, electric field, and electric potential.

24.7 Numerical Methods: Relaxation (Optional)

Gauss's law and the other equations connecting charge density, electric field, and electric potential are called *field equations*. For simple charge configurations of high symmetry the field equations can be solved with relative ease. Often, we can model actual configurations using simple disks, cylinders, and spheres and obtain answers with adequate accuracy. For complicated configurations where high accuracy is required numerical solutions of the field equations become indispensable. We introduce one technique, **the method of relaxation,** to give an example of the kinds of calculations that can be done and to provide you with a tool for examining some nonsymmetrical geometries.

To keep the mathematics tractable we consider only two-dimensional problems, allowing no variation in, say, the z-direction. We fix the potentials on certain surfaces, usually conductors, because this is often done in the real world. The laws of electrostatics then determine the potential at all other points. Indeed this is the very meaning of *field equations*: fields have values at all points in space; field equations connect those values.

We model our space as composed of small cubic building blocks with dimensions Δx, Δy, Δz. The imaginary walls of a typical cube act as a Gaussian surface as shown in Figure 24.22. Our goal is to relate the potential in the center of a cube to the values of the potentials of the neighboring cubes. The final relation is almost incredibly simple. To get to that final result we proceed in two steps:

STEP 1: First we use Gauss's law to obtain a relation between the electric fields on the walls of the cube under study.

STEP 2: Next we use $E = -\Delta V/\Delta d$ to relate the fields at the cube walls to the potential of adjacent cubes.

In step (1) we apply Gauss's law to the representative cube shown in Figure 24.23. For ease of identification we denote the walls as F, B, R, L, O, U for Front, Back, Right, Left, Over, and Under. The outward-pointing unit vectors for these walls are $\hat{\mathbf{k}}$, $-\hat{\mathbf{k}}$, $\hat{\mathbf{i}}$, $-\hat{\mathbf{i}}$, $\hat{\mathbf{j}}$, and $-\hat{\mathbf{j}}$, respectively. Because the areas of all of the walls are equal, we call the value simply ΔA. The total flux on the left-hand side of Gauss's law is the sum of six integrals, one over each face. We now make an important approximation. Although the field can have different values on different faces, we assume that the faces are small enough so that E *is constant on any one face*. With this simplification the six integrals become six products of the (constant) $\mathbf{E}$ at each face times the respective area ΔA at each face.

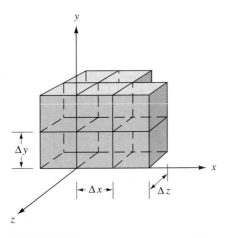

FIGURE 24.22

Space is subdivided into an array of cubic Gaussian surfaces of sides $\Delta x = \Delta y = \Delta z$.

$$\oint \mathbf{E} \cdot d\mathbf{A} = \int_{\text{Front}} \mathbf{E} \cdot d\mathbf{A} + \int_{\text{Back}} \mathbf{E} \cdot d\mathbf{A} + \int_{\text{Right}} \mathbf{E} \cdot d\mathbf{A} + \int_{\text{Left}}$$

$$\mathbf{E} \cdot d\mathbf{A} + \int_{\text{Over}} \mathbf{E} \cdot d\mathbf{A} + \int_{\text{Under}} \mathbf{E} \cdot d\mathbf{A}$$

$$\approx \mathbf{E}(F) \cdot (\hat{\mathbf{k}} \Delta A) + \mathbf{E}(B) \cdot (-\hat{\mathbf{k}} \Delta A) + \mathbf{E}(R) \cdot (\hat{\mathbf{i}} \Delta A) + \mathbf{E}(L) \cdot (-\hat{\mathbf{i}} \Delta A)$$

$$+ \mathbf{E}(O) \cdot (\hat{\mathbf{j}} \Delta A) + \mathbf{E}(U) \cdot (-\hat{\mathbf{j}} \Delta A)$$

The arguments of the electric field are meant to symbolize the face at which the field should be evaluated. For example $\mathbf{E}(L)$ means to evaluate the electric field at the left face of the cube. The electric field at any face can be written as $\mathbf{E} = E_x\hat{\mathbf{i}} + E_y\hat{\mathbf{j}} + E_z\hat{\mathbf{k}}$. Taking the indicated dot products, we obtain for the flux

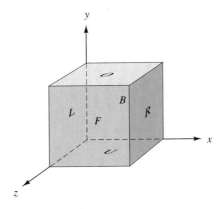

FIGURE 24.23

The Front, Back, Left, Right, Over, and Under sides of a typical cube are labeled F, B, L, R, O, and U, respectively.

$$\oint \mathbf{E} \cdot d\mathbf{A} \approx \Delta A[E_z(F) - E_z(B) + E_x(R) - E_x(L) + E_y(O) - E_y(U)]$$

Because we have restricted our problem to two dimensions, we assume no variation in the y-direction. Consequently, $E_y(O) = E_y(U)$. This result means that the last two terms of the above expression for the flux cancel. The right-hand side of Gauss's law is q/ϵ_o. If we restrict our problem to regions of space where there are no charges, then the right-hand side of Gauss's law is zero, which in turn means the previous expression for the flux is equal to zero. Consequently, dividing the flux expression by ΔA and dropping the last two terms, we reach our first milestone in the derivation:

$$E_z(F) - E_z(B) + E_x(R) - E_x(L) = 0 \tag{24.22}$$

Step (2) of our solution is to approximate the electric fields E in the above equation using $E = -\Delta V/\Delta d$. Because the problem is now fully two-dimensional, let's change our perspective and look down on our cube and its neighbors. As shown in Figure 24.24, we now label the row and column of our cube with indices m and n. We designate the potential in the center of the (m, n) cube as $V(m, n)$. As you can see from Figure 24.24, the left face of our cube lies between cube $(m - 1, n)$ and cube (m, n). Consequently, we can approximate $E_x(L) \approx -\Delta V/\Delta x = -[V(m, n) - V(m - 1, n)]/\Delta x$. Similarly, $E_x(R) \approx -[V(m + 1, n) - V(m, n)]/\Delta x$. We leave it as an exercise for you to find the equivalent expressions for $E_z(R)$ and $E_z(L)$ and then substitute them into Equation (24.22) to produce

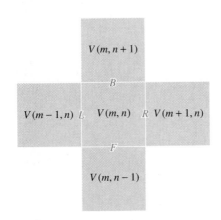

FIGURE 24.24

Top view of cell (m, n) and its neighbors. The potential in the center of cell (m, n) is $V(m, n)$. The left face (L) is between cell $(m - 1, n)$ and cell (m, n).

$$4V(m, n) - V(m - 1, n) - V(m + 1, n) - V(m, n - 1) - V(m, n + 1) = 0$$

from which we immediately obtain our sought-after connection between the potentials of adjacent cubes

$$V(m, n) = \frac{V(m - 1, n) + V(m + 1, n) + V(m, n - 1) + V(m, n + 1)}{4} \quad (24.23)$$

This result is remarkably simple! The potential at the center of a cube is just the average of the potential of its four neighboring cubes. Wouldn't it be nice if all of electrostatics were this simple?

We can easily implement Equation (24.23) to solve two-dimensional potential problems using a spreadsheet. We simply let each spreadsheet cell represent one of our cubes. We define a closed border of cells of any shape by putting a fixed number in each of them to represent our boundary potential. Equation (24.23) is copied into each cell internal to the border. Then we instruct the spreadsheet to calculate. Of course, the answer doesn't appear all at once. For example, suppose all the internal cells originally contain zero and the border cells all contain 10 V. After one calculation a cell near the center still has zero in it because all of its neighbors were zero before the calculation. Information about the potential on the border creeps into the interior cells only after many cycles of calculation. Eventually, the interior cells relax to their steady-state values. This process gives these types of techniques their generic name, **relaxation methods.** The particular algorithm of Equation (24.23) dates back to the previous century and is officially known as **Jacobi's method.** In Example 24.15 we demonstrate the use of the spreadsheet template contained on the diskette that accompanies this text and a simple way to display equipotential lines using a spreadsheet.

In parting, we should mention that Jacobi's method is rather slow to converge. For an $N \times N$ array of cells the number of iterations required for a given accuracy goes as N^2. If you get serious about such techniques and use large arrays, you will want more efficient algorithms, such as the method of **simultaneous over-relaxation,** or SOR. (We admit the name is reminiscent of spring break!) The idea of SOR is to overcorrect at each step, using the expression

$$V(m, n) =$$

$$(1 - w)V(m, n) + w\frac{V(m - 1, n) + V(m + 1, n) + V(m, n - 1) + V(m, n + 1)}{4} \quad (24.24)$$

where w is called the **over-relaxation parameter.** If you choose $w = 1$, this becomes Jacobi's method. For convergence you must choose $1 < w < 2$. The rate of convergence depends on the proper choice of w. When properly chosen, the rate of convergence goes as N rather than N^2. Selecting w is a subject of study in numerical analysis, although it is often chosen by trial and error. For a 25×25 spreadsheet area, we've found a value of 1.8 works reasonably well. To pursue these matters here would take us too far afield. If you are interested, you should look in a text on numerical analysis or numerical computing.[3]

EXAMPLE 24.15   Learn to Relax and Enjoy Electrostatics

Set up a spreadsheet to apply simultaneous over-relaxation to model the two-dimensional potential distribution interior to the rectangular region shown in Figure 24.25 if the entire boundary is held at 10.0 V, and the diamond-shaped region in the interior is held at 0.00 V.

SOLUTION   A portion of our spreadsheet template is shown in Figure 24.26. We put numbers symbolizing the boundary potential around the border cells A9–S9,

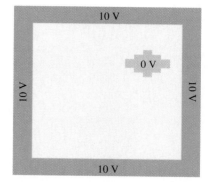

FIGURE 24.25

We wish to calculate the potential within the rectangular area. The border is held at a fixed potential of 10 V. A small region in the interior is held at 0 V.

---

[3] One of our favorites is T.A. Press, B.R. Flannery, S.A. Teulkolsky, and W.T. Vetterling, *Numerical Recipes . . .* (Cambridge University Press, NY, 1988).

	A	B	C	D	E	F	G	H	I	J	K	L	M	N	O
1	═══
6	Constants:					SOR parameter: w =			1.80		(1<=w<=2 required)				
7	(w=1 is Jacobi's method)							1-w=	-0.8			w/4 =0.45				
8	═══
9	10.0	10.0	10.0	10.0	10.0	10.0	10.0	10.0	10.0	10.0	10.0	10.0	10.0	10.0	10.0
10	10.0	9.9	9.8	9.7	9.5	9.4	9.2	9.1	8.9	8.8	8.7	8.6	8.7	8.9	9.1
11	10.0	9.8	9.6	9.3	9.1	8.8	8.5	8.1	7.8	7.5	7.2	7.2	7.3	7.7	8.1
12	10.0	9.7	9.4	9.0	8.6	8.2	7.7	7.2	6.6	6.1	5.7	5.5	5.8	6.4	7.1
13	10.0	9.6	9.2	8.7	8.2	7.7	7.0	6.3	5.5	4.6	3.8	3.3	4.0	4.9	6.0
14	10.0	9.5	9.0	8.5	7.9	7.2	6.4	5.5	4.4	3.1	1.7	0.0	1.8	3.4	4.9
15	10.0	9.5	8.9	8.3	7.7	6.9	6.0	4.8	3.4	1.6	0.0	0.0	0.0	1.8	4.0
16	10.0	9.4	8.9	8.2	7.5	6.7	5.7	4.5	2.8	0.0	0.0	0.0	0.0	0.0	3.3
17	10.0	9.4	8.8	8.2	7.5	6.7	5.8	4.6	3.2	1.5	0.0	0.0	0.0	1.7	3.8
18	10.0	9.4	8.9	8.3	7.6	6.9	6.0	5.0	3.9	2.7	1.5	0.0	1.6	3.1	4.6
19	10.0	9.5	8.9	8.4	7.8	7.1	6.4	5.6	4.7	3.9	3.2	2.7	3.4	4.3	5.4
20	10.0	9.5	9.0	8.5	8.0	7.4	6.8	6.2	5.5	5.0	4.6	4.4	4.8	5.4	6.2

B10: (F1) +$J$7*B10+$M$7*(B9+B11+A10+C10)

FIGURE 24.26

Partial spreadsheet template for simultaneous over-relaxation. See Example 24.15.

S9–S28, S28–A28, and A28–A9. In cell B10 interior to the border we placed the formula for Equation (24.24):

B10:   +$J$7*B10+$M$7*(B9+B11+A10+C10)

On our spreadsheet program the $ character is used to "anchor" cells. When formulas are copied, cells that are anchored are not changed. We then copied the formula in B10 to all the other cells interior to the boundary. (Note that we calculated $w/4$ and $1 - w$ in cells M7 and J7 and then referenced these cells in our formula. This procedure saves computation time because these expressions do not have to be recalculated for each cell.) After the SOR formula was copied to the interior, we modeled the diamond-shaped region by overwriting the interior fixed-voltage cells with 0. Once this task was completed, it was a simple matter to hold down the function key that makes the spreadsheet RECALCULATE until the numbers in the cells stopped changing.

Just gazing at the numbers that result from our calculation doesn't give us much "feeling" for the way the potential changes throughout the region between the boundary and the area of fixed 0-V potential. It would be nice if we had a way to generate equipotential lines. In Figures 24.27(a) and (b) we show a couple of simple ways to simulate the effect of contour lines. In each case we made an image in A29–S48 of the potential map (A9–S28) using some criterion to decide whether to print an integer that indicates the voltage or to leave the image cell blank. We did this by using the IF statement in our spreadsheet. The syntax for your spreadsheet may not be identical, but you probably have a similar function. For our spreadsheet the syntax is

@IF(condition, true, false)

The @ sign means that IF is a function built into our spreadsheet. The *condition* is some statement to be tested as true or false. If the *condition* is true, the value following the first comma is printed into the cell. If the *condition* is false, the statement following the second comma is printed into the cell. Cell A29 of Figure 24.27 con-

	A	B	C	D	E	F	G	H	I	J	K	L	M	N	O	P	Q	R	S
29	10	10	10	10	10	10	10	10	10	10	10	10	10	10	10	10	10	10	10
30	10							8	8	8	8	8	8						10
31	10				8	8	8								8	8			10
32	10				8	8		6	6					6			8		10
33	10			8	8		6		4				4				8		10
34	10			8		6		4				0			4	6		8	10
35	10		8	8		6		4			0	0	0					8	10
36	10		8	8		6		4	2	0	0	0	0	0				8	10
37	10		8	8		6		4			0	0	0					8	10
38	10		8	8		6	6			2		0			4	6		8	10
39	10		8	8			6		4			2		4		6		8	10
40	10			8			6	6		4	4	4	4		6		8		10
41	10			8	8		6	6					6	6			8		10
42	10			8	8	8			6	6	6	6	6				8	8	10
43	10				8	8	8									8	8	8	10
44	10					8	8	8	8	8		8	8	8	8	8			10
45	10						8	8	8	8	8	8	8	8	8				10
46	10																		10
47	10																		10
48	10	10	10	10	10	10	10	10	10	10	10	10	10	10	10	10	10	10	10

A29:   @IF(2*@INT(@INT(A9)/2)=@INT(A9),@INT(A9), " ")

	A	B	C	D	E	F	G	H	I	J	K	L	M	N	O	P	Q	R	S
29	10	10	10	10	10	10	10	10	10	10	10	10	10	10	10	10	10	10	10
30	10						9	9							9	9			10
31	10			9			8			7	7				8		9		10
32	10		9		8		7		6					7				9	10
33	10	9		8		7	6									7	8	9	10
34	10	9			7			3			0								10
35	10									0	0	0							10
36	10		8						0	0	0	0	0			7			10
37	10		8					3			0	0	0			7			10
38	10		8			6	5					0		3		6			10
39	10				7					3									10
40	10		9				6							6	7	8	9		10
41	10		9	8		7											9		10
42	10		9			8								7		8			10
43	10			9			8							8					10
44	10			9	9				8	8		8	8					9	10
45	10			9	9											9			10
46	10							9	9	9	9	9	9	9	9				10
47	10																		10
48	10	10	10	10	10	10	10	10	10	10	10	10	10	10	10	10	10	10	10

B30:   @IF(@ABS(B10-@INT(B10))<0.3,@INT(B10), " ")

FIGURE 24.27

Two methods for generating approximate equipotential plots.

tains the formula

A29:   @IF(2*@INT(@INT(A9)/2)=@INT(A9),INT(A9), " ")

The @INT function takes the greatest integer of a number; for example @INT(8.623) = 8. If you examine our *condition* you will see that we are testing to see if the integer part of the voltage is even or odd. If it is even, we print the integer. If it is odd, we print a blank, which we made using an empty pair of quotes. The results of this computation are shown in Figure 24.27(a).

In Figure 24.27(b) we show the results of an alternative approach that generates pseudoequipotential lines. Here we plot the @INT value only if the actual voltage is within some small neighborhood of a whole number, in this case ±0.3:

```
B30:    @IF((@ABS(B10)-@INT(B10))<0.3, @INT(B10), " ")
```

where @ABS is the absolute-value function.

You can now connect the equal integers in the spreadsheet map and at least approximate the position of equipotential lines. You may wish to try other schemes, including "ANDing" our two conditions. The whole business of drawing computer-generated equipotential lines is an interesting subject in itself. We leave further exploration to the computer scientists among you. ◀

In Example 24.15 we surrounded the entire region of interest with a fixed potential. The potential need not be constant around the border, but if you have jumps in the potential, the potentials generated in the immediate neighborhood of the jump will not be accurate. You can simulate structures that are periodic in one direction by using **periodic boundary conditions** instead of fixed voltages along all or part of the border. For example, you could replace the top border with a row of cell formulas with a minor modification of Equation (24.24): replace the reference to the cell immediately above with the address of the cell in the same column but on the lower border. Similarly, on the lower border, change the cells so that instead of referencing the cell below, they reference the cell directly above on the top border. This wrapping around process simulates a structure that repeats periodically.

24.8 Summary

The electric potential energy for two point-charges q and q_t is

$$U(r) = \frac{1}{4\pi\epsilon_o} \frac{qq_t}{r} \qquad (24.6)$$

The electric potential is a scalar field defined as the potential energy per charge

$$\Delta V = \frac{\Delta U}{q_t} \qquad (24.8)$$

and its unit is a volt = joule/coulomb. The electric potential is defined only to within an arbitrary constant, and for a point-charge this constant is usually taken as zero at an infinite distance from the charge. We can compute the electric potential change from the electric field **E**:

$$\Delta V = V_f - V_o = -\int_o^f \mathbf{E} \cdot d\mathbf{s} \qquad (24.9)$$

For a point-charge, an electric potential function can be defined

$$V(r) = \frac{1}{4\pi\epsilon_o} \frac{q}{r} \qquad (24.7)$$

and for a continuous charge distribution

$$V = \frac{1}{4\pi\epsilon_o} \int_{\substack{\text{all} \\ \text{charges}}} \frac{dq}{r}$$ (24.17)

where again, the potential is assigned a value of zero at an infinite distance from the charge distribution.

In one dimension the electric field can be computed from the negative derivative of the electric potential:

$$E(x) = -\frac{dV}{dx}$$ (24.18)

(Optional) In three dimensions the electric field is the negative gradient of the electric potential:

$$\mathbf{E}(x, y, z) = -\boldsymbol{\nabla}V(x, y, z)$$ (24.20)

Equipotential surfaces are composed of points that all have the same electric potential. No work is required to move a charge from one position to another on the same equipotential surface. The surfaces of an ideal conductor are equipotential. All points within and on the surface of a conductor are at the same electric potential.

(Optional) Numerical methods, such as the **method of relaxation,** can be used to solve for potential distributions in complicated geometries containing little or no symmetry.

PROBLEMS

24.1 Electric Potential Energy

1. A proton is fixed at the origin. Compute both the gravitational and electrostatic potential energies due to (a) another proton placed at $x = 2.00$ m, (b) an electron placed at $x = 2.00$ m, and (c) an electrically neutral 1.00-g mass charged to 1.00 C placed at $x = 2.00$ m.

2. An alpha particle consists of two protons and two neutrons. Compute the electrostatic potential energy between an alpha particle and a gold nucleus when the alpha particle is 2.00×10^{-10} m from the gold nucleus. The positive charge in the nucleus of an atom is Ze, where Z is the atomic number.

3. Two protons are released from rest when they are separated by 1.00 m. What are their velocities when they are separated by 4.00 m? Assume the protons are in a region of space where any force other than the electrostatic force is negligible.

4. Two electrons are separated by a distance such that when the separation is increased by 1.00 m their potential energy decreases to 0.250 times its original value. What is the original distance between the electrons?

5. Three charges are located along the x-axis. A 6.00-μC charge is located at the origin, a -3.00-μC charge is located at $x = 1.50$ m, and a 4.00-μC charge is located at $x = 5.00$ m. (a) What is the electrostatic potential energy of this system of charges? (As usual, take the zero of potential energy to occur when the charges are infinitely far apart.) (b) If the 6.00-μC charge is released and the others are held fixed, what is its kinetic energy when it is infinitely far away? (c) What if the 4.00-μC charge is released instead?

6. Three charges have the following (x, y) locations: 2.00 μC at (2.00, 0.00) m, -4.00 μC at (0.00, 3.00) m, and 3.00 μC at (2.00, 3.00) m. What is the electrostatic potential energy of this system of charges? (Take the potential energy to be zero when the charges are inifinitely far apart.)

7. Compute the work required to assemble four, 1.00-μC charges on the corners of a square 10.0 cm on a side if the charges are initially at rest and separated from each other by an infinite distance.

8. (a) Compute the work required to bring the following three charges from rest at an infinite separation to the folowing (x, y) locations: $q_1 = +2.00$ μC at (0.00, 0.00) m, $q_2 = +3.00$ μC at (1.00, 0.00) m, and $q_3 = +4.00$ μC at (0.00, 2.00) m. Assume the charges are initially at rest and separated by an infinite distance. (b) After the charges are assembled the 2.00-μC charge is released from rest. What is its kinetic energy when it is infinitely far from the other two?

24.2 The Electric Potential

9. The uniform electric field between two parallel plates separated by 5.00 cm has a magnitude of 2.40×10^3 N/C. What is the potential difference between the plates?

10. Two parallel plates are separated by a distance of 4.00 cm. A proton placed between the plates experiences a force of 1.28×10^{-16} N. (a) What is the potential difference between the plates? (b) How much work must be done to move the proton from the negative plate to the positive plate?

11. A 1.50-V battery is used to establish a uniform electric field between two parallel plate conductors. (a) How much work must be done to move an electron from the positive plate to the negative plate? (b) How much work must be done to move 1.00 mol of electrons from the positive plate to the negative plate?

12. An electron is traveling due east with a kinetic energy of 6.40×10^{-14} J when it enters a uniform electric field. After traveling through a distance of 10.0 cm parallel to the electric field, the electron comes momentarily to rest. (a) What potential difference did the proton travel through from where it entered the E-field to where it came to rest? (b) What is the magnitude and direction of the electric field?

13. A proton is traveling in a region of space where there is a uniform electric field. At a particular location (A) its velocity is 2.50×10^5 m/s, and at another location (B) 8.00 cm along a line parallel to the electric field its velocity is 1.80×10^5 m/s. What is the potential difference between points A and B?

14. A uniform electric field is given by $\mathbf{E} = -(1200 \text{ V/m})\hat{\mathbf{i}}$. Find the potential difference between the locations $(1.00, 6.00)$ m and $(5.00, 2.00)$ m.

15. A uniform electric field is given by $\mathbf{E} = (2400 \text{ V/m})\hat{\mathbf{j}}$. Find the potential difference between the locations $(2.00, 3.00)$ m and $(-4.00, -2.00)$ m.

16. In the region of space shown in Figure 24.P1 there is an electric field given by $\mathbf{E} = (250. \text{ V/m})\hat{\mathbf{i}}$. By direct application of Equation (24.9) evaluate the potential difference between the points A and C (a) along the path $A \rightarrow B \rightarrow C$, (b) along the path $A \rightarrow D \rightarrow C$, and (c) along the diagonal path $A \rightarrow C$.

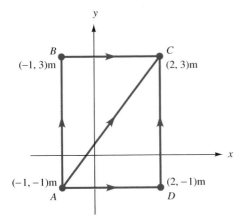

FIGURE 24.P1   Problems 16, 17, and 18

17. In the region of space shown in Figure 24.P1 there is an electric field given by $\mathbf{E} = (600. \text{ V/m})\hat{\mathbf{i}} + (800. \text{ V/m})\hat{\mathbf{j}}$. By direct application of Equation (24.9) evaluate the potential difference between the points A and C (a) along the path $A \rightarrow B \rightarrow C$, (b) along the path $A \rightarrow D \rightarrow C$, and (c) along the diagonal path $A \rightarrow C$.

18. In the region of space shown in Figure 24.P1 there is an electric field given by $\mathbf{E} = (120 \text{ V/m})\hat{\mathbf{i}} + (90.0 \text{ V/m})\hat{\mathbf{j}}$. By direct application of Equation (24.9) evaluate the potential difference between the points A and C (a) along the path $A \rightarrow B \rightarrow C$ and (b) along the diagonal path $A \rightarrow C$.

19. Two thin, concentric conducting spheres have radii R_1 and R_2 ($R_1 < R_2$). The inner sphere is charged with a uniform area charge density $+\sigma$, whereas the outer sphere carries a uniform charge density $-\sigma$. Compute the potential difference between the two spheres.

20. A long, hollow conducting cylinder carrying charge per length $-\lambda$ surrounds a smaller diameter cylinder of charge with linear density $+\lambda$. The axes of the two cylinders coincide. If the radius of the inner cylinder is R_1 and that of the outer is R_2, compute the potential difference between the two charged cylinders. This arrangement is used in the Geiger-Müller radiation detector.

24.3 Computing the Electric Potential from a Charge Distribution

21. At what distance from a 0.020-μC point-charge is the electric potential equal to 100. V if the potential is zero at an infinite distance from the point-charge?

22. The electric potential at an infinite distance from a point-charge is zero, and it is 250. V at a distance of 2.00 m from the charge. What is the electric potential at a distance of 0.500 m from the charge?

23. A 12.0-μC charge is located at (x, y) coordinates $(2.00, 0.00)$ m. Compute the potential difference between the two points $(4.00, 0.00)$ m and $(0.00, 2.00)$ m due to this charge.

24. A charge of $+q$ is located at the origin, and a charge $-2q$ is located at $x = d$. (a) Find the location along the x-axis where $V = 0$. (b) Does $E = 0$ at this location? Explain.

25. A 6.00-μC charge is located at the origin, and a -8.00-μC charge is positioned at $(10.0, 0.00)$ cm. (a) Find the electric potential at $(0.00, 12.0)$ cm due to these charges. (b) What is the potential energy of a 24.0-μC charge placed at $(0.00, 12.0)$ cm?

26. A 24.0-μC charge is located at the origin, and a -36.0-μC charge is located at $(12.0, 16.0)$ cm. Compute the potential difference between the points $(0.00, 8.00)$ cm and $(12.0, 0.00)$ cm.

27. Sodium metal ions occupy the corners and the center of a cube 4.20×10^{-10} m on an edge. Assume the charge of each ion is $+e$ and (a) compute the electric potential at the cube center due to the eight ions at the cube's corners. (b) What is the potential energy of the ion at the cube's center?

28. The charges in Figure 24.P2 are: $q_1 = q_2 = +4.00$ μC, and $q_3 = -6.00$ μC. The distances $a = 10.0$ cm and $b = 15.0$ cm. (a) Find the electric potential at point A. (b) What is the potential energy of a $+8.00$-μC charge placed at point A?

29. In Figure 24.P2 the charges $q_1 = q_2 = +Q$ and $q_3 = -Q$. Find an expression for the electric potential at points along the line from A up to (but not including) the position of charge q_3. Your answer should be expressed in terms of the charges $\pm Q$, the distances a and b, and the distance x along the line from point A.

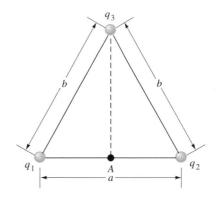

FIGURE 24.P2   Problems 28, 29, and 40

30. A linear charge distribution along the x-axis between $x = 0$ and $x = 24.0$ cm varies according to the charge density $\lambda = +(2.00 \ \mu C/m^3)x^2$. Compute the electric potential at the position $x = 1.00$ m.

31. A line of charge of length a with uniform density λ is placed along the x-axis as shown in Figure 24.P3. Compute the electric potential at point P at distance y from the origin along the y-axis.

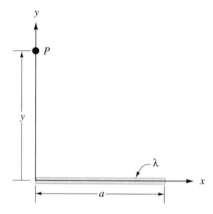

FIGURE 24.P3   Problems 31, 32, and 41

32. A line of charge of length a with nonuniform density $\lambda = Ax$ lies along the x-axis as shown in Figure 24.P3. A is a constant. Compute the electric potential at point P at distance y along the y-axis.

33. A line of charge of length a with uniform density λ is placed along the x-axis so that its end nearest the origin is at distance c from the origin (Fig. 24.P4). Compute the electric potential at point P at distance y along the y-axis.

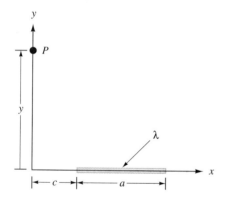

FIGURE 24.P4   Problems 33 and 42

34. A line of charge with uniform density λ lies along the x-axis from $x = -a/2$ to $x = +a/2$. Compute the electric potential at point P a distance y from the origin along the y-axis.

24.4 Computing the Electric Field from the Electric Potential

35. The electric potential of a field directed parallel to the x-axis is given by

$$V(x) = (5.00 \ \text{V/m}^3)x^3 - (60.0 \ \text{V/m}^2)x^2 - (240. \ \text{V/m})x$$

(a) Find the electric field $E(x)$. (b) Locate the position on the x-axis where the force on a charge $q = +1.20 \times 10^{-3}$ C is zero.

36. The electric potential function for an E-field directed parallel to the x-axis is graphed in Figure 24.P5. (a) Construct a graph of the electric field strength $E(x)$. (b) If the x-axis passes through a conductor, where is the conductor located?

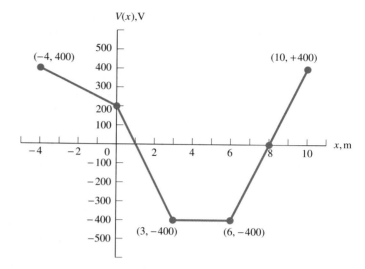

FIGURE 24.P5   Problem 36

37. The electric potential function for an E-field directed parallel to the y-axis is graphed in Figure 24.P6. Construct a graph of the electric field strength $E(y)$.

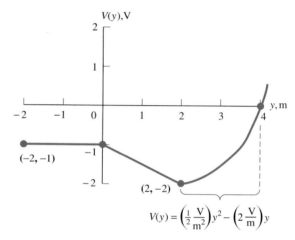

$$V(y) = \left(\tfrac{1}{2}\tfrac{\text{V}}{\text{m}^2}\right)y^2 - \left(2\tfrac{\text{V}}{\text{m}}\right)y$$

FIGURE 24.P6   Problem 37

38. Three point-charges are arranged as shown in Figure 24.P7. (a) Compute the electric potential at point A, at distance x from the $+2Q$-charge and along the line on which the three charges lie. (b) From your result in part (a) find the electric field strength at point A.

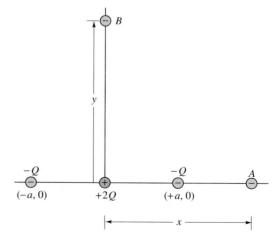

FIGURE 24.P7   Problems 38 and 39

39. Three point-charges are arranged as shown in Figure 24.P7. (a) Compute the electric potential at point B, at distance y from the $+2Q$-charge and perpendicular to the line on which the three charges lie. (b) From your result in part (a) find the electric field strength at point B.

40. From the electric potential function computed in Problem 29 find the electric field strength along the line between point A and charge q_3 in Figure 24.P2.

41. (a) Compute the y-component of the electric field strength at point P for the charge distribution given in Problem 31 shown in Figure 24.P3. (b) Why is it not possible to find the x-component of the field from the result of Problem 31?

42. (a) From the electric potential function computed in Problem 33 find the y-component of the electric field strength at a distance y from the origin along the y-axis in Figure 24.P4. (b) Why is it not possible to find the x-component of the field from the result of Problem 33?

43. (a) From the electric potential function computed in Problem 34 find the y-component of the electric field strength at point P a distance y from the origin along the y-axis. (b) Why is it not possible to find the x-component of the field from the result of Problem 33?

24.5 Equipotential Surfaces

44. Two conducting spheres are connected by a long wire. The spheres' radii are 6.00 cm and 9.00 cm, and they are separated by 3.00 m. A total charge of 1.00 μC is transferred to the spheres. (a) Compute the charge on each sphere. (b) Compute the electric potential of the spheres.

45. Two conducting spheres of 4.00 cm and 12.0 cm radii are separated by 1.50 m and are connected by a wire. Charge is transferred to the smaller sphere until its potential reaches 400. V. (a) Estimate the surface charge density on each sphere. (b) Estimate the electric field strength just outside each sphere.

46. A metallic sphere of radius R is charged to a potential $+V$. An alpha particle of charge $+2e$ and mass m_α is initially at a distance $r_o > R$ from the sphere's center and is traveling with velocity v_o toward the sphere's center. (a) Find an expression for the velocity of the alpha particle at distance r ($R \le r \le r_o$) from the sphere's center. (b) If r_o is sufficiently large that it can be set to ∞, what must v_o be for the alpha particle to just touch the surface of the sphere?

24.6 The Big Picture

47. Figure 24.P8 is a matching exercise. The first row shows four hollow metal spheres each with internal radius a and external radius b. You are to match each charge distribution in the first row with its corresponding E-field graph and V graph from rows two and three. As shown in Figure 24.P8 a point-charge of $\pm Q$ is present at the center of some spheres. The charge indicated on the spherical shell itself is the *net* charge on the shell, that is, any induced charge distribution is not shown. Also, the label for the net charge on a conducting sphere does not necessarily indicate the actual position of the charge on or within the conductor.

48. Redraw each of the four conducting hollow spheres shown in the top row of Figure 24.P8. (a) Carefully label the surface of each sphere with the net charge found on that surface; include induced charges. (If no net charge resides on the surface, label the surface with "no charge.") (b) On each drawing show electric field lines in the appropriate regions. Make your drawing qualitative in the sense that the number of lines of E starting (or ending) on a surface is proportional to the total charge on that surface.

Numerical Methods and Computer Applications

49. An electric dipole is oriented along the y-axis with its charges $+q$ and $-q$ located at $(0, +d/2)$ and $(0, -d/2)$, respectively. Write a program or develop a spreadsheet template to compute and plot the resulting electric potential from $x = -10d$ to $x = +10d$ along lines parallel to the x-axis given by an input value for y. Use the program or spreadsheet template for $q = 4.00$ μC, $d = 1.00$ mm, and (a) $y = 0.250$ mm, (b) 0.750 mm, (c) 2.00 mm, and (d) 10.0 mm.

50. Write a program or implement a spreadsheet template to graph the solution to Example 24.5 with $R = 0.200$ m in the region $0.00 \le r \le 0.600$ m.

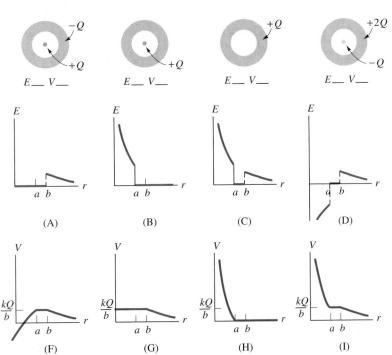

FIGURE 24.P8

Problems 47 and 48

51. In Example 24.8 a linear charge distribution of 0.400 μC/m extends along the x-axis from $x = 0.0$ to $x = 10.0$ cm. (a) Write a program or develop a spreadsheet template that approximates the line of charge by four, 0.100-μC charges placed at locations $x = 0, 2.50, 5.00,$ and 10.0 cm, and compute the electric potential at any given position $x > 10.0$ cm. Compare the electric potential your program predicts for the location $x = 25.0$ cm with the exact result of Example 24.8. (b) Modify your program or spreadsheet template to approximate the line of charge by 25 identical point-charges with total charge (0.400 μC/m)(0.10 m) = 0.040 μC, and compare your new answer with that of Example 24.8. (c) Use your last program to compute $V(x)$ for $x = 24.5$ cm and $x = 25.5$ cm, and use the result to estimate the electric field at $x = 25.0$ cm. How does the result compare to the exact solution of Example 22.5?

52. In Problem 34, a line of charge with uniform density λ lies along the x-axis from $x = -a/2$ to $x = +a/2$. Suppose $\lambda = 60.0$ μC/m and $a = 10.0$ cm. (a) Write a program or develop a spreadsheet template that approximates the line of charge by four equally spaced point-charges, and compute the electric potential to any given position y along the y-axis. (b) Modify your program or spreadsheet template to approximate the line of charge by 37 identical point-charges. How does your answer compare to the exact answer obtained by the analytic expression derived in Problem 34? (c) Use your last program or spreadsheet template to compute $V(y)$ for $y = 29.5$ cm and $y = 30.5$ cm, and use the result to estimate the electric field at $y = 30.0$ cm. How does the result compare with the exact solution for $E(y)$ at this location predicted by the result of Problem 43?

53. A uniform distribution of constant linear charge density λ lies along the x-axis from $x = 0$ to $x = L$. (a) Derive an expression for the electric potential at a general point (X_o, Y_o) in the xy-plane. (b) Develop a spreadsheet template or write a program to compute the potential at this point. (c) Use the program or spreadsheet template to locate points of equipotential, and plot equipotential lines through these points on a sheet of graph paper. On the graph indicate the line of charge from $x = 0$ to $x = L$. Try $\lambda = 36.0$ pC/m and $L = 20.0$ cm.

24.7 Numerical Methods: Relaxation (Optional)

54. Modify the spreadsheet template SOR.WK1 accompanying this text to find the potential within a 25 $\times$ 25 cell region bounded by a constant potential of 10.0 V when there are two interior cells at 0 V: one cell located near the upper left and one cell near the lower right. (Put the 0-V cells at least five cells inward from the border.) Hand in a printout of the potential values and draw a set of equipotential lines for 2, 4, 6, and 8 V directly on your printout. In a different color than that used to draw equipotentials draw in at least five electric field lines in different parts of the plot.

55. Create a computer program or spreadsheet template (or modify SOR.WK1) to calculate the potential within a 26 $\times$ 26 cell region. Split the region down the center and make the boundary of the] shaped border on the right 10.0 V. Make the other half of the border 0 V. Hand in a printout of the potentials calculated in the interior, and draw equipotential lines for 2, 4, 6, and 8 V. In a different color than that used to draw equipotentials draw in at least five electric field lines in different parts of the plot.

56. Create a spreadsheet template (or modify SOR.WK1) to calculate the potential within an isosceles triangle with a height of 25 cells and base of 25 cells. Make the potential of the border cells 10.0 V. Place one cell of 0 V within the interior, not at the center, but at least five cells from any border. Hand in a printout of the potentials calculated in the interior, and draw equipotential lines for 2, 4, 6, and 8 V. In a different color than that used to draw equipotentials draw in at least five electric field lines in different parts of the plot.

★**57.** Write a computer program (or use a symbolic mathematics software package) to implement the SOR algorithm. Demonstrate its capabilities by producing a plot for the potential problem in Example 24.15.

General Problems

58. Four charges, two of +2.00 μC and two of -4.00 μC, are located on the corners of a square 15.0 cm on a side. Charges with the same sign are positioned diagonally opposite each other. Compute the work required to move one of the +2.00-μC charges to infinity.

59. At large distances from a dipole charge distribution the electric field is given by $\mathbf{E}(r) = -A/x^3\,\widehat{\mathbf{i}}$, where A is a constant. Compute the electric potential function $V(x)$ and require $V = 0$ at $x = \infty$.

60. The interior of a sphere of radius R has a charge density $\rho = Ar^2$. (A is a constant with units of C/m^5.) (a) Find the electric potential at a radial position r outside the sphere, where $V_{out} = 0$ at $r \to \infty$. (b) Find the electric potential at a radial position r inside the sphere. (*Hint:* Begin by using Gauss's law to find E both outside and inside the sphere.)

61. Consider an electric dipole consisting of a charge $+q$ located in the xy-plane at $(0, a/2)$ and $-q$ located at $(0, -a/2)$. (a) Find the electric potential at any point (x, y). (b) Use this expression for electric potential to determine the electric field components E_x and E_y at the point (x, y).

62. Find the electric potential on the axis of a disk of radius R that carries a charge density that varies with distance r from the disk center according to $\sigma = Ar$. A is constant.

63. A disk with a hole in its center has inner radius R_1 and outer radius R_2 and carries a uniform charge density σ as shown in Figure 24.P9. Compute the electric potential on the axis of the disk a distance x from the center of the disk.

64. From the electric potential function derived in Problem 63 compute the electric field strength on the axis of a charged disk with a hole in its center (Fig. 24.P9).

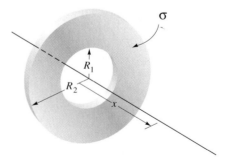

FIGURE 24.P9   Problems 63 and 64

65. In a region of space where there is a nonuniform charge density the electric potential is given by the function

$$V(x, y, z) = (2.00 \text{ V/m})x - (4.00 \text{ V/m}^2)y^2 - (1.00 \text{ V/m})z$$

(a) Compute the potential difference between the points (1.00, 1.00, 3.00) m and (0.00, 1.00, 2.00) m. (b) What is the magnitude of the electric field at (2.00, 2.00, 3.00) m?

66. In a region of the xy-plane there exists a nonuniform charge distribution that gives rise to an electric potential of the form

$$V(x, y) = -(1.00 \text{ V/m}^4)x^3 y$$

(a) Compute the magnitude of the electric field at the point (2.00, 3.00) m. (b) What is the magnitude of the force on an electron placed at (2.00, 3.00) m?

67. In a Van de Graaff generator (Fig. 24.P10) a moving belt made from an insulating material transfers charge from a metallic brush (point A) to another brush (point B). The latter is connected to a conducting sphere. Because the E-field inside the sphere is zero, charge is easily removed by the brush at B and travels to the outer surface of the conducting sphere. Air is not an ideal insulator, and a spark will travel through air from a conductor when the electric field strength reaches about 3.00×10^6 V/m. (a) What is the maximum electric potential that can be created on the conducting sphere of a Van de Graaff generator if the radius of the sphere is 20.0 cm? (b) What must the radius of the sphere be for its maximum potential to be one million volts?

★68. In a region of the xy-plane the electric field is given by $\mathbf{E} = (2.00 \text{ V/m}^5)xy^3\hat{\mathbf{i}} + (3.00 \text{ V/m}^5)x^2y^2\hat{\mathbf{j}}$. (a) Find the potential difference between the points $(-2.00, 1.00)$ m and $(+4.00, 1.00)$ m. (b) Find the potential difference between the points $(1.00, -4.00)$ m and $(1.00, +2.00)$ m. (c) Find the potential difference between the points $(0.00, 0.00)$ m and $(3.00, 3.00)$ m.

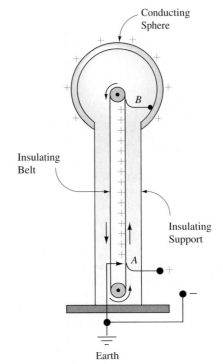

FIGURE 24.P10

Problem 67. A Van de Graaff generator. Charge is transferred to the moving belt by discharge from a metallic brush to a grounded plate (A). At point B the charge is removed to the metallic sphere.

CHAPTER

25

Capacitors and Dielectrics

In this chapter you should learn

- the definition of capacitance and how to calculate this quantity for a variety of geometrical configurations.

- to compute the equivalent capacitance for combinations of capacitors.

- to find the energy stored in a capacitor and in the electric field.

- how the presence of a dielectric affects capacitance.

- (optional) about the displacement field, the polarization, the permittivity constant, and the electric susceptibility.

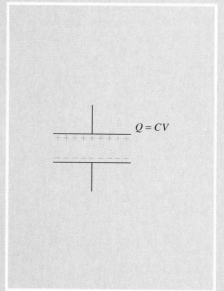

$Q = CV$

In the earliest days of the study of electricity, before the invention of the battery, ways of storing electric charge were eagerly sought. Among the most successful was the Leyden jar named after its town of origin (Leyden, Holland) in 1745. The original Leyden jar consisted of a glass jar filled with water. The jar had a cork stopper with a nail through it to contact the water. After repeatedly charging the nail, one then grasped the outside of the jar and the nail to unleash the electrical "genie."[1] By the year 1746, John Bevis of London had formed a prototype of what we presently call capacitors. He had eliminated the water and lined the inside and outside of the jar with two separate sheets of lead. This chapter describes the electrostatics of the Leyden jar and its many modern counterparts.

In the previous chapter we used the arrangement of two parallel, charged conductors to create a uniform electric field in the region between the plates. In this chapter we will examine other configurations of two charged conductors. From our analysis we will discover how to create *E*-fields of other geometries and how electric energy can be stored with various conductor arrangements. This ability to store energy and charge and later get them back turns out to be very important and is used from the smallest micron-sized integrated circuits to megawatt-scale industrial electrical facilities.

Left: A Leyden jar. (Photo courtesy of Tim Thornberry.)

Right: This electronic flash unit uses a capacitor to store energy from a battery. The battery by itself is not capable of providing energy at a rate fast enough to produce the discharge in the flash tube. It typically takes several seconds for a battery to store energy on the capacitor, but the capacitor can discharge in a very small fraction of a second. (Photo courtesy of Tim Thornberry.)

We begin with some terminology. A **capacitor** is any device that consists of two conductors (with *any* geometry) isolated from each other so that they can be given and hold equal, but opposite, charges. Some "old-timers" occasionally refer to capacitors as *condensers* as they were named by Alessandro Volta (1745–1827) in 1782.[2] Even though the charge on one of the conductors is negative, *the* charge q of a capacitor is taken to be the *positive* value. The conductors of a capacitor are often called **plates.** This name is left over from the parallel-plate arrangement, which is now the exception rather than the rule. Nonetheless, whether they are spherical, cylindrical, or even rolled-up sheets separated by an insulator, the conductors are still called plates.

---

[1] There are many hair-raising accounts of early attempts to trap the electrical fluid. It is reported that one Mr. Winckler (about whom we know little else) "... the first time he tried the Leyden experiment, he found great convulsions in his body; and that it put his blood into great agitation; so, that he was afraid of ardent fever, and was obliged to use refrigerating medicines. He also felt a heaviness in his head, as if a stone lay upon it. Twice he says it gave him a bleeding at the nose to which he was not inclined. His wife (whose curiosity, it seems, was stronger than her fears) received the shock twice, and found herself so weak, that she could hardly walk; and a week after, upon recovering courage to receive another shock, she bled at the nose after taking it only once." (From Joseph Priestly, *The History and Present State of Electricity*, 3rd ed. (London, 1775), p. 107, Vol. 1, as quoted in M. Brotherton, *Capacitors* (D. Van Nostrand, New York, 1946), p. 3.)

[2] The old-timers don't have to be that old! Capacitors were known as condensers until after World War II.

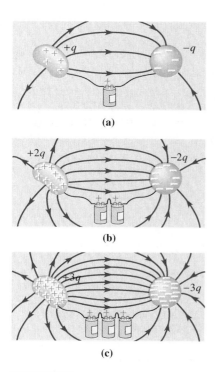

(a)

(b)

(c)

FIGURE 25.1

A capacitor consists of two conductors that are electrically isolated from each other. The magnitude of the charge that accumulates on the conductors is proportional to the applied potential difference. In (a) the voltage difference is ΔV, and the charge on the capacitor is q. In (b) and (c) the potential differences and charges are $2\Delta V$ and $2q$, and $3\Delta V$ and $3q$, respectively.

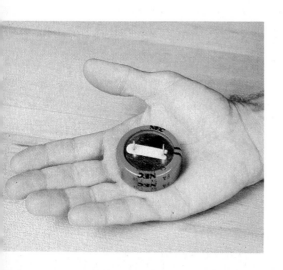

A one farad capacitor. (Photo courtesy of Tim Thornberry.)

25.1 Capacitance

When we apply a potential difference ΔV between the two conductors of a capacitor (perhaps by connecting them to the opposite terminals of a battery), we find that charge flows from the source of the electric potential to the conductors. We also discover that for a certain potential difference ΔV, a fixed amount of charge q accumulates on each conductor ($+q$ on one plate and $-q$ on the other). Next, as illustrated in Figure 25.1, we experiment a little. When the potential difference between the conductors is doubled so that it is $2\Delta V$, we find that the magnitude of the charge on each conductor becomes $2q$. When we triple the potential difference the charge on the capacitor becomes $3q$. In short, we find that the magnitude of the charge on each conductor is proportional to the potential difference between them. The proportionality constant is called the **capacitance** C, and it is defined by the relation

$$q = C\,\Delta V \qquad (25.1)$$

Implicit in this equation is a sign convention. The positively charged plate is always at a higher potential than the negatively charged plate. Therefore, q is positive when ΔV is the change of the potential when going from negative plate to positive plate.

The capacitance depends on the particular geometry of the two conductors constituting the capacitor and does *not* depend on the charge q nor on the potential difference ΔV. The capacitance C is a useful number because, if you know C, then knowledge of either q or ΔV permits you to find the other. As we shall also see, many applications of capacitors require us to know both q and ΔV.

From Equation (25.1) we can see that the unit of capacitance is coulombs per volt. This ratio is called the **farad** and is symbolized F:

$$1\,\text{F} = 1\,\frac{\text{C}}{\text{V}}$$

The term *farad* is derived from the name of the physicist Michael Faraday (1791–1867), one of the greatest experimental physicists of all time.

Although 1-F capacitors are commercially available, the farad is a rather large unit for most purposes. (See Problem 9 at the end of this chapter.) The electronics industry labels capacitors using the microfarad (1 μF = 10^{-6} F) or picofarad (1 pF = 10^{-12} F) almost exclusively. By the way, sometimes in older literature the picofarad unit is written $\mu\mu$F, micro-microfarad. Pushed by the desire for miniaturization, great strides are being made in the reduction of the size of capacitors. Indeed, present day microelectronic technology is such that a 1-F capacitor can be constructed small enough to fit in the palm of your hand.

There are three steps we can apply to compute the capacitance of an arrangement of two conductors. These steps require us to employ two equations we hope you have already learned to apply in a variety of situations.

Problem Solving Strategy

To compute the capacitance of an arrangement of two conductors:

1. Assume charges $\pm q$ are placed on the two conductors, and use Gauss's law to compute the electric field E between them as a function of the charge q.
2. Substitute the expression for E from step 1 into the master equation $\Delta V = -\int \mathbf{E} \cdot d\mathbf{l}$ to find a relation between q and ΔV. (Remember to always integrate from low potential to high potential when choosing the direction of $d\mathbf{l}$, so that ΔV is positive.)
3. Solve for $C = q/\Delta V$.

The following three examples illustrate how to apply these three steps. In the first example we will compute the capacitance of what we call an *ideal*, parallel-plate capacitor. In this idealized model of a true parallel-plate capacitor we assume the electric field to be perfectly uniform between the plates, and that the field drops to zero discontinuously outside the plates. The *E*-field of an actual parallel-plate capacitor exhibits a transition region near the edge of the plates where the *E*-field weakens and field lines bow outward in the so-called **fringe field.** In our ideal parallel-plate model we ignore fringing of the *E*-field near the edges of the plates. The model is illustrated in Figure 25.2. It will be our approach to ignore fringing effects in almost all of our analyses. After reading Example 25.1, try to solve the problems posed in Examples 25.2 and 25.3 before you look at our work. After all, you have already mastered the techniques of steps (1) and (2) in the previous two chapters!

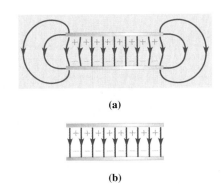

(a)

(b)

FIGURE 25.2

A cross-sectional view of a parallel-plate capacitor. (a) In a true capacitor with this geometry the *E*-field lines exhibit fringing near the edges of the plates. (b) In the ideal model the *E*-field lines are uniform between the plates, and $E = 0$ in the region outside the capacitor.

EXAMPLE 25.1 *Capacitance of an Ideal, Parallel-Plate Capacitor*

Find the capacitance of two parallel, conducting plates each of area A separated by distance d.

SOLUTION

1. Figure 25.3 shows an ideal parallel-plate capacitor with charges $\pm q$ on the two plates. To apply Gauss's law we imagine a rectangular, box-shaped Gaussian surface with a bottom that is between the plates where **E** is constant and a top that is within the conductor where $\mathbf{E} = 0$. The sides of the Gaussian surface do not contribute to the electric flux because the electric field is perpendicular to the area vectors for these surfaces. Because $\mathbf{E} = 0$ through the top surface, only the bottom surface contributes to the flux integral of Gauss's law. The total charge enclosed by the Gaussian surface is that on the top plate $(+q)$, so Gauss's law

$$\oint \mathbf{E} \cdot d\mathbf{A} = \frac{q_{\text{encl}}}{\epsilon_o}$$

becomes

$$EA = \frac{q}{\epsilon_o}$$

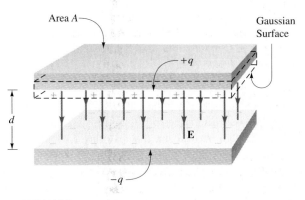

FIGURE 25.3

An ideal, parallel-plate capacitor. Charges $+q$ and $-q$ are imagined to reside on the top and bottom plates, respectively. A Gaussian surface is constructed with a top surface that is inside the upper conductor. The *E*-field lines pass perpendicularly through the lower surface, and $E = 0$ over the surface that is inside the conductor. See Example 25.1.

which can be solved for the electric field to give

$$E = \frac{q}{A\epsilon_o}$$

Notice that E is a function of q.

2. We compute the potential difference between the plates by assigning the bottom plate a position of 0 and a potential $V_1 = 0$. When $d\mathbf{l}$ is directed from the bottom plate to the top plate (where $V_2 = V$), $\mathbf{E}$ and $d\mathbf{l}$ are in opposite directions:

$$\mathbf{E} \cdot d\mathbf{l} = -E\,dl$$

Hence,

$$\Delta V = -\int_0^d \mathbf{E} \cdot d\mathbf{l}$$

becomes

$$\Delta V = +\int_0^d E\,dl = \int_0^d \frac{q}{A\epsilon_o}\,dl$$

$$\Delta V = \frac{qd}{A\epsilon_o}$$

Now we have ΔV as a function of q.

3. From our previous result we compute the ratio $C = q/\Delta V$.

$$C = \frac{A\epsilon_o}{d} \qquad (25.2)$$

There are no charges q or potential differences ΔV left over in Equation (25.2). The capacitance of an ideal, parallel-plate capacitor depends only on the geometrical factors A and d. In fact, the capacitance for any geometry *always* turns out to be ϵ_o times a combination of geometrical factors with the overall unit of length. For this reason you will often see ϵ_o written with units of farads per meter. Equation (25.2) is an important result to which we refer on several occasions. ◀

Concept Question 1
Would the capacitance of a real parallel-plate capacitor be greater or less than that of our ideal capacitor? (*Hint:* What is the effect of fringing?)

EXAMPLE 25.2   *A Spherical Capacitor*

Find the capacitance of two thin, concentric, conducting spheres with radii R_1 and R_2 ($R_1 < R_2$).

SOLUTION

1. This arrangement of conductors is shown in Figure 25.4(a), and a cut-away view is shown in Figure 25.4(b). We assume charges of $-q$ on the inner sphere and $+q$ on the outer sphere. Figure 25.4(b) shows that the resulting E-field is radially inward between the spheres.

   As usual, we imagine a spherical Gaussian surface of radius r between the spheres. The charge enclosed is $-q$ and, for a given radius r, the E-field is constant and antiparallel to the area vectors at all points on the Gaussian surface. Hence, Gauss's law

$$\oint \mathbf{E} \cdot d\mathbf{A} = \frac{q_{\text{encl}}}{\epsilon_o}$$

becomes

$$-E(4\pi r^2) = \frac{-q}{\epsilon_o}$$

which results in the magnitude of E being given by

$$E = \frac{1}{4\pi\epsilon_o}\frac{q}{r^2}$$

a familiar expression!

2. We designate the potential of the negative inner surface as $V_1 = 0$ and that of the positive outer surface as $V_2 = V$. Hence, at point P as shown in Figure 25.4(c), $d\mathbf{l}$ is directed from the inner surface to the outer surface; that is $d\mathbf{l} = d\mathbf{r}$, because $\mathbf{r}$ points radially outward from the center of the sphere. Thus, $\mathbf{E} \cdot d\mathbf{l} = \mathbf{E} \cdot d\mathbf{r} = E\, dr \cos(180°) = -E\, dr$. Therefore, the master equation

$$\Delta V = -\int_{R_1}^{R_2} \mathbf{E} \cdot d\mathbf{l}$$

can be written in terms of r

$$\Delta V = +\int_{R_1}^{R_2} E\, dr = \frac{q}{4\pi\epsilon_o}\int_{R_1}^{R_2}\frac{dr}{r^2}$$

resulting in a potential difference between spheres of

$$\Delta V = \frac{q}{4\pi\epsilon_o}\left(\frac{1}{R_1} - \frac{1}{R_2}\right)$$

3. From this result we compute C:

$$C = \frac{q}{\Delta V} = 4\pi\epsilon_o\left(\frac{R_1 R_2}{R_2 - R_1}\right)$$

Again, we note that this result depends only on the geometry through R_1 and R_2, and has the form ϵ_o times a geometrical factor with overall units of length. ◀

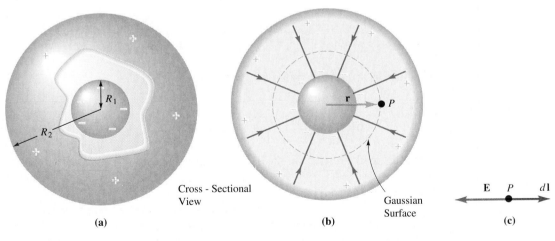

(a)   (b)   (c)

FIGURE 25.4

(a) A spherical capacitor consisting of two concentric spherical conductors of radii R_1 and R_2. (b) A cut-away view of the capacitor. Charges $+q$ and $-q$ reside on the outer and inner conductors, respectively, and a spherical Gaussian surface of radius r surrounds the inner sphere. See Example 25.2.

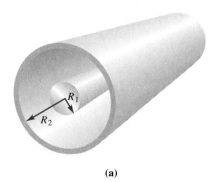

(a)

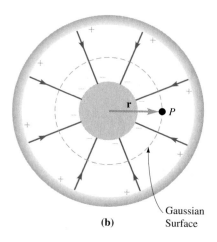

(b)   Gaussian
                          Surface

(c)

FIGURE 25.5

A cylindrical capacitor is constructed
from two concentric, conducting
cylinders. A cylindrical Gaussian
surface of radius r surrounds the inner
cylinder. See Example 25.3.

EXAMPLE 25.3   A Cylindrical Capacitor

Find the capacitance of an ideal, cylindrical capacitor comprised of two concentric cylindrical conductors of length L. The inner radius is R_1 and the outer radius is R_2 as shown in Figure 25.5.

SOLUTION

1. We assume a charge $-q$ on the inner cylinder and $+q$ on the outer one. For the ideal case we ignore fringing of the electric field at the ends of the cylinders and suppose **E** to be directed radially inward at all points between the cylinders. Applying Gauss's law to a cylindrical Gaussian surface of radius r (Fig. 25.5), Gauss's law

$$\oint \mathbf{E} \cdot d\mathbf{A} = \frac{q_{encl}}{\epsilon_o}$$

    becomes for the present case

$$-E(2\pi r L) = \frac{-q}{\epsilon_o}$$

    where we have recognized that E is perpendicular to the ends of the cylindrical Gaussian surface and is antiparallel to the area vectors on the curved portion of the surface. Hence, the E-field has magnitude

$$E = \frac{q}{2\pi \epsilon_o r L}$$

2. Because we have (quite arbitrarily) chosen to charge the outer sphere positively, **E** is directed radially inward. As we recall from Chapter 24, the electric field always points to the lower potential, so we must integrate from R_1 to R_2 in order for ΔV to be positive. For this case we once again have $d\mathbf{l} = d\mathbf{r}$ and $\mathbf{E} \cdot d\mathbf{l} = \mathbf{E} \cdot d\mathbf{r} = E\, dr \cos(180°) = -E\, dr$. Thus, the master equation

$$\Delta V = -\int_{R_1}^{R_2} \mathbf{E} \cdot d\mathbf{l}$$

    becomes

$$\Delta V = +\int_{R_1}^{R_2} E\, dr = \frac{q}{2\pi \epsilon_o L} \int_{R_1}^{R_2} \frac{dr}{r}$$

    resulting in a potential difference given by

$$\Delta V = \frac{q}{2\pi \epsilon_o L} \ln\!\left(\frac{R_2}{R_1}\right)$$

3. Solving for $C = q/\Delta V$, we have

$$C = \frac{2\pi \epsilon_o L}{\ln(R_2/R_1)}$$

    which, as you have no doubt noticed, is the product of ϵ_o and a combination of geometrical factors with the overall unit of length. The geometry of coaxial cables consists of two concentric cylindrical conductors as described in this example. It is common to specify their *capacitance per length* $C/L = 2\pi\epsilon_o/\ln(R_2/R_1)$.   ◄

25.2 Combinations of Capacitors

The symbol for a capacitor is shown in Figure 25.6. This schematic depicts a cross-sectional view of a parallel-plate capacitor, but you should realize that it represents a capacitor of unspecified geometry. Often the value of the capacitance is provided next to the symbol. In some applications we may want a capacitance that is not available commercially. However, capacitors can be connected together to form a combination that can have a capacitance closer to our desired value. There are two ways that two capacitors can be connected together. Once we understand how to compute the equivalent capacitance for each of these two configurations, it is possible to compute, in a straightforward manner, the equivalent capacitance of many combinations of three or more capacitors.

It is important that you not confuse the symbol for a capacitor, Figure 25.7(a), with that of a source of electric potential, such as a cell or a battery of cells as shown in Figures 25.7(b). Notice that for the source of electric potential one vertical line is longer than the other, and the shorter line is thicker. It is customary to represent the higher potential side of the source by the longer line and designate it as the positive terminal. The symbol shown in Figure 25.7(c) is also that of a source of electric potential but one with a potential difference that is significantly larger than that of Figure 25.7(b).

Capacitors in Parallel

Figure 25.8(a) shows two parallel-plate capacitors connected to each other in an arrangement referred to as **parallel.** The capacitors are also connected to a battery. In Figure 25.8(b) we show the circuit diagram for this capacitor combination. In more com-

FIGURE 25.6

The circuit symbol for a capacitor.

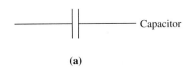

(a)

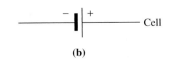

(b)

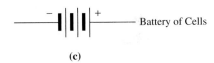

(c)

FIGURE 25.7

Circuit symbols for (a) a capacitor and sources of electric potential difference: (b) a cell and (c) a battery of cells.

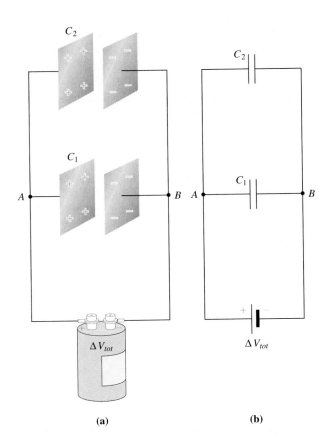

(a)   (b)

FIGURE 25.8

(a) Two capacitors in a parallel arrangement are connected to the terminals of a battery. Notice that the potential difference ΔV across each capacitor is the same. (b) The circuit diagram for this parallel configuration of capacitors.

plex circuits, students sometimes have trouble distinguishing between a parallel configuration of two capacitors in the circuit and the **series** arrangement that we will analyze soon hereafter. For this reason we discuss the topology of series and parallel connections in the special boxed section: *Series and Parallel Circuit Connections*. Here is the important feature you should remember:

When two capacitors are connected in parallel the potential difference ΔV across each capacitor is the same.

Series and Parallel Electrical Connections

In schematic drawings of electrical circuits, the batteries, capacitors, resistors, inductors, and other devices are shown as they are to be connected by conductors on circuit boards (usually by either wires or by thin copper strips called *traces*). The point where two or more elements are connected is called a **node.** In this box we describe how combinations of the generic circuit element shown in Figure 25.9 can be arranged. This circuit element might symbolize a battery, capacitor, or even combinations of other circuit elements. Two special ways of connecting the elements together are called **series** and **parallel** connections. Most of the circuit elements we will encounter in this text are called **two-terminal elements,** meaning that the combination of elements can be connected to the outside world by just two wires, one on each end of the element. When the two terminals of one circuit element are connected to the two terminals of a second element as shown in Figure 25.10, the elements are in parallel. Note that the elements must be *directly* connected by a wire only. That is, in order to say that the elements are connected in parallel, you must be able to trace a wire *directly* from one element to the other *without going through any other element*. It doesn't matter if other elements are also connected at the node. Furthermore, the elements must be directly connected on *both* ends to be in parallel. In *good* circuit diagrams designers deliberately draw elements to emphasize that they are in parallel, but it is not the fact that the elements are geometrically parallel that counts; it is the way they are electrically connected.

In a *series* connection only *one end* of the two-terminal devices are *directly* connected. The critical feature of this connection is that *no other element is connected* at the node where two series elements are joined.

Figure 25.11 shows examples of both parallel and series connections of generic circuit elements drawn only as rectangles. Notice that the three circuit elements in Figure 25.11(d) and (e) are neither in series nor parallel. It is worth a few minutes of your time to trace the connections between elements to ensure that you can tell whether the elements are in series or parallel (or neither).

FIGURE 25.9

A generic circuit element that represents a battery, capacitor, resistor, inductor, or other device.

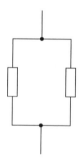

FIGURE 25.10

A parallel arrangement of two circuit elements.

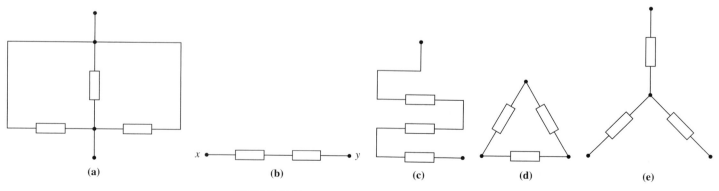

FIGURE 25.11

Various combinations of circuit elements: (a) parallel; (b) series; (c) series; (d) neither; (e) neither.

In Figure 25.8 the left plates of each capacitor are connected to each other by a wire. This connection makes the wire and the capacitor plates on the left side all one, continuous conductor. Because the surface of a conductor in equilibrium is an equipotential surface, both left-side capacitor plates must be at the same potential. In fact, because these plates are connected to the positive terminal of the battery and they are in electrostatic equilibrium, the plates must both be at the potential of that battery terminal.

We can make a similar argument about the capacitor plates on the right side in Figure 25.8. The plates and wire form a continuous conductor and are, therefore, at the same potential, namely, that of the negative terminal of the battery.

Let's compute the equivalent capacitance of these two capacitors connected in parallel. That is, suppose we conceal these two capacitors (connected as shown) in a box with only the wires at points A and B protruding from holes in the ends of the box. Moreover, imagine we try to deceive you by reporting that there is only one capacitor in the box. What capacitance do we have to claim is in the box?

We begin by noting that the potential difference across each capacitor is the same and equal to the total potential difference ΔV_{tot} across the battery:

$$\Delta V_1 = \Delta V_2 = \Delta V_{\text{tot}}$$

Next, we see that the total charge on the A side of *whatever* is in the box is

$$Q_{\text{tot}} = Q_1 + Q_2 \qquad \qquad (25.3)$$

where Q_1 and Q_2 are the charges on capacitors C_1 and C_2, respectively. Now, from the definition of capacitance,

$$C_1 = \frac{Q_1}{\Delta V_1} = \frac{Q_1}{\Delta V_{\text{tot}}} \qquad \text{and} \qquad C_2 = \frac{Q_2}{\Delta V_2} = \frac{Q_2}{\Delta V_{\text{tot}}}$$

or

$$Q_1 = C_1 \, \Delta V_{\text{tot}} \qquad \text{and} \qquad Q_2 = C_2 \, \Delta V_{\text{tot}}$$

Substituting these results into Equation (25.3), we have

$$Q_{\text{tot}} = C_1 \Delta V_{\text{tot}} + C_2 \Delta V_{\text{tot}} = (C_1 + C_2)\Delta V_{\text{tot}}$$

Solving for the total capacitance C_{tot} between points A and B, we have

$$C_{\text{tot}} = \frac{Q_{\text{tot}}}{\Delta V_{\text{tot}}} = C_1 + C_2 \qquad \qquad (25.4)$$

That is, for two capacitors in a parallel arrangement the equivalent capacitance is equal to the sum of the individual capacitances. We can extend our result to N capacitors in parallel:

$$C_{\text{tot}} = \sum_{i=1}^{N} C_i \qquad \text{(Capacitors in parallel)} \qquad \qquad (25.5)$$

It should be obvious to you that the equivalent capacitance of several capacitors connected in parallel is *larger* than any of the individual capacitors. Therefore, if we want a capacitance that is larger than any we have available, we need to connect the capacitors we have in a parallel arrangement. What do we do if we want to build a small capacitance out of a bunch of large capacitors? Read on!

Capacitors in Series

In Figure 25.12(a) we show the terminals of a battery attached to two parallel-plate capacitors that are connected in a *series* arrangement. We have also included the corresponding circuit diagram. Here is the feature that characterizes this type of circuit configuration:

> When two capacitors are connected in series the *charge* on each capacitor is the same.

We can see how this situation arises by considering what happens when we attach the arrangement of capacitors between the points A and B shown in Figure 25.12 to the battery. The capacitors are initially neutral, and when connected together the bottom plate of capacitor C_1 and the top plate of capacitor C_2 form an *isolated* conductor. Therefore, any charge that accumulates on the bottom plate of C_1 must necessarily come from the top plate of C_2. Hence, the magnitude of the charge on capacitor C_2 must be exactly equal to the magnitude of the charge on capacitor C_1. (Remember, when we refer to *the* charge on a capacitor we always refer to the positive value only.) It should be clear from Figure 25.12 that both top plates have the same positive charge and both bottom plates have the same negative charge. Mathematically, we are saying

$$Q_1 = Q_2 = Q_{tot}$$

where Q_{tot} is the charge on the equivalent capacitance of this two-capacitor combination. (If you like, put a box over C_1 and C_2 with points A and B again sticking out.) To compute the equivalent capacitance we also note that the total potential difference between points A

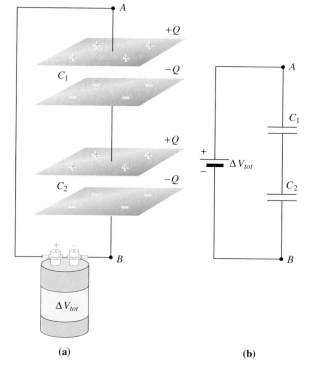

(a)   (b)

FIGURE 25.12

(a) Two capacitors in a series arrangement are connected to the terminals of a battery. The positive charge $+Q$ on the top plate of C_1 attracts electrons with a total charge $-Q$ from the top plate of C_2, leaving the top plate of C_2 with a charge $+Q$. (b) The circuit diagram for this series configuration of capacitors.

and B, which we have called ΔV_{tot}, must be the sum of the potential differences across the individual capacitors:

$$\Delta V_{\text{tot}} = \Delta V_1 + \Delta V_2 \qquad (25.6)$$

But from the definition of capacitance

$$\Delta V_1 = \frac{Q_1}{C_1} = \frac{Q_{\text{tot}}}{C_1} \qquad \text{and} \qquad \Delta V_2 = \frac{Q_2}{C_2} = \frac{Q_{\text{tot}}}{C_2}$$

Substituting this result into Equation (25.6), we have

$$\Delta V_{\text{tot}} = \frac{Q_1}{C_1} + \frac{Q_2}{C_2} = Q_{\text{tot}}\left(\frac{1}{C_1} + \frac{1}{C_2}\right)$$

That is,

$$\frac{\Delta V_{\text{tot}}}{Q_{\text{tot}}} = \frac{1}{C_1} + \frac{1}{C_2}$$

But, for our fictitious equivalent capacitor we must also have $\Delta V_{\text{tot}}/Q_{\text{tot}} = 1/C_{\text{tot}}$, so

$$\frac{1}{C_{\text{tot}}} = \frac{1}{C_1} + \frac{1}{C_2} \qquad (25.7)$$

For N capacitors arranged in series,

$$\frac{1}{C_{\text{tot}}} = \sum_{i=1}^{N} \frac{1}{C_i} \qquad \text{(Capacitors in series)} \qquad (25.8)$$

For capacitors in *series*, the equivalent capacitance must always be *less* than any of the individual capacitances.

Concept Question 2
A thin sheet of metal is inserted between and parallel to the plates of a parallel-plate capacitor. What effect does this have on the capacitance? Does the position of the sheet matter? Explain.

EXAMPLE 25.4   *Finding Equivalent Capacitance*

Find the equivalent capacitance (between the points A and B) of the three capacitors shown in Figure 25.13(a).

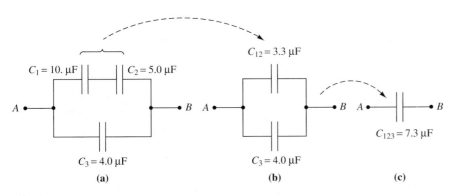

FIGURE 25.13

A combination of three capacitors. See Example 25.4.

SOLUTION You must train your eyes to search for the simplest combination of *either* a parallel or series combination in a circuit. The simplest arrangement in Figure 25.13(a) is the series combination of C_1 and C_2. (Look back to Figure 24.13 on page 699.) These two capacitors form the equivalent capacitor shown schematically as C_{12} in Figure 25.13(b). From Equation (25.7) we have

$$\frac{1}{C_{12}} = \frac{1}{C_1} + \frac{1}{C_2}$$

or

$$C_{12} = \left(\frac{1}{C_1} + \frac{1}{C_2}\right)^{-1} = \left(\frac{1}{10.0\ \mu\text{F}} + \frac{1}{5.00\ \mu\text{F}}\right)^{-1} = 3.33\ \mu\text{F}$$

(Practice using your calculator so you can minimize the number of keystrokes needed to calculate this result.)

From Figure 25.13(b) it is easy to recognize that C_{12} and C_3 are in parallel. Applying Equation (25.8), we have

$$C_{123} = C_{12} + C_3 = 3.33\ \mu\text{F} + 4.00\ \mu\text{F} = 7.33\ \mu\text{F} \qquad \blacktriangleleft$$

In Example 25.5 we illustrate the steps necessary to compute the charge on one of the capacitors in a combination.

EXAMPLE 25.5 *All Charged Up and No Place to Go*

(a) Find the equivalent capacitance between points A and B in Figure 25.14. (b) If a potential difference of 12.0 V is applied across the points A and B, what is the charge on C_2?

SOLUTION (a) The sequence we apply to reduce this combination to one equivalent capacitor is shown in Figure 25.15. Capacitors C_2 and C_3 in Figure 25.14 are in parallel so

$$C_{23} = C_2 + C_3 = 2.00\ \mu\text{F} + 1.00\ \mu\text{F} = 3.00\ \mu\text{F}$$

The equivalent circuit segment between points A and B is shown in Figure 25.15(a). From this figure we see that capacitors C_1 and C_{23} are in series:

$$\frac{1}{C_{123}} = \frac{1}{C_1} + \frac{1}{C_{23}}$$

or

$$C_{123} = \left(\frac{1}{C_1} + \frac{1}{C_{23}}\right)^{-1} = \left(\frac{1}{6.00\ \mu\text{F}} + \frac{1}{3.00\ \mu\text{F}}\right)^{-1} = 2.00\ \mu\text{F}$$

Finally, from Figure 25.15(b), C_{123} and C_4 are in parallel, so

$$C_{\text{tot}} = C_{123} + C_4 = 2.00\ \mu\text{F} + 4.00\ \mu\text{F} = 6.00\ \mu\text{F}$$

(b) In order to find the charge on C_2 we work our way back from the equivalent capacitance C_{tot} to C_2, keeping in mind that capacitors in series have equal charges and capacitors in parallel have equal voltage drops. The potential difference across C_4 and C_{123} must be $\Delta V_{AB} = 12.0$ V because these two capacitances are in parallel (see

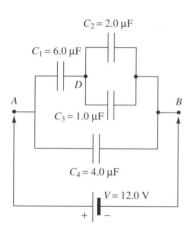

FIGURE 25.14

A combination of four capacitors. See Example 25.5.

Fig. 25.15(b)). We can, therefore, find the charge on C_{123}:

$$Q_{123} = C_{123} \, \Delta V_{AB} = (2.00 \ \mu C)(12.0 \ V) = 24.0 \ \mu C$$

From Figure 25.15(a), we see that this charge must in fact be on the left plate of capacitor C_1. Hence, $Q_1 = 24.0 \ \mu C$. However, C_1 is in series with C_{23}, so C_{23} must have this same charge:

$$Q_{23} = 24.0 \ \mu C$$

This result allows us to find the potential difference across C_{23}:

$$\Delta V_{DB} = \frac{Q_{23}}{C_{23}} = \frac{24.0 \ \mu C}{3.00 \ \mu F} = 8.00 \ V$$

Because C_2 and C_3 are in parallel ΔV_{DB} is also the potential difference across each of these capacitors. In particular, the potential difference across C_2 is 8.00 V, and we can at last find the charge on C_2:

$$Q_2 = C_2 \, \Delta V_{DB} = (2.00 \ \mu F)(8.00 \ V) = 16.0 \ \mu C$$

Can you find the charges on C_1 and C_3? Answers: $Q_1 = 8.0 \ \mu C$ and $Q_3 = 48.0 \ \mu C$. ◀

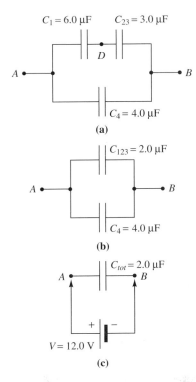

(a)

(b)

(c)

FIGURE 25.15

Reducing the capacitor combination of Figure 25.14 to a single, equivalent capacitance. (a) Capacitors C_2 and C_3 of Figure 25.14 are in parallel and are equivalent to C_{23}. (b) C_1 and C_{23} in (a) are in series and are represented by C_{123}. (c) C_{123} and C_4 in (b) are in parallel and are equivalent to C_{tot}.

25.3 Energy Storage in a Capacitor

Work is required to charge a capacitor because charge already accumulated on one of the capacitor plates repels any additional charge we try to add to that plate. We can view this work as contributing to potential energy stored in the capacitor.

Suppose we take a capacitor already holding a charge q so that it has a potential difference ΔV across its terminals. From the definition of capacitance,

$$\Delta V = \frac{q}{C}$$

Now let's move a small bit of charge dq from the negative terminal to the positive. We have to do work in the amount $\Delta V \, dq$ on this charge to move it against the E-field, and this increases the potential energy stored by the capacitor by the same amount.

$$dU = \Delta V \, dq$$

Replacing ΔV with q/C, we find

$$dU = \frac{1}{C} q \, dq$$

To find the total energy stored by the capacitor we integrate these changes starting from when the capacitor was uncharged until it holds charge Q. If we choose the potential energy to be zero when the capacitor is uncharged, we have

$$\int_0^U dU = \frac{1}{C} \int_0^Q q \, dq$$

Concept Question 3
Show that the circuit given below is equivalent to that in Figure 25.14.

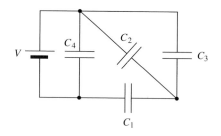

Concept Question 4
Because $Q = CV$, Equation (25.9) can be written $U = QV/2$. However, the work required to move a charge Q through a potential difference V is $W = QV$. Explain this difference in the W and U.

where Q is the final charge accumulated on the capacitor. Integrating,

$$U = \frac{Q^2}{2C} \tag{25.9}$$

We can find another useful expression for U by substituting $Q = C\,\Delta V$ into Equation (25.9):

$$U = \frac{(C\,\Delta V)^2}{2C}$$

$$U = \frac{1}{2}C(\Delta V)^2 \tag{25.10}$$

Equations (25.9) and (25.10) are useful when we want to determine how much energy we can expect to get out of a charged capacitor. Equation (25.10) is particularly valuable because we need only know the values of the capacitance and the voltage of the battery (or other source) that we use to charge the capacitor.

There is another important result we can obtain from Equation (25.10). We will derive this result for the special case of an ideal, parallel-plate capacitor, although the result turns out to be true in general. In Example 25.1 we found the capacitance of a parallel-plate capacitor to be

$$C = \frac{A\epsilon_o}{d}$$

where A is the area of the plates and d is their separation. Substituting this result into Equation (25.10), we have

$$U = \frac{A\epsilon_o(\Delta V)^2}{d}$$

Now, we have also observed that the uniform electric field between the plates is given by $E = \Delta V/d$ (see Eq. (24.13)), or $\Delta V = Ed$. Making this substitution, we have

$$U = \frac{(Ad)\epsilon_o E^2}{2}$$

You should recognize (Ad) as the volume enclosed between the plates of the parallel-plate capacitor. We define the **energy density** u as the energy per unit volume. For the present case $u = U/(Ad)$:

$$u = \frac{1}{2}\epsilon_o E^2 \tag{25.11}$$

In SI, u has the units of joules per cubic meter. Equation (25.11) suggests that we can associate energy with the electric field; that is, energy is stored in the electric field. Although we have derived this result for the special case of an ideal, parallel-plate capacitor in which the E-field is uniform and entirely between the plates, keep in mind that for any E-field (that can change with position) we can focus our attention to a smaller and smaller region so that the *E*-field becomes closer and closer to uniformity in this region. That is, as the size of the region of our interest approaches a point, the spatial changes in the *E*-field approach zero. Equation (25.11) provides the energy density *at a point* and is true in general.

EXAMPLE 25.6   A Storage Box for Energy

A 3.00-μF and a 2.00-μF capacitor are separately charged with a 24.0-V battery. They are then connected as shown in Figure 25.16. What is the energy stored on each capacitor (a) before switch *S* is closed, and (b) after switch *S* is closed?

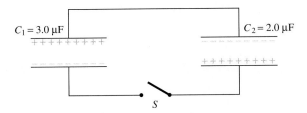

FIGURE 25.16

Two capacitors are charged, and the plates of opposite polarity are connected. When switch *S* is closed, charge is redistributed, but the new final voltage is the same across each capacitor. See Example 25.6.

SOLUTION   (a) The charge on the capacitors remains constant when the capacitors are connected at one end only because the charges on the unconnected ends have no place to go. The Coulomb attraction from the trapped charges on the unconnected ends holds the charges on the connected ends in place, and keeps them from combining. Thus, each capacitor has the same potential difference and charge that it had when originally connected to the battery. The potential difference across each capacitor is 24.0 V, so we can use Equation (25.10) to compute the initial energy storage for each capacitor:

$$U_1 = \frac{1}{2}C_1(\Delta V)^2 = \frac{1}{2}(3.00 \times 10^{-6}\text{ F})(24.0\text{ V})^2 = 8.64 \times 10^{-4}\text{ J}$$

$$U_2 = \frac{1}{2}C_2(\Delta V)^2 = \frac{1}{2}(2.00 \times 10^{-6}\text{ F})(24.0\text{ V})^2 = 5.76 \times 10^{-4}\text{ J}$$

The total energy stored by this system of capacitors is

$$U_{\text{tot}} = U_1 + U_2 = 1.44 \times 10^{-3}\text{ J}$$

(b) When the switch *S* is closed the charges are no longer fixed in place. The charges on the top plates of the capacitors can now combine, provided the charges on the bottom plates behave similarly. Only the net charge is preserved. The magnitude of the initial charge on each capacitor is

$$Q_1 = C_1\,\Delta V = (3.00\ \mu\text{F})(24.0\text{ V}) = 72.0\ \mu\text{C}$$

$$Q_2 = C_2\,\Delta V = (2.00\ \mu\text{F})(24.0\text{ V}) = 48.0\ \mu\text{C}$$

The positive plate of C_1 has been connected to the negative plate of C_2. Therefore, when switch *S* is closed, the magnitude of the combined total charge on both top plates is the *difference* between charges Q_1 and Q_2. (The total charge on the two bottom plates together is also the difference, but has the opposite sign.) Let's compute the total charge on the top plates:

$$Q_{net} = Q_1 + Q_2 = 72.0 \ \mu C - 48.0 \ \mu C = 24.0 \ \mu C$$

Notice that, after the switch is closed, the two capacitors are in parallel. Consequently, the potential differences across the two capacitors are equal to each other (but no longer equal to 24.0 V). Thus, because $\Delta V = Q/C$ for each of the capacitors,

$$\frac{Q_1}{C_1} = \frac{Q_2}{C_2}$$

Solving the last two equations for Q_1 and Q_2, we find

$$Q_1 = 14.4 \ \mu C \qquad \text{and} \qquad Q_2 = 9.60 \ \mu C$$

The potential difference across each capacitor is now

$$\Delta V = \frac{Q_1}{C_1} = \frac{Q_2}{C_2} = \frac{14.4 \ \mu C}{3.0 \ \mu F} = \frac{9.60 \ \mu C}{2.0 \ \mu F} = 4.80 \ V$$

Finally, we apply Equation (25.9) to find the energy now stored on each capacitor:

$$U' = \frac{Q^2}{2C}$$

$$U_1' = \frac{(14.4 \times 10^{-6} \ C)^2}{2(3.00 \times 10^{-6} \ F)} = 3.46 \times 10^{-5} \ J$$

$$U_2' = \frac{(9.60 \times 10^{-6} \ C)^2}{2(2.00 \times 10^{-6} \ F)} = 2.30 \times 10^{-5} \ J$$

The total energy U_{tot}' is now

$$U_{tot}' = U_1' + U_2' = 5.76 \times 10^{-5} \ J$$

Work was required to move the charges between the plates when switch S was closed. This work was supplied by the electric field and accounts for the decrease in energy calculated in parts (a) and (b):

$$\Delta U = U_{tot}' - U_{tot} = -1.38 \times 10^{-3} \ J. \qquad \blacktriangleleft$$

EXAMPLE 25.7   More Than One Way to Skin a Capacitor

The spherical capacitor of Example 25.2 is charged such that the inner sphere carries charge $-Q$ and the outer sphere carries charge $+Q$. (a) By direct integration of Equation (25.11) find the electric potential energy stored on the capacitor. (b) Verify this result from Equation (25.9).

SOLUTION   (a) The charged capacitor is shown in Figure 25.17. In Example 25.2 we found that the electric field between the spheres is radially inward with magnitude given by

$$E = \frac{1}{4\pi\epsilon_o} \frac{Q}{r^2}$$

The energy dU within a volume dV enclosed by a spherical shell of inner radius r and

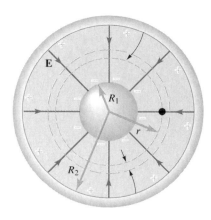

FIGURE 25.17

The electric field between the conductors of a spherical capacitor. A small-volume element of thickness dr forms a spherical shell of volume $dV = 4\pi r^2 \ dr$. See Example 25.7.

thickness dr is

$$dU = u \, d\mathcal{V} = u(4\pi r^2 \, dr)$$

or

$$U = \int_{R_1}^{R_2} u(4\pi r^2 \, dr)$$

where u is the energy density given by Equation (25.11):

$$u = \frac{1}{2}\epsilon_o E^2 = \frac{1}{2}\epsilon_o \left(\frac{1}{4\pi\epsilon_o}\frac{Q}{r^2}\right)^2$$

Substituting into the integral for u, we have

$$U = \frac{1}{2}\epsilon_o \int_{R_1}^{R_2} E^2 (4\pi r^2) \, dr = \frac{Q^2}{8\pi\epsilon_o} \int_{R_1}^{R_2} \frac{dr}{r^2}$$

$$U = \frac{Q^2}{8\pi\epsilon_o}\left(\frac{1}{R_1} - \frac{1}{R_2}\right)$$

(b) To employ Equation (25.9) we use the capacitance calculated in Example 25.2:

$$U = \frac{Q^2}{2C} = \frac{Q^2}{2[4\pi\epsilon_o R_1 R_2/(R_2 - R_1)]}$$

$$U = \frac{Q^2}{8\pi\epsilon_o}\left(\frac{1}{R_1} - \frac{1}{R_2}\right)$$

This result agrees with our answer to part (a).   ◀

25.4 Dielectrics

We now investigate what happens when we place a nonconducting material between the conductors of a capacitor. These materials, called **dielectrics,** can be grouped into one of two broad categories: **polar** and **nonpolar** substances. Polar materials are composed of molecules that have permanent electric dipole moments, whereas nonpolar materials contain molecules that do not have permanent dipole moments. However, in some cases nonpolar molecules can acquire significant induced moments when they are placed in an applied E-field. These latter types of materials may become **polarized** when the external field is applied, because the positive and negative charge distributions that make up the molecules experience forces in opposite directions. (See the discussion at the end of Section 22.4.)

Dielectric materials, such as mylar, silicone oil, metallic oxides, or simply air may be used to fill the space between capacitor plates. For example, a capacitor with capacitance in the range from 0.001–10 μF can be constructed from a mylar film with aluminum deposited on its two surfaces to form the plates. The film is then rolled up into a cylindrical configuration. For a further rundown on types of capacitors, see the special boxed section on capacitors. Dielectrics sometimes provide a mechanical support to keep the conductors of a capacitor separated. More importantly, dielectric materials increase the capacitance of a particular plate geometry above the value that exists when there is only vacuum between the conductors. One reason this result is desirable can be found by looking at Equation (25.10). This equation shows that for a given potential difference ΔV, a larger value of C permits a greater amount of energy to be stored in the capacitor. A second reason, important in this age of miniaturization, is that with a dielectric in place, any given capacitance can be obtained with smaller dimensions.

Capacitor Types and Characteristics

TYPE	RANGE	COMMENT
Air	1–5000 pF	
Ceramic	10 pF–1 μF	Inexpensive, small, and common
Electrolytic	0.1 μF–2 F	Polarized, so you must put them in the circuit with the correct orientation; terrible stability; usually used only in power supplies
Glass	10–1000 pF	Long-lasting and stable
Mica	1 pF–0.01 μF	Dependable and stable
Mylar	0.001–50 μF	Poor temperature stability, but inexpensive and popular
Oil	0.1–20 μF	Long life; used in high-voltage power supplies
Polystyrene	10 pF–2 μF	Bulky but reliable, exceptionally low leakage
Porcelain	100 pF–0.1 μF	Long-term stability and inexpensive
Tantalum	0.1–500 μF	Popular for large capacitance requirements; polarized; poor stability

Let's try to discover what effect a dielectric has when it is inserted between the conductors of a capacitor. We take as our example a dielectric with permanent dipoles; similar results hold for dielectrics with induced dipoles. Figure 25.18(a) shows a representation of a polar material when it is *not* subject to an applied E-field; the polar molecules are randomly oriented and the net charge on all of the dielectric's surfaces is zero. When the dielectric slab is inserted between the plates of a charged capacitor (see Fig. 25.18(b)), the polar molecules experience a torque and tend to align with the E-field. The strong

FIGURE 25.18

(a) In the absence of an applied E-field the polar molecules of a dielectric are randomly oriented. (b) When the dielectric slab is inserted between the charged plates of a capacitor, the dipole moments tend to align so that the surface of the dielectric closest to the positive plate acquires a negative charge density. The induced surface charge on the dielectric surface closest to the negative plate is positive.

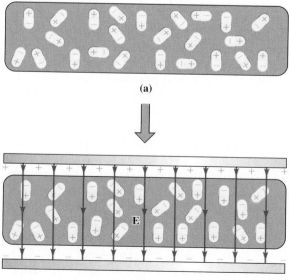

(a)

(b)

intermolecular bonds and the thermal motion of the molecules prevent these molecules from ever reaching complete alignment. As the applied field strength is increased (perhaps by increasing the potential difference applied across the capacitor plates), the molecular alignment increases. Eventually, the molecules can reach the state of maximum alignment allowed by the molecular structure. Even in this case, thermal vibrations of the molecules limits the degree of alignment, so as you might expect, temperature plays an important role in the overall, average orientation of the molecules. You should realize that Figure 25.18 presents the simplest model of a polar material. Real dielectrics always let some charge "leak" through; that is, they are not perfect insulators. We shall ignore these effects in our perfect-dielectric model.

Notice that the overall effect of the dipole alignment is to make the top surface of the dielectric negative and the bottom surface positive. This **induced surface charge** density σ_i creates an electric field $\mathbf{E}_i$ that opposes the field $\mathbf{E}_o$ produced by the **free charge** density σ_o on the surface of the capacitor plates. Both these charge densities and resulting E-fields are illustrated in Figure 25.19. It should be clear that the induced surface charge is not free because it is *bound* to the dipole moments that constitute the dielectric. Sometimes the free charge on the capacitor is called **real charge.** This term is somewhat unfortunate because both free charge and bound charge are quite real in a physical sense. Whichever term you use, it is important for you to realize that the free or real charge is that which you place on the conducting capacitor plates by some external device (such as a battery).

Suppose the charged capacitor is not connected to a source of additional charge when the dielectric is inserted between its plates. Then the creation of the induced surface charge density σ_i means that the electric field strength E between the capacitor plates (and within the dielectric material) is reduced to a lower value E from its value E_o when no dielectric is present. One way to characterize how strongly the dielectric responds to the presence of the applied field E_o is to define the **relative dielectric constant** κ based on this reduction of field strength:

$$\kappa = \frac{E_o}{E} \qquad (25.12)$$

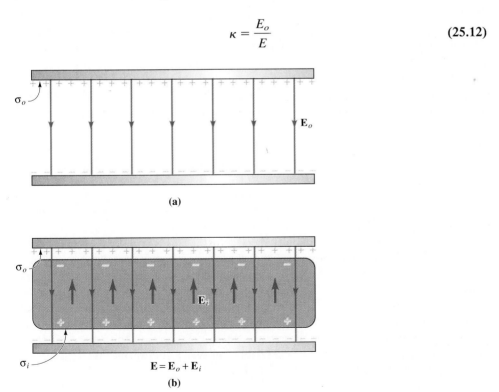

(a)

(b)

FIGURE 25.19

(a) The field $\mathbf{E}_o$ produced by a parallel-plate capacitor carrying free surface charge density σ_o.
(b) When a dielectric slab is inserted between the plates, the induced surface charge density σ_i creates a field $\mathbf{E}_i$ that opposes the field $\mathbf{E}_o$. The field within the dielectric is the vector sum of the fields $\mathbf{E}_o$ and $\mathbf{E}_i$.

Because $E < E_o$, the dielectric constant is a number greater than 1 (except for a vacuum where $E = E_o$ and $\kappa_{vac} = 1$). The dielectric constants of some common substances are listed in Table 25.1.

There is a maximum electric field strength that can be applied across any dielectric. When this field strength is exceeded, the insulating properties of the material fail and the material begins to conduct. The maximum electric field a material can withstand is called the **dielectric strength** of that material. This dielectric strength is a function of temperature, the rate of change of the electric field, and in some cases the thickness of the material. Nonetheless, "typical" values are often quoted for comparison. For air, a typical dielectric strength is about 3×10^6 V/m. Typical values for some materials are included in Table 25.1. Notice that $\kappa_{air} \approx \kappa_{vac} = 1$. Hence, no serious error resulted when we treated air-filled capacitors as if they were "vacuum-filled" in earlier sections of this chapter.

TABLE 25.1   The Dielectric Constants of Several Substances

MATERIAL	DIELECTRIC CONSTANT $\kappa = \epsilon/\epsilon_o$	DIELECTRIC STRENGTH $(10^6$ V/m)
Air	1.00054	3
Amber	2.8	
Bakelite	4.9	24
Barium titanate	1200	5
Fused quartz	3.9	8
Mica	5.4	100
Paper	3.5	15
Plexiglass (Lucite)	3.4	40
Polystyrene	2.5	24
Porcelain	7.0	6
Potassium dihydrogen phosphate (near its tricritical point)	400 000	
Pyrex glass	4.5–5.5	15
Silicon	11.9 (300 K)	semiconductor
Silicon dioxide	3.9	
Strontium titanate	310	
Tantalum oxide	26	500
Teflon	2.0	60
Titanium dioxide	90	6
Water	80.4 (20°C)	

Let's now find how the potential difference across a capacitor and its capacitance are affected by the presence of a dielectric. Although we are analyzing a parallel-plate capacitor, our results turn out to be valid for any geometry. Let's begin with two parallel plates with a capacitance C_o when only a vacuum is between them. We place a charge Q_o on the capacitor by connecting it across the terminals of a battery that maintains a potential difference ΔV_o between its terminals. Then we *remove the battery so that the capacitor is isolated*. The relation between the capacitance, charge, and potential difference is, therefore, $C_o = Q_o/\Delta V_o$. We have seen that the relation between a uniform electric field and the potential difference ΔV_o across a separation d is

$$E_o = \frac{\Delta V_o}{d} \tag{25.13}$$

where E_o is the electric field strength in the vacuum between the plates. When we slide a dielectric material snugly between the plates, the induced surface charge on the dielectric causes the field strength to be reduced to E and, from Equation (25.12),

$$E = \frac{E_o}{\kappa}$$

Now this change in E implies that the potential difference between the plates must also have changed. However, the electric field is still uniform and related to the potential difference by $E = \Delta V/d$. Therefore,

$$\frac{\Delta V}{d} = \frac{E_o}{\kappa}$$

Substituting for E_o from Equation (25.13), we have

$$\frac{\Delta V}{d} = \frac{(\Delta V_o/d)}{\kappa}$$

or

$$\Delta V = \frac{\Delta V_o}{\kappa} \qquad \text{(Isolated capacitor)} \qquad (25.14)$$

The potential difference across the isolated capacitor is reduced by the presence of the dielectric. It is important to remember that the capacitor is *isolated*. If you connect the plates of any capacitor to the terminals of a 1.5-V battery, the potential difference across the conductors is 1.5 V no matter what is between the plates (as long as it's not a conductor). If, however, you connect the terminals of a 1.5-V battery to a capacitor with plates that are separated by only a vacuum, *remove the battery from the capacitor*, and then slide a dielectric between the plates, the potential difference always decreases.

How is the capacitance of our isolated capacitor affected by a dielectric? Before the dielectric is inserted, we know that $C_o = Q_o/\Delta V_o$. After the dielectric is inserted the new capacitance is

$$C = \frac{Q_o}{\Delta V}$$

where the free charge Q_o on the capacitor remains unchanged because the capacitor is isolated. Substituting for ΔV from Equation (25.14), we have

$$C = \frac{Q_o}{(\Delta V_o/\kappa)} = \kappa\left(\frac{Q_o}{\Delta V_o}\right)$$

$$C = \kappa C_o \qquad (25.15)$$

The capacitance is increased by the presence of a dielectric.

We emphasize that Equation (25.15) is true in general. For example, we have already found (Example 25.1) that the capacitance of a parallel-plate capacitor is $C_o = A\epsilon_o/d$. For a parallel-plate capacitor with a dielectric between the plates, the capacitance is

$$C = \frac{A\kappa\epsilon_o}{d}$$

The energy stored in a capacitor is given by Equation (25.9):

$$U = \frac{Q^2}{2C}$$

When a dielectric is placed between the conductors of our isolated capacitor, $C = \kappa C_o$ and the energy stored becomes

$$U = \frac{Q^2}{2(\kappa C_o)} = \frac{1}{\kappa}\left(\frac{Q^2}{2C_o}\right)$$

$$U = \frac{U_o}{\kappa} \qquad \text{(Isolated capacitor)}$$

For a fixed charge the energy stored in the capacitor is reduced. This energy is expended as the capacitor does work pulling the dielectric into the space between the plates.

We can also find the relation between the induced surface charge σ_i on the dielectric and the free charge σ_o on the capacitor plates. We have seen that the relation between the uniform electric field E_o just outside a conductor with a uniform surface charge density σ_o is (see Eq. (23.7)) $E_o = \sigma_o/\epsilon_o$. Moreover, the induced field E_i is a result of the uniform polarization surface charge density σ_i and is given by[3]

$$E_i = \frac{\sigma_i}{\epsilon_o}$$

Because the field due to free charge and the field due to bound charge are opposite in direction, the magnitude of the total field is

$$E = E_o - E_i = \frac{\sigma_o}{\epsilon_o} - \frac{\sigma_i}{\epsilon_o} \tag{25.16}$$

However, by Equation (25.12),

$$E = \frac{E_o}{\kappa} = \frac{\sigma_o}{\epsilon_o}\frac{1}{\kappa}$$

Substituting this result into Equation (25.16), we have

$$\frac{\sigma_o}{\epsilon_o}\frac{1}{\kappa} = \frac{\sigma_o}{\epsilon_o} - \frac{\sigma_i}{\epsilon_o}$$

Solving for σ_i, we obtain

$$\sigma_i = \left(\frac{\kappa - 1}{\kappa}\right)\sigma_o \tag{25.17}$$

Suppose instead of isolating a charged capacitor we leave it connected to the battery so that the potential difference ΔV *remains constant* across the plates as the dielectric is placed between them. Because $\Delta V = Q_o/C_o$ and C_o increases to κC_o, the charge on the capacitor must change:

---

[3] Because all the field lines exit from only one side of the sheet of charge, the result is the same as for a conductor. Recall that for a sheet of charge with equal electric fields on both sides, the electric field strength is $E = \sigma/2\epsilon_o$.

$$\Delta V = \frac{Q_o}{C_o} = \frac{Q}{\kappa C_o}$$

Solving for the new charge Q on the capacitor, we have

$$Q = \kappa Q_o \qquad (\Delta V \text{ held constant})$$

We can use Equation (25.9) to find how the energy stored by the capacitor has changed:

$$U = \frac{Q^2}{2C} = \frac{(\kappa Q_o)^2}{2(\kappa C_o)} = \kappa \left(\frac{Q_o^2}{2C_o} \right)$$

$$U = \kappa U_o \qquad (\Delta V \text{ held constant})$$

The energy storage capacity has increased by a factor of κ.

Let's summarize what we have learned about the presence of a dielectric between the conductors of a capacitor:

ΔV HELD CONSTANT (BATTERY ATTACHED)	Q HELD CONSTANT (CAPACITOR ISOLATED)
$Q = \kappa Q_o$	$Q = Q_o$
$\Delta V = \Delta V_o$	$\Delta V = \frac{1}{\kappa}\Delta V_o$
$E = E_o$	$E = \frac{1}{\kappa}E_o$
$U = \kappa U_o$	$U = \frac{1}{\kappa}U_o$

Finally, we want to clear up the mysteries about the "o" subscript on the constant ϵ_o and the name of this constant. We confine our attention to isotropic dielectrics and define the **permittivity** ϵ (without the "o" subscript) of a material with dielectric constant κ as

$$\epsilon = \kappa \epsilon_o$$

When the material is no material at all (that is, a vacuum), $\kappa = 1$ and the permittivity is that of free space. If you look back to Chapter 22, you will discover that we first defined the constant ϵ_o as the *permittivity of free space*. Hence, the permittivity constant is a number that, like κ, characterizes how polarizable a material is. If there is no material, ϵ becomes ϵ_o. We should remark here that if one applies Gauss's law inside a material medium, then ϵ for that medium must be used, rather than ϵ_o. A derivation of this result is given in Section 25.5.

Concept Question 5
What might be the reason that water is not used in capacitors even though it has a high dielectric constant?

Concept Question 6
Explain what effect raising the temperature of a dielectric material has on its dielectric constant.

EXAMPLE 25.8 A Partially Filled Capacitor

A parallel-plate capacitor of area A and plate separation d is partially filled with a dielectric slab of constant κ. As shown in Figure 25.20(a), the thickness of the slab is $d/4$. (a) Find the equivalent capacitance of this arrangement. (b) If the plate area of the capacitor is 6.00 cm^2, the separation is 2.00 mm, the dielectric constant is $\kappa = 3.00$, and a potential difference of 24.0 V is maintained across the plates, find the induced surface charge on the dielectric.

SOLUTION   (a) The given capacitor arrangement is equivalent to the series combination of the two capacitors shown in Figure 25.20(b). We can understand this equivalence by noting that the electric potential is the same at all points on the flat, bottom

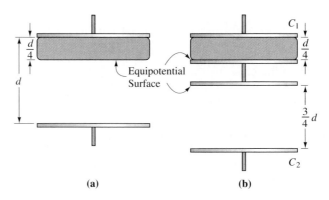

FIGURE 25.20

A parallel-plate capacitor partially filled with a dielectric slab. See Example 25.8.

surface of the dielectric. We long ago established that all points on the surface of a conductor are equipotential: this result is therefore also true for all points on the I-shaped conductor connecting the two capacitors in Figure 25.20(b). Because the surface bounding all regions in both versions of the capacitor are identical, the capacitors are electrically equivalent. The capacitances of the two partial capacitors are

$$C_1 = \frac{A\kappa\epsilon_o}{(d/4)} \quad \text{and} \quad C_2 = \frac{A\epsilon_o}{(3d/4)}$$

Applying Equation (25.7) for two capacitors in series

$$\frac{1}{C} = \frac{1}{C_1} + \frac{1}{C_2} = \frac{(d/4)}{A\kappa\epsilon_o} + \frac{(3d/4)}{A\epsilon_o}$$

and solving for C, we obtain

$$C = \frac{4A\epsilon_o}{d}\left(\frac{3\kappa + 1}{\kappa}\right)$$

(b) The charge on the top capacitor in Figure 25.20(b) is

$$Q = C_1 \, \Delta V = \frac{A\kappa\epsilon_o}{(d/4)}\Delta V = \frac{4A\kappa\epsilon_o \, \Delta V}{d}$$

And the free charge density $\sigma_o = Q/A$ is

$$\sigma_o = \frac{4\kappa\epsilon_o \, \Delta V}{d}$$

We find the induced charge density from Equation (25.17)

$$\sigma_i = \left(\frac{\kappa - 1}{\kappa}\right)\sigma_o = \left(\frac{\kappa - 1}{\kappa}\right)\frac{4\kappa\epsilon_o \, \Delta V}{d} = \frac{4\epsilon_o \, \Delta V}{d}(\kappa - 1)$$

$$= \frac{4(8.854 \times 10^{-12} \text{ C}^2/(\text{N}\cdot\text{m}^2))(24.0 \text{ V})}{(2.00 \times 10^{-3} \text{ m})}(3.00 - 1)$$

$$= 8.50 \times 10^{-7} \text{ C/m}^2 = 0.850 \ \mu\text{C/m}^2$$

25.5 Gauss's Law and the Electric Field Vectors

In this section we provide a more complete description of a dielectric's behavior when it is subjected to an external electric field. We continue to use the ideal, parallel-plate capacitor as a specific case, but our results are quite general.

We begin by deriving an auxiliary form of Gauss's law. We don't do this to confuse you; this form of the law proves to be easier to apply when one or more dielectric media are present. For reference we restate Gauss's law in its familiar form:

$$\oint \mathbf{E} \cdot d\mathbf{A} = \frac{q_{\text{encl}}}{\epsilon_o} \tag{25.18}$$

Figure 25.21 shows a cross-section of our familiar parallel-plate capacitor with free charge $\pm Q_o$ on the conductors and induced charge $\pm Q_i$ on the surfaces of the dielectric. Figure 25.21(a) shows the capacitor with no dielectric present, and the resulting field E_o. Figure 25.21(b) displays the same capacitor with the dielectric between the plates.

In each case, a cross section of a Gaussian surface is also designated. The top side of each Gaussian surface is within the conductor so that $E = 0$ through this portion of the surface. The electric field is perpendicular to the area vectors of the front, back, and sides of the rectangular, box-shaped Gaussian surface. Therefore, there is no contribution to the electric flux $\int \mathbf{E} \cdot d\mathbf{A}$ from these four sides of the Gaussian surface in either Figures 25.21(a) or (b). In fact, only the bottom surface contributes to the flux and through this surface both $\mathbf{E}_o$ in the left capacitor and $\mathbf{E}$ in the right capacitor are parallel to the area vectors (that is, perpendicular to the plane of the bottom Gaussian surface). In

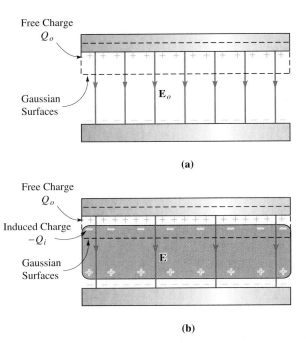

(a)

(b)

FIGURE 25.21

(a) A cross section of a parallel-plate capacitor with no dielectric. The top of the Gaussian surface is within the conductor, and the uniform field $\mathbf{E}_o$ passes perpendicularly through the bottom portion of the surface. (b) The capacitor and the same Gaussian surface with a dielectric present. The total charge enclosed by the surface is $Q_o - Q_i$, and the uniform field passing through the bottom portion of the Gaussian surface is $\mathbf{E}$.

Figure 25.21(a) the charge enclosed is Q_o, so Gauss's law gives

$$\oint \mathbf{E} \cdot d\mathbf{A} = \int_{\substack{\text{bottom} \\ \text{area } A}} E_o \, dA = E_o A = \frac{Q_o}{\epsilon_o}$$

or

$$E_o = \frac{Q_o}{A\epsilon_o} \tag{25.19}$$

In Figure 25.21(b), the total charge enclosed is $Q_o - Q_i$:

$$\oint \mathbf{E} \cdot d\mathbf{A} = \frac{Q_o - Q_i}{\epsilon_o} \tag{25.20}$$

$$\int_{\substack{\text{bottom} \\ \text{area } A}} E \, dA = EA = \frac{Q_o - Q_i}{\epsilon_o}$$

or

$$E = \frac{Q_o - Q_i}{A\epsilon_o} \tag{25.21}$$

Now, $\kappa = E_o/E$. If in this expression for κ we replace E_o and E with the expressions from Equations (25.19) and (25.21), we have

$$\kappa = \frac{Q_o/A\epsilon_o}{(Q_o - Q_i)/A\epsilon_o}$$

or

$$Q_o - Q_i = \frac{Q_o}{\kappa} \tag{25.22}$$

Finally, we put this result back into the form of Gauss's law when the dielectric is present, and Equation (25.20) becomes

$$\oint \mathbf{E} \cdot d\mathbf{A} = \frac{Q_o}{\kappa\epsilon_o} \tag{25.23}$$

This form of Gauss's law is convenient to use when dielectrics are present. It is very important to remember that Q_o is the free charge. Notice when there is no dielectric present $\kappa = 1$ (the value for a vacuum), and Equation (24.23) reduces to the original form of Gauss's law (Eq. (25.18)).

Polarization and the Displacement Field (Optional)

Let's return to Equation (25.21) for the net electric field within the dielectric material shown in Figure 25.21(b). Rearranging this expression, we can write

$$\frac{Q_o}{A} = \epsilon_o E + \frac{Q_i}{A} \tag{25.24}$$

We define the **free surface charge density** D as

$$D = \frac{Q_o}{A} \tag{25.25}$$

and the **induced surface charge density** P as

$$P = \frac{Q_i}{A} \tag{25.26}$$

With these substitutions Equation (25.24) becomes

$$D = \epsilon_o E + P \tag{25.27}$$

In addition to the surface charge induced on the dielectric, P has another interpretation. If we multiply both the numerator and denominator of our expression for P by the thickness d of the dielectric, we have

$$P = \frac{Q_i d}{Ad} \tag{25.28}$$

The numerator in Equation (25.28) is the product of the magnitude of the induced charge times the separation distance between these positive and negative charges. This is just the net dipole moment induced across the dielectric. The denominator is the volume of the dielectric. Hence, P is the *dipole moment per volume* and is called the **polarization** of the dielectric.

We again focus our attention to Equation (25.27). Because the electric field E is a vector, it is useful to define P and D so that they are also vectors. Defining P as a vector is quite natural because a dipole moment is itself a vector that is directed from the negative charge to the positive charge. Accordingly, **P** is defined as a vector directed from the negative induced surface charge to the positive induced surface charge. Its magnitude is the dipole moment per volume or, equivalently, the magnitude of the induced surface charge density on the dielectric.

The vector **D** is the vector sum of ϵ_o**E** and **P** and is called the **displacement field.** Its magnitude is the free surface charge density σ_o. A schematic representation of the relation between **P, E,** and **D** is shown in Figure 25.22.

We again emphasize that although Equation (25.27) was obtained for a parallel-plate capacitor, it is a quite general result and valid for dielectrics of any shape. However, the vector fields **E, D,** and **P** can change magnitude and direction within a dielectric of complicated geometry.

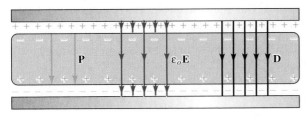

$$P + \varepsilon_o E = D$$

FIGURE 25.22

The relation between the field vectors **P, E,** and **D.** Note that **P** is associated with the induced surface charge, and **D** is related to the free surface charge.

Gauss's Law for the Displacement Field (Optional)

With our definition of the displacement field (Eq. (25.25)) we can simplify Equation (25.23) for Gauss's law. We have

$$\oint \mathbf{E} \cdot d\mathbf{A} = \frac{Q_o}{\kappa \epsilon_o}$$

or, because $\kappa \epsilon_o$ is a constant

$$\oint \kappa \epsilon_o \mathbf{E} \cdot d\mathbf{A} = Q_o \qquad (25.29)$$

From Equation (25.19) the E-field between the parallel plates, in the absence of a dielectric, is

$$E_o = \frac{Q_o}{A \epsilon_o}$$

When a dielectric is present, the E-field is

$$E = \frac{E_o}{\kappa} = \frac{Q_o}{A \kappa \epsilon_o} = \frac{Q_o}{A} \frac{1}{\kappa \epsilon_o} = \frac{D}{\kappa \epsilon_o}$$

Hence,

$$\mathbf{D} = \kappa \epsilon_o \mathbf{E} \qquad (25.30)$$

With this result we can rewrite Equation (25.29):

$$\oint \mathbf{D} \cdot d\mathbf{A} = Q_o \qquad (25.31)$$

Equation (25.31) emphasizes the fact that the displacement field $\mathbf{D}$ is associated with the free charge Q_o and it is also a very compact form of Gauss's law. However, this equation does require the additional relation of Equation (25.30) to complete the relation between the free charge Q_o and the electric field $\mathbf{E}$. Indeed, it is important to remember that although $\mathbf{D}$ and $\mathbf{P}$ have useful interpretations related to the surface charge densities and the dipole moment per volume, $\mathbf{E}$ is still the field we require for many important physical properties. The force on a charge q placed in the dielectric is still $\mathbf{F} = q\mathbf{E}$, and the torque on a dipole with moment $\mathbf{p}$ placed in the dielectric is still $\boldsymbol{\tau} = \mathbf{p} \times \mathbf{E}$.

EXAMPLE 25.9   P, D, and E

A capacitor consists of two, 1.00 cm^2 parallel plates separated by 0.500 mm. A dielectric ($\kappa = 2.60$) is placed between the plates, and a total charge of 6.00×10^{-9} C is transferred to the capacitor. Compute the displacement and electric fields between the plates and the polarization of the dielectric.

SOLUTION   The magnitude of the displacement field is given by the surface density of the free charge, Equation (25.25):

$$D = \sigma_o = \frac{Q_o}{A} = \frac{6.00 \times 10^{-9} \text{ C}}{(0.010 \text{ m})^2} = 60.0 \ \mu\text{C/m}^2$$

From Equation (25.31)

$$E = \frac{D}{\kappa \epsilon_o} = \frac{60.0 \times 10^{-6} \text{ C/m}^2}{(2.60)(8.854 \times 10^{-12} \text{ C}^2/(\text{N} \cdot \text{m}^2))} = 2600 \text{ kV/m}$$

We can find the polarization of the dielectric from Equation (25.27):

$$P = D - \epsilon_o E$$
$$= 60.0 \times 10^{-6} \text{ C/m}^2 - (8.854 \times 10^{-12} \text{ C}^2/(\text{N} \cdot \text{m}^2))(2.60 \times 10^6 \text{ V/m})$$
$$= 37.0 \text{ } \mu\text{C/m}^2$$

25.6 Numerical Application (Optional)

The spreadsheet QLINES.WK1, which is available on the disk accompanying this text, computes and plots the *magnitude* of the electric field along a line that lies between, parallel to, and in the plane of two parallel lines of charge. These charged lines are finite in length and carry equal but opposite uniform charge densities $\pm\lambda$. The E-field magnitude is computed along lines, such as l_1, l_2, or l_3, shown in Figure 25.23.

FIGURE 25.23

The electric field magnitude is computed along lines, such as l_1, l_2, or l_3.

In Figure 25.24 we show the geometry used to compute the components E_x and E_y of the electric field at a general point (X_o, Y_o). The length of the charge distribution is L, and it extends from $x = 0$ to $x = L$. If you worked Problem 50 in Chapter 22 you have already solved this portion of the problem.

The electric field at location (X_o, Y_o) generated by charge $dq = \lambda \, dx$ is

$$dE = \frac{1}{4\pi\epsilon_o} \frac{\lambda \, dx}{(X_o - x)^2 + Y_o^2}$$

and the components of dE are

$$dE_x = dE \cos(\theta)$$

$$dE_y = dE \sin(\theta)$$

where

$$\sin(\theta) = \frac{Y_o}{\sqrt{(X_o - x)^2 + Y_o^2}}$$

$$\cos(\theta) = \frac{X_o - x}{\sqrt{(X_o - x)^2 + Y_o^2}}$$

The expressions for the field components become

$$E_y = -\frac{\lambda Y_o}{4\pi\epsilon_o} \int_0^L \frac{dx}{[(X_o - x)^2 + Y_o^2]^{3/2}}$$

$$E_x = \frac{\lambda}{4\pi\epsilon_o} \int_0^L \frac{(X_o - x)dx}{[(X_o - x)^2 + Y_o^2]^{3/2}}$$

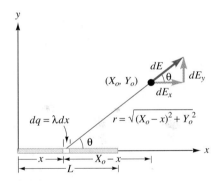

FIGURE 25.24

Computing the E-field at point (X_o, Y_o) from a line of charge.

Integrating, we find

$$E_y = \frac{\lambda}{4\pi\epsilon_o Y_o}\left(\frac{X_o}{\sqrt{X_o^2 + Y_o^2}} - \frac{X_o - L}{\sqrt{(X_o - L)^2 + Y_o^2}}\right) \qquad (25.32a)$$

$$E_x = \frac{\lambda}{4\pi\epsilon_o Y_o}\left(\frac{1}{\sqrt{(X_o - L)^2 + Y_o^2}} - \frac{1}{\sqrt{X_o^2 + Y_o^2}}\right) \qquad (25.32b)$$

We must next compute the components E_x' and E_y' for the other line of charge (with charge density $-\lambda$) parallel to the one shown in Figure 25.24. This line is displaced a distance D along the y-axis. Therefore, E_x' and E_y' are given by Equations (25.32) with Y_o replaced with $D - Y_o$ and λ replaced by $-\lambda$. We then add the corresponding components:

$$E_{x_{tot}} = E_x + E_x'$$

$$E_{y_{tot}} = E_y + E_y'$$

The magnitude of the E-field at (X_o, Y_o) is

$$E_{tot} = \sqrt{E_{x_{tot}}^2 + E_{y_{tot}}^2}$$

The input data section of the spreadsheet QLINES.WK1 is shown in Figure 25.25. In addition to λ, L, Y_o, and D, we specify a distance ΔX for the step size along the line for which we wish to compute the E-field magnitude. We have also computed the value of $\lambda/4\pi\epsilon_o$ for use in the equations we generate.

The X-coordinates of the points along the line are placed in column A. These values increase from the value placed in cell A10 by increments of ΔX. Columns B, D, L, and N are used to compute distances from $(X, 0)$ or (X, D) to (X_o, Y_o). The quantities E_x, E_y, E_x', E_y', $E_{x_{tot}}$, and $E_{y_{tot}}$ are in the designated columns (F through V). The magnitude of the electric field is tabulated in column X, and the resulting field for the data shown in Figure 25.25 is displayed graphically in Figure 25.26.

	A	B	C	D	E	F	G
1		E-Field Magnitude for Two Parallel Charged Lines					
2							
3	Lambda =	2.4E-11	!	1/4(pi)eo =	9E+09		!
4	L =	0.4	! Line Separation =		0.05		!
5	Yo =	0.025	! (lambda)/4(pi)eo =	2.16E-01			!
6	delta-X =	0.02	!				!
7							
8	X	[(Xo-L)^2 + Yo^2]^1/2	!	[Xo^2 + Yo^2]^1/2		Ey	!
9							
10	-0.3	7.00E-01		3.01E-01		2.43E-02	
11	-0.28	6.80E-01		2.81E-01		2.84E-02	
12	-0.26	6.60E-01		2.61E-01		3.35E-02	
13	-0.24	6.40E-01		2.41E-01		3.99E-02	
14	-0.22	6.21E-01		2.21E-01		4.82E-02	
15	-0.2	6.01E-01		2.02E-01		5.92E-02	
16	-0.18	5.81E-01		1.82E-01		7.41E-02	
17	-0.16	5.61E-01		1.62E-01		9.50E-02	
18	-0.14	5.41E-01		1.42E-01		1.25E-01	
19	-0.12	5.21E-01		1.23E-01		1.72E-01	
20	-0.1	5.01E-01		1.03E-01		2.47E-01	

FIGURE 25.25

A portion of the spreadsheet QLINES.WK1. Columns H through X are also used but are not shown.

Notice that for these 40.0-cm long lines with a separation of 5.00 cm the field is fairly uniform within about the center one-third of the length L. Within the first five or so centimeters in the region to the left and right outside of the charged lines the field magnitude is still very significant. Beyond 20.0 cm from the edge of the lines the field strength is less than 10% of its maximum value. In all of the models for our examples in this chapter we ignored fringing effects. Indeed, a "nonfringing" model of the E-field for our two parallel lines of charge would have assumed **E** to be uniform between the lines in Figure 25.23 and zero in the region to their left and right. We performed our calculation in the plane of the lines, so we did not investigate the fringing of the E-field outside of this plane. We hope, however, that this numerical example has given you some "feel" for fringing effects.

Do not forget that **E** is a vector, and we have only plotted its magnitude. You will find it interesting and instructive to plot the individual components of the electric field.

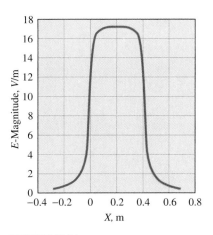

E-Field Magnitude Between Two Parallel Lines

FIGURE 25.26

The magnitude of the E-field for the parameters shown in Figure 25.25.

25.7 Summary

The **capacitance** of an arrangement of two conductors depends only on the geometry of the arrangement and is defined by

$$C = \frac{q}{\Delta V}$$

where q is the magnitude of the charge on one of the conductors (sometimes called plates), and ΔV is the electric potential difference between them.

For N capacitors in a **parallel** arrangement the equivalent capacitance is

$$C_{\text{tot}} = \sum_{i=1}^{N} C_i \qquad \qquad (25.5)$$

and for N capacitors in a **series** combination the equivalent capacitance is

$$\frac{1}{C_{\text{tot}}} = \sum_{i=1}^{N} \frac{1}{C_i} \qquad \qquad (25.8)$$

The energy stored in a capacitor C with potential difference ΔV across its conductors and carrying charge Q is

$$U = \frac{1}{2}C(\Delta V)^2 = \frac{Q^2}{2C}$$

The **energy density** at a point in an electric field E is

$$u = \frac{1}{2}\epsilon_o E^2 \qquad \qquad (25.11)$$

When an insulator of **dielectric constant** κ is inserted between the conductors of a capacitor, its capacitance is increased by a factor of κ over its value C_o for a vacuum between the plates:

$$C = \kappa C_o \qquad \qquad (25.15)$$

When the dielectric is inserted into an *isolated* capacitor, the electric field between the conductors is reduced and the potential difference across an isolated capacitor also decreases. The reduction of the field strength between the conductors arises from an opposing E-field generated by the **induced surface charge** on the dielectric. This induced charge originates from the alignment of the dipoles that comprise the dielectric material.

When a dielectric is inserted between the conductors of a capacitor where the potential difference between the conductors is held constant, the charge on the capacitor increases: $Q = \kappa Q_o$.

The **permittivity constant** ϵ is related to the dielectric constant by

$$\epsilon = \kappa\epsilon_o$$

Gauss's law can be written in the form

$$\oint \mathbf{E} \cdot d\mathbf{A} = \frac{Q_o}{\kappa\epsilon_o} \tag{25.23}$$

(Optional) Gauss's law can also be rewritten in the form

$$\oint \mathbf{D} \cdot d\mathbf{A} = Q_o \tag{25.31}$$

where **D** is the **displacement field** and Q_o is the **free charge** enclosed by the Gaussian surface. The magnitude of **D** at a surface is equal to the surface density of the free charge on that surface. The **polarization P** is the dipole moment per volume within a dielectric, and its magnitude at the dielectric surface is also the surface charge density of the induced charge on the dielectric. The relation between **D, P,** and the electric field **E** within the dielectric is

$$\mathbf{D} = \epsilon_o\mathbf{E} + \mathbf{P} \tag{25.27}$$

PROBLEMS

25.1 Capacitance

1. Show that the unit farad is equivalent to $C^2 \cdot s^2/(kg \cdot m^2)$.
2. What charge resides on the plates of a 100.-pF capacitor when there is a potential difference of 24.0 V across its conductors?
3. The magnitude of the charge carried by 1.00 mol of electrons is called a **faraday.** How many faradays of charge reside on a 0.0100-μF capacitor with a potential difference of 1200. V applied across it?
4. A certain electrolytic capacitor has a capacitance of 4700. μF. What is the magnitude of the charge on its plates when there is a potential difference of 250. V between the plates?
5. If we wish to store a charge of 60.0 μC and the maximum potential difference we have available is 2400. V, what is the minimum value of capacitance we can use?
6. The charge on one plate of a 100.-pF capacitor is 0.0300 μC. What is the potential difference between the plates of this capacitor?
7. If an air-filled 1.00-F, parallel-plate capacitor is to be constructed with plate separation of 1.00 mm, what must be the area of its plates?
8. When the potential difference across a certain capacitor is reduced by 120. V, the charge on the capacitor changes from 360. μC to 120. μC. What is the capacitance of the capacitor?
9. Show that the capacitance of an isolated, spherical conducting surface of radius R is $4\pi\epsilon_o R$ and use this result to compute the capacitance of a sphere the size of the earth. (*Hint:* Suppose the other conductor is also a sphere located at an infinite radius from the center of the actual sphere.)

10. A parallel-plate capacitor is constructed with a plate separation of 0.500 mm, a plate area of 36.0 cm^2, and air between the plates. Charge is transferred from one plate to the other until a potential difference of 400. V is established across the plates. Find (a) the capacitance and (b) the magnitude of the charge on each plate.
11. The radius of the outer sphere of a spherical capacitor is four times that of the inner sphere. What must be the radius of a sphere that is to replace the inner sphere if the new capacitance of the capacitor is to be half the original value?
12. The length of a cylindrical capacitor is tripled. By what factor must the ratios of the cylinders' radii (R_2/R_1) be changed for the new capacitance to be equal to the original capacitance?
13. The variable capacitor shown in Figure 25.P1 is used in the circuit

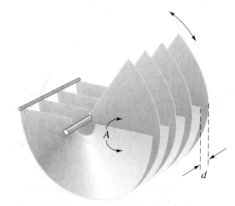

FIGURE 25.P1   Problem 13

of a radio to "tune-in" a particular station. This capacitor is constructed with alternating plates, each of area A and separation d. If there are N plates in the capacitor, show that its maximum capacitance is

$$C = \frac{(N-1)\epsilon_o A}{d}$$

14. Two conductors carry charges $+Q$ and $-Q$. They create a nonuniform electric field as shown in Figure 25.P2. A coordinate system is placed on the positive conductor, and the electric field strength along the y-axis is found to vary according to the function

$$E(y) = QA\epsilon_o\left(1 + \frac{y^2}{B}\right)$$

where $A = 4.70 \times 10^{17}$ V/(F $\cdot$ C) and $B = 2.00 \times 10^{-5}$ m^2. The distance d is 6.80 mm. Compute the capacitance for this arrangement of conductors.

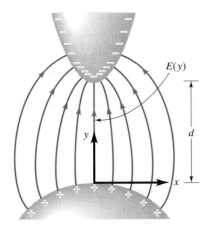

FIGURE 25.P2   Problem 14

25.2 Combinations of Capacitors

15. What are the maximum and minimum values of capacitance you can construct from a 6.00-μF capacitor and a 12.0-μF capacitor?
16. How many 4.00-μF capacitors do you need to connect in series to create an equivalent capacitance of 2.00 μF?
17. What is the equivalent capacitance of four capacitors with values of 2.00, 4.00, 6.00, and 8.00 μF connected in (a) series and (b) parallel?
18. Two capacitors, 10.0 pF and 24.0 pF, are connected in series, and a potential difference of 240. V is applied across the combination. What is the charge on each and the potential difference across each?
19. Three capacitors (2.00 μF, 4.00 μF, and 6.00 μF) are connected in series, and a potential difference of 120. V is applied across the combination. Compute the charge on each capacitor and the potential difference across each.
20. Two capacitors, 4.00 pF and 12.00 pF, are connected in parallel, and a potential difference of 48.0 V is applied across the combination. Compute the charge on each capacitor.
21. Three capacitors (12.0 μF, 18.0 μF, and 24.0 μF) are connected in parallel. Compute the charge on each capacitor when 110. V is applied across the combination.
22. An 8.00-pF capacitor and a 12.0-pF capacitor are connected in series, and a potential difference of 24.0 V is applied across the combination. The potential difference source is then removed, and the two free wires from the two capacitors are connected together. Find the charge on each capacitor and new potential difference across each.
23. Both a 0.50-μF capacitor and a 1.50-μF capacitor are (separately) charged by a 24.0-V battery. The battery is removed, and the positive plate of the 0.50-μF capacitor is connected by a wire to the negative plate of the 1.50-μF capacitor. Also, the negative plate of the 0.500-μF capacitor is connected by a wire to the positive plate of the 1.50-μF capacitor. Find the resulting charge on each capacitor and the potential difference across each.
24. Find the equivalent capacitance between points A and B and the charge on each capacitor for the circuit segment shown in Figure 25.P3. Use $C_1 = 4.00$ μF, $C_2 = 6.00$ μF, $C_3 = 3.00$ μF, and $\Delta V = 12.0$ V.

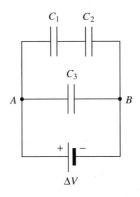

FIGURE 25.P3   Problem 24

25. Find the equivalent capacitance between points A and B and the charge on each capacitor for the circuit segment shown in Figure 25.P4 if $C_1 = 6.00$ μF, $C_2 = 12.0$ μF, $C_3 = 4.00$ μF, and $\Delta V = 240.$ V.

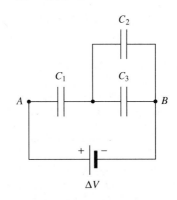

FIGURE 25.P4   Problem 25

26. For the circuit segment shown in Figure 25.P5 find the equivalent capacitance between points A and B if $C_1 = 1.00$ μF, $C_2 = 2.00$ μF, $C_3 = 4.00$ μF, $C_4 = 6.00$ μF, and $C_5 = 3.00$ μF.

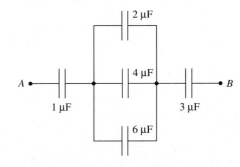

FIGURE 25.P5   Problem 26

27. Compute the equivalent capacitance between points A and B in Figure 25.P6.

28. For the circuit segment shown in Figure 25.P6 compute the charges on the 5.00-pF capacitor and the 2.00-pF capacitor when a potential difference of 72.0 V is applied across points A and B.

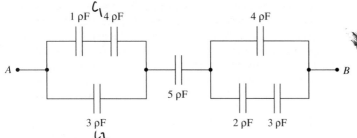

FIGURE 25.P6   Problems 27 and 28

29. A 3.00-μF capacitor and a 6.00-μF capacitor are connected in series, and their free ends are then connected to a 24.0-V battery. After charging, the capacitors are disconnected from the battery and each other without discharging either. (a) If the two capacitors are now connected positive side to positive side and negative to negative, what is the resulting potential difference across the pair? (b) What would be the potential difference across the pair if they had been connected negative to positive?

25.3 Energy Storage in a Capacitor

30. Find the energy stored in a 10.0-μF capacitor when it is charged with a potential difference of 110. V.

31. A tantalum capacitor has a capacitance of 100. μF. What potential difference do you need in order to store enough energy in this capacitor to lift a 1.00-kg mass vertically through 1.00 m at the earth's surface?

32. A potential difference of 240. V is applied across the plates of an ideal, parallel-plate capacitor. The plate separation is 2.00 mm, and each plate has an area of 2.00 cm^2. (a) Find the electric field strength and energy density between the capacitor plates. (b) Compute the total energy stored in the capacitor.

33. A 10.0-pF parallel-plate capacitor is charged to a potential difference of 240. V, and the charging source is then removed. The area of each plate is 10.0 cm^2. The plates are pulled apart until their separation is double the original value. (a) Compute the new potential difference across the plates. (b) Compute the original energy stored in the capacitor before the plates were separated. (c) Find the work required to separate the plates.

34. Compute the energy density at the surface of an isolated metal sphere that is charged to a potential difference of 5.00 kV if the radius of the sphere is 12.0 cm.

35. (a) Compute the potential energy stored in a parallel-plate capacitor with plates that have area A, are separated by a distance x, and carry charges $+Q$ and $-Q$. (b) From your answer to part (a) show that the force exerted by one plate on the other is $Q^2/2A\epsilon_o$.

36. The plates of a parallel-plate capacitor have an area of 8.00 cm^2 and are separated by 12.0 cm. A potential difference of 96.0 V is maintained across the capacitor while a copper slab 6.00 cm thick is inserted midway between the plates. How much work is required to insert the copper slab?

37. The plates of an ideal parallel-plate capacitor have an area of 12.0 cm^2 and are separated by 8.00 cm. Charge is transferred to the plates by connecting them to the terminals of a 48.0-V battery. After charging, the battery is removed, and a 2.00-cm thick copper slab is inserted midway between the plates. How much work is required to insert the copper slab?

25.4 Dielectrics

38. A metal-oxide-semiconductor (MOS) diode used in integrated circuits consists of a sandwich of metal-SiO$_2$-Si. This layered structure acts as a capacitor with the metal and silicon as the conductors, and the silicon dioxide layer as the dielectric ($\kappa = 3.9$). What is the capacitance per unit area (C/A) for a typical MOS diode with a SiO$_2$ layer 25.0 nm thick?

39. When a parallel-plate capacitor with a certain dielectric material between its plates is connected to a battery, the charge on its plates is 400. μC. When the dielectric is removed while the battery remains connected, the charge on the capacitor is reduced to 110. μC. What is the dielectric constant of the material?

40. The plates of a parallel-plate capacitor are separated by a dielectric material and are charged by connecting them to a 6.00-V battery. The battery is then removed. When the dielectric material is also removed, the potential difference across the plates drops to 24.0 V. What is the dielectric constant of the material?

41. What is the maximum potential difference that can be applied across a parallel-plate capacitor with a 0.500-cm slab of Pyrex glass ($\kappa = 5.00$) separating the plates?

42. When the plates of a parallel-plate capacitor are separated by a slab of fused quartz, its capacitance is 150. pF. When the fused quartz is replaced with Pyrex glass ($\kappa = 5.00$), what is the new capacitance?

43. The plates of a parallel-plate capacitor have area A and are separated by a distance d. Two different dielectric materials of constants κ_1 and κ_2 but each with equal thickness sit one on top of the other and fill the space between the horizontal plates. Show that the equivalent capacitance of this arrangement is

$$C = \frac{2\epsilon_o A}{d}\left(\frac{\kappa_1 \kappa_2}{\kappa_1 + \kappa_2}\right)$$

44. Between the plates of the parallel-plate capacitor of area A shown in Figure 25.P7 a dielectric material of constant $\kappa = 2.00$ sits on a slab of copper. (a) Compute the equivalent capacitance of this arrangement. (b) What fraction of the total energy of this arrangement is stored in the field of the dielectric?

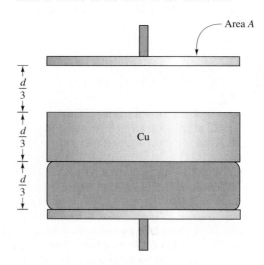

FIGURE 25.P7   Problem 44

45. The parallel-plate capacitor shown in Figure 25.P8 is constructed from plates of area $A = 3.60$ cm^2 and separation $d = 2.40$ mm. Two different dielectric materials of equal size occupy the space between the plates and have constants $\kappa_1 = 5.40$ and $\kappa_2 = 4.20$. Compute the equivalent capacitance of this system.

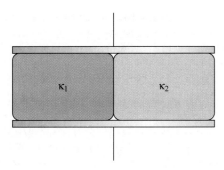

FIGURE 25.P8   Problems 45 and 73

46. The parallel-plate capacitor arrangement shown in Figure 25.P9 consists of plates of area $A = 4.00$ cm^2 separated by a distance $d = 1.80$ mm. The three different dielectric materials have constants $\kappa_1 = 3.60$, $\kappa_2 = 4.80$, and $\kappa_3 = 7.20$. Find the capacitance of this arrangement.

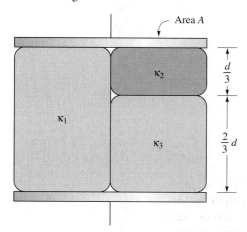

FIGURE 25.P9   Problem 46

47. What is the capacitance of a capacitor constructed from two rolls of aluminum foil each with area of 1.00 ft × 20.0 yd, separated by wax paper of the same area. Assume the thickness of the wax paper to be 0.125 mm, and use $\kappa = 3.5$ for its dielectric constant.

48. A parallel-plate capacitor has a plate separation of 3.00 mm and an area of 2.40 cm^2. A dielectric of constant $\kappa = 4.80$ fits snugly between the plates. A battery is used to charge the capacitor to a potential difference of 500. V, and then the battery is removed. Compute the amount of work required to remove the dielectric slab from between the plates.

49. A parallel-plate capacitor has a plate separation of 2.00 mm and an area of 4.00 cm^2. A dielectric of constant $\kappa = 5.60$ fits snugly between the plates. Find the work required to remove the dielectric slab from the capacitor if a potential difference of 250. V is maintained across the plates.

Numerical Methods and Computer Applications

50. Write a computer program to calculate the electric field magnitude along a line just as performed in the spreadsheet example of Section 25.6. Test your program with the data from that example.

51. A parallel-plate capacitor consists of two thin plates of length L and width W, separated by a distance d. With no dielectric between them, the plates are charged by a battery of potential difference ΔV, and then the battery is removed. A dielectric of constant κ is partially inserted between the plates so that a cross section of the capacitor appears as in Figure 25.P10 *without the voltage source present*. (a) Write a program or develop a spreadsheet to

compute the potential energy of this system as a function of x. Increment x in steps of Δx from $-L$ to $+L$. (b) Try out your program or spreadsheet for the following data: $L = W = 10.0$ cm, $d = 2.00$ mm, $\Delta x = 0.500$ cm, $\Delta V = 2.50$ kV, and $\kappa = 5.00$. If the dielectric slab is released, what type of motion does the shape of the potential energy function suggest? (c) Investigate the effect of different values of κ on the shape of the potential energy function.

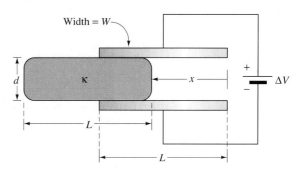

FIGURE 25.P10   Problems 51 and 72

52. The capacitor shown in Figure 25.P11 consists of two alternating layers of plates, separated by air. The width and length of each plate are L and W, respectively, and the distance between plates is d. Write a program or construct a spreadsheet to compute and plot the capacitance as a function of overlap distance x. Try out your program for the following data: $L = 2.00$ cm, $W = 1.00$ cm, and $d = 0.200$ mm. Compute x at intervals of $\Delta x = 0.005$ cm.

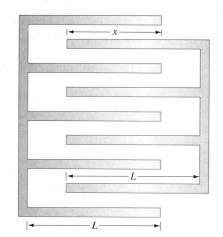

FIGURE 25.P11   Problem 52

53. Use the relaxation method introduced in Chapter 24 to explore the fringing field of a parallel-plate capacitor. Use as large a rectangular area as your spreadsheet allows, and set the border cells to a constant potential of 0.00 V. Then insert two parallel plates at ±10.0 V. Make the distance from the ends of the plates to the border at least twice the length of the plates. For starters you might try a 50 × 50 cell area with plates ten cells long separated by five cells. Draw equipotentials at 2-V intervals, and then sketch in the electric field lines by hand. (Because the capacitor is symmetric about both vertical and horizontal center lines, you can save space and computation time by displaying only one-quarter of the solution. You must, however, think carefully about the boundary conditions on your border; they are no longer simply zero along two edges.)

54. Suppose a parallel-plate capacitor with area 0.400 m^2 has an inhomogeneous dielectric that varies linearly between $\kappa = 1.00$ and

$\kappa = 2.00$ between the two plate surfaces, which are separated by distance $d = 1.00$ mm. Develop a computer program or spreadsheet to calculate the capacitance of such a capacitor. Begin by dividing the distance between plates into two layers of thickness $d/2$ with dielectric constants of 1.25 and 1.75, and calculate the value of the series equivalent. Then divide the dielectric into three layers with dielectric constants 1.17, 1.50, and 1.83 each of thickness $d/3$, and calculate the equivalent capacitance. Continue this process for more and more layers until the equivalent capacitance becomes constant to within at least 0.1%. Produce a plot of effective capacitance versus the number of layers.

General Problems

55. Show that if the separation between the two concentric spheres of a spherical capacitor is small in comparison to their radii, the capacitance approaches that of a parallel-plate capacitor with an area equal to that of one of the spheres.

56. Use the approximation that $\ln(1 + x) \approx x$ when $x \ll 1$ to show that the capacitance of a cylindrical capacitor approaches that of a parallel-plate capacitor when $R_1 \rightarrow R_2$. (*Hint:* Write $R_2 = R_1 + d$ where $d \ll R_1$.)

57. While helping your physics instructor rummage through the stock room, you discover a 0.500-m long, solid cylinder and a solid sphere. Both have 2.00-cm radii. You also discover two conducting, hemispherical shells that can be placed together to form a spherical shell with a diameter of 10.0 cm. (a) What is the capacitance of a spherical capacitor constructed from a concentric arrangement of the solid sphere and the shell? (b) Is it practical to search for a 0.500-m long pipe to use concentrically with the solid cylinder to build a capacitor with the same capacitance as the spherical arrangement? (*Hint:* What is the required radius of the pipe?)

58. The capacitor shown in Figure 25.P12 consists of two, nonparallel, square plates. By dividing the plates into small, parallel strips (with lengths that run into the page), show that for a small angle θ the capacitance of this arrangement is

$$C = \frac{\epsilon_o l^2}{d}\left(1 - \frac{l\theta}{2d}\right)$$

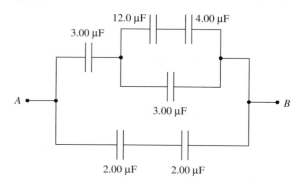

FIGURE 25.P12   Problem 58

59. You discover a large box of 2.00-μF capacitors each with a voltage rating of 10.0 V (i.e., a potential difference of no greater than 10.0 V can safely be placed across any one capacitor). Draw a diagram showing how you can connect these capacitors to produce an equivalent capacitor of 1.00 μF that can be connected across 100. V.

60. A 1.00-μF capacitor and a 3.50-μF capacitor are (separately) charged by a 36.0-V battery. The battery is removed, and the positive plate of the 1.00-μF capacitor is connected by a wire to the negative plate of the 3.50-μF capacitor. Also, the negative plate of the 1.00-μF capacitor is connected by a wire to the positive plate of the 3.50-μF capacitor. Find the resulting charge on each capacitor and the potential difference across each.

61. A 2.50-μF capacitor and a 4.00-μF capacitor are connected in series, and their free ends are then connected to a 12.0-V battery. After the charging, the capacitors are disconnected from the battery and each other without discharging either. (a) If the two capacitors are now connected positive side to positive side and negative to negative, what is the resulting potential difference across the pair? (b) What is the potential difference across the pair if they are connected negative to positive?

62. Find the charge on each capacitor in the circuit shown in Figure 25.P13 if $C_1 = 2.00$ μF, $C_2 = 3.00$ μF, $C_3 = 4.00$ μF, $C_4 = 6.00$ μF, and $\Delta V = 24.0$ V.

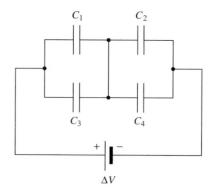

FIGURE 25.P13   Problem 62

63. Compute the equivalent capacitance between points A and B in Figure 25.P14.

64. Determine the charge on the 4.00-μF capacitor in Figure 25.P14 when a potential difference of 120. V is applied across points A and B.

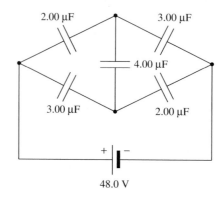

FIGURE 25.P14   Problems 63 and 64

★65. Determine the potential difference and charge on the 4.00-μF capacitor in the circuit shown in Figure 25.P15.

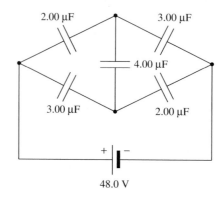

FIGURE 25.P15   Problem 65

66. The cylinders of a cylindrical capacitor have inner and outer radii of 0.500 cm and 2.50 cm, respectively, and are 10.0 cm in length. By direct integration find the energy stored in the region $0.500 \text{ cm} \leq r \leq 1.50 \text{ cm}$ when a potential difference of 1.20 kV is maintained between the cylinders. Assume there is only a vacuum between the cylinders.

67. The heat of combustion of gasoline leads to an available energy density of $35.0 \times 10^{10} \text{ J/m}^3$. (a) What electric field magnitude is necessary to store electric energy at an equivalent density? (b) Compare your answer to (a) the dielectric strength of some typical materials as listed in Table 25.1. What do you conclude about the possibility of an electric car that uses electrostatic energy as a power source?

68. When electronic circuits operate in high-pressure chambers or in high vacuum, it is sometimes desirable to be able to change the capacitance in an electric circuit without actually touching the capacitor. Such a "varicap" can be made by joining a p-type and an n-type semiconductor. At the junction between them an insulating dielectric "depletion layer" is formed between the two (semi)conductors. The width of this layer (and hence the capacitance of the device) can be controlled by the potential difference across the pn-junction. If one of the semiconductors is much more highly conducting than the other, the width of the depletion layer is given by

$$W = \sqrt{\frac{2\kappa\epsilon_o(V_{bi} - V)}{eN_b}}$$

where κ is the dielectric constant of the semiconductor, V_{bi} is the so-called built-in voltage, e is the magnitude of the charge on an electron, and N_b is the impurity concentration of the more poorly conducting semiconductor. If the junction has an area $A = 0.010$ mm^2 and potential differences from $V = 0$ to $V = -20.0$ V can be applied to the device, what range of capacitances can be achieved? Use values typical for silicon: $\kappa = 11.9$, $V_{bi} = 0.800$ V, and $N_b = 1.00 \times 10^{16}$ atoms/cm^3.

69. A spherical capacitor consists of two concentric, metallic spheres of radii R_1 and R_2 ($R_1 < R_2$) separated by a material of dielectric constant κ. Apply the steps presented in Section 25.1 to compute the capacitance of the arrangement. Be sure to consider the induced charge on the surface of the dielectric.

70. A spherical capacitor consists of two concentric, metallic spheres of radii R_1 and R_3 ($R_1 < R_3$). Between the spheres are two concentric, spherical dielectrics of constants κ_1 and κ_2. As shown in Figure 25.P16 the radius of the boundary between the two dielectrics is R_2 ($R_1 < R_2 < R_3$). Apply the steps presented in Section 25.1 to compute the capacitance of the arrangement. Be sure to consider the induced charge on the surface of the dielectric.

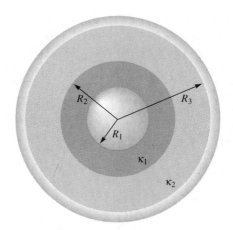

FIGURE 25.P16   Problem 70

71. A cylindrical capacitor consists of two concentric, metallic cylinders of length L separated by two concentric dielectric material of constants κ_1 and κ_2. As shown in Figure 25.P17, the radii of the inner and outer conductors is R_1 and R_3, respectively. The radius of the boundary between the dielectrics is R_2. Apply the steps presented in Section 25.1 to compute the capacitance of the arrangement. Be sure to consider the induced charge on the surface of the dielectric.

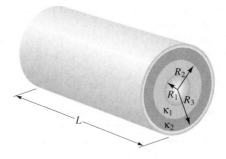

FIGURE 25.P17   Problem 71

72. A parallel-plate capacitor consists of two thin plates of length L and width W separated by a distance d. A dielectric material of constant κ is inserted a distance x between the plates as shown in Figure 25.P10. A potential difference ΔV is maintained across the plates. (a) Compute the total energy stored in the capacitor. (b) Find the magnitude and direction of the force on the dielectric slab.

73. Find the electric field strength and the induced surface charge on each of the dielectrics described in Problem 45 when a potential difference of 64.0 V is applied across its terminals.

CHAPTER

26

Electric Current and Resistance

In this chapter you should learn

- the definitions of current and current density.

- how the current density is related to an applied electric field.

- the definition of resistance and how it is related to a resistor's shape.

- about Ohm's law and how to apply it.

- how to calculate the rate of energy dissipation in a resistor.

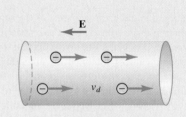

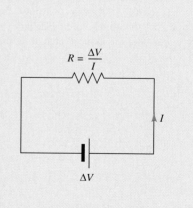

To this point we have considered only systems in *electrostatic equilibrium.* This condition led us to make statements, such as "The electric field is zero inside an ideal conductor," and "The excess charge of a charged conductor resides on its surface." We are now ready to relax this static equilibrium restriction and consider what happens when charges are moving. In this nonstatic case, free charge does not necessarily reside on a conductor's surface and can move throughout its volume.

The laws for electric currents that you will study in this chapter apply to devices from the smallest microcircuit to the largest power-generating systems. The flow of charge dominates our modern industrialized lives. If there is any doubt in your mind that the study of electric currents is important to modern scientists and engineers, try counting up the number of times you have used electricity today already. Start with your toast in the morning; remember the electric starter on your car, and don't overlook the calculator with which you do your homework. You likely will be amazed at how dependent your life style is on moving electric charges.

Our first order of business is to define the flow of charges. As usual, we find it convenient to begin in one dimension, but we quickly realize the necessity for a broader, three-dimensional description of charge flow. The good news is that the three-dimensional description has a mathematical form that is completely analogous to that of a concept we developed in Chapter 23, namely, flux.

26.1 Electric Current and Current Density

The easiest way to describe the flow of charge is to state the rate at which charge passes through a reference surface; the **average electric current** is defined as

$$I_{av} = \frac{\Delta q}{\Delta t} \qquad \text{(Average current)} \qquad (26.1)$$

where charge Δq flows through an imaginary plane in space in time Δt. The unit of current is the coulombs per second, and is defined as an **ampere** after the French physicist André Marie Ampère (1775–1836).

When the amount of charge Δq changes in time, we define the **instantaneous current** as

$$I = \lim_{\Delta t \to 0} \frac{\Delta q}{\Delta t}$$

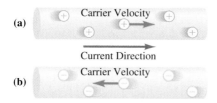

(a) Carrier Velocity

Current Direction

(b) Carrier Velocity

$$I = \frac{dq}{dt} \qquad \text{(Instantaneous current)} \qquad (26.2)$$

By convention, the direction of the current flow is that of the *positive* charge. This convention is illustrated in Figure 26.1. In metals, the moving charges are negative because the charge carriers are electrons. Hence, the electron flow is in the opposite direction from the current flow; when an electron moves toward the left in Figure 26.1(b), the result is the same as if a charge $+e$ moves toward the right as in Figure 26.1(a).

Most common conductors are metals. We don't know about you, but it would have been more comfortable for us if electrons had somehow ended up positive, the atomic nuclei negative and, consequently, positive charges really were flowing along the direction of the current. However, when he made the charge assignments that eventually led to negatively charged electrons, Benjamin Franklin knew nothing about the nature of the

FIGURE 26.1

The current direction is that of the positive charge carriers. In (a) the charge carriers are positive, so the current is in the same direction as the carriers' velocities. In (b) the carriers are negative, so the current direction is in the opposite direction from the actual charge flow. The situation for metals is that of (b) because the charge carriers are electrons.

An aerial view of the Fermi Lab accelerator.

charge carriers in metals. His assignments were based on electrostatic experiments and were quite arbitrary.

It is not necessary to have a wire present to have a current of electrons. In the picture tube of a conventional television set, a beam of electrons flows from the back of the tube to the screen. Our convention for current direction means there is a current flowing from the front of the TV toward the back. Particle physicists use accelerators to create beams of protons where the current of a beam is in the direction of the protons' velocities. In electrolytic solutions the charge carriers consist of both positive and negative ions traveling in opposite directions; both ion types contribute to a current in the *same* direction.

Most of the currents we will consider throughout this and the next chapter are carried by the electrons in metals. For this reason we devote a fair amount of space to develop a model for this conduction process. Initially our microscopic picture of a metal is that of a lattice of positive ions with a sea of virtually free electrons (one or two per ion) dashing around with random velocities with average magnitudes determined by the metal's temperature. You might even think of the electrons as behaving in many ways like an **electron gas.** It is a rather amazing fact deduced from quantum mechanics that a *perfectly* ordered lattice is transparent to the moving electrons. Only when the lattice is distorted in some way do the electrons interact with the lattice. As illustrated in Figure 26.2, free electrons can undergo collisions with impurities, imperfections, and vibrations in the ionic lattice structure. The smallest vibrational waves of the lattice are known as **phonons.** When electrons scatter from these elementary vibrations, the conservation of energy and momentum can be applied to the interaction by treating these phonons as if they were in fact themselves particles. Thus, when describing the interaction of electrons with lattice vibrations, we speak of electron scattering by phonons.

In order to establish a current in a conductor we must apply an electric field $\mathbf{E}$. Free charges q experience forces $\mathbf{F} = q\mathbf{E}$ and accelerate. If the charge carriers are electrons in a metal, they accelerate in a direction opposite that of the applied field. The velocities these electrons achieve due to this acceleration is superposed on their random thermal motions. The electron's velocity after any collision is random in magnitude and direction, hence its ''memory'' of the acceleration is erased. However, immediately after a collision the electron once again experiences the applied E-field and, therefore, begins to accelerate again. The result is a sequence of acceleration–collision–acceleration–collision, and so on as illustrated in Figure 26.2(d). Notice, although there are many random collisions, there is

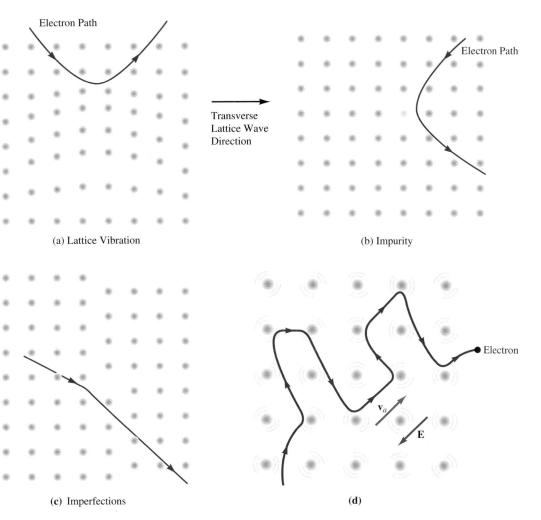

(a) Lattice Vibration

(b) Impurity

(c) Imperfections

(d)

FIGURE 26.2

Some mechanisms by which electrons are scattered in solids. Over most temperatures scattering by lattice vibrations is most dominant. Here an electron is scattered by a transverse distortion wave propagating to the right. At low temperatures most scattering occurs from impurities (b) and lattice imperfections (c). (d) When an electric field is present, the electron is accelerated in a direction opposite to the E-field so that over a number of collisions it exhibits a net displacement with an average velocity v_d. The curved paths are a result of the electron's acceleration.

a net displacement of the carrier in the direction opposite to the field. Consequently, we can model the electron's average motion as a constant velocity called the **drift velocity** v_d, equal to the average drift displacement per unit time.

The model we described above for conduction in a metal is also applicable to electrolytic solutions, but in this case both positive and negative ions are accelerated (in opposite directions, of course). These ions collide with the randomly moving molecules of the solution, and the overall result is again that the charge carriers drift with the same average velocity. The models for the conduction process in other materials, such as semiconductors, plasmas, and superconductors, are complicated by quantum mechanical effects and some of these are described in (optional) Section 26.4.

Now that we have described drift velocity, let's establish the relationship between the drift velocity v_d of the charge carriers and the current I. The question we wish to answer is this: How is I related to the number of charge carriers per volume n, their drift velocity v_d, and the cross-sectional area A through which they pass? For simplicity we take the sign of the charge carriers as positive so that we don't continually have to refer to a direction reversal between current and velocity.

In Figure 26.3 we show a cylindrical volume of positive charges traveling toward the right through a fixed surface of area A. We assume that the average velocity component of

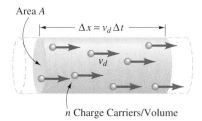

FIGURE 26.3

A uniform density of n charge carriers per volume each with drift velocity v_d move through a cross-sectional area A. In time Δt the carriers travel through a distance $\Delta x = v_d \, \Delta t$.

all the charge carriers is v_d and assign this value to each charge. In time Δt these carriers travel an average distance $\Delta x = v_d \Delta t$, and all charges within the volume $\Delta x A = v_d \Delta t A$ pass the fixed reference plane. If n is the number of carriers per volume, then the total number of carriers that pass the reference plane in time Δt is $nAv_d \Delta t$. Moreover, if each carrier has charge q, the total charge moving through the reference surface is

$$\Delta q = qnAv_d \Delta t$$

and the current is

$$I = \frac{\Delta q}{\Delta t} = nqv_d A \tag{26.3}$$

This result is adequate if the charge flow is uniform as we have depicted in Figure 26.3. In many circumstances, however, the charge flow through an area varies with position across that area. This situation is depicted in Figure 26.4 where we have drawn lines to indicate the path of individual charge carriers through an imaginary surface S. There are more lines per area passing through the area near the top of the surface than toward the bottom. You already have some experience with definitions that describe how much of something passes through a surface, namely, the idea of *flux*. In fact, if you look back to Figure 23.2 in which we first defined *electric* flux, you will discover that Figures 26.4 and 23.2 are almost identical! It seems natural, therefore, to define the flow of charge in terms of current per unit area, which we designate with the symbol J:

$$J = \frac{I}{A} = nqv_d \tag{26.4}$$

If we assign the direction of J to be that of the positive charge flow, J becomes a vector called the **current density** that has units of amperes per square meter. From knowledge of $\mathbf{J}$ we can compute the current *through* a surface. Figure 26.5 shows the current density vector passing through a small area element ΔA. This area element is chosen small enough that $\mathbf{J}$ can be taken as constant over its surface. The current through ΔA is

$$I = J \cos(\theta) \, \Delta A$$

The component of charge flow parallel to a surface does not contribute to the current through the surface. We may write this last equation as

$$I = \mathbf{J} \cdot \Delta \mathbf{A}$$

Now, we suspect you know what is coming next; to obtain the current through the entire surface we must sum the contributions of $\mathbf{J}$ through all the surfaces that together compose surface S:

$$I = \int_S \mathbf{J} \cdot d\mathbf{A} \tag{26.5}$$

This prescription for calculating total current is analogous to the definition of electric flux. Whereas the electric flux is the sum of the normal component of the electric field through a surface, the current is the sum of the normal component of the current density through a surface.

FIGURE 26.4

Charge flow of nonuniform density through a surface. The paths of the charge carriers are indicated by the arrows. More charge carriers per area pass through the upper portion of the surface than through the lower region.

FIGURE 26.5

The current density $\mathbf{J}$ through a small area element $\Delta \mathbf{A}$.

EXAMPLE 26.1 *Charge Ahead!*

The conducting bar shown in Figure 26.6 is made of a nonuniform material so that the current density varies across the bar according to the relation

$$\mathbf{J} = \alpha(1 - \beta y^2)\hat{\mathbf{k}}$$

The values of the constants are $\alpha = 1.20 \times 10^4$ A/m^2 and $\beta = 9.00 \times 10^3$ m^{-2}. The bar has a square cross section, and the length of one of its sides is $L = 1.00$ cm. Find the total current I flowing along the bar.

SOLUTION As shown in Figure 26.6 a small area element through which $\mathbf{J}$ is approximately constant is given by

$$d\mathbf{A} = L\,dy\,\hat{\mathbf{k}}$$

Substituting the expression for $\mathbf{J}$ and this area differential into Equation (26.5), we have

$$I = \int_0^L [\alpha(1 - \beta y^2)\hat{\mathbf{k}}] \cdot (L\,dy\,\hat{\mathbf{k}}) = \alpha L \int_0^L (1 - \beta y^2)\,dy = \alpha L \left(y - \frac{\beta}{3}y^3\right)_0^L$$

$$= \alpha L^2 \left(1 - \frac{\beta}{3}L^2\right)$$

Substituting for α, β, and L, we have

$$I = (1.20 \times 10^4 \text{ A/m}^2)(0.010 \text{ m})^2 \left[1 - \frac{(9.00 \times 10^3 \text{ m}^{-2})}{3}(0.010 \text{ m})^2\right] = 0.840 \text{ A}$$

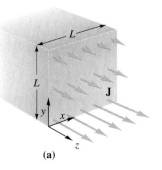

(a)

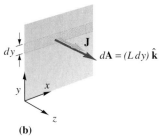

(b)

FIGURE 26.6

A nonuniform current through a conductor of square cross-section.

Concept Question 1
If the radius of a cylindrical conductor doubles, by what factor must the current density change in order to maintain the same current along the axis of the cylinder?

EXAMPLE 26.2 *As Bad as Waiting for Godot*

Assume each atom in a copper wire contributes one conduction electron. Compute the drift velocity of an electron if the wire carries a current of 15.0 A through a cross-sectional area of 3.30×10^{-6} m^2. This area is that of the 12-gauge wire commonly used in most household wiring. The density of copper is $\rho = 8.96$ g/cm^3.

SOLUTION The atomic mass of copper is $m = 63.5$ g/mol, and the number of atoms in this mass is Avogadro's number $N_A = 6.02 \times 10^{23}$. The volume $\mathcal{V}$ occupied by N_A atoms is

$$\mathcal{V} = \frac{m}{\rho} = \frac{6.35 \times 10^{-2} \text{ kg}}{8.96 \times 10^3 \text{ kg/m}^3} = 7.09 \times 10^{-6} \text{ m}^3$$

For one free electron per atom, the number of electrons per unit volume is

$$n = \frac{N_A}{\mathcal{V}} = \frac{6.02 \times 10^{23}}{7.09 \times 10^{-6} \text{ m}^3} = 8.49 \times 10^{28} \text{ electrons/m}^3$$

We assume the electron flow is uniform, so from Equation (26.5) we have

$$I = \int_S \mathbf{J} \cdot d\mathbf{A} = JA$$

Substituting from Equation (26.4), $J = nev_d$, we have

$$v_d = \frac{I}{neA} = \frac{15.0 \text{ A}}{(8.49 \times 10^{28} \text{ electrons/m}^3)(1.60 \times 10^{-19} \text{ C})(3.30 \times 10^{-6} \text{ m}^2)}$$
$$= 3.34 \times 10^{-4} \text{ m/s}$$

At this rate, an electron needs almost 51 min to travel 1 m!

Concept Question 2
Explain why we have been considering nonzero E-fields within conductors in this section whereas in the previous chapters we required E to be zero within a conductor.

The solution to Example 26.2 may have you wondering why you don't have to wait several hours for the light to come on when you throw the switch. Although the drift velocity of the electrons that constitute a current is small, their effect is felt almost instantaneously. In order to start the electrons moving we must apply an electric field. (We have often supposed that a battery has supplied the necessary potential difference to establish this field.) In Chapter 34 we will discover that the effect of an applied electric field travels at the speed of light, 3×10^8 m/s. Hence, the electrons at the position of the light bulb feel the applied electric field almost immediately after you flip the switch. The situation is analogous to a garden hose full of water; as soon as you turn on the faucet, water starts to run out the other end.

There is another startling feature about the small value of v_d. The magnitude of the random velocities of the virtually free electrons is on the order of 10^{+4} m/s. Hence, the magnitude of the drift velocity is very much smaller than the thermal velocity magnitudes. It may seem remarkable to you that such a small additional velocity as that of Example 26.1 can result in anything at all happening, especially a light bulb illuminating when you turn on the switch. Keep in mind, however, that there are a vast number of electrons all with this average drift velocity.

26.2 Resistivity and Resistance

When a potential difference $V_b - V_a = \Delta V$ is applied across a small segment of a conductor, an electric field **E** is established in that conductor (Fig. 26.7). Under the influence of this field the charge carriers within the conductor begin to move, and a current density is established. For many conductors we find by experiment that the resulting current density **J** is proportional to the applied field **E**. This result is called **Ohm's law,** after the German physicist Georg Ohm (1787–1854), and can be written

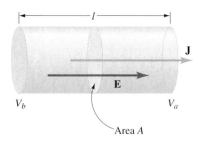

FIGURE 26.7

When a potential difference $\Delta V = V_b - V_a$ is applied across a conductor, the resulting electric field causes a current to begin to flow in the conductor. For a uniform current I through cross-sectional area A the current density is $J = I/A$.

$$\mathbf{J} = \sigma\mathbf{E} \tag{26.6}$$

The proportionality constant σ is called the **conductivity** of the conductor. Be sure not to confuse this use of the symbol σ with our previous definition of surface charge density. It is common practice to use σ for both definitions, so you must be certain of the context in which this symbol is employed.

Materials that obey Equation (26.6) are called **ohmic.** Not all materials obey the linear relation between **J** and **E** expressed by Equation (26.6). Materials with current densities not directly proportional to the magnitude of the applied field **E** are called **nonohmic.**

Sometimes, it is more convenient to restate Equation (26.6) in terms of the **resistivity** ρ, which is defined as the reciprocal of the conductivity:

$$\rho = \frac{1}{\sigma}$$

With this definition Equation (26.6) becomes $\mathbf{J} = 1/\rho\,\mathbf{E}$. You guessed it; here comes another caution! Be careful not to confuse the symbol ρ for resistivity with the same symbol we use for volume charge density; again, it's tradition to use this same symbol for both. The SI unit of ρ is $(V/m)/(A \cdot m^2) = (V/A)$ m. The ratio V/A is defined as an **ohm** and designated by the symbol Ω:

$$1\,\Omega = 1\,\frac{V}{A}$$

Hence, the SI unit of resistivity is the ohm meter ($\Omega \cdot$ m). Table 26.1 lists the resistivity for some common materials. Notice that the values of resistivity extend over a factor of about

10^{25}. That this range is very large is understandable if you remember that a perfect conductor has a zero resistivity and a perfect insulator has an infinite resistivity. Materials are often classified as conductors or insulators depending on the value of their resistivities. **Semiconductors** are materials with resistivities in a range of values intermediate to that of conductors and insulators.

TABLE 26.1   Resistivities of Some Materials at 20°C

MATERIAL	RESISTIVITY ($\Omega \cdot$ m)	TEMPERATURE COEFFICIENT ($°C^{-1}$)
Conductors:		
Aluminum	2.8×10^{-8}	3.9×10^{-3}
Copper	1.7×10^{-8}	3.9×10^{-3}
Gold	2.4×10^{-8}	3.4×10^{-3}
Iron	$10. \times 10^{-8}$	5.0×10^{-3}
Lead	$22. \times 10^{-8}$	3.9×10^{-3}
Mercury	$96. \times 10^{-8}$	0.9×10^{-3}
Nichrome (a Ni-Cr alloy)	100×10^{-8}	0.4×10^{-3}
Platinum	$10. \times 10^{-8}$	3.0×10^{-3}
Silver	1.6×10^{-8}	3.8×10^{-3}
Tungsten	5.6×10^{-8}	4.5×10^{-3}
Carbon	3.5×10^{-5}	-0.5×10^{-3}
Semiconductors:		
Germanium	0.45	-4.8×10^{-2}
Silicon	640	-0.5×10^{-2}
Insulators:		
Glass	10^{10}–10^{14}	
Quartz (fused)	7.5×10^{17}	
Rubber	10^{13}–10^{16}	
Sulfur	10^{15}	

Resistance

Equation (26.6) is an empirical relation between the current density **J** through an ohmic conductor and the electric field **E** within that conductor. In many applications it is more useful to know the relation between the potential difference ΔV applied across a conductor, such as the conducting segment shown in Figure 26.7, and the resulting current I. The relation between the ΔV and **E** is given by the master equation of Chapter 25:

$$V_b - V_a = \Delta V = -\int_a^b \mathbf{E} \cdot d\mathbf{l} \qquad (26.7)$$

If the segment of the conductor is uniform in nature, the E-field is constant. On several occasions (see Eq. (24.13), for example) we have seen that for constant **E** Equation (26.7) yields

$$\Delta V = E(b - a) = El \qquad (26.8)$$

where l is the length of the segment. In terms of ΔV the expression for current density (Equation (26.6)) becomes

$$J = \sigma \frac{\Delta V}{l}$$

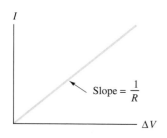

FIGURE 26.8

The current versus voltage for a resistor constructed from an ohmic material. The curve is linear.

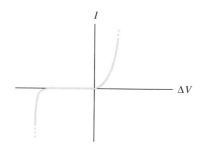

FIGURE 26.9

The current versus voltage for a zener diode. These devices are often used to establish a constant voltage somewhere in an electronic circuit. From this I–V curve you should be able to see that when a sufficient negative voltage is applied across the zener, the current flows easily through it.

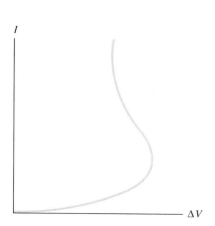

FIGURE 26.10

A typical current-versus-voltage curve for a thermistor. This device is often used in temperature-measuring instruments.

or

$$J = \frac{1}{\rho} \frac{\Delta V}{l} \tag{26.9}$$

Because the electric field **E** is uniform, the current density $J = (1/\rho)E$ is also uniform and equal to I/A, where A is the cross-sectional area of the conducting segment. When we substitute this result into Equation (26.9), we have

$$\frac{I}{A} = \frac{1}{\rho} \frac{\Delta V}{l}$$

or

$$\Delta V = \left(\frac{\rho l}{A}\right) I \tag{26.10}$$

We define the **resistance** R of a particular conductor as

$$R = \frac{\rho l}{A} \tag{26.11}$$

so that Equation (26.10) becomes

$$\Delta V = IR \tag{26.12a}$$

or

$$I = \frac{\Delta V}{R} \tag{26.12b}$$

This equation is a consequence of the empirical relation Equation (26.6), Ohm's law. Notice that the unit of resistance $R = \Delta V/I$ is the ohm: $V/A = \Omega$. Equations (26.12) tell us that a potential difference of 1 V across a resistance of 1 Ω gives rise to a current of 1 A. It is important to realize that Equations (26.12) are valid even for nonohmic materials: the ratio of an applied potential difference ΔV to the resulting current I is the resistance R *for that potential difference*. Also notice from Equation (26.11) that the resistance R of a material with resistivity ρ depends only on the magnitude of ρ for that material and the geometrical factors of the length l and the area A.

Unfortunately, it has become customary in electric circuit analysis to call Equations (26.12) Ohm's law. We do not follow this practice for two reasons: First, the equation $\Delta V = IR$ is sometimes used to *define* R even for nonohmic devices. That is, for a current I through a nonohmic material with a potential difference ΔV applied across it, the resistance for that potential difference is defined to be $\Delta V/I$. Second, Ohm's "law" is really an empirical description of how *some* materials behave when subjected to an electric field.

Figure 26.8 shows the current through a long copper wire as a function of applied potential difference. The slope of this line is constant and equal to $1/R$. If the relation between the current flow through a material and the applied potential difference is not a linear function, the material can, in fact, still be useful in electric circuits. Figure 26.9 shows the I-versus-ΔV curve for a zener diode often used to create a reference potential. Figure 26.10 shows I versus ΔV for a thermistor, a device with a resistance that is particularly sensitive to temperature variations. Notice from Figure 26.10 that for a range of potential differences *two* different currents are possible!

EXAMPLE 26.3   *Die Hard Music*

A certain car stereo system draws a current of 400. mA when connected to a 12.0-V battery. (a) What is the resistance of the stereo system? (b) The stereo is left playing from the battery for several hours while the engine is off. Finally, the battery voltage begins to drop. The radio can continue to operate until the current drops to 320. mA. At what battery voltage does the stereo stop playing?

SOLUTION   (a) We model the radio as a resistance R. If we know the current I through a resistance R when a potential difference ΔV is applied across it we can use Equation (26.12) to compute the resistance:

$$R = \frac{\Delta V}{I} = \frac{12.0 \text{ V}}{400. \times 10^{-3} \text{ A}} = 30.0 \ \Omega$$

(b) If we assume the resistance of the radio remains 30.0 Ω, when 320. mA flows through it, the applied potential difference must be

$$\Delta V = IR = (320. \times 10^{-3} \text{ A})(30.0 \ \Omega) = 9.60 \text{ V}$$

So in this simple model the radio continues to play until the potential difference across the battery terminal drops below 9.60 V.   ◀

EXAMPLE 26.4   *It's Not Music, But It Is Hard Rock*

Geophysicists use the experimental arrangement shown in Figure 26.11 to measure the resistivity of a mineral. Often, the electrodes are constructed by dipping the ends of the mineral sample into molten solder. For a given current I the potential difference is measured across a distance L. The diameter of a cylindrical sample of volcanic rock is 5.00 cm, and the length L is 4.00 cm. When the voltmeter indicates a potential difference $\Delta V = 200.$ V, the current is 3.30 mA. What is the resistivity of the volcanic rock?

SOLUTION   From Equation (26.11) the resistivity is given by

$$\rho = R \frac{A}{L}$$

But $R = \Delta V/I$ so

$$\rho = \frac{\Delta V A}{IL} = \frac{\Delta V \pi (D/2)^2}{IL} = \frac{(200. \text{ V})\pi(0.025 \text{ m})^2}{(3.30 \times 10^{-3} \text{ A})(0.040 \text{ m})} = 2.97 \times 10^3 \ \Omega \cdot \text{m}$$
$$= 2.97 \text{ k}\Omega \cdot \text{m}$$ ◀

In electric circuits it is common to use **carbon film resistors.** This type of resistor is made by depositing a thin film of carbon over a cylindrical insulator. Wires are attached at the ends so that current flows through the film parallel to the axis of the cylinder. Such resistors can be made with resistance values ranging from a few ohms to several million ohms. The joule heating of such resistors is generally limited to $\frac{1}{4}$ or $\frac{1}{2}$ W. Power dissipated much larger than the rated value causes the resistor film to deteriorate (often releasing smoke and an unpleasant odor). For applications where higher power dissipation is required, wire-wound and **composition resistors** are used. Composition resistors are made from carbon granules bonded together by resin and encased in a cylindrical phenolic jacket. Wire-wound and composition resistors may have power ratings of several watts.

Concept Question 3
When a constant potential difference ΔV is applied across a cylindrical resistance, what effect does doubling the radius of the cylinder have on the current flowing through it if its length remains unchanged?

Concept Question 4
Explain the difference between *resistance* and *resistivity*.

Concept Question 5
Two wires made from the same material have the same resistance. However, one wire has twice the diameter as the other. How do their lengths compare?

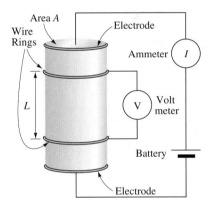

FIGURE 26.11

Measuring the resistivity of a cylindrical sample of rock.

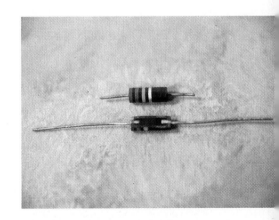

The inside of a composition resistor. The carbon granules are encased in a phenolic housing.

There is a color-code scheme from which the particular resistance of a resistor is indicated. Three colored stripes around the resistor are sufficient to designate its resistance. Each color has an associated digit in the sequence of digits that form the value of the resistance. The first colored stripe gives you the first digit, the second colored stripe gives the second digit, and the third colored stripe provides the **multiplier.** This multiplier is just the power of ten you use to multiply the two-digit number you have determined from the first two stripes. Table 26.2 provides the digits associated with each color, and an example is shown in Figure 26.12. There are various mnemonic devices used to remember this code. Our favorite is ''Bad Bees Ravage Our Yellow Geraniums But Violets Grow Wildly.''

The fourth band on a resistor indicates its tolerance, that is, its accuracy. The most commonly used resistors are reliable to within ±5% of their indicated value.

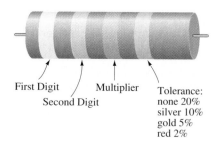

First Digit / Multiplier
Second Digit

Tolerance:
none 20%
silver 10%
gold 5%
red 2%

FIGURE 26.12

A 4700-Ω resistor with a 10% tolerance. The yellow stripe designates the first digit as a 4, the violet stripe indicates that the second digit is a 7, and the third color is red. The color red stands for a 2, so we multiply 47 by 10^2 to obtain 4700 Ω.

TABLE 26.2   The Resistor Color Code

DIGIT	COLOR	MULTIPLIER
	silver	10^{-2} (when used as the third band)
	gold	10^{-1} (when used as the third band)
0	black	1
1	brown	10^1
2	red	10^2
3	orange	10^3
4	yellow	10^4
5	green	10^5
6	blue	10^6
7	violet	10^7
8	gray	10^8
9	white	10^9

Tolerance: ±10% silver band (when used as the fourth band)
              ±5% gold band (when used as the fourth band)

Temperature Dependence of Resistivity (Optional)

The vibrational energy of the lattice increases with temperature; that is to say, the number of phonons increases. As a result the number of electron collisions with phonons increases with increasing temperature. Therefore, the resistivity of metals increases with temperature. For limited temperature variations the resistivity may often be approximated by a linear function of temperature. Such a function can be written in the form

$$\rho = \rho_o[1 + \alpha_{av}(T - T_o)] \tag{26.13}$$

where ρ_o is the resistivity at temperature T_o. This functional form is illustrated in Figure 26.13. The empirically found coefficient α_{av} is the **mean temperature coefficient of resistivity:**

$$\alpha_{av} = \frac{1}{\rho_o}\frac{\Delta\rho}{\Delta T}$$

Because the resistance of a conductor is proportional to its resistivity (Eq. (26.11)), the temperature dependence of a resistor over a limited temperature range can be written

$$R = R_o[1 + \alpha_{av}(T - T_o)] \tag{26.14}$$

where R_o is the resistance at temperature T_o.

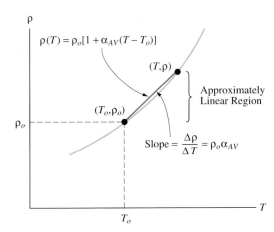

FIGURE 26.13

Over a limited range of temperatures the resistivity can be approximated by a linear function.

Over extended temperature ranges Equation (26.13) does not accurately describe the temperature variation of ρ. For example, Figure 26.14 shows the temperature dependence of ρ for copper. As the temperature approaches zero Kelvin, the resistivity approaches a finite value ρ_i. At low temperatures few phonons exist within the metal, and electrons are scattered primarily by impurity atoms and static defects within the metal lattice. At any temperature the total resistivity is the sum of that due to electron scattering from the phonons ρ_p and a contribution from impurity and defect scattering ρ_i: $\rho = \rho_p + \rho_i$. The value of ρ_i depends on the purity and mechanical treatment of the metal. There are models in *solid-state physics* that correctly predict the temperature dependence of the resistivity for most metals over a wide range of temperatures.

The density of charge carriers for *semiconductors* increases with temperature faster than the effect of scattering due to phonons, so that these materials generally decrease in resistance when their temperature is raised. Hence, in Equation (26.13) $\alpha_{av} < 0$; the mean temperature coefficient is negative for semiconductors. Certain types of impurities can play a dominant role in determining the resistivity of a semiconductor. The resistivity of a semiconductor can be controlled quite accurately by the deliberate addition of particular impurities that alter the number of charge carriers. This phenomenon is described in Section 26.4.

When the temperature of a **superconducting** material is brought below a certain temperature T_c, its resistivity completely vanishes. These types of materials are also described in Section 26.4.

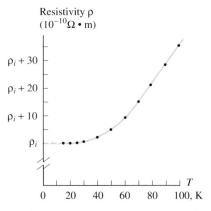

FIGURE 26.14

The resistivity of copper for temperatures below 100 K. At low temperatures the resistivity approaches a value ρ_i that depends on the nature and concentration of the impurities in the particular specimen of copper.

26.3 Energy Dissipation

In this section we describe a model for what happens when we attach a resistance across the terminals of a battery. Rather than considering the forces of the charge carriers we utilize the energy picture. The symbol for a resistor is shown in Figure 26.15, and the circuit diagram for our battery and resistor is shown in Figure 26.16. For now we assume that the resistances of the wires connecting the resistor to the battery are zero. The electric potential of the negative battery terminal is arbitrarily assigned a value of zero. By chemical means the battery takes charge dq entering at the low-potential terminal and performs work on it so that upon reaching the higher potential (positive) terminal the charge dq has gained potential energy dU in the amount

$$dU = \Delta V \, dq \qquad (26.15)$$

where ΔV is the electric potential of the battery. From this potential energy, on leaving the positive terminal the charge carriers are accelerated by the electric field and gain kinetic energy. However, as the carriers move through the resistor they collide with phonons, impurities, or defects within the resistor material, giving up part of their kinetic energy to lattice vibrations (phonons). This increase in lattice vibrational energy manifests itself as

FIGURE 26.15

The circuit symbol for a resistor.

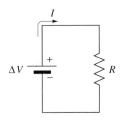

FIGURE 26.16

A simple circuit consisting of a battery of voltage ΔV and a resistor of resistance R. Current flows in the clockwise direction.

an increase in the temperature of the resistor. Hence, through a two-step process, the chemical potential energy of the battery is converted to kinetic energy of the charge carriers and then to energy of the lattice. The net effect is that the energy supplied by the battery is transformed into internal energy of the resistor. Because this increase in internal energy causes the resistor to become warmer, we occasionally refer to this internal energy as thermal energy.

When the charge carriers exit from the resistor, they are at the site of the low-potential terminal of the battery and have lost all the potential energy they had at the high-potential terminal. But, as long as the battery has chemical energy available, the process can be continued at a steady rate. Therefore, for a steady current through the resistor, the rate at which energy is supplied by the battery is equal to the rate at which the thermal energy is generated in the resistor. From Equation (26.15) we have

$$\frac{dU}{dt} = \Delta V \frac{dq}{dt}$$

Now, $dq/dt = I$, the current through the resistor, and the rate at which energy is supplied by the battery is equal to the rate it is converted to internal energy in the resistor. This latter rate is just the power loss $\mathcal{P} = dU/dt$, so our equation becomes

$$\mathcal{P} = I \Delta V \qquad (26.16)$$

We may apply $\Delta V = IR$ to obtain two alternative forms of Equation (26.16):

$$\mathcal{P} = I^2 R = \frac{\Delta V^2}{R} \qquad (26.17)$$

Concept Question 6
The relation $\mathcal{P} = (\Delta V^2)/R$ appears to suggest that increasing the resistance of a resistor decreases the power dissipated in it, whereas the relation $\mathcal{P} = I^2 R$ seems to suggest the opposite. Reconcile these two equations.

In Equations (26.16) and (26.17) I is in amperes, ΔV is in volts, R is in ohms, and the SI unit of power is the watt (W). Finally, we should mention that this power loss is sometimes called **joule heating,** or I^2R-loss.

EXAMPLE 26.5   An Enlightening Example

What current do you expect to flow through a 6.00-W automobile taillight bulb if you connected it to a 12.0-V battery? What is the resistance of the bulb?

SOLUTION   From Equation (26.16) we have

$$I = \frac{\mathcal{P}}{\Delta V} = \frac{6.00 \text{ W}}{12.0 \text{ V}} = 0.500 \text{ A}$$

and from Equation (26.17) we find

$$R = \frac{\Delta V^2}{\mathcal{P}} = \frac{(12.0 \text{ V})^2}{6.00 \text{ W}} = 24.0 \text{ } \Omega$$

This resistance is much higher than the resistance of a cold light bulb filament, which is a fraction of an ohm. The resistance of a light bulb has a huge temperature dependence and, therefore, increases dramatically as the bulb heats up. The filament of the light bulb is definitely nonohmic, yet the relation $\Delta V = IR$ may be applied. ◀

26.4 Microscopic Models of Resistance (Optional)

We will present two complementary models of conductivity. The first, the valence-bonding model, is intended to provide a visualization of how charge carriers actually move through a lattice of ions. The second, the band model, employs the energy approach and is important because from it we can immediately see what makes one material a conductor, another an insulator, and still others semiconductors. We begin with the valence-bonding picture.

Valence-Bonding Model of Conduction

We have represented a metallic conductor as a lattice of positive ions with about one or two virtually-free electrons per atom traveling around within the material. In this model there are two important ways that the electrical conductivity is influenced. We have already noted that electron collisions with phonons transform the kinetic energy of the electrons into internal energy of the lattice. The picture you should have in your mind is a fairly high density of electrons (again, one or two per ion) dashing around with average thermal velocities on the order of 10^4 m/s. It's not hard to see that we should expect the electrons to collide frequently with phonons. However, it also seems logical that the electrons should collide with each other. At room temperature electron–electron collisions turn out to have little influence on electrical conductivity because of the very same principle that dictates how the electron energy levels (described in the next section) are filled, namely, *the Pauli exclusion principle.* This rule places sufficient restrictions on the behavior of the most energetic electrons that the mean distance traveled by an electron between electron–electron collisions is about ten times that between electron–phonon collisions. Hence, in metals electron–phonon collisions dominate the electrical conductivity at room temperature. At low temperatures electron–electron collisions become more important, but, as we described in Section 26.3, collisions with impurity atoms dominate conductivity and result in a finite value for ρ as $T \rightarrow 0$ K.

Let's now turn to semiconductors. In **semiconductors,** such as silicon and germanium, all of the valence electrons are involved in creating bonds between neighboring atoms. This situation is similar to the case for an **insulator** in that, based on the number of chemical bonds, there are no free electrons. In an insulator the bonds are strong enough that there is essentially no electronic conduction. In the case of a semiconductor, however, the bonds are somewhat weaker, and a very few electrons (perhaps only 1 in 10^{12} at room temperature) are shaken loose from their bonds by thermal vibrations and roam about the

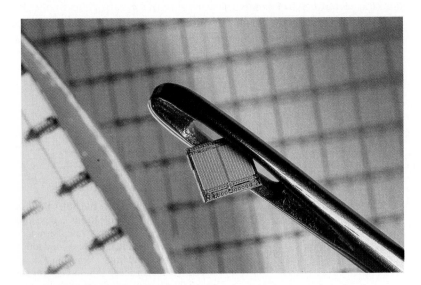

An integrated circuit capable of performing millions of electronic functions fits through the eye of a needle. Such circuits function due to the electronic properties of semiconductors such as silicon.

lattice. A two-dimensional picture of this process is illustrated in Figure 26.17 for a lattice of silicon atoms. You should realize that this is only a schematic representation because, not only are silicon atoms not flat, but their bonds are not restricted to two dimensions! We have squashed all the bonds into a plane to provide you with a rough idea of what is happening in three dimensions.

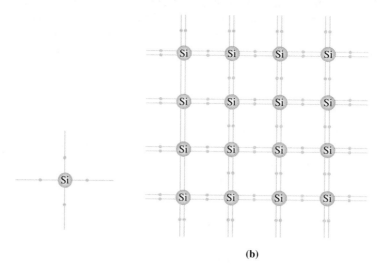

(b)

FIGURE 26.17

(a) A two-dimensional representation of a silicon ion (Si) and the bonds formed by its valence electrons —•. (b) A two-dimensional portrayal of a silicon lattice. Two electrons complete each of four bonds to neighboring ions.

Each silicon atom has four valence electrons to form bonds with the valence electrons of four neighbors. When a thermal vibration of the lattice kicks an electron out of one of these bonds, it is free to travel through the lattice as a conduction electron. However, this electron leaves behind a vacancy in the bond structure. This vacancy is called a **hole.** As illustrated in Figure 26.18, a neighboring valence electron can, by a slight nudge from the thermal vibrations of the lattice, hop over into the vacant position. By this process the hole can move around the lattice. In fact, because the presence of a hole leaves its associated ion with a net positive charge, hole motion is equivalent to a conduction process by a positive charge!

To illustrate this point imagine that we apply an electric field toward the upper right in Figure 26.18. If the magnitude of this field is sufficient to remove an electron from its valence state, the electron is accelerated toward the lower left of the figure. But other electrons also feel the force to the lower left and are attracted toward the position of the hole. When an electron moves to the lower left to fill the hole, the hole moves toward the

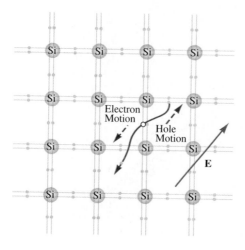

FIGURE 26.18

The motion of electrons and holes through a silicon lattice. A few electrons can be knocked out of bonding positions by the thermal energy of the lattice, providing electrical conduction by both electrons and holes. When an E-field is applied to the lattice, holes and electrons move parallel and antiparallel, respectively, to the field.

upper right, along the field direction. The net effect of all this motion is a small current consisting of negatively charged electrons going in one direction and positively charged holes traveling in the other.

The process of **doping** can greatly enhance the conductivity of semiconductors. Arsenic ions have five valence electrons as illustrated in Figure 26.19(a). When an arsenic ion is introduced into the silicon lattice, one electron is left over from the bonding. This electron is fairly easy to remove from the vicinity of the arsenic atom and becomes a conduction electron. Notice that there is no hole left behind. This means that the conductivity of an arsenic-doped silicon semiconductor is dominated by electrons. Such materials are called ***n*-type** semiconductors.

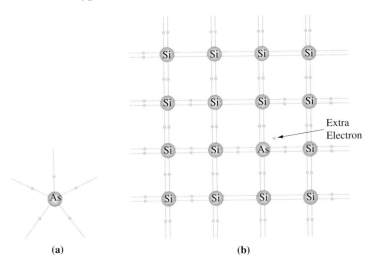

(a)   (b)

FIGURE 26.19

(a) A two-dimensional representation of an arsenic atom with its five valence electrons. (b) When an arsenic atom replaces a silicon atom, the extra electron contributes to the conduction process.

Aluminum atoms have three valence electrons. As shown in Figure 26.20, when aluminum is used as the doping atom, an extra hole is introduced into the silicon lattice. Materials for which the conductivity is dominated by hole carriers are called ***p*-type** semiconductors because the holes act as positive carriers. The transistor and other semiconductor devices are constructed by putting into contact various arrangements of *n*- and *p*-type semiconductors.

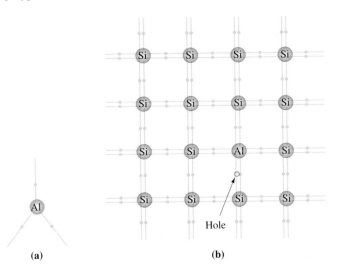

(a)   (b)

FIGURE 26.20

(a) A two-dimensional representation of an aluminum atom with its three valence electrons.
(b) When an aluminum atom replaces a silicon atom, one bond is left with a hole.

The Band Model of Conductivity

We hope you are impressed with the enormous range of the values for the resistivities listed in Table 26.1. The physical theory required to explain this immense range is remarkable. Indeed, a detailed application of **quantum mechanics** is required to provide a comprehensive model of the conductivity of all materials. Although you may not have a detailed knowledge of quantum mechanics, you probably know that the energy levels of the electrons in an atom are quantized. You may even have learned what "1s, 2s, 2p," and so on, are all about. If so, read on; we think you will find the next few paragraphs interesting.

Quantum mechanics places major restrictions on the behavior of electrons. They cannot roam wildly about the lattice "doing their own thing." In fact, only a carefully prescribed set of energies is available from which an electron can choose. Each characteristic energy level corresponds to a particular **quantum-mechanical state.** These levels are "allowed" by the solutions to the quantum-mechanical equations that describe the atom. One way to construct a chunk of a pure material is to assemble it from its constituent atoms. When you bring together two atoms of the same type to form a simple molecule, quantum mechanics predicts that each of the original (and identical) energy levels of the two atoms are split into two slightly different levels. Figure 26.21 provides a schematic

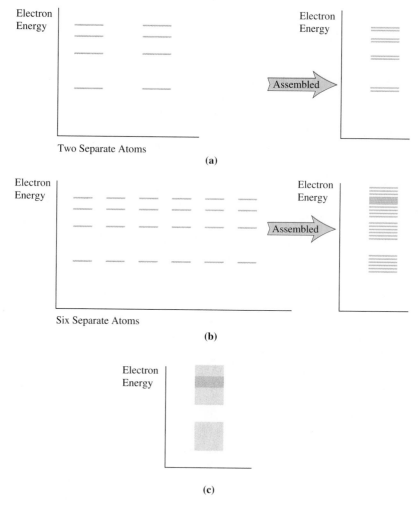

FIGURE 26.21

(a) When two atoms are brought together to form a molecule, the original electron energy levels of each atom split into two, closely spaced levels. (b) When six atoms of an element are assembled, each level splits into six levels. Some levels may overlap. (c) For $\sim 10^{23}$ atoms the levels form a virtual continuum of allowed electron energy states.

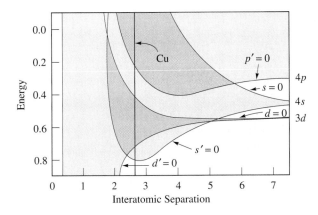

FIGURE 26.22

The band structure of copper as a function of interatomic distance d. The vertical line indicates the equilibrium separation of atoms in a copper crystal at room temperature and 1 atm of pressure. At this spacing several bands overlap, and there are no band gaps.

representation of this process for a two-atom collection, a group of six atoms, and the result for an N-atom collection. The important point to note is that the energy levels form regions on the energy plots where the separation (in energy units) between levels is very small. Hence, in a real solid ($N \approx 10^{23}$ atoms) there can be many energy levels so closely spaced that they form a virtual continuum of energy states called an **energy band.** At other energies, no energy states exist, creating **band gaps** between the bands. Another critical point to remember about energy bands is that the number of states in a band, although large, is finite and *exactly* countable. If N atoms come together, there might, for example, be exactly $4N$ states created in a particular band.

Figure 26.22 shows how the energy levels of copper are perturbed into bands as atoms are assembled to form a crystal of copper. We should mention that in addition to the process illustrated in Figure 26.22 there are other methods by which the band structure of a solid can be determined. These techniques begin with a lattice of ions and apply the methods of quantum mechanics to compute the allowed energies of the valence (outermost) electrons. Solid-state physicists have successfully explained many of the important and useful properties of materials based on their band structure. Furthermore, these scientists are able to speculate about the properties of materials before they are actually produced.

From knowledge of how electrons are distributed in the energy levels within the bands we are able to understand why a material is a conductor, an insulator, or a semiconductor. Electrons fill up the levels in the same manner they do in atoms: from the lowest energy upward. A fundamental tenet of quantum mechanics called the **Pauli exclusion principle** provides the reason the levels are filled in this manner. The exclusion principle states that only two electrons, each with an oppositely directed intrinsic angular momentum (called *spin*), can occupy the same energy level. Sometimes, the last filled level lies within a band, such as illustrated in Figure 26.23(a). If this is the case, the highest energy electrons can move to even higher energy states with very little increase in the electron's energy. On the other hand if the band is completely filled, no electron can move to a higher energy state unless it gains enough energy to jump the relatively large band gap.

In order to generate a current in a conductor, we establish an electric field within the conductor. From the classical model we presented in Section 26.1 an electron experiences a force of magnitude $F = eE$, is accelerated, and over time achieves an average drift velocity v_d. However, quantum mechanics tells us that the electron can be accelerated to a new kinetic energy level only if there are higher energy states available. When the highest energy band is only partially filled, this is no problem because unoccupied states above those of the most energetic electrons are available for occupancy. Hence, materials in which the highest energy bands are only partially filled are conductors.

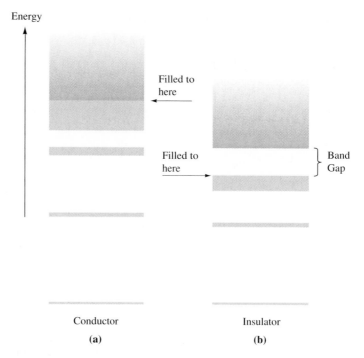

Conductor

(a)

Insulator

(b)

FIGURE 26.23

(a) In a conductor the most energetic electrons only partially fill an energy band and, therefore, can easily move to higher energy states. (b) In an insulator the electrons just complete the filling of a band, and there is a large energy gap between the top of this band and the band containing unfilled energy states. Notice in each case that lower energies, which correspond to those of electrons deep in the metallic ions, still appear as narrow energy levels.

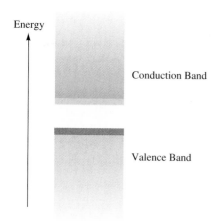

FIGURE 26.24

Energy bands of a semiconductor. Electrons at the top of the valence band are thermally excited into the conduction band. This transfer makes higher energy states available to the conduction electrons and leaves vacant states at the top of the valence band.

If the last energy state filled by the electrons of a material is at the top of a band and there is a substantial band gap between this band and the next higher one, the material is an insulator. This situation is illustrated in Figure 26.23(b). Electric fields of ordinary magnitudes are insufficient to give an electron at the top of the filled band enough energy to jump the energy gap in to the higher energy band.

Semiconductors are materials, such as silicon and germanium, that are similar to insulators in that their electrons just complete the filling of a band. However, the band gap is small enough that the kinetic energy of a few electrons is sufficient to push them from this **valence band** into the next higher band, called the **conduction band.** These electrons then act as conduction electrons because they have available to them the higher energy states of the conduction band. The band structure for a semiconductor is illustrated in Figure 26.24.

The number of electrons able to jump across the band gap from the valence band into the conduction band depends on the size of the band gap relative to the available thermal energy. In silicon, for example, the band gap is 1.12 eV, whereas at room temperature the available average energy is only 0.025 eV.[1] Consequently, in *intrinsic* silicon only about 3 out of every 10^{12} atoms contributes an electron, resulting in a free electron density of $1.45 \times 10^{10}/cm^3$.

This free electron density can be changed dramatically by the process of **doping.** For example, a very small fraction of arsenic atoms can be introduced into a silicon lattice. As we showed in Figure 26.19, the arsenic atoms contain five valence electrons instead of the four like silicon, and the "extra" electrons are not used in the lattice bonds. Moreover, these electrons are only weakly bonded to the arsenic atoms and form energy levels only 0.054 eV below the conduction band. The energy levels of these electrons are sufficiently

---

[1] Recall in Section 24.2 we calculated 1.60×10^{-19} J = 1.0 eV.

close to the conduction band that the 0.025-eV thermal energy from the lattice is sufficient to ionize nearly all of the added arsenic atoms.

The atomic density of silicon is 5×10^{22} atoms/cm^3. Thus, if we add only *one* arsenic atom per *million* silicon atoms, we add 5×10^{16} atoms/cm^3. Because virtually all of these arsenic atoms contribute an electron to the conduction band, the electrons from the arsenic far, far outnumber the 1.45×10^{10} electrons contributed by the silicon itself. This vast increase in conduction electrons reduces the resistivity of the silicon by over a factor of a million!

A simple one-dimensional analogy might help your understanding the electron politics involved in band structures. Let's imagine that the electrons are people and that the quantum-mechanical states are robotic cars traveling along a highway. Each car can hold two people (but only if one sits upside down). Let's further suppose that there are exactly 100 cars and that the laws of quantum mechanics dictates that these cars can travel only at velocities that are *exactly* multiples of 1 mph. Hence, our 100 cars have velocities ± 1, ± 2, . . . , ± 50 mph.

The analog of a conductor is to have only 100 people riding on our highway (band). We can put these 100 people into the 50 slowest moving cars. There is no net flow of people (no current) because as many are traveling to the left as to the right. If we want to create a flow of people to the right, we can with relative ease, move a few folks to some of the 50 empty cars moving to the right at speeds of $+51$, $+52$ mph, and so forth, thus creating our flow of people.

The analog of an insulator is to have 200 people on our highway. In this case every car is full. There is simply no way to create an excess of people going to the right, because there are no vacant cars and we can't make any given car go faster.

To model a semiconductor, we must imagine a second highway parallel to our full highway. This second highway might contain 100 empty cars traveling at velocities of ± 100, ± 101, . . . , ± 150 mph. Now it is possible to obtain an excess people flow to the right, if we can get one of our left-moving passengers from the full highway (valence band) to gain enough kinetic energy to hop into one of the empty cars (conduction band). If one of our passengers makes it, we will have an extra person moving to the right, *and one less person moving to the left*. Thus, the difference between people moving right and people moving left is $51 - 49 = 2$. The empty car is the analog of a hole.

Now, you shouldn't take this somewhat whimsical analogy too seriously, but perhaps it will aid your intuition when thinking about the strange quantum-mechanical world experienced by electrons inside of a crystalline solid.

Superconductors

In 1911, the Dutch physicist Heike Kamerlingh-Onnes (1853–1926) discovered that below a certain very low temperature the resistivity of mercury becomes immeasurably small (Fig. 26.25). Since that time, researchers have discovered many metals and alloys with resistivities that completely vanish when their temperature is reduced below a temperature called the **transition** (or **critical**) **temperature** T_c. The transition temperatures for several metals and alloys are shown in Table 26.3. Superconductors are of great technological interest because they hold the promise of many new devices and applications, such as very fast computer switches; low-cost, powerful magnets for a whole range of uses; and more efficient transmission of electricity through superconducting power lines, to name just a few. At present the problem with using the materials listed in Table 26.3 for these applications is that we must use liquid helium to cool them into the superconducting state. Liquid helium is expensive and supplies are limited.

A theory that is able to explain the superconducting properties of the materials listed in Table 26.3, as well as many other superconductors, was provided by John Bardeen (1908–), Leon Cooper (1930–), and Robert Schrieffer (1931–) in 1957. This theory, now known as the BCS theory, is based on the quantum-mechanical behavior of the conduction electrons as they travel through the metallic lattice. There is no classical explanation of superconductivity. We can tell you that a small band gap is required and the primary

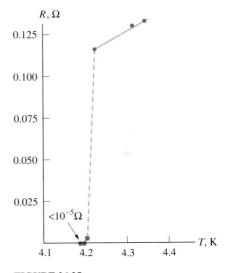

FIGURE 26.25

The resistance of mercury near its critical temperature T_c.

TABLE **26.3**   The Transition Temperature for Some Superconducting Materials

MATERIAL*	T_c (K)
Al	1.20
Pb	7.19
Hg	4.15
Nb	9.26
Sn	3.72
Zn	0.85
Nb$_3$Ge	23.2
Nb$_3$Sn	18.05
NbSn$_2$	2.60
PbTl$_{0.27}$	6.43

*There are many other superconducting elements and alloys that are not listed in this table.

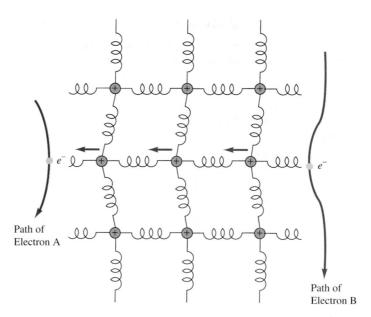

FIGURE 26.26

The origin of an attractive force between two electrons. The Coulomb attraction causes an ion to be displaced slightly toward a passing electron (electron A). Via the interatomic bonds, this displacement is transferred through the lattice and eventually another electron (electron B) is pulled slightly toward electron A as a result of the nearby ion's displacement.

interaction that is absolutely necessary for superconductivity to take place (according to BCS theory) is an *attractive* interaction between pairs of electrons (called **Cooper pairs**).

You might ask how it is possible for electrons to attract one another. Keep in mind that the electrons travel within a regular array of positive ions. As we illustrate in Figure 26.26, when an electron moves past an ion in one region of the lattice its coulomb attraction for a nearby ion can cause that ion to move slightly toward the electron's path. Through the bonds that connect that ion to its neighbors these neighboring ions feel this movement as a tug. The resulting slight displacement is transferred through the lattice to a point where the movement of another positive ion can cause a nearby electron to vary its path in the direction of the displacement. Notice that the path change of this latter electron is toward the original electron. Hence, via a vibration of the ionic lattice there arises an attractive force between electrons, and Cooper pairs can be established.

From the description in the previous paragraph it should be clear to you that as the temperature of the superconductor increases the random motions of the ionic lattice also grow, and the delicate correlated motion between the lattice and Cooper pairs is upset. The result, of course, is the disappearance at a certain temperature (T_c) of the associated superconductivity. On the other hand, if the interatomic bonds are sufficiently weak that the ionic displacements are easily transmitted across regions of the material, it is easier for Cooper pairs to be formed and remain at higher temperatures. Indeed, it turns out that the interatomic bonds of almost all superconductors show softening at certain vibrational frequencies. This observation, however, has led many solid-state physicists to be pessimistic about the possible discovery of superconductors with transition temperatures that are much above those in Table 26.3; for high-temperature superconductivity (on the order of room temperature) the bonds might have to be so weak that the material would simply fall apart. This thought is just another way of saying that the material does not exist.

In 1987, a superconductor with a transition temperature of 98 K was discovered. This temperature is remarkably higher than any of those in Table 26.3. The structure of this material is complex and we will not describe it here. Since 1987, several other materials with similar structures and even higher transition temperatures have been discovered. Even these new materials have transition temperatures below room temperature. Nonetheless, these discoveries are significant because liquid nitrogen has a 77 K temperature at 1 atm of pressure, and it may be used to cool these superconductors below the transition temperature. Liquid nitrogen is much cheaper and more abundant than the liquid helium required to cool conventional superconductors into the superconducting state.

At this time solid-state physicists are not certain of the mechanism that gives rise to the superconductivity in these high-T_c materials. It may be the electron–lattice–electron interaction we described previously, or it may be some completely different mechanism. Hopefully, when physicists discover the source of this type of superconductivity, they can design new materials with transition temperatures much closer or even above room temperature.

26.5 Summary

The electric current is the rate at which charge flows through a plane. When charge Δq travels past a point in time Δt, the average current is

$$I_{av} = \frac{\Delta q}{\Delta t} \tag{26.1}$$

The instantaneous current is given by

$$I = \frac{dq}{dt} \tag{26.2}$$

The unit of current is the ampere (A) and is defined as a coulomb per second. The direction of the current is along the direction of positive charge flow. In metals the charge carriers are electrons so that the current is in the opposite direction from the average electron velocity. The current density is the current per area flowing through a surface. If there are n charge carriers per volume, each carrying charge q, and the average drift velocity of the carriers is v_d, the magnitude of the current density is given by

$$J = nqv_d \tag{26.4}$$

The direction of the current density is also along the direction of the positive charge flow. The relation between the current and the current density is

$$I = \int_S \mathbf{J} \cdot d\mathbf{A} \tag{26.5}$$

The relation between the current density $\mathbf{J}$ and the applied electric field $\mathbf{E}$ is

$$\mathbf{J} = \sigma \mathbf{E} \tag{26.6}$$

The constant σ is called the **conductivity.** Materials for which σ is a constant are said to be *ohmic* and obey Ohm's law. The **resistivity** ρ is the reciprocal of the conductivity; that is, $\rho = 1/\sigma$.

For a uniform cross-sectional area and length l the **resistance** of a material is given by

$$R = \frac{\rho l}{A} \tag{26.11}$$

The resistance of a material is also the ratio of the potential difference across it to the current that flows through it:

$$\Delta V = IR \tag{26.12a}$$

The unit of resistance is the ohm: $1 \, \Omega = 1 \, \text{V/A}$.

The rate at which electric energy is dissipated as heat within a conductor that carries current I when a potential difference ΔV is maintained across it is

$$\mathcal{P} = I \, \Delta V \qquad (26.16)$$

For resistors of resistance R

$$\mathcal{P} = I^2 R = \frac{\Delta V I^2}{R} \qquad (26.17)$$

(Optional) The resistivity of a material changes with temperature. For small temperature changes where the relation between the resistivity and the temperature can be approximated by a linear function

$$\rho = \rho_o[1 + \alpha_{av}(T - T_o)] \qquad (26.13)$$

where α_{av} is the **mean temperature coefficient of resistivity** for the material.

PROBLEMS

26.1 Electric Current and Current Density

1. How many electrons pass a cross section through a metallic wire if the wire carries a current of 3.00 A for 5.00 min?
2. Find the current in a wire if 6.25×10^{16} electrons pass a point in the wire in 1.00 s.
3. The electron beam in the cathode-ray tube of a certain oscilloscope is to provide 2.00×10^{16} electrons to strike the screen in 1.00 min. What current must be supplied to form the electron beam?
4. In a certain electrolytic cell silver ions Ag^+ are deposited on one of the electrodes at a rate such that 0.100 g of silver accumulates during 1.00 h. What current flows through the electrolytic cell? (The mass number of silver is 108.)
5. Compute the drift velocity of electrons traveling through an 18-gauge copper wire (diameter = 1.02 mm) when the wire carries the maximum safe current of 3.00 A.
6. Current flows through the constricted conductor shown in Figure 26.P1. The diameter D_1 is 1.00 mm, and the current density to the left of the constriction is $J = 1.27 \times 10^6$ A/m². (a) What current flows into the constriction? (b) If the current density is doubled as it emerges from the right side of the constriction, what is the diameter D_2?

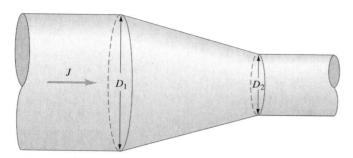

FIGURE 26.P1   Problem 6

7. The total charge that passes perpendicularly through a circular surface area of 1.00 cm diameter after a time t is given by $q(t) = (1.00 \text{ C/s}^2)t^2 - (2.00 \text{ C/s})t + 3.00$ C. (a) What is the instantaneous current through the circular area at $t = 2.00$ s? (b) What is the current density at this same instant?
8. A current density given by $\mathbf{J} = (2.00 \text{ A/m}^4)xy\hat{\mathbf{i}} + (3.00 \text{ A/m}^5)x^2y\hat{\mathbf{k}}$ flows through an area of the xy-plane bounded by the x- and y-axes and the lines $x = 8.00$ cm and $y = 12.0$ cm. Compute the current through this area.

26.2 Resistivity and Resistance

9. A cylindrical conductor with a diameter of 0.250 mm has a resistivity of 4.00×10^{-5} $\Omega \cdot$m. A uniform current of 0.500 A flows along the axis of the cylinder. What is the electric field within the conductor?
10. A cylindrical conductor with a length of 2.00 m and a diameter of 0.500 cm is found to carry a current of 4.60 A when 60.0 V is applied between its ends. (a) What is the resistance of this conductor? (b) What is the resistivity of the material from which it is made?
11. Copper wire with a diameter of 1.02 mm is wound around an insulating cylinder in a single layer with 300. turns. If the diameter of the cylinder is 2.40 cm, what is the resistance of the coil of copper wire?
12. When the voltage across a resistor increases by 20.0 V, the current through it increases by 5.00 A. What is the resistance of the resistor?
13. Two pieces of 14-gauge wire (diameter = 0.163 cm) each are 50.0 cm in length. One wire is made of aluminum, the other of tungsten; both are maintained at a temperature of 20.0°C. The wires are joined together at one end to form a single wire of 1.00-m length. A potential difference of 12.0 V is applied across the length of this compound wire. (a) What is the resistance of each 0.500-m segment? (b) What is the current in each? (c) What is the potential difference across each segment? (d) What is the electric field in each segment? (e) What is the current density in each?
14. A high-voltage transmission cable ($\rho = 2.2 \times 10^{-8}$ $\Omega \cdot$m) is 10.0 km in length and has a diameter of 2.00 cm. (a) What is the resistance of the cable? (b) What length of aluminum cable of the same diameter has the same resistance?

15. A 6.00-m long wire with a diameter of 1.60 mm has a resistance of 0.170 Ω. (a) What is the resistivity of this material? (b) Can you identify the material from Table 26.1? (c) What current flows through the wire when a potential difference of 6.00 V is applied across its ends?

16. A silver wire and an aluminum wire are both 12.0 m in length. The silver wire is 0.500 mm in diameter. What must be the diameter of the aluminum wire for it to have the same resistance as the silver wire? Assume both wires are at 20°C.

17. A specimen of the mineral molybdenite (MoS_2) has a resistivity of 2000. $\Omega \cdot m$. A cylinder of this material with a 2.40-cm radius is prepared for the experimental arrangement shown in Figure 26.11. What must the length L be for a current of 2.00 mA to flow through this mineral when the potential difference between the wire rings is 250. V?

18. A prankster borrows a spool of insulated copper wire, unwinds it, tangles it into a hopeless mess, and returns it to the physics lab. Because the insulation is a thin coat of lacquer paint of negligible mass, your physics teacher says that the wire's length can be found and instructs you to find it. With an *ohmmeter* attached to the two bare ends of the wire you find its total resistance to be 0.760 Ω. With a balance you find that the mass of the wire is 500. g. What is the wire's length?

19. By how much must the temperature of an iron wire be raised above 20.0°C for its resistance to be increased by 1.00%?

20. The resistance of a platinum wire at 20.0°C is 120. Ω. In order to measure the temperature of a hot plate the wire is placed between the plate and a piece of thermal insulation. The new (equilibrium) resistance of the wire is 156 Ω. What is the temperature of the hot plate?

21. By how much must the temperature of nichrome be increased above 20.0°C for its resistance to change by 0.500%?

22. Through what temperature change must you raise a length of tungsten wire to double its resistance?

23. When placed in an ice and water mixture at 0.0°C the resistance of a wire is 84.0 Ω, and when the wire is immersed in boiling water at 100.°C its resistance changes to 118 Ω. What is the mean temperature coefficient of resistivity for this material in this temperature range?

24. A 100.-W light bulb has a resistance of 10.0 Ω at a temperature of 20.0°C when it is not illuminated. When the bulb is illuminated, its resistance is about 150. Ω. Estimate the temperature of the bulb's filament by assuming its temperature coefficient is constant and equal to 5.00×10^{-3}°C^{-1}.

26.3 Energy Dissipation

25. What is the maximum potential difference you should apply across a 10.0-kΩ resistor rated at 0.500 W?

26. What resistance must you place between the terminals of a 6.00-V battery for 1.00 mW of power to be dissipated by the resistor?

27. How much heat is dissipated in a 500.-Ω resistor in 1.00 h when a current of 2.00 A is maintained through the resistor?

28. The resistance wire of an electric heater dissipates 1.50 kW of power. (a) What current flows through the wire if a potential difference of 120. V is maintained across the resistance wire? (b) What is the resistance of the heater wire?

29. A large resistor is placed in an insulated container filled with 200. g of water at 15.0°C, and a potential difference of 50.0 V is maintained across the resistor. After 6.00 min the temperature of the water has increased to 25.0°C. (a) What current has been flowing through the resistor? (b) What is the resistance of the resistor?

30. Two wires of identical length but of cross-sectional areas $A_1 = A$ and $A_2 = A/2$ and resistivities $\rho_1 = 4\rho$ and $\rho_2 = \rho$ are connected side by side to the terminals of the same battery. What is the ratio $\mathcal{P}_1/\mathcal{P}_2$ of the power dissipated in the wires?

31. What value resistor (in ohms) do you need to raise the temperature of 500. g of water 30.0°C in 5.00 min if you maintain a potential difference of 120. V across the resistor while it is immersed in the water?

32. What is the cost of leaving a 100.-W light on for a month (30 days) if electricity costs 6.00 cents/kW $\cdot$ h?

26.4 Microscopic Models of Resistance

33. A bar of silicon is constructed with a square cross section 1.00 mm on a side and a length of 1.00 cm. The silicon is doped so that the charge carrier density is 2.00×10^{18} electrons/m^3, making its resistivity 4.00×10^{-3} $\Omega \cdot m$. As shown in Figure 26.P2 a 3.00-V battery is placed across the ends of a silicon bar. (a) Calculate the resistance of the bar, the current and electric field in the bar, and the drift velocity of the electrons in the bar. (b) Suppose the doping concentration is doubled. This causes the resistivity to be halved. What effect does this have on each quantity you calculated in part (a)?

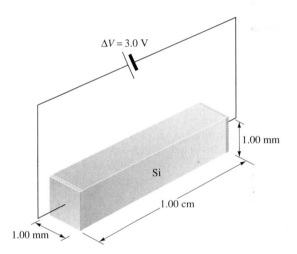

$\Delta V = 3.0$ V

1.00 mm

Si

1.00 cm

1.00 mm

FIGURE 26.P2   Problem 33

34. Examine the positions of P, B, Tl, and Sb in the periodic table. Which can be used to dope silicon to create an n-type semiconductor? A p-type?

35. The resistivity of an n-type semiconductor is $\rho = 1/(en\mu_n)$, where n, the number density of conduction band electrons, is equal to the N_D, the number density of dopant atoms in atoms per cubic centimeter. The parameter μ_n is the **mobility** of the electrons. For impurity concentrations below 10^{16} atoms/cm^3 the mobility of electrons in silicon is approximately 1500 cm^2/(V $\cdot$ s). A semiconductor manufacturer measures the resistivity of one of its As-doped silicon wafers to be 100. $\Omega \cdot cm$. What is the concentration of As atoms?

Numerical Methods and Computer Applications

36. The following table provides the resistivities of copper at various temperatures. Assume you are given a piece of copper with resistance R_o at temperature T_o. Use this table to write a program that

computes the resistance of this piece of copper at any temperature between 15.0 K and 295 K.

T (K)	ρ ($10^{-10}\,\Omega \cdot m$)
15	0.01
20	0.08
25	0.25
30	0.63
40	2.2
50	5.0
60	9.5
70	15.
80	21.
90	28.
100	35.
120	49.
140	63.
160	77.
180	92.
200	106.
220	120.
250	140.
273	155.
295	170.

37. A student applies various voltages across the filament of a flash-light bulb and measures the current through the bulb at each voltage. The table below summarizes her results:

V (V)	I (mA)
0.5	46
1.0	60
1.5	72
2.0	84
2.5	94
3.0	100
3.5	105
4.0	110
4.5	112
5.0	113

Write a program or develop a spreadsheet template to compute the resistance of the bulb as a function of applied voltage and the power dissipated $\mathcal{P}$ as a function of voltage. From a plot of $\mathcal{P}$ versus V estimate the power dissipated if a potential difference of 6.00 V is applied across the bulb.

38. The *dynamic resistance* of a nonohmic circuit element is defined as the slope of the V-versus-I curve at some particular value of I. This dynamic resistance is useful for determining the element's effect when small changes of current occur near some steady current (often called the *bias current*). Find the dynamic resistance of

the filament described in Problem 37 when the average current is 2.5 mA.

General Problems

39. The total charge that passes a certain point in a conductor after a time t is given by $q(t) = (60.0\ \mu C)e^{-t/3.00s}$. Find the current in the conductor at (a) $t = 1.00$ s, (b) $t = 3.00$ s, and (c) $t = 6.00$ s.

40. The conducting bar shown in Figure 26.6 carries a current density given by $\mathbf{J} = J_o e^{\alpha x}\hat{\mathbf{k}}$, where $J_o = 0.360\ A/m^2$ and $\alpha = 1.50\ cm^{-1}$. The width L of the bar is 2.00 cm. Find the current that flows through the cross-sectional area of the bar.

41. Two concentric, cylindrical conductors are each 60.0 cm in length. Their radii are $R_1 = 4.00$ cm and $R_2 = 12.0$ cm. The space between the cylinders is filled with a different conducting material, and a 2.00-A current flows radially outward between the cylinders, perpendicular to their common axis. Compute the current density at radius R, where (a) $R = 6.00$ cm and (b) $R = 10.0$ cm.

42. For proper balance in some small aircraft the 12.0-V engine battery must be located in the tail of the plane. If the aluminum cable between the positive pole of the battery and the starter motor is 6.10 m in length, has a diameter of 8.25 mm, and carries a current of 125 A, what is the potential drop between the cable ends?

43. An advertisement for the 12.0-V battery in one of the author's cars claims that the battery can provide 405 "cold-cranking" amperes. The copper cable to the starter motor is 0.500 m in length and 5.20 mm in diameter. (a) What is the resistance of the cable? (b) What is the potential drop along the cable if it carries 405 A? (c) What is the potential drop across the cable during a normal engine start-up when the cable carries a current of 85.0 A?

44. A resistor is made from a material of resistivity ρ. It is formed in the shape of a hollow cylinder with inner radius R_1, outer radius R_2, and length L (Fig. 26.P3). Compute the resistance of this resistor if current is to flow *parallel* to the axis of the cylinder.

*45. A resistor is made from a material of resistivity ρ. It is formed in the shape of a hollow cylinder with inner radius R_1, outer radius R_2, and length L (Fig. 26.P3). Show that the resistor's resistance to current flow *perpendicular* to the cylinder's axis is $R = (\rho/2\pi L) \ln(R_2/R_1)$.

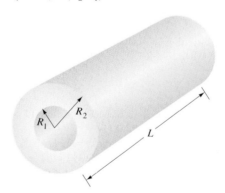

FIGURE 26.P3   Problems 44 and 45

46. An RG-58/u coaxial cable is formed from an inner copper wire with a diameter that is 0.0320 in and a concentric copper cylinder 0.116 in in diameter. The conductors are separated by polyethylene with a room temperature resistivity of $2.00 \times 10^{11}\ \Omega \cdot m$. What current do you expect to leak between the two conductors in a 1.00-m length of cable if the maximum operating voltage of 1900. V is maintained between them? (*Hint:* See Problem 45.)

47. A resistor is formed in the shape of a hollow, quarter-cylinder from a material of resistivity ρ. Find the resistance of this resistor between faces A and B shown in Figure 26.P4.

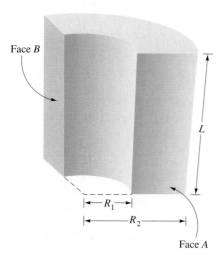

Face B

L

R_1

R_2

Face A

FIGURE 26.P4   Problem 47

*48. The region between two spherical, conducting shells of radii R_1 and R_2 ($R_1 < R_2$) is filled with a material of resistivity ρ. Find the total resistance between the two spherical conductors.

*49. A material of resistivity ρ is formed in the shape of a truncated, right-circular cone as shown in Figure 26.P5. Assume that the current density is uniform through any cross section perpendicular to the axis of the cone, and show that the resistance between the two ends is $R = (\rho/\pi)(L/R_1R_2)$.

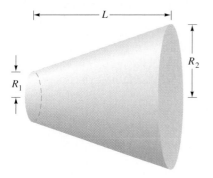

L

R_2

R_1

FIGURE 26.P5   Problem 49

50. In the *classical* theory of electrical conduction, charge carriers of mass m and charge q acquire a constant acceleration a from the force $F = qE$. If T is the average time between carrier collisions with the lattice vibrations, (a) use a kinematic equation to show that the *average* carrier velocity is $v_d = qET/2m$. (b) Apply Equations (26.4) and (26.6) to show that this theory predicts $\sigma = nq^2T/2m$.

51. A 1.40-kW heater uses a 4.80-m long nichrome wire as the heating element. If the heater requires a potential difference of 110. V to be applied across the wire, what must be the wire's diameter?

52. The volume of a certain house is 14 000 ft^3. In order to keep the house heated during the winter, each hour the electric furnace must take in this amount of air at an average temperature of 15.0°C and pass it through wire heating resistors so that its temperature is raised to 22°C. (a) If the heater is to run continuously, what power must be dissipated as heat by the wires? The heat capacity of air is about 700. J/(kg°C) and its density is 1.20 kg/m^3. (b) If a potential difference of 120. V is maintained across the resistance wires, what current flows through the wires?

CHAPTER

27

Direct-Current Circuits

In this chapter you should learn

- how sources of emf are used in direct-current (DC) circuits.

- how to compute the equivalent resistance for series and parallel combinations of resistors.

- how to apply Kirchhoff's laws to analyze DC circuits.

- how to construct a voltmeter or ammeter from a galvanometer.

- how a Wheatstone bridge and potentiometer work.

- to analyze circuits that consist of a series combination of a resistor and a capacitor.

- (optional) how to use a spreadsheet or electronic calculator to solve for the currents from a series of linear equations.

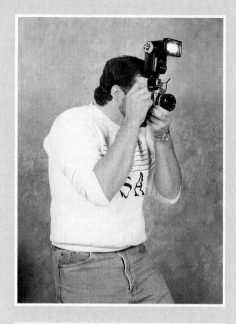

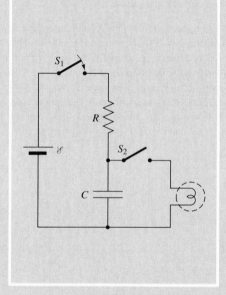

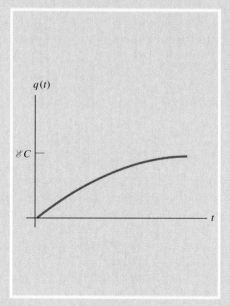

From a photographic flash unit to the intricate structure internal to a computer, electronic circuits play a major role in our lives. In this chapter we will present the fundamentals of circuit analysis. We can't promise that you will be able to build a stereo system, repair a VCR, or wire a house when you complete this chapter. Indeed, for a lot of good reasons most of us shouldn't even attempt any of these activities! However, we assure you that you will meet many of the fundamental concepts that are applicable to all electrical systems. We begin by taking a more detailed look at the sources of an electric potential difference.

27.1 Electromotive Force

On numerous occasions we have found it convenient to use the concept of a battery to establish a potential difference between two points. A battery is only one example of a type of electric element called a source of **electromotive force,** or **emf.** A source of emf is any device that can do work on a charge to increase that charge's potential energy. Generators and solar cells are other sources of emf. The name *electromotive force* has a historical origin and is really not an appropriate term to describe what an emf is all about; as we will show in more detail below, it certainly is *not* a force. We will follow tradition, however, and use the term *emf* and hope you will soon forget that the *f* once stood for force.

Our interest in sources of emf does not extend to the details of their internal workings. Rather, we will consider their external electrical properties and how they are used in electrical circuits. It is sufficient to know that a battery converts chemical energy into electrical energy, a generator converts mechanical energy into electrical energy, and a solar cell converts the energy of light into electrical energy. Sometimes it is convenient to think of a source of emf as a kind of pump that acts on charge to bring it to a higher potential energy. When an infinitesimal charge dq enters the low-potential terminal (designated with a minus sign), by chemical, mechanical, or other means, the source of emf performs work dW on that charge and brings the charge to a higher potential at the other terminal (designated by a plus sign). The mathematical statement that the emf $\mathscr{E}$ is the work dW done per charge dq is

$$\mathscr{E} = \frac{dW}{dq} \qquad (27.1)$$

From Equation (27.1) we see that the units of emf are those of work per unit charge. Hence, the SI unit of emf is the volt.

Figure 27.1(a) shows a simple circuit loop consisting of a source of emf connected to an external resistance R so that current I flows through the circuit. For simplicity, we assume the wires of this circuit have no resistance. A consequence of this approximation is that no work is required to move the charge dq from the right side of resistance R back to the negative terminal of the source of emf. In reality, all wires have some resistance and, therefore, a small potential difference remains across the wires to drive charge dq through them. However, we usually ignore this effect and model all wires in circuit diagrams as if they had no resistance.

All real sources of emf have some **internal resistance.** In Figure 27.1(b) we model the internal resistance of the emf in our circuit as if it were a separate resistance r. In practice this is not the case. The internal resistance is an *integral part* of the emf source itself. For example, the same lead and sulfuric acid that gives rise to the emf of an automobile battery simultaneously gives rise to a resistance through which charge must travel. Although one cannot actually separate the emf from this internal resistance, for ease of computation it is convenient to model the complete source of emf as a series combination of an **ideal (resistanceless) emf** $\mathscr{E}$ and a separate internal resistance r.

A positive charge entering the negative terminal of the emf at point l of Figure 27.1(b) has its electric potential increased to $\mathscr{E}$ at point m. However, as the charge moves through the resistance r of the emf (from point m to point n), its electric potential is *reduced* by an

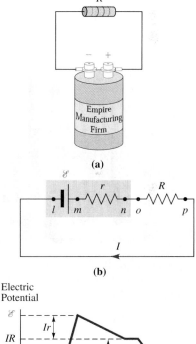

FIGURE 27.1

(a) A resistor R connected to a source of emf. (b) The circuit diagram. The internal resistance is designated by a small resistance r in series with the emf $\mathscr{E}$. (c) The electric potential from point a to point e in the circuit. The terminal potential depends on the current I and the internal resistance: $\Delta V_{lm} = \mathscr{E} - Ir$.

amount Ir. The **terminal potential** is defined as the potential difference between the terminals of the emf source. In this case the terminals are l and m, and the terminal potential ΔV_{lm} is

$$\Delta V_{lm} = \mathscr{E} - Ir \qquad (27.2)$$

Equation (27.2) is important because it tells us that the potential difference we observe across the terminals of the source of emf is not really $\mathscr{E}$ but instead is a smaller potential with a value that depends on the current I. Only when the current is zero does the terminal potential equal the emf $\mathscr{E}$. The **load resistance** R is connected to the emf terminals, and thus the potential difference across R is the terminal potential ΔV_{ln}. That is,

$$\Delta V_{ln} = IR$$

Substituting this result into Equation (27.2), we have

$$IR = \mathscr{E} - Ir$$

or

$$I = \frac{\mathscr{E}}{R + r} \qquad (27.3)$$

It should be clear to you from Equation (27.3) that the current in a circuit depends not only on the visible external load resistance but also on the hidden internal resistance r of the emf. In many of the circuits we analyze, the load resistance R is sufficiently larger than the internal resistance r so that we can ignore the latter. In this approximation Equation (27.3) becomes $I = \mathscr{E}/R$. It's a good idea, however, to always keep in the back of your mind that anytime the load resistance is small you may need to consider the effect of the internal resistances of the emf sources present in your circuit.

From Equation (27.1) we find the work done by the source of emf to move charge $dq = I\,dt$ through potential difference $\mathscr{E}$:

$$dW = \mathscr{E}\,dq = \mathscr{E}(I\,dt)$$

Hence, the power $\mathscr{P}$ delivered by the source of emf is

$$\mathscr{P} = \frac{dW}{dt} = I\mathscr{E}$$

When we solve Equation (27.3) for $\mathscr{E}$ and multiply both sides of the resulting equation by the current I, we have

$$I\mathscr{E} = I^2R + I^2r \qquad (27.4)$$

This result tells us that the power provided by the source of emf is dissipated as joule heating in both the load resistance R and the internal resistance r.

Concept Question 1
Could the terminal potential of a battery in a circuit ever exceed its emf? Explain.

Concept Question 2
Why would a circuit designer want to connect two sources of emf in series? In parallel?

Concept Question 3
Explain the difference between potential difference and emf.

EXAMPLE 27.1   *Power for a Cold Start*

While starting a certain automobile engine a 12.0-V battery delivers 90.0 A to a starter motor. If the battery has an internal resistance of 0.0400 Ω (a) what is the terminal potential of the battery while delivering this current? (b) If the engine requires 2.50 s to start, what electric energy is delivered *to the starter motor* during this time? (Assume no energy losses in the electric cables.)

SOLUTION (a) The terminal potential is given by Equation (27.2):

$$\Delta V = \mathcal{E} - Ir = 12.0 \text{ V} - (90.0 \text{ A})(0.0400 \ \Omega) = 8.40 \text{ V}$$

(b) The power provided to the starter motor by the battery is $\mathcal{P}_{del} = I \Delta V$, and in time Δt the energy delivered is

$$\Delta W = \mathcal{P}_{del} \Delta t = I \Delta V \Delta t = (90.0 \text{ A})(8.40 \text{ V})(2.50 \text{ s}) = 1.89 \text{ kJ}$$

The total power delivered by the source of emf is $\mathcal{P} = I\mathcal{E}$. Can you show that the energy dissipated in the battery during this same time is 810 J? ◀

27.2 Combinations of Resistors

Most electric circuits consist of combinations of various circuit elements, many of which are resistors. Later in this chapter we will learn what happens when we construct a circuit loop that consists of a sequence of a resistor, a capacitor, and a source of emf. For now, we consider only combinations of resistors. As usual, we continue to assume that all wires connecting the resistors to each other and also to sources of emf are ideal and have no resistance. We also assume that the internal resistances of the sources of emf are sufficiently small compared to the resistances in the external circuits that they also can be ignored.

Resistors in Series

Figure 27.2 shows two resistors connected in a **series** combination to a source of emf, and the schematic drawing for this circuit. Let's compute the resistance R_{tot} of a single, equivalent resistor with which we can replace this combination of resistors and still maintain the same current flow I from the source of emf. The first thing you should notice is that any current coming from the emf must pass through both resistors; this current has no place else to go! This feature is a general characteristic of circuit elements arranged in series,[1] so

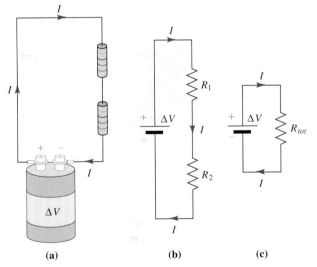

(a)   (b)   (c)

FIGURE 27.2

(a) A series combination of two resistors R_1 and R_2 connected to a source of emf ΔV. (b) The schematic diagram for this circuit. The total voltage drop across the resistors is $\Delta V = \Delta V_1 + \Delta V_2$, while the same current I flows through each. (c) A single, equivalent resistor must have a value such that $R_{tot} = \Delta V / I$.

---

[1] See Section 25.2 to review the meaning of *series* and *parallel* circuit connections.

we emphasize this point:

> At every instant of time all circuit elements connected in a series combination carry the same current.

Next, notice that the *total* potential drop across both resistors must be ΔV, the potential across the terminals of the emf. Moreover, the potential drops across resistors R_1 and R_2 are IR_1 and IR_2, respectively, so the total potential drop across the series combination is

$$\Delta V = IR_1 + IR_2 = I(R_1 + R_2) \tag{27.5}$$

From Figure 27.2(c) we see that the potential drop across the equivalent resistance R_{tot} is

$$\Delta V = IR_{tot}$$

Comparing this result with Equation (27.5), we see

$$R_{tot} = R_1 + R_2 \tag{27.6}$$

For N resistors connected in a series combination, the equivalent resistance is

$$R_{tot} = R_1 + R_2 + R_3 + \cdots + R_N = \sum_{i=1}^{N} R_i \qquad \text{(Resistors in series)} \tag{27.7}$$

It should be clear from Equation (27.7) that *the equivalent resistance of a series combination of resistances is always larger than any of the individual resistances.*

Notice what happens in series circuits when one element fails. Suppose your physics teacher connects several resistances in series, but one of the resistors has a power rating too low for the current that flows through the combination. As you might anticipate (perhaps with some excitement), a portion of that resistor is destined to become smoke! When that happens, an opening forms in the circuit. An **open circuit** is equivalent to a circuit that somewhere contains an infinite resistance, and hence no current flows through any of the resistors in series with the damaged one. Some less-expensive Christmas tree lights are built this way; it's a dismal chore to try to find the burned-out bulb!

A **fuse** is a resistor with a resistance that is normally small. A fuse is designed to vaporize when the current passing through it exceeds a certain value. In homes and other buildings **circuit breakers** are often used in place of fuses. Both fuses and circuit breakers are placed in a series combination with the other circuit elements they are designed to protect.

Resistors in Parallel

A **parallel** combination of resistors connected to a source of emf and the equivalent circuit diagram are shown in Figure 27.3. We again compute the equivalent resistance R_{tot}, which may replace resistors R_1 and R_2 and still permit the flow of the same current I from the source of emf. In Figure 27.3 the left-hand side of each resistor is connected to the positive terminal of the emf. Likewise, the right-hand side of each of these resistors is connected to the emf's negative terminal. Therefore, the potential drop across each resistor must be the same and equal to ΔV. This equivalence of potential difference is a general characteristic of any collection of circuit elements connected in parallel:

> At every instant of time the potential difference across all circuit elements connected in parallel is the same.

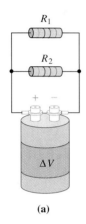

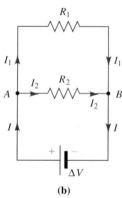

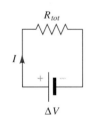

FIGURE 27.3

(a) A parallel combination of two resistors R_1 and R_2 connected to a source of emf ΔV. (b) The schematic diagram for this circuit. The total current I splits into two currents at junction A such that $I = I_1 + I_2$, but the voltage drop across each resistor is the same. (c) A single, equivalent resistor must have a value $R_{tot} = \Delta V/I$.

Notice from Figure 27.3(b) that at junction A the current I splits into two currents I_1 and I_2 such that

$$I = I_1 + I_2 \tag{27.8}$$

At junction B the currents I_1 and I_2 recombine to again total to current I. The potential drop across resistor R_1 is I_1R_1, and across resistor R_2 the potential drop is I_2R_2. Both these potential changes are equal to ΔV:

$$\Delta V = I_1 R_1 = I_1 R_2$$

or

$$I_1 = \frac{\Delta V}{R_1} \quad \text{and} \quad I_2 = \frac{\Delta V}{R_2}$$

Substituting these expressions for I_1 and I_2 into Equation (27.8), we have

$$I = \Delta V \left(\frac{1}{R_1} + \frac{1}{R_2} \right) \tag{27.9}$$

From Figure 27.3(c) we see that the current through the equivalent resistance R_{tot} is

$$I = \frac{\Delta V}{R_{\text{tot}}}$$

Comparing this result with Equation (27.9), we have

$$\frac{1}{R_{\text{tot}}} = \frac{1}{R_1} + \frac{1}{R_2} \tag{27.10}$$

For N resistors in a parallel combination a similar analysis yields

$$\frac{1}{R_{\text{tot}}} = \frac{1}{R_1} + \frac{1}{R_2} + \frac{1}{R_3} + \cdots + \frac{1}{R_N} = \sum_{i=1}^{N} \frac{1}{R_i} \qquad \text{(Resistors in parallel)} \tag{27.11}$$

From Equations (27.10) and (27.11) you should be able to see that *the total resistance of any parallel combination of resistors must be less than any one of the individual resistances.* Also, notice from Figure 27.3 that if either resistor R_1 or R_2 should fail, causing that segment of the circuit to be open, current still flows through the other resistor.

All circuit elements in homes and other buildings are arranged in groups, each consisting of several parallel circuits. When you plug in a lamp and turn it on, you are putting a resistance in parallel to other devices on the same circuit. This way, if a device in one of these parallel circuits should fail and result in an open circuit (perhaps the light bulb in your lamp burns out), all the other devices continue to function. Moreover, each device is designed to work at the same potential difference, a condition that is satisfied when all are connected in a parallel arrangement. Each group of circuit elements is connected in series to a fuse or circuit breaker. The proper place for a fuse in Figure 27.3(b) is between point A and the positive terminal of the emf or, equivalently, between point B and the negative terminal. If either resistor should *short circuit* (that is, have its resistance drop to zero), the resulting large current would vaporize the resistive material of the fuse and produce an open circuit.

Concept Question 4
Why might a circuit designer connect two resistors with small resistances in parallel rather than use one equivalent resistor?

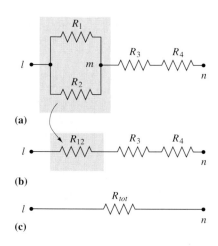

(a)

(b)

(c)

FIGURE 27.4

(a) The parallel combination R_1 and R_2 are replaced by a single equivalent resistor R_{12}. (b) The three series resistors R_{12}, R_3, and R_4 are replaced by a single resistor R_{tot}. (c) An ohmmeter placed between points l and m will read the equivalent resistance R_{tot}. See Example 27.2.

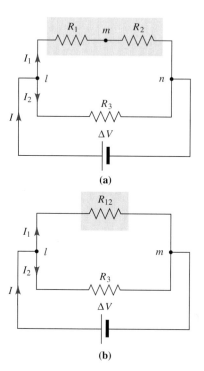

(a)

(b)

FIGURE 27.5

(a) A series combination R_1 and R_2 are replaced by a single equivalent resistor R_{12}. (b) The resistors R_{12} and R_3 are replaced by a single resistor R_{tot}.

EXAMPLE 27.2   "The Intelligence of Congressmen Adds Like Resistors in Parallel."—Anonymous

Find the equivalent resistance between the point A and B for the combination of resistors shown in Figure 27.4(a). Take $R_1 = 3.00\ \Omega$, $R_2 = 6.00\ \Omega$, $R_3 = 4.00\ \Omega$, and $R_4 = 2.00\ \Omega$.

SOLUTION   We begin by focusing our attention on the two resistors R_1 and R_2, which are in parallel. We can find their equivalent resistance $R_{tot} = R_{12}$ from Equation (27.10):

$$\frac{1}{R_{12}} = \frac{1}{R_1} + \frac{1}{R_2}$$

Solving for R_{12}, we have

$$R_{12} = \left(\frac{1}{R_1} + \frac{1}{R_2}\right)^{-1} \tag{27.12}$$

Equation (27.12) is a convenient way to write our answer because most calculators have a $\boxed{1/x}$ function key that allows this computation to be performed with a small number of keystrokes. (Try it!) Substituting, we find

$$R_{12} = 2.00\ \Omega$$

Notice that R_{12} is smaller than either R_1 or R_2. An equivalent circuit is shown in Figure 27.4(b). All these resistors are in series, so

$$R_{tot} = R_{12} + R_3 + R_4 = 2.00\ \Omega + 4.00\ \Omega + 2.00\ \Omega = 8.00\ \Omega \qquad \blacktriangleleft$$

EXAMPLE 27.3   More Practice with Series and Parallel Resistors

For the combination of resistors shown in Figure 27.5(a) compute (a) the current I from the source of emf and (b) the potential drop across R_2. Take $R_1 = 1.00\ \Omega$, $R_2 = 3.00\ \Omega$, $R_3 = 12.0\ \Omega$, and $\Delta V = 24.0$ V.

SOLUTION   (a) In order to find the current I, we must compute the equivalent resistance for this combination of resistors; for then, as you can see from Figure 27.5(c), $I = \Delta V_{ln}/R_{tot}$. We begin by trying to find the simplest looking part of the circuit to analyze. Resistors R_1 and R_2 are in series, so

$$R_{12} = R_1 + R_2 = 1.00\ \Omega + 3.00\ \Omega = 4.00\ \Omega$$

An equivalent circuit is shown in Figure 27.5(b) where we recognize that R_{12} and R_3 are in a parallel arrangement. Applying Equation (27.12) to the resistors in Figure 27.5(b), taking care to use the proper subscripts, we obtain

$$R_{tot} = \left(\frac{1}{R_{12}} + \frac{1}{R_3}\right)^{-1} = \left(\frac{1}{4.00\ \Omega} + \frac{1}{12.0\ \Omega}\right)^{-1} = 3.00\ \Omega$$

The current through R_{tot} is

$$I = \frac{\Delta V_{ln}}{R_{tot}} = \frac{24.0\text{ V}}{3.00\ \Omega} = 8.00\text{ A}$$

(b) To compute the potential drop ΔV_{mn} across the resistor R_2 we must determine the current I_2 through this resistor. For then $\Delta V_{mn} = I_1 R_2$. We begin by noting from Figure 27.5(b) that resistors R_{12} and R_3 are in parallel so that the potential drop across

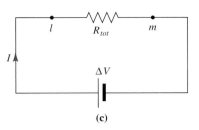

each is the same, namely, 24.0 V. Hence, the current I_2 through resistor R_3 must be

$$I_2 = \frac{\Delta V_{ln}}{R_3} = \frac{24.0 \text{ V}}{12.0 \text{ V}} = 2.00 \text{ A}$$

In part (a) we found that the total current $I = 8.00$ A. Therefore, the current $I_1 = I - I_2 = 8.00$ A $- 2.00$ A $= 6.00$ A. The potential drop across R_2 is

$$\Delta V_{mn} = I_1 R_2 = (6.00 \text{ A})(3.00 \text{ } \Omega) = 18.0 \text{ V}$$

Can you explain why the potential drop across resistor R_1 is 6.00 V? ◀

FIGURE 27.5 Continued

(c) An ohmmeter placed between points l and m will read the equivalent resistance R_{tot}. See Example 27.3.

27.3 Multiple-Loop Circuits: Kirchhoff's Laws

In Example 27.3 we found that by applying the rules for resistors in parallel or series and the relation $\Delta V = IR$ we could compute the potential changes and currents in circuits that consist of a combination of resistors with a source of emf. A crucial step in our analysis was the reduction of the arrangement of resistors to a single, equivalent resistance. Circuits for which this type of analysis is possible are known as **simple circuits.** In many useful circuits, however, it is often impossible to perform this reduction; circuit elements can be connected so that simple substitutions for series or parallel equivalents are not possible. Such circuits are called **nonsimple circuits.**[2] There are, however, two laws, based on the conservation of charge and the conservation of energy, respectively, that can be applied to both simple and nonsimple circuits. When applied to any circuit, the laws provide us with a set of simultaneous equations from which we can solve for the current through any resistor present (or any other circuit element present, for that matter). Here are **Kirchhoff's laws:**

1. At any circuit junction the sum of the currents entering the junction must equal the sum of the currents leaving the junction.

2. The sum of the potential changes across all the circuit elements found when traversing in one direction around any closed circuit-loop must be zero.

The first law is simply an expression of charge conservation; as current flows through various branches of a circuit, none piles up nor disappears at any junction. The second law requires a charge to always have the same energy value every time it passes a particular place in the circuit. That is, as a charge dq moves around a complete circuit loop, its energy gains and losses must total zero. In the next three paragraphs we will describe how to apply these laws to a complex circuit. Examples 27.4 and 27.5 illustrate the application of these laws.

As a practical matter, you will always have to begin by assigning current names, along with their directions, to each branch of the complex circuit. It is not always easy to correctly guess the current direction. It is no disaster, however, if you choose the wrong direction. When you have finished the problem, if the numerical value for the current comes out negative, this result indicates that this particular current is in the opposite direction from your original guess.

Once you have assigned currents to all the circuit branches you must pick a point and begin to walk (mentally) around a circuit loop. As you traverse each resistor or emf you must write down the potential change for that element. The signs of the potential changes

---

[2] The classification of simple and nonsimple circuits does not refer to the complexity of the circuit.

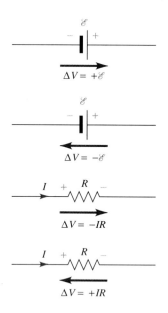

FIGURE 27.6

Rules for computing the potential change when sources of emf and resistors are traversed while writing circuit equations. The blue arrows indicate the direction in which the circuit element is traversed.

Concept Question 5
Why is it not necessary to always assign an electric potential of 0.00 V to the negative terminal of the sources of emf in circuits?

are very important. Here are four rules to help you write down the correct potential change:

> 1. If you traverse through a source of emf from the negative terminal to the positive terminal, the change in potential is $+\mathcal{E}$.
> 2. If you traverse through a source of emf from the positive terminal to the negative terminal, the change in potential is $-\mathcal{E}$.
> 3. If a resistor R is traversed in the same direction as the assumed current flow, the change in potential is $-IR$.
> 4. If a resistor R is traversed in the direction opposite from the assumed current flow, the change in potential is $+IR$.

These rules are illustrated in Figure 27.6. In each case the blue arrow indicates the direction you traverse the particular circuit element. There is a limit to the number of times you can apply each of Kirchhoff's laws. This limit is not governed by your personal exhaustion or distaste for things electrical, but rather by the number of independent equations that can be generated from the circuit loops. If you exceed this number, the new equations you generate turn out to be combinations of the equations you have already written. In general, you can apply the first rule one less time than the number of circuit junctions present. By the way, a right- or left-angle bend does *not* constitute a junction. A junction has three or more wires connected at a point. You can apply the second of Kirchhoff's laws as long as you introduce a new current or another circuit element (emf or resistor) into each new equation.

More advanced treatments of complex circuits utilize matrix algebra to solve the system of equations that is generated by Kirchhoff's laws. Also, there are elaborate computer programs that can quickly analyze very complex circuits. The computations performed by these programs are all based on Kirchhoff's two laws, so even if you eventually use such programs, it is worthwhile for you to understand the basic nature of the computations they are performing. You should become intimately familiar with Examples 27.4 and 27.5 so that you can eventually work each without referring to this text for help at any step.

EXAMPLE 27.4   *With All the Details!*

Find the current in each branch of the circuit shown in Figure 27.7(a) if $\mathcal{E}_1 = 4.00$ V, $\mathcal{E}_2 = 16.0$ V, $\mathcal{E}_3 = 6.00$ V, $R_1 = 4.00\ \Omega$, $R_2 = 8.00\ \Omega$, and $R_3 = 10.0\ \Omega$.

SOLUTION   In Figure 27.7(b) we have redrawn the circuit, arbitrarily assigned currents (and directions) in each circuit branch, and labeled the two junctions A and B.

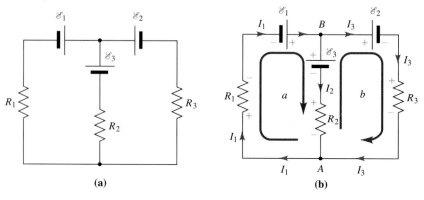

FIGURE 27.7

(a) A circuit with three resistors and three emf sources, (b) the same circuit redrawn with arbitrarily assigned currents and directions and two nodes A and B.

We have also shown the path we will take as we mentally walk around the two circuit loops with centers labeled ''*a*'' and ''*b*.'' Before we begin, some comments are in order. First of all, the paths are both clockwise. This choice is completely arbitrary and at some point you should verify that an application of Kirchhoff's laws for anticlockwise paths yields equations equivalent to those we obtain below. (They are the negative of our equations.) Our second point pertains to the current assignments. The preassigned directions are also arbitrary. Also note carefully that the *current* does *not* change magnitude when it passes through a resistor or battery or just because a branch changes direction. Finally, there are three unknown currents, so we must generate three independent equations.

Let's begin by applying the first of Kirchhoff's laws to junction *A*. The currents entering this junction are I_2 and I_3, and the current leaving is I_1. Hence,

$$I_2 + I_3 = I_1 \qquad\qquad (27.13)$$

Next, we follow the path clockwise around loop *a* starting from junction *A*. The first circuit element we encounter is resistance R_1, and we traverse it in the same direction as the current I_1. Hence, we assign a potential change of $-I_1R_1$ to this circuit element. Next, we encounter emf $\mathcal{E}_1$ and our path takes us in the direction of this emf, so we assign it a potential change of $+\mathcal{E}_1$. After junction *B* we traverse emf $\mathcal{E}_3$ in the opposite direction of this emf. Therefore, we allot a $-\mathcal{E}_3$ potential change for this circuit element. Finally, to return to junction *A* we traverse resistor R_2 in the direction of current I_2. Its potential change is $-I_2R_2$. We have returned to our starting point, so by Kirchhoff's second law all these potential changes[3] must total zero:

$$-I_1R_1 + \mathcal{E}_1 - \mathcal{E}_3 - I_2R_2 = 0 \qquad\qquad (27.14)$$

For loop *b* we also begin at point *A*. As we follow the clockwise path, the potential changes in the order in which we encounter them, are $+I_2R_2$, $+\mathcal{E}_3$, $-\mathcal{E}_2$, and $-I_3R_3$. The second Kirchhoff law gives

$$+I_2R_2 + \mathcal{E}_3 - \mathcal{E}_2 - I_3R_3 = 0 \qquad\qquad (27.15)$$

If you apply Kirchhoff's second law to the outer loop, once again starting at point *A*, you obtain the equation $-I_1R_1 + \mathcal{E}_1 - \mathcal{E}_3 - I_3R_3 = 0$. You should verify that this equation does not constitute any new information by showing that it is simply the sum of Equations (27.14) and (27.15). We rearrange Equations (27.13), (27.14), and (27.15) to summarize our three equations in the three unknowns I_1, I_2, and I_3:

$$I_1 \quad - I_2 \quad - I_3 \quad = 0 \qquad\qquad (27.16a)$$

$$I_1R_1 + I_2R_2 \qquad = \mathcal{E}_1 - \mathcal{E}_3 \qquad\qquad (27.16b)$$

$$I_2R_2 - I_3R_3 = \mathcal{E}_2 - \mathcal{E}_3 \qquad\qquad (27.16c)$$

At this point the easiest approach is to substitute for R_1, R_2, R_3, $\mathcal{E}_1$, $\mathcal{E}_2$, and $\mathcal{E}_3$ to get

$$I_1 = I_2 + I_3 \qquad\qquad (27.17a)$$

---

[3] In electrical engineering circuit courses, it is conventional to sum potential *drops* rather than potential *changes*. A potential *change* of $-IR$ is equivalent to a potential *drop* of $+IR$. Consequently, your electrical engineering professor might write for exactly the same loop:

$$+I_1R_1 - \mathcal{E}_1 + \mathcal{E}_3 + I_2R_2 = 0$$

The difference in convention causes all kinds of unnecessary confusion. You can see that one equation is merely the negative of the other. The difference has nothing at all to do with electron current versus positive current or anything like that. The only difference is whether one chooses to sum *potential changes* or *potential drops*. Unfortunately, there seems to be little hope in changing either physicists or electrical engineers, so be forewarned.

$$(4.00 \ \Omega)I_1 + (8.00 \ \Omega)I_2 = 4.00 \ \text{V} - 16.0 \ \text{V} = -12.0 \ \text{V} \qquad \textbf{(27.17b)}$$

$$(8.00 \ \Omega)I_2 - (10.0 \ \Omega)I_3 = 16.0 \ \text{V} - 6.00 \ \text{V} = 10.0 \ \text{V} \qquad \textbf{(27.17c)}$$

(If you have a calculator that can solve simultaneous equations, you may wish to consult the instruction manual for directions on how to solve equations like those above. Also see optional Section 27.7 at the end of this chapter.)

If you (i) substitute the expression for I_1 of Equation (27.17a) into Equation (27.17b), (ii) solve the resulting equation for I_2, (iii) substitute this result into Equation (27.17c), and solve this equation for I_3, you will find

$$I_3 = -0.895 \ \text{A}$$

followed by

$$I_2 = +0.132 \ \text{A}$$

and

$$I_1 = -0.763 \ \text{A}$$

Negative values for I_1 and I_3 indicate that these currents actually flow in the opposite direction from those we assigned. If it looks like a rainy weekend, you may want to boost your confidence in your analytical skills by solving the system of Equations (27.6) for I_1, I_2, and I_3 in terms of the resistances and emfs. The results are

$$I_1 = \frac{(R_2 + R_3)\mathscr{E}_1 - R_2\mathscr{E}_2 - R_3\mathscr{E}_3}{R_1R_2 + R_1R_3 + R_2R_3} \qquad \textbf{(27.18a)}$$

$$I_2 = \frac{R_3\mathscr{E}_1 + R_1\mathscr{E}_2 - (R_1 + R_3)\mathscr{E}_3}{R_1R_2 + R_1R_3 + R_2R_3} \qquad \textbf{(27.18b)}$$

$$I_3 = \frac{R_2\mathscr{E}_1 - (R_1 + R_2)\mathscr{E}_2 + R_1\mathscr{E}_3}{R_1R_2 + R_1R_3 + R_2R_3} \qquad \textbf{(27.18c)}$$

Equations (27.18) are valid only for the special case of the circuit shown. Whatever you do, don't try to memorize them. The whole point is to learn how to create such equations as Equations (27.16) for any circuit you might encounter.

Want more practice? Try working this problem again by drawing the circuit on a sheet of paper and assigning these values: $R_1 = 2.00 \ \Omega$, $R_2 = 4.00 \ \Omega$, $R_3 = 6.00 \ \Omega$, $\mathscr{E}_1 = 12.0 \ \text{V}$, $\mathscr{E}_2 = 4.00 \ \text{V}$, and $\mathscr{E}_3 = 3.00 \ \text{V}$. (It will do you little good to just substitute into the above equations. Start from scratch!)
Answers: $I_1 = 1.95 \ \text{A}$, $I_2 = 1.27 \ \text{A}$, $I_3 = 0.68 \ \text{A}$.

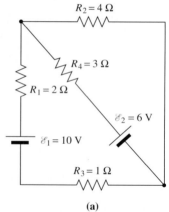

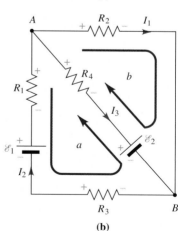

FIGURE 27.8

(a) A circuit with four resistors and two emf sources. (b) The same circuit showing loops a and b.

EXAMPLE 27.5   *More Kirchhoff, But You Supply the Details*

Find the currents in each of the segments of the circuit shown in Figure 27.8(a), where $\mathscr{E}_1 = 10.0 \ \text{V}$, $\mathscr{E}_2 = 6.00 \ \text{V}$, $R_1 = 2.00 \ \Omega$, $R_2 = 4.00 \ \Omega$, $R_3 = 1.00 \ \Omega$, and $R_4 = 3.00 \ \Omega$.

SOLUTION   Don't let the angled branch bother you. There are just three segments each in both loops a and b of Figure 27.8(b). We also show our current assignments and paths for these two loops. Applying the first Kirchhoff law to junction A yields

$$I_2 = I_1 + I_3$$

The counterclockwise path around loop a from junction A gives

$$+I_2R_1 - \mathscr{E}_1 + I_2R_3 + \mathscr{E}_2 + I_3R_4 = 0$$

The clockwise path around loop b from junction A gives

$$-I_1R_2 + \mathcal{E}_2 + I_3R_4 = 0$$

Substituting the values for the resistors and sources of emf into these equations, we have

$$I_2 = I_1 + I_3$$

$$(3.00\ \Omega)I_2 + (3.00\ \Omega)I_3 = 8.00\ \text{V}$$

$$(4.00\ \Omega)I_1 - (3.00\ \Omega)I_3 = 4.00\ \text{V}$$

Solving these three equations for I_1, I_2, and I_3, we find

$$I_1 = 1.45\ \text{A}$$

$$I_2 = 2.06\ \text{A}$$

$$I_3 = 0.606\ \text{A}$$

27.4 Potential Difference, Current, and Resistance Measurements

There are a variety of potential difference and current-measuring devices. Analog meters consist of a needle attached to a coil that is able to rotate in the field of a magnet.[4] As we shall see in the next chapter, when a small current passes through the coil, it experiences a torque that causes the needle-coil combination to rotate. The combination of the needle, the coil, the magnet, and a meter face is called a **galvanometer.** Additional resistors are added to make voltmeters and ammeters. You need to know just two basic features about a galvanometer to be able to construct a potential difference or current-measuring instru-

Three types of meters. The analog meter employs a needle that can rotate up to a maximum angle. A digital meter provides a numerical readout. An oscilloscope can be used to measure voltage as a function of time; the vertical axis is a selected range of voltage, whereas a time scale provides the horizontal axis. (Photo courtesy of Tim Thornberry.)

---

[4] Digital meters employ integrated circuits to generate a numerical readout; the principles of their operation are discussed in Section 27.6.

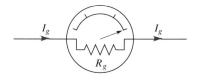

FIGURE 27.9

Schematic representation of a galvanometer. At this point it is only important to know that a galvanometer provides a full-scale reading when a current I_g flows through it and that a galvanometer has an internal resistance R_g.

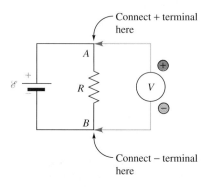

FIGURE 27.10

A voltmeter is connected in parallel to the circuit element across which it measures the potential difference. The positive terminal of the voltmeter is always attached to the higher potential side of the circuit element. Because a voltmeter provides a path for the flow of a small amount of current, its presence modifies the circuit.

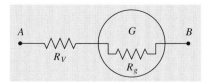

FIGURE 27.11

A schematic representation of a voltmeter. The voltmeter circuit consists of a large resistor R_v in series with the galvanometer resistance R_g.

ment from one:

A galvanometer has
1. a small internal resistance R_g and
2. a characteristic current I_g for which it gives a full-scale deflection of the needle.

These features of a galvanometer are illustrated in Figure 27.9. Notice when a current I_g flows through the galvanometer, its display shows the maximum reading, a full-scale deflection for an analog device. The small internal resistance plays an important role in the following subsection where we discover how to make voltmeters and ammeters from a galvanometer.

Voltmeters

A **voltmeter** is used to measure the potential difference between two points. In order to measure the potential difference across the resistor shown in Figure 27.10 we connect the terminals of the voltmeter to the two ends of the resistor. That is, we connect the voltmeter in *parallel* to the device across which we want to measure the potential difference. The terminals of a voltmeter are designated as $\oplus$ and $\ominus$, so we must be careful to connect the $\oplus$ terminal to the high-potential side of the resistor and the $\ominus$ terminal to the low-potential side.

When we connect the voltmeter in parallel with the resistor in Figure 27.10, we change the circuit because the voltmeter requires a small current to operate. Remember, a voltmeter is constructed, in part, from a galvanometer and as such has a small internal resistance R_g. If we use a galvanometer alone as the voltmeter in Figure 27.10, this small resistance would dramatically affect the circuit because most of the current would go through the small R_g rather than the much larger load resistance R. In fact, we would probably find ourselves with the smoking remnants of a has-been galvanometer because only a few microamperes are required for most galvanometers to give a full-scale reading! The solution to this problem is to place a very large resistance in series with the galvanometer:

An analog voltmeter consists of a large resistance in *series* with a galvanometer.

The voltmeter, therefore, consists of a series arrangement of a galvanometer and a large resistor R_V as shown in Figure 27.11. Because the galvanometer itself has a resistance R_g, the voltmeter appears to the circuit shown in Figure 27.10 as a large resistance in parallel to the resistor R. If the resistance of the voltmeter is very much larger than R, the current through the voltmeter is very small and the presence of the voltmeter has a very small effect on the circuit. In fact, an **ideal voltmeter** has an *infinite* resistance. The best analog voltmeters have internal resistances of about 10^6 Ω, and typical digital voltmeters reach internal resistances in excess of 10^8 Ω. In Example 27.6 we show how to determine the size of the resistor R_V required to construct a voltmeter to measure a given maximum potential difference ΔV_{max}.

EXAMPLE 27.6   *Building a Voltmeter*

You locate an old galvanometer in the physics stockroom. The label on its back says that this galvanometer has an internal resistance of 50.0 Ω, it provides a full-scale deflection for a current $I_g = 300.$ μA, and it was built in 1924. (a) What resistance R_V must you place in series with the galvanometer to have a voltmeter that reads a maximum potential difference of $\Delta V_{max} = 15.0$ V? (b) When you connect the terminals of your newly constructed voltmeter to the terminals of a battery, you find that the needle on the galvanometer deflects 40% of full scale. What is the potential difference supplied by this battery?

SOLUTION   (a) As illustrated in Figure 27.12, when we place a potential difference $\Delta V_{\max}$ across the terminals A and B of our voltmeter, we want a current I_g to flow through the two resistors R_V and R_g; for then it reads full scale. Because the resistors R_V and R_g are in series, the total resistance through which I_g flows is $R_V + R_g$. Now $\Delta V = IR$, so the potential drop across this resistance is

$$\Delta V_{\max} = I_g(R_V + R_g)$$

Solving for R_V, we have

$$R_V = \frac{\Delta V_{\max}}{I_g} - R_g = \frac{15.0 \text{ V}}{300. \times 10^{-6} \text{ A}} - 50.0 \ \Omega = 49\,950 \ \Omega \approx 50.0 \text{ k}\Omega$$

(b) A needle deflection of 40% of full scale corresponds to the potential difference

$$V = (0.40)(15.0 \text{ V}) = 6.00 \text{ V} \qquad \blacktriangleleft$$

Ammeters

An **ammeter** is used to measure the current through a circuit element and consequently must be placed in *series* with that element (Fig. 27.13). Normally, the currents we wish to measure are much greater than that required to cause a full-scale reading on a galvanometer. Hence, an alternative route is required through which most of the current in a circuit can flow:

> An analog ammeter is constructed by placing a small resistance in *parallel* with a galvanometer.

Such an arrangement is shown in Figure 27.14. During operation, current entering terminal A splits into two currents at junction N so that a current less than or at most equal to I_g flows through R_g. These currents recombine at junction N' and exit the meter at terminal B. Recall that for resistors in parallel the equivalent resistance is smaller than any of the individual resistances. However small the equivalent resistance, when an ammeter is placed in a circuit, the ammeter's resistance is introduced and the original circuit is therefore modified. Thus, an **ideal ammeter** has *zero* resistance.

EXAMPLE 27.7   *Building an Ammeter*

Using the galvanometer described in Example 27.6 ($I_g = 300.$ μA, $R_g = 50.0 \ \Omega$), find the resistance R_A necessary to construct an ammeter that reads a maximum current of 5.00 A.

SOLUTION   In Figure 27.15 we show an ammeter with a current $I_{\max}$ entering the device through terminal A. At junction N the current splits so that current I_g enters the galvanometer and current $I_{\max} - I_g$ passes through resistor R_A. The two resistors R_g and R_A are parallel to each other so the potential difference across each is the same:

$$\Delta V_{R_g} = \Delta V_{R_A}$$

$$(I_{\max} - I_g)R_A = I_g R_g$$

$$R_A = \frac{I_g R_g}{I_{\max} - I_g} = \frac{(300. \times 10^{-6} \text{ A})(50.0 \ \Omega)}{5.00 \text{ A} - 300. \times 10^{-6} \text{ A}} = 3.00 \times 10^{-3} \ \Omega$$

Can you use the techniques described in the previous chapter to show that this R_A value is the resistance of about 60.0 cm of 12-gauge (2.06-mm diameter) copper wire? $\qquad \blacktriangleleft$

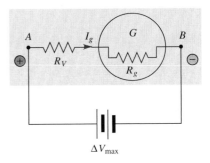

FIGURE 27.12

Constructing a voltmeter that can read full scale for a voltage of 15 V. See Example 27.6.

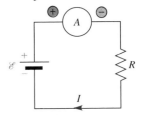

FIGURE 27.13

An ammeter is placed in series with the circuit element through which the measured current flows. The positive side of the ammeter is always connected to the high-potential side in the circuit. Because an ammeter has a small internal resistance, its presence modifies the circuit.

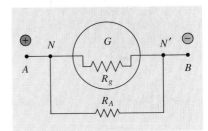

FIGURE 27.14

A schematic representation of an ammeter. The ammeter circuit consists of a small resistance R_A in parallel with the galvanometer resistance R_g.

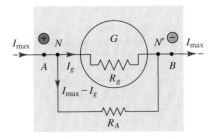

FIGURE 27.15

Constructing an ammeter that can read full scale for a current of 5 A. See Example 27.7.

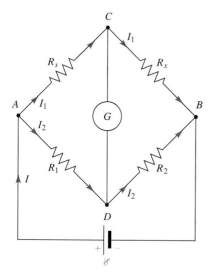

FIGURE 27.16

The Wheatstone bridge is used to measure the value of an unknown resistance. When resistors R_1 and R_2 are adjusted so that no current flows through the galvanometer G, the potential drop across R_s is equal to that across R_1. Moreover, the potential drop across R_x is equal to that across R_2.

The Wheatstone Bridge

We next describe a resistance-measuring device that employs a technique called the **bridge method.** The bridge method has the advantage that it does not require a calibrated meter. Moreover, most ordinary sources of emf work as a source of potential difference. Figure 27.16 shows the circuit diagram for a resistance-measuring device known as a **Wheatstone bridge.** Current flows from the source of emf into junction A where it divides into currents I_1 and I_2. The galvanometer segment of this circuit between junctions C and D forms the bridge. Resistors R_1 and R_2 are variable and are adjusted until no current flows through the galvanometer. When this condition is satisfied, the electric potential at point C must be equal to that at point D (otherwise current would flow through the galvanometer). Hence, the potential drop across resistor R_s must be equal to the potential drop across resistor R_1:

$$I_1 R_s = I_2 R_1 \tag{27.19}$$

Moreover, the potential drop across R_x must be equal to the potential drop across R_2:

$$I_1 R_x = I_2 R_2 \tag{27.20}$$

Dividing these two equations and solving for R_x, we have

$$R_x = \frac{R_2}{R_1} R_s \tag{27.21}$$

Can you see the advantage of this bridge technique? When the circuit is properly adjusted so that Equation (27.21) is applicable (the bridge is said to be *balanced*), no current flows through the galvanometer. Hence, it has no effect on the circuit. Because of practical limitations on the adjustment of resistances R_1 and R_2, the Wheatstone bridge is difficult to apply to unknown resistances much larger than about $10^5 \ \Omega$. Ohmmeters that employ circuits constructed from a special type of transistor (called *field-effect transistors* or FETs) can measure resistances as large as $10^{12} \ \Omega$.

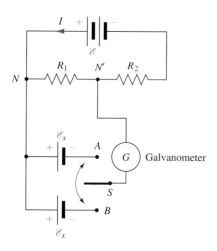

FIGURE 27.17

The schematic of a potentiometer. Resistors R_1 and R_2 can be adjusted but in a way such that the sum $R_1 + R_2$ remains constant. Therefore, the current in the upper loop remains a fixed value I. When switch S is in position A, and R_1 and R_2 are adjusted so that no current flows through the galvanometer, the potential drop across $R_1 = R_s$ is $\mathscr{E}_s$. When switch S is in position B, and R_1 and R_2 are adjusted so that no current flows through the galvanometer, the potential drop across $R_1 = R_x$ is $\mathscr{E}_x$.

The Potentiometer

The final analog instrument we wish to describe to you is the **potentiometer.** This device is used to find the value of an unknown emf $\mathscr{E}_x$ by comparison with a known emf $\mathscr{E}_s$, sometimes called a **standard emf.** The circuit diagram for a potentiometer is shown in Figure 27.17. Switch S can be moved to either position A (where $\mathscr{E}_s$ is connected to the galvanometer) or position B (where $\mathscr{E}_x$ is connected to the galvanometer). The resistors R_1 and R_2 can be varied in such a way that their total resistance remains constant. (How this is accomplished is explained in Problem 36 at the end of this chapter.) Just as in the case of the Wheatstone bridge, R_1 and R_2 are adjusted so that no current flows through the galvanometer. In this case, the current I in the upper loop of Figure 27.17 always has the same value because at junction N no current enters the loop NAN'. Suppose switch S is placed in position A, and R_S is the value of the resistance R_1 for which no current flows through the galvanometer. Then the potential drop across R_s equals the potential $\mathscr{E}_s$:

$$IR_s = \mathscr{E}_s$$

Next, the switch S is changed to position B, and resistances R_1 and R_2 are again adjusted so that the current through the galvanometer is zero (while the sum $R_1 + R_2$ remains constant). If R_x is the value of R_1 for no current through the galvanometer, the potential drop across R_x must equal that across $\mathscr{E}_x$ and

$$IR_x = \mathscr{E}_x$$

Dividing these last two equations and solving for $\mathscr{E}_x$, we have

$$\mathscr{E}_x = \frac{R_x}{R_s}\mathscr{E}_s$$

Therefore, by measuring the values of the two resistances R_x and R_s we can compute the unknown emf $\mathscr{E}_x$ from the known value $\mathscr{E}_s$. Because no current is drawn from the source of emf $\mathscr{E}_x$, the potentiometer provides a more accurate measurement of the emf than simply placing a voltmeter across the terminals. As we saw above, a voltmeter requires some small current in order to operate. This current reduces the measured terminal potential below the true emf of the source (see Section 27.1).

27.5 RC Circuits

To this point we have considered only DC circuits, that is circuits in which currents and potential differences do not change in time. We have ignored what happens immediately after a source of emf is attached to a circuit; we have performed our analysis of the circuit only after it has reached a condition of *steady state*. Although the important applications of circuits that operate in the steady state are too long to list, it is also impossible to tally the uses of circuits with currents and potential differences that change in time. One of the most important time-varying circuits is that of a series combination of a resistor and a capacitor. This circuit forms the basis for many practical applications ranging from variable-speed windshield wipers to the timer delay on electric lights that go dark minutes after you shut them off. We refer to circuits that contain only a resistor, a capacitor, and perhaps a battery, as **RC circuits.**

An *RC* circuit is shown in Figure 27.18. At time $t = 0$ s switch S is closed, current I starts to flow and, as illustrated in Figure 27.20(b), the capacitor begins to charge. Let's

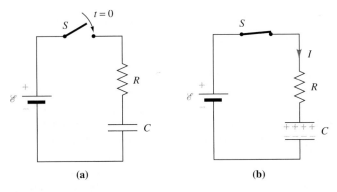

(a)   (b)

FIGURE 27.18

(a) An *RC* circuit. At $t = 0$ s the switch S is closed. (b) As current I flows, the capacitor C charges.

apply Kirchhoff's second law to this circuit. If we begin at the lower-potential terminal of the source of emf and walk clockwise around the circuit, we have

$$+\mathscr{E} - IR - \Delta V_c = 0 \qquad (27.22)$$

where we have recognized that the electric potential decreases across the capacitor as we traverse it from the positive plate to the negative plate. If q is the charge on the capacitor, $\Delta V_c = q/C$ and Equation (27.22) becomes

$$IR + \frac{q}{C} = \mathscr{E} \qquad (27.23)$$

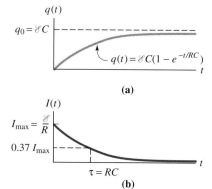

FIGURE 27.19

(a) The charge $q(t)$ on the capacitor of an RC circuit. (b) The current $I(t)$ in the RC circuit decays exponentially from an initial value of $\mathscr{E}/R = I_{max}$.

Now, it is important to recognize that the current I is just the rate at which charge is flowing through the circuit and, therefore, this current is also the rate at which charge is accumulating on the capacitor: $I = dq/dt$. Hence, Equation (27.23) becomes

$$\frac{dq}{dt} + \frac{1}{RC}q = \frac{\mathscr{E}}{R} \qquad (27.24)$$

Equation (27.24) is a differential equation with solution

$$q(t) = \mathscr{E}C(1 - e^{-t/RC}) \qquad (27.25)$$

This result describes the charge $q(t)$ on capacitor C at any time t. You should verify by direct substitution that Equation (27.25) is indeed a solution to Equation (27.24) (see Problem 37). A graph of $q(t)$ is shown in Figure 27.19(a); at $t = 0.00$ s the charge on the capacitor is 0.00 C. This charge $q(t)$ increases, and after a long time it approaches the final value $q_c = \mathscr{E}C$. We can compute the current in the circuit:

$$I(t) = \frac{dq}{dt} = \frac{d}{dt}[\mathscr{E}C(1 - e^{-t/RC})] = \mathscr{E}C\left[0 - \left(-\frac{1}{RC}\right)e^{-t/RC}\right]$$

$$I(t) = \frac{\mathscr{E}}{R}e^{-t/RC} \qquad (27.26)$$

At time $t = 0.00$ s the current in the RC circuit has the value $\mathscr{E}/R$ but decays exponentially toward zero as the capacitor becomes charged. We might expect this result; as charge accumulates on the capacitor plate, more work is required to overcome its growing Coulomb repulsion for the charge carried by the current I. A graph of $I(t)$ is shown in Figure 27.19(b). From Equation (27.26) we note that at time $t = RC$

$$I = \frac{\mathscr{E}}{R}e^{-RC/RC} = \frac{1}{e}\frac{\mathscr{E}}{R}$$

$$I = 0.368I_{max} \qquad \text{(Current after one time constant)}$$

The time $\tau = RC$ is called the **time constant,** and it is the time required for the current to drop to about 36.8% of its maximum (initial) value. The only thing special about 36.8% is that $1/e = 0.3678794\cdots$. That is, the time constant RC is an indicator of how quickly the current in the RC circuit decays. Equivalently, it is also a measure of how quickly the charge $q(t)$ accumulates on the capacitor. Most students (and some physics teachers) think it would be nice if $1/e$ were to equal 0.5 so that $\tau = RC$ would indicate the time for the current to decrease to half its original value. Sometimes physics is just not as simple as we might like. In fact, the time $T_{1/2}$ for the current to decrease to half its original value is $T_{1/2} = 0.693\tau$ (see Problem 68). Figure 27.20 shows the current-versus-time graph for two circuits with different time constants. In Figure 27.20(a) the product RC is a small number so the time constant is also small and the current decays very quickly. In Figure 27.20(b) the values of R and C are such that $\tau = RC$ is a relatively large number (compared to that for Fig. 27-20(a)) and the current decays toward zero very slowly.

The potential differences across the capacitor and the resistor can be computed from Equations (27.25) and (27.26) and are given by

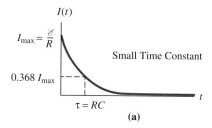

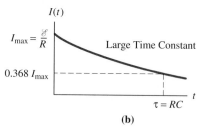

FIGURE 27.20

(a) When the time constant $\tau = RC$ is small, the current decays quickly.
(b) When the time constant is large, the current decays slowly.

$$\Delta V_C(t) = \frac{q(t)}{C} = \mathscr{E}(1 - e^{-t/RC}) \qquad (27.27)$$

$$\Delta V_R(t) = I(t)R = \mathscr{E}e^{-t/RC} \qquad (27.28)$$

If you sum these expressions for $\Delta V_R(t)$ and $\Delta V_C(t)$, you will obtain $\mathscr{E}$ regardless of the time t (try it!), as you must because this is the emf supplied to the series combination of the resistor and capacitor. A graph of $\Delta V_R(t)$, $\Delta V_C(t)$, and their sum is shown in Figure 27.21. The time $T_{1/2}$ in Figure 27.21 is that at which the potential differences across the resistor and capacitor are equal (see Problem 38).

Let's next consider a discharging RC circuit. Such a circuit is shown in Figure 27.22, in which the initial charge on the capacitor is taken to be $q_o = \mathscr{E}C$. When switch S is closed at time $t = 0.00$ s, the current $I(t)$ starts to flow through the resistor as charge begins to leave the capacitor. When we apply Kirchhoff's second law to an anticlockwise walk around this circuit from switch S, we have

$$\frac{q(t)}{C} - I(t)R = 0 \qquad (27.29)$$

However, charge is *leaving* the capacitor, so dq/dt is negative. Thus, to obtain a positive current in the direction shown, we must take $I(t) = -dq/dt$. With this substitution, Equation (27.29) can be rewritten

$$\frac{dq}{dt} + \frac{1}{RC}q = 0$$

$$\frac{dq}{q} = -\frac{1}{RC}\,dt$$

We can integrate this expression by remembering that at time $t = 0.00$ s the charge on the capacitor is q_o, whereas at time t the charge is q:

$$\int_{q_o}^{q} \frac{dq}{q} = -\frac{1}{RC}\int_0^t dt$$

$$\ln\!\left(\frac{q}{q_o}\right) = -\frac{1}{RC}t$$

$$q(t) = q_o e^{-t/RC} \qquad (27.30)$$

The current in the discharging circuit is given by

$$I(t) = -\frac{dq}{dt} = +\frac{q_o}{RC}e^{-t/RC} = \frac{\mathscr{E}C}{RC}e^{-t/RC}$$

$$I(t) = I_o e^{-t/RC} \qquad (27.31)$$

where we have designated the initial current as $\mathscr{E}/R = I_o$. Equations (27.30) and (27.31) show that both the charge on the capacitor and the current in the circuit decay exponentially. Once again, the time constant $\tau = RC$ is a measure of how quickly $q(t)$ and $I(t)$ decrease.

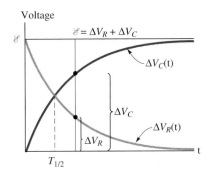

FIGURE 27.21

At any instant the voltage across the resistor ΔV_R and the voltage across the capacitor ΔV_C must sum to $\mathscr{E}$.

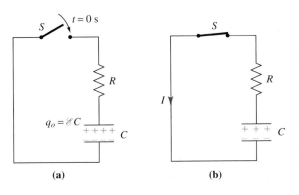

FIGURE 27.22

A discharging RC circuit. (a) Before switch S is closed a charge $q_o = \mathscr{E}C$ is placed on the capacitor by charging the circuit for a long time with a source of emf $\mathscr{E}$. (b) When the capacitor discharges, the current I in the circuit equals the rate of discharge $-dq/dt$ from the capacitor.

Concept Question 9
What effect does the internal resistance of the source of emf have on the time constant of an *RC* circuit? Explain.

Concept Question 10
What effect does the internal resistance of the source of emf have on the final charge on the capacitor of a charging *RC* circuit? Explain.

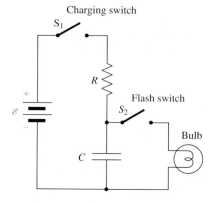

Charging switch

S_1

R

Flash switch

$\mathscr{E}$

S_2

Bulb

C

FIGURE 27.23

An *RC* circuit used to charge an electronic flash unit. Switch S_1 is closed to charge capacitor *C*. After a sufficient charging interval, switch S_2 is closed, and the capacitor is discharged through the flashbulb. See Example 27.8.

Example 27.8 provides an illustration of a useful *RC* circuit. We will, however, reserve the most interesting part of this example for you to work out (see Problem 39).

EXAMPLE 27.8   *An Enlightening Example*

A photographer's electronic flash unit consists of a 50.0-kΩ resistor in series with a 140.-μF capacitor and a 9.00-V source of emf (Fig. 27.23). The flashbulb is placed in parallel with the capacitor so that when a sufficient charge is stored on the capacitor switch S_2 can be closed and the capacitor quickly discharges through the small resistance of the bulb, causing it to flash. Suppose switch S_2 remains open and at $t = 0$ s switch S_1 is closed. (a) How much time elapses before the current through the resistor drops to $1/e$ of its initial value? (b) Determine the functions $q(t)$ and $I(t)$ for this circuit. (c) What charge is on the capacitor at $t = \tau/2$?

SOLUTION   (a) The time constant for this circuit is

$$\tau = RC = (50.0 \times 10^3 \ \Omega)(140. \times 10^{-6} \ \text{F}) = 7.00 \ \text{s}$$

(b) The charge and current functions are given by Equations (27.25) and (27.26):

$$q(t) = \mathscr{E}C(1 - e^{-t/RC})$$

$$I(t) = \frac{\mathscr{E}}{R}e^{-t/RC}$$

Now, $\mathscr{E}C = (9.00 \ \text{V})(140. \times 10^{-6} \ \text{F}) = 1.26 \times 10^{-3} \ \text{C}$ and $\mathscr{E}/R = (9.00 \ \text{V})/(50. \times 10^3 \ \Omega) = 1.8 \times 10^{-4} \ \text{A}$. Hence,

$$q(t) = (1.26 \ \text{mC})(1 - e^{-t/(7.00 \ \text{s})})$$

$$I(t) = (0.180 \ \text{mA})e^{-t/(7.00 \ \text{s})}$$

These functions are graphed in Figure 27.24.

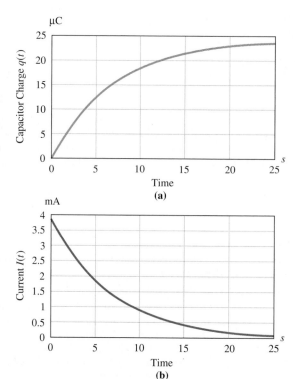

FIGURE 27.24

(a) The charge $q(t)$ on the capacitor for Example 27.8. (b) The current $I(t)$ through this *RC* circuit.

(c) At $t = \tau/2 = (7.00 \text{ s})/2 = 3.50$ s

$$q = (1.26 \text{ mC})(1 - e^{-(3.50 \text{ s})/(7.00 \text{ s})})$$

$$= 0.496 \text{ mC}$$

Can you determine the potential differences across the resistor and the capacitor at time $\tau/2$? Answers: $V_R = 5.46$ V and $V_C = 3.54$ V.   ◀

Electronic Calculus: Differentiating and Integrating Circuits (Optional)

With the proper selection of the resistance and capacitance values in an RC circuit this circuit can perform some electronic calculus for us. There are numerous circuit applications that reduce to a desire for the time derivative or the integral over time of a particular potential difference function that itself is changing in time. Sometimes, time-varying potential difference functions are called electric *signals*. Look first at Figure 27.25, where $V_{in} = \Delta V_{AB}$ is the potential difference across terminals A and B. (We are taking conductor BE as a 0.00-V reference potential.) Similarly $\Delta V_{DE} = V_{out}$ is the potential at point D relative to the 0.00-V reference and is equal to the potential drop across the resistor. The potential difference ΔV_C across the capacitor is $V_{in} - V_{out}$, and the current in the circuit can be calculated from the rate of change of charge on the capacitor:

$$I = \frac{dq}{dt} = C\frac{d}{dt}V_C = C\frac{d}{dt}(V_{in} - V_{out})$$

$$I = C\left(\frac{dV_{in}}{dt} - \frac{dV_{out}}{dt}\right)$$

The current is also given by

$$I = \frac{V_{out}}{R}$$

Equating these two expressions for I, we have

$$\frac{dV_{in}}{dt} - \frac{dV_{out}}{dt} = \frac{V_{out}}{RC} \qquad (27.32)$$

If we choose RC to be small, the right-hand side of Equation (27.32) is large and $dV_{in}/dt \gg dV_{out}/dt$. If dV_{out}/dt is small enough compared to dV_{in}/dt to be ignored, Equation (27.32) becomes

$$V_{out} = RC\frac{dV_{in}}{dt} \qquad (27.33)$$

The output voltage signal is the derivative of the input voltage signal, multiplied by the constant RC. For example, if the input voltage function is that shown in Figure 27.26(a), the output voltage function is that shown in Figure 27.26(b).

Next, consider the circuit shown in Figure 27.27. If R and C are selected so that the product RC is large enough that $V_{in} \gg V_{out}$ and the latter can be ignored, then (see Problem 48(a))

$$V_{out} = \frac{1}{RC}\int V_{in}\, dt + \text{constant} \qquad (27.34)$$

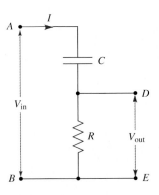

FIGURE 27.25

For sufficiently small RC the output voltage V_{out} is the time derivative of V_{in}.

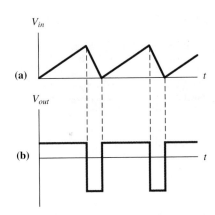

FIGURE 27.26

V_{out} for the given V_{in}.

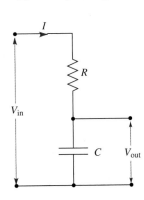

FIGURE 27.27

For sufficiently large RC the output voltage V_{out} is the integral over time of V_{in}.

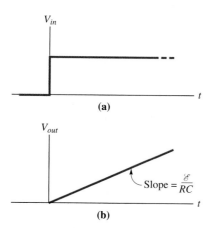

FIGURE 27.28

As t increases from $t = 0$, the integral of the constant function $V_{in} = \mathcal{E}$ is the ramp with slope $\mathcal{E}/RC$.

The output voltage signal is the integral of the input signal, multiplied by the constant $1/RC$. Because V_{out} is the potential difference across the capacitor, you may be wondering why Equation (27.34) appears so different from the equation we first derived for this potential difference, Equation (27.27). In the derivation of Equation (27.27) our input voltage began at $t = 0$ s and had a constant magnitude $V_{in} = \mathcal{E}$. This V_{in} signal is shown in Figure 27.28(a). When we substitute $V_{in} = \mathcal{E}$ into Equation (27.34) and make the constant zero to correspond to this value at $t = 0$ s, we have

$$V_{out} = \frac{1}{RC} \int_0^t \mathcal{E}\, dt$$

$$V_{out} = \frac{\mathcal{E}}{RC} t \tag{27.35}$$

Equation (27.35) is a linear equation and is graphed in Figure 27.28(b). This particular output signal is called a *voltage ramp* and its slope (the time rate of voltage increase) is $\mathcal{E}/RC$. Keep in mind that this result is contingent on RC being large enough that $V_{in} \gg V_{out}$. We leave it as an exercise for you to show that for large RC, Equation (27.27) gives Equation (27.35) (see Problem 48(b)).

27.6 Digital Voltmeters (Optional)

Modern digital voltmeters make use of RC circuits and internal clocks to measure potential differences. One type, known as a **dual slope**, integrating meter, charges an internal capacitor through an internal series resistor. To make a potential difference measurement two distinct processes occur after the voltmeter is connected to the external potential difference: (1) the discharged internal capacitor is first charged by the external potential difference *for a fixed time*; (2) then the internal capacitor is discharged by a *fixed current*. Let's examine these two processes to see how they can be used to calculate the unknown potential difference.

We have seen that a discharged capacitor connected to an external emf $\mathcal{E}$ charges according to Equation (27.25)

$$q(t) = \mathcal{E}C(1 - e^{-t/RC})$$

In the voltmeter's first cycle, t is a fixed interval known as the charging time, which we call T_C. The charging time is chosen to be short compared to the time constant of the voltmeter's internal R and C. For small T_C/RC the exponential in Equation (27.25) can be expanded by the binomial theorem to give

$$e^{-T_c/RC} \approx 1 - \frac{T_C}{RC} + \cdots$$

Consequently, we have

$$q(T_C) \approx \mathcal{E}C\frac{T_C}{RC} = \frac{T_C}{R}\mathcal{E}$$

The point here is that the charge q stored on the capacitor during the fixed charging time is directly proportional to the unknown voltage $\mathcal{E}$.

After the time T_C, the capacitor is disconnected from the external circuit by a transistor switch and then connected to an internal discharge circuit. This discharge circuit is a combination of transistors, which guarantees that *the capacitor discharge current I_D is constant*. This means that the capacitor is completely discharged in a time T_D given by

$$T_D = \frac{q}{I_D}$$

where I_D is the magnitude of the discharge current. Putting the expression for the charge q accumulated during the charging process into the relation above, we find

$$T_D = \left(\frac{T_C}{I_D R}\right)\mathscr{E}$$

The point to notice here is that the discharge time is directly proportional to the unknown potential difference being measured. Notice, too, that once R, C, and I_D are chosen, the process of measuring a potential difference is reduced to measuring a time, and measuring time is something that digital circuits do very well by counting the "ticks" from a fixed-frequency oscillator.

27.7 A Matrix Method for Complex Circuits (Optional)

In this section we show you how to use the matrix power of a spreadsheet to solve the system of linear equations generated by the application of Kirchhoff's laws to a circuit. You can use the methods described below even if you have never before encountered a matrix. You need to understand that a matrix is a carefully arranged rectangular array of numbers and that matrices can be multiplied using rules we will not describe here. (See Appendix 1 if you are interested.)

Perhaps you wondered about the strange way we chose to present Equations (27.16) in Example 27.4. We wrote these equations in a form such that the symbols for the currents occupied the same position in each equation. For convenience we rewrite the equations here

$$I_1 + I_2 + I_3 = 0 \qquad (27.36a)$$

$$R_1 I_1 + R_2 I_2 + (0)I_3 = \mathscr{E}_1 - \mathscr{E}_3 \qquad (27.36b)$$

$$(0)I_1 + R_2 I_2 - R_3 I_3 = \mathscr{E}_2 - \mathscr{E}_3 \qquad (27.36c)$$

Notice that this time we have filled in the currents missing from the second and third equations by including zero coefficients. We can represent Equations (27.36) by a matrix equation as shown here with a symbolic definition of each matrix given below each.

$$\underbrace{\begin{bmatrix} 1 & -1 & -1 \\ R_1 & R_2 & 0 \\ 0 & R_2 & -R_3 \end{bmatrix}}_{\{R\}} \underbrace{\begin{bmatrix} I_1 \\ I_2 \\ I_3 \end{bmatrix}}_{\{I\}} = \underbrace{\begin{bmatrix} 0 \\ \mathscr{E}_1 - \mathscr{E}_2 \\ \mathscr{E}_2 - \mathscr{E}_3 \end{bmatrix}}_{\{E\}} \qquad (27.37)$$

If you know how to multiply matrices, you should verify that when the matrix multiplication $\{R\}\{I\}$ is performed the result is a matrix the rows of which appear as the left side of Equations (27.36). The matrix equation for Equation (27.27) is

$$\{R\}\{I\} = \{E\} \qquad (27.38)$$

If we can find the inverse of matrix $\{R\}$, called $\{R\}^{-1}$, we can multiply both sides of Equation (27.38) by $\{R\}^{-1}$ to obtain

$$\{R\}^{-1}\{R\}\{I\} = \{R\}^{-1}\{E\} \qquad (27.39)$$

Now, the product $\{R\}^{-1}\{R\}$ is the identity matrix $\{\mathscr{I}\}$ (ones down the diagonal and zeros everywhere else), and when $\{\mathscr{I}\}$ multiplies the current matrix $\{I\}$ the result is just $\{I\}$; that is, $\{R\}^{-1}\{R\}\{I\} = \{\mathscr{I}\}\{I\} = \{I\}$. Hence,

$$\{I\} = \{R\}^{-1}\{E\} \qquad (27.40)$$

The rows of the matrix $\{I\}$, computed by performing the product of $\{R\}^{-1}$ and $\{E\}$, contain the unknown currents I_1, I_2, and I_3. We hasten to remind you that matrices do not commute. Therefore, you must always multiply $\{R\}^{-1}$ and $\{E\}$ *in the order shown by Equation* (27.40).

To solve for the unknown currents in Equation (27.37) our first step must be to find $\{R\}^{-1}$ because then we can perform the multiplication given by Equation (27.39). The straightforward technique for computing the inverse of a matrix is sometimes tedious and is illustrated in most linear algebra texts. We will not describe this process but simply urge you to listen to your advisor when she or he advocates your enrollment in a linear algebra course. Instead we will take advantage of the matrix inversion and multiplication routines provided by most spreadsheet programs.[5]

Here is a general description of how you can use a spreadsheet to solve for the currents in the matrix $\{I\}$. Below Figure 27.29 we have included a list of spreadsheet commands that should, if executed in the order given, perform the operations we are about to describe. You must first enter the elements of the matrices $\{R\}$ and $\{E\}$ in two ranges of cells. To invert the matrix $\{R\}$ the usual procedure with most spreadsheets is to first select the DATA (or, perhaps ADVANCED) submenu from the main menu. Next choose the MATRIX submenu followed by the INVERT option. At this point there are usually just two steps: (1) designate the range of cells that contain the matrix $\{R\}$ and (2) designate a place on the spreadsheet to put the inverted matrix. Once these steps are completed, the spreadsheet computes the matrix $\{R\}^{-1}$ and places it in the set of cells you designated to hold the inverse.

The next step is to multiply $\{R\}^{-1}$ times the matrix $\{E\}$. After you select the matrix multiplication option from the MATRIX submenu, the multiplication procedure usually involves three steps: (1) designate the first matrix ($\{R\}^{-1}$ in our case), (2) designate the second matrix $\{E\}$, and (3) designate the place to put the result of the multiplication. It is important that you keep the order of matrix multiplication correct because matrices do not commute: $\{R\}^{-1}$ is the first matrix and $\{E\}$ is the second. At this point our task is complete; this last matrix contains one column and the currents I_1, I_2, $\cdots$ are the elements in rows 1, 2, $\cdots$, respectively.

On the disk accompanying this text there is a spreadsheet called KIRCHHOF.WK1, which allows you to find up to five unknown currents from the equations generated by an application of Kirchhoff's laws. You must always write these equations in the form illustrated in Equations (27.36) so that you will obtain the proper coefficients in the proper cells. That is, $I_1, I_2, I_3, \cdots$ must appear in just this order in each of the equations. Below we illustrate how KIRCHHOF.WK1 can be used to solve the set of equations generated in Example 27.4. In this case we have only three equations and three unknowns. The general procedure here is to include the equations $I_4 = 0$ and $I_5 = 0$ to our set of equations by simply putting ones on the diagonal of the remaining rows and columns of the $\{R\}$ matrix, and to put zeros in the other positions of the fourth and fifth rows and columns. Zeros are also placed in the fourth and fifth row of the $\{E\}$ matrix. The matrix equation that represents Equation (27.37) with the values from Example 27.4 becomes

$$\underbrace{\begin{bmatrix} 1 & -1 & 1 & 0 & 0 \\ 0 & 3 & 3 & 0 & 0 \\ 4 & 0 & -3 & 0 & 0 \\ 0 & 0 & 0 & 1 & 0 \\ 0 & 0 & 0 & 0 & 1 \end{bmatrix}}_{\{R\}} \underbrace{\begin{bmatrix} I_1 \\ I_2 \\ I_3 \\ I_4 \\ I_5 \end{bmatrix}}_{\{I\}} = \underbrace{\begin{bmatrix} 0 \\ 8 \\ 4 \\ 0 \\ 0 \end{bmatrix}}_{\{E\}}$$

---

[5]Many modern scientific calculators also contain routines for solving simultaneous equations using matrix notation. Often the operation is as simple as keying in the $\{R\}$ matrix and the $\{E\}$ matrix (sometimes called a vector) from Equation (27.37) and then hitting the "divide" key on your calculator. Check the owner's manual to see if your calculator has this capability. If it does, it's well worth half an hour of your time to learn to use it!

Notice that the equations that correspond to the fourth and fifth row are $I_4 = 0$, and $I_5 = 0$, respectively.

The spreadsheet KIRCHHOF.WK1 appears as:

	A	B	C	D	E	F	G	H	I	J
1										
2			Matrix Solution for Circuit Equations							
3										
4				R—Matrix					E—Matrix	
5		1	−1	1	0	0			0	
6		0	3	3	0	0			8	
7		4	0	−3	0	0			4	
8		0	0	0	1	0			0	
9		0	0	0	0	1			0	
10										
11										
12				R—Inverse—Matrix					I—Values	
13		0.273	0.091	0.182	0.000	0.000			1.455	
14		−0.364	0.212	0.091	0.000	0.000			2.061	
15		0.364	0.121	−0.091	0.000	0.000			0.606	
16		0.000	0.000	0.000	1.000	0.000			0.000	
17		0.000	0.000	0.000	0.000	1.000			0.000	
18										
19										
20										

FIGURE 27.29

The spreadsheet KIRCHHOF.WK1.

The $\{R\}$ matrix is contained in cells B5 through F9, and the elements of matrix $\{E\}$ are in cells I5 through I9. The computed inverse matrix $\{R\}^{-1}$ is in cells B13 through F17, and the results of the multiplication of $\{R\}^{-1}\{E\}$ are in cells I13 through I17. After the elements of $\{R\}$ and $\{E\}$ are entered in their appropriate cells, a typical sequence of spreadsheet commands to compute $\{I\}$ might be (with our explanation contained in brackets):

/	[obtain the main menu]
D	[select DATA submenu] *or* A [select ADVANCED submenu]
M	[select MATRIX submenu]
I	[select matrix inversion]
B5 . . . F9	[designate matrix to be inverted]
B13	[designate where the inverted matrix is to be placed]

At this point $\{R\}^{-1}$ is computed and placed in cells B13 through B17. To perform the matrix multiplication follow these steps:

/	[obtain the main menu]
D	[select DATA submenu] *or* A [select ADVANCED submenu]
M	[select MATRIX submenu]
M	[select matrix multiplication]
B13 . . . F17	[designate first matrix]
I5 . . . I9	[designate second matrix]
I13	[designate where the result is to be placed]

After you instruct the spreadsheet to perform the matrix multiplication the solution currents should appear in cells I13 through I17. The answers in the worksheet shown in Figure 27.29 are those we obtained in Example 27.4. If you have some experience with macros you may want to create one to perform the steps given above.

27.8 Summary

Any device that can do work on a charge to increase that charge's potential energy is called a **source of emf.** The magnitude of the emf is given by

$$\mathcal{E} = \frac{dW}{dq} \tag{27.1}$$

Common sources of emf include batteries, generators, and solar cells. All real sources of emf have an **internal resistance** r so that the **terminal potential** depends on the current delivered by the emf.

The current through all circuit elements connected in a series arrangement is the same. For N resistors connected in series, the equivalent resistance is given by

$$R_{\text{tot}} = R_1 + R_2 + R_3 + \cdots + R_N = \sum_{i=1}^{N} R_i \tag{27.7}$$

The change in electric potential is the same across all circuit elements connected in a parallel arrangement. The equivalent resistance for N resistors connected in a parallel arrangement is given by

$$\frac{1}{R_{\text{tot}}} = \frac{1}{R_1} + \frac{1}{R_2} + \frac{1}{R_3} + \cdots + \frac{1}{R_N} = \sum_{i=1}^{N} \frac{1}{R_i} \tag{27.11}$$

Kirchhoff's laws can be applied to write circuit equations from which the currents through circuits containing one or more branches can be found. These laws are:

1. At any circuit junction the sum of the currents entering the junction must equal the sum of the currents leaving the junction.
2. The sum of the potential changes across all the circuit elements around any closed circuit loop must be zero.

A **galvanometer** is a current-measuring device that has an internal resistance and requires a small current for a maximum reading. A **voltmeter** can be constructed from a large resistance in series with a galvanometer, and an **ammeter** is built by placing a small resistance in parallel with a galvanometer. A **Wheatstone bridge** is a device that can be used to measure an unknown resistance, and a **potentiometer** can be used to measure the magnitude of an unknown emf.

When a source of emf is attached to a series combination of a resistor R and a capacitor C, the charge on the capacitor and the current in the circuit vary in time according to the relations

$$q(t) = \mathcal{E}C(1 - e^{-t/RC}) \tag{27.25}$$

$$I(t) = \frac{\mathcal{E}}{R}e^{-t/RC} \tag{27.26}$$

The **time constant** $\tau = RC$ is a measure of how quickly the current $I(t)$ in the circuit decays or, equivalently, how fast the charge $q(t)$ accumulates on the capacitor. For a capacitor discharging through a resistor (with no source of emf involved) the charge on the capacitor and the current in the circuit are given by

$$q(t) = q_o e^{-t/RC} \tag{27.30}$$

$$I(t) = I_o e^{-t/RC} \tag{27.31}$$

PROBLEMS

27.1 Electromotive Force

1. Show that the power dissipated in the load resistance of Figure 27.1 is a maximum when the value of the load resistance R is equal to the internal resistance r.

2. When the terminals of a battery are connected by a thick wire of negligible resistance, a current of 40.0 A flows through the wire. If the battery has an emf of 8.00 V what is its internal resistance?

3. Compute the potential difference across an 8.00-Ω resistor that is connected across the terminals of a 10.0-V emf with an internal resistance of 0.250 Ω.

4. A resistance R is connected across the terminals of a battery that has an emf of 12.0 V and an internal resistance r of 0.400 Ω. If the current through the load resistor is 3.00 A, what is R?

5. When a load resistance of 8.00 Ω is placed between the terminals of a source of emf, the potential difference across the emf is found to be 22.0 V. The manufacturer claims that the emf of the source is 24.0 V. What is its internal resistance?

6. If each of the headlights of an automobile draws a current of 2.00 A, how much energy is used from a 12.0-V battery when the headlights are left on for 1.00 h while the engine is not running?

7. When a battery with an emf of 6.00 V is connected to a resistor R, the power dissipated in this load resistor is 8.20 W. If the terminal potential of the battery is 5.70 V, find the load resistance R and the internal resistance r.

27.2 Combinations of Resistors

8. Find the equivalent resistance between points A and B of Figure 27.P1.

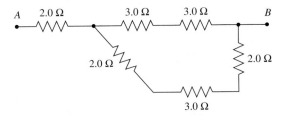

FIGURE 27.P1   Problem 8

9. Find the equivalent resistance between points A and B of Figure 27.P2.

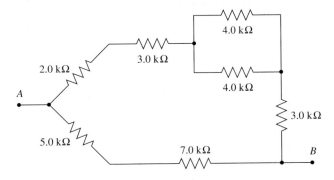

FIGURE 27.P2   Problem 9

10. A 72.0-V battery has an internal resistance of 6.00 Ω. (a) A 24.0-Ω resistor and a 36.0-Ω resistor are connected in series across the terminals of the battery. What is the potential drop across the 24.0-Ω resistor? (b) What is the potential drop across the 24.0-Ω resistor when it and the 36.0-Ω resistor are connected in parallel across the battery terminals?

11. Find the equivalent resistance between points A and B of Figure 27.P3.

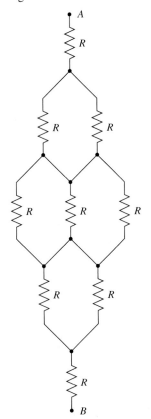

FIGURE 27.P3   Problem 11

12. What is the equivalent resistance between points A and B in Figure 27.P4?

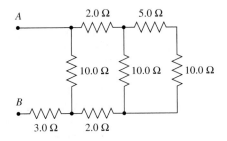

FIGURE 27.P4   Problems 12 and 17

13. For the circuit shown in Figure 27.P5, $R_1 = 4.00\ \Omega$, $R_2 = 8.00\ \Omega$, $R_3 = 10.0\ \Omega$, $R_4 = 4.00\ \Omega$, $R_5 = 6.00\ \Omega$, and $\mathscr{E} = 36.0$ V. Find

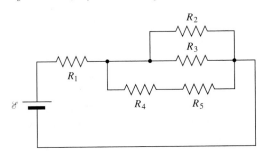

FIGURE 27.P5   Problem 13

the equivalent resistance between the terminals of the source of emf, the current in R_4, and the potential difference across R_2.

14. For the circuit shown in Figure 27.P6, $R_1 = 9.00\ \Omega$, $R_2 = 6.00\ \Omega$, $R_3 = 4.00\ \Omega$, $R_4 = 2.00\ \Omega$, $R_5 = 3.00\ \Omega$, and $\mathcal{E} = 12.0$ V. Find the equivalent resistance between the terminals of the 12.0-V source of emf and the power dissipated in R_2.

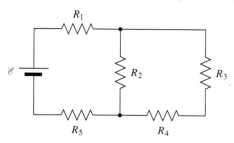

FIGURE 27.P6   Problems 14 and 15

15. Find the potential difference across each resistor in Problem 14.
16. Find the equivalent resistance across the terminals of the 24.0-V source of emf in Figure 27.P7.

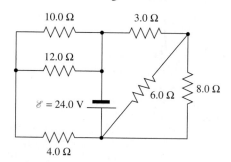

FIGURE 27.P7   Problems 16 and 18

17. If the current in the 5.00-Ω resistor of Figure 27.P4 is 0.200 A, what is the potential difference between points A and B?
18. What is the current through each resistor in Figure 27.P7?

27.3 Multiple-Loop Circuits: Kirchhoff's Laws

19. For the circuit shown in Figure 27.P8 $R_1 = 4.00\ \Omega$, $R_2 = 8.00\ \Omega$, $R_3 = 2.00\ \Omega$, $\mathcal{E}_1 = 12.0$ V, and $\mathcal{E}_2 = 4.00$ V. Find the current in each resistor.

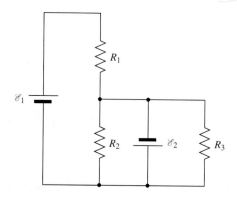

FIGURE 27.P8   Problem 19

20. For the circuit shown in Figure 27.P9, $R_1 = 4.00$ kΩ, $R_2 = 3.00$ kΩ, $R_3 = 2.00$ kΩ, $R_4 = 1.00$ kΩ, $\mathcal{E}_1 = 10.0$ V, and $\mathcal{E}_2 = 8.00$ V. Find the power supplied by $\mathcal{E}_2$ and the potential drop across R_3.

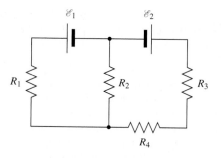

FIGURE 27.P9   Problem 20

21. Find (a) the current through each of the resistors in the circuit shown in Figure 27.P10, and (b) the potential difference between points A and B. Which point is at the higher potential? Take $R_1 = 4.00$ kΩ, $R_2 = 4.00$ kΩ, $R_3 = 2.00$ kΩ, $\mathcal{E}_1 = 12.0$ V, and $\mathcal{E}_2 = 8.00$ V.

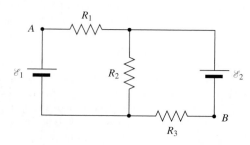

FIGURE 27.P10   Problem 21

22. Analyze the circuit shown in Figure 27.P11 for $R_1 = 5.00\ \Omega$, $R_2 = 2.00\ \Omega$, $R_3 = 6.00\ \Omega$, $R_4 = 5.00\ \Omega$, $\mathcal{E}_1 = 8.00$ V, $\mathcal{E}_2 = 16.0$ V, and $\mathcal{E}_3 = 20.0$ V. Find (a) the current through each resistor and (b) the potential difference between points A and B.

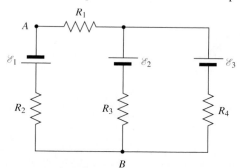

FIGURE 27.P11   Problem 22

23. Find the current in each resistor of the circuit shown in Figure 27.P12.

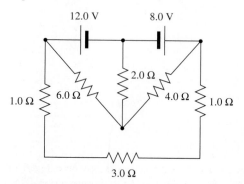

FIGURE 27.P12   Problem 23

24. Compute the potential drop across the 2.0-kΩ resistor and the power dissipated in the 4.0-kΩ resistor for the circuit shown in Figure 27.P13.

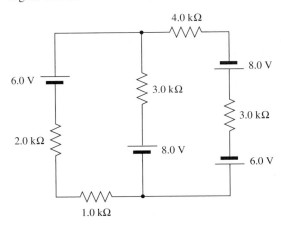

FIGURE 27.P13   Problem 24

25. For the circuit shown in Figure 27.P14, R_1 = 6.00 kΩ, R_2 = 8.00 kΩ, R_3 = 2.00 kΩ, R_4 = 4.00 kΩ, $\mathcal{E}_1$ = 12.0 V, $\mathcal{E}_2$ = 4.00 V, and $\mathcal{E}_3$ = 8.00 V. Find the power dissipated in R_4, the power supplied by $\mathcal{E}_2$, and the potential drop across R_1.

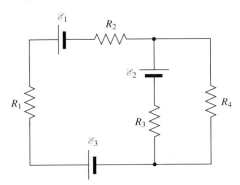

FIGURE 27.P14   Problem 25

27.4 Potential Difference, Current, and Resistance Measurements

26. A galvanometer has an internal resistance of 40.0 Ω and provides a full-scale reading when a current of 200. μA flows through it. Describe how to make this galvanometer into the following meters: (a) a 5.00-V voltmeter; (b) a 50.0-V voltmeter; (c) a 2.00-mA ammeter; (d) a 2.00-A ammeter.

27. A galvanometer has an internal resistance of 60.0 Ω and provides a full-scale reading when a current of 0.50 mA flows through it. Describe how to make this galvanometer into the following meters: (a) a 10.0-A ammeter; (b) a 1.00-A ammeter; (c) a 20.0-V voltmeter; (d) a 200.-V voltmeter.

28. When a resistor is connected to a 24.0-V source of emf a current of 2.00 A flows through the resistor. What does an ammeter with an internal resistance of 0.150 Ω read when it is placed in the circuit?

29. A 6.00-V source of emf with an internal resistance of 15.0 Ω is connected to a series combination of a 30.0-Ω and a 50.0-Ω resistor. A voltmeter with an internal resistance of 2.00 kΩ is used to measure the potential difference across the 50.0-Ω resistor. What is the percentage error in the potential difference indicated by the voltmeter?

30. An ideal source of emf is connected across a series combination of an ammeter with an internal resistance of 0.0400 Ω and a 5.000-Ω resistor. The ammeter indicates the current to be 3.175 A. What current flows through the resistor when the ammeter is removed from the circuit and only the resistor is connected to the source of emf?

31. Consider the circuit shown in Figure 27.P15. (a) What potential difference does a voltmeter with an internal resistance of 20.0 kΩ read between points A and B? (b) What current does an ammeter with an internal resistance of 100. Ω read if it is connected between points A and B?

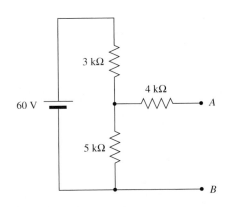

FIGURE 27.P15   Problem 31

32. A Wheatstone bridge is used to measure the resistance of an unknown resistor R_x. The standard resistor is 10.00 m of 18-gauge copper wire (diameter = 1.02 mm). The bridge is balanced when the resistances in Figure 27.16 are such that R_1 = 1.80 R_2. What is the value of R_x?

33. When a Wheatstone bridge is used to measure the resistance of a long, insulated segment of wire, the ratio R_2/R_1 is found to be 3.60. The wire is cut into three segments of equal length. The three segments are laid out parallel to each other, and their ends are connected together so that they form three resistors of equal value in a parallel arrangement. This parallel combination is then reattached as the unknown resistor in the Wheatstone bridge. What is the new ratio of R_2/R_1?

34. In terms of R_1, R_2, R_x, R_s, and the emf supply $\mathcal{E}$, find the current through the galvanometer of an *unbalanced* Wheatstone bridge.

35. When a certain potentiometer is used to measure an unknown emf, the standard emf has a value of 1.2076 V. When the standard emf is in the circuit and resistances R_1 and R_2 are adjusted so that no current flows through the galvanometer, the values of R_1 and R_2 are 2.42 Ω and 4.84 Ω, respectively. When the unknown emf is in the circuit, the value of R_2 is 3.66 Ω. What is the value of the unknown emf?

36. A slide-wire potentiometer is shown in Figure 27.P16. In this device the resistances R_1 and R_2 of Figure 27.17 are actually a uniform wire 1.000 m in length. Hence, the resistance R_1 is proportional to length L_1, and the resistance R_2 is proportional to the length L_2. The galvanometer junction N' in Figure 27.17 is a sharp-edged key that can slide along the wire and be made to make contact with it at any position along its length. When a standard cell with an emf of 1.0218 V is used and the switch S is in position A, the length L_1 is 42.05 cm for no current to flow through the galvanometer. When the switch is moved to position B, the length L_1 must be changed to 64.10 cm for the galvanometer to carry no current. What is the value of the emf $\mathcal{E}_x$?

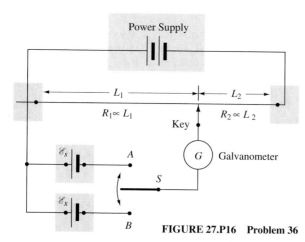

FIGURE 27.P16   Problem 36

27.5 RC Circuits

37. By direct substitution verify that Equation (27.25) is a solution to the differential equation, Equation (27.24).

38. Show that if an uncharged capacitor is attached to an emf through a series resistance R at time $t = 0$, then at time $t = RC \ln(2)$ the potential differences across the resistor and the capacitor of a charging RC circuit are equal.

39. The potential difference across the capacitor in Example 27.8 necessary for the flashbulb to work is 6.00 V. How much time elapses after switch S_1 is closed before the capacitor reaches this potential difference?

40. How many time constants must pass before a charging capacitor reaches 99.0% of the potential difference applied across it?

41. A capacitor is charged to a potential of 10.0 V and is then connected across a voltmeter with a resistance of 1.00 MΩ (1.00 megaohm). After 8.00 s the voltmeter reads 5.00 V. What is the capacitance of the capacitor?

42. A 5.00-μF capacitor is completely charged with a 24.0-V source of emf. It is then connected across a 2.00-MΩ resistor. (a) How long does it take for the potential difference across the capacitor to reach 6.00 V? (b) What is the current in the circuit at this time?

43. At $t = 0$ s the switch in Figure 27.P17 is closed. After 160. s the charge on the capacitor is 8.00 mC. (a) What is the value of the resistance R? (b) What is the potential difference across R at $t = 120$. s?

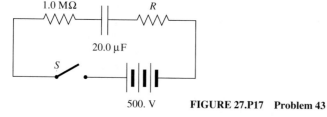

500. V   **FIGURE 27.P17   Problem 43**

44. The RC circuit shown in Figure 27.18 consists of a 20.0-V source of emf, a 10.0-μF capacitor, and an unknown resistance. When the switch S is closed, the current is found to drop to $1/e$ of its initial value in 5.00 s. (a) What is the resistance R? (b) What is the potential drop across the resistor 3.00 s after the switch is closed?

45. A 1.00-μF capacitor is connected in series to a 5.00-MΩ resistor. With the capacitor discharged the combination is attached to a 12.0-V source of emf for 5.00 s. (a) How much charge is on the capacitor at $t = 5.00$ s? (b) At the end of 5.00 s the source of emf is removed, and the ends of the series combination of R and C are connected together for 5.00 s. How much charge remains on the capacitor after the 5.00-s connection?

46. What percentage of the energy supplied by a source of emf to a charging RC circuit becomes stored on the capacitor?

47. The RC circuit shown in Figure 27.18 consists of a 12.0-V source of emf, a 1.00-kΩ resistor, and a 2.00-μF uncharged capacitor. How long after the switch is closed does it take for the capacitor to charge from one third of its final potential difference to two-thirds of its final potential difference?

48. (a) Show for $t \ll RC$ in Figure 27.27, Equation (27.34) provides V_{out}. (b) Show for large RC, $V_{\text{in}} \gg V_{\text{out}}$ in Figure 27.27 and V_{out} is given by Equation (27.35).

27.6 Digital Voltmeters

49. You wish to design a digital voltmeter with a charging time T_C of 100. ms and an input resistance of 1.00 MΩ. (a) What size capacitor should you use? (Use the criterion that $\tau = 10 T_C$ to ensure that $T_C \ll \tau$.) (b) If the voltmeter is used to measure a potential difference of 5.00 V, how much charge accumulates on the capacitor? (c) During the discharge cycle a current of 2.00 μA is used. How long does it take the capacitor to discharge?

27.7 A Matrix Method for Complex Circuits

50. Use the spreadsheet technique of matrix inversion and multiplication to find five different currents for the circuit shown in Figure 27.P18.

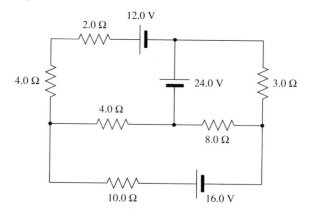

FIGURE 27.P18   Problem 50

51. Apply the matrix inversion and multiplication technique with a spreadsheet to find the current through each of the resistors in Figure 27.P19.

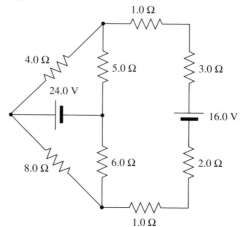

FIGURE 27.P19   Problem 51

52. Use the spreadsheet technique of matrix inversion and multiplication to find the current through each resistor in the circuit shown in Figure 27.P20.

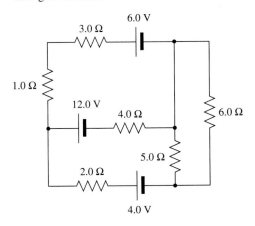

FIGURE 27.P20 Problem 52

53. Modify the spreadsheet KIRCHHOF.WK1 to accommodate eight equations with eight unknown currents and solve for the current in each resistor of Figure 27.P21.

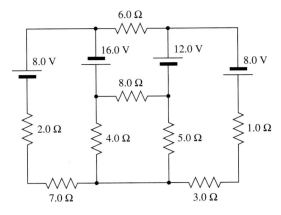

FIGURE 27.P21 Problem 53

Numerical Methods and Computer Applications

54. The internal resistance of a certain battery is $r = 0.500$ Ω. The emf of the battery is 12.0 V. Write a program or construct a spreadsheet template to plot the power dissipation in a load resistance R as a function of R in the range $0 \le R \le 2.00$ Ω. Increment R in steps of 0.100 Ω. At what value of R is the power dissipation maximum? Change r to 1.00 Ω and verify that the power dissipation is a maximum when $R = r$.

55. Write a program or construct a spreadsheet template to plot the $q(t)$ and $I(t)$ curves for an RC circuit. Test your program with the R, C, and $\mathscr{E}$ values from (a) Example 27.8 and (b) Problem 45.

56. Real switches dissipate energy. One model for a switch employs the following function for the resistance of a switch as a function of time [See S. D. Harper, "The energy dissipated by a switch," *American Journal of Physics*, **56**, 886 (1988).]: $r = R_s(T_s/t - 1)$. Notice when $t = 0$ the switch resistance is infinite and at $t = T_s$ the resistance is zero. The *switch resistance* R_s characterizes how quickly the resistance approaches zero; you should verify that at $T = \frac{2}{3}T_s$, $r = R_s/2$. The expression for the power dissipated by the switch when it is placed in a circuit consisting of a series combination of a potential difference source V and a load resis-

tance R is

$$\mathscr{P} = \frac{V^2}{R}K\left(\frac{t/T_s - (t/T_s)^2}{[1 - (t/T_s)(1 - K)]^2}\right)$$

where $K = R/R_s$, and t is the time 0 s $\le t \le T_s$. Write a program or develop a spreadsheet template to compute and plot the power at 51 times between $t = 0$ s and $t = T_s$. The energy dissipated by the switch is the area under the $\mathscr{P}$-versus-t curve. Numerically compute this area. Verify that for a given load resistance R the power dissipated approaches a maximum when R_s gets close to the load resistance R (but it's best not to let $R_s = R$. Why?). You may want to compare your calculated value of the energy dissipated with the exact expression given in the reference.

General Problems

57. You find an old battery in the physics stockroom. When you place a 10.0-Ω resistor between its terminals, the current through the resistor is 1.50 A. When the 10.0 Ω-resistor is replaced with a 5.00-Ω resistor, the current changes to 2.80 A. What is the battery's emf and internal resistance?

58. For the circuit shown in Figure 27.P22 $R_1 = 5.00$ Ω, $R_2 = 9.00$ Ω, $\mathscr{E}_1 = 8.00$ V, $\mathscr{E}_2 = 6.00$ V, and $\mathscr{E}_3 = 4.00$ V. Find (a) the current in each resistor and (b) the potential difference between points A and B.

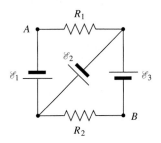

FIGURE 27.P22 Problem 58

59. For the circuit shown in Figure 27.P23 find the current flowing from the 16.0-V source of emf and the potential difference across the 6.00-Ω resistor.

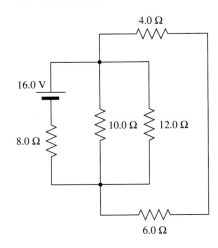

FIGURE 27.P23 Problem 59

★60. Six 10.0-Ω resistors are connected together so that they form a tetrahedron. What is the equivalent resistance between any two of the vertices?

★**61.** Twelve 1.00-Ω resistors are connected together so that each lies along the edge of a cube as shown in Figure 27.P24. What is the equivalent resistance between two diagonally opposite corners of the cube?

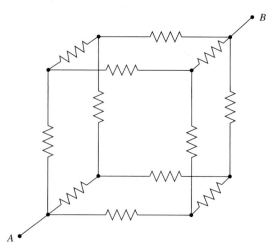

FIGURE 27.P24   Problem 61

★**62.** An infinite chain of identical resistors R are arranged as shown in Figure 27.P25. What is the equivalent resistance between points A and B? [*Hint:* You need not sum an infinite series to find the solution.]

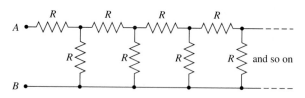

FIGURE 27.P25   Problem 62

63. For the circuit shown in Figure 27.P26 find the current through each resistor.

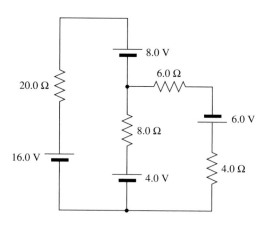

FIGURE 27.P26   Problem 63

64. A voltmeter is used to measure the terminal potential of a source of emf that is known to have an internal resistance of 1.500 Ω. If the internal resistance of the voltmeter is 6.000 kΩ and it indicates a potential difference of 5.994 V, what is the emf of the battery when no current flows from it?

65. (a) Compute the time constant for the circuit shown in Figure 27.P27. (b) What are the potential differences across the capacitor and the 60.0-kΩ resistor at $t = 0.500$ s after switch S is closed? (c) At what time after the switch S is closed does the potential difference across the capacitor equal that across the parallel combination of resistors?

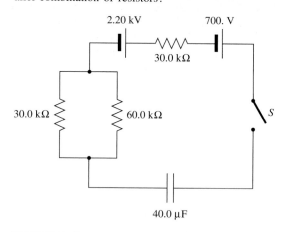

FIGURE 27.P27   Problem 65

66. At $t = 0.00$ s the switch shown in Figure 27.P28 is closed. At $t = 5.00$ s the current supplied by the source of emf is found to be 0.330 μA. (a) What is the capacitance C? (b) What is the current through the 30.0-MΩ resistor at $t = 3.00$ s?

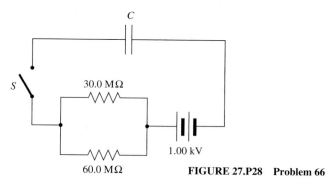

FIGURE 27.P28   Problem 66

67. The circuit shown in Figure 27.P29 is connected long enough so that the capacitor carries a constant charge. (a) Compute the charge on the capacitor. (b) How long after the source of emf is removed from the circuit does it take for the charge on the capacitor to drop to one-quarter of its original value?

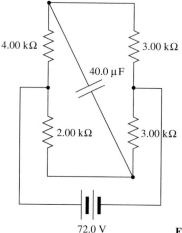

FIGURE 27.P29   Problem 67

68. Show that if at $t = 0$ a switch is thrown so that a capacitor of capacitance C is discharged through a resistor R, the charge remaining on the capacitor falls to half the original value in a time given by $t = 0.693RC$.

69. The face of analog voltmeters are often labeled with a phrase such as "100 000 ohms per volt." This is a way of designating the size of the resistance that is switched in series with the galvanometer for each voltage range on the meter. That is, if you use the voltmeter on a scale with full-scale range of 20 V, then the internal resistance of the meter is (20 V) (100 000 Ω/V) = 2.0 MΩ. (Note you multiply by the *full-scale* reading, not the particular reading you make.) (a) Assume an inexpensive voltmeter marked "10 000 ohms per volt" is used on the 5.00-V full-scale range to measure the potential difference across an unknown resistor connected in a series circuit with other resistors and an ammeter. The ammeter shows that the current in the series circuit is 12.1 mA, and the voltmeter across the unknown resistor reads 4.06 V. What is the approximate resistance of the unknown resistor? (b) What is the resistance of the voltmeter? (c) What is the actual resistance of the unknown resistor?

CHAPTER
28

The Magnetic Field

In this chapter you should learn

- how to compute the magnetic force on a moving charge.

- how to compute the magnetic force on a current-carrying conductor.

- the definition of a magnetic dipole.

- how to compute both the torque on a magnetic dipole and its potential energy in a uniform magnetic field.

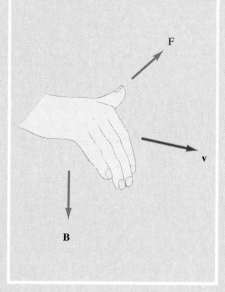

$$\mathbf{F} = q(\mathbf{E} + \mathbf{v} \times \mathbf{B})$$

28.1 The Magnetic Field

Diet Smith, a character in the Dick Tracy comic strip, was noted for his firm belief that, "The country that controls magnetism will control the world." We are not willing to make a claim quite this strong, but there is no denying that magnetism plays a pervasive role in modern civilization. Understanding magnetism has resulted in such diverse applications as data storage on computer diskettes, the development of electric motors and generators, and measurements of the fundamental properties of solids and elementary particles. (We should also not fail to mention the all-important task of attaching millions of notes, papers, and pictures to refrigerator doors throughout the civilized world!)

Your own introduction to magnetism probably started when you first played with a pair of bar magnets and learned about their north and south "poles." You no doubt quickly discovered simple rules of attraction and repulsion: like poles repel and unlike attract. Pictures like Figure 28.1 showing iron filings sprinkled around the poles of magnets bring to mind lines of magnetic field that fill the space around a magnet. We wish now to formalize this intuitive picture of the magnetic field.

One way to define the magnetic field is to follow a procedure analogous to that used to define the electric field. We could define a magnetic pole q_m of a certain strength and polarity, say, north. Then we plunk this pole down at any point in space and measure the magnetic force on it. The magnetic field vector **B** could then be defined using the force on our standard north pole. The analogy with the electric field definition is complete:

$$\mathbf{E} = \frac{\mathbf{F}}{q} \quad \text{(Electric field)} \qquad \mathbf{B} = \frac{\mathbf{F}}{q_m} \quad \text{(Magnetic field)}$$

Historically, exactly this kind of procedure was followed. It leads, however, to a quite complicated mixture of units and unnecessary complications. More importantly, it tends to obscure the fundamental nature of the magnetic field. For this and other reasons[1] we will take the more modern, but unfortunately less intuitive, approach and define the *magnetic* field in terms of its effect on *electric* charges. Nonetheless, it is still perfectly all right to think of the magnetic field lines as being parallel to the direction of the force on a magnetic north pole of a very small magnetic compass arrow placed in the field. The direction of the magnetic field is parallel to the small compass needle and in the direction indicated by the north pole of the compass.

We will defer the question of how a magnetic field is produced until Chapter 29 and assume for now that the field is given in a region of space. Magnetic field lines have some important similarities to E-field lines. A tangent to the magnetic field line gives the direction of the magnetic field at the point of tangency. Also, the number of field lines per unit area that pass at right angles through a surface is proportional to the strength of the magnetic field in that region of space.

The force due to a magnetic field acts on electric charges in a peculiar manner. If you simply place an electric charge in a magnetic field nothing happens; when an electric charge is at rest in a constant magnetic field, it experiences no magnetic force. *Electric* charges have to *move* to experience a force from a *magnetic* field.

Imagine the north pole of a large magnet near the left edge of this book and the south pole of a similar magnet near the right edge. These magnets create a magnetic field in the region between the poles. The direction of the field is from the left to right (along the direction you are reading). Next, imagine that a positive charge springs directly toward you, straight out of this page through this funny-looking symbol ⊙. Even though the

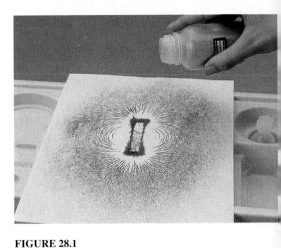

FIGURE 28.1

Iron filings sprinkled around a bar magnet reveal magnetic field lines. (Courtesy of Central Scientific Company.)

---

[1] There is another problem with this approach. If you try to purchase a standard north pole, you undoubtedly will find the bookstore doesn't stock them. As of this writing no one has yet been able to unambiguously detect an isolated magnetic pole, that is, a **magnetic monopole.** Magnetic poles always seem to come in pairs. If you break a bar magnet in half, each of the halves has a north and a south pole on opposite ends. Indeed, even if you continue subdividing the magnet until only a single electron is left, this electron too has the equivalent of both a north and a south pole.

charge was aimed right between your eyes, it missed! The charge flew over the top of your head; during its motion between this page and your face there was a magnetic force on the charge and the direction of this force was from the bottom of the page toward the top. "Turn-about is fair play," so let's suppose you shoot a positive charge back toward the book (perpendicular to, and into, the page) through the symbol ⊗. Go ahead, fire away. Oops! You nearly shot yourself in the foot. This time the magnetic force on the moving charge was directed downward from the top of the page toward the bottom. In other words, for each case the charge experienced a force that was at a right-angle to its velocity; the direction of the force depends on the direction of the velocity.

Later on, we will write a mathematical description of this peculiar magnetic force and use it to define the magnetic induction. First, however, we need a symbol to represent the magnetic field. The **magnetic induction field** is designated by $\mathbf{B}(x, y, z)$. Properly known as the magnetic induction, $\mathbf{B}$ is often simply called *"the" magnetic field* or, less frequently, the *magnetic flux density*.[2] Once again, remember that for now, we take $\mathbf{B}$ as given; we will show you how to compute $\mathbf{B}$ from its sources in the next chapter.

We can describe the magnetic force on a charge q that travels with velocity $\mathbf{v}$ through a magnetic induction field $\mathbf{B}$, and the properties described above, by the expression

$$\mathbf{F} = q\mathbf{v} \times \mathbf{B} \tag{28.1}$$

Because, in principle at least, $\mathbf{F}$, q, and $\mathbf{v}$ are all independently measurable, this equation serves to define the magnetic induction $\mathbf{B}$ at the point where q is located.

Equation (28.1) also determines the units of the magnetic induction $\mathbf{B}$. The magnitude of the magnetic force is

$$F = qvB \sin(\theta)$$

where θ is the angle between $\mathbf{v}$ and $\mathbf{B}$ when the vectors are placed tail to tail. This equation can be rewritten

$$B = \frac{F}{qv \sin(\theta)}$$

In SI units, F is in newtons, q is in coulombs, v is in meters per second, and $\sin(\theta)$ is unitless. Hence, the units of the magnetic field $\mathbf{B}$ must be

$$\frac{N}{C \, (m/s)} = \frac{N}{m \, (C/s)} = \frac{N}{A \cdot m}$$

This unit is called a tesla (T):

$$1 \, \frac{N}{A \cdot m} \equiv 1 \, T$$

The gauss (G) is also a common, but non-SI, unit and is related to the tesla by

$$1 \, T = 10^4 \, G = 10 \, kG$$

When given the magnetic field in gauss, you must change the unit to tesla before performing calculations using the formulas in this text because we use SI units.

---

[2] Another magnetic field designated by $\mathbf{H}$ is sometimes defined. The current practice in more advanced physics texts is to call $\mathbf{H}$ "the" magnetic field and $\mathbf{B}$ the magnetic induction. For now we follow the modern practice in introductory texts of calling $\mathbf{B}$ "the" magnetic field.

Returning to the force law, Equation (28.1), there are several important features of the magnetic force that we wish to emphasize:

---

1. The force on a charge q is proportional to the magnitude of the charge and its speed v. When the charge is not moving, there is no magnetic force on it.
2. The direction of the force is perpendicular to the direction of both the velocity $\mathbf{v}$ and the magnetic induction field $\mathbf{B}$. This result, of course, is a feature of the cross product in Equation (28.1).
3. When the velocity of the charge is parallel to the magnetic field, the force is zero: $F = |\mathbf{F}| = qvB \sin(0°) = 0$. (Also a feature of the cross product in Equation (28.1).)

In Figure 28.2 we illustrate how to apply the right-hand rule to compute the direction of the magnetic force on a moving charge. Figure 28.2(a) shows a *positive* charge moving with velocity $\mathbf{v}$ in the region of a magnetic induction field $\mathbf{B}$. The velocity vector makes an angle θ with the magnetic field. Application of the cross product yields a force of magnitude $F = qvB \sin(\theta)$. This vector is perpendicular to and upward from the plane of the vectors $\mathbf{v}$ and $\mathbf{B}$. In Figure 28.2(b) the vector $\mathbf{v} \times \mathbf{B}$ is also upward and perpendicular to the plane, but this time the moving charge is negative. Therefore, the resulting magnetic force is perpendicular but *downward* from the plane.

Figure 28.3 illustrates the direction of the magnetic force on charges $\pm q$ when the magnetic field is perpendicular to and out of the plane of the page. (This direction is represented by the symbol $\odot$.) You should verify these directions on your own and also confirm the directions of the magnetic force for the examples we presented above in which a fictitious charge traveled between this text and your face. First, however, if there is anyone around you, you might want to explain to him or her that you are applying the right-hand rule to determine the direction of a would-be magnetic force. Otherwise, you might look rather silly to your roommate as you appear to be waving your fingers at your face!

You might be surprised to learn that no work is performed by the magnetic force on a moving charge. This result is easy to obtain from the definition of work:

$$\mathcal{W} = \int \mathbf{F} \cdot d\mathbf{r} = \int q\mathbf{v} \times \mathbf{B} \cdot d\mathbf{r}$$

The charge q has velocity $\mathbf{v}$, so $d\mathbf{r} = \mathbf{v}\, dt$ and

$$\mathcal{W} = \int q\mathbf{v} \times \mathbf{B} \cdot (\mathbf{v}\, dt)$$

$$= \int q(\mathbf{v} \times \mathbf{B}) \cdot \mathbf{v}\, dt$$

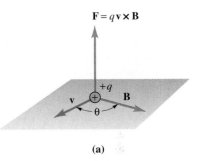

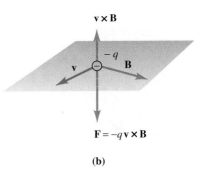

FIGURE 28.2

The magnetic force on a charge moving with velocity $\mathbf{v}$. (a) The force on a moving charge $+q$ is perpendicular to the plane of the velocity $\mathbf{v}$ and the magnetic field $\mathbf{B}$. The direction of the magnetic force $\mathbf{F} = q\mathbf{v} \times \mathbf{B}$ is given by the right-hand rule. (b) The magnetic force on a negative charge $-q$ is directed opposite from the direction of $\mathbf{v} \times \mathbf{B}$.

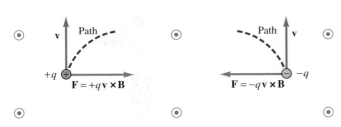

FIGURE 28.3

The magnetic force on two charges $+q$ and $-q$, each with initial velocity $\mathbf{v}$, when the charges enter a region of uniform magnetic field $\mathbf{B}$ that is directed upward, out of the page.

But the vector resulting from the cross product $\mathbf{v} \times \mathbf{B}$ is perpendicular to $\mathbf{v}$. Consequently, $(\mathbf{v} \times \mathbf{B}) \cdot \mathbf{v} = 0$ because the dot product of two perpendicular vectors is zero. Thus, we see that

$$\mathcal{W} = 0$$

This result, combined with the work-energy theorem first introduced in Chapter 7 implies that the particle's kinetic energy is not changed by the magnetic field.

Finally, if there is both an electric field $\mathbf{E}$ and a magnetic field $\mathbf{B}$ in a region of space, the force on a charge q moving with velocity $\mathbf{v}$ is (assuming any gravitational force to be negligible):

$$\boxed{F = q(\mathbf{E} + \mathbf{v} \times \mathbf{B})} \tag{28.2}$$

This expression is called the **Lorentz force law,** and it provides the total electromagnetic force on a charge q.

Concept Question 1
Is it possible for a uniform magnetic field to change the speed of an electron? a proton?

Concept Question 2
If an electron is moving along a straight line in a region where we know there is a magnetic field, can we necessarily conclude that it is moving parallel to the *B*-field lines? Explain.

Concept Question 3
Is the magnetic force conservative? Carefully explain your answer.

EXAMPLE 28.1   *Reviewing Your Cross Product Skills*

A 64.0-μC charge traveling with velocity $\mathbf{v} = (4.00 \text{ m/s})\hat{\mathbf{i}} + (2.00 \text{ m/s})\hat{\mathbf{j}}$ enters a region where there is a magnetic field $\mathbf{B} = -(5.00 \text{ kG})\hat{\mathbf{i}} + (3.00 \text{ kG})\hat{\mathbf{j}}$. Compute the force on the charge.

SOLUTION   The vectors $\mathbf{v}$ and $\mathbf{B}$ are shown in Figure 28.4. Because both $\mathbf{v}$ and $\mathbf{B}$ lie in the xy-plane, we suspect that $\mathbf{F}$ is parallel to the z-axis. We use the determinant method explained in Chapter 2 to compute the cross product indicated by Equation (28.1):

$$\mathbf{F} = q\mathbf{v} \times \mathbf{B} = (64.0 \ \mu\text{C}) \begin{vmatrix} \hat{\mathbf{i}} & \hat{\mathbf{j}} & \hat{\mathbf{k}} \\ 4.00 \text{ m/s} & 2.00 \text{ m/s} & 0 \\ -5.00 \text{ kG} & 3.00 \text{ kG} & 0 \end{vmatrix}$$

$$= (64.0 \times 10^{-6} \text{ C})[(12.0 + 10.0) \text{ kG} \cdot \text{m/s } \hat{\mathbf{k}}]\left(\frac{1 \text{ T}}{10 \text{ kG}}\right)$$

$$= 1.41 \times 10^{-4} \text{ N } \hat{\mathbf{k}}$$

As we expected, $\mathbf{F}$ is parallel to the z-axis.   ◀

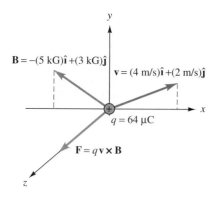

FIGURE 28.4

A charge q traveling with velocity $\mathbf{v}$ enters magnetic field $\mathbf{B}$. See Example 28.1.

EXAMPLE 28.2   *The Velocity Selector*

A velocity selector is used in many particle-scattering experiments that require charged particles with well-known velocities. This device provides perpendicular electric and magnetic fields as shown in Figure 28.5. Compute the velocity of parti-

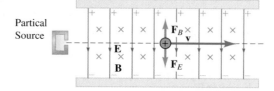

FIGURE 28.5

The velocity selector. Two oppositely charged, parallel plates create a uniform, downward electric field $\mathbf{E}$. A uniform magnetic field $\mathbf{B}$ is directed into the page, perpendicular to $\mathbf{E}$. A charge q is undeflected when $\mathbf{F}_B = \mathbf{F}_E$. See Example 28.2.

cles with charge q that pass undeflected through the velocity selector if the electric field magnitude is E and the magnetic field strength is B.

SOLUTION For the configurations of the fields shown in Figure 28.5, the magnetic force on charge q is

$$F_B = |q\mathbf{v} \times \mathbf{B}| = qvB \sin(90°) = qvB \qquad \text{(Upward)}$$

and the electric force on q is

$$F_E = qE \qquad \text{(Downward)}$$

For a particle to pass through the velocity selector undeflected

$$F_B = F_E$$

or

$$qvB = qE$$

which can be solved for v giving

$$v = \frac{E}{B}$$

Particles with velocities greater than this value are deflected upward by the stronger magnetic force, while those with smaller velocities are deflected downward by the stronger electric force. In practice, a small aperture is placed on the far right end of the selector to allow only undeviated charges through. The field strengths B and E are then adjusted to select the desired particle velocity. ◀

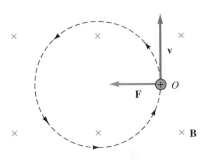

(a)

Moving Charges in Uniform Magnetic Fields

Consider a charge $+q$ traveling with a velocity $\mathbf{v}$ in a uniform magnetic field $\mathbf{B}$ perpendicular to $\mathbf{v}$. The resulting particle trajectory has the following three important aspects: (1) If the **B**-field is perpendicular to the velocity, the charge exhibits uniform circular motion. (2) Unlike satellite orbits in an inverse-square gravitational field, the radius of a charged particle's orbit in a constant magnetic field depends on the particle's mass. (3) The period of the particle's orbit is independent of particle velocity. Let's examine these results in order.

1. In Figure 28.6 we show the charge q, its velocity $\mathbf{v}$, the magnetic field $\mathbf{B}$, and the resultant force $\mathbf{F}$. The direction of the force $\mathbf{F} = q\mathbf{v} \times \mathbf{B}$ is perpendicular to the velocity. In Chapter 6 we learned that any time a constant magnitude force remains perpendicular to a mass's velocity, that mass exhibits uniform circular motion. Notice that as the charge moves from point O in Figure 28.6(a) to point P in (b) the force remains directed perpendicular to the velocity. The magnitude of the magnetic force is

$$F = q|\mathbf{v} \times \mathbf{B}| = qvB \sin(90°) = qvB$$

Because q, v, and B are all constants, the magnitude of the centripetal force F is also constant and, therefore, the charge q exhibits *uniform* circular motion.

2. To find the radius of the particle's orbit, we recall Newton's second law $F_{net} = ma$ and that for circular motion the magnitude of the centripetal acceleration $a = v^2/r$. In

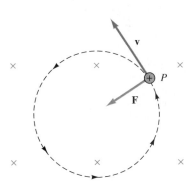

(b)

FIGURE 28.6

The motion of a charge q with velocity $\mathbf{v}$ perpendicular to a uniform magnetic field $\mathbf{B}$. The force $\mathbf{F}$ remains perpendicular to the velocity, and uniform circular motion results.

our case, however, $F_{net} = qvB$. Therefore,

$$qvB = \frac{mv^2}{r}$$

or

$$r = \frac{mv}{qB}$$

$$r = \frac{v/B}{q/m}$$

For a certain velocity v and a fixed magnetic field strength B, a charge q with mass m has a fixed radius of orbit determined by this expression. Because ionized atoms or molecules have fixed charge-to-mass ratios q/m, we can use this result to establish a technique to identify these atoms and molecules. Indeed, this last result forms the basis of the **mass spectrometer.** In Problem 12 we explore this device further.

3. In order to compute the particle's angular velocity ω, we recall that angular velocity and tangential velocity are related by the radius r of the circular motion:

$$v = r\omega$$

or

$$\omega = \frac{v}{r}$$

Substituting our expression for the radius r obtained from Newton's second law into the above expression, we have

$$\omega = \frac{v}{mv/qB}$$

$$\omega = \frac{q}{m}B \qquad \text{(Cyclotron frequency)} \qquad (28.3)$$

Notice that the angular velocity is *independent of the charge's speed v* and depends only on the charge-to-mass ratio q/m and the magnetic field strength B. For a particle (such as a proton or an electron) q/m is fixed, and by selecting a magnetic field strength B we can trap these particles by causing them to move in a circular path. Moreover, if during this circular motion we also accelerate the charged particle through a potential difference, we can dramatically increase its energy. This process is the basis of such accelerators as cyclotrons and synchrotrons. In fact, the expression we have derived for ω is called the **cyclotron frequency.** The operating principle of a cyclotron is shown in Figure 28.7.

When the velocity of a charge is not perpendicular to the direction of the uniform magnetic field, the path traced out by the charge's motion is that of a helix as shown in

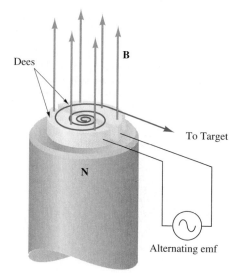

FIGURE 28.7

The cyclotron consists of two evacuated D-shaped containers (called "dees," of course) between the poles of a magnet. (Only the north magnetic pole is shown.) After a charged particle enters the region between the dees, an alternating emf accelerates the charge back and forth between the dees. As the charge's velocity increases so does its radius until the charge emerges with a very high energy.

Concept Question 4
What is the sign of the charged particle in the cyclotron shown in Figure 28.7?

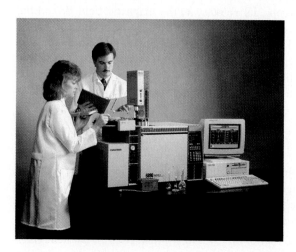

A modern mass spectrometer. (Photo courtesy of Hewlett-Packard Co.)

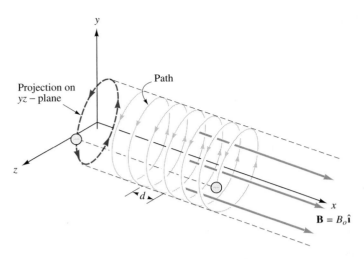

FIGURE 28.8

When a charge's velocity is not entirely perpendicular to the B-field direction, the motion is that of a helix. The distance the charge travels parallel to the magnetic field lines during one revolution is called the pitch d.

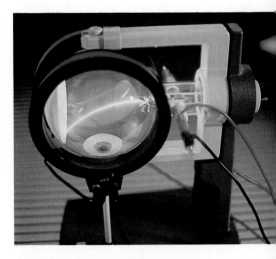

This electron beam is bent into a curved path from a magnetic field produced by the circular coils. The beam is visible because this tube contains low-pressure helium that glows when struck by electrons. (Photo courtesy of Central Scientific Company)

Figure 28.8. The component of the velocity that is parallel to the B-field is unaffected by the magnetic force. That is, $a_x = 0$ in Figure 28.8 (see Problem 14). The components of the velocity that are in the plane perpendicular to $\mathbf{B} = B_o\hat{\mathbf{i}}$ cause the charge to circulate. The projection of the charge's motion on the yz-plane is that of uniform circular motion. Because there is no component of B in the yz-plane, the magnitude of the velocity of the motion projected on the yz-plane is constant and is given by $v_\perp = \sqrt{v_y^2 + v_z^2}$. The charge's angular velocity is given by the relation for ω in Equation (28.3). The distance d along the B-field direction that the charge travels during one revolution is called the **pitch.**

Moving Charges in Nonuniform Magnetic Fields

When the B-field is *nonuniform,* the motion of a charged particle within it is usually complicated. Nonuniform magnetic fields are used in nuclear fusion research, the goal of which is to produce a clean and inexpensive source of energy. In order to achieve nuclear fusion, however, ionized gases with temperatures measured in millions of kelvins are necessary. Magnetic fields provide the only practical method for containing such states of matter called **plasmas.** One field orientation, called the **magnetic bottle,** is shown in Figure 28.9. We will see in the next chapter that two parallel, current-carrying coils can be used to create such a field. In the following paragraphs we will see that many ions that spiral toward one end of the magnetic bottle are reflected by the increased magnetic field strength at that end. Magnetic bottles are not perfect, however, and they leak sufficient numbers of ions that fusion experiments must be performed in a small fraction of a second.

In Figure 28.10 we show a charged particle's orbit in a region of space where the magnetic field strength increases slightly along the positive x-axis. Consider a charged particle moving in this field with a velocity that has a small component along the x-axis. If the magnetic field were uniform, the particle's trajectory would be a helix as described previously. Let's assume that the change in magnetic field is small enough that the particle must move through many turns of the helix before there is a noticeable change in the magnetic field strength. In this case we can think of the convergence of the field lines as producing a small perturbation to the particle's helical orbit. What effects can we expect to see due to the increasing magnetic field strength? First, because the radius of the helix is approximately the same as that for circular orbits ($r = mv/qB$), we expect the radius of the orbit to decrease as the field strength B increases. Second, if we apply the right-hand rule

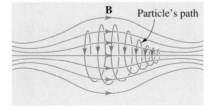

FIGURE 28.9

The magnetic field lines of a magnetic bottle. Charges spiraling toward regions of higher field strength can be reflected back into the bottle.

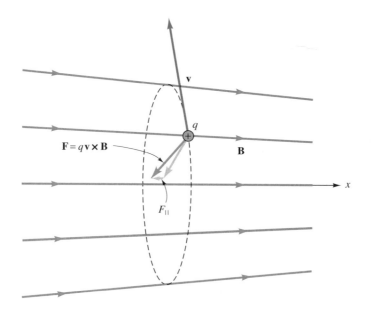

FIGURE 28.10

Motion of a charge $+q$ in a converging B-field. Because the direction of $\mathbf{B}$ is tilted slightly toward the x-axis, a component $F_\parallel$ of $\mathbf{F} = q\mathbf{v} \times \mathbf{B}$ is directed along the *negative x*-axis.

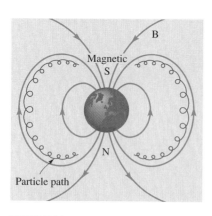

FIGURE 28.11

Charged particles, called cosmic rays, are trapped in a region of the earth's magnetic field known as the Van Allen belt.

and draw the resulting magnetic force, we see, as in Figure 28.10, that throughout the particle's orbit, the magnetic force has a small component in the negative x-direction. You should be sure to observe that this small component $F_\parallel$ arises *not* because the charge has an x-component to its velocity but rather because the B-field lines converge slightly. This component of the magnetic force accelerates the particle in the negative x-direction, ultimately reversing its direction. Note, however, that the magnitude of the particle's total velocity cannot change because the magnetic force is always perpendicular to the velocity. Nonetheless, the magnetic force can, of course, change the direction of the particle's velocity.

The earth's magnetic field deflects and sometimes traps charged particles, mostly protons and electrons, in a region called the Van Allen belt (Fig. 28.11). A few of these particles originate in distant stars but most come from our sun; all are called **cosmic rays.** When these particles spiral around the earth's magnetic field lines, they occasionally collide with air molecules in the upper atmosphere, causing them to emit light. This phenomenon can be observed in the latitudes near the north and south poles. In the north, this phenomenon is called the aurora borealis, or northern lights, and in the south it is known as the aurora australis.

In fusion experiments, large superconducting magnets are employed to temporarily confine plasmas whose temperatures, for a very short period, reach millions of kelvins.

The northern lights.

The Hall Effect (Optional)

The **Hall effect** provides a direct method by which the sign of the charge carriers in a conductor can be determined. Moreover, the drift velocity and density of the charge carriers can also be found. A schematic of the experiment is shown in Figure 28.12(a). A source of emf is attached to a thin sample of the conductor so that a current flows along its length in a known direction. Next, a magnetic field **B** is applied at right angles to the face of the conductor.

Let's first suppose that the charge carriers are positive as illustrated in Figure 28.12(b). In this case the velocity of the carriers is along the direction of the current I. The resulting magnetic force **F** is toward the top of the conductor and, therefore, as the charge carriers move through the conductor they also drift toward its top side as shown in Figure 28.12(b). The accumulation of charge from this drift leaves the bottom side of the conductor with a net negative charge. Moreover, this charge separation results in a potential difference between the top and bottom sides of the conductor; this potential difference

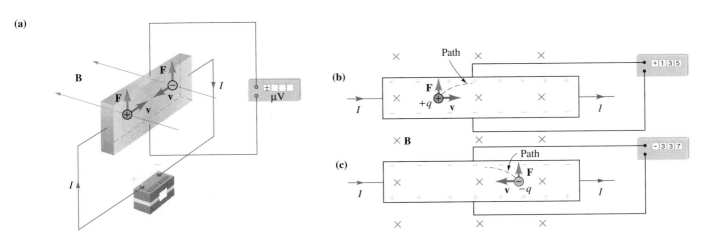

FIGURE 28.12

The Hall effect. (a) A magnetic field **B** is applied perpendicular to the current direction through a slab of conducting material. Because their velocity directions are opposite for a current in the same direction, both positive and negative charge carriers experience the magnetic force in the same direction. (b) When the majority of charge carriers are positive, the top of the slab attains a higher potential than the bottom of the slab. (c) When the majority of charge carriers are negative, the bottom of the slab is at the higher potential.

across the *width* of the conductor is called the **Hall voltage** and can be measured with a voltmeter. As shown in Figure 28.12(b), this voltmeter indicates that the top of the conductor is at a higher potential than the bottom side.

What happens when the charge carriers are negative? As illustrated in Figure 28.12(c), the negative carriers move toward the left because the applied electric field is toward the right. Because the carriers are indeed negative, the magnetic force $\mathbf{F} = -|q|\mathbf{v} \times \mathbf{B}$ is still toward the top of the conductor. Hence, the negative charge carriers drift upward in Figure 28.12(c), and our voltmeter tells us that the top of the conductor is at a *lower* potential than the bottom. This result is the opposite voltmeter polarization for the case of positive carriers. Isn't this a slick tool from which the sign of the carriers can be determined?

From the polarity of the Hall voltage the sign of the charge carriers can be determined. From the magnitude of the Hall voltage the concentration of the charge carriers can be determined as follows. All that charge buildup on the top and bottom surfaces of the conductor in the Hall experiment creates an electric field $\mathbf{E}$ within the conductor. In Figure 28.13 we show this situation for positive charge carriers. The result of the following analysis is the same for negative carriers. Because the E-field is directed downward, the

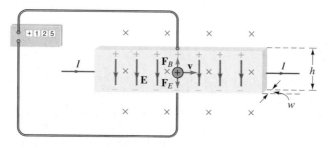

FIGURE 28.13

The Hall effect through a slab of material of width w and height h. When sufficient charge has accumulated on the top and bottom surfaces of the conducting slab, the magnetic and electric forces balance each other.

resulting *electric* force $\mathbf{F}_E$ opposes the *magnetic* force $\mathbf{F}_B$. When sufficient charge has accumulated on the surfaces of the conductor so that the magnitude of the electric force just equals the magnitude of the magnetic force, the resultant force on the carriers is zero, and the charge carriers' velocities are unaffected. The situation is a sort of self-imposed velocity selector! The electric force on charge q is $F_E = qE$, and the magnetic force is $F_B = qvB \sin(90°)$. Therefore, when $F_B = F_E$, we have

$$v_d B = qE$$

or, the drift velocity of the charge carriers is

$$v_d = \frac{E}{B}$$

It is easy to determine the electric field strength E if we know the height h of the conductor. Our voltmeter provides us with the **Hall voltage** V_H between the top and bottom sides of the conductor, and in Chapter 24 (Eq. (24.13)) we found that the magnitude of the uniform E-field is just $E = \Delta V/h$. Hence,

$$v_d = \frac{V_H}{Bh}$$

The current density through the conductor is $J = I/A$, where A is the cross-sectional area of the slab through which current I flows (see Fig. 28.13). In our case, the area is the product of the width and the height, $A = wh$. In Chapter 26, Equation (26.4), we found that the current density is related to the charge carrier drift velocity by $J = nqv$. Therefore,

$$J = \frac{I}{wh} = nqv_d = nq \frac{V_H}{Bh}$$

or

$$V_H = \frac{1}{nq} \frac{IB}{w} \qquad (28.4)$$

The constant $1/nq$ is called the **Hall coefficient** R_H. Because we can measure all the other quantities in Equation (28.4), we can use the Hall effect to measure the charge density of the carriers; the magnitude and sign of R_H provide the density of the charge carriers and their sign.

About 100 years after Hall's discovery, solid-state physicists discovered a rather remarkable variation of the classical Hall effect, called the **quantum Hall effect.** This phenomenon takes place within two-dimensional conductors in strong magnetic fields (15–20 T) and at temperatures only a few degrees above absolute zero. In a two-dimensional conductor all of the charge carriers travel in planes and have parallel velocity vectors. For example, in a metal-oxide-semiconductor (MOS) there is a thin layer of oxide between the semiconducting material and a metallic surface. Within the semiconductor, near the oxide layer, electrons exhibit two-dimensional behavior. In 1980, Klaus von Klitzing (1943–   ) reported that both the Hall voltage (Fig. 28.14) and the resistance of the conductor exhibit a periodic behavior. In particular, as the applied magnetic field strength is increased, the resistance drops to zero at regular field intervals. Moreover, at the field strengths with zero resistance the Hall voltage remains constant over a limited range of B-field strengths. The difference between the Hall voltages for adjacent plateaus is exactly equal to

$$\Delta V_H = \frac{h}{e^2} I \qquad (28.5)$$

where h is **Planck's constant,** a fundamental constant of quantum mechanics that you will learn more about in Chapter 40. Solid-state theorists have been able to satisfactorily explain these phenomena with quantum theory. Since the discovery of the quantum Hall effect, however, a new phenomenon, called the **fractional quantum Hall effect,** has been observed. Experiments in very strong magnetic fields have revealed a limited number of certain fractional values for ΔV_H.

The Hall effect provides a useful mechanism by which we can probe the microscopic behavior of solids. From a practical standpoint, the quantum Hall effect may one day provide a way to define the ohm. Measurements of such combinations of fundamental constants as e^2/h provide important cross checks on the precision to which the constants are known. Experimentally, it has been found that $h/e^2 = 25\,813\ \Omega$.

The Hall effect was first observed in 1879 by Edwin H. Hall (1855–1938). When Hall first asked to be accepted as a graduate student at Harvard University he was discouraged from entering the physics program by the chairman of that department. He was told that most of the important discoveries of physics had already been made. We all know now that this notion (held by many physicists at that time) could not have been less accurate. Indeed, both the theories of relativity and quantum mechanics lay just on the other side of 1900. It is, therefore, somewhat ironic that even the Hall effect requires a quantum mechanical treatment to completely explain it.

Magnetohydrodynamics (Optional)

The magnetic force on moving particles can also be exploited on a large scale for the generation of electric power. The process is known by the rather grand name of **magneto-**

V_H

B

FIGURE 28.14

Unlike the classical Hall effect, the quantum Hall voltage exhibits plateaus in the V_H versus B plot.

Concept Question 5
Why do you expect good conductors, such as copper and silver, to have smaller Hall voltages than poor conductors, such as germanium and silicon, if each has the same conductor dimensions, applied B-field, current, and orientation?

Concept Question 6
Some current-carrying ionic solutions transport positive and negative charges of equal magnitude in opposite directions. Do you expect to observe the Hall effect in such a solution? Explain.

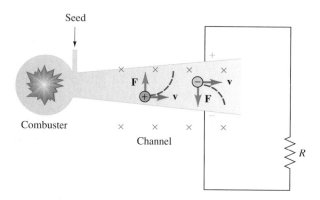

FIGURE 28.15

Schematic diagram for MHD electric power generation. A high-temperature gas is accelerated to high velocity and enters the channel from the left. A "seed" composed of a low-ionization-potential alkali metal is added to create a conducting plasma in the channel. Moving charged particles experience a magnetic force due to the large magnetic field directed into the page of this figure. Positive charges are forced toward the top of the channel and negative charges toward the bottom, creating a source of emf that can drive a current through the external load symbolized by the resistor R.

hydrodynamics, or, more simply, MHD. As shown in Figure 28.15, the process starts with a heat source used to raise the temperature of a "working fluid," usually a gas. Any high-temperature energy source, for example burning fossil fuels, such as gas or coal, can be used. The working fluid is usually "seeded" to increase its conductivity. The seed is typically an alkali metal with a low ionization potential, such as potassium or cesium. When introduced into the high-temperature gas, the seed atoms ionize and the gas becomes a conducting **plasma.** This plasma is then allowed to expand into the MHD channel where the magnitude of its velocity can be on the order of the speed of sound.

On either side of the MHD channel are the opposite poles of a magnet, which creates a large magnetic field perpendicular to the gas flow. The direction of the magnetic force on positive seed ions is opposite that on negative electrons. Consequently, the positive and negative charges separate. The result, as shown in Figure 28.15, is that the upper and lower plates of the channel become charged positively and negatively, respectively. This process creates a giant battery with the magnetic force on the plasma supplying the emf. In practice, the sides of the channels where the electric power is taken off are often constructed of many segmented electrodes to prevent current from flowing down the walls of the channel.

The exhaust gas exiting from an MHD channel is still hot enough to run a conventional steam-turbine electric power plant. Thus, any electric energy extracted by the MHD process adds to the total electric power output resulting from the energy of the heat source. Moreover, all of this occurs without any moving parts! The disadvantage of the MHD process is that the resulting electricity is direct current (DC) and must be converted to alternating current (AC) before being placed on commercial power-transmission lines. In addition, as you might imagine, there are nontrivial problems associated with the materials used to build a channel that can survive a 3000 K plasma traveling at nearly the speed of sound down its interior. Nonetheless, much progress has been made in this area and both gas-fired and coal-fired MHD generators are being tested. If successful, this technology may add greatly to the electric energy available from our limited fossil fuel reserves.

It is also possible to reverse the sequence described in the preceding paragraphs and use the MHD process to push on the working fluid. The idea is to replace the electrical load (represented by a resistor symbol in Fig. 28.15) with an external DC voltage source. This external source is connected with polarity so as to drive an electric current backward across the MHD channel. If you use the right-hand rule, you can convince yourself that the positive and negative charges that contribute to such a current both experience a magnetic force in the forward direction. These charged particles collide with the molecules of the working gas and accelerate it.

An MHD channel for coal-fired magnetohydrodynamics in the test bay at the Component Development and Integration Facility at MSE, Inc., in Butte, Montana. (Photo courtesy of CDIF, MSE, Inc.)

The MHD accelerator process described above is used to pump corrosive high-temperature fluids, such as the liquid sodium used to cool some types of nuclear reactors. In such applications the benefit of an external pump with no moving parts is obvious. This same principle has been employed to accelerate the gas in high-velocity wind tunnels used to test aircraft designs. High-temperature superconductors may provide sufficiently intense magnetic fields so that submarine propulsion using seawater as the working fluid may become viable.

28.2 Force on a Current-Carrying Conductor

We often want to determine the magnetic force on charges with motions that can most easily be described by a current I. Therefore, we want to modify Equation (28.1) in some manner such that q and v are replaced with a reference to the current I. Figure 28.16 shows a segment of a conductor carrying a current I in a uniform magnetic field $\mathbf{B}$. We focus our attention on a small element of charge dQ, which in time dt travels through displacement $d\mathbf{l}$. If $\mathbf{v}_d$ is the drift velocity of this charge, then the distance through which the charge flows in this time is $d\mathbf{l} = \mathbf{v}_d\, dt$. From Equation (28.1), the force $d\mathbf{F}$ on the charge dQ is given by

$$d\mathbf{F} = dQ\, \mathbf{v}_d \times \mathbf{B} \qquad (28.6)$$

But $dQ = I\, dt$ so that

$$d\mathbf{F} = (I\, dt)\mathbf{v}_d \times \mathbf{B} = I(\mathbf{v}_d\, dt) \times \mathbf{B} = I\, d\mathbf{l} \times \mathbf{B}$$

$$\boxed{d\mathbf{F} = I\, d\mathbf{l} \times \mathbf{B}} \qquad (28.7)$$

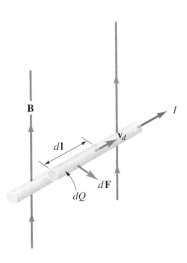

FIGURE 28.16

The current I gives rise to charge flow $dQ = I\, dt$. In the presence of the magnetic field $\mathbf{B}$, the force and charge carriers with velocity $\mathbf{v}$ is $dF = dQ\mathbf{v} \times \mathbf{B}$.

It is important for you to remember that $d\mathbf{l}$ is along the direction of the current. To obtain the total force on the wire we must sum all the force contributions $d\mathbf{F}$ by integrating Equation (28.7). If the current is uniform along a straight segment of a conductor, and the magnetic field is a constant, then both can be removed from the integral, making the integration trivial:

$$\int d\mathbf{F} = \int (I \, d\mathbf{l} \times \mathbf{B})$$

$$= \left(I \int d\mathbf{l} \right) \times \mathbf{B}$$

$$\mathbf{F} = I \, \mathbf{l} \times \mathbf{B} \qquad \text{Magnetic force on a straight current-carrying wire} \qquad (28.8)$$

EXAMPLE 28.3   *"You Take the High Road, I'll Take the Low Road" Takes a New Turn*

A wire formed in the shape of a semicircular loop of radius R carries a clockwise current I as shown in Figure 28.17(a). A uniform magnetic field $\mathbf{B}$ is directed downward, perpendicular to the plane of the loop. Compute the magnetic force on (a) the straight segment of current between points A and B, and (b) the semicircular segment between these two points.

SOLUTION   (a) Because this segment is a straight line, we can use Equation (28.8) to compute the force. The right-hand rule tells us the force is downward, toward the bottom of the page. The length of this segment is $2R$, so

$$F_{\text{line}} = |I \, \mathbf{l} \times \mathbf{B}| = I \, lB \sin(90°) = I(2R)B$$

$$F_{\text{line}} = 2IRB \qquad \text{(Downward)}$$

(b) Because this wire is not straight we cannot use Equation (28.8). Instead, we must use the differential form in Equation (28.7) and perform a summation. To begin our computation of the magnetic force on the curved segment, we focus our attention on the small element $d\mathbf{l}$ shown in Figure 28.17(b). In this figure we have designated the position of the element $d\mathbf{l}$ by its angle ϕ from the horizontal. The element dl subtends an angle $d\phi$ so that its length is $Rd\phi$. Because the magnetic field is perpendicular to the plane of the current, the magnitude of the force on segment dl is

$$dF = I|d\mathbf{l} \times \mathbf{B}| = I \, dl \, B \sin(90°) = I(Rd\phi)B = IRB \, d\phi$$

We must be careful when we integrate this expression to obtain the total force F_{arc} on the curved segment because, although the magnitude of dF is the same on each dl segment, each $d\mathbf{F}$ is pointed in a different direction. Consequently, before we integrate we must resolve $d\mathbf{F}$ into its horizontal and vertical components as shown in Figure 28.17(c). On the left side of this figure we show the components of the force for the $d\mathbf{F}$ shown in Figure 28.17(b). It is important to realize in this example that for every $d\mathbf{l}$ and $d\mathbf{F}$ at angle ϕ on the left side of the curve, there is a corresponding $d\mathbf{l}$ and $d\mathbf{F}$ at $(180° - \phi)$ on the right side. Corresponding vectors $d\mathbf{l}$ and $d\mathbf{F}$ are also shown in Figure 28.17(c). From this figure we see that the horizontal components of the forces are in opposite directions and, therefore, cancel. However, the vertical components are both in the same direction, upward. Hence, the total force on the curved segment of the conductor can be computed by summing (integrating) only the vertical components of dF:

$$dF_{\text{arc}} = dF \sin(\phi) = (IRB \, d\phi)\sin(\phi)$$

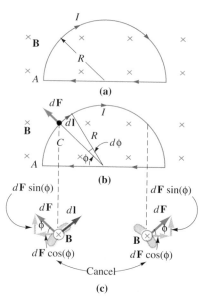

FIGURE 28.17

(a) A wire in a semicircular loop of radius R carries a clockwise current I. (b) The force $d\mathbf{F}$ on a small segment $d\mathbf{l}$ with length $dl = R \, d\phi$. (c) For every element $d\mathbf{l}$ on the left side of the curve there is a $d\mathbf{l}$ on the right side such that the horizontal components cancel. The vertical component $dF \sin(\phi)$ contributes to the net force and is summed by integrating ϕ from $\phi = 0$ rad to $\phi = \pi$ rad.

or

$$F_{arc} = IRB \int_0^{\pi} \sin(\phi)d\phi = IRB\Big[-\cos(\phi)\Big]_0^{\pi}$$

$$= 2IRB \qquad \text{(Upward)}$$

The net force on the curved segment is exactly equal in magnitude but opposite in direction from the net force on the straight segment. Hence, this current loop does not move translationally in the presence of this B-field. If we tilt the plane of the loop so that it is no longer perpendicular to the magnetic field **B**, the loop still does not move. However, as we will see in the next section, there is a net torque on the loop. ◀

Concept Question 7
What effect does a magnetic field have on the ions of a conducting material due to their thermal motion? Explain your answer.

Concept Question 8
Even when a current flows through a conductor the net charge within any segment of the conductor is zero. How can such a current-carrying conductor experience a magnetic force when placed in a B-field?

28.3 Current-Carrying Loops in a Uniform Magnetic Field

In this section we compute the torque τ on a current loop placed in a uniform magnetic field **B**, and the potential energy U of this loop. Although we obtain expressions for τ and U using a rectangular loop, the results we obtain are valid for a loop of any shape. Two practical applications of the material in this section include electric motors and the mechanical galvanometer used to construct analog meters. However, the concepts we develop here play an important role in many other electromagnetic phenomena.

In Figure 28.18(a) we show a rectangular current-carrying loop of width a and length b. The loop is tilted with respect to a uniform B-field so that a line perpendicular to the plane of the rectangle (the normal) makes an angle θ with **B.** When a current I flows through the loop, each straight segment experiences a force as shown in the figure. The first feature to recognize about these forces is that $\mathbf{F}_2$ and $\mathbf{F}_4$ are equal in magnitude, opposite in direction, and directed along the same line. Therefore, these two forces cause no acceleration of the loop. Applying Equation (28.8), we find

$$F_2 = F_4 = IbB \sin(90°) = IbB$$

Forces $\mathbf{F}_1$ and $\mathbf{F}_3$ are also equal in magnitude and opposite in direction. However, as shown in Figure 28.18(b), they do not act along the same line and, therefore, these forces

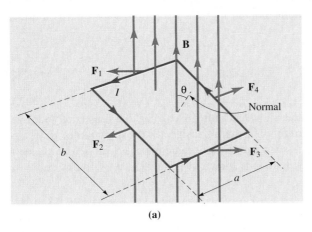

(a)

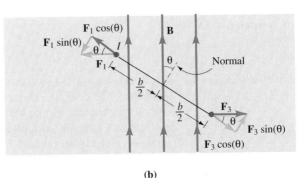

(b)

FIGURE 28.18

A rectangular current-carrying loop in a uniform magnetic field **B.** (a) The forces $\mathbf{F}_2$ and $\mathbf{F}_4$ cause no translational motion of the loop. However, although forces $\mathbf{F}_1$ and $\mathbf{F}_3$ cause no translation motion, they do exert a net torque on the loop. (b) An end view. The net torque is $\tau = IabB \sin(\theta)$, and its direction is such as to tend to align the normal to the loop area parallel to the field **B.**

exert a torque on the loop. The magnitudes of $\mathbf{F}_1$ and $\mathbf{F}_3$ are

$$F_1 = F_3 = IaB \sin(90°) = IaB$$

The rotation axis passes through the center of the loop so that the radial distance from this axis to the point of application of the magnetic force is $b/2$. The net torque on the loop is

$$\tau = \frac{b}{2}[F_1 \sin(\theta)] + \frac{b}{2}[F_3 \sin(\theta)] = \frac{b}{2}(IaB + IaB)\sin(\theta)$$

$$= I(ab)B \sin(\theta) \tag{28.9}$$

The product ab is the area A of the loop. Hence, we can write Equation (28.9) as

$$\tau = I[AB \sin(\theta)] \tag{28.10}$$

It is once again convenient to represent the area A as a vector with a magnitude equal to the area and a direction that is normal to the surface. To determine in which direction normal to the surface the area vector points, wrap the fingers of your right hand around the loop in the direction of the current flow. Your extended thumb indicates the proper direction for the A vector. Because the angle θ is between $\mathbf{A}$ and $\mathbf{B}$, we can represent the term in the parentheses of Equation (28.10) by the vector cross product:

$$\boldsymbol{\tau} = I\mathbf{A} \times \mathbf{B} \tag{28.11}$$

You should verify from Figure 28.18(b) that the order of $\mathbf{A}$ and $\mathbf{B}$ in this cross product is correct; the torque is out of the page because it tends to rotate the loop anticlockwise, and the direction of $\mathbf{A} \times \mathbf{B}$ is also out of the page. Although we have derived Equation (28.11) for a rectangular current loop, this result is valid for a planar loop of any shape.

The Magnetic Dipole Moment

In Chapter 22 we found that the torque on an electric dipole moment $\mathbf{p}$ placed in a uniform electric field $\mathbf{E}$ is given by

$$\boldsymbol{\tau} = \mathbf{p} \times \mathbf{E}$$

We can make Equation (28.11) look similar to this expression if we define the **magnetic dipole moment** $\boldsymbol{\mu}$ of a single current loop as $\mu = IA$. If the loop has N turns the total magnetic moment is

$$\boldsymbol{\mu} = NI\mathbf{A} \tag{28.12}$$

The direction of $\boldsymbol{\mu}$ is given by another right-hand rule; if you curl the fingers of your right hand along the direction of the current, your thumb designates the direction of $\boldsymbol{\mu}.$ The SI unit of the magnetic dipole moment is amperes times square meter $(\text{A} \cdot \text{m}^2)$. The magnetic moment of a circular, current-carrying loop is shown in Figure 28.19. With this definition Equation (28.11) becomes

$$\boldsymbol{\tau} = \boldsymbol{\mu} \times \mathbf{B} \tag{28.13}$$

Both the electric and the magnetic dipole moments experience no net force when placed in *uniform* electric and magnetic fields, respectively. However, if the fields are nonuniform, each dipole moment experiences a net force.

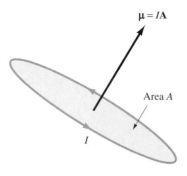

$\mu = IA$

Area A

I

FIGURE 28.19

A magnetic dipole moment is composed of current I circulating around a loop of area $\mathbf{A}.$ The direction of the magnetic moment vector $\boldsymbol{\mu} = I\mathbf{A}$ is given by the right-hand rule.

Direction convention for magnetic dipole moment

In Chapter 22 we also found that the potential energy of an electric dipole is given by

$$U_E = -\mathbf{p} \cdot \mathbf{E}$$

We leave it as an exercise (Problem 43) for you to show that the potential energy of a magnetic dipole moment $\boldsymbol{\mu}$ in the field $\mathbf{B}$ is

$$U_B = -\boldsymbol{\mu} \cdot \mathbf{B} \qquad (28.14)$$

where $U_B = 0$ when $\theta = 90°$. The potential energy function U_B has all the peculiarities that we warned you about for U_E. In particular,

when
$$\begin{aligned}
\theta &= 180° & U_B &= U_{B_{max}} = +\mu B \\
\theta &= 90° & U_B &= 0 \\
\theta &= 0° & U_B &= U_{B_{min}} = -\mu B
\end{aligned}$$

In Example 28.4 we compute the torque on a current-carrying loop placed in a magnetic field. Electric motors convert electric energy into mechanical energy. Although electric motors have much more complex designs to maximize efficiency, the basic principles of motor operation are embodied in this illustration.

EXAMPLE 28.4 Magnetic Dipole in a Uniform Magnetic Field

The rectangular coil shown in Figure 28.20 consists of 20.0 turns of wire, each turn carrying a current of 50.0 mA. The dimensions of the coil are $a = 1.00$ cm and $b = 1.50$ cm, and it is placed in a uniform magnetic field of 0.0200 T. (a) What torque must be applied to the coil to rotate it at a constant rate from equilibrium to an orientation in which the normal to the plane of the coil makes an angle of 70.0° with the direction of the B-field? (b) What is the change in potential energy of the coil when rotated to this orientation?

SOLUTION (a) Because there are $N = 20.0$ loops, each carrying $I = 50.0$ mA, the magnitude of the magnetic moment

$$\mu = NIA = NI(ab)$$

The magnitude of the torque acting on the loop when its normal is at angle θ is

$$\tau = |\boldsymbol{\mu} \times \mathbf{B}| = \mu B \sin(\theta) = NIabB \sin(\theta)$$

In order to rotate the coil at a constant rate we must apply a torque that just balances the magnetic torque at each angle as we turn the coil. As indicated by the preceding equation

$$NIabB = (20.0)(50.0 \times 10^{-3}\,\text{A})(0.0100\,\text{m})(0.0150\,\text{m})(0.0200\,\text{T})$$
$$= 3.00 \times 10^{-6}\,\text{N} \cdot \text{m}$$

That is,

$$\tau = (3.00 \times 10^{-6}\,\text{N} \cdot \text{m}) \sin(\theta)$$

When the coil is in equilibrium $\theta = 0°$, and initially the required torque is zero. At 70°

Concept Question 9
When a magnetic moment μ is antiparallel to a magnetic field $\mathbf{B}$, is the moment in a stable or unstable equilibrium? Explain.

Concept Question 10
Explain how the maximum *change* in potential energy of a dipole of moment μ in a magnetic field $\mathbf{B}$ can be $2\mu B$ when its maximum potential energy is μB.

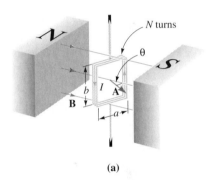

(a)

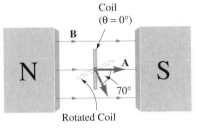

Top View

(b)

FIGURE 28.20

A rectangular, current-carrying loop with N turns, each carrying current I. (a) The loop is rotated about a vertical axis so that a normal to its surface area makes an angle θ with the direction of the uniform magnetic field $\mathbf{B}$. (b) A top view of the rectangular loop between the magnet pole faces. The equilibrium ($\theta = 0°$) and rotated ($\theta = 70°$) positions are depicted. See Example 28.4.

the torque required increases to

$$\tau = (3.0 \times 10^{-6}\,\text{N} \cdot \text{m}) \sin(70.0°) = 2.82 \times 10^{-6}\,\text{N} \cdot \text{m}$$

Between these two positions we must apply a changing torque given by $\tau = (3.0 \times 10^{-6}\,\text{N} \cdot \text{m}) \sin(\theta)$.

(b) The potential energy is given by Equation (28.13): $U_B = -\boldsymbol{\mu} \cdot \mathbf{B} = -(NIab)B\cos(\theta)$. The change in potential energy is, therefore,

$$\Delta U_B = U_f - U_o = -(NIab)B[\cos(\theta_f) - \cos(\theta_o)]$$

$$= -(3.00 \times 10^{-6}\,\text{J})[\cos(70.0°) - \cos(0.0°)] = +1.97 \times 10^{-6}\,\text{J}$$

The work we do while turning the coil increases the potential energy of the coil-magnetic field system. Note that the work done to rotate the coil can also be found from the result of part (a):

$$\mathcal{W} = \int_{0°}^{70°} \tau \, d\theta = \int_{0°}^{70°} (3.00 \times 10^{-6}\,\text{N} \cdot \text{m}) \sin(\theta) \, d\theta$$

$$= (3.00 \times 10^{-6}\,\text{N} \cdot \text{m})\Bigl(-\cos(\theta)\Bigr)\Big|_{0°}^{70°}$$

$$= (3.00 \times 10^{-6}\,\text{N} \cdot \text{m})[\cos(0°) - \cos(70°)] = 1.97 \times 10^{-6}\,\text{J} \quad \blacktriangleleft$$

EXAMPLE 28.5   *Magnetic Moments to Remember*

A rotating rod is pivoted about one end and carries a total charge Q uniformly distributed along its length L (Figure 28.21(a)). If the rod rotates with angular velocity ω, compute its magnetic moment μ.

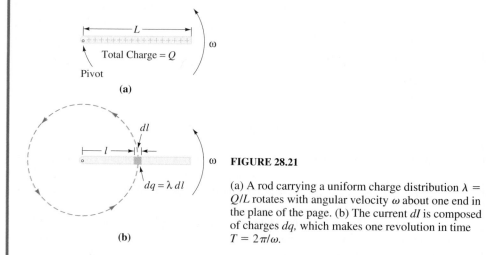

FIGURE 28.21

(a) A rod carrying a uniform charge distribution $\lambda = Q/L$ rotates with angular velocity ω about one end in the plane of the page. (b) The current dI is composed of charges dq, which makes one revolution in time $T = 2\pi/\omega$.

SOLUTION   We visualize the rotating rod of charge as a series of concentric current loops, each carrying a current dI. The charge per length λ along the rod is

$$\lambda = \frac{Q}{L}$$

so that in a segment of the rod dl in length

$$dq = \lambda \, dl$$

The time for charge dq to complete one revolution is the period of the rotation $T = 2\pi/\omega$. Hence, the current dI is given by

$$dI = \frac{dq}{(2\pi/\omega)} = \frac{\omega}{2\pi}\, dq = \frac{\omega}{2\pi}\lambda\, dl$$

The magnetic moment of this differential current loop is the product of the current dI and the area of this loop πl^2:

$$d\mu = dI\,(\pi l^2) = \left(\frac{\omega}{2\pi}\lambda\, dl\right)\pi l^2 = \frac{\omega\lambda}{2}l^2\, dl$$

To find the total magnetic moment of the rotating rod we integrate:

$$\mu = \frac{\omega\lambda}{2}\int_0^L l^2\, dl = \frac{\omega\lambda L^3}{6}$$

Substituting for λ, we have

$$\mu = \frac{Q\omega L^2}{6} \qquad \blacktriangleleft$$

28.4 Summary

The **magnetic induction field** $\mathbf{B}(x, y, z)$ is defined by the force it exerts on a moving electric charge q through the equation

$$\mathbf{F} = q\mathbf{v} \times \mathbf{B} \qquad (28.1)$$

where $\mathbf{v}$ is the velocity of the moving charge. The magnetic force on a moving electric charge is perpendicular to both the magnetic field direction and that of the velocity. The magnetic force does no work.

The SI unit of the magnetic field is the tesla (T):

$$1\,\text{T} = 1\,\frac{\text{N}}{\text{A}\cdot\text{m}}$$

When a charge q moves with velocity $\mathbf{v}$ in the presence of both an electric field $\mathbf{E}$ and a magnetic field $\mathbf{B}$, the total force on the charge is given by the **Lorentz force law:**

$$\mathbf{F} = q(\mathbf{E} + \mathbf{v} \times \mathbf{B}) \qquad (28.2)$$

When a charge q with mass m moves in a uniform magnetic field $\mathbf{B}$ with its initial velocity $\mathbf{v}$ perpendicular to the field, it travels in a circle of radius

$$r = \frac{mv}{qB}$$

The angular frequency of its uniform circular motion (the **cyclotron frequency**) is

$$\omega = \frac{q}{m}B \qquad (28.3)$$

The force on a current I that is directed along a length $d\mathbf{l}$ is given by

$$d\mathbf{F} = I\,d\mathbf{l} \times \mathbf{B} \tag{28.7}$$

The **Hall effect** can be used to determine the sign of charge carriers, their density, and drift velocity.

The **magnetic moment** $\boldsymbol{\mu}$ is defined in terms of the current I that circulates around an area $\mathbf{A}$:

$$\boldsymbol{\mu} = NI\mathbf{A} \tag{28.12}$$

The torque on a magnetic moment in a uniform magnetic field $\mathbf{B}$ is given by

$$\boldsymbol{\tau} = \boldsymbol{\mu} \times \mathbf{B} \tag{28.13}$$

and its potential energy is

$$U_B = -\boldsymbol{\mu} \cdot \mathbf{B} \tag{28.14}$$

PROBLEMS

28.1 The Magnetic Field

1. The poles of a strong magnet are oriented so that its 2.40-T field is directed from the east toward the west. What is the force (magnitude and direction) on an alpha particle (the nucleus of a He atom) when it enters the region of the field with a velocity of 1.80×10^5 m/s, northward?

2. An electron experiences a force of 4.60×10^{-14} N when it travels with a velocity of 4.00×10^5 m/s through a uniform, 1.20-T magnetic field. What is the angle between the electron's velocity and the magnetic field?

3. An electron travels with a velocity $\mathbf{v} = 1.20 \times 10^6$ m/s along the positive x-axis. If there is a uniform magnetic field $B = 0.0500$ T along the positive z-axis, determine the magnitude and direction of the electric field necessary to keep the electron's motion along the positive x-axis.

4. The magnetic field in a region of space is given by $\mathbf{B} = (0.80\hat{\mathbf{i}} - 1.20\hat{\mathbf{j}})$ T. (a) Compute the magnetic force on an electron that has velocity $\mathbf{v} = (2.40\hat{\mathbf{i}} - 3.60\hat{\mathbf{j}} - 1.20\hat{\mathbf{k}}) \times 10^5$ m/s. (b) What is the magnitude of this force?

5. An electron traveling with velocity $\mathbf{v} = (4.20\hat{\mathbf{i}} + 1.60\hat{\mathbf{j}} - 2.80\hat{\mathbf{k}}) \times 10^4$ m/s enters a region of space where a magnetic field $\mathbf{B} = (25.0\hat{\mathbf{i}} + 15.0\hat{\mathbf{j}} + 20.0\hat{\mathbf{k}})$ kG. (a) Compute the magnitude of the magnetic force on the electron. (b) What angle does this force make with the positive z-axis?

6. What potential difference must protons be accelerated through for them to pass undeflected through a velocity selector that has electric field strength of 9.60×10^5 V/m and a magnetic field strength of 0.600 T?

7. What is the magnetic force on an electron when it travels due north with a speed of 2.00×10^6 m/s in a region where the earth's magnetic field is 60.0 μT and is directed downward (toward the center of the earth)?

8. A proton with 1.50 MeV of kinetic energy encounters a region of the Van Allen belt where the B-field strength is 0.200 G. If the proton exhibits circular motion, (a) what is its minimum radius of orbit and (b) how long does it take to complete a revolution for this orbit radius?

9. A charge q travels with velocity $\mathbf{v} = v_x\hat{\mathbf{i}} + v_y\hat{\mathbf{j}} + v_z\hat{\mathbf{k}}$ in the magnetic field $\mathbf{B} = B\hat{\mathbf{j}}$. By directly computing the force $\mathbf{F} = q\mathbf{v} \times \mathbf{B}$ show that the y-component of the charge's acceleration $a_y = 0$.

10. There is a magnetic field in a certain region of space where the field lines are all parallel to the xy-plane (that is, $B_z = 0$). When a 2.00-μC charge enters this region of space with a velocity $\mathbf{v} = (4.00\hat{\mathbf{i}} - 2.00\hat{\mathbf{j}} + 3.00\hat{\mathbf{k}}) \times 10^5$ m/s, it encounters a force $\mathbf{F} = (-0.90\hat{\mathbf{i}} + 1.50\hat{\mathbf{j}} + 2.20\hat{\mathbf{k}})$ N. Compute the magnitude and direction of the magnetic field $\mathbf{B}$.

11. A charge $+q$ travels along the x-axis with velocity $\mathbf{v} = v_o\hat{\mathbf{i}}$. At the instant it passes through the origin a uniform magnetic field $\mathbf{B} = -B\hat{\mathbf{k}}$ is turned on. Show that the components of the charge's acceleration depend on its instantaneous velocity components v_x and v_y, and in particular,

$$a_x = -\frac{qv_y}{m}B \quad \text{and} \quad a_y = \frac{qv_x}{m}B$$

where m is the mass of the charge.

12. In a **mass spectrometer** ions (formed in an oven or made by contact with a hot wire) are accelerated through a potential difference ΔV and enter a region of uniform magnetic field strength B as shown in Figure 28.P1. The magnetic force causes the ions to move through a semicircular path where they are observed by a detector. If y is the lateral displacement of the ion, show that its charge-to-mass ratio is given by

$$\frac{q}{m} = \frac{8\,\Delta V}{y^2 B^2}$$

13. Suppose the accelerating potential ΔV in the mass spectrometer described in Problem 12 is replaced with a velocity selector so

that the ions exit the velocity selector at the same point. If E and B' are the electric and magnetic fields strengths of the selector that determine the velocity $\mathbf{v}$ shown in Figure 28.P1, show that

$$\frac{q}{m} = \frac{2E}{BB'y}$$

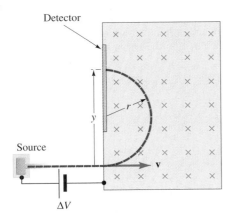

Detector

Source

ΔV

FIGURE 28.P1   Problems 12 and 13

14. A charge $+q$ travels in the yz-plane with velocity $\mathbf{v} = v_{y_o}\widehat{\mathbf{j}} + v_{z_o}\widehat{\mathbf{k}}$. At the instant it passes through the origin a uniform magnetic field $\mathbf{B} = B\widehat{\mathbf{j}}$ is turned on. Show that the y-component of the charge's acceleration is zero.

15. A deuteron consists of a proton and a neutron bound together. A proton and a deuteron are both accelerated through the same potential difference ΔV and then enter a region where there is a uniform magnetic field perpendicular to their paths. What is the ratio of the radii of their circular paths, $r_{\text{prot}}/r_{\text{deut}}$?

16. (a) What velocity must a proton have to be confined to circular motion with a 1.00-cm radius in a 1.00-T magnetic field? (b) Through what potential difference must the electron be accelerated to obtain this velocity?

17. Compute the kinetic energy of a charge Q of mass M that travels in a circle of radius R while in a uniform magnetic field B.

18. What is the cyclotron frequency of a proton that orbits in a plane perpendicular to a 1.00-T magnetic field?

19. A **bubble chamber** consists of a container filled with a liquid (often hydrogen) near its boiling point. When charged particles travel through the liquid, their disturbances produce bubbles that indicate the path of the particle. An alpha particle traveling with a velocity of 5.08×10^5 m/s in a circle with a 6.40-mm radius is observed in a bubble chamber. What is the magnitude of the magnetic field perpendicular to the particle's motion?

20. Each electron in a beam of electrons travels with the same velocity $v = 1.50 \times 10^7$ m/s. The beam is bent 90.0° through an arc of 4.25 mm by a uniform magnetic field perpendicular to the beam path. (a) Find the magnetic field strength B. (b) Through what potential difference must each electron have been accelerated to obtain this velocity? (c) How much time is required for an electron to travel through the 90.0° arc?

21. An alpha particle is accelerated through a potential difference ΔV and then enters a region of space perpendicular to a uniform magnetic field $\mathbf{B}$. The alpha particle travels in a circle of radius R. In terms of R, what is the radius of a proton's circular motion if it is accelerated through the same potential difference ΔV and then enters perpendicular to the same magnetic field?

Note: Problems 22 through 27 refer to thin conducting slabs and ribbons. The convention for the terms *length, width,* and *thickness* are shown in Figure 28.P2.

22. A thin slab of metal 1.50 mm wide is placed in a uniform 1.20-T magnetic field. The direction of the magnetic field is perpendicular to the surface of the slab, and a Hall voltage of 1.25 μV is observed across its height when a constant current of 10.0 A is maintained along the slab's length. What is the density n of the conduction electrons in this metal? See Figure 28.P2.

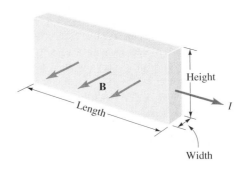

Height

$\mathbf{B}$

Length

Width

I

FIGURE 28.P2   Problems 22–27

23. A uniform magnetic field is applied perpendicular to the face of a thin slab of conducting material, and a constant current I flows along the slab's length. A Hall voltage of 36.0 μV is observed across this slab's height. This slab is removed from the magnetic field, and a slab of identical height but made from a different conductor is placed in the same magnetic field. The magnetic field is also perpendicular to this slab's face. The Hall coefficient of the new slab is twice that of the original slab, and the width of the new slab is one-third that of the original. What Hall voltage appears across the height of the new slab when the same current I flows along its length? See Figure 28.P2.

24. A certain semiconductor is found to have a density of charge carriers $n = 1.80 \times 10^{20}$ carriers/m³. A current of 5.00 mA is made to flow along the length of a 0.850-mm wide ribbon of this material. What magnetic field strength must be applied perpendicular to the face of this ribbon for Hall voltages of (a) 0.100 mV, (b) 1.00 mV, and (c) 10.0 mV to be across the height of the ribbon? See Figure 28.P2.

25. A thin ribbon of copper is found to have a Hall coefficient of -0.600×10^{-10} m³/C. The width of the ribbon is 0.250 mm, and a Hall voltage across the height of the ribbon is 5.80 μV when a current of 15.0 A flows along its length. What is the magnitude of the magnetic field that has a direction perpendicular to the plane of the ribbon? See Figure 28.P2.

26. A thin ribbon of silver 0.240 mm thick carries a current of 18.0 A in a Hall-effect experiment. When a 2.20-T magnetic field is applied perpendicular to the plane of the ribbon, a Hall voltage of -16.5 μV is observed across the height of the ribbon. (a) What is the Hall coefficient for silver? (b) What is the density of charge carriers? (c) If the height of the ribbon is 1.20 cm, what do you predict for the drift velocity of the charge carriers? See Figure 28.P2.

27. Probes that utilize the Hall effect are often used to measure magnetic field strengths. A constant current of 48.0 mA is maintained through a small slab of conducting material. When a magnetic field of 0.250 T is applied perpendicular to the plane of the slab, a 24.0-μV Hall voltage is observed across the height of the slab.

(a) What is the magnetic field strength when the Hall voltage rises to 144. μV? (b) Suppose this Hall probe is sensitive to only the component of the magnetic field perpendicular to the plane of the conducting slab. What is the magnetic field strength when a normal to the plane of the slab makes a 40.0° angle with the magnetic field direction and the Hall voltage is 125. μV?

28. An important parameter that describes the electrons in the plasma of an MHD channel (see Section 28.1) is known as the **Hall parameter** $\omega\tau$ (not to be confused with the Hall coefficient R_H). The angular frequency ω is the cyclotron frequency, and τ is the mean free time between collisions for the electron. The Hall parameter is thus the number of radians an electron moves in its circular orbit before it is interrupted by a collision. (a) What is the electron Hall parameter for an electron with a velocity 2.00×10^5 m/s if the electron moves in magnetic field of 32 500 G and experiences a collision every 1.00×10^{-11} s on the average? (b) What fraction of a full circle does the electron typically complete between collisions?

28.2 Force on a Current-Carrying Conductor

29. A uniform magnetic field is given by $\mathbf{B} = (0.300\text{ T})\hat{\mathbf{j}} + (0.400\text{ T})\hat{\mathbf{k}}$. What is the magnitude and direction of the force on a 15.0-A current-carrying wire that lies along the x-axis between $x = 0.00$ m and $x = 2.50$ m?

30. A uniform 1.20-T magnetic field has no component along the y-direction. However, the field makes a 60.0° angle from the z-axis toward the positive x-axis. Compute the force per length on a 24.0-A current-carrying wire if the direction of the current is along the positive y-axis.

31. When a wire carries a current of 3.60 A along the positive y-axis and perpendicular to a magnetic field, it experiences a force per unit length of 1.80 N/m along the positive z-axis. What is the magnitude and direction of the magnetic field?

32. A 2.20-T magnetic field is directed along a line 60.0° from the z-axis in the yz-plane. What is the force (magnitude and direction) on a 1.50-m segment of a wire that carries an 8.00-A current along the positive x-axis?

33. The direction of a magnetic field is 40.0° from the z-axis toward the positive x-axis in the xz-plane. A wire carrying a current of 12.0 A along the positive x-axis experiences a magnetic force per length of 14.7 N/m toward the negative y-axis. What is the magnitude of the magnetic field?

34. Two conducting rails are connected to a source of emf and form an incline as shown in Figure 28.P3. A bar of mass m slides without friction down the incline through a vertical magnetic field **B**. If the length of the bar is L, what current I must the emf provide for the bar to slide at a constant velocity?

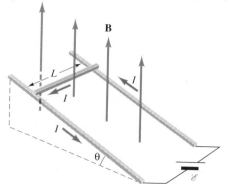

FIGURE 28.P3   Problem 34

35. In a trapeze-shaped structure, two rigid wires of negligible mass support a conducting bar of mass m and length L as shown in Figure 28.P4. A source of emf is applied to the wires so that a current I flows through the bar. A uniform magnetic field **B** is perpendicular to the plane of the wires and bar. (a) Compute the current that the source of emf must provide for there to be no tension in the wires. (b) If the current is reduced to half the value computed in (a) and the plane of the structure is moved through an angle θ, compute the tension in the wires and the magnitude of the net unbalanced force on the bar at the instant it is released from this angle.

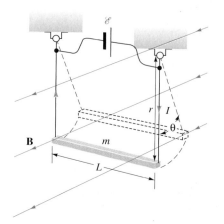

FIGURE 28.P4   Problem 35

36. One segment of a conductor is Z-shaped as shown in Figure 28.P5. The conductor carries a current $I = 8.00$ A. The uniform magnetic field $B = 1.60$ T is in the plane of the page but is directed at a 30.0° angle below the horizontal. Compute the magnitude and direction of the net force on this Z-shaped segment.

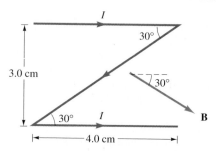

FIGURE 28.P5   Problem 36

37. A superconducting ring has a radius of 1.40 cm and a mass of 30.0 g. The ring carries a constant current I and is placed in a 0.500-T magnetic field with field lines that are tilted at a 20.0° angle, outward from the vertical at every location around the ring (Fig. 28.P6). What must the current I be for the ring to float in the magnetic field?

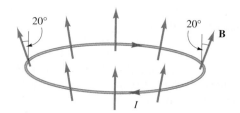

FIGURE 28.P6   Problem 37

38. The pie-shaped current loop shown in Figure 28.P7 subtends an angle of $\pi/6$ rad and lies in the xy-plane. The radius $R = 40.0$ cm and the current I is 6.00 A. The uniform magnetic field **B** is parallel to the positive z-axis and has a magnitude of 0.750 T. (a) Compute the magnetic force (magnitude and direction) on the segment ab. (b) Compute the magnetic force on the segment bc. (c) Compute the magnetic force on the segment ca, and show that the net force on the current loop is zero.

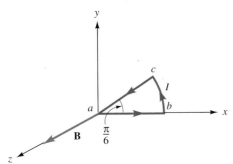

FIGURE 28.P7   Problem 38

28.3 Current-Carrying Loops in a Uniform Magnetic Field

39. For the magnetic moment $\boldsymbol{\mu} = (2.00 \times 10^3 \text{ A} \cdot \text{m}^2)\hat{\mathbf{i}} - (3.00 \times 10^3 \text{ A} \cdot \text{m}^2)\hat{\mathbf{j}}$ in the magnetic field $\mathbf{B} = (3.00 \text{ T})\hat{\mathbf{j}} - (2.00 \text{ T})\hat{\mathbf{k}}$, find (a) the magnitude of the torque on the dipole, and (b) the potential energy of the dipole in the field **B**.

40. A circular loop of wire has 50.0 turns, and each loop carries a current of 20.0 mA. (a) Compute the magnetic moment of this loop if it has a radius of 6.00 cm. (b) How much work is required to rotate the loop from an orientation in which the direction of its magnetic moment makes a 30.0° angle with that of an applied 2.20-T B-field to an orientation in which the moment makes a 135° angle with the B-field direction? (c) What is the torque on the moment when it is oriented at the 135° angle?

41. A single-turn, rectangular loop carries a 2.00-A current as shown in Figure 28.P8. A uniform, 1.60-T magnetic field is directed along the z-axis. (a) Using the $\hat{\mathbf{i}}, \hat{\mathbf{j}},$ and $\hat{\mathbf{k}}$ notation, write the magnetic moment of the current-carrying loop. (b) What is the magnitude of this torque on the loop? (c) What is the potential energy of the loop in the field **B**?

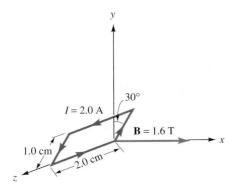

FIGURE 28.P8   Problem 41

42. The single, rectangular loop shown in Figure 28.P9 carries a 3.00-A current in the uniform magnetic field $B = 4.00$ T. (a) Use the $\hat{\mathbf{i}}, \hat{\mathbf{j}},$ and $\hat{\mathbf{k}}$ notation to write the magnetic moment of the loop. (b) What is the magnitude of the torque that the B-field exerts on the loop? (c) What is the potential energy of the loop?

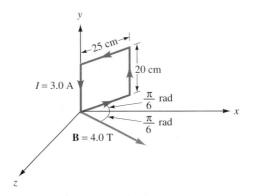

FIGURE 28.P9   Problem 42

43. Show that the potential energy U of a magnetic dipole moment $\boldsymbol{\mu}$ in a uniform magnetic field **B** is $U = -\boldsymbol{\mu} \cdot \mathbf{B}$, where $U = 0$ when the angle between $\boldsymbol{\mu}$ and **B** is 90°.

44. When it lies along the x-axis with one end at the origin, a rod of length L carries a nonuniform linear charge density λ described by $\lambda = Cx^2$, where C is a constant. If the rod is pivoted about an axis through the origin perpendicular to the x-axis and rotates with angular velocity ω, compute its magnetic moment $\boldsymbol{\mu}$.

45. A total charge Q is uniformly distributed over the surface of a disk with radius R. If the disk rotates about an axis through its center (and perpendicular to the plane of its surface) with angular velocity ω, show that the magnetic moment is $\mu = Q\omega R^2/4$.

46. Atomic nuclei have magnetic moments. In **nuclear magnetic resonance** experiments (NMR) the environment surrounding particular atomic nuclei may be studied by measuring the energy necessary to flip these moments in an applied magnetic field. If the magnetic moment of a proton is $1.40 \times 10^{-26} \text{ A} \cdot \text{m}^2$, what energy is necessary to flip 1.00 mol of protons from a parallel alignment with a 0.180-T field to an antiparallel orientation with this B-field?

Numerical Methods and Computer Applications

47. Create a spreadsheet template such that a user enters (1) a particle's charge, (2) the three components of a particle's velocity in three adjacent cells in one row (or column), and (3) three magnetic field components in another set of three adjacent cells. Have the spreadsheet then produce the three components of magnetic force on the particle in three adjacent cells and the magnitude of the force in another cell. Document your template using clear labels. Verify your spreadsheet template by using it to solve Problems 4 and 5.

★48. Write a computer program (or modify the RK2D.BAS) to use the fourth-order Runge-Kutta method to compute the trajectory of a charged particle traveling in the xy-plane in a nonuniform B-field $B(x, y)\hat{\mathbf{k}}$. Test your program by letting B equal a constant and see if that trajectory is a circle. Once you have satisfied yourself that the program is working, launch the particle with an x-velocity into a field given by $B(x, y) = B_o x/L\,\hat{\mathbf{k}}$. Use the same initial velocity and magnetic field B_o that gave you acceptable circles and let L equal a distance comparable to one-third the screen width.

General Problems

49. In 1897, J. J. Thomson measured the charge-to-mass ratio e/m of an electron by using a device similar to that illustrated in Figure 28.P10. Electrons from the heated filament A are accelerated by a potential difference ΔV through a small opening O. The electrons pass through a region where perpendicular electric and magnetic fields can be applied; these electrons eventually collide with a fluorescent screen, where they are observed. Thomson measured the vertical deflection y at the screen that a certain known field E produced, and then he measured the magnetic field strength B that returned the beam to its undeflected position. If L is the length of the deflecting plates, show that

$$\frac{e}{m} = \frac{2yE}{B^2 L^2}$$

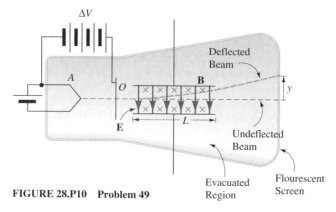

FIGURE 28.P10   Problem 49

50. A charge q enters the region of space where there is a uniform magnetic field of magnitude B along the x-axis. The charge enters this region at the origin while traveling in the xy-plane. If its velocity $\mathbf{v}$ initially makes an angle θ with the x-axis in the xy-plane, compute the pitch d and the radius of helical motion.

51. The electrons in a TV picture tube are accelerated horizontally through a 6.50-kV potential difference. The vertical component of the earth's magnetic field is 48.0 μT downward, and the electrons travel 30.0 cm before hitting the screen. (a) What is the acceleration of an electron due to this magnetic field? (b) What is the horizontal deflection of an electron beam due to the earth's magnetic field?

52. A charge Q of mass m is accelerated through a potential difference ΔV. It then enters a region of space perpendicular to a uniform magnetic field where it travels in a circle of radius R. If a charge $2Q$ with mass $m/2$ is accelerated through the same potential difference ΔV and then enters the same magnetic field region, what is the radius of this charge's circular motion?

53. What must be the diameter of a cyclotron that is to accelerate protons to 30.0 MeV in a magnetic field of 5.00 T?

54. In a certain cyclotron designed to accelerate protons, the electric potential alternates with a frequency of 7.20 MHz. What must be the radius of the cyclotron for the protons to reach an energy of (a) 0.200 MeV, (b) 2.00 MeV, and (c) 20.0 MeV?

55. A 1.50-m diameter cyclotron is to accelerate protons to an energy of 10.5 MeV. (a) What must be the magnetic field strength for this cyclotron? (b) At what frequency must this cyclotron operate?

56. A 15.0-g conductor 40.0 cm long is suspended from two springs that are attached to a 72.0-V source of emf. The springs, connecting wires, and conductor have a total resistance (represented by the resistor in Fig. 28.P11) of 2.00 Ω. When the bar is attached to the springs, they are observed to stretch 0.600 cm. When a magnetic field with a direction that is perpendicular to the plane of Figure 28.P11 is turned on, the springs are found to compress 0.800 cm from their equilibrium positions. What is the magnitude and direction of the applied magnetic field?

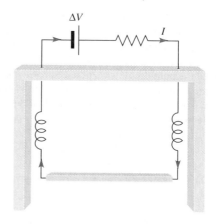

FIGURE 28.P11   Problem 56

★57. A nonuniform magnetic field is parallel to the y-axis and increases in magnitude with distance x from the origin according to the relation $\mathbf{B} = [(0.250 \text{ T/m})x + (0.100 \text{ T})]\hat{\mathbf{j}}$. A rectangular loop carries a constant current $I = 6.00$ A as shown in Figure 28.P12. The dimensions of the loop are given by $a = 30.0$ cm and $b = 20.0$ cm, and the distance c is 10.0 cm. Compute the net force on the current loop.

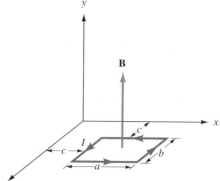

FIGURE 28.P12   Problem 57

58. As young lads, both authors constructed simple electric motors with basic designs similar to that shown in Figure 28.P13. (Despite the outcome, they both decided to go into physics, anyway!) As a coil of N turns rotates under the influence of a torque $\boldsymbol{\mu} \times \mathbf{B}$, the direction of the current flow is reversed by the action of the commutator and brushes. Hence, the torque is always in one direction. (a) Show for a current I in a magnetic field B, the magnitude of the torque on the coil is $\tau = NIBab|\sin(\theta)|$, where a and b are the length and width of the coil. (b) Suppose $a = 2.00$ cm, $b = 2.50$ cm, $B = 0.200$ T, $I = 0.250$ A, and $N = 20.0$. What is the average torque over one complete rotation?

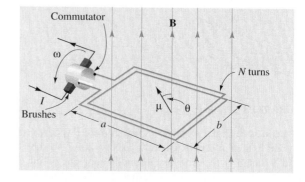

FIGURE 28.P13   Problem 58

59. The mechanism of a galvanometer is shown in Figure 28.P14. A coil of N turns is attached to a cylindrical core that is centered in a *radial* magnetic field **B.** The effect of the radial field is to keep the resultant B-field parallel to the plane of the coil. A spring of torsion constant κ causes a torque that acts in the opposite direction from that created when a current I flows through the galvanometer coil. Suppose the area of the coil is 2.50 cm^2, it has 50.0 turns, and the magnetic field has a magnitude of 0.300 T. Moreover, when a current of 10.0 μA flows through each turn of the coil the needle is deflected through 45.0°. (a) What is the torsion constant of the spring? (b) What angle is the galvanometer needle deflected through when a current of 3.50 μA flows through the coil?

★60. A sphere of radius R carries a uniform charge density σ over its surface. If the sphere rotates about a diameter with angular velocity ω, show that the magnitude of its magnetic moment is

$$\mu = \frac{4\pi\sigma\omega R^4}{3}$$

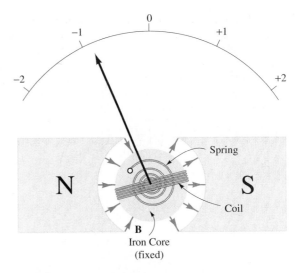

FIGURE 28.P14   Problem 59

Sources of Magnetic Fields

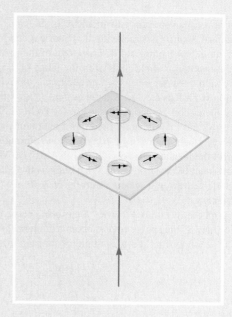

In this chapter you should learn

- how to apply the Biot-Savart law to compute the magnetic field **B.**

- how the units ampere and coulomb are defined.

- how to apply Ampère's law to compute the magnetic field for certain high-symmetry current distributions.

- about magnetic flux and Gauss's law for magnetism.

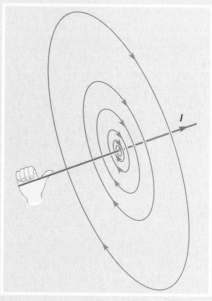

$$\oint \mathbf{B} \cdot d\mathbf{l} = \mu_o I_{\text{encl}}$$

In the previous chapter we assumed that we could produce a magnetic field in a region of space when we needed it. We used bar magnets as a way of creating magnetic fields to get us started. However, before we can understand how the magnetic fields of bar magnets are created, we must look at the fundamental mechanism by which magnetic fields are produced. The set of four Maxwell equations toward which we are building can be thought of as a set of recipes for creating electric and magnetic fields. One of these recipes tells us that magnetic fields can be created either by *changing* electric fields or by *moving* charges. In this chapter we concentrate on the production of magnetic fields by moving charges in the form of electric currents.

It may interest you to know that the discovery that currents create magnetic fields took place in a classroom during a physics lecture demonstration by Hans Christian Oersted (1777–1851) in 1819. Shortly after Oersted's discovery, Jean Baptiste Biot (1774–1862) and Félix Savart (1791–1841) were able to establish the properties (magnitude and direction) of the current-generated magnetic field and from these properties write down a mathematical relation that can be used to compute the field **B** from a current I. We present this expression, the Biot-Savart law, in the first section of this chapter and illustrate its use. Next we introduce a second form of the relation between currents and magnetic fields: Ampère's law. In the final section of the chapter we present the Maxwell equation that describes for magnetic fields a rule that is analogous to Gauss's law for electric fields.

The directions of vectors remain crucial in this chapter, so we urge you to read carefully every statement about the right-hand rule and anything else that involves a vector direction. It is best *not* to read any of this chapter casually; every time we describe the directions of vectors and the outcome of a vector cross product you should stop and use your right hand to verify what we have said.

29.1 The Biot-Savart Law

In Chapter 22 we found that the *electric* field **E** generated by a charge distribution can be computed from the expression

$$\mathbf{E} = \frac{1}{4\pi\epsilon_o} \int \frac{dq}{r^2}\hat{\mathbf{r}}$$

where the integration is carried out over the entire charge distribution, and $\hat{\mathbf{r}}$ is a unit vector that points from the infinitesimal charge element dq to the point in space where we want to find **E.** The Biot-Savart law is an analogous expression from which we can compute the *magnetic* field **B.** Steady currents, rather than static charges, are the sources of the magnetic fields we consider in this chapter. The Biot-Savart law is written in a form that utilizes the $I\,d\mathbf{l}$ term described in the previous chapter; recall that $d\mathbf{l}$ is an infinitesimal length parallel to the current I. You should keep in mind, however, that in this chapter we are not focusing our attention on what happens *to* moving charges (currents) in a magnetic field, but rather on the properties (magnitude and direction) of the field *produced by* moving charges. The Biot-Savart law is a mathematical description of the magnetic field $d\mathbf{B}$ that arises from a current I flowing along an infinitesimal path element $d\mathbf{l}$. Before looking at the mathematics, we describe the field's properties. Refer to Figure 29.1 as you read these properties.

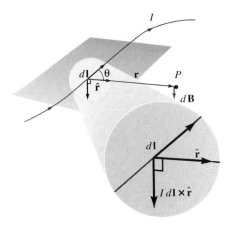

FIGURE 29.1

The magnetic field $d\mathbf{B}$ at point P is given by the Biot-Savart law. The direction of $d\mathbf{B}$ is parallel to $I\,d\mathbf{l} \times \hat{\mathbf{r}}$.

1. The magnetic field grows weaker as you move farther from its source. In particular, the magnitude of the magnetic field dB is inversely proportional to r^2, the square of the distance from the current source $I\,d\mathbf{l}$.
2. The larger the electric current, the larger the magnetic field. In particular, the magnitude of the magnetic field dB is proportional to the current I.
3. If you stay a fixed distance from the $I\,d\mathbf{l}$ source, the magnetic field (due to this element only) is stronger off to the side of the current element than it is directly in front of it or behind it. To be precise, the magnitude of the magnetic field dB is

proportional to $\sin(\theta)$, where θ is the angle between the current direction $d\mathbf{l}$ and a unit vector $\hat{\mathbf{r}}$ that *points from the current element to the point in space where dB is evaluated*.

4. The direction of the B-field is *not* radially away from its source as the gravitational field and the electric field are from theirs. In fact, the direction of $d\mathbf{B}$ is perpendicular to both $d\mathbf{l}$ and the unit vector $\hat{\mathbf{r}}$.

The Biot-Savart law summarizes all these properties in the mathematical form

$$dB = \frac{\mu_o}{4\pi} \frac{I\,d\mathbf{l} \times \hat{\mathbf{r}}}{r^2} \qquad (29.1)$$

As usual, $\hat{\mathbf{r}}$ is a unit vector that points from the current element to the point in space where $d\mathbf{B}$ is to be evaluated. The constant μ_o is called the **permeability of free space,** or more simply, the **permeability constant.** Its value is defined to be

$$\mu_o = 4\pi \times 10^{-7} \text{ T} \cdot \text{m/A}$$

In the next section we will explain why we are free to define this constant.

The direction of the magnetic field element $d\mathbf{B}$ as given by the Biot-Savart law is shown in Figure 29.1. The first step in applying Equation (29.1) is to put your pencil down (unless you are left-handed) and pick up your right hand and perform the cross product $I\,d\mathbf{l} \times \hat{\mathbf{r}}$. As illustrated in Figure 29.1, $d\mathbf{B}$ is parallel to this direction. You can calculate the magnitude of $\mathbf{B}$ by employing the magnitude of the cross product in Equation (29.1):

$$dB = \frac{\mu_o}{4\pi} \frac{I\,dl\,\sin(\theta)}{r^2} \qquad (29.2)$$

For a conducting path of finite length, it is important for you to realize that the total magnetic field $\mathbf{B}$ at any given point in space is computed by summing the contributions from each small element of the current along its total path. In practice, this is done by integrating Equation (29.1) along the current path made up of $d\mathbf{l}$ vectors.

As a practical matter, you should always begin by checking on the direction of $d\mathbf{B}$ caused by several $I\,d\mathbf{l}$ elements along the current path. When you do this, one of two cases occurs, each requiring a slightly different strategy:

CASE 1: The $d\mathbf{B}$ contributions from each $d\mathbf{l}$ all point in the same direction (or at worst in opposite directions). In this happy circumstance, you can integrate immediately using the magnitude expression in Equation (29.2).

CASE 2: The $d\mathbf{B}$ contributions from $I\,d\mathbf{l}$ sources located on different parts of the current path point in different directions. In this case, before you integrate you *must* resolve the $d\mathbf{B}$ contributions into components. The equations you derive for the components dB_x, dB_y, and dB_z are then ready to be integrated.

Concept Question 1
Describe how the Biot-Savart law accounts for the four properties of the magnetic field listed just before Equation (29.1).

Case 1 occurs whenever the lines of current and the observation point lie in the same plane. We begin with Example 29.1, which is of the case-1 type. Example 29.2 explores a case-2 type problem.

EXAMPLE 29.1   *What Goes Around Comes Around*

A straight wire segment carries a constant current $I = 1.00$ A along the x-axis between x_o and x_f. Compute the magnetic field $\mathbf{B}$ at point P with coordinates $(x, y) = (0, 1.00)$ cm for (a) $x_o = -5.00$ cm and $x_f = +3.00$ cm, and (b) at this same point P for a very long wire.

SOLUTION   We make the approximation that the distance y is large compared to the diameter of the wire. In such a case we ignore its finite diameter and treat the wire as a line. In Figure 29.2 we show a current of magnitude I along the x-axis with one end located at $x = x_o$ and the other at $x = x_f$. We wish to compute the magnetic field $d\mathbf{B}$ at point P, a distance y from the current line. Notice that we have chosen an *arbitrary* segment $d\mathbf{l}$ that is along the direction of the current. (Avoid choosing a $d\mathbf{l}$ at either end or the middle.) Also note that the unit vector $\hat{\mathbf{r}}$ points from $d\mathbf{l}$ to the point P. By using the right-hand rule you should verify that the direction of $d\mathbf{l} \times \hat{\mathbf{r}}$ is perpendicular to and out of the page; this direction is also that of $d\mathbf{B}$. It is important to realize that *every* element $d\mathbf{l}$ along this wire contributes a $d\mathbf{B}$ at P that is in this same direction. (Verify this last statement for a segment $d\mathbf{l}$ located on a different segment of the current line.) The magnitude of the field is given by Equation (29.2):

$$dB = \frac{\mu_o}{4\pi} \frac{I \, dx \, \sin(\theta)}{r^2}$$

where we have recognized that $dl = dx$. The integral we need to compute is

$$B = \frac{\mu_o}{4\pi} \int \frac{I \, dx \, \sin(\theta)}{r^2}$$

Because I is a constant, it can be removed from the integration; however, both r and θ change as x changes. One way to integrate over the entire current line is to allow θ to go from $\theta = \theta_o$ (where $x = x_o$) to $\theta = \theta_f$ (where $x = x_f$). Be sure to see that θ_f is an obtuse angle in Figure 29.2. To make the change of variable to θ we note

$$\sin(\theta) = \frac{y}{r}$$

or

$$\frac{1}{r^2} = \frac{\sin^2(\theta)}{y^2}$$

Also, $\tan(\theta) = y/x$ or $x = y \cot(\theta)$ so that

$$dx = y \csc^2(\theta) \, d\theta = \frac{y}{\sin^2(\theta)} \, d\theta$$

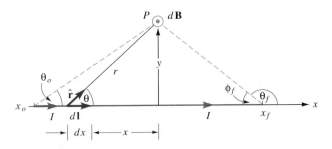

FIGURE 29.2

A finite current segment carries a current I along the x-axis between x_o and x_f. The magnetic field $d\mathbf{B}$ at point P is out of the page and is generated by the current along the length element $dl = dx$. Notice that both r and θ depend on x. See Example 29.1.

Substituting these last two relations into our expression for dB, we have

$$dB = \frac{\mu_o I}{4\pi} \left(\frac{\sin^2(\theta)}{y^2} \right) \left[\frac{y}{\sin^2(\theta)} \, d\theta \right] \sin(\theta) = \frac{\mu_o I}{4\pi} \frac{1}{y} \sin(\theta) \, d\theta$$

Integrating,

$$B = \frac{\mu_o I}{4\pi} \frac{1}{y} \int_{\theta_o}^{\theta_f} \sin(\theta) \, d\theta$$

$$= \frac{\mu_o I}{4\pi y} \left[-\cos(\theta) \right]_{\theta_o}^{\theta_f} = \frac{\mu_o I}{4\pi y} [\cos(\theta_o) - \cos(\theta_f)]$$

Often, it is convenient to write this expression in terms of the interior angles of the triangle formed by the wire and the observation point P. In the notation of Figure 29.2, $\theta_f = \pi - \phi_f$. Thus, we have

$$-\cos(\theta_f) = -\cos(\pi - \phi_f) = +\cos(\phi_f)$$

The final result can be written

$$B = \frac{\mu_o I}{4\pi y} [\cos(\theta_o) + \cos(\phi_f)]$$

For a wire of finite length with $x_o = -5.00$ cm and $x_f = +3.00$ cm we find $\theta_o = 11.3°$ and $\phi_f = 18.4°$, and with $y = 1.00$ cm $= 0.0100$ m we have

$$B = \frac{(4\pi \times 10^{-7} \text{ T} \cdot \text{m/A})(1.0 \text{ A})}{4\pi(0.0100 \text{ m})} [\cos(11.3°) + \cos(18.4°)] = 1.93 \times 10^{-5} \text{ T}$$

(b) As the wire length becomes large relative to y, both θ_o and ϕ_f approach zero. In such cases it is often adequate to model the wire as "infinitely long" and let $\theta_o = \phi_f = 0$. Substituting these values into our expression for B, this approximation yields

$$B = \frac{\mu_o I}{2\pi y}$$

For the "infinitely long wire model," when $I = 1.00$ A and $y = 1.00$ cm, we have

$$B = \frac{(4\pi \times 10^{-7} \text{ T} \cdot \text{m/A})(1.0 \text{ A})}{2\pi(0.0100 \text{ m})} = 2.00 \times 10^{-5} \text{ T}$$

Thus, we see that, even for a relatively short wire, the infinitely long wire model differs from the exact answer by less than 4%. ◀

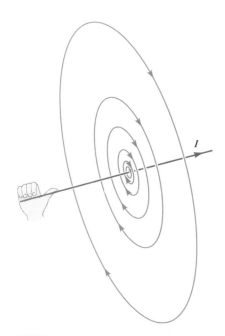

FIGURE 29.3

Applying the right-hand rule to determine the direction of the magnetic field **B** from a line of current. When the thumb of the right hand is placed along the direction of the current, the fingers curl around the line in the direction of the B-field.

As illustrated in Figure 29.2, we found the direction of the magnetic field computed in Example 29.1 to be out of the page. There is nothing unique about the plane of the page in Example 29.1; if we take the position of point P in Figure 29.2 to be directly above both the line and the plane of the page, the direction of **B** at this position (as determined by the direction of $d\mathbf{l} \times \hat{\mathbf{r}}$) is toward the bottom of the page and has the exact same magnitude. This result suggests that the B-field lines make circles in planes perpendicular to the current direction as shown in Figure 29.3. Although you are free to draw any of the circular B-field lines you choose, we show those at positions where the field magnitude decreases by a constant amount; close to the current I the magnitude of **B** is large and there is a high density of circles, whereas farther from I the B-field magnitude decreases and the spacing between the circles increases.

Finally, also notice from Figure 29.3, that a way to find the direction that the B-field circles around a current is to place the thumb of your right hand along the direction of the current. Your fingers curl around this current in the direction of **B**.

EXAMPLE 29.2   A Superloop

There is a fascinating phenomenon exhibited by a circular loop of superconducting material. If you start a current in the loop while the material is in its superconducting state, the current persists without perceptibly diminishing even with no source of emf in the loop. Find the magnetic field **B** a distance x along the axis of a superconducting loop of radius R if the loop carries a current I.

SOLUTION   There are three angles to keep your eyes on in this problem and we show all three in Figure 29.4(a) (α, θ, and the angle between $d\mathbf{l}$ and $\hat{\mathbf{r}}$). The good news is that for any element $d\mathbf{l}$ on the loop, $\hat{\mathbf{r}}$ is perpendicular to $d\mathbf{l}$. Hence, the magnitude of each dB, as given by the Biot-Savart law, is constant and equal to

$$dB = \frac{\mu_o}{4\pi} \frac{I \, dl \, \sin(90°)}{r^2} = \frac{\mu_o}{4\pi} \frac{I \, dl}{r^2}$$

The bad news is that the directions of the $d\mathbf{B}$ vectors are all different. As you can see, $d\mathbf{B}$ is not perpendicular to the loop axis but, as shown in Figure 29.4(a), makes an angle α with that axis. (Verify this result with your right hand!) In principle, this means we must resolve the $d\mathbf{B}$ vectors into components and integrate each component separately. However, in this example we can use the cylindrical symmetry of the problem to deduce (without actually doing an integral!) that two of the three components integrate to zero. Now for the symmetry: for every $d\mathbf{l}$, such as the element shown in Figure 29.4(a), there is another $d\mathbf{l}$ on the opposite side of the loop (Fig. 29.4(b)) so that, as we sum the contributions to $d\mathbf{B}$, the components parallel to the yz-plane, each given by $dB \sin(\alpha)$, cancel in pairs. However, all x-components $dB \cos(\alpha)$ are parallel and contribute to the total field **B**. Hence, the total B-field at P can be calculated by integrating the x-components of dB only:

$$dB_x = dB \cos(\alpha)$$

$$= \frac{\mu_o}{4\pi} \frac{I \, dl}{r^2} \cos(\alpha)$$

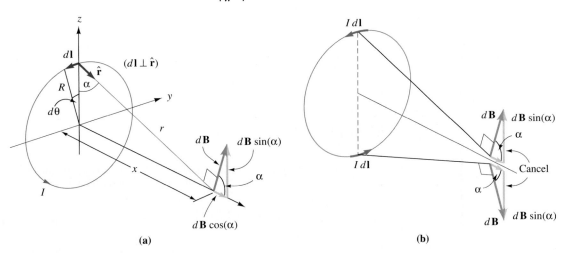

(a)   (b)

FIGURE 29.4

Computing the B-field at distance x along the axis of a conducting ring from its center. (b) Notice that for every length element $d\mathbf{l}$ such as that shown, there is an equivalent element on the opposite side of the ring so that the components $dB \sin(\alpha)$ sum to zero. See Example 29.2.

But $r = \sqrt{x^2 + R^2}$ and

$$\cos(\alpha) = \frac{R}{r} = \frac{R}{\sqrt{x^2 + R^2}}$$

Now, $dl = R\,d\theta$, so our expression for dB_x becomes

$$dB_x = \frac{\mu_o}{4\pi} \frac{IR\,d\theta}{(x^2 + R^2)} \frac{R}{\sqrt{x^2 + R^2}}$$

We need only integrate this expression around the entire loop from $\theta = 0$ rad to $\theta = 2\pi$ rad:

$$B_x = \frac{\mu_o I}{4\pi} \frac{R^2}{(x^2 + R^2)^{3/2}} \int_0^{2\pi} d\theta$$

$$B_x = \frac{\mu_o I}{2} \frac{R^2}{(x^2 + R^2)^{3/2}} \qquad \text{(For points on the loop axis)}$$

$$B_y = B_z = 0 \qquad \text{(From symmetry, for points on the loop axis)}$$

Notice that for $x = 0$ we find the field at the center of the loop to be

$$B_x = \frac{\mu_o I}{2R} \qquad \text{(Field at the center of a circular loop)}$$

You might wonder what the magnetic field lines do off the axis of the loop. From Example 29.1 we know that if we consider a position that is very close to the conducting loop, the magnetic field lines are shaped as circles around the loop. As the distance from the loop increases toward its center, the *B*-field lines change shape as shown in Figure 29.5. For distances far from the loop the field is identical to that of the *magnetic dipole* that we will investigate further in this and the next chapter.

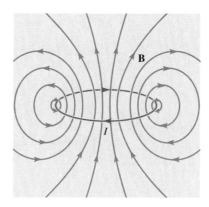

FIGURE 29.5

The magnetic field lines of a current loop. Close to the current the lines approach circles. Far from the loop the *B*-field lines are identical to those of a magnetic dipole.

The equation for the *B*-field due to a short wire segment as derived in Example 29.1 is a useful pocket tool. Another helpful equation for *B* (also computed from the Biot-Savart law) is that for a current-carrying arc of radius *R* (see Problem 62). Many current configurations can be built up from segments of straight wires and portions of circular loops. We summarize these results here and show one example of their application in Example 29.3.

Magnetic Fields for Two Useful Special Cases

1. A point P near a straight wire of any length:

$$B = \frac{\mu_o I}{4\pi y} [\cos(\theta) + \cos(\phi)]$$

where θ and ϕ are the angles made by the wire and lines connecting the wire ends to P and y is the perpendicular distance from the wire to point P.

2. A point P at the vertex of a circular arc that subtends an angle θ:

$$B = \frac{\mu_o I}{R} \frac{\theta}{4\pi}$$

where R is the radius of the circular arc.

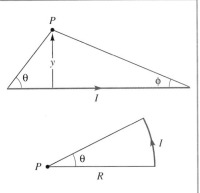

The magnetic field at point P can be calculated using the "special case" formulas at left.

EXAMPLE 29.3   *The Magnetooptic Spatial Light Modulator (MOSLM)*

A magnetooptic spatial light modulator (MOSLM) is composed of a rectangular array of small square picture elements called pixels. Each pixel is covered by a transparent iron garnet film. The light-transmitting properties of the film over each pixel can be changed by altering its magnetic state with an applied magnetic field. Pixels are separated by thin metallic strips that run both vertically and horizontally. The conducting strips, called address lines, make small partial loops in the corner of each pixel as shown in Figures 29.6(a) and (b). When current is sent through both the vertical wire and the horizontal wire nearest the corner of a particular pixel, the

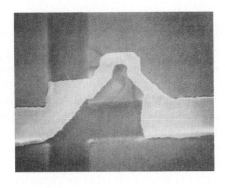

(a)

(b)

(c)

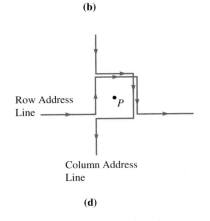

Row Address Line

Column Address Line

(d)

FIGURE 29.6

(a) A picture created with a scanning electron microscope reveals a partial current loop formed by a horizontal conductor. A similar loop created by a vertical conductor on the reverse side of the board is fairly visible. (b) The current loops of figure (a) are located in the corners of square pixels. (c) A 48 × 48 array of pixels is used to "write" a visible pattern. (Missing portions are due to defects in this particular MOSLM.) (d) Simplified model of current flow near the corner of the MOSLM pixel. Current flowing in one vertical wire and one horizontal wire (located on opposite sides of the circuit board) create a large magnetic field inside the square area. See Example 29.3.

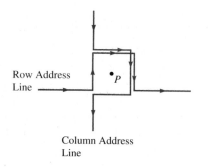

Row Address Line

Column Address Line

FIGURE 29.6(d)

magnetic field from the combined currents is sufficient to change the magnetic state of the material in the pixel's corner. This changed magnetic state then spreads to the rest of the pixel, altering the pixel's light-transmission characteristic.

Model the region near the pixel's corner by square half-loops as shown in Figure 26.6(d), and calculate the magnetic field at the center point P of the model. Take the sides of the square to be 50.0 nm in length, and assume the current in each of the address lines is 100. mA.

SOLUTION   We ignore the field created by the long undeviated portions of the address lines and assume the field at point P of Figure 29.6(d) is created primarily by six straight line segments in the surrounding square area. (See Problem 66 for an evaluation of this approximation.) Because each of the six wire segments contributes equally to the field, we can calculate the contribution from any one of them and multiply by six. For one side of the square we use the formula from special case number 1:

$$B = \frac{\mu_o I}{4\pi y} [\cos(\theta) + \cos(\phi)]$$

with $\theta = \phi = 45°$, $I = 100.$ mA, and $y = 25.0$ nm.

$$B = \frac{(4\pi \times 10^{-7}\ \text{T} \cdot \text{m/A})(100. \times 10^{-3}\ \text{A})}{4\pi(25.0 \times 10^{-9}\ \text{m})} [\cos(45°) + \cos(45°)]$$

$$= 0.566\ \text{T}$$

The total field is six times larger than this result because the B-field contributions from each segment are parallel. The total field is quite strong:

$$B = 6(0.566\ \text{T}) = 3.39\ \text{T} \qquad \blacktriangleleft$$

29.2 Parallel Wires, Amperes, and Coulombs

In the previous section we *defined* the constant μ_o to have a value of $4\pi \times 10^{-7}\ \text{T} \cdot \text{m/A}$. In fact, we were simultaneously defining the unit of current, the ampere, which in turn defines the coulomb. This domino effect occurs because of interrelationships among the three new concepts defined in our study of electromagnetism; namely, charge, current, and magnetic field. If we had defined these three concepts completely independently from one another, then the *physical* relations between them would have contained three proportionality constants, say C_1, C_2, and C_3.[1] One choice for the three basic relations (ignoring the vector complications for the moment) are listed below.

$$Q = C_1 It$$

$$F = C_2 IBl$$

and

$$B = \frac{C_3 I}{2\pi y}$$

---

[1] To a large extent, this is what happened historically: four major unit systems evolved each with different values for C_1, C_2, and C_3. To say that the situation was confusing is an understatement.

In the SI system of units the choices are $C_1 = C_2 = 1$ and $C_3 = 4\pi \times 10^{-7}$ T·m/A $\equiv \mu_o$. Once these three, free choices are made, the magnitudes of each of the units for charge, current, and magnetic field are determined by physical relations implied by the three equations. In this section we put all the pieces together, and describe an operational method from which the magnitude of the ampere can be determined.

The experimental determination of an ampere's magnitude is based on the magnetic force of attraction between two, parallel current-carrying conductors. Two such conductors are shown in Figure 29.7, where currents I_1 and I_2 are in the same direction and are separated by distance d. In Figure 29.7(a) we show only a few B-field lines generated by current I_1. Be sure to realize that the B-field created by I_1 acts on wire 2 at all points along its length. Moreover, wire 1 experiences a force along its entire length due to the magnetic field created by current I_2. Let's focus our attention on the force *on* wire 2 *by* wire 1, which we call F_{21} and show in Figure 29.7(b). From Example 29.1 we know that the magnetic field due to wire 1 at distance d from this wire is

$$B_1 = \frac{\mu_o I_1}{2\pi d}$$

Also notice that the direction of $\mathbf{B}_1$ is perpendicular to wire 2 so that the magnitude of the force on a length l of wire 2 by this field is

$$F_{21} = I_2|\mathbf{l} \times B_1| = I_2 l B_1 \sin(90°)$$

$$= I_2 l \left(\frac{\mu_o I_1}{2\pi d}\right)$$

$$F_{21} = \frac{\mu_o I_1 I_2}{2\pi d} l \qquad (29.3)$$

By applying the right-hand rule you should verify that the direction of F_{21} is toward wire 1. Also, you should perform an analysis similar to that above to convince yourself that the force F_{12} *on* wire 1 *by* wire 2 is equal in magnitude to F_{21} but opposite in direction, as it must be by Newton's third law. We may use Equation (29.3) to determine the ampere experimentally. If both currents are equal, $I_1 = I_2 = I$, we have

$$F_{12} = \frac{\mu_o I^2}{2\pi d} l$$

Suppose the separation distance between the wires is 1.00 m, and they each carry a current of 1.00 A. From our equation, the force per length on a segment of wire 1 is

$$\frac{F_{12}}{l} = \frac{\mu_o I^2}{2\pi d} = \frac{(4\pi \times 10^{-7} \text{T·m/A})(1.00 \text{ A})^2}{2\pi (1.00 \text{ m})}$$

$$= 2.00 \times 10^{-7} \text{ N/m}$$

This result is useful because we can *experimentally determine* the magnitude of one ampere. As you can see, when one ampere flows through each of two parallel wires, separated by one meter, the force of attraction per unit length on each is 2.00×10^{-7} N/m. In practice, *current balances* are used to measure the force between parallel conductors so that standard currents can be calibrated.

Once we have established the magnitude of an ampere, the coulomb is also defined through $Q = It$. When a conductor carries a current of one ampere, one coulomb of charge flows through a cross section of that conductor in one second. Furthermore, the tesla is defined by the relation $B = F/Il$. The magnetic field strength one meter away from a very long wire carrying one ampere of current is 2.00×10^{-7} T.

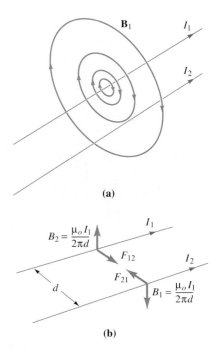

(a)

(b)

FIGURE 29.7

(a) Two parallel, current-carrying wires each experience the magnetic field of the other. Here we show only the magnetic field $\mathbf{B}_1$ due to current I_1.
(b) Forces on the two current-carrying, parallel wires, separated by distance d. The force on a segment of wire 2 by the magnetic field of wire 1 is $F_{21} = I_2 l B_1$, where $B_1 = \mu_o I_1 / 2\pi d$. When the currents are in the same direction, the wires are attracted.

Concept Question 2
Two perpendicular, current-carrying wires pass very close to each other. Describe the force that one wire exerts on the other.

André Marie Ampère (1775–1836) was a French philosopher, mathematician, and chemist. Although he is properly credited with the law that bears his name, Ampère also made many other significant contributions to the theory of electromagnetism and mathematics, particularly in the areas of partial differential equations and game theory.

29.3 Ampère's Law

In Chapters 22 and 23, we found that the electric field could be related to its sources through either the *differential* relationship of Coulomb's law or the *integral* relationship of Gauss's law:

$$d\mathbf{E} = \frac{1}{4\pi\epsilon_o} \frac{dq}{r^2} \hat{\mathbf{r}}, \qquad \text{Coulomb's law}$$

$$\epsilon_o \oint \mathbf{E} \cdot d\mathbf{A} = q_{\text{encl}}, \qquad \text{Gauss's law}$$

Gauss's law provided us with some powerful insights into the behavior of electric fields that were far from obvious when we looked at Coulomb's law alone. In addition, when the charge distribution has a sufficiently high symmetry, Gauss's law can be used to deduce the magnitude of the electric field. The Biot-Savart law, Equation (29.1), provides a *differential* relation for the magnetic field. One might reasonably ask if there is a corresponding *integral* relation between the magnetic field **B** and its current source *I*. Such an integral relation between the magnetic field and *steady* currents does exist and is known as **Ampère's law:**

$$\oint \mathbf{B} \cdot d\mathbf{l} = \mu_o I_{\text{encl}} \tag{29.4}$$

We will explain the meaning of I_{encl} below, but for the moment let's consider the left side of Equation (29.4). The left-hand side of Ampère's law is a line integral *around a closed path*. You have encountered line integrals previously when you studied the work concept; recall that the work done by a force **F** between points *o* and *f* is $\mathcal{W} = \int_o^f \mathbf{F} \cdot d\mathbf{l}$. For our present case the integral $\oint \mathbf{B} \cdot d\mathbf{l} = \oint B \cos(\theta)\, dl$ simply means to add up (integrate) the product of the incremental path element dl and the component of the magnetic field **B** that is parallel to this path element ($B \cos(\theta)$). The only restriction on the path is that you must end up where you start out. That is, the path must be closed. Now, we intend to inflict this operation on you in only the most symmetrical of cases for which you can perform the integral using simple multiplication. Below we are going to show you how easy it is, but first let's clear up the meaning of I_{encl}.

Ampère's law relates the magnetic field **B** to the total current that flows *through* the area enclosed by the closed path we have been considering. The closed path can take any shape, but, if it is a hoop, you can say that the current has to jump through the hoop in order to count. Hence, I_{encl} represents the steady current that is actually *enclosed* within the path over which $\oint \mathbf{B} \cdot d\mathbf{l}$ is evaluated. The enclosed current need not be a single current but can indeed be several steady current sources. If there are currents that flow in opposite directions from each other while passing through the closed loop, we must assign opposite signs to currents with opposite directions.

Ampère's law can be applied to any current distribution and any closed path. But, in actual fact, there are only a few simple current configurations where Ampère's law can be used to calculate the magnitude of the *B*-field without encountering quite messy integrals. Even in these simple cases you must be careful to choose a convenient path. Let's apply Ampère's law to a case for which we already know the answer. Figure 29.8 shows a long, straight line of current and some of the *B*-field lines circling it. Suppose we choose one of these circular *B*-field lines as our path. Then, as shown in Figure 29.8, at every point on the path not only is the magnitude of **B** a constant, but the direction of **B** and $d\mathbf{l}$ are parallel so that

$$\oint \mathbf{B} \cdot d\mathbf{l} = \oint B \cos(0°)\, dl = B \oint dl = B(2\pi y) \tag{29.5}$$

where we have recognized $\oint dl$ as the distance around the circular path.

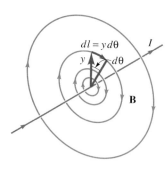

FIGURE 29.8

Application of Ampère's law to a long line of current *I*. The magnetic field **B** along any path element $dl = y\, d\theta$ is constant in magnitude and parallel to $d\mathbf{l}$.

The current enclosed by the circular path of Figure 29.8 is simply the current I, so that the right-hand side of Equation (29.4) is $\mu_o I$. Therefore, using this result for the right-hand side of Ampère's law and Equation (29.5) for the left-hand side, we have

$$B(2\pi y) = \mu_o I$$

or

$$B = \frac{\mu_o I}{2\pi y} \tag{29.6}$$

This result is exactly the same as that we obtained in Example 29.1 where we applied the Biot-Savart law (and exerted a lot more effort).

Infinite Current Sheet

As our next application of Ampère's law let's consider a sheet of current confined to the xz-plane, with a direction parallel to the z-axis. This current distribution is illustrated in Figure 29.9(a). We can think of the sheet as being made up of a large number of parallel wires all carrying current in the same direction. If you place the thumb of your right hand along any one wire's current, the field due to that wire forms a circular B-field with the wire at the center. Above the sheet (say along the positive y-axis) the **B** field from each wire element has an x-component and a y-component. The B_y components from wires that cross the positive x-axis are negative (pointing toward the sheet as illustrated in Fig. 29.9(b)). The B_y components from wires that cross the negative x-axis are positive (pointing away from the sheet as illustrated in Fig. 29.9(c)). Because the contributions from the positive x-axis and the negative x-axis are equal, they must cancel. We conclude that the B_y component is zero along the y-axis. But the y-axis could be anywhere along this infinite sheet. This observation leads us to the conclusion that the B-field everywhere above the sheet has a negative x-component only. Moreover, you will also discover that for any point below the sheet ($-y$ values) the B-field is parallel to the positive x-axis (see Fig. 29.9(d)). Hence, for this symmetry, we see for positive y-values $\mathbf{B} = -B\hat{\mathbf{i}}$ and for negative y-values $\mathbf{B} = +B\hat{\mathbf{i}}.$

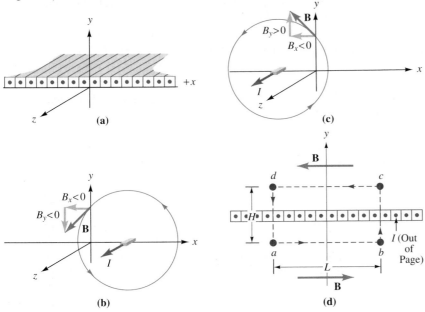

FIGURE 29.9

(a) A model of an infinite current sheet. The sheet lies in the xy-plane, and the direction of the current is parallel to the positive x-axis. The current per length (along the y-axis) is given by $\mathcal{J}_l$.
(b) **B** on the z-axis from a current that crosses the positive y-axis. (c) **B** on the z-axis from a current that crosses the negative y-axis. (d) For this symmetry the magnetic field above the sheet is uniform and directed toward the negative y-axis, whereas the field below the sheet is also uniform but directed parallel to the positive y-axis.

We can characterize the flow of the current along the sheet by $\mathscr{J}_l$ its *current per length* along the x-axis. In this case the total current that flows past a region of length L is $I = (\mathscr{J}_l L)$. To apply Ampère's law we consider a rectangular patch of length L and height H as shown in Figure 29.9(d). Applying Equation (29.4), we have

$$\oint \mathbf{B} \cdot d\mathbf{l} = \mu_o I_{\text{encl}}$$

$$\int_a^b \mathbf{B} \cdot d\mathbf{l} + \int_b^c \mathbf{B} \cdot d\mathbf{l} + \int_c^d \mathbf{B} \cdot d\mathbf{l} + \int_d^a \mathbf{B} \cdot d\mathbf{l} = \mu_o(\mathscr{J}_l L) \qquad (29.7)$$

The path directions $b \rightarrow c$ and $d \rightarrow a$ are both perpendicular to the direction of $\mathbf{B}$ so that for each of these paths $\mathbf{B} \cdot d\mathbf{l} = 0$ and

$$\int_b^c \mathbf{B} \cdot d\mathbf{l} = \int_d^a \mathbf{B} \cdot d\mathbf{l} = 0$$

Moreover, along both paths $a \rightarrow b$ and $c \rightarrow d$ the field $\mathbf{B}$ is parallel to the path direction and constant in magnitude so that Equation (29.7) becomes

$$\int_a^b \mathbf{B} \cdot d\mathbf{l} + \int_c^d \mathbf{B} \cdot d\mathbf{l} = \mu_o \mathscr{J}_l L$$

$$B \int_a^b dl + B \int_c^d dl = \mu_o \mathscr{J}_l L$$

$$BL + BL = \mu_o \mathscr{J}_l L$$

$$B = \frac{\mu_o \mathscr{J}_l}{2}$$

Notice that this magnitude is independent of the distance from the current sheet. Hence, not only is the magnetic field $\mathbf{B}$ parallel to the x-axis, but it is also uniform in the regions above and below the infinite sheet.

It is interesting to think about the resultant fields if a second parallel sheet of current is placed a distance d from the first with the current running in the opposite direction. You should be able to convince yourself that in the region between the two sheets the fields add giving $B = \mu_o \mathscr{J}_l$, while in the regions exterior to the sheets the fields exactly cancel. This result is the magnetic analog of the electric field between oppositely charged conductors of an ideal, parallel-plate capacitor.

It would be difficult (to say the least) to actually construct two infinite sheets of current such as described in the previous paragraphs. A more practical arrangement which leads to a similar field configuration is the solenoid described next.

The Solenoid

A **solenoid** is a coil of wire wound on a right-circular cylinder. Each turn of the current-carrying wire is very close to a current loop and contributes a field much like that shown in Figure 29.5. Actually, no matter how closely wound, the wire is really helical in shape. However, in our model of an *ideal solenoid* we ignore this helix and only worry about the field generated by current loops. You can apply the right-hand rule to any part of the loosely wound coil of Figure 29.11(a) to verify the direction shown for **B.** Notice that although the B-field lines diverge at one end of the solenoid and converge at the other end, all the lines are closed (even though we couldn't show this feature for all the lines drawn in this figure).

When we wind the current turns closer together and make the length of the solenoid very long, Figure 29.11(b) illustrates that the field within the solenoid becomes stronger and more uniform. Figure 29.11(b) also implies that the B-field magnitude in the region outside of the solenoid, but far from its ends (point P), is very weak. At this point, you

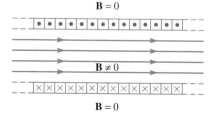

FIGURE 29.10

The magnetic field between two parallel, infinite sheets of current is uniform, whereas the field above the upper sheet and below the lower sheet is zero.

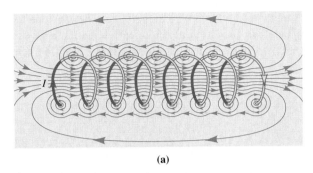

(a)

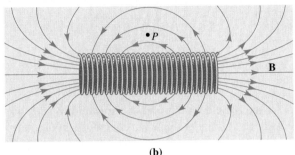

(b)

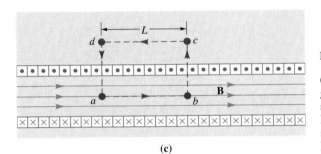

(c)

FIGURE 29.11

(a) The magnetic field lines generated by a loosely wound solenoid. (b) The magnetic field lines for a long, tightly wound solenoid. (c) The magnetic field lines for the ideal-solenoid model.

have no way of predicting that the field is weak outside the coil. Indeed, looking at Figure 29.11(a) you might reasonably conclude that the field just outside the solenoid might be quite large because the circles of B all add on the outside just as they do on the inside. In fact, however, the field outside the solenoid *is* weak for the following reason: Exterior to the coil, the magnetic field from the nearer parts of the coil is opposed by the magnetic field generated by the portion of the coil on the opposite side. The cancellation, although by no means obvious, is similar to the two parallel sheets of oppositely directed currents described in the previous subsection. To really see that the fields exactly cancel for our infinitely long ideal-solenoid model requires some rather complex integrals if we use the Biot-Savart law. Instead, we use Ampère's law to deduce the field outside of the coil.

Before we can apply Ampère's law to determine the field outside the coil, we need the value of the B-field inside the solenoid along the coil axis. This equation can be readily obtained from the result for a single current loop that we determined in Example 29.2:

$$B_x = \frac{\mu_o I}{2} \frac{R^2}{(x^2 + R^2)^{3/2}}$$

We can think of the field at the center of the solenoid as the superposition of an infinite number of single coils extending from $x = -\infty$ to $x = +\infty$. Thus, B_x in the above equation is actually just part of the field dB that is due to part of the current dI:

$$dB = \frac{\mu_o(dI)}{2} \frac{R^2}{(x^2 + R^2)^{3/2}}$$

If n is the number of loops per unit length along the x-axis, then the current dI is due to dN loops of wire, where $dN = n\, dx$. Therefore, the circumferential current within a length dx of the solenoid is $dI = (dN)I = (n\, dx)I$, where I is the current in a *single* turn. We can now write

$$B = \int dB_x = \int \frac{\mu_o dI}{2}\frac{R^2}{(x^2 + R^2)^{3/2}} = \int_{-\infty}^{+\infty} \frac{\mu_o In\, dx}{2}\frac{R^2}{(x^2 + R^2)^{3/2}}$$

or simply

$$B = \frac{\mu_o In R^2}{2}\int_{-\infty}^{+\infty}\frac{dx}{(x^2 + R^2)^{3/2}} = \frac{\mu_o In R^2}{2}\left[\frac{x}{R^2(x^2 + R^2)^{1/2}}\right]_{-\infty}^{+\infty}$$

$$B = \frac{\mu_o nI}{2}\left[\frac{1}{(1 + R^2/x^2)^{1/2}}\right]_{-\infty}^{+\infty} = \frac{\mu_o nI}{2}(1 + 1) = \mu_o nI$$

Now that we know the magnitude of the B-field on the axis of a solenoid, we can use Ampère's law to show that the field outside an infinitely long ideal solenoid is in fact zero. Consider an ideal-solenoid model such as that illustrated in Figure 29.11(c). To apply Ampère's law we consider the rectangular path $a \to b \to c \to d \to a$ of length L shown in this figure. If the solenoid contains n turns per length, then the total current enclosed by this path is $(nL)I$, where I is the current in each turn of the wire. Applying Equation (29.4), we have

$$\oint \mathbf{B} \cdot d\mathbf{l} = \mu_o I_{\text{encl}}$$

$$\int_a^b \mathbf{B} \cdot d\mathbf{l} + \int_b^c \mathbf{B} \cdot d\mathbf{l} + \int_c^d \mathbf{B} \cdot d\mathbf{l} + \int_d^a \mathbf{B} \cdot d\mathbf{l} = \mu_o(nLI)$$

Now, the second and fourth integrals along the paths $b \to c$ and $d \to a$ are zero because, from symmetry, $\mathbf{B}$ has no vertical component. (See the arguments in Example 29.2.) Thus, $\mathbf{B}$ is perpendicular to $d\mathbf{l}$ so that $\mathbf{B} \cdot d\mathbf{l} = 0$. We can directly evaluate the first integral using the result for the B-field on the solenoid axis as calculated in the previous paragraph.

$$\int_a^b \mathbf{B} \cdot d\mathbf{l} = \int_a^b B\, dl = \int_a^b (\mu_o nI)\, dl = (\mu_o nI)L$$

Hence, Ampère's law becomes

$$(n\mu_o I)L + 0 + 0 + \int_c^d \mathbf{B} \cdot d\mathbf{l} = \mu_o(nLI)$$

or

$$\int_c^d \mathbf{B} \cdot d\mathbf{l} = 0$$

But $\mathbf{B}$ is a constant and parallel to $d\mathbf{l}$ along cd so

$$B \int_0^L dl = 0$$

$$BL = 0$$

which implies

$$B = 0, \qquad \text{outside of the solenoid.}$$

Now that we know that B is zero everywhere *outside* our solenoid model, we can easily find B at positions other than the central axis *inside* the solenoid. We have merely to repeat the application of Ampère's law around a loop for which side ab is parallel to the solenoid's central axis but located some distance from it. If you do this you will quickly discover that the B-field has the same magnitude everywhere on the interior of our solenoid model:

$$B = \mu_o nI \qquad (29.8)$$

You should keep in mind that Equation (29.8) is only an approximation of the true field within a physical solenoid. However, Equation (29.8) works well for points near the center of a long solenoid. In the last section of this chapter we provide a numerical computation of the B-field on the axis of a solenoid of finite length.

The Toroid

A **toroid** consists of N turns of wire wound around a doughnut-shaped core as shown in Figure 29.12(a). As was the case with the solenoid, the shape of each turn of the current-carrying wire is very close to a current loop and contributes a field much like that shown in Figure 29.5. An ideal toroid consists of such closely spaced current turns that the B-field is completely confined to the interior of the doughnut, and these lines make circles as illustrated in Figure 29.12(b). The current path of a real toroid forms a helix that closes on itself rather than a set of current loops. In our model of the ideal toroid we ignore this helical characteristic.

We can easily apply Ampère's law if we choose one of these circles with radius r as our path. For then, not only is the magnitude of $\mathbf{B}$ a constant, but $\mathbf{B}$ and $d\mathbf{l}$ are parallel along the entire circular path. Hence, Ampère's law gives

$$\oint \mathbf{B} \cdot d\mathbf{l} = \mu_o I_{encl}$$
$$B \oint dl = B(2\pi r) = \mu_o I_{encl} \qquad (29.9)$$

The current enclosed by this circle consists of the N currents each of magnitude I, which in Figure 29.12(b) go downward through the center of the doughnut. Hence, Equation (29.8) becomes

$$B(2\pi r) = \mu_o(NI)$$

or

$$B = \frac{\mu_o NI}{2\pi r} \qquad (a \le r \le b)$$

Notice that the magnetic field strength varies as $1/r$ within a toroid. For this ideal model it is easy to show that the B-field exterior to the toroid is zero. We consider a circular path with a radius again measured from the center of the doughnut but with a length that is either smaller than the inner radius of the doughnut ($r \le a$) or larger than the outer radius ($r \ge b$). In either of these cases the *net* current enclosed is zero so that Equation (29.4) gives $\oint \mathbf{B} \cdot d\mathbf{l} = B(2\pi r) = \mu_o(0)$ and, therefore, $\mathbf{B} = 0$.

Solenoids and toroids have many practical applications. Solenoids are used to create small and useful magnetic fields in devices that range from doorbells to components in electric circuits. The field of a toroid is used in fusion research to confine plasmas.

EXAMPLE 29.4   *The Field Within*

A long straight wire of radius R carries a uniform current I. Compute the magnetic field in the region $0 \le r \le R$, where r is measured from the center of the wire.

SOLUTION   We intend to use Ampère's law. Before we begin we must identify a suitable path. From our solution for the B-field outside a long wire we expect the

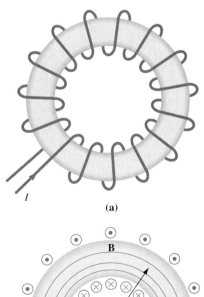
(a)

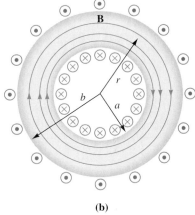
(b)

FIGURE 29.12

(a) A toroid consists of a current-carrying wire wrapped around a doughnut-shaped object. (b) A cut-away drawing of the toroid with inner radius a and outer radius b. The magnetic field lines are circles of radius r, where $a \le r \le b$.

The Tokamak, an experimental device used for the magnetic confinement of plasmas in fusion research.

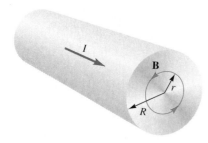

FIGURE 29.13

The magnetic field lines within a long, straight wire carrying a uniform current I are concentric circles. Ampère's law is applied along a circular path of radius r. See Example 29.3.

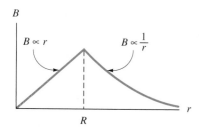

FIGURE 29.14

The magnetic field strength B as a function of the distance r from the center of a long, straight wire carrying a uniform current I. The radius of the wire is R.

current near the axis of the wire to create magnetic field lines that are also circular but this time within the cross section of the wire. Therefore, we choose a circular path of radius $r \leq R$ as shown in Figure 29.13.

Having chosen our path we are left with two calculations to perform. The first problem is to calculate the line integral on the left-hand side of Ampère's law. The second problem is to calculate the amount of current passing through our Amperian loop. We begin with the line integral. Along the circular path shown in Figure 29.13, **B** is constant in magnitude and parallel to every $d\mathbf{l}$ constituting the path. Thus, the line integral is trivial to calculate

$$\oint \mathbf{B} \cdot d\mathbf{l} = B \oint dl = B(2\pi r)$$

Now we must calculate the amount of current that threads this loop. For a uniform current I passing through a circular cross-sectional area πR^2 the current density is $\mathcal{J} = I/\pi R^2$. Hence, the current enclosed by the circular path of Figure 29.13 is

$$I_{\text{encl}} = \mathcal{J}(\pi r^2) = \left(\frac{I}{\pi R^2}\right)\pi r^2 = I\frac{r^2}{R^2}$$

Substituting the expression for the line integral and the expression for I_{encl}, Ampère's law

$$\oint \mathbf{B} \cdot d\mathbf{l} = \mu_o I_{\text{encl}}$$

becomes

$$B(2\pi r) = \mu_o\left(I\frac{r^2}{R^2}\right)$$

From which we find

$$B = \frac{\mu_o I}{2\pi R^2}r \qquad (0 \leq r \leq R)$$

We see that the magnetic field strength increases linearly in r for $0 \leq r \leq R$ and is zero at $r = 0$. Notice also that this expression and that which we found earlier for $r \geq R$ ($B = \mu_o I/2\pi r$) agree at $r = R$. The B-field is plotted as a function of r in Figure 29.14. ◀

29.4 Magnetic Flux and Gauss's Law for Magnetism

Concept Question 3
Does Ampère's law hold true if the magnetic field is not constant along the path over which $\oint \mathbf{B} \cdot d\mathbf{l}$ is evaluated? Explain.

Concept Question 4
A current runs along a copper water pipe. What is the magnetic field strength inside the pipe? Explain your answer.

Concept Question 5
Explain how B can be zero outside a toroid but must be nonzero outside a solenoid of finite length.

In Chapter 23 we defined the electric flux through a surface S as a measure of the number of E-field lines passing perpendicularly through that surface:

$$\Phi_E = \int_S \mathbf{E} \cdot d\mathbf{A}$$

The magnetic flux Φ_B is defined in an analogous fashion as a measure of the number of B-field lines that pass perpendicularly through a surface. As illustrated in Figure 29.15, when the magnetic field **B** passes through an area element $d\mathbf{A}$, the magnetic flux $d\Phi_B$ through that area element is $\mathbf{B} \cdot d\mathbf{A}$. Over the entire surface the magnetic flux is defined to be

$$\Phi_B = \int_S \mathbf{B} \cdot d\mathbf{A} \qquad \textbf{(29.10)}$$

In Chapter 23 we discovered from Gauss's law that the total electric flux through a *closed* surface was equal to the charge enclosed within that surface divided by the constant ϵ_o. Only when no charge is enclosed by the surface is the net electric flux through that surface zero. Experimentally, we find a very different situation for the magnetic field. Gauss's law for magnetism states that the net magnetic flux through a closed surface is *always* zero:

$$\oint \mathbf{B} \cdot d\mathbf{A} = 0 \qquad (29.11)$$

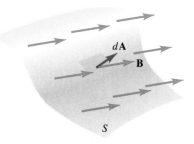

FIGURE 29.15

The magnetic flux Φ_B is computed by integrating the perpendicular component of $\mathbf{B}$ over all the area elements $d\mathbf{A}$ that compose surface S.

Equation 29.11 is another of Maxwell's equations, and it is therefore worthwhile to make some additional comments about its meaning. As is the case with the rest of Maxwell's equations, this law is based on experimental observations. In particular, isolated magnetic poles (sometimes called **monopoles**) have never been observed and may, in fact, not exist. Unlike electric field lines that begin or terminate at their sources (positive or negative charges) magnetic field lines form closed loops.

The magnetic field lines of a current ring are shown in Figure 29.16, along with two arbitrarily shaped closed surfaces. Notice that for these or any other *closed* surface you might imagine, the number of B-field lines entering the surface equals the number leaving. This feature is very different from that of the *electric* flux through a closed surface. When the surface encloses a net charge, the net electric flux through the surface is not zero. We have seen that currents are the sources of magnetic fields. But even for the closed surface S_2 of Figure 29.16, which does have a current passing through it (inward through one portion of the surface and outward through another), the net magnetic flux through this surface is zero.

Concept Question 6
The cross-sectional area of a certain solenoid is a circle that has a diameter equal to that of the solenoid's interior. If the current supplied to an ideal solenoid is halved, by what factor does flux through its cross section change?

29.5 Field on the Axis of a Solenoid: A Numerical Application (Optional)

In our analysis of the solenoid we noted that, to a good approximation, the current distribution can be considered as a set of closely spaced current loops. In this section we calculate the B-field on the axis of a solenoid constructed in this manner. The derivation follows the same procedure we applied in the previous section to compute B at the center of a very long solenoid. From Example 29.2 we found that the field on the axis of a single loop of current is given by

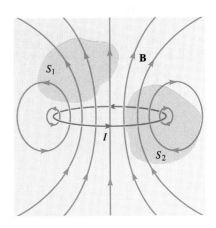

FIGURE 29.16

Magnetic field lines from a ring of current passing through two closed surfaces S_1 and S_2. The net flux through both surfaces is zero, even for surface S_2, which also has the current I passing into and then out of it.

$$B_x = \frac{\mu_o I}{2} \frac{R^2}{(x^2 + R^2)^{3/2}} \qquad (29.12)$$

In Figure 29.17 we show a solenoid of length L and radius R composed of a series of N current loops each of which carries a current I. We assume the number of turns per length $n = N/L$ is large so that in a segment of length dx the number of turns is $n\, dx$.

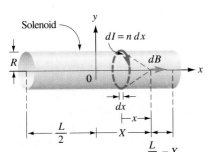

FIGURE 29.17

Computing the field on the axis of a solenoid of length L composed of N rings each carrying current I. An origin is placed at the center of the solenoid, and distances from this origin along the solenoid axis are given by X.

Because I is the current per turn, we can write the current in the element dx as $dI = I(n\,dx)$. From Equation (29.12) the magnetic field generated by this current element is

$$dB = \frac{\mu_o}{2} \frac{R^2}{(x^2 + R^2)^{3/2}} dI = \frac{\mu_o}{2} \frac{R^2}{(x^2 + R^2)^{3/2}} (In\,dx)$$

where it is important to remember (from Example 29.2) that x is measured from the loop center to the point at which B is to be computed. As shown in Figure 29.17, we place an origin at the center of the solenoid and label the point along its axis where we want to compute B with the capital letter X. In this case the solenoid loops extend from $x = -(X + L/2)$ to $x = +(L/2 - X)$. Hence, at point X the B-field is given by

$$B = \frac{\mu_o nIR^2}{2} \int_{-(X+L/2)}^{+(L/2-X)} \frac{dx}{(x^2 + R^2)^{3/2}}$$

$$= \frac{\mu_o nI}{2} \left(\frac{1}{\sqrt{x^2 + R^2}} \right)_{-(X+L/2)}^{+(L/2-X)}$$

$$B(X) = \frac{\mu_o nI}{2} \left[\frac{(L/2 - X)}{\sqrt{(L/2 - X)^2 + R^2}} + \frac{(X + L/2)}{\sqrt{(X + L/2)^2 + R^2}} \right] \qquad (29.13)$$

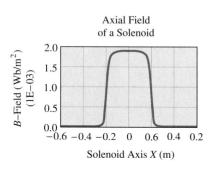

Axial Field
of a Solenoid

FIGURE 29.18

A plot of the magnetic field magnitude along the axis of a solenoid composed of current-carrying rings.

The spreadsheet SOLENOID.WK1 on the disk accompanying this text provides a calculation of this function for a solenoid of length $L = 40.0$ cm, radius $R = 2.00$ cm, current $I = 5.00$ A, and $n = 300$ turns/m. Figure 29.18 shows the resulting graph for this solenoid. You should vary the length and radius and observe how the field uniformity near the center of the solenoid changes. Also note how quickly the field magnitude decreases to zero outside the solenoid. In Problems 40 and 41 we investigate further the B-field magnitudes at the ends and center of the solenoid.

29.6 Summary

The magnetic field $d\mathbf{B}$ generated by a current I flowing along a direction $d\mathbf{l}$ is given by the Biot-Savart law:

$$d\mathbf{B} = \frac{\mu_o}{4\pi} \frac{Id\mathbf{l} \times \hat{\mathbf{r}}}{r^2} \qquad (29.1)$$

where r is the distance from $d\mathbf{l}$ to the point at which the magnetic field is evaluated, and $\hat{\mathbf{r}}$ is a unit vector that points in this direction. The constant $\mu_o = 4\pi \times 10^{-7}$ T·m/A is called the **permeability of free space.** The magnetic field lines for a long line of current I form concentric circles. The magnitude of the field at distance r from the line is

$$B = \frac{\mu_o I}{2\pi r}$$

The force F_{21} of attraction or repulsion on either of two parallel wires carrying currents I_1 and I_2, separated by distance d is given by

$$F_{21} = \frac{\mu_o I_1 I_2}{2\pi d} l \qquad (29.3)$$

where l is the length of the wires. If the currents run in the same direction, the wires attract each other, and if the currents run in opposite directions the wires repel each other. When one **ampere** of current flows through each of two parallel wires separated by one meter, the force of attraction between the wires is 2.00×10^{-7} N/m. The **coulomb** is the amount of charge that flows in one second past a point in a wire that carries a current of one ampere.

Ampère's law relates the line integral of the magnetic field **B** along a closed path to the net, steady current I_{encl} enclosed by that path:

$$\oint \mathbf{B} \cdot d\mathbf{l} = \mu_o I_{encl} \tag{29.4}$$

This law is easiest to apply in cases in which the current distribution is symmetrical. In particular, for the ideal-**solenoid** model where n turns per length of wire each carry a current I, the magnetic field within the solenoid is given by $\mu_o n I$. For an ideal **toroid** of inner radius a and outer radius b, with N turns, the magnetic field strength is $\mu_o N I / 2\pi r$, where $a \le r \le b$.

Gauss's law for magnetism states that *the net magnetic flux through a closed surface is always zero*:

$$\Phi_B = \oint \mathbf{B} \cdot d\mathbf{A} = 0 \tag{29.11}$$

Another statement of this law often given is that isolated magnetic poles (called monopoles) do not exist.

PROBLEMS

29.1 The Biot-Savart Law

1. A long, straight wire carries a current $I = 5.00$ A. (a) Compute the magnetic field strength at a distance of 2.00 cm from the wire. (b) To what value must the current be increased for the magnetic field at a distance of 4.00 cm from the wire to be equal to that computed in part (a)?

2. In the **Bohr model of the hydrogen atom** the electron orbits the nucleus with a radius of 5.29×10^{-11} m and an angular velocity of 4.13×10^{16} rad/s. Compute the magnetic field strength at the nucleus from the current formed by the orbiting electron.

3. A total charge $+Q$ is distributed uniformly over a thin disk of radius R. The disk rotates with angular velocity ω about a perpendicular axis through its center. Compute the magnetic field strength **B** at the center of the disk.

4. A current $I = 1.00$ A flows along a small current element $\Delta \mathbf{l} = (2.50 \text{ mm})\hat{\mathbf{k}}$ at the origin. Compute the magnetic field $\Delta \mathbf{B}$ at the points (a) $(0, 0, 2.00)$ m, (b) $(0, 2.00, 0)$ m, (c) $(2.00, 2.00, 0)$ m, and (d) $(2.00, 0, 2.00)$ m.

5. A current $I = 2.00$ A flows along a small current element $\Delta \mathbf{l} = (3.00 \text{ mm})\hat{\mathbf{j}}$ at the origin. Compute the magnetic field $\Delta \mathbf{B}$ at the points (a) $(0, 2.00, 5.00)$ m, (b) $(2.00, 4.00, 0)$ m, and (c) $(2.00, 0, 3.00)$ m.

6. Apply the Biot-Savart law to compute the magnetic field strength at point P generated by the current flowing along the segment $abcd$ shown in Figure 29.P1. The radius of the curved segment is R.

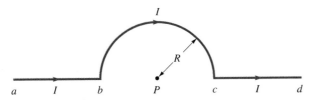

FIGURE 29.P1   Problem 6

7. Show that the magnetic field strength at point P in Figure 29.P2 is given by

$$B = \frac{\mu_o I \phi}{4\pi} \left(\frac{1}{R_2} - \frac{1}{R_1} \right)$$

where I is the current in the loop.

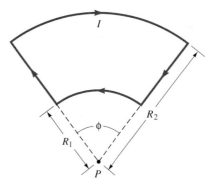

FIGURE 29.P2   Problem 7

★8. A square loop of side length l carries a current I. Show that the magnetic field strength a distance x along the axis of the loop from its center is given by

$$B = \frac{4\mu_o I l^2}{\pi(4x^2 + l^2)(4x^2 + 2l^2)^{1/2}}$$

9. A wire of length L is bent so as to form a square. (a) Compute the magnetic field at its center if a current I is made to flow around this square loop. (b) If the same wire, carrying the same current I, is next made into a circle, compute the new B-field at its center. (c) Which field is larger, that of part (a) or part (b)?

10. A long, insulated wire is bent into the shape shown in Figure 29.P3. If the wire carries a current I, compute the magnetic field strength at point P, the center of the circular loop.

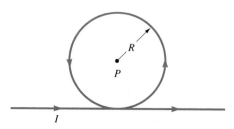

FIGURE 29.P3   Problem 10

11. Show that the magnetic field strength at the center of a rectangular current loop of length l and width w is given by

$$B = \frac{4\mu_o I}{\pi} \frac{(l^2 + w^2)^{1/2}}{lw}$$

where I is the current in the loop.

12. A wire forms a square the sides of which are 20.0 cm long. If a current of 4.00 A flows around the square, what is the magnetic field strength B at the center of the square?

13. A very long wire is bent so that the current I it carries makes a right angle as shown in Figure 29.P4. Compute the magnetic field **B** at point P shown in this figure. Your answer should be in terms of the distance y and the current I.

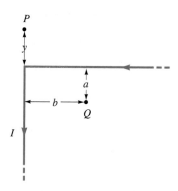

FIGURE 29.P4   Problems 13 and 14

14. A very long wire is bent so that the current I it carries makes a right angle as shown in Figure 29.P4. Apply the Biot-Savart law to find the magnetic field at point Q shown in this figure. Your answer should be in terms of the distances a and b and the current I.

15. **Helmholtz coils** are constructed from two circular coils of wire, each perpendicular to the same axis, each carrying the same current in the same direction. As shown in Figure 29.P5, the coils are separated by the distance R, which is also the radius of each coil. Use the result of Example 29.2 to find an expression for the magnetic field strength on the axis of the coils at point P located a distance x from the center of the arrangement.

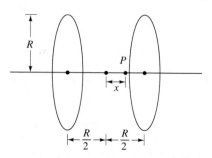

FIGURE 29.P5   Helmholtz coils (Problems 15 and 16)

16. Show that the magnetic field **B** for the Helmholtz coils described in Problem 15 is uniform by proving that both dB/dx and d^2B/dx^2 are zero at $x = 0$.

29.2 Parallel Wires, Amperes, and Coulombs

17. Two long, parallel wires separated by 50.0 cm carry currents $I_1 = 5.00$ A and $I_2 = 10.0$ A in opposite directions. (a) Draw a figure and verify that the wires repel each other. (b) Compute the force by wire 1 on a 1.50-m segment of wire 2.

18. Two parallel wires carry currents $I_1 = 2.00$ A and $I_2 = 6.00$ A in opposite directions. The wires are separated by 30.0 cm. Compute the magnetic field strength in the plane of the wires at positions (a) 10.0 cm from wire 1 on the opposite side from wire 2, and (b) 10.0 cm from wire 1 on the same side as wire 2.

19. Two parallel wires separated by 20.0 cm carry currents $I_1 = 3.00$ A and $I_2 = 5.00$ A. Compute the magnetic field strength in the plane of the wires at a point half-way between them if (a) the currents flow in the same direction, and (b) the currents flow in opposite directions.

20. Two parallel lines of current flow along the positive x-direction in the xy-plane. Current $I_1 = 4.00$ A flows along the x-axis, and current $I_2 = 2.00$ A flows along this same direction but passes through the point $y = 20.0$ cm. Compute the magnetic field **B** at the point (0, 20.0, 10.0) cm.

21. Two parallel lines of current flow along the positive x-direction in the xy-plane. Current $I_1 = 1.00$ A flows along the x-axis, and current $I_2 = 2.00$ A flows along this same direction but passes through the point $y = 30.0$ cm. Compute the magnetic field **B** at the point (0, 15.0, 15.0) cm.

22. A long, straight wire carries a current $I_1 = 8.00$ A, and a rectangular loop carries a current $I_2 = 6.00$ A as shown in Figure 29.P6. The current loop and the line of current both lie in the same plane. If $d_1 = 5.00$ cm, $d_2 = 7.00$ cm, and $l = 15.0$ cm, find the net force by the current loop on the straight line of current.

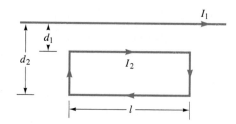

FIGURE 29.P6   Problem 22

23. The four parallel currents shown in Figure 29.P7 flow perpendicular to the page. Current $I_1 = 2.00$ A, $I_2 = 3.00$ A, $I_3 = 1.00$ A,

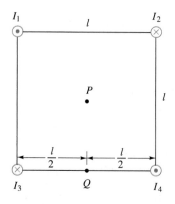

FIGURE 29.P7   Problems 23 and 24

and $I_4 = 2.00$ A. Compute the magnetic field strength at the center of the square (point P) if $l = 15.0$ cm.

24. The four parallel currents shown in Figure 29.P7 flow perpendicular to the page. Current $I_1 = 1.00$ A, $I_2 = 2.00$ A, $I_3 = 3.00$ A, and $I_4 = 2.00$ A. Compute the magnetic field strength at point Q if $l = 15.0$ cm.

25. Three long, parallel wires each carry a current of 2.00 A and are located at the corners of a 30.0 cm by 40.0 cm by 50.0 cm right triangle as shown in Figure 29.P8. The wire at the vertex of the right angle carries current into the page, whereas the other two wires carry currents out of the page. (a) Find the magnetic field (magnitude and direction) at point P. (b) If a fourth wire is now located at point P and carries 3.00 A into the page, what is the magnitude and direction of the force on a 1.00-m length of this new wire?

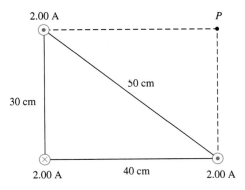

FIGURE 29.P8   Problem 25

26. Two long, parallel wires separated by $d = 2.54$ cm each carry a current $I = 5.00$ A into the plane of the page as shown in Figure 29.P9. (a) Find the magnitude and direction of the magnetic field produced by these wires at a point P that forms an equilateral triangle with the points where the wires cross the plane of the page. (b) If an electron passes through point P with a velocity 650. m/s parallel to the two wires and directed into the page, what is the magnitude and direction of the force on the electron?

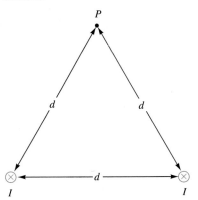

FIGURE 29.P9   Problem 26

29.3 Ampère's Law

27. Estimate the magnetic field strength produced by a 12 000-A lightning bolt at a distance of 100 m from the bolt.

28. Two large, flat conductors are separated by a distance d and are parallel to the xy-plane. One conductor carries a uniform current per transverse width $\mathcal{J}_1 = (I/l)\hat{\mathbf{i}}$, and the other carries $\mathcal{J}_2 = -(I/l)\hat{\mathbf{i}}$. What is the B-field magnitude, (a) between the conductors, and (b) above or below the conductors?

29. Two large, flat conductors are separated by a distance d and are parallel to the xy-plane. One conductor carries a uniform current per transverse width $\mathcal{J}_1 = (I/l)\hat{\mathbf{i}}$, and the other carries $\mathcal{J}_2 = (I/l)\hat{\mathbf{j}}$. What is the B-field magnitude, (a) between the conductors, and (b) above or below the conductors?

30. Lead has a **superconducting transition** temperature of 7.20 K. However, when below this temperature its superconducting state can be destroyed if a B-field greater than about 0.0803 T exists at its surface. What is the maximum current that a long, straight, superconducting lead wire with a 1.00-mm radius can carry and remain in the superconducting state?

31. A long, flat conductor of negligible thickness but width w carries a uniform current I as shown in Figure 29.P10. By using the result of Example 29.1 for the field of a long line of current, compute the magnetic field **B** at a point P a distance y directly over the center of the strip.

32. A long, flat conductor of negligible thickness but width w carries a uniform current I as shown in Figure 29.P10. By using the result of Example 29.1 for the field of a long line of current, compute the magnetic field **B** at a point Q in the plane of the strip a distance x from its center.

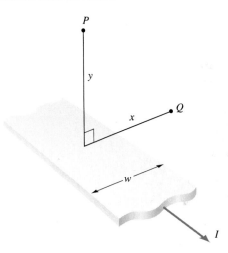

FIGURE 29.P10   Problems 31 and 32

33. The axis of a long, cylindrical conductor of radius a is along the y-axis. The current density carried by the conductor is not uniform but varies with distance r from the axis and is given by $\mathbf{J} = (br)\hat{\mathbf{j}}$, where b is a constant. (a) What are the units of b? (b) What is the total current carried by the conductor? (c) Apply Ampère's law to find the B-field magnitude at locations $r \le a$. (d) What is the magnetic field strength at locations $r > a$?

34. A long, cylindrical conductor of radius R_o lies along the x-axis and carries a current density $\mathbf{J} = J_o(1 - r^2/R_o^2)\hat{\mathbf{i}}$, where J_o is a constant and r is the radial distance from the conductor's axis. (a) Compute the total current flowing in the conductor. (b) Apply Ampère's law to find the B-field strength for points within the conductor at distances $r \le R_o$. (c) Apply Ampère's law to find the B-field strength at locations $r \ge R_o$.

35. A long, cylindrical rod with a radius $R = 2.00$ mm carries a uniform current $I = 4.00$ A. (a) Compute the magnetic field strength at a distance $r = 1.00$ cm from the center of the rod. (b) At what distance from the center of the rod, but within it, does the B-field have this same magnitude?

36. A coaxial cable consists of an inner cylindrical conductor of radius R_1 and a concentric, hollow cylindrical conductor with inner radius R_2 and outer radius R_3 (Fig. 29.P11). A current I travels down the inner conductor, and a uniform current of the same

FIGURE 29.P11   Problem 36

magnitude travels in the opposite direction along the outer conductor. Apply Ampère's law to compute the magnetic field strength in the regions (a) $0 \leq r \leq R_1$, (b) $R_1 \leq r \leq R_2$, (c) $R_2 \leq r \leq R_3$, and (d) $r \geq R_3$.

37. The long, hollow, cylindrical conductor with inner radius a and outer radius b shown in Figure 29.P12 carries a uniform current I. Apply Ampère's law to find the magnetic field strength **B** in the regions (a) $r < a$, (b) $a \leq r \leq b$, and (c) $r > b$.

FIGURE 29.P12   Problem 37

38. A 15.0-cm long solenoid has a radius of 2.00 cm and 250 turns of wire. (a) If the wire carries a current of 3.00 A, what is the magnetic field strength at the center of the solenoid? (b) How much current does a single loop of wire with the same radius have to carry to produce a magnetic field with the same magnitude at its center?

39. A solenoid is to be constructed to produce a magnetic field of 0.0500 T at its center. The available wire can safely carry a maximum current of 5.00 A and at its tightest winding can have 100. turns/cm. Can the solenoid be constructed from this wire?

40. In Section 29.5, Equation (29.13) was derived for the magnetic field on the axis of a solenoid of length L and radius R. Verify for such a solenoid that when the length is such that $R/L \ll 1$, the magnetic field strength at the solenoid ends is $\mu_o nI/2$.

41. Show from Equation (29.13) that the magnetic field strength at the center of a solenoid of length L and radius R approaches that of an ideal solenoid when $L \gg R$.

42. A 200.-m long copper wire is to be wound around a 30.0-cm long plastic tube with a 5.00-cm diameter. Assuming the coil extends the total 30.0-cm length, what current must flow through the wire to produce a magnetic field of magnitude 0.100 T at the center of the tube?

43. A coil wrapped around a toroid has an inner radius of 20.0 cm and an outer radius of 25.0 cm. If the wire wrapping makes 800. turns and carries a current of 12.0 A, what are the minimum and maximum values of the magnetic field within the toroid?

29.4 Magnetic Flux and Gauss's Law for Magnetism

44. The rectangular loop shown in Figure 29.P13 has dimensions $a = 4.00$ cm and $b = 6.00$ cm. If the uniform magnetic field is given by $\mathbf{B} = (0.500 \text{ T})\hat{\mathbf{i}}$, compute the magnetic flux through the rectangular surface for (a) $\theta = 20.°$, (b) $\theta = 50.°$, and (c) $\theta = 90.°$.

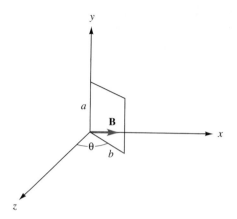

FIGURE 29.P13   Problem 44

45. A long, straight wire carries a current $I = 6.00$ A parallel to one of the sides of a rectangular loop as shown in Figure 29.P14. If $a = 2.00$ cm, $b = 5.00$ cm, and $w = 6.00$ cm, compute the net magnetic flux Φ_B through the rectangular loop.

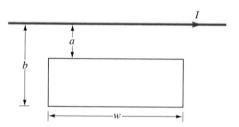

FIGURE 29.P14   Problem 45

46. A long solenoid with a radius of 2.50 cm carries a current of 3.00 A. The solenoid is wound with 84.0 turns/cm. As shown in Figure 29.P15 the solenoid is oriented so that its axis is along the x-axis of a coordinate system the origin of which is placed at the center of the solenoid. Apply the ideal-solenoid model (a) to compute the flux Φ_B through a circular surface in the yz-plane that has a radius equal to that of the solenoid, and (b) to compute the flux Φ_B through a circular surface in the yz-plane with a radius of 5.00 cm.

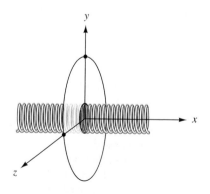

FIGURE 29.P15   Problem 46

✳ **47.** As shown in Figure 29.P16, a solenoid passes through a ring of radius $R = 6.00$ cm tilted in such a way that the axis of the solenoid makes an angle of θ with the perpendicular to the ring. The solenoid is 70.0 cm long, has a radius of 2.00 cm, and is composed of 300. turns of wire, each carrying a current of 2.00 A. What is the flux Φ_B through the ring if (a) $\theta = 30.0°$, (b) $\theta = 60.0°$, and (c) $\theta = 90.0°$?

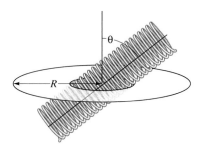

FIGURE 29.P16   Problem 47

48. A magnetic field $\mathbf{B} = (-4.00\hat{\mathbf{i}} + 3.00\hat{\mathbf{j}} + 2.00\hat{\mathbf{k}})$T passes through the surfaces of the rectangular solid shown in Figure 29.P17. If $a = 10.00$ cm, $b = 6.00$ cm, and $c = 8.00$ cm, compute the magnetic flux through each of the six surfaces and show that they sum to zero.

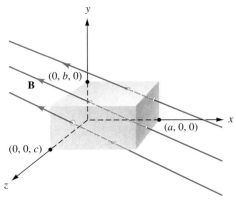

FIGURE 29.P17   Problem 48

★ **49.** A *rectangular* toroid consists of N turns of wire, each carrying current I. The inner radius of the toroid is a and its outer radius is b. The height of each rectangular loop is h. (a) Find an expression for the *total* flux through the N turns of the toroid. (b) Compute this flux for $N = 300.$, $I = 4.00$ A, $a = 10.0$ cm, $b = 12.0$ cm, and $h = 4.00$ cm.

Numerical Methods and Computer Applications

50. Figure 29.P5 shows an arrangement of two current-carrying loops separated by a distance just equal to their radius. These coils are called Helmholtz coils (see Problem 15). Write a program or construct a spreadsheet template to plot the magnetic field strength $\mathbf{B}$ along the axis of the coils.

51. Two wires carrying currents I_1 and I_2 lie in the xy-plane and are parallel to the x-axis. Current I_1 crosses the y-axis at $y = -d/2$, and current I_2 crosses the y-axis at $y = +d/2$. (a) Write a program or construct a spreadsheet template to plot both the x- and y-components as well as the total magnitude of the magnetic field B along the y-axis at a distance $z = h$ above the wires. (b) Try out your program or spreadsheet for $I_1 = 1.00$ A, $I_2 = 2.00$ A, $d = 10.0$ cm, and $h = 5.00$ cm. (c) Change the direction of I_2 by assigning it a value of -2.00 A.

52. Use the results from Example 29.4 to write a program or construct a spreadsheet template to plot the magnetic field strength as a function of distance from the center of a wire of radius R that carries uniform current I (similar to Fig. 29.14). Try out your program or spreadsheet with $I = 2.00$ A and $R = 2.00$ mm by plotting B from $r = 0$ to $r = 6.00$ mm in increments of 0.25 mm.

53. Use the results of Example 29.1 to construct a spreadsheet template or write a computer program to plot a graph of the percentage error between the B-field magnitude of an infinitely long, current-carrying wire and one of finite length.

General Problems

54. (a) Compute the magnetic field strength at point P in Figure 29.P18 if the current in the loop is I. (b) What is the magnetic dipole moment of this current loop?

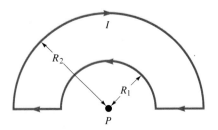

FIGURE 29.P18   Problem 54

55. (a) Compute the magnetic field strength at point R within the current loop shown in Figure 29.P19. (b) What is the magnetic dipole moment of this current loop?

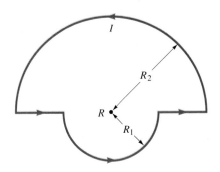

FIGURE 29.P19   Problem 55

56. Four parallel wires, each separated by distance R from the next and located in the same plane, carry current I. Find an expression for the net force per unit length on each wire. Sketch the four wires and draw near each wire a vector the length of which is proportional to the net force per unit length on that wire.

57. A long, cylindrical conductor of radius b has a cylindrical cavity of radius a running parallel to its axis. The axis of the cavity is located a distance d from the axis of the cylinder as shown in Figure 29.P20. If the conductor carries a uniform current density J, compute the magnetic field strength within the cavity and verify that it is uniform.

58. A long, cylindrical conductor of radius b has a cylindrical cavity of radius a running parallel to its axis. The center of the hole is located a distance d along the negative y-axis from the center of the cylinder as shown in Figure 29.P20. If the conductor carries a uniform current I, compute the magnetic field strength at (a) $(0, 3b/2)$, and (b) $(3b/2, 0)$.

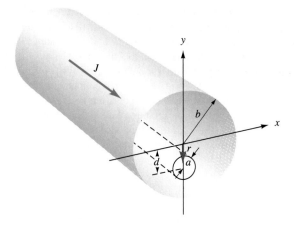

FIGURE 29.P20　Problems 57 and 58

59. A long, cylindrical conductor of radius R has two cylindrical cavities of diameter R running parallel to its axis as shown in Figure 29.P21. The conductor carries a uniform current density J. Compute the magnetic field strength **B** at position P along the y-axis where $y > R$.

60. A long, cylindrical conductor of radius R has two cylindrical cavities of diameter R running parallel to its axis as shown in Figure 29.P21. The conductor carries a uniform current density J. Compute the magnetic field strength B at position Q along the x-axis where $x > R$.

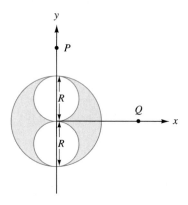

FIGURE 29.P21　Problems 59 and 60

61. A toroid is to be constructed so that the variation in the B-field magnitude from its value at the inner radius r_1 to its value at the outer radius r_2 is given by the fraction $f = \Delta B/B$. Here, B is the field magnitude at the average distance from the toroid center, where $r = (r_1 + r_2)/2$, and ΔB is the change in B from radial position r_1 to radial position r_2. Show that $f = \frac{1}{2}(r_1/r_2 - r_2/r_1)$.

62. Show that the magnetic field strength at the vertex of an arc segment that subtends an angle θ is given by

$$B = \frac{\mu_o I}{R}\frac{\theta}{4\pi}$$

where R is the radius of the arc.

63. Apply the results of Example 29.1 to find the magnetic field strength at point P in Figure 29.P22 in terms of the length a and the current I.

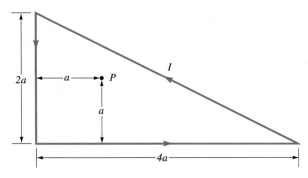

FIGURE 29.P22　Problem 63

64. Apply the result of Problem 62 and that of Example 29.1 to compute the magnetic field strength at point P shown in Figure 29.P23 in terms of the radius a, the length b, and the current I.

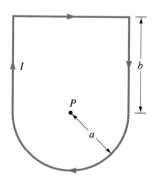

FIGURE 29.P23　Problem 64

65. Apply the result of Problem 62 and that of Example 29.1 to compute the magnetic field strength at point P shown in Figure 29.P24 in terms of the radius a and the current I.

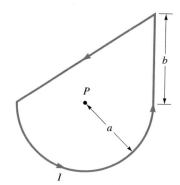

FIGURE 29.P24　Problem 65

66. In Example 29.3, the magnetic field contribution of the long portions of the address lines of the MOSLM were ignored. Although the address lines are less than a half centimeter in length, we can take them as infinitely long compared to the 50.0-nm length of the sides of the square in our model pictured in Figure 29.6(d). (a) How does the direction of the magnetic field from the long portions of the address lines compare with that of the field from the sides of the square loop? (b) Calculate the magnitude of the magnetic field at point P due to the long portions of the address lines that do not form part of the square loop in Figure 29.6(d).

67. The parallel address lines of a MOSLM (see Example 29.3) are 0.500 cm long and separated by 228 nm. If two adjacent address lines both carry a current of 100. mA in the same direction, (a) what is the magnitude of the force between them? (b) Is the force attractive or repulsive? (c) In actual operation an external bias field of 30.0 G is directed into the plane of the MOSLM.

What is the magnitude of the force on a current-carrying address line due to the bias field? (d) If the external bias field is created by a 500.-mA current flowing in a circular coil of mean diameter 0.950 cm, how many turns of wire are in this coil? Assume the field is 30.0 G at the center of the coil.

Faraday's Law and Induction

In this chapter you should learn

- how to apply Faraday's law to find the magnitude of the induced emf in a circuit that experiences a changing magnetic flux.

- how to apply Lenz's law to find the direction of the induced emf in a circuit.

- how to calculate Maxwell's displacement current.

- to recognize Maxwell's equations.

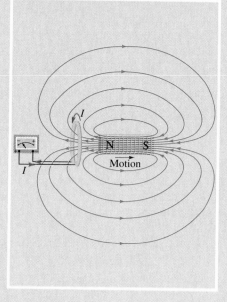

$$\mathcal{E} = -\frac{d\Phi_B}{dt}$$

When driving under a power line have you ever been annoyed by harsh radio static that drowns out your favorite song? Have you ever wondered how the detectors used at airports to screen passengers for metal objects work? You might be surprised to learn that the basic physics behind these two phenomena is the same as that behind the creation of electricity in hydroelectric, nuclear, and coal-fired electric-generating plants. On a much smaller scale the same phenomenon allows physicists, chemists, and medical technicians to ''listen'' as the nuclei of atoms broadcast information about their magnetic environment. Such ''broadcasts'' are known as nuclear magnetic resonance (NMR). Nuclear magnetic resonance has wide applications, which range from unraveling the chemical bonding of molecules to whole-body imaging of soft tissues in humans.

This chapter discusses what, from a technological point of view, may be the most important of the laws of electromagnetism: a prescription for turning mechanical energy of motion into electrical energy. While exploring this prescription, an unexpected bonus results: we learn how to move electric power from one place to another *without interconnecting wires*. Once again, we discover that a single law of physics is able to explain phenomena that occur in applications ranging from the nuclei of atoms to multimegawatt electric power plants.

The powerful law alluded to in the previous paragraphs is the fourth, and last, of Maxwell's equations. Thus far we have described only one way that electric fields can be created, namely, by electric charges. Experimentally, it has been found that it is also possible to create an electric field from a changing magnetic flux. The law we state in the first section of this chapter was first deduced experimentally by both Michael Faraday (1791–1867) in England and Joseph Henry (1797–1878) in the United States at about the same time (1831). Henry did not publish his findings, so the law bears the name of Faraday, but the physical unit associated with what we shall come to call ''inductance'' is named in honor of Joseph Henry.

As we discuss Faraday's law, you will see that it is a *changing* magnetic flux that gives rise to an electric *field*. The magnetic flux $d\Phi_B$ is computed from the vector dot product of the magnetic field **B** and an area element $d\mathbf{A}$, $d\Phi_B = B \, dA \cos(\theta)$. Consequently, in the examples and illustrations that follow, you should be alert for the changes in any of the three quantities (1) magnetic field strength B, (2) the area element dA, (3) the angle (θ) between these vectors, or any combination of changes in these quantities. If a conductor is present, the electric field produced by a changing flux can cause currents to flow through that conductor. In fact, the detection of these currents is one way we can infer that an electric field has been generated.

30.1 The Laws of Faraday and Lenz

To introduce the final Maxwell equation we describe two experiments that illustrate the ways by which a changing magnetic flux can cause a current to flow through a conductor. In these first two cases the changing flux is caused by a changing magnetic field. The first experiment, illustrated in Figure 30.1(a), shows a bar magnet that can slide back and forth along the axis of a conducting ring. The ring has been cut so that it does not make a complete circle, and a galvanometer has been attached between the two ends of the gap to complete the circuit. When the magnet is at rest, no current is observed to pass through the galvanometer. However, when the magnet moves toward the ring, a current flows through the ring in the direction shown in Figure 30.1(b). If the magnet is brought to rest when it is closer to the ring, no current is observed while the magnet is not in motion, a point illustrated in Figure 30.1(c). Even if the magnet were to be inside the ring, no current would be observed so long as the magnet remained at rest. When the magnet is moved toward the right, away from the ring, a current is again observed in the ring *while the magnet is moving*. This time, however, as illustrated in Figure 30.1(d), the direction of the current is opposite its previous direction when the magnet was moving toward the left.

Notice from Figures 30.1(a) and (b) that the number of B-field lines that pass through the area of the ring increases as the magnet moves toward the left. Hence, during this motion the magnetic flux through the ring increases. Our experiment has shown that this

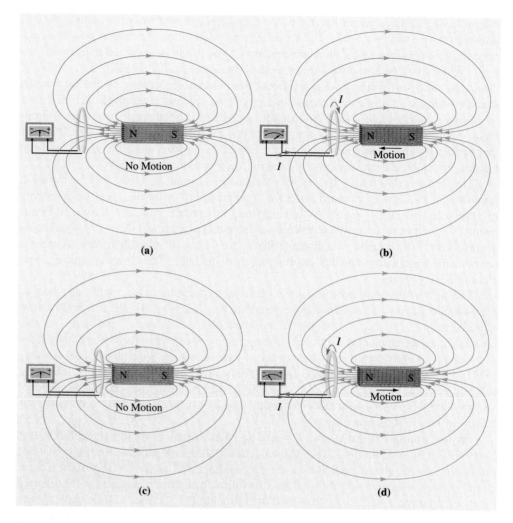

FIGURE 30.1

(a) The magnetic flux from a bar magnet passes through a conducting ring attached to a galvanometer. When the magnet is at rest, no current is induced in the ring. (b) As the magnet moves toward the left the flux increases through the ring, and the flow of current through the galvanometer indicates that an emf has been induced around the ring. (c) When the magnet is at rest, no emf is induced around the ring. (d) As the magnet moves toward the right the magnetic flux through the ring decreases and an induced emf drives the current around the ring in the direction opposite that shown in part (b).

increasing flux causes an emf to be *induced* around the ring and this **induced emf** causes a current to flow in the ring. The magnitude of the **induced current** caused by the induced emf depends on both the magnitude of the emf and the resistance of the ring. Only when the flux is *changing* is there an induced emf; when the magnet is at rest, no current is observed in the ring. By the way, if the magnet is turned around and the motion described above repeated, we again observe induced emfs in the ring, but this time the current directions are reversed. We will have more to say about the induced current directions shortly, but for now, don't worry too much about the directions; just remember what condition causes a current to flow. That condition is, of course, a changing magnetic flux. And don't miss the exciting fact that we have been able to make current flow *without any batteries.* We have, in fact, changed the mechanical motion of the magnet into electric current in the loop.

Let's look at a second experiment. This time we do employ a battery but end up generating a current in a circuit loop that is *not* connected to the battery. Figure 30.2(a)

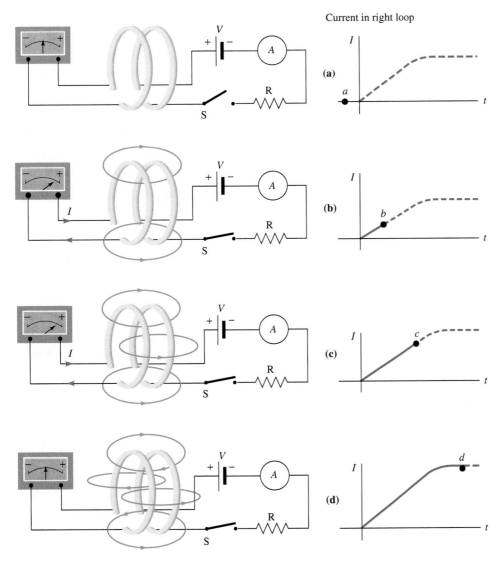

Current in right loop

FIGURE 30.2

(a) Two conducting rings with a common axis. Notice that the graphs show the current in the *battery-powered circuit*. (b) When switch S is closed, a current begins to flow through the right ring, creating a magnetic field with lines that pass through the left ring. (c) As the magnetic field increases, the number of B-field lines linking the left loop increases. (d) Although the flux through the left ring is large, it is no longer changing. Hence, the induced current in the left circuit loop is zero.

shows two conducting rings with a common axis. Across the gap in the ring on the right is a series arrangement of a battery, a resistor R, a switch S, and an ammeter. The loop on the left is attached to a galvanometer just as in our previous experiment. When switch S is closed, a current grows from zero to some final, finite value through the ring on the right. The graphs down the right-hand column of Figure 30.2 indicate the time dependence of the current read by the ammeter. Here's what happens in the left ring. Just after the switch is closed the galvanometer deflects, indicating that an emf has been induced in the left loop, causing a current to flow in this circuit. However, the galvanometer needle returns to the zero-current position quite quickly (Fig. 30.2(d)). This lack of current implies that the emf no longer exists around the left loop, even though we know that a steady current flows through the coil on the right. If the switch in the right circuit is once again opened, the galvanometer needle once again *momentarily* deflects but, as you probably have already guessed, this deflection is in the opposite direction from the deflection that occurred when we first closed the switch.

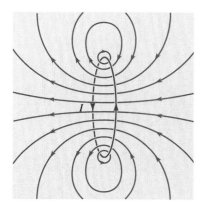

FIGURE 30.3

The magnetic field lines generated by a circular current loop.

Can you suggest an explanation Faraday or Henry might have used to describe what is happening in this experiment? (The experiment we just described is similar to the ones these two scientists performed.) Remember, it is a *changing* flux that causes the induced emf. When we close the switch and the current begins to flow, a magnetic field is generated by the current in the right loop. Figure 30.3 shows the manner in which the *B*-field fills the space around a current-carrying loop. In Figure 30.2(b) the growing *B*-field lines created by the right-hand loop pass through the conducting ring on the left and, therefore, the magnitude of the magnetic flux through the left-hand ring increases in the moments after the switch is closed (Figs. 30.2(b) and (c)). This increasing flux induces the emf in the left loop, and a current flows in this loop. When the current through the right-hand ring reaches its constant final value, the magnetic flux through the left ring is constant and there is no longer an induced emf in this loop (Fig. 30.2(d)). Just after the switch *S* is reopened, the current in the right loop begins to decrease rapidly, and the magnetic field strength caused by this current decreases. The resulting decrease in the magnitude of the flux through the left-hand ring induces an emf that drives current through the galvanometer.

Faraday's Law

From the two experiments described above it is clear that a changing magnetic flux can induce an emf. The formal statement of this result constitutes **Faraday's law of induction:**

> The induced emf in a circuit is proportional to the time rate of change of magnetic flux through a surface enclosed by that circuit.

In mathematical form, we write

$$\mathcal{E} = -\frac{d\Phi_B}{dt} \tag{30.1}$$

where $d\Phi_B = \mathbf{B} \cdot d\mathbf{A}$, the magnetic flux passing through the surface. The minus sign in Faraday's law pertains to the direction of the induced emf. (The minus sign is significant and is even enshrined in a separate law: Lenz's law. We will discuss this law later, but for now we concentrate on the magnitudes in Equation (30.1).) Because the magnetic flux is the dot product of the magnetic field **B** and the area $d\mathbf{A}$, we can list four ways that the emf $\mathcal{E}$ can be induced in a circuit:

1. The magnetic field strength *B* can vary in time.
2. The area enclosed by the circuit can vary in time.
3. The angle between the magnetic field and circuit area can change in time.
4. Any combination of the events listed in 1, 2, or 3 can take place.

If the circuit consists of *N* loops, each with the same area, and ϕ_B is the *flux through a single loop* so that $\Phi_B = N\phi_B$ is the flux through all *N* loops, Equation (30.1) becomes

$$\mathcal{E} = -N\frac{d\phi_B}{dt} \tag{30.2}$$

Before reading about the use of Lenz's law to determine the direction of induced currents in circuit loops, you should carefully read Example 30.1, which illustrates how to apply Faraday's law to compute the *magnitude* of the emf $\mathcal{E}$.

EXAMPLE 30.1 *Look Ma, No Batteries!*

A circular coil of wire consists of exactly 250 turns with a 5.00-cm radius. A magnetic field **B** is directed through the loop, perpendicular to its plane as shown in Figure 30.4. The magnetic field is increased in strength at a rate of 0.600 T/s. If the total resistance of the coil is 8.00 Ω, find the current induced in the one strand of the wire constituting the coil.

SOLUTION   The magnitude of the induced emf is given by Equation (30.2):

$$\left| \mathscr{E} \right| = \left| -N \frac{d\phi_B}{dt} \right| = N \left| \frac{d\phi_B}{dt} \right|$$

Now, the flux through one turn of the coil is

$$\phi_B = \int \mathbf{B} \cdot d\mathbf{A} = \int B(t)\cos(0°)\, dA$$

Although the *B*-field changes in time, at any given instant it is constant over the coil area and can therefore be removed from the area integral to give

$$\phi_B = B \int dA = B\pi r^2$$

Thus, on taking the time derivative

$$\frac{d\phi_B}{dt} = \pi r^2 \frac{dB}{dt}$$

The total emf $\mathscr{E}$ due to N turns of the coil is

$$\left| \mathscr{E} \right| = N\pi r^2 \frac{dB}{dt} = (250)\pi(0.0500 \text{ m})^2(0.600 \text{ T/s}) = 1.18 \text{ V}$$

The current is given by

$$I = \frac{\left| \mathscr{E} \right|}{R} = \frac{1.18 \text{ V}}{8.00 \ \Omega} = 0.147 \text{ A}$$

◀

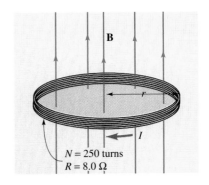

FIGURE 30.4

An increasing magnetic field **B** is directed through the loop, perpendicular to its plane. See Example 30.1.

Concept Question 1
Suppose a bar magnet is held vertically above a circular loop of wire that rests in a horizontal plane. If the north pole of the magnet is at the lower end, describe the current that is induced in the circular loop as the magnet is released from rest and falls through the loop.

Concept Question 2
Describe what happens when a magnet is dropped down a long conducting tube.

Lenz's Law

We can determine the direction of the induced emf (and therefore the direction of any resulting induced current) from **Lenz's law,** named after the German physicist Heinrich Lenz (1804–1865):

> The direction of the induced emf is such as to oppose the change in magnetic flux that causes the induced emf.

Read this statement again and pay careful attention to the words " . . . oppose the *change* in magnetic *flux* . . . " The words *change* and *flux* are very important, so be sure you understand exactly what they mean. First consider *change*. Lenz's law says that when the flux is increasing, the induced emf tries to make it decrease; when the flux is decreasing the induced emf tries to make it increase. Also notice that Lenz's law does *not* say the induced emf opposes the magnetic *field,* but rather it says it opposes the *change* in magnetic *flux*. First-time users of Lenz's law often have a tendency to simply make the

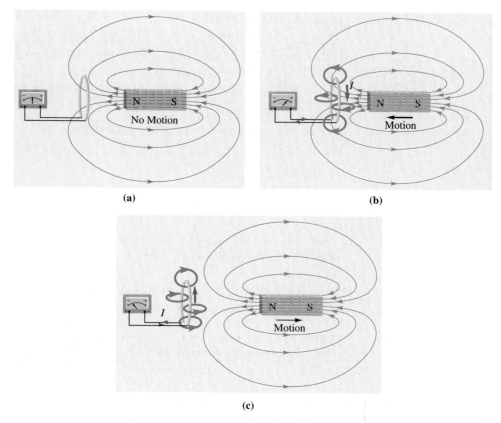

FIGURE 30.5

(a) When the magnet is at rest, there is no change in flux through the ring. (b) When the magnet moves toward the ring, the induced emf creates a current that gives rise to a magnetic flux that opposes the increase in flux through the ring. (c) When the magnet moves away from the ring, the induced emf creates a current that gives rise to a magnetic flux that opposes the decrease in the flux through the ring.

induced emf oppose the magnetic field. Regrettably, it's not that easy. Sometimes the induced emf does oppose the B-field, but sometimes it does not!

Let's illustrate Lenz's law by reexamining our magnet and conducting-ring experiment. You should keep in mind that flux is a *scalar,* so although it is appropriate to state the direction of a magnetic field that creates a flux, it is not correct to refer to the direction of a flux. For clarity we call the original field from the magnet the **primary** field. In Figure 30.5(a) the bar magnet is oriented so that the primary magnetic field through the conducting ring is directed toward the left. When the magnet moves toward the ring (Fig. 30.5(b)), the strength of the left-pointing magnetic field through the ring increases and, therefore, the magnitude of the magnetic flux *increases.* Lenz's law says that an induced emf is generated around the ring in such a direction as to *oppose this increase* in flux. To oppose the increase in flux, the induced field attempts to *decrease* the magnitude of the total magnetic flux. Because the flux is increasing due to a left-pointing field, clearly the way to decrease the flux is to create a right-pointing field. When an induced current flows through the ring in the direction indicated in Figure 30.5(b), this current creates its own magnetic field, and the direction of this induced B-field (given by the right-hand rule) opposes the increasing field from the bar magnet. Hence, the net field through the ring is reduced. The current resulting from the induced emf has created a magnetic flux that opposes the change in the flux from the moving magnet. *In this case,* the B-field from the induced current creates a B-field that opposes the B-field of the moving magnet. Got it? If not, it's a good idea to read this paragraph again!

Now consider what happens when the bar magnet is moved toward the right, away from the conducting ring. From Figure 30.5(c) we see that the magnetic flux from the bar magnet is still due to a primary field directed toward the left. This primary flux is *decreas-*

ing in magnitude as the magnet slides toward the right. Hence, the induced emf creates a current so as to *increase* the magnitude of the magnetic flux. To increase the flux created by a left-directed field, the left-directed magnetic field through the ring must be strengthened. This can be accomplished by a current flow in the direction shown. Here is a situation where the induced magnetic field is in the same direction as the primary field. We are reminded that it is not the magnetic field that the induced emf tends to oppose, but rather it is the *change in the magnetic flux* that it opposes.

As you no doubt have deduced from the above discussion, the rule for the direction of the induced field is quite simple:

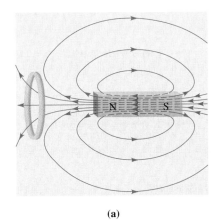

(a)

---

1. If the magnitude of the flux through a circuit is *increasing,* the induced (secondary) *B*-field is in the *opposite* direction of the primary field.
2. If the magnitude of the flux through a circuit is *decreasing,* the induced (secondary) field is in the *same* direction as the primary field.

Lenz's law is required by the law of conservation of energy. When the magnet in Figure 30.6 is moved toward the conducting ring at the left, the induced current in the ring creates an induced field $\mathbf{B}_{ind}$. Notice that the field lines of this induced *B*-field are in the opposite direction from those of the magnet. Hence, the field lines of $\mathbf{B}_{ind}$ are such that the left side of the current loop is analogous to a south magnetic pole and the right side to a north magnetic pole. Because like poles repel each other, we must do work to move the magnet toward the ring. Indeed, this work is the source of energy that goes into creating the induced emf in the ring that drives the current around it. The energy generated by this work is eventually dissipated as Joule heating (I^2R loss) in the ring or radiated into space.

Imagine the violation of the conservation of energy that would result if the induced emf were in such a direction as to *enhance* the change in flux, that is, opposite to that stated by Lenz's law. In such a case the induced *B*-field would be in the opposite direction from that shown in Figure 30.6, and the right side of the ring would become a south magnetic pole. Because unlike poles attract each other, the north pole of the magnet would be drawn toward the ring. Hence, by giving the magnet a small initial motion toward the ring it would be attracted by the induced *B*-field of the ring and accelerate toward it. We would then have a system that could generate more energy without any additional work being put into it: a clear violation of the conservation of energy principle!

In Example 30.1 and in our magnet-ring illustrations we demonstrated how a changing magnetic field strength through a constant cross-sectional area can generate an induced emf (item (1) in our list of three ways to change a magnetic flux). Examples 30.2 and 30.3 illustrate how items (2) and (3) in our list can also generate an induced emf. We begin with an example of a cross-sectional area that changes as a result of a conducting bar sliding through a uniform *B*-field (Example 30.2). Any emf induced from the motion of any conductor through a magnetic field is called **motional emf.** In Example 30.3 we illustrate how the change in the angle between the *B*-field and the area vector can produce an induced emf. This last example is important because the physics of the rotating current loop in a uniform *B*-field forms the basis for alternating-current generators on which we all depend for the electric power in our homes, offices, and residence halls.

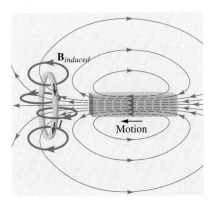

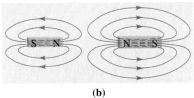

(b)

FIGURE 30.6

As the magnet moves closer to loop, the loop is linked by a larger number of field lines. Thus, the magnetic flux through the ring increases, and the induced *B*-field opposes that of the magnet. Similar to two magnets in which like poles repel each other (insert) so do the ring and the magnet.

EXAMPLE 30.2 *Riding the Rails in Search of Motional emf*

A conducting rod of length *l* slides on two parallel conducting rails as shown in Figure 30.7. The rods are connected at one end by a resistance *R*. The conducting bar slides along the rails with velocity of magnitude *v* toward the open end. A uniform magnetic field **B** is normal to the plane of the rod and rails and points into the page. Compute the emf across the rod and determine the magnitude and direction of the induced current in the circuit.

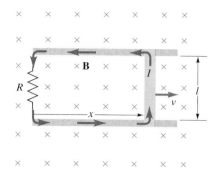

FIGURE 30.7

A conducting bar slides on two parallel conducting rails connected by a resistor R. See Example 30.2.

Concept Question 3
Explain why an applied force is required to keep the bar described in Example 30.2 moving at a constant velocity.

SOLUTION In this example the flux increases because the area of the circuit increases in time. If x is the length of the area enclosed by the rod and rails at some instant, then the flux through the circuit is

$$\Phi_B = \int \mathbf{B} \cdot d\mathbf{A} = \int B \cos(0°) \, dA = B \int dA = Blx$$

where we have recognized that because the constant B-field is normal to the surface, it is parallel to the area vector. The magnitude of the induced emf is given by Faraday's law:

$$|\mathcal{E}| = \frac{d\Phi_B}{dt} = \frac{d}{dt}(Blx) = Bl\frac{dx}{dt} = Blv$$

because $dx/dt = v$, the velocity of the bar.
    The current in the circuit is

$$I = \frac{|\mathcal{E}|}{R} = \frac{Blv}{R}$$

Now for the real fun; we must apply Lenz's law to find the direction of the induced current. We realize you have already looked at Figure 30.7 and, therefore, know the correct direction. So here is our reasoning that led us to this result. As the bar slides to the right the area enclosed by the loop increases, and therefore the magnitude of the flux due to the field directed into the page also increases. Lenz's law tells us that the induced current is in such a direction as to decrease the magnitude of the flux due to the field directed into the page. To accomplish this, a flux must be generated from a magnetic field directed *out of the page*. Hence, the current must flow in the anticlockwise direction. Then, an application of the right-hand rule shows that the B-field resulting from this current is out of the page, and the magnitude of the net flux is decreased. This example is an illustration of rule 1: If the magnitude of the flux is *increasing*, the induced (secondary) B-field is in the *opposite* direction from that of the primary field. ◀

You might be wondering what happens if the rails and resistor in Figure 30.7 are removed while the conducting bar still travels with speed v in the plane perpendicular to the B-field. In this case an emf is still induced across the rod, but in equilibrium, no current flows across it. To find the magnitude of this induced emf let's apply an alternative approach, one based on what we learned in Chapter 28 about the magnetic force on moving charges. As shown in Figure 30.8(a), the electrons in the conducting bar experience a force given by $\mathbf{F}_B = -e\mathbf{v} \times \mathbf{B}$ and, therefore, move to the lower end of the bar. This charge displacement leaves the upper end of the bar positive. The positive charge at the top of the bar and a negative charge at its bottom create an *electric* field directed from the top toward the bottom of the bar (see Fig. 30.8(b)). In equilibrium, the magnitudes of the electric force $F_E = eE$ and the magnetic force $F_B = evB \sin(90°) = evB$ on electrons between the ends of the bar are equal, and no additional charge accumulates on either end:

$$evB = eE$$

or after simplifying

$$vB = E$$

Now, we have often seen that a uniform E-field acting over a distance l creates a potential difference ΔV given by $\Delta V = El$. This potential difference ΔV is referred to as a **motional**

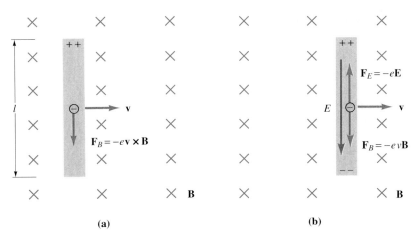

FIGURE 30.8

(a) Motional emf is established in a conducting bar moving in a plane perpendicular to an applied *B*-field. (b) In equilibrium no additional charge moves because the magnetic force is balanced by the force of the induced *E*-field.

emf $\mathcal{E}$. Therefore,

$$vB = \frac{\mathcal{E}}{l}$$

Solving for the motional emf induced across the sliding bar, we find

$$\mathcal{E} = vlB$$

As you can see, the faster you slide a conducting bar in a plane perpendicular to a magnetic field, the greater the emf you can establish between its ends. Just be sure to also keep the length of the bar perpendicular to the *B*-field direction.

Concept Question 4
Suppose the bar shown in Figure 30.8 slides to the left. Deduce the direction of the force on the electrons in the bar and determine the direction of the resultant induced *E*-field.

EXAMPLE 30.3 *The Generator*

A circular loop of area *A* and resistance *R* rotates with angular velocity ω about an axis through its diameter as shown in Figure 30.9. The plane of the loop is initially perpendicular to a constant magnetic field *B*. Find the induced current in the loop $I(t)$.

SOLUTION We begin by computing the magnetic flux through the loop when the area vector makes an angle θ with the *B*-field:

$$\Phi_B = \int \mathbf{B} \cdot d\mathbf{A} = \int B\cos(\theta)\, dA = B\cos(\theta)\int dA = BA\cos(\theta)$$

Now when the loop rotates at angular speed ω, the angle θ is given by $\theta = \omega t$, so

$$\Phi_B = BA\cos(\omega t)$$

and from Faraday's law

$$\mathcal{E} = -\frac{d\Phi_B}{dt} = -\frac{d}{dt}[BA\cos(\omega t)]$$

$$\mathcal{E}(t) = BA\omega\sin(\omega t)$$

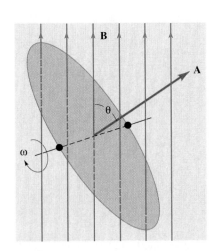

FIGURE 30.9

A circular loop of area *A* and resistance *R* rotates with angular velocity ω about an axis through its diameter. See Example 30.3.

The current $I(t)$ is given by

$$I(t) = \frac{\mathscr{E}}{R} = \frac{BA\omega}{R}\sin(\omega t)$$

Both the induced emf and the current vary sinusoidally in time. The amplitude of the emf $\mathscr{E}(t)$ is given by the coefficient $BA\omega$, and the amplitude of the current $I(t)$ is $BA\omega/R$. Electric generators are constructed from loops of many turns that are rotated in a magnetic field. The addition of N turns, each of area A, increases area to a total of NA. The energy source for the rotational motion can be falling water, high-pressure steam, or even wind. ◀

30.2 Induced Electric Fields

In the opening paragraphs of this chapter we stated that a changing magnetic flux gives rise to an *electric field*. Indeed, if this changing flux can cause the charges of a current to move through a conductor, an electric field must be present. It is this induced *E*-field that provides the induced emf. Even if the conductor does not form a closed path so that no current can flow, a changing magnetic flux still gives rise to an induced *E*-field and the associated induced emf. Let's consider a uniform magnetic field **B** out of the page and the circular conducting path in the plane of the page as shown in Figure 30.10. If the magnetic field out of the page increases in magnitude, Faraday's law tells us that an emf is established around the circular path. However, Faraday's law says nothing about conductors and, in fact, *even if no conductor is present,* an electric field **E** is established in this region of space as long as the magnetic field **B** is changing in time. If in place of the conducting ring shown in Figure 30.10 we place a single·charge q, this charge experiences a force $\mathbf{F} = q\mathbf{E}$ from the induced *E*-field. Electric field lines describing *induced E*-fields are quite different from those that describe fields that are produced from static charges. Field lines for induced electric fields form closed paths because there are no charges present from which the *E*-field lines originate or terminate.

To compute the emf between two locations we need only compute the potential difference between those locations. Hence, the emf along a path is evaluated from the line integral of $\mathbf{E} \cdot d\mathbf{l}$ between the starting and ending points of that path. For a closed path

$$\mathscr{E} = \oint \mathbf{E} \cdot d\mathbf{l}$$

and from Faraday's law

$$\oint \mathbf{E} \cdot d\mathbf{l} = -\frac{d\Phi_B}{dt}$$

Applications of the physics described by Faraday's law can be found in many useful devices such as magnetic tape recorders, hydroelectric generators, and power generating windmills.

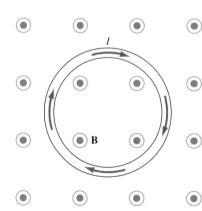

FIGURE 30.10

A region of space where a magnetic field **B** out of the plane of the page is increasing in intensity. The induced current I is in a direction so as to oppose the increasing field.

Writing Φ_B explicitly, we have

$$\oint \mathbf{E} \cdot d\mathbf{l} = -\frac{d}{dt} \int \mathbf{B} \cdot d\mathbf{A} \qquad\qquad (30.3)$$

From Equation (30.3) we can see that the *induced electric field* **E** *is nonconservative*. In order for **E** to be conservative the line integral around the closed loop must be zero (refer back to Chapter 8, Section 1 for the definition of a conservative quantity, such as a conservative force). Clearly, from Equation (30.3) the line integral of the induced *E*-field is not zero but equals the negative rate of change of the magnetic flux. Induced electric fields are very different from electrostatic fields.

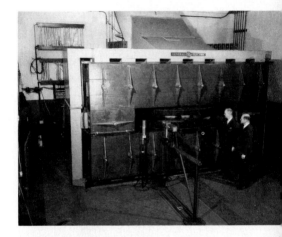

EXAMPLE 30.4   Blasted Betatron

A betatron (Fig. 30.11) is an instrument used to accelerate electrons (beta particles) to high velocities. As illustrated in Figure 30.12, electrons are injected into a horizontal, toroidal vacuum chamber where a vertical magnetic field B_1 causes them to move in circular orbits. Within the region that is interior to the circular paths, the magnetic field is varied, and the resulting induced emf accelerates the electrons. If B_2 is the average field strength over the area within the circle, find the relation between B_1 and B_2.

SOLUTION   We apply Equation 30.3 where E is the electric field strength around the circular path of radius R:

$$\oint \mathbf{E} \cdot d\mathbf{l} = -\frac{d\Phi_B}{dt}$$

$$E(2\pi R) = -\frac{d}{dt} \int \mathbf{B} \cdot d\mathbf{A}$$

We assume B_2 is constant over the circular area $A = \pi R^2$, so that (working with magnitudes only) we have

$$E(2\pi R) = \frac{d}{dt}[B_2(\pi R^2)] = \pi R^2 \frac{dB_2}{dt}$$

This leads to an electric field of magnitude

$$E = \frac{1}{2} R \frac{dB_2}{dt}$$

The magnetic field $\mathbf{B}_1$ at the orbit location is responsible for the circular paths of the electrons. Hence, the magnetic force $F = |-e\mathbf{v} \times \mathbf{B}_1| = evB\sin(90°)$ provides the centripetal force $m_e v^2/R$ on the electron:

$$evB_1 = m_e \frac{v^2}{R}$$

This last condition implies that the momentum p of the electron of mass m_e is

$$m_e v = eRB_1$$

FIGURE 30.11

The betatron, an instrument used to accelerate electrons.

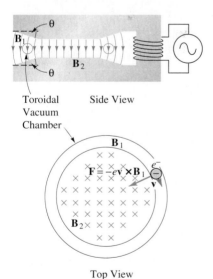

Toroidal   Side View
Vacuum
Chamber

Top View

FIGURE 30.12

The magnetic fields of a betatron. The alternating field $\mathbf{B}_1$ experienced by an electron as it travels through the evacuated toroidal chamber is related to the alternating field $\mathbf{B}_2$ in the central region by the relation $\mathbf{B}_1 = \frac{1}{2}\mathbf{B}_2$.

The force on the electron that accelerates it around the circular path is given by the tangential components of Newton's second law

$$(F_{net})_t = \left(\frac{dp}{dt}\right)_t$$

$$eE = \frac{d}{dt}(m_e v)$$

Substituting for $(m_e v)$ and E from our expressions above, we have

$$e\left(\frac{1}{2}R\frac{dB_2}{dt}\right) = \frac{d}{dt}(eRB_1)$$

from which we find

$$\frac{dB_1}{dt} = \frac{1}{2}\frac{dB_2}{dt}$$

Integrating both sides of this equation with respect to time we obtain

$$\int_0^t \frac{dB_1}{dt}\,dt = \frac{1}{2}\int_0^t \frac{dB_2}{dt}\,dt$$

$$\int_0^{B_1} dB_1 = \frac{1}{2}\int_0^{B_2} dB_2$$

$$B_1 = \frac{1}{2}B_2$$

where we have assumed that both fields are zero at $t = 0$. This result is called the *betatron condition*. This feature is achieved by designing the pole faces to have the proper angle θ in the vicinity of the toroidal vacuum chamber (see Fig. 30.12).

◀

Eddy Currents

When a changing magnetic flux is applied to a bulk piece of conducting material, circulating charge motions called **eddy currents** are induced. Because the resistance of the bulk conductor is usually low, eddy currents often have large magnitudes. A popular demonstration of this phenomenon is shown in Figure 30.13. A pendulum is constructed by attaching a flat piece of nonmagnetic metal, such as copper or aluminum, to the end of a stiff rod; this plate is then allowed to swing between the poles of a strong magnet. When the plate enters the B-field, the magnetic flux through it increases causing eddy currents to be induced. As we saw previously, these currents dissipate internal energy within the conductor. The source of the dissipated energy is the kinetic energy of the moving metallic plate, so its kinetic energy is reduced and the plate slows down. For a very strong magnet the pendulum can come to an abrupt stop.

Eddy currents are often undesirable. To reduce their effects slots are sometimes cut into moving metallic parts of machinery. These slots interrupt the conducting paths and decrease the magnitudes of the induced currents. If, for example, the pendulum plate in Figure 30.13 is replaced with one having parallel gaps running partially along its length, the pendulum swings more freely through the magnet. Most transformers (discussed in Chapter 33) and motors have parts constructed from alternating layers of conducting and nonconducting substances. Such laminations serve to interrupt the conducting paths and reduce the conversion of electrical energy to thermal energy due to I^2R heating caused by eddy currents.

Concept Question 5
If the direction of the magnetic field were reversed could a betatron be used to accelerate protons? Explain.

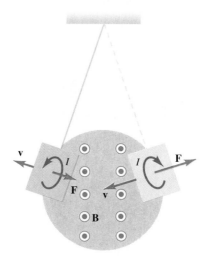

FIGURE 30.13

A pendulum constructed from a light rigid rod with a flat, nonmagnetic plate at its free end is allowed to swing between the pole faces of a magnet. Eddy currents are induced from the changing magnetic flux through the conducting plate. The resulting magnetic force on the moving charges slows the pendulum. Notice that the directions of the eddy currents depend on whether the plate is entering or leaving the magnetic field region.

Not all eddy currents are undesirable. Some rapid-transit train cars use them in braking mechanisms. When braking is desired, electromagnets on the cars near the steel rails are activated, and the eddy currents created in the rails provide the braking force on the cars. Because the magnitude of the force decreases as the car slows, these types of brakes tend to produce a smooth stop. Eddy currents are also used as damping mechanisms in such instruments as mechanical and electronic balances.

30.3 The Displacement Current and Maxwell's Equations

Throughout this and the previous eight chapters you have studied electromagnetic phenomena and the theories that describe them. Periodically, we have also attempted to convey to you our excitement about Maxwell's equations and how they form the heart of this theory. By this time you are probably getting tired of our implications that "we're almost there," but really, we're almost there. We have only one more detail to bring to your attention and then we will be able to write down Maxwell's equations in their complete splendor. There is one more fundamental way that electromagnetic systems can behave that we haven't told you about. This particular phenomenon leads to a modification of Ampère's law:

$$\oint \mathbf{B} \cdot d\mathbf{l} = \mu_o I \qquad (30.4)$$

Indeed, this law cannot be correct as it stands. To see the flaw in Equation (30.4) consider the charging capacitor shown in Figure 30.14. We imagine that the current I is kept constant as the capacitor charges by suitable adjustment of the emf source supplying the current. According to Ampère's law, the line integral of $\mathbf{B}$ around the curve $\mathscr{C}$ must be the product of μ_o and the current I passing through a surface bounded by $\mathscr{C}$. Here's the problem: If we consider the surface to be the one labeled S_1 in Figure 30.14, then the current through this surface is I and therefore $\oint \mathbf{B} \cdot d\mathbf{l} = \mu_o I$. But Ampère's law must hold for *any* surface bounded by the curve $\mathscr{C}$. A contradiction appears to occur, if we consider curve $\mathscr{C}$ to bound surface S_2. For then $\oint \mathbf{B} \cdot d\mathbf{l} = 0$ because no current flows through S_2. To reconcile the apparent contradiction we must remember that, because a steady current is flowing, charge is continuously accumulating on the capacitor plates, thus increasing the strength of the electric field between these plates. James Clerk Maxwell's insight was to recognize this increasing electric field strength as equivalent to an electric current for purposes of applying Ampère's law.

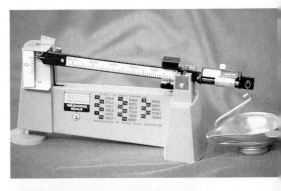

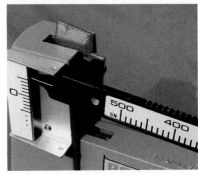

Using the same mechanism illustrated in Figure 30.13, eddy currents provide the damping mechanism for this reloading powder balance.

Concept Question 6
Explain the induced current directions and resulting force directions (and magnitudes) in Figure 30.13.

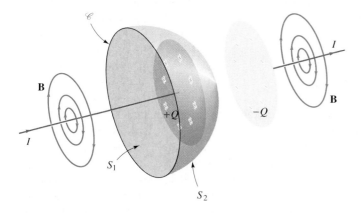

FIGURE 30.14

The curve $\mathscr{C}$ bounds both surfaces S_1 and S_2. The current I, however, only passes through surface S_1.

Maxwell showed that Ampère's law can be cast in a correct form by including another term called the **displacement current:**

$$I_d = \epsilon_o \frac{d\phi_E}{dt}$$ (30.5)

where, as you probably remember, ϕ_E is the electric flux. With the addition of this term Ampère's law becomes

$$\oint \mathbf{B} \cdot d\mathbf{l} = \mu_o(I + I_d)$$

or, in its fullest splendor,

$$\oint \mathbf{B} \cdot d\mathbf{l} = \mu_o I + \mu_o \epsilon_o \frac{d\phi_E}{dt}$$ (30.6)

It is easy to understand how the displacement current behaves for the ideal, parallel plate capacitor shown in Figure 30.14. For plates of area A separated by distance d the capacitance $C = Q/V$ is given by $C = A\epsilon_o/d$, and the magnitude of the electric field between the plates is

$$E = \frac{\Delta V}{d} = \frac{Q}{dC} = \frac{Q}{d(A\epsilon_o/d)} = \frac{Q}{A\epsilon_o}$$

The electric flux is, therefore, given by

$$\phi_E = EA = \frac{Q}{\epsilon_o}$$

from which we can discover the magnitude of the displacement current

$$I_d = \epsilon_o \frac{d\phi_E}{dt} = \epsilon_o \frac{d}{dt}\left(\frac{Q}{\epsilon_o}\right) = \frac{dQ}{dt} = I$$

The displacement current magnitude is exactly equal to the magnitude of actual current I passing through the surface S_1.

Equation (30.6) is the most general form of Ampère's law, and it has an important implication for such systems as our charging, parallel-plate capacitor. In the region between the plates the current I is zero. However, the electric flux is increasing (as long as the capacitor is charging), and for this region Ampère's law becomes

$$\oint \mathbf{B} \cdot d\mathbf{l} = \mu_o \epsilon_o \frac{d\phi_E}{dt}$$

This equation says that the displacement current, generated by the changing electric flux $d\phi_E/dt$, gives rise to a magnetic field $\mathbf{B}$ just as real currents do! As shown in Figure 30.15 the B-field lines circulate around the area of changing electric flux much the same way that B-field lines circulate around a normal current flow. However, you must remember that displacement currents exist only when the electric flux changes. Consequently the magnetic field shown in Figure 30.15 is present *only when the flux ϕ_E changes in time.* When the charging current I is zero, the electric field between the capacitor plates is *constant in time* and no induced B-field lines are present.

The inclusion of the displacement current in Ampère's law, Equation (30.6), provides a kind of conceptual symmetry between this law and Faraday's law, Equation (30.3). The

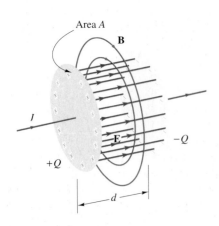

FIGURE 30.15

While the flux $\phi_E = EA$ changes in time an induced B-field circulates around the region of changing flux as well as around the actual current-carrying wire.

latter says that a changing *magnetic* flux produces an *electric* field, whereas the extended Ampère's law now says that a changing *electric* flux produces a *magnetic* field.

Maxwell's Equations

We are now ready to summarize Maxwell's equations. We have made so much to-do about them because they form the empirical foundation for the description of all classical electromagnetic phenomena. Indeed, when setting out to solve some problem of electromagnetism, scientists invariably begin with these equations. We will also take this approach in Chapter 33 where we shall see what Maxwell's equations have to say about electromagnetic waves.

Here, at last, are Maxwell's equations:

Maxwell's Equations

Gauss's law for **E**:	$$\oint \mathbf{E} \cdot d\mathbf{A} = \frac{q_{encl}}{\epsilon_o}$$	(30.7)
Gauss's law for **B**:	$$\oint \mathbf{B} \cdot d\mathbf{A} = 0$$	(30.8)
Faraday's law:	$$\oint \mathbf{E} \cdot d\mathbf{l} = -\frac{d\phi_B}{dt}$$	(30.9)
Ampère's law:	$$\oint \mathbf{B} \cdot d\mathbf{l} = \mu_o I + \mu_o \epsilon_o \frac{d\phi_E}{dt}$$	(30.10)
(as modified by Maxwell)		

In addition, we are compelled to write down the Lorentz force law, from which the force on any charge q can be found once **E** and **B** have been determined from Maxwell's equations:

$$\mathbf{F} = q(\mathbf{E} + v \times \mathbf{B}) \qquad (30.11)$$

In one sense we can think of Maxwell's equations as a set of recipes that tell us how to create electric and magnetic fields. Gauss's law says that if you want to create an electric field that diverges in all directions never to return, find yourself an electric charge. Gauss's law for magnetism says that if you wish to make a magnetic field that diverges in all directions, you're out of luck (at least so far). Faraday's law tells us how to make electric fields that circle back on themselves in closed loops. This is done, we learn, by changing the strength of a magnetic field somewhere in space. To make circling loops of magnetic field, Ampère's law tells us we have two choices, either use an electric current or simply change the strength of an electric field somewhere in space.

With these thoughts as an intuitive guide, let's now summarize the meaning of each of Maxwell's equations more precisely. First, Gauss's law establishes the relationship between a charge distribution and its electric field. Equation (30.7) states that the net electric flux through any closed surface equals the total charge enclosed by that surface divided by ϵ_o. It is important to keep in mind that, except for those that wander off to infinity, electric field lines originate on positive charges and terminate on negative charges; otherwise they must form closed loops as magnetic field lines do.

Gauss's law for **B**, Equation (30.8), states that the net magnetic flux through a closed surface is zero; the number of *B*-field lines that enter a surface equals the number that leave it. As a consequence, magnetic field lines are always closed and do not originate or terminate at any point. Equation (30.8) states that magnetic monopoles do not exist.

Equation (30.9) is Faraday's law, which establishes the relationship between a changing magnetic flux and the resulting electric field (and its related emf). Precisely, this law states that the line integral of the electric field **E** around a closed path is equal to the negative of the time rate of change of magnetic flux enclosed by that path.

Finally, Ampère's law provides a relation between a magnetic field **B** and its sources: a current I or a changing electric flux $d\phi_E/dt$. Equation (30.10) says that the line integral of the magnetic field **B** around any closed path is determined by the current I that passes through that path and the time rate of change of electric flux that passes through any surface that is bounded by that path.

30.4 Summary

Faraday's law states that a changing magnetic flux induces an emf:

$$\mathcal{E} = -\frac{d\Phi_B}{dt} \tag{30.1}$$

The most general form of Faraday's law is

$$\oint \mathbf{E} \cdot d\mathbf{l} = -\frac{d}{dt} \int \mathbf{B} \cdot d\mathbf{A} \tag{30.3}$$

where **E** is a nonconservative electric field. **Lenz's law** is effectively the minus sign in Faraday's law and states that the induced emf is in such a direction as to oppose the change in magnetic flux that produces it.

When bulk conductors are subjected to a changing magnetic flux, the induced emf can generate **eddy currents** in the conductors.

Time-varying electric flux gives rise to the **displacement current:**

$$I_d = \epsilon_o \frac{d\phi_E}{dt} \tag{30.5}$$

The displacement current is a source of the magnetic field **B**, and the complete form of Ampère's law is

$$\oint \mathbf{B} \cdot d\mathbf{l} = \mu_o I + \mu_o \epsilon_o \frac{d\phi_E}{dt} \tag{30.6}$$

This law states that both currents I and changing electric flux $d\phi_E/dt$ produce magnetic fields.

Maxwell's equations form the foundation of classical electromagnetic theory. They are

Gauss's law for **E**: $\oint \mathbf{E} \cdot d\mathbf{A} = \dfrac{q_{\text{encl}}}{\epsilon_o}$ (30.7)

Gauss's law for **B**: $\oint \mathbf{B} \cdot d\mathbf{A} = 0$ (30.8)

Faraday's law: $\oint \mathbf{E} \cdot d\mathbf{l} = -\dfrac{d\phi_B}{dt}$ (30.9)

Ampère's law: $\oint \mathbf{B} \cdot d\mathbf{l} = \mu_o I + \mu_o \epsilon_o \dfrac{d\phi_E}{dt}$ (30.10)

(as modified by Maxwell)

PROBLEMS

30.1 The Laws of Faraday and Lenz

1. A square coil of wire lies in the plane of this page, is 8.00 cm on a side, and contains 36.0 turns. If a magnetic field is directed perpendicular to the plane of the coil (into the page) and increases from 0.000 T to 0.800 T in 1.50 s, (a) what is the magnitude of the induced emf in the coil? (b) Is the induced current in the coil moving in the clockwise or anticlockwise sense?

2. A rubber band coated with flexible conducting paint makes a circle and lies in the plane of this page. It has a diameter of 6.00 cm and a total resistance of 12.0 Ω. A magnetic field of intensity $B = 0.750$ T is directed into the page, normal to the plane of the coil. If the diameter of the coil is made to suddenly decrease to half its original value in 0.200 s while the coil's resistance remains fixed, (a) what is the induced emf in the loop? (b) What is the induced current in the coil?

3. A square coil of wire lying in the plane of this page is 12.0 cm on a side and has 60 turns. If the total resistance of the coil is 6.00 Ω, at what rate must the magnitude of a magnetic field $B(t)$ directed perpendicular to the surface of the coil change to induce a current of 0.0500 A in the wire of the coil?

4. A 500-turn, circular coil of wire has a diameter of 36.0 cm and is held in a position so that the direction of the earth's magnetic field is perpendicular to the plane of the coil. When the coil is flipped over (one-half of a complete rotation) in 0.200 s, an average emf of 25.0 mV is observed between the ends of the coil. What is the strength of the earth's magnetic field at the location of the coil?

5. A coil of wire with a square cross section, 10.0 cm on a side, is placed in a uniform magnetic field that makes a 40.0° angle with the normal to the plane of the coil. The magnetic field initially has a magnitude of 1.20 T. This field is then reduced at a constant rate to 0.00 T in 2.40 s. As the field is reduced an emf of 2.00 V is induced between the ends of the coil. What is the total length of the wire that composes the coil?

6. A circular coil of wire lying in the plane of this page has 20.0 turns, a radius of 6.00 cm, and a total resistance of 4.00 Ω. (a) At what rate must a magnetic field vary to produce a current of 250. mA in the wire of the coil if the magnetic field is directed into this page but at an angle of 60.0° from a perpendicular to the plane of the coil? (b) In what sense (clockwise or anticlockwise) does the induced current circulate if the B-field is increasing in magnitude?

7. A circular coil lying in the plane of this page has a 5.00-cm radius, 50.0 turns, and a total resistance of 2.00 Ω. A magnetic field directed into this page perpendicular to the plane of the loop varies in time according to the relation $B(t) = at^2 + bt + c$, where $a = 20.0$ mT/s², $b = 30.0$ mT/s, and $c = 50.0$ mT. (a) Find the induced current in the coil at $t = 2.00$ s. (b) In what sense does the induced current circulate, clockwise or anticlockwise?

8. A circular coil lies in the plane of this page and has 25.0 turns, a radius of 6.00 cm, and a total resistance of 1.80 Ω. A magnetic field is directed into this page at an angle of 40.0° from a perpendicular to the plane of the coil. If the magnitude of the B-field varies in time according to the relation $B(t) = at^2 + bt$, where $a = 50.0$ mT/s² and $b = 20.0$ mT/s, (a) find the induced current in the coil at $t = 0.500$ s. (b) In what sense does the induced current circulate, clockwise or anticlockwise?

9. The 1500-turn solenoid shown in Figure 30.P1 is 80.0 cm in length and has a diameter of 9.00 cm. Around the solenoid is wrapped a 6-turn coil attached to a resistance $R = 6.00$ Ω. If the current I to the solenoid increases at a uniform rate from 0.500 A to 6.00 A in 1.50 s, what is the magnitude and direction of the induced current through the resistor R?

10. The 1200-turn solenoid shown in Figure 30.P1 is 60.0 cm in length and has a diameter of 8.00 cm. Around the solenoid is wrapped a 6-turn coil attached to a resistance $R = 40.0$ Ω. If the current in the solenoid is reduced at a uniform rate from 4.00 A to 1.50 A in 0.500 s, what is the magnitude and direction of the induced current through the resistor R?

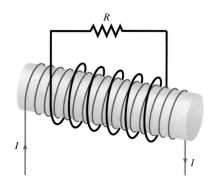

FIGURE 30.P1   Problems 9 and 10

11. The single loop illustrated in Figure 30.P2 symbolizes a 300-turn current loop that has a radius of 4.00 cm and rotates at 500. rpm. What is the maximum current that flows through the resistance $R = 36.0$ Ω if the constant magnetic field into the page is of magnitude $B = 0.200$ T?

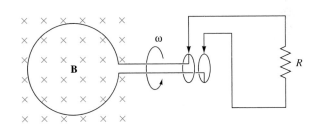

FIGURE 30.P2   Problem 11

12. A 50.0-turn rectangular loop of wire has a width of 8.00 cm and a length of 12.0 cm. A magnetic field applied along the direction perpendicular to the plane of the loop varies in time according to the relation $B(t) = B_o \sin(\omega t)$, where $B_o = 0.0400$ T and $\omega = 120.\pi$ rad/s. (a) What is the induced emf $\mathcal{E}(t)$ in the loop of wire? (b) If the total resistance of the loop is 24.0 Ω, what is the current in the loop at $t = 0.680$ s?

★13. A cut-away view of a toroid with a rectangular cross section is shown in Figure 30.P3. The toroid has $N_T = 800.$ turns, radii $R_1 = 6.00$ cm and $R_2 = 12.0$ cm, and height $h = 4.00$ cm. The rectangular conducting loop also shown in the figure has $N_L = 24.0$ turns, and the plane of this loop coincides with a rectangular cross section of the toroid. If the current in the toroid changes at a constant rate from 1.00 A to 6.00 A in 0.250 s, what is the induced emf around the rectangular loop?

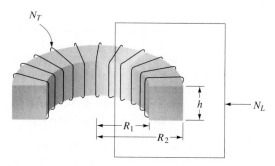

FIGURE 30.P3   Problems 13 and 31

★14. The long, straight wire shown in Figure 30.P4 carries a current $I(t) = I_o e^{-t/\tau}$, where $I_o = 12.0$ A and $\tau = 4.00$ s. If $x = 0.500$ cm, $w = 6.00$ cm, and $l = 12.0$ cm, find the induced emf around the rectangular loop at $t = 2.00$ s.

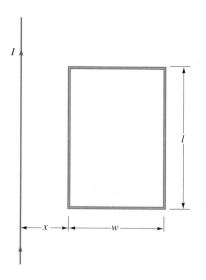

FIGURE 30.P4   Problems 14, 36, and 37

15. A small aircraft has a wing span of 15.0 m (from the tip of one wing to the tip of the other) and flies straight and level at 78.0 m/s in a region where the vertical component of the earth's magnetic field is 5.00×10^{-5} T. What is the induced emf between the ends of the wings?

16. The conducting rod shown in Figure 30.P5 moves with a constant velocity $v = dx/dt$. Show that the magnitude of the emf induced across the rod is $|\mathscr{E}| = (\mu_o/2\pi)(vIl/x)$.

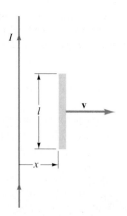

FIGURE 30.P5   Problem 16

17. If, in Figure 30.7, $B = 0.600$ T, $l = 12.0$ cm, and $R = 36.0$ Ω, at what velocity must the bar slide for 2.00 mW of power to be dissipated in the resistor?

18. Suppose the current generated through the resistor R shown in Figure 30.7 is 20.0 mA when the resistance is 500. Ω and the bar's velocity is 10.0 cm/s. Compute the constant force that must be applied to the sliding bar to maintain this current.

19. A pair of conducting, parallel rails are separated by 12.0 cm and are connected at one end by a resistance $R = 8.00$ Ω as shown in Figure 30.P6. A conducting bar is pulled along the frictionless rails with a velocity of 3.60 m/s. A uniform magnetic field $B = 0.750$ T is directed out of the page. (a) What power is dissipated in the resistor? (b) What is the direction of the induced current and the magnetic force on the bar? (c) At what rate does the pulling force do work on the bar?

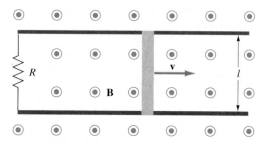

FIGURE 30.P6   Problem 19

20. A conducting bar moves in a direction parallel to a constant current $I = 64.0$ A as shown in Figure 30.P7. If $d = 0.500$ cm, $l = 12.0$ cm, and $v = 8.00$ m/s, compute the induced emf between the ends of the bar.

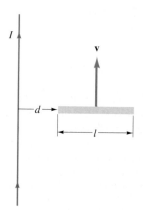

FIGURE 30.P7   Problem 20

21. A circular coil of wire consists of 250. turns and has a radius of 8.40 cm. At what angular velocity must the coil rotate about a diameter in a uniform 2.40-mT field to produce a maximum emf of 6.00 V between the open ends of the wire?

22. The coil of a certain generator produces a maximum emf of 100. V when rotating at 60.0 rev/s. If the area of the coil is 240. cm^2 and it rotates in a uniform B-field of 0.800 T, how many turns must the coil have?

30.2 Induced Electric Fields

23. A 1200.-turn solenoid is coaxial with a single-turn conducting ring as shown in Figure 30.P8. The length l of the solenoid is 20.0 cm, and the current applied to it varies in time according to

the relation $I(t) = I_o(1 - e^{-t/\tau})$, where $I_o = 3.00$ A and $\tau = 2.00$ s. The diameter of the solenoid is 4.00 cm, whereas that of the conducting ring is 8.00 cm. (a) Compute the induced emf in the ring at $t = 3.00$ s. (b) If the resistance of the ring is 0.500 Ω, find the current induced in the ring at $t = 3.00$ s.

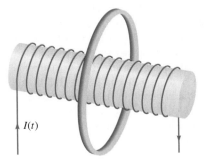

$I(t)$

FIGURE 30.P8   Problem 23

24. A square, flat loop of wire with side length l has N turns and a total resistance R. As shown in Figure 30.P9, the loop moves with velocity **v** into a region of constant magnetic field **B**. Find the magnitude and direction of the total magnetic force on the loop (a) as it enters the field region, (b) while it is completely within the field region, and (c) as it leaves the field region.

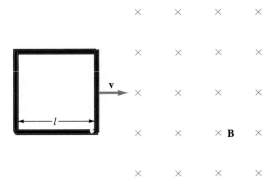

FIGURE 30.P9   Problem 24

25. Figure 30.P10 shows a crude model for B-field lines between the pole faces of a magnet. Notice that, for this model, there is no fringing of the B-field lines. Apply Ampère's law to the rectangular path shown in the figure to show that fringing must occur.

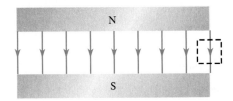

FIGURE 30.P10   Problem 25

26. A uniform magnetic field is confined to a cylindrical region of space with a radius $R = 20.0$ cm. As shown in Figure 30.P11 the field is into the page, and its magnitude is decreasing at a rate of 0.0500 T/s. Use Faraday's law to compute the line integral $\oint \mathbf{E} \cdot d\mathbf{l}$ around the closed circular path of radius (a) $r = 8.00$ cm, and (b) $r = 16.0$ cm.

27. A uniform magnetic field is confined to a cylindrical region of space with a radius of $R = 30.0$ cm. As shown in Figure 30.P11 the field is into the page, and its magnitude is decreasing according to the relation $B(t) = (1.20$ T$)e^{-t/(2.00 \, s)}$. Use Faraday's law to

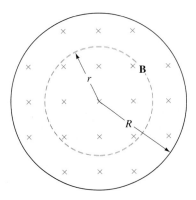

FIGURE 30.P11   Problem 26

compute the line integral $\oint \mathbf{E} \cdot d\mathbf{l}$ at time $t = 1.00$ s around the closed circular path of radius (a) $r = 20.0$ cm, and (b) $r = 32.0$ cm.

30.3 The Displacement Current and Maxwell's Equations

28. Within a cylindrical volume of space of radius $R = 1.28$ cm an electric field exists that is directed parallel to the cylinder axis. The magnitude of the electric field can be approximately modeled as $E = at$ for $r < R$ ($a = 250$ (V/m)/s and t is in seconds) and $E = 0$ for $r > R$. (a) What is the displacement current in the region $r < R$? Give your result in amperes. (b) What is the magnitude of the magnetic field along a circle of radius R concentric with the cylinder? If the changing electric field is directed into the page of this book, in what sense do magnetic field lines circulate around the cylinder: clockwise or anticlockwise? (c) Suppose a long, straight wire carries a real current the magnitude of which is equal to the magnitude of displacement current calculated in part (a). What is the magnitude of the magnetic field at a distance of R from the wire?

29. Consider the electric field between two capacitor plates of circular geometry each with radius $r = 2.00$ cm and separated by a distance of 1.00 mm. (a) If the capacitor is being charged by a constant current source supplying 1.50 mA of current, what is the rate at which the electric field between the plates is increasing? (b) What is the magnetic field strength along a circle of radius 1.00 cm midway between and concentric with the capacitor plates?

30. (a) From which of Maxwell's equations can Coulomb's law be derived? (b) Which of Maxwell's equations can be used to calculate the magnetic field strength around a long current-carrying wire? (c) Which equations must be modified if magnetic monopoles are discovered? (d) Which equation describes how the alternator in a car generates electricity by rotating a coil of wire in a magnetic field? (e) Identify the term that Maxwell added to this collection of equations.

Numerical Methods and Computer Applications

31. In Problem 13 the toroid of rectangular cross section and the rectangular loop shown in Figure 30.P3 are described. Suppose the current does not increase steadily as given in that problem, but instead is defined numerically by the $I(t)$ function provided on the spreadsheet RTOROID.WK1 on the disk accompanying this text. From the data given in Problem 13, construct a plot of (a) the current $I(t)$, and (b) the emf induced around the rectangular loop $\mathcal{E}(t)$. What is the maximum magnitude of the emf induced around the loop? Do you note any unusual changes in the emf the pres-

ence of which you might not have guessed from an examination of the $I(t)$ plot?

32. Two long, parallel wires are separated by a distance d and carry currents I_1 and I_2. Figure 30.P12 shows an xy-coordinate system in a plane perpendicular to the currents, with the origin midway between them on a line connecting the currents. Write a program or construct a spreadsheet template to plot the energy density u_B as a function of x along the line $y = h$ from $x = -3d/2$ to $x = +3d/2$. Try out your program or spreadsheet for $I_1 = 2.0$ A, $I_2 = -3.0$ A, $d = 1.0$ cm, and $h = 1.5$ cm. Try various other values for the current magnitudes and directions and for values of d and h.

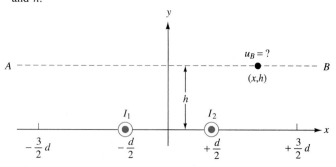

FIGURE 30.P12   Problem 32

★33. If one attempts to use an electromagnetic launcher to accelerate projectiles, the induced back emf acts to counter the currents used to accelerate the projectile. To model this effect, consider a projectile that is a copper bar of mass $m = 20.0$ g, length $l = 1.00$ cm, and resistance $R_o = 0.0100$ Ω. Let this bar ride on parallel copper rails with a resistance per unit length of rail $\lambda = 0.0100$ Ω/cm. Power the launcher with an ideal emf source of $V = 10.0$ V capable of supplying currents in excess of 1000 A. Now suppose a magnetic field of 3.00 T exists normal to the plane of the rails. The force on the projectile is given by $F = BIl$. The current I is given by $I = (V - \mathcal{E})/R$, where R is the total resistance in the circuit after the projectile has traveled distance x down the rail and is given by $R = R_o + 2\lambda x$. The induced back emf $\mathcal{E}$ has a magnitude $\mathcal{E} = lBv$, where v is the instantaneous velocity of the projectile (see Example 30.2).

The position versus time for the projectile can be found by integrating $x = \int a\, dt$, using, for example, the fourth-order Runge-Kutta method as outlined in Chapter 6. Combine the preceding equations to produce an expression for F as a function of x and v. Use the computational aid of your choice (spreadsheet, computer program, symbolic mathematics software) to find x, v, and F as a function of t. Notice that as the velocity of the projectile increases the available force decreases owing to the induced back emf. This same phenomenon occurs in rotating electric motors as their angular speed increases.

General Problems

34. A 36-turn loop of wire has a cross-sectional area of 24.0 cm². A magnetic field applied along the direction perpendicular to the plane of the loop varies in time according to the relation

$$B(t) = B_o e^{-t/\tau} \sin(\omega t)$$

where $B_o = 0.0800$ T, $\tau = 0.500$ s, and $\omega = 30.0\pi$ rad/s. (a) What is the induced emf in the loop at $t = 0.360$ s? (b) If the total resistance of the loop is 12.0 Ω, what is the current in the loop at $t = 0.360$ s?

35. When a metal blade, such as an airplane or helicopter propeller, rotates in the earth's magnetic field, an emf can be induced between the ends of the blade. Suppose a blade 1.40 m in length is rotating at 500. rpm in an orientation such that the component of the earth's magnetic field perpendicular to the plane of the rotation is 4.00×10^{-5} T. What is the induced emf between the tip and center of the blade (i.e., the point of rotation)?

★36. A straight wire carries a current $I(t) = I_o \cos(\omega t)$ in a direction parallel to two sides of the rectangular conducting loop shown in Figure 30.P4. The parameters shown in this figure are $x = 1.00$ cm, $w = 5.00$ cm, and $l = 10.0$ cm. If $I_o = 12.0$ A and $\omega = 9.00\pi$ rad/s, compute the induced emf in the rectangular loop at $t = 1.80$ s.

★37. The long, straight wire shown in Figure 30.P4 carries a constant current $I = 24.0$ A, and the dimensions of the rectangular conducting loop are $l = 12.0$ cm and $w = 5.00$ cm. The rectangular loop moves away from the line of current in a direction perpendicular to the line so that the distance x is given by the relation $x(t) = (16.0 \text{ m/s})t + 0.0500$ m. (a) Compute the induced emf in the rectangular loop at $t = 0.500$ s. (b) In what direction, clockwise or anticlockwise, does the induced current travel in the conducting loop?

38. A thin, conducting disk of radius $R = 24.0$ cm rotates with angular velocity $\omega = 36.0$ rad/s about an axis through its center and perpendicular to the plane of the disk. A 0.500-T uniform magnetic field is directed parallel to the disk's rotation axis. Compute the potential difference between the center and the edge of the disk.

39. A constant current I flows along a long, straight wire as shown in Figure 30.P13. A bar slides parallel to the current along a frictionless U-shaped conductor. The bar, conductor, and current I are all in the same plane. If $I = 24.0$ A, $d = 0.500$ cm, $l = 4.00$ cm, and $v = 6.00$ m/s what is the induced emf in the loop formed from the bar and the U-shaped conductor?

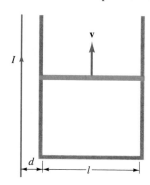

FIGURE 30.P13   Problem 39

40. A thin, conducting bar slides down a ramp consisting of two parallel, frictionless rails of negligible resistance as shown in Figure 30.P14. The uniform magnetic field **B** is directed vertically upward, and the angle of the incline is θ. If the mass of the bar is m and its length l, show that the terminal velocity of the bar is

$$v = \frac{mgR}{B^2 l^2} \frac{\sin(\theta)}{\cos^2(\theta)}$$

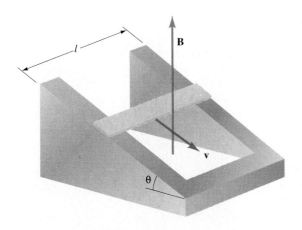

FIGURE 30.P14   Problem 40

41. If the radius of the conducting loop shown in Figure 30.9 is r and the magnetic field strength is changing at a rate dB/dt, show that the electric field induced around the conducting loop is given by

$$E = -\frac{r}{2}\frac{dB}{dt}$$

42. A spy taps the serial link between two computers by wrapping a small coil around the current-carrying wire connecting the computers. The current versus time for the transmission between the two computers is a digital signal as shown in Figure 30.P15. Draw a sketch of the induced emf signal versus time as detected by the spy.

FIGURE 30.P15   Problem 42

CHAPTER

31

Inductance

In this chapter you should learn

- how to compute the self and mutual inductance for various circuit arrangements.

- about the energy stored in an inductor.

- about the energy stored in a magnetic field.

- how the charge and current vary in *LC* circuits.

- how current and potential vary across an inductor in an *LR* circuit.

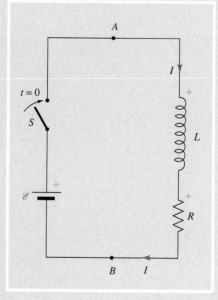

$$\mathcal{E} = -L\frac{dI}{dt}$$

If you have ever peeked inside a radio or television set you may have noticed small coils of wire resembling the solenoids described in the previous chapter. When used in electric circuits these solenoids are known as *inductors*. An inductor's electrical effect in a circuit is measured by its *inductance*. In many respects our discussion of inductors and inductance unfolds in a manner parallel to that of capacitors and capacitance. In Chapter 25 we found that a capacitor can be used to store energy in an *electric* field, and in this chapter we will find that an inductor can be used to store energy in a *magnetic* field. You probably remember that the capacitance of a capacitor depends on its geometrical characteristics only. Similarly, we will discover that this quantity, called the inductance, also depends on geometrical factors only. If your instructor chooses to include the optional section on *LC* circuits, you will find that energy stored in this type of circuit can oscillate back and forth between the *E*-field of the capacitor and the *B*-field of the inductor. As usual, we begin with definitions.

31.1 Induction

In Chapter 27 we first studied simple electric circuits, such as those shown in Figure 31.1. In these circuits we ignored what happens immediately after a switch is closed and applied Kirchhoff's rules to determine the steady-state values of the currents and voltage changes around the circuits. Having studied Faraday's law we might consider ourselves somewhat wiser now and able to determine what happens during the moments just after the switch is closed. We know, for instance, that when the switch S_1 is closed in the upper circuit of Figure 31.1, a current I_1 begins to flow around the circuit and through the resistor R_1. This current creates a magnetic field, and thereby increases the magnetic flux through this same circuit loop. Lenz's laws tell us that this change in flux induces an emf that opposes the current I_1 in this circuit because the current is increasing after the switch is closed. We have good reason to suspect, therefore, that the current does not immediately jump to a value of $I_1 = \mathcal{E}_1/R_1$, but rather increases gradually toward this steady-state value.

The effects of the initial flow of current can extend beyond this self-induced emf in circuit 1. If there is another circuit in close vicinity to circuit 1, such as circuit 2 (the one shown below circuit 1 in Fig. 31.1), the changing magnetic field generated by changing current I_1 can also produce a changing magnetic flux through circuit 2. Hence, an emf will be induced in circuit 2 by a changing current in circuit 1. Conversely, if the current I_2 in circuit 2 changes in time, perhaps by opening or closing switch S_2, an emf will be induced in circuit 1 due to the variation of the current I_2.

We know from the Biot-Savart law that the magnitude of the magnetic field created by current I_1 is proportional to I_1. Hence, the magnetic flux due to this current is also proportional to I_1. Similarly, the magnetic flux through circuit 1 that results from the current in circuit 2 is proportional to I_2. The total magnetic flux through circuit 1 is the sum of the flux from both circuits. Including the proportionality constants, we can therefore write the total magnetic flux in circuit 1 as

$$\Phi_{B_1} = L_1 I_1 + M_{12} I_2 \tag{31.1}$$

The constant L_1 is called the **self-inductance** of circuit 1, and the constant M_{12} is the **mutual inductance** on circuit 1 by circuit 2. The self-inductance L_1 depends on the geometrical arrangement of circuit 1 only, whereas the mutual inductance M_{12} depends on the shape and relative orientations of both circuits. In a similar fashion, the flux through circuit 2 can be written

$$\Phi_{B_2} = L_2 I_2 + M_{21} I_1 \tag{31.2}$$

We won't prove it here, but it turns out that

$$M_{12} = M_{21}$$

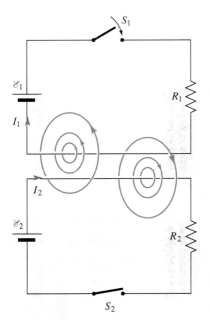

FIGURE 31.1

Two circuits close to each other. When switch S_1 is closed and current I_1 begins to flow around the upper circuit, magnetic flux increases through the area enclosed by this circuit, creating a self-induced emf. This magnetic flux can extend into the lower circuit as well. Moreover, any change in the current I_2 flowing in the lower circuit can create a changing magnetic flux that can influence the upper circuit.

The induced emf in circuit 1 can be found by application of Faraday's law to Equation (31.2):

$$\mathscr{E}_{LM1} = -\frac{d}{dt}\Phi_{B_1} = -L_1\frac{dI_1}{dt} - M_{12}\frac{dI_2}{dt} \tag{31.3}$$

If there is only one circuit present, the mutual inductance is zero. Even when there are two or more circuits present, it is sometimes possible to ignore the mutual inductance term; for example, when there is a relatively large distance between the circuits or some other feature of their orientation makes their mutual inductance negligible. When the mutual inductance is small enough that only the self-inductance of a circuit is important, we can write Equation (31.3) as

$$\mathscr{E}_L = -L\frac{dI}{dt} \tag{31.4}$$

We can compute its self-inductance from Equation (31.4) if we can determine the induced emf $\mathscr{E}$ in a circuit when the rate of change in current through the circuit is dI/dt:

$$L = -\frac{\mathscr{E}_L}{dI/dt} \tag{31.5}$$

We can use Equation (31.5) to discover the units of inductance. Clearly, the SI unit of inductance is $V/(A/s) = V \cdot s/A$, which is called a **henry:**

$$1\,H = 1\,\frac{V \cdot s}{A}$$

When we substitute Faraday's law, $\mathscr{E} = -d\Phi_B/dt$, into Equation (31.4), we obtain

$$-L\frac{dI}{dt} = -\frac{d\Phi_B}{dt}$$

Integrating this expression with respect to time, we have

$$L\int_0^t \frac{dI}{dt}\,dt = \int_0^t \frac{d\Phi_B}{dt}\,dt$$

Assuming the magnetic flux through the circuit is zero when the current is zero, the integrals become

$$L\int_0^I dI = \int_0^{\Phi_B} d\Phi_B$$

$$L\,I = \Phi_B$$

or

$$L = \frac{\Phi_B}{I} \tag{31.6}$$

The following example illustrates how to apply Equation (31.6) to compute the self-inductance of a solenoid.

EXAMPLE 31.1   *Inductance of an Ideal Solenoid*

Find the inductance of an ideal solenoid of length l, cross-sectional area A, and N turns.

SOLUTION   We found previously that the magnetic field strength in an ideal solenoid carrying current I is $B = \mu_o(N/l)I$. Hence, the magnetic flux through *one turn* of the solenoid is

$$\phi_B = BA = \mu_o \frac{N}{l} IA$$

The total flux through the circuit that comprises the entire solenoid (all N turns) is

$$\Phi_B = N\phi_B = \frac{\mu_o N^2 A}{l} I$$

Applying Equation (31.6), we have

$$L = \frac{\mu_o N^2 A}{l}$$

Notice that the inductance L depends only on geometrical factors: length l, area A, and the number of turns N. The geometric factors are always found in combinations such that the overall unit is length. For this reason you will often find μ_o written with units of henrys per meter.   ◄

Concept Question 1
How does the inductance *per unit length* at the center of a *real* solenoid compare to its inductance per unit length near its ends? Explain.

Concept Question 2
How would you construct a coil with a large resistance but almost negligible self-inductance?

EXAMPLE 31.2   *An Induction Coil*

Apply the ideal-solenoid model to compute the rate of current change required to induce an emf of 1.00 kV through a solenoid with 250. turns, a 20.0-cm length, and a 5.00-cm^2 area.

SOLUTION   We begin by computing the inductance of the solenoid. From Example 31.1, we have

$$L = \frac{\mu_o N^2 A}{l} = \frac{(4\pi \times 10^{-7} \text{ N/A}^2)(250)^2(5.00 \times 10^{-4} \text{ m}^2)}{(0.200 \text{ m})} = 1.96 \times 10^{-4} \text{ H}$$

$$= 0.196 \text{ mH}$$

Next, we solve Equation (31.5) for dI/dt:

$$\left| \frac{dI}{dt} \right| = \frac{|\mathscr{E}_L|}{L} = \frac{1.00 \times 10^3 \text{ V}}{1.96 \times 10^{-4} \text{ H}} = 5.09 \times 10^6 \text{ A/s}$$

Such rapid changes can be brought about by suddenly opening a switch in a current-carrying circuit. In older cars this was accomplished by "breaker points"; but these days the switching is done with transistors. In both cases the resulting large emf from the induction coil is subsequently applied to the spark plugs of the automobile engine.   ◄

When there are two circuits present and their geometry is such that the mutual inductance is important, we can compute $M_{12} = M_{21} = M$ in a manner similar to the self-inductance L. From Equation (31.3) the emf induced in circuit 2 from the mutual inductance only of circuit 1 is

$$\mathscr{E}_{M2} = -M\frac{dI_1}{dt}$$

However, from Faraday's law this emf must come from a change in magnetic flux through circuit 2, $\mathscr{E}_{M2} = -d\Phi_{B_2}/dt$ so that

$$\frac{d\Phi_{B_2}}{dt} = M\frac{dI_1}{dt} \tag{31.7}$$

Integrating this equation with respect to time, we have

$$M\int_0^t \frac{dI}{dt}\,dt = \int_0^t \frac{d\Phi_{B_2}}{dt}\,dt$$

leading to

$$M\int_0^I dI = \int_0^{\Phi_{B_2}} d\Phi_{B_2}$$

Resulting in a mutual inductance

$$M = \frac{\Phi_{B_2}}{I_1} \tag{31.8}$$

where Φ_{B_2} is only that flux through circuit 2 generated by the circuit in circuit 1. We illustrate the computation of mutual inductance in Example (31.3).

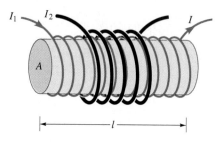

FIGURE 31.2

An ideal solenoid of length l and cross-sectional area A is surrounded by a second coil. A changing current in one coil causes an induced current in the other. See Example 31.3.

EXAMPLE 31.3   *Cozy Coils*

An ideal solenoid of length l and cross-sectional area A has N_1 turns and carries a current I_1. Wrapped snugly around this solenoid is another coil with N_2 turns as shown in Figure 31.2. Find the mutual inductance M of this system of solenoids.

SOLUTION   The magnetic field within the solenoid from current I_1 is

$$B = \mu_o \frac{N_1}{l} I_1$$

so that the flux ϕ_{B_2} through a *single* turn of the outer coil's winding with cross-sectional area A is

$$\phi_{B_2} = BA = \frac{\mu_o N_1 I_1 A}{l}$$

Through N_2 turns of the secondary the *total* flux Φ_{B_2} is

$$\Phi_{B_2} = N_2\phi_{B_2} = \frac{\mu_o N_1 N_2 I_1 A}{l}$$

The mutual inductance M is given by Equation (31.8):

$$M = \frac{\Phi_{B_2}}{I_1} = \frac{\mu_o N_1 N_2 A}{l}$$

31.2 *LR* Circuits

Example 31.2 illustrated that a changing current through a solenoid produces an emf across the solenoid. Circuit elements, such as solenoids, that can provide induced emfs in a circuit when the current is changing are called **inductors.** The circuit symbol for an inductor is shown in Figure 31.3. You should realize that the actual inductor may or may not be a simple coil of wire. Although many inductors are formed by winding a length of insulated wire around a cylinder, any conducting arrangement that can produce a significant induced emf can rightly be called an inductor.

Consider a simple circuit in which a source of emf $\mathcal{E}$ is placed in series with a switch S, an inductor of inductance L, and a resistance R as shown in Figure 31.4. We call such circuits *LR* circuits. Because most inductors consist of a coil made from a long wire that itself has some resistance, we must always assume that there is some resistance in an *LR* circuit even if we were to place an inductor only across the emf $\mathcal{E}$. We assume any resistance of the inductor is included in the value of the resistance R. When switch S is closed, current I begins to flow clockwise through the circuit. As the current begins to flow through the inductor, however, an emf is induced across it so as to oppose the flow of this current. In order to oppose this current, the direction of the induced emf must be as indicated by the plus and minus signs in Figure 31.4. Don't forget that emfs drive current *from* their positive terminal around the *external* circuit *toward* their negative terminal. Hence, the emf provided by the inductance L in Figure 31.4 opposes the current I. In order to find the current $I(t)$ through the circuit we apply Kirchhoff's law by starting at the source of emf $\mathcal{E}$ and "walking" around the circuit clockwise, summing potential changes as we go. (Consider the switch S closed and don't forget that the magnitude of the induced emf across an inductor is $L\,dI/dt$):

$$\mathcal{E} - L\frac{dI}{dt} - IR = 0$$

Putting this in more standard form, we have

$$\frac{dI}{dt} + \frac{R}{L}I = \frac{\mathcal{E}}{L} \tag{31.9}$$

Equation (31.9) is a linear differential equation with constant coefficients and, in fact, if you look back to Chapter 27, Equation (27.24), you will discover that this differential equation is analogous to the one we derived for a simple *RC* circuit. In the present case, however, we are solving for the current $I(t)$ rather than the charge on the capacitor $q(t)$. As before, we simply state the solution:

$$I(t) = \frac{\mathcal{E}}{R}(1 - e^{-t/\tau_L}) \tag{31.10}$$

where we define the **inductive time constant** τ_L as:

$$\tau_L = \frac{L}{R}$$

The inductive time constant is the time required for the current to reach the fraction $(1 - e^{-1}) \approx 0.63$ of its final value of $\mathcal{E}/R$:

$$I(\tau_L) = \frac{\mathcal{E}}{R}(1 - e^{-1}) \approx 0.63\,\frac{\mathcal{E}}{R}$$

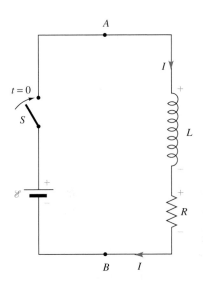

FIGURE 31.3

The circuit symbol for an inductor.

FIGURE 31.4

An *LR* circuit with a switch that closes at time $t = 0$. The induced emf across the inductor opposes the current flow.

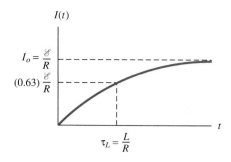

FIGURE 31.5

The current $I(t)$ in the LR circuit of Figure 31.4. At later times the emf across the inductor decreases so that I approaches its steady-state value $\mathcal{E}/R$.

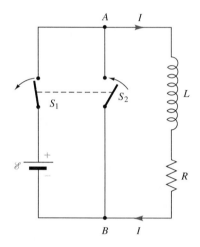

FIGURE 31.6

An LR circuit with two linked switches S_1 and S_2. When S_1 opens, S_2 simultaneously closes so that at $t = 0$ the external emf in the right-hand loop becomes $\mathcal{E} = 0$.

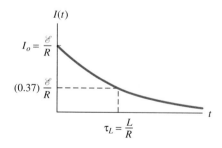

FIGURE 31.7

$I(t)$ for the circuit shown in Figure 31.6. The current decays exponentially to zero.

A typical plot of $I(t)$ is shown in Figure 31.5. You should verify that the units of L/R work out to be seconds. Notice from Figure 31.5 that the induced emf is 100% effective at opposing the current from the source of emf $\mathcal{E}$ only at $t = 0$. As time increases, the current through the circuit increases and $I(t)$ approaches a steady-state value of $I_o = \mathcal{E}/R$ when t becomes very large.

Let's next consider the LR circuit shown in Figure 31.6, in which we added a switch S_2 between the terminals A and B of Figure 31.4. We will suppose that switch S_1 has been closed for a long time so that the current is $I_o = \mathcal{E}/R$. At $t = 0$ we close switch S_2 while simultaneously opening switch S_1. The effect of this switching is to set $\mathcal{E} = 0$ in the circuit as well as in Equation (31.9), which we derived from Kirchhoff's law. Our differential equation becomes

$$\frac{dI}{dt} + \frac{R}{L} I = 0 \qquad (31.11)$$

The solution to this differential equation is

$$I(t) = I_o e^{-t/\tau_L} = \frac{\mathcal{E}}{R} e^{-t/\tau_L} \qquad (31.12)$$

The graph of this function, Figure 31.7, shows that the current decreases in time, as we expect because there is no longer any source of constant emf. From this figure we also see that the emf across the inductor, $\mathcal{E}_L = -L(dI/dt)$, is positive because the slope dI/dt is always negative. Physically, this positive induced emf is a result of the decreasing flux through the inductor; Lenz's law states that the emf opposes this flux decrease, thereby providing an emf in the direction that tries to keep the current flowing. Also shown in Figure 31.7 is the time τ_L at which the current drops to $1/e \approx 0.37$ of its maximum value.

EXAMPLE 31.4 An Inductive Time Constant

A solenoid with an inductance of 240. mH and a resistance of 48.0 Ω is connected across a 12.0-V source of emf. What is the inductive time constant τ_L for this circuit and what is the current through the solenoid 1.00 ms after it is first connected to the source of emf?

SOLUTION The inductive time constant is

$$\tau_L = \frac{L}{R} = \frac{0.240\ \text{H}}{48.0\ \Omega} = 5.00 \times 10^{-3}\ \text{s} = 5.00\ \text{ms}$$

We apply Equation (31.10) to find the current at $t = 1.00$ ms:

$$I(t) = \frac{\mathcal{E}}{R}(1 - e^{-t/\tau_L})$$

$$I(1.00\ \text{ms}) = \frac{12.0\ \text{V}}{48.0\ \Omega}(1 - e^{-(1.00\ \text{ms})/(5.00\ \text{ms})}) = 0.0453\ \text{A} = 45.3\ \text{mA} \quad \blacktriangleleft$$

31.3 Energy and the Magnetic Field

In this section we compute the energy stored in a magnetic field. Our approach is similar to that in Section 25.3 where we found that the energy density (energy per unit volume) of an electric field is $u_E = \epsilon_o E^2/2$. To obtain this result we considered the rate at which a source of emf supplied energy to an RC circuit. Part of this energy was dissipated in the resistor, and the remainder was stored in the uniform electric field between the parallel plates of the capacitor. In the present case we look back at Figure 31.4 where (when switch S is closed) a source of emf supplies energy to the inductor L and the resistor R. When we applied Kirchhoff's laws to this circuit, we obtained Equation (31.9). If we rearrange this equation and multiply through by the current I, we obtain

$$\mathscr{E}I = LI\frac{dI}{dt} + I^2R$$

We saw previously that the rate at which the source of emf supplies energy to the circuit is $\mathscr{E}I$ and the rate at which energy is dissipated in the resistor is I^2R. From the conservation of energy for this closed system, it is easy to identify the remaining term as the rate at which energy is stored in the inductor L. That is

$$\frac{dU}{dt} = LI\frac{dI}{dt}$$

We integrate this expression with respect to time subject to the initial condition that $U = 0$ when $I = 0$:

$$\int_0^t \frac{dU}{dt}\,dt = \int_0^t LI\frac{dI}{dt}\,dt$$

$$\int_0^U dU = L\int_0^I I\,dI$$

$$U = \frac{1}{2}LI^2 \qquad\qquad (31.13)$$

In Chapter 25 we learned that the energy stored by a capacitor C with a potential difference ΔV between its conductors is $C\,\Delta V^2/2$. Equation (31.13) provides us with the energy stored by an inductor L carrying current I. If you believe in the conservation of energy, then about the only place you can "visualize" this energy being stored is in the magnetic field created by the inductor. By all experiments to date this view is indeed a valid way to account for the energy. Let's relate U to the B-field by taking the inductor to be an ideal solenoid of length l and cross-sectional area A. In this case, we saw in Example 31.1 that the inductance L could be calculated from

$$L = \frac{\mu_o N^2 A}{l} = \mu_o\left(\frac{N}{l}\right)^2(Al)$$

With this expression for L and the number of turns per length written $n = N/l$, Equation (31.13) can be written

$$U = \frac{1}{2}[\mu_o n^2(Al)]I^2 = \frac{\mu_o^2 n^2 I^2}{2\mu_o}(Al)$$

The magnetic field of an ideal solenoid is $B = \mu_o nI$, and the volume $\mathscr{V}$ of the solenoid is Al. Hence,

$$U = \frac{B^2}{2\mu_o}(Al)$$

Concept Question 3
For an *LR* circuit, such as that shown in Figure 31.4, is it possible for the induced emf in the inductor to be greater than that of the source of emf? Explain your answer.

Concept Question 4
When a large electric appliance, such as a furnace, stove, or refrigerator, first comes on, it is common for the lights to momentarily dim. Explain this phenomenon.

and we can write the energy density $u_B = U/V = U/(Al)$ as

$$u_B = \frac{B^2}{2\mu_o}$$

(31.14)

Although we have derived Equation (31.14) for the special case of the uniform B-field of an ideal solenoid, it turns out that this expression is valid for any magnetic field B. Notice that both energy densities $u_B = B^2/2\mu_o$ and $u_E = \epsilon_o E^2/2$ are proportional to the square of the respective field strengths.

EXAMPLE 31.5   A Coaxial Cable

A long coaxial cable consists of two concentric, cylindrical conductors of radii R_1 and R_2, where $R_1 < R_2$. The inner cylinder carries a uniform current I, and the outer cylinder carries the same current in the reverse direction as shown in Figure 31.8. Compute the energy stored in the magnetic field between the two cylinders of length l.

SOLUTION   The magnetic field between the two conductors a distance r from the cylinders' axes is (see Problem 29.36)

$$B = \frac{\mu_o I}{2\pi r}$$

The energy density at radius r is, therefore,

$$u_B = \frac{B^2}{2\mu_o} = \frac{1}{2\mu_o}\left(\frac{\mu_o I}{2\pi r}\right)^2 = \frac{\mu_o I^2}{8\pi^2 r^2}$$

Now, the energy density can be written $u_B = dU_B/dV$, where dV is a volume element between the cylinders. Hence,

$$dU_B = u_B\, dV$$

$$U_B = \int \frac{\mu_o I^2}{8\pi^2 r^2}\, dV$$

Because u_B is cylindrically symmetric (depending only on the radius r), we can write the volume element as a cylindrical shell of length l and a radius which extends from r

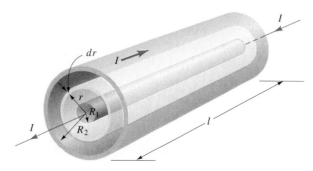

FIGURE 31.8

Computing the energy stored in the magnetic field between the conductors of a coaxial cable. See Example 31.5.

to $r + dr$:

$$d\mathcal{V} = (2\pi r)l\, dr$$

The total energy between the cylinders is, therefore,

$$U_B = \int_{R_1}^{R_2} \frac{\mu_o I^2}{8\pi^2 r^2}(2\pi r)l\, dr = \frac{\mu_o I^2 l}{4\pi}\int_{R_1}^{R_2}\frac{dr}{r}$$

$$= \frac{\mu_o I^2 l}{4\pi}\ln\left(\frac{R_2}{R_1}\right)$$

◀

Concept Question 5
If the cross-sectional area of a solenoidal inductor decreases to half its original value (while its number of turns and length remain constant), by what factor must the current in the inductor change for the solenoid to store the same amount of energy?

31.4 *LC* Circuits (Optional)

In the first part of this section we analyze an idealized model for a circuit containing only an inductor and a capacitor (an *LC* circuit). The model is ideal because we ignore any resistance in the circuit. After we describe the physics behind the ideal *LC* circuit, we will show you the effect of adding resistance to the circuit.

Figure 31.9 shows a capacitor *C* connected in series with a switch *S* and an inductor *L*. We assume that the capacitor plates already hold opposite charges $\pm Q_o$ and at $t = 0$ the switch *S* is closed. At any instant, the total energy U_{tot} stored in the circuit is the sum of that stored on the capacitor and that stored on the inductor:

$$U_{\text{tot}} = \frac{1}{2}LI^2 + \frac{Q^2}{2C} \tag{31.15}$$

where *I* is the current through the inductor when the charge on the capacitor is *Q*. If no energy is converted to heat or transferred to the external world by some other means, the electrical energy U_{tot} must remain constant and

$$\frac{dU_{\text{tot}}}{dt} = 0$$

That is,

$$\frac{d}{dt}\left(\frac{1}{2}LI^2 + \frac{Q^2}{2C}\right) = 0$$

Now, both *C* and *L* are constants, whereas *I* and *Q* change in time. Hence, we have

$$LI\frac{dI}{dt} + \frac{1}{C}Q\frac{dQ}{dt} = 0$$

But

$$\frac{dQ}{dt} = I \quad \text{and} \quad \frac{dI}{dt} = \frac{d^2Q}{dt^2}$$

so our equation becomes, after canceling a factor of *I*,

$$\frac{d^2Q}{dt^2} + \frac{1}{LC}Q = 0 \tag{31.16}$$

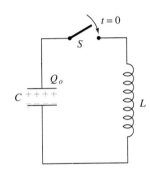

FIGURE 31.9

An ideal *LC* circuit. At $t = 0$ charge Q_o resides on the capacitor, and the switch *S* is closed.

We hope you recognize this linear differential equation. It has the very same form as that which we first derived for a simple harmonic oscillator composed of mass m attached to a spring of stiffness constant k and vibrating with angular frequency $\omega_o = \sqrt{k/m}$:

$$\frac{d^2x}{dt^2} + \omega_o^2 x = 0 \tag{31.17}$$

Back in Chapter 14 we found that the solution to Equation (31.17) is $x(t) = A\cos(\omega_o t + \delta)$, where A is the amplitude and δ is a phase constant determined by the initial conditions of the oscillator. Therefore, we are led to write the solution to Equation (31.16) as

$$Q(t) = A\cos(\omega_o t + \delta) \tag{31.18}$$

By comparing differential Equation (31.16) and (31.17), we also recognize the angular frequency

$$\omega_o = \frac{1}{\sqrt{LC}} \tag{31.19}$$

Moreover, the current is given by

$$I(t) = \frac{dQ}{dt} = -\omega_o A\sin(\omega_o t + \delta) \tag{31.20}$$

Now, at $t = 0$ the current is zero:

$$I(0) = -\omega_o A\sin(\delta) = 0$$

Thus, $\delta = 0$, and Equation (31.18) becomes

$$Q(t) = A\cos(\omega_o t)$$

We also have the initial condition $Q(0) = Q_o$ so that $A = Q_o$, leading to the final solution

$$Q(t) = Q_o\cos(\omega_o t) \tag{31.21}$$
$$I(t) = -\omega_o Q_o\sin(\omega_o t) \tag{31.22}$$

Figure 31.10 shows graphs of $Q(t)$ and $I(t)$. The charge on the capacitor oscillates between $\pm Q_o$, while the current through the inductor varies between $\mp \omega_o Q_o$. The period of the harmonic oscillations is $T = 2\pi/\omega_o = 2\pi\sqrt{LC}$.

To find the energy stored in the inductor we substitute the current from Equation (31.22) into the expression for the energy of the inductor and note that $\omega_o^2 = 1/LC$:

$$U_B = \frac{1}{2}LI^2 = \frac{1}{2}L\omega_o^2 Q_o^2 \sin^2(\omega_o t)$$

$$U_B = \frac{Q_o^2 \sin^2(\omega_o t)}{2C} \tag{31.23}$$

Similarly, Equation (31.21) can be used to find the energy stored by the capacitor at any time t:

$$U_E = \frac{Q^2}{2C} = \frac{Q_o^2 \cos^2(\omega_o t)}{2C} \tag{31.24}$$

$Q(t)$

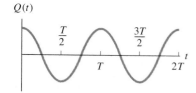

$I(t)$

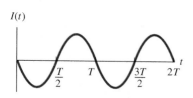

FIGURE 31.10

The charge $Q(t)$ and current $I(t)$ in an ideal LC circuit. The charge leads the current by $\pi/2$ rad.

The total energy U_{tot} of the LC circuit is the sum of $U_B + U_E$:

$$U_{\text{tot}} = \frac{Q_o^2 \sin^2(\omega_o t)}{2C} + \frac{Q_o^2 \cos^2(\omega_o t)}{2C} = \frac{Q_o^2}{2C}[\sin^2(\omega_o t) + \cos^2(\omega_o t)]$$

$$U_{\text{tot}} = \frac{Q_o^2}{2C}$$

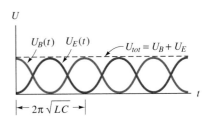

FIGURE 31.11

The energies $U_B(t)$ and $U_E(t)$ for an ideal LC circuit. The total energy at all times is a constant $U_B + U_E$.

This result for the total energy is just as expected because it is equal to the initial energy of the capacitor. The energies U_B, U_E, and U_{tot} are plotted in Figure 31.11. We see that the energy sloshes back and forth between the inductor and the capacitor.

The *RLC* Circuit

When a resistor is present in the *LC* circuit the total energy of the circuit is no longer constant but decreases in time. We refer to circuits such as that shown in Figure 31.12 as *RLC* circuits. As before, we assume an initial charge Q_o on the capacitor and that switch S is closed at $t = 0$. Applying Kirchhoff's laws to this circuit, we obtain

$$L\frac{dI}{dt} + \frac{Q}{C} + IR = 0$$

or, with $I = dQ/dt$,

$$L\frac{d^2Q}{dt^2} + R\frac{dQ}{dt} + \frac{1}{C}Q = 0$$

This differential equation is analogous to Equation (14.29) for a damped harmonic oscillator:

$$m\frac{d^2x}{dt^2} + b\frac{dx}{dt} + kx = 0$$

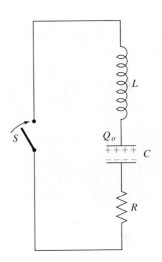

FIGURE 31.12

An *RLC* circuit with an initial charge Q_o on the capacitor C.

In fact, we see that in the present case the inductance L corresponds to the mass m, the resistance R corresponds to the damping constant b, and the reciprocal capacitance $1/C$ corresponds to the spring constant k. Therefore, we expect the charge $Q(t)$ to behave in a fashion completely analogous to the solution $x(t)$ for the damped harmonic oscillator. In Chapter 14 we found that the form of the solution $x(t)$ depends on the relative size of the damping constant b. Hence, in our case the form of $Q(t)$ depends on the relative size of the resistance R. When the resistance is sufficiently small, the charge on the capacitor exhibits damped harmonic motion:

$$Q(t) = Q_o e^{-(R/2L)t} \cos(\omega t) \qquad (31.25)$$

where

$$\omega = \sqrt{\frac{1}{LC} - \left(\frac{R}{2L}\right)^2} \qquad (31.26)$$

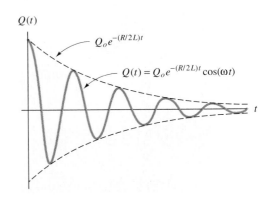

FIGURE 31.13

For relatively small values of the resistance the charge $Q(t)$ in the *RLC* circuit exhibits damped harmonic motion. The exponential amplitude envelope $Q_o e^{-(R/2L)t}$ is also shown.

A graph of $Q(t)$ is shown in Figure 31.13. Because $I = dQ/dt$, the current in the circuit also exhibits damped harmonic motion. Notice from Equation (31.26) that when $1/LC \gg (R/2L)^2$ or equivalently, $R \ll \sqrt{4L/C}$, the angular frequency ω is very near that of the undamped oscillator $\omega_o = 1/\sqrt{LC}$. When the resistance is sufficiently large, we find that the charge and current damp out quickly. In fact, for resistances $R > \sqrt{4L/C}$ no oscillations occur at all, and the system is said to be **overdamped.** When $R = \sqrt{4L/C}$, the charge and current are **critically damped.**

EXAMPLE 31.6   *Watching the Energy Slosh Back and Forth*

A 10.0-μF capacitor is initially charged with a 12.0-V battery, and then the ends of a 4.00-mH inductor are connected to its plates (without the battery). Compute the frequency of oscillation and the maximum current through the inductor.

SOLUTION   The frequency of the oscillation is related to the angular frequency by

$$f = \frac{\omega_o}{2\pi} = \frac{1}{2\pi\sqrt{LC}} = \frac{1}{2\pi\sqrt{(4.00 \times 10^{-3}\text{ H})(10.0 \times 10^{-6}\text{ F})}} = 796 \text{ Hz}$$

We can see from Equation (31.22) that the maximum current is

$$I_{max} = \omega_o Q_o = \frac{Q_o}{\sqrt{LC}}$$

where the charge Q_o is given by

$$Q_o = CV_o = (10.0 \times 10^{-6}\text{ F})(12.0 \text{ V}) = 1.20 \times 10^{-4}\text{ C}$$

Therefore,

$$I_{max} = \frac{1.20 \times 10^{-4}\text{ C}}{\sqrt{(4.00 \times 10^{-3}\text{ H})(10.0 \times 10^{-6}\text{ F})}} = 0.600 \text{ A}$$

Electric circuits in security systems sense the change of inductance of wire coils when metal is brought near them.

The material in this chapter is put to practical use in the metal detectors that have become common place at airports and government buildings. Similar devices are used by amateur treasure hunters seeking coins on vacant lots and by professional geophysicists searching for mineral deposits. The common element in all of these applications is an inductor in the form of a coil. In some gateway metal detectors at airports, you actually walk between two inductors located on opposite sides of the doorway. An external power source creates an alternating current in one of the coils called the sender. As a consequence a small current is induced in the opposite coil, called the receiver. When a metal object is present between the two coils currents are induced in the metal by the sender coil. These currents greatly increase the currents induced in the receiver coil thereby setting off an alarm.

Geophysicists exploring for minerals also use sender and receiver coils. However, the loops may be separated by distances of a few meters to a few kilometers. These geophysicists measure the mutual inductance between the coils as a function of their separation. Intervening conducting mineral deposits act to increase the mutual inductance.

Nuclear magnetic resonance (NMR) is a technique used to measure the electric signals induced by the magnetic moments of atomic nuclei. The nuclei of many atoms have a permanent magnetic dipole moment. That is, they act like tiny bar magnets. When placed in a large external magnetic field (a few tenths of a tesla to several tesla) these nuclear bar magnets tend to align with that field. Quantum-mechanical effects prevent the nuclei from completely aligning with the external field, and the nuclear bar magnets precess rapidly (with typical frequencies of many megahertz) about the field direction, rather like the precession of a spinning toy top when set not quite upright in the earth's gravitational field. If these nuclei are placed inside a tightly wound coil (an inductor), changes in the alignment of the nuclei can be sensed in the circuit connected to the coil.

Nuclear magnetic resonance (NMR) spectrometers sense the emf induced in coils by atomic nuclei rotating in the external field of a large magnetic. The rotation rate of the nuclei gives information about the magnetic field created by neighboring atoms.

In one type of NMR experiment known as the pulse-echo technique, a strong oscillating current is applied to the coil at a frequency that just matches the nuclei's precession rate. At this resonance condition the nuclei feel the oscillating magnetic field of the coil as a steady field in their rotating frame of reference. The nuclei precess around the new total field produced by the coil and the external magnet. If the coil field is shut off at just the right moment during their precession, the nuclei are caught ''upside down'' in the large

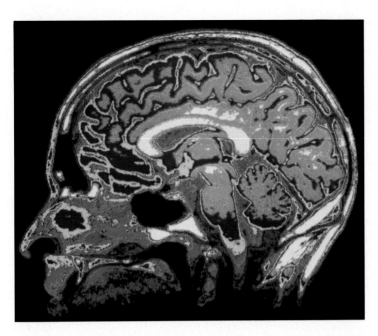

Magnetic resonance imaging (MRI) uses electric currents induced by rotating nuclei to distinguish the location of atoms with different local magnetic environments. A color enhanced MRI image of a human head. (Photo courtesy of Siemens Medical Systems, Inc.)

field of the external magnet. Like a compass needle pointing the wrong way in the earth's magnetic field, the nuclear bar magnets soon return to their original aligned directions. As they do so, the rapid precession of the nuclear moments, as they flip over, induce an emf in the surrounding coil. This induced emf is the so-called echo of the pulse that first tipped the nuclei over. The emf can be amplified and displayed on an oscilloscope. Detailed analysis of the rate at which the nuclei return to equilibrium can reveal much about their atomic environments. Similar nuclei in different local environments have different resonant frequencies. By mapping the spatial location of nuclei with different environments, scientists can produce maps of the location of certain types of molecules. This technique is exploited in NMR whole-body imaging for medical diagnosis (now often called *magnetic resonance imaging* MRI).

31.5 Summary

When the current in a coil or other circuit element, including the circuit itself, changes in time, a self-induced emf is generated and is given by

$$\mathcal{E} = -L\frac{dI}{dt} \qquad (31.3)$$

where L is the **self-inductance** of the circuit element. The inductance of a coil of wire, such as a solenoid or toroid, is given by

$$L = \frac{\Phi_B}{I} \qquad (31.6)$$

where Φ_B is the *total* flux through the coil. If the coil has N turns, each with a flux ϕ_B, the total flux is $\Phi_B = N\phi_B$. The inductance of a solenoid with N turns, cross-sectional area A, and length l is given by

$$L = \frac{\mu_o N^2 A}{l}$$

When two circuits or circuit elements are close to each other, a changing current in one can induce an emf in the other. The emf induced in circuit element 2 from circuit element 1 is

$$\mathcal{E}_2 = -M_{21} \frac{dI_1}{dt}$$

where M_{21} is the **mutual inductance** on circuit element 2 by circuit element 1. The mutual inductance depends on the relative orientations of each circuit element as well as their own geometry. Moreover, $M_{12} = M_{21} = M$.

When an inductor, a resistor, a switch, and a source of emf are connected together in series and the switch is closed at time $t = 0$, the current in the LR circuit is given by

$$I(t) = \frac{\mathcal{E}}{R}(1 - e^{-t/\tau_L}) \tag{31.10}$$

where $\tau_L = L/R$ is the **inductive time constant.** If the source of emf and the switch are removed from the circuit and the free ends of the inductor and resistor are connected together, the current in the circuit is given by

$$I(t) = I_o e^{-t/\tau_L} = \frac{\mathcal{E}}{R}e^{-t/\tau_L} \tag{31.12}$$

The energy stored by an inductor L carrying current I is

$$U = \frac{1}{2}LI^2 \tag{31.13}$$

The energy per volume stored in a magnetic field B is

$$u_B = \frac{B^2}{2\mu_o} \tag{31.14}$$

The charge on the capacitor and the current through the inductor of an ideal (resistanceless) LC circuit oscillate harmonically with angular frequency

$$\omega_o = \frac{1}{\sqrt{LC}} \tag{31.19}$$

The charge and current are given by

$$Q(t) = Q_o \cos(\omega_o t) \tag{31.21}$$

and

$$I(t) = -\omega Q_o \sin(\omega_o t) \tag{31.22}$$

For small resistance R the current through the circuit and the charge on the capacitor of an RLC circuit exhibit damped harmonic motion, analogous to an underdamped harmonic oscillator.

PROBLEMS

31.1 Inductance

1. Show that the inductance of an ideal solenoid is proportional to the product of its volume and the square of the number of turns per length.

2. The current to a 2.50-H inductor increases from 1.20 A to 3.60 A in 2.00 s. What is the induced emf that appears across the inductor during this time?

3. (a) Compute the inductance of a 360-turn solenoid that has a 2.50-cm radius and is 12.0 cm in length. (b) What emf appears across the solenoid during the time that the current through it changes at a rate of 0.500 A/s?

4. A 500-turn solenoid has a radius of 4.00 cm and is 24.0 cm in length. At what rate must the current to the solenoid change for there to be a 16.0-V potential difference across it?

5. What is the total flux through a 400-turn solenoid when it carries a current of 360. mA if the self-inductance of the solenoid is 2.00 mH?

6. Show that the two equations for inductance L have the same units:

$$L = \frac{\Phi_B}{I} \quad \text{and} \quad L = -\frac{\mathcal{E}}{dI/dt}$$

7. At the same instant that the current through a 360-turn solenoid is 1.50 A and the induced emf across it is 20.0 mV, the rate of change of current is 4.20 A/s. What is the flux through *one turn* of the solenoid's wire?

8. Two inductors L_1 and L_2 are connected together in a series arrangement but are separated by a large enough distance that their mutual inductance is negligible. Show that the equivalent inductance for this arrangement is $L_{\text{tot}} = L_1 + L_2$.

9. Two inductors L_1 and L_2 are connected together in a parallel arrangement but are separated by a large enough distance that their mutual inductance is negligible. Show that the equivalent inductance for this arrangement is $1/L_{\text{tot}} = 1/L_1 + 1/L_2$.

10. A current $I(t) = I_o \sin(\omega t)$ flows through an 8.00-mH inductor. (a) If $I_o = 2.00$ A and $\omega = 60.0\pi$ rad/s, what is the induced emf across the inductor at $t = 0.280$ s? (b) What is the phase of $\mathcal{E}(t)$ with respect to $I(t)$?

★11. A toroidal coil with a square cross section consists of N turns of wire. If the inner and outer radii of the toroid are R_1 and R_2, respectively, show that the self-inductance of this toroid is

$$L = \frac{\mu_o N^2 h}{2\pi} \ln\left(\frac{R_2}{R_1}\right)$$

12. Two long, parallel wires each have radius R. Their centers are separated by a distance d. Assuming equal currents flow in opposite directions through the wires, compute the inductance per length of this pair of wires.

13. When the current in a solenoid is increased linearly from 2.00 A to 6.00 A in 0.500 s, an induced emf of 24.0 mV is observed in another solenoid. What is the mutual inductance between the two solenoids?

14. Two nearby conducting loops have a common axis. When the current changes in one loop at a rate of 3.60 A/s, an induced emf of 48.0 mV is observed in the other loop. What is the mutual inductance between the two loops?

15. Two coils have a 64.0-mH mutual inductance. If the current in one coil is given by $I(t) = at^2 - bt + c$, where $a = 5.00$ A/s^2,

$b = 2.00$ A/s, and $c = 3.00$ A, what is the induced emf in the other coil at $t = 0.500$ s?

16. In terms of x, w, and l what is the mutual inductance of the arrangement shown in Figure 31.P1.?

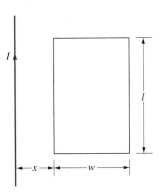

FIGURE 31.P1   Problem 16

31.2 *LR* Circuits

17. Show that the unit of L/R is the second.

18. By direct substitution show that Equation (31.10) is a solution to the differential equation Equation (31.9).

19. By direct substitution show that Equation (31.12) is a solution to the differential equation Equation (31.11).

20. The maximum current in an *LR* circuit like that of Figure 31.4 is 500. mA, and its time constant is 10.0 ms. If the source of emf in the circuit is 120. V, what is the inductance in the circuit?

21. Find the inductance in an *LR* circuit like that of Figure 31.4 if the potential drop across the 2.40 Ω resistor reaches half its final value 1.50 s after the switch is closed.

22. What resistance must be placed in series with a 24.0-V source of emf and a 2.50-H inductor that itself has a resistance of 20.0 Ω if the current is to reach 63.2% of its final magnitude in 0.0500 s?

23. An inductance $L = 270.$ mH and a resistance $R = 9.00$ Ω are connected in a series arrangement. The combination is connected across a 6.00-V source of emf. (a) What is the current through the resistor 15.0 ms after the combination is connected to the source of emf? (b) What is the current in the circuit 10 min after the combination is connected to the source of emf?

24. How much time passes after a series combination of a 12.0-H inductor and a 24.0-Ω resistor is connected to a 72.0-V source of emf until the current reaches (a) 50.0% of its final value, and (b) 80.0% of its final value?

25. For the *LR* circuit shown in Figure 31.4 suppose $L = 6.00$ H, $R = 12.0$ Ω, and $\mathcal{E} = 36.0$ V. (a) Find the current in the circuit at $t = 18.0$ ms after switch S is closed. (b) At what time is the potential difference across the resistor one-fourth of its maximum value?

26. How many time constants must pass before the current in the circuit shown in Figure 31.4 reaches 95.0% of its steady-state value?

27. The resistor in Figure 31.4 is 460. Ω, and the current in the circuit requires 84.0 μs to reach 90.0% of its final value. What is the inductance in the circuit?

28. (a) From Equation (31.10) find an expression for the *time rate of change* of the current in an *LR* circuit. If a 6.00-H inductor is connected in series with a 4.00-Ω resistor and a 36.0-V source of

emf, what is the potential difference across the inductor at (b) $t = 0.500$ s, and (c) $t = 2.00$ s?

29. For the LR circuit shown in Figure 31.4, $L = 3.00$ mH, $R = 12.0$ kΩ, and $\mathscr{E} = 24.0$ V. (a) Compute the emf across the inductor when the voltage drop across the resistor is 6.00 V. (b) What power is dissipated in the resistor at the instant when the emf across the inductor is 6.00 V?

30. The inductor L shown in Figure 31.4 has an inductance of 12.0 mH, and the resistor has a resistance $R = 16.0$ Ω. The emf is $\mathscr{E} = 24.0$ V, and at $t = 0$ the switch S is closed. (a) What is the current in the circuit at $t = 500.$ μs? (b) What power is dissipated in the resistor at $t = 1.00$ ms? (c) How much time passes until the current in the circuit reaches 90.0% of its final value?

31. For the circuit shown in Figure 31.P2 the inductance $L = 24.0$ mH, the resistance $R = 60.0$ Ω, and the emf $\mathscr{E} = 120.$ V. (a) What is the time constant for this circuit? (b) What is the steady-state value of the current? (c) After S_1 is closed for several minutes it is opened and simultaneously switch S_2 is closed. What is the current in the circuit 150. μs after switch S_2 is closed?

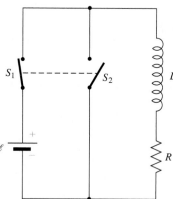

FIGURE 31.P2   Problems 31, 32, 33, 34, 35, 39, 40, and 42

32. In the circuit shown in Figure 31.P2 the inductance $L = 8.00$ H, the resistance $R = 36.0$ Ω, and the emf $\mathscr{E} = 12.0$ V. Switch S_1 is closed, and 10.0 min later switch S_2 is closed and simultaneously switch S_1 is opened. (a) What is the time constant for this circuit? (b) How long after switch S_2 is closed does the current reach the value it had 0.50 s after switch S_1 was initially closed?

33. For the circuit shown in Figure 31.P2, $L = 250.$ mH, $R = 5.00$ Ω, and $\mathscr{E} = 12.0$ V. At $t = 0$, switch S_1 is closed. At $t = 150.$ ms switch S_1 is opened and simultaneously switch S_2 is closed. What current passes through the resistor R at a time of 100. ms after switch S_2 is closed?

34. For the circuit shown in Figure 31.P2, $L = 24.0$ mH, $R = 8.00$ Ω, and $\mathscr{E} = 24.0$ V. At $t = 0.00$ s switch S_1 is closed. At $t = 1.50$ ms switch S_1 is opened and simultaneously switch S_2 is closed. What current passes through the resistor R at a time of 5.00 ms after switch S_2 is closed?

35. For the circuit shown in Figure 31.P2, $L = 240.$ mH, $R = 16.0$ Ω, and $\mathscr{E} = 6.40$ V. At $t = 0.0$ ms switch S_1 is closed. At $t = 32.0$ ms switch S_1 is opened and simultaneously switch S_2 is closed. How long does it take for the current to decay to 250. mA?

31.3 Energy and the Magnetic Field

36. Compute the energy stored in the B-field of a 250-turn ideal solenoid that is 30.0 cm in length if it has a cross-sectional area of 24.0 cm^2 and carries a current of 2.00 A.

37. What current must flow through a 500-turn ideal solenoid of 24.0-cm length and 2.00-cm radius for it to store 1.00 J of energy?

38. A 360-turn ideal solenoid carries a current $I = 2.40$ A that results in a magnetic flux of 5.40×10^{-3} T $\cdot$ m^2 through each turn of the solenoid. What energy is stored in the magnetic field of this solenoid?

39. Suppose $L = 2.20$ mH, $R = 10.0$ kΩ, and $\mathscr{E} = 12.0$ V in the LR circuit shown in Figure 31.P2. (a) What energy is stored by the inductor at $t = 0.100$ μs after switch S_1 is closed? (b) What energy is stored by the inductor when the current reaches its steady-state value?

40. In the circuit shown in Figure 31.P2, $L = 360$ mH, $R = 4.00$ Ω, and $\mathscr{E} = 6.00$ V. At time $t = 0.00$ ms the switch S_1 is closed. At $t = 50.0$ ms, find (a) the instantaneous power dissipated in the resistor, (b) the rate at which energy is stored by the inductor, and (c) the instantaneous rate of power supplied by the source of emf.

41. A 460.-mH inductor and a 20.0-Ω resistor are connected in series, and a 12.0-V source of emf is applied across the combination. What energy is stored in the inductor when the current reaches half its final value?

42. In the circuit shown in Figure 31.P2 the inductance $L = 3.00$ mH, the resistance $R = 12.0$ Ω, and the emf is $\mathscr{E} = 24.0$ V. (a) What energy is stored by the inductor at the instant the current reaches half its steady-state value? (b) After a sufficient time for the current in the circuit to reach its steady-state value switch S_1 is opened and simultaneously switch S_2 is closed. What energy is stored by the inductor at the instant when switch S_2 has been closed 0.150 ms?

★43. A long, straight wire with a 0.250-mm radius carries a steady current of 2.50 A. (a) What is the energy density of the magnetic field at the surface of the wire? (b) Compute the total energy stored in the magnetic field in a hollow, cylindrical region 36.0 cm long between the surface of the wire and a distance of 1.00 cm from the wire's axis.

44. Superconducting magnets can produce magnetic fields on the order of 10.0 T. Suppose you could fill a volume of space similar to your car's gas tank, say 0.500 m^3, with such a field. (a) How much energy could you store? (b) The heat of combustion of gasoline leads to an available energy density of 3.5×10^{10} J/m^3. How much gasoline would it take to store energy equivalent to your answer in part (a)?

45. A certain laboratory superconducting magnet creates a uniform magnetic field of 15.0 T in a volume of 20.0 cm^3. (a) How much magnetic energy is stored in this region of the field? (b) How much gasoline (heat of combustion 3.5×10^{10} J/m^3) would release an equivalent amount of energy?

31.4 LC Circuits (Optional)

46. An LC circuit consists of a 2.00-mH inductor and a 10.0-μF capacitor. (a) What is the period T of the oscillations? (b) If at $t = 0.00$ s the current is zero and the charge on the capacitor is 150. μC, what is the current in the circuit at $t = 0.250$ s?

47. What capacitance must be used in an LC circuit with a 20.0-mH inductor to cause current oscillations in the circuit with a frequency of 500. kHz?

48. The capacitance of a certain variable air capacitor can be changed from 120. pF to 360. pF. Over what range of broadcast frequencies can the resulting LC circuit be made to oscillate if a 0.250-mH inductor is connected to this variable capacitor?

49. A 36.0-V source of emf is used to charge a 100.-μF capacitor. The source of emf is removed, and at $t = 0.00$ s the ends of a 3.00-mH inductor are connected to the capacitor. (a) What charge is on the capacitor at $t = 0.500$ s? (b) What is the current in the capacitor at $t = 0.500$ s? (c) What is the energy stored by the capacitor at $t = 0.500$ s? (d) What energy is stored on the inductor at $t = 0.500$ s?

50. A 500.-μF capacitor is charged with a 120.-V source of emf. After the source of emf is removed the capacitor is connected to a 0.240-mH inductor. Assume an ideal LC circuit model. (a) What is the maximum current in the circuit? (b) What is the maximum energy stored on the inductor? (c) What is the emf induced across the inductor 50.0 s after the capacitor is connected to the inductor?

51. A 1.00×10^3 μF capacitor is charged with a 36.0-V source of emf. The capacitor is then removed from the source of emf, and its ends are connected to a 24.0-mH inductor. Compute the following quantities at $t = 250$. ms after the connection is made: (a) the charge on the capacitor; (b) the current through the inductor; (c) the energy stored by the capacitor; (d) the emf induced across the inductor.

52. A 96.0-V source of emf is used to charge a 25 000-pF capacitor. After the source of emf is removed the capacitor is connected to a 30.0 mH-inductor. Compute the following quantities at $t = 7.50$ μs after the connection is made: (a) the energy stored on the capacitor; (b) the energy stored on the inductor; (c) the emf induced across the inductor.

53. At $t = 0$ the charge Q_o on the capacitor of an ideal LC circuit is a maximum. (a) If the circuit oscillates with a frequency f, find the time at which the energy stored on the inductor is equal to that stored on the capacitor. (b) What charge is stored on the capacitor at this time?

54. What is the maximum resistance an RLC circuit can have and exhibit damped oscillations (and, therefore, not be overdamped) if $L = 3.0$ μH and $C = 100$ μF?

55. What is the frequency of the damped oscillations for the circuit shown in Figure 31.12 if $R = 450.$ Ω, $L = 24.0$ mH, and $C = 0.100$ μF?

Numerical Methods and Computer Applications

56. Write a program or construct a spreadsheet template that plots $I(t)$ for a time duration of $3\tau_L$ of an LR circuit. Try out your program or spreadsheet with the values of L, R, and $\mathcal{E}$ from Example 31.4.

57. Write a program or construct a spreadsheet template that plots graphs, such as those of Figures 31.10 and 31.11, for given values of L, C, and Q_o. Try out your program or spreadsheet with the values of L, C, and Q_o given in Example 31.6.

General Problems

58. Compute the self-inductance of a length l of a coaxial cable if the inner conductor is a thin, cylindrical conductor of radius R_1 and the outer conductor is a thin cylindrical shell of radius R_2. [*Hint:* Find the energy stored and equate it to $LI^2/2$.]

★59. Compute the mutual inductance between the toroid and rectangular loop described in Problem 30.13 and pictured in Figure 31.P3.

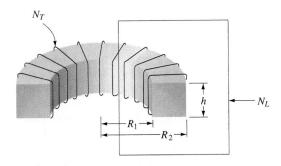

FIGURE 31.P3   Problems 59 and 68

60. An ideal 24.0-cm long solenoid is wound with 500. turns around a radius of 2.00 cm (the primary). A second coil (the secondary) is wrapped around the solenoid and has 80.0 turns. (a) What is the mutual inductance of the coils? (b) If the current is changing at a rate of 4.80 A/s in the primary, what emf is induced across the secondary? (c) If the current is changing at a rate of 4.80 A/s through the secondary, what is the emf induced across the primary?

61. Near the earth's north magnetic pole the average strength of the magnetic induction field is about 6.0×10^{-5} T, whereas in a region of the Pacific Ocean the average B is as small as 2.5×10^{-5} T. (a) What energy is stored in one cubic meter of the earth's magnetic field in each of these two regions? (b) The local magnetic field near the earth's surface can be as high as $30. \times 10^{-5}$ T due to local geological formations. What is the energy density stored in the B-field in such a region?

62. Figure 31.P4 shows the current $I(t)$ through a 0.0500-H ideal (resistanceless) inductor. Construct a graph of the voltage across the inductor $\mathcal{E}(t)$. Is the inductor a "differentiating circuit?" Explain.

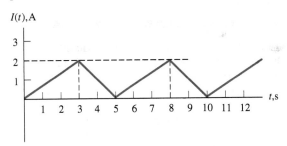

FIGURE 31.P4   Problem 62

63. Show that a toroid with N turns, a mean radius R, and a cross-sectional area A with a diameter that is very small compared to the radius R, has an approximate self-inductance given by

$$L \approx \frac{\mu_o N^2 A}{2\pi R}$$

(*Hint:* Assume the B-field inside the toroid can be approximated by that of a solenoid.)

64. Two adjacent circuits have a mutual inductance of 0.240 mH. The current in one circuit is given by $I(t) = I_o \sin[(3.00\pi \text{ rad/s})t]$. For proper operation, the maximum induced emf in the second circuit, which is caused by the varying current in the first circuit, cannot exceed 800. μV. What maximum current I_o can be permitted in the first circuit?

65. A toroidally shaped inductor has a mean radius of R and a cross-sectional area A. The shape is such that $B \approx \mu_o I/(2\pi R)$ is approximately constant throughout the volume of the inductor. It is wound with two coils, one with N_1 turns and the other on top of the first with N_2 turns. Find the mutual inductance between the two windings.

66. Two resistors, an inductor and a source of emf are connected with two switches as shown in Figure 31.P5. Their values are $R_1 = 4.00$ kΩ, $R_2 = 2.00$ kΩ, $L = 18.0$ mH, and $\mathscr{E} = 36.0$ V. At time $t = 0.00$ μs switch S_1 is closed. (a) What is the current in the circuit at $t = 4.00$ μs? (b) After a sufficient time for the current in the circuit to reach its steady-state value, switch S_1 is opened while switch S_2 is simultaneously closed. What is the current through the resistor R_2 10.0 μs after switch S_2 is closed?

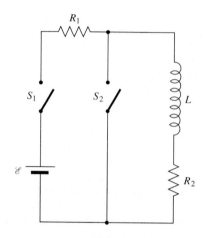

FIGURE 31.P5   Problems 66 and 67

67. For the circuit shown in Figure 31.P5, $R_1 = 8.00$ kΩ, $R_2 = 16.0$ kΩ, $L = 48.0$ mH, and $\mathscr{E} = 96.0$ V. At time $t = 0.00$ μs switch S_1 is closed. (a) What is the current in the circuit at $t = 1.50$ μs? (b) If at $t = 1.50$ μs switch S_1 is opened while switch S_2 is simultaneously closed, what is the potential difference across the resistor 2.00 μs after switch S_2 is closed?

★68. The rectangular toroid shown in Figure 31.P3 has 360 turns, an inner radius $R_1 = 6.00$ cm, an outer radius $R_2 = 12.0$ cm, a height $h = 4.00$ cm, and carries a uniform current of 2.50 A. (a) Apply Equation (31.13) and the result of Problem 11 to compute the total energy stored by this inductor. (b) By a direct integration of Equation (31.14) verify that the result to part (a) is indeed the total energy stored in the magnetic field B of the toroid.

★69. A long, straight copper rod of 1.00-cm diameter carries a current of 16.0 A. Assume the current density within the bar is uniform and compute the total energy stored in the magnetic field *within* the volume of a 1.00-m length of the bar (that is, within a 1-m long region that is bounded by the surface of the bar).

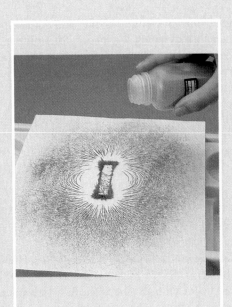

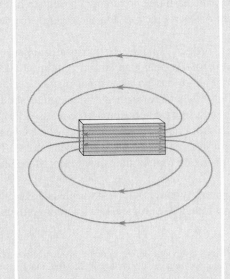

Magnetic Properties of Materials (Optional)

In this chapter you should learn

- about the atomic sources of the permanent and induced dipole moments.

- the definitions of magnetic field intensity and magnetization.

- the magnetic characteristics of materials in the diamagnetic, paramagnetic, and ferromagnetic phases.

For many of us our first introduction to magnetism came while playing with bar magnets. The apparent action-at-a-distance that magnets display holds fascination for children and adults and has provided a basis for numerous "executive toys" found in many offices. Magnetic materials play vital roles in modern society. These roles range from pole pieces for electric motors and generators to coatings that record information on audio tapes and floppy disks. We will discuss magnetic materials from two distinct points of view. On one hand we will characterize materials according to their macroscopic magnetic properties without regard to the causes of the magnetism. At the other extreme we will look at the atomic level to discover the fundamental causes for the different macroscopic magnetic behavior of different materials.

32.1 Overview of the Magnetic Properties of Matter

To understand the magnetic properties of bulk matter we must understand a long and sometimes intricate chain of interactions that begin with the electron. In a fundamental sense all the important bulk magnetic properties of matter come from electrons.[1] Next in the chain come atoms or molecules with magnetic properties that represent the combined effects of their many electrons. When molecules combine to form a solid, the solid's magnetic properties arise from combinations of the magnetic properties of the constituent molecules. The solid material can be uniform throughout. However, in materials with strong magnetic properties, it is more typical for the solid to form small microcrystalline regions, each with its magnetic field oriented in a different direction. In such cases the macroscopic behavior of the object itself depends on the relative sizes and orientations of all its microcrystalline regions. This magnetic structure is influenced by many factors, including the shape and the magnetic history of the object. As you can see, to do justice to the subject of magnetic properties of matter would take a course in itself. We content ourselves with a description of a few relatively simple models that account for the most common magnetic behavior in materials.

To reiterate an important point, electrons are the primary source of the magnetic properties of matter. Electrons contribute to the magnetic properties of a material in two ways: (1) each electron itself has an **intrinsic magnetic dipole moment**; that is, the electron acts as if it were a tiny bar magnet. This intrinsic magnetic moment is proportional to the electron's intrinsic angular momentum, the so-called *spin* of the electron. In addition, the intrinsic magnetic dipole moment of the electron and the electron's intrinsic angular momentum vector are exactly antiparallel. (2) Any electron that has a nonzero value of orbital angular momentum about the nucleus creates an electric current and hence creates an **orbital magnetic dipole moment**. The net magnetic moment of the atom is the vector sum of the *intrinsic* magnetic moment of each electron and the *orbital* magnetic moment of each electron.

In some atoms all of the electrons are "paired" so that the intrinsic moments of electrons in similar orbits are antiparallel. If the total orbital angular momentum of all the electrons of this type of atom is also zero, the atom has no permanent magnetic dipole moment. On the other hand, some atoms have what are termed "incomplete shells" containing several unpaired electrons with aligned spins. Such atoms have relatively large **permanent** magnetic dipole moments. When atoms bond into molecules, the magnetic moment of the molecule may be different from that of any of the constituent atoms. The point is that atoms and molecules may have magnetic dipole moments, or not, depending on their atomic structure.

The difference in magnetic behavior of materials composed of molecules with permanent magnetic dipoles and those without is profound. When molecules with a permanent

---

[1] The nuclei of some atoms also carry a net magnetic moment, resulting from the intrinsic magnetic moments of neutrons and protons. However, nuclear magnetic moments are generally about one thousand times smaller than the moments due to the electrons.

dipole moment are placed in an external magnetic field, the dipoles tend to align with that field much as the magnetic dipole of a compass needle aligns with the earth's field. However, when a molecule with no permanent dipole moment is placed in an external field, the molecule *develops* a magnetic moment. Such a moment is called an **induced magnetic dipole moment**. What is most significant about this behavior is that this induced magnetic dipole is *antialigned* with the external field. (Note that this behavior is exactly opposite from that of induced electric dipoles that are created aligned with the external electric field.) Hence, there are two fundamental types of magnetic dipoles, permanent and induced, which exhibit opposite behavior in externally applied fields. The macroscopic magnetic behavior of materials depends on the type of dipoles it contains and the degree of interaction between nearby dipoles.

When neighboring magnetic dipoles of a material do not interact strongly, the magnetic behavior is relatively simple. In the case of permanent dipoles with small magnetic dipole moments, thermal motions cause the moments to be randomly oriented. The material exhibits no bulk magnetism in the absence of an externally applied field. In the presence of an applied magnetic field these dipole moments exhibit weak alignment with the applied field. This alignment tends to make the *B*-field inside the material greater than it would be if the material were not present. As the temperature of a material is decreased the degree of dipole alignment increases, but the overall magnetic effect remains relatively small. This type of behavior is known as **paramagnetism**, and the material is said to be paramagnetic, or more precisely in a **paramagnetic phase**.

When there are no permanent dipoles in the material, clearly there is no bulk magnetic effect in the absence of an externally applied field. When an external field is imposed, the induced dipoles created by the external field are antialigned with the external field, causing the *B*-field inside the material to be weaker than it would be if the material were not present. It is as if the applied magnetic field were being driven from the inside of the material. This behavior is known as **diamagnetism**. Although all atoms exhibit diamagnetism, a material is said to be diamagnetic only in those cases in which the diamagnetism is not overshadowed by paramagnetic or ferromagnetic effects.

All of the materials we normally refer to as "magnets" are, in fact, **ferromagnets**. Such materials contain strongly interacting permanent magnetic dipoles. The strong interaction between neighboring dipoles causes a spontaneous ordering of the dipoles even in the absence of an external field. In the typical case not every dipole throughout the entire

This 180-ton superconducting magnet, built at Argonne National Laboratory, is one of the world's largest. The magnet is used for advanced energy research in magnetohydrodynamics (see Section 28.1). A thorough understanding of the magnetic properties of materials is required for the design of such magnets. (Photo courtesy of Argonne National Laboratory.)

macroscopic object becomes aligned. Rather, dipoles within small regions of the material become aligned. These small regions are known as **domains**. The direction of alignment is different in neighboring domains. Even within a domain the thermal motion prevents the alignment from becoming complete. We will have more to say about domains in Section 4 of this chapter.

In some cases the lack of alignment between dipoles in neighboring domains results in the cancellation of the overall magnetic effect. In this case a bulk sample of ferromagnetic material may not act like a bar magnet at all. By placing such a ferromagnetic material in an external field it is possible to cause domains with dipole alignments in one direction to grow in size at the expense of antialigned domains. In this way the bulk sample can be made to resemble a large dipole; that is, act like a bar magnet. If, however, you repeatedly strike an ordinary iron bar magnet with a hammer, it "loses" its magnetism. In fact, its domains are still magnetized, but the domains become random in size and orientation.

As we have pointed out, if a material contains permanent dipoles, it can exist in either a paramagnetic or a ferromagnetic state. Which of the two is the actual state depends on a competition between magnetic forces tending to align the dipoles and thermal motion tending to randomize the dipole alignment. As you might expect, at higher temperatures thermal motions ultimately win, and a ferromagnet becomes paramagnetic. This transition from ferromagnet to paramagnet is often abrupt, the phase transition occurring at a special temperature known as the **Curie temperature**. The Curie temperature is different for different materials and depends on the detailed nature of the interactions of the magnetic dipoles.

You should be aware that there are many other more exotic forms of magnetic ordering in solids than simple alignment of adjacent dipoles. Some crystals exhibit an **antiferromagnetic** phase wherein adjacent dipole moments point in *opposite* directions. Some substances, like the rare-earth elements, crystallize into structures with many different kinds of unusual magnetic behaviors.[2] Depending on temperature, the magnetic moments in these crystals can perform such feats as switching directions between planes of atoms or spiraling along the surface of imaginary cones. Moreover, changing the temperature can cause them to suddenly change orientations. But enough about all that. We must content ourselves with a description of the three basic magnetic phases: diamagnetism, paramagnetism, and ferromagnetism. We begin with a closer look at the dipole moments associated with the electron.

32.2 The Source of Magnetism in Materials

Concept Question 1
If the electron had no intrinsic magnetic dipole moment, what would be the dominant type of magnetic material?

As we mentioned previously, the magnetic properties of matter stem primarily from the electrons of atoms. We do not attempt a detailed treatment of the quantum mechanics of electrons in solid materials here. Instead, we rely on a few fundamental ideas about the quantum mechanics of atoms, such as you have no doubt encountered in your chemistry classes. For our purposes the important result we need from quantum mechanics is that certain dynamical variables, such as energy and angular momentum, are not continuous but rather come only in certain discrete values. Quantum mechanics tells us that the state of an electron in an atom can be characterized by a set of quantum numbers n, l, m_l, s, and m_s. The principal quantum number n is associated with the total electron energy and can take on integer values from $n = 1$ to infinity. Electrons with the same principal quantum number are said to occupy the same **shell**. The quantum number l is the orbital angular momentum quantum number that can take on integer values $0 \geq l \geq n - 1$. The orbital

---

[2] The rare-earths are those elements in the next to the last line of the periodic table but which are always shown at the bottom; they run from lanthanum to lutetium and are also referred to as the lanthanide series. For this series of elements electrons are being added to the $4f$ ($n = 4$, $l = 4$) subshell.

angular momentum L_{orb} of the electron depends on the quantum number l through the relation

$$L_{\text{orb}} = \sqrt{l(l + 1)}\, \frac{h}{2\pi}$$

where h is called **Planck's constant**, and its value is

$$h = 6.626 \times 10^{-34}\,\text{J} \cdot \text{s}$$

The combination $h/2\pi$ appears quite frequently in quantum-mechanical applications and is abbreviated by the symbol $\hbar$, pronounced "h-bar,"

$$\hbar = \frac{h}{2\pi} = \frac{6.626 \times 10^{-34}\,\text{J} \cdot \text{s}}{2\pi} = 1.055 \times 10^{-34}\,\text{J} \cdot \text{s}$$

Thus, the magnitude of the orbital angular momentum can take on only values

$$L_{\text{orb}} = \sqrt{l(l + 1)}\, \hbar$$

One of the more remarkable predictions of quantum mechanics is that the direction of the angular momentum is also quantized. In fact when a measurement of one component of angular momentum is made, say along the z-direction, the only possible results have values

$$L_{\text{orb}_z} = m_l \hbar$$

where the so-called **magnetic quantum number** m_l takes on integer values $-l \leq m_l \leq +l$. The symbol m_l is known as the magnetic quantum number because measured components of the magnetic dipole moment μ_{orb_z} due to the electron's orbital motion can take on only values

$$\mu_{\text{orb}_z} = m_l \mu_B \tag{32.1}$$

where μ_B is called the **Bohr magneton**:

$$\mu_B = \frac{e\hbar}{2m_e} = 9.274 \times 10^{-24}\,\text{A} \cdot \text{m}^2$$

Here m_e is the mass of the electron, and e is the magnitude of the charge of the electron. The Bohr magneton is named after Niels Bohr (1885–1962), who in 1913 proposed the first successful model of the atom employing quantized angular momentum. By adding a quantum condition to a simple planetary model of an electron orbiting the nucleus, Bohr's model predicts that the orbital angular momentum of the electron gives rise to a magnetic moment of exactly the size indicated by Equation (32.1). (See Problem 1.)

The final quantum numbers s and m_s are associated with the electron itself. The electron has an intrinsic angular momentum of magnitude

$$L_{\text{spin}} = \sqrt{s(s + 1)}\, \hbar$$

where s is a quantum number the only possible value of which for the electron is $\frac{1}{2}$. The intrinsic angular momentum is often referred to as the **spin** of the electron. The name comes from imagining the electron as a small sphere rotating on its axis. This mental picture is sometimes helpful but is in fact much too naive. As of this writing no one has yet detected a finite radius for the electron; it may be a true point-particle. Like the orbital angular momentum, only one component of the spin angular momentum can be measured.

Allowed values are

$$L_{\text{spin}_z} = m_s \hbar$$

where $m_s = \pm\frac{1}{2}$. The two values $+\frac{1}{2}$ and $-\frac{1}{2}$ are often referred to as ''spin up'' and ''spin down.'' The quantum number m_s is also associated with an intrinsic magnetic dipole moment of the electron through

$$\mu_{\text{spin}_z} = 2.0023 m_s \mu_B$$

It is perhaps worthy of note that, because $m_s = \pm\frac{1}{2}$, the magnetic dipole moment due to electron spin is approximately the same magnitude as the orbital dipole moment of an electron with $m_l = 1$.

We now see that there is an intimate connection between the angular momentum and the magnetic dipole moment for both the orbital motion and the intrinsic spin of the electron. We can summarize this connection in a single expression

$$\boldsymbol{\mu}_z = -g\left(\frac{\mu_B}{\hbar}\right)\mathbf{L}_z \tag{32.2}$$

which relates the measurable component of the magnetic moment to the measurable values of the angular momentum. The negative sign in this vector equation is due to the negative charge of the electron and reminds us that the magnetic moment and the angular momentum point in opposite directions. The factor g is known as the **gyromagnetic ratio**, or **Landé g factor**. For orbital motion $g = 1$; for electron spin g is approximately 2. It is possible to write a proportion between angular momentum and magnetic moment for an entire atom in which the orbital angular momenta and spins of all the atom's electrons combine, often in a quite complicated fashion. In such cases the gyromagnetic ratio varies between 1 and 2 and depends on the relative importance of orbital and spin moments.

In order to calculate the total magnetic dipole moment of an atom, we must combine the orbital and spin magnetic moments of all the atom's electrons. Unfortunately, for many-electron atoms this vector addition often leads to a rather involved calculation using quantum mechanics. The calculation is simplified by the fact that the magnetic moments of electrons filling closed shells of an atom sum to zero. Thus, only electrons in partially filled shells contribute to the atom's permanent magnetic dipole moment. The results of such computations, as well as experiments, show that atomic magnetic moments are generally on the order of a few Bohr magnetons. For example, the magnetic moment of an iron atom is about 2.2 μ_B. The magnetic moments of the other elements with permanent dipole moments are given in Table 32.1.

Concept Question 2
What does it mean to say a quantity is quantized?

Concept Question 3
If the quantum of angular momentum were macroscopic, what would it feel like to set a toy gyroscope spinning by pulling on a string wrapped around its axle? What effect would you notice when you set one end of the gyroscope axle down on a tabletop?

TABLE 32.1   Magnetic Dipole Moments of Selected Elements

ELEMENT	MOMENT (IN UNITS OF μ_B)
Iron	2.2
Cobalt	1.7
Nickel	0.61
Gadolinium	7.1
Terbium	5.0
Dysprosium	10.0
Holmium	8.5

32.3 The Magnetic Field Vectors

In the previous section we concentrated on the atomic origins of magnetic dipole moments. Throughout the remainder of this chapter we are more concerned with the *macroscopic* behavior of magnetic materials, rather than their *microscopic* origin. To begin we need to establish some macroscopic way to describe the magnetic properties of a material.

The magnetic state of a bulk piece of material is characterized by its **magnetization M**, which is defined as the *magnetic dipole moment per unit volume*. Physically, the magnetization depends on both the magnitude of the individual atomic dipole moments and their degree of alignment. Magnetization is large when atoms with dipole moments are present and these dipoles exhibit a large degree of alignment. The definition of magnetization implies that the magnetic moment $d\boldsymbol{\mu}$ associated with a volume element dV of a material with a magnetization $\mathbf{M}$ is

$$d\boldsymbol{\mu} = \mathbf{M}\,dV$$

so that the average dipole moment $\boldsymbol{\mu}_{av}$ over some volume v is

$$\boldsymbol{\mu}_{av} = \int_v \mathbf{M} \, d\mathcal{V} \tag{32.3}$$

When we apply a magnetic field to a material, the *net* magnetic field within the material depends on both the applied field and the magnetization of the material. To discover how to describe the magnetic field within a material let's consider a very long solenoid with cross-sectional area A that contains a cylindrically shaped piece of material also with a cross-section A and magnetization $\mathbf{M}$. In order to avoid any complications from effects near the ends of the solenoid we suppose that our material is isotropic and that it completely fills the solenoid, which is quite long compared to its diameter. With these assumptions we can be confident that the B-field inside the solenoid is uniform. Figure 32.1 shows our material in the solenoid and also provides a model for the individual magnetic dipole moments, which combine to form the magnetization of the material. Each dipole moment represents the average moment over some volume $\Delta\mathcal{V}$, which is large compared to individual atoms but small compared to the size of the solenoid. Thus, each dipole moment we speak of here is the local average of a good many atoms so that we have smoothed out the graininess due to the discrete nature of atomic dipoles. We can think of these fictitious dipole moments as if each were created by a small current loop i_m as illustrated in Figure 32.1. Focus your attention on two adjacent current loops on the interior of the material. Note that where the currents of adjacent loops are closest together the currents move in opposite directions and therefore cancel. The beauty of this model is that, in order to compute the magnetization $\mathbf{M}$, we need only consider the net current I_M around the outermost curved surface (where, of course, individual currents do not cancel). The magnetic moment created by the (fictitious) outer current loop is calculated as usual by taking the product of the current and the area of the loop. If the cross-sectional area of the solenoid is A, the magnitude of the net magnetic dipole moment is $I_M A$. Let's consider

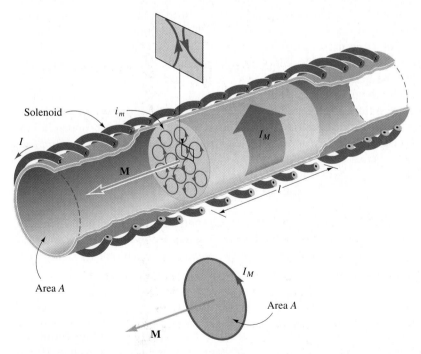

FIGURE 32.1

A long solenoid of cross-sectional area A is filled with a material in the paramagnetic or ferromagnetic phase with magnetization $\mathbf{M}$ parallel to the solenoid axis. We look at a small section of length l near the center of the solenoid. Within the material, adjacent microscopic currents i_m cancel so that the magnetization $\mathbf{M}$ can be computed from the net surface current I_M. This magnetization current I_M is distinct from any real current flowing through the wires surrounding the material.

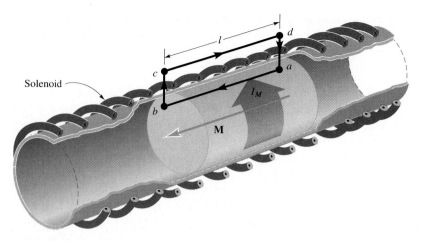

FIGURE 32.2

Consider a rectangular, closed path *abcda* of length *l*, parallel to the axis of the solenoid. The total current passing through the closed path is the real current from the solenoid wires plus the magnetization current I_M.

a slice of the material perpendicular to the axis of the solenoid. If the length of the slice is *l*, then the volume of the material is *Al* so the magnitude of the magnetization is

$$M = \frac{Dipole\ moment}{Volume} = \frac{I_M A}{Al} = \frac{I_M}{l} \qquad (32.4)$$

You can verify from an application of the right-hand rule to Figure 32.1 that the direction of this magnetization is along the axis of the solenoid.

In the preceding paragraphs we modelled the magnetic field due to the internal atomic dipoles of the material filling the solenoid as if it were instead due to a fictitious current circulating around the outside surface of the material. This fictitious current I_M is known as the **magnetization current**, or **Ampèrian current**. Thus, we reduce the problem to one we already know how to solve. In order to find the magnetic field inside the filled solenoid, we calculate the field as if the solenoid were empty but surrounded by two currents: (1) the real current flowing in the wires of the actual solenoid and (2) the fictitious magnetization current flowing around the material surface representing the average effect of the dipoles internal to the material. To calculate the field we apply Ampère's law, Equation (30.10). Because there is no *electric* field inside the solenoid, the term $\epsilon_o(d\phi_E/dt)$ in Ampère's law is zero,[3] leaving us with

$$\oint \mathbf{B} \cdot d\mathbf{l} = \mu_o I_{encl}$$

As shown in Figure 32.2 we apply this law to the rectangular loop of arbitrary length *l*. In accordance with our plan, we take the total current enclosed by the loop *abcda* as the sum of the *real* current of the solenoid I_r and the *magnetization* current I_M. Thus, Ampère's law becomes

$$\oint \mathbf{B} \cdot d\mathbf{l} = \mu_o(I_r + I_M) \qquad (32.5)$$

This form of Ampère's law is not particularly useful as it contains the fictitious current I_m. To do better we must replace I_m with an expression involving the magnetization *M*. For example, we can solve Equation (32.4) for I_M to obtain

$$I_M = Ml \qquad (32.6)$$

---

[3] You may remember that this term is often called the *displacement current*. The displacement "current" is not literally a current. The term is merely to remind us that changing electric fields have the same effect as real current. The fictitious displacement current has nothing to do with the fictitious magnetization current.

Notice that the units of **M** in SI must be amperes per meter. We derived Equation (32.6) for the special case in which **M** points along the axis of the solenoid. For the general case it can be shown that

$$I_M = \oint \mathbf{M} \cdot d\mathbf{l} \tag{32.7}$$

You should verify that Equation (32.7) indeed reduces to Equation (32.6) for the path *abcda*. (Remember, **M** = 0 outside the material and $\mathbf{M} \perp d\mathbf{l}$ along the paths $b \rightarrow c$ and $d \rightarrow a$.)

In principle, if we know the material's magnetization, we can combine Equation (32.6) with Ampère's law to obtain the *B*-field inside the material. It is not particularly helpful to do this because the magnetization is not a fundamental property of the material. In fact, the magnetization depends on *B* itself, among other things. Instead, what we desire is to find some relation for the magnetic field which depends on the material only. To this end we substitute Equation (32.7) into Ampère's law, as written in Equation (32.5), to obtain

$$\oint \mathbf{B} \cdot d\mathbf{l} = \mu_o \left(I_r + \oint \mathbf{M} \cdot d\mathbf{l} \right)$$

Because both integrals are taken over the same path, we can combine them to write

$$\oint \left(\frac{\mathbf{B} - \mu_o \mathbf{M}}{\mu_o} \right) \cdot d\mathbf{l} = I_r \tag{32.8}$$

We can write Equation (32.8) in a simpler form if we define a new vector called the **magnetic field intensity H**

$$\mathbf{H} = \frac{\mathbf{B} - \mu_o \mathbf{M}}{\mu_o} \tag{32.9}$$

With this definition Ampère's law becomes

$$\oint \mathbf{H} \cdot d\mathbf{l} = I_r \tag{32.10}$$

This alternative form of Ampère's law states that the magnetic field intensity **H** is associated with only the *real* current I_r. The magnetic field intensity is also called the **magnetic field strength**. You need to keep the names of **H** and **B** straight. Our old friend **B** is the more fundamental magnetic field and is called the **magnetic flux density**, or the **magnetic induction**. To keep things clear we refer to **B** as *the* magnetic field and refer to this new field simply as the **H**-field. We can solve Equation (32.9) for the magnetic field **B**:

$$\mathbf{B} = \mu_o \mathbf{H} + \mu_o \mathbf{M} \tag{32.11}$$

The *B*-field is associated with *all* the current sources, real and magnetization. It is still this field we use if we want to compute the force on a moving charge or the torque on a rotated magnetic dipole. The introduction of the **H**-field is nonetheless convenient because it is determined by the external current I_r, a current we can often control.

For materials in the paramagnetic or diamagnetic phases we find experimentally that the magnetization is approximately proportional to the magnetic field intensity **H**, as long as the magnitude of **H** is not too large. The factor relating **M** and **H** is called the **magnetic susceptibility** χ:

$$\mathbf{M} = \chi \mathbf{H} \tag{32.12}$$

Concept Question 4
We know that *B*-field lines always form continuous, unbroken loops. What can you say about lines of *M*? What about lines of *H*?

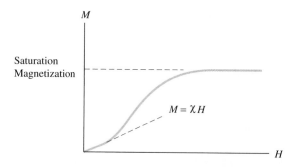

FIGURE 32.3

The internal magnetization of a ferromagnetic material as a function of the *H*-field created by real currents. Note that for large *H* the magnetization reaches a maximum value known as the saturation magnetization.

TABLE **32.2**   Magnetic Susceptibilities of Some Materials at 300 K

MATERIAL	χ
Paramagnetic	
Aluminum	2.3×10^{-5}
Calcium	1.9×10^{-5}
Chromium	2.7×10^{-4}
Lithium	2.1×10^{-5}
Magnesium	1.2×10^{-5}
Niobium	2.6×10^{-4}
Oxygen*	2.1×10^{-6}
Platinum	2.9×10^{-4}
Tungsten	6.8×10^{-5}
Diamagnetic	
Bismuth	-1.7×10^{-5}
Copper	-9.8×10^{-6}
Diamond	-2.2×10^{-5}
Gold	-3.6×10^{-5}
Lead	-1.7×10^{-5}
Mercury	-3.2×10^{-5}
Nitrogen*	-5.0×10^{-9}
Silver	-2.6×10^{-5}
Sodium	-2.4×10^{-7}
Silicon	-4.2×10^{-7}

*at STP

When the material is in a paramagnetic phase, χ is positive and **M** is parallel to **H**. When the material is in a diamagnetic phase, χ is negative and **M** is directed opposite from **H**. The magnetic susceptibility χ provides a quantitative measure of how susceptible a material is to being magnetized when placed in an external field of intensity **H**. Magnetic susceptibility is a function of temperature. Values of susceptibility for several materials at 300 K are given in Table 32.2. For a material in a ferromagnetic phase M is not simply proportional to H, so that Equation (32.12) is not particularly useful; χ is not a constant. When the magnitude of **H** is so large that all the moments within a material are aligned, no additional increase in H can contribute to the magnitude of the magnetization as shown in Figure 32.3. The material is said to be **saturated**, and the magnetization is called the **saturation magnetization**. In addition to its usual dependence on temperature, χ has different values depending on H itself and on the history of the material. By "history" we mean such things as how the material was prepared and the magnitude and direction of previous applications of **H**. We will have more to say about this situation later in Section 32.4.

It is often useful to relate the actual *B*-field inside our material directly to the *H*-field that we create with external currents. We can accomplish this by eliminating **M** from Equation (32.11) by substituting its equivalent from Equation (32.12):

$$\mathbf{B} = \mu_o(\mathbf{H} + \mathbf{M}) = \mu_o(\mathbf{H} + \chi\mathbf{H}) = \mu_o(1 + \chi)\mathbf{H} \qquad (32.13)$$

It is convenient to define the **permeability** μ of the material:

$$\mu = \mu_o(1 + \chi)$$

With this definition Equation (32.13) becomes simply

$$\mathbf{B} = \mu\mathbf{H} \qquad (32.14)$$

From Table 32.2 it should be clear that μ is always very nearly equal to μ_o for materials in the paramagnetic or diamagnetic phases. For substances in a ferromagnetic phase μ also depends not only on the material and temperature but also on the history of the material.

Concept Question 5
Equation (32.14) implies that $B = 0$ whenever $H = 0$. This is certainly not true for permanent magnets. What step in the derivation of Equation (32.14) does not apply to permanent magnets?

32.4 Diamagnetic, Paramagnetic, and Ferromagnetic Phases

In this section we describe the properties of the three magnetic phases: diamagnetism, paramagnetism, and ferromagnetism.

Diamagnetism

When placed in an applied magnetic field, materials in the diamagnetic phase have induced magnetic moments with directions opposing that of the applied field. We can understand how such a moment can be induced by considering the simple model shown in

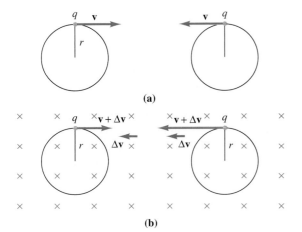

(a)

(b)

FIGURE 32.4

A model of diamagnetism. Two charges q each travel in circular paths of radius r with velocity v. (a) In the absence of an applied magnetic field the net moments cancel. (b) Lenz's law tells us that during the time when a magnetic field into the page is increasing, an E-field is induced such that $\oint \mathbf{E} \cdot d\mathbf{l}$ is positive when calculated around the orbits in the anticlockwise direction. The torque on the charge pictured in the left orbit decreases that charge's angular momentum while the angular momentum of the charge on the right increases.

Figure 32.4. This figure shows two positive charges q orbiting with circular paths each with the same radius r but with opposite orbital directions. Although electrons provide the induced moment in atoms, we choose to consider positive charges in our model so as to make the induced fields simpler to follow. When no external magnetic field is present, the magnetic moments of these two orbits cancel. When a magnetic field is applied into the page (Fig. 32.4(b)) Lenz's law tells us that in this case an induced current is generated in a direction so as to oppose the increase in magnetic field directed. An application of the right-hand rule shows that in both cases the direction of this induced current is anticlockwise. (When your right thumb points out of the page to oppose the increasing magnetic field, your fingers curl anticlockwise.) In the case of the charge already circulating anticlockwise the increased current can be created by the charge speeding up. If the charge moving clockwise slows down, the magnetic effect is the same as an increase in anticlockwise current.

Here's the point: The smaller clockwise current produced by the charge on the left results in a smaller magnetic moment into the page. The greater anticlockwise current created by the charge on the right results in a larger magnetic moment out of the page. In each case the *change* in the magnetic moment is opposite to the applied field and is directed out of the page. Hence, the two oppositely directed magnetic moments of these charges no longer cancel, and instead a net magnetic moment that opposes the applied field has been induced. You should verify that the induced moment still opposes the applied field if the direction of this field is opposite that shown in Figure 32.4(b); that is, out of the page.

Atoms that have filled electronic shells have no net angular momentum and, therefore, no permanent magnetic moment. However, the orbital motion of all the electrons contributes to diamagnetism, even those in filled shells. Thus, all materials are diamagnetic. For substances in the ferromagnetic phase the magnetic behavior of the material is dominated by the interactions of the large permanent magnetic dipole moments, and any diamagnetism is negligible. Other materials with permanent magnetic moments can be either paramagnetic or diamagnetic, depending on which effect is stronger.

Some types of superconductors are perfect diamagnets; that is, their internal B-field remains zero in the presence of an applied field. When a magnetic field is applied to one of these superconductors, currents are induced in it that have the effect of exactly canceling the magnetic field internal to the superconducting material. The diamagnetism of the superconductor is due to actual currents caused by electrons flowing freely within the material, rather than the diamagnetism of individual atoms. Such behavior, called the **Meissner effect**, permits the superconducting material to float above a permanent magnet (Fig. 32.5).

Concept Question 6
If the charge traveling anticlockwise in Figure 32.4 has its speed increased by the changing magnetic field, it will require a larger centripetal force to stay in an orbit of the same radius. What is the origin of this additional centripetal force?

FIGURE 32.5

An illustration of the Meissner effect. A superconducting disk expels magnetic flux from its interior and floats a small permanent magnet above it.

In strong magnetic fields even the weak paramagnetism of liquid oxygen causes the oxygen to be attracted to the poles of the magnet.

Paramagnetism

Materials in the paramagnetic phase are composed of atoms that have permanent magnetic dipole moments. These moments interact only weakly, and in the absence of a magnetic field the dipole directions are randomly oriented due to thermal motion. When an external magnetic field is applied to a substance in the paramagnetic phase, its moments tend to align with the field. However, the weak interactions between the moments compete with the thermal motions of the atoms, which tend to disrupt alignment. Hence, the degree to which paramagnetic moments align with an applied magnetic field depends on both the applied field strength and the temperature. Experimentally, it is found that for moderate field strengths and temperatures not too low, the magnetization of a material in the paramagnetic phase follows the **Curie law**:

$$M = C\frac{B}{T} \tag{32.15}$$

This behavior was discovered by Pierre Curie (1859–1906), and the constant C is called the **Curie constant**. Clearly, larger B-fields tend to order more moments, thereby increasing M; when $B = 0$, Equation (32.15) says $M = 0$, implying that all the moments are randomly oriented when no external field is applied. According to the Curie law, a high temperature T results in a small value for the magnetization M. Microscopic theories, using a formalism of quantum mechanics applicable to many-body systems, correctly predict the value of C for the paramagnetic phases of many materials.

We can estimate how large an influence the thermal motions have on the magnetic dipole alignments in the paramagnetic phase. In Chapter 21, we found that the average energy of an atom or molecule in a gas at temperature T is about $k_B T$, where k_B is Boltzmann's constant. In Chapter 28, we found that the difference between the highest and lowest potential energy of a magnetic dipole μ_B in a magnetic field **B** is $\Delta U_B = 2\mu_B B$ (see Section 28.3); this energy is the amount necessary to move a dipole from complete alignment with the B-field to an orientation directly opposite that of the field. For a magnetic moment of 1.00 μ_B in a field of 1.00 T

$$\Delta U_B = 2(9.27 \times 10^{-24} \text{ A} \cdot \text{m}^2)(1.00 \text{ T}) = 1.85 \times 10^{-23} \text{ J}$$

Now, at room temperature $T = 300$ K, the average kinetic energy of a typical dipole due to thermal motion is about

$$k_B T = (1.38 \times 10^{-23} \text{ J/K})(300 \text{ K}) = 4.14 \times 10^{-21} \text{ J}$$

As you can see, the magnitude of this average energy is about 200 times the size required to rotate the magnetic dipole toward antialignment with the applied B-field. It is easy to conclude that under these circumstances most of the dipoles are randomly oriented.

Ferromagnetism

For temperatures below a certain value, called the **Curie temperature** T_C, materials, such as iron, cobalt, and nickel, are in the ferromagnetic phase. The Curie temperature for iron is 1043 K. Thus, iron is in its ferromagnetic phase at room temperature. Crystals of rare-earth elements that exhibit ferromagnetic phases include gadolinium, terbium, dysprosium, and holmium. Table 32.3 lists the Curie temperatures for these elements. Alloys and some compounds of these elements also exhibit ferromagnetic phases; several examples are also given in Table 32.3.

In the ferromagnetic phase atomic magnetic moments of large magnitude tend to line up parallel to each other throughout microscopic regions of the material. If you have ever played with two bar magnets, you know that, because unlike poles attract, the magnets have a tendency to line up *antiparallel* to each other. Therefore, it may come as somewhat of a surprise to you that strong magnetic dipoles that are located close to each other should

TABLE **32.3** The Curie Temperature for Ferromagnetic Phases of Selected Substances*

ELEMENT	T_C (K)	SATURATION MAGNETIZATION M_S (10^6 A/m)
Iron	1043	1.75
Cobalt	1404	1.45
Nickel	631	0.51
Gadolinium	289	2.00
Terbium	230	1.44
Dysprosium	85	2.01
Holmium	20	2.55
$Nd_2Fe_{14}B$	585	1.27
Alnico	1123	0.87–0.95
$SmCo_5$	993	0.77

*From David Jiles, *Introduction to Magnetism and Magnetic Materials* (Chapman and Hall, New York, 1991.)

want to line up in *parallel* fashion. The reason these atomic moments tend to align has to do with the electrostatic repulsion between the electrons. When the electrons are anti-aligned, their symmetry properties as described by quantum mechanics allows them to be close together on the average. Electron–electron Coulomb repulsion then raises their electrostatic energy. When the electron spins are aligned, the electrons must spend more time farther apart, thereby lowering their electrostatic energy.[4] In ferromagnets the overall effect is that the total energy (magnetic plus electrostatic) is lower when the spins are aligned. This behavior is a quantum mechanical effect and has no classical analog.

The spatial regions of a material throughout which all the atomic magnetic moments exhibit strong parallel alignment are the *domains* we spoke of in Section 1 of this chapter. A material in the ferromagnetic phase with no net magnetic moment has many domains with differing magnetic orientations resulting in an average magnetization of zero (Fig. 32.6). The magnetic domains themselves are usually microscopic in size and often contain a millionth to a hundredth of a mole of magnetic moments. As illustrated in Figure 32.7 each moment in the narrow region between domains (called the **domain wall**) is rotated slightly from its neighbor. When a magnetic field is applied to a ferromagnetic material,

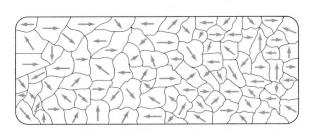

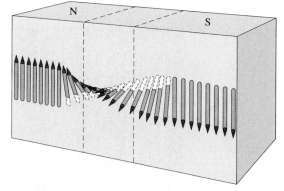

FIGURE 32.6

A model for an unmagnetized material in the ferromagnetic phase. Atomic magnetic moments create a strong magnetization within each domain, yet the overall magnetization averaged over many domains is zero.

FIGURE 32.7

In the region between domains, known as the domain wall, the magnetic moments are gradually rotated from one orientation to the other.

[4]This situation is analogous to the Pauli exclusion principle for single atoms. Recall that two electrons with aligned spins cannot be in exactly the same orbital configuration. That is, if their spins are aligned, at least one of their other quantum numbers (n, l, m_l) must be different.

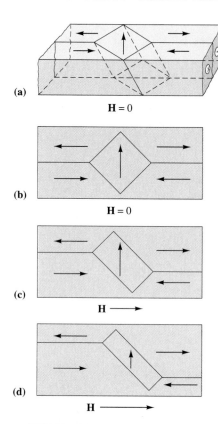

(a)

$H = 0$

(b)

$H = 0$

(c)

$H \longrightarrow$

(d)

$H \longrightarrow$

FIGURE 32.8

A microscopic view of domain wall motions as a magnetic field **H** is applied to a material in the ferromagnetic phase. (a) A three-dimensional schematic of the domains with arrows to indicate the orientation of the magnetization of each block. (b) **H** = 0. (c) Small **H** applied to the right. (d) Large **H** applied to the right.

the moments tend to align along the field direction. Microscopically, this process takes place by the movement of domain walls; domains with moments aligned along the direction of the applied field grow, and domains with moments in other directions decrease in size. It is possible to use a microscope to observe domain walls and their motion by placing a liquid containing a fine powder of ferromagnetic substance on the material to be observed. The powder is suspended in the liquid and tends to collect in the regions where the domain walls meet the surface of the ferromagnet. Figure 32.8 shows a small region of a ferromagnetic material and its behavior as a field **H** is applied.

The interaction between ferromagnetic dipoles is so strong that when a magnetic field is applied to a material in this phase and then removed, the material can remain magnetized; that is, the material can acquire a net magnetization. Hence, ferromagnetic materials are used to create permanent magnets. One way to remove this residual magnetization is to raise the temperature of the material above the Curie temperature T_C. When the temperature is above T_C, the average energy of the dipoles is large compared to the energy necessary to align adjacent moments; the moments become randomly oriented, and the material is in a paramagnetic state.

Great quantities of data can be stored using a film of ferromagnetic material on the surface of the hard disk of a personal computer.

The magnetic pattern representing binary data is written by the recording head shown here.

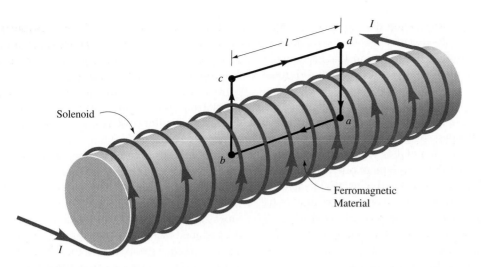

FIGURE 32.9

A portion of the material in the ferromagnetic phase near the center of a long solenoid wrapped with n turns per length of wire that carries a current I.

Another way to return a magnetized ferromagnetic substance to an unmagnetized state is to apply a magnetic field in the reverse direction. To illustrate such a process and the macroscopic behavior of a material in the ferromagnetic phase let's consider a cylindrical ferromagnet filling a long solenoid (Fig. 32.9). We begin with a sample of material with no net magnetization. We apply Ampère's law in the form of Equation (32.10), $\oint \mathbf{H} \cdot d\mathbf{l} = I_r$, so that we know the magnetic field strength H within the sample. Using the path *abcda* shown in Figure 32.9, we have for a length l within the sample and a solenoid of n turns per unit length carrying current I:

$$\oint \mathbf{H} \cdot d\mathbf{l} = (nl)I$$

$$\int_a^b \mathbf{H} \cdot d\mathbf{l} + \int_b^c \mathbf{H} \cdot d\mathbf{l} + \int_c^d \mathbf{H} \cdot d\mathbf{l} + \int_d^a \mathbf{H} \cdot d\mathbf{l} = nlI$$

$$Hl + 0 + 0 + 0 = nlI$$

$$H = nI \qquad \textbf{(32.16)}$$

The point of this computation was to show you that we really can know the magnetic field strength $\mathbf{H}$ from knowledge of the number of turns per unit length of the solenoid and the current I flowing through each turn.[5] The magnetic induction $\mathbf{B}$ within the ferromagnet depends on both the applied magnetic field strength $\mathbf{H}$ and the magnetization $\mathbf{M}$ within the sample. We can control $\mathbf{H}$ by controlling the current in the solenoid; $\mathbf{M}$ is then determined by the material. Recall from Equation (32.11),

$$\mathbf{B} = \mu_o \mathbf{H} + \mu_o \mathbf{M}$$

Let's start by assuming that our sample has no initial bulk magnetization. When $H = 0$, it has a magnetic induction B also equal to zero. The relation between H and B for a ferromagnetic material is shown in Figure 32.10. As we increase H from zero, domains begin to orient along the direction of $\mathbf{H}$, and the net magnetization in the sample increases. Consequently, the magnetic induction $\mathbf{B}$ within the material also increases as illustrated by the curve $O \rightarrow a$ in Figure 32.10. In the region approaching point a (characterized by the decreasing slope of B versus H) the domains along the direction of $\mathbf{H}$ have grown to the point where nearly all of the material is composed of domains directed along $\mathbf{H}$. If we

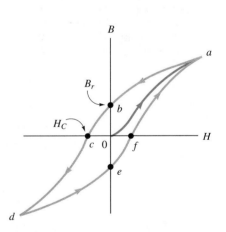

FIGURE 32.10

A hysteresis curve for a material in the ferromagnetic phase. The H-field is created by the real current of an external solenoid. Arrows around the curve indicate the history of the sample. At point b the external field is reduced to zero, but the ferromagnetic material has a nonzero B-field within it. At point c the external field is reversed from its original direction, coercing the B-field in the ferromagnetic material to fall to zero.

---

[5] In this derivation we are ignoring the depolarizing H-field originating at the ends of the material.

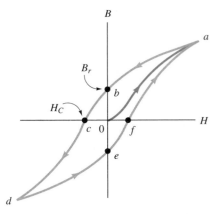

FIGURE 32.10

A hysteresis curve for a material in the ferromagnetic phase.

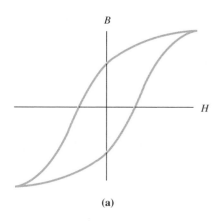

(a)

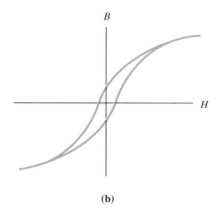

(b)

FIGURE 32.11

(a) The hysteresis loop for a hard ferromagnetic material. (b) The hysteresis curve for a soft ferromagnetic material.

Concept Question 7
Discuss the desirable magnetic properties of the magnetic material bonded to the reverse side of a bank card. Should the material be magnetically hard or soft? What about the Curie temperature?

continue to increase the magnitude of **H** to a point where the material is essentially one domain, the material becomes **saturated** and its magnetization is the **saturation magnetization** M_s. The saturation magnetization is also listed for the materials in Table 32.3. When a material is saturated, no additional increase in the magnitude of **H** will increase the magnitude of **M**.

If after reaching point a (which may or may not represent saturation) we begin to decrease the magnetic field intensity, we find when the magnitude of **H** reaches zero at point b the magnetic induction within the material is not zero but has a finite value B_r. The magnetization of the material at this point is called the **remnant magnetization**. A majority of the domains remain oriented along the direction that **H** had been applied. In order to reduce the field **B** in the material to zero we must increase the magnetic intensity **H** in the opposite direction by applying current in the reverse direction to the solenoid. The magnitude of the magnetic intensity H at which B becomes zero is known as the **coercive field**[6] H_C and is identified by point c in Figure 32.10. A continued increase in the current causes the path to proceed toward point d. If we then once more reduce the current in the solenoid to zero so that **H** again becomes zero, a remnant magnetic field represented by point e in Figure 32.10 remains in the material. At this point the remnant magnetic field is in the opposite direction from that represented by point b. In order to bring the magnetization of the material to zero (point f) current to the solenoid must be increased in the original direction so that a magnetic intensity **H** is also applied along its original direction. By further increasing the magnitude of **H** along this same direction the material can be returned to point a where it can be cycled through these same magnetization reversals.

The graph shown in Figure 32.10 is called a **hysteresis curve**, and the process that this curve describes is called **magnetic hysteresis**. It is not necessary that a hysteresis curve be closed. For example, we need not return all the way to point a from point f before we begin to reduce the magnitude of **H**. In this case the hysteresis curve is not closed. If the curve is closed, it is referred to as a **hysteresis loop**. The shape and size of a hysteresis loop depends on the strength of the applied field **H** and the magnetic properties of the material. "Hard" ferromagnetic materials have large coercivity and are characterized by broad hysteresis loops such as that shown in Figure 32.11(a). These materials are used to make permanent magnets. Materials that have coercivity of small magnitude (below 1 kA/m), such as that depicted in Figure 32.11(b), are called "soft" ferromagnets.

From our previous description of the process it should be clear to you that the hysteresis loop represents an irreversible process. In Chapter 20 we came to associate irreversible processes with an increase in entropy and the dissipation of heat to the environment. It turns out that the area enclosed by a hysteresis curve is proportional to the electrical energy dissipated as heat. Soft ferromagnets dissipate less energy when they are cycled through a hysteresis loop. The work performed during this process appears as thermal energy, and the temperature of the material increases with each cycle. Ferromagnetic materials are used in devices, such as transformers, but because these materials are subjected to alternating fields, soft ferromagnets are desirable so that the electric power wasted in heating the core is small.

EXAMPLE 32.1   To B or Not to B? That Is the Question. (Permeability Is the Answer.)

A 15.0-cm long, narrow solenoid with 240 turns is filled with bismuth. (a) Ignoring end effects, compute the magnetic intensity H within the bismuth when the solenoid carries a current of 2.00 A. (b) Show that the magnetization contribution to the magnetic induction within the bismuth is negligible. (c) If the bismuth is replaced with an iron alloy with a permeability of $400\mu_o$, what is the magnetic induction within the iron?

---

[6] A distinction is often made between the *coercivity* that is required by the H-field to reduce the magnetization to zero starting from the saturation magnetization and the *coercive field* required to reduce the magnetization to zero starting with any magnetization. The coercivity, a property of the material, is the maximum value of the coercive field.

SOLUTION   (a) Applying Equation (32.16)

$$H = nI = \frac{NI}{l} = \frac{(240)(2.00\ A)}{0.150\ m} = 3200\ A/m$$

(b) For a paramagnet, such as bismuth, the magnetization is given by Equation (32.12):

$$M = \chi H = (-1.66 \times 10^{-5})(3200\ A/m) = -0.0531\ A/m$$

The magnitude of the magnetization M is less than 0.002% of the magnetic intensity H, indicating that the magnetic effect of the bismuth is negligible. The magnetic induction is

$$B = \mu_o(H + M) = \mu_o(3200\ A/m - 0.0531\ A/m)$$
$$= (4\pi \times 10^{-7}\ H/m)(3200\ A/m) = 4.02 \times 10^{-3}\ T$$

(c) For the iron-alloy core we are given the permeability, hence we calculate the magnetic induction inside the solenoid using Equation (32.14):

$$B = \mu H = 400\mu_o H = (400)(4\pi \times 10^{-7}\ H/m)(3200\ A/m) = 1.61\ T$$

Can you show that (ignoring end effects) the magnetization of the soft iron is 1.28×10^6 A/m?

◀

32.5 Summary

The magnetic dipole moments associated with atoms are the sources of the magnetism in materials. The magnetic moment of an atom arises from both its orbital motion about the nucleus and the intrinsic spin of the electron. A convenient unit of the atomic magnetic moment is the **Bohr magneton**:

$$\mu_B = \frac{e\hbar}{2m_e} = 9.274 \times 10^{-24}\ A \cdot m^2$$

The **magnetization M** is defined as the magnetic dipole moment per volume within a material. The magnetization can be thought of as arising from **Ampèrian currents** I_m, the sources of which are microscopic magnetic dipoles. The **magnetic field intensity H** is related to real current sources and is defined in terms of the magnetic induction **B** and the magnetization **M**:

$$\mathbf{H} = \frac{\mathbf{B} - \mu_o\mathbf{M}}{\mu_o} \tag{32.9}$$

For linear materials the magnetization is related to the magnetic field intensity by the relation

$$\mathbf{M} = \chi\mathbf{H} \quad \text{(Diamagnets and paramagnets only)} \tag{32.12}$$

where χ is the **magnetic susceptibility**. The permeability of a material is related to its susceptibility by the relation

$$\mu = \mu_o(1 + \chi)$$

so that the relation between the magnetic field intensity and the magnetic induction is

$$\mathbf{B} = \mu\mathbf{H} \tag{32.14}$$

Concept Question 8
At the mid-Atlantic ridge the molten rock from deep within the earth flows out of the boundary between continental plates in a phenomenon known as sea floor spreading. This molten rock cools beneath the water continually adding new crustal material at the boundary and forcing older material outward in both directions from the crack. Geologists studying the expanding ocean floor in the neighborhood of such boundaries have discovered that the direction of magnetization of ferromagnetic domains presently being formed is preferentially aligned parallel to the earth's magnetic field. They find, however, that the magnetization of older rock farther from the boundary reverses itself periodically from parallel to the earth's field to antiparallel and back again. These zones of reversal occur on both sides of the boundary in pairs. Geologists interpret these reversals as evidence that the earth's magnetic field has reversed itself many times over the geologic past. Discuss this phenomenon in terms of the magnetic behaviors described in this chapter.

The **diamagnetic phase** is characterized by a negative magnetic susceptibility. When an external magnetic field is applied to a material in this phase, the induced magnetization opposes the applied field so that its magnitude is slightly reduced. In the **paramagnetic** and **ferromagnetic** phases the material is composed of permanent magnetic dipoles. In the absence of a magnetic field the magnetic moments of a paramagnet are randomly oriented. The susceptibility of a paramagnet is positive but small, and in the presence of an applied magnetic field a substance in the paramagnetic phase slightly enhances the field strength. In the ferromagnetic phase the moments of the material interact strongly to create regions called **domains** in which the moments are aligned. When an external magnetic field is applied to a substance in the ferromagnetic phase, the domains with moments aligned along the direction of the applied field grow in size at the expense of those domains with moments oriented in other directions. When the applied field is removed, a ferromagnetic material may remain magnetized.

PROBLEMS

32.2 The Sources of Magnetism in Materials

1. Derive the magnetic moment of the hydrogen atom as predicted by the Bohr model by performing the following: (a) Write an expression for the angular momentum of an electron of mass m_e with velocity v orbiting the nucleus in a circle of radius r as shown in Figure 32.P1. (b) Write an expression for the magnetic moment created by the electron's orbital motion. (*Hint:* The effective current created by the electron's motion is e/T, where e is the magnitude of the electron's charge and T is the period of the orbital motion.) (c) Write your answer for the magnetic moment found in (b) in terms of the angular momentum found in part (a). Eliminate both v and T from your result. (d) Show if, as Bohr assumed, angular momentum comes in integer bundles of $\hbar$ only, that the magnetic moment due to the orbital motion of the electron must be an integer multiple of a Bohr magneton.

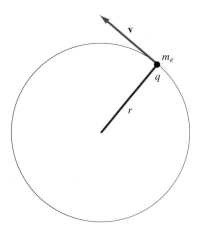

FIGURE 32.P1   Problem 1

2. *Nuclear* magnetic moments are usually stated in terms of the *nuclear magneton* $\mu_N = e\hbar/2M_p$, where M_p is the mass of a proton. How many times larger is the Bohr magneton than the nuclear magneton?

3. When the magnetic moments of a sample of iron are all aligned (the saturation condition), its magnetization is 1.75×10^6 A/m. If the number density of iron atoms is 8.5×10^{28} atoms/m^3, show that the effective magnetic moment per atom is 2.2 μ_B.

★4. Calculate the gyromagnetic ratio for a uniform circular disk of mass m, radius r, and charge e spinning about a perpendicular axis through its center. (Assume both the mass density and the charge density are uniform throughout the disk.)

★5. Calculate the gyromagnetic ratio for a uniform solid sphere of mass m and radius r. Assume the sphere has a total charge e uniformly distributed over its surface and that the sphere is spinning about a diameter.

★6. Assume that the electron is a small sphere of radius R, mass m_e, with a total charge $q = -e$ distributed uniformly throughout its volume. By dividing the charge distribution into infinitesimal current loops and summing the contributions from each, show that its magnetic moment is $\mu_{\text{spin}} = qL/2m_e$, where L is the angular momentum of the spinning electron. This classical result is about half the observed value for μ_{spin}.

32.3 The Magnetic Field Vectors

7. A long solenoid is wound with 2500 turns/m and carries a current of 2.00 A. A cylinder of copper fits snugly inside the solenoid completely filling it. (a) Ignoring end effects, compute the magnetic intensity **H** in the copper cylinder. (b) Compute the magnetization **M** within the copper. (c) By what percentage is the total magnetic induction different inside the copper cylinder from the value it would have if the copper were not present?

8. A long solenoid is wound with 800 turns/m and carries a current of 5.00 A. A cylinder of chromium completely fills the solenoid. (a) Ignoring end effects, compute the magnetic intensity **H** inside the chromium cylinder. (b) Compute the magnetic induction **B** inside the cylinder. (c) What is the magnetization of the chromium?

9. The magnetic induction B inside a particular material is 2.00 T when its magnetization is 1.50×10^6 A/m. Compute the magnetic field intensity **H** inside the material.

10. When a solenoid is filled with a material with permeability $1800\mu_o$, the magnetic induction **B** within the material has a magnitude of 6.80 T. (a) If the material is removed and the solenoid is then filled with bismuth, what is the new magnitude of the magnetic induction within the solenoid? Assume the current to the solenoid remains constant. (b) If the bismuth is replaced by aluminum, what is the magnitude of the magnetic induction?

11. A 24.0-cm long solenoid is filled with a material that has a permeability of $1200\mu_o$. If a current of 1.50 A flows through each turn of the solenoid, how many turns must it have for the magnetic induction B to be 2.00 T within the solenoid?

12. A 36.0-cm long solenoid is filled with a material with a permeability of $2200\mu_o$. If the diameter of the solenoid is 10.0 cm, and it is wound with 250 turns of wire, what current must flow through the wire to produce a magnetic flux of 0.0320 Wb through a cross section of the solenoid?

13. The average radius of a toroidal coil is R (Fig. 32.P2). (a) Apply Equation (32.10) to show that at radius R, $H = IN/2\pi R$, where N is the total number of wire turns, each of which carry current I. (b) If the toroid is filled with a material of permeability μ, find an expression for the magnetization M at radius R.

FIGURE 32.P2   Problem 13

14. An air-core solenoid is wound with 250 turns on a thin, hollow cylinder that is 24.0 cm long and 2.00 cm in radius. The cylinder is filled with a material that has a permeability of $1500\mu_o$. Ignoring end effects, compute the self-inductance of the solenoid.

15. An air-core solenoid is wound with 300 turns and has a self-inductance of 1.80 mH. If the solenoid is filled with soft iron that has a permeability of $900\mu_o$, how many turns are required to maintain the same value of self-inductance?

32.4 Diamagnetic, Paramagnetic, and Ferromagnetic Phases

16. Show that the product of the magnetic induction B and the magnetic field intensity H has the units of joules per cubic meter in SI.

17. Consider the hysteresis loop for the material shown in Figure 32.P3. (a) If a solenoid of volume 15.6 cm^3 is filled with this material, how much electrical energy does an external AC power

source supply to the material during each cycle around the loop? (b) If the external circuit drives the solenoid at a frequency of 60.0 Hz, what is the power dissipated as heat?

Numerical Methods and Computer Applications

18. The table below provides the permeability $\mu(H)$ of a particular type of unmagnetized steel at increasing magnetic intensities H applied to the steel. Compute the work required to increase H from zero to 11.0×10^{-4} A/m in a solenoid that has a volume of 1.00 L filled with this material.

H (A/m)	μ (H/m)
0.60×10^{-4}	3300
0.90×10^{-4}	4400
1.1×10^{-4}	5500
1.5×10^{-4}	5300
2.3×10^{-4}	4300
3.9×10^{-4}	2800
11.0×10^{-4}	1100

General Problems

19. Estimate the saturation magnetization for iron as follows: (a) Use the density of iron and its atomic mass to determine the number of atoms per cubic centimeter. (b) Determine the magnetization of iron, assuming that all the atomic magnetic moments are completely aligned. (c) Compare your answer from part (b) with the measured value of the saturation magnetization for iron in Table 32.3.

20. The iron in a long cylindrical magnet has a magnetization of 850. A/m near its center. (a) Express this magnetization in Bohr magnetons per cubic centimeter. (b) If the magnet has a radius of 0.550 cm, what is the magnitude of the (fictitious) Ampèrian current circulating around the circumference of a 1.00-cm length of the magnet?

21. Derive an expression for the self-inductance of a solenoid of length L and radius R that is wrapped with n turns of wire per unit length if the solenoid is filled with a material of susceptibility χ. Ignore end effects.

★22. Consider the model of diamagnetism illustrated in Figure 32.4. Suppose that the electric force holding the right-hand particle in its original orbit has magnitude $F_E = mR\omega_o^2$. Let the new angular speed after the magnetic field has been turned on be ω. Assuming that the radius of the orbit remains constant, show that if $\Delta\omega = \omega - \omega_o$ is small, then $\Delta\omega = qB/2m$, where q and m are the charge and mass of the particle respectively. (Hint: See concept question 32.6.)

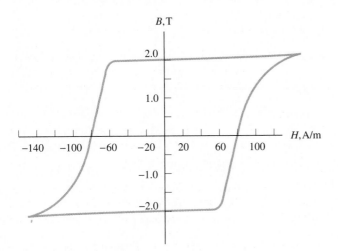

FIGURE 32.P3   Problem 17

*G*UEST *ESSAY*

Geophysical Applications of Electromagnetic Induction

by WILLIAM R. SILL

The use of metal detectors by "treasure hunters" in the search for lost coins makes use of the same principles exploited by geophysicists in the search for certain types of metallic ore deposits. The principles of electromagnetic induction are also put to work in the search for groundwater resources, the inves-tigation of rock structures near dams, and in the study of the deep interior of the earth.

You might not ordinarily think of rocks in the earth as being particularly good conductors since most common rock-forming minerals are insulators at room temperature. However, rocks and soils near the surface contain some pore water with various dis-solved substances. These dissolved ions can move through the pore pas-sages under the influence of an elec-tric field and produce the most com-mon form of rock conductivity. In some cases, the presence of metallic ore minerals can dramatically increase rock conductivity. These metallic ore minerals are mostly sulfides and ox-ides of metals such as iron, copper,

WILLIAM R. SILL
Montana College of Mineral Science and Technology

William R. Sill is Department Head and Professor of Geophysical Engineering at Mon-tana College of Mineral Science and Technology in Butte, Montana. He received his B.S. degree in Geophysics from Michigan State University in 1960 and his M.S. and Ph.D. degrees from MIT in 1963 and 1969. In the early seventies, while working for Bellcomm, Inc., on projects associated with the Apollo Program, he applied electro-magnetic techniques to the problem of interaction of the moon with the solar wind magnetic field and plasma. At the University of Utah, in the mid-seventies, he was part of an interdisciplinary group that conducted a multifrequency, radar-sounding experiment on the Apollo 17 mission and was involved with measurements of the electrical properties of lunar rocks and soils. At the conclusion of the Apollo pro-gram, he became interested in more earthly pursuits such as the uses of electrical techniques in the exploration for geothermal resources in southwestern U. S. In 1983 he went to "Montana Tech" to pursue an interest in teaching and the development of geophysical techniques for groundwater and engineering applications.

lead, and silver, and they are usually classified as semiconductors. Deep in the earth, below 100 kilometers, the temperature is high enough that even many common rock-forming minerals are in a semiconducting state. The iron/nickel core of the earth, encountered at a depth about halfway to the center, is, of course, a very good conductor.

Geophysicists have devised a variety of induction techniques to investigate these geologic conductors. These applications also illustrate the large range of scale in the use of electromagnetic induction, from the scale size of coins (centimeters) to the scale of the earth (millions of meters).

One type of geophysical induction technique is shown in Figure 1(a). The transmitter (T_x) consists of a loop of wire carrying a current I_1, which, in turn, generates a magnetic field B_1. The time varying magnetic field (B_1) induces subsurface eddy currents whose strength depends on the conductivity of the rocks. One such eddy current loop (I_2) is shown circulating in a horizontal slab of rock just below the transmitter loop. This eddy current is the source of the secondary magnetic field B_2. The resulting magnetic fields, B_1 and B_2, induce a voltage in the receiver coil shown on the right. If all of the rocks in the subsurface are poor conductors, the eddy currents and the secondary magnetic fields are small. On the other hand, if some of the rocks are good conductors, the eddy currents will be large and the secondary magnetic field may be almost as large as the primary field.

All of this probably looks and sounds complicated, but an electrical engineer might characterize the system by the simplified circuit in Figure 1(b). The transformer M_{13} is the mutual inductance between the transmitter and receiver, M_{12} is the mutual inductance between the transmitter and the conductive slab of rock, and M_{23} is the mutual inductance between the conductive rock and the receiver. These mutual inductances are just a function of the geometry of the transmitter, receiver and rock slab. Information about the conductive rock slab is contained in the self-inductance of the slab (L_2) and the resistance (R_2). The self-inductance is determined by the size and thickness of the rock slab (geometry), and the resistance is determined by the geometry and conductivity.

Use of this circuit provides a handy way to analyze the response of this measurement system. Let us look at the case where the current in the transmitter loop has been on for a long time at a steady value and is suddenly turned off by opening a switch.

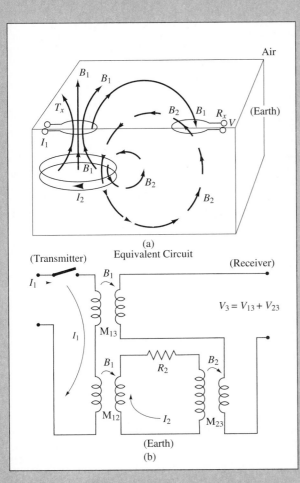

(a)

Equivalent Circuit

(b)

FIGURE 1

(a) Geometry of a loop-loop induction measurement on the surface of the earth. T_x is the transmitter which generates the primary magnetic field (B_1) and R_x is receiver loop. The disk represents a conductive slab of rock below the surface which has eddy currents (I_2) induced in it by the primary magnetic field. (b) The equivalent circuit for the induction problem in (a). The loops and their magnetic fields are replaced by their mutual inductances. The earth with its eddy current (I_2) is represented by the part of the circuit containing the resistance R_2 and the self inductance L_2. The receiver measures voltages induced by coupling through the air (M_{13}) and through the earth (M_{12} and M_{23}).

Deep in the earth, . . . the temperature is high enough that many common rock-forming minerals are in a semiconducting state.

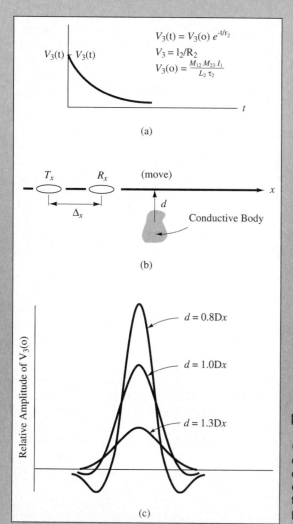

(a)

(b)

(c)

■■■■■■ **FIGURE 2**

(a) Decay of the voltage in the receiver coil when current in the transmitter coil is turned off at $t = 0$. (b) Geometry of moving coil measurement. (c) Relative amplitude of $V_3(0)$ as a function of position over conductor.

Prior to opening the switch, the steady current produces constant magnetic field (B_1), and no voltages will be induced through M_{13}, M_{12} and M_{23}. When the switch is opened, I_1 (and B_1) will fall rapidly to zero and induce voltage pulses through M_{13} and M_{12}. The voltage pulse through M_{12} will induce a current in the conductive rock (I_2). The rock current I_2 flows through a series circuit containing L_2 and R_2 and as Section 31.2 shows, this current will decay exponentially with a time constant given by $\tau = L_2/R_2$. Since I_2 and B_2 decrease exponentially with time, an exponential voltage decay will be induced in the receiver loop through the mutual inductance M_{23}. The form of the decay is shown in Figure 2(a).

The initial value of the voltage (V_3 at $t = 0$) and the time constant of the decay can be used to determine the resistance (R_2) and inductance (L_2) of the conductive rock slab. The job remaining for the geophysicist is to interpret these values in terms of the slab size and rock conductivity. In general, this may not be an easy task without some additional information.

The task of locating a conducting body is much easier and it involves traversing the surface with the two loops at a fixed distance (M_{13} is constant). If a conductor is below the traverse, M_{12} and M_{23} will change rapidly as the transmitter and receiver pass over the conductive rocks. Figure 2(c) shows a plot of the relative amplitude

of the decay voltage as a function of position along a traverse passing over a small conductive body. The peaks on the curves show the horizontal location of the center of the conductor. The three curves show the relative amplitudes for bodies at increasing depths where the depth is measured in terms of the separation between the transmitter loop and the receiver loop. Not surprisingly, the response is largest when the body is shallow. When a body depth is more than a few times the separation, the response is so small that the body is essentially undetectable. In order to see deeper into the ground with this system, the separation between the transmitter and receiver must be increased, a principle that applies to many geophysical techniques.

For typical near-surface rocks ($\sigma \sim 10^{-2}$ s/m) and a length scale of 100 m, the time constant is 10^{-4} s. On the other hand, a massive sulfide ore body ($\sigma \sim 10^{2}$ s/m) with a length scale of only 10 m has a time constant of 10^{-2} s, which is one hundred times longer than a typical rock body. The value for the ore body is about the same as that for a steel ball bearing with a radius of 1 centimeter. If we really want to go to extremes, we can consider a very large sphere like the iron-nickel core of the earth. In this case ($\sigma \sim 10^{7}$ s/m and $\ell \sim 3 \times 10^{6}$ m), the time constant is 10^{13} s, or one third of a million years. As you may know, currents in the core generate the earth's magnetic field. The fact that the earth's magnetic field has existed for billions of years indicates that there must be some source regenerating the currents, or otherwise the field would have decayed away a long time ago.

Alternating-Current Circuits

In this chapter you should learn

- the phase relations between the sinusoidal current and emfs in circuits that contain resistors, capacitors, and inductors.

- the definitions of the capacitive reactance, the inductive reactance, and the impedance, and why they are important in circuits containing *RLC* elements.

- to compute the root-mean-square current and potential changes.

- how the power dissipated in a series *RLC* circuit depends on the values of *R*, *L*, and *C*.

- about transformers.

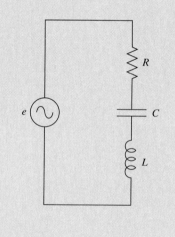

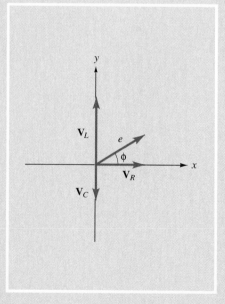

Thus far, the time-dependent emfs we have imposed on circuit elements have been very simple; by closing or opening a switch we turned the emf either on or off. In this chapter we will investigate what happens when we apply a sinusoidally varying emf or current source to a resistor, capacitor, inductor, or combination of these circuit elements. The electric power delivered to your home arrives as alternating current (AC) with a frequency of 60 Hz. Your radio, television, stereo system, and microwave oven all make use of alternating current. We won't be building any stereo systems, but the material in this chapter is fundamental to all AC circuit applications. Our strategy is to ease our way into AC circuits by first considering how each circuit element (resistor, capacitor, or inductor) behaves when separately subjected to an alternating-current source. We will then demonstrate a method, called the *phasor* technique, that allows us to determine how a circuit responds when all three of these elements are present.

33.1 Circuit Elements in AC Circuits

When the emf applied across a linear circuit element varies sinusoidally between two extreme values, the current through that circuit element also varies between two extreme values. Such circuits are referred to as alternating current (AC) circuits. In AC circuits we have to consider emfs and currents that are time-dependent; we also have to refer to the time-independent values for these emfs and currents. In order to avoid any confusion, we adopt the convention that lowercase letters e, v, and i represent *time-dependent* quantities and capital letters $\mathscr{E}$, V, and I represent *time-independent* quantities. For example, the emf source graphed in Figure 33.1(a) has a frequency f and period T related by $f = 1/T$ and can be represented by the function

$$e(t) = \mathscr{E}_{max} \sin(\omega t)$$

where $\omega = 2\pi f = 2\pi/T$. When discussing the time-dependent potential drop v across the terminals of a particular circuit element, we write

$$v(t) = V_{max} \sin(\omega t)$$

The potential amplitude V_{max} does not vary in time. When it is necessary to distinguish between potential drops across different elements, such as a resistor R and capacitor C, we add an additional subscript, such as v_R or $V_{C_{max}}$. The alternating current $i(t)$ graphed in Figure 33.1(b) varies between $+I_{max}$ and $-I_{max}$ and is represented by the function:

$$i(t) = I_{max} \cos(\omega t)$$

For sinusoidally varying emfs and currents with more general starting points we must include a phase constant ϕ within the argument of the trigonometric function. For example,

$$e(t) = \mathscr{E}_{max} \sin(\omega t + \phi) \qquad \text{or} \qquad i(t) = I_{max} \cos(\omega t + \phi)$$

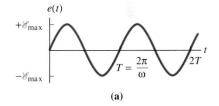

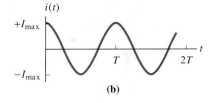

FIGURE 33.1

(a) A time-dependent emf $e(t)$ varies sinusoidally with an amplitude $\mathscr{E}_{max}$.
(b) A time-dependent current $i(t)$ with an amplitude I_{max}.

EXAMPLE 33.1   *Didn't We Do All This Back in Chapter 14?*

An emf source $e(t)$ varies sinusoidally in time, reaching a maximum of 0.500 V with a frequency of 6.366 Hz. If at $t = 0.00$ s the emf is 0.300 V, find an expression for $e(t)$.

SOLUTION   For a sinusoidal time dependence we can write

$$e(t) = \mathscr{E}_{max} \sin(\omega t + \phi)$$

The angular frequency is given by

$$\omega = 2\pi f = 2\pi (6.366 \text{ s}^{-1}) = 40.0 \text{ rad/s}$$

Therefore, with $\mathscr{E}_{max} = 0.500$ V we have

$$e(t) = (0.500 \text{ V}) \sin[(40.0 \text{ rad/s})t + \phi]$$

But, $e(0) = 0.300$ V so

$$0.300 \text{ V} = (0.500 \text{ V}) \sin(0 + \phi)$$

Solving for ϕ we find

$$\phi = \sin^{-1}\left(\frac{0.3 \text{ V}}{0.5 \text{ V}}\right) = 0.644 \text{ rad}$$

We must be quite circumspect here. The answer 0.644 rad given by our calculator is not unique. A quick sketch of the sine function should convince you that the sine takes on a value of 0.644 at two different angles during a single cycle. The other angle is in the second quadrant and is given by $\pi - 0.644$ rad $= 2.498$ rad. From the information given in the problem statement for this example, there is no way of knowing which is appropriate. In order to determine which angle is correct we must know whether the potential is increasing ($\phi = 0.644$ rad) or decreasing ($\phi = 2.498$ rad) at the moment in question. (We encountered a similar ambiguity when we tried to determine the phase constant for an object undergoing simple harmonic motion. For that case a knowledge of both x and v was required to determine ϕ.) Here we assume the potential is increasing. Hence,

$$e(t) = (0.500 \text{ V}) \sin[(40.0 \text{ rad/s})t + 0.644 \text{ rad}] \qquad \blacktriangleleft$$

We now investigate what happens when we place a sinusoidally varying emf source across each of three circuit elements (resistor, capacitor, and inductor), separately. Later, we combine these elements into a series circuit. Recall the model for an ideal emf source is one in which a power source provides a predetermined potential difference, regardless of the amount of current it must supply to do this. Conversely, an ideal-current source is one that regulates the current in a circuit at a predetermined value regardless of the emf required to drive this current. In the development that follows we have chosen to apply an emf source rather than a current source. The results we obtain are general, however, and apply for either time-dependent emf sources or time-dependent current sources. Later in this chapter we will find it convenient to apply a sinusoidally varying current source.

A Resistor in an AC Circuit

We first investigate what happens when we place an alternating emf source across a resistance R. The circuit symbol for an AC emf source is ⊙, and our simple circuit is shown in Figure 33.2. In order to keep the mathematics simple, we write the applied emf $e(t)$ as a sine function with given angular frequency ω and phase constant $\phi = 0$. The applied emf across our resistor is, therefore,

$$e(t) = \mathscr{E}_{max} \sin(\omega t)$$

Now, at every instant the sum of the potential changes around the circuit in Figure 33.2 must be zero. We must establish a sign convention before proceeding. We take the emf $e(t)$ to be positive when the potential of terminal A of the oscillating emf source is more positive than terminal B. We designate the current $i(t)$ as positive when the current flows in the clockwise direction through the circuit in Figure 33.2. As in the case of direct-current (DC) circuits we know that the potential decreases as the current fights its way through the resistor R. Hence, we label the upper side of the resistor with a plus sign to remind us that the upper end is at a more positive potential than the lower end *at this*

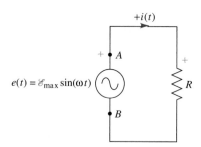

FIGURE 33.2

A sinusoidal voltage source $e(t)$ applied to a single resistor R. The polarities indicated at the source of emf and the resistor are for a particular instant of time.

particular instant of time when the current is flowing in the positive direction as shown. You should realize that we place the plus signs in Figure 33.2 only for the convenience of writing a circuit equation; as $e(t)$ changes in time the potential differences across both $e(t)$ and across the resistor oscillate in time. Traversing the circuit in the clockwise direction and summing potential changes, we have, according to Kirchhoff's potential law:

$$e(t) - i(t) R = 0$$

or solving for i,

$$i(t) = \frac{e(t)}{R} = \frac{\mathscr{E}_{max}}{R} \sin(\omega t) \qquad (33.1)$$

From Equation (33.1), we see that the maximum current is

$$I_{max} = \frac{\mathscr{E}_{max}}{R} \qquad (33.2)$$

and Equation (33.1) can be rewritten

$$i(t) = I_{max} \sin(\omega t) \qquad (33.3)$$

At any instant, the potential *drop* across the resistor,[1] which we denote as v_R, is

$$v_R(t) = i(t)R$$

Substituting the expression for current Equation (33.2), we find

$$v_R(t) = I_{max}R \sin(\omega t) = \mathscr{E}_{max} \sin(\omega t) \qquad (33.4)$$

Both Equation (33.3) for the current $i(t)$ and Equation (33.4) for the potential drop $v_R(t)$ across the resistor vary in time according to the function $\sin(\omega t)$:

In an AC circuit the current through a resistor is exactly in phase with the potential drop across its terminals.

A graph of this relationship between $v_R(t)$ and $i(t)$ is shown in Figure 33.3. Although it is usually not a good idea to plot two quantities with different units on the same graph, doing so here helps us represent the potential and current relationships when other circuit elements are present. To emphasize that the current amplitude I_{max} and the emf amplitude $\mathscr{E}_{max}$ have different units we indicate the potential scale on the left side of the graph and the current scale on the right side of the graph in Figure 33.3.

A **phasor diagram** provides a useful means by which we can graphically represent the phase relationships between the current $i(t)$ flowing through a circuit and the potential changes $v(t)$ throughout a circuit. A **phasor** is a vector-like directed line segment that rotates anticlockwise in a Cartesian plane with angular velocity ω. The initial point of the phasor (its "tail" if you will) remains fixed to the origin. The phasor's length represents

---

[1] Note the words being used here. If we agree to traverse the circuit in the clockwise direction when we apply Kirchhoff's law, then the *change* in potential across the resistor is $\Delta v_R = -iR$. Here, as everywhere in this text, the change, denoted by Δ, is "final minus original." This *change* $(-iR)$ is a *drop* of $+iR$. When discussing potential drops across circuit elements, we use the symbol v.

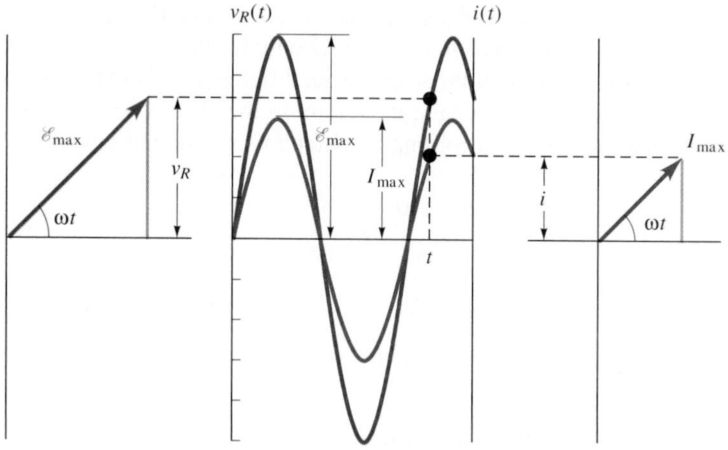

$v_R(t)$ $i(t)$

$\mathscr{E}_{max}$

v_R

ωt

$\mathscr{E}_{max}$

I_{max}

t

I_{max}

i

ωt

FIGURE 33.3

The potential drop $v_R(t)$ and current $i(t)$ for a resistor in an AC circuit are in phase. As the phasors rotate anticlockwise with angular frequency ω the projections of $\mathscr{E}_{max}$ and $\mathbf{I}_{max}$ onto the y-axis provide the values of $i(t)$ and $v_R(t)$.

the *amplitude* of either the current or the potential drop. The phasor diagram for our AC resistor circuit is shown in Figure 33.3(b). Because $i(t)$ and $v_R(t)$ are in phase, both phasors $\mathbf{I}_{max}$ and $\mathscr{E}_{max}$ are parallel and, at time t, make an angle ωt with the positive x-axis. Notice also from Figure 33.3 that the projections on the y-axis of the phasors $\mathbf{I}_{max}$ and $\mathscr{E}_{max}$ yield the functions $i(t)$ and $v_R(t)$, respectively. Because the angle ωt increases in time t, the phasors rotate in the anticlockwise direction, and the curves $i(t) = I_{max} \sin(\omega t)$ and $v_R(t) = \mathscr{E}_{max} \sin(\omega t)$ shown in Figure 33.3 are traced out by the projection of the phasors $\mathbf{I}_{max}$ and $\mathscr{E}_{max}$ on the y-axis. We must admit that this phasor representation is not too useful for resistors, but we will find it very useful when capacitors and inductors are added to the circuit.

A Capacitor in an AC Circuit

A circuit with a single capacitor connected to a sinusoidal emf source is shown in Figure 33.4. Once again we place plus signs to indicate our sign convention for positive values of the applied emf e, the current i, the charge q on the capacitor, and the potential drop v_C. (Please remember that the plus signs on the figure merely indicate our convention for positive values of e, i, q, and v_C. The actual variables alternate sign continually. When i is negative, the current is flowing in the opposite sense from that indicated in the figure. When e is negative, terminal B in the figure is more positive than terminal A.) When we apply Kirchhoff's law to this circuit, we obtain

$$e(t) - \frac{q}{C} = 0$$

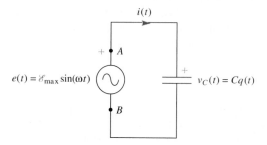

$i(t)$

$+$ A

$e(t) = \mathscr{E}_{max} \sin(\omega t)$

$+$

$v_C(t) = Cq(t)$

B

FIGURE 33.4

A sinusoidal voltage source $e(t)$ applied to a single capacitor C. The polarities indicated at the source of emf and the capacitor are for a particular instant of time.

from which we can determine the charge on the capacitor at any time t

$$q(t) = Ce(t) = C\mathcal{E}_{max} \sin(\omega t)$$

Now, by definition, the current in the circuit is given by

$$i(t) = \frac{dq}{dt} = (\omega C)\mathcal{E}_{max} \cos(\omega t) \qquad (33.5)$$

But for any angle ωt, $\cos(\omega t) = \sin(\omega t + \pi/2)$, so we can write Equation (33.5) as

$$i(t) = (\omega C)\mathcal{E}_{max} \sin\left(\omega t + \frac{\pi}{2}\right) \qquad (33.6)$$

The potential drop across the capacitor is given by

$$v_C(t) = \frac{q(t)}{C} = \mathcal{E}_{max} \sin(\omega t)$$

Plots of $v_C(t)$ and $i(t)$ are shown in Figure 33.5. Comparing the potential drop $v_C(t) = \mathcal{E}_{max} \sin(\omega t)$ across the capacitor with Equation (33.6), we note the following rule:

The potential drop across the capacitor lags the current in the AC circuit by $\pi/2$ radians.

Therefore, in the phasor diagrams of Figure 33.5 we show the phasor $\mathcal{E}_{max}$ representing the potential drop across the capacitor lagging the current phasor $\mathbf{I}_{max}$ by $\pi/2$ radians; when $\mathcal{E}_{max}$ makes an angle of ωt with the positive x-axis, the phasor $\mathbf{I}_{max}$ makes an angle $(\omega t + \pi/2)$ with this axis. It is also correct to say that *the current through the circuit leads the potential drop across the capacitor by $\pi/2$ radians*.

In Figure 33.5 we also show that the projections of the phasors $\mathbf{I}_{max}$ and $\mathcal{E}_{max}$ on the y-axis again represent the functions $i(t)$ and $v_C(t)$, respectively. As the time t increases, these phasors rotate in the anticlockwise direction and their y-components yield the curves shown in Figure 33.5.

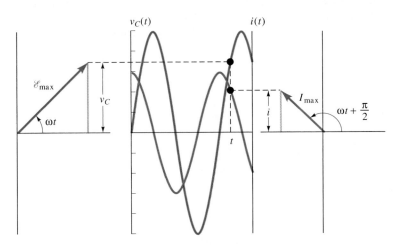

FIGURE 33.5

The potential drop $v_C(t)$ and current $i(t)$ for a capacitor in an AC circuit. The current leads the potential by $\pi/2$ rad. The projections of $\mathcal{E}_{max}$ and $\mathbf{I}_{max}$ on the y-axis give the instantaneous values of $v_C(t)$ and $i(t)$, respectively.

In our analysis of the resistor in the AC circuit we found a familiar looking equation for the current amplitude I_{max} in terms of the resistance R and the emf amplitude $\mathscr{E}_{max}$; namely, Equation (33.2), which says $I_{max} = \mathscr{E}_{max}/R$. It would be nice if we had a similar equation to apply to a capacitor in an AC circuit so we could write I_{max} equals $\mathscr{E}_{max}$ over "something." From Equation (33.6) we see that the amplitude of the current is given by

$$I_{max} = (\omega C)\mathscr{E}_{max}$$

Hence, if we define the **capacitive reactance** X_C as

$$X_C = \frac{1}{\omega C} \tag{33.7}$$

the "something" becomes X_C, and our equation for I_{max} is

$$I_{max} = \frac{\mathscr{E}_{max}}{X_C} \tag{33.8}$$

From Equation (33.8) we can see that the unit of X_C must be the ohm. In an AC circuit, a capacitor limits the amplitude of the alternating current in a manner similar to that of a resistor in a DC circuit. However, Equation (33.7) says that the capacitive reactance X_C is inversely proportional to both the capacitance C and the angular frequency ω. Therefore, for a given capacitance C, the number of ohms that impede the flow of current is a function of frequency. An emf source $e(t) = \mathscr{E}\sin(\omega t)$ with a small angular frequency ω results in a large capacitive reactance X_C, and according to Equation (33.8) the current amplitude is small. On the other hand, if the emf source has a large angular frequency ω, the capacitive reactance X_C is small and the current amplitude $I_{max} = \mathscr{E}_{max}/X_C$ is large.

Concept Question 1
As $\omega \to 0$ an AC circuit becomes more and more like a DC circuit. What happens to the capacitive reactance as $\omega \to 0$? Does this make sense? Can current flow through a capacitor in a steady-state DC circuit?

An Inductor in an AC Circuit

A sinusoidal emf source connected to a single inductor is shown in Figure 33.6. As before, we consider an instant when terminal A of our sinusoidal emf source is positive and define a positive current direction as that shown in the figure. We adopt the convention that a positive potential drop v_L across the inductor means that its upper terminal is positive with respect to its lower terminal. When we apply Kirchhoff's law to this circuit, we find

$$e(t) - L\frac{di}{dt} = 0$$

Solving for di/dt and integrating with respect to time

$$\int \frac{di}{dt}\,dt = \frac{1}{L}\int e(t)\,dt = \frac{\mathscr{E}_{max}}{L}\int \sin(\omega t)\,dt$$

resulting in[2]

$$i(t) = -\frac{\mathscr{E}_{max}}{\omega L}\cos(\omega t) \tag{33.9}$$

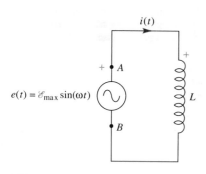

FIGURE 33.6

A sinusoidal voltage source $e(t)$ applied to a single inductor L. The polarities indicated at the source of emf and the inductor are for a particular instant of time.

---

[2] A nonzero integration constant would represent a time-independent DC current flowing in our circuit. This possibility is an artificial consequence of the fact that we have no resistance in our circuit model. We set the integration constant equal to zero because we have no interest in this possibility.

Now, $-\cos(\omega t) = +\sin(\omega t - \pi/2)$, so Equation (33.9) can be written

$$i(t) = \frac{\mathscr{E}_{max}}{\omega L} \sin\left(\omega t - \frac{\pi}{2}\right) \qquad (33.10)$$

Comparing Equation (33.10) with the potential drop across the inductor, $v_L(t) = e(t) = \mathscr{E}_{max} \sin(\omega t)$, we observe the following rule:

The potential drop across the inductor leads the current in the AC circuit by $\pi/2$ radians.

It is also correct to say that *the current through the inductor lags the potential drop across the inductor by $\pi/2$ radians.* Figure 33.7 shows the potential drop $v_L(t)$ and current $i(t)$ for our AC inductive circuit and the related phasor diagrams. As the $\mathscr{E}_{max}$ and $\mathbf{I}_{max}$ phasors rotate anticlockwise the phasor representing the potential drop across the inductor maintains a $\pi/2$-rad lead over the current phasor. The projections of $\mathscr{E}_{max}$ and $\mathbf{I}_{max}$ on the y-axis provide the functions $v_L(t)$ and $i(t)$.

From Equation (33.10) we can identify the amplitude of the current

$$I_{max} = \frac{\mathscr{E}_{max}}{\omega L} \qquad (33.11)$$

We are again motivated to get I_{max} to equal $\mathscr{E}_{max}$ over "something." We define the **inductive reactance** X_L as

$$X_L = \omega L \qquad (33.12)$$

so that Equation (33.11) can be written

$$I_{max} = \frac{\mathscr{E}_{max}}{X_L} \qquad (33.13)$$

Concept Question 2
As $\omega \to 0$ an AC circuit becomes more and more like a DC circuit. What happens to the reactance of an inductor as $\omega \to 0$? Does this make sense? What must an inductor seem like to a steady-state DC current?

Concept Question 3
What happens to the magnitude of the reactance of a capacitor as $\omega \to \infty$? What happens to the magnitude of the reactance of an inductor as $\omega \to \infty$? Can you make physical arguments that make these results plausible?

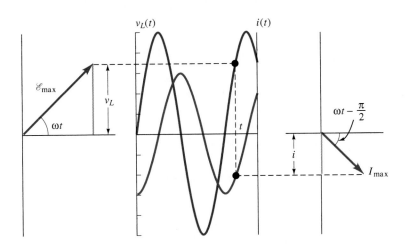

FIGURE 33.7

The potential drop $v_L(t)$ and current $i(t)$ for an inductor in an AC circuit. The current lags the potential by $\pi/2$ rad. The projections of $\mathscr{E}_{max}$ and $\mathbf{I}_{max}$ on the y-axis give the instantaneous values of $v_L(t)$ and $i(t)$.

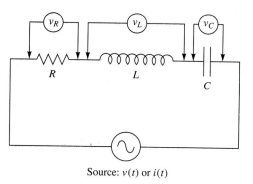

Source: $v(t)$ or $i(t)$

A series RLC circuit showing how the potential drop across each element is defined.

The unit of the inductive reactance is the ohm, and X_L is directly proportional to the inductance L and the angular frequency ω. The inductive reactance limits the maximum current flow in AC circuits that contain an inductor only.

The relationships between the current $i(t)$ and the potential drop $v_X(t)$ for AC circuits containing a single resistor, capacitor, or inductor are summarized in Table 33.1.

TABLE 33.1   Relationships between $i(t)$ and $v_X(t)$ for an AC Circuit

AT ANY INSTANT	AMPLITUDE RELATION	PHASE RELATION
$v_R(t) = i(t)R$	$V_{R_{max}} = I_{max}R$	$v_R(t)$ and $i(t)$ are in phase
$v_C(t) = i(t)X_C$	$V_{C_{max}} = I_{max}X_C$	$v_C(t)$ lags $i(t)$ by $\pi/2$ rad
$v_L(t) = i(t)X_L$	$V_{L_{max}} = I_{max}X_L$	$v_L(t)$ leads $i(t)$ by $\pi/2$ rad

Example 33.2 is a straightforward application of the above mentioned results to a capacitor that is subjected to an AC emf source. In Example 33.3, however, we demonstrate the phase relationship between $i(t)$ and $v_L(t)$ for a circuit containing an inductor in which a **current source** (rather than an emf source) is applied. We represent current sources by the symbol ⬆.

EXAMPLE 33.2   A Reactionary Problem

A 10.0-μF capacitor is connected to a 10.0-Hz sinusoidal emf source shown in Figure 33.8. What is the current in the circuit at $t = 0.280$ s?

$$v(t) = (80.0 \text{ V}) \sin[(20\pi \text{ rad/s})t] \quad \sim \quad \mathrel{\vert\vert} \quad C = 10.0 \, \mu F$$

FIGURE 33.8

A 10.0-μF capacitor connected to a 10.0-Hz sinusoidal emf source of 80.0 V amplitude. See Example 32.2.

SOLUTION   The angular frequency is $\omega = 2\pi f = 62.8$ rad/s so the capacitive reactance is

$$X_C = \frac{1}{\omega C} = \frac{1}{(62.8 \text{ rad/s})(10.0 \times 10^{-6} \text{ F})} = 1.59 \times 10^3 \, \Omega$$

The emf amplitude is $\mathcal{E}_{max} = 80.0$ V, so amplitude of the current is

$$I_{max} = \frac{\mathcal{E}_{max}}{X_C} = \frac{80.0 \text{ V}}{1.59 \times 10^3 \, \Omega} = 0.0503 \text{ A}$$

The current through the capacitor leads the potential drop across it by $\pi/2$ rad (see Eq. 33.6)) and is given by

$$i(t) = I_{max} \sin\left(\omega t + \frac{\pi}{2}\right)$$

$$= (0.0503 \text{ A}) \sin\left[(62.8 \text{ rad/s})t + \frac{\pi}{2}\right]$$

Remembering to put our calculator in radian mode, we find at $t = 0.280$ s

$$i(0.280 \text{ s}) = (0.0503 \text{ A}) \sin\left[(62.8 \text{ rad/s})(0.280 \text{ s}) + \frac{\pi}{2}\right] = 0.0155 \text{ A} = 15.5 \text{ mA}$$

◄

EXAMPLE 33.3   *A Current Source*

A 25.0-mH inductor is connected to a 60.0-Hz sinusoidal *current source* of amplitude 100. mA as shown in Figure 33.9. Compute the potential drop across the inductor at $t = 8.40$ ms.

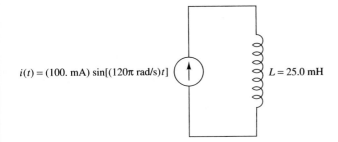

$i(t) = (100. \text{ mA}) \sin[(120\pi \text{ rad/s})t]$ $L = 25.0 \text{ mH}$

FIGURE 33.9

A 25.0-mH inductor connected to a 60.0-Hz sinusoidal current source of amplitude 100. mA. See Example 32.3.

SOLUTION   We must find the function $v_L(t)$. The angular frequency is 377 rad/s, so the inductive reactance is

$$X_L = \omega L = (377 \text{ rad/s})(25.0 \times 10^{-3} \text{ H}) = 9.42 \text{ } \Omega$$

The current amplitude $I_{max} = 100.$ mA, and amplitude of the potential drop is given

$$V_{max} = I_{max}X_L = (0.100 \text{ A})(9.42 \text{ } \Omega) = 0.942 \text{ V}$$

Now for the most subtle part of the problem. From Table 33.1 we know that the *potential drop across the inductor leads the current by $\pi/2$ radians*. The current is given by the simple sine function $i(t) = I_{max} \sin(\omega t)$. Hence, for the potential drop to lead the current by $\pi/2$ radians, its functional form must be

$$v_L(t) = V_{max} \sin\left(\omega t + \frac{\pi}{2}\right)$$

or

$$v_L(t) = (0.942 \text{ V}) \sin\left[(377 \text{ rad/s})t + \frac{\pi}{2}\right]$$

At $t = 8.40$ ms

$$v_L = (0.942 \text{ V}) \sin\left[(377 \text{ rad/s})(8.40 \times 10^{-3} \text{ s}) + \frac{\pi}{2}\right] = -0.942 \text{ V}$$

If the inductor in Figure 33.8 is replaced by a 1.20-μF capacitor, can you show that the potential drop across this capacitor is given by $v_C(t) = (221 \text{ V}) \sin[(377 \text{ rad/s})t - \pi/2]$?   ◄

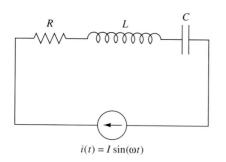

FIGURE 33.10

An *RLC* circuit consisting of a series combinations of a resistance *R*, an inductance *L*, and a capacitance *C* driven by a current source.

33.2 *RLC* Circuits

Now we are ready to have some fun! In this section we pull together the analyses of the three separate circuits from the previous section. Our objective is to discover how the potential and current behave for a series combination of a resistance *R*, an inductance *L*, and a capacitance *C*. The analysis is simplified by using a sinusoidal *current source* (rather than an emf source) as shown in Figure 33.10. Our final result, however, is applicable to either type of source. In order to more clearly distinguish the case of a current source from that of an emf source, we shall call the terminal potential of the current source $v(t)$ rather than $e(t)$. Keep in mind that $v(t)$ is applied by the current source across the *entire* series of circuit elements and is therefore equal in magnitude and sign to the total potential *drop* across all elements in series with the current source.

In Section 33.1, we found for a circuit that contains either a capacitor only or an inductor only, the potential drop either lags or leads the current by $\pi/2$ radians. It is reasonable, therefore, to assume that $v(t)$ across a series combination of *R*, *L*, and *C* is not in phase with the current $i(t)$. In particular, if the driving current from our source is given by

$$i(t) = I_{max} \sin(\omega t)$$

we assume that the terminal potential of the current source is given by

$$v(t) = V_{max} \sin(\omega t + \phi) \tag{33.14}$$

where V_{max} is the maximum emf applied by the current source to the series combination of *R*, *L*, and *C*, and the **phase angle** between $i(t)$ and $v(t)$ is represented by ϕ. Our objective is to find V_{max} and ϕ in terms of *R*, *L*, *C*, I_{max}, and ω.

As you will see, the phasor description provides a convenient method by which we can find V_{max} and ϕ. We begin by noting that the circuit elements are in series, so that at any instant the current $i(t) = I_{max} \sin(\omega t)$ must be the same at all points in the circuit. The phasor diagrams for *R*, *L*, and *C* are shown separately in Figure 33.11. The current limiting abilities of *L* and *C* are determined by their inductive and capacitive reactance, respectively. For brevity we refer to both *L* and *C* as reactive elements. Because the current is the same through each circuit element, we place the $\mathbf{I}_{max}$ phasor along a common direction; for simplicity we choose the positive *x*-axis for this direction. You should understand that this is the correct direction for $\mathbf{I}_{max}$ only at a particular instant because, as time *t*

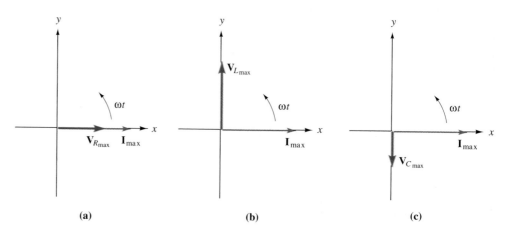

(a)   **(b)**   **(c)**

FIGURE 33.11

The phasor diagrams for each element in a series *RLC* circuit. The current phasor $\mathbf{I}_{max}$ is aligned along the same direction in each diagram because the current must be the same in each circuit element. We show the particular instant when $\mathbf{I}_{max}$ is directed along the positive *x*-axis. (a) The resistance phasor diagram. (b) The inductance phasor diagram. (c) The capacitance phasor diagram.

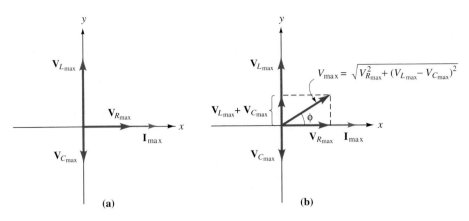

FIGURE 33.12

(a) A combined phasor diagram for an *RLC* circuit. (b) The maximum potential drop V_{max} across the entire circuit and the phase angle ϕ are computed from the vector sum of the phasors ($\mathbf{V}_{L_{max}}$ + $\mathbf{V}_{C_{max}}$) and $\mathbf{V}_{R_{max}}$.

increases, the phasor rotates about the origin with angular velocity ω. However, as the phasors representing current in each diagram rotate, at any instant they always point in a common direction.

When the three phasor diagrams from Figure 33.11 are combined into a single figure, we obtain the diagram shown in Figure 33.12(a). Although the $\mathbf{I}_{max}$, $\mathbf{V}_{R_{max}}$, $\mathbf{V}_{C_{max}}$, and $\mathbf{V}_{L_{max}}$ phasors rotate in the anticlockwise direction, at any instant they maintain their relative phase relationships. Notice, therefore, that at any instant the potential drops across the capacitor and the inductor are of opposite sign; these two phasors are always π radians out of phase. As shown in Figure 33.12(b), the $\mathbf{V}_{L_{max}}$ and $\mathbf{V}_{C_{max}}$ phasors combine to yield a phasor $\mathbf{V}_X = \mathbf{V}_{L_{max}} + \mathbf{V}_{C_{max}}$ of magnitude $|V_{L_{max}} - V_{C_{max}}|$, which is the amplitude of net potential drop across the two reactive elements. If $V_{L_{max}} > V_{C_{max}}$, this contribution to the total potential drop across the series *LC* combination is inductive in nature, and $\mathbf{V}_X$ leads the current phasor $\mathbf{I}_{max}$ by $\pi/2$ radians. If $V_{L_{max}} < V_{C_{max}}$, the resultant phasor $\mathbf{V}_X$ is capacitive in nature and lags the current phasor $\mathbf{I}_{max}$ by $\pi/2$ radians. In either case, the potential drop $V_{R_{max}}$ across the resistor is $\pi/2$ radians out of phase with the phasor $\mathbf{V}_X$, which represents the potential drop across the series inductor and capacitor. In order to find the phasor for the total potential drop $\mathbf{V}_{max}$ across the series combination of all three elements we must compute the vector sum of $\mathbf{V}_{R_{max}}$ and $\mathbf{V}_X$. From Figure 33.12(b) we see that the magnitude of the phasor $\mathbf{V}_{max}$ is

$$V_{max} = \sqrt{V_{R_{max}}^2 + V_X^2} = \sqrt{V_{R_{max}}^2 + (V_{L_{max}} - V_{C_{max}})^2} \qquad (33.15)$$

Moreover, the phase angle ϕ between the phasor $\mathbf{V}_{max}$ and the current phasor $\mathbf{I}_{max}$ is readily determined from Figure 33.12(b):

$$\tan(\phi) = \frac{V_{L_{max}} - V_{C_{max}}}{V_{R_{max}}}$$

so that

$$\phi = \arctan\left(\frac{V_{L_{max}} - V_{C_{max}}}{V_{R_{max}}}\right) \qquad (33.16)$$

Now, we make the substitutions

$$V_{R_{max}} = I_{max} R$$

$$V_{L_{max}} = I_{max} X_L$$

$$V_{C_{max}} = I_{max} X_C$$

into Equation (33.15) to obtain

$$V_{max} = \sqrt{(I_{max}R)^2 + (I_{max}X_L - I_{max}X_C)^2}$$

$$V_{max} = I_{max}\sqrt{R^2 + (X_L - X_C)^2} \qquad \textbf{(33.17)}$$

We can make Equation (33.17) into an equation that looks like V_{max} equals I_{max} times "something" by defining the impedance Z

$$Z = \sqrt{R^2 + (X_L - X_C)^2} \qquad \textbf{(33.18)}$$

leading to,

$$V_{max} = I_{max}Z \qquad \textbf{(33.19)}$$

Because $X_L = \omega L$ and $X_C = 1/\omega C$, Equations (33.18) and (33.19) are the relations we seek in order to determine V_{max} from the given values of R, L, C, I_{max}, and ω. Moreover, we can make the same substitutions for $V_{R_{max}}$, $V_{L_{max}}$, and $V_{C_{max}}$ in Equation (33.16) so that the phase angle can be written

$$\phi = \arctan\left(\frac{I_{max}X_L - I_{max}X_C}{I_{max}R}\right)$$

$$\phi = \arctan\left(\frac{X_L - X_C}{R}\right) \qquad \textbf{(33.20)}$$

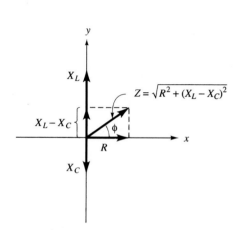

FIGURE 33.13

A simplified diagram from which the impedance Z and phase angle ϕ can be computed. It is important to maintain the proper phase relations in this diagram: always place X_L along the positive y-axis, X_C along the negative y-axis, and R along the positive x-axis.

The unit of the impedance Z is the ohm. In a direct current (DC) circuit a constant current I results in a steady potential drop V across a total resistance R_{tot}. We learned in Chapter 26 that the relation between the DC quantities V and I is "$V = IR_{tot}$." Equation (33.19) is the equivalent expression for a series of elements subject to a sinusoidally varying emf or current source. Often the first step in solving an AC-circuit problem is to compute the impedance Z and the phase angle ϕ. To this end we can make a simplification in the phasor diagram. Equation (33.20) says that we need only consider the resistance R and the reactances X_L and X_C in order to compute the phase constant ϕ. If the correct phase relations are maintained from the phasor diagram, Figure 33.12(b), a diagram of R, X_L, and X_C is sufficient to compute Z and ϕ. Such a drawing is known as an **impedance diagram** and is shown in Figure 33.13.

Recall that the phase angle ϕ indicates by how much the *applied potential leads the current*:

$$i(t) = I_{max}\sin(\omega t)$$

$$v(t) = V_{max}\sin(\omega t + \phi)$$

In fact, however, $v(t)$ can lead, lag, or be in phase with the current. Which of these relationships holds depends on the relative sizes of X_L and X_C. From Figure 33.13 you should verify each case presented in Table 33.2.

TABLE 33.2 Phase Relations between $v(t)$ and $i(t)$ for an *RLC* Circuit

When				
	$X_L > X_C$	$\phi > 0$	and	$v(t)$ leads $i(t)$ by ϕ
	$X_L < X_C$	$\phi < 0$	and	$v(t)$ lags $i(t)$ by $|\phi|$
	$X_L = X_C$	$\phi = 0$	and	$v(t)$ and $i(t)$ are in phase

It is important to remember that $v(t)$ is the potential difference applied across the *entire RLC* circuit. In our analysis we assumed that the current source was used so that $i(t) = I_{max} \sin(\omega t)$ is imposed on the circuit. If this is how $i(t)$ is given in a problem, then $v(t)$ is given by $v(t) = V_{max} \sin(\omega t + \phi)$ where $V_{max} = I_{max}Z$. As described previously, Z is given by Equation (33.18), and ϕ is positive if $X_L > X_C$, negative if $X_L < X_C$, or zero if $X_L = X_C$.

On the other hand, sometimes an emf source is used so that $e(t) = \mathcal{E}_{max} \sin(\omega t)$ is given. In this case, we still find the impedance Z from Equation (33.18) and the phase angle ϕ from Equation (33.20) just as in the current source case. In fact, you can even write

$$i(t) = I_{max} \sin(\omega t)$$

$$e(t) = \mathcal{E}_{max} \sin(\omega t + \phi)$$

and similar relations just as in the current source case. Now, however, you must add one small final step: you must replace every occurrence of ωt with $\omega t - \phi$. As simply as that you solved the emf source case. (You must of course be careful when ϕ is a negative number. Remember "minus a minus is a plus.")

TABLE 33.3   Summary of *RLC* Circuit Equations

Given: $i(t) = I_{max} \sin(\omega t)$ **then** $v(t) = V_{max} \sin(\omega t + \phi)$

$\quad\quad v_R(t) = V_{R_{max}} \sin(\omega t)$ $\quad\quad\quad\quad (v_R(t)$ is in phase with $i(t))$
$\quad\quad v_L(t) = V_{L_{max}} \sin(\omega t + \pi/2)$ $\quad\quad (v_L(t)$ leads $i(t)$ by $\pi/2$ rad)
$\quad\quad v_C(t) = V_{C_{max}} \sin(\omega t - \pi/2)$ $\quad\quad (v_C(t)$ lags $i(t)$ by $\pi/2$ rad)

Given: $e(t) = \mathcal{E}_{max} \sin(\omega t)$ **then** $i(t) = I_{max} \sin(\omega t - \phi)$

$\quad\quad v_R(t) = V_{R_{max}} \sin(\omega t - \phi)$ $\quad\quad\quad\quad (v_R(t)$ is in phase with $i(t))$
$\quad\quad v_L(t) = V_{L_{max}} \sin(\omega t - \phi + \pi/2)$ $\quad (v_L(t)$ leads $i(t)$ by $\pi/2$ rad)
$\quad\quad v_C(t) = V_{C_{max}} \sin(\omega t - \phi - \pi/2)$ $\quad (v_C(t)$ lags $i(t)$ by $\pi/2$ rad)

In either case you may use

$$Z = \sqrt{R^2 + (X_L - X_C)^2} \tag{33.18}$$
$$\mathcal{E}_{max} = I_{max}Z \quad\text{or}\quad V_{max} = I_{max}Z \tag{33.19}$$

and

$$V_{R_{max}} = I_{max}R$$
$$V_{L_{max}} = I_{max}X_L$$
$$V_{C_{max}} = I_{max}X_C$$

EXAMPLE 33.4   *A Current Source*

Consider a series *RLC* circuit driven at 5.00 Hz by an alternating-current source of amplitude 0.200 A as shown in Figure 33.14. For $R = 250.\ \Omega$, $L = 20.0$ H, and $C = 100.\ \mu$F find: (a) the impedance Z, (b) the phase angle ϕ, and (c) $v(t)$, $v_R(t)$, $v_L(t)$, and $v_C(t)$ at $t = 0.260$ s.

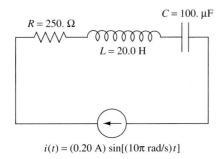

$$i(t) = (0.20\ \text{A}) \sin[(10\pi\ \text{rad/s})t]$$

FIGURE 33.14

A series *RLC* circuit driven at 5.00 Hz by an AC source of amplitude 0.200 A. See Example 32.4.

SOLUTION   The angular frequency ω is 10.0π rad/s so that

$$X_L = \omega L = (31.4 \text{ rad/s})(20.0 \text{ H}) = 628 \text{ } \Omega$$

$$X_C = \frac{1}{\omega L} = \frac{1}{(31.4 \text{ rad/s})(100. \times 10^{-6} \text{ F})} = 318 \text{ } \Omega$$

(a) The impedance is given by Equation (33.18):

$$Z = \sqrt{R^2 + (X_L - X_C)^2} = \sqrt{(250 \text{ } \Omega)^2 + (628 \text{ } \Omega - 318 \text{ } \Omega)^2} = 398 \text{ } \Omega$$

(b) The phase constant is given by Equation (33.20):

$$\phi = \arctan\left(\frac{X_L - X_C}{R}\right) = \arctan\left(\frac{628 \text{ } \Omega - 318 \text{ } \Omega}{250 \text{ } \Omega}\right) = 0.892 \text{ rad} = 51.1°$$

The impedance diagram for this circuit is shown in Figure 33.15.

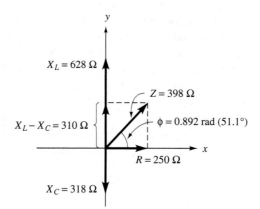

FIGURE 33.15

The impedance diagram for the circuit in Figure 33.15.

(c) The phase constant ϕ is greater than zero, so the potential drop $v(t)$ across the *RLC* combination leads the current:

$$v(t) = V_{\text{max}} \sin(\omega t + \phi) = V_{\text{max}} \sin[(31.4 \text{ rad/s})t + 0.892 \text{ rad}]$$

The amplitude V_{max} of the applied potential can be computed from Equation (33.19):

$$V_{\text{max}} = I_{\text{max}}Z = (0.200 \text{ A})(398 \text{ } \Omega) = 79.6 \text{ V}$$

Therefore,

$$v(t) = (79.6 \text{ V}) \sin[(31.4 \text{ rad/s})t + 0.892 \text{ rad}]$$

We compute the functions $v_R(t)$, $v_L(t)$, and $v_C(t)$ and then substitute $t = 0.260$ s. First, we compute the amplitudes of each potential drop:

$$V_{R_{\text{max}}} = I_{\text{max}}R = (0.200 \text{ A})(250 \text{ } \Omega) = 50.0 \text{ V}$$

$$V_{L_{\text{max}}} = I_{\text{max}}X_L = (0.200 \text{ A})(628 \text{ } \Omega) = 126 \text{ V}$$

$$V_{C_{\text{max}}} = I_{\text{max}}X_C = (0.200 \text{ A})(318 \text{ } \Omega) = 63.6 \text{ V}$$

Notice that the algebraic sum of amplitudes is nearly 240 V, although previously we found that the maximum total potential across the circuit is 79.6 V. Therefore, it should be abundantly clear that the potential drops across *R*, *L*, and *C* are not in phase.

The point is that in AC circuits *you cannot merely add the amplitudes of the potential drops*. To find the amplitude of the total potential drop you must add the phasors, exactly as you do in the case of vector addition.

Once the amplitudes of the individual potential drops are known, the instantaneous potential drop across any one of these elements at time t is given by $v_R(t)$, $v_L(t)$, or $v_C(t)$. Unlike the amplitudes, these instantaneous potential drops *do* add as simple (signed) numbers. The phase of each potential drop is determined relative to $i(t)$. In particular $v_R(t)$ is in phase with $i(t)$, $v_L(t)$ leads $i(t)$ by $\pi/2$ rad, and $v_C(t)$ lags $i(t)$ by $\pi/2$ rad. Thus, because $i(t)$ is given by $I_{max} \sin(\omega t)$, we have

$$v_R(t) = V_{R_{max}} \sin(\omega t)$$

$$v_L(t) = V_{L_{max}} \sin\left(\omega t + \frac{\pi}{2}\right)$$

$$v_C(t) = V_{C_{max}} \sin\left(\omega t - \frac{\pi}{2}\right)$$

or

$$v_R(t) = (50.0 \text{ V}) \sin[(31.4 \text{ rad/s})t]$$

$$v_L(t) = (126 \text{ V}) \sin\left[(31.4 \text{ rad/s})t + \frac{\pi}{2}\right]$$

$$v_C(t) = (63.6 \text{ V}) \sin\left[(31.4 \text{ rad/s})t - \frac{\pi}{2}\right]$$

Previously we found that the total applied potential difference across the circuit (which is, of course, equal to the total potential drop across the circuit) at time t is given by

$$v(t) = (79.6 \text{ V}) \sin[(31.4 \text{ rad/s})t + 0.892 \text{ rad}]$$

At $t = 0.210$ s:

$$v(0.26 \text{ s}) = (79.6 \text{ V}) \sin[(31.4 \text{ rad/s})(0.260 \text{ s}) + 0.892 \text{ rad}] = 28.4 \text{ V}$$

We can compare this to the instantaneous potential drop across the individual elements

$$v_R(0.26 \text{ s}) = (50.0 \text{ V}) \sin[(31.4 \text{ rad/s})(0.260 \text{ s})] = 47.6 \text{ V}$$

$$v_L(0.26 \text{ s}) = (126 \text{ V}) \sin\left[(31.4 \text{ rad/s})(0.260 \text{ s}) + \frac{\pi}{2}\right] = -38.9 \text{ V}$$

$$v_C(0.26 \text{ s}) = (63.6 \text{ V}) \sin\left[(31.4 \text{ rad/s})(0.260 \text{ s}) - \frac{\pi}{2}\right] = 19.7 \text{ V}$$

Notice that the sum $v_R(0.26 \text{ s}) + v_L(0.26 \text{ s}) + v_C(0.260 \text{ s}) = 28.4 \text{ V} = v(0.260 \text{ s})$; all is well! ◀

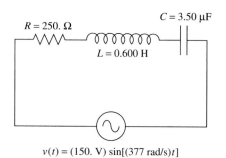

FIGURE 33.16

An *RLC* circuit of resistance 250. Ω, capacitance 3.50 μF, and inductance 0.600 H. See Example 32.5.

EXAMPLE 33.5 *An emf Source*

For the *RLC* circuit shown in Figure 33.16 find: (a) the impedance Z, (b) the phase angle ϕ, (c) the current in the circuit $i(t)$ and the potential drops $v_R(t)$, $v_L(t)$, and $v_C(t)$ at $t = 1.24$ s.

SOLUTION The angular frequency ω is 377 rad/s, so that

$$X_L = \omega L = (377 \text{ rad/s})(0.600 \text{ H}) = 226 \; \Omega$$

$$X_C = \frac{1}{\omega L} = \frac{1}{(377 \text{ rad/s})(3.50 \times 10^{-6} \text{ F})} = 758 \; \Omega$$

(a) The impedance is given by Equation (33.18):

$$Z = \sqrt{R^2 + (X_L - X_C)^2} = \sqrt{(250 \; \Omega)^2 + (226 \; \Omega - 758 \; \Omega)^2} = 588 \; \Omega$$

(b) The phase constant is given by Equation (33.20):

$$\phi = \arctan\left(\frac{X_L - X_C}{R}\right) = \arctan\left(\frac{226 \; \Omega - 758 \; \Omega}{250 \; \Omega}\right) = -1.13 \text{ rad} = -64.8°$$

The phasor diagram for this circuit is shown in Figure 33.17.

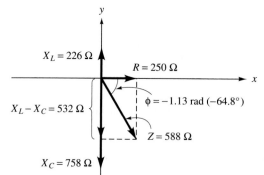

FIGURE 33.17

The phasor diagram for the circuit in Figure 33.16.

(c) The phase constant ϕ is negative, so the applied emf $e(t)$ lags the current, or, equivalently, the current *leads* the potential drop by $|\phi|$. The emf is given as $e(t) = \mathscr{E}_{max} \sin(\omega t)$, thus

$$i(t) = I_{max} \sin(\omega t - \phi)$$

$$= I_{max} \sin[(377 \text{ rad/s})t + 1.13 \text{ rad}]$$

Be sure to notice that we obtained $+1.13$ rad in this expression when we substituted $\phi = -1.13$ rad. We can find the current amplitude I_{max} from Equation (33.19):

$$I_{max} = \frac{\mathscr{E}_{max}}{Z} = \frac{150 \text{ V}}{588 \; \Omega} = 0.255 \text{ A}$$

Therefore,

$$i(t) = (0.255 \text{ A}) \sin[(377 \text{ rad/s})t + 1.13 \text{ rad}]$$

At $t = 1.24$ s

$$i(1.24 \text{ s}) = (0.255 \text{ A}) \sin[(377 \text{ rad/s})(1.24) + 1.13 \text{ rad}] = -0.125 \text{ A}$$

We now compute the functions $v_R(t)$, $v_L(t)$, and $v_C(t)$. First, we find the amplitudes for each potential drop:

$$V_{R_{max}} = I_{max}R = (0.255 \text{ A})(250 \text{ }\Omega) = 63.8 \text{ V}$$

$$V_{L_{max}} = I_{max}X_L = (0.255 \text{ A})(226 \text{ }\Omega) = 57.6 \text{ V}$$

$$V_{C_{max}} = I_{max}X_C = (0.255 \text{ A})(758 \text{ }\Omega) = 193 \text{ V}$$

Now, $i(t)$ is given by $\sin(\omega t - \phi)$ and $v_R(t)$ is in phase with $i(t)$, $v_L(t)$ leads $i(t)$ by $\pi/2$ rad and $v_C(t)$ lags $i(t)$ by $\pi/2$ rad. Thus,

$$v_R(t) = V_{R_{max}} \sin(\omega t - \phi)$$

$$v_L(t) = V_{L_{max}} \sin(\omega t - \phi + \pi/2)$$

$$v_C(t) = V_{C_{max}} \sin(\omega t - \phi - \pi/2)$$

or

$$v_R(t) = (63.8 \text{ V}) \sin[(377 \text{ rad/s})t + 1.13 \text{ rad}]$$

$$v_L(t) = (57.6 \text{ V}) \sin\left[(377 \text{ rad/s})t + 1.13 \text{ rad} + \frac{\pi}{2}\right]$$

$$v_C(t) = (193 \text{ V}) \sin\left[(377 \text{ rad/s})t + 1.13 \text{ rad} - \frac{\pi}{2}\right]$$

At $t = 1.24$ s:

$$v_R(1.24 \text{ s}) = (63.8 \text{ V}) \sin[(377 \text{ rad/s})(1.24 \text{ s}) + 1.13 \text{ rad}] = -31.3 \text{ V}$$

$$v_L(1.24 \text{ s}) = (57.6 \text{ V}) \sin\left[(377 \text{ rad/s})(1.24 \text{ s}) + 1.13 \text{ rad} + \frac{\pi}{2}\right] = -50.2 \text{ V}$$

$$v_C(1.24 \text{ s}) = (193 \text{ V}) \sin\left[(377 \text{ rad/s})(1.24 \text{ s}) + 1.13 \text{ rad} - \frac{\pi}{2}\right] = 168 \text{ V}$$

Both the sum of these potential drops and the expression for $e(t)$ in Figure 33.16 yield a total of 87 V at $t = 1.24$ s. ◀

33.3 The Root-Mean-Square Potential and Current

Bare wires can be dangerous. If you have ever accidentally touched a "live" wire, you no doubt received a painful shock. The power supplied to your home comes in the form of a sinusoidal emf with an amplitude of about 170 V. It probably would not have made the shock you received any less painful if you had known that the average emf applied to your body and the average current through it was zero! In this chapter we have been considering sinusoidal emfs and currents, such as those given by the equations

$$v(t) = V_{max} \sin(\omega t) \tag{33.21a}$$

or

$$i(t) = I_{max} \sin(\omega t) \tag{33.21b}$$

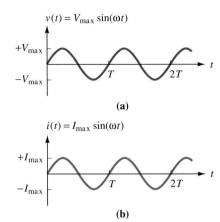

$v(t) = V_{max} \sin(\omega t)$

(a)

$i(t) = I_{max} \sin(\omega t)$

(b)

FIGURE 33.18

(a) The sinusoidal potential function $v(t) = V_{max} \sin(\omega t)$. Because the potential is maintained in one direction for the same amount of time that it is in the other, the average potential is zero. (b) Similarly, the average of the current $i(t) = I_{max} \sin(\omega t)$ is zero.

These functions are shown in Figure 33.18. It is clear from these graphs that the average value of $v(t)$ is zero and the average value of $i(t)$ is also zero; both $v(t)$ and $i(t)$ are positive for the same length of time that they are negative. If we look at the power delivered to a resistor, we recall that the expression $\mathcal{P} = i^2 R$ depends on the square of the current. Thus, the power is positive regardless of the direction of current flow. When the current follows a sine wave of frequency f, the power rises to a maximum and falls to zero with a frequency $2f$. Seldom are we concerned with the rapid bursts with which power is delivered. More often, the average power over many cycles is all that is of interest. In order to facilitate such average power calculations it is useful to define a peculiar type of average based on the squares of the current and potential drops. In this section we describe this average, called the root-mean-square (rms) average. In the section following this one we will apply these averages to power calculations in which the utility of the rms average will quickly become apparent.

The **root mean square** of the potential is defined as:

$$V_{rms} = \sqrt{<v(t)^2>_{av}} \tag{33.22}$$

where $< \quad >_{av}$ means take the average over a whole number of cycles. If T is the period over which a periodic function $f(t)$ repeats itself, then the time average of $f(t)$ is defined to be

$$<f(t)>_{av} = \frac{\displaystyle\int_0^T f(t)\,dt}{T} \tag{33.23}$$

Notice that Equation (33.22) says to first square $v(t)$, then find the average of $v^2(t)$ and, finally, take the square root of this result. Similarly the root mean square of the current is defined as the square root of the average of the square of $i(t)$:

$$I_{rms} = \sqrt{<i(t)^2>_{av}} \tag{33.24}$$

It is not difficult to use Equation (33.23) to compute $<v(t)^2>_{av}$ and $<i(t)^2>_{av}$ for various waveforms (see Problems 38 through 41). However, when $v(t)$ and $i(t)$ are *sinusoidal* functions we can find expressions for V_{rms} and I_{rms} without actually performing the integration indicated by Equation (33.24). We use a technique that only works for sinusoidally varying potential and current sources.

Let's look at $v(t)$ given by Equation (33.21a). First, we square $v(t)$:

$$v^2(t) = V_{max}^2 \sin^2(\omega t)$$

Next, we take the average over time of $v^2(t)$. The amplitude V_{max} does not change in time and, therefore, is not affected by the averaging process:

$$<v^2(t)>_{av} = <V_{max}^2 \sin^2(\omega t)>_{av} = V_{max}^2 <\sin^2(\omega t)>_{av} \tag{33.25}$$

Now, $\sin^2(\omega t) + \cos^2(\omega t) = 1$ for all t, so it is certainly also true on the average

$$<\sin^2(\omega t)>_{av} + <\cos^2(\omega t)>_{av} = 1 \tag{33.26}$$

In Figure 33.19 we show the functions $\sin^2(\omega t)$ and $\cos^2(\omega t)$. It is clear that both these functions have the same average over a whole number of periods; after all, one function is just the other translated by one-fourth period along the t-axis. Hence,

$$<\cos^2(\omega t)>_{av} = <\sin^2(\omega t)>_{av}$$

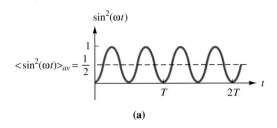

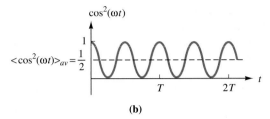

FIGURE 33.19

(a) A graph of the functions $\sin^2(\omega t)$ and $\cos^2(\omega t)$. The average of each function is $\frac{1}{2}$.

and Equation (33.26) becomes

$$<\sin^2(\omega t)>_{av} + <\sin^2(\omega t)>_{av} = 1$$

$$2 < \sin^2(\omega t)>_{av} = 1$$

$$<\sin^2(\omega t)>_{av} = \frac{1}{2}$$

Substituting this result into Equation (33.25), we have

$$<v^2(t)>_{av} = \frac{1}{2}\,(V_{max})^2 \qquad\qquad (33.27)$$

Finally, when we substitute Equation (33.27) into Equation (33.22) we obtain

$$V_{rms} = \frac{V_{max}}{\sqrt{2}} \qquad\qquad (33.28)$$

In a similar manner, it is easy to show for the sinusoidal current given by Equation (33.21b)

$$I_{rms} = \frac{I_{max}}{\sqrt{2}} \qquad\qquad (33.29)$$

Many of the equations we will obtain in the next section will look quite similar to their DC circuit counterparts when written in terms of rms values.

EXAMPLE 33.6 *Just an Average Problem*

The root-mean-square emf from a household wall outlet is 120. V. (a) What is the amplitude of this sinusoidal emf source? (b) What is the root-mean-square current through a 150.-Ω resistor connected to this household source?

SOLUTION (a) From Equation (33.28)

$$V_{max} = \sqrt{2}\ V_{rms} = \sqrt{2}(120.\ \text{V}) = 170.\ \text{V}$$

(b) The relation between the current amplitude and the amplitude of the potential drop for an AC circuit containing a resistance only is

$$I_{max} = \frac{V_{max}}{R}$$

Dividing both sides of this equation by $\sqrt{2}$, we have

$$\frac{I_{max}}{\sqrt{2}} = \frac{V_{max}}{\sqrt{2}} \frac{1}{R}$$

or

$$I_{rms} = \frac{V_{rms}}{R} = \frac{120.\ V}{150.\ \Omega} = 0.800\ A \qquad \blacktriangleleft$$

33.4 Power and Resonance in AC Circuits

In this section we show you how the power dissipated in a series RLC circuit depends critically on the phase constant ϕ of the circuit. We begin with an observation about energy dissipation we think will interest you: *In the steady state, on the average, no electrical energy is dissipated as heat by either an ideal capacitor or an ideal inductor in an RLC circuit.* By "ideal" we mean that the capacitor does not leak charge between its plates, and the inductor has no resistance. Here's a microscopic model for what happens at the capacitor of an RLC circuit that is subjected to a sinusoidal current or emf source: As charge begins to accumulate on the capacitor C, the potential drop $v_C(t)$ across its plates increases and so does the electric field $\mathbf{E}$ between these conductors. At any instant, the energy stored by the capacitor is $U_C = \frac{1}{2}Cv_C^2(t)$, and when the potential drop across the capacitor reaches a maximum $V_{C_{max}}$, the energy stored is $U_C = \frac{1}{2}CV_{C_{max}}^2$. At this instant of maximum charge, energy is stored in the electric field between the plates of the capacitor. As the cycle continues, the capacitor begins to discharge and does work on the charges flowing in the circuit. When the capacitor is completely discharged, it has returned an energy of $\frac{1}{2}CV_{C_{max}}^2$ to the rest of the circuit. The other half of the cycle commences as the capacitor's plates begin to be charged with charge of polarity opposite from the first half of the cycle. You should see that twice during each cycle an energy of $\frac{1}{2}CV_{C_{max}}^2$ is stored on the capacitor and then returned to the other parts of the circuit.

A similar sequence of events takes place for the inductor. This time, however, energy is stored in the magnetic field $\mathbf{B}$ established by the inductor as the current through it varies sinusoidally. In an inductor of inductance L the energy stored when the current reaches its amplitude I_{max} is $U_L = \frac{1}{2}LI_{max}^2$. Twice each cycle this amount of energy is stored in the B-field of the inductor and then returned to the remainder of the circuit. The only place in an ideal RLC circuit that energy is dissipated is in the resistor R, and this energy is dissipated as Joule heat in the same manner as it is dissipated in steady-current circuits.

In Chapter 26 we learned that the power provided to a circuit by a source of emf e that delivers current i is

$$\mathcal{P} = ie \qquad\qquad (33.30)$$

In the case of sinusoidally varying sources of emf this expression provides the *instantaneous* power. However, it is usually more useful to know something about the *average* power $\mathcal{P}_{av}$ delivered to the circuit:

$$\mathcal{P}_{av} = <i(t)e(t)>_{av}$$

In general, the emf $e(t)$ and current $i(t)$ can differ in phase. Hence, if the emf is given by

$$e(t) = \mathscr{E}_{max} \sin(\omega t) \qquad (33.31)$$

then the current can be written

$$i(t) = I_{max} \sin(\omega t - \phi) \qquad (33.32)$$

Substituting Equations (33.31) and (33.32) into Equation (33.30), we have

$$\mathscr{P}_{av} = <I_{max}\mathscr{E}_{max} \sin(\omega t - \phi) \sin(\omega t)>_{av} = I_{max}\mathscr{E}_{max} <\sin(\omega t - \phi) \sin(\omega t)>_{av} \quad (33.33)$$

where we have removed the product $I_{max}\mathscr{E}_{max}$ from the indicated averaging process because it is a constant. We can make use of the trigonometric identity $\sin(\omega t - \phi) = \sin(\omega t) \cos(\phi) - \sin(\phi) \cos(\omega t)$ in Equation (33.33) and also recognize that $\sin(\phi)$ and $\cos(\phi)$ are both constants:

$$\mathscr{P}_{av} = I_{max}\mathscr{E}_{max}<[\sin(\omega t) \cos(\phi) - \sin(\phi) \cos(\omega t)] \sin(\omega t)>_{av}$$
$$= I_{max}\mathscr{E}_{max}[<\sin^2(\omega t)>_{av} \cos(\phi) - \sin(\phi) <\sin(\omega t) \cos(\omega t)>_{av}]$$

We have seen that $<\sin^2(\omega t)>_{av} = \frac{1}{2}$, and it is easy to apply Equation (33.23) (see Problem 44) to show that

$$<\sin(\omega t) \cos(\omega t)>_{av} = 0$$

Hence,

$$\mathscr{P}_{av} = I_{max}\mathscr{E}_{max}\left[\frac{1}{2} \cos(\phi) + 0\right] = \frac{I_{max}\mathscr{E}_{max}}{2} \cos(\phi) = \frac{I_{max}}{\sqrt{2}} \frac{\mathscr{E}_{max}}{\sqrt{2}} \cos(\phi)$$

$$\mathscr{P}_{av} = I_{rms}\mathscr{E}_{rms} \cos(\phi) \qquad (33.34)$$

The term $\cos(\phi)$ is called the **power factor** and can be determined from the impedance diagram shown in Figure 33.20. From this figure we see

$$\cos(\phi) = \frac{R}{Z} = \frac{R}{\sqrt{R^2 + (X_L - X_C)^2}} \qquad (33.35)$$

When $\cos(\phi) = R/Z$ is substituted into Equation (33.34), an alternative form for $\mathscr{P}_{av}$ can be derived:

$$\mathscr{P}_{av} = I_{rms}\mathscr{E}_{rms}\frac{R}{Z} = I_{rms}\left(\frac{\mathscr{E}_{rms}}{Z}\right)R$$

$$\mathscr{P}_{av} = I_{rms}^2 R \qquad (33.36)$$

Equation (33.34) tells us that the average power delivered to the resistor in the series RLC circuit is a maximum when $\cos(\phi) = 1$. From Equation (33.35) we can see that this is the case when

$$X_L = X_C \qquad (33.37)$$

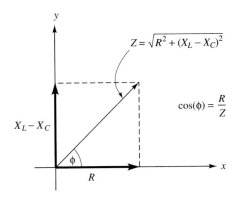

FIGURE 33.20

Computing the power factor $\cos(\phi) = R/Z$.

for then

$$\cos(\phi) = \frac{R}{Z} = \frac{R}{\sqrt{R^2 + 0^2}} = 1 \tag{33.38}$$

When the condition given by Equation (33.37) is satisfied, the series RLC circuit is said to be in **resonance.** In this case, the energy is being dissipated in the resistor at the maximum rate possible for fixed values of $\mathcal{E}_{max}$, R, L, and C. Notice from Equation (33.38) that at resonance the impedance of the circuit Z is just equal to the resistance R:

$$Z_{res} = R$$

Therefore, $I_{rms} = \mathcal{E}_{rms}/Z = \mathcal{E}_{rms}/R$, and Equation (33.34) becomes

$$\mathcal{P}_{av} = \frac{\mathcal{E}_{rms}^2}{R} \qquad \text{(At resonance)} \tag{33.39}$$

Resonance occurs at only one particular angular frequency $\omega = \omega_{res}$. Using the definitions for X_L and X_C, Equation (33.37) can be used to find an expression for ω_{res}:

$$\omega_{res}L = \frac{1}{\omega_{res}C}$$

or

$$\omega_{res} = \frac{1}{\sqrt{LC}} \tag{33.40}$$

Suppose you apply an emf source $e(t) = \mathcal{E}_{max} \sin(\omega t)$ to a series combination of a resistance R, an inductance L, and a capacitance C. Moreover, suppose the angular frequency ω is variable and you increase its magnitude beginning from $\omega = 0$. The amplitude of the current in the circuit is given by

$$I_{max} = \frac{\mathcal{E}_{max}}{Z} = \frac{\mathcal{E}_{max}}{\sqrt{R^2 + (X_L - X_C)^2}}$$

As ω approaches ω_{res} the inductive reactance X_L approaches the capacitive reactance X_C and not only does the power near a maximum, but the current amplitude I_{max} also grows toward a maximum. At resonance $X_L = X_C$ and

$$I_{max} = \frac{\mathcal{E}_{max}}{R} \qquad \text{(At resonance)} \tag{33.41}$$

This expression for peak current can be combined with Equation (33.36) to give the average power dissipated in the resistor as a function of the frequency of the emf source (see Problem 76):

$$\mathcal{P}_{av} = \frac{\mathcal{E}_{rms}^2 R \omega^2}{R^2 \omega^2 + L^2(\omega^2 - \omega_{res}^2)^2}$$

Concept Question 4
Make a drawing of an impedance diagram of a certain *RLC* circuit for which $R = X_C = X_L$ at resonance. Draw qualitative sketches showing the impedances for $\omega < \omega_{res}$ and at $\omega > \omega_{res}$.

The dependencies on ω of the power factor $\cos(\phi)$, the current amplitude I_{max}, and the average power $\mathcal{P}_{av}$ dissipated in the circuit are shown in Figure 33.21. From Equations (33.39) and (33.41) you should see that the resonant peaks for I_{max} and P_{av} are higher for smaller values of the resistance R. Incidentally, you need not worry about the

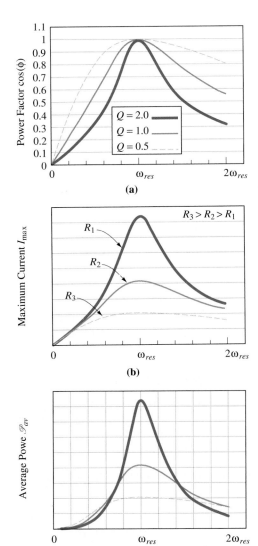

FIGURE 33.21

The behavior of (a) the power factor $\cos(\phi)$, (b) the current amplitude I_{max}, and (c) the average power $\mathcal{P}_{av}$ as a function of the angular frequency ω of the source. The peaks occur at $\omega = \omega_{res} = 1/\sqrt{LC}$.

current or power reaching infinity if $R = 0$. In any real series *RLC* circuit, these peaks never diverge even in the absence of a resistor R because the inductor and the wires connecting the capacitor and inductor to the $i(t)$ or $v(t)$ source always have some nonzero resistance.

A parameter Q is often defined to describe the sharpness of the resonance peaks of both electrical and mechanical oscillating systems. The "Q" of a system is defined by the relation $Q \equiv 2\pi U_{max}/\Delta U$, where U_{max} is the maximum energy stored in the system and ΔU is the energy dissipated by the system during a single cycle. In the case of the series *RLC* circuit Q is easily calculated. We recall from our discussion of oscillating circuits that energy is alternately stored in the inductor and the capacitor. If we choose to examine, say, the inductor, we know that the maximum energy stored is $U = LI_{max}^2/2$. We also know that it is the resistor only that dissipates energy in the circuit and that it does so at the average rate $\mathcal{P} = I_{rms}^2 R = I_{max}^2 R/2$. Thus, during a single cycle the electrical energy dissipated is $\Delta U = I_{max}^2 RT/2$, where T is the period of the oscillation. Combining the expressions for U and ΔU, we find that for the series *RLC* circuit

$$Q = 2\pi \frac{U}{\Delta U} = 2\pi \frac{(LI_{max}^2/2)}{(I_{max}^2 RT/2)} = \frac{\omega_{res}L}{R}$$

where in the previous step we noted $2\pi/T = 2\pi f_{res} = \omega_{res}$. As a practical matter when Q is greater than 3 or 4, its value can be determined from a graph of power versus frequency.

For values of $Q > 4$, $Q \approx f_{res}/\Delta f$, where Δf is the difference between the two frequencies on either side of the resonance peak at which the dissipated power falls to one-half the value at the resonance peak.

For such a case it can be shown (see Problem 81) that

$$Q \approx \frac{f_{res}}{\Delta f}$$

where Δf is the **full width at half maximum** (FWHM) of the resonance peak.

Example 33.7 provides another opportunity for you to exercise your knowledge of the phase relations between the current and potential functions of a series *RLC* circuit.

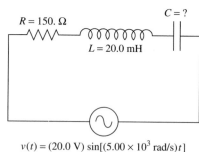

$$v(t) = (20.0 \text{ V}) \sin[(5.00 \times 10^3 \text{ rad/s})t]$$

FIGURE 33.22

An *RLC* circuit with $R = 150.\ \Omega$, $L = 20.0$ mH and an unknown capacitance C. See Example 32.7.

EXAMPLE 33.7　*A Series of Calculations for a Series Circuit*

For the series *RLC* circuit shown in Figure 33.22: (a) Find the value of the capacitance C that causes the circuit to be in resonance. (b) Find I_{rms} and $\mathscr{E}_{rms}$ at resonance. (c) If the angular frequency of the emf source $e(t)$ is changed to 2.00×10^3 rad/s, find the power dissipated and the equations for $i(t)$, $v_L(t)$, and $v_C(t)$.

SOLUTION　(a) We must find the capacitance for the resonant angular frequency $\omega_{res} = 5.00 \times 10^3$ rad/s. By Equation (33.40),

$$C = \frac{1}{L\omega_{res}^2} = \frac{1}{(20.0 \times 10^{-3} \text{ H})(5.00 \times 10^3 \text{ rad/s})^2} = 2.00 \ \mu\text{F}$$

(b) The root-mean-square emf is given by

$$\mathscr{E}_{rms} = \frac{\mathscr{E}_{max}}{\sqrt{2}} = \frac{20.0 \text{ V}}{\sqrt{2}} = 14.1 \text{ V}$$

At resonance $Z = R$, so the current amplitude is

$$I_{max} = \frac{\mathscr{E}_{max}}{Z} = \frac{\mathscr{E}_{max}}{R}$$

or

$$I_{rms} = \frac{\mathscr{E}_{rms}}{R} = \frac{14.1 \text{ V}}{150.\ \Omega} = 0.0940 \text{ A}$$

(c) At $\omega = 2.00 \times 10^3$ rad/s

$$X_L = \omega L = (2.00 \times 10^3 \text{ rad/s})(20.0 \times 10^{-3} \text{ H}) = 40.0 \ \Omega$$

$$X_C = \frac{1}{\omega C} = \frac{1}{(2.00 \times 10^3 \text{ rad/s})(2.00 \times 10^{-6} \text{ F})} = 250.\ \Omega$$

$$\phi = \arctan\left(\frac{X_L - X_C}{R}\right) = \arctan\left(\frac{40.0\ \Omega - 250.\ \Omega}{150.\ \Omega}\right) = -0.951\ \text{rad}$$

The impedance Z is

$$Z = \sqrt{R^2 + (X_L - X_C)^2} = \sqrt{(150.\ \Omega)^2 + (40.0\ \Omega - 250.\ \Omega)^2} = 258\ \Omega$$

The current amplitude is

$$I_{max} = \frac{\mathscr{E}_{max}}{Z} = \frac{20.0\ \text{V}}{258\ \Omega} = 0.0775\ \text{A}$$

The average power dissipated is given by Equation (33.34):

$$\mathscr{P}_{av} = I_{rms}\mathscr{E}_{rms}\cos(\phi) = \frac{I_{max}\mathscr{E}_{max}}{2}\cos(\phi)$$

$$= \frac{(0.0775\ \text{A})(20.0\ \text{V})}{2}\cos(-0.951\ \text{rad}) = 0.450\ \text{W}$$

The phase constant is negative, so the current $i(t)$ leads the applied emf $e(t)$ by $|\phi|$:

$$i(t) = I_{max}\sin(\omega t) = (0.0775\ \text{A})\sin[(2.00 \times 10^3\ \text{rad/s})t + 0.951\ \text{rad}]$$

The potential drop across the inductor *leads* the current $i(t)$ by $\pi/2$ radians:

$$v_L(t) = V_{L_{max}}\sin\left(\omega t - \phi + \frac{\pi}{2}\right)$$

Now,

$$V_{L_{max}} = I_{max}X_L = (0.0775\ \text{A})(40.0\ \Omega) = 3.10\ \text{V}$$

so

$$v_L(t) = (3.10\ \text{V})\sin\left[(2.00 \times 10^3\ \text{rad/s})t + 0.951\ \text{rad} + \frac{\pi}{2}\right]$$

The potential drop across the capacitor *lags* the current by $\pi/2$ rad:

$$v_C(t) = V_{C_{max}}\sin\left(\omega t - \phi - \frac{\pi}{2}\right)$$

Now,

$$V_{C_{max}} = I_{max}X_C = (0.0775\ \text{A})(250\ \Omega) = 19.4\ \text{V}$$

so that

$$v_C(t) = (19.4\ \text{V})\sin\left[(2.00 \times 10^3\ \text{rad/s})t + 0.951\ \text{rad} - \frac{\pi}{2}\right]$$ ◀

Electric energy is transmitted from one point to another along high-voltage transmission lines.

33.5 Transformers (Optional)

Have you ever wondered why power lines are designed to carry electric power at very high potentials? These transmission lines are usually miles long and, therefore, have a significant amount of resistance. The average power dissipated in a wire of resistance R is given by Equation (33.36): $\mathcal{P}_{\text{dis}_{av}} = I_{\text{rms}}^2 R$. Therefore, if we require that the Joule heat transferred to the environment be small, we need to keep the current magnitude small during transmission. However, the purpose of the transmission line is to deliver large amounts of power to distant loads. The total delivered power is given by Equation (33.34): $\mathcal{P}_{\text{tot}_{av}} = I_{\text{rms}} V_{\text{rms}} \cos(\phi)$. This equation reminds us that to transmit a useful average power $\mathcal{P}_{av}$ with a small current I_{rms} we need to make V_{rms} large. In practice, it is common for V_{rms} to be about 350 kV, but in some cases it may be as high as 750 kV. Now, we don't know about your home, but 350 kV is not very useful around ours. In fact, there are many devices, such as radios, televisions, and stereo systems, that have electronic components that are designed to operate with potentials significantly lower than the root-mean-square 120 V that is available at a common wall socket.

A **transformer** is a device used to lower or raise the amplitude of AC potentials and currents. A simple transformer consists of two separate coils wound around a common soft-iron core as shown in Figure 33.23. The coil on the left has N_1 turns, is attached to the emf source $v(t)$, and is called the **primary.** The **secondary** coil is on the right in Figure 33.23 and its N_2 turns provide the output of the transformer. The impedance Z_2, placed across the secondary when switch S is closed, is called a **load** impedance. The load may consist of your stereo system, your toaster, or the combination of all the impedances connected to your household wiring that, in turn, is attached to the power company's transformer out on your street.

When switch S in Figure 33.23 is open, there is no current in the secondary and so the power source to the transformer acts like a simple inductor L_1. In this case the primary current of an actual transformer is made up of two components: the **magnetizing current** required to establish the magnetic field in the core and the **core-loss current** due to hysteresis and eddy current losses in the core. The magnetizing current is the I in $-L_1(dI/dt)$ and, as we have seen, it lags the potential applied to the primary by 90°. The core-loss current is in phase with the applied potential and the resulting power loss is like having an extra resistor in the primary circuit. The phasor sum of these currents constitutes the **energizing current** of the transformer.

As we have done many times, we examine a model as a means to understanding the fundamental processes at work. Our **ideal transformer** has coils of zero resistance and a core of very high permeability μ. We assume that all lines of B that pass through the

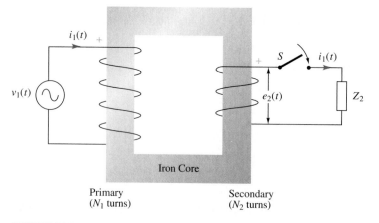

FIGURE 33.23

A model of an ideal transformer. An AC potential $v_1(t)$ with amplitude $V_{1_{\text{max}}}$ is applied to the primary with N_1 turns of wire. An output emf of amplitude $\mathcal{E}_{2_{\text{max}}}$ is induced in the secondary, which has N_2 turns.

primary coil are confined within the core and thus also pass through the secondary. In the discussions of the ideal transformer that follow we assume that the transformer impedance is much larger than any other impedance in the problem so that the magnetizing current is very small and can be ignored in comparison to other currents of interest. Furthermore we assume there are no hysteresis losses so that the core-loss current is also zero. In short, in the ideal-transformer model we take the total energizing current to be negligible. Figure 33.23 shows our ideal-transformer model and indicates the direction of positive applied potential difference v_1, positive induced emf e_2, and positive current flow i_1 in the primary and i_2 in the secondary.

We first consider the transformer in Figure 33.23 when the switch S is open. Because the coils of our ideal transformer have no resistance, we can model the circuit on the left side of Figure 33.23 as a pure inductor of magnitude L_1 so that the power factor phase constant ϕ is 90°. Thus, the average power delivered to the primary by the source $v_1(t)$ is zero.

The iron core helps to confine the magnetic flux lines generated in the primary to the core so that the magnetic flux through the secondary is as large as possible. In the ideal case all of the magnetic field lines are maintained within the iron core, so that the flux through each turn of the secondary is equal to the flux through each turn of the primary. Denoting the *total* flux through the primary and secondary coils as Φ_1 and Φ_2, respectively, we can write the equality of the flux per turn as:

$$\frac{\Phi_2}{N_2} = \frac{\Phi_1}{N_1}$$

Because the number of turns in each coil is constant, the time derivatives of both sides of this equation yield

$$\frac{1}{N_2}\frac{d\Phi_2}{dt} = \frac{1}{N_1}\frac{d\Phi_1}{dt}$$

Faraday's law requires the *induced* emfs across the primary and secondary to be, respectively,

$$e_1 = -\frac{d\Phi_1}{dt} \qquad \text{and} \qquad e_2 = -\frac{d\Phi_2}{dt}$$

The previous two equations taken together imply

$$\frac{e_1}{N_1} = \frac{e_2}{N_2} \qquad\qquad (33.42)$$

Transformers used to convert electric power into convenient potentials and currents come in all sizes.

For the case of a sinusoidal applied potential and ideal inductors, the induced emfs $e_2 = \mathcal{E}_{2_{max}} \sin(\omega t)$ and emf $e_1 = \mathcal{E}_{1_{max}} \sin(\omega t)$ are in phase so that

$$\frac{\mathcal{E}_{2_{max}}}{N_2} = \frac{\mathcal{E}_{1_{max}}}{N_1}$$

For our ideal transformer the applied potential difference v_1 is exactly equal and opposite to the induced back emf $e_1 = -L_1(dI_1/dt)$ so that $\mathcal{E}_{1_{max}} = V_{1_{max}}$. Similarly, for our ideal transformer, the potential drop v_2 across any load connected to the terminals of the secondary is exactly equal and opposite to the induced emf e_2 so that $\mathcal{E}_{2_{max}} = V_{2_{max}}$. We can therefore rewrite the previous equation as

$$V_{2_{max}} = \frac{N_2}{N_1} V_{1_{max}} \qquad\qquad (33.43)$$

If $N_2 < N_1$, the output amplitude $V_{2_{max}}$ of the secondary is less than $V_{1_{max}}$ and we have a **step-down** transformer. On the other hand, when $N_2 > N_1$, then $V_{2_{max}}$ is greater than $V_{1_{max}}$, and the device is called a **step-up** transformer.

When the switch S in Figure 33.23 is closed, the induced emf e_2 in the secondary is imposed on a load impedance Z_2. The result is a current i_2 flowing in the secondary. The magnitude of this current can be found in the usual way:

$$I_{2_{max}} = \frac{V_{2_{max}}}{Z_2}$$

As we expect, the potential drop v_2 across the load in the secondary leads the current i_2 by a phase angle $\phi_2 = \arctan[(X_{L_2} - X_{C_2})/R_2]$.

The induced current i_2 in the secondary has a rather dramatic effect. The current i_2 itself *attempts* to change the magnetic field inside the transformer core. But for our ideal-transformer model the B-field in the core cannot change; it is fixed by the applied potential $v_1(=-e_1)$ through Equation (33.42). Before the switch was closed the power source connected to the primary had to supply only a very small polarizing current. Now, however, this source must supply an additional current i_1' of exactly the correct magnitude and phase to cancel the flux change that the current i_2 in the secondary is attempting to create.

The magnitude of the magnetic field change ΔB that i_2 creates is given by Ampère's law

$$\oint \Delta \mathbf{B} \cdot d\mathbf{l} = \mu N_2 i_2$$

where the magnetic loop is taken around the transformer core and μ is the permeability of the core material. The primary must now supply a new current i_1' of exactly the size required to cancel the magnetic circulation created by the secondary current:

$$\oint \Delta \mathbf{B} \cdot d\mathbf{l} = \mu N_1 i_1'$$

This requirement implies

$$\mu N_1 i_1' = \mu N_2 i_2$$

In our ideal-transformer model the energizing current in the primary is small so that i_1' is essentially the only current in the primary; therefore we call it simply i_1. The previous equation must hold at each moment in time. This equality implies that the currents i_1 and i_2 have the same frequency and phase, so that

$$N_1 I_{1_{max}} \sin(\omega t) = N_2 I_{2_{max}} \sin(\omega t)$$

or

$$I_{2_{max}} = \frac{N_1}{N_2} I_{1_{max}} \tag{33.44}$$

Of course, a similar relation holds for rms values of the currents. Comparing this expression with Equation (33.43) we see that for a step-up transformer the price we pay for an increase in the emf is a decrease in current. On the other hand a step-down transformer has a larger current in the secondary than in the primary.

Let's now compare the power delivered to the load impedance in the secondary with the power output by the emf source in the primary. The power delivered to the load Z_2 is

$$\mathcal{P}_2 = I_{2_{rms}} \mathcal{E}_{2_{rms}} \cos(\phi)$$

We first note that because v_1 and v_2 are in phase and the currents i_1 and i_2 are in phase, the phase constant ϕ between the emf and the current is the same for both primary and secondary of our ideal transformer. Thus, using the rms equivalent of Equations (33.43) and (33.44), we can write

$$\mathscr{P}_2 = \left(\frac{N_1}{N_2}I_{1_{rms}}\right)\left(\frac{N_2}{N_1}\mathscr{E}_{1_{rms}}\right)\cos(\phi) = I_{1_{rms}}\mathscr{E}_{1_{rms}}\cos(\phi) = \mathscr{P}_1$$

We see that our *ideal* transformer conserves energy perfectly. Real transformers have efficiencies usually in excess of 90% (sometimes as high as 99%).

Let's now look to see what effective load is observed by the external potential source driving the primary. The magnitude of the effective load impedance in the primary is given by

$$Z_1 = \frac{V_{1_{max}}}{I_{1_{max}}} = \frac{\left(\frac{N_1}{N_2}\right)V_{2_{max}}}{\left(\frac{N_2}{N_1}\right)I_{2_{max}}} = \left(\frac{N_1}{N_2}\right)^2\frac{V_{2_{max}}}{I_{2_{max}}} = \left(\frac{N_1}{N_2}\right)^2 Z_2$$

Hence the effective impedance seen by the source driving the primary is the actual load impedance in the secondary multiplied by the square of the turns ratio $n = N_1/N_2$. The circuit symbol for a transformer is shown in Figure 33.24. In Table 33.4 we summarize these results for a transformer and show the circuit models for both the primary and the secondary.

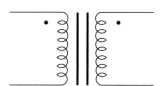

FIGURE 33.24

The circuit symbol for a transformer.

TABLE 33.4   Circuit Models for the Ideal Transformer

Let $n = N_2/N_1$ so that $n > 1$ is a step-up transformer

potential transformation: $v_2 = nv_1$

current transformation: $i_2 = \dfrac{i_1}{n}$

For a secondary with load impedance Z_2, the *apparent* load Z' in the primary is

$$Z' = \frac{Z_2}{n^2}$$

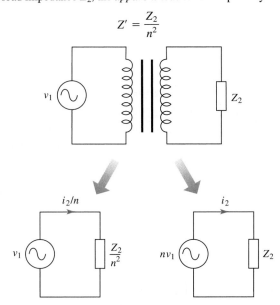

FIGURE 33.25

A source and a load (Z_2) coupled by a transformer can be modeled as two distinct circuits as shown.

When a real emf source with internal resistance r is attached directly to a load resistance R, we have seen previously (Problem 27.1) that the maximum power is delivered to the load when $r = R$. The same holds true for AC circuits. Often, however, both R and r

Concept Question 5
How is a lever with a fulcrum
nearer one end than the other
end like a transformer? Describe
the analog for both the force at
each end of the lever and for the
distance traveled by each end of
the lever.

are determined by other design constraints and cannot be changed. In such cases the load can be coupled to the power source using an **impedance-matching transformer** with turns ratio chosen so that the *apparent load* seen by the power source driving the primary is equal to the internal resistance of that source. For example, an output amplifier with internal resistance of 800 Ω can efficiently drive an 8-Ω speaker if these two devices are coupled via a transformer with a turns ratio $n = N_2/N_1 = 1/10$.

EXAMPLE 33.8   *Higher Potential, Lower Current*

A generator delivers a power of 100. MW at an rms potential of 22.0 kV. The emf is stepped up to 500. kV rms for transmission along a power line with total resistance 50.0 Ω. Assume the transformer is ideal and compute the rms current in the transmission line and the percentage power loss during transmission. (Assume a phase angle of zero in the secondary.)

SOLUTION   For an ideal transformer the power delivered by the secondary is equal to that provided to the primary. Hence, in the secondary

$$\mathscr{P}_{av} = I_{2_{rms}} V_{2_{rms}} \cos(\phi)$$

Assuming $\phi = 0°$, we find

$$I_{2_{rms}} = \frac{\mathscr{P}_{av}}{V_{2_{rms}}} = \frac{1.00 \times 10^8 \text{ W}}{5.00 \times 10^5 \text{ V}} = 200. \text{ A}$$

The electric power lost to Joule heating of the transmission line is

$$\mathscr{P}_{lost} = I_{2_{rms}}^2 R = (200. \text{ A})^2(50.0 \text{ } \Omega) = 2.00 \text{ MW}$$

The percentage power lost is

$$\frac{\mathscr{P}_{lost}}{\mathscr{P}_{av}} \times 100\% = \frac{2.00 \text{ MW}}{100. \text{ MW}} \times 100\% = 2.00\%$$ ◀

EXAMPLE 33.9   *Transforming the Load*

The primary coil of an ideal transformer is driven by an ideal potential source with an rms value of 125. V at a frequency of 60.0 Hz. The secondary is connected to a load resistor of 900. Ω in series with an inductor of 450. mH as shown in Figure 33.26. The primary has 500 turns, and the secondary has 1500 turns. (a) Find the power delivered to the load. (b) Write expressions for the current $i_2(t)$ in the secondary and the output $v_2(t)$ of the secondary. (c) Determine the effective load as seen by the power source connected to the primary.

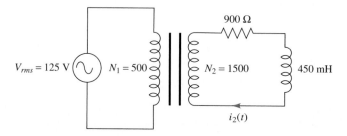

FIGURE 33.26

The secondary is connected to a load resistor of 900. Ω in series with an inductor of 450. mH. See Example 32.9.

SOLUTION   (a) The turns ratio $n = N_2/N_1 = 3$, leading to an induced emf of 375 V rms across the secondary. The reactance of the load inductor in the secondary is

$$X_L = \omega L = 2\pi fL = 2\pi(60.0 \text{ Hz})(450. \text{ mH}) = 170. \ \Omega$$

Thus, the impedance of the load in the secondary is

$$Z_2 = \sqrt{X_L^2 + R^2} = \sqrt{(170. \ \Omega)^2 + (900. \ \Omega)^2} = 916 \ \Omega$$

The rms current in the secondary is

$$I_{2_{rms}} = \frac{V_{2_{rms}}}{Z_2} = \frac{375 \text{ V}}{916 \ \Omega} = 0.409 \text{ A}$$

The phase angle between v_2 and the i_2 in the secondary is given by

$$\phi = \arctan\left(\frac{X_{L_2}}{R_2}\right) = \arctan\left(\frac{170. \ \Omega}{900. \ \Omega}\right) = 0.187 \text{ rad}$$

Thus, the power delivered to the load resistor is

$$\mathcal{P}_2 = I_{2_{rms}} V_{2_{rms}} \cos(\phi) = (0.409 \text{ A})(375 \text{ V}) \cos(0.187 \text{ rad}) = 151 \text{ W}$$

You may wish to check that this same value results from $\mathcal{P}_2 = (I_{2_{rms}})^2 R_2$.
   (b) We take the induced emf in the secondary as our reference phase and write

$$v_2(t) = V_{2_{max}} \sin(\omega t) = \sqrt{2}V_{2_{rms}} \sin(\omega t) = \sqrt{2} \ (375 \text{ V}) \sin[2\pi(60.0 \text{ Hz})t]$$
$$= (530. \text{ V}) \sin[(377 \text{ rad/s})t]$$

The current lags v_2 by ϕ, so that

$$i_2(t) = I_{2_{max}} \sin(\omega t - \phi)$$
$$= \sqrt{2} \ I_{2_{rms}} \sin(\omega t - \phi) = \sqrt{2}(0.409 \text{ A}) \sin[2\pi(60.0 \text{ Hz})t - 0.187 \text{ rad}]$$
$$= (5.78 \text{ A}) \sin[(377 \text{ rad/s})t - 0.187 \text{ rad}]$$

(c) The effective load seen by the driving potential v_1 is an impedance

$$Z' = \frac{Z_2}{n^2} = \frac{916 \ \Omega}{(3)^2} = 102 \ \Omega$$

Thus, the current in the primary is

$$I_{1_{rms}} = \frac{V_{1_{rms}}}{Z'} = \frac{125 \text{ V}}{102 \ \Omega} = 1.22 \text{ A}$$

It is a convenient check on our work to note that $I_{1_{rms}} = nI_{2_{rms}} = (3)(0.409 \text{ A}) = 1.23$ A and that $\mathcal{P}_1 = I_{1_{rms}} V_{1_{rms}} \cos(\phi) = (1.22 \text{ A})(125 \text{ V}) \cos(0.187 \text{ rad}) = 150$ W, equal (to within round-off error) to the power delivered to the load. ◀

33.6 Filter Circuits (Optional)

A filter circuit can be constructed with a series combination of a resistor and a capacitor. Filters are used to eliminate unwanted periodic potentials at certain frequencies. For example, undesirable emfs can be induced in audio equipment by its 60-Hz power sources. If

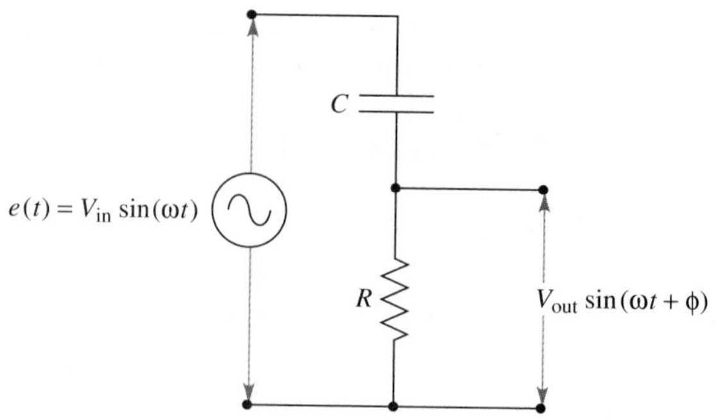

$e(t) = V_{in} \sin(\omega t)$

C

R $V_{out} \sin(\omega t + \phi)$

FIGURE 33.27

A simple high-pass RC filter circuit. For proper choices of R and C the resistor passes periodic signals with high frequencies and attenuates signals with low frequencies.

these emfs are not eliminated, they manifest themselves as an annoying hum from the speaker. There are many electronic applications in which it is desirable to eliminate either high-frequency or low-frequency noise from a periodic signal. To discover how we might accomplish this task let's take a look at the circuit shown in Figure 33.27.

The sinusoidal emf source $e(t) = V_{in} \sin(\omega t)$ represents the input signal to the RC circuit. We first investigate the amplitude of the potential drop across the resistor. The *maximum input* V_{in} is related to the maximum input current I by Equation (33.19):

$$V_{in} = IZ = I\sqrt{R^2 + X_C^2} = I\sqrt{R^2 + (1/\omega C)^2}$$

We take the maximum potential drop across the resistor to be the *output potential* V_{out} of the filter:

$$V_{out} = I_{max}R$$

The *gain* of the filter is the ratio of the potential drop across the output resistor to the emf applied across both elements:

$$\frac{V_{out}}{V_{in}} = \frac{R}{\sqrt{R^2 + (1/\omega C)^2}} \qquad (33.45)$$

Notice from Equation (33.45) that when the angular frequency ω is large, the term $1/\omega C$ is small and the gain V_{out}/V_{in} is close to unity. However, when ω is small, the denominator in Equation (33.45) is large and the gain is much less than 1. For proper choices of R and C we can, therefore, construct a **high-pass filter.** A plot of a typical gain curve for a high-pass filter is shown in Figure 33.28.

Next, we reverse the role of the resistor and the capacitor so that our output is equal to the potential drop across the capacitor as shown in Figure 33.29. In this case

$$V_{out} = I_{max}X_C = \frac{I_{max}}{\omega C}$$

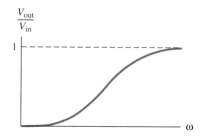

$\dfrac{V_{out}}{V_{in}}$

1

ω

FIGURE 33.28

The gain V_{out}/V_{in} for a high-pass filter.

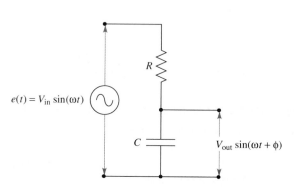

$e(t) = V_{in} \sin(\omega t)$

R

C $V_{out} \sin(\omega t + \phi)$

FIGURE 33.29

A low-pass filter. For proper choices of R and C the capacitor passes periodic signals with low frequencies and attenuates signals with high frequencies.

The gain of this new filter is

$$\frac{V_{\text{out}}}{V_{\text{in}}} = \frac{1/\omega C}{\sqrt{R^2 + (1/\omega C)^2}} \tag{33.46}$$

This gain is plotted as a function of frequency in Figure 33.30. Notice at $\omega = 0$, the gain $V_{\text{out}}/V_{\text{in}}$ is unity. However, as ω increases, the gain decreases toward zero. At low frequencies the potential drop across the capacitor is large and is available across the output terminals, whereas for high frequencies the potential drop across the capacitor tends to zero. This arrangement of R and C is known as a **low-pass filter.** Methods for determining R and C to produce a given response are explored in the problems. (See Problems 66 and 80.)

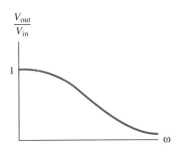

FIGURE 33.30

The gain $V_{\text{out}}/V_{\text{in}}$ for a low-pass filter.

Concept Question 6
Show that the gain of a high-pass filter approaches 1 as $\omega \rightarrow \infty$.
Show that the gain of a low-pass filter approaches 1 as $\omega \rightarrow 0$.

33.7 Summary

When a sinusoidal emf or current source is applied to a resistor, the current through the resistor and the potential drop across it are in phase. When one of these sources is applied to a capacitor, the current $i(t)$ leads the potential drop $v_C(t)$ across the capacitor by $\pi/2$ rad. Finally, when a sinusoidal emf or current source is applied to an ideal (resistanceless) inductor, the current through the inductor lags the potential drop across the inductor by $\pi/2$ rad.

The **capacitive reactance** and the **inductive reactance** are defined, respectively, as

$$X_C = \frac{1}{\omega C} \tag{33.7}$$

$$X_L = \omega L \tag{33.12}$$

where ω is the angular frequency of the current or emf source.

The **impedance** Z of an AC circuit containing an inductance L, a capacitance C, and a resistance R in series is given by

$$Z = \sqrt{R^2 + (X_L - X_C)^2} \tag{33.18}$$

and the relation between the current amplitude I_{max} and the amplitude V_{max} of the potential drop across the entire circuit is

$$V_{\text{max}} = I_{\text{max}}Z \tag{33.19}$$

In general, when a sinusoidal current or emf source is applied to a series RLC circuit, the potential drop across the circuit and the current through it are not necessarily in phase. The phase angle between the total potential change $v(t)$ and the current $i(t)$ is given by

$$\phi = \arctan\left(\frac{X_L - X_C}{R}\right) \tag{33.20}$$

When this **phase angle** ϕ is positive, the total potential change across the series RLC combination leads the current by ϕ, and when the phase angle is negative, the potential change lags the current by $|\phi|$.

The root-mean-square potential V_{rms} and current I_{rms} are related to the potential and current amplitudes V_{max} and I_{max} by

$$V_{\text{rms}} = \frac{V_{\text{max}}}{\sqrt{2}} \tag{33.28}$$

and

$$I_{rms} = \frac{I_{max}}{\sqrt{2}} \tag{33.29}$$

The **average power** delivered to a series RLC circuit by an emf or current source is

$$\mathcal{P}_{av} = I_{rms} V_{rms} \cos(\phi) = I_{rms}^2 R$$

where $\cos(\phi)$ is called the **power factor** and is given by

$$\cos(\phi) = \frac{R}{Z} = \frac{R}{\sqrt{R^2 + (X_L - X_C)^2}} \tag{33.36}$$

When the inductive reactance X_L equals the capacitive reactance X_C, the phase constant is zero, the power factor is unity, and the RLC circuit is said to be in **resonance.** The frequency at which resonance occurs is

$$\omega_{res} = \frac{1}{\sqrt{LC}} \tag{33.40}$$

The current in a series RLC circuit and the power dissipated in the circuit are both maximum at resonance. Large values of the quantity $Q = \omega_{res} L/R$ indicate a sharp resonance peak.

A transformer is a device that can be used to lower or raise an AC potential. It consists of a **primary** with N_1 turns of wire and a secondary with N_2 turns both wound on a common iron core. When an emf amplitude $V_{1_{max}}$ is applied across the primary of an ideal transformer, the emf across the secondary $V_{2_{max}}$ is given by

$$V_{2_{max}} = \frac{N_2}{N_1} V_{1_{max}} \tag{33.43}$$

In an ideal transformer the power provided to the primary is delivered by the secondary so that

$$I_{1_{rms}} V_{1_{rms}} \cos(\phi) = I_{2_{rms}} V_{2_{rms}} \cos(\phi)$$

PROBLEMS

33.1 Circuit Elements in AC Circuits

1. The emf source shown in Figure 33.2 has a frequency of 40.0 Hz and an emf amplitude of 15.0 V. (a) If $R = 2.20$ kΩ what is the amplitude of the current in the circuit? (b) Compute the current in the circuit at $t = 0.64$ s and at $t = 1.44$ s.

2. The emf source described by $e(t) = \mathcal{E}_{max} \sin(\omega t)$ and shown in Figure 33.2 has an amplitude of 36.0 V. At $t = 0.150$ s the potential drop across the resistor is -36.0 V for the first time since $t = 0$. The resistance R is 12.0 Ω. (a) What is the angular frequency ω of $e(t)$? (b) What is the current amplitude I_{max}? (c) What is the current in the circuit at $t = 0.0750$ s?

3. The amplitude of the emf source $e(t) = \mathcal{E}_{max} \sin(\omega t)$ shown in Figure 33.2 is 12.5 V. At $t = 10.0$ ms the potential drop across the resistor $R = 30.0$ Ω is 10.0 V and increasing. This instant is the first time since $t = 0$ that this voltage has occurred. (a) Compute the angular frequency ω of $e(t)$. (b) Find the earliest time when the potential drop across the resistor is -10.0 V.

4. The amplitude of the emf source $e(t) = \mathcal{E}_{max} \sin(\omega t)$ shown in Figure 33.2 is 24.0 V. At $t = 15.0$ ms the potential drop across the resistor $R = 72.0$ Ω is 19.4 V and decreasing. This instant is the first time since $t = 0$ that this voltage has occurred. (a) Compute the angular frequency ω of $e(t)$. (b) What is the current in the circuit at $t = 64.0$ ms?

5. A 0.0100-μF capacitor is connected to a sinusoidal emf source with a 24.0-V amplitude and a 60.0-Hz frequency. (a) What is the capacitive reactance of the capacitor in this circuit? (b) What is the current amplitude I_{max}? (c) What is the magnitude of the maximum charge acquired by the capacitor?

6. A sinusoidal emf source with an amplitude of 36.0 V is connected across a 20.0-μF capacitor. What must be the frequency (in hertz) of the source if the resulting current amplitude is 135 mA?

7. A power supply provides a sinusoidal emf with a frequency of 560. Hz. What must be the amplitude of the emf for the source to provide a current with a 0.500-A amplitude when connected across a 4.70-μF capacitor?

8. The emf source $e(t) = \mathcal{E}_{max} \sin(\omega t)$ shown in Figure 33.4 has an amplitude of 18.0 V and a frequency of 440. Hz. A 3.70-μF capacitor is connected to the source. (a) Compute the current amplitude and the maximum charge on the capacitor. (b) What is the potential drop across the capacitor and the current in the circuit at $t = 360.$ μs?

9. The emf source $e(t) = \mathcal{E}_{max} \sin(\omega t)$ shown in Figure 33.4 has an amplitude of 24.0 V, whereas its frequency can vary. The capacitor shown in the figure has a capacitance of 0.0100 μF. (a) What frequency must be selected for a current amplitude of 15.0 mA? (b) For the frequency you found in part (a), what is the potential drop across the capacitor and the current in the circuit at $t = 20.0$ μs?

10. The frequency of the emf source $e(t) = \mathcal{E}_{max} \sin(\omega t)$ shown in Figure 33.4 is 480. Hz, and its amplitude is 72.0 V. The capacitor shown in the figure carries a maximum charge of 3.60 μC. (a) What is the capacitive reactance X_C of the capacitor in this circuit? (b) Compute the potential drop across the capacitor and the current in the circuit at $t = 84.0$ ms.

11. A 0.0200-μF capacitor is connected to a sinusoidal current source $i(t) = (3.00 \text{ mA}) \sin[(1.20 \times 10^3 \text{ rad/s})t]$. (a) Compute the capacitive reactance X_C for this circuit. (b) What is the amplitude of the potential drop across the capacitor? (c) Write the expression for the potential drop $v_C(t)$ across the capacitor and evaluate it at $t = 2.50$ s.

12. A sinusoidal emf source $e(t) = (2.00 \text{ V}) \sin[(540. \text{ rad/s})t + \pi/3]$ is connected to a 4.70-μF capacitor. (a) Find the capacitive reactance X_C for this circuit. (b) What is the amplitude I_{max} of the current in the circuit? (c) Find an expression for the current $i(t)$ in the circuit and evaluate it at $t = 4.00$ s.

13. A power supply provides a sinusoidal emf with an amplitude of 6.00 V. (a) What frequency must the power supply provide to a 2.20-H inductor for the current amplitude to be 360. μA? (b) What is the current amplitude if the frequency is half of that computed in part (a)?

14. The amplitude of a sinusoidal current source is maintained at 0.720 A. When the source is attached to a 0.250-H inductor, the amplitude of the potential drop across the inductor is 36.0 V. What is the amplitude of the potential drop when the frequency is reduced to half its original value?

15. A 50.0-mH inductor is attached to a sinusoidal current source that provides a peak current of 2.50 A at a frequency of 76.0 Hz. (a) Compute the inductive reactance X_L for this circuit. (b) Find the peak potential drop across the inductor. (c) For the same current what must the frequency be for the peak potential drop to be 1.00 kV?

16. A sinusoidal emf source $e(t) = (48.0 \text{ V}) \sin[(60.0\pi \text{ rad/s})t]$ is attached to a 0.500-H inductor. (a) Compute the inductive reactance X_L of the inductor. (b) What is the current amplitude I_{max} in the circuit? (c) Write down the expression for the current in the circuit $i(t)$ and compute its value at $t = 160.$ ms.

17. A sinusoidal current source $i(t) = (1.20 \text{ A}) \sin[(1000 \text{ rad/s})t]$ is attached to a 60.0-mH inductor. (a) Find the inductive reactance X_L for this circuit. (b) Find the peak potential drop across the inductor. (c) Write down the expression for the potential drop across the inductor and compute its value at $t = 135$ ms.

18. A 1.50-mH inductor is attached to a sinusoidal emf source $e(t) = (6.00 \text{ V}) \sin[(3200 \text{ rad/s})t + \pi/4]$. (a) Compute the inductive reactance X_L for this circuit. (b) What is the amplitude of the current in the circuit? (c) Write down the expression for the current $i(t)$ through the circuit and compute its value at $t = 1.80$ s.

33.2 *RLC* Circuits

19. A series *RLC* circuit consists of a 0.0500-μF capacitor, a 64.0-mH inductor, and a resistance R. The combination is connected to a sinusoidal emf $e(t)$ that has a frequency $f = 1500$ Hz. (a) Does the current lead or lag the applied emf? (b) Find the resistance R such that the current through the circuit $i(t)$ is $\pi/6$ rad out of phase with the applied emf $e(t)$. (c) What is the impedance Z of the circuit you designed in part (b)?

20. A 1200-Hz sinusoidal emf is applied to a series *RLC* circuit that has an impedance of 6.00 kΩ. The circuit includes an inductance $L = 500.$ mH and a resistance $R = 5.50$ kΩ. (a) Find the magnitude of the capacitor in the circuit. (b) Compute the phase angle ϕ. Does the current through the circuit lead or lag the applied emf?

21. The *RCL* circuit of Figure 33.P1 has component values $R = 310.$ Ω, $L = 1.60$ mH, and $C = 0.0250$ μF. The frequency is such that the current amplitude is $I_{max} = 215$ mA when the amplitude of the applied emf is $\mathcal{E}_{max} = 96.0$ V. If the current $i(t)$ leads the emf $e(t)$, find (a) the impedance of the circuit, (b) the frequency f of the sinusoidal source, and (c) the phase constant ϕ.

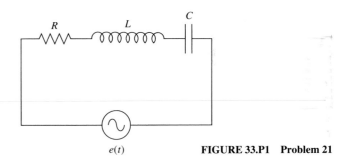

$e(t)$ **FIGURE 33.P1   Problem 21**

22. A wire coil with a resistance $R = 180.$ Ω has an inductance $L = 96.0$ mH. A current source $i(t) = (240 \text{ mA}) \sin[(2400 \text{ rad/s})t]$ is applied to the coil. (a) Compute the impedance and phase angle for this circuit. (b) Find an expression for the potential drop $v_L(t)$ across the inductor and evaluate it at $t = 1.50$ s. (*Hint*: You can model the real inductor as a series combination of an ideal inductor and a resistor. Remember, however, that the potential across *the* inductor is the potential across both elements of your model, that is, the potential drop across the ideal inductor alone has no physical meaning.)

23. A series *RLC* circuit consists of a 4.70-kΩ resistor, a 360.-mH inductor, and a variable capacitor C connected in series to a sinusoidal current source with frequency $f = 5.00$ kHz. (a) To what value must the capacitance be set for the current through the circuit to lead the applied potential by $\pi/6$ rad? (b) Compute the impedance of this circuit when the capacitance is that which you found in part (a).

24. A sinusoidal emf source $e(t) = (8.00 \text{ V}) \sin[(8000 \text{ rad/s})t]$ is applied to a series combination of a resistance $R = 3.00$ kΩ and a capacitance $C = 0.0240$ μF. (a) Find the impedance and phase angle for this circuit. (b) Find an expression for the current $i(t)$ in the circuit and evaluate it at $t = 0.300$ s.

25. A source of emf $e(t) = (120. \text{ V}) \sin[(7600 \text{ rad/s})t]$ is applied to a series combination of a resistance $R = 350.$ Ω and an inductance $L = 36.0$ mH. (a) Compute the impedance and phase angle for this circuit and plot the phasor diagram. (b) Find an expression for the current in the circuit and evaluate it at $t = 150.$ ms. (c) Find expressions for the potential drop across the resistor and the inductor, and evaluate them at $t = 150.$ ms.

26. A 6.00-μF capacitor is connected in series to an 800.-Ω resistor, and a current source described by (50.0 mA) sin[(500. rad/s)t] is applied across the combination. (a) Find the impedance and phase angle for this combination and plot the phasor diagram. (b) Find an expression for the potential drop across the capacitor and resistor (together) and evaluate it at $t = 3.50$ s. (c) Compute the potential drop across each circuit element at $t = 3.50$ s and verify that their sum equals that computed in part (b).

27. An *RLC* circuit consists of a series combination of a resistance $R = 15.0$ Ω, an inductance $L = 0.300$ H, and a capacitance $C = 100.$ μF. A current source $i(t) = (5.00$ A) sin[(200. rad/s)t] is applied across the series combination. (a) Find the impedance Z and the phase angle ϕ, and state whether the current leads or lags the total potential drop across the circuit. (b) What is the potential drop across the entire circuit at $t = 0.300$ s? (c) Compute the potential drop across the inductor and across the capacitor at $t = 0.300$ s.

28. A series *RLC* circuit consists of a resistance $R = 40.0$ Ω, an inductor $L = 0.400$ H, and a capacitor $C = 50.0$ μF, connected to a sinusoidal emf $e(t) = (180.$ V) sin[(200. rad/s)t]. (a) Find the expression for the current $i(t)$ in the circuit and evaluate it at $t = 0.700$ s. (b) Does the applied emf $e(t)$ lead or lag the current $i(t)$? (c) Find the potential drops across the resistor, capacitor, and inductor at $t = 0.700$ s.

29. A series combination of a resistance $R = 180.$ Ω, an inductance $L = 2.50$ H, and a capacitance $C = 0.150$ μF is connected to a sinusoidal current source. The emf $e(t)$ across the entire circuit is found to lead the current $i(t)$ by $\pi/3$ rad. Find the frequency f of the source and the impedance of the circuit.

30. The *RLC* circuit shown in Figure 33.P1 has component values $L = 2.50$ H, $R = 200$ Ω, and $C = 30.0$ μF. For an applied emf described by $e(t) = (10.0$ V) sin[(100 rad/s)t] find (a) the impedance of the circuit, (b) the phase constant ϕ, and (c) the expression for the current $i(t)$ in the circuit. (d) Make a plot of the phasor diagram for this circuit. (e) Evaluate the current in the circuit and the potential drops across the inductor and capacitor at $t = 1.50$ s.

31. A series *RLC* circuit consisting of a resistance $R = 250.$ Ω, an inductance $L = 6.00$ H, and a capacitance $C = 15.0$ μF is connected to an emf source $e(t) = (300.$ V) sin[(50π rad/s)t]. (a) Find the impedance and phase angle for this circuit. Does the current through the circuit lead or lag the applied emf? (b) Compute the applied emf and the current in the circuit at $t = 215$ ms. (c) Compute the potential drop across each circuit element at $t = 215$ ms and show that their sum equals the applied emf found in part (b).

32. A resistance $R = 150.$ Ω, an inductance $L = 1.00$ H, and a capacitance $C = 20.0$ μF are connected in series to a sinusoidal emf source described by $e(t) = (100.$ V) sin[(60π rad/s)t]. (a) Find the impedance and phase angle, and plot the phasor diagram for this circuit. (b) Find an expression for the current in the circuit and evaluate it at $t = 1.11$ s. (c) Evaluate the potential drop across each circuit element at $t = 1.11$ s and verify that their sum is equal to $e(t)$ at this same time.

33. For the series *RLC* circuit shown in Figure 33.P2 with $R = 400.$ Ω, $C = 0.600$ μF, $L = 100.$ mH driven by the current source $i(t) = (5.00$ A) sin[(3.00 × 10³ rad/s)t] find (a) the impedance and the phase angle and (b) the expression for the potential drop $v(t)$ across the entire circuit. (c) Plot the phasor diagram. (d) Compute the potential drop across each circuit element at $t = 1.50$ s and verify that their sum equals the total potential drop calculated in part (b) at this same time.

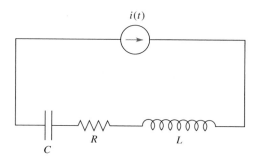

FIGURE 33.P2   Problem 33

34. A series *RLC* circuit has components with the following values: $R = 50.0$ Ω, $L = 200.$ mH, and $C = 10.0$ μF. The circuit contains an emf source $e(t) = (200.$ V) sin[(500. rad/s)t]. (a) Compute the impedance and phase angle for this circuit and plot a phasor diagram. (b) Find an expression for the current $i(t)$ in the circuit. (c) Evaluate the potential drop across each circuit element at $t = 0.700$ s and verify that their sum totals to $e(t)$ at this same time.

33.3 The Root-Mean-Square Potential and Current

35. What is the potential amplitude for an AC emf source if its root-mean-square potential is 115 V?

36. Alternating-current (AC) meters are used to measure the current I_{rms} through a resistor and the potential V_{rms} across the resistor. If $R = 76.0$ Ω and the potential drop is found to be $V_{rms} = 32.0$ V, find (a) the peak potential drop across the resistor and (b) the current I_{rms}.

37. By directly evaluating the integral given in Equation (33.23) show that the $<v^2(t)>_{av} = V^2_{max}/2$ and $<i^2(t)>_{av} = I^2_{max}/2$, where $v(t)$ and $i(t)$ are given by Equations (33.21a) and (33.21b). [*Hint*: You may wish to use the trigonometric identity $\sin^2(\theta) = [1 - \cos(\theta)]/2$.]

38. Compute V_{rms} for the potential function $v(t)$ shown in Figure 33.P3.

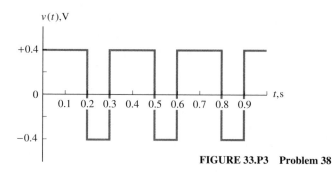

FIGURE 33.P3   Problem 38

39. Compute V_{rms} for the potential function $v(t)$ shown in Figure 33.P4.

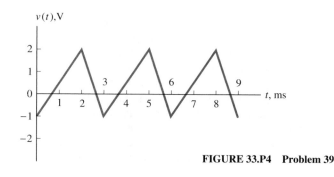

FIGURE 33.P4   Problem 39

40. Compute V_{rms} for a symmetrical square wave of amplitude V_{max} shown in Figure 33.P5.

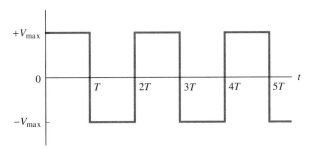

FIGURE 33.P5   Problem 40

41. Compute V_{rms} for the symmetrical triangle wave shown in Figure 33.P6.

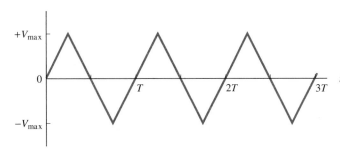

FIGURE 33.P6   Problem 41

42. (a) Compute the root-mean-square current in the circuit described in Problem 24. Compute the root-mean-square potential drop across (b) the capacitor and (c) the resistor in Problem 24.

43. (a) Compute the root-mean-square current in the circuit described in Problem 25. Compute the root-mean-square potential drop across (b) the inductor and (c) the resistor described in Problem 25.

44. Apply Equation (33.23) to show that $<\sin(\omega t)\cos(\omega t)>_{av} = 0$.

33.4 Power and Resonance in AC Circuits

45. A generator provides energy to an *RLC* circuit so that $V_{rms} = 48.0$ V and $I_{rms} = 2.40$ A. If the phase constant for the circuit is $52.9°$, compute (a) the reactance $(X_L - X_C)$ and (b) the resistance of the circuit R.

46. Compute the resistance of a series *RCL* circuit for which $I_{rms} = 240$. mA when the power dissipated in the circuit is 36.0 W.

47. When connected to a 60.0-Hz emf source where $V_{rms} = 120$. V, a certain electric motor behaves like a 15.0-Ω resistor in series with an inductive reactance of 4.00 Ω. (a) Compute the power factor for this circuit. (b) At what average rate is energy dissipated by the motor? (c) What is the magnitude of the peak current?

48. A generator delivers a sinusoidal emf with $V_{rms} = 75.0$ V to an *RLC* circuit with a 360. Ω impedance. If the resistance in the circuit is $R = 120$. Ω, compute the power dissipated in the circuit.

49. Compute the power delivered to the resistor in the circuit described in Problem 25.

50. For the circuit described in Problem 27 find (a) the power factor and (b) the average power delivered to the load resistor.

51. For the circuit described in Problem 28 find (a) the power factor, (b) the average power delivered to the resistor.

52. A series *RLC* circuit consists of a 5.40-kΩ resistor, a 3.80-mH inductor, and a 120.-pF capacitor connected in series to an 11.0-V

(rms) emf source. The frequency of the source is 2.00 MHz. (a) Compute the power factor for this circuit. (b) What power is dissipated by each of the *RLC* elements?

53. An *RLC* circuit consists of a 3.60-μF capacitor, a 4.70-kΩ resistor, and a 150.-mH inductor connected in series to a sinusoidal emf source. Compute the frequency f at which resonance occurs.

54. The tuning circuit of a radio consists of an 18.0-Ω resistor, a 0.0150-mH inductor, and a variable capacitor C in series. The circuit is to be set to a resonant frequency of 92.9 MHz. What capacitance should be selected?

55. For the circuit described in Problem 52 (a) to what value must the frequency be changed for resonance to occur? (b) What is the rms current in this circuit at resonance? (c) What power is dissipated at resonance?

56. When a 120.-V (rms) ideal emf source with a frequency of 60.0 Hz is applied across a series *RLC* circuit, an rms current of 16.0 mA is observed in the circuit. The current is found to lag the applied potential across the circuit by 35.0°. (a) What circuit element (capacitor or inductor) must be added in series to the circuit to make the phase constant zero? (b) What is the magnitude of the capacitor or inductor?

33.5 Transformers (Optional)

57. A sinusoidal emf $e(t) = (180.$ V$) \sin[(1.40 \times 10^3$ rad/s$)t]$ is applied to the primary of a transformer. The primary and secondary have 360 turns and 1200 turns, respectively. Compute the rms potential drop across the secondary.

58. A sinusoidal emf source provides 120. V (rms) to the primary of a transformer with 120. turns. How many turns must the secondary have to provide an output of 1.50 kV?

59. The primary of a certain transformer has 800. turns, and the rms potential applied to it is 120. V. For an output potential of 6.00 V, how many turns must the secondary have?

60. A transformer has 1500 turns in the primary and 500 turns in the secondary. If the secondary root-mean-square current and potential are 3.50 A and 110. V, respectively, what are the rms current and potential in the primary?

61. A 24.0-V (rms) source is attached across a 2000-turn primary of a transformer. The secondary has 100 turns. If the impedance of the load in the secondary is 1.50 kΩ (a) what is the impedance seen by an emf source attached to the primary? (b) What are I_{rms} and V_{rms} in the secondary?

62. A transformer with the circuit symbol shown in Figure 33.P7 has $N_1 = 800$ turns. An rms potential $V_1 = 120$. V is applied to this primary. The secondary is *tapped* in three places as indicated in the figure. Compute the number of turns N_2 and N_3 for the rms potential outputs from the secondary to be $V_2 = 12.0$ V and $V_3 = 18.0$ V.

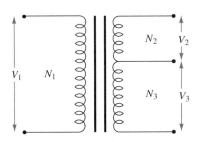

FIGURE 33.P7   Problem 62

33.6 Filter Circuits (Optional)

63. For the low-pass filter shown in Figure 33.29 take $R = 400.$ Ω and $C = 1.00$ μF. Compute the gain V_{out}/V_{in} for (a) a frequency of $f = 150.$ Hz and (b) a frequency $f = 1.60$ kHz.

64. The high-pass filter shown in Figure 33.27 consists of a resistance $R = 600.$ Ω and a capacitor $C = 0.100$ μF. Compute the gain V_{out}/V_{in} for (a) a frequency $f = 400.$ Hz and (b) a frequency $f = 5.00$ kHz.

65. The capacitance in the low-pass filter shown in Figure 33.29 is 1.00 μF. (a) If the gain V_{out}/V_{in} at an angular frequency of $\omega = 40.0 \times 10^3$ rad/s is to have a value of 0.100, compute the required resistance R. (b) For this resistance in the circuit compute the gain at $\omega = 400.$ rad/s.

66. Show that for a low-pass filter with resistance R and gain $V_{out}/V_{in} = G$ at the angular frequency ω the required capacitance is given by

$$C = \frac{\sqrt{1 - G^2}}{\omega G R}$$

67. A high-pass filter is designed with a 0.500-μF capacitor and is to provide a gain of 0.100 at an angular frequency of 1.00×10^3 rad/s. (a) What resistance R must be used in the circuit? (b) What is this circuit's gain at $\omega = 40.0 \times 10^3$ rad/s?

Numerical Methods and Computer Applications

68. For the RLC circuit given in Problem 34 write a program or construct a spreadsheet template to make plots of $i(t)$, $v(t)$, $v_R(t)$, $v_L(t)$, and $v_C(t)$ from $t = 0$ to $t = 3T$, where T is the period of the sinusoidal source.

69. Write a program or construct a spreadsheet template to plot the average power $\mathcal{P}_{av}$ as a function of ω as given in Problem 76. Input should consist of R, L, C, and the emf V_{rms}. Verify that the peak in $\mathcal{P}_{av}$ decreases in height with increasing values of R.

70. Use a spreadsheet to graph the power transferred to a resistor in a series RLC circuit as a function of angular frequency for several different values of resistors R. Use $C = 2.50$ nF; $L = 6.00$ μH; and resistances of 2.00, 4.00, and 8.00 Ω. Choose your frequency range to exhibit the resonance peaks clearly.

71. Figure 33.21 was originally created using the spreadsheet template RLC_RES.WK1 found on the diskette accompanying this text. (a) Change the value of $R1$ in this template to create a power resonance curve with $Q = 10$. (b) Determine the values of ω_+ and ω_- on either side of ω_{res} where the power drops to one-half of its value at the peak. Define $\Delta\omega = \omega_+ - \omega_-$ and then compute $\omega_{res}/\Delta\omega$ and compare this value to Q. [*Hint:* If you need finer resolution, both the initial frequency and the frequency scale can be changed simply by altering the cells following the labels "w start = " and "w scale = ", respectively.]

General Problems

72. The emf amplitude provided by a certain power supply is 5.00 V, and its frequency is 60.0 Hz. (a) What is the range of current amplitude delivered to a variable capacitor connected to the power supply if the capacitor can be changed from 0.0200 μF to 0.0600 μF? (b) If the frequency of the power supply is increased to 90.0 Hz, to what capacitance must the capacitor be set to produce the same maximum current as when the supply frequency was 60.0 Hz and the capacitance was 0.0600 μF?

73. A 10.0-mH inductor is connected to the terminals of a sinusoidal emf source that provides a peak potential of 12.0 V at a frequency of $480.$ Hz. (a) What is the inductive reactance X_L for this circuit?

(b) Compute the peak current in the circuit. (c) For the same applied emf what must the frequency be for the peak current to be 0.500 A?

74. A resistance $R = 10.0$ Ω, an inductance $L = 0.100$ H, and a capacitance $C = 100.$ μF are connected in series to a sinusoidal current source $i(t) = (2.00\text{ A}) \sin[(400.\text{ rad/s})t]$. (a) What are the impedance and phase angle for this circuit? Does the potential applied across the circuit lead or lag the current through it? (b) Find an expression for the potential $v(t)$ across the entire circuit and evaluate it at $t = 0.750$ s. (c) Compute the potential drop across each circuit element at $t = 0.750$ s. Verify that the sum of the potential drops across these circuit elements equals the total that you found in part (b). (d) Plot a phasor diagram.

75. A wire coil has a resistance $R = 240.$ Ω and an inductance $L = 2.50$ mH. A current source $i(t) = (780.\text{ m/A}) \sin[(1.50 \times 10^3\text{ rad/s})t]$ is applied to the coil. Compute the root-mean-square current I_{rms} through the coil and the root-mean-square potential V_{rms} across this coil.

76. Show that the power delivered to a series RLC circuit by a sinusoidal potential source with angular frequency ω is

$$\mathcal{P}_{av} = \frac{\mathcal{E}_{rms}^2 R \omega^2}{R^2\omega^2 + L^2(\omega^2 - \omega_{res}^2)^2}$$

where $\omega_{res} = 1/\sqrt{LC}$.

77. A common transformer used to step down the $120.$ V (rms) line potential has rms ratings for the secondary of 1.00 A at 6.30 V (rms). (a) What is the ratio of secondary turns to primary turns? (b) When operating at full rating what is the load impedance attached to the secondary? (c) What load impedance is seen by an emf source attached to the primary under the conditions of part (b)?

78. An impedance-matching transformer is used to connect an amplifier with an output resistance of $800.$ Ω to an 8.00-Ω speaker. Model the amplifier output as an ideal emf of 12.0 V (rms) in series with a real resistance of $800.$ Ω. (a) Find the secondary-to-primary turns ratio that causes the 8.00-Ω load on the secondary to appear as an additional $800.$-Ω load in the primary. (b) Assume the magnetizing current in the primary is quite small so that the primary circuit can be modeled as an ideal emf in series with two resistors—one real, and the other representing the effect of the secondary. Find the current in the primary and the secondary. (c) Find the power delivered to the load. (d) Now double the turns ratio calculated in part (a) and determine the power delivered to the load. What do you conclude?

79. A series RLC circuit consisting of a 2.00-H inductor, a 2.00-μF capacitor, and a 20.0-Ω resistor in series is driven by a variable frequency generator with an amplitude of 10.0 V. Model the generator as an ideal emf source and (a) find the root-mean-square current I_{rms} when the angular frequency is $400.$ rad/s. (b) Compute the phase angle ϕ at this same frequency. (c) Does the current lead or lag the potential drop across the combined RLC elements? (d) Find the resonant frequency and answer the questions posed in parts (a) through (c) if the circuit is in resonance.

80. The frequency at which the capacitive reactance is equal to the resistance in a low-pass or high-pass filter is called the **corner frequency**. (a) Derive an expression for the corner frequency for a high-pass filter. (b) What is the potential gain of a high-pass filter at its corner frequency?

★81. Start from the expression for power given in Problem 76 and show that, for large Q, the circuit $Q = \omega_{res} L/R$ is given approximately by $Q = f_{res}/\Delta f$, where Δf is the difference between the two frequencies on either side of f_{res} where the power falls to one-half its peak value.

GUEST ESSAY

Magnetic Resonance

by JOHN E. DRUMHELLER

Magnetic resonance is a widely used technique that results from the interaction of the intrinsic magnetic moment of an ion or nucleus with a magnetic field. As you will see, this interaction is very much like that of a spinning top in the earth's gravitational field. The applications of magnetic resonance provide important information from the microscopic world of the nucleus to the macroscopic world of medicine.

Fundamental particles such as electrons and nucleons have a quality called spin, that is, they behave as if they were very small moving charges. Therefore, each particle possesses an intrinsic magnetic moment, μ, which makes each particle like a very tiny magnet. When in a magnetic field, this intrinsic magnet moment will interact with that field in the same way that a loop of current carrying wire will interact in a magnetic field: there will be a torque on the magnetic moment according to

$$\tau = \mu \times \mathbf{B}$$

The torque causes the magnetic moment to precess about the field in a manner such as is shown in Figure 1 and with a frequency called the Larmor frequency. By integrating this torque over the space angle, an energy results which is given by

$$U = -\mu \cdot \mathbf{B}$$

This energy of interaction is called the Zeeman effect.

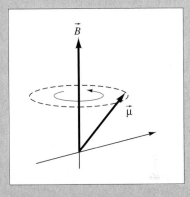

FIGURE 1

The magnetic moment precessing in a magnetic field with the Larmor frequency ν.

JOHN E. DRUMHELLER
Montana State University

John E. Drumheller received his B.S. in Physics from Washington State University, his M.S. and Ph.D. degrees in Physics from the University of Colorado, and Post Doctoral in Physics from the University of Zurich. He has held positions at the University of Colorado, the University of Zurich, and Montana State University and was a 1st Lt. in the U.S. Air Force—Special Weapons from 1954 to 1956. The research area that he is primarily interested in is "Experimental Condensed Matter Physics." Drumheller is a member of several professional societies and fellowships including the American Association of Physics Teachers, Phi Kappa Phi, and the American Physical Society. He has also served as a manuscript reviewer for several professional journals, as an expert witness for a law firm and a drilling company, and on the Thesis Advisory Committee for the Indian Institute of Technology. He has received many awards and honors, including the "Outstanding Teaching Award" from the College of Letters and Science at Montana State University. He currently is Dean, College of Letters and Science at Montana State University.

In classical physics the magnetic moment can precess about the field at virtually any angle. However, in the world of quantum physics, which governs the behavior of the fundamental particles, the magnetic moment is restricted to only a very few angles according to the amount of "spin" quantum number of the particle. The electron and the proton each have a spin of $\frac{1}{2}$ so that the moment of either of them will only precess at two possible angles with respect to the field: "along" the field or "opposite" the field. This reflects the fact that for a given spin quantum number, S, there are $2S + 1$ possible "magnetic quantum numbers. It is important to note that the magnetic moment is given by

$$\mu = (eh/2m)\, S$$

and therefore proportional to the spin. The expression in parentheses is called the Bohr magneton where e is the charge on the electron, h is Planck's constant and m is the mass of the particle.

For the electron (or the proton) with spin $\frac{1}{2}$ the magnetic quantum numbers are $+\frac{1}{2}$ and $-\frac{1}{2}$, which in turn prescribes the possible directions in space that it may have in a magnetic field. Since each direction has a specific energy associated with it, there are two possible energy levels for the electron in a magnetic field. The electron can either precess along the magnetic field—this is called the spin "up" state—which is its lower energy state,

or opposite the field—spin "down"— in a higher energy state as shown in Figure 2. This limitation to just two states for these spin $\frac{1}{2}$ particles is the remarkable consequence of the quantum world and permits the experimental application called magnetic resonance.

The resonance phenomenon is the transition from one energy state to the other. That is, when the electron (or proton) is in one of the energy states and can make a transition or "jump" to the other state. There is a specific energy difference between these two states so that the transition from the lower energy state to the higher state requires a specific amount of energy or a "quantum" of energy to be absorbed. Hence the name quantum physics. Conversely, a quantum of energy is released when the particle makes the transition from the higher energy state to the lower one. This quantum is called a photon. Einstein showed us that the energy of a photon is given by $E = h\nu$ where ν is the frequency of the photon. The transition from higher energy levels to lower levels in certain atomic cases, such as in the hydrogen atom, releases quanta with energy and, therefore, frequency in the optical portion of the electromagnetic spectrum. This is how light is produced. If you have a small optical spectrograph or grating, turn it toward a neon bulb and the discrete lines of the quantum transitions can be seen. Conversely, if photons of the same frequency as the preces-

sional frequency—the Larmor frequency—impinge on a two-level system, the photon will have just the right energy to cause a transition to the higher state. That is, the photon will resonate with the quantum system and "flip" the spin to the higher state. This gives rise to the famous *resonance equation,* which says that the photon energy must equal the energy level separation, i.e.,

$$h\nu = \mu B$$

This resonance phenomenon is much like exciting a mechanical system or electric circuit at its resonant frequency causing it to oscillate. If the quantum system happens to be in the higher state to begin with, the photon will cause the quantum system to jump to the lower state giving up another photon in the process. However, Boltzmann has shown us that the lower energy levels will be more highly populated according to a distribution law that says the population of the upper state will be reduced from that of the lower state by a factor of $\exp(-\mu B/kT)$. Here k is the Boltzmann constant and T is the absolute temperature. The net result of transitions going both ways will be the absorption of energy.

Classically the absorption of a photon and the transition to a higher state can be seen in the following way. Since the photon is the quantum of the electromagnetic field, it is made up of electric and magnetic fields. The

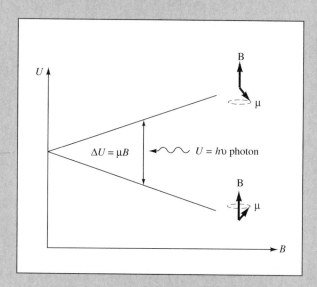

■■■■■ FIGURE 2

The Zeeman effect: a splitting of the two-level system in a magnetic field. A transition from one state to the other is caused by a photon with energy $h\nu$.

magnetic field portion is rotating at the same frequency as the precession of the particle so that when the photon is appropriately superimposed on the precessing particle, the particle, as it precesses about the original B-field, will "see" the small magnetic field of the photon, which in the rotating frame of the particle appears to be stationary to it. The magnetic moment will then also try to precess about this new small field. Since it is much smaller, the "precession" about it will be slower but it has the result that the particle will rotate to the other direction. This is the classical equivalence of "flipping" the particle.

[In MRI] . . . the patient is the sample to be studied . . .

There are many applications of this resonance phenomenon but two are most common: electron paramagnetic resonance (EPR) and nuclear magnetic resonance (NMR). The first of these uses the fact that electrons in atoms, ions, or molecules, if they are "unpaired" or not oppositely aligned with one another, create a magnetic atom, ion, or molecule. The iron ion, for example, has five inner electrons that all align with one another together giving a spin of $\frac{5}{2}$ and therefore a strong magnetic moment. When the magnetic ions are introduced into a material in very small amounts as an impurity the slight amount of paramagnetism can be studied by resonance techniques. Such an impurity ion will behave differently in different environments so that these ions serve as effective probes to study the microscopic surroundings within materials. There are a wide variety of magnetic ions and a wide variety of materials into which to put them so that EPR is a rich lode of information. Because the

magnetic moment is inversely proportional to the mass and the mass of the electron is very small, the experimental frequency in these experiments is very high, typically in the microwave region for convenient laboratory magnetic fields.

The nuclear masses utilized in NMR on the other hand are very much larger than the electron so that the frequency is proportionately lower—typically in the radio frequency (RF) region. Again, the environment of the nucleus will determine how the energy levels split so that the nuclei can be used as effective probes for studying the most fundamental aspects of materials. The first NMR experiments were done by Felix Bloch at Stanford and Edward Purcell at Harvard for which they were given the 1952 Nobel prize.

One of the most common applications of NMR is seen in many hospitals today: magnetic resonance imaging or MRI. In this remarkable experiment, the patient is the sample to be studied and usually it is the protons of his or her body that are the probes. In order to describe this unique apparatus it is important to briefly discuss how the transitions of the protons, or other nuclei, are detected in a NMR experiment.

When the external magnetic field is applied there will be more protons precessing along the field—that is, in the lower state—than against it owing to the Boltzmann factor. This results in a measurable net macroscopic magnetic moment, or magnetization vector, along the direction of the field. When the radio frequency is applied in an appropriate way to make the transition, it will tip this magnetization vector to the opposite direction if left on long enough. If, however, it is left on only long enough to bring the magnetization vector perpendicular to the applied field (this is called a 90° pulse) and then turned off, the system will try to "relax" back to its original state. The relaxation takes two forms. First, the individual spins (magnetic moments) will interact with the local environment and will try to go back to their original orientation before the RF was applied. The result is that the macroscopic magnetization vector tries to return to its original direction along the field. The individual magnetic

moments and therefore the magnetization vector will relax with different times for different environments—a proton behaves differently if it is in bone than if it is in fatty tissue—providing a nonintrusive probe of the body.

The second type of relaxation was utilized by Irwin Hahn of Berkeley in a remarkable experiment called "spin echo." Here the moments of the protons also see other magnetic moments and will interact with them causing individual moments to precess at different rates. This is called transverse relaxation and has the results that the spins become disengaged from one another losing their "coherence." This can be detected by applying another RF pulse after a given time, this time a 180° pulse, that tips the magnetization vector completely over so that the different precession rates of the individual protons now has the effect that they come back together again. As they do so, the magnetization vector is reformed causing the "echo" which can be detected. Again, different environments will cause different transverse relaxations, another independent probe of the body.

There are two more tricks important in MRI. Normally in NMR a single uniform field is applied to the whole sample. There is no way then to determine where a given proton with its signatures of relaxation is located within the sample. However, we saw above that the frequency of the resonance is directly proportional to the applied field. Therefore, if a linearly varying field is applied along an axis of the patient to be examined, each point along that axis requires its own resonant frequency. In this way a frequency scale can be transformed into a distance scale. In addition two different axes can be distinguished by using different times for the echo pulses. Using an appropriate combination of these techniques, a computer can separate out the effects. As a result, a very well-defined, high-resolution internal picture of the patient can be formed. For the development of the two-dimensional NMR techniques necessary for MRI, the Nobel prize was awarded to Richard Ernst in 1991. MRI has become the new standard for diagnostic and preventive medicine.

CHAPTER
34

Electromagnetic Waves

In this chapter you should learn

- the form of the wave equation satisfied by electro-magnetic waves.
- how to compute the energy transported by electro-magnetic waves.
- how to find the pressure exerted by an electromagnetic wave when it is incident on an object.
- about the sources of electromagnetic waves.
- about the electromagnetic spectrum.

$$\frac{\partial^2 E}{\partial x^2} = \frac{1}{c^2}\frac{\partial^2 E}{\partial t^2}$$

The warmth you experience from a heat lamp and the bright display you see from a poster under a novelty shop's "black light" have something in common. Both, in fact, share this similarity with the X-rays used by your dentist and the signals from your favorite radio station. All are a result of **electromagnetic waves.** Because these waves are an electro-magnetic phenomenon, their behavior is governed by Maxwell's equations. Put another way, one important consequence of Maxwell's equations is that they permit wavelike solutions for **E** and **B**. In this chapter we will explore what Maxwell's equations have to say about electromagnetic waves. We begin in the first section by taking a quantitative look at the description of electromagnetic waves. In the second section we briefly describe dipole radiation, one of the important ways by which electromagnetic waves are produced. In the remainder of the chapter we describe the classification of electromagnetic waves over the broad range of wavelengths and frequencies they exhibit.

34.1 The Prediction of Waves from Maxwell's Equations

Let's begin our study of electromagnetic waves by trying to understand how electric and magnetic fields can break free of charges and currents and propagate through space by themselves. The heart of the matter is this: the displacement term in Ampère's law pre-dicts that a changing electric field produces a magnetic field; Faraday's law predicts that a changing magnetic field produces an electric field. Clearly, if changing E-fields cause B-fields, and changing B-fields cause E-fields, one can imagine that the process might sustain itself and continue forever. This is indeed the case. What is perhaps most signifi-cant is that this self-perpetuating chain of fields propagates through space at exactly the speed of light. In what follows we verify that our intuitive guess regarding propagating electromagnetic waves holds up under more rigorous analysis.

In a region of free space, void of any currents, Ampère's law (Equation (30.10)) can be written

$$\oint \mathbf{B} \cdot d\mathbf{l} = \mu_o \epsilon_o \frac{d\phi_E}{dt} \tag{34.1}$$

This equation tells us that a changing electric flux $d\phi_E/dt$ produces a magnetic field **B**. Moreover, the line integral $\oint \mathbf{B} \cdot d\mathbf{l}$ of the magnetic field **B** around any closed path always equals the product of $\mu_o\epsilon_o$ times the rate of change of electric flux $d\phi_E/dt$ through an area enclosed by the closed path. Similarly, Faraday's law (Eq. (30.9))

$$\oint \mathbf{E} \cdot d\mathbf{l} = -\frac{d\phi_B}{dt} \tag{34.2}$$

indicates that a changing magnetic flux $d\phi_B/dt$ produces an electric field **E**. In this case the line integral $\oint \mathbf{E} \cdot d\mathbf{l}$ is the negative of the rate of change of magnetic flux $d\phi_B/dt$ enclosed by the path of the integral. An illustration of the phenomena described by these two equations is shown in Figure 34.1. When an electric field toward the right, such as that shown in Figure 34.1(a) increases in magnitude, the electric flux through planes perpen-dicular to the field increases in magnitude. As suggested by Figure 34.1(a), a **B**-field is produced that circulates around the region of increasing **E**. Similarly, as Figure 34.1(b) suggests, when the strength of a magnetic field toward the right increases, the increased magnitude of the magnetic flux produces an E-field that circulates around the region of increasing **B**. Note that the directions in which **B** and **E** circulate are determined by Equations (34.1) and (34.2). You should verify the directions indicated in the figure by justifying them with the right-hand rule and the signs of $d\phi_E/dt$ and $d\phi_B/dt$. The circular geometry of the field lines of the induced **B**- and **E**-fields in Figure 34.1 is meant to suggest the types of induced fields that result in a nonzero line integral around a closed path. The exact shape of the induced field depends on the spatial geometry of the inducing field. Other possibilities are shown in Figures 34.2 and 34.3.

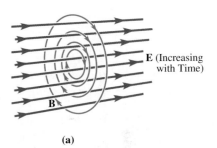

(a)

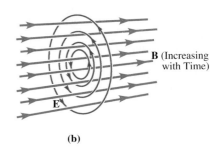

(b)

FIGURE 34.1

(a) If the magnitude of a cylindrically symmetric E-field changes in time, an associated B-field forms closed loops around the E-field lines. The circulation sense of the B-field is shown here for an increasing electric field in the direction shown. (b) Similarly, if the magnitude of a cylindrically symmetric B-field changes in time, an associated E-field forms closed loops around the B-field lines. The sense of the circulation of the E-field is shown for the B-field increasing.

In order to keep the mathematics relatively simple we examine only a special case of electromagnetic waves. We assume that the electric field has only one component, say E_y, and that this component varies in only one direction, say x. This choice makes E_y a function of x and t only and so we may write $\mathbf{E}_y = E_y(x, t)\hat{\mathbf{j}}$. Similarly, we assume that $\mathbf{B} = B_z(x, t)\hat{\mathbf{k}}$. The waves described in this special case are known as **plane waves.** Even spherical wavefronts approximate plane waves when one is far from the source and examines only a small region of space over which the curvature of the wavefront can be ignored. We are not going to prove that such waves exist; that is for experiment to do. We will, however, show that Maxwell's equations predict that such fields, if they exist, satisfy the familiar wave equation

$$\frac{\partial^2 \psi}{\partial x^2} = \frac{1}{v^2}\frac{\partial^2 \psi}{\partial t^2}$$ (34.3)

that we first encountered in Chapter 15. In Section 15.2 we saw that this equation admitted solutions in the form of waves described by

$$\psi = A \sin(kx - \omega t + \phi)$$

We therefore look to see if the electric and magnetic fields described by Maxwell's equations obey an equation similar to Equation (34.3). What we desire is an equation that relates the *spatial* change in E to the *temporal* change in B. To proceed let's look at the fields as shown in Figure 34.2. In this figure we focus our attention on two adjacent E-field vectors that are separated by a small distance Δx. Because we assume that the electric and magnetic fields have only one component, we drop the subscripts on E_y and B_z for convenience. In Figure 34.2 we draw a rectangular path two sides of which coincide with the two closely spaced E-field lines. As shown in this figure, the width of the rectangle is Δx and its height is l. We are going to apply Faraday's law to this rectangular path and eventually take the limit as both l and Δx approach zero.

Let's now apply Faraday's law to the loop *abcda* shown in Figure 34.2:

$$\oint \mathbf{E} \cdot d\mathbf{l} = -\frac{d\phi_B}{dt}$$

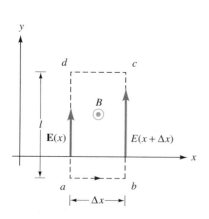

FIGURE 34.2

Faraday's law applied to a rectangular region in the plane of an electric field E_y. This region of space contains no electric charge, yet the magnitude of the field E_y increases with increasing x. Therefore, the magnetic field through the loop *abcd* increases with time.

For the left-hand side we decompose the closed path into four segments

$$\oint \mathbf{E} \cdot d\mathbf{l} = \int_a^b \mathbf{E} \cdot d\mathbf{l} + \int_b^c \mathbf{E} \cdot d\mathbf{l} + \int_c^d \mathbf{E} \cdot d\mathbf{l} + \int_d^a \mathbf{E} \cdot d\mathbf{l}$$ (34.4)

The first and third integrals on the right-hand side of this equation are zero because in both cases $\mathbf{E}$ is perpendicular to the path element $d\mathbf{l}$ along ab and cd. From b to c, $\mathbf{E}$ is constant, parallel to $d\mathbf{l}$, and has a magnitude $E(x + \Delta x) = E_f$ so that the dot product reduces to scalar multiplication. Similarly, from d to a, $\mathbf{E}$ is *anti*parallel to $d\mathbf{l}$ and has a constant magnitude of $E(x) = E_o$, thus again we have a simple product but with a negative sign due to $\cos(180°)$ appearing in the dot product. Hence, Equation (34.4) can be written

$$\oint \mathbf{E} \cdot d\mathbf{l} = 0 + E_f l + 0 - E_o l = (E_f - E_o)l = \Delta E l$$ (34.5)

We now calculate the flux term on the right-hand side of Faraday's law. Because the B-field is perpendicular to the plane of the loop, the area vector is parallel to $\mathbf{B}$ so that $\mathbf{B} \cdot d\mathbf{A} = B \, dA$. We define the average value of B within the rectangle *abcd* as

$$B_{av} \equiv \frac{\int B \, dA}{A}$$

Consequently, we can write the time derivative of the flux through the rectangle *abcd* as

$$-\frac{d\phi_B}{dt} = -\frac{d}{dt}\int \mathbf{B}\cdot d\mathbf{A} = -\frac{d}{dt}(B_{av}A) = -\frac{d}{dt}(B_{av}\,l\,\Delta x) = -l\,\Delta x\frac{\partial B_{av}}{\partial t} \quad \textbf{(34.6)}$$

where we use a partial derivative $\partial B_{av}/\partial t$ because B_{av} depends on both time t and position x. Substituting the results of Equations (34.5) and (34.6) into Faraday's law, we have

$$\Delta El = -l\,\Delta x\frac{\partial B_{av}}{\partial t}$$

or

$$\frac{\Delta E}{\Delta x} = -\frac{\partial B_{av}}{\partial t}$$

as $l \rightarrow 0$ and $\Delta x \rightarrow 0$ the average B-field approaches $B(x)$, and the fraction $\Delta E/\Delta x$ becomes the partial derivative $\partial E/\partial t$. Hence, in this limit the previous equation becomes

$$\frac{\partial E}{\partial x} = -\frac{\partial B}{\partial t} \quad \textbf{(34.7)}$$

Concept Question 2
Look at Figure 34.2. If $E(x + \Delta x) =$ 3.00 μV/mm, $E(x) = 2.00$ μV/mm, $l = 2.00$ mm, and $\Delta x = 0.500$ mm, what is the magnitude of $\oint \mathbf{E}\cdot d\mathbf{l}$ around the path *abcd*?

Equation (34.7) expresses a relationship that exists between the spatial change in an electric field $\mathbf{E}$ and the temporal change in the associated magnetic field $\mathbf{B}$. We can obtain a similar expression relating the spatial change in the magnetic field $\partial B/\partial x$ to the temporal change in the electric field $\partial E/\partial t$ by applying Ampère's law to a segment of the electromagnetic fields in Figure 34.3. As shown in this figure, the area of the rectangle is again $l\,\Delta x$. This time, we apply Ampère's law to the path *abcda*:

$$\oint \mathbf{B}\cdot d\mathbf{l} = \mu_o\epsilon_o\frac{d}{dt}\int \mathbf{E}\cdot d\mathbf{A}$$

Applying reasoning similar to that which led us to Equations (34.4) and (34.6) to the left-hand side and right-hand side of Ampère's law respectively, we find

$$\int_b^a \mathbf{B}\cdot d\mathbf{l} + \int_b^c \mathbf{B}\cdot d\mathbf{l} + \int_c^d \mathbf{B}\cdot d\mathbf{l} + \int_d^a \mathbf{B}\cdot d\mathbf{l} = \mu_o\epsilon_o\frac{d}{dt}(E_{av}A)$$

Because $\mathbf{B}$ is parallel to $d\mathbf{l}$ along the path *ab*, antiparallel to $d\mathbf{l}$ along *cd*, and perpendicular to $d\mathbf{l}$ along both segments *bc* and *da*, we have

$$B(x)l + 0 - B(x + \Delta x)l + 0 = \mu_o\epsilon_o\frac{d}{dt}E_{av}(l\,\Delta x) = \mu_o\epsilon_o(l\,\Delta x)\frac{\partial E_{av}}{\partial t}$$

leading to

$$-\frac{\Delta B}{\Delta x} = \mu_o\epsilon_o\frac{\partial E_{av}}{\partial t}$$

In the limit as $\Delta x \rightarrow 0$ and $l \rightarrow 0$, we have

$$\frac{\partial B}{\partial x} = -\mu_o\epsilon_o\frac{\partial E}{\partial t} \quad \textbf{(34.8)}$$

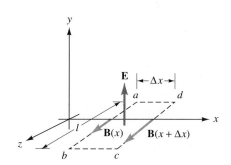

FIGURE 34.3

Ampère's law applied to a rectangular region in the plane of a magnetic field B_z. This region of space contains no electric currents, yet the magnitude of B_z increases with increasing x. Therefore, the electric field through the loop *abcd* increases with time.

Equation (34.8) expresses the relation between the spatial change in the magnetic field B and the temporal change in the associated electric field E. We can combine the results of Equations (34.7) and (34.8) by first differentiating Equation (34.7) with respect to x

$$\frac{\partial^2 E}{\partial x^2} = -\frac{\partial^2 B}{\partial x \, \partial t} \tag{34.9}$$

and then differentiating Equation (34.8) with respect to t

$$\frac{\partial^2 B}{\partial t \, \partial x} = -\mu_o \epsilon_o \frac{\partial^2 E}{\partial t^2} \tag{34.10}$$

If we can assume B to be a continuous function, then the order of differentiation is unimportant and $\partial^2 B/\partial x \, \partial t = \partial^2 B/\partial t \, \partial x$. As a consequence the left-hand side of Equation (34.9) and the right-hand side of Equation (34.10) are equal:

$$\frac{\partial^2 E}{\partial x^2} = \mu_o \epsilon_o \frac{\partial^2 E}{\partial t^2} \tag{34.11}$$

This is the one-dimensional wave equation we first derived for mechanical waves in Chapter 15. From comparison of this equation with Equation (34.3) we recognize the velocity of the wave to be

$$v = \frac{1}{\sqrt{\mu_o \epsilon_o}} = \frac{1}{\sqrt{(4\pi \times 10^{-7} \text{ H/m})(8.854 \times 10^{-12} \text{ F/m})}} = 3.00 \times 10^8 \text{ m/s} = c$$

Concept Question 3
The next time a local event is televised nationally, listen to the event on the TV and at the same time on a local radio station. A noticeable time difference between the sound from the TV and from the radio will be apparent. Why does the local station broadcast arrive first? (*Hint:* Think about how television signals are relayed over long distances in this age of satellite communications.)

Thus we find that the electromagnetic waves predicted by Maxwell's equations travel at the speed of light!

In 1888, eight years after Maxwell's death, Heinrich Hertz (1857–1894) published the results of his extensive proof for the existence of long-wavelength electromagnetic waves. The verification of Maxwell's predictions was complete. This result, that the velocity of electromagnetic plane waves in free space is precisely that measured for light, was regarded near the end of the nineteenth century as compelling evidence that light is an electromagnetic phenomenon.

To complete our discussion we rewrite Equation (34.11) as

$$\frac{\partial^2 E}{\partial x^2} = \frac{1}{c^2} \frac{\partial^2 E}{\partial t^2} \tag{34.12}$$

We can also derive the wave equation for B by differentiating Equations (34.7) and (34.8) with respect to t and x, respectively, and then equating expressions for $\partial^2 E/\partial t \, \partial x$. The result is

$$\frac{\partial^2 B}{\partial x^2} = \mu_o \epsilon_o \frac{\partial^2 B}{\partial t^2}$$

or

$$\frac{\partial^2 B}{\partial x^2} = \frac{1}{c^2} \frac{\partial^2 B}{\partial t^2} \tag{34.13}$$

Equations (34.12) and (34.13) are consequences of Maxwell's equations. The solutions to these partial differential equations describe electromagnetic waves. Recall that the E-field and B-field in these equations are perpendicular and represent the special case called plane waves. In the next section we investigate this important special case further.

34.2 Sinusoidal Electromagnetic Waves

When we studied mechanical waves in Chapter 15 we found that equations such as Equations (34.12) and (34.13) have sinusoidal solutions. In the present context we write these solutions as

$$E(x, t) = E_o \sin(kx - \omega t) \tag{34.14}$$

$$B(x, t) = B_o \sin(kx - \omega t) \tag{34.15}$$

where $k = 2\pi/\lambda$ is the **wave number** of a wave with wavelength λ, and $\omega = 2\pi\nu$ is the angular frequency of a wave with frequency ν.[1] Despite these similarities, there is an important difference between electromagnetic wave phenomena and the mechanical waves we studied previously. Although the mathematical description of each is governed by the solutions to the same form of wave equation, electromagnetic waves do not require any type of material medium in which to propagate. Electromagnetic waves can propagate through empty space as evidenced by the light reaching us from distant galaxies.

In Chapter 15 we learned that the product of the wave frequency ν and the wavelength λ is the phase velocity of the wave. Therefore, for electromagnetic waves

$$c = \lambda\nu \tag{34.16}$$

Because an electromagnetic plane wave in free space always has a velocity $c = 3.00 \times 10^8$ m/s, knowledge of ν is sufficient to determine λ, and vice versa.

When electromagnetic waves are discussed, it is common to focus attention on the *electric field only* because the electric field dominates the waves' interaction with charged particles. We can see why this is true by calculating the magnitudes of the amplitudes E_o and B_o of the electric and magnetic fields. Substituting the expressions for E and B from Equations (34.14) and (34.15) into Equation (34.7), we find

$$\frac{\partial}{\partial x}[E_o \sin(kx - \omega t)] = -\frac{\partial}{\partial t}[B_o \sin(kx - \omega t)] \tag{34.17}$$

$$kE_o \cos(kx - \omega t) = +\omega B_o \cos(kx - \omega t)$$

$$\frac{E_o}{B_o} = \frac{\omega}{k} = \frac{(2\pi\nu)}{(2\pi/\lambda)} = \lambda\nu = c$$

$$\frac{E_o}{B_o} = c \tag{34.18}$$

As illustrated in Example 34.1, the relation $B_o = E_o/c$ results in an extremely small magnetic field.

---

[1] It is conventional to designate the frequency of electromagnetic waves as ν rather than f as used in mechanical oscillations and sound waves.

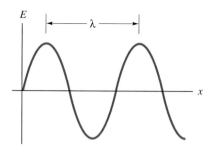

FIGURE 34.4

One representation of an electromagnetic plane wave. (See Example 34.1.)

EXAMPLE 34.1   A Radio Wave

A sinusoidal electromagnetic wave, such as that shown in Figure 34.4, has a wavelength of 2.88 m, and the maximum electric field strength E_o of this wave is 48.0 mV/m. (a) At what frequency must you set your FM radio to receive this wave? (b) What is the maximum magnetic field strength B_o of this electromagnetic wave? (c) Write equations for the orthogonal components $E(x, t)$ and $B(x, t)$ for this wave.

SOLUTION   (a) To find the frequency of this wave we use Equation (34.16):

$$\nu = \frac{c}{\lambda} = \frac{3.00 \times 10^8 \text{ m/s}}{2.88 \text{ m}} = 1.04 \times 10^8 \text{ Hz} = 104 \text{ MHz}$$

(b) We apply Equation (34.18) to compute B_o:

$$B_o = \frac{E_o}{c} = \frac{48.0 \times 10^{-3} \text{ V/m}}{3.00 \times 10^8 \text{ m/s}} = 1.60 \times 10^{-10} \text{ T}$$

(c) In order to use Equations (34.14) and (34.15) we must compute the wave number k and the angular frequency ω for these expressions:

$$k = \frac{2\pi}{\lambda} = \frac{2\pi}{2.88 \text{ m}} = 2.18 \text{ m}^{-1}$$

$$\omega = 2\pi\nu = 2\pi(1.04 \times 10^8 \text{ Hz}) = 6.53 \times 10^8 \text{ rad/s}$$

Hence, Equations (34.14) and (34.15) give

$$E(x, t) = E_o \sin(kx - \omega t) = (48.0 \text{ mV/m}) \sin[(2.18 \text{ m}^{-1})x - (6.53 \times 10^8 \text{ rad/s})t]$$

$$B(x, t) = B_o \sin(kx - \omega t)$$

$$= (1.60 \times 10^{-10} \text{ T}) \sin[(2.18 \text{ m}^{-1})x - (6.53 \times 10^8 \text{ rad/s})t] \quad \blacktriangleleft$$

34.3 Energy Transport by Electromagnetic Waves

In this section we examine the energy transported by electromagnetic waves. Because the ratio of the wave amplitudes E_o to B_o is $c = 3.00 \times 10^8$ m/s (Eq. (34.18)), you might expect that most of the energy carried by an electromagnetic wave is related to its E-field component. However, this assumption is erroneous: Recall that the expressions for the energy densities (energy per unit volume) of electric and magnetic fields (see Sections 25.3 and 31.3) are given by

$$u_E = \frac{1}{2}\epsilon_o E^2 = \frac{1}{2}\epsilon_o E_o^2 \sin^2(kx - \omega t) \qquad (34.19)$$

and

$$u_B = \frac{B^2}{2\mu_o} = \frac{B_o^2}{2\mu_o} \sin^2(kx - \omega t) \qquad (34.20)$$

Using Equation (34.18), we now rewrite the B-field in terms of the E-field, obtaining

$$\frac{1}{\mu_o}B_o^2 = \frac{1}{\mu_o}\frac{E_o^2}{c^2} = \frac{1}{\mu_o}(\epsilon_o \mu_o)E_o^2 = \epsilon_o E_o^2$$

so that the magnetic energy density in Equation (34.20) can be written

$$u_B = \frac{1}{2}\epsilon_o E_o^2 \sin^2(kx - \omega t),$$

which is the same as Equation (34.19). Therefore, the energy carried by the electromagnetic wave is shared equally by the $\mathbf{E}(x, t)$ and $\mathbf{B}(x, t)$ components. Thus, the total energy carried by the electromagnetic wave has density

$$u_T = u_E + u_B = 2u_E$$

$$u_T = \epsilon_o E_o^2 \sin^2(kx - \omega t)$$

The **energy flux** S, that is, the *energy per unit area per unit time,* is proportional to the product of the energy density and the wave speed. (This result is analogous to fluid flux for which the mass per area per time is the product of the mass density and the fluid velocity and is given by ρv.)

$$S = u_T c = c\epsilon_o E_o^2 \sin^2(kx - \omega t)$$

$$= [E_o \sin(kx - \omega t)][\epsilon_o c E_o \sin(kx - \omega t)]$$

We now reverse our earlier substitution and return the B-field to the previous equation by writing $E_o = cB_o$ and $c^2 = 1/\mu_o\epsilon_o$ to obtain

$$S = [E_o \sin(kx - \omega t)]\left[\frac{B_o}{\mu_o} \sin(kx - \omega t)\right] \tag{34.21}$$

However, we note that $E_o \sin(kx - \omega t) = E(x, t)$ and $B_o \sin(kx - \omega t) = B(x, t)$ so that Equation (34.21) can also be written

$$S = \frac{1}{\mu_o}EB \tag{34.22}$$

The energy flux S is proportional to the product of E and B. This energy is transferred along the direction of wave propagation. Recalling that E and B here represent orthogonal vectors that are each perpendicular to the direction of wave propagation, we can assign a direction to S using the cross product of $\mathbf{E}$ and $\mathbf{B}$ with the order chosen to give the correct direction of energy flow:

$$\mathbf{S} = \frac{1}{\mu_o}\mathbf{E} \times \mathbf{B} \tag{34.23}$$

The vector $\mathbf{S}$ is called the **Poynting vector,** and its magnitude is the time rate of energy flow per area; that is, the power per area. If we rewrite Equation (34.21) in the form

$$S = E_o\frac{B_o}{\mu_o} \sin^2(kx - \omega t) \tag{34.24}$$

we see that the energy flux is a rapidly varying function of time. Seldom are we interested in these rapid fluctuations occurring at a frequency 2ν. What is usually of interest is the *average power* delivered to a fixed region in space over relatively long times, seconds for example. The average power per area transported by the wave is just the time average of S. We designate this time average by angular brackets $\langle \; \rangle$. Thus, the time average of S is designated $\langle S \rangle$. We compute this quantity for Equation (34.24):

$$\langle S \rangle = (E_o)\left(\frac{B_o}{\mu_o}\right)\langle \sin^2 (kx - \omega t)\rangle$$

We have seen (Section 33.3) that, for fixed x, $\langle \sin^2 (kx - \omega t)\rangle = \frac{1}{2}$. Hence,

$$\langle S \rangle = \frac{1}{\mu_o} E_o B_o\left(\frac{1}{2}\right) = \frac{1}{\mu_o}\frac{E_o}{\sqrt{2}}\frac{B_o}{\sqrt{2}}$$

$$\langle S \rangle = \frac{1}{\mu_o} E_{\text{rms}} B_{\text{rms}} \qquad\qquad (34.25)$$

where $E_{\text{rms}} = E_o/\sqrt{2}$ is the root-mean-square value of the oscillating field. (Look back to Section 33.3 if you wish to review the root-mean-square value of an oscillating quantity.) When studying sound waves (Section 17.4), we used the word *intensity* to describe the average energy per unit area per unit time. However, the proper SI term **irradiance** and its symbol I are commonly used in the science of optics to designate this quantity, and we follow this modern convention throughout the remainder of this text when referring to the average energy flux of electromagnetic waves.[2] We will find, particularly when dealing with optical fields, that an alternative expression to describe the energy flux in terms of the amplitude of the oscillating electric field is often useful. From Equation (34.25)

$$\langle S \rangle = \frac{1}{\mu_o}\frac{E_o B_o}{2} = \frac{1}{\mu_o}\frac{E_o}{2}\frac{E_o}{c}$$

or

$$\langle S \rangle = I = \frac{1}{2c\mu_o}E_o^2 \qquad\qquad (34.26)$$

The point not to be missed here is that the energy flux, that is, the irradiance, is proportional to the *square* of the amplitude of the electric field.

EXAMPLE 34.2   *Let's Put a Little Light on the Subject*

A small flashlight bulb filament emits 10.0 W of electromagnetic radiation uniformly in all directions. (a) What is the irradiance at a distance of 1.00 m from the source? (b) What is the maximum energy density in the electric field at a point 1.00 m from the source?

SOLUTION   (a) The power emitted from the source is $\mathcal{P} = 10.0$ W. At a distance of $r = 1.00$ m from the source, the irradiance (the average magnitude of the Poynting vector) is the average power divided by the surface area of a sphere over which that power is spread

$$I = \langle S \rangle = \frac{\mathcal{P}}{4\pi r^2} = \frac{10.0 \text{ W}}{4\pi(1.00 \text{ m})^2} = 0.796 \text{ W/m}^2$$

(b) The E-field contribution to the energy density at a point in space is

$$u_E = \frac{1}{2}\epsilon_o E^2$$

---

[2] The intensity of a source is defined in SI as the power radiated by the source divided by the solid angle into which the source radiates.

The maximum electric field strength is E_o, so that

$$u_{E_{\max}} = \frac{1}{2}\epsilon_o E_o^2$$

From Equation (34.26)

$$I = \frac{1}{2}\frac{E_o^2}{c\mu_o}$$

or

$$E_o^2 = 2c\mu_o I$$

Therefore, at a distance of 1.00 m,

$$u_{E_{\max}} = \frac{1}{2}\epsilon_o(2c\mu_o I) = \frac{I}{c} = \frac{0.796 \text{ W/m}^2}{3.00 \times 10^8 \text{ m/s}} = 2.65 \times 10^{-9} \text{ J/m}^3 \qquad \blacktriangleleft$$

34.4 Radiation Pressure

In addition to transporting energy, electromagnetic waves exert a force on objects they encounter. The force is not large; you aren't likely to be knocked over when someone shines a flashlight beam on you. On the other hand the force is critically important to our existence: The force of the radiation from the interior of the sun is in large part the reason the sun does not collapse under its own gravity. On a smaller scale, it is in part the pressure of the sunlight that pushes the tail of a comet away from the sun. With modern lasers, focused beams can be used to suspend macroscopic particles as in Figure 34.5. As we will see below, the amount of force electromagnetic waves exert depends on whether the wave is reflected or absorbed by the material it encounters.

Although a rigorous derivation is beyond the scope of this text, Maxwell was able to show that the average pressure P exerted on a surface by an electromagnetic wave that is completely absorbed is given by

$$P = \frac{I}{c} \qquad \text{(Complete absorption)} \qquad (34.27)$$

This pressure is consistent with assigning to the electromagnetic wave a momentum p given by

$$p = \frac{U}{c} \qquad (34.28)$$

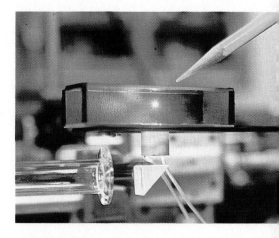

FIGURE 34.5

A glass sphere is suspended in midair by the radiation pressure from an upwardly directed 250-mW laser beam. (See Problem 43.)

where U is the *total* energy carried by the wave that is absorbed. This point of view has interesting implications regarding the effect of an electromagnetic wave on a reflecting surface. In this case the original momentum of the electromagnetic wave is reversed, that is, the *change* in the wave's momentum is $2U/c$. Because the momentum change is twice as great for a reflected wave as it is for an absorbed wave, the pressure exerted on a perfectly reflecting surface is

$$P = \frac{2I}{c} \qquad \text{(Complete reflection)} \qquad (34.29)$$

This difference between the pressure from an absorbed wave and a reflected wave has a direct analog in the mechanics of particle collisions: when a particle collides with a wall in a perfectly elastic collision, it delivers twice the impulse to the wall as it does if the collision is perfectly inelastic.

EXAMPLE 34.3   *About the Same as Three Snow Flakes*

The sun provides an average irradiance of about 1.00×10^3 W/m² to the earth's surface. (a) If the roof dimensions of a building are as shown in Figure 34.6, calculate the total power incident on the roof. (b) Find the radiation pressure and total force on the black (perfectly absorbing) roof.

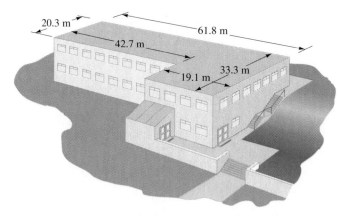

FIGURE 34.6

The pressure of the sunlight falling on this roof is calculated in Example 34.3.

SOLUTION   (a) The total area is $A = (20.3 \text{ m})(61.8 \text{ m}) + (13.0 \text{ m})(19.1 \text{ m}) = 1.50 \times 10^3$ m². The irradiance is $I = 1.00 \times 10^3$ W/m² so that

$$\text{Power} = \mathscr{P} = (1.00 \times 10^3 \text{ W/m}^2)(1.50 \times 10^3 \text{ m}^2) = 1.50 \times 10^6 \text{ W}$$

(b) The pressure is given by Equation (34.27):

$$P = \frac{I}{c} = \frac{1.00 \times 10^3 \text{ W/m}^2}{3.00 \times 10^8 \text{ m/s}} = 3.33 \times 10^{-6} \text{ N/m}^2$$

The total force on the roof is

$$F = PA = (3.33 \times 10^{-6} \text{ N/m}^2)(1.50 \times 10^3 \text{ m}^2) = 5.00 \times 10^{-3} \text{ N}$$ ◄

34.5 Sources of Electromagnetic Waves

According to Maxwell's equations, when any electrically charged particle undergoes an acceleration, it radiates energy. The form in which this energy leaves the charged particle is known as **electromagnetic radiation.** The acceleration of the charges can be created by a variety of mechanisms. On a large scale they can be produced by alternating currents in a radio or television antenna. On the atomic scale, radiation can result from rearrangements of the outer electrons of atoms. In both cases the important features of the process by which electromagnetic waves are created can be modeled as an oscillating electric dipole. Figure 34.7 shows the *E*-field lines in the plane of the page of an oscillating electric dipole. In this model the negative charge is exhibiting simple harmonic motion about a stationary positive charge of equal magnitude.

At $t = 0$, Figure 34.7(a) shows the *E*-field of a static dipole. As the charge $+q$ moves toward the charge $+q$ (Fig. 34.7(b)) the field lines are slightly distorted from those of the static moment. After $-q$ passes $+q$ (Fig. 34.7(c)) the direction of the *E*-field lines near the dipole are reversed (Fig. 34.7(d)) from those shown in Figures 34.7(a) and (b). In the region of space very near the dipole the *E*-field is essentially that of a static dipole. A little farther from the dipole, Maxwell's equations show that the field is quite complicated, and we do not concern ourselves with its form in this intermediate region. However, *far from*

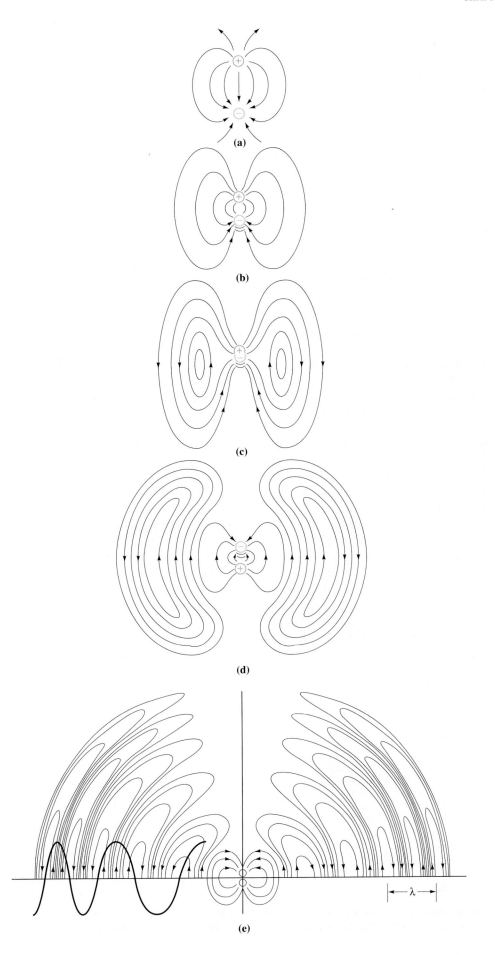

(a)

(b)

(c)

(d)

(e)

FIGURE 34.7

The electric field lines of an oscillating dipole. With each oscillation the field breaks loose from the dipole and travels through space at the speed of light. Most of the energy is radiated along the direction perpendicular to the dipole axis. For waves propagating perpendicular to the dipole axis the *E*-field oscillations are polarized parallel to the dipole axis.

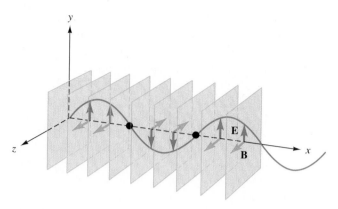

FIGURE 34.8

In the far field, many wavelengths away from the source, the radiation from a harmonically oscillating dipole can be modeled as a plane electromagnetic wave. Note that the electric and magnetic fields are perpendicular to each other and to the direction of propagation and that the electric field is polarized parallel to the dipole axis.

the dipole Maxwell's equations predict that the electric field becomes a transverse wave with a well-defined wavelength (Fig. 34.7(e)). In addition to the *E*-field there is an associated *B*-field that is also transverse. Moreover, at any instant and at any point in space both **E** and **B** are perpendicular to each other and are also perpendicular to the direction of wave propagation. Associated with the *E*-field lines shown in Figure 34.7(e) there are *B*-field lines that alternate in direction, in and out of the plane of the page.

Suppose the negative charge in our dipole model exhibits simple harmonic motion. We orient a coordinate system so that the *x*-axis is along the propagation direction of the electromagnetic wave at a distance far from the wave source. If we look at a small region of space at a sufficiently large distance from the source, the *E*-field lines approach straight lines and the associated perpendicular *E*- and *B*-field lines lie in common planes as illustrated in Figure 34.8. In this case the wave closely resembles the electromagnetic plane-wave model introduced earlier. Another representation of the plane-wave model that emphasizes its sinusoidal wave nature is shown in Figure 34.9. There we have removed the planes and show the orthogonal relation between **E** and **B**. You should realize that Figure 34.9 is a "snapshot" of the electromagnetic wave and that it propagates along the positive *x*-axis. The plane containing the Poynting vector and the electric field direction (the *xy*-plane of Fig. 34.9) is called the **plane of polarization** of the electromagnetic wave.

We wish to emphasize two important features regarding the electromagnetic waves originating from a dipole source. The first point is that the majority of the energy leaves

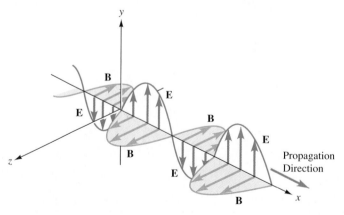

FIGURE 34.9

An alternative representation of a sinusoidal electromagnetic wave. This figure represents a snapshot of the wave that is traveling along the *x*-axis in the direction of increasing *x*.

the dipole in directions nearly perpendicular to the dipole axis, and none of the radiation leaves along the dipole axis. Figure 34.7 shows this feature; recall that the *E*-field lines are closest together where the electric field is strongest. (To get a complete picture you should imagine rotating the radiation pattern about the dipole axis.) The second point we wish to emphasize is that, for the part of the wave propagating in a direction perpendicular to the dipole axis, the electric field oscillates in a direction parallel to the dipole axis. These two statements together allow us to create a particularly simple model for the dominant features of dipole radiation: the electromagnetic wave leaves the dipole at right angles to the dipole axis, and the plane of polarization is parallel to the dipole axis. This simple model is useful for understanding a wide variety of phenomena from the orientation of radio antennas to the polarization of light from the sky. (Polarization phenomena involving visible light are discussed in detail in Section 35.5.)

When we wish to understand the detection of electromagnetic waves, the arguments contained in the previous paragraphs related to the production of such waves can be reversed. When an electromagnetic wave passes a receiving antenna, its electric field induces currents in that conductor. When the conductor of the antenna is parallel to the direction of the *E*-field oscillations of the electromagnetic wave, the electrons along the length of the conductor are free to move resulting in maximum current as the wave passes the antenna. An electronic circuit attached to the antenna can detect the resulting periodic current and ampify it into a useful signal so that we might hear the news from Lake Wobegon or enjoy our favorite television program.

34.6 The Electromagnetic Spectrum

Maxwell's equations predict that when any charged particle undergoes an acceleration, it radiates energy. All electromagnetic waves are essentially the same regardless of wavelength. However, because of the different means by which they are produced and detected, different names have evolved to designate electromagnetic waves within various wavelength intervals. The named regions, which overlap to some extent, range from wavelengths that are less than 10^{-15} m to greater than 10^{+9} m. A map relating the names of the regions to their wavelengths is shown in Figure 34.10. This diagram displays part of what is called the **electromagnetic spectrum.** There is neither an upper nor a lower limit to the electromagnetic spectrum. The electromagnetic wave frequencies, as computed from the relation $\lambda\nu = c$, are also given in Figure 34.10.

The range of wavelengths most important to humans extends from about 400 nm to about 750 nm. This range is the visible region of the electromagnetic spectrum to which our eyes are sensitive. As we are sure you are aware, electromagnetic waves with wavelengths falling within this range are called **light** waves. Actually, the average human eye is most sensitive to light with a wavelength of about 550 nm. Our eyes identify this wavelength by its yellow-green color. As shown in Figure 34.11, at both higher and lower

Concept Question 4
Find the direction of the Poynting vector of the *EM* wave shown in Figure 34.3.

Concept Question 5
Make a simple sketch of an oscillating dipole and a sine wave indicating the oscillations of the electric field of the predominant radiation from an electric dipole at a large distance from the dipole.

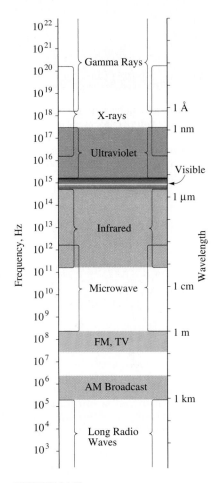

FIGURE 34.10

The electromagnetic spectrum. Notice that in some regions the names of the electromagnetic radiation overlap.

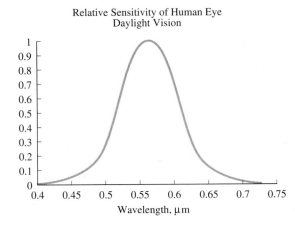

FIGURE 34.11

The spectral sensitivity of the human eye in daylight.

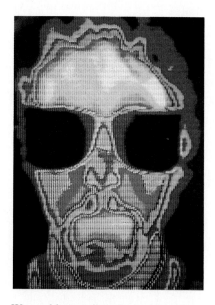

Warm objects radiate electromagnetic radiation. Warmer areas on the person's face emit more infrared radiation. In this photo color has been artificially added to emphasize the temperature differences.

Concept Question 6
The electromagnetic radiation given off by room temperature objects is in the far infrared, and the maximum emitted radiation has a wavelength around 10 μm. In this wavelength region we all glow in the dark. If our eyes were sensitive to this wavelength of radiation, what advantages would we have? What disadvantages?

wavelengths the eye's sensitivity decreases until at 430 nm (violet) and 690 nm (red) it is about 1% of that at 550 nm. Electromagnetic waves with wavelengths just beyond these limits can be detected by the eye if the light is sufficiently bright.

Electromagnetic waves in the **infrared** region extend from the longest visible region to a few millimeters in wavelength. Most objects, such as our bodies, can absorb infrared waves. We detect them with our skin as warmth. The warm sensation felt from a heat lamp is due to infrared radiation. Infrared waves have applications in physical therapy, but are also used in infrared spectroscopic experiments to study interatomic bonding in molecules. With infrared imaging techniques waves emitted by warm objects, such as humans or poorly insulated houses, can be observed by cameras sensitive to these wavelengths.

You can tell from Figure 34.10 that there are regions of the electromagnetic spectrum for which an electromagnetic wave can belong to either of two defined regions. For example, the infrared region overlaps somewhat with the **microwave** band. In the region of overlap which term is used usually depends on the mechanism by which the wave is produced or on its application. The term *microwave* is generally used for waves generated using electronic means, whereas *infrared* refers to waves produced using thermal sources or molecular vibrations and rotations. Microwaves have wavelengths that range from a few millimeters to a few centimeters. Beyond microwaves are the electromagnetic waves that constitute the wavelengths assigned by regulatory statutes to broadcast enterprises, such as television and radio stations, and to amateur radio enthusiasts. Such waves are known generically as *radio waves.* Waves beyond 1 km in wavelength are simply called **long waves.**

Electromagnetic waves with wavelengths slightly shorter than those visible to the human eye are called **ultraviolet** waves. Even when only moderately intense, these waves carry enough energy to damage living cells, especially the cornea and lens of the eye, which absorb ultraviolet radiation strongly. Along with visible and infrared waves, our sun emits ultraviolet waves. Fortunately, molecules in our upper atmosphere (most notably ozone, O_3) absorb most of this harmful radiation. In the early 1990s much concern evolved about the possible depletion of the ozone layer of the earth's atmosphere when holes in the ozone layer were detected near the south and north poles.

X-rays are very energetic electromagnetic waves with wavelengths extending from about 10 nm down to about 10^{-4} nm. Some X-rays are produced spontaneously as the electrons of an atom rearrange following a change in the atomic number of the atom's nucleus as a result of naturally occurring radioactivity. For application purposes, however, X-rays are usually created by the abrupt deceleration of electrons as they are propelled onto the surface of a metallic target. These X-rays are used for medical diagnosis as well as investigations of the crystal structure of solids. Because they are so energetic, X-rays can easily damage living tissue and care must be exercised in any production or application of X-rays.

Gamma rays are extremely energetic electromagnetic waves that are often produced by both natural and artificially induced nuclear reactions. Because of their high energy, they are potentially lethal to living cells, making them useful when carefully directed, as in cancer therapy, but dangerous in the case of uncontrolled exposure. Thick shielding must be employed to absorb these electromagnetic waves.

34.7 Summary

The orthogonal components of electric and magnetic fields of an electromagnetic plane wave traveling in free space obey wave equations that are a consequence of Maxwell's equations:

$$\frac{\partial^2 E}{\partial x^2} = \frac{1}{c^2}\frac{\partial^2 E}{\partial t^2} \tag{34.12}$$

$$\frac{\partial^2 B}{\partial x^2} = \frac{1}{c^2}\frac{\partial^2 B}{\partial t^2} \tag{34.13}$$

where the velocity of the waves through empty space is

$$c = \frac{1}{\sqrt{\mu_o \epsilon_o}} = 3.00 \times 10^8 \text{ m/s}$$

The frequency ν and wavelength λ of electromagnetic waves in free space are related by

$$c = \lambda \nu \tag{34.16}$$

and at any instant at any point in space the magnitudes of E_o and B_o of an electromagnetic wave are related by

$$\frac{E_o}{B_o} = c \tag{34.17}$$

The energy transported by an electromagnetic wave is shared equally by its electric and magnetic field components. The energy per unit area per unit time carried by the electromagnetic wave is described by the **Poynting vector:**

$$\mathbf{S} = \frac{1}{\mu_o} \mathbf{E} \times \mathbf{B} \tag{34.23}$$

The average power per area transported by the wave is known as the **irradiance** I. For sinusoidal waves the irradiance is:

$$I = \langle S \rangle = \frac{1}{\mu_o} E_{\text{rms}} B_{\text{rms}} \tag{34.25}$$

where E_{rms} and B_{rms} are the root-mean-square values of E and B at a particular point in space. The irradiance is proportional to the square of the amplitude of the electric field:

$$I = \frac{1}{2c\mu_o} E_o^2 \tag{34.26}$$

The pressure that an electromagnetic wave applies to an object that completely absorbs that wave is given by

$$P = \frac{I}{c} \tag{34.27}$$

whereas for a wave that is perfectly reflected

$$P = \frac{2I}{c} \tag{34.29}$$

An electromagnetic wave of total energy U carries a momentum of magnitude

$$p = U/c \tag{34.28}$$

According to Maxwell's equations, accelerating charges provide the sources of electromagnetic waves. An oscillating dipole provides a useful model for the production of some types of electromagnetic waves. The radiation from oscillating dipoles is directed primarily at right angles to the dipole axis, and for radiation emitted along this direction the electric field is parallel to the dipole axis.

The **electromagnetic spectrum** (in order of decreasing wavelength and increasing frequency) includes long radio waves, AM broadcast waves, FM and television waves, microwaves, infrared waves, the visible spectrum, ultraviolet waves, X-rays, and gamma rays.

PROBLEMS

34.1 The Prediction of Waves from Maxwell's Equations

1. Verify that $1/\sqrt{\mu_o \epsilon_o}$ has the units of velocity.
2. When a time-dependent electric field, such as that of an electromagnetic wave, is applied to a transparent medium, the dielectric constant κ of that medium depends on the frequency of the E-field variations. The speed of an electromagnetic wave that propagates through a transparent medium is given by $v = 1/\sqrt{\kappa\mu_o\epsilon_o}$. Find the speed of light of wavelength 589 nm in (a) water ($\kappa = 1.78$), (b) diamond ($\kappa = 5.85$), and (c) salt (NaCl) ($\kappa = 2.38$).
3. Show that the expression $E(x, t) = E_o e^{i(kx - \omega t)}$ where $i = \sqrt{-1}$ is a solution to the wave equation Equation (34.12).
4. Verify that any function $E(x, t) = f(x \pm ct)$ is a solution to the wave Equation (34.12). (*Hint:* Make the substitution $u = x \pm ct$.)
5. Consider an electric field, such as that shown in Figure 34.2. (a) If the electric field is increasing at a rate of (2.45 V/m)/m along the x-axis, what is the time rate of change of the magnetic field in this region? (b) In what direction is the change in the magnetic field? (c) Can you deduce the actual direction of the magnetic field from the data given?
6. Consider an electric field, such as shown in Figure 34.3. (a) If the electric field is growing in the sense of the positive y-axis at a rate of (2.89 mV/m)/s, what is the change in magnetic field strength between two points 1.00 cm apart along the x-axis, that is, two points located on the x-axis such that $\Delta x = 1.00$ cm? (b) What is the difference in magnetic field between two points with identical x-coordinates but separated by $\Delta y = 1.00$ cm?

34.2 Sinusoidal Electromagnetic Waves

7. A sinusoidal electromagnetic wave traveling through empty space has an electric field amplitude of 248 μV/m. What is the amplitude of the magnetic field associated with this wave?
8. Compute the amplitude of the magnetic field component of a sinusoidal electromagnetic wave with electric field amplitude of 18.5 mV/m.
9. What must the amplitude of the electric field component of a sinusoidal electromagnetic wave be for the magnetic field amplitude to be 1.00 G?
10. (a) Verify by direct substitution that Equations (34.14) and (34.15) are solutions to Equations (34.12) and (34.13), respectively. (b) Verify by direct substitution that $E(x, t) = E_o \cos(kx + \omega t)$ and $B(x, t) = B_o \cos(kx + \omega t)$ are also solutions to Equations (34.12) and (34.13).
11. A sinusoidal electromagnetic wave propagating through empty space along the positive x-direction has a frequency of 3.86×10^{14} Hz and an amplitude of 540 mV/m. (a) Compute the angular frequency ω and the wave number k for this wave. (b) Write equations for $E_y(x, t)$ and $B_z(x, t)$ for this wave.
12. A sinusoidal electromagnetic wave propagates through empty space along the x-axis in the direction of increasing x with a wavelength of 12.6 cm. (a) Compute the frequency of this wave. (b) If the amplitude of $E_y(x, t)$ for this wave is 12.0 N/C, compute the magnitude and direction of the associated magnetic field at a point in space at the instant when the E-field is one-third its maximum value and directed along the negative y-axis.

13. The E-field component of a sinusoidal electromagnetic plane wave is given by the equation

$$E_y(x, t) = (0.240 \text{ N/C}) \sin[(1.26 \text{ m}^{-1})x - (3.78 \times 10^8 \text{ rad/s})t]$$

(a) In what region of the electromagnetic spectrum is this wave classified? (b) Verify that the wave is electromagnetic by computing its propagation velocity. (c) Write an equation for the associated magnetic field $B(x, t)$. Be sure to indicate the direction of this component.

14. The E-field component of a sinusoidal electromagnetic plane wave is given by the equation

$$E_y(x, t) = (28.5 \text{ } \mu\text{V/m}) \sin[(1.05 \times 10^8 \text{ m}^{-1})x + (3.15 \times 10^{16} \text{ rad/s})t]$$

(a) In what region of the electromagnetic spectrum is this wave classified? (b) Verify that the wave is electromagnetic by computing its propagation velocity. (c) Is the wave propagating toward the direction of the positive or negative x-axis? (d) Write an equation for the associated magnetic field $B(x, t)$.

15. The B-field component of a sinusoidal electromagnetic plane wave is given by the equation

$$B(x, t) = (8.00 \times 10^{-8} \text{ T}) \cos[(3.14 \times 10^8 \text{ m}^{-1})x + (9.42 \times 10^{16} \text{ rad/s})t]$$

(a) In what region of the electromagnetic spectrum is this wave classified? (b) Verify that the wave is electromagnetic by computing its velocity. (c) Is the wave propagating toward the direction of increasing or decreasing x? (d) Write an equation for the associated electric field $E(x, t)$.

16. The B-field component of a sinusoidal electromagnetic plane wave is given by the equation

$$B(x, t) = (2.50 \times 10^{-12} \text{ T}) \sin[(15.7 \text{ m}^{-1})x + (4.71 \times 10^9 \text{ rad/s})t]$$

(a) In what region of the electromagnetic spectrum is this wave classified? (b) Verify that the wave is electromagnetic by computing its velocity. (c) Is the wave propagating toward the direction of the positive or negative x-axis? (d) Write an equation for the associated electric field $E(x, t)$.

34.3 Energy Transport by Electromagnetic Waves

17. How far must you stand from a light bulb that emits 60.0 W of radiant energy uniformly in all directions for the E-field amplitude to be 10.0 V/m?
18. A point source emits an electromagnetic wave with spherical symmetry. What power must the source radiate for the electric field amplitude E_o to be 10.0 V/m at a distance of 12.0 m from the source?

19. An incandescent light bulb radiates 75.0 W with spherical symmetry. Compute the rms values of the E-field and B-field at a distance of 12.0 m from the bulb.

20. A radio station's broadcast antenna radiates electromagnetic waves with spherical symmetry so that at a distance of 500. m from the antenna the amplitude of the electric field is 1.50 V/m. Compute the amplitude of the electric field at a distance from the antenna of (a) 1.00 km, (b) 10.0 km, and (c) 20.0 km.

21. Compute the average magnitude of the Poynting vector at a distance of 3.22 km (2.00 miles) from a radio broadcast antenna if the antenna emits electromagnetic waves with spherical symmetry and a power of 100. kW.

22. A point source emits electromagnetic waves isotropically so that at a distance of 125 m from the source the rms electric field is 2.40 V/m. At a distance of 500. m from the source compute (a) the rms magnetic field strength, (b) the average value of the Poynting vector, (c) the energy density of the electric field, and (d) the energy density of the magnetic field.

34.4 Radiation Pressure

23. The average irradiance of the sun on the earth's upper atmosphere is about 1.4 kW/m². What pressure would be exerted on the atmosphere at this altitude if the radiation were completely absorbed?

24. Assume that a 100.-W light bulb radiates uniformly in all directions and compute the pressure it exerts on a perfectly absorbing surface located (a) 0.500 m, (b) 1.00 m, and (c) 2.00 m from the light bulb.

25. A plane electromagnetic wave has an irradiance of 500. W/m² and falls at normal incidence on a flat material of 0.500-m² area. Half of the energy carried by the wave is reflected by the material. (a) Compute the total energy absorbed by the material in 5.00 min. (b) What is the magnitude of the total impulse delivered to the material during this time?

26. Assume 5.00×10^2 W of light are incident on a screen for 10.0 min. Compute the total linear momentum transferred to the screen if (a) it completely absorbs the radiation, (b) it completely reflects the radiation, (c) it absorbs 30.0% of the radiation and reflects the remainder.

27. Compute the force on a 4.00-m² perfectly absorbing screen if the electromagnetic wave described in Problem 13 is incident on the screen.

28. Compute the rate at which momentum is delivered to a 3.00-m² perfectly reflecting screen if the electromagnetic wave described in Problem 14 is incident on the screen.

29. Compute the force on a 3.00-m² perfectly absorbing screen if the electromagnetic wave described in Problem 15 is incident on the screen.

30. Compute the rate at which momentum is delivered to a 4.00-m² perfectly absorbing screen if the electromagnetic wave described in Problem 16 is incident on the screen.

34.6 The Electromagnetic Spectrum

31. The human eye is most sensitive to light with a wavelength of 555 nm. What is the frequency of light with this wavelength?

32. A helium–neon laser emits electromagnetic radiation with wavelengths of 1152.3 nm, 3391.2 nm, and 632.8 nm. (a) What are the frequencies of these waves? (b) In what region of the electromagnetic spectrum are each of these waves categorized? (c) Which is the familiar red light observed in the laboratory?

33. The most intense emissions from an argon laser have wavelengths of 488.0 nm and 514.5 nm. What are the frequencies of these waves?

34. The frequencies of the AM radio band extend from 540 kHz to 1600 kHz. What is the range of wavelengths covered by this band?

35. The frequencies of the FM radio band extend from 88 MHz to 108 MHz. What is the range of wavelengths covered by this band?

36. The FM radio band extends from 88 MHz to 108 MHz. Stations are separated by 0.2 MHz. (a) What is the difference in wavelength between two stations broadcasting at 88.1 MHz and 88.3 MHz? (b) What is the difference in wavelength between stations at 107.7 MHz and 107.9 MHz?

37. What wavelength does the National Weather Service broadcast marine weather forecasts if the frequency of these radio waves is 162.5 MHz?

38. Show that for electromagnetic waves in free space that a small change $\Delta\lambda$ in wavelength from some initial wavelength λ_o produces a change in frequency $\Delta\nu$ given by $\Delta\nu = -c\,\Delta\lambda/\lambda_o^2$.

General Problems

39. At a certain place on the earth's surface the rms value of the electric field due to the sun's irradiance is 400. V/m. Compute (a) the magnetic field amplitude, (b) the magnitude of the Poynting vector, and (c) the average energy density.

★40. Because of absorption by air, water vapor, and other molecules, the sun's irradiance increases with altitude above the earth's surface. Assume that the earth's atmosphere is 35.0 km in height and at this altitude the irradiance provided by the sun is 1.35×10^3 W/m², whereas at the earth's surface the irradiance is reduced to 1.00×10^3 W/m². (a) By assuming a linear increase in irradiance with altitude, derive an equation for the irradiance as a function of altitude h from the earth's surface. (b) Use the linear model you developed in part (a) to compute the energy content of a column of air with a 1.00-m² cross section that is 35.0 km high.

★41. (a) Use the result of Ampère's law applied at the surface of a wire of radius r and length l, which carries a steady current I to compute B and then (b) show that the magnitude of the Poynting vector at the wire's surface is given by $\langle S \rangle = (IR^2)/(2\pi l r)$, where R is the resistance of the wire.

★42. Suppose that two electromagnetic plane waves with angular frequencies ω_1 and ω_2, both with their electric fields E_1 and E_2 parallel to the y-axis, travel along the x-axis and fall on a screen placed in the yz-plane at $x = 0$. In this case the total electric field at the screen is given by $E_{tot} = E_1 \sin(\omega_1 t) + E_2 \sin(\omega_2 t)$. (a) Show that the total irradiance on the screen is given by $I_{tot} = I_1 + I_2 + 2\sqrt{I_1 I_2}\cos[(\omega_1 - \omega_2)t]$, where I_1 and I_2 are the irradiances caused by each wave separately. (b) Using the answer to (a) show that when the frequencies of the waves are quite different (incoherent waves), the time average irradiance is the simple sum of the individual irradiances. (c) Using the result from part (a), determine the irradiance that results when two waves of equal amplitude and identical frequencies (perfect coherence) are added.

43. The glass bead suspended in midair by the laser beam pictured in Figure 34.5 has a diameter of approximately 1.00×10^{-3} in. Estimate the power of the laser required to accomplish this feat. (For simplicity assume all of the radiation is absorbed by the glass. This is clearly not the case, but you obtain the correct order of magnitude nevertheless.)

44. What is the minimum time it takes for an electromagnetic wave to travel from the surface of the earth to a geosynchronous communications satellite and back to earth? (Geosynchronous satellites orbit above the equator at an altitude where their period is 24.00 h.)

PART

IV

OPTICS

In the following four chapters we discuss several classical models for light. In Chapter 35 we begin with a model of light as an electromagnetic wave. Much of the introductory material draws heavily on the discussion of electromagnetic waves in Chapter 34, which we assume you have already studied. A number of important and fundamental phenomena are described along with several applications. Two classical models are then each discussed in more detail. Chapter 36 contains a description of the behavior of light rays and the way they travel through lenses and reflect from mirrors. These topics are traditionally known as *geometrical optics*. Applications are well known and range from eyeglasses to astronomical telescopes. Later in Chapters 37 and 38 the wave nature of light is examined more thoroughly. The study of wave phenomena of interference and diffraction is traditionally known as *physical optics*. Since the advent of the laser these phenomena are now exploited in a large number of practical devices ranging from the compact-disc player to holography. Chapter 38 ends with an optional section describing X-ray phenomena; this final section reminds us again that light is a special case of electromagnetic radiation that takes many forms.

More modern aspects of the phenomena associated with light (special relativity and quantization) are found in Part V of this text.

Reflection, Refraction, and Polarization of Light

In this chapter you should learn:

- the basic laws for the propagation, reflection, and refraction of light.

- how to apply Snell's law to describe the refraction of light.

- about total internal reflection and optical fibers.

- how light becomes polarized.

- how to apply the law of Malus to polarized light.

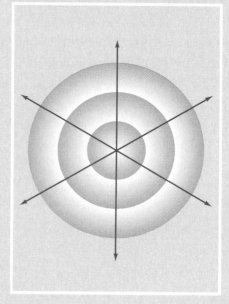

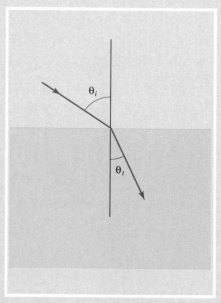

The study of light is without a doubt one of the most interesting topics in physics. Light is both beautiful and confounding. It is responsible for rainbows and sunburns. It provides one of our primary sensors of the world around us. Our view of the world is, literally, colored by the way light interacts with matter. Light provides means of communication ranging from your ability to read this page to fiber-optic cables that transport hundreds of telephone conversations over a single thread of glass. Humankind's attempt to understand the detailed nature of light has resulted in the creation (and discard) of many models of both light and matter. Although we do not attempt to present a detailed history of the evolution of theories of light, we do find it impossible to resist the temptation to include occasionally the historical context in which current models evolved.

35.1 Particles and Waves: A Tale of Two Models

As we described in Chapter 34, one of the triumphs of electromagnetic theory was the prediction of electromagnetic waves. Maxwell's equations brought together X-rays, visible light, radio waves, and radiant heat under a single description. This synthesis was perhaps as profound as Newton's synthesis of terrestrial and celestial motion. At that time, the prediction (1865) and experimental production (1887) of electromagnetic waves was viewed as the final victory in a long series of battles between two models of light: the particle model and the wave model.

In *La'Dipotrique* (1637) René Descartes suggested that light was a longitudinal compression wave, and from this (incorrect) model he deduced the law of refraction discovered empirically by Willebrod Snell (1591–1626) in 1621. (Refraction, which we will describe in detail shortly, is the bending of light as it passes obliquely from one medium, such as air, into another medium, such as glass.) Robert Hooke (of Hooke's law spring fame) observed the bending of light around small obstacles and the colors that appear in a thin film of oil floating on water. Hooke proposed in 1665 that light was a high-speed vibration. Christiaan Huygens, working at about the same time, extended the wave theory to explain polarization.

Isaac Newton, as you might expect, also studied the nature of light. He performed now famous experiments wherein he broke white light into a spectrum of colors using a prism and then collected the colored light, sending it through a second prism to create white light again. Newton generally favored the particle model. However, his earlier work contains interesting references to both ''corpuscles'' and vibrations. Perhaps because of the inability to reconcile the straight-line motion of light with the spreading behavior observed in waves, Newton ultimately came down on the side of the particle model. Given Newton's reputation it is no surprise that, as Eugene Hecht writes, ''The great weight of Newton's opinion hung like a shroud over the wave theory during the eighteenth century, all but stifling its advocates.''[1]

Throughout the early nineteenth century proponents of the wave theory began to enjoy more and more success using the wave model to describe and predict the behavior of light. The battle was not without unpleasantness, occasionally becoming little more than name-calling. Nonetheless, in the game of physics Nature is the ultimate referee, and adherents of the wave model could not be denied. The remainder of this chapter and the next two contain what might be called classical optics, albeit with many modern applications. We will use the wave model almost exclusively, although from time to time we will point out how the same phenomena might be described using a particle model.

Lest you think the battle ended in the middle of the 1800s, we should point out that it was rejoined in the early 1900s when new phenomena were discovered that, seemingly, were impossible to explain using a wave model. Indeed, only a particle model seemed to fit! That story is described in the final chapters of this book where we introduce so-called

---

[1] Eugene Hecht, *Optics,* 2nd ed. (Addison-Wesley, Reading, MA, 1987), p. 5.

modern physics, a name that has outlived its descriptive power. Suffice it to say that Nature does not seem compelled to fit light neatly into either the classical wave or the classical particle models. Indeed, here at the end of the twentieth century our description of light, presented in Chapter 40, sounds remarkably like Newton's strange combination of wavelike and particlelike adjectives.

35.2 Properties of the Wave Model of Light

In Chapter 34 we described light as a transverse electromagnetic wave. Light waves are fully three-dimensional waves, that is to say they generally spread out in space in three dimensions. This property sometimes makes them difficult to visualize and even more difficult to draw. In Figure 35.1 we show a representation of one of the simplest of light waves, a monochromatic plane wave. On each of the rectangular sheets of Figure 35.1 the electric field has a constant magnitude. As one progresses from sheet to sheet, the electric field is different, oscillating as a sine wave between $+E_{max}$ and $-E_{max}$. The electric field exists everywhere within a given sheet, not just where the E-vector arrows are drawn. The field also exists between adjacent sheets, not just on the sheets that are drawn in the figure.

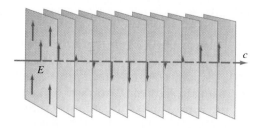

FIGURE 35.1

A monochromatic plane wave. The electric field of the wave is constant in magnitude and direction throughout an infinite plane, symbolized here by a finite sheet. A given sheet in the figure above moves to the right at the speed of light. At any given moment of time, the electric field is different in planes located at different positions along the propagation direction.

Wavefronts and Rays

Drawings, such as Figure 35.1, sometimes contain too much information and are needlessly complicated. To simplify our representation we follow tradition and pick a particular phase of the wave oscillation and draw only those sheets over which the electric field has that particular phase. If we choose the phase where E has its maximum value and is pointed, say, upward, we can simplify the representation of Figure 35.1 to that of Figure 35.2. The rectangular sheets of Figure 35.2 represent **wavefronts.** That is, on each sheet the wave has an equivalent phase. As you move from sheet to sheet the phase increases (or decreases) by 2π rad. Figures 35.3 and 35.4 show partial wavefronts for cylindrical and spherical waves, respectively. Note that for cylindrical and spherical wavefronts the mag-

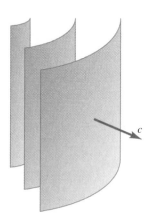

FIGURE 35.3

Cylindrical wavefronts of light. Such wavefronts might be created when monochromatic light passes through a narrow (compared with the wavelength of light) slit.

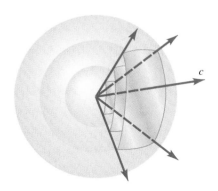

FIGURE 35.4

Spherical wavefronts. Such wavefronts are created when light passes through a small (compared with the wavelength of light) circular opening.

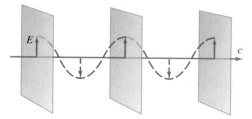

FIGURE 35.2

A simplified representation of plane waves shows wavefronts where the electric field has a particular value.

nitude of the electric field is the same everywhere within a given sheet but differs from sheet to sheet. When the light contains a single wavelength it is called **monochromatic** (literally "one color"). In the case of monochromatic light the distance between wavefronts is equal to one wavelength. Sometimes the wavefront representation is used for light containing many different wavelengths, that is **polychromatic** light. In such cases the drawings should be thought of merely as generic representations of the shape of typical wavefronts.

Often it is desirable to reduce the dimensionality of our representation of wavefronts even further. We can do so by drawing only the intersection of the wavefronts with some plane, the plane of this page for example. In this case we display lines representing the wavefronts. As you can see from Figure 35.5, the information content is greatly reduced. For example, we cannot distinguish between spherical wavefronts and cylindrical wavefronts with this representation. The advantage of this simple line representation is that we can draw the intersections of multiple wavefronts—a task that would be difficult, if not impossible, in more complete representations.

Huygens' Principle

In Chapter 34 we indicated how Maxwell's equations lead to a wave equation and how that wave equation leads to plane waves in an isotropic medium. The wave equation can also be solved for more complicated geometries and nonisotropic media. The mathematics quickly becomes more involved than we wish to present here. In fact, scientists today are employing high-speed digital computers to aid in the solution of these equations for complicated geometries required by fabrication techniques for modern micron-scaled integrated electronic circuits. On the other hand, a great deal of insight into wave propagation can be gained from a quite simple model proposed by the Dutch physicist Christiaan Huygens in 1678.

Huygens' model is quite general in that it describes the behavior of both longitudinal and transverse waves and does not at all depend on the electromagnetic character of light. The essence of the model is contained in a geometric construction described by **Huygens' principle:** *Consider each point on a wavefront as a source of secondary spherical waves, each spreading with the same speed. A short time later, the new wavefront is a surface tangent to the secondary wavelets.* In Figure 35.6 we show how the secondary waves, often called **Huygens' wavelets,** can be used to predict the propagation of different shapes of wavefronts. Note that in Figure 35.6 Huygens' principle implies that light, indeed any wave, can "bend" around obstacles. This is in fact the case, and the phenomenon is known as **diffraction.** The study of diffraction and other wave properties of light is known as **physical optics.** We will return to Huygens' principle in Chapter 37 when we concentrate on the various wave phenomena exhibited by light.

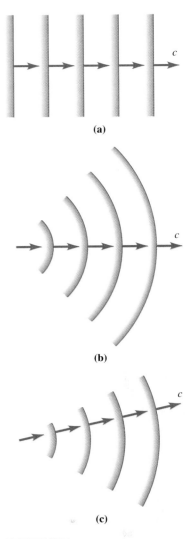

(a)

(b)

(c)

FIGURE 35.5

Simplified representations of (a) planar, (b) cylindrical, and (c) spherical wavefronts. In such simplified representations it is not possible to distinguish between cylindrical and spherical wavefronts.

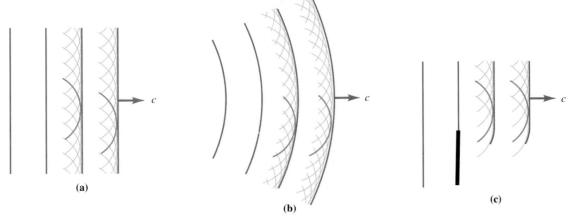

(a)

(b)

(c)

FIGURE 35.6

Huygens' principle. Each point on a wavefront is considered the source of secondary spherical wavelets. The location of the subsequent wavefront is the line of tangency to the wavelets.

The Ray Model

As we picture the wavefronts described in the previous sections, we need to keep in mind that they always move at the speed of light. Therefore, the wavefronts travel a distance longer than three football fields during each one millionth of a second. When we want to consider the motion of the wavefront through space, a different view is more convenient. In this second representation we concentrate our attention on a small segment of a wavefront and the flow of energy through space. The direction of energy flow is given by the Poynting vector as discussed in Chapter 34. A line tracing the flow of radiant energy through space is called a **ray.** In an *isotropic* medium, rays are perpendicular to wavefronts. Figure 35.7 shows rays superposed on several wavefront representations. From Figure 35.7 we can make some useful associations between our wavefront picture and our ray picture for isotropic media:

1. Parallel rays indicate plane-wave wavefronts.
2. Diverging rays indicate expanding wavefronts.
3. Converging rays indicate collapsing wavefronts.

When one considers only the ray paths through an optical system, ignoring diffraction and related wave effects, the model is known as **geometrical optics.** In Chapter 36 we will concentrate on this model.

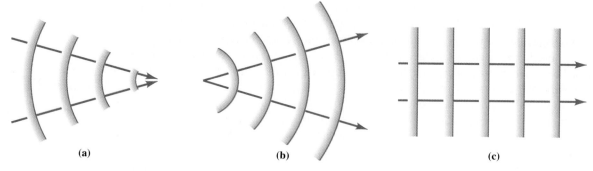

(a)   (b)   (c)

FIGURE 35.7

In isotropic media, rays (lines indicating the direction of energy flow) are perpendicular to wavefronts. Parallel rays indicate plane waves; diverging rays indicate expanding wavefronts; converging rays indicate collapsing wavefronts.

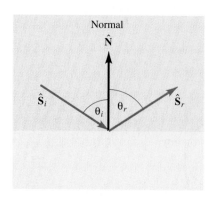

FIGURE 35.8

In specular reflection the incoming ray, reflected ray, and a normal to the surface all lie in the same plane. The angle of incidence θ_i equals the angle of reflection θ_r.

35.3 Reflection

Light reflects from surfaces in two distinct ways. As you read these words, the light entering your eyes has reflected from this page. The light reflecting from this page scatters quite differently from the way light scatters from a mirror or the surface of a smooth pond. In the former case the page is rough and an incident wavefront is scrambled when it strikes the surface; rays are scattered more or less randomly in many different directions. This type of reflection is known as **diffuse reflection.** On the other hand, when light strikes an optically smooth surface (a surface with irregularities of a size much less than the wavelength of light), the reflection of the wavefront is quite regular. To describe this regularity we let $\hat{\mathbf{S}}$ represent a unit vector pointed in the direction of energy propagation. Then $\hat{\mathbf{S}}_i$ is in the direction of the incident ray that makes an angle θ_i with a normal $\hat{\mathbf{N}}$ drawn to the reflecting surface. The plane formed by $\hat{\mathbf{S}}_i$ and $\hat{\mathbf{N}}$ is called the **plane of incidence.** The **law of reflection** states that $\hat{\mathbf{S}}_r$ for the reflected ray lies in the plane of incidence and makes an angle $\theta_r = \theta_i$ on the other side of the normal as shown in Figure 35.8.

$$\theta_i = \theta_r \qquad \text{Law of reflection}$$

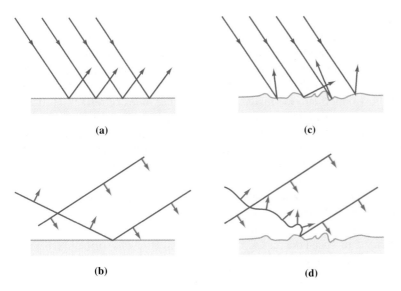

FIGURE 35.9

Specular reflection as depicted by (a) the ray model and (b) the wavefront model. Diffuse reflection as depicted in the (c) ray model and (d) the wavefront model.

This type of reflection is called **specular reflection.** The law of reflection is a direct consequence of a continuity requirement for the components of certain electric and magnetic fields at the reflecting interface. These continuity requirements are a consequence of Maxwell's equations. In Figure 35.9 we contrast diffuse reflection and specular reflection; each are represented with both the wavefront and the ray pictures. We are concerned primarily with specular reflection in this text. For convenience, when we use the word *reflection* we mean "specular reflection" unless we specifically indicate diffuse reflection.

35.4 Refraction

You have probably heard that "light travels in straight lines." Such a statement is an oversimplification. One might better say that "in a uniform, isotropic medium ray trajectories are straight lines." Actually, light bends into the shadow region when the wavefront is sheared (broken) by an obstacle or an aperture. This phenomenon is known as **diffraction** and is discussed in Chapter 38. The direction of a light wave is also altered when light obliquely crosses the boundary into a different medium, moving from air into glass, for example. It is convenient to describe this bending, called **refraction,** in terms of the ray picture of geometrical optics. We let θ_1 represent the angle between the incoming ray and a normal to the interface between the two media. Then θ_2 represents the angle between the normal and the transmitted ray in the second medium. The transmitted ray also lies in the plane of incidence. The geometry is shown in Figure 35.10. The rule that describes the connection between the angles θ_i and θ_t is known as Snell's law after Willebrod Snell, who discovered the law empirically in 1621.

$$n_i \sin(\theta_i) = n_t \sin(\theta_t) \qquad \text{Snell's law}$$

The n_i and n_t are numbers that characterize the respective media. In general, n is known as the **index of refraction** of the medium; the subscripts i and t indicate the incident and the transmitted rays respectively. Table 35.1 lists the index of refraction for some common materials. Notice that when a light ray enters a medium with a higher index of refraction, the ray is bent toward the normal. When the ray enters a medium of lower index of refraction, it is bent away from the normal.

Concept Question 1
Draw an *S*-shaped curve on a large sheet of paper. Construct Huygens' wavelets along one side of the curve. (If you don't have a drawing compass you can do pretty well with a penny. Remember to keep the center of the penny on the old wavefront.) Draw the new wavefront. Repeat the process starting from your new wavefront. Now repeat the process one more time. What happened to the convex portion of the original wavefront? What happened to the concave portion?

Concept Question 2
If you haven't done so already, complete the construction described in Concept Question 1. Now draw a few rays on your diagram starting from points on the original *S*-shaped wavefront. What do the rays do on the convex portion of the wavefront? What do they do on the concave portion?

Concept Question 3
Explain the law of reflection based on a particle model for light. Explain the law of reflection using Huygens' construction for the wave model of light.

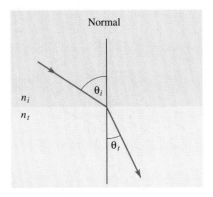

FIGURE 35.10

A ray of light is refracted at the boundary between two different dielectric media. For isotropic media, the incident ray, refracted ray, and a normal to the boundary all lie in a plane. The angle of incidence θ_i and angle of refraction θ_t are related by Snell's law.

TABLE 35.1   Indices of Refraction for Selected Isotropic Media at Sodium D Wavelength 589 nm

SUBSTANCE	n
Gases at 0°C and 1 atm	
Air	1.000293
Helium	1.000036
Hydrogen	1.000132
Carbon dioxide	1.00045
Liquids (at 20°C unless stated otherwise)	
Liquid nitrogen	1.205 (−190°C)
Water	1.333
Acetone	1.357
Ethanol	1.361
Carbon tetrachloride	1.461
Glycerine	1.473
Benzene	1.501
Carbon disulfide	1.628
Solids (at room temperature unless stated otherwise)	
Ice	1.31 (0°C)
Fluoride	1.392
Fused silica	1.458
Glass (typical)	1.50
Gelatin	1.516–1.514
Sodium chloride	1.544
Diamond	2.417

Snell's law can be deduced from either a particle model or a wave model. However, the two models predict *different* changes for the speed for light as it enters a medium of higher index! When light enters a medium of high index, such as water, from one of lower index, such as air, the particle model predicts that the light will speed up, whereas the wave model predicts it will slow down. In the particle model we can speculate that the light particle is somehow pulled into the material with a higher index of refraction. If this pull occurs, the component of the light's velocity vector normal to the interface increases, whereas the component parallel to the interface remains unchanged. Such an effect would cause the ray to bend toward the normal. The important point is that the particle model predicts that light should *speed up* when entering a material with a higher index of refraction.

In the wave model, the rays bend toward the normal when entering the medium with the higher index of refraction if the wave *slows down*. As shown in Figure 35.11, we imagine a series of plane wavefronts incident on a boundary. As each portion of a wavefront crosses the interface, its speed decreases. It follows from Huygens' principle that the portion of the wavefront that encounters the boundary first, falls farther and farther be-

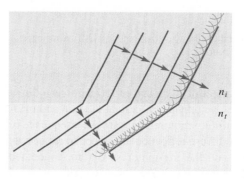

FIGURE 35.11

A construction based on Huygens' principle indicates that if waves slow down on entering the medium with the lower index of refraction, their line of propagation is bent toward the normal.

hind, causing the wavefront to bend at the interface. The direction of the bending is such as to cause the ray to bend toward the normal if the wave slows down.

During the course of his Ph.D. thesis work in 1850, Jean Bernard Léon Foucault (1819–1868) demonstrated that the speed of light in water ($n = 1.33$) is less than the speed of light in air ($n = 1.00$). Although the electromagnetic character of the waves would not be shown by Maxwell for another 15 years, Foucault's measurements added to the growing acceptance of the wave model over the particle model.

Using the wave model we can discover how the index of refraction of a material is directly related to the speed of light in the material. From Figure 35.12, we calculate the length of segment AB from the two right triangles ABC and ABD.

$$AB = \frac{\sin(\theta_1)}{\lambda_1} = \frac{\sin(\theta_2)}{\lambda_2}$$

From this result it follows that

$$\frac{\sin(\theta_1)}{\sin(\theta_2)} = \frac{\lambda_1}{\lambda_2}$$

The frequency ν of the light is the same in both media. (If this fact at first seems strange, notice that if, say, 100 wavefronts hit the interface each second, then 100 wavefronts must also leave the interface each second. Otherwise, wavefronts must either be created or destroyed at the interface.) Also recall that for all waves, the phase velocity V is given by $V = \lambda\nu$. Consequently, we find

$$\frac{\sin(\theta_1)}{\sin(\theta_2)} = \frac{\lambda_1}{\lambda_2} = \frac{\lambda_1\nu}{\lambda_2\nu} = \frac{V_1}{V_2}$$

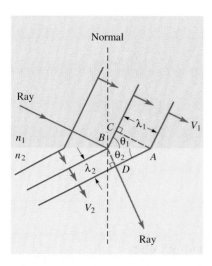

FIGURE 35.12

Wavefronts encounter the boundary between two isotropic dielectric media. Note that the angle between the boundary and the wavefront is equal to the angle between the ray and the normal to the surface.

From Snell's law we also have

$$\frac{\sin(\theta_1)}{\sin(\theta_2)} = \frac{n_2}{n_1}$$

Hence,

$$\frac{V_1}{V_2} = \frac{n_2}{n_1}$$

If medium 1 is a vacuum for which $n_1 \equiv 1$ and $V_1 = c$, we have

$$n_2 = \frac{c}{V_2} \qquad\qquad (35.1)$$

Thus, we see that the wave model provides us with an interpretation of Snell's index of refraction: the index of refraction of a material is equal to the ratio of the speed of light in a vacuum to its speed in the material:

$$n = \frac{c}{V} \qquad\qquad (35.2)$$

Concept Question 4
Does light travel faster in water or in glass?

EXAMPLE 35.1   A Fishy Example

A ray of light enters the side of a water-filled glass aquarium making an angle of 30.0° with a normal to the aquarium face. Calculate the angle made by the ray within the glass and within the water.

SOLUTION   We trace the ray through three media as shown in Figure 35.13. We begin by applying Snell's law at the first interface

$$n_1 \sin(\theta_1) = n_2 \sin(\theta_2) \qquad \text{or} \qquad \sin(\theta_2) = \frac{n_1 \sin(\theta_1)}{n_2}$$

Taking numerical values for the refraction indices from Table 35.1, we calculate

$$\sin(\theta_2) = \frac{1.00 \sin(30.0°)}{1.50} = 0.333$$

$$\theta_2 = \sin^{-1}(0.333) = 19.5°$$

Assuming that the sides of the glass are parallel as shown in Figure 35.13, the angle of incidence at the glass–water interface is θ_2. Hence, we can write Snell's law at the second interface as

$$n_2 \sin(\theta_2) = n_3 \sin(\theta_3)$$

From which

$$\theta_3 = \sin^{-1}\left(\frac{n_2 \sin(\theta_2)}{n_3}\right) = \sin^{-1}\left[\frac{(1.50)(0.333)}{1.33}\right] = 22.1°$$

Can you show that this angle is the same as would be obtained had the ray entered the water directly?

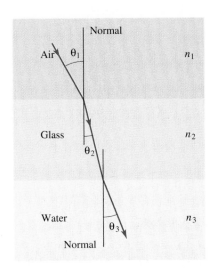

FIGURE 35.13

A ray of light passes through three dielectrics.

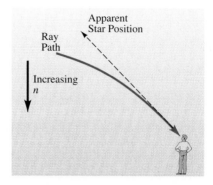

FIGURE 35.14

When a light ray encounters a region with a nonuniform index of refraction, the ray deviates toward the higher index of refraction.

We should also point out that this bending of the light as it travels from one medium to another need not occur all in a single step. If the medium gradually changes its index of refraction, a light ray is gradually bent toward the medium with the higher index as shown in Figure 35.14. This type of bending occurs when light from a star enters the earth's atmosphere. As a result stars appear out of position when viewed near the horizon, and the sun is actually already below the horizon when we observe its disk about to set. (See also Fig. 35.P14 and Problem 43b.)

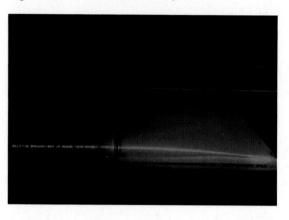

Light doesn't always travel in straight lines! The density of this sugar solution increases toward the bottom. The ray is bent toward the more dense medium. Light from stars and the sun is similarly bent when entering the earth's atmosphere.

Total Internal Reflection

Generally, whenever light makes an abrupt transition between two media of different indices of refraction, part of the energy is transmitted and part is reflected. If the light is moving from a medium of low index to high index, such as air to glass, the reflection is called **external reflection.** If the light moves from a high-index material to a low-index material, for example glass to air, the reflection is called **internal reflection.**

When traveling from a more dense to a less dense medium, the direction of a light ray is bent away from the normal to the interface. There is a limit to how far from the normal the ray can be bent. Clearly the angle of refraction cannot exceed 90°. The angle of incidence where the angle of refraction becomes 90° is known as the **critical angle.** For angles of incidence greater than the critical angle the light cannot escape the medium with the higher index of refraction. Instead, the light is reflected back into the medium in accordance with the law of reflection. This phenomenon is known as **total internal reflection** and is illustrated in Figure 35.15.

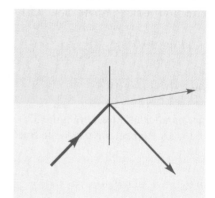
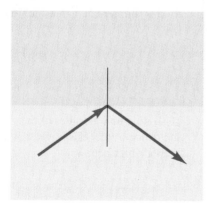

FIGURE 35.15

In the case of internal reflection transmitted light is bent away from the normal. If the angle of incidence is greater than the critical angle, the light is totally reflected.

EXAMPLE 35.2   *The Critical Angle for Total Internal Reflection*

Determine the critical angle for total internal reflection between glass and air.

SOLUTION   When the angle of incidence is equal to the critical angle, the refracted ray makes an angle of 90° with the normal. In Figure 35.15, air is above the horizontal boundary and a typical glass is below it. We write Snell's law for this glass–air interface

$$n_g \sin(\theta_g) = n_a \sin(\theta_a)$$

and solve for the critical angle

$$\sin(\theta_g) = \frac{n_a \sin(\theta_a)}{n_g} = \frac{1.00 \sin(90.0°)}{1.50} = 0.667$$

$$\theta_{\text{crit}} = \theta_g = \sin^{-1}(0.667) = 41.8°$$

If the angle of incidence in glass is greater than 41.8°, the light is totally reflected and unable to escape from the glass into the air. Can you show that if the glass in this example is replaced with water, the critical angle for the water–air interface is 48.8°? ◄

Notice from Figure 35.15 that the transition to total internal reflection is a continuous occurrence. At every angle less than the critical angle (including normal incidence) part of the light energy is transmitted (Fig. 35.16(a)) and part is reflected (Fig. 35.16(b)). As the critical angle is approached the transmitted fraction falls continuously to zero as shown in Figure 35.16(b).

It is also true that, when light is transmitted from a medium with low index of refraction to a medium of high index of refraction, part of the light energy is reflected as well as transmitted. The fraction of light reflected increases with the angle of incidence. Thus, even clear window glass makes a reasonably good mirror when the light strikes it at a grazing angle. (The direction of oscillation of the E-field becomes quite important for large angles of incidence. See Section 35.5.)

By applying boundary conditions that require the continuity of the components of certain electric and magnetic fields at the interface between two dielectric media, one can obtain expressions for the ratio of reflected and transmitted energy as a function of indices of refraction and angle of incidence. These derivations are beyond the scope of this text; however, the results are graphed in Figure 35.16. For *normal* incidence they reduce to

$$R = \frac{\text{Reflected power}}{\text{Incident power}} = \left(\frac{n_2 - n_1}{n_2 + n_1}\right)^2 \qquad \text{(35.3a)}$$

Concept Question 5
Can total *external* reflection ever occur, for example, when light travels from air into water?

Concept Question 6
A diver with a face mask looks up at the water surface from her position at the bottom of an indoor swimming pool. She is able to see the lights on the ceiling over the pool through a circular area directly overhead. However, if she looks at the water surface to the side of the overhead circular area, she cannot see the ceiling lights, but instead she sees a nearly uniform bright surface that has the same color as the bottom of the pool. Explain her observations.

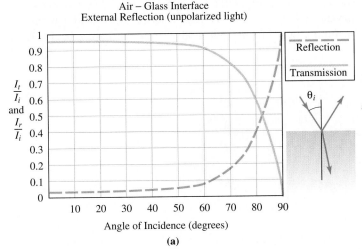

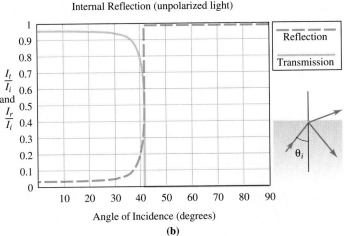

(a)

(b)

FIGURE 35.16

The fraction of light energy that is reflected at the boundary between the two dielectrics depends on the angle of incidence. I_t and I_r are the transmitted and reflected irradiances, respectively, and I_i is the incident irradiance. The fractions calculated here are for glass ($n = 1.50$) and air ($n = 1.00$).

$$T = \frac{\text{Transmitted power}}{\text{Incident power}} = \frac{4n_1 n_2}{(n_2 + n_1)^2} \qquad \text{(35.3b)}$$

which define the **reflection coefficient** R and **transmission coefficient** T at normal incidence. Of interest to eyeglass wearers is the fact that, for normal incidence, about 4% of the light is reflected at *each* surface of your eyeglasses. Hence, only 92% of the light energy striking the front of your glasses makes it to your eye. (Formulas for other angles are given in Problem 41.)

Optical Fibers (Optional)

The phenomenon of total internal reflection raises the possibility of creating optical waveguides. Imagine a thin slab of, say, glass with light entering at one end. If the ray makes a small enough angle with the sides of the glass slab, the energy cannot escape because of total internal reflection. The light is confined to the slab until it reaches its opposite end. Waveguides can also be made in the form of long thin cylinders known as **optical fibers.** These fiber waveguides are strands of glass, often thinner than a human hair. Advanced technology has created low-attenuation optical fibers that are kilometers in length. Hundreds of telephone conversations can be transmitted over a single, high-purity fiber. Optical fibers can be used to transmit light into otherwise inaccessible places. Orderly bundles of fibers, which maintain their relative position from one end of the bundle to the other, are used to transmit images for applications ranging from electronic typesetting to medical endoscopy.

Although a simple strand of glass will conduct light, modern optical fibers are actually formed from two layers of glass. As shown in Figure 35.17, the fiber **core** is surrounded by a concentric layer of lower index glass known as the **cladding.** The cladding is surrounded by a protective layer of a material, such as silicone or a polymer epoxy acrylate. The total internal reflection occurs at the core–cladding interface. In fibers designed for high-speed telecommunications the core is only a few microns in diameter, not much larger than the wavelength of the light used. In such cases, the full electromagnetic wave picture must be used to describe the propagation of the light. However, when the highest data transmission rates are not required, fibers with a "large" core of perhaps a hundred microns or more are used. Such fibers are known as **multimode** fibers. For multimode fibers the ray picture is adequate to describe the behavior of the light.

Two types of rays can propagate in a multimode fiber as shown in Figure 35.18. **Meridional rays** bounce back and forth across the fiber, always passing through the central axis of the fiber. **Skew rays** on the other hand careen down the fiber in a corkscrew fashion and never pass through the fiber axis. We can describe the meridional rays in multimode fiber using the simple two-dimensional model shown in Figure 35.19. Let's

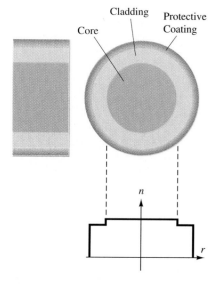

FIGURE 35.17

A cross section of a modern multimode optical fiber. The core has a slightly higher refractive index than the surrounding cladding.

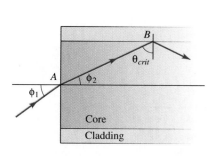

FIGURE 35.19

Cross-sectional view of a meridional ray striking the core–cladding interface at the critical angle for total internal reflection.

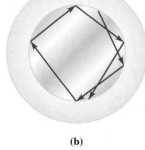

FIGURE 35.18

An end view of ray paths in a multimode fiber. (a) As a meridional ray propagates down an optical fiber it bounces back and forth between opposite sides of the fiber, passing through the central fiber axis after each bounce. (b) A skew ray does not pass through the fiber axis as it travels down the length of the fiber.

look at a ray that hits the core–cladding boundary at θ_{crit}, the critical angle for total internal reflection. Tracing this ray back to the point where it entered the face of the fiber, we find it makes an angle of $\phi_2 = 180° - \theta_{\text{crit}}$ with a normal to the tip of the fiber. Outside the fiber this ray makes a larger angle ϕ_1 with the normal as dictated by Snell's law. All rays with entrance angles less than ϕ_1 strike the core–cladding interface at an angle greater than the critical angle and are guided by the fiber. Rays with entrance angles greater than ϕ_1 strike the core–cladding boundary at less than the critical angle. These rays are only partially reflected at the interface, and their energy is quickly lost from the core after a few centimeters of travel (see Problem 25).

The angle ϕ_1 described above defines the **acceptance cone** for the fiber. The sine of ϕ_1 is an important engineering parameter for the fiber and is known as the **numerical aperture** NA. In Example 35.3 we show that the numerical aperture can be written

$$NA = \sqrt{n_{\text{core}}^2 - n_{\text{clad}}^2} \qquad (35.4)$$

Light is guided by optical fibers because of total internal reflection.

EXAMPLE 35.3   *Numerical Aperture of a Multimode Fiber*

(a) Show that $NA = \sqrt{n_{\text{core}}^2 - n_{\text{clad}}^2}$. (b) Use this result to find the half-angle of the acceptance cone of a multimode fiber with a core index of $n_{\text{core}} = 1.486$ and a cladding index such that $\Delta \equiv (n_{\text{core}} - n_{\text{clad}})/n_{\text{core}} = 0.010$.

SOLUTION   We write Snell's law for the total internal reflection occurring at the critical angle between the core and the cladding at point B of Figure 35.19:

$$n_{\text{core}} \sin(\theta_{\text{crit}}) = n_{\text{core}} \sin(90° - \phi_2) = n_{\text{clad}} \sin(90°)$$

Because $\sin(90° - \phi_2) = \cos(\phi_2) = \sqrt{1 - \sin^2(\phi_2)}$, the above expression for Snell's law becomes

$$n_{\text{core}}\sqrt{1 - \sin^2(\phi_2)} = n_{\text{clad}}$$

or

$$\sin(\phi_2) = \sqrt{\frac{n_{\text{core}}^2 - n_{\text{clad}}^2}{n_{\text{core}}^2}}$$

We now substitute this result for $\sin(\phi_2)$ into the expression for Snell's law describing the refraction occurring at point A in Figure 35.19:

$$n_1 \sin(\phi_1) = n_{\text{core}} \sin(\phi_2)$$

$$n_1 \sin(\phi_1) = n_{\text{core}} \sqrt{\frac{n_{\text{core}}^2 - n_{\text{clad}}^2}{n_{\text{core}}^2}}$$

By definition $NA = \sin(\phi_1)$. If the fiber is in air, then $n_1 = 1.00$ so that

$$\sin(\phi_1) = \sqrt{n_{\text{core}}^2 - n_{\text{clad}}^2}$$

or

$$NA = \sqrt{n_{\text{core}}^2 - n_{\text{clad}}^2}$$

(b) The expression for Δ can be rewritten as $n_{\text{clad}} = n_{\text{core}}(1 - \Delta)$, so that we can write

$$n_{\text{clad}}^2 = n_{\text{core}}^2(1 - 2\Delta + \Delta^2)$$

Because Δ is small, we can ignore Δ^2 compared to 2Δ in the above expression. Substituting this result into the previous expression for $\sin(\phi_1)$, we obtain

$$NA = \sin(\phi_1) = \sqrt{n_{\text{core}}^2 - n_{\text{core}}^2(1 - 2\Delta)} = n_{\text{core}}\sqrt{2\Delta}$$

Substituting numerical values, we have

$$NA = n_{core}\sqrt{2\Delta} = 1.486\sqrt{2(0.010)} = 0.210$$

With this result the half-angle of the acceptance cone is

$$\phi_1 = \sin^{-1}(0.210) = 12.1°$$

Dispersion

Although we have discussed *the* index of refraction of a material, in reality, a material does not have a single index value for all colors of light. For some types of glass the change in index is nearly 10% over the visible portion of the spectrum, varying as shown in Figure 35.20. For most transparent dielectrics, like those shown in Figure 35.20, the index of refraction gradually decreases as the wavelength progresses from shorter visible wavelengths (blue light) to the longer visible wavelengths (red light). Thus, if white light, composed of a combination of wavelengths, is refracted at an air–glass boundary, the blue light is bent more toward the normal than the red light. A prism, such as that shown in Figure 35.21, can be used to separate the wavelengths into a spectrum in which the colors are ordered according to wavelength. The order of the colors is often remembered by referring to the letters in the name of the famous, but fictitious, ROY G. BIV. When an index of refraction n decreases with increasing wavelength, the phenomenon is known as **normal dispersion.**

White light is dispersed into colors because the index of refraction is different for different wavelengths.

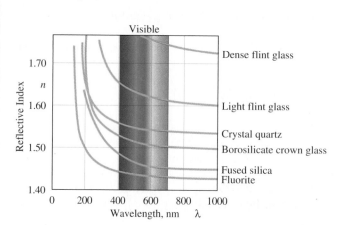

FIGURE 35.20

The variation of index of refraction with wavelength is known as dispersion.

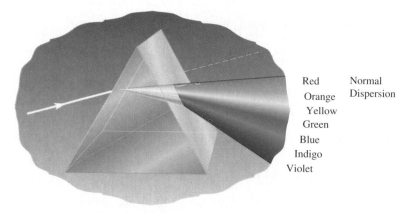

FIGURE 35.21

Normal dispersion of white light by a prism.

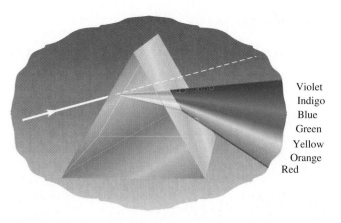

FIGURE 35.22

Anomalous dispersion of white light by a prism.

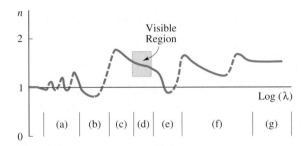

FIGURE 35.23

Dielectric materials absorb electromagnetic radiations strongly at many different wavelengths, depending on the molecular structure of the dielectric. These absorption bands are indicated by dashed lines. Near each of these resonances, regions of anomalous and normal dispersion occur. (a) X-rays, (b) far ultraviolet, (c) near ultraviolet, (d) visible, (e) near infrared, (f) far infrared, (g) radio waves.

Over certain ranges of wavelength the index of refraction for a substance increases with increasing wavelength. In such a case rays of light with longer wavelengths are deviated more than rays with shorter wavelengths. This phenomenon is known as **anomalous dispersion.** Figure 35.22 shows how anomalous dispersion appears in the visible spectrum. All dielectric materials have regions of anomalous dispersion; however, rarely does it occur in the visible spectrum. Consequently, when first observed in the mid 1800s it was considered an anomaly. Anomalous dispersion occurs in a range of wavelengths near any particular wavelength at which the radiation is strongly absorbed by the medium. Figure 35.23 shows the complete dispersion curve throughout the electromagnetic spectrum for a typical substance. Some liquids that strongly absorb light in the middle of the visible spectrum exhibit anomalous dispersion for shorter wavelengths and normal dispersion at the longer wavelengths. Imagine how the ''rainbow'' for a hollow prism filled with such a substance would appear.

Prism Geometry (Optional)

Because of dispersion, a thick piece of glass with a triangular cross section can be used to separate the wavelengths of light that are originally combined in a single beam. A typical geometry for a triangular prism is shown in Figure 35.24. Using the index of refraction for air $n_1 = 1$ and that for the prism $n_2 = n$, we can write Snell's law for the refractions occurring at each of the two faces:

$$(1) \sin(\theta_1) = n \sin(\theta_2) \qquad (35.5)$$

$$n \sin(\theta_1') = (1) \sin(\theta_2') \qquad (35.6)$$

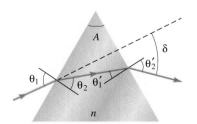

FIGURE 35.24

A cross section of a triangular prism used to separate different wavelengths contained in the incoming ray.

You can also show from plane geometry that

$$\delta = \theta_1 + \theta_1' - A \tag{35.7}$$

and

$$A = \theta_2 + \theta_2' \tag{35.8}$$

Using these last four equations and the calculus, it is possible to show that the deviation δ is a minimum for the symmetric case

$$\theta_1 = \theta_1' = \frac{\delta + A}{2}, \qquad \theta_2 = \theta_2' = \frac{A}{2}$$

For this symmetric case the index of refraction is given by

$$n = \frac{\sin[(\delta + A)/2]}{\sin(A/2)} \tag{35.9}$$

EXAMPLE 35.4   *Doing Time in Prism*

By what angle is light from a helium-neon laser operating at 633 nm deviated when it passes through a fused silica glass prism of apex angle $A = 60.0°$, if the light enters the prism at the angle leading to minimum deviation?

SOLUTION   From the graph of Figure 35.20 we estimate the index of refraction of the glass to be $n = 1.46$. Equation 35.9 can be solved for the deviation δ

$$\delta = 2 \sin^{-1}\left[n \sin\left(\frac{A}{2}\right) \right] - A$$

$$= 2 \sin^{-1}[1.46 \sin(30.0°)] - 60.0°$$

$$= 2 \sin^{-1}(0.73) - 60° = 2(46.9°) - 60° = 93.8° - 60° = 33.8° \quad \blacktriangleleft$$

35.5 Polarization

In Section 33.2 we showed that our model of an electromagnetic wave is one in which an electric field oscillates. In free space the direction of this electric field is perpendicular to the propagation direction of the wave. When the plane of the oscillating electric field is stationary in space, we say that the wave is **linearly polarized.** Such a wave is shown in Figure 35.25. Note that the **plane of polarization** in Figure 35.25 contains the electric field $\mathbf{E}$ and the vector $\hat{\mathbf{S}}$ that indicates the direction of propagation. Several simplified representations of a wave's polarization are also given in Figure 35.25. The magnetic field oscillates in a direction perpendicular to the plane of polarization.

Ordinary white light from an incandescent source, such as a light bulb, is said to be unpolarized. Such light is made up of a great many linearly polarized wave trains with their planes of polarization oriented at random angles. The net effect of these random orientations is the same as if none of the light waves were polarized.

In our classical model, electromagnetic waves are created by accelerating charges. One very useful model is that of the oscillating dipole. For electromagnetic waves with wavelengths corresponding to visible light, the dipoles are atoms or molecules. Recall that when a dipole oscillates, the strength of the radiation is greatest in the direction perpendicular to the dipole, dropping to zero in the direction of the dipole axis. In the direction of maximum strength the radiation is polarized parallel to the dipole axis. Thus, we can understand that the unpolarized nature of the light from an ordinary light bulb is due to the random orientation of the atomic dipoles in the filament of the bulb.

Polarized light can be produced in several ways: scattering, reflection, transmission through a wire-grid polarizer, or transmission through a birefringent crystal. Let's look at each in turn.

Polarization by Scattering

When light strikes small particles in its path, the light is scattered in different directions. This phenomenon occurs, for example, when light from the sun travels through the earth's atmosphere. One simple descriptive model for this process considers the scattering centers as electric dipoles. The scattering mechanism can then be thought of as a two-step process. First, the electric field of the incoming light pushes the opposite charges of the dipole in opposite directions, and the dipole starts to oscillate. By this mechanism energy is transferred from the light wave to the dipole. Next, the oscillating dipole radiates a new light wave. We know that an oscillating dipole emits primarily in the directions perpendicular to its axis. Furthermore, that radiation is polarized with the electric field parallel to the dipole axis. With these simple facts in mind let's consider Figure 35.26.

In Figure 35.26 we look at unpolarized light striking dipoles from three different directions. Dipole A has its axis parallel to the direction of the incoming light. This dipole cannot be excited by the incoming light because the electric field of all the incoming light is at right angles to the dipole axis. The electric fields do not push the dipole charges toward or away from each other. Dipole B is excited by the incoming light, but when it radiates, the resulting light is emitted into the horizontal plane and not down toward our eye. (One of many possible waves traveling in this horizontal plane is illustrated in Figure 35.26(b).) Finally, consider dipole C. This dipole is excited by the incoming light. It radiates light into the vertical plane. Some of the resulting light travels toward our eye, and the resulting radiation is polarized parallel to the dipole axis.

Of course not all of the molecules in the atmosphere have one of the three orientations of Figure 35.26. However, our simple model reveals the underlying source of polarization by scattering. The actual result is complicated by averages over all directions and the translational and rotational motions of the molecules. The effect is closest to our model when the sun is low on the horizon and we look straight upward. If you have a pair of Polaroid sunglasses, try rotating them in front of your eyes while looking straight up when

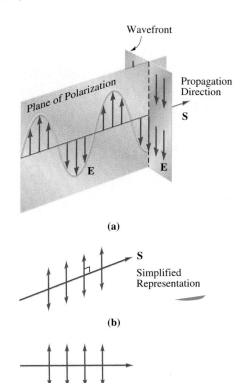

(a)

(b)

(c)

(d)

FIGURE 35.25

Plane-polarized light. (a) The plane of polarization is defined by the direction of the electric field and the direction of propagation. (b) A simplified representation of the polarization shows only double-ended arrows. The sole purpose of the arrows is to show the polarization direction. (c) A schematic representation of a ray with the plane of polarization lying in the plane of the page. (d) A ray with the plane of polarization perpendicular to the plane of the page.

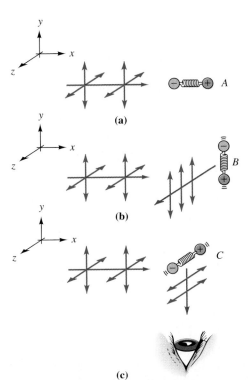

(a)

(b)

(c)

FIGURE 35.26

Polarization of light by scattering. (a) A ray of unpolarized light travels along the x-axis, striking a dipole oriented with its axis parallel to the x-axis. Because the electric field oscillates in the yz-plane, the dipole is not caused to oscillate. (b) A dipole with its axis parallel to the y-axis is caused to oscillate by the incoming light. Light is radiated by the dipole primarily into the xz-plane and is thus not seen by an observer below (in the negative y-direction). (c) A dipole with its axis oriented parallel to the z-axis is set into oscillation by the incoming wave. The oscillating dipole radiates primarily into the yz-plane. The radiated light is polarized primarily along the z-direction and is seen by an observer below.

the sun is near the horizon. The changing light intensity incident on your eyes will reveal significant polarization from the light scattered by air molecules.

Incidentally, our simple model also gives a clue as to why the sky looks blue. The dipole oscillations we are talking about are essentially oscillations of the electron clouds around the nuclei of the molecules that constitute the air. The natural frequency of this oscillation is higher than the frequency of visible light. We know from our study of resonance in mechanics that the closer the frequency of a driving force is to the natural frequency of an oscillating system, the more efficient the driving force is at setting the system into vibration. (See Section 14.5.) The higher frequency of the blue light excites larger amplitude vibration in the dipoles, and thus more blue light is "scattered." This same scattering is also the reason the sun looks red when it is near the horizon. Under this condition we observe light that was scattered *least*, namely, the lower frequencies.

The two polarizers have their polarization axes perpendicular. One polarizer appears darker than the other, indicating the polarization of sky light.

Polarization by Reflection

As we mentioned above, when light falls on the boundary between two dielectric media, it is partially reflected and partially refracted. How much of the light is reflected and how much is refracted depends on the polarization of the light relative to the dielectric interface. For light incident at a boundary between air and water, Figure 35.27 shows the

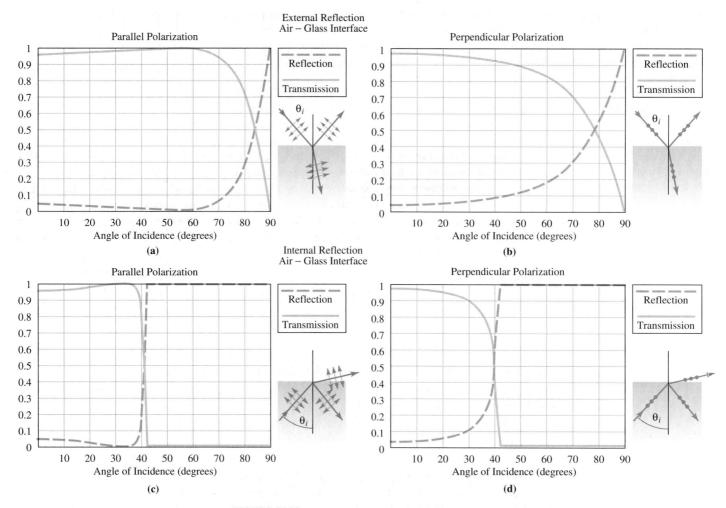

FIGURE 35.27

The fraction of incident light power transmitted and reflected at the boundary between air and glass for internal ((a) and (b)) and external ((c) and (d)) reflection. Note that, when the plane of polarization is the same as the plane of incidence ((a) and (d)), there is one angle where no light is reflected. The angle where no light of parallel polarization is reflected is called Brewster's angle. Brewster's angles for internal and external reflection are complementary.

percentage of power reflected and the percentage transmitted for (a) light polarized parallel to the plane of incidence ($E_\parallel$) and (b) light polarized perpendicular to the plane of incidence ($E_\perp$). Figures 35.27 (c) and (d) show similar results for internal reflection. The results shown in these figures follow directly from continuity requirements for the components of certain electric and magnetic fields at the interface. As we mentioned earlier, these continuity requirements are a consequence of Maxwell's equations.

A close examination of Figure 35.27(a) indicates a rather remarkable result. At one particular angle, no light with polarization in the plane of incidence is reflected. Put another way, all the reflected light is polarized perpendicular to the plane of incidence. (We hasten to add, however, that the refracted light contains both polarizations.) In general, this special case comes about when the reflected and refracted rays form a 90° angle as shown in Figure 35.28. The angle of incidence in this case is known as the **polarizing angle.**

It is possible to derive a simple expression for the polarizing angle using Snell's law

$$n_1 \sin(\theta_1) = n_2 \sin(\theta_2)$$

and the law of reflection. Examining Figure 35.28 we note that when θ_1 equals the polarizing angle θ_p, we have $\theta_p + \theta_2 + 90° = 180°$. Hence, $\theta_2 = 90° - \theta_p$. Using the trigonometric identity for the sine of the sum of two angles, we find

$$\sin(\theta_2) = \sin(90° - \theta_p) = \sin(90°)\cos(\theta_p) - \sin(\theta_p)\cos(90°) = \cos(\theta_p)$$

With this substitution, Snell's law becomes

$$n_1 \sin(\theta_p) = n_2 \cos(\theta_p)$$

which leads to

$$\tan(\theta_p) = \frac{n_2}{n_1} \qquad \text{Brewster's law}$$

This expression is known as **Brewster's law** after Sir David Brewster (1781–1868), who discovered the law experimentally. The polarization angle θ_p is often called **Brewster's angle.**

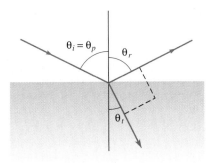

FIGURE 35.28

When the angle of incidence is equal to the polarizing angle (Brewster's angle), the reflected and refracted rays form a right angle.

EXAMPLE 35.5 *It Seems Like More When the Glare Is in Your Eyes*

When light traveling in air strikes a glass surface at the polarizing angle, what fraction of the light is reflected?

SOLUTION We use Figure 35.27 to estimate the answer. The polarization angle is 56.3°. At this angle, none of the polarization components parallel to the plane of incidence are reflected. From the graph in Figure 35.27(b) we estimate that about 15% of the perpendicularly polarized light is reflected. If we assume the original light was an equal mixture of both polarizations, then 7.5% of the original light is reflected. ◀

As we mentioned above, when light is incident on the boundary between two dielectrics at the polarizing angle, all the reflected light is polarized in the direction perpendicular to the plane of incidence. The transmitted light, however, still contains both polarizations. Suppose that the transmitted light subsequently reenters the original medium through a face parallel to the first boundary. In general, the light is again both transmitted and reflected. However, it can be shown (see Problem 33) that the exiting ray strikes the second interface at the polarizing angle for internal reflection. Hence, once again, the reflected light is completely polarized perpendicular to the plane of incidence. By employ-

Light reflected from the water surface is strongly polarized. (a) The polarizer in front of the camera lens has its polarization axis horizontal. (b) The polarizer in front of the camera lens has its polarization axis vertical, thus blocking the majority of the reflected light.

(a)

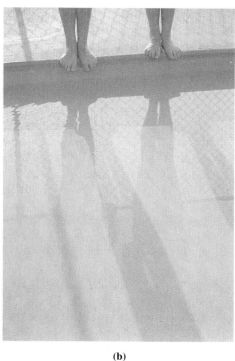

(b)

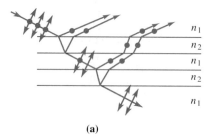

(a)

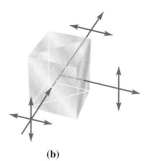

(b)

FIGURE 35.29

By using several layers of alternating dielectrics, incident unpolarized light can be separated into reflected and transmitted beams of light with orthogonal polarizations.

ing several alternating layers of the two dielectrics, more and more light polarized perpendicular to the plane of incidence can be removed from the transmitted beam. The situation is shown in Figure 35.29(a).

Alternating layers of dielectrics are used to construct **polarizing beam splitters.** Such a beam splitter in the shape of a cube is shown in Figure 35.29(b). The cube is made up of two triangular prisms joined along the hypotenuse. Between the two halves are a large number of thin alternating layers of two dielectrics. When unpolarized light enters the cube, two beams emerge at right angles to one another. The two beams have orthogonal polarizations. If the beam entering is already polarized perpendicular to the plane of incidence, it is reflected at the boundary where the half-cubes join. If the incoming beam is polarized parallel to the plane of incidence, it is transmitted. Such a cube can be used to steer light in different directions depending on its polarization. This property has many practical applications. Later in this section we will describe how such cubes are used in a compact-disc player.

Wire-Grid Polarizers

The type of polarizer with which you are most likely to be familiar is that often found in sunglasses. We refer to this type as a "wire-grid" polarizer because the underlying mechanism is identical to that of a type of polarizer that is literally made of a wire grid.

For wavelengths greater than a few microns, a polarizer can be made by arranging fine strands of wire closely spaced and parallel to one another. Such a **wire-grid polarizer** is sketched in Figure 35.30. The idea behind this polarizer is straightforward. If the plane of polarization of an incident electromagnetic wave is parallel to the conducting wires, then the electric field of that wave pushes on free electrons in the direction parallel to the wire. Because the electrons can move, they do. That is, the electric field energy of the wave is transferred to kinetic energy of the electrons, and much of that energy is then transferred to the atoms that make up the wire, manifesting itself as an increase in wire temperature. In short, such waves are absorbed.

On the other hand, electrons pushed by the electric field of waves with a plane of polarization perpendicular to the wires can move only across the diameter of the wire.

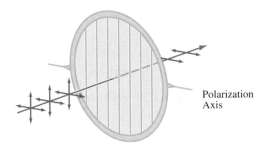

FIGURE 35.30

When unpolarized electromagnetic waves are incident on an array of closely spaced wires, electric field components parallel to the wires are selectively absorbed.

Concept Question 7
An analogy is sometimes made between the transverse electric field oscillations of an electromagnetic wave passing through a wire-grid polarizer and the oscillations of a vibrating rope passing through a space in a picket fence. Why is this analogy misleading?

Such electrons absorb far less energy from the incoming wave. As shown in Figure 35.30, the result is that light polarized parallel to the wire is absorbed, whereas light polarized perpendicular to the wire is transmitted. The plane of polarization of the transmitted light defines the **polarization axis** of the polarizer. The polarization axis is perpendicular to the direction of the grid wires.

When a wave polarized at an angle θ to the wire direction is incident on the polarizer, only the component perpendicular to the wire is transmitted. As shown in Figure 35.31, this component is proportional to $\cos(\theta)$. The irradiance is proportional to the square of the electric field. Hence, the transmitted irradiance is proportional to $\cos^2(\theta)$. This relation is known as the **law of Malus.** It states that when a completely linearly polarized wave of irradiance I_o is incident on a polarizer, the transmitted irradiance I varies according to

$$I = I_o \cos^2(\theta) \qquad \text{Law of Malus}$$

Note that when an *unpolarized* wave falls on a polarizer, the transmitted power is cut in half because the average value of $\cos^2(\theta)$ averaged over all angles is $\frac{1}{2}$ (see Chapter 33, Section 3).

As a practical matter, the wire-grid polarizer described above works only for electromagnetic waves with wavelengths greater than very long infrared wavelengths due to the limitation on the size and spacings of the wires.[2] It is possible to make polarizers with more finely spaced grids by using photolithographic techniques similar to those used to make integrated circuits. Using such techniques, wire-grid polarizers can be made with the "wires" separated by a fraction of a micron. Such polarizers are useful throughout the infrared region of the electromagnetic spectrum.

For visible wavelengths, polarizers are created from plastics the molecules of which are formed in long polymer chains. In one type of polarizer, iodine atoms are added to the plastic. Each iodine atom bonds to a polymer chain and releases an electron that is free to travel along the length of the chain. By stretching the plastic, the polymer chains can be preferentially aligned, creating an atomic-scale wire grid. There are also crystals that, because of their structure, selectively absorb one polarization more strongly than another. For historical reasons these types of polarizers are known as **dichroic** polarizers. Dichroic literally means "two-color." The name stems from the fact that certain materials of the type used to make sheet polarizers exhibit two colors when subject to polarized white light.

If you have a pair of sunglasses with polarizing lenses, you should not miss the chance to observe polarized light from the sky or reflecting surfaces by tilting your head or, better still, rotating the glasses in front of your eye.

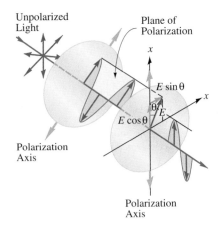

FIGURE 35.31

The first polarizer transmits light with a single polarization. When this polarized light strikes a second polarizer, we resolve its electric field into components parallel to and perpendicular to the polarization axis of the second polarizer. Only the electric field component parallel to the polarization axis is transmitted.

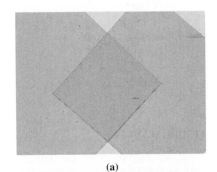

(a)

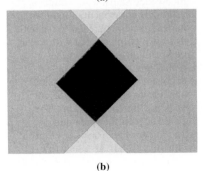

(b)

(a) Overlapping polarizers with parallel polarization axes. (b) Overlapping polarizers with crossed polarization axes.

---

[2] Wire-grid polarizers also work well and are quite convenient at microwave frequencies for which wavelengths are on the order of millimeters to centimeters.

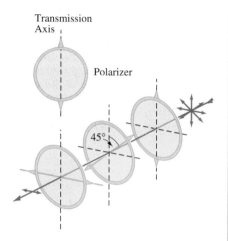

FIGURE 35.32

A double application of the law of Malus as described in Example 35.6.

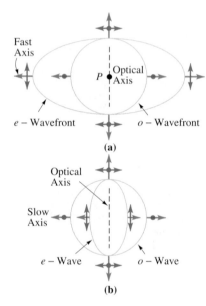

FIGURE 35.33

Imagine a small source of unpolarized light embedded inside of a birefringent crystal. The component of light with polarization perpendicular to the optic axis (o-wave) spreads out with equal speed in all directions, forming a circular light pulse. The light component with polarization parallel to the optic axis (e-wave) has different speeds in different directions. The forward edge of the spreading extraordinary light forms an ellipsoid within the crystal. (a) A negative uniaxial crystal where $(n_e - n_o) < 0$. (b) A positive uniaxial crystal with $(n_e - n_o) > 0$.

EXAMPLE 35.6   *With Malus Aforethought*

An unpolarized beam of light passes through three ideal polarizers. The first has its polarizing axis vertical, the second 45° from the first, and the third 90° from the first. Find the irradiance of the beam after passing through each polarizer.

SOLUTION   Polarizer 1: We call the irradiance of the incoming beam I_o. After passing through the first polarizer the irradiance is reduced to $I_1 = I_o/2$ as explained in the text.

Polarizer 2: The beam incident on polarizer 2 is polarized vertically as shown in Figure 35.32. Polarizer 2 has its polarization axis at an angle of 45° from the vertical so that from the law of Malus we have

$$I_2 = I_1 \cos^2(\theta) = I_1 \cos^2(45°) = I_1\left(\frac{1}{2}\right) = \frac{1}{4}I_o$$

Polarizer 3: The beam exiting the second polarizer has its polarization plane oriented at 45° from the axis of polarizer 3, so once again applying the law of Malus we have

$$I_3 = I_2 \cos^2(\theta) = I_2 \cos^2(45°) = I_2\left(\frac{1}{2}\right) = \frac{1}{8}I_o$$

Notice that if polarizer 2 were not present, the transmitted irradiance would be zero! ◀

Birefringence (Optional)

Liquids, gases, amorphous solids, such as glass, and crystals with cubic atomic arrangements, such as salt, can be characterized by a single index of refraction. Not all materials are that simple, however. Many crystals, such as calcite and quartz, must be characterized by two indices of refraction.[3] Such crystals are called **birefringent.** When unpolarized light enters a birefringent crystal it breaks into two component light waves with perpendicular polarizations. One of the waves behaves in a quite ordinary way. In fact, it is called the **ordinary wave.** The ordinary wave travels with the same speed in all directions within the crystal and this speed is characterized by an index n_o. If we could embed a small light bulb in the center of the crystal, ordinary waves would spread out in all directions with spherical wavefronts. The second wave, however, is quite extraordinary.[4] The speed of the **extraordinary wave** is different in each direction within the crystal. Wavefronts of the extraordinary wave form three-dimensional ellipsoids as shown in Figure 35.33. Let's assume for the moment that the speed of the extraordinary wave is generally greater than that of the ordinary wave as shown in Figure 35.33(a). The direction of fastest propagation (known as the *fast axis*) is characterized by an index of refraction n_e. In one direction within the crystal the speeds of the ordinary wave and extraordinary wave are equal. This direction defines the **optic axis** of the crystal.[5] When an extraordinary wave travels

---

[3] Crystals that belong to the hexagonal, tetragonal, and trigonal crystal groups are known as uniaxial and are characterized by two principal indices of refraction. Crystals belonging to the orthorhombic, monoclinic, and triclinic groups are biaxial; they have two optical axes and require three principal indices.

[4] Extraordinary indeed! These rays do not obey Snell's law and, in fact, may not even lie in the plane of incidence.

[5] This is true only for the class of birefringent crystals known as **optically inactive.** For optically **active** crystals the e-wave and o-wave never have the same speed in any direction within the crystal. For optically active crystals there is, however, one direction in which the difference between the e-wave speed and o-wave speed is a *minimum*. This direction defines the optic axis for optically active crystals. Calcite is an example of an optically inactive crystal; quartz is optically active.

through a crystal in a direction other than the optic axis or the fast axis, the effective index of refraction is between n_o and n_e. As shown in Table 35.2, n_o can be greater or less than n_e. Thus, the extraordinary wave can, in general, travel either faster or slower than the ordinary wave. Crystals for which $(n_e - n_o) < 0$ are known as **negative uniaxial,** whereas if $(n_e - n_o) > 0$ the crystals are said to be **positive uniaxial.** For positive uniaxial crystals the fast axis is, of course, replaced by a *slow axis.*

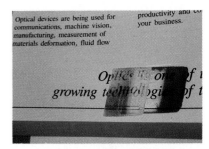

(a)

TABLE **35.2**   Refractive Indices for Some Birefringent Crystals

CRYSTAL	n_o	n_e	$\Delta n = n_e - n_o$
Calcite	1.6584	1.4864	−0.1720
Ice	1.309	1.313	+0.0040
Quartz	1.5443	1.5534	+0.0091
Sodium nitrate	1.5854	1.3369	−0.2485
Rutile	2.616	2.903	+0.2870
Tourmaline	1.669	1.638	−0.0310

(b)

Generally, when an unpolarized wave enters a birefringent crystal, the ordinary and extraordinary waves separate and follow distinct paths within the crystal. This separation causes a double image to form when an object is viewed through such a crystal. Figure 35.34 shows the strong birefringence of a calcite crystal. When this double image is viewed through a polarizer, the individual images can be blocked one at a time by rotating the polarizer as shown in Figures 35.34(b) and (c). The mechanism by which the image caused by the extraordinary ray comes to be separated from that of the ordinary rays is shown in Figure 35.35.

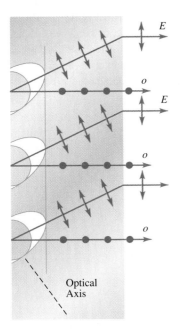

FIGURE 35.34

(a) Separation of the ordinary and extraordinary waves in a calcite crystal creates a double image. (b) and (c) The images are produced by light with orthogonal polarization.

FIGURE 35.35

Wavefronts and rays in a birefringent crystal. When one invokes Huygens' principle in a birefringent crystal, the wavelets for the extraordinary wave are elliptical in cross section. Wavefronts are formed by the line of tangency to the ellipsoids. The ray direction for the extraordinary waves is found by joining the center of the ellipse to the point of tangency of the wavefront. Note that the ray direction for the extraordinary wave is not perpendicular to the wavefront.

Half-Wave and Quarter-Wave Plates (Optional)

Imagine that we cut a negative uniaxial birefringent crystal so that two opposite faces are perpendicular to the fast axis. If an unpolarized wave enters such a crystal normal to this face, the ordinary and extraordinary rays take the same path straight through the crystal. Because $n_e < n_o$, the wave with ordinary wave polarization (the o-wave) falls farther and farther behind the e-wave as the waves progress through the crystal. If the thickness of the crystal is such that the ordinary wave falls behind by exactly one-half wavelength, the crystal is said to be a **half-wave plate.** As a practical matter such plates are sometimes made so that the delay is $3\lambda/2$, $5\lambda/2$, or more so that the relative phase change is the same, but the plate is thicker and therefore stronger.

Half-wave plates are useful for rotating the plane of polarization of a linearly polarized wave. In Figure 35.36(a) we show the electric field of a wave with its plane of polarization making a 45° angle with the x- and y-axes of a Cartesian coordinate system. If we like, we can resolve the electric field vector into components along these axes. If we now imagine that this wave travels toward us in time, the electric field shrinks to zero and then grows in the opposite direction as shown in the sequence of drawings of Figure 35.36(a). Note that each of the two components also shrinks and grows, while the vector sum is, of course, the resultant total **E** vector.

Imagine now that we send this wave through a half-wave plate oriented so that the polarization component parallel to the y-axis becomes the e-wave and the component with polarization parallel to the x-axis becomes the o-wave. After traveling through the half-wave plate, the x-axis component is 180° behind in its oscillation relative to the y-axis component. This situation is shown in the sequence of drawings in Figure 35.36(b). Note carefully that both components oscillate exactly as before, except for the phase of the x-component. The effect, as you can see, is to rotate the plane of polarization of the resultant by 90° relative to the wave that entered the half-wave plate.

A **quarter-wave plate** is half as thick as a half-wave plate. If the wave in Figure 35.36(a) is incident on a quarter-wave plate, the result is as shown in Figure 35.36(c). Note that once again both E-field components oscillate from positive to negative just as before. This time, however, the resultant E-field is quite different. Instead of its *length* changing as it oscillates in a fixed direction, its magnitude stays constant and the *direction* of the E-field vector rotates. Light with this type of behavior is said to be **circularly**

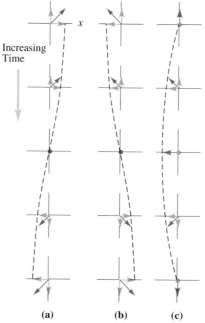

FIGURE 35.36

(a) A sequence of sketches showing that as the magnitude of the electric field of a polarized wave oscillates in time, its components along the x- and y-axes also oscillate in time. (b) The electric field components oscillate as in (a) except that the x-component is delayed by one-half oscillation. The result is that the resultant E-field is rotated by 90° relative to that in (a). (c) When the x-component is delayed by one-quarter wave relative to that in (a), the resultant E-field has a constant magnitude but rotates.

(a)   (b)   (c)

polarized. The *E*-field pictured in Figure 35.36(c) rotates anticlockwise as the light comes toward you. Such light is *left* circularly polarized.

The uses for half-wave and quarter-wave plates in optics are many and varied. You may own a quarter-wave plate and not even know it. In the next section we describe how one is cleverly used in the design of many compact-disc players.

Optics of the Compact-Disc Player (Optional)

The musical sounds produced by a compact disc are encoded in a digital format. That is, they are represented by a binary sequence, essentially a string of 1's and 0's, such as 00100011101011000. . . . Electronic circuits within the compact-disc player convert the digital pattern into sound. On the compact disc itself, the string of 1's and 0's is encoded by a series of bumps arranged in a long spiral as shown in Figure 35.37. The source of the light used to "read" the compact disc is a small diode laser. This laser produces a narrow beam of light that is directed at the reflecting surface within the disc. After reflecting, the light falls on a phototransistor the electrical output of which varies with the irradiance of the light striking it. Thus, the optical signal is transformed into an electric signal. The presence or absence of a bump is detected by monitoring the reflection of the focused laser beam from the disc. The size of the spot illuminated by the laser beam is about twice the size of the bump. When no bump is present, the reflected beam is bright. The height of the bumps is about one-quarter of the wavelength of the laser light used to read the pattern. When laser light reflects from an area with a bump, the reflected irradiance is diminished by interference between the light reflected from the bump and light reflected from the surrounding flat area.[6] The change from high reflected irradiance to low reflected irradiance is interpreted as a 1; the return to brighter reflected irradiance is interpreted as a 0. The parts of the compact disc of interest to us here are the optics by which the beam is steered from the laser to the detector.

The two key optical elements, shown in Figure 35.38, are the polarizing beam splitter and the quarter-wave plate, both of which we discussed earlier. Light from the laser is polarized as it passes through the beam splitter on the way to the compact disc. The transmitted light (now polarized) then passes through the quarter-wave plate, becoming circularly polarized. After reflecting from the disc the returning light passes once again through the quarter-wave plate. Two passes through the quarter-wave plate are equivalent

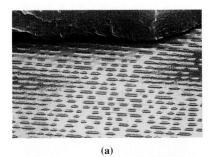

(a)

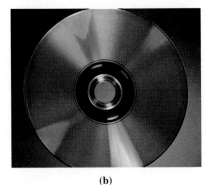

(b)

FIGURE 35.37

(a) The surface of a compact disc is covered by lines of small bumps that encode information. (b) The lines of bumps form a long spiral on the surface of the compact disc.

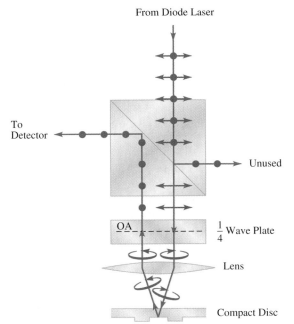

From Diode Laser

To Detector

Unused

OA $\frac{1}{4}$ Wave Plate

Lens

Compact Disc

FIGURE 35.38

The light path in a compact-disc player. A polarizing beam splitter transmits polarized light through a quarter-wave plate, creating circularly polarized light. After reflecting from the compact disc, the light once again passes through the quarter-wave plate becoming linearly polarized in a direction perpendicular to the incoming light. The polarizing beam splitter now reflects the light to the detector.

---

[6]Interference effects are discussed in detail in Chapter 37.

to one pass through a half-wave plate. The net effect is to rotate the plane of polarization of the original plane-polarized light by 90°. Thus, when the returning beam strikes the polarizing beam splitter, it is reflected rather than transmitted. Consequently, the returning light is directed to the detector rather than back into the laser source.

35.6 Summary

When light strikes a reflective surface, the angle θ_i between the incident ray and a normal to the reflecting surface and the angle θ_r between the reflected ray and the normal lie in the **plane of incidence** and are related by

$$\theta_i = \theta_r \qquad \text{Law of specular reflection}$$

When light strikes the interface between two dielectric media, it is generally both reflected and transmitted. The direction of the transmitted ray and the incident ray are related by their respective angles with a normal to the surface by the equation

$$n_1 \sin(\theta_1) = n_2 \sin(\theta_2) \qquad \text{Snell's law}$$

Partial reflection always occurs at the interface between two dielectric media. If the light is traveling from a substance with a lower index of refraction to a substance with a higher index of refraction, the refracted ray is bent toward the normal and the reflected ray is said to have undergone **external reflection.** If the light is traveling from a substance with a higher index of refraction to a substance with lower index of refraction, the refracted ray is bent away from the normal and the reflected ray is said to have undergone **internal reflection.** When the angle of incidence is equal to the **critical angle,** the angle of refraction is equal to 90°. For angles of incidence greater than the critical angle **total internal reflection** occurs.

An optical fiber is composed of concentric cylinders of glass. The outer **cladding** layer has a lower index of refraction than the inner **core.** The half-angle θ of the acceptance cone is given by the **numerical aperture** NA

$$NA = \sin(\theta) = \sqrt{n_{\text{core}}^2 - n_{\text{clad}}^2}$$

The index of refraction of a material varies with the wavelength of the incident light—a phenomenon known as **dispersion.** Dispersion is exploited in **prisms** to separate the various wavelengths in a beam of light.

Light can be modeled as a transverse electromagnetic wave. The plane of oscillation of the electric field is known as the **plane of polarization.** Waves with a single plane of polarization are known as **plane-polarized waves.** Plane-polarized waves can be created by scattering, reflection, wire-grid type polarizers, and birefringent crystals. When light that is already polarized passes through a **wire-grid polarizer,** the transmitted irradiance is related to the incident irradiance by

$$I = I_o \cos^2(\theta) \qquad \text{Law of Malus}$$

where θ is the angle between the plane of polarization of the incident wave and the polarization plane of the polarizer.

When light is reflected at the boundary between two dielectric media, all of the reflected light is polarized perpendicular to the plane of incidence when the angle of incidence is equal to the **polarizing angle** (Brewster's angle), which is given by

$$\tan(\theta_p) = \frac{n_2}{n_1} \qquad \text{Brewster's law}$$

(Optional) **Birefringent** materials can be used to create **half-wave plates,** which rotate the plane of polarization of plane-polarized waves, and **quarter-wave plates,** which change linearly polarized light into **circularly polarized light.**

PROBLEMS

35.3 Reflection

1. Two mirrors are set at right angles as shown in Figure 35.P1. Prove, using the law of reflection and plane geometry, that a light ray coming in at any angle is reflected back parallel to itself after striking both mirrors. (A three-dimensional version of this arrangement is called a *corner reflector* and is made up of three mirrors at right angles to each other. Notice the shape of the small indentations on the taillights of a car or the reflectors on the side of a highway.)

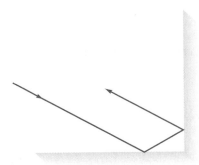

FIGURE 35.P1   Problem 1

2. **Fermat's principle** states that light travels in such a way as to take the least time when traveling between two points. Figure 35.P2 shows a light ray leaving point A and traveling to point B after reflecting from mirror M. (a) Show that the time it takes the light to travel from A to B is given by

$$t = \frac{\sqrt{D^2 + x^2} + \sqrt{(L - x)^2 + d^2}}{c}$$

(b) Show that this expression for t is a minimum when

$$\frac{x}{\sqrt{D^2 + x^2}} = \frac{L - x}{\sqrt{(L - x)^2 + d^2}}$$

(c) Explain how the expression in part (b) implies the law of reflection.

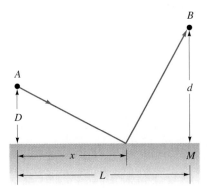

FIGURE 35.P2   Problem 2

3. Prove that if a mirror is rotated through an angle θ, then the reflected ray is deviated by 2θ from its original direction.
4. A scanner used to read bar codes is composed of a collimated laser beam, which strikes an eight-sided scanning mirror as shown in Figure 35.P3. Over what range of angles does the reflected beam scan when the scanning mirror is rotated about its center?

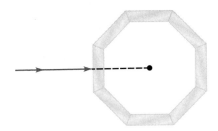

FIGURE 35.P3   Problem 4

35.4 Refraction

5. A helium-neon laser has a wavelength of 632.8 nm in air. If light from this laser enters water (a) what is its speed in water? (b) its wavelength? (c) its frequency?
6. Light from an argon-ion laser has a wavelength in air of 488 nm. The speed of the light emitted by this laser in a certain block of glass is found to be 2.07×10^8 m/s. (a) What is the index of refraction of the glass? (b) What is the wavelength of the light while in the glass? (c) What is the frequency of the light in the glass? In air?
7. Two rays of light with wavelength (in air) of 632.8 nm travel parallel to each other. One ray encounters a 1.00-mm thick layer of glass with refractive index $n = 1.50$. (a) How much more time does it take the ray traveling through the glass to go through the 1.00 mm of glass than it takes the ray traveling in air to go this same distance? (b) Which ray completes more oscillations while traveling this 1.00-mm distance? (c) What is the actual difference in number of oscillations made by the two waves over the 1.00-mm distance?
8. A ray of light traveling in air is incident on the smooth surface of a lake at an angle of 36.87° with the normal. Find the angle (from the normal) the transmitted ray makes in the water.
9. A ray of light from a submerged diver's flashlight reaches the surface of the lake, making an angle of 75° with the water's smooth *surface*. What angle does the emerging ray make in air with the water's *surface?*
10. A ray of light traveling in air strikes a parallel-sided dielectric layer with index of refraction n and makes an angle of θ with a normal to the dielectric surface as shown in Figure 35.P4. After exiting the dielectric the ray travels in a direction parallel to its

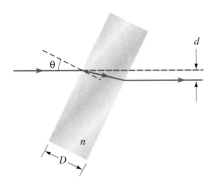

FIGURE 35.P4   Problem 10

original path but is displaced laterally. (a) Derive an expression for the lateral displacement d of the beam for a dielectric layer of thickness D. (b) Show that for small angles θ your expression reduces to $d = D\theta(n - 1)/n$.

11. **Fermat's principle** states that light travels in such a way as to take the least time to travel between two points. Figure 35.P5 shows a light ray leaving point A and traveling to point B after being refracted at point M. The velocity of the light in the upper medium is V_1, and the velocity in the lower medium is V_2. (a) Show that the total time it takes the light to travel from A to B is given by

$$t = \frac{\sqrt{d^2 + x^2}}{V_1} + \frac{\sqrt{(L - x)^2 + D^2}}{V_2}$$

(b) Show, using the calculus, that t is a minimum when

$$\frac{1}{V_1} \frac{x}{\sqrt{d^2 + x^2}} = \frac{1}{V_2} \frac{(L - x)}{\sqrt{(L - x)^2 + D^2}}$$

(c) Show that the expression in part (b) reduces to Snell's law.

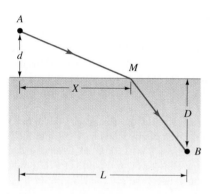

FIGURE 35.P5   Problem 11

12. A slender pole extends straight up from the bottom of a clear pond 2.00 m deep. The pole extends 1.20 m above the surface of the pond. The sun's light creates a shadow of the pole on the surface of the water. This shadow is 0.800 m long. What is the length of the pole's shadow on the bottom of the pond?

13. A small ball floats on the surface of a swimming pool that is 3.00 m deep. On the bottom of the pool, what is the distance between the point directly below the ball and the ball's shadow when the sun is 30.0° above the horizon?

14. What is the critical angle for total internal reflection at the interface between air and dense flint glass?

15. It is found that the critical angle for total internal reflection between air and a certain substance is 24.4°. (a) What is the index of refraction of the substance? (b) What might the substance be?

16. (a) What is the critical angle for total internal reflection between water and dense flint glass? (b) In which direction can total internal reflection occur, as light goes from glass to water or water to glass?

17. A ray of light is traveling in glass ($n = 1.50$) and strikes an internal surface at $\theta = 60.0°$ as shown in Figure 35.P6. If a small drop of liquid is placed on the surface of the glass as shown in the figure, what is the maximum index of refraction of the liquid such that the ray suffers total internal reflection?

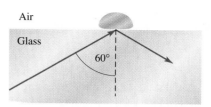

FIGURE 35.P6   Problem 17

18. Verify the claim made in Section 35.2 that at normal incidence 4% of the light energy is reflected at each surface of a pair of eyeglasses. That is, using $n = 1.50$ for the glass, show that $R \approx 0.04$.

19. Follow the ray through the prism ($n = 1.45$) shown in Figure 35.P7 until it has struck three faces. At each face describe the angle made by both the reflected ray and the transmitted ray (if any). You will need a protractor to make a careful diagram of the light path.

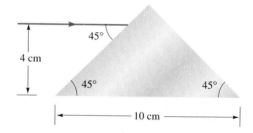

FIGURE 35.P7   Problems 19 and 21

20. Follow the ray through the prism ($n = 1.47$) shown in Figure 35.P8 until it has struck three faces. At each face describe the angle made by both the reflected ray and the transmitted ray (if any). You will need a protractor to make a careful diagram of the light path.

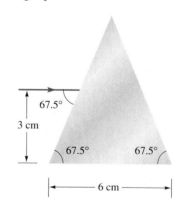

FIGURE 35.P8   Problems 20 and 22

21. Repeat the exercise described in Problem 19, except assume the prism is made of air and the surrounding medium is water.

22. Repeat the exercise described in Problem 20, except assume the prism is made of air and the surrounding medium is water.

23. (a) Calculate the numerical aperture of a multimode optical fiber (Fig. 35.P9) with core index $n = 1.450$ and cladding index $n = 1.435$. (b) If a screen is placed 10.0 cm in front of the tip of a light-filled fiber, what is the diameter of the circle of light on the screen? (Assume the circle is formed only by light exiting the fiber at angles within the acceptance cone.)

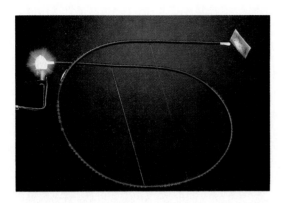

FIGURE 35.P9 **The sine of the half-angle of the cone of light exiting the fiber indicated the fiber's numerical aperture. See Problem 24.**

24. A certain step-index multimode fiber has a core index of $n = 1.465$. When the fiber is filled with light and a screen is held 15.0 cm in front of the fiber, a circle of light with diameter 6.20 cm is created by light escaping from the fiber. (a) Calculate the cladding index of the fiber. (b) Express the difference between the core and cladding indices in terms of the fiber parameter $\Delta = (n_{core} - n_{clad})/n_{core}$.

25. Consider a meridional ray that is introduced into a step-index multimode fiber at an angle too large to be confined to the core by total internal reflection at the core–cladding boundary. (a) If a ray makes an angle of 10.0° with the central axis of an optical fiber, how many reflections at the core–cladding boundary does the ray make when traveling 10.0 cm down the length of an optical fiber? Take the core diameter of the fiber to be 100. μm. (b) If only 99.0% of the light is reflected each time, what fraction of the light remains in the core after it travels this 10.0-cm distance?

26. Consider meridional rays in a step-index multimode fiber with core index $n = 1.453$, $\Delta = 0.0100$, and core diameter $a = 100.$ μm. (a) What is the largest angle (inside the fiber) that a ray can make with the fiber axis and still suffer total internal reflection at the core–cladding boundary? Such a ray corresponds to the **highest-order mode** in the fiber. (b) The **lowest-order mode** in a multimode fiber corresponds to a ray that travels straight down the center of the fiber core. Find the difference in time required for rays corresponding to the lowest-order and highest-order modes to travel through 1.00 km of fiber. (This time difference is referred to as **modal dispersion.** It is responsible for the spreading (in time) of short pulses of light as the pulses traverse a fiber. Modal dispersion is the major limitation of high-speed data communications in multimode fiber.)

27. For light of wavelength $\lambda = 589$ nm the angle of minimum deviation for a prism with a vertex angle of 60.0° is found to be 37.2°. (a) What is the index of refraction of the prism? (b) If the prism is replaced with a thin hollow prism of the same vertex angle but filled with glycerine ($n = 1.473$), what is the angle of minimum deviation?

28. Light from a discharge lamp containing hydrogen gas is composed of a mixture of light with many *distinct* wavelengths. Brightest among them are 656.3 nm, 486.1 nm, and 410.2 nm. Suppose a narrow beam of this light strikes a light flint glass prism which has an apex angle $A = 60.0°$. If the beam is incident at an angle $\theta = 48.0°$ from the normal as shown in Figure 35.P10, calculate the angles at which 656.3-nm and 410.2-nm beams emerge from the prism. (*Hint:* Use Fig. 35.20 to estimate the index of refraction of the prism at these wavelengths.)

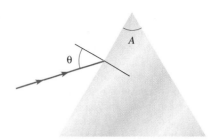

FIGURE 35.P10 **Problem 28**

29. (a) Use the geometry of Figure 35.P11 to show that if a ray strikes one face of a wedge at normal incidence, the angle of deviation is given approximately by $\delta = A(n - 1)$ for small A. (Assume that A is small enough that the approximation $\sin(A) \approx A$ is valid.) (b) Show that Equation (35.9), which relates n, A, and δ for the minimum deviation of a prism, reduces to this same expression for small A. (c) If the maximum allowable deviation of a ray passing through a fused silica plate is 2.00 min of arc, what is the maximum allowable angle between the faces of the plate?

FIGURE 35.P11 **Problem 29**

35.5 Polarization

30. The polarizing angle for light traveling in air when reflecting off fluorite (CaF_2) is 55.1°. What is the critical angle for total internal reflection for fluorite and air?

31. Unpolarized light falls on a polarizing sheet with its polarization axis vertical. Behind this polarizer is a second polarizing sheet (called the analyzer) with its polarizing angle at 30° from the first. (a) What fraction of the energy of the original beam is transmitted through the first polarizer? (b) What fraction of the *original* intensity is transmitted through both polarizers?

32. Two sheet polarizers are placed together with their polarization directions offset by an unknown angle θ. When placed into a beam of unpolarized light, the transmitted light power is 20.0% of the original power. What is the angle θ between the two polarization axes?

33. (a) Show that the polarizing angle for external reflection and internal reflection are complementary. (b) Show geometrically that a ray entering a parallel-sided dielectric layer at the external polarizing angle strikes the opposite face at the internal polarizing angle. The important consequence of this result is that all the reflected light is polarized regardless of whether it is reflected from the front surface or the back surface.

34. Four ideal polarizers are placed in a row. The transmission axis of the second is 20° from the first, the third is 20° from the second, and the fourth is 20° from the third (making the fourth 60° from the first). If a beam of unpolarized light of intensity I_o falls on this combination, what fraction is transmitted through all four?

35. Consider a race between the ordinary wave (*o*-wave) and the extraordinary wave (*e*-wave) traveling along a direction perpendicular to the optic axis in a calcite crystal. (a) Show that the total number of wavelengths that "fit" into a thickness *d* of the crystal is nd/λ_o, where λ_o is the wavelength in air, and *n* is the appropriate index of refraction. The quantity *nd* is known as the **optical path length.** (b) A half-wave plate is formed when the optical path length of the *o*-wave and the *e*-wave differ by an odd multiple of $\lambda_o/2$. Calculate the minimum thickness of a half-wave plate of calcite designed for use with light having a free-space wavelength of 589 nm.

36. The thicknesses required for half-wave and quarter-wave plates are impracticably small for some crystals (see Problem 35). Mica on the other hand can be easily cleaved such that the indices for light transmission at right angles to the cleavage plane are 1.6049 and 1.6117 for light with a free-space wavelength of 589.3 nm. (a) Calculate the minimum thickness of a mica half-wave plate cleaved in this fashion. (b) What is the fractional wavelength retardation if such a plate is used at 633 nm?

37. An ideal polarizing sheet and a quarter-wave plate are cemented together such that the optical axis of the quarter-wave plate makes an angle of 45° with the transmission axis of the polarizer. The combination is placed on top of a highly polished twenty-five-cent piece. (a) If the quarter-wave plate is closest to the coin, can you see the coin? Explain why or why not. (b) If the combination is reversed so that the plane polarizer is placed next to the coin's surface, can you see the coin? Explain why or why not.

Numerical Methods and Computer Applications

38. Construct a spreadsheet template that uses Equations (35.5) through (35.8) to calculate the deviation of a ray that strikes a prism at angle θ_1 as in Figure 35.24. Take the refractive index of the prism to be 1.65 and the apex angle to be 60.0°. Have your spreadsheet calculate the angle of deviation for angles of incidence (in 1° increments) between 45° and 75°. Graph these results and compare the minimum deviation with that predicted by Equation (35.9).

39. When light propagates in a medium where the index of refraction changes gradually, its direction of propagation changes slowly. The path of the light is thus curved. Model this effect by using 20 layers of dielectrics with indices of refraction varying from *n* = 1 to *n* = 3 in steps of 0.1 (i.e., *n* = 1.0 1.1, 1.2 1.3, . . .). Let the ray enter at a 5° angle (as measured from the interface surface) in the least dense medium. Construct a spreadsheet template or computer program to calculate the Snell's law result at each interface. (a) Determine the angle of the ray in the densest medium. (b) Produce a graph showing the trajectory of the ray through the layered medium, assuming each layer is 1.0 mm thick.

40. Two polarizing sheets are placed with polarization axes at right angles to one another so that no light can be transmitted through the pair. A third sheet is introduced such that its polarizing direction makes a 45.0° angle with the first. As explained in Example 35.6 one-eighth of the irradiance that is incident on the first polarizer passes through all three. (a) Suppose now that two sheets of polarizer are inserted between the original two. Let each of these two sheets make an angle of 30° with the previous polarizer (Fig. 35.P12). (b) What fraction of light incident on the first polarizer now exits the system? (c) Next insert three polarizers between the original two, each at 22.5° from the previous. Calculate the fraction of transmitted irradiance. (d) Make a spreadsheet or computer program to continue this process, adding *N* sheets of polar-

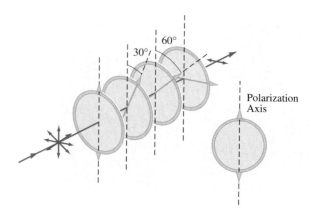

FIGURE 35.P12   Problem 40

izer each at an angle of $90°/(N + 1)$ from the previous. Let *N* vary from 1 to 30. Produce a graph of fractional transmitted power versus *N*. Is the result what you expected?

41. The equations that describe the transmission and reflection of light at the boundary between two dielectrics are known as the **Fresnel equations.** The ratios of the transmitted power to the incident power for light polarized in the plane of incidence and for light polarized perpendicular to the plane of incidence are denoted $T_\parallel$ and $T_\perp$, respectively. Similarly, the ratios of reflected power to incident power for the two polarizations are denoted $R_\parallel$ and $R_\perp$. These ratios are given by

$$R_\perp = \left(\frac{N_1}{D_1}\right)^2$$

$$R_\parallel = \left(\frac{N_2}{D_1}\right)^2$$

$$T_\perp = \frac{N_3}{(D_1)^2}$$

$$T_\parallel = \frac{N_3}{(D_2)^2}$$

where $N_1 = n_1 \cos(\theta_1) - \rho$, $N_2 = n_2 \cos(\theta_1) - (n_1/n_2)\rho$, $N_3 = 4n_1 \cos(\theta_1)\rho$, $D_1 = n_1 \cos(\theta_1) + \rho$, $D_2 = n_2 \cos(\theta_1) + (n_1/n_2)\rho$, and $\rho = \sqrt{n_2^2 - n_1^2 \sin^2(\theta_1)}$. Use a spreadsheet or a computer program to produce graphs like those shown in Figure 35.27 for an air–diamond interface.

General Problems

42. Calculate the fiber parameter Δ for a multimode fiber with a core index of refraction 1.450 if light from the fiber forms a circle of radius 3.00 cm on a screen held 10.0 cm from the fiber's tip.

43. Draw ray diagrams and explain the appearance of Figures 35.P13 and 35.P14. (a) How it is possible in Figure 35.P13 to see the reflection of the sun but not the sun directly? (b) Why does the image of the sun in Figure 35.P14 appear to be smeared into the hot (low-density) exhaust of the plane's engines?

44. A small mirror is placed in a bowl of water, making an angle $\phi = 10.5°$ with the water surface as shown in Figure 35.P15. A beam of light strikes the water at an angle $\theta = 33.3°$ from its surface and subsequently reflects from the mirror. At what angle from the water surface does the ray emerge from the water?

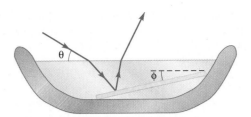

FIGURE 35.P15   Problem 44

45. Consider a slab of glass of refractive index n covered by a thin sheet of wet paper. When a beam of light strikes the paper at normal incidence, a distinctive circular pattern is formed as shown in Figure 35.P16(a). As shown in Figure 35.P16(b) the dark region around the central area occurs because scattered light from the front surface is able to escape the glass at the back surface. However, light striking the back surface at greater than the critical angle is reflected back to the front surface where it escapes into the wet paper and is scattered, forming the brighter surrounding area. (a) Show that the radius of the bright ring is given by $R = 2D/\sqrt{n^2 - 1}$. (b) Calculate the index of refraction of a glass plate of thickness 0.50 cm if the *diameter* of the bright ring is 2.42 cm.

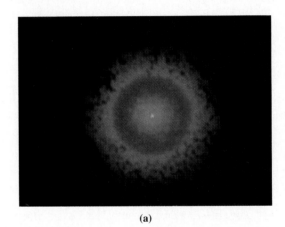

(a)

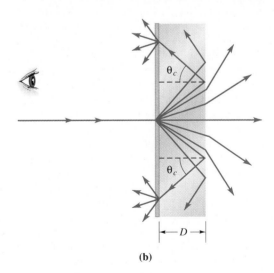

(b)

FIGURE 35.P16   Problem 45

FIGURE 35.P13   Notice that the sun is visible in the reflection but not in the directly viewed scene. See Problem 43.

FIGURE 35.P14   Note the image of the sun in the exhaust from the plane's engines. See Problem 43.

Geometrical Optics

In this chapter you should learn

- how images are formed.
- the difference between real and virtual images.
- how to predict the size and location of images formed by flat and curved mirrors.
- how to predict the size and location of images formed by refracting surfaces and lenses.

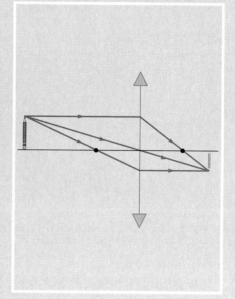

$$\frac{1}{o} + \frac{1}{i} = \frac{1}{f}$$

Images are all around us. Some images we see when light is reflected to us. Bathroom mirrors, rearview mirrors, shiny rings, the front and back of metal spoons are all familiar places where we see the images we call reflections. Other images we see when light is refracted before entering our eyes. When you look through eyeglasses or contact lenses, through a magnifying lens, a microscope, or even simply at the bottom of a swimming pool, the images you see have been affected by refraction. In this chapter we study how such images are formed and learn how to predict their size and location.

36.1 Images

In this chapter we assume that light travels in isotropic media so that ray paths are straight lines except at a reflecting surface and at a boundary between dielectric media. We will use the ray model for light almost exclusively.

Before beginning, we need to remind ourselves how we "see" in the first place. Here we don't mean a detailed description of the eye, such as that presented later in the chapter, but rather a quite naive picture of how we see. Young children often have the incorrect notion that we see things because something comes *out* of our eyes and falls on external objects (an unfortunate mistake reinforced by certain comic book heroes whose magical vision comes from rays emanating from their eyes). In actual fact you know, of course, that we see because something *enters* our eyes, namely, light. It is critical to remember that the *only* information we have about the objects we see is due to the light entering our eye.

Figure 36.1(a) is a representation of some of the light rays that have been diffusely reflected from a pencil point. Some of these rays are directed toward an observer's eye. It is the small cone of divergent light rays entering our pupil that causes us to believe the

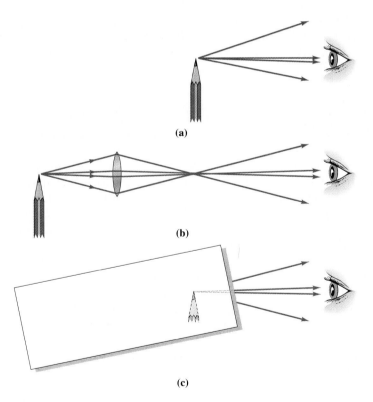

(a)

(b)

(c)

FIGURE 36.1

(a) Light reflected from the tip of a pencil enters the pupil of an observer's eye. (b) The pencil is moved and a lens interposed between the pencil and the observer. Rays from the pencil tip eventually enter the observer's eye. (c) The rays carry information about the tip of the pencil. When asked to determine the location of the pencil, the innocent eyeball backtracks the rays to their apparent point of origin. This apparent point of origin is the location of the image of the pencil.

pencil is located at the position where we see it.[1] However, there is no need for the pencil to actually be in the position shown. In Figure 36.1(b) we show another optical configuration (don't worry about the details) that produces exactly the same pattern of light falling on the observer's eye. *To this observer the pencil appears to be in exactly the same place as it was in Figure 36.1(a).* We emphasize this point in Figure 36.1(c) by masking the actual light path. The only information the observer has about the pencil is contained in the rays entering the eye. In essence, the process we call vision is a matter of our brain ''backtracking'' along the direction of the rays entering the eye to determine where they must have come from.

In the discussions that follow we will continually return to this ''innocent-eyeball'' model of image formation. When employing this model, we will cover the intermediate ray paths as in Figure 36.1(c) and ask, ''Where do we, looking with this innocent eyeball that knows only about the rays reaching it, *think* the object is located?'' The point to which the innocent eyeball backtracks the rays is the point where we believe the object is situated. But, as we have seen, this point is not necessarily the object's position. In fact we define that apparent point of origin to be the **image** position.

> Consider a set of rays that originate at a single point on some object. If, after traversing an optical system, an arbitrarily large number of these rays can be extrapolated backward to a common crossing point, an image is said to be formed. The position of that crossing point is the image location.

You need to be able to distinguish between two types of images called *real* and *virtual*. When the rays entering the viewer's eye are backtracked to their apparent point of origin, it may turn out that all the rays forming the image actually did pass through this image point (although they may not have originated there). In such a case the image is said to be **real.** On the other hand, it may be that the rays that *appear* to have come from this image point never *actually* passed through the point at all.[2] This type of image is said to be **virtual.** The distinction between these two types of images is best illustrated by specific examples. Let's start with mirrors.

36.2 Images Formed by Plane Mirrors

By a **plane mirror** we mean a flat mirror. Most everyday plane mirrors have a ''silvered'' side behind a layer of glass so that the reflecting surface is protected. We ignore this layer of glass and model all mirrors as if they were silvered on the front face. High-quality mirrors for precision optical applications are made this way and must be cleaned with great care to avoid scratching the reflecting surface. The only fundamental ''law'' we need is the law of specular reflection, which tells us that the angle of incidence and the angle of reflection, as measured from a normal to the surface, are equal.

Figure 36.2(a) shows an observer looking in a mirror at the reflection of a pencil. We have traced a few of the many rays that leave the pencil tip and fall on the observer's eye. Note that these rays originated from a light source somewhere in the room. These rays strike the pencil tip and are diffusely reflected. We show the scattered rays only after they have left the pencil tip. In Figure 36.2(b) we employ our innocent-eyeball model to find the point from which the rays appear to come. As the figure suggests, the image is located somewhere behind the mirror. No light actually ever travels on the back side of the mirror, so the image is said to be virtual.

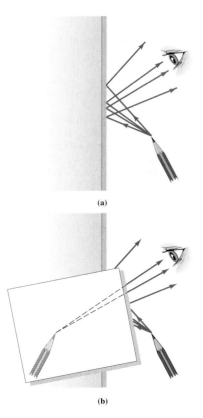

(a)

(b)

FIGURE 36.2

(a) Some of the rays scattered by the tip of the pencil strike the mirror and enter the observer's pupil. (b) The innocent eyeball believes the rays to have originated at a point behind the mirror.

---

[1] A full sense of three-dimensional depth perception requires divergent cones of light to enter both eyes. The brain then translates the parallax between the two images into a single three-dimensional image. See Section 36.4 for a discussion of parallax.

[2] In some cases a single ray may have passed through the image point, but all others did not. Such an image is still virtual. To be classified as real, an arbitrarily large number of rays must actually pass through the image.

The image of the pencil tip is actually located exactly as far behind the mirror as the real pencil is in front of the mirror. To prove this fact it is convenient to redraw our picture with different rays. The rays we draw (shown in Fig. 36.3) are chosen to make the geometrical proof simple. The only rule required to construct the rays is the law of reflection. Now, it is certainly the case that these rays probably are *not* the actual rays that enter the viewer's eyes. Nonetheless, they allow us to locate the image conveniently. We will use such rays selected for convenience continually throughout this chapter.

In Figure 36.3, ray number 1 proceeds from O to an arbitrary point B on the mirror. This ray reflects according to the law of reflection such that $\theta_i = \theta_r$. Ray number 2 leaves point O and strikes the mirror normally at point A, and reflects back on itself. We wish to prove that triangle ABO is congruent to triangle ABI. This result is easy to demonstrate. Angle $\theta_i = \theta_r$ by the law of reflection and $\theta_m = \theta_r$ because opposite angles at intersecting lines are equal. As a result, $\alpha = \beta$ because, if two angles are equal, their complements are also equal. Because right triangles ABO and ABI also have leg AB in common, they must be congruent.

There is nothing unique about the tip of the pencil; an identical proof could be used for any point on the pencil. Consequently, the congruency of triangles ABO and ABI in Figure 36.3 implies the congruency of the image with the pencil itself. Two important facts follow. First, because $OA = IA$, the image is as far behind the mirror as the object is in front of it. Second, the height of the image is equal to the height of the object. The ratio of image height to object height is often expressed as the **lateral magnification,** the *magnitude* of which is defined by

$$|m| = \frac{Image\ height}{Object\ height}$$

For a plane mirror we see that the magnification $m = 1$. A sign convention for lateral magnification is given in Section 36.3.

A final interesting property of plane mirrors is that they reverse the direction of images front to back. As shown in Figure 36.4, a Cartesian coordinate system placed with its z-axis pointing away from the mirror surface has an image that appears to have a z-axis also pointing away from the surface but with the direction reversed from the actual coordinate system. This inversion of a single axis normal to the mirror plane has the effect of changing the "handedness" of the coordinate system. That is, when you view your right hand in a mirror, it appears as a left hand. Have a friend try to orient her right hand to make it look (when viewed without the mirror) like the image of your right hand in the mirror. It can't be done. However, your friend is able to make her actual left hand match the image of your right hand.

To summarize, the relevant properties of an image formed by a plane mirror are:

1. The image is located as far behind the mirror as the object is in front of it.
2. The image is upright (as opposed to upside down), virtual, and has magnification $m = 1$.
3. Normal inversion causes a change in the "handedness" of the image.

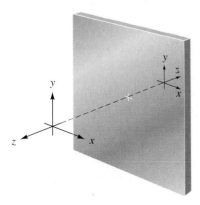

FIGURE 36.4

When a coordinate system is viewed in the mirror, the axis normal to the mirror is inverted in direction. This inversion changes the "handedness" of the coordinate system.

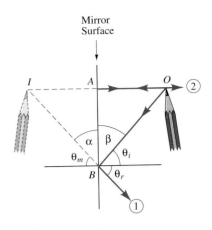

FIGURE 36.3

Construction rays 1 and 2 can be used to prove that the image is congruent to the object and located as far behind the mirror as the object is in front of it.

Concept Question 1
Place a clean sheet of paper on your desktop. Prop up a small mirror in the center of the paper so that the mirror is perpendicular to the desk; draw a line on the paper along the base of the mirror. Stand a small object, such as a marking pen, upright a couple of inches in front of the mirror and mark its position on the paper. Now move your head down near the level of desktop and to the right of the pen. With one eye look at the image of the pen in the mirror. Draw a line on the paper from the position of your eye toward the *image* of the pen. If you sight along this line it should look like it is heading straight toward the image. Now move your eye to the left-hand side of the mirror. Again draw a line from the position of your eye toward the position of the image. Extend the two sight lines you have drawn to the point where they cross. This crossing point is the position of the image of the pen. How does the image position compare with the actual position of the object?

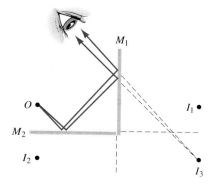

FIGURE 36.5

Three images of point O are formed by the two perpendicular mirrors M_1 and M_2. Example 36.1 describes how to determine their locations.

EXAMPLE 36.1   *Reflect on This Example a While*

Two mirrors are placed at right angles to each other as shown in Figure 36.5. Find the location of all images formed from point O.

SOLUTION   The *method* of solution is quite simple. Understanding *why* the method works is a bit more difficult but provides us with some insight into the formation of images. Here's how to find the answer when multiple mirrors are involved:

1. First find the direct image of the object in each of the mirrors separately. Do this by using the simple rule that the image lies along the line drawn from the object normal to the mirror and is as far behind the mirror as the object is in front of it.
2. Next find the image of the image in each of the other mirrors *or the imaginary extension of them*. This process can continue indefinitely unless you reach a point where the images become coincident.

Figure 36.5 illustrates this method. Images I_1 and I_2 are formed in mirrors M_1 and M_2 in the usual manner. To see how we located image I_3, extend mirror M_1 downward and then find the image of I_2 in the extension of mirror M_1. Similarly, you can extend mirror M_2 and locate the image of I_1 in it. In this case both of the secondary images fall at the same point labeled I_3.

Also in Figure 36.5 we draw the actual path taken by the rays forming image I_3. Try to extend the pair of rays backward after they undergo only a single reflection (from mirror M_2). Can you verify that they come from image I_2? ◀

EXAMPLE 36.2   *How Much Cheaper Is That Smaller Mirror?*

Many department stores have full-length mirrors for their customers. Show that to view a full-length image a mirror only half the size of the customer is required.

SOLUTION   Figure 36.6 shows the geometry of rays traveling from the feet to the eye and from the top of the head to the eye. We can imagine the customer viewing his image behind the mirror as shown in Figure 36.6. From the angular equalities required by the law of reflection we determine that triangle $I_1 I_2 E$ and triangle abE are similar triangles. Furthermore, the mirror is halfway between the image and the object. Thus, side $\overline{ab}$ equals half of side $\overline{I_1 I_2}$. Notice that rays originating anywhere on the person but striking the mirror below point b or above point a do not reach the observer's eyes. ◀

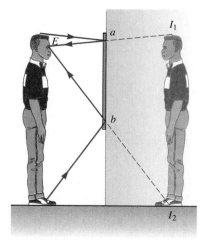

FIGURE 36.6

A man can view his entire body in a mirror that is only half his height.

36.3 Images Formed by Curved Mirrors

Imagine taking a parabola and rotating it about its axis of symmetry. The surface swept out by such a process is called a *paraboloid of revolution* and is pictured in Figure 36.7(a). A mirror with its reflecting surface shaped like a paraboloid of revolution is called a **parabolic mirror.** The point where the axis of symmetry intersects the paraboloid is called the **vertex** of the mirror. Rays traveling parallel to the axis of revolution are reflected by the interior surface in such a way that, after reflection, they all converge and travel through a single point as shown in Figure 36.7(b). This point is known as the **focal point** of the mirror. The converging rays are said to be focused.

As a practical matter, parabolic mirrors are expensive and difficult to fabricate compared to mirrors with surfaces that are simply spherical. A mirror formed from part of a spherical surface is called a **spherical mirror.** The **axis** of a spherical mirror is an axis of symmetry that passes through both the center of the mirror and the center of the sphere of which the mirror is a part. Spherical mirrors do not focus light to a single point. As illustrated in Figure 36.8, however, rays that initially travel close to the mirror axis are quite nearly focused at a single point. Rays that travel close to the mirror axis and make small angles with it are called **paraxial rays.**

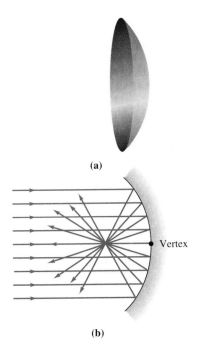

(a)

(b)

Vertex

FIGURE 36.7

(a) A parabolic mirror formed by rotating a parabola about its axis of symmetry. (b) Parabolic mirrors focus rays parallel to the axis of symmetry to a single point called the focal point.

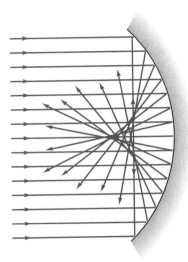

FIGURE 36.8

A spherical mirror does an imperfect job of focusing parallel rays. The farther a ray is from the mirror axis, the greater is the error in focus. The resulting lack of focus is known as spherical aberration.

It is easy to see why spherical mirrors behave similarly to parabolic mirrors for paraxial rays. Figure 36.9 shows a parabola and the closest fitting circle. The equation for the parabola is

$$y^2 = 4fx \qquad \text{(Equation of a parabola)}$$

where f is the distance from the vertex of the parabola to the focal point of the parabola. This distance is known as the **focal length.** The equation for the circle is $(x - R)^2 + y^2 = R^2$, where R is the radius of the circle. On simplification this equation becomes

$$y^2 = 2Rx - x^2 \qquad \text{(Equation of a circle)}$$

Notice that when $x \ll 2R$, corresponding to points near the vertex of the parabola, we can ignore the x^2 term compared to $2Rx$. In this case, the equation for the parabola and the equation for the circle are identical if we take $R = 2f$. Notice also that paraxial rays strike the parabolic–circular mirror at points described by small values of y corresponding to small values of x for which the curves are similar.

To summarize the above discussion:

1. Spherical mirrors bring paraxial rays traveling parallel to the mirror axis to an approximate focus at a point on the mirror axis.
2. The focal length of a spherical mirror is equal to half the radius of curvature of the mirror.

In the remainder of this section we are concerned with images formed by paraxial rays reflecting from spherical mirrors. Image distortion caused by the use of spherical mirrors is known as **spherical aberration.** As a practical matter, in many of our illustrations we use objects that are large compared to the radius of curvature of the mirrors shown. Therefore, you may find that image locations computed by graphical methods may disagree slightly from the results found by the algebraic formulas we present later. The graphical methods we describe later on are wonderful aids to your intuition and invaluable as "sanity checks" on calculated answers. However, drawings using spherical mirrors may not exhibit perfect image formation because of significant spherical aberration.

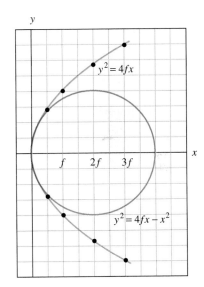

FIGURE 36.9

A circle and a parabola are nearly identical for small values of x and y. As a consequence, spherical mirrors focus rays provided that the rays strike the mirror near the optic axis. The focal length of the spherical mirror is half its radius of curvature.

Concave Mirrors

Figure 36.10 shows an object in front of a concave spherical mirror. Along the axis the center of the sphere is marked C and, halfway between the center and the vertex V, the focal point is labeled F. To find the location of the image, we examine four rays from the tip of the object. These rays are chosen only because their directions are simple to determine after reflection. You should realize that an arbitrarily large number of rays could be drawn originating at the object and reflecting from the mirror.

> Ray 1 is chosen parallel to the axis; it is reflected back through the focal point as shown in Figure 36.10(a).
> Ray 2 passes through the focal point on the way to the mirror; it is reflected back parallel to the mirror axis as shown in Figure 36.10(b).

These two rays are all that are really needed to locate the image. However, two other simple rays are available. We strongly urge you to always use at least one of them as a check.

FIGURE 36.10

Convenient construction rays used to locate the image formed by spherical mirrors. (a) A ray parallel to the optic axis is reflected through the focal point F. (b) A ray passing through the focal point is parallel to the optic axis after reflection. (c) A ray passing through the center of curvature of the mirror reflects directly back on itself. (d) A ray that strikes the vertex of the mirror is reflected so that its path makes equal angles with the optic axis. Therefore, this ray passes the object location at a point as far below the optic axis as it was above the axis when it left the object.

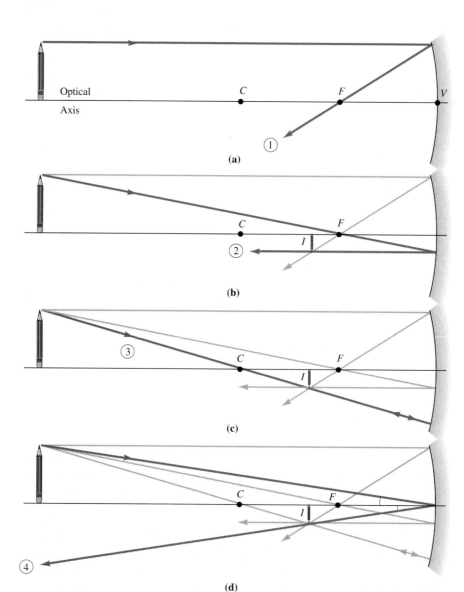

Ray 3 passes through the center of the circle and thus strikes the mirror at right angles and, therefore, reflects back on itself as shown in Figure 36.10(c).

Ray 4 strikes the vertex of the mirror and reflects at an equal angle on the other side of the mirror axis. (Ray 4 can be conveniently constructed by drawing a line from the vertex to the tip of an inverted copy of the object.)

Please trace the path of each of these rays yourself, and be sure you know why they travel as they do.

Using our innocent-eyeball model of vision, we find that an observer believes the pencil tip to be at the image location I in the figure. Notice in Figure 36.10(d) that all the rays forming the image really do pass through the image point. Hence, we say that the image is real. If the object is a small light bulb, we can cast an image of its filament on a sheet of paper held at position I. Notice too that the image is inverted compared to the object.

At this point we have identified the position of the tip of the pencil, which acts as our object. You should realize that if we instead pick the midpoint of the pencil as our object point, nothing in the procedure for locating the image changes. The location of the image of the midpoint of the pencil falls halfway between the image of the pencil tip and the optic axis. On the other hand we cannot use our four special rays to locate the image of the pencil eraser located precisely on the optic axis. Our procedure fails only because the four special rays we chose all happen to fall on top of one another so that we cannot locate the image using only these rays. Rest assured that the image of the eraser is still formed by many other rays and that the image of the eraser lies on the optic axis directly below the pencil tip when the object pencil is vertical.

In Figure 36.11(a) we have moved the object closer so that it is between the focal point F and the mirror. Ray 1 is drawn as in the previous example. However, if we draw ray 2 from the object to the focus, it does not hit the mirror; instead it goes the wrong way! Nonetheless, by extending a straight line that connects the focal point and the tip of the object, we can draw a ray that strikes the mirror *as if it had come through the focus*. This ray is reflected in a direction parallel to the mirror axis. Rays 3 and 4 can be drawn as

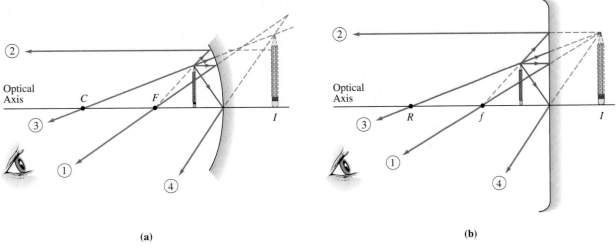

(a)   **(b)**

FIGURE 36.11

(a) Four construction rays used to locate the image when the object is between the focal point and the mirror. Note that ray 2 strikes the mirror at an angle such that it appears to have come through the focal point F. A virtual image is located behind the mirror. Using a large object near the mirror results in noticeable spherical aberration because the backtracked rays do not cross at a single point. (b) Spherical aberration in ray drawing can be reduced by drawing the mirror surface as if it were flat. The construction rules for rays 1 through 4 must still be used, however.

described previously. We advise that you trace each ray in Figure 36.11(a), paying particular attention to ray 2.

Once again we now backtrack the four rays to find where they appear to have originated. The point of apparent origin is behind the mirror. This point is where the image is located. No light rays ever pass through the image, and, therefore, the image is classified as virtual. The image is also upright and enlarged. Such mirrors are often used in bathrooms as shaving or makeup mirrors.

Notice that, when the rays of Figure 36.11(a) are traced back to their apparent origin, they fail to meet at exactly one point. This is not a defect in the figure but rather a manifestation of the spherical aberration we warned you about earlier. Please note that rays 3 and 4 best locate the image. Ray 3 is independent of mirror size because its direction is determined by the center of curvature of the mirror not the mirror itself. Ray 4 is unaffected by spherical aberration because it strikes the mirror at the vertex. The spherical aberration exhibited by the drawing can be reduced by using an object of much smaller height. This choice often makes drawings inconveniently small. An alternative approach is to draw the mirror surface as if it were flat, the way it appears to an extremely small object. Of course, even though the mirror surface is *drawn* flat you must still use the appropriate spherical mirror rules to determine the paths for the various rays as in Figure 36.11(b). Our opinion is that it probably isn't worth worrying about eliminating the spherical aberration in your drawings because the results are only approximate anyway. However, if you like your images sharply focused, draw your spherical mirrors flat.

Figure 36.12 shows two photographs of a teenager seated in front of the same mirror. In Figure 36.12(a) the young woman is seated more than one focal length from the mirror; in (b), she is less than one focal length from the mirror. In order to record the image, the photographer must focus on the images labeled *I* in Figures 36.10 and 36.11 and *not* on the mirror. Pencils are placed along a line parallel to the mirror axis to indicate where the camera is focused. The location of her image in the mirror is the same as that of the pencil that is in focus. (Note the red pencils are behind the mirror and the green pencils are in front of the mirror.)

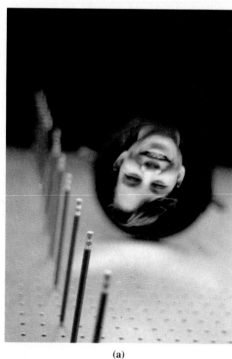 

(a)   (b)

FIGURE 36.12

The image locations predicted by Figures 36.10 and 36.11 can be verified experimentally by determining where one must focus a camera in order to obtain a sharp image. (a) In order to focus on the young woman's image the photographer must focus on a point in front of the mirror, when she is outside the spherical mirror's focal point. (b) When her face is between the focal point and the mirror the photographer must focus on a point behind the mirror.

It is possible to calculate numerical values for the position of images using algebraic equations deduced from simple geometry. Before deriving these equations, however, we must establish a sign convention for the distances we will use. We measure all distances from the vertex of the mirror. For convenience we call the distance from the vertex to the object the **object distance** and denote it by the symbol o. Similarly, the distance from the image to the vertex we call the **image distance** i. Following the nearly universal convention, we assume light enters the system from the left on all diagrams. When rays *diverge* from an object to the left of the mirror, we designate the *object distance* as *positive*. When rays *converge* to form a real image in front of the mirror, we specify the *image distance* as *positive*. Thus, the situation in Figure 36.10 is our standard configuration with both object and image distances positive. These sign conventions are summarized in Table 36.1.

If the *image* is formed by *diverging rays* only, we take the image distance i to be a *negative* number. For example, i is negative for the image formed behind the mirror depicted in Figure 36.11. (It is also possible to have a "virtual object" formed by rays converging toward a point behind the mirror. In such a case we designate the object distance o as negative. You needn't worry about how this might happen yet, but keep the possibility in mind.)

Finally we must also assign an algebraic sign to the radius of curvature of the mirror. The convention is that mirrors that are concave toward the incoming rays are assigned positive values for the radius of curvature. If the incoming rays strike a convex surface, the radius of that surface is assigned a negative value.

TABLE 36.1   **Sign Conventions for Spherical Mirrors**

f and R are positive for concave mirrors
f and R are negative for convex mirrors
o is positive when rays diverge from a real object
o is negative when rays converge toward a virtual object
i is positive for real images formed by converging rays
i is negative for virtual images created by diverging rays only

To obtain an algebraic relation between object position and image position, consider a point-object on the mirror axis as shown in Figure 36.13. A line from the center of curvature C of the mirror to point a is normal to the mirror surface. The ray from the object point O strikes the mirror at point a and reflects according to the law of reflection, intersecting the optic axis at image point I. Consequently, lines $\overline{Oa}$ and $\overline{Ia}$ make equal angles θ with line $\overline{Ca}$. We recall from plane geometry that the exterior angle of a triangle is equal to the sum of the opposite interior angles. Applied to triangles OCa and OIa we find that $\beta = \alpha + \theta$ and $\gamma = \alpha + 2\theta$. Eliminating θ from these two equations, we discover

$$2\beta = a + \gamma$$

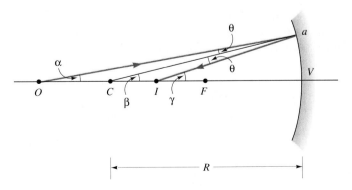

FIGURE 36.13

Geometry for the derivation of the mirror equation.

We now calculate each angle in radians as the subtended arc length divided by the corresponding radius:

$$2\frac{\overline{Va}}{R} = \frac{\overline{Va}}{o} + \frac{\overline{Va}}{i}$$

This last equation is only approximately correct. Clearly, only angle β really subtends the circular arc of the mirror surface from a to V. The angles α and γ subtend other circular arcs that do not exactly follow the mirror surface between a and V. Consequently, the approximation is good only for small angles α and γ. But this restriction to small angles is precisely the restriction made earlier for paraxial rays. Thus, for paraxial rays we have

$$\frac{2}{R} = \frac{1}{o} + \frac{1}{i}$$

Recalling that $f = R/2$, we can write the expression as

$$\frac{1}{f} = \frac{1}{o} + \frac{1}{i}$$ The mirror equation

This equation is the desired algebraic relation between object position and image position for the mirror.

The **lateral magnification** is the ratio of the image height to the object height. In Figure 36.14 we show a ray diagram locating object and image. Direct your attention to the triangles formed by the dark lines in the figure. From similar triangles $OO'V$ and $II'V$ in Figure 36.14 we see that $h_i/i = h_o/o$, and thus

$$\frac{h_i}{h_o} = \frac{i}{o}$$

It is conventional to define the lateral magnification so that inverted images have negative lateral magnification. Thus, we calculate the lateral magnification using

$$m = -\frac{i}{o}$$ Lateral magnification

In Examples 36.3 and 36.4 we show how the mirror equation can be applied to predict the location and character of images.

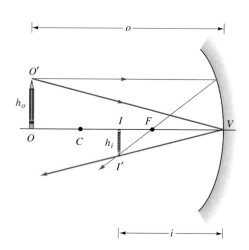

FIGURE 36.14

An object and its image formed by a concave mirror. The type 4 construction ray is darkened for emphasis. Similar triangles $OO'V$ and $II'V$ indicate that the lateral magnification m can be calculated by $m = -i/o$. See text for discussion of the sign convention.

EXAMPLE 36.3   A Concave Mirror Forming a Real Image

Find the location of the image formed from a pencil placed 20.0 cm in front of a concave spherical mirror of focal length $f = 15.0$ cm. State whether the image is real or virtual, erect or inverted, and give its lateral magnification.

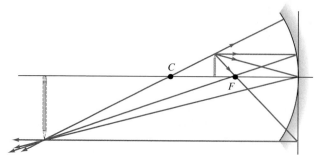

FIGURE 36.15

Ray diagram for the image formed of a pencil placed farther than one focal length in front of a concave spherical mirror. See Example 36.3.

SOLUTION   The given parameters are

$$f = +15.0 \text{ cm}$$

$$o = +20.0 \text{ cm}$$

We solve for *i* from the image equation

$$i = \left(\frac{1}{f} - \frac{1}{o} \right)^{-1} = \left(\frac{1}{15.0 \text{ cm}} - \frac{1}{20.0 \text{ cm}} \right)^{-1} = 60.0 \text{ cm}$$

The lateral magnification is

$$m = -\frac{i}{o} = -\frac{60.0 \text{ cm}}{20.0 \text{ cm}} = -3.00$$

Because the image distance is positive, the image is real and is formed in front of the mirror. The image is enlarged by a factor of 3. The lateral magnification is negative, indicating that the image is inverted. A quick sketch as shown in Figure 36.15 serves to check our conclusions. ◀

Concept Question 2
On Figure 36.15 draw three additional rays that form the image.

EXAMPLE 36.4   *A Concave Mirror Forming a Virtual Image*

The pencil described in Example 36.3 is moved to a position 10.0 cm in front of the 15.0-cm focal length concave mirror. Characterize the new image.

SOLUTION   Once again we use the mirror equation and find

$$i = \left(\frac{1}{f} - \frac{1}{o} \right)^{-1} = \left(\frac{1}{15.0 \text{ cm}} - \frac{1}{10.0 \text{ cm}} \right)^{-1}$$

$$= -30.0 \text{ cm}$$

The lateral magnification is

$$m = -\frac{i}{o} = -\frac{-30.0 \text{ cm}}{10.0 \text{ cm}} = +3.00$$

Concept Question 3
On Figure 36.16 draw three additional rays that form the image.

The image is formed 30.0 cm behind the mirror, is virtual, upright, and magnified by a factor of 3. The sketch in Figure 36.16 verifies our calculation and interpretation.

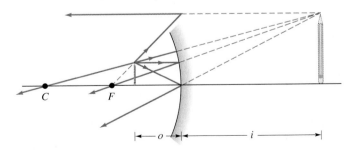

FIGURE 36.16

Ray diagram for the magnified image of a pencil closer than one focal length in front of a concave mirror. See Example 36.4. ◀

Convex Mirrors

When the outwardly curved surface of a spherical mirror is used to reflect light rays, the mirror is said to be convex. To accommodate such cases in our mirror equation we have only to use negative values for the focal length f. A few minor alterations are also necessary to adopt our ray-tracing rules to such mirrors. Referring to Figure 36.17 we note that ray 1 incoming parallel to the mirror axis now reflects in such a direction that it *appears* to have come from the focal point located behind the mirror. Similarly, ray 2, which is heading toward the focal point behind the mirror, is reflected so that it becomes parallel to the mirror axis. Ray 3 heads toward the center of curvature C and reflects directly back on itself. Finally, ray 4 strikes the mirror vertex, and its reflected ray makes an equal angle with the mirror axis, as in the concave case. We illustrate these alterations in Example 36.5. (Once again some spherical aberration can be observed in Fig. 36.17(e).)

In order to focus on the image formed by the convex mirror the camera must be focused on a point well behind the mirror as indicated by the sharper image of pencils located behind the mirror.

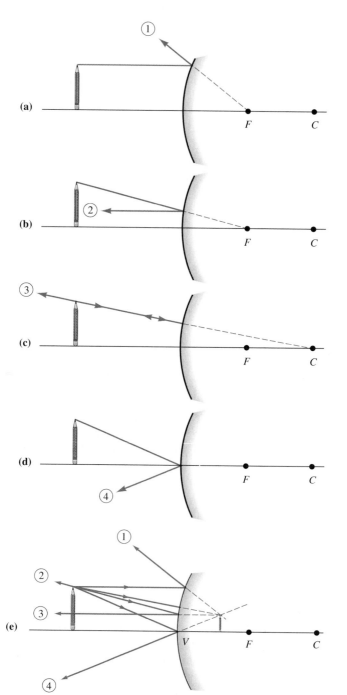

FIGURE 36.17

(a)–(d) Four construction rays used to locate the image formed by a convex mirror. (e) The location of the virtual image formed by the mirror is located by backtracking the four construction rays to their apparent point of origin. Some spherical aberration may be noticed when large objects are used in the drawing.

EXAMPLE 36.5   *"Caution: Objects in This Mirror Are Closer Than They Appear"*

A wide-angle convex mirror used on the passenger side of an automobile has a 4.00-m radius of curvature. If a bus is located 6.00 m from this mirror, where is the image located?

SOLUTION   We employ the mirror equation once again, using $f = -2.00$ m.

$$i = \left(\frac{1}{f} - \frac{1}{o}\right)^{-1} = \left(\frac{1}{-2.00\text{ m}} - \frac{1}{6.00\text{ m}}\right)^{-1} = -1.50\text{ cm}$$

The lateral magnification is

$$m = -\frac{i}{o} = -\frac{-1.50\text{ m}}{6.00\text{ m}} = +0.25$$

The image is virtual, upright, and made smaller by a factor of 0.25. The ray diagram in Figure 36.18 confirms these conclusions.

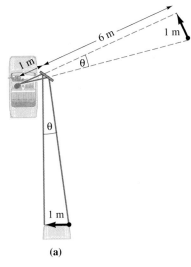

(a)

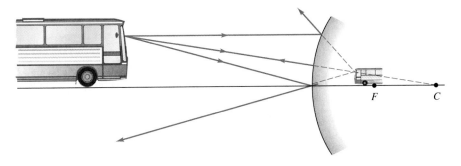

FIGURE 36.18

Ray diagram for the image of a bus reflected in a wide-angle convex mirror on an automobile. See Example 36.5.

(b)

FIGURE 36.19

Comparison of the images formed by a plane mirror (a) and a convex mirror (b) when used as a sideview mirror. The convex mirror produces a smaller image that is closer to the observer. The angle subtended by the image is smaller in the case of the convex mirror, causing the driver to think that the object is further away than it actually is.

We note in passing that we judge the apparent distance of objects by the angle they subtend at our eye. Suppose we are seated 1.0 m from a *plane* side mirror and looking at a 1.0-m wide feature on the bus behind us. Then, as diagrammed in Figure 36.19, the image is 7 m from our eye and subtends an angle of $\theta = 1$ m/7 m = 1/7 rad. In the *convex* mirror described previously, the image is only 1.5 m behind the mirror, which is itself 1.0 m from the viewer's eye. The size of the image in this mirror is 0.25 m. Thus, the angle subtended by the image is 0.25 m/2.5 m = 1/10 radian. The smaller angular size of the trailing bus creates the illusion that it is farther behind us than it really is. ◀

36.4 Images Formed by Refracting Surfaces

The images fashioned by eyeglasses, microscopes, and magnifying glasses are formed because of the refraction of light at the surfaces of glass. Have you ever jumped into a swimming pool and been surprised when it turned out to be deeper than you thought? The apparent shallow depth of objects submerged beneath a layer of water is due to refraction at the water's surface. In this section we study how these images are formed. We will start by considering a single curved refracting surface, and then we will look at a flat surface as a special case. Finally, we will combine two curved surfaces to form a lens.

Refraction at Curved Surfaces

We ultimately wish to find an equation that describes the focusing properties of a thin lens. We begin by looking at the way in which a single spherical surface refracts light. Figure 36.20 shows such a boundary between two media with indices of refraction n_1 and n_2.

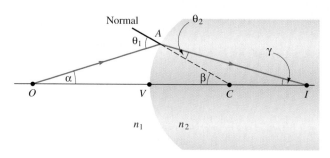

FIGURE 36.20

Refraction of a ray at a curved surface separating media with different indices of refraction.

We construct a ray diverging from an object point located at distance o from the vertex V of the boundary between the media. This ray strikes the curved surface at point A. A normal to the surface is found by drawing a radial line from the center of curvature located at point C. The incoming ray is refracted at the boundary according to Snell's law:

$$n_1 \sin(\theta_1) = n_2 \sin(\theta_2)$$

Once again we restrict ourselves to paraxial rays so that we can use the approximation $\sin(\theta) \approx \theta$. In this approximation Snell's law becomes

$$n_1 \theta_1 = n_2 \theta_2 \tag{36.1}$$

The external angle of a triangle is equal to the sum of the opposite two internal angles. Thus, from Figure 36.20 we see that

$$\beta = \gamma + \theta_2$$

and

$$\theta_1 = \alpha + \beta$$

We can combine the last three equations in such a way as to eliminate θ_1 and θ_2. The result can be written

$$(n_2 - n_1)\beta = n_2\gamma + n_1\alpha \tag{36.2}$$

We calculate the magnitude of each of the angles α, β, and γ in radians as the arc length $\overline{AV}$ divided by the corresponding radius:

$$\beta = \frac{\overline{AV}}{R}, \qquad \gamma = \frac{\overline{AV}}{i}, \qquad \alpha = \frac{\overline{AV}}{o}$$

Of these relations, only the expression for β is exact. The expressions for α and γ are approximately correct in the limit of small angles (our paraxial approximation). Substituting these expressions for α, β, and γ into Equation (36.2) and canceling the common factor $\overline{AV}$, we obtain our master equation for refraction at a curved boundary

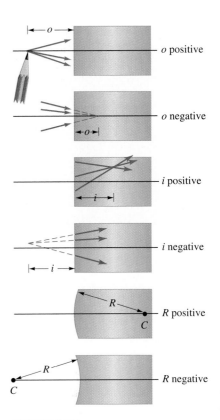

$$\frac{(n_2 - n_1)}{R} = \frac{n_1}{o} + \frac{n_2}{i}$$ Master equation for refraction   (36.3)

This equation relates the object and image distances for our standard configuration shown in Figure 36.20. As we did with mirrors, we must establish a sign convention to deal with situations that differ from this standard. The underlying principle for our sign convention remains the same as for the mirror equations. When rays *diverge from a real object*, we take the object distance o as *positive*. When rays *converge to a real image*, we take the image distance i as *positive*. The rule for the sign of the radius R, however, is opposite from that for mirrors. When the refracting surface is *convex* toward the incoming rays, we take R to be positive. This rule is applicable to our standard situation shown in Figure 36.20. Conversely, when rays *converge to form a virtual object*, we take o as *negative*. (This seemingly improbable case occurs when multiple lenses are present. We will deal with it more fully later.) When rays *diverge to form a virtual image*, the image distance i is *negative*. Finally, when rays encounter a *concave* curved surface, the radius R is counted *negative*. These various cases are summarized in Figure 36.21 and Table 36.2.

FIGURE 36.21

A pictorial summary of the sign conventions for the master equation for refraction at a curved boundary. (See Table 36.2.)

TABLE 36.2   Sign Conventions for Refraction at a Spherical Interface Between Two Dielectrics

R is positive if the surface is convex toward the incoming ray	the radius convention
R is negative if the surface is concave toward the incoming ray	is *opposite* that for mirrors

o is positive when rays diverge from a real object	the object and image
o is negative when rays converge to form a virtual object	distance convention
i is positive for real images formed by converging rays	are the *same* as for mirrors
i is negative for virtual images created by diverging rays	

Apparent Depth

It is interesting to examine the predictions of our master equation (Eq. (36.3)) when the boundary between the two media is flat. The surface of a curved boundary becomes flat as the radius of curvature R becomes infinite. If we let R approach infinity in Equation (36.3), we find

$$0 = \frac{n_1}{o} + \frac{n_2}{i}$$

or

$$i = -\frac{n_2}{n_1} o \qquad (36.4)$$

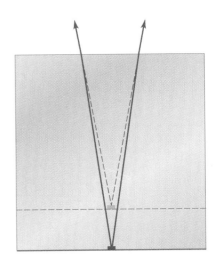

FIGURE 36.22

The depth of a pool of water appears less than it actually is due to the refraction of rays at the water's surface.

From this relation we can understand why swimming pools appear more shallow than they actually are. When the light rays begin from the bottom of the pool ($n_1 = 1.33$) and enter the air ($n_2 = 1.00$), they are bent away from the normal. As shown in Figure 36.22, the bottom appears closer to the observer than it actually is.

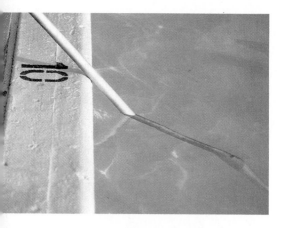

The pole appears bent due to the refraction of light at the water's surface.

Locating Images by Parallax

Parallax is the apparent movement of an object due to the actual movement of the observer. Close one eye and hold two fingers in front of you at different distances from your open eye. Now move your head from side to side. Note that the finger nearest you appears to move in the *opposite* direction from the motion of your head. Or, relative to the near finger, the far finger moves in the *same* direction as your head. Obviously, if you place your two fingers at the same distance from your eye, they exhibit no relative motion when you move your head. The absence of relative movement (parallax) can be used to locate the position of an image.

Equation (36.4) can be used to determine a liquid's index of refraction. Consider a column of liquid with known height and a small object at its bottom. The observer can locate the height of the image by finding the point where the image and a pointer *outside the liquid* exhibit no parallax. Figure 36.23 shows how this can be done.

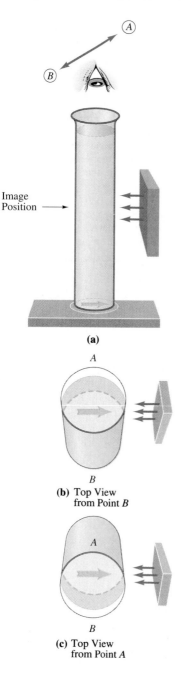

Image Position →

(a)

A

B

(b) Top View from Point B

A

B

(c) Top View from Point A

FIGURE 36.23

Image location by parallax. (a) An arrow-shaped object is placed at the bottom of a column of liquid and viewed from above. Because of refraction, the image of the arrow at the bottom of the liquid is located closer to the eye than the actual object. (See Fig. 36.22.) We compare the position of the image with three pointers attached to a small block of wood on the outside of the liquid column. When the eye is moved between positions *B* and *A*, no parallax is observed between the middle pointer and the image, because they are the same distance from the eye. (b) A view of the liquid column from the point of view of the eye when located at position *B*. Note the middle arrow on the pointer block does not move relative to the image. (c) A top view from position *A*. In practice the image can be located using a single pointer. For example, a pencil tip can be moved up and down the outside of the tube until a position is found where the image of the submerged object and the external pencil exhibit no parallax as the eye is moved back and forth between positions *B* and *A*.

EXAMPLE 36.6   *Still Waters Run Deeper Than They Appear*

If a fish is 40.0 cm below the surface of a pond, how deep does the fish appear to be to an observer directly above?

SOLUTION   We use Equation (36.4) with $i = x$ and $o = d$: $|x| = (n_2/n_1) d$ because light comes from the fish in the water ($n_1 = 1.33$) to an observer in air ($n_2 = 1.00$):

$$|x| = \frac{1.00}{1.33} d = 0.752 \, d = (0.752)(40.0 \text{ cm}) = 30.1 \text{ cm}$$

Concept Question 4
Perform the experiment shown in Figure 36.23 using a glass of water. Because $n_w = 4/3$, where do you expect the point of zero parallax to be if the water is 12 cm deep?

It is perhaps interesting to note that the fish observing a mayfly hovering above the water sees the image of the mayfly as being higher above the water than the mayfly actually is.   ◄

36.5 Images Formed by Lenses

Now that we have developed our master equation for refraction at a curved surface (Eq. (36.3)), and settled on a sign convention for its use, let's proceed to develop an expression for the focal length of a lens.

The Lens-Maker's Formula

We begin our development of an equation for the focal length of a lens by following a ray past two refracting surfaces, such as in Figure 36.24. The application of our master equation to the first boundary yields

$$\frac{(n_2 - n_1)}{R_1} = \frac{n_1}{o_1} + \frac{n_2}{i_1}$$

Note that for the case shown in Figure 36.24 our sign convention implies that i_1 is a negative number. The image from the first refraction acts as an object for the second refraction. This second object is at a distance $o_2 = |i_1| + t$. The distance o_2 is positive because, although there is not actually an object at o_2, rays are *diverging* from this point as if a *real* object were present. Applying our master equation to the second surface, we obtain

$$\frac{(n_1 - n_2)}{R_2} = \frac{n_2}{|i_1| + t} + \frac{n_1}{i_2}$$

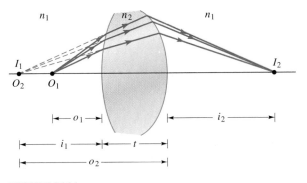

FIGURE 36.24

Rays of light are deviated by a thick lens. When the distance t approaches zero, the thin-lens model results. Note carefully that for this lens the radius of curvature R_1 of the first surface is a positive number, whereas the radius of curvature R_2 of the second surface is a negative number.

Because at the second surface rays are traveling from a medium with index of refraction n_2 to a medium of index of refraction n_1, the indices are reversed relative to the first surface. If desired, these last two equations can be used in sequence to find the location of the image created by a thick lens. However, our intent is to model the behavior of thin lenses only. Hence, we take thickness t to be zero. The previous two equations can then be added to obtain

$$(n_2 - n_1)\left(\frac{1}{R_1} - \frac{1}{R_2}\right) = \frac{n_1}{o_1} + \frac{n_2}{i_1} + \frac{n_2}{|i_1|} + \frac{n_1}{i_2}$$

Recalling that i_1 is a negative number, we see that the second and third terms on the right-hand side of this equation cancel leaving us with

$$(n_2 - n_1)\left(\frac{1}{R_1} - \frac{1}{R_2}\right) = n_1\left(\frac{1}{o_1} + \frac{1}{i_2}\right) \quad (36.5)$$

In the case of a lens of index $n_2 = n$ in air ($n_1 = 1$), this expression reduces to

$$(n - 1)\left(\frac{1}{R_1} - \frac{1}{R_2}\right) = \frac{1}{o_1} + \frac{1}{i_2}$$

When the object distance approaches infinity, the rays entering the lens are essentially parallel. According to the expression above, such rays come to a focus at i_2, as shown in Figure 36.25. Thus, for $o_1 \to \infty$, we are led to identify i_2 as the **focal length** f of the lens. Consequently, we have

$$\boxed{(n - 1)\left(\frac{1}{R_1} - \frac{1}{R_2}\right) = \frac{1}{f}} \quad \text{The len's-makers formula} \quad (36.6)$$

which is known as **the lens-maker's formula.** The lens-maker's formula provides the relationship between a lens's optical properties (f) and its physical properties (n, R_1, R_2). Lenses are made in a variety of forms as shown in Figure 36.26. Please note carefully the

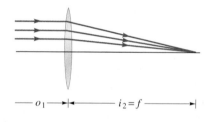

FIGURE 36.25

When the object recedes to infinity, rays coming from points on it are parallel when they reach the lens. The rays converge to an image point. For this case the image distance is, by definition, the focal length of the lens.

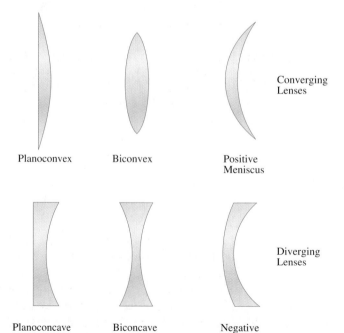

Converging Lenses

Planoconvex   Biconvex   Positive Meniscus

Diverging Lenses

Planoconcave   Biconcave   Negative Meniscus

FIGURE 36.26

A variety of shapes are possible for lenses. Lenses thicker in the middle are converging; lenses thinner in the middle are diverging.

sign convention that must be used when assigning numerical values to the radii. Those lenses that are thicker in the middle converge parallel rays and have positive focal lengths. If a lens is thicker at the edges, its focal length is negative, and parallel rays diverge after passing through the lens. These observations lead us to a simple statement that may be worth remembering, ''Rays are always bent toward the thicker part of the lens.'' The lens-maker's formula is accurate only for paraxial rays and thin lenses. In particular, the thickness of the lens must be much less than any other dimension in the equation.

Concept Question 5
Give the sign for each surface of each lens in Figure 36.26. Write the signs assuming light is coming from the left. List lenses in the top row first proceeding from left to right and then the second row from left to right.

EXAMPLE 36.7 *Watch the Signs of the Radii in the Lens-Maker's Formula*

Suppose that the lens in Figure 36.24 is constructed such that the radii of the first and second surfaces are 20.0 cm and 10.0 cm, respectively, and that the glass has an index of refraction of 1.500 for light at 589 nm. What is the focal length of this lens for light of 589 nm?

SOLUTION We solve the lens-maker's formula (Eq. 36.6) for the focal length f

$$f = \frac{[(1/R_1) - (1/R_2)]^{-1}}{(n-1)} = \frac{[(1/20.0 \text{ cm}) - [1/(-10.0 \text{ cm})]^{-1}}{(1.500 - 1.000)} = \frac{6.67 \text{ cm}}{0.500} = 13.3 \text{ cm}$$

Please note carefully that the radius R_2 was entered as a negative number in accordance with the sign convention illustrated in Figure 36.21. ◄

The Thin-Lens Formula

Consider a bundle of rays that are parallel to each other but make an angle of θ with the lens axis. These rays are also brought to a focus as shown in Figure 36.27. For *paraxial rays* and *thin lenses* such parallel rays are focused in a plane that is parallel to the lens and passes through the focal point. This plane is known as the **focal plane** of the lens.

Having identified the right-hand side of Equation (36.6) with the focal length, we can now rewrite it in a form identical to the mirror equation:

$$\frac{1}{o} + \frac{1}{i} = \frac{1}{f} \qquad \text{Thin-lens equation} \qquad (36.7)$$

Note that we dropped the subscripts on the object and image distances because we now treat the lens as if all the bending occurs at a single plane. This plane is often referred to as the **principal plane** of the thin lens. To emphasize that this expression applies only to what we might call the **thin-lens model,** we draw lenses as single lines, such as those shown in Figure 36.28. A thin lens is reversible. That is, rays parallel to the optic axis are focused at the same distance f from the lens regardless of which side of the lens they strike. For this reason it is common to show thin lenses with two foci located equal

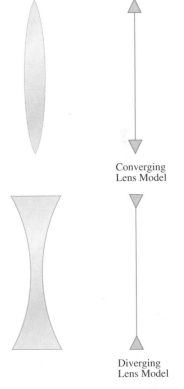

Converging
Lens Model

Diverging
Lens Model

FIGURE 36.28

The thin-lens model. We treat thin lenses as if all of the refraction occurs in a single plane. To emphasize this approximation we draw lenses as lines indicating their converging or diverging nature with small triangles located at the ends of the lines.

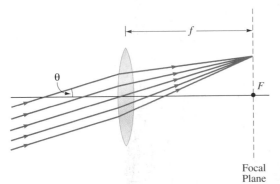

Focal
Plane

FIGURE 36.27

For an ideal thin lens, parallel rays making an angle θ with the optic axis are brought to a focus at a point in a plane that makes right angles with the optic axis and passes through the focal point.

distances f on either side of the lens. The focal point on the side of incoming rays is called the **primary** focal point. The focal point on the side of the outgoing rays is called the **secondary** focal point. The sign convention used with the thin-lens formula is similar to that for mirrors and is summarized in Table 36.3.

TABLE 36.3   Sign Conventions for the Thin-Lens Equation

f is positive for converging lenses
f is negative for diverging lenses
o is positive when rays diverge from a real object
o is negative when rays converge toward a virtual object
i is positive for real images formed by converging rays
i is negative for virtual images created by diverging rays

As was the case for mirrors, it is helpful to precede numerical calculations with a graphical solution. Again, there are several convenient rays we can use to plot the location of images.

> 1. A ray going through the center of the lens is undeviated.
> 2. A ray incoming parallel to the axis of a converging lens passes through the secondary focus. (For a diverging lens, a ray parallel to the axis is bent outward in a direction such that it *appears* to have come from the primary focus.)
> 3. A ray striking the lens after passing through the primary focus of a converging lens exits the lens parallel to the lens axis. (For a diverging lens, a ray headed toward the secondary focus before striking the lens emerges parallel to the lens axis after refraction by the lens.)

Note that these special construction rays can be used to locate the image *even if the lens is too small to allow the actual rays drawn to pass through it.* For small lenses, merely extend the line that represents the lens and pretend the construction rays are deflected as described by the rules above. Applications of both the graphical and algebraic methods appear in Examples 36.8 to 36.12.

EXAMPLE 36.8   *Converging Lens, Real Object, Real Image*

An object 1.00 cm in height is placed 30.0 cm from a converging lens with focal length 10.0 cm. Find the location of the image and describe its characteristics.

SOLUTION   The configuration is diagrammed in Figure 36.29. We estimate the image location using the three special rays described previously. Because rays are

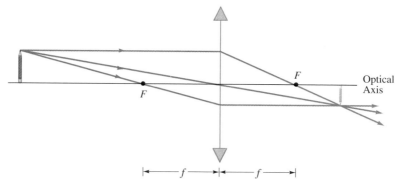

FIGURE 36.29

Ray diagram for Example 36.8.

diverging from a real object, we take the object distance as $o = +30.0$ cm. The lens is a converging type so that $f = +10.0$ cm. We solve the thin-lens equation for i and substitute these values to obtain

$$\frac{1}{o} + \frac{1}{i} = \frac{1}{f}$$

$$i = \left(\frac{1}{f} - \frac{1}{o}\right)^{-1} = \left(\frac{1}{10.0 \text{ cm}} - \frac{1}{30.0 \text{ cm}}\right)^{-1} = 15.0 \text{ cm}$$

The image distance is positive, indicating that the image is real. The lateral magnification is

$$m = -\frac{i}{o} = -\frac{15.0 \text{ cm}}{30.0 \text{ cm}} = -0.500$$

The image is half the size of the object and thus is 0.500 cm in length. The negative lateral magnification implies that the image is inverted. ◀

EXAMPLE 36.9   *Converging Lens, Real Object, Virtual Image*

An object 1.00 cm high is placed 6.00 cm from a converging lens of focal length 10.0 cm. Locate and characterize the image.

SOLUTION   We once again employ the lens equation with positive object distance and focal length:

$$i = \left(\frac{1}{f} - \frac{1}{o}\right)^{-1} = \left(\frac{1}{10.0 \text{ cm}} - \frac{1}{6.00 \text{ cm}}\right)^{-1} = -15.0 \text{ cm}$$

$$m = -\frac{i}{o} = -\frac{-15.0 \text{ cm}}{+6.00 \text{ cm}} = +2.50$$

The image is located 15.0 cm on the object side of the lens. It is virtual, upright, and 2.50 cm high. This result is confirmed by the construction in Figure 36.30. Notice that the principal plane of the lens is extended so that the ray that appears to come from the primary focus can be used to locate the image.

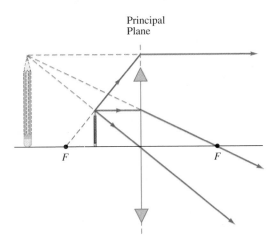

FIGURE 36.30

Ray diagram for Example 36.9. ◀

EXAMPLE 36.10   *Diverging Lens, Real Object, Virtual Image*

An object 1.00 cm in height is placed 15.0 cm from a diverging lens with a 10.0-cm focal length. Find the location of the image and describe its characteristics.

SOLUTION   In accordance with our sign convention, the focal length of the diverging lens must be $f = -10.0$ cm when used with the lens equation:

$$i = \left(\frac{1}{f} - \frac{1}{o}\right)^{-1} = \left(\frac{1}{-10.0 \text{ cm}} - \frac{1}{15.0 \text{ cm}}\right)^{-1} = -6.0 \text{ cm}$$

$$m = -\frac{i}{o} = -\frac{-6.0 \text{ cm}}{+15.0 \text{ cm}} = +0.40$$

The image is 0.40 cm high and upright. It is virtual and located 6.0 cm in front of the lens. Figure 36.31 is a graphical construction illustrating these conclusions.

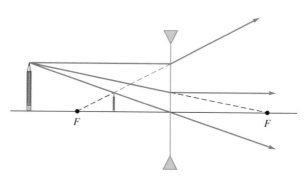

FIGURE 36.31

Ray diagram for Example 36.10.

Concept Question 6
On each of the Figures 36.29 through 36.31 draw three additional rays that form the image.

EXAMPLE 36.11   *Two Converging Lenses, Virtual Object for the Second Lens*

An object 1.00 cm in height is placed 15.0 cm from a converging lens with focal length 10.0 cm. A distance 20.0 cm behind the first lens is a second converging lens with a 15.0-cm focal length. Find the location of the final image formed by the second lens and describe its characteristics.

SOLUTION   The configuration is diagrammed in Figure 36.32(a). We divide the problem into two parts. First, we find the image of the first lens *completely ignoring the existence of the second lens.*

$$i = \left(\frac{1}{f_1} - \frac{1}{o}\right)^{-1} = \left(\frac{1}{10.0 \text{ cm}} - \frac{1}{15.0 \text{ cm}}\right)^{-1} = 30.0 \text{ cm}$$

We thus find that the first lens is *attempting* to form a real image 30 cm behind it. However, before this image can be formed, these rays are intercepted by the second lens located 20 cm behind the first. These converging rays are trying to form an *object* for the second lens. This "would-be" object is known as a **virtual object** and is located at the position of the "would-be" image of the first lens. Thus, for the second lens we write $o_2 = 20.0$ cm $- 30.0$ cm $= -10.0$ cm. In accordance with our sign convention, we sign this object distance as negative because *converging* rays strike the second lens. Algebraically, we have

$$i_2 = \left(\frac{1}{f_2} - \frac{1}{o_2}\right)^{-1} = \left(\frac{1}{15.0 \text{ cm}} - \frac{1}{-10.0 \text{ cm}}\right)^{-1} = 6.0 \text{ cm}$$

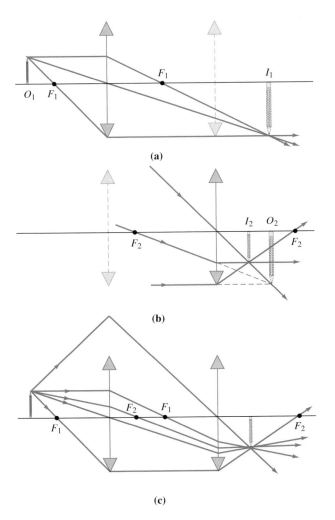

FIGURE 36.32

Ray diagram for Example 36.11. (a) If the second lens were not present, the first lens would form a real image at position I_1. (b) The image that would be formed by the first lens acts as a virtual object for the second lens. (c) Complete paths of all construction rays. All rays forming the image of the pencil tip originated at the pencil tip.

This result indicates that a real image is formed 6.0 cm behind the second lens. The lateral magnification of the final image is the product of the two individual lateral magnifications

$$m_1 m_2 = \left(-\frac{i}{o}\right)\left(-\frac{i'}{o'}\right) = \left(-\frac{30.0}{15.0}\right)\left(-\frac{6.0}{-10.0}\right) = -1.2$$

To verify these results graphically, in Figure 36.32(b) we trace three rays that strike the second lens while converging to form the real image from the first lens. These rays need not be the same construction rays used to locate the image from the first lens. Once we know where the image from the first lens is located we can pick any rays we like that are converging toward it. In particular, in Figure 36.32(b) we choose a ray going through the center of the second lens, a ray traveling parallel to the lens axis, and a ray traveling through the primary focus of the second lens. Each of these rays is deviated at the second lens according to our usual rules. The image formed by these construction rays verifies our algebraic calculation. In Figure 36.32(c) we show the complete paths taken by all construction rays. ◄

EXAMPLE 36.12   Two Thin Lenses in Contact

Two thin lenses of focal lengths f_1 and f_2 are placed in contact. Show that the combined focal length is given by

$$\frac{1}{f_c} = \frac{1}{f_1} + \frac{1}{f_2}$$

SOLUTION   We follow the reasoning used in Example 36.11 and find the image location of a real object formed by the first lens only:

$$\frac{1}{i_1} + \frac{1}{o_1} = \frac{1}{f_1}$$

This image acts as a (virtual) object for the second lens, thus we can write for the second lens

$$\frac{1}{i_2} + \frac{1}{o_2} = \frac{1}{f_2}$$

$$\frac{1}{i_2} + \frac{1}{-|i_1|} = \frac{1}{f_2}$$

Adding this last equation to the equation describing the first lens, we have

$$\frac{1}{i_1} + \frac{1}{o_1} + \frac{1}{i_2} + \frac{1}{-|i_1|} = \frac{1}{f_1} + \frac{1}{f_2}$$

which reduces to

$$\frac{1}{o_1} + \frac{1}{i_2} = \frac{1}{f_1} + \frac{1}{f_2}$$

This expression relates the *original* object position to the *final* image position in such a way that we are led to identify

$$\frac{1}{f_1} + \frac{1}{f_2} = \frac{1}{f_c}$$

as the combined focal length f_c of the lens pair.   ◀

(a)

(b)

FIGURE 36.33

(a) The real image of a light bulb filament is cast on a ground glass screen using a lens. (b) The image of a light bulb is completely formed on a viewing screen even though half of the lens is covered. Compare this result with that of Figure 36.34.

When predicting the location of images formed by lenses, it is quick and convenient to use the three special rays (one through the center, one through the focus, and one parallel to the axis). We must be careful, however, not to fall into the trap of believing that these are *the* rays that create the image. In fact, a multitude of rays diverge from each point on the object. Many of these rays are collected by the lens and then focused in its focal plane. The complete formation of a real image in this plane, however, does not require that all of the rays be used. Indeed, we can cover half the lens and yet *the entire image nonetheless forms on a screen in the image plane.* Because many people find this result surprising, we have photographed an experiment demonstrating this result (Fig. 36.33). As you can see, the entire image is present even when only half the lens is used. The image is only half as bright to be sure, but it's all there.

The reason many people are surprised to learn that half of a lens can form an entire image is that this result contradicts another observation that many of us have made when *looking through* a lens; namely, if we cover half the lens, half the image is no longer visible. This situation is shown in Figure 36.34. It is natural, but *incorrect,* to conclude from this experiment that the top half of the lens makes one half the image and the bottom half of the lens makes the other half of the image. To resolve the apparent contradiction between these two experiments we must take a careful look at a ray diagram that shows how we actually see an image formed by a magnifying lens. To do this we once more invoke our innocent-eyeball model to remind us that we see images only by virtue of light from that image entering our eye.

In Figure 36.35(a) we show a single lens forming a real image of an arrow. We apologize for having so many rays on the diagram, but they are important for emphasizing that rays from the top of the arrow are collected by the entire lens and focused to an image. The same is true for the bottom of the arrow. Notice, however, despite this multitude of rays, only a comparative few within a narrow cone intercepted by the eye's pupil are actually used by our innocent eyeball to "see" the real image. Now look at Figure 36.35(b). Here we have blocked the top half of the lens. Notice two critical things. First, the entire real image is still formed. Second, rays forming the image of the bottom of the arrow do not enter the observer's eye; therefore, the observer cannot see this image. With this realization the apparent paradox between our two experiments is resolved. The entire image is formed even when only part of the lens is used. Nonetheless, this image may not be observable if rays creating the image are not intercepted by the viewer's eye. In order to see the entire image we must insert a diffuse scatterer of some sort, perhaps a piece of paper or a ground glass screen. In Figure 36.35(c), we show the effect of placing such a scattering screen at the position of the real image. When light forming the image strikes the screen, the rays are scattered in all directions. Observers at many positions can now view the image because rays from the entire image are scattered toward their eyes.

(a)

(b)

FIGURE 36.34

(a) The real image of a light bulb filament is viewed directly through a lens. (b) When we attempt to view the image of a light bulb through a half-covered lens, only part of the image is visible. Compare this result to that of Figure 36.33.

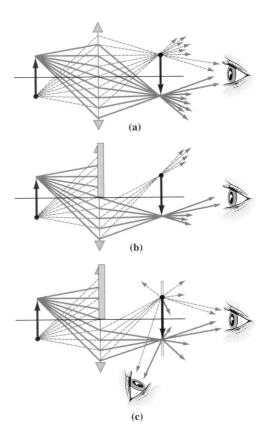

(a)

(b)

(c)

FIGURE 36.35

Image formation by a thin lens. (a) Rays forming an image of the bottom of the arrow pass through the entire lens. (b) When the top half of the lens is covered, an image of the bottom of the arrow is still formed. Looking along the optic axis, we cannot see the bottom of the arrow because the rays that form it are directed upward and do not enter our eye. (c) If a glass plate with a roughened surface is placed at the image plane, the entire image can be seen from any position because rays are scattered in many directions.

Concept Question 7
Why is it that in Figure 36.35(b) the entire image is formed but you see only half of it?

36.6 The Eye and Simple Optical Instruments (Optional)

In this section we examine basic principles behind magnifying glasses, microscopes, and telescopes, which are used to magnify images for our viewing. To do this we must first look at how images are formed in the eye.

The Eye

Figure 36.36 shows a simplified view of the eye with typical dimensions. Entering light is refracted first at the air–cornea boundary. As shown in Figure 36.37(a) about 75% of the focusing occurs at this boundary. The final focusing results from the refraction of the

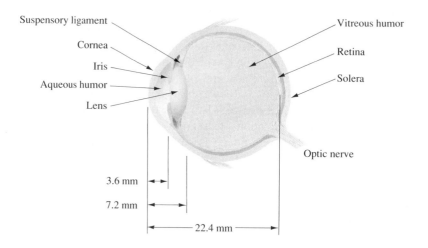

FIGURE 36.36

A simplified schematic of the human eye with approximate dimensions.

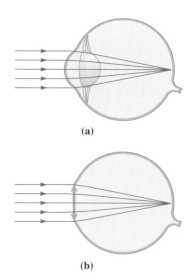

(a)

(b)

FIGURE 36.37

Refraction of light by the human eye. (a) Most of the bending of light rays occurs at the air–cornea boundary. The lens of the eye provides the final focusing of the image on the retina. (b) A thin-lens model of the eye.

Concept Question 8
Is the image formed inside the eye real or virtual?

eye's lens situated between the vitreous and aqueous humors. For our purposes it is sufficient to model the eye as if it had a single lens that forms an image on the light-sensitive retina of the eye as shown in Figure 36.37(b). Light falling on the retina stimulates electric signals that are sent to the brain and intercepted. (Yes, the image on the retina is inverted. Our brain inverts the image again.) It is the size of the image falling on the retina that determines the perceived size of an object. The size of the image on the retina is determined by the angle subtended by the object being viewed. The eye is different from the systems we have looked at thus far in that the focal length of its lens varies (within limits) to ensure that the image is always formed at the retina. A normal eye can focus objects located at positions from infinity (parallel rays) to some nearest point. This point of closest focus is known, naturally enough, as the **near point.** In young people the lens of the eye is quite flexible and the near point can be as close as 7–10 cm. As one ages the lens hardens and the near point recedes, requiring that objects be held farther away to be in focus. (You may know older folks who claim that their eyes are fine, but that their arms have grown too short!) In order to define magnifications in a standard way it is typical to define the **standard near point** as 25 cm from the eye.

The focal length of the eye changes to ensure that the image is focused on the retina. In Figure 36.38 we locate this image in the usual manner by ray construction. If you look at Figure 36.38 for a moment you will realize that *because the image is always focused on the retina,* we really need only a single construction ray to locate the image, namely, the one that passes undeviated through the center of the lens. This ray is the darker ray shown in Figure 36.38. Notice that *the image height is determined solely by the angle θ subtended by the object.* One obvious way you can make an object appear larger is to move the object closer to your eye. For a standard eye, the best we can do is to move the object to the near point, 25 cm away. To do better we need a magnifier. In order to standardize the magnification of a magnifier we use as a benchmark the angle subtended by the object when it is located at the near point of the unaided eye. To quantify how much larger the object

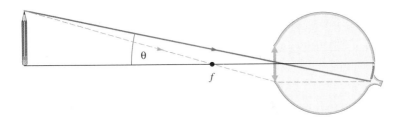

FIGURE 36.38

Because the image is formed on the retina of the eye, the image distance is fixed. Thus, the apparent size of an object is governed solely by the angle the object subtends at the eye.

appears to us we define a second type of magnification based on the *angle* subtended at the eye. This **angular magnification** is defined as

$$M_\theta \equiv \frac{\textit{Angle subtended by the image of an object}}{\textit{Angle subtended by the same object at the near point}} \equiv \frac{\theta'}{\theta_{np}}$$

We will soon see that the *angular* magnification depends on both the image size and how far the image is from our eye, whereas the *lateral* magnification depends only on the ratio of the image size to the object size.

EXAMPLE 36.13   *The Range of Focal Lengths for a Standard Eye*

Calculate the range of focal lengths attained by the lens of an eye that can focus on objects between infinity and the standard near point. Model the eye's cornea and lens as a single thin lens and assume a typical value of 20.0 mm for the distance from this model lens to the retina.

SOLUTION   With the eye focused at infinity the object distance o is infinite and hence the image distance is the focal length. Thus, the focal length is the lens-to-retina distance 20.0 mm.

When the eye is focused at the near point, we have $o = 250.$ mm and $i = 20.0$ mm. Solving the thin-lens equation, we have

$$f = \left(\frac{1}{o} + \frac{1}{i}\right)^{-1} = \left(\frac{1}{250.\text{ mm}} + \frac{1}{20.0\text{ mm}}\right)^{-1} = 18.5\text{ mm}$$

We see that the change in effective focal length required for a normal eye is quite small.   ◀

The Simple Magnifying Glass

Perhaps the first optical instrument you ever used was a magnifying glass. You no doubt enjoyed looking at the back of your hand, the letters on a printed page, and every insect you could find. (If you *haven't* done this, you should treat yourself to a magnifying glass the next time you're in the bookstore.) Let's start by analyzing how this simple, single-lens, magnifying glass made these enlarged images. Suppose our unaided "standard eye" is observing an object of height h, placed at our point of nearest focus. The angle subtended by our object in this case is, as shown in Figure 36.39(a),

$$\theta_{np} = \frac{h}{25\text{ cm}}$$

When we use a magnifying glass, we place the object inside its focal point and thus create an enlarged virtual image of height h' as shown in Figure 36.39(b). The angle subtended

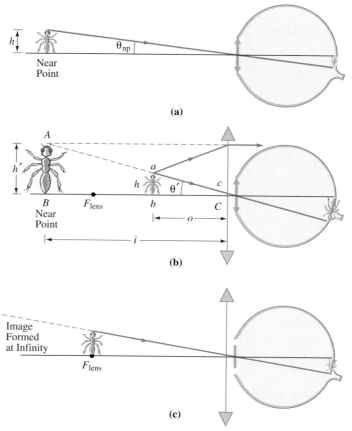

FIGURE 36.39

(a) An insect viewed at the near point of the eye. (b) A simple magnifier is used to create a larger, virtual image at the near point of the eye. (c) As the object approaches the focal point of the magnifying glass the image becomes slightly smaller and recedes toward infinity.

by this image is

$$\theta' = \frac{h'}{|i|}$$

where we have added absolute values signs because i is a negative number for this virtual image. If we concentrate on the ray from the tip of the object through the center of the lens, we see that this ray appears to come undeviated from the tip of the image. Triangles ABC and abc formed by this ray are similar. Consequently, we have $h'/|i| = h/o$ and the expression for the angular magnification becomes

$$M_\theta = \frac{\theta'}{\theta_{np}} = \frac{h/o}{h/25 \text{ cm}} = \frac{25 \text{ cm}}{o}$$

In this last form for the angular magnification we see, not surprisingly, that the closer we move the object toward the lens while still being able to focus on the image, the greater the magnification. The best we can do in this regard is to place the object so that the image formed by the lens is at our near point as in Figure 36.39(b). For a lens of focal length f the thin-lens equation gives

$$\frac{1}{o} = \frac{1}{f} - \frac{1}{i} = \frac{1}{f} - \frac{1}{-25 \text{ cm}} = \frac{1}{f} + \frac{1}{25 \text{ cm}}$$

leading to an angular magnification of

$$M_\theta = \frac{25 \text{ cm}}{o} = 25 \text{ cm}\left(\frac{1}{o}\right) = 25 \text{ cm}\left(\frac{1}{f} + \frac{1}{25 \text{ cm}}\right)$$

or

$$M_\theta = 1 + \frac{25 \text{ cm}}{f} \qquad \text{(Image at near point)} \qquad (36.8)$$

When the object is placed at the focus of the lens, the image recedes to infinity as shown in Figure 36.39(c). In this case $o = f$, and the angular magnification is thus

$$M_\theta = \frac{25 \text{ cm}}{f} \qquad \text{(Image at infinity)} \qquad (36.9)$$

In practice the observer usually adjusts the position of the object until the image, which falls somewhere between infinity and the near point, is comfortable to observe. The magnification is then somewhere between the values predicted by Equations (36.8) and (36.9). For standardization purposes it is traditional to use Equation (36.9) (image at infinity) when quoting the angular magnification.

As a practical matter, magnifications greater than 10 cannot be obtained without unacceptable distortion resulting from the violation of the paraxial approximation. (See the descriptions of lens aberrations at the end of this chapter for a more detailed discussion.) To achieve magnifications greater than 10, a two-lens magnifier must be employed. Possible configurations are explored in the problem set at the end of this chapter.

The Compound Microscope

If we wish to observe small objects with magnifications greater than those available from simple magnifiers we can use a microscope. Modern high-quality microscopes employ sophisticated optical systems to produce sharp images and flat fields of view. However, the basic principle behind these instruments is straightforward. We use one lens, called the **objective,** to form a real image of the object. We then view this image with a magnifying glass, now called the **eyepiece.** The objective of focal length f_{obj} and eyepiece of focal length f_{ep} are arranged so that their nearest foci are separated by a distance L called the **tube length.** The geometry is shown in Figure 36.40. When you focus the microscope, you are changing the object-to-objective distance until the image formed by the objective is located at the focus of the eyepiece.

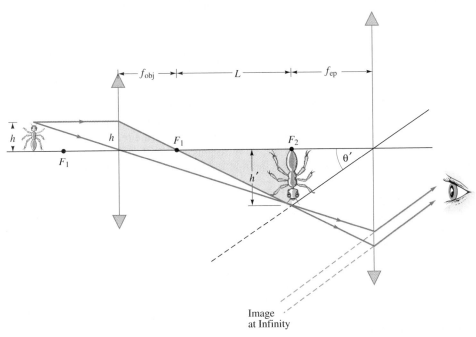

FIGURE 36.40

A simple microscope. The front objective creates a real image that is then examined with a magnifying glass.

Concept Question 9
At first it may be difficult to understand how an image can be formed at infinity. In particular it is difficult to imagine how an image infinitely far away can look big! To help understand this phenomenon trace a drawing of the magnifying lens and optic axis of Figure 36.39(b) onto a sheet of paper. Draw a bug the same size as ours onto your paper but place it halfway between its position in Figure 36.39(b) and the focal point of the lens. Now locate the image of your bug by drawing two construction rays similar to the ones we have shown. How big is your bug? Where is it located? (Now you know why we didn't include this drawing in the text.) What about the angle θ' for your drawing; is it larger or smaller than ours? Remember, it is the angle subtended by the image that determines the size of the image on the retina. If you're still not comfortable with the image-at-infinity idea, try the drawing one more time. This time move the bug half again the distance to the focus. (Be sure you have lots of paper!)

The overall angular magnification of the microscope is

$$M_\theta = \frac{\theta'}{\theta_{np}} = \frac{h'/f_{ep}}{h/25 \text{ cm}} = \left(\frac{h'}{h}\right)\left(\frac{25 \text{ cm}}{f_{ep}}\right)$$

Written this way, it is clear that the overall magnification is the product of the lateral magnification of the objective and the angular magnification of the eyepiece. From the similar triangles in Figure 36.40 we see that $h'/h = L/f_{obj}$, giving us the final result

$$M_\theta = -\left(\frac{L}{f_{obj}}\right)\left(\frac{25 \text{ cm}}{f_{ep}}\right) \tag{36.10}$$

where we have again inserted a minus sign in keeping with the convention that inverted images are denoted by negative magnification.

Telescopes

Telescopes constructed by using only lenses are known as refracting telescopes, or often simply "refractors." The basic design of a simple refracting telescope is similar to a microscope in the sense that a real image is formed by an objective lens and that image is then viewed through an eyepiece. The angular magnification of a telescope, however, is defined somewhat differently from that of a magnifying glass and a microscope. It is no longer useful to compare the angle subtended by the image to the angle θ_{np} subtended by the object when it is placed at the near point. (It's unlikely that you are ever going to get many celestial objects 25 cm from your eye!) Instead, we compare the angle subtended by the image to the angle θ_o subtended by the distant object *at its present position.* For example, the sun and the moon subtend nearly the same angle (about 0.5°) even though they have vastly different diameters.

An **astronomical telescope** intended for viewing celestial objects is shown in Figure 36.41. The diagram is similar to that of a microscope except that the tube length L is zero. Rays from celestial objects are essentially parallel when they enter the telescope, and hence the image is formed at the common focus of the objective and eyepiece. The eyepiece then forms an image also at infinity, but this image subtends a much greater angle. The angular magnification of the astronomical telescope is

$$M_\theta = \frac{\theta'}{\theta_o} = \frac{-h'/f_{ep}}{h'/f_{obj}} = -\frac{f_{obj}}{f_{ep}}$$

When a telescope is used for terrestrial observations, the inverted image can be quite bothersome. The image can be inverted again by employing a diverging lens for the eyepiece as is often used in "opera glasses." Alternatively, a third lens can be inserted to invert the inverted image. This latter possibility requires that the telescope be lengthened by at least four times the focal length of the third lens (see Problem 60). Another alternative is to reinvert the image using total internal reflections inside of erecting prisms, such

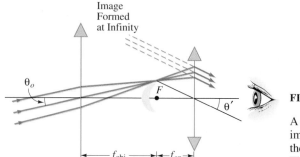

FIGURE 36.41

A simple astronomical telescope. A real image of a distant object is formed at the focal point of a magnifying glass. The observer examines the image with the magnifying glass.

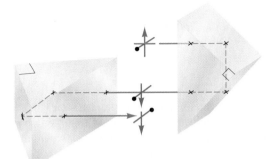

FIGURE 36.42

A pair of Porro (right angle) prisms can be used to invert an image. Such prisms are common in binoculars.

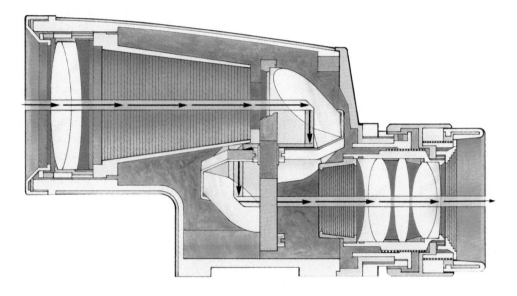

A cut-away view of binoculars shows the Porro prisms used to invert the image.

as the pair of Porro prisms shown in Figure 36.42. This approach provides a compact instrument and is the choice used in binoculars.

Newton was troubled by the chromatic dispersion of refracting telescopes, and so instead of an objective lens he used a concave mirror to form the real image. A small plane mirror diverts the real image formed by the concave mirror to the side of the telescope where it can be viewed by the eyepiece. The geometry of such a **Newtonian telescope** is shown in Figure 36.43. Telescopes that use a mirror or mirrors to provide the real image are known as "reflectors" even though the eyepiece is a refractive element. The expression for the angular magnification of a reflector is the same as for a refractor except the focal length of the objective f_{obj} is replaced by the focal length of the mirror.

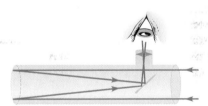

FIGURE 36.43

A Newtonian telescope.

36.7 Optical Aberrations (Optional)

Throughout this chapter we make use of the paraxial ray approximation. In this approximation we require that rays travel near to the optical axis and make small angles with it. We have repeatedly applied the approximation $\sin(\theta) \approx \theta$. We recognize that this approximation is merely the first term in the power series expansion for the sine function:

$$\sin(\theta) = \theta - \frac{\theta^3}{3!} + \frac{\theta^5}{5!} - + \cdots$$

The theory we presented makes use of only the first term in this power series and is consequently known as a **first-order** model. When the next term $\theta^3/3!$ is included, departures from the perfect focusing implied by our first-order model arise. These departures are known as **third-order** aberrations.[3] A detailed study of such aberrations is beyond the scope of this text; however, we briefly describe some first-order and third-order aberrations.

Chromatic Aberration

As discussed in Section 35.4, the index of refraction of common glasses varies as a function of wavelength. This variation means that a lens designed to have a focal length f for one particular wavelength has a different focal length for rays of another wavelength. For example, if the lens has a focal length of 20.0 cm for yellow light of $\lambda = 589$ nm, then its focal length is longer than 20.0 cm for red light and shorter for blue light. Thus, objects illuminated by white light (a mixture of many wavelengths) have many images formed at different distances from the lens, and each is a different color. This phenomenon is known as **chromatic aberration** and is a first-order effect.

In order to minimize chromatic aberration lenses are often made as **doublets.** These doublets consist of a combination of converging and diverging lenses, often cemented together. Each of the constituent lenses is made from a different glass type. One glass is chosen because it has a large deviation but small dispersion (Fig. 36.44). The other lens has a small average deviation but a large dispersion. By combining two such lenses it is possible to obtain a doublet with minimal chromatic aberration. Such doublets are known as **achromatic.**

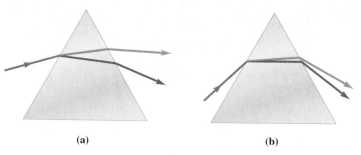

(a)   (b)

FIGURE 36.44

Comparison of prisms made from glasses with different dispersive properties. Prism (a) has a smaller average deviation than (b). However, prism (a) has larger dispersion than prism (b). A pair of lenses (one converging, the other diverging) can create a doublet lens with low chromatic aberration.

Third-Order Aberrations

The aberrations described by the third-order corrections to the paraxial approximation were studied by the German mathematician Phillip Ludwig von Seidel (1821–1896) in the 1850s. These aberrations are known as **third-order aberrations,** or **Seidel aberrations.** Seidel identified five such aberrations resulting from five distinct mathematical terms in his corrections. Three of the aberrations result in a lack of clarity of the image. These aberrations are called **spherical aberration, coma,** and **astigmatism.** The other two aberrations result in an overall deformation of the image and are designated as **field curvature** and **distortion.** Examples of these aberrations are shown in Figure 36.45.

Spherical aberration describes a lens's inability to focus rays originally parallel to the optical axis to a single point. Typically rays traveling farthest from the optic axis miss the focal point by the largest distance (Fig. 36.45(a)). **Coma** (Fig. 36.45(b)) describes a lens's inability to focus parallel rays that originally are not traveling parallel to the optic axis.

---

[3]The second restriction in the paraxial approximation is that the distance y between the ray and the optic axis must be small. Thus, terms such as θy^2 are also included as third-order aberrations.

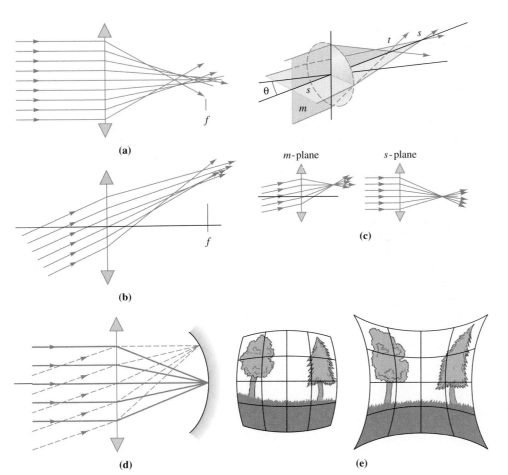

FIGURE 36.45

(a) Spherical aberration. Rays far from the optic axis are not focused to the focal point. (b) Coma. Rays not parallel to the optic axis are not focused to a point. (c) Astigmatism. Rays in different planes are focused to different points. (d) Curvature of field. The focal plane is actually a curved surface. (e) Distortion.

When an object point lies far from the optic axis, a cone of rays from the point strikes the lens asymmetrically. The lens has a shorter focal length for rays that lie in a plane that contains both the optic axis and the object point than it does for rays lying in a plane perpendicular to this one. This aberration is known as **astigmatism** (Fig. 36.45(c)).

Minimizing chromatic, Seidel, and even higher-order distortions is the challenging task of the modern lens designer. Lens design often involves intricate tradeoffs between the reduction of one type of aberration at the cost of increasing another. Often, the best results are obtained by employing **aspherical** lenses the surfaces of which differ from perfect spheres. Today's designers are aided by powerful computer programs that can perform *exact* ray trace diagrams to evaluate image quality even before the lens is created.

36.8 Numerical Methods for Paraxial Ray Tracing (Optional)

When a paraxial ray travels through an optical system composed of mirrors and thin lenses, it is possible to describe its path with only two parameters: the ray's distance y from the optic axis and the angle θ it makes with the optic axis. As the ray travels from optical element to optical element its *distance* from the optic axis changes. (For convenience we refer to the distance of the ray from the optic axis as the "height" of the ray.) As the ray passes through (or is reflected by) an optical "element" the *angle* that it makes with the optic axis changes. In either case the new height y' and the new angle θ' can be calculated from the previous height and angle by simple linear equations that we can write generically as

$$y' = Ay + B\theta \qquad (36.11)$$

$$\theta' = Cy + D\theta \qquad (36.12)$$

The four parameters A, B, C, and D depend on the type of optical element. For example, when the ray is merely traveling a distance d in air, its height changes but its angle does not. As shown in Figure 36.46, the new height can be obtained from the old height from the relation

$$y' = y + d \tan(\theta)$$

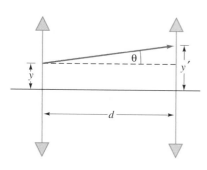

FIGURE 36.46

A ray travels in the space between two lenses. The new distance y' from the optic axis is related to the old distance y from the optic axis by $y' = y + d \tan(\theta)$.

Using the paraxial approximation $\tan(\theta) \approx \theta$, we see that this equation has the form of Equation (36.11) with $A = 1$ and $B = d$. Because the angle the ray makes with the optic axis does not change as it travels, we have $C = 0$ and $D = 1$ in Equation (36.12). Thus, we can describe the transformation of a ray traveling between elements by the $ABCD$ parameters: 1, d, 0, 1.

When a ray strikes a thin lens, its direction is changed but its position is fixed. (Remember, "thin lens" means zero thickness.) The thin-lens equation predicts

$$\frac{1}{i} = \frac{1}{f} - \frac{1}{o}$$

We multiply this expression by y to obtain

$$\frac{y}{i} = \frac{y}{f} - \frac{y}{o}$$

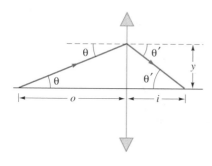

FIGURE 36.47

At a lens the ray direction is deviated from θ to θ'.

Examining Figure 36.47, we find $y/o = \tan(\theta) \approx \theta$ and $y/i = \tan(\theta') \approx -\theta'$. Note that we have introduced a negative sign in front of θ' because we wish θ' to be a negative number when the ray has a negative slope as in Figure 36.47. With these substitutions the thin-lens equation becomes $\theta' = -y/f + \theta$. Because the ray height is unchanged we have $y' = y$. Consequently, for a ray deviated by a thin lens we have $ABCD$ parameters: 1, 0, $-1/f$, 1.

In a similar way all of the rules for paraxial rays derived in this chapter can be written in terms of $ABCD$ parameters. The results are summarized in Table 36.4. To trace a ray through an optical system we need only operate on the ray height y and angle θ repeatedly with Equations (36.11) and (36.12). Each time we update y' and θ', we can plot the new ray height at the appropriate position along the axis.

TABLE 36.4   *ABCD* Parameters

ELEMENT TYPE	A	B	C	D
Translation	1	L	0	1
Refraction (plane surface)	1	0	0	$\dfrac{n_1}{n_2}$
Refraction (curved surface)	1	0	$\dfrac{n_1 - n_2}{Rn_2}$	$\dfrac{n_1}{n_2}$
Thin lens	1	0	$-\dfrac{1}{f}$	1
Spherical mirror	1	0	$\dfrac{2}{R}$	1

A Spreadsheet Template for the *ABCD* Parameters

Repetitive arithmetic procedures, such as the one just described, are made to order for computers. In Example 36.14 we show a spreadsheet organized so that rays can be plotted as ordered pairs of (x, y). Here, x is the position along the optic axis starting from $x = 0$ at the object and y is the ray height.

EXAMPLE 36.14   *A Spreadsheet Template to Trace Paraxial Rays*

Create a spreadsheet template to trace rays through an optical system with up to three thin lenses.

SOLUTION   One possible spreadsheet template organization using the *ABCD* parameters is shown in Figure 36.48. The spreadsheet template RAY__TRAC.WK1 on the diskette accompanying this text allows six rays to be traced through three lenses. Tabular listings for two of the six rays are shown in Figure 36.48. The *ABCD* parameters are calculated from the formulas summarized in Table 36.4 using values of positions and focal lengths input by the user in rows 6 and 7.

	A	B	C	D	E	F	G	H	I	J
	A	B	C	D	E	F	G	H	I	J
4										
5	Lens 1		Lens 2		Lens 3		End of trace			
6	x = 10		x = 15		x = 25		x = 35			
7	f = 5		f = 10		f = 15					
8										
9							Ray 1		Ray 2	
10		position	A	B	C	D	height	angle	height	angle
11		(cm)					(cm)	(rad)	(cm)	(rad)
12										
13	object	0.0					1.00	0.00	1.00	0.05
14	space	10.0	1.00	10.00	0.00	1.00	1.00	0.00	1.50	0.05
15	lens 1	10.0	1.00	0.00	−0.20	1.00	1.00	−0.20	1.50	−0.25
16	space	15.0	1.00	5.00	0.00	1.00	−0.00	−0.20	0.25	−0.25
17	lens 2	15.0	1.00	0.00	−0.10	1.00	−0.00	−0.20	0.25	−0.28
18	space	25.0	1.00	10.00	0.00	1.00	−2.00	−0.20	−2.50	−0.28
19	lens 3	25.0	1.00	0.00	−0.07	1.00	−2.00	−0.07	−2.50	−0.11
20	space	35.0	1.00	10.00	0.00	1.00	−2.67	−0.07	−3.58	−0.11

A13:	[W8]	'object		H14:	(F2)	[W7] +$E14*G13+$F14*H13
B13:	(F1)	[W8] 0		I14:	(F2)	[W7] +$C14*I13+$D14*J13
G13:	(F2)	[W7] 1		J14:	(F2)	[W7] +$E14*I13+$F14*J13
H13:	(F2)	[W7] 0		A15:	(F3)	[W8] 'lens 1
I13:	(F2)	[W7] 1		B15:	(F1)	[W8] +B6
J13:	(F2)	[W7] 0.05		C15:	(F2)	[W7] 1
A14:	(F3)	[W8] 'space		D15:	(F2)	[W7] 0
B14:	(F1)	[W8] +B6		E15:	(F2)	[W7] −1/B7
C14:	(F2)	[W7] 1		F15:	(F2)	[W7] 1
D14:	(F2)	[W7] +B15−B13		G15:	(F2)	[W7] +$C15*G14+$D15*H14
E14:	(F2)	[W7] 0		H15:	(F2)	[W7] +$E15*G14+$F15*H14
F14:	(F2)	[W7] 1		I15:	(F2)	[W7] +$C15*I14+$D15*J14
G14:	(F2)	[W7] +$C14*G13+$D14*H13		J15:	(F2)	[W7] +$E15*I14+$F15*J14

FIGURE 36.48

Spreadsheet template for paraxial ray tracing. See Example 36.14.

Initial ray heights and angles must be entered in row 13. We use an initial ray height of 1 cm so that magnifications are easy to determine from the plot. Ray angles less than 0.1 rad generally guarantee paraxial rays. (Note the *x*- and *y*-axes have different scales.)

In order to calculate the updated positions and angles as the rays proceed through the system only two equations ((36.11) and (36.12)) are needed throughout. These formulas were typed only once into cells G14 and H14 in the form

$$+\$C14*G13+\$D14+H13 \quad \text{and} \quad +\$E14*G13+\$F14*H13$$

Once typed, the two formulas were then copied using the spreadsheet's copy command into the remainder of their respective columns. Next, the two columns for Ray 1 were copied into two columns for ray 2 and so on. Note the $ in front of the column designator in $C14, $D14, $E14, and $F14. For our spreadsheet this $ symbol "anchors" the cell address parameter following the $ so that when the formula is copied, other rows and columns in the formula are updated, but columns C, D, E, and F are always used for the *ABCD* parameters. For a given ray, each row merely updates the *y* and *θ* values using the values from the previous row and the corresponding *A*, *B*, *C*, or *D* values.

Figure 36.49 shows a ray trace created by this spreadsheet. We have added the object, image, and lenses to the figure for clarity.

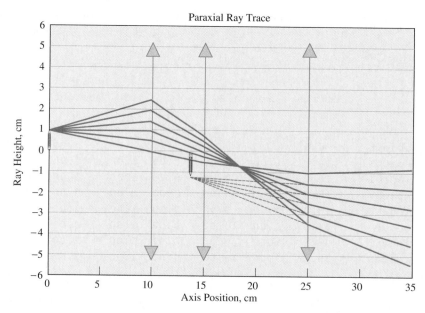

FIGURE 36.49

Ray tracing produced by the spreadsheet in Example 36.14. Lenses, object, and image (including dotted lines) have been added by hand for clarity.

We hope you find the spreadsheet template of Example 36.14 fun to play with, and we also hope you try to design your own systems (or perhaps just check your answers to other problems). Various adaptations are explored in the problems at the end of this chapter.

The *ABCD* Matrix (Optional)

Equations (36.11) and (36.12) describing the height and angle of paraxial rays can be written conveniently in matrix form

$$\begin{bmatrix} y' \\ \theta' \end{bmatrix} = \begin{bmatrix} A & B \\ C & D \end{bmatrix}\begin{bmatrix} y \\ \theta \end{bmatrix}$$

The rules for multiplication of the column "vector"

$$\begin{bmatrix} y \\ \theta \end{bmatrix}$$

with the matrix

$$\begin{bmatrix} A & B \\ C & D \end{bmatrix}$$

are discussed in Appendix 2 and are illustrated in Section 27.7, which deals with matrix techniques for solving electric circuits.

Using matrix algebra, we can reduce many of the derivations of this chapter to simple multiplication. For example, consider refraction by a thick lens. As Figure 36.50 reminds us, the process occurs as a refraction at a spherical surface, followed by a displacement, followed by another refraction at a second spherical surface. Referring to Table 36.4, we see that the process can be expressed as the product of three matrices

$$\begin{bmatrix} 1 & 0 \\ \dfrac{n-1}{R_2} & n \end{bmatrix}\begin{bmatrix} 1 & t \\ 0 & 1 \end{bmatrix}\begin{bmatrix} 1 & 0 \\ \dfrac{1-n}{R_1 n} & \dfrac{1}{n} \end{bmatrix} \qquad (36.13)$$

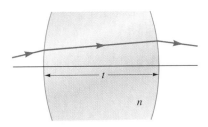

FIGURE 36.50

Refraction of a ray by a thick lens results from refraction at the front surface followed by a displacement through the lens, and finally another refraction at the second surface.

Note that the first surface encountered is represented by the rightmost matrix so that it multiplies the original y and θ first. When the lens is thin ($t = 0$), the middle matrix reduces to the identity matrix and can thus be omitted. The result of multiplying the two remaining matrices is

$$\begin{bmatrix} 1 & 0 \\ -(n-1)\left(\dfrac{1}{R_2} - \dfrac{1}{R_1}\right) & 1 \end{bmatrix}$$

By comparing this result with Table 36.4, we recognize this result as the *ABCD* matrix for a thin lens with a focal length given by

$$f = \left[(n-1)\left(\frac{1}{R_2} - \frac{1}{R_1}\right)\right]^{-1}$$

which is, of course, the lens-maker's formula we found previously.

The power of the *ABCD* matrix formalism should now be evident. The techniques are made even more powerful when combined with modern computers. Many spreadsheets and calculators allow for the multiplication of matrices when entered in numerical form. Modern symbolic algebra programs permit matrix multiplication of symbolic forms.

36.9 Summary

Images are formed by light rays entering the eye. The apparent location of images can be found by backtracking rays from a single point to their apparent point of origin. Images are **real** if many light rays from the entire object actually pass through the image point. The image is **virtual** if rays do not pass through the image.

The images formed by a plane mirror are virtual and upright, exhibit opposite handedness, and have a magnification of 1.

Paraxial rays are rays that travel near the **optic axis** of a lens and/or mirror system and make small angles with it.

Paraxial rays are focused by spherical mirrors according to the mirror equation

$$\frac{1}{f} = \frac{1}{o} + \frac{1}{i}$$

where the **focal length** f is half the **radius of curvature** R of the mirror. The magnitude of the **object distance** o is the distance from the object to the vertex of the mirror; the magnitude of the **image distance** i is the distance from the image to the vertex of the mirror. The following sign convention holds for the mirror equation:

> f and R are positive for concave mirrors
> f and R are negative for convex mirrors
> o is positive when rays diverge from a real object
> o is negative when rays converge toward a virtual object
> i is positive for real images formed by converging rays
> i is negative for virtual images created by diverging rays

The **lateral magnification** of an image can be calculated from

$$m = -\frac{i}{o}$$

where a negative magnification denotes an inverted image.

When a ray is refracted by a spherical boundary of radius R between two dielectric materials, the object and image distances are related by

$$\frac{(n_2 - n_1)}{R} = \frac{n_1}{o} + \frac{n_2}{i}$$

where n_1 and n_2 are the indices of refraction of the initial and final material, respectively. The sign convention for distances o and i is the same as for curved mirrors. However, the convention for the radius R is opposite that for curved mirrors. That is, for refraction at a spherical boundary

> R is positive if the surface is convex toward the incoming ray
> R is negative if the surface is concave toward the incoming ray

The focal length of a thin lens can be deduced from the index of refraction of the lens material and the radius of curvature of the two surfaces by the **lens-maker's formula**

$$(n - 1)\left(\frac{1}{R_1} - \frac{1}{R_2}\right) = \frac{1}{f}$$

This formula holds only for paraxial rays and assumes the rays enter from air into a lens fabricated from material of refractive index n. The radii are signed numbers following the convention described previously for refraction at a spherical boundary.

For thin lenses the object and image distances o and i are related to the focal length of the lens by the **thin-lens equation**

$$\frac{1}{f} = \frac{1}{o} + \frac{1}{i}$$

The sign conventions for the thin-lens equation are similar to those for the mirror equation:

> f is positive for converging lenses
> f is negative for diverging lenses
> o is positive when rays diverge from a real object
> o is negative when rays converge toward a virtual object
> i is positive for real images formed by converging rays
> i is negative for virtual images created by diverging rays

For two thin lenses in contact the effective focal length f_c of the combination is given by

$$\frac{1}{f_c} = \frac{1}{f_1} + \frac{1}{f_2}$$

(Optional) The magnifying power of optical instruments is usually stated in terms of the **angular magnification**

$$M_\theta = \frac{Angle\ subtended\ by\ the\ image\ of\ an\ object}{Angle\ subtended\ by\ the\ same\ object\ at\ some\ reference\ point} = \frac{\theta'}{\theta_{ref}}$$

For simple magnifiers and microscopes the reference position is taken as the standard **near point** of normal vision, 25 cm from the observer's eye. In this case, the reference angle is given by the object height h divided by 25 cm:

$$\theta_{ref} = \theta_{np} = \frac{h}{25\ cm}$$

The angular magnification of a magnifying glass of focal length f is

$$M_\theta = -\frac{25\ cm}{f}$$

when the image is formed at infinity.

The angular magnification of a compound microscope is found from the product of the lateral magnification of the **objective** lens and the angular magnification of the **eyepiece** lens.

$$M_\theta = -\left(\frac{L}{f_{obj}}\right)\left(\frac{25\ cm}{f_{ep}}\right)$$

(Optional) **Chromatic aberration** results from variation of a material's index of refraction with wavelength. Monochromatic aberrations arise when the paraxial approximation fails to hold. The primary image defects created by small deviations from the paraxial approximation are called **third-order aberrations.**

(Optional) Paraxial rays can be traced through optical systems using a set of linear equations of the ray height and ray angle with the optic axis. This set of equations involves four parameters $ABCD$ that characterize each segment of the system. This method lends itself to computerized computation.

PROBLEMS

36.2 Images Formed by Plane Mirrors

1. An automobile rearview mirror is located 60.0 cm from a driver's eyes and 2.20 m from the back window, which is 1.30 m wide. Assume the normal to the mirror is tilted by 20° from the center line of the car as shown in Figure 36.P1. (a) Make a copy of Figure 36.P1 and add to it a careful drawing of the image of the rear window where it appears as seen in the mirror. (b) Use your sketch to estimate the minimum width of the rearview mirror for the driver to be able to see the image of the entire rear window. (Measuring with a rule on a scale drawing provides sufficient accuracy.)

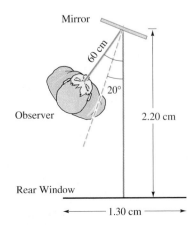

FIGURE 36.P1   Problem 1

2. Draw a sketch showing exactly how several rays bounced off the surface of the lake to form the images displayed in Figure 36.P2.

FIGURE 36.P2   Problem 2

3. Two plane mirrors are joined at one edge so that their reflecting surfaces form an angle of 60°. A small object is placed between the mirrors along a line bisecting the angle between the mirrors. (a) How many images of the object are formed? (b) What happens if the object is moved off the bisector?

4. Show your understanding of Example 36.1 by drawing the images of the extended object formed by the mirrors shown in Figure 36.P3.

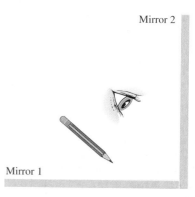

Mirror 2

Mirror 1

**FIGURE 36.P3
Problem 4**

36.3 Images Formed by Curved Mirrors

5. An object 1.00 cm high is placed on the axis 90.0 cm from the vertex of a spherical concave mirror with radius of curvature 75.0 cm. (a) Construct a scale drawing (horizontal and vertical scales need not be equal) showing the path of at least three of the four special construction rays described in the text. Use these rays to locate the image formed. On your sketch also show a small cone of rays that might be intercepted by an eye viewing the image. (b) Locate the position of the image formed using the mirror equation. (c) Describe the image by telling whether it is real or virtual, erect or inverted, and give its magnification.

6. An object 1.00 cm high is placed on the axis 50.0 cm from the vertex of a spherical concave mirror with radius of curvature 75.0 cm. Characterize the image by carrying out the activities described in parts (a), (b), and (c) of Problem 5.

7. An object 2.00 cm high is placed on the axis 25.0 cm from the vertex of a spherical concave mirror with radius of curvature 75.0 cm. Characterize the image by carrying out the activities described in parts (a), (b), and (c) of Problem 5.

8. An object 1.50 cm high is placed on the axis 90.0 cm from the vertex of a spherical convex mirror with radius of curvature 75.0 cm. Characterize the image by carrying out the activities described in parts (a), (b), and (c) of Problem 5.

9. An object 0.50 cm high is placed on the axis 50.0 cm from the vertex of a spherical convex mirror with radius of curvature 75.0 cm. Characterize the image by carrying out the activities described in parts (a), (b), and (c) of Problem 5.

10. An object 3.00 cm high is placed on the axis 25.0 cm from the vertex of a spherical convex mirror with radius of curvature 75.0 cm. Characterize the image by carrying out the activities described in parts (a), (b), and (c) of Problem 5.

11. Find the location and character of the image formed from an object located in front of a spherical mirror with radius of curvature of 10.0 cm for each of the following object positions: (a) 25.00 from a concave mirror, (b) 1.25 cm from a concave mirror, (c) 5.00 cm from a convex mirror, (d) 20.00 cm from a convex mirror. In each case obtain numerical values using an appropriate formula, and verify your answers using a ray diagram.

12. A meterstick lies along the optic axis of a concave mirror with radius of curvature 60.0 cm. If the closer end of the meterstick is 40.0 cm from the mirror, how long is the image of the meterstick?

13. A meterstick lies along the optic axis of a convex mirror of focal length 40.0 cm. If the closer end of the meterstick is 10.0 cm from the mirror, how long is the image of the meterstick?

36.4 Images Formed by Refracting Surfaces

14. A gem-quality diamond has a small flaw located a distance of 1.00 mm directly below one face. How far below the face does this flaw appear to an observer?

15. A caddis fly hovers 10.0 cm above the surface of a stretch of flat water (Fig. 36.P4). How far above the surface does the insect appear to be to a hungry rainbow trout in the water directly below?

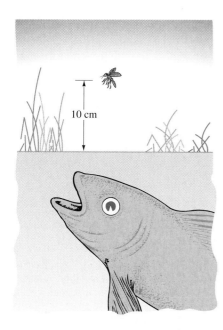

FIGURE 36.P4
Problem 15

16. A small bubble is located in the exact center of a transparent marble of radius 2.54 cm. If the index of refraction of the glass marble is 1.52, how far from the surface does the bubble appear to be to an observer?

17. A hollow glass sphere of radius 5.00 cm is filled with glycerin. When shaken, bits of white plastic simulate snow falling on a small figure of Santa Claus inside the sphere. If the Santa figure is 4.26 cm from the outside surface of the sphere, how close to the surface does he appear to be to someone looking in at him? (Assume the sphere has the same index of refraction as glycerine.)

36.5 Images Formed by Lenses

18. A thin biconvex lens is formed of borosilicate crown glass, which has an index of refraction of 1.510 for light of wavelength 400 nm. The radii of curvature of the front and back spherical surfaces are 0.500 m and 0.600 m, respectively. (a) What is the focal length of this lens for 400-nm light? (b) For light of 680 nm the index of refraction of this particular glass is 1.500. What is the focal length of this lens for 680-nm light?

19. A thin biconcave lens is formed of borosilicate crown glass, which has an index of refraction of 1.510 for light of wavelength 400 nm. The radii of curvature of the front and back spherical surfaces are 0.500 m and 0.600 m, respectively. (a) What is the focal length of this lens for 400-nm light? (b) For light of 680 nm the index of refraction of this particular glass is 1.500. What is the focal length of this lens for 680-nm light?

20. A thin planoconvex lens is formed of light flint glass, which has an index of refraction of 1.655 for light of wavelength 400 nm. The focal length of this lens is to be 30.0 cm at 400 nm. (a) What should be the radius of curvature of the spherical surface? (b) For light of 680 nm the index of refraction of this particular glass is 1.615. What is the focal length of this lens for 680-nm light?

21. A thin planoconcave lens is formed of light flint glass, which has an index of refraction of 1.655 for light of wavelength 400 nm. The focal length of this lens is to be −25.0 cm at 400 nm. (a) What should be the radius of curvature of the spherical surface? (b) For light of 680 nm the index of refraction of this particular glass is 1.615. What is the focal length of this lens for 680-nm light?

22. The two radii of curvature of a biconvex lens are measured with a spherometer (Fig. 36.P5) to be 30.0 cm and 60.0 cm. The focal length of this lens is measured to be 26.0 cm at 589 nm. What is the index of refraction of this glass for light of 589 nm?

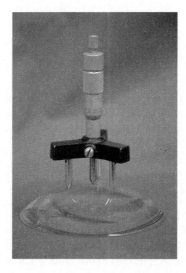

FIGURE 36.P5 A spherometer is a device used to measure the radius of curvature of a spherical surface. See Problems 22 and 59.

23. An object is placed at the following distances from a converging lens with focal length 35.0 cm: (a) 50.0 cm, (b) 30.0 cm, (c) 15.0 cm. In each case state the image location, magnification, and character (real or virtual), and tell whether it is upright or inverted. Verify your numerical calculations with ray diagrams.

24. An object is placed at the following distances from a diverging lens with focal length 35.0 cm: 50.0 cm, 30.0 cm, 15.0 cm. In each case state the image location, magnification, and character (real or virtual), and tell whether it is upright or inverted. Verify your numerical calculations with ray diagrams.

25. An object is placed at a distance of 25.0 cm from several lenses with the following focal lengths: (a) +50.0 cm, (b) +10.0 cm, (c) −15.0 cm, (d) −30.0 cm. In each case state the image location, magnification, and character (real or virtual), and tell whether it is upright or inverted. Verify your numerical calculations with ray diagrams.

26. An object is placed at a distance of 15.0 cm from several lenses with the following focal lengths: (a) −10.0 cm, (b) +10.0 cm, (c) −15.0 cm, (d) +20.0 cm. In each case state the image location, magnification, and character (real or virtual), and tell whether it is upright or inverted. Verify your numerical calculations with ray diagrams.

27. When an object is placed 5.00 cm to the left of a particular lens, the image, observed from the right side of the lens, is found to be 7.50 cm to the left of the lens. (a) What is the focal length of the lens? (b) Is it a converging or diverging lens? (c) If the object is moved to a location 20.0 cm to the left of the lens, where is the image? Characterize the image.

28. (a) Using Figure 36.P6, show that the *Gaussian* form of the thin lens equation

$$\frac{1}{o} + \frac{1}{i} = \frac{1}{f}$$

can be written in the *Newtonian* form $xx' = f^2$ where x is the distance from the object to the front focus of the lens, and x' is the distance from the back focal point to the image. (b) Show that the lateral magnification can be written $m = -f/x = -x'/f$.

FIGURE 36.P6   Problem 28

29. Use the Newtonian form of the thin-lens equation (see Problem 28) to find the location of the image formed by a diverging lens of focal length -10.0 cm, which is 25.0 cm from an object. What problem occurs if we try to use the Newtonian form for a diverging lens?

30. The glowing filament of a small, clear light bulb acts as an object. A lens of focal length f is used to create a real image of the filament on a screen a distance $L > 4f$ away. (a) If the distance L is held fixed, there are two positions of the lens for which a real image is formed. Show that these two lens positions are separated by $l = \sqrt{L(L-4)}$. (b) Show that the focal length of the lens is given by $f = (L^2 - l^2)/4L$. Application of this result is known as *Bessel's method* for finding the focal length of a lens.

31. To apply *Abbe's method* for finding the focal length of a lens we measure the lateral magnifications m_1 and m_2 of the lens with an object located at two different positions o_1 and o_2. Show that the focal length is given by

$$f = \frac{o_2 - o_1}{(m_1)^{-1} - (m_2)^{-1}}$$

32. The *longitudinal magnification m_l* of a lens describes the apparent shortening or lengthening of an object in the direction *along the optic axis*. (a) Show that the longitudinal magnification $m_l \equiv di/do$ is given by

$$m_l = -m^2 = -\left(\frac{f}{o-f}\right)^2$$

The negative sign in this result is often ignored. What is its significance? (b) Draw a scale diagram showing the image formed of an insect 2.00 cm long (yuk!) lying on the optic axis and having its head positioned 10.0 cm in front of a converging lens of focal length 6.00 cm. Be sure to show the position of the head of the insect on both the object and the image. What is the actual longitudinal magnification of the insect? (c) What is the lateral magnification of the insect's head? of its tail? Show this distortion in your sketch. (d) The equation derived in part (a) is valid for objects with dimensions small relative to f. If you use the average position of the insect in part (b) as o, what percentage error does the formula of part (a) give?

33. Two converging lenses with a common optic axis and the secondary focal point of the first coincident with the primary focal point of the second is known as an *afocal* arrangement. A pair of afocal lenses can be used as a *beam expander* to change the diameter of a collimated beam of light entering along the optic axis. If the lenses have focal lengths f_1 and f_2 and the entering beam has diameter d, find the expression for the diameter D of the exiting beam.

34. A converging lens of focal length f is to be used to create a real image of a real object. Using the calculus, find the minimum distance between the object and the image.

35. A converging lens ($f_1 = 12.0$ cm) is separated by 50.0 cm from a diverging lens ($f_2 = -10.0$ cm) as shown in Figure 36.P7. An object 1.0 cm high is placed 25.0 cm to the left of the first lens. Find the position of the final image and its height. Complete a ray diagram and characterize the image.

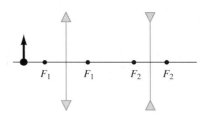

$\vert\blacktriangleleft$25 cm$\blacktriangleright\vert\blacktriangleleft$——50 cm——$\blacktriangleright\vert$ **FIGURE 36.P7   Problem 35**

36. A diverging lens ($f_1 = -10.0$ cm) is separated by 14.0 cm from a converging lens ($f_2 = 8.00$ cm). An object 1.0 cm high is placed 15.0 cm to the left of the first lens. Find the position and height of the final image as viewed from the right of the second lens. Complete a ray diagram and characterize the image.

37. A diverging lens ($f_1 = -10.0$ cm) is separated by 5.00 cm from a converging lens ($f_2 = 15.0$ cm). An object 1.0 cm high is placed 15.0 cm to the left of the first lens. Find the position and height of the final image as viewed from the right of the second lens. Complete a ray diagram and characterize the image.

36.6 The Eye and Simple Optical Instruments (Optional)

38. A lens of focal length $f = 10.0$ cm is used as a magnifying glass to examine the fine print of a legal contract. (a) What are the largest and smallest angular magnifications that can be obtained using this lens? (b) How far from the page is the lens in each of these two cases?

39. A fisherman tying flies desires a magnifier that produces an angular magnification of 2.00 when used with eyes in the relaxed (unaccommodated) state. What focal length lens should be used? How close to the fly does the sportsman's eyes have to be?

40. A microscope objective with focal length 4.00 mm is used in a simple microscope with tube length L of 10.0 cm and a $5.00\times$ eyepiece. (The designation $5.00\times$ indicates that the eyepiece has an angular magnification of 5.00.) (a) What is the overall magnification of the microscope? (b) What is the focal length of the eyepiece? (c) What is the *working distance* of the objective? (The working distance is defined as the distance from the objective to the object when the object is in focus.)

41. A student in physics lab has two converging lenses with focal lengths 5.00 cm and 10.0 cm at her lab station. She decides to make an impromptu microscope to examine a ladybug crawling across the lab bench. She quickly finds that putting the 5 cm lens closer than 6.00 cm to the bug results in unacceptable aberration. (a) With the ladybug 6.00 cm from the first lens, what is the maximum magnification possible if she positions the two lenses as in a simple "microscope"? (b) If she places the two lenses in contact and then places the combination 6.00 cm from the ladybug and views the resulting image, what is the magnification?

42. A student decides to use the lenses described in Problem 41 to make a telescope. She looks out the window at a clocktower on the far side of campus. (a) What is the magnification of the telescope with the 5.00 cm lens used as the eyepiece? (b) How far apart are the lenses when the distant clock is in focus? (c) To what values do the answers to parts (a) and (b) change when the order of the lenses is changed?

43. The planet Jupiter subtends an angle of about 50 arc seconds when viewed from earth. (a) How far away from your eye would a 0.50-mm (diameter) pencil lead be if it were to subtend the same angle? (b) What is the angular size of Jupiter if viewed through a telescope with an objective having focal length of 75.0 cm with an eyepiece of focal length 3.0 cm? (c) How far away would the pencil lead be if it had the angular size calculated in part (b) when viewed with the naked eye?

44. An elderly person's near point has receded to 1.10 m. (a) What focal length must a contact lens have to restore the near point to 25 cm? Model the cornea and lens of the eye as a single lens located 20.0 mm from the retina. Assume this model eye lens is in contact with the contact lens. (b) If this person's eyes focus light from infinity exactly on to the retina when no external lens is used, how far away is the most distant object on which the person is able to focus sharply when wearing the contacts? Are such contacts a viable substitute for reading glasses?

45. When sound waves are introduced into the earth, perhaps by an explosion of dynamite, they obey the same laws of reflection and refraction as do light rays. *Seismic surveys* a linear array of *geophones* (seismic wave detectors) positioned at regular intervals (Fig. 36.P8). Usually, the velocity of sound v_1 in the unconsolidated region immediately below the earth's surface is much lower than the sound velocity v_2 in the rock layers below. Figure 36.P8 shows two (of many) sound rays that travel from a dynamite source and eventually are detected by the same geophone. Ray *SRD* is a direct reflection. Ray *SABD* represents the ray that encounters the high-velocity layer at the critical angle so that it is refracted along the boundary. As this critically refracted ray travels along this boundary it produces waves that return to the surface, one of which is *BD*. (a) Show that if the depth to the bedrock is d, the time for the ray to travel the critically refracted path *SABD* is

$$t = \frac{x}{v_2} + \frac{2d}{v_1}\cos(\theta_c)$$

(b) Show that for

$$x > 2d\sqrt{\frac{v_1 + v_2}{v_2 - v_1}}$$

the critically refracted ray arrives before the directly reflected ray.

36.8 Numerical Methods for Paraxial Ray Tracing (Optional)

46. The thin-lens equation for converging lenses and convex mirrors

$$\frac{1}{o} + \frac{1}{i} = \frac{1}{f}$$

can be "normalized" by dividing all distances by the focal length f. This procedure results in an equation of the form

$$\frac{1}{u} + \frac{1}{v} = 1$$

where u is the object distance and v is the image distance, both measured in units of f. Use a spreadsheet or computer program to calculate the value of v for values of u ranging from $u = -3$ to $+3$ in steps of 0.1. Plot a graph with v on the vertical axis and u on the horizontal. Describe the character (real or virtual) of the object and image in each quadrant of your graph. Calculate and display on your graph the transverse magnification for all integer values of u and for all integer values of $v > -3$ when $0 < u < 1$. (See Albert A. Bartlett, "Image formation in lenses and mirrors," *The Physics Teacher,* May (1976).)

47. The thin-lens equation for diverging lenses and concave mirrors

$$\frac{1}{o} + \frac{1}{i} = \frac{1}{f}$$

can be "normalized" by dividing all distances by the magnitude of focal length $|f|$. This procedure results in an equation of the form

$$\frac{1}{u} + \frac{1}{v} = -1$$

where u is the object distance and v is the image distance, both measured in units of $|f|$. Use a spreadsheet or computer program to calculate the value of v for values of u ranging from $u = -3$ to $+3$ in steps of 0.1. Plot a graph with v on the vertical axis and u on the horizontal. Describe the character (real or virtual) of the object and image in each quadrant of your graph. Calculate and display on your graph the transverse magnification for all integer values of u and for all integer values of $v > -3$ when $0 < u < 1$. (See Albert A. Bartlett, "Image formation in lenses and mirrors," *The Physics Teacher,* May (1976).)

48. Alter the input data of the spreadsheet template of Example 36.14 to trace the rays from a 0.5-cm high object positioned in front of three lenses with focal lengths of 4.0 cm, 6.0 cm, and −10.0 cm located at distances of 5.0 cm, 15.0 cm, and 25.0 cm, respectively, from the object. On the ray tracing sketch an object and all real images (if any). Also, using dotted lines, backtrack the final rays and locate the image seen by an observer located to the right of all three lenses.

49. Using formulas from Table 36.4, alter the spreadsheet template of Example 36.14 to trace rays through a thick lens. Demonstrate your finished template by tracing rays from a 1-cm high object located 5.0 cm from a convex surface of radius 40.0 cm. Let the second surface be located 4.00 cm from the first surface and also be convex outward with a radius of 50.0 cm. Use 1.60 as the index of refraction of the thick lens. Using dotted lines, backtrack the final rays to locate the image. Sketch the object and the image on your printed output.

50. Write a computer program to produce ray-tracing diagrams similar to those produced in the spreadsheet template of Example 36.14.

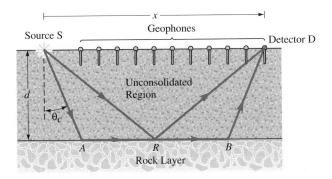

FIGURE 36.P8 Seismic waves used for geophysical exploration obey the same laws of reflection and refraction as do light waves. See Problem 45.

51. A more realistic model for the eye can be constructed from the following parameters:

SURFACE OR SPACE	DISTANCE FROM CORNEA VERTEX (mm)	RADIUS OF CURVATURE (mm)	INDEX OF REFRACTION	COMMENT
Cornea	0	+8	—	Assumed infinitely thin
Aqueous humor			1.336	Fills space from cornea to lens
Lens			1.42	
front surface	3.6	+10		For relaxed (unaccommodated) eye
front surface	3.6	+6		Fully accommodated eye
back surface	7.2	−6		
Vitreous humor			1.336	Fills space from lens to retina
Retina	22.4			

Modify the spreadsheet template of Example 36.14 to trace rays through the model of the eye described above. Show at least four rays at different heights parallel to the optic axis incoming from air. Does this model eye focus parallel rays at the retina for the unaccommodated radius of curvature given? Where does the fully accommodated eye focus parallel rays? (Model adapted from Matthew Alpern, "The eyes and vision," *Handbook of Optics*, Table 1, Section 12 (McGraw-Hill, NY, 1978).)

General Problems

52. Use the data from the table in Problem 51 to calculate the focal length of the lens of a fully accommodated eye in vivo, that is, situated as it is between the vitreous and aqueous humors.

53. A goldfish inside a spherical bowl sees a Cheshire cat through the side of the bowl. If the bowl has a diameter of 25.0 cm and the cat's eye is 10.0 cm from the bowl, where is the image of the cat's eye as seen by the goldfish? Ignore the bowl's thin layer of glass.

54. A novelty aquarium is built so that one glass face ($n = 1.50$) is actually a planoconvex lens. The convex face with radius of curvature 10.0 cm is on the outer side of the aquarium, which is filled with water ($n = 1.33$). To a viewer on the outside, what is the apparent length of a 1.00-cm long tetra swimming parallel to and 25.0 cm from the inside face of the aquarium? [Hint: When two different media of refractive indices n_1 and n_3 are on opposite sides of a lens of index n_2, the lens maker's formula becomes

$$\frac{n_1}{o} + \frac{n_3}{i} = \frac{n_2 - n_1}{R_1} + \frac{n_3 - n_2}{R_2}$$

55. A compound magnifier is constructed of two planoconvex lenses each with focal length 24.0 cm and separated by 18.0 cm. (a) Where should an object be located so that rays exiting the magnifier are parallel? (b) What is the magnification of the magnifier? (c) Is the image upright or inverted?

56. A **Cassegrain** telescope has a geometry like that shown in Figure 36.P9. Suppose the primary mirror has radius of curvature of

4.00 m, and the secondary mirror has a 130. cm radius of curvature and is located 1.50 m in front of the primary mirror. (a) What is the overall focal length of this mirror combination? The overall focal length of the mirror combination is the distance (measured along the optic axis) that parallel rays entering the telescope travel after stiking the primary mirror before being brought to a focus. (b) If the image is now viewed with an eyepiece of 10.0-cm focal length, what is the angular magnification of the telescope?

57. If you hold this text at a distance of 25.0 cm from your eye, how large is the image of the letter "I" on your retina? (Use the model for the eye described in Example 36.13.)

58. A simple camera focuses images on the film plane by varying the distance between the lens and the film. (a) If a camera with a lens of focal length 50.0 mm is to focus at various times on objects located between infinity and 65 cm, what must be the range of distances over which the lens must move? (b) A 35-mm camera uses film that is 35 mm wide, but the image size is about 24 mm × 36 mm. What is the maximum angular field of view of the camera when focused at infinity? when focused at 65 cm? (c) What are the answers to the questions in part (b) if the standard 50-mm lens is replaced with a 35.0-mm focal length wide-angle lens?

59. A spherometer is a device used to measure the radius of curvature R of spherical surfaces (see Fig. 36.P5). It consists of three legs positioned in an equilateral triangle with sides of length y. A center probe is raised or lowered by means of a micrometer drive. The spherometer is first placed on a flat surface and the zero position of the micrometer noted. The instrument is then placed on a spherical surface and the micrometer is adjusted until the center probe just touches the surface. The displacement z of the micrometer between the two readings is known as the *sagitta*. (a) Make a simple argument to show that $y^2 + (z - R)^2 = R^2$ so that

$$R = \frac{z^2 + y^2}{2z}$$

(b) A student records the following micrometer positions in millimeters: flat surface 11.222, convex face 8.716, concave face 12.473, when using a spherometer with y given by the manufacturer as 50.0 mm. The student subsequently measures the focal length of the lens to be 186 cm. What is the index of refraction of the material from which the lens is constructed?

60. Verify the claim in the text that, in order to invert the image of a celestial telescope using a converging lens of focal length f, the length of the telescope must be lengthened by at least $4f$.

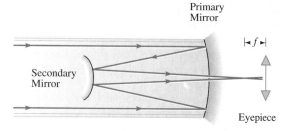

FIGURE 36.P9   A Cassegrain telescope. See Problem 56.

★**61.** A biconvex lens ($f_1 = 10.0$ cm) is placed 40.0 cm in front of a concave mirror ($f_2 = 7.50$ cm) as shown in Figure 36.P10. An object 2.00 cm high is placed 20.0 cm to the left of the lens. Using the lens and mirror equations and a ray diagram, find the locations of three images: (a) the image formed by the lens as rays travel to the right, (b) the image formed after rays reflect from the mirror, and (c) the final image after the leftward-traveling rays once again pass through the lens. Complete the ray diagram and characterize each image and object.

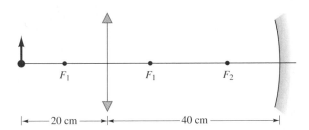

FIGURE 36.P10   Problem 61

★**62.** A biconvex lens ($f_1 = 25.0$ cm) is placed 105. cm in front of a convex mirror ($f_2 = -25.0$ cm) as shown in Figure 36.P11. A 2.00-cm high object is placed 35.0 cm to the left of the lens. Using the lens and mirror equations and a ray diagram, find the location of three images: (a) the image formed by the lens as rays travel to the right, (b) the image formed after rays reflect from the mirror, and (c) the final image after the leftward-traveling rays once again pass through the lens. Complete the ray diagram and characterize each image and object.

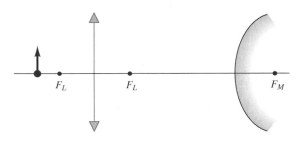

FIGURE 36.P11   Problem 62

★**63.** A biconcave lens ($f_1 = -25.0$ cm) is placed 20.0 cm in front of a concave mirror ($f_2 = 5.00$ cm) as shown in Figure 36.P12. A 2.00-cm high object is placed 15.0 cm to the left of the lens. Using the lens and mirror equations and a ray diagram, find the location of three images: (a) the image formed by the lens as rays travel to the right, (b) the image formed after rays reflect from the mirror, and (c) the final image after the leftward-traveling rays once again pass through the lens. Complete the ray diagram and characterize each image and object.

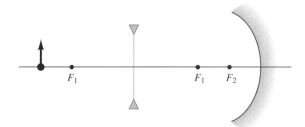

FIGURE 36.P12   Problem 63

★**64.** A biconcave lens ($f_1 = -25$ 0 cm) is placed 60.0 cm in front of a concave mirror ($f_2 = 45.0$ cm) as shown in Figure 36.P13. Note that the primary focus of the mirror is to the left of the secondary focus of the lens. A 2.00-cm high object is placed 35.0 cm to the left of the lens. Using the lens and mirror equations and a ray diagram, find the location of three images: (a) the image formed by the lens as rays travel to the right, (b) the image formed after rays reflect from the mirror, and (c) the final image after the leftward-traveling rays once again pass through the lens. Complete the ray diagram and characterize each image and object.

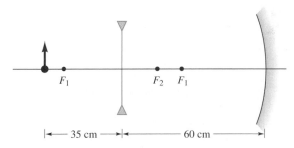

FIGURE 36.P13   Problem 64

65. Figure 36.P14 shows six lens each having a different focal length. The focal length values are ± 167 mm, ± 333 mm, and ± 500 mm. List the focal length of each lens in the photograph in order from left to right.

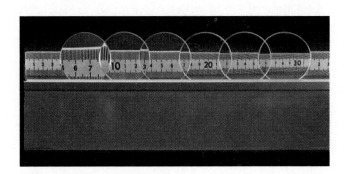

FIGURE 36.P14   Problem 65

Interference of Light

In this chapter you should learn

- how light waves interfere.
- how antireflection coatings work.
- the difference between coherent and incoherent light.
- the details of two-source interference patterns.

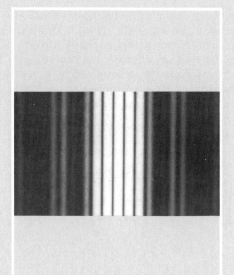

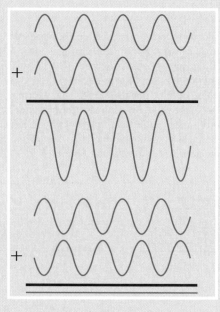

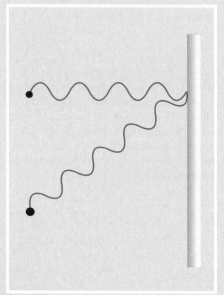

In this chapter we will focus on the wave model for light. The fact that light behaves like a wave is not at all obvious to the casual observer. We all know that sound is a wave. One noticeable characteristic of sound waves is their ability to bend around obstacles. This bending enables us to hear sound from sources that are around the corner from us. So, if light is a wave, why can't we *see* around corners? The answer to this question is related to the fact that the wavelength of visible light is about a million times smaller than the wavelength of sound. If we are to discover the wave properties of light we must look closely at the edges of objects and consider openings that are very small.

Nonetheless, clues that light is wavelike can be found if we keep our eyes open. Take your index finger and middle finger and put them quite close together to form as narrow a slit as you can. Hold the slit right next to your eye and look at a distant light bulb or fluorescent lamp. If you look very carefully, you will see fine dark lines in the bright slit. We will see that such lines are evidence for the wave nature of light. If you have ever looked at a pool of water with a thin layer of oil or gasoline on its surface, you have likely noticed bands of swirling colors indicating the oil's presence. Such colors are also evidence for the wave properties of light. In this chapter we will explore these and other manifestations of light's wave nature.

37.1 Interference

One of the hallmarks of wave behavior is summarized in the **principle of superposition.** This principle reminds us that the wave equation (Eq. (34.12)) is linear, so that the sum of any two solutions is also a solution. The physical consequence of this mathematical property is that when two light waves overlap, the electric field **E** of the resultant wave is the sum of the E-fields due to the two individual waves. A key point to remember here is that it is the *electric field* that obeys the principle of superposition. The irradiance, however, is proportional to the square of the resultant electric field. (See Eq. (34.26).) Thus, to determine the irradiance resulting from two overlapping light waves, we need first to add their electric fields and then to calculate the irradiance by squaring the resultant field. In the paragraphs below we will follow this plan and see where it takes us.

Two-Source Interference

Consider two waves with the same amplitude, frequency, and phase. As shown in Figure 37.1(a), when the two waves are added in accordance with the principle of superposition, the result is a wave with twice the original amplitude. Two such waves are said to interfere constructively or result in **constructive interference.** In Chapter 34 we saw that the irradiance of a light wave is proportional to the square of its electric field amplitude. As a consequence, the irradiance of the resultant wave is *four* times that of either wave alone.

On the other hand, consider the same two waves when their relative phase differs by π radians, that is, waves that are out of phase such as shown in Figure 37.1(b). When these two waves are added, the resulting electric field is zero—a phenomenon called **destructive interference.** The implication is rather marvelous. It suggests that we can shine two bright lights on the same place and the result will be darkness! This result is certainly counterintuitive and its observation gave great impetus to the wave model for light. In the remainder of this chapter we will introduce experiments whereby constructive and destructive interference can be observed. In the remainder of this section we will restrict ourselves to obtaining relations useful for predicting the location of bright spots and dark spots. In later sections of this chapter we will examine the wave equations in greater detail and obtain descriptions for the irradiance at all points in the observation plane.

Figure 37.2 shows the result of shining light from two monochromatic point sources onto a screen. The pattern of light and dark bands is called an **interference pattern.** The bands themselves are called **fringes.** One speaks of bright fringes and dark fringes. A method for producing an interference pattern such as that shown in Figure 37.2 is known as **Young's double-slit experiment.** Monochromatic light from a small source is incident

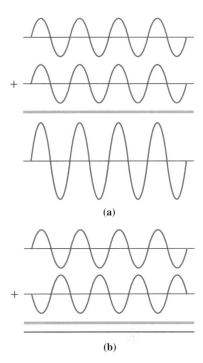

(a)

(b)

FIGURE 37.1

(a) Constructive interference. The two waves have the same wavelength and amplitude and are in phase. When added, they result in a wave with twice the amplitude of either wave alone. (b) Destructive interference. Waves of equal amplitude and wavelength that are out of phase. When such waves are added together, the sum is zero.

FIGURE 37.2

Part of the irradiance pattern produced when two monochromatic point-sources both shine on the same screen. The bands are known as fringes. The pattern of fringes is called an interference pattern.

Concept Question 1
Imagine you have two ropes, each painted with alternating red and blue stripes. The length of a single stripe on each rope is 0.50 m. You now carefully tie each rope to a different fence post in an identical fashion so that a red stripe is just starting as the rope leaves the post. You now hold both ropes in one hand and back away to a point 6.25 m from post A and 4.25 m from post B. If both ropes are stretched taut, what color are the sections of rope in each hand? What does this thought experiment have to do with the interference of light? Describe how you could model destructive interference using the ropes.

on two narrow slits separated by a distance d as shown in Figure 37.3. The interference pattern similar to that in Figure 37.1 appears on a screen located at distance L from the slits. The light passing through each slit is spread in all directions as predicted by Huygens' principle (Chapter 35, Section 2). In the ray model we picture a multitude of rays emanating from each slit in every direction.

It is easy to understand the origin of the fringes by applying the principle of superposition to two of these rays that converge to point P a distance y from the screen's center as shown in Figure 37.4. We have drawn a line from slit A to point C in such a way that distances AP and CP are equal. We see that the ray from slit B must travel an extra distance Δl equal to the distance from B to C. When light waves leave slits A and B, they are in phase. If the path difference Δl is an integer multiple of the light's wavelength λ, the waves meet in phase and combine to form an irradiance maximum as in Figure 37.1(a). That is, the condition

$$m\lambda = \Delta l, \qquad m = 0, 1, 2, 3, \ldots$$

results in constructive interference at point P. On the other hand, if the path length difference is equal to an odd multiple of $\lambda/2$, the situation at point P resembles Figure 37.1(b), and destructive interference results.

We wish to obtain a relation that predicts where constructive and destructive interference occurs on the screen. To find this relation we approximate angle ACB of Figure 37.4 as a right angle so that we can write $\Delta l = d \sin(\theta)$. This approximation is quite accurate for the usual situation where $d \ll L$. For small θ we can see from Figure 37.4 that $\sin(\theta) \approx \tan(\theta) = y/L$. Hence, the conditions for constructive and destructive interference can be written

$$m\lambda = d \sin(\theta) = d\frac{y}{L}, \qquad m = 0, 1, 2, 3, \ldots \tag{37.1a}$$

(Irradiance maximum, Young's double slit)

$$\left(m - \frac{1}{2}\right)\lambda = d \sin(\theta) = d\frac{y}{L}, \qquad m = 1, 2, 3, \ldots \tag{37.1b}$$

(Irradiance minimum, Young's double slit)

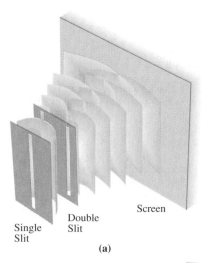

Single Slit   Double Slit   Screen

(a)

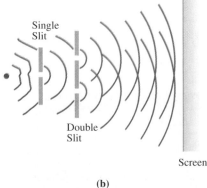

Single Slit

Double Slit

Screen

(b)

FIGURE 37.3

One method for producing two monochromatic sources with a constant phase relationship is to illuminate a pair of slits with a small monochromatic source. (a) Cylindrical wavefronts are formed by the slits. (b) Top view.

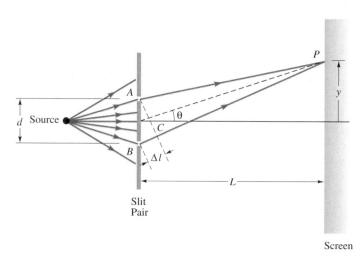

Slit Pair

Screen

FIGURE 37.4

Two of the many rays emanating from slits A and B are pictured meeting at point P on a screen located distance L from the slits. Whether P is bright or dark depends on Δl the difference in path lengths of the two waves from A and B.

Note that the expressions in Equations (37.1) describe maxima and minima for positive y only; however, the interference pattern is symmetrical about $y = 0$.

Our analysis for Young's double-slit experiment is limited because it doesn't describe the irradiance at points between these maxima and minima. The expression for the complete irradiance pattern is developed in Section 37.2.

EXAMPLE 37.1 *Young's Double Slit*

The expanded beam from a helium-neon laser is allowed to fall on two equidistant small slits separated by 0.250 mm. If a screen is placed 3.00 m from the slits, find the distance from the center bright spot of the pattern to the third bright spot from the center. (Count the center spot as $m = 0$.)

SOLUTION The wavelength of the He-Ne laser is 633 nm. (a) According to Equation (37.1a) the third maximum to the right of the center bright spot occurs when

$$d\frac{y}{L} = m\lambda, \qquad m = 3$$

Solving for y, we obtain

$$y = \frac{m\lambda L}{d} = \frac{3(633 \times 10^{-9}\ \text{m})(3.00\ \text{m})}{(0.250 \times 10^{-3}\ \text{m})} = 2.28 \times 10^{-2}\ \text{m} = 2.28\ \text{cm} \quad \blacktriangleleft$$

(a)

(b)

FIGURE 37.5

The reflection of (a) white light and (b) monochromatic light from a soap film.

Thin-Film Interference

You are likely to have seen the bands of colors due to a thin layer of oil or gas floating on a puddle of water or the colored bands in a soap film like those shown in Figure 37.5. Such patterns result from the constructive and destructive interference of waves reflected from the front and back surfaces of the film layer. In Figure 37.6 we show the series of reflections of a single ray encountering a thin layer of index of refraction n bounded on both sides by a dielectric of index n_o. This figure might represent the cross section of a soap film in air for example. In Chapter 35 we saw that a ray is partially reflected and partially transmitted each time it encounters the boundary between two dielectrics with different refractive indices. For the process shown in Figure 37.6 the result is an infinite number of rays reflecting upward. The first reflection is from the top surface. The remaining rays exit the top surface after an odd number of reflections (1, 3, 5, . . .) inside the film. When the reflection coefficient is small (see Section 35.4), the majority of the energy in the reflected beam can be accounted for by examining only two waves: one reflected from the top surface and the other ray exiting after a single internal reflection. We begin by examining this *two-reflection model*. As in the previous section we discover that combining two similar waves can lead to constructive and destructive interference. In Section 37.2 we will describe corrections to this result when multiple reflections are taken into account.

Let's look at the conditions leading to constructive and destructive interference between the two waves reflected from the front and back of a thin film. Because the two reflected waves started as one, they were originally in phase. There are two ways these rays can get out of step before recombining: (1) a difference in path lengths traveled between separation and recombination or (2) a phase change occurring on reflection. We examine the second possibility first.

Solutions to Maxwell's equations show that when a wave is reflected, its phase is affected differently depending on whether the reflection is internal or external. When an electromagnetic wave is partially reflected on trying to enter a medium of higher index of refraction (external reflection), the reflected wave undergoes a 180° change of phase as

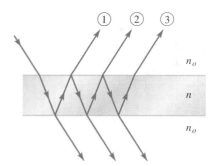

FIGURE 37.6

When a ray of light strikes a dielectric film, it reflects an infinite number of times. The majority of the reflected power is contained in the first two reflected rays, which are of nearly equal power. For a film of refractive index $n = 1.33$, ray 1 reflected from the upper surface contains about 2.01% of the power of the incoming ray. Ray 2 is reflected once from the lower surface and contains $(98\%)(2.01\%) \approx 1.97\%$ of the power of the incoming ray. Ray 3, which suffers three reflections, contains only $(98\%)(2.01\%)(2.01\%)(2.01\%) \approx 0.000\ 8\%$ of the power of the incoming ray.

Waves reflected at the boundary into a dielectric medium with higher index of refraction or a metallic surface undergo a 180° = π rad change of phase on reflection. Waves reflected at the boundary of a dielectric medium with lower index of refraction undergo no change of phase when the angle of incidence is less than the critical angle. The rule is sometimes remembered using the phrase, "low to high, a change of pi."

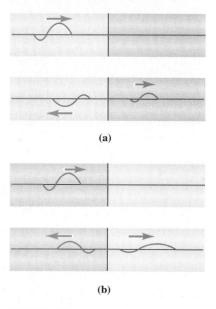

(a)

(b)

FIGURE 37.7

(a) When an electromagnetic wave strikes a metallic surface or the boundary leading to a dielectric medium of higher refractive index, the reflected portion undergoes a phase inversion. (b) When an electromagnetic wave strikes the boundary leading to a dielectric medium of lower refractive index, the phase of the reflected wave is not inverted.

Concept Question 2
A certain monochromatic light has a wavelength of 0.50 μm in air. A thin film has a thickness of 3.00 μm and an index of refraction of 2.00. As light travels in one direction through the film, how many wavelengths of the light fit between the surfaces of the film?

shown in Figure 37.7(a). Contrarily, a wave partially reflected while trying to enter a medium of lower index of refraction does not undergo a phase change as illustrated in Figure 37.7(b). This is a classical wave phenomenon and, as discussed in Chapter 15, analogous properties hold for mechanical waves.

Now let's return to the discussion of reflections from a thin dielectric film. To be concrete (and because "dielectric film" sounds stuffy) we consider a thin film of soap solution stretched across the face of a circular ring. Let's suppose that the film thickness is much less than a wavelength of light. In this case the two reflected waves have a phase difference $\Delta\phi_{\text{refl}} = 180° = \pi$ rad when they recombine. Thus, the total power in the reflected wave is zero. This result is evident at the top of the soap bubble photographed in reflected light as shown in Figure 37.5. The reflected waves are completely out of phase, and destructive interference results.

If the thin film is "thick" compared to the wavelength of the light used, an additional phase difference $\Delta\phi_{\text{path length}}$ arises due to the extra distance traveled by the wave that enters the film. For convenience we continue to refer to the wave that reflects from the first surface as the "first wave" and the wave that enters the film and reflects from the second surface as the "second wave." To see the effect of film thickness, consider a monochromatic light wave that strikes a film of thickness equal to one-quarter of the wavelength of light. If the light strikes the film at normal incidence, the second wave traverses the film thickness twice and thus travels farther than the first wave by a distance equivalent to one-half wavelength. When they reunite, the total phase difference between the two waves is 180°(reflection) + 180°(path length) = 360° = 2π rad. That is, the two waves are *in phase,* and constructive interference results.

One subtlety must be emphasized. The film in the discussion above was one-quarter of a wavelength thick. But the relevant wavelength is *not* the wavelength λ_o of the light in air. Rather, it is the wavelength λ_{film} of the light *in the soap film.* For the particular case discussed, we require that the film thickness $t = 0.250\lambda_{\text{film}} = 0.250(\lambda_o/n_{\text{film}})$. This same relationship can be written as $n_{\text{film}}t_{\text{film}} = 0.25\lambda_o$. The product nt is known as the **optical path length.** The optical path length is the equivalent distance the wave would have to travel in vacuum so as to achieve the same phase change that results from traveling a distance t in a medium with refractive index n.

As a general rule the relative phase change between two waves is proportional to the *difference* in their optical path lengths. In fact, a simple proportion can be used to calculate the phase difference $\Delta\phi_{\text{path length}}$

or

$$\frac{\Delta\phi_{\text{path length}}}{2\pi} = \frac{OPD}{\lambda_o}$$

$$\Delta\phi_{\text{path length}} = 2\pi\frac{OPD}{\lambda_o} \qquad (37.2)$$

where the *OPD* is the **o**ptical **p**ath **d**ifference, that is, the difference in optical path lengths nt of the two waves. Once again λ_o is the free-space wavelength of the light. The total phase difference $\Delta\phi_{\text{tot}}$ between the two waves is the sum of that due to reflection and that due to optical path-length difference.

We can now understand the sequence of colors in the soap film pictured in Figure 37.5. Near the top of the figure, the film is so thin that the reflected waves are out of phase due to the opposite phase behavior of the waves on internal and external reflection. But the thickness of the soap film increases with distance down the film (toward the bottom of the figure). When the optical thickness becomes equal to one-quarter the wavelength of blue light, reflected beams of that color are in phase and create a bright reflection. Slightly farther down, the film reaches a greater optical thickness equal to one-quarter of the wavelength of yellow light. At that point yellow light is strongly reflected. This process continues on down the film and through the spectrum to the longest visible wavelength, red. Eventually, the process repeats as the film thickness reaches three-quarters of the wavelength for blue light.

Concept Question 3
Figure 37.5 records light reflected off a soap film. If one were looking through the soap film back toward the light source, how would the topmost portion of the film appear, that is, the portion of the film that is black in Figure 37.5?

EXAMPLE 37.2 *Is This What Vampires Use?*

A layer of magnesium fluoride ($n_1 = 1.38$) is to be applied as an antireflection coating to a camera lens with refractive index $n_2 = 1.53$. What thicknesses of MgF$_2$ results in minimum reflection for waves of $\lambda = 550.$ nm?

SOLUTION In order to produce destructive interference between the reflected waves we require that $\Delta\phi_{\text{tot}} = m\pi$, where $m = 1, 3, 5, \ldots$. Because $n_{\text{air}} < n_1 < n_2$, the waves that reflect from the front and the back of the MgF$_2$ coating *both* undergo external reflection. Consequently, both waves undergo a π phase change. However, the *relative* phase change due to these reflections is $\Delta\phi_{\text{refl}} = 0$. The ray reflected from the back surface of the film travels through its thickness t twice so that we are left with the requirement

$$\Delta\phi_{\text{tot}} = \Delta\phi_{\text{refl}} + \Delta\phi_{\text{path length}} = 0 + 2\pi\frac{n_1(2t)}{\lambda_o} = m\pi$$

which leads to thicknesses

$$t = \frac{m\lambda_o}{4n_1} = \frac{m(550.\ \text{nm})}{4(1.38)} = m(99.6\ \text{nm}), \qquad m = 1, 3, \ldots$$

We see that any odd multiple of one-quarter of the wavelength of light in the MgF$_2$ film, that is, any odd multiple of 99.6 nm, will work. In practice, such a film can reduce reflections from 4% to an average of 1% over the visible spectrum. ◀

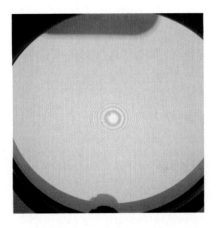

Newton's rings. A glass plate with a spherical surface floats on a thin layer of air above an optically flat surface. An interference pattern is formed between rays reflected from the two surfaces.

EXAMPLE 37.3 *Newton's Rings*

A planoconvex lens with a radius R is placed on an optical flat and illuminated by monochromatic light of wavelength λ. Relate the radius of the ring pattern formed to the sphere's radius of curvature.

SOLUTION A pattern of concentric rings is formed as shown in Figure 37.8. Figure 37.9 shows a cross-sectional view of the spherical and flat surfaces. The interference pattern is formed from light waves that undergo internal reflection at the spherical surface and superpose with those that undergo external reflection at the flat surface. In this case the thin film is a film of air sandwiched between two layers of glass. Because one ray undergoes an internal reflection and the other ray undergoes an external reflection, we conclude $\Delta\phi_{\text{refl}} = \pi$.

We describe the geometry using a coordinate system with the origin at the bottom edge of the sphere, so that the cross section of the spherical surface in the xy-plane is a circle described by $x^2 + (y - R)^2 = R^2$. The thickness of the air film is equal to the distance y. Simplifying the equation of the circle, we find

$$x^2 = (2R - y)y$$

Because $2R \gg y$, we approximate $2R - y \approx 2R$. The previous equation therefore leads to a film thickness $y = x^2/2R$. Assuming that the flat and the lens are nearly in contact at the center, the center spot is dark because $\Delta\phi_{\text{refl}} = \pi$. A new dark ring is formed each time y increases to a multiple of $\lambda/2$:

$$\frac{x^2}{2R} = m\frac{\lambda}{2}, \qquad m = 1, 2, 3, \ldots$$

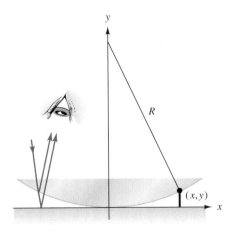

FIGURE 37.9

Geometry showing an air wedge forming the variable thickness film that results in the interference pattern known as Newton's rings. The angle from vertical of the sample light rays has been exaggerated for clarity.

Thus, the radius of the mth dark ring is

$$x = \sqrt{m\lambda R}$$

Note, when Newton's rings are formed by "floating" a lens on an optical flat, the actual order of any given fringe is not usually known because the curved surface seldom actually touches the reference flat. Therefore, the proper way to calculate the radius of curvature of a spherical surface from the ring pattern is to plot x^2 versus any set of consecutive integers. The straight line formed has a slope equal to $R\lambda$ regardless of the actual values assigned to the orders m. (See Problem 35.) ◄

EXAMPLE 37.4 Lloyd's Mirror

Figure 37.10 shows how a monochromatic point-source and a front-surface mirror can be used to create a double source that can then be used to form an interference pattern. (a) Describe the effect on the interference pattern caused by the fact that a π phase change occurs when a wave is reflected from a metallic surface. (b) If the point-source emits light of wavelength 514.5 nm and is located 1.00 mm above the mirror, what is the spacing between the fringe maxima on a screen 1.50 m away?

SOLUTION (a) The interference pattern is formed between the actual point source of wavelength 514.5 nm and its mirror image located 1.00 mm below the mirror surface. The effect of the reflection is to introduce a π phase difference between the real and the image sources. This phase difference results in an interchange of the maxima and minima relative to the interference pattern created when the two sources are in phase. Therefore, the positions of *bright* spots are given by

$$d\frac{y}{L} = \left(m - \frac{1}{2}\right)\lambda, \qquad m = 1,\ 2,\ 3,\ \ldots$$

(b) Two consecutive bright spots at locations y' and y satisfy the relations

$$d\frac{y'}{L} = \left[(m+1) - \frac{1}{2}\right]\lambda \qquad \text{and} \qquad d\frac{y}{L} = \left(m - \frac{1}{2}\right)\lambda$$

On subtracting these relations, we have for $\Delta y = y' - y$

$$d\frac{\Delta y}{L} = \lambda$$

leading to

$$\Delta y = \frac{\lambda L}{d} = \frac{(514.5 \times 10^{-9}\ \text{m})(1.50\ \text{m})}{2.00 \times 10^{-3}\ \text{m}} = 3.86 \times 10^{-4}\ \text{m} = 0.386\ \text{mm} \quad ◄$$

37.2 Irradiance for Simple Interference Patterns

In the previous section we investigated the conditions for irradiance maxima and minima created by two-source and thin-film interference. In this section we take a second look at these same phenomena to obtain a more complete description of the full interference pattern. In doing so we obtain insights that allow us to understand details of the patterns created by multiple slits.

In Chapter 34 we learned that light is an electromagnetic wave. In free space such waves consist of oscillating electric and magnetic fields that are perpendicular to each

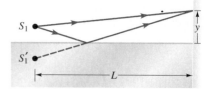

FIGURE 37.10

Lloyd's mirror. The actual source and its mirror image form an interference pattern on the screen. The areas of light and dark are reversed from those created by Young's experiment (see Example 37.1). The image-source appears to be exactly out of phase with the actual source because waves that appear to come from the image-source had their phases reversed when they reflected from the metallic mirror surface.

Concept Question 4
Reread Concept Question 1. If you were going to modify the rope arrangement described in that question so that it became analogous to the Lloyd's mirror experiment of Example 37.4, what would you change?

other and to the wave propagation direction. At a particular point in space (say, $x = 0$, for simplicity) the plane waves of Chapter 34 can be described by $\mathbf{E}(t) = \mathbf{E}_o \sin[k(0) - \omega t] = -\mathbf{E}_o \sin(\omega t)$. The $\mathbf{E}$ vector and the direction of propagation define the plane of polarization as described in Chapter 34. Our eyes, and indeed most optical detectors, cannot respond to the E-field directly. Instead, our eyes are sensitive to the average power transmitted by the wave. The important quantity for our purposes is the power per unit area known as the irradiance I. We saw in Chapter 34 (Eq. (34.26)) that, for a single wave in free space, the irradiance was given by the time-averaged value of the Poynting vector

$$I = \frac{1}{2c\mu_o} E_o^2 = \frac{c\epsilon_o}{2} E_o^2$$

where the speed of light $c = 1/\sqrt{\mu_o \epsilon_o}$. The key point here is that the irradiance is proportional to the *square* of the amplitude of the electric field.

Without worrying about the source, consider once again two light waves overlapping in some region of space. We designate the E-fields of the two waves as $\mathbf{E}$ and $\mathbf{E}'$

$$\mathbf{E} = \mathbf{E}_o \sin(\omega t)$$

$$\mathbf{E}' = \mathbf{E}_o' \sin(\omega' t + \phi)$$

These waves are quite general and may have different polarizations, frequencies, and amplitudes. By including ϕ we even allow for a possible difference in phase at time $t = 0$. We now apply the principle of superposition to find the total E-field.

$$\mathbf{E}_{\text{tot}} = \mathbf{E}_o \sin(\omega t) + \mathbf{E}_o' \sin(\omega' t + \phi)$$

To find the total irradiance I_{tot} that results from the superposition of these two waves, we follow the procedures of Chapter 34, Section 3, and calculate

$$I_{\text{tot}} = c\epsilon_o \langle \mathbf{E}_{\text{tot}} \cdot \mathbf{E}_{\text{tot}} \rangle$$

where the angle brackets $\langle \ \rangle$ indicate the time average. Substituting the expression for the total electric field, we find that we must calculate

$$I_{\text{tot}} = c\epsilon_o \langle [\mathbf{E}_o \sin(\omega t) + \mathbf{E}_o' \sin(\omega' t + \phi)] \cdot [\mathbf{E}_o \sin(\omega t) + \mathbf{E}_o' \sin(\omega' t + \phi)] \rangle$$

Preforming the indicated dot products, leaves us with

$$I_{\text{tot}} = c\epsilon_o [E_o^2 \langle \sin^2(\omega t) \rangle + E_o'^2 \langle \sin^2(\omega' t + \phi) \rangle + 2(\mathbf{E}_o \cdot \mathbf{E}_o') \langle \sin(\omega t) \sin(\omega' t + \phi) \rangle]$$

Remember that the frequencies of visible light are enormous. A typical frequency is $\nu > 10^{14}$ Hz, making $\omega > 10^{15}$ rad/s. No detector, including our eye, can respond to variations this rapid. Thus, we detect only the time averages of $\sin^2(\omega t)$ and $\sin^2(\omega' t + \phi)$, which are each $\frac{1}{2}$. As a result, we have

$$I_{\text{tot}} = \frac{c\epsilon_o}{2} E_o^2 + \frac{c\epsilon_o}{2} E_o'^2 + 2c\epsilon_o (\mathbf{E}_o \cdot \mathbf{E}_o') \langle \sin(\omega t) \sin(\omega' t + \phi) \rangle$$

The first two terms on the right-hand side of this equation are the irradiances I and I' that would be produced by each wave separately.

$$I_{\text{tot}} = I + I' + 2c\epsilon_o (\mathbf{E}_o \cdot \mathbf{E}_o') \langle \sin(\omega t) \sin(\omega' t + \phi) \rangle$$

The third term on the right-hand side is called the **interference term,** and that's what this chapter is all about. In what follows we will examine several special cases in detail. First

Concept Question 5

In the expression $\mathbf{E}_o \sin(\omega t)$ representing the electric field at a point $x = 0$ what symbol represents the frequency of the field oscillation? What symbol represents the amplitude of the oscillation? What in this expression indicates the wave's polarization direction?

let's suppose that the polarization directions of the two E-fields are orthogonal. In that case the dot product $\mathbf{E}_o \cdot \mathbf{E}_o' = 0$. Consequently, the total irradiance of the two beams taken together is the simple sum of the irradiance of each beam separately: $I_{\text{tot}} = I + I'$. We hereby declare that case uninteresting. Two beams with orthogonal polarizations do not interfere. Hence, in all that follows in this chapter we assume that the two beams are polarized in the same plane.[1] With this the case, we can drop the vector notation and simply write $\mathbf{E}_o \cdot \mathbf{E}_o' = E_o E_o'$. While we are at it, we may as well use the fact (see Eq. 34.26) that

$$E_o = \sqrt{\frac{2I}{c\epsilon_o}}$$

to write the remaining electric fields in terms of the irradiances caused by each individual source acting alone:

$$I_{\text{tot}} = I + I' + 4\sqrt{II'}\langle \sin(\omega t) \sin(\omega' t + \phi) \rangle$$

As a final assault on this equation, we use the trigonometric identity

$$\sin(A)\sin(B) = \frac{1}{2}[\cos(A - B) - \cos(A + B)]$$

to write

$$I_{\text{tot}} = I + I' + 2\sqrt{II'}\{\langle \cos[\omega t - (\omega' t + \phi)] \rangle - \langle \cos[\omega t + (\omega' t + \phi)] \rangle\}$$

or

$$I_{\text{tot}} = I + I' + 2\sqrt{II'}\{\langle \cos[(\omega - \omega')t - \phi] \rangle - \langle \cos[(\omega + \omega')t + \phi] \rangle\}$$

The second term in the curved braces { } once again describes oscillations with extremely high frequencies $(\omega + \omega')$ for which detectors can register only an average. Because the cosine is not squared, the time average of the term $\cos(\omega t + \omega' t + \phi)$ is zero. This result leaves us with what we call our *master interference equation:*

Master interference equation

$$I_{\text{tot}} = I + I' + 2\sqrt{II'}\langle \cos[(\omega - \omega')t - \phi] \rangle \tag{37.3}$$

Now the fun begins. We look at several special cases for the frequencies ω and ω'.

Coherence

Suppose that the frequencies ω and ω' of the two overlapping light waves described in Equation (37.3) differ greatly. Because each is quite large, their difference is also a large frequency. Thus, the observable time average is again zero, and we recover the simple relation $I_{\text{tot}} = I + I'$. Waves with greatly different frequencies are said to be **incoherent.** Most of the light you encounter each day is made up of incoherent waves. Incandescent light bulbs are incoherent sources. Waves of many different frequencies are emitted by the atoms that constitute the bulb's filament. The irradiances from such sources add in a simple way. If you turn on two identical light bulbs positioned symmetrically relative to this book, this page of the book receives twice the radiant power it would receive from either source alone.

---

[1] Unpolarized light beams can interfere. The point is that the overall interference results from each polarization component interfering with light of the same polarization.

Now let's imagine the opposite extreme. Consider two waves with identical frequencies ω and ω'. For $\omega = \omega'$ our master interference equation becomes

$$I_{\text{tot}} = I + I' + 2\sqrt{II'}\langle\cos(\phi)\rangle \tag{37.4}$$

We now see that the phase difference ϕ between the two waves is critical. If the two waves have the same frequency, *and* their relative phase is constant in time, the waves are said to be **coherent.** Real light sources can never be perfectly coherent.[2] The length of time during which they maintain a stable phase relationship is known as their **coherence time** t_c. The distance the waves travel during the coherence time is known as the longitudinal **coherence length** $l_c = ct_c$. For incoherent light the coherence length is less than one wavelength, that is a fraction of a micron (1 μm $= 10^{-6}$ m). On the other hand, for a stabilized carbon dioxide laser the coherence length can be kilometers.

Optical Beats (Optional)

Examination of the time-dependent term in Equation (37.3) raises the interesting possibility of observing optical beats. That is, if ω and ω' are sufficiently close, is it possible that we might observe a slow sinusoidal variation of irradiance at a frequency of $(\omega - \omega')$? The answer is yes and no. No, if by "observe" we mean see with our eyes using light sources found in most college laboratories. The human eye cannot detect temporal fluctuations much faster than about 30 Hz. Thus, the two angular frequencies ω and ω' could differ by no more than about 200 rad/s. This difference requires that the individual frequencies be monochromatic to at least this extent. That is, the waves would have to have identical frequencies to an accuracy of 200 parts in 10^{14}. Such an accuracy is now possible using especially stabilized lasers available in only a few research laboratories in the world.

On the other hand, electronic detectors are able to observe variations of intensity occurring with frequencies of tens of megahertz. This capability lowers the stability requirements on our two waves to a few parts in 10^7, a level well within reach of even inexpensive lasers. Example 37.5 shows how beat frequencies can be used to determine the number of modes in a helium-neon laser. Beats between Doppler-shifted waves reflected from a moving object and an unshifted wave can be used to determine the velocity of the moving object. A similar technique employing radar—and more recently, laser beams—is used by law enforcement agencies to monitor the speed of automobiles. These applications are explored further in the problems at the end of this chapter.

EXAMPLE 37.5 *The Beat of a Really Different Drummer*

Light from a polarized helium-neon laser falls on a photodiode detector able to respond to high frequencies. The electrical output of the detector is input to a spectrum analyzer used to measure the beat frequencies contained in the original light beam. If the laser light contains waves from five different modes separated by 300. MHz as shown in Figure 37.11, what beat frequencies are detected? (A laser "mode" is a wave with an *extremely* narrow range of frequencies, typically 1 MHz for simple He-Ne lasers. These modes are standing electromagnetic waves created between the mirrors on opposite ends of the laser cavity. They are analogous to the standing waves of a vibrating string fixed at both ends.)

SOLUTION We expect beats at frequencies that are the difference between optical frequencies in the laser beam. The lowest beat frequency occurs between adjacent modes and is $\nu = 300.$ MHz. In addition, modes differing in frequency by 2ν also beat against one another. Similarly, beats are found between modes separated by 3ν.

---

[2] The coherence of real sources is limited by their finite spatial extent and a finite spread in the wavelengths emitted. The spread in wavelengths may be due to collisions between atoms, motion of the emitting atoms, and the fundamental uncertainty in the atomic energy levels themselves.

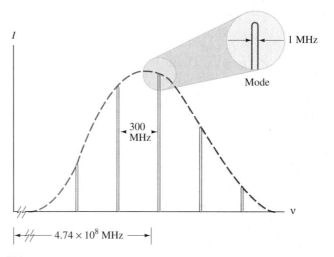

FIGURE 37.11

The light from a typical 632.8-nm wavelength helium-neon laser consists of several narrow-band modes. To distinguish the wavelengths of the modes we have to quote the wavelength to a precision of 0.000 000 5 nm.

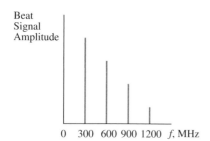

FIGURE 37.12

A spectrum analyzer connected to a fast optical detector can detect beat frequencies between laser modes. A beat frequency of 300 MHz implies that the light from the laser is "blinking" off and on 300 million times each second.

The largest separation between any two modes is 4ν. These four beat frequencies can be observed in the spectrum shown in Figure 37.12. The mode structure described here is typical of that found in He-Ne lasers common to undergraduate physics laboratories. (Note that in unpolarized He-Ne lasers adjacent modes have opposite polarizations and hence do not beat against one another.) ◀

Thin-Film Interference

In our "first look" at thin-film interference in Section 37.1 we analyzed thin-film reflections using a model in which we assumed only two reflected waves contributed to the reflected power. We assumed that the two reflected waves had identical frequencies. This two-reflection model describes actual thin-film reflection so long as the reflectance of the two surfaces is small and the path differences between the waves is less than the coherence length. For this model $\omega = \omega'$ so that the master interference equation (Eq. (37.3)) reduces to

$$I_{\text{tot}} = I + I' + 2\sqrt{II'}\langle\cos(\phi)\rangle$$

where ϕ is the phase difference between the two waves when they recombine. If we assume that the waves are coherent, ϕ is constant so that the time average $\langle\cos(\phi)\rangle$ is simply $\cos(\phi)$ itself. Let's assume that because these two beams have each been reflected once, they have nearly equal magnitudes and, therefore, we can take $I = I'$ in our interference equation:

$$I_{\text{tot}} = 2I[1 + \cos(\phi)]$$

Recall that I is the irradiance due to either *reflected* ray alone. The trigonometric identity $1 + \cos(\phi) = 2\cos^2(\phi/2)$ can be used to simplify the previous expression. This result can be combined with our knowledge of ϕ as described in Section 37.1. To summarize, the irradiance of normally incident light reflected equally from opposite sides of a thin film is given by

$$I_{\text{tot}} = 4I\cos^2\left(\frac{\phi}{2}\right) \qquad \text{(Low-reflectance model)} \qquad (37.5)$$

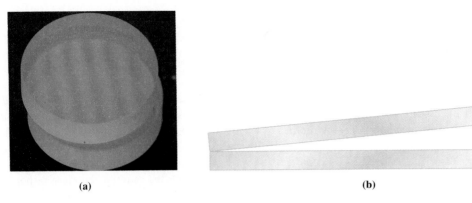

(a)   (b)

FIGURE 37.13

(a) A thin-film interference pattern created when a wedge of air between two optical flats is viewed in monochromatic light. (b) Geometry leading to the interference pattern in (a). The irradiance pattern is described by Equation (37.4).

where

$$\phi = \Delta\phi_{\text{path length}} + \Delta\phi_{\text{refl}} \qquad (37.6)$$

and the contributions to the phase difference ϕ can be calculated from

$$\Delta\phi_{\text{path length}} = 2\pi\frac{n\Delta l}{\lambda_o}, \qquad (37.7)$$

where Δl = physical path length difference and $\Delta\phi_{\text{refl}} = 0$ or π rad depending on whether reflections are both internal, both external, or one internal and one external.

When films of varying thickness are viewed in monochromatic light, an alternating pattern of constructive and destructive interference can be observed as shown in Figure 37.13. Such fringes are known as **fringes of equal thickness.**

High-Reflectance Result for Thin-Film Interference (Optional)

Our discussion of thin-film interference is incomplete because our model for the reflected light included only two rays, one reflected from the front surface and one reflected from the back. However, if we are to accurately describe the irradiance due to light reflected from a thin film, Figure 37.6 reminds us that, in principle, the irradiances of an infinite number of rays must be added. This summation results in an infinite series that is easily totaled.[3] The resulting formula for the reflected irradiance is

$$I = I_o\frac{F \sin^2(\Delta\phi_{\text{path length}}/2)}{1 + F \sin^2(\Delta\phi_{\text{path length}}/2)} \qquad (37.8)$$

where I_o is the irradiance of the incident beam. The parameter F is called the **coefficient of finesse**

$$F \equiv \frac{4R}{1 - R^2}$$

Concept Question 6
Compared to the irradiance I of one reflected beam alone, what is the maximum irradiance occurring in a thin-film interference pattern described by the two-wave model? What is the minimum irradiance for this case?

---

[3] See, for example, Eugene Hecht, *Optics,* 2nd ed. (Addison-Wesley, Reading, MA, 1987) pp. 363–367.

Reflected Irradiance
Thin Film

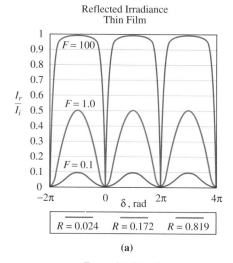

$R = 0.024$ $R = 0.172$ $R = 0.819$

(a)

Transmitted Irradiance
Thin Film

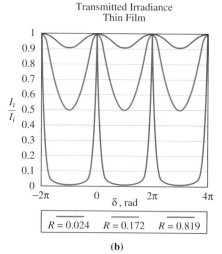

$R = 0.024$ $R = 0.172$ $R = 0.819$

(b)

FIGURE 37.14

Three thin-film interference patterns formed from films with differing values of the power reflection coefficient R. (a) For large values of R the reflected light exhibits sharp, dark fringes. (b) The transmitted light exhibits sharp, bright fringes as R becomes larger.

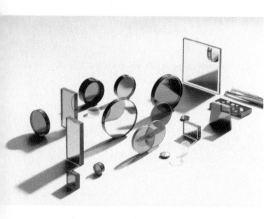

FIGURE 37.16

Examples of interference filters.

where R is the power reflection coefficient for a single reflection as defined in Section 35.4. The phase angle $\Delta\phi_{\text{path length}}$ includes only that portion of the phase difference due to the optical path length difference as defined in Equation (37.2). Figure 37.14(a) shows a graph of the relative irradiance of the reflected light, whereas Figure 37.14(b) shows the complementary pattern for transmitted light.

The first thing to notice regarding the more exact calculation is that the values of $\Delta\phi$ for maximum and minimum irradiances are the same as those predicted by our model in which we considered only two reflected waves. For small values of the reflection coefficient R, Equation (37.8) approaches the irradiance expression obtained for the two-wave model. (See Problem 34.) In the case of reflected light the minima in the interference pattern occur at the same positions for all values of R, but the minima become sharper as the reflection coefficient becomes larger.

We have seen that a thin-film interference pattern can be sharpened when the number of reflections is increased by raising the reflection coefficient of a single film. An alternative method for increasing the number of reflections is to increase the number of layers of film. Figure 37.15 shows a stack of dielectric layers that can be used to create an **interference filter.** When polychromatic light is incident on such a stack, only light within a narrow band of wavelengths satisfies the interference condition represented by the narrow

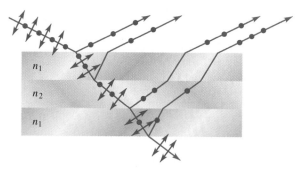

FIGURE 37.15

A multilayer system of alternating dielectrics. Each layer has an optical thickness nt equal to one-quarter of the design wavelength.

peaks in Figure 37.14(b). As a consequence, the transmitted light is nearly monochromatic. Figure 37.16 shows a variety of such filters constructed for specific wavelengths.

Two-Source Interference

We now return to the case of two-source interference, such as the Young's double-slit interference of Section 37.1. Our goal this time is to calculate the irradiance throughout the interference pattern. Consider once again a model in which two point-sources are arranged in front of a screen as shown in Figure 37.17. The irradiance follows directly from the master interference equation (Eq. (37.3)) once we calculate the phase difference ϕ. In this case the phase difference arises directly from the path length difference $\Delta l = l - l'$ in Figure 37.17. Both l and l' can be found from the Pythagorean theorem:

$$l' = L\sqrt{1 + \left(\frac{y + (d/2)}{L}\right)^2} \tag{37.9}$$

$$l = L\sqrt{1 + \left(\frac{y - (d/2)}{L}\right)^2} \tag{37.10}$$

We look at only the **far-field** case in which L is much, much larger than y and d. In this case we can use the binomial expansion $\sqrt{1 + x} = 1 + x/2 + \cdots$ to approximate both distances. On subtracting, we find (see Problem 24),

$$l' - l = \Delta l = \frac{yd}{L} \qquad (37.11)$$

The resulting phase difference follows from Equation (37.2)

$$\phi = 2\pi\left(\frac{yd}{L\lambda}\right)$$

This expression for the phase can be used with our irradiance expression Equation (37.3):

$$I_{tot} = I + I' + 2\sqrt{II'}\cos(\phi)$$

to compute the irradiance throughout the Young's double-slit pattern.

For the special case in which the irradiances I and I' due to our two points-sources are equal, we obtain

$$I_{tot} = 2I[1 + \cos(\phi)]$$

Using the trigonometric identity $1 + \cos(\phi) = 2\cos^2(\phi/2)$, the irradiance can be rewritten

$$I_{tot} = 4I\cos^2\left(\frac{\phi}{2}\right)$$

$$\phi = 2\pi\left(\frac{yd}{L\lambda}\right) \qquad (37.12)$$

Figure 37.18 shows a graph of this irradiance as a function of ϕ. Figure 37.19 shows a photograph of an actual interference pattern between two coherent point-sources. We note that maxima occur whenever ϕ is an even multiple of π, whereas minima occur when ϕ is an odd multiple of π. These conditions lead to

$$d\frac{y}{L} = m\lambda, \qquad m = 0, 1, 2, 3, \ldots \qquad \text{(Irradiance maxima)} \quad (37.13a)$$

$$d\frac{y}{L} = \left(m - \frac{1}{2}\right)\lambda, \qquad m = 1, 2, 3, \ldots \qquad \text{(Irradiance minima)} \quad (37.13b)$$

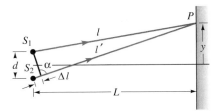

FIGURE 37.17

Two of the many rays giving rise to the complete two-source interference pattern. The distances from S_1 to P and from S_2 to P differ by Δl. When L is much larger than the slit separation d, the angle α is approximately a right angle.

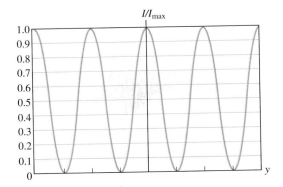

FIGURE 37.18

A graph of the irradiance pattern predicted by Equation (37.11) showing "cosine-squared" type fringes.

FIGURE 37.19

A photograph of the central portion of the irradiance pattern produced by two-source interference. The irradiance varies as in Figure 37.18.

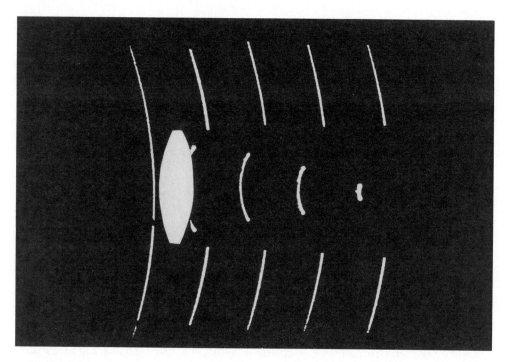

FIGURE 37.20

Photograph of holographic images showing how the shape of wavefronts are changed by a converging lens. Note that the entire original wavefront arrives at the point of focus at the same time. (The photograph shows only a few representative wavefront shapes. Actual wavefronts are separated by fractions of a micron. The bending of light around the lens due to diffraction is too small to be seen on this scale; remember the actual wavelength is much smaller than that implied by the few representative wavefronts shown here.)

just as predicted by our simpler treatment in Section 37.1. The derivation of the preceding expressions for the combined irradiance from two coherent point sources has assumed that $L \gg y + d/2$. As we can see from Figure 37.17, large L implies that rays that eventually join to create, say, a maximum in the interference pattern are nearly parallel. Indeed, under the circumstances that our approximations become exact, the rays are precisely parallel and L is infinite. This requirement can sometimes be a burden on a short laboratory bench. However, we have seen that a lens focuses parallel rays into a focal plane located a distance f behind the lens. Perhaps a lens could be used to focus our ideal interference pattern in its focal plane. But we must be quite circumspect here. After all, our interference pattern is a delicate creation because it depends on exact optical path-lengths. We might, for example, worry about the fact that the wavelength of the light changes while the light is inside the glass lens. Furthermore, each ray passes through a different thickness of the lens. Surely there is great potential for introducing extra phase differences between waves from our two sources. To investigate this matter we have to digress for a moment to examine the focusing properties of a lens from the wave-model point of view.

Figure 37.20 is a remarkable photograph showing representative wavefronts of very short pulses of light as they travel through a lens. This figure shows very clearly that the wavefronts are bent as they pass through the lens, causing them to converge to the lens focus. This process is quite a different view of focusing than we had in our geometrical optics ray models. From our present point of view the important thing to notice is that every point on the original wavefront entering the lens arrives at the focus at the same instant. In order for this simultaneous arrival to occur, the optical path lengths *from any point* on the original wavefront to the focal point must be the same. Figure 37.21 shows how this photograph might appear if the original wavefronts were plane waves and had entered at angle θ with respect to the optic axis. Again, the important fact is that all points on a plane wavefront entering the lens maintain a constant phase relationship until they reach the focus.

FIGURE 37.21

Portions of a plane wavefront striking a lens at a nonzero angle all reach the focal plane at the same time.

Returning now to our two-source interference case, Figure 37.22 shows how spherical waves from our two coherent point-sources combine (by the principle of superposition) to form composite wavefronts. In Figure 37.23 we see how to determine the path-length difference between our two point-sources. We extend the line AB from one source to a ray path from the second source in such a way that line AB is perpendicular to both rays. Energy traveling along parallel rays from points A and B arrives at the same point in the lens's focal plane simultaneously. (Remember, because both rays are refracted through the lens, optical path lengths ACP and BDP are equivalent.) From triangle ABO we conclude that the path-length difference $\Delta l = d\sin(\theta)$. By drawing a construction ray passing undeviated through the center of the lens, we find that $\tan(\theta) = y/f$. Combining these relations with Equation (37.7) for the phase difference accruing due to the path length difference between our point sources, we find that, when a lens is used to focus a two-point interference pattern,

$$\phi = 2\pi\left[\frac{d\sin(\theta)}{\lambda}\right] \tag{37.14}$$

$$d\sin(\theta) = m\lambda, \qquad m = 0, 1, 2, 3, \ldots \qquad \text{(Irradiance maxima)} \quad \textbf{(37.15a)}$$

$$d\sin(\theta) = \left(m - \frac{1}{2}\right)\lambda, \qquad m = 1, 2, 3, \ldots \qquad \text{(Irradiance minima)} \quad \textbf{(37.15b)}$$

with

$$\tan(\theta) = \frac{y}{f} \tag{37.16}$$

Even when a lens is not used, the above equations (with $f = L$) give accurate results, providing the distance between the slits and the screen is large compared to the distance between slits. In fact Equations (37.15a) and (37.15b) reduce to Equations (37.13a) and (37.13b) for this case. (See Problem 26.)

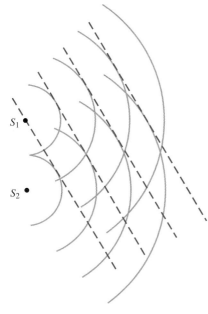

FIGURE 37.22

Spherical waves from two point-sources combine to form plane wavefronts indicated by lines of tangency. These wavefronts correspond to the parallel rays indicated in Figure 37.23. The straight-line wavefronts shown correspond to the first-order ($m = 1$) bright spot. Can you sketch in the straight lines corresponding to wavefronts that give rise to the $m = 0$ and $m = 2$ maxima?

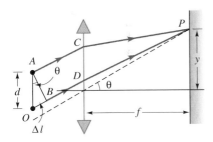

FIGURE 37.23

Parallel rays from the two point-sources interfere in the focal plane of the lens. The optical path length from point A to point P is the same as the optical path length from point B to point P. (See Figs. 37.20 and 37.21.)

EXAMPLE 37.6   *Focusing on Young's Double Slit*

Reconsider the Young's double-slit experiment of Example 37.1 in which the expanded beam from a He-Ne laser is allowed to fall on two equidistant small slits separated by 0.250 mm. These slits then reradiate the light as two coherent point-sources. If a lens of focal length 40.0 cm is placed behind the slits and a screen is placed in its focal plane, where is the third bright spot located?

SOLUTION   The wavelength of the He-Ne laser is 633 nm. Because we are now using an imaging geometry, we employ Equation (37.15a)

$$d\sin(\theta) = m\lambda \qquad \text{with } m = 3$$

and find

$$\sin(\theta) = \frac{m\lambda}{d} = \frac{3(633 \times 10^{-9}\text{ m})}{(0.250 \times 10^{-3}\text{ m})} = 7.60 \times 10^{-3}$$

This angle is sufficiently small that we can employ the approximation $\tan(\theta) \approx \sin(\theta)$ so that from Equation (37.16) we have

$$y = f\tan(\theta) = (40.0\text{ cm})(7.60 \times 10^{-3}) = 0.304\text{ cm}$$

37.3 Multiple-Source Interference

We now imagine that we have an array of N coherent point-sources located a distance d from one another. Let's assume they all have the same intensity and calculate the interference pattern created in the back focal plane of a lens placed behind the sources. Our task is to sum the electric fields from each of the point-sources. These fields are given by

$$E_1 = E_o \sin(\omega t)$$

$$E_2 = E_o \sin(\omega t + \phi)$$

$$E_3 = E_o \sin(\omega t + 2\phi)$$
$$\vdots$$
$$E_N = E_o \sin[\omega t + (N - 1)\phi]$$

where the phase difference ϕ between adjacent sources is given by

$$\phi = 2\pi \left[\frac{d \sin(\theta)}{\lambda} \right]$$

as in the two-slit case. Not surprisingly, the maxima occur at the same angles θ as for the two-source case. From Figure 37.24 it is clear that if the path-length difference between the first and second sources is λ, then the path-length difference between rays from the second source and third source is also λ. Thus, waves from all three sources are in phase when they meet. This argument can be extended to show that waves from all the remaining sources are also in phase when they meet. That is, the condition for an intensity maximum for a multisource, such as that shown in Figure 37.24, is

$$m\lambda = d \sin(\theta) \qquad m = 0, 1, 2, 3, 4, \ldots \qquad (37.17)$$

(Irradiance maxima, multiple sources)

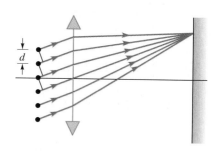

FIGURE 37.24

When multiple sources are used, the path-length differences between adjacent sources are all equal.

The situation is not quite as simple for the location of the minima. Suppose the path-length difference between the first and second sources is one-half wavelength. Then the E-fields from these two sources cancel when they overlap at the focal plane. Similarly, the E-fields from the third and fourth sources cancel, and so on. But whether a completely dark spot occurs depends on whether there are an even or an odd number of total sources. If the number of sources is even, cancellation is complete. If the number of sources is odd, there is still a small uncanceled electric field at an otherwise dark spot.

We can find the exact irradiance values for this model by direct calculation. We have performed this computation for several values of N, and the results are shown in Figure 37.25, in which the calculated irradiances are compared with actual multiple-slit interference patterns. When comparing multiple-source interference patterns to two-source interference patterns, the important point not to miss is that the *locations* of the maxima do not change, but the *width* of each maximum becomes increasingly narrow as the number of sources is increased.

Phasors

To aid our visualization it is helpful to picture in a graphic fashion the addition of the electric fields from our multiple sources. We represent each electric field by a directed line segment known as a **phasor**. The amplitude E_o of the electric field is represented by the length of the phasor. We orient each phasor so that the phasor makes an angle with the horizontal axis equal to the argument of the sine function in the corresponding expression for the electric field. In Figure 37.26 we show a phasor diagram for the vector addition of the E-fields for two point-sources $E_1 = E_o \sin(\omega t)$ and $E_2 = E_o \sin(\omega t + \phi)$. The magnitude of each electric field at any moment is the projection of the phasor on the vertical

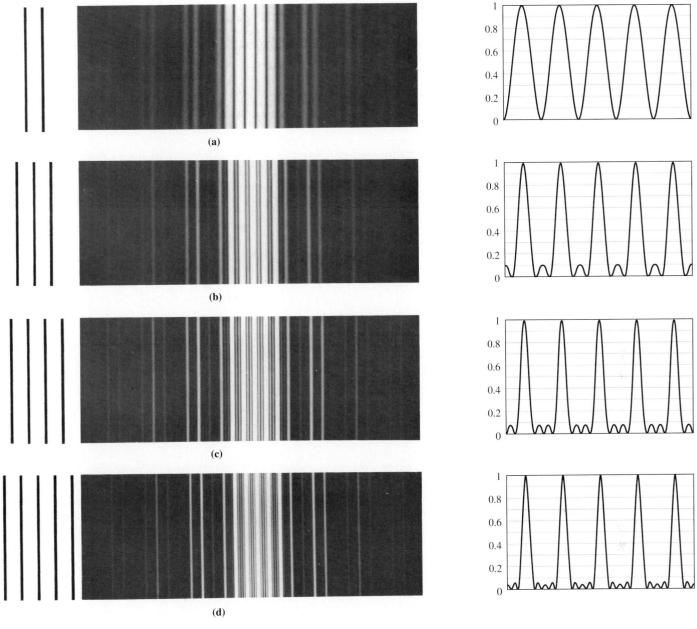

(a)

(b)

(c)

(d)

FIGURE 37.25

Comparison of measured and calculated irradiance for multiple-source interference. Coherent light falling on the slit patterns shown in the left-hand column create complex interference patterns shown in the center column. In the right-hand column are calculated irradiances which describe the central portion of the interference patterns. (The fainter fringes on either side of the bright central regions result from the finite width of the slits. This effect is discussed in Chapter 38.) Notice that as the number of sources increases the bright lines become narrower. When the number of sources is even, the midpoint between two bright fringes is completely dark. When the number of sources is odd, a small irradiance from one uncanceled source remains.

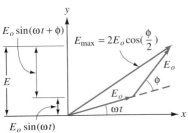

FIGURE 37.26

The addition of phasors representing two electric fields with phases that differ by ϕ. The total field at any time t is given by the projection of the phasors on the vertical axis. As the entire phasor diagram rotates about the origin, the maximum projection on the y-axis is given by the length of the resultant.

axis. Now we see that the sum $E = E_o \sin(\omega t) + E_o \sin(\omega t + \phi)$ is the sum of two projections on the vertical axis. However, we can also obtain this sum by first adding the two phasors *and then* taking the projection. The sum of the projections is the projection of the phasor sum, providing that the two phasors make the correct angle with each other. Phasor addition is formally equivalent to the addition of vectors, and the resultant phasor can be calculated in the same manner as for resultant vectors. Keep in mind, however, that the entire diagram is rotating about the origin approximately 10^{14} times each second (Fig.

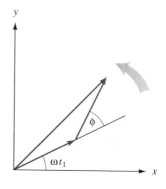

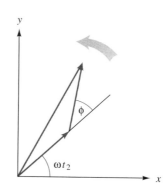

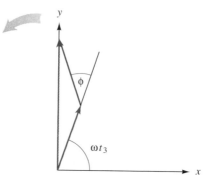

FIGURE 37.27

As time progresses the entire phasor diagram rotates about the origin. At any instant of time the actual electric field is the projection of the resultant phasor onto the y-axis.

Concept Question 7
Draw a phasor diagram similar to Figure 37.26, showing the addition of two phasors that are exactly in phase. Show four more sketches indicating the addition of two phasors that differ in phase by 45°, 90°, 135°, and 180°. In each case draw the phasor sum representing the resultant electric field amplitude. What optical path-length difference (measured in wavelengths) between waves from two coherent sources is necessary to produce each of these relative phase differences?

37.27). The largest electric field value that is ever projected on the vertical axis is the magnitude of the resultant phasor that connects the origin to the tip of the last phasor. If we freeze the picture at a convenient time, say $t = 0$, we can calculate the *amplitude* of the resultant electric field as the length of the resultant phasor. In the case of two phasors of equal length, the resultant amplitude can be calculated using the law of cosines (see Appendix 1) to obtain

$$E^2 = E_o^2 + E_o^2 - 2E_o^2 \cos(\pi - \phi)$$

Employing the trigonometric identities $\cos(A + B) = \cos(A) \sin(B) + \sin(A) \cos(B)$ and $1 + \cos(A) = 2 \cos^2(A/2)$, we can write our equation for E^2 as

$$E^2 = 2E_o^2[1 + \cos(\phi)] = 4E_o^2 \cos^2\left(\frac{\phi}{2}\right)$$

Recalling that in free space the irradiance is given by $I = (c\epsilon_o/2)E_o^2$, we can write

$$I = 4I_o \cos^2\left(\frac{\phi}{2}\right)$$

where I_o is the irradiance produced by a single source alone. This result is exactly Equation (37.12) obtained previously by direct calculation of the time-averaged E-fields.

Our phasor picture is easily extended to three or more sources. In Figure 37.28 we show the configurations of five phasors corresponding to several points of the graph in Figure 37.25(d). In the next chapter we will use this phasor picture to model the far-field diffraction pattern of a single slit.

37.4 Summary

The conditions for constructive and destructive interference in **Young's double-slit experiment** can be written

$$m\lambda = d \sin(\theta) = d\frac{y}{L}, \qquad m = 1, 2, 3, \ldots$$

(Irradiance maximum, Young's double slit)

$$\left(m - \frac{1}{2}\right)\lambda = d \sin(\theta) = d\frac{y}{L} \qquad m = 1, 2, 3, \ldots$$

(Irradiance minimum, Young's double slit)

The length of time during which two waves maintain a nearly constant relative phase is known as the **coherence time** t_c. During the coherence time the waves travel a distance $l_c = c\, t_c$ known as the longitudinal **coherence length.** Waves with long coherence lengths relative to distances of interest are said to be **coherent.** Waves with short coherent lengths are **incoherent.**

When electromagnetic waves overlap, their electric fields obey the **principle of superposition. Irradiance** is found from the square of the resulting electric field by

$$I_{tot} = c\epsilon_o\langle \mathbf{E}_{tot} \cdot \mathbf{E}_{tot}\rangle$$

The irradiance resulting from the superposition of two **incoherent** waves is the sum of the irradiance from the individual waves:

$$I_{tot} = I + I'$$

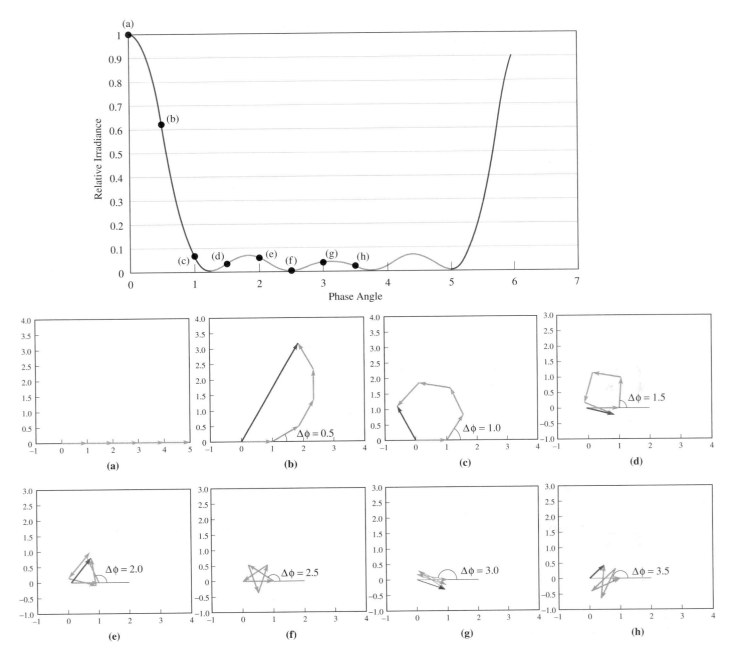

FIGURE 37.28

Magnified view of Figure 27.25(d) showing the irradiance as a function of ϕ for five slits. (a)–(h) Five electric fields with phases of 0, ϕ, 2ϕ, 3ϕ, and 4ϕ are added by means of phasors for different values of ϕ. The resultant represents the amplitude and phase of the sum. Compare the length of the resultant to the irradiance at corresponding values of ϕ.

The irradiance resulting from the superposition of two **coherent** waves is given by the master interference equation

$$I_{\text{tot}} = I + I' + 2\sqrt{II'}\langle \cos[(\omega - \omega')t - \phi]\rangle \tag{37.3}$$

The **interference term** $2\sqrt{II'}\langle\cos[(\omega - \omega')t - \phi]\rangle$ results in a number of optical phenomena, including optical beats, thin-film interference, and single-slit and multiple-slit diffraction patterns.

Fluctuations of irradiance known as **optical beats** occur when two highly monochromatic waves with angular frequencies ω and ω' are superposed. The beat frequencies, $(\omega - \omega')/(2\pi)$, of tens and hundreds of megahertz can be detected by electronic means.

Electromagnetic waves undergo a phase change of π when they reflect from a metallic surface or from the boundary of a dielectric with a higher value of refractive index. When an electromagnetic wave reflects from a thin film, the total reflected wave is canceled if the partial waves reflecting from the front and back surface of the film have a phase difference given by

$$\Delta\phi_{tot} = m\pi, \qquad m = 1, 3, 5, \ldots$$

The phase difference may be due either to reflection or optical path-length difference

$$\Delta\phi_{tot} = \Delta\phi_{path\ length} + \Delta\phi_{refl} \tag{37.6}$$

where

$$\Delta\phi_{path\ length} = 2\pi\frac{n\,\Delta l}{\lambda_o}, \qquad \text{and} \qquad \Delta l = \text{physical path-length difference} \tag{37.7}$$

$$\Delta\phi_{refl} = 0 \text{ or } \pi, \qquad \text{depending on whether reflections are internal or external}$$

The **optical path length** $n\,\Delta l$ is the distance a wave would have to travel in a vacuum in order to undergo the same phase change that it does traveling distance Δl in a medium of refractive index n.

When waves from two in-phase coherent sources are incident on a distant screen, the resulting interference pattern is described by

$$I_{tot} = 4I\cos^2\left(\frac{\phi}{2}\right) \tag{37.12}$$

$$\phi = 2\pi\left(\frac{yd}{L\lambda}\right)$$

where L is the distance from the slits to the screen, d is the slit separation, and y is the distance from the center of the pattern on the screen measured along a direction parallel to a line joining the two sources.

When the interference pattern is formed on a screen located at the back focal plane of a lens with focal length f placed after the slits, the phase ϕ is calculated from

$$\phi = 2\pi\left[\frac{d\sin(\theta)}{\lambda}\right]$$

where $\tan(\theta) = y/f$.

The location of the maxima in multiple-source interference patterns is the same as that for two-source patterns; however, the width of each maximum decreases as the number of interfering sources increases.

The addition of oscillating electric fields with different relative phases can be illustrated graphically by means of **phasor diagrams.**

PROBLEMS

37.1 Interference

1. Two coherent point-sources emitting light of wavelength 633 nm form an interference pattern on a screen 1.35 m away as shown in Figure 37.P1. If the point-sources are located 1.25 mm apart, calculate the distances from the central bright spot to the locations where the first three interference maxima are located.

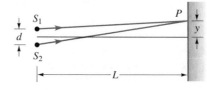

FIGURE 37.P1   Problem 1

2. Two coherent point-sources emit light of 589 nm to form an interference pattern on a wall directly in front of them and 3.45 m away. If the bright spots in the central region of the interference pattern are separated by 1.30 cm, how far apart are the point-sources?

3. Young's double-slit experiment is performed with the slits illuminated by a laser diode emitting light of wavelength 680. nm. If the second-order bright fringe is 3.50 cm from the center line and the slits are located 2.00 m from the observation screen, find (a) the slit separation and (b) the position of the second dark fringe from the center.

4. A Young's double-slit experiment is set up using a slit separation of 1.00 mm and a screen 4.00 m away. The third dark fringe is located at a distance of 6.60 mm from the center of the screen directly across from the slits. (a) What is the wavelength of the incident monochromatic light? (b) What should be the new slit separation if the second bright fringe of a new pattern is to be located where the third dark fringe of the previous pattern was located?

5. When two point-sources separated by 50.0 μm emit monochromatic light, the interference pattern shown in Figure 37.P2 is observed on a screen 1.20 m away. If the distance between the center of adjacent bright fringes is measured to be 1.52 cm, what is the wavelength of the light?

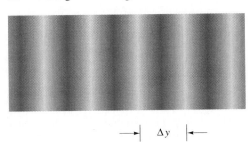

$$\longrightarrow| \quad \Delta y \quad |\longleftarrow$$

FIGURE 37.P2   Problem 5

6. Two slits are illuminated by a nearly monochromatic source of wavelength 546 nm. Because of the interference between the light passing through the two slits, a pattern is formed on a screen 0.500 m away. If the distance between the first- and second-order maxima is 1.00 mm, what is the separation of the two slits?

7. Light of wavelength 633 nm is used in a Lloyd's mirror arrangement to produce an interference pattern on a screen located 1.60 m from the source. What is the distance from the point of intersection of the mirror plane and the screen of the third dark fringe if the source is 1.00 mm above the mirror? (Count the center as the zeroth dark fringe.)

8. Consider the soap film in Figure 37.5. How thick is the film at the point where the second band of yellow appears in the reflected light? Take the refractive index of the soap to be 1.40.

9. A film of oil ($n = 1.24$) floats on top of a puddle of water. What is the thinnest layer of oil that results in a strong reflection of red light with wavelength 680. nm?

10. A thin layer of transparent material of thickness 260. nm and refractive index 1.40 is used as an antireflection coating on the surface of glass with index of refraction 1.50. What reflected, visible wavelengths are suppressed by this coating?

11. An antireflection coating of MgF_2 ($n = 1.38$) is to be applied onto a lens formed from glass of refractive index 1.52. (a) If the coating is to reduce reflections at 550 nm, what should be its thickness? Give possibilities in order of increasing thickness starting with the thinnest. (b) If the coating is to be made of CeF_3 ($n = 1.65$), what are the three thinnest possible coatings?

12. A thin wedge of air is formed between two flat glass plates in contact along one edge and separated by a thread at the other end. The distance between two ends of the plate is 8.00 cm. An observer looking at reflected monochromatic light of wavelength 546 nm sees 40 dark interference fringes when viewing the plate from above. (a) What is the change in vertical separation of the two plates between two adjacent dark fringes? (b) What is the thickness of the thread?

13. Two optical flats are placed in near contact at one edge and are separated by a human hair at the other edge as shown in Figure 37.P3. (a) At the thin edge of the air wedge the separation of the glass surfaces is much less than the wavelength of visible light. Does this point appear bright or dark when viewed by reflected light? (b) If filtered light from a mercury vapor source ($\lambda = 546$ nm) is used, exactly nine bright fringes are located between the edges in contact and the position of the hair. How thick is the hair?

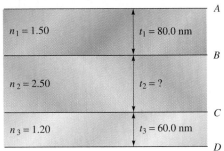

FIGURE 37.P3   Problem 13

14. Monochromatic visible light falls at nearly normal incidence on a three-layer dielectric film surrounded by air as shown in Figure 37.P4. (a) If a minimum in reflected irradiance is observed from light reflected from surfaces A and B, what is the wavelength of the incident light? (b) What is the minimum thickness t_2 if an irradiance minimum is observed due to reflections from surfaces B and C?

FIGURE 37.P4   Problem 14

15. A lens with a slightly curved surface is placed on an optical flat generating a pattern of circular rings when viewed in reflected monochromatic light. The curvature is sufficiently slight that it is not possible to tell by simple observation whether the surface is convex or concave. (a) If one pushes gently on the center of the lens, circular fringes move toward the center or away from the center depending on whether the lens surface is concave or convex (Fig. 37.P5). Which way do they move in the convex case?

(b) An alternative technique for distinguishing concave from convex is to view the interference pattern using white light. In that case the order of the colors from center out is opposite for the two cases. What is the order of the colors (from the center out) for the convex case?

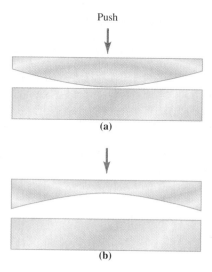

FIGURE 37.P5 Problem 15

16. A spherical convex lens is placed on an optical flat and the pattern of Newton's rings is viewed by reflected light of wavelength 546 nm. The radius of the fifth dark ring is 1.26 cm. Assume that the center of the lens and the optical flat are separated by much less than one wavelength of light. What is the radius of curvature of the lens surface?

37.2 Irradiance for Simple Interference Patterns

★17. Consider two waves, one of frequency $\nu_1 = 5.455 \times 10^{15}$ Hz, the other with $\nu_2 = 5.450 \times 10^{15}$ Hz. If these two waves start exactly in phase at $t = 0$, (a) how long is it before they are exactly out of phase? [*Hint:* The number of oscillations made by a wave in time t is νt. You are looking for a particular time T when the number of oscillations made by the first wave is 0.5 greater than the number made by the second.] (b) How far has each wave traveled in this time?

18. Consider two waves with frequencies ν_1 and ν_2. After traveling distance d the first wave has made one-half oscillation more than the second wave. Show that the frequencies are related by $\Delta \nu = \nu_1 - \nu_2 = V/2d$, where V is the speed of the wave.

19. Show that if two waves differ in frequency by an amount $\Delta \nu$ that is much smaller than either frequency, then the waves differ in wavelength by $\Delta \lambda = -(\Delta \nu / \nu) \lambda$.

★20. A particular helium-neon laser used in a physics lab supports two modes. That is, it emits light with essentially two wavelengths. The longitudinal coherence length of the laser is 25.0 cm, i.e. when waves due to the two modes start out exactly in phase, they are exactly out of phase after traveling 25.0 cm. [*Hint:* See problem 18.] What is the frequency difference between the two modes?

21. The wavelengths of two modes of a He-Ne laser (average wavelength 632.8 nm) differ by only 2 parts in 10^8. What beat frequency is detected between these two modes?

22. (a) Use the relativistic Doppler shift (see Section 17.6) to calculate the change in the frequency of light with wavelength 632.8 nm that is reflected from a reflector moving 2.00 m/s away from the light source. (Don't forget both shifts!) (b) If the original light wave is combined with the Doppler-shifted light, what beat frequency is observed? (c) If the optical detector is able to respond to irradiance changes occurring no faster than 750. MHz, what is the fastest reflector speed that can be measured?

23. Two identical waves with free-space wavelength λ_o start in phase and travel parallel routes for a distance d. The first wave travels distance d in air ($n_o = 1.00$), and the second wave travels distance d in a medium of refractive index n. Show that after traveling the distance d the waves have a phase difference $\Delta \phi = 2\pi[(n-1)d/\lambda_o]$.

24. Expand the expression for l and l' in Equations (37.9) and (37.10) using the binomial expansion and obtain Equation (37.11).

25. Two slits separated by 0.200 mm are to be used in Young's double-slit experiment. Immediately behind the slits is a lens of focal length 0.500 m used to form the interference pattern on a screen located in the focal plane of the lens. What is the wavelength of a monochromatic light source used to illuminate the slits if adjacent maxima of the interference pattern are separated by 1.00 mm?

26. Show that in the limit of small θ, Equations (37.15a) and (37.15b) reduce to Equations (37.13a) and (37.13b).

37.3 Multiple-Source Interference

27. Draw a phasor diagram showing six phasors each of the same length when the phase lag between adjacent phasors is 15°. Determine the comparative length of the resultant phasor.

★28. Draw a phasor diagram showing six phasors each with the same length when the phase lag between adjacent phasors is (a) 0°, and (b) 45°. (c) What should be the phase angle between the adjacent phasors for the resultant phasor to have zero length?

29. (a) Draw the phasor diagram representing four point-sources of equal strength when the phase lag between adjacent phasors is 0°, 45°, 90°, 135°, and 180°. (b) Calculate the amplitude of the resultant phasor for each case. Express your answer in terms of E_o the electric field amplitude of each individual source. (c) Calculate the resulting irradiance in each case. Express your answer as a ratio of the actual irradiance to the maximum possible irradiance ($I_{max} = c\epsilon_o(4E_o)^2/2$).

★30. Without performing any calculations, sketch the irradiance pattern you expect for the far-field diffraction pattern of seven point-sources emitting monochromatic light. Assume all of the sources are emitting in phase. Present your sketch in the form used in Figure 37.25.

Numerical Methods and Computer Applications

31. Spreadsheet template PHASOR.WK1 on the diskette accompanying this text can be used to create phasor diagrams, such as those shown in Figure 37.28 for five point-sources. (a) Change the phase difference $\Delta \phi$ between adjacent phasors to 0.750 rad. Print the graph, add arrow heads to the individual phasors, and draw the resultant phasor. (b) Modify the spreadsheet to show the addition of six (total) point-sources. Calculate the $\Delta \phi$ needed for zero resultant and produce a graph showing your result. Add arrow heads to your graph.

32. (a) Create a spreadsheet template or computer program or use symbolic mathematics software to produce a graph similar to those in Figure 37.25 showing the interference pattern created by seven coherent point-sources. (b) Increase the number of point-sources to 25 and obtain a graph of the expected irradiance pattern. Is this what you expected?

General Problems

★**33.** A Fresnel biprism can be used together with a point-source located distance d behind the center line of the prism to create a two-source interference pattern. As shown in Figure 37.P6 the two images P_1' and P_2' formed by refraction at the prism act as coherent sources. (a) Show that the separation a of the images is given approximately by $a = 2d\alpha(n-1)$. [*Hint:* Assume that the prism angle is small ($\alpha < 1°$) and that the image is formed in the same plane as the source. Trace a ray striking the prism such that it is deviated by the angle of minimum deviation.] (b) Calculate the fringe separation on a screen located a distance $s = 1.750$ m from the back face of a prism of refractive index $n = 1.500$ and vertex angle $1.00°$ if the point–source has wavelength $\lambda = 632.8$ nm and is located a distance $d = 0.750$ cm behind the back face of the prism.

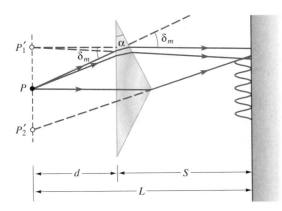

FIGURE 37.P6   Problem 33

★**34.** Show that the exact equation of reflective irradiance of a thin film (Eq. (37.8)) approaches the expression deduced from the two-wave model when the reflection coefficient R is much less than 1. In addition to the simplification of Equation (37.7) for small R, you have to reconcile the appearance of $4RI_o$ in one expression with $4I$ in the other. Finally, show how the sine squared appears in one while the cosine squared appears in the other.

35. The following are measured diameters of the observed dark Newton's rings formed when a concave lens is placed on an optical flat and observed in reflected light of wavelength 546 nm: 9.23, 10.7, 11.9, 13.0, 14.1 mm. Follow the method suggested at the end of Example 37.3 and calculate the radius of curvature of the lens face.

36. When light of wavelength λ falls on two narrow slits separated by 0.0200 mm, an interference pattern is observed on a screen located several meters from the slits. When the wavelength of the light is changed to λ', the position of the original third bright fringe (from the central bright fringe) becomes occupied by the fifth dark fringe. (a) Find the relation between the wavelengths λ and λ'. (b) If $\lambda = 620.$ nm, what is λ'?

37. A Young's double-slit experiment is observed in air. When the whole apparatus (source, slits, and screen) is submerged into a fluid, the screen position originally occupied by the mth bright fringe becomes the location of the m'th dark fringe. In terms of m and m', find an expression for the index of refraction n of the fluid.

38. Light of 600.-nm wavelength is normally incident on a thin, wedge-shaped film of 1.55 refractive index. How much does the film thickness change over a distance in which there are eight bright and seven dark fringes?

39. Monochromatic light of variable wavelength falls normally incident on a thin, uniform film of oil ($n = 1.25$) that rests on a glass plate ($n = 1.55$). The reflected light exhibits destructive interference for wavelengths of 432 nm and 720 nm but at no other wavelengths between these two. What is the film thickness?

40. Reflected white light is observed from above a convex-shaped oil drop ($n = 1.24$) floating on water ($n = 1.33$). (a) What is the oil film thickness where the third red fringe (from the outer edge of the drop) is observed? (b) Is the outermost edge of the oil drop dark or bright? Explain your answer. (Take 700. nm as the wavelength of red light.)

41. A scuba diver observes an oil slick ($n = 1.22$) from beneath. If the film is 480 nm thick, what wavelengths of light appear brightest to the diver?

42. Bargain-brand jewelry usually advertised on your favorite home shopping channel is often made of SiO-coated ($n = 1.20$) glass ($n = 1.50$) to increase its reflectance. What should the thickness of the coating be to enhance reflectance at a wavelength of 550 nm?

43. Find the sum of the following (a) using phasors and (b) using trigonometric identities:

$$E_1 = (5 \text{ V/m}) \sin(\omega t)$$

$$E_2 = (3 \text{ V/m}) \sin\left(\omega t + \frac{\pi}{3}\right)$$

44. Find the sum of the following using the phasor method:

$$E_1 = (4 \text{ V/m}) \sin(\omega t)$$

$$E_2 = (8 \text{ V/m}) \sin\left(\omega t - \frac{\pi}{6}\right)$$

$$E_3 = (6 \text{ V/m}) \sin\left(\omega t + \frac{\pi}{4}\right)$$

45. In Young's double-slit experiment a thin layer of mica is placed directly behind one of the slits. The thickness of the mica is such that the optical path length of the light in the mica is $0.500\,\lambda$. (a) Assuming no power is absorbed from the light passing through the mica, describe the change in the far-field diffraction pattern that occurs when the mica is inserted behind one slit. (b) If the wavelength of the light used is 632.8 nm, what is the actual thickness of the mica? (Assume the mica is oriented so that its index of refraction is 1.552.)

Diffraction of Light

In this chapter you should learn

- about single-slit diffraction.
- about combined single-slit and double-slit diffraction.
- about diffraction gratings (optional).
- the Bragg condition for X-ray diffraction (optional).

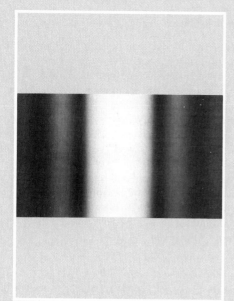

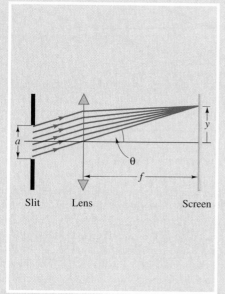

$$a \sin(\theta) = m\lambda$$

In the previous chapter we examined the interference patterns caused when light from two or more coherent sources falls on a distant screen. When predicting the resulting pattern of light and dark we modeled each of the individual sources as a point-source. We know, of course, that in any real case each source must have a finite spatial extent. In this chapter we examine the effects that occur when parts of wavefronts are obstructed by apertures and obstacles. When parts of a wavefront are sheared off by the obstruction, the wavefront bends into the shadow region of the obstacle. This behavior is known as **diffraction.**

If you use a magnifying glass to examine a photograph as reproduced in a printed document, such as a newspaper or this text, you will discover that the reproduction is made up of a great many small dots. You cannot distinguish these dots from normal viewing distances. This is not necessarily due to any defect in your vision. Indeed even if your eyes were optically perfect, you would not be able to distinguish the dots. Ultimately the ability of our eyes and other optical instruments to resolve fine detail is limited by diffraction due to the wave nature of light itself.

In order to model the diffraction effect we will make frequent use of Huygens' wavelets as introduced in Chapter 35. That is, we will model the wavefront as if it were composed of a multitude of point-sources. Thus, our discussion of diffraction takes up where the previous chapter left off, looking at the interference pattern created by a large number of point-sources.

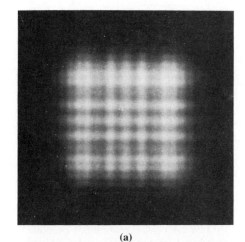

(a)

38.1 Single-Slit Diffraction

Huygens' principle (Chapter 35, Section 2) implies that when a wavefront is sheared by passing an obstruction, the remaining wavefront spreads out into the shadow region. If we carefully examine the "shadow" created by coherent light falling on an opaque object, we notice, as in Figure 38.1, that the shadow is not sharp; some light is **diffracted** into the geometric shadow region. In addition, the bright region near the edge shows alternating light and dark bands. We can now understand this phenomenon in terms of interference. In Figure 38.2 we show the shadow cast by a rectangular aperture illuminated by plane waves from a He-Ne laser. In each picture the aperture used is the same size. However, the distance between the aperture and the screen increases for each photo. Note that initially the character of the interference pattern changes rapidly when the screen is near the aperture. But when the aperture–screen distance becomes large compared to the aperture dimensions, the pattern remains essentially the same except for a size increase.

When the screen's distance from the aperture is comparable to the aperture dimensions, the interference patterns formed are known as **near-field,** or **Fresnel,** diffraction patterns, after Augustin Jean Fresnel (1788–1827). Fresnel was a brilliant proponent of the wave model for light. The **far-field** patterns created when the screen distance is large compared to the aperture size are known as **Fraunhofer** diffraction patterns, after Joseph von Fraunhofer (1787–1826). We describe only Fraunhofer diffraction in this text. As a

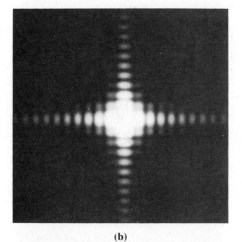

(b)

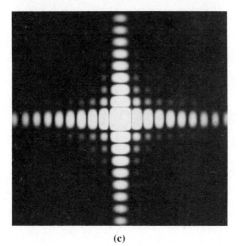

(c)

FIGURE 38.2

The transition between near-field (Fresnel) diffraction and far-field (Fraunhofer) diffraction.

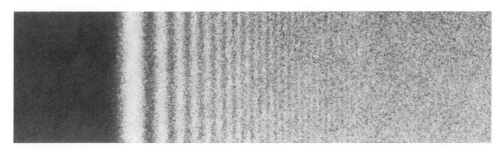

FIGURE 38.1

A close-up view of the shadow cast by a straight edge in coherent light shows a diffraction pattern due to the wave nature of light.

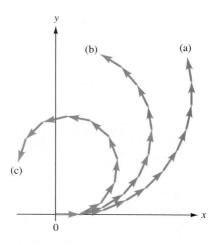

FIGURE 38.4

Phasors from ten sources are added. The angles between each phasor are (a) 0.2 rad, (b) 0.3 rad, (c) 0.5 rad. When the number of sources is large the summation created by chaining together phasors resembles a circular arc as in Figure 38.5.

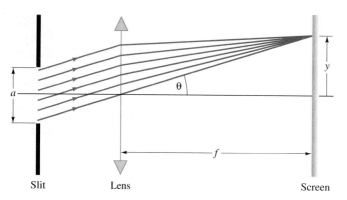

FIGURE 38.3

Geometry for the analysis of single-slit diffraction.

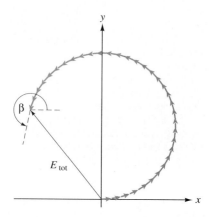

FIGURE 38.5

A large number of phasors approaches a circular arc. Compare with Figure 38.4.

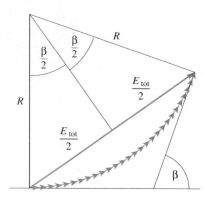

FIGURE 38.6

Geometric construction for calculating the chord length of the circular arc. The chord length E_{tot} is the resultant electric field amplitude. The arc length E_{max} is the maximum electric field that results if all the constituent phasors are parallel.

rule of thumb, for an aperture with dimension a and an aperture-to-screen distance L, the diffraction pattern is of the far-field type when $L > a^2/\lambda$.

Let's look at the diffraction pattern caused by a narrow slit. In Figure 38.3 we show a cross section of a slit with narrow dimension a. We can be certain we are observing the far-field pattern by using a lens to project the pattern in the focal plane of the lens. By applying Huygens' principle we model the wavefront passing through the slit as if it were composed of a very large number of point–sources. Figure 38.4 shows the phasor diagram for such a collection of sources. In Figure 38.5 we extend the phasor picture to represent a continuum of point–sources. The resultant electric field E_{tot} is the chord connecting the initial and final points of the arc. The angle β is the phase-angle difference between one ray from the top of the slit and one from the bottom of the slit. The arc length itself is the maximum electric field E_{max} that would occur if $\beta = 0$. To aid our calculation of E_{tot} we consider an arc of shorter length as shown in Figure 38.6. Notice that β is again the angle between the phasors of the bottom and top rays. We have also constructed a triangle and labeled angles pertinent to our analysis. From the definition of radian measure, the chord length is $E_{max} = \beta R$, and from the definition of the sine we have $E_{tot}/2 = R \sin(\beta/2)$. Eliminating R between these two expressions yields

$$E_{tot} = E_{max} \frac{\sin(\beta/2)}{\beta/2} \tag{38.1a}$$

The combination $\sin(x)/x$ occurs often in optics and electrical engineering and has been given a functional name of its own, **sinc(x).** Recalling that the irradiance we observe on the screen is proportional to the square of the electric field, we can write the irradiance compactly as

$$I_{tot} = I_{max} \operatorname{sinc}^2\left(\frac{\beta}{2}\right) \tag{38.1b}$$

The expression for the maximum phase angle β between a ray from the top of the slit and a ray from the bottom of the slit is analogous to our expression for the phase angle between two point-sources (see Eq. (37.14)), except that the distance between sources is replaced with the slit width a:

$$\beta = 2\pi\left[\frac{a\sin(\theta)}{\lambda}\right] \tag{38.2}$$

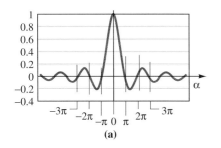
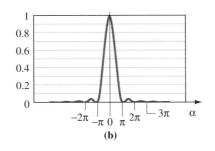
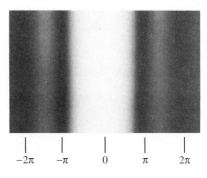

FIGURE 38.7

Comparison of (a) sinc(α), (b) sinc$^2(\alpha)$, (c) the irradiance from a single-slit interference pattern.

(c)

The expression for calculating the ray direction angle θ depends on whether we are in a lensless geometry with the screen at distance L, or in an imaging geometry using a lens of focal length f

$$\tan(\theta) = \frac{y}{L}, \quad \left(\text{No lens, screen at distance } L > \frac{a^2}{\lambda} \right)$$

$$\tan(\theta) = \frac{y}{f}, \quad \text{(Screen in focal-plane of a lens)}$$

From Equation (38.1b) we see that minima occur in the diffraction pattern when $\sin(\beta/2) = 0$ or $\beta = m(2\pi)$, where m is a nonzero integer. From Equation (38.2) we find that this result implies

$$a \sin(\theta) = m\lambda, \qquad m = \pm 1, \pm 2, \ldots \qquad \text{(Irradiance minima)} \qquad \textbf{(38.3)}$$

You must be wary here. The above expression for *minima* in a single-slit diffraction pattern is quite similar to the expression for *maxima* in multiple-slit interference. Figure 38.7 shows a graph of the calculated single-slit irradiance compared with a photograph of the pattern from an actual slit.

Diffraction-Limited Optics

One of the most important features of the single-slit diffraction pattern is the width of the central bright maximum. Taking $m = 1$ in Equation (38.3), we find that the angle between the center bright spot to the first complete irradiance minimum is given by $a \sin(\theta) = \lambda$. In most cases of practical interest the angle θ is small enough that $\sin(\theta) \approx \theta$, so to an excellent approximation we can write

$$\theta = \frac{\lambda}{a}$$

If we were to perform a similar calculation for a circular aperture, we would obtain a similar relation. The resulting diffraction pattern for the irradiance involves **cylinder functions,** which are similar in many respects to sines and cosines, although more difficult to calculate (at least with today's calculators). Nonetheless, the resulting cylindrically symmetric irradiance pattern resembles the sinc-squared pattern closely as you can see from Figure 38.8. Furthermore, the angle between the center and the first minimum in the

Concept Question 1
Draw a single slit of width a and place the screen a distance L away from it. Designate the top-most and bottommost ray paths to an irradiance maximum on the screen. From an analysis similar to that for Figure 37.23 verify Equation (38.2). Make use of the fact that every time the path difference for these two rays changes by one wavelength the phase difference between the corresponding waves changes by 2π.

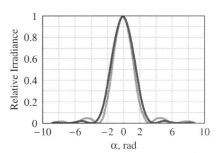

FIGURE 38.8

Comparison of the relative irradiance cross section from a rectangular slit of width a (lighter line) and the relative irradiance from a circular aperture with diameter equal to a (darker curve). The patterns are quite similar.

Concept Question 2
When the width of a single slit is doubled, the radiant energy that passes through the slit in a given time also doubles. Explain why the irradiance of the central maximum of the far-field interference pattern increases by a factor of *four*.

pattern can be found by

$$\theta = 1.22\frac{\lambda}{a} \tag{38.4}$$

This relation defines the smallest spot into which a lens of diameter a can focus parallel light rays; the finite aperture of the lens itself causes diffraction. Even when all the aberrations discussed in Chapter 36 are reduced to an unmeasurable amount, diffraction remains.

Resolution

If you have ever driven on a long stretch of straight highway at night, you may have noticed that the two headlights of an approaching car appear as only one spot of light. Only when the car gets closer are you eventually able to distinguish the two headlights. At the point where you can distinguish both headlights, the lights are said to be **resolved.** The resolution of an optical system determines our ability to see closely spaced detail. With sufficient effort (and money) the optical aberrations described in Chapter 36 can be reduced to arbitrarily small amounts. Ultimately, however, our ability to resolve objects is limited by the diffraction caused by the fundamental wave nature of light itself. Optical systems for which resolution is limited by diffraction are known, logically enough, as **diffraction-limited** systems. Figure 38.9 shows the diffraction-limited image of two point–sources separated by differing amounts. In Fig. 38.9(a) the points are clearly resolved; it's easy to tell that there are two points. Eventually, as the two points get closer and closer, it becomes impossible to tell that there are really two such points. Somewhere between these two extremes one must make a judgment that the points are "just barely" resolved. The criterion for resolution is a value judgment, and not everyone would agree on exactly what constitutes two resolved objects. A number of proposals have been made to standardize the criterion. One of the most popular statements is known as the **Rayleigh criterion,** which assumes that light from the two points is incoherent. According to the Rayleigh criterion the points are just resolved when the first minimum of one of the point's diffraction pattern falls on the maximum of the adjacent point's pattern. Figure 38.9(b) shows irradiance of two points just resolved according to the Rayleigh criterion.

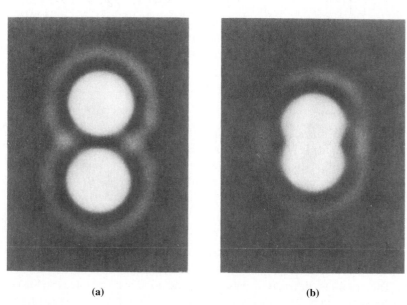

(a)   (b)

FIGURE 38.9

Diffraction patterns of two incoherent sources: (a) well resolved and (b) just resolved by the Rayleigh criterion.

EXAMPLE 38.1   *Be Sure to Wave When You Go out the Front Door*

Assume a reconnaissance satellite uses a diffraction-limited telescope with a mirror diameter comparable to the Hubble telescope (2.40 m). Is it possible to read the numbers on the license plate of your car if the satellite is in an orbit 160. km above the earth?

SOLUTION   According to the Rayleigh criterion two points are just resolved when their angular separation is equal to angle θ, at which the first diffraction minimum occurs as given by Equation (38.4):

$$\theta = 1.22\frac{\lambda}{a}$$

We can see from Figure 38.10 that the angular separation between the two points on

FIGURE 38.10

The resolution of a large telescope mirror on a reconnaissance satellite is limited by diffraction.

the license plate is $\theta = x/R$, where x is their linear separation and R is the altitude of the satellite. Equating these two expressions for θ, we find

$$\frac{x}{R} = 1.22\frac{\lambda}{a}$$

which leads to

$$x = 1.22R\frac{\lambda}{a}$$

Using $\lambda = 550$ nm as the average value for white light we have

$$x = 1.22(1.60 \times 10^5 \text{ m})\frac{550 \times 10^{-9} \text{ m}}{2.40 \text{ m}} = 4.47 \text{ cm}$$

We find that, although one can distinguish the license plates, the numbers are not quite resolved.   ◀

38.2 Effect of Finite Slit Width on Double-Slit Interference Patterns (Optional)

In Section 37.1 we described Young's double-slit experiment using a model based on two point-sources. In reality the two sources are actually slits of finite width. Consequently, the actual diffraction pattern has irradiance variations corresponding to both single-slit and double-slit interference. Consider the geometry of Figure 38.11 in which a lens is placed to the right of the double slits, and the diffraction pattern is formed in the focal-plane of the lens. If only the right-hand slit is uncovered, a single-slit diffraction pattern is formed as described in Section 38.1. If only the left-hand slit is uncovered, again a single-slit pattern is formed. What is important to note is that both of these patterns are formed in *exactly* the same position on the screen. It may at first seem surprising that the patterns fall in exactly the same spot. We are accustomed to thinking about the *imaging* properties of lenses and, therefore, we may be tempted to think that the two single-slit patterns are offset laterally by a distance equal to that between the slits. They are not. In fact for this case the image distance i is equal to f, the focal length of the lens. Consequently, from the point of view of image formation, the "object" is located at infinity. Recall that all parallel rays are focused to the same point in the back focal plane of an ideal lens. It does not matter *where* the rays originated; only their orientation in space is critical. In our analysis of the single-slit diffraction pattern, we calculated the irradiance by tracing rays that left the slit at the same angle.

From the discussion in the preceding paragraph we know that either slit alone creates a single-slit diffraction pattern on the screen, and that these two patterns overlap exactly. However, when we open both slits simultaneously, the resulting irradiance is not simply the sum of the irradiance from each slit. Because the waves from the two slits are coherent, the electric fields, rather than the irradiance values, add at each point. As we saw in the discussion of a two-source interference pattern, the electric fields from the two sources are sometimes out of phase and thus add to zero. At other points the electric fields are exactly in phase and combine constructively to give an irradiance *four* times greater than the irradiance when only one slit is open. The resulting total irradiance pattern is given by the product of the single-slit and double-slit irradiances

$$I = I_{max} \text{sinc}^2\left(\frac{\beta}{2}\right)\cos^2\left(\frac{\phi}{2}\right)$$

(38.5)

where

$$\beta = \frac{2\pi a}{\lambda}\sin(\theta) \quad \text{and} \quad \phi = \frac{2\pi d}{\lambda}\sin(\theta)$$

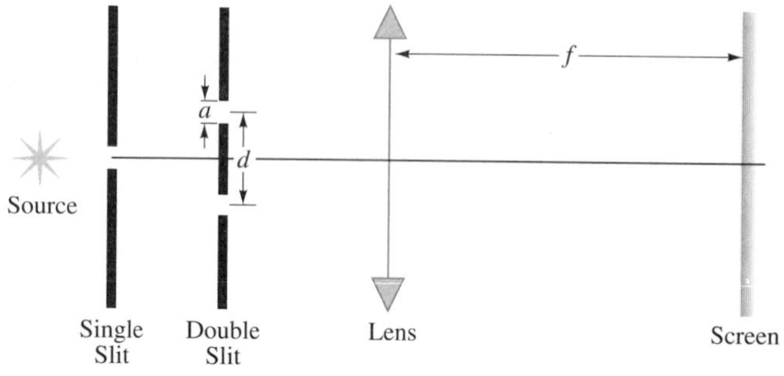

FIGURE 38.11

Geometry for Young's double-slit experiment with the interference pattern projected on a screen located in the focal plane of the lens. The finite width of the slits causes a modulation of the cosine-squared fringes predicted by the point-source model.

The visual appearance of such patterns is shown in Figure 38.12. The broad features in the pattern are due to the width of the individual slits. The finer lines are due to the interferences between the two slits. The central broad maximum of the single-slit pattern is called the **diffraction envelope,** or **diffraction halo.** The effect of slit separation and slit width is shown in Figure 38.13. As differences in the rows of Figure 38.13 remind us, narrow slits give rise to a broad diffraction envelope, whereas broader slits give rise to a narrower envelope. Differences between patterns in different columns of Figure 38.13 are due to the interference between slits, with more widely spaced slits giving rise to more closely spaced fringes.

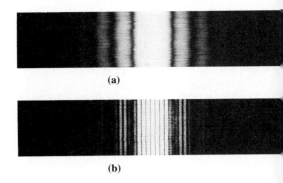

(a)

(b)

FIGURE 38.12

(a) Single-slit interference pattern. (b) Double-slit interference pattern created by two slits each with the same width as in (a). The fine fringes are cosine-squared fringes due to double-source interference. The overall modulation is a result of the finite width of the individual slits.

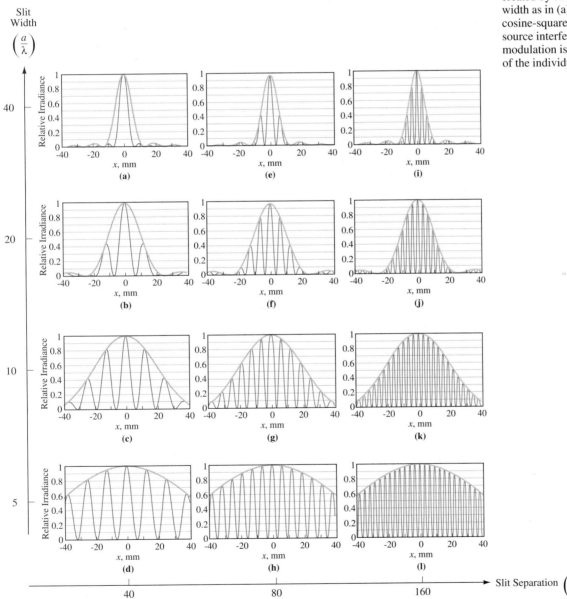

FIGURE 38.13

Double-slit interference patterns showing the effect of slit width and slit separation. Relative slit width, shown along the vertical scale at the left, increases toward the top. Relative slit separation increases to the right as indicated on the horizontal scale at the bottom.

Concept Question 3
Carefully distinguish the meanings of β, θ, and ϕ in Equation (38.5).

Concept Question 4
In Young's double-slit experiment is the location of the intensity maxima changed if the double slits are of unequal width? Describe the observed pattern.

38.3 Diffraction Gratings (Optional)

In Chapter 35 we saw that prisms can be used to separate the various wavelengths composing a beam of light. In the case of prisms this dispersion occurs because the index of refraction of the prism material varies as a function of wavelength. In this chapter we have seen that constructive interference caused by diffraction also depends on wavelength. We can employ this wavelength dependence of diffraction to disperse different wavelengths of light into different regions of space. Light from sources, such as the sun and incandescent light bulbs, provide a continuous rainbow of colors known as a spectrum. Other sources, such as gas-discharge lamps, produce light that is composed of a series of very narrow wavelength bands known as **spectral lines.** The analysis of these spectral lines gives important information regarding the atomic structure and composition of the emitting source. (See Chapter 40.) In this section we describe the principle behind one type of diffractive element used to separate optical spectra.

We begin our discussion by returning to our analysis of diffraction patterns created by multiple slits. If you look back at Figure 37.25 a general feature is clearly evident: the greater the number of slits, the narrower the maxima, and, therefore, the dark areas between the maxima are broader. When the number of slits becomes extremely large, say several hundred, the pattern for monochromatic light consists of many narrow lines, one for each value of m in Equation (37.17). The line corresponding to each value of m is known as a **diffraction order.** If light composed of two wavelengths is incident on such an array of slits, each wavelength forms a diffraction pattern. In accord with $d \sin(\theta) = m\lambda$, the bright spots corresponding to different wavelengths occur at different angles θ and, therefore, are separated on any screen where the pattern falls. In other words, we have a means by which we can sort out in space the different wavelengths that make up a single beam of light in a fashion similar to the dispersion effect that occurs in prisms. In the case of a prism the dispersion is due to the wavelength dependence of the refractive index of the prism material. In the case of the diffraction grating the dispersion is caused by diffraction and interference.

A periodic array of slits used to separate light into its spectral components is known as a **diffraction grating.** In principle, any periodic array of diffracting elements can serve as a diffraction grating. If the elements actually block light at regular intervals, changing the amplitude of the wavefront, the grating is known as an **amplitude grating.** Some gratings are completely transparent, but have thickness that varies in a periodic fashion. The alternating thickness causes Huygens' wavelets to have different phases when they emerge from the grating. Such gratings are known as **phase gratings.** Because nearly all of the light striking a phase grating passes through it, these gratings produce brighter interference patterns than amplitude gratings. A grating can be constructed so that light passes through it or so that light is reflected from its surface. These gratings are known as **transmission** and **reflection gratings,** respectively. The machines used to create the rulings on master diffraction gratings are among the most precise ever created by humans. They operate while on vibration-isolated platforms in temperature-controlled rooms. Moreover, interference patterns are used to control the motion of the cutting tool as the grating is fabricated. Less expensive copies, known as **replica gratings,** are often made by pressing master gratings into softened plastic.

Two important parameters are used to characterize diffraction gratings. One parameter, **resolving power,** describes how close two wavelengths can be and still be spatially separated by the grating. The second parameter, the **free spectral range,** describes how different two wavelengths can be before different diffraction orders begin to overlap. Let's examine each parameter in more detail.

Resolving Power of a Diffraction Grating

Our ability to separate the diffraction maxima due to nearly equal wavelengths depends on two factors: the width of the bright spot formed by each wavelength and the spatial separation of the two bright spots. The width of each bright spot depends on the number of slits illuminated; the separation of the bright spots depends on the distance between adjacent slits. First we find the angular width of each maximum. Equation (37.17) reminds us that the bright spots for a diffraction grating occur when $d \sin(\theta_m) = m\lambda$. In this case we know that at the observation screen the phasors representing the electric field from each slit are all aligned as in Figure 38.14(a). In order to reach a minimum, the chain of phasors must bend into a circular arc that closes on itself, thus creating a zero resultant. To accomplish this geometry each phasor must turn by an angle $\Delta\phi = 2\pi/N$ as shown in Figure 38.14(b). This change in phase angle $\Delta\phi$ occurs when the rays turn their direction in space from angle θ_m to angle $\theta_m + \Delta\theta$. We can relate the phase ϕ to the direction angle θ from the relation

$$\phi = 2\pi\frac{OPD}{\lambda}$$

$$\phi = 2\pi\frac{d \sin(\theta)}{\lambda}$$

To find the effect of small changes in θ, we differentiate this expression, obtaining

$$\frac{d\phi}{d\theta} = 2\pi\frac{d \cos(\theta)}{\lambda}$$

For small changes we write $\Delta\phi = (d\phi/d\theta)\,\Delta\theta$ so that near $\theta = \theta_m$

$$\Delta\phi = 2\pi\frac{d \cos(\theta_m)\,\Delta\theta}{\lambda}$$

As we mentioned previously in our discussion of phasors, a minimum occurs in the interference pattern when the phase difference $\Delta\phi$ between electric fields contributed by adjacent slits is equal to $2\pi/N$. Hence, we can solve the equation above for the angular width $\Delta\theta_{\text{width}}$ of the spectral line when the phase change is $\Delta\phi = 2\pi/N$:

$$\frac{2\pi}{N} = 2\pi\frac{d \cos(\theta_m)\,\Delta\theta_{\text{width}}}{\lambda}$$

or

$$\Delta\theta_{\text{width}} = \frac{\lambda}{Nd \cos(\theta_m)} \tag{38.6}$$

The second quantity we need is the angular separation of two wavelengths that differ by a small amount $\Delta\lambda$. To find this separation we begin by differentiating with respect to λ the equation that describes the location of bright spots:

$$d \sin(\theta) = m\lambda$$

$$d \cos(\theta)\frac{d\theta}{d\lambda} = m$$

The quantity $d\theta/d\lambda$ is known as the **dispersion** $\mathcal{D}$ of the grating, and from the preceding equation we can see that

$$\mathcal{D} = \frac{m}{d \cos(\theta)}$$

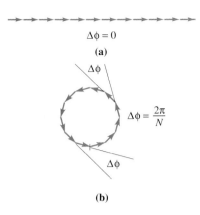

FIGURE 38.14

(a) The maximum resultant amplitude occurs when all the phasors have the same phase. (b) Zero resultant amplitude occurs when the chain of N phasors wraps around and closes on itself. This complete cancellation occurs when the phase difference between adjacent phasors is $2\pi/N$.

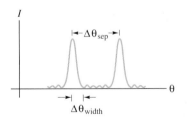

The critical factors in determining resolution are the angular width of the individual bright spots and the angular separation between the centers of the bright spots belonging to different wavelengths.

Near a bright spot where $\theta = \theta_m$ the angular separation $\Delta\theta_{\text{sep}}$ of two wavelengths separated by $\Delta\lambda$ is given by

$$\Delta\theta_{\text{sep}} = \frac{d\theta_m}{d\lambda}\,\Delta\lambda = \mathcal{D}\,\Delta\lambda = \frac{m\,\Delta\lambda}{d\cos(\theta_m)} \tag{38.7}$$

If we employ the Rayleigh criterion for resolution, two lines are just resolved when the maximum from one wavelength falls on the first minimum of the second wavelength's diffraction pattern. We achieve this result by setting $\Delta\theta_{\text{width}}$ from Equation (38.6) equal to $\Delta\theta_{\text{sep}}$ from Equation (38.7):

$$\frac{\lambda}{Nd\cos(\theta_m)} = \frac{m\,\Delta\lambda}{d\cos(\theta_m)}$$

or

$$\frac{\Delta\lambda}{\lambda} = \frac{1}{mN}$$

The quantity $\Delta\lambda/\lambda$ is the fractional change in wavelength that can be resolved by our grating. The smaller this fraction, the better the resolution of the grating. It is traditional, however, to use a "bigger-is-better" parameter $\lambda/\Delta\lambda$ known as the **resolving power** $\mathcal{R}$

$$\mathcal{R} = \frac{\lambda}{\Delta\lambda} = mN$$

We see that the greater the number of slits illuminated and the higher the diffraction order, the greater the resolving power.

EXAMPLE 38.2   The Sodium Doublet

When rock salt is placed on a Meker-type burner, the flame becomes bright yellow. The yellow light results from two closely spaced wavelengths (the sodium doublet) at 589.00 nm and 589.59 nm. Light from the flame passes through a slit and is collimated by a lens. The resulting beam then strikes a diffraction grating with 400 lines/mm at normal incidence. If the third-order spectrum is used, how much of the grating must be illuminated to narrow the lines sufficiently so that the angular separation between the lines is equal to twice that required by the Rayleigh criteria?

SOLUTION   We require twice the resolving power required by the Rayleigh criterion $2\mathcal{R} = 2\lambda/\Delta\lambda = 2(589.3\ \text{nm})/(0.59\ \text{nm}) = 2.0 \times 10^3$. Because $\mathcal{R} = mN$, we must illuminate

$$N = \frac{\mathcal{R}}{m} = \frac{2.0 \times 10^3}{3} = 670\ \text{slits}$$

which requires 670 slits/(400 slits/mm) = 1.7 mm of the grating to be illuminated, that is, only a fairly narrow beam is required.   ◄

Free Spectral Range

The equation describing the location of maxima in multiple-source interference patterns (Equation 37.17) tells us that longer wavelengths are diffracted through larger angles. We can anticipate that eventually the first-order pattern from one wavelength will overlap the second-order pattern from a lower wavelength. That is, if

$$d\sin(\theta_1) = m\lambda_1$$

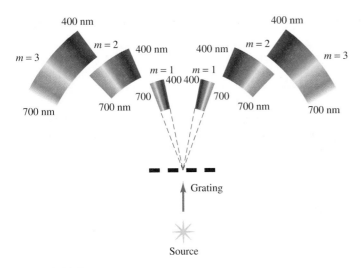

FIGURE 38.15

When white light is incident normally on a diffraction grating, the wavelengths are separated into a spectrum. A separate spectrum is produced for each order m. Notice that the angular width of the spectrum increases with order. Also note that the spectra from different orders can overlap.

and

$$d \sin(\theta_2) = (m + 1)\lambda_2$$

then, when $\theta_1 = \theta_2$,

$$m\lambda_1 = (m + 1)\lambda_2$$

The difference between the two overlapping wavelengths is known as the **free spectral range**

$$\Delta\lambda_{\text{fsr}} = \lambda_1 - \lambda_2 = \frac{\lambda_1}{m + 1} = \frac{\lambda_2}{m}$$

The phenomenon of overlapping spectra from different orders is shown in Figure 38.15.

EXAMPLE 38.3   Seeing Double

A student intends to use a diffraction grating to observe the bright lines in the Balmer series spectrum of hydrogen. The visible wavelengths of this spectrum are 656.7 nm (red), 486.1 nm (blue-green), 432.1 nm (blue), and 410.2 nm (violet). If the student observes the second-order diffraction spectrum, what does she see?

SOLUTION   The free spectral range of the grating near $\lambda_1 = 656.7$ nm is

$$\Delta\lambda_{\text{fsr}} = \frac{\lambda_1}{m + 1} = \frac{656.7 \text{ nm}}{2 + 1} = 218.9 \text{ nm}$$

This means that the third-order maximum of a wavelength $\lambda_2 = m\,\Delta\lambda_{\text{fsr}} = 2(218.9 \text{ nm}) = 437.8$ nm overlaps this line, that is, $3(437.8) = 2(656.7)$. We conclude that two of the shorter wavelengths repeat before the longest wavelength is observed. Perhaps the easiest way to see this is to calculate the angle at which each line appears for each of the first three orders. Because the order of appearance of spectral lines is independent of d, we can pick any convenient value, say $d =$

3000 nm. Applying $d \sin(\theta) = m\lambda$, we find

COLOR	WAVELENGTH (nm)	1ST ORDER (DEGREES)	2ND ORDER (DEGREES)	3RD ORDER (DEGREES)
Red	656.7	12.6	26.0	41.0
Blue-green	486.1	9.3	18.9	29.1
Blue	432.1	8.3	16.7	25.6
Violet	410.2	7.9	15.9	24.2

As the angle of observation is increased from 15° the following colors are seen (the diffraction order is shown in parentheses): violet (2), blue (2), blue-green (2), violet (3), blue (3), red (2).

38.4 X-Ray Diffraction (Optional)

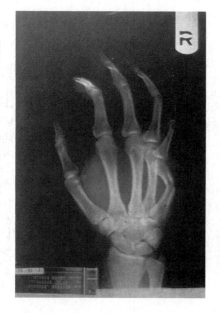

Medical radiographs show the shadows of dense bones which shield the film from X-rays.

The principles of diffraction and interference apply to all electromagnetic waves, not just the visible spectrum. **X-rays** are electromagnetic waves with wavelengths ranging from 0.01 nm to 10 nm, about 5000 to 50 times shorter than those of visible light. X-rays were discovered in 1895 by Wilhelm Röntgen (1845–1923) and are known as Röntgen rays in most European countries. These waves interact strongly with electrons in matter and are absorbed within distances that range from a fraction of a millimeter to several centimeters in most materials. Because of their ability to penetrate matter, X-rays are used to create images of structures, such as broken bones, which are inaccessible to visible light. The images formed for medical and dental applications are actually simple shadowgraphs. No lenses are used in the image formation. X-ray-sensitive film is placed on one side of the structure of interest, and X-rays irradiate the structure from the opposite side. Bones and teeth are visible in the shadowgraph because they absorb X-rays more strongly than does surrounding soft tissue, leaving shadows on the photographic film.

It is difficult to create optical systems that use X-rays. Because X-rays are strongly absorbed by matter, refractive optical elements are not employed. Reflective elements can be used but only when the angle of incidence is large. Figure 38.16 shows a mirror used to focus short-wavelength X-rays. Diffractive optical elements appear promising as

FIGURE 38.16

A mirror used to focus high-energy X-rays. Note that X-rays traveling near the axis of the cylinder strike the mirror at quite large angles of incidence.

FIGURE 38.17

The zone plate is a diffractive optical element used to focus X-rays. Only rays that reach the focal point in phase are allowed to pass through transparent zones of the plate. The rings that block the unwanted X-rays are gold rings supported on a silicon nitride membrane 120 nm thick. The plate was fabricated by Erik Anderson. (From Malcolm Howells, Jano Kirz, and David Sayre, "X-ray Microscopes," *Scientific American,* **264**(2), 89 (February 1991).)

"lenses" for X-rays. The technology that was originally developed to fabricate integrated microcircuits with dimensions on the order of tens of nanometers can, in fact, be used to create ultrafine diffraction gratings. These gratings can be used to focus long-wavelength X-rays. Many of the diffractive elements used for X-rays do not resemble the parallel-ruled lines of traditional gratings. Patterns of concentric rings with varying radii and thickness, such as that shown in Figure 38.17, are known as zone plates. Zone plates can be used as lenses to focus X-rays.[1] Other special-purpose focusing elements have complicated patterns derived from diffraction equations.

Bragg Diffraction

The regular arrangement of atoms that occurs in crystals forms natural diffraction gratings. One can deduce the spacing between slits in an optical grating by an analysis of the diffraction pattern it causes when illuminated by light. (See Problem 25.) In 1913, Max von Laue (1879–1960) proposed that one could, in a similar fashion, determine the interatomic spacing of atoms by an analysis of the pattern created when X-rays are diffracted by crystals. His prediction proved correct, and X-ray diffraction experiments have become one of the primary means by which crystal structures are determined.

One experimental arrangement for X-ray diffraction is shown in Figure 38.18(a). A collimated beam of X-rays containing a broad range of wavelengths is incident on a single crystal. Constructive interference occurs only at particular angles and results in a diffraction pattern that consists of many bright spots similar to those in Figure 38.18(b). Such an

---

[1] For more details regarding this emerging technology see Malcolm R. Howells, Janos Kirz, and David Sayre, "X-ray microscopes," *Scientific American,* **264**(2), 88–94 (February, 1991).

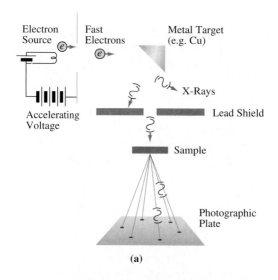

(a)

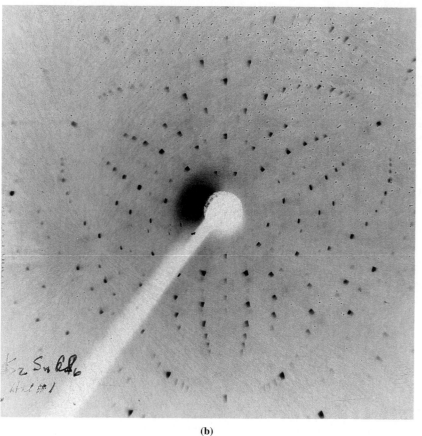

(b)

FIGURE 38.18

(a) Schematic diagram showing collimated X-rays shining on a crystal sample. (b) A Laue diffraction pattern. (Laue pattern courtesy of H. G. Smith.)

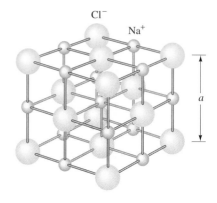

FIGURE 38.19

The cubic arrangement of atoms in a crystal of salt (NaCl). The structure shown is a unit cell with edge length 0.5627 nm.

interference pattern is known as a **Laue diffraction pattern.** To illustrate the principles involved we consider a cubic crystal structure. Figure 38.19 shows a structure model of a salt crystal. The Na^+ and Cl^- ions occupy alternate positions in a cubic array of points called the crystal **lattice.** In this ionic crystal the electrons with which the X-rays interact are closely bonded to the ionic sites, so it is sufficient to consider the ions themselves. Notice that the ions lie in planes separated by a constant distance.

The ions shown in Figure 38.19 constitute a **unit cell.** A unit cell is the smallest subset of atoms that, if repeated in each of three directions, reproduces the entire crystal. The shortest distance a unit cell must be translated to become coincident with the next adjacent unit cell is called the **lattice constant.** For noncubic unit cells there can be up to three lattice constants. The unit cell of Figure 38.19 is said to contain four chlorine ions. This number results from considering that only one-eighth of each of the chlorine ions located at the corners belongs to this particular unit cell. In addition only one-half of each of the chlorine ions located in the center of the cell faces are inside the unit cell. Thus, one calculates $8 \times (1/8) + 6 \times (1/2) = 4$ chlorine ions in this unit cell. Crystals of organic molecules can contain hundreds of atoms per unit cell.

In Figure 38.20 we show a cross section of several layers of unit cells. Notice that it is traditional in X-ray diffraction to indicate the direction of rays with the **grazing angle** θ *measured from the crystal plane* rather than a normal to the plane. To describe X-ray diffraction we defer to Sir William Lawrence Bragg (1890–1971), who in 1912 suggested the mathematical form of the law that describes the conditions of constructive interference:

> The easiest way to approach the optical problem of X-ray diffraction is to consider the X-ray waves as being reflected from sheets of atoms inside the crystal. When a beam of monochromatic (uniform wavelength) X-rays strike the crystal, the wavelets scattered by atoms in each sheet combine to form a reflected wave. If the path difference for waves reflected from different sheets is a whole number of wavelengths, the wave trains will combine to produce a strong reflected beam.[2]

As discussed in Problem 47, when X-rays are incident on the crystal, wavelets reflected from adjacent atoms within a given plane combine through the mechanism of constructive interference to form a strongly diffracted wavefront at many possible angles. However, only when the angle of incidence θ equals the angle of diffraction θ' is the condition for constructive interference independent of both wavelength and interatomic spacing within the plane. Because $\theta = \theta'$, one speaks of reflections from Bragg planes. Let's now look for the conditions under which such specularly reflected waves from adjacent planes combine constructively.

Figure 38.20 shows two rays that are partially reflected by adjacent crystal planes. The two rays travel equal distances until arriving at points a and b. Furthermore, after reaching points c and b they again travel equal distances to the detector. As a consequence, the extra distance traveled by the lower ray is the distance along the path *aoc*, which is $2\,\Delta l$ in Figure 38.20. Because $\angle abo$ is equal to θ, we can write that $\Delta l = d \sin(\theta)$. When the extra path length traveled is equal to an integer number m of wavelengths, the diffracted rays are in phase when they reach the detector:

$$2d \sin(\theta) = m\lambda \qquad \text{(Constructive interference)} \qquad (38.8)$$

The two conditions (angle of incidence equal to angle of reflection and Eq. (38.8)) for constructive interference are known as **Bragg's law.**[3] Regarding Equation (38.8) Bragg modestly wrote, "I first stated the diffraction condition in this form in my initial adventure into research in a paper presented to the Cambridge Philosophical Society in 1912, and it has come to be known as Bragg's law. It is, I have always felt, a cheaply earned honor, because the principle had been well known for some time in the optics of visible light."[4] (Incidentally, in 1915 Sir William Lawrence Bragg (for whom the law is named) was awarded the Nobel prize jointly with his father, Sir William Henry Bragg (1862–1942), for their study of crystal structures using X-rays.) Bragg's law is similar in form to the bright-spot condition for double-slit interference. There are, however, two important differences. Note that the "2" does not appear in the double-slit formula (Eq. (37.15(a)).

[2] Sir Lawrence Bragg, "X-ray crystallography," *Scientific American,* **218**(6), 58–70 (July 1968) p. 58.

[3] F. K. Richtmyer and E. H. Kennard, *Introduction to Modern Physics,* 4th ed., 3rd Impression. (McGraw-Hill, New York, 1947), p. 472.

[4] Sir Lawrence Bragg, "X-ray crystallography," *Scientific American,* **218**(6), 58–70 (July 1968) p. 58.

Concept Question 5
How many sodium ions are there in the unit cell shown in Figure 38.19?

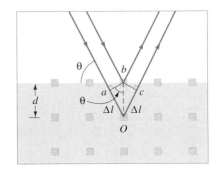

FIGURE 38.20

A two-dimensional cross section of a cubic lattice. The dots represent the centers of unit cells not the location of individual atoms. The unit cells can be thought of as lying in any of a great number of parallel planes with differing orientations. Strong reflections can occur if the path difference between X-rays scattered from identical, parallel planes equals an integral number of wavelengths of the X-rays used.

Moreover, θ in Bragg's law is measured from the crystal plane, while θ in the double-slit formula is measured from the normal.

Notice that, as shown in Figure 38.20, many crystal planes can be defined within a crystal even when it has cubic symmetry. Constructive interference can result when the X-rays satisfy the Bragg condition for any set of planes within the crystal. We should also note that Bragg's law describes conditions for interference maxima but says nothing about the intensity of these ''bright spots.'' It may well happen that spots allowed by interference conditions do not appear because the details of the electron distribution within the unit cell result in low reflectivity from a particular plane. Sorting out such details and unraveling the underlying crystal structure is the business of X-ray crystallography.

EXAMPLE 38.4   *Reflecting on Bragg Diffraction*

Consider Bragg reflections of X-rays with wavelength $\lambda = 1.544$ Å from the mineral aragonite ($CaCO_3$) the unit cell of which is orthorhombic (box-shaped) with sides $a = 5.72$ Å, $b = 7.94$ Å, and $c = 4.94$ Å. (The Angstrom (Å) is a unit of length commonly used in X-ray studies and equals 10^{-10} m.) Find the angles for the first four diffracted orders permitted by Bragg's law from the family of planes shown in Figure 38.21(a). Note carefully that the small dots in Figure 38.21(a) represent the position of the centers of the unit cells and not the location of individual atoms.

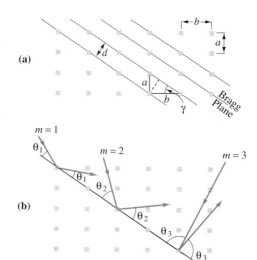

FIGURE 38.21

(a) Location of unit cells in a rectangular lattice. See Example 38.4. (b) Directions of three orders of the Bragg diffraction equation.

SOLUTION   We first determine the interplane spacing d. From the right triangle formed by connecting lattice points with the Bragg plane as the hypotenuse in Figure 38.21(a) we see that $d = b \sin(\gamma)$ where $\tan(\gamma) = a/b$. Substituting numerical values, we find

$$d = b \sin\left[\tan^{-1}\left(\frac{a}{b}\right)\right] = (7.94 \text{ Å}) \sin\left[\tan^{-1}\left(\frac{5.72}{7.94}\right)\right] = (7.94 \text{ Å}) \sin(35.8°) = 4.64 \text{ Å}$$

We can now find the grazing angles permitted by Bragg's law $d \sin(\theta) = m\lambda$ through

$$\theta = \sin^{-1}\left(m\frac{\lambda}{d}\right) = \sin^{-1}\left(m\frac{1.544 \text{ Å}}{4.64 \text{ Å}}\right) = \sin^{-1}(m\,0.333)$$

resulting in values of

m:	1	2	3	4
θ:	19.4°	41.7°	86.6°	not permitted as it requires $\sin(\theta) > 1$

Directions for the first three orders are shown in Figure 38.21(b).   ◀

EXAMPLE 38.5   *The Origin of Bragg Planes*

Consider Bragg scattering from a lattice in which the unit cells are arranged in a rectangular lattice. (a) Find the condition such that X-rays scattered from atoms separated by distance a in the surface of a crystal are in phase. (b) Find the condition such that X-rays scattered from atoms on the surface and atoms located a distance b below the surface are in phase. (c) Show that these two conditions taken together can be described as specular reflection from Bragg planes.

SOLUTION   (a) Figure 38.22(a) shows X-rays incident on a rectangular lattice making angle α with the crystal surface. If the path length difference between the two rays is a whole number i of wavelengths, the scattered waves are in phase. We find the path-length difference Δl_1 in the usual way by connecting the two rays with a perpendicular line segment at the position of each atom. The geometry of the right triangles thus formed leads to

$$\Delta l_1 = a \cos(\alpha) - a \cos(\beta) = i\lambda$$

(b) In a similar fashion we examine two rays scattered by atoms separated vertically by a distance b in Figure 38.22(b). The path-length differences Δl_2 between these two rays must also be equal to a whole number j of wavelengths:

$$\Delta l_2 = b \sin(\alpha) + b \sin(\beta) = j\lambda$$

(c) If we take the ratio of the two path-length conditions in parts (a) and (b), we can write

$$\frac{\cos(\alpha) - \cos(\beta)}{\sin(\alpha) + \sin(\beta)} = \frac{i/a}{j/b}$$

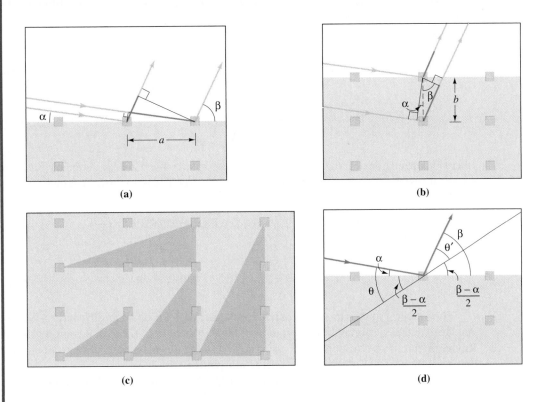

(a)   (b)

(c)   (d)

FIGURE 38.22

Coherent superposition of X-rays reflected from within a plane and between planes can be thought of as reflections from Bragg planes. See Example 38.5.

Using the trigonometric identities for the differences of the sines and the cosines of two angles (see Appendix 1), this expression can be written

$$\tan\left(\frac{\beta - \alpha}{2}\right) = \frac{ib}{ja}$$

Recall that the tangent of an angle is often thought of as the "opposite side over the adjacent side" in a right triangle containing the angle. From the previous equation we see that the angle $(\beta - \alpha)/2$ can be found in a right triangle with sides the lengths of which are equal to integral numbers of atomic spacings. As shown in Figure 38.22(c) the hypotenuse of such angles always connects lines of atoms within the crystal. These lines are the edges of planes described by Bragg.

The significance of these Bragg planes of atoms can be seen if we look at the angle θ between the incoming X-rays and the Bragg plane at angle $(\alpha - \beta)/2$ and compare θ to the angle θ' between the diffracted beam and the same Bragg plane. Figure 38.22(d) shows the geometry for the case in which $i = j = 1$. We see that

$$\theta = \alpha + \frac{\beta - \alpha}{2} = \frac{\alpha + \beta}{2}$$

$$\theta' = \beta - \frac{\beta - \alpha}{2} = \frac{\alpha + \beta}{2}$$

This result holds regardless of the values of i and j. We have thus discovered that $\theta = \theta'$, that is, the incoming and diffracted beams can always be thought of as making equal grazing angles with some set of parallel Bragg planes.   ◀

38.5 Summary

The **far-field**, or **Fraunhofer**, diffraction pattern from a single slit of width a has an irradiance pattern described by

$$I_{\text{tot}} = I_{\text{max}} \, \text{sinc}^2\left(\frac{\beta}{2}\right) \tag{38.1b}$$

$$\beta = 2\pi\left(\frac{a \sin(\theta)}{\lambda}\right) \tag{38.2}$$

where θ depends on whether the diffraction takes place in a lensless geometry with the screen at distance L, or in an imaging geometry using a lens of focal length f:

$$\tan(\theta) = \frac{y}{L}, \qquad \left(\text{No lens, screen at distance } L > \frac{a^2}{\lambda}\right)$$

$$\tan(\theta) = \frac{y}{f}, \qquad \text{(Screen in focal plane of lens)}$$

The smallest spot into which parallel rays can be focused by a **diffraction-limited** lens of diameter a subtends an angle θ at the lens, where $\theta = 1.22\lambda/a$. According to the **Rayleigh criterion** two points are just resolved when the maximum of one falls on the first minimum of the other point's diffraction pattern.

(Optional) The irradiance pattern caused by the interference between two slits of finite size is given by

$$I = I_{\text{max}} \, \text{sinc}^2\left(\frac{\beta}{2}\right) \cos^2\left(\frac{\phi}{2}\right) \tag{38.5}$$

where

$$\beta = \frac{2\pi a}{\lambda} \sin(\theta) \quad \text{and} \quad \phi = \frac{2\pi d}{\lambda} \sin(\theta)$$

(Optional) A **diffraction grating** can be used to separate different wavelengths that compose a beam of light. Bright spots for a given wavelength occur when

$$d \sin(\theta_m) = m\lambda$$

where d is the slit separation. The **resolving power** $\mathcal{R} = \lambda/\Delta\lambda$ of a diffraction grating is given by

$$\mathcal{R} = mN$$

(Optional) Diffractive optical elements can be used to focus longer wavelength X-rays. The atomic structure of crystals can be determined from patterns formed by X-rays reflected from the crystal. Constructive interference occurs when the distance d between atomic planes in the crystal satisfies the **Bragg condition**

$$2d \sin(\theta) = m\lambda \qquad\qquad (38.8)$$

where θ is the **grazing angle** measured from the crystal plane, and d is the separation of reflecting planes.

PROBLEMS

38.1 Single-Slit Diffraction

1. The width of a narrow slit is to be measured using its far-field diffraction pattern. Collimated light of wavelength 633 nm from a helium-neon laser illuminates the slit, and the diffraction pattern is observed on a screen 1.50 m from it. The distance between the first-order dark spots on either side of the central maximum is 2.78 cm. What is the width of the slit?

2. A collimated beam of light from a sodium-vapor lamp (589 nm) is incident on a slit of width 25.4 μm. A lens of focal length 50.0 cm is placed behind the slit and the diffraction pattern observed on a screen in the focal plane of the lens. What is the width of the diffraction halo, that is, the distance between the first-order dark spots on either side of the central maximum?

3. Light of 488-nm wavelength is incident on a single slit with 0.500 mm width. How far from the slit must the screen be located if the first diffraction minimum is to be 0.600 mm from the center of the central maximum?

4. Light of 633 nm is incident on a 0.250-mm wide slit. Compute the width of the central maximum on a screen located 2.50 m from the slit.

5. A collimated light beam containing only two wavelengths illuminates a single slit. In the far-field diffraction pattern it is found that the fourth-order minimum of one color falls exactly on the third-order minimum of the other color. What is the ratio of the two wavelengths?

6. The eye can be modeled as a thin lens of 20.0-mm focal length. (See Section 36.6.) The eye is most sensitive to light with a free-space wavelength of 555 nm. (a) What is the smallest spot to which this model eye can focus a 555-nm beam of collimated light on a bright day when the pupil diameter is 2.00 mm? [*Hint:*

It is the wavelength of the light in the vitreous humor ($n = 1.336$) filling the eye that must be used.] (b) What is the smallest spot to which this model eye can focus a 555-nm beam of collimated light in dim light when the pupil diameter is 8.00 mm? (c) Compare these spot sizes to the diameter of the light-sensitive cells in the eye: rods (2 μm) used for night vision and cones (6 μm) used in bright light.

7. A 4.00-cm diameter lens with a 50.0-cm focal length is used to focus a collimated beam of nearly monochromatic light of wavelength 546 nm. What is the diameter of the smallest diffraction-limited spot to which the beam can be focused?

8. The headlights of a certain automobile are separated by 1.80 m. How far away is the automobile when an observer can just resolve the two headlights? Assume an average wavelength of 550. nm and a nighttime pupil diameter of 6.00 mm.

9. A certain night-vision camera is sensitive to wavelengths of 10.6 μm. The effective diameter of the diffraction-limited germanium telephoto lens is 12.0 cm. At what distance can the camera just resolve two people who are standing 3.00 m apart?

10. Optical fibers are formed by drawing a long thin fiber from the heated end of a large glass cylinder known as a preform. In the drawing tower the diameter of the fiber is often monitored by using diffraction. A collimated laser beam is directed perpendicular to the length of the fiber, and the position of diffraction minima are monitored using detector arrays. The analysis makes use of the fact that the locations of diffraction minima for an opaque strip are the same as those for an equivalent-width slit in an otherwise opaque screen. If light of 633 nm is directed at a fiber of outer diameter 140. μm, how far apart are the two first-order minima on a detector array located 0.75 m from the fiber?

11. A police radar gun emits electromagnetic waves at a frequency of

10.525 GHz. (a) If the emitting antenna has an effective diameter of 25.0 cm, how large a spot does the radar gun illuminate when the beam has reached 60.0 m from the gun? (b) Out to what distance does the beam spread no more than the width of a single oncoming car, say about 2.00 m?

12. The binary stars in Kruger 60 in the constellation Cepheus have an angular separation of about 2.5 seconds of arc. What is the minimum diameter of the primary mirror of a telescope that could just resolve this binary? (Assume the telescope is diffraction-limited and just resolves the binary pair according to the Rayleigh criterion at a wavelength of 555 nm.)

13. An 8.20-m diameter mirror is in preparation for the European Southern Observatory in Chile's Atacama Desert. (a) If such a mirror is used in a diffraction-limited telescope what is the angular separation θ between two stars that are just resolved (according to the Rayleigh criterion) at a wavelength of 555 nm? (b) Calculate the resolving power $R = \theta^{-1}$ of this telescope. As usual θ must be in radians. (c) Compare this resolving power with that of the 200-in diameter Mount Palomar telescope.

14. Light of 514-nm wavelength is incident on a single slit of 0.500-mm width. A diffraction pattern is formed on a screen 1.60 m from the slit. Find the fraction of the intensity maximum (I_{tot}/I_{max}) at a distance of 2.00 mm from the center of the principal maximum.

15. A diffraction pattern is formed on a screen 2.50 m from a slit of 0.125 mm-width. If the wavelength of the incident light is 488 nm, find the fraction of the intensity maximum (I_{tot}/I_{max}) at a distance of 1.80 mm from the center of the principal maximum.

16. Light from a He-Ne laser is directed through a slit of width 0.254 mm. The diffraction pattern is created on a screen in the focal plane of a lens with focal length 20.0 cm. (a) Sketch a picture of the geometry of the experiment and a qualitative drawing of the diffraction pattern. (b) What is the distance between the two first-order dark spots on either side of the central maximum? (c) What is the relative irradiance (I_{tot}/I_{max}) at a distance of 0.745 mm from the center of the pattern?

17. The diffraction pattern of a single slit of width 0.125 mm is formed using a lens of focal length 35.0 cm and light (633 nm) from a helium-neon laser. Calculate the relative irradiance (I_{tot}/I_{max}) at points located 0.500 mm, 1.00 mm, and 1.50 mm from the middle of the central maximum.

38.2 Effect of Finite Slit Width on Double-Slit Interference Patterns

18. Show that, in principle, the number of bright fringes contained within the first-order diffraction halo for the Fraunhofer diffraction pattern of a double slit is $2[\![d/a]\!] + 1$ where $[\![\ ]\!]$ denotes the greatest-integer function. Note that this formula assumes when the final double-slit interference maxima fall exactly at the position of the single-slit diffraction minima that the maxima will actually be observed. Because in this case the final maxima are not actually observed, the formula is often written $2(d/a) - 1$. (Other authors note that although the full maxima are not observed, slight increases in irradiance can be seen just inside the edges of the diffraction halo. These authors count the missing bright spots as $\frac{1}{2}$ each and thus obtain $2(d/a)$. We prefer to count *observable* fringes only.)

19. (a) How many observable fringes appear between the two first-order minima of the diffraction halo for a double-slit pattern if $\lambda = 550.$ nm, the slit separation is 0.250 mm, and the slit width is 0.100 mm? [*Hint:* See Problem 18.] (b) What is the ratio of the irradiance of the second bright fringe lying to the right of the center line to that of the central maximum?

✳20. A helium-neon laser of wavelength 632.8 nm illuminates a double slit with slit separation 0.325 mm. If the $m = \pm 3$ order bright spots of the double-slit pattern fall exactly on the first-order minima of the single-slit diffraction envelope, what is the width of the slits?

21. Light from a mercury-vapor lamp is filtered to produce a beam of light with wavelength 546.1 nm. This light falls on a double-slit arrangement where the two slits each have a width of 0.150 mm. In the resulting interference pattern exactly seven fringes are visible within the central diffraction envelope. The next pair of fringes are not seen because they fall exactly on the minima of the single-slit pattern. What is the separation of the two slits?

38.3 Diffraction Gratings (Optional)

22. Light composed of two dominant wavelengths illuminates a diffraction grating. When the incoming light strikes the grating at normal incidence, the third-order maximum of one wavelength falls on the second-order maximum of the other wavelength. What is the ratio of the two wavelengths?

23. A diffraction grating produces a second-order maximum for light from a sodium-vapor lamp ($\lambda = 589$ nm) at 30.0°. For what wavelength is the second-order maximum at an angle 1.00% greater?

✳24. A diffraction grating is 3.50 cm wide and has a slit spacing of 2.00 μm. What is the grating's resolving power in the third order?

25. A certain diffraction grating produces a deviation of 30.0° in the second order when illuminated by light of wavelength 600. nm at normal incidence. (a) What is the slit separation for this grating? (b) If 2.54 cm of this grating is illuminated, what is the resolving power of the grating?

✳26. Two spectral lines from an argon ion laser have wavelengths 488 nm and 514 nm. (a) What is the angular separation of these two lines in the second-order spectrum produced by passing the laser beam through a diffraction grating of 520. lines/mm? (b) Is there any order m such that the mth order of the 488-nm line occurs at a smaller angle than the $(m - 1)$th order of the 514-nm line?

27. The H_α spectral line of the hydrogen atom has a wavelength of 656.30 nm. When the nucleus of the atom contains an extra neutron, the atom (now called deuterium) produces a wavelength of 656.48 nm for the same transition. Using a grating of 450 lines/mm what width of the grating must be illuminated in order that these two lines be resolved in second order?

38.4 X-Ray Diffraction

28. A beam of monochromatic X-rays is incident on a single crystal. If the first-order Bragg reflection maximum is observed at an angle of 4.10° from a set of crystal planes separated by 0.234 nm, (a) what is the wavelength of the X-rays being used? (b) At what angle should the second-order diffracted beam be expected?

29. X-rays can be used to identify the mineral content of ores. The mineral galena (PbS) has the rock salt structure with a lattice constant of 0.5936 nm. If a Bragg reflection maximum is detected at a grazing angle of 11.7°, what is the wavelength of the X-rays being used? (Assume $m = 2$. The odd orders have zero intensity for cubic crystals of the rock salt structure. See Problem 48.)

30. If a beam of X-rays with wavelength 0.150 nm falls on a set of crystal planes with spacing 0.275 nm, what is the highest order Bragg reflection possible?

31. A monochromatic beam of copper K_α X-rays with wavelength 0.154 nm are incident on crystal planes in silicon separated by 0.313 nm. Calculate all the possible grazing angles that satisfy the Bragg condition.

32. One way to produce a monochromatic beam of X-rays is to reflect a multiwavelength beam from a crystal and block reflected X-rays except those that pass through a narrow slit. Suppose a beam of X-rays is directed at a NaCl crystal (lattice spacing 0.563 nm) so that the only rays exiting this spectrometer must reflect at an angle of 60.0° with a normal to the crystal face. What is the wavelength of the X-rays the second-order Bragg reflection of which is passed? (Note that for crystals with cubic unit cells such as NaCl reflections with odd orders have zero intensity. Can you see why this is the case? See Problem 48.)

33. Lithium fluoride crystals are sometimes used for the analysis of X-rays diffracted by other crystals. For one particular orientation of the LiF crystal an interplanar spacing of 0.1424 nm is used. (a) If a first-order Bragg reflection occurs at a grazing angle of 35.4° with these planes, what is the wavelength of the X-rays being analyzed? (b) If instead, an orientation of the LiF crystal such that planes with a spacing of 0.2014 nm are used, at what angle is the first-order Bragg reflection found?

Numerical Methods and Computer Applications

34. Using the Bessel function of order 1 denoted J_1, the far-field irradiance pattern due to diffraction of a circular slit is described by the equation

$$I = I_o \left[\frac{2J_1(x)}{x} \right]^2$$

where $x = ka \sin(\theta)$, $k = 2\pi/\lambda$, and a is the diameter of the circular aperture. Spreadsheet template J1(X).WK1 calculates $J_1(x)$ using the power series

$$J_1(x) = \frac{x}{2}\left[1 - \frac{1}{1!2!}\left(\frac{x}{2}\right)^2 + \frac{1}{2!3!}\left(\frac{x}{2}\right)^4 - \frac{1}{3!4!}\left(\frac{x}{2}\right)^6 + - \cdots \right]$$

Column A of the spreadsheet contains x values. Column B contains the $J_1(x)$ values calculated by summing columns D through W. Columns D through W each contain one additional term in the power series contained in the [ ] brackets above. Each term a_N is generated from the previous term using $a_N = -\{(a_{N-1})/[N(N-1)]\} (x/2)^2$. (a) Starting with the spreadsheet given, create a column of values for I/I_o for the diffraction of a circular aperture and plot a graph of I/I_o versus x. (b) With an accuracy of at least three significant figures find the value of x for which the first zero of I/I_o occurs. (c) In an actual diffraction pattern $x = ka \sin(\theta)$, where a is the diameter of the aperture, k is the wave number $2\pi/\lambda$, and θ is the angle subtended from the aperture by a line on the screen joining the center of the pattern and the point of observation a distance r away. Use these definitions and your answer from part (b) to show that, for small angles, the radial distance r to the first dark ring is given by $r = 1.22\lambda/\theta$.

35. Spreadsheet template SLITS.WK1 contains a spreadsheet template used to calculate interference patterns, such as those shown in Figure 38.13. Modify this spreadsheet by adding three additional slits. Produce graphs that are the equivalent of Figure 38.13(a) and Figure 38.13(f) for five slits in each case.

36. If you have not already done so, perform the modification of SLITS.WK1 as described in Problem 35. Now add the following second modification. Increase the original phase of the wave exit-ing each slit by $n\beta$, where the extra phase increment is β and the slits are numbered $n = 0, 1, 2, 3, 4. \ldots$ Produce graphs for $\beta = 0, 0.50$, and 1.00 radian. Describe what is happening. (This same effect can be used to "steer" the beam from a line of fixed radio antennas.)

37. (a) Modify spreadsheet template PHASOR.WK1 (see Problem 37.31) to show the addition of ten phasors each with a phase difference of $\Delta\phi = 0.250$ rad from the next. Produce a graph and add arrow heads to the phasors. (b) Using values from the spreadsheet, calculate the length of the resultant for the case described in (a). (c) Compare your result from part (b) to that predicted by Equation (38.16b) for the continuous case.

★38. Use symbolic mathematics software or numerical integration to determine the fraction of the total power in a single-slit diffraction pattern contained between the two first-order minima.

General Problems

39. Photographs in newspapers and books (including this one) are made up of a very large number of closely spaced small dots. (a) How far apart should these dots be so that they are not quite resolved (according to the Rayleigh criterion) when the printed picture is held at the near point of normal vision (25.0 cm)? Use a pupil diameter of 2.00 mm and an average wavelength of 555 nm. (b) How does this compare with 300 dots/in spacing used in many laser printers?

40. Monochromatic light falls at normal incidence on a 10 000 line/cm diffraction grating that is submerged in a liquid with an $n_1 = 1.40$ index of refraction. A first-order maximum is produced at 31.2°. The grating is now submerged in a liquid of unknown refractive index n_2, and the first-order maximum occurs at 38.6°. What is the free-space wavelength of the light and the refractive index n_2?

41. Show that if collimated light is incident on a diffraction grating at an angle ψ with respect to the normal as shown in Figure 38.P1, the grating formula for bright spots becomes $d [\sin(\theta) + \sin(\psi)] = m\lambda$.

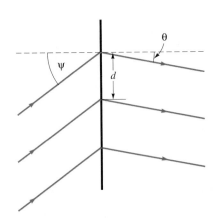

FIGURE 38.P1 Problem 41

42. Compute the angular resolution of the 2.40-m Hubble telescope for light of wavelength 555 nm.

43. Compute the smallest diameter required of a telescope that can resolve the two stars of a binary system if their angular separation is 2.00 seconds of an arc. Assume a visible wavelength of 555 nm.

44. The energy distribution across the wavefront of a laser beam is not uniform but instead varies like a bell-shaped curve known as "Gaussian." A detailed analysis of the angular width of a circular Gaussian beam measured in the far-field shows that the angular half-width of the beam is given by $\theta = 2\lambda/(\pi a)$, where a is the minimum spot size to which the beam is focused. (a) Compare the three equations relating the beam angle, diffraction spot size, and wavelength for the following cases: plane waves through a single-slit, plane wave through a circular aperture, and Gaussian beam of circular cross section. Aside from a numerical factor on the order of 1, all these equations indicate that the spreading of a beam is dependent on a fundamental ratio. What is that ratio? (b) If a He-Ne laser with a diameter of 0.500 mm is aimed at the moon, how large is the diffraction-limited spot?

45. A typical edge-emitting solid-state laser emits light of wavelength 720 nm from a rectangular area measuring 2.00 μm by 10.0 μm. (a) What is the approximate shape and dimensions of the spot from such a laser at a distance of 2.50 m? (A simple estimate based on single-slit diffraction is all that is required.) (b) Draw a simple sketch clearly showing how the long axis of the diode's emitting area is related to the long axis of the expanded beam.

46. Light from a helium-neon laser illuminates a double slit. On a screen 3.65 m away the interference pattern shown in Figure 38.P2 is created to the actual scale shown. Make appropriate measurements on the photograph and calculate (a) the slit separation and (b) the slit width.

FIGURE 38.P2   Problem 46

47. (a) Show that the two X-ray beams reflected from adjacent points in a plane shown in Figure 38.P3 have a path-length difference given by

$$AB - CD = d[\cos(\alpha) - \cos(\beta)]$$

(b) The two reflected rays are in phase if the path length difference is $n\lambda$, where $n = 0, 1, 2, \ldots$. If the rays consist of many wavelengths, for what value(s) of n are the rays in phase for all wavelengths? What does this condition on n imply about α and β?

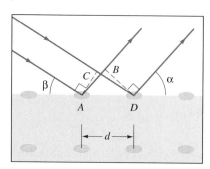

FIGURE 38.P3
Problem 47

48. Consider Bragg diffraction from the particular set of Bragg planes of the rectangular lattice shown in Figure 38.P4(a). In this figure we do not follow the conventional practice of showing points that represent the centers of unit cells. Instead we show the position of the atoms themselves and outline the boundary of a typical unit cell with a dotted line. Figure 38.P4(b) shows a crystal lattice identical to that of Figure 38.P4(a) except that an additional atom is present on the center of each rectangular face. These new atoms lie in planes parallel to the planes of the original lattice and are separated from the original lattice by $d/2$. (Note that we changed the *structure* of the unit cell but not the spacing of the cells. On a conventional diagram showing only the position of the centers of unit cells, both lattices would appear identical.) Show that if the Bragg condition is fulfilled for odd m using the original set of planes, that the reflections from the newly introduced set of

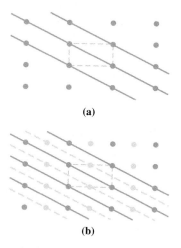

(a)

(b)

FIGURE 38.P4   Problem 48

planes are exactly out of phase with the waves reflected from the original planes.

49. Diffraction effects occur for all waves including sound. Estimate the angular width of a 20.0 cm-diameter speaker emitting sound of frequency 1.00×10^4 Hz.

50. In a certain experiment X-rays are obtained by bombarding copper with electrons. The wavelengths of the resulting X-rays are measured using a Bragg diffraction from an analyzing crystal. Diffraction peaks are found at Bragg angles of 63.86° and 54.22°. If the peak at the larger angle corresponds to a wavelength of 0.15405 nm (the K_{α_1} line), what is the wavelength corresponding to the smaller angle?

MODERN PHYSICS

At the close of the 1800s the branches of physics known as mechanics, thermodynamics, and electricity and magnetism were well established. Mechanics founded on Newton's three laws of motion had passed innumerable tests. Sophisticated mathematical techniques had been brought to bear on complex problems with great success. The laws of thermodynamics had proved successful in explaining a vast number of phenomena from chemical processes to steam engines. Maxwell's equations successfully accounted for electromagnetic radiation and laid the foundation for numerous practical inventions from the electric motor to the wireless telegraph.

Newton's laws together with the conservation of momentum and energy, the three laws of thermodynamics, and Maxwell's equations constitute the cornerstones of what is now known as classical physics. The first decade of the 1900s brought with it new and unexpected results that required modification of many of the ideas of classical physics. The most significant of these are special relativity and quantum theory. These topics are traditionally referred to as *modern physics* and are the subject of the final chapters of this text. For those of us living at the close of the twentieth century it may seem strange to call work done during the first decades of the 1900s

"modern" physics. Nonetheless, the then-new theories of relativity and quantum physics changed our way of thinking about the universe so dramatically that every theory since has had to incorporate their implications. It is in this sense that one speaks of "modern physics" as distinguished from "classical physics."

Although these ideas are now nearly a hundred years old, you are likely to find them new and surprising because they lie outside most of our common experience. To notice significant results of special relativity, objects must travel at speeds near that of light. Few of us knowingly come into contact with such objects in our daily lives. Although the results of quantum mechanics affect everything we do, we seldom observe fundamental quantum mechanical effects directly. For exactly this reason the study of twentieth-century physics is great fun and full of surprises and new insights. If you sometimes have difficulty believing the results of some of the theories we discuss, don't feel alone. Some professional physicists too, especially those educated during the classical era, had great difficulty accepting some of these new models. However, experiment is not to be denied. And it appears that the universe is a very strange place indeed . . .

Special Theory of Relativity

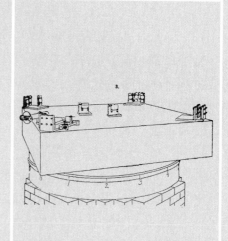

In this chapter you should learn

- about the search for ether.
- the postulates of Einstein's special theory of relativity.
- how to apply the Lorentz transformations.
- formulas for time dilation and length contraction.
- how to apply the relativistic Doppler shift.
- relativistically correct equations for momentum and kinetic energy.
- how to apply $E_o = m_o c^2$.

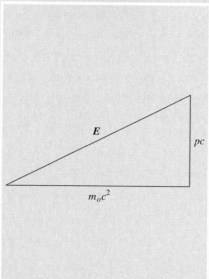

The physics described in this chapter is not intuitive. Indeed, we are going to have to ask you to lay aside many of your preconceptions about the way things are. If you have some trouble believing that the results discussed in this chapter are true, don't feel bad. An entire generation of physicists had similar difficulty coming to grips with the implications of the experiments we describe here. But Nature is stubborn. The universe is the way it is, not the way we think it should be.

The drama begins innocently enough. Even before Maxwell's description of light as an electromagnetic wave, a search began to find the medium in which light propagated. Ocean waves travel on water. Sound waves require air to propagate. So what is the medium of propagation for light? The mysterious medium came to be called luminiferous ether. This ether must fill the entire vacuum of space in order to allow light to propagate to us from distant stars. The hope was that this ether would act as an absolute reference frame for the entire universe. As we describe in the remainder of the chapter, the search for ether failed. Light, it appears, needs no medium for propagation and stubbornly maintains a constant speed regardless of the observer's motion. This universal constancy of the speed of light ultimately forced a rethinking of the very nature of space and time.

39.1 The Speed of Light

One of the earliest attempts to measure the speed of light was reported by Galileo in his *Dialogues Concerning Two New Sciences* published in 1638. In this monumental book the character Sagredo, representing Galileo, describes an experiment in which two observers stand a great distance apart while holding shuttered lanterns. The first observer uncovers his lantern. On seeing the first lantern, the second observer is to uncover his own lantern. The first observer notes the time interval between first uncovering his own lantern and seeing light from the second lantern. This time interval is the sum of the time light takes to make the round trip plus, of course, the reaction time of the distant observer. If we speculate that the observers were 1 km apart, the time for light to travel the 2-km round trip is about 7 μs, a value completely lost in the typical human reaction time of 0.2 s. Galileo was led to conclude regarding the speed of light, " . . . if not instantaneous it is extraordinarily rapid."[1]

The first quantitative measurements of the speed of light were made by Danish astronomer Olaus Roemer (1644–1710) in 1675 while observing the rotation of the moons of Jupiter. The period of rotation of the moons could be measured accurately by watching many cycles. From knowledge of the rotational period the expected time when a given moon would pass into Jupiter's shadow could be predicted accurately. Roemer noted that as the earth went through its yearly orbit there was a difference between the predicted time of an eclipse of a Jovian moon and the observed time. When the earth was farthest from Jupiter, the eclipse occurred later than expected, whereas when the earth was closest to Jupiter the eclipse occurred early. Roemer concluded that the difference was due to the time taken by light to travel the diameter of the earth's orbit, about 17 min. From his data Roemer estimated the speed of light to be about 2×10^8 m/s.

By the end of the nineteenth century a number of terrestrial measurements of the speed of light had been made with the result of 2.99×10^8 m/s that matches today's accepted value to three significant figures.

Luminiferous Ether

By 1880, there seemed little doubt that the wave model of light was correct. Earlier that century Young, Fresnel, Huygens, and others had successfully described interference and diffraction phenomena in terms of a wave model. In 1865, Maxwell showed that the laws of electromagnetism predicted waves traveling at the speed of light. At that time it must

---

[1] Galileo Galilei, *Dialogues Concerning Two New Sciences,* translated by Henry Crew and Alfonso De Salvio (Northwestern University Press, Evanston, IL, 1968), p. 44.

have seemed as if all that remained for physicists to do was to identify the medium of propagation for light waves. Just as sound travels in air, it was thought that light must travel in something, and that something was eventually named the **luminiferous ether.** Because light reaches us from distant stars, the ether must fill the entire universe. Surely such a ubiquitous and important substance was worth investigating.

When we say that sound travels at 340 m/s, we are describing the speed of sound relative to the air through which it propagates. If we drive 30 m/s through still air toward an oncoming sound wave, the sound approaches us at 370 m/s. If we drive away from a sound source at 30 m/s (relative to the air) the sound catches up with us but at a speed that we observe to be 310 m/s. Clearly, the speed of sound *relative to us* varies with our speed *relative to the air.* Such calculations are simple and obvious. In an exactly similar way we can imagine that as we sit on the surface of the earth we are hurtling through space as our planet circles the sun. Consequently, it is reasonable to expect that the speed of light that we measure relative to us varies with our speed relative to the ether.

The speed of the earth around the sun is about 30 km/s. Although this is a rather astounding speed it is only about 0.01% of the speed of light. Therefore, the maximum change in the velocity of light that we might hope to observe is this same fraction. Unfortunately, this velocity change was also the same size as the experimental uncertainty in the measurement of light velocity around 1880. What was needed to overcome this dilemma was a method to measure only the *changes* in the speed of light rather than the total speed itself. Of course, the only thing to which one can easily compare the speed of a light beam is that of another light beam. One convenient arrangement for comparing the speed of two light beams taking different paths is known as a Michelson interferometer.

Michelson Interferometer

In 1907, Albert Abraham Michelson (1852–1931) became the first American to receive the Nobel Prize in science. He was cited for the invention of precision optical instruments and measurements carried out with them. One of these instruments, known as the **Michelson interferometer,** is shown schematically in Figure 39.1. Light from a single monochromatic source is split by a partially silvered mirror called a beam splitter. The two beams travel along perpendicular paths. Each beam is reflected by a mirror and returned to

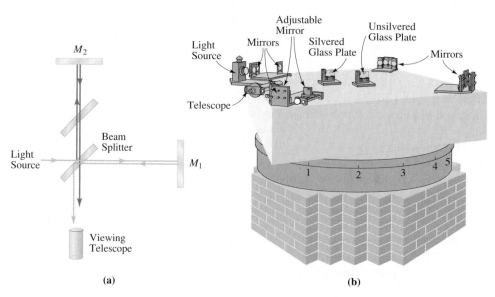

(a)　　　　　　　　　　　　　　(b)

FIGURE 39.1

(a) A schematic diagram of the Michelson interferometer. Light from the source strikes the beam splitter where part of the beam is reflected to mirror M_2 and part passes to mirror M_1. After reflection from the mirrors the beams are reunited at the beam splitter and viewed in the telescope. Rays reflected at the mirrors have been offset for clarity. (b) Rendering of Michelson and Morley's apparatus.

the beam splitter. At the beam splitter, portions of each beam are recombined and travel together to a telescope. In the telescope, interference fringes similar to those described in Chapter 37 are observed. Because the light sources available to Michelson had quite short coherence lengths (see Section 37.2), the path lengths in each of the perpendicular paths (called arms) had to be made as identical as possible. The extra glass plate introduced into one arm guarantees that both beams travel through equal thicknesses of glass. This arrangement ensures that any interaction between the ether and the glass occurs symmetrically in both arms.

The Experiment of Michelson and Morley

Let's see how Michelson's interferometer could be used to detect our motion through the ether. Imagine that the interferometer is oriented such that one arm is parallel to our velocity V through the ether as shown in Figure 39.2. For each arm we need to calculate the time it takes for light to make a round trip from the beam splitter to the fully silvered mirror and back again to the beam splitter. Because we assume all velocities V are constant, we'll make repeated use of the formula for the relation between time t and the distance d: $t = d/V$.

Let's first look at light traveling parallel to our motion through the ether, the x-axis of Figure 39.3. Remember, the *assumption* is that light travels at its characteristic speed c relative to the ether, which is stationary in our picture. Light moving rightward from the beam splitter to the mirror chases after the moving mirror with a relative speed of $c - V$. The time t_r for light to travel the distance L to the mirror is thus

$$t_r = \frac{L}{c - V}$$

After reflecting from the mirror the light travels to the left and runs into the oncoming beam splitter. The light and beam splitter approach with a relative speed $c + V$. Thus, the time t_l for the light to return to the beam splitter is

$$t_l = \frac{L}{c + V}$$

Thus, the total time $t_{\parallel}$ for the beam traveling parallel to our motion is

$$t_{\parallel} = t_r + t_l = \frac{L}{c - V} + \frac{L}{c + V} = \frac{2Lc}{c^2 - V^2}$$

It will turn out to be convenient to write this last expression for $t_{\parallel}$ in terms of the time it would ordinarily take light to travel a distance $2L$ multiplied by a correction factor. Factoring out the ordinary time $2L/c$, we have

$$t_{\parallel} = \frac{2L}{c} \left(\frac{1}{1 - V^2/c^2} \right) \tag{39.1}$$

We now need to perform a similar calculation for the beam traveling at right angles to our assumed motion. Remember that our point of view is that we are at rest relative to the ether and the entire apparatus moves past us at velocity V. As illustrated in Figure 39.4, the mirror M_2 and the beam splitter move to the right as the light makes its journey. We see that in order to strike the mirror in the center and return through the center of the moving beam splitter, the light must travel along the diagonal lines indicated. If we denote the time between the moment when the light leaves the beam splitter and when it strikes the mirror as t_u, we can calculate the length of the hypotenuse from the Pythagorean theorem

$$(ct_u)^2 = L^2 + (Vt_u)^2$$

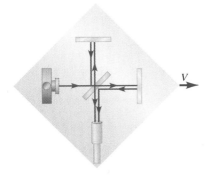

FIGURE 39.2

We imagine the entire Michelson interferometer to be traveling through the ether at speed V along the direction of one interferometer arm.

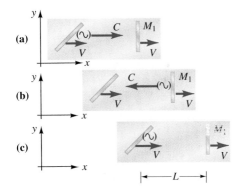

FIGURE 39.3

(a) Light traveling through the ether at speed c chases the mirror M_1, which is receding at speed V. (b) After reflection, the light travels back toward the approaching beam splitter. (c) The light arrives at the beam splitter.

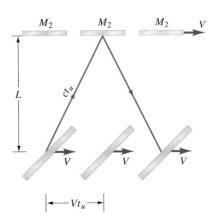

FIGURE 39.4

Light traveling along the arm of the Michelson interferometer that is perpendicular to the direction of the presumed motion must take the diagonal path shown.

From this expression we can solve for the time of travel

$$t_u = \sqrt{\frac{L^2}{c^2 - V^2}}$$

From symmetry we see that the total time of travel $t_\perp$ for the round trip in the direction perpendicular to our motion is twice this value. We then write $t_\perp = 2t_u$ in a form similar to Equation (39.1):

$$t_\perp = \frac{2L}{c} \sqrt{\frac{1}{1 - V^2/c^2}} \tag{39.2}$$

As you can see, the two beams traveling in different directions relative to our motion through the ether take different times for their respective trips. Before calculating the time difference it is traditional to expand the expression for these two time intervals in a power series in V^2/c^2 using the binomial expansion. (See Appendix 1.) The expression V/c occurs so frequently in relativity that it is also traditional to give this velocity ratio its own symbol β. Thus, for our two times we have

$$t_\parallel = \frac{2L}{c}\left(\frac{1}{1 - V^2/c^2}\right) = \frac{2L}{c}\left(\frac{1}{1 - \beta^2}\right) = \frac{2L}{c}(1 + \beta^2 + \beta^4 + \cdots)$$

and

$$t_\perp = \frac{2L}{c}\sqrt{\frac{1}{1 - V^2/c^2}} = \frac{2L}{c}\sqrt{\frac{1}{1 - \beta^2}} = \frac{2L}{c}\left(1 + \frac{1}{2}\beta^2 + \frac{3}{8}\beta^4 + \cdots\right)$$

For velocities on the order of the earth's orbital velocity around the sun we expect the velocity ratio β to be on the order of 10^{-4} so that $\beta^2 \approx 10^{-8}$. Clearly, we need not worry about β^4. Hence, we can find the time difference Δt between the travel time for these two beams by keeping track of the power series out to terms of order β^2:

$$\Delta t = t_\parallel - t_\perp = \frac{2L}{c}\left(\frac{1}{2}\beta^2\right) \quad \text{(Time difference predicted by ether model)} \tag{39.3}$$

In Michelson's first interferometer the distance L was 1.20 m. For $\beta = 10^{-4}$ we obtain $\Delta t = 4 \times 10^{-17}$ s. How could Michelson ever hope to measure such a short time? The answer is: by observing the interference pattern created when the light beams recombine.

Let's compute the phase difference between the two beams that would be created by the time delay calculated from Equation (39.3). If the delay were exactly equal to the oscillation period T of the light, the phase delay would be 2π rad. Thus, the actual phase delay $\Delta\phi$ caused by a delay of time Δt can be calculated from a simple proportion

$$\frac{\Delta\phi}{2\pi} = \frac{\Delta t}{T}$$

Substituting the expression for Δt from Equation (39.3) and recalling that $cT = \lambda$, we find

$$\frac{\Delta\phi}{2\pi} = \left(\frac{L}{\lambda}\right)\beta^2 \tag{39.4}$$

Using visible light of $\lambda = 550$ nm and $L = 1.2$ m, we obtain the fractional phase shift $\Delta\phi/2\pi = 0.02$. Michelson's plan was to rotate his interferometer about a vertical axis so that the two arms alternately reversed roles. Thus, during an entire rotation the total

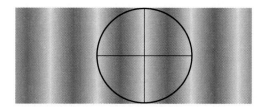

FIGURE 39.5

When the mirrors M_1 and M_2 of the Michelson interferometer are precisely perpendicular, but the arm lengths are unequal, fringes form a circular bull's-eye pattern. If one of the mirrors is tilted from its perpendicular orientation, the center of the bull's-eye moves off to one side, and a nearly straight-line fringe pattern is observed as shown here. A cross hair in the viewing telescope can be used to determine small changes in position of the fringes.

change is twice this value. Michelson was looking at fringe patterns similar to that shown in Figure 39.5. A fractional phase change of 0.500 would move the pattern exactly one-half fringe, moving a dark fringe to precisely where a bright fringe was previously located. The fringe change of 0.04 that Michelson was looking for was small indeed, but Michelson was confident that he could detect changes even smaller. In addition, these calculations based on the orbital velocity of the earth around the sun were a "worst-case" scenario based on the improbable premise that the sun just happened to be at rest in the ether. It seemed much more likely that the sun itself was traveling through the ether, perhaps with a quite large velocity.

When Michelson carried out his experiment in 1881 the results were startling. No fringe shift of the expected size was observed! With his apparatus Michelson was certain, in fact, that no shift larger than 0.02 fringes occurred. The null result was so surprising that Michelson was quickly encouraged to repeat the experiment with greater sensitivity. In 1887, Michelson, in collaboration with E. W. Morley (1838–1923), published the results of another trial for which the distance L was effectively increased by a factor of 10 through the use of multiple reflections from several mirrors. The *expected* fringe shift was now an easily detectable 0.4 fringe. Once again, the result was contrary to expectations. Michelson and Morley placed the upper limit of the fringe shift at no greater than 0.01 fringe.

The Michelson-Morley experiment has been repeated many times always with the same null result.[2] Numerous proposals were put forward to explain the null result. Dutch physicist H. A. Lorentz (1853–1928), for example, proposed that the ether might be partially dragged along with the earth in its motion. In 1892, G. F. FitzGerald (1851–1901) and, independently, H. A. Lorentz suggested that the null results could be explained if the arm oriented parallel to the velocity through the medium were shortened by a factor of $\sqrt{1 - \beta^2}$. It is clear from Equations (39.1) and (39.2) that this effect would indeed account for the lack of any time difference, but the suggestion seemed rather ad hoc.

In 1905, Albert Einstein (1879–1955) suggested that the search for the ether was in fact both fruitless and misleading. If the ether could not be found, it was because it did not exist! Einstein's insight was to suggest that the fundamental assumptions should be discarded if they led to conclusions that contradicted experiment. Einstein challenged the fundamental assumptions regarding the existence of an absolute frame of reference in space and even our fundamental notion about the nature of time itself. As described by

---

[2] See the review paper "New Analysis of the Interferometer Observations of Dayton C. Miller," R. S. Shankland, S. W. McCuskey, F. C. Leone, and G. Kuerti, *Reviews of Modern Physics,* **27,** 167–178 (1955).

Newton, ''Absolute, true, and mathematical time, of itself and from its own nature, flows equably without relation to anything external, . . .''[3] We shall soon see that this is not the case.

39.2 Postulates of Special Relativity

The null result of the Michelson-Morley experiment suggested that no absolute reference frame could be found. Rather than trying to explain away this unexpected result, Einstein raised it to the level of a postulate and looked for its consequences. He suggested that all reference frames moving at constant velocity relative to an intertial frame were equally good, and that the laws of physics should be identical in each of them. He argued, for example, that when a magnet causes a current to be induced in a coil, it doesn't matter which actually moves, the magnet or the coil. He then set out the simple yet profound basis for what is now called his special theory of relativity:

> Examples of this sort, together with unsuccessful attempts to discover any motion of the earth relatively to the ''light medium,'' suggest that the phenomena of electrodynamics as well as the mechanics possess no properties corresponding to the idea of absolute rest. They suggest rather that . . . the same laws of electrodynamics and optics will be valid for all frames of reference for which the equations of mechanics hold good. We will raise this conjecture . . . to the status of a postulate, and also introduce another postulate, which is only apparently irreconcilable with the former, namely, that light is always propagated through empty space with a definite velocity c which is independent of the state of motion of the emitting body.[4]

The collective results following from these two simple postulates are now known as **Einstein's special theory of relativity.** The theory is called ''special'' because of its restriction to '' . . . frames of reference for which the equations of mechanics hold good.'' These days, such frames of reference are known as inertial frames of reference. In 1915, Einstein published his general theory of relativity wherein the restriction to inertial frames was removed.

For convenience of reference we number Einstein's postulates as follows:

Postulate 1: The laws of physics are the same in all inertial reference frames.

Postulate 2: The velocity of light in free space is a constant c regardless of the state of motion of the source.

Because all reference frames moving with constant velocity relative to each other are equivalent, the second postulate implies that the velocity of light is independent of the velocity of the observer as well as the source. Be sure you appreciate the remarkable nature of these statements. They suggest that if you were to run forward at half the speed of light and turn on a flashlight, the light coming out travels at c, no faster than when you stand still and turn on the light. Furthermore, if your friend shines a light in your direction and you run toward your friend at $0.5c$, you measure the speed of the light coming toward you as c, not $1.5c$ as you might expect. In fact, the implication is that observers in constant straight-line motion relative to each other *all* measure the very same beam of light as traveling at c regardless of their individual velocities. To say the implications are profound

Concept Question 1
Suppose two observers charge toward each other each traveling at half the speed of light as measured relative to this book. If one observer turns on a flashlight, how fast does the second observer see the light approaching?

---

[3] From Newton's preface to the first edition of the *Principia* as quoted in *Newton's Philosophy of Nature,* edited by H. S. Thayer (Hafner Publishing, New York, 1960), p. 17. The editor indicates the translation appearing in this collection is Florian Cajori's revision of Motte's translation of the *Principia.*

[4] A. Einstein, *Annalen der Physik,* **17,** 891 (1905); translated by W. Perrett and G. B. Jeffery in *The Principle of Relativity* (Dover, New York, 1923).

is an understatement. Even our intuitive notion of two events occurring simultaneously is destroyed when the events are separated in space.

The Downfall of Simultaneity

When we say that two events are simultaneous, we mean, of course, that they happen at the same time. If you are on one side of campus and a friend is on the other, you might agree to close your physics books simultaneously at 2:00 A.M., exactly. Seemingly, there is no problem provided your watches are synchronized. How might you synchronize your watches? One possible way is shown in Figure 39.6. Here we imagine a third party standing exactly halfway between you and your friend. This third party sets off a bright flash. When you and your friend see the flash you both start your clocks immediately. Because the third party is exactly midway between the two of you and because light travels at the same speed c in both directions, you both see the flash at the same instant and start your clocks synchronously.[5] If all this seems trivial and quite obvious, it may surprise you to learn that a moving observer (perhaps a jet pilot flying by) will say, quite accurately, that you and your friend have in fact failed to synchronize your clocks at all!

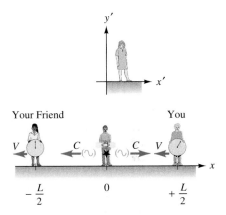

FIGURE 39.7

An observer in the $x'y'$-frame of reference watches an attempt to synchronize clocks in a frame of reference moving to the left at speed V. A flash of light is set off midway between two clocks. The clocks are started when they receive the light pulse. Because the right-hand clock moves toward the light, it starts before the left-hand clock, which is moving away from the light.

When clocks separated by distance are synchronized in their rest frame, they are not synchronized in a frame of reference moving parallel to a line joining the clocks. To an observer who sees the clocks moving, the clock leading in position is trailing in time.

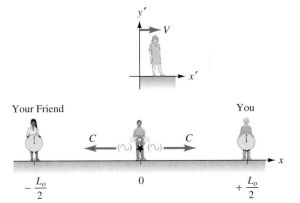

FIGURE 39.6

One method for synchronizing clocks. A light pulse is created midway between two clocks. When the light reaches the clocks, they start simultaneously.

Imagine an observer flying at speed V toward you from your friend's position. From the flyer's frame of reference, it is you and your friend who are moving at speed V as shown in Figure 39.7. Remember, all inertial frames of reference are equivalent, and the flyer's point of view is every bit as valid as your own. The pilot observes the flash set off by the person between you and your friend. The pilot sees that the light travels outward toward you and toward your friend at exactly the same speed, namely, c. This point is critical and we repeat it for emphasis. According to postulate 2, the flyer measures both light beams traveling at exactly c. As a consequence, from the flyer's point of view, you are rushing toward the oncoming light beam while your friend is moving away from the beam directed toward her. Thus, the beam traveling toward you reaches you first and you start your clock before your friend. In the flyer's frame of reference your clocks are not synchronized. Clearly, even time itself is *relative* and does not flow " . . . equably without relation to anything external."

39.3 Relativistic Kinematics

To further illustrate the rather remarkable implications of postulate 2 we describe another thought experiment. Imagine you and your friend each standing at the origin of different coordinate systems. We call yours the S-frame and your friend's the S'-frame. Let the x- and y-axes of your frame be parallel to the respective x'- and y'-axes of your friend's frame. Finally, let your friend's frame move at speed V along your common x-axis as

Concept Question 2
Consider a train with an engineer at the front and a conductor in the rear. This train is coming straight toward a crossing where you are waiting. If the engineer and the conductor start their clocks when they receive a light pulse from a flashbulb set off at the midpoint of the train, which do you see start his clock first?

---

[5] If you're worried about differences in reaction time, we can always arrange for the clocks to be started by identical light-sensitive photodetectors manufactured to arbitrarily close specifications.

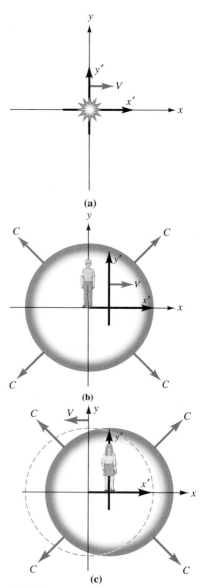

FIGURE 39.8

The S'-coordinate system is traveling along the x-axis of the S-frame of reference at speed V. (a) When the two origins are coincident, a bright flash at the origin creates a spherical pulse of light. (b) The observer in the stationary frame sees light move outward in all directions at exactly c and is therefore in the center of the spherical pulse of light. (c) The observer in the moving frame sees light spread outward in all directions at exactly c and is therefore in the center of the same spherical pulse of light.

shown in Figure 39.8(a). At the moment when the origins of the two coordinate systems are coincident, you set off a light flash at the common origin and you both also start your identical clocks. (There is no problem with simultaneity here, as you are together at the same spot. In all equations and problems in this chapter we assume that clocks are set so that $t = t' = 0$ when the S and S' origins are coincident.) Now from your point of view, light from the flash heads off with speed c in all directions, forming a spherical pulse with your location at the center as in Figure 39.8(b). You might describe this light flash by writing the equation of spherical pulse with radius ct as

$$(ct)^2 = x^2 + y^2 + z^2$$

To keep life a little simpler, we ignore the z-direction out of the page and look only at the circle in the plane of the page. Nothing is lost; in any of the equations that follow, z behaves exactly as y. For this simplified case we can write

$$(ct)^2 = x^2 + y^2 \qquad \textbf{(39.5)}$$

According to postulate 2, your friend also sees the light spreading out at exactly the same rate in all directions as in Figure 39.8(c). Hence, your friend believes that she, too, is in the center of a spherical pulse of light which intersects the $x'y'$-plane in a circle. Therefore, she writes

$$(ct')^2 = x'^2 + y'^2 \qquad \textbf{(39.6)}$$

But, contrary to the drawing of Figure 39.8(c), there are *not* two light pulses. The flash went off just once and both observers are describing one and the same pulse of light. How can you both think that you are at the center of one circle of light? Furthermore, how can you both be right? To resolve these difficulties we must give up some of our preconceived notions about space and time. One of the notions we must relinquish is our classical view of relative motion.

Lorentz Transformations

The rules that determine the relationships between the positions and times given in two different coordinate systems, such as yours and your friend's in the previous paragraphs, are known as **coordinate transformations.** In particular, the transformations we have used throughout this text since Chapter 3 are known as **Galilean transformations.** When something occurs at a well-defined place and time, the "something" is called an **event.** A flashbulb going off is a good example of an event. In Figure 39.9 we show an event, which in your frame of reference, the S-frame, has coordinates (x, y, t). In your friend's frame, the S'-frame, the event has coordinates (x', y', t'). Using a Galilean transformation we expect the coordinates to transform between your frame and your friend's frame according to

$$y' = y$$
$$t' = t \qquad \text{Galilean transformations} \qquad \textbf{(39.7)}$$
$$x' = x - Vt$$

The transformation of the x- and y-positions is clear from Figure 39.7 and, of course, it is obvious that time is the same for all observers—well—at least it used to be obvious. In fact, the transformations of Equation (39.7) are inconsistent with the observations represented by Equations (39.5) and (39.6). That is, if you make the substitutions indicated by Equation (39.7) into Equation (39.6), you will not arrive at Equation (39.5). We are now in a rather uncomfortable position. Either we must give up the Galilean transformations, including the absolute nature of time, or we must deny the evidence of experiment. This choice is by no means clear. History records many instances in which persons of good intent denied the evidence of their eyes rather than give up cherished beliefs. But the whole of modern science is built on the conviction that Nature and experiment are not to be denied. If our intuition is at odds with the results of careful experiments, we must give up our assumptions. Otherwise we are not doing science but something altogether different.

Einstein set out to find a new set of transformation equations that would transform Equation (39.6) into Equation (39.5). We leave the details as a problem and simply present the results here.

$$y' = y$$
$$t' = \gamma\left(t - \frac{Vx}{c^2}\right)$$
$$x' = \gamma(x - Vt)$$

Lorentz transformations   (39.8)

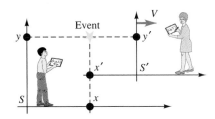

FIGURE 39.9

Two observers witness the same event. In the S-frame the event has space-time coordinates (x, y, t). In the S'-frame moving at speed V the event has coordinates (x', y', t'). The Lorentz transformations must be used to transform between the two sets of coordinates.

where

$$\gamma = \frac{1}{\sqrt{1 - V^2/c^2}} = \frac{1}{\sqrt{1 - \beta^2}}$$

The transformations of Equations (39.8) were discovered by both H. A. Lorentz and A. Einstein. However, they are usually referred to simply as the **Lorentz transformations** in recognition of the fact that Lorentz deduced them in 1904 (a year before Einstein's independent discovery) while attempting to reconcile electromagnetic theory with the null result of the Michelson-Morley experiment. Lorentz's treatment, however, fell far short of the insights of Einstein's elegant theory.

Using the substitutions indicated by the Lorentz transformations, Equation (39.6) does become Equation (39.5). Notice that although we are quite accustomed to seeing time appear in distance equations, such as in the term $x - Vt$, it is rather startling to see position occurring in the equation for time. Putting numerical values into the Lorentz transformations, it is easy to see why you and I don't notice relativistic effects in our everyday comings and goings. When $V \ll c$, the Lorentz transformations reduce exactly to the Galilean transformations with which we are familiar. Thus, we need not abandon our familiar Galilean transformations but merely note their proper place as perfectly adequate low-velocity approximations to the more correct Lorentz transformations.

The factor γ appearing in the Lorentz transformations is a good indicator of the importance of relativistic effects. In Figure 39.10 we show values of γ as a function of the velocity ratio β. For low speeds $\gamma \approx 1$. In particular, taking γ to be 1 introduces an error of less than 1% for speeds up to 4×10^7 m/s.

According to postulate 1, two observers moving with relative speed V find that the same laws of physics hold in each of their reference frames. It is not surprising, therefore, that if you solve the Lorentz transformations of Equations (39.8) for the unprimed variables, the inverse transformations are identical in form except that V is replaced by $-V$ as we would expect:

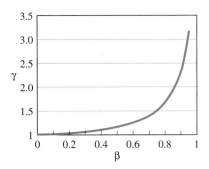

FIGURE 39.10

The factor $\gamma = \sqrt{1 - \beta^2}$ is a measure of the importance of relativistic effects. When γ is near 1, classical Galilean relativity can be used.

$$y = y'$$
$$t = \gamma\left(t' + \frac{Vx'}{c^2}\right)$$
$$x = \gamma(x' + Vt')$$

Inverse Lorentz transformations   (39.9)

To change the Lorentz transformations to their inverse, change all primed variables (x', y', t') to their unprimed counterparts and all unprimed variables to primed. Then change all occurrences of V to $-V$.

We now examine some of the consequences of the Lorentz transformations, which may surprise you even further.

Time Dilation

To explore this business of the relativity of time, let's compare the time interval between two events occurring at the same place in your frame of reference. These two events might be the start and end of a class that takes place at some fixed point in our coordinate system. We could list the space-time coordinates (x, y, t) of a wristwatch resting on the corner of your desk at the beginning and the end of class as $(x_b, 0, t_b)$ and $(x_e, 0, t_e)$, respectively. If you don't move the position of your watch during the class, then $x_b = x_e$. Because the class takes place in your frame of reference and you measure its duration with a clock resting next to you, we refer to the time interval $t_e - t_b$ as the **proper time** T_o for the duration of the class. That is,

$$T_o = t_e - t_b$$

Because the clock is at rest in your frame of reference, it is traditional and logical to call this frame of reference the **rest frame.**

We now imagine that your friend in the moving S'-frame also measures the duration of the class. Your friend finds that the class lasts some time T given naturally enough as

$$T = t'_e - t'_b$$

To compare this interval to times in your frame we use the Lorentz transformations listed in Equations (39.8).

$$T = t'_e - t'_b = \gamma\left(t_e - \frac{Vx_e}{c^2}\right) - \gamma\left(t_b - \frac{Vx_b}{c^2}\right) = \gamma(t_e - t_b) - \frac{\gamma V}{c^2}(x_e - x_b)$$

Because you didn't move your watch, its x-coordinate is constant. Thus, $x_e - x_b = 0$. Consequently, the previous equation reduces to simply

$$T = \gamma T_o \qquad\qquad (39.10)$$

As mentioned earlier γ is always greater than 1 when your friend's frame of reference is in motion relative to yours. Thus, the duration of the class as measured by your friend is longer than the duration you measure in your frame. Put another way, the duration of an event is shorter in the rest frame of the event than in any other frame. It is this time T_o, measured by a single stationary clock located at the position of the event, that is called the proper time.

Let's make up some numbers for illustration. Suppose you measure your class period to be 50 min long. Your friend might find that the class lasted 55 min. If you ask your friend what is going on, she might reply, "Well, your class lasted 55 min. Your watch ticked off only 50 min, so your watch must be running slow." From your friend's point of view, your clock was in motion and that moving clock ran slow.

Moving clocks run slow.

EXAMPLE 39.1   Can Your Friend Get an Extra Five Minutes of Test Time This Way?

How fast does your friend have to be moving to measure the duration of a test as 55.0 min if that test lasted 50.0 min in the frame of the clock on the wall?

SOLUTION   We find your friend's speed through the γ factor in the time dilation formula Equation (39.10):

$$\gamma = \frac{T}{T_o} = \frac{55.0 \text{ min}}{50.0 \text{ min}} = 1.10$$

Recalling the definition of γ

$$\gamma = \frac{1}{\sqrt{1 - \beta^2}}$$

we solve for β to find

$$\beta = \sqrt{1 - \frac{1}{\gamma^2}} = \sqrt{1 - \frac{1}{(1.10)^2}} = 0.417$$

Your friend must travel at a speed $V = 0.417c = 1.25 \times 10^8$ m/s. ◀

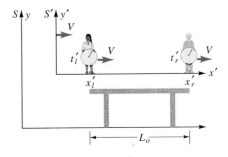

Concept Question 3
As a certain nucleus travels relative to us with a speed such that $\gamma = 2$ it exists for a time of 0.5 s from creation to decay as measured in our frame. How long did the particle exist as measured in its own frame?

Length Contraction

Suppose we set out to perform the relatively (if you'll excuse the expression) simple task of measuring the length of your desk. In your frame of reference the task is straightforward. Because the desk is at rest you have merely to set the desk next to the x-axis of your coordinate system and read its length from the scale. This length measured when the object is at rest in your frame of reference is known as the object's **proper length,** which we denote by L_o. When you record the right and left positions of the desk as x_r and x_l, respectively, this proper length is simply

$$L_o = x_r - x_l$$

Your friend at rest in the S'-frame of Figure 39.11 does not have quite as easy a time of it when she wishes to measure the length of your desk. To her, the desk is in motion. She can determine the length by reading the position of the edges of your desk against her x'-axis. However, because the desk is moving, it is critical that she note the position of both desk edges *at the same moment of time* as measured in the S'-frame. She might make this measurement, for example, by enlisting the aid of an assistant with whom she has synchronized watches so that she can read the position of the right-hand edge of the desk at the same instant that the assistant reads the position of the left-hand edge. Denoting the coordinates of the right-hand and left-hand edges of the desk as x'_r and x'_l, respectively, the length she and her assistant measure is simply

$$L = x'_r - x'_l$$

This length is correct if $t'_r = t'_l$, where these times are the instants when each of the two workers records the position of the edges of your desk. To see how this length compares with the proper length we might again use the Lorentz transformation of Equations (39.8). When we try this, we find

$$L = x'_r - x'_l = \gamma(x_r - Vt_r) - \gamma(x_l - Vt_l) = \gamma(x_r - x_l) - \gamma V(t_r - t_l)$$

$$= \gamma L_o - \gamma V(t_r - t_l)$$

Unfortunately, this equation does *not* give us the result we desire. The reason is that, although she has taken care to ensure that $t'_r = t'_l$, we have no guarantee that $t_r = t_l$. The unprimed times t_r and t_l here are the times at which your friend recorded the positions of the edges of your desk *as you see those times in your frame.* We have already noted in Section 39.2 that events that are simultaneous in one frame are not simultaneous in another frame moving relative to the first when the events occur at different spatial positions.

In order to relate the lengths measured in the two frames let's use the *inverse* Lorentz transformations as described by Equations (39.9). Although we may not be sure of the times t_r and t_l *when* your friend measured your desk, we do know *where* the edges of the desk were in your frame when she measured them in hers; namely, at x_r and x_l. So we transform these back to her frame.

FIGURE 39.11

Observers in the moving S'-frame measure the length of a desk at rest in the S-frame. The observers measure the correct length of the desk if they both record the position of the desk edges at the same instant of time. However, the length so measured is shorter than the length of the desk as measured in the S-frame.

$$L_o = x_r - x_l = \gamma(x_r' + Vt_r') - \gamma(x_l' + Vt_l') = \gamma(x_r' - x_l') + \gamma V(t_r' - t_l')$$

For the case we've been discussing, if your friend is careful to measure the position of the desk edges simultaneously in her frame, then $t_r' - t_l' = 0$ and $x_r' - x_l' = L$, the length of your desk in your friend's frame. Thus, the previous equation reduces to

$$L_o = \gamma L$$

Or, as it is more traditionally written,

$$L = \frac{L_o}{\gamma} \tag{39.11}$$

Objects that are moving appear to be shorter along their direction of motion.

Because γ is always greater than 1, we see that your friend finds that the desk's width L is shorter than you measured it to be. Objects that are moving appear to be shortened along their direction of motion. This phenomenon is known as **length contraction.**

EXAMPLE 39.2   *My Desk Isn't Big Enough as It Is*

Your desk is 90.0 cm wide. If your friend travels at $V = 0.417c$ parallel to the direction of the width, how wide does your friend measure the desk to be?

SOLUTION   The γ factor for this speed is

$$\gamma = \frac{1}{\sqrt{1 - \beta^2}} = \frac{1}{\sqrt{1 - (0.417)^2}} = 1.10$$

The length observed in the moving frame is given by Equation (39.11)

$$L = \frac{L_o}{\gamma} = \frac{90.0 \text{ cm}}{1.10} = 81.8 \text{ cm}$$ ◀

Concept Question 4
A bit of rock from another solar system flies near the earth with a speed such that $\gamma = 2$. If the rock has a length of 0.5 m as measured from the earth, what is its length in its own frame of reference?

Experimental Tests of Relativity

The predictions of special relativity are so counterintuitive that one can hardly help but ask, "Yes, but do these things really happen?" The answer appears to be, unequivocally, yes. Special relativity's predictions have been verified many times. We describe only one such experiment here. In this experiment the "clocks" used are charged particles called μ mesons. (That's pronounced "mu mesons" or simply "muons," if you're on a first-name basis.) These muons are created in the upper atmosphere by cosmic rays and they rain down onto the earth with a speed nearly equal to the speed of light. The muons are unstable particles and fall apart, decaying into an electron and two other elementary particles known as a neutrino and an antineutrino. We call the muons "clocks" because, although the decay of any particular muon is a random event, the decay rate of a large collection of muons is quite predictable. Given a large number of muons, half of them decay in 2.2 μs. In another 2.2 μs half of those remaining decay, leaving only one-quarter of the original number. After another 2.2 μs half of the remaining again have decayed, leaving only one-eighth of the original. Although there are some statistical fluctuations, this rule is quite accurate. The time 2.2 μs is called the **half-life** of the muons. Traveling at a speed greater than $0.99c$ these muons take about 6.5 μs to reach the earth from the upper atmosphere where they are created. This time is nearly three half-lives, so we should expect only about one-eighth (12%) of the muons originally created to reach the earth. When an experiment was performed,[6] scientists discovered that nearly 70% of the muons actually reach the earth.

---

[6]For a more detailed description see D. H. Frisch and J. H. Smith, "Measurement of The Relativistic Time Dilation Using μ-Mesons," *American Journal of Physics*, **31**, 342–355 (1963). Also see the film *Time Dilation—An Experiment with μ-Mesons* (Education Development Center, Newton, Mass., 1963) by the same authors.

What could account for the fact that 70% of the muons survive to reach the earth when only 12% are expected to arrive based on the rest-frame half-life of 2.2 μs? According to the time dilation predicted by the special theory of relativity, moving clocks run slowly. The observed 70% survival rate of the muons can be accounted for if only 0.8 μs elapsed on the moving muon's clock during the 6.5 μs that elapsed on our clock as the muon traveled to earth.

Similar experimental results using muons created by accelerators at CERN (European Council for Nuclear Research) were reported in 1976. In the CERN experiments muons with well-known speeds ($V = 0.9994c$) circulated in a storage ring as they decayed. The moving muons exhibited half-lives about 30 times longer than stationary muons. These results agree with the predictions of the special theory of relativity to better than 0.2%.

EXAMPLE 39.3   *On the Other Hand*

A muon created in the upper atmosphere is traveling toward the earth at a speed such that $\gamma = 8.00$. (a) How long does it take the muon to travel 2.00 km to the earth's surface as measured from the earth's frame of reference? (b) How long is this travel time as viewed by someone in the muon's frame of reference? (c) How does the person in the muon's frame account for the difference in time intervals?

SOLUTION   (a) The muon's speed can be calculated from the definition of γ:

$$\beta = \sqrt{1 - \frac{1}{\gamma^2}} = \sqrt{1 - \frac{1}{(8.00)^2}} = 0.992$$

so that

$$V = \beta c = 0.992c$$

At this speed the muon reaches earth in a time calculated by

$$t = \frac{d}{V} = \frac{2.00 \times 10^3 \text{ m}}{0.992 \ (3.00 \times 10^8 \text{ m/s})} = 6.72 \ \mu\text{s}$$

(a)

(b) We wish to know the time registered in the muon's frame. The muon is the moving clock, thus the time measured in its frame is the proper time T_o. The time in our frame is the time T of Equation (39.10). Therefore, we have

$$T_o = \frac{T}{\gamma} = \frac{6.72 \ \mu\text{s}}{8.00} = 0.840 \ \mu\text{s}$$

Note from our point of view, the muon's clock is ticking too slowly.
(c) From the point of view of someone in the muon's frame, her own clocks are running just fine. However, the distance to the earth is not 2.00 km. The 2.00 km is the proper distance L_o measured by an observer in the earth's frame of reference. To an observer riding along with the muon the distance to the earth is L where, due to length contraction,

$$L = \frac{L_o}{\gamma} = \frac{2.00 \times 10^3 \text{ m}}{8.00} = 250. \text{ m}$$

(b)

Moving objects appear shortened along their direction of motion.

According to a person in the muon's frame, the earth is approaching at $0.992c$, and thus it should take

$$t = \frac{d}{V} = \frac{250. \text{ m}}{0.992 \ (3.00 \times 10^8 \text{ m/s})} = 0.840 \ \mu\text{s}$$

for the earth to arrive. This time is exactly that which the muon-frame observer sees on her watch. The results are completely consistent with her expectations. ◀

Relativistic Doppler Shift

In Chapter 17 we saw that in the case of sound waves, motion of either the source or the receiver causes a shift in the frequency of the sound. This shift is not symmetric, that is, different frequency shifts occur for the same relative velocity depending on which is moving, source or receiver. For example, if the source and receiver are receding from one another, one finds for the shifted frequency ν_r measured by the receiver in terms of the unshifted frequency ν_s of the source (see Eq. (17.27), in which we used the symbol f for frequency of sound):

$$\nu_r = \nu_s\left[\frac{1}{1 + (V_s/V)}\right] \qquad \text{(Doppler shift for sound; source moving away from observer)} \qquad \textbf{(39.12)}$$

$$\nu_r = \nu_s\left(1 - \frac{V_r}{V}\right) \qquad \text{(Dopper shift for sound; observer moving away from source)} \qquad \textbf{(39.13)}$$

where V is the velocity of sound, and V_r and V_s are the speeds of the receiver and the source, respectively. Even when $V_r = V_s$, these equations do not give the same frequency. This result is not a problem for sound waves, because there really is a preferred frame of reference, namely, the frame of the medium through which the sound propagates. Let's see what these equations look like in the context of electromagnetic waves where $V = c$, $V_r/c = \beta_r$, and $V_s/c = \beta_s$. With these substitutions we have

$$\nu_r = \nu_s\left(\frac{1}{1 + \beta_s}\right) \qquad \text{(Incorrect equation)}$$

and

$$\nu_r = \nu_s(1 - \beta_r) \qquad \text{(Incorrect equation)}$$

Now we have a problem. According to the special theory, there is no special frame of reference (no ether), so we should not be able to distinguish which object is moving, source or observer. This problem is solved if we correct Equations (39.12) and (39.13) to take into account time dilation.

To examine the first case, we place ourselves in the rest frame S of the stationary observer. To be concrete let's assume that the source is receding as shown in Figure 39.12. We call the source's moving frame of reference the S'-frame. Let's assume that the frequency of the source is such that a new wave is generated with each tick of the clock in the S'-frame. From Figure 39.12 we can note one reason we receive a lower frequency than the source is emitting: The effective wavelength is longer because the movement of the source away from us stretches out the distance interval between the front and the end of the wave. This feature is precisely the effect described by Equation (39.12). But we now know that there is a second effect due to relativity, namely, "moving clocks run slow." From our point of view in the observer's frame, the source's clocks are ticking more slowly. Because a new wave starts with each tick of the source's clock, we certainly expect to receive a lower frequency from the source. The proper time interval between ticks in the S'-frame is T_o, resulting in a proper frequency of $\nu_o = 1/T_o$. In our frame the source's clock ticks at γT_o, resulting in a source frequency of $\nu_s = 1/(\gamma T_o) = \nu_o/\gamma$. With this substitution, Equation (39.12) becomes

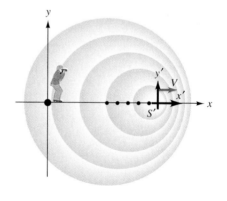

FIGURE 39.12

When observing a receding light source, the frequency of the source appears lower than the frequency measured when the source is at rest. The frequency is lower for two reasons: (1) the wavelength is stretched out as shown in the figure, and (2) moving clocks tick slowly and the oscillations producing the light are essentially a clock.

$$\nu_r = \nu_s\left(\frac{1}{1 + \beta_s}\right) = \frac{\nu_o}{\gamma}\left(\frac{1}{1 + \beta_s}\right) = \nu_o\left(\frac{\sqrt{1 - \beta_s^2}}{1 + \beta_s}\right) = \nu_o\left(\frac{\sqrt{1 - \beta_s}\,\sqrt{1 + \beta_s}}{\sqrt{1 + \beta_s}\,\sqrt{1 + \beta_s}}\right)$$

$$= \nu_o\left(\frac{\sqrt{1 - \beta_s}}{\sqrt{1 + \beta_s}}\right)$$

or

$$\nu_r = \nu_o \sqrt{\frac{1 - \beta_s}{1 + \beta_s}} \tag{39.14}$$

where here ν_o is the source frequency measured in the source frame of reference. In this case (the point of view of the receiver looking at a moving source) we corrected the frequency of the moving source, because to us moving clocks run slowly. When we look at this same situation from the reference frame of the source, we observe the receiver receding and, therefore, we must use Equation (39.13). However, we (with the stationary source) see the receiver's clocks running slow. Consequently, the receiver's frequency must be corrected on the *left-hand* side of Equation (39.13). Performing this calculation (see Problem 21) leads to

$$\nu_o = \nu_s \sqrt{\frac{1 - \beta_r}{1 + \beta_r}} \tag{39.15}$$

where here ν_o is the frequency measured by the receiver in the receiver's frame. As you can see, Equations (39.13) and (39.14) both describe exactly the same result; namely,

$$\nu_r = \nu_s \sqrt{\frac{1 - \beta}{1 + \beta}} \qquad \text{Relativistic Doppler shift} \tag{39.16}$$

where ν_r is the frequency measured by the receiver in the receiver's frame and ν_s is the frequency of the source as measured in the frame of reference of the source. In the case of the receiver and source approaching, the signs of β reverse.

The Twin Paradox

The time dilation factor is a symmetrical effect. If you are moving relative to your friend, you measure her clock to be running slow and she measures yours to be running slow. This situation seems paradoxical indeed, as the following thought experiment illustrates. Consider twins Sara and Teresa both of whom are 11 years old. Sara jumps in a space ship and travels outward at a speed such that $\gamma = \frac{5}{3}$, while Teresa stays on earth. After traveling outward for 25 years (measured on Teresa's clock), Sara turns around and returns to earth at the same speed. Teresa claims that Sara's clocks are running slow by a factor of $\frac{3}{5}$. Thus, during the 50 years of Sara's trip Teresa believes that Sara has aged only $\frac{3}{5}$ (50 years) = 30 years, and when the ship returns, 61-year-old Teresa expects to meet her 41-year-old astronaut sister stepping off the ship! On the other hand, Sara makes the same argument. From the point of view of Sara's frame of reference, it is the earth that is moving away and hence clocks back on earth are running slow. Thus, when Sara returns at age 61 she expects Teresa to be only 41. Which of the twins is correct?

While Sara is on her outward trip, the symmetry between the twins is exact. Both are correct in believing the other to be aging more slowly. This statement presents no real physical problem because the twins are separated. The statement "Sara turned 21 at the same time that Teresa turned 17" may be true in one reference frame, and the statement "Teresa turned 21 at the same time Sara turned 17" true in another frame. The paradox becomes real only when we bring the twins back together. But, to do this one or the other twin must turn around. Which twin actually turns around is easily determined from the measurements of the acceleration. Thus, the symmetry of the situation is broken. It is no longer paradoxical that the twins who have undergone different motions should have different ages on being reunited. (You might still find it hard to accept, but it is not paradoxical.) A careful treatment of the space-time history as presented in Example 39.4 reveals that it is the twin who actually turns around that ends up younger.

EXAMPLE 39.4   *Movie Stars Who Would Remain Young Must Live Very Fast Lives*

Sara and Teresa are 11-year-old twins. Teresa stays on earth while Sara travels at $(\frac{4}{5})c$ to a space station $L_o = 20.0$ light-years away as measured in the earth's frame of reference. (A light-year is the *distance* light travels in 1.00 year. We symbolize this distance as $1.00\ c \cdot \text{yr}$.) On reaching the station Sara immediately turns around and returns to earth. Every year on her own birthday each twin sends the other a greeting in the form of a light pulse. (a) Describe the situation from Sara's point of view, and calculate the number of birthday greetings she receives from Teresa. (b) Describe the situation from Teresa's point of view, and calculate the number of greetings she receives from Sara.

SOLUTION[7]   (a) We first calculate γ from

$$\gamma = \frac{1}{\sqrt{1 - \beta^2}} = \frac{1}{\sqrt{1 - (4/5)^2}} = \frac{5}{3}$$

From Sara's point of view it is the earth and the space station that are moving at $(\frac{4}{5})c$. Furthermore, due to length contraction, Sara sees the distance between the earth and the space station as

$$L = \frac{L_o}{\gamma} = \frac{20.0\ c \cdot \text{yr}}{(5/3)} = 12.0\ c \cdot \text{yr}$$

Thus, Sara calculates that it will take her

$$t = \frac{d}{V} = \frac{12.0\ c \cdot \text{yr}}{(4/5)c} = 15.0\ \text{yr}$$

to make the outward trip. As a consequence, Sara expects to age 30 years during the round trip and thus be 41 years old on returning.

During this time Sara knows that Teresa will send a greeting each time she (Teresa) becomes a year older. Sara calls this source frequency $\nu_s = 1$ b/yr, that is, one birthday per year as measured by Teresa in Teresa's frame. However, as shown in Figure 39.13(a), during her outward journey Sara receives Teresa's greeting at a lower frequency ν_r related to ν_s by the relativistic Doppler formula of Equation (39.15).

$$\nu_r = \nu_s \sqrt{\frac{1 - \beta}{1 + \beta}} = \left(1\ \frac{\text{b}}{\text{yr}}\right)\sqrt{\frac{1 - (4/5)}{1 + (4/5)}} = \left(1\ \frac{\text{b}}{\text{yr}}\right)\sqrt{\frac{(1/5)}{(9/5)}} = \frac{1}{3}\ \text{b/yr}$$

So during her 15-year outward journey Sara receives $(15\ \text{yr})(\frac{1}{3}\ \text{b/yr}) = 5$ b, that is, notification of 5 of Teresa's birthdays. However, as shown in Figure 39.13(b), the instant Sara reverses direction, she starts receiving greetings at a higher rate ν_r given by

$$\nu_r = \nu_s \sqrt{\frac{1 + \beta}{1 - \beta}} = \left(1\ \frac{\text{b}}{\text{yr}}\right)\sqrt{\frac{1 + (4/5)}{1 - (4/5)}} = \left(1\ \frac{\text{b}}{\text{yr}}\right)\sqrt{\frac{(9/5)}{(1/5)}} = 3\ \text{b/yr}$$

The result is that during the 15-year return trip Sara receives $(15\ \text{yr})(3\ \text{b/yr}) = 45$ b greetings. In total then, on meeting her sister Sara has received $5 + 45 = 50$ birthday greetings and expects to find her sister 61 years old, which is indeed the case.

---

[7]The solution in this example is patterned after that of C. G. Darwin, *Nature,* **180,** 976 (1957).

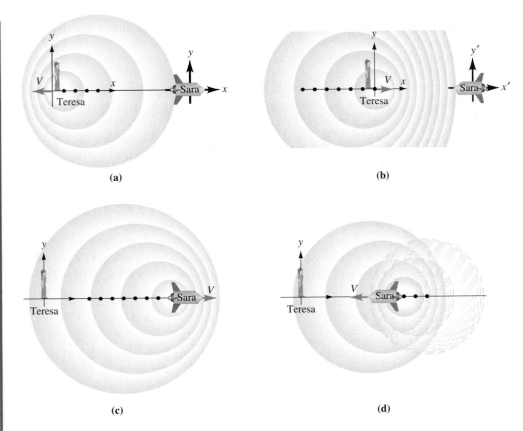

(a) (b)

(c) (d)

FIGURE 39.13

Sara leaves her twin sister, Teresa, on earth as she travels to a distant space station. (a) From Sara's point of view, it is Teresa who is moving. As Teresa moves away the signals she sends are received by Sara at a lower rate. (b) The instant Sara turns around Teresa's signals increase in frequency. (c) From Teresa's point of view it is Sara who is moving. Teresa receives Sara's greetings at a lower frequency as Sara moves away. (d) Because it is Sara who actually turns around, Teresa does not start receiving higher frequencies the instant Sara turns around. Instead, all of the lower frequency pulses must first reach Teresa before the high-frequency waves start arriving.

(b) From earthbound Teresa's point of view, the distance to the space station is 20.0 $c \cdot$ yr. At a speed of $(\frac{4}{5})c$ it will take Sara

$$ t = \frac{d}{V} = \frac{20.0 \ c \cdot \mathrm{yr}}{(4/5)c} = 25.0 \ \mathrm{yr} $$

each way. During Sara's outward trip, shown in Figure 39.13(c), Teresa receives Sara's birthday messages at the lower rate of $\frac{1}{3}$ b/yr as we previously calculated. However, when Sara turns around, Teresa does *not* immediately start receiving the greetings at a higher frequency. You can see why by looking at Figure 39.13(d) in which we show the situation shortly after Sara has turned around. Notice that there are a large number of greetings still "in flight" on their way back to earth. Not until the last of the lower frequency greetings has reached earth does Teresa start receiving the higher-rate flashes. How long does it take the last low-frequency message to get back to earth? Well, the message travels at the speed of light and has a distance of 20 $c \cdot$ yr to travel. Hence, it takes 20 years. Thus, Teresa receives the low-frequency greetings for a total of 25 yr + 20 yr = 45 yr. During this time the number of greetings received is

$$ (45 \ \mathrm{yr})\left(\frac{1}{3} \ \mathrm{b/yr}\right) = 15 \ \mathrm{b} $$

Now, only 5 years are left before Sara's return and during this time Teresa receives greetings at the higher rate of 3 b/yr, and thus receives 15 more greetings.[8] On Sara's return Teresa has received notice that Sara has had 30 birthdays and hence expects her to be 41 years old, consistent with Sara's own calculation.

We thus see that both twins expect Sara to be 41 and Teresa to be 61, although their explanations of how this came about are different. In all of this explanation don't miss the essential point of asymmetry: it *really is* Sara who turns around. As a consequence, it is Sara for whom the Doppler-shifted frequencies immediately change from low to high. For Teresa there is a time lag after Sara's turn around before the higher frequencies are received. ◀

Velocity Transformations

Suppose you are driving down the highway at $V = 80$ km/h and a truck traveling at $u = 100$ km/h passes you. As you look out your side window you see the truck pass your vehicle at $u' = 20$ km/h. In this circumstance we applied a simple rule of velocity subtraction:

$$u' = u - V$$

In words, the truck's velocity u' relative to you is the difference between the truck's velocity u relative to the road and your velocity V relative to the road. In the parlance of this chapter, we have transformed the truck's velocity from the S-frame (the road) to the moving S'-frame (your vehicle). Given what we now know of special relativity, it is perhaps not surprising that this simple rule of velocity subtraction is not accurate when the speeds involved are near those of light. We find, in fact, that this subtraction is but a low-velocity approximation of the relativistically correct law for determining relative velocity.

To determine the correct velocity-transformation law, we begin with the Lorentz transformation laws from Equations (39.8), which we repeat here as written for small changes in position

$$\Delta y' = \Delta y$$
$$\Delta x' = \gamma(\Delta x - V\,\Delta t) \tag{39.17}$$
$$\Delta t' = \gamma\!\left(\Delta t - \frac{V\,\Delta x}{c^2}\right)$$

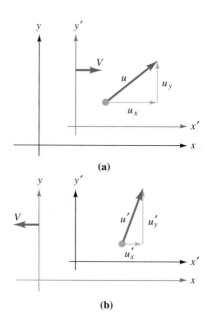

FIGURE 39.14

The S'-frame is moving at speed v parallel to the x-axis of the S-frame of reference. (a) In the S-frame a particle has velocity given by $\mathbf{u} = u_x\hat{\mathbf{i}} + u_y\hat{\mathbf{j}}$. (b) As seen from the S'-frame this same particle has velocity $\mathbf{u}' = u_x'\hat{\mathbf{i}}' + u_y'\hat{\mathbf{j}}'$. The Lorentz transformations must be used to relate the two velocity expressions.

The situation we wish to describe is shown in Figure 39.14 in which an object is depicted moving through space. Relative to a coordinate system S, which we take as stationary relative to the book page, the object has velocity $\mathbf{u}$ with components $u_x = \Delta x/\Delta t$ and $u_y = \Delta y/\Delta t$. A second frame of reference S', oriented as shown in Figure 39.14, travels along the x-axis of the S-frame with a constant velocity V. As viewed from the S'-frame the object has a velocity $\mathbf{u}'$ with components $u_x' = \Delta x'/\Delta t'$ and $u_y' = \Delta y'/\Delta t'$. Our goal is to find equations such that when we know the velocity components in one frame, we can find the velocity components in the other.

We can obtain the equations that transform velocities by taking the ratio of the equations that transform position and time. For example,

$$\frac{\Delta x'}{\Delta t'} = \frac{\gamma(\Delta x - V\,\Delta t)}{\gamma(\Delta t - V\,\Delta x/c^2)} = \frac{\Delta x/\Delta t - V}{1 - V\,(\Delta x/\Delta t)/c^2}$$

where we obtained the final fraction by dividing numerator and denominator by Δt and

---

[8] You might have expected 25 greetings to "pile up" in front of Sara during her 25-year return trip. But remember from Teresa's point of view Sara's clocks are ticking slow by a factor of $\tfrac{3}{5}$.

canceling the common factor of γ. Recalling the definitions of u_x' and u_x, we find the above expression is

$$u_x' = \frac{u_x - V}{1 - Vu_x/c^2} \qquad (39.18)$$

We leave it as an exercise for you to calculate u_y' from $\Delta y'/\Delta t'$ to obtain (see Problem 29)

$$u_y' = \frac{u_y}{\gamma(1 - Vu_x/c^2)} \qquad (39.19)$$

Note two features of the y-velocity component transformation: (1) There is a factor of γ in the denominator that does not appear in the x-component equation. (2) The *x-component* of velocity appears in the denominator of the y-transformation. As usual, to invert the transformation change the u' components to unprimed u components, change unprimed u components to u' components, and substitute $-V$ for V throughout:

$$u_x = \frac{u_x' + V}{1 + Vu_x'/c^2} \qquad (39.20)$$

$$u_y = \frac{u_y'}{\gamma(1 + Vu_x'/c^2)} \qquad (39.21)$$

We illustrate the application of the velocity-transformation equations in Examples 39.5 and 39.6.

EXAMPLE 39.5   *Now I c*

A US astronaut at rest in the S-frame of reference fires a signal laser toward a distant Russian colleague moving toward him at a speed of $0.800c$. What is the speed of the oncoming laser pulse as measured by the Russian?

SOLUTION   We adopt the point of view of the Russian and assign to him the S'-frame, which travels to the right at $V = 0.800c$ as shown in Figure 39.15. In the

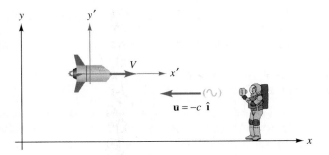

FIGURE 39.15

As seen from the stationary S-frame, a rocket moves to the right at speed V while a light pulse moves to the left at speed c. As seen from the rocket the light does not approach at $1.8c$ but rather at exactly c. The phenomenon is discussed in Example 39.5.

S-frame of the US astronaut the laser pulse is traveling at $u_x = -c$. This velocity, as measured by the Russian, is u'_x and is given by Equation (39.18):

$$u'_x = \frac{u_x - V}{1 - Vu_x/c^2} = \frac{(-c) - (0.800c)}{1 - (0.800c)(-c)/c^2} = \frac{-1.8c}{1.8} = -c$$

The Russian astronaut sees the light pulse advancing at precisely c. Although you may still find this result disquieting, it can hardly be a surprise. After all, the Lorentz transformation equations were designed precisely to fulfill the requirement that all observers measure the speed of light in free space as exactly c. ◀

EXAMPLE 39.6   *The Aberration of Starlight*

An astronomer stationary in the S-frame sees light from a distant star traveling toward him exactly down the y-axis of his coordinate system. A second astronomer is traveling at a velocity of $0.800c$ parallel to the x-axis of the S-frame as shown in Figure 39.16(a). At what angle does the moving astronomer see the light, and how fast is the light from that star traveling, relative to her?

SOLUTION   The components of the velocity of the starlight as seen in the S-frame are $u_x = 0$ and $u_y = -c$. The S'-frame is traveling at velocity $V = 0.800c$. Using velocity transformations Equations (39.18) and (39.19), we find

$$u'_x = \frac{u_x - V}{1 - Vu_x/c^2} = \frac{(0) - (0.800c)}{1 - (0.800c)(0)/c^2} = -0.800c$$

and

$$u'_y = \frac{u_y}{\gamma(1 - Vu_x/c^2)} = \frac{(-c)}{(5/3)[1 - (0.800c)(0)/c^2]} = \frac{3}{5}(-c) = -0.600c$$

The total speed of the light beam as measured in the S'-frame is (of course!)

$$u' = \sqrt{u'^2_x + u'^2_y} = \sqrt{(-0.800c)^2 + (-0.600c)^2} = c$$

In the S'-frame the light makes an angle

$$\phi' = \arctan\left(\frac{u'_y}{u'_x}\right) = \arctan\left(\frac{-0.600c}{-0.800c}\right) = 36.9°$$

with the x'-axis as shown in Figure 39.16(b).

The apparent change in angle of the position of a star due to the motion of the observer is known as the aberration of starlight. The phenomenon was recorded by British astronomer James Bradley (1693–1762) while observing γ-Draconis from 1727 to 1728.[9] Bradley found an angle change of $40''$ of arc associated with the earth's orbital motion around the sun. ◀

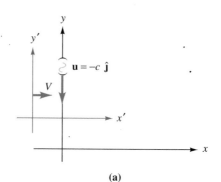

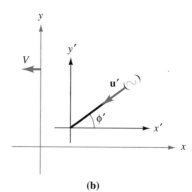

FIGURE 39.16

The S'-coordinate system moves to the right at speed V parallel to the x-axis of the S-frame. (a) As seen from the S-frame a pulse of light travels straight downward parallel to the y-axis. (b) As seen from the moving frame the light makes an angle with the y'-axis. This phenomenon is known as the aberration of starlight.

39.4 Relativistic Dynamics

In Section 39.3 we described the rather far-reaching implications of the Lorentz transformations on the kinematics of particles. It is perhaps not surprising to find an equally profound effect on dynamics. In this section we explore the implications of special relativity for momentum and energy.

Concept Question 5
In the Lorentz transformations $y = y'$. Why is it then that $u_y \neq u'_y$?

---

[9] See the description by A. Stewart, "The discovery of stellar aberration," *Scientific American,* **210,** 100 (1964).

Relativistic Momentum

The conservation of momentum is one of the most cherished and powerful laws of physics. The classical (nonrelativistic) form for momentum is of course $\mathbf{p} = m\mathbf{u}$ where $\mathbf{u}$ is the particle velocity. When momentum is written this way, the conservation of momentum is "preserved under Galilean transformations." What this statement means is that, if we make up a thought experiment and adjust parameters so that momentum is conserved in the experiment, then another observer in a moving frame of reference who looks at this same experiment agrees that momentum is conserved, *providing we transform between frames using the Galilean transformations.* As a trivial example consider a perfectly elastic collision between two spheres of identical mass. Let one move vertically along the y-axis with a velocity $\mathbf{u}_1 = +u_o\hat{\mathbf{j}}$. Let the second sphere have a velocity given by $\mathbf{u}_2 = +V\hat{\mathbf{i}} - u_o\hat{\mathbf{j}}$. The two spheres are launched so that they collide as shown in Figure 39.17. Because this is a perfectly elastic collision between identical spheres, the conservation of momentum implies that the y-components of each sphere are reversed as shown in the figure. Thus, the change in momentum of sphere 1 is $\Delta\mathbf{p}_1 = \mathbf{p}_f - \mathbf{p}_o = -2mu_o\hat{\mathbf{j}}$. Similarly, the change of momentum of sphere 2 is $\Delta\mathbf{p} = +2mu_o\hat{\mathbf{j}}$. The total change is thus zero. Momentum is conserved.

If we view this same collision in an S'-frame moving at velocity V to the right, the picture appears as in Figure 39.18. Providing we use the Galilean transformations to obtain the new velocities, we find that $\mathbf{u}_1' = -V\hat{\mathbf{i}}' + u_o\hat{\mathbf{j}}'$ and $\mathbf{u}_2' = -u_o\hat{\mathbf{j}}'$. After the collision the velocities are once again reversed. If we calculate the changes in momentum, we once again find them equal and opposite so that momentum is conserved in the S'-frame too.

The calculations of the previous paragraphs illustrate what we mean when we say that, for $\mathbf{p} = m\mathbf{u}$, the law of momentum conservation is preserved under Galilean transformation. By now you probably see the problem. We learned that the Galilean transformations are merely approximations to the Lorentz transformations, appropriate to low velocities. We leave it as an exercise for you to show that if the Lorentz transformations are used for the collision described above, momentum $m\mathbf{u}$ is not conserved. (See Problem 43.) Even without doing the detailed calculation, it is easy to see the difficulty. In the S-frame we made the y-velocity components equal so that on reversal of these velocities, the changes in momenta are equal. This equality of the y-velocity components is not preserved by the Lorentz velocity transformations of Equations (39.18) and (39.19) because the x-component of sphere 2's velocity alters its u_y' velocity component. Thus, in the S'-frame the y-components of the velocity of the two spheres are no longer equal. This inequality results in the nonconservation of momentum.

Thus, special relativity now forces us to look for some modification of the definition of momentum such that

1. momentum is conserved under Lorentz transformations and
2. the definition reduces to $m\mathbf{u}$ for low velocity.

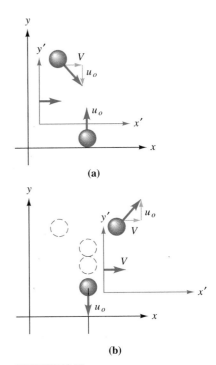

FIGURE 39.17

A perfectly elastic collision between two identical spheres. The vertical components of velocity are equal in the stationary S-frame of reference, guaranteeing that classical momentum $p = m_o u$ is conserved.

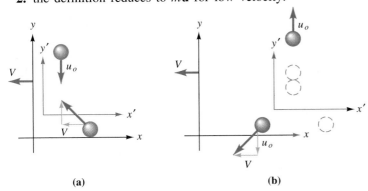

FIGURE 39.18

The same collision as that in Figure 39.17 seen this time from the point of view of the S'-frame moving to the right relative to the S-frame. If Galilean transformations are used, classical momentum $m_o u$ is conserved in the S'-frame as well as the S-frame. However, if the Lorentz transformations are used, classical momentum $m_o u$ is not conserved in both frames.

The modified definition of momentum satisfying these requirements is

$$\mathbf{p} = \frac{m_o \mathbf{u}}{\sqrt{1 - u^2/c^2}} \qquad \text{Relativistically correct momentum equation} \quad \textbf{(39.22)}$$

The symbol m_o is the mass of the particle measured in a frame in which the mass is at rest. For this reason m_o is referred to as the **rest mass.** (You may sometimes see $m = m_o/\sqrt{1 - u^2/c^2}$ defined as the ''relativistic'' mass. We will not use the relativistic mass concept but retain the notation m_o for emphasis.) Using this relativistically correct definition and the Lorentz transformations, momentum is conserved in all reference frames. (See Problem 44.)

Kinetic Energy

As you may recall, the work–energy theorem of Chapter 7 suggests that the work done by the net force is equal to the change in a particle's kinetic energy. We arrived at this conclusion by actually calculating the work done by the net force $\mathbf{F}_{\text{net}}$, which we set equal to $d\mathbf{p}/dt$. Let's repeat this procedure now using the relativistically correct expression for the momentum. We assume the particle is traveling along the x-axis with a velocity $u = dx/dt$. Thus, the integral we must calculate is

$$\mathcal{W} = \Delta K = \int \frac{dp}{dt}\, dx$$

Rather than tackling this integral directly we employ the chain rule to write

$$\frac{dp}{dt}\, dx = \frac{dp}{dx}\frac{dx}{dt}\, dx = \frac{dp}{dx}\, u\, dx = u\,\frac{dp}{dx}\, dx = u\, dp$$

We can solve Equation (39.22) for the velocity u:

$$u = \frac{p}{\sqrt{m_o^2 + p^2/c^2}}$$

With these two substitutions the integral we desire becomes

$$\int \frac{dp}{dt}\, dx = \int u\, dp = \int \frac{p\, dp}{\sqrt{m_o^2 + p^2/c^2}}$$

Let's take the specific case of a particle starting from rest and acquiring kinetic energy as work is done on it. Accordingly, we calculate the particle's kinetic energy as

$$K = \int_0^{p_f} \frac{p\, dp}{\sqrt{m_o^2 + p^2/c^2}}$$

With the substitution $\xi = m_o^2 + p^2/c^2$ this last integral can be put into a form with the following solution

$$\int \xi^{-1/2}\, d\xi = 2\xi^{1/2}$$

The final result (see Problem 45) is

$$K = \sqrt{(m_o c^2)^2 + (p_f c)^2} - m_o c^2$$

The form of this result suggests that we define the **total energy** of a body as

$$E = \sqrt{(m_o c^2)^2 + (pc)^2}$$ Relativistic total energy (39.23)

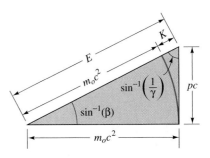

This triangle is a handy mnemonic for remembering the relativistic energy–mass–momentum relations. All geometric relations true for the triangle are valid relations.

Equation (39.23) gives the total energy of a particle traveling with a momentum p. Notice that even when the particle has zero momentum and, hence, zero speed, it has nonzero energy in the amount

$$E_o = m_o c^2$$ (39.24)

This energy, which the particle has by virtue of its very existence, is called the **rest mass energy** of the particle. Thus, we can think of the kinetic energy as the difference between the total energy and the rest mass energy

$$K = E - m_o c^2$$ (39.25)

As an exercise (see Problem 46), we leave it for you to show that for a particle traveling with velocity parameter $\beta = u/c$,

$$K = \left(\frac{1}{\sqrt{1 - \beta^2}} - 1\right) m_o c^2$$ (39.26)

This last expression for kinetic energy hardly looks like our old friend $\frac{1}{2}mu^2$. However, if we perform a binomial expansion on the factor $(1 - \beta^2)^{-1/2}$, the kinetic energy can be written

$$K = \left[\left(1 + \frac{1}{2}\beta^2 + \frac{3}{8}\beta^4 + \cdots\right) - 1\right] m_o c^2 = \frac{1}{2} m_o u^2 + \frac{3}{8} m_o u^2 \left(\frac{u^2}{c^2}\right) + \cdots$$

As you can see, the classical expression for the kinetic energy is recovered for low speeds when $u/c \ll 1$.

Concept Question 6
For a given speed, which gives the larger result for kinetic energy: the classical formula or the relativistically correct expression?

39.5 Binding Energy and Mass

When two particles are held together by an attractive force, work must be done by an external agent to separate them. The amount of work that must be done is known as the **binding energy** of the joined particles. The nuclei of atoms are composed of protons and neutrons. The protons are particles with positive charge equal in magnitude to the charge of an electron. Neutrons are slightly more massive than protons and have no electric charge. When we don't wish to bother distinguishing between protons and neutrons, we use their generic name, **nucleons.** Nuclei are often represented in a notation, such as

$$_{Z}^{A}X$$

where Z is the number of protons, A is the total number of nucleons, and X represents the chemical symbol for an atom with this particle as a nucleus. For example, $_{2}^{4}\text{He}$ is the symbol for the nucleus of a helium atom composed of two protons and $A - Z = 2$ neutrons. This particle is also referred to as an **alpha particle.**

The protons and neutrons are held together in the nuclei of atoms by attractive short-range, but strong, nuclear forces. As an example, consider a single proton and single neutron that combine to form a composite nucleus called a deuteron. When the proton and neutron approach closely enough, the strong nuclear force draws them together. As they fall toward each other, they gain kinetic energy. If they can dispose of this kinetic energy, perhaps by passing it to some third particle, then the proton and neutron "stick," creating the deuteron. The energy disposed of during the deuteron creation is the binding energy. If we wish to separate the two particles again, we have to supply this same amount of energy, perhaps by striking them with another particle.

One of the laws used often in chemistry is the law of conservation of mass. We might reasonably expect that matter can neither be created nor destroyed. Let's check this law for the previous example of deuteron formation. The mass of the proton, neutron, and the deuteron have been carefully measured.

mass of proton	$1.672\ 623 \times 10^{-27}$ kg
mass of neutron	$1.674\ 929 \times 10^{-27}$ kg
sum of masses	$3.347\ 552 \times 10^{-27}$ kg
mass of deuteron	$3.343\ 586 \times 10^{-27}$ kg

As you can see from the computation above, mass is not conserved! The whole is less than the sum of its parts. Let's recount how this came about. The proton and neutron came together, gave up some energy, and thus became bound. Yet, the composite deuteron has less mass than the proton and neutron masses summed. Where could the mass have gone? The answer seems inescapable. The only quantity to leave our pair of particles was energy! It appears that the missing mass was somehow converted to energy, which then left the system. Could it be that the relationship $E_o = m_o c^2$ has a more profound meaning than we originally thought? The answer is unequivocally, yes. Because, if we multiply the missing mass by the square of the speed of light, the energy calculated is exactly the magnitude of the binding energy that escaped.

When performing calculations involving nuclear binding energies and masses, it is inconvenient to use the SI system. Instead, the MeV is chosen as an energy unit of convenience. The MeV stands for million electron volts, and thus is equal to

$$1 \text{ MeV} = 10^6 \text{ eV} = (10^6)(1.602\ 177 \times 10^{-19} \text{ J/eV}) = 1.602\ 177 \times 10^{-13} \text{ J}$$

For example, if we calculate the rest mass energy of an electron, we find

$$E = m_o c^2 = (9.109\ 390 \times 10^{-31} \text{ J})(2.997\ 924 \times 10^8 \text{ m/s})^2$$
$$= 8.187\ 108 \times 10^{-14} \text{ J} = 0.511 \text{ MeV}$$

This type of conversion is performed so frequently in nuclear physics that you sometimes see masses written in a form so that the conversion is already done. For example, we could quote the rest mass of the electron as

$$m_o = \frac{E}{c^2} = 0.511 \text{ MeV}/c^2$$

Another mass unit common in nuclear physics is the unified atomic mass unit u. A handy conversion to have available is

$$u = 931.494\ 32 \text{ MeV}/c^2$$

Example 39.7 illustrates the calculation of binding energy. Be sure to read the note of caution at the end of the example.

EXAMPLE 39.7 *Watch Out for Electrons*

Determine the binding energy of an alpha particle.

SOLUTION An alpha particle is the nucleus of a helium atom and is represented by $^4_2$He. We can determine the mass of the helium *atom* from a table of atomic masses, such as in Appendix 6.

$$\text{atomic mass of helium} \quad 4.002\ 60\ \text{u}$$

This value is to be compared to the total mass of the two protons and two neutrons that constitute the alpha particle. We use the masses of the hydrogen *atom* and the neutron:

$$\text{mass of constituents} \quad 2(1.007\ 97) + 2(1.008\ 665) = 4.033\ 27\ \text{u}$$

The **mass defect,** as the mass difference is sometimes called, is

$$\Delta m = 4.033\ 27 - 4.002\ 60 = 0.030\ 67\ \text{u}$$

Converting this mass value to energy units, we have

$$\Delta m = (0.030\ 67\ \text{u})[931.494\ 32\ (\text{MeV}/c^2)/\text{u}] = 28.6\ \text{MeV}/c^2$$

So that the binding energy is 28.6 MeV. It is often of interest to calculate the average binding energy per nucleon. In this case, we find

$$Binding\ energy\ per\ nucleon = \frac{28.6\ \text{MeV}}{4} = 7.14\ \text{MeV}$$

Note of caution: In this calculation we used *atomic* masses for helium and hydrogen. These masses include the mass of the electron(s) in the surrounding shells. Strictly speaking, we desire the mass of the nucleus only. However, ultimately in this type of calculation we are interested only in the *differences* between the whole and its parts. When we used the mass of the helium atom we included two extra electrons. When we used the mass of the hydrogen atom we included one extra electron but this number was doubled because there were two protons. When we subtracted, the masses of the extra electrons exactly canceled so that we made no error. We caution you therefore when doing such calculations to always use the mass of hydrogen when calculating "proton" masses. Furthermore should you ever need the mass of the bare nucleus of an atom you must subtract the mass of Z electrons from the masses found in Appendix 6. ◀

Concept Question 7
If a proton and a neutron bind together to form a deuteron, is the deuteron mass larger or smaller than the sum of the proton and neutron masses?

39.6 Summary

The two postulates of Einstein's special theory of relativity are

POSTULATE 1. The laws of physics are the same in all inertial reference frames.
POSTULATE 2. The velocity of light in free space is a constant c regardless of the state of motion of the source.

These postulates require that Galilean relativity be abandoned in favor of the Lorentz transformation equations

$$y' = y$$

$$t' = \gamma\left(t - \frac{Vx}{c^2}\right) \tag{39.8}$$

$$x' = \gamma(x - Vt)$$

$$u'_x = \frac{u_x - V}{1 - Vu_x/c^2} \tag{39.18}$$

$$u'_y = \frac{u_y}{\gamma(1 - Vu_x/c^2)} \tag{39.19}$$

where $\gamma = 1/\sqrt{1 - \beta^2}$. To invert the transformation equations, interchange primed and unprimed quantities, and replace V by $-V$.

As a consequence of the Lorentz transformations one finds that moving clocks run slower. This phenomenon is called **time dilation.** The time interval T measured in the moving frame is related to the **proper time** interval T_o measured in the frame in which the clock is stationary by

$$T = \gamma T_o \tag{39.10}$$

Length contraction results when a moving observer measures the length of an object along the direction of motion. The contracted length is given by

$$L = \frac{L_o}{\gamma} \tag{39.11}$$

where L_o is the **proper length** of the object when measured in a frame in which it is at rest.

When a source and receiver are separating, the source frequency ν_s is detected by the receiver as ν_r. The two frequencies are related by the **relativistic Doppler shift** for electromagnetic waves given by

$$\nu_r = \nu_s \sqrt{\frac{1 - \beta}{1 + \beta}} \tag{39.16}$$

regardless of whether it is the source or observer that is in motion. When the source and receiver approach, the $+$ and $-$ signs in this equation are interchanged.

The correct expression for the **momentum** of an object of rest mass m_o traveling at speed u is

$$\mathbf{p} = \frac{m_o\mathbf{u}}{\sqrt{1 - u^2/c^2}} \tag{39.22}$$

The **total energy** of such an object is

$$E = \sqrt{(m_oc^2)^2 + (pc)^2} \tag{39.23}$$

The **kinetic energy** K of such an object is

$$K = E - m_oc^2 \tag{39.25}$$

A particle has **rest mass energy** $E_o = m_oc^2$ simply by virtue of its existence.

When two or more particles are bound together by an attractive force, the energy needed to separate them is the **binding energy.** Energy equal to the binding energy escapes when the particles bind. This energy loss results in the mass of the composite particle being less than the sum of the masses of the constituent particles. The amount of the mass loss is given by

$$\Delta m_o = \frac{\Delta E}{c^2}$$

where ΔE is the binding energy.

PROBLEMS

39.1 Speed of Light

1. Using data from Appendix 8, calculate the time it takes for light to travel across the diameter of the earth's orbit.
2. Satellites in geosynchronous orbits stay directly above a point on the earth's equator at an altitude of 3.58×10^7 m above the earth's surface. Such satellites are used to relay television signals. How long does it take an electromagnetic signal to travel directly up to the satellite and back down?
3. How long does it take light from the sun to reach the earth?
4. A light-year is defined as the distance light travels in one year. How many meters are there in a light-year?
5. It is now known that our sun is moving relative to the center of the Milky Way galaxy at a speed of 250 km/s. If there were an ether, and it were stationary relative to the center of the Milky Way, what size fringe shift would be expected from the Michelson-Morley experiment? Assume an arm length of $L = 10.0$ m in the interferometer and a wavelength of 550. nm.
6. A river 0.800 km wide flows with a speed of 6.00 km/h to the right. A power boat is able to travel at 10.0 km/h in still water. (a) How long does it take the power boat to complete a round trip to a point on the bank straight downstream a distance of 0.800 km? (b) The boat now wishes to travel *directly* across the river to the opposite shore. At what angle upstream must the boat be pointed to guarantee that it travels directly across the river? (c) How long does it take the boat to make a round trip to a point directly across the river when traveling as described in part (b)? (d) What has this problem to do with the Michelson-Morley experiment?

39.3 Relativistic Kinematics

7. A baby is born. The space-time coordinates (x, y, t) of this blessed event are (10.0 m, 5.00 m, 155 s) in the S-frame of reference. If birth announcements are sent out in an S'-frame of reference traveling at $V = 0.800c$ parallel to the x-axis of the S-frame of reference, what space-time coordinates should be listed for the birth?
8. In the S-frame of reference the space-time coordinates (x, t) of two events A and B are given by A:(0, 0) and B:(300. km, 2.00 μs). With what velocity and in what direction must a second frame of reference S' move so that the events are simultaneous in the S'-frame?
9. Carbon-14, often used for dating archeological artifacts, has a half-life of 5730 yr. What is the apparent half-life of carbon-14 as measured by an observer traveling at $0.994c$?
10. A certain astronaut has a heart rate such that the time between beats is 1.00 s. If this astronaut is traveling at $0.990c$ relative to an observer on earth, what is the time interval between beats measured by an earthbound observer?

11. How fast does a clock have to travel for the duration of one tick to be twice that of an identical stationary clock?
12. If the half-life of a moving muon is found to be 9.00 times greater than a stationary muon, what is the speed of the muon?
13. You measure the half-life of a certain radioactive isotope as 1.35 s when it has a velocity of $0.896c$ relative to you. What is the half-life of the isotope when it is at rest?
14. A certain movie appears to last 18.0 h (by your clock) when shown in a spaceship traveling at $0.995c$ relative to you. How long does the movie last if you see it in a theater at rest relative to you?
15. What is the radius of the earth's orbit around the sun as seen by a traveler approaching our solar system at 2.70×10^8 m/s? Assume the traveler's velocity is parallel to the diameter of the earth's orbit.
16. An astronaut is 1.85 m tall. If you observe this astronaut traveling at $0.800c$ along a direction parallel to her height, how tall does she appear to be?
17. How fast does a meterstick have to travel past you to appear 0.250 m long? Express your answer both as a fraction of the speed of light and in meters per second.
18. How fast must an object travel before its length, measured by a stationary observer, is 1.00% less than its proper length?
19. A spaceship has a length of 5.67 m when it travels past you at $0.999c$. What is its proper length?
20. What is the proper length of a meteorite when it is observed to have a length of 0.455 m as it travels at $0.600c$?
21. Perform the substitution $\nu_r = 1/(\gamma T_o) = \nu_o/\gamma$ as described in the text to obtain Equation (39.15) from Equation (39.13).
22. A certain source emits light of frequency 5.67×10^{14} Hz. (a) What frequency is detected by an observer approaching the source at $0.994c$? (b) What frequency is detected by a receiver receding from the source at $0.994c$?
23. A source is receding from an observer at $0.800c$. The observer detects light of frequency 6.23×10^{14} Hz coming from the source. What frequency is the source emitting as measured in its own frame?
24. At rest, a helium-neon laser emits light of wavelength 632.8 nm. (a) If the laser is approaching you at $0.965c$, what wavelength do you detect? (b) If the laser is stationary and you are approaching it at $0.965c$, what wavelength do you detect?
25. The speed of rotation of Saturn's rings has been measured by means of the Doppler shift. The velocity of the outer boundary of the A-ring has a velocity of 20 km/s relative to the planet surface. (a) Ignoring the relative motion of Saturn and the earth and assuming a measurement wavelength of 555 nm, what percentage change in wavelength is measured for that part of the ring moving toward an observer on earth? (b) What percentage change in

wavelength is detected from the side of the ring moving away from the earth?

26. How fast do you have to travel toward a red light ($\lambda = 620.$ nm) for the light to appear green ($\lambda = 550.$ nm)?

27. The nearest star to our sun is Alpha Centauri, which is "only" $4.20 \ c \cdot \text{yr}$ away. If you travel to Alpha Centauri and back at a speed of $0.900c$ how much younger are you on return than if you had stayed on earth?

28. How long do you have to fly in an airplane traveling at 800. km/h to be one second younger when you land than you would have been had you stayed at the airport?

29. Follow procedures similar to those used to obtain Equation (39.18) to derive Equation (39.19) for the velocity transformation of a velocity component perpendicular to the relative velocity between two frames of reference.

30. In an attempt to once and for all go faster than the speed of light a student designs two rockets. Each has sufficient power to reach a speed of $0.750c$ relative to the frame of reference in which it is launched. The two rockets are placed end to end and launched from a reference platform. Powered by the first rocket, the pair reach $0.750c$. At that moment the second rocket is fired and attains a speed of $0.750c$ relative to the first rocket. What is the speed of the second rocket relative to the reference platform?

31. A pizza delivery truck is traveling at $u'_x = +0.800c$ down the x'-axis of an S'-coordinate system, which is itself moving at $V = +0.600c$ down the x-axis of a stationary S-frame of reference. How fast is the pizza truck traveling relative to the S frame of reference?

32. In his quest to go faster than the speed of light a student builds two rockets each of which will reach a speed of $0.750c$ relative to its launch point. The student boards one rocket and launches himself ahead at $v = 0.750c$ relative to a space station. His assistant launches the second rocket directly toward the first rocket at speed $u_x = -0.750c$ relative to the space station. At what velocity u'_x does the student riding in the first rocket see the second rocket approach?

33. In a colliding-beam experiment two protons travel toward each other, each with a speed of $0.99000c$ relative to the laboratory frame of reference. In a frame of reference attached to one of the protons, at what speed does the other approach?

34. Rocket A travels at a speed $u_y = +0.800c$ parallel to the y-axis of a fixed coordinate system. What is the velocity (magnitude and direction) of rocket A as observed from rocket B, which is traveling at $V = +0.600c$ parallel to the x-axis of the fixed reference frame?

35. As observed in the stationary S-frame of reference a particle has velocity $\mathbf{u} = 0.600c\hat{\mathbf{i}} + 0.400c\hat{\mathbf{j}}$. What is the velocity (in unit-vector form) of this particle as viewed from a frame of reference S', which is itself moving with a velocity $\mathbf{V} = 0.600c\hat{\mathbf{i}}$?

36. A light pulse is traveling with velocity $\mathbf{u} = 0.800c\hat{\mathbf{i}} + 0.600c\hat{\mathbf{j}}$ relative to a fixed frame of reference. Calculate the velocity components of this light pulse as observed from a frame of reference traveling with a velocity $\mathbf{V} = 0.800c\hat{\mathbf{i}}$ relative to the fixed frame.

39.4 Relativistic Dynamics

37. (a) If a particle is traveling at $0.400c$, by what factor does its momentum increase if its speed doubles? (b) If a particle is traveling at $0.400c$, by what factor must its speed increase for its momentum to double?

38. (a) For a given speed, which is greater: a particle's classical momentum (mu) or its relativistic momentum? (b) What speed must

a particle have for its relativistically correct momentum to deviate from its classical value $m_o u$ by 1.00%?

39. When performing relativistic calculations on nuclear particles, it is often convenient to measure momentum in units of MeV/c, where 1 MeV = one million electron volts = 1.602×10^{-13} J, and c is the velocity of light in vacuum. (a) Show that MeV/c has the units of momentum. (b) What is 1.00 MeV/c in kilogram meters per second?

40. What is the momentum of an electron traveling at $0.900c$? Express your answer in both SI units and in MeV/c (see Problem 39).

41. What is the momentum of a proton traveling at $0.800 \ c$? Express your answer in both SI units and in MeV/c (see Problem 39).

42. What should be the speed of a proton so that it has the same momentum as an alpha particle with a speed of $0.500c$?

43. Consider a perfectly elastic collision between two identical spheres of mass m_o as discussed in Section 39.4 of the text and shown in Figure 39.P1. As viewed in the S-frame the velocity of the lower ball is $\mathbf{u}_A = +0.400c\hat{\mathbf{j}}$ before the collision and $\mathbf{u}_A = -0.400c\hat{\mathbf{j}}$ after the collision. The upper ball has velocity $\mathbf{u}_B = +0.500c\hat{\mathbf{i}} - 0.400c\hat{\mathbf{j}}$ before the collision and $\mathbf{u}_B = +0.500c\hat{\mathbf{i}} + 0.400c\hat{\mathbf{j}}$ after the collision. (a) Find the change in the classical momentum $m_o \, \Delta\mathbf{u}$ of each ball. What is the total change for both balls together? (b) Now, using the Lorentz velocity transformations, transform all four velocities into the S'-frame, which is moving to the right at speed $\mathbf{V} = 0.500c\hat{\mathbf{i}}$. Sketch the appearance of the collision in the S'-frame. (c) Using your answers from part (b) calculate the total change in classical momentum for both spheres as seen in the S'-frame. Is classical momentum conserved?

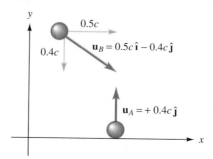

FIGURE 39.P1   Problem 43

44. Consider a perfectly elastic collision between two identical spheres of mass m_o as discussed in Section 39.4 and shown in Figure 39.P2. As viewed in the S-frame the velocity of the lower ball is $\mathbf{u}_A = +0.4619c\hat{\mathbf{j}}$ before the collision and $\mathbf{u}_A = -0.4619c\hat{\mathbf{j}}$ after the collision. The upper ball has velocity $\mathbf{u}_B = +0.500c\hat{\mathbf{i}} - 0.400c\hat{\mathbf{j}}$ before the collision and $\mathbf{u}_B = +0.500c\hat{\mathbf{i}} + 0.400c\hat{\mathbf{j}}$ after

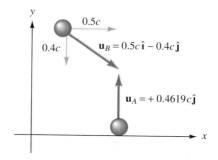

FIGURE 39.P2   Problem 44

the collision. (a) Find the change in the relativistic momentum of each ball. Show that momentum is conserved (to three significant figures) in this collision. (b) Now transform all four velocities into the S'-frame, which is moving to the right at speed $\mathbf{V} = 0.500c\hat{\mathbf{i}}$. Sketch the appearance of the collision in the S'-frame. (c) Using your answers from part (b) calculate the total change in relativistic momentum for both spheres as seen in the S'-frame. Is relativistic momentum conserved?

45. Perform the integration leading to the expression for kinetic energy in Equation (39.23).

46. Combine the energy–momentum relation of Equation (39.23) with the expression for kinetic energy in Equation (39.25) to obtain the alternative expression for kinetic energy given in Equation (39.26).

47. (a) What is the rest mass energy (in MeV) of a deuteron with mass 3.343×10^{-27} kg? (b) What is the mass in kilograms of a J/ψ particle with a rest mass energy of 3100 MeV? What atom has a mass close to this value?

48. (a) The three heavy leptons are the electron, the muon, and the tau particle with rest mass energies of 0.511 MeV, 105.66 MeV, and 1784 MeV, respectively. What are the rest masses of these particles in unified atomic mass units? (b) What is the rest mass energy (in MeV) of 1.00 mg of matter?

49. Find the momentum (in MeV/c) and kinetic energy (in MeV) of a proton traveling at $0.672c$.

50. Find the momentum (in MeV/c) and kinetic energy (in MeV) of an electron traveling at $0.900c$.

51. In nuclear and particle physics it is common to describe the motion of a particle by quoting its kinetic energy. For example, to describe an electron with a kinetic energy of 2.00 MeV, one simply says, "a 2.00-MeV electron." Find the speed (as a fraction of c) and the momentum (in MeV/c) of a 2.00-MeV electron.

52. Find the momentum (in MeV/c) and the speed (as a fraction of c) of a proton the kinetic energy of which is equal to its rest mass energy.

53. A pi-zero (π^0) meson has a rest mass of 135 MeV/c^2 and a momentum of 250. MeV/c. Find the particle's speed (as a fraction of c) and its kinetic energy in MeV.

54. The lambda particle (Λ^0) is a baryon similar to a neutron in that it has zero charge and spin of 1/2. Its rest mass of 1115.6 MeV/c^2 is somewhat greater than the neutron's, but its lifetime is only 2.62×10^{-10} s. If the Λ^0 particle has a momentum of 3000 MeV/c, what is its speed (as a fraction of c) and kinetic energy in MeV?

39.5 Binding Energy and Mass

55. (a) Calculate the binding energy of the lithium nucleus $^7_3$Li. (b) What is the binding energy per nucleon for $^7_3$Li?

56. Iron has the highest binding energy per nucleon of any nucleus. (a) Calculate the binding energy of ^{56}Fe. (b) Calculate the binding energy per nucleon of ^{56}Fe.

57. The mass defect due to binding energy loss holds for chemical binding as well as nuclear binding. One might wonder why in chemical reactions the mass difference between the resultant atom and the sum of its parts was never noticed. The binding energy between an electron and the proton in the hydrogen atom is 13.6 eV. (a) How much less massive is the hydrogen atom than the separate electron and proton? (b) If you have a mole of hydrogen atoms what is the total "missing mass"? Can you detect this with a laboratory balance accurate to 0.1 μg?

58. The sun's rate of energy output is 3.76×10^{26} W. This energy is produced through the release of nuclear binding energy. How much mass is converted into energy each second? What fraction of the sun's total mass is this?

59. An electron e^- and its antiparticle e^+ can annihilate each other, leaving only electromagnetic energy behind. (a) What is the minimum energy of the electromagnetic radiation given off when these antiparticles annihilate? (Hint: This minimum results when the kinetic energy of both particles is zero at the time of annihilation.) (b) When the electron and antielectron annihilate each other as described in part (a) two identical electromagnetic waves are given off in opposite directions. What fundamental law of physics would be violated if only one wave were created? (c) What are the frequencies of the two electromagnetic waves described in part (b)? In what part of the electromagnetic spectrum do these waves fall?

Numerical Methods and Computer Applications

60. (a) Use a spreadsheet or computer program to produce a graph of γ as a function of β. (b) How close to $\beta = 1$ can you get before your computer gives an overflow error indication?

61. Use a spreadsheet to produce a graph showing both classical kinetic energy and relativistic kinetic energy as a function of β. Use your graph to find the speed where the error in the classical approximation becomes 10%.

62. The classical relation between momentum and kinetic energy can be written as $K = p^2/2m_o$. Use a spreadsheet to compare the classical expression with the relativistically correct Equation (39.25). For convenience use units such that $m_o = 1$ and $c = 1$. On the same graph make two plots of K versus p for values of p from 0 to 10, one graph using the classical expression, the other using Equation (39.25).

General Problems

★63. The earth and a space station X are separated by $L_o = 40$ light-years ($c \cdot$ yr). Moreover, they are at rest relative to each other and have synchronized their clocks. The space transport *Messenger* is to make a trip from the earth to station X traveling at a speed of $V = 0.800c$. Clocks on the earth and the transport both read 0.00 year the instant the transport leaves the earth. Take the earth as the origin of the S-frame of reference as shown in Figure 39.P3. (a) Describe this trip completely from the point of view of someone in the earth's (S) frame of reference. In particular, state the departure time as registering on all three clocks (one on earth, one on station X, and one on the transport). Give the arrival times as shown on each of the three clocks. Explain (from the earth's point of view) why station X and the transport show different times on

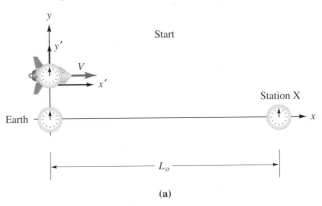

(a)

FIGURE 39.P3   Problem 63

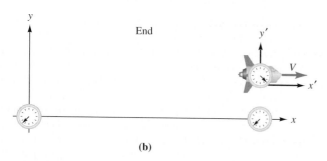

(b)

FIGURE 39.P3   Problem 63

their clocks when the transport arrives at X. (b) Now describe the trip completely from the transport's point of view: How far apart are the earth and space station X? What time is showing on each of the three clocks when the transport leaves? What times are shown on each of the three clocks when the transport arrives? How does a passenger on board the *Messenger* explain the difference between the clock on the transport and the clock on station X when he arrives?

64. Consider two events (x, t) in the S-frame of reference: $(0, 0)$ and (L, T). Show that there is an S' reference frame moving at speed $v < c$ relative to the S-frame in which the second event occurs prior to the first event only if $L/T > c$. Because nothing can travel faster than light, the first event can cause the second only if $L/T < c$. Thus, if the first event caused the second, the order of the events cannot be reversed in any reference frame.

65. As seen in the S' coordinate system, a light pulse travels in the $x'y'$-plane at an angle of $45°$ with respect to the x'-axis. The S'-frame is moving at $0.600c$ parallel to the x-axis of a second frame of reference S. The x- and x'-axes are parallel as shown in Figure 39.P4. Calculate the velocity components, the speed, and the angle of the light pulse as observed from the S-frame.

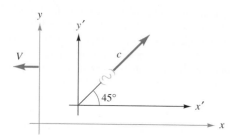

FIGURE 39.P4   Problem 65

66. In yet another attempt to make an object go faster than the speed of light, a student fires two identical rockets from the origin of a fixed coordinate system. One rocket is fired directly up the y-axis with a speed of $0.750c$. The second rocket is fired along the x-axis at $0.750c$. (a) Find the velocity components of the first rocket as measured by an observer in the second rocket, which is traveling along the x-axis. (b) What is the total speed of the first rocket as measured by the observer in the second?

67. Neon atoms in a helium-neon laser have typical speeds on the order of 600 m/s. One of the electromagnetic waves emitted by these atoms has a wavelength (in the rest frame of the molecule) of 632.8 nm. What is the approximate width (in nanometers) of this spectral line due to Doppler broadening? (*Hint:* Calculate the largest and smallest wavelengths that result from the moving atoms. The difference in these wavelengths is the approximate wavelength width of the spectral line.)

68. It has been suggested that police use light from an infrared laser instead of radar to detect speeders. If the light has a wavelength of 830. nm and an automobile is approaching at 150. km/h, what is the beat frequency between the original light and reflected light from the oncoming car? (*Hint:* The reflected light has been Doppler-shifted twice.)

69. Among the lightest of elementary particles is the neutrino ("little neutral-one"). It may be that the neutrino in fact has zero rest mass. (a) What expression does the relativistic energy–momentum equation predict for the relation between energy and momentum for a particle with zero rest mass? (b) Assuming the neutrino does have zero rest mass, what is the momentum of a neutrino with a kinetic energy of 1.50 MeV? (c) If the neutrino actually has a rest mass of $20 \text{ eV}/c^2$, what is the answer to the question posed in part (b)?

70. Show that the relativistic Doppler shift written in terms of wavelength change $\Delta\lambda = \lambda_r - \lambda_s$ is given by

$$\frac{\Delta\lambda}{\lambda_s} = \sqrt{\frac{1 + V/c}{1 - V/c}} - 1$$

for a source and receiver receding at speed V.

71. When calcium atoms are at rest, their absorption spectrum shows a strong line at a wavelength of 393.3 nm. When the spectrum of a moving celestial source is analyzed, this absorption line has a wavelength of 379.2 nm. (a) If the motion of the source is in the radial direction, is the source moving toward or away from the observer? (b) At what speed is the source moving?

72. Assume two identical clocks are located at $x = +L_o/2$ and $x = -L_o/2$. A flashbulb is set off at the origin $x = 0$ and $t = 0$. When light from the bulb reaches each of the clocks, they start ticking at $t = t_o = L_o/(2c)$. The start of each clock is an event with space-time coordinates (x, t). (a) Use the Lorentz transformations to transform the two events $(-L_o/2, t_o)$ and $(+L_o/2, t_o)$ into a coordinate system that is traveling parallel to the x-axis at a speed V. (b) What is the time difference between the "synchronized" clocks as seen from the moving frame?

73. The expression for the radius R of the circular orbit of a particle of charge q moving in a plane perpendicular to a magnetic field B was obtained in Chapter 28 as $R = p/qB$. This relation holds true for relativistic velocities as well. Calculate the radius of curvature of the orbit of an electron with kinetic energy 1.55 MeV traveling in a magnetic field of 2.34 T.

74. Show that the Lorentz transformation of Equations (39.8) does indeed transform the equation $(x')^2 + (y')^2 = (ct')^2$ into $x^2 + y^2 = (ct)^2$ as asserted in Section 39.3. (See Fig. 39.8.)

The Birth of Quantum Physics

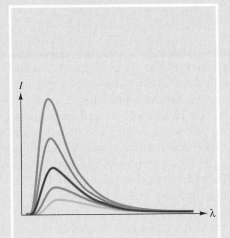

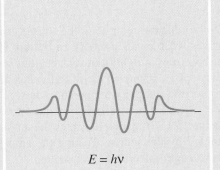

$E = h\nu$

In this chapter you should learn

- about blackbody radiation.
- Planck's quantum hypothesis.
- about the photoelectric effect.
- to calculate the energy of a photon.
- to calculate the de Broglie wavelength of an electron.
- about electron diffraction.
- about Rutherford scattering.
- about line spectra of atoms.
- about Bohr's model for the atom.

The world as we know it is a pretty smooth place. As we observe the motion of objects around us, we observe velocities that change continuously from one speed to another. We can lift masses in the earth's gravitational field and stop at any point we wish. The water that runs through our hands seems completely continuous. Yet you have been told from an early age that water is in fact made up of discrete molecules. You have probably accepted the existence of atoms that give an underlying graininess to all matter if we look closely enough. Perhaps far more surprising is the fact that momentum and energy share this grainy nature and can be parceled out only in small bundles. One striking example of bundled energy is light, which we previously modeled as a continuous wave. In this chapter we review evidence that suggests that light, too, has a particlelike side to its nature. Most startling of all is the realization that not only does light have a particle nature, but matter, which we consider to be particlelike, has a wave nature. We are reminded in a dramatic way that waves and particles are mental models that we construct to describe our macroscopic experience. Nature seems little compelled to restrict herself to our pre-conceptions. We are thus led to develop a new model to describe the behavior of both light and matter. The name given to this new model is *quantum,* and this chapter describes the evolution of this new concept during the first several decades of the 1900s.

40.1 The Particle Model for Light Revisited

In Section 35.1 we described the competition between the wave and particle models for light. At the close of the nineteenth century it must surely have seemed that the wave model for light had triumphed. The wave model had accounted for the many optical phenomena described in Chapters 35 through 38. Heinrich Hertz had demonstrated the reality of electromagnetic waves predicted by James Clerk Maxwell. It must also have seemed that all the important laws of physics had been discovered and that there was only a little clean-up work to be done to solve a few final, stubborn problems. In this section we describe some of these problems and the failure of classical physics to resolve them. To adequately explain these experiments, old and cherished beliefs had to be set aside. To compensate for this loss a new and rich field of physics was born. Central to this new physics is the concept of **quantization.** To say that a quantity is quantized is to say that each time we measure that quantity we find only one of a discrete set of possible values. We already encountered one example of quantization when we dealt with electric charge. The only values for electric charge measured to date have been multiples of the magnitude of the charge on an electron. Thus, we say that electric charge is quantized. The following experiments suggest that the energy of light is also quantized.

Blackbody Radiation

Figure 40.1 shows a kiln used to melt glass. Notice that the color of all the objects inside is uniform. Light from such furnaces approximates what is called **cavity radiation.** Physicists like to study cavity radiation because the spectrum of light from such cavities is quite smooth, showing none of the telltale frequency spikes characteristic of individual atoms. In fact, the spectrum of cavity radiation is a function of cavity temperature only and not the substance from which the cavity is made. The spectrum of radiation can be conveniently described in terms of **spectral excitance** $I_\lambda(\lambda, T)$, which indicates the amount of power per unit area per unit wavelength emitted by the source. To find the amount of power per unit area ΔI radiated within a wavelength interval $\Delta\lambda$ we write

$$\Delta I(\lambda, T) = I_\lambda(\lambda, T)\,\Delta\lambda$$

Figure 40.2 shows the spectral excitance of cavity radiation for several different temperatures. A body that emits light with the spectrum of cavity radiation is known as a **blackbody.** A small hole in a heated cavity that is in thermal equilibrium is an excellent

FIGURE 40.1

Light emerging from the opening in this kiln approximates blackbody radiation. The color is due to cavity radiation, not individual objects in the interior.

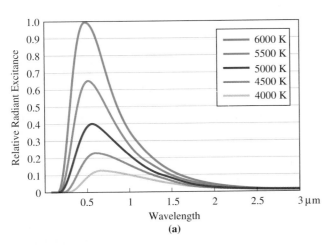

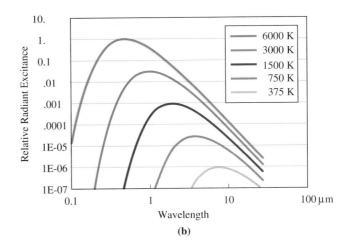

FIGURE 40.2

The spectral excitance of blackbodies at different temperatures. (a) Note the rapid decrease in total emission (area under the curve) with decreasing temperature. (b) Because of the large changes in spectral excitance with temperature, it is necessary to display radiant excitance for widely different temperatures using logarithmic scales.

approximation to a blackbody. The name *blackbody* stems from the fact that the hole acts as a perfect absorber as shown in Figure 40.3.

Several regularities in the blackbody spectral excitance had been noted by the late 1800s and two empirical laws describing these regularities emerged:

1. **Wien's displacement law:** The peak in the spectral excitance occurs at a wavelength λ_m, which decreases with increasing temperature of the cavity according to

$$\lambda_m = \frac{2898 \ \mu\text{m} \cdot \text{K}}{T} \qquad (40.1)$$

2. **Stefan-Boltzmann law:** The total power emitted per unit area of the cavity opening (**total radiant excitance**) increases rapidly with temperature according to

$$I(T) \equiv \int_0^\infty I_\lambda(\lambda, T) \ d\lambda = \sigma T^4 \qquad (40.2)$$

where the Stefan-Boltzmann constant $\sigma = 5.670 \times 10^{-8} \ \text{W} \cdot \text{m}^{-2} \cdot \text{K}^{-4}$.

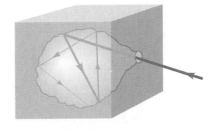

FIGURE 40.3

A small opening in an enclosed cavity acts as a blackbody. The hole is a perfect absorber. Radiation entering the hole is absorbed by the cavity walls before it can be reflected back out the hole.

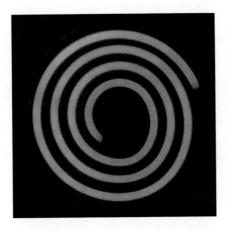

Even for objects that are not perfect blackbodies, the light emitted when they are heated indicates their temperature.

EXAMPLE 40.1   Surface Temperature of the Sun

The sun's spectrum, as measured from above the earth's atmosphere, exhibits a peak in spectral excitance at 465 nm. What is the temperature of a blackbody with this same spectral peak? (The surface of the sun is actually about 400 K cooler than this equilibrium blackbody model predicts.)

SOLUTION   We solve Wien's displacement law from Equation (40.1) for the temperature T

$$T = \frac{2898 \ \mu\text{m} \cdot \text{K}}{\lambda_m} = \frac{2898 \ \mu\text{m} \cdot \text{K}}{0.465 \ \mu\text{m}} = 6230 \ \text{K} \qquad \blacktriangleleft$$

By 1900, two partially successful attempts had been made to deduce the detailed shape of the measured spectral radiance function from classical physics. Wilhelm Wien (1864–1928), using an analogy between the radiation in the cavity and the distribution of speeds of molecules in a gas, had proposed

$$I_\lambda(\lambda, T) = 8a\left(\frac{c}{2\lambda}\right)^5 e^{-bc/\lambda T}$$

where a and b are constants to be determined. Lord Rayleigh (John William Strutt (1842–1919)) derived an expression for the spectral excitance later modified by Sir James Jeans (1877–1946). Their derivation was based on purely classical theories. This **Rayleigh-Jeans law,** as it is known, predicted

$$I_\lambda(\lambda, T) = \frac{32\pi k_B T}{c^3}\left(\frac{c}{2\lambda}\right)^4$$

As you can see from Figure 40.4 neither of these descriptions fits the data well over the entire spectrum. At short wavelengths the relation proposed by Wien is accurate; at long wavelengths the Rayleigh-Jeans law describes the data. In October of 1900, Max Planck (1858–1947) proposed a solution that fit the blackbody spectrum throughout its entire range. What Planck did was to find a functional form that reduced to the Wien description for short wavelengths and to the Rayleigh-Jeans law for long wavelengths. (See Problem

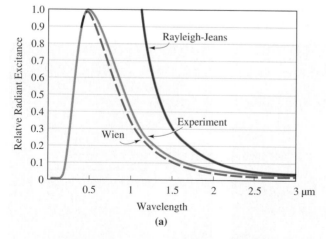

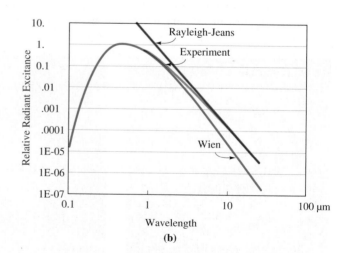

FIGURE 40.4

Blackbody radiation is approximated by the Wien distribution at small wavelengths and the Rayleigh-Jeans distribution at long wavelengths. (a) Linear scales. (b) Logarithmic scales.

2.) The functional form proposed by Planck was

$$I_\lambda(\lambda, T) = 8a\left(\frac{c}{2\lambda}\right)^5 \frac{1}{e^{bc/\lambda T} - 1}$$

At this stage Planck's formula was empirical only, and a and b were again constants chosen to fit the data. However, two (no doubt labor-filled) months later Planck had produced a theoretical model that predicted a spectral excitance of

$$I_\lambda(\lambda, T) = \left(\frac{2\pi hc^2}{\lambda^5}\right) \frac{1}{e^{hc/\lambda k_B T} - 1} \qquad (40.3)$$

where k is Boltzmann's constant, and h is an entirely new constant chosen to fit the data. Today this constant is known as **Planck's constant,** and its value has been determined to be

$$h = 6.626 \times 10^{-34}\,\text{J} \cdot \text{s} = 4.136 \times 10^{-15}\,\text{eV} \cdot \text{s}$$

To see the significance of Planck's new law we need to see why the classical derivation of the Rayleigh-Jeans law goes astray and what Planck did to correct it. The classical calculation models the cavity radiation as standing waves with a node at each cavity wall. The mental picture for a cubic cavity is shown in Figure 40.5. For a cavity length L, the standing wave condition is satisfied for $\lambda = 2L/n$, where $n = 1, 2, 3, \ldots$. Waves with a wavelength corresponding to a particular value of n are said to be of the same **mode.** The situation reminds us of pictures of standing waves on a string. According to the classical equipartition theorem, each of these modes should receive $3k_B T$ worth of energy. (In classical thermodynamics each degree of freedom is assigned $k_B T/2$. Each standing wave has three possible directions and two possible polarizations leading to $3 \times 2 \times k_B T/2 = 3k_B T$.) By looking at Figure 40.5 you can see the problem: there are a great many modes with short wavelengths! In fact the wavelengths have no shortest value. Furthermore, the wavelengths of standing waves become quite similar for short wavelengths. (For example, if $L = 1.000$ m, then $\lambda_1 = 2.000$ m and $\lambda_2 = 1.000$ m differ by 100%, but $\lambda_{1000} = 2.000$ mm and $\lambda_{1001} = 1.998$ mm differ by only 0.1%.) Thus, the number of waves satisfying $\lambda = 2L/n$ in any given $\Delta\lambda$-interval increases dramatically as λ becomes small. Now, if we give each of these waves the same energy $3k_B T$, the result is just what you expect: there is an enormous amount of energy concentrated at short wavelengths. You can see this clearly in Figure 40.4 where the Rayleigh-Jeans spectral density shoots off the graph at the low wavelengths. This situation is often referred to as the **ultraviolet catastrophe.**

To avoid the ultraviolet catastrophe Planck proposed a radical new idea. He proposed that a mode cannot have an arbitrary amount of energy. Instead, he hypothesized that the oscillators in the walls of the cavity that formed the blackbody could emit and absorb energy E only in bundles of the amount

$$E = h\nu$$

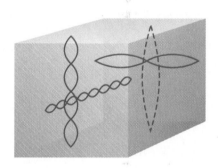

FIGURE 40.5

A pictorial view suggesting the formation of standing electromagnetic waves in three directions. The waves are actually plane waves filling the cavity interior.

Thus, a radiation mode in the cavity could have $h\nu, 2h\nu, 3h\nu, \ldots$ amounts of energy, but nothing in between. Within the cavity an average energy on the order of $k_B T$ is all that is available for any oscillator. If a given mode requires substantially more energy than this average, chances are good that it never becomes excited. The effect of this low probability is to "freeze out" the high-frequency (short-wavelength) modes.

Because of the small size of Planck's constant, we seldom see evidence for this quantization of energy on the macroscopic scale. If we did, the experience might be like the following illustration. Imagine a guitar with quantized string vibrations. For simplicity let's assume that each string vibrates only in its fundamental mode. The frequency of this

Concept Question 1
Which is more energetic, a photon corresponding to blue light or one corresponding to red light?

fundamental is determined by the string's length, tension, and mass per unit length. Thus, like the atomic oscillator, the guitar string frequency has a characteristic value determined by its physical parameters. But this is not the type of quantization that Planck meant. For although the frequency of the string is fixed, in the classical case, the string can vibrate with any amplitude large or small. Let's think what it would be like if the *amplitude* of the string's vibration were quantized: You strum the guitar lightly and nothing happens; the strings don't vibrate. You pluck the bass string forcefully and still it doesn't vibrate. Striking the string still more forcefully, you finally get it to vibrate at a low amplitude that remains constant. You strike the bass string hard as before and now it vibrates with greater amplitude, in fact with twice the original energy. Next, you try a similar experiment on the highest pitched string. Pound as you may, however, you cannot get this string to vibrate. You just don't have enough energy available to supply the larger $h\nu$ required to excite this string. Finally, you give up and settle back to some vigorous strumming. The lowest frequency string vibrates vigorously, the next highest frequency string not quite as much, and the next highest still less. And, of course, the highest pitched string remains stubbornly fixed. We should note too that our quantized guitar is generally silent! In a real guitar the string vibration amplitude gradually decreases as energy is slowly transferred to the surrounding air and carried off in sound waves in a fairly continuous fashion after the string is plucked. Because the vibration amplitudes of the strings of our quantum guitar are quantized, sound is emitted only in short tone bursts when the vibration amplitudes change discontinuously. Improbable as this illustration sounds, it is precisely what Planck was proposing for the behavior of the atomic oscillators in the cavity walls.

As you might suspect, acceptance of Planck's hypothesis was slow in coming. It took a great many confirming experiments before the idea of the quantization of energy was accepted. Nonetheless, it is fair to say that Planck's paper presented in December 1900 opened the new era of quantum physics.

Heat Capacity of Solids

Another of the problems with which physicists wrestled in the late 1800s was the molar heat capacity of solids. You may recall from Chapter 18 that the molar heat capacity is the heat required to raise the temperature of one mole of a substance by one degree Celsius. When the heat capacity is measured at constant volume, the added heat increases the internal energy of the solid. A classical argument suggests that each atom should have an average energy of $k_B T/2$ for each degree of freedom (see Chapter 21, Section 3). Each atom can move in three dimensions and has both kinetic and potential energy giving a total of $3 \times 2 \times k_B T/2 = 3k_B T$ average energy per atom. To find the energy per mole we multiply by Avogadro's number N_A and obtain $U = 3N_A k_B T = 3RT$, where $R = 8.314$ J/mol/K is the universal gas constant first introduced in Chapter 18. The molar heat capacity is given by $c_V = dU/dT = 3R$, where the subscript V denotes that the volume is held constant (see Eq. (19.8)). If we examine the molar heat capacity of several metals as in Figure 40.6, the good news is that at high temperatures the metals do approach the values of $3R$ as predicted by classical thermodynamics. The bad news is that nothing in the classical theory predicts any temperature dependence at all!

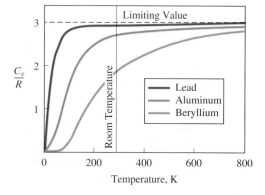

FIGURE 40.6

Molar heat capacity at constant volume for several metals. Note that the general shape is the same for each, but only at high temperatures does the heat capacity approach its classical value of $3R$.

Einstein was the first to point out (in 1907) that Planck's quantization condition could explain the major features of the temperature dependence of these specific heats. Einstein quantized the energy of the individual atoms in the metal. Once again, this restriction had the effect of not permitting the higher frequency modes to be excited at lower temperatures. The fewer number of modes leads to lower specific heat values. Einstein's result is

$$c_V = \left(\frac{3N_A h^2 \nu^2}{k_B T^2}\right)\frac{e^{h\nu/k_B T}}{(e^{h\nu/k_B T} - 1)^2} \qquad (40.4)$$

By defining an appropriately chosen **Einstein temperature** $T_E = h\nu/k_B$ for each type of metal, one finds that the heat capacity of a great variety of substances all have the same shape and, for all but the lowest temperatures, that shape is described by

$$c_V = 3R\left(\frac{T_E}{T}\right)^2 \frac{e^{T_E/T}}{(e^{T_E/T} - 1)^2} \qquad (40.5)$$

Despite this success, Einstein's theory did not quite fit the data at the lowest temperatures. Five years later (in 1912) Dutch physicist Peter Debye (1884–1966) modified Einstein's model. Instead of quantizing individual atoms, Debye treated the elastic vibrations within the solid as standing waves created between the sides of the metal. He quantized these standing waves in the same fashion that Planck had quantized the light waves in cavity radiation. This modification resulted in an expression for the molar heat capacity that was in complete accord with measured values.[1] Figure 40.7 shows a comparison of measured values of molar heat capacities for several materials compared with Debye's quantum theory. The agreement is impressive and gives added credence to the reality of Planck's quanta.

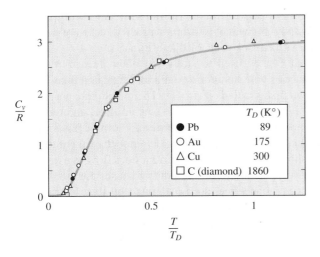

FIGURE 40.7

The Debye model for the molar heat capacity shows agreement with measured values for a large range of substances, giving further experimental evidence of quantization of energy. The Debye temperature T_D is a constant for each material and accounts for differences in masses and vibration constants between materials.

Photoelectric Effect

In 1905, Einstein made a bold conjecture concerning the nature of light. Planck had argued that the individual oscillators in the walls of a blackbody cavity emitted and absorbed light in quantized amounts. Einstein took the argument one step further and suggested that light itself is quantized into localized bundles, which we now call **photons.** He proposed that his hypothesis be tested by using the photoelectric effect, which we will describe shortly.

---

[1] Strictly speaking measured values of the heat capacity are heat capacities at constant pressure rather than constant volume. The difference is small for solids and the correction relatively easy to compute. In addition, measured heat capacities of metals contain contributions from free electrons, whereas Debye's theory accounted only for the heat capacity due to the lattice. Again, the correction is straightforward, and the corrected lattice heat capacity agrees with Debye's theory.

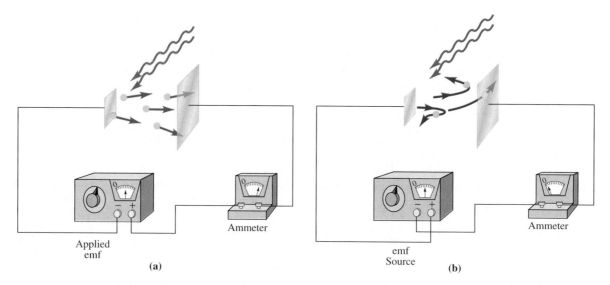

FIGURE 40.8

An apparatus for demonstrating the photoelectric effect. (a) Light falling on the emitter causes electrons to be ejected. These electrons are attracted by the positively charged collector and are accelerated to it, thereby completing the circuit. When the circuit is thus completed, the ammeter deflection indicates the electron current across the photocell. (b) When leads on the emf source are reversed, electrons ejected from the emitter by the light are repelled from the collector. Only the fastest electrons make it across. As the reverse emf is increased, the current in the ammeter eventually falls to zero, indicating that no electrons are crossing the gap. The total potential difference between the emitter and collector that just stops the fastest electrons is called the stopping potential.

It was not until much later, in 1916, that Robert Millikan (of the Millikan oil-drop fame) published results of an extensive study in which Einstein's predictions were verified completely, apparently much to the surprise of even Millikan himself.

Figure 40.8 shows a circuit that can be used to study the photoelectric effect. A variable emf source is connected in series with an ammeter and an evacuated tube. Inside the tube the circuit is connected to two plates called electrodes. The metal plate nearest the negative terminal of the emf source in Figure 40.8(a) is called the **emitter,** and the other plate is called the **collector.** As you no doubt can tell, ordinarily no current flows in this circuit because there is no way for charge to move across the vacuum between emitter and collector. However, when a light is shined on the emitter, a current does flow as evidenced by the deflection of the ammeter. The light falling on the emitter causes electrons to be ejected from the metal. These electrons, once free, are accelerated by the electric field between the plates and travel to the collector, thus completing the circuit.

What is interesting about this effect is that even if we reverse the polarity of the applied emf as in Figure 40.8(b), some electrons are still able to make the journey from emitter to collector despite the opposing electric field. These electrons have been ejected from the metal with sufficient kinetic energy to climb the potential hill to the collector. However, if a sufficiently large electric field is applied, even the most energetic electrons are forced back to the emitter before they reach the collector. The potential difference required to bring the current to zero is called the **stopping potential.**[2] We define the stopping potential V_s as the potential difference between the emitter and the collector when the kinetic energy K_m of the most energetic electrons is just equal to the potential energy

---

[2] When we discuss the potential difference required to stop electrons, this potential difference may not equal the applied emf $\mathscr{E}$. In fact, when the emitter and collector are made of different metals (which is usually the case in practice), a potential difference exists between them even in the absence of an applied emf. This so-called *contact potential* ϕ arises because of the different electron affinities of the two metals. The stopping potential is then the sum of the applied emf and the contact potential. In our examples and problems we assume the emitter and collector are made of the same material in order to avoid this complication.

of the collector relative to the emitter. That is, when

$$K_m = \frac{1}{2}mv_m^2 = eV_s \qquad (40.6)$$

A typical graph of the current through the circuit as a function of the potential difference V between the electrodes is shown in Figure 40.9.

In his experiments Millikan varied a number of parameters associated with this phenomenon. The most important results are summarized here:

1. When the irradiance of the light is increased, the magnitude of the current increases, but the stopping potential remains constant. That is, it takes no greater potential to stop electrons ejected by bright light than those ejected by dim light.
2. When the frequency of the light is increased, the stopping potential increases in a linear fashion.
3. There exists a cutoff frequency below which no electrons are ejected no matter how intense the light. This cutoff frequency is different for different metals.
4. The ejection of photoelectrons is immediate, even when the light is made exceedingly dim.

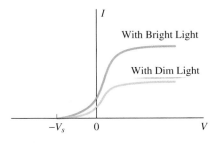

FIGURE 40.9

The current in the ammeter of Figure 40.8 as a function of the total potential difference between emitter and collector. Higher irradiance on the emitter causes a larger current to flow, but the stopping potential V_s required to stop the electrons is no greater. Different metals have different values for the stopping potential.

The significance of these results becomes clear when we compare them to the predictions of classical wave theory and to Einstein's radical proposal. Result 1 suggests that the energy of photoejected electrons does not depend on the irradiance of the illuminating light. This result seems strange from a classical point of view because we know that the irradiance is proportional to the square of the electric field of the incident light wave. Thus, a higher irradiance means stronger electric fields. Why are the electrons not given greater kinetic energy by the larger force due to the larger electric fields? Results 2 and 3 are still more difficult to reconcile with the classical view. They tell us that the maximum kinetic energy, indeed even the existence, of ejected electrons depends on the frequency of the incoming light. Yet we found in Chapter 34 that the energy of a light wave is proportional to the square of the amplitude of the electric field *and is completely independent of the frequency of the light.* Result 4 is surprising, too. From the classical point of view, the energy of an electromagnetic wave is spread out smoothly over the entire wavefront. In order to escape from an atom, an electron must acquire an energy at least equal to its binding energy, something on the order of a few electron volts. We might reasonably expect that this amount of energy would have to fall on an area the size of an individual atom before the electron could accumulate a sufficient amount to escape. When the illuminating beam is made extremely dim, the predicted time for the absorption of sufficient energy can be increased to several seconds. Yet Millikan observed no delay whatsoever. The instant light hits the emitter, electrons are ejected regardless of how dim the light is made.

The inability of the classical wave model of light to explain the results of the photoelectric effect stands in sharp contrast to the predictions of Einstein. According to Einstein's explanation, the energy of the light wave is already localized in discrete bundles (the photons), each with energy $h\nu$. When a photon strikes an atom, the photon's energy can be delivered to an electron that is knocked out of the metal surface. A certain amount of work must be done to liberate even the most loosely held electrons. This minimum amount of work is known as the **work function** W of the metal. Thus, the maximum kinetic energy K_m of the most loosely held electrons is given by

$$K_m = h\nu - W \qquad \text{Einstein's photoelectric equation} \qquad (40.7)$$

Equation (40.7) is known as **Einstein's photoelectric equation.** This equation has the form of a straight line $y = mx + b$ if we plot the maximum kinetic energy as a function of the frequency of the illuminating light. The maximum kinetic energy is deduced from

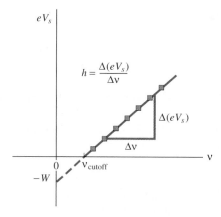

FIGURE 40.10

Results of a typical photoelectric experiment. The slope of the straight line is Planck's constant. The smallest frequency of light for which electrons are ejected is called the cutoff frequency. The extrapolated intercept with the energy axis is the work function W.

TABLE **40.1**   **Photoelectric Work Function of Selected Elements (Polycrystalline)[3]**

ELEMENT	W (eV)
Al	4.28
Cs	2.14
Cu	4.65
Hg	4.49
K	2.30
Na	2.75
Ni	5.15
Pb	4.25
Pt	5.65

Concept Question 3
The work function of potassium is larger than that of cesium. Each is to be used with different monochromatic light in the photoelectric effect. It is desired to achieve the same maximum kinetic energy for photoejected electrons from both surfaces. Should the light shined on the potassium be bluer or redder than the light shined on the cesium?

the stopping potential through Equation (40.6). Figure 40.10 shows a comparison of the photoelectric equation and typical measurements. Not only is the straight-line nature of the relation confirmed, but the slope of the straight line is indeed equal to Planck's constant. Here, in an entirely different context, is further evidence for the reality of the quantization of energy.

EXAMPLE 40.2   *Cesium Has the Lowest Work Function of Any Stable Element*

The photoelectric work function for cesium is 2.14 eV. (a) What is the cutoff frequency for cesium? (b) If the photocurrent is brought to zero by a stopping potential of 0.600 V, what is the wavelength of the illuminating light?

SOLUTION   (a) Cutoff occurs when the energy of incoming photons falls below the work function

$$h\nu \leq W$$

so that for frequencies

$$\nu \leq \frac{W}{h} = \frac{2.14 \text{ eV}}{4.135 \times 10^{-15} \text{ eV} \cdot \text{s}} = 5.18 \times 10^{14} \text{ Hz}$$

no electrons are ejected.

(b) When the maximum kinetic energy $K_{\max}$ of the photoejected electrons is equal to the potential energy eV_s the photocurrent falls to zero. Thus, we can write Einstein's photoelectric equation as

$$eV_s = h\nu - W$$

We substitute $\nu = c/\lambda$ and solve for the wavelength λ

$$\lambda = \frac{hc}{eV_s + W} = \frac{(4.135 \times 10^{-15} \text{ eV} \cdot \text{s})(2.998 \times 10^8 \text{ m/s})}{(0.600 \text{ eV} + 2.14 \text{ eV})} = 452 \text{ nm} \quad \blacktriangleleft$$

---

[3] Taken from Robert E. Weast, editor in chief, *Handbook of Chemistry and Physics,* 1st Student Edition. (CRC Press, Boca Raton, FL, 1988), pp. E78–79.

Compton Scattering

If Einstein's photons really represent localized electromagnetic energy, one can imagine treating them as particles. We can imagine then collision experiments between these photons and other types of particles, say electrons. Moreover, we should be able to analyze these collisions just as we did particle collisions back in Chapter 9. To study such collisions we applied the conservation of momentum. Provided we keep track of the (relativistically correct) energy of electrons and photons, we can also apply the energy conservation principle to collisions between photons and electrons. Even without performing calculations, we can anticipate what happens. If a photon gives some of its energy to an electron it strikes, the photon's energy must decrease. Because the photon energy is $h\nu$, this loss of energy implies the scattered photon has lower frequency. Lower frequency implies longer wavelength. Thus, we expect to see photons that strike electrons come away from the collision with longer wavelengths. Our goal is to compute the wavelength change of these scattered photons.

Before we can proceed to find this wavelength change, we need to find an expression for the momentum of Einstein's photons. The energy–momentum relation from special relativity is (see Eq. (39.23))

$$E^2 = (m_o c)^2 + (pc)^2$$

But what do we take as the rest mass m_o of a photon, this "particle" that is never at rest?! One hint comes from classical electromagnetic theory. In Chapter 34 we found that the wave model of light based on Maxwell's equations led us to conclude in Equation (34.28) that $U = pc$. Where, in the notation of that chapter, U was the total energy in the electromagnetic wave; this same energy is symbolized by E in the energy–momentum relation. Once we realize the equivalence of U and E we see that the classical wave model and Einstein's photon model result in the same energy–momentum relation if we take the rest mass of the photon as zero. In this regard at least, the classical electromagnetic wave model and the photon model agree: light carries both energy and momentum but no rest mass.

We can now continue with our program to find the wavelength change that results when a photon collides with a traditional particle. Treating the photons as particles with zero rest mass leads to

$$E = pc$$

When we combine this equation with the relation $E = h\nu$, we are led to an expression for the **momentum of a photon**:

$$p = \frac{h\nu}{c} = \frac{h}{\lambda} \qquad \text{Momentum of a photon} \qquad (40.8)$$

We now imagine a collision between a photon and a stationary free electron of rest mass m_o as diagrammed in Figure 40.11. First, we equate the x-components of total momentum before and after the collision:

$$\frac{h}{\lambda} = \frac{h}{\lambda'}\cos(\phi) + \frac{m_o u}{\sqrt{1 - (u/c)^2}}\cos(\theta) \qquad (40.9)$$

Next, we equate the y-components of total momentum before and after the collision to find

$$0 = \frac{h}{\lambda'}\sin(\phi) + \frac{m_o u}{\sqrt{1 - (u/c)^2}}\sin(\theta) \qquad (40.10)$$

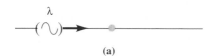

(a)

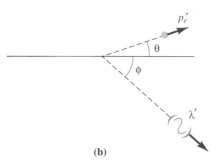

(b)

FIGURE 40.11

Compton scattering. (a) A photon of wavelength λ strikes a stationary free electron. (b) The photon is scattered at an angle ϕ with respect to its original direction and has a longer wavelength λ than the original photon.

The conservation of energy requires that

$$h\nu + m_o c^2 = h\nu' + \frac{m_o c^2}{\sqrt{1 - (u/c)^2}} \qquad (40.11)$$

Equations (40.9) through (40.11) can be solved (after a considerable bit of algebra) for the wavelength shift $\Delta\lambda$, known as the **Compton shift,** of the scattered photon

$$\Delta\lambda = \lambda' - \lambda = \frac{h}{m_o c}[1 - \cos(\phi)] \qquad \text{Compton shift} \qquad (40.12)$$

The quantity $h/(m_o c)$ has the dimensions of length and is known as the **Compton wavelength** λ_C. The Compton wavelength indicates the scale of the wavelength changes that can be expected when the photon collides with a free particle of mass m_o. When m_o is the mass of the electron, this quantity is the **Compton wavelength of the electron,** which has a magnitude of 0.00243 nm = 2.43 pm.

Because the Compton wavelength of the electron is quite small, photons of very short wavelength must be used to produce a sizable fractional change in wavelength. Figure 40.12 shows a schematic diagram of the apparatus used by Arthur Compton (1892–1962) in 1923 to irradiate a graphite sample with monochromatic X-rays. Figure 40.13 shows Compton's results. The shift in the wavelength of the scattered photons as a function of angle dramatically verified Einstein's photon hypothesis.

As the concepts of energy quantization and the photon gained acceptance, the major players in the drama were recognized by Nobel Prizes for their contributions to the formation of the new physics: Max Planck (1918), energy quanta; Albert Einstein (1921), theoretical physics, especially the photoelectric effect; Robert Millikan (1923), experimental measurement of the charge of an electron and the photoelectric effect; Arthur Compton (1927), discovery of the Compton effect.

Concept Question 4
Which has greater momentum, a photon of wavelength corresponding to blue light or one corresponding to red light?

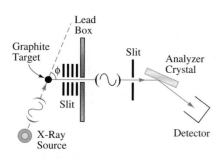

FIGURE 40.12

A schematic diagram of the Compton scattering experiment. Monochromatic X-rays strike a graphite target. X-rays scattered from free electrons in the graphite pass through a series of collimating slits into a lead box. The X-ray wavelengths are determined by diffraction from an analyzer crystal. (After A. H. Compton, "The scattering of x-rays as particles," *American Journal of Physics,* **29**(12), 817–820 (Dec. 1961).)

EXAMPLE 40.3   Momentum of a Photon

(a) What is the momentum of a photon from a helium-neon laser ($\lambda = 633$ nm)? (b) How fast does a hydrogen atom have to travel in order to have this same momentum?

SOLUTION   Equation (40.8) can be used to calculate the photon momentum. The combination hc occurs so often in the calculation of photon energy and momentum that it is convenient to treat it as a single constant with value $hc = 1240$ eV · nm.

$$p = \frac{h}{\lambda} = \frac{hc}{c\lambda} = \frac{1240 \text{ eV} \cdot \text{nm}}{c(633 \text{ nm})} = 1.96 \text{ eV}/c$$

In SI units the momentum is

$$p = 1.96 \text{ eV}/c = (1.96 \text{ eV})\frac{1.602 \times 10^{-19} \text{ J/eV}}{2.998 \times 10^8 \text{ m/s}} = 1.05 \times 10^{-27} \text{ kg} \cdot \text{m/s}$$

(b) We assume that the hydrogen atom moves slowly, so we can use the classical expression for momentum. Thus, to find the speed of an atom with the same momentum value as the photon of part (a) we use $p = m_o v$:

$$v = \frac{p}{m_o} = \frac{1.05 \times 10^{-27} \text{ kg} \cdot \text{m/s}}{1.67 \times 10^{-27} \text{ kg}} = 0.627 \text{ m/s} \qquad \blacktriangleleft$$

EXAMPLE 40.4   Compton Shift at 90°

By what percentage is the wavelength of a 35.0-keV photon shifted if a collision with an electron results in the photon being scattered at 90°?

SOLUTION   The wavelength change is given by the Compton scattering formula of Equation (40.12):

$$\Delta\lambda = \lambda_C[1 - \cos(\phi)] = (2.43 \text{ pm})[1 - \cos(90°)] = 2.43 \text{ pm}$$

For scattering at 90° the shift, which is independent of the incoming photon wavelength, is exactly equal to the Compton wavelength of the electron.

The wavelength of the incoming photon can be found from its energy $E = h\nu = hc/\lambda$

$$\lambda_o = \frac{hc}{E} = \frac{1240 \text{ eV} \cdot \text{nm}}{35.0 \times 10^3 \text{ eV}} = 3.54 \times 10^{-2} \text{ nm} = 35.4 \text{ pm}$$

This result leads to a percentage change

$$\frac{\Delta\lambda}{\lambda_o} = \frac{2.43 \text{ pm}}{35.4 \text{ pm}} = 0.0686 \approx 7\%$$

◄

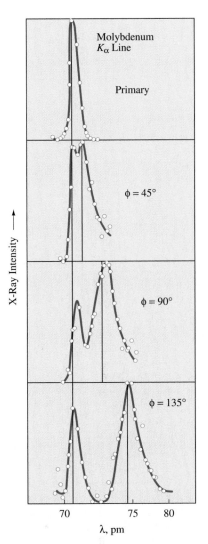

FIGURE 40.13

Results of Compton scattering. The peak on the right (in the bottom three frames) represents X-rays scattered by free electrons. The wavelength shift increases with increasing angle as predicted by the Compton scattering formula. (After A. H. Compton, ''The scattering of x-rays as particles,'' *American Journal of Physics,* **29**(12), 817–820 (Dec. 1961).)

40.2 Wave Model for Particles

In the previous section we outlined experiments that present convincing evidence for the particlelike nature of electromagnetic radiation. Clearly, that which we previously thought to be waves can exhibit properties like particles. In 1924, Louis de Broglie (1892–1987) in his Ph.D. thesis proposed that the tables could be turned and that entities that we think of as particles have a wave nature. His suggestion, known as the **de Broglie hypothesis,** was that the expression of Equation (40.8) for the momentum of a photon could be used to predict the effective wavelength of a particle, such as the electron

$$\lambda = \frac{h}{p} \tag{40.13}$$

This **de Broglie wavelength** λ is a function of the momentum and, hence, the energy of the particle. We show in Example 40.5 that the wavelength associated with macroscopic objects is too small to ever be detected. On the other hand, for electrons with modest energies, the de Broglie wavelength is on the order of tenths of a nanometer.

EXAMPLE 40.5   Wavelength of a Golf Ball and an Electron

Calculate (a) the de Broglie wavelength of a golf ball traveling at 60.0 m/s and (b) an electron with kinetic energy of 100. eV.

SOLUTION   (a) Employing the de Broglie relation of Equation (40.13), we find for the golf ball of mass 0.0459 kg

$$\lambda = \frac{h}{p} = \frac{6.626 \times 10^{-34} \text{ J} \cdot \text{s}}{(0.0459 \text{ kg})(60.0 \text{ m/s})} = 2.41 \times 10^{-34} \text{ m}$$

This length is about 10^{-19} times the radius of a proton, quite beyond measurement!

(b) Because $K \ll m_o c^2$ for this electron, we can use the classical expression $K = p^2/2m_o$ to write

$$\lambda = \frac{h}{p} = \frac{h}{\sqrt{2m_o K}} = \frac{6.626 \times 10^{-34}\,\text{J} \cdot \text{s}}{\sqrt{2\,(9.109 \times 10^{-31}\,\text{kg})(100.\ \text{eV})(1.602 \times 10^{-19}\,\text{J/eV})}}$$
$$= 0.123\ \text{nm}$$

This result represents a small wavelength to be sure, but it is on the order of the wavelength of the electromagnetic waves we call X-rays. ◄

When we studied light waves, we discovered that their wave nature could be revealed through interference and diffraction patterns. To create such patterns we found that the slits and grating lines had to have separations comparable to the wavelength of the radiation. Although no physical *slits* can be constructed with spacings on the order of tenths of a nanometer, this length is comparable to the spacings between atoms in solids. It is reported that in reply to questions from a skeptical examining committee during his Ph.D. thesis defense, de Broglie proposed the wave nature of electrons might be detected using diffraction from crystalline solids in the same manner that the wave nature of X-rays was first demonstrated.[4] Two confirming experiments using low-energy electron beams quickly followed, one performed by C. J. Davisson (1881–1958) and L. H. Germer at Bell Telephone Laboratories in the United States. The other experiment was performed by Sir George Thomson (1892–1975) at the University of Aberdeen in Scotland. J. V. Hughes, and others, later carried out experiments with electrons moving at high speeds where relativistic corrections were required.[5]

The Davisson-Germer Experiment

In 1925, C. J. Davisson and L. H. Germer tested de Broglie's hypothesis literally by accident. They were measuring the electron scattering from a polycrystalline nickel sample. The explosion of a liquid air bottle broke the vacuum system while the nickel sample was hot, causing the sample to oxidize. In order to remove the nickel oxide the workers reheated the sample in hydrogen and then in vacuum. In the process, the polycrystalline sample recrystallized into several relatively large single crystals. This process produced exactly the experimental conditions required to test de Broglie's hypothesis! In this experiment a beam of electrons is accelerated toward a sample of crystalline nickel as shown in Figure 40.14. A detailed analysis of Bragg scattering for electrons that strike the surface at *normal incidence* reveals that constructive interference occurs at angles ϕ, such that

$$D \sin(\phi) = m\lambda \tag{40.14}$$

where D is the spacing of adjacent rows of atoms in the surface of the crystal and m is an integer.[6] Figure 40.15 shows the results of Davisson and Germer's experiment. Note the appearance of a sharp diffraction maximum at an accelerating potential of 54 V and an angle of 50°. In Example 40.6 we show that this is precisely the electric potential and angle at which such a peak should occur according to the diffraction equation (Eq. (40.14)) for electron waves.

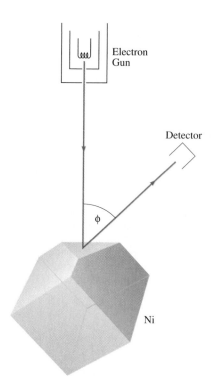

Electron Gun

Detector

ϕ

Ni

FIGURE 40.14

Schematic diagram of the Davisson-Germer experiment. A collimated electron beam strikes a nickel crystal at normal incidence. If electrons have wave properties, diffraction peaks are expected in the scattered intensity at certain angles.

[4] A. P. French and Edwin F. Taylor, *An Introduction to Quantum Physics* (Norton, New York, 1978), p. 61.

[5] A more complete list of experimental tests is given in H. R. Brown and R. de A. Martins, "De Broglie's relativistic phase waves and wave groups," *American Journal of Physics,* **52,** 1130 (1984).

[6] Although Equation (40.14) can be obtained from the Bragg scattering condition $2d \sin(\theta) = m\lambda$, the angle ϕ is not the Bragg angle and D is not the perpendicular distance between Bragg planes. The relation between the two sets of variables is discussed in Problem 63 and shown in Figure 40.P6.

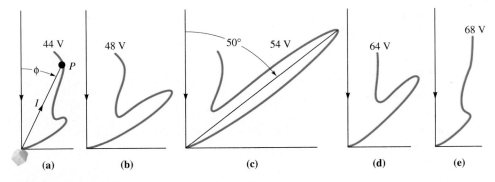

FIGURE 40.15

Polar plots of the scattered electron current in the Davisson-Germer experiment. The distance from the origin to the curve is proportional to the number of electrons scattered at angle ϕ measured from the vertical axis. A quite strong peak occurs at just the angle and electron energy predicted by the de Broglie hypothesis.

EXAMPLE 40.6   *Diffraction of Electrons*

Calculate the angular position of the diffraction peak for electrons with kinetic energy 54.0 eV when diffracted from a nickel crystal with a spacing $D = 2.15$ Å between rows of atoms. (The Angstrom Å is a non-SI unit of length traditionally used in X-ray work. 1 Å = 0.1 nm.)

SOLUTION   The wavelength of the electrons is

$$\lambda = \frac{h}{p} = \frac{h}{\sqrt{2m_o k}} = \frac{6.626 \times 10^{-34} \text{ J} \cdot \text{s}}{\sqrt{2(9.109 \times 10^{-31} \text{ kg})(54.0 \text{ eV})(1.602 \times 10^{-19} \text{ J/eV})}}$$
$$= 0.167 \text{ nm}$$

According to the interference formula of Equation (40.14) the first-order ($m = 1$) maximum occurs when

$$D \sin(\phi) = \lambda$$

so that the diffraction angle is

$$\phi = \arcsin\left(\frac{\lambda}{D}\right) = \arcsin\left(\frac{0.167 \text{ nm}}{0.215 \text{ nm}}\right) = \arcsin(0.777) = 51.0°$$

Incidentally, there is a change in wavelength of the electron matter waves as they enter the nickel and, consequently, the electron waves are refracted as they leave the surface of the nickel. It can be shown that *for the case of normal incidence* the effect of wavelength change and refraction at the surface cancel each other and the free-space wavelength of the electron can be used in Equation (40.14).[7]   ◀◀◀

The Experiment of G. P. Thomson

In 1927, Sir George P. Thomson performed an experiment analogous to the powder diffraction experiments carried out with X-rays. A thin polycrystalline foil was illuminated with an electron beam. The resulting diffraction patterns were similar to patterns that are

---

[7] See, for example, A. P. French and Edwin F. Taylor, *An Introduction to Quantum Physics* (Norton, New York, 1978), pp. 71–72.

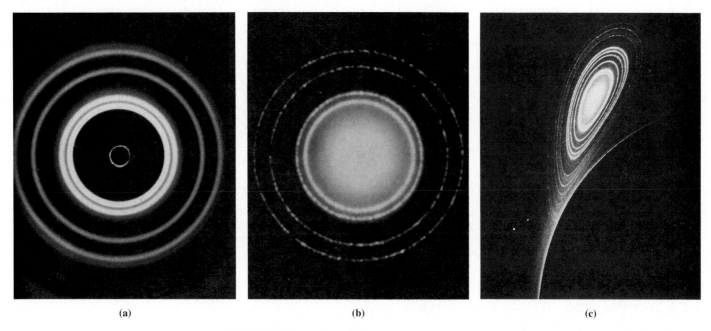

(a)   (b)   (c)

FIGURE 40.16

Electron and X-ray diffraction. (a) Diffraction pattern of monochromatic X-rays striking an aluminum foil at normal incidence. (b) Diffraction pattern created by electrons with de Broglie wavelength similar to that of the X-rays in (a). (c) A small magnet placed near the path of scattered electrons shows that the diffraction pattern is due to electrons. (The path of X-rays would not be deviated by the magnet.)

created when X-rays strike the foil. Figure 40.16 shows a more recent comparison of a diffraction pattern created by X-rays and by an electron beam. Examining these remarkable photographs, one can hardly deny the reality of the electron's wave nature. In 1937, Sir George P. Thomson shared the Nobel Prize with Clinton Davisson for their experimental demonstrations of the wave nature of the electron. Thirty-one years earlier George Thomson's father, Joseph J. Thomson (1856–1940), was awarded the Nobel Prize for the discovery of the electron and his measurements of *e/m* for this *particle*.

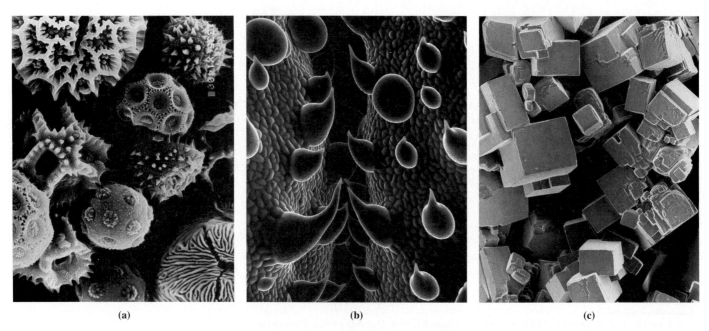

(a)   (b)   (c)

Scanning electron micrograph (SEM) images of (a) pollen grains, (b) a marijuana leaf, and (c) salt crystals.

40.3 The Spectra of Atoms

We turn our attention now to the light given off by atoms in a low-pressure gas. Heating such gases or passing an electric current through them can cause the gases to glow. When light from the glowing gas is passed through a narrow slit and then through a prism or a diffraction grating, images of the slit formed by different wavelengths of the light are spread out spatially, giving rise to a characteristic spectrum. Unlike the case in which the light originates from blackbody radiation, these slit images are not spread out into a continuous rainbow of color. Instead, a series of discrete colored images is observed. These slit images, each formed by light of a particular wavelength, are often referred to as **spectral lines.** As evidenced by Figure 40.17 the series of spectral lines is unique for each different type of atom or molecule that make up the gas. The pattern of the spectral lines acts like a fingerprint that can be used to identify elements present in a gas or vaporized sample.

When white light composed of a continuous spread of wavelengths is passed through a cool gas, that gas absorbs exactly the same wavelengths that it emits when heated. Radiation from stars resembles blackbody radiation. However, when this radiation passes through the relatively cool outermost layers of a star, atoms in those layers selectively absorb certain characteristic wavelengths. These **absorption spectra,** as they are known, are routinely used to identify the elements present in the outer layers of distant stars. In fact, the element helium was first discovered not here on earth but on our own sun. Looking at the absorption spectrum of the sun, it was noticed that lines appeared that did not correspond to any element yet found on earth. The new element was named helium after *helios,* the Greek word for sun.

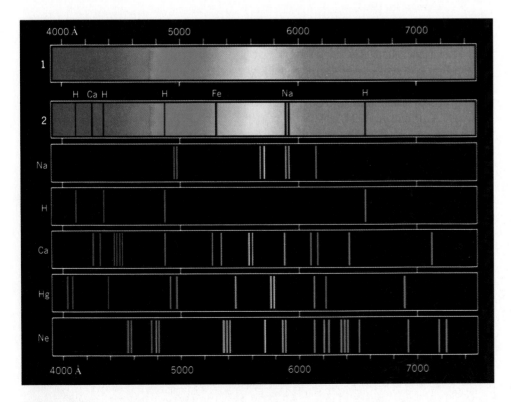

FIGURE 40.17

A comparison of spectra. (a) A continuous spectrum, such as that produced by a blackbody. (b) Fraunhofer lines in the solar spectrum. The spectrum of the sun shows dark lines where certain frequencies have been selectively absorbed by the cooler gases of the outer layers of the sun. (c) The emission spectrum of sodium. (d) Emission spectrum of hydrogen. The regular spacing of the lines is described by the Balmer formula. (e) The emission spectrum of calcium (Ca), (f) mercury (Hq), (g) neon (Ne).

The spectral lines of hydrogen are particularly interesting because of their simplicity and regularity. Fascinated by this regularity a Swiss schoolteacher, Johann Balmer (1825–1898), found in 1885 an empirical formula that predicted the frequencies of the lines in the visible hydrogen spectrum that were then known:

$$\lambda_n = 364.6 \frac{n^2}{n^2 - 4} \text{ nm}, \qquad n = 3, 4, 5, \ldots$$

Balmer's formula was later recast in the following form by Johannes R. Rydberg (1854–1919):

$$\frac{1}{\lambda} = R_H\left(\frac{1}{2^2} - \frac{1}{n^2}\right), \qquad n = 3, 4, 5, \ldots \qquad \textbf{(40.15)}$$

The modern value of the **Rydberg constant** R_H is 10 967 758.5 m^{-1}. Since the time of Balmer other spectra have been observed, the wavelengths of which are described by the more general expression

$$\frac{1}{\lambda} = R_H\left(\frac{1}{m^2} - \frac{1}{n^2}\right), \qquad \begin{array}{l} m = 1, 2, 3, \ldots \\ n = m + 1, m + 2, \ldots \end{array} \qquad \textbf{(40.16)}$$

Today, we speak of these series as

the Lyman series, $m = 1$, $n \geq 2$, with wavelengths in the ultraviolet (reported by T. Lyman in 1914)
the Balmer series, $m = 2$, $n \geq 3$, with wavelengths in the visible
the Paschen series, $m = 3$, $n \geq 4$, with wavelengths in the infrared (reported by F. Paschen in 1908)
the Brackett series, $m = 4$, $n \geq 5$, with wavelengths in the far infrared (reported by F. Brackett in 1922)
the Pfund series, $m = 5$, $n \geq 6$, with wavelengths in the far infrared (reported by H. A. Pfund in 1924).

Equation (40.16) also holds for ions when a single electron orbits a nucleus of charge Ze, however, R_H must be multiplied by Z^2.

EXAMPLE 40.7   The Long and Short of It

Find (a) the longest and (b) the shortest wavelengths present in the Balmer series of the hydrogen spectrum.

SOLUTION   (a) We make use of the Rydberg formula of Equation (40.15)

$$\frac{1}{\lambda} = R_H\left(\frac{1}{2^2} - \frac{1}{n^2}\right), \qquad n = 3, 4, 5, \ldots$$

The longest wavelength occurs for $n = 3$:

$$\frac{1}{\lambda} = R_H\left(\frac{1}{2^2} - \frac{1}{3^2}\right) = R_H\left(\frac{5}{36}\right)$$

So that the wavelength of the H_α line, as it is called, is

$$\lambda = \frac{36}{5(10\,967\,758.5 \text{ m}^{-1})} = 656 \text{ nm}$$

a red line clearly visible to the human eye.

(b) The shortest wavelength, known as the series limit, occurs as $n \to \infty$

$$\frac{1}{\lambda} = R_H\left(\frac{1}{2^2} - \frac{1}{\infty^2}\right) = R_H\left(\frac{1}{4}\right)$$

resulting in

$$\lambda = \frac{4}{(10\ 967\ 758.5\ \text{m}^{-1})} = 365\ \text{nm}$$

This wavelength lies in the ultraviolet, just outside the range of sensitivity of the human eye. ◄

40.4 The Rutherford-Bohr Model of the Atom

At the end of the 1800s physicists knew that matter was composed of atoms. Moreover, it was clear that atoms were composed of equal and opposite charges with the positive charge being nearly 2000 times more massive than the negative charge. However, at that time, a model that satisfactorily described how these charges combined to give stable atoms had not been formulated. A complete model would ultimately have to account for the structure of the periodic table and the line spectra of the elements. Some early models of the atom resembled a raisin muffin.[8] The much more massive positive charge was presumed to also occupy more space. Thus, the mental picture was one of a large muffin of positive charge with small negatively charged raisins within. Attempts to predict line spectra from the oscillations of the electrons within the surrounding positive charge proved unsuccessful.

Rutherford Scattering

In an attempt to probe the positive charge of the atom, Ernest Rutherford (1871–1937) and his collaborators Hans Geiger (1882–1945) and Ernest Marsden (1889–1970) along with a host of students performed a series of experiments whereby energetic alpha particles from radioactive sources were directed toward thin foils of gold.[9] These experiments employed a radioactive source of particles called **alpha particles** that we now know are helium nuclei. After scattering from the gold atoms, the positively charged alpha particles collided with a phosphorescent screen with tiny flashes of light indicating at what angle the alpha particles had been scattered (Fig. 40.18). By comparing the statistical average of scattered alpha-particle trajectories with various models for the atom, Rutherford hoped to reveal the size and structure of the atom's charge distribution.

The basic idea behind Rutherford's analysis is fairly simple. At large distances the positively charged alpha particles and the positive charge of the atom are expected to repel one another according to Coulomb's law.[10] The statistical distribution of trajectories for randomly aimed alpha particles can be calculated. As long as the alpha particle and positive charge of the gold atoms don't touch, the trajectories predicted from Coulomb's law result. If, on the other hand, the alpha particles have sufficient energy to overcome the Coulomb repulsion and actually collide with the positive charge of the gold atom, differ-

---

[8] The most famous of these models, proposed by J. J. Thomson, Cavendish Professor at Cambridge University, is fondly known as the ''plum pudding'' model, testifying to its English heritage.

[9] For a delightful, first-hand account of the difficulties of visual alpha-particle counting by those who worked with Rutherford as students see T. H. Osgood and H. S. Hirst, ''Rutherford and his alpha particles,'' *American Journal of Physics,* **32**(9), 681–686 (1964).

[10] The negative charge has little effect on the alpha-particle's trajectory because the alpha particles are about 8000 times more massive than the negative charges. If you throw a handful of BBs at a moving bowling ball, you don't expect to change the bowling ball's direction much!

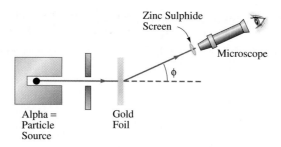

FIGURE 40.18

Schematic diagram of an alpha-particle scattering experiment as performed by Geiger and Marsden. A collimated beam of alpha particles from a radioactive source strikes a gold foil. When the alpha particle strikes a zinc sulfide screen, a small flash of light results. Observers looking through microscopes counted the flashes of light. In practice many observers surrounded the gold foil simultaneously, sitting in a darkened room for hours at a time.

Concept Question 5
Suppose that an alpha particle is acted on by a Coulomb force of magnitude F_o when it is a distance $3r_o$ from a nucleus of radius $2r_o$. (a) If the radius of the nucleus shrinks to $1r_o$, how does the force on the alpha particle change? (*Answer:* It doesn't.) (b) In view of the answer to part (a), why is it that a smaller nuclear radius implies that the alpha particle can experience a larger force from the nucleus?

ent scattering statistics apply. The approximate size of gold atoms can be calculated from knowledge of the atomic mass and density of gold. These calculations predict a diameter for the gold *atom* to be on the order of 0.25 nm.

The results of the scattering experiments by Geiger and Marsden were dramatically different from what anyone expected. Deviations from Coulomb scattering were not noted when alpha particles and gold atoms approached to within 0.25 nm. Nor within 0.025 nm. Nor within even 0.0025 nm. Rutherford was quick to realize that his data implied that the positive charge of the atom was *not* spread throughout the entire volume of the atom but instead concentrated in a minute nucleus. In fact, results of modern experiments indicate that deviations do not occur until the alpha particles and positive charge of the gold nucleus are on the order of 0.000 006 nm apart! Figure 40.19 shows how the concentration of positive charge changes the shape of the electric potential experienced by the alpha particles as they traverse the gold foil.

Bohr Model of the Atom

In 1913, thirteen years after Planck's quantum hypothesis and two years after Rutherford's paper demonstrating that the alpha-particle scattering experiments could be explained based on a small nucleus of positive charge, Niels Bohr (1885–1962) published a description of a model for the atom that united both ideas. Bohr's model achieved dramatic success by explaining the observed line spectra of hydrogen and gave great impetus to the new quantum physics, which was still in its infancy. Although Bohr's model was superseded by more correct and accurate models, the mental picture of Bohr's planetary atom provides a useful aid when thinking about the radiation and absorption of light by atoms.

Bohr envisioned the hydrogen atom as composed of a massive, but small, nucleus of positive charge with an electron orbiting this nucleus, much as the moon orbits the earth. The Coulomb attraction of the negatively charged electron for the positively charged nucleus provided the centripetal force necessary to hold the electron in its orbit. From the classical point of view, this simple planetary model of the atom is completely untenable. Maxwell's equations predict and many experiments have verified that accelerating charges radiate electromagnetic energy. The centripetal acceleration of the electron in its orbit combined with classical electromagnetic theory implies that the orbiting electron should radiate away its kinetic energy and fall into the nucleus in a minute fraction of a second. Bohr's dramatic step was to suggest that this classical law of radiation simply does not apply to atoms and that, instead, a new quantum law of radiation should be used.

Bohr proposed that there exists for the orbiting electron certain **stationary states** such that, when an electron is in one of these states it does not radiate electromagnetic waves. Instead, he theorized, the atom emits or absorbs electromagnetic energy only when electrons make transitions between these stationary states. We outline Bohr's ideas in a simplified model in which we take the massive nucleus with charge Ze to be completely stationary and assume the electron is in a circular orbit with velocity V small enough that we can

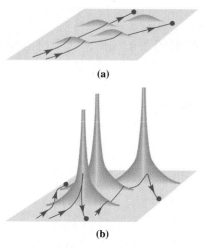

(a)

(b)

FIGURE 40.19

Two-dimensional representation of the scattering of alpha particles (small dark spheres) by atoms in a gold foil. Height of the hills represents the electric potential due to the positive charge of the gold atom. (a) In the "plum-pudding" model of the atom, positive charge is smeared out over large areas, and hence the potential hills are low. Energetic alpha particles can easily traverse them, and their trajectories are changed only slightly. (b) In the Rutherford model, positive charge was concentrated in a minute nucleus. The $1/R$ Coulomb's-law behavior of the electric potential results in enormously high peaks from which alpha particles can be scattered at quite large angles.

ignore relativistic effects. For this case we can apply Newton's second law to the orbiting electron shown in Figure 40.20. The force of attraction is given by Coulomb's law, and the acceleration is the centripetal acceleration $a_c = V^2/r$.

$$F_{net} = ma_c$$

$$\frac{1}{4\pi\epsilon_o}\frac{Ze^2}{r^2} = m\frac{V^2}{r} \qquad (40.17)$$

To this purely classical argument Bohr added a quantization condition that described his hypothesized stationary states. Bohr's assumption was that angular momentum was quantized in bundles of magnitude $\hbar = h/(2\pi)$, so that

$$mVr = n\hbar, \qquad n = 1, 2, 3, \ldots \qquad (40.18)$$

(The symbol $\hbar$ is spoken "h bar.")

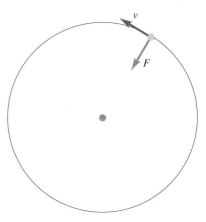

FIGURE 40.20

Planetary model of the hydrogen atom. A massive positive charge is nearly stationary in the center. An electron travels in a circular orbit. The electron is held in its orbit by the Coulomb's-law attraction of the positive nucleus and the negative electron.

Assumptions of the Bohr Model

1. Electrons move in circular orbits about a positive nucleus with the Coulomb force providing the centripetal acceleration.
2. The electrons undergo transitions between stable orbits and emit (or absorb) electromagnetic radiation only during transitions between these orbits.
3. The energy of the emitted (or absorbed) electromagnetic radiation is given by

$$h\nu = |E_i - E_f|$$

where E_i and E_f are the energies of the initial and final stable electron states, respectively.
4. The angular momentum of the electrons is quantized and given by

$$mrV = n\hbar$$

where m is the electron mass, r is its radius of orbit, V is its velocity, and n is a positive integer.

With only the two ideas embodied in Equations (40.17) and (40.18) as input, a remarkably accurate picture of the hydrogen atom emerges. We first look at the energy levels of Bohr's stationary states. The energy of the atom is comprised of the kinetic energy of the electron and the electrostatic potential energy of the nucleus–electron system:

$$E = K + U = \frac{1}{2}mV^2 - \frac{1}{4\pi\epsilon_o}\frac{Ze^2}{r} \qquad (40.19)$$

We expect this total energy to be negative when the electron is bound to the nucleus. Equations (40.17), (40.18), and (40.19) constitute three relations between the variables E, r, V, and the **quantum number** n. Our goal is to untangle these equations and arrive at an expression for each variable in terms of fundamental constants and n only.

Multiplying both sides of Equation (40.17) by $r/2$ reveals that

$$\frac{1}{2}mV^2 = \frac{1}{2}\left(\frac{1}{4\pi\epsilon_o}\frac{Ze^2}{r}\right)$$

Substituting this relation into Equation (40.19) allows us to eliminate V and write the total energy of the atom as

$$E = -\frac{1}{8\pi\epsilon_o}\frac{Ze^2}{r} \tag{40.20}$$

or to eliminate r and write the total energy as

$$E = -\frac{1}{2}mV^2$$

Bohr's quantization condition, Equation (40.18), can be solved for V yielding $V = n\hbar/mr$. This expression for V can then be substituted into the expression for the total energy E to yield a second equation for E as a function of r:

$$E = -\frac{n^2\hbar^2}{mr^2} \tag{40.21}$$

Equations (40.20) and (40.21) now involve only E, r, and the quantum number n. By equating these expressions for E

$$-\frac{1}{8\pi\epsilon_o}\frac{Ze^2}{r} = -\frac{n^2\hbar^2}{mr^2}$$

we can solve for the radius r

$$r_n = \frac{4\pi\epsilon_o\hbar^2}{Ze^2m}n^2 \tag{40.22}$$

where we have subscripted the radius r with the integer n to remind us that only certain discrete radii are allowed by Bohr's quantum condition. The quantity $(4\pi\epsilon_o\hbar^2)/(e^2m)$ is the radius of the smallest orbit of the electron in the hydrogen atom. It is commonly known as the **Bohr radius** a_o

$$a_o = \frac{4\pi\epsilon_o\hbar^2}{e^2m}$$

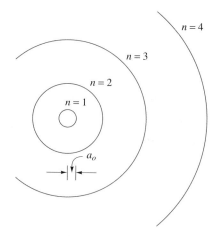

which has a numerical value of 0.0529 nm. In terms of the Bohr radius, Equation (40.22) for the allowed radii becomes

$$r_n = \frac{a_o}{Z}n^2$$

In Figure 40.21 we show the radii of the first few allowed states in the Bohr model.
We now substitute this expression for r into the Equation (40.20) for the total energy to obtain

$$E_n = -\left(\frac{1}{8\pi\epsilon_o}\frac{Z^2e^2}{a_o}\right)\frac{1}{n^2} \tag{40.23}$$

for the allowed energy levels of the Bohr atom. The lowest energy state is called the **ground state** and occurs for $n = 1$. For the hydrogen atom ($Z = 1$) the magnitude E_o of the ground-state energy is given by

$$E_o = \frac{1}{8\pi\epsilon_o}\frac{e^2}{a_o} \tag{40.24}$$

FIGURE 40.21

Scale drawing of the circular excited states of the electron in the Bohr model of the hydrogen atom. The radius of the lowest energy state is known as the Bohr radius a_o of the hydrogen atom.

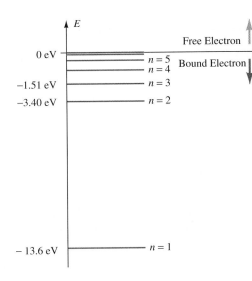

FIGURE 40.22

Energy level diagram for the Bohr model of the hydrogen atom. The vertical axis represents energy. The (arbitrary) zero of energy is taken as the energy of a stationary electron, infinitely far from the positive nucleus. The lowest energy level ($n = 1$) is known as the ground state.

and has a value of 13.6 eV. In these terms the allowed energy levels of hydrogenlike atoms become

$$E_n = -E_o \frac{Z^2}{n^2}, \qquad n = 1, 2, 3, \ldots \qquad \textbf{(40.25)}$$

In Figure 40.22 we show an **energy level diagram** that gives a pictorial representation of allowed energy levels of the hydrogen atom. Note that the levels become more closely spaced as the total energy approaches zero. When the total energy of the atom is zero, the electron is no longer bound to the nucleus, that is to say, the atom has been ionized. Consequently, E_o is often referred to as the **ionization energy** of the atom.

The triumph of Bohr's model is its ability to account for the observed spectral lines of the Balmer series. According to Bohr's model, electromagnetic energy is emitted only when an electron makes transitions between the stationary states described by Equations (40.22) and (40.23). Making use of Einstein's photon concept, Bohr suggested that the energy $h\nu$ of the emitted radiation is just that required by the conservation of energy. Hence, the relationship between the initial-state energy E_i and the final-state energy E_f when a photon is emitted by the atom is just

$$E_i = E_f + h\nu$$

Denoting the quantum numbers of the atom's initial and final states by n_i and n_f, respectively, the resulting emitted photon energy is

$$h\nu = E_i - E_f = E_o \left(\frac{1}{n_f^2} - \frac{1}{n_i^2} \right)$$

Written in terms of the photon wavelength, we find the Bohr model predicts photon energies such that

$$\frac{1}{\lambda} = \frac{E_o}{hc} \left(\frac{1}{n_f^2} - \frac{1}{n_i^2} \right) \qquad \textbf{(40.26)}$$

The form of this expression is exactly that of the Rydberg formula, which we repeat here for convenience

$$\frac{1}{\lambda} = R_H \left(\frac{1}{m^2} - \frac{1}{n^2} \right), \qquad \begin{array}{l} m = 1, 2, 3, \ldots \\ n = m + 1, m + 2, \ldots \end{array} \qquad \textbf{(40.27)}$$

Concept Question 6
Kepler's third law of planetary motion indicates that the orbital period T and average distance R from the sun for planets in our solar system are such that T^2/R^3 is a constant. Does this law suggest that the orbits of planets are quantized?

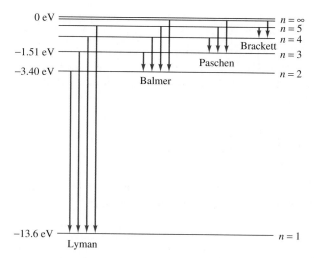

FIGURE 40.23

Light is emitted from the hydrogen atom only when the electron makes transitions between stationary states. The Balmer series of spectral lines, for instance, results when electrons from higher energy levels fall into the $n = 2$ level, releasing their energy as a single photon.

Furthermore, Bohr's model predicted that the Rydberg constant could be calculated from other fundamental constants. Substituting expressions for E_o and a_o, Bohr's model predicts a Rydberg constant of

$$R_\infty \equiv \left(\frac{1}{4\pi\epsilon_o}\right)^2 \frac{me^4}{4\pi c\hbar^3} = 1.097\ 373\ 18 \times 10^7\ \text{m}^{-1} \qquad \text{(Infinitely heavy nucleus)}$$

for our model, which assumed an infinitely heavy nucleus for the hydrogen atom. When corrections are made for the finite mass of the hydrogen nucleus (a proton), the resulting prediction is $R_H = 1.096\ 775\ 85 \times 10^7\ \text{m}^{-1}$, which agrees with the experimentally measured value to six significant figures! (See Problem 42.)

In addition to understanding the origin of the Rydberg constant, we now have a model that allows us to interpret the integers in the various spectral series observed for hydrogen. The Balmer series for example results from electronic transitions into the $n_f = 2$ state from some higher energy state for which $n_i = 3, 4, \ldots$. Figure 40.23 is a diagram depicting electron transitions responsible for several of the known spectral series for hydrogen.

Bohr's model was a marvelous success. However, it does have a number of shortcomings: (1) It does not account for the complex spectra emitted by multielectron atoms, nor that of molecules. (2) It does not allow one to predict the variation in intensity observed for different spectral lines. (3) His theory does not explain splittings that occur in some of the spectral lines when the excited atoms are subjected to a magnetic field. (4) One must also wonder about the reason for the quantization condition for the angular momentum. Nonetheless, the partial success of Bohr's model did demonstrate that victory ultimately would be found in the development of the quantization concept first introduced by Planck.

The formalism that ultimately evolved from Bohr's early model is known as **quantum mechanics.** As the name suggests, quantum mechanics is a set of rules for performing calculations to describe these new quantum waves. The theories of quantum mechanics have proven dramatically successful in accounting for the line spectra of atoms, the structure of the periodic table, and the behavior of electrons in solids. Despite these calculational successes, physicists today, perhaps like you, continue to ponder and debate the significance of the wave properties of matter.

Concept Question 7
Describe what it might be like to ride a bicycle if Planck's constant had a macroscopic value of, say, 10 J · s?

40.5 Summary

A perfect absorber is known as a **blackbody.** The electromagnetic radiation emitted by blackbodies can be characterized by a spectra excitance described by **Planck's radiation formula**

$$I_\lambda(\lambda, T) = \left(\frac{2\pi hc^2}{\lambda^5}\right)\frac{1}{e^{hc/\lambda kT} - 1}$$

where h is a new fundamental constant of nature called **Planck's constant.** This distribution correctly predicts other observed characteristics of blackbody radiation, in particular, **Wien's displacement law:**

$$\lambda_m = \frac{2898 \ \mu\text{m} \cdot \text{K}}{T} \tag{40.1}$$

and the **Stefan-Boltzmann law,**

$$I(T) = \int_0^\infty I_\lambda(\lambda, T) \, d\lambda = \sigma T^4 \tag{40.2}$$

where the **Stefan-Boltzmann constant** $\sigma = 5.670 \times 10^{-8} \ \text{W} \cdot \text{m}^{-2} \cdot \text{K}^{-4}$.

When light falls on a metal surface, electrons are ejected in a process known as the **photoelectric effect.** Albert Einstein correctly predicted that the results of this experiment can be explained by postulating that the energy in electromagnetic waves is concentrated in bundles called **photons,** each with energy

$$E = h\nu$$

With this assumption the results of the photoelectric effect are described by **Einstein's photoelectric equation**

$$K_m = h\nu - W \tag{40.7}$$

where W, the **work function** of the particular metal, is the minimum energy required to free an electron from the metal's surface.

Einstein's **photons have a momentum** given by

$$p = \frac{h}{\lambda}$$

When these photons collide with free particles, the process is known as **Compton scattering.** Photons scattered at angle ϕ have their wavelengths increased by $\Delta\lambda$ as given by the **Compton scattering formula**

$$\Delta\lambda = \lambda_C \left[1 - \cos(\phi)\right]$$

where $\lambda_C = h/(m_o c)$ is the **Compton wavelength.**

Material particles have associated with them a **de Broglie wavelength** λ, related to the particle's momentum by the relation

$$\lambda = \frac{h}{p} \tag{40.13}$$

Electron scattering experiments by Davisson and Germer and by G. P. Thomson, as well as many later experiments, have confirmed the wave properties of electrons.

The spectra of gaseous elements at low pressure exhibit spectral lines at discrete wavelengths rather than a continuum. The spectrum of hydrogen is particularly simple and described by the **Rydberg formula**

$$\frac{1}{\lambda} = R_H\left(\frac{1}{m^2} - \frac{1}{n^2}\right), \qquad n = m + 1, m + 2, \ldots \qquad (40.16)$$

For $n = 2$, the series is partly in the visible spectrum and is known as the **Balmer spectrum.**

Lord Rutherford's analysis of the scattering of alpha particles from gold foils indicates that the positive charge of atoms is concentrated in a **nucleus** much smaller than the size of the atom itself. Niels Bohr proposed a planetary model of the atom with electrons orbiting this small nucleus. Bohr proposed that electrons in **stationary states** do not radiate their energy. He characterized these stationary states by requiring that the electrons of the atom have angular momentum only in bundles of $\hbar = h/(2\pi)$

$$mVr = n\hbar, \qquad n = 1, 2, 3, \ldots \qquad (40.18)$$

This assumption leads to a set of discrete energy levels in hydrogenlike atoms as given by

$$E_n = -\left(\frac{1}{8\pi\epsilon_o}\frac{Z^2 e^2}{a_o}\right)\frac{1}{n^2} \qquad (40.23)$$

where $a_o = 0.0529$ nm is the first Bohr radius.

PROBLEMS

40.1 The Particle Model for Light Revisited

1. One method for estimating the temperature of stars is to model them as blackbodies and compare their spectral excitance to the Planck function, choosing the temperature that gives the best fit. However, measuring the entire spectrum is time-consuming. In order to shorten the process, the excitance is often measured at only two wavelengths, and the ratio is then taken. This ratio can then be used to indicate the temperature of the surface of the star. One common ratio, called the "$B - V$ color index," is calculated as $B - V = -2.5 \log_{10}[I_\lambda(\lambda_1, T)/I_\lambda(\lambda_2, T)]$, where $\lambda_1 = 440$ nm and $\lambda_2 = 550$ nm. Calculate the color index for a blackbody with temperature $T = 4000$ K.

2. (a) Show that for small wavelengths the Planck function for spectral excitance reduces to the Rayleigh-Jeans law. (b) Show that for long wavelengths the Planck function reduces to Wien's expression for spectral excitance (not to be confused with Wien's law).

3. At what wavelength is the peak in the spectral excitance of a blackbody with temperature of 30 000 K? (This is the approximate surface temperature of the bluish white star Naos.)

4. Stars known as red giants (Fig. 40.P1) have a surface temperature around 3200 K. At what wavelength is the peak spectral excitance of a blackbody with this same temperature?

5. The burners of an electric range can get "red hot." Estimate the temperature of such a burner by modeling it as a blackbody with a peak in spectral excitance that occurs at 2400 nm.

6. The universe appears to be filled with a background electromag-

netic radiation that precisely fits the Planck function for radiant excitance. According to one model, this radiation is left over from the "big bang" that occurred at the creation of the universe. The blackbody temperature corresponding to this radiation is 2.74 K. (a) At what wavelength is the peak of the spectral excitance for such a blackbody? (b) In what part of the electromagnetic spectrum is radiation of this wavelength?

7. A *greybody* is defined as an object with a spectral excitance lower than that of a blackbody by a constant factor $\epsilon \leq 1$. This factor ϵ is known as the emissivity of the greybody. Calculate the total radiant power emitted by a spherical greybody of radius 2.54 cm and emissivity $\epsilon = 0.100$ when it is at a temperature of 330°C.

8. The average power radiated by the sun is 3.75×10^{26} W. Using data from Appendix 8, estimate the surface temperature of the sun based on a blackbody model and the Stefan-Boltzmann law.

9. (a) What must be the temperature of a spherical blackbody of radius 1.00 m for it to radiate energy at a rate of 2.50 kW? (b) At what wavelength does the peak of spectral excitance $I_\lambda(\lambda, T)$ occur?

10. Consider two blackbodies exchanging heat by radiation as shown in Figure 40.P2. Blackbody 1 is at temperature T_1 and blackbody 2 is at temperature T_2. If $T_1 = T_2 + \Delta T$, show that for small ΔT the rate of heat exchange between the two bodies is *linear* in ΔT.

11. Calculate the energy (in eV) of a photon for the following radiation types: (a) a radio signal with frequency 97.7 MHz, (b) the light from a helium-neon laser ($\lambda = 632.8$ nm), (c) the copper K_α X-ray of wavelength 0.154 nm.

12. For photons with the following energies, calculate their wave-

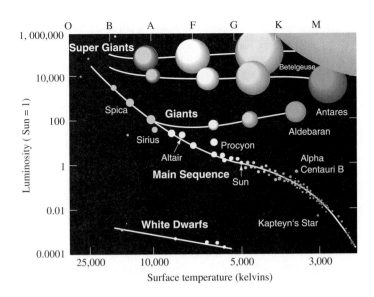

O   B   A   F   G   K   M

Luminosity (Sun = 1)

1,000,000

10,000

100

1

0.01

0.0001

Super Giants

Betelgeuse

Spica   **Giants**

Sirius

Altair   Procyon

Main Sequence   Sun

Antares

Aldebaran

Alpha
Centauri B

White Dwarfs   Kapteyn's Star

25,000   10,000   5,000   3,000

Surface temperature (kelvins)

FIGURE 40.P1   A Hertzsprung-Russell diagram shows the variation of brightness (vertical scale) and temperature (horizontal scale) for stars. Red giants are in the upper right-hand corner. See Problem 4.

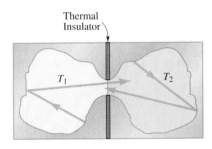

Thermal
Insulator

T_1 T_2

FIGURE 40.P2   Two objects exchange energy by radiation only. See Problem 10.

lengths and identify the region of the electromagnetic spectrum in which they fall: (a) 0.100 eV, (b) 1.00 eV, (c) 1.00 keV, (d) 1.00 MeV.

13. How many photons per second arrive at a target irradiated by the beam from a 7.50-mW helium-neon laser ($\lambda = 632.8$ nm)?

14. Calculate the number of photons per second in a 1.00-mW beam from (a) an argon laser (488 nm) and (b) a carbon dioxide laser (10 600 nm).

15. If a page of this book is illuminated by 135 W/m² of light with wavelength 555 nm, how many photons strike each square centimeter each second?

16. One thousand billion (10^{12}) photons of light with wavelength 546 nm arrive at a target each second. What is the incident power on the target?

17. (a) What is the longest wavelength light that ejects electrons from metallic sodium? (b) If light of wavelength 410. nm illuminates sodium, what is the maximum kinetic energy of the photoejected electrons?

18. The alkali metals are often used as coatings on photocells because of the low magnitudes of their work functions. To see the difficulty in using more common metals, calculate the longest wavelength that can be used to eject electrons from a copper surface using the photoelectric effect. In what part of the electromagnetic spectrum is this wave?

19. Light from the 488-nm line of an argon laser is used in the photoelectric effect. When this light strikes the emitter, the stopping

potential of photoejected electrons is 0.380 V. What is the work function of the material from which the emitter is made? (Assume emitter and collector are made of the same material.)

20. (a) Light of wavelength 546 nm from a filtered mercury lamp falls on a cesium metal surface in a photoelectric experiment. What is the stopping potential for this arrangement? (b) If this same light falls on a potassium-coated surface, what is the stopping potential?

21. From which of the materials with photoelectric work functions listed in Table 40.1 can you build a photocell that operates with visible light (400–750 nm)?

22. When light of wavelength 390. nm falls on a certain metal surface, the photoelectrons require a stopping potential of 0.931 V. When a wavelength of 488 nm is used on the same surface, the stopping potential is 0.282 V. (a) What is the work function of the metal? (b) What value of Planck's constant is implied by these results? What is the percentage difference between this experimental value and the accepted value?

23. (a) If X-rays with wavelength 0.100 nm undergo Compton scattering from electrons, what is the wavelength of photons scattered through an angle of 135° from the direction of the incident X-rays? (b) How much kinetic energy is transferred to the electron that such a photon hits?

24. A muon is an elementary particle like the electron but with a larger mass (105.66 MeV/c^2). (a) Calculate the Compton wavelength of the muon. (b) What is the wavelength shift of a photon that scatters from a muon at an angle of 45.0° from its original direction. (c) For photons, in what region of the electromagnetic spectrum is this shift a 10% change?

25. Show that when $\Delta\lambda \ll \lambda_o$, the maximum energy transferred to an electron by a Compton collision with a photon of wavelength λ_o and energy E_o is $E = 2E_o(\lambda_C/\lambda_o)$.

26. A photon of original wavelength 0.0750 nm strikes a stationary electron in a Compton collision. If the photon scatters at an angle of 30.0° from its original direction, find the magnitude and direction of the electron's velocity after the collision.

27. An X-ray photon of wavelength 0.115 nm collides with a stationary electron. After the collision the electron has a kinetic energy of 187 eV. Calculate (a) the wavelength and (b) the direction of the scattered photon.

40.2 The Wave Model for Particles

28. Calculate the de Broglie wavelength of (a) an electron with a kinetic energy of 25.5 eV, and (b) a neutron with a kinetic energy of 1/40 eV.

29. An electron has a de Broglie wavelength equal to 0.223 nm. What is the electron's kinetic energy? What is its speed?

30. (a) A beam of fast neutrons from nuclear reactions can be used in diffraction experiments with atomic nuclei as targets. Calculate the de Broglie wavelength of a neutron with kinetic energy 4.76 MeV. (b) If the neutron's kinetic energy is reduced by half, by what factor is its wavelength changed?

31. Show that the de Broglie wavelength (in nanometers) for nonrelativistic electrons with kinetic energy K (in electron volts) can be calculated from $\lambda = 1.226/\sqrt{K}$.

32. Suppose the Davisson-Germer experiment is repeated using a beam of neutral helium atoms traveling at a speed of 1640 m/s. Moreover, suppose these atoms impinge on a lithium fluoride crystal with atomic spacing $D = 2.014$ Å. At what angle is the diffracted beam expected?

33. At what angle does one expect the diffraction peak in Davisson and Germer's experiment if the electrons incident on the nickel sample have a kinetic energy of 60.0 eV?

40.3 Spectra of Atoms

34. Calculate the longest and shortest wavelengths that result from the Lyman series for atomic hydrogen.

35. Calculate the wavelengths of the three longest wavelengths in the Balmer spectrum ($n_f = 2$) of singly ionized helium.

36. Suppose that a stationary unbound electron was captured by a proton to form a hydrogen atom. If this electron falls all the way to the $n_f = 1$ level and the energy left the electron–nucleus system as a single photon, what is the wavelength of that photon? In what region of the electromagnetic spectrum does such a photon lie?

37. Consider a doubly ionized lithium atom. Find at least one transition that results in visible radiation. What wavelength is emitted? What are the n_i and n_f values for this line?

40.4 The Rutherford-Bohr Model of the Atom

38. Calculate the (a) energy, (b) wavelength, and (c) frequency of a photon emitted when the electron of a hydrogen atom makes a transition from the $n_i = 5$ level to the $n_f = 4$ level.

39. Calculate (a) the ionization potential and (b) the first Bohr radius for a doubly ionized lithium atom.

40. (a) Create a scale drawing that shows the energy levels of the five lowest energy states of a singly ionized helium atom. (b) Draw circles (or parts thereof) to scale designating orbits for these five states.

41. A muonic atom consists of a proton nucleus and a negatively charged mu meson orbiting about their common center of mass. The mu meson (sometimes called a muon) is a particle of charge $-e$ and mass 207 times that of an electron. Find the ground-state energy and the first Bohr radius of a muonic atom. [*Hint:* When the approximation that the nucleus has infinite mass is not valid, the equations for the Bohr atom must be modified by replacing the electron mass with the **reduced mass** $\mu = (mM)/(m + M)$, where m and M are the masses of the two bound particles that form the atom.]

42. Deuterium is an isotope of hydrogen the nucleus of which contains one proton and one neutron. The spectrum of deuterium is identical to that of hydrogen except for the correction for the finite mass of the nucleus in each case. See the hint in Problem 41 for the finite-mass correction. (a) Calculate the Rydberg constant for both hydrogen and deuterium. (b) Calculate the wavelength difference between the H_β line of the Balmer series ($n_i = 4$, $n_f = 2$) for hydrogen and deuterium. This difference has been measured spectroscopically. (You should carry at least six significant figures for all fundamental constants.)

Numerical Methods and Computer Applications

43. Starting with the Planck formula for spectral excitance, show that Wien's law can be written

$$\lambda_m T = \frac{hc}{4.965 k_B}$$

[*Hint:* After differentiating to find the maximum, you have to find the root of the transcendental equation $x = 5(1 - e^{-x})$. This equation can easily be solved by iteration, as long as your initial guess is not too far off. Guess an x-value and compute the right-hand side of this equation. Use whatever answer you obtain as the next value for x and once again calculate the right-hand side. Repeat this operation until your values for x remain stable to five decimal places.] Quote your answer to five decimal places.

44. For computational purposes it is convenient to write Planck's formula for the radiant excitance of a blackbody in the form

$$I_\lambda(\lambda, T) = \left(\frac{c_1}{\lambda^5}\right) \frac{1}{e^{c_2/\lambda T} - 1}$$

where c_1 and c_2 are known as the first and second radiation constants. Write a spreadsheet template or computer program to create a graph of I_λ for wavelengths between 100 nm and 10 000 nm for temperatures of 10 000 K and 8000 K. Plot both functions on the same graph. (*Hint:* The range of values is quite extreme. You have to calculate the radiation constants in a system of units such that your computer does not underflow or overflow.) If your spreadsheet allows, repeat the graphs using a common log scale for both axes.

45. Calculate the $B - V$ color index (see Problem 1) for blackbodies with temperatures between 3000 K and 30 000 K in increments of 1000 K. Use your table to estimate the surface temperature of the star Betelgeuse, which has a $B - V$ color index of 1.85.

46. In order to derive the Stefan-Boltzmann law it is necessary to integrate the Planck function for spectral excitance over the wavelength range from zero to infinity. (a) Show that this integral can be cast in the form of several constants multiplying the integral

$$\int_0^\infty \frac{x^3 \, dx}{e^x - 1}$$

This integral cannot be evaluated by techniques of elementary calculus. (b) Use Simpson's rule to numerically evaluate this integral. [*Hint:* You need to choose a lower limit greater than zero and a finite upper limit. Choose each so that the value of the integral is accurate to at least three significant figures. Also, see the hint in Problem 44 regarding units.] (c) Using advanced techniques in calculus, the integral can be shown to be equal to $\pi^5/15$. What is the resulting expression for the Stefan-Boltzmann constant in terms of other fundamental constants?

★**47.** Write a computer program or spreadsheet template to calculate the fraction of total energy radiated by a blackbody that falls between two wavelengths. The blackbody temperature and wavelength range should be selected by the user. Test your program by determining the fraction of the total radiant energy emitted by a blackbody with temperature 5780 K that falls in the visible region. (The temperature of the sun's photosphere is 5780 K.) (See hints for Problems 44 and 46.)

48. According to the Debye model the lattice heat capacity of solids is given by

$$c_V = 9R\left(\frac{T}{T_D}\right)^3 \int_0^{T_D/T} \frac{x^4 e^x}{(e^x - 1)^2}\, dx$$

where T_D is a substance-dependent parameter known as the Debye temperature. Write a spreadsheet template or computer program that evaluates the indicated integral numerically for any T and T_D as input. Test your program by calculating the molar heat capacity of copper ($T_D = 300$ K) at room temperature ($T = 300$ K).

★49. Write a computer program or spreadsheet template to calculate the molar heat capacity given by the Debye model as a function of temperature T. (See Problem 48.) Plot a universal graph of C_V/R versus T/T_D for values of T/T_D from 0.01 to 1.5.

50. Using the least-squares regression function of your spreadsheet or calculator, compute Planck's constant from the following data for the photoelectric effect:

Frequency (10^{14} Hz)	54.5	69.0	74.0	82.0	96.5	118.5
Stopping emf (V)	-2.07	-1.50	-1.26	-0.90	-0.40	$+0.53$

These data are estimated from the graph appearing in R. A. Millikan's article in *Physical Review*, **7**, 355 (1916). Millikan's value was about 5% lower than the currently accepted value. The stopping emf graphed here does not contain the contact potential between the emitter and the collector, hence the positive value of the stopping emf. See footnote 2 of this chapter.

General Problems

51. (a) Calculate the wavelength of maximum spectral excitance for a blackbody with a temperature of 30.0°C. (b) Because of water content, the emissivity of human skin is equal to 1.0 regardless of skin color in the range of infrared wavelengths near the peak you calculated in part (a). (See Problem 7.) Estimate the surface area of the human body as 2.00 m^2 and calculate the rate at which our bodies radiate heat to the surroundings. (Our surroundings also radiate heat to us so that the net heat exchange is much less. See Problem 10.)

52. The red star Antares A and the bluish white star Antares B have surface temperatures of 3000 K and 15 000 K, respectively. These companion stars are bound gravitationally and revolve around their common center of mass. Because of the great distance to each, they are both essentially the same distance from the earth. (a) Calculate the ratio of the total radiant excitance of Antares A to Antares B. (b) Although Antares B emits more power per square meter of surface than Antares A, Antares A is actually brighter by a factor of 40 as seen from earth. Calculate the ratio of the diameter of Antares A to that of Antares B. (The "brightness" of a star is the product of its total radiant excitance and the area of the flat circular disk with a diameter equal to that of the star.)

53. In the photoelectric experiment light of a certain wavelength falls on cesium, resulting in a maximum kinetic energy $K_m = 0.401$ eV for photoejected electrons. When the same light illuminates another coated surface, $K_m = 0.241$ V. What is the wavelength of the light and what material is the other surface?

54. The production of X-rays can be likened to the photoelectric effect in reverse. That is, a target is irradiated with high-energy electrons and, as each electron is brought to rest in the target, it may give up its kinetic energy in the form of a single photon. (Radiation created by the deceleration of charges is known as **bremsstrahlung,** the German word meaning literally "braking radiation.") The process is complicated and actually gives rise to a continuum of wavelengths, often with strong peaks, providing information about the atomic energy levels of the atoms constituting the target as shown in Figure 40.P3. Nonetheless, the *maximum* energy of X-rays produced in this manner is equal to the kinetic energy of a single incoming electron. Use this fact to calculate the **short-wavelength limit** for X-rays produced by bombardment of a copper target with electrons having kinetic energy of 35.0 keV.

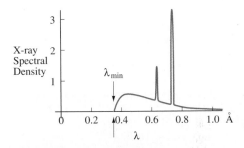

FIGURE 40.P3   When electrons of kinetic energy K strike a metal target, X-rays are produced. The spectrum of the X-rays has a minimum wavelength for any given K, independent of target material. Sharp peaks in the spectrum are due to the internal energy levels of the atoms of the particular target. See Problem 54.

55. Thermal neutrons are neutrons having kinetic energies on the order of kT, where the temperature T is around room temperature. Such neutrons have de Broglie wavelengths on the order of 0.1 nm and are used in diffraction experiments to reveal the structure of crystals. Calculate the de Broglie wavelength of a neutron with kinetic energy $K = k_B T$ for $T = 300$ K.

56. An electron is to have a de Broglie wavelength equal to the wavelength of an X-ray photon of energy 50.0 keV. Through what potential difference should the electron be accelerated assuming it starts from rest?

57. Show that the condition of a whole number of electron de Broglie wavelengths fitting into the circumference of a circle of radius r produces the same relationship as does the Bohr quantization condition for angular momentum.

58. Young's double-slit experiment has been performed using neutral helium atoms. (a) The de Broglie wavelength of these atoms was 0.103 nm. What was the velocity of the atoms? (b) The slits used were 1.0 μm wide and 8.00 μm apart. In the resulting fringe pattern the maxima were located 8.00 μm apart. How far was the detector plane from the plane of the double slits? (The experiment was performed at the University of Konstanz.[11] For a description of other interference experiments using neutral atoms see *Physics Today*, **44**(7), July 1991, pp. 17–20.)

59. A 15.0-mW beam from a helium-neon laser falls on a completely absorbing screen. (a) How many photons strike the screen each second? (b) What is the momentum of each photon? (c) What force is exerted on the screen by the beam?

---

[11]O. Carnal and J. Mlynek, *Physical Review Letters*, **66**, 181 (1991).

60. A photon of wavelength 34.5 nm strikes a hydrogen atom in the ground state. An inelastic collision results whereby the electron in the hydrogen atom is excited to the $n = 2$ state. Calculate the wavelength of the scattered photon. (Ignore the kinetic energy of the recoiling hydrogen atom.)

61. Electrons in semiconductors are allowed to take on energies within certain bands that are separated in energy by **band gaps.** These band gaps represent values of energy where there are no allowed states in which electrons can reside. The energy band gap between the conduction band and the valence band in gallium arsenide is 1.43 eV. When an electron falls from the conduction band to the lower energy valence band, the electron may give up this energy as a single photon. This is the mechanism for the emission of light in light-emitting diodes (LEDs). What is the wavelength of light given off by a GaAs LED?

62. When observed under a "black light" many posters, articles of clothing, teeth, and minerals seem to glow (Fig. 40.P4). The phenomenon is called **fluorescence.** The black light emits photons with wavelengths in the ultraviolet region of the spectrum to which our eyes are not sensitive. These photons excite to a higher energy level electrons in the molecules contained in certain dyes.

FIGURE 40.P4 Some minerals seem to glow under a "black light," that is, illumination by ultraviolet light to which the human eye is not sensitive: powellite (yellow), willenite (green), schealite (blue), and calcite (red).

Visible light is emitted when the electrons fall to lower energy levels but not all the way back to the ground state. From these lower levels the electrons eventually return to the ground state, losing energy by nonradiative means. Figure 40.P5 is an energy

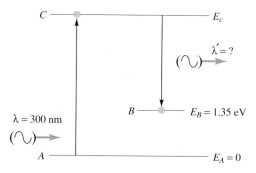

FIGURE 40.P5   A three-level model for fluorescence. See Problem 62.

level diagram for a typical process. (a) Calculate the energy level of the uppermost state (C) on the diagram. (b) Find the wavelength of the photon emitted when the electron makes a transition from state C to state B.

63. Relate the diffraction in the Davisson-Germer experiment to the Bragg scattering. Figure 40.P6 shows the geometry of the Davisson-Germer experiment and that of the underlying Bragg planes. Show that the Bragg condition $2d \sin(\theta) = m\lambda$ leads to the Equation (40.14) $D \sin(\phi) = m\lambda$.

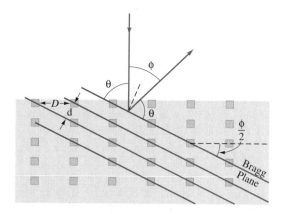

FIGURE 40.P6   Problem 63

APPENDIXES

APPENDIX 1   Mathematics Summary

$\pi = 3.141\ 592\ 654\ldots$

$e = 2.718\ 281\ 829\ldots$

A1.1 Geometry

Area

Triangle of base b and height h:	$\dfrac{1}{2}bh$
Circle of radius r:	πr^2
Right circular cylinder of radius r and height h:	$2\pi rh$
Sphere of radius r:	$4\pi r^2$

Volume

Cylinder of radius r and length l:	$\pi r^2 l$
Sphere of radius r:	$\dfrac{4}{3}\pi r^3$

A1.2 Trigonometry

Definitions

$$\sin(\theta) = \frac{y}{r} \qquad \cos(\theta) = \frac{x}{r}$$

$$\tan(\theta) = \frac{\sin(\theta)}{\cos(\theta)} = \frac{y}{x}$$

$$\csc(\theta) = \frac{1}{\sin(\theta)} \qquad \sec(\theta) = \frac{1}{\cos(\theta)} \qquad \operatorname{ctn}(\theta) = \frac{1}{\tan(\theta)}$$

$$\theta\ (\text{in radians}) = \frac{s}{r}$$

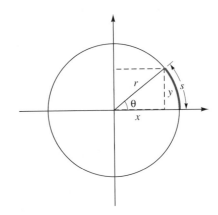

Symmetry and Periodicity

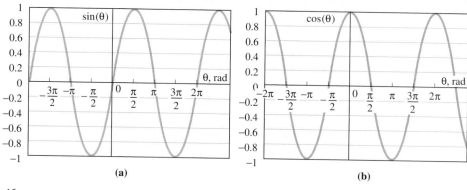

(a)

(b)

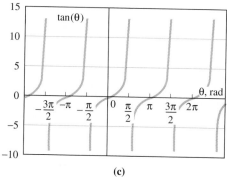

(c)

$$\sin(-\theta) = -\sin(\theta)$$
$$\cos(-\theta) = \cos(\theta)$$
$$\tan(-\theta) = -\tan(\theta)$$

$$\sin(90° - \theta) = \cos(\theta)$$
$$\cos(90° - \theta) = \sin(\theta)$$
$$\tan(90° - \theta) = \text{ctn}(\theta)$$

Identities

$$\sin^2(\theta) + \cos^2(\theta) = 1$$
$$\cos^2(\theta) - \sin^2(\theta) = \cos(2\theta)$$
$$1 + \sec^2(\theta) = \tan^2(\theta)$$
$$1 + \text{ctn}^2(\theta) = \csc^2(\theta)$$
$$\sin^2(\theta) = \frac{1 - \cos(2\theta)}{2}$$
$$\cos^2(\theta) = \frac{1 + \cos(2\theta)}{2}$$
$$\sin(\alpha \pm \beta) = \sin(\alpha)\cos(\beta) \pm \cos(\alpha)\sin(\beta)$$
$$\cos(\alpha \pm \beta) = \cos(\alpha)\cos(\beta) \mp \sin(\alpha)\sin(\beta)$$
$$\sin(\alpha) \pm \sin(\beta) = 2 \sin\left(\frac{\alpha \pm \beta}{2}\right)\cos\left(\frac{\alpha \mp \beta}{2}\right)$$
$$\cos(\alpha) + \cos(\beta) = 2 \cos\left(\frac{\alpha + \beta}{2}\right)\cos\left(\frac{\alpha - \beta}{2}\right)$$
$$\cos(\alpha) - \cos(\beta) = -2 \sin\left(\frac{\alpha + \beta}{2}\right)\sin\left(\frac{\alpha - \beta}{2}\right)$$
$$\tan(\alpha + \beta) = \frac{\tan(\alpha) + \tan(\beta)}{1 - \tan(\alpha)\tan(\beta)}$$
$$\tan(\alpha - \beta) = \frac{\tan(\alpha) - \tan(\beta)}{1 + \tan(\alpha)\tan(\beta)}$$
$$\tan\left(\frac{\alpha - \beta}{2}\right) = \frac{\sin(\alpha) - \sin(\beta)}{\cos(\alpha) + \cos(\beta)}$$

A1.3 Series Expansions

Exponential:
$$e^x = 1 + x + \frac{x^2}{2!} + \frac{x^3}{3!} + \cdots$$

Natural logarithm:
$$\ln(1 + x) = x - \frac{x^2}{2} + \frac{x^3}{3} - + \cdots$$

Sine (x in radians):
$$\sin(x) = x - \frac{x^3}{3!} + \frac{x^5}{5!} - + \cdots$$

Cosine (x in radians):
$$\cos(x) = 1 - \frac{x^2}{2!} + \frac{x^4}{4!} - + \cdots$$

Binomial ($|x| < 1$):
$$(1 + x)^n = 1 + nx + \frac{n(n - 1)}{2!}x^2 + \frac{n(n - 1)(n - 2)}{3!}x^3 + \cdots$$

A1.4 Calculus

Derivatives

The derivative with respect to x of a function $y(x)$ is defined as

$$\frac{dy}{dx} = \lim_{\Delta x \to 0} \frac{\Delta y}{\Delta x} = \lim_{\Delta x \to 0} \frac{y(x + \Delta x/2) - y(x - \Delta x/2)}{\Delta x}$$

and algebraically is the rate of change Δy in the function $y(x)$ with respect to a change Δx in x. Graphically, dy/dx represents the slope of a tangent line to the curve $y(x)$ at coordinate x. Some useful derivatives are listed below:

$$\frac{d}{dx}(ax) = a \qquad (a = \text{constant})$$

$$\frac{d}{dx}x^n = nx^{n-1}$$

$$\frac{d}{dx}e^x = e^x$$

$$\frac{d}{dx}\ln(x) = \frac{1}{x}$$

$$\frac{d}{dx}\log_{10}(x) = \frac{\log_{10}(e)}{x}$$

$$\frac{d}{dx}[u(x)v(x)] = u\frac{dv}{dx} + v\frac{du}{dx}$$

$$\frac{d}{dx}\sin(ax) = a\cos(ax)$$

$$\frac{d}{dx}\cos(ax) = -a\sin(ax)$$

$$\frac{d}{dx}\tan(ax) = a\sec^2(ax)$$

$$\frac{d}{dx}\csc(ax) = -a\,\text{ctn}(ax)\csc(ax)$$

$$\frac{d}{dx}\sec(ax) = a\tan(ax)\sec(ax)$$

$$\frac{d}{dx}\text{ctn}(ax) = -a\csc^2(ax)$$

Partial Derivatives

Partial derivatives may look imposing, but they are quite easy to use (assuming you can perform ordinary derivatives with ease). If $U(x, y)$ is a function of the two variables x and y, the partial derivative of U with respect to x is written $\partial U/\partial x$, and while carrying out the differentiation we treat all terms containing the variable x just as we would during an ordinary differentiation. However, we treat the other variable y just as we would a *constant*. Similarly, the partial derivative of U with respect to y is written $\partial U/\partial y$. While performing this derivative we treat x as if it were a constant and perform ordinary derivative operations with respect to y.

Examples:

For
$$U(x, y) = Ax^m y^n \qquad \frac{\partial U}{\partial x} = mAx^{m-1}y^n \qquad \frac{\partial U}{\partial y} = nAx^m y^{n-1}$$

For $U(x, y, z) = xy^2 + Ay\cos(\kappa z)$, where A and κ are constants

$$\frac{\partial U}{\partial x} = y^2 \qquad \frac{\partial U}{\partial y} = 2xy + A\cos(\kappa z) \qquad \frac{\partial U}{\partial z} = -\kappa yA\sin(\kappa z)$$

Integrals

The *definite integral* of a function $f(x)$ between the integration limits a and b is equal to

$$\int_a^b f(x)\, dx = \lim_{N \to \infty} \sum_{i=1}^{N} f(z_i)\, \Delta x, \quad \Delta x = x_i - x_{i-1} = \frac{b-a}{N}, \qquad x_{i-1} \le z_i \le x_i$$

Some useful integrals are listed below.

$$\int dx = x + C$$

$$\int x^n\, dx = \frac{x^{n+1}}{n+1} + C \qquad (n \ne -1)$$

$$\int \frac{dx}{x} = \ln|x| + C$$

$$\int \frac{dx}{a+bx} = \frac{1}{b}\ln(a+bx) + C$$

$$\int \frac{dx}{a^2+x^2} = \frac{1}{a}\arctan\left(\frac{x}{a}\right) + C$$

$$\int \frac{dx}{a^2-x^2} = \frac{1}{2a}\ln\left(\frac{a+x}{a-x}\right) + C \qquad |a| > |x|$$

$$\int \frac{dx}{x^2-a^2} = \frac{1}{2a}\ln\left(\frac{x-a}{x+a}\right) + C \qquad |x| > |a|$$

$$\int \frac{x\,dx}{a^2 \pm x^2} = \pm\frac{1}{2}\ln(a^2 \pm x^2) + C$$

$$\int \frac{dx}{\sqrt{x^2 \pm a^2}} = \ln(x + \sqrt{x^2 \pm a^2}) + C$$

$$\int \frac{dx}{\sqrt{a^2-x^2}} = \arcsin\left(\frac{x}{a}\right) + C \qquad |a| > |x|$$

$$\int \frac{x\,dx}{\sqrt{a^2-x^2}} = -\sqrt{a^2+x^2} + C$$

$$\int \frac{dx}{(x^2+a^2)^{3/2}} = \frac{x}{a^2\sqrt{x^2+a^2}} + C$$

$$\int \frac{x\,dx}{(x^2 + a^2)^{3/2}} = \frac{1}{a^2\sqrt{x^2 + a^2}} + C$$

$$\int \sin(ax)\,dx = -\frac{1}{a}\cos(ax) + C$$

$$\int \cos(ax)\,dx = \frac{1}{a}\sin(ax) + C$$

$$\int \tan(ax)\,dx = \frac{1}{a}\ln|\sec(ax)| + C$$

$$\int \sin^2(ax)\,dx = \frac{1}{2}x - \frac{1}{4a}\sin(2ax) + C$$

$$\int \cos^2(ax)\,dx = \frac{1}{2}x + \frac{1}{4a}\sin(2ax) + C$$

$$\int \tan^2(ax)\,dx = \frac{1}{a}\tan(ax) - x + C$$

$$\int e^{ax}\,dx = \frac{1}{a}e^{ax} + C$$

$$\int xe^{-ax}\,dx = -\frac{1}{a^2}(ax + 1)e^{-ax} + C$$

$$\int \ln(ax)\,dx = x\ln(ax) - x + C$$

A1.5 Matrix Multiplication

A matrix is a rectangular array of numbers like the one shown below

$$\{M\} = \begin{bmatrix} A & B & C \\ D & E & F \\ G & H & I \end{bmatrix}$$

We designate matrices by using { } brackets as above. Vectors may also be represented in a similar form:

$$\mathbf{r} = x\hat{\mathbf{i}} + y\hat{\mathbf{j}} + z\hat{\mathbf{k}} = \begin{bmatrix} x \\ y \\ z \end{bmatrix}$$

If the number of columns of one matrix is equal to the number of rows of a second matrix, the two can be multiplied. The product of two matrices is another matrix. The product matrix has as many rows as the first matrix and as many columns as the second matrix. For example, the product of a (3×3) matrix $\{M\}$ and a (3×1) vector $\mathbf{r}$ is a (3×1) vector the components of which can be calculated using the simple rule demonstrated below:

$$\{M\}\,\mathbf{r} = \begin{bmatrix} A & B & C \\ D & E & F \\ G & H & I \end{bmatrix}\begin{bmatrix} x \\ y \\ z \end{bmatrix} = \begin{bmatrix} Ax + By + Cz \\ Dx + Ey + Fz \\ Gx + Hy + Iz \end{bmatrix}$$

As you can see from the multiplication above, when we multiply our 3×3 matrix times a vector, the result is another vector. The pattern of the multiplication is easy to remember. To see this pattern place the index finger of your left hand on A and the index finger of your right hand on x. The first contribution to the x-component of the answer vector is Ax. Now move your left finger to B and your right finger to y. These letters are members of the second product By. Now move to C and z. Sum these three products and you have the x-component of the answer. Repeat the same kind of finger movement for the three members of the sum that represent the y-component of the answer. The pattern should now be clear to you. Try it on the z-component.

Introduction to Numerical Methods Useful in Physics

Numerical techniques have been introduced in a number of optional sections throughout this text. The choice of which techniques to introduce has been a compromise. Techniques that are intuitive and obvious extensions of fundamental definitions are to be preferred. Unfortunately, the most straightforward methods do not always give reliable results even in simple cases. On the other hand, the origins of advanced and robust numerical techniques are often obscure. Our strategy has been to include the simplest techniques available that give reasonable results with data of the quality taken in introductory laboratories.

A2.1 Solving Equations

Sometimes complex physical models result in equations that cannot be readily solved using simple algebra. This is especially true in the case of transcendental equations involving trigonometric or exponential functions. In such cases it is handy to have a method for obtaining a numerical value of the unknown. Three possibilities follow:

Binary search Suppose you are given a function $f(x)$ and you wish to find the value of x so that $f(x) = 0$. If you can find (by trial and error, for example) two values of x such that $f(x_h) > 0$ and $f(x_l) < 0$, then (if $f(x)$ has only a single root between x_h and x_l) it is possible to find the root of $f(x) = 0$ in the following systematic way. First calculate $x_{\text{test}} = (x_h + x_l)/2$ and then calculate $f(x_{\text{test}})$. If $f(x_{\text{test}}) > 0$, then replace x_h by x_{test}. If $f(x_{\text{test}}) < 0$, then replace x_l by x_{test}. Repeat the process over and over until values of x_{test} repeat themselves to the desired number of significant figures.

Iteration If you can write the equation you wish to solve in the form $x = g(x)$, it is often possible to determine the solution by iteration. Guess a value for x and calculate $g(x)$. Use this value of $g(x)$ as the new x and recalculate $g(x)$. Repeat the process over and over. The iteration is convergent if the values of x converge on a self-consistent value. Beware however; the values of x sometimes run off to infinity or simply oscillate about two or more values that are not close.

Brute force Although frowned on in polite circles, there's really nothing wrong with brute-force calculation for one-time problems. Such calculations are especially convenient if you have a spreadsheet handy. Write your equation in the form $h(x) = 0$. In column A of your spreadsheet write 100 equally spaced values that seem reasonable for x. (Use the FILL command of your spreadsheet.) Then in column B, write the function $g(x)$ in the 100 cells next to the x-values in column A. (Use the COPY command, of course!) Now just scan column B and look for the place(s) where $g(x)$ changes sign. Refill column A with new x-values starting with a value just less than that for which $g(x)$ changed signs and using a smaller incremental Δx-value in the FILL command. Once again scan the $g(x)$ column for the sign change and repeat the process until you find a value of x for which $g(x) = 0$ to sufficient accuracy. Inelegant as it is, this method is fairly quick and just the ticket when you're not feeling clever.

A2.2 Differentiation

Consider a function $y(t)$ represented by a set of data points (y_i, t_i) labeled with index $i = 1 \ldots N$. When a parabola is fit through three consecutive evenly spaced data points, the derivative of the function $y(t)$ at the middle point of the three can be approximated by

$$\left(\frac{dy}{dt}\right)_i = \frac{y_{i+1} - y_{i-1}}{t_{i+1} - t_{i-1}} \tag{3.23}$$

The second derivative of a function approximated by five equally spaced data points can be approximated at the middle of the five points by

$$\left(\frac{d^2y}{dt^2}\right)_i = \frac{1}{12\,(\Delta t)^2}(-y_{i+2} + 16y_{i+1} - 30y_i + 16y_{i-1} - y_{i-2}) \tag{3.25}$$

Implementation of these formulas can be found on the spreadsheet template AFIST.WK1 and computer program AFIST.BAS on the diskette accompanying this text. Also see Example 3.14 for further documentation.

A2.3 Integration

If a function $v(t)$ is approximated by N data points (v_i, t_i), then the integral $\int v\,dt$ can be approximated by the **trapezoidal rule:**

$$\int_{t_1}^{t_N} v\,dt = \frac{1}{2}(v_1 + v_2)\,\Delta t_1 + \frac{1}{2}(v_2 + v_3)\,\Delta t_2 + \cdots \frac{1}{2}(v_{N-1} + v_N)\,\Delta t_{N-1}$$

where $\Delta t_i = t_{i+1} - t_i$ and the t_i need not be evenly spaced. The integral calculated by the trapezoidal rule is the actual integral of a function obtained by connecting the data points with straight lines. See Examples 3.12 and 3.13 for additional discussion of spreadsheet and computer programs implementing this technique.

If an odd number N of evenly spaced data points (v_i, t_i) are available, then the integral $\int_{t_1}^{t_N} v\,dt$ can be approximated by **Simpson's rule:**

$$\int_{t_1}^{t_N} v\,dt = \frac{\Delta t}{3} \sum_{i=1}^{\frac{N-1}{2}} (v_{2i-1} + 4v_{2i} + v_{2i+1}) \tag{3.22}$$

Simpson's rule results in an integral equivalent to joining three consecutive data points by a best-fit parabola. See Section 3.5 for a more detailed description.

A2.4 Differential Equations

An equation that relates a function, say $x(t)$, and one or more of its derivatives is called a differential equation. Although a relation between $x(t)$ and its derivatives may be known, the form of the function itself may not be known. "Solving" a differential equation means finding the function $x(t)$ itself. Analytical techniques that produce an actual equation for $x(t)$ are available to solve many classes of differential equations. Numerical techniques are also possible. Numerical techniques do not result in an equation for $x(t)$ but rather a series of values for (x, t). In some complicated cases numerical solutions may be the only ones available.

Differential equations often occur in mechanics when Newton's second law of motion, $F_{net} = ma$, is applied. The acceleration a is the second derivative of position with respect to time d^2x/dt^2. The net force F_{net} may be a function of position x, time t, or velocity $v =$

dx/dt. Thus, in one dimension Newton's second law can be thought of as a differential equation for the unknown function $x(t)$:

$$\frac{d^2x}{dt^2} = \frac{F(t, x, v)}{m}$$

A particularly useful numerical technique for solving differential equations is known as the fourth-order **Runge-Kutta method.** Starting with velocity v_{old} at an initial point (x_{old}, t_{old}), the coordinates of the next point (x_{new}, t_{new}) and a new velocity v_{new} are predicted from knowledge of the force acting on the particle. To implement the Runge-Kutta method you must calculate a series of four intermediate numbers called k_1, k_2, k_3, and k_4. From these numbers the new position and velocity can be calculated. The relevant equations for motion in one dimension are

$$k_1 = F(t_{old}, x_{old}, v_{old}) \frac{\Delta t}{m}$$ (6.23)

$$k_2 = F\left(t_{old} + \frac{\Delta t}{2}, x_{old} + \frac{v_{old} \Delta t}{2} + \frac{k_1 \Delta t}{8}, v_{old} + \frac{k_1}{2}\right) \frac{\Delta t}{m}$$

$$k_3 = F\left(t_{old} + \frac{\Delta t}{2}, x_{old} + \frac{v_{old} \Delta t}{2} + \frac{k_1 \Delta t}{8}, v_{old} + \frac{k_2}{2}\right) \frac{\Delta t}{m}$$

$$k_4 = F\left(t_{old} + \Delta t, x_{old} + v_{old} \Delta t + \frac{k_3 \Delta t}{2}, v_{old} + k_3\right) \frac{\Delta t}{m}$$

$$x_{new} = x_{old} + \left(v_{old} + \frac{(k_1 + k_2 + k_3)}{6}\right) \Delta t$$

$$v_{new} = v_{old} + \frac{(k_1 + 2 k_2 + 2 k_3 + k_4)}{6}$$

$$t_{new} = t_{old} + \Delta t$$

A brief discussion of the rationale behind the Runge-Kutta method is given in Section 6.6. A spreadsheet template entitled RK1D.WK1, which implements these equations for one-dimensional motion, is included with the diskette accompanying this textbook and is discussed in Example 6.10. The source code for a computer program implementing the two-dimensional version of this method is entitled RK2D.BAS and is discussed in Example 6.11.

A2.5 Simultaneous Overrelaxation

One of the most important equations in physics is a partial differential equation known as Laplace's equation:

$$\nabla^2 V = 0$$

$$\frac{\partial^2 V}{\partial x^2} + \frac{\partial^2 V}{\partial y^2} + \frac{\partial^2 V}{\partial z^2} = 0$$

where the function $V(x, y, z)$, which satisfies both Laplace's equation and certain boundary conditions, is to be found. In this text we encounter Laplace's equation in the study of electrostatics in which the function $V(x, y, z)$ represents the electric potential in a certain charge-free region of space.

When the problem is two-dimensional, V is not a function of one of the variables, say z, for example. A common type of two-dimensional problem is one in which the boundary conditions are given as fixed values of $V(x, y)$ around the periphery of a region of space. In such cases the region can be divided into a rectangular grid of points labeled with indices (m, n). The value of $V(m, n)$ at any grid point is then related to adjacent points by

$$V(m, n) = \frac{V(m - 1, n) + V(m + 1, n) + V(m, n - 1) + V(m, n + 1)}{4} \quad \textbf{(24.23)}$$

except for the grid points on the border the values of which are determined by the boundary conditions. In practice one fills the interior cells with the formula of Equation (24.23) and the cells constituting the boundary with fixed numbers that represent the boundary conditions. The CALCULATE command is then given to the spreadsheet or computer program over and over until the numerical values in the interior reach stable values. This procedure is an example of the numerical technique known as **relaxation.** The particular algorithm of Equation (24.23) dates back to the previous century and is known as **Jacobi's method.**

Jacobi's method is rather slow to converge, that is, it takes many calculation cycles before stable answers appear in the interior. For an $N \times N$ array of points the number of iterations required for a given accuracy goes as N^2. A more efficient algorithm is known as **simultaneous overrelaxation,** or SOR. The idea of SOR is to overcorrect at each step using the expression

$$V(m, n) = (1 - w)V(m, n) + \quad \textbf{(24.24)}$$

$$w\frac{V(m - 1, n) + V(m + 1, n) + V(m, n - 1) + V(m, n + 1)}{4}$$

where w is called the **overrelaxation parameter.** If you choose $w = 1$ Equation (24.24) becomes Jacobi's method. For convergence you must choose $1 < w < 2$. The rate of convergence depends on the choice of w. When properly chosen, the rate of convergence goes as N rather than N^2. Selecting w is a subject of study in numerical analysis, although it is often chosen by trial and error. For a 25×25 spreadsheet area, we've found a value of 1.8 works reasonably well. The SOR method is suitable for program or spreadsheet use. A detailed application of SOR can be found in Example 24.15, which also illustrates two techniques for generating contour maps of equal V values using spreadsheets.

APPENDIX 3   List of Symbols

A	area	c	speed of light
A	apex angle of prism	D	drag force
A	amplitude of simple harmonic motion	d	slit separation
a	acceleration	d	distance between Bragg planes
a	slit width	$đQ$	a small quantity of heat
a_c	centripetal acceleration	$đW$	a small quantity of work
a_g	free-fall acceleration	E	energy
a_o	Bohr radius	E	Young's modulus
a_T	tangential acceleration	$\mathbf{E}$	electric field strength
B	bulk modulus (isothermal)	$e(t)$	time-varying emf
B_{ad}	bulk modulus (adiabatic)	$\mathscr{E}$	emf
$\mathbf{B}$	magnetic induction field	$\mathbf{F}$	force
C	capacitance	$\mathbf{F}_{fric}$	force of friction
C	heat capacity	F	finesse
C	Curie constant	F_g	gravitational force
C_D	drag coefficient	f	focal length of lens
c	molar heat capacity	f	frequency of oscillation

G	universal gravitational constant
G	voltage gain
g	local gravitational field strength
g	gyromagnetic ratio
$\mathbf{H}$	magnetic intensity
h	Planck's constant
$\hbar$	$h/(2\pi)$
I	rotational inertia (moment of inertia)
I	electric current
I	intensity of sound
I	irradiance of light
i	image distance
$i(t)$	time-varying current
$\hat{\mathbf{i}}$	Cartesian unit vector in $+x$-direction
$\mathbf{J}$	impulse
$\mathbf{J}$	current density
$\hat{\mathbf{j}}$	Cartesian unit vector in $+y$-direction
$\mathscr{J}$	rotational impulse
K	kinetic energy
k	spring constant
k	wave number
k_B	Boltzmann's constant
$\hat{\mathbf{k}}$	Cartesian unit vector in $+z$-direction
L	length
L	latent heat
L	inductance
L_l	sound intensity level
$\mathbf{L}$	angular momentum of a solid
l	length
l	orbital quantum number
$\mathbf{l}$	angular momentum of a point-particle
$\mathbf{M}$	magnetization
M_s	saturation magnetization
M	mutual inductance
m	mass
m	lateral magnification
m	an integer indicating diffraction order
m_e	mass of the electron
m_o	rest mass
$\mathscr{M}$	mass ratio
$\mathbf{N}$	normal force
N	number of turns of wire
N_A	Avogadro's number
NA	numerical aperture
$\hat{\mathbf{n}}$	a unit vector
n	index of refraction
n	number of turns per unit length
n	number of moles
n	principal quantum number
o	object distance
P	pressure
$\mathbf{p}$	linear momentum
$\mathbf{p}$	electric dipole moment
$\mathscr{P}$	power
Q	electric charge
Q	Q-value of resonance
q	electric charge
$\mathscr{Q}$	quantity of heat
R	universal gas constant
R	power reflection coefficient
R	resistance
R	radius
Re	Reynolds number
r	radius, radial distance
$\mathbf{S}$	Poynting vector
S	shear modulus
S	entropy
s	distance
s	arc length
T	tension force
T	period of oscillation or rotation
T	temperature
T	transmission coefficient
T_C	Curie temperature
T_D	Debye temperature
t	time
U	potential energy
U	internal energy
u	energy density
u	velocity
V	velocity
V	electric potential
v	velocity
v_D	drift velocity
v_T	tangential velocity
$v(t)$	time-varying potential difference
$\mathscr{V}$	volume
W	work function of a metal
$\mathscr{W}$	work
X	reactance
x	a Cartesian coordinate
x	a generic unknown quantity
y	a Cartesian coordinate
Y	Young's modulus
Z	impedance
z	a Cartesian coordinate
α	an angle
α	coefficient of thermal expansion
β	an angle, phase angle
β	coefficient of volume expansion
β	the ratio v/c in relativity
γ	an angle
γ	parameter in relativity
Δ	fiber-optic parameter
Δ	when in front of a symbol, signifies the change in that quantity calculated as final minus initial
δ	deviation angle of ray through prism
ϵ	coefficient of restitution
ϵ	permittivity
ϵ_o	permittivity of free space
η	viscosity (dynamic)
θ	an angle
κ	relative dielectric constant
κ	thermal conductivity

κ	torsion constant		π	3.1415 . . .	
λ	wavelength		ρ	volume density of mass or charge	
λ	linear density of charge or mass		ρ	resistivity	
λ_C	Compton wavelength		σ	area density of mass or charge	
μ	reduced mass		τ	torque	
μ	linear mass density (waves only)		Φ_E	electric flux	
$\boldsymbol{\mu}$	magnetic dipole moment		Φ_B	magnetic flux	
μ_B	Bohr magneton		ϕ	an angle	
μ_M	nuclear magneton		χ	magnetic susceptibility	
μ	permeability		ψ	generic wave function	
μ_o	permeability of free space		ω	angular frequency	
ν	frequency of light		Ω_p	angular frequency of precession	

APPENDIX 4   Greek Letters

Alpha	A	α	Eta	H	η	Nu	N	ν	Tau	T	τ
Beta	B	β	Theta	Θ	θ	Xi	Ξ	ξ	Upsilon	Υ	υ
Gamma	Γ	γ	Iota	I	ι	Omicron	O	o	Phi	Φ	ϕ
Delta	Δ	δ	Kappa	K	κ	Pi	Π	π	Chi	X	χ
Epsilon	E	ϵ	Lambda	Λ	λ	Rho	P	ρ	Psi	Ψ	ψ
Zeta	Z	ζ	Mu	M	μ	Sigma	Σ	σ	Omega	Ω	ω

APPENDIX 5   Conversion Factors

Length

1 in = 2.54 cm
1 m = 39.37 in = 3.281 ft
1 yd = 0.9144 m
1 km = 0.621 mi
1 mi = 1.609 km
1 mi = 5280 ft
1 μm = 10^{-6} m
1 Å = 10^{-10} m = 10^{-4} μm

Mass

1 slug = 14.59 kg
1 u = $1.660\,540 \times 10^{-27}$ kg
1000 kg = 1 t (metric ton)

Time

1 h = 3600 s
1 day = 86 400 s
1 yr = $365\frac{1}{4}$ days = 3.16×10^7 s

Velocity

1 m/s = 3.281 ft/s = 2.237 mi/h
1 mi/h = 0.4470 m/s = 1.467 ft/s
60 mi/h = 88 ft/s

Acceleration

1 m/s^2 = 3.281 ft/s^2
1 ft/s^2 = 0.3048 m/s^2

Force

$1 \text{ N} = 10^5 \text{ dyn} = 0.2248 \text{ lbf}$
$1 \text{ lbf} = 4.448 \text{ N}$
$1 \text{ dyn} = 10^{-5} \text{ N} = 2.248 \times 10^{-6} \text{ lbf}$
$1 \text{ ton} = 2000 \text{ lbf}$
$1 \text{ oz} = 1/16 \text{ lbf}$

Pressure

$1 \text{ Pa} = 1 \text{ N/m}^2 = 1.45 \times 10^{-4} \text{ lbf/in}^2$
$1 \text{ bar} = 10^5 \text{ N/m}^2 = 14.5 \text{ lbf/in}^2$
$1 \text{ atm} = 14.7 \text{ lbf/in}^2 = 1.013 \times 10^5 \text{ N/m}^2$
1 atm is equivalent to 760 mm of Hg

Energy

$1 \text{ J} = 0.738 \text{ ft} \cdot \text{lb} = 10^7 \text{ ergs}$
$1 \text{ kW} \cdot \text{h} = 3.60 \times 10^6 \text{ J}$
$1 \text{ cal} = 4.184 \text{ J}$
$1 \text{ Btu} = 252 \text{ cal} = 1.054 \times 10^3 \text{ J}$
$1 \text{ eV} = 1.60 \times 10^{-19} \text{ J}$

Power

$1 \text{ W} = 1 \text{ J/s} = 0.738 \text{ ft} \cdot \text{lbf/s}$
$1 \text{ hp} = 550 \text{ ft} \cdot \text{lb/s} = 745.7 \text{ W}$
$1 \text{ W} = 3.413 \text{ Btu/h}$

Volume

$1 \text{ m}^3 = 10^6 \text{ cm}^3 = 6.102 \times 10^4 \text{ in}^3$
$1 \text{ L} = 10^{-3} \text{ m}^3 = 10^3 \text{ cm}^3$
$1 \text{ L} = 0.0353 \text{ ft}^3 = 1.0576 \text{ qt}$
$1 \text{ ft}^3 = 2.832 \times 10^{-2} \text{ m}^3 = 7.481 \text{ gal(US)}$
$1 \text{ gal(US)} = 3.786 \text{ L}$
$1 \text{ oil barrel} = 42 \text{ gal(US)} = 0.159 \text{ m}^3$
$1 \text{ gal (British)} = 4.55 \text{ L}$
$1 \text{ fluid ounce} = 1/128 \text{ gal (US)} = 2.96 \times 10^{-2} \text{ L}$

Area

$1 \text{ m}^2 = 10^4 \text{ cm}^2 = 10.76 \text{ ft}^2 = 1550 \text{ in}^2$
$1 \text{ ft}^2 = 9.29 \times 10^{-2} \text{ m}^2$
$1 \text{ acre} = 4.356 \times 10^4 \text{ ft}^2 = 4047 \text{ m}^2$
$1 \text{ barn} = 10^{-28} \text{ m}^2$

Magnetism

$1 \text{ T} = 1 \text{ Wb/m}^2 = 10^4 \text{ G} = 10^9 \gamma$

Periodic Table of the Elements

GROUP IA	IIA	IIIA	IVA	VA	VIA	VIIA	VIIIA			IB	IIB	IIIB	IVB	VB	VIB	VIIB	VIII
1 1.0079 **H** Hydrogen																	2 4.00260 **He** Helium
3 6.941 **Li** Lithium	4 9.01218 **Be** Beryllium											5 10.81 **B** Boron	6 12.011 **C** Carbon	7 14.0067 **N** Nitrogen	8 15.9994 **O** Oxygen	9 18.998403 **F** Fluorine	10 20.179 **Ne** Neon
11 22.98977 **Na** Sodium	12 24.305 **Mg** Magnesium											13 26.98154 **Al** Aluminum	14 28.0855 **Si** Silicon	15 30.97376 **P** Phosphorus	16 32.06 **S** Sulfur	17 35.453 **Cl** Chlorine	18 39.948 **Ar** Argon
19 39.0983 **K** Potassium	20 40.08 **Ca** Calcium	21 44.9559 **Sc** Scandium	22 47.90 **Ti** Titanium	23 50.9415 **V** Vanadium	24 51.996 **Cr** Chromium	25 54.9380 **Mn** Manganese	26 55.847 **Fe** Iron	27 58.9332 **Co** Cobalt	28 58.70 **Ni** Nickel	29 63.546 **Cu** Copper	30 65.38 **Zn** Zinc	31 69.72 **Ga** Gallium	32 72.59 **Ge** Germanium	33 74.9216 **As** Arsenic	34 78.96 **Se** Selenium	35 79.904 **Br** Bromine	36 83.80 **Kr** Krypton
37 85.4678 **Rb** Rubidium	38 87.62 **Sr** Strontium	39 88.9059 **Y** Yttrium	40 91.22 **Zr** Zirconium	41 92.9064 **Nb** Niobium	42 95.94 **Mo** Molybdenum	43 (98) **Tc** Technetium	44 101.07 **Ru** Ruthenium	45 102.9055 **Rh** Rhodium	46 106.4 **Pd** Palladium	47 107.868 **Ag** Silver	48 112.41 **Cd** Cadmium	49 114.82 **In** Indium	50 118.69 **Sn** Tin	51 121.75 **Sb** Antimony	52 127.60 **Te** Tellurium	53 126.9045 **I** Iodine	54 131.30 **Xe** Xenon
55 132.9054 **Cs** Cesium	56 137.33 **Ba** Barium	57 138.9055 **La** Lanthanum	72 178.49 **Hf** Hafnium	73 180.9479 **Ta** Tantalum	74 183.85 **W** Tungsten	75 186.207 **Re** Rhenium	76 190.2 **Os** Osmium	77 192.22 **Ir** Iridium	78 195.09 **Pt** Platinum	79 196.9665 **Au** Gold	80 200.59 **Hg** Mercury	81 204.37 **Tl** Thallium	82 207.2 **Pb** Lead	83 208.9804 **Bi** Bismuth	84 (209) **Po** Polonium	85 (210) **At** Astatine	86 (222) **Rn** Radon
87 (223) **Fr** Francium	88 226.0254 **Ra** Radium	89 227.0278 **Ac** Actinium	104 (261) **Unq** (Unnilquadium)	105 (262) **Unp** (Unnilpentium)	106 (263) **Unh** (Unnilhexium)												

★
58 140.12 **Ce** Cerium	59 140.9077 **Pr** Praseodymium	60 144.24 **Nd** Neodymium	61 (145) **Pm** Promethium	62 150.4 **Sm** Samarium	63 151.96 **Eu** Europium	64 157.25 **Gd** Gadolinium	65 158.9254 **Tb** Terbium	66 162.50 **Dy** Dysprosium	67 164.9304 **Ho** Holmium	68 167.26 **Er** Erbium	69 168.9342 **Tm** Thulium	70 173.04 **Yb** Ytterbium	71 174.967 **Lu** Lutetium

★★
90 232.0381 **Th** Thorium	91 231.0359 **Pa** Protactinium	92 238.029 **U** Uranium	93 237.0482 **Np** Neptunium	94 (244) **Pu** Plutonium	95 (243) **Am** Americium	96 (247) **Cm** Curium	97 (247) **Bk** Berkelium	98 (251) **Cf** Californium	99 (252) **Es** Einsteinium	100 (257) **Fm** Fermium	101 (258) **Md** Mendelevium	102 (259) **No** Nobelium	103 (260) **Lr** Lawrencium

NOTES:
(1) Physical state at room temperature and atmospheric pressure
    Black—solid.
    Red—gas.
    Blue—liquid.
    Outline—synthetically prepared.
(2) Based upon carbon-12. ( ) indicates most stable or best known isotope.
(3) Entries marked with asterisks refer to the gaseous state at 273 K and 1 atm and are given in units of g/l.

Z	ELEMENT	SYMBOL	A	ATOMIC MASS	ABUNDANCE OR DECAY SCHEME	HALF-LIFE
0	(Neutron)	n	1	1.008 665	β^-	10.6 min
1	Hydrogen	H	1	1.007 825	99.985%	
	Deuterium	D	2	2.014 102	0.015%	
	Tritium	T	3	3.016 049	β^-	12.33 yr
2	Helium	He	3	3.016 029	0.00014%	
			4	4.002 603	~100%	
			5	5.012 22		
			6	6.018 886	β^-	0.808 s
3	Lithium	Li	6	6.015 121	7.5%	
			7	7.016 003	92.5%	
4	Beryllium	Be	7	7.016 930	$\beta^-_{capture}$, γ	53.3 days
			8	8.005 305	2α	6.7×10^{-17} s
			9	9.012 182	100%	
5	Boron	B	10	10.012 937	19.8%	
			11	11.009 305	80.2%	
6	Carbon	C	11	11.011 433	β^+, $\beta^-_{capture}$	20.4 min
			12	12.000 000	98.89%	
			13	13.003 355	1.10%	
			14	14.003 242	β^-	5730 yr
7	Nitrogen	N	13	13.005 739	β^+	9.96 min
			14	14.003 074	99.63%	
			15	15.000 109	0.37%	
8	Oxygen	O	15	15.003 065	β^+, $\beta^-_{capture}$	122 s
			16	15.994 915	99.759%	
			18	17.999 159	0.204%	
9	Fluorine	F	19	18.998 403	100%	
10	Neon	Ne	20	19.992 435	90.51%	
			21	20.993 843	0.21%	
			22	21.991 384	9.22%	
11	Sodium	Na	22	21.994 435	β^+, $\beta^-_{capture}$, γ	2.6 yr
			23	22.989 767	100%	
			24	23.990 961	β^-, γ	15.0 h
12	Magnesium	Mg	24	23.985 042	78.99%	
13	Aluminum	Al	27	26.981 541	100%	
14	Silicon	Si	28	27.976 928	92.23%	
			29	28.976 495	4.67%	
			31	30.975 364	β^-, γ	2.60 h
15	Phosphorus	P	31	30.973 763	100%	
			32	31.973 908	β^-	14.3 days
16	Sulfur	S	32	31.972 072	95.0%	
			35	34.969 033	β^-	87.4 days
17	Chlorine	Cl	35	34.968 853	75.77%	
			37	36.965 903	24.23%	
18	Argon	Ar	40	39.962 383	99.60%	
19	Potassium	K	39	38.963 798	93.26%	
			40	39.964 000	β^-, β^- capture, γ, β^+	1.28×10^9 yr
20	Calcium	Ca	40	39.962 591	96.94%	
21	Scandium	Sc	45	44.955 914	100%	
22	Titanium	Ti	48	47.947 947	73.7%	
23	Vanadium	V	51	50.943 963	99.75%	
24	Chromium	Cr	52	51.940 510	83.79%	
25	Manganese	Mn	55	54.938 046	100%	
26	Iron	Fe	54	53.939 612	5.8%	
			56	55.934 939	91.8%	
			57	56.935 396	2.2%	

Z	ELEMENT	SYMBOL	A	ATOMIC MASS	ABUNDANCE OR DECAY SCHEME	HALF-LIFE
27	Cobalt	Co	59	58.933 198	100%	
			60	59.933 820	β^-, γ	5.27 yr
28	Nickel	Ni	58	57.935 347	69.3%	
			60	59.930 789	26.1%	
			64	63.927 968	0.91%	
29	Copper	Cu	63	62.929 599	69.2%	
			64	63.929 766	β^-, β^+	12.7 h
			65	64.927 792	30.8%	
30	Zinc	Zn	64	63.929 145	48.6%	
			66	65.926 035	27.9%	
31	Gallium	Ga	69	68.925 581	60.1%	
32	Germanium	Ge	72	71.922 080	27.4%	
			74	73.921 179	36.5%	
33	Arsenic	As	75	74.921 596	100%	
34	Selenium	Se	80	79.916 521	49.8%	
35	Bromine	Br	79	78.918 336	50.7%	
36	Krypton	Kr	84	83.911 506	57.0%	
			89	88.917 563	β^-	3.2 min
37	Rubidium	Rb	85	84.911 800	72.17%	
38	Strontium	Sr	86	85.909 273	9.8%	
			88	87.905 625	82.6%	
			90	89.907 746	β^-	28.8 yr
39	Yttrium	Y	89	88.905 856	100%	
40	Zirconium	Zr	90	89.904 708	51.5%	
41	Niobium	Nb	93	92.906 378	100%	
42	Molybdenium	Mo	98	97.905 405	24.1%	
43	Technetium	Tc	98	97.907 210	β^-, γ	4.2×10^6 yr
44	Ruthenium	Ru	102	101.904 348	31.6%	
45	Rhodium	Rh	103	102.905 50	100%	
46	Palladium	Pd	106	105.903 48	27.3%	
47	Silver	Ag	107	106.905 095	51.8%	
			109	108.904 754	48.2%	
48	Cadmium	Cd	114	113.903 361	28.7%	
49	Indium	In	115	114.903 88	95.7%, β^-	5.1×10^{14} yr
50	Tin	Sn	120	119.902 199	32.4%	
51	Antimony	Sb	121	120.903 824	57.3%	
52	Tellurium	Te	130	129.906 23	34.5%, β^-	2×10^{21} yr
53	Iodine	I	127	126.904 477	100%	
			131	130.906 118	β^-, γ	8.04 days
54	Xenon	Xe	132	131.904 15	26.9%	
			136	135.907 22	8.9%	
55	Cesium	Cs	133	132.905 43	100%	
56	Barium	Ba	137	136.905 82	11.2%	
			138	137.905 24	71.7%	
			144	143.922 673	β^-	11.9 s
57	Lanthanum	La	139	138.906 36	99.9%	
58	Cerium	Ce	140	139.905 44	88.5%	
59	Praseodymium	Pr	141	140.907 66	100%	
60	Neodymium	Nd	142	141.907 73	27.2%	
61	Promethium	Pm	145	144.912 75	$\beta^-_{\text{capture}}, \alpha, \gamma$	17.7 hr
62	Samarium	Sm	152	151.919 74	26.6%	
63	Europium	Eu	153	152.921 24	52.1%	

Z	ELEMENT	SYMBOL	A	ATOMIC MASS	ABUNDANCE OR DECAY SCHEME	HALF-LIFE
64	Gadolinium	Gd	158	157.924 11	24.8%	
65	Terbium	Tb	159	158.925 35	100%	
66	Dysprosium	Dy	164	163.929 18	28.1%	
67	Holmium	Ho	165	164.930 33	100%	
68	Erbium	Er	166	165.930 31	33.4%	
69	Thulium	Tm	169	168.934 23	100%	
70	Ytterbium	Yb	174	173.938 87	31.6%	
71	Lutecium	Lu	175	174.940 79	97.4%	
72	Hafnium	Hf	180	179.946 56	35.2%	
73	Tantalum	Ta	181	180.948 01	99.99%	
74	Tungsten	W	184	183.950 95	30.7%	
75	Rhenium	Re	187	186.955 77	62.60%, β^-	4×10^{10} yr
76	Osmium	Os	191	190.960 94	β^-, γ	15.4 days
			192	191.961 49	41.0%	
77	Iridium	Ir	191	190.960 60	37.3%	
			193	192.962 94	62.7%	
78	Platinum	Pt	195	194.964 79	33.8%	
79	Gold	Au	197	196.966 56	100%	
80	Mercury	Hg	202	201.970 63	29.8%	
81	Thallium	Tl	205	204.974 41	70.5%	
			210	209.990 056	β^-	1.3 min
82	Lead	Pb	204	203.973 044	β^-, 1.48%	1.4×10^{17} yr
			206	205.974 46	24.1%	
			207	206.975 89	22.1%	
			208	207.976 64	52.3%	
			210	209.984 16	α, β^-, γ	22.3 yr
			211	210.988 74	β^-, γ	36.1 min
			212	211.991 88	β^-, γ	10.6 h
			214	213.999 80	β^-, γ	26.8 min
83	Bismuth	Bi	209	208.980 39	100%	
			211	210.987 26	α, β^-, γ	2.15 min
84	Polonium	Po	210	209.982 86	α, γ	138.4 days
			214	213.995 19	α, γ	164 μs
85	Astatine	At	218	218.008 70	α, β^-	~2 s
86	Radon	Rn	222	222.017 574	α, γ	3.82 days
87	Francium	Fr	223	223.019 734	α, β^-, γ	21.8 min
88	Radium	Ra	226	226.025 406	α, γ	1.60×10^3 yr
			228	228.031 069	β^-	5.76 yr
89	Actinium	Ac	227	227.027 751	α, β^-, γ	21.8 yr
90	Thorium	Th	228	228.028 73	α, γ	1.91 yr
			232	232.038 054	100%, α, γ	1.41×10^{10} yr
91	Protactinium	Pa	231	231.035 881	α, γ	3.28×10^4 yr
92	Uranium	U	232	232.037 14	α, γ	72 yr
			233	233.039 629	α, γ	1.59×10^5 yr
			235	235.043 925	0.7%; α, γ	7.038×10^8 yr
			236	236.045 563	α, γ	2.34×10^7 yr
			238	238.050 786	99.3%; α, γ	4.47×10^9 yr
			239	239.054 291	β^-, γ	23.5 min
93	Neptunium	Np	239	239.052 932	β^-, γ	2.35 days
94	Plutonium	Pu	239	239.052 158	α, γ	2.41×10^4 yr
95	Americium	Am	243	243.061 374	α, γ	7.37×10^3 yr
96	Curium	Cm	245	245.065 487	α, γ	8.5×10^3 yr
97	Berkelium	Bk	247	247.070 03	α, γ	1.4×10^3 yr

Z	ELEMENT	SYMBOL	A	ATOMIC MASS	ABUNDANCE OR DECAY SCHEME	HALF-LIFE
98	Californium	Cf	249	249.074 849	α, γ	351 yr
99	Einsteinium	Es	254	254.088 02	α, γ, β^-	276 days
100	Fermium	Fm	253	253.085 18	$\beta^-_{\text{capture}}, \alpha, \gamma$	3.0 days
101	Mendelevium	Md	255	255.091 1	$\beta^-_{\text{capture}}, \alpha$	27 min
102	Nobelium	No	255	255.093 3	$\beta^-_{\text{capture}}, \alpha$	3.1 min
103	Lawrencium	Lr	257	257.099 8	α	~35 s
104	Unnilquadium	Rf	261	261.108 7	α	1.1 min
105	Unnilpentium	Ha	262	262.113 760	α	34 s
106	Unnilhexium		263	263.118 4	α	0.9 s
107	Unnilseptium		261	261	α	1–2 ms
109						5 ms

APPENDIX 7 Physical Constants[1]

QUANTITY	SYMBOL	VALUE[2]
Atomic mass unit (unified)	u	$1.660\,540\,2(10) \times 10^{-27}$ kg
Avogadro's number	N_A	$6.022\,136\,7(36) \times 10^{23}$ mol^{-1}
Bohr magneton	$\mu_B = e\hbar/2m_e$	$9.274\,015\,4(31) \times 10^{-24}$ J/T
Bohr radius	$a_o = \hbar^2/m_e e^2 k$	$0.529\,177\,249(24) \times 10^{-10}$ m
Boltzmann's constant	$k = R/N_A$	$1.380\,658(12)$ J/K
Electron mass	m_e	$9.109\,389\,7(54) \times 10^{-31}$ kg
Electron charge	e	$1.602\,177\,33(49) \times 10^{-19}$ C
Gas constant	R	$8.314\,510(70)$ J/(mol·K)
Gravitational constant	G	$6.672\,59(85) \times 10^{-11}$ N·m^2/kg^2
Neutron rest mass	m_n	$1.674\,928\,6(10) \times 10^{-27}$ kg
Nuclear magneton	μ_N	$5.050\,786\,6(17) \times 10^{-27}$ J/T
Permeability of free space	$\epsilon_o = 1/(\mu_o c^2)$	$8.854\,187\,817\ldots \times 10^{-12}$ F/m (exact)
Permittivity of free space	μ_o	$4\pi \times 10^{-7}$ N/A^2 (exact)
Planck's constant	h	$6.626\,075\,5(40) \times 10^{-34}$ J·s
Proton mass	m_p	$1.672\,623\,1(10) \times 10^{-27}$ kg
Rydberg constant	R_∞	$1.097\,373\,153\,4(13) \times 10^7$ m^{-1}
Speed of light in a vacuum	c	$2.997\,924\,58 \times 10^8$ m/s (exact)

WATER (at 20°C and 1 atm of pressure unless noted otherwise)

Density		1.00×10^3 kg/m^3
Specific heat capacity		4186 J/kg·K
Melting point		273 K
Boiling point		373 K
Heat of fusion		3.34×10^5 J/kg
Heat of vaporization		2.257×10^6 J/kg

[1] From E. R. Cohen and B. N. Taylor, *Reviews of Modern Physics* **59**, 1121, 1987.
[2] Values given in parentheses are the uncertainties (one standard deviation) for the last two digits given.

Free-Fall Acceleration at Various Locations[3]

	a_g (m/s^2)
National Institute of Standards and Technology Gaithersburg, Maryland	$9.801\ 023\ 94 \pm 0.000\ 000\ 55$
National Physical Laboratory Teddington, England	$9.811\ 819\ 30 \pm 0.000\ 000\ 42$
Bureau International des Poids et Mesures Sèvres, France	$9.809\ 259\ 60 \pm 0.000\ 000\ 41$
Air Force Cambridge Research Laboratories Bedford, Massachusetts	$9.803\ 786\ 71 \pm 0.000\ 000\ 42$
Geophysics Institute, University of Alaska Fairbanks, Alaska	$9.822\ 349\ 53 \pm 0.000\ 000\ 42$
Universidad Nationale de Colombia Bogota, Colombia	$9.773\ 900\ 15 \pm 0.000\ 000\ 87$
University of Denver Denver, Colorado	$9.795\ 977\ 08 \pm 0.000\ 000\ 42$
Wesleyan University Middletown, Connecticut	$9.803\ 053\ 06 \pm 0.000\ 000\ 41$

[3] From J. A. Hammond and J. E. Faller, "Laser-interferometer system for the determination of the acceleration of gravity," *IEEE Journal of Quantum Electronics,* **3,** 597 (1971).

APPENDIX 8   Astronomical Table

Earth

Mass	5.98×10^{24} kg
Mean radius	6.378×10^6 m
Mean density	5.5×10^3 kg/m^3
Surface gravitational field strength	9.80 N/kg
Mean earth–sun distance	1.5×10^{11} m

Sun

Mass	1.99×10^{30} kg
Mean radius	6.96×10^8 m
Mean density	1.4×10^3 kg/m^3
Surface gravitational field strength	274 N/kg

Moon

Mass	7.35×10^{22} kg
Mean radius	1.74×10^6 m
Mean density	3.3×10^3 kg/m^3
Surface gravitational field strength	1.62 N/kg
Mean earth–moon distance	3.85×10^8 m

Planets of the Solar System

	EQUATORIAL RADIUS (km)	MASS (10^{24} kg)	SURFACE GRAVITY (Earth = 1)	PERIOD OF REVOLUTION	ORBITAL PERIOD (Years)
Mercury	2440	0.33	0.38	58.65 days	0.241
Venus	6050	4.87	0.91	243.01 days	0.615
Earth	6378	5.98	1.00	23 h 56 min 4.1 s	1.000
Mars	3394	0.64	0.39	24 h 37 min 22.6 s	1.88
Jupiter	71 400	1900	2.74	9 h 50.5 min	11.86
Saturn	60 000	569	1.17	10 h 14 min	29.46
Uranus	25 050	87	0.94	17 h	84.01
Neptune	27 700	103	1.15	17 h 50 min	164.8
Pluto	1100	0.01	0.03	6.39 days	248.6

APPENDIX 9　Properties of Sporting Balls and Related Information

SPORT	MASS (kg)	RADIUS (cm)	COMMENT
Baseball	0.142–0.149	3.64–3.74	
Basketball	0.600–0.650	11.9–12.4	
Bowling	7.26 (max)	10.9 (max)	
Bowling pins	1.64	6.06 (max)	38.1 cm height
Clay pigeons	0.105 ± 0.005	5.4	
Discus	2.00 (min)	21.9–22.1 (dia.)	4.4–4.6 cm thick in flat center
Field hockey	0.156–0.163	3.56–3.74	
Golf	0.0459	2.13	
Hammer	7.26	5.50–6.50	1.175–1.215 m from grip to side of head
Hockey	0.156–0.170	3.81	2.54 cm thick
Lacrosse	0.142–0.149	3.13–3.23	
Ping Pong	0.00240–0.00253	1.86–1.91	
Polo	0.120–0.124	4.44	
Pool	0.156–0.170	2.86	
Racquet ball	0.0397	2.86	
Shotput (men's)	7.26	5.50–6.50	
Shotput (women's)	4.00	4.75–5.50	
Soccer	0.397–0.454	10.9–11.3	
Softball	0.184–0.198	4.80–4.90	
Squash	0.0233–0.0246	1.98–2.08	
Snooker		2.62	
Tennis	0.0567–0.0585	3.18–3.33	
Volleyball	0.260–0.280	10.3–10.7	court 18.3 m × 9.14 m; net height: 2.43 m (men's), 2.29 m (women's)
Water polo	0.400–0.450	10.8–11.3	

Spreadsheets provide a powerful tool for data analysis. You can perform quick and elegant computations, plot graphs, and then make observations or decisions based on what your spreadsheet tells you. This appendix is intended to provide an overview of spreadsheets for those who are not already familiar with them. It is not intended to be a guide for the use of a particular spreadsheet software. Indeed, because the procedures necessary to carry out a sequence of spreadsheet operations is software-dependent, you must consult your operator's manual for the required command sequence to carry out such operations as copying, moving, selecting a graph type, and so forth. Nonetheless, there are features common to all spreadsheets, and it should help you to learn quickly to apply your own spreadsheet software if you spend a few minutes with this appendix.

The basic idea of a spreadsheet is to divide the workpage (actually the computer screen and then some) into a collection of cells, each labeled by its row and column. Each cell may contain a number, letters of a label, or even a mathematical formula. Although a typical spreadsheet can contain as many as 256 columns and 8192 rows of cells, at any one time only 20 rows of cells are displayed. The number of columns visible on the screen depends on the width chosen for an individual column. Columns are labeled by letters A, B, C, . . . , Z, AA, BB, . . . , and rows are labeled with integers 1, 2, 3, . . . In addition to the cells, the screen also displays some information that relates to the contents of a particular cell and some additional information about the entire spreadsheet. A spreadsheet template (or worksheet) with cells that do not yet contain any information is shown in the figure on the facing page.

Spreadsheets have a distinct advantage over an ordinary sheet of paper: they can perform sums, subtractions, multiplications, averages, take sines, cosines, and tangents, and even make decisions. In fact, you can define almost any mathematical relation you like in or between cells. Furthermore, you can graph the contents of a group of cells.

You can move from one cell to the next by using the arrow keys $\rightarrow$ $\uparrow$ $\leftarrow$ $\downarrow$ (and the (PgDn), (PgUp) keys) to move a **pointer** around the spreadsheet. The label (A1, G13, H39, etc.) and the contents of the cell highlighted by the pointer are displayed on the screen in the **descriptor line.** Below are a few general rules and procedures for working with a spreadsheet. It will be helpful if you are familiar with these concepts before you try to use a spreadsheet.

Entering Information into a Cell

To enter a label, number, or formula into a cell, position the pointer to the desired cell and then type the desired contents. As you type, the information appears at a position on the screen, called the **input line.** The information you type is not entered into the cell until you press (Enter) or use one of the arrow keys to move the pointer. You can edit the contents of a cell by positioning the cursor to that cell and then pressing the function key (F2). Use the right and left arrow keys to move through the text, and the (Del) and (BkSp) keys to delete characters. Press (Enter) when your editing is completed.

Labels

Any string of characters that begins with a letter is assumed to be a label. When you type such a label, an apostrophe (') is automatically placed in the first column of that cell to designate what follows as a label. Sometimes you may wish a number to be a label or the first letter of a label. To distinguish this type of string as a label just begin it with an apostrophe. The contents of a particular cell is not limited in length by the width of the cell. If the contents is a label and its length is wider than the cell width, the label covers the cell (or cells) to its immediate right as long as those cells are empty. (By ''empty'' we mean the

	A	B	C	D	E	F	G	H
1								
2								
3								
4								
5								
6								
7								
8								
9								
10								
11								
12								
13								
14								
15								
16								
17								
18								
19								
20								

cell does not even contain "blanks"!) However, if these cells do contain any information (even blanks), they cover the label.

Numbers

Numbers can be entered in a cell by moving the pointer to the desired cell and typing the number. Many of your applications will require only the fixed number format (a decimal number). However, it is possible to change the way a number is displayed to scientific form or even to currency notation. The procedure necessary to change the **format** depends on your particular software package. A word of warning. *Do not precede numbers in a cell by an apostrophe('), quotation marks('') or a caret(^).* These marks tell the spreadsheet that what follows is a *label,* not a number.

Formulas

When a formula is entered into a cell, the spreadsheet does not display it in the cell. Rather, the spreadsheet software performs the computation indicated by the formula and displays the result in the worksheet cell. To view the formula in a cell, move the cursor to the cell and note the descriptor line. The operations of formulas are usually the same as that for most computer languages:

- *   multiplication
- /   division
- +   addition
- −   subtraction
- ^   exponentiation (to the power of)

Moreover, logical formulas can use the symbols:

=	equal
<>	not equal
<	less than
>	greater than
<=	less than or equal
>=	greater or equal

Formulas must begin with one of the following characters:

$$0\ 1\ 2\ 3\ 4\ 5\ 6\ 7\ 8\ 9\ +\ -\ .\ (\ @\ \#\ \$$$

They must begin with a plus sign (+) if the first part of the formula is a cell address (column and row), otherwise the spreadsheet interprets the contents as a label. Equations can be up to 240 characters in length and spaces must not be used. As an example, suppose we want the contents of cell C5 to contain the result of multiplying the contents of cell A1 with the sum of cells B1, B2, B3, and B4. The formula we enter in cell C5 is:

$$\text{+A1*(B1+B2+B3+B4)}$$

After this formula is entered into cell C5, the numerical result is displayed in this cell.

Spreadsheets have a number of built-in functions similar to most computer languages. However, unlike their language counterparts, spreadsheet functions begin with a special character, typically the @ sign. Below is a list of the more useful built-in functions:

@ABS(x)	@LN(x)	@TAN(x)
@ATAN(x)	@LOG(x)	@MAX(list)
@COS(x)	@ROUND(x)	@MIN(list)
@EXP(x)	@SIN(x)	@SUM(list)
@INT(x)	@SQRT(x)	

This list is by no means complete. Furthermore, function names vary slightly from one spreadsheet brand to another, so you should consult your spreadsheet operating manual for the functions it provides and the exact designation. There is a shorthand method employed to specify a range of cells. The double period (..) is often used to specify a cell range. For example, to specify all the cells in the range from C3 to C14, one need only type C3..C14. Hence, if we employ the @SUM function given above, the equation of our previous example can be written:

$$\text{+A1*@SUM(B1..B4)}$$

One final note about formulas: We have often observed students typing the same number numerous times in formulas that occupy different cells. An important style feature you should get used to is placing the values of frequently used parameters in a particular cell and then referencing that cell in formulas. You may even want to ''anchor'' that cell reference so that it doesn't change if you copy or move the formula. Anchoring is described in the following section.

Cell Addressing

The ability to enter a formula into a cell is one indication of how useful a spreadsheet can be when applied to data analysis. However, there is another feature that makes the spreadsheet even more powerful. Unless you specify otherwise, when you enter a cell label in a formula, the spreadsheet software assumes that it is a **relative** address. For example, suppose the pointer is positioned at cell C2 and the formula 2*A1 is entered into this cell. The software interprets this instruction as: take the contents of the cell two columns to the left and one row up, multiply it by 2, and put the result in this cell C2. The cell A1 is two columns to the left and one row above the cell C2. If you COPY or MOVE the contents of cell C2 to another location (a procedure you will eventually learn), the A1 in the equation is updated by the software to refer to the corresponding cell, which is two columns to the left and one row above the cell to which you copy or move this formula.

It is possible to prevent the software from changing the cell label (address) in a formula, a process called **anchoring.** To anchor a cell one need only prefix each part of the label (both column and row) with a special symbol. Many spreadsheets use a dollar sign ($). For example, if the equation 2*$A$1 is moved from cell C2 to any other cell, it has *exactly* this same form. It is possible to anchor only a row or a column by placing the $ sign by either the row only or column only designation of the cell reference. For example, if the contents of cell C2 is 2*A$1 and it is moved to cell E3, the new formula in cell E3 is 2*C$1. Explanation: the A of the formula in cell C2 refers to two columns to the left. When this formula is moved to cell E3, "two columns to the left" means column C. The $ sign before the row 1 ($1) prevents row 1 from being changed when this formula is moved (or copied).

Menus

In order to perform operations with a spreadsheet it is necessary to obtain the **main menu** of the spreadsheet. This "pop-up" menu is obtained by pressing the slash (/) key or using a mouse. The main menu is actually the first of a whole "tree structure" of menus that allow you to perform many operations on your spreadsheet. To move back to the previous menu (or completely remove the main menu) press either the escape key (Esc) or, if the particular menu has the option "Quit" available, press Q.

When you display a menu on the screen, you will also find a short, single-sentence description of the particular menu item that is presently being highlighted (reversed video). To make a menu selection, use the (↑) or (↓) key to move the highlight to the desired selection and press (Enter) or just type the first letter of your selection.

Importing Files

If your particular software package is different from ours, it is still possible for you to use the spreadsheet templates on the disk accompanying this text. All spreadsheet software packages provide an option (usually under the FILE submenu) that allows you to **import** our spreadsheet templates into your software. We have successfully imported our files into the QUATTRO, LOTUS-123, and the EXCEL software packages.

Graphing

You should consult your particular software package for the commands necessary to plot a graph. (It's easy, and takes far less time than a pencil and a sheet of graph paper!) You should be aware that most of our applications require an XY-type graph. This type is not the default for most spreadsheets and we have noticed that students sometimes mistakenly designate a LINE-type graph. Avoid this error because a LINE graph does not provide a correct horizontal scale for your graph unless the independent variable points are evenly spaced.

APPENDIX 11   List of Tables

1.1 SI units used in mechanics
1.2 Some approximate lengths
1.3 Some approximate time intervals
1.4 Some approximate masses
1.5 Several atomic masses
1.6 Powers of ten prefixes
3.1 Data from ''Galileo's experiment''
3.2 Trained hammerfist strike data
3.3 Author's hammerfist data
4.1 Kinematic equations specialized for projectile motion
4.2 Special conditions in projectile-motion problems
5.1 Units of force, mass, and acceleration in several systems
6.1 Typical coefficients of static and kinetic friction
6.2 Dynamic viscosity of common substances at 24°C and atmospheric pressure
6.3 Force models
9.1 Coefficient of restitution model
10.1 The big picture (momentum and kinetic energy)
11.1 Analogy between rotational kinematics and straight-line kinematics
11.2 Relations between angular and linear quantities for constant radius R
11.3 Comparison of linear and rotational equations for the constant acceleration model
11.4 Rotational inertia for some common shapes
12.1 Energy division between translational motion of the center of mass and rotational motion about the center of mass for objects that roll without slipping
16.1 A classification scheme for the states of matter
16.2 Elastic moduli for various substances
16.3 Densities for some substances
16.4 Dynamic viscosities of some fluids
17.1 The velocity of sound in various materials
17.2 Sound intensity levels for some common sources
18.1 Expansion coefficients for some materials
18.2 Thermal conductivities
18.3 R-values for some typical building materials
18.4 Specific heats and molar specific heat capacities of some common substances
18.5 Van der Waals constants for some gases
19.1 Summary of thermodynamic processes
20.1 Carnot cycle
21.1 Specific heat capacities for selected gases at 15°C
22.1 Charge and mass of the proton, electron, and neutron
22.2 Electric dipole moments
22.3 E-field magnitudes for various symmetric geometries
24.1 Analogy between relations for force and potential energy and electric field and electric potential
25.1 The dielectric constants of several substances
26.1 Resistivities of some materials at 20°C
26.2 The resistor color code
26.3 The transition temperature for some superconducting materials
32.1 Magnetic dipole moments of selected elements

32.2 Magnetic susceptibilities of some materials at 300 K
32.3 The Curie temperature for ferromagnetic phases of selected substances
33.1 Relationships between $i(t)$ and $v_X(t)$ for an AC circuit
33.2 Phase relationships between $v(t)$ and $i(t)$ for an RCL circuit
33.3 Summary of RCL circuit equations
33.4 Circuit models for the ideal transformer
35.1 Indices of refraction for selected isotropic media at sodium D wavelength 589 nm
35.2 Refractive index of some birefringent crystals
36.1 Sign convention for spherical mirrors
36.2 Sign conventions for refraction as a spherical interface between two dielectrics
36.3 Sign convention for the thin-lens equation
36.4 The $ABCD$ parameters
40.1 Photoelectric work function of selected elements (polycrystalline)

APPENDIX 12 Nobel Prize Winners in Physics

1901 Wilhelm Roentgen (1845–1923):
for the discovery of X-rays

1902 Hendrick Antoon Lorentz (1853–1928) and Pieter Zeeman (1865–1943):
for predicting and observing the splitting of spectral lines by a magnetic field

1903 Antoine Henri Becquerel (1852–1908):
for the discovery of radioactivity

Pierre Curie (1859–1906) and Marie Sklodowska-Curie (1867–1934):
for their studies of radioactivity

1904 Lord Rayleigh (John William Strutt) (1842–1919):
for his study of gas densities and the discovery of argon

1905 Philipp Eduard Anton von Lenard (1862–1947):
for his studies of cathode rays

1906 Sir Joseph John Thomson (1856–1940):
for his studies of conduction through gas discharge tubes and the discovery of the electron

1907 Albert Abraham Michelson (1852–1931):
for his optical instrument inventions and the measurement of the speed of light

1908 Gabriel Lippmann (1845–1921):
for the application of interference techniques to produce the first color photographic plate

1909   Guglielmo Marconi (1874–1937) and Carl Ferdinand Braun (1850–1918):

for inventing wireless telegraphy

1910   Johannes Diderik van der Waals (1837–1923):

for his studies of the equations of states for liquids and gases

1911   Wilhelm Wien (1864–1928):

for his discovery of laws of governing blackbody radiation

1912   Nils Gustaf Dalén (1869–1937):

for his invention of gas regulators for the lighting mechanisms in light-houses and buoys

1913   Heike Kamerlingh Onnes (1853–1926):

for liquefying helium and the discovery of superconductivity

1914   Max T. F. von Laue (1879–1960):

for his studies of X-ray diffraction by crystals

1915   Sir William Henry Bragg (1862–1942) and his son Sir William Lawrence Bragg (1890–1971):

for their contributions to the study of crystal structures by X-rays

1917   Charles Glover Barkla (1877–1944):

for his studies of atoms by X-ray emissions

1918   Max Planck (1858–1947):

for the discovery of energy quanta

1919   Johannes Stark (1874–1957):

for the discovery that spectral lines can be split by an electric field

1920   Charles Édouard Guillaume (1861–1938):

for the development of Invar, a nickel–steel alloy with a very small expansion coefficient

1921   Albert Einstein (1879–1955):

for his contributions to theoretical physics and, in particular, his explanation of the photoelectric effect

1922   Niels Bohr (1885–1962):

for his atomic model and its explanation of spectral emissions

1923   Robert Andrews Millikan (1868–1953):

for the measurement of the charge of an electron

1924   Karl Manne Georg Siegbahn (1886–1978):

for his research in X-ray spectroscopy

1925   James Franck (1882–1964) and Gustav Hertz (1887–1975):

for their discovery of the *Franck-Hertz* effect for collisions of electrons with atoms

1926   Jean Baptiste Perrin (1870–1942):

　　　　for his studies of the discontinuous structure of matter

1927   Arthur Holly Compton (1892–1962):

　　　　for his discovery of the *Compton effect* for X-ray scattering by electrons

　　　　Charles Thomson Rees Wilson (1869–1959):

　　　　for his method of making the paths of electrically charged particles visible
　　　　　　by condensation of vapor

1928   Sir Owen Willans Richardson (1879–1959):

　　　　for his studies of thermionic effects and electron emissions by hot metals

1929   Louis Victor de Broglie (1892–1987):

　　　　for the discovery of the wave nature of electrons

1930   Sir Chandrasekhara Venkata Raman (1888–1970):

　　　　for his studies of the *Raman effect,* the wavelength changes undergone by
　　　　　　light scattered by atoms and molecules

1932   Werner Heisenberg (1901–1976):

　　　　for the development of quantum mechanics

1933   Erwin Schrödinger (1887–1961):

　　　　for the development of wave mechanics

　　　　Paul Adrien Maurice Dirac (1902–1984):

　　　　for the development of relativistic quantum mechanics

1935   Sir James Chadwick (1891–1974):

　　　　for the discovery of the neutron

1936   Victor Franz Hess (1883–1964):

　　　　for the discovery of the positron

　　　　Carl David Anderson (1905–1991):

　　　　for the discovery of cosmic rays

1937   Clinton Joseph Davisson (1881–1958) and Sir George Paget Thomson (1892–
　　　　　　1975):

　　　　for their discovery of electron diffraction by crystals

1938   Enrico Fermi (1901–1954):

　　　　for his discovery of transuranic elements by neutron–neutron irradiation

1939   Ernest Orlando Lawrence (1901–1958):

　　　　for the invention of the cyclotron and its applications

1943   Otto Stern (1888–1969):

　　　　for his development of molecular beam techniques and their use for the
　　　　　　discovery of the magnetic moment of the proton

1944   Isidor Issac Rabi (1898–1988):

　　　　for the discovery of nuclear magnetic resonance in molecular and atomic
　　　　　　beams

1945   Wolfgang Pauli (1900–1958):

> for the discovery of the exclusion principle

1946   Percy Williams Bridgman (1882–1961):

> for the development of high-pressure containers and discoveries made with such vessels

1947   Sir Edward Victor Appleton (1892–1965):

> for his studies of the ionosphere

1948   Patrick Maynard Stuart Blackett (1897–1974):

> for the development of the cloud chamber and discoveries made by it in nuclear and cosmic-ray physics

1949   Hideki Yukawa (1907–1981):

> for the prediction of the existence of mesons

1950   Cecil Frank Powell (1903–1969):

> for the development of photographic emulsion techniques and their applications to studies of cosmic rays and mesons

1951   Sir John Douglas Cockroft (1897–1967) and Ernest Thomas Sinton Walton (1903–):

> for the transmutation of nuclei by accelerated particles

1952   Felix Bloch (1905–1983) and Edward Mills Purcell (1912–):

> for their discoveries of nuclear magnetic resonances in liquids and gases

1953   Frits Zernike (1888–1966):

> for his development of the phase-contrast method and its application to the invention of the phase-contrast microscope

1954   Max Born (1882–1970):

> for the statistical interpretation of quantum mechanics

Walther Bothe (1891–1957):

> for the development of the coincidence method for studying subatomic particles

1955   Willis Eugene Lamb, Jr. (1913–1993):

> for the discovery of the fine structure of the hydrogen atom

Polykarp Kusch (1911–):

> for the determination of the magnetic moment of the electron

1956   William B. Shockley (1910–1989), John Bardeen (1908–1991), and Walter Houser Brattain (1902–1987):

> for the invention of the transistor

1957   Tsung-Dao Lee (1926–) and Chen Ning Yang (1922–):

> for their prediction that parity is not conserved in beta decay

1958   Pavel Alexeyevich Cerenkov (1904–):

> for the discovery of Cerenkov radiation

Ilya Mikhailovich Frank (1908–1990) and Igor Yevgenevich Tamm (1895–1971):

> for the explanation of Cerenkov radiation

1959   Emilio Gino Segrè (1905–1989) and Owen Chamberlain (1920–):

> for the discovery of the antiproton

1960   Donald Arthur Glaser (1926–):

> for the invention of the bubble chamber

1961   Robert Hofstadter (1915–1990):

> for discovering the internal structure of the proton and neutron

Rudolf Ludwig Mössbauer (1929–):

> for discovering the Mössbauer effect

1962   Lev Davydovich Landau (1908–1968):

> for his theoretical investigations of liquid helium and other forms of condensed matter

1963   Eugene Paul Wigner (1902–):

> for his applications of symmetry principles to elementary particle theory

Maria Goeppert Mayer (1906–1972) and J. Hans D. Jensen (1906–1973):

> for their studies of the nuclear shell model

1964   Charles Hard Townes (1915–), Nicolai Gennediyevich Basov (1922–), and Aleksandr Mikhailovich Prochorov (1916–):

> for developing lasers and masers

1965   Sin-itiro Tomonaga (1906–1979), Julian Seymour Schwinger (1918–), and Richard Phillips Feynman (1918–1988):

> for developing the theory of quantum electrodynamics

1966   Alfred Kastler (1902–1984):

> for the development of optical methods for studying atomic energy levels

1967   Hans Albrecht Bethe (1906–):

> for his theory of nuclear reactions related to energy production in stars

1968   Luis Walter Alvarez (1911–1988):

> for his contributions to elementary particle physics

1969   Murray Gell-Mann (1929–):

> for his contributions to the classifications of elementary particles

1970   Hannes Olof Gösta Alfvén (1908–):

> for his development of magnetohydrodynamic theory

Louis Eugène Félix Néel (1904–):

> for his discoveries of ferromagnetism and antiferromagnetism

1971   Dennis Gabor (1900–1979):

> for his development of holography

1972   John Bardeen (1908–1991), Leon N. Cooper (1930–), and John Robert Schrieffer (1931–):

for their theory of superconductivity

1973   Leo Esaki (1925–):

for the discovery of tunneling in semiconductors

Ivar Giaever (1929–):

for the discovery of tunneling in superconductors

Brian David Josephson (1940–):

for predicting tunneling of electron pairs in superconductors

1974   Antony Hewish (1924–):

for the discovery of pulsars

Sir Martin Ryle (1918–1984):

for his contributions to radio interferometry

1975   Aage N. Bohr (1922–), Ben Roy Mottelson (1926–), and Leo James Rainwater (1917–1986):

for their development of the collective model of the nucleus

1976   Burton Richter (1931–) and Samuel Chao Chung Ting (1936–):

for the discovery of the J/ψ particle

1977   John Hasbrouck Van Vleck (1899–1980), Sir Nevill Francis Mott (1905–), and Philip Warren Anderson (1923–):

for their contributions to the quantum theory of solids

1978   Arno A. Penzias (1933–) and Robert Woodrow Wilson (1936–):

for their discovery of the cosmic background radiation

Pjotr Leonidovich Kapitza (1894–1984):

for his contributions to low-temperature physics

1979   Sheldon Lee Glashow (1932–), Abdus Salam (1926–), and Steven Weinberg (1933–):

for their theory of the unification of the weak and electromagnetic forces

1980   Val Logsdon Fitch (1923–) and James Watson Cronin (1931–):

for their discovery of charge–parity violation

1981   Nicolaas Bloembergen (1920–) and Arthur Leonard Schawlow (1921–):

for their contributions to laser spectroscopy

Kai Manne Börje Siegbahn (1918–):

for his development of high-resolution electron spectroscopy

1982   Kenneth Geddes Wilson (1936–):

for his theories of phase transitions to study critical phenomena

1983   William Alfred Fowler (1911–):

for his theoretical studies of astrophysical chemical formation

Subrahmanyan Chandrasekhar (1910–):

for his theoretical studies of stellar evolution

1984   Carlo Rubbia (1934–) and Simon van der Meer (1925–):

for their contributions to elementary particle physics

1985   Klaus von Klitzing (1943–):

for his discovery of the quantum Hall effect

1986   Ernst Ruska (1906–1988):

for the invention of the electron microscope

Gerd Binnig (1947–) and Heinrich Rohrer (1933–):

for their invention of the scanning-tunneling electron microscope

1987   Karl Alex Müller (1927–) and Johannes George Bednorz (1950–):

for the discovery of high-temperature superconductivity

1988   Leon M. Lederman (1922–), Melvin Schwartz (1932–), and Jack Steinberger (1921–):

for the first use of a neutrino beam to study the weak nuclear force

1989   Norman Ramsey (1915–):

for the development of the cesium clock and techniques for atomic structure studies

Hans Dehmelt (1922–) and Wolfgang Paul (1913–):

for the development of the ion-trap method of separating charged particles

1990   Jerome Friedman (1930–), Henry Kendall (1926–), and Richard Taylor (1929–):

for their pioneering investigations concerning deep inelastic scattering of electrons on protons and bound neutrons

1991   Pierre de Gennes (1932–):

for discovering that methods developed for studying order phenomena in simple systems can be generalized to more complex forms of matter; in particular to liquid crystals and polymers

1992   George Charpak (1924–):

for his invention of fast electronic detectors for high-energy particles

Chapter 1

1. (a) 1 microphone, (b) 1 Megaphone, (c) 2 kilomockingbirds, (d) 1 picoboo, (e) 1 terabulls
3. 9.00 m   **5.** 2.83×10^{22} atoms
7. (a) 5.10×10^{14} m^2, (b) 5.10×10^{18} cm^2,
(c) 6.10×10^{14} yd^2 **9.** (a) 8.36×10^3 m^2,
(b) 8.36×10^7 cm^2, (c) 3.23×10^{-3} mi^2 **11.** 2.59 km^2
13. 1.00×10^3 kg/m^3 **15.** $(\alpha, \beta) =$ (a) (m/s^2, m/s^3),
(b) (m/s, m/s^2), (c) (m, m^{-1}), (d) (m, s^{-2}), (e) (m, s^{-1}),
(f) (m/s, s^{-2})   **19.** (a) 4, (b) 6, (c) 3, (d) 4
21. (a) 2, (b) 2, (c) 4, (d) 2, (e) 3, (f) 2
23. 6.707×10^8 mi/h   **25.** 5.70 m^3 ± 0.23 m^3
27. 3.34×10^{25} water molecules, 55.5 mol, 6.69×10^{25}
hydrogen atoms   **29.** (a) 10.3 m^2, (b) 9.68×10^{-2} mm
31. (a) $m = 0$, (b) $m = 1$, $n = -2$, (c) $m = 3$, $n = -1$,
(d) $m = 2$ **33.** (a) 4.82×10^{-5}, (b) 4.71×10^5,
(c) 9.48×10^4, (d) 6.7, (e) 137, (f) 700

Chapter 2

1. (6.60 m, 58°)   **3.** (−8.03 m, −9.58 m)
5. (a) $(x_1, y_1) = $ (4.00 m, 3.00 m),
$(x_2, y_2) = $ (−5.04 m, 4.86 m), (b) 9.22 m,
(c) (9.22 m, 168°)   **7.** 34.1 km, 13.1° N of E
9. (3.6, 214°)   **11.** (−16.1 m/s, 9.30 m/s)
13. (−10.6 m, −22.7 m)   **15.** (8.14 m, 152°)
17. $A_x = 4.0$, $A_y = 4.0$, $B_x = -2.0$, $B_y = -8.0$, $C_x = -5.0$,
$C_y = 2.0$, $R_x = -3.0$, $R_y = -2.0$ **19.** (a) (−24.0, 120),
(b) (−120, −24.0), (c) (−96.0, 168°)
21. (a) $A_x = 64.3$, $A_y = 54.0$, $B_x = -28.2$, $B_y = 53.0$
(b) (148.9, −104.9) (c) (182.2, 325°)
23. (a) $A_x = 10.1$ m, $A_y = 7.35$ m, $B_x = -5.74$ m,
$B_y = 8.19$ m, $C_x = -1.94$ m, $C_y = -7.24$ m,
(b) $D_x = 24.0$ m, $D_y = -0.741$ m, (c) (24.0 m, −1.8°)
25. (a) (−4.80 m, −3.60 m), (6.00 m, 217°),
(b) (20.4 m, 0), (20.4 m, 0°), (c) (−15.6 m, 3.60 m),
(16.0 m, 167°)   **27.** (a) $19.0\hat{\mathbf{i}} - 5.00\hat{\mathbf{j}} - 5.00\hat{\mathbf{k}}$,
(b) $-2.0\hat{\mathbf{i}} - 14.0\hat{\mathbf{j}} + 37.0\hat{\mathbf{k}}$, (c) $-17.0\hat{\mathbf{i}} + 19.0\hat{\mathbf{j}} - 32.0\hat{\mathbf{k}}$
29. (a) 8.00, (b) 77.5   **31.** 115°   **33.** (a) 162°,
(b) 72.4°, (c) 180°   **39.** (a) $-26.0\hat{\mathbf{k}}$, (b) $+26.0\hat{\mathbf{k}}$,
(c) $-52.0\hat{\mathbf{k}}$, (d) $-52.0\hat{\mathbf{k}}$
45. (a) $6.00\hat{\mathbf{i}} + 1.50\hat{\mathbf{j}} + 2.50\hat{\mathbf{k}}$, (b) 78.6°   **47.** 44.7
49. (141, −137, 30)   **51.** −234   **53.** (a) (3.53 m E,
4.71 m N), (b) (5.89 m, 53.1° N of E)   **55.** (11.4, 199°)
57. $l = 1.5$, $m = 2.9$, $m = -3.8$ **59.** (a) $8.00\hat{\mathbf{k}}$, (b) 0,
(c) $-16.0\hat{\mathbf{i}} - 8.00\hat{\mathbf{j}}$

Chapter 3

1. 4.83 km/h = 1.34 m/s   **3.** 29.5 s   **5.** (a) 2.00 mph,
(b) 2.67 mph   **7.** 21.4 km/h/s = 6.70 m/s/s
9. 2.68 m/s/s
11. 3.0 cm/s

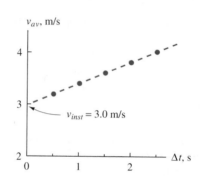

15. (a) $v(t) = (12.0 \text{ m/s}^3)t^2 - 5.00$ m/s, (b) $a(t) = $
$(24.0 \text{ m/s}^3)t$, (c) 5.00 m/s in − x direction, (d) 23.0 m/s
17. (a) m/s^3, (b) $A = -0.5$ m/s^3, $B = 2.5$ m/s^2,
(c) 3.5 m/s^2 **19.** (a) 4 s, (b) 2 s, (c) never
21. (a) 5.7×10^{-10} m/s, 1.8 cm/y, (b) 54 cm/y
23. (a) 8 s, (b) 11 s ≤ t ≤ 13 s, and 16 s ≤ t ≤ 18 s
25. (a) 2 s and 19 s, (b) 10 s ≤ t ≤ 13 s   **27.** (a) −1.5 s,
(b) 2.0 m/s   **29.** (a) 14.5 m/s, (b) 13.0 m/s
31. (a) ~13 s, (b) first, (c) ~$0.55g$
33. $\mathscr{A} = 1.67$ m/s^2

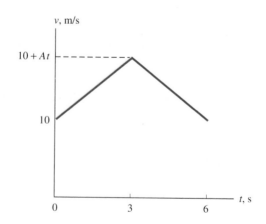

35. (a) 8.00 m/s^2, (b) 2.50 m/s, (c) 0.277 s   **37.** 51.0 m
39. (a) 2.00 m/s^2, (b) 1.00 s   **41.** 139 m/s
43. 6.21 km

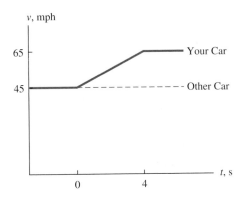

45. (b) 3.79 s (other car 76.1 m), 93.0 m (you are 3.1 m behind), (c) 0.91 s, (d) 119 m

47. 0.178 s   **49.** $v_o/2$ **51.** (a) $v(t) = (6.00 \text{ m/s}^3)t^2 + (5.00 \text{ m/s}^2)t - 6.00 \text{ m/s}$, (b) $x(t) = (2.00 \text{ m/s}^3)t^3 + (2.5 \text{ m/s}^2)t^2 - (6.00 \text{ m/s})t - 3.00 \text{ m}$

53. (a) $x(t) = (4.00 \text{ m/s}^3)t^3 - (4.00 \text{ m/s}^2)t^2 + (4.00 \text{ m/s})t - 27.0 \text{ m}$, (b) 12.0 m/s, (c) 64.0 m/s²   **55.** (a) 12.1 s, (b) +x direction, (c) 77.4 m, (d) 8.60 m/s, (e) 8.60 m/s

63. (a)

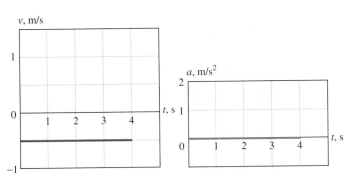

63. (b)

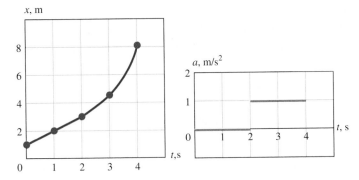

65. (a) 3.3 m/s², (b) 10 m/s at 5 s, (c) ~70 m
67. (a) $0.57g$ = 12.5 mph/s, (b) 67.0 m/s, (c) 191 m
69. (a) 15.8 s, (b) 250 m, (c) 70.7 mph (yes!)
71. (a) 1.28 s, (b) 2.03 m, (c) 2.43 m   **73.** 6th floor
75. (a) $x(t) = (10.0 \text{ m})(1 - e^{-t/\tau})$

Chapter 4

1. (a) 0.672 m/s, (b) 0.746 m/s   **3.** (a) 0.833 m/s at 36.9°S of E, (b) 1.17 m/s   **5.** (a) $\mathbf{v}(t) = (2.00 \text{ m/s})\hat{\mathbf{i}} -$

$(6.00 \text{ m/s}^2)t\hat{\mathbf{j}}$, (b) $(-6.00 \text{ m/s}^2)\hat{\mathbf{j}}$, (c) 12.2 m/s, (d) 6.00 m/s² (magnitude)   **7.** (a) $\mathbf{v}(t) = (9.00 \text{ m/s}^3)t^2\hat{\mathbf{i}} + (4.00 \text{ m/s}^2)t\hat{\mathbf{j}}$, (b) 57.1 m/s, (c) $\mathbf{a}(2.50 \text{ s}) = (45.0 \text{ m/s}^2)\hat{\mathbf{i}} + (4.00 \text{ m/s}^2)\hat{\mathbf{j}}$

9. (a)

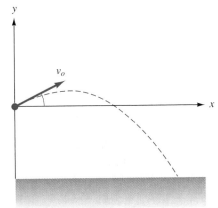

9. (b) v_{o_x} = 10.9 m/s, v_{o_y} = 5.07 m/s, (d) 1.46 s, (e) 15.9 m   **11.** 3.32 m/s   **13.** (a) 1.54 m, (b) 5.16 m/s   **15.** (a) 1.75 m, (b) 5.64 m/s
17. (a) 5.34 m/s, (b) 5.30 m/s $\leq v_o \leq$ 5.38 m/s
19. (a) 1.74 m, (b) 7.17 m/s   **21.** (a) 39.4 s, (b) 5280 m = 3.28 mile   **23.** (b) if θ = 45° then $R_{max} = v_o^2/a_g$, (c) $y_{max} = [v_o^2/(2a_g)] \sin^2(\theta)$ **25.** (a) 11, (b) 1.40 m from the side it rolled in on
29. (a) 46.4 m/s² = $4.74a_g$, (b) 1.97 rad/s
31. (a) 3.37×10^{-2} m/s², (b) 5.93×10^{-3} m/s²
33. (a) 1.52×10^{-16} s, (b) 9.07×10^{22} m/s², (c) 4.13×10^{16} rad/s   **35.** 5.44×10^3 m = 3.38 mile
37. 60.0 m   **39.** (a) 4.08 s, (b) 25.0 m, (c) 0.500 rad = 28.6°, (d) 4.24 m/s²
55. (a) $\mathbf{v}(t) = (10.0 \text{ m/s})\hat{\mathbf{i}} - (10.0 \text{ m/s}) \times \sin[(0.500 \text{ rad/s})t]\hat{\mathbf{j}}$, (b) $\mathbf{a}(t) = (-5.0 \text{ m/s}) \times \cos[(0.500 \text{ rad/s})t]\hat{\mathbf{j}}$, (c) 11.1 m/s, 14.1 m/s, (d) 334° anticlockwise from the +x-axis, 315° anticlockwise from +x-axis
57. (a)

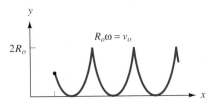

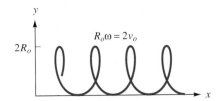

57. (b) $[v_o - \omega R_o \sin(\omega t)]\hat{\mathbf{i}} + [\omega R_o \cos(\omega t)]\hat{\mathbf{j}}$, (c) $v_o + \omega R_o$, (d) $v_o - \omega R_o$ **59.** 0.73 m when θ = 76.2°   **61.** (a) 0.236 m over net, (b) 0.671 s
65. (a) $\omega_1 R_1 + \omega_2 R_2$, (b) $-\omega_1 R_1 + \omega_2 R_2$, (c) $\omega_1^2 R_1 + \omega_2^2 R_2$

Chapter 5

1. (a) 11.5 N, 269°, (b) no, it could have a constant velocity   **5.** (a) 8.25 N, 166° anticlockwise from the positive x-axis, (b) 2.06 m/s^2 at 346° anticlockwise from positive x-axis   **7.** (a) 4.17 m/s^2, (b) 0.833 m/s^2, (c) 89.8°   **9.** (a) 9.00 N, (b) 3.75 m/s^2 **11.** (a) 3.11 slug = 45.4 kg, (b) 49.0 N = 11.0 lbf   **13.** 141 ft/s^2, (b) 29.4 N   **15.** (a) 60.0 kg, (b) 226 N = 50.8 lbf
17. (a) 60.0 N/m, (b) 27.0 N
19.

(a)

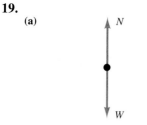

(b)

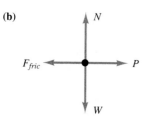

(c)

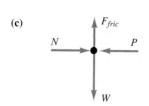

(d)

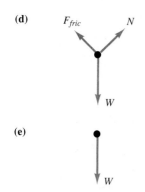

(e)

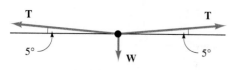

21. (a) 10.0 N, (b) 6.88 cm   **23.** (a) 345 N/m, (b) 17.6 cm   **25.** (a) −1.50 m/s^2, (b) 8.30 cm (total), (c) 9.80 m/s^2 downward
27. (a)

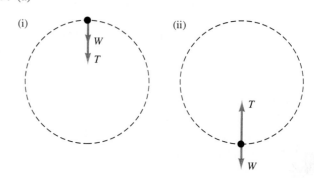

27. (b) 2860 N   **29.** 598 N, 343 N   **31.** (a) 29.4 cm, (b) 30.4 cm, (c) 0.600 cm   **33.** 0.545 m/s^2, 1390 N
35. (a) 630 N, (b) 420 N, (c) 210 N
37. (a) 2.16 m/s^2, 11.5 N, (b) 4.38 m/s^2, 8.14 N

39. 2.04 m/s^2, 38.8 N, 54.3 N   **41.** (a) 9.80 N, (b) 9.80 N, (c) 9.80 N   **43.** (a) a downward force on hands by barbell, (b) a horizontal force on the tire by the road, (c) a backward pull by the trailer on the bumper, (d) the upward force of the floor on the heel, (e) the upward gravitational force on the earth by your body
45. (a)

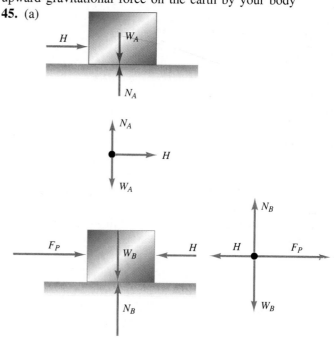

45. (b) 4.00 N, 2.00 N   **47.** (a) 0.452 s, (b) 735 N, (c) 735 N, (d) 1.23 × 10^{-22} m/s^2, (e) 1.26 × 10^{-23} m
49. (a) 1.96 m/s^2, (b) 2.94 m/s^2 **51.** 36.2 N
53. (a) 1.96 m/s^2, 11.8 N, 15.7 N, (b) 0.960 m/s^2, 10.8 N, 17.7 N, (c) 2.96 m/s^2, 12.8 N, 13.7 N   **55.** 0.0290 m/s^2, 24.4 N

Chapter 6

1. 0.800   **3.** (a) $\theta = \arctan(\mu_s)$, (b) 0.400 m/s^2, (c) 0.700   **5.** 6.37 m/s^2 **7.** (a) 0.993 m, (b) no
9. 0.500   **11.** $[m_A m_B/(m_A + m_B)](\mu_A - \mu_B)g \cos(\theta)$
13. 2.41 cm   **15.** 39.2°   **17.** the car does not slide down even if at rest
19. (a)

(i)   (ii)

19. (b) 3.43 m/s, (c) 29.4 N   **21.** (a) 0.35%,
(b) 9.766 m/s^2 **23.** (a) 3.61 × 10^{-47} N,
(b) 8.32 × 10^{-8} N, the gravitational force is a factor of
2.3 × 10^{39} times too small   **25.** $GmM/[a(a + L)]$
27. 5.07 × 10^3 s = 1.41 hr   **29.** (a) 28.0g, (b) 274 N
31. 1.98 × 10^{20} N, 4.31 × 10^{20} N   **33.** (a) 5.00 × 10^4,
(b) 3.13 × 10^5, (c) 3.13   **37.** (a) ~800, (b) ~5 m/s
39. (a) 3.66 × 10^{-6} m, (b) 1.91 × 10^{-12} N
49. 0.033 m/s^2, 2.48 N   **51.** $\theta = \arctan(\mu_k)$
53. (a) 7.45 m/s^2, 7.06 N, 2.35 N, (b) 4.31 m/s^2, 16.5 N,
5.49 N   **55.** (a) 232 N, (b) 2.05 × 10^3 N
57. 3.59 × 10^7 m   **59.** 45.3 h

Chapter 7

1. 160. J   **3.** (a) 73.5 J, 0.0 J, −24.5 J, 49.0 J,
(b) −73.5 J, 0.0 J, 24.5 J, −49.0 J   **5.** 6.00 J, 12.0 J,
−12.0 J, 0.0 J, −6.00 J, 0.0 J   **7.** 199 N
9. (a) 8.00 J, (b) 19.3 J   **11.** 0.450 J   **13.** (a) 4.25 J,
(b) −4.25 J   **15.** (a) $ax^2/2 + bx^4/4$, (b) 4.52 J
17. −2.50 J   **19.** (a) 0.0 m < x < 0.50 m,
(b) 0.50 m < x < 5.50 m, (no comment), (c) 20.0 J,
(d) −20.0 J
21. 32.0 J

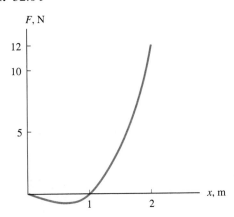

25. 155 ft · lbf   **27.** 7.52 × 10^{-15} J, 3.88 × 10^{-5} m/s
29. (a) 13.7 J, (b) 1.37 × 10^3 N, (c) 54.8 N
31. (a) 18.0 m/s, (b) 11.2 m/s   **33.** (a) 35.0 J,
(b) 37.4 m/s   **35.** 9.18 m/s   **37.** (a) −6.00 J,
(b) 0.0 J, (c) 14.7 J, (d) −5.66 J, (e) 3.05 J, 1.10 m/s
39. (a)

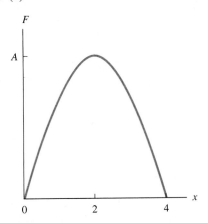

39. (b) 1.07 × 10^4 J, (c) 1.46 × 10^3 m/s   **41.** (a) 22.2
(23 total 20 m long passes required), 1.32 × 10^4 J,
(b) 11.0 W = 1.47 × 10^{-2} hp   **43.** (a) 1.95 × 10^{-6}
cents/J, (b) 31.9 km = 19.8 mi   **45.** (a) 1.50 MW,
(b) 20.0 mW   **51.** $\sqrt{gR}$ **53.** (a) ~−98 J, ~28 m/s,
(b) ~38 J, ~17 m/s   **55.** 28.0 N   **57.** (a) 0.429 J,
(b) 1.88 m/s
59. (a) $A(v_{max} − v)v$, (b) $v_{max}/2$

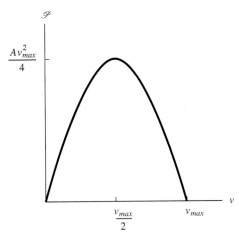

Chapter 8

1. may be conservative, but test is not conclusive
3. (a) 14.0 J, (b) 14.0 J, (c) this fact alone does not prove
the force is conservative   **5.** (a) 30.0 J, (b) 30.0 J,
(c) 30.0 J, (d) no; the results are consistent with the force
being conservative, but alone are not conclusive
7. (a) −20.8 J, (b) −20.8 J, (c) −20.8 J
9. (a) $B(x_f^{-1} − x_i^{-1})$, (b) −B/x, (c) $U = \infty$ at $x = 0$,
(d) $U = 0$ at $x = \pm\infty$
11.

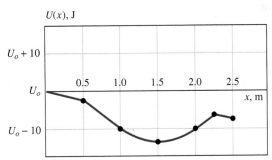

13.

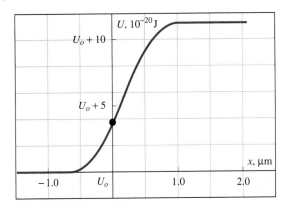

15. (a)

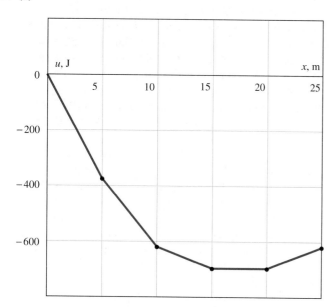

17. (a) 5.42 m/s, (b) 1.40 m   **19.** (a) 7.79 m/s,
(b) 0.171 m   **21.** 2030 N/m   **23.** 1.26 m/s
25. 9.18 m/s   **27.** (a) 10.0 m/s, (b) 8.14 m
29. (a)

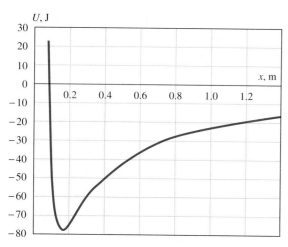

29. (b) $(4.00)x^{-3} - (25.0)x^{-2}$, (c) 0.160 m, (d) stable
31. $-A[2xz\hat{\mathbf{i}} + 2yz\hat{\mathbf{j}} + (x^2 + y^2)\hat{\mathbf{k}}]$
33. $-A\cos(z)[y\hat{\mathbf{i}} + x\hat{\mathbf{j}} - xy\tan(z)\hat{\mathbf{k}}]$
35.

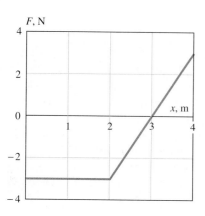

37. (a) 6.18×10^5 m/s, (b) 1.12×10^4 m/s,
(c) 2.37×10^3 m/s   **39.** $2GM/c^2$,
(b) $U = -GMm/(Rs) = -(1/2)mc^2$
41. $3R_o$ **43.** (a) conservative, (b) nonconservative
45. $(x^2 + y^2) + f(z)$, where $f(z)$ is an arbitrary function of z
only   **51.** $v = \sqrt{2(2D - L)g}$, (b) $3L/5$ **53.** 48.2°,
$h = R/3$ **55.** (a) $MgL/2$,
(b) $L/2$ **57.** (a) 5.42 m/s, (b) 3.13 m/s, (c) 0.746 m,
(d) 3.51 m to the left of point C **59.** (a) 0.858 m,
(b) 0.6125 m (total spring stretch 0.245 m), (c) 0.3675 m
(total spring stretch 0.6125 m), (d) next rebound is
0.1225 m (total spring stretch 0.490 m). Force balance
indicates mass does not move if it stops with total spring
stretch h such that 0.4288 m $\leq h \leq$ 0.5513 m. Thus the
masses stop with total stretch at 0.490 m   **61.** (a) 1.00,
(b) $a_A/a_E = m_E/m_A$, (c) $d_A/d_E = m_E/m_A$, (d) $K_A/K_E = m_E/m_A$

Chapter 9

1. (a) $16.0 \, \text{N} \cdot \text{s}\,\hat{\mathbf{i}}$, (b) $2.21 \, \text{N} \cdot \text{s}$, (c) $5.00 \, \text{N} \cdot \text{s}$ at 36.87°
from the positive y-axis   **3.** $(8.00\hat{\mathbf{i}} - 5.00\hat{\mathbf{j}}) \, \text{N} \cdot \text{s}$
5. (a) $K = p^2/(2m)$, (b) 0.417 m/s   **7.** (a) 5.48 s,
(b) 27.4 N, (c) 13.7 N   **9.** (a) $1.87 \times 10^4 \, \text{kg} \cdot \text{m/s}$,
(b) 4.15×10^3 N
9. (c)

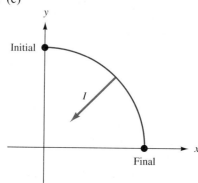

11. $368 \, \text{N}\hat{\mathbf{j}}$ **13.** (a) $\Delta\mathbf{p} = (-0.627\hat{\mathbf{i}} - 0.134\hat{\mathbf{j}}) \, \text{kg} \cdot \text{m/s}$,
(b) 257 N at 12.1° above the wall normal to the right
15. 4.50 N   **17.** 2.47 m/s   **19.** 2.78 m from the
fence which the 60-kg mass hits   **21.** (a) 47.1 m/s,
(b) 2.56×10^3 J   **23.** (a) 0.286 m/s, (b) 0.578 m/s
25. (a) 1.88 m/s, (b) 62.5%   **27.** (a) 3.98, (b) alpha
1.20×10^6 m/s, neutron 1.80×10^6 m/s, (c) 64%
29. (a) 6.25 m/s, 17.5 m/s, (b) 14.6%   **31.** (a) 7.0 m/s,
10.0 m/s, (b) 31.3%, (c) 60.0 N   **33.** $\mathcal{M} = 1$, $\epsilon = 0.40$
35. (a) -27.4 m/s, -5.43 m/s   **37.** 2.34 m/s at 46.2° on
the opposite side of the incoming ball's line of travel
39. warheads 0.40°, decoys 5.00°   **41.** 1.00
43. (a) 4.96 m/s at 23.8° from the positive x-axis,
(b)

(c) inelastic, $K_o = 40$ J, $K_f = 16.1$ J   **45.** *Hint:* use the law of cosines   **61.** (a) 7.07×10^{12} N/s^3, (b) 8.38×10^3 N, (c) 4.71×10^3 N   **63.** (a) 2.79 ft/s, (b) Jake must be launched at 44 ft/s. Assuming a double charge doubles the recoil momentum, there is still less than half the momentum required to launch Jake this fast even if he absorbed it all. But it still makes a good yarn.
65. $0.727 \leq \epsilon \leq 0.803$ **67.** (a) 11.9 cm, (b) 1.71 m/s, 5.71 m/s   **69.** 0.2%

Chapter 10

1. (a) lesser mass ($p^2/2m$), (b) greater mass ($\sqrt{2mK}$), (c) false, (d) true   **3.** (a) $p = 5.00$ kg·m/s for both, $K = 12.5$ J for the 1-kg mass, 6.25 J for the 2-kg mass, (b) $K = 2.00$ J for both, $p = 2.00$ kg·m/s for the 1-kg mass, the momentum of the 2-kg mass is larger by a factor of $\sqrt{2}$ **5.** (a) 0.1165 m/s by larger mass, 0.0686 by the smaller mass, (b) larger mass 1.50 s, smaller mass 0.866 s   **7.** 0.408 m   **9.** 0.075 m   **11.** 45.2° (*Note:* length of string is not needed.)   **13.** −0.667 cm
15. (−1.07 cm, 12.9 cm)   **17.** (a) (−2, 1), (0.750, 1.25), (0, −2.07)   **19.** 9L/20   **21.** (0, 4R/(3π))
23. (a) 1.875 m/s, (b) 62.5% in the lab frame
25. alpha 1.20×10^6 m/s, proton -1.80×10^6 m/s
27. (a) −0.250 m/s, 8.75 m/s, (b) 180 J, (c) 105 J, (d) 75 J, (e) $K_{lab} = K_{cm} + m_{total}v_2^2/2$ **29.** (a) 8.25 m/s, 10.5 m/s, (b) approach 9.00 m/s, recede 2.25 m/s, (c) $2.25/9.00 = \epsilon$ **31.** 3.38 m/s, 7.38 m/s
33. (a) $m_{total}v_{cm}^2/2$, (b) the maximum kinetic energy that can be "lost" in a collision is $\frac{1}{2}\Sigma m_i u_i^2$ **35.** (a) 3.50 N, (b) 0.969 m/s^2, (c) 10.3 m/s   **37.** (a) The rocket does not push off against the earth or against the atmosphere. It "pushes off" against its own exhaust; that is, the rocket recoils forward when it propels exhaust backward. (b) The revolver recoils backward from the expelled gasses even in a vacuum. The experiment is exactly analogous to rocket propulsion.   **39.** $580 + 550 + 1530 = 2660$ m/s
41. (b) 0.445 m/s is the asymptotic limit as $N \rightarrow \infty$ in Problem 9-49   **51.** 125 J   **53.** 482 m/s
55. (a) 0.378 m, (b) the same distance to three significant figures, (c) twice as far   **57.** (a) 1500 N, (b) 18 000 J, (c) 9000 J, (d) the work done by friction is dissipated as heat   **61.** for example, $x = (1.00 \text{ m}^{1/2})\sqrt{4.00 \text{ m} - y}$, $I_1 = (2\text{m}^{1/2})\sigma \int \sqrt{4.00 \text{ m} - y}\, dy = \sigma(256 \text{ m}^3)/15$
63. 0.582L

Chapter 11

1. (a) 0.644 rad, (b) 3.49 rad/s, (c) 157 rad, (d) 7.27×10^{-5} rad/s, (e) 1.45×10^{-4} rad/s^2
3. (a) $\omega(t) = \alpha_o\tau[1 - \exp(-t/\tau)]$, $\theta(t) = \alpha_o\tau\{t - \tau[1 - \exp(-t/\tau)]\}$, (b) 1.74×10^{-2} rad/s^2, (c) 9.17 rev, (d) 0.961 rad/s   **5.** (a) 255 rad, (b) 6.28 rad/s^2 **7.** 0.222 rad/s   **9.** (a) 10.6 cm,

(b) 1.75×10^{-3} rad/s, (c) 1.85×10^{-2} cm/s
11. (a) 1.88 m/s, (b) 1.26 m/s, (c) 0.628 m/s
13. $\sqrt{1/\alpha}$ **15.** (a) 2.30×10^5 s, (b) 1.53×10^8 rev, (c) -3.63×10^{-2} rad/s^2 **17.** (a) 0.133 rad/s^2, (b) 4.77 rev   **19.** (a) 107 s, (b) 3.59 rev, 96.4 rev
21. (a) 0.125 kg·m^2, (b) 0.0417 kg·m^2, (c) 0.125 kg·m^2, (d) in (a) and (c) all the mass is the same distance from the axis of rotation, while in (b) the mass is distributed between $x = 0$ and $x = L/2$ **23.** $L = 1.00$ m, $W = 0.500$ m
25. 13.4 kg·m^2 **27.** (a) $\lambda_o L/3$, (b) $3ML^2/5$
29. $MR^2/2$ **31.** (a) 2.57×10^{29} J, (b) 4.24×10^{32} J, that is, earth–sun is factor of 1.65×10^3 larger
33. (a) 3.11 m/s, (b) More mass concentrated on the outer edge causes the moment of inertia to be larger. For the same angular speed, the kinetic energy is larger, requiring a larger translational speed.   **35.** 1.83 m/s   **37.** 5.36 m/s
39. −1.08 N·m   **41.** $LF \sin(\phi)$ **43.** 100. rad/s
47. 1.67 m/s^2, 3.44 N, 4.07 N   **49.** (a) 0.327 m/s^2, 9.47 N, 9.31 N, (b) $a = 0$, 9.80 N, 7.80 N
51. 2.87×10^{-3} km·m^2 **53.** 5.40 rad/s
55. (a)

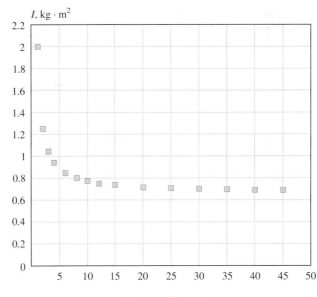

Number of Point Masses

61. (a) 0.500 m/s^2, (b) 225 m/s^2, (c) 50.0 rad/s^2, (d) 150. rad/s   **67.** (a) 9.38 N·m, (b) 312 rad/s^2

Chapter 12

1. (a) 1.57 N·m, (b) 0.213 m, 1.57 N·m, (c) the torques are the same   **3.** 376 N, 604 N in cable nearest painter
5. 3.00 kg   **7.** 1.52 m from right end   **9.** 213 lbf, 250. lbf at 31.4° from horizontal   **11.** $T = 1.054$ W, $V = 0$, $H = W/3$ **13.** (a) $V = Mg$, $N = H = Mgd \tan(\theta)/L$, (b) $d \tan(\theta)/L$ **15.** (a) order of reaching bottom: disk, sphere, hoop, (b) 0.171 s
17. $[(M + 4m/3)v^2]/[(M + 2m)g] + R$ **19.** (a) 0.459 m, (b) 0.643 m   **21.** 0.943 m/s   **25.** $[k/(k + 1)]\tan(\theta)$
27. (a) 2.22×10^3 N, (b) 60.0 N

31. $2\{[2m - M\sin(\theta)]/[(k + 1)M + 4m]\}g$
33. 188 rad/s   **35.** 194 ms   **37.** $v = v_o/(1 + k)$
39. 0.107   **47.** $\sqrt{[2(R/h) - 1]/[(R/h) - 1]^2}W$
49. (a) 14.5°, (b) $T = 12.6$ N each, $H = 6.33$ N
51. disk: $v = 0.750\,R\omega$, sphere: $v = 0.286\,R\omega$, hoop:
$v = 0.500\,R\omega$ **53.** (a) $\rho LgD^2/2$, (b) $\rho LgD^3/3$, (c) $2D/3$

Chapter 13

1. (a) 7.00 N·m $(-\hat{\mathbf{i}} + \hat{\mathbf{j}} + \hat{\mathbf{k}})$, (b) 0.00 N·m (that is
$\boldsymbol{\tau} \perp \hat{\mathbf{n}}$)   **3.** (a) $(-11.0\hat{\mathbf{i}} + 6.00\hat{\mathbf{j}} + 7.00\hat{\mathbf{k}})$ N·m,
(b) 7.00 N·m   **5.** (a) $(-8.00\hat{\mathbf{i}} + 5.00\hat{\mathbf{k}})$ N·m,
(b) $(-8.00\hat{\mathbf{i}} + 9.00\hat{\mathbf{k}})$ N·m   **7.** $\boldsymbol{\tau} = (r_x F_y - r_y F_x)\hat{\mathbf{k}}$
11. $\boldsymbol{\omega} = (\mathbf{r} \times \mathbf{v})/r^2$ **13.** (a) $(5.00 \times 10^{-4}$ kg·m²/s$) \times$
$(-\hat{\mathbf{i}} + \hat{\mathbf{j}} + \hat{\mathbf{k}})$, (b) $(+1.50\hat{\mathbf{i}} - 2.50\hat{\mathbf{j}} - 4.50\hat{\mathbf{k}}) \times$
10^{-3} kg·m²/s   **15.** $(-0.250\hat{\mathbf{i}} + 0.100\hat{\mathbf{j}} - 0.800\hat{\mathbf{k}})$
kg·m²/s
19. (a) $-(2.83$ rad/s$)\hat{\mathbf{k}}$, (b) $-(2.83$ rad/s$)\hat{\mathbf{k}}$,
(c)

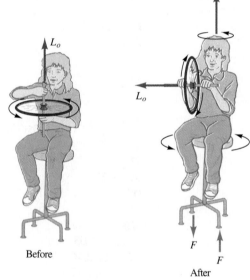

Before   After

23.

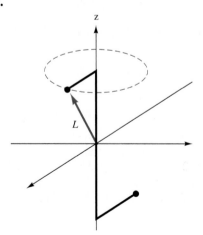

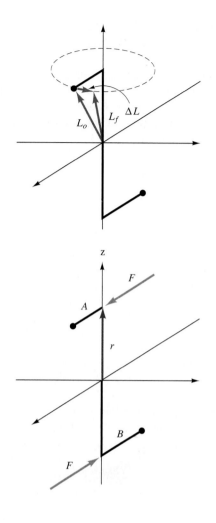

Chapter 14

1. 24.0 ms, 262 rad/s   **3.** (a) 0.040 s, 157 rad/s,
(b) $x(t) = (2.50$ cm$)\cos[(157$ rad/s$)t - 2.35$ rad$]$,
(c) -2.21 cm   **5.** (a) 35.0 cm, 3.59 rad/s,
(b) $y(t) = (35.0$ cm$)\cos[(3.59$ rad/s$)t]$, (c) -31.5 cm
7. (a) 2.00 ms, 500. Hz, (b) 0.0225 mm, (c) 466 mm/s
9. (a) $x(t) = (0.6$ mm$)\cos[(2.76 \times 10^3$ rad/s$)t - 1.05$ rad$]$,
$v(t) = -(1.66 \times 10^3$ mm/s$)\sin[(2.76 \times 10^3$ rad/s$)t - $
1.05 rad$]$, $a(t) = -(4.59 \times 10^6$ mm/s²$) \times$
$\cos[(2.76 \times 10^3$ rad/s$)t - 1.05$ rad$]$, (b) 0.342 mm,
-1.36×10^3 mm/s, -2.62×10^6 mm/s²
11. (a) 20.0 cm, 0.167 s, 6.00 Hz, (b) 753 cm/s,
2.84×10^4 cm/s², (c) 750. cm/s   **13.** (a) 6.00 cm,
4.00 s, 1.57 rad/s, (b) 1.23 N/m,
(c) $x(t) = (6.00$ cm$)\cos[(\pi/2$ rad/s$)\,t - 1.23$ rad$]$,
(d) -3.14 cm/s, -14.0 cm/s²   **15.** $3k$
17. (a) 24.5 N/m, (b) 1.58 Hz   **19.** (a) 0.507 kg,
(b) 80.0 N/m   **21.** 7.14×10^3 N/m
23. $2\pi\sqrt{m/(k_1 + k_2)}$ **25.** 6.12 s   **27.** -1.97 m/s²
29. $\theta(t) = (0.0873$ rad$)\cos[(2.86$ rad/s$)t]$ **31.** 0.286 kg
33. 2.71 s   **35.** $2\pi\sqrt{(m_1 l_1^2 + m_2 l_2^2)/(m_2 g l_2 - m_1 g l_1)}$
37. 0.580 kg   **39.** 1.82 s
41. (a) $2\pi\sqrt{(l^2 + 12d^2)/(12gd)}$, (b) $(0.211)l$
43. (a) 0.183 m, (b) 3.16 m/s, (c) 0.363 s

45. (a) 205 N/m, (b) 2.31 J **47.** $A/\sqrt{2}$
49. (a) 17.3 rad/s, (b) 9.68 cm, (c) 11.5 cm
61. (a) $x(t) = (20.0 \text{ cm}) \cos[(8.00 \text{ rad/s})t - 2.40 \text{ rad}]$,
(b) 0.156 kg m/s, (c) 0.300 J **63.** 2.05 ms
65. (a) $(mg/k) \sin(\theta)$, (b) $2\pi\sqrt{m/k}$
67. (a) 1.12×10^{-6} N m/rad, (b) 1.00×10^{-11} kg m^2
69. (a) $-2T(x/l)$, (c) $(15/16)(TA^2/l)$ **71.** $\sqrt{k/(m + 3M)}$
75. 376 m/s

Chapter 15

1. $(3.00 \text{ cm})/\{4 + (2.00 \text{ cm}^{-2})[x - (150 \text{ cm/s})t]^2\}$
3. (a) 0.050 m, (b) 0.300 m, (c) 60.0 cm/s, (d) 2.00 Hz
5. 4.00 m **7.** (a) $A = 4.00$ cm, $f = 1.91$ Hz,
$\lambda = 2.00$ cm, (b) 1.53 cm, (c) 3.82 m/s along $-x$ direction,
(d) -44.4 cm/s **11.** (a) $y(x, t) = (6.00 \text{ cm}) \times$
$\sin[(3.49 \text{ m}^{-1})x + (83.8 \text{ rad/s})t]$, (b) 2.05 cm, -472 cm/s,
-1.44×10^4 cm/s^2 **13.** (a) 3.00 Hz, 5.24 m^{-1},
(b) $y(x, t) = (2.40 \text{ cm}) \sin[(5.236 \text{ m}^{-1})x - (18.85 \text{ rad/s})t]$,
(c) -0.500 cm, -44.3 cm/s **15.** (a) 0.178 m,
$y(x, t) = (1.50 \text{ cm}) \sin[(35.3 \text{ m}^{-1})x - (180\pi \text{ rad/s})t]$,
(b) 0.574 cm, 784 cm/s **17.** (a) 450 cm/s, (b) $y(x, t) =$
$(3.20 \text{ cm}) \sin[(0.209 \text{ cm}^{-1})x - (30\pi \text{ rad/s})t]$ **19.** 2.33 N
21. 2.88 N **23.** (a) $A = 2.00$ cm, $\lambda = 3.60$ m, $f =$
17.6 Hz, (b) $y(x, t) = (2.00 \text{ cm}) \sin[(1.75 \text{ m}^{-1})x -$
(110. rad/s)t] **25.** 1560 W **27.** 0.321 kg/m
29. (a) 15.0 m/s, (b) 18.8 Hz, (c) 1.03 W **31.** 8.00 m/s
33. 1.35 cm **35.** (a) $7\pi/12$ rad, (b) $y(x, t) =$
$(6.09 \text{ cm}) \sin[(12.0 \text{ m}^{-1})x - (180. \text{ rad/s})t + \pi/24]$,
(c) 4.67 cm **37.** 1.80 cm, 45.7 m/s, 255 Hz, 0.180 m
39. $y(x, t) = (6.00 \text{ cm}) \sin[(5\pi \text{ m}^{-1})x - (314 \text{ rad/s})t -$
1.23 rad] **41.** 104 N **43.** 25.0 cm **45.** 168 cm
47. $1.18f$ **61.** (a) 17.3 cm,
(b) $y(x, t) = (1.50 \text{ cm}) \sin[(36.3 \text{ m}^{-1})x + (126 \text{ rad/s})t]$,
(c) -1.27 cm, -101 cm/s, (d) 30.8 N **63.** 933 Hz
65. 1.60 m **67.** 16.3 Hz

Chapter 16

1. (a) 3.54×10^7 N/m^2, (b) 3.04 mm **3.** between
2.00 mm and 2.63 mm **7.** 308 N **9.** 8.27×10^4 N/m^2
11. copper **13.** (a) 7.16×10^{15} kg/m^3,
(b) 7.02×10^{10} N $= 7.89 \times 10^6$ tons!
15. 9.28×10^{17} N **17.** (a) 5.03×10^7 Pa, (b) larger
19. 7.93×10^3 m **25.** 7.22 mm
27. (a) 2.45×10^3 N, (b) 0.250 m^3 **29.** 1.73×10^3 N
31. aluminum **33.** $\rho_{\text{wood}} = 6.00 \times 10^2$ kg/m^3,
$\rho_{\text{fluid}} = 706$ kg/m^3 **35.** $V_{\text{cav}} = 2.59 \times 10^{-4}$ m^3
37. (a) 31.3 m/s, (b) 2.67×10^5 Pa **39.** (a) 6.25 cm/s,
(b) 1.00×10^5 Pa **41.** (a) 0.500 m/s, (b) 160. N
43. (a) 0.211 m, (b) 782 s **45.** 4.60×10^3 N
(additional lift is required) **49.** (a) 1.23×10^{-7} m^3,
(b) 4.93×10^{-3} m **51.** $\rho g l H^3/6$
53. (b) $T = 2\pi\sqrt{\rho_c L/\rho_\omega g}$ **55.** (b) 16.4 km
57. 1.60 m **59.** 1.40 m/s
61. $h = [\rho_a(r_2^2 - r_1^2)\omega^2]/(2\rho_f g)$

Chapter 17

1. 4.64×10^3 m/s, 2.10×10^3 m/s **5.** 1.72×10^{-2} m,
17.2 m **7.** 18.6 m **9.** 4.55×10^3 m/s **11.** $2.78 \times$
10^{-4} N/m^2 **13.** 0.105 Pa **15.** (a) $X(x, t) =$
$(1.20 \times 10^{-10} \text{ m}) \sin[(91.6 \text{ m}^{-1})x - (1000\pi \text{ rad/s})t]$,
(b) 1.16×10^{-10} m **17.** (a) 0.670 m,
(b) 2.81×10^{-6} m **19.** $X(x, t) =$
$(3.60 \times 10^{-7} \text{ m}) \cos[(36.6 \text{ m}^{-1})x - (4000\pi \text{ rad/s})t]$
21. 100. dB **23.** (a) 83.0 dB, (b) 93.0 dB, (c) 103 db,
(d) 113 dB **25.** 4.52×10^{-10} W
27. (a) 2.97×10^{-2} Pa, (b) 29.7 Pa
29. (a) 0.796 W/m^2, 4.77×10^{-6} m,
(b) 7.96×10^{-3} W/m^2, 4.77×10^{-7} m
31. 1.99×10^4 Hz **33.** 784 Hz, 1780 Hz
35. 0.434 m **37.** (a) 286 Hz, 572 Hz, 858 Hz,
(b) 143 Hz, 429 Hz, 715 Hz **39.** 338 m/s
41. 331 m/s **43.** (a) 70.4 Hz, 211 Hz, (b) 3.17 Hz
45. 518 Hz **47.** 363 Hz, 336 Hz **49.** 352 Hz
51. parked driver hears 7.4 beats/s, moving driver hears
7.2 beats/s **53.** (a) 617 m/s, (b) 33.7° **55.** 34.9°
59. 85.8 m **61.** 100 **63.** (a) 0.655 m, (b) 254 Hz
67. 36.0 m/s, 499 Hz **69.** 7.68 m/s
71. 2.66×10^8 m/s

Chapter 18

1. (a) -65.6°C, (b) 208 K **3.** (a) $T_{He} = -452.16$°F,
$T_O = -297.35$°F, $T_{Cu} = 1982$°F, $T_{Cs} = 83.3$°F,
(b) $T_{He} = 4.17$ K, $T_O = 90.17$ K, $T_{Cu} = 1356$ K,
$T_{Cs} = 301.65$ K **5.** (a) $T = \frac{6}{5}T_C + 30.00°$,
(b) $T = \frac{2}{3}T_F + 26/3$ **7.** -40.0°C **9.** three
11. (a) 4.09 cm Hg, (b) 6.59 cm Hg **13.** 12.01 m
15. 0.0385% **17.** 50.02 m **19.** 23.8 cm^2
21. 0.573 cm^2 **23.** 4.81 cm^3 **27.** 7.03 kcal
29. 0.235°C **31.** -27.9°F **33.** 8.32×10^7 J
35. 2.08×10^2 J/s **37.** 121.°C **39.** 48.8°C
41. 54.5°C **43.** 59.3°C **45.** 6.30 kcal
47. -71.4°C **49.** 283 K **59.** 243 lbf
61. 2.74 mm **63.** 1.06×10^{-4}°C^{-1} **67.** 2.35 kcal

Chapter 19

1. (a) 600. J, (b) 400. J **3.** 560. J **5.** (a) 1.22 J,
(b) 0.743 J **7.** 138 J **9.** 329 J **11.** 5.79×10^4 Pa
13. 85.5 cal **15.** 935 J **17.** (a) 400. cal,
(b) -1000. cal, (c) 150. cal, (d) -850 cal
19. (a) 2780 J, (b) 4130 J **21.** (a) 0.450 J, (b) 0.926 J
23. (b) 250. J, (c) 150. J, (d) 81.1 J **25.** (b) 210. J,
(c) -534 J, -382 J **27.** (a) 4.47×10^5 Pa, 121 K,
(b) -3.22×10^3 J
29.

	ΔU	Q	W
AB	295 J	491 J	196 J
BC	-435 J	-435 J	0
CA	140	0	-140 J
ABCA	0	56.0 J	56.0 J

31.

	ΔU	Q	W
AB	0	368 J	368 J
BC	−334 J	−572 J	−228 J
CA	334 J	334 J	0
ABCA	0	140 J	140 J

37.

	ΔU	Q	W
AB	68.0 J	296 J	228 J
BC	−152 J	−152 J	0
CA	84.0 J	0	−84.0 J
ABCA	0	144 J	144 J

(b) 144 J, (c) 68.0 J, (d) −152 J

39.

	ΔU	Q	W
AB	0	55.5 J	55.5 J
BC	−45.1 J	−75.1 J	−30.0 J
CA	45.1 J	45.1 J	0
ABCA	0	25.5 J	25.5 J

41.

	ΔU	Q	W
AB	−135 J	0	135 J
BC	0	2930 J	2930 J
CA	−2910 J	−4840 J	−1940 J
DA	3040 J	3040 J	0
ABCDA	0	1130 J	1130 J

43. (a) $W_{\text{isotherm}} = P_o V_o \ln(V_f/V_o)$, $W_{\text{isobar}} = P_o(V_f - V_o)$,
(b) isobaric   **45.** (a) 2.60×10^3 J, (b) 6.65×10^3 J,
(c) 3.65×10^3 J   **47.** $T_A = T_B = 546$ K, $T_C = 273$ K,
$T_D = 406$ K, $V_B = 4.48$ L, $V_D = 1.66$ L,
(c)

	ΔU	Q	W
AB	0	315 J	315 J
BC	−568 J	−568 J	0
CA	276 J	0	−276 J
DA	292 J	409 J	117 J
ABCDA	0	156 J	156 J

Chapter 20

1. (a) 0.300, (b) 4.46 kJ, (c) 0.080   **3.** (a) 35.0 kJ,
(b) 0.714, (c) 1100°C   **5.** no   **7.** (a) 3.66, (b) 273 J,
(c) 1.27×10^3 J   **9.** (a) 15.4, 163 J, (b) 2660 J
11. (a) 302 J/s, (b) 2.10 kW   **13.** 0.114   **15.** 0.253
17. (a) 3.53×10^3 K, (b) 32.4 kJ   **19.** (a) 0.541,
54.1 kJ, (b) $T_B = 812$ K, $P_B = 2.29 \times 10^6$ Pa, $T_C = 5620$ K,
$P_C = 1.58 \times 10^7$ Pa, $T_D = 2580$ K, $P_D = 1.04 \times 10^6$ Pa
23. 9.13 J/K   **25.** (a) −5.80 J/K, (b) no
27. (a) −30.8 J/K, (b) 35.1 J/K, (c) 4.29 J/K
29. 9.64 J/K   **31.** 767 J/K   **35.** $\Delta S_{AB} = 8.4$ J/K,
$\Delta S_{BC} = -8.4$ J/K, $\Delta S_{CA} = 0$ **37.** $\Delta S_{AB} = 1.15$ J/K,
$\Delta S_{BC} = -2.88$ J/K, $\Delta S_{CA} = 1.73$ J/K   **39.** (a) 2.18 kW,
(b) 1.83 kW   **45.** 0.31   **51.** 5.76 J/K
53. $\Delta S_{AB} = 5.72$ J/K, $\Delta S_{BC} = -4.45$ J/K,
$\Delta S_{CA} = -1.26$ J/K, $\Delta S_{DA} = 0$ **55.** (a) -1.12×10^6 J,
(b) 1.28×10^6 J, (c) 21.7 W

Chapter 21

1. 1.24 N/m²   **3.** (a) 5.42 m/s, (b) 6.01 m/s
5. (a) 1.93×10^3 m/s (b) 484 m/s   **7.** 3.74 kJ

9. 7.72×10^{-21} J, 719 m/s   **11.** 7.39×10^{-21} J
13. 56.7°C   **15.** $\Delta T_{H_2} = -72.5$ K, $\Delta T_A = -109$ K
17. 8.75×10^4 Pa

19.

METAL	c_p (cal/mole-K)	c_p (J/mole-K)	a
Theor	5.96	24.9	3.00
Ag	6.06	25.4	3.05
Al	5.82	24.4	2.93
Au	6.08	25.4	3.06
Cu	5.84	24.4	2.94
Fe	5.99	25.1	3.01
Hg	6.69	28.0	3.37
Pb	6.36	26.6	3.20

25. 2.25×10^{19} m   **31.** 1.01   **33.** 4.97×10^{-16} kg
35. (a) monatomic, (b) 13.7 L, (c) 1.21×10^3 J
37. 5.18×10^{-5} Pa

Chapter 22

1. One possibility:

	+	−	★
+	R	A	A
−	A	R	A
★	A	A	R

3. 9.63×10^4 C   **5.** (a) -0.382μC, (b) -0.270μC
7. $Q_+ = 0.386 \mu$C, $Q_- = -0.579 \mu$C
9. $8.38 \times 10^{-2} \mu$C   **11.** (a) 0.122μC,
(b) $3.80 \times 10^{-3} \mu$C   **13.** 4.80×10^{-12} m
15. (a) $(1.29 \times 10^{-2}$ N$)\hat{\mathbf{i}}$, (b) $-(2.64 \times 10^{-2}$ N$)\hat{\mathbf{i}}$
17. (a) $(Q/\pi\epsilon_o)[b/(a^2/2 + b^2)^{3/2}$, (b) $(10.2$ N$)\hat{\mathbf{j}}$
19. $-(1.08 \times 10^{-27}$ N$)\hat{\mathbf{i}} - (4.77 \times 10^{-27}$ N$)\hat{\mathbf{j}}$
21. 4.00×10^5 N/C   **23.** 9.36×10^{-14} kg
25. (a) 3.30×10^6 N/C at 221° anticlockwise from the
+x-axis, (b) 6.61 N at 40.9° anticlockwise from the
+x-axis   **27.** 21.8 m
29. (a) $(2q/4\pi\epsilon_o x^2)[1/(1 + d^2/x^2)^{3/2} - 1]$
31. $(1.97 \times 10^4$ N/C$)\hat{\mathbf{i}}$ **33.** $(\lambda/4\pi\epsilon_o y)(-\hat{\mathbf{i}} + \hat{\mathbf{j}})$
35. $-(\lambda/2\pi\epsilon_o a)\sin(\theta)\hat{\mathbf{j}}$ **37.** 1.66×10^7 N/C
39. $\{\lambda a/[2\epsilon_o(a^2 + x^2)^{3/2}]\}\sqrt{x^2/4 + a^2/\pi^2}$ at
$\arctan(-2a/\pi x)$ **41.** (a) 1.29 N m, (b) −0.400 J,
(c) +1.35 J, (d) 2.70 J   **51.** (a) 0,
(b) -2.26×10^5 N/C, (c) -3.39×10^5 N/C, (d) 0
55. 2.09×10^{-7} C or 3.47×10^{-7} C
57. (a) 3.07×10^{11} m/s², (b) 0.102 m, (c) 8.14×10^{-7} s
59. (a) 844 ns, (b) 3.38×10^{-2} m, (c) $v_y = 2.67 \times 10^5$ m/s, $v = 2.70 \times 10^5$ m/s at 81.6° from the +x-axis
61. (a) 0.488 m, (b) 1.20×10^6 m/s at 78.4°

Chapter 23

3. (a) 3.00 N m²/C, (b) 6.00 N m²/C   **5.** (a) 476 N/C,
(b) 37.4°   **7.** 4.07×10^5 N m²/C   **9.** 4.52×10^5 N m²/C   **11.** 18.1 N m²/C   **13.** (a) $(\sigma/\epsilon_o)\hat{\mathbf{j}}$, (b) 0
15. (a) 1.12×10^7 N/C, (b) 9.99×10^6 N/C   **17.** (a) 0,

(b) 2.50×10^4 N/C, (c) 5.39×10^4 N/C,
(d) 5.62×10^4 N/C   **19.** (a) 0, (b) $(\rho/3\epsilon_o)(r - R_1^3/r^2)$,
(c) $(\rho/3\epsilon_o)(R_2^3 - R_1^3)/r^2$ **21.** (a) 0,
(b) $(\rho/2\epsilon_o)(r - R_1^2/r)$,
(c) $(\rho/2\epsilon_o)(R_2^2 - R_1^2)/r$ **23.** 4.51×10^5 C
25. 5.97×10^{-6} C   **27.** $(2.88 \times 10^6$ N/C$)\hat{\mathbf{r}}$, 0,
$-(1.44 \times 10^6$ N/C$)\hat{\mathbf{r}}$, (b) -3.20×10^{-6} N/C
37. (a) $CR^5/5\epsilon_o r^2$, (b) $Cr^3/5\epsilon_o$
41. (a) $(\rho/2\epsilon_o)[r + R_2^2/(d - r)]$, (b) $\rho d/2\epsilon_o$,
(c) $(\rho/2\epsilon_o)[r - R_2^2/(d - r)]$, (d) $(\rho/2\epsilon_o)[R_1^2/r - R_2^2/(d - r)]$
43. (a) $\rho r/2\epsilon_o$, $(\rho/2\epsilon_o)R_1^2/r$, 0, $(\rho/2\epsilon_o)R_1^2/r$, (b) $-\pi R_1^2\rho$,
$+\pi R_1^2\rho$

Chapter 24

1. (a) 9.30×10^{-65} J, 1.15×10^{-28} J, (b) 5.07×10^{-68} J,
-1.15×10^{-28} J, (c) 5.57×10^{-41} J, 7.20×10^{-10} J
3. 0.322 m/s   **5.** (a) 8.06×10^{-2} J, (b) 0.111 J,
(c) 0.113 J   **7.** 0.487 J   **9.** 120. V
11. (a) 2.40×10^{-19} J, (b) 1.44×10^5 J   **13.** 157. V
15. 12 000 V   **17.** (a) -5000 V, (b) -5000 V,
(c) -5000 V,   **19.** $|\Delta V| = (\sigma R_1/\epsilon_o R_2)(R_2 - R_1)$
21. 1.80 m   **23.** 15.8 kV   **25.** (a) -1.09×10^4 V,
(b) -0.262 J   **27.** (a) 31.7 V, (b) 5.07×10^{-18} J
29. $(Q/4\pi\epsilon_o)\{2/\sqrt{x^2 + (a/2)^2} - 1/(\sqrt{b^2 - (a/2)^2} - x)\}$
31. $(\lambda/4\pi\epsilon_o)\ln[(a + \sqrt{y^2 + a^2})/y]$
33. $(\lambda/4\pi\epsilon_o)\ln[(c + a + \sqrt{y^2 + (c + a)^2})/(c + \sqrt{y^2 + c^2})]$
35. (a) 240. V/m $+ (120.$ V/m$^2)x - (15.0$ V/m$^3)x^2$,
(b) 9.66 m and -1.66 m
39. (a) $(Q/2\pi\epsilon_o)(1/y - 1/\sqrt{y^2 + a^2})$,
(b) $(Q/2\pi\epsilon_o)(1/y^2 - y/(y^2 + a^2)^{3/2})$
41. (a) $\lambda a/(4\pi\epsilon_o y\sqrt{y^2 + a^2})$, (b) the expression for V is
not a function of x **43.** (a) $\lambda a/(4\pi\epsilon_o y\sqrt{y^2 + (a/2)^2})$,
(b) the expression for V is not a function of x
45. (a) $\sigma_1 = 88.5$ nC/m^2, $\sigma_2 = 29.5$ nC/m^2,
(b) $E_1 = 10.0$ kV/m, $E_2 = 3.33$ kV/m   **47.** (B, H), (C, J),
(A, G), (D, F)   **59.** $-(A/2)/x^2$ **61.** (a) $V = (q/4\pi\epsilon_o) \times$
$[1/\sqrt{x^2 + (y - a/2)^2} - 1/\sqrt{x^2 + (y + a/2)^2}]$, (b) $E_x =$
$(qx/4\pi\epsilon_o)(1/[x^2 + (y - a/2)^2]^{3/2} - 1/[x^2 + (y + a/2)^2]^{3/2}$,
$E_y = (q/4\pi\epsilon_o)((y - a/2)/[x^2 + (y - a/2)^2]^{3/2} -$
$(y + a/2)/[x^2 + (y + a/2)^2]^{3/2})$
63. $(\sigma/2\epsilon_o)(\sqrt{R_2^2 + x^2} - \sqrt{R_1^2 + x^2})$ **65.** (a) 1.00 V,
(b) 16.2 V/m   **67.** (a) 600. kV, (b) 33.3 cm

Chapter 25

3. 1.24×10^{-10} Faradays   **5.** 25.0 nF   **7.** 113 km^2
9. 708 μF   **11.** $4R_1/7$ **15.** $C_{max} = 18.0$ μF,
$C_{min} = 4.00$ μF   **17.** (a) 0.960 μF, (b) 20.0 μF
19. $Q = 131$ μF, $\Delta V_{2\mu F} = 65.6$ V, $\Delta V_{4\mu F} = 32.7$ V,
$\Delta V_{6\mu F} = 21.8$ V   **21.** $Q_{12\mu F} = 1.32$ mC,
$Q_{18\mu F} = 1.98$ mC, $Q_{24\mu F} = 2.64$ mC   **23.** 24.0 μC,
12.0 V   **25.** 4.36 μF, $Q_1 = 1.05$ mC, $Q_2 = 0.785$ mC,
$Q_3 = 0.262$ mC   **27.** 1.53 pF   **29.** (a) 10.7 V, (b) 0
31. 443 V   **33.** (a) 480. V, (b) 2.88×10^{-7} J,

(c) 2.88×10^{-7} J   **35.** $(Q^2x)/(2A\epsilon_o)$
37. -3.82×10^{-11} J   **39.** 3.64   **41.** 75.0 kV
45. 6.37 pF   **47.** 1.38 μF   **49.** -0.254 μJ
57. (a) 2.78 pF, (b) $R_2 = 438$ m (too large to be practical)
59. five parallel rows of capacitors, each with ten
capacitors in series   **61.** (a) 5.68 V, (b) 0
63. 3.00 μF   **65.** 3.69 V, 14.8 μC
67. 2.81×10^{11} V/m   **69.** $4\pi\kappa\epsilon_o R_1 R_2/(R_2 - R_1)$
71. $2\pi\epsilon_o\kappa_1\kappa_2 L/[\kappa_2 \ln(R_2/R_1) + \kappa_1 \ln(R_3/R_2)]$
73. $E_1 = E_2 = 2.67 \times 10^4$ V/m, $Q_{1i} = 1.87 \times 10^{-10}$ C,
$Q_{2i} = 1.36 \times 10^{-10}$ C

Chapter 26

1. 5.63×10^{21} electrons   **3.** 53.3 μA   **5.** 0.270 mm/s
7. (a) 2.00 A, (b) 2.55×10^4 A/m^2 **9.** 407 V/m
11. 0.471 Ω **13.** (a) $R_{Al} = 0.671$ Ω, $R_W = 1.34$ Ω,
(b) 5.96 A, (c) $\Delta V_{Al} = 4.00$ V, $\Delta V_W = 8.00$ V,
(d) $E_{Al} = 0.080$ V/m, $E_W = 0.160$ V/m,
(e) 2.86×10^6 A/m^2 **15.** (a) 5.70×10^{-8} $\Omega \cdot$m,
(b) Tungsten, (c) 35.3 A   **17.** 11.3 cm
19. $\Delta T = 2.00°$C   **21.** $\Delta T = 12.5°$C
23. 4.05×10^{-3} °C^{-1} **25.** 70.7 V   **27.** 7.20×10^6 J
29. (a) 0.465 A, (b) 108 Ω **31.** 68.8 Ω
33. (a) 40.0 Ω, 75.0 mA, 300. V/m, 2.34×10^5 m/s,
(b) 20.0 Ω, 150. mA, 300. V/m, 2.34×10^5 m/s
35. 4.17×10^{13} atoms/cm^3 **39.** (a) 14.3 μA,
(b) 7.36 μA, (c) 2.71 μA   **41.** (a) 8.84 A/m^2,
(b) 5.31 A/m^2 **43.** (a) 4.00×10^{-4} Ω, (b) 0.162 V,
(c) 0.0340 V   **47.** $\pi\rho/[2L \ln(R_2/R_1)]$ **51.** 0.841 mm

Chapter 27

3. 9.70 V   **5.** 0.727 Ω **7.** 3.96 Ω, 0.209 Ω
9. 5.45 kΩ **11.** $3.33R$ **13.** 7.08 Ω, 1.57 A, 15.7 V
15. $\Delta V_1 = 7.20$ V, $\Delta V_2 = 2.40$ V, $\Delta V_3 = 1.60$ V,
$\Delta V_4 = 0.80$ V, $\Delta V_5 = 2.40$ V   **17.** 8.00 V
19. $I_1 = 4.00$ A, $I_2 = 0.50$ A, $I_3 = 2.00$ A
21. (a) $I_1 = 1.25$ mA, $I_2 = 1.75$ mA, $I_3 = 0.50$ mA,
(b) 13.0 V, point A **23.** $I_{1\Omega} = I_{5\Omega} = 0.800$ A,
$I_{6\Omega} = 1.27$ A, $I_{4\Omega} = 0.909$ A, $I_{2\Omega} = 2.18$ A
25. 1.09 mW, 3.83 mW, 2.61 V   **27.** (a) 0.003 Ω in
parallel with the galvanometer, (b) 0.030 Ω in parallel with
the galvanometer, (c) 40.0 kΩ in series with the
galvanometer, (d) 400. kΩ in series with the galvanometer
29. 1.17%   **31.** (a) 29.0 V, (b) 6.28 mA   **33.** 0.400
35. 1.80 V   **39.** 7.69 s   **41.** 11.5 μF
43. (a) 3.97 MΩ, (b) 350. V   **45.** (a) 7.59 μC,
(b) 2.79 μC   **47.** 1.39 ms   **49.** (a) 1.00 μF,
(b) 500. nC, (c) 250. ms   **57.** $r = 0.769$ Ω, $\mathcal{E} = 16.2$ V
59. 1.39 A, 2.94 V   **61.** $(5/6)R$ **63.** $I_{20\Omega} = 0.345$ A,
$I_{6\Omega} = I_{4\Omega} = 0.709$ A, $I_{8\Omega} = 0.354$ A   **65.** (a) 2.00 s,
(b) $\Delta V_C = 641$ V, $\Delta V_R = 903$ V, (c) 0.673 s
67. (a) 494 μC, (b) 166 ms   **69.** (a) 336 Ω,
(b) 50.0 kΩ, (c) 338 Ω

Chapter 28

1. 1.38×10^{-13} N (upward)   3. $(6.00 \times 10^4$ V/m$)\hat{\mathbf{j}}$
5. (a) 2.76×10^{-14} N, (b) $82.2°$ 7. 1.92×10^{-17} N
east   15. $r_{prot}/r_{deut} = \sqrt{2}/2$ 17. $Q^2R^2B^2/2M$
19. 1.65 T   21. $R_p = (0.709)R_H$ 23. $216\,\mu$V
25. 1.61 T   27. (a) 1.50 T, (b) 1.70 T   29. 18.8 N at
$53.1°$ below the z-axis in the yz-plane   31. $(-0.5\,$T$)\hat{\mathbf{i}}$
33. 1.60 T   35. (a) mg/LB, (b) $T = (mg/2)\cos(\theta)$,
$F_{net} = (mg/2)\sin(\theta)$ 37. 19.5 A
39. (a) 9.38×10^3 N $\cdot$ m, (b) 9.00 kJ
41. (a) $\boldsymbol{\mu} = (3.46 \times 10^{-4}$ A $\cdot$ m$^2)\hat{\mathbf{i}} - (2.00 \times 10^{-4}$ A $\cdot$ m$^2)\hat{\mathbf{j}}$,
(b) 3.20×10^{-4} N $\cdot$ m, (c) -5.54×10^{-4} J
51. (a) 4.03×10^{14} m/s^2, (b) 7.94 mm   53. 31.7 cm
55. (a) 0.351 T, (b) 5.35 MHz   57. $(0.0900$ N$)\hat{\mathbf{i}}$
59. (a) 4.77×10^{-8} N $\cdot$ m/rad, (b) 0.275 rad

Chapter 29

1. (a) 5.00×10^{-5} T, (b) 10.0 A   3. $\mu_o Q\omega/2\pi R$
5. (a) $(1.92 \times 10^{-11}$ T$)\hat{\mathbf{i}}$, (b) $-(1.34 \times 10^{-11}$ T$)\hat{\mathbf{k}}$,
(c) $(3.84 \times 10^{-11}$ T$)\hat{\mathbf{i}} - (2.56 \times 10^{-11}$ T$)\hat{\mathbf{k}}$
9. (a) $16\mu_o I/\sqrt{2}\pi L$, (b) $\pi\mu_o I/L$, (c) square loop
13. $\mu_o I/4\pi y$ 15. $(\mu_o NIR^2/2)\{1/[(x + R/2)^2 + R^2]^{3/2} + 1/[(x - R/2)^2 + R^2]^{3/2}\}$ 17. 3.00×10^{-5} N
19. (a) 4.00×10^{-6} T, (b) 1.60×10^{-5} T   21. $2.11\,\mu$T
at $18.4°$ below the negative y-axis   23. $3.77\,\mu$T
25. (a) $0.926\,\mu$T at $22.9°$ clockwise from $\mathbf{B_3}$, (b) $2.78\,\mu$N
at $22.8°$ CW from the direction of $\mathbf{B_1}$ 27. $24.0\,\mu$T
29. (a) $\mu_o I\sqrt{2}/2l$, (b) $\mu_o I\sqrt{2}/2l$
31. $(\mu_o I/\pi\omega)\arctan(\omega/2y)$ 33. (a) A/m^3, (b) $2\pi a^3 b/3$,
(c) $\mu_o br^2/3$, (d) $\mu_o a^3 b/3r$ 35. (a) $80.0\,\mu$T,
(b) 0.400 mm   37. (a) 0,
(b) $(\mu_o I/2\pi r)(r^2 - a^2)/(b^2 - a^2)$, (c) $\mu_o I/2\pi r$ 39. yes
43. 7.68 mT $\le B \le 9.60$ mT   45. 6.60×10^{-8} T $\cdot$ m^2
47. (a) 1.17×10^{-6} T $\cdot$ m^2, (b) 6.67×10^{-7} T $\cdot$ m^2, (c) 0
49. 1.75×10^{-6} T $\cdot$ m^2 55. (a) $(\mu_o I/4)(1/R_1 + 1/R_2)$,
(b) $(\pi I/2)(R_1^2 + R_2^2)$ out of the page   57. $\mu_o Jd/2$
59. $(\mu_o R^2 J/2y)(R^2 - 2y^2)/(4y^2 - R^2)$
63. $(7.41 \times 10^{-7}$ T $\cdot$ m/A$)(I/a)$
65. $(7.97 \times 10^{-7}$ T $\cdot$ m/A$)I/a$ 67. (a) 4.39×10^{-5} N,
(b) attractive, (c) 1.50×10^{-6} N, (d) 45.4 turns

Chapter 30

1. (a) 0.123 V, (b) anticlockwise   3. 0.347 T/s
5. $160.$ m   7. (a) 21.6 mA, (b) anticlockwise
9. $55.0\,\mu$A right to left   11. 0.439 A   13. 2.13 mV
15. 58.5 mV   17. 3.73 m/s   19. (a) 13.1 mW,
(b) clockwise, left, (c) 13.1 mW   21. 4310 rpm
23. (a) $3.17\,\mu$V, (b) $6.34\,\mu$A   27. (a) 45.7 mV,
(b) 103 mV   29. (a) 1.35×10^{11} V $\cdot$ s/m,
(b) 7.50×10^{-9} T   35. 2.05 mV
37. (a) 7.07×10^{-9} V, (b) clockwise   39. $63.3\,\mu$V,
anticlockwise

Chapter 31

1. $L = \mu_o n^2(Al)$ 3. (a) 2.66 mH, (b) 1.33 mV
5. 7.20×10^{-4} Wb   7. 1.98×10^{-5} Wb
13. 3.00 mH   15. 0.192 V   21. 5.19 H
23. (a) 0.262 A, (b) 0.667 A   25. (a) 0.106 A,
(b) 0.144 s   27. 16.8 mH   29. (a) 18.0 V,
(b) 27.0 mW   31. (a) 0.400 ms, (b) 2.00 A, (c) 1.38 A
33. 0.309 A   35. 5.16 ms   37. 34.9 A
39. (a) 0.211 nJ, (b) 1.58 nJ   41. 20.7 mJ
43. (a) 1.59 J/m^3, (b) $0.830\,\mu$J   45. (a) 1790 J,
(b) 1.33×10^{-5} gal   47. 5.07 pF   49. (a) -0.850 mC
(polarity reverse that of original charge), (b) 6.39 A,
(c) 3.61 mJ, (d) 61.2 mJ   51. (a) 22.6 mC, (b) 5.71 A,
(c) 0.255 J, (d) 22.6 V   53. (a) $1/(8f)$, (b) $0.707\,Q_o$
55. 2.89 kHz   59. $(\mu_o N_1 N_2 h/2\pi) \ln(R_2/R_1)$
61. (a) 1.43 mJ, 0.249 mJ, (b) 35.8 mJ
65. $(\mu_o/2\pi)N_1 N_2(A/R)$ 67. (a) 2.11 mA, (b) 1.08 mA
69. 6.40×10^{-6} J

Chapter 32

1. (a) $l = mvr$, (b) $\mu = e\pi r^2/T$ 5. $5e/(6m)$
7. (a) 5.00×10^3 A/m, (b) -49×10^{-3} A/m,
(c) -0.00098% 9. 9.16×10^4 A/m   11. 212
13. $[(\mu/\mu_o) - 1][NI/(2\pi R)]$ 15. 10 turns
17. (a) 10.6 mJ, (b) 0.636 W   19. (a) 8.48×10^{28}
atoms/m^3, (b) 1.73×10^6 A/m, (c) agrees with measured
value to two significant figures   21. $\mu_o(1 + \chi)n^2 Al$

Chapter 33

1. (a) 6.82 mA, (b) $i(0.64$ s$) = -4.0$ mA,
$i(1.44$ s$) = -4.0$ mA   3. (a) 92.7 rad/s, (b) 43.9 ms
5. (a) 265 kΩ, (b) $90.5\,\mu$A, (c) $0.240\,\mu$C   7. 30.2 V
9. (a) 9.95 kHz, (b) 22.8V, 4.73 mA (this result is very
sensitive to significant figures)   11. (a) 41.7 kΩ,
(b) 125 V, (c) $v_C(t) = (125$ V$) \sin[(1.20 \times 10^3$ rad/s$)t - \pi/2]$, $v_C(2.50$ s$) = 122$ V   13. (a) 1.21 kHz,
(b) $720.\,\mu$A   15. (a) $23.9\,\Omega$, (b) 59.7 V, (c) 1.27 kHz
17. (a) $60.0\,\Omega$, (b) 72.0 V, (c) -71.7 V   19. (a) i
leads e, (b) 2.63 kΩ, (c) 3.04 kΩ 21. (a) $446\,\Omega$,
(b) 13.8 kHz, -0.803 rad   23. (a) 2.27×10^{-9} F,
(b) 5.43 kΩ 25. (a) $444\,\Omega$, 0.664 rad, (b) 0.236 A,

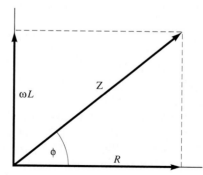

(c) −36.0 V   **27.** (a) 18.0 Ω, 0.588 rad, current leads applied emf, (b) −70.4 V, (c) $\Delta V_C = 238$ V, $\Delta V_L = -286$ V   **29.** 270. Hz, 360. Ω   **31.** (a) 575 Ω, 1.12 rad, current lags emf, (b) 212 V, 0.492 A, (c) $\Delta V_L = 162$ V, $\Delta V_C = -73.1$ V, $\Delta V_R = 123$ V, the total is 212 V = e(0.215 s)   **33.** (a) 475 Ω, −0.568 rad, (b) e(t) = (2380 V) sin[(3000 rad/s)t − 0.568 rad], (c)

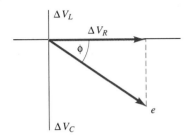

(d) 1480 V, $\Delta V_R = 1890$ V, $\Delta V_L = 488$ V, $\Delta V_C = -905$ V, the sum of the potential drops equals the applied emf
35. 163 V   **39.** 1.00 V   **41.** $V_{max}/\sqrt{3}$
43. 0.191 A, 66.8 V, 52.2 V   **45.** (a) 15.9 Ω, (b) 12.1 Ω
47. (a) 0.966, (b) 896 W, (c) 10.9 A   **49.** 12.8 W
51. (a) 0.894, (b) 324 W   **53.** 217 Hz
55. (a) 236 kHz, (b) 2.04 mA, (c) 22.5 mW   **57.** 424 V
59. 40 turns   **61.** (a) 600. kΩ, (b) 1.20 V, 800. μA
63. (a) 0.935, (b) 0.241   **65.** (a) 24.9 Ω, (b) 1.00
67. (a) 201 Ω, (b) 0.970   **73.** (a) 30.2 Ω, (b) 0.398 A, (c) 382 Hz   **75.** (a) 552 mA, (b) 2.07 V   **77.** (a) 1/19, (b) 6.3 Ω, (c) 2.19 kΩ   **79.** (a) 15.7 mA, (b) −1.53 rad, (c) current leads applied emf, (d) 79.6 Hz, 354 mA, 0.0°, current in phase with applied emf

Chapter 34

5. (a) −2.45 T/s, (b) in the negative $\hat{\mathbf{k}}$ direction (into page), (c) no, only the change is determined
7. 8.27×10^{-13} T   **9.** 3.00×10^4 V/m
11. (a) 2.42×10^{15} rad/s, 8.08×10^6 m^{-1}, (b) $E_y(x, t) = (540$ mV/m) $\times$ sin[$(8.08 \times 10^6$ m$^{-1})x - (2.42 \times 10^{15}$ rad/s)t], $B_z(x, t) = (1.80 \times 10^{-9}$ T) sin[$(8.08 \times 10^6$ m$^{-1})x - (2.42 \times 10^{15}$ rad/s)t]
13. (a) 6.02×10^7 Hz, just below the fm band, (b) 3.00×10^8 m/s, (c) $B_z(x, t) = (8.00 \times 10^{-10}$ T) $\times$ sin[$(1.26$ m$^{-1})x - (3.78 \times 10^6$ rad/s)t], parallel or antiparallel to z-axis depending upon x and t
15. (a) 1.50×10^{16} Hz, ultraviolet light, (b) 3.00×10^8 m/s, (c) wave propagates in the direction of decreasing x, (d) (24.0 V/m) cos[$(3.14 \times 10^8$ m$^{-1})x + (9.42 \times 10^{16}$ rad/s)t]   **17.** 6.00 m   **19.** 3.95 V/m, 1.32×10^{-8} T
21. 7.67×10^{-4} W/m^2 **23.** 4.67×10^{-6} N/m^2
25. (a) 37.5 kJ, (b) 3.75×10^{-4} N·s
27. 1.02×10^{-12} N   **29.** 7.64×10^{-9} N
31. 5.40×10^{14} Hz   **33.** 6.15×10^{14} Hz, 5.83×10^{14} Hz
35. 3.41 m, 2.78 m   **37.** 1.85 m
39. (a) 1.89×10^{-6} T, (b) 212 W/m^2, (c) 1.42×10^{-6} J/m^3 **43.** ≈60 mW

Chapter 35

5. (a) 2.25×10^8 m/s, (b) 475 nm, (c) 4.74×10^{14} Hz
7. (a) 1.67×10^{-12} s, (b) ray in glass makes more oscillations, (c) 790   **9.** 69.9°   **13.** 2.57 m
15. 2.42, diamond   **17.** 1.30   **19.** $\phi_1 = 29.2°$, $\phi_2 = 60.8°$, $\phi_3 = 15.8°$, $\theta_3 = 23.3°$

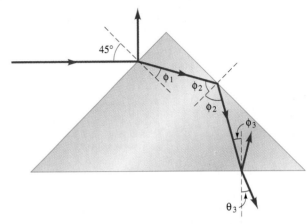

21. $\phi_1 = 70.1°$, $\phi_2 = 19.9°$, $\theta_2 = 14.8°$, $\phi_3 = 25.1°$, $\theta_3 = 18.6°$

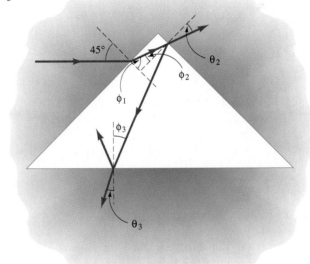

23. (a) 0.208, (b) 4.25 cm   **25.** (a) 176 collisions, (b) 17.0%   **27.** (a) 1.50, (b) 34.9°   **29.** (c) 4.37 minutes of arc   **31.** (a) 1/2, (b) 37.5%
35. (b) 1.71 μm   **37.** (a) no, (b) yes
43.

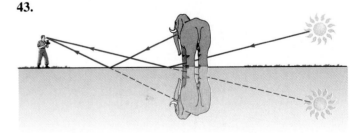

(a)

(Continued on next page)

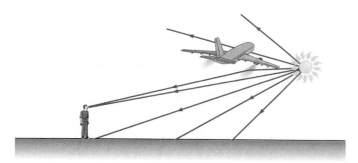

45. 1.30   **(b)**

Chapter 36

1. (a)

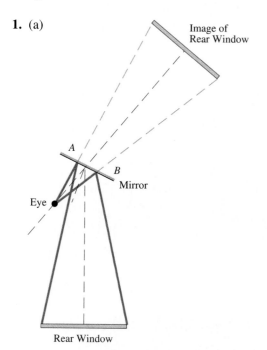

1. (b) $AB = 33$ cm   **3.** (a) 6, (b) 6
5. (a)

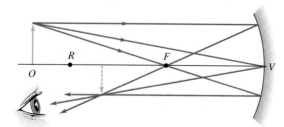

5. (b) 64.3 cm, (c) −0.714, real, inverted
7. (a)

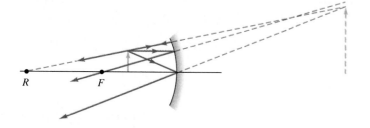

7. (b) −75.0 cm, (c) +3.00, virtual, erect
9. (a)

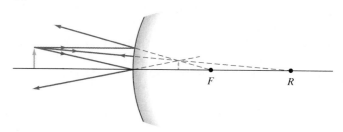

9. (b) −21.4 cm, (c) +0.429, virtual, erect

(a)

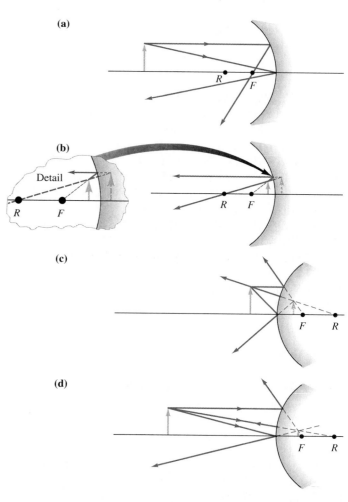

(b)

Detail

(c)

(d)

11. (a) 6.25 cm, −0.250, real, inverted, (b) −1.67 cm,
1.33, virtual, erect, (c) −2.50 cm, 0.50, virtual, erect,
(d) −4.00 cm, +0.200, virtual, erect   **13.** 21.3 cm
15. 13.3 cm   **17.** −3.98 cm   **19.** (a) −0.535 m,
(b) −0.546 m   **21.** (a) 16.4 cm, (b) −26.6 cm
23.
(a)

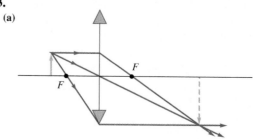

(b)

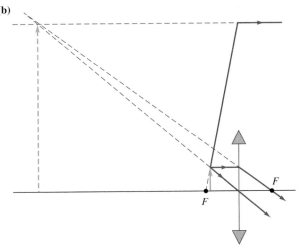

(c)

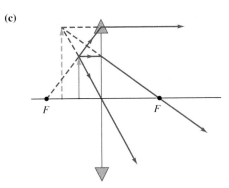

23. (a) 117 cm, −2.33, real, inverted, (b) −210 cm, 7.00, virtual, erect, (c) −26.2 cm, 1.75, virtual, erect
25.

(a)

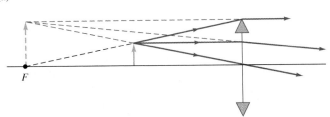

(b)

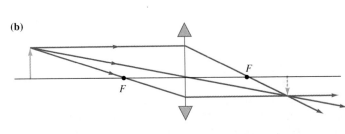

(c)

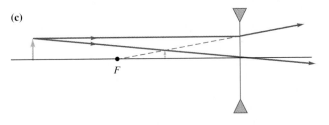

(d)

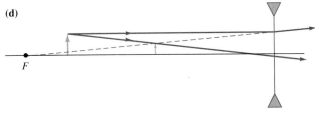

25. (a) −50.0 cm, 2.00, virtual, erect, (b) 16.7 cm, −0.667, real, inverted, (c) −9.38 cm, 0.375, virtual, erect, (d) −13.6 cm, 0.546, virtual, erect
27. (a) 15.0 cm, (b) converging, (c) 60.0 cm to right of lens   **29.** +6.67 cm, an incorrect answer because we can't tell the difference between converging and diverging lenses   **33.** $D = (f_2/f_1)d$
35.

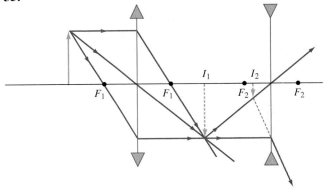

35. −7.29 cm, to the left of the second lens, $m = -0.250$, virtual, inverted, $h_2 = 0.250$ cm
37.

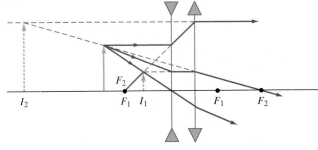

37. −41.2 cm, $h = 1.50$ cm, virtual, erect   **39.** 12.5 cm
41. (a) 12.5, (b) 1.25   **43.** (a) 2.06 m, (b) 20.8 minutes of arc, (c) 8.26 cm   **53.** −18.1 cm
55. 4.80 cm, 1.3   **57.** a factor of 0.080 times the actual height of the I   **59.** (b) 1.538
61.

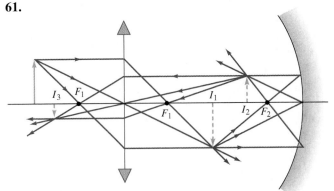

61. 20 cm, −1, real, inverted; +12 cm, −0.6, real, erect (that is, reinverted); 15.6 cm, −0.556, real, inverted; overall magnification 0.333

63.

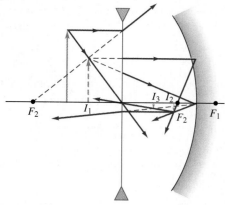

63. −9.375 cm, +0.625, virtual, erect; 6.026 cm, −0.205, real, inverted; −8.96 cm, 0.641; final image is 8.96 cm to the right of the lens, −0.0822, virtual, inverted

Chapter 37

1.

n	y (mm)
1	0.684
2	1.37
3	2.05

3. (a) 77.7 μm, (b) 2.62 cm   **5.** 633 nm   **7.** 1.52 mm
9. 0.274 μm   **11.** (a) 0.0996 μm, 0.299 μm, 0.498 μm,
(b) 0.167 μm, 0.333 μm, 0.500 μm   **13.** (a) dark,
(b) 2.32 μm   **15.** (a) convex move out, (b) from blue
to red   **17.** (a) 1.00×10^{-13} s, (b) 30.0 μm
21. 9.48 MHz   **25.** 400. nm
27.

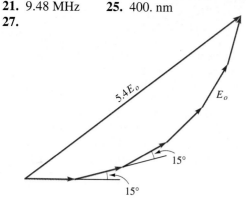

29. (a)

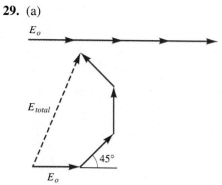

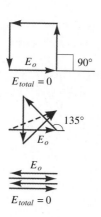

29. (b) $4.00E_o$, $2.61E_o$, $0.00E_o$, $1.08E_o$, $0.00E_o$,
(c) $1.00\, I_{max}$, $0.426\, I_{max}$, $0.00\, I_{max}$, $0.0732\, I_{max}$, $0.00\, I_{max}$

33. (b) 8.52 mm   **35.** 52.0 m   **37.** $n = \left(m' - \dfrac{1}{2}\right)/m$

39. 432 nm **41.** 585.6 nm, 390.4 nm **43.** (a) (7.00 V/m) ×
$\sin(\omega t + 21.8°)$, (b) (6.5 V/m) $\sin(\omega t)$ + (2.6 V/m) ×
$\cos(\omega t)$ **45.** (a) pattern is shifted by one-half fringe:
dark to light, light to dark, (b) 204 nm

Chapter 38

1. 68.3 μm   **3.** 61.5 cm   **5.** 4/3   **7.** 16.6 μm
9. 27.8 km   **11.** (a) 8.34 m, (b) 7.19 m
13. (a) 0.0826 μrad, (b) 1.21×10^7, (c) $R_A = 1.61R_P$
15. 0.893   **17.** 0.764, 0.306, 0.305×10^{-2}
19. (a) 5, (b) 0.0547   **21.** 0.600 nm   **23.** 594 nm
25. (a) 2.40 μm, (b) 2.12×10^4 **27.** 4.05 mm
29. 0.120 nm   **31.** 14.5°, 29.5°, 47.6°, 79.8°
33. (a) 0.165 nm, (b) 24.2°   **39.** (a) 84.6 μm,
(b) 84.7 μm (approximately equal)   **43.** 7.00 cm
45. 1.80 m, 0.360 m, the longer dimension of the diffraction
rectangle is perpendicular to the longer dimension of the
emitter   **47.** (b) only for $n = 0$, therefore $\alpha = \beta$
49. $2\theta = 23.8°$

Chapter 39

1. 16.7 min   **3.** 8.33 min   **5.** 12.6 fringes
7. $(-6.20 \times 10^{10}$ m, 5.00 m, 258 s)   **9.** 52 400 yr
11. 0.866c   **13.** 0.599 s   **15.** $r_{\parallel} = 6.54 \times 10^{10}$ m,
$r_{\perp} = 1.50 \times 10^9$ m   **17.** 0.968c = 2.90×10^8 m/s
19. 127 m   **23.** 1.87×10^{15} Hz   **25.** $-6.67 \times 10^{-3}\%$,
$+6.67 \times 10^{-3}\%$ **27.** 5.27 yr   **31.** 0.946c
33. 0.99995c   **35.** $0\hat{\mathbf{i}} + 0.5c\hat{\mathbf{j}}$ **37.** (a) 3.06, (b) 1.64
39. (b) 5.34×10^{-22} kg·m/s   **41.** 6.68×10^{-19} kg·m/s,
1250 MeV/c   **43.** (a) ball A: $-0.8m_o c\hat{\mathbf{j}}$, ball B:
$+0.8m_o c\hat{\mathbf{j}}$, total change = 0, (b) before $u'_{Ay} = 0.346c$, after
$u'_{Ay} = -0.346c$, before $u'_{Bx} = 0$, after $u'_{Bx} = 0$, before
$u'_{By} = -0.462c$, after $u'_{By} = +0.462c$, (c) change in

classical momentum $-0.692\,m_o c$ for A, $+0.924\,m_o c$ for B, $\Delta p_{\text{total}} \neq 0$, classical momentum is not conserved
47. (a) 1880 MeV, (b) 5.52×10^{-27} kg = 3.33 u, similar to ^{3}He and T   **49.** 851 MeV/c, 329 MeV   **51.** 0.979c, 2.46 MeV/c **53.** 0.880c, 149 MeV   **55.** (a) 39.2 MeV, (b) 5.61 MeV/nucleon   **57.** (a) 1.46×10^{-8} u = 2.42×10^{-35} kg, (b) 1.46×10^{-11} kg, (c) 0.0146 μg, no, a factor of 10 too small   **59.** (a) 1.022 MeV total, (b) conservation of momentum, (c) 1.24×10^{20} Hz, gamma rays   **63.** times given in years as (earth clock, space station X clock, *Messenger* clock), from earth & SSX point of view: departure (0, 0, 0), arrival (50, 50, 30), from *Messenger* point of view:
departure (0, 32, 0), arrival (18, 50, 30)   **65.** 0.9177c, 0.3972c, 1.00c, 23.4°   **67.** 2.53×10^{-3} nm
69. (a) $E = pc$, (b) 1.50 MeV/c, (c) larger by a factor of 1.000 013   **71.** toward observer, 0.0365c **73.** 2.83 nm

Chapter 40

1. 0.567   **3.** 96.6 nm   **5.** 1210 K = 934°C
7. 6.08 W   **9.** 243 K   **11.** (a) 4.04×10^{-7} eV,
(b) 1.96 eV, (c) 8060 eV   **13.** 2.39×10^{16} photons/s
15. 3.77×10^{16} photons/s/cm^2 **17.** (a) 0.451 μm,
(b) 0.28 eV   **19.** 2.16 eV   **21.** Cs, K, Na
23. (a) 0.104 nm, (b) 495 eV   **27.** (a) 117 nm,
(b) 80.5°   **29.** 30.2 eV, 0.0109c **33.** 47.4°
35. 164.2 nm, 121.6 nm, 108.6 nm   **37.** one possibility is $n_i = 26$, $n_f = 25$, $\lambda = 537$ nm   **39.** (a) 122.4 eV,
(b) 0.0176 nm   **41.** 0.284 pm, 2.53 keV
51. (a) 9.54 μm, (b) 956 W   **53.** (a) 488 nm,
(b) 0.24 eV (potassium)   **55.** 0.178 nm
59. (a) 4.79×10^{16} photons/s, (b) 1.96 eV/c = 1.04×10^{-27} kg·m/s, (c) 5.01×10^{-11} N   **61.** 867 nm (infrared)

Index

Page numbers in italics indicate figures; page numbers followed by t indicate tables.

ABCD parameters
matrix for, 1016t, 1018–1019, *1019*
for paraxial ray tracing, 1016, *1016,*
1016t, 1021
spreadsheet for, 1017–1018
example of, *1017,* 1017–1018, *1018*
Aberration of starlight, example of, 1098,
1098
Absolute pressure, 430
Absolute temperature, speed-distribution
function for molecules and, 576,
576
Absolute temperature scale
Kelvin scale and, 560
third law of thermodynamics and, 548–
549, *549*
Absolute value notation, 23
Absorption spectra, 1125
Acceleration
along curved path, components of, 94
average, 87
centripetal, 87
change of velocity vector and, 87
direction of, 87
computing from position, 62t, 66, 66t,
67
computing from velocity, 66
displacement and, in simple harmonic
motion, 361, 362
due to gravity, magnitude of, 57
graphical methods for computing, 68
instantaneous, 87
magnitude of, displacement and, 373
magnitudes of forces and, 110
normal and tangential components of,
91–93, *91–93*
positive or negative, Newton's second
law of motion and, 117
sign of, 55–57, *56*
of system, connected masses example
of, *297,* 297–299, *298*
tension and, 109
Acceleration curve, area under (change in
velocity), 60
Acceleration function, for simple
harmonic motion, 363, 366, 380
Acceleration graphs, velocity from, *51,*
51–52, *52*
Acceleration vector(s)
components of, 89, 90, *90*
velocity vectors and, 77–79

Acceptance cone, 963
determining half-angle of, *962,* 963–
964
Achromatic doublets, 1014, *1014*
Acoustic pressure, impulse-momentum
theorem and, 454
Acoustics, 477
Action, reaction and, in Newton's third
law of motion, 117–119, *119*
Actual frictional force, 130
"Adaptive step size" in Runge-Kutta
programs, 152
Additive inverse of vector, 24, *25*
Adiabat
in example of heat cycle problem, 531
of ideal gas, 525, 526
Adiabatic bulk modulus, 454, 473
Adiabatic process(es), 517, 532
in thermodynamics, 525–528, *526,*
526t
Air columns as source of sound, *465,*
466, 466–467
Air hockey (example of two-dimensional
collision), *240,* 240–241
Air resistance, correcting trajectory for,
145
example of, 145–146
Airplane cables (example of Young's
modulus), 420–421
Airstream method of holding Indycars on
track, 160–161
Alcohol in thermometer, 486
Alpern, Matthew, 1026
Alpha particles, 1101, 1127
motion in two dimensions and, *207,*
207–208
example of, *207,* 208
Alternating current (AC), 786
Alternating-current (AC) circuits, 890–
924
circuit elements in, *891,* 891–899
capacitor, *894,* 894–896, *895*
inductor, *896,* 896–899, *897,* 898t
resistor, *892,* 892–894, *894*
power and resonance in, 910–915,
911, 913
relationships in, 898t
RLC circuits, *900–902,* 900–907,
902t, 903t
root-mean-square potential and current
in, 907–910, *908, 909*

transformers and, *916,* 916–921, *919,*
919t
See also Circuit(s); Direct-current
(DC) circuits
Aluminum atoms
in doping, conductivity of
semiconductors and, 731, *731*
valence electrons in, 731, *731*
Ammeter, 755, *755*
Ampere(s), 717, 737, 818
experimental determination of
magnitude of, 809, *809*
parallel wires, coulombs and, 808–
809, *809*
Ampère, André Marie, 717
Ampère's law, *810,* 810–816, 819, 842,
874
applied to infinite current sheet, *811,*
811–812, *812*
electromagnetic waves and, 933, *933,*
935, 935
Maxwell's equations and, 841, 842
modification of, displacement current
and, *839,* 839–840
solenoid and, *806,* 812–815, *813*
toroid and, *806, 815,* 815–816
Ampèrian current (magnetization
current), 874, 883
calculating, 874–875
of transformer, 916
Amplitude, 408
of composite sound wave, 467
displacement and, 380
of light wave, energy and, 1117
periodic spatial variation in standing
waves, 400
of simple harmonic motion, 357
variation in, with position, *399,* 399–
400
of a wave, 388
Amplitude grating, 1060
Amplitude-modulated waves, *399,* 399–400
Amplitude of motion in simple harmonic
motion, 366
Analog meter, *753*
Anchor toss (example of relative
velocities and conservation of
momentum), 231, *231*
Angle(s)
between force and displacement in
work, 169

Angle(s) (*Cont.*)
  resolving forces into components and, 115, *115*, 116
Angle of incidence, *961*, 961–962
Angular acceleration, 278
  displacement and, in simple harmonic motion, 368
  distinguished from linear acceleration, 281
  electric motor example of, 283
  in force analysis of rotation and translation, 319
  moment of inertia and, for given torque, 297
  of torsion pendulum, 369
Angular acceleration function, in example of angular velocity, 280
Angular acceleration vector, 341–344, *342, 343*
Angular displacement, 277
Angular displacement function in example of angular velocity, 280
Angular frequency, 358, 374, 380, 408
  expressions for, in common systems, 380t
  for position equation of sinusoidal wave train, 393
  of simple harmonic motion, 358
Angular magnification, 1009, 1010, 1021
  of magnifying glass, *1010*, 1010–1011
  overall, of compound microscope, *1011*, 1012
Angular momentum
  conservation of, 300–303
  external torques and, 348
  magnetic dipole moment and, 871–872
  merry-go-round example of, *301*, 301–302
  for rotation about fixed axis, 299–303
Angular momentum vector, *344*, 344–349
  conservation of vector angular momentum and, 346
  Newton's second law and, *345*, 345–346
  precession of earth's axis of rotation and, 348–349, *349*
  precession of gyroscope and, *346*, 346–348, *347*
Angular position, 277
Angular position function in example of angular velocity, 280
Angular separation of wavelengths, 1061
Angular velocity, 86, 278
  distinguished from linear velocity, 281
  gyroscope example of, 279–281
  infinitesimal rotation in, 342–343
  of particle in uniform magnetic field, 780
  in simple harmonic motion, 362, 368
  tangential velocity and, for circular motion, 352

units for, 278
vector definition of, circular motion and, 343
Angular velocity vector, 341–344, *342, 343*
Angular width of spectral line, 1061
Anomalous dispersion, 965, *965*
Anticlockwise direction
  of current in diamagnetism, 877, *877*
  in torque, 293
Antiferromagnetic phase, 870
Antineutrino, 1090
Antinode(s), 401, 409
Antinode boundary condition, 466
Antiparallel alignment in ferromagnetism, 878–879
Antiparallel direction of intrinsic moments of electrons, 868
Antiparallel force components, 293, *293*
Antiparallel geometry, 88
Apparent depth, *997*, 997–999, *998*
Apparent weight of astronauts, free fall and, 141–142
Applied electric field
  effect on drift velocity, 722
  ideal metallic conductor in, 633, *633*
Applied force, distortion and, 418
Applied potential difference for ideal transformer, 917
Applied stress, strain and, proportional relationship of, 418
Arc length, 277
  floppy disk drive example of, 278, *278*
Archimedes' principle, *427*, 427–428, 440, 498–499
  floating cylinder example of, 428, *428*
Area(s), computing, 61–64, *62*, 62t, *64, 65*
Area charge density, in computing electric field from disk charge, 607
Area expansion, 490t, 491, *491*
  of isotropic substance, 508
Area mass density function, 255
Area under a curve, 60
  karate hammerfist strike example of, 62, *62*, 62t
  parabola in computing, *62*, 63
  trapezoids in computing, 62, *62*
Argon (example of root-mean-square speed of ideal-gas molecules), 569–570
Armenti, Angelo, Jr., 157
Arsenic ions, in doping, conductivity of semiconductors and, 731, *731*
''Artificial gravity,'' 136
Aspherical lenses, 1014
Astigmatism, 1014, *1015*
Astronaut's pencil (example of recoil), 228, *228*
Astronomical telescopes, 1012, *1012*
Astronomy, Doppler effect in, 471
Asymmetrical rotor (example of vector relations), 352, *352*

Asymmetry of speed distribution curve in molecular motion, 577, *577*
Atmosphere, 422
Atmospheric pressure, local, 430
Atom(s), 586
  Bohr model of, 1128–1132, *1129–1132*
  Rutherford-Bohr model of, 1127–1132
  Rutherford scattering and, *1127*, 1127–1128, *1128*
  spectra of, *1125*, 1125–1127
Atomic clocks (cesium clocks), 7, 7–8
Atomic particles, 587
  charge and mass of, 587t
Atomic physics, perfectly elastic collisions and, 240
Atomizers, as applications of Bernoulli's equation, 438, *438*
Atwood machine (example of Newton's second law of motion), 116–117, *117*
Aurora borealis (aurora australis), 782
Auto battery (example of electromotive force), 744–745
Average acceleration, 43–44, 68, 87, 93
  angular, in example of angular velocity, 280
  instantaneous acceleration and, 48–49
  from velocity-versus-time graph, 48–49
Average current, 717, 737
Average electron velocity, 737
Average magnetism, 879, *879*
Average power, 178, 924, 939, 1035
  in AC circuits, 910, 911
  dissipated, 912, *913*, 915
  over many cycles, 908
Average pressure, 422
Average speed, 68, 93
  average velocity and, 42–43, *43*
Average translational kinetic energy
  of gas molecule, 578
  of ideal-gas molecules, 569
  macroscopic variable pressure and, 568
Average velocity, 68, 93
  average speed and, 42–43, *43*
  of gas in disturbed region, sound production and, 453
  graphical interpretation of, 46–47, *47*
Average wave number, 399
Avogadro's number, 9, 15, 500
Axis(es), 19, *19*
  of spherical mirror, 986, *987*

''BAC-CAB'' rule to simplify double cross product, 350
Bad bunt (example of projectile motion), *83*, 83–85
Balance, Wheatstone bridge and, 756
Balance point, center of mass as, 253
Balanced molecule (example of center of mass), 253, *253*

Balanced rod (example of center of mass), 255–256

Ballet dancers, center of mass and, 259, *259*

Ballistic pendulum (example of conservation of momentum and mechanical energy), 250–252, *251*

Balloon
  example of angle coordinates, *20*, 20–21
  example of electric field, 634–635, *635*

Balmer, Johann, 1126

Balmer series of spectra, 1126, 1131, 1132, 1134
  long and short example of, 1126–1127

Band gaps, 733

Band model of conductivity, 723t, *731–734*, 732–735

Band structure of semiconductors, 734, *734*

Bar, 422

Bardeen, John, 735

Bartlett, Albert A., 446–449, 1025

Basic units, 5

Battery, 651
  effects on capacitance, 700–701
  effects on energy stored by capacitor, 701
  example of change in potential energy, *651*, 651–652
  example of electric field strength, 651, *651*

Battery-powered horn (example of Doppler shift), 471

Beam splitter in Michelson interferometer, *1080*, 1080–1081

Beat(s), 400, 409
  resonance and, as source of sound, 467, *467*

Beat frequency(ies), 400, 1047
  example of, 1037–1038, *1038*
  uses of, 1037

Bell, Alexander Graham, 461

Bernoulli effect, 160, 161

Bernoulli's equation, 440
  in fluid dynamics, *434*, 434–438
  garden hose example of, 435–436, *436*
  Pitot tube example of, *436*, 436–437
  qualitative applications of, 438, *438*
  Torricelli's law and, 437–438

Betatron condition, 838

Betatron (example of induced emf), *837*, 837–838

Bevis, John, 679

Biaxial birefringent crystals, 972

Bicycle wheel (example of rotational impulse-angular momentum theorem), 319t, *325*, 325–326

Billiard ball (example of impulse of force), 221, *221*

Binding energy, 1104
  calculation of, 1102
    example of, 1103
  mass and, 1101–1103
  per nucleon, 1103

Biomechanical loading of human body, 335–338, *336–338*

Biomechanics, 335

Biot, Jean Baptiste, 801

Biot-Savart law, *801*, 801–808, *803*, *804*, 810, 818, 849

Birefringence, 972–973, *973*, 973t, *974*, 976
  polarization and, 972–973, *973*, 973t, *974*

Birefringent crystals, 972–973, *973*
  refractive indices for, 973t

Birefringent materials, 976

Bismuth (example of permeability and magnetic induction), 882–883

Blackbody, 1110–1111, *1111*, 1133
  sun temperature example of, 1112

Blackbody radiation, *1110–1113*, 1110–1114, 1125

Bloch, Felix, 931

"Blue sky", polarization by scattering and, 968

Bohr, Niels, 871, 1128, 1134

Bohr magneton, 871, 872, 883, 930

Bohr model of hydrogen atom, 1128–1132, *1129–1132*
  assumptions of, 1129
  shortcomings of, 1132

Bohr radius, 1130, 1134

Boltzmann factor, 931

Boltzmann's constant, 569, 878, 930

Bound charges, 697

Boundary conditions, 466
  in example of computing electric potential, 654
  for standing waves, 402, *402*

Boyle, Robert, 504

Brackett, F., 1126

Brackett series of spectra, 1126

Bradley, James, 1098

Bragg, Sir William Henry, 1067

Bragg, Sir William Lawrence, 1067, 1070

Bragg condition, 1071

Bragg diffraction, X-ray diffraction and, 1065–1070, *1066*, *1067*

Bragg planes, 1067
  example of, *1069*, 1069–1070

Bragg reflections (example of Bragg's law), *1068*, *1068*

Bragg scattering
  in Davisson-Germer experiment, 1122
  in example of Bragg planes, *1069*, 1069–1070

Bragg's law, 1067
  Bragg reflections example of, 1068, *1068*

Brahe, Tycho, 139

"Breaker points," 851

Breaking point, 420, *420*

Brewster, Sir David, 969

Brewster's angle (polarizing angle), 969, 976
  glare example of, *968*, 969

Brewster's law, of polarization by reflection, *968*, 969

Bridge method of measuring resistance, 756, *756*

British Engineering system, 10, 42, 106, 120
  unit of heat in (BTU), 494, 547
  units of torque in, 293

British thermal unit (BTU), 494, 547

Brody, Howard, *183*, 326

Brotherton, M., 679

Brown, H. R., 1122

BTU (*see* British thermal unit)

Building insulation, transfer of heat and energy and, 497–498, 498t

Bulk modulus(i), 415, *419*, 419–421, *420*, 420t, 440
  adiabatic, 454
  isothermal, 454
  velocity and, 458

Buoyant force, 440
  Archimedes' principle of, *427*, 427–428

Calculators
  significant figures and, 13–15
  solving for phase constant with, 360
  solving simultaneous equations with, 752

Calorie, 493

Capacitance, *680*, 680–684, *681*, 709
  equivalent
    for capacitors in parallel arrangement, 687
    for capacitors in series arrangement, 789
  increased by dielectric materials, 695
  parallel-plate capacitor example of, *681*, 681–682
  problem solving strategy for computing, 680
  spherical capacitor example of, 682–683, *683*
  units of (coulombs per volt; farad), 680

Capacitive reactance, 896, 923

Capacitors
  in AC circuits, *894*, 894–896, *895*
    "reactionary problem" example of, *898*, 898–899
  charge on, 895, *895*
  combinations of, *685*, 685–691
    example of computing charge, *690*, 690–691

Capacitors (*Cont.*)
   in parallel, *685*, 685–687
   in series, *688*, 688–691
   cylindrical, example of, 684, *684*
   dielectrics and, 678–710
   energy storage in, 691–695
   microscopic model for, 910
   parallel-plate, example of, *681*, 681–682
   potential drop across, 895, *895*
   spherical, example of, 682–683, *683*, *694*, 694–695
   symbol for, distinguished from other symbols, 685
   total energy of, LC circuits and, 859, *859*
Car on banked curve (example of friction and circular motion), 132–133, *132–133*
Carbon dioxide (example of specific heat), 502
Carbon film resistors, 725
Carnal, O., 1137
Carnot, Lazare, 542
Carnot, Nicolas Léonard Sadi, 542
Carnot cycle, *542*, 542–545, 544t, 560
   efficiency of, 560
   ideal efficiency of, 544
   net work obtained from, 544, 544t
   volumes at corners of, 543
Carnot engine, 542
   "downhill" heat flow example of, 545
   efficiency of, 542
Carnot refrigerator, 546, *546*
   coefficient of performance of, 546
   "uphill" heat flow example of, 547–548
Carnot's theorem, 545, 551, 560
   proof of, *552*, 552–553
Cartesian components of vector, 26
Cartesian coordinate(s), polar coordinates and, 19–20
Cartesian coordinate system, 19, *19*, 35, 41
   for computing electric field from point charges, 598
   for kinematics of rotational motion, 277, *277*
Cavendish, Lord Henry, 138
Cavendish balance
   determining proportionality constant G by, *138*, 138–139
   "weighing" sun and, 139–140
Cavity radiation, 1110, *1110*
Celsius scale, *485*, 485–486, 508
   thermometer example of, 485–486
Center of gravity, 106, 259, *259*
Center of mass, 252–257, 268–269, 312
   balanced molecule example of, 253, *253*
   balanced rod example of, 255–256
   ballet dancers and, 259, *259*

for collection of point-objects, 252–254, *253*
energy and momentum and, 257–263
   gravitational potential energy and, 258
   momentum conservation in, 259–262, *260*, *261*
   Newton's second law and, 258–259, *259*
first moment to calculate, 286
in force analysis of rotation and translation, 319
homecoming float example of, *256*, 256–257
pitcher and glasses example of, *253*, 253–254
rotation about, 326
for solid objects
   by integration, 254–257, *255*, *256*
   symmetrical, 254, *254*
of system of particles, 269
torque about, 325
Center-of-mass (CM) frame, 260, 261
   one-dimensional collisions from, *260*, 260–261
   example of, *261*, 261–262
   velocities relative to, 262
   total kinetic energy of particles and, 262
   total momentum in, 262
Center-of-mass velocity, 317, *317*
Center of percussion, linear and rotational analysis of, *326*, 326–327, *327*
Centimeter-gram-second (cgs) system of measurement, 10
   units of energy (ergs) in, 493
Central bright maximum, width of, in single-slit diffraction, 1055
Centrifugal, 87
Centripetal acceleration, 87–88, 93, 94, 282
   cause of, 130
   in circle of constant radius, 343, *343*
   in circular motion, 135
   magnitude of, 87, *87*
Centripetal force, 130
   in Indianapolis 500, 161
Cerenkov radiation, 472, *472*
CERN (European Council for Nuclear Research), 1091
Cesium, example of photoelectric work function, 1118
Cesium clocks (atomic clocks), 7, 7–8
cgs system (centimeter-gram-second) of measurement, 10
   units of energy (ergs) in, 493
Chain rule (calculus), 171
Change in potential energy
   battery example of, *651*, 651–652
   work performed and, 612, 648

Changing magnetic flux
   electric field and, 827
   induced emf and, 828
Charge, 615
   computing, combined capacitors example of, *690*, 690–691
Charge carriers
   in Hall effect
      negative, *783*, 784
      positive, 783, *783*
   per volume, 737
Charge density(ies), 588
   in electrostatic configuration, 666, *666*
Charge distribution, computing electrical potential from, 655–658
   disk example of, *657*, 657–658
Charge per area, 616
Charge per length, 616
Charge per volume, 616
Charge separation, Hall effect and, 783–784
Charge-to-mass ratios of ionized atoms or molecules, 780
Checkout lines at stores, continuity equation and, 447, *447*
Child's hoop (example of physical pendulum), 369, *369*
Chromatic aberration, 1014, *1014*, 1021
Circuit(s)
   elements in, *891*, 891–899
      capacitor, *894*, 894–896, *895*
      inductor, *896*, 896–899, *897*, 898t
      resistor, *892*, 892–894, *894*
   junctions in, 750
   See also Alternating-current (AC) circuits; Direct-current (DC) circuits
Circuit breakers, 746
Circular arc, magnetic field for, 807
Circular motion, *85*, 85–90, *87*, 130–135, *131–134*
   angular velocity and tangential velocity for, 352
   example of, *131*, 131–132
      coefficient-of-friction model and, 132–133, *132–133*
   tangential and normal components, 89–90, *90*
   vector definition of angular velocity and, 343
Circularly polarized light, 974–975, *975*, 976
Cladding of optical fiber, 962, *962*
Classical mechanics, 1
Classical physics, 4
Clausius inequality, 553–556
   efficiencies of real engines and, 551–553, *552*
Clausius statement, 559
   of second law of thermodynamics, 540
Closed integral in electric flux for closed surface, 625

Closed path, Ampère's law applied to, *810,* 810–811

Closed surface
  electric flux for, 625
    example of, *625,* 625–626
  magnetic flux through, Gauss's law for
    magnetism and, 817, *817*

Closed systems, 262
  change in momentum of
    calculating, 265–266
    Newton's second law and, 266

CM-frame (*see* Center-of-mass frame)

Coaxial cables
  as cylindrical conductors, 684
  example of energy stored in magnetic
    field, *856,* 856–857

Coefficient of dynamic viscosity, 439

Coefficient of finesse, 1039

Coefficient of friction, typical values for, 128, 128t

Coefficient-of-friction model, 127–130, 128t, *129*
  circular motion and, example of, 132–133, *132–133*
  for kinetic friction, 152
  for static friction, 152

Coefficient of kinetic friction, 127, 152

Coefficient of linear expansion, 508

Coefficient of performance, 546, 560

Coefficient of restitution, 236, 242
  collisions classified by, 242t
  gliders example of, 237
  in one-dimensional collisions, 261
  racquetball example of, 236

Coefficient of static friction, 127, 152

Coefficient of thermal expansion
  for common materials, 490, 490t
  road median example of, *490,* 491

Coefficient of volume expansion, 490t, 491–492

Coercive field, *881,* 882

Coercivity, 882

Coherence, 1036–1037

Coherence length, 1037, 1038

Coherence time, 1037, 1046

Coherent waves, 1037, 1046

Collector for photoelectric effect, 1116, *1116*

Collision(s), 242
  of molecules in ideal-gas model, 565
    pressure and molecular motion in, *565,* 565–568, *567*
  in one dimension, 232–238
    from center-of-mass frame, *260,* 260–261
    partially elastic, 236t, 236–238, *238*
    perfectly elastic, 234–235, *235*
    perfectly inelastic, 233–234
  in two dimensions, 238–241, *239*
    air hockey example of, *240,* 240–241
    hockey players example of, 239, *239*

partially elastic, 241
perfectly elastic, 240–241
perfectly inelastic, 239, *239*

Collision approximation, 233, 242
  rotational equivalent of, in angular momentum, 301–302

Color-code scheme, for resistance of resistors, 726, *726,* 726t

Coma, 1014, *1015*

Common logarithm, 461

Commutative law, vectors and, 341, *342*

Compact disc digital audio, sound reproduction of, 480

Compact-disc player, optics of, polarization and, *975,* 975–976

Complete absorption
  of electromagnetic waves, 947
  of radiation pressure, 941

Complete constructive interference
  between two one-dimensional waves, 408, *408*

Complete cycle of heat engine, 540–541, *541*

Complete reflection
  of electromagnetic waves, 947
  of radiation pressure, 941

Complex circuits, matrix method for, 763–765, *765*

Component, 26
  of a torque, finding, 341

Component method of vector addition, 26–30, *28, 29*

Component vector, 26

Composite sound wave, amplitude of, 467

Composite wavefronts, principle of superposition in, 1043, *1043*

Composition resistors, 725

Compound bow
  example of force versus extension, *176,* 176–177, *177*
  example of numerical calculations of work, *176,* 176–177, *177*

Compound microscope, *1011,* 1011–1012
  angular magnification of, 1021

Compound slab
  example of conductivity, *496,* 496–497
  example of R-value, 498, 498t

Compressibility, 419

Compressible-fluid model, 428–430, *429*
  Granite Peak air pressure example of, 429–430

Compression, 166, 451, *451*
  along a tube, sound production and, 453

Compton, Arthur, 1120

Compton scattering, 1118–1121, *1119–1121,* 1133

Compton scattering formula, 1133

Compton shift at 90°, example of, 1121

Compton wavelength, 1120, *1120,* 1133
  of electron, 1120, *1120*

Computer(s), in calculating areas, 69

Computer programs
  BASIC, to compute displacement-versus-time, 61t, 63–64, *64*
  to calculate velocity and acceleration of golf club, 91–93, *91–93*
  to calculate velocity-versus-time record, 62t, 66t, *67*
  for complex circuit analysis, 750, 763–765, *765*
  for complex circuits, 763–765, *765*
  for current distribution in solenoid, 818
  for displacement-versus-time record, 61t, 63–64, *64*
  for electric field application, 707–709, *707–709*
  for example of force versus extension, *176,* 176–177, *177*
  example of superposition of one-dimensional waves, *407,* 407–408, *408*
  to implement trapezoidal rule, 62t, 64, *65*
  for paraxial ray tracing, 1017–1018
  for Runge-Kutta method
    one-dimensional motion and, 148–149, *149*
    two-dimensional motion and, 149, *150–151,* 151–152
  for simple harmonic motion
    equation of motion in, 378
    in large-amplitude oscillations, 380, *380*
    velocity and acceleration, 379, *379*
  for superposition of one-dimensional waves, *407,* 407–408, *408*
  for van der Waals gas, 507–508

Concave mirror(s)
  example of mirror equation, 993, *993*
  example of virtual image, 993–994, *994*
  images formed by, 988–992, 988–994, 991t

Concentric contraction, 338

Concert halls, *479,* 479–480

Condensers, 679

Condition, in spreadsheet example of simultaneous overrelaxation, 669–670, *670*

Conducting bar (example of computing total current), 721, *721*

Conducting bodies, locating, exponential voltage decay and, 887–888, *888*

Conducting rings, changing flux and, induced emf in, *829,* 829–830, *830*

Conducting rod (example of motional emf), 833–834, *834*

Conduction
  for transfer of heat and energy, *494,* 494–497, *495,* 496t

Conduction (*Cont.*)
  valence-bonding model of, 729–731, *730, 731*
Conduction band, 734, *734*
Conduction process, models for, complications in, 719
Conductivity, 722, 737
  band model of, 723t, *731–734, 732–*735
  compound slab example of, *496,* 496–497
Conductors, 587
  analog of, 735
  band structure of, 733
  current-carrying
    force on, *787, 787–789, 788*
    wire loop example of force on, *788,* 788–789
  electric field lines and, 634, *634*
  relation of current density and electric field in, *722,* 723
  surface of, equipotential, 663
  See also Semiconductor(s)
Connected mass(es)
  example of acceleration and tension, *297,* 297–299, *298*
  example of Newton's second law of motion with non-constant tension, *112,* 112–113
Connected-mass problems, nonzero rotational inertia and, 297
Conservation of angular momentum, 300–303
  law of, 353
Conservation of charge, Kirchhoff's first law and, 749
Conservation of energy, 186–211, 1119
  Bernoulli's equation and, 434
  conservative forces in, 188–202, *200, 201*
    potential energy and, 187–188
    two and three dimensions, 206–210
  first law of thermodynamics and, 520
  ideal transformer and, 918–919
  Kirchhoff's second law and, 749
  mechanical, 190–196, *193, 194*
  Pascal's principle and, 426
  potential energy in, 188–190
    for Newtonian gravity, 202–206, *203–205*
  systems with nonconservative forces, 196–199, *198*
Conservation-of-energy theorem, kinetic and potential energy and, 201, *201*
Conservation of fluid-mass, 433
Conservation of kinetic energy, "working equations" in, 234
Conservation of mass
  binding energy and, 1102
  equation of continuity and, 434
Conservation of matter, 446

Conservation of mechanical energy
  ballistic pendulum example of, 250–252, *251*
  in energy analysis of rotation and translation, *317,* 317–318
  in isolated systems, 195–196, 211
  problem-solving strategy for, 192
Conservation of mechanical energy theorem, 191
  egg-dropping example of, 193
Conservation of momentum, 226–232
  astronaut's pencil example of recoil and, 228, *228*
  ballistic pendulum example of, 250–252, *251*
  law of, 241–242
  for linear motion, 304t
  recoil and, 227–230, *228, 229*
  relative motion and, 230–232, *231*
  relativistic momentum and, *1099,* 1099–1100
  for rotational motion, 304t
  "working equations" in, 234
Conservation of motion, using relative velocities in, 264
Conservation of *vis viva*, 238
Conservative force(es), 188
  as function of position, 200, 210
  motion in three dimensions and, 208–210
  motion in two dimensions and, *207,* 207–208
  negative derivative of potential energy, 199–202, *200, 201*
    equilibrium and, *201,* 201–202, *202*
  in one dimension, 188
  partial derivative test for, 208–210
  in Piñata example of conservation of mechanical energy, 195
  potential energy and, 662, 662t
    in conservation of energy, 187–188
  potential energy function and, 206, 210, 643
  three-derivative test and, 210
  total mechanical energy and, 191–192
  in two and three dimensions, 206–210
Constant acceleration equations, 93
  applicability in position and velocity, 59
Constant-acceleration model, 68
  in correcting for air resistance, 145–146
  of motion, *52,* 52–59, *54*
    projectile, *79,* 79–80
    rotational, 282–283
Constant angular velocity, with uniform circular motion, 374, *374*
Constant direction, in work
  constant force and, *164,* 164–165, *165*
  variable force and, 165–169, *166–169*
Constant-energy curves, isotherms of ideal gas as, 523

Constant force, constant direction and, in work, *164,* 164–165, *165*
Constant phase relationship of points on plane wavefront, 1042, *1042*
Constant pressure
  in isobaric processes, 523–524
  in volume expansions, 492
Constant radius in uniform circular motion, 85
Constant speed in uniform circular motion, 85
Constant temperature in isothermal processes, 524, *524*
Constant velocity, 49–50, 395
  forces and, 103
    frictional, 129
Constant-volume gas thermometer, 486–488, *487*
Constant volume in isochoric processes, 523
Constantly varying mass, conveyer belt example of, *267,* 267–268
Construction rays
  to locate image formed by spherical mirror, *988,* 988–989, *989*
  to plot image locations, 1002
Constructive interference, *398,* 398–399, 1071
  Bragg's law for, 1067
  two-source, 1029, *1029,* 1030
Contact forces, 109, *109*
Continuity equation in everyday life, 446–449
Continuous charge distributions
  computing electric potential from, *656,* 656–658
  electric field and, *601,* 601–608, 616
  electric potential function for, 672
  electrical field and, 588–590, *589, 590*
Continuous function, electric potential as, 647
Convection, radiation and, for transfer of heat and energy, 498–499
Converging lens(es)
  real object, real image and, example of, *1002,* 1002–1003
  real object, virtual image and, example of, 1003, *1003*
  virtual object for second lens and, example of, *1005,* 1005–1006
Conversion(s), between systems of units, 11
Conversion factors, 11
Convex mirrors, images formed by, *994,* 994–995, *995*
Conveyer belt (example of system with varying mass), *267,* 267–268
Cooper, Leon, 735
Cooper pairs, 736, *736*
Coordinate(s), 19, *19*
Coordinate systems, *19,* 19–21, *20*
  in finding torque, 341

for force analysis of rotation and translation, 320
in problem-solving strategy for conservation of mechanical energy, 192
relative velocities and, 231
for two-mass system in simple harmonic motion, 370
using Newton's second law of motion, 110
Atwood machine example if, 116
Coordinate transformations, 1086
Copper, energy levels of, 733, *733*
Copper cube (example of volume expansions), 492–493
Copper wire (example of drift velocity), 721
Core-cladding boundary, 962, *962*, 963
Core-loss current of transformer, 916
Corridors and stairs, flow of persons through, continuity equation and, 447–448
Cosmic rays, 782
Coulomb, 587, 646, 648, 776, 818
amperes, parallel wires, and, 808–809, *809*
defining by determination of magnitude of ampere, 809
Coulomb, Charles, 590
Coulomb attraction, in system of capacitors, 693
Coulomb force, magnitude of, 591
Coulomb force vector, sign convention for, 591
Coulomb per second (ampere), 737
Coulomb repulsion for a charge, 758, *758*
Coulomb scattering, 1128
Coulomb's constant, 590
Coulomb's law, *590*, 590–596, *591*, 615, 810
in Bohr model of hydrogen atom, 1127, 1129
electric field lines and, 610
electric potential and, 646
from Gauss's law, *627*, 627–633, *632*
hydrogen atom example of, 592, *592*
Critical angle, 976
for total internal reflection, 960, *960*
example of, *960*, 960–961
Critical temperature, 506–507
for superconducting materials, 735, 735t
Critically damped motion, *376*, 376–377
Critically damped system, RLC circuits and, 859
Cross product, 31–32, *32*, 34
in calculating torque, 341
double, ''BAC-CAB'' rule to simplify, 350
example of force on a charge, 778, *778*
for magnitude of magnetic force, 777

Cross-sectional area, pressure independent of, 425, *425*
Crummett, W. P., 404, 510
Crystal(s), dichroic polarizers made from, 971
Crystal lattice, in salt crystal, *1066*, 1066–1067
Crystal planes, Bragg's law and, *1067*, 1067, 1068
Cubic Gaussian surface, 632, *632*
Cue ball (example of speed after impact), 222
Curie constant, 878
Curie law, 878
Curie temperature, 870, 878
''Curious'' heat ratio fraction for arbitrary cycles, 553–554
Current, 923
properties of magnetic field and, 801, *801*
root-mean-square potential and, 907–910, *908*, *909*
Current balances, 809
Current-carrying conductor(s)
force on, *787*, 787–789, *788*
wire loop example of, *788*, 788–789
in uniform magnetic field, torque on, 789–793, *790*
Current change, induction coil example of, 851–852
Current decrease in time, 854, *854*
Current density, 737
electric current and, *717*, 717–722, *719*, *720*
through conductor in Hall effect, *784*, 784–785
Current density vector, 720, *720*
Current distribution
Ampère's law applied to, *810*, 810–811
in solenoid, 817
Current flow, direction of, 717, *717*
Current source, 898
example of potential drop across inductor, 899, *899*
example of RLC circuit, *903*, 903–905, *904*
Current-versus-time graph, for different time constants, 758, *758*
Curvature, charge density and, 665
Curve ball, as application of Bernoulli's equation, 438, *438*
Curved surfaces, refraction at, *996*, 996–997, *997*, 997t
Cutoff frequency, in photoelectric effect, 1117
Cyclic permutation, 34
Cyclic processes, 533
general, *549*, 549–554, *550*
reversible, analyzing, 549–550
in thermodynamics, 528–532, *529*
Cyclotron, operating principle of, 780, *780*

Cyclotron frequency, 780, 794
Cylinder functions, 1055–1056, *1056*
Cylindrical symmetry, 652, 806
in coaxial cable example, of energy stored in magnetic field, 856
computing electric field by negative derivative in, 658
of irradiance pattern in single-slit diffraction, 1055, *1056*

Dam (example of fluid pressure), 425, *425*
Damped harmonic motion, 862
in RLC circuits, 859, *859*
Damped harmonic oscillator, similarity of RLC circuits to, 859, *859*
Damped oscillations, *375*, 375–377, *376*
critically damped motion, *376*, 376–377
driven oscillations and, 375–378
overdamped motion, 376, *376*
underdamped motion, 375, *376*
Damping, 381
Damping out selected harmonics, 463
Darwin, C. G., 1094
DATs (Digital audio tape recorders), 480
Davisson, Clinton J., 1122, 1124, 1134
Davisson-Germer experiment, *1122*, 1122–1123, *1123*
dB (decibel), 461, 473
de Broglie, Louis, 1121, 1122
de Broglie hypothesis, 1121
de Broglie wavelength, 1121, 1133
golf ball example of, 1121–1122
Debye, Peter, 1115
Debye T-cubed law, 575
Debye temperature, 575
Debye's quantum theory, 1115
Decibel (dB), 461, 473
Decibel scale, *460*, 460–462, 461t
Decomposition into components, 80–81, 81t
Defective toaster (example of work-energy theorem), 172
Deformation(s), elastic moduli to describe, 415
Deformation waves in solids, longitudinal or transverse, 456t, 457
Degree of alignment of dipoles, magnetization and, 872
Degree of freedom, in equipartition-of-energy theorem, 570–571, *571*
Density, 5, 422t, 440
pressure and, *422*, 422t, 423
proportional, 428
Denver tire pressure (example of gauge pressure change), 430–431
Depth, variation of pressure with, *423*, 423–424
Descartes, René, 238, 953

Desk width (example of length contraction), 1090
Desktop toy (example of perfectly elastic collision), 238, *238*
Destructive interference, 399, 1029, *1029*
  magnesium fluoride example of, 1033
Determinant(s), 34
  example of, 34–35
Deuteron, 1102
Dialogues Concerning Two New Sciences (Galileo), 57, 1079
Diamagnetic phase, 884
Diamagnetism, 869, 876–877, *877*
  induced magnetic moments in, 876–877
  Meissner effect in, 877
Diatomic ideal gas(es), specific heats of, 573, 574t
Diatomic ideal gas models, equipartition-of-energy theorem for, 578–579, 579t
Diatomic molecules
  two-mass system as model of, 370
  vibrational energy levels of, 574t, 574–575
Dichroic polarizers, 971
"Die hard music" (example of resistance), 725
Dielectric(s), 587, 695–702, *696, 697,* 698t
  alternating layers of, polarizing by reflection and, 969–970, *970*
  behavior of, *703,* 703–707
    Gauss's law for displacement field, 706–707
    polarization and displacement field, *703,* 704–705, *705*
  Gauss's law for displacement field and, 706–707
    example of, 706–707
  in isolated capacitor
    effects on capacitance, 698–699
    effects on energy stored, 699–700
    effects on potential across capacitor, 699
  in parallel-plate capacitors
    example of, 701–702, *702*
    Gaussian surfaces and, 703, *703*
Dielectric constant, 698, 698t, 709
Dielectric film (soap film), 1032
Dielectric strength, 698, 698t
Diesel cycle, 550
Differential equations, 145
  for RLC circuits, 859
Differentiation with respect to time, in problems of varying mass, 268
Diffraction, 955, 957, 1052–1071, *1053*
  diffraction gratings in, 1060–1064
    free spectral range and, 1062–1064, *1063*
    resolving power of, 1060–1062, *1061*

finite slit width, double-slit interference patterns and, *1058,* 1058–1059, *1059*
  single-slit, *1053–1055,* 1053–1057
    diffraction-limited optics and, 1055–1056, *1056*
    resolution and, *1056,* 1056–1057
  X-ray, *1064,* 1064–1070, *1065*
    Bragg diffraction and, 1065–1070, *1066, 1067*
Diffraction envelope (diffraction halo), 1059
Diffraction grating(s), 1060–1064, 1071
  resolving power of, 1060–1062, *1061*
Diffraction halo (diffraction envelope), 1059
Diffraction-limited lens, 1070
Diffraction-limited optics, 1055–1056, *1056*
Diffraction-limited systems, 1056, *1056*
Diffraction of electrons, example of, 1123
Diffraction order, 1060
Diffraction patterns, experiment in, 1123–1124, *1124*
Diffuse reflection, 956
Digital audio tape recorders (DATs), 480
Digital data, 480
Digital meter, *753*
Digital voltmeters, 762–763
Dimension(s), 12, 15
  components of a vector and, 26
Dimensional analysis, 12
  example of, 12
  unit conversion and, 10–13
Dimensional consistency, 12
Dimensionality of wave propagation, 388
Dipole, equipotential surfaces for, 662, *663*
Dipole moment
  per volume, 705
  water molecule example of, 613–614
Dipotrique, La (Descartes), 953
Direct current (DC), 786
Direct-current (DC) circuits, 742–766
  combinations of resistors and, 745–749
    in parallel, *746,* 746–749
    in series, *745,* 745–746
  complex, matrix method for, 763–765, *765*
  digital voltmeters and, 762–763
  electromotive force and, *743,* 743–745
  measurements in, 753–757, *754*
    ammeters, 755, *755*
    potentiometer, *756,* 756–757
    voltmeters, *754,* 754–755
    Wheatstone bridge, 756, *756*
  multiple loop, Kirchhoff's laws and, 749–753, *750*
  RC circuits, *757–759,* 757–762
    differentiating and integrating, *761,* 761–762, *762*

See also Alternating-current (AC) circuits; Circuit(s)
Direct sound in concert halls, 479
Directed line segment, 21, *21*
Direction
  associated with angular momentum, 344, *344*
  as critical part of motion, 220
  differences in, in wave interference, 400–402, *401, 402*
  rules of, 21, 22
  of waves, 388
Discharging RC-circuits, 759, *759*
Discontinuities in electric potential function, 647
Disk charge, electric field and, example of, *607,* 607–608
Disorder, entropy and, 559
Dispersion, 976
  of grating, 1061
  in large-amplitude waves, 390
  refraction and, *964,* 964–965, *965*
Displacement, 21, 41–42, *42,* 68, 93
  acceleration and, in simple harmonic motion, 361, 362
  amplitude and, 380
  area under velocity curve and, 68
  force and, in work, 169
  magnitude of acceleration and, 373
  on velocity-versus-time graph, *50,* 50–51, *51*
  vertical component of, change in gravitational potential energy and, 190
Displacement amplitude, pressure amplitude and, in harmonic waves, 458
Displacement current, 840, 842, 874
  Maxwell's equations and, *839,* 839–842, *840*
Displacement field, 710
  free surface charge density and, 705
  Gauss's law for, 706–707
Displacement node of sound wave, 466, *466*
Displacement vector, 22, *22*
Distance, properties of magnetic field and, 801, *801*
Distortion, 1014
  applied force and, 418
Distortion waves
  propagating in elastic media, 451
  propagation speeds for, 455, 456t
Distribution of energy in conservative system, 192
Distribution of molecular speeds, 576–578
  Maxwell speed-distribution function and, *576,* 576–577, *577*
  mean free path and, *577,* 577–578
Distributive law, vector cross product and, 32

Diverging lens, real object, virtual image and, example of, 1004, *1004*

Diving board (example of simple harmonic motion), 360–361, *361*

"Doing time in prism" (example of index of refraction), *964*, 966

Domains
  in ferromagnetism, *879*, 879–880, *880*
    observing microscopically, 880
  in ferromagnets, 870
  in magnetism, 884

Door, example of rotational inertia, *286*, 286–287
    with perpendicular axis, *287*, 287–288, 289t, 290

Doping
  conductivity of semiconductors and, 731, *731*
  free electron density and, *731*, 734–735

Doppler, Christian Johann, 468

Doppler effect, 455t, *468*, 468–472, *469*, 470t
  shock waves and, *468*, *469*, 472, *472*

Doppler equation, relativistic, 471

Doppler shift, 473
  battery-powered horn example of, 471
  relativistic, 1104
  for sound, 1092

Doppler-shifted frequency(ies), 470t, 1096
  computing, *468*, 468–469

Dot product, 31, *31*, 33
  of constant force and direction in work, 164

Double-slit interference patterns, effect of finite slit width on, *1058*, 1058–1059, *1059*

Doublets, 1014, *1014*

Doughnut models of mass of sun and moon, 349, *349*

"Downhill" heat flow (example of Carnot engine), 545

Downward acceleration, horizontal motion and, 79, *79*

Drag coefficient, 142
  as function of Reynolds number, 143, *143*

Drag force, 142, 152–153

Drift velocity
  copper wire example of, 721
  of electrons, 719
  magnitude of, thermal velocity magnitude and, 722

Driven oscillations, *377*, 377–378, *378*
  damped oscillations and, 375–378
  in simple harmonic motion, 381

Driver safety at Indianapolis 500, 162

"Dropping a perpendicular," 26

Drumheller, John E., 929–931

Dual slope digital voltmeters, 762

Dynamic(s), 103

Dynamic quantities
  linear, 304t
  rotational, 304t

Dynamic viscosity(ies), *438*, 438–440, *439*, 440t
  of common fluids, 439, 440t

Earth model, torque generated on, 349

Earth's axis of rotation, precession of, 348–349, *349*

Eccentric contraction, 338

Eddy current(s), 842
  as braking mechanisms, 839
  effects of, 838
  in induced electric fields, *838*, 838–839

Eddy current loop, 887

Edgerton, Harold, 91

Effective load in ideal transformer, example of, *920*, 920–921

Efficiency
  of Carnot cycle, 560
  of heat engine, 541

Egg-dropping (example of conservation of mechanical energy), 193

Egg-toss, momentum statement of Newton's second law and, 223–224

Einstein, Albert, 119, 202, 1083–1084, 1087, 1115, 1120, 1133

Einstein temperature, 1115

Einstein's general theory of relativity, 136

Einstein's photoelectric equation, 1117–1118, 1133

Einstein's postulates, 1084–1085

Ejection of photoelectrons in photoelectric effect, 1117

Elastic collisions of molecules in ideal-gas model, *565*, 565–568, *567*

Elastic limit, 418, 420, *420*

Elastic moduli
  to describe deformations, 415
  stress, strain and, 416–421, *517*

Elastic response, 417

Electric charge, 585–590, 587t
  continuous charge distributions in, 588–590, *589*, *590*
  magnetic field and, 775–776

Electric current, 737
  current density and, *717*, 717–722, *719*, *720*
  resistance and, 716–738
  sign convention in, *719*, 719–720

Electric dipole, 598, 616
  example of computing electric field, *598*, 598–599
  in nonuniform fields, *614*, 614–615

Electric dipole moment, 598
  defined as vector, 611, *611*

Electric field, *595*, 595–611, 615
  amplitude of, law of cosines and, 1046
  applied, ideal metallic conductor in, 633, *633*
  balloon example of, 634–635, *635*
  changing magnetic flux and, 827
  computing
    from continuous charge distributions, 602
    disk charge example of, *607*, 607–608
    electric dipole example of, *598*, 598–599
    from electric potential, *658*, 658–662
    from infinite line of charge (example of), *603*, 603–604
    multiple point-charge example of, 599–601, *600*
    from ring charge (example of), *605*, 605–606
  continuous charge distribution and, *601*, 601–608, 616
  electric field lines in, 608, *609*, 610–611
  in electrostatic configuration, *666*, 666
  importance to electromagnetic waves, 937
  infinite line of charge example of, 630–631, *631*
  magnitude of, spreadsheet application for, 707–709, *707–709*
  magnitudes of, for various symmetric geometries, 616t
  negative derivative of electric potential and, 672
  negative gradient of electric potential and, 672
  "nonfringing" model of, 709, *709*
  nonuniform, electric dipole in, *614*, 614–615
  numerical application for, 707–709, *707–709*
  obtaining from electric potential, example of, 659
  point-charge distributions in, 596–601, *597*
  point charge examples of, *636*, 636–637
  point-charge examples of, 660
  precise definition of, 596
  relation with potential function in three dimensions, 660–661
  stationary charges and, 584–616
  strength of
    battery example of, 651, *651*
    example of, *665*, 665–666
    infinite plane example of, *632*, 632
    spherical charge distribution example of, *628*, 628–630, *630*
  three-dimensional nature of, 610
  uniform, electric dipole in, *611–613*, 611–614

Electric field lines, 608, *609*, 610–611, 616, 630
  conducting surface and, 634, *634*
  equipotential surfaces perpendicular to, 662, *663*
  rules for, 608, *609*, 610
  thin insulator and, 634, *634*
Electric field units (volts per meter), 650
Electric flux, 637, 720, *720*
  changing, magnetic field and, 933
  for closed surface, 625
    example of, *625*, 625–626
  Gauss's law and, *623–625*, 623–626
  magnetic flux and, 816
  magnitude of displacement current and, 840
  surface shape in, 624, *624*
Electric motor (example of angular acceleration), 283
Electric potential, 642–672
  on axis of charged ring, example of, 659, *659*
  computing electric field from, *658*, 658–663
    relation to force and potential energy, 662, 662t
    in three dimensions, 660–661
  computing from charge distribution, 655–658
    continuous, *656*, 656–658
    point-charge, 655–656
  as continuous function, 647
  cubic surfaces, problem of, 666–668, *667*
  differentiated from electric potential energy, 646
  "electrifying" triangle and, 666, *666*
  electromotive force and, *743*, 743–744
  in electrostatic configuration, 666, *666*
  equipotential surfaces and, 662–666, *663*, *664*
  example of, 652–654, *653*, *654*
  of isolated point-charge, example of, 649
  numerical methods for, 666–671, *667*
  potential energy and, 643–646, *644*, *645*
  as scalar field, 647, 671
  unit of (volt), 671
  using gradient operator to find, example of, 661
Electric potential change
  due to uniform electric field, *649*, 649–652
  in nonuniform electric field, 652–654
Electric potential energy
  spherical capacitor example of, *694*, 694–695
  for two point-charges, 671
Electric potential function
  for continuous charge distribution, 672
  finding form of, 652

master equation for computing, 647
  for point-charge, 671
Electric signals, 761, *761*
Electrolytic solutions, 719
Electromagnetic induction, geophysical applications of, 886–889, *887–889*
Electromagnetic radiation, 942
  oscillating electric dipole model of, 942, *943*, 944
  plane-wave model of, 944, *944*
  sinusoidal nature of, 944, *944*
Electromagnetic source for RLC circuit, example of, *906*, 906–907
Electromagnetic spectrum, *945*, 945–946
Electromagnetic waves, 388, 932–948
  detection of, 945
  energy transport by, 938–941
  frequency of
    convention for, 937
    example of, 938, *938*
    wavelength and, 947
  material medium unnecessary for, 937
  oscillating dipole model of, 947
  prediction from Maxwell's equations, *933–935*, 933–937
  radiation pressure of, *941*, 941–942
  sinusoidal, 937–938
  sources of, 942, *943*, *944*, 944–945
  spectrum of, *945*, 945–946
Electromagnetism in radiation, 499
Electromotive force (emf)
  auto battery example of, 744–745
  direct-current circuits and, *743*, 743–745
  magnitude of, 766
Electron(s), 586, 868
  velocity after collision, 718, *719*
Electron gas, 718
Electron paramagnetic resonance (EPR), 931
Electron scattering by phonons, 718
Electron volt, 648
Electronic calculus, differentiating and integrating circuits, *761*, 761–762, *762*
Electrostatic configuration, 666, *666*
Electrostatic equilibrium, 638, 717
  ideal-conductor model in, Gauss's law and, *633*, 633–637, *634*
Electrostatic fields, induced dipole moment and, 613
Electrostatic force on torsion balance, 590, *590*
Electrostatic potential energy, computing from point-charge distribution, example of, 656, *656*
Electrostatic repulsion between electrons in ferromagnetism, 879
Emf (*see* Electromotive force)
Emitter, for photoelectric effect, 1116, *1116*

Empirical law, 140
Energizing current of transformer, 916
Energy
  differentiated from momentum, 250
  of light wave, amplitude and, 1117
  magnetic field and, *853*, 854–857
  per unit area per unit time (energy flux), 939
  per volume, stored in magnetic field, 862
  simple harmonic motion and, example of, 373
  storage in capacitor, 691–695
    work required for, 691
  storage in system of capacitors, storage box example of, *693*, 693–694
  stored in magnetic field, coaxial cable example of, *856*, 856–857
Energy band, 733
"Energy checkbook" (example of internal energy change), *521*, 521–522
Energy conservation condition, Bernoulli's equation and, 435
Energy density, 692, 709
Energy dissipation, *727*, 727–728, 910
Energy flux, 939
Energy level diagram, of Bohr model of hydrogen atom, 1131, *1131*
Energy-momentum relation, from special relativity, 1119
Energy transfer
  mechanisms of, 493–499
  time rate of, 174–175, 178
Energy transport, by electromagnetic waves, 938–941
Entropy, 555–556, 560
  changes in
    for irreversible processes, 555–558
    principle of increasing entropy and, *555*, 555–558
  disorder and, 559
  ideal monatomic gas example of, 555
  of isolated system, 557
  reversible change of phase example of, 556
  as state variable, 554–555
  water mixture example of, 558
EPR (Electron paramagnetic resonance), 931
Equation of continuity, 440
  in fluid dynamics, *432*, 432–434, *433*
Equation(s) of motion
  for electric charge, point-charge example of, *610*, 610–611
  in simple harmonic motion, 362, 368
  for transverse, one-dimensional wave, 393–394
Equation of spherical pulse in relativity, 1086
Equation of state, *504–506*, 504–508, 509

for ideal gas, 504
lack of applicability to irreversible processes, 517
Equilibrium, 357
  infinitesimal deviation from, in reversible processes, 516–517
  surface of conductor in, *685*, 687
Equipartition-of-energy theorem, 570–571, 578–579, 579t
  ideal polyatomic-gas models and, 571, *571*
Equipotential surfaces, 662–666, *663, 664,* 672
  of conductor, 663
Equivalence of inertial frames of reference, 1085, *1085*
Equivalent resistance, 766
  parallel combination of resistors example of, 748, *748*
  for resistors in parallel, 747
  for resistors in series, *745*, 746
Ergonomics, 338
Ergs, 493
Ernst, Richard, 931
Escalators, continuity equation and, 448
Escape velocity, 206
European Council for Nuclear Research (CERN), 1091
Events, 1086, *1087*
Experiment of G. P. Thomson, in wave model for particles, 1123–1124, *1124*
Experimental tests of relativity, 1090–1091
Experimental value, of gas constant, 572
Exponential notation, 7
Exponential voltage decay, locating conducting bodies and, 887–888, *888*
Extended objects as point-particles, 344
External field, dipoles aligned with, 614
External forces
  changing momentum of closed system and, 266
  on system of particles, 227
External reflection, 960, 976
External torques, angular momentum and, 348
"Extra five minutes" (example of time dilation), 1088–1089
Extraordinary wave, 972–973, *973*
Eye, *1008,* 1008–1009, *1009*
Eyepiece, of compound microscope, 1011
Eyepiece lens, 1021

Fabric softeners, static electricity and, 585
Face-centered cubic array (example of rotational motion), 285, *285*
Fahrenheit scale, 486, 508

Falling objects, 57t, 57–59, 68
  gravitational field strength, mass and, 119–120
Far-field case, of two-source interference patterns, 1040–1043, *1041–1043*
Far-field diffraction patterns, 1053, 1070
Farad, 680
Faraday, Michael, 680, 827
Faraday's law, 842, 849
  in current change, 852
  electromagnetic waves and, *933, 933–935, 934*
  in inductance, 850
  of induction, 830–831
  Lenz's law and, 827–836, *828–830*
  Maxwell's equations and, 841–842
  transformers and, 917
  using to compute magnitude of induced emf, example of, 831, *831*
Fast axis in birefringence, 973
Feld, M. S., 62
Ferromagnet(s), 869–870
Ferromagnetic phase, 884
Ferromagnetism, 878–883, 879t, *879–882*
  domains in, *879,* 879–880, *880*
FETs (field-effect transistors), 756, *756*
Fiber core of optical fiber, 962, *962*
Fiddling (example of harmonics), 465
Field curvature aberrations, 1014
Field-effect transistors (FETs), 756, *756*
Field equations, 666
Field lines in electric flux, with constant field and flat area, 623
Field strength, electric field lines and, 608
Figure skaters, high angular velocity and, 302–303
Film thickness, effect in thin-film interference, 1032
Filter
  gain of, 922
  high-pass, 922, *922*
  low-pass, *922,* 922–923, *923*
Filter circuits, 921–923, *922, 923*
Final velocity(ies)
  intergalactic probe example of, 267
  range of values for, 237
  of a rocket, 269
Finite change, irreversible processes and, 517
Finite slit width, effect on double-slit interference patterns, *1058,* 1058–1059, *1059*
Fire hose (example of average force), 226, *226*
First harmonic (fundamental), 402, 463, 466
First law of thermodynamics, *520,* 520–522, 533

First moment
  to calculate center of mass, 286
  of mass distribution, 256
First-order model of paraxial ray approximation, 1013
First overtone (second harmonic), 402, 465, 466
"Fishy" (example of index of refraction), 959, *959*
FitzGerald, G. F., 1083
Fixed axis, rotation about
  angular momentum for, 299–303
  Newton's second law for, 296–299
Fixed current, 762–763
Fixed points of scale, 485
Fixed step-size in Runge-Kutta programs, advantages of, 152
Fixed time, 762
Flannery, B. R., 668
Flash unit (example of RC-circuit), *760,* 760–761
Flashlight bulb (example of irradiance), 940–941
Floating cylinder (example of Archimedes' principle), 428, *428*
Floppy disk drive (example of arc length), 278, *278*
Flow of persons or vehicles, continuity equation and, 446–449
Flow rate, continuity equation and, in fluid dynamics, 434
Fluid(s), 416, 416t
Fluid displacement, dynamic viscosity and, 438, *439*
Fluid dynamics, 432–438
  Bernoulli's equation in, *434,* 434–438
    qualitative applications of, 438, *438*
  equation of continuity, *432,* 432–434, *433*
  ideal-fluid model, 432
Fluid elements, 432
Fluid-mass, conservation of, 433
Fluid statics, 423–430
  Archimedes' principle, *427,* 427–428
  compressible-fluid model, 428–430, *429*
  incompressible-fluid model, 424–425, *424–426*
  Pascal's principle, 426, *426*
  variation of pressure with depth, *423,* 423–424
Flux, 720, *720*
  See also Electric flux; Magnetic flux
Focal length(s), 987, *987,* 1019
  of lens, 1000, *1000*
  for one thin lens, 1020
  range of, example of, 1009
  for two thin lenses, 1020
Focal plane of lens, 1001
Focal point of parabolic mirror, 986, *987*
"Folded-axis" sign convention, 298

"Follow your nose" (example of frictional forces in rotational motion), 321, *321*

Foot, 6

Foot-pounds, 293

Football plays (example of component method of vector addition), 27–29, *28*

Football trainer (example of power), 175

Force(s), 105–109, 120
  added as vectors, 104
  application of, 417
  classes of, 187
  direction of, direction of velocity and, 776
  displacement and, in work, 169
  of elastic molecular collisions in ideal-gas model, 566–567, *567*
  frictional, 109, *109*
  magnitude of, work and, 165–169, *166–169*
  normal, 109, *109*
  spring, 108, *108*
  tension, 109
  torque and, in static equilibrium, 313–314
  vector sum of, constant velocity and, 103
  weight, 106–108, 107t

Force analysis
  in example of static equilibrium, 316
  in rotation and translation with no slipping, 319–320, *320*, 322–323

Force-at-a-point model, 417

Force equation, in example of force and torque, 323

Force functions
  defining, in Runge-Kutta method, 151–152
  negative integral of, 199
  one-dimensional potential energy function and, 211
  partial derivative test on, example of, 209

Force models, summary of, 146, 146t

Force of attraction on parallel wires, 818

Force of repulsion on parallel wires, 818

Force on a charge, cross product example of, 778, *778*

Force per unit charge in electric field, 596, 597

Force per unit mass in gravitational field, 595

Force plate, 337

Force-versus-extension curve for Non-Hooke's-law spring, example of, 168–169, *169*

Force-versus-time curve, 224, *224*

Forced convection, 499

Forward frictional force in rotational motion, 321, *321*

Forward orbital motion, gravitational force and, in apparent weightlessness, 142

Foucault, Jean Bernard Léon, 958

Fourier, Jean Baptiste Joseph, 403

Fourier series, 404, *404*, 409

Fourier's theorem, 403–404, *404*

Fourth-order polynomial(s), in computing acceleration, 66

Fourth-order Runge-Kutta method (*see* Runge-Kutta method)

Fractional quantum Hall effect, 785

Frame of reference, 19, 35
  "muons" example of, 1091
  "Now I c" example of, *1097*, 1097–1098

Franklin, Benjamin, 585, 717–718

Fraunhofer, Joseph von, 1053–1054

Fraunhofer diffraction patterns, 1053–1054, 1070

Free-body diagram, 110–112, *111*, *112*, 115, *115*, 128, *129*, 298
  in biomechanics, 335, 336, *336*, *337*
  for example of static equilibrium, *315*, 315, 316
  for wave equation, 404

Free charge, 710
  induced surface charge and, relation between, 700

Free charge density, 697

Free electron(s), 587

Free electron density, doping and, *731*, 734–735

Free fall, 57
  zero apparent weight of astronauts and, 141–142

Free space, speed of light in, 1098

Free-space irradiance, 1046

Free-space wavelength, 1032

Free spectral range, 1060
  diffraction gratings and, 1062–1064, *1063*
  "seeing double" example of, 1063–1064

Free surface charge density, 705

Freely falling bodies, gravitational field strength and, 120

Freely-falling-body model of projectile motion, 80

French, A. P., 1122, 1123

Frequency, 358
  differences in, in wave interference, *399*, 399–400
  in example of twin paradox, 1094–1095, *1095*
  for position equation of sinusoidal wave train, 392
  of sinusoidal wave, 394
  of sound, 460, *460*
  of wave motion, 396

Frequency of oscillation in simple harmonic motion, 368

Frequency spectrum, 464, *464*

Fresnel, Augustin Jean, 1053, 1079

Fresnel diffraction patterns, 1053

Friction, 187
  in conveyer belt example of system with varying mass, 268
  in wave motion, 390

Frictional force(s), 109, *109*, 120
  describing, 127–128
  example of, 128–130, *129*
  in example of circular motion, 133, *133*
  impulse approximation and, *221*, 222–223
  Newton's third law of motion and, 118
  in rotational motion, 321, *321*
  in simple harmonic motion, 381
  work done by, 165

Frictional losses, energy transported by wave and, 396

Frictionless mass (example of tangential velocity), 289t, 292, *292*

Fringe(s), 1029, *1029*, 1030
  of equal thickness, 1039

Fringe field of parallel-plate capacitor, 681, *681*

Fringe patterns, 1082–1083, *1083*

Frisch, D. H., 1090

Frost (example of latent heat of vaporization), 503

Fuel cell of Indycars, 162

Full width at half maximum (FWHM), of resonance peak, 914

Fundamental (first harmonic), 402, 463, 465, 466

Fuse, 746

Fusion research, toroid used in, 815

FWHM (full width at half maximum), of resonance peak, 914

Gain of filter, 922

Galilean transformations, 1086, 1099, 1103
  Lorentz transformations and, 1087, *1087*

Galileo (Galileo Galilei Linceo), 57, 1079

Galvanometer, 753–754, *754*, 766
  ammeter in parallel with, 755, *755*
  potentiometer and, 756
  voltmeter in series with, 754, *754*
  Wheatstone bridge and, 756, *756*

Gamma rays, *945*, 946

Garden hose (example of Bernoulli's equation), 435–436, *436*

Gas(es), 416, 416t
  models for sound waves in, *451*, 451–452, *452*

Gas constant, experimental value of, 572

Gas expansion (examples of work done), 519, *519*

Gas temperature scale, 487
Gasoline can (example of volume
  expansions), 492
Gasoline engine (example of Otto cycle),
  550–551
Gauge pressure, 430
  change in, Denver tire pressure
    example of, 430
Gaussian surface(s)
  charge enclosed within, 629
  cubical, 632, *632*
  cylindrical, 630–631, *631*
    for planar charge surface, 632, *632*
  of cylindrical capacitor, 684, *684*
  dielectrics and, in parallel-plate
    capacitors, 703, *703*
  Gauss's law and, *626*, 626–627, *627*
  orientation of, 628
  of parallel-plate capacitor, 681, *681*
  spherical, *628*, 629
  of spherical capacitor, 682–683, *683*
  for symmetrical charge distributions,
    626
  walls of cube as, 667
Gauss's law, 622–638, 701, 810
  alternate statements of, 710
  applied to problem of electric
    potential, 667, *667*
  Coulomb's law from, *627*, 627–633,
    632
  for cylindrical capacitor, 684, *684*
  differential form of, in electrostatic
    configuration, 666
  electric flux and, *623–625*, 623–626
  Gaussian surfaces and, *626*, 626–627,
    627
  ideal-conductor model in electrostatic
    equilibrium and, *633*, 633–637,
    634
  for magnetism, 819
    magnetic flux and, 816–817, *817*
  Maxwell's equations and, 841, 842
  for parallel-plate capacitor, *681*, 681–
    682
    with dielectrics, *703*, 703–704
  problem-solving strategy for, 628
  for spherical capacitor, 682–683, *683*
Geiger, Hans, 1127, 1128
Geiger-Müller tubes, 604
General Conference on Weights and
  Measures, 5, 6
General cyclic processes, *549*, 549–554,
  550
  Clausius inequality in, 551–553, *552*
  ''curious'' heat ratio fraction for
    arbitrary cycles, 553–554
General theory of relativity (Einstein),
  119
Generator (example of induced current in
  loop), *835*, 835–836
Geologic conductors, induction
    techniques to investigate, 887, *887*

Geometrical optics, 951, 956, 982–1021
  curved mirrors, 986–995, *987*
    apparent depth, *997*, 997–999, *998*
    concave, *988–992*, 988–994, 991t
    convex, *994*, 994–995, *995*
    refraction at curved surfaces, *996*,
      996–997, *997*, 997t
  images, *983*, 983–984
    formed in eye, *1008*, 1008–1009,
      1009
  lenses
    lens-maker's formula, *100*, 997, *999*,
      999–1001
    thin-lens formula, *1001*, 1001–1007,
      1002t, *1006*, *1007*
  numerical methods for paraxial ray
    tracing, 1015–1019, *1016*, 1016t
    ABCD matrix, 1016t, 1018–1019,
      1019
    spreadsheet template for ABCD
      parameters, 1017–1018
  optical aberrations, 1013–1015
    chromatic, 1014, *1014*
    third-order, 1014–1015, *1015*
  plane mirrors, *984*, 984–986, *985*
  simple optical instruments
    compound microscope, *1011*, 1011–
      1012
    magnifying glass, 1009–1011,
      1010
    telescope, *1012*, 1012–1013, *1013*
Geometries, symmetric, electric field
    magnitudes for, 616t
Geophysical applications of
    electromagnetic induction, 886–
    889, *887–889*
Germanium, 729
Germer, L. H., 1122, 1134
Glare (example of polarizing angle), *968*,
    969
Gliders
  example of coefficient of restitution,
    237
  example of perfectly elastic collisions,
    235, *235*
Gold atoms, in Rutherford scattering,
    1127–1128
Golf ball (example of de Broglie
    wavelength), 1121–1122
Golf swing (example of velocity and
    acceleration), 91–93, *91–93*
Goodyear Rubber Company, 161
Gossert, Dana, 98
Gradient, 207
Gradient operator, example of using to
    find electric potential, 661
Granite Peak air pressure (example of
    compressibility), 429–430
Gravitation
  as inverse square of distance, 137
  potential energy function for, 210
Gravitational field, 595

Gravitational field strength, 106, 107
  falling objects, mass and, 119–120
  freely falling bodies and, 120
  magnitude of, in example of force and
    torque, 324
Gravitational force, forward orbital
    motion and, in apparent
    weightlessness, 142
Gravitational mass, equivalence to
    inertial mass, 119–120
Gravitational potential energy, 291
  center of mass and, 258
  near earth's surface, *188*, 188–190,
    189
  negative, 206
  of system of particles, 269, 317
Gravity, 187
  acceleration and, 89
  finding rate of change in, 140
  work done by, final speed and, 173
Grazing angles, 1067, 1071
  in example of Bragg's law, 1068
Great Newtonian synthesis, 137
Ground reaction forces, 337
Ground state, 1130
Guest essays
  biomechanical loading of human body,
    335–338, *336–338*
  continuity equation in everyday life,
    446–449
  geophysical applications of
    electromagnetic induction, 886–
    889, *887–889*
  magnetic resonance, *929*, 929–931,
    930
  musical acoustics, 477–481
  racing at the Indianapolis 500, 160–
    162
Guitar string (illustration of quantization
    of energy), 1113–1114
Gyromagnetic ratio (Landé *g* factor), 872
Gyroscope
  example of angular velocity, 279–281
  example of rate of precession, 347–
    348
  precession of, *346*, 346–348, *347*

Hahn, Irwin, 931
Half-angle of acceptance cone,
    determining, *962*, 963–964
Half-life of muons, 1090
Half-wave plates, 976
  polarization and, *974*, 974–975
Hall, D. E., 481
Hall, Edwin H., 785
Hall coefficient, 785
Hall effect, 794
  in magnetic field, 783–785, *783–785*
Hall voltage, 783, *783*
Halley, Edmond, 137

Hanging sign (example of static equilibrium), *315*, 315–316, *316*
"Hard rock" (example of resistivity), 725, *725*
Harmonic oscillation in inductance, 862
Harmonic oscillator
  example of power of sinusoidal wave, 397
  simple, energy of, 371–373, *372*
  undamped, total mechanical energy of, 372
Harmonic waves in air, *452*, *453*, *457*, 457–458
Harper, S. D., 771
Heat, 493, 508
  energy dissipated as, 910
  work extracted from, 539
Heat capacity, 508
  latent heat and, 500–505, 501t
    heat of transformation, 502–504, *503*
  of solids, *1114*, 1114–1115, *1115*
Heat cycle, helium cylinder example of, 523–532, *529*, *530*
Heat dumping in thermodynamic process, 539
Heat engines
  efficiency of, 541
  second law of thermodynamics and, *540*, 540–541, *541*
Heat of transformation, 502–504, *503*
  ice to steam example of, 503–504
Heat pumps
  coefficient of performance of, 546
  refrigerators and, *546*, 546–547
Heat ratios, "curious" fraction in, 547
Heat reservoir, 515
Heat transfer
  mechanisms of, 493–499
  as path variable, 520
Hecht, Eugene, 953, 1039
Helium cylinder (example of heat cycle), 523–532, *529*, *530*
Helix, motion of charge in, 780–781, *781*
Hemingway, Ernest, 484
Henry, 850
Henry, Joseph, 827
Hertz, Heinrich, 358, 936, 1110
Hertz (Hz), 358
High Flux Isotope Reactor (HFIR), 472, *472*
High jumpers, center of mass and, 259, *259*
High-pass filter, 922, *922*
High-reflectance result, for thin-film interference, *1031*, 1039–1040, *1040*
"High tea" (example of work and kinetic energy), 172
Higher harmonics, 463

Hirst, H. S., 1127
Hockey players (example of two-dimensional collision), 239, *239*
Hoffman, Banish, 22
Hole, 730, *730*
Hole motion, *730*, 730–731
Holographic interferometry for studying vibrational modes, 478, *478*
Holton, G., 137
Homecoming float (example of center of mass), *256*, 256–257
Homogeneous objects, density and, 422
Hooke, Robert, 953
Hooke's law, 398, 418
  rotational, in torsion pendulum, 369
  of spring forces, 108
  in two-mass system, 370
  in wave equation, 406
  work-energy theorem and, 172
Hooke's-law spring, 197–199, *198*
  example of Runge-Kutta method, 148–149, *149*
  magnitude of force exerted by, 166
  model of motion, 488
  negative work done by, 167, *167*
  positive work done on, 166–167, *167*
  potential energy function for, 210
  potential energy of, 190, *190*
  proportionality constant for, 418
  work done by, in piñata example of work, 167–168
Hooke's-law spring force, 120
Hooke's-law spring model, 415
Hoop, rotational inertia for, 288, 289t
Horizontal motion, downward acceleration and, 79, *79*
Horse power (hp), 174, 175
Howells, Malcolm R., 1065
Howes, Ruth H., 160–162
Hp (horse power), 174, 175
Hughes, J. V., 1122
Huygens, Christiaan, 238, 953, 955, 1079
Huygens' principle, 958, 1030, 1053, 1054, *1054*
  in wave model of light, 955, *955*
Huygens' wavelets, 955, *955*, 1053, 1060
Hydraulic lift, Pascal's principle and, 426, *426*
Hydrogen atom
  Bohr's stationary states and, 1129
  example of Coulomb's force law for two-point charges, 592, *592*
Hyperbolic tangent, 145
Hyperelastic collisions, 233
Hypothesis, 3
Hysteresis curve, *881*, 882
Hysteresis loop, 882, *882*
Hysteresis losses, core-loss current and, in ideal-transformer model, *916*, 916–917

Ice point, 485, 487
Ice skater (example of velocity and acceleration), 90, *90*
Ice to steam (example of heat of transformation), 503–504
Ideal ammeter, 755
Ideal Carnot efficiency, 544
Ideal conductor, 638
Ideal-conductor model, in electrostatic equilibrium, Gauss's law and, *633*, 633–637, *634*
Ideal-current source, 892
Ideal emf (resistanceless), 743
Ideal emf source, 892
Ideal-fluid model, 440
  in fluid dynamics, 432
Ideal gas
  kinetic theory of, 565–570
  summary of thermodynamic processes for, 526t
Ideal-gas equation of state, 509, 524, *524*
Ideal-gas law, 504
Ideal-gas model
  to illustrate techniques of thermodynamics, 522
  kinetic theory of ideal gas and, 565
Ideal metallic-conductor model, 633, *633*
Ideal monatomic gas (example of entropy), 555
Ideal solenoid, 812
  example of inductance, 851
  magnetic field of, 855–856
Ideal-solenoid model, 819
Ideal transformer, *916*, 916–917, 924
  circuit models for, 910t, *919*
  effective load in, example of, *920*, 920–921
  power loss from, example of, 920
Ideal voltmeter, 754
Idealized projectile motion, 90
Identity matrix, 763
Image
  locating
    plane mirror example of, 986, *986*
    wide-angle convex mirror example of, 995, *995*
  position of, 984
  object position and, 991, *991*
Image(s), 1019
  formed by curved mirrors, 986–995, *987*
    concave, *988–992*, 988–994, 991t
    convex, *994*, 994–995, *995*
  formed by lenses, 999–1007
    lens-maker's formula, *997*, *999*, 999–1001, *1000*
    thin-lens formula, *1001*, 1001–1007, 1002t, *1006*, *1007*
  formed by plane mirrors, 984, 984–986, *985*
  formed by refracting surfaces, 995–999

in geometric optics, *983*, 983–984
locating by parallax, 998
Image distance, 991, 1019, 1020
Image distortion (*see* Spherical aberration)
Imaging geometry (with lens), 1055, 1070
Impedance, 923
for RLC circuit, 904, 906
Impedance diagram, 902, *902*
power factor determined from, *911*
for RLC circuit, 904, *904*
Impedance-matching transformer, 919–920
Impulse, 220–221, *221*, 241
of force, example of, 221, *221*
linear momentum and, 219–242
momentum and, *221*, 221–223, *223*
Impulse approximation, *221*, 222–223
Impulse-momentum theorem, 222, 241
acoustic pressure and, 454
Impurities, resistivity of semiconductors and, 727
Incandescent source of unpolarized light, 966
Incoherent waves, 1036, 1046
Incompressibility in ideal-fluid model, 432
Incompressible-fluid model, 424–425, *424–426*, 440
Increased fuel, benefits of, in rocket propulsion, 267, *267*
Incremental work done, in rope swing example of work, 170
Index of refraction, 957
"doing time in prism" example of, *964*, 966
"fishy" example of, 959, *959*
gradual change in, 960, *960*
of initial and final moments, 1020
of liquid, 998, *998*
for selected isotropic media, 957, 958t
speed of light and, 958
"still waters" example of, 999
Indianapolis 500, 160–162
Induced charge density in parallel-plate capacitor, 702
Induced current, 828
effect of, in secondary coil of transformer, 918
in loop, generator example of, *835*, 835–836
Induced dipole(s), 614
in dielectric material, 695, 696
Induced dipole moment, 613
Induced electric fields, *836*, 836–839
eddy currents in, *838*, 838–839
field lines in, 836
nonconservative nature of, 837
Induced emf, 827–828
betatron example of, *837*, 837–838

direction of
in Faraday's law, 830
Lenz's law for, 831–836, *832*, *833*, *835*
finding magnitude of, 834–835, *835*
example of Faraday's law for, 831, *831*
ways of producing, 830
Induced field, rules for direction of, 833
Induced magnetic dipole moment, 869
Induced magnetic moments in diamagnetism, 876–877
Induced surface charge, 710
density of, 697, 705
free charge and, 700
Inductance, 848–862
energy, magnetic field and, *853*, 854–857
ideal solenoid example of, 851
LC circuits and, *857–859*, 857–861
RLC circuits, *859*, 859–860
LR circuits and, *853*, 853–854, *854*
process of induction and, *849*, 849–852
Induction
Faraday's Law and, 826–842
Faraday's law and, statement of, 830–831
Induction coil (example of current change), 851–852
Induction techniques, to investigate geologic conductors, 887, *887*
Inductive reactance, 897, 923
Inductive time constant, 853, 862
solenoid example of, 854
Inductor(s), 849, 853, *853*
in AC circuits, *896*, 896–899, *897*, 898t
energy stored by, 862
magnetic field created by, energy stored in, 855
microscopic model for, 910
use in metal detectors, 860
Indycars
fuel cell of, 162
kinetic energy in collision, 162
methods of holding on track, 160–161
tires of, 161–162
Inelastic collisions, 233
Inertial frames of reference, equivalence of, 1085, *1085*
Inertial mass, 105
equivalence to gravitational mass, 119–120
Infinite current sheet
Ampère's law applied to, *811*, 811–812, *812*
electric analog of, 812
Infinite line of charge
example of computing electric field from, *603*, 603–604

example of electric field, 630–631, *631*
Infinite plane (example of electric field strength), 632, *632*
Infinite resistance in ideal voltmeter, 754
Infinite resistivity of perfect insulator, 723
Infinitely heavy nucleus, Rydberg constant for, 1132
Infinitely long wire model for computing magnetic field, 804
Infinitesimal rotations as vectors, 342
Infrared region of electromagnetic spectrum, *945*, 946
Infrasonic waves, 452
Initial displacement, 360
Initial velocity vector in projectile motion, 80, *80*
Inner product (*see* Scalar product)
Inner tube
example of work, 165, *165*
example of work-energy theorem, *165*, 172–173
"Innocent-eyeball" model of image formation, 984, 989, 1007
Input voltage signal, output voltage signal and, 761, *761*
Instantaneous acceleration, 45–46, 68, 87, 93
from velocity-versus-time graph, 48–49
Instantaneous current, 717, 737
Instantaneous potential drop for RLC circuit, 905
Instantaneous power, 178, 910
calculating, 174
expended by torque, 296
Instantaneous speed, 93
Instantaneous velocity, 44–45, *45*, 68, 93
graphical interpretation of, *47*, 47–48, *48*
Insulation for adiabatic processes, 525
Insulator(s), 517, 587, 729
analog of, 735
band structure of, 734, *734*
electric field lines and, 634, *634*
electrical properties of surfaces, 587–588
Integral calculus, to find position and velocity, 69
Integral calculus to find position and velocity, 60–61
Intensity
of sound, sound level and, *453*, 458–463
of sound waves, 940
Interacting phase (collision), conservation of momentum and, 232
Interference filters, 1040, *1040*
Interference of light, 1028–1048
thin-film, *1031*, 1031–1034, *1032*
two-source, *1029*, 1029–1031, *1030*

Interference of waves, superposition and, 397–404
Interference pattern(s), 1029, *1029*, 1034–1043
  coherence, 1036–1037
  double-slit, effect of finite slit width on, *1058,* 1058–1059, *1059*
  Lloyd's mirror example of, 1034, *1034*
  master interference equation in, 1036
  optical beats, 1037–1038
  thin-film, 1038–1039, *1039*
   high-reflectance result, *1031,* 1039–1040, *1040*
  two-source, 1040–1043, *1041–1043*
Interference term, 1035, 1047
Intergalactic probe (example of final velocity), 267
Intermolecular bonds in dielectric material, 697
Internal energy (thermal energy), 493, 508, 520, 533
  change as state variable, 520
   example of, 554
  changes in, "energy checkbook" example of, *521,* 521–522
Internal forces, 118
  on system of particles, 227
Internal friction of a medium, viscosity as, 142
Internal reflection, 960, 976
Internal resistance, 766
  of sources in electromotive force, 743, *743,* 744
Internal torques, cancellation in angular momentum, *345,* 345–346
International Bureau of Weights and Measures (France), 5
International Steam Table Conference, 494
International watt-hour, 494
Intrinsic angular momentum (spin) of electron, 868, 871–872
Intrinsic elastic constant, 418
Intrinsic magnetic dipole moment, 868
Intrinsic magnetic moment, 929
Inverse Lorentz transformations, 1087, 1089
Inverse of matrix, 763
Irradiance, 940, 947, 1035, 1046, 1054
  flashlight bulb example of, 940–941
  in low-reflectance model, 1038
  in photoelectric effect, 1117
  reflected, 1039
Irradiance maximum(a), 1041, 1043, 1055, *1055*
  multiple sources, 1044, *1044, 1045*
  Young's double-slit experiment, 1046
Irradiance minimum(a)
  multiple sources, 1044, *1045*
  Young's double-slit experiment, 1030, 1046

Irradiance pattern of double-slit interference, 1070–1071
Irregularly shaped conductor
  charge density on, 664, *664*
  charge distribution on, 664, *664*
Irreversible process(es), 517, 532
  entropy changes for, 555–558
   principle of increasing entropy and, *556,* 556–558
  represented by hysteresis loop, 882
Isentropic processes, 555
Isobar(s), 505, *505*
  in example of heat cycle problem, 531
Isobaric process(es), 532
  isothermal processes and, "two-step" example of, *527,* 527–528
  in thermodynamics, 523–524
Isochor(s), 505, *505*
  in example of heat cycle problem, 531
Isochoric process(es), *518,* 518–520, 532
  in thermodynamics, 523
Isolated capacitor(s), 709–710
  effects of dielectric materials in
   on capacitance, 698–699
   on energy stored, 699–700
   on potential difference across capacitor, 699
  Gaussian surface in, 633, *633*
  in series arrangement of capacitors, 688
Isolated point-charge, example of electric potential of, 649
Isolated system(s)
  conservation of mechanical energy in, 195–196, 211
  nonconservative forces in, 211
  principle of increasing entropy and, *556,* 556–557
Isotherm(s), 505, *505*
  in example of heat cycle problem, 531
  of ideal gas as constant-energy curves, 523
Isothermal bulk modulus, 454
Isothermal process(es), 532
  isobaric processes and, "two-step" example of, *527,* 527–528
  in thermodynamics, 524, *524*
Isotropic media in ray model of light, 956, *956*
Isotropic substance, area expansion of, 490t, 491, *491,* 508

Jacobi's method, 668
Jeans, Sir James, 1112
Jefferson, Thomas, 189
Jiles, David, 879
Joule, James Prescott, 165, 494
Joule heating, 728
Joules, 174, 493, 646, 648
Junction, in circuit, 750

"Junk", square root of, in simple harmonic motion, 365

Kamerlingh-Onnes, Heike, 735
Karate hammerfist strike, examples of
  computing acceleration from position, 62t, 66, 66t, *67*
  computing areas under a curve, 62, *62,* 62t
Kelvin, Lord (William Thompson), 488, *488*
Kelvin-Planck statement, 559
  of second law of thermodynamics, 540, 552
Kelvin scale, 487, *487,* 508
  absolute temperature scale and, 548, 560
Kelvin temperature, 569
Kelvins, 487
Kennard, E. H., 1067
Kepler, Johannes, 139–140
Kepler's third law, 139–140
Kilogram, 8, 15, 105, 120
Kilogram-meters per second, 222
Kinematic(s), 103
  of rotational motion, *277,* 277–283
Kinematic quantities, linear and rotational, 304t
Kinetic energy, 171, 178, 192, 1100–1101, 1104
  center of mass and, 262–263
  of Indycars in collision, 162
  in inner tube example of work-energy theorem, 173
  of moving fluid element, 435
  in partially elastic collisions, 241
  in perfectly elastic collisions, 240
  of rotation, 296
   of rigid body, 284
  of rotation and translation, 327
  rotational inertia and, *283,* 283–292
  as scalar quantity, 220
  in simple harmonic motion, 373
  of simple harmonic oscillator, 381
  of system of particles, 269, 317
  work and, 163–178, *165,* 170–173
   "high tea" example of, 172
   inner tube example of, *165,* 172–173
   rope swing example of, 173
Kinetic friction, negative work done by, 197
Kinetic frictional force, 127–128
Kinetic theory of gases, 451
Kinetic theory of ideal gas, 565–570
  ideal-gas model in, 565
  pressure and molecular motion in, *565,* 565–568, *567*
  temperature and molecular motion in, 568–570

Kirchhoff's first law, 751
  applied to circuit segments, example
      of, *752*, 752–753
Kirchhoff's law(s), 766, 853, 893
  capacitors and, 894
  inductors and, 896
  multiple-loop circuits and, 749–753,
      750
  potential change and, example of, *750*,
      750–752
Kirchhoff's rules, 849
Kirchhoff's second law, 751–753
  RC-circuits and, 757–758
      discharging, 759, *759*
Kirz, Janos, 1065
Klitzing, Klaus von, 785
Krypton 86 lamp (standard meter), 6, *6*
Kuerti, G., 1083–1084

Lab-frame velocities, center-of-mass
      frame velocities and, 262
Lack of turbulence, in ideal-fluid model,
      432
Laminar body(ies), 290
  vector relations and, 351
Laminations, reducing eddy currents by,
      838, *838*
Landé *g* factor (gyromagnetic ratio), 872
Larmor frequency, *929*, 929, 930
Latent heat, 502, 509
  heat capacity and, 500–505, 501t
      heat of transformation, 502–504,
      503
Latent heat of fusion, 503
Latent heat of vaporization, 503
Lateral adhesion, 160
  increasing, 161
Lateral magnification, 985, 992, *992*,
      1020
Lattice, in current flow, 718
Lattice constant, 1066–1067
Lattice kinetic energy, resistor
      temperature and, 727–728
Laue, Max von, 1065
Laue diffraction pattern, 1066, *1066*
Law(s), 3
Law of Charles and Gay-Lussac, 504
Law of conservation of angular
      momentum, 353
Law of conservation of energy, Lenz's
      law and, 833, *833*
Law of conservation of momentum, 227,
      231, 241–242
Law of cosines, amplitude of electric
      field and, 1046
Law of Dulong and Petit, 575
Law of inertia, 103–104, *104*
  example of, 104, *104*
Law of Malus, 971, 976
Law of reflection, 956, 985

Law of specular reflection, 956–957,
      957, 976
LC circuits, *857–859*, 857–861
  ideal, charge and current in, 858, *858*
  RLC circuits and, 859, *859*
Leibniz, Gottfried Wilhelm, 238
Length, 5–6, 7t
  units of, 5–6, 7t
Length contraction, *1089*, 1089–1090,
      1104
  desk width example of, 1090
Lens(es)
  forms of, *1000*
  images formed by, 999–1007
      lens-maker's formula, *997*, *999*, 999–
      1001, *1000*
      thin-lens formula, *1001*, 1001–1007,
      1002t, *1006*, *1007*
Lens-maker's formula, *997*, *999*, 999–
      1001, *1000*, 1020
  radii example of sign convention in,
      997, *999*, 1001
Lensless geometry, 1055, 1070
Lenz, Heinrich, 831
Lenz's law, 831–836, *832*, *833*, *835*,
      842, 849
  conducting rod example of motional
      emf and, 833–834, *834*
  Faraday's law and, 827–836, *828–830*
  opposition to magnetic flux in, 831–
      832
Leone, F. C., 1083–1084
Lever
  geometry of, biomechanical loading
      and, 335–336, *336*
  movement of, 336, *336*
Lever arm of force, 293
"License plate" (example of Rayleigh
      criterion), 1057, *1057*
Light, 952–976
  diffraction of (*see* Diffraction)
  frequency of, in photoelectric effect,
      1117
  interference of, 1028–1048
  particle model of, 953–954
  polarization of, *966*, 966–976
      birefringence, 972–973, *973*, 973t,
      974
      half-wave and quarter-wave plates
      in, *974*, 974–975
      optics of compact-disc player and,
      975, 975–976
      by reflection, 968–970, *968–970*
      by scattering, *967*, 967–968
      wire-grid polarizers in, 970–972,
      971
  rays of, in geometric optics, *983*, 983–
      984
  reflection of, *956*, 956–957, *957*
  refraction of, *957*, 957–966, *958*, 958t,
      960

dispersion and, *964*, 964–965, *965*
  optical fibers and, *962*, 962–964
  prism geometry and, *965*, 965–966
  total internal reflection and, *960*,
      960–962, *961*
speed of, 1081
wave model of, 953–954
  properties of, *954*, 954–956
See also Polarization; Refraction
Light waves, *945*, 945–946
  energy of, amplitude and, 1117
Line of action of force, 293
Linear acceleration, vectors associated
      with, 282, *282*
Linear charge distribution
  computing electrical potential from,
      example of, 656–657, *657*
  magnitude of force of point-charge,
      example of, *594*, 594–595
Linear combinations of vectors, 25
Linear differential equations
  in finding current through circuit, 853
  for LC circuits, 857–858
  See also Differential equations
Linear expansion, 508
  thermal, *490*, 490t, 490–491
Linear expansion coefficient, 491
Linear impulse-momentum theorem, 325
Linear mass density function, 255
Linear momentum, impulse and, 219–
      242
Linear objects, center of mass for, 255,
      255
Linear quantities, related to rotational
      quantities, by vector statements,
      343–344
Linear velocity, distinguished from
      angular velocity, 281
Linearly polarized wave, 966
Lines of force (electric field lines), 608,
      609, 610–611
Liquid(s)
  index of refraction of, 998, *998*
  velocity distribution function for
      molecular speeds of, 577
Liquid crystal displays (LCDs), 613
Liquid helium in superconductors, 735
Lloyd's mirror (example of interference
      pattern), 1034, *1034*
Load impedance of transformer, 916
Load resistance, electromotive force and,
      744
Local atmospheric pressure, 430
Local model, for gravitational potential
      energy, 204
  incorrect solution to example using,
      204–205
Lone canoeist (example of momentum),
      260, *260*
Long and short (example of Balmer
      series), 1126–1127

Long-wavelength electromagnetic waves, proof of existence of, 936
Long waves, *945*, 946
Longitudinal coherence length, 1046
Longitudinal waves, *388*, 388–389, 408
  one-dimensional, 394–395
  sinusoidal transverse wave example of, 394–395
Lorentz, H. A., 1083, 1087
Lorentz force law, 778, 794
  Maxwell's equations and, 841
Lorentz transformation equations, 1103
Lorentz transformation laws, 1096
Lorentz transformations, *1085*, 1086–1087, *1087*, 1099
  Galilean transformations and, 1087, *1087*
  inverse, 1087, 1089
  in time dilation, 1088
Loudness, 481, *481*
Low-attenuation optical fibers, 962
Low-pass filter, *922*, 922–923, *923*
Low-reflectance model, of irradiance, 1038
LR circuits, *853*, 853–854, *854*
Luminiferous ether, 1079–1080
Lyman, T., 1126
Lyman series of spectra, 1126
Lynch, Rosemary, 157

McCuskey, S. W., 1083–1084
Mach number in shock waves, 472
McInerney, Michael, 98
McNair, R. E., 62
Macroscopic object, 1
Magnesium fluoride (example of destructive interference), 1033
Magnet and conducting ring experiment in induced emf
  Lenz's law and, *832*, 832–833
  production of current in, 827–828, *828*
Magnetic bottle, 781, *781*
Magnetic dipole moment, *791*, 791–793
  angular momentum and, 871–872
  example of torque in, 792, *792*
  per unit volume (magnetization), 872
Magnetic field(s), 774–794, *775*, *777*
  changing electric flux and, 933
  electric charges and, 775–776
  energy and, *853*, 854–857
  energy stored in, coaxial cable example of, *856*, 856–857
  of ferromagnetic substances, reversing, 881, *881*
  force on current-carrying conductor and, *787*, 787–789, *788*
  Hall effect in, 783–785, *783–785*
  inside solenoid, determining value of, 813–814
  magnetohydrodynamics and, 785–787, *786*

nonuniform, moving charges in, *781*, 781–782, *782*
  properties of, *801*, 801–802
  sources of, 800–819
    Ampère's law and, *810*, 810–816
    Biot-Savart law and, *801*, 801–808, *803*, *804*
    field on axis of solenoid, *817*, 817–818, *818*
    magnetic flux and, 816–817, *817*
    parallel wires, amperes, and coulombs, 808–809, *809*
    superloop example of, *805*, 805–806, *806*
  in toroid, example of, 815–816, *816*
  for two special cases, 807
  uniform
    current-carrying loops in, 789–793, *790*
    magnetic dipole moment and, *791*, 791–793
    moving charges in, 779–781, *779–781*
Magnetic field elements
  direction in same plane, 802
    example of, 802–804, *803*
  direction not in same plane, 803–805, 803–806
    example of, *805*, 805–806, *806*
Magnetic field intensity, 875, 883
Magnetic field lines, 775, *775*
  of current loop, 806, *806*
  in transformer, 917
Magnetic field strength, 875
  increasing, effects in nonuniform magnetic fields, 781–782, *782*
Magnetic field vectors, 872–876, *873*, *874*, 876t
Magnetic flux
  change in, Lenz's law and, 831–832
  Gauss's law for magnetism and, 816–817, *817*
  in transformer, 917
Magnetic flux density, 875
Magnetic force
  direction of, 777, *777*
  features of, 777
  on moving electric charge, 793
Magnetic hysteresis, *881*, 882
Magnetic induction, 875
  permeability and, bismuth example of, 882–883
Magnetic induction field, 776, 793
Magnetic moment, 794
  "moments to remember" example of, 793, *793*
Magnetic monopole, 775
Magnetic properties of materials, 867–884
  magnetic field vectors and, 872–876, *873*, *874*, 876t
  overview of, 868–870

phases of, 876–883
  diamagnetism, 876–877, *877*
  ferromagnetism, 878–883, 879t, *879–882*
  paramagnetism, 878
  source of, 870–872, 872t
Magnetic quantum number(s), 871, 930
Magnetic resonance, *929*, 929–931, *930*
Magnetic resonance imaging (MRI), 861, *861*, 931
Magnetic susceptibility, 875
  in paramagnetic or diamagnetic phase, 875–876, 876t
  permeability and, 883
Magnetization, 872, 883
Magnetization current (Ampèrian current), 874, 883
  calculating, 874–875
  of transformer, 916
Magnetohydrodynamics (MHD), 785–787, *786*
Magnetooptic spatial light modulator (MOSLM), example of, *807*, 807–808
Magnifying glass, 1009–1011, *1010*
  angular magnification of, 1021
Magnifying power of optical instruments, 1021
Magnitude(s)
  of angular momentum, 300
  of component vector, 26
  of field, 818
  of forces, 110
  of induced emf, determining, 834–835, *835*
  of total linear acceleration, 282
  of a vector, 21, 22
"Malus aforethought" (example of transmitted irradiance), 972, *972*
Manometer, *430*, 430–431, 486, *487*
Marching soldiers (example of resonance phenomena), 377
Marsden, Ernest, 1127, 1128
Martins, R. de A., 1122
Mass, 8–9, 9t
  binding energy and, 1101–1103
  differentiated from weight, 106
  gravitational field strength, falling objects and, 119–120
  net force on, 259
Mass defect, 1103
Mass densities, in wave reflection, 403, *403*
Mass distribution(s)
  first moment of, 256
  moments of, 286
Mass element(s), 451
  of string in sinusoidal wave, 396, *396*
Mass of constituents, 1103
Mass of fluid displaced in Archimedes' principle, 427

Master equation
  for computing electric potential
      function, 647
  for refraction, 997
Master interference equation, 1036
  coherence and, 1047
  incoherence and, 1037, 1046
Materials
  magnetic properties of, 867–884
  magnetic susceptibility of, 875–876,
      876t
  sources of magnetism in, 870–872,
      872t
Matrix algebra, 1016t, 1019
Matrix method for complex circuits, 763–
      765, 765
Matrix multiplication, 763
Matter
  mechanical properties of, 414–440
  states of, 415–416, 416t
Maximum acceleration for simple
      harmonic motion, 380
Maximum current through a resistor,
      potential drop across terminals
      and, 893, 894
Maximum frictional force, 130
Maximum phase angle in single-slit
      diffraction, 1054
Maximum velocity for simple harmonic
      motion, 380
Maxwell, James Clerk, 476, 583, 839,
      958, 1079, 1110
Maxwell speed-distribution function, 479
  in molecular speeds, 576, 576–577,
      577
Maxwell's equations, 583, 946, 953,
      1119
  in Bohr model of hydrogen atom,
      1128
  displacement current and, 839, 839–
      842, 840
  prediction of electromagnetic waves
      from, 933–935, 933–937
  specular reflection and, 957
  summarized, 841–842
Mean free path, 479, 577, 577–578
  oxygen tank example of, 578
Mean temperature coefficient of
      resistivity, 726, 727, 738
Measurement, 2–15
  models, 3–5
  significant figures, 13–15
  systems of units, 5t, 5–10
  unit conversion and dimensional
      analysis, 10–13
Measurement system, for investigating
      geologic conductors, 887
Measuring devices, 753–754, 754
  ammeters, 755, 755
  potentiometer, 756, 756–757
  voltmeters, 754, 754–755
  Wheatstone bridge, 756, 756

Mechanical energy, conservation of, 190–
      196, 193, 194
Mechanical properties of matter, 414–
      440
Mechanical waves, 388
Meissner effect in diamagnetism, 877
Mercury
  millimeters of, 431
  in thermometer, 485–486
Mercury barometer, 423, 431, 431–432
  straw example and, 431–432
Merging highway traffic, continuity
      equation and, 448–449
Meridional rays, 962, 962
Merry-go-round
  example of angular momentum, 301,
      301–302
  example of torque vector, 340, 340–
      341
Metal(s), as conductors, 717–718
Metal detectors, 886
  inductors in, 860
Metal-oxide-semiconductor in quantum
      Hall effect, 785
Meter, 5, 15
Meter-kilogram-second (mks) system of
      measurement, 10
Method of relaxation, 672
  in field equations, 666–671, 667
Metric system, 10, 10t
MHD (magnetohydrodynamics), 785–
      787, 786
Michelson, Albert Abraham, 1080–1082
Michelson interferometer, 1080, 1080–
      1084, 1081, 1083
Michelson-Morley experiment, 1081,
      1081–1084, 1083, 1087
Microcrystalline regions, 868
Microfarad, 680
Microwave region of electromagnetic
      spectrum, 945, 946
Miller, Dayton C., 1083–1084
Millikan, Robert, 1116, 1117, 1120
Miniature golf (example of circular
      motion), 134, 134–135
Mirror, to focus short-wavelength X-rays,
      1064, 1064
Mirror equation, 992, 1019
  concave mirror example of, 993, 993
  sign convention for, 1020
mks system (meter-kilogram-second) of
      measurement, 10
Mlynek, J., 1137
Mode, wavelength and, 1113
Models, 3–5, 15
  construction of, 4–5
Modern physics, 1077 (see Twentieth-
      century physics)
Modulus(i), 418
  proportionality constant as inverse of,
      418
  values for, 419t

Molar heat capacity(ies), 500, 509
  for common substances, 501t
  ratio of, in adiabatic processes, 525
Molar specific heat
  at constant pressure, 522
  at constant volume, 522
  stating processes for, 522
Molar specific heat capacity, 524
Mole(s), 9, 15, 500
Molecular mass (M), 9
Molecular motion
  pressure and
    in ideal-gas model, 565, 565–568,
        567
    in kinetic theory of ideal gas, 565,
        565–568, 567
  temperature and
    in ideal-gas model, 568–579
    in kinetic theory of ideal gas, 568–
        570
Molecules, in ideal-gas model, 565
Moment-arm method
  of finding net torque, 294, 294–295
  for force and torque, in static
      equilibrium, 314
  for torques due to weight force, 313
Moment-arm of force, 293
Moment of inertia, 284, 286
  Angular acceleration and, for given
      torque, 297
  in example of force and torque, 323
  in example of rotational inertia, 287
  in example of tangential velocity, 289t,
      291
  mass of door and, 288
Moment of inertia integral, in rotational
      inertia example, 287
Moments of inertia, in no-slip condition,
      317
Moments of mass distributions, 286
"Moments to remember" (example of
      magnetic moment), 793, 793
Momentum, 241, 1104
  center of mass and, 259–262, 260, 261
  conservation of, 226–232
  definition of, 221–222
  differentiated from energy, 250
  impulse and, 221, 221–223, 223
  lone canoeist example of, 260, 260
  of photons, 1119, 1133
    example of, 1120
  redefining, relativity and, 1099–1100
  reducing to zero, Newton's second law
      and, 223–224
  as vector, 220, 222
Momentum equation, relativistically
      correct, 1100
Momentum statement of Newton's
      second law of motion, 223–226,
      224, 226
Monatomic ideal gas(es), specific heats
      of, 572

Monatomic ideal gas models, equipartition-of-energy theorem for, 578–579, 579t
Monochromatic aberration, 1021
Monochromatic light, 955
Monochromatic plane wave, light as, 954, *954*
Monopoles, 817
Montgomery, Carla W., *499*
Morley, E. W., 1083
MOSLM (Magnetooptic spatial light modulator), example of, *807*, 807–808
Motion
  along a straight line, 40–69
    applying integral calculus to find position and velocity, 60–61
    constant-acceleration model, *52*, 52–59, *54*
    graphical interpretation of, 46–52
    handling real data, 61–67
    position, velocity, and acceleration, 41–46, *42*
  direction as critical part of, 220
  Newton's first law of, 103–104, *104*
  Newton's second law of, 104–105
    force and acceleration in, weight and tension example, 116–117, *117*
    friction example, *115*, 115–116
    spring force example, *111*, 111–112
    tension and friction examples, *112*, 112–115, *114*
    units in, 105
  Newton's third law of, 117–119, *119*
  through resistive medium, 142t, 142–146, *143*
  in two dimensions, 76–94
    circular, *85*, 85–90, *87*
    numerical techniques in, 90–93, *91–93*
    projectile, 79–85
    velocity and acceleration vectors, 77–79
Motional emf, 833
  conducting rod example of, 833–834, *834*
Mountain climbers' lunch (example of position and velocity), *58*, 58–59
Moving charges
  Gauss's law valid for, 626
  in nonuniform magnetic fields, *781*, 781–782, *782*
  in uniform magnetic fields, 779–781, *779–781*
Moving dresser (example of frictional forces), 128–130, *129*
Moving object, power delivered by, 178
MRI (magnetic resonance imaging), 861, *861*, 931
μ mesons ("muons"), 1090, 1091

Multimode optical fiber(s), 962, *962*
  numerical aperture of, 963
    example of, *962*, 963–964
Multiple-loop circuits, Kirchhoff's laws and, 749–753, *750*
Multiple point-masses, vector relations and, *351*, 351–352
Multiple-source interference, *1044*, 1044–1046, *1045*
  phasors in, 1044–1047, *1045–1047*
Multiple wavefronts, 955, *955*
Multiplication by scalar, 25
Multiplier, 726, *726*, 726t
"Muons," 1090
  example of frame of reference, 1091
Muscular force in biomechanical loading, 336
Musical acoustics, 477–481
  musical sound in (*see* Musical sound)
Musical instruments, resonance in, 378
Musical sound
  perception of, 480–481, *481*
  production of, 478, *478*
  transmission to listener, 479–480
    in concert halls, *479*, 479–480
    by sound recording and reproduction, 480
Mutual inductance, 862, 887
  of circuits, 849–850
  solenoid system example of, 852, *852*
Mutual time dependence in projectile motion, 79

National Institute of Standards and Technology (NIST), 6, *6*, 8, 106, 494
Natural convection, 498–499
Natural frequency, 377
Near-field diffraction patterns, 1053
Near point, 1008, *1010*
  of normal vision, 1021
Negative angular acceleration, 280
Negative average velocity, 43
Negative charge, 585
  electric field of, 597, *597*
Negative derivative
  of electric potential, electric field and, 658, 672
  of potential energy function, 199
Negative gradient of electric potential, electric field and, 672
Negative gravitational potential energy, 206
Negative integral of force function, 199
Negative potential energy, 189, 204, 645
Negative uniaxial birefringent crystals, 973
Negative work, 164–165
  done by kinetic friction, 197

Net force, 104
  calculating, 226–232
  definition of impulse and, 221
  momentum and, 226
Net magnetization in ferromagnetism, 880
Net torque
  on electric dipole, *611*, 611–612
    right-hand rule and, 612
  wheel example of, *294*, 294–295
Net work performed
  in cyclic thermodynamic process, 529, *529*
  in piñata example, 168
Neutral equilibrium, 201
Neutrino, 1090
Neutrons, 586
Newton, Isaac, 87, 137–138, 140, 141, 202, 238, 349, 488, 953, 1084
Newton-meters, 293
Newton (N), 105, 120, 776
Newton-second, 220
Newtonian gravity
  potential energy for, 202–206, *203–205*, 211
  potential energy function for, 203, 204, *204*
Newtonian mechanics, 1
Newtonian telescopes, 1013, *1013*
Newton's first law of motion, 103–104, *104*, 136, *136*
Newton's rings
  calculating radius of curvature in, 1034
  example of, *1033*, 1033–1034
Newton's second law of motion, 104–105, 241
  for angular momentum, *345*, 345–346
  applications of, 110–117, *111*, *112*, *114–117*, 119–120
  applied to wave pulse, 395
  center of mass and, 258–259, *259*
  centripetal acceleration and, 131
  circular motion and, 134, 135
  for linear motion, 304t
  momentum statement of, 223–226, *224, 226*
  numerical methods for, 146t, 146–152, *147, 149–151*
  recast in terms of momentum, 224
  for rotational motion (vector form), 304t, 346
    change in angular momentum and, 347
  in simple harmonic motion
    rotational form, 368
    for two-mass system, 370
  in work and kinetic energy, 170–171
Newton's third law of motion, 117–119, *119*
  average force and, 225
  cancellation of internal forces and, 227

Coulomb's law and, 591–592
in wave reflection, 403
Newton's universal law of gravitation
(NULG), 136–142, 152
Cavendish balance in, *138*, 138–139
for spherical masses, 202
"weighing" the sun and, 139–140
weight and universal gravitation and,
140–142, *141*
90° pulse, 931
NIST (National Institute of Standards
and Technology), 494
NMR (*see* Nuclear magnetic resonance)
No-slip condition, 317, 327
in energy analysis of rotation and
translation, 318
in example of force and torque, 323
in force analysis of rotation and
translation, 320, 322
in tangential acceleration, 298
Node(s), 400–401, 409, 686
Non-Hooke's-law spring
example of work, 168–169, *169*
example of work done by external
force, 168–169, *169*
"Nonconservation" condition, entropy
as, 557–558
Nonconservative forces
conservation of energy and, 187
in isolated system, 211
systems with, 196–199, *198*
slingshot example of, 197–199, *198*
Noncyclic permutation, 34
Noninteracting phase before collision,
conservation of momentum and, 232
Nonisolated systems, application of
conservation of mechanical energy
to, 196
Nonohmic materials, 722
validity of resistance equation for, 724
Nonpolar dielectric materials, 695–696,
696t
Nonsimple circuits, 749
Nonslipped disk (example of force and
torque), *323*, 323–324
Nonsymmetrical solid objects, center of
mass for, 254–257, *255, 256*
Nonuniform electric field(s)
electric dipole in, *614*, 614–615
electric potential change in, 652–654
Nonuniform linear charge distribution,
example of, 589, *589*
Nonuniform magnetic fields, moving
charges in, *781*, 781–782, *782*
Nonuniform volume charge distribution,
example of, 589–590, *590*
Nonviscosity in ideal-fluid model, 432
Nonzero integration constant, 896
Nonzero resistance, 913
Nonzero rotational inertia, special
precautions for, 297

Normal contact force in circular motion,
135
Normal conversation (example of sound
intensity level), 461t, 461–462
Normal dispersion, 964
Normal forces, 109, *109*, 120
work done by, 165
Normal frequencies, 409
of a string, 402
Normal modes
displacement waves for, *466*, 466–467
frequencies of, 467
wavelengths of, 402, *402*, 467
Normal stress, 417
"Now I c" (example of frame of
reference), *1097*, 1097–1098
*N*th moment of mass distribution for
advanced mechanics, 286
n-type semiconductors, 731, *731*
Nuclear magnetic resonance (NMR), 827,
860, 931
Nucleons, 1101
Nucleus of atom, 586, 1134
NULG (*see* Newton's universal law of
gravitation)
Number density, elastic molecular
collisions in ideal-gas model and,
567
Numerical aperture, 976
of multimode fiber, 963
example of, *962*, 963–964
Numerical calculations for periodic
motion, 378–380
equation of motion, 378
large-amplitude pendulum oscillations,
380, *380*
velocity and acceleration, 379, *379*
Numerical integration techniques, *176*,
176–178, *177*
Numerical methods
in field equations, 666–671, *667*
for handling real data, 69
for Newton's second law, 146t, 146–
152, *147, 149–151*
for paraxial ray tracing, 1015–1019,
1016, 1016t
Numerical techniques, in two dimensions,
90–93, *91–93*, 98–99

Oak Ridge National Laboratory, 472
Object(s), motion of, work done and,
170, 171
Object distance, 991, 1019, 1020
Object position, image position and, 991,
991
Objective lens, 1021
of compound microscope, 1011
Observer, frequency change due to
motion of, 468
Oersted, Hans Christian, 801

Ohm, Georg, 722
Ohmic materials, 722, 737
Ohms, 722, 737
as unit of impedance, 902
as units of capacitive reactance, 896
as units of inductive reactance, 898
Ohm's law, 722, 724, 737
One dimension
collisions in, 232–238
obtaining force from potential energy
in, 199
One-dimension, collisions in, from center-
of-mass frame, *260*, 260–261
One-dimensional fields, computing
electric field by negative
derivative in, 658
One-dimensional motion, applying Runge-
Kutta method to, 148–149, *149*
One-dimensional potential energy curve,
201, *201*
One-dimensional potential energy
function, 200
force function and, 211
One-dimensional wave(s), 387–409
particle behavior in time and, 389–
390, *390*
particle motion with respect to wave
direction in, *388*, 388–389
sinusoidal, energy transported by, *396*,
396–397
superposition and interference of, 397–
404
differing in direction only, 400–402,
401, 402
differing in frequency only, *399*,
399–400
differing in phase only, *398*, 398–
399
Fourier's theorem, 403–404, *404*
spreadsheet calculations for, 407–
408
wave reflection, *402*, 402–403, *403*
traveling on string, 390–395, *391*
longitudinal, 394–395
sinusoidal wave train, *392*, 392–395
velocity of, *395*, 395–396
wave pulses, 390–392, *391*
wave dimension in, 389, *389*, 390
wave equation and, 404–407, *405*
analysis of: wave velocity, 406–407
One-dimensional wave equation, 409
Open circuit, 746
Opera glasses, diverging lens in, 1012
Optic axis, 1019
of birefringent crystal, 973
Optical aberrations, 1013–1015
chromatic, 1014, *1014*
third-order, 1013–1015, *1015*
Optical beats, 1037–1038, 1047
Optical fibers, refraction and, *962*, 962–
964

Optical path length(s), 1032
  phase differences and, 1047
  two-source interference patterns and, 1042
Optical systems, using X-rays, *1064,* 1064–1065
Optical wave-guides, 962
Optically inactive birefringent crystals, 973
Optics of compact-disc player, polarization and, *975,* 975–976
Orbital magnetic dipole moment, 868
Ordinary wave in birefringence, 972, 973, *973*
Ore bodies, 888–889
Origin, 19, *19*
Oscillating dipole model
  of electromagnetic radiation, 942, *943,* 944
  of electromagnetic waves, 947
    unpolarized light and, 966
  for polarization by scattering, 967, *967*
Oscillating field, root-mean-square value of, 940
Oscillating particle (example of simple harmonic motion), 362–363
Oscillating piston to generate sound, 452, *452*
Oscillation(s), 356–381
  damped, *375,* 375–377, *376*
  driven, *377,* 377–378, *378*
  dynamics of simple harmonic motion and, 363–371
    mass attached to string, 363–366, *364*
    physical pendulum, *368,* 368–369
    simple pendulum, *366,* 366–368
    torsion pendulum, *369,* 369–370
    two-mass system, *370,* 370–371
  energy of simple harmonic oscillator, 371–373, *372*
  kinematics of simple harmonic motion and, *357,* 357–363, *359*
    velocity and acceleration for, 361–363, *362*
  numerical calculations for periodic motion, 378–380
    equation of motion for simple harmonic motion, 378
    large-amplitude pendulum oscillations, 380, *380*
    velocity and acceleration for simple harmonic motion, 379, *379*
  of sound wave in gas, 451
  uniform circular motion and simple harmonic motion, 374, *374*
Oscilloscope, *753*
Osgood, T. H., 1127
Otto cycle, 550, *550*
  gasoline engine example of, 550–551
Ounce-inches, 293
Outdoor concert speaker (example of sound intensity level), 461t, 462

Outer product (*see* Vector product)
Output voltage signal, input voltage signal and, 761, *761*
Over-relaxation parameter, 668
Overdamped motion, 376, *376*
Overdamped system, RLC circuits and, 859
Overlapping spectra from different orders, 1063, *1063*
Oxygen tank (example of mean free path), 578

Parabola
  in computing areas under a curve, *62,* 63
  in computing slope, 65, *65*
Parabolic mirror, 986, *987*
Parabolic trajectory
  equations of motion for electric charge and, 611, 611t
  in projectile motion, 83, *83*
Paraboloid of revolution, 986, *987*
Parallax, locating images by, 998
Parallel arrangement of capacitors, *685,* 685–687, 709
  circuit diagram for, *685*
  equivalent capacitance of, 687
  series arrangement and, 686, *686*
Parallel-axis theorem, 289t, 290
Parallel combination of resistors, *746,* 746–749
  example of equivalent resistance, 748, *748*
  example of potential drop, 748–749, *749*
Parallel connection of voltmeter, 754–755, *755*
Parallel force components, 293, *293*
Parallel particle motion in longitudinal waves, 388
Parallel-plate capacitor
  energy density in, 692–693
  example of capacitance, *681,* 681–682
  example of dielectrics, 701–702, *702*
  ideal, 681, *681*
  magnetic analog of, *811,* 811–812, *812*
Parallel-plate model, in example of electric field strength, *665,* 665–666
Parallel wires
  amperes, coulombs and, 808–809, *809*
  in experimental determination of magnitude of ampere, 809
Paramagnetic phase, 869, 884
Paramagnetism, 869, 878
Paraxial ray(s), 1019
  spherical mirror and, 986, *987*
  tracing, 1021
  numerical methods for, 1015–1019, *1016,* 1016t

Paraxial ray approximation, 1016
  first-order model of, 1013
Partial derivative(s)
  combined with unit vectors (gradient), 207
  for pressure change, 458
  transverse velocity of sinusoidal wave as, 394–395
Partial derivative test, 208–210
  on force function, example of, 209
Partial reflection, 976
Partially elastic collisions, 236t, 236–238, *238*
Particle, wavelength of, momentum of photon and, 1121–1122
Particle behavior in time, 389–390, *390*
  sinusoidal wave train, 390, *390*
  wave pulse, 389, *390*
Particle model
  of light, 953–954, 1110–1121
    blackbody radiation, *1110–1113,* 1110–1114
    Compton scattering, 1118–1121, *1119–1121*
    heat capacity of solids, *1114,* 1114–1115, *1115*
    photoelectric effect, 1115–1118, *1116–1118,* 1118t
    speed of light in, 958
  of potential energy function, example of, 200–201, *201*
Particle motion
  Runge-Kutta method for determining, *147,* 147–152, *149–151*
  wave direction and, *388,* 388–389
Particle trajectory in uniform magnetic field, aspects of, *779,* 779–780
Pasaschoff, Jay, 185
Pascal, Blaise, 425
Pascal (Pa), 420, 420t, 422, 440
Pascal's principle, 426, *426,* 440
Paschen, F., 1126
Paschen series of spectra, 1126
Path, work dependent on, *518,* 518–520
Path length
  conditions in example of Bragg planes, *1069,* 1069–1070
  of thin-film interference, 1031–1032
Path variables, 519
  work and heat as, 520
Pauli exclusion principle, 729, 733, *734*
Peak current, 912
Peastrel, Mark, 157
Percussion instruments, 478
Perfect conductor, zero resistivity of, 723
Perfect insulator, infinite resistivity of, 723
Perfectly elastic collision(s)
  desktop toy example of, 238, *238*
  gliders example of, 235, *235*
  in one dimension, 234–235, *235*
  in production of sound, 453, *453*

relativistic momentum and, *1099,* 1099–1100
Perfectly inelastic collision(s), 233
  ballistic pendulum as, 251, *251*
  in one dimension, 233–234
Period, 358
  of planet's orbit, 139
  for position equation of sinusoidal wave train, 392
  of sinusoidal wave, 394
  of wave motion, 396
Periodic boundary conditions, 671
Periodic motion, numerical calculations for, 378–380
  equation of motion, 378
  large-amplitude pendulum oscillations, 380, *380*
  velocity and acceleration, 379, *379*
Permanent dipole moment(s), 868
  paramagnetic and ferromagnetic states and, 870
  in paramagnetism, 878
  polarity and, 613, *613*
Permanent dipoles
  in dielectric material, 695, 696
  of water molecule, solubility and, 615
Permeability
  magnetic induction and, bismuth example of, 882–883
  magnetic susceptibility and, 883
Permeability constant, 802
Permeability of free space, 802, 818
Permittivity, 701
Permittivity constant, 591, 710
Permittivity of free space, 591, 615, 701
Perpendicular-axis theorem, 289t, 290, *290*
Perpendicular particle motion in transverse waves, 388
Pfund, H. A., 1126
Pfund series of spectra, 1126
Phase
  differences in, wave interference and, *398,* 398–399
  of two-source interference pattern, 1058
Phase angle, 900, 923
  change in, 1061, *1061*
Phase change
  in reflected wave, 403, *403*
  of thin-film interference, 1031–1032
Phase constant, 358, 380
  dependence on initial position and velocity, 359, *359*
  for RLC circuit, 904, 906
  in wave interference, *398,* 398–399
Phase difference in Michelson interferometer, 1082
Phase gratings, 1060
Phase relations for RLC circuit, 902t
Phase velocity
  of sinusoidal wave, 394, 395

of sound waves in pipes, 456
of wave, 390
Phasor(s), 893–894, *894,* 1061
  in multiple-source interference, 1044–1046, *1045–1047*
Phasor addition, 1045
Phasor description for RLC circuits, *900,* 900–901, *901*
Phasor diagram(s), 893–894, *894,* 1047
  of capacitor, 895, *895*
  for RLC circuit, 906, *906*
  for RLC circuits, *900,* 900–901, *901*
Phasor sum of currents in transformer, 916
Phonograph records, sound reproduction of, 480
Phonons, 718
Photoelectric effect, 1115–1118, *1116–1118,* 1118t, 1133
  parameters associated with, 1117
Photoelectric equation, 1117–1118, 1133
Photoelectric work function, cesium example of, 1118
Photon(s), 1115, 1133
  example of momentum of, 1120
  momentum of, 1119
Physical optics, 951, 955
Physical pendulum
  child's hoop example of, 369, *369*
  simple harmonic motion and, 368, 368–369
Picofarad, 680
Piñata
  example of conservation of mechanical energy, 194–195
  example of work, 167–168, *168*
Pitch, 481, 781
Pitcher and glasses (example of center of mass), *253,* 253–254
Pitot tube (example of Bernoulli's equation), *436,* 436–437
Pivot point, 346
  for torques in example of static equilibrium, 316
Pixels, altering light-transmission characteristic of, example of, *807,* 807–808
Planar objects, center of mass for, 255, *255*
Planck, Max, 1112, 1113, 1115, 1120, 1128, 1132
Planck's constant, 785, 871, 930, 1113, 1118, 1133
Planck's radiation formula, 1133
Plane mirrors
  image location example, 986, *986*
  images formed by, *984,* 984–986, *985*
  images reversed in, 985, *985*
  small mirror example of, 986, *986*
Plane of circular motion, vector perpendicular to, 343, *343*
Plane of incidence, 956, 976

Plane of polarization, *944,* 944–945, 974, 976
Plane-polarized waves, 976
Plane wave(s), 451, *451,* 934
Plane-wave model, of electromagnetic radiation, 944, *944*
Plane wavefronts, sound waves with, 457
Plasma(s), 781, 786
Plastic(s), wire-grid polarizers made from, 971
Plastic behavior, 420, *420*
Plastic deformation, 416–417
Plates, 679, 709
"Plum pudding" model of atom, 1127, 1128
Pohlmann, K. C., 481
Point-charge(s)
  computing electric field and, *597,* 597–598
    problem-solving strategy for, 598
  electric potential function for, 671
  equipotential surfaces of, 662, *663*
  example of electric field, *636,* 636–637
  example of equations of motion, *610,* 610–611
  multiple (in example of computing electric field), 599–601, *600*
Point-charge distribution(s)
  computing electrical potential from, 655–656
    example of, 655, *655*
  computing electrostatic potential energy from, example of, 656, *656*
  in electric field, 596–601, *597*
Point-mass(es)
  multiple, vector relations and, *351,* 351–352
  net torque on, 345
  in rope swing example of work, 170
  single, vector relations and, *350,* 350–351
  time derivative of angular momentum of, 345
Point-objects, collection of, center of mass for, 252–254, *253*
Point of apparent origin, finding, *988–990,* 990–991
Point of contact in perfectly elastic collisions, 240
Point-particle(s), 269
  extended objects as, 344
  rotating about fixed axis
    angular momentum for, 299–300
    in merry-go-round example, *301,* 302
Point-particle model(s)
  analysis of, center-of-mass motion and, 327
  applied to circular motion, 134
  center of mass and, 257

Point-sources in wavefront model, 1054, *1054*
Polar coordinates, 19, *19*
  Cartesian coordinates and, 19–20
Polar dielectric materials, 695–696, 696t
Polarity, permanent dipole moment and, 613, *613*
Polarization, 710, *966*, 966–976
  birefringence and, 972–973, *973*, 973t, *974*
  of dielectric, 705
  displacement field and, Gauss's law and, *703*, 704–705, *705*
  half-wave and quarter-wave plates and, *974*, 974–975
  of nonpolar dielectric materials, 695
  optics of compact-disc player and, *975*, 975–976
  by reflection, 968–970, *968–970*
  by scattering, *967*, 967–968
    ''blue sky'' and, 968
  by wire-grid polarizers, 970–972, *971*
Polarization axis, 971, *971*
Polarizing angle (Brewster's angle), 969, *969*, 976
  glare example of, *968*, 969
Polarizing beam splitter(s), 970, *970*
  in compact-disc player, *975*, 975
Polychromatic light, 955
Porro prisms, 1012–1013, *1013*
Position, 41–42, *42*, 68, 93
  computing acceleration from, 62t, 66, 66t, *67*
  conservative forces as function of, 200, 210
  velocity and, example of, *58*, 58–59
  from velocity graphs, 49–51, *50*, *51*
Position equation for sinusoidal wave train, 392
Position function
  of sinusoidal wave, 396
  in wave equation, 404, 406
Position-versus-time graph
  average velocity from, 47, *47*
  instantaneous velocity from, *47*, 47–48, *48*
Positive charge, 585
  electric field of, 597, *597*
Positive charge flow, 737
Positive uniaxial birefringent crystals, 973
Postulates of special theory of relativity, 1084–1085
  simultaneity and, 1085, *1085*
Potential change
  Kirchhoff's laws and, example of, *750*, 750–752
  rules for, 750, *750*
Potential drop, 915, 923
  across inductor, current source example of, 899, *899*
  across terminals, maximum current through a resistor and, 893, *894*

parallel combination of resistors
  example of, 748–749, *749*
  for RLC circuit, 904
    amplitudes of, 907
Potential energy, 187–190, 794
  conservative forces and, 662, 662t
    in conservation of energy, 187–188
  of electric dipole, 616
  gravitational, near earth's surface, *188*, 188–190, *189*
  of Hooke's-law spring, 190, *190*
  importance of changes in, 189
  incremental change in, 188
  of moving fluid element, 435
  negative, 189
  negative derivative of, 199–202, *200*, *201*
  for Newtonian gravity, 202–206, *203–205*, 211
  in simple harmonic motion, 373
  of simple harmonic oscillator, 381
  of two-point charges, 643–646, *644*, *645*
Potential energy curve, lack of symmetry in thermal expansion, 489, *489*
Potential energy function(s), 192–193
  conservative force and, 206, 643
  electric dipole and, 612, *612*
  of Hooke's-law spring, 200
  impossibility of defining for nonconservative forces, 197
  negative derivative of, 199
  for Newtonian gravity, 203, 204, *204*
  particle model of, example of, 200–201, *201*
  relationship between objects and, 196
  in thermal expansion, 488–489, *489*
  of two-mass system, 370
Potential energy per charge, electric potential and, 646
Potentiometer, *756*, 756–757, 766
Pound force, 120
Power, 174–175, 178
  applied to rotating object, 296
  football trainer example of, 175
  resonance and, in AC circuits, 910–915, *911*, *913*
  in RLC circuit, example of, *914*, 914–915
  sound intensity level and, 462
  as time rate of energy transfer, 459
  transferred per unit area, intensity as, 459
  transmitted by sinusoidal wave, 397
Power factor, 911, 924
  determined from impedance diagram, *911*
Power loss from ideal transformer, example of, 920
Power per unit area (*see* Irradiance)
Poynting vector, 939, 940, 947
Precedence effect in concert halls, 479

Precession
  of Earth's axis of rotation, 348–349, *349*
  of gyroscope, *346*, 346–348, *347*
  rate of, 347
    gyroscope example of, 347–348
Precession period, 347
Precessional angular frequency, 347
Press, T. A., 668
Pressure, 419, *419*, 422, 440
  density and, *422*, 422t, 428, 432–423
  of fluid, dam example of, 425, *425*
  of gas, 578
  molecular motion and, in kinetic theory of ideal gas, *565*, 565–568, *567*
  variation with depth, *423*, 423–424
Pressure amplitude, in harmonic waves, 458
Pressure-measuring devices, 422, *422*, 430–432
  manometer, *430*, 430–431
  mercury barometer, *431*, 431–432
Pressure variation, pressure wave as, 457, *457*
Pressure wave amplitude, sound intensity and, 459
Priestly, Joseph, 679
Primary coil of transformer, 916, *916*, 924
Primary field, 832
Primary focal point, 1002
Primary vibrating system, music and, 478
Principal plane of thin lens, 1001
Principia (Newton), 137
Principle axes, 352, 353
Principle of increasing entropy, *556*, 556–558
Principle of superposition, 397, 592, 1029, 1046
  in composite wavefronts, 1043, *1043*
Principles of Philosophy (Descartes), 238
Prism(s), 976, 1060
Prism geometry, refraction and, *965*, 965–966
Probability, 14
Problem-solving strategy
  for analyzing thermodynamic cycles, 526t, 529
  to compute capacitance, 680
  for computing electric field from continuous charge distributions, 602
  for computing electric field from point charges, 598
  for conservation of mechanical energy, 192
  for constant-acceleration model, 53
    example of, 53–54, *54*
  for force analysis of rotation and translation, 323
  for Gauss's law, 628

using Newton's second law, 110
for vector addition, 27
Process, 532
in thermodynamics, 515
"Product rule," in adiabatic processes, 525
Projectile(s), 79
Projectile motion, 79–85, 93
constant-acceleration model in, *79*, 79–80
examples of, *82*, 82–85, *83*
Propagation
of sound wave in gas, 451
of wave, 396
Proper distance, frame of reference and, 1091
Proper frequency, 1092
Proper length, 1089, 1104
Proper time, 1088, 1104
Proper time interval, 1092
Proportional limit, 420, *420*
Proportionality constant, 137–138, 418, 490, 490t
for thermal conductivity, 495
Protons, 586
"Pseudo" forces, 136, *136*, 152
Pseudoequipotential lines, 671
Psychoacoustics, 479, 480
p-type semiconductors, 731, *731*
Pulse, 389–390, *390*
Pulse-echo technique, in nuclear magnetic resonance experiments, 860–861
Purcell, Edward, 931
Pythagorean theorem, 26, 1081
in two-source interference patterns, 1040

"Q" of a system, resonance peaks and, 913–914, *914*
Quadratic model, 152–153
in calculating Reynolds number, 144
errors in correcting for air resistance and, 146
of fluid resistance, 143
Quantization, 1110
of energy
guitar string illustration of, 1113–1114
rotational and vibrational, 574–575
Quantization condition, in Bohr model of hydrogen atom, 1130
Quantized charge, 586
Quantum Hall effect, 785
Quantum mechanical effects in vibrating diatomic ideal gas, 573
Quantum-mechanical state, energy levels and, 731
Quantum mechanics, 1132
applied to paramagnetism, 878
model of material conductivity, 731

Planck's constant in, 785
sources of magnetism in materials and, 870–872
Quantum number, 1129
Quantum physics, 4, 1109–1134
magnetic moment and, 930
particle model for light and, 1110–1121
blackbody radiation, *1110–1113*, 1110–1114
Compton scattering, 1118–1121, *1119–1121*
heat capacity of solids, *1114*, 1114–1115, *1115*
photoelectric effect, 1115–1118, *1116–1118*, 1118t
Rutherford-Bohr model of hydrogen atom, 1127–1132
Bohr model and, 1128–1132, *1129–1132*
Rutherford scattering and, *1127*, 1127–1128, *1128*
spectra of atoms, *1125*, 1125–1127
wave model for particles and, 1121–1124
Davisson-Germer experiment, *1122*, 1122–1123, *1123*
experiment of G. P. Thomson, 1123–1124, *1124*
Quarks, 587
Quarter-wave plate(s), 976
in compact-disc player, 975, *975*
polarization and, *974*, 974–975
Quasi-static process, 516

R-value
for building insulation, common materials used, 498t
compound slab example of, 498, 498t
Racquetball (example of coefficient of restitution), 236
Rad, 85, 86
Radian(s), 85
Radian measure for angular position, 277
Radians per second, 278
Radiation, convection and, for transfer of heat and energy, 498–499
Radiation pressure
of electromagnetic waves, *941*, 941–942
"snow flakes" example of, 942, *942*
Radii (example of sign convention in Lens-maker's formula), *997*, *999*, 1001
Radio frequency region, nuclear magnetic resonance and, 931
Radio wave (example of frequency of electromagnetic waves), 938, *938*
Radius of curvature, 1019
calculating in Newton's rings experiment, 1034

Radius of orbit of particle in uniform magnetic field, 779–780
Rafting (example of force and torque in static equilibrium), *314*, 314–315
Railroad yards, flow of vehicles through, continuity equation and, 448
Random motion of molecules in ideal-gas model, 565
Range of focal lengths, 1009
Range of validity of theory(ies), 4
Rare-earth elements, unusual magnetic behaviors of, 870
Rarefactions, 451, *451*
Raspberry jello (example of shear modulus), 421
Ray(s), 956, *956*
wavefronts and, in wave model of light, *954*, 954–955, *955*
Ray direction angle, calculating, 1055
Ray model of light, 956, *956*
Rayleigh, Lord (Strutt, John William), 1112
Rayleigh criterion, 1056, *1056*, 1070
"license plate" example of, 1057, *1057*
for resolution, 1062
sodium doublet example of, 1062
Rayleigh-Jeans law, 1112
Rayleigh waves, 389
RC-circuits, *757–759*, 757–762
charging, *757*, 757–759, *758*
flash unit example of, *760*, 760–761
R-DAT (rotating heads), 480
Reaction, action and, in Newton's third law of motion, 117–119, *119*
"Reactionary problem" (example of capacitor in AC circuit), *898*, 898–899
Real charge, 697
Real image(s), 984, 1019
complete formation of, *1006*, 1006–1007, *1007*
converging lens, real object and, example of, *1002*, 1002–1003
Real object
converging lens, real image and, *1002*, 1002–1003
converging lens, virtual image and, 1003, *1003*
diverging lens, virtual image and, 1004, *1004*
Recoil, 227–230, *228*, *229*, 242
Recording tape, sound reproduction of, 480
Rectangular coordinates, 19, *19*
"Red-shifted" light caused by Doppler effect in astronomy, 471
Reduced mass, 371
Reducibility of theory(ies), 4
Reference angle, 1021
Reflected irradiance, 1039

Reflection, *956,* 956–957, *957*
    polarization by, 968–970, *968–970*
        Brewster's law of, *968,* 969
            percentage of power reflected, *968,* 968–969
Reflection coefficient, 962
Reflection gratings, 1060
Refracting surfaces, images formed by, 995–999
Refracting telescopes (refractors), 1012
Refraction, 957–966, 995
    at curved surfaces, *996,* 996–997, *997,* 997t
    dispersion and, *964,* 964–965, *965*
    master equation for, 997
    optical fibers and, *962,* 962–964
    prism geometry and, *965,* 965–966
    at spherical boundary, 1020
    at spherical interface, sign convention for, *997,* 997, 997t
    total internal reflection and, *960,* 960–962, *961*
Refractors (refracting telescopes), 1012
Refrigerators, heat pumps and, *546,* 546–547
Related bond strength of resonance frequency, 378
Relative amplitudes of component frequencies, as sources of sound, 464, *465*
Relative density (*see* Specific gravity)
Relative dielectric constant, 697
Relative motion, 141–142, 1089, *1089*
    conservation of momentum and, 230–232, *231*
    relinquishing classical view of, 1086
Relative velocities, 241
    combining, 230–231
    in conservation of motion, 264
    coordinate system and, 231
Relative velocity expression, 232
Relative velocity subscript rule, 265
Relativistic Doppler formula, 1094
Relativistic Doppler shift, *1092,* 1092–1093, 1104
Relativistic dynamics, 1098–1101
    kinetic energy and, 1100–1101
    relativistic momentum and, *1099,* 1099–1100
Relativistic kinematics, 1085–1098, *1086*
    experimental tests of relativity, 1090–1091
    length contraction, *1089,* 1089–1090
    Lorentz transformations, *1085,* 1086–1087, *1087*
    relativistic Doppler shift, *1092,* 1092–1093
    time dilation, 1088–1089
    twin paradox, 1093–1096
    velocity transformations, *1096,* 1096–1098
Relativistic momentum, *1099,* 1099–1100

Relativistic physics, 4
Relativistically correct momentum equation, 1100
Relativity, special theory of, 1078–1105
Relaxation methods, 668
Relaxation of magnetic moments in MRI, 931
Remnant magnetization, 881–882
Replica gratings, 1060
Resistance, *722,* 723–726, *724, 725,* 737
    defined, 724
    "die hard music" example of, 725
    electric current and, 716–738
    energy dissipation and, *727,* 727–728
    microscopic models of, 729–737
        band model of conductivity, 723t, *731–734,* 732–735
        superconductors, *735,* 735t, 735–737, *736*
        valance-bonding model of conduction, 729–731, *730, 731*
    resistivity and, 722–727
    of resistors, color-code scheme for, 726, *726,* 726t
    taillight example of, 728
    temperature and, 726–727, *727,* 738
Resistivity, 722–723, 723t, 737
    "hard rock" example of, 725, *725*
    resistance and, 722–727
    temperature and, 726–727, *727*
Resistor(s)
    in alternating-current (AC) circuits, *892,* 892–894, *894*
    carbon film, 725
    combinations of, 745–749
        in parallel, *746,* 746–749
        in series, *745,* 745–746
    composition, 725
    resistance of, color-code scheme for, 726, *726,* 726t
Resolution, *1056,* 1056–1057
    criterion for, 1056, *1056*
Resolving a vector, 26, *26*
Resolving power, 1060, 1062
    of diffraction grating, 1060–1062, *1061,* 1071
Resonance, 377, 912, 924
    beats and, as source of sound, 467, *467*
    power and, in AC circuits, 910–915, *911, 913*
    in RLC circuit, example of, *914,* 914–915
    in simple harmonic motion, 381
Resonance equation, 930
Resonance frequency, 377
Resonance peak(s), 912–913, *913*
    full width at half maximum (FWHM) of, 914
    sharpness of, 924
        "Q" of a system and, 913–914, *914*

Resonance phenomenon(a), 377–378, 930–931
    applications of, 931
Resonant frequencies, 467
Rest frame, 1088
    of stationary observer, 1092, *1092*
Rest mass, 1100
Rest mass energy, 1101, 1104
Restoring force, 455
Resultant vector, 24, *24*
Return to noninteracting phase after collision, conservation of momentum and, 232
Reverberant sound in concert halls, 479, *479*
Reverberation time in concert halls, 479
Reversible change of phase (example of entropy), 556
Reversible processes, 516, *516, 517,* 532
    cyclic, heat transfer and, 528
Revolutions per minute (rpm), 278
Reynolds, Osborne, 143
Reynolds number, 143, 146
    calculating, example of, 144
    drag coefficient and, 143, *143*
Richtmyer, F. K., 1067
Right-hand rule, 36, 340, *340,* 344, *344*
    for direction of magnetic field elements, 803
    for direction of magnetic force, 777, *777*
    for magnetic dipole moment, 791
    in magnetohydrodynamics, 786
    for net torque on electric dipole, 612
    for torque on current-carrying loops, 790
    in vector multiplication, 31–32, *32*
        example of, *32,* 32–33
Rigid body, 283
Rigid diatomic ideal gas
    degrees of freedom for, 571, *571*
    specific heats of, 573
    total kinetic energy for, 571
Ring charge, example of computing electric field from, *605,* 605–606
RLC circuit(s), *900–902,* 900–907, 902t, 903t
    current source example of, *903,* 903–905, *904*
    example of power and resonance in, *914,* 914–915
    LC circuits and, 859, *859*
    summary of equations for, 903t
Road median (example of coefficient of thermal expansion), *490,* 491
Rock conductivity, 886–887
Rocket propulsion (example of system with changing mass), 264–267, *265, 267*
Roemer, Olaus, 1079
Roller skaters (example of conservation of momentum), *229,* 229–230

Rolling objects, static equilibrium and, 311–328

Röntgen, Wilhelm, 1064

Röntgen rays (X-rays), 1064

Root-mean-square, example of, 909–910

Root-mean-square emf, 914

Root-mean-square potential, 923–924
current and, 907–910, *908, 909*

Root-mean-square (rms) average, 908

Root-mean-square speed of ideal-gas molecules, 569
argon example of, 569–570

Root-mean-square values, irradiance and, 947

Rope swing
example of work, 170, *170*
example of work-energy theorem, 173

Rossing, Thomas D., 477–481, *478, 481*

Rotating heads (R-DAT), 480

Rotation
about fixed axis, 276–304
angular momentum for, 299–303
Newton's second law for, 296–299
system of equations for, 303, 304t
kinematics of rotational motion and, *277,* 277–283
constant-acceleration model of, 282–283
kinetic energy and rotational inertia in, *283,* 283–292
with no slipping, 316–324
energy analysis in, *317,* 317–319, 319t
force analysis in, 319–320, *320,* 322–323
problem-solving strategy for force analysis of, 323
with slipping, 324–327
angular impulse-angular momentum theorem and, 324–326
center of percussion and, *326,* 326–327, *327*
torque and, *292,* 292–296, *293*
work done by, *295,* 295–296

Rotational impulse, 325, 327
of torque, 324

Rotational impulse-angular momentum theorem, 324–326
scalar version of, 324

Rotational impulse-momentum theorem, 327

Rotational inertia, 284, 286
door example of, *286,* 286–287
with perpendicular axis, *287,* 287–288, 289t, 290
in energy analysis of rotation and translation, 318
kinetic energy and, *283,* 283–292
slender rod example of, 286
of solid object, 285

Rotational kinematics, straight-line kinematics and, 279, 279t

Rotational kinetic energy, 290

Rotational motion
constant-acceleration model for, 282–283
in energy analysis of rotation and translation, 318, 319t
face-centered cubic array example of, 285, *285*
frictional forces in, 321, *321*
of gas molecules, quantum mechanical treatment of, 574
vector descriptions of, 339–353
angular momentum, *344,* 344–349
angular velocity and acceleration, 341–344, *342, 343*
Newton's second law for, 346, 347
torque, *340,* 340–341
vector relations and, 349–352

Rotational quantities, linear quantities and, 343–344

Rotational speed, spinning teacher example of, 303, *303*

Royal Society of London for Improving of Natural Knowledge, 238

Rpm (revolutions per minute), 278

"Rug shuffle" (example of electric field strength), *665,* 665–666

Runge-Kutta method, 152
applied to large-amplitude pendulum oscillations, 380, *380*
applied to two-dimensional motion, 149, *150–151,* 151–152
for determining particle motion, *147,* 147–152, *149–151*

Running, biomechanical loading and, *337,* 337–338

Rutherford, Ernest, 1127, 1128, 1134

Rutherford, F. J., 137

Rutherford-Bohr model of hydrogen atom, 1127–1132
Bohr model and, 1128–1132, *1129–1132*
Rutherford scattering and, *1127,* 1127–1128, *1128*

Rutherford scattering, *1127,* 1127–1128, *1128*

R-value, for building insulation, 497

Rydberg, Johannes R., 1126

Rydberg constant, 1126, 1132

Rydberg formula, 1131, 1134

Satellites, determining mass by, 139

Saturation, 876, 881–882

Saturation magnetization, 876, 881–882

Savart, Félix, 801

Sayre, David, 1065

Scalar(s), 21, 35
energy as, 250
multiplication by, 25
vectors and, *21,* 21–23, *22*

Scalar field, electric potential as, 647, 671

Scalar product(s), 36
in unit-vector representation, 33–35
in vector multiplication, 31, *31*

Scalar quantity, kinetic energy as, 220

Scattering, polarization by, *967,* 967–968
oscillating dipole model of, 967, *967*

Schreiffer, Robert, 735

Scientific notation, 15

Second, 7, 15

Second derivatives for time and position, in wave equation analysis, 406, 406t

Second harmonic (first overtone), 402, 465, 466

Second law of thermodynamics, 538–560
Carnot cycle in, *542,* 542–545, 544t
Clausius statement of, 540
entropic statement of, 559, 560
heat engines and, *540,* 540–541, *541*
Kelvin-Planck statement of, 540
refrigerators and heat pumps and, *546,* 546–548
"curious" heat ratio fraction and, 548
statistical mechanical statement of, 559

Second moment (rotational inertia), 286

Second-order differential equation
partial, wave equation as, 404, 406
in simple harmonic motion, 364

Second overtone (third harmonic), 402, 465, 466

Secondary coil of transformer, 916, *916,* 924
effect of induced current in, 918

Secondary focal point, 1002

Secretariat (example of unit conversion), 11

"Seeing double" (example of free spectral range), 1063–1064

Seidel, Phillip Ludwig von, 1014

Seidel aberrations, 1014–1015, *1015*

Seismic waves, 456t, 457

Self-induced emf, 861

Self-inductance, 861, 887
of circuit, 849–850

Semiconductor(s), 587, 727, 729
analog of, 735
bonding process in, 729–730, *730*
conductivity of, doping and, 731, *731*
resistivity of, impurities and, 727
See also Conductors

Sequence of colors in soap film, thin-film interference and, 1032–1033

Series arrangement of capacitors, *688,* 688–691, 709
charge on each capacitor in, 688
equivalent capacitance of, 689
isolated conductors in, 688
parallel arrangement and, 686, *686*

Series combination
of resistor and capacitor (RC-circuit), 757
of resistors, *745,* 745–746

Series connection of ammeter, 755, *755*
Seven-day clock, 189
Shadowgraphs, dental X-rays as, 1064
Shear modulus, 415, 419, *419*, 440, 456
  raspberry jello example of, 421
Shear strain in dynamic viscosity, 438–439, *439*
Shear stress, 417, 419, *419*
SHM (*see* Simple harmonic motion)
Shock waves, 473
  Doppler effect and, *468, 469, 472, 472*
Short circuit, 747
SI (*see Système Internationale*)
SI units (*see Système Internationale* (SI) units)
Sign convention(s)
  for applied emf, *892, 892*–893
  for capacitance, 680
  in Carnot refrigeration cycle, 547
  in complete cycle of heat engine, 541
  for Coulomb force vector, 591
  for electric charges, 648
  in electric current, *719*, 719–720
  for focal length of lens, 999
  folded-axis, 298
  importance in force and torque equations, 322
  for mirror equation, 1020
  nonzero rotational inertia and, 297
  for object and image distances, 1020
  in radii example of lens-maker's formula, *997, 999,* 1001
  for refraction at spherical interface, 997, *997,* 997t
  for spherical mirrors, 991, 991t
  in thermodynamics, 520
  for thin-lens equation, 1002, 1002t, 1020
  for translational and rotational motion, 315
Sign of acceleration, 55–57, *56*
Significant figures, 13–15
  rule of thumb for, 14
Silicon, 729
Silicon atom, valence electrons in, 730
Sill, William R., 886–889
Simple circuits, 749
Simple harmonic motion curve, sketching, 360
Simple harmonic motion (SHM), 357
  diving board example of, 360–361, *361*
  dynamics of, 363–371
  energy and, example of, 373
  kinematics of, *357,* 357–363, *359*
  mass attached to string, 363–366, *364*
  oscillating particle example of, 362–363
  physical pendulum, *368,* 368–369
  simple pendulum, *366,* 366–368
  sinusoidal wave train as, 392
  torsion pendulum, *369,* 369–370

two-mass system, *370, 370*–371
  uniform circular motion and, 374, *374*
  velocity and acceleration for, 361–362, *362*
    maximum values of, 361
Simple harmonic oscillator
  energy of, 371–373, *372*
  similarity of LC circuit equations to, 857–858
Simple optical instruments, 1008–1013
Simple pendulum model, of simple harmonic motion, *366,* 366–368
  example of, 368
  frequency of vibrations in, 367
Simpson's rule, 64, 69, 176, 178
  example of, 61t, 63–64, *64*
Simultaneity
  frame of reference and, 1089, *1089*
  special theory of relativity and, 1085, *1085*
Simultaneous arrival of wavefronts, in two-source interference patterns, 1042, *1042*
Simultaneous over-relaxation, 668
  spreadsheet example of, 668–670, 668–671
Sinc(x), 1054, 1070
Single point-masses, vector relations and, *350,* 350–351
Single-slit diffraction, *1053–1055,* 1053–1057
  diffraction-limited optics and, 1055–1056, *1056*
  resolution and, *1056,* 1056–1067
Single-slit diffraction pattern, 1058
Sinusoidal current source for RLC circuits, 900, *900*
Sinusoidal electromagnetic waves, 937–938
Sinusoidal time dependence, example of, 891–892
Sinusoidal wave(s), 408
  energy transported by, *396,* 396–397
  example of, 396
  harmonic oscillator example of power of, 397
  transverse (example of longitudinal waves), 394–395
Sinusoidal wave train, 390, *390, 392,* 392–395
Skate-board (example of Newton's second law of motion), resolving forces, 115, *115*
Skew rays, 962, *962*
Slender rod (example of rotational inertia), 286
Slingshot (example of conservation of energy), 197–199, *198*
Slit images, 1125
Slope(s), 47, 64–65, *65*
  parabola in computing, 65, *65*
Slow axis in birefringence, 973

Slug, 120
Small diode laser in compact-disc player, 975, *975*
Small mirror (example of plane mirrors), 986, *986*
Smith, J. H., 1090
Smokestack "scrubbers"
  mechanism of, 604
  static electricity and, 585
"Snapshot" technique, differences in, 399
"Snapshot" view of sound wave, 451, *451*
Snell, Willebrod, 953, 957
Snell's law, 957, 958, 976
  in prism geometry, 965
  for total internal reflection, *962,* 963
"Snow flakes" (example of radiation pressure), 942, *942*
Soap film (dielectric film), 1032
Sodium doublet (example of Rayleigh criterion), 1062
Solenoid(s), *806,* 812–815, *813,* 849
  Ampère's law and, *813,* 813–814
  computing magnetization in, *873,* 873–874
  example of inductive time constant, 854
  field on axis of, numerical application for, *817,* 817–818, *818*
Solenoid system (example of mutual inductance), 852, *852*
Solid(s), 416, 416t
  molar heat capacity of, *575,* 575–576
  as thermal conductors, 495
  thermal expansion of, 488
Solid disk, rotational inertia for, 288, 289t
Solid objects
  center of mass for, by integration, 254–257, *255, 256*
  rotational inertia of, 285
Solid-state physics, models of temperature dependence of resistivity in, 727
"Sonic boom," 472
Sound, 450–473
  Doppler effect and, 455t, *468,* 468–472, *469,* 470t
    shock waves, *468, 469, 472, 472*
  harmonic waves in air, *452, 453, 457,* 457–458
  intensity and intensity level of, *453,* 458–463
    decibel scale, *460,* 460–462, 461t
  models for sound waves in gas, *451,* 451–452, *452*
  sources of, 463–467
    air columns, *465, 466,* 466–467
    resonance and beats, 467, *467*
    vibrating strings, 463–465, *463–465*

velocity of, *452,* 452–457, *453,* 455t
  other distortion waves and, 455–457, 456t
Sound decay rate in concert halls, 479–480
Sound intensity, 460
  examples of
    normal conversation, 461t, 461–462
    outdoor concert speaker, 461t, 462
    threshold of pain, 460
Sound level, sound intensity and, *453,* 458–463
Sound power, 458
Sound pressure level, 481
Sound recording and reproduction, 480
Sound waves, 473
  frequency of, Doppler shift and, 469
  intensity of, 473
  pressure variation and gas-layer displacement for, 457
Source(s)
  of electromagnetic waves, 942, *943, 944,* 944–945
  of emf, 766
  frequency change due to motion of, 468
  frequency of, 1092
Soutas-Little, Robert, 99, 335–338
Spatial change in magnetic field, temporal change in electric field and, 934–936
Special theory of relativity, 1078–1105
  binding energy and mass, 1101–1103
  energy-momentum relation from, 1119
  postulates of, 1084–1085
    simultaneity and, 1085, *1085*
  relativistic dynamics and, 1098–1101
    kinetic energy, 1100–1101
    relativistic momentum, *1099,* 1099–1100
  relativistic kinematics and, 1085–1098, *1086*
    experimental tests of relativity, 1090–1091
    length contraction, *1089,* 1089–1090
    Lorentz transformations, *1085,* 1086–1087, *1087*
    relativistic Doppler shift, *1092,* 1092–1093
    time dilation, 1088–1089
    twin paradox, 1093–1096
    velocity transformations, *1096,* 1096–1098
  speed of light and, 1079–1084
    luminiferous ether in, 1079–1080
    Michelson interferometer and, *1080,* 1080–1081
    Michelson-Morley experiment and, *1081,* 1081–1084, *1083*
Specific gravity (relative density) of substances, 422

Specific heat(s), 500, 572–576
  for common substances, 501t
  of diatomic ideal gases, 573, 574t
  of monatomic ideal gases, 572
  quantum mechanical effects in, 574t, 574–575
  spoon in thermos example of, 501t, 501–502
Specific heat capacity, 508
  of solids, *575,* 575–576
Spectral excitance, 1110
Spectral lines, 1060, 1125
  angular width of, 1061
  in Bohr model of hydrogen atom, 1131
  of hydrogen, 1126
Spectrum(a), 1060
  of atoms, *1125,* 1125–1127
Specular reflection, 956–957, *957*
Speed, angular velocity and, of particle in uniform magnetic field, 780
Speed of light, 6, 958, 1079–1084
  in free space, 1098
  luminiferous ether in, 1079–1080
  Michelson interferometer and, *1080,* 1080–1081
  Michelson-Morley experiment and, *1081,* 1081–1084, *1083*
Speed of mass in simple harmonic motion, 373
Spencer, R. L., *663*
Spherical aberration, 987, *987, 989, 990*
  as third-order aberration, 1014, *1015*
Spherical capacitor
  example of capacitance, 682–683, *683*
  example of electric potential energy, *694,* 694–695
Spherical charge distribution (example of electric field strength), *628,* 628–630, *630*
Spherical conductor, charged, electric potential and field of, 663, *663*
Spherical masses, Newton's universal law of gravitation (NULG) for, 202
Spherical mirrors, 986, *987*
  paraxial rays and, 1019
Spherical objects, Newton's universal law of gravitation (NULG) and, 138
Spherical symmetry, computing electric field by negative derivative in, 658
Spherical wave(s), 451, *451,* 473
Spherical wavefronts, plane waves and, 934
Spin "down" state, of electron, 930, *930*
Spin (intrinsic angular momentum), 733, *734*
  of electron, 868, 871–872
  magnetic resonance and, 929
Spin quantum number, 930
Spin "up" state, of electron, 930
Spinning teacher (example of rotational speed), 303, *303*

Spoon in thermos (example of specific heat), 501t, 501–502
Sports activities, *337,* 337–338, *338*
Spreadsheets (*see* Computer programs)
Spring, potential energy of, 371
Spring constant, 108, 109
  as bond in resonance, 378
Spring-distortion distance, in two-mass system, 371
Spring forces, 108, *108*
  example of, 108, *111,* 111–112
Spring-mass system, simple harmonic motion and, 363–366, *364*
  example of, 365–366
Square wave, 404, *404*
Stable equilibrium, 201
Stagnation pressure, 437
Standard emf, 756
Standard mass, 8, *8*
Standard meter bar, 6, *6*
Standard meter (krypton 86 lamp), 6, *6*
Standard near point, 1008
Standard units, 5
Standards, 15
Standing sound waves, 473
Standing waves, 400–402, *401, 402,* 409
  quantization of, 1115
State of thermodynamic system, 504
State variables, 504, 509, 519, 532
  entropy as, 554–555
  formal definition of, 554
  internal energy change as, 520
  zeroth law of thermodynamics and, 515
Static charges, Coulomb's law valid for, 626
Static electricity, 585
Static equilibrium, 312–316, 327
  in biomechanical loading, 337
  force and torque in, 313–314
    hanging sign example of, *315,* 315–316, *316*
    rafting example of, *314,* 314–315
  rolling objects and, 311–328
  torques due to weight force and, *312,* 312–313
Static fluid pressure, behavior of force due to, 423
Static frictional force in example of circular motion, 133, *133*
Stationary center of mass, 259
Stationary charges, electric field and, 584–616
Stationary states, 1134
  in Bohr model of hydrogen atom, 1128, 1129
Statistical mechanics, statement of second law of thermodynamics, 559
Steady flow in ideal-fluid model, 432
Steady state, 757
  in thermodynamics, 514
Steady-state condition, 377, *377*

Steam engines, 174
Steam point, 485, 487
Stefan-Boltzmann constant, 1111, 1133
Stefan-Boltzmann law, 1111, 1133
Step size, reducing, in Runge-Kutta method, 148
Step-up transformer, 918
Stepdown transformer, 918
Stewart, A., 1098
Stiffness constant of a spring, 108
"Still waters" (example of index of refraction), 999
Stokes, Sir George G., 143
Stokes' law, 143, 152
  in calculating Reynolds number, 144
Stopping potential, in photoelectric effect, 1116, *1116*
Storage box (example of energy stored in system of capacitors), *693*, 693–694
Straight-line kinematics, rotational kinematics and, 279, 279t
Straight wire, magnetic field for, 807
Strain, 417–418
  applied stress and, 418
Straw (example of pressure-measuring devices), 431–432
Streamline, 432, *432*
Stress, 417
  magnitude of, 417
Stringed instruments
  sound produced by, 473
  standing waves and, 401–402
Strutt, John William, Lord Rayleigh, 1112
Subjective loudness, 481, *481*
Subjective pitch, 481
"Sudden approximation," 251–252, 268
  in ballistic pendulum example, 251
Summary tables, 528
  in example of heat cycle problem, 531
  importance in solving thermodynamic problems, 522
Sun temperature (example of blackbody), 1112
Sunglasses, wire-grid polarizers in, 970
Superconducting materials, 727
Superconductors, *735*, 735t, 735–737, *736*
  as diamagnets, 877
  high-temperature, in magnetohydrodynamics, 787
  transition temperature for, 735, 735t, 736
Superloop (example of magnetic field), *805*, 805–806, *806*
Superposition
  of one-dimensional waves, *407*, 407–408, *408*
  principle of, 1029, 1046
  of selected harmonics, 463
  vector addition example of, 592–594, *593*

of waves
  interference and, 397–404
  irradiance from, 1035
Superposition principle, 409
Surface charge distribution, 630
Surface distortion, compression waves and, 455–456
Surface integral
  calculating, shape of Gaussian surface and, 632, *632*
  for computing electric flux, 624–625
Surface shape, electric flux and, 624, *624*
Symbolic notation for SI units, 13
Symbols, confusing, in resistivity and resistance, 722, 723
Symmetrical solid objects, center of mass for, 254, *254*
Symmetry
  cylindrical, 652, 658, 806, 856
    of irradiance pattern, 1055, *1056*
  of electric field, charge distribution and, 627, *627*, 628
  lack of, in thermal expansion, 489, *489*
  in twin paradox, 1093
Symon, K. R., 346
Synchronization, 1085, *1085*
System
  in thermodynamics, 504
  work done by, distinguished from work done on, *518*, 518–520
System(s)
  of units, 5t, 5–10
    length, 5–6, 7t
    mass, 8–9, 9t
    metric system, 10, 10t
    time, 7–8, 8t
  of variable mass, unbalanced external forces on, 263–268, *265*, *267*
System of particles
  center of mass of, 269
  with changing mass, Newton's second law for, 269
  gravitational potential energy of, 269
  kinetic energy of, 262, 269
System of the World, The (Newton), 141
Système Internationale (SI), 5t, 5–10, 42
Système Internationale (SI) units
  coulombs, 776
  of electric charge (coulomb), 587
  of electric potential (volt), 646
  of emf (volt), 743
  for energy density (joules per cubic meter), 692
  of energy (joule), 493
  of force, 120
  of frequency (Hz), 358
  of inductance (henry), 850
  for magnetic dipole moment (amperes per square meter), 791
  of mass, 120
  of momentum, 222

newtons, 776
  of potential energy (joule), 646
  of power (watt), 728
  for pressure (pascal), 420, 420t, 422
  of resistivity (ohmmeter), 722
  tesla, 776, 793
  of torque, 293
Systems of particles in conservation of momentum, 226–227

Tabular procedure, for solving thermodynamics problems, 527, 543, 544t
Tabular procedure for solving thermodynamics problems, 521
Tachometer, 278–279
Tacoma Narrows Bridge Disaster (example of resonance phenomena), 378, *378*
Taillight (example of resistance), 728
Taking a limit, 45
Tangent line, 47
Tangential acceleration, 93
  in circular motion, 134
  of particle moving in circle of constant radius, 343, *343*
Tangential force, torque and, 295
Tangential velocity
  angular velocity and, for circular motion, 352
  frictionless mass example of, 289t, 292, *292*
  wall example of, 289t, 291, *291*
Tank of fluid (example of Torricelli's Law), *437*, 437–438
Taylor, Edwin F., 1122, 1123
Taylor, John R., 485
Telescopes, *1012*, 1012–1013, *1013*
Temperature, 485–488
  bulk modulus and, 454–455
  Celsius and Fahrenheit scales, *485*, 485–486
  constant-volume gas thermometer, 486–488, *487*
  effect on velocity distribution function, 577, *577*
  of a gas, 578
  molecular motion and, in kinetic theory of ideal gas, 568–570
  resistance and, 738
  speed of sound waves as function of, 456
Temperature dependence of resistivity, 726–727, *727*
Temperature gradient in conduction, 495–496
Temperature scale, 485
Temporal change in electric field, spatial change in magnetic field and, 934–936

Tennis ball (example of average force), 225

Tensile stress, 418–419, *419*

Tension, 109, 120, 166
  acceleration and, 109
  connected masses example of, *297*, 297–299, *298*
  non-constant, *114*, 114–115
  wave motion and, 396

Tension equation for Newton's second law of motion, 117

Tension force, magnitude of, in wave equation, 404

Terminal potential, 766
  electromotive force and, 744

Terminal velocity, 145

Tesla, 776, 793

Teulkolsky, S. A., 668

Theory(ies), 3

Theory of Light and Color (Newton), 137

Thermal conductivity, 495, *495*, 508
  for common solids and gases, 496t

Thermal energy (internal energy), 493, 508
  kinetic energy and, in resistors, 727–728

Thermal expansion, 488–493, *489*
  area, 490t, 491, *491*
  linear, *490*, 490t, 490–491
  potential energy function in, 488–489, *489*
  volume, 490t, 491–493

Thermal motions
  magnetic dipole moments and, in paramagnetism, 878
  of molecules in dielectric material, 697
  orientation of dipole moments and, 869
    within domains, 870

Thermal velocity, magnitude of, drift velocity magnitude and, 722

Thermistor, 724, *725*

Thermodynamic cycles, problem-solving strategy for analyzing, 526t, 529

Thermodynamic equilibrium, 514

Thermodynamic internal energy, 520

Thermodynamic processes, 522–528
  summary of, 526t

Thermodynamic variables
  finite changes in, in real processes, 549
  in ideal-gas law, 505, *505*

Thermodynamics, 504
  cyclic processes in, 528–532, *529*
  distribution of molecular speeds in, 576–578
    Maxwell speed-distribution function, *576*, 576–577, *577*
    mean free path and, *577*, 577–578
    equipartition-of-energy theorem in, 570–571

ideal polyatomic-gas models and, 571, *571*
  first law of, *520*, 520–522
  kinetic theory of ideal gas, 565–570
    ideal-gas model and, 565
    pressure and molecular motion in, *565*, 565–568, *567*
    temperature and molecular motion in, 578–580
  microscopic connections to, 564–579
  processes and first law, 513–533
  second law of, 538–560
  specific heats in, 572–576
    diatomic ideal gases and, 573, 574t
    heat capacity of solids and, *575*, 575–576
    monatomic ideal gases and, 572
    quantum mechanical effects and, 574t, 574–575
  specific processes in, 522–528
    adiabatic, 525–528, *526*, 526t
    isobaric, 523–524
    isochoric, 523
    isothermal, 524, *524*
  third law of, 560
    absolute temperature scale and, 548–549, *549*
  work and, *518*, 518–520
  zeroth law of, 514–517, *516*, *517*

Thermometer (example of Celsius scale), 485–486

Thin-film interference, 1031–1034, 1038–1039, 1047
  pattern of, interference filters and, 1040, *1040*

Thin lens(es), 1020
  in contact, example of, 1006

Thin-lens equation, 1010, 1020
  in paraxial ray tracing, 1016
  sign conventions for, 1002, 1002t, 1020

Thin-lens formula, *1001*, 1001–1007, 1002t, *1006*, *1007*

Third harmonic (second overtone), 402, 465, 466

Third-order aberrations, 1021
  from paraxial ray approximation, 1013–1015, *1015*

Thompson, Benjamin, 493–494

Thompson, William, Lord Kelvin, 488, *488*

Thomson, Joseph J., 798, 1124, 1127

Thomson, Sir George P., 1122–1124, 1134

Three dimension(s)
  conservative force in, 211
  relation between electric field and potential function in, 660–661

Three-dimensional forces, conservative, 211

Three-dimensional objects, center of mass for, 255, *255*

Three-dimensional waves, 389, *390*

Threshold of pain for rap (example of sound intensity), 460

Thrust, 264, 266

Timbre, 481
  of percussion instruments, 478

Time, 7–8, 8t
  amount of work as function of, 174
  behavior of wave in, 388
  current decrease in, 854, *854*

Time constant, 758, 766

Time-dependent quantities, 891

Time derivatives, 258, 259
  of work-energy theorem, 296

Time difference predicted by ether model, Michelson interferometer and, 1082

Time dilation, 1088–1089, 1104
  ''extra five minutes'' example of, 1088–1089

Time-independent quantities, 891

Tires of Indycars, 161–162

Toll gates on highways, flow of vehicles through, continuity equation and, 447

Toroid
  Ampère's law and, *806*, *815*, 815–816
  computing magnetic field in, example of, 815–816, *816*
  varying magnetic field strength in, 815

Torque, *292*, 292–296, *293*, 616
  in biomechanical loading, 337
  on current-carrying loops in uniform magnetic field, 789–793, *790*
  definition of, 352
  due to weight force, static equilibrium and, *312*, 312–313
  in force analysis of rotation and translation, 319, 320
  force and, in static equilibrium, 313–314
  magnetic dipole moment and, example of, *792*, *792*
  on magnetic moment, 794
  of polar molecules, in dielectric material, 696, *696*
  sign conventions for, 298
  units of, 293
  as vector, 293
  work done by, *295*, 295–296

Torque equation
  in example of force and torque, 323
  vector nature of, 313

Torque vector, *340*, 340–341
  merry-go-round example of, *340*, 340–341

Torricelli's law, tank of fluid example of, *437*, 437–438

Torsion balance (Coulomb), 590, *590*

Torsion constant
  Cavendish balance and, 139
  in torsion pendulum, 369

Torsion pendulum in simple harmonic motion, *369*, 369–370
Total acceleration, 91
Total angular momentum, 300
　axis of rotation and, 351, *351*
Total charge
　nonuniform linear charge distribution and, 589, *589*
　nonuniform volume charge distribution and, 589–590, *590*
Total charge enclosed in parallel-plate capacitors, Gauss's law and, 704
Total current, calculating
　conducting bar example of, 721, *721*
　definition of electric flux and, 720
Total electromagnetic force on a charge (Lorentz force law), 778
Total energy, 1101, 1104
　of motion, 290
Total internal reflection, 976
　critical angle for, 960, *960*
　　example of, *960*, 960–961
　refraction and, *960*, 960–962, *961*
Total irradiance pattern in double-slit interference, 1058–1059, *1059*
Total kinetic energy
　in energy analysis of rotation and translation, 318
　of rigid body, 284
Total linear acceleration, magnitude of, 282
Total mechanical energy
　in simple harmonic motion, 381
　of simple harmonic oscillator, 371–373, *372*
　of a system, 191
Total momentum, 227
Total potential drop, in RLC circuits
　amplitude of, 904–905
　phasor for, 901–902
Total resistance of resistors in parallel, *746*, 747
Total system mass (zeroth moment), 286
Total work done in rope swing example of work, 170
Touchdown (example of perfectly inelastic collision), 233–234
Tow chain (example of Newton's second law of motion), *114*, 114–115
Toy gun (example of projectile motion), *82*, 82–83
Trajectories, computing with Runge-Kutta method, *148*, 148–149
Transfer mechanisms for heat and energy, 493–499
　building insulation, 497–498, 498t
　conduction, *494*, 494–497, *495*, 496t
　convection and radiation, 498–499
Transformation equations, inverting, 1104
Transformer(s), *916*, 916–921, *919*, 919t
　impedance-matching, 919–920

primary and secondary coils of, 924
　stepdown and step-up, 918
Transformer impedance, 917
Transistors, 851
Transition temperature for superconducting materials, 735, 735t, 736
Translation
　with no slipping, 316–324
　　energy analysis in, *317*, 317–319, 319t
　　force analysis in, 319–320, *320*, 322–323
　problem-solving strategy for force analysis of, 323
　with slipping, 324–327
　　angular impulse-angular momentum theorem and, 324–326
　　center of percussion and, *326*, 326–327, *327*
Translation theorem, 390, 392
Translational kinetic energy
　in energy analysis of rotation and translation, 318, 319t
　of a particle, 570
Translational motion
　of gas molecules, 574
　nonzero rotational inertia and, 297
　in simple pendulum model of simple harmonic motion, 366–367
Translational velocity, final, 326
Transmission coefficient, 962
Transmission gratings, 1060
Transmitted irradiance
　law of Malus and, 971
　"Malus aforethought" example of, 972, *972*
Transverse electromagnetic wave, light as, 954
Transverse relaxation of magnetic moments in MRI, 931
Transverse (shear) deformation waves in elastic media, 456t, 457
Transverse velocity
　differentiated from phase velocity, 405
　of sinusoidal wave, 394–395
Transverse waves, 388, *388*, 408
　speed of, 456t, 457
Trapezoid(s), in computing areas under a curve, 62, *62*
Trapezoidal rule, example of spreadsheet for, 62t, 64, *65*
Trigonometric identity, 400
　in wave interference, *398*, 398–399
Triple point, 506, *506*
　of water, 487
Truck tire (example of circular motion), 89
Tsai, J., *478*
Tube length of compound microscope, 1011

Tug-of-war (example of action and reaction), Newton's third law of motion and, 118, *119*
Tuning forks, 467, *467*
Twentieth-century physics, 4
Twin paradox, 1093–1096
　example of, 1094–1096, *1095*
Two-dimensional conductor in quantum Hall effect, 785
Two-dimensional motion, applying Runge-Kutta method to, 149, *150–151*, 151–152
Two-dimensional NMR techniques, 931
Two-dimensional problems, numerical methods for, 666–671, *667*
Two-dimensional waves, 389, *389*
Two-fluid theory of electric charge, 585
Two-mass system in simple harmonic motion, *370*, 370–371
Two-point charges
　Coulomb's force law for, hydrogen atom example of, 592, *592*
　potential energy function for, example of, 645–646
　potential energy of, 643–646, *644*, *645*
Two-reflection model of thin-film interference, 1031
Two-source interference of light, *1029*, 1029–1031, *1030*
　patterns of, 1040–1043, *1041–1043*
"Two-step" (example of isobaric and isothermal processes), *527*, 527–528
Two-terminal elements, 686

Ultrafine diffraction gratings, to focus long-wavelength X- rays, 1065
Ultrasonic waves, 452
Ultraviolet catastrophe, 1113
Unbalanced force, 104
Uncertainty
　optimistic analysis of, 14
　significant figures and, 13–15
　worst-case analysis of, 13–14
Underdamped harmonic oscillator, 862
Underdamped motion, 375, *376*
Uniaxial birefringent crystals, 972
Unified atomic mass unit, 9, 9t, 1102
Uniform charge flow, electric current and, 720, *720*
Uniform circular motion, 85, *85*, 90
　of particle in uniform magnetic field, 779, *779*
　radian measure in, 94
　simple harmonic motion and, 374, *374*
Uniform distribution of total charge, calculating charge densities from, 588–589
Uniform electric field
　conductor in, 663, *663*

electric dipole in, *611–613,* 611–614

electric potential change due to, *649,* 649–652

Uniform magnetic field, moving charges in, 779–781, *779–781*

Unit cell in salt crystal, *1066,* 1066–1067

Unit conversion, dimensional analysis and, 10–13

Unit vector(s), *25,* 25–26, *26,* 35
   combined with partial derivatives (gradient), 207
   creating, 90

Unit-vector notation, 36

United States Auto Club (USAC), 160

Universal gas constant, 504

Universal gravitation, weight and, 140–142, *141*

Universal model, for gravitational potential energy, correct solution to example using, *205,* 205–206

Universal model of gravitational potential energy, 205

Unmagnetized state, returning ferromagnetic substance to, 881, *881*

Unpolarized light, from incandescent source, 966

Unstable equilibrium, 201

"Uphill" heat flow (example of Carnot refrigerator), 547–548

USAC (United States Auto Club), 160

Valence band, 734, *734*

Valence-bonding model of conduction, 729–731, *730, 731*

Valence electrons
   in aluminum atoms, 731, *731*
   in silicon atom, 730

Van Allen belt, 782, *782*

Van de Graff generator, *594*

van der Waals gas, *506,* 507t, 507–508
   constants for, 507t
   isobaric process for, 524

van der Waals model, 507

Vapor, 507

Variable(s)
   of integration, time limits of integral and, 221
   relations between, 3

Variable direction, variable force and, in work, 169–170, *170*

Variable force
   constant direction and, in work, 165–169, *166–169*
   variable direction and, in work, 169–170, *170*

Variable mass, systems of, 263–268, *265, 267*

Vector(s), 21, 35
   additive inverse of, 24, *25*
   angular momentum as, 300
   associated with linear accelerations, 282, *282*
   components of, 26, *26,* 35
   conservation of momentum and
      in polar form, *229,* 229–230
      in rectangular form, 229, *229*
   infinitesimal rotations as, 342
   integrating, example of, 220
   linear combinations of, 25
   magnetic field, 872–876, *873, 874,* 876t
   momentum as, 220, 222, 250
   perpendicular to plane of circular motion, 343, *343*
   scalars and, *21,* 21–23, *22*

Vector addition, 23–30, *24, 25*
   associative nature of, 24, *24*
   commutative nature of, 24, *24*
   component method, 26–30, *28, 29*
   example of superposition, 592–594, *593*
   examples of, 27–30, *28, 29*
   multiplication by scalar, 25
   sum of forces, constant velocity and, 103
   unit vectors, *25,* 25–26, *26*
   vector components, 26, *26*

Vector algebra, 18–36
   coordinate systems in, *19,* 19–21, *20*
   rules of, 22
   scalars and vectors, *21,* 21–23, *22*

Vector angular momentum, conservation of, 346

Vector descriptions of rotational motion, 339–353
   angular momentum, *344,* 344–349
   angular velocity and acceleration, 341–344, *342, 343*
   torque, *340,* 340–341
   vector relations and, 349–352

Vector dot product notation for electric flux, 624

Vector multiplication, 30–35
   of constant force and direction in work, 164
   product, 31–33, *32*
   scalar product of, 31, *31*
   unit-vector representation, 33–35

Vector product(s), 36
   in unit-vector representation, 33–35
   in vector multiplication, 31–33, *32*

Vector quantities, symbols for, 22–23

Vector relations, 349–352
   asymmetrical rotor example of, 352, *352*

Vector translation, 29, *29*

Velocity(ies)
   from acceleration graphs, *51,* 51–52, *52*

computing acceleration from, 66

direction of, direction of force and, 776

of electrons after collision, 718, *719*

graphical methods for computing, 68

inability to measure, 44–45, *45*

magnitude of, in circular motion, 88

position and, example of, *58,* 58–59

ratio of, one-dimensional collisions and, 261

of sinusoidal wave, 395

of sound, *452,* 452–457, *453,* 455t
   magnitude of, 453, *453*
   in various materials, 455t

Velocity change, area under acceleration curve and, 68

Velocity curve, area under (displacement), *50,* 50–51, *51*

Velocity distribution function, effect of temperature on, 577, *577*

Velocity function, for simple harmonic motion, 362–363, 366, 380

Velocity gradient in dynamic viscosity, 439

Velocity graphs, position from, 49–51, *50, 51*

Velocity of light, velocity of observer and, in Einstein's second postulate, 1084

Velocity of observer, velocity of light and, in Einstein's second postulate, 1084

Velocity ratio in relativity, 1082

Velocity selector
   example of, *778,* 778–779
   Hall effect as, 784

Velocity subtraction rule, 1096

Velocity transformations, *1096,* 1096–1098
   aberration of starlight and, 1098
   equations for, 1096–1097

Velocity vectors
   acceleration vectors and, 77–79
   change of direction in, 87
   in circular motion, 88
   tangential component of, 89, 90, *90*
   vector perpendicular to, 90

Velocity-versus-time graph, average and instantaneous acceleration from, 48–49

Vertex of parabolic mirror, 986, *987*

Vetterling, W. T., 668

Vibrating diatomic gas molecules, quantum theory for, 579

Vibrating diatomic ideal gas
   average energy of, 571
   degrees of freedom for, 571, *571*
   specific heats of, 573

Vibrating strings as source of sound, 463–465, *463–465*

Virtual image(s), 984, 1019
   concave mirror example of, 993–994, *994*

Virtual image(s) (*Cont.*)
converging lens, real object and, example of, 1003, *1003*
diverging lens, real object and, example of, 1004, *1004*
Virtual object for second lens, example of, *1005*, 1005–1006
Viscosity, 142, 142t
Visible spectrum of electromagnetic wavelengths, *964*, 964–965, *965*
Volleyball (example of velocity following impulse), 223, *223*
Volt(s), 646
per meter, as electric field units, 650
Volta, Alessandro, 679
Voltage ramp, 762, *762*
Voltmeters, *754*, 754–755, 766, 783, *783*
digital, 762–763
example of building, 754–755, *755*
Volume density function, 255
Volume expansions, 490t, 491–493, 508
copper cube example of, 492–493
gasoline can example of, 492
Volume flux, equation of continuity and, in fluid dynamics, 434
Volume of fluid displaced in Archimedes' principle, 427

Wall (example of tangential velocity), 289t, 291, *291*
Wallis, John, 238
Water balloon (example of air resistance), 145–146
Water mixture (example of entropy increase), 558
Water molecule (example of dipole moment), 613–614
Watson, F. G., 137
Watt, James, 174, 175
Watt-second, 174
Watterson, Bill, 415
Watts, 174
Wave(s)
amplitude-modulated, *399*, 399–400
amplitude of, 388
direction of, particle motion and, *388*, 388–389
one-dimensional, 387–409
properties of, 388
superposition and interference of, 397–404
traveling on string, 390–395, *391*
longitudinal, 394–395
pulses, 390–392, *391*
sinusoidal wave train, *392*, 392–394
types of, 388
Wave dimension, 389, *389*, *390*
one-dimensional waves, 389, *389*
three-dimensional waves, 389, *390*
two-dimensional waves, 389, *389*

Wave equation, *405*, 405–407
Wave equation analysis, wave velocity as, 406–407
Wave function, 391
for transverse, one-dimensional wave, 393–394
wave pulse example of, *391*, 391–392, *392*
Wave-guide dispersion, 456
Wave model
of light, 953–954
properties of, *954*, 954–956
Snell's law and, 959
speed of light in, 958
for particles, 1121–1124
Davisson-Germer experiment, *1122*, 1122–1123, *1123*
experiment of G. P. Thomson, 1123–1124, *1124*
Wave number, 393, 408, 937
Wave number difference, 399
Wave pulses, 389–390, *390*, 390–392, *391*, 391–392, *392*
Wave reflection, *402*, 402–403, *403*
Wave train, 390, *390*
Wave velocity
of sinusoidal wave train, 393
on string, *395*, 395–396
wave equation analysis as, 406–407
Wavefront(s), 452, *452*
force on compressed region and, in sound production, 453–454
rays and, in wave model of light, *954*, 954–955, *955*
Wavefront emissions of sound waves, 469–470
Wavelength(s)
angular separation of, 1061
differing, of sound wave and string, 463
for position equation, of sinusoidal wave train, 393
of sinusoidal wave, 394
of sound waves, Doppler shift and, 469
Weast, Robert E., 1118
Weight, 120
mass and, example of differentiating, 107–108
universal gravitation and, 140–142, *141*
Weight and mass, 106
Weight force, 106–108, 107t
distributed nature of, 312
magnitude of, 106
torques due to, static equilibrium and, *312*, 312–313
work done by, 165, *165*
in piñata example of work, 167, *168*
"Weight-loss" (example of gravitational field strength), 141

Weight of fluid displaced in Archimedes' principle, 427
"Weightlessness" (example of gravitational field strength), *141*, 141–142
Western, A. B., 494
Wheatstone bridge, 756, *756*, 766
Wheel (example of net torque), 294, 294–295
Wheeler, G. F., 404
Wide-angle convex mirror (example of image location), 995, *995*
Wien, Wilhelm, 1112
Wien's displacement law, 1111, 1133
Wilk, S. R., 62
Wind instruments, 478
sound produced by, 473
Wing mounted on car, as method of holding Indycars on track, 161
Wire-grid polarizers, 970–972, *971*, 976
Wire loop (example of force on current-carrying conductor), *788*, 788–789
Wolfson, Richard, 185
Work, 177, *518*, 518–520
with constant force and angle, 177
with constant force and direction, *164*, 164–165, *165*
inner tube example of, 165, *165*
extraction from heat, 539
general definition of, 169
kinetic energy and, 163–178, *165*, 170–173
"high tea" example of, 172
inner tube example of, *165*, 172–173
rope swing example of, 173
negative, 164–165
numerical calculations of, 176–177
compound bow example of, *176*, 176–177, *177*
with variable force and constant direction, 165–169, *166–169*
non-Hooke's-law spring example of, 168–169, *169*
piñata example of, 167–168, *168*
with variable force and direction, 169–170, *170*
rope swing example of, 170, *170*
Work differential, in rope swing example of work, 170
Work-energy theorem, 171–173, 178, 187, 191, 296
for linear motion, 304t
for rotational motion, 304t
Work function, 1117, 1133
Work per charge, 648
Work performed
in adiabatic process, 525–526, *526*
change in potential energy and, 648
on electric dipole to change orientation, 612

gas expansion examples of, 519, *519*
in isothermal processes, 524, *524*
as path variable, 520
by system, *518,* 518–520
by torque, *295,* 295–296
on two-point charges, potential energy
    function and, 645–646
Work required
    to charge capacitors, energy storage
      and, 691
    to move charges in system of
      capacitors, 694
"Working equations," 234
    for partially elastic collisions, 236
"Working fluid", in
      magnetohydrodynamics, 786–787
Wren, Christopher, 238

X-ray(s), *945,* 946
X-ray diffraction, *1064,* 1064–1070,
    1065
    Bragg diffraction and, 1065–1070,
      1066, 1067
    experimental arrangement for, 1065–
      1066, *1066*

Yard, 6
Young's double-slit experiment, 1029–
    1031, *1030,* 1046
    example of, 1031, 1043
Young's double-slit pattern for
    computing irradiance, 1041
Young's modulus, 415, 418–419, *419,*
    440
    airplane cables example of, 420–421

Zeeman effect, 929, *930*
Zener diode, 724, *724*
Zeno of Elea, 44
Zeno's paradox, 44
Zero momentum frame in one-
      dimensional collision, 262
Zero resistance in ideal ammeter, 755
Zero resistivity of perfect conductor, 723
Zeroth law of thermodynamics, 514–517,
    516, 517, 532
Zeroth moment (total system mass), 286
Zone plates to focus X-rays, 1065
Zypman, F. R., 346

Credits

Conversion Factors

Length

1 in = 2.54 cm
1 m = 39.37 in = 3.281 ft
1 yd = 0.9144 m
1 km = 0.621 mi
1 mi = 1.609 km
1 mi = 5280 ft
1 μm = 10^{-6} m
1 Å = 10^{-10} m = 10^{-4} μm

Mass

1 slug = 14.59 kg
1 u = 1.660 540 × 10^{-27} kg
1000 kg = 1 t (metric ton)

Time

1 h = 3600 s
1 day = 86 400 s
1 yr = $365\frac{1}{4}$ days = 3.16 × 10^7 s

Velocity

1 m/s = 3.281 ft/s = 2.237 mi/h
1 mi/h = 0.4470 m/s = 1.467 ft/s
60 mi/h = 88 ft/s

Acceleration

1 m/s^2 = 3.281 ft/s^2
1 ft/s^2 = 0.3048 m/s^2

Force

1 N = 10^5 dyn = 0.2248 lbf
1 lbf = 4.448 N
1 dyn = 10^{-5} N = 2.248 × 10^{-6} lbf
1 ton = 2000 lbf
1 oz = 1/16 lbf

Pressure

1 Pa = 1 N/m^2 = 1.45 × 10^{-4} lbf/in^2
1 bar = 10^5 N/m^2 = 14.5 lbf/in^2
1 atm = 14.7 lbf/in^2 = 1.013 × 10^5 N/m^2
1 atm is equivalent to 760 mm of Hg

Energy

1 J = 0.738 ft · lb = 10^7 ergs
1 kW · h = 3.60 × 10^6 J
1 cal = 4.184 J
1 Btu = 252 cal = 1.054 × 10^3 J
1 eV = 1.60 × 10^{-19} J

Power

1 W = 1 J/s = 0.738 ft · lbf/s
1 hp = 550 ft · lb/s = 745.7 W
1 W = 3.413 Btu/h

Volume

1 m^3 = 10^6 cm^3 = 6.102 × 10^4 in^3
1 L = 10^{-3} m^3 = 10^3 cm^3
1 L = 0.0353 ft^3 = 1.0576 qt
1 ft^3 = 2.832 × 10^{-2} m^3 = 7.481 gal(US)
1 gal(US) = 3.786 L
1 oil barrel = 42 gal(US) = 0.159 m^3
1 gal (British) = 4.55 L
1 fluid ounce = 1/128 gal (US) = 2.96 × 10^{-2} L

Area

1 m^2 = 10^4 cm^2 = 10.76 ft^2 = 1550 in^2
1 ft^2 = 9.29 × 10^{-2} m^2
1 acre = 4.356 × 10^4 ft^2 = 4047 m^2
1 barn = 10^{-28} m^2

Magnetism

1 T = 1 Wb/m^2 = 10^4 G = $10^9 \gamma$